Praise for *Taguchi's Quality Engineering Handbook*

"Classic! This handbook is an excellent reference for engineers and quality professionals in any field."

Kevin W. Williams
Vice President, Quality
General Motors North America

"ITT has learned from experience that Dr. Taguchi's teachings can prevent variation before it happens. This book is a great demonstration of this powerful approach and how it can make a meaningful difference in any type of business. It takes a dedicated engineering approach to implement, but the pay back in customer satisfaction and growth is dramatic."

Lou Giuliano
Chairman, President and CEO
ITT Industries

"Dr. Taguchi's enduring work is of great importance to the field of quality."

Dr. A.V. Feigenbaum
President and CEO
General Systems Company, Inc.

"This handbook represents a new bible for innovative technical leadership."

Saburo Kusama
President
Seiko Epson

"It is a very well thought out and organized handbook, which provides practical and effective methodologies to enable you to become the best in this competitive world for product quality."

Sang Kwon Kim
Chief Technical Officer
Hyundai Motor Company & Kia Motors Corporation

"For those who see the world as connected, this landmark book offers significant insights in how to design together, build together, and lead together."

Dr. Bill Bellows
Associate Technical Fellow
The Boeing Company

"Dr. Taguchi has been an inspiration and has changed the paradigm for quality with his concept of loss function. Robust Engineering is at the heart of our innovation process and we are proud to be associated with this work."

Donald L. Runkle
Vice Chairman and CTO
Delphi Corporation

"The elegantly simple Taguchi theory is demonstrated through many successful case studies in this excellent book."

Don Dees
Vice President, Manufacturing
DaimlerChrysler

"We have and are practicing the methods and teachings in this valuable handbook; it simply works."

Dr. Aly A. Badawy
Vice President, Engineering
TRW Automotive

"To have all of the theories and application of the lifetime of learning from these three gentlemen in one book is amazing. If it isn't in here, you probably don't need it."

Joseph P. Sener, P.E.
Vice President, Business Excellence
Baxter Healthcare

"*Taguchi's Quality Engineering Handbook* is a continuation of the legacy by Dr. Taguchi, Subir Chowdhury and Yuin Wu in Robust Engineering, which is increasingly being recognized in the automotive industry as a 'must have' tool to be competitive.

Dan Engler
Vice President, Engineering and Supply Management
Mark IV Automotive

"Taguchi's principles, methods, and techniques are a fundamental part of the foundation of the most highly effective quality initiatives in the world!"

Rob Lindner
Vice President, Quality, CI and Consumer Service
Sunbeam Products, Inc.

"The value of Dr. Genichi Taguchi's thinking and his contributions to improving the productivity of the engineering profession cannot be overstated. This handbook captures the essence of his enormous contribution to the betterment of mankind."

John J. King
Engineering Methods Manager, Ford Design Institute
Ford Motor Company

"I am glad to see that *Taguchi's Quality Engineering Handbook* will now be available for students and practitioners of the quality engineering field. The editors have done a great service in bringing out this valuable handbook for dissemination of Taguchi Methods in quality improvement."

C. R. Rao, Sc.D., F.R.S.
Emeritus Eberly Professor of Statistics and Director of the Center for Multivariate Analysis
The Pennsylvania State University

"This book is very practical and I believe technical community will improve its competence through excellent case studies from various authors in this book."

Takeo Koitabashi
Executive Vice President
Konica-Minolta

"Digitization, color, and multifunction are the absolute trend in imaging function today. It is becoming more and more difficult for product development to catch up with market demand. In this environment, at Fuji Xerox, Taguchi's Quality Engineering described in this handbook is regarded as an absolutely necessary tool to develop technology to meet this ever-changing market demand."

Kiyoshi Saitoh
Corporate Vice President, Technology & Development
Fuji Xerox Co., Ltd.

"In the today's IT environment, it is extremely critical for technical leaders to provide an effective strategic direction. We position Taguchi Methods of Robust Engineering as a trump card for resolving difficulty during any stage of technology development. Technology is advancing rapidly and competition is getting global day by day. In order to survive and to win in this environment, we use the Taguchi Methods of Robust Engineering described in this handbook to continue developing highly reliable products one product after another product."

Haruo Kamimoto
Executive Vice President
Ricoh

"I was in a shock when I have encountered Taguchi Methods 30 years ago. This book will provide a trump card to achieve improved 'Product Cost' and 'Product Quality' with less 'R&D Cost' simultaneously."

Takeshi Inoo
Executive Director
Isuzu Motor Co.
Director
East Japan Railroad Co.

Taguchi's Quality Engineering Handbook

Genichi Taguchi
Subir Chowdhury
Yuin Wu

Associate Editors
Shin Taguchi and Hiroshi Yano

WILEY
John Wiley & Sons, Inc.

ASI Consulting Group, LLC
Livonia, Michigan

Copyright © 2005 by Genichi Taguchi, Subir Chowdhury, Yuin Wu. All rights reserved.

Published by John Wiley & Sons, Inc., Hoboken, New Jersey
Published simultaneously in Canada

For general information on our other products and services or for technical support, please contact our Customer Care Department within the United States at (800) 762-2974, outside the United States at (317) 572-3993 or fax (317) 572-4002.

Wiley also publishes its books in a variety of electronic formats. Some content that appears in print may not be available in electronic books. For more information about Wiley products, visit our web site at www.wiley.com.

Library of Congress Cataloging-in-Publication Data:

Taguchi, Genichi, 1924–
    Taguchi's quality engineering handbook / Genichi Taguchi, Subir Chowdhury, Yuin Wu.
        p. cm.
    Includes bibliographical references and index.
    ISBN 0-471-41334-8
   1. Quality control—Statistical methods. 2. Production management—Quality control. 3. Engineering design. I. Chowdhury, Subir. II. Wu, Yuin. III. Title.
    TS156.T343   2004
    658.5′62—dc22

                                        2004011335

Printed in the United States of America.

10 9 8 7

To our colleagues, friends, and families
for their enormous support over the years

Contents

Preface   xxv
Acknowledgments   xxix
About the Authors   xxxi

SECTION 1
THEORY

Part I
Genichi Taguchi's Latest Thinking   3

1
The Second Industrial Revolution and Information
Technology   5

2
Management for Quality Engineering   25

3
Quality Engineering: Strategy in Research and
Development   39

4
Quality Engineering: The Taguchi Method   56

Part II
Quality Engineering: A Historical Perspective   125

5
Development of Quality Engineering in Japan   127

6
History of Taguchi's Quality Engineering in the United
States   153

Part III
Quality Loss Function   **169**

7
Introduction to the Quality Loss Function   171

8
Quality Loss Function for Various Quality Characteristics   180

9
Specification Tolerancing   192

10
Tolerance Design   208

Part IV
Signal-to-Noise Ratio   **221**

11
Introduction to the Signal-to-Noise Ratio   223

12
SN Ratios for Continuous Variables   239

13
SN Ratios for Classified Attributes   290

Part V
Robust Engineering   **311**

14
System Design   313

15
Parameter Design   318

16
Tolerance Design   340

17
Robust Technology Development   352

18
Robust Engineering: A Manager's Perspective   377

19
Implementation Strategies   389

Part VI
Mahalanobis–Taguchi System (MTS)   **395**

20
Mahalanobis–Taguchi System   397

Part VII
Software Testing and Application   **423**

21
Application of Taguchi Methods to Software System
Testing   425

Part VIII
On-Line Quality Engineering   **435**

22
Tolerancing and Quality Level   437

23
Feedback Control Based on Product Characteristics   454

24
Feedback Control of a Process Condition   468

25
Process Diagnosis and Adjustment   474

Part IX
Experimental Regression   **483**

26
Parameter Estimation in Regression Equations   485

Part X
Design of Experiments   **501**

27
Introduction to Design of Experiments   503

28
Fundamentals of Data Analysis   506

29
Introduction to Analysis of Variance   515

30
One-Way Layout   523

31
Decomposition to Components with a Unit Degree of
Freedom   528

32
Two-Way Layout   552

33
Two-Way Layout with Decomposition   563

34
Two-Way Layout with Repetition   573

35
Introduction to Orthogonal Arrays   584

36
Layout of Orthogonal Arrays Using Linear Graphs   597

37
Incomplete Data   609

38
Youden Squares   617

SECTION 2
APPLICATION (CASE STUDIES)

Part I
Robust Engineering: Chemical Applications   **629**

Biochemistry

Case 1
Optimization of Bean Sprouting Conditions by Parameter
Design   631

Case 2
Optimization of Small Algae Production by Parameter
Design   637

Chemical Reaction

Case 3
Optimization of Polymerization Reactions   643

Case 4
Evaluation of Photographic Systems Using a Dynamic
Operating Window   651

Measurement

Case 5
Application of Dynamic Optimization in Ultra Trace
Analysis   659

Case 6
Evaluation of Component Separation Using a Dynamic
Operating Window   666

Case 7
Optimization of a Measuring Method for Granule
Strength   672

Case 8
A Detection Method for Thermoresistant Bacteria   679

Pharmacology

Case 9
Optimization of Model Ointment Prescriptions for In Vitro
Percutaneous Permeation   686

Separation

Case 10
Use of a Dynamic Operating Window for Herbal
Medicine Granulation   695

Case 11
Particle-Size Adjustment in a Fine Grinding Process for a
Developer   705

Part II
Robust Engineering: Electrical Applications   **715**

Circuits

Case 12
Design for Amplifier Stabilization   717

Case 13
Parameter Design of Ceramic Oscillation Circuits   732

Case 14
Evaluation Method of Electric Waveforms by Momentary
Values   735

Case 15
Robust Design for Frequency-Modulation Circuits   741

Electronic Devices

Case 16
Optimization of Blow-off Charge Measurement Systems   746

Case 17
Evaluation of the Generic Function of Film Capacitors   753

Case 18
Parameter Design of Fine-Line Patterning for IC
Fabrication   758

Case 19
Minimizing Variation in Pot Core Transformer Processing   764

Case 20
Optimization of the Back Contact of Power MOSFETs   771

Electrophoto

Case 21
Development of High-Quality Developers for
Electrophotography   780

Case 22
Functional Evaluation of an Electrophotographic Process   788

Part III
Robust Engineering: Mechanical Applications   **793**

Biomechanical

Case 23
Biomechanical Comparison of Flexor Tendon Repairs   795

Machining

Case 24
Optimization of Machining Conditions by Electrical
Power   806

Case 25
Development of Machining Technology for High-
Performance Steel by Transformability   819

Case 26
Transformability of Plastic Injection-Molded Gear   827

Material Design

Case 27
Optimization of a Felt-Resist Paste Formula Used in
Partial Felting   836

Case 28
Development of Friction Material for Automatic
Transmissions   841

Case 29
Parameter Design for a Foundry Process Using Green
Sand   848

Case 30
Development of Functional Material by Plasma Spraying   852

Material Strength

Case 31
Optimization of Two-Piece Gear Brazing Conditions   858

Case 32
Optimization of Resistance Welding Conditions for
Electronic Components   863

Case 33
Tile Manufacturing Using Industrial Waste   869

Measurement

Case 34
Development of an Electrophotographic Toner Charging
Function Measuring System   875

Case 35
Clear Vision by Robust Design   882

Processing

Case 36
Optimization of Adhesion Condition of Resin Board and
Copper Plate   890

Case 37
Optimization of a Wave Soldering Process   895

Case 38
Optimization of Casting Conditions for Camshafts by
Simulation   900

Case 39
Optimization of Photoresist Profile Using Simulation   904

Case 40
Optimization of a Deep-Drawing Process   911

Case 41
Robust Technology Development of an Encapsulation
Process   916

Case 42
Gas-Arc Stud Weld Process Parameter Optimization
Using Robust Design   926

Case 43
Optimization of Molding Conditions of Thick-Walled
Products   940

Case 44
Quality Improvement of an Electrodeposited Process for
Magnet Production   945

Case 45
Optimization of an Electrical Encapsulation Process
through Parameter Design   950

Case 46
Development of Plastic Injection Molding Technology by
Transformability   957

Product Development

Case 47
Stability Design of Shutter Mechanisms of Single-Use
Cameras by Simulation   965

Case 48
Optimization of a Clutch Disk Torsional Damping System
Design   973

Case 49
Direct-Injection Diesel Injector Optimization   984

Case 50
Optimization of Disk Blade Mobile Cutters   1005

Case 51
D-VHS Tape Travel Stability   1011

Case 52
Functionality Evaluation of Spindles   1018

Case 53
Improving Minivan Rear Window Latching   1025

Case 54
Linear Proportional Purge Solenoids   1032

Case 55
Optimization of a Linear Actuator Using Simulation   1050

Case 56
Functionality Evaluation of Articulated Robots   1059

Case 57
New Ultraminiature KMS Tact Switch Optimization   1069

Case 58
Optimization of an Electrical Connector Insulator Contact
Housing   1084

Case 59
Airflow Noise Reduction of Intercoolers   1100

Case 60
Reduction of Boosting Force Variation of Brake Boosters 1106

Case 61
Reduction of Chattering Noise in Series 47 Feeder
Valves   1112

Case 62
Optimal Design for a Small DC Motor   1122

Case 63
Steering System On-Center Robustness   1128

Case 64
Improvement in the Taste of Omelets   1141

Case 65
Wiper System Chatter Reduction   1148

Other

Case 66
Fabrication Line Capacity Planning Using a Robust
Design Dynamic Model   1157

Part IV
Mahalanobis–Taguchi System (MTS)   **1169**

Human Performance

Case 67
Prediction of Programming Ability from a Questionnaire
Using the MTS   1171

Case 68
Technique for the Evaluation of Programming Ability
Based on the MTS   1178

Inspection

Case 69
Application of Mahalanobis Distance for the Automatic
Inspection of Solder Joints   1189

Case 70
Application of the MTS to Thermal Ink Jet Image
Quality Inspection   1196

Case 71
Detector Switch Characterization Using the MTS   1208

Case 72
Exhaust Sensor Output Characterization Using the MTS   1220

Case 73
Defect Detection Using the MTS   1233

Medical Diagnosis

Case 74
Application of Mahalanobis Distance to the
Measurement of Drug Efficacy   1238

Case 75
Use of Mahalanobis Distance in Medical Diagnosis   1244

Case 76
Prediction of Urinary Continence Recovery among
Patients with Brain Disease Using the Mahalanobis
Distance   1258

Case 77
Mahalanobis Distance Application for Health
Examination and Treatment of Missing Data   1267

Case 78
Forecasting Future Health from Existing Medical
Examination Results Using the MTS   1277

Product

Case 79
Character Recognition Using the Mahalanobis Distance   1288

Case 80
Printed Letter Inspection Technique Using the MTS   1293

Part V
Software Testing and Application   **1299**

Algorithms

Case 81
Optimization of a Diesel Engine Software Control
Strategy   1301

Case 82
Optimizing Video Compression   1310

Computer Systems

Case 83
Robust Optimization of a Real-Time Operating System
Using Parameter Design   1324

Software

Case 84
Evaluation of Capability and Error in Programming   1335

Case 85
Evaluation of Programmer Ability in Software Production 1343

Case 86
Robust Testing of Electronic Warfare Systems   1351

Case 87
Streamlining of Debugging Software Using an
Orthogonal Array   1360

Part VI
On-Line Quality Engineering   **1365**

On-Line

Case 88
Application of On-Line Quality Engineering to the
Automobile Manufacturing Process   1367

Case 89
Design of Preventive Maintenance of a Bucket Elevator
through Simultaneous Use of Periodic Maintenance and
Checkup   1376

Case 90
Feedback Control by Quality Characteristics   1383

Case 91
Control of Mechanical Component Parts in a
Manufacturing Process   1389

Case 92
Semiconductor Rectifier Manufacturing by On-Line
Quality Engineering   1395

Part VII
Miscellaneous   **1399**

Miscellaneous

Case 93
Estimation of Working Hours in Software Development   1401

Case 94
Applications of Linear and Nonlinear Regression
Equations for Engineering   1406

SECTION 3
TAGUCHI'S METHODS VERSUS OTHER
QUALITY PHILOSOPHIES

39
Quality Management in Japan   1423

40
Deming and Taguchi's Quality Engineering   1442

41
Enhancing Robust Design with the Aid of TRIZ and
Axiomatic Design   1449

42
Testing and Quality Engineering   1470

43
Total Product Development and Taguchi's Quality
Engineering   1478

44
Role of Taguchi Methods in Design for Six Sigma   1492

APPENDIXES

A
Orthogonal Arrays and Linear Graphs: Tools for Quality
Engineering   1525

B
Equations for On-Line Process Control   1598

C
Orthogonal Arrays and Linear Graphs for Chapter 38   1602

Glossary   **1618**
Bibliography   **1625**
Index   **1629**

Preface

To compete successfully in the global marketplace, organizations must have the ability to produce a variety of high-quality, low-cost products that fully satisfy customers' needs. Tomorrow's industry leaders will be those companies that make fundamental changes in the way they develop technologies and design and produce products.

Dr. Genichi Taguchi's approach to improving quality is currently the most widely used engineering technique in the world, recognized by virtually any engineer, though other quality efforts have gained and lost popularity in the business press. For example, in the last part of the twentieth century, the catch phrase "total quality management" (TQM) put Japan on the global map along with the revolution in automotive quality. Americans tried with less success to adopt TQM into the American business culture in the late 1980s and early 1990s. Under Jack Welch's leadership, GE rediscovered the Six Sigma Quality philosophy in the late 1990s, resulting in corporations all over the world jumping on that bandwagon.

These methods tend to be but a part of a larger, more global quality effort that includes both upstream and downstream processes—espoused by Dr. Taguchi since the 1960s. It is no surprise that the other methods use parts of his philosophy of Robust Engineering in what they try to accomplish, but none is as effective.

Dr. Taguchi strongly feels that an engineering theory has no value unless it is effectively applied and proven to contribute to society. To that end, a Taguchi study group led by Dr. Taguchi has been meeting on the first Saturday of every month in Nagoya, Japan, since 1953. The group has about one hundred members—engineers from various industries—at any given time. Study group members learn from Dr. Taguchi, apply new ideas back on the job, and then assess the effectiveness of these ideas. Through these activities, Dr. Taguchi continues to develop breakthrough approaches. In Tokyo, Japan, a similar study group has been meeting monthly since 1960. Similar activities continue on a much larger scale among the more than two thousand members of the Quality Engineering Society.

For the past fifty years, Dr. Taguchi has been inventing new quality methodologies year after year. Having written more than forty books in the Japanese language, this engineering genius is still contributing profoundly to this highly technical field, but until now, his work and its implications have not been collected in one single source. *Taguchi's Quality Engineering Handbook* corrects that omission. This powerful handbook, we believe, will create an atmosphere of great excitement in the engineering community. It will provide a reference tool for every engineer and quality professional of the twenty-first century. We believe that this book could remain in print for one hundred years or more.

The handbook is divided into three sections. The first, "Theory," consists of ten parts:

- ❏ Genichi Taguchi's Latest Thinking
- ❏ Quality Engineering: A Historical Perspective
- ❏ Quality Loss Function
- ❏ Signal-to-Noise Ratio
- ❏ Robust Engineering
- ❏ Mahalanobis–Taguchi System (MTS)
- ❏ Software Testing and Application
- ❏ On-Line Quality Engineering
- ❏ Experimental Regression
- ❏ Design of Experiments

After presenting the cutting-edge ideas Dr. Taguchi is currently working on in Part I, we provide an overview of the quality movement from earliest times in Japan through to current work in the United States. Then we cover a variety of quality methodologies, describe them in detail, and demonstrate how they can be applied to minimize product development cost, reduce product time-to-market, and increase overall productivity.

Dr. Taguchi has always believed in practicing his philosophy in a real environment. He emphasizes that the most effective way to learn his methodology is to study successful and failed case studies from actual applications. For this reason, in Section 2 of this handbook we are honored to feature ninety-four case studies from all over the world from every type of industry and engineering field, mechanical, electrical, chemical, electronic, etc. Each of the case studies provides a quick study on subject areas that can be applied to many different types of products and processes. It is recommended that readers study as many of these case studies as they can, even if they are not ones they can relate to easily. Thousands of companies have been using Dr. Taguchi's methodologies for the past fifty years. The benefits some have achieved are phenomenal.

Section 3 features an explanation of how Dr. Taguchi's work contrasts with other quality philosophies, such as Six Sigma, Deming, Axiomatic Design, TRIZ, and others. This section shows readers ways to incorporate Taguchi Methods into existing corporate activities.

In short, the primary goals of this handbook are

- ❏ To provide a guidebook for engineers, management, and academia on the Quality Engineering methodology.
- ❏ To provide organizations with proven techniques for becoming more competitive in the global marketplace.
- ❏ To collect Dr. Taguchi's theories and practical applications within the confines of one book so the relationships among the tools will be properly understood.
- ❏ To clear up some common misinterpretations about Taguchi's Quality Engineering philosophy.

❑ To spread the word about Quality Engineering throughout the world, leading to greater success in all fields in the future.

We strongly believe that every engineer and engineering manager in any field, every quality professional, every engineering faculty member, research and development personnel in all types of organizations, every consultant, educator, or student, and all technical libraries will benefit from having this book.

Subir Chowdhury
Yuin Wu
April 14, 2004

Acknowledgments

The authors gratefully acknowledge the efforts of all who assisted in the completion of this handbook:

To all of the contributors and their organizations for sharing their successful case studies.

To the American Supplier Institute, ASI Consulting Group, LLC, and its employees, Board of Directors, and customers for promoting Dr. Taguchi's works over the years.

To our two associate editors, Shin Taguchi of the United States and Hiroshi Yano of Japan, for their dedication and hard work on the manuscript development.

To two of the finest teachers and consultants, Jim Wilkins and Alan Wu, for nonstop passionate preaching of Dr. Taguchi's philosophy.

To our colleagues and friends, Dr. Barry Bebb, Jodi Caldwell, Ki Chang, Bill Eureka, Craig Jensen, John King, Don Mitchell, Gopi Palamaner, Bill Sutherland, Jim Quinlan, and Brad Walker, for effectively promoting Dr. Taguchi's philosophy.

To Emy Facundo for her very hard work on preparing the manuscript.

To Bob Argentieri, our editor at Wiley, for his dedication and continuous support on the development of this handbook.

To Rebecca Taff for her dedicated effort toward refining the manuscript.

Finally, this handbook never would have been possible without the continuous support of our wonderful wives, Kiyo Taguchi, Malini Chowdhury, and Suchuan Wu.

About the Authors

DR. GENICHI TAGUCHI is Executive Director of American Supplier Institute (ASI). Over the past three decades, he has promoted his continually evolving philosophy of quality engineering (called Taguchi Methods™ in the United States) through the Japanese Society for Quality Control, Japanese Standards Association (JSA), Japanese Union of Scientists and Engineers (JUSE), and Central Japan Quality Control Association (CJQCA). In 1982, the American Supplier Institute first introduced Dr. Taguchi and his methods to the United States. Since that time his methods, including techniques for engineering, business, and management, have saved American companies millions of dollars in product development and warranty costs.

Dr. Taguchi has been awarded the coveted Deming Prize on three separate occasions for his contributions to the field of quality engineering. In 1986, he received the Willard F. Rockwell Medal for combining engineering and statistical methods to achieve rapid improvements in cost and quality by optimizing product design and manufacturing processes. He received the Blue Ribbon Award from the Emperor of Japan in 1990 for his contribution to industry and the Shewhart Medal from the American Society for Quality (ASQ) in 1996. In 1997, Dr. Taguchi was elected as an Honorary Member of the ASQ, the highest honor of the ASQ, and is only the third Japanese to be inducted into the Automotive Hall of Fame in Dearborn, Michigan.

DR. SUBIR CHOWDHURY is chairman and CEO of American Supplier Institute (ASI) Consulting Group—the world leader on Six Sigma, Design For Six Sigma and Quality Leadership implementation consulting and training. A respected quality strategist, Dr. Chowdhury's clients include global *Fortune* 100 companies as well as small organizations in both the public and private sectors. Hailed by the *New York Times* as a "leading quality expert," Dr. Chowdhury is the author of eleven books, including international bestseller *The Power of Six Sigma*, which has been translated in more than 20 languages and has sold more than a million copies worldwide. Dr. Chowdhury's *Design for Six Sigma* (DFSS) is the *first* book on the topic and is credited with popularizing DFSS philosophy in the United States and around the globe. His *Management 21C* was selected as the Best Business Book of 1999 by Amazon.com in U.K. and translated into more than ten languages. His most recent book is *Next Generation Business Handbook* (Wiley, September 2004).

Dr. Chowdhury has received numerous international awards for his leadership in quality management and his major contributions to various industries world-wide. In addition to being honored by the Automotive Hall of Fame, the Society of Automotive Engineers awarded him its most prestigious recognition, the *Henry Ford II Distinguished Award for Excellence in Automotive Engineering*. He also received *Honorable U.S. Congressional Recognition* and the Society of Manufacturing Engineers' *Gold Medal*. The American Society for Quality honored him with the first *Philip Crosby Medal* for authoring the most influential business book on Six Sigma. He is also an honorary member of the World Innovation Foundation and International Technology Institute (ITI). In 2004 Dr. Chowdhury was inducted into the Hall of Fame for Engineering, Science and Technology (World Level) and the ITI honored him with its most prestigious *Rockwell Medal for Excellence in Technology*.

Dr. Chowdhury has an undergraduate degree in aerospace engineering from the Indian Institute of Technology (IIT), Kharagpur, India, a graduate degree in industrial management from Central Michigan University (CMU), and an honor-ary doctorate in engineering from the Michigan Technological University. In 2003, in its golden anniversary, IIT featured Dr. Chowdhury as one of its top 15 eminent alumni and CMU awarded him with its Distinguished Alumni Award (which has been bestowed upon only 22 alumni in its 100-plus year history). Most engineering schools and business schools throughout the world include his engineering and management books in undergraduate and graduate programs. Dr. Chowdhury lives with his wife and daughter in Northville, Michigan.

YUIN WU received a B.S. in chemical engineering from Cheng Kung University, taught as a professor at several institutions in Taiwan, and held senior management positions with industrial firms in Taiwan as well. Mr. Wu was a past Executive Di-rector of ASI.

Mr. Wu was active in quality control and improvement. He became acquainted with Dr. Taguchi in 1966 while in Japan on a study sponsored by the Taiwan gov-ernment. Mr. Wu made some of the first English (and Chinese) translations of Dr. Taguchi's works. He is also credited with conducting the first Taguchi Methods experiments in the United States while working with private industry in California.

He worked with the American Supplier Institute (and its predecessor, the Ford Supplier Institute) since 1982, and also provided consultation and training in many diverse Taguchi Methods applications. He was active as a consultant in North America as well as many countries in Europe, South America, and Asia. He was trained for the automotive, computer, and defense companies and also mechani-cal, electrical, chemical, and food industries. He wrote numerous publications on the subject of Taguchi Methods.

Section 1

Theory

Part I

Genichi Taguchi's Latest Thinking

The Second Industrial Revolution and Information Technology

1

1.1. **Elements of Productivity**   5
     Economy and Productivity   5
     Quality and Productivity   7
     Quality and Taxation   8
1.2. **Developing Productivity**   10
     Corporate Organization and Two Types of Quality   10
     Product Quality Design for a Financial System   10
     What Is Productivity?   11
     Product Planning and Production Engineering   12
1.3. **Risks to Quality**   13
     Risk Management   13
     Countermeasures against Risks   13
1.4. **Management in Manufacturing**   16
     Ford's Strategy   16
     Determining Tolerance   16
     Evaluation of Quality Level and Consumer Loss   17
     Balance of Cost and Quality Level   18
1.5. **Quality Assurance Department**   23
     Responsibility of Each Department for Quality   23
     Responsibilities of the Quality Assurance Department   23
     References   24

1.1. Elements of Productivity

In this chapter we discuss the second Industrial Revolution, led by information technology and management's role regarding quality.

To lead a better life we need various kinds of products and services, which are produced by human work and shared by all people. Although products are produced by manufacturers in most cases, one that is difficult to produce is land. Services include quite a few things (e.g., raising babies, nursing bedridden people) that we cannot produce efficiently because they cannot be automated by current

Economy and Productivity

5

technology and need direct human work. On the other hand, productivity of tele-communication, transportation, or banking services can be improved through equipment innovation or automation.

To produce more products and services with the same number of or fewer workers or to improve their quality results in *productivity improvement*. Two hundred years ago, 90% of the U.S. population were farmers. At present, fewer than 2% of Americans are farmers; however, they can produce crops for twice as many people as the entire population of the United States. This implies that their productivity has improved about 110-fold. Yet even if agricultural productivity increased, if 88% among the remaining 90% of U.S. citizens do not produce anything, overall national productivity would not improve. U.S. managers have offered other jobs, such as building railways, making automobiles, and producing telephones or television sets, to many people no longer employed in the agricultural field. In other words, overall productivity has been improved by the production of new products and services. The total amount of products manufactured and services performed is called the *gross national product* (GNP). However, the GNP does not include household services such as raising children, preparing meals, doing laundry, or cleaning. In addition, land and environment are not included in the GNP. This means that the GNP does not reflect real productivity, so the discussion that follows deals only with limited products and services.

Now imagine a case in which a certain company produces the same amount with half the workforce by improving work productivity. If the company keeps half of its employees idle, paying them the same amount in wages, its overall productivity does not increase. When all employees produce twice as much as before, and society needs the products, the company is said to have doubled its productivity. Otherwise, it causes other companies' productivity to deteriorate. In sum, society's overall productivity does not improve. This holds true when one country exports products to another country, causing the loss of many jobs in that country. If there is no need for quantity, increased productivity because of quantity leads to un-employment. Therefore, to improve productivity in the true sense, we should provide jobs to people unemployed as a result of productivity improvement. As companies develop new products, new jobs are created. While productivity improvement such as producing double the products with the same workforce results in partial realization, new jobs, producing new products and services, must also be available to workers newly unemployed. An entity focusing mainly on creating new products and services is a research and development (R&D) department.

We could also have another type of productivity improvement: the creation of twice as valuable a product using the same number of workers. In this case, if our costs are doubled and we sell the products at double the former price, the company's productivity increases twofold. If every company could achieve this goal, all people's incomes would double, they would buy doubly priced products, and as a consequence, the overall standard of living would double. On the other hand, if the company produced products of double value at the same labor cost and sold the products at the same prices, consumers would enjoy twice as high a standard of living in terms of these products. In other words, if all companies offered products of double value at the same price by improving their productivity with the same number of employees, our living standard would double, even if our wages and salaries remained the same. Although, in this case, we did not need to develop a totally new product, we needed to develop production technology that would allow us to produce products of twice the value using the same workforce.

Therefore, wages and living standards are not important from a national point of view. What is important is that we increase the number of products by improving productivity; that once we achieve it, we develop new products, create new industries, and offer jobs to unemployed people; and that once we enhance the quality of products and services, we sell them at the same prices as before. In fact, the price of a color television set is decreasing more than the price of a black-and-white TV set, so although the former is superior in quality, the cost of large-scale integrated circuits remains the same in the end, despite the high degree of integration. These are typical examples of developed technologies that have enabled us to produce higher-quality products at the same cost.

Although Japan's standard of living is lower than that of the United States, Japan's income is higher. An exchange rate is determined by competitive products that can be exported. On the one hand, Japan is highly productive because of several high-level technologies. On the other hand, Japan's productivity in terms of other products and services is low. On balance, Japan's overall living standard remains low. This is because a standard of living depends totally on productivity and allocation of products and services. To raise Japan's overall standard of living, quantitative productivity in agriculture, retail trade, and the service industry should be increased. Otherwise, qualitative productivity should be improved. As a result, Japan should invest in R&D to improve productivity. This is regarded as the most essential role of top management in society. The objective of quality engineering is to enhance the productivity of R&D itself.

As noted earlier, there are two different types of productivity improvement, one a quantitative approach, the other qualitative. As for the latter, the word *qualitative* can be replaced by *grade*, although in this section we focus on quality rather than grade. In the case of an automobile, quality indicates losses that a product imposes on society, such as fuel inefficiency, noise, vibration, defects, or environmental pollution. An auto engine at present has only 25% fuel efficiency; that is, to gain the horsepower claimed, four times as much gasoline is consumed as would seem to be required. Now, if we double fuel efficiency, we can expect to halve noise, vibration, and environmental pollution as well as fuel consumed. It is obvious that the excess fuel consumption is converted into vibration, noise, and exhaust, thereby fueling environmental pollution. Also, it can be theorized that if the fuel efficiency of an engine doubles, Japan's trade deficit with the United States would disappear and global environmental pollution would be reduced by 50%. In contrast, sales in oil-producing nations and petroleum companies would be decreased by half. Even if these companies doubled the price of gasoline to maintain the same number of workers, the global standard of living would be improved, owing to decreased pollution because fuel efficiency would not have changed. Therefore, instead of reducing economically unimportant defects by 0.1% in the production process, engineers should focus on more substantial quality improvement, such as in engine efficiency. That is, since more than 75% of gasoline fuel is wasted, thus incurring an enormous economic loss, to improve fuel efficiency is desirable.

Indeed, petroleum companies might cause joblessness because they cannot maintain their doubled prices; in actuality, however, there are quite a few people who wish to purchase a car. In the United States, many families need a car for each adult. Therefore, if the price of a car and that of fuel decline by half, the demand for cars would increase, and eventually, fuel consumption would decrease

Quality and Productivity

less than expected. In quality engineering, improvements in functionality that set up fuel input as a signal and output as mechanical energy are important because we believe that such functionality would mitigate any type of environmental pollution. Thus, quality engineering does not deal basically with improvements using data such as quality characteristics expressed in consumer interests such as fuel efficiency, vibration, pollution, or defects. In short, we in quality engineering define an essential function as a *generic function*. Thus, quality engineering facilitates quality improvement for consumers using a generic function.

Take the function of an electric lamp, whose input for lighting is electric power and whose output is light. Because its efficiency is less than 1%, there is much that might be improved. Suppose that we can prolong the lamp's life threefold, keeping its power consumption the same: Sales of lamp manufacturers would possibly decrease to one-third. Some people would be reluctant to do research that would lead to such a longer-lived lamp. Yet this would not hold true for companies with a smaller market share because they could expand their market share and increase their sales, whereas companies with a larger market share would lose sales. That is, quality improvement could lead to unemployment. Although both qualitative and quantitative improvement could lead to jobless people, qualitative improvement often mitigates environmental pollution or consumption of resources. If a lamp lasts three times as long as before, only one third of material resources are consumed. In that case, quality improvement is regarded as playing a more important role in living standards than does quantitative improvement. Nevertheless, both bring about unemployment.

Other examples: If car breakdowns decrease, we will need fewer repair shops, and if a tire's life is extended, we will need fewer resources. Accordingly, tire manufacturers' sales will also decline, and as a result, manufacturers will need to offer different jobs to the unemployed or to reduce wages by cutting working hours. Since many people do not usually accept reduced wages, employers should provide new jobs aimed at improving productivity. Then not only would top management have a more positive attitude toward investing in new product development, but developing the capability of R&D engineers would become crucial. As discussed above, a number of people are in need of products and services to subsist. Some of them wish to have higher-quality housing, and some are not satisfied with the current environment. Handicapped or bedridden persons seek services such as nursing whose demand is regarded as unlimited. Technological development to offer such products and services at lower prices is anticipated. Of course, medical treatment to rehabilitate such patients is also crucial, together with preventive measures. From a technical point of view, these fields are so difficult that many more resources should be invested in their research and development. If we do not see any prospect of progress in these fields, additional public investment and national research organizations will be required.

Quality and Taxation   In nations that have a wider variety of products than that in developing countries, qualitative improvement is preferable to quantitative from a social perspective. This is because the qualitative approach tends to reduce pollution and resource consumption. I believe that taxation is more desirable than pollution regulation to facilitate quality improvement. This would include a consumption tax for natural resources (including water) and a pollution tax. The resource consumption tax would stimulate recycling.

For lands not leading to higher productivity (as with housing), proprietors should be forced to pay taxes for their occupancy. In contrast, for public lands or places where many people can visit (e.g., department stores, amusement parks, ski areas), no or only a small amount of tax is needed. This will help to make lands public and supply housing. Occupancy should be allowed; however, equivalent prices should be paid.

Next, except for special cases such as exhaust or noise pollution, environmental pollution should not be regulated but be taxed. Some examples of currently taxed items are cigarettes and alcohol. Despite their harm to human health, they can be dealt with not by prohibition but by taxation.

In sum, even if certain issues are detrimental to society, we should not regulate them but impose taxes on their use. The underlying idea is that to whatever degree we can broaden individual freedom and take action by our free will, a higher level of culture is represented. That is, a real level of production, or real productivity, represents a large sum of individual freedoms. Freedom contains not only products and services or an affluent lifestyle where we can obtain goods freely, but also health and a free life not restrained by others. Indeed, freedom from authority is important; however, that each person not interfere with others is more essential. If we have no other choice but to restrain others' freedom, we need to pay a sufficient price as well as to obtain their consent. For example, the reason that we can receive meals or services from those with whom we are unfamiliar is that we pay a price equivalent to their worth.

So the sum of individual freedoms is regarded as important, even though it is quite difficult to measure. In this regard, the social value of a person should be evaluated by how much freedom he or she can offer. To support this, the government, as an authoritative organization, should not restrain but should protect individual freedom. We might safely say that the government is obligated to regulate those who transgress others' freedom. But here we discuss only productivity improvement that can be calculated economically.

The first Industrial Revolution relieved us from physical labor by mechanizing machining operations within manufacturing processes. After this mechanization, the major jobs of operators were production control, such as preparation of raw material, transportation of in-process products, or machine setup; and quality control, such as machining diagnosis and control and inspection. Because automation requires proper equipment, investment in automation must be augmented.

To reduce the investment in manufacturing processes, process stability needs to be improved in R&D functions. More specifically, speeding-up the production or development of products that have high value, large-scale accumulation, or complicated functions should be studied. All of them require technological development in laboratories to enhance machining performance. In particular, production speedup is effective for reducing the total cost of indirect labor, except in a manufacturing department. As we discuss later, a manufacturing department is responsible for only a small percentage of total complaints in the market because its principal activity is cost reduction. By improving manufacturing cycle time, it contributes considerably to company-wide cost reduction. To achieve this, we need to measure process stability. Primarily, we detail the variability of product quality.

1.2. Developing Productivity

Corporate Organization and Two Types of Quality

To allocate managers and leaders in an organization is one of the primary jobs of top management. To assess the business performance of each department and each person is also one of top management's roles. The performance of top management is judged by a balance sheet. In turn, top managements needs to assess the productivity of managers of each department. Among all departments, an R&D department may be the most difficult but most important to evaluate, as an R&D department should always improve organizational productivity.

Top management is charged with planning business strategies, determining the types of products to be produced, and allocating managers and budgets of engineering departments that design products (R&D and design departments). They must also take responsibility for the results of these departments in business competition in accordance with the balance sheet or profit and loss statement.

In corporate management, product development is considered a major tactic for profitability. It consists of the following two aspects:

1. *Product quality:* what consumers desire (e.g., functions or appearance)
2. *Engineering quality:* what consumers do not want (e.g., functional variability, running cost, pollution)

Engineers, especially those in product design, are required to design in such a way as to improve engineering quality as much as possible in parallel with product quality. If we compete with other companies, our level of quality is expected to exceed those of the other companies. This type of evaluation by management, dubbed *benchmarking*, consists of comparing one's own company with competitors on the two types of quality noted above. That is, an engineering department (in charge of tactics) assesses both product quality and engineering quality as well as production cost against others' results. In addition, the evaluation of production cost is one of the most essential tasks of a production engineering department.

In corporate management, a battle often occurs between manufacturing and sales, both of which are front-line departments. For evaluation of a manufacturing department, see Section 1.4.

Product Quality Design for a Financial System

An engineer is charged with responsibility to design a product with an objective function. For example, a bank's objective function consists of supplying a vast amount of money collected from both capitalists and the public. In the twenty-first century, when electronic money is supposed to prevail, the design of such a financial system does not require gigantic buildings and a large number of bank clerks. In designing a system to gather money, what is most important is whether a moneylender trusts a borrower. When a bank lends money to a company, whether the company continues to pay the interest on the loan is of significance. Such considerations are tied closely to credit.

Checks and credit cards have long been used in Europe and the United States. During my days at Princeton University as a visiting professor, shortly after I opened a bank account, four checkbooks were sent to me. They seemed to say, "You can use freely." However, I did not have any money in the account. Even when a payment due date arrived, I did not need to deposit any money. This was equivalent to a nonmortgage loan. While the bank paid for me, I would be charged interest on the money without a mortgage.

Credit means that money can be lent without security and that credit limits such as \$50,000 or \$10,000 are determined based on each person's data. How we judge individual credibility is a technological issue. In designing a system, how we predict and prevent functional risk is the most critical design concern. This is about how we offer credit to each person in a society. How to collect credit information and how much credit to set for a given person constitute the most important details in the era of information. In the field of quality engineering, an approach to minimizing risk of deviation, not from function design per se, but from functionality or an ideal state is called *functionality design*. In the twenty-first century, this quality engineering method is being applied to both hardware and software design.

An organization collecting money from a number of people and corporations and lending to those in need of money is a bank. Its business is conducted in the world of information or credit. Such a world of reliance on information (from plastic money to electronic money) has already begun. The question for the customer is whether a bank pays higher interest than others do, or runs a rational business that makes a sufficient profit. What is important is whether a bank will be able to survive in the future because its information system is sufficiently rationalized and its borrowers are properly assessed, even if it keeps loan interest low to borrowers, and conversely, pays somewhat higher interest to money providers. After all, depositors evaluate a bank. A credible bank that can offer higher interest is regarded as a good and highly productive bank. Therefore, we consider that a main task of a bank's R&D department is to rationalize the banking business and lead to a system design method through rationalization of functional evaluation in quality engineering.

Unfortunately, in Japan, we still lack research on automated systems to make a proper decision instantaneously. Because a computer itself cannot take any responsibility, functional evaluation of a system based on software plays a key role in establishing a company. After a company is established, daily routine management of software (update and improvement of the database) becomes more important. Globalization of information transactions is progressing. A single information center will soon cover all the world and reduce costs drastically. Soon, huge bank buildings with many clerks will not be required.

What Is Productivity?

We define productivity as follows: *Total social productivity* (GDP) is the sum of individual freedoms. Freedom includes situations where we can obtain what we want freely, that is, without restraint of individual freedom by others. As discussed earlier, when electronic money systems are designed, numerous people become unemployed because many bank clerks are no longer needed. Once only 100 bank clerks can complete a job that has required 10,000 people, the bank's productivity is regarded as improved 100-fold. Nevertheless, unless the 9900 redundant people produce something new, total social productivity does not increase.

To increase productivity (including selling a higher-quality product at the same price) requires technological research, and engineers designing a productive system constitute an R&D laboratory. An increase in productivity is irreversible. Of course, a bank does not itself need to offer new jobs to people unemployed due to improved productivity. It is widely said that one of the new key industries will be the leisure industry, including travel, which has no limit and could include space travel.

The government is responsible for incrementing GDP at a certain rate. In fact, a nominal increase in GDP, for example 3%, is quite an easy goal to reach because we can attain it by hiking a nominal amount of wages by 3% on a yearly basis. The question is: How much is equivalent to 3% of GDP? This does not indicate a 3% improvement in the standard of living (real GDP), because this is a nominal figure. In elementary schools, where productivity is difficult to improve, we could decrease the number of students in a class, thus employing more teachers and reducing the number of jobless teachers. Raising salaries by 3%, reducing the number of students in a class, or preparing a larger number of academic courses causes more than a 3% increase in cost in the form of additional employment. This is an improvement in the overall social standard of living. Rather than keeping unemployed people doing nothing and paying an unemployment allowance, we should pay unemployed people a certain amount of money to let them work, such as reducing the number of students in a class or preparing a larger number of academic courses. This is an important action to take when attempting to solve the unemployment issue. We will not discuss here whether the necessary expenses would be borne by government or a part of them shared by families of students. Instead, we should determine the improved living standard over a 20-year school life. The best chance to test this is, when a number of classrooms are empty because the number of students is declining.

The debate over whether an enormous expenditure is needed to improve the standard of living of the elderly has heated up. A key point at issue is whether older people can continue to lead healthy lives without a loss of mental acuity. Developing a medicine to prevent such disability is one of the most significant technologies today. On the other hand, organizing a group of people to talk with the elderly before they show signs of senility is more essential than inventing a robot to assist and talk with them.

What is important is to develop practical means of achieving the goal and evaluating it. We should create a specialized laboratory. In fact, there are quite a few people who age without becoming senile who could be studied. Using the Mahalanobis–Taguchi system (MTS), regarded as a key method in quality engineering, we should study how they maintain their health. This also holds true for "medical checkups" for corporate management. We discuss later how a company comes to be considered healthy.

Some corporations run an active and sound business at all times. How we evaluate and predict sound and unsound management is a major issue in business strategy, which is a totally different evaluation system from that of financial accounting, which assesses only results.

Product Planning and Production Engineering

Assuming that each consumer's taste and standard of living (disposable income) is different for both hardware and software that he or she wishes to buy, we plan a new product. Toyota is said to be able to deliver a car to a customer within 20 days of receiving an order. A certain watch manufacturer is said to respond to 30 billion variations of dial plate type, color, and size and to offer one at the price of $75 to $125 within a short period of time. An engineer required to design a short-cycle-time production process for only one variation or one product is involved in a *flexible manufacturing system* (FMS), used to produce high-mix, low-volume products. This field belongs to production engineering, whose main interest is production speed in FMS.

Production engineering thus focuses on improving production speed to reduce the cost of indirect departments, including sales, administration, and development. To achieve the goal by taking advantage of quality engineering, we should stabilize the production process and drastically increase production speed. Because a manufacturing department can improve market quality by a few percent only, it does not need to take that responsibility. Its most important task is to manufacture efficiently products planned and designed by other departments.

1.3. Risks to Quality

In Japan, we have recently had some unexpected events, such as a missile launched over our archipelago, a large-scale earthquake, and a prime minister's sudden resignation due to illness. The media splashed articles about our inappropriate preparation for these risks all over the front pages. In quality engineering we call such risks either *signals* or *noises*. To make preparations in anticipation of extraordinary events is called *strategic planning* in the field of management. One example is an air force that prepares sufficient weapons against enemy attack. The best possible result is that the enemy hesitates to attack for fear of such weapons.

Risk Management

An order to an army to "Get the mountain," meaning to occupy the mountain efficiently, is not strategy but *tactics*. What officers and soldiers on the front line use in battle is not strategy but tactics. What investments in various fields require, especially those in R&D, is *strategy*. Strategy should include generic techniques and advanced technologies that are useful in many fields. Quality engineering in R&D aims at designing robustness (sturdiness, functionality).

On the other hand, quality engineering recommends that we evaluate uncountable noises in the market with only two noise factors. Because market noises are generated by users and are due to their conditions of use, the effects evaluated in their study would be minimal or nonexisting. Take as an example an earthquake-proof building. Being "earthquake-proof" does not mean that the building will not break down at all; it means that the effects of an earthquake will be minimized. Therefore, we do not assess a building using the point on the seismic intensity scale at which it will collapse. Using the signal-to-noise (SN) ratio, we evaluate its robustness to noises at a seismic intensity scale of about 4, for example. In addition, as a countermeasure for human life, earthquake prediction is important as well as earthquake-proof and safety studies. This is because our current houses may not be sufficiently earthquake-proof. Further, a robust house is not economical in the face of an enormous earthquake.

We usually have the following noises:

1. Noises due to erroneous or careless use

2. Noises due to the environment

3. Intentional noises such as jamming radio waves

Countermeasures against Risks

HUMAN ERRORS

Now let's look at noises from number 1 in the list above from the standpoint of quality engineering. Among common countermeasures against such noises are the training of users to head off misuse and the prevention of subsequent loss and damage by, for example, the design of easy-to-use products. In Europe and the United States, the term, *user friendly* is often used for designs whose goal is to prevent misuse of software or medical errors.

Of course, all error cannot be designed out of a product. For example, there are a vast number of automobile accidents every year because of mistakes in driving. Since human errors are inevitable, it is essential to design sensors and alarms to let us know our mistakes or to design a system to avoid a car accident automatically. In developing an integrated sensing system that can judge as human beings do, the MTS process in quality engineering may be instrumental. In other examples, such as handling radioactive substances, human errors cannot be prevented completely. And whereas the incidence of fire per person in Japan is one-third that in the United States, the loss incurred by fire is said to 100 times as much. The reason is that there are no effective measures against fire for such household structures as shoji screens, fusuma sliding doors, and curtains, whereas we are strict with regard to the safety of automobile carpets.

If certain human errors do not lead to such important results as sustaining human life or extremely valuable property, we do not need to take technical countermeasures. For example, if we were to drop a piece of porcelain on the floor and break it, we tend to discipline ourselves so as not to repeat the mistake. On the other hand, risk management handles noises that jeopardize human life, important property, or national treasures. In terms of hardware failures, there are some measures that can be used, such as redundant systems, daily routine checkups, or preventive maintenance. Such rational design is called *on-line quality engineering* [1, Chaps. 11 and 12].

ENVIRONMENTAL NOISES

There are two types of environment: natural and artificial. The natural environment includes earthquakes and typhoons. From an economic point of view, we should not design buildings that can withstand any type of natural disaster. For an earthquake, for which point on the seismic intensity scale we design a building is determined by a standard in tolerance design. If we design the robustness of a building using the quality engineering technique, we select a certain seismic intensity, such as 4, and study it to minimize the deformation of the building. However, this does not mean that we design a building that is unbreakable even in a large-scale earthquake.

To mitigate the effects of an earthquake or typhoon on human life, we need to forecast such events. Instead of relying on cause-and-effect or regression relationships, we should focus on prediction by pattern recognition. This technique, integrating multidimensional information obtained to date, creates Mahalanobis space (see Section 4.7) using only time-based data with a seismic intensity scale below 0.5, at which no earthquake happens. The Mahalanobis distance becomes 1, on average, in the space. Therefore, distance D generated in the Mahalanobis space remains within the approximate range 1 ± 1. We assume that the Mahalanobis space exists as unit space only if there is no earthquake. We wish to see how the Mahalanobis distance changes in accordance with the SN ratio (forecasting accuracy) proportional to the seismic intensity after we calculate a formal equation of the distance after an earthquake. If the Mahalanobis distance becomes large enough and is sufficiently proportional to its seismic intensity, it is possible to predict an earthquake using seismic data obtained before the earthquake occurs.

The same technique holds for problems of the elderly. In actuality, there are quite a few 80-year-olds who are still healthy and alert. We would collect information from their youth, such as how many cigarettes they smoked. What sorts of information should be gathered is a matter of the design of the information system. For n different-aged persons belonging to a unit space, we create the Mahalanobis space for their information by collecting either quantitative or qualitative data, such as professions. This Mahalanobis space is a unit space. For this information we calculate a distance for a single person who is senile or cannot lead a normal life. If the distance becomes great and at the same time matches the degree of how senile or bedridden a person is, we may be able to forecast and change the futures of some elderly people.

For the most part, some items in the list are not helpful. MTS can also play a significant role in improving prediction accuracy using an orthogonal array, and the SN ratio may be useful in earthquake forecasting and senility in the elderly. To learn more about MTS, see Chapter 21 of this book and other books specializing in MTS.

CRIMINAL NOISES

Many social systems focus on crimes committed by human beings. Recently, in the world of software engineering, a number of problems have been brought about by hackers. Toll collection systems for public telephones, for example, involve numerous problems. Especially for postpaid phones, only 30% of total revenue was collected by the Nippon Telegraph and Telephone Public Corporation. Eventually, the company modified its system to a prepaid basis. Before you call, you insert a coin. If your call is not connected, the coin is returned after the phone is hung up. Dishonest people put tissue paper in coin returns to block returning coins because many users tend to leave a phone without receiving their change. Because phone designers had made the coin return so small that a coin could barely drop through, change collectors could fill the slot with tissue paper using a steel wire and then burn the paper away with a hot wire and take the coins. To tackle this crime, designers added an alarm that buzzes when coins are removed from a phone; but change collectors decoy people in charge by intentionally setting off alarms at some places, in the meantime stealing coins from other phones.

A good design would predict crimes and develop ways to know what criminals are doing. Although the prepaid card system at first kept people from not paying, bogus cards soon began to proliferate. Then it became necessary to deal with counterfeiting of coins, bills, and cards. Another problem is that of hackers, who have begun to cause severe problems on the Internet. These crimes can be seen as intentional noises made by malicious people. Education and laws are prepared to prevent them and the police are activated to punish them. Improvement in people's living standard leads to the prevention of many crimes but cannot eliminate them.

No noises are larger than the war that derives from the fact that a national policy is free. To prevent this disturbance, we need international laws with the backing of the United Nations, and to eradicate the noise, we need United Nations' forces. At the same time, the mass media should keep check on UN activities to control the highest-ranked authority. Although the prevention of wars around the world is not an objective of quality engineering, noises that accompany businesses should be handled by quality engineering, and MTS can be helpful by designing ways to detect counterfeit coins, for example, and to check the credibility of borrowers. Quality engineering can and should be applied to these fields and many others, some examples of which are given below.

1.4. Management in Manufacturing

Ford's Strategy   The first process to which Ford applied the quality engineering method was not product design but the daily routine activity of quality control (i.e., on-line quality engineering). We reproduce below a part of a research paper by Willie Moore.

Quality Engineering at Ford

Over the last five years, Ford's awareness of quality has become one of the newest and most advanced examples. To date, they have come to recognize that continuous improvement of products and services in response to customers' expectation is the only way to prosper their business and allocate proper dividends to stockholders. In the declaration about their corporate mission and guiding principle, they are aware that human resources, products, and profits are fundamentals for success and quality of their products and services is closely related to customer satisfaction. However, these ideas are not brand-new. If so, why is it considered that Ford's awareness of quality is one of the newest and advanced? The reason is that Ford has arrived at the new understanding after reconsidering the background of these ideas. In addition, they comprehend the following four simple assertions by Dr. Taguchi:

1. Cost is the most important element for any product.
2. Cost cannot be reduced without any influences on quality.
3. Quality can be improved without cost increase. This can be achieved by the utilization of the interactions with noise.
4. Cost can be reduced through quality improvement.

Historically, the United States has developed many quality targets, for example, zero-defect movement, conformity with use, and quality standards. Although these targets include a specific definition of quality or philosophy, practical ways of training to attain defined quality targets have not been formulated and developed. Currently, Ford has a philosophy, methods, and technical means to satisfy customers. Among technical means are methods to determine tolerances and economic evaluation of quality levels. Assertion 4 above is a way of reducing cost after improving the SN ratio in production processes. Since some Japanese companies cling too much to the idea that quality is the first priority, they are losing their competitive edge, due to higher prices. Quality and cost should be well balanced. The word *quality* as discussed here means market losses related to defects, pollution, and lives in the market. A procedure for determining tolerance and on- and off-line quality engineering are explained later.

Determining   Tolerance   Quality and cost are balanced by the design of tolerance in the product design process. The procedure is prescribed in JIS Z-8403 [2] as a part of national standards. JIS Z-8404 is applied not only to Ford but also to other European companies. Balancing the cost of quality involves how to determine targets and tolerances at shipping.

Tolerance is determined after we classify three quality characteristics:

1. *Nominal-the-best characteristic:* a characteristic that incurs poorer quality when it falls below or exceeds its target value m (e.g., dimension, electric current). Its tolerance Δ, where a standard is $m \pm \Delta$, is calculated as follows:

$$\Delta = \sqrt{\frac{A}{A_0}} \, \Delta_0 \tag{1.1}$$

where A_0 is the economic loss when a product or service does not function in the marketplace, A the manufacturing loss when a product or service does not meet the shipping standard, and Δ_0 the functional limit. Above $m + \Delta_0$ or below $m - \Delta_0$, problems occur.

2. *Smaller-the-better characteristic:* a characteristic that should be smaller (e.g., detrimental ingredient, audible noise). Its tolerance Δ is calculated as follows:

$$\Delta = \sqrt{\frac{A}{A_0}} \, \Delta_0 \tag{1.2}$$

3. *Larger-the-better characteristic:* a characteristic that should be larger (e.g., strength). Its tolerance Δ is calculated as follows:

$$\Delta = \sqrt{\frac{A_0}{A}} \, \Delta_0 \tag{1.3}$$

Since a tolerance value is set by the designers of a product, quite often production engineers are not informed of how tolerance has been determined. In this chapter we discuss the quality level after a tolerance value has been established.

Evaluation of Quality Level and Consumer Loss

A manufacturing department is responsible for producing specified products routinely in given processes. Production engineers are in charge of the design of production processes. On the other hand, hardware, such as machines or devices, targets, and control limits of quality characteristics to be controlled in each process, and process conditions, are given to operators in a manufacturing department as technical or operation standards. Indeed, the operators accept these standards; however, actual process control in accordance with a change in process conditions tends to be left to the operators, because this control task is regarded as a calibration or adjustment. Production cost is divided up as follows:

$$\text{production cost} = \text{material cost} + \text{process cost} + \text{control cost} + \text{pollution cost} \tag{1.4}$$

Moreover, control cost is split up into two costs: production control costs and quality control costs. A product design department deals with all cost items on the right side of equation (1.4). A production engineering department is charged with the design of production processes (selection of machines and devices and setup of running conditions) to produce initially designed specifications (specifications at shipping) as equitably and quickly as possible. This department's responsibility covers a sum of the second to fourth terms on the right-hand side of (1.4). Its particularly important task is to speed up production and attempt to reduce total cost, including labor cost of indirect employees after improving the stability of production processes. Cost should be regarded not only as production cost but as companywide cost, which is several times that of production cost as expressed in (1.4).

Since product designers are involved in all cost items, they are responsible not for considering process capability but for designing product reliability in such a way that a product functions sufficiently over its life span (how many years a product endures) under various environmental conditions. In designing a product, the *parameter design method*, which is a way of stabilizing product function, helps to broaden a product's tolerances so that it can be manufactured easily. For example, to improve stability twofold means that a target characteristic of a product never changes, even if all possible factors of its variability double. A production engineering department needs to design production processes to reduce total corporate cost as much as possible as well as to satisfy the initial specifications given by a design department. This means that the design and production processes are to be studied until the variability of characteristics of the product produced actually match the allowable total cost, including the labor cost of indirect employees.

No matter how stabilized production processes are, if we do not control them, many defective products are produced in the end. Moreover, we should design stable processes, speed up production speed, and reduce production cost. Speedup of production usually leads to increased variability. In this case, management should play the role of building up a system so that production cost and loss due to quality variability are balanced automatically.

In production processes, the variability of objective characteristics should be restrained at an appropriate level by process control (feedback control of quality and process conditions). This is equivalent to balancing cost in equation (1.4) and loss due to quality or inventory. If cost is several times as important as loss due to quality or increased inventory, cost should be calculated as being several times as great as actual cost. According to Professor Tribus, former director of the Center for Advanced Engineering Studies at MIT, Xerox previously counted the price of a product as four times that of its unit manufacturing cost (UMC). Since the UMC does not include the cost of development, sales, and administration, if the company did not sell a product for four times the UMC, it would not make a profit.

Top management needs to determine a standard for how many times as much as actual cost the quality level should be balanced. Offering such a standard and an economic evaluation of quality level is regarded as a manufacturing strategy and is management's responsibility.

Balance of Cost and Quality Level

Since price basically gives a customer the first loss, production cost can be considered more important than quality. In this sense, balance of cost and quality is regarded as a balance of price and quality. Market quality as discussed here includes the following items mentioned earlier:

1. Operating cost
2. Loss due to functional variability
3. Loss due to evil effects

The sum of 1, 2, and 3 is the focus of this chapter. Items 1 and 3 are almost always determined by design. Item 1 is normally evaluated under conditions of standard use. For items 2 and 3, we evaluate their loss functions using an average sum of squared deviations from ideal values. However, for only an initial specification (at the point of shipping) in daily manufacturing, we assess economic loss due to items 2 and 3, or the quality level as defined in this book, using the average sum of squared deviations from the target. In this case, by setting loss when an initial characteristic of a product falls below a standard to A dollars, its squared deviation from the target value to σ^2, and the characteristic's tolerance to Δ, we evaluate its quality level by the following equation:

$$L = \frac{A}{\Delta^2} \sigma^2 \qquad \text{(for the nominal-the-best or smaller-the-better characteristic)}$$

(1.5)

Then, σ^2 is calculated as follows, where n data are $y_1, y_2, \cdots, y_n$:

$$\sigma^2 = \frac{1}{n} [(y_1 - m)^2 + (y_2 - m)^2 + \cdots + (y_n - m)^2] \qquad (1.6)$$

$$\sigma^2 = \frac{1}{n} (y_1^2 + y_2^2 + \cdots + y_n^2) \qquad (1.7)$$

Equation (1.6) is used to calculate σ^2 for the nominal-the-best characteristic, and equation (1.7) for the smaller-the-better characteristic.

$$L = A\Delta^2\sigma^2 \qquad \text{(for the larger-the-better characteristic)} \qquad (1.8)$$

Now

$$\sigma^2 = \frac{1}{n} \left(\frac{1}{y_1^2} + \frac{1}{y_2^2} + \cdots + \frac{1}{y_n^2} \right) \qquad (1.9)$$

We evaluate daily activity in manufacturing as well as cost by clarifying the evaluation method of quality level used by management to balance quality and cost.

QUALITY LEVEL: NOMINAL-THE-BEST CHARACTERISTIC
A manufacturing department evaluates quality level based on tolerances as clarified in specifications or drawings. Instead of a defect rate or approval rate, we should assess quality using loss functions. For the nominal-the-best characteristic, equations (1.5) and (1.6) are used.

❏ Example

A certain plate glass maker whose shipping price is $3 has a dimensional tolerance of 2.0 mm. Data for differences between measurements and target values (m's) regarding a product shipped from a certain factory are as follows:

0.3	0.6	−0.5	−0.2	0.0	1.0	1.2	0.8	−0.6	0.9
0.0	0.2	0.8	1.1	−0.5	−0.2	0.0	0.3	0.8	1.3

Next we calculate the quality level of this product. To compute loss due to variability, after calculating an average sum of squared deviations from the target, mean-squared error σ^2, we substitute it into the loss function. From now on, we call mean-squared error σ^2 *variance*.

$$\sigma^2 = \tfrac{1}{20} (0.3^2 + 0.6^2 + \cdots + 1.3^2) = 0.4795 < mm^2 \qquad (1.10)$$

Plugging this into equation (1.5) for the loss function gives us,

$$L = \frac{A}{\Delta^2} \sigma^2 = \frac{300}{2^2} (0.4795) = 36 \text{ cents} \qquad (1.11)$$

As an average value of plate glass, 0.365 mm is regarded as somewhat large. To check this, we can create an analysis of variance (ANOVA) table.

$$S_T = 0.3^2 + 0.6^2 + \cdots + 1.3^2 = 9.59 \qquad (f = 20) \quad (1.12)$$

$$S_m = \frac{(0.3 + 0.6 + \cdots + 1.3)^2}{20} = \frac{7.3^2}{20} = 2.66 \qquad (f = 1) \quad (1.13)$$

$$S_e = S_T - S_m = 9.59 - 2.66 = 6.93 \qquad (1.14)$$

Variance V is equal to the value of variation S divided by degrees of freedom f. Pure variation S' is subtraction of error variance multiplied by its degrees of freedom from variation S. By dividing pure variation S' by total variation S_T, we can obtain degrees of contribution ρ, which represents the quality level. As a result, we have Table 1.1, which, reveals that since the average is much greater than the standard, mean-squared error increases accordingly and thereby leads to enlarging loss due to variability. In a plant, adjusting an average to a target is normally regarded to be easy.

Once the average gets close to the target, variance σ^2 can be changed to approximately match the error variance V_e (Table 1.1). As a result, the following value of the loss function is obtained:

$$L = \frac{300}{2^2} (0.365) = (75)(0.365) = 27 \text{ cents} \qquad (1.15)$$

As compared to equation (1.11), for one sheet of plate glass, this brings us a quality improvement of

$$36.0 - 27.4 = 8.6 \text{ cents} \qquad (1.16)$$

If 500,000 sheets are produced monthly, we can obtain $43,000 through quality improvement.

No special tool is required to adjust an average to a target. We can cut plate glass off with a ruler after comparing an actual value with a target value. If the

Table 1.1
ANOVA table

Factor	f	S	V	S'	$\rho(\%)$
m	1	2.66	2.66	2.295	23.9
e	19	6.93	0.365	7.295	76.1
Total	20	9.59	0.4795	9.590	100.0

time interval of comparison and adjustment is varied, a variance will be changed as well as an average. This is regarded as an issue of calibration cycle in quality engineering. At a factory, we can determine an optimal calibration cycle using the procedure detailed in Chapter 2.

QUALITY LEVEL: SMALLER-THE-BETTER CHARACTERISTIC

Smaller-the-better characteristics should be nonnegative, and the most desirable value is zero. The quality level of smaller-the-better characteristics is given as the loss function L.

☐ Example

Now suppose that a tolerance standard of roundness is less than 12 μm, and if a product is disqualified, loss A costs 80 cents. We produce the product with machines A_1 and A_2. Roundness data for this product taken twice a day over a two-week period are as follows:

Morning:	0	5	4	2	3	1	7	6	8	4
Afternoon:	6	0	3	10	4	5	3	2	0	7

The unit used here is micrometers. $\Delta = 12$ μm and $A = 80$ cents. The quality level using the smaller-the-better characteristic is calculated as

$$\sigma^2 = \tfrac{1}{20}(y_1^2 + y_2^2 + \cdots + y_{20}^2)$$

$$= \tfrac{1}{20}(0^2 + 5^2 + 4^2 + \cdots + 7^2)$$

$$= 23.4 \ \mu m^2 \tag{1.17}$$

$$L = \frac{A}{\Delta^2}\sigma^2 = \frac{80}{12^2}(23.4) = 13 \text{ cents} \tag{1.18}$$

QUALITY LEVEL: LARGER-THE-BETTER CHARACTERISTIC

The larger-the-better characteristic should be nonnegative, and its most desirable value is infinity. Even if the larger the better, a maximum of nonnegative heat efficiency, yield, or nondefective product rate is merely 1 (100%); therefore, they are not larger-the-better characteristics. On the other hand, amplification rate, power, strength, and yield amount are larger-the-better characteristics because they do not have target values and their larger values are desirable.

□ Example

Now we define y as a larger-the-better characteristic and calculate a loss function. Suppose that the strength of a three-layer reinforced rubber hose is important:

K_1: adhesion between tube rubber and reinforcement fiber

K_2: adhesion between reinforcement fiber and surface rubber

Both are crucial factors for rubber hose. For both K_1 and K_2, a lower limit for Δ is specified as $\Delta = 5.0$ kgf. When hoses that do not meet this standard are discarded, the loss is \$5 per hose. In addition, its annual production volume is 120,000. After prototyping eight hoses for each of two different-priced adhesives, A_1 (50 cents) and A_2 (60 cents), we measured the adhesion strengths of K_1 and K_2, as shown in Table 1.2.

Compare quality levels A_1 and A_2. We wish to increase their averages and reduce their variability. The scale for both criteria is equation (1.9), an average of the sum of squared reciprocals. By calculating A_1 and A_2, respectively, we compute loss functions expressed by (1.8). For A_1,

$$\sigma_1^2 = \frac{1}{16}\left(\frac{1}{10.2^2} + \frac{1}{5.8^2} + \cdots + \frac{1}{16.5^2}\right) = 0.02284 \qquad (1.19)$$

$$L_1 = A_0\Delta_0\sigma^2 = (500)(5^2)(0.02284) = \$2.86 \qquad (1.20)$$

For A_2,

$$\sigma_2^2 = \frac{1}{16}\left(\frac{1}{7.6^2} + \frac{1}{13.7^2} + \cdots + \frac{1}{10.6^2}\right) = 0.01139 \qquad (1.21)$$

$$L_2 = (500)(5^2)(0.01139) = \$1.42 \qquad (1.22)$$

Therefore, even if A_2 is 10 cents more costly than A_1 in terms of adhesive cost (sum of adhesive price and adhesion operation cost), A_2 is more cost-effective than A_1 by

$$(50 + 285.5) - (60 + 142.4) = \$1.33 \qquad (1.23)$$

In a year, we can save \$159,700.

Table 1.2
Adhesion strength data

A_1	K_1	10.2	5.8	4.9	16.1	15.0	9.4	4.8	10.1
	K_2	14.6	19.7	5.0	4.7	16.8	4.5	4.0	16.5
A_2	K_1	7.6	13.7	7.0	12.8	11.8	13.7	14.8	10.4
	K_2	7.0	10.1	6.8	10.0	8.6	11.2	8.3	10.6

1.5. Quality Assurance Department

A *strategy* is a system that stimulates employees to rationalize a corporate business process or to endeavor voluntarily to improve productivity. To encourage employees voluntarily to enhance a management system, in particular, it is essential to make them strategically predict and assess the quality level (i.e., part of the quality level: only loss due to manufacturing variability, or A/A_0 in the total loss) of a product at the point of shipping. An individual management system (e.g., preventive quality management system, preventive maintenance) is not strategy but *tactics*, whose design is detailed in Section 2.4. In this section we demonstrate the management of unknown items. The countermeasure to unknown items is a strategy.

Responsibility of Each Department for Quality

The main objective of quality engineering is to offer a common effective procedure for use by an R&D, design, or manufacturing department to improve productivity. That is, quality engineering takes no responsibility for quality, quality improvement, or productivity. Similarly, a computer is in charge of information processing but it cannot solve problems directly and is merely a tool for the person attempting to solve a problem. A computer cannot itself solve a problem.

Quality engineering is also merely an instrument for people to use to solve problems such as improvement of technological development, product design, process design, process control, or product control. If selection of control factors or calculation of SN ratios is inappropriate, eventual improvement of quality or productivity obviously results in failure. On the contrary, to shed light on how to determine improper control factors and SN ratios is one of the roles of quality engineering. In this regard, quality engineering has a feature regarded as evaluation of technological research, or self-management.

Some companies have a quality assurance department. What responsibilities are its members supposed to take? Quality assurance is not just a department that takes responsibility for complaints in the marketplace and tries to compensate customers and solve problems. Quality assurance is not simply a department that apologizes to customers. That is, it does not have the sole responsibility for quality assurance.

A general role of management is to encourage employees to do a good job. To do this, management needs to have authority over personnel affairs and allocation of monetary resources. Managers of the quality assurance department do not have this kind of authority. In the next section we discuss how one deals with the quality assurance personnel's responsibilities and clarifies their duties.

The quality assurance department should be responsible for unknown items and complaints in the marketplace. Two of the largest-scale incidents of pollution after World War II were the arsenic poisoning of milk by company M and the PCB poisoning of cooking oil by company K. (We do not here discuss Minamata disease because it was caused during production.) The arsenic poisoning incident led to many infants being killed after the intake of milk mixed with arsenic. If company M had been afraid that arsenic would be mixed with milk and then had produced the milk and inspected for arsenic, no one would have bought its products. This is because a company worrying about mixtures of arsenic and milk cannot be trusted.

Responsibilities of the Quality Assurance Department

The case of PCB poisoning is similar. In general, a very large pollution event is not predicable, and furthermore, prediction is not necessary. Shortly after World

War II, Fuji Film decimated the ayu (sweetfish) population in the Sakawa River by polluting through industrial drainage. Afterward, it decided to keep fish in the drainage ponds rather than inspecting and analyzing the drainage. By examining the ponds several times a day, company personnel could determine whether the fish were swimming actively. Instead of checking up on what kinds of harmful substances were drained, they made sure that the fish were alive. Since then the company has never caused any pollution as a result of industrial drainage.

Checking on whether something is harmful by using organisms rather than inspecting harmful substances themselves is a practice used widely in Europe and the United States. Since we do not need to specify and measure detrimental substances, specific technologies for measuring them are also not necessary. Thus, we believe that a quality assurance department should be responsible for unknown items. *Their mission is not to inspect to assure that harmful substances are not released, but to inspect whether they have been released and to block them from reaching society when detected.*

For unknown items, specialized technologies do not function at all. Therefore, it is not unreasonable that a quality assurance department take responsibility. We hope that plants all over Japan will adopt such a quality assurance system. In sum, a test for unknown items should be conducted by feeding organisms with substances in question or by using products under actual conditions. More detailed procedures are described in Section 2.5.

References

1. Genichi Taguchi, 1989, *Quality Engineering in Manufacturing*. Tokyo: Japanese Standards Association.
2. Japanese Industrial Standards Committee, 1996. *Quality Characteristics of Products*. JIS Z-8403. Tokyo: Japanese Standards Association.

2 Management for Quality Engineering

2.1. **Management's Role in Research and Development**   **25**
Planning Strategies   25
Design of Product Quality: Objective Function   26
Selection of Systems: Concepts   26
2.2. **Evaluation of Functionality**   **28**
Functionality (Performance) Evaluation, Signals, and Noise   29
2.3. **Design Process**   **30**
Tools for Designing   30
Evaluation of Reproducibility   30
2.4. **Automated Process Management**   **31**
2.5. **Diagnosis to Prevent Recall Problems**   **33**
System Design for Recall Prevention   35
Functions of the Quality Assurance Department   38
References   **38**

2.1. Management's Role in Research and Development

Planning Strategies

The principal job of top management in an R&D or engineering department is planning strategy for technological development, classified into four general groupings. These are not specialized technologies but generic approaches that contribute to a wide range of technical fields in the long term.

Sun Tzu on the Art of War [1] is regarded as strategy, whereas technology represents tactics. Each specialized technology offers a concrete solution (design or means). New products or technologies are concrete results obtained through engineering research. However, since new products can survive only a few years without continuous creation of new products and technologies, we can be defeated by competitors. Conceptualization and selection of systems are creative jobs often conducted by engineers. Because determining parameters is only routine design work, it should be both rationalized and computerized.

Unlike the truth, technology is not in pursuit of what lasts eternally. However, strategic technology should be used as widely and as long as possible. R&D investment is supposed to be ongoing continually to ensure the success of a corporation. As a part of corporate strategy, some percentage of total sales should be invested on a continuing basis. Through technical activities, people in charge of R&D need to develop new products and technologies that are competitive enough to maintain a company and help it to grow. To date, Japanese financial groups have invested very little in R&D organizations. Although Mitsubishi Research Institute and Nomura Research Institute have research departments, little research has been conducted on R&D itself.

A key task of an R&D department is to rationalize and streamline a broad range of technical activities directed toward the development of new products and technologies. A means that can often be used in most fields of technological development is technological strategy per se, called *generic technology*.

One of the principal generic technologies is the streamlining of measurement and evaluation technologies. The reason is not that development engineers do not have good ideas but that they need to use most development time and resources for experimenting, prototyping, and testing to evaluate their ideas. A key rationalization of experiments and tests, including simulation, is to develop a product that functions well under conditions of mass production or various markets (including targeted design life) and causes little pollution and few difficulties when small-scale research is being conducted or test pieces are being used. In fact, it does not mean that engineers are not creative. However, they do not attempt to proceed to the next ideas until their current ideas have clearly failed. Then, rational evaluation, especially the accurate prediction of ideas, is required.

Strategy planning and personnel affairs are management roles in R&D. Top management is charged with planning business strategies, determining types of products to be produced, and allocating managers and budgets of engineering departments to design products (R&D and design departments). Quite often they proceed with their duties without knowing whether or not their decisions are correct; however, they need to take responsibility for their results in business competition in accordance with a balance sheet or profit and loss statement.

Design of Product Quality: Objective Function

Manufacturers plan products in parallel with variations in those products. If possible, they attempt to offer whatever customers wish to purchase: in other words, made-to-order products. Toyota is said to be able to deliver a car to a customer within 20 days after receipt of an order, with numerous variations in models, appearance, or navigation system. For typical models, they are prepared to deliver several variations; however, it takes time to respond to millions of variations. To achieve this, a production engineering department ought to design production processes for the effective production of high-mix, low-volume products, which can be considered rational processes.

On the other hand, there are products whose functions only are important: for example, invisible parts, units, or subsystems. They need to be improved with regard to their functions only.

Selection of Systems: Concepts

Engineers are regarded as specialists who offer systems or concepts with objective functions. All means of achieving such goals are artificial. Because of their artificiality, systems and concepts can be used exclusively only within a certain period protected by patents. From the quality engineering viewpoint, we believe that the more complex systems are, the better they become. For example, a transistor was originally a device invented for amplification. Yet, due to its simplicity, a transistor itself cannot reduce variability. As a result, an amplifier put into practical use in a circuit, called an *op-amp*, is 20 times more complicated a circuit than that of a transistor.

As addressed earlier, a key issue relates to what rationalization of design and development is used to improve the following two functionalities: (1) design and development focusing on an objective function under standard conditions and (2) design and evaluation of robustness to keep a function invariable over a full period of use under various conditions in the market. To improve these functionalities, as many parameters (design constants) should be changed by designers as possible. More linear and complex systems can bring greater improvement. Figure 2.1 exemplifies transistor oscillator design by Hewlett-Packard engineers, who in choosing the design parameters for a transistor oscillator selected the circuit shown in the figure. This example is regarded as a functional design by simulation based on conceptualization. Through an L_{108} orthogonal array consisting of 38 control factors, impedance stability was studied.

Top management is also responsible for guiding each project manager in an engineering department to implement research and development efficiently. One practical means is to encourage each manager to create as complicated a system as possible. In fact, Figure 2.1 includes a vast number of control factors. Because a complex system contains a simpler one, we can bring a target function to an ideal one.

The approach discussed so far was introduced in a technical journal, *Nikkei Mechanical*, on February 19, 1996 [2]. An abstract of this article follows.

"LIMDOW (Light Intensity Modulation Direct Overwrite) Disk," developed by Tetsuo Hosokawa, chief engineer of the Business Development Department, Development Division, Nikon Corp., together with Hitachi Maxell, is a typical example that has successfully escaped from a vicious cycle of tune-up by taking advantage of the Taguchi Method. LIMDOW is a next-generation magneto-optical disk (MO) that can be both read and rewritten. Its writing speed is twice as fast as a conventional one because we

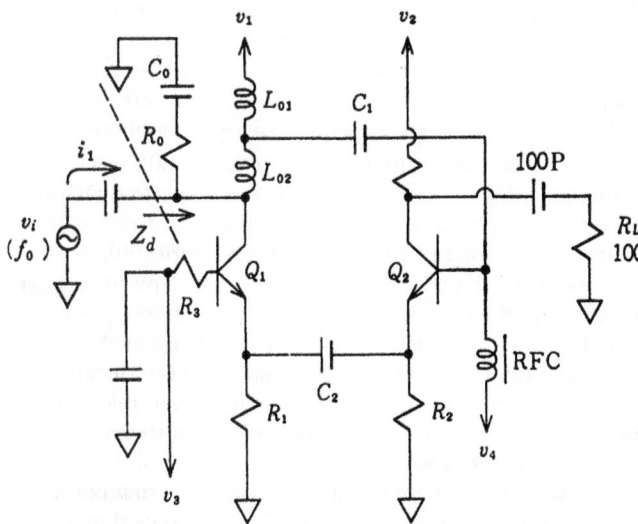

Figure 2.1
Transistor oscillator

can directly rewrite new data on an old record. For a conventional MO, old data needs to be erased before new ones are recorded.

To achieve this, we have to form at least six and at most nine magnetic layers on a LIMDOW disk, whereas only one layer is formed on a conventional one. Because of unstable production processes, Nikon faced a number of technical difficulties. There were approximately ten design parameters to form only a single layer in process; therefore, they had to adjust over ninety parameters in total for nine layers.

According to Hosokawa, this problem with parameters was so complex that they almost lost track of the direction in development. In addition, interaction between each layer happens once a multi-layer structure is designed. That is, if a certain optimal manufacturing condition is changed, other conditions are accordingly varied. Moreover, since optimal conditions depended greatly on which evaluation characteristics for a product were chosen, a whole development department fell into chaos.

Six years passed with prototyping and experimentation repeated. Despite six years spent, they could not obtain a satisfactory functional prototype. To break this deadlock, they introduced the Taguchi Method. Three years afterward, they stabilized forming processes, and finally in 1995, they found a prospect for mass production. A LIMDOW-type MO disk is already established as an international standard of the International Standardization Organization (ISO) and being developed by other major disk manufacturers. Nikon Corp. was the first company to succeed in mass production and is still monopolizing the market.

Currently, photographic film comprises about 20 layers. Since some of the layers could be unnecessary, engineers at photographic filmmakers need to evaluate which layers are required as control factors, using parameter design. A development approach whereby a complicated system is attempted after a simple one fails does not lead to effective optimal design of robustness but instead, results in technical improvement by tolerance adjustment. Designers' philosophy at Hewlett-Packard can help your understanding (see Figure 2.1). In quality engineering, a system to be selected by specialized engineers should be more complicated because we have the flexibility to adjust functional robustness to an objective function due to many control factors (design constants that can be selected at the designers' discretion).

2.2. Evaluation of Functionality

Technological development is different from general personnel management in that unlike human beings, new products and technologies or hardware and software cannot operate voluntarily. In the development and design phases, an R&D department needs to predict accurately how many problems a product (including software) they have developed will cause in the marketplace throughtout its life span. There are various kinds of uses expected in the marketplace, some of which cannot be predicted. In the chemical industry, how to evaluate whether a product developed in a laboratory can be produced on a large scale without adjustment (in most cases, in chemical reaction time) is quite important. A key issue in technological development is to predict functionality "downstream," in contrast to "upstream," in the development stage.

Up to this point, where we are dealing with technical procedures in technological development, each engineer's ability is highly respected and final evaluation of a product is assessed by other departments, such as reliability engineering or quality assurance. This procedure is similar to that of financial accounting in terms of checking up only on final results. However, it cannot function well as technological management. Each engineer is required to provide not only predictability of results but functionality (reliability) in the development stage.

The reason that the consequence approach fails in technological management is that in most technological research, engineers are limited in the following ways:

1. Since they aim to research as simple a system for an objective function as possible, they often fail to improve functional stability (robustness) in a market that involves various conditions, and to perform tune-ups for an objective function under standard conditions. This is because they are not trained to become creative enough to select a complicated system in the first place. For example, *Parameter Design for New Product Development* [3] shows complex systems designed by Hewlett-Packard. Because, in general, circuit elements (especially integrated-circuit and large-scale integrated elements) are quite cheap, to develop a larger-scale, more complex circuit helps to improve functionality; at the same time, encouraging developers to do so is considered a job for management.

2. Researching specialized characteristics in engineering books does not help us avoid technical problems in the marketplace involving numerous unknown factors. Since there are only two types of factors, *signal factors* (without them, a product becomes useless) and *noise factors* (the smaller they become, the better), few researchers have not utilized an SN ratio that measures functional robustness while maintaining signal factor effects. Therefore, after technological development or design research is complete, conventional evaluation methods have not been able to predict unexpected problems caused by various conditions regarding mass production and customer use, which are different from those in the laboratories. We discuss functionality evaluation in Chapter 3.

3. Since quality and cost are predicted according to economic evaluation in both the design and production stages, they are not well balanced.

It is a essential managerial task in engineering departments to change the foregoing paradigms (schemes) of thinking used by engineers. In Europe and the United States, a change in thinking, termed a *paradigm shift*, is attained in two ways. For all three above, cases, we train research and development engineers to change their way of thinking. Especially for case 2, by altering functional test procedures in a design department, we lead engineers to use SN ratios for functional robustness assessment. In this section we explain case 1 and 2.

Functionality (Performance) Evaluation, Signals, and Noise

For both hardware and software, to evaluate how well a product functions in the marketplace (when designed or developed), we need to consider signal factors, which represent consumers' use conditions, and noise (error) factors for both hardware and software. Signal factors are of two types, active and passive, and noise also comprises two types, indicative and true. An *active signal factor* is a variable that a person uses actively and repeatedly: for example, stepping down on an accelerator pedal. On the other hand, a *passive signal factor* is a sensed or observed value that is used passively for processing of measurement or judgment. In an actual system, there are a number of these factors. For an entire range of signal, we should predict how well a function works. That is, by clarifying an ideal function for a signal we need to evaluate how close to the ideal function the signal factor effects can come.

A *true noise factor* is a factor whose noise effects should be smaller: for instance, an environmental condition or deterioration in a life span. If initial signal factor effects of a product never change under standard conditions or over any length of time, we say that it has good functionality or robust design. *Indicative factors* whose effects are regarded as not vital are selected to prove that a product can function well under any of them. For example, although an automobile's performance does not need to be the same at low, middle, and high speeds, it should be satisfactory at all of the speeds. An indicative factor is a noise condition that is used to evaluate a product's performance for each condition in a large environment.

We detail each field of functionality evaluation in Chapter 3.

2.3. Design Process

Tools for Designing   Among tools (procedures, techniques) to streamline design research are the following:

1. *Generic tools:* computer, orthogonal arrray
2. *Specialized tools:* finite element method software, circuit calculation method software
3. *Measurement standard*

We describe an *orthogonal array* as a tool specialized for quality engineering. Although many other tools are also important related to quality engineering, they play an essential role in information processing in all engineering fields, not just in quality engineering. An orthogonal array is regarded as special in the quality engineering field because it not only deals with difference equation calculation but also evaluates reproducibility of functionality for "downstream" conditions.

Evaluation of   To evaluate functionality under conditions of mass production or various appli-
Reproducibility   cations by means of test pieces, downsized prototypes, or limited flexibilities (a life test should be completed in less than one day, a noise test is limited to only two conditions) in a laboratory, signals and noises expected to occur in the market, as discussed above, should be taken into account.

However, by changing design constants called *control factors*, which are not conditions of use but parameters that can freely be altered by designers (including selection of both systems and parameter levels), designers optimize a system. Even if we take advantage of functionality evaluation, if signal factors, noises, and measurement characteristics are selected improperly, optimal conditions are sometimes not determined. Whether optimal conditions can be determined depends on the evaluation of reproducibility in the downstream (conditions regarding mass production or various uses in the market). Therefore, in the development phase, development engineers should design parameters using an orthogonal array to assess reproducibility. Or at the completion point of development, other evaluation departments, as a management group, should assess functionality using benchmarking techniques. The former approach is desirable; however, the latter is expected to bring a paradigm shift that stimulates designers to use SN ratios.

From a quality engineering viewpoint, a field that derives formulas or equations to explain target output characteristics, including reliability, is not considered engineering but physics. This is because any equation does not include economic factors, no matter how well it predicts the result. Physics deals with creating theories and formulas to account for natural phenomena. These theories are scientifically quite important, however, they are not related to designing products that are artificial. Since engineers design products or production processes that do not exist in nature, they are allowed exclusive use of their own inventions. Truth cannot be accepted as a patent because it is ubiquitous. Design is judged based on market factors such as cost and quality or productivity when used by customers. Productivity discussed here is productivity from a *user's* perspective. If a product of a certain manufacturer is much cheaper and has many fewer failures and much less pollution in each market segment than in others, it can increase its own market share. Good market productivity means that a product not only has good quality under standard conditions but also low production costs. That is, manufacturers that have good market productivity can sell at lower prices products of better technical quality (fewer defects, lower running cost, or lower social costs, such as pollution).

A key issue is whether or not we can predict and improve production cost or technical quality in the market prior to mass production or shipping. Market productivity, including product quality, is a way for a corporation to make a profit. Means for improving market productivity include design and manufacturing. Can we predict such market productivity accurately in the design and development phase? To predict market productivity, in addition to the marketability of a product, production cost and functionality must be forecast. This forecast is conducted through design evaluation. Although design is evaluated after development has been completed, in order to pursue optimal design, it should also be evaluated in the development stage when design can still be changed flexibly. Therefore, we need to find optimal design under laboratory conditions (e.g., using test pieces, small-scale studies, limited test conditions). We should evaluate whether factor effects hold true under downstream conditions (e.g., actual products, large-scale production processes; conditions of use, including various product lives). This is regarded as an issue of reproducibility in the downstream.

For product design to have reproducibility in downstream, evaluation characteristics need to be scrutinized as well as signals and noises. To make optimal design become optimal downstream, we should (1) realize that all conditions of use in the marketplace belong to either signals and noises, or do not; (2) determine levels of signals and noises; and (3) select characteristics to calculate rational SN ratios. In short, the use of rational SN ratios is a key factor. Thus, the reason that only the main effects of control factors are assigned to an orthogonal array is to determine whether they have additivity for SN ratios.

2.4. Automated Process Management

The first Industrial Revolution relieved humans of much physical labor by mechanizing machining operations in manufacturing processes. Currently, the major jobs of operators are production control, such as preparation of raw material, transportation, and fixturing of in-process products; machine setup and quality control, such as machining diagnosis and control; and inspection. Rationalization and mechanization of management operations are keys in today's second Industrial Revolution.

To rationalize and automate management operations, we are urged to establish theoretical economic fundamentals for them. Reference 4 interprets basic formulas and applications of daily management jobs in on-line departments such as manufacturing. Its distinguishing point from other guidelines is that the theory rests on system design and economic calculation.

Manufacturing departments are responsible for productivity, including cost. There are seven approaches to this [4]. The contents are explained in detail in Chapters 23 through 25.

1. *Quantitative quality evaluation of a shipped product.* Since defective products are not to be shipped and because that does not affect consumers, defect problems are not quality issues but cost issues.

2. *Product quality standard that is important in manufacturing.* By detailing how to determine tolerances, we show ways not only to estimate significance quantitatively when quality should come close to an ideal or target value but also ways to determine tolerances in negotiation or contracts. A distinctive feature is a new way of selecting safety factors, which has not been well defined. See JIS Z-8403 [5], which details how to determine standards.

3. *Feedback control in process. Process control* checks product characteristics or process conditions at a certain interval of time. If the values are within a limit, it determines whether or not to continue production; conversely, if they are beyond a limit, process conditions are adjusted. Chapters 3 to 5 of Reference 4 detail optimal system design for feedback control in a machining process. Refer to Chapter 23 through 25 for more details. This method hinges on economic system design that aims to balance checkup and adjustment cost and the economic quality level of a shipped product. Chapter 5 of Reference 4 covers calibration system design of measurement errors, that is, an assessment of the optimal number of operators.

4. *Process maintenance design.* Chapters 6 to 8 of Reference 4 offer ways to design process management when we can obtain only qualitative values, such as soldering characteristics instead of quantitative values, or when we can perform checkups only by inspecting gauges in lieu of management-designed gauges whose shapes are matched to products. To emphasize preventive maintenance during processing (preventive quality control), basic formulas and various applications, including preventive maintenance methods, are elucidated.

5. *Feedforward control.* In feedback control in process, we investigate the characteristics of a product and return it to its process. *Feedforward control* is a method of adjusting final characteristics to target values by changing process conditions. For example, in manufacturing film or iron, after the gelatin or ore received is inspected, optimal treatments or reaction conditions corresponding to those raw materials are selected for production. Feedforward control methods according to environment or adaptive control is detailed in Chapter 9 of Reference 4.

6. *Design of inspection systems.* For each product, quality characteristics are measured, their differences to target values are adjusted, and if they cannot be adjusted, each product is discarded. Although this procedure is considered

inspection in a broader sense, unlike approaches 3, 4, and 5, each product is inspected in this procedure. See Chapter 10 of Reference 4.

7. *Maintenance system design for a shipped product.* We should manage systems rationally in manufacturing, telecommunication, and traffic for products and services. When products and services will be produced, maintenance system design for the production system is crucial. In fact, availability management systems in production processes belong in this category.

These control, management, and maintenance systems are included in management activities based on information fed back. Currently, frontline operators are in charge of process management. Details of management design are discussed in Reference 4.

2.5. Diagnosis to Prevent Recall Problems

☐ Example

When we visited a Rolls-Royce helicopter engine plant in 1975, we saw that several-minute bench tests for all engines were being run, as well as life tests for every 25 engines. Since the annual engine production volume at that time was approximately 1000, they ran the life test approximately once a week. The life test took 160 hours to run, which was equivalent to the time before the first overhaul, and cost $25,000. If the test frequency could be reduced from every 25 engines to every 50, an annual reduction of $500,000 in the cost of inspections would be achieved.

We were asked whether the life test might be changed to every 50 engines. When I asked why they conducted the life test, they answered that they wished to find unexpected failures. That is, this is failure for unknown reasons, as discussed at the end of Chapter 1. Since it was almost impossible to investigate quality characteristics for each of thousands of parts, they substituted a life test as a simpler method. Because they could not prevent problems due to unknown items, inspection was the only solution. If inspection found defects and failures, human lives would not be lost.

To find such a serious problem that an engine stops is an example of technical management using function limits. In this case, a quality assurance department could be responsible for inspection because this inspection checks only whether or not an engine stops in the life test. We asked them for the following three parameters:

A: loss when a product does not function

B: inspection cost

$\bar{u}$: mean failure interval

Loss when a product does not function could be regarded as plant loss when an engine stops before its first overhaul. In this case we selected an engine replacement cost of about $300,000. Because the life test was conducted once a week and took almost a week to complete (this is called *time lag*), in the meantime a certain number of engines were mounted on various helicopters. However, since the life test at a plant can detect problems much faster than customers in the market can, we did not consider problems of loss of life due to crashes.

As noted earlier, the inspection expense is $25,000 and the mean failure interval (including failures both in the marketplace and at a plant), $\bar{u}$ was estimated to be "once in a few years." For the latter, we judged that they had one failure for every 2500 engines, and so set $\bar{u}$ to 2500 units. (Even if parameters deviate from actual values, they do not have any significant impact on inspection design.) For $\bar{u} = 2500$, we can calculate the optimal inspection interval using the equation that follows. The proof of this equation is given in Chapter 6 of Reference 4. For the sake of convenience, as a monetary unit, $100 is chosen here.

$$n = \sqrt{\frac{2\bar{u}B}{A}}$$

$$= \sqrt{\frac{(2) \times (2500) \times (250)}{3000}}$$

$$\approx 20.4 \text{ engines} \qquad (2.1)$$

This happens to be consistent with their inspection frequency. For a failure occurring once in three years, a life test should be done weekly. We were extremely impressed by their method, fostered through long-time experience.

We answered that the current frequency was best and that if it were lowered to every 50, the company would lose more. We added that if they had certain evidence that the incidence declined from once in two or three years to once in a decade, they could conduct the life test only once for every 50 engines.

In fact, if a failure happens once in a decade, the mean failure interval $\bar{u} = 10{,}000$ units:

$$n = \sqrt{\frac{(2)(100)(250)}{3000}}$$

$$= 41 \text{ engines} \qquad (2.2)$$

This number has less than a 20% difference from 50 units; therefore, we can select $n = 50$.

On the other hand, when we keep the current $\bar{u}$ unchanged and alter n from 25 to 50, we will experience cost increases. Primarily, in case of $n = 25$, loss L can be calculated as follows:

$$L = \frac{B}{n} + \frac{n+1}{2}\frac{A}{\bar{u}} + \frac{IA}{\bar{u}}$$

$$= \frac{250}{25} + \left(\frac{25+1}{2}\right)\left(\frac{3000}{2500}\right) + \frac{(25)(3000)}{2500}$$

$$= 10 + 15.6 + 30.0$$

$$= 55.6 \text{ cents } (\times \$100) \qquad (2.3)$$

Then, total annual cost, multiplied by an annual production volume of 1000, amounts to $5,560,000.

Similarly, in case of $n = 50$, L is computed as follows:

$$L = \frac{250}{25} + \left(\frac{50 + 1}{2}\right)\left(\frac{3000}{2500}\right) + \frac{(25)(3000)}{2500}$$

$$= 5 + 30.6 + 30.0$$

$$= 65.6 \text{ cents} \tag{2.4}$$

Thus, total annual cost amounts to $6,560,000. As a consequence, the company will suffer another $1 million loss.

Based on long-time technical experience, they had balanced inspection cost and the cost of problems following sale of the product.

Occasionally, problems create a sensation. This is because, unlike Rolls-Royce, some manufacturers have a management system that cannot detect in-house failures. A quality assurance department can take responsibility for recall due to unknown items. If there is no quality assurance department, a quality control section in a manufacturing department should take responsibility.

System Design for Recall Prevention

❏ Example

For one particular product, failures leading to recall are supposed to occur only once a decade or once a century. Suppose that a manufacturer producing 1 billion units with 600 product types yearly experiences eight recalls in a year. Its mean failure interval $\bar{u}$ is

$$\bar{u} = \frac{\text{total annual production volume}}{\text{annual number of failures}}$$

$$= \frac{1,000,000,000}{8}$$

$$= 125,000,000 \text{ units} \tag{2.5}$$

This is equivalent to *mean time between recalls* if no life test is conducted.

Whereas $\bar{u}$ is calculated for all products, parameters of A and B are computed for each product. For example, a certain product has an annual production volume of 1.2 million units and is sold at a price of $4. In addition, a life test for it costs $50 per unit and takes one week. In sum, $A = \$4$ and $B = \$50$. Given that mean failure interval $\bar{u}$ is equal to the value in equation (2.5),

$$n = \sqrt{\frac{2\bar{u}B}{A}}$$

$$= \sqrt{\frac{(2)(125,000,000)(50)}{4}}$$

$$\approx 56,000 \text{ units} \qquad (2.6)$$

Now by taking into account the annual production volume of 1.2 million and assuming that there are 48 weeks in a year, we need to produce the following number of products weekly:

$$\frac{1,200,000}{48} = 25,000 \qquad (2.7)$$

Therefore, 56,000 units in (2.6) means that the life test is conducted once every other week. The mean failure interval of 125,000,000 indicates that a failure occurs almost once in

$$\frac{125,000,000}{1,200,000} = 104 \text{ years} \qquad (2.8)$$

In short, even if a failure of a certain product takes place only once a century, we should test the product biweekly.

What happens if we cease to run such a life test? In this case, after a defect is detected when a product is used, it is recalled. "Inspection by user" costs nothing. However, in general, there should be a time lag of a few months until detection. Now supposing it to be about two months and nine weeks, the, time lag l is

$$l = (25,000)(9)$$

$$= 225,0000 \text{ units} \qquad (2.9)$$

Although, in most cases, loss A for each defective product after shipping is larger that that when a defect is detected in-house, we suppose that both are equal. The reasoning behind this is that A will not become so large on average because we can take appropriate technical measures to prevent a larger-scale problem in the marketplace immediately after one or more defective products are found in the plant.

If we wait until failures turn up in the marketplace, $B = 0$ and $n = 1$ because consumers check all products; this can be regarded as screening or 100% inspection. Based on product recall, loss L_0 is

$$L_0 = \frac{B}{n} + \frac{n+1}{2}\frac{A}{\bar{u}} + \frac{lA}{\bar{u}} \qquad (2.10)$$

Substituting $n = 1$, $B = 0$, $A = \$4$, $l = 225,000$ units, and $\bar{u} = 125,000,000$ units into (2.11), we obtain

$$L_0 = \frac{0}{1} + \left(\frac{1+1}{2}\right)\left(\frac{4}{125,000,000}\right) + \frac{(225,000)(400)}{125,000,000}$$

$$= 0.0072 \text{ cent} \qquad (2.11)$$

$$(0.0072)(1,200,000) = \$864 \qquad (2.12)$$

If we conduct no inspection, we suffer an annual loss of $8640 from recalls.

On the other hand, loss L when we run a life test once every two weeks is computed when we use $n = 50,000$ units, $B = \$50$, $A = \$4$, $I = 25,000$ units, and $\bar{u} = 125,000,000$ units:

$$L = \frac{50}{50,000} + \left(\frac{50 + 1}{2}\right)\left(\frac{4}{125,000,000}\right) + \frac{(25,000)(4)}{125,000,000}$$

$$= 0.001 + 0.0008 + 0.0008$$

$$= 0.0026 \text{ cent} \qquad (2.13)$$

For an annual production volume of 1.2 million, the annual total loss is $3120:

$$(0.0026)(1,200,000) = \$3120 \qquad (2.14)$$

Comparing this with (2.12), we can reduce total cost by

$$8640 - 3120 = \$5520 \qquad (2.15)$$

If we can expect this amount of gain in all 600 models as compared to the case of no life test, we can save

$$(5520)(600) = \$3,312,000 \qquad (2.16)$$

That is, $3.3 million can be saved annually.

In this case we assume that cost A for the case of recall is equal to loss A, the cost for a defective product detected in-house. In most cases, this assumption holds true. On the other hand, once a defective product is shipped, if the product threatens human life or enormous loss of property, the loss should be increased, for example, to $2 million per product. However, no matter how many products are defective, we need to take measures. Thus, we assume that only one product puts human life at risk.

If the sum of other product costs is equal to A, for its original value of $4, the average of A should be increased by

$$\frac{2,000,000}{225,000} = \$8.89 \qquad (2.17)$$

Defining loss A in the case of recall as A_0, we have

$$A_0 = 8.89 + 4 = \$12.89 \qquad (2.18)$$

As for the value of a life test, only loss in the case of no life test is increased. Putting $A_0 = \$12.89$ into A of equation (2.10), the loss L_0 is calculated as to be 0.0232 cent. Comparing this loss with the loss using a life test, $L = 0.0026$ cent, the loss increases by

$$0.0232 - 0.0026 = 0.0206 \text{ cent} \qquad (2.19)$$

This is equivalent to an annual loss of

$$(0.0206)(1,200,000) = \$24,720 \qquad (2.20)$$

Therefore, unlike A when a defect is detected in-house, A_0 in the case of no life test (in general, a test only to check functions) is a loss for each defective product recalled.

Functions of the Quality Assurance Department

We propose the following two items as functions of a quality assurance department: (1) to test functions of a product (benchmarking test) in terms of design quality, and (2) to conduct inspection to prevent pollution and recall in terms of manufacturing quality. A design quality test is a functional test for environment or deterioration, more specifically, a test based on the dynamic SN ratio. Originally a design department is responsible for this test; however, they do not usually test the functions of parts purchased from outside suppliers. Therefore, a quality assurance department should implement a functional test for an objective function to assist a purchasing department. A quality assurance department ought to take full responsibility for both items 1 and 2. When unexpected problems occur in the marketplace, a quality assurance department takes full responsibility in terms of design quality, and when a product is recalled, so does the quality control department. Regarding item 2, it is more desirable to use SN ratios than to conduct life tests.

References

1. Sun Tzu, 2002. *The Art of War,* Dover Publications.
2. *Nikkei Mechanical,* February 19, 1996.
3. Genichi Taguchi, 1984. *Parameter Design for New Product Development.* Tokyo: Japanese Standards Association.
4. Genichi Taguchi, 1989. *Quality Engineering Series,* Vol. 2. Tokyo: Japanese Standards Association.
5. Japanese Industrial Standards Committee, 1996. *Quality Characteristics of Products.* JIS Z-8403.

3 Quality Engineering: Strategy in Research and Development

3.1. Research and Development Cycle Time Reduction   39
3.2. Stage Optimization   40
      Orthogonal Expansion for Standard SN Ratios and Tuning Robustness   40
      Variability (Standard SN Ratio) Improvement: First Step in Quality Engineering   41

3.1. Research and Development Cycle Time Reduction

Functionality evaluation, which facilitates all research and development activities, is an issue for R&D managers. Their major task is not as specialists but as, those who develop new technologies or products. Streamlining of R&D tasks is called *generic technology* in Japan and *technological strategy* in Europe and the United States.

The main job of top management in an R&D or engineering department is to plan strategy for technological development, classified into four general groupings.

1. *Selection of technical themes.* Fundamental research for creative products prior to product planning is desirable. Testing of current and new products, downsizing, and simulations without prototyping are included in this process.

2. *Creation of concepts and systems.* Parameter design is conducted through a complex system to enhance reliability. The more complicated a system becomes, the more effectively robust design must be implemented.

3. *Evaluation for parameter design.* This procedure involves functional evaluation and checkup. The former rests on the SN ratio and the latter is a checkup of additivity based on an orthogonal array.

4. *Preparation of miscellaneous tools.* The finite element and circuit calculation methods should be used in addition to computers. The difference calculation method by orthogonal array, which we introduced in the United States, is applied in numerous fields. An orthogonal array is a generic tool for difference calculation.

Quality engineering is related to all four of the strategic items above. Although item 2 does not seem to be concerned with quality engineering, it must be included because in quality engineering we do not take measures once problems occur but design a complex system and parameters to prevent problems. Quality engineering gives guidelines only; it does not include detailed technical measures.

3.2. Stage Optimization

Orthogonal Expansion for Standard SN Ratios and Tuning Robustness

In a manufacturing industry, product and process designs affect a company's future significantly, but the design of production systems to produce services is an essential role in a service industry such as telecommunications, traffic, or finance. The design process comprises the following two stages: (1) synthesis and (2) analysis.

Quality engineering classifies the stage of synthesis into another two phases: (1) system (concept) selection and (2) selection of nominal system parameters (design constants). For the former, designer creativity is desirable, and if a designer invents a new method, it can be protected with a patent. The latter is related to development cycle time and to quality engineering. How a designer determines levels of design constants (nominal system parameters), which can be selected at a designer's discretion, changes the functional stability of a product under various conditions of use. A method of minimizing deviation of a product function from an ideal function under various conditions by altering nominal design constants is called *parameter design*. To balance functional robustness and cost and to enhance productivity (minimize quality and cost) by taking advantage of the loss function after robustness is improved is also an issue in quality engineering. Here, only the strategy for parameter design is illustrated.

Research and design consist primarily of system selection and parameter design. Although creation of new systems and concepts is quite important, until parameter design is complete, it is unclear whether a system selected can become competitive enough in the market. Therefore, it is essential in the short term to design parameters effectively for the system selected. In case of a large-scale or feedback control system, we should divide a total system into subsystems and develop them concurrently to streamline whole system design. Proper division of a system is the design leader's responsibility.

To improve functional robustness (reliability) in such a way that a product can function well in the market is the basis of product design. This means that the research and design parameters should be drawn up so that a product can work under various conditions in the marketplace over the product life span. Current simulation design approaches an objective function. After improving robustness (the standard SN ratio), we should approach the objective function using only standard conditions. Reducing variability is crucial. Parameters should be dispersed around standard values in simulations. All parameters in a system are control and noise factors. If simulation calculation takes a long time, all noise factors can be compounded into two levels. In this case we need to check whether compounded noises have an original qualitative tendency. To do this we investigate compounded noise effects in an orthogonal array to which only noises are assigned under initial design conditions, in most cases the second levels of control factors. (A design example in Chapter 4 shows only compounded noise effects.)

Not all noise factors need to be compounded. It is sufficient to use only the three largest. After improving robustness, a function curve is quite often adjusted to a target curve based on a coefficient of linear term β_1 and a coefficient of quadratic term β_2. (See the example in Chapter 4.) Since it is impossible to conduct the life test of a product under varied marketplace conditions, the strategy of using an SN ratio combined with noise, as illustrated in this chapter, has fundamentally changed the traditional approaches. The following strategies are employed:

1. Noise factors are assigned for each level of each design parameter and compounded, because what happens due to the environment and deterioration is unpredictable. For each noise factor, only two levels are sufficient. As a first step, we evaluate robustness using two noise levels.

2. The robust level of a control factor (having a higher SN ratio) does not change at different signal factor levels. This is advantageous in the case of a time-consuming simulation, such as using the finite element method by reducing the number of combinations in the study and improving robustness by using only the first two or three levels of the signal factor.

3. Tuning to the objective function can be made after robustness is improved using one or two control or signal factors. Tuning is done under the standard condition. To do so, orthogonal expansion is performed to find the candidates that affect linear coefficient β_1 and quadratic coefficient β_2.

Variability (Standard SN Ratio) Improvement: First Step in Quality Engineering

In quality engineering, all conditions of use in the market can be categorized into either a signal or a noise. In addition, a signal can be classified as active or passive. The former is a variable that an engineer can use actively, and it changes output characteristics. The latter is, like a measurement instrument or receiver, a system that measures change in a true value or a signal and calculates output. Since we can alter true values or oscillated signals in research, active and passive signals do not need to be differentiated.

What is most important in two-stage optimization is that we not use or look for cause-and-effect relationships between control and noise factors. Suppose that we test circuits (e.g., logic circuits) under standard conditions and design a circuit with an objective function as conducted at Bell Labs. Afterward, under 16 different conditions of use, we perform a functional test. Quality engineering usually recommends that only two conditions be tested; however, this is only because of the cost. The key point is that we should not adjust (tune) a design condition in order to meet the target when the noise condition changes. For tuning, a cause-and-effect relationship under the standard condition needs to be used.

We should not adjust the deviation caused by the usage condition change of the model or regression relationship because it is impossible to dispatch operators in manufacturing to make adjustments for uncountable conditions of use. Quality engineering proposes that adjustments be made only in production processes that are considered standard conditions and that countermeasures for noises should utilize the interactions between noise and control factors.

TESTING METHOD AND DATA ANALYSIS
In designing hardware (including a system), let the signal factor used by customers be M and its ideal output (target) be m. Their relationship is written as $m - f(M)$. In two-stage design, control factors (or indicative factors) are assigned to an orthogonal array. For each run of the orthogonal array, three types of outputs are obtained either from experimentation or from simulation: under standard conditions N_0, under negative-side compounded noise conditions N_1, and under

positive-side compounded noise conditions, N_1. The output under N_0 at each level of the signal factor (used by customers) are now "redefined" as signal factor levels: $M_1, M_2, \ldots, M_k$. Table 3.1 shows the results.

First stage: To reduce variability, we calculate SN ratios according to Table 3.1. However, in this stage, *sensitivities are not computed.* $M_1, M_2, \ldots$, and M_k are equal to output values under N_0 for each experiment. From the inputs and outputs of the table, an SN ratio is calculated. Such an SN ratio is called a *standard SN ratio.* The control factor combination that maximizes a standard SN ratio is the optimal condition. However, it is useless to compute sensitivities because we attempt to maximize standard SN ratios in the first stage. Under the optimal conditions, N_0, N_1, and N_2 are obtained to calculate the SN ratio. This SN ratio is used to check the reproducibility of gain.

Second stage: After robustness is optimized in the first stage, the output is tuned to the target. The signal used by customers (original signal, M) is tuned so that the output, Y, may meet the target, m. $y_1, y_2, \ldots, y_k$ represent the outputs under N_0 of optimal conditions, and $m_1, m_2, \ldots, m_k$ represent the targets under the conditions of the original signal, M. Objective function design is to bring output value y's close to target value m's by adjusting the signal used by the user, M. In most cases, the proportionality

$$y = \beta M \qquad (3.1)$$

is regarded as an ideal function except for its coefficient. Nevertheless, the ideal function is not necessarily a proportional equation, such as equation (3.1), and can be expressed in various types of equations. Under an optimal SN ratio condition, we collect output values under standard conditions N_0 and obtain the following:

M (*signal factor*)	M_1	M_2	...	M_k
m (*target value*):	m_1	m_2	...	m_k
y (*output value*):	y_1	y_2	...	y_k

Although the data in N_1 and N_2 are available, they are not needed for adjustment because they are used only for examining reproducibility.

If a target value m is proportional to a signal M for any value of β, it is analyzed with level values of a signal factor M. An analysis procedure with a signal M is the same as that with a target value m. Therefore, we show the common method of adjustment calculation using a target value m, which is considered the Taguchi method for adjustment.

Table 3.1
Data for SN ratio analysis

N_0	$M_1, M_2, \ldots, M_k$	Linear Equation
N_1	$y_{11}, y_{12} \ldots, y_{1k}$	L_1
N_2	$y_{21}, y_{22}, \ldots, y_{2k}$	L_2

If outputs $y_1, y_2, \ldots, y_k$ under the standard conditions N_0 match the target value m's, we do not need any adjustment. In addition, if y is proportional to m with sufficient accuracy, there are various ways of adjusting β to the target value of 1. Some of the methods are the following:

1. We use the signal M. By changing signal M, we attempt to match the target value m with y. A typical case is proportionality between M and m. For example, if y is constantly 5% larger than m, we can calibrate M by multiplying it by 0.95. In general, by adjusting M's levels, we can match m with y.

2. By tuning up one level of control factor levels (sometimes, more than two control factor levels) in a proper manner, we attempt to match the linear coefficient β_1 with 1.

3. Design constants other than the control factors selected in simulation may be used.

The difficult part in adjustment is not for coefficients but for deviations from a proportional equation. Below we detail a procedure of adjustment including deviations from a proportional equation. To achieve this, we make use of orthogonal expansion, which is addressed in the next section. Adjustment for other than proportional terms is regarded as a significant technical issue to be solved in the future.

FORMULA OF ORTHOGONAL EXPANSION

The formula of orthogonal expansion for adjustment usually begins with a proportional term. If we show up to the third-order term, the formula is:

$$y = \beta_1 m + \beta_2 \left(m^2 - \frac{K_3}{K_2} m \right)$$

$$+ \beta_3 \left(m^3 + \frac{K_3 K_4 - K_2 K_5}{K_2 K_4 - K_3^2} m + \frac{K_3 K_5 - K_4^2}{K_2 K_4 - K_3^2} \right) + \cdots \qquad (3.2)$$

Now, $K_1, K_2, \ldots$ are expressed by

$$K_i = \frac{1}{k} (m_1^i + m_2^i + \cdots + m_k^i) \qquad (i = 2, 3, \ldots) \qquad (3.3)$$

$K_2, K_3, \ldots$ are constant because m's are given. In most cases, third- and higher-order terms are unnecessary; that is, it is sufficient to calculate up to the second-order term. We do not derive this formula here.

Accordingly, after making orthogonal expansion of the linear, quadratic, and cubic terms, we create an ANOVA (analysis of variance) table as shown in Table 3.2, which is not for an SN ratio but for tuning. If we can tune up to the cubic term, the error variance becomes V_e. The loss before tuning,

$$L_0 = \frac{A_0}{\Delta_0^2} V_T$$

is reduced to

$$L = \frac{A_0}{\Delta_0^2} V_e$$

Table 3.2
ANOVA table for tuning

Source	f	S	V
β_1	1	$S_{\beta 1}$	
β_2	1	$S_{\beta 2}$	
β_3	1	$S_{\beta 3}$	
e	$k - 3$	S_e	V_e
Total	k	S_T	V_T

after tuning. When only the linear term is used for tuning, the error variance is expressed as follows:

$$V_e = \frac{1}{k-1}\,(S_{\beta 2} + S_{\beta 3} + S_e)$$

Since the orthogonal expansion procedure in Table 3.2 represents a common ANOVA calculation, no further explanation is necessary.

❏ Example

This design example, introduced by Oki Electric Industry at the Eighth Taguchi Symposium held in the United States in 1990, had an enormous impact on research and design in the United States. However, here we analyzed the average values of N_1 and N_2 by replacing them with data of N_0 and the standard SN ratio based on dynamic characteristics. Our analysis is different from that of the original report, but the result is the same.

A function of the color shift mechanism of a printer is to guide four-color ribbons to a proper position for a printhead. A mechanism developed by Oki Electric Industry in the 1980s guides a ribbon coming from a ribbon cartridge to a correct location for a printhead. Ideally, rotational displacement of a ribbon cartridge's tip should match linear displacement of a ribbon guide. However, since there are two intermediate links for movement conversion, both displacements have a difference. If this difference exceeds 1.45 mm, misfeeds such as ribbon fray (a ribbon frays as a printhead hits the edge of a ribbon) or mixed color (a printhead hits a wrong color band next to a target band) happen, which can damage the function. In this case, the functional limit Δ_0 is 1.45 mm. In this case, a signal factor is set to a rotational angle M, whose corresponding target values are as follows:

Rotational angle (rad):	1.3	2.6	3.9	5.2	6.5
Target value (mm):	2.685	5.385	8.097	10.821	13.555

For parameter design, 13 control factors are selected in Table 3.3 and assigned to an L_{27} orthogonal array. It is preferable that they should be assigned to an L_{36} array. In addition, noise factors should be compounded together with control factors other than design constants.

Since variability caused by conditions regarding manufacturing or use exists in all control factors, if we set up noise factors for each control factor, the number of noise factors grows, and accordingly, the number of experiments also increases. Therefore, we compound noise factors. Before compounding, to understand noise factor effects, we check how each factor affects an objective characteristic of displacement by fixing all control factor levels at level 2. The result is illustrated in Table 3.4.

By taking into account an effect direction for each noise factor on a target position, we establish two compounded noise factor levels, N_1 and N_2, by selecting the worst configuration on both the negative and positive sides (Table 3.5). For some cases we do not compound noise factors, and assign them directly to an orthogonal array.

According to Oki research released in 1990, after establishing stability for each rotational angle, the target value for each angle was adjusted using simultaneous equations. This caused extremely cumbersome calculations. By "resetting" an output value for each rotational angle under standard conditions as a new signal M, we can improve the SN ratio for output under noises. Yet since we cannot obtain a program for the original calculation, we substitute an average value for each angle at N_1 and N_2 for a standard output value N_0. This approach holds true only for a case in which each output value at N_1 and N_2 mutually has the same absolute value with a different sign around a standard value. An SN ratio that uses an output value under a standard condition as a signal is called a *standard SN ratio*, and one substituting an average value is termed a *substitutional standard SN ratio* or *average standard SN ratio*.

Table 3.3
Factors and levels[a]

Factor	Level 1	Level 2	Level 3	Factor	Level 1	Level 2	Level 3
A	6.0	6.5	7.0	H	9.6	10.6	11.6
B	31.5	33.5	35.5	I	80.0	82.0	84.0
C	31.24	33.24	35.24	J	80.0	82.0	84.0
D	9.45	10.45	11.45	K	23.5	25.5	27.5
E	2.2	2.5	2.8	L	61.0	63.0	65.0
F	45.0	47.0	49.0	M	16.0	16.5	17.0
G	7.03	7.83	8.63				

[a]The unit for all levels is millimeters, except for F (angle), which is radians.

Table 3.4
Qualitative effect of noise factor on objective function

Factor	Effect	Factor	Effect
A'	+	H'	−
B'	−	I'	−
C'	−	J'	+
D'	+	K'	+
E'	+	L'	−
F'	+	M'	+
G'	−		

Calculation of Standard SN Ratio and Optimal Condition

For experiment 1 in the orthogonal array (L_{27} in this case), each sliding displacement for each rotational angle under the standard conditions, N_1 and N_2 is shown in Table 3.6. Decomposition of the total variation is shown in Table 3.7.

Now we obtain the SN ratio:

$$\eta = 10 \log \frac{(1/2r)(S_\beta - V_e)}{V_N} = 3.51 \text{ dB} \qquad (3.4)$$

For other experimental runs, by resetting sliding displacement under standard conditions as a signal, we should compute SN ratios. Each signal level value is the average of output values at both N_1 and N_2 for each experiment. Therefore, even if we use this approximation method or the normal procedure based on standard output values, the signal value for each of experiments 1 to 18 is different.

Tables 3.8 and 3.9 and Figure 3.1 show factor assignment and SN ratios, level totals for SN ratios, and response graphs, respectively. The optimal condition is $A_1B_3C_3D_1E_1F_1G_3H_1I_1J_1K_1L_3M_3$, which is the same as that of Oki Electric Industry. Table 3.10 summarizes estimation and confirmation. This simulation calculation strongly influenced U.S. engineers because it involved strategic two-stage optimi-

Table 3.5
Two levels of compounded noise factor

Factor	N_1	N_2	Factor	N_1	N_2
A'	−0.1	+0.1	H'	+0.1	−0.1
B'	+0.1	−0.1	I'	+0.1	−0.1
C'	+0.1	−0.1	J'	−0.15	+0.15
D'	−0.1	+0.1	K'	+0.15	−0.15
E'	−0.05	+0.5	L'	+0.15	−0.15
F'	−0.5	+0.5	M'	−0.15	+0.15
G'	+0.1	−0.1			

Table 3.6
Data of experiment 1

Rotational Angle:	1.3°	2.6°	3.9°	5.2°	6.5°	Loss Function
Standard Condition:	3.255	6.245	9.089	11.926	14.825	
N_1	2.914	5.664	8.386	11.180	14.117	463.694309
N_2	3.596	6.826	9.792	12.672	15.533	524.735835

zation and noise selection in simulation. However, quite often in the United States, instead of compounding noise factors, they assign them to an outer orthogonal array. That is, after assigning three levels of each noise factor to an L_{27} orthogonal array, they run a simulation for each combination (direct product design) formed by them and control factor levels.

Analysis for Tuning
At the optimal SN ratio condition in simulation, shown in the preceding section, data at N_0 and the average values at both N_1 and N_2 are calculated as follows:

Signal factor M (deg):	1.3	2.6	3.9	5.2	6.5
Target value m (mm):	2.685	5.385	8.097	10.821	13.555
Output value y (mm):	2.890	5.722	8.600	11.633	14.948

These data can be converted to Figure 3.2 by means of a graphical plot. Looking at Figure 3.2, we notice that each (standard) output value at the optimal condition digresses with a constant ratio from a corresponding target value, and this tendency makes the linear coefficient of output differ from 1. Now, since the coefficient β_1 is more than 1, instead of adjusting β_1 to 1, we calibrate M, that is, alter a signal M to M*:

Table 3.7
Decomposition of total variation

Source	f	S	V
β	1	988.430144	988.430144
$N\beta$	1	3.769682	3.769682
e	8	0.241979	0.030247
N	9	4.011662	0.445740
Total	10	992.441806	

Table 3.8
Factor assignment and SN ratio based on average value

No.	A	B	C	D	E	F	G	H	I	J	K	L	M	η (dB)
1	1	1	1	1	1	1	1	1	1	1	1	1	1	3.51
2	1	1	1	1	2	2	2	2	2	2	2	2	2	2.18
3	1	1	1	1	3	3	3	3	3	3	3	3	3	0.91
4	1	2	2	2	1	1	1	2	2	2	3	3	3	6.70
5	1	2	2	2	2	2	2	3	3	3	1	1	1	1.43
6	1	2	2	2	3	3	3	1	1	1	2	2	2	3.67
7	1	3	3	3	1	1	1	3	3	3	2	2	2	5.15
8	1	3	3	3	2	2	2	1	1	1	3	3	3	8.61
9	1	3	3	3	3	3	3	2	2	2	1	1	1	2.76
10	2	1	2	3	1	2	3	1	2	3	1	2	3	3.42
11	2	1	2	3	2	3	1	2	3	1	2	3	1	2.57
12	2	1	2	3	3	1	2	3	1	2	3	1	2	0.39
13	2	2	3	1	1	2	3	2	3	1	3	1	2	5.82
14	2	2	3	1	2	3	1	3	1	2	1	2	3	6.06
15	2	2	3	1	3	1	2	1	2	3	2	3	1	5.29
16	2	3	1	2	1	2	3	3	1	2	2	3	1	3.06
17	2	3	1	2	2	3	1	1	2	3	3	1	2	-2.48
18	2	3	1	2	3	1	2	2	3	1	1	2	3	3.86
19	3	1	3	2	1	3	2	1	3	2	1	3	2	5.62
20	3	1	3	2	2	1	3	2	1	3	2	1	3	3.26
21	3	1	3	2	3	2	1	3	2	1	3	2	1	2.40
22	3	2	1	3	1	3	2	2	1	3	3	2	1	-2.43
23	3	2	1	3	2	1	3	3	2	1	1	3	2	3.77
24	3	2	1	3	3	2	1	1	3	2	2	1	3	-1.76
25	3	3	2	1	1	3	2	3	2	1	2	1	3	3.17
26	3	3	2	1	2	1	3	1	3	2	3	2	1	3.47
27	3	3	2	1	3	2	1	2	1	3	1	3	2	2.51

Table 3.9
Level totals of SN ratio

	1	2	3		1	2	3
A	34.92	27.99	20.01	H	29.35	27.23	26.34
B	24.26	28.55	30.11	I	28.67	27.21	27.07
C	10.62	27.33	44.97	J	37.38	28.48	17.06
D	32.92	27.52	22.48	K	32.94	26.59	23.39
E	34.02	28.87	20.03	L	16.10	27.78	39.04
F	35.40	27.67	19.85	M	22.06	26.63	34.23
G	24.66	28.12	30.14				

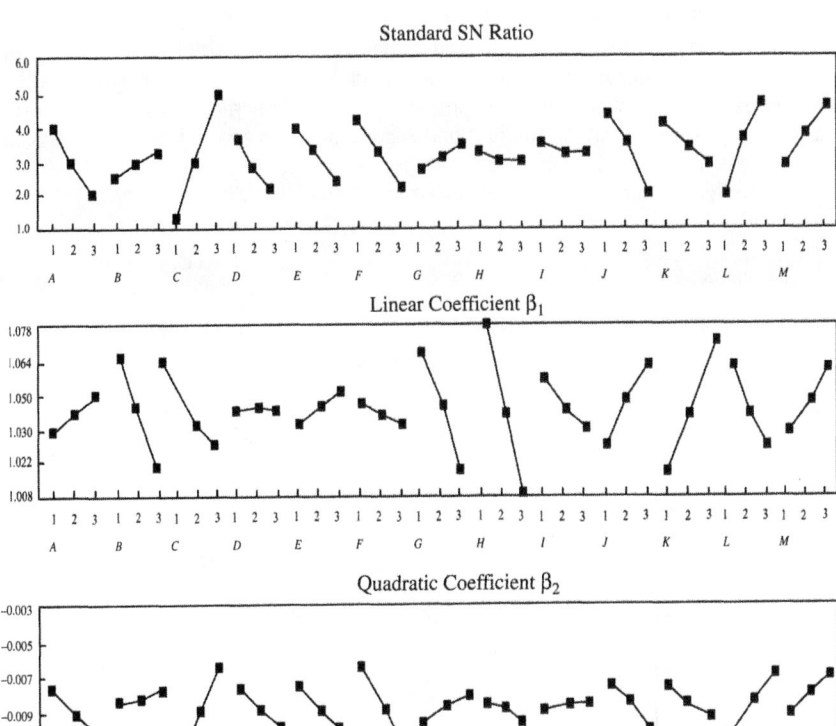

Figure 3.1
Factor effect plots of standard SN ratio, linear coefficient, and quadratic coefficient

Table 3.10
Estimation and confirmation of optimal SN ratio (dB)

	Estimation by All Factors	Confirmatory Calculation
Optimal condition	12.50	14.58
Initial condition	3.15	3.54
Gain	9.35	11.04

$$M^* = \frac{1}{\beta_1} M \tag{3.5}$$

However, in some cases we cannot correct M to M^* in accordance with (3.5). Although, normally in the case of $\beta_1 \neq 1$, we calibrate using a signal factor, we change β_1 to 1 using control factors on some occasions. Quality engineering has not prescribed which ways should be selected by designers because tuning is regarded as straightforward. We explain a procedure by analyzing data using control factors, whereas we should keep in mind that a calibration method using a signal is more common.

Looking at the data above, there is a gap between a target value m and an output value y under the standard conditions of N_0. To minimize this gap is called *adjustment*, where accuracy is our primary concern. To achieve this, using an output value y, we decompose the first and second terms in the orthogonal expansion equation (3.5).

Primarily, we calculate the total output variation for S_T with 5 degrees of freedom as follows:

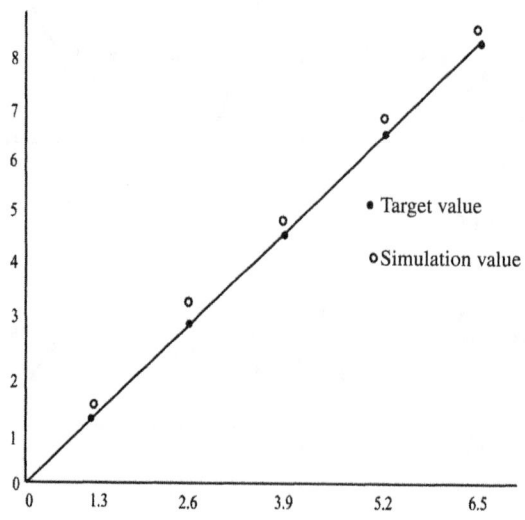

Figure 3.2
Standard output value
at optimal S/N ratio
condition

$$S_T = 2.890^2 + \cdots + 14.948^2 = 473.812784 \qquad (f = 5) \qquad (3.6)$$

Secondarily, the variation of a linear term S_{β_1} is computed as follows:

$$
\begin{aligned}
S_{\beta_1} &= \frac{(m_1 y_1 + m_2 y_2 + \cdots + m_5 y_5)^2}{m_1^2 + m_2^2 + \cdots + m_5^2} \\[2mm]
&= \frac{(2.685)(2.890) + \cdots + (13.555)(14.948)^2}{2.685^2 + \cdots + 13.555^2} \\[2mm]
&= 473.695042 \qquad\qquad\qquad (3.7)
\end{aligned}
$$

On the other hand, a linear term's coefficient β_1 is estimated by the following equation:

$$
\begin{aligned}
\beta_1 &= \frac{m_1 y_1 + m_2 y_2 + \cdots + m_5 y_5}{m_1^2 + m_2^2 + \cdots + m_5^2} \\[2mm]
&= \frac{436.707653}{402.600925} \\[2mm]
&= 1.0847 \qquad\qquad\qquad (3.8)
\end{aligned}
$$

In this case, a calibration method using a signal is to multiply a signal of angle M by $1/1.0847$. In contrast, in a calibration procedure by control factors, we select one control factor that significantly affects β_1.

As a next step for analyzing the quadratic term, we derive the second-order equation after calculating constants K_1 and K_2:

$$
\begin{aligned}
K_2 &= \tfrac{1}{5}(m_1^2 + m_2^2 + \cdots + m_5^2) \\[2mm]
&= \tfrac{1}{5}(2.685^2 + \cdots + 13.555^2) \\[2mm]
&= 80.520185 \qquad\qquad\qquad (3.9)
\end{aligned}
$$

$$
\begin{aligned}
K_3 &= \tfrac{1}{5}(m_1^3 + m_2^3 + \cdots + m_5^3) \\[2mm]
&= \tfrac{1}{5}(2.685^3 + \cdots + 13.555^3) \\[2mm]
&= 892.801297 \qquad\qquad\qquad (3.10)
\end{aligned}
$$

Thus, the second-order term can be calculated according to the expansion equation (3.2) as follows:

$$
\begin{aligned}
\beta_2\left(m^2 - \frac{K_3}{K_2}m\right) &= \beta_2\left(m^2 - \frac{892.801297}{80.520185}m\right) \\[2mm]
&= \beta_2(m^2 - 11.0897m) \qquad (3.11)
\end{aligned}
$$

Next, to compute β_2 and variation in the second-order term, we need to calculate the second-order term's coefficients $w_1, w_2, \ldots, w_5$ for $m_1, m_2, \ldots, m_5$ using the following formula:

$$w_i = m_i^2 - 11.0897 m_i \qquad (i = 1, 2, \ldots, 5) \qquad (3.12)$$

Now we obtain

$$w_1 = 2.685^2 - (11.0897)(2.685) = -22.567 \qquad (3.13)$$

$$w_2 = 5.385^2 - (11.0897)(5.385) = -30.720 \qquad (3.14)$$

$$w_3 = 8.097^2 - (11.0897)(8.097) = -24.232 \qquad (3.15)$$

$$w_4 = 10.821^2 - (11.0897)(10.821) = -2.9076 \qquad (3.16)$$

$$w_5 = 13.555^2 - (11.0897)(13.555) = 33.417 \qquad (3.17)$$

Using these coefficients $w_1, w_2, \ldots, w_5$, we compute the linear equation of the second-order term L_2:

$$L_2 = w_1 y_1 + w_2 y_2 + \cdots + w_5 y_5$$

$$= -(22.567)(2.890) + \cdots + (33.477)(14.948)$$

$$= 16.2995 \qquad (3.18)$$

Then, to calculate variation in the second-order term $S_{\beta 2}$, we compute r_2, which denotes a sum of squared values of coefficients $w_1, w_2, \ldots, w_5$ in the linear equation L_2:

$$r_2 = w_1^2 + w_2^2 + \cdots + w_5^2$$

$$= (-22.567)^2 + \cdots + 33.477^2$$

$$= 3165.33 \qquad (3.19)$$

By dividing the square of L_2 by this result, we can arrive at the variation in the second-order term $S_{\beta 2}$. That is, we can compute the variation of a linear equation by dividing the square of L_2 by the number of units (sum of squared coefficients). Then we have

$$S_{\beta 2} = \frac{16.2955^2}{3165.33} = 0.083932 \qquad (3.20)$$

Finally, we can estimate the second-order term's coefficient β_2 as

$$\beta_2 = \frac{L_2}{r_2} = \frac{16.2995}{3165.33} = 0.005149 \qquad (3.21)$$

Therefore, after optimizing the SN ratio and computing output data under the standard conditions N_0, we can arrive at Table 3.11 for the analysis of variance to compare a target value and an output value. This step of decomposing total variation to compare target function and output is quite significant before adjustment (tuning) because we can predict adjustment accuracy prior to tuning.

Economic Evaluation of Tuning

Unlike evaluation of an SN ratio, economic evaluation of tuning, is absolute because we can do an economic assessment prior to tuning. In this case, considering that the functional limit is 1.45 mm, we can express loss function L as follows:

Table 3.11
ANOVA for tuning (to quadratic term)

Source	f	S	V
β_1	1	473.695042	
β_2	1	0.083932	
e	3	0.033900	0.011300
Total	5	473.812874	
$\beta_2 + e$	(4)	(0.117832)	(0.029458)

$$L = \frac{A_0}{\Delta_0^2}\sigma^2 = \frac{A_0}{1.45^2}\sigma^2 \tag{3.22}$$

A_0 represents loss when a product surpasses its function limit of 1.45 mm in the market and is regarded as cost needed to deal with a user's complaint (e.g., repair cost). We assume that $A_0 = \$300$. σ^2 includes tuning and SN ratio errors, each of which is mutually independent.

It is quite difficult to evaluate an SN ratio on an absolute scale because absolute values of errors due to environment, deterioration, or variability among products are unknown. In short, SN ratio gain is based on a relative comparison of each loss. In contrast, for adjustment errors we can do an absolute assessment by computing the error variance according to the following procedure. When adjusting only the linear term, the error variance σ_1^2 is:

$$\sigma_1^2 = \frac{S_{\beta2} + S_e}{4} = \frac{0.083932 + 0.033930}{4} = 0.029458 \tag{3.23}$$

When adjusting up to the quadratic term, the error variance σ_2^2 is

$$\sigma_2^2 = \frac{S_e}{3} = \frac{0.033930}{3} = 0.001130 \tag{3.24}$$

Plugging these into (3.22), we can obtain the absolute loss. For the case of tuning only the linear term, the loss is

$$L_1 = \frac{300}{1.45^2}(0.029458) = \$4.20 \tag{3.25}$$

For the case of adjusting the linear and quadratic terms,

$$L_2 = \frac{300}{1.45^2}(0.001130) = 16 \text{ cents} \tag{3.26}$$

Thus, if we tune to the quadratic term, as compared to tuning only the linear term, we can save

$$420.3 - 16.1 = \$4.04 \tag{3.27}$$

Suppose a monthly production volume of only 20,000 units, approximately $80,000 can be saved.

Tuning Procedure

To tune in a practical situation, we generally use control factors that affect β_1 and β_2 significantly. Since only β_1 can be adjusted by a signal M, we sometimes use *only control factors that affect β_2*. Traditionally, we have used two tuning methods, one to change individual control factors and the other to use various factors at a time based on the least-squares method. In both cases, tuning has often failed because the selection of control factors for adjustment has depended on each designer's ability.

   To study adjustment procedures, we calculate coefficients of the first- and second-order terms at each level of the L_{27} orthogonal array shown in Table 3.6. Table 3.12 summarizes the level average of coefficients, and Figure 3.1 plots the factorial effects. Even if we do not adjust the linear term's coefficient, to 1, quite often we can do tuning by changing a range of rotational angles. Since in this case each linear term's coefficient is 8% larger, we can adjust by decreasing the rotational angle by 8%. This implies that elimination of quadratic terms is a key point in adjustment. Since the quadratic term's coefficient $\beta_2 = 0.005$, by referring to the response graphs of the quadratic term, we wish to make $\beta_2 = 0$ without worsening the SN ratio. While B, G, H, and I have a small effect on SN ratios, none of

Table 3.12

Level average of linear and quadratic coefficients

Factor	Linear Coefficient			Quadratic Coefficient		
	1	2	3	1	2	3
A	1.0368	1.0427	1.501	−0.0079	−0.0094	−0.0107
B	1.0651	1.0472	1.0173	−0.0095	−0.0094	−0.0091
C	1.0653	1.0387	1.0255	−0.0124	−0.0091	−0.0064
D	1.0435	1.0434	1.0427	−0.0083	−0.0093	−0.0104
E	1.0376	1.0441	1.0478	−0.0081	−0.0094	−0.0105
F	1.0449	1.0434	1.0412	−0.0077	−0.0093	−0.0109
G	1.0654	1.0456	1.0185	−0.0100	−0.0092	−0.0088
H	1.0779	1.0427	1.0089	−0.0090	−0.0092	−0.0098
I	1.0534	1.0430	1.0331	−0.0094	−0.0093	−0.0093
J	1.0271	1.0431	1.0593	−0.0076	−0.0093	−0.0111
K	1.0157	1.0404	1.0734	−0.0083	−0.0094	−0.0103
L	1.0635	1.0405	1.0256	−0.0113	−0.0095	−0.0073
M	1.0295	1.0430	1.0570	−0.010	−0.0095	−0.0084

them has a great effect on the quadratic term. Therefore, even if we use them, we cannot lower the quadratic term's coefficient. As the next candidates, we can choose K or D. If the K' level is changed from 1 to 2, the coefficient is expected to decrease to zero. Indeed, this change leads to increasing the linear term's coefficient from the original value; however, it is not vital because we can adjust using a signal. On the other hand, if we do not wish to eliminate the quadratic term's effect with a minimal number of factors, we can select K or D. In actuality, extrapolation can also be utilized. The procedure discussed here is a generic strategy applicable to all types of design by simulation, called the *Taguchi method* or *Taguchi quality engineering* in the United States and other countries.

4 Quality Engineering: The Taguchi Method

4.1. **Introduction**   57
4.2. **Electronic and Electrical Engineering**   59
  Functionality Evaluation of System Using Power (Amplitude)   61
  Quality Engineering of System Using Frequency   65
4.3. **Mechanical Engineering**   73
  Conventional Meaning of Robust Design   73
  New Method: Functionality Design   74
  Problem Solving and Quality Engineering   79
  Signal and Output in Mechanical Engineering   80
  Generic Function of Machining   80
  When On and Off Conditions Exist   83
4.4. **Chemical Engineering**   85
  Function of an Engine   86
  General Chemical Reactions   87
  Evaluation of Images   88
  Functionality Such as Granulation or Polymerization Distribution   91
  Separation System   91
4.5. **Medical Treatment and Efficacy Experimentation**   93
4.6. **Software Testing**   97
  Two Types of Signal Factor and Software   97
  Layout of Signal Factors in an Orthogonal Array   97
  Software Diagnosis Using Interaction   98
  System Decomposition   102
4.7. **MT and MTS Methods**   102
  MT (Mahalanobis–Taguchi) Method   102
  Application of the MT Method to a Medical Diagnosis   104
  Design of General Pattern Recognition and Evaluation Procedure   108
  Summary of Partial MD Groups: Countermeasure for Collinearity   114
4.8. **On-line Quality Engineering**   116
  References   123

4.1. Introduction

The term *robust design* is in wide spread use in Europe and the United States. It refers to the design of a product that causes no trouble under any conditions and answers the question: What is a good-quality product? As a generic term, *quality* or *robust design* has no meaning; it is merely an objective. A product that functions under any conditions is obviously good. Again, saying this is meaningless. All engineers attempt to design what will work under various conditions. The key issue is not design itself but how to evaluate functions under known and unknown conditions.

At the Research Institute of Electrical Communication in the 1950s, a telephone exchange and telephone were designed so as to have a 40- and a 15-year design life, respectively. These were demanded by the Bell System, so they developed a successful crossbar telephone exchanger in the 1950s; however, it was replaced 20 years later by an electronic exchanger. From this one could infer that a 40-year design life was not reasonable to expect.

We believe that design life is an issue not of engineering but of product planning. Thinking of it differently from the crossbar telephone exchange example, rather than simply prolonging design life for most products, we should preserve limited resources through our work. No one opposes the general idea that we should design a product that functions properly under various conditions during its design life. What is important to discover is how to assess proper functions.

The conventional method has been to examine whether a product functions correctly under several predetermined test conditions. Around 1985, we visited the Circuit Laboratory, one of the Bell Labs. They developed a new circuit by using the following procedure; first, they developed a circuit of an objective function under a standard condition; once it could satisfy the objective function, it was assessed under 16 different types of conditions, which included different environments for use and after-loading conditions.

If the product did not work properly under some of the different conditions, design constants (parameters) were changed so that it would function well. This is considered parameter design in the old sense. In quality engineering (QE), to achieve an objective function by altering design constants is called *tuning*. Under the Taguchi method, functional improvement using tuning should be made under standard conditions only after conducting stability design because tuning is improvement based on response analysis. That is, we should not take measures for noises by taking advantage of cause-and-effect relationships.

The reason for this is that even if the product functions well under the 16 conditions noted above, we cannot predict whether it works under other conditions. This procedure does not guarantee the product's proper functioning within its life span under various unknown conditions. In other words, QE is focused not on response but on interaction between designs and signals or noises. Designing parameters to attain an objective function is equivalent to studying first-order moments. Interaction is related to second-order moments. First-order moments involve a scalar or vector, whereas second-order moments are studied by two-dimensional tensors. Quality engineering maintains that we should do tuning only under standard conditions after completing robust design. A two-dimensional tensor does not necessarily represent noises in an SN ratio. In quality engineering, noise effects should have a continual monotonic tendency.

Quality engineering says that we should adjust a function to a target value or curve under standard conditions only after reducing variability. The idea is that, first, robust design or functional improvement for certain conditions of use needs to be completed, following which tuning (a method of adjusting a response value to a target value or curve) is done. Yet people in charge of designing a receiver sometimes insist that unless tuning is done first, output characteristics cannot be attained. In this case, what can we do? This is the reason that quality engineering requires us to acquire knowledge of measurement and analysis technologies. Several different measurement methods should be investigated.

A study to attain an objective value or function under 16 types of noise conditions by altering parameters can be regarded as one parameter design method; however, this is different from quality engineering. Indeed, it is true that quality characteristics have been studied by changing system parameters in a number of design of experiments cases, utilized broadly in the Japanese engineering field, and consequently, contributed to improving product quality. However, in terms of efficacy, these methods have not been satisfactory. We do not describe them in detail here because many publications do so. Improving objective characteristics by parameters, termed *design of experiments* (DOE), has the following two purposes: (1) to check out effects on objective characteristics and improve through tuning, and (2) to check out effects on output characteristics and conduct tolerance design. The objective of this chapter is to compare quality engineering and conventional methods.

Quality engineering has been used in the United States since the 1980s; however, it differs from the quality engineering detailed here, which is called the Taguchi method, the Taguchi paradigm, or Taguchi quality engineering in Europe and the United States. Quality engineering is robust design based on the following three procedures: (1) orthogonal array, (2) SN ratio, and (3) loss function. These three procedures are not robust design per se because they are used to evaluate technical means or products. To design robustness, we are required to understand the meaning of robust design before considering technical or management means. Those who consider practical means are engineers and production people.

Robust design (including product and process design) generally means designing a product that can function properly under various conditions of use. If this is its lone objective, we can study reliability or fraction nondefective, whose goal is 100%. Yet a corporation is merely a means of making a profit. Increased cost caused by excessive quality design leads to poor profitability and potentially to bankruptcy. The true objective of design and production is to earn a corporate profit.

The essence of the Taguchi method lies in measurement in functional space (to calculate a signal's proximity to an ideal function of output using the reciprocal of the average sum of squared deviations from an ideal function), overall measurement in multidimensional space from a base point, and accuracy. Quality engineering involves a research procedure for improving functionality and accuracy of diagnosis or prediction by taking advantage of measurement methods in functional space and in multidimensional space. Each improvement is an individual engineer's task. In the Taguchi method for hardware design, downstream reproducibility is important. For this purpose, it is desirable to collect data whose input signal and output characteristic are energy-related. Also, analysis can be conducted using the square root of the data. On the other hand, in case of diagnosis and prediction, as long as multidimensional data have sufficient accuracy, the equation of Mahalanobis distance in homogeneous or unit space is used. To measure the validity of use, an SN ratio is used by calculating the Mahalanobis distance from objects that obviously do not belong to the homogeneous space. The SN ratio is used to examine the effect of control factors. In the case of hardware study, control factors include design constants, and in the case of multidimensional data, control vectors include the selection of items. The reproducibility of control factor effects is checked using orthogonal arrays.

In this chapter we explain the Taguchi method in the field of quality engineering by explaining how to use it in several typical technological fields. Since the concepts apply for general technological fields, the Taguchi method is classified as a generic technology. In this chapter we briefly explain its application to each of the following fields: (1) electronic and electrical engineering, (2) mechanical engineering, (3) chemical engineering, (4) medical treatment and efficacy experimentation, (5) software testing, (6) MT and MTS Methods, and (7) on-line quality engineering.

Sections 4.2 to 4.5 cover quality engineering for hardware. Section 4.4 includes engine function as a generic function. In Section 4.6 we discuss debugging tests in terms of software quality. Section 4.7 highlights a new quality engineering method, which includes pattern recognition in regard to diagnosis and prediction. Section 4.8 addresses the system design of feedback quality management in the category of the on-line quality engineering. All formulas used in this chapter are given in Chapter 3 of Reference 1.

4.2. Electronic and Electrical Engineering

In general, the electronic and electrical engineering field emphasizes the following two functions: to supply power and to supply information. Power includes not only electric power but also output, such as current and voltage. However, these functions are almost the same as those in mechanical system design, as discussed in the next section. Thus, as a unique function in electronic and electrical engineering (and in mechanical engineering), we explain experimentation and evaluation of functionality in cases where both input signal and output data are expressed in complex numbers. The electronic and electrical technology regarding information systems generally consists of the following three types:

1. *Systems using power (amplitude)*: power supply or AM (amplitude modulation) system

2. *Systems using frequency:* measurement or FM (frequency modulation) system

3. *Systems using phase:* PM (phase modulation) system

After elucidating the significance of two-stage optimization, which is common to all categories, we provide details. Even a system classified as 2 or 3 above needs stability and functionality to power. In contrast, a receiving system (or transmitting and receiving system) focuses more on accuracy in receiving information than on stability of power. Therefore, we should develop a system by considering each functionality carefully. Energy difference for frequency difference is too small to measure directly. As for phase, energy difference for phase difference is so small that it cannot be measued. We need to calculate the SN ratio and sensitivity by

measuring frequency or phase itself for 2 or 3. Therefore, a procedure is needed
for studying functionality by computing the frequency or phase data in system 2
or 3, as explained later.

Quality engineering proposes that for all three categories above, after creating
as complicated a circuit or system as possible, we implement parameter design to
determine optimal levels of design parameters in the first stage and adjust them
to target values in the second stage. This is called *two-stage optimization*.

An ideal function shows a relationship between a signal and an output characteristic under certain conditions of use created by engineers. Indeed, no ideal
function exists in nature; however, the process of studying and designing a function
so as to match it with an ideal function is research and development. In quality
engineering, we conduct research by following two steps:

1. Reduce the variability of a function under various conditions.

2. Bring the function close to an ideal function under standard conditions.

The first step is evaluation by the SN ratio; the second, called *adjustment* or
tuning, brings a function close to an ideal by using traditional methods such as
least squares. Most conventional research has been conducted in reverse, that is,
the second step, then the first. In 1984, we visited the Circuit Laboratory, one of
the Bell Labs, for the first time. They were developing a new circuit. As a first step,
they developed it such that its function as a signal could match an objective function under standard conditions. Next, they tested whether its function could work
under 16 different types of conditions (environmental conditions or noise conditions, including a deterioration test). If it functioned well under the 16 conditions,
its development was completed. However, if not, they altered design constants
(parameters) so that the objective function would work under all 16 conditions.
Now, setting design constants to A, B, ... , and the 16 noise conditions to N_1, N_2,
... , N_{16}, we obtain

$$f(A, B, \ldots , N_i) = m \qquad (i = 1, 2, \ldots , 16) \tag{4.1}$$

In this case, m is a target value (or target curve). If we have 16 or more control
factors (called *adjusting factors* in quality engineering), we can solve equation (4.1).
If not, we determine A, B, ... , to minimize a sum of squared differences between
both sides of equation (4.1). In other words, we solve the following:

$$\min_{A,B,\ldots} \sum_{i=1}^{16} [f(A,B, \ldots , N_i) - m]^2 \tag{4.2}$$

This method is termed *least squares*. Although functional improvement based
on this method has long been used, the problem is that we cannot predict how
the function behaves under conditions other than the preselected 16 conditions.
It should be impossible to adjust the function to be objective function under a
variety of conditions of use.

Tuning or adjustment in quality engineering, adjust a certain value to a target
value or curve, prescribes that tuning must be done only under standard conditions. What type of characteristic or target curve a product should have under
standard conditions is a user's decision and an ideal function in design. Although
a target value or curve of a product can be selected by a designer before shipping,
it is predetermined by customers of the product or parent companies. In both
cases, the target value or curve needs to be studied under standard conditions.

This is because a production engineering or manufacturing department should produce products with a target value or function when shipped. Even if a product meets the target function at the point of shipment, as conditions of use vary or the product itself deteriorates when used, it sometimes malfunctions. This market quality issue is regarded as a functionality problem for which a design department should take full responsibility. The main objective of quality engineering is to take measures for functional variability under various conditions of use in the market. Indeed, we can study an ideal function for signals under standard conditions; however, we need to improve functionality for noises in the market when designing a product. When there is trouble in the market, no operator of a production plant can adjust the functions in that market. Thus, a designer should not only maintain effects for signals but also minimize functional variability caused by noises.

EFFECTIVE AND INEFFECTIVE ELECTRIC POWER

Functionality Evaluation of System Using Power (Amplitude)

In quality engineering, to assess functionality, we regard a sum of squared measurements as total output, decompose it into effective and harmful parts, and improve the ratio of the effective part to the harmful part by defining it as an SN ratio. To decompose measurements properly into quadratic forms, each measurement should be proportional to the square root of energy. Of course, the theory regarding quadratic form is related not to energy but to mathematics. The reason that each measurement is expected to be proportional to a square root of energy is that if total variation is proportional to energy or work, it can be decomposed into a sum of signals or noises, and each gain can be added sequentially.

In an engineering field such as radio waves, electric power can be decomposed as follows:

$$apparent\ power = effective\ power + reactive\ power$$

The effective power is measured in watts, whereas the apparent power is measured in volt-amperes. For example, suppose that input is defined as the following sinusoidal voltage:

$$input = E \sin(\omega t + \alpha) \tag{4.3}$$

and output is expressed by the following current:

$$output = I \sin(\omega t + \beta) \tag{4.4}$$

In this case, the effective power W is expressed as

$$W = \frac{EI}{2} \cos(\alpha - \beta) \tag{4.5}$$

$\cos(\alpha - \beta)$ is the *power factor*. On the other hand, since the

$$apparent\ power = \frac{EI}{2} \tag{4.6}$$

the reactive power is expressed by

$$reactive\ power = \frac{EI}{2} [1 - \cos(\alpha - \beta)] \tag{4.7}$$

Since in the actual circuit (as in the mechanical engineering case of vibration), phases α and β vary, we need to decompose total variation into parts, including not only effective energy but also reactive energy. This is the reason that variation decomposition by quadratic form in complex number or positive Hermitian form is required.

OUTPUT DECOMPOSITION BY HERMITIAN FORM

We set the signal to M, measurement to y, and the proportionality between both to

$$y = \beta M \tag{4.8}$$

If we know an ideal theoretical output, y should be selected as loss. If M, y, and β are all complex numbers, how can we deal with these? When y is a real number, defining the total variation of measurement y as total output S_T, we decompose S_T into useful and harmful parts:

$$S_T = y_1^2 + y_2^2 + \cdots + y_n^2 \tag{4.9}$$

Using the Hermitian form, which deal with the quadratic form in complex number, we express total variation as

$$S_T = y_1\bar{y}_1 + y_2\bar{y}_2 + \cdots + y_n\bar{y}_n \tag{4.10}$$

Now $\bar{y}$ is a complex conjugate number of y. What is important here is that S_T in equation (4.10) is a positive (strictly speaking, nonnegative) real number which reflects not only real number parts but also imaginary number parts.

The proportionality coefficient β is estimated by

$$\beta = \frac{\overline{M}_1 y_1 + \overline{M}_2 y_2 + \cdots + \overline{M}_n y_n}{M_1\overline{M}_1 + M_2\overline{M}_2 + \cdots + M_n\overline{M}_n} \tag{4.11}$$

In addition, the complex conjugate number of β is expressed by

$$\bar{\beta} = \frac{M_1\bar{y}_1 + M_2\bar{y}_2 + \cdots + M_n\bar{y}_n}{M_1\overline{M}_1 + M_2\overline{M}_2 + \cdots + M_n\overline{M}_n} \tag{4.12}$$

The variation of proportional terms is calculated as

$$S_\beta = \beta\bar{\beta}(M_1\overline{M}_1 + M_2\overline{M}_2 + \cdots + M_n\overline{M}_n)$$
$$= \frac{(\overline{M}_1 y_1 + \cdots + \overline{M}_n y_n)(M_1\bar{y}_1 + \cdots + M_n\bar{y}_n)}{M_1\overline{M}_1 + M_2\overline{M}_2 + \cdots + M_n\overline{M}_n} \quad (f = 1) \tag{4.13}$$

Therefore, the error variation is computed as

$$S_e = S_T - S_\beta \tag{4.14}$$

Since, in this case, a compounded noise factor is not included,

$$V_N = V_e = \frac{S_e}{n - 1} \tag{4.15}$$

The SN ratio and sensitivity are calculated as

$$\eta = 10 \log \frac{(1/r)\,(S_\beta - V_e)}{V_e} \tag{4.16}$$

$$S = 10 \log \frac{1}{r}\,(S_\beta - V_e) \tag{4.17}$$

Now r (effective divider), representing a magnitude of input, is expressed as

$$r = M_1\overline{M_1} + M_2\overline{M_2} + \cdots + M_n\overline{M_n} \tag{4.18}$$

In cases where there is a three-level compounded noise factor N, data can be tabulated as shown in Table 4.1 when a signal M has k levels. L_1, L_2, and L_3 are linear equations computed as

$$L_1 = \overline{M_1}\,y_{11} + \overline{M_2}\,y_{12} + \cdots + \overline{M_k}\,y_{1k}$$

$$L_2 = \overline{M_1}\,y_{21} + \overline{M_2}\,y_{22} + \cdots + \overline{M_k}\,y_{2k}$$

$$L_1 = \overline{M_1}\,y_{31} + \overline{M_2}\,y_{32} + \cdots + \overline{M_k}\,y_{3k} \tag{4.19}$$

Based on these conditions, we can find the variation and variance as follows:

Total variation:

$$S_T = y_{11}\bar{y}_{11} + y_{12}\bar{y}_{12} + \cdots + y_{3k}\bar{y}_{3k} \qquad (f = 3k) \tag{4.20}$$

Variation of proportional terms:

$$S_\beta = \frac{(L_1 + L_2 + L_3)(\overline{L_1} + \overline{L_2} + \overline{L_3})}{3r} \qquad (f = 1) \tag{4.21}$$

Effective divider:

$$r = M_1\overline{M_1} + M_2\overline{M_2} + \cdots + M_k\overline{M_k} \tag{4.22}$$

Variation of sensitivity variability:

$$S_{N\beta} = \frac{L_1\overline{L_1} + L_2\overline{L_2} + L_3\overline{L_3}}{r} - S_\beta \qquad (f = 2) \tag{4.23}$$

Error variation:

$$S_e = S_T - S_\beta - S_{N\beta} \qquad (f = 3k - 3) \tag{4.24}$$

Table 4.1
Input/output data

Noise	Signal				Linear Equation
	M_1	M_2	...	M_k	
N_1	y_{11}	y_{12}	...	y_{1k}	L_1
N_2	y_{21}	y_{22}	...	y_{2k}	L_2
N_3	y_{31}	y_{32}	...	y_{3k}	L_3

Total noise variation:

$$S_N = S_e + S_{N\beta} \qquad (f = 3k - 1) \tag{4.25}$$

Error variance for correction:

$$V_e = \frac{S_e}{3k - 3} \tag{4.26}$$

Noise magnitude including nonlinearity (total error variance):

$$V_N = \frac{S_N}{3k - 1} \tag{4.27}$$

SN ratio:

$$\eta = 10 \log \frac{(1/3r)\ (S_\beta - V_e)}{V_N} \tag{4.28}$$

Sensitivity:

$$S = 10 \log \frac{1}{3r} (S_\beta - V_e) \tag{4.29}$$

We have thus far described application of the Hermitian form to basic output decomposition. This procedure is applicable to decomposition of almost all quadratic forms because all squared terms are replaced by products of conjugates when we deal with complex numbers. From here on we call it by a simpler term, *decomposition of variation,* even when we decompose variation using the Hermitian form because the term *decomposition of variation by the Hermitian form* is too lengthy.

Moreover, we wish to express even a product of two conjugates y and $\bar{y}$ as y^2 in some cases. Whereas y^2 denotes itself in the case of real numbers, it denotes a product of conjugates in the case of complex numbers. By doing this, we can use both the formulas and degrees of freedom expressed in the positive quadratic form. Additionally, for decomposition of variation in the Hermitian form, decomposition formulas in the quadratic form hold true. However, as for noises, we need to take into account phase variability as well as power variability because the noises exist in the complex plane. Thus, we recommend that the following four-level noise N be considered:

N_1:   negative condition for both power and phase

N_2:   negative condition for power, positive for phase

N_3:   positive condition for power, negative for phase

N_4:   positive condition for both power and phase

Change of phase is equivalent to change of reactance. By studying noises altering reactance, we set up positive and negative levels.

Except for noise levels, we can calculate the SN ratio in the same manner as that for the quadratic form. After maximizing the SN ratio, we aim at matching sensitivity β (a complex number) with a target value. To adjust both real and imaginary parts to target values, instead of sensitivity β^2 we need to consider control factor effects for β itself.

If we compute a theoretical output value in case of no loss, by redefining a deviation from a theoretical output value as an output datum y, we analyze with SN ratio and sensitivity. Although the calculation procedure is the same, sensitivity with loss should be smaller. In some cases, for a smaller-the-better function, we use a smaller-the-better SN ratio using all data.

HOW TO HANDLE FREQUENCY

Setting frequency to f and using Planck's constant h, we obtain frequency energy E as follows:

$$E = hf \tag{4.30}$$

When E is expressed in joules (J), h becomes the following tiny number:

$$h = 6.62 \times 10^{-34} \text{ J/s} \tag{4.31}$$

In this case, E is too small to measure. Since we cannot measure frequency as part of wave power, we need to measure frequency itself as quantity proportional to energy. As for output, we should keep power in an oscillation system sufficiently stable. Since the system's functionality is discussed in the preceding section, we describe only a procedure for measuring the functionality based on frequency.

If a signal has one level and only stability of frequency is in question, measurement of time and distance is regarded as essential. Stability for this case is nominal-the-best stability of frequency, which is used for measuring time and distance. More exactly, we sometimes take a square root of frequency.

STABILITY OF TRANSMITTING WAVE

The stability of the transmitting wave is considered most important when we deal with radio waves. (A laser beam is an electromagnetic wave; however, we use the term *radio wave*. The procedure discussed here applies also for infrared light, visible light, ultraviolet light, x-rays, and gamma rays.) The stability of radio waves is categorized as either stability of phase or frequency or as stability of power. In the case of frequency-modulated (FM) waves, the former is more important because even if output fluctuates to some degree, we can easily receive phase or frequency properly when phase or frequency is used.

In Section 7.2 of Reference 2, for experimentation purposes the variability in power source voltage is treated as a noise. The stability of the transmitting wave is regarded as essential for any function using phase or frequency. Since only a single frequency is sufficient to stabilize phase in studying the stability of the transmitting wave, we can utilize a nominal-the-best SN ratio in quality engineering. More specifically, by selecting variability in temperature or power source voltage or deterioration as one or more noises, we calculate the nominal-the-best SN ratio. If the noise has four levels, by setting the frequency data at N_1, N_2, N_3, and N_4 to y_1, y_2, y_3, and y_4, we compute the nominal-the-best SN ratio and sensitivity as follows:

$$\eta = 10 \log \frac{\frac{1}{4}(S_m - V_e)}{V_e} \tag{4.32}$$

$$S = 10 \log \tfrac{1}{4}(S_m - V_e) \tag{4.33}$$

Quality Engineering of System Using Frequency

Now

$$S_m = \frac{(y_1 + y_2 + y_3 + y_4)^2}{4} \qquad (f = 1) \qquad\qquad (4.34)$$

$$S_T = y_1^2 + y_2^2 + y_3^2 + y_4^2 \qquad (f = 4) \qquad\qquad (4.35)$$

$$S_e = S_T - S_m \qquad (f = 3) \qquad\qquad (4.36)$$

$$V_e = \tfrac{1}{3} S_e \qquad\qquad\qquad\qquad (4.37)$$

After determining the optimal SN ratio conditions, we perform a confirmatory experiment under those conditions and compute the corresponding frequency. When dealing with a clock, its frequency can be set to 1 second. If we have a target value for a frequency, we can bring the frequency close to the target by using control factors affecting sensitivity S greatly or changing the effective inductance or capacitance. Since tuning is relatively easy for radio waves, sensitivity data are often of no importance.

To generalize this procedure, we recommend using multiple frequencies. We select K levels from a certain band and measure data at two compounded noises, N_1 and N_2, as illustrated in Table 4.2. For signal M, k levels can be selected arbitrarily in a certain band. For example, when we obtain k levels by changing the inductance, we do not need to consider an ideal relationship between inductance and M. If an equation representing such a relationship exists, it does not matter so much, and we do not have to include a deviation from the equation in errors. Therefore, we can decompose total variation as shown in Table 4.3. The effect of signal M is the effect of all k levels.

$$S_M = \frac{(y_{11} + y_{12})^2 + \cdots + (y_{k1} + y_{k2})^2}{2} \qquad (f = k) \qquad\qquad (4.38)$$

$$S_N = \frac{[(y_{11} + \cdots + y_{k1}) - (y_{12} + \cdots + y_{k2})]^2}{2k} \qquad\qquad (4.39)$$

$$S_e = S_T - S_M - S_N \qquad\qquad (4.40)$$

S_T is now the total variation of output.

Table 4.2
Cases where signal factor exists

	Noise		
Signal	N_1	N_2	Total
M_1	y_{11}	y_{12}	y_1
M_2	y_{21}	y_{22}	y_2
⋮	⋮	⋮	⋮
M_k	y_{k1}	y_{k2}	y_k

Table 4.3
Decomposition of total variation

Factor	f	S	V
M	k	S_M	V_M
N	1	S_N	
e	$k-1$	S_e	V_e
Total	$2k$	S_T	

The SN ratio and sensitivity are as follows:

$$\eta = 10 \log \frac{\frac{1}{2}(V_M - V_e)}{V_N} \tag{4.41}$$

$$S = 10 \log \tfrac{1}{2}(V_M - V_e) \tag{4.42}$$

Now

$$V_M = \frac{S_M}{k} \tag{4.43}$$

$$V_e = \frac{S_e}{k-1} \tag{4.44}$$

$$V_N = \frac{S_N + S_e}{k} \tag{4.45}$$

Since we consider stabilizing all frequencies in a certain range of frequency, this procedure should be more effective than a procedure of studying each frequency.

THE CASE OF MODULATION FUNCTION
When we transmit (transact) information using a certain frequency, we need frequency modulation. The technical means of modulating frequency is considered to be the signal. A typical function is a circuit that oscillates a radio wave into which we modulate a transmitting wave at the same frequency as that of the transmitting wave and extract a modulated signal using the difference between two waves. By taking into account various types of technical means, we should improve the following proportionality between input signal M and measurement characteristic y for modulation:

$$y = \beta M \quad (y \text{ and } M \text{ take both positive and negative numbers}) \tag{4.46}$$

Thus, for modulation functions, we can define the same SN ratio as before. Although y as output is important, before studying y we should research the frequency stability of the transmitting wave and internal oscillation as discussed in the preceding section.

When we study the relationship expressed in equation (4.46) by setting a difference between a transmitting wave and an internally oscillated radio wave to output, it is necessary for us to measure data by changing not only conditions regarding the environment or deterioration as noises but also frequency (not a modulated signal but a transmitting wave). If we select two levels for noise N, three levels for frequency F, and k levels for modulated signal M, we obtain the data shown in Table 4.4. Frequency F is called an *indicative factor*. According to this, we decompose total variation as shown in Table 4.5.

Now r represents the total variation of the input modulated signal. The calculation procedure is as follows:

$$S_\beta = \frac{(L_1 + L_2 + \cdots + L_6)^2}{6r} \qquad (f = 1) \qquad (4.47)$$

$$S_{F\beta} = \frac{(L_1 + L_4)^2 + \cdots + (L_3 + L_6)^2}{2r} - S_\beta \qquad (f = 2) \qquad (4.48)$$

$$S_{N(F)\beta} = \frac{L_1^2 + \cdots + L_6^2}{r} - S_\beta - S_{F\beta} \qquad (f = 3) \qquad (4.49)$$

$$S_e = S_T - (S_\beta + S_{F\beta} + S_{N(F)\beta}) \qquad (F = 6k - 6) \qquad (4.50)$$

Using these results, we calculate the SN ratio and sensitivity:

$$\eta = 10 \log \frac{(1/6r)(S_\beta - V_e)}{V_N} \qquad (4.51)$$

$$S = 10 \log \frac{1}{6r}(S_\beta - V_e) \qquad (4.52)$$

Now

$$V_N = \frac{S_{N(F)\beta} + S_e}{6k - 3} \qquad (4.53)$$

FUNCTIONALITY OF SYSTEM USING PHASE
In the case of analog modulation, modulation technology has evolved from AM and FM to PM. Indeed, digital systems also exist; however, in quality engineering

Table 4.4
Case where modulated signal exists

Noise	Frequency	Modulated Signal				Linear Equation
		M_1	M_2	$\cdots$	M_k	
N_1	F_1	y_{11}	y_{12}	$\cdots$	y_{1k}	L_1
	F_2	y_{21}	y_{22}	$\cdots$	y_{2k}	L_2
	F_3	y_{31}	y_{32}	$\cdots$	y_{3k}	L_3
N_2	F_1	y_{41}	y_{42}	$\cdots$	y_{4k}	L_4
	F_2	y_{51}	y_{52}	$\cdots$	y_{5k}	L_5
	F_3	y_{61}	y_{62}	$\cdots$	y_{6k}	L_6

Table 4.5
Decomposition of total variation

Factor	f	S	V
β	1	S	
$F\beta$	2	$S_{F\beta}$	
$N(F)\beta$	3	$S_{N(F)\beta}$	
e	$6(k-1)$	S_e	V_e
Total	$6k$	S_T	

we should research and design only analog functions because digital systems are included in all AM, FM, and PM, and its only difference from analog systems is that its level value is not continuous but discrete: that is, if we can improve SN ratio as an analog function and can minimize each interval of discrete values and eventually enhance information density.

❏ Example

A phase-modulation digital system for a signal having a 30° phase interval such as 0, 30, 60, ... , 330° has a σ value that is expressed in the following equation, even if its analog functional SN ratio is −10 dB:

$$10 \log \frac{\beta^2}{\sigma^2} = -10 \tag{4.54}$$

By taking into account that the unit of signal M is degrees, we obtain the following small value of σ:

$$\sigma = 31.6\beta = 3.16° \tag{4.55}$$

The fact that σ representing RMS is approximately 3° implies that there is almost no error as a digital system because 3° represents one-fifth of a function limit regarded as an error in a digital system. In Table 4.6 all data are expressed in radians. Since the SN is 48.02 dB, σ is computed as follows:

$$\sigma = \sqrt{10^{-4.802}}\,\beta = 0.00397 \text{ rad}$$
$$= 0.22° \tag{4.56}$$

The ratio of this σ to the function limit of ±15° is 68. Then, if noises are selected properly, almost no error occurs.

In considering the phase modulation function, in most cases frequency is handled as an indicative factor (i.e., a use-related factor whose effect is not regarded as noise.) The procedure of calculating its SN ratio is the same as that for a system

Table 4.6
Data for phase shifter (experiment 1 in L_{18} orthogonal array; rad)

Temperature	Frequency	Voltage		
		V_1	V_2	V_3
T_1	F_1	$77 + j101$	$248 + j322$	$769 + j1017$
	F_2	$87 + j104$	$280 + j330$	$870 + j1044$
	F_3	$97 + j106$	$311 + j335$	$970 + j1058$
T_2	F_1	$77 + j102$	$247 + j322$	$784 + j1025$
	F_2	$88 + j105$	$280 + j331$	$889 + j1052$
	F_3	$98 + j107$	$311 + j335$	$989 + j1068$

using frequency. Therefore, we explain the calculation procedure using experimental data in *Technology Development for Electronic and Electric Industries* [3].

☐ Example

This example deals with the stability design of a phase shifter to advance the phase by 45°. Four control factors are chosen and assigned to an L_{18} orthogonal array. Since the voltage output is taken into consideration, input voltages are selected as a three-level signal. However, the voltage V is also a noise for phase modulation. Additionally, as noises, temperature T and frequency F are chosen as follows:

Voltage (mV): $V_1 = 224$, $V_2 = 707$, $V_3 = 2234$

Temperature (°C): $T_1 = 10$, $T_2 = 60$

Frequency (MHz): $F_1 = 0.9$, $F_2 = 1.0$, $F_3 = 1.1$

Table 4.6 includes output voltage data expressed in complex numbers for 18 combinations of all levels of noises V, T, and F. Although the input voltage is a signal for the output voltage, we calculate the phase advance angle based on complex numbers because we focus only on the phase-shift function:

$$y = \tan^{-1} \frac{Y}{X} \tag{4.57}$$

In this equation, X and Y indicate real and imaginary parts, respectively. Using a scalar Y, we calculate the SN ratio and sensitivity.

As for power, we compute the SN ratio and sensitivity according to the Hermitian form described earlier in the chapter. Because the function is a phase advance (only 45° in this case) in this section, the signal has one level (more strictly, two phases, 0 and 45°). However, since our focal point is only the difference, all that we need to do is to check phase advance angles for the data in Table 4.6. For the sake of convenience, we show them in radians in Table 4.7, whereas we can also use degrees as the unit. If three phase advance angles, 45, 90, and 135°, are selected, we should consider dynamic function by setting the signal as M.

Table 4.7
Data for phase advance angle (rad)

Temperature	Frequency	Voltage			Total
		V_1	V_2	V_3	
T_1	F_1	0.919	0.915	0.923	2.757
	F_2	0.874	0.867	0.876	2.617
	F_3	0.830	0.823	0.829	2.482
T_2	F_1	0.924	0.916	0.918	2.758
	F_2	0.873	0.869	0.869	2.611
	F_3	0.829	0.823	0.824	2.476

Since we can adjust the phase angle to 45° or 0.785 rad (the target angle), here we pay attention to the nominal-the-best SN ratio to minimize variability. Now T and V are true noises and F is an indicative factor. Then we analyze as follows:

$$S_T = 0.919^2 + 0.915^2 + \cdots + 0.824^2 = 13.721679$$

$$(f = 18) \tag{4.58}$$

$$S_m \frac{15.701^2}{18} = 13.695633$$

$$(f = 1) \tag{4.59}$$

$$S_F = \frac{5.515^2 + 5.228^2 + 4.958^2}{6} - S_m = 0.0258625$$

$$(f = 2) \tag{4.60}$$

$$S_e = S_T - S_m - S_F = 0.0001835$$

$$(f = 15) \tag{4.61}$$

$$V_e = \frac{S_e}{15} (0.00001223) \tag{4.62}$$

Based on these results, we obtain the SN ratio and sensitivity as follows:

$$\eta = 10 \log \frac{\frac{1}{18}(13.695633 - 0.00001223)}{0.00001223}$$

$$= 47.94 \text{ dB} \tag{4.63}$$

$$S = 10 \log \tfrac{1}{18}(13.695633 - 0.00001223)$$

$$= -1.187 \text{ dB} \tag{4.64}$$

The target angle of 45° for signal S can be expressed in radians as follows:

$$S = 10 \log \frac{45}{180} (3.1416) = -1.049 \text{ dB} \qquad (4.65)$$

As a next step, under optimal conditions, we calculate the phase advance angle and adjust it to the value in equation (4.65) under standard conditions. This process can be completed by only one variable.

When there are multiple levels for one signal factor, we should calculate the dynamic SN ratio. When we design an oscillator using signal factors, we may select three levels for a signal to change phase, three levels for a signal to change output, three levels for frequency F, and three levels for temperature as an external condition, and assign them to an L_9 orthogonal array. In this case, for the effective value of output amplitude, we set a proportional equation for the signal of voltage V as the ideal function. All other factors are noises. As for the phase modulation function, if we select the following as levels of a phase advance and retard signal,

Level targeted at $-45°$: $M_1' - M_2' = m_1$

Level targeted at $0°$: $M_2' - M_2' = m_2$

Level targeted at $+45°$: $M_3' - M_2' = m_3$

the signal becomes zero-point proportional. The ideal function is a proportional equation where both the signal and output go through the zero point and take positive and negative values. If we wish to select more signal levels, for example, two levels for each of the positive and negative sides (in total, five levels, M_1, M_2, M_3, M_4, and M_5), we should set them as follows:

$$M_1 = -120°$$

$$M_2 = -60°$$

$$M_3 = M_4 = 0 \quad \text{(dummy)}$$

$$M_5 = +60°$$

$$M_6 = +120°$$

Next, together with voltage V, temperature T, and frequency F, we allocate them to an L_{18} orthogonal array to take measurements. On the other hand, by setting $M_1 = -150°$, $M_2 = -90°$, $M_3 = -30°$, $M_4 = +30°$, $M_5 = +90°$, and $M_6 = +150°$, we can assign them to an L_{18} array. After calibrating the data measured using $M_0 = 0°$, they are zero proportional.

DESIGN BY SIMULATION

In electronic and electric systems, theories showing input/output relationships are available for analysis. In some cases, approximate theories are available. Instead of experimenting, we can study functionality using simulation techniques on the computer. About half of the electrical and electronic design examples in References 1 and 4 are related to functionality (reliability) design by simulation, even though there are not as many electrical and electronic examples. Even if a theory is imperfect or partial, functionality design often goes well. This is because in quality engineering we study basic functionality first and then adjust the outputs to target values with a few parameters on an actual circuit or machine.

Chapter 16 of Reference 4 illustrates Hewlett-Packard's simulation design using complex numbers, in which real and imaginary parts are dealt with separately for parameter design. Ideally, it would be better to design parameters for power according to the procedure discussed earlier and to study phase stability based on the technique shown earlier. It was necessary to allocate all control factors selected to an L_{108} orthogonal array. In place of analyzing real and imaginary parts separately, parameter design should be performed using the data of effective amplitude and phase angle. However, decomposition of total variation by the Hermitian form was not known at that time.

If simulation calculation does not take long, after creating three noise factor levels before and after each control factor level, we should assign them to an orthogonal array. Then, to compute the SN ratio, we measure data for the noise-assigned orthogonal array (called the *outer array*) corresponding to each experiment in the orthogonal array.

In the twenty-first century, simulation will play a major role in R&D. In this situation, partially established theories or approximation methods suffice for studying functional variability because such techniques can help improve robustness. Tuning the outputs to the objective functions may be performed using hardware.

4.3. Mechanical Engineering

A research way of reducing variability only because an objective function is varying is regarded as *whack-a-mole*, or unplanned. Although this problem-solving method should not be used, engineers have been trying to prevent problems with a new product caused by current research and design. Ideally, we should prevent quality problems in the manufacturing process or the market through optimization at the design stage.

Conventional Meaning of Robust Design

Although the Nippon Telegraph and Telephone Public Corporation (NTT) designed exchangers or telephones in the 1950s, those who produced these products were private companies such as NEC, Hitachi, Fujitsu, and Oki. Therefore, without knowing how to produce them, NTT designed them in such a way that they could prevent quality problems in the market.

While the system realization department conceptualized future transmission systems and requested devices, units, materials, and parts with new functions needed at that time, the device realization department developed new products and technologies to meet the requests. That is, the telecommunication laboratory did not have a design department. The results from research and development were drawings and specifications, and the manufacturers of communication devices noted above produced products according to these specs.

A design department is responsible for avoiding problems in the manufacturing process or the marketplace before mass production or shipping is begun. Therefore, in the design phase, manufacturers had to improve quality (functional variability) in the manufacturing process or the marketplace without knowing the type of manufacturing process that would be used or how quality would be controlled. In fact, no problems occurred. Researchers knew the objective functions of products they had developed themselves. In fact, they knew not only the objective functions but also the generic functions that were the means to attaining the objective functions. The mistake that researchers in Europe and the United States made in evaluating objective or generic functions was not related to the creation of technical means.

Although these researchers are creative and talented, their research procedure was first to develop and design a function under standard conditions, then test functionality under those conditions and study countermeasures against the problems through design changes. Then, by modifying product design, they take countermeasures for functional variability (functionality) caused by conditions of production or use after discovering functional problems through tests under various conditions. Needless to say, this can be regarded as "whack-a-mole" research.

Quality engineering recommends that we modify product design using an L_{18} orthogonal array before finding problems. Researchers quite often say that they cannot consider technical measures without looking at actual problems. The conventional design procedure comprises the following steps:

1. Under standard conditions, we design a product or system that each of us considers optimal.

2. To find design problems, we test the product or system under various conditions.

3. We change the design of the product or system (called *tuning*, a countermeasure based on cause-and-effect relationships) such that it functions well even under conditions where the original design does not work properly.

Quite often the second and third steps are repeated. We call this procedure "whack-a-mole." Quality engineering insists that such research methods be stopped.

New Method: Functionality Design

Quality engineering recommends improving functionality before a quality problem (functional variability) occurs. In other words, before a problem happens, a designer makes improvements. Quite a few researchers or designers face difficulties understanding how to change their designs. Quality engineering advises that we check on changes in functionality using SN ratios by taking advantage of any type of design change. A factor in design change is called a *control factor*. Most control factors should be assigned to an L_{18} orthogonal array. A researcher or designer should have realized and defined both the objective and generic functions of the product being designed.

❏ Example

Imagine a vehicle steering function, more specifically, when we change the orientation of a car by changing the steering angle. The first engineer who designed such a system considered how much the angle could be changed within a certain range of steering angle. For example, the steering angle range is defined by the rotation of a steering wheel, three rotations for each, clockwise and counterclockwise, as follows:

$$(-360)(3) = -1080°$$

$$(360)(3) = +1080°$$

In this case, the total steering angle adds up to 2160°. Accordingly, the engineer should have considered the relationship between the angle above and steering curvature y, such as a certain curvature at a steering angle of $-1080°$, zero curvature at 0, and a certain curvature at $+1080$.

For any function, a developer of a function pursues an ideal relationship between signal M and output y under conditions of use. In quality engineering, since any function can be expressed by an input/output relationship of work or energy, essentially an ideal function is regarded as a proportionality. For instance, in the steering function, the ideal function can be given by

$$y = \beta M \tag{4.66}$$

In this equation, y is curvature (the reciprocal of a turning radius) and M is a steering angle range. European or U.S. researchers (especially inventors of functions) who develop a product or system know that a relationship exists between its signal and output characteristic y. They make efforts to design functions and invent a number of techniques in order to match a function with an ideal function under standard conditions. This matching procedure known as *tuning*, utilizes the least-squares method in mathematical evaluation.

Quality engineering maintains that we first improve functionality (functional variability) under conditions of use by deferring until later the adjustment of a function to the ideal function under standard conditions. Conditions of use are various environmental conditions under which a product or system is used, including both environmental conditions and deterioration. For the case of the steering function, some environmental conditions are road conditions and a tire's frictional condition, both of which are called *noises*. In quality engineering we compound all noises into only two levels, N_1 and N_2, no matter how many there are.

Now, we consider compounding noises for the steering function. For the output, curvature of radius, we define two levels of a compounded error factor N:

N_1: conditions where the proportionality coefficient β becomes smaller

N_2: Conditions where the proportionality coefficient β becomes larger

In short, N_1 represents conditions where the steering angle functions well, whereas N_2 represents conditions where it does not function well. For example, the former represents a car with new tires running on a dry asphalt road. In contrast, the latter is a car with worn tires running on a wet asphalt or snowy road.

By changing the vehicle design, we hope to mitigate the difference between N_1 and N_2. Quality engineering actively takes advantage of the interaction between control factors and a compounded error factor N to improve functionality. Then we measure data such as curvature, as shown in Table 4.8 (in the counterclockwise turn case, negative signs are added, and vice versa). Each M_i value means angle.

Table 4.8 shows data of curvature y, or the reciprocal of the turning radius for each signal value of a steering angle ranging from zero to almost maximum for both clockwise and counterclockwise directions. If we turn left or right on a normal road or steer on a highway, the range of data should be different. The data in Table 4.8 represent the situation whether we are driving a car in a parking lot or steering in a hairpin curve at low speed. The following factor of speed K, indicating different conditions of use, called an *indicative factor*, is often assigned to an outer orthogonal array. This is because a signal factor value varies in accordance with each indicative factor level.

$$K_1: \text{low speed (less than 20 km/h)}$$

$$K_2: \text{middle speed (20 to 60 km/h)}$$

$$K_3: \text{high speed (more than 60 km/h)}$$

The SN ratio and sensitivity for the data in Table 4.8 are computed as follows. First, we calculate two linear equations, L_1 and L_2, using N_1 and N_2:

$$L = M_1 y_{11} + M_2 y_{12} + \cdots + M_7 y_{17} \tag{4.67}$$
$$L_2 = M_1 y_{21} + M_2 y_{21} + M_2 y_{22} + \cdots + M_7 y_{27} \tag{4.68}$$

Next, we decompose the total variation of data in Table 4.8.

$$S_T = \text{total variation} \quad (f = 14) \tag{4.69}$$

Variation of proportional terms:

$$S_\beta = \frac{(L_1 + L_2)^2}{2r} \tag{4.70}$$

where

$$r = M_1^2 + M_2^2 + \cdots + M_7^2$$
$$= (-720)^2 + (-480)^2 + \cdots + 720^2$$
$$= 1{,}612{,}800 \tag{4.71}$$

Table 4.8
Curvature data

	M_1 -720	M_2 -480	M_3 -240	M_4 0	M_5 240	M_6 480	M_7 720
N_1	y_{11}	y_{12}	y_{13}	y_{14}	y_{15}	y_{16}	y_{17}
N_2	y_{21}	y_{22}	y_{23}	y_{24}	y_{25}	y_{26}	y_{27}

Although we choose $1°$ as a unit of magnitude of signal r, we can choose $120°$ and eventually obtain $r = 112$. That is, designers can define a unit of signal as they wish. On the other hand, the difference of proportionality coefficient for signal M (difference between linear equations N_1 and N_2), $S_{N\beta}$, is calculated

$$S_{N\beta} = \frac{(L_1 - L_2)^2}{2r} \tag{4.72}$$

Therefore, the error variation

$$S_e = S_T - (S_\beta + S_{N\beta}) \tag{4.73}$$

As a result of decomposing total variation, we obtain Table 4.9.
Finally, we can compute SN ratio and sensitivity as follows:

$$\eta = 10 \log \frac{(1/2r)\,(S_\beta - V_e)}{V_N} \tag{4.74}$$

$$\hat{\beta} = S = \frac{L_1 + L_2}{2r} \tag{4.75}$$

$$V_e = \frac{S_e}{12} \tag{4.76}$$

$$V_N = \frac{S_{N\beta} + S_e}{13} \tag{4.77}$$

The reason that sensitivity S is calculated by equation (4.75) instead of using S_β is because S_β always takes a positive value. But sensitivity is used to adjust β to target, and β may be either positive or negative, since the signal (M), takes both positive and negative values.

A major aspect in which quality engineering differs from a conventional research and design procedure is that QE focuses on the interaction between signal and noise factors, which represent conditions of design and use. As we have said

Table 4.9
Decomposition of total variation

Source	f	S	V
β	1	S_β	
$N\beta$	1	$S_{N\beta}$	$V_e \brace V_N$
e	12	S_e	
Total	14	S_T	

previously, in quality engineering, as a first step, we improve functionality, and second, do tuning (adjustment) on a difference between a function at an optimal SN ratio condition and an ideal function under standard conditions.

ADVANTAGES AND DISADVANTAGES OF THE NEW METHOD

Since our new method focuses on functions, we do not need to find problems through a number of tests, and setting up noise conditions does not have to be perfect. The new method has five major advantages:

1. Before product planning, it is possible to study subsystems (including over-looked items) or total system functionality. In other words, the study of functional variability at the product design stage is unnecessary, leading to earlier delivery.

2. When we improve functionality using generic functions, such developed common technologies are applicable to a variety of products. This helps not only the current project plan, but the next and consequently streamlines the research process.

3. Instead of considering all possible noise conditions such as in the traditional reliability study, only a few noise conditions selected from conditions of use are needed for analysis.

4. Since only the type of functionality is sufficient for each signal factor type, we can attain optimal design simply and quickly. Even if there are multiple types of signal factors, we can study them at the same time.

5. When systems are going to be connected, only timing is needed, without the study on variability.

The disadvantages of our new method are as follows:

1. It is necessary to persuade engineers who have never thought about the concept of signal factors to acknowledging the importance of selecting them. However, this might also be regarded as an advantage or key role of our new method because we can teach them engineering basics. That may stop engineers using easy ways to measure something related to an objective and let them look back to the essence of engineering. Some engineers work extremely hard and long hours but still often experience unexpected problems in the marketplace. The necessity of preventing such problems forces engineers to consider signal factors and to evaluate functionality using SN ratios under a variety of noise conditions because this method of improving design is quite effective.

2. Engineers need to calculate SN ratios, but this does not mean that they must do the actual calculation. Calculation is not a burden to engineers, since computer software is available. What they must do is to think about the engineering meaning of an SN ratio, then determine what type of output generated by the signal factor level must be collected. Engineers must understand the new evaluation method: Use only one SN ratio to express a function.

3. In quality engineering, when we consider a generic function, both input signal and output characteristic are not objective related but are work and energy related. However, since we have difficulties measuring work and energy in many cases, we have no other choice but to select an objective func-

tion. For example, the steering function discussed earlier is not a generic but an objective function. When an objective function is chosen, we need to check whether there is an interaction between control factors after allocating them to an orthogonal array, because their effects (differences in SN ratio, called *gain*) do not necessarily have additivity. An L_{18} orthogonal array is recommended because its size is appropriate enough to check on the additivity of gain. If there is no additivity, the signal and measurement characteristics we used need to be reconsidered for change.

4. Another disadvantage is the paradigm shift for orthogonal array's role. In quality engineering, control factors are assigned to an orthogonal array. This is not because we need to calculate the main effects of control factors for the measurement characteristic *y*. Since levels of control factors are fixed, there is no need to measure their effects for *y* using orthogonal arrays. The real objective of using orthogonal arrays is to check the reproducibility of gains, as described in disadvantage 3.

In new product development, there is no existing engineering knowledge most of the time. It is said that engineers are suffering from being unable to find ways for SN ratio improvement. In fact, the use of orthogonal arrays enables engineers to perform R&D for a totally new area. In the application, parameter-level intervals may be increased and many parameters may be studied together. If the system selected is not a good one, there will be little SN ratio improvement. Often, one orthogonal array experiment can tell us the limitation on SN ratios. This is an advantage of using an orthogonal array rather than a disadvantage.

We are asked repeatedly how we would solve a problem using quality engineering. Here are some generalities that we use.

Problem Solving and Quality Engineering

1. Even if the problem is related to a consumer's complaint in the marketplace or manufacturing variability, in lieu of questioning its root cause, we would suggest changing the design and using a generic function. This is a technical approach to problem solving through redesign.

2. To solve complaints in the marketplace, after finding parts sensitive to environmental changes or deterioration, we would replace them with robust parts, even if it incurs a cost increase. This is called *tolerance design*.

3. If variability occurs before shipping, we would reduce it by reviewing process control procedures and inspection standards. In quality engineering this is termed *on-line quality engineering*. This does not involve using Shewhart control charts or other management devices but does include the following daily routine activities by operators:

   a. *Feedback control.* Stabilize processes continuously for products to be produced later by inspecting product characteristics and correcting the difference between them and target values.

   b. *Feedforward control.* Based on the information from incoming materials or component parts, predict the product characteristics in the next process and calibrate processes continuously to match product characteristics with the target.

   c. *Preventive maintenance.* Change or repair tools periodically with or without checkup.

      d. *Inspection design.* Prevent defective products from causing quality problems in downstream processes through inspection.

These are daily routine jobs designed to maintain quality level. Rational design of these systems is related to on-line quality engineering and often leads to considerable improvement.

Signal and Output in Mechanical Engineering

In terms of the generic function representing mechanical functionality,

$$y = \beta M \tag{4.78}$$

both signal M and output characteristic y should be proportional to the square root of energy or work. That is, we need to analyze using its square root because total output can be expressed as follows:

$$S_T = y_1^2 + \cdots + y_n^2 \tag{4.79}$$

By doing so, we can make S_T equivalent to work, and as a result, the following simple decomposition can be completed:

$$\text{total work} = (\text{work by signal}) + (\text{work by noise}) \tag{4.80}$$

This type of decomposition has long been used in the telecommunication field. In cases of a vehicle's acceleration performance, M should be the square root of distance and y the time to reach. After choosing two noise levels, we calculate the SN ratio and sensitivity using equation (4.80), which holds true for a case where energy or work is expressed as a square root. But in many cases it is not clear if it is so. To make sure that (4.80) holds true, it is checked by the additivity of control factor effects using an orthogonal array. According to Newton's law, the following formula shows that distance y is proportional to squared time t^2 under constant acceleration:

$$y = \frac{b}{2} t^2 \tag{4.81}$$

From a quality engineering viewpoint, it is not appropriate to use (4.81). Instead, taking ease of calculation into account, we should use

$$t = \sqrt{\frac{2}{b}} \sqrt{y}$$

$$t = \beta M \qquad \beta = \sqrt{\frac{2}{b}} \qquad \sqrt{y} = M \tag{4.82}$$

Generic Function of Machining

The term *generic function* is used quite often in quality engineering. This is not an objective function but a function as the means to achieve the objective. Since a function means to make something work, work should be measured as output. For example, in case of cutting, it is the amount of material cut. On the other hand, as work input, electricity consumption or fuel consumption can be selected. When

Ford Motor Co. experimented on functionality improvement of an engine during idling in 1995, they chose a signal, output, and ideal function as follows:

> *Signal M:* amount of fuel flow
>
> *Output y:* indicated mean effective pressure (IMEP)
>
> *Ideal function:* $y = \beta M$

Now, output y indicates average pressure to rotate a crankshaft, which is regarded as a force for work. However, the generic function of an engine is a chemical reaction, discussed in the next section.

Quality engineering maintains that what is picked up as signal, what is measured, and what types of noises are chosen depends on the theme selected by researchers. Since the output of work is a request from users (consumers or operators in manufacturing), work should be selected as signal, and the amount needed for the work (in case of machinery, used power) should be measured as data. In the experiment on cutting conducted by JR in 1959 (known as National Railway at that time), the power needed for cutting was measured as net cutting power in the unit kilowatthours.

In an actual experiment, we should take measurements as follows: Selecting test pieces and cutting them for different durations, such as $T_1 = 10$ s, $T_2 = 20$ s, and $T_3 = 30$ s, we measure the amount of cut M (grams) and electricity consumption y, as shown in Table 4.10. N indicates a two-level noise. According to what these data show, we perform two different analyses.

1. *Analysis of cutting performance.* This analysis is based on the following ideal function:

$$y = \beta M \tag{4.83}$$

Since in this case, both signal of amount of cut M and data of electricity consumption y are observations, M and y have six levels and the effects of N are not considered. For a practical situation, we should take the square root of both y and M. The linear equation is expressed as

$$L = M_{11}y_{11} + M_{12}y_{12} + \cdots + M_{23}y_{23} \tag{4.84}$$

$$r = M_{11}^2 + M_{12}^2 + \cdots + M_{23}^2 \tag{4.85}$$

Table 4.10
Data for machining

	T_1	T_2	T_3
N_1	M_{11}	M_{12}	M_{13}
	y_{11}	y_{12}	y_{13}
N_2	M_{21}	M_{22}	M_{23}
	y_{21}	y_{22}	y_{23}

Regarding each variation, we find that

$$S_T = y_{11}^2 + y_{12}^2 + \cdots + y_{23}^2 \qquad (f = 6) \tag{4.86}$$

$$S_\beta = \frac{L^2}{r} \qquad (f = 1) \tag{4.87}$$

$$S_e = S_T - S_\beta \qquad (f = 5) \tag{4.88}$$

$$V_e = \frac{S_e}{5} \tag{4.89}$$

Using these results, we obtain the SN ratio and sensitivity:

$$n = 10 \log \frac{(1/r)\ (S_\beta - V_e)}{V_e} \tag{4.90}$$

$$S = 10 \log \frac{1}{r}\ (S_\beta - V_e) \tag{4.91}$$

Since we analyze based on a root square, the true value of S is equal to the estimation of β^2, that is, consumption of electricity needed for cutting, which should be smaller. If we perform smooth machining with a small amount of electricity, we can obviously improve dimensional accuracy and flatness.

2. *Improvement of work efficiency.* This analysis is not for machining performance but for work efficiency. In this case, an ideal function is expressed as

$$y = \beta T \tag{4.92}$$

The SN ratio and sensitivity are calculated as

$$S_T = y_{11}^2 + y_{12}^2 + \cdots + y_{23}^2 \qquad (f = 6) \tag{4.93}$$

$$S_\beta = \frac{(L_1 + L_2)^2}{2r} \qquad (f = 1) \tag{4.94}$$

$$S_{N\beta} = \frac{(L_1 - L_2)^2}{2r} \qquad (f = 1) \tag{4.95}$$

Now

$$L_1 = T_1 y_{11} + T_2 y_{12} + T_3 y_{13} \tag{4.96}$$

$$L_2 = T_1 y_{21} + T_2 y_{22} + T_3 y_{23} \tag{4.97}$$

$$r = T_1^2 + T_2^2 + T_3^2 \tag{4.98}$$

$$\eta = 10 \log \frac{(1/2r)\ (S_\beta - V_e)}{V_N} \tag{4.99}$$

$$S = 10 \log \frac{1}{2r}\ (S_\beta - V_e) \tag{4.100}$$

Then

$$V_e = \tfrac{1}{4}(S_T - S_\beta - S_{N\beta}) \tag{4.101}$$

$$V_N = \tfrac{1}{5}(S_T - S_\beta) \tag{4.102}$$

In this case, β^2 of the ideal function represents electricity consumption per unit time, which should be larger. To improve the SN ratio using equation (4.99) means to design a procedure of using a large amount of electricity smoothly. Since we minimize electricity consumption by equations (4.90) and (4.91), we can improve not only productivity but also quality by taking advantage of both analyses.

The efficiency ratio of y to M, which has traditionally been used by engineers, can neither solve quality problems nor improve productivity. On the other hand, the hourly machining amount (machining measurement) used by managers in manufacturing cannot settle noise problems and product quality in machining.

To improve quality and productivity at the same time, after measuring the data shown in Table 4.10 we should determine an optimal condition based on two values, SN ratio and sensitivity. To lower sensitivity in equation (4.91) is to minimize electric power per unit removal amount. If we cut more using a slight quantity of electricity, we can cut smoothly, owing to improvement in the SN ratio in equation (4.90), as well as minimize energy loss. In other words, it means to cut sharply, thereby leading to good surface quality after cutting and improved machining efficiency per unit energy.

When On and Off Conditions Exist

Since quality engineering is a generic technology, a common problem occurs for the mechanical, chemical, and electronic engineering fields. We explain the functionality evaluation method for on and off conditions using an example of machining.

Recently, evaluation methods of machining have made significant progress, as shown in an experiment implemented in 1997 by Ishikawajima–Harima Heavy Industries (IHI) and sponsored by the National Space Development Agency of Japan (NASDA). The content of this experiment was released in the Quality Engineering Symposium in 1998. Instead of using transformality [setting product dimensions input through a numerically controlled (NC) machine as signal and corresponding dimensions after machined as measurement characteristic], they measured input and output utilizing energy and work by taking advantage of the concept of generic function.

In 1959, the following experiment was conducted at the Hamamatsu Plant of the National Railway (now known as JR). In this experiment, various types of cutting tools (JIS, SWC, and new SWC cutting tools) and cutting conditions were assigned to an L_{27} orthogonal array to measure the net cutting power needed for a certain amount of cut. The experiment, based on 27 different combinations, revealed that the maximum power needed was several times as large as the minimum. This finding implies that excessive power may cause a rough surface or variability in dimensions. In contrast, cutting with quite a small amount of power means that we can cut sharply. Consequently, when we can cut smoothly and power effectively, material surfaces can be machined flat and variability in dimension can be reduced. In the Quality Engineering Symposium in June 1998, IHI released research on improvement in machining accuracy of a space rocket engine, entitled "Optimization of T_i High-Speed Machining Process" by Kenji Fujmoto and two others from Ishikawajima–Harima Heavy Industries, Kazuyuki Higashihara from the National Space Development Agency of Japan, and Kazuhito Takahashi from the University of Electro-Communications.

This research did not adopt transformality, which defines an ideal function as the following equation by setting the input signal of the NC machine N to signal M, and a dimension of a machined test piece y to output:

$$y = \beta M \tag{4.103}$$

The paradigm of the testing method in quality engineering maintains that we should select energy or work as input signal M and output y. The ideal function in equation (4.103) is not expressed by an equation proportional to energy, as IHI clearly enunciated in the forum. Since cutting itself is work in the cutting process, the amount of cut is regarded as work. Therefore, output is work. To make something work, we need energy or power. IHI originally wished to input electricity consumption. However, since it changes every second, they considered it inappropriate for use as signal. They conducted an experiment based on Table 4.11, in which T, M, and y indicate the number of tool paths, measured amount of cut, and electricity consumption, respectively. In addition, as noise factor levels, two different types of thickness of a flat plate N_1 and N_2, each of which also had different Young's modulus, were chosen.

The reason that they chose the amount of cut as signal M (as did the JR experiment in 1959) is that the amount of cut is regarded as required work. The ideal function was defined as follows, after they took the square root of both sides of the original equation:

$$\sqrt{y} = \beta \sqrt{M} \tag{4.104}$$

The only difference between (4.103) and (4.104) is that the former is not a generic but an objective function, whereas the latter is a generic function representing the relationship between work and energy. Since y represents a unit of energy, a square of y does not make sense from the viewpoint of physics. Thus, by taking the square root of both sides beforehand, we wish to make both sides equivalent to work and energy when they are squared.

On the other hand, unless the square of a measurement characteristic equals energy or work, we do not have any rationale in calculating the SN ratio based on a generic function. After taking the square root of both sides in equation (4.104),

Table 4.11

Cutting experiment

		T_1	T_2	T_3
N_1	M: amount of cut y: electricity consumption	M_{11} y_{11}	M_{12} y_{12}	M_{13} y_{13}
N_2	M: amount of cut y: electricity consumption	M_{21} y_{21}	M_{22} y_{22}	M_{23} y_{23}

we compute the total variation of output S_T, which is a sum of squares of the square root of each cumulative electricity consumption:

$$S_T = (\sqrt{y_{11}})^2 + (\sqrt{y_{12}})^2 + \cdots + (\sqrt{y_{23}})^2 = y_{11} + y_{12} + \cdots + y_{23} \qquad (f = 6)$$

(4.105)

This is equivalent to a total sum of electricity consumptions. Subsequently, total consumed energy S_T can be decomposed as follows:

$$S_T = \text{(Electricity consumption used for cutting)}$$
$$+ \text{(electricity consumption used for loss or variability in cutting)}$$

(4.106)

While an ideal relationship is defined by energy or work, we use the square root of both sides of the relationship in analyzing the SN ratio. After we take the square root of both sides, the new relationship is called a *generic function*.

We can compute the power used for cutting as a variation of proportional terms:

$$S_\beta = \frac{L_1 + L_2}{r_1 + r_2}$$

(4.107)

Now L indicates a linear equation using square roots and is calculated as

$$L_1 = \sqrt{M_{11}}\,\sqrt{y_{11}} + \sqrt{M_{12}}\,\sqrt{y_{12}} + \sqrt{M_{13}}\,\sqrt{y_{13}}$$
$$L_2 = \sqrt{M_{21}}\,\sqrt{y_{21}} + \sqrt{M_{22}}\,\sqrt{y_{22}} + \sqrt{M_{23}}\,\sqrt{y_{23}}$$

(4.108)

$$r_1 = (\sqrt{M_{11}})^2 + (\sqrt{M_{12}})^2 + (\sqrt{M_{13}})^2$$
$$= M_{11} + M_{12} + M_{13}$$

$$r_2 = M_{21} + M_{22} + M_{23}$$

(4.109)

Variation of sensitivity fluctuations $N\beta$,

$$S_{N\beta} \frac{L_1 - L_2}{r_1 + r_2}$$

(4.110)

According to Table 4.11, a general performance evaluation, or test of work quantity and quality, is simple to complete. All that we need to do in designing a real product is tuning, regarded as a relatively simple procedure. For any machine function, after defining an ideal function of signal and output based on work and energy, we should calculate the SN ratio and sensitivity by taking the square root of both sides of the functional equation.

4.4. Chemical Engineering

As a function regarding chemical reaction or molecules, often we can measure only percentage data of molecular conditions. Although we conduct R&D based on experiments on as small a scale as possible (e.g., flasks), the ultimate goal is optimal design of large-scale production processes. Quality engineering in chemical reactions deals with a technical method of studying functional robustness on a small scale. Robustness of chemical reactions revolves around control of re-action speed.

Function of an Engine

The genetic function of an engine is a chemical reaction. Our interest in oxygen aspirated into an engine is in how it is distributed in the exhaust.

❏ *Insufficient reaction.* We set the fraction of oxygen contained in CO_2 and CO to p.

❏ *Sufficient reaction.* We set the fraction of oxygen contained in CO_2 to q.

❏ *Side reaction* (e.g., of N_{ox}). We set the fraction of oxygen contained in side-reacted substances to $1 - p - q$.

Based on this definition, ideally the chemical reaction conforms to the follow-ing exponential equations:

$$p = e^{-\beta_1 T} \tag{4.111}$$

$$p + q = e^{-\beta_2 T} \tag{4.112}$$

Here T represents the time needed for one cycle of the engine and p and q are measurements in the exhaust. The total reaction rate β_1 and side reaction rate β_2 are calculated as

$$\beta_1 = \frac{1}{T} \ln \frac{1}{p} \tag{4.113}$$

$$\beta_2 = \frac{1}{T} \ln \frac{1}{p + q} \tag{4.114}$$

It is desirable that β_1 become greater and β_2 be close to zero. Therefore, we compute the SN ratio as

$$\eta = 10 \log \frac{\beta_1^2}{\beta_2^3} \tag{4.115}$$

The sensitivity is

$$S = 10 \log \beta_1^2 \tag{4.116}$$

The larger the sensitivity, the more the engine's output power increases, the larger the SN ratio becomes, and the less effect the side reaction has.

Cycle time T should be tuned such that benefits achieved by the magnitude of total reaction rate and loss by the magnitude of side reaction are balanced optimally.

Selecting noise N (which has two levels, such as one for the starting point of an engine and the other for 10 minutes after starting) as follows, we obtain the data shown in Table 4.12. According to the table, we calculate the reaction rate for N_1 and N_2. For an insufficient reaction,

$$\beta_{11} = \frac{1}{T} \ln \frac{1}{p_1} \tag{4.117}$$

$$\beta_{12} = \frac{1}{T} \ln \frac{1}{p_2} \tag{4.118}$$

Table 4.12
Function of engine based on chemical reaction

	N_1	N_2
Insufficient reaction p	p_1	p_2
Objective reaction q	q_1	q_2
Total	$p_1 + q_1$	$p_2 + q_2$

For a side reaction,

$$\beta_{21} = \frac{1}{T} \ln \frac{1}{p_1 + q_1} \tag{4.119}$$

$$\beta_{22} = \frac{1}{T} \ln \frac{1}{p_2 + q_2} \tag{4.120}$$

Since the total reaction rates β_{11} and β_{12} are larger-the-better, their SN ratios are as follows:

$$\eta_1 = -10 \log \frac{1}{2} \left(\frac{1}{\beta_{11}^2} + \frac{1}{\beta_{12}^2} \right) \tag{4.121}$$

Since the side reaction rates β_{21} and β_{22} are smaller-the-better, their SN ratio is

$$\eta_2 = -10 \log \tfrac{1}{2}(\beta_{21}^2 + \beta_{22}^2) \tag{4.122}$$

Finally, we compute the total SN ratio and sensitivity as follows:

$$\eta = \eta_1 + \eta_2 \tag{4.123}$$

$$S = \eta_1 \tag{4.124}$$

General Chemical Reactions

If side reactions barely occur and we cannot trust their measurements in a chemical reaction experiment, we can separate a point of time for measuring the total reaction rate $(1 - p)$ and a point of time for measuring the side reaction rate $(1 - p - q)$. For example, we can set T_1 to 1 minute and T_2 to 30 minutes, as illustrated in Table 4.13. When $T_1 = T_2$, we can use the procedure described in the preceding section.

Table 4.13
Experimental data

	T_1	T_2
Insufficient reaction p	p_{11}	p_{12}
Objective reaction q	q_{11}	q_{12}
Total	$p_{11} + q_{11}$	$p_{12} + q_{12}$

If the table shows that $1 - p_{11}$ is more or less than 50% and $1 - (p_{12} + q_{12})$ ranges at least from 10 to 50%, this experiment is regarded as good enough. p_{11} and $(p_{12} + q_{12})$ are used for calculating total reaction and side reaction rates, respectively:

$$\beta_1 = \frac{1}{T} \ln \frac{1}{p_{11}} \qquad (4.125)$$

$$\beta_2 = \frac{1}{T_2} \ln \frac{1}{p_{12} + q_{12}} \qquad (4.126)$$

Using the equations above, we can compute the SN ratio and sensitivity as

$$\eta = 10 \log \frac{\beta_1^2}{\beta_1^2} \qquad (4.127)$$

$$S = 10 \log \beta_1^2 \qquad (4.128)$$

Evaluation of Images   An *image* represents a picture such as a landscape transformed precisely on each pixel of a flat surface. Sometimes an image of a human face is whitened compared to the actual one; however, quality engineering does not deal with this type of case because it is an issue of product planning and tuning.

In a conventional research method, we have often studied a pattern of three primary colors (including a case of decomposing a gray color into three primary colors, and mixing up three primary colors into a gray color) as a test pattern. When we make an image using a pattern of three primary colors, a density curve (a common logarithm of a reciprocal of permeability or reflection coefficient) varies in accordance with various conditions (control factors). By taking measurements from this type of density curve, we have studied an image. Although creating a density curve is regarded as reasonable, we have also measured $D_{\max}$, $D_{\min}$, and gamma from the curve. In addition to them (e.g., in television), resolution and image distortion have been used as measurements.

Because the consumers' demand is to cover a minimum density difference of three primary colors (according to a filmmaker, this is the density range 0 to 10,000) that can be recognized by a human eye's photoreceptor cell, the resolving power should cover up to the size of the light-sensitive cell. However, quality engineering does not pursue such technical limitations but focuses on improving imaging technology. That is, its objective is to offer evaluation methods to improve both quality and productivity.

For example, quality engineering recommends the following procedure for taking measurements:

1. *Condition M_1*. Create an image of a test pattern using luminosity 10 times as high and exposure time one-tenth as high as their current levels.

2. *Condition M_2*. Create an image of a test pattern using as much luminosity and exposure time as their current levels.

3. *Condition M_3*. Create an image of a test pattern using luminosity one-tenth as high and exposure time 10 times as high as their current levels.

At the three sensitivity curves, we select luminosities corresponding to a permeability or reflection coefficient of 0.5. For density, it is 0.301. For a more practical experiment, we sometimes select seven levels for luminosity, such as 1000, 100, 10, 1, 1/10, 1/100, and 1/1000 times as much as current levels. At the same time, we choose exposure time inversely proportional to each of them.

After selecting a logarithm of exposure time E for the value 0.301, the SN ratio and sensitivity are calculated for analysis.

Next, we set a reading of exposure time E (multiplied by a certain decimal value) for each of the following seven levels of signal M (logarithm of luminosity) to $y_1, y_2, \dots, y_7$:

$$M_1 = -3.0, \quad M_2 = -2.0, \quad \dots, \quad M_7 = +3.0 \qquad (4.129)$$

This is a procedure for calculating a necessary image-center luminosity from a sensitivity curve by taking into account a range of consumer uses. This procedure is considered appropriate from the standpoint not only of quality engineering but also the performance evaluation of photo film.

The ASA valve of a photo film indicates a necessary luminosity at the center of an image for each film. For a permeability or reflection coefficient, a point with value 0.5 is easiest to measure and the most appropriate for evaluation of an image, whereas we do not create images at D_{max} and D_{min}. Quality engineering is a strategic technology used to evaluate performance by taking consumer uses into account. Consumer uses are all signals and noises. In the case of an image, a signal is color density. To study an image by breaking down each of the miscellaneous color densities into a pattern of three primary colors is a rationalization of the measurement method.

The reason for measuring data at the center (at the point where the reflection or permeability coefficient is 50%) is that we wish to avoid inaccurate measurement, and this method has long been used in the telecommunication field. In addition, the reason for taking luminosity and exposure time in inverse proportion to this to see the density range is that we wish to check the linearity of color density of an image for luminosity.

This linearity is totally different from the study of images by pixels. That is, in dealing with digital information, we can select only the accuracy of each dot as a measurement because total image performance improves as the accuracy of each dot increases.

Among consumer uses are not only signals but also noises, which should be smaller to enhance performance. For a photographic film, conditions regarding film preservation and environment when a picture is taken are included in noises. We select the following levels of film preservation conditions as two noise levels:

N_1: good preservation condition, where the density change is small

N_2: poor preservation condition, where the density change is large

In this case, a small difference between the two sensitivity curves for N_1 and N_2 represents a better performance. A good way to designing an experiment is to compound all noise levels into only two levels. Thus, no matter how many noise levels we have, we should compound them into two levels.

Since we have three levels, K_1, K_2, and K_3, for a signal of three primary colors, two levels for a noise, and seven levels for an output signal, the total number of data is $(7)(3)(2) = 42$. For each of the three primary colors, we calculate the SN ratio and sensitivity. Now we show only the calculation of K_1. Based on Table 4.14,

Table 4.14
Experimental data

		M_1	M_2	...	M_7	Linear Equation
K_1	N_1	y_{11}	y_{12}	...	y_{17}	L_1
	N_2	y_{21}	y_{22}	...	y_{27}	L_2
K_2	N_1	y_{31}	y_{32}	...	y_{37}	L_3
	N_2	y_{41}	y_{42}	...	y_{47}	L_4
K_3	N_1	y_{51}	y_{52}	...	y_{57}	L_5
	N_2	y_{61}	y_{62}	...	y_{67}	L_6

we proceed with the calculation. Now, by taking into account that M has seven levels, we view $M_4 = 0$ as a standard point and subtract the value of M_4 from M_1, M_2, M_3, M_5, M_6, and M_7, each of which is set to y_1, y_2, y_3, y_5, y_6, and y_7.

By subtracting a reading y for $M_4 = 0$ in case of K_1, we obtain the following linear equations:

$$L_1 = -3y_{11} - 2y_{12} - \cdots + 3y_{17} \tag{4.130}$$

$$L_2 = -3y_{21} - \cdots + 3y_{27} \tag{4.131}$$

Based on these, we define $r = (-3)^2 + (-2)^2 + \cdots + 3^2 = 28$.

$$S_\beta = \frac{(L_1 + L_2)^2}{2r} \tag{4.132}$$

$$S_{N\beta} = \frac{(L_1 - L_2)^2}{2r} \tag{4.133}$$

$$S_T = y_{11}^2 + y_{12}^2 + \cdots + y_{27}^2 \tag{4.134}$$

$$V_e = \tfrac{1}{12}(S_T - S_\beta - S_{N\beta}) \tag{4.135}$$

$$V_N = \tfrac{1}{13}(S_T - S_\beta) \tag{4.136}$$

$$\eta_1 = 10 \log \frac{(1/2r)\,(S_\beta - V_e)}{V_N} \tag{4.137}$$

$$S_1 = 10 \log \frac{1}{2r}\,(S_\beta - V_e) \tag{4.138}$$

As for K_2 and K_3, we calculate η_2, S_2, η_3, and S_3. The total SN ratio η is computed as the sum of η_1, η_2, and η_3. To balance densities ranging from low to high for K_1, K_2, and K_3, we should equalize sensitivities for three primary colors, S_1, S_2, and S_3, by solving simultaneous equations based on two control factors. Solving them such that

$$S_1 = S_2 = S_3 \tag{4.139}$$

holds true under standard conditions is not a new method.

The procedure discussed thus far is the latest method in quality engineering, whereas we have used a method of calculating the density for a logarithm of luminosity log E. The weakness of the latter method is that a density range as output is different at each experiment. That is, we should not conduct an experiment using the same luminosity range for films of ASA 100 and 200. For the latter, we should perform an experiment using half of the luminosity needed for the former. We have shown the procedure to do so.

Quality engineering is a method of evaluating functionality with a single index η. Selection of the means for improvement is made by specialized technologies and specialists. Top management has the authority for investment and personnel affairs. The result of their performance is evaluated based on a balance sheet that cannot be manipulated. Similarly, although hardware and software can be designed by specialists, those specialists cannot evaluate product performance at their own discretion.

Functionality Such as Granulation or Polymerization Distribution

If you wish to limit granulation distribution within a certain range, you must classify granules below the lower limit as excessively granulated, ones around the center as targeted, and ones above the upper limit as insufficiently granulated, and proceed with analysis by following the procedure discussed earlier for chemical reactions. For polymerization distribution, you can use the same technique.

Separation System

Consider a process of extracting metallic copper from copper sulfide included in ore. When various types of substances are mixed with ore and the temperature of a furnace is raised, deoxidized copper melts, flows out of the furnace, and remains in a mold. This is called *crude copper ingot*. Since we expect to extract 100% of the copper contained in ore and convert it into crude copper (product), this ratio of extraction is termed *yield*. The percentage of copper remaining in the furnace slag is regarded as the loss ratio p. If $p = 0$, the yield becomes 100%. Because in this case we wish to observe the ratio at a single point of time during reaction, calibration will be complicated.

On the other hand, as the term *crude copper* implies, there exist a considerable number (approximately 1 to 2% in most cases) of ingredients other than copper. Therefore, we wish to bring the ratio of impurity contained in copper ingot, q^*, close to zero. Now, setting the mass of crude copper to A (kilograms), the mass of slag to B (kilograms) (this value may not be very accurate), the ratio of impurity in crude copper to q^*, and the ratio of copper in slag to p^*, we obtain Table 4.15 for input/output. From here we calculate the following two error ratios, p and q.

$$p = \frac{Bp^*}{A(1 - q^*) + Bp^*} \tag{4.140}$$

$$q = \frac{Aq^*}{Aq^* + B(1 - p^*)} \tag{4.141}$$

The error ratio p represents the ratio of copper molecules that is originally contained in ore but mistakenly left in the slag after smelting. Then $1 - p$ is called the *yield*. Subtracting this yield from 1, we obtain the error ratio p. The error ratio q indicates the ratio of all noncopper molecules that is originally contained in the furnace but mistakenly included in the product or crude copper. Both ratios are calculated as a mass ratio. Even if the yield $1 - p$ is large enough, if the error

Table 4.15

Input/output for copper smelting

Input	Output		Total
	Product	Slag	
Copper	$A(1 - q^*)$	Bp^*	$A(1 - q^*) + Bp^*$
Noncopper	Aq^*	$B(1 - p^*)$	$Aq^* + B(1 - p^*)$
Total	A	B	$A + B$

ratio q is also large, this smelting is considered inappropriate. After computing the two error ratios p and q, we prepare Table 4.16.

Consider the two error ratios p and q in Table 4.15. If copper is supposed to melt well and move easily in the product when the temperature in the furnace is increased, the error ratio p decreases. However, since ingredients other than copper melt well at the same time, the error ratio q rises. A factor that can decrease p and increase q is regarded as an adequate variable for tuning. Although most factors have such characteristics, more or less, we consider it real technology to reduce errors regarding both p and q rather than obtaining effects by tuning. In short, this is smelting technology with a large functionality.

To find a factor level reducing both p and q for a variety of factors, after making an adjustment so as not to change the ratio of p to q, we should evaluate functionality. This ratio is called the *standard SN ratio*. Since gain for the SN ratio accords with that when $p = q$, no matter how great the ratio of p to q, in most cases we calculate the SN ratio after p is adjusted equal to q. Primarily, we compute p_0 as follows when we set $p = q = p_0$:

$$p_0 = \frac{1}{1 + \sqrt{[(1/p) - 1][(1/q) - 1]}} \tag{4.142}$$

Secondarily, we calculate the standard SN ratio η as

$$\eta = 10 \log \frac{(1 - 2p_0)^2}{4p_0(1 - p_0)} \tag{4.143}$$

The details are given in Reference 5.

Once the standard SN ratio η is computed, we determine an optimal condition according to the average SN ratios for each control factor level, estimate SN ratios

Table 4.16

Input/output expressed by error ratios

Input	Output		Total
	Product	Slag	
Copper	$1 - p$	p	1
Noncopper	q	$1 - q$	1
Total	$1 - p + q$	$1 + p - q$	2

for optimal and initial conditions, and calculate gains. After this, we conduct a confirmatory experiment for the two conditions, compute SN ratios, and compare estimated gains with those obtained from this experiment.

Based on the experiment under optimal conditions, we calculate p and q. Unless a sum of losses for p and q is minimized, timing is set using factors (such as the temperature in the furnace) that influence sensitivity but do not affect the SN ratio. In tuning, we should gradually change the adjusting factor level. This adjustment should be made after the optimal SN ratio condition is determined.

The procedure detailed here is even applicable to the removal of harmful elements or the segregation of garbage.

4.5. Medical Treatment and Efficacy Experimentation

A key objective of quality engineering is to rationalize functional evaluation. For example, for medical efficacy, it proposes a good method of evaluating great effect and small side effects in a laboratory. Evaluation in a laboratory clarifies whether we can assess both main effects and side effects without clinical tests on patients at the same time. We discuss monitoring of individual patients later in this section.

☐ Example

First, we consider a case of evaluating the main effects on a cancer cell and the side effects on a normal cell. Although our example is an anticancer drug, this analytic procedure holds true for thermotherapy and radiation therapy using supersonic or electromagnetic waves. If possible, by taking advantage of an L_{18} orthogonal array, we should study the method using a drug and such therapies at the same time.

Quality engineering recommends that we experiment on cells or animals. We need to alter the density of a drug to be assessed by h milligrams (e.g., 1 mg) per unit time (e.g, 1 minute). Next, we select one cancer cell and one normal cell and designate them M_1 and M_2, respectively. In quality engineering, M is called a *signal factor*. Suppose that M_1 and M_2 are placed in a certain solution (e.g., water or a salt solution), the density is increased by 1 mg/minute, and the length of time that each cell survives is measured. Imagine that the cancer and normal cells die at the eighth and fourteenth minutes, respectively. We express these data as shown in Table 4.17, where 1 indicates "alive" and 0 indicates "dead." In addition, M_1 and

Table 4.17

Data for a single cell

	T_1	T_2	T_3	T_4	T_5	T_6	T_7	T_8	T_9	T_{10}	T_{11}	T_{12}	T_{13}	T_{14}	T_{15}
M_1	1	1	1	1	1	1	1	0	0	0	0	0	0	0	0
M_2	1	1	1	1	1	1	1	1	1	1	1	1	1	0	0

M_2 are cancer and normal cells, and T_1, T_2, ... indicate each lapse of time: 1, 2, ... minutes.

For digital data regarding a dead-or-alive (or cured-or-not cured) state, we calculate LD_{50} (lethal dose 50). In this case, the LD_{50} value of M_1 is 7.5 because a cell dies between T_7 and T_8, whereas that of M_2 is 13.5. LD_{50} represents the quantity of drug needed until 50% of cells die.

In quality engineering, both M and T are signal factors. Although both signal and noise factors are variables in use, a signal factor is a factor whose effect should exist, and conversely, a noise is a factor whose effect should be minimized. That is, if there is no difference between M_1 and M_2, and furthermore, between each different amount of drug (or of time in this case), this experiment fails. Then we regard both M and T as signal factors. Thus, we set up X_1 and X_2 as follows:

$$X_1 = LD_{50} \text{ of a cancer cell} \tag{4.144}$$

$$X_2 = LD_{50} \text{ of a normal cell} \tag{4.145}$$

In this case, the smaller X_1 becomes, the better the drug's performance becomes. In contrast, X_2 should be greater. Quality engineering terms the former the *smaller-the-better characteristic* and the latter *larger-the-better characteristic*. The following equation calculates the SN ratio:

$$\eta = 20 \log \frac{X_2}{X_1} \tag{4.146}$$

In this case, since there is no noise, to increase η is equivalent to enlarging the ratio X_2/X_1.

For a more practical use, we often test N_1, N_2, ... , N_k as k cancer cells, and N_1', N_2', ... , N_k' as k' normal cells. Suppose that $k = 3$ and $k' = 3$. N_1, N_2, and N_3 can be selected as three different types of cancer cells or three cancer cells of the same type. Or we can choose two cancer cells of the same type and one cancer cell of a different type.

Next, we obtain dead-or-alive data, as illustrated in Table 4.18, after placing three cells for each of M_1 and M_2. Table 4.19 shows the LD_{50} values of N_1, N_2, and N_3 and N_1', N_2', and N_3'.

Table 4.18

Dead-or-alive data

		T_1	T_2	T_3	T_4	T_5	T_6	T_7	T_8	T_9	T_{10}	T_{11}	T_{12}	T_{13}	T_{14}	T_{15}	T_{16}	T_{17}	T_{18}	T_{19}	T_{20}
M_1	N_1	1	1	1	1	1	0	0	0	0	0	0	0	0	0	0	0	0	0	0	0
	N_2	1	1	1	0	0	0	0	0	0	0	0	0	0	0	0	0	0	0	0	0
	N_3	1	1	1	1	1	1	1	1	1	1	1	0	0	0	0	0	0	0	0	0
M_2	$N_{1'}$	1	1	1	1	1	1	1	1	1	1	1	1	1	1	1	0	0	0	0	0
	$N_{2'}$	1	1	1	1	1	1	1	1	0	0	0	0	0	0	0	0	0	0	0	0
	$N_{3'}$	1	1	1	1	1	1	1	1	1	1	1	1	1	1	1	1	1	1	1	0

Table 4.19

LD_{50} data

M_1	5.5	3.5	11.5
M_2	14.5	8.5	19.5

Since the data for M_1 should have a smaller standard deviation as well as a smaller average, after calculating an average of a sum of squared data, we multiply its logarithm by 10. This is termed the *smaller-the-better SN ratio*:

$$\eta_1 = -10 \log \tfrac{1}{3}(5.5^2 + 3.5^2 + 11.5^2) = -17.65 \text{ dB} \qquad (4.147)$$

On the other hand, because the LD_{50} value of a normal cell M_2 should be infinitesimal, after computing an average sum of reciprocal data, we multiply its logarithm by 10. This is *larger-the-better SN ratio*:

$$\eta_2 = -10 \log \frac{1}{3} \left(\frac{1}{14.5^2} + \frac{1}{8.5^2} + \frac{1}{19.5^2} \right) = 21.50 \text{ dB} \qquad (4.148)$$

Therefore, by summing up η_1 and η_2, we calculate the following η, which is the total SN ratio integrating both main and side effects:

$$\eta = \eta_1 + \eta_2$$
$$= -17.65 + 21.50$$
$$= 3.85 \text{ dB} \qquad (4.149)$$

For example, given two drugs A_1 and A_2, we obtain the experimental data shown in Table 4.20. If we compare the drugs, N_1, N_2, and N_3 for A_1 should be consistent with N'_1, N'_2, and N'_3 for A_2. Now suppose that the data of A_1 are the same as those in Table 4.19 and that a different drug A_2 has LD_{50} data for M_1 and M_2 as illustrated in Table 4.20. A_1's SN ratio is as in equation (4.149). To compute A_2's SN ratio, we calculate the SN ratio for the main effect η_1 as follows:

$$\eta_1 = -10 \log \tfrac{1}{3}(18.5^2 + 11.5^2 + 20.5^2) = -24.75 \text{ dB} \qquad (4.150)$$

Table 4.20

Data for comparison experiment

A_1	M_1	5.5	3.5	11.5
	M_2	14.5	8.5	19.5
A_2	M_1	18.5	11.5	20.5
	M_2	89.5	40.5	103.5

Next, as a larger-the-better SN ratio, we calculate the SN ratio for the side effects η_2 as

$$\eta_2 = -10 \log \frac{1}{3}\left(\frac{1}{89.5^2} + \frac{1}{40.5^2} + \frac{1}{103.5^2}\right) = 35.59 \qquad (4.151)$$

Therefore, we obtain the total SN ratio, combining the main and side effects, by adding them up as follows:

$$\eta = -24.75 + 35.59 = 10.84 \qquad (4.152)$$

Finally, we tabulate these results regarding comparison of A_1 and A_2 in Table 4.21, which is called a *benchmarking test*. What we find out from Table 4.21 is that the side effect of A_2 is larger than A_1's by 14.09 dB, or 25.6 times, whereas the main effect of A_2 is smaller than A_1's by 7.10 dB, or 1/5.1 times. On balance, A_2's effect is larger than A_1's by 6.99 dB, or 5 times. This fact reveals that if we increase an amount of A_2, we can improve both main and side effects by 3.49 dB compared to A_1's. That is, the main effect is enhanced 2.3 times, and the side effect is reduced 1/2.3 times. By checking the SN ratio using the operating window method, we can know whether the main and side-effects are improved at the same time.

When we study therapy such as using heat or electromagnetic waves, we select as many control factors as possible, assign them to an orthogonal array L_{18}, increase the power at a certain interval, and measure data as shown in Table 4.18. Then, after preparing a table such as Table 4.19 for each experiment in the orthogonal array, we calculate the SN ratios. As for remaining calculations, we can follow the normal quality engineering procedures.

When we wish to measure temperature itself when dealing with thermotherapy, we should choose a temperature difference from a standard body temperature. For example, suppose that we set the starting temperature to a body temperature (e.g., 36°C) when we raise the temperature of a Schale or petri dish containing one cancer cell M_1 and one normal cell M_2 by 1° each 10 minutes (or 5 minutes). Indeed, it is possible to raise the temperature quickly; however, by increasing it slowly, we should monitor how many times we need to raise the temperature until a cell dies. In this case, the temperature T is regarded as a signal whose standard

Table 4.21

Comparison of drugs A_1 and A_2

	Main Effect	Side Effect	Total
A_1	−17.65	21.50	3.85
A_2	−24.75	35.59	10.84
Gain	−7.10	14.09	6.99

point is 36°C. If the cancer cell dies at 43°C, the LD_{50} value is 6.5°C, the difference from the standard point of 36°C. If the normal cell dies at 47°C, the LD_{50} value is 10.5°C. When a sonic or electromagnetic wave is used in place of temperature, we need to select the wave's power (e.g., raise the power by 1 W/minute).

4.6. Software Testing

Quality engineering classifies a signal type into two categories, the quality of an active signal (in the case of software that a user uses actively) and the quality of the MTS (Mahalanobis–Taguchi system), dealing with a passive signal. In this section we detail the former, and in the next section we discuss the case for the latter.

Quality engineering supposes that all conditions of use necessarily belong to a signal or noise factor. Apart from a condition where a user uses it actively or passively, the effects of a signal factor are those that are essential. When we plan to design software (software product) using a system with computers, the software is considered a user's *active signal factor*. On the other hand, when we conduct inspection, diagnosis, or prediction using research data (including various sensing data), the entire group of data is regarded as consisting of *passive signal factors*. Signal factors should not only have a common function under various conditions of use but also have small errors.

Two Types of Signal Factor and Software

In this section we explain how to use the SN ratio in taking measures against bugs when we conduct a functional test on software that a user uses actively. Software products have a number of active signal factors and consist of various levels of signal factors. In this section we discuss ways to measure and analyze data to find bugs in software.

In testing software, there is a multiple-step process, that is, a number of signals. In quality engineering, we do not critque software design per se, but discuss measures that enable us to find bugs in designed software. We propose a procedure for checking whether software contains bugs or looking for problems with (diagnosis of) bugs.

Layout of Signal Factors in an Orthogonal Array

As we discussed before, the number of software signal factors is equivalent to the number of steps involved. In actuality, the number of signal factors is tremendous, and the number is completely different in each factor level. With software, how large an orthogonal array we can use should be discussed even when signal factors need to be tested at every step or data can be measured at a certain step. To use an L_{36} orthogonal array repeatedly is one of the practical methods we can use, as shown next.

❑ *Procedure 1*. We set the number of multilevel signal factors to k. If k is large, we select up to 11 signal factors of two levels and up to 12 factors of three levels. Therefore, if $k \leq 23$, we should use an L_{36} orthogonal array. If k is more than 23, we should use a greater orthogonal array (e.g., if $24 \leq k \leq 59$, L_{108} is to be used) or use an L_{36} orthogonal array repeatedly to allocate all factors. We should lay out all signal factors as an array after reducing the number of each factor level to two or three.

❏ *Procedure 2.* We conduct a test at 36 different conditions that are laid out on an orthogonal array. If we obtain an acceptable result for each experiment in the orthogonal array, we record a 0; conversely, if we do not obtain an acceptable result, we record a 1. Basically by looking at output at the final step, we judge by 0 or 1. Once all data (all experiments in case of L_{36}) are zero, the test is completed. If even one datum of 1 remains in all experimental combination, the software is considered to have bugs.

❏ *Procedure 3.* As long as there are bugs, we need to improve the software design. To find out at what step a problem occurs when there are a number of bugs, we can measure intermediate data. In cases where there are a small number of bugs, we analyze interactions. Its application to interactions is discussed in Section 4.7.

Software Diagnosis Using Interaction   For the sake of convenience, we explain the procedure by using an L_{36} orthogonal array; this also holds true for other types of arrays. According to procedure 3 in the preceding section, we allocate 11 two-level factors and 12 three-level factors to an L_{36} orthogonal array, as shown in Table 4.22. Although one-level factors are not assigned to the orthogonal array, they must be tested.

We set two-level signal factors to A, B, ... , and K, and three-level signal factors to L, M, ... , and W. Using the combination of experiments illustrated in Table 4.23, we conduct a test and measure data about whether or not software functions properly for all 36 combinations. When we measure output data for tickets and changes in selling tickets, we are obviously supposed to set 0 if both outputs are correct and 1 otherwise. In addition, we need to calculate an interaction for each case of 0 and 1.

Now suppose that the measurements taken follow Table 4.23. In analyzing data, we sum up data for all combinations of A to W. The total number of combinations is 253, starting with AB and ending with VW. Since we do not have enough space to show all combinations, for only six combinations of AB, AC, AL, AM, LM, and LW, we show a two-way table, Table 4.23. For practical use, we can use a computer to create a two-way table for all combinations.

Table 4.23 includes six two-way tables for AB, AC, AL, AM, LM, *and* LW, and each table comprises numbers representing a sum of data 0 or 1 for four conditions of A_1B_1 (Nos. 1–9), A_1B_2 (Nos. 10–18), A_2B_1 (Nos. 19–27), and A_2B_2 (Nos. 28–36). Despite all 253 combinations of two-way tables, we illustrate only six. After we create a two-dimensional table for all possible combinations, we need to consider the results.

Based on Table 4.23, we calculate combined effects. Now we need to create two-way tables for combinations whose error ratio is 100% because if an error in software exists, it leads to a 100% error. In this case, using a computer, we should produce tables only for L_2 and L_3 for A_2, M_3 for A_2, M_2 for L_3, W_2 and W_3 for L_2, and W_2 and W_3 for L_3. Of 253 combinations of two-dimensional tables, we output only 100% error tables, so we do not need to output those for AB and AC.

We need to enumerate not only two-way tables but also all 100% error tables from all possible 253 tables because bugs in software are caused primarily by combinations of signal factor effects. It is software designers' job to correct errors.

Table 4.22

Layout and data of L_{36} orthogonal array

No.	A 1	B 2	C 3	D 4	E 5	F 6	G 7	H 8	I 9	J 10	K 11	L 12	M 13	N 14	O 15	P 16	Q 17	R 18	S 19	T 20	U 21	V 22	W 23	Data
1	1	1	1	1	1	1	1	1	1	1	1	1	1	1	1	1	1	1	1	1	1	1	1	0
2	1	1	1	1	1	1	1	1	1	1	1	2	2	2	2	2	2	2	2	2	2	2	2	1
3	1	1	1	1	1	1	1	1	1	1	1	3	3	3	3	3	3	3	3	3	3	3	3	1
4	1	1	1	1	1	2	2	2	2	2	2	1	1	1	1	2	2	2	2	3	3	3	3	0
5	1	1	1	1	1	2	2	2	2	2	2	2	2	2	2	3	3	3	3	1	1	1	1	0
6	1	1	1	1	1	2	2	2	2	2	2	3	3	3	3	1	1	1	1	2	2	2	2	0
7	1	1	2	2	2	1	1	1	2	2	2	1	1	2	3	1	2	3	3	1	2	2	3	0
8	1	1	2	2	2	1	1	1	2	2	2	2	2	3	1	2	3	1	1	2	3	3	1	0
9	1	1	2	2	2	1	1	1	2	2	2	3	3	1	2	3	1	2	2	3	1	1	2	0
10	1	2	1	2	2	1	2	2	1	1	2	1	1	3	2	1	3	2	3	2	1	3	2	0
11	1	2	1	2	2	1	2	2	1	1	2	2	2	1	3	2	1	3	1	3	2	1	3	1
12	1	2	1	2	2	1	2	2	1	1	2	3	3	2	1	3	2	1	2	1	3	2	1	0
13	1	2	2	1	2	2	1	2	1	2	1	1	2	3	1	3	2	1	3	2	2	1	3	0
14	1	2	2	1	2	2	1	2	1	2	1	2	3	1	2	1	3	2	1	3	3	2	1	0
15	1	2	2	1	2	2	1	2	1	2	1	3	1	2	3	2	1	3	2	1	1	3	2	0
16	1	2	2	2	1	2	2	1	2	1	1	1	2	3	2	1	1	3	2	3	3	2	1	0
17	1	2	2	2	1	2	2	1	2	1	1	2	3	1	3	2	2	1	3	1	1	3	2	0
18	1	2	2	2	1	2	2	1	2	1	1	3	1	2	1	3	3	2	1	2	2	1	3	1
19	2	1	2	2	1	1	2	2	1	2	1	1	2	1	3	3	3	1	2	2	3	1	2	1
20	2	1	2	2	1	1	2	2	1	2	1	2	3	2	1	1	1	2	3	3	1	2	3	1
21	2	1	2	2	1	1	2	2	1	2	1	3	1	3	2	2	2	3	1	1	2	3	1	0
22	2	1	1	2	2	2	2	1	1	1	2	1	2	2	3	3	1	2	1	1	3	3	2	1
23	2	1	1	2	2	2	2	1	1	1	2	2	3	3	1	1	2	3	2	2	1	1	3	0
24	2	1	1	2	2	2	2	1	1	1	2	3	1	1	2	2	3	1	3	3	2	2	1	1

Table 4.22 (*Continued*)

No.	A 1	B 2	C 3	D 4	E 5	F 6	G 7	H 8	I 9	J 10	K 11	L 12	M 13	N 14	O 15	P 16	Q 17	R 18	S 19	T 20	U 21	V 22	W 23	Data
25	2	1	1	2	2	2	1	2	2	1	1	1	3	2	1	2	3	3	1	3	1	2	2	0
26	2	1	1	2	2	2	1	2	2	1	1	2	1	3	2	3	1	1	2	1	2	3	3	1
27	2	1	1	2	2	2	1	2	2	1	1	3	2	1	3	1	2	2	3	2	3	1	1	1
28	2	2	2	1	1	1	1	2	2	1	2	2	1	3	3	3	2	2	1	3	1	2	1	1
29	2	2	2	1	1	1	1	2	2	1	2	2	1	3	3	3	2	2	1	3	1	2	1	1
30	2	2	2	1	1	2	1	2	2	2	2	3	2	1	1	1	3	3	2	2	2	3	2	0
31	2	2	1	2	2	1	1	1	1	2	1	1	3	3	3	2	3	2	2	1	2	1	1	1
32	2	2	1	2	2	2	1	1	1	2	1	2	1	1	1	3	1	3	3	2	3	2	2	1
33	2	2	1	2	2	2	2	1	1	2	1	3	3	2	2	3	2	1	1	3	1	3	3	1
34	2	2	1	2	2	1	2	1	2	2	1	1	1	1	2	2	2	3	1	2	2	3	1	1
35	2	2	1	1	2	2	2	2	2	2	1	2	1	2	3	1	3	1	2	3	3	1	2	1
36	2	2	1	1	2	2	2	2	2	2	1	3	2	3	1	2	1	2	3	1	1	2	3	1

Table 4.23
Supplemental tables

(1) *AB* two-way table

	B_1	B_2	Total
A_1	4	4	8
A_2	6	8	14
Total	10	12	22

(2) *AC* two-way table

	C_1	C_2	Total
A_1	4	4	8
A_2	7	7	14
Total	11	11	22

(3) *AL* two-way table

	L_1	L_2	L_3	Total
A_1	0	4	4	8
A_2	2	6	6	14
Total	2	10	10	22

(4) *AM* two-way table

	M_1	M_2	M_3	Total
A_1	1	2	5	8
A_2	6	4	4	14
Total	7	6	9	22

(5) *LM* two-way table

	M_1	M_2	M_3	Total
L_1	0	0	2	2
L_2	4	2	4	10
L_3	3	4	3	10
Total	7	6	9	22

(6) *LW* two-way table

	W_1	W_2	W_3	Total
L_1	1	0	1	2
L_2	2	4	4	10
L_3	2	4	4	10
Total	5	8	9	22

Once they modify 100% errors, they perform a test again based on an L_{36} orthogonal array. If 100% error combinations remain, they correct them again. Although this procedure sounds imperfect, it quite often streamlines the debugging task many times over.

System Decomposition

When, by following the method based on Table 4.22, we find it extremely difficult to seek root causes because of quite a few bugs, it is more effective for debugging to break signal factors into small groups instead of selecting all of them at a time. Some ways to do so are described below.

1. After splitting signal factors into two groups, we lay out factors in each of them to an L_{18} orthogonal array. Once we correct all bugs, we repeat a test based on an L_{36} orthogonal array containing all signal factors. We can reduce bugs drastically by two L_{18} tests, thereby simplifying the process of seeking root causes in an L_{36} array with few bugs.

2. When we are faced with difficulties finding causes in an L_{36} array because of too many bugs, we halve the number of signal factors by picking alternative factors, such as A, C, E, G, ... , W, and check these bugs. Rather than selecting every other factor, we can choose about half that we consider important. The fact that the number of bugs never changes reveals that all bugs are caused by a combination of about half the signal factors. Then we halve the number of factors. Until all root causes are detected, we continue to follow this process. In contrast, if no bug is found in the first-half combinations of signal factors, we test the second half. Further, if there happens to be no bug in this test, we can conclude that interactions between the first and second halves generate bugs. To clarify causes, after dividing each of the first and second halves into two groups, we investigate all four possible combinations of the two groups.

3. From Table 4.22, by correcting 100% errors regarding A, L, and M based on Table 4.23, we then check for bugs in an L_{36} orthogonal array. Once all bugs are eliminated, this procedure is complete.

We have thus far shown some procedures for finding the basic causes of bugs. However, we can check them using intermediate output values instead of the final results of a total system. This is regarded as a method of subdividing a total system into subsystems.

4.7. MT and MTS Methods

MT (Mahalanobis–Taguchi) Method

We cannot describe human health with a single measurement. The MT method represents human health by integrating measuring characteristics. We can apply this method to a medical examination or patient monitoring. The concept of distance expressing interrelationships among multidimensional information was introduced by Mahalanobis. However, since his method calculates a distance for each population (group) of samples, it is regarded as an application of deviation value in a one-dimensional space to a multidimensional space. Taguchi proposes introducing distance in a multidimensional space by prescribing the Mahalanobis distance only in the unit cluster of a population and defining variable distance in a population in consideration of variable interrelationships among items.

The MTS method also defines a group of items close to their average as a unit cluster in such a way that we can use it to diagnose or monitor a corporation. In fact, both the MT and MTS methods have started to be applied to various fields because they are outstanding methods of pattern recognition. What is most important in utilizing multidimensional information is to establish a fundamental database. That is, we should consider what types of items to select or which groups to collect to form the database. These issues should be determined by persons expert in a specialized field.

In this section we detail a procedure for streamlining a medical checkup or clinical examination using a database for a group of healthy people (referred to subsequently as "normal" people). Suppose that the total number of items used in the database is k. Using data for a group of normal people—for example, the data of people who are examined and found to be in good health in any year after annual medical checkups for three years in a row (if possible, the data of hundreds of people is preferable)—we create a scale to characterize the group.

As a scale we use the distance measure of P. C. Mahalanobis, an Indian statistician, introduced in his thesis in 1934. In calculating the Mahalanobis distance in a certain group, the group needs to have homogeneous members. In other words, a group consisting of abnormal people should not be considered. If the group contains people with both low and high blood pressure, we should not regard the data as a single distribution.

Indeed, we may consider a group of people suffering only from hepatitis type A as being somewhat homogeneous; however, the group still exhibits a wide deviation. Now, let's look at a group of normal people without hepatitis. For gender and age we include both male and female and all adult age brackets. Male and female are denoted by 0 and 1, respectively. Any item is dealt with as a quantitative measurement in calculation. After regarding 0 and 1 data for male and female as continuous variables, we calculate an average m and standard deviation σ. For all items for normal people, we compute an average value m and standard deviation σ and convert them below. This is called *normalization*.

$$Y = \frac{y - m}{\sigma} \tag{4.153}$$

Suppose that we have n normal people. When we convert $y_1, y_2, \ldots, y_n$ into $Y_1, Y_2, \ldots, Y_n$ using equation (4.153), $Y_1, Y_2, \ldots, Y_n$ has an average of 0 and a standard deviation of 1, leading to easier understanding. Selecting two from k items arbitrarily, and dividing the sum of normalized products by n or calculating covariance, we obtain a correlation coefficient.

After forming a matrix of correlation coefficients between two of k items by calculating its inverse matrix, we compute the following square of D representing the Mahalanobis distance:

$$D^2 = \frac{1}{k}\left(\sum_{ij} a_{ij} Y_i Y_j\right) \tag{4.154}$$

Now a_{ij} stands for the (i, j)th element of the inverse matrix. $Y_1, Y_2, \ldots, Y_n$ are converted from $y_1, y_2, \ldots, y_n$ based on the following equations:

$$Y_1 = \frac{y_1 - m_1}{\sigma_1}$$

$$Y_2 = \frac{y_2 - m_2}{\sigma_2}$$

$$\vdots$$

$$Y_k = \frac{y_k - m_k}{\sigma_k} \qquad (4.155)$$

In these equations, for k items regarding a group of normal people, set the average of each item to $m_1, m_2, \ldots, m_k$ and the standard deviation to $\sigma_1, \sigma_2, \ldots, \sigma_n$. The data of k items from a person, $y_1, y_2, \ldots, y_k$, are normalized to obtain $Y_1, Y_2, \ldots, Y_k$. What is important here is that "a person" whose identity is unknown in terms of normal or abnormal is an arbitrary person. If the person is normal, D^2 has a value of approximately 1; if not, it is much larger than 1. That is, the Mahalanobis distance D^2 indicates how far the person is from normal people.

For practical purposes, we substitute the following y (representing not the SN ratio but the magnitude of N):

$$y = 10 \log D^2 \qquad (4.156)$$

Therefore, if a certain person belongs to a group of normal people, the average of y is 0 dB, and if the person stays far from normal people, y increases because the magnitude of abnormality is enlarged. For example, if y is 20 dB, in terms of D^2 the person is 100 times as far from the normal group as normal people are. In most cases, normal people stay within the range of 0 to 2 dB.

Application of the MT Method to a Medical Diagnosis

Although, as discussed in the preceding section, we enumerate all necessary items in the medical checkup case, we need to beware of selecting an item that is derived from two other items. For example, in considering height, weight, and obesity, we need to narrow the items down to two because we cannot compute an inverse of a matrix consisting of correlation coefficients.

When we calculate the Mahalanobis distance using a database for normal people, we determine the threshold for judging normality by taking into account the following two types of error loss: the loss caused by misjudging a normal person as abnormal and spending time and money to do precise tests; and the loss caused by misjudging an abnormal person as normal and losing the chance of early treatment.

☐ Example

The example shown in Table 4.24 is not a common medical examination but a special medical checkup to find patients with liver dysfunction, studied by Tatsuji

Table 4.24

Physiological examination items

Examination Item	Acronym	Normal Value
Total protein	TP	6.5–7.5 g/dL
Albumin	Alb	3.5–4.5 g./dL
Cholinesterase	ChE	0.60–1.00 ΔpH
Glutamate oxaloacetate transaminase	GOT	2–25 units
Glutamate pyruvate transaminase	GPT	0–22 units
Lactatdehydrogenase	LDH	130–250 units
Alkaline phosphatase	ALP	2.0–10.0 units
γ-Glutamyltranspeptidase	γ-GTP	0–68 units
Lactic dehydrogenase	LAP	120–450 units
Total cholesterol	TCh	140–240 mg/dL
Triglyceride	TG	70–120 g/dL
Phospholipases	PL	150–250 mg/dL
Creatinine	Cr	0.5–1.1 mg/dL
Blood urea nitrogen	BUN	5–23 mg/dL
Uric Acid	UA	2.5–8.0 mg/dL

Kanetaka at Tokyo Teishin Hospital [6]. In addition to the 15 items shown in Table 4.24, age and gender are included.. The total number of items is 17.

By selecting data from 200 people (it is desirable that several hundred people be selected, but only 200 were chosen because of the capacity of a personal computer) diagnosed as being in good health for three years at the annual medical checkup given by Tokyo's Teishin Hospital, the researchers established a database of normal people. The data of normal people who were healthy for two consecutive years may be used. Thus, we can consider the Mahalanobis distance based on the database of 200 people. Some people think that it follows an F-distribution with an average of 1 approximate degree of freedom of 17 for the numerator and infinite degrees of freedom for its denominator on the basis of raw data. However, since its distribution type is not important, it is wrong. We compare the Mahalanobis distance with the degree of each patient's dysfunction.

If we use decibel values in place of raw data, the data for a group of normal people are supposed to cluster around 0 dB with a range of a few decibels. To minimize loss by diagnosis error, we should determine a threshold. We show a simple method to detect judgment error next.

Table 4.25 demonstrates a case of misdiagnosis where healthy people with data more than 6 dB away from those of the normal people are judged not normal. For 95 new persons coming to a medical checkup, Kanetaka analyzed actual diagnostic error accurately by using a current diagnosis, a diagnosis using the Mahalanobis distance, and close (precise) examination.

Category 1 in Table 4.25 is considered normal. Category 2 comprises a group of people who have no liver dysfunction but had ingested food or alcohol despite being prohibited from doing so before a medical checkup. Therefore, category 2 should be judged normal, but the current diagnosis inferred that 12 of 13 normal people were abnormal. Indeed, the Mahalanobis method misjudged 9 normal people as being abnormal, but this number is three less than that obtained using the current method.

Category 3 consists of a group of people suffering from slight dysfunctions. Both the current and Mahalanobis methods overlooked one abnormal person. Yet both of them detected 10 of 11 abnormal people correctly.

For category 4, a cluster of 5 people suffering from moderate dysfunctions, both methods found all persons. Since category 2 is a group of normal persons, combining categories 1 and 2 as a group of liver dysfunction − , and categories 3 and 4 as a group of liver dysfunction + , we summarize diagnostic errors for each method in Table 4.26. For these contingency tables, each discriminability (0: no discriminability, 1: 100% discriminability) is calculated by the following equations (see "Separation System" in Section 4.4 for the theoretical background). A_1's discriminability:

$$\rho_1 = \frac{[(28)(15) - (51)(1)]^2}{(79)(16)(29)(66)} = 0.0563 \qquad (4.157)$$

A_2's discriminability:

$$\rho_2 = \frac{[(63)(15) - (16)(1)]^2}{(76)(16)(64)(31)} = 0.344 \qquad (4.158)$$

Table 4.25

Medical checkup and discriminability

Category[a]	A_1: Current Method		A_2: Mahalanobis Method		Total
	Normal	Abnormal	Normal	Abnormal	
1	27	39	59	7	66
2	1	12	4	9	13
3	1	10	1	10	11
4	0	5	0	5	5
Total	29	66	64	31	95

[a] 1, Normal; 2, normal but temporarily abnormal due to food and alcohol; 3, slightly abnormal; 4, moderately abnormal.

Table 4.26
2 × 2 Contingency table

A_1: Current method

Diagnosis: Liver dysfunction	Normal	Abnormal	Total
− (Normal)	28	51	79
+ (Abnormal)	1	15	16
Total	29	66	95

A_2: Mahalanobis method

Diagnosis: Liver dysfunction	Normal	Abnormal	Total
− (Normal)	63	16	79
+ (Abnormal)	1	15	16
Total	64	31	95

Each of the equations above consists of a product of peripheral frequencies for the denominator and a squared difference of diagonal products for the numerator. Whereas the discriminability of the current method is 5.63%, that of the Mahalanobis distance is 34.4%, approximately six times as large as that of the current method. This index is the concept of squared correlation coefficient and equivalent to what is called *contribution*.

On the other hand, in expressing discriminability, the following SN ratios are regarded as more rational. A_1's SN ratio:

$$\eta_1 = 10 \log \frac{\rho_1}{1 - \rho_1}$$

$$= 10 \log \frac{0.0563}{0.9437} = -12.2 \text{ dB} \tag{4.159}$$

A_2's SN ratio:

$$\eta_2 = 10 \log \frac{0.344}{0.656} = -2.8 \text{ dB} \tag{4.160}$$

Therefore, a medical examination using the Mahalanobis distance is better by 9.7 dB or 8.7 times than the current item-by-item examination. According to Table 4.25, both methods have identical discriminability of abnormal people. However, the current method diagnosed 51 of 79 normal people to be or possibly be abnormal, thereby causing a futile close examination. whereas the Mahalanobis method judges only 16 of 79 to be abnormal. That is, the latter enables us to eliminate such a wasteful checkup for 35 normal people.

In calculating the SN ratio, we should use the dynamic SN ratio, as explained below.

Design of General Pattern Recognition and Evaluation Procedure

MT METHOD

There is a large gap between science and technology. The former is a quest for truth; the latter does not seek truth but invents a means of attaining an objective function. In mechanical, electronic, and chemical engineering, technical means are described in a systematic manner for each objective function. For the same objective function, there are various types of means, which do not exist in nature.

A product planning department considers types of functions. The term *product planning*, including hardware and software, means to plan products to be released in the marketplace, most of which are related to active functions. However, there are quite a few passive functions, such as inspection, diagnosis, and prediction. In evaluating an active function, we focus on the magnitude of deviation from its ideal function. Using the SN ratio, we do a functional evaluation that indicates the magnitude of error.

In contrast, a passive function generally lies in the multidimensional world, including time variables. Although our objective is obviously pattern recognition in the multidimensional world, we need a brand-new method of summarizing multidimensional data. To summarize multidimensional data, quality engineering (more properly, the Taguchi method) gives the following paradigms:

- ❑ *Procedure 1:* Select items, including time-series items. Up to several thousand items can be analyzed by a computer.
- ❑ *Procedure 2:* Select zero-point and unit values and base space to determine a pattern.
- ❑ *Procedure 3:* After summarizing measurement items only from the base space, we formulate the equation to calculate the Mahalanobis distance. The Mahalanobis space determines the origin of multidimensional measurement and unit value.
- ❑ *Procedure 4:* For an object that does not belong to the base space, we compute Mahalanobis distances and assess their accuracy using an SN ratio.
- ❑ *Procedure 5:* Considering cost and the SN ratio, we sort out necessary items for optimal diagnosis and proper prediction method.

Among these specialists are responsible for procedures 1 and 2 and quality engineering takes care of procedures 3, 4, and 5. In quality engineering, any quality problem is attributed to functional variability. That is, functional improvement will lead to solving any type of quality problem. Therefore, we determine that we should improve functionality by changing various types of design conditions (when dealing with multiple variables, and selecting items and the base space).

Since the Mahalanobis distance (variance) rests on the calculation of all items selected, we determine the following control factors for each item or for each group of items and, in most cases, allocate them to a two-level orthogonal array when we study how to sort items:

- ❑ *First factor level.* Items selected (or groups of items selected) are used.
- ❑ *Second factor level.* Items selected (or groups of items selected) are not used.

We do not need to sort out items regarded as essential. Now suppose that the number of items (groups of items) to be studied is l and an appropriate two-level orthogonal array is L_N. When $l = 30$, L_{32} is used, and when $l = 100$, L_{108} or L_{124} is selected.

Once control factors are assigned to L_N, we formulate an equation to calculate the Mahalanobis distance by using selected items because each experimental condition in the orthogonal array shows the assignment of items. Following procedure 4, we compute the SN ratio, and for each SN ratio for each experiment in orthogonal array L_N, we calculate the control factor effect. If certain items chosen contribute negatively or little to improving the SN ratio, we should select an optimal condition by excluding the items. This is a procedure of sorting out items used for the Mahalanobis distance.

MTS METHOD

Although an orthogonalizing technique exists that uses principal components when we normalize data in a multidimensional space, it is often unrelated to economy and useless because it is based too heavily on mathematical background. Now we introduce a new procedure for orthogonalizing data in a multidimensional space, which at the same time reflects on the researchers' objective.

We select $X_1, X_2, \ldots, X_k$ as k-dimensional variables and define the following as n groups of data in the Mahalanobis space:

$$X_{11}, X_{12}, \ldots, X_{1n}$$

$$X_{21}, X_{22}, \ldots, X_{2n}$$

$$\vdots$$

$$X_{k1}, X_{k2}, \ldots, X_{kn}$$

All of the data above are normalized. That is, n data, $X_1, X_2, \ldots, X_n$ have a mean of zero and a variance of 1. $X_1, X_2, \ldots, X_k$ represent the order of cost or priority, which is an important step and should be determined by engineers. In lieu of $X_1, X_2, \ldots, X_k$, we introduce new variables, $x_1, x_2, \ldots, x_k$ that are mutually orthogonal:

$$x_1 = X_1 \tag{4.161}$$

x_1 is equal to X_1 and has a mean of zero and a variance of 1. Variable x_2 is the part of variable X_2, that remains after removing the part related to, except for the part involving variable x_1 (or X_1). We consider the following equation, where X_2 is expressed by x_1:

$$X_2 = b_{21}x_1 \tag{4.162}$$

In this case b_{12} is not only the regression coefficient but also the correlation coefficient. Then the remaining part of X_2, excluding the part related to x_1 (regression part) or the part independent of x_1, is

$$x_2 = X_2 - b_{21}x_1 \tag{4.163}$$

Orthogonality of x_1 and x_2 indicates that a sum of products of x_{1j} and x_{2j} amounts to zero in the base space:

$$\sum_{j=1}^{n} x_{1j} x_{2j} = \sum_{j=1}^{n} X_{1j}(X_{2j} - b_{21}x_{1j}) = 0 \qquad (4.164)$$

Thus, x_1 and x_2 become orthogonal. On the other hand, whereas x_1's variance is 1, x_2's variance σ_2^2, called *residual contribution*, is calculated by the equation

$$\sigma_2^2 = 1 - b_{21}^2 \qquad (4.165)$$

The reason is that if we compute the mean square of the residuals, we obtain the following result:

$$\frac{1}{n} \sum (X_{2j} - b_{21}x_{1j})^2 = \frac{1}{n} \sum X_{2j}^2 - \frac{2}{n} b_{21} \sum X_{2j} x_{ij} + \frac{1}{n} (-b_{21})^2 \sum x_{1j}^2$$

$$= 1 - \frac{2n}{n} b_{2l}^2 + \frac{n}{n} b_{2l}^2$$

$$= 1 - b_{2l}^2 \qquad (4.166)$$

We express the third variable, X_3, with x_1 and x_2:

$$X_3 = b_{31}x_1 + b_{32}x_2 \qquad (4.167)$$

By multiplying both sides of equation (4.167) by x_1, calculating a sum of elements in the base space, and dividing by n, we obtain

$$b_{31} = \frac{1}{nV_1} \sum X_{3j} x_{1j} \qquad (4.168)$$

$$V_1 = \frac{1}{n} \sum x_{1j}^2 \qquad (4.169)$$

Similarly, by multiplying both sides of equation (4.167) by x_2, we have b_{32}

$$b_{32} = \frac{1}{nV_2} \sum X_{3j} x_{2j} \qquad (4.170)$$

$$V_2 = \frac{1}{n} \sum x_{2j}^2 \qquad (4.171)$$

The orthogonal part of x_3 is thus a residual part, so that X_3 cannot be expressed by both X_1 and X_2 or both x_1 and x_2. In sum,

$$x_3 = X_3 - b_{31}x_1 - b_{32}x_2 \qquad (4.172)$$

In the Mahalanobis space, x_1, x_2, and x_3 are orthogonal. Since we have already proved the orthogonality of x_1 and x_2, we prove here that of x_1 and x_3 and that of x_2 and x_3. First, considering the orthogonality of x_1 and x_3, we have

$$\sum_{j} x_{1j}(X_{3j} - b_{31}x_{1j} - b_{32}x_{2j}) = \sum_{j} x_{1j}X_{3j} - b_{31} \sum x_{1j}^2$$

$$= 0 \qquad (4.173)$$

This can be derived from equation (4.167) defining b_{31}. Similarly, the orthogonality of x_2 and x_3 is proved according to equation (4.170), defining b_{32} as

$$\sum x_{2j}\,(X_{3j} - b_{31}x_{1j} - b_{32}x_{2j}) = \sum x_{2j}X_{3j} - b_{32}\sum x_{2j}^2$$
$$= 0 \qquad (4.174)$$

For the remaining variables, we can proceed with similar calculations. The variables normalized in the preceding section can be rewritten as follows:

$$x_1 = X_1$$
$$x_2 = X_2 - b_{21}x_1$$
$$x_3 = X_3 - b_{31}x_1 - b_{32}x_2$$
$$\vdots$$
$$x_k = X_k - b_{k1}x_1 - b_{2k}x_2 - \cdots - b_{k(k-1)}x_{k-1} \qquad (4.175)$$

Each of the orthogonalized variables $x_2, \ldots, x_k$ does not have a variance of 1. In an actual case we should calculate a variance right after computing n groups of variables, $x_1, x_2, \ldots, x_n$ in the Mahalanobis space. As for degrees of freedom in calculating a variance, we can regard n for x_1, $n-1$ for x_2, $n-2$ for x_3, $\ldots$, $n-k+1$ for x_k. Instead, we can select n degrees of freedom for all because $n \gg k$ quite often.

$$V_1 = \frac{1}{n}\,(x_{11}^2 + x_{12}^2 + \cdots + x_{1n}^2)$$
$$V_2 = \frac{1}{n-1}\,(x_{21}^2 + x_{22}^2 + \cdots + x_{2n}^2)$$
$$\vdots$$
$$V_k = \frac{1}{n-k+1}\,(x_{k1}^2 + x_{k2}^2 + \cdots + x_{kn}^2) \qquad (4.176)$$

Now setting normalized, orthogonal variables to $y_1, y_2, \ldots, y_k$, we obtain

$$y_1 = \frac{x_1}{\sqrt{V_1}}$$
$$y_2 = \frac{x_2}{\sqrt{V_2}}$$
$$\vdots$$
$$y_k = \frac{x_k}{\sqrt{V_k}} \qquad (4.177)$$

All of the normalized $y_1, y_2, \ldots, y_k$ are orthogonal and have a variance of 1. The database completed in the end contains m and σ as the mean and standard deviation of initial observations, $b_{21}, V_2, b_{31}, b_{32}, V_2, \ldots, b_{k1}, b_{k2}, \ldots, b_{k(k-1)}, V_k$ as the coefficients and variances for normalization. Since we select n groups of data, there are k means of $m_1, m_2, \ldots, m_k$, k standard deviations of $\sigma_1, \sigma_2, \ldots, \sigma_k$, $(k-1)k/2$ coefficients, and k variances in this case. Thus, the number of necessary memory items is as follows:

$$k + k + \frac{(k-1)k}{2} + k = \frac{k(k+5)}{2} \tag{4.178}$$

The correlation matrix in the normalized Mahalanobis space turns out to be an identity matrix. Then the correlation matrix R is expressed as

$$R = \begin{pmatrix} 1 & 0 & \cdots & 0 \\ 0 & 1 & \cdots & 0 \\ \vdots & \vdots & \ddots & \vdots \\ 0 & 0 & \vdots & 1 \end{pmatrix} \tag{4.179}$$

Therefore, the inverse matrix of R, A is also an identity matrix:

$$A = R^{-1} = \begin{pmatrix} 1 & 0 & \cdots & 0 \\ 0 & 1 & \cdots & 0 \\ \vdots & \vdots & \ddots & \vdots \\ 0 & 0 & \vdots & 1 \end{pmatrix} \tag{4.180}$$

Using these results, we can calculate the Mahalanobis distance D^2 as

$$D^2 = \frac{1}{k}(y_1^2 + y_2^2 + \cdots + y_k^2) \tag{4.181}$$

Now, setting measured data already subtracted by m and divided by σ to $X_1, X_2, \ldots, X_k$, we compute primarily the following $x_1, x_2, \ldots, x_k$:

$$x_1 = X_1$$

$$x_2 = X_2 - b_{21}x_1$$

$$x_3 = X_3 - b_{31}x_1 - b_{32}x_2$$

$$\vdots$$

$$x_k = X_k - b_{k1}x_1 - \cdots - b_{k(k-1)} \tag{4.182}$$

Thus, $y_1, y_2, \ldots, y_k$ are calculated as

$$y_1 = x_1$$

$$y_2 = \frac{x_2}{\sqrt{V_2}}$$

$$y_3 = \frac{x_3}{\sqrt{V_3}}$$

$$\vdots$$

$$y_k = \frac{x_k}{\sqrt{V_k}} \tag{4.183}$$

When for an arbitrary object we calculate y_1, y_2, ... , y_k and D^2 in equation (4.181), if a certain variable belongs to the Mahalanobis space, D^2 is supposed to take a value of 1, as discussed earlier. Otherwise, D^2 becomes much larger than 1 in most cases.

Once the normalized orthogonal variables y_1, y_2, ... , y_k are computed, the next step is the selection of items.

A_1: Only y_1 is used

A_2: y_1 and y_2 are used

$\vdots$

A_k: y_1, y_2, ... , y_k are used

Now suppose that values of signal factor levels are known. Here we do not explain a case of handling unknown values. Although some signals belong to the base space, l levels of a signal that is not included in the base space are normally used. We set the levels to M_1, M_2, ... , M_l. Although l can only be 3, we should choose as large an l value as possible to calculate errors.

In the case of A_k, or the case where all items are used, after calculating the Mahalanobis distances for M_1, M_2, ... , M_l by taking the square root of each, we create Table 4.27. What is important here is not D^2 but D per se. As a next step, we compute dynamic SN ratios. Although we show the case where all items are used, we can create a table similar to Table 4.27 and calculate the SN ratio for other cases, such as the case when partial items are assigned to an orthogonal array.

Based on Table 4.27, we can compute the SN ratio η as follows:

Total variation:

$$S_T = D_1^2 + D_2^2 + \cdots + D_l^2 \qquad (f = l) \qquad (4.184)$$

Variation of proportional terms:

$$S_\beta = \frac{(M_1 D_1 + M_2 D_2 + \cdots + M_l D_l)^2}{r} \qquad (f = 1) \qquad (4.185)$$

Effective divider:

$$r = M_1^2 + M_2^2 + \cdots + M_l^2 \qquad (4.186)$$

Error variation:

$$S_e = S_T - S_\beta \qquad (f = l - 1) \qquad (4.187)$$

Table 4.27
Signal values and Mahalanobis distance

Signal-level value	M_1	M_2	$\cdots$	M_k
Mahalanobis distance	D_1	D_2	$\cdots$	D_l

SN ratio:

$$\eta = 10 \log \frac{(1/r)(S_\beta - V_e)}{V_e} \tag{4.188}$$

On the other hand, for the calibration equation, we calculate

$$\beta = \frac{M_1 D_1 + M_2 D_2 + \cdots + M_l D_l}{r} \tag{4.189}$$

and we estimate

$$M = \frac{D}{\beta} \tag{4.190}$$

In addition, for A_1, A_2, ... , A_{k-1}, we need to compute SN ratios, η_1, η_2, ... , η_{k-1}. Table 4.28 summarizes the result. According to the table, we determine the number of items by balancing SN ratio and cost. The cost is not the calculation cost but the measurement cost for items. We do not explain here the use of loss functions to select items.

Summary of Partial MD Groups: Countermeasure for Collinearity

To summarize the distances calculated from each subset of distances to solve collinearity problems, this approach can be widely applied. The reason is that we can select types of scale to use, as what is important is to select a scale that expresses patients' conditions accurately no matter what correlation we have in the Mahalanobis space. The point is to be consistent with patients' conditions diagnosed by doctors.

Now let's go through a new construction method. For example, suppose that we have data for 0 and 1 in a 64 × 64 grid for computer recognition of handwriting. If we use 0 and 1 directly, 4096 elements exist. Thus, the Mahalanobis space formed by a unit set (suppose that we are dealing with data for n persons in terms of whether a computer can recognize a character "A" as "A": for example, data from 200 sets in total if 50 people write a character of four items) has a 4096 × 4096 matrix. This takes too long to handle using current computer capability.

How to leave out character information is an issue of information system design. In fact, a technique for substituting only 128 data items consisting of 64 column sums and 64 row sums for all data in a 64 × 64 grid has already been proposed. Since two sums of 64 column sums and 64 row sums are identical, they have collinearity. Therefore, we cannot create a 128 × 128 unit space by using the relationship 64 + 64 = 128.

In another method, we first create a unit space using 64 row sums and introduce the Mahalanobis distance, then create a unit space using 64 column sums and

Table 4.28

Signal values and Mahalanobis distances

Number of items	1	2	3	...	k
SN ratio	η_1	η_2	η_3	...	η_k
Cost	C_1	C_2	C_3	...	C_k

calculate the Mahalanobis distance. For a few alphabets similar to "B," we calculate Mahalanobis distances. If 10 persons, N_1. N_2, ... , N_{10} write three letters similar to "B," we can obtain 30 signals. After each of the 10 persons writes the three letters "D," "E," and "R," we compute the Mahalanobis distances for all signals from a unit space of "B." This is shown in Table 4.29.

We calculate a discriminability SN ratio according to this table. Because we do not know the true difference among M's, we compute SN ratios for unknown true values as follows:

Total variation:

$$S_T = D_{11}^2 + D_{12}^2 + \cdots + D_{3,10}^2 \qquad (f = 30) \qquad (4.191)$$

Signal effect:

$$S_M = \frac{D_1^2 + D_2^2 + D_3^2}{10} \qquad (f = 3) \qquad (4.192)$$

$$V_M = \frac{S_M}{3} \qquad (4.193)$$

$$S_e = S_T - S_M \qquad (f = 27) \qquad (4.194)$$

$$V_e = \frac{S_e}{27} \qquad (4.195)$$

By calculating row sums and column sums separately, we compute the following SN ratio η using averages of V_M and V_e:

$$\eta = 10 \log \frac{\frac{1}{10}(V_M - V_e)}{V_e} \qquad (4.196)$$

The error variance, which is a reciprocal of the antilog value, can be computed as

$$\sigma^2 = 10^{-1.1\eta} \qquad (4.197)$$

The SN ratio for selection of items is calculated by setting the following two levels:

❏ *Level 1:* item used

❏ *Level 2:* item not used

We allocate them to an orthogonal array. By calculating the SN ratio using equation (4.196), we choose optimal items.

Table 4.29
Mahalanobis distances for signals

Signal	Noise				Total
	N_1	N_2	$\cdots$	N_{10}	
M_1 (D)	D_{11}	D_{12}	$\cdots$	$D_{1,10}$	D_1
M_2 (E)	D_{21}	D_{22}	$\cdots$	$D_{2,10}$	D_2
M_3 (R)	D_{31}	D_{32}	$\cdots$	$D_{3,10}$	D_3

Although there can be some common items among groups of items, we should be careful not to have collinearity in one group. Indeed, it seems somewhat unusual that common items are included in several different groups, even though each common item should be emphasized; however, this situation is considered not so unreasonable because the conventional single correlation has neglected other, less related items completely.

A key point here is that we judge whether the summarized Mahalanobis distance corresponds to the case of using the SN ratio calculated by Mahalanobis distances for items not included in the unit space. Although we can create Mahalanobis spaces by any procedure, we can judge them only by using SN ratios.

4.8. On-line Quality Engineering

None of us cares about causes of errors when we set a clock. We sometimes calibrate the time by comparing it with a time signal from radio or TV. This is the case with almost all characteristics in the production process. Let's say that according to company S, the cost of comparing the time with the time signal is approximately $1.50. In addition, assuming that calibration takes 5 seconds, it will cost around 10 cents. A quality control system in the production process is responsible for handling eventual clock errors, given $1.50 for comparison of the time with the time signal and 10 cents for calibration. If we predict such errors, we can determine optimal checkup intervals or calibration limits using loss functions.

A technical procedure for designing optimal checkup intervals and adjustment limits is called *on-line quality engineering* or *quality control in process*. In on-line quality engineering, we should clarify three system elements used to design a system: process, checkup procedure (comparison with a standard such as a time signal and generally called *measurement*), and correction method (in the case of calibration of a measuring machine called *calibration*, and in the case of process called *correction* or *adjustment*). In this section we discuss a quality control system using a simple example.

❏ Example

Although automobile keys are generally produced as a set of four identical keys, this production system is regarded as dynamic because each set has a different dimension and shape. Each key has approximately 9 to 11 cuts, and each cut has several different dimensions. If there are four different dimensions with a 0.5-mm step, a key with 10 cuts has $4^{10} = 1,048,576$ variations.

In actuality, each key is produced such that it has a few different-dimension cuts. Then the number of key types in the market will be approximately 10,000. Each key set is produced based on the dimensions indicated by a computer. By using a master key we can check whether a certain key is produced, as indicated by the computer. Additionally, to manage a machine, we can examine particular-shaped keys produced by such machines at certain intervals. Now we consider a case where we conduct process quality control by measuring differences between dimensions of cuts of the last key in one set and those shown by a computer.

For the sake of convenience, the standard of key dimensions is ± 30 μm for each target value. Whether or not this ± 30 μm tolerance is rational depends on whether the safety factor of 8 is appropriate for the function limit of ± 250 μm. (In this case, since the height of a cut is designed using a 500-μm step, if a certain dimension deviates by 250 μm more or less than its design dimension, it can be taken for other designs. Thus, the function limit is 250 μm. In this case, is a safety factor of 8 appropriate? If we suppose the price of a key set shipped from a plant to be $1.90, the eventual loss in the market when a key does not work becomes 64 times as great as the price, or $1.20. We consider the tolerance appropriate; that is, when the malfunction of the key costs $120, ± 30 μm makes sense.

A manufacturing department's task is not to determine tolerances for a product but to produce a product with dimensions that lie within a range of tolerances. To achieve this, how should the manufacturing department design a quality control system? For example, in an inspection room, once an hour we check the dimensions of a key shortly after it has been machined. When we inspect every n products produced, we call the number n the *inspection interval* (also called *diagnosis, checkup,* or *measurement interval*). If we produce 300 sets in an hour, the current inspection interval N_0 is 300 sets. Now the inspection cost B when we pick up a product in the production process and check it in the inspection room (including labor cost) amounts to $12. In addition, the loss A for one defective set is already assumed to be $1.90. On the other hand, the number of key sets produced while the key selected is inspected for 10 minutes is called the *time lag of inspection.* In this case we regard this as 50 sets. At present, we are controlling the production process within a range of ± 20 μm, which is two-thirds of the current tolerance, $\Delta = 30$ μm. Then the current adjustment limit D_0 is 20 μm. Currently, one of 19,560 products goes beyond the adjustment limit of ± 20 (μm). Since we stop the production line and take countermeasures such as an adjustment or tool change when a product exceeds the limit, we need an adjustment cost C of $58. In sum, we enumerate all preconditions for designing a process control system:

- ❏ *Dimensional tolerance:* $\Delta = 30$ μm
- ❏ *Loss of defective product:* $A = \$1.90$ per set
- ❏ *Inspection cost:* $B = \$12$
- ❏ *Time lag between inspection and judgment:* $I = 50$ sets
- ❏ *Adjustment cost:* $C = \$58$
- ❏ *Current inspection interval:* $n_0 = 300$ sets
- ❏ *Current adjustment limit:* $D_0 = 20$ μm
- ❏ *Observed mean adjustment interval:* $u_0 = 19{,}560$ sets

When we inspect a set in an hour and control the production process with the adjustment limit $D_0 = 20$ μm, the only observation is how many inspections happen from one to another adjustment. In this case, this value, u_0, indicating process stability, is 19,560.

Design of Feedback Control System

To design all process control systems in an economical manner is the most crucial task for managers. The cost of the current control system for one set is

$$\text{control cost} = \frac{B}{n_0} + \frac{C}{u_0} \tag{4.198}$$

In addition, we need a monetary evaluation of the quality level according to dimensional variability. When the adjustment limit is $D_0 = \pm 20 \ \mu m$, the dimensions are regarded as being distributed uniformly within the limit because they are not controlled in the range. The variance is as follows:

$$\frac{D_0^2}{3} \tag{4.199}$$

On the other hand, the magnitude of dispersion for an inspection interval n_0 and an inspection time lag of l is

$$\left(\frac{n_0 + 1}{2} + l\right)\frac{D_0^2}{u_0} \tag{4.200}$$

Consequently, we obtain the following loss function:

$$\frac{A}{\Delta^2}\left[\frac{D_0^2}{3} + \left(\frac{n_0 + 1}{2} + l\right)\frac{D_0^2}{u_0}\right] \tag{4.201}$$

Adding equations (4.198) and (4.201) to calculate the following economic loss L_0 for the current control system, we have 33.32 cents per product:

$$L_0 = \frac{B_0}{n_0} + \frac{C}{u_0} + \frac{A}{\Delta^2}\left[\frac{D_0^2}{3}\left(\frac{n_0 + 1}{2} + l\right)\frac{D_0^2}{u_0}\right]$$

$$= \frac{12}{300} + \frac{58}{19,560} + \frac{1.90}{30^2}\left[\frac{20^2}{3} + \left(\frac{301}{2} + 50\right)\frac{20^2}{19,560}\right]$$

$$= 0.04 + 000.30 + 0.2815 + 0.87$$

$$= 33.32 \text{ cents} \tag{4.202}$$

Assuming that the annual operation time is 1600 hours, we can see that in the current system the following amount of money is spent annually to control quality:

$$(33.32)(300)(1600) \approx \$160,000 \tag{4.203}$$

If an error σ_m accompanies each measurement, another loss of measurement is added to the loss above:

$$\frac{A}{\Delta^2}\sigma_m^2 \tag{4.204}$$

It is an optimal control system that improves the loss in equation (4.203). In short, it is equivalent to determining an optimal inspection interval n and adjustment limit D, both of which are calculated by the following formulas:

$$n = \sqrt{\frac{2u_0 B}{A}\frac{\Delta}{D_0}} \qquad (4.205)$$

$$= \sqrt{\frac{(2)(19,560)(12)}{1.90}}\left(\frac{30}{20}\right)$$

$$= 745$$

$$\approx 600 \text{ (twice an hour)} \qquad (4.206)$$

$$D = \left(\frac{3C}{A} \times \frac{D_0^2}{u_0}\Delta^2\right)^{1/2} \qquad (4.207)$$

$$= \left[\frac{(3)(58)}{1.90}\left(\frac{20^2}{19,560}\right)\Delta^2\right]^{1/4}$$

$$= 6.4 \qquad (4.208)$$

$$\approx 7.0 \ \mu m \qquad (4.209)$$

Then, by setting the optimal measurement interval to 600 sets and adjustment limit to $\pm 7.0\ \mu m$, we can reduce the loss L:

$$L = \frac{B}{n} + \frac{C}{u} + \frac{A}{\Delta^2}\left[\frac{D^2}{32} + \left(\frac{n+1}{2} + l\right)\frac{D^2}{u}\right] \qquad (4.210)$$

Using this equation, we estimate the mean adjustment interval u:

$$u = u_0 \frac{D^2}{D_0^2}$$

$$= (19,560)\left(\frac{7^2}{20^2}\right)$$

$$= 2396 \qquad (4.211)$$

Therefore, we need to change the adjustment interval from the current level of 65 hours to 8 hours. However, the total loss L decreases as follows:

$$L = \frac{12}{600} + \frac{58}{2396} + \frac{1.80}{30^2}\left[\frac{7^2}{3} + \left(\frac{601}{2} + 50\right)\frac{7^2}{2396}\right]$$

$$= 0.02 + 0.0242 + 3.45 + 1.51$$

$$= 9.38 \text{ cents} \qquad (4.212)$$

This is reduced from the current level by

$$33.32 - 9.38 = 23.94 \text{ cents} \qquad (4.213)$$

On a yearly basis, we can expect an improvement of

$$(23.94)(300)(1600) = \$114{,}912 \tag{4.214}$$

Management in Manufacturing
The procedure described in the preceding section is a technique of solving a balancing equation of checkup and adjustment costs, and necessary quality level, leading finally to the optimal allocation of operators in a production plant. Now, given that checkup and adjustment take 10 and 30 minutes, respectively, the work hours required in an 8-hour shift at present are:

(10 min) × (daily no. inspections) + (30 min) × (daily no. adjustments)

$$= (10)\frac{(8)(300)}{n} + (30)\frac{(3)(300)}{u} \tag{4.215}$$

$$= (10)\left(\frac{2400}{600}\right) + (30)\left(\frac{2400}{2396}\right)$$

$$= 40 + 30.1 = 70 \text{ minutes} \tag{4.216}$$

Assuming that one shift has a duration of 8 hours, the following the number of workers are required:

$$\frac{70}{(8)(60)} = 0.146 \text{ worker} \tag{4.217}$$

If there are at most six processes,

$$(0.146)(6) = 0.88 \text{ worker} \tag{4.218}$$

is needed, which implies that one operator is sufficient. As compared to the following workers required currently, we have

$$(10)\frac{(8)(300)}{300} + (30)\frac{(3)(300)}{19{,}560} = 83.6 \text{ minutes} \tag{4.219}$$

so we can reduce the number of workers by only 0.028 in this process. On the other hand, to compute the process capability index C_p, we estimate the current standard deviation σ_0 using the standard deviation of a measurement error σ_m:

$$\sigma_0 = \sqrt{\frac{D_0^2}{3} + \left(\frac{n_0+1}{2} + 1\right)\frac{D_0^2}{u_0} + \sigma_m^2} \tag{4.220}$$

Now setting $\sigma_m = 2$ μm, we obtain σ_0:

$$\sigma_0 = \sqrt{\frac{20^2}{3} + \left(\frac{301}{2} + 50\right)\frac{20^2}{19560} + 2^2}$$

$$= 11.9 \text{ μm} \tag{4.221}$$

Thus, the current process capability index C_p is computed as follows:

$$C_p = \frac{2\Delta}{6\sigma_0}$$

$$= \frac{(2)(30)}{(6)(11.9)}$$

$$= 0.84 \qquad (4.222)$$

The standard deviation in the optimal feedback system is

$$\sigma = \sqrt{\frac{7^2}{3} + \left(\frac{601}{2} + 50\right)\frac{7^2}{2396} + 2^2}$$

$$= 5.24 \ \mu m \qquad (4.223)$$

Then we cannot only reduce the required workforce by 0.028 worker (0.84 times) but can also enhance the process capability index C_p from the current level to

$$C_p = \frac{(30)(2)}{(6)(5.24)} = 1.91 \qquad (4.224)$$

Manufacturing Strategy: Balance of Production Measurements and Quality Control System

The most essential strategy planned by manufacturing managers is not to balance quality and cost but to improve productivity. The difference in profitability between corporations is primarily the difference in production speed. Quite a few managers insist that we should continue to produce only a certain volume because even if the production speed is doubled and the production volume is doubled, the increased volume tends to remain unsold in most cases. Nevertheless, we believe that a key point is to develop a technology to double the production speed to prepare for such a demand. As an extreme idea, after doubling the production speed, we can stop after half a day. In this case, an R&D department could offer new jobs to idle workers, or after lowering the retail price by two-thirds, could double sales.

Unless a company sells a product at a retail price several times as high as a plant price (UMC or production cost), it cannot survive. This is because labor and running expenses in sales, administration, and R&D departments account for a few percent of its total cost. In other words, the retail price needs to include the total running cost and profit. When a product is sold at a price three times as high as a plant price, if the production volume is doubled, the plant price, excluding material cost, will be cut in half. Therefore, if the material cost accounts for 30% of the plant price, in cases of double production speed, the retail price can be reduced to an amount 1.65 times as great as the plant cost:

$$[0.3 + \tfrac{1}{2}(0.7)] + \tfrac{1}{2}(2) = 1.65 \text{ times} \qquad (4.225)$$

As compared to 3 times, 1.65 times implies that we can lower the retail price 0.55-fold. This estimation is grounded on the assumption that if we sell a product at 45% off the original price, the sales volume doubles. This holds true when we offer new jobs to half the workers, with the sales volume remaining at the same level. Now, provided that the mean adjustment interval decreases by one-fourth due to increased variability after the production speed is raised, how does the cost eventually change? When frequency of machine breakdowns quadruples, u decreases from its current level of 2396 to 599, one-fourth of 2396. Therefore, we alter the current levels of u_0 and D_0 to the values $u_0 = 599$ and $D_0 = 7$ μm. By taking these into account, we consider the loss function. The cost is cut by 0.6-fold and the production volume is doubled. In general, as production conditions change, the optimal inspection interval and adjustment limit also change. Thus, we need to recalculate n and D here. Substituting $0.6A$ for A in the formula, we obtain

$$n = \sqrt{\frac{2u_0B}{0.6A}} \frac{\Delta}{D_0}$$

$$= \sqrt{\frac{(2)(599)(12)}{0.6 \times 190}} \left(\frac{30}{7}\right)$$

$$= 481 \rightarrow 600 \quad \text{(once in an hour)} \tag{4.226}$$

$$D = \left[\frac{(3)(58)}{(0.6)(1.90)} \left(\frac{7^2}{599}\right) (30^2)\right]^{1/4}$$

$$= 10.3 \rightarrow 10.00 \tag{4.227}$$

On the other hand, the mean adjustment interval is as follows:

$$u = (599)\left(\frac{10^2}{7^2}\right) = 1222 \tag{4.228}$$

Hence, the loss function is computed as follows:

$$L = \frac{B}{n} + \frac{C}{u} + \frac{0.6A}{30^2}\left[\frac{D^2}{3} + \left(\frac{n+1}{2} + 2l\right)\frac{D^2}{u}\right]$$

$$= \frac{12}{600} + \frac{58}{1222} + \frac{(0.6)(1.90)}{30^2}\left[\frac{10^2}{3} + \left(\frac{601}{2} + 100\right)\frac{10^2}{1222}\right]$$

$$= 0.02 + 0.0475 + 4.22 + 4.15$$

$$= 15.12 \text{ cents} \tag{4.229}$$

As a consequence, the quality control cost increases by 5.74 cents (= 15.12 − 9.38). However, the total loss including cost is

$$\$1.90 + 9.38 \text{ cents} = \$1.99 \tag{4.230}$$

$$(0.6)(190 \text{ cents}) + 15.12 \text{ cents} = \$1.29 \tag{4.231}$$

Suppose that the production volume remains the same, $(300)(1600) = 480,000$ sets. We can save the following amount of money on an annual basis:

$$(1.99 - 1.29)(48) = \$33,600,000 \qquad (4.232)$$

References

1. Genichi Taguchi, 1987. *System of Experimental Design*. Dearborn, Michigan: Unipub/American Supplier Institute.
2. Genichi Taguchi, 1984. *Reliability Design Case Studies for New Product Development*. Tokyo: Japanese Standards Association.
3. Genichi Taguchi et al., 1992. *Technology Development for Electronic and Electric Industries*. Quality Engineering Application Series. Tokyo: Japanese Standards Association.
4. Measurement Management Simplification Study Committee, 1984, *Parameter Design for New Product Development*. Japanese Standards Association.
5. Genichi Taguchi et al., 1989. *Quality Engineering Series*, Vol. 2. Tokyo: Japanese Standards Association.
6. Tatsuji Kenetaka, 1987. *An application of Mahalanobis distance. Standardization and Quality Control*, Vol. 40, No. 10.

Part II

Quality Engineering: A Historical Perspective

5 Development of Quality Engineering in Japan

5.1. Origin of Quality Engineering   127
5.2. Conceptual Transition of the SN Ratio and Establishment of Quality Engineering   131
5.3. On-line Quality Engineering and Loss Function   133
5.4. SN Ratio in Machining   135
5.5. Evaluation Method Using Electric Power in Machining   138
5.6. SN Ratio of Transformability   139
5.7. SN Ratio in Electric Characteristics   140
5.8. Chemical Reaction and Pharmacy · 142
5.9. Development of the Mahalanobis–Taguchi System   142
5.10. Application of Simulation   144
5.11. Summary   147
       References   150

5.1. Origin of Quality Engineering

As an explanation of the origin of quality engineering in the post–World War II rehabilitation period in Japan, Genichi Taguchi said the following about his experiences at the Musashino Telecommunication Laboratory of the Nippon Telegraph and Telephone Public Corporation in his book, *My Way of Thinking* [1]:

It was in 1948 when the Telecommunication Laboratory was founded in the Ministry of Communications (later Ministry of Telecommunications). The Electrotechnical Laboratory, which had supported Japanese communication and electrical research so far, was dissolved and developed into the Telecommunication Laboratory, focused on studying telecommunications. Behind the scenes, there existed a strong intention of the Common Communication Support (CCS) of the General Headquarters (GHQ).

As soon as The CSS entered into Japan, it started to investigate the communicational situation in Japan. This must have been because they considered reinforcement of a communication network essential for the U.S. occupation policy. At that time, the telecommunication infrastructure in Japan was poorly established insomuch as a phone in Japan was ridiculed as a "bone." While the CSS recommended a drastic improvement in Japanese telecommunication research, it requested foundation of a research organization such as the Bell Labs in the U.S. As a result, the Japanese government founded the Telecommunication Laboratory using 2.2 percent of the national budget.

It was in 1950 when the laboratory completed a full-scale research organization by renewing a disaster-stricken building into a new building in the ex-Nakajima airplane factory in Musashino-shi. In those days, the Japanese economy was accelerating its rehabilitation speed due to the Korean War that broke out in June of 1950. Accordingly, demand for subscribed phones began to skyrocket. Therefore, to build the telecommunication infrastructure was one of the most urgent national projects.

During this period, I entered the Telecommunication Laboratory. Because it completely followed the system of the Bell Labs, I noticed that it was an R&D laboratory, much more Americanized than expected in Japan at that time and filled with a liberal atmosphere.

At this point, I would like to mention its peculiarity. In line with inauguration of the Nippon Telegraph and Telephone Public Corporation in 1952 (currently known as NTT), the Telecommunication Laboratory also came under its control. The special characteristics of the Telecommunication Laboratory were equivalent to those of the Nippon Telegraph and Telephone Public Corporation, different from other common companies.

What are required for telecommunication are telephone machines, telecommunication networks, and exchangers. Although we conducted research and development regarding hardware and system design at that time, we did not produce actual products. By receiving design drawings and specifications from us, private manufacturers such as Nippon Electric or Fujitsu manufactured products based on them. After the Nippon Telegraph and Telephone Corporation purchased all of them, exchangers and cables were used by ourselves, and telephone machines were lent to subscribers. That is, we were "consumers."

Once exchangers broke down or cables malfunctioned, we had to squander a tremendous amount of labor force and expense. When telephone machines went out of order, we had to repair them free of charge or replace them with new ones. Those who got into trouble when these failures happened were ourselves as consumers. However, since we designed them on our own, there were no other people to make complaints against.

On the other hand, usually manufacturers sell their products produced. In other extreme words, "Everything is done once a product is sold." Thus, they do not need to seriously consider the issues of quality or cost of products after customers purchase them. In fact, quality management implemented by many of the manufacturers deals not with customers' issues but with production issues in most cases. From the viewpoint of economy, these approaches make a great difference.

If a product breaks down, we suffer from a loss—human beings are quite sensitive to their own interests. Therefore, we have no other choice but to consider how we can design a product before mass production of it, so that it does not cause failures and troubles.

Then, how we can prevent such failures and problems or what types of test should be implemented to do so becomes a critical issue. The design of experiments that I instructed was a method of conducting a test and make an improvement.

So to speak, this can be regarded as the origin of quality engineering. That is, the quality engineering defines the following: "Quality of a product is a loss given to a society after it is shipped."

This definition is seen in the journal *Standardization and Quality Management* (1965) [2]. When "the Taguchi method" was discussed in the Japanese Quality Control Society in 1986 [3], it was discussed whether the above definition agreed with the one defined by the quality control group, which said: "an entity of unique characteristics or performances to be evaluated for determining whether a product or service satisfies its objective in use." JIS Z-9011, *Single Sampling Inspection Plans by Attribute with Adjustment* (1963), regarding sampling inspection created under this new definition, was not often recognized by experts in sampling inspection. However, it proposed an epoch-making idea that loss is economically balanced based on the break-even point.

While I was watching many researchers doing their jobs in the Musashino Telecommunication Laboratory, it was noticed that most of the time spent was not on researching the idea itself but on the evaluation of the idea, such as spending their time on calculation, experimentation, prototype preparation, or testing. In order to rationalize and make the evaluation efficiently, the book *Design of Experiment* [4] was published. This book is considered to have originated from the educational textbook for engineers of the laboratory and related companies and had a strong impact on development of the design of experiments in Japan. When it was updated from the original edition of 1957 to the second one of 1962, the outlook of the design of experiments as a technical procedure was fostered from the conventional one as a phenomenal analysis. In addition, when its third edition was published, the *Design of Experiments* clarified its position as a methodology of technological development by introducing the concept of SN ratio. While it is still regarded as a heresy by experts in the statistical design of experiments, it is highly evaluated as the most creative method. However, the concept of SN ratio was already proposed in a different form in the first edition of *Design of Experiments;* Vol. 2. That is, the fact that there already existed a methodology from the beginning has perplexed many people after its new proposal was offered.

For example, the objective of using orthogonal arrays in the design of experiment is not for optimizing experimental conditions but for finding faults of experiments by the utilization of interactions. Such an idea existed from the very beginning. This contradicts the statisticians' view that interactions are neglected by Taguchi. Conversely, the new definition of quality leads to the idea that we should predict *before* product design whether or not the quality of a product will be sufficient after it has debuted in the market. From the current standpoint of quality engineering, we take it for granted that in the 1960s, when quite a few people argued that an orthogonal array should be used because of its effectiveness in technical optimization or that it could not be used because of the existence of interactions, the idea that we should take advantage of the fact that an orthogonal experiment tends to fail more frequently than a normal one could not be comprehended by ordinary engineers. [5] Quality engineering advocates that it is worthwhile to discover failure in the early phase of each experiment.

Quality engineering is often said to be difficult to understand. One of the reasons is related to its concept, another to its analysis procedure. Following is a summary of the history of change and development of quality engineering:

- ❑ *1957:* Genichi Taguchi's *Design of Experiments* was published [4]; Taguchi proposed the idea that the orthogonal array should be used to check experiments [5]; Taguchi proposed the prototype of the *SN ratio.*

- ❑ *1966: Statistical Analysis* was published [6]; *Process Adjustment* was published; Taguchi proposed the SN ratio.

- ❑ *1967: Production Process Control* was published [5]; a design quality control system during production was introduced.

- ❑ *1972: SN Ratio Manual* was published [7]; experiments on vehicle drivability, plastic injection molding, and cutting were performed.

❑ *1975:* Correspondence course on the SN ratio was begun [8]; application of the SN ratio to design and machining was accelerated; The term *quality engineering* was born.

❑ *1980:* The term *Taguchi method* began to be used in the United States; Taguchi succeeded in experiments at Bell Labs [9].

❑ *1984: Parameter Design for New Product Development* was published [10]; the ASI Taguchi Method Symposium began; 13 serial articles, "Evaluation of SN Ratio in Machining," appeared in *Machine Technology* (Nikkan Kogyo) [11].

❑ *1985:* A research group was dispatched to Japan by Bell Labs; 12 serial articles, "Design of Experiments for Quality Engineering," appeared in *Precision Machine* [13]; Taguchi proposed decomposition of $N\beta$.

❑ *1986:* JIS K-7109, *Rules of Dimensional Tolerances of a Plastic Part* was established; Taguchi–Fujikoshi Award for Measurement in Management was founded; Taguchi insisted that research on product development should be stopped.

❑ *1988: Quality Engineering Series* began to be published [14]; research on quality design systems in next-generation production systems was implemented by the Advanced Machining Tool Technology Promotion Association [15]; Taguchi proposed that research on technological development should be started.

❑ *1989:* Taguchi proposed *transformability* [16].

❑ *1990:* Taguchi proposed *generic function.*

❑ *1991:* The Quality Engineering Symposium at Ueno Ikenohata Culture Hall resulted in success; Taguchi's *Quality Engineering for Technological Development* was published [17]; JIS Z-9090, "*Measurement: General Rules for Calibration System* was added; Taguchi stated that to get quality, don't measure quality"; Taguchi proposed electrical characteristics, chemical reaction, and the MTS method [12].

❑ *1993:* Quality Engineering Forum was founded [18]; Quality Engineering Award was founded by the Precision Metrology Promotion Foundation.

❑ *1995:* Taguchi proposed an evaluation method based on electric power [19].

❑ *1996:* JIS Z-8403, *Product Characteristics: General Rules for Tolerancing* was added; 9 serial articles, "Quality Engineering for Mechanical Engineers," appeared in *Machine Technology* (Nikkan Kogyo) [20].

❑ *1997:* ISO standardization of quality engineering began to be discussed [21]; Genichi Taguchi was elected to the American Automotive Hall of Fame [22].

❑ *1998:* The Quality Engineering Forum was renamed the Quality Engineering Society; the Quality Engineering Taguchi Award was founded [23].

❑ *1999: Technological Development in Chemistry, Pharmacy, and Biology* was published [24].

❑ *2000: Technological Development in Electric and Electronic Engineering* was published [25]; Taguchi was elected "Quality Champion in the 20th Century"; Taguchi made a new proposal for applying quality engineering to simulation [26].

❑ *2001:* Taguchi proposed quality engineering for the 21st century [27]; *Technological Development in Machinery, Material and Machining* was published [28].

We can see that in most cases its difficult data analysis procedure stands out because it was founded on the design of experiments. However, investigating why Taguchi tackled the design of experiments, we can see that this is because rationalization and streamlining of experiments were essential for technological development; that is, they comprised management of technological development per se [29]. Therefore, without understanding the origin of the concept, we are mired in methodological difficulty.

The origin of the concept of quality engineering arises in nature, and because of this, the concept is considered difficult to understand. That is, although we can think something, we need to spend an enormous amount of energy to prove it concretely and to realize it practically. Quality engineering attempts to rationalize and streamline this cumbersome process, thereby enriching our lives in the time saved. There are a number of scholars who think of the idea abstractly, yet engineers are obligated to use it in a concrete manner. To achieve this, we always need to combine an idea and a methodology. However, since each reader can come to the book *Design of Experiments* only within his or her capability of understanding, many readers miss its overall concepts.

5.2. Conceptual Transition of the SN Ratio and Establishment of Quality Engineering

According to Masanobu Kawamura, ex-president of the Japanese Standards Association, who was a behind-the-scenes promoter of quality engineering, Genichi Taguchi often says something 10 years ahead of the natural course of development. And there is another reason that his ideas are sometimes obscure: Although developed extremely early, they are always evolving; thus, it is quite difficult to distinguish between old and new proposals. For example, the concept of the SN ratio in quality engineering appeared in Chapter 26 of Reference 4, published in 1958. Because this seemed to be based on the variance ratio of noises and effective elements, some people believe that it can be regarded as being completely different from the current SN ratio. However, its essence has not changed at all. Although it was devised a long time ago, it took a long time to be understood by many people and applied to many practical uses. Finally, mixed development of its concept and methodology made it difficult to comprehend.

Although publication of the *SN Ratio Manual* [7] in 1972 clarified the SN ratio in measurement, the idea that it is important to evaluate measurements based on the SN ratio in the measurement world, even in the case when true values are unknown, is still not comprehended, even by some measurement engineers today. This is because they focus on results instead of making an effort to understand their essence.

Since 1973, the SN ratio has most often been applied to design and machining. A typical example is evaluation of driving performance by Takeshi Inao, the ex-CEO of Isuzu Motors [30]. Initiated at Taguchi's proposal, this research dealt with changes in driving performance according to the characteristics of truck tires. By setting the rotational angle of a steering wheel, M to a signal factor, they calculated the SN ratio using a linear equation between the rotational angle and turning radius when a truck's turning radius arrived at the state shown in Figure 5.1. Although we can take advantage of a zero-point proportional equation to calculate the SN ratio if this were analyzed today, up to 1987, linear equations were used in

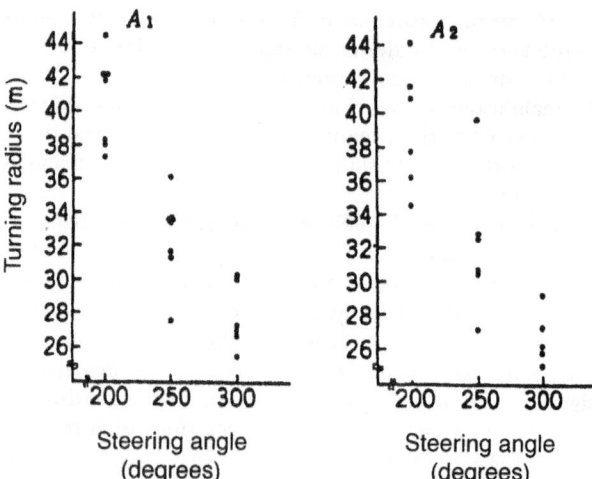

Figure 5.1
Relationship between
steering angle and
turning radius

the majority of SN ratio applications. In some cases the main effect of a signal factor when its true value was not known was analyzed using S_M instead of S_β.

Although this Isuzu experiment is considered a monumental achievement in research on the dynamic SN ratio, shortly after this experiment, to obtain the Deming Prize, Isuzu stopped utilizing the quality engineering technique, because the panel of judges disliked the technique. It is well known that afterward, Isuzu became affiliated with General Motors. Some authors, including Soya Tokumaru [31, 32], harshly criticized this symbolic incident as doing harm to a specific company and to the Japanese concept of total quality control. In contrast, Tokumara highly praised the Taguchi method.

Since 1975, a correspondence course on the SN ratio has been available from the Japanese Standards Association. The primary field in which the SN ratio was utilized in Japan at that time was quality control; therefore, the course was targeted to metrology and quality control engineers. When Hiroshi Yano began to work at the National Research Laboratory of Metrology of the Agency of Industrial Science and Technology in the Ministry of International Trade and Industry (currently known as the National Research Laboratory of Metrology of the National Institute of Advanced Industrial Science and Technology in the Ministry of Economy, Trade and Industry), he attempted to introduce the SN ratio in the field of metrology. Yano studied under Genichi Taguchi as visiting researcher at the laboratory between 1972 and 1976 [33].

The SN ratio of formability was applied by Yano to the evaluation of plastic injection molding at that time and led to transformability in 1989 [34] after almost a 15-year lapse. On the other hand, adaptation to the cutting process led to the current evaluation of machining based on electric power [35]. This change took 20 years to accomplish.

Uses of the SN ratio before and after the 1970s was related to test, analysis, and measurement. Around 1975, when this was well established, quality engineering began to be applied to design and production engineering for the purpose of using dynamic characteristics. According to the latest classification, the SN ratio used for measurement is categorized as a passive dynamic SN ratio, and the ratio used for design and machining is seen as an active dynamic ratio. Using it as an active dynamic SN ratio, Yano attempted to apply quality engineering to plastic injection molding and cutting processes. Here's what he said [16]:

> For the SN ratio in plastic injection molding, we began our research by considering what to select as signal factors. As a result, setting holding pressure to the input signal and dimension to the output, we used the SN ratio expressed in a linear equation with dimensions of a formed product. Many achievements obtained by this method eventually led to proposing transformability of setting die dimensions to signals, which was considered a revolutionary change in 1989.
>
> As for a cutting process, we started to argue whether to select a linear or zero-point proportional equation by choosing the depth of cut as a signal and amount of cut as an output value. Finally, we failed to come to convince machining experts.

A basic understanding of quality engineering was decided by the publication of Reference 14 early in 1988. In particular, Volume 1 of the *Quality Engineering Series* contributed much to fostering the concept of quality engineering. However, this is a revised edition of the textbook used in the quality engineering seminars offered by the Japanese Standards Association.

Quality Engineering for Technological Development [17], published in 1994, a revised edition of "Introduction to Quality Engineering," which appeared in serial form in a technical journal, *Standardization and Quality Control*, posed a new problem after the Quality Engineering Series appeared. Its focus on semiconductor, electrical characteristics, chemical reactions, and the Mahalanobis–Taguchi system (MTS) especially pioneered a new phase of quality engineering.

At about that time, manufacturers of electric appliances, many of which are concentrated in the Kansai region in western Japan, began research using quality engineering. One piece of systematic research was carried out by Yano Laboratory of the University of Electro-Communications and Yoshishige Kanemoto of Clarion [25]. This fundamental research developed into the concept of selecting a direct-current pulse for an alternating-current circuit as a signal, and furthermore, frequency itself as output.

In the area of chemical reactions, Tosco and Sampo Chemical began experimenting on chlorella and bean sprouts; later, Toagosei and Tsumura & Co. announced research on use of the SN ratio in chemical reactions. It is not well known that at that time, Konica, under the guidance of Genichi Taguchi, led the way by evaluating the performance of photographic film using quality engineering. Because in chemical reactions an individual reaction cannot be quantified and only an overall reaction rate can be estimated, a logarithmic SN ratio based on the dynamic operating window method began to be used [24].

As transformability developed, the concept of a generic function SN ratio and a functional evaluation SN ratio, which are used to develop materials based on proportionality between load and deformation or masses in the water and air, spread in the late 1990s. Around 2000, more and more people began to acknowledge that the SN ratio plays a key role in the MTS.

5.3. On-line Quality Engineering and Loss Function

Around 1975, the study of expressing improvement effects using the loss function was heated up. Although the special research program subsidized by the Ministry of International Trade and Industry motivated corporations, their understanding of the study was not sufficient. Indeed, they had a strong interest in the concept per se but did not comprehend how to put the loss function to practical use. Even if they understood it, the concept of reducing consumer loss was difficult to understand in a corporate-oriented culture, positioning a company itself in the center of business. In those days, the idea that reduction of in-house defects would lead to quality improvement in terms of quality control was dominant in many industries, whereas the phrase *customer satisfaction* is commonly used now. Therefore, the technique of expressing quality through the loss function was considered only an idea.

An article about quality problems in color television sets manufactured by Sony, titled "Japanese Company in America," appeared in *Asahi Shimbun* on April 17, 1979. As illustrated in Figure 5.2, the fact that U.S. buyers preferred Sony products manufactured in Japan to those made in the United States, even if the design of the sets was exactly the same, gave some implications of the quality issue and proved the appropriateness of the quality engineering definition of quality; that is, the quality of a product should be measured not by an acceptance rate but by the deviation from a design target value.

The reason that the evaluation method based on the loss function strongly attracted those engineers who promoted quality engineering was that it could express technological improvements by quality losses or economic effects. Additionally, aggressive promotion of this concept by ASI (the American Supplier Institute) influenced Japan significantly. On the other hand, in Japan, the Metrology Management Association, supervised by the Ministry of International Trade and Industry, emphasized the loss function when they taught quality engineering. However, in the 1990s the loss function concept began to be recognized by design engineers. Although the amount of improvement, estimated to be $1 million, is not regarded as surprising among quality engineering experts, this amount was viewed as questionable by many at that time.

Figure 5.2
Quality characteristic
distribution of Japanese
and U.S. products

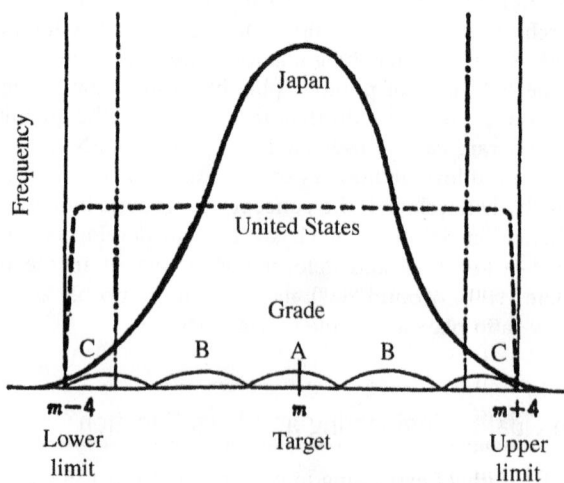

Color concentration

The original objective of the loss function is related to optimization of manufacturing process management and can be regarded as most significant. This field, called *On-line quality engineering*, was introduced earlier than off-line quality engineering, which had been studied since the 1960s as a successor of the design of experiments for manufacturing process management. The studies began with managerial problems in the manufacturing process at Fuji Film and were established through research on managerial issues regarding manufacturing processes at Toyota and Nippon Denso. As a consequence, in the late 1980s, on-line quality engineering came to be well known. In addition, after being introduced under the name *statistical process control* (SPC) in the United States, it was again recognized in Japan.

Because the initial managerial problems in manufacturing processes were complex, it was considered reasonable that many could not understand them completely [35]. Although this issue is closely related to Taguchi's overall achievements, the most essential idea is gradual improvement of the method itself after identifying the essence of any problems. In other words, a key to understanding the method is to define the range of applications through concrete results. This process of transition and development is regarded as the history of quality engineering. Therefore, quality engineering evolves continually and remains unfinished at all times.

On-line quality engineering has been applied to a wider range of fields since the latter half of the 1980s. However, the number of case studies involving on-line quality engineering is still much smaller than those related to off-line quality engineering. To reduce CEO costs, doubling of the production speed was proposed in the 1990s [36].

A major concept of on-line quality engineering is that we need only parameter study, with no experiments. Therefore, we proceed with analysis only by investigating current problems or through calculation. Based on these calculations, we can then grasp solutions to current problems. As a result, we can use calculations to study the possibility of doubling production speed. We can identify problems in a targeted manufacturing process and clarify problems to be solved technically. Yet because this concept is concerned heavily with the loss function, top management or supervisors who have considered the quality issues to focus on the number of in-process defects cannot easily understand the concept. Around the year 2000, Mazda began to tackle this problem positively [37].

5.4. SN Ratio in Machining

According to quality engineering guidelines—that an output characteristic, y, is proportional to an input signal, M—an experimental cutting process, reported in 1978, was grounded on the idea of proportionality of the amount of cut, y, with respect to the depth of cut, M [38]. In those days, if the experimental results were not consistent with conventional technical knowledge, the experimental procedure was questioned; even if the results conformed with the knowledge, the new experimental methodology was not appreciated. As shown in Figure 5.3, because the mainstream calculation method was based on a linear equation, most analyses focused on a comparison of the linear and proportional equations.

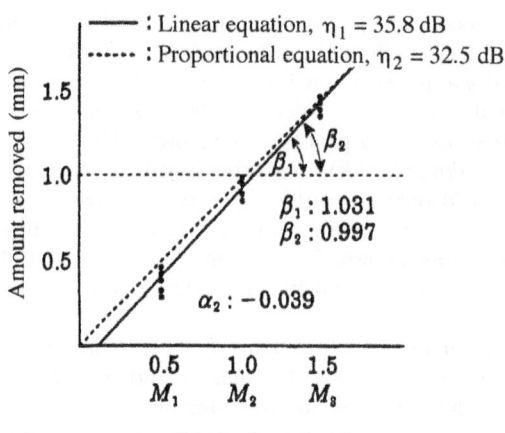

Figure 5.3
Depth of cut and
amount removed

Depth of cut (mm)

In 1981, once the subcommittee on the machining precision SN ratio was organized by Hidehiko Takeyama, the chairman, and Hiroshi Yano, the chief secretary, in the Japan Society for Seikigakkai (currently known as the Japan Society for Precision Engineering), research on the application of quality engineering to various machining technologies began to be conducted, primarily by local public research institutes.

These research results were reported in the research forum of the Japan Society for Seikigakkai and in the technical journals *Machine Technology,* published by Nikkan Kogyo Shimbun, and *Machine and Tool,* published by Kogyo Chosakai. Finally, a round-table talk titled "Expected Evaluation Method Based on SN Ratio Applicable to Design and Prototyping Such as Machine Tools" was reported in *Machine Technology* in August 1986 [11]. The participants were four well-known experts in each field: Hidehiko Takeyama (Tokyo University of Agriculture and Technology), Sadao Moritomo (Seiko Seiki), Katsumi Miyamoto (Tokyo Seimitsu), and Yoshito Uehara (Toshiba Tungaloy), together with Hiroshi Yano, whose job it was to explain quality engineering.

Following are the topics discussed by the round table:

1. SN ratio (for evaluation of signal and noise effects)

2. Error-ridden experiments (regarded as more reliable)

3. Importance of successful selection of control factors

4. New product development and design of experiments

5. Possible applicability to CAD systems

6. Effectiveness in evaluation of prototypes

7. Possible applicability of method to tools

8. High value of method when used properly

9. Significance of accumulation of data

Although the phrase *design of experiments* is used, all contents were sufficiently well understood. However, there were some questions about why it takes time for the SN ratio to be applied for practical uses.

Selecting depth of cut as input, the zero point becomes unclear. To clarify this, an experiment on electrode machining using the machining center was conducted by Sanjo Seiki in 1988. In this case, as illustrated in Figure 5.4, they chose dimensions indicated by the machining center as signal factors and machined dimensions as output. This concept is considered as the origin of transformability [39]. It was in 1987 when the zero-point proportionality displaced the linear equation, and thereafter, in the field of machining, the zero-point proportionality has been in common use.

In 1988 a study subcommittee of the Advanced Machining Tool Technology Promotion Association encouraged application of quality engineering to the mechanical engineering field [15], and a quality engineering project by the IMS in 1991 accelerated it further. When an experiment implemented by Kimihiro Wakabayashi of Fuji Xerox was published in the journal *Quality Engineering* in 1995 [19], Genichi Taguchi, the chairman of the judging group, advised them to calculate the relationships between time and electric power and between work and electric power. An experiment that reflected his advice was conducted at the University of Electro-Communication in 1996 [40], and later, Ishikawajima-Harima Heavy Industries [41] and Nachi-Fujikoshi Corp. [42] continued the experiment.

Setting electric power y to output, and time T and work M to signal factors, we define the following as generic functions:

$$Y = \begin{cases} \beta_1 T & \beta_1 \text{ large} & (5.1) \\ \beta_z M & \beta_z \text{ small} & (5.2) \end{cases}$$

Since electric power is energy, considering that each equation is expressed in terms of energy, we take the square root of both sides of each equation. If β_2 decreases, we can obtain a great deal of work with a small amount of electric power. That is, we can improve energy efficiency. Converting into a relationship between T and M, we obtain

$$M = (\beta_1/\beta_2) T \qquad (5.3)$$

As a result, the time efficiency can be improved. Looking at the relationship between T and Y, we see that momentary electric power is high. This idea can be applied to a system that makes use of energy conversion.

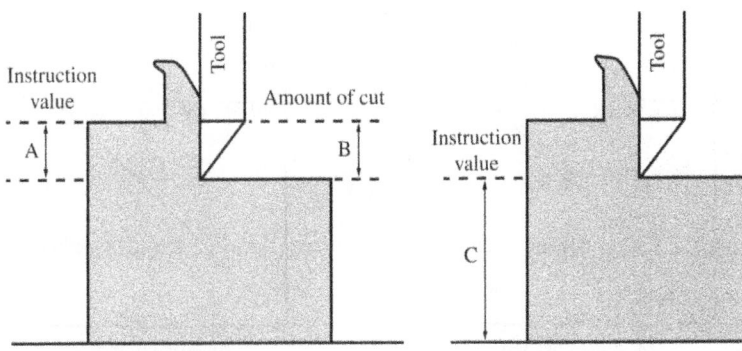

(a) Cut by target amount of cut

(b) Cut up to target dimension

Figure 5.4
Depth and amount of cut (1978)

5.5. Evaluation Method Using Electric Power in Machining

Quality engineering advocates the invariable principle that "to get quality, don't measure quality." In fact, this idea is difficult to comprehend because the concept that "quality is everything" has been widespread. However, according to Taguchi, cost should be prioritized.

When the experiment on cutting was implemented at the University of Electro-Communications in 1996, good reproducibility was obtained using electric power evaluation, and quality characteristics were also improved. However, not everything went well from the beginning. What was regarded as a problem at first was fluctuation of electric power during idling. We barely succeeded in calibrating actual electric power by taking advantage of its average. Yet when a different machine was used in the next experiment, we could not obtain sufficient reproducibility. After reconsidering the data shown in Figures 5.5 to 5.7, we finally fixed the problem in 1998 [43].

In the meantime, since evaluation of electric power during idling was questioned due to poor reproducibility in an experiment conducted by Ishikawajima Harima Heavy Industries (IHI), IHI and Nachi-Fujikoshi Corp. undertook various studies. Among them was a new proposal of the concept of *on* and *off* functions offered by Taguchi [44]. In case of machining, the state of machining is regarded as an on function, whereas the state of idling is an off function.

However, in much of the research on machining, machining conditions and tools were selected as control factors. That is, from the standpoint of a user of a machine tool, machining conditions and tools can be chosen as control factors. Therefore, we needed to conduct research on a manufacturer of machine tools from the user's point of view. In 2000, Matsuura Machinery Corp. implemented parameter design of the main axis of a high-speed machining center. Although they had a problem with heat generated in the main axis turning at high speeds, they clarified that it was possible to reduce heat in the main axis only by evaluating electric power during idling [45]. This symbolized the invariable principle: "To get quality, don't measure quality."

There is a tendency to evaluate the machining process quite often, for example, as a method of measuring stress on a tool. Although we wished to do this based on our desire to explain a phenomenon, this method looks at a tool only in place of a machine. Taguchi insists that signal and noise factors are consumer conditions. If we relate this idea to the fact that the SN ratio is energy conversion, it is sufficient

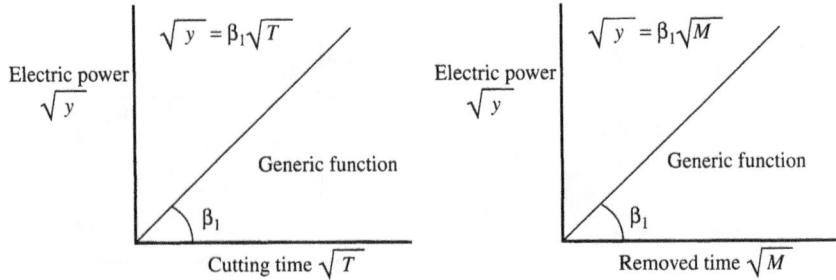

Figure 5.5
Relationship between instructive values and amount of cut (1980)

1. Idling   2. Cutting run 1

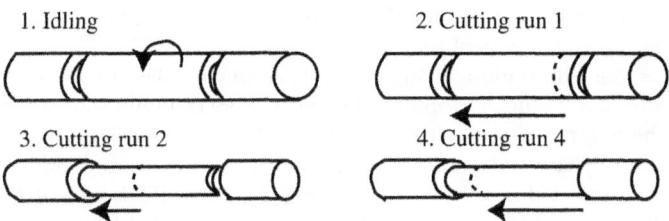

3. Cutting run 2   4. Cutting run 4

Figure 5.6
Evaluation by electric power

to evaluate only a result instead of a process. However, in the case of technological elements in lieu of final products, we face difficulties deciding what to do.

Yet from the viewpoint of original technological functions, evaluation of results is totally different. We have a strong interest in whether quality characteristics are improved or not when evaluated based on electric power. In this case, variability in dimensions and surface roughness of the machined product under optimal conditions are reduced to 1/4 and 1/1.3. That is, the principle "to get quality, don't measure quality" holds true. The fact that electric consumption is small for the amount of cut indicates that a tool cuts well and its life span will be prolonged. Therefore, we can dispense with a design life test.

This method of evaluating machining based on electric power rather than the dimensions of a product implies a new possibility. As mentioned earlier, this method is widely applicable not only to machining but also to almost all systems that hinge on energy conversion.

5.6. SN Ratio of Transformability

Although the dynamic SN ratio was commonly used in the 1980s for practical uses such as in design or machining areas, it was not a standardized practice, especially when in the late 1980s, the SN ratio was classified into dynamic and nondynamic characteristics. Smaller-the-better, nominal-the-best, and larger-the-better SN ratios were defined as nondynamic SN ratios. To standardize the idea of the SN ratio, we digressed considerably from the original objective of using the dynamic SN ratio. Since the nondynamic SN ratio was easily understood by statisticians, many statistical discussions took place. In fact, Kenzo Ueno of the quality engineering

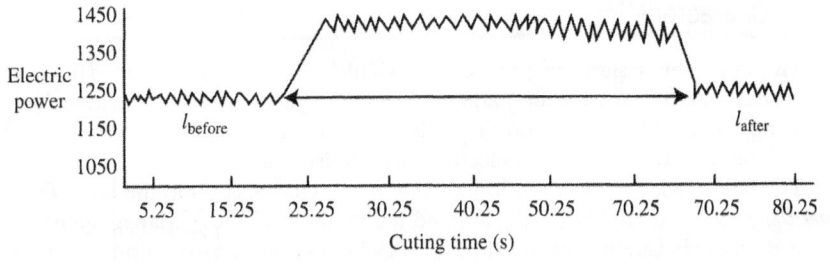

Figure 5.7
Power changes of cutting run 1

promotion department in the Nissan Technical Center prohibited the use of non-dynamic SN ratios because of the power of dynamic SN ratios.

Study of the application of quality engineering to plastic injection molding began in the plastic study group of the Japan Society of Plastic Technology, led by Yano, in the latter half of the 1970s, and significant achievements were obtained regarding the dynamic SN ratio based on a linear equation or nominal-the-best SN ratio. Eventually, this activity led to JIS K-7109, *Rules of Dimensional Tolerances of a Plastic Part,* the first JIS related to quality engineering, in 1986.

Also, The National Research Laboratory of Metrology triggered the research and development of CAMPS (Computer-Aided Measurement and Process Design and Control System), which aimed to optimize production engineering and to design and manage production processes. Its first focus was plastic injection molding; however, later the focus changed to ceramic injection molding. Still, the study of dynamic SN ratios was insufficient [46].

In the Quality Engineering Seminar on plastic injection molding given by the Japan Society of Plastic Technology in March 1989, Genichi Taguchi proposed a new idea: that "the function in plastic injection molding is the relationship between the dimension of a die and the corresponding dimension of the product." Because of much difficulty measuring plastic dimensions in this case, this idea was not accepted immediately, but it had a great impact. This led to compilation of a book, *Technological Development of Transformability* [16].

Nevertheless, the common idea that each engineer's technical skill in designing a die is linked directly to the ability to forecast product shrinkage was prevalent in the field of plastic injection molding at that time, so the proposal of transformability provoked many plastic molding engineers, and eventually, the idea was criticized as a "pedant's nonsense." The SN ratio of transformability was applied to a broad range of other engineering fields, such as forming, photocopying, printing, and cutting. What was most important was that the discussion encouraged engineers to consider what the SN ratio was, or what its generic function was.

Because Genichi Taguchi regarded transformability not as a generic function but as an objective function, transformability in plastic injection molding evolved into a relationship between load and deformation of a product, or proportionality between masses in the air and water.

On a different note, about the time of the Great Hanshin Earthquake in 1995, the concept of SN ratio for shape retentivity was suggested. Although it should function well for cases such as earthquakes, no effective application has been reported to date.

5.7. SN Ratio in Electric Characteristics

When parameter design was proposed by Genichi Taguchi in the late 1970s, the initial example focused on the parameter design of the Wheatstone bridge, based on a nominal-the-best SN ratio. Therefore, the concept of parameter design is considered to have begun with electric characteristic simulation.

The first major achievement in Japanese parameter design was the book *Parameter Design for New Product Development,* edited by the Metrology Management Simplification Study Committee of the Metrology Management Association (chairman:

Genichi Taguchi; chief secretary: Hiroshi Yano) [10]. This book was initiated to collect examples of parameter design, most of which were characterized as applications of electric characteristic simulation by measurement instrument manufacturers. The complex-number example of Yokokawa Hewlette Packard transistor development, based on an L_{108} orthogonal array, highly evaluated by Genichi Taguchi, was introduced in this book.

However, since the analysis of simulation was thought to be a commonplace method using nominal-the-best SN ratio with no excitement of using such an SN ratio, and regarded as mere application of simulation using the software existing in Japan, the topics of parameter design shifted from simulation to experimentation.

For these reasons, only a small number of examples focused on electric characteristics, whereas mechanical engineering examples accounted for the majority. However, this trend changed when parameter design was applied to semiconductors for the first time. Patterning of semiconductors started in the field of transformability. Although many Japanese companies kept this application extremely confidential and almost no examples were reported publicly, Sumitomo Electric Industries released their experimental result of patterning transformability because the company was a follower in the semiconductor industry.

Based on the idea that a generic function was essential in the 1990s, utilization of Ohm's law led to its application to electric characteristics. At first, in the semiconductor field, Sumitomo Electric Industries obtained a significant achievement by measuring electric characteristics in place of patterning transformability [47]. Next, Nachi-Fujikoshi succeeded in using a procedure to calculate voltage–current characteristics instead of observing soldering defects as a conventional metric. However, at first, these researchers did not mention the fact that a key to success is to utilize four-terminal circuits. It was later unveiled in other experimental reports [48].

Whereas the example of sensor welding studied by the University of Electro-Communications and IHI utilized electric characteristics in lieu of the original function, research on electric circuits themselves was begun by the University of Electro-Communications and Clarion. In the latter research, various types of experiments were attempted to evaluate the reproducibility of gains in the parameter design of a dc amplification circuit. Although a number of problems with the experimental method were identified successfully, inconsistency of a few decibels caused Genichi Taguchi dissatisfaction with the result, which led to a new method of utilization of pulse to evaluate ac circuits. From this study, Yoshishige Kanemoto of Clarion began to devote more time to research using frequency characteristics as the output. Indeed, it can be regarded as a follow-up to research on a nominal-the-best SN ratio in a ceramic oscillating circuit conducted by Sharp Corp. in 1995 [49]; it successfully clarified a formulated method of evaluating electric circuits [25].

Although Genichi Taguchi has made many new proposals, including applications of amplitude, frequency, and pulse and new circuits and research using simulation, up to the present time, few significant achievements based on his proposals have been reported. However, because the book *Technological Development in Electric and Electronic Engineering* was published in 2000 [25], research in this area can be expected to flourish.

5.8. Chemical Reaction and Pharmacy

In the chemical reaction area, quality engineering began to be applied in 1993 after Genichi Taguchi proposed the use of natural logarithms in the technical journal *Standardization and Quality Control.* Thereafter, this analysis procedure was formulated as the *dynamic operating window method.* This was subdivided into two methodologies, the *speed difference method* and the *speed ratio method* [24].

Since it is quite difficult to measure chemical reaction or cell division directly as a molecular-level reaction, we need to measure it on a scale such as a percentage or fraction. That is, instead of a generic function, an objective function is used. In reference to the relationship between time and the sensitivity curve, by taking a natural logarithm of the fraction and linearizing it, we can calculate a dynamic SN ratio.

This idea was also applied to the cultivation of bean sprouts and led to the achievement obtained by Fushimi Yoshino of Sampo Chemical [24]. Afterward, the use of natural logarithms attracted many people and was also applied to cases where a functional curve saturates, as in a capacitor [25].

The method of using natural logarithms is not limited to chemical reactions. However, Toagosei and Tsumura & Co. accomplished significant results by applying the technique to chemical synthesis and particle granulation, respectively. Toagosei's synthesis is an example of the speed difference method [51], and Tsumura's granulation is based on the speed ratio method [52]. Both of them are aimed at stabilizing the reaction speed using two curves of main and side reactions without noise factors in the studies. In the future, setting of noise factors may become a subject of study.

For the pharmaceutical industry, Genichi Taguchi proposed a method using cells looking at main and side effects [44]; however, no significant example has been reported thus far. Yet for the evaluation of medical efficacy, the results of experiments on the permeability of ointment into human skin and medical efficacy of an herbal medicine using mice have been reported [24]. Although government regulations allow the use of statistical methods, including using more animals for performing such experiments, the new method enabled researchers to eliminate this science-oriented research attitude and attain productive achievements, according to Genichi Taguchi.

Especially for the latter case, where it had been regarded as impossible to obtain approval of a new medicine in the field of herbal medicine, by taking advantage of parameter design, they proved its medical efficacy and obtained official approval. This event is considered as epoch-making in that it succeeded in advancing the phase of experimentation on medical efficacy [53]. In other words, quality engineering demonstrated its effectiveness in the field of herbal medicine in a situation that Western medical analysis could not deal with.

5.9. Development of the Mahalanobis–Taguchi System

Since the quality engineering methodology developed by Genichi Taguchi utilizes conventional statistical methods in a different manner, at first the method was quite often misunderstood and criticized as being an existing statistical method.

Similarly, the SN ratio was criticized as mere ANOVA (analysis of variance), and the zero-point proportional equation was misunderstood as regression analysis. In addition, the orthogonal array was considered to be an already existing method. However, even though the orthogonal array existed at that time, the concept of the L_{18} orthogonal array was new and had a different objective.

In addition, the MTS method was misunderstood as discriminant analysis because it takes advantage of the Mahalanobis distance. Whereas discriminant analysis deals with the distinction between two different groups, the MTS method takes the position that there is a single group, but distances are compared by deviations from the base space, which is the origin. This is identical to the concept of a zero-point proportional equation defining a zero point and unit amount, and it can be regarded as a problem of SN ratio and parameter design in a multidimensional space.

The origin of the MTS method dates back to the research on reliability improvement of medical checkups undertaken by Tatusji Kanetaka in 1987. Despite the fact that Kanetaka attained a significant number of results, the new method did not spread to technological fields because its application had only a limited scope in medicine. After Genichi Taguchi proposed application of the MTS method to a fire alarm in the technical journal *Quality Engineering* in 1996 [51], an experiment at the University of Electro-Communications fueled its application to technological fields.

The essential idea in the MTS method is to create a base space. In the case of a disease, we create a space of "being healthy," and in case of fire, a space of "no fire." Since this concept is regarded as a paradox or as mere rhetoric, it often cannot be understood, and additionally, there existed no data for the base space. That is, even if there are data for the disease or fire, we have no other choice but to prepare or seek the data as to "being healthy" or "no fire" to create a base space. This is the uniqueness and difficulty of the MTS method.

Further, it was difficult to set a signal factor level to express the degree of disease or the degree of fire from a specialized standpoint. The SN ratio was dealt with as a larger-the-better characteristic in most cases because it was difficult to set such a signal. For the study of setup of signals in the 1970s, there were similar problems when the SN ratio was proposed. In particular, in the case of measurement engineering for the case when true values are unknown, many attempts were made to set up signals; the MTS method still has a similar serious technical problem.

In any case, there is no doubt that MTS brought about numerous achievements through a growing number of applications to technological and medical fields in the late 1990s. However, when we apply the method to practical uses, we sometimes face troubles with multiple correlation caused by calculation of the inverse matrix of coefficients or the problems caused by the order of items in selecting them when Schmidt's orthogonal expansion is used. To solve this, Genichi Taguchi has suggested bridging MTS methods for the tenty-first century. This application is regarded as our future task [27].

The MTS software released by Oken, based on the software developed based on the University of Electro-Communication's research on a fire alarm, is considered easy to use and has, in fact, contributed greatly to application of the MTS method by taking advantage of Genichi Taguchi's hands-on coaching.

5.10. Application of Simulation

Quality engineering relies heavily on Genichi Taguchi's creativity and powerful coaching style in all aspects, including Design of experiments in the 1960s, the SN ratio in the 1970s, the Taguchi method in the 1980s, and the transformability and generic function in the 1990s. The fact that his achievements have been supported by various groups that have applied his ideas to practical uses should not go unnoted. What should we do next in the transitional period? One anticipated issue is innovation of the MTS method; another is a new proposal for the SN ratio in simulation. One typical example of this is described below. It shows how Genichi Taguchi's proposal has developed.

As discussed thus far, simulation was used in some cases of use of the SN ratio in the 1970s. However, since the nominal-the-best SN ratio was generally used (reflecting the fact that many cases of simulation are based on a linear relationship), most engineers did not have a strong interest in calculating SN ratios in simulation. Therefore, application of the SN ratio to simulation gradually became excluded from the mainstream in quality engineering. Yet in the Judging Subcommittee in 2000, Genichi Taguchi mapped out a new policy of vigorous evaluation of parameter design in simulation, as there existed few good examples of simulation at that time.

When the simulation session was added in the 8th Quality Engineering Research Symposium in 2000, it was noted that many simulation experts avoided using parameter design in simulation because results obtained through simulation often disagreed with actual results. In addition, it was thought that with regard to the finite element method, if a "finer mesh" is used (meaning that if more combinations are studied), if many more meshes are created in order to match the simulation results with the actual results, an enormous amount of time would be needed and would thus discourage the experts to use the method.

In response to this concern, Genichi Taguchi vehemently argued that because simulation results do not need to agree with actual results and we should stress only stability (evaluated by the SN ratio) and also argued that it is not necessary to increase the fineness of a net (more knots by more combinations) in a study.

This argument was posed in the Taguchi Method Symposium held by ASI in November 2000. In fact, the nonlinearity simulation on a switch implemented by ITT Cannon triggered Genichi Taguchi's new proposal, which was presented at the Quality Engineering Symposium given by the Central Japan Quality Management Association in Nagoya in the same month. In this symposium, a new method of calculating the SN ratio was suggested, with the practical example of "Optimization of Dimensional Parameters in Mechanical Design" studied by the Oki Electric Industry [26].

More specifically, the following was argued as a general discussion [28]: In case of a conventional simulation, on assumption of no nonlinearity effect against signal factors, nominal-the-best SN ratios have been applicable. However, at the ASI fall symposium in 2000, an example regarding a nonlinear simulation was presented which triggered a new approach. The calculations are as follows:

Signal factor:
Design target of signal output:

$$M_1^* \quad M_2^* \quad \cdots \quad M_k^*$$

$$m_1 \quad m_2 \quad \cdots \quad m_k$$

Output under standard condition (N_0): Output under N_0 as signal:

$$Y_{01} \quad Y_{02} \quad \cdots \quad Y_{0k}$$
$$M_1 \quad M_2 \quad \cdots \quad M_k$$

Output value under noise condition:

$$Y_{11} \quad y_{12} \quad \cdots \quad Y_{1k}$$

N_1 (negative side), N_2 (positive side)

$$Y_{21} \quad Y_{22} \quad \cdots \quad Y_{2k}$$

Total variation:

$$S_T = y_{11}^2 + y_{12}^2 + \cdots + y_{2k}^2 \qquad (f = 2k) \tag{5.4}$$

Variation of proportional terms:

$$S_\beta = \frac{(L_1 + L_2)^2}{2r} \qquad (f = 1) \tag{5.5}$$

Effective divider:

$$r = M_1^2 + M_2^2 + \cdots + M_k^2$$
$$= y_{01}^2 + y_{02}^2 + \cdots + y_{0k}^2 \tag{5.6}$$

Linear equations:

$$L_1 = M_1 y_{11} + M_2 y_{12} + \cdots + M_k y_{1k}$$
$$L_2 = M_1 y_{21} + M_2 y_{22} + \cdots + M_k y_{2k} \tag{5.7}$$

Variation of differences of proportional terms:

$$S_{N\beta} = \frac{(L_1 - L_2)^2}{2r} \qquad (f = 1) \tag{5.8}$$

Error variation:

$$S_e = S_T - S_\beta - S_{N\beta} \qquad (f = 2k - 2) \tag{5.9}$$

Error variance:

$$V_e = \frac{S_e}{2(k - 1)} \tag{5.10}$$

Total noise variance:

$$V_N = \frac{S_{N\beta} + S_e}{2k - 1} \tag{5.11}$$

SN ratio:

$$\eta = 10 \log \frac{(1/2r)\,(S_\beta - V_e)}{V_N} \tag{5.12}$$

Sensitivity (proportional term):

$$\beta = \frac{M_1^* m_1 + M_2^* m_2 + \cdots + M_k^* m_k}{M_1^{*2} + M_2^{*2} + \cdots + M_k^{*2}} \tag{5.13}$$

First, we select control factor levels to maximize the SN ratio; that is, we make an adjustment to match the sensitivity with the target.

The standard SN ratio, which we renamed the SN ratio based on this idea, was proposed in the chapter on reliability testing in the third edition of *Design of Experiments*, Vol. 2 [4]. As the SN ratio of the relationship between on–off output and input in the quality engineering of a digital system, this idea is explained on the following pages. [See the calculations preceding equation (5.4) and the explanation that follows.]

The standardized error rate is

$$\rho_0 = \frac{1}{1 + \sqrt{[(1/p) - 1][(1/q) - 1]}} \tag{5.14}$$

Assuming the standardized error rate above, we obtain Table 5.1 as the input/output table. The contribution for this case is called the *standard degree of contribution,* which is calculated as follows:

$$\begin{aligned} \rho_0 &= [(1 - p_0)^2 - p_0^2]^2 \\ &= (1 - 2p_0)^2 \end{aligned} \tag{5.15}$$

By taking into account noise's contribution of $1 - \rho_0$ based on the result above, we can calculate the standardized SN ratio η_0 as follows:

$$\begin{aligned} \eta_0 &= 10 \log \frac{\rho_0}{1 - \rho_0} \\ &= -10 \log \left(\frac{1}{\rho_0} - 1 \right) \end{aligned} \tag{5.16}$$

This SN ratio, calculated in the standard error rate after leveling, is called the *standard SN ratio.*

The standard SN ratio above is constructed with on–off (or digital) data. The new SN ratio, which is the same concept but is constructed with continuous variables, is still called the standard SN ratio. Genichi Taguchi considers the standard

Table 5.1

Standardized input/output table

Input	Output 0	Output 1	Total
0	$1 - p_0$	p_0	1
1	p_0	$1 - p_0$	1
Total	1	1	2

SN ratio with continuous variables to be the quality engineering of the twenty-first century.

The key is to stabilize variability first, then tune the output to the target. This is the philosophy of two-stage optimization: that the new standard SN ratio attempts to perform thoroughly. This idea is regarded as the basis of quality engineering and a clue to the future development of quality engineering. That is, it will be taken advantage of in all fields.

We can see that each of various problems considered in the initial phase of the design of experiments is being investigated through the succeeding research process. In actuality, we are so occupied in working on practical studies through evolution of the design of experiments that we do not have enough time to deliberate each of the issues that was originally brought out. Therefore, even if each problem posed at each time is seen as standing alone, all of them are interrelated continuously.

5.11. Summary

Most of Genichi Taguchi's achievements have been announced in the technical journal *Standardization and Quality Control,* published by the Japanese Standards Association, which is not a commercial magazine but an organ of the foundation. The journal contents are often related to standardization (JIS) and quality control or statistical methods. Thus, even if a certain report focused on quality engineering, its title ended with the words "for quality management" until the 1980s, which meant that Taguchi's achievements appeared in the organ only because it was in the field of quality control. In the late 1980s the uniqueness of quality engineering as distinct from "quality control" was finally accepted. Since 1986, the journal regularly features quality engineering in a special article and has sometimes covered activities in the United States. The insight shown by the Japanese Standards Association, which continued to use his reports as articles, cannot be overlooked.

In general, these reports were regarded as academic achievements that were supposed to appear in academic journals. The reason that his reports were instead, systematized in *Standardization and Quality Management* is because each academic discipline in Japan is usually focused on achievements in a specialized field. Therefore, the universal concept of quality engineering did not fit a journal dealing with a specific discipline.

Quality engineering as formed in the foregoing context took full shape after 1990 and led to establishment of the Quality Engineering Society. The theme of quality engineering in the 1990s was its proposal of a new perspective in each technological field, which began with a mention in the publication *Quality Engineering for Technological Development* [17].

When we look back to the history of quality engineering, we notice that a new proposal is initiated every time an achievement is accomplished. The field is not static. So what is the newest proposal? This is what quality engineering positions as *functionality evaluation* or *technological development methodology* as a managerial issue. Unfortunately, one problem is that the majority of those who attempt to comprehend and promote quality engineering are not strategic but tactical people. Thus, the functionality evaluation has often been seen not as a strategy but as a tactic.

To alter this perspective, *technological development using simulation* has been suggested as a strategy for the twenty-first century. In addition, *functionality evaluation in trading* was proposed for businesses. This has already been published using the term *evaluation technique for optimization* and is under discussion as a topic regarding ISO or JIS standardization of quality engineering.

Standardization of quality engineering, a project founded on a budget of $700,000, was begun as research on the optimization design and evaluation technique of quality engineering in the Global Standard Creation Research and Development Program supported by the Ministry of International Trade and Industry. Although this project accomplished a number of significant results through technological research that commonly would not be attempted, Genichi Taguchi's R&D methodology was judged not to be standardized, and subsequently, a standardization draft called *functionality evaluation,* limited to business trading, was written up. In addition, a group was dispatched to the United States to promote quality engineering. Unfortunately, it has been difficult to understand for engineers familiar with reliability or design life test in conventional quality control. Since quality engineering is not a concept lodged in traditional statistical methods, it cannot be acknowledged easily by statistical experts who are assigned to be responsible for standardization. Quality engineering has offered a ubiquitous methodology of technological development centering on parameter design. By using the evaluation scale of the SN ratio, we can easily assess the technical possibilities. Nevertheless, determination of a SN ratio is dependent on specialized or individual technology. This fact brings up another difficulty with quality engineering.

The essence of this problem is that to determine the evaluation scale of an SN ratio or generic function requires technological creativity. In short, a generic function is not a ready-made product but a custom-made invention. Therefore, if we obtain poor reproducibility of gain in parameter design, we have no other choice but to reconsider a new measurement characteristics. Furthermore, if this idea is extended, we come back to Genichi Taguchi's insistence that "we should discover the failure of experiments as soon as possible," typified by an experiment conducted by IHI in which, despite the fact that masses of drills were damaged in the experiment, an efficient machining process was attained under optimal conditions. However, this experimental process is not easy for many engineers to accept.

Between the 1960s and 1970s, Genichi Taguchi's achievements were widely accepted by automobile manufacturers and related companies, especially in the Nagoya area. Based on the judgment that his ideas would not become more widespread in Japan, his accomplishments at Bell Labs were "imported" back from the United States to Japan and had a considerable impact. As quite often happens, a certain thing that has been evaluated in Japan to only a limited degree comes to be highly valued once it succeeds in the United States. This is regarded as a traditional Japanese custom, which some people believe derived from ancient history, as when Japan introduced Chinese culture.

Genichi Taguchi's main activity to promote his methodology has been lectures and case studies of examples in the Quality Control Research Group (QCRG) of the Central Japan Quality Control Association and the QCRG of the Japanese Standards Association. Whereas the former was begun in 1951 and held primarily in the Nagoya area, the latter originated in 1963 and continued around Tokyo.

Through an annual symposium held at an alternative location, the technical exchange of both associations was facilitated.

In the research forum, lectures based on Genichi Taguchi's books were given, and practical applications were reportedly discussed by participants from many corporations. In 1989, the QCRG was renamed the Quality Engineering Research Group (QRG). In line with these research forums, the two associations held a seminar on the design of experiments (a 20-day course) led by Genichi Taguchi, which is considered to have contributed a great deal to promoting the design of experiments and quality engineering. This corresponds to the expert course offered by ASI in the United States.

Although the design of experiments seminar gained an extremely favorable reception in the Japanese Standards Association, Genichi Taguchi changed the content of the seminar as the design of experiments developed into quality engineering. Although it was not always well understood by many, it is still highly valued.

After his success at Bell Labs renewed public awareness of quality engineering, its evolution not only gave rise to enthusiastic believers but also kept many from shying away from it, due to its being overly complicated. Its successful application by Fuji Xerox triggered technological competition based on quality engineering among Japanese photocopy manufacturers. The peak of this is recognized to be the panel discussion "Competition on Quality Engineering Applications Among Copy Machine and Printer Manufacturers" at the 7th Annual Symposium of the Quality engineering Society in 1999.

As with the standard SN ratio in simulation, quality engineering has many methods that are not taken advantage of fully. The Technical Committee of the Quality Engineering Society has planned to sponsor seminars on important problems since 2001.

Now we close with two remarkable points. One is *management by total results,* an idea created by Genichi Taguchi that any achievement can be calculated based on the extent of loss given to "downstream" or following departments as the "degree of trouble" when the tasks of each department are interrelated. In fact, it seems that this idea, developed by Genichi Taguchi in the late 1960s, was put forward too early. However, now is considered to be an excellent time to make the most of this idea, as more people have become receptive to it.

The other point related to *experimental regression analysis,* which is related to the analysis of business data. Although this supplements regression analysis from a methodological viewpoint, it is considered a good example of encouraging engineers to be aware of a way of thinking grounded on technologies. Unfortunately, technical experts do not make full use of this method; however, it is hoped that it will be utilized more widely in the future.

Since the Taguchi methods are effective to elucidate phenomena as well as to improve technological analyses, it is expected to take much longer for these methods to move further into the scientific world from the technological field.

It is roughly estimated that the number of engineers conducting quality engineering–related experiments is several tens of times the current 2000 members in the Quality Engineering Society. We hope the trend will continue. In Chapter 6 we cover the introduction of Taguchi methods in the United States.

References

1. Genichi Taguchi, 1999. *My Way of Thinking*. Tokyo: Keizaikai.
2. Genichi Taguchi, 1965. Issue of adjustment. *Standardization and Quality Management*, Vol. 18, No. p. 10.
3. Hitoshi Kume, Akira Takeuchi and Genichi Taguchi, 1986. QC by Taguchi method. *Quality*, Vol. 16, No. 2.
4. Genichi Taguchi, 1957–1958. *Design of Experiments*, Vols. 1 and 2. Tokyo: Maruzen.
5. Shigeru Mizuno, Genichi Taguchi, Shin Miura, et al., 1959. Production plant experiments based on orthogonal array. *Quality Management*, Vol. 10, No. 9.
6. Genichi Taguchi, 1966. *Statistical Analysis*. Tokyo: Maruzen.
7. Shozo Konishi, 1972. *SN Ratio Manual*. Tokyo: Japanese Standards Association.
8. Genichi Taguchi, et al., 1975. *Evaluation and Management of Measurement*. Tokyo: Japanese Standards Association.
9. L. E. Katz, and M. S. Phadake. 1985. Macro-quality with macro-money. *Record*.
10. Genichi Taguchi, 1984. *Parameter Design for New Product Development*. Tokyo: Japanese Standards Association.
11. Hidehiko Takeyama, 1987. Evaluation of SN ratio in machining. *Machine Technology*, Vol. 3, No. 9.
12. Hiroshi Yano, 1995. *Introduction to Quality Engineering*. Tokyo: Japanese Standards Association.
13. Genichi Taguchi, 1985–1986. Design of experiments for quality engineering. *Precision Machine*, Vol. 51, No. 4 to Vol. 52, No. 3.
14. Genichi Taguchi, 1988–1990. *Quality Engineering Series*, Vols. 1 to 7. Tokyo: Japanese Standards Association.
15. Study Subcommittee for Quality Design System in Next-Generation Production System, 1989. *Research on Production Machining System Using Analytic Evaluation, Measurement, and Inspection*. Advanced Machining Tool Technology Promotion Association.
16. Ikuo Baba, 1992. *Technological Development of Transformability*. Tokyo: Japanese Standards Association.
17. Genichi Taguchi, 1994. *Quality Engineering for Technological Development*. Tokyo: Japanese Standards Association.
18. Genichi Taguchi, 1993. Inaugural speech as a chairman. *Quality Engineering*, Vol. 1, No. 1.
19. Kimihiro Wakabayashi, Yoshio Suzuki, et al., 1995. Study of precision machining by MEEC/IP cutting. *Quality Engineering*, Vol. 3, No. 6.
20. Genichi Taguchi, 1996. What does quality engineering think? *Machine Technology*, Vol. 44, No. 1.
21. Katsuichiro Hata, 1997. New JIS and ISO based on quality engineering concept. *Standardization and Quality Management*, Vol. 50, No. 5.
22. Genichi Taguchi, 1998. Memorial lecture: quality engineering technology for 21st century. *Quality Engineering*, Vol. 6, No. 4.
23. Taguchi Award Planning Subcommittee. 1999. Prescription of Quality Engineering Society Taguchi Award (draft). *Quality Engineering*, Vol. 7, No. 4.
24. Akimasa Kume, 1999. *Quality Engineering in Chemistry, Pharmacy, and Biology*. Tokyo: Japanese Standards Association.

25. Kazuhiko Hara, 2000. *Technological Development in Electric and Electronic Engineering.* Tokyko: Japanese Standards Association.

26. Genichi Taguchi, 2001. Robust design by simulation: standard SN ratio. *Quality engineering,* Vol. 9, No. 2.

27. Genichi Taguchi, 2001. Quality engineering for the 21st century. *Collection of Research Papers for the 91st Quality Engineering Research Forum.*

28. Hiroshi Yano, 2001. *Technological Development in Machinery, Material, and Machining.* Tokyo: Japanese Standards Association.

29. Genichi Taguchi, and Hiroshi Yano. 1994. *Management of Quality Engineering.* Tokyo: Japanese Standards Association.

30. Takeshi Inao, 1978. Vehicle drivability evaluation based on SN ratio. *Design of Experiments Symposium '78.* Tokyo: Japanese Standards Association.

31. Soya Tokumaru, 1999. *Ups and Downs of Japanese-Style Business Administration.* Tokyo: Diamond.

32. Soya Tokumaru, 2001. *Novel: Deming Prize.* Tokyo: Toyo Keizai Inc.

33. Taguchi Genichi, Tokyo: Coronasha.

34. Hiroshi Yano, 1976. Technical trial for evaluation of measurement and moldability in plastic injection molding. *Plastics,* Vol. 22, No. 1.

35. Genichi Taguchi, 1967. *Process Adjustment.* Tokyo: Japanese Standards Association.

36. Genichi Taguchi, 1995. Reduce CEO's cost. *Standardization and Quality Management,* Vol. 48, No. 6.

37. Teizo Toishige, and Hiroshi Yano, 1978. Consideration of cutting error evaluation by dynamic SN ratio. *Collection of Lectures at Seikigakkai Spring Forum.*

38. Keiko Yamamura, and Tsukako Kuroiwa, 1990. Parameter design of electrode machining using machining center. *Plastics,* Vol. 36, No. 12.

39. Kazuhito, Takahashi, Hiroshi Yano, et al, 1997. Study of dynamic SN Ratio in Cutting," *Quality Engineering,* Vol. 5, No. 3.

40. Kenji Fujioto, Kiyoshi Fujikake, et al., 1998. Optimization of Ti high-speed cutting. *Quality Engineering Research Forum.*

41. Masaki Toda, Shinichi Kazashi, et al., 1999. Research on technological development and productivity improvement of drilling by machining energy. *Quality Engineering,* Vol. 7, No. 1.

42. Kazuhito Takahashi, Hiroshi Yano, et al., 1999. Optimization of cutting material using electric power. *Quality Engineering,* Vol. 8, No. 1.

43. Genichi Taguchi, 2000. *Robust Design and Functionality Evaluation.* Tokyo: Japanese Standards Association.

44. Tsuguo Tamamura, and Hiroshi Yano, 2001. Research on functionality evaluation of spindle. *Quality Engineering,* Vol. 8, No. 3.

45. Hiroshi Yano, 1987. Precision machining and metrology for plastics. *Polybile,* Vol. 24, No. 7.

46. Susumu Sakano, 1994. *Technological Development of Semiconductor Manufacturing.* Tokyo: Japanese Standards Association.

47. Genichi Taguchi, 1994. Issues surrounding quality engineering (1). *Quality Engineering,* Vol. 2, No. 3.

48. Koya Yano, 2001. *Quality Engineering,* Vol. 9, No. 3.

49. Genichi Taguchi, 1995. Total evaluation and SN ratio using multi-dimensional information, and Design of multi-dimensional machining system. *Quality Engineering,* Vol. 3, No. 1.

50. Tatusji Kanetaka, 1997. Medical checkup judgment by applying Mahalanobis distance. *Quality Engineering*, Vol. 5, No. 2.

51. Takashi Kamoshita, Kazuto Tabata, Hiroshi Yano, et al., 1996. Optimization of multidimensional information system using multi-dimensional information: case of fire alarm. *Quality Engineering*, Vol. 4, No. 3.

52. Genichi Taguchi, 2001. Robust design by simulation: standard SN ratio. *Quality Engineering*, Vol. 9, No. 2.

53. Genichi Taguchi, 2000. *Evaluation Technique for Optimal Design: Standardization and World-wide Promotion of Quality Engineering.* Tokyo: Japanese Standards Association.

6 History of Taguchi's Quality Engineering in the United States

6.1. Introduction   153
      ECL and India in the 1950s   154
      Princeton University, Bell Laboratory, and Aoyama Gakuin University   155
6.2. The Beginning, 1980–1984   156
      Bell Laboratories   156
      Ford Motor Company   157
      Xerox   158
6.3. Second Period: Applications to Quality Problem Solving,
      1984–1992   158
6.4. Third Period: Applications to Product/Process Optimization,
      1992–2000   161
6.5. Fourth Period: Institutionalization of Robust Engineering,
      2000–Present   164

6.1. Introduction

Genichi Taguchi was born in 1924 in Tokamachi, Niigata prefecture, 120 miles north of Tokyo. His father died when he was 10 years old, leaving him with a hard-working mother, Fusaji, and three younger brothers. Tokamachi has been very famous for centuries for the production of Japanese kimonos. Just about every household had a family business with a weaving machine to produce kimonos. As the oldest boy in the family, Taguchi's intention was to succeed with the family business of kimono production.

Genichi's grandmother, Tomi Kokai, was known for her knowledge of how to raise silkworms to produce the high-quality silk used in kimono production. In the old days in Japan, it was very unusual for a woman to be a well-respected technical consultant. It is interesting that her grandson later became one of the most successful consultants in product development, traveling internationally.

Genichi attended Kiryu Technical College, majoring in textile engineering. However, by the time he returned to his hometown in 1943, the Japanese government had requisitioned all the weaving machines in Tokamachi for use in weapons production, as World War II was to be in progress for two more years. He thus found no family business in which to work.

He was then conscripted into the armed forces and assigned to the astronomical department of the Navy Navigation Institute. As he was charged with tabulating astronavigational data, he encountered and became interested in the *least squares method*, used to estimate a vessel's position from the data available. He then began a self-study of statistics at the Congressional Library in Tokyo.

When, the war ended, the Allied Forces established their general headquarters (GHQ) in Tokyo and set out to help reconstruct Japanese economics and its infrastructure. The U.S. military found that the quality and reliability of the Japanese telecommunication system were very poor. (Some people suggested that it might be better to use smoke signals or carrier pigeons.) GHQ was desperate to improve telecommunications so that they could perform proper intelligence work. GHQ forced the Japanese government to establish the Electrical Communication Laboratory (ECL), on which it spent almost 2% of its national budget. ECL was the first pure research and development organization in Japanese history, modeled after Bell Laboratories in the United States. ECL began operations in 1949 with 1400 people.

ECL and India in the 1950s

After the war, Taguchi joined the Japanese Ministry of Public Health and Welfare, where he worked on collecting data and conducting analysis. While working there, he encountered M. Masuyama of the University of Tokyo, the author whose book on statistics he had studied during the war. Masuyama guided him to develop further his knowledge of statistics and sent Taguchi to industries as a consultant in place of himself. In 1948, Taguchi became a member of the Institute of Statistical Mathematics, part of the Ministry of Education. By the age of 26, Taguchi had, by applying statistical methods, generated numerous successful applications. In 1950, the newly established ECL acquired Taguchi's services. He was assigned to a section responsible to assist design engineers to put together test plans and to conduct data analysis.

For whatever reason, ECL executives decided to compete with Bell Laboratories in developing a cross-bar switching system, which was the state-of-the-art telecommunication system at that time. Table 6.1 compares ECL and Bell Laboratories. In 1957, ECL completed development, meeting all the very difficult requirements, before Bell Labs, and ECL won the contract.

This was the first success story by a Japanese company in product development following World War II. Of course, we cannot conclude that this success was due entirely to Taguchi's ideas. However, during those six years, he developed and applied his industrial *design of experiment*. He developed not only efficient use of such tools as orthogonal arrays and linear graphs, but also developed and applied strategies for product and process optimization, which became the essence of what is today known as *robust design*. Those strategies include:

Table 6.1

ECL and Bell Labs compared

	Budget	No. of People	No. of Years	Result
AT&T Bell Labs	50	5	7	Base
NT&T ECL	1	1	6	Superior

❑ Interaction between control factors and noise factors to achieve robustness

❑ Inner and outer arrays

❑ Two-step optimization

❑ A focus on additivity and reproducibility

During development of the cross-bar switching system, ECL had optimized over 2000 design and process parameters. ECL was not a manufacturing company, and its contribution was complete with the design and specifications. Contractors produced the system and subsystems based on the design created by ECL. Then ECL purchased the system from contractors and leased it to users, such as telephone companies and end users. Leases included a 100% warranty for 40 years for the exchanging system and 15 years for telephone sets. In other words, ECL had to repair or replace, at no charge, a failure or defect. Therefore, it was very critical for ECL to assure performance, quality, and reliability at the stage of product design. This is why ECL was forced to think of robustness before aging and the customer's environment.

During the development phase, Taguchi also consulted contractor companies, such as Toshiba, Hitachi, and NEC, to optimize their processes. On weekends, Taguchi continued to consult industries outside telecommunications. A famous study at Ina Seito, a ceramic tile manufacturing company, was conducted in 1953.

In 1955, Taguchi visited the Indian Statistical Institute in Calcutta, India, for almost a full year as a visiting professor in place of Masuyama. In India, he met P. C. Mahalanobis, the founder of ISI, and Shewhart from Bell Labs. After he came back, he published the first edition of a two-volume book, *System of Experimental Design* (Tokyo: Maruzen). In 1962, Taguchi was granted a Ph.D. degree in science from Kyushu University for his work in developing the industrial design of experiments. In 1960 he was awarded the Deming Prize from the Japanese Union of Scientists and Engineers.

In 1963, Taguchi went to the United States for the first time and spent one year at Princeton University and Bell Labs. He faced a culture shock but fell in love with the American way of life and thinking, especially with its rational, pragmatic, and entrepreneurial spirit. During that time, his work was concentrated in the area of the SN ratio for digital systems.

Princeton University, Bell Laboratory, and Aoyama Gakuin University

After he returned from the United States, he was invited by Aoyama Gakuin University to act as a professor in its newly established engineering college. He taught industrial and management engineering for 17 years, until his retirement in 1981. During that time, he lectured at other universities, including the University of Tokyo and Ochanomizu University.

Taguchi leads the Quality Control Research Group (QRG) at CJQCA, the Central Japan Quality Control Association. The organization includes Toyota Motor Company, its group companies, and many others. The group has been meeting monthly since the 1950s. He has led a similar research group at the Japanese Standard Association (JSA) in the Tokyo area since 1961. This group has also been meeting monthly since its start.

While teaching at Aoyama, Taguchi has continued to teach seminars at the Japanese Union of Scientists and Engineers, CJQCA, and JSA and to consult with numerous companies from various industries. He says that he must have consulted on over 150,000 projects over the years. He has written a monthly article for JSA's publication, *Standardization and Quality Control,* for over 40 years. He has authored literally hundreds of articles and published many, many books on his techniques and strategies.

In addition to concepts developed during the ECL era, he has continued to develop new tools and strategies, such as:

- ❑ Quality loss function
- ❑ Parameter design
- ❑ Tolerance design
- ❑ Business data analysis (experimental regression analysis)
- ❑ Dynamic SN ratio for measurement systems
- ❑ SN ratio for digital data
- ❑ On-line quality engineering
- ❑ Management by total results
- ❑ Nondynamic SN ratios
- ❑ More sophisticated two-step optimization
- ❑ Parameter design with computer simulation
- ❑ Dynamic SN ratios

His methodologies were introduced to the United States in the early 1980s and spread throughout U.S. industries. This process can be divided into four periods.

6.2. The Beginning, 1980–1984

It was very much by coincidence that three organizations, Bell, Ford, and Xerox, encountered Taguchi's methodologies within a one-year time frame, around 1981. All three experienced a culture shock, then promoted the methodologies.

Bell Laboratories   In 1980, Taguchi wrote a letter to Bell Labs indicating that he was interested in visiting them while on sabbatical. He visited there on his own expense and lectured on his methodology to those who were involved with design of experiment. These people were interested but still skeptical. They included Madhave Phadke, Rague Kakar, Anne Shoemaker, and Vijayan Nair, who were students of the famous statistician George Box at the University of Wisconsin or of Stu Hunter at Princeton.

Taguchi asked them to bring him the most difficult technical problem that they were facing. They brought the problem of 256k photolithography. The process was to generate over 150,000 windows on a ceramic chip. The window size had to meet a requirement of 3.00 ± 0.25 μm. The process yielded only 33% after several years of research. Taguchi designed an experiment with nine process parameters assigned to an inner array of L_{18} and measured the window size as its response. Noise factors in the outer array were simply positioned between chips on a wafer and within chips. Then the response was treated as a nominal-the-best response and the now famous two-step optimization was applied. The yield was thus improved to an amazing 87%. People at Bell Labs were impressed and asked him to keep it secret for a few years, although the study was published in the May 1983 issue of the *Bell System Technical Journal.* Taguchi continued to visit Bell Labs frequently until its huge reorganization later in the 1980s.

In 1979, Edward Deming was featured in an NBC White Paper 90-minute TV program titled "If Japan Can, Why Can't We?" Deming was introduced as the father of quality in Japan. In 1950, because of his statistical competence, Deming visited Japan to help development of a national census. During his visit, the JUSE had asked Deming to conduct seminars with Japanese top executives. He had preached to Japanese industrial leaders the importance of statistical quality control (SQC) if the Japanese were interested in catching up to Western nations. Because Japanese top executives were good students and had nothing to lose, they practiced what Deming preached. Later, JUSE established the Deming prizes for industries and individuals. Based on his teaching, Japan had developed unique total quality management systems, together with various practical statistical tools, including Taguchi methods.

Ford Motor Company

In the United States, it became apparent in the late 1970s that made-in-Japan automobiles were starting to eat into the market share of the U.S. Big 3, starting with markets in California. U.S. industries were puzzled as to how the Japanese could design and produce such high-quality products at such low cost. Then the TV program featuring Deming was broadcast. Ford's CEO Don Peterson had hired Deming and started Ford's movement called "Quality Is Job #1" under Deming's guidance. Deming suggested that Ford establish the Ford Supplier Institute (FSI) to train and help its suppliers in Deming's philosophy with statistical tools such as Shewhart's process control chart. FSI conducted seminars, and thousands of Ford and supplier people were trained in SPC and Deming's quality management philosophy. Deming also suggested that they send missions to Japan to study what was being practiced there.

During the mission in spring 1982, Ford and supplier executives visited Nippon Denso. They found that Denso's engineering standards, in their thick blue binders, were of very high quality and of a type they had never seen. They asked how such excellent technical knowledge had been developed as part of corporate memory. They mentioned various statistical techniques and emphasized Taguchi's design of experiment for product/process optimization. Taguchi happened to be consulting at Denso that day. The leader of the mission, Larry Sullivan, immediately asked Taguchi to visit Ford and give lectures.

Taguchi visited Ford in October 1982 and gave a five-day seminar with a very small English handout. It was somewhat difficult for both sides because of translation problems, but the power of his ideas was understood by some participants. FSI decided to include Taguchi's design of experiment in their curriculum. Yuin Wu, a long-time associate of Taguchi's, was hired by FSI. Wa translated training material and began to conduct seminars with FSI. Wu, who is from Taiwan, was very knowledgeable in Japanese quality control and its culture, especially in Taguchi's methodologies. He had immigrated to the United States in the mid-1970s and is credited with conducting the very first Taguchi-style optimization study in the United States.

In May 1983, Shin Taguchi, a son of Taguchi's, joined the FSI. Shin is a graduate of the University of Michigan in industrial and operations engineering. Wu, along with Shin and other professors, taught Taguchi's design of experiment (DoE) to Ford suppliers week after week. Then, to qualify as a Q1 Ford supplier, Ford made it mandatory to practice DoE in addition to SPC. Those supplier companies included huge conglomerates such as ITT, GE, United Technology, and GTE. Because of that, Taguchi DoE became well known outside the automotive industry. Willie Moore of Ford claimed that Ford had contributed greatly to promoting Taguchi's DoE in U.S. industries. In 1984, FSI became ASI, the American Supplier Institute, one of its goals being to conduct training and implementation assistance for companies other than Ford and its suppliers. The first Taguchi Symposium was held by FSI in 1984, and it has been held annually since then. The Taguchi symposia contributed greatly to promoting the latest and greatest of Taguchi's methodologies.

Xerox   At Xerox Corporation, after its xerography patent had expired, management could not help but notice that many Japanese office imaging companies were shipping small and medium-sized copy machines to the United States which produced high-quality copies at low cost, taking market share from Xerox. They were especially anxious to protect their large-machine market share. They had no choice but to conduct a benchmark study. Xerox had a sister company called Fuji Xerox, owned 50–50 by Rank Xerox of Europe and Fuji Film of Japan. Xerox found out that Fuji Xerox could sell a copy machine at the manufacturing cost to Xerox and still make a handsome profit.

Fuji Xerox was established in the mid-1960s and had no technical know-how regarding copy machines. Yet, after 15 years, Fuji Xerox had developed the technical competence to threaten Xerox Corporation. They found that Fuji Xerox was practicing total quality management and various powerful techniques, such as quality function deployment (QFD) and Taguchi's methodologies for design optimization. Xerox found that Taguchi had been a consultant to Fuji Xerox since its establishment and suspected that Fuji's competence had something to do with implementing Taguchi's methodologies.

Xerox invited Taguchi to visit in 1981, and he began to provide lectures and consultations. Xerox executive Wayland Hicks and his people, including Barry Bebb, Maurice Holmes, and Don Clausing, were to take leadership in implementing QFD and Taguchi methods. Clausing, the creator of the concept of *robustness* called "operating window," gave the method *Taguchi methods* to Taguchi's approach to design optimization. Clausing later became a professor at MIT and taught total product development, in which QFD and Taguchi methods played a major role.

6.3. Second Period: Applications to Quality Problem Solving, 1984–1992

In 1980, there were fewer than 10 consulting companies in the area of quality in North America. By 1990, the number has grown to more than 300. During the 1980s, U.S. industries introduced TQM, policy (Hoshin) management, just-in-time, the Toyota production system, TPM (total productive maintenance), 7 basic tools, 7 management tools, poka yoke, SPC, FMEA, QFD, design of experiment, Taguchi methods, and many others. Many people were calling these endless initiatives "flavors of the month," "alphabet soup," or "the kitchen sink." However, Japanese total quality management was the thing to practice during the 1980s.

During this period, most applications of Taguchi methods focused on *firefighting*, solving *existing* manufacturing quality problems. Naturally, the responses measured were nondynamic, typically smaller-the-better or larger-the-better response. For example, a typical objective was to minimize defects such as voids, shrinkage, cracks, leakage, audible noise, and so on, or to maximize strength, time to failure, pressure to leak, and so on. Two-step optimization with nominal-the-best was mentioned but not emphasized enough, and it was not practiced as much as it should have been. In reality, Taguchi methods in the 1980s in the United States were nothing but fractional factorial design of experiment to solve manufacturing quality problems. Many successful studies were reported, with cost saving in the millions of dollars, and Taguchi methods were becoming more and more popular. When these tremendous cost reductions were reported, typical Japanese practitioners were saying: "Why is so much cost reduction possible? In Japan, that kind of cost reduction is unheard of, because we reduce cost from the beginning." I believe that part of the reason that even simple design of experimentation was not used in typical U.S. corporations is that it was practiced only by statisticians in R&D environments.

The Department of Defense also promoted Taguchi methods in the late 1980s. The U.S. Air Force published the *Reliability and Maintainability 2000 Program* in 1988, which encouraged contractors to practice Taguchi's quality loss function and parameter design.

As soon as the Bell case study on photolithography was published in 1983, Taguchi's approach to industrial design of experiment was recognized by the U.S. statistical community. Since that time, many articles on Taguchi methods have appeared in magazines and journals such as *Quality Progress, Quality, Journal of Quality Technology, Technometrics, Quality Engineering*, and *Quality Observer*. By then, Taguchi methods had caused tremendous debate and controversy. Some people were supportive and very enthusiastic, some were against use of the methods and antagonistic, and others were neutral. The positive responses can be summarized from these articles as follows:

1. The quality loss function and its philosophy are valuable concepts.
2. The methods are useful to analyze and optimize variability.
3. Inner and outer arrays are useful designations.
4. Two-step optimization is a valuable process.
5. Getting engineers to conduct balanced/orthogonal experimentation is a great contribution.
6. DoE tools are easy for engineers to use.
7. This is a wake-up call to the quality movement in U.S. industries.
8. Prediction and confirmation are very consistent with the PDCA cycle.

The following criticisms were made:

1. Taguchi ignores interactions among factors.
2. Taguchi ignores interactions between control factors.
3. The validity of accumulation analysis for attribute data is questionable.

4. The validity of the SN ratio for nominal-the-best response is questionable.

5. The validity of signal-to-noise ratio for smaller-the-better response and larger-the-better response is questionable.

6. There is a lack of statistical testing in parameter design.

7. There is no use of distribution theory.

8. Taguchi does not provide a model.

9. Taguchi does not credit statisticians.

10. Taguchi has made contributions, but there is a better way to do what he is trying to do.

Meanwhile, Taguchi was coming to the United States in May and October every year, visiting Xerox, Ford, ASI, Bell, ITT, General Motors, and many other companies. He made numerous presentations at conferences and symposia. He also visited universities and professional societies. As he visited them, he emphasizing that his method was misused and that he prefered to see it used for design optimization for robustness rather than for quality problem solving. Quality problem solving focuses only on a symptom of poor functioning. It works on a nonoptimized design to solve problems. If the design were optimized, there is a good chance the problem would not have existed. Robust optimization focuses on optimizing the energy transformation of a system to get rid of all symptoms. It is important to recognize that any hardware function is essentially a series of energy transformations and that defects and failures are nothing but symptoms of energy transformation variability.

The idea of energy transformation, its ideal functioning, and the dynamic SN ratio became better understood in the late 1980s. At the 1987 ASI Taguchi Symposium, Taguchi said: "From now on, every case study must use dynamic response based on the ideal function of energy transformation." The theme of the Taguchi Symposium in 1988 was: "To get quality, don't measure quality"; but it was the least well attended symposium. Taguchi methods were very successful, saving millions of dollars by solving problems related to defects and failures, but it was not easy to change the emphasis from firefighting to design optimization—it has taken several years to make this transition.

The first U.S. case study using dynamic response was conducted by Flex Technology and presented at the 1988 Taguchi Symposium. ASI had begun to change its training material from an emphasis on firefighting to one of fire prevention. However, many U.S. industries had just begun their quality movement, and most companies were not ready to digest Taguchi's upstream thinking.

By 1990, an initial quality survey of the number of things that go wrong in the first three months in service reported that Ford cars were as good as many Japanese-made automobiles. It was a 400% improvement over the early 1980s, when typically three things were going wrong per car. It was a great improvement; however, Ford was still having problems in the area of warranty and resale value. It indicated the necessity to improve not only initial quality but also reliability and durability. Ford executives recognized that this was not just a manufacturing problem but that product design engineering had to be involved to develop products that were robust against customers' environments and aging.

6.4. Third Period: Applications to Product/Process Optimization, 1992–2000

By 1992, TQM had become less popular because of its vagueness and its spiritual nature. In place of TQM, *six sigma* was popularized by Mikel Harry, Larry Bossidy (Allied Signal CEO), and Jack Welch (GE CEO). On the other hand, Taguchi methods had evolved from a quality problem-solving tool to a system of robust engineering for fire prevention.

In Japan, the Quality Engineering Society was formed and began to publish a bimonthly technical journal in 1993, its mission being to research and advance applications of Taguchi methods in various industries. Originally simply a forum with 600 members, today it is an official society with a membership of more than 2000. Taguchi was the first chairman, and the chief editor for the technical journal was Hiroshi Yano. The journal has contributed tremendously to accomplishing the society's mission.

Ford Motor Company established the Ford Design Institute in 1992 and began to generate robust design case studies. With FDI, more than 100 projects were completed from 1993 to 1995. In addition to Ford Motor Company, ITT, Kodak, Xerox, Delphi, the Bobcat Division of Ingersoll-Rand, and many other companies were starting to apply robust design in product development, using dynamic response based on energy transformation.

Since Taguchi began to emphasize dynamic response based on energy transformation in the late 1980s, Taguchi methods have advanced further. These new ideas and techniques include:

❏ SN ratio used in conjunction with operating window response (originated by Xerox's Don Clausing)
❏ Dynamic SN ratio based on the ideal function of energy transformation
❏ Dynamic SN ratio in chemical and biochemical systems
❏ Dynamic SN ratio using complex numbers
❏ Dynamic SN ratio in software function
❏ Noise compounding
❏ New ideas on noise strategies
❏ Double signal dynamic response
❏ Triple signal dynamic response
❏ Dynamic operating window response
❏ Speed ratio method
❏ Mahalanobis–Taguchi system for diagnostic/pattern recognition system optimization
❏ Robust software testing (software debugging using orthogonal array)
❏ Standard SN ratio for continuous variable

Figure 6.1 summarizes the development of robust engineering in the United States.

The six sigma technique became popular in the mid-1990s. Successful implementation by Jack Welch of General Electric contributed greatly to its popularity. For instance, GE reported a large annual saving in 1998. GE scheduled 50,000 six sigma and design for six sigma (DFSS) projects in 2002.

Figure 6.1
Development of robust
engineering

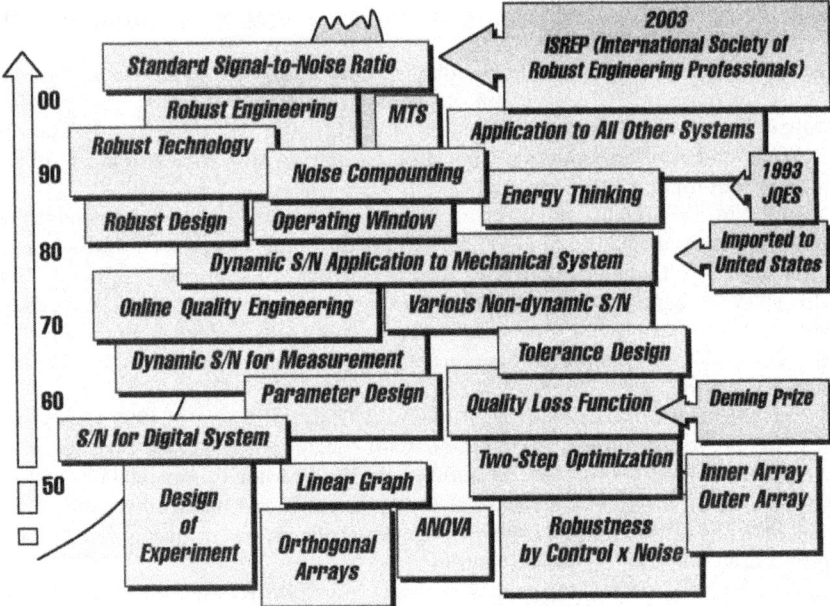

While the DMAIC (define–measure–analyze–improve–control) process of six sigma typically deals with solving current problems and reducing waste, DFSS's IDDOV (identify opportunity–define requirement–develop concept–optimize concept–verify and launch) process optimizes design of product and processes to prevent future problems. Robust engineering plays a major role in DFSS, especially at the optimization stage. Corporations such as Delphi, Caterpillar, Hyundai-Kia Motor, TRW, and many others are integrating robust engineering into their DFSS deployment. Figure 6.2 shows the DFSS and RE deployment process, and Figure 6.3 shows the monetary savings expected from DFSS and RE activities. DFSS is an excellent vehicle to use to integrate robust engineering into new product introduction for pure "prevention."

During this period, many excellent studies were developed and presented in Japan, the United States, and Europe. The Quality Engineering Society in Japan hosts an annual symposium in June every year, with over 900 people participating. ASI continues to host an annual Taguchi symposium in October with approximately 200 participants. ITT has its own annual Taguchi symposium, with around 200 people participating. In 1995, ASI started a 16-day Taguchi expert course; more than 360 engineers have graduated at this point.

Taguchi was inducted to the Automotive Hall of Fame in 1997 for his contribution to Big 3's quality improvement. He also received an honorary membership in the ASQ in 1996 and the same honor from ASME in 199X. These awards are evidence of the contributions made by Taguchi methods in U.S. industries.

Additionally, new ideas developed in recent years were published in the late 1990s. *Robust Engineering* by Taguchi, Chowdhury, and Taguchi was published by McGraw-Hill in 2000. *Taguchi Methods for Robust Design* by Wu and Wu was published

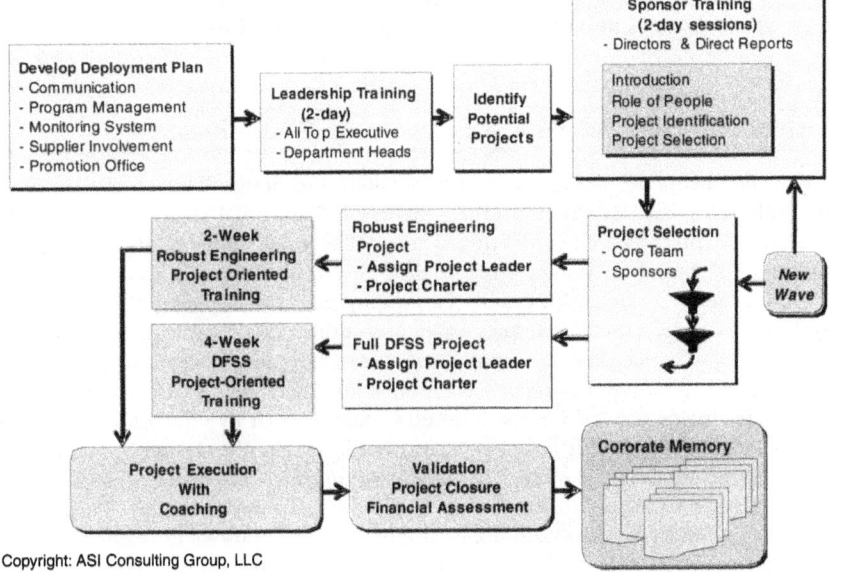

Figure 6.2
DFSS and RE deployment model

Copyright: ASI Consulting Group, LLC

by ASME Press in 2000. *Mahalanobis–Taguchi System*, also published by McGraw-Hill, appeared in 2001.

Despite all these publications, there remain many misconceptions regarding Taguchi methods. Most people still take them to be quality problem-solving tools and do not recognize their application to fire prevention. Only those companies trained by ASI seem to be using Taguchi methods for design optimization.

	Robust Engineering	Full DFSS	Total
Number of projects	120	60	180
Saving/project	$0.5M	$1.0M	$120M

(1) *Robust engineering project involves optimizing existing design concept.*

(2) *Full DFSS project involves new design where concept is not determined.*

(3) *Saving includes savings in "hard" and "soft" dollars.*

(4) *Existing new product development processes need to be integrated into DFSS curriculum.*

Figure 6.3
DFSS and RE implementation

Moreover, even within companies trained by ASI, it has been difficult to adapt Taguchi methods as an engineering strategy to prevent fires.

6.5. Fourth Period: Institutionalization of Robust Engineering, 2000–Present

Today, engineers are doing their best to rotate the design–build/simulate–Test–fix cycle. They are very busy rotating this cycle. Figure 6.4 shows what I call the "traditional product development process." The cycle starts by developing "requirements" based on vision, mission, and corporate strategies, and the voice of the customer. The house of quality for quality function deployment (Figures 6.5 and 6.6) provides an effective process to accomplish this. Then concepts are generated and selected. Concepts may have been developed previously during *seed technology*. Stuart Pugh's concept selection is an excellent process for this stage. Then a prototype is developed and tested by hardware or by simulation. It is tested against requirements; if it does not meet requirements, the design is studied and analyzed heavily. Based on what is learned, a design change takes place. Then the new design is tested again to see if it meets the requirements now. That is how the design–build/simulate–test–fix cycle goes, using iterations as needed.

This traditional process is still missing a critical step: parameter design or robust optimization. Once the design is optimized for robustness based on its energy transformation, you have the opportunity to meet all the requirements at once: that is, to kill all the birds with one stone. Moreover, chances are that you may exceed requirements by far and therefore are left with plenty of opportunities to reduce cost. On the other hand, when you optimize the design (based on its energy transformation, of course) and find that it does not meet the requirements, you have a strong indication of a poor concept. You must either reject the design

Figure 6.4
Traditional product development

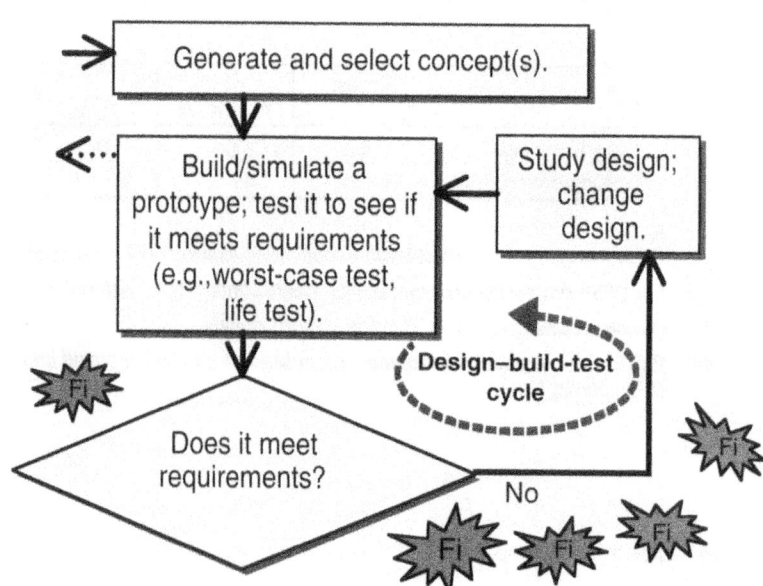

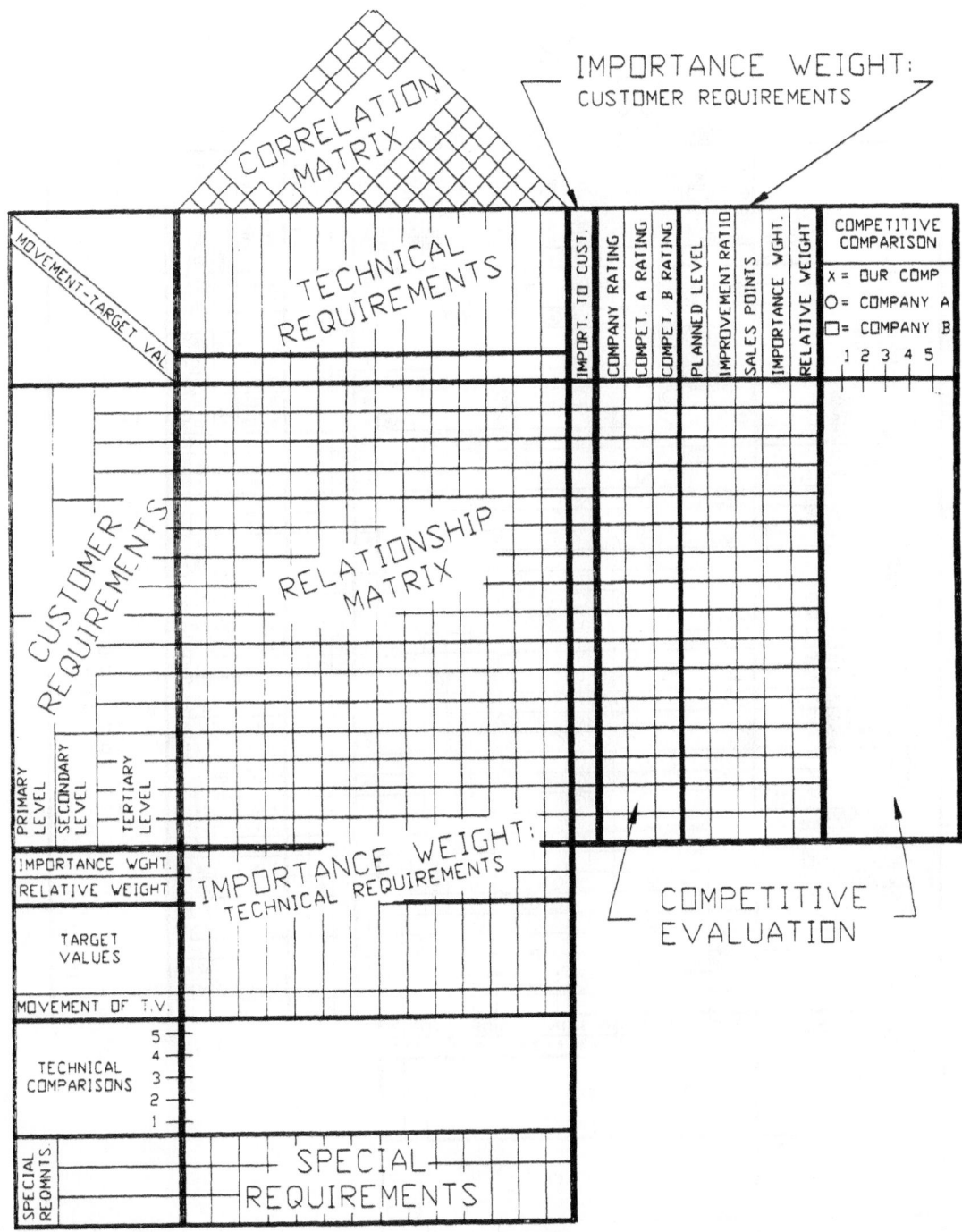

Figure 6.5
House of quality

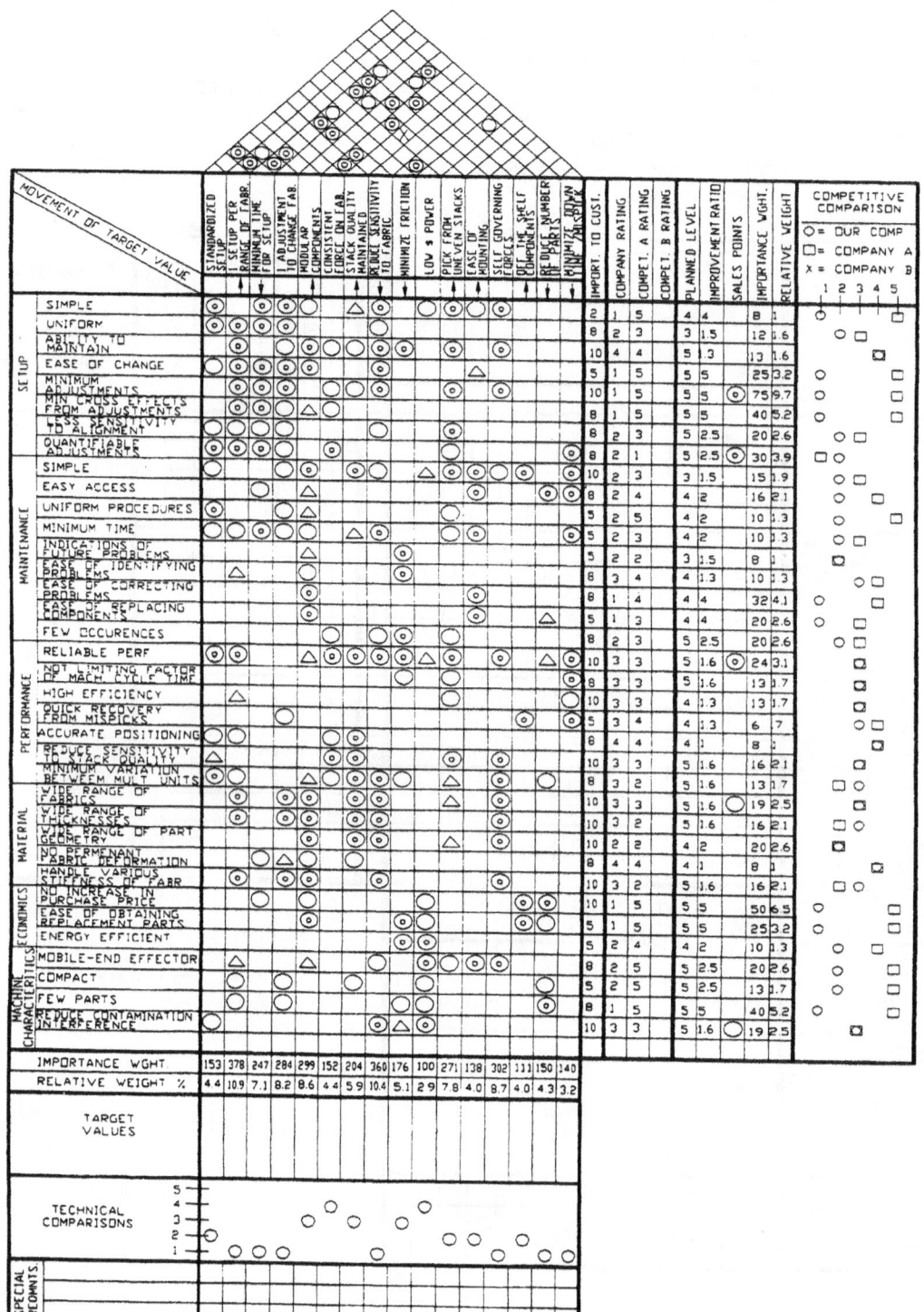

Figure 6.6
House of quality for pickup device

and change the concept, or go to the next step of tolerance design for further improvement with a minimum increase in cost. Recognizing that robust optimization seeks for perfection of the conceptual design will prevent a poor concept from going downstream, thus avoiding the need for firefighting.

In a typical company in the United States, it is said that 75% of engineers' time is spent on fighting fires. Successful firefighters are recognized as good engineers and are promoted. When one *prevents* fires, he or she will not be recognized. This is a serious issue that only top executives can resolve. Just ask anyone which is more effective, firefighting or fire prevention and I am sure that the majority of people will say that fire prevention is more effective and is the right thing to do. However, in a typical company, engineers developing technology and products are not asked to optimize design for robustness. They are getting paid to meet "requirements." Therefore, they rotate through the design–build–test–fix cycle to test against requirements. Or their job description does not spell out "optimization"; that is, nobody is asking them to optimize the design.

Figure 6.7 shows product development with robust engineering from 30,000 feet above. This process includes robust optimization of technology elements in the R&D phase. With energy thinking, technology elements can be optimized for robustness without having the requirements for a specific product. For instance, a small motor can be optimized without knowing the detailed requirements. This concept of technology robustness is highly advantageous in achieving quality, reliability, low cost, and shorter development time.

Although it has been 20 years since Taguchi methods were introduced into the United States, we cannot say that any company has changed its culture successfully from firefighting to fire prevention. Some got close, but they did not sustain the momentum. Many are working on it today, but struggling. For the sake of reducing the "loss to society," robust engineering should be "imprinted in the DNA" of engineers, so to speak. To accomplish this vision, we need to do the following:

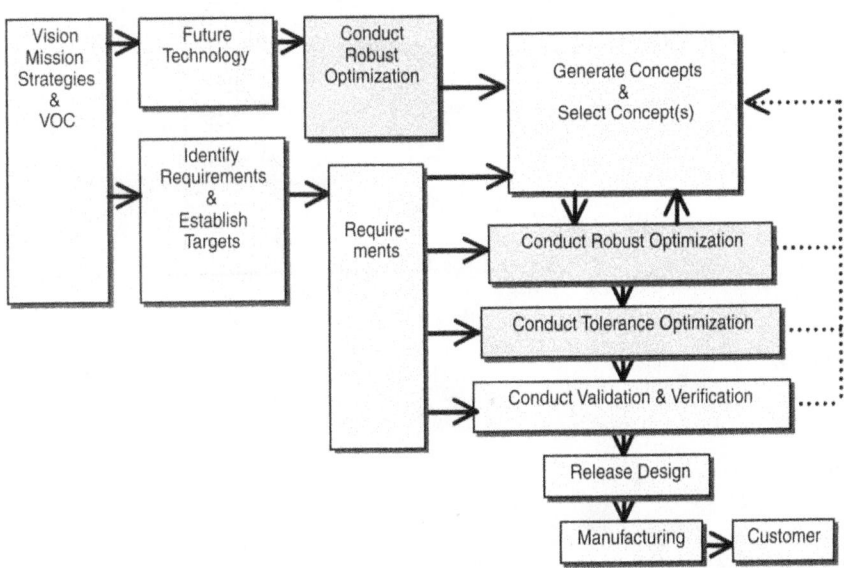

Figure 6.7
Product development with robust engineering

1. Introduce fire prevention as a corporate strategy. It is absolutely necessary to have top executives' leadership to establish a robust engineering culture.

2. Integrate robust optimization and tolerance optimization in R&D and in the product development process. Prioritize and provide appropriate resources for optimization.

3. Convince several major engineering schools to establish a solid course on robust engineering. This should not be simply one to two hours of coverage within a class on design of experiments, as many universities provide today, but at least 4 credits (about 50 hours) at the undergraduate level and a master's degree in robust engineering at the graduate level.

4. Develop networks among practitioners internationally. The Quality Engineering Society in Japan is well established. A tremendous amount of research has been accomplished by society in the last 10 years. Congratulations to all of you, but today there is no effective mechanism to exchange ideas internationally.

5. In 2003, under the leadership of Subir Chowdhury, the American Supplier Institute created the first International Society for Robust Engineering Professionals (*www.isrep.com*). Taguchi launched the society site at ASI's annual Taguchi Symposium in September 2003 in Dearborn, Michigan.

We look forward to continued growth in the area and to companies recognizing the importance of fire prevention.

Part III

Quality Loss Function

7 Introduction to the Quality Loss Function

7.1. Introduction   171
Definition of Quality   171
Quality Loss Function and Process Capability   173
7.2. Steps in Product Design   176
System Design   176
Parameter Design: Two-Step Optimization   177
Tolerance Design   179

7.1. Introduction

Definition of Quality

A product is sold by virtue of its product species and its price. *Product species* relate to product function and market size. *Product quality* relates to loss and market size. Quality is often referred to as *conformance to specifications*. However, Taguchi proposes a different view of quality, one that relates it to cost and loss in dollars, not just to the manufacturer at the time of production but to the consumer and to society as a whole.

Loss is usually thought of as additional manufacturing costs incurred up to the point that a product is shipped. After that, it is society, the customer, who bears the cost for loss of quality. Initially, the manufacturer pays in warranty costs. After the warranty period expires, the customer may pay for repair on a product. But indirectly, the manufacturer will ultimately "foot the bill" as a result of negative customer reaction and costs that are difficult to capture and account for, such as customer inconvenience and dissatisfaction, and time and money spent by customers. As a result, the company's reputation will be damaged, and eventually the market share will be lost.

Real growth comes from the market, cost, and customer satisfaction. The money the customer spends for a product and the perceived loss due to poor quality ultimately come back as long-term loss to the manufacturer. Taguchi defines *quality* as "the loss imparted by the product to the society from the time the product is shipped." The objective of the quality loss function is quantitative evaluation of loss caused by functional variation of a product.

☐ Example

Once a serious problem occurred concerning the specification of the thickness of vinyl sheets used to build vinyl houses for agricultural production. Figure 7.1 shows the relationship between thickness, cost, and quality versus loss. Curve L shows the sum of cost and quality. The midvalue of specification, or target value, is the lowest point on curve L. Assume that the specification given by the Vinyl Sheet Manufacturing Association is

$$1.0 \pm 0.2 \text{ mm} \qquad (7.1)$$

A vinyl sheet manufacturer succeeded in narrowing the variation of thickness down to ± 0.02 mm after a series of quality improvement efforts. To reduce material cost, the manufacturer changed its production specification to

$$0.82 \pm 0.02 \text{ mm} \qquad (7.2)$$

Although the production specification above still meets the one given by the association, many complaints were lodged after the products within the specification of equation (7.2) were sold.

When the thickness of vinyl sheets is increased, the production cost (including material cost, processing cost, and material-handling cost) increases as shown by curve C. On the other hand, when the thickness is decreased, sheets become easy to break and farmers must replace or repair them; thus, the cost caused by inferior quality increases following curve Q.

Initially, the midvalue of specification (denoted by m) is the lowest point on curve L, which shows the sum of the two curves. When the average thickness is lowered from $m = 1.0$ mm to 0.82 mm, the manufacturing cost is cut by B dollars,

Figure 7.1
Objective value and its
limits

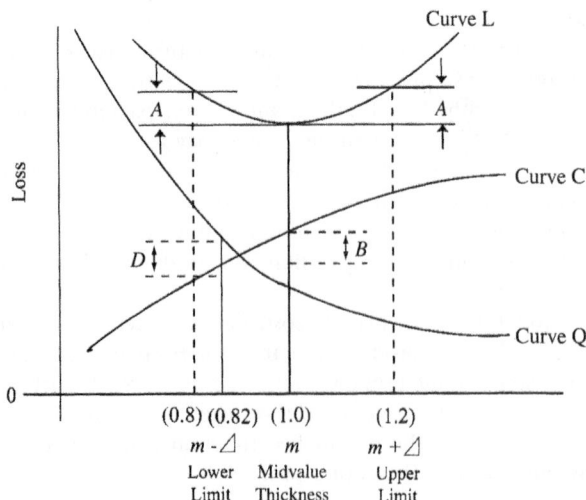

but consumer loss is increased by D dollars, which is much larger than B dollars. Also, if the thickness varies by 0.2 mm on either side of the target value (m), the manufacturer incurs a total loss of A dollars.

Because of many complaints, the manufacturer had to adjust the midvalue of thickness back to the original value of 1.0 mm. The association still maintained the original specification after adding a clause that the average thickness must be 1.0 mm.

Traditionally, the core of quality control is the fraction defective and its counter-measures. If defective products are shipped, it creates quality problems. If defectives are *not* shipped, it causes loss to the manufacturer. To avoid damage to the company's reputation, it is important to forecast the quality before products are shipped. Since the products within specifications are shipped, it is necessary to forecast the quality level of nondefective products. For this purpose, the *process capability index* has been used. This index is calculated from the tolerance divided by 6σ, which is quite abstract. (In general, root mean square is used instead of sigma. This is described in a later chapter.) Since there is no economic issue and no standard to follow a particular value, such as 0.8 or 1.3 or 2.0, it is difficult for management to understand the concept.

Quality Loss Function and Process Capability

The loss function is calculated from the square of the reciprocal of the process capability index after multiplying a constant related to economy. It is an economic forecasting value that is imparted to the customer in the market.

The process capability index, C_p, is calculated by the following equation:

$$C_p = \frac{\text{tolerance}}{6(\text{standard deviation})} \tag{7.3}$$

The loss function is given as

$$L = k(y - m)^2 \tag{7.4}$$

where L is the loss in dollars when the quality characteristic is equal to y, y the value of the quality characteristic (i.e., length, width, concentration, surface finish, flatness, etc.), m the target value of y, and k a constant to be defined later. As shown in Figure 7.2, this quadratic representation of the loss function $L(y)$ is

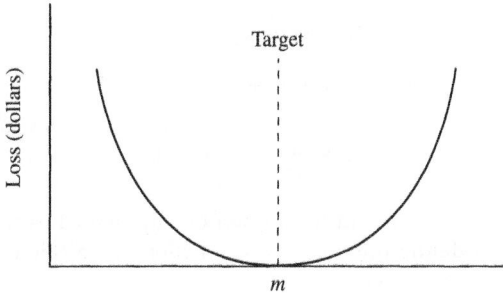

Figure 7.2
Quality loss function

minimum at $y = m$, increases as y deviates from m, and is expressed in monetary units.

However, one might still question whether equation (7.4) is a reasonable representation. The loss function L can be expanded in a Taylor series around the target value m:

$$L = L(m + y - m)$$

$$= L(m) + \frac{L'(m)}{1!}(y - m) + \frac{L''}{2!}(y - m)^2 + \cdots \qquad (7.5)$$

Because L is a minimum at $y = m$, $L'(m) = 0$. $L(m)$ is always a constant and is ignored since its effect is to raise or lower the value of $L(y)$ uniformly at all values of y. The $(y - m)^2$ term is the dominant term in equation (7.5) (higher-order terms are neglected). Therefore, we use the following equation as an approximation:

$$L = \frac{L''}{2!}(y - m)^2$$

$$= k(y - m)^2 \qquad (7.6)$$

In reality, for each quality characteristic there exists some function that uniquely defines the relationship between economic loss and the deviation of the quality characteristic from its target value. The time and resources required to obtain such a relationship for each quality characteristic would represent a considerable investment. Taguchi has found the quadratic representation of the quality loss function to be an efficient and effective way to assess the loss due to deviation of a quality characteristic from its target value (i.e., due to poor quality). The concept involved in Taguchi methods is that useful results must be obtained quickly and at low cost. Use of a quadratic, parabolic approximation for the quality loss function is consistent with this philosophy.

For a product with a target value m, from most customers' point of view, $m \pm \Delta_0$ represents the deviation at which functional failure of the product or component occurs. When a product is manufactured with its quality characteristic at the extremes, $m + \Delta_0$ or at $m - \Delta_0$, some countermeasure must be undertaken by the average customer (i.e., the product must be discarded, replaced, repaired, etc.). The cost of the countermeasure is A_0 since the quality loss function is

$$L = k(y - m)^2 \qquad (7.4)$$

at

$$y = m + \Delta_0 \qquad (7.7)$$

$$A_0 = k(m + \Delta_0 - m)^2 \qquad (7.8)$$

$$k = \frac{A_0}{\Delta_0^2} \qquad (7.9)$$

This value of k, which is constant for a given quality characteristic, and the target value, m, completely define the quality loss function curve (Figure 7.3).

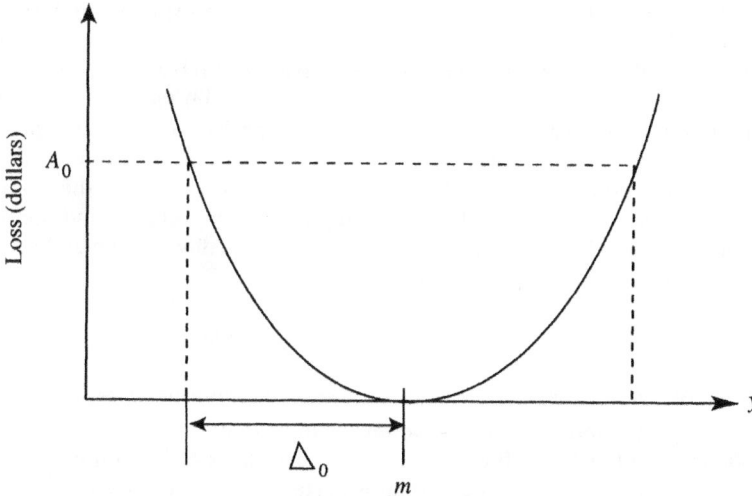

Figure 7.3
Quality loss function

Up to this point, the quality loss function is explained with one piece of product. In general, there are multiple pieces of products. In that case, the loss function is given by

$$L = k\delta^2 \tag{7.10}$$

Equation (7.3) can be written as

$$C_p = \frac{\text{tolerance}}{6(\text{standard deviation})} = \frac{2\Delta_0}{6\sigma} \tag{7.11}$$

Therefore,

$$L = \frac{A}{\Delta_0^2}\sigma^2 = \frac{A}{9}\frac{1}{C_p^2} \tag{7.12}$$

It is seen that the loss function can be expressed mathematically by the process capability index. But the former has dollars as a unit and the latter has no unit. For example, the process capability index of last month was 0.9, and it is 0.92 this month. This fact does not explain the difference from the economic or productivity viewpoint. Using the loss function, monetary expression can be made and productivity can be compared.

There are two aspects to the quality loss function. The first is the *justice-related* aspect. Take, for example, a product specified to have a 100-g content. If the amount is 97 g instead of 100, the customer loses 3 g. If the price is 10 cents a gram, the customer loses 30 cents. In this case, it is easy to understand.

If the content is 105 g, the customer may be happy to have a free 5 g of product. However, it might be a loss to the customer. If the manufacturer aimed for 105 g in a package, there would be a 5% production loss. If a package is supposed to contain 100 g, a planned production of 10,000 packages would result in producing only 9500 packages in the case of 105-g packages. The manufacturer normally would calculate the manufacturing cost based on 9500 units; then overhead and profit would be added. The product selling price is a few times the manufacturing cost. Selling a 100-g package based on a 105-g package cost calculation is simply cheating.

The second aspect is a *quality* issue. Assume that a package of medicine specified to contain 100 g actually contains 105 g. If the customer measured the quantity accurately and used it as prescribed, there would be a 5-g waste. But if the customer used the package of 105 g as 100 g, there might be some problems and the problems could be fatal. It is therefore important that the quality level be at the target from both the legal and quality aspects.

7.2. Steps in Product Design

The objective function of a TV set's power circuit is to convert an ac input of 100 V into a dc output of 115 V. If the power circuit maintains a 115-V output anytime, anywhere, the quality of its power circuit is perfect as far as voltage is concerned. The variables causing the voltage to move away from the target of 115 V, called *noise variables*, are classified as follows:

1. *Outer noise:* variation of environmental conditions (e.g., temperature, humidity, dust, input voltage)

2. *Inner noise:* deterioration of elements or materials comprising the circuit (e.g., the resistance of a resistor increases by 10% in 10 years, or the parameter of a transistor changes with time)

3. *Between-product noise:* manufacturing imperfection resulting in different outputs

To minimize the effects caused by these noise sources, some countermeasures may be considered. Of these countermeasures, the most important is the countermeasure by design. It consists of (1) system design, (2) parameter design, and (3) tolerance design.

System Design   Knowledge from a specialized field is applied to system design: from electronics, research into the type of circuit used to convert alternating current into direct current; and from the chemical industry, studies concerning the specific chemical reaction needed to obtain the best process given the available technology.

Automatic controlling systems are included in the scope of this step. For example, an output voltage is measured from time to time. When there is a deviation from the target voltage, a parameter in the circuit is changed automatically. However, it is difficult to control the deterioration or variation of an automatic controlling system. Not only that but there is an additional cost for the automatic control mechanism. Therefore, it is not the design of a stable circuit at a low cost. Our goal is to design a stable and low-cost circuit by introducing a drastic change to the system. However, we must rely on knowledge from specialized fields. This is the specialist's territory, and neither quality control nor the design of experiments can help.

Many systems are conceivable for a certain function. We select a system that matches our objective from countless possible systems. It is desirable to select a new system that does not exist anywhere else in the world. Patents protect new systems. System selection is often determined at a meeting of the persons in charge, where opinions are exchanged regarding the advantages and disadvantages of each item in the functions, price, and life that are set at the planning stage. Judgment of each item is only quantitative, but this is important in order to convince executives to go along with it. When the development section has the resources, two or three systems may be developed simultaneously.

If social loss is great in case a function fails, considerations of safety are also a part of system selection. Safety design for improvement of productivity is an important factor in system selection. One of the most important duties of a system designer is to include a safety mechanism that does not endanger human life or cause significant financial loss if the function fails, rather than simply continuing to increase reliability.

Once the system design is finished, the next step is to determine the optimum level of individual parameters of the system in both quality and cost. *Parameter design* is a method used to reduce the influence of sources of variation. It has been called *utilization of nonlinearity* or *utilization of interaction between control and noise factors*. In the technical world, analytical methods have been used for parameter design.

Parameter Design: Two-Step Optimization

Research on this has been neglected, especially in developing countries, where researchers look for information from literature, designing power circuits that "seem to be the best." A prototype is made, then charged with 100 V ac current. If the output voltage equals the target value, it is labeled a success; and if the result is unsatisfactory, the value of a parameter that seemed effective in the circuit is changed. This approach is dated, reminiscent of the dyer of old. Dyers had color samples and tried to create dyes to match. If a dyer got a color identical to a sample, he succeeded; if not, he changed either the mixing ratio or the dyeing conditions and tried again.

This is not a design, merely operational work, called a *modification* or *calibration*. "Designs" in developing countries are similar to the dyers' methods; their engineers proceed in a similar fashion.

◻ Example

An element transistor A and resistor B in a power circuit affect the output voltage as shown in Figure 7.4.

Suppose that a design engineer forecasts the midvalues of the elements in a circuit, assemblies these to get a circuit, and puts 100 V ac into it. If the output voltage so obtained is only 100 V instead of 115 V, he changes resistance A from $A_1 = 200 \ \Omega$ to $A' = 250 \ \Omega$ to adjust the 15-V deviation from the target value. From the standpoint of quality control, this is very poor methodology. Assume that the resistance used in the circuit is the cheapest grade. It either varies or deteriorates to a maximum range of $\pm 10\%$ during its service period as the power source of a TV set. From Figure 7.4, we see that the output voltage varies within a range of ± 6 V. In other words, the output voltage varies by the influence of noise, such as the original variation in the resistance itself, or due to deterioration. If resistance

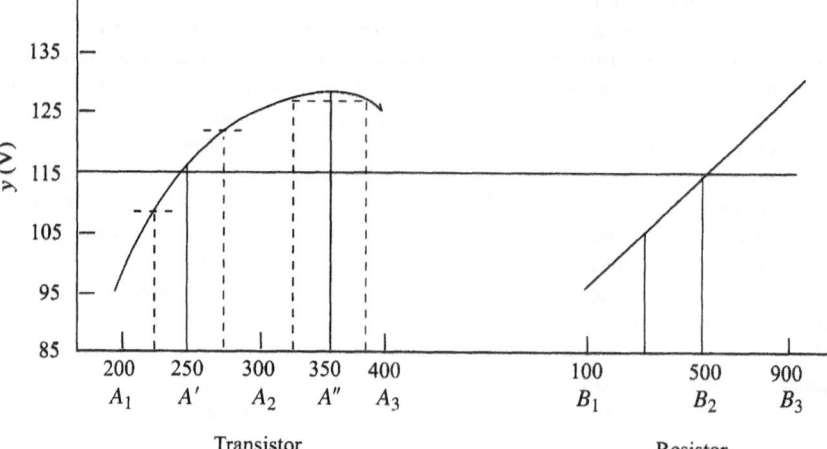

Figure 7.4
Relationship between
factors *A* and *B* and the
output voltage

A″, which has a midvalue of 350 Ω, is used instead of *A*′, its influence on the output voltage is only ±1 V, even if its variation is ±10% (35 Ω). However, the output voltage goes up about 10 V, as shown in the figure. Such a deviation can be adjusted by a factor like *B* with influence on the output voltage almost linear. In this case, 200 Ω is selected instead of 500 Ω for level *B*. A factor like *B*, with a differential coefficient to the output voltage that is nearly constant no matter what level is selected, is not useful for a high-level quality control purpose; it is used merely to adjust a deviation from the target value.

Parameter design is the most important step in developing stable and reliable products or manufacturing processes. With this technique, nonlinearity may be utilized positively. (Factor *A* has this property.) In this step we find a combination of parameter levels that are capable of damping the influences not only of inner noise but of all noise sources, while keeping the output voltage constant. At the heart of research lives a conflict: to design a product that is reliable within a wide range of performance conditions but at the lowest price. Naturally, elements or component parts with a short life and wide tolerance variation are used. The aim of design of experiments is the utilization of nonlinearity. Most important in applying design of experiments is to cite factors or to select objective characteristics with this intention.

There is a difference between the traditional quality improvement approach and parameter design. In the traditional approach, the first step is to try to hit the target, then reduce variability. The countermeasures of reducing variability depend on the tolerance design approach: improving quality at high cost increase. In parameter design, quality is improved by selecting control factor combinations. After the robustness of a product is improved, the target is adjusted by selecting a control factor whose level change affects the average but affects variability minimally. For this reason, parameter design is also called *two-step optimization*.

Once parameter design is completed, we have the midvalues of factors comprising   **Tolerance Design**
the system elements. The next step is to determine the tolerances of these factors,
or to rationally select the grades of parts and materials. Once the system design is
complete, we have the midvalues of the factors comprising the system elements.
The next step is to determine the tolerances of these factors. To make this determination, we must consider environmental conditions as well as the system elements. The midvalues and varying ranges of these factors and conditions are
considered as noise factors and are arranged in orthogonal arrays so that the
magnitude of their influence on the final output characteristics may be determined. A narrower allowance will be given to those noise factors, imparting a large
influence on the output. Cost considerations determine the allowance. Different
from parameter design, this step results in "cost-up" by controlling noise in narrower ranges; therefore, quality-controlling countermeasures should be achieved
through parameter design analysis. If this is not an option, tolerance design is the
only countermeasure left that will allow for influencing factors with a narrow
range.

8 Quality Loss Function for Various Quality Characteristics

8.1. **Classification of Quality Characteristics**   180
8.2. **Nominal-the-Best Characteristic**   180
    More Than One Piece   183
8.3. **Smaller-the-Better Characteristic**   186
8.4. **Larger-the-Better Characteristic**   188
8.5. **Summary**   189

8.1. Classification of Quality Characteristics

Quality loss function is used for the nominal-the-best, smaller-the-better, larger-the-better characteristics. The *nominal-the-best* characteristic is the type where there is a finite target point to achieve. There are typically upper and lower specification limits on both sides of the target. For example, the plating thickness of a component, the length of a part, and the output current of a resistor at a given input voltage are nominal-the-best characteristics.

A *smaller-the-better* output response is the type where it is desired to minimize the result, with the ideal target being zero. For example, the wear on a component, the amount of engine audible noise, the amount of air pollution, and the amount of heat loss are smaller-the-better output responses. Notice that all these examples represent things that we do not want, not the intended system functions. In the smaller-the-better characteristic, no negative data are included.

The *larger-the-better* output response is the type where it is desired to maximize the result, the ideal target being infinity. For example, strength of material, and fuel efficiency are larger-the-better output responses. Percentage yield seems to be the larger the better, but it does not belong to the larger-the-better category in quality engineering, since the ideal value is 100%, not infinity. In the larger-the-better characteristic, negative data are not included.

8.2. Nominal-the-Best Characteristic

To demonstrate the criteria for quality evaluation, consider Figure 8.1, which contains the frequency distribution curves for TV set color density. One curve

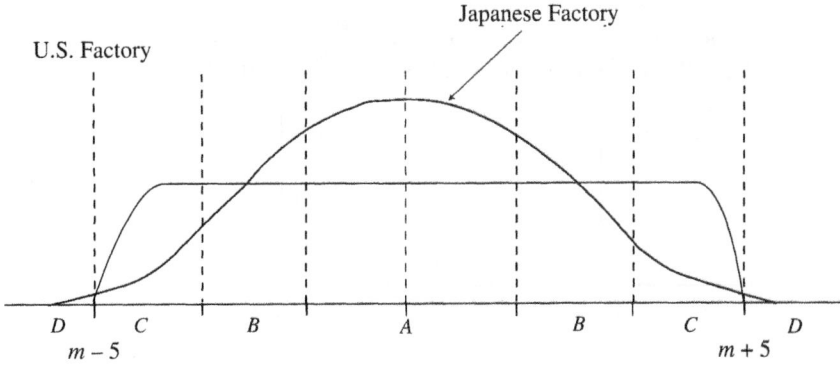

Figure 8.1
Distribution of color density in TV sets

represents the color density frequency distribution associated with sets built in Japan, and the other curve represents the same distribution for sets built in the United States. The two factories belong to the same manufacturing company.

In 1979, an article appeared in *The Asahi* (a newspaper) relative to the preference of American consumers for TV sets built by a company in Japan. Apparently, identical subsidiary plants had been built in the United States and in Japan. Both facilities were designed to manufacture sets for the U.S. market and did so using identical designs and tolerances. However, despite the similarities, the American consumer displayed a preference for sets that had been made in Japan.

Referring again to Figure 8.1, the U.S.-built sets were defect-free; that is, no sets with color density out of tolerance were shipped from the San Diego plant. However, according to the article and as can be seen in Figure 8.1, the capability index, C_p, of the Japan-built sets was $C_p = 1.0$ and represented a defect rate of 3 per 1000 units. Why, then, were the Japanese sets preferred?

Before answering that question, it is important to note that:

❑ Conformance to specification limits is an inadequate measure of quality or of loss due to poor quality.

❑ Quality loss is caused by customer dissatisfaction.

❑ Quality loss can be related to product characteristics.

❑ Quality loss is a financial loss.

Traditional methods of satisfying customer requirements by controlling component and/or subsystem characteristics have failed. Consider Figure 8.2, which represents the inspection-production-oriented traditional concept, whereby all products or processes that exist or function within some preestablished limits are considered to be equally good, and all products or processes that are outside these limits are considered to be equally bad. The fallacy in this type of judgment criteria was illustrated in the example where although all of the sets manufactured in the United States were within specifications, the customer requirements were apparently better satisfied by the sets built in Japan.

Taguchi compared specification limits to pass/fail criteria often used for examinations. In school, for example, 60% is generally considered to be a passing grade, but the difference between a 60% student who passes the course and a 59%

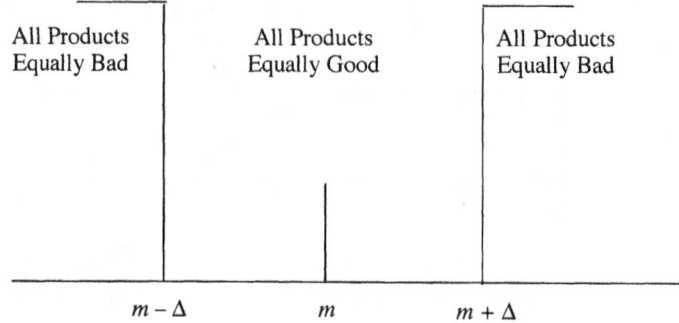

Figure 8.2
Conformance to
requirements

QUALITY CHARACTERISTIC

student who fails is nil. If 100 points represent perfection, the student with 60 points is more like the student with 59 than he or she is like the perfect student. Similarly, the manufactured product that barely conforms to the specification limits is more similar to a defective part than to a perfect part.

Specification limits only provide the criteria for acceptance and/or rejection. A product that just barely conforms to some preestablished limits functions relative to optimum just as the 60% student functions relative to perfection. For some customers, such deviation from optimum is not acceptable. In Figure 8.1, the quality characteristic has been graded depending on how close its value is to the target or best value denoted by m. The figure shows that the majority of the sets built in Japan were grade A or B, whereas the sets built in the United States were distributed uniformly in categories A, B, and C.

The grade-point average of the TV sets built in Japan was higher than that of the U.S.-built sets. This example rejects the quality evaluation criteria that characterize product acceptance by inspection with respect to specification limits. How, then, can quality be evaluated? Since perfection for any student is 100%, loss occurs whenever a student performs at a lower level. Similarly, associated with every quality characteristic is a best or target value: that is, a best length, a best concentricity, a best torque, a best surface finish, and so on. Quality loss occurs whenever the quality characteristic deviates from this best value. Quality should, therefore, be evaluated as a function of deviation of a characteristic from target.

If a TV color density of m is best, quality loss must be evaluated as a function of the deviation of the color density from this m value. This new definition of product quality is the uniformity of the product or component characteristics around the target, not the conformity of these characteristics to specification limits. Specification limits have nothing to do with quality control. As quality levels increase and the need for total inspection as a tool for screening diminishes, specifications become the means of enunciating target values. Associated with every product characteristic, however, there exists some limit outside of which the product, as viewed by 50% of the customers, does not function. These limits are referred to as the *customers' tolerance.*

The rejection of quality assurance by means of inspection is not new. The Shewhart concept also rejects quality assurance by inspection and instead tries to assure quality by control methods. From an economic standpoint, however, this concept has, a weakness. If the cost for control is larger than the profit realized from the resulting reduced variation, we should, of course, do nothing. In other words, it is the duty of a production department to reduce variation while maintaining the profit margin required.

Suppose that four factories are producing the same product under the same engineering specifications. The target values desired are denoted by m. The outputs are as shown in Figure 8.3. Suppose further that the four factories carry out 100% inspection and ship out only pieces within specification limits.

More Than One Piece

Although the four factories are delivering products that meet specifications, factory 4 offers more products at or near the desired target value and exhibits less piece-to-piece variability than the other factories. Factory 4 is likely to be selected as the preferred vendor. If a person selects factory 4, it is probably because he or she believes in the loss function but being within specifications is not the entire story.

The loss function, L, as described previously, is used for evaluating 1 unit of product:

$$L = k\,(y - m)^2 \tag{7.4}$$

where k is a proportionality constant and y is the output. More often, the qualities of all pieces of products are evaluated. To do this, the average of $(y - m)^2$, which is called the *mean-squared deviation* (MSD), is used. When there is more than one piece of product, the loss function is given by

$$L = k(\text{MSD}) \tag{8.1}$$

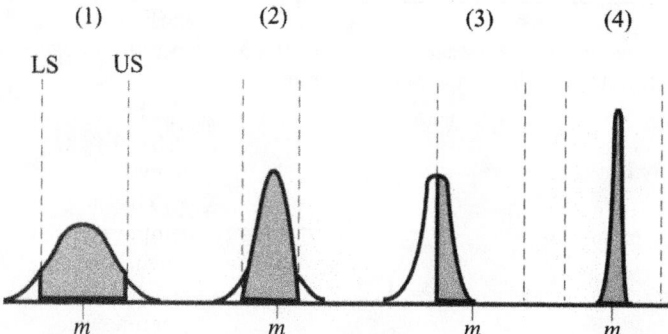

Figure 8.3
Outputs for four factories

LS: Lower specification limit
US: Upper specification limit

For n pieces of products with output y_1, y_2, ... , y_n, the average loss is

$$L = \frac{k(y_1 - m)^2 + k(y_2 - m)^2 + \cdots + k(y_n - m)^2}{n}$$

$$= k \frac{(y_1 - m)^2 + (y_2 - m)^2 + \cdots + (y_n - m)^2}{n} \qquad (8.2)$$

$$\text{MSD} = \frac{1}{n} \sum_{i=1}^{n} (y_i - m)^2$$

$$= \frac{1}{n} \sum_{i-1}^{n} (y_i - \bar{y})^2 + (\bar{y} - m)^2$$

$$= \frac{(y_1 - \bar{y})^2 + \cdots + (y_n - \bar{y})^2}{n} + (\bar{y} - m)^2$$

$$= \sigma^2 + (\bar{y} - m)^2 \qquad (8.3)$$

where $\bar{y}$ is the average of y. The loss function for more than one piece then becomes

$$L = k[\sigma^2 + (\bar{y} - m)^2] \qquad (8.4)$$

where σ^2 is the variability around the average and $\bar{y}$ is the deviation of the average from the target.

We can now evaluate the quality of all our outputs. To reduce the loss, we must reduce the MSD. This can be accomplished with:

δ^2:   reducing the variability around the average

$(y - m)^2$:   adjusting the average to the target

The quality loss function gives us a quantitative means of evaluating quality. Let us reexamine our comparison of the four factories with different output distributions (Table 8.1).

☐ Example

The following analysis involves a sample size of 13 pieces each. Suppose that $k = 0.25$. Then, using the nominal-the-best format:

$$L = k[\sigma^2 + (\bar{y} - m)^2] = k(\text{MSD}) \qquad (8.5)$$

For factory 3:

$$\bar{y} = \frac{112 + 113 + \cdots + 114}{13}$$

$$= 113$$

$$(\bar{y} - m)^2 = (113 - 115)^2$$

$$= 4 \qquad (8.6)$$

Table 8.1
Data for the four factories

Factory	Data			MSD	Loss per Piece
1	115	113	113	2.92	$0.73
	114	114	115		
	115	116	116		
	117	117	115		
	118				
2	113	114	114	1.08	$0.27
	114	115	115		
	115	115	115		
	116	116	116		
	113				
3	112	113	112	4.92	$1.23
	113	112	113		
	114	115	112		
	113	114	112		
	114				
4	114	115	116	0.62	$0.15
	114	115	116		
	114	115	116		
	114	115	116		
	115				

$$\sigma^2 = \frac{(y_1 - \bar{y})^2 + (y_2 - \bar{y})^2 + \cdots + (y_n - \bar{y})^2}{n}$$

$$= \frac{(112 - 113)^2 + (113 - 113)^2 + \cdots + (114 - 113)^2}{n} = 0.92 \quad (8.7)$$

$$MSD = \sigma^2 + (\bar{y} - m)^2$$

$$= 0.92 + 4$$

$$= 4.92 \quad (8.8)$$

$$L = k(MSD)$$

$$= 0.25(4.92)$$

$$= \$1.23 \text{ per piece} \quad (8.9)$$

Losses for factories 1, 2, and 4 are calculated in the same way. The results are summarized in Figure 8.4. The 73-cent loss for factory 1, for example, is interpreted as follows: As a rough approximation, one randomly selected product shipped from factory 1 is, on average, imparting a loss of 73 cents. Somebody spends the 73 cents: a customer, the company itself, an indirect consumer, and so on. Does factory 4 still appear to be the best choice?

Notice that in all cases, the smaller the MSD, the less the average loss to society. *Our job, to obtain high quality at low cost, is to reduce the MSD.* This can be accomplished through use of parameter design and tolerance design.

The loss function offers a way to quantify the benefits achieved by reducing variability around the target. It can help to justify a decision to invest $20,000 to improve a process that is already capable of meeting specifications.

Suppose, for example, that you are an engineer at factory 2 and you tell your boss that you would like to spend $20,000 to raise the quality level of your process to that of factory 4. What would your boss say? How would you justify such an investment?

Let's assume that monthly production is 100,000 pieces. Using the loss function, the improvement would account for a savings of $(0.27 − 0.15)(100,000) = $12,000 per month. The savings would represent customer satisfaction, reduced warranty costs, future market share, and so on.

8.3. Smaller-the-Better Characteristic

The loss function can also be determined for cases when the output response is a smaller-the-better response. The formula is a little different, but the procedure is much the same as for the case of nominal-the-best (Figure 8.5). For the case of smaller-the-better, where the target is zero, the loss function becomes

$$L = ky^2 \qquad k = \frac{A_0}{y_0^2} \tag{8.10}$$

$$L = k(\text{MSD}) \qquad \text{MSD} = \sum_{i=1}^{n} \frac{y_i^2}{n} = \sigma^2 + \bar{y}^2 \tag{8.11}$$

where A_0 is the consumer loss and y_0 is the consumer tolerance.

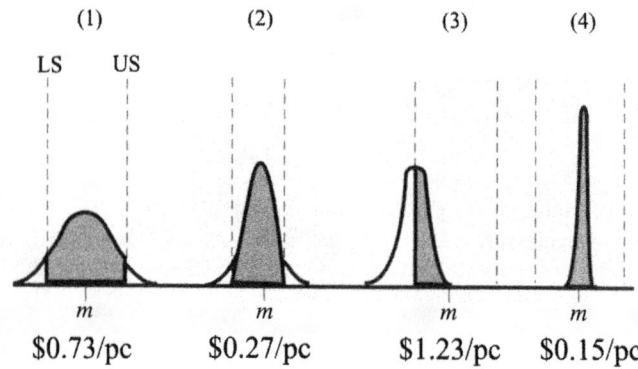

Figure 8.4
Evaluating four factories

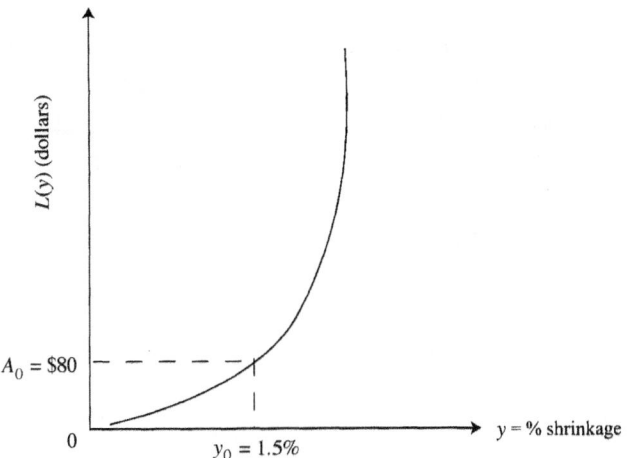

Figure 8.5
Smaller-the-better curve
(% shrinkage)

❏ Example

The smaller-the-better case is illustrated for the manufacturing of speedometer cable casing, where the output response is

$$y = \% \text{ shrinkage of speedometer casing}, \qquad y \geq 0$$

When y is 1.5%, the customer complains about 50% of the time and brings the product back for replacement. The replacement cost is $80.

$$A_0 = \$80$$

$$y_0 = 1.5$$

Thus,

$$k = \frac{80}{(1.5)^2} = 35.56 \qquad (8.12)$$

The loss function can then be written as

$$L(y) = 35.56y^2 \qquad (8.13)$$

The data in Table 8.2 are percent shrinkage values of the casings made from two different materials. The losses were computed using the formula

$$L = 35.56 \, (\bar{\delta}^2 + \bar{y}^2) \qquad (8.14)$$

While the shrinkage measurements from both materials meet specifications, the shrinkage from material type B is much less than that of type A, resulting in a much smaller loss. If both materials cost the same, material type B would be the better choice.

Table 8.2
Smaller-the-better data (% shrinkage)

Type of Material	Data		$\widetilde{\delta}^2$	$\bar{y}$	MSD	L
A	0.28	0.24	0.00227	0.0729	0.0751	$2.67
	0.33	0.30				
	0.18	0.26				
	0.24	0.33				
B	0.08	0.12	0.00082	0.00439	0.0052	$0.19
	0.07	0.03				
	0.09	0.06				
	0.05	0.03				

8.4. Larger-the-Better Characteristic

For a larger-the-better output response where the target is infinity, the loss function is (Figure 8.6)

$$L = k\frac{1}{y^2} \qquad k = A_0 y_0^2 \tag{8.15}$$

$$L = k(\text{MSD}) \qquad \text{MSD} = \frac{1}{n}\sum_{i=1}^{n}\frac{1}{y_1^2}$$

$$= \frac{1}{n}\left(\frac{1}{y_1^2} + \frac{1}{y_2^2} + \cdots + \frac{1}{y_n^2}\right) \tag{8.16}$$

Figure 8.6
Larger-the-better curve

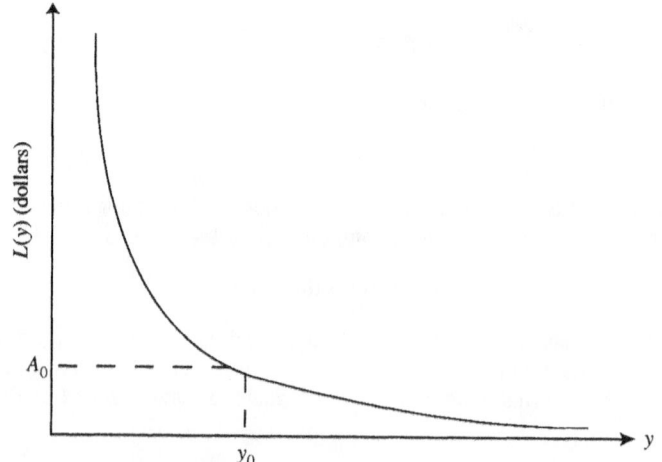

□ **Example**

A company wants to maximize the weld strength of its motor protector terminals. When the weld strength is 0.2 lb/in², some welds have been known to break and result in an average replacement cost of $A_0 = \$200$.

1. Find k and set up the loss function.
2. If the scrap cost at production is $2 per unit, find the manufacturing tolerance.
3. In an experiment that is conducted to optimize the existing semimanual welding process (Table 8.3), compare the "before" and "after."

8.5. Summary

The equations to calculate the loss function for different quality characteristics are summarized below.

1. Nominal-the-best (Figure 8.7)

$$L = k(y - m)^2 \qquad k = \frac{A_0}{\Delta_0^2}$$

$$L = k(\text{MSD}) \qquad \text{MSD} = \frac{1}{n} \sum_{i=1}^{n} (y_1 - m)^2$$

$$= \sigma^2 + (\bar{y} - m)^2$$

2. Smaller-the-better (Figure 8.8)

$$L = ky^2 \qquad k = \frac{A_0}{y_0^2}$$

Table 8.3
Larger-the-better data (weld strength)

	Data			MSD	L
Before experiment	2.3 2.0 1.9 1.7 2.1 2.2 1.4 2.2 2.0 1.6			0.28529	$2.28
After experiment	2.1 2.9 2.4 2.5 2.4 2.8 2.1 2.6 2.7 2.3			0.16813	$1.35

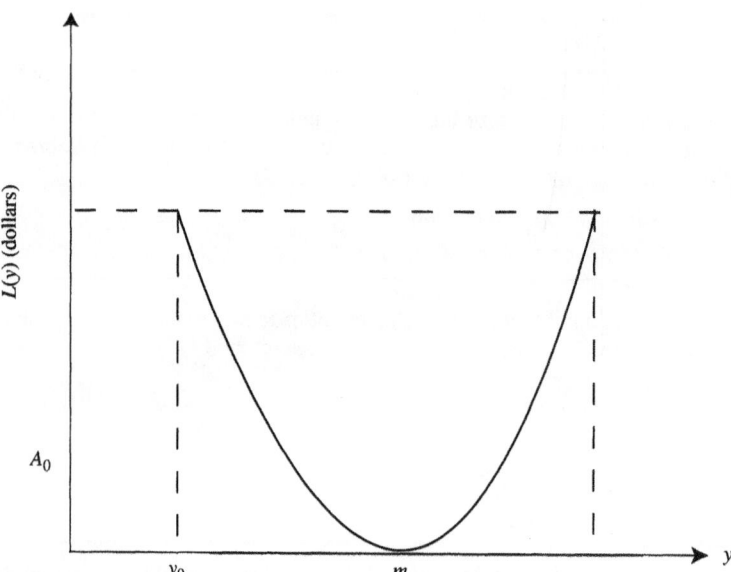

Figure 8.7
Nominal-the-best curve

$$L = k(\text{MSD}) \qquad \text{MSD} = \sum_{i=1}^{n} \frac{1}{y_i^2}$$

$$= \sigma^2 + \bar{y}^2$$

3. Larger-the-better (Figure 8.9)

$$L = k\frac{1}{y^2} \qquad k = A_0 y_0^2$$

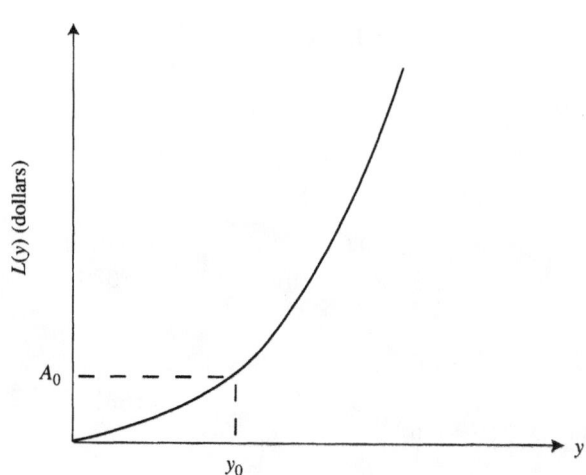

Figure 8.8
Smaller-the-better curve

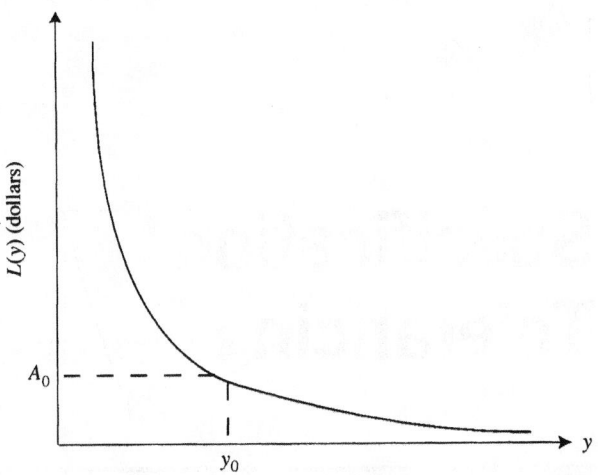

Figure 8.9
Larger-the-better curve

$$L = k(\text{MSD}) \qquad \text{MSD} = \frac{1}{n} \sum_{i=1}^{n} \frac{1}{y_i^2}$$

$$= \frac{1}{n} \left(\frac{1}{y_i^2} + \frac{1}{y_2^2} \cdots + \frac{1}{y_n^2} \right)$$

9 Specification Tolerancing

9.1. Introduction   192
9.2. Safety Factor   195
9.3. Determination of Low-Rank Characteristics   201
9.4. Distribution of Tolerance   204
9.5. Deteriorating Characteristics   205

9.1. Introduction

In traditional quality control, the fraction defective of a manufacturing process is considered to be the quality level of the product. When a defective product is discovered, it is not shipped. So if all products are within specifications, are these products perfect? Obviously they are not, because many of the products shipped, considered as being good or nondefective, have problems in the marketplace. Problems occur due to lack of robustness against a user's conditions. The only responsibility that a manufacturer has is to ship nondefective products, that is, products within specifications. Defective products cause a loss to the company. In other words, it is not a quality problem but a cost problem. It is important to realize that the tolerances, the upper and lower limits, stated in the drawings are for inspection purposes, *not* for quality control.

Again, as in school, any mark above 60 is a passing grade, with a full mark being 100. A student passing with a grade of 60 is not perfect and there is little difference between a student with a grade of 60 and one with a grade of 59. It is clear that if a product is assembled with the component parts marginally within tolerance, the assembled product cannot be a good one.

When troubles occur from products that are within tolerance, the product design engineer is to be blamed. If robust designs were not well performed and specifications were not well studied, incorrect tolerances would come from the design department.

Should the tolerance be tightened? If so, the loss claims from the market could be reduced a little. But then manufacturing cost increases, giving the manufacturer a loss and ending up with an increased loss to society.

Although the determination of tolerance is important, there have been no methods with a sound basis. A safety factor such as 4 has been widely used but not for sound reasons. From Taguchi's definition, *quality is the loss that a product imparts to society after the product is shipped.* The quality problem for a manufacturing process is in taking countermeasures to minimize the loss to society.

However, it is difficult to actually measure the societal loss associated with a product. We might produce 1000 units of product, distribute them to various type of customers, and let the customers use the product for the designed life, measure the loss due to problems, and take the average. Such an approach is realistically impossible. Even if it were possible, after the designed year is passed, the product would become obsolete. It is therefore necessary to forecast the quality level. The quality loss function provides a basis for the determination of tolerance.

In quality engineering, tolerance is not the deviation between products. Tolerance is defined as *a deviation from target.* Tolerance is determined so that the loss caused by the manufacturer and the one caused by the customer are balanced.

❏ Example

To determine a specification, it is necessary to determine two values: functional tolerance and customer loss. For every product characteristic we can find a value at which 50% of customers view the product as not functioning. This value represents an *average customer viewpoint* and is referred to as the *functional tolerance* or LD_{50} (denoted as Δ_0). The average loss occurring at LD_{50} is referred to as the *customer loss, A_0. The functional tolerance and consumer loss are required to establish a loss function.*

Let us set up the loss function for a color TV power supply circuit where the target value of y (output voltage) is $m = 115$ V. Suppose that the average cost for

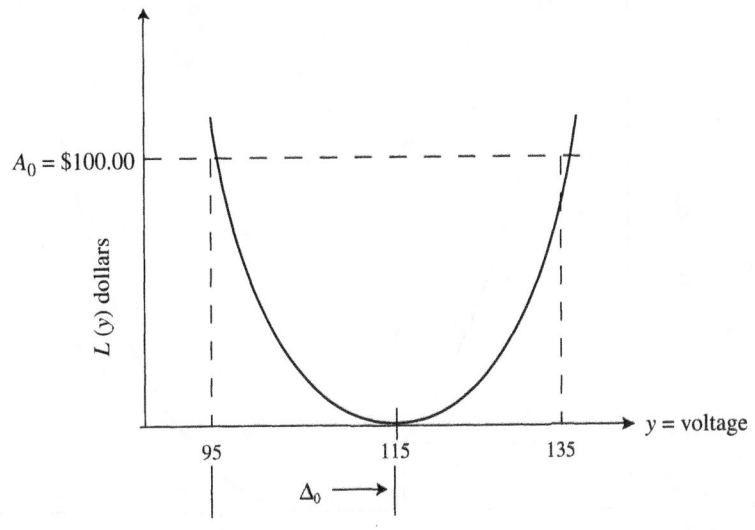

Figure 9.1
Loss function for TV
power supply

repairing the color TV is $100. This occurs when y goes out of the range $115 \pm$ 20 V in the hands of the consumer (Figure 9.1). We see that $L = \$100$ when y = 95 or 135 V, consumer tolerance $\Delta_0 = \pm 20$ V, and consumer loss $A_0 = \$100$. Then we have

$$L(y) = k(y - m)^2 \tag{7.4}$$

$$A_0 = k\Delta_0^2 \tag{9.1}$$

$$k = \frac{A_0}{\Delta_0^2} = \frac{\$100}{(20 \text{ V})^2} = \$0.25 \text{ per volt}^2 \tag{9.2}$$

The loss function can now be rewritten as

$$L = 0.25(y - 115)^2 \quad \text{dollars/piece} \tag{9.3}$$

When the output voltage becomes 95 or 135 V, somebody is paying $100. As long as the output is 115 V, society's financial loss is minimized.

With the loss function established, let's look at its various uses. Suppose that a circuit was shipped with an output of 110 V. It is imparting a loss of

$$L = \$0.25(110 - 115)^2 = \$6.25 \tag{9.4}$$

This means that on the average, someone is paying $6.25. This figure is a rough approximation of loss imparted to society due to inferior quality (Figure 9.2).

With information about the after-shipping or consumer tolerance, we can calculate the prior-to-shipping or manufacturing tolerance. *The manufacturing tolerance is the economical break-even point for rework for scrap.*

Figure 9.2
Loss function for 110-V circuit

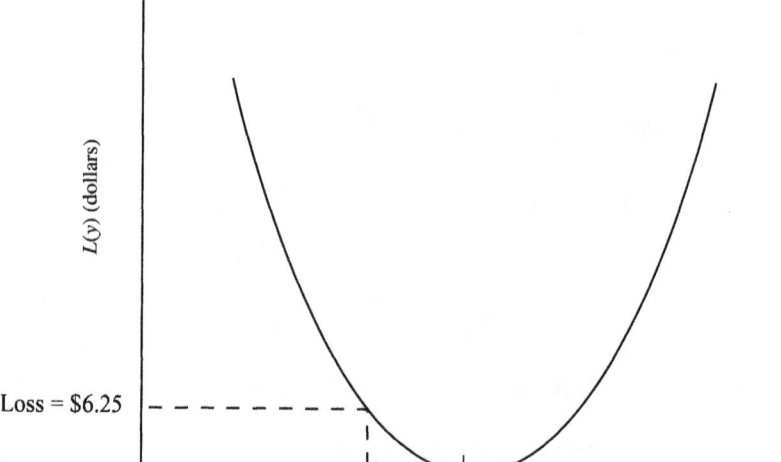

Suppose that the output voltage can be recalibrated to target at the end of the production line at a cost of $2. What is the manufacturing tolerance? Stated differently, at what output voltage should the manufacturer spend $2 to fix each set?

The manufacturing tolerance, Δ, is obtained by setting L at $2.

$$L(y) = 0.25(y - m)^2 \tag{9.5}$$

$$\$2 = \$0.25(y - 115)^2 \tag{9.6}$$

$$y = 115 \pm \sqrt{\frac{2.00}{0.25}}$$

$$= 115 \pm \sqrt{8}$$

$$= 115 \pm 2.83$$

$$\approx 115 \pm 3\ \text{V} \tag{9.7}$$

$$A = k\Delta^2$$

$$= \frac{A_0}{\Delta_0^2}\Delta^2 \tag{9.8}$$

or

$$\frac{A}{A_0}\Delta_0^2 = \Delta^2 \tag{9.9}$$

$$\Delta = \sqrt{\frac{A}{A_0}}\Delta_0 \tag{9.10}$$

As long as y is within 115 ± 3, the factory should not spend $2 for rework because the loss without the rework will be less than $2. The manufacturing tolerance sets the limits for shipping a product. It represents a break-even point between the manufacturer and the consumer. Either the customer or the manufacturer can spend $2 for quality (Figure 9.3).

Suppose that a circuit is shipped with an output voltage of 110 V without rework. The loss is

$$L(y) = \$0.25(110 - 115)^2 = \$6.25 \tag{9.11}$$

The factory saves $2 by not reworking, but it is imparting a loss of $6.25 to society. This loss becomes apparent to the manufacturer through customer dissatisfaction, added warranty costs, consumer's expenditure of time and money for repair, damaged reputation, and long-term loss of market share.

9.2. Safety Factor

As described before, a safety factor of 4 or 5 has often been used in industry. In most cases it is determined by considering the technological possibilities or the frequencies of actual problem occurrence. More specifically, special research was conducted when a high safety factor was needed in an especially demanding area

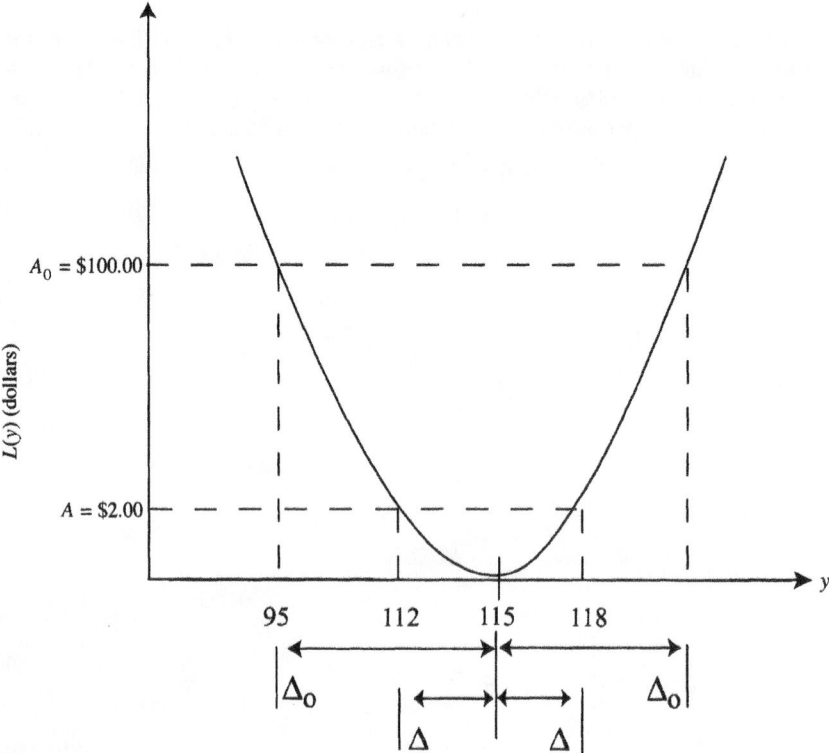

Figure 9.3
Tolerancing using the
loss function

(military, communication), and a higher level of safety was attained through technological advancement.

In advanced countries where such high-precision products can be produced, those new technologies were applied for general products, and tolerances for general products have become more and more stringent. In Japan, industrial standards (JIS) are reviewed every five years, and of course, tolerances are included in the list of reviews.

Again, let us denote the function limit as Δ_0. Δ_0 represents the point at which the function actually fails. It is determined as the point where the function fails due to the deviation of y by Δ_0 from m, assuming that other characteristics and operating conditions are in the standard (nominal value) state.

Determination of the function limit, assuming that the other conditions are in the standard state, means that the other conditions are in the normal state. The normal state means the median or mode condition, or where half of the conditions are more severe and the other half are less severe. It is abbreviated as LD_{50} (lethal dosage). In most cases, a test for determining the function limit may be conducted appropriately only under standard conditions.

The value of Δ_0 can easily be determined. For example, the LD_{50} value is determined by finding the voltage at which ignition fails by varying the voltage from the mean value of 20 kV of the ignition voltage of an engine. Suppose that the ignition fails at 8 kV when the voltage is lowered under the standard operating conditions; the lower function limit is 8 kV − 20 kV = −12 kV. On the other hand, the upper function limit is determined where problems begin to occur by increasing the output voltage. When the voltage is high, failures occur due to corona discharge, and so on. Let's assume that the function limit is 20 kV, for example. When function limits are different above and below the standard conditions, different tolerance values may sometimes be given. The tolerances in this case are as follows:

$$m \begin{matrix} -\Delta_1 \\ \\ +\Delta_2 \end{matrix} \qquad (9.12)$$

Tolerance values Δ_1 and Δ_2 are determined separately for above and below the upper and lower function limits, Δ_{01} and Δ_{02}, with separate safety factors, ϕ_1 and ϕ_2, respectively.

$$\Delta_i = \frac{\Delta_{0i}}{\phi_i} \qquad (i = 1,2) \qquad (9.13)$$

The function limit Δ_0 is often determined by experiments or calculations. However, the determination of a safety factor is based on experience and is unclear. Therefore, recently some people have begun to determine the tolerance by introducing the concept of probability. The author believes that the probability method is not a good approach. One of the reasons is that a safety factor of about 1000 with respect to LD_{50} is adopted in cases of side effects of drugs and harmful contents in food. How can the people who deal with probability theory determine the probability? It is likely that probability theory is useless in such cases. The form of the distribution is unknown, not to mention probability calculations.

The following equation was proposed for the safety factor ϕ, which is now beginning to be accepted:

$$\phi = \sqrt{\frac{\text{average loss when the function limit is exceeded}}{\text{loss at the factory when the factory standard is not met}}} \qquad (9.14)$$

A_0 and A are used for the numerator and denominator under the square root in equation (9.14), where A_0 is the average loss when the function limit is exceeded (after shipping) and A is the loss at the factory when the factory standard is not met (sale price or adjustment cost).

Therefore, equation (9.14) is written as

$$\phi = \sqrt{\frac{A_0}{A}} \qquad (9.15)$$

Let us explain this reason first. Let the characteristic value be y and the target value be m. The loss due to deviation of y from the target value m is considered here. The actual economic loss is $L(y)$ when a product with a characteristic value, y, is shipped and used under various conditions. N is the size of the total market where the product can potentially be used. $L_i(t,y)$ is the actual loss occurring at location i after t years.

$L_i(t,y)$ is equal to zero in most cases. If the function fails suddenly, an actual loss is incurred. The average loss $L(y)$ caused by using the product at all locations during the entire period of the design life is

$$L(y) = \frac{1}{N} \sum_{i=1}^{n} \int_0^T L_i(t,y) \, dt \qquad (9.16)$$

Although each $L_i(t,y)$ is a discontinuous function, $L(y)$ becomes approximately a continuous function when it is averaged for the losses of many individuals. As mentioned before, the loss function above is approximated by a quadratic equation and expressed by the following equations using the function limit Δ_0 and the loss A_0 caused by failure of the function.

$$L(y) = \frac{A_0}{\Delta_0^2} (y - m)^2 \qquad (9.17)$$

$(y - m)$ in equation (9.17) is the deviation from the target value. Assuming that the tolerance level of the deviation is the tolerance Δ and A is the value of the point where quality and cost balance in the tolerance design, equation (9.17) yields

$$A = \frac{A_0}{\Delta_0^2} \Delta^2 \qquad (9.18)$$

$$\Delta = \sqrt{\frac{A}{A_0}} \, \Delta_0 = \frac{\Delta_0}{\phi} \qquad (9.19)$$

$$\phi = \sqrt{\frac{A_0}{A}} \qquad (9.20)$$

$\phi = \sqrt{A_0/A}$ is called the *safety factor*.

☐ Example

On January 5, 1988, an illumination device weighing 1.6 tons fell in a disco-dancing hall in Tokyo. Three youths died and several were injured. The device was hung with six wires, but these wires could stretch easily, so when the chain used for driving the device broke, the device fell.

The design was such that the wires stretched and the device fell when the chain broke. The main function of the chain was to move the device up and down. Because of its design, when the chain broke and stopped functioning, human lives were threatened. The tensile strength of the chain had a safety factor of 4 with respect to the weight of the illuminating device, 1.6 tons. Two chains with a tensile strength of 3.2 tons had been used. A safety factor of 4 to 5 is adopted in many corporations in the United States and quite a few Japanese companies have adopted the same value.

Because the weight of the illuminating device was 1.6 tons, the function limit, Δ_0, of the chain was apparently 1.6 tons,. On average, several people are located in the area underneath the illuminating device at any time. Assuming that the loss caused by the death of one person is $1,550,000 and that four people died, the loss caused by losing the suspension function is

$$A_0 = (1,550,000)(4) = \$6,200,000 \qquad (9.21)$$

By applying a larger-the-better characteristic, the loss function for strength y of the chain is as follows:

$$L = \frac{A_0 \Delta_0^2}{y^2} = \frac{(\$6{,}200{,}000)(1.6^2)}{y^2} \tag{9.22}$$

When the safety factor is 4, $y = (1.6)(4) = 6.4$ tons,. This leads to a poor quality level, as follows:

$$L = \frac{(\$6{,}200{,}000)(1.6^2)}{6.4^2}$$

$$= \$387{,}500 \tag{9.23}$$

If the price of the chain were \$387,500, the price and quality would have been balanced. Actually, a chain costs about \$1000, or \$2000 for two. Therefore, the loss of quality that existed in each chain was about 190 times the price. If deterioration of the chain were considered, the loss of quality would be even larger. Therefore, the problem existed in selection of the safety factor.

In this case, the graver error was the lack of safety design. Safety design is not for improvement of the functioning rate or reliability. It is for reducing the loss to merely the repair cost in the case of malfunction. Some examples of safety design: At JR (Japan Railroad), design is done so that a train will stop whenever there is any malfunction in the operation of the train. At NTT, there is a backup circuit for each communication circuit, and when any type of trouble occurs, the main circuit is switched to the backup circuit; customers feel no inconvenience. Instead of allowing a large safety factor for prevention of a breakdown of the chain, the device in the disco-dancing hall should have had shorter wires so that it would have stopped above the heads of the people if the chain did break. In this case, the safety design could have been done at zero cost. If no human lives were lost, the loss caused by the breakdown of the chain would probably be only about \$10,000 for repair. The loss in this case would be

$$L = \frac{(\$10{,}000)(1.6^2)}{6.4^2}$$

$$= \$625 \tag{9.24}$$

This is less than the price of the chains, \$2000. Therefore, the quality of chains having 6.4 tons, strength is excessive, and even cheaper chains may be used.

Table 9.1 compares the loss with and without safety design. For example, when there is no safety design and one chain is used, the safety factor is equal to 2. Since this is the case of a larger-the-better characteristic, the equality is calculated as

$$L = \frac{A_0 \Delta_0^2}{y^2} = \frac{(\$6{,}200{,}000)(1.6^2)}{(1.6 \times 2)^2} = \$1{,}550{,}000 \tag{9.25}$$

Table 9.1

Loss function calculation of chain ($\times$\$10)[a]

Number of chains	Price	No Safety Design		With Safety Design	
		Quality	Total	Quality	Total
1	100	155,000	155,100	250	350
2	200	38,750	38,950	60	*260
3	300	17,220	17,520	30	330
4	400	9,690	10,090	20	420
6	500	4,310	4,910	7	607
8	800	2,420	3,220	4	804
10	1,000	1,550	2,550	3	1,003
12	1,200	1,080	2,280	2	1,202
14	1,400	790	*2,190		
15	1,500	610	2,200		

[a] Asterisks denote an optimum solution.

Since the cost of a chain is \$1000, the total loss is

$$1,550,000 + 1000 = \$1,551,000 \qquad (9.26)$$

When there is safety design using one chain, the loss, A_0, is equal to a repair cost of \$10,000. The quality is calculated as

$$A_0 \Delta_0^2 = (10,000)(1.6^2)$$

$$L = \frac{A_0 \Delta_0^2}{[(1.6)(2)]^2} = \frac{(10,000)(1.6^2)}{[(1.6)(2)]^2} = \$2500 \qquad (9.27)$$

Adding the cost of the chain, the total loss is

$$2500 + 1000 = \$3500 \qquad (9.28)$$

This calculates the sum of price and quality for cases with and without safety design corresponding to the number of chains (1,2,...). In actual practice, a factor 2 or 3 is often given to the purchase price, considering the ratio of purchase price to the sales price and the interest, because the price is the first expenditure for a consumer. Regarding the deterioration of quality, it is only necessary to think about the mode condition, or the most possible condition. Although it is usually better to consider the deterioration rate in the mode condition, it is ignored here because the influence of deterioration is small.

As we can see, the optimum solution for the case without a safety design is 14 chains and $14,000 for the price of the chains. For the case with a safety design, the solution is two chains and $2000 as the cost of the chains. The total loss that indicates productivity is $21,900 and $2600 for the optimum solutions, respectively. Therefore, the productivity of the design with a safety design included is 8.4 times higher.

9.3. Determination of Low-Rank Characteristics

Usually, the purchaser gives the specifications for parts and materials to the parts manufacturers or materials manufacturers. For finalproduct manufacturers, the planning section usually gives the standards of a product. Regarding these objective characteristics, parameter design and tolerance design consider how to determine the standard for the characteristics of the cause. Parameter design shows how to obtain the level of the parameter for the cause of low-rank characteristics. That is explained following Chapter 10, where the method of determining the central value is discussed. Now we discuss how to determine tolerance assuming that the central value of the low-rank characteristics has been determined.

When the shipping standards of characteristics of a product are given, the shipping standards are the high-rank characteristics with respect to the characteristics of subsystems and parts used in the product. When the characteristic values of parts and materials influence the characteristic values of a subsystem, the level of the characteristic values of the subsystem is high with respect to the characteristic values of the parts and materials. If the characteristic values of the product shipped from a company influence the characteristic values of the worked products of a purchaser, the characteristic values of the purchaser are high-rank characteristic values.

❒ Example

A stamped product is made from a steel sheet. If the shape after press forming is defective, an adjustment is needed at a cost, A_0, of $12. The specification for a pressed product dimension is $m \pm 300$ μm. The dimension of the product is affected by the hardness and thickness of the steel sheet. When the hardness of steel sheet changes by a unit quantity (in Rockwell hardness H_R), the dimension changes by 60 μm. When the thickness of the steel sheet changes by 1 μm, the dimension changes by 6 μm.

Let's determine the tolerance for the hardness and thickness of the steel sheet. When either the hardness or thickness fails to meet the specifications, the product is scrapped. The loss is $3 for each stamped product.

The method of determining the tolerance of the cause or the low-rank characteristics is to establish an equation for a loss function for the high-rank characteristics and to convert it to a formula for the characteristics of the cause. When the standard of the high-rank characteristic is $m_0 + \Delta_0$ and the loss for failing to meet the standard is A_0, the loss function is given by the following equation, where the high-rank characteristic is y.

$$L = \frac{A_0}{\Delta_0^2}(y - m_0)^2 \qquad (9.29)$$

For a low-rank characteristic x, the right-hand side of equation (9.29) becomes the following, due to the linear relationship between y and x, where b is the influence of a unit change of x to the high-rank characteristic y:

$$\frac{A_0}{\Delta_0^2}[b(x - m)]^2 \qquad (9.30)$$

where m is the central value of the low-rank characteristic x. Substituting this expression into the right-hand side of equation (9.29), and substituting the loss, A dollars, due to rejection by the low-rank characteristic for L on the left-hand side, we get

$$A = \frac{A_0}{\Delta_0^2}[b(x - m)]^2 \qquad (9.31)$$

By solving this equation for $\Delta = x - m$, the tolerance, Δ, for the low-rank characteristic, x, is given by

$$\Delta = \sqrt{\frac{A}{A_0}}\frac{\Delta_0}{b} \qquad (9.32)$$

The parameters are as follows:

A_0:   loss due to high-rank characteristics (objective characteristics) failing to meet the standard

Δ_0:   tolerance for the high-rank characteristics

A: loss due to the low-rank characteristics failing to meet the standard

b: the influence of a unit change in the low-rank characteristics to the high-rank characteristics

Both A_0 and A are losses for a product failing to meet the standard found by the purchaser or in the factory of the company. The cost for inspection is not included here. A_0 is the loss for adjustment in the case when the shipping standard is not satisfied in the company. In the case of high-rank characteristics of the purchaser, the loss occurs when the standard is not satisfactory to the purchaser.

In the case of hardness, A_0 is the loss when the purchaser finds a defective shape.

$$A_0: \quad \$12$$

$$A: \quad \$3$$

$$\Delta_0: \quad 300 \ \mu m$$

$$b: \quad 60 \ \mu m/H_R$$

Therefore, the tolerance for hardness is

$$\Delta = \sqrt{\frac{3}{12}} \left(\frac{300}{60}\right) = 2.5 H_R \tag{9.33}$$

Accordingly, the tolerance of hardness for shipping is $m \pm 2.5H_R$. Similarly, the tolerance, Δ, of the thickness is derived as follows, where A_0, A, and Δ_0 are the same as before and $b = 6 \ \mu m/\mu m$:

$$\Delta = \sqrt{\frac{3}{12}} \left(\frac{300}{6}\right) = 25.0 \ \mu m \tag{9.34}$$

Therefore, the tolerance of the thickness of the product shipped is $\pm 25.0 \ \mu m$. In this case, the average of the thickness is assumed to agree with the target value of the dimension after press forming. In this example, the purchaser or the press former is the one who determines the tolerance of the dimension.

In many cases, the relationship between the low- and high-rank characteristics can be approximated by a linear function. When this is not possible, a graphic approach is employed. For example, the influence of the change of a low-rank characteristic to the objective characteristic is found to be as shown in Figure 9.4. To study this relationship, it is important to keep other low-rank characteristics and environmental conditions constant.

The function limits, Δ_{10} and Δ_{20}, of the low-rank (cause) characteristic, x, are determined from the point where the high-rank (objective) characteristic, y, crosses the upper and lower limits. These are the points where the objective characteristic, y, exceeds the tolerance under the condition that all the other low-rank characteristics are in the standard state. When the upper and lower tolerance limits of x are to be determined separately, the lower tolerance limit, Δ_1, and the upper tolerance limit, Δ_2, are

$$\Delta_1 = \frac{\Delta_{10}}{\phi} \qquad \phi_1 = \sqrt{\frac{A_0}{A_1}} \tag{9.35}$$

$$\Delta = \frac{\Delta_{20}}{\phi} \qquad \phi_1 = \sqrt{\frac{A_0}{A_2}} \tag{9.36}$$

A_0 is the loss when the high-rank characteristic, y, exceeds the tolerance, A_1 is the loss when the low-rank characteristic, x, does not satisfy the lower limit for x

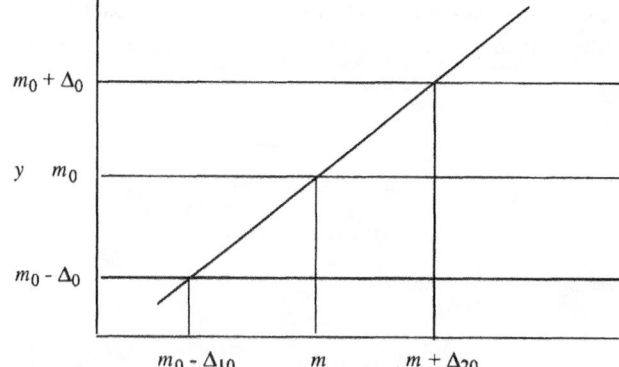

Figure 9.4
Relationship between
low- and high-rank
characteristics

(usually, parts and materials cannot be repaired; the loss is their price) A_2 is the loss when x exceeds the upper limit for x. ϕ_1 and ϕ_2 are safety factors. Because it is cumbersome to give different tolerances for above and below the standard, generally

$$\Delta_x = \min(\Delta_1, \Delta_2) \tag{9.37}$$

is used and the specification for x is given by $m \pm \Delta_x$, where Δ_x is the tolerance for x.

9.4. Distribution of Tolerance

Let's assume that there are k low-rank characteristics of parts and materials that influence the objective characteristics of assembled products. When the price of the ith part (more exactly, the loss when the i-th characteristic fails to meet the specification) is A_i $(i = 1, 2, ..., k)$ and the tolerance for each part is Δ_i, the variability range, Δ, of the objective characteristic, y, is given by equation (9.38). This is so because Δ_i can be calculated from equation (9.32) by assuming the additivity of variability.

$$\Delta^2 \frac{A_1 + A_2 + \cdots + A_k}{A_0} \Delta_0^2 \tag{9.38}$$

There are different relationships between A_0, which is the loss caused by the characteristics of an assembled product being unable to meet the tolerance, and the total sum of the price of component parts $(A_1 + \cdots + A_k)$.

$$\frac{A_1 + A_2 + \cdots + A_k}{A_0} \ll 1 \tag{9.39}$$

In the case of equation (9.39), if an assembled product fails to meet the standard, it must be discarded. Usually, the loss caused by rejection of an assembled product is several times the total of the price of the component parts. If it is four times, the process capability index of the characteristics of the assembled product will be twice that of the component parts. Of course, there is no rejection in such a case.

$$\frac{A_1 + A_2 + \cdots + A_k}{A_0} >> 1 \tag{9.40}$$

With equation (9.40), the assembled product can be fixed or adjusted at a low cost. There may be many rejections, but it is all right because the cost for adjustment is low.

$$\frac{A_1 + A_2 + \cdots + A_k}{A_0} \approx 1 \tag{9.41}$$

The case of equation (9.41) seldom occurs. The cost for adjustment is just equal to the total sum of the price of parts. Distribution of the tolerance is justified only in this case.

Generally speaking, it is wrong to distribute the tolerance to each characteristic of a part so that the characteristic of the assembled product falls within the tolerance limit around the target value. The best way is to determine tolerances separately for each part, as shown in this chapter. The most common situations are equations (9.39) and (9.40).

9.5. Deteriorating Characteristics

By definition, when the standard value of the high-rank characteristic fails to meet the standard at $m_0 \pm \Delta_0$, the loss is A_0 dollars.

The tolerance, Δ, of the initial value of the low-rank characteristic, x, which affects the high-rank characteristic, y, at a coefficient b for a change of unit value is given by the following equation, as was also shown in equation (9.32):

$$\Delta = \sqrt{\frac{A}{A_0}} \frac{\Delta_0}{|b|} \tag{9.42}$$

where A is the loss due to the low-rank (part) characteristic failing to meet the specification, A_0 the loss due to the high-rank characteristic failing to meet the specification, Δ_0 the tolerance for the high-rank characteristic, and $|b|$ is the change in the high-rank characteristic due to a unit change in the low-rank characteristic.

The loss function for deterioration per year is obtained by calculating the variance due to deterioration. The variance, δ^2, of the objective characteristic is obtained by considering the deterioration of the low-rank characteristic in T years:

$$\sigma^2 = \frac{1}{T} \int_0^T (b\beta t)^2 \, dt$$

$$= \frac{1}{T} \left(\frac{b^2 \beta^2 t^3}{3} \right)_0^t$$

$$= \frac{b^2 \beta^2 T^2}{3} \tag{9.43}$$

where β is the coefficient of deterioration. Therefore,

$$L = \frac{A_0}{\Delta_0^2} \frac{b^2 \beta^2 T^2}{3} \qquad (9.44)$$

If the tolerance for the deterioration in one year is Δ^*,

$$\Delta^* = \sqrt{\frac{3A^*}{A_0} \frac{\Delta_0}{bT}} \qquad (9.45)$$

by substituting Δ^* in β. A_0 and Δ_0 are the same as above, and A^* and T are defined as follows:

    A^*:   loss due to the low-rank characteristic failing to meet the standard

    T: design life (years)

☐ Example

A quality problem occurs when a luminance changes by 50 lx, and the social loss for repair is $150:

$$\Delta_0: \quad 50 \text{ lx}$$

$$A_0: \quad \$150$$

When the luminous intensity of a lamp changes by a unit amount of 1 cd in the manufacturing process, the illuminance changes by 0.8 lx. When the initial luminous intensity of the lamp fails to meet the specification, it can be adjusted by a cost of $A - \$3$. When the rejection is due to deterioration, the lamp is discarded, and the loss, A^*, is $32.

If the design life is 20,000 hours,

$$b: \quad 0.8 \text{ lx/cd}$$

$$A: \quad \$3$$

$$A^*: \quad \$32$$

$$T: \quad 20,000 \text{ hours}$$

Using equations (9.42) and (9.45), the tolerance Δ for the initial luminous intensity and the tolerance Δ^* for deterioration are given as follows:

$$\Delta = \sqrt{\frac{A}{A_0} \frac{\Delta_0}{|b|}} = \sqrt{\frac{3}{150}} \left(\frac{50}{0.8} \right) = 8.8 \text{ cd} \qquad (9.46)$$

$$\Delta^* = \sqrt{\frac{3A^*}{A_0} \frac{\Delta_0}{|\beta|} \frac{1}{T}}$$

$$= \sqrt{\frac{(3)(32)}{150}} \left(\frac{50}{0.8} \right) \left(\frac{1}{20,000} \right) = 0.00225 \text{ cd} \qquad (9.47)$$

Therefore, the tolerance for the initial luminous intensity of the lamp is +8.8 cd, and the tolerance for the coefficient of deterioration is less than 0.00225 cd/h. Even if the lamp has sufficient luminous intensity, the loss due to cleaning must be considered if the luminance is lost rapidly because of stains on the lamp.

In Chapter 10, we discuss tolerance design in more detail.

10 Tolerance Design

10.1. Introduction 208
10.2. Tolerance Design for Nominal-the-Best and Smaller-the-Better
Characteristics 208
10.3. Tolerance Design for Larger-the-Better Characteristics 210
10.4. Tolerance Design for Multiple Factors 212

10.1. Introduction

In most cases of parameter design, quality can be improved without cost increases. In tolerance design, quality is improved by upgrading raw materials or component parts that increase cost. In other words, tolerance design is used to compare the total loss caused by quality and cost.

10.2. Tolerance Design for Nominal-the-Best and Smaller-the-Better Characteristics

After the system and parameter designs have been completed, tolerance design is conducted to complete the process. At the product design stage, tolerance design must include noise factors associated with deterioration and environmental conditions.

☐ Example

Values of the linear thermal coefficient, b (percentage of expansion per 1°C), and wear, β, per year (percentage of wear per year) of three materials, A_1, A_2, and A_3, are as shown in Table 10.1. If the dimension changes by 6%, there will be a problem in the market, and the loss, A_0, in this case is \$180. Among A_1, A_2, and

Table 10.1
Material characteristics

Material	b (%)	β (%)	Price
A_1	0.08	0.15	$1.80
A_2	0.03	0.06	3.50
A_3	0.01	0.05	6.30

A_3, which is the best material? The standard deviation, σ_x, of the temperature condition x at which the material is used is 15°C, and the design life is 20 years.

Because the dimension of the product at the time of shipment is equal to the target value, m, at the standard temperature, the variance, σ^2, of the sum of variability due to temperature and variability due to deterioration is given by

$$\sigma^2 = b^2\sigma_x^2 + \frac{T^2}{3}\beta^2 \tag{10.1}$$

The second term of equation (10.1) is determined because the variance is given by the following equation, where the deterioration per year is β and the design life is T years:

$$\sigma^2 = -\frac{1}{T}\int_0^T (\beta t)^2\, dt = \frac{\beta^2}{3}T^2 \tag{10.2}$$

By substituting $\sigma_x = 15°C$ and $T = 20$ years in equation (10.1), the variances, σ^2, of materials A_1, A_2, and A_3 are

$$A_1: \quad \sigma^2 = (0.08^2)(15^2) + \frac{20^2}{3}(0.15^2)$$

$$= 4.440 \tag{10.3}$$

$$A_2: \quad \sigma^2 = (0.03^2)(15^2) + \frac{20^2}{3}(0.06^2)$$

$$= 0.6825 \tag{10.4}$$

$$A_3: \quad \sigma^2 = (0.01^2)(15^2) + \frac{20^2}{3}(0.05^2)$$

$$= 0.3558 \tag{10.5}$$

Table 10.2 was obtained in this way, and the total loss is the sum of price and quality. The optimum solution is material A_2.

The quality level L is obtained by the equation

$$L = \frac{A_0}{\Delta_0^2}\sigma^2 = \frac{180}{6^2}\sigma^2 \tag{10.6}$$

Table 10.2
Tolerance design

Material	b (%)	β (%)	Price	σ^2	Quality Level	Total Loss
A_1	0.08	0.15	$1.80	4.44	$22.20	$24.00
A_2	0.03	0.06	3.50	0.6825	$3.41	$6.91
A_3	0.01	0.05	6.30	0.3558	$1.78	$8.08

Because materials and parts cannot be adjusted later, the optimum solution is approximately at the point where the quality level and the price balance. The optimum solution, A_2, is used to determine equation (10.7), or more generally, the safety factor. It is important to balance quality and cost in advance for rational determination of the tolerance.

Minimizing the sum of production cost and quality is tolerance design for selection of production methods and production tools. It is important to minimize the sum of the quality evaluation level, Q, and the product cost, P. The values of loss, A, due to failure are the cost of materials, parts, and products after balancing quality and cost in the tolerance design. In this case, the safety factor, ϕ, is given by

$$\phi = \sqrt{\frac{A_0}{A}} \tag{10.7}$$

10.3. Tolerance Design for Larger-the-Better Characteristics

The loss function of larger-the-better characteristics is given by

$$L = \frac{A_0 \Delta_0^2}{y^2} = A_0 \Delta_0^2 \sigma^2 \tag{10.8}$$

where σ^2 is the square average of the inverse of the larger-the-better characteristic values.

☐ Example

Let's assume that both the strength and price of a pipe are proportional to the cross section of the pipe. A resin pipe is broken at a stress of 5000 kg, and the loss in this case is $300,000. The strength of a unit area b is set at 80 kg/mm² and the price per unit area a is set at \$40 (per/mm²). For a cross-sectional area of x mm², the sum L_T of the price and the quality is given as

$$L_T = ax + A_0 \Delta_0^2 \frac{1}{(bx)^2} \qquad (10.9)$$

Setting the differential of L_T with x equal to zero and solving for x yields the x that minimizes this value:

$$x = \left(\frac{2A_0 \Delta_0^2}{ab^2} \right)^{1/3} \qquad (10.10)$$

By substituting $a = \$40$, $b = 80$, $A_0 = \$300,000$, and $\Delta_0 = 5000$, we obtain

$$x = \left[\frac{(2)(300,000)(5000^2)}{(40)(80^2)} \right]^{1/3}$$

$$= 388 \text{ mm}^2 \qquad (10.11)$$

Therefore, the price is

$$(388)(40) \approx \$15,520 \qquad (10.12)$$

and the quality level is

$$(300,000) \cdot (5000^2) \left(\frac{1}{(80)(388)^2} \right)$$

$$\approx \$7800 \qquad (10.13)$$

The total loss is

$$L = 15,520 + 7800 = \$23,320 \qquad (10.14)$$

Because the strength deteriorates in an actual situation, the loss is determined using the following variability σ^2:

$$\sigma^2 = \frac{1}{(bx)^2} \frac{1}{T} \int_0^T e^{2\beta t} \, dt$$

$$= \frac{1}{(bx)^2} \frac{1}{2\beta T} (e^{2\beta T} - 1) \qquad (10.15)$$

where T is the design life and β is the deterioration in one year. That is not calculated here. If the characteristic values fluctuate when tested under various noise conditions, the average of the square of the inverse of these values is taken as the target value of larger-the-better characteristics.

The calculation above was made from one type of resin. The same calculation can be made for other types, and the one that gives the minimum total loss will be selected.

10.4. Tolerance Design for Multiple Factors

In the preceding sections we described the tolerance design for one factor, such as material type (A_1, A_2, and A_3) or a certain resin pipe. In general, many factors can be used simultaneously to conduct tolerance design in any one case.

In parameter design, optimum control factor levels are determined to assure that the system output becomes least sensitive to noise factors. This means that even if wider tolerances are used around these factor levels, the system output will still yield minimal variability. Once this testing is done, the proper tolerances around these factor levels need to be fine-tuned further, since some tolerances may further reduce output variability. Tolerances that do not greatly affect variability should remain the same.

Determining which tolerances to tighten in tolerance design should be a conscious trade-off between quality improvement (variability reduction) and the cost of upgrading. Use analysis of variance to estimate the quality improvement as the tolerance of a factor is tightened, and use the quality loss function to translate the quality improvement into monetary units. Then compare the improvement to the cost of upgrading to decide whether the tolerance should be tightened to "add value." If the improvement is greater than the cost increase, use the tighter tolerance; otherwise, allow the tolerance to remain as is.

The primary steps in tolerance design are:

1. Conduct an experiment using existing tolerances.

2. Perform an analysis of variance (ANOVA) on the experimental data, and obtain the current total variance and the percent contribution from each factor.

3. Establish the loss function for the system output response, and calculate the current total loss.

4. For each factor, calculate the existing monitory loss using the loss function.

5. For each factor, calculate the new monitory loss and the quality improvement (in loss) using the upgraded tolerance; then compare to the cost increase to decide whether or not the upgraded tolerance should be used.

6. For factors to be upgraded, calculate the total quality improvement (in loss) and the total upgrade cost to determine the total net gain.

□ Example

In the design of an engine control circuit, the output response is y = number of *on* signals per minute. The target is 600, and the specification is ± 60. If the output is out of specification, the average repair cost is \$2.50. The optimum nominal values of the factors obtained through parameter design are shown in Table 10.3 with the existing tolerances. The tolerances of resistors, transistors, and condensers can be upgraded through the respective cost increases shown in Table 10.4.

The steps for tolerance design are implemented as follows.

Table 10.3

Factors, nominal values, and tolerances

Tolerance Factor		Nominal Value	σ
P	Resistor R_1	2200	5%
Q	Resistor R_2	470	5%
R	Resistor R_3	100k	5%
S	Resistor R_4	10k	5%
T	Resistor R_5	1500	5%
U	Resistor R_6	10k	5%
V	Transistor T_1	180 (h_{FE})	50
W	Transistor T_2	180 (h_{FE})	50
X	Condenser C_1	0.68	20%
Y	Condenser C_2	10.00	20%
Z	Voltage	6.6 V	0.3

Step 1. Conduct an experiment using the existing tolerances. Based on the nominal values and existing tolerances, the factor levels for the experiment are set as shown in Table 10.5. In two-level tolerance design experiments, the factor levels are established as

$$\text{level } 1 = \text{nominal} - \sigma$$

$$\text{level } 2 = \text{nominal} + \sigma \qquad (10.16)$$

Table 10.4

Upgrade cost and tolerance

Component	Grade	Cost	σ
Resistor	Low	Base	5%
	High	$2.75	1%
Transistor	Low	Base	50
	High	$2.75	25
Condenser	Low	Base	20%
	High	$5.50	5%

Table 10.5
Factor levels

Tolerance Factor		Level 1	Level 2
P	Resistor R_1	2090	2310
Q	Resistor R_2	446.5	493.5
R	Resistor R_3	95k	10.5k
S	Resistor R_4	9.5k	10.5k
T	Resistor R_5	1425	1575
U	Resistor R_6	9.5k	10.5k
V	Transistor T_1	130	230
W	Transistor T_2	130	230
X	Condenser C_1	0.544	0.816
Y	Condenser C_2	8.00	12.00
Z	Voltage	6.3	6.9

Table 10.6
Experimental layout

L_{12}	P 1	Q 2	R 3	S 4	T 5	U 6	V 7	W 8	X 9	Y 10	Z 11	No. of *on* Signals min
1	1	1	1	1	1	1	1	1	1	1	1	588
2	1	1	1	1	1	2	2	2	2	2	2	530
3	1	1	2	2	2	1	1	1	2	2	2	597
4	1	2	1	2	2	1	2	2	1	1	2	637
5	1	2	2	1	2	2	1	2	1	2	1	613
6	1	2	2	2	1	2	2	1	2	1	1	630
7	2	1	2	2	1	1	2	2	1	2	1	584
8	2	1	2	1	2	2	2	1	1	1	2	617
9	2	1	1	2	2	2	1	2	2	1	1	601
10	2	2	2	1	1	1	1	2	2	1	2	621
11	2	2	1	2	1	2	1	1	1	2	2	579
12	2	2	1	1	2	1	2	1	2	2	1	624

where as in three-level experiments, the factor levels are established as

$$\text{level } 1 = \text{nominal} - \sqrt{3/2}\,\sigma$$

$$\text{level } 2 = \text{nominal} \tag{10.17}$$

$$\text{level } 3 = \text{nominal} + \sqrt{3/2}\,\sigma$$

Use an L_{12} orthogonal array for the experiment to obtain the data, as shown in Table 10.6.

Step 2. Perform the analysis of variance on the experimental data, and obtain the current total variance and the percent contribution from each factor. Use the 12 data points from the experiment for the ANOVA calculations. First calculate the total sum of squares and the total variance. Then, for each factor, calculate the sum of squares, variance, pure sum of squares, and percent contribution. Start by calculating, the correction factor, CF:

$$CF = \frac{(\text{sum of all data points})^2}{\text{total number of data points}}$$

$$= \frac{(588 + 530 + \cdots + 624)^2}{12}$$

$$= 4{,}345{,}236.75 \tag{10.18}$$

The total sum of squares, S_T, is

$$S_T = y_1^2 + y_2^2 + \cdots + y_{12}^2 - CF$$

$$= 588^2 + 530^2 \cdots + 624^2 - 4{,}345{,}236.75$$

$$= 4{,}354{,}695 - 4{,}345{,}236.75$$

$$= 9458.25 \tag{10.19}$$

Calculate the current total variance, V_T, as follows:

$$V_T = \frac{S_t}{f_T} \tag{10.20}$$

where f_T shows total degrees of freedom; in this case,

$$f_T = (\text{number of experiments}) - 1$$

$$= 12 - 1$$

$$= 11 \tag{10.21}$$

Therefore,

$$V_T = \frac{9458.25}{11}$$

$$= 859.84 \tag{10.22}$$

Calculate the level sums of all the factors as shown in Table 10.7. Calculate the sum of squares for factor P:

$$S_P = P_1^2 + \frac{P_2^2}{n} - CF$$

$$= 3595.00^2 + \frac{3626.00^2}{6} - 4,345,236.75$$

$$= 80.08 \tag{10.23}$$

where n is the number of data points per each factor level, in this case, 6. The degrees of freedom of factor P are:

$$f_P = (\text{number of levels}) - 1$$

$$= 2 - 1$$

$$= 1 \tag{10.24}$$

The variance of factor P, V_P, is

$$V_P = \frac{S_P}{f_P}$$

$$= \frac{80.08}{1}$$

$$= 80.08 \tag{10.25}$$

Perform similar calculations for the other factors. The results are shown in the ANOVA table, Table 10.8.

Table 10.7
Response table: Level sums

Factor	Level 1	Level 2	Factor	Level 1	Level 2
P	3595.00	3626.00	V	3599.00	3622.00
Q	3517.00	3704.00	W	3635.00	3586.00
R	3559.00	3662.00	X	3618.00	3603.00
S	3593.00	3628.00	Y	3694.00	3527.00
T	3532.00	3689.00	Z	3640.00	3581.00
U	3651.00	3570.00			

Table 10.8
ANOVA table

Source	f	S	V
P	1	80.08	80.08
Q	1	2914.08	2914.08
R	1	884.08	884.08
S	1	102.08	102.08
T	1	2054.08	2054.08
U	1	546.75	546.75
V	1	44.08	44.08
W	1	200.08	200.08
X	1	18.75	18.75
Y	1	2324.08	2324.08
Z	1	290.08	290.08
Total	11	9458.25	859.84

Since the effects (sum of squares) of factors P, V, and X are relatively small, "pool" them together to represent $S_{(e)}$, the pooled error sum of squares:

$$S_{(e)} = S_P + S_V + S_x$$

$$= 80.08 + 44.08 + 18.75$$

$$= 142.91 \tag{10.26}$$

The pooled degrees of freedom, $f_{(e)}$, are

$$f_{(e)} = f_P + f_V + f_x$$

$$= 1 + 1 + 1$$

$$= 3 \tag{10.27}$$

Thus, the pooled error variance, $V_{(e)}$, is

$$V_{(e)} = \frac{S_{(e)}}{f_{(e)}}$$

$$= \frac{142.91}{3}$$

$$= 47.64 \tag{10.28}$$

Next, calculate the pure sum of squares, S', and the percent contribution, $\rho(\%)$, for each factor. For example:

$$S'_Q = S_Q - V_{(e)} f_Q$$

$$= 2914.08 - (47.64)(1)$$

$$= 2866.44 \qquad (10.29)$$

$$\rho(\%)_Q = \frac{S'_Q}{S_T}$$

$$= \frac{2866.44}{9458.25}$$

$$= 30.31 \qquad (10.30)$$

The rearranged ANOVA table is shown in Table 10.9.

Step 3. Establish the loss function for the system output response, and calculate the current total loss. For the output response, y = (number of *on* signals)/minute, the specification Δ_0 is 60, and the average repair cost A_0 is \$250. Calculate the proportional constant k in the loss function as

$$k = \frac{A_0}{\Delta_0^2}$$

$$= \frac{250}{60^2}$$

$$= 0.0694 \qquad (10.31)$$

The loss function can now be established as

$$L = 0.0694 V_T \qquad (10.32)$$

The current total loss (Table 10.10) is

$$L_{T(current)} = (0.0694)(859.84)$$

$$= \$59.67 \text{ per circuit} \qquad (10.33)$$

Step 4. For each factor, calculate the existing monitory loss using the loss function. Calculate the loss due to each factor, for example, Q:

$$L_{Q(current)} = L_{T(current)} \rho(\%)_Q$$

$$= (59.67)(30.31\%)$$

$$= \$18.09 \text{ per circuit} \qquad (10.34)$$

Make similar calculations for the other factors.

Step 5. For each factor, calculate the new monitory loss, the quality improvement (in loss) using the upgraded tolerance, then compare to the cost increase to decide if the upgraded tolerance should be used. Recall the tolerances and cost increases shown in Table 10.4. If we consider upgrading the tolerance of factor Q,

Table 10.9
Rearranged ANOVA table

Source	f	S	V	S'	ρ (%)
P	1	80.08	80.08		
Q	1	2914.08	2914.08	2866.4	30.31
R	1	884.08	884.08	836.44	8.84
S	1	102.08	102.08	54.44	0.58
T	1	2054.08	2054.08	2006.44	21.21
U	1	546.75	546.75	499.11	5.28
V	1	44.08	44.08		
W	1	200.08	200.08	152.44	1.61
X	1	18.75	18.75		
Y	1	2324.08	2324.08	2276.44	24.07
Z	1	290.08	290.08	242.44	2.56
(e)	3	142.91	47.64	524.03	
Total	11	9458.25	859.84		100.00

the improvement ratio is $(1\%/5\%)^2$. Then the new loss due to factor Q after upgrading is

$$L_{Q(new)} = L_{Q(current)}\left(\frac{1\%}{5\%}\right)^2$$

$$= (18.09)(0.04)$$

$$= \$0.72 \text{ per circuit} \qquad (10.35)$$

Table 10.10
Current loss

Factor	ρ (%)	$L_{current}$
Q	30.31	18.09
R	8.84	5.27
S	0.58	0.35
T	21.21	12.66
U	5.28	3.15
W	1.61	0.96
Y	24.07	14.36

Therefore,

$$\text{quality improvement} = L_{Q(current)} - L_{Q(new)}$$

$$= 018.09 - 0.72$$

$$= \$17.37 \text{ per circuit} \qquad (10.36)$$

The upgrade cost for factor Q is $2.75. Thus, the

$$\text{net gain} = (\text{quality improvement}) - (\text{cost increase})$$

$$= 17.37 - 2.75$$

$$= \$14.62 \text{ per circuit} \qquad (10.37)$$

In this case, it does pay off to use the upgraded tolerance for factor Q. Similar calculations and justification are done for the other factors as shown in Table 10.11.

Step 6. *For factors to be upgraded, calculate the total quality improvement (in loss), the total upgrade cost, and therefore the total new gain.* Since factors Q, R, T, U, and Y are to be upgraded,

$$\text{total quality improvement} = 17.37 + 5.06 + 12.15 + 3.02 + 13.46$$

$$= \$51.06 \text{ per circuit} \qquad (10.38)$$

The total upgrade cost for these factors is

$$\text{total upgrade cost} = 2.75 + 2.75 + 2.75 + 2.75 + 5.50$$

$$= \$16.50 \text{ per circuit} \qquad (10.39)$$

Therefore,

$$\text{total net gain} = (\text{total quality improvement}) - (\text{total upgrade cost})$$

$$= 51.06 - 16.50$$

$$= \$34.56 \text{ per circuit} \qquad (10.40)$$

Table 10.11
Upgrading decision making

Factor	$L_{current}$	L_{new}	Quality Improvement	Upgrade Cost	Net Gain	Upgrade Decision
Q	18.09	0.72	$17.37	$2.75	$14.62	Yes
R	5.27	0.21	5.06	2.75	2.31	Yes
S	0.35	0.07	0.34	2.75	−2.41	No
T	12.66	0.51	12.15	2.75	9.40	Yes
U	3.15	0.13	3.02	2.75	0.27	Yes
W	0.96	0.24	0.72	2.75	−2.03	No
Y	14.36	0.90	13.46	5.50	7.96	Yes

Part IV

Signal-to-Noise Ratio

11 Introduction to the Signal-to-Noise Ratio

11.1. **Definition**   224
11.2. **Traditional Approach for Variability**   224
11.3. **Elements of the SN Ratio**   224
11.4. **Benefits of Using SN Ratios**   225
   Direct Relationship with Economy   225
   Simplification of Interaction Calculations   227
   Efficient Robust Design Achievement   227
   Efficient Adjustment or Calibration   227
   Efficient Evaluation of Measuring Systems   228
   Reduction of Product/Process Development Cycle Time   228
   Efficient Research for a Group of Products: Robust Technology Development   228
11.5. **Various Ideal Functions**   228
   Machining Technology   229
   Injection Molding Process   229
   Fuel Delivery System   229
   Engine Idle Quality   230
   Wiper System   230
   Welding   230
   Chemical Reactions   230
   Grinding Process   232
   Noise Reduction for an Intercooler   232
   Wave Soldering   233
   Development of an Exchange-Coupled Direct-Overwrite Magnetooptical Disk   233
   Equalizer Design   233
   Fabrication of a Transparent Conducting Thin Film   233
   Low-Pass Filter Design   233
   Development of a Power MOSFET   234
   Design of a Voltage-Controlled Oscillator   234
   Optimization of an Electrical Encapsulant   234
   Ink Formulation   234
11.6. **Classification of SN Ratios**   234
   Continuous Variables and Classified Attributes   234
   Classification Based on Intention   236

Fixed and Multiple Targets   236
Classification Based on Input and Output   236
Other Ways of Classifying SN Ratios   237
References   237

11.1. Definition

The signal-to-noise ratio (SN ratio) is a measurement scale that has been used in the communication industry for nearly a century. A radio measures the signal or the wave of voice transmitted from a broadcasting station and converts the wave into sound. The larger the voice sent, the larger the voice received. In this case, the *magnitude of the voice* is the input signal, and the *voice received* is the output. Actually, the input is mixed with the audible noise in the space. Good measuring equipment catches the signal and is not affected by the influence of noise.

The quality of a measurement system is expressed by the ratio of signal and noise. SN ratio is expressed in decibels (dB). For example, 40 dB means that the magnitude of the output is 10,000 times the magnitude of noise. For example, if the SN ratio of a transistor radio is 45 dB, the power of the input is about 30,000 times the magnitude of noise. The larger the SN ratio, the better the quality.

Taguchi has generalized the concept of SN ratio as used in the communication industry and applied it for the evaluation of measurement systems as well as for the function of products and processes.

11.2. Traditional Approach for Variability

Traditionally, the analysis of data for variability has been traditionally studied by decomposing data into deviation and variation. Such an approach is applicable only when one object is to be measured. Using a watch as an example, say that we want to measure the variability after 24 hours. But that is not the best way to measure. It is important to determine the error within a certain range. We would want to measure the variation of the watch from actual time at any time. This same principle is true for a scale, which has a smaller error when objects of different weight are measured. The quality of any measurement system must be evaluated using the dynamic characteristics of the items measured.

In the automobile industry, the quality of measurement systems is traditionally evaluated by studying repeatability, reproducibility, and stability. *Repeatability* is the variability when the *same person* measures the *same sample* repeatedly. *Reproducibility* is the variability when *different persons* measure the *same sample*. *Stability* is the variability when the *same person* measures the same sample repeatedly at *different times*. But those three types of variability belong to one category: *noise*. Using quality engineering, there is no necessity to study these three types of variability separately. It is only necessary to conduct a study including signal.

11.3. Elements of the SN Ratio

Conceptually, the SN ratio is the ratio of signal to noise in terms of power. Another way to look at it is that it represents the ratio of sensitivity to variability.

In a measurement system, the input-to-output relationship is studied. The true value of the object is the input, and the result of measurement is the output. In a good measuring system, the result of measurement must be proportional to the true value; thus the input/output relationship must be linear. A good measurement system must also be sensitive to different inputs (whatever different objects are measured); thus, the slope showing the input/output relationship must be steep. Of course, the variability must be small. When an SN ratio is used to evaluate a measuring system, the three elements sensitivity, slope, and variability are combined into a single index. As a result, engineers can easily evaluate and improve a system by using SN ratios. Figure 11.1 shows the three elements in the SN ratio.

There are two modes in the SN ratio: dynamic and nondynamic. The measurement system described above belongs to the *dynamic* mode. In the design of a product for a fixed target, there is no need to adjust the target from time to time. For example, we may want to design an electric circuit having an output of 110 V. Once the product is designed and put into production, it is not necessary to change the output voltage as long as we want to manufacture and sell this particular product only. In a nondynamic mode, therefore, mean is considered as being equivalent to sensitivity and is placed in the numerator of the SN ratio. In this case, the SN ratio is the ratio of average to variability in square terms. But in a dynamic system, such as a control system, there is always a need to adjust the output to the target by varying a certain input signal. In such a case, *adjustability* becomes critical for the design. Here it is important that the input/output relationship be proportional, or linear. In other words, linearity becomes critical for adjusting systems. The better the linearity and the steeper the input/output relationship, the better the adjustability. Therefore, the slope is used as being equivalent to sensitivity and is placed in the numerator of the SN ratio.

The concept of the dynamic-type SN ratio evolved further in the area of technology development. The activity is defined by the development of technology for a group of products instead of developing a particular product with a fixed target, so as to avoid redundant research and thereby reduce research and development cycle time. (For technology development, see Chapter 14.)

11.4. Benefits of Using SN Ratios

There are many benefits to using SN ratios in quality engineering, as discussed below.

The SN ratio is defined as follows: **Direct Relationship
                                                                                          with Economy**

$$\text{SN ratio} = \frac{\text{power of signal}}{\text{power of noise}} = \frac{(\text{sensitivity})^2}{(\text{variability})^2} = \frac{\beta^2}{\sigma^2} \qquad (11.1)$$

As seen from equation (11.1), the SN ratio is the ratio of sensitivity to variability squared. Therefore, its inverse is the variance per unit input. In the loss function, the loss is proportional to variance. Therefore, monetary evaluation is possible.

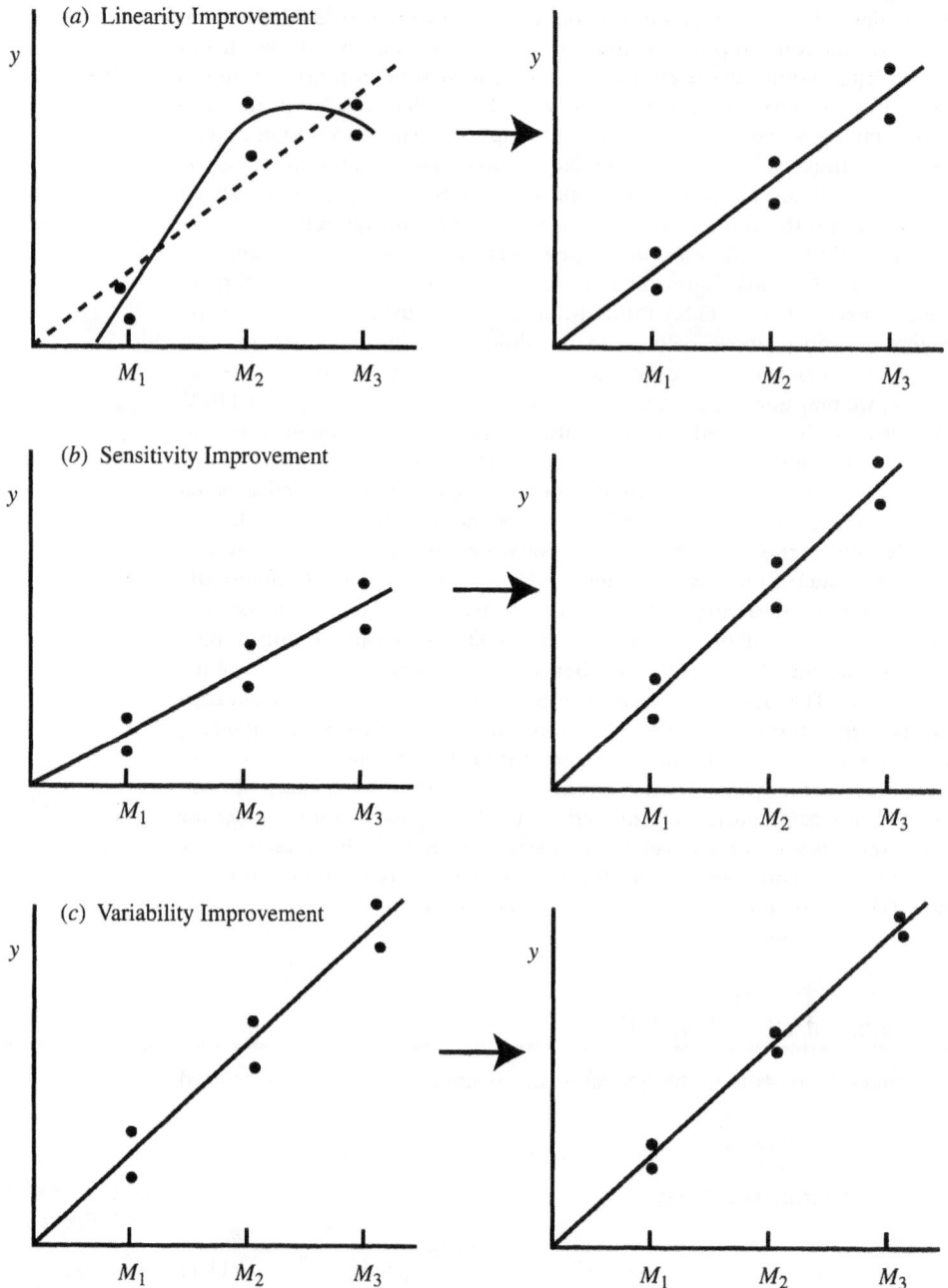

Figure 11.1
Three elements of SN ratio

In the early days of quality engineering applications, research for robustness was conducted by discovering the interactions between control and noise factors. When there is a significant interaction between a control and a noise factor, it shows that different levels of a control factor behave differently under different noise factors. If this is true, there is a possibility that robustness can be improved. Such a study method was called *direct product design,* where the experiment was laid out in such a way that every interaction between a control and a noise factor may be calculated. But such calculations were so tedious that many people had to work full-time on them. By using the SN ratio, such calculations are no longer necessary. Instead, a comparison of SN ratios between control factor levels is made, and the one with a higher SN ratio is selected as the robust condition.

Simplification of Interaction Calculations

This is one of the most important benefits of using the SN ratio in product or process design. Using traditional methodology, engineers tend to design a product or a process by meeting the target first instead of by maximizing robustness first. This "hitting target first" approach is very inefficient, for the following reasons. After the target is met, the engineer has to make sure that the product works under noise conditions: for example, temperature.

Efficient Robust Design Achievement

If the product does not work at a certain extreme temperature, the engineer has to change or adjust the design parameters to meet the target. But while studying the effects of temperature, the engineer needs to keep other noise factors, such as humidity, fixed. If the product does not work in a certain humidity environment, the engineer tries to vary some parameter levels and change parameter levels again to meet the target. In the first trial, where the aim was to meet the target at a certain temperature, the control factor levels selected would probably be different from the control factor levels selected in the second trial, where the aim was to meet the target at a certain humidity condition.

To meet the target under a certain temperature as well as under a certain humidity condition, the study must be started again from the beginning. The engineer has to do the same for any other noise conditions that ought to be considered. Generally, there are a lot of noise factors. Not only is this type of approach tedious and time consuming, it is not always successful. Designing in this fashion is similar to solving many simultaneous equations, not by calculation but by trial-and-error experimentation on hardware.

In contrast, use of the SN ratio enables the engineer to maximize robustness simply by selecting the levels of control factors that give the largest SN ratio, where many noise factors are usually compounded into one. After robustness is achieved, the next step is to adjust the mean to the target by adjusting the level of a control factor, the factor having a maximum effect on sensitivity and a minimum effect on robustness. In this way it is possible that meeting the target will have scant effect on robustness. Very often, the efficiency of research can be improved by a factor of 10, 100, or 1000.

Linearity, one of the three elements in an SN ratio discussed earlier, is very important for simplifying adjustment in product or process design and for calibration in measurement systems. Because the relationship between the value of an SN ratio and linearity is so simple and clear, adjusting the output of control systems and calibration of measurement systems is easier than when traditional methods are used. When the input/output relationship is not linear, deviation from linearity

Efficient Adjustment or Calibration

is evaluated as the error after the decomposition of variation. Therefore, the SN ratio becomes smaller.

Efficient Evaluation of Measuring Systems

In a measuring system, there are two types of calibration: zero-point calibration and slope calibration. *Zero-point calibration* is used to calibrate the output to zero when the input is equal to zero (when nothing is measured). *Slope calibration* is used to calibrate the input/output ratio (slope) to 1. Since the slope of the input/output becomes 1, the inverse of the SN ratio is equal to the variance of the measuring system. *Such estimation is possible without conducting physical calibration.* This greatly improves the efficiency of measurement system optimization.

Reduction of Product/Process Development Cycle Time

An SN ratio is written and calculated based on the ideal function of a product or process. When the input/output relationship of the ideal function is related to energy, the effects of control factors are cumulative (additive).

While interaction between a control factor and a noise factor enables us to find a robust condition, interaction between control factors shows the inconsistency of conclusions. If the conclusions are inconsistent, the conclusion obtained from a study might be different when control factors change. How to deal with or avoid interactions between control factors is one of the most important issues in quality engineering.

Various case studies in the past have shown that using SN ratios based on the ideal function that relates to input/output energy transformation enables us to avoid interactions between control factors. This means that the conclusions obtained in a small-scale laboratory study can be reproduced downstream in a large-scale manufacturing process and in the marketplace. That is why using dynamic SN ratios can reduce product/process development cycle time.

Efficient Research for a Group of Products: Robust Technology Development

Traditionally, research is conducted every time a new product is planned. But for a group of products having the same function within a certain output range, it would be wasteful to conduct research for each product. Instead, if we could establish an appropriate SN ratio based on the ideal function of a group of products and maximize it, the design of a specific product with a certain output target would be obtained easily by adjusting the input signal. This approach, called *robust technology development*, is a breakthrough in quality engineering. It is believed that such an approach will soon become the core of quality engineering.

11.5. Various Ideal Functions

The traditional problem-solving approach generally brings about quality improvement. In the problem-solving approach, root-cause analysis is important, and the engineer analyzes whatever data are measured (such as symptoms or defect rates, called *responses*). There is no consideration of what type of data should be avoided and what type of data must be analyzed. In the Taguchi Methods Symposium held by the American Supplier Institute in 1989, the theme was: "To get quality, don't measure quality." This meant that to improve quality, do not measure and analyze "response," because the conclusions obtained from such analyses are not reproducible in most cases.

Instead, we should try to think about the *ideal function* and use for analysis the dynamic SN ratio based on that ideal function. It is not an easy task to determine the ideal function, and it takes a lot of discussion. The ideal function is different from product to product, from one system to another, and there are no generalizable approaches or solutions. Ideal functions must be discussed case by case.

Generally, it is desirable that the ideal function be identified based on the product function, which is energy related, since energy has additivity. This type of ideal function is called a *generic function*. A generic function can be explained in physics, for example, by Ohm's law or Hooke's law. In some cases, however, the data to be collected for the study of such relationships are not available, due to lack of measurement technology. In such cases, a function that describes an objective may be used. This is called the *objective function*. Many published cases are available for engineers to use as a reference to establish an ideal function. Following are some examples.

Machining Technology

In a machining process, the dimensions of an actual product are traditionally measured for analysis. To reduce error and to optimize the process to cover a range of the products the machine can process, it is recommended that test pieces be used for study.

In a study entitled "NC Machining Technology Development [1], the concept of *transformability* was developed. The ideal function in this case was that product dimension be proportional to programmed dimension:

$$y = \beta M,$$

where y is the product dimension, β a constant, and M the programmed dimension. The function of machining was studied using different ideal functions [2]. There were two functions in the study: function 1, $y = \beta M$, where y is the square root of total power consumption and M is the square root of time (per unit product); and function 2, $y = \beta M$, where y is the square root of total power consumption and M is the square root of total amount removed. The objective of the first function is to improve productivity, and the second is to improve machining quality.

Injection Molding Process

In the injection molding process, quality characteristics such as reject rate, flash, or porosity have traditionally been measured. In a study of a carbon fiber—reinforced plastic injection molding process [3], the following function was studied: $y = \beta M$, where y is the product dimension and M is the mold dimension. In a study of an injection molding experiment [4], the function above was used together with another function, $y = \beta M$, where y is the deformation and M is the load. In this case study, the former function was the objective-related function, whereas the latter belonged to the generic function.

Fuel Delivery System

An automobile's fuel delivery system must provide a consistent supply of liquid fuel in the expected operating range, regardless of external (environmental) conditions. Traditionally, symptoms such as difficulty of starting, rolling engine idles, and engine stumbles have been measured. In a study conducted by Ford Motor Company [5], the ideal function was defined as follows. The pump efficiency is defined as

$$\text{pump efficiency} = \frac{QP}{VI}$$

where Q is the fuel pump flow rate, P the fuel system back pressure, V the fuel pump voltage, and I the fuel pump current. Letting the fuel pump flow rate be y, the system input signal be $M \, (= VI)$, and the adjustment signal be $M^* \, (= P)$, the output response, y, is rewritten as

$$y = \frac{\beta M}{M^*}$$

Engine Idle Quality   Idle quality is a major concern on some motor vehicles. It contributes to the vibration felt by vehicle occupants. A commonly used process for evaluating engine idle quality involves running an engine at idle for several cycles and collecting basic combustion data, such as burn rates, cylinder pressure, air/fuel ratio, and actual ignition timing.

In a study conducted by Ford Motor Company [6], it was considered that *indicated mean effective pressure* (**IMEP**) is an approximately linear function of fuel flow at idle when air/fuel ratio, ignition timing, and some of the engine combustion parameters were held constant under steady operation: $y = \beta M$, where y is the IMEP and M is the fuel flow.

Wiper System   The function of a wiper system is to clear precipitation (rain, snow, etc.) and foreign material from a windshield to provide a clear zone of vision. An annoying phenomenon that affects the performance of a windshield wiper system is wiper chatter. Chatter occurs when the wiper blade does not take the proper set and "skips" across the windshield during operation, potentially causing deterioration in both acoustic and wiping quality.

The characteristics measured traditionally to improve the quality of a wiper system are chatter, clear vision, and uniformity of wiper pattern, quietness, and life. The ideal function in a study [7] was defined as "the actual time for a wiper to reach a fixed point on the windshield for a cycle should be the same as the theoretical time (ideal time) for which the system was designed," expressed as

$$y = \beta M,$$

where y is the actual time and M is the theoretical time (1/rpm of the motor).

Welding   In welding processes, traditional measurements include appearance, maximum strength or reject rate, and so on. A study conducted for a welding process [8] defined the ideal function as $y = \beta M M^*$, where y is the force, M the displacement, and M^* the length. In a study of spot welding conditions [9], the ideal function $y = \beta M$, where y is the current and M is the voltage, was used. This means that if the welding is performed ideally, the welded unit is as uniform as the material itself; therefore, the voltage–current relationship must be proportional.

Chemical Reactions   In the manufacture of chemical products, the determination of conditions favoring synthesis reactions greatly affects quality and cost. Traditionally, yield of product has been used as the quality characteristic. Such digital-type characteristics are not recommended in quality engineering. One of the most important issues in chemical reactions is to control reaction speed.

Figure 11.2 shows the case of a chemical reaction with side reactions. In the figure we have

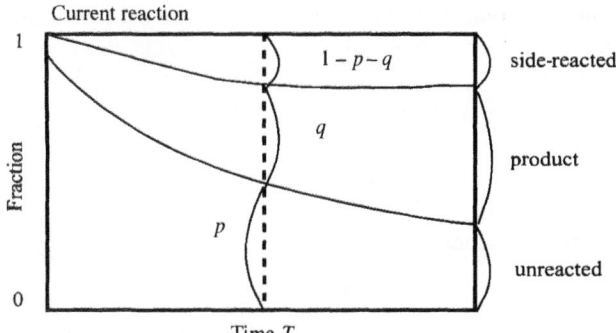

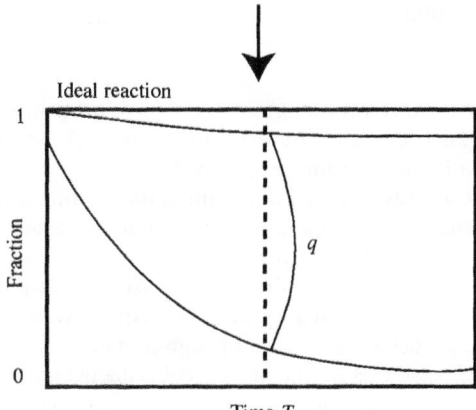

Figure 11.2
Dynamic operating
window

p: fraction unreacted

q: fraction of reacted product

T: time

Letting

        β_1:   total reaction speed

        β_2:   side reaction speed

the fraction of total reaction is written as

$$p = e^{-\beta_1 T}$$

Setting

$$M_1 = \ln \frac{1}{p}$$

the function is written as

$$M_1 = \beta_1 T$$

When there are side reactions, the fraction of side-reacted products is written as

$$1 - p - q = e^{-\beta_2}T$$

Setting

$$M_2 = \frac{1}{1 - p - q}$$

the fraction of side reactions is written as

$$M_2 = \beta_2 T$$

In a study by Toa Gosei Chemical Company [10], the equations above were used for calculation of the SN ratio. As a result, the reaction was increased 1.8-fold above the current condition, which means that the reaction time can be cut approximately in half.

Grinding Process

The dry developer used for electrophotographic copiers and printers consists of toner and carrier. The particle size of the developer is one of the physical properties that has a strong influence on image quality.

Size distribution typically was analyzed as classified data. Quite often, yield was used for analysis. In a study of developing a technique for improving uniformity and adjusting particle size conducted by the Minolta Camera Company [11], the ideal function was $y = \beta/M$, where y is the particle size and M is the grinding energy. In a later study [12], conducted by the same company, a different ideal function was used. Since particles are ground by high-pressure air, the collision of particles in the grinding process is similar to the collision between molecules in chemical reactions.

In this case, the portion with a particle size larger than the upper specification limit was treated as unreacted, and the portion with a particle size smaller than the lower specification limit was treated as overreacted. The ideal functions, similar to the chemical reaction (i.e., the total reaction and side reactions) were used to optimize the grinding process.

Noise Reduction for an Intercooler [13]

Intercoolers are used with turbochargers to increase the output power of car engines. The output power of an engine is determined by the amount of mixed gas that is sucked into the engine cylinder. The more gas sucked in, the higher the feeding efficiency by compressing the mixed air and consequently, raising the air pressure above the atmospheric pressure.

When the airflow resistance inside the cooler is high, the cooling efficiency is reduced and noise is often noted. Before this study, the problem-solving approach has been to detect causes by measuring audible noise level, but without success. From the viewpoint of the product function, it is required that the airflow resistance be low and the flow rate be uniform. The engine output power varies with driving speed. The airflow changes as the accelerator is pushed down. Therefore, it is ideal that the airflow rate in the intercooler be proportional to the amount of air. As the rpm rate of the turbocharger varies under different driving conditions, it is also ideal for the airflow rate to vary proportional to the rpm rate of the turbocharger. The ideal function was defined as $y = \beta M M^*$, where y is the airflow speed of the intercooler, M the amount of airflow, and M^* the rpm rate of the turbocharger.

In the manufacture of high-density printed circuit boards, quality characteristics such as bridge or nonsoldering have been used to improve quality. The yields of these characteristics were studied. In the early stage of quality engineering, nondynamic characteristics such as smaller-the-better and larger-the-better SN ratios were used. These two characteristics were combined to be the *nondynamic operating window SN ratio.*

Wave Soldering [14]

In this study, voltage and current relationships were used as the ideal function. The relationship between current and cross-sectional area was also studied. Thus, there were two signal factors. To maximize the SN ratio of the voltage–current relationship is to optimize the product function. To maximize the SN ratio of the current and cross-sectional area is to optimize manufacturability. In this way, product- and manufacturing-related functions can be studied simultaneously using only one index: an SN ratio with double signals. It is truly a *simultaneous engineering* or *concurrent engineering* approach, as shown by $y = \beta MM^*$, where y is the current, M the voltage, and M^* the cross-sectional area.

Magnetooptical disks are used as computer memory components. In early versions of the products, information already recorded had to be erased before rerecording, taking a long time. To save on recording time, developers tried to develop direct-overwrite MO disks. However, disks consisted of multiple complicated layers. After years of work, there was little success.

Development of an Exchange-Coupled Direct-Overwrite Magnetooptical Disk [15]

Using the quality engineering approach, the ideal function was defined as follows: "The length of magnetic mark is proportionally transformed to the time of light emitted from the laser." In the past, interactions between control factors had been studied. By use of the SN ratio based on the ideal function as defined, the time required for development was greatly reduced.

One of the most important quality items for amplifiers is that the gain of frequency is flat. However, it is not quite flat for amplifiers with a wide frequency range. To make the response flat, a circuit called an *equalizer* is connected to compensate for the declining gain.

Equalizer Design [16]

Using two-stage optimization, the robustness of the equalizer was optimized; the control factors that least affect SN ratio but were highly sensitive were used to tune them so that the gain became flat. The ideal function is that the output voltage (complex number) is proportional to the input voltage.

Transparent conducting flat films are used as flat displaying devices for computers. This type of film is used because of its high conductivity and good pattern quality following etching.

Fabrication of a Transparent Conducting Thin Film [17]

In manufacturing, an electron-beam-heating vacuum deposition process was used. Traditionally, resistance and transparency were used for quality measurement. In this study, the ideal function was defined as $y = \beta MM^*$, where y is the current, M the film thickness, and M^* the voltage. The result showed that the standard deviation was reduced by the fraction 1.83, and conductivity was increased 1.16-fold.

In the design of ac circuits, root mean squares have traditionally been used. By using complex numbers in parameter design, the variability of both the amplitude and phase of the output can be reduced. Therefore, a system whose input and output are expressed by sine waves should use complex numbers.

Low-Pass Filter Design [18]

In the study, amplitude was varied under a fixed frequency, and the proportionality between the input and the output in complex numbers was measured to optimize the stability. After optimization, a 3.8-dB gain was confirmed.

Development of a Power MOSFET [19]

A power MOSFET is used as a reliable switching device in automobile electronics. It is required that the *on* resistance be low so that the unit may be miniaturized. Traditionally, quality characteristics such as film thickness or bonding strength were measured. In this study, the ideal function was defined as $y = \beta M$, where y is the voltage drop and M is the current. From the results, the standard deviation was reduced by a factor of 8, and the resistance was reduced by a factor of 7.

Design of a Voltage-Controlled Oscillator [20]

Voltage-controlled oscillators are used in wireless communication systems. The performance of those oscillators is affected by environmental conditions and the variability of component parts in printed circuit boards. Traditionally, quality characteristics such as transmitting frequency or output power were used for evaluation. In this study, the function $y = \beta MM^*$, where y is the output (complex number), M the dc voltage, and M^* the ac voltage, was used. A 7.27-dB gain was obtained after tuning.

Optimization of an Electrical Encapsulant [21]

An encapsulant is used to isolate the elements in a night-vision image intensifier. In the past, cycle testing was used to evaluate reliability. In this study, the following ideal function, $y = \beta M/M^*$, where y is the current leakage, M the applied voltage, and M^* the electrode spacing, was used. The SN ratio of the encapsulation system was improved by 3.8 dB compared to the standard conditions, and the slope was reduced to 80%.

Ink Formulation [22]

In the development of ink used in digital printers, characteristics such as permeability, stability, flow value, and flexibility have commonly been used. But the results of the use of these characteristics were not always reproducible.

In a digital printer, a document is scanned and the image is converted to electric signals. The signals are then converted to heat energy to punch holes on the resin film, which forms the image on paper. The ideal function for the ink development is that the dot area of the master is proportional to the dot area printed.

11.6. Classification of SN Ratios

Continuous Variables and Classified Attributes

There are two types of data: continuous variables and classified attributes. Voltage, current, dimension, and strength belong to the *continuous variables* category. In quality engineering, *classified attributes* are divided into two groups: *two-class* classified attributes and *three-or-more-class* classified attributes. The SN ratio equations are different for different types.

From the quality engineering viewpoint, it is recommended that one avoid using attribute data, for the followings reasons. First, attribute data are less informative. For example, if a passing score in school is 60, students having scores of 60 and 100 are classified equally as having passed, whereas a student with a grade of 59 is classified as having failed. There is a big difference between the two students who passed, but little difference between a student with a score of 60 and one with a score of 59.

Second, attribute type data are inefficient. In athletic competition, it is difficult to draw a conclusion from the result of two teams playing only one game. We do not conclude that the winning team is going to have a 100% chance to beat the defeated team. In tennis, for example, a champion is commonly declared after one player and wins two of three sets or three of five sets.

The third, and most serious, problem encountered when using attribute data is the possibility of developing interactions. In one study of a welding process, yield was studied. Yield looks like a continuous variable, but it is the fraction represented by the number of nondefective pieces divided by the total number of pieces. A nondefective piece may be counted as 1 and a defective as 0. For example, a one-factor-at-a-time experiment was conducted to compare two control factors, A and B.

$$A_1B_1: \quad \text{yield} = 40\%$$

$$A_2B_1 \quad \text{yield} = 70\%$$

$$A_1B_1 \quad \text{yield} = 40\%$$

$$A_1B_2 \quad \text{yield} = 80\%$$

From the results, one might conclude that the best condition would be A_2B_2 and the yield might be over 90%. But one might be surprised if a confirmatory experiment under condition A_2B_2 resulted in less than 40%. This shows the existence of an interaction between A and B. Let's assume the following levels of A and B:

A_1: low current

A_2: high current

B_1: short welding time

B_2: long welding time

Condition A_1B_1 was low current and short welding time; therefore, too little energy was put into the system, and the result was: no weld. Conditions A_2B_1 and A_1B_2 have more energy going in and the yields are higher. But under condition A_2B_2, high current and long welding time, the yield was low because too much energy was put into the system and the result was: burnthrough. It is seen that if 0–1 data were used, the two extremely opposite conditions are considered equally bad. It is also seen that the superiority and inferiority of factor A is inconsistent, depending on the condition of another factor, B, and vice versa. The interactions between control factors indicate inconsistency of conclusions. We should avoid using 0–1 data. But in some cases, continuous data are unavailable because of a lack of technology. In such cases it is necessary to convert the data so that interactions may be avoided.

Following is an example of data collection. In the case of welding, samples are collected from each welding condition. From each sample, load and deformation are considered as the input and output, respectively. The SN ratio is calculated from the relationship above. There will be SN ratios for A_1B_1, A_2B_1, A_1B_2, and A_2B_2. The effect of A is compared by the average SN ratios of (A_1B_1, A_1B_2) and $(A_2B_1,$

A_2B_2), and the effect of B is compared by the averages of (A_1B_1, A_2B_1) and (A_1B_2, A_2B_2).

Classification Based on Intention

There are two types of SN ratios, based on intention: passive and active. The SN ratio used in measurement is called *passive*. An object, the input signal, is measured, and from the result, the true value of the input signal is estimated. That is why the SN ratio is called passive. For the passive type we have:

❏ *Input:* true value to be measured

❏ *Output:* measured value

In the *active* type, a signal is put into a system to change the output:

❏ *Input:* intention

❏ *Output:* result

For example, a steering wheel is turned to change the direction of a car. The output is changed by intention, so it is an active output. In control systems or product/process design, the parameters for tuning are of the active type. The difference is in the way that the signal factor is considered. Although the intentions are different, the equations and calculation of the SN ratio used are the same.

Fixed and Multiple Targets

Dynamic SN ratios are used to study either multiple targets or a certain range of targets, whereas nondynamic SN ratios are used to study a single target. The former includes three aspects: linearity, sensitivity, and variability. The latter includes two aspects: sensitivity and variability.

In earlier stages of quality engineering, nondynamic characteristics were commonly used. There are three types of nondynamic characteristics: nominal-the-best, smaller-the-better, and larger-the-better. *Nominal-the-best characteristics* are used to hit the target after reducing variability. The output is ideally equal to the target. *Smaller-the-better characteristics* are used to reduce both average and variability. The output is ideally equal to zero. *Larger-the-better characteristics* are used to maximize average and minimize variability. Ideally, the output is equal to infinity.

Dynamic SN ratios are more powerful and efficient; therefore, it is recommended that one use them as much as possible.

Classification Based on Input and Output

As described earlier, quality characteristics are classified as continuous variables, and discrete or classified attributes, the latter also being referred to as digital. SN ratios are classified into the following four cases:

Case	Input	Output
1	Continuous variables	Continuous variables
2	Continuous variables	Digital
3	Digital	Continuous variables
4	Digital	Digital

Case 1 is the type that occurs most frequently. Driving a car, for example, the steering wheel angle change is the input and the turning radius is the output.

Case 2 occurs in digital functions, such as an on–off system using temperature as the signal. For case 3, such as a transmitter in a communication system, the input is digital (mark and space) and the output is voltage. Case 4 occurs often in such chemical processes as separation, purification, and filtration.

SN ratios can be classified based on the number of signal factors: single signal factors, double signal factors, multiple signal factors, and no signal factors. SN ratios with single signal factors occur most frequently. These without signal factors are of the nondynamic type. In Chapter 12, we illustrate the most important and frequently applied case: the SN ratio for continuous variables.

Other Ways of Classifying SN Ratios

References

1. K. Ueno et al., 1992. NC machining technology development. Presented at the American Supplier Institute Symposium.
2. Kazuyoshi Ichikawa et al., 2000. Technology development and productivity improvement of a machining process through the measurement of energy consumed during on and off mode. Presented at the American Supplier Institute Symposium.
3. K. Ueno et al., 1992. Stabilization of injection molding process for carbon fiber reinforced plastic. Presented at the American Supplier Institute Symposium.
4. A. Obara et al., 1997. A measurement of casting materials for aerocraft using quality engineering approaches. *Quality Engineering Forum*, Vol. 5, No. 5.
5. J. S. Colunga et al., 1993. Robustness of fuel delivery systems. Presented at the American Supplier Institute Symposium.
6. Waheed D. Alashe et al., 1994. Engine idle quality robustness using Taguchi methods. Presented at the American Supplier Institute Symposium.
7. M. Deng, et al., 1996. Reduction of chatter in a wiper system. Presented at the American Supplier Institute Symposium.
8. Mitsugi Fukahori, 1995. Parameter design of laser welding for lap edge joint piece. *Journal of Quality Engineering Forum*, Vol. 3, No. 2.
9. Shinichi Kazashi et al., 1996. Optimization of spot welding conditions by generic function. *Journal of Quality Engineering Forum*, Vol. 4, No. 2.
10. Yoshikazu Mori et al., 1995. Optimization of the synthesis conditions of a chemical reaction. *Journal of Quality Engineering Forum*, Vol. 3, No. 1.
11. Hiroshi Shibano et al., 1994. Establishment of particle size adjusting technique for a fine grinding process for developer. *Journal of Quality Engineering Forum*, Vol. 2, No. 3.
12. Hiroshi Shibano et al., 1997. Establishment of control technique for particle size in roles of fine grinding process for developer. *Journal of Quality Engineering Forum*, Vol. 5, No. 1.
13. Hisahiko Sano et al., 1997. Air flow noise reduction of inter-cooler system. Presented at the Quality Engineering Forum Symposium.
14. S. Kazashi et al., 1993. Optimization of a wave soldering. *Journal of Quality Engineering*, Vol. 1, No. 3.
15. Tetsuo Hosokawa et al., 1994. Application of QE for development of exchange-coupled direct-overwrite MO disk. *Journal of Quality Engineering Forum*, Vol. 2, No. 2.

16. Yoshishige Kanemoto, 1994. Robust design and tuning for an equalizer. *Journal of Quality Engineering Forum*, Vol. 2, No. 5.

17. Yoshiharu Nakamura et al., 1995. Fabrication of transparent conducting thin films. *Journal of Quality Engineering Forum*, Vol. 3, No. 1.

18. Satoshi Takahashi, 1995. Parameter design of a low pass filter using complex numbers. *Journal of Quality Engineering Forum*, Vol. 3, No. 4.

19. K. Makabe et al., 1996. Technology development of drain electrode process for power MOSFET. *Journal of Quality Engineering Forum*, Vol. 4, No. 2.

20. Fumikazu Harazono, 1995. Robust design of voltage controlled oscillator. *Journal of Quality Engineering Forum*, Vol. 3, No. 4.

21. Lapthe Flora et al., 1995. Optimization of electrical encapsulant. Presented at the American Supplier Institute Symposium.

22. Takeshi Nojima et al., 1996. Optimization of ink prescription. *Journal of Quality Engineering Forum*, Vol. 4, No. 1.

12 SN Ratios for Continuous Variables

12.1. Introduction   239
12.2. Zero-Point Proportional Equation   241
12.3. Reference-Point Proportional Equation   247
12.4. Linear Equation   250
12.5. Linear Equation Using a Tabular Display of the Orthogonal Polynomial
       Equation   254
12.6. When the Signal Factor Level Interval Is Known   256
12.7. When Signal Factor Levels Can Be Set Up   260
12.8. When the Signal Factor Level Ratio Is Known   262
12.9. When the True Values of Signal Factor Levels Are Unknown   264
12.10. When There Is No Signal Factor (Nondynamic SN Ratio)   264
       Nominal-the-Best, Type I   265
       Nominal-the-Best, Type II   267
       Smaller-the-Better   268
       Larger-the-Better   269
       Nondynamic Operating Window   269
12.11. Estimation of Error   271
12.12. Double Signals   275
12.13. Split Analysis: When There Is No Noise   284
       Split Analysis for Two-Level Orthogonal Array   284
       Split Analysis for Mixed Orthogonal Arrays   286
       References   289

12.1. Introduction

From a quality engineering viewpoint, continuous variables are far more important then classified attributes, as described in Chapter 11. The SN ratios of continuous variables are used much more frequently than those of classified attributes. SN ratios for continuous variables are classified as follows:

1. When the true values of the signal factor are known
   a. Zero-point proportional equation
   b. Reference-point proportional equation
   c. Linear equation
2. When the level interval values are known
3. When the ratio of the signal factor level values is known
4. When the true values of the signal factor levels are unknown
5. When there is no signal factor (nondynamic characteristics)
   a. Nominal-the-best characteristic
   b. Smaller-the-better characteristic
   c. Larger-the-better characteristic
   d. Nondynamic operating window

The following explanation is based on an experiment done under the assumption that at each level of the control factor, there is one signal factor and the remaining factors are noise factors. It is assumed that the data shown in Table 12.1 are given. That is, for true signal value, M_i ($i = 1, ... , k$), r_0 experiments are repeated and data y_{ij} ($j = 1, ... , r_0$) are obtained. y_i is the sum of data for signal M_i.

The value of the SN ratio obtained here is given in terms of the antilog value, and to apply it as a characteristic value for analysis of the effect of control factors, it is transformed into a decibel value, which is equal to 10 times the common log of the SN ratio in terms of the antilog value: $10 \log \eta$ (dB). Selection of the correct SN ratio for a study is based on engineering objectives and viewpoints. It is important for an engineer to understand the function of a product or process and his or her objectives for the study. Once a proper SN ratio is selected, the calculation can be made by using the equations in this chapter or by inputting the data to the software. To simplify the explanation, only one control factor, and usually, one level of the factor, is used in the examples.

When the true values of a signal factor are known, there are basically three types: zero-point proportional equation, reference-point proportional equation, and linear equation. In such cases, the linear effect of the signal factor can be calculated accurately. *The error variance is estimated by including the deviation from the linearity (or nonlinear portion) into the error term.* Different equations are used depending on the ideal functional relationship considered to exist between the input

Table 12.1
Signal factor and data

Signal	M_1	M_2	M_3	...	M_k
Data	y_{11}	y_{21}	y_{31}	...	y_{k1}
	y_{12}	y_{22}	y_{32}	...	y_{k2}
	$\vdots$	$\vdots$	$\vdots$	$\vdots$	$\vdots$
	y_{1r_0}	y_{2r_0}	y_{3r_0}	...	y_{kr_0}
Total	y_1	y_2	y_3	...	y_k

(signal factor) and the output (response data). Following are typical functional relations.

1. *Zero-point proportional equation.* In this case, $y = 0$ when $M = 0$:

$$y = \beta M$$

2. *Reference-point proportional equation.* In this case, the ideal relationship is expressed by the proportional equation that goes through a particular point, $M = M_s$:

$$y - \bar{y}_s = \beta(M - M_s)$$

3. *Linear equation.* In this case, the ideal relationship is expressed by the linear equation between M and y:

$$y = m + \beta(M - \overline{M})$$

12.2. Zero-Point Proportional Equation

A zero-point proportional equation is used most frequently. In this case the input is equal to zero and the output "must" pass through the origin based on the situation considered by the engineer. It must be noted that the selection of an SN equation is not based on "how the response looks."

For example, Figure 12.1*a* is an example of the voltage–current relationship of a dc motor study in a company. Since the regression line does not pass through the origin, as can be seen from Figure 12.1*b*, to improve quality the engineers tried to use a linear equation. The engineers said that based on their design, there is no current until the voltage reaches 1.5 V. However, this conclusion is misleading because, ideally, even when voltage is as low as 0.1 V, current is flowing.

Therefore, the zero-point proportional equation must be used for this case, as shown in Figure 12.1*c*. Any deviation from the regression line in the figure then becomes an error.

Table 12.2 shows the inputs and outputs of a measurement experiment. The inputs are the samples whose true values are known; these are the levels of the signal factors. The outputs are the results measured.

An explanation of symbols used is as follows:

$M_1, M_2, \ldots, M_k$: true values of samples

$y_{11}, y_{12}, \ldots, y_1 r_0$: results of measuring M_1

$y_{21}, y_{22}, \ldots, y_2 r_0$: results of measuring M_2

$y_1, y_2, \ldots, y_k$: total of measurements at M_i

where r_0 is the number of data points in each signal level.

There are kr_0 observations in Table 12.2. Reading from y_{11} to $y_1 r_0$ shows the results of measuring sample M_1. Variation among kr_0 pieces of data is caused by different samples and repetitions. Variation caused by different samples includes the following components: a linear portion, a quadratic portion, and a higher-order portion.

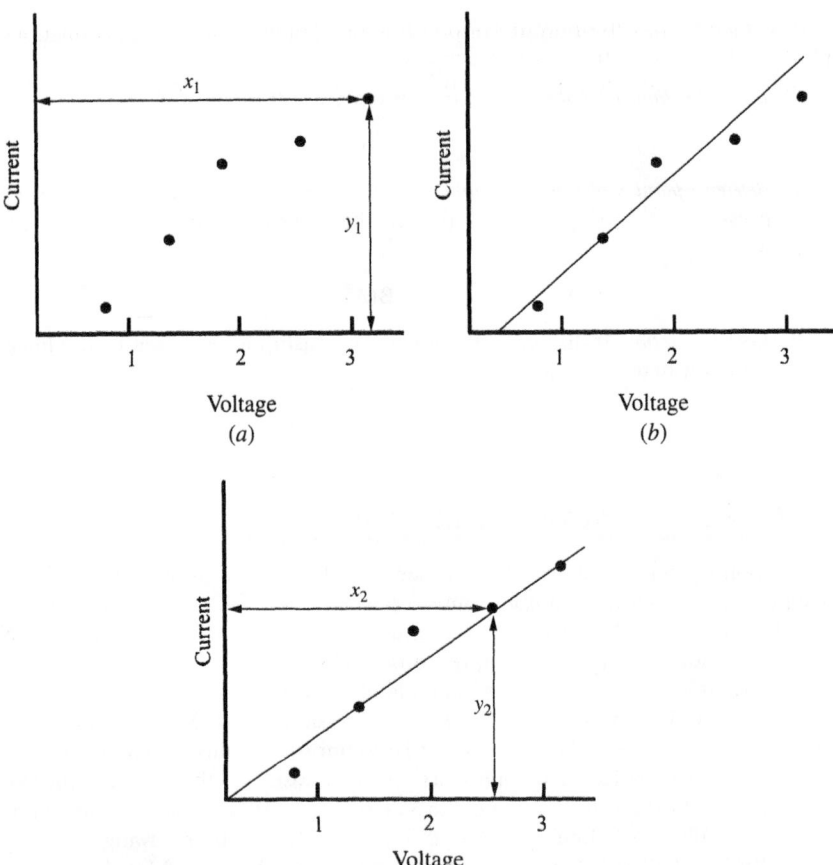

Figure 12.1
Voltage–current
relationship

Variation caused by repetitions is a type of noise factor. In many studies, a compounded noise factor is set and observed rather than just observing simple repetitions. Since an SN ratio is determined from two parts, a desirable portion and a harmful portion, the data must be decomposed as follows:

$$\text{total variation} = \text{variation caused by linear portion} + \text{variations}$$
$$\text{caused by other portions}$$

Table 12.2
Reading from a measurement system

Signal Factor	M_1	M_2	...	M_k
Reading	y_{11}	y_{21}	...	y_{k1}
	y_{12}	y_{22}	...	y_{k2}
	$y_{1\gamma_0}$	$y_{2\gamma_0}$	...	$y_{k\gamma_0}$
Total	y_1	y_2	...	y_k

In the case of a zero-point proportional equation, when the signal is zero, the output must be zero. In a measurement system, if a sample to be measured contains no ingredient, the measured result must read zero. Here the ideal function is expressed as

$$y = \beta M \tag{12.1}$$

where y is the output, β the slope of the response line or the regression coefficient, and M the signal.

In reality, there are errors in observations, and the data in Table 12.2 are expressed by

$$y_{ij} = \beta M_i + e_{ij} \tag{12.1a}$$

where $i = 1, 2, ... , k$ and $j = 1, 2, ... , r_0$.

Equation (12.1a) is expressed as a series of simultaneous equations with one unknown, β. The value of β is selected so as to minimize the total of the squares of differences between the right and left sides of the equation (the least squares method):

$$S_e = (y_{11} - \beta M_1)^2 + (y_{12} - \beta M_1)^2 + \cdots + (y_{kr_0} - \beta M_k)^2 \tag{12.2}$$

The value of β that minimizes S_e is given by

$$\beta = \frac{1}{r_0 r} (M_1 y_1 + M_2 y_2 + \cdots + M_k y_k) \tag{12.3}$$

where

$$r = M_1^2 + M_2^2 + \cdots + M_k^2 \tag{12.4}$$

and r_0 is the total number of measurements under M_i.

To decompose the data in Table 12.2, S_T, the total variation, is first calculated from the squares of kr_0 pieces of data. The symbol f denotes the degrees of freedom:

$$S_T = y_{11}^2 + y_{12}^2 + \cdots + y_{kr_0}^2 \quad \text{and} \quad f = kr_0 \tag{12.5}$$

The variation caused by the linear effect, denoted by S_β, is given by

$$S_\beta = \frac{1}{r_0 r} (M_1 y_1 + M_2 y_2 + \cdots + M_k y_k)^2 \quad \text{and} \quad f = 1 \tag{12.6}$$

The error variation S_e, including the deviation from linearity, V_e, is

$$S_e = S_T - S_\beta \quad \text{and} \quad f = kr_0 - 1 \tag{12.7}$$

The error variance V_e is the error variation divided by its degrees of freedom:

$$V_e = \frac{S_e}{kr_0 - 1} \tag{12.8}$$

The definition of this SN ratio is

$$\eta \rightarrow 10 \log \frac{\beta^2}{\sigma^2} \tag{12.9}$$

To calculate the SN ratio of equation (12.9), the error variance V_e calculated in equation (12.8) is used to estimate σ^2 in equation (12.9). To estimate β^2, a correction is necessary. Statistically, S_β, the variation caused by the linear effect, has the following content:

$$E(S_\beta) = r_0 r \beta^2 + \sigma^2 \qquad (12.10)$$

where E is the expected value as defined in statistics.

When data are repeatedly collected in the same way for an infinite number of times and their S_β value is calculated each time, the average of an infinite number of S_β values is denoted by $E(S_\beta)$. The term σ^2 designates the error contained in the data, such as experimental error or measurement error. What is needed to calculate the SN ratio is the use of β^2 as the numerator. Therefore, the SN ratio in decibel units (dB) is calculated as

$$\eta = 10 \log \frac{(1/r_0 r)(S_\beta - V_e)}{V_e} \qquad \text{dB} \qquad (12.11)$$

The sensitivity in decibel units is calculated as

$$S = 10 \log \frac{1}{r_0 r}(S_\beta - V_e) \qquad (12.12)$$

☐ Example 1 [1]

In this example, the SN ratio of a newly developed displacement gauge is calculated. The value of displacement that was used as the signal is measured with another displacement gauge, which is believed to be accurate. The readings given in Table 12.3 were obtained from the measurement calculated twice by different testing personnel. A zero-point proportional equation was assumed, since the output, y, of the displacement gauge is zero when the displacement, M, is zero. The ANOVA is done as follows:

$$S_T = 65^2 + 74^2 + \cdots + 197^2 = 131{,}879 \qquad (f = 6) \quad (12.13)$$

$$r_0 r = (2)(30^2 + 60^2 + 90^2) = 25{,}200 \qquad (12.14)$$

$$S_\beta = \frac{[(30)(139) + (60)(283) + (90) \times (405)]^2}{r_0 r}$$

$$= \frac{57{,}600^2}{25{,}200} = 131{,}657.14 \qquad (f = 1) \quad (12.15)$$

$$S_e = S_T - S_\beta = 131{,}879 - 131{,}657.14 = 221.86 \qquad (f = 5) \quad (12.16)$$

$$V_e = \frac{S_e}{5} = \frac{221.86}{5} = 44.37 \qquad (12.17)$$

The results are summarized in Table 12.4

Table 12.3
Displacement and displacement gauge data

Reading (mV)	Displacement M (μm)		
	30	60	90
R_1	65	136	208
R_2	74	147	197
Total	139	283	405

The SN ratio is

$$\eta = \frac{(1/25,220)\,(131,657.14 - 44.37)}{44.37} = 0.1177 \qquad (12.18)$$

This was transformed into a decibel value:

$$\eta = 10 \log(0.1177) = -9.29 \text{ dB} \qquad (12.19)$$

The estimated value of the sensitivity coefficient was

$$\hat{\beta} = \frac{57,600}{25,200} = 2.285 \qquad (12.20)$$

and the relationship between M and y was

$$y = 2.285M \qquad (12.21)$$

☐ Example 2 [2]

In Example 1, the SN ratio of one product (or one condition) is calculated . When there is a control factor with two levels (two conditions), two SN ratios are calculated for comparison. In the present example two quantitative cadmium microanalysis methods, dithizone extraction–atomic absorption spectrophotometry and

Table 12.4
ANOVA table for zero-point proportional equation: Displacement gauge

Factor	F	S	V	$E(V)$
β	1	131,657.14	131,657.14	$\sigma^2 + \gamma_0\gamma\beta^2$
e	5	221.86	44.37	σ^2
Total	6	131,879.00		

DDTC extraction–atomic absorption spectrophotometry, were compared. Control factor levels A_1 and A_2 denote the two methods, respectively.

❑ *Signal factor:* cadmium content, M

M_1: 1 ppm
M_2: 2 ppm
M_3: 3 ppm
M_4: 4 ppm
M_5: 5 ppm

❑ *Noise factor:* measuring personnel, R_1 and R_2

Table 12.5 shows the results of measurement.

1. *SN ratio of A_1:*

$$S_T = 12.0^2 + 11.0^2 + \cdots + 54.0^2 = 12{,}907.50 \qquad (f = 10) \qquad (12.22)$$

$$S_\beta = \frac{[(1)(23.0) + (2)(45.5) + (3)(67.0) + (4)(86.5) + (5)(106.0)]^2}{(2)(1^2 + 2^2 + 3^2 + 4^2 + 5^2)}$$

$$= \frac{1191.0^2}{110} = 12{,}895.28 \qquad (f = 1) \qquad (12.23)$$

$$S_e = 12{,}907.50 - 12{,}895.28 = 12.22 \qquad (f = 9) \qquad (12.24)$$

From this result, Table 12.6 is obtained.

$$\eta = 10 \log \frac{\frac{1}{10}(12{,}895.28 - 1.36)}{1.36} = 10 \log(86.2) = 19.36 \text{ dB} \qquad (12.25)$$

2. *SN ratio of A_2:* A similar calculation is done and Table 12.7 is obtained.

$$\eta = 10 \log \frac{\frac{1}{10}(1{,}539.38 - 0.35)}{0.35} = 10 \log(39.97) = 16.02 \text{ dB} \qquad (12.26)$$

Compared to the DDTC extraction method (A_2), the dithizone extraction method (A_1) is better by 3.34 dB.

Table 12.5
Quantitative cadmium microanalysis data

		M_1	M_2	M_3	M_4	M_5	Total
A_1	R_1	12.0	23.0	33.0	42.5	52.0	
	R_2	11.0	22.5	34.0	44.0	54.0	
	Total	23.0	45.5	67.0	86.5	106.0	328.0
A_2	R_1	3.5	7.0	10.5	14.5	18.0	
	R_2	4.0	7.5	12.0	15.5	19.5	
	Total	7.5	14.5	22.5	30.0	37.5	112.0

Table 12.6
ANOVA table of quantitative cadmium microanalysis data of A_1

Factor	f	S	V	$E(V)$
β	1	12,895.28	12,895.28	$\sigma^2 + (2)(1^2 + 2^2 + \cdots + 5^2)$
e	9	12.22	1.36	σ^2
Total	10	12,907.50		

12.3. Reference-Point Proportional Equation

In a measuring system, there are two types of calibration: point calibration and slope calibration. For example, a watch measures time. We correct our watches by the standard time signal. Another example is a bathroom scale, on which an adjusting knob can be used to calibrate the reading to zero when nobody steps on the scale. These are examples of *point calibration.*

In old-style (mechanical and nonquartz) watches, there was a pace-adjusting slide inside the watch. The slide was used to adjust the pace. This is called *slope calibration* (adjustment).

In point calibration, the reference point could be either the zero point or some other point. However, it often happens that a small section in the entire range of measurement is used for a particular application and that range is quite far from the zero point. In such a case, it would be better to select a reference point within the range for calibration rather than using the zero point, because by doing so, the error after calibration becomes smaller.

In reference-point calibration, a standard, M_s, is used as the reference point. Measurements are conducted and the average is calculated to be $\bar{y}_s$. The ideal relationship between the readings and the signals is

$$y - \bar{y}_s = \beta(M - M_s) \qquad (12.27)$$

Table 12.8 shows the data after reference-point calibration. As noted above, first the average of the r_0 readings at the reference point, $y_{s1}, y_{s2}, \ldots, y_{s r_0}$ is calculated:

$$\bar{y}_s = \frac{y_{s1} + y_{s2} + \cdots + y_{s r_0})}{r_0} \qquad (12.28)$$

Table 12.7
ANOVA table of quantitative cadmium microanalysis data of A_2

Factor	f	S	V	$E(V)$
B	1	1539.38	1539.38	$\sigma^2 + 100\beta^2$
e	9	3.12	0.35	σ^2
Total	10	1542.50		

Table 12.8
Data after reference-point calibration

Signal	$M_1 - M_s$	$M_2 - M_s$	...	$M_k - M_s$
Reading	$y_{11} - \bar{y}_s$ $y_{12} - \bar{y}_s$ ... $y_{1r_0} - \bar{y}_s$	$y_{21} - \bar{y}_s$ $y_{22} - \bar{y}_s$ ... $y_{2r_0} - \bar{y}_s$	... ...	$y_{k1} - \bar{y}_s$ $y_{k2} - \bar{y}_s$ ... $y_{kr_0} - \bar{y}_s$
Total	y_1	y_2	...	y_k

The slope or the linear coefficient, β, is also called the *sensitivity coefficient*. It is given as:

$$\beta = \frac{1}{r_0 r} [y_1(M_1 - M_s) + y_2(M_2 - M_s) + \cdots + y_k(M_k - M_s)] \quad (12.29)$$

where

$$r = (M_1 - M_s)^2 + (M_2 - M_s)^2 + \cdots + (M_k - M_s)^2 \quad (12.30)$$

The proportional equation is

$$y - \bar{y}_s = \beta(M - M_s) \quad (12.31)$$

For calculation of the SN ratio, the total variation, S_T, is

$$S_T = \text{total of square of } y_{ij} - \bar{y}_s = \sum_{i=1}^{k} \sum_{j=1}^{r_0} (y_{ij} - \bar{y}_s)^2 \quad (f = kr_0) \quad (12.32)$$

The linear effect of the signal, S_β, is

$$S_\beta = \frac{1}{r_0 r} [y_1(M_1 - M_s) + y_2(M_2 - M_s) + \cdots + y_k(M_k - M_s)]^2 \quad (f = 1) \quad (12.33)$$

The error variation, S_e, is

$$S_e = S_T - S_\beta \quad (f = kr_0 - 1) \quad (12.34)$$

☐ Example [1]

The SN ratio of an analyzer that measures olefin density was calculated. Standard solutions of four different densities were used. The analyzer was used after the reference-point proportional equation, with 5% standard solution as the reference point. The data in Table 12.9 were obtained after measurement was conducted by two different testing personnel.

Table 12.9
Measured olfefin density data

Reading	Density M(%)			
	5	10	15	20
R_1	5.2	10.3	15.4	20.1
R_2	5.0	10.1	15.5	20.3

First, preparatory data processing was performed. The average of data at $M_s = 5\%$ is

$$\bar{y}_s = \frac{5.2 + 5.0}{2} = 5.1 \qquad (12.35)$$

and after subtracting this value from each datum, reference-point calibration was done. The result is shown in Table 12.10.
The ANOVA is done as follows:

$$S_T = 0.1^2 + (-0.1)^2 + \cdots + 15.2^2 = 722.35 \qquad (f = 8) \qquad (12.36)$$

$$\gamma_0\gamma = (2)(0^2 + 5^2 + 10^2 + 15^2) = 700 \qquad (12.37)$$

$$S_\beta = \frac{[(0)(0) + (5)(10.2) + (10)(20.7) + (15)(30.2)]^2}{700}$$

$$= \frac{711^2}{700} = 722.1729 \qquad (f = 1) \qquad (12.38)$$

$$S_e = S_T - S_\beta = 722.35 - 722.1729 = 0.1771 \qquad (f = 7) \qquad (12.39)$$

$$V_e = \frac{S_e}{7} = 0.1771.7 = 0.0253 \qquad (12.40)$$

Table 12.10
Olefin density data after reference-point calibration

	Density M(%)			
	5	10	15	20
$M - M_s$	0	5	10	15
Reading after calibration				
R_1	0.1	5.2	10.3	15.0
R_2	−0.1	5.0	10.4	15.2
Total	0.0	10.2	20.7	30.2

The result is summarized in Table 12.11.
   The SN ratio is

$$\eta = \frac{(1/700)\,(722.1729 - 0.0523)}{0.0253} = 40.76 \qquad (12.41)$$

The decibel transformation of the SN ratio is

$$\eta = 10\,\log(40.76) = 16.10\ \text{dB} \qquad (12.42)$$

The sensitivity, S_β, is

$$\eta = 10\,\log\frac{1}{700}\,(722.1729 - 0.0523) = 10\,\log(1.0316) \qquad (12.43)$$

The estimate of the linear coefficient is

$$\hat{\beta} = \sqrt{1.0316} = 1.015 \qquad (12.44)$$

The function of M and y is determined as

$$y - 5.1 = 1.015(M - 5) \qquad (12.45)$$

12.4. Linear Equation

This is the case in which a linear relationship without a particular restriction is expected between the signal factor and the output responses. In zero-point proportional equations, the regression line must pass through the origin. In reference-point proportional equations, the line passes through the reference point and the reference data, as shown in Figure 12.2.

Which case should be used depends on the objective of the study or on the ideal function of the product or process to be studied. When there is no specific restriction on the input/output relationship shown by a regression line, a linear equation may be used. In a study of an injection molding process, we are not interested in the area in which the injection pressure is equal to zero, and whether the regression line passes through the origin is not our concern. In such a case, a linear equation may be used.

Table 12.11
ANOVA table of the reference-point proportional equation of olefin density

Factor	f	S	V	E(V)
β	1	722.1729	722.1729	$\sigma^2 + \gamma_0\gamma\beta^2$
e	7	0.1771	0.0253	σ
Total	8	722.3500		

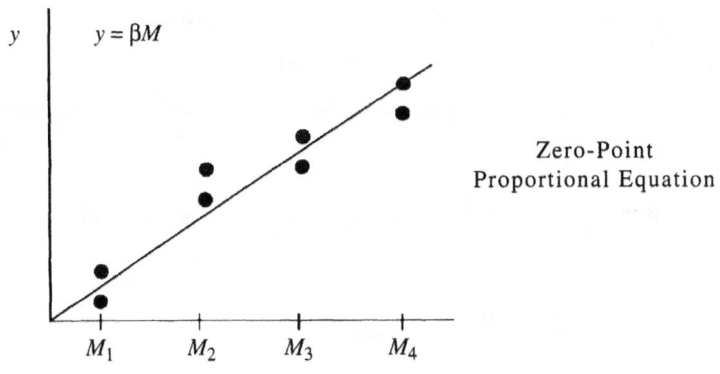

Zero-Point
Proportional Equation

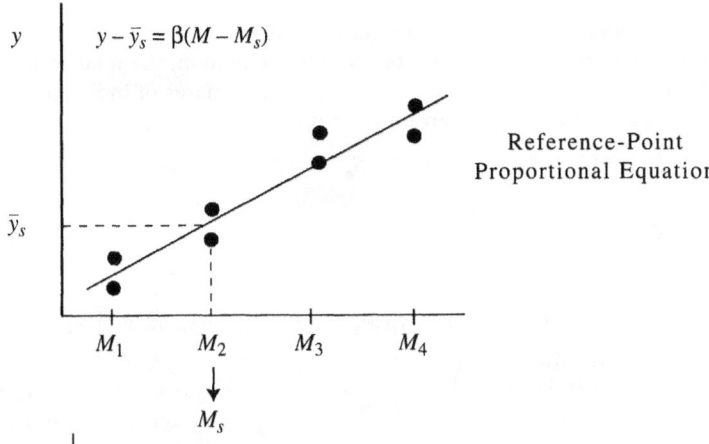

Reference-Point
Proportional Equation

Figure 12.2
Three different cases of
equations

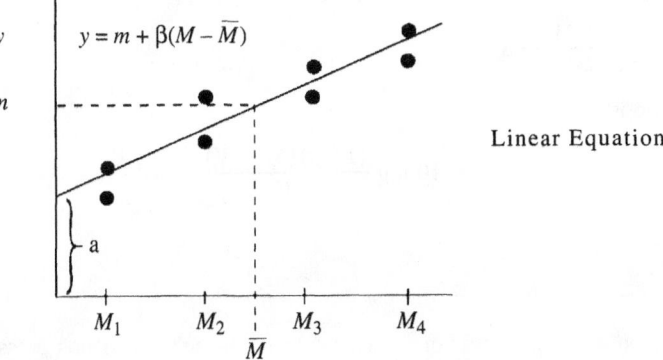

Linear Equation

From the results in Table 12.2, a linear equation is given as

$$y - m + \beta(M - \overline{M}) + e \tag{12.46}$$

where m is the average and e is the error. Parameters m and β are estimated by

$$m = \bar{y} \tag{12.47}$$

$$\beta = \frac{1}{r_0 r} [y_1(M_1 - \overline{M}) + y_2(M_2 - \overline{M}) + \cdots + y_k(M_k - \overline{M})] \tag{12.48}$$

where

$$\overline{M} = \frac{M_1 + M_2 + \cdots + M_k}{k} \tag{12.49}$$

$$r = (M_1 - \overline{M})^2 + (M_2 - \overline{M})^2 + \cdots + (M_k - \overline{M})^2 \tag{12.50}$$

and r_0 is the number of repetitions under each signal factor level.

It must be noted that in the case of the linear equation, the total variation, S_T, is calculated by subtracting S_m from the total of the squares of individual data. In this case, S_T shows the variation around the mean:

$$S_T = y_{11}^2 + y_{12}^2 + \cdots + y_{kr_0}^2 - S_m \qquad (f = kr_0 - 1) \tag{12.51}$$

$$S_m = \frac{(\Sigma y_{ij})^2}{kr_0} \tag{12.52}$$

$$S_\beta = \frac{1}{r_0 r} [y_1(M_1 - \overline{M}) + y_2(M_2 - \overline{M}) + \cdots + y_k(M_k - \overline{M})]^2$$

$$(f = 1) \tag{12.53}$$

$$S_e = S_T - S_\beta \qquad\qquad (f = kr_0 - 2) \tag{12.54}$$

$$V_e = \frac{1}{kr_0 - 2} S_e \tag{12.55}$$

The SN ratio is

$$\eta = 10 \log \frac{(1/r_0 r)(S_\beta - V_e)}{V_e} \tag{12.56}$$

❏ Example [1]

In an injection molding process of a plastic product, using the injection pressure as the signal factor controls the dimensions of the plastic product. To obtain the SN ratio for such a system, the molding injection pressure was altered, and the data given in Table 12.12 were obtained by measuring the dimensions of two molded products. In this case, a linear relationship with no particular restriction was assumed.

Table 12.12
Dimension measurement data for plastic injection molding

	Pressure (kg$_f$/cm²)			
	30	40	50	60
$M - \overline{M}$	−15	−5	5	15
Data (mm)				
R_1	4.608	4.640	4.682	4.718
R_2	4.590	4.650	4.670	4.702
Total	9.198	9.290	9.352	9.420

Table 12.12 also shows the value of $M_i - \overline{M}$ that was used in the ANOVA. Here $\overline{M} = 45$. ANOVA was performed as follows:

$$S_T = 4.608^2 + 4.590^2 + \cdots + 4.702^2 = 173.552216 \qquad (f = 8) \qquad (12.57)$$

$$S_m = \left(\frac{9.198 + \cdots + 9.420}{8}\right)^2 = \frac{37.26^2}{8}$$

$$= 173.538450 \qquad (f = 1) \qquad (12.58)$$

$$\gamma_0\gamma = (2)[(-15)^2 + (-5)^2 + 5^2 + 15^2] = 1000 \qquad (12.59)$$

$$S_\beta = \frac{[(-15)(9.198) + \cdots + (15)(9.420)]^2}{1000}$$

$$= \frac{3.64^2}{1000} = 0.0132496 \qquad (f = 1) \qquad (12.60)$$

$$S_e = S_T - S_m - S_\beta = 173.552216 - 173.538450$$

$$- 0.0132496 = 0.00051564 \qquad (f = 6) \qquad (12.61)$$

$$V_e = \frac{S_e}{6} = \frac{0.005164}{6} = 0.0000861 \qquad (12.62)$$

The calculation results are summarized in Table 12.13.
The SN ratio and the results of decibel transformation are

$$\eta = \frac{(1/1000)(0.0132496 - 0.0000861)}{0.0000861} = 0.1529 \qquad (12.63)$$

$$= 10 \log(0.1529) = -8.155 \text{ dB} \qquad (12.64)$$

254   12. SN Ratios for Continuous Variables

Table 12.13
ANOVA table: Linear equation for plastic injection molding

Factor	f	S	V	E(V)
m	1	173.538450	173.538450	$\sigma^2 + (4)(2\sigma_m^2)$
β	1	0.0132496	0.0132496	$\sigma^2 + \gamma_0\gamma\beta^2$
e	6	0.0005164	0.0000861	σ^2
Total	8	173.552216		

The estimates of the mean value, m, and sensitivity coefficient, β, are

$$\hat{m} = \frac{37.26}{8} = 4.6575 \tag{12.65}$$

$$\hat{\beta} = \frac{3.64}{1000} = 0.00364 \tag{12.66}$$

The equation relating M and y is

$$y = 4.6575 + 0.00364 \quad (M - 45) \tag{12.67}$$

12.5. Linear Equation Using a Tabular Display of the Orthogonal Polynomial Equation

When the intervals for signal factor level are set equal, the SN ratios can easily be calculated. Table 12.14 is part of a tabular display of Chebyshev's orthogonal polynomial equation. Although the table can be used to calculate linear, quadratic, or higher-order terms, only the linear term is used in the SN ratio calculation. (*Note:* This table is used only when level intervals are equal.) In the table,

$$b_1 = \text{linear term}$$

$$b_2 = \text{quadratic term}$$

$$b_i = \frac{W_1 A_1 + \cdots + W_k A_k}{r_0 \lambda S h^i} \tag{12.68}$$

$$h = \text{level interval}$$

$$n_e \text{ of } b_i = r_0 S h^{2i} \tag{12.69}$$

$$S_{bi} = \frac{(W_1 A_1 + \cdots + W_k A_k)^2}{r_0 \lambda^2 S} \tag{12.70}$$

Table 12.14

Orthogonal polynomial equation[a]

No. of Levels:	$k = 2$	$k = 3$		$k = 4$			$k = 5$			
Coefficients:	b_1	b_1	b_2	b_1	b_2	b_3	b_1	b_2	b_3	b_4
W_1	-1	-1	-1	-3	1	-1	-2	2	-1	1
W_2	1	0	-2	-1	-1	3	-1	-1	2	-4
W_3		1	1	1	-1	-3	0	-2	0	6
W_4				3	1	1	1	-1	-2	-4
W_5							2	2	1	1
$\lambda^2 S$	2	2	6	20	4	20	10	14	10	70
λS	1	2	2	10	4	6	10	14	12	24
S	$1/2$	2	$2/3$	5	4	$9/5$	10	14	$72/5$	$288/35$
λ	2	1	3	2	1	$10/3$	1	1	$5/6$	$35/12$

[a] Level intervals must be equal.

From Table 12.2 the SN ratio and the linear coefficient are calculated as follows:

$$S_T = y_{11}^2 + y_{12}^2 + \cdots + y_{km_0}^2 - S_m \qquad (f = kr_0 - 1) \qquad (12.71)$$

$$S_m = \frac{(y_{11} + y_{12} + \cdots + y_{km_0})^2}{kr_0} \qquad (f = 1) \qquad (12.72)$$

$$S_\beta = \frac{(W_1 y_1 + W_2 y_2 + \cdots + W_k y_k)^2}{r_0 \lambda^2 S} \qquad (f = 1) \qquad (12.73)$$

$$\beta = \frac{W_1 y_1 + W_2 y_2 + \cdots + W_k y_k}{r_0 \lambda S h} \qquad (12.74)$$

$$S_e = S_T - S_\beta \qquad (f = kr_0 - 2) \qquad (12.75)$$

$$V_e = \frac{S_e}{kr_0 - 2} \qquad (12.76)$$

$$\eta = 10 \log \frac{(1/r_0 \lambda^2 S h^2)(S_\beta - V_e)}{V_e} \qquad (12.77)$$

To calculate S_β and β, the values of W_1, W_2, ... , W_k are found from Table 12.6 under column b_1 corresponding to the signal factor level, k. h is the level interval of the signal factor. A_1, A_2, ... , A_k in Table 12.14 correspond to y_1, y_1, ... , y_k in Table 12.2.

☐ Example

The example in Table 12.12 can be calculated easily by use of Table 12.14. From the table,

$$k = 4$$
$$r_0 = 2$$
$$h = 5$$
$$W_1 = -3$$
$$W_2 = -1$$
$$W_3 = 1$$
$$W_4 = 3$$
$$\lambda^2 S = 20$$

(12.78)

S_β is calculated by

$$S_\beta = \frac{(-3y_1 - y_2 + y_3 + 3y_4)^2}{20\gamma_0}$$

$$= \frac{[(3)(9.198) - 9.290 + 9.352 + (3)(9.420)]^2}{(20)(2)}$$

$$= 0.0132496$$

(12.79)

and

$$\eta = 10 \log \frac{(1/r_0(\lambda^2 S)h^2 (S_\beta - V_e)}{V_e}$$

$$= 10 \log \frac{[1/(2)(20)(5^2)](0.0132496 - 0.0000861)}{0.0000861}$$

$$= 10 \log(0.1529) = -8.155 \text{ dB}$$

(12.80)

12.6. When the Signal Factor Level Interval Is Known

If the signal factor level interval is known, the linear effect can be calculated without knowing the signal factor's true values. In a chemical analysis, we try to set the levels of the content of a certain ingredient.

Suppose that there is a sample containing a certain ingredient, such as the alcohol content in soybean sauce, but its true value is unknown. To set up levels as signal, the following samples are prepared: M_1: unknown sample containing x

percent alcohol, $M_2 : M_1$ plus 3% alcohol, $M_3 : M_1$ plus 6% alcohol, and $M_4 : M_1$ plus 9% alcohol. Thus, the signal-factor levels M_1, M_2, M_3, and M_4 have equal level intervals. Chebyshev's table of the orthogonal polynomial equation can be used to calculate the linearity and thus the SN ratios.

From Table 12.15, the SN ratio of this case is calculated as follows:

$$S_T = y_{11}^2 + y_{12}^2 + \cdots + y_{34}^2 - S_m \qquad (f = 11) \qquad (12.81)$$

$$S_m = \frac{(y_{11} + y_{12} + \cdots + y_{34})^2}{12} \qquad (12.82)$$

and the following definitions apply:

❏ *Level interval: $h = 3$*
❏ *Number of levels: $k = 4$*
❏ *Number of repetitions: $r_0 = 3$*

From Table 12.14,

$$W_1 = -3 \qquad W_2 = -1 \qquad W_3 = 1 \qquad W_4 = 3$$

$$A_1 = y_1 \qquad A_2 = y_2 \qquad A_3 = y_3 \qquad A_4 = y_4$$

$$\lambda^2 S = 20$$

so

$$S_\beta = \frac{(W_1 A_1 + W_2 A_2 + W_3 A_3 + W_4 A_4)^2}{r_0(\lambda^2 S)}$$

$$= \frac{(-3y_1 - y_2 + y_3 + 3y_4)^2}{(3)(20)} \qquad (f = 1) \qquad (12.83)$$

$$S_e = S_T - S_\beta \qquad (f = 10) \qquad (12.84)$$

$$V_e = \frac{S_e}{10} \qquad (12.85)$$

$$\eta = 10 \log \frac{(1/r_0\lambda^2 Sh^2)(S_\beta - V_e)}{V_e} \qquad (12.86)$$

Table 12.15
Equal level interval

Sample	$M_1 = x$	$M_2 = x + 3\%$	$M_3 = x + 6\%$	$M_4 = x + 9\%$
Reading	y_{11} y_{21} y_{31}	y_{12} y_{22} y_{32}	y_{13} y_{23} y_{33}	y_{14} y_{24} y_{34}
Total	y_1	y_2	y_3	y_4

☐ Example [3]

Table 12.16 shows the measuring results of an electronic balance for chemical analysis. Each raw data point is multiplied by 1000 and M_1 is used as a reference point. Signal factor levels are set as follows:

$$
\begin{aligned}
M_1 &= x & M_1 - M_1 &= 0 \\
M_2 &= x + h & M_2 - M_1 &= h \\
&\ \vdots & &\ \vdots \\
M_k &= x + (k - 1)h & M_k - M_1 &= (k - 1)h
\end{aligned}
\tag{12.87}
$$

Since M_1 was used as a reference point (M_s), M_1 is subtracted from $M_1, M_2, \ldots, M_k$ to obtain the signal factor levels in Table 12.17b. The average of reading corresponding M_1, denoted by $\bar{y}_s$, is calculated as

$$
\bar{y}_s = \frac{120{,}585.7 + 120{,}584.6 + 120{,}585.9}{3} = 120{,}585.4 \tag{12.88}
$$

The following calculations are made for the zero-point proportional equation.

Total variation:

$$
S_T = 0.3^2 + (-0.8)^2 + \cdots + 38.4^2 = 7.132.19 \qquad (f = 15) \tag{12.89}
$$

Table 12.16
Measuring results of an electronic balance

(a) Raw Data

	Signal				
Noise	M_1 x	M_2 x + h	M_3 x + 2h	M_4 x + 3h	M_5 x + 4h
N_1	120.5857	120.5938	120.6018	120.6132	120.6206
N_2	120.5846	120.5933	120.6026	120.6117	120.6213
N_3	120.5859	120.5914	120.6041	120.6095	120.6238

(b) Tranformed Data

	Signal					
Noise	0 $M_1 - M_s$	h $M_2 - M_s$	2h $M_3 - M_s$	3h $M_4 - M_s$	4h $M_5 - M_s$	Linear Equation
N_1	0.3	8.4	16.4	27.8	35.2	$L_1 = 265.4h$
N_2	−0.8	7.9	17.2	26.3	35.9	$L_2 = 264.8h$
N_3	0.5	6.0	18.7	24.1	38.4	$L_3 = 269.3h$
Total	$y_1 = 0.0$	$y_2 = 22.3$	$y_3 = 52.3$	$y_4 = 78.2$	$y_5 = 108.9$	

Note: $L_1 = (0)(0.3) + h(8.4) + 2h(16.4) + 3h(27.8) + 4h(35.2) = 265.4h$.

Table 12.17
COD values of wastewater

(a) Raw Data

Sample	Signal				
	M_1	M_2	M_3	M_4	M_5
R_1	9.3	25.0	43.3	61.9	81.5
R_2	11.3	27.0	44.5	63.1	80.6
R_3	10.4	27.7	45.2	62.0	79.4

(b) Transformed Data

Sample	Signal				
	M_1 $-0.5(\beta - \alpha)$	M_2 $-0.25(\beta - \alpha)$	M_3 0	M_4 $0.25(\beta - \alpha)$	M_5 $0.5(\beta - \alpha)$
R_1	-35.0	-19.3	-1.0	17.6	37.2
R_2	-33.0	-17.3	0.2	18.8	36.3
R_3	-33.9	-16.6	0.9	17.7	35.1
Total	-101.9	-53.20	0.1	54.1	108.6

Effective divider:

$$r = 0^2 + h^2 + (2h)^2 + (3h)^2 + (4h)^2 = 30h^2 \qquad (12.90)$$

Variation of linear term:

$$S_\beta = \frac{1}{\gamma_0 \gamma} [y_1(M_1 - \overline{M}_s) + \cdots + y_5(M_5 - \overline{M}_s)]^2$$

$$= \frac{(L_1 + L_2 + L_3)^2}{(3)(30h^2)} = \frac{(799.5h)^2}{(3)(30h^2)} = 7102.225 \qquad (f = 1) \quad (12.91)$$

Variation of linear term due to noise:

$$S_{NB} = \frac{L_1^2 + L_2^2 + L_3^2}{30h^2} - S_\beta$$

$$= \frac{(264.4h)^2 + (265.8h^2) + (269.3h)^2}{30h^2} - S_\beta$$

$$= 7102.623 - S_\beta = 0.398 \qquad (f = 2) \qquad (12.92)$$

For the calculation of S'_{NB}, see Section 12.11.)

Error variation:

$$S_e = S_T - S_\beta - S_{NB} = 29.567 \qquad (f = 12) \qquad (12.93)$$

Error variance:

$$V_e = \frac{29.567}{12} = 2.464 \tag{12.94}$$

Pooled total error variance:

$$V_N = \frac{0.398 + 29.567}{14} = 2.140 \tag{12.95}$$

The linear coefficient is calculated from equation (12.29):

$$\beta = \frac{1}{r_0 r} [y_1(M_1 - \overline{M}_s) + y_2(M_2 - \overline{M}_s) + \cdots + y_5(M_5 - \overline{M}_5)]$$

$$= \frac{(L_1 + L_2 + L_3)h}{(3)(30h^2)} = \frac{8.883}{h}$$

$$\approx \frac{8.883}{10} = 0.8883 \tag{12.96}$$

where $h = 0.01$ (unit: grams) is used. The following SN ratio is calculated using $h = 10$ (unit: milligrams):

$$\eta = 10 \log \frac{[1/(3)(30)(10^2)](7102.225 - 2.140)}{2.140}$$

$$= 10 \log (0.369) = -4.33 \text{ dB} \tag{12.97}$$

Sensitivity is given by

$$S = 10 \log \frac{1}{(3)(30)(10)^2} (7102.225 - 2.140)$$

$$= 10 \log(0.789) = -1.03 \text{ dB} \tag{12.98}$$

In this example, pooled error variance (V_N) is used as the dominator in the calculation of SN ratio instead of using unpooled error variance. (This is discussed further in Section 12.11.)

12.7. When Signal Factor Levels Can Be Set Up

Although the true values of signal factor levels are unknown, their intervals can be set up correctly. Consider creating a level with a particular chemical content. Two samples, A and B, of content α and β percent, respectively, are prepared, and samples with different content ratios are prepared by mixing these two samples. If the levels are taken by setting $A : B = 0.2 : 0.8$ for M_1, $A : B = 0.4 : 0.6$ for M_2, and so on, the following will be obtained:

$$M_1 = 0.2\alpha + 0.8\beta = (1 - 0.8)\alpha + 0.8\beta = \alpha + 0.8(\beta - \alpha)$$

$$M_2 = \alpha + 0.6(\beta - \alpha)$$

$$M_3 = \alpha + 0.4(\beta - \alpha) \qquad\qquad (12.99)$$

$$M_4 = \alpha + 0.2(\beta - \alpha)$$

$$M_5 = \alpha$$

The levels of such a signal factor have the same interval, $0.2(\beta - \alpha)$. Even if the values of α and β are not known accurately, the equal-interval relationship between the adjacent levels is maintained. A level setting of this type is effective in experiments that check sample measurement errors.

❑ Example [3]

Table 12.17 shows the measuring results of chemical oxygen demand (COD) values from the wastewater of a factory. Two different water samples were mixed to set the following signed factor levels:

$$M_1 = \alpha$$
$$M_2 = 0.75\alpha + 0.25\beta$$
$$M_3 = 0.50\alpha + 0.50\beta \qquad\qquad (12.100)$$
$$M_4 = 0.25\alpha + 0.75\beta$$
$$M_5 = \beta$$

The results of the table were transformed using M_3 as the reference point. The average at M_3 is calculated as

$$\bar{y}_s = \frac{43.3 + 44.5 + 45.2}{3} = 44.3 \qquad\qquad (12.101)$$

Total variation:

$$S_T = (-35.0)^2 + (-33.0)^2 + \cdots + 35.1^2 = 9322.43 \qquad (f = 15) \quad (12.102)$$

Signal factor level interval:

$$h = 0.25(\beta - \alpha) \qquad\qquad (12.103)$$

[The table of orthogonal polynomial equation (Table 12.14) is needed for calculation.]

Effect divider:

$$r = (-2)^2 + (-1)^2 + 0^2 + 1^2 + 2^2 = 10 \qquad\qquad (12.104)$$

Variation of linear effect:

$$S_\beta = \frac{[(-2)(-101.9) - (-53.20) + 54.1 + (2)(108.6)]^2}{3r}$$

$$= \frac{528.3^2}{(3)(10)} = 9303.3630 \qquad (f = 1) \tag{12.105}$$

Error variation:

$$S_e = S_r - S_\beta = 9322.43 - 9303.3630 = 19.067 \qquad (f = 14) \tag{12.106}$$

Error variance:

$$V_e = \frac{19.067}{14} = 1.3619 \tag{12.107}$$

SN ratio:

$$\eta = 10 \log \frac{(1/3r)(S_\beta - V_e)}{V_e} = \frac{[1/(3)(10)](9303.3630 - 1.3619)}{1.3619}$$

$$= 10 \log(227.67) = 23.57 \text{ dB} \tag{12.108}$$

Sensitivity:

$$S = 10 \log \frac{1}{(3)(10)}(9303.3630 - 1.3619)$$

$$= 10 \log(310.07) = 24.91 \text{ dB} \tag{12.109}$$

12.8. When the Signal Factor Level Ratio Is Known

There are times when the absolute value of signal factor levels is not accurately known, but the ratio of the level values can be established accurately. In such a case, the base value of the comparison does not need to be known as long as the proportions to the base value are known (e.g., a geometrically progressing level case or a case where M_1 is a certain times larger than M_2). This is because the linear effect of the signal factor can be calculated accurately as long as the relation among signal factor levels is known.

For example, when the SN ratio is calculated for a method measuring the content ratio of a chemical compound, diluting the sample of certain content by $\frac{1}{2}$, $\frac{1}{4}$, and $\frac{1}{8}$ can create levels with a common ratio of $\frac{1}{2}$. Further, if a scale is used, it is possible to divide a mass into $\frac{1}{2}$, and by repeating this, the levels with a common ratio of $\frac{1}{2}$ can be arranged.

Although the absolute values of levels are also unknown in this case, it is not practical to assume a reference-point proportional equation between M and y, since the distance between the zero point and the maximum level point, M_2, is divided into ratio intervals to create the levels. The following two cases are applicable in practice: (1) a zero-point proportional equation or (2) a linear equation. In either case, the signal-factor-level values are transformed into x times the value M_s, which is used as the base:

$$M_i' = \frac{M_i}{M_s}$$

Specifically, the transformation is done as follows:

$$M_1 \rightarrow M_1' M_s$$

$$M_2 \rightarrow M_2' M_s$$

$$\vdots$$

$$M_k \rightarrow M_k' M_s$$

The value M_s, which is used as the base, can be chosen arbitrarily. The maximum level value is generally used. The transformed level value, M', is used for the remaining calculation.

In the case of a zero-point proportional equation, it is known that data $y = 0$ if signal $M = 0$. It is assumed that the signal factor, M, and data, y, are proportional. In other words, the relationship between M and y is assumed to be

$$y = \beta M \tag{12.110}$$

The distance between levels is not equal, and the calculation formulations are the same as the case of a zero-point proportional equation. The linear effect of the signal factor is calculated using the transformed signal level value, M'. For example, if the geometrically progressive levels have a common ratio of $\frac{1}{2}$ and the number of levels $k = 4$, the following will result:

$$M_1' = \tfrac{1}{8} \qquad M_2' = \tfrac{1}{4} \qquad M_3' = \tfrac{1}{2} \qquad M_4' = 1 \tag{12.111}$$

$$S_\beta = \frac{1}{\gamma_0 \gamma} \left(\frac{1}{8} y_1 + \frac{1}{4} y_2 + \frac{1}{2} y_3 + y_4 \right)^2 \tag{12.112}$$

$$r = \left(\frac{1}{8} \right)^2 + \left(\frac{1}{4} \right)^2 + \left(\frac{1}{2} \right)^2 + 1^2 \tag{12.113}$$

The alternative SN ratio, η', which is based on M_s, is given by

$$\eta' = \frac{(1/\gamma_0 \gamma)(S_\beta - V_e)}{V_e} \tag{12.114}$$

This SN ratio, which is calculated assuming that $M_s = 1$, is not the minus square of the unit of the signal factor. However, this alternative SN ratio is sufficient if it is used only for comparison.

To obtain the SN ratio corresponding to the unit of the signal factor, we substitute M_s with the value expressed in terms of the signal factor unit and transform it according to the equation

$$\eta = \frac{\eta'}{M_s^2} \tag{12.115}$$

12.9. When the True Values of Signal Factor Levels Are Unknown

When the signal factor level values are not known accurately, the variation of samples with unknown values is used as the signal. Signal factor variation S_M is calculated, and the residual of the total variation, S_T, after subtracting S_M, is considered the error term. This is an expedient measure which assumes that a linear relationship holds between the signal M and the data y.

The method of decomposing the total variation is different depending on what is expected with the linear relationship, as in the following two cases:

1. *Proportional equation:* $S_T = S_M + S_e$
2. *Linear equation:* $S_T = S_m + S_M + S_e$

However, these two cases share a basic concept: Calculation of the effect, S_M, of the signal factor, M, and the general mean, S_m, as well as the degrees of freedom is different. The proportional equation is explained below.

The total variation is calculated similarly to the case in which the true value is known. The linear effect cannot be calculated since the signal factor true value is unknown. Therefore, the total effect, S_M, of M is considered the signal variation and is obtained as follows:

$$S_T = y_{11}^2 + y_{12}^2 + \cdots + y_{kr_0}^2 \qquad (f = kr_0) \qquad (12.116)$$

$$S_M = \frac{y_1^2 + y_2^2 + \cdots + y_k^2}{r_0} \qquad (f = k) \qquad (12.117)$$

In the case of a proportional equation, the degrees of freedom of the signal variation, S_M, is k. The error variation is calculated as follows:

$$S_e = S_T - S_M \qquad (12.118)$$

Since the degrees-of-freedom expression of the error term is

$$f_e = kr_0 - k \qquad (12.119)$$

the error variance becomes

$$V_e = \frac{1}{kr_0 - k} S_e \qquad (12.120)$$

The variance of signal factor is obtained as

$$V_M = \frac{1}{k} S_M \qquad (12.121)$$

The SN ratio is determined by

$$\eta = 10 \log \frac{(1/r_0)(V_M - V_e)}{V_e} \qquad (12.122)$$

12.10. When There Is No Signal Factor (Nondynamic SN Ratio)

One of the most important applications of the dynamic SN ratio is to improve the robustness of product or process functions within a certain output range. In a group of products, each product has a different output. A process can produce different outputs for different products or components. It is therefore important to use dynamic SN ratios to develop a group of products or a process for manufacturing different components so that different outputs can easily be obtained.

Traditionally, only one output of a product or only one output in a process has been studied in research and development. Even today, most of the product or process development is being performed this way. In such a case, a nondynamic SN ratio may be used to improve robustness for only a specific or fixed target instead of the entire output range.

For example, a study was conducted to plate gold on the contact point of an electronic component. The target was to plate to a thickness of 50 μin. on the surface, also to reduce variability around 50 μin. using a parameter design approach. This is a nondynamic approach. The target may be 65 μin. for another component. Then another parameter design has to be conducted. But using a dynamic SN ratio, this plating process can be optimized within a certain range of thickness. After the completion of a dynamic approach, any plate thickness within that range can be manufactured without repeating the study for a different thickness every time. Thus, it is easy to understand that the dynamic approach is far more powerful and efficient than the nondynamic approach.

In a nondynamic SN ratio, there are two issues: (1) to reduce variability and (2) to adjust the average to the target. Once variability is reduced and adjusted to the target, there will be no more issues of moving or adjusting the target. It is seen that a nondynamic approach is useful for a specific product with a fixed output, in contrast to a dynamic approach, where the output can easily be adjusted within a range.

There are four types of nondynamic output response applications: nominal-the-best, smaller-the-better, larger-the-better, and nondynamic operating window. In nominal-the-best, there are two types, making a total of five.

Nominal-the-Best, Type I

Recall that in nominal-the-best, there is a fixed target: for example, the plating thickness of a component, the resistance of a resistor, or various dimensions of a product. There are typically upper and lower specification limits. Type I is the case where data are all nonnegative. In type II, data are a mixture of positive and negative values.

In two-step optimization for a nominal-the-best application, the first step is to maximize the SN ratio and thereby minimize variability around the average. The second step is to adjust the average to the target.

The nominal-the-best SN ratio is described as

$$SN = \frac{\text{desired output}}{\text{undesired output}}$$

$$= \frac{\text{effect of the average}}{\text{variability around the average}}$$

In nominal-the-best applications, the data are not negative. This is especially true if the output response is energy related, where zero input to the system should produce zero output; thus, there are no negative data.

Calculations for a nominal-the-best SN ratio are as follows. There are n pieces of data:

$$y_1, y_2, \dots, y_n \quad (f = n) \tag{12.123}$$

The average variation, S_m, is

$$S_m = \frac{(y_1 + y_2 + \dots + y_n)^2}{n} \quad (f = 1) \tag{12.124}$$

The error variation, S_e, is

$$S_e = S_T - S_m \quad (f = n - 1) \tag{12.125}$$

The error variance, V_e, is the error variation divided by its degrees of freedom:

$$V_e = \frac{S_e}{n - 1} \tag{12.126}$$

The SN ratio, denoted by η, is given by

$$\eta = 10 \log \frac{(1/n)\ (S_m - V_e)}{V_e} \tag{12.127}$$

The sensitivity, denoted by S, is given by

$$S = 10 \log \frac{1}{n} (S_m - V_e) \tag{12.128}$$

◻ Example

In a tile manufacturing experiment, samples were taken from different locations in the tunnel kiln, with the following results: 10.18 . . . 10.18 . . . 10.12 . . . 10.06 . . . 10.02 . . . 9.98 . . . 10.20

Total variation:

$$S_T = y_1^2 + y_2^2 + \dots + y_7^2$$

$$= 10.18^2 + 10.18^2 + \dots + 10.20^2$$

$$= 714.9326 \quad (f = 7) \tag{12.129}$$

Average variation:

$$S_m = \frac{(y_1 + y_2 + \dots + y_7)^2}{7}$$

$$= \frac{(10.18 + 10.18 + \dots + 10.20)^2}{7}$$

$$= 714.8782 \quad (f = 1) \tag{12.130}$$

Error variation:

$$S_e = S_T - S_m$$
$$= 714.9326 - 714.8782 \qquad (12.131)$$
$$= 0.05437 \quad (f = 6)$$

Error variance:

$$V_e = \frac{S_e}{6} = \frac{0.05437}{6} = 0.007562 \qquad (12.132)$$

SN ratio:

$$\eta = 10 \log \frac{(1/n)(S_m - V_e)}{V_e}$$
$$= 10 \log \frac{(1/7)(714.8782 - 0.007562)}{0.007562}$$
$$= 10 \log(13504.9427) = 41.37 \text{ dB} \qquad (12.133)$$

Sensitivity:

$$S = 10 \log \frac{1}{n}(S_m - V_e)$$
$$= 10 \log \tfrac{1}{7}(714.8782 - 0.007562)$$
$$= 10 \log(102.1244) = 20.09 \text{ dB} \qquad (12.134)$$

When the results of an experiment include negative values, the positive and negative values cancel each other, which makes the calculation of S_m meaningless. Therefore, the information regarding sensitivity, or average, cannot be obtained. The SN ratio calculated for this case shows variability only. It is less informative than type I. **Nominal-the-Best, Type II**

Conceptually, the SN ratio gives the following information:

$$\frac{SN = 1}{\text{variability around the average}}$$

The SN ratio is calculated as follows where there are n pieces of data: $y_1, y_2, \ldots, y_n$. The error variation is

$$S_e = y_1^2 + y_2^2 + \cdots + y_n^2 - \frac{(y_1 + y_2 + \cdots + y_n)^2}{n} \quad (f = m - 1) \quad (12.135)$$

The error variance is

$$V_e = \frac{S_e}{n - 1} \qquad (12.136)$$

The SN ratio is denoted by

$$\eta = 10 \log \frac{1}{V_e}$$

$$= -10 \log V_e \qquad (12.137)$$

□ **Example**

For data points 1.25, −1.48, −2.70, and 0.19, the SN ratio is calculated as follows:

$$S_e = 1.25^2 + (-1.48)^2 + (-2.70)^2 + 0.19^2$$

$$-\frac{(1.25 - 1.48 - 2.70 + 0.19)^2}{4}$$

$$= 9.2021 \qquad (f = 3) \qquad (12.138)$$

$$V_e = \frac{S_e}{n-1} = \frac{9.2021}{3} = 3.0674 \qquad (12.139)$$

$$\eta = -10 \log(3.0674)$$

$$= -4.87 \text{ dB} \qquad (12.140)$$

Smaller-the-Better Smaller-the-better obtains when the objective is to minimize the output value and there are no negative data. In such a case, the target is zero. For example, we want to minimize the audible engine noise level or to minimize the number of impurities. These examples are easy to understand; however, they are used without success in many cases because such outputs are "symptoms." They are called *downstream quality characteristics*. As explained in Chapter 14, using symptoms as the output quality characteristics in a study is not recommended. It is better, instead, to use this type of SN ratio together with a larger-the-better SN ratio to construct an SN ratio for a nondynamic operating window.

A smaller-the-better SN ratio is calculated as

$$\eta = -10 \log \frac{1}{n} (y_1^2 + y_2^2 + \cdots + y_n^2) \qquad (12.141)$$

□ **Example**

For data points 0.25, 0.19, and 0.22, the SN ratio is calculated as

$$\eta = -10 \log \tfrac{1}{3}(0.25^2 + 0.19^2 + 0.22^2)$$

$$= 13.10 \text{ dB} \qquad (12.142)$$

Larger-the-better obtains when the intention is to maximize the output and there are no negative data. In such a case, the target is infinity. For example, we use this if we want to maximize the power of an engine or the strength of material. Similar to smaller-the-better, such quality characteristics are symptoms: downstream quality characteristics. A study of downstream quality characteristics will exhibit interactions between control factors; therefore, conclusions from the study cannot be reproduced in many cases. It is recommended that this type of SN ratio be used together with the smaller-the-better type to construct an SN ratio of a nondynamic operating window.

Larger-the-Better

The calculation of a larger-the-better SN ratio is given by

$$\eta = -10 \log \frac{1}{n} \left(\frac{1}{y_1^2} + \frac{1}{y_2^2} + \cdots + \frac{1}{y_n^2} \right) \tag{12.143}$$

❐ Example

For data points 23.5, 43.1, and 20.8, the SN ratio is calculated as

$$\eta = -10 \log \frac{1}{3} \left(\frac{1}{23.5^2} + \frac{1}{43.1^2} + \frac{1}{20.8^2} \right)$$

$$= 28.09 \text{ dB} \tag{12.144}$$

As described for smaller-the-better and larger-the-better, the approaches tend to be unsuccessful if the output quality characteristic is a symptom. However, if we can express a function using either smaller-the-better or larger-the-better characteristics, the results have a better chance of being reproduced. A nondynamic operating window combines smaller-the-better and larger-the-better SN ratios to express a function that moves in two opposite directions.

Nondynamic Operating Window

For example, the carriage of a copy machine needs to send the top sheet of paper on a pile to the next process. When the carriage fails to feed a sheet of paper, it is called a *misfeed;* when it feeds two or more sheets of paper, it is called a *multifeed.* The ideal function is that there be no misfeed or multifeed under any conditions of paper quality or use environment. The authors of this book have been promoting a method for checking the effect of noise using the signal if the input is a continuous variable.

The evaluation of carriage function can be done as follows: When friction between the paper and the carriage roller, which is pressed by a spring toward the paper, is used to feed paper, different spring strengths create different levels of pressure and thus different levels of paper-feeding force. If the pressure is zero, paper cannot be fed (Figure 12.3). As the pressure level is increased, the carriage starts feeding paper. If only one sheet of paper is fed, the pressure level is the threshold value of misfeed. Let x denote that value. A further increase in the pressure makes the carriage feed two or more sheets of paper. That pressure level, which is denoted as y, is the threshold value of multifeed. If two sheets of paper are fed as soon as the carriage starts feeding paper, $x = y$. Such a case is called a *zero operating window.* It suggested that a design that gives a large operating window

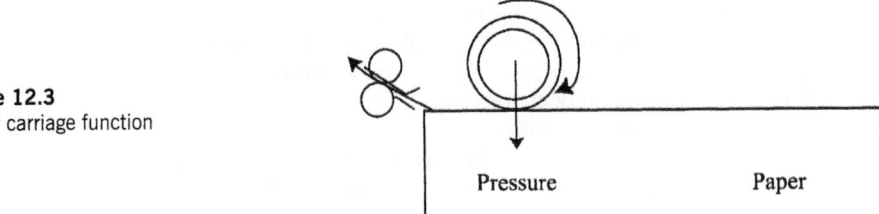

Figure 12.3
Paper carriage function

is a good design and has developed a method of stability design called the *operating window method.*

It is possible to have $x = 0$ or $y = \infty$, but these are not a problem. The important point is that the values of x and y change depending on the paper and environmental conditions. Based on noise factors such as paper size, weight, static electricity, water content, wrinkle size, room temperature, humidity, and power voltage, the following compounded noise factor is created using technical know-how and preparatory experiments (which are done under only one design condition).

N_1:   standard condition

N_2:   condition that is most likely to create misfeed

N_3:   condition that is most likely to create multifeed

Under these three conditions, misfeeding threshold, value x, and double-feeding threshold, value y, are obtained. According to the data in Table 12.18, the following two SN ratios are obtained by treating x as a smaller-the-better characteristic and y as a larger-the-better characteristic.

$$\eta_x = -10 \log \tfrac{1}{3}(x_2 + x_2^2 + x_3^2) \tag{12.145}$$

$$\eta_y = -10 \log \frac{1}{3}\left(\frac{1}{y_1^2} + \frac{1}{y_2^2} + \frac{1}{y_3^2}\right) \tag{12.146}$$

Two SN ratios are obtained for each design (e.g., every condition of an inner orthogonal array to which control factors are assigned). By considering η_x and η_y as outside data at K_1 and K_2, in contrast to the design condition, the main effect of the control factor is obtained from the sum of K_1 and K_2. The key point is to maximize the sum, $\eta_x + \eta_y$.

Table 12.18
Two threshold values x and y

Characteristic Noise	Misfeed	Multifeed
N_1	x_1	y_1
N_2	x_2	y_2
N_3	x_3	y_3

□ Example

In a paper-feeding experiment, the responses selected were:

    x: spring force required to begin feeding one piece of paper

    y: spring force required to begin feeding two or more pieces of paper

The experimental objective was defined as:

   ❏ *Minimize x:* x was treated as a smaller-the-better response.

   ❏ *Maximize y:* y was treated as a larger-the-better response.

   Table 12.19 shows the results of one run in the L_{18} experiment. The SN ratio is calculated as

$$\eta = \eta_x + \eta_g$$

$$= -10 \log \tfrac{1}{3}(x_1^2 + x_2^2 + x_3^2)$$

$$-10 \log \frac{1}{3}\left(\frac{1}{y_1^2} + \frac{1}{y_2^2} + \frac{1}{y_3^2}\right)$$

$$= -10 \log \frac{1}{3}(30^2 + 50^2 + 50^2) - 10 \log \frac{1}{3}\left(\frac{1}{50^2} + \frac{1}{80^2} + \frac{1}{100^2}\right)$$

$$= 3.66 \text{ dB} \tag{12.147}$$

12.11. Estimation of Error

Table 12.20 shows the results of a dynamic study. For the zero-point proportional equation, the SN ratio is given by

$$\eta = 10 \log \frac{(1/r_0 r)(S_\beta - V_e)}{V_e} \tag{12.148}$$

Table 12.19
Results of experiment

	x	y
N_1	30	50
N_2	50	80
N_3	50	100
	N_x	N_y

Table 12.20
Inputs and outputs

	M_1	M_2	$\cdots$	M_k	L
N_1	y_{11}	y_{21}	$\cdots$	y_{k1}	L_1
N_2	y_{12}	y_{22}	$\cdots$	y_{k2}	L_2
$\cdots$	$\cdots$	$\cdots$	$\cdots$	$\cdots$	$\cdots$
N_l	y_{1l}	y_{2l}	$\cdots$	y_{kl}	L_l
Total	y_1	y_2	$\cdots$	y_k	

Based on the concept of SN ratio, the total variation is decomposed as

$$\text{total variation } (S_T) = \text{useful part} + \text{harmful part}$$
$$= \text{proportional part } (S_\beta) + \text{remaining part } (S_e) \quad (12.149)$$

V_e in equation (9.148) is calculated from S_e in equation (12.149).

Table 12.20 includes various sources of variation. In addition to the proportional part of M, there are high-order terms of M, the main effect of N, and interactions between M and N. It is obvious that using the remaining part in equations (12.149) to calculate V_e is an overestimate of V_e to be used in the numerator in equation (12.148). Also, the V_e in the denominator should have a different notation to distinguish the V_e in the numerator.

In Table 12.20, L_1, L_2, ... , L_l denote the linear effects at noise factor conditions N_1, N_2, ... , N_l respectively. The difference between L's is the variation caused by different noise factor conditions, or the interaction between L and N. The main interaction, the variation of linear effect caused by noise, is denoted by βN.

To avoid overestimation of V_e, the total variation in Table 12.20 is decomposed as shown in Figure 12.4. Following is the information for the decomposition of total variation, S_T, into S_β, S_N, and S_e from the results in Table 12.19.

Total variation:

$$S_T = y_{11}^2 + y_{12}^2 + \cdots + y_{kl}^2 \quad (f = kl) \quad (12.150)$$

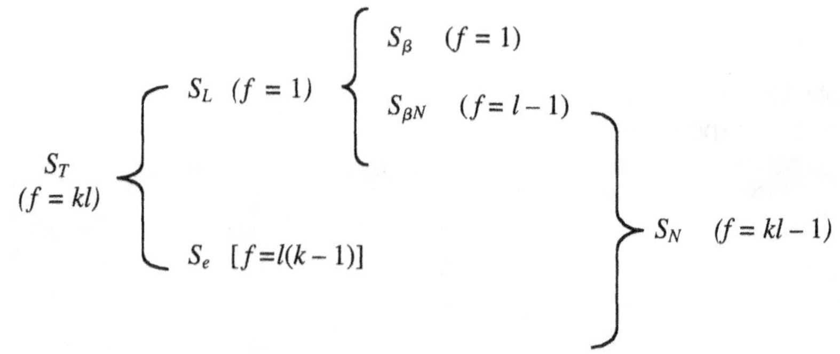

Figure 12.4
Decomposition of variation

Linear effect:

$$L_1 = M_1 y_{11} + M_2 y_{21} + \cdots + M_k y_{k1}$$

$$L_2 = M_1 y_{12} + M_2 y_{22} + \cdots + M_k y_{k2} \qquad (12.151)$$

$$\vdots$$

$$L_l = M_1 y_{1l} + M_2 y_{2l} + \cdots + M_k y_{kl}$$

Variation of linear effect:

$$S_\beta = \frac{(M_1 y_1 + M_2 y_2 + \cdots + M_k y_k)^2}{\gamma_0 \gamma}$$

$$= \frac{(L_1 + L_2 + \cdots + L_l)^2}{\gamma_0 \gamma} - S_\beta \qquad (f = 1) \qquad (12.152)$$

where

$$\gamma_0 = l \qquad (12.153)$$

$$\gamma = M_1^2 + M_2^2 + \cdots + M_k^2 \qquad (12.154)$$

Variation of L:

$$S_L = \frac{L_1^2 + L_2^2 + \cdots + L_l^2}{\gamma} \qquad (f = l) \qquad (12.155)$$

Variation of β due to N:

$$S_{\beta N} = \frac{L_1^2 + L_2^2 + \cdots + L_l^2}{\gamma} - S_\beta \qquad (f = l - 1) \qquad (12.156)$$

Variation of error:

$$S_e = S_T - S_\beta - S_{\beta N} \qquad [f = l(k - 1)] \qquad (12.157)$$

Error variance:

$$V_e = \frac{S_e}{l(k - 1)} \qquad (12.158)$$

Variation of the harmful part:

$$S_N = S_T - S_\beta \qquad (f = kl - 1) \qquad (12.159)$$

Variance of the harmful part:

$$V_N = \frac{S_N}{kl - 1} \qquad (12.160)$$

SN ratio:

$$\eta = 10 \log \frac{(1/\gamma_0 \gamma)(S_\beta - V_e)}{V_N} \qquad (12.161)$$

Table 12.21 shows the results of an experiment for a car brake. It is ideal that the input signal, the line pressure, is proportional to the output, the braking torque.

For this case, two noise conditions, labeled N_1 and N_2, are determined by compounding a few important noise factors, so that

N_1: noise condition when the response tends to be lower
N_2: noise condition when the response tends to be higher

The noise factors are compounded into two levels:

N_1: 60°F, wet, 80% worn pad

N_2: 360°F, dry, 10% worn pad

Another noise factor, Q, is torque variation:

Q_1: maximum torque during braking

Q_2: minimum torque during braking

$$S_T = 4.8^2 + 0.9^2 + 5.8^2 + 0.8^2 + 8.5^2 + \cdots + 34.5^2 = 7342.36 \quad (12.162)$$

$$L_1 = (0.008)(4.8) + (0.016)(8.5) + (0.032)(20.4) + (0.064)(36.9)$$
$$= 3.1888$$

$$L_2 = (0.008)(0.9) + (0.016)(6.5) + (0.032)(13.2) + (0.064)(32.7)$$
$$= 2.6264$$

$$L_3 = (0.008)(5.8) + (0.016)(11.5) + (0.032)(25.0) + (0.064)(43.5)$$
$$= 3.8144$$

$$L_4 = (0.008)(0.8) + (0.016)(6.8) + (0.032)(16.2) + (0.064)(34.5)$$
$$= 2.8416 \quad\quad (12.163)$$

Table 12.21
Results of experiment

	$M_1 = 0.008$	$M_2 = 0.016$	$M_3 = 0.032$	$M_4 = 0.064$	L
N_1Q_1	4.8	8.5	20.4	36.9	L_1
N_1Q_2	0.9	6.5	13.2	32.7	L_2
N_2Q_1	5.8	11.5	25.0	43.5	L_3
N_2Q_2	0.8	6.8	16.2	34.5	L_4
Total	12.3	33.3	74.8	147.6	

$$S_\beta = \frac{(L_1 + L_2 + L_3 + L_4)^2}{\gamma_0 \gamma}$$

$$= \frac{(3.1888 + 2.6264 + 3.8144 + 2.8416)^2}{(4)(0.00544)}$$

$$= 7147.5565 \quad (f = 16) \tag{12.164}$$

$$\gamma_0 = 4$$

$$\gamma = 0.008^2 + 0.016^2 + 0.032^2 + 0.064^2 = 0.00544 \tag{12.165}$$

$$S_{\beta N} = \frac{L_1^2 + L_2^2 + L_3^2 + L_4^2}{\gamma} - S_\beta$$

$$= \frac{3.1888^2 + 2.6264^2 + 3.8144^2 + 2.8416^2}{0.00544} - S_\beta$$

$$= 148.5392 \quad (f = 3) \tag{12.166}$$

$$S_e = S_T - S_\beta - S_{\beta N}$$

$$= 7342.36 - (7147.5565 + 148.5392)$$

$$= 46.2643 \quad (f = 12) \tag{12.167}$$

$$V_e = \frac{S_e}{(3)(4)} = \frac{46.2643}{12} = 3.8544 \tag{12.168}$$

$$S_N = S_T - S_\beta = 7342.36 - 7147.5565$$

$$= 194.8035 \quad (f = 15) \tag{12.169}$$

$$V_N = \frac{S_N}{(4)(4) - 1} = \frac{194.8035}{15} = 12.9869 \tag{12.170}$$

$$\eta = 10 \log \frac{(1/\gamma_0 \gamma)(S_\beta - V_e)}{V_N}$$

$$= 10 \log \frac{[1/(4)(0.00544)](7147.5565 - 3.8554)}{12.9869}$$

$$= 44.03 \text{ dB} \tag{12.171}$$

12.12. Double Signals

In the dynamic SN ratio study, the number of signal factors may not always be equal to 1. For example, in a study of fine patterning for integrated-circuit fabrication, the ideal function is that current be proportional to voltage. It is also ideal that current be proportional to line width. The former is a product-related function, whereas the latter is manufacturing related. The study would become efficient if these two ideal functions were put together for study at the same time. This is called *concurrent engineering* or *simultaneous engineering:* to develop product and process at the same time. The use of double signals makes it possible.

The ideal function for double signals would be

$$y = \beta M M^* \tag{12.172}$$

where y is the current, β the regression coefficient, M the voltage, and M^* the line width.

Calculation of the SN ratio is done similarly by considering the combinations of M and M^* as one signal. Table 12.22 shows the layout of input/output for double signals with noise. By considering the product of M and M^* as one signal factor, L_1 and L_2, the linear equations of the proportional term at N_1 and N_2, respectively, are calculated as follows:

$$L_1 = M_1^* M_1 y_{11} + M_1^* M_2 y_{12} + \cdots + M_3^* M_k y_{5k}$$
$$L_2 = M_1^* M_1 y_{21} + M_1^* M_2 y_{22} + \cdots + M_3^* M_k y_{6k} \tag{12.173}$$

Total variation, S_T, is decomposed as shown in Figure 12.5

$$S_T = y_{11}^2 + y_{12}^2 + \cdots + y_{6k}^2 \qquad (f = 6k) \tag{12.174}$$

$$S_\beta = \frac{(L_1 + L_2)^2}{2r} \qquad (f = 1) \tag{12.175}$$

$$r = (M_1^* M_1)^2 + (M_1^* M_2)^2 + \cdots + (M_3^* M_k)^2 \tag{12.176}$$

$$S_{\beta N} = \frac{L_1^2 + L_2^2}{\gamma} - S_\beta \qquad (f = 1) \tag{12.177}$$

$$S_e = S_T - S_\beta - S_{\beta N} \qquad (f = 6k - 2) \tag{12.178}$$

$$V_e = \frac{S_e}{6k - 2} \tag{12.179}$$

$$V_N = \frac{S_N}{6k - 1} \tag{12.180}$$

$$\eta = 10 \log \frac{(1/2r)(S_\beta - V_e)}{V_N} \tag{12.181}$$

$$S = 10 \log \frac{1}{2r}(S_\beta - V_e) \tag{12.182}$$

❑ Example [5]

In a laser welding experiment, the ideal function was defined as follows:

$$y = \beta M M^*$$

Table 12.22
Input/output table for double signals

		M_1	M_2	$\cdots$	M_k
M_1^*	N_1	y_{11}	y_{12}	$\cdots$	y_{1k}
	N_2	y_{21}	y_{22}	$\cdots$	y_{2k}
M_2^*	N_1	y_{31}	y_{32}	$\cdots$	y_{3k}
	N_2	y_{41}	y_{42}	$\cdots$	y_{4k}
M_3^*	N_1	y_{51}	y_{52}	$\cdots$	y_{5k}
	N_2	y_{61}	y_{62}	$\cdots$	y_{6k}

where y is the force, M the deformation, and M^* the welding length. The force at each deformation level was measured for each test piece of different welding length. Table 12.23 shows the results of one of the runs.

$$L_1 = M_1^* M_1 y_{11} + M_1^* M_2 y_{12} + \cdots + M_3^* M_2 y_{52}$$

$$= (2)(20)(149.0) + (2)(40)(146.8) + \cdots + (6)(40)(713.2)$$

$$= 338840 \tag{12.183}$$

$$L_2 = M_1^* M_1 y_{21} + M_1^* M_2 y_{22} + \cdots + M_3^* M_2 y_{62}$$

$$= (2)(20)(88.6) + (2)(40)(87.0) + \cdots + (6)(40)(304.4)$$

$$= 152196 \tag{12.184}$$

$$S_T = y_{11}^2 + y_{12}^2 + \cdots + y_{62}^2$$

$$= 149.0^2 + 146.8^2 + \cdots + 304.4^2$$

$$= 1,252,346.16 \qquad (f = 12) \tag{12.185}$$

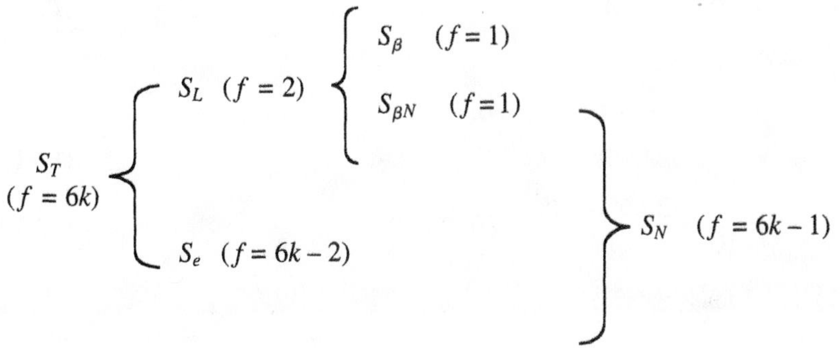

Figure 12.5
Decomposition of variation

Table 12.23
Results of experiment

		M_1 (20)	M_2 (40)
M_1^* (2 mm)	N_1	149.0	146.8
	N_2	88.6	87.0
M_2^* (4 mm)	N_1	300.4	481.1
	N_2	135.7	205.5
M_3^* (6 mm)	N_1	408.0	713.2
	N_2	207.5	304.4

$$S_\beta = \frac{(L_1 + L_2)^2}{2r}$$

$$= \frac{(338,840 + 152,196)^2}{(2)(112,000)} = 1,076,412.292 \qquad (f = 1) \qquad (12.186)$$

$$\gamma_0 = 2 \qquad\qquad (12.187)$$

$$\gamma = (M_1^* M_1)^2 + (M_1^* M_2)^2 + \cdots + (M_3^* M_2)^2$$

$$= [(2)(20)]^2 + [(2)(40)]^2 + \cdots + [(6)(40)]^2$$

$$= 112,000 \qquad\qquad (12.188)$$

$$S_{\beta N} = \frac{L_1^2 + L_2^2}{\gamma} - S_\beta$$

$$= \frac{338,840^2 + 152,196^2}{112,000} - 1,076,412.292$$

$$= 155,517.78 \qquad (f = 1) \qquad (12.189)$$

$$S_e = S_T - S_\beta - S_{\beta N}$$

$$= 1,252,436.16 - 1,076,412.292 - 155,517.78$$

$$= 20,506.088 \qquad (f = 10) \qquad (12.190)$$

$$S_N = S_T - S_\beta$$

$$= 1,252,436.16 - 1,076,412.292$$

$$= 176,023.868 \qquad (f = 11) \qquad (12.191)$$

$$V_e = \frac{S_e}{10} = \frac{20,506.088}{10} = 2050.6088 \qquad\qquad (12.192)$$

$$V_N = \frac{S_N}{11} = \frac{176{,}023.868}{11} = 16{,}002.1698 \tag{12.193}$$

$$\eta = 10 \log \frac{(1/2r)(S_\beta - V_e)}{V_N}$$

$$= 10 \log \frac{[1/(2)(112{,}000)](1{,}076{,}412.295 - 2050.6088)}{16{,}002.1698}$$

$$= -35.23 \text{ dB} \tag{12.194}$$

$$S = 10 \log \frac{1}{2r}(S_\beta - V_e)$$

$$= 10 \log \frac{1}{(2)(112{,}000)}(1{,}076{,}412.292 - 2050.6088)$$

$$= 0.68 \text{ dB} \tag{12.195}$$

When there are two signals, the ideal function could be as follows:

$$y = \beta \frac{M}{M^*} \tag{12.196}$$

where y is the deformation, M the load, and M^* the cross-sectional area. In a material development, for example, the deformation of test pieces must be linearly proportional to the load added and also inversely proportional to the cross-sectional area. Table 12.24 shows the deformation of test pieces at each load and cross-sectional area under noise conditions N_1 and N_2.

Table 12.24
Load, cross section, and deformation

		M_1^*	M_2^*	$\cdots$	M_k^*
M_1	N_1	y_{11}	y_{12}	$\cdots$	y_{1k}
	N_2	y_{21}	y_{22}	$\cdots$	y_{2k}
M_2	N_1	y_{31}	y_{32}	$\cdots$	y_{3k}
	N_2	y_{41}	y_{42}	$\cdots$	y_{4k}
M_3	N_1	y_{51}	y_{52}	$\cdots$	y_{5k}
	N_2	y_{61}	y_{62}	$\cdots$	y_{6k}

The linear equations at different noise factor levels are calculated by

$$L_1 = \frac{M_1}{M_1^*} y_{11} + \frac{M_1}{M_2^*} y_{12} + \cdots + \frac{M_3}{M_k^*} y_5 k$$

$$L_2 = \frac{M_1}{M_1^*} y_{21} + \frac{M_1}{M_2^*} y_{22} + \cdots + \frac{M_3}{M_k^*} y_{6k} \qquad (12.197)$$

The total variation is

$$S_T = y_{11}^2 + y_{12}^2 + \cdots + y_{6k}^2 \qquad (f = 6k) \qquad (12.198)$$

The total variation is decomposed as shown in Figure 12.6.

Variation of L:

$$S_L = \frac{L_1^2 + L_2^2}{r} \qquad (f = 2) \qquad (12.199)$$

Variation of the proportional term:

$$S_\beta = \frac{(L_1 + L_2)^2}{2r} \qquad (f = 1) \qquad (12.200)$$

where

$$r = \left(\frac{M_1}{M_1^*}\right)^2 + \left(\frac{M_1}{M_2^*}\right)^2 + \cdots + \left(\frac{M_3}{M_k^*}\right)^2 \qquad (12.201)$$

Variation of L due to noise:

$$S_{\beta N} = S_L - S_\beta$$

$$= \frac{L_1^2 + L_2^2}{r} - \frac{(L_1 + L_2)^2}{2r} = \frac{(L_1 - L_2)^2}{2r} \qquad (f = 1) \qquad (12.202)$$

Variation of error:

$$S_e = S_T - S_L = S_T - S_\beta - S_{\beta N} \qquad (f = 6k - 2) \qquad (12.203)$$

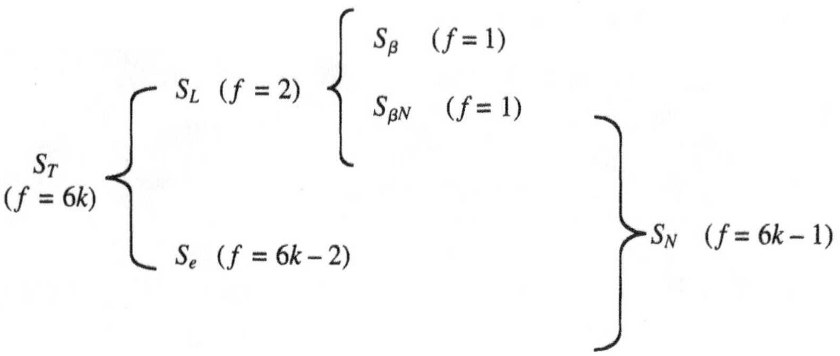

Figure 12.6
Decomposition of
variation

Variation of noise:

$$S_N = S_T - S_\beta \qquad (f = 6k - 1) \tag{12.204}$$

Variance of error:

$$V_e = \frac{S_e}{6k - 2} \tag{12.205}$$

Variance of noise:

$$V_N = \frac{S_N}{6k - 1} \tag{12.206}$$

SN ratio:

$$\eta = 10 \log \frac{(1/2r)(S_\beta - V_e)}{V_N} \tag{12.207}$$

Sensitivity:

$$S = 10 \log \frac{1}{2r}(S_\beta - V_e) \tag{12.208}$$

❑ Example [6]

A process used to apply an encapsulant, which provides electrical isolation between elements on the outer surface of a night-vision image intensifier tube, was optimized using double signal factor SN ratio. The ideal function was identified as

$$y = \beta \frac{M}{M^*} \tag{12.196}$$

where y is the current leakage, M the applied voltage ($M_1 = 2$, $M_2 = 4$, $M_3 = 6$, $M_4 = 8$, $M_5 = 10$), and M^* the electrode spacing ($M_1^* = 0.016$, $M_2^* = 0.030$, $M_3^* = 0.062$), Table 12.25 shows the input/output relationship of this study, and Table 12.26 shows the results of one run of the experiment.

Calculations for SN ratio and sensitivity are made as follows:

$$L_1 = \frac{M_1}{M_1^*} y_{11} + \frac{M_1}{M_2^*} y_{12} + \cdots + \frac{M_5}{M_3^*} y_{53}$$

$$= \frac{2}{0.016}(9) + \frac{2}{0.030}(5) + \cdots + \frac{10}{0.062}(34)$$

$$= 217{,}781.7 \tag{12.209}$$

$$L_2 = 266{,}009.9$$

$$L_3 = 3{,}687{,}281$$

$$L_4 = 3{,}131{,}678$$

Table 12.25
Input/output table

		M_1	M_2	M_3	M_4	M_5
N_1	$M_1^\star$	y_{11}	y_{12}	y_{13}	y_{14}	y_{15}
	$M_2^\star$	y_{21}	y_{22}	—	—	y_{25}
	$M_3^\star$	y_{31}	y_{32}	—	—	y_{35}
N_2	$M_1^\star$	y_{41}	y_{42}	—	—	y_{45}
	$M_2^\star$	y_{51}	y_{52}	—	—	y_{55}
	$M_3^\star$	y_{61}	y_{62}	—	—	y_{65}
N_3	$M_1^\star$	y_{71}	y_{72}	—	—	y_{75}
	$M_2^\star$	y_{81}	y_{82}	—	—	y_{85}
	$M_3^\star$	y_{91}	y_{92}	—	—	y_{95}
N_4	$M_1^\star$	$y_{10\cdot1}$	$y_{10\cdot2}$	—	—	$y_{10\cdot5}$
	$M_2^\star$	$y_{11\cdot1}$	$y_{11\cdot2}$	—	—	$y_{11\cdot5}$
	$M_3^\star$	$y_{12\cdot1}$	$y_{12\cdot2}$	—	—	$y_{12\cdot5}$

$$S_T = y_{11}^2 + y_{12}^2 + \cdots + y_{12\cdot5}^2$$
$$= 9^2 + 26^2 + \cdots + 361^2$$
$$= 20{,}974{,}160 \quad (f = 60) \tag{12.210}$$

$$S_L = \frac{L_1^2 + L_2^2 + L_3^2 + L_4^2}{\gamma}$$
$$= \frac{217{,}781.7^2 + 266{,}009.9^2 + 3{,}687{,}281^2 + 3{,}131{,}678^2}{1{,}161{,}051}$$
$$= 20{,}258{,}919 \tag{12.211}$$

$$\gamma_0 = 4 \tag{12.212}$$

$$\gamma = \left(\frac{M_1}{M_1^\star}\right)^2 + \left(\frac{M_1}{M_2^\star}\right)^2 + \cdots + \left(\frac{M_5}{M_3^\star}\right)^2$$
$$= \left(\frac{2}{0.016}\right)^2 + \left(\frac{2}{0.030}\right)^2 + \cdots + \left(\frac{10}{0.062}\right)^2 = 1{,}161{,}051 \tag{12.213}$$

$$S_\beta = \frac{(L_1 + L_2 + L_3 + L_4)^2}{4\gamma}$$
$$= \frac{(217{,}781.7 + 266{,}009.9 + 3{,}687{,}281 + 3{,}131{,}678)^2}{(4)(1{,}161{,}051)}$$
$$= 11{,}483{,}166 \quad (f = 1) \tag{12.214}$$

Table 12.26
Results of experiment

		M_1	M_2	M_3	M_4	M_5	L
N_1	M_1^*	9	26	53	92	152	L_1
	M_2^*	5	13	24	42	62	
	M_3^*	5	9	14	23	34	
N_2	M_1^*	13	36	66	100	170	L_2
	M_2^*	8	23	44	62	90	
	M_3^*	6	14	24	37	45	
N_3	M_1^*	209	517	915	1452	2266	L_3
	M_2^*	165	374	622	892	1186	
	M_3^*	110	235	385	515	655	
N_4	M_1^*	177	457	897	1321	2113	L_4
	M_2^*	100	234	408	621	895	
	M_3^*	50	104	171	262	361	

$$S_{\beta N} = S_L - S_\beta$$

$$= 20{,}258{,}919 - 11{,}483{,}166$$

$$= 8{,}775{,}753 \qquad (f = 3) \qquad (12.215)$$

$$S_e = S_T - S_L = S_T - S_\beta - S_{\beta N}$$

$$= 20{,}974{,}160 - 20{,}258{,}919$$

$$= 715{,}241 \qquad (f = 56) \qquad (12.216)$$

$$S_N = S_T - S_\beta = 20{,}974{,}160 - 11{,}483{,}166$$

$$= 9{,}490{,}994 \qquad (f = 59) \qquad (12.217)$$

$$V_e = \frac{S_e}{56} = \frac{715{,}241}{56} = 12{,}772 \qquad (12.218)$$

$$V_N = \frac{S_N}{59} = \frac{9{,}490{,}994}{59} = 160{,}864 \qquad (12.219)$$

$$\eta = 10 \log \frac{(1/\gamma_0 \gamma)(S_\beta - V_e)}{V_N}$$

$$= 10 \log \frac{\dfrac{1}{(4)(1{,}161{,}051)}(11{,}483{,}166 - 12{,}772)}{160{,}864}$$

$$= -48.138 \text{ dB} \qquad (12.220)$$

$$S = 10 \log \frac{1}{\gamma_0 \gamma} (S_\beta - V_e)$$

$$= 10 \log \frac{1}{(4)(1,161,051)} (11,483,166 - 12,772)$$

$$= 3.93 \text{ dB} \hspace{4cm} (12.221)$$

12.13. Split Analysis: When There Is No Noise

The objective of parameter design is to reduce the effect of noise factors by evaluating the magnitude of their SN ratio. Therefore, noise factors must be assigned in an experiment. But sometimes it is difficult to assign noise factors. In experiments with chemical reactions, for example, taking extra data points means conducting one more experiment for a run, which doubles the time and cost of the entire experiment. Therefore, only one experiment is conducted for one run of the orthogonal array in many cases.

In such cases it is necessary to provide some type of noise for SN ratio calculation. When dynamic characteristics are used, there are signal factors. In such cases we want the input signal and the output response to be proportional. Therefore, the deviation from linearity can be used as the noise. But the number of signal factor levels is limited and the degrees of freedom for noise are limited. Generally, it is better to have more degrees of freedom for noise. Split design is used for this purpose. It is a method of calculating the SN ratio from the results obtained from an orthogonal array experiment without providing noise factors.

Split Analysis for Two-Level Orthogonal Array

To simplify explanation, an L_8 orthogonal array is used. (In practice, however, it is recommended that one use L_{12}, L_{18}, or L_{36} arrays.) Two-level control factors A, B, C, D, E, F, and G are assigned to the inner array. A three-level signal factor M is assigned to the outer array. Table 12.27 shows the layout.

Assume that zero-point proportional equations are used for M and y. The SN ratio of A_1 is calculated from the data for runs 1, 2, 3, and 4. The data for A_1 are shown in Table 12.28.

The 12 data points in Table 12.28 are decomposed as follows:

$$\begin{matrix} S_T \\ (f = 12) \end{matrix} \begin{cases} S_{\text{row}} & (f = 3) \\ S_\beta & (f = 1) \\ S_e & (f = 8) \end{cases}$$

S_{row} is the variation between rows 1, 2, 3, and 4. It is the variation caused by control factors B, C, D, E, F, and G. This variation belongs neither to the signal nor to the noise, and therefore is separated out.

Total variation:

$$S_T = y_{11}^2 + y_{12}^2 + \cdots + y_{43}^2 \hspace{1cm} (f = 1) \hspace{2cm} (12.222)$$

Table 12.27
Results without noise factors

No.	A 1	B 2	C 3	D 4	E 5	F 6	G 7	M_1	M_2	M_3	Total
1	1	1	1	1	1	1	1	y_{11}	y_{12}	y_{13}	y_1
2	1	1	1	2	2	2	2	y_{21}	y_{22}	y_{23}	y_2
3	1	2	2	1	1	2	2	y_{31}	y_{32}	y_{33}	y_3
4	1	2	2	2	2	1	1	y_{41}	y_{42}	y_{43}	y_4
Total								$y_{.1}$	$y_{.2}$	$y_{.3}$	
5	2	1	2	1	2	1	2	y_{51}	y_{52}	y_{53}	y_5
6	2	1	2	2	1	2	1	y_{61}	y_{62}	y_{63}	y_6
7	2	2	1	1	2	2	1	y_{71}	y_{72}	y_{73}	y_7
8	2	2	1	2	1	1	2	y_{81}	y_{82}	y_{83}	y_8

Totals of M:

$$y_{.1} = y_{11} + y_{21} + y_{31} + y_{41}$$

$$y_{.2} = y_{12} + y_{22} + y_{32} + y_{42}$$

$$y_{.3} = y_{13} + y_{23} + y_{33} + y_{43} \qquad (12.223)$$

$$y_{.4} = y_{14} + y_{24} + y_{34} + y_{44}$$

Variation of linear term:

$$S_\beta = \frac{1}{\gamma_0 \gamma} (M_1 y_{.1} + M_2 y_{.2} + M_3 y_{.3})^2 \qquad (f = 1) \qquad (12.224)$$

where γ_0 is the number of data points under M:

$$\gamma_0 = 4 \qquad (12.225)$$

Table 12.28
Results of A_1

No.	1 A	2 B	3 C	4 D	5 E	6 F	7 G	M_1	M_2	M_3	Total
1	1	1	1	1	1	1	1	y_{11}	y_{12}	y_{13}	y_1
2	1	1	1	2	2	2	2	y_{21}	y_{22}	y_{23}	y_2
3	1	2	2	1	1	2	2	y_{31}	y_{32}	y_{33}	y_3
4	1	2	2	2	2	1	1	y_{41}	y_{42}	y_{43}	y_4
Total								$y_{.1}$	$y_{.2}$	$y_{.3}$	

Effective divider:

$$\gamma = M_1^2 + M_2^2 + M_3^2 \tag{12.226}$$

Variation between rows:

$$S_{\text{row}} = \frac{1}{k}(y_1^2 + y_2^2 + y_3^2 + y_4^2) - \frac{(y_1 + y_2 + y_3 + y_4)^2}{\gamma_0 k} \qquad (f = \gamma_0 - 1) \tag{12.227}$$

where k is the number of data points in each run:

$$k = 3 \tag{12.228}$$

Error variation:

$$S_e = S_T - S_\beta - S_{\text{row}} \qquad (f = 8) \tag{12.229}$$

Error variance:

$$V_e = \frac{S_e}{8} \tag{12.230}$$

SN ratio:

$$\eta = 10 \log \frac{(1/\gamma_0 \gamma)(S_\beta - V_e)}{V_e} \tag{12.231}$$

The SN ratio of A_2 is calculated similarly from the results for A_2. The SN ratio of B_1 is calculated from the results for B_1, as shown in Table 12.29.

Split Analysis for Mixed Orthogonal Arrays

Calculation of the SN ratios for a three-level orthogonal array is exactly the same as that for a two-level orthogonal array. An L_{18} orthogonal array is used as an example to explain the calculations in Table 12.30.

Table 12.31 shows the layout of an experiment. A two-level control factor, A, and seven three-level control factors, B, C, D, E, F, G, and H, are assigned to an L_{18} orthogonal array. The signal factor M is assigned to outside the L_{18} array. There are no noise factors.

First, the calculation of the SN ratio of A_1 is explained. The results under A_1 are shown in Table 12.31.

Total variation:

$$S_T = y_{11}^2 + y_{12}^2 + \cdots + y_{9k}^{20} \qquad (f = 9k) \tag{12.232}$$

Table 12.29
Results of B_1

No.	M_1	M_2	M_3	Total
1	y_{11}	y_{12}	y_{13}	y_1
2	y_{21}	y_{22}	y_{23}	y_2
5	y_{51}	y_{52}	y_{53}	y_5
6	y_{61}	y_{62}	y_{63}	y_6
Total	$y_{\cdot 1}$	$y_{\cdot 2}$	$y_{\cdot 3}$	

Table 12.30
Results without noise factors

No.	A 1	B 2	C 3	D 4	E 5	F 6	G 7	H 8	M_1	M_2	...	M_k	Total
1	1	1	1	1	1	1	1	1	y_{11}	y_{12}	...	y_{1k}	y_1
2	1	1	2	2	2	2	2	2	y_{21}	y_{22}	...	y_{2k}	y_2
3	1	1	.	.	.	.	.	.	.	.	...	.	.
4	1	2	.	.	.	.	.	.	.	.	...	.	.
5	1	2	.	.	.	.	.	.	.	.	...	.	.
6	1	2											
7	1	3											
8	1	3											
9	1	3											
10	2	1											
11	2	1											
12	2	1											
13	2	2											
14	2	2											
15	2	2											
16	2	3											
17	2	3											
18	2	3	3	2	1	2	3	1	$y_{18\cdot 1}$	$y_{18\cdot 2}$	...	y_{18k}	y_{18}

Table 12.31
Results of A_1

No.	M_1	M_2	...	M_k	Total
1	y_{11}	y_{12}	...	y_{1k}	y_1
2	y_{21}	y_{22}	...	y_{2k}	y_2
3	.				
4	.				
5	.				
6	.				
7					
8					
9	y_{91}	y_{92}	...	y_{9k}	y_9
Total	$y_{\cdot 1}$	$y_{\cdot 2}$	...	$y_{\cdot k}$	

Column totals:

$$y_{.1} = y_{11} + y_{21} + \cdots + y_{91}$$

$$y_{.2} = y_{12} + y_{22} + \cdots + y_{92} \qquad (12.233)$$

$$\vdots$$

$$y_{.k} = y_{1k} + y_{2k} + \cdots + y_{9k}$$

Variation of linear term:

$$S_\beta = \frac{1}{\gamma_0 \gamma} (M_1 y_{.1} + M_2 y_{.2} + \cdots + M_k y_{.k})^2 \qquad (f = 1) \qquad (12.234)$$

where γ_0 is the number of runs under M_1:

$$\gamma_0 = 9 \qquad (12.235)$$

Effective divider:

$$\gamma_0 \gamma = 9(M_1^2 + M_2^2 + \cdots + M_k^2) \qquad (12.236)$$

Variation between rows:

$$S_{\text{row}} = \frac{1}{k} (y_1^2 + y_2^2 + \cdots + y_9^2) - \frac{1}{\gamma_0 k} (y_1 + y_2 + \cdots + y_9)^2$$

$$(f = \gamma_0 - 1 = 8) \qquad (12.237)$$

where k is the number of data points in each run.

Error variation:

$$S_e = S_T - S_\beta - S_{\text{row}} \qquad (f = 9k - 9) \qquad (12.238)$$

Error variance:

$$V_e = \frac{S_e}{9k - 9} \qquad (12.239)$$

SN ratio:

$$\eta = 10 \log \frac{(1/\gamma_0 \gamma)(S_\beta - V_e)}{V_e} \qquad (12.240)$$

The SN ratio of B_1 is calculated from the results under B_1, as shown in Table 12.32.

$$S_T = y_{11}^2 + y_{12}^2 + \cdots + y_{12 \cdot k}^2 \qquad (f = 6k) \qquad (12.241)$$

$$y_{.1} = y_{11} + y_{21} + y_{31} + y_{10 \cdot 1} + y_{12 \cdot 1}$$

$$y_{.2} = y_{12} + y_{22} + y_{32} + y_{10 \cdot 2} + y_{11 \cdot 2} + y_{12 \cdot 2}$$

$$\vdots$$

$$y_{.k} = y_{1k} + y_{2k} + y_{3k} + y_{10 \cdot 2} + y_{11 \cdot k} + y_{12 \cdot k} \qquad (12.242)$$

$$S_\beta = \frac{1}{\gamma_0 \gamma} (M_1 y_{.1} + M_2 y_{.2} + \cdots + M_k y_{.k})^2 \qquad (12.243)$$

Table 12.32
Results of B_1

No.	M_1	M_2	$\cdots$	M_k	Total
1	y_{11}	y_{12}	$\cdots$	y_{1k}	y_1
2	y_{21}	y_{22}	$\cdots$	y_{2k}	y_2
3	y_{31}	$\cdots$			$\cdot$
10	$y_{10\cdot1}$	$\cdots$			$\cdot$
11	$y_{11\cdot1}$	$\cdots$	$\cdots$		$\cdot$
12	$y_{12\cdot1}$	$y_{12\cdot2}$	$\cdots$	y_{12k}	y_{12}
Total	$y_{\cdot1}$	$y_{\cdot2}$	$\cdots$	$y_{\cdot k}$	

where

$$\gamma_0 = 6 \tag{12.244}$$

$$\gamma = M_1^2 + M_2^2 + \cdots + M_k^2 \tag{12.245}$$

$$S_{\text{row}} = \frac{1}{k}(y_1^2 + y_2^2 + \cdots + y_{12}^2) - \frac{(y_1 + y_2 + \cdots + y_{12})^2}{\gamma_0 k}$$

$$(f = \gamma_0 - 1 = 5) \tag{12.246}$$

$$S_e = S_T - S_\beta - S_{\text{row}} \qquad (f = 9k - 9) \tag{12.247}$$

$$V_e = \frac{S_e}{9k - 9} \tag{12.248}$$

$$\eta = 10 \log \frac{(1/\gamma_0\gamma)(S_\beta - V_e)}{V_e} \tag{12.249}$$

References

1. Genichi Taguchi, 1991. *Quality Engineering Series*, Vol. 3. Tokyo: Japanese Standards Association/ASI Press.

2. Kazuyoshi Ishikawa et al., 1972. *SN Ratio Manual*, Tokyo: Japanese Standards Association.

3. Hiroshi Yano, 1998. *Introduction to Quality Engineering Calculation*. Tokyo: Japanese Standards Association.

4. Shin Taguchi, 1998. *ASI Workshop Manual*. Livonia, Michigan: American Supplier Institute.

5. Mitsugi Fukahori, 1995. Optimization of an electrical encapsulant. *Journal of Quality Engineering Forum*, Vol. 3, No. 2.

6. Lapthe Flora et al., 1995. Parameter design of laser welding for lap edge joint piece. Presented at the ASI Symposium.

13 SN Ratios for Classified Attributes

13.1. **Introduction**   **290**
13.2. **Cases with Two Classes, One Type of Mistake   291**
13.3. **Case With Two Classes, Two Types of Mistake   293**
13.4. **Chemical Reactions without Side Reactions   296**
   Nature of Chemical Reactions   296
   Basic Function of Chemical Reactions   296
   SN Ratio of Chemical Reactions without Side Reactions   297
13.5. **Chemical Reactions with Side Reactions: Dynamic Operating Window   298**
   Reaction Speed Difference Method   298
   Reaction Speed Ratio Method   301
   References   309

13.1. Introduction

Classified data consist of 0 and 1. For example, the number of defects can be expressed by 1, and nondefects can be expressed by 0. The yield calculated from a mixture of defects and nondefects looks like a continuous variable, but actually, it is calculated from 0 and 1 data. Even in the case of chemical reactions, yield is the fraction of substance reacted within a whole substance in units of molecules or atoms.

Classified attributes are used very widely for data collection and analysis, but they are not good-quality characteristics from a quality engineering viewpoint. Using such characteristics only, a lot of information cannot be picked up, so the study would be inefficient and there would be interactions between control factors.

There are strategies to avoid such interactions. First, avoid 0–1 data. Second, when continuous-variable data cannot be measured due to lack of technology or for other reasons, try to provide more than two classes, such as good, normal, and bad, or large, medium, and small. The more levels that are classified, the more information that is revealed. In this way, the trend toward control factor change can be observed and incorrect conclusions can be avoided to some extent. The third and best strategy is to use an SN ratio based on the function. This is discussed in Chapter 14.

When 0 and 1 are used to classify attribute data, there are only two classes. In this case, there are two situations: one type of mistake and two types of mistakes. Both types of situation are discussed below.

13.2. Cases with Two Classes, One Type of Mistake

When using a vending machine, for example, we put coins into the machine and expect to get a drink. If no drink is delivered, there is one type of mistake. Another type of mistake is that a drink comes out of the machine when no coins were put in. Normally, vending machines are designed in such a way that the second type of mistake rarely occurs. In a reliability study of vending machine, only the first type of mistake is studied.

As we said earlier, the use of 0–1 data will reveal interactions between control factors. Interaction is synonymous with nonadditivity, inconsistency, and nonreproducibility. In other words, good plus good cannot be better. For example, in a manufacturing process, the fraction defective of producing a component part, denoted by p, is 0.10. Three countermeasures were taken separately to obtain the following results:

A: material

A_1:   current material $(B_1 C_1)$; $p = 0.10$

A_2:   new material $(B_1 C_1)$; $p = 0.02$

B: machine

B_1:   current machine $(A_1 C_1)$; $p = 0.10$

B_2:   new machine $(A_1 C_1)$; $p = 0.04$

C: method

C_1:   current method $(A_1 B_1)$; $p = 0.10$

C_2:   new method $(A_1 B_1)$; $p = 0.02$

Such a method is called the *one-factor-at-a-time approach*. If arithmetic additivity is assumed, the fraction defective under $A_2 B_2 C_2$ is calculated as

$$p = 0.10 + (0.02 - 0.10) + (0.04 - 0.10) + (0.02 - 0.10)$$

$$= -0.12 \tag{13.1}$$

A negative fraction is unrealistic. To avoid such a problem, it is recommended that one use the omega transformation. Following are the results of a test: $y_1, y_2, \ldots, y_n$. y values are either 0 or 1. Assume that 0 and 1 are good and defective products, respectively.

Letting the fraction defective be p, we have

$$p = \frac{y_1 + y_2 + \cdots + y_n}{n} \tag{13.2}$$

To calculate the SN ratio, data are decomposed as follows. The total variation is given by

$$S_T = y_1^2 + y_2^2 + \cdots + y_n^2 = np \tag{13.3}$$

The effect of the signal is calculated as the variation of p. Since equation (13.2) is a linear equation, denoted by L, its variation, denoted by S_p, is given by

$$S_p = \frac{L^2}{D} \tag{13.4}$$

where D is the number of units of equation (13.2):

$$D = \left(\frac{1}{n}\right)^2 + \left(\frac{1}{n}\right)^2 + \cdots + \left(\frac{1}{n}\right)^2 = \left(\frac{1}{n}\right)^2 (n) = \frac{1}{n} \tag{13.5}$$

Putting (13.2) and (13.5) into equation (13.4), we get

$$S_p = \frac{p^2}{1/n} = np^2 = S_m \tag{13.6}$$

The error variation, S_e, is calculated by subtracting S_p from S_T:

$$S_e = S_T - S_p$$
$$= np - np^2 = np(1 - p) \tag{13.7}$$

The SN ratio of digital data is calculated as the ratio of the variation of the signal to the variation of error in decibel (dB) units:

$$\eta = 10 \log \frac{S_p}{S_e} = 10 \log \frac{np^2}{np(1 - p)}$$
$$= 10 \log \frac{p}{1 - p} = -10 \log \left(\frac{1}{p} - 1\right) \tag{13.8}$$

This is the same equation as that used in the omega transformation.

Using the omega transformation, the omega value of each condition is calculated:

$$A_1 B_1 C_1 (p = 0.10) = -10 \log \left(\frac{1}{p} - 1\right) = -10 \log \left(\frac{1}{0.10} - 1\right) = -9.54 \tag{13.9}$$

$$A_2 B_1 C_1 (p = 0.02) = -10 \log \left(\frac{1}{0.02} - 1\right) = -16.90 \tag{13.10}$$

$$A_1 B_2 C_1 (p = 0.04) = -10 \log \left(\frac{1}{0.04} - 1\right) = -13.80 \tag{13.11}$$

$$A_1 B_1 C_2 (p = 0.02) = -10 \log \left(\frac{1}{0.02} - 1\right) = -16.90 \tag{13.12}$$

Letting current condition be T, condition $A_2B_2C_2$ is estimated by

$$-10 \log \left(\frac{1}{p} - 1\right) = T + (A_2 - T) + (B_2 - T) + (C_2 - T)$$

$$= A_2 + B_2 + C_2 - 2T$$

$$= -16.90 - 13.80 - 16.90 - (2)(-9.54)$$

$$= -28.52 \qquad (13.13)$$

p in the fraction is

$$p = 0.0014 \qquad (13.14)$$

13.3. Case with Two Classes, Two Types of Mistake

To explain this case, a copper smelting process is used as an example. From copper ore (copper sulfide), metal copper is extracted to produce crude copper. It is desirable that all copper in the ore be included in the crude copper and that all noncopper materials be included in the slag (waste). Actually, some impurities are included in the crude copper, and some copper is included in the slag.

Table 13.1 shows the input/output in the copper smelting process. In the table, p^* is the fraction of copper included in slag, and q^* is the fraction of impurities in the product. See also Table 13.2 There are two types of mistakes:

p: fraction of copper included in the slag

q: the fraction of impurities included in crude copper ingot

From the table, the two types of mistakes are calculated as follows:

$$p = \frac{Bp^*}{A(1 - q^*) + Bp^*} \qquad (13.15)$$

$$q = \frac{Aq^*}{Aq^* + B(1 - p^*)} \qquad (13.16)$$

p is the fraction of copper mistakenly included in the slag, and $(1 - p)$ is called *yield*. q is the fraction of noncopper materials included in the product. Both fractions are calculated by the fraction in weight.

To compare p and q, when the furnace temperature is high, more copper melts to be the product; therefore, p becomes smaller. But at the same time, more

Table 13.1
Input/output of the copper smelting process

Input	Output		
	Product	Slag	Total
Copper	$A(1 - q^*)$	Bp^*	$A(1 - q^*) + Bp^*$
Noncopper	Aq^*	$B(1 - p^*)$	$Aq^* + B(1 - p^*)$
Total	A	B	$A + B$

Table 13.2
Input/output using p and q

Input	Output		Total
	Product	Slag	
Copper	$1 - p$	p	1
Noncopper	q	$1 - q$	1
Total	$1 - p + q$	$1 + p - 1$	2

noncopper materials melt into the product, which increases q. The factor, which reduces p but increases q, is called a *tuning factor*. The tuning factor does not improve the separating function of the process. A good separating function is to reduce both p and q at the same time.

To find the control factors that can reduce p and q together, it is important to evaluate the function after tuning p and q on the same basis. For example, we want to compare which level of control factor, A_1 and A_2, has a better separating function. Assume that the p and q values for A_1 and A_2 are p_1, q_1 and p_2, q_2, respectively. As described before, we cannot compare $(p_1 + q_1)$ with $(p_2 + q_2)$. We need to transform the data so that the comparison can be made on the same basis.

For this purpose, the standard SN ratio is used. The SN ratio is calculated after tuning to reach the point when $p = q$. This fraction is denoted by p_0, calculated as

$$p_0 = \frac{1}{1 + \sqrt{[(1/p) - 1][(1/q) - 1]}} \tag{13.17}$$

The standard SN ratio is calculated by

$$\eta = 10 \log \frac{(1 - 2p_0)^2}{4p_0(1 - p_0)} \tag{13.18}$$

For a derivation, see Reference 1.

☐ Example [1]

In a uranium concentrating process, the aim was to separate ^{235}U from the mixture of ^{235}U and ^{238}U. The separating functions of two conditions were compared. Table 13.3 shows hypothetical results after separation. To calculate the standard SN ratio, the results in the table are converted into fractions to obtain Table 13.4.

Condition A_1:

$$p = \frac{3975}{5000} = 0.79500 \tag{13.19}$$

Table 13.3
Results of experiment

	Input	Output Product	Output Slag	Total
A_1	^{235}U	1,025	3,975	5,000
	^{238}U	38,975	156,025	195,000
	Total	40,000	160,000	200,000
A_2	^{235}U	1,018	3,782	4,800
	^{238}U	38,982	256,218	195,200
	Total	40,000	160,000	200.000

$$q = \frac{38,975}{195,000} = 0.19987 \tag{13.20}$$

$$p_0 = \frac{1}{1 + \sqrt{[(1/p) - 1][(1/q) - 1]}}$$

$$= \frac{1}{1 + \sqrt{[(1/0.79500) - 1][(1/0.19987) - 1]}}$$

$$= 0.49602 \tag{13.21}$$

$$\eta = 10 \log \frac{(1 - 2p_0)^2}{4p_0(1 - p_0)}$$

$$= 10 \log \frac{[1 - (2)(0.49602)]^2}{(4)(0.49602)(1 - 0.49602)}$$

$$= -41.981 \text{ dB} \tag{13.22}$$

Table 13.4
Results in fraction

	Input	Output Product	Output Slag	Total
A_1	^{235}U	0.20500	0.79500	1.00000
	^{238}U	0.19987	0.80013	1.00000
	Total	0.40487	1.59513	2.00000
A_2	^{235}U	0.21208	0.78792	1.00000
	^{238}U	0.19970	0.80030	1.00000
	Total	0.41178	1.58822	2.00000

Condition A_2:

$$\eta = -34.449 \text{ dB} \qquad (13.23)$$

Therefore, A_2 is better than A_1 by $(-34.449) - (-41.981) = 7.532$ dB

13.4. Chemical Reactions without Side Reactions

Nature of Chemical Reactions

The function of a chemical reaction is to change the combining condition of molecules or atoms. In most cases, it is impossible to measure the behavior of individual molecules or atoms. Instead, only the unit, such as yield, can measure the behavior of an entire group. The yield used in chemical reactions sounds like a continuous variable, but it is a fraction of the numbers of molecules or atoms. Therefore, it should be classified as two classified attributes (i.e., the data with 0 and 1).

Basic Function of Chemical Reactions

Conducting a chemical reaction is to change the combining state of molecules or atoms. When substances A and B react to form substance C, A and B collide and the molecules of A gradually decrease. In this case it is ideal that "reaction speed is proportional to the concentration of A."

$$A + B \rightarrow C \qquad (13.24)$$

Letting the initial amount of main raw material A be Y_0 and the amount at time T be Y, the fraction of the amount of change by time T is Y/Y_0. The reaction speed is the fraction of the reduced amount differentiated by time:

$$\frac{d(Y/Y_0)}{dT} = \beta \left(\frac{1 - Y}{Y_0} \right) \qquad (13.25)$$

The amount shown in the parentheses on the right side of (13.25) is the concentration of A remaining, denoted by p_0:

$$p_0 = 1 - \frac{Y}{Y_0} \qquad (13.26)$$

From equation (13.25),

$$\frac{Y}{Y_0} = 1 - e^{-\beta T} \qquad (13.27)$$

The concentration of A remaining is then

$$p_0 = e^{-\beta T} \qquad (13.28)$$

$$-\ln p_0 = \beta T \qquad (13.29)$$

Setting

$$-\ln p_0 = y \qquad (13.30)$$

in Equation (13.29), the ideal function is expressed by zero-point proportion equation.

Letting time, T, be the signal factor, the fractions remaining at different times, p_1, $p_2, \ldots, p_k$ are shown in Table 13.5.

The ideal relationship between the fraction remaining and time is shown in Figure 13.1.

From the results in Table 13.5, the variation of $y_1, y_2, \ldots, y_k$ is decomposed as follows.

Total variation:

$$S_T = y_1^2 + y_2^2 + \cdots + y_k^2 \quad (f = k) \tag{13.31}$$

The total variation is decomposed into the variation due to the generic function and the variation caused by the deviation from the generic function.

Variation of proportional term:

$$S_\beta = \frac{L^2}{\gamma} \quad (f = 1) \tag{13.32}$$

Linear equation:

$$L = y_1 T_1 + y_2 T_2 + \cdots + y_k T_k \tag{13.33}$$

Effective divider:

$$\gamma = T_1^2 + T_2^2 + \cdots + T_k^2 \tag{13.34}$$

Error variation:

$$S_e = S_T - S_\beta \quad (f = k - 1) \tag{13.35}$$

Table 13.6 shows the ANOVA table.

SN ratio:

$$\eta = 10 \log \frac{(1/\gamma)(S_\beta - V_e)}{V_e} \tag{13.36}$$

Table 13.5
Fraction remaining at different times

Time	T_1	T_2	...	T_k
Fraction of Remaining Main Raw Material	p_1	p_2	...	p_k
$\ln \dfrac{1}{p_0}$	y_1	y_2	...	y_k

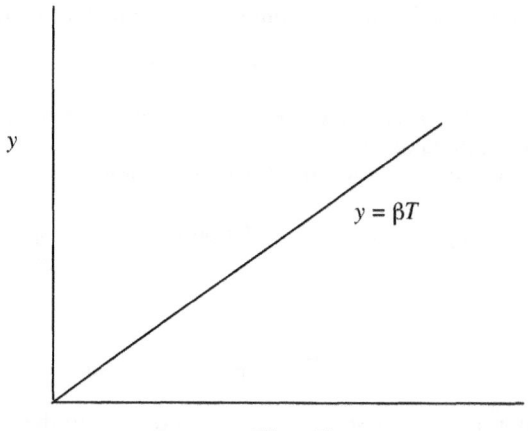

Figure 13.1
Fraction of remained
material and time

η is calculated to estimate the following:

$$\eta \rightarrow 10 \log \frac{\beta^2}{\sigma^2} \tag{13.37}$$

Sensitivity:

$$S = 10 \log \frac{1}{\gamma} (S_\beta - V_e) \tag{13.38}$$

S is calculated to estimate the following:

$$S \rightarrow 10 \log \beta^2 \tag{13.39}$$

13.5. Chemical Reactions with Side Reactions: Dynamic Operating Window

Reaction Speed
Difference Method

When substances A and B react to form the objective substance C and side-reacted substance D, it is desirable to maximize the reaction speed for producing C and minimize the one for D:

$$A + B \rightarrow C + D \tag{13.40}$$

Table 13.6
ANOVA table for chemical reactions

Source	f	S	V	E(V)
β	1	S_β	—	$\sigma^2 + \gamma\beta^2$
e	$k - 1$	S_e	V_e	σ^2
Total	k			

To do this, the following fractions are measured at time T as shown in Table 13.7.

p_0: faction of unreacted raw material, A

p_1: fraction of objective product, C

p_2: fraction of side-reacted substance, D

From p_0, p_1, and p_2, y_1 and y_2 are calculated to show two situations, M_1 and M_2, respectively. For total reaction M_1 and reaction speed β_1,

$$y_1 = \beta_1 T \qquad (13.41)$$

$$y_1 = \ln \frac{1}{p_0} \qquad (13.42)$$

For side reaction M_2 and reaction speed β_2,

$$y_2 = \beta_2 T \qquad (13.43)$$

$$y_2 = \ln \frac{1}{p_0 + p_1} \qquad (13.44)$$

Since the objectives are to maximize total reaction M_1 and minimize side reaction M_2, the operating window is shown in Figures 13.2 and 13.3. This is called the *speed difference method.*

The SN ratio is calculated from the following decomposition.

Total variation:

$$S_T = y_{11}^2 + y_{12}^2 + \cdots + y_{2k}^2 \qquad (f = 2k) \qquad (13.45)$$

Variation of proportional terms:

$$S_\beta = \frac{(L_1 + L_2)^2}{2\gamma} \qquad (f = 1) \qquad (13.46)$$

Table 13.7
Data collection for dynamic operating window

Time	T_1	T_2	...	T_k
Unreacted Raw Material	p_{01}	p_{02}	...	p_{0k}
Objective Product	p_{11}	p_{12}	...	p_{1k}
Side-Reacted Substance	p_{21}	p_{22}	...	p_{2k}
M_1: $\ln \dfrac{1}{p_0}$	y_{11}	y_{12}	...	y_{1k}
M_2: $\ln \dfrac{1}{p_0 + p_1}$	y_{21}	y_{22}	...	y_{2k}

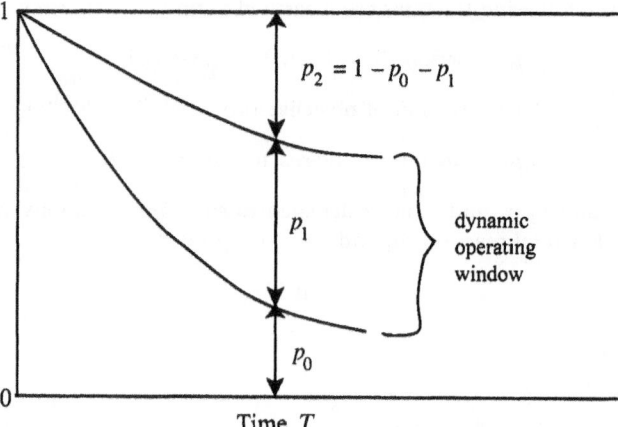

Figure 13.2
Dynamic operating
window (shown in a
fraction)

Effective divider:

$$\gamma = T_1^2 + T_2^2 + \cdots + T_k^2 \tag{13.47}$$

Linear equations:

$$L_1 = y_{11}T_1 + y_{12}T_2 + \cdots + y_{1k}T_k \tag{13.48}$$

$$L_2 = y_{21}T_1 + y_{22}T_2 + \cdots + y_{2k}T_k \tag{13.49}$$

Variation of the difference between proportional terms:

$$S_{M\beta} = \frac{(L_1 - L_2)^2}{2\gamma} \quad (f = 1) \tag{13.50}$$

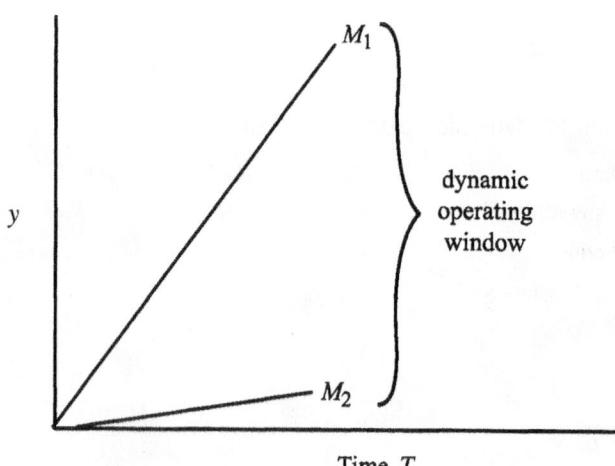

Figure 13.3
Dynamic operating
window (after data
transformation)

Error variation:

$$S_e = S_T - S_\beta - S_{M\beta} \qquad (f = 2k - 2) \qquad (13.51)$$

Table 13.8 shows the ANOVA table for a dynamic operating window.
SN ratio of the speed difference method:

$$\eta = 10 \log \frac{(1/2\gamma)(S_{M\beta} - V_e)}{V_e} \qquad (13.52)$$

Sensitivity of the speed difference method:

$$S = 10 \log \frac{1}{2\gamma} (S_{M\beta} - V_e) \qquad (13.53)$$

Sensitivity of total reaction:

$$S* = 10 \log \frac{1}{2\gamma} (S_\beta - V_e) \qquad (13.54)$$

In chemical reactions, noise factors are not included in the experiment because of the cost increase. When there are side reactions, the reaction speed of side-reacted products, β_2, may be considered as noise. Therefore, its SN ratio is written as

Reaction Speed Ratio Method

$$\eta = \frac{\beta_1^2}{\beta_2^2} \qquad (13.55)$$

Since we want to maximize the total reaction speed, β_1, denoted by η_1, it is calculated using a larger-the-better characteristic. We also want to minimize the side-reacting speed, β_2, denoted by η_2. The SN ratio of β_2 is calculated using a smaller-the-better characteristic. The two SN ratios are added together to obtain the overall SN ratio, denoted by η:

$$\eta = \eta_1 + \eta_2 \qquad (13.56)$$

In the decibel scale,

$$\eta = 10 \log \frac{\beta_1^2}{\beta_2^2} \qquad (13.57)$$

Table 13.8
ANOVA table for a dynamic operating window

Source	f	S	V	E(V)
β	1	S_β		$\sigma^2 + 2\gamma\beta^2$
$M\beta$	1	$S_{M\beta}$		$\sigma^2 + 2\gamma\sigma_{M\beta}^2$
e	$2k - 2$	S_e	V_e	σ^2
Total	$2k$	S_T		

This is called the *speed ratio method*. A sample of the data collection is shown in Table 13.9.

$$\beta_{11} = \frac{y_{11}}{T_1}$$

$$\beta_{12} = \frac{y_{12}}{T_2}$$

$$\vdots$$

$$\beta_{1k} = \frac{y_{1k}}{T_k} \tag{13.58}$$

$$\beta_{21} = \frac{y_{21}}{T_1}$$

$$\beta_{22} = \frac{y_{22}}{T_2}$$

$$\vdots$$

$$\beta_{2k} = \frac{y_{2k}}{T_k} \tag{13.59}$$

SN ratio for M_1:

$$\eta_1 = -10 \log \frac{1}{k}\left(\frac{1}{\beta_{11}^2} + \frac{1}{\beta_{12}^2} + \cdots + \frac{1}{\beta_{1k}^2}\right) \tag{13.60}$$

SN ratio for M_2:

$$\eta_2 = -10 \log \frac{1}{k}(\beta_{21}^2 + \beta_{22}^2 + \cdots + \beta_{2k}^2) \tag{13.61}$$

SN ratio of the speed ratio method:

$$\eta = \eta_1 + \eta_2 \tag{13.62}$$

Table 13.9
Data collection for the speed ratio method

	T_1	T_2	...	T_k
M_1	y_{11}	y_{12}	...	y_{1k}
M_2	y_{21}	y_{22}	...	y_{2k}
M_1	β_{11}	β_{12}	...	β_{1k}
M_2	β_{21}	β_{22}	...	β_{2k}

☐ Example [2]

In a chemical reaction, the primary raw material, F_0, reacts with another raw material to produce F_3 after forming intermediate products F_1 and F_2:

$$F_0 \rightarrow F_1 \rightarrow F_2 \rightarrow F_3 \tag{13.63}$$

In the reaction, side-reacted product was also produced. The fraction of side-reacted product is calculated by

$$\text{fraction of side-reacted product} = 1 - (F_0 + F_1 + F_2 + F_3) \tag{13.64}$$

Table 13.10 shows the fraction changes of constituents at different time segments.

SN Ratio Using the Speed Difference Method
The function of unreacted raw material, denoted by p, at time T is given by

$$p = e^{-\beta T} \tag{13.65}$$

Equation (13.65) is rewritten as

$$\ln \frac{1}{p} = \beta T \tag{13.66}$$

The fraction of unreacted raw material, p, is transformed to y as follows:

$$y = \ln \frac{1}{p} \tag{13.67}$$

From equations (13.66) and (13.67), the ideal function is written as a zero-point proportional equation:

$$y = \beta T \tag{13.68}$$

The objectives are to: maximize the reaction speed of the objective product (or maximize the reducing speed of raw material) and to minimize the reaction speed of the side-reacted product.

In Table 13.10, F_0 is raw material, F_1 and F_2 are intermediate products, and F_3 is the objective product. Therefore, the fraction of unreacted raw material is $F_0 + F_1 + F_2$. Let the total of F_0, F_1, and F_2 be p. Also, the fraction of the objective product is q. The fraction of side-reacted product is then equal to $1 - p - q$. Table 13.11 is calculated from Table 13.10.

The values of y to be used for the SN ratio are calculated as follows. At time T_3, for example, the total reaction M_1 is

$$y_{13} = \ln \frac{1}{p_3} = \ln \frac{1}{0.786} \tag{13.69}$$

Table 13.10
Fractions of constituents at time levels

	Time (h)										
	T_0 0.0	T_1 1.0	T_2 2.0	T_3 3.0	T_4 4.0	T_5 5.0	T_6 6.0	T_7 7.0	T_8 8.0	T_9 9.0	T_{10} 10.0
F_0	100.00	36.88	5.52	0.94	0.39	0.11	0.04	0.03	0.01	0.0001	0.001
F_1	0.00	51.86	46.04	19.08	9.59	2.85	1.13	0.64	0.37	0.32	0.12
F_2	0.00	8.55	42.28	58.56	51.80	39.13	27.45	20.22	15.64	11.83	8.28
F_3	0.00	2.71	5.80	20.42	36.21	53.72	65.21	71.93	76.19	79.80	83.39
Side-reacted product	0.00	0.00	0.36	1.00	2.01	4.20	6.17	7.18	7.79	8.05	8.21
Total	100.00	100.01	100.00	100.00	100.00	100.00	100.00	100.00	100.00	100.00	100.00

Table 13.11
Fractions at time levels

	Time (h)									
	T_1 1.0	T_2 2.0	T_3 3.0	T_4 4.0	T_5 5.0	T_6 6.0	T_7 7.0	T_8 8.0	T_9 9.0	T_{10} 10.0
$M_1 : p$	0.973	0.938	0.786	0.618	0.421	0.286	0.209	0.160	0.122	0.084
q	0.027	0.06	0.204	0.362	0.537	0.652	0.719	0.768	0.798	0.834
$1 - p - q$	0.000	0.004	0.010	0.020	0.042	0.062	0.072	0.078	0.081	0.082
$M_2 : p + q$	1.000	0.944	0.990	0.980	0.958	0.938	0.928	0.922	0.920	0.918

and the side reaction M_2 is

$$y_{23} = \ln \frac{1}{p_3 + q_3} = \ln \frac{1}{0.786 + 0.204} = \ln \frac{1}{0.990} \qquad (13.70)$$

Table 13.12 shows the results of calculation. The dynamic operating window is shown in Figure 13.4.

The SN ratio is calculated from Table 13.11. There are 10 time levels. Two linear equations, L_1 and L_2, are calculated.

$$L_1 = (0.0275)(1.0) + (0.0636)(2.0) + \cdots$$
$$+ (2.477)(10.0) = 83.990 \qquad (13.71)$$
$$L_2 = (0.000)(1.0) + (0.0036)(2.0) + \cdots$$
$$+ (0.0857)(10.0) = 3.497 \qquad (13.72)$$

Effective divider:

$$\gamma = 1^2 + 2^2 + \cdots + 10^2 = 385 \qquad (13.73)$$

Table 13.12
Results after conversion

	Time (h)									
	T_1 1.0	T_2 2.0	T_3 3.0	T_4 4.0	T_5 5.0	T_6 6.0	T_7 7.0	T_8 8.0	T_9 9.0	T_{10} 10.0
$M_1 : y_1$	0.0275	0.0636	0.2410	0.4815	0.8654	1.2510	1.5661	1.8313	2.1078	2.4770
$M_2 : y_2$	0.000	0.0036	0.0100	0.0203	0.0429	0.0636	0.0746	0.0811	0.0839	0.0857

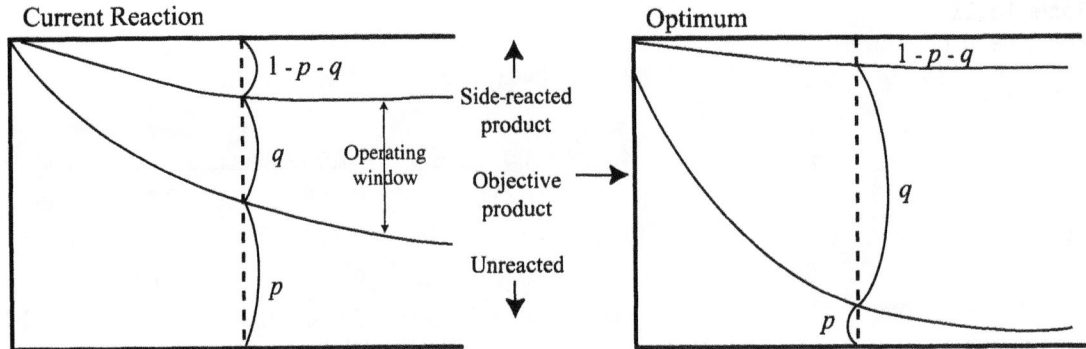

Figure 13.4
Dynamic operating window

Variation of linear term:

$$S_\beta = \frac{(L_1 + L_2)^2}{2\gamma} = \frac{(83.990 + 3.497)^2}{(2)(385)} = 9.940 \qquad (f = 1) \qquad (13.74)$$

Variation of the difference between M_1 and M_2:

$$S_{M\beta} = \frac{(L_1 - L_2)^2}{2\gamma} = \frac{(83.990 - 3.497)^2}{(2)(385)} = 8.4144 \qquad (f = 1) \qquad (13.75)$$

Total variation:

$$S_T = 0.0275^2 + 0.0636^2 + \cdots + 2.4770$$
$$+ 0.000^2 + 0.0036^2 + \cdots + 0.0857^2$$
$$= 19.026 \qquad (f = 20) \qquad\qquad (13.76)$$

Error variation:

$$S_e = S_T - S_\beta - S_{M\beta}$$
$$= 19.0262 - 4.940 - 8.4144$$
$$= 5.6715 \qquad (f = 18) \qquad\qquad (13.77)$$

Error variance:

$$V_e = \frac{S_e}{18} = \frac{5.6715}{18} = 0.0373 \qquad\qquad (13.78)$$

Table 13.13 shows the ANOVA table.

Table 13.13
ANOVA table

Source	f	S	V
β	1	9.9403	9.9403
Mβ	1	8.4144	8.4144
e	18	0.6715	0.0373
Total	20	19.0262	

The magnitude of the objective reaction is shown by $M\beta$, which corresponds to the difference between β_1 and β_2. Therefore, the SN ratio is calculated by the following equation:

$$\eta = 10 \log \frac{(1/2\gamma)(S_{M\beta} - V_e)}{V_e}$$

$$= 10 \log \frac{[1/(2)(385)](8.4144 - 0.0373)}{0.0373}$$

$$= -5.35 \text{ dB} \tag{13.79}$$

The sensitivity corresponding the difference between β_1 and β_2, denoted by S, is calculated by

$$S = 10 \log \frac{1}{2r}(S_{M\beta} - V_e) = 10 \log \frac{1}{(2)(385)}(8.4144 - 0.0373)$$

$$= -19.6 \tag{13.80}$$

The sensitivity corresponding to the total reaction, or corresponding to the average of β_1, and β_2, denoted by S^*, is calculated by

$$S^* = 10 \log \frac{1}{2\gamma}(S_\beta - V_e)$$

$$= 10 \log \frac{1}{(2)(385)}(9.940 - 0.0373)$$

$$= -17.6 \text{ dB} \tag{13.81}$$

SN Ratio Using the Speed Ratio Method
Since the generic function of chemical reactions is defined as

$$y = \beta T \tag{13.82}$$

where

$$y = \ln \frac{1}{p} \tag{13.83}$$

β is estimated from the responses of M_1 and M_2 in Table 13.12.

$$\beta = \frac{y}{T} \tag{13.84}$$

The values of β estimated under M_1 are larger-the-better, and the ones under M_2 are smaller-the-better. Therefore, the larger-the-better SN ratio, η_1, is calculated from the former β_1 values, and the smaller-the-better SN ratio, η_2, is calculated from the latter β_2 values. The SN ratio of the operating window using the speed ratio method is calculated by

$$\eta = \eta_1 + \eta_2 \tag{13.85}$$

Table 13.14 shows the estimates of β_1 (under M_1) and β_2 (under M_2). From Table 13.14, η_1 and η_2 are calculated:

$$\eta_1 = -10 \log \frac{1}{10} \left(\frac{1}{\beta_{11}^2} + \frac{1}{\beta_{12}^2} + \cdots + \frac{1}{\beta_{1 \cdot 10}^2} \right)$$

$$= -10 \log \frac{1}{10} \left(\frac{1}{0.275^2} + \frac{1}{0.0318^2} + \cdots + \frac{1}{0.2477^2} \right)$$

$$= -25.17 \text{ dB} \tag{13.86}$$

$$\eta_2 = -10 \log \frac{1}{10} \left(\beta_{21}^2 + \beta_{22}^2 + \cdots + \beta_{2 \cdot 10}^2 \right)$$

$$= -10 \log \frac{1}{10} \left(0^2 + 0.0018^2 + \cdots + 0.0086^2 \right)$$

$$= 42.57 \text{ dB} \tag{13.87}$$

Therefore,

$$\eta = \eta_1 + \eta_2$$

$$= -25.17 + 42.57 = 17.40 \tag{13.88}$$

Table 13.14

Estimates of proportional constant

	Time (h)									
	T_1 1.0	T_2 2.0	T_3 3.0	T_4 4.0	T_5 5.0	T_6 6.0	T_7 7.0	T_8 8.0	T_9 9.0	T_{10} 10.0
M_1: β_1	0.2750	0.0318	0.0803	0.1204	0.1731	0.2085	0.2237	0.2289	0.2342	0.2477
M_2: β_2	0.0000	0.0018	0.0033	0.0051	0.0086	0.0106	0.0107	0.0101	0.0093	0.0086

References

1. Genichi Taguchi, 1987. *System of Experimental Design.* Livonia, Michigan: Unipub/American Supplier Institute.
2. Genichi Taguchi et al., 1999. *Technology Development for Chemical, Pharmaceutical and Biological Industries.* Tokyo: Japanese Standards Association.

Part V

Robust Engineering

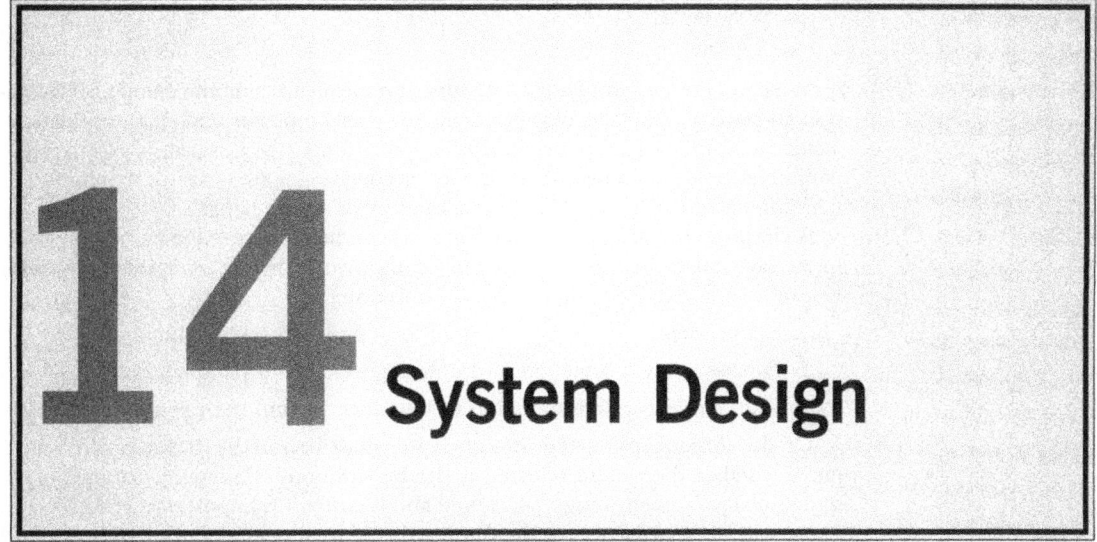

14 System Design

14.1 Design and System Selection   313
14.2. System Functionality and Evaluation   314
      References   317

14.1. Design and System Selection

The core of quality engineering is to improve productivity by improving functionality and by reducing cost. For any given objective function, there are several technical means to achieve it. For example, the paper feeder for a copy machine can use belts, air, or rollers. Selection of the mean for this objective function is called *system design*. The department that is responsible for determining an objective function is the product planning department. Selection of the function directly relates to future sales of the company. Traditionally, only the objective function has been studied. To distinguish it from an objective function, the function of the mean is called a *generic function*. In the case of the paper feeder for a copy machine, for example, the objective function is no misfeed or no multifeed. If a roller were used as the technical mean, the generic function would be the relationship between the roller-rotating angle and the paper travel distance.

To be successful in system selection, it is better to select a complex system rather than a simple one. The more complex the system being studied, the more chance of the system being improved. For example, in a development of a transistor transmitter by Yokogawa Hewlett-Packard, 38 control factors were studied by simulation [1]. In the development of LIMDOW direct-overwrite disks, it involved over 90 control factors [2]. In the early stage of this development, a traditional approach was used for six years without their being able to obtain a satisfactory prototype. After Taguchi methods were applied, the formation of multilayers was stabilized in just three years.

A large system may be divided into several subsystems. The objective function of each subsystem and its generic function must be defined. An example of determining an objective function is the selection of product types to satisfy customers. Engineers have to select the generic function based on the objective function selected, and the generic function selected must be optimized.

Selecting the system of a generic function means to combine elements or component parts so that the objective function can be performed. If the system selected does not exist, a patent can be applied for. Once a system is selected, the parameter levels of individual elements or component parts are determined. Depending on which levels are selected, the functionality changes.

To improve functionality, studies must be conducted by varying parameter levels to see how much the functionality can be improved. This is called *parameter design*. In product or process design, there are three steps:

1. *System design:* selection of a system for a given objective function
2. *Parameter design:* selection of the optimum levels of parameters
3. *Tolerance design:* determination of tolerance around each parameter level

Of the three steps, system design is the most important, but it is difficult to judge whether the system selected is the best without conducting parameter design. Since parameter design takes time and is costly, it is important to be efficient. As we can see, parameter design is the core of quality engineering. It is used to improve a system and to find the ultimate functionality of a system.

One example of the use of quality engineering is that by the Asahi Industrial Company in Japan, which developed the products used as the standards for Rockwell hardness testers. Prior to these studies, there had been only one source in the world for such hardness standards. To develop the standards, a joint research effort was undertaken by several large steel companies in Japan. But they did not succeed, despite several years of research efforts.

Assisted by the Japanese Bearing Inspection Institute and the National Research Laboratory of Metrology, using the robust technology development approach, Asahi successfully developed standards of quality superior to those existing. The development was completed in a year and a half.

14.2. System Functionality and Evaluation

The *quality* discussed in quality engineering is the variation related to (1) the loss caused by deviation from the ideal function of an objective function, and (2) the loss caused by harmful items, including the cost when the product is used. We are not going to study an objective function itself here, but will study how the function is close to the ideal function. Quality engineering is the evaluation of functionality. In quality engineering, a measure called the *signal-to-noise ratio* (SN ratio) is used for this purpose. Conceptually, it is the ratio of the magnitude of energy consumed for the objective function divided by the magnitude of energy consumed for variability. The total output is decomposed into the following two parts:

$$\text{total output} = (\text{output of useful part}) + (\text{output of harmful part}) \quad (14.1)$$

The SN ratio is the first term on the right side of the equation divided by the second term. Actual calculation is different from one generic function to another. That is the main focus of quality engineering.

Many systems exist that perform a specific objective function. If the wrong system were selected, its functionality could not be improved. As a result, expensive tolerance design would have to be performed; thus, the company would lose its competitiveness in the market.

How much improvement can be accomplished depends not only on the new and old of the system but on the system's complexity. If a system is not complicated enough, it is difficult to improve its functionality by parameter design. It seems that selecting and improving a complicated system would be costly. But if the gain (improvement in SN ratio) were 10 dB, the trouble caused in the market or during the production stage could be reduced to one-tenth through the process. If the gain were 20 dB, half of it (10 dB) could be consumed for quality improvement, and the other half could be used to loosen up tolerance or increase production speed.

If a simple system were selected, there would be little possibility for improvement of functionality, and the improvement would need to be done by conducting tolerance design, controlling the causes of variation that increases cost.

❏ Example [3,4]

The resistance of a resistor can be measured by connecting the two ends, applying a certain voltage, and reading the current. Letting the resistance, voltage, and current be R, V, and I, respectively, R is given by

$$R = \frac{V}{I} \qquad (14.2)$$

This system is shown in Figure 14.1, the parameters are unknown resistance, R, power source, V, and ammeter, X. Since it is so simple, it seems to be a good system from the cost viewpoint. However, parameter design cannot be conducted to improve the function for such a simple system. In this case it cannot be performed to improve the precision of measurement. Because there is no complexity in the system, there are no control factors to be studied for parameter design. In Figure 14.1 there is only one control factor: power voltage. When there are variations either in voltage or ammeter, there is no way to reduce their effects and to reduce the measuring error of current. The relative error of resistance is given approximately by

$$\frac{\Delta R}{R} \approx \frac{\Delta V}{V} - \frac{\Delta I}{I} \qquad (14.3)$$

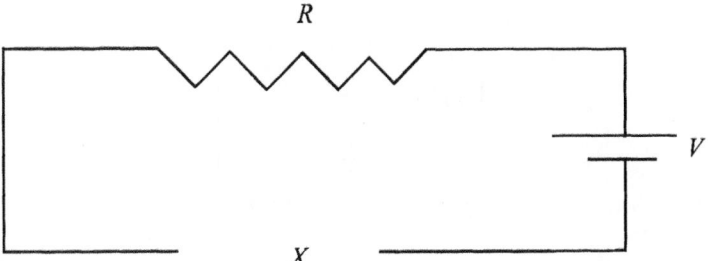

R

V

X

Figure 14.1
Simplest system to measure a resistance

In a simple system such as Figure 14.1, the only way to reduce measurement error is to reduce the causes of variation: in this case, either to reduce the variation of input voltage or to use a more precise ammeter.

To provide a constant voltage, we need to use a long-life power source with a small voltage variation. Or we need to develop a constant-voltage power-supplying circuit. For the former attempt, there is no such ideal power source. For the latter attempt, a complicated circuit is necessary. A precise ammeter must be a large unit and expensive. To improve quality, use of a simple circuit would end up being too expensive. It is why a more complicated circuit such as the Wheatstone bridge (Figure 14.2) is used. In the figure, X is an ammeter, B is a rheostat (continuously variable resistor), C and D are resistors, E is power voltage, A is the resistance of the ammeter, and F is the resistance of the power source.

A resistor of unknown resistance, y, is measured as follows: y is connected between a and b. Adjust the rheostat, B, until the reading of the ammeter, X, becomes zero.

$$y = \frac{BD}{C} \tag{14.4}$$

Even when the ammeter indicates zero, there is actually a small amount of current flowing. There are also some variations of resistance in resistors C, D, and F. There might be variation in the rheostat. In such cases,

$$y = \frac{BD}{C} - \frac{X}{C^2 E} [A(D + C) + D(B + C)][B(C + D) + F(B + C)] \tag{14.5}$$

Compared with Figure 14.1, Figure 14.2 looks complicated, but by varying the values of control factors C or D, the impacts of the variability caused by the variations of input power voltage or ammeter can be reduced substantially.

Parameter design is to reduce the variation of an objective function by varying the levels of control factors without trying to reduce the source of variation, which

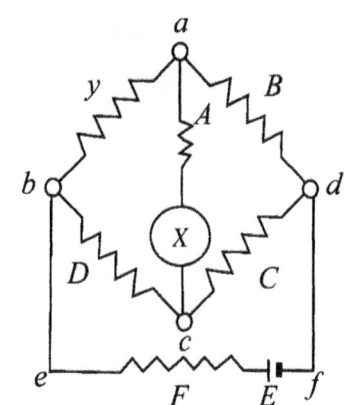

Figure 14.2
Wheatstone bridge

is costly and difficult to do. The more control factors in the system, the more possibility for improvement. The more complications between the input signal (the unknown resistance to be measured) and the output (the reading of rheostat), the higher the possibility for improvement.

Using a Wheatstone bridge, the power supply voltage variation has scarce effect-to-measurement error. Variation caused by the ammeter can often be reduced to a fraction of 100. The effect of variation caused by an inexpensive battery that is difficult to improve can also be minimized to nearly zero.

But the variation caused by resistors C, D, and B cannot be improved through parameter design, so higher-grade resistors have to be used. How high the grade should be is determined by the last step of product design: tolerance design.

References

1. Genichi Taguchi, 1984. *Parameter Design for New Product Development.* Tokyo: Japanese Standards Association.
2. *Nikkei Mechanical,* February 19, 1996.
3. Genichi Taguchi et al., 1996. *Management of Technology Development.* Tokyo: Japanese Standards Association.
4. Yuin Wu et al., 1986. *Quality Engineering: Product and Process Design Optimization.* Livonia, Michigan: American Supplier Institute.

15 Parameter Design

15.1. Introduction   318
15.2. Noise   319
15.3. Parameter Design of a Wheatstone Bridge   320
15.4. Parameter Design of a Water Feeder Valve   330
        References   339

15.1. Introduction

After a system is selected, a prototype is made for experimentation, or simulation without experimentation is conducted. If the prototype functions, or the results of the simulation, are satisfactory under certain conditions (normal or standard), the system design is complete. *Prototype functions* indicates that it hits the objective target value "under certain conditions."

The most important objective after the completion of system design is the determination of parameter midvalues (levels). This study is conducted by using the lowest-grade, least expensive raw materials and component parts. Tolerance design, which is to tighten tolerances to secure the function and better product quality, is accompanied by cost increase. Therefore, *tolerance design should be conducted after parameter design is complete.*

In many cases of traditional product design, drawings and specifications are made for production immediately after the prototype functions only under certain conditions. No further studies are made until problems are reported or complaints occur in the marketplace. The product is then reviewed, adjusted, or redesigned, which is called *firefighting.* Firefighting is needed when there has not been robust design.

Good product design engineers study robustness after system design. The prototype is tested under some customer's usage conditions. Suppose that there are several extreme usage conditions (noise factors). Using the traditional approach, an engineer tests a product by varying one of the noise conditions. If the output response deviates from the target, the engineer adjusts one of the design parameters to hit the target. Then the second noise factor is varied. The output deviates again. The engineer then varies another or other control factors to adjust output to the target. Such procedures are repeated again and again until the target is hit under all extreme conditions. But this is not a product design; it is merely operational work, called *modification, adjusting,* or *tuning.* It is extremely tedious, since this approach is similar to trying to solve several simultaneous equations—not mathematically, but through hardware.

The foregoing approach is misguided because the engineer tries to hit the target first and reduce variability last. In parameter design, robustness must be improved first, and the target is adjusted last. This is called two-step optimization.

15.2. Noise

Variables that cause product functions are called *noise.* There are three types of noise:

1. *Outer noise:* variation caused by environmental conditions (e.g., temperature, humidity, dust, input voltage)
2. *Inner noise:* deterioration of elements or materials in the product
3. *Between-product noise:* piece-to-piece variation between products

Parameter design is used to select the best control-factor-level combination so that the effect of all of the noise above can be minimized.

❏ Example [1]

The objective of a TV set's power circuit is to convert 100-V ac input into 115-V dc output. If the power circuit maintains a 115-V output anytime, anywhere, the quality of this power circuit is perfect as far as voltage is concerned. For example, element resistor (A) and the h_{FE} value of a transistor (B) in a power circuit affect the output voltage as shown in Figure 15.1.

Suppose that a design engineer forecasts the midvalues of the elements in the circuit, assembles these to get a circuit, and puts 100 V of ac into it. If the output voltage so obtained is only 100 V instead of 115 V, he then changes resistance A from $A_1 = 200\ \Omega$ to $A' = 250\ \Omega$ to adjust the 15-V deviation from the target value. From the standpoint of quality control, this is very poor methodology. Assume that the resistance used in the circuit is the cheapest grade. It either varies or deteriorates to a maximum range of $\pm 10\%$ during its service period as the power source of a TV set. From Figure 15.1, we see that the output voltage varies within a range of ± 6 V. In other words, the output voltage varies by the influence of inner noise, such as the original variation of the resistance itself, or due to deterioration. If resistance A", which has a midvalue of 350 Ω, is used instead of A', its influence on the output voltage is only ± 1 V, even if its variation is $\pm 10\%$ (35 Ω). However, the output voltage goes up about 10 V, as seen from the figure. Such a deviation can be adjusted by a factor such as B with an almost linear influence on the output voltage. In this case, 200 Ω is selected instead of 500 Ω for level B. A factor such

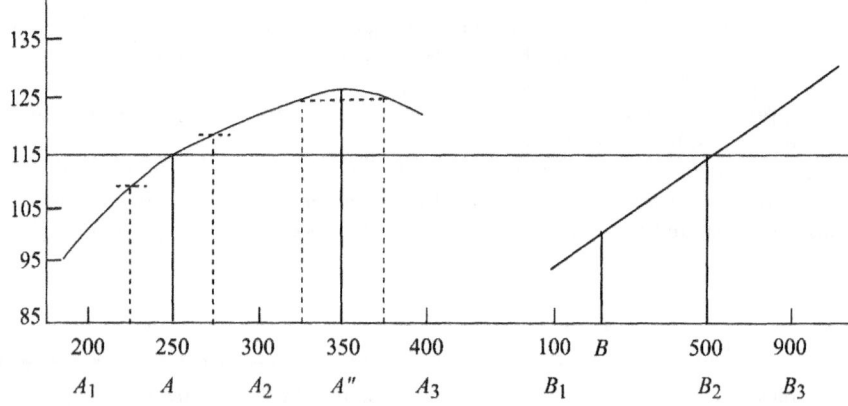

Figure 15.1
Relationship between
factors *A* and *B* and the
output voltage

as *B*, with a differential coefficient to the output voltage that is nearly constant no matter what level is selected, is not useful to reduce variability. It is used merely for the purpose of adjusting the deviation from a target value.

Parameter design is the most important step in developing stable and reliable products or manufacturing processes. With this technique, nonlinearity may be utilized positively. (Factor *A* has this property.) In this step we find a combination of parameter levels that are capable of damping the influences not only of inner noise, but also of all noise sources, while keeping the output voltage constant. At the heart of research lives a conflict: to design a product that is reliable within a wide range of performance conditions but at the lowest price. Naturally, elements or component parts with a short life and wide tolerance variation are used. The aim of the design of experiments is the utilization of nonlinearity.

15.3. Parameter Design of a Wheatstone Bridge [2]

☐ Example

The purpose of a Wheatstone bridge is to measure a resistance denoted by *y*. Figure 15.2 shows the circuit diagram for the Wheatstone bridge. The task is to select the midvalues of parameters *A*, *C*, *D*, *E*, and *F*.

To measure resistor *y*, adjustable resistor *B* is adjusted to bring the reading of the ammeter, *X*, to zero. *B* is therefore not a control factor, but factors *A*, *C*, *D*, *E*, and *F* are. The levels of these control factors are selected as shown in Table 15.1.

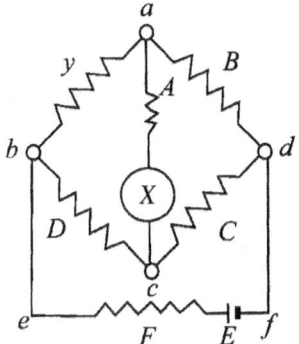

The purpose of parameter design is to investigate the overall variation caused by inner and outer noise when the levels of the control factors are allowed to vary widely. The next step is to find a stable or robust design that is essentially unaffected by inner or outer noise. Therefore, the most likely types of inner and outer noise factors must be identified and their influence must be investigated.

Consider the variations of the elements of the bridge. First, the low-priced elements are used. The varying ranges of these characteristics are shown in the three levels of the seven noise factors in Table 15.2. Since there may be a reading error, the error factors B', adjustable resistor, and X', the ammeter, are cited in the table.

Since it is believed that the power source voltage has a minimal effect on the measurement error, the lowest-priced battery on the market is used, and the three levels of battery error are set at -5%, 0, and $+5\%$. Regardless of whether the research concerns a measurement technique for minimizing error or a design for product stability, it is advisable first to consider the lower-priced component parts and materials.

In parameter design, wide intervals between factor levels are used to increase the possibility of finding a parameter level combination at which the variability of the product quality characteristic is reduced. Therefore, it would be wasteful to use high-priced components or materials at the early stage of product design.

Table 15.1
Three levels of control factors

Factor	Level 1	Level 2	Level 3
A (Ω)	20	100	500
C (Ω)	2	10	50
D (Ω)	2	10	50
E (V)	1.2	6	30
F (Ω)	2	10	50

Table 15.2
Three levels of noise factors

Factor	Level 1	Level 2	Level 3
A' (%)	−0.3	0	0.3
B' (%)	−0.3	0	0.3
C' (%)	−0.3	0	0.3
D' (%)	−0.3	0	0.3
E' (%)	−5.0	0	5.0
F' (%)	−0.3	0	0.3
X' (mA)	−0.2	0	0.2

Parameter design is the first priority in the improvement of measuring precision, stability, and/or reliability. When parameter design is completed, tolerance design is used to further reduce error factor influences.

In a Wheatstone bridge measurement, an unknown resistor, y, is connected between points a and b (Figure 15.2). Resistor B is adjusted to the point where no current, denoted by X, flows through the ammeter. The unknown resistance, y, is calculated by

$$y = \frac{BD}{C} \tag{15.1}$$

To investigate the error of measurement, assume that when the ammeter reads zero, there is a small amount of current actually flowing. In this case, the resistance, y, is not calculated by equation (15.1) but by

$$y = \frac{BD}{C} - \frac{X}{C^2 E}[A(D + C) + D(B + C)][B(C + D) + F(B + C)] \tag{15.2}$$

Control factors A, C, D, E, and F, shown in Table 15.1, are assigned in columns 1, 3, 4, 5, and 6 of orthogonal array L_{36}. An orthogonal array where control factors are assigned is called an *inner orthogonal array*. Orthogonal array L_{36} is the most popular array for use in parameter design and tolerance design when there are equations for the research. This is true because, in most instances, three levels are used for control and noise factors and the interactions between control factors are not considered. Since interactions are not calculated, the interaction effects are treated as errors. It is advantageous to have the effects of these interactions distributed uniformly in all columns. Orthogonal array L_{36} is the one array with interactions between any columns distributed uniformly among all columns. Of course, there are situations where orthogonal arrays other than the L_{36} array are used to assign control factors. In actual experiments, the adoption of the L_{36} array requires an enormous amount of time and expense. Therefore, many other sophisticated layout techniques have been developed that can be used.

Layout and Data Analysis

In Table 15.3 there are 36 combinations of the control factors A, C, D, E, and F. These represent 36 combinations of the midvalues of these factors and form the inner array. Prepare another orthogonal array, L_{36}, the *outer array*, to assign the error factors of A', B', C', D', E', F', and X'. There are 36×36, or 1296, combinations, and the values of y are calculated from these 1296 combinations. Such a layout is called the *direct product layout* of the inner orthogonal array (L_{36}) and the outer orthogonal array (L_{36}). Table 15.3 displays the direct product layout.

There are nine combinations between a control factor and its noise factor. For example, actual levels of the nine combinations between control factor A and noise factor A' are given in Table 15.4. In the outer array, noise factors A', B', C', D', E', F', and X' are assigned to columns 1, 2, 3, 4, 5, 6, and 7, respectively. Table 15.5 shows the three levels of noise factors for level 2 configuration (condition) of control factors in the inner array. The combination of experiment 2 of the inner array is $A_2C_2D_2E_2F_2$. The actual levels are: resistors $A_2 = 100\ \Omega$, $C_2 = 10\ \Omega$, $D_2 = 10\ \Omega$, battery $E_2 = 6\ V$, and resistor $F_2 = 10\ \Omega$. Around these control factor

Table 15.3
Direct product layout

		Outer Array					
	Col. No.	1	2	3	4	⋯	36
	A'	1	1	2	3	1	⋯	3
	B'	2	1	2	3	1	⋯	2
	C'	3	1	2	3	1	⋯	3
	D'	4	1	2	3	1	⋯	1
	E'	5	⋮	⋮	⋮	⋮	⋯	⋮
	F'	6	⋮	⋮	⋮	⋮	⋯	⋮
	X'	7	⋮	⋮	⋮	⋮	⋯	⋮
		⋮	⋮	⋮	⋮	⋮	⋯	⋮

Inner Array

Col No.	A 1	e 2	C 3	D 4	E 5	F 6	⋯	e 13	13						η
1	1	1	1	1	·	·	⋯	1		$y_{1.1}$	$y_{1.2}$	$y_{1.3}$	$y_{1.4}$	⋯ $y_{1.36}$	η_1
2	2	2	2	2	·	·	⋯	1		$y_{2.1}$	$y_{2.2}$	$y_{2.3}$	$y_{2.4}$	⋯ $y_{2.36}$	η_2
3	3	3	3	3	·	·	⋯	1		$y_{3.1}$	$y_{3.2}$	$y_{3.3}$	$y_{3.4}$	⋯ $y_{3.36}$	η_3
4	1	1	1	1	·	·	⋯	1		$y_{4.1}$	$y_{4.2}$	$y_{4.3}$	$y_{4.4}$	⋯ $y_{4.36}$	η_4
⋮	⋮	⋮	⋮	⋮	·	·	⋯	·		⋮	⋮	·	·	⋯ ⋮	⋮
36	3	2	3	1	·	·	⋯	3		$y_{36.1}$	$y_{36.2}$	·	·	⋯ $y_{36.36}$	η_{36}

Table 15.4
Actual levels of resistor A

	Noise Factor A'		
Control Factor A	Level 1 A'_1	Level 2 A'_2	Level 3 A'_3
Level 1, A_1	19.94	20.00	20.06
Level 2, A_2	99.70	100.00	100.30
Level 3, A_3	498.50	500.00	501.50

levels, three-level noise factors are prepared, that is, $\pm 0.3\%$ around the foregoing level of each resistor and $\pm 5\%$ around the 6 V for the battery. The midvalue of resistor B is always equal to 2 Ω, and the midvalue of the ammeter reading is always equal to zero. The three levels of these two factors of experiment 2 of the inner orthogonal array are shown in Table 15.5.

For example, the actual levels of noise factors of experiment 2 of the inner array and experiment 1 of the outer array are shown in the first column of Table 15.5, such as $A = 99.7\ \Omega$, $B = 1.994\ \Omega$, ... , $X = -0.0002$ A. Putting these figures in equation (15.2), y is calculated as

$$
\begin{aligned}
y &= \frac{(1.994)(9.97)}{9.97} - \frac{-0.0002}{(9.97^2)(5.7)} \\
&\quad \times [(99.7)(9.97 + 9.97) + (9.97)(1.994 + 9.97)] \\
&\quad \times [(1.994)(9.97 + 9.97) + (9.97)(1.994 + 9.97)] \\
&= 2.1123
\end{aligned}
\tag{15.3}
$$

Table 15.5
Three levels of noise factors of experiment 2 of inner array

Factor	1	2	3
$A\ (\Omega)$	99.7	100.0	100.3
$B\ (\Omega)$	1.994	2.0	2.006
$C\ (\Omega)$	9.97	10.0	10.03
$D\ (\Omega)$	9.97	10.0	10.03
$E\ (V)$	5.7	6.0	6.3
$F\ (\Omega)$	9.97	10.0	10.03
$X\ (A)$	-0.0002	0	0.0002

Assume that the true value of y is 2 Ω. From this value the true value, 2 Ω, must be subtracted to obtain an error, which is shown in column 1 of the results of Table 15.6.

$$2.1123 - 2 = 0.1123 \tag{15.4}$$

Similar calculations are made for the other 35 configurations of the outer array to get the data of results column 1 of Table 15.6. Since the configuration of experiment 2 represents current levels (before parameter design), current variability is calculated from the experiment 2 data. The total variation of errors of these 36 pieces of data, denoted by S_T, is

$$S_T = 0.1123^2 + 0.0000^2 + \cdots + (-0.0120)^2 \tag{15.5}$$

$$= 0.31141292 \quad (f = 36) \tag{15.6}$$

The number shown in equation (15.6) was calculated using a computer, with the control factors at the second level and with the varying noise factor ranges as given in Table 15.2, the total error variance, V_T, is

$$V_T = \frac{S_T}{36} = \frac{0.31141292}{36} = 0.00865036 \tag{15.7}$$

With a different combination of control factors, and the noise factors varying according to the levels in Table 15.2, the error variance of measurements changes. From experiment 1 through 36, combinations of the inner array, the sum of squares of errors, S_e, and the sum of squares of the general mean, S_m, are calculated:

$$S_e = \text{(total of the squares of 36 errors)}$$

$$- \text{(correction factor, } S_m) \quad (f = 35) \tag{15.8}$$

$$V_e = \frac{S_e}{35} \tag{15.9}$$

$$S_m = \frac{[(2)(36) + \text{(total of 36 errors)}]^2}{36} \tag{15.10}$$

The SN ratio, η, is

$$\eta = \frac{\frac{1}{36}(S_m - V_e)}{V_e} \tag{15.11}$$

For example, η of experiment 2 is calculated as

$$S_e = 0.1123^2 + 0.0000^2 + (-0.1023)^2 + \cdots + (-0.0120)^2 \ - S_m$$

$$= 0.31140718 - \frac{0.0006^2}{36} = 0.31140717 \tag{15.12}$$

$$V_e = \frac{0.31140717}{35} = 0.008897347 \tag{15.13}$$

Table 15.6
Layout of noise factors and the data of measurement error

No.	A' 1	B' 2	C' 3	D' 4	E' 5	F' 6	X' 7	8	9	10	11	12	13	(1) Expt. 2	(2) Improved Condition
1	1	1	1	1	1	1	1	1	1	1	1	1	1	0.1123	−0.0024
2	2	2	2	2	2	2	2	2	2	2	2	2	1	0.0000	0.0000
3	3	3	3	3	3	3	3	3	3	3	3	3	1	−0.1023	0.0027
4	1	1	1	1	2	2	2	2	3	3	3	3	1	−0.0060	−0.0060
5	2	2	2	2	3	3	3	3	1	1	1	1	1	−0.1079	−0.0033
6	3	3	3	3	1	1	1	1	2	2	2	2	1	0.1252	0.0097
7	1	1	2	3	1	2	3	3	1	2	2	3	1	−0.1188	−0.0036
8	2	2	3	1	2	3	1	1	2	3	3	1	1	0.1009	−0.0085
9	3	3	1	2	3	1	2	2	3	1	1	2	1	0.0120	0.0120
10	1	1	3	2	1	3	2	3	2	1	3	2	1	−0.0120	−0.0120
11	2	2	1	3	2	1	3	1	3	2	1	3	1	−0.1012	0.0086
12	3	3	2	1	3	2	1	2	1	3	2	1	1	0.1079	0.0033
13	1	2	3	1	3	2	1	3	3	2	1	2	2	0.0950	−0.0087
14	2	3	1	2	1	3	2	1	1	3	2	3	2	0.0120	0.0120
15	3	1	2	3	2	1	3	2	2	1	3	1	2	−0.1132	−0.0035
16	1	2	3	2	1	1	3	2	3	3	2	1	2	−0.1241	−0.0096
17	2	3	1	3	2	2	1	3	1	1	3	2	2	0.1317	0.0215
18	3	1	2	1	3	3	2	1	2	2	1	3	2	−0.0120	−0.0120
19	1	2	1	3	3	3	1	2	2	1	2	3	2	0.1201	0.0153
20	2	3	2	1	1	1	2	3	3	2	3	1	2	0.0000	0.0000
21	3	1	3	2	2	2	3	1	1	3	1	2	2	−0.1250	−0.0154
22	1	2	2	3	3	1	2	1	1	3	3	2	2	0.0060	0.0060
23	2	3	3	1	1	2	3	2	2	1	1	3	2	−0.1247	−0.0096
24	3	1	1	2	2	3	1	3	3	2	2	1	2	0.1138	0.0035
25	1	3	2	1	2	3	3	1	3	1	2	2	3	−0.1129	−0.0035
26	2	1	3	2	3	1	1	2	1	2	3	3	3	0.0951	−0.0087
27	3	2	1	3	1	2	2	3	2	3	1	1	3	0.0120	0.0120
28	1	3	2	2	2	1	1	3	2	3	1	3	3	0.1186	0.0095
29	2	1	3	3	3	2	2	1	3	1	2	1	3	−0.0060	−0.0060
30	3	2	1	1	1	3	3	2	1	2	3	2	3	−0.1197	−0.0036
31	1	3	3	3	2	3	2	2	1	2	1	1	3	0.0060	0.0060
32	2	1	1	1	3	1	3	3	2	3	2	2	3	−0.1133	−0.0093
33	3	2	2	2	1	2	1	1	3	1	3	3	3	0.1194	0.0036
34	1	3	1	2	3	2	3	1	2	2	3	1	3	−0.0957	0.0087
35	2	1	2	3	1	3	1	2	3	3	1	2	3	0.1194	0.0036
36	3	2	3	1	2	1	2	3	1	1	2	3	3	−0.0120	−0.0120

$$S_m = \frac{[(2)(36) + 0.0006)]^2}{36} = 144.00240000 \qquad (15.14)$$

$$\eta = \frac{\frac{1}{36}(144.0024 - 0.008897347)}{0.008897347} = 449.552 \qquad (15.15)$$

On the decibel scale,

$$\eta = 10 \log \frac{\frac{1}{36}(S_m - V_e)}{V_e} = 10 \log(449.552) = 26.7 \text{ dB} \qquad (15.16)$$

Analysis of SN Ratio

From the 36 data for each of the 36 conditions of the inner orthogonal array, the SN ratios are calculated using equation (15.11). Table 15.7 shows the SN ratios converted to decibels. The SN ratio is the reciprocal measurement of error variance. Accordingly, the optimum design is a combination of levels, which maximizes the decibel value. For this purpose, the decibel is used as the objective characteristic unit, and data analysis is made based on that value for the inner orthogonal array. The method is the same as the ordinary analysis of variance. For example, the main effect of A is (result from nontruncated data):

$$S_A = \frac{A_1^2 + A_2^2 + A_3^2}{12} - S_m$$

$$= \frac{378.7^2 + 225.4^2 + 80.9^2}{12} - \frac{685.0^2}{36} \qquad (15.17)$$

$$= 3700.21 \qquad (f = 2) \qquad (15.18)$$

Other factors—S_C, S_D, S_E, S_F, and S_T—are calculated similarly. Table 15.8 is the ANOVA table for the SN ratio.

$$S_T = 32.2^2 + 26.7^2 + \cdots + 8.0^2 - \frac{685.0^2}{36} = 11{,}397.42 \qquad (f = 35)$$

$$(15.19)$$

When the SN ratio is the objective characteristic, it is important to investigate how it varies with respect to control-factor-level changes. Factor-level selection should be made based on the respective SN ratios, no matter how small differences between ratios may be. Cost differences between levels should always be taken into account.

Table 15.9 shows the average of each significant control factor level. These effects are also shown in Figure 15.3. From Figure 15.3, the combination that gives the largest SN ratio is A_1, C_3, D_2, E_3, and F_1. The forecasting of the gain from the midlevel combination $A_2C_2D_2E_2F_2$ is

$$\text{gain} = (31.56 - 18.78) + (21.42 - 21.10) + (21.24 - 21.24)$$

$$+ (32.89 - 18.52) + (27.58 - 19.68)$$

$$= 12.78 + 0.32 + 0 + 14.37 + 7.90$$

$$= 35.37 - 14.54 = 20.83 \qquad (15.20)$$

Table 15.7
Layout of control factors (inner array) and SN ratio

No.	A 1	B 2	C 3	D 4	E 5	F 6	e 7	e 8	e 9	e 10	e 11	e 12	e 13	Results (dB)
1	1	1	1	1	1	1	1	1	1	1	1	1	1	32.2
2	2	2	2	2	2	2	2	2	2	2	2	2	1	26.7[a]
3	3	3	3	3	3	3	3	3	3	3	3	3	1	15.9
4	1	1	1	1	2	2	2	2	3	3	3	3	1	36.4
5	2	2	2	2	3	3	3	3	1	1	1	1	1	28.6
6	3	3	3	3	1	1	1	1	2	2	2	2	1	7.2
7	1	1	2	3	1	2	3	3	1	2	2	3	1	16.5
8	2	2	3	1	2	3	1	1	2	3	3	1	1	13.0
9	3	3	1	2	3	1	2	2	3	1	1	2	1	28.0
10	1	1	3	2	1	3	2	3	2	1	3	2	1	15.0
11	2	2	1	3	2	1	3	1	3	2	1	3	1	16.4
12	3	3	2	1	3	2	1	2	1	3	2	1	1	25.5
13	1	2	3	1	3	2	1	3	3	2	1	2	2	43.8
14	2	3	1	2	1	3	2	1	1	3	2	3	2	−8.3
15	3	1	2	3	2	1	3	2	2	1	3	1	2	14.6
16	1	2	3	2	1	1	3	2	3	3	2	1	2	29.0
17	2	3	1	3	2	2	1	3	1	1	3	2	2	6.9
18	3	1	2	1	3	3	2	1	2	2	1	3	2	14.7
19	1	2	1	3	3	3	1	2	2	1	2	3	2	21.5
20	2	3	2	1	1	1	2	3	3	2	3	1	2	17.4
21	3	1	3	2	2	2	3	1	1	3	1	2	2	17.4
22	1	2	2	3	3	1	2	1	1	3	3	2	2	46.5
23	2	3	3	1	1	2	3	2	2	1	1	3	2	5.5
24	3	1	1	2	2	3	1	3	3	2	2	1	2	−8.2
25	1	3	2	1	2	3	3	1	3	1	2	2	3	27.3
26	2	1	3	2	3	1	1	2	1	2	3	3	3	43.4
27	3	2	1	3	1	2	2	3	2	3	1	1	3	−20.9
28	1	3	2	2	2	1	1	3	2	3	1	3	3	44.1
29	2	1	3	3	3	2	2	1	3	1	2	1	3	39.3
30	3	2	1	1	1	3	3	2	1	2	3	2	3	−17.0
31	1	3	3	3	2	3	2	2	1	2	1	1	3	23.0
32	2	1	1	1	3	1	3	3	2	3	2	2	3	44.2
33	3	2	2	2	1	2	1	1	3	1	3	3	3	−0.9
34	1	3	1	2	3	2	3	1	2	2	3	1	3	43.4
35	2	1	2	3	1	3	1	2	3	3	1	2	3	−7.7
36	3	2	3	1	2	1	2	3	1	1	2	3	3	8.0

[a] Example calculated.

Table 15.8
ANOVA table for the SN ratio

Source	f	S	V
A	2	3,700.21	1,850.10
C	2	359.94	179.97
D	2	302.40	151.20
E	2	4,453.31	2,226.65
F	2	1,901.56	950.97
e	25	680.00	27.20
Total	35	11,397.42	

The next step is to attempt to confirm the foregoing conclusion by an actual error calculation. In the neighborhood of the optimum condition, three levels of error factors A', B', C', D', E', F', and X' are prepared according to Table 15.2, and the errors are calculated.

Results are shown in the last column of Table 15.6. The sum of squares of error, S_T, is

$$S_T = (-0.0024)^2 + 0.0000^2 + \cdots + (-0.0120)^2$$

$$= 0.00289623 \tag{15.21}$$

The error variance including the general mean is

$$V_T = \frac{0.00289623}{36} = 0.00008045 \tag{15.22}$$

Compared with the result in equation (15.7), this error variance, under the optimum condition, is reduced by 1/107.5, a gain of 20.32 dB. Such an improvement is much greater than the improvement obtainable by tolerance design when the varying range of each element is reduced by $\frac{1}{10}$ (probably at a huge cost increase). In

Table 15.9
Estimate of significant factors: Average of each level (decibels)

Level	A	C	D	E	F
1	31.56	14.56	20.91	5.66	27.58
2	18.78	21.10	21.24	18.52	19.68
3	6.73	21.42	14.93	32.89	9.81

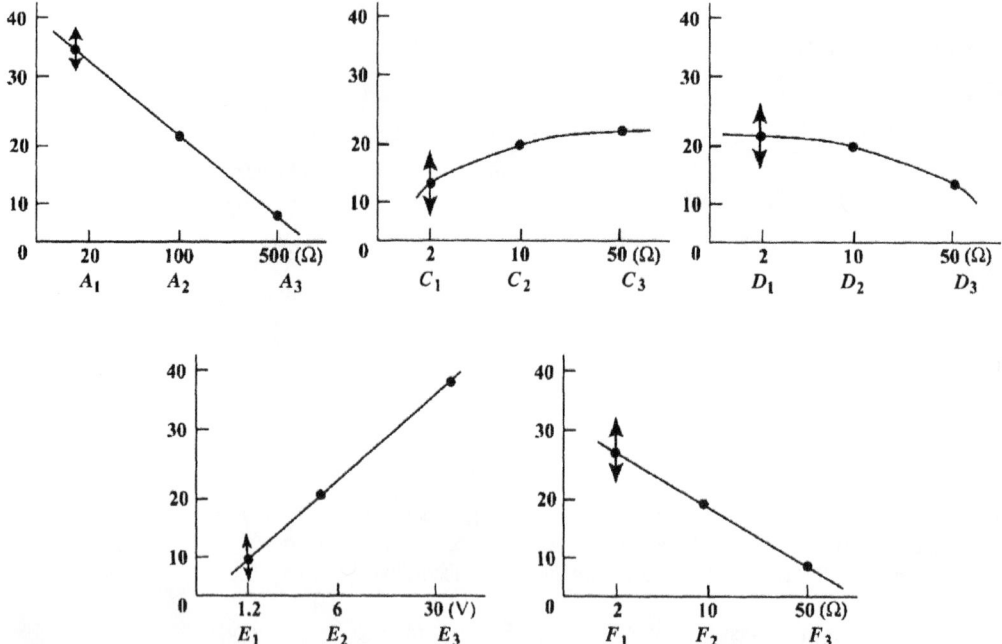

Figure 15.3
Effects of significant factors

parameter design, all component parts used in the circuit are low-priced, with wide varying ranges. In fact, the role of parameter design is to reduce measurement error, achieve higher stability, and create a significant improvement of quality by using components and/or materials that have a widely varying range and that are inexpensive. In the case of product or process design, a significant quality improvement is expected by adopting this same method.

15.4. Parameter Design of a Water Feeder Valve [3]

❏ Example

McDonnell & Miller Company produces hot-water boiler control systems. There was a problem from the field: chattering of water feeder valves. Chattering manifested itself when a feeder valve was slightly open or slightly closed. In such a case, a system harmonic was created, resulting in a loud, rolling noise through the heating pipes. The noise was considered by customers to be a significant nuisance that had to be minimized or eliminated. Figure 15.4 shows the configuration of the cartridge in the valve. A traditional one-factor-at-a-time approach was tried. The solutions, in particular the quad ring and the round poppet, caused ancillary difficulties and later had to be reversed.

Based on the Taguchi methods concept that "to get quality, don't measure quality," the engineers tried not to measure quality, that is, symptoms such as vibration and audible noise. Instead, the product function was discussed to define a generic function.

The generic function was defined as (Figure 15.5)

$$y = \frac{\beta M}{M*} \tag{15.23}$$

where y is the flow rate, M the cross-sectional flow area, and $M*$ the square root of the inlet pressure. The idea was that if the relationship above was optimized, problems such as vibration or audible noise should be eliminated. For calculation of the SN ratio, zero-point proportional equation was used.

Experimental Layout

Figure 15.6 illustrates the parameter diagram that shows the control, noise, and signal factors. Table 15.10 explains the control, noise, and signal factors and their levels.

Orthogonal array L_{18} was used to assign control factors. Noise and signal factors were assigned to outside the array. Table 15.11 shows the layout, calculated SN ratio, and sensitivity. From Tables 15.11 and 15.12, response tables and response

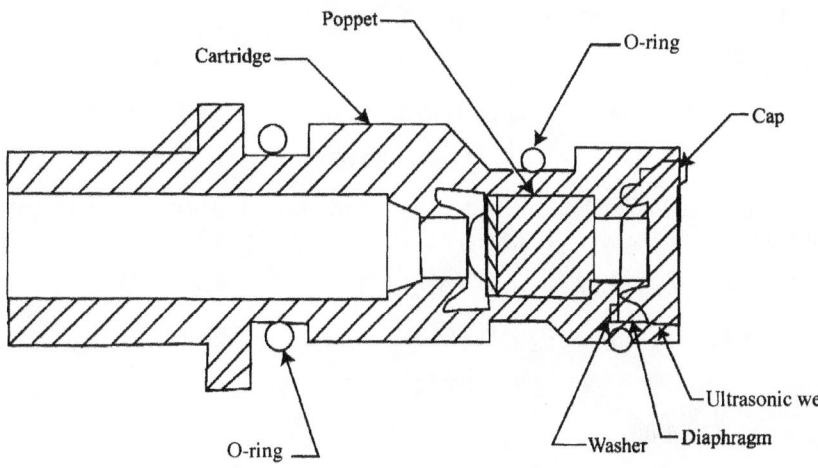

Figure 15.4
Cartridge configuration

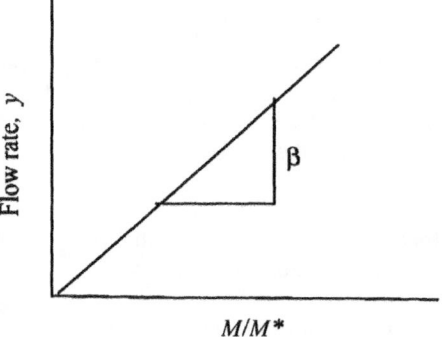

Figure 15.5
Ideal function

graphs for the SN ratio and sensitivity were prepared. Table 15.13 shows the response tables for the SN ratio and sensitivity, respectively. Figure 15.7 shows the response graph of the SN ratio.

Optimization and Confirmation
The optimum condition was selected as

$$A_2 \, B_2 \, C_3 \, D_2 \, E_2 \, F_2$$

The optimum condition predicted for the SN ratio was calculated by adding together to larger effects: A, C, D, and E, as follows:

$$\eta_{opt} = \bar{T} + (\bar{A}_2 - \bar{T}) + (\bar{C}_3 - \bar{T}) + (\bar{D}_2 - \bar{T}) + (\bar{E}_2 - \bar{T})$$
$$= \bar{A}_2 + \bar{C}_3 + \bar{D}_2 + \bar{E}_2 - 3\bar{T}$$
$$= -5.12 - 4.94 - 5.24 - 4.87 - (3)(-5.59)$$
$$= -3.40 \text{ dB} \tag{15.24}$$

The initial condition was purposely set as trial 1:

$$A_1 \, B_1 \, C_1 \, D_1 \, E_1 \, F_1$$

Therefore, its SN ratio (actual or predicted) is found from Table 15.13 as

$$\eta_{initial} = -8.27 \text{ dB} \tag{15.25}$$

The predicted optimum conditions of sensitivity was calculated using the larger effects on sensitivity, A, B, D, E, and F.

$$S_{opt} = \bar{T} + (\bar{A}_2 - \bar{T}) + (\bar{B}_2 - \bar{T}) + (\bar{D}_2 - \bar{T}) + (\bar{E}_2 - \bar{T}) + (\bar{F}_2 - \bar{T})$$
$$= \bar{A}_2 + \bar{B}_2 + \bar{D}_2 + \bar{E}_2 + \bar{F}_2 - 4\bar{T}$$
$$= -6.27 - 6.24 - 6.44 - 7.80 - 6.60 - (4)(-6.45)$$
$$= -7.55 \text{ dB} \tag{15.26}$$

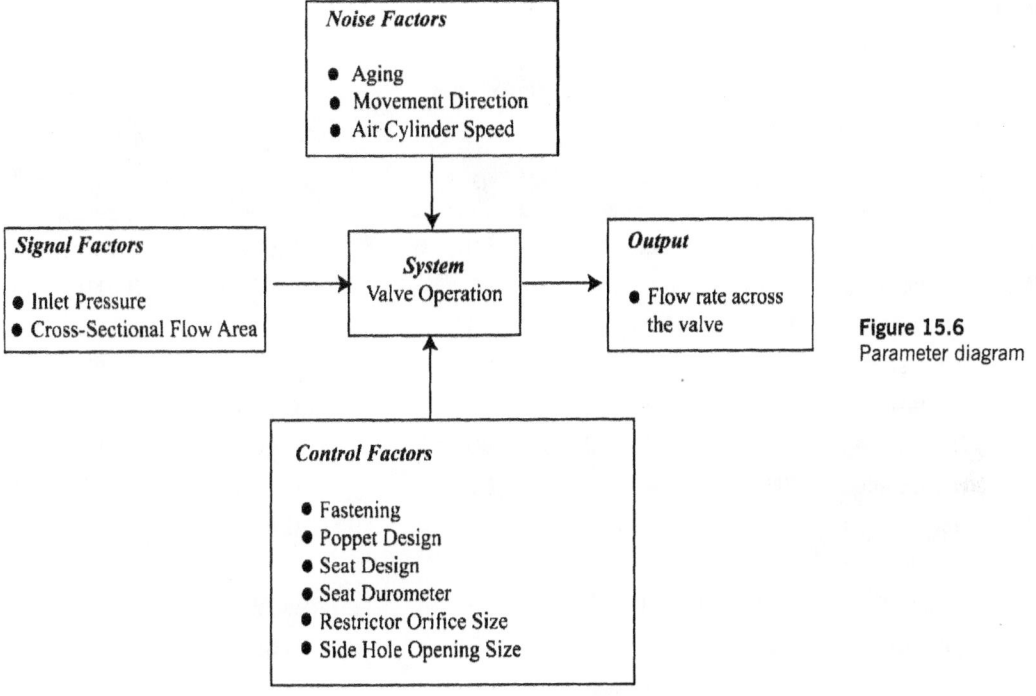

Figure 15.6
Parameter diagram

Table 15.10
Factors and levels

Factor	Description	Level 1	Level 2	Level 3
Control Factors				
A	Fastening	Not fastened	Fastened	
B	Poppet design	Hex	Square	
C	Seat design	Flat	Conical	Spherical
D	Durometer	80	60	90
E	Restrictor size (in.)	$\frac{1}{2}$	$\frac{1}{8}$	$\frac{1}{16}$
F	Cartridge side hole size (in.)	0.187	0.210	
Noise Factors				
N	Aging	New	Aged	
Flow rate	Air cylinder speed (in./sec)	0.002	0.001	
Drain rate	Air cylinder direction	Down (as valve closing)	Up (as valve opening)	
Signal factors				
M	Inlet pressure (psig)	40	150	80
M*	Cross-sectional flow area	Flow area calculated as valve closing or opening		

Table 15.11
Layout of experiments[a]

				Inner Array				
Trial No.	A Fastening	B Poppet Design	C Seat Design	D Durometer	e	E Restrictor Size	F Cartridge Slide Hole Size	e
1	Not fastened	Hex	Flat	80		0.500	0.187	
2	Not fastened	Hex	Conical	60		0.125	0.210	
3	Not fastened	Hex	Spherical	90		0.063	*0.187	
4	Not fastened	Square	Flat	80		0.125	*0.187	
5	Not fastened	Square	Conical	60		0.063	0.187	
6	Not fastened	Square	Spherical	90		0.500	0.210	
7	Not fastened	*Hex	Flat	60		0.063	0.210	
8	Not fastened	*Hex	Conical	90		0.500	*0.187	
9	Not fastened	*Hex	Spherical	80		0.125	0.187	
10	Fastened	Hex	Flat	90		0.125	0.210	
11	Fastened	Hex	Conical	80		0.063	*0.187	
12	Fastened	Hex	Spherical	60		0.500	0.187	
13	Fastened	Square	Flat	60		0.500	*0.187	
14	Fastened	Square	Conical	90		0.125	0.187	
15	Fastened	Square	Spherical	80		0.063	0.210	
16	Fastened	*Hex	Flat	90		0.063	0.187	
17	Fastened	*Hex	Conical	80		0.500	0.210	
18	Fastened	*Hex	Spherical	60		0.125	*0.187	

[a] Cells marked with asterisks.

Again, the initial condition of sensitivity was trial 1:

$$S_{initial} = -1.12 \text{ dB} \tag{15.27}$$

The gain predicted for the SN ratio is calculated from equations (15.24) and (15.25) as

$$gain(\eta) = -3.40 - (-8.27) = 4.87 \text{ dB} \tag{15.28}$$

The predicted gain in sensitivity is calculated from equations (15.26) and (15.27) as

$$gain(S) = -7.55 - (-1.12) = -6.43 \text{ dB} \tag{15.29}$$

Table 15.12
SN ratio and sensitivity

Trial No.	SN Ratio	Sensitivity $(S = 10 \log \beta^2)$
1	−8.27	−1.12
2	−4.73	−7.96
3	−3.97	−18.32
4	−6.86	−7.62
5	−4.33	−17.60
6	−5.85	−1.54
7	−5.86	−18.21
8	−9.74	−0.62
9	−4.83	−8.92
10	−3.96	−9.02
11	−4.26	−18.01
12	−6.15	−1.82
13	−5.86	−0.48
14	−4.29	−6.56
15	−4.30	−6.56
16	−4.56	−18.25
17	−8.21	−0.28
18	−4.54	−6.99

Table 15.14 compares the gains between the optimum and initial conditions.
   Based on the new design, products were manufactured and shipped to the field. In the first year there were no problems of chattering. This success encouraged the company to start applying such robust design approaches to other projects.

Discussion of the Case
In quality engineering it is recommended that noise factors be compounded to simplify experimentation. But in this particular case, all combinations of noise factors were tested. Also, as a rule, one data point for each noise factor level should be enough for analysis. In this case, a multitude of data points were collected, and the number of data points for different noise factor levels were different. The conclusions were made based on these methods of data collection.

Table 15.13
Response tables of SN ratio and sensitivity

	SN Ratio	Sensitivity ($S = 10 \log \beta^2$)
A_1	−6.05	−6.63
A_2	−5.12	−6.27
B_1	−5.76	−6.55
B_2	−5.25	−6.24
C_1	−5.90	−6.50
C_2	−5.93	−5.94
C_3	−4.94	−6.93
D_1	−6.12	−6.40
D_2	−5.24	−6.44
D_3	−5.39	−6.51
E_1	−7.35	−0.96
E_2	−4.87	−7.80
E_3	−4.55	−10.59
F_1	−5.64	−6.38
F_2	−5.48	−6.60
Average	−5.59	−6.45

Figure 15.7
Response graph of SN ratio

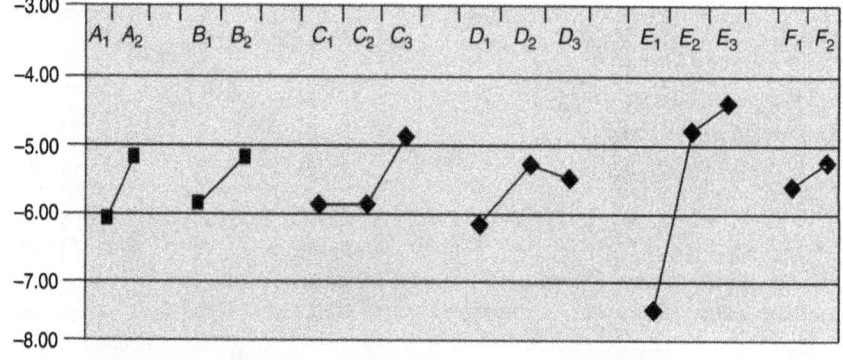

Table 15.14
Results from the confirmation run

	SN Ratio (dB)		Sensitivity (dB)	
	Predicted	Actual	Predicted	Actual
Initial	−8.27	−8.27	−1.12	−1.12
Optimum	−3.40	−4.25	−7.55	−8.64
Gain	4.87	4.02		

After the success of using quality improvement methods, their methods were studied further to see the influence of reducing the number of data points for analysis. First, six data points for each noise factor condition were used for calculation. The results of optimization were the same. In other words, comparing with the case when all data points were used for calculation, there was no difference in the selection of the optimum level for an individual control factor. Next, two data points, maximum and minimum, of each noise factor condition were used for calculation. The results of optimization were again the same. These comparisons can be seen from Figures 15.8 and 15.9.

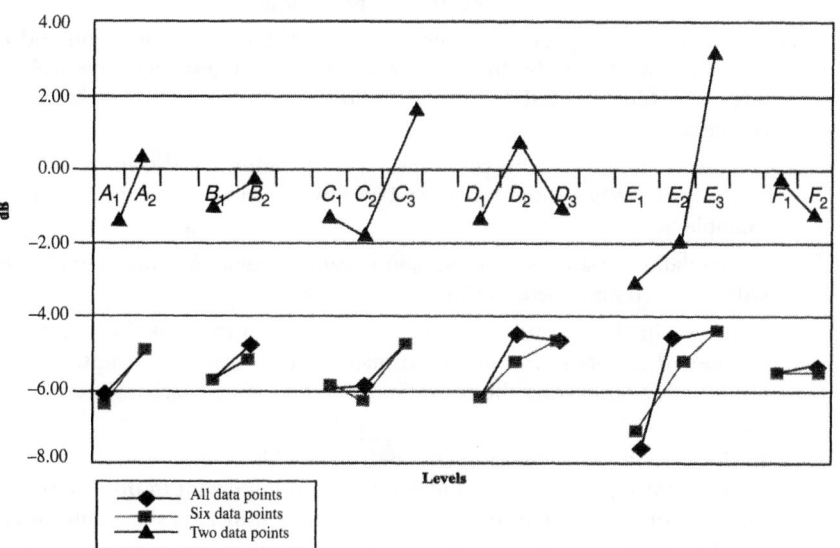

Figure 15.8
SN ratio comparison

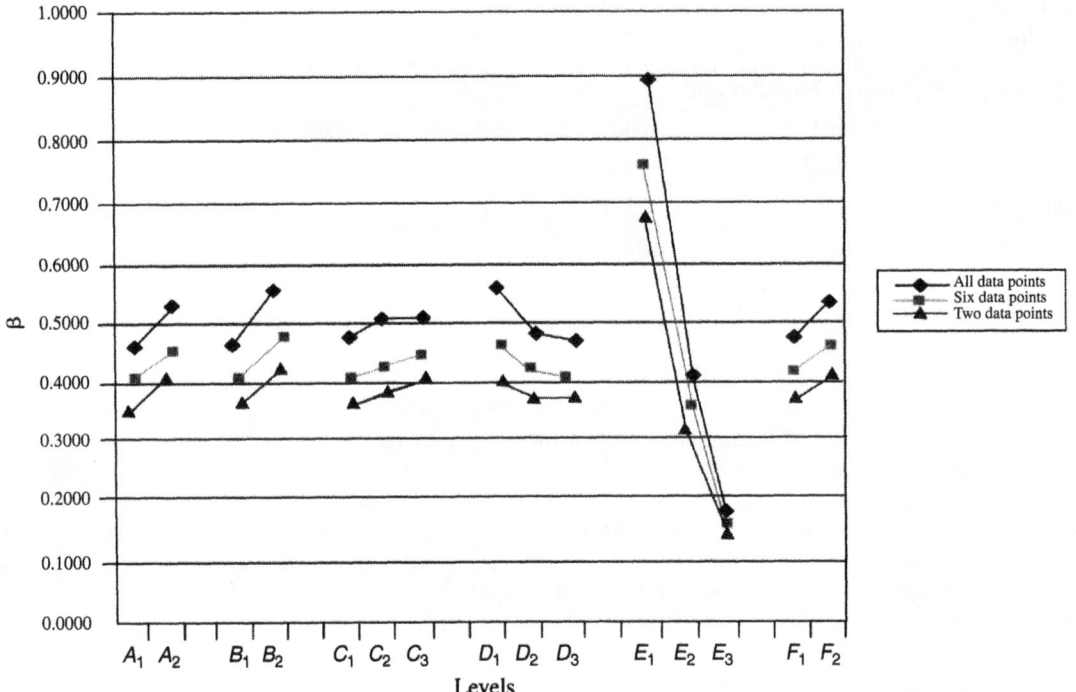

Figure 15.9
Sensitivity comparison

There are some important points in parameter design.

1. From the foregoing two examples, it is noted that cause detection and its removal were not made. Instead, the midvalues of parameters (control factors) were widely varied and the SN ratio was calculated from each of their combinations.

2. Parameter design can be performed by simulation, as shown in the first example of a Wheatstone bridge and also by experimentation in the second example.

3. In simulation, mathematical equations can be used. A more complicated system is recommended, as shown in the first example.

4. As shown in the second example, the generic function of the system was established for SN ratio calculation instead of studying the symptom, chattering. After optimizing the SN ratio, chattering disappeared without studying it.

5. In the second example, too many data were collected. However, the study showed that repetitions were unnecessary. It is important to provide proper noise factors rather than repetitions. The study indicated that noise factors could be compounded without affecting conclusions.

References

1. Genichi Taguchi et al., 1979. *Introduction to Off-Line Quality Control.* Nagoya, Japan: Central Japan Quality Control Association.
2. Yuin Wu et al. 1986. *Quality Engineering: Product and Process Optimization.* Livonia, Michigan: American Supplier Institute.
3. Amjed Shafique, 2000. Reduction of chatter noise in 47-feeder valve. Presented at ASI's 18th Annual Taguchi Methods Symposium.

16 Tolerance Design

16.1. Introduction   340
16.2. Analysis of Variance   342
16.3. Tolerance Design of a Wheatstone Bridge   346
16.4. Summary of Parameter and Tolerance Design   349
       Reference   351

This chapter is based on Yuin Wu et al., *Quality Engineering: Product and Process Optimization.* Livonia, Michigan. American Supplier Institute, 1986.

16.1. Introduction

Two specific sets of characteristics determine the quality level of a given product. These are the characteristics of the product's subsystems and/or component parts, called *low-rank characteristics,* and the characteristics of the product itself, called *high-rank characteristics.* Often, specifications and characteristic come from different sources. The end-item manufacturer often provides the specifications for component parts, while the marketing or product planning activity might specify the end item itself. In any case, one function is responsible for determining the specification of low-rank quality characteristics (or cause parameters), which will eventually affect high-rank quality characteristics. Determination of these low-rank specifications includes parameter design and tolerance design.

Parameter design, which was the subject of Chapter 15, is the determination of optimum component or parameter midvalues in such a way that variability is reduced. If, after these midvalues are determined, the level of variability is still not acceptable, components and materials must be upgraded (usually at a higher cost) to reduce variability further. The determination of acceptable levels of variability after midvalues have been established is *tolerance design,* the subject of this chapter.

☐ Example

A 100-V, 50-Hz power supply across a resistance A in series with an inductor B results in a current of y amperes.

$$y = \frac{100}{\sqrt{A^2 + [(2\pi)(50B)]^2}} \qquad (16.1)$$

The customer's tolerance of this circuit is $y = 10.0 \pm 4.0$ A. When the current is out of tolerance, the loss (including after services), denoted by A, is \$150. The annual production of this circuit is 200,000 units.

Currently, second-grade resistors, A, and inductors, B, are used. For resistor A (Ω), we have

Mean: $\qquad m_A = 9.92$
Standard deviation: $\quad \sigma_A = 0.300$

For inductor B (mH), we have

Mean: $\qquad m_B = 4.0$
Standard deviation: $\quad \sigma_B = 0.80$

The variation in the components due to 10 years of deterioration is included in the standard deviations σ_A and σ_B. The problem is to reduce the variability of the current by upgrading the resistor, the inductance coil, or both. If first-grade components were used, one-half would reduce the varying ranges, including the deterioration. The cost increases would be 12 cents for a resistor and \$1 for an inductor. Taylor's expansion and Monte Carlo simulation have frequently been used to solve such problems of component upgrade to reduce variability.

Here the experimental design approach will be discussed. The noise factor levels are set in the following way:

Three-Level Factors	Two-Level Factors
First level = second level $-\sqrt{\tfrac{3}{2}}\,\sigma$	First level = nominal $-\sigma$
Second level = nominal	Second level = nominal $+\sigma$
Third level = second level $+\sqrt{\tfrac{3}{2}}\,\sigma$	

The three levels of these components are set as follows:

$$A_1 = 9.920 - \sqrt{\tfrac{3}{2}}(0.300) = 9.552$$

$$A_2 = 9.920 \qquad\qquad (16.2)$$

$$A_3 = 9.920 + \sqrt{\tfrac{3}{2}}(0.300) = 10.288$$

$$B_1 = 0.00400 - \sqrt{\tfrac{3}{2}}(0.00080) = 0.00302$$

$$B_2 = 0.00400 \qquad\qquad (16.3)$$

$$B_3 = 0.00400 + \sqrt{\tfrac{3}{2}}(0.00080) = 0.00498$$

The units for B are changed from millihenries to henries. Using the nine possible combinations of A and B, the output current is computed from equation (16.1). For example, the current that results when the resistor/inductor combination A_1B_1 is used is

$$y_{11} = \frac{100}{A_1^2 + [(2\pi)(50B_1)]^2} = \frac{100}{9.552^2 + [(314.16)(0.00302)]^2}$$

$$= 10.42 \tag{16.4}$$

Similar computations are made for each configuration to obtain Table 16.1. In the table, the target value of 10 A has been subtracted from each observation.

16.2. Analysis of Variance

From equations (16.2) and (16.3) the quantitative contribution of each of the two component parts to the variability of output current is difficult to assess, so an analysis of variance is performed.

☐ Example [1]

From Table 16.1, the following factorial effects (constituents of the orthogonal polynomial equation) are computed:

$$m = \text{general mean}$$

$$A_l = \text{linear term of } A$$

$$A_q = \text{quadratic term of } A$$

$$B_l = \text{linear term of } B$$

$$B_q = \text{quadratic term of } B$$

$$A_1B_1 = \text{interaction of linear terms of } A \text{ and } B$$

$$e = \text{error term (including all higher-order terms)}$$

Sum of squares of these terms are calculated as follows:

$$S_m = \frac{0.06^2}{9} = 0.00040 \tag{16.5}$$

Table 16.1
Results of two-way layout

	B_1	B_2	B_3	Total
A_1	0.42	0.38	0.33	1.13
A_2	0.03	0.00	−0.04	−0.01
A_3	−0.32	−0.35	−0.39	−1.06
Total	0.13	0.03	−0.10	0.06

S_{Al}, and S_{Aq}, and so on, are calculated.

$$S_{A_l} = \frac{(W_1A_1 + W_2A_2 + W_3A_3)^2}{r(\lambda^2 S)} = \frac{[(-1)A_1 + (0)A_2 + (1)A_3]^2}{(3)(2)}$$

$$= \frac{(-1.13 - 1.06)^2}{(3)(2)} = 0.79935 \tag{16.6}$$

$$S_{A_q} = \frac{[(1)(1.13) - (2)(-0.01) + (1)(-1.06)]^2}{(3)(6)} = 0.00045 \tag{16.7}$$

$$S_{B_l} = \frac{[(-1)(0.13) + (1)(-0.10)]^2}{(3)(2)} = 0.00882 \tag{16.8}$$

$$S_{B_q} = \frac{[(1)(0.13) - (2)(0.03) + (1)(-0.10)]^2}{(3)(6)} = 0.00005 \tag{16.9}$$

To obtain the interaction $A_1 \times B_1$, the linear effects of A at each level of B are written using coefficients $W_1 = -1$, $W_2 = 0$, and $W_3 = 1$.

$$L(B_1) = (-1)(0.42) + (0)(0.03) + (1)(-0.32) = -0.74$$
$$L(B_2) = (-1)(0.38) + (0)(0.00) + (1)(-0.35) = -0.73 \tag{16.10}$$
$$L(B_3) = (-1)(0.33) + (0)(-0.04) + (1)(-0.39) = -0.72$$

and

$$S_{A_lB_l} = \frac{[W_1L(B_1) + W_2L(B_2) + W_3L(B_3)]^2}{r(\lambda^2 S)_A(\lambda^2 S)_B}$$

$$= \frac{[(-1)(-0.74) + (0)(-0.73) + (1)(-0.72)]^2}{(1)(2)(2)}$$

$$= 0.00010 \tag{16.11}$$

The total variation, S_T, is

$$S_T = 0.42^2 + 0.38^2 + \cdots + -0.39^2 = 0.8092 \tag{16.12}$$

The error sum of squares is

$$S_e = S_T - (S_m + S_{A_i} + \cdots + S_{A_iB_i})$$

$$= 0.8092 - (0.00040 + 0.79935 + \cdots + 0.00010)$$

$$= 0.00003 \tag{16.13}$$

Table 16.2 is the ANOVA table. From an ANOVA table, a design engineer can decide what should be considered as the error variance. It is exactly the same as the case of the expansion of the exponential function. In the latter case, one can include terms up to those of a certain order as factors and assign the remainder as the error. However, it is important to evaluate the magnitude of the error base on such a consideration. In other words, there is freedom to select a model, but the error of the model has to be evaluated.

After comparing the variances, B_q and A_iB_i are pooled with error to get (e). The pooled error sum of squares is 0.00018, with 5 degrees of freedom. The pooled error variance is

$$V_{(e)} = \frac{0.00018}{5} = 0.000036 \tag{16.14}$$

This variance is used to calculate the contribution percentages; for example,

$$\rho_m = \frac{S_m - V_e}{S_T}(100) = \frac{0.00040 - 0.000036}{0.80920}(100) = 0.045\% \tag{16.15}$$

The total variance when second-grade component parts are used is

$$V_T = \frac{S_T}{9} = \frac{0.89020}{9} = 0.08991 \tag{16.16}$$

Table 16.2
ANOVA table

Source		f	S	V	ρ (%)
m		1	0.00040	0.00040	0.045
A	l	1	0.79935	0.79935	98.778
	q	1	0.00045	0.00045	0.051
B	l	1	0.00882	0.00882	1.086
	q	1	0.00005	0.00005°	—
A_iB_i		1	0.00010	0.00010°	—
e		3	0.00003	0.00001	—
(e)		(5)	(0.00018)	(0.000036)	0.040
Total		9	0.80920	0.08991	100.00

Resulting Loss
Tolerance design is done at the design stage where the grades of the individual component parts are determined. For this purpose, it is necessary to acquire information regarding the loss function and the magnitude of the variability of each component part (standard deviation including deterioration). In the current example, the loss function is

$$L = k(y - m)^2$$

$$= \frac{150}{4.0^2} \sigma^2 = \$9.38(\sigma^2) \tag{16.17}$$

Letting $V_T = \sigma^2$ in equation (16.17), the loss is

$$L = 9.38V_T = (9.38)(0.08991) = 84 \text{ cents} \tag{16.18}$$

The annual loss is $(0.84)(200,000) = \$1680$. When first-grade component parts were used, the varying range would be reduced to one-half, that is, the standard deviation would be reduced to one-half. The variance would be reduced to $(\frac{1}{2})^2$ or $\frac{1}{4}$ for a linear term and to $(\frac{1}{2})^4$ or $\frac{1}{16}$ for a quadratic term.

The variance after upgrading component parts is calculated using the equation

$$V_N = V_C\left[\rho_{A_l}\left(\frac{H_N}{H_C}\right)^2 + \rho_{A_q}\left(\frac{H_N}{H_C}\right)^4 + \rho_{B_l}\frac{H_N'}{H_C'} + \rho_e\right] \tag{16.19}$$

where

V_N = variance after upgrading (new)

V_C = variance before upgrading (current)

ρ_{A_l} = percent contribution of linear term of A

ρ_{A_q} = percent contribution of quadratic term of A

ρ_{B_l} = percent contribution of linear term of B

H_N = varying range of new component A

H_C = varying range of current component A

H_N' = varying range of new component B

H_C' = varying range of current component B

In the case of component A, the loss due to using second-grade resistors is

$$L = \frac{9.375V_T(\rho_{A_l} + \rho_{A_q})}{100}$$

$$= (9.375)(0.08991)(0.98778 + 0.00051)$$

$$= \$0.8330 \tag{16.20}$$

The loss due to using first-grade resistors is

$$L = (9.375)(0.08991)[(0.98778)(\tfrac{1}{4}) + (0.00051)(\tfrac{1}{16})]$$

$$= \$0.2082 \tag{16.21}$$

The quality improvement is

$$0.8330 - 0.2082 = \$0.6248 \tag{16.22}$$

Since there is a 12-cent cost increase for upgrading, the net gain would be

$$\$0.6248 - 12 = \$0.5048 \tag{16.23}$$

The annual improvement would be

$$(0.5048)(200,000) = \$100,000 \tag{16.24}$$

Similar calculations can be made for the inductor.
Table 16.3 is the summary of the calculation results.
   The first-grade resistor and second-grade inductor should be used. In this example, the tolerances of only two factors are explained. Normally, there are many system elements in a product, and orthogonal arrays are used to facilitate analysis.

16.3. Tolerance Design of a Wheatstone Bridge

After parameter design is completed, further research may be conducted by widening the ranges of levels or by citing additional control factors and conducting additional experiments.

☐ Example

Consider the parameter design of the Wheatstone bridge example described in Chapter 15. It is assumed that the final optimum configuration is $A_1C_3D_2E_3F_1$. To start the tolerance design, lower-price elements are again considered, and the three

Table 16.3
Summary of tolerance design

Element	Grade	Price	Quality	Total
Resistor	Second	Base	83.30	83.30
	First	+12	20.82	32.82
	Second	Base	0.92	0.92
Inductor	First	+100	0.23	100.23

levels for each element are set closest to the optimum condition (with variations such as those shown in Table 15.2). In this step we usually cite more error factors than in the parameter design stage. In this example, however, the levels and error factors shown in Table 15.2 are investigated. Let the midvalue and the standard deviation of each element be m and σ, respectively. The three levels are prepared as follows:

$$\text{First level:} \quad m - \sqrt{\tfrac{3}{2}}\,\sigma$$

$$\text{Second level:} \quad m \qquad\qquad (16.25)$$

$$\text{Third level:} \quad m + \sqrt{\tfrac{3}{2}}\,\sigma$$

These are assigned in the orthogonal array L_{36}. For this optimum configuration, the measurement errors of the 36 combinations of the outer array factors are given in the last column of Table 15.6. To facilitate the analysis of variance, multiply each of the 36 data points and the percent contribution of each is calculated. The main effects of these error factors are decomposed into linear and quadratic components using the orthogonal polynomial equation.

Table 16.4
ANOVA for tolerance design

Source		f	S	V
m		1	0°	
A	l	1	1°	1
	q	1	0°	
B	l	1	86,482	86,482
	q	1	2°	1
C	l	1	87,102	87,102
	q	1	0°	
D	l	1	87,159	87,159
	q	1	0°	
E	l	1	0°	
	q	1	0°	
F	l	1	0°	
	q	1	1°	1
X	l	1	28,836	28,836
	q	1	0°	
e		21	41°	1.95
Total		36	289,623	
(e)		32	44	1.38

°: Pooled into error.
(e): Error after pooling.

From column 2 of Table 15.6, each datum is multiplied by 10,000.

$$A_1 = -24 - 60 + \cdots + 87 = -3$$

$$A_2 = 0 - 33 + \cdots + 36 = 3 \qquad\qquad (16.26)$$

$$A_3 = 27 + 97 + \cdots - 120 = 3$$

$$S_m = \frac{(A_1 + A_2 + A_3)^2}{36} = \frac{(-3 + 3 + 3)^2}{36} = \frac{9}{36} = 0.25 \qquad (f = 1) \quad (16.27)$$

$$S_{A_l} = \frac{(W_1 A_1 + W_2 A_2 + W_3 A_3)^2}{r(\lambda^2 S)} = \frac{(-A_1 + A_3)^2}{r(\lambda^2 S)}$$

$$= \frac{[-(-3) + 3]^2}{(12)(2)} = \frac{36}{24} = 1.5 \qquad (f = 1) \qquad\qquad (16.28)$$

$$S_{A_q} = \frac{(A_1 - 2A_2 + A_3)^2}{r(\lambda^2 S)} = \frac{[(-3) - (2)(3) + 3]^2}{(12)(6)} = \frac{36}{72}$$

$$= 0.5 \qquad (f = 1) \qquad\qquad (16.29)$$

$\lambda^2 S$ is the sum of squares of W_i, which can be obtained from the orthogonal polynomial table.

Other factors are calculated similarly. Table 16.4 shows the results. Notice that the four linear effects, B, C, D, and X, are significantly large. Pool the small effects with the error to obtain Table 16.5. Because B, C, and D have large contributions, higher-quality (less varying) resistors must be used for these elements. Assume that premium-quality resistors with one-tenth of the original standard deviations are used for B, C, and D. For the ammeter, the first-grade product with one-fifth of the original reading error is used. As a result, the percents of contribution of B, C, and D are one one-hundredth of, and X is one twenty-fifth of, the original. The error variance, V_e, is given as (see Table 16.5)

$$V_e = V_T[(\tfrac{1}{10})^2(\rho_B + \rho_C + \rho_D)] + (\tfrac{1}{5})^2 \rho_X + \rho_e$$

$$= (8045.1)[(0.01)(0.29860 + 0.30074 + 0.30093)]$$

$$+ (0.04)(0.09956) + 0.00017$$

$$= (8045.1)(0.013155)$$

$$= 105.8 \qquad\qquad (16.30)$$

After reconverting the measurement, the error variance, σ^2, of equation (12.7) is now reduced by 1/8176:

$$\sigma^2 = 0.000001058 \qquad\qquad (16.31)$$

The allowance, which is three times that of the standard deviation, becomes

$$\pm 3\sigma = \pm 3(\sqrt{0.000001058})$$

$$= \pm 0.0031 \qquad\qquad (16.32)$$

Table 16.5
Rearranged ANOVA table

Source	f	S	V	ρ (%)
B	1	86,482	86,482	29.86
C	1	87,102	87,102	30.07
D	1	87,159	87,159	30.09
X	1	28,836	28,836	9.96
e	32	44	1.38	0.02
Total	36	289,623	8045.1	100.000

If only resistors B, C, and D are upgraded, the resulting error variance will be

$$V_e = (8045)[(0.01)(0.29860 + 0.30074 + 0.30093)$$

$$+ 0.09956 + 0.00017]$$

$$= (8045)(0.108733) = 874.8 \qquad\qquad (16.33)$$

Converting back to the original units (dividing by $1/10^6$) gives 0.00000875.

16.4. Summary of Parameter and Tolerance Design

In the Wheatstone bridge example, the midvalues of the parameters first considered were those in column (1) of Table 16.6. When these midvalues were widely changed as shown in Table 15.1, the optimum midvalues were obtained and are shown in column (2) of Table 16.6. The information in the table also reveals that the error variance was reduced to less than 1/100 of the original value without upgrading components.

Through parameter design investigation, a method of increasing the measurement precision using low-priced and widely varying elements was sought. This method is effective and applicable to all product design situations. On the other hand, the tolerance design method demands that higher-priced and less-varying elements be used whenever these elements have a large influence on measurement results. Therefore, it is necessary to evaluate carefully the profitability of variation reduction. The countermeasures taken to minimize error effects due to resistors A, C, D, and ammeter X must be evaluated for economic impact.

In the preceding section, the conclusion was to use the highest-grade resistors for A, C, and D and the first-grade ammeter X. However, such a conclusion might not be correct from an economic viewpoint.

☐ Example

Wheatstone bridge is used in a resistor manufacturing plant to control the quality of resistors. The plant is producing 120,000 resistors annually. The tolerance of manufactured resistors is $2 \pm 0.5\ \Omega$. When a resistor is out of specification, the loss is $3. At the stage of parameter design, there is no cost increase for optimization. In the tolerance design stage, the cost for upgrading components is $4 for a resistor and $20 for an ammeter. The annual cost for interest and depreciation combined is 50% of the upgrading cost.

Losses are calculated using the following:

1. Before parameter design [column (1) of Table 16.6]
2. After parameter design [column (2) of Table 16.6]
3. After tolerance design by upgrading only resistors to the premium grade
4. After tolerance design by upgrading resistors to the premium grade and the ammeter to the first grade

The loss due to measurement before parameter design is

$$L = \frac{3}{0.5^2}\, \sigma^2 = (12)(0.00865036) = \$0.103804 \qquad (16.34)$$

The loss after parameter design is

$$L = \frac{3}{0.5^2}\, (0.00008045) = \$0.000965 \qquad (16.35)$$

The loss after tolerance design by upgrading only three resistors to the premium grade is

$$L = \frac{(4)(0.5)(3)}{120,000} + \frac{3}{0.5^2}\, (0.00000875) = \$0.000155 \qquad (16.36)$$

Table 16.6
Initial and optimum midvalues and variance of parameters

	(1) Initial Midvalue	(2) Optimum Midvalue
$A\ (\Omega)$	100	20
$C\ (\Omega)$	10	50
$D\ (\Omega)$	10	10
$E\ (V)$	6	30
$F\ (\Omega)$	10	2
Error variance	0.00865036	0.00008045
3σ	0.2790	0.0269

Table 16.7

Product design stages and costs

Design Stage	σ^2 Out	Quality L	Upgrading Cost	Total	Annual Loss
System design only	0.00805636	$0.103804	(base)	$0.103804	$12,456.00
After parameter design	0.00008045	0.00965	+0	0.000965	115.80
Tolerance design Premium-grade resistors only	0.00000875	0.000105	$0.000050	0.000155	18.60
Premium-grade resistors only plus first-grade ammeter	0.00000106	0.000013	0.000883	0.000896	107.52

The loss after tolerance design by upgrading only resistors to the premium grade and the ammeter to the first grade is

$$L = \frac{(4)(0.5)(3) + (20)(0.5)}{120,000} + (1200)(0.000001058)$$

$$= \$0.0000896 \tag{16.37}$$

The best solution is therefore (3) above, that is, equation (16.36), as shown in Table 16.7.

Reference

1. Genichi Taguchi, 1987. *System of Experimental Design.* Livonia, Michigan: Unipub/ American Supplier Institute.

17 Robust Technology Development

17.1. **Introduction**   352
17.2. **Concepts of Robust Technology Development**   353
    Two-Step Optimization   353
    Selection of What to Measure   354
    Objective Function versus Generic Function   355
    SN Ratio   355
    Orthogonal Array   356
17.3. **Advantages of Using Robust Technology Development**   358
    Technology Readiness   358
    Flexibility   358
    Reproducibility   359
17.4. **How to Apply Robust Technology Development**   359
    Paradigm Shift   359
    Identification of the Generic Function   360
    SN Ratio and Sensitivity   360
    Use of Test Pieces   360
    Compounding Noise Factors   360
    Use of Orthogonal Arrays   361
    Calculation of Response Tables   361
    Optimization and Calculation of Gains   361
    Confirmatory Experiment   361
    References   376

17.1. Introduction

Robust technology development is a revolutionary approach to product design. The idea is to develop a family of products so that development efforts do not have to be repeated for future products in the same family. This is possible because the generic function of the entire family of products is the same. Once the robustness of this generic function is maximized, all that remains to be done for a newly planned product is to adjust the output.

The optimization of a generic function can be started before product planning in the research laboratory. Since there are no actual products, a generic function can be studied using test pieces, which are easier to prepare and less expensive.

Robust design is the design of the product that will be least affected by user conditions. Noise conditions must therefore be considered in the development. Since the development is conducted in a small-scale research laboratory, the conclusions obtained must be reproducible downstream, that is, in large-scale production and under a customer's conditions. To check reproducibility, the concepts SN ratio and orthogonal array are used.

The most direct benefit of robust technological development is that the product development cycle time can be reduced dramatically. At the Quality Engineering Symposium in Nagoya, Japan, in 1992, the theme was: Can we reduce R&D cycle time to one-third? Case studies presented at the symposium offered convincing proof that it is possible to do so. If a product was shipped one year earlier than a competitor's, the profit potential would be enormous.

17.2. Concepts of Robust Technology Development

Some key concepts in robust technology development are two-step optimization, selection of what to measure, generic function, SN ratio, and orthogonal arrays. These concepts are described below.

There are two common paradigms among engineers: (1) Hit the target first at the design stage, then make it robust; and (2) to improve quality, higher-grade raw materials or component parts must be used.

Two-Step Optimization

Because of competition, there is always a deadline for the completion of product development. In many cases, therefore, drawings and specifications are made right after the target for the product to be developed is hit, and production starts immediately without any study of robustness. After receiving warrantee returns or complaints from the market, firefighting is begun.

A good product design engineer considers robustness after the target is hit. Assume that there are 10 or 15 extreme customer conditions. The test under the first extreme condition shows a deviation from the target. Therefore, the engineer changes some design parameters and the target is hit. The test under the second extreme condition deviates again. The engineer changes the same or other design parameters to hit the target to satisfy the first and second extreme conditions. In this way, the engineer tries to satisfy all extreme conditions. This is very difficult and time-consuming work. It is similar to solving 10 or 15 simultaneous equations through hardware experimentation. It is difficult, even when the study is conducted by simulation.

Using Taguchi methods, the followings are new paradigms:

1. Robustness is first, adjusting average is last.

2. To improve quality, parameter design is first, tolerance design is last.

In golf, for example, the most important thing is to reduce variability in flight distance and direction, not to improve the average distance. Using a driver, one may hit the ball to a respectable average of 200 yards, with a range of ±50 yards. Using the parameter design approach, by changing the levels of control factors such as stance, grip, or swing, variability can be reduced. After variability is reduced, adjusting the average distance is accomplished simply by selecting the proper club. This parameter design approach is now widely recognized as *two-step optimization.*

Selection of What to Measure

In the 1989 ASI Taguchi Methods Symposium, the following was selected as the theme: To get quality, don't measure quality! The second *quality* here refers to the *symptom* or subjects measured in firefighting. Typical examples are audible noise level and chattering or vibration of a machine or a car.

In the case of a company manufacturing thermosetting and steel laminated sheets, cracking was measured before shipment [1]. No cracking was detected in the final product inspection, but cracking was noticed two or three years after shipment at the rate of 200 ppm in the market. According to the method of testing before shipment, the problem would not have been noticed, even after an 80-hour test.

Parameter design was conducted by measuring the following generic function of the product: *Load is proportional to deformation.*

In the study, evaluation of cracking was made using an 18-hour test instead of 80 hours without observing cracks. A 4.56-dB gain was confirmed from the study.

Quality is classified into the following four levels:

1. Downstream quality (customer quality)
2. Midstream quality (specified quality)
3. Upstream quality (robust quality)
4. Origin quality (functional quality)

Downstream quality refers to the type of quality characteristics that are noticed by customers. Some examples in the auto industry are gasoline mileage, audible noise, engine vibration, and the effort it takes to close a car door. Downstream quality is important to management in an organization. However, it is of limited value to engineers in determining how to improve the quality of a product. Downstream quality serves to create a focus on the wrong thing and is the worst type of quality characteristic to use for quality improvement. Downstream quality, however, is easy to understand. If such quality characteristics were used for quality improvement, the statement would be: To improve quality, measure that quality.

Midstream quality is also called *specified quality*, since it is specification-related quality, such as dimension, strength, or the contents of impurities in a product. Midstream quality is important for production engineers, since it is essential to "make the product to print." But many engineers today have begun to realize that making to specifications (print) does not always mean that good quality is achieved. It is slightly better to measure these downstream quality characteristics— but not much.

Upstream quality is expressed by the nondynamic SN ratio. Since the nondynamic SN ratio relates to the robustness of a fixed output, upstream quality is the second-best quality characteristic to measure. It can be used to improve the robustness of a particular product instead of a group or a family of products. However, the concept of SN ratio is not easily understood by engineers.

Origin quality is expressed by the dynamic SN ratio. This is the best and most powerful type of quality and is the heart of robust technology development. In contrast to nondynamic SN ratios that are used to improve the robustness of the average, the output of a particular product, the dynamic SN ratio is used to improve the *generic* function of a product, the outputs of a *group* of products. By using this type of quality characteristics, we can expect the highest probability of reproducing the conclusions from a research laboratory to downstream, that is, to large-scale production or market. The use of origin quality therefore improves the efficiency of R&D. However, this type of quality level is hardest for engineers to understand.

Why is downstream quality the worst, midstream the next worst, upstream the next best, and origin quality the best? This concept is based on philosophy as well as on actual experiments over the years. Origin quality can be described from the viewpoint of interactions between control factors, which we try to avoid, because the existence of such interactions implies that there is no reproducibility of conclusions. When downstream quality is used, these interactions occur most frequently. Such interactions occur next most frequently when midstream quality is used, less in upstream quality, and least in origin quality.

The most important quality, origin quality, is the SN ratio of dynamic characteristics, the relationships between an objective function and a generic function.

Objective Function versus Generic Function

An *objective function* is the relationship between the signal factor input used by a customer and the objective output. In an injection molding process, for example, the user (production engineer) measures the relationship between mold dimension and product dimension. Mold dimension is the input signal, and product dimension is the objective output. It is the relationship between the user's intention and the outcome.

A *generic function* is the relationship between the signal factor input and output of the technical mean (method) the engineer is going to use. In the case of an injection molding process, the process is the mean (material) used to achieve a certain objective function. The engineer expects the material to have a certain physical property, such as the relationship between load and deformation.

In another example, the objective function of a robot is the relationship between the programmed spot the robot is supposed to travel to and the spot actually reached. If an engineer used an electric motor to move the arm of the robot, this is the technical means. The generic function would be the relationship between the input rpm and the angle change of the robot's arm.

Generic function is related to physics. It is energy-related; therefore, there is additivity or reproducibility. We want additivity for reaching conclusions before product planning. Conclusions from a small-scale study should be reproducible in large-scale manufacturing and under the customer's conditions. Therefore, generic functions are better to study than are objective functions.

In some cases, generic function cannot be used for reasons such as lack of technology to measure the input or output. In such a case, objective functions may be used, but it is always preferable to study generic functions.

SN Ratio

The SN ratio has been used since the beginning of the last century in the communications industry. For example, a radio measures the signal from a broadcasting station and reproduces the original sound. The original sound from the broadcasting station is the *input*, and the sound reproduced from the radio is the *output*. But there is noise mixed with the sound reproduced. A good radio catches mostly original sound and is least affected by noise. Thus, the quality of a radio is expressed by the ratio of the power of the signal to the power of the noise. The unit decibel (dB) is used as the scale. For example, 40 dB indicates that the magnitude of the signal is 10,000 and the magnitude of noise is 1; 45 dB indicates that the power of the signal is approximately 30,000 times the power of noise. It is important to remember that the higher the SN ratio, the better the quality. In quality engineering, every function is evaluated by the SN ratio: the power of the signal to the power of the noise.

The objective of design of experiments is to search for the relationship between various factors and the objective characteristic (response). In design of experiments, it is important to find the correct relationship (model) precisely and efficiently. When the interactions between factors (causes) are significant, these terms must be included.

The primary uses of parameter design in quality engineering are (1) to introduce the SN ratio, the measure of functionality; and (2) to use orthogonal arrays to check the significance of the interactions between control factors. If such interactions are significant, the reproducibility of conclusions is questionable.

The SN ratio gives the interactions between a control factor, the signal factor, and noise factors. Introduction of the SN ratio enables one to avoid interactions between control factors. However, it is not known whether or not the interactions between control factors are significant just from the introduction of SN ratio, so orthogonal arrays are used to check the existence of significant interactions.

Traditionally, technology has been developed to find the following equations:

$$y = \beta M \tag{17.1}$$

$$y = f(M, x_1, x_2, \ldots, x_n) \tag{17.2}$$

where y is the response, M the signal, and $x_1, x_2, \ldots, x_n$ are noise factors. In such a study it is important to find an equation that expresses the relationship precisely. A product with good functionality means has a large portion of equation (17.1) included in equation (17.2). In quality engineering, the responses of equation (17.2) are decomposed into the useful part: equation (17.1), and the harmful part, the rest: equation (17.2) − equation (17.1). The latter is the deviation from the useful part:

$$f(M, x_1, x_2, \ldots, x_n) = \beta M + [f(M, x_1, x_2, \ldots, x_n) - \beta M] \tag{17.3}$$

In traditional design of experiments, there is no distinction between control and noise factors. Error is assumed to be random, and its distribution is discussed seriously. In quality engineering, neither random error nor distribution is considered. Thus, quality engineering differs entirely from the traditional design of experiments in this aspect.

Orthogonal Array In technology development, studies are conducted in research laboratories using test pieces. This method is used to improve and forecast actual product quality: for both large-scale manufacturing and for quality in the market. Therefore, it is extremely important that the conclusions from the laboratory be reproducible.

Reproducibility does not mean that an effect is reproduced under the same conditions, as are repetitions in a laboratory. Instead, it means the effects can be reproduced in the following situations:

1. The conclusions from the test piece study are reproduced in the actual product.

2. The conclusions from a small-scale study are reproduced in large-scale manufacturing.

3. The conclusions from limited conditions are reproduced under various other customers' conditions.

Since conditions in the laboratory are different from those downstream (large-scale manufacturing or the market), the output response measured in the laboratory is different from the targeted response in manufacturing or in the market. Such differences must be adjusted later in the product design so that the output hits the target. This is the second-stage tuning process in two-step optimization. At the first stage, only functionality is improved.

Suppose that there are two different designs (such as different control factor levels or different conditions: initial and optimum). The SN ratios of these designs are calculated, and their difference, the gain, is calculated. Since the SN ratio is the measure of the stability of a function, it is expected that the gain may be reproduced downstream. However, there is no guarantee, and it is necessary to check reproducibility. That is why orthogonal arrays are used.

For experimentation, orthogonal array L_{18} is used most of the time. For two-level control factors, L_{12} is used. For simulation, L_{36} is used. Using orthogonal arrays does not improve the efficiency of experiments. Indeed, it can take longer. In the arrays above, the interactions between control factors are distributed to other columns and confounded with main effects. It is important to conduct an experiment in such a fashion that interactions between control factors are purposely confounded with control factors. When interactions are found between control factors, main effects are affected and deviate. Therefore, the gain between estimation from an orthogonal array experiment and the confirmatory experiment changes. If the gains between the two do not change significantly, we can expect good reproducibility.

The one-factor-at-a-time approach has been used widely in experimentation. Using this method, however, one cannot know whether the main effect will be reproduced under other conditions. Therefore, the one-factor-at-a-time method is successful only when there are no interactions.

The same is true with assigning only main effects to an orthogonal array. The experiment will be successful only if there are no interactions. Therefore, the possibility of success using this approach is the same as that for using the one-factor-at-a-time approach. In other words, if there are no interactions between control factors, it does not matter whether an orthogonal array or a one-factor-at-a-time approach is used.

The reason for using an orthogonal array without assigning interactions is to check the existence of interactions by checking for reproducibility. In other words, orthogonal arrays are used to see if the experiment is a success or a failure. It is the same as inspecting product quality during manufacturing. If the product passes the test, the inspection was a waste, since the product is a good one without inspection. Inspection is beneficial or valuable only when a defective product is found. Inspection is not useful to improve quality, but it can prevent the problems caused by quality.

Similarly, the objective of using orthogonal arrays is to find bad experiments. It is to prevent improper technology or a bad design from being transferred forward and causing problems downstream.

If an orthogonal array is used and the gain shows good reproducibility, it was a waste to use the array. The benefit is when the gain shows no reproducibility. Then we should be thankful for the discovery.

To be successful using orthogonal arrays for experimentation, it is necessary to find a characteristic to be measured that has minimum interactions between control factors, that is, the SN ratio derived from functionality. From the viewpoint of quality engineering, the dynamic SN ratio of a generic function is superior to the dynamic SN ratio of an objective function, and the dynamic SN ratio of an objective function is superior to the nondynamic SN ratio of an objective characteristic. Of course, these are still much superior to midstream quality and downstream quality characteristics.

17.3. Advantages of Using Robust Technology Development

There are three main features in using robust technology development: technology readiness, flexibility, and reproducibility.

Technology Readiness
Problem solving or firefighting, on which engineers spend most of their time, is conducted based on customer complaints or dissatisfaction. But it is too late by then. *Quality must be built into a product before the product design stage by first studying the function of the product.*

This has been a distinctive advantage that U.S. engineers have had over Japanese engineers, who have done most research by studying actual products. Research conducted based on the ideal product function enables manufacturers to be technologically ready and be able to bring new products to the market ahead of competitors.

In the development of a soldering process, for example, research has been done on test pieces in U.S. laboratories. The process was developed without actual products. Once this process is optimized, the technology is ready to produce future products.

Flexibility
The nondynamic SN ratio was used widely and successfully in the 1970s to optimize the robustness of one target or one particular product. But it would be much better and more powerful if multiple targets were optimized in one study. Dynamic SN ratios are used for this purpose. The dynamic SN ratio can improve linearity, improve sensitivity, and reduce variability, as we have said.

In the case of an injection molding process, the objective function is that the input dimension be proportional to the output dimension. If linearity, sensitivity, and variability are all improved, the change is as shown in Figure 17.1 from *a* to *b*.

After optimization, the process can produce any product having dimensions within the range of the outputs studied. Therefore, one study is good enough to optimize a group of the products. In other words, one dynamic SN ratio is used to evaluate all three aspects for the products within the ranges in the study.

A generic function may be used to optimize the same injection molding process. For example, the input and output may be load and deformation, respectively; or the input and output could be the weight of the product measured inside and the one outside the water. What type of generic function should be used depends on

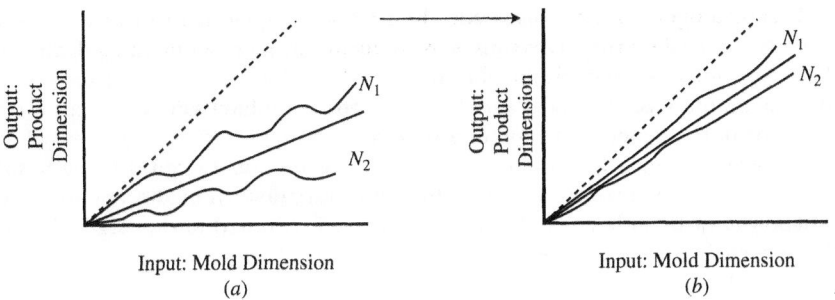

Figure 17.1
Input/output for an objective function

the reproducibility of conclusions. If the linearity, sensitivity, and variability are improved, the change is as shown in Figure 17.2 from *a* to *b*.

In this way, one-shot optimization enables engineers to utilize the database over and over without repeating similar research from product to product. This means a potentially huge savings in time, work power, and capital.

Reproducibility

As shown earlier, interaction (between control factors) is synonymous with poor reproducibility and also with lack of additivity, because downstream quality gives us the best chance of having interactions and origin quality the least. If an appropriate generic function is selected and the system is optimized, a minimum chance of having interactions, that is, maximum chance of downstream reproducibility, can be expected.

17.4. How to Apply Robust Technology Development

Here are some guidelines to follow for the application of robust technology development, based on the explanations presented earlier.

Paradigm Shift

Quality control activities have traditionally been performed within the manufacturing organization. Quality engineers, production engineers, and statisticians have played a major role in quality improvement. In contrast, R&D and product design engineers have been less involved in quality activities. It is said that a definition of R&D in the United States has wrongly excluded the issue of quality.

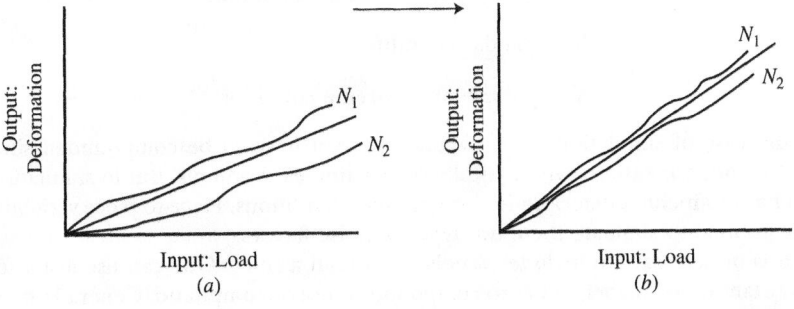

Figure 17.2
Input/output for a generic function

This paradigm has to be changed. The traditional approach to quality control issues has largely been synonymous with problem solving or firefighting. Utilizing the robust technology development approach, one does not focus on the symptoms or on the basic causes of variation. Instead, as we have said, the focus is on the study of the generic function of a product.

Another paradigm for product development or product design is trying to hit the target first, then considering how to reduce variation. In quality engineering, variation must be reduced first, then the average is adjusted to the target. This is the concept of two-step optimization.

Identification of the Generic Function

Problems cannot be solved by observing symptoms. In one case in the auto industry, for example, the audible noise level of a brake system was observed to solve the problem of squealing, but this did nothing to solve the problem in the long term. In another case, symptoms such as wear and audible noise level were observed in a study of timing belts. Those problems disappeared completely after dynamic SN ratios were used for evaluation rather than inspection and observation.

Ideally, the generic function would be used to calculate the dynamic SN ratio. If technology is lacking, the objective function of a system may be used instead. For computer simulation, nondynamic SN ratio may be used, since nonlinearity is not reflected in the software in most cases.

SN Ratio and Sensitivity

Maximizing the SN ratio is to maximize robustness. Sensitivity is analyzed to adjust slope in the case of dynamic characteristics or to adjust the average in the case of nondynamic characteristics.

Use of Test Pieces

Since robust technology development is applied before product planning, there are no products yet, and test pieces must be used for development. The advantages of using test pieces are low cost and shorter time frame.

Compounding Noise Factors

To conduct a study efficiently, noise factors are compounded to set two or three conditions, such as:

N_1:   negative-side extreme condition

N_2:   positive-side extreme condition

or

N_1:   negative-side extreme condition

N_2:   standard condition

N_3:   positive-side extreme condition

In the case of simulation, noise factors may not have to be compounded since calculations are easier than physically conducting experiments. But in simulation, it is hard to include deterioration or customer conditions. Piece-to-piece variability may be used to simulate these two types of noise factors.

It is unnecessary to include *all* noise factors in a study. One can use just a few important noise factors, either to compound or not to compound. Generally, noise factors related to deterioration and customer conditions are preferable for experimentation rather than the factors related to piece-to-piece variation.

Orthogonal arrays are used to check the existence of interactions. It is recommended that one use L_{12} and L_{18} for experimentation or L_{36} arrays for simulation.

Use of Orthogonal Arrays

Response tables for the SN ratio and sensitivity are used for two-step optimization. No ANOVA tables are necessary for selection of the optimum levels of control factors.

Calculation of Response Tables

The optimum condition is determined from the two response tables: tables of SN ratio and sensitivity. The SN ratios of the current condition and the optimum condition are calculated. The predicted gain is then calculated from the difference. The same is done for sensitivity.

Optimization and Calculation of Gains

Under the current and optimum conditions, confirmatory experiments are run, along with their SN ratios and sensitivity; then the gain is calculated. The gain must be close enough to the predicted gain to show good reproducibility. How close it should be is determined from the engineering viewpoint rather than by calculating percent deviation. If the two gains are not close enough, it indicates poor reproducibility, and the quality characteristic used must be reexamined.

Confirmatory Experiment

❒ Example [2]

To save energy for copy machines and printers, a new fusing system was developed using a resistant heater that warms up in a short period of time. This project was scheduled to be completed in one year. In the midst of development, there were design changes both mechanically and electrically that affected the progress. By the use of robust technology development approaches, however, development was completed within the scheduled time. From past experience it was estimated that the development would have taken two years if the traditional approaches had been used.

The electrophotographic system requires a high-temperature roller that fixes resin toner to the paper by heat fusion. In this system, 90% of the power is consumed for idle time (the time to wait when the machine is not running). To save energy, it would be ideal that no heating be applied during waiting and heat applied only when needed. But heating takes time that makes users wait.

To solve this problem, a new system, which heats up the system to a high temperature in a short period of time, was to be developed in one year. This system was named the Minolta advanced cylindrical heater (MACH).

Traditional Fixing System
Figures 17.3 and 17.4 show the structure of the traditional and the MACH heat roller, respectively. The differences between the two systems are in the heat source and the method of power transformation. In the former system, heat is transferred

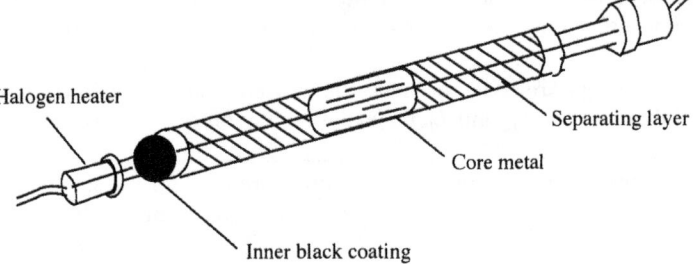

Figure 17.3
Traditional roller

by the radiation from a halogen heater. In the latter case, it is transferred directly by a resistor. To heat the heat roller surface to 180°C requires that the surface temperature of the halogen lamp reach 350°C. But using the resistor heater, its surface temperature needs only to be 180°C.

Functions to Be Developed

Figure 17.5 shows the functions of the system. Of those functions, the following four were selected to develop:

- ❑ Power supply function
- ❑ Heat-generating function
- ❑ Separating function
- ❑ Temperature-measuring function

These functions were developed independently and concurrently because (1) there was a short period of time for development, (2) several suppliers would be involved, (3) any delay of one supplier affects the entire project, and (4) no technological accumulations was available.

Development of the Power Supply Function

This system was constructed by a carbon brush and power receiving parts. It was considered to be ideal that the power be transformed without loss.

Figure 17.4
MACH heat roller

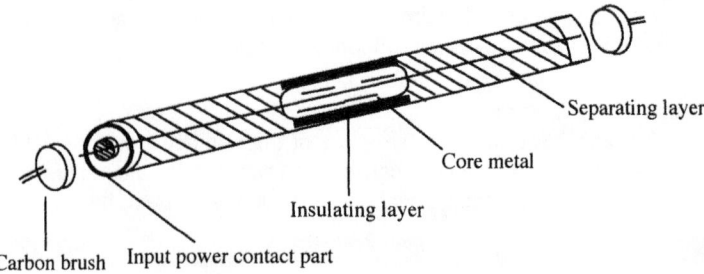

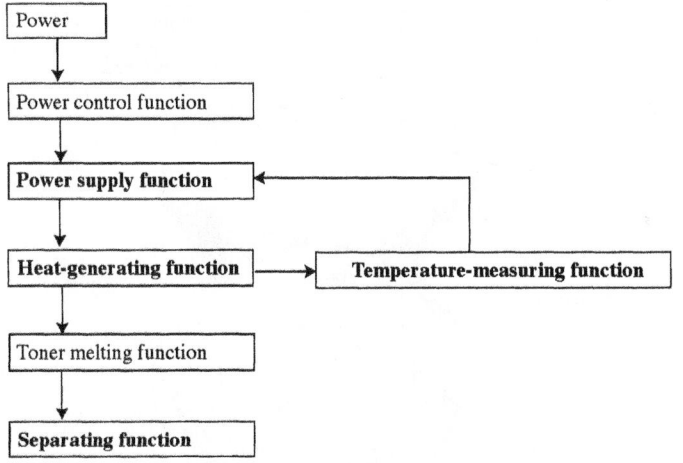

Figure 17.5
Functions of the system

- *Input:* square root of input power (five levels)
- *Output:* square root of output power
- *Noise:* not compounded (eight levels)
- *Duration time:* 0, 750 hours
- *Temperature:* 20, 200°C
- *Contact pressure:* small, large

Figure 17.6 shows the ideal function, Figure 17.7 shows the test piece prepared for the study, and Table 17.1 shows the control factors and their levels. Orthogonal array L_{18} was used to assign control factors. The SN ratios and sensitivity of the 18

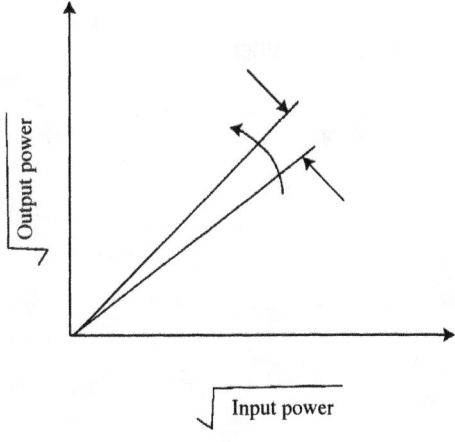

Figure 17.6
Ideal function of power supply

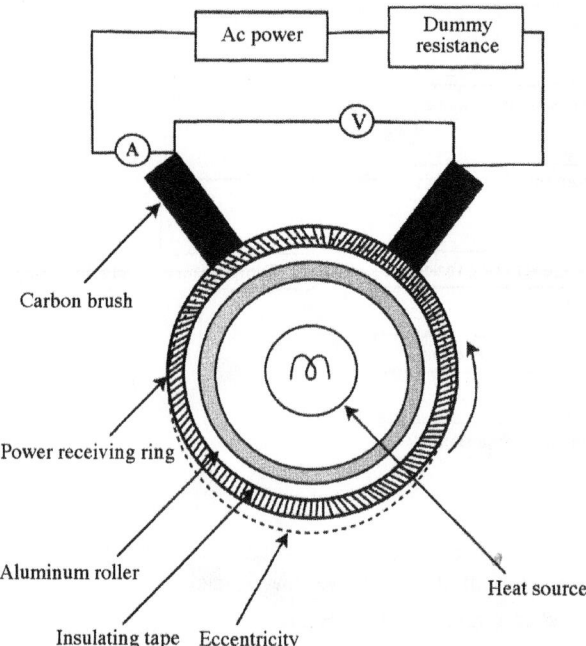

Figure 17.7
Test piece for power
supply function

runs were calculated, and response curves were drawn as shown in Figures 17.8
and 17.9.

   The levels with "o" marks in the figures were selected as the optimum conditions. Table 17.2 shows the results of the confirmatory experiment. Factors A, C,
and E were used for estimation. The gain of the SN ratio was fairly reproduced.

Table 17.1
Control factors for power supply function

	Factor	Level 1	Level 2	Level 3
A	Temperature during deterioration	Low	High	—
B	Brush shape	I	II	III
C	Pressure	Weak	Medium	Strong
E	Brush material	A	B	C
F	Lead wire cross-sectional area	Small	Medium	Large
G	Bush area	Small	Medium	Large
H	Holder distance	Small	Medium	Large

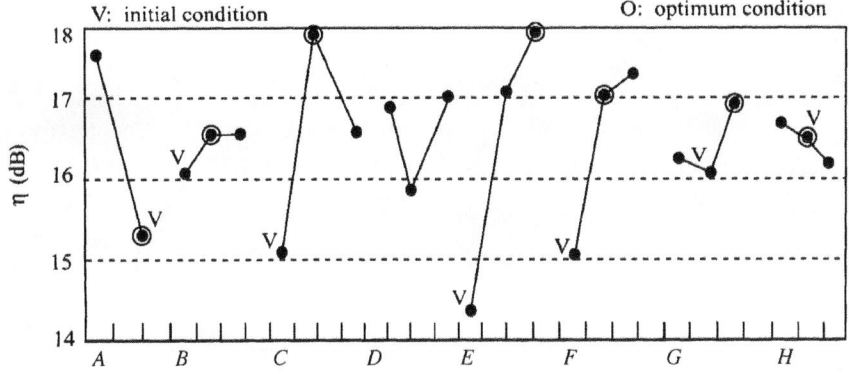

Figure 17.8
SN ratio of power supply function

The gain of sensitivity was small, but the power loss at the brush was reduced by half.

Figures 17.10 and 17.11 are the results of confirmation under current and optimum conditions where twice the range of signal factor in the orthogonal array experiment was used. It shows a higher power transformation efficiency and a smaller variation due to noise.

Development of the Heat-Generating Function
The heat-generating part is constructed by a lamination of core metal, insulating layer, and heat-generating element. The ideal function is to transfer power to heat efficiently and uniformly.

❑ *Input:* square root of input power (three levels)

❑ *Output:* square root of temperature rise

❑ *Noise:* four measuring positions and one non-heat-generating position (five levels)

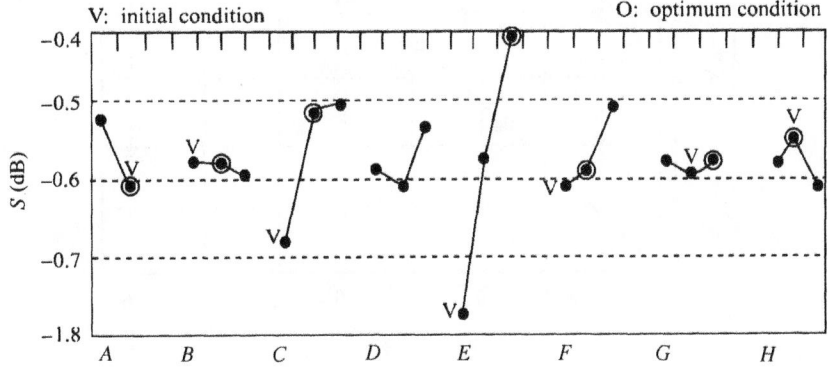

Figure 17.9
Sensitivity of power supply function

Table 17.2
Results of confirmation

	SN Ratio		Sensitivity	
	Estimation	**Confirmation**	**Estimation**	**Confirmation**
Initial condition	12.18	6.16	−0.89	−0.93
Optimum condition	18.22	10.98	−0.40	−0.81
Gain	6.04	4.82	0.49	0.12

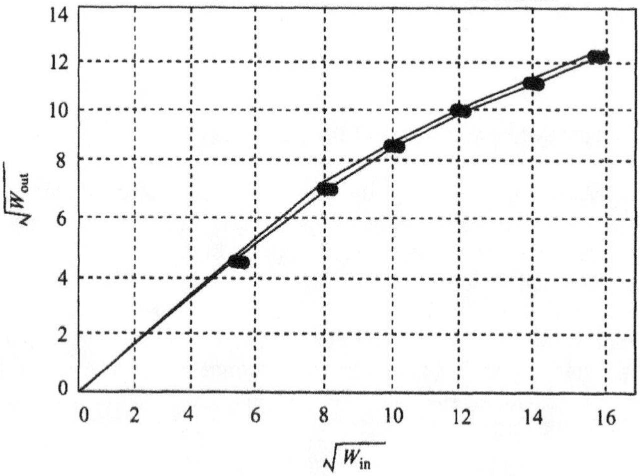

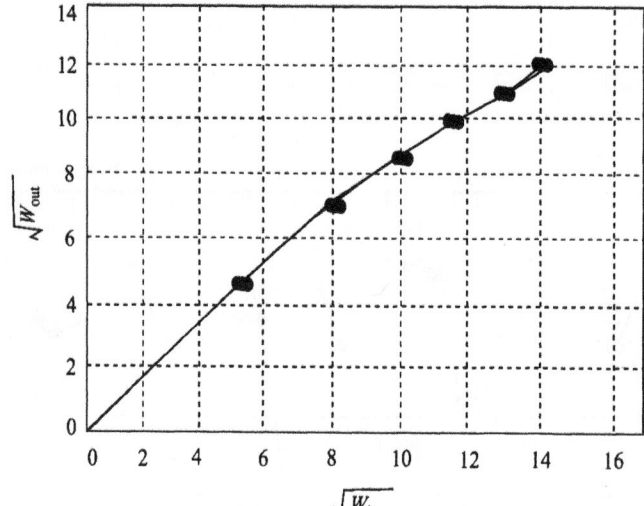

Figure 17.10
Input/output of initial
condition

Figure 17.11
Input/output of
optimum condition

Figure 17.12 shows the ideal function, and Figure 17.13 shows the test piece for heat-generating function. Heat-generating positions and materials were cited as control factors as shown in Table 17.3. From the results of SN ratio and sensitivity, response graphs were plotted as shown in Figures 17.14 and 17.15. Factors A, B, C, D, and F were used to estimate the SN ratio and sensitivity of the optimum condition. The results of confirmatory experiments are shown in Table 17.4.

The input/output characteristics of the initial and optimum conditions are plotted in Figures 17.16 and 17.17. From the figures, it can be seen that the temperature variation of the optimum condition was improved significantly compared with the initial condition.

Development of the Peel-Off Function

The peel-off unit is constructed by a core metal coated by primer and fluorine resin layers. It is ideal that the peel-off force of melted toner from the unit is small, as shown in Figure 17.18. For test pieces, plain peel-off layers were used for simplicity and easy preparation. The plates were heated, and unfixed toner layer was pressed and melted, then peel-off force was measured. Figure 17.19 shows the test piece.

Table 17.5 shows the control factors. Since the time to complete development is limited, the control factors related to fluorine resin were not studied and only manufacturing-related control factors were studied. The optimum condition was determined from response graphs in Figures 17.20 and 17.21.

Results of the confirmatory experiment are shown in Table 17.6. They show poor reproducibility. But from Figures 17.22 and 17.23, one can see that the

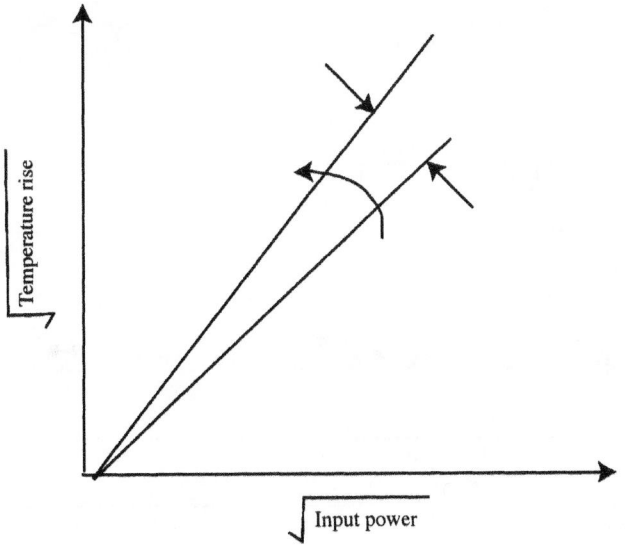

Figure 17.12
Ideal function of heat-generating function

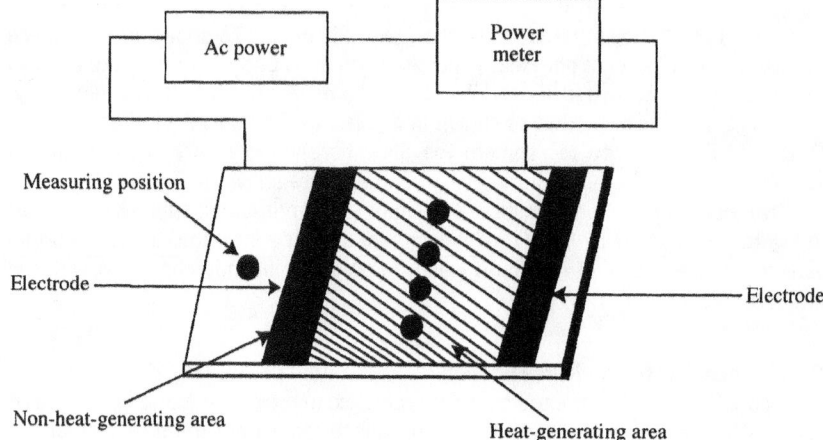

Figure 17.13
Test piece for heat-generating function

Table 17.3
Control factors for heat-generating function

	Factor	Level 1	Level 2	Level 3
A	Position of heat-generating element	Face	Reverse	—
B	Material I	A	B	C
C	Material II	Small	Medium	Large
D	Material III	Small	Medium	Large
F	Material IV	X	Y	Z

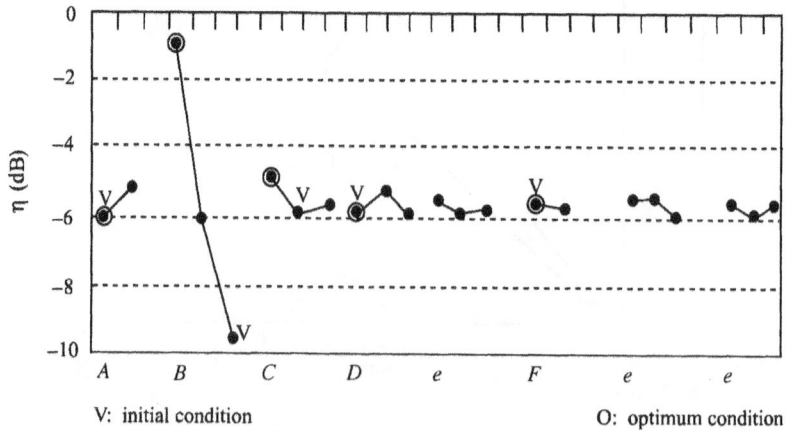

Figure 17.14
SN ratio of heat-generating function

V: initial condition   O: optimum condition

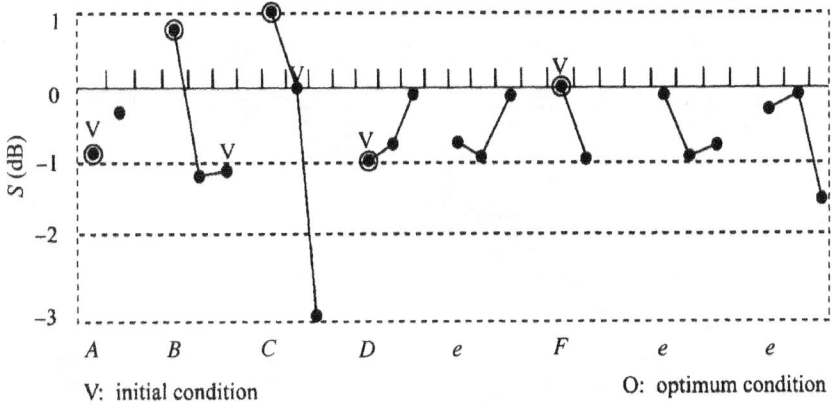

Figure 17.15
Sensitivity of heat-generating function

V: initial condition   O: optimum condition

peel-off force under the optimum condition became smaller. As a result, a peel-off force that would satisfy the product requirement was obtained. The poor reproducibility was probably caused by uneven setting of noise factor conditions in the experiment.

Development of the Temperature-Measuring Function
A unit constructed by a thermistor and a sponge measures temperature (Figure 17.24). The ideal function is that the temperature of the object measured be proportional to the reading, as shown in Figure 17.25. Temperature reading was transformed from the reading of voltage. The construction of thermistor, resistance, and so on, was studied as control factors, as shown in Table 17.7.

From the response graphs shown in Figures 17.26 and 17.27, the optimum conditions were selected from the SN ratio. Factor B was selected based on the situation of self-heat generation. Factors G and H was selected based on the cost aspect.

Table 17.8 shows the results of confirmation. All factors were used to estimate both the SN ratio and sensitivity. Both gains showed good reproducibility. From Figures 17.28 and 17.29, the optimum condition has a better linearity. The study

Table 17.4
Results of confirmation

	SN Ratio		Sensitivity	
	Estimation	**Confirmation**	**Estimation**	**Confirmation**
Initial condition	−10.32	0.01	−0.02	−0.05
Optimum condition	−0.79	10.94	1.70	3.26
Gain	9.53	10.93	1.72	3.31

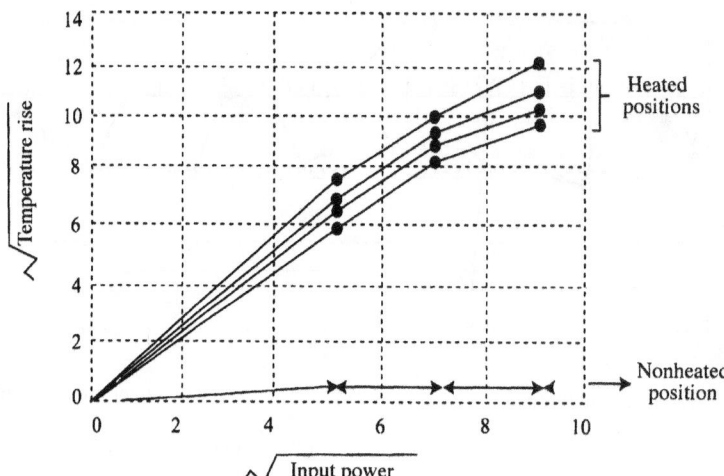

Figure 17.16
Input/output of initial
condition

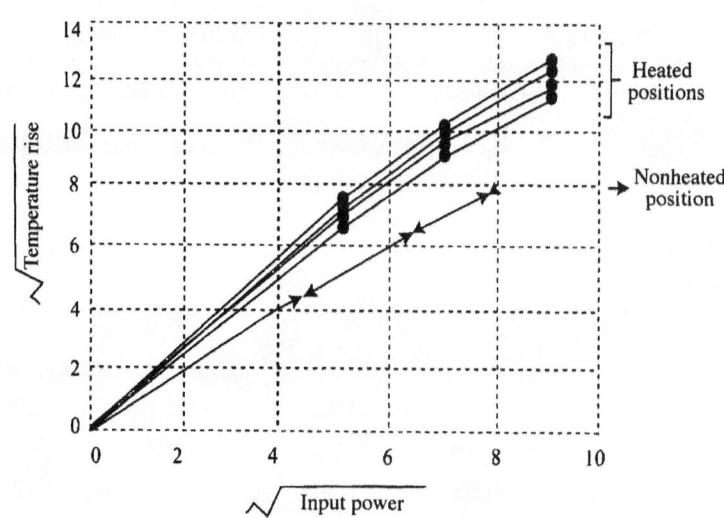

Figure 17.17
Input/output of
optimum condition

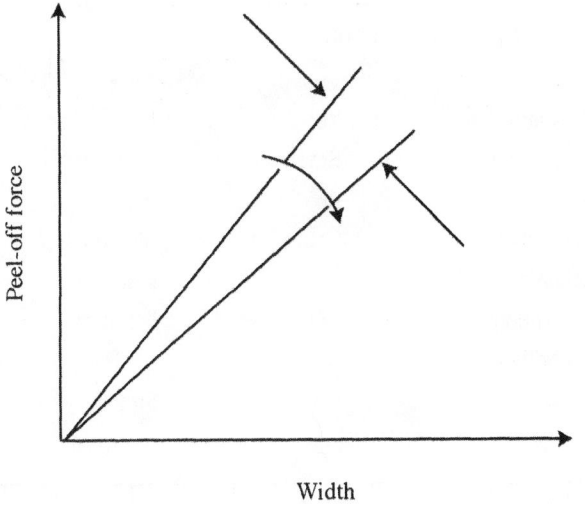

Figure 17.18
Peel-off function

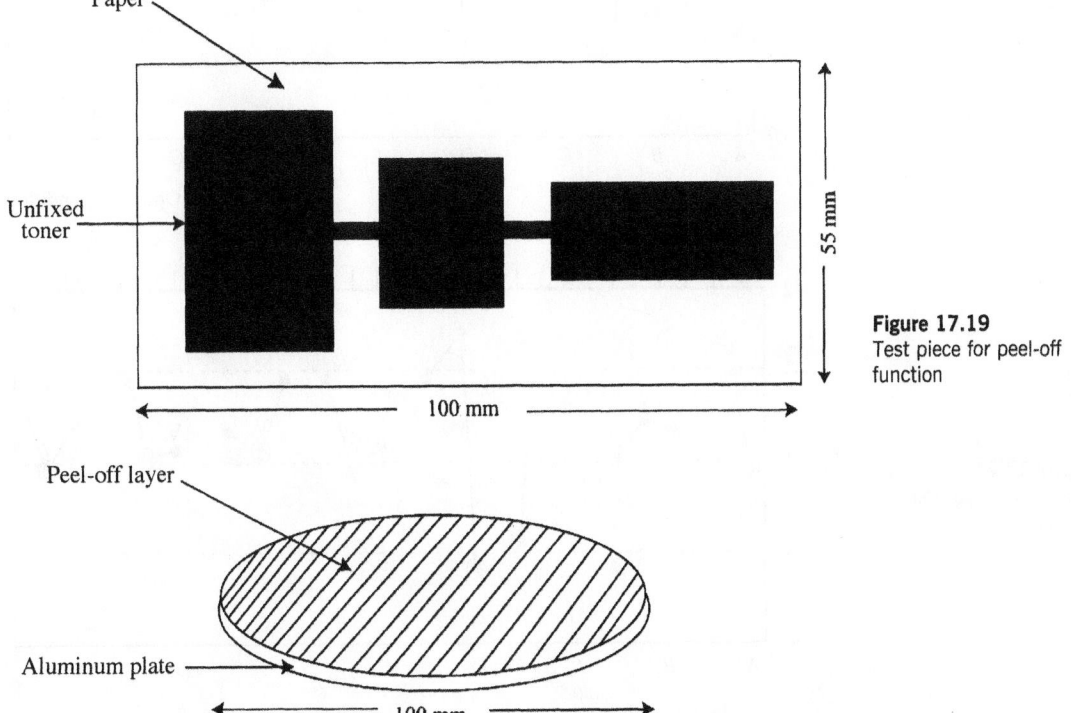

Figure 17.19
Test piece for peel-off function

Table 17.5
Control factors for peel-off function

	Factor	Level 1	Level 2	Level 3
A	Heat treatment	Yes	No	—
B	Baking condition 1	Small	Medium	Large
C	Material	A	B	C
D	Baking condition 2	Small	Medium	Large
E	Film thickness	Small	Medium	Large
F	Baking condition 3	Small	Medium	Large
G	Baking condition 4	Small	Medium	Large
H	Pretreatment	No	Small	Large

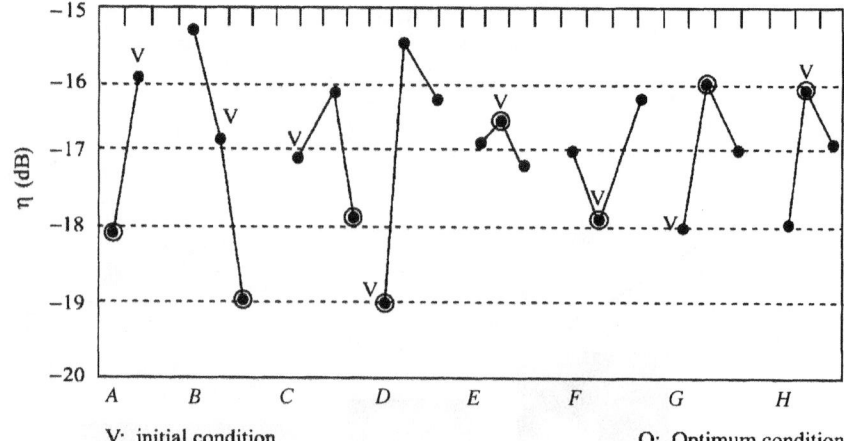

Figure 17.20
SN ratio of peel-off function

V: initial condition   O: Optimum condition

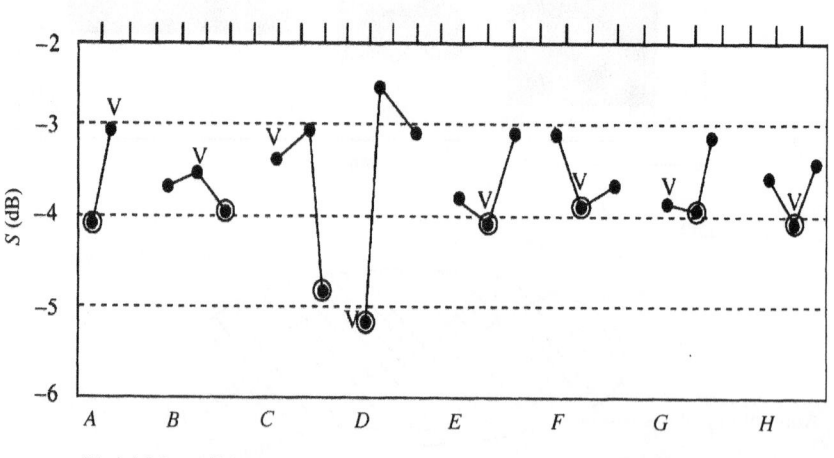

Figure 17.21
Sensitivity of peel-off function

V: initial condition   O: Optimum condition

Table 17.6
Results of confirmation

	SN Ratio		Sensitivity	
	Estimation	**Confirmation**	**Estimation**	**Confirmation**
Initial condition	−14.03	−21.25	−4.51	0.23
Optimum condition	−22.22	−24.66	−7.00	−8.84
Gain	−8.19	−3.41	−2.49	−9.07

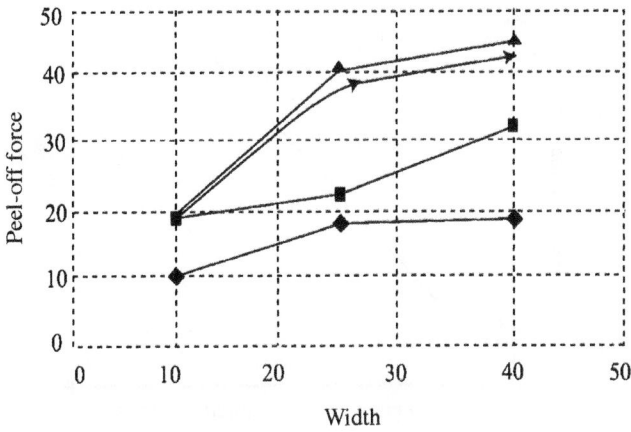

Figure 17.22
Input/output of initial condition

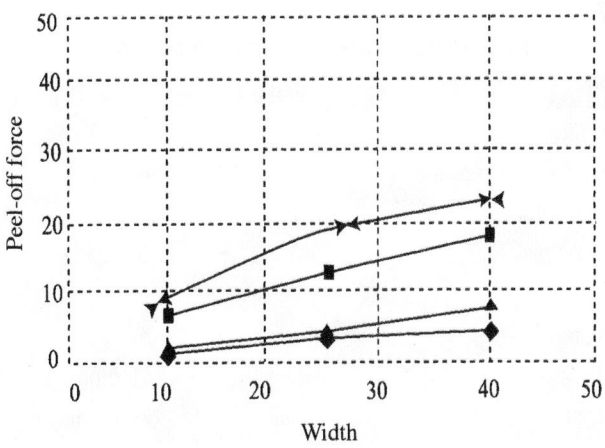

Figure 17.23
Input/output of optimum condition

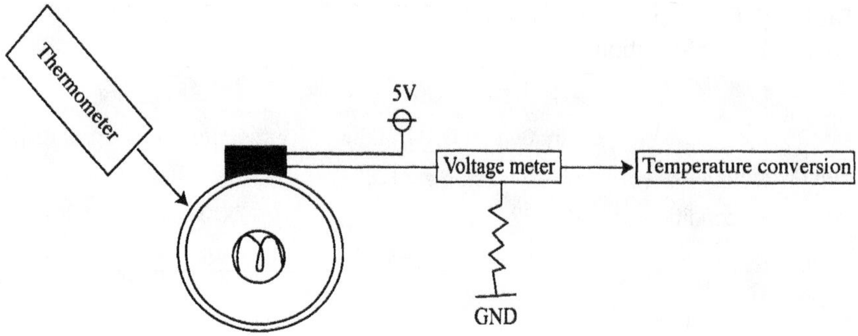

Figure 17.24
Test piece for
temperature-measuring
function

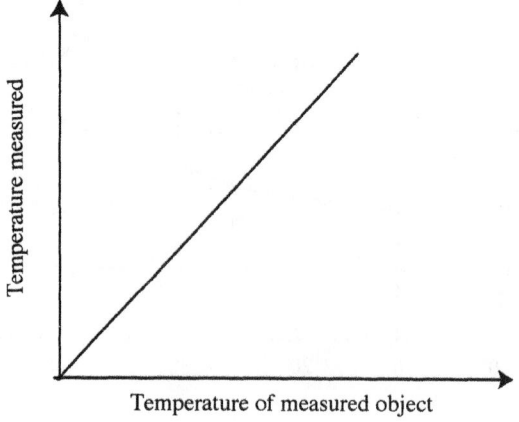

Figure 17.25
Ideal function of
temperature
measurement

Table 17.7
Control factors for the temperature-measuring function

	Factor	Level 1	Level 2	Level 3
A	Surface film	A	B	C
B	Pressure-dividing resistor	2 kΩ	4 kΩ	8 kΩ
C	Thermistor chip	I	II	III
D	Thermistor resistor	0.3 kΩ	0.55 kΩ	1 kΩ
E	Sponge	X	Y	Z
F	Plate material	a	b	c
G	Plate thickness	Small	Medium	Large
H	Plate area	Small	Medium	Large

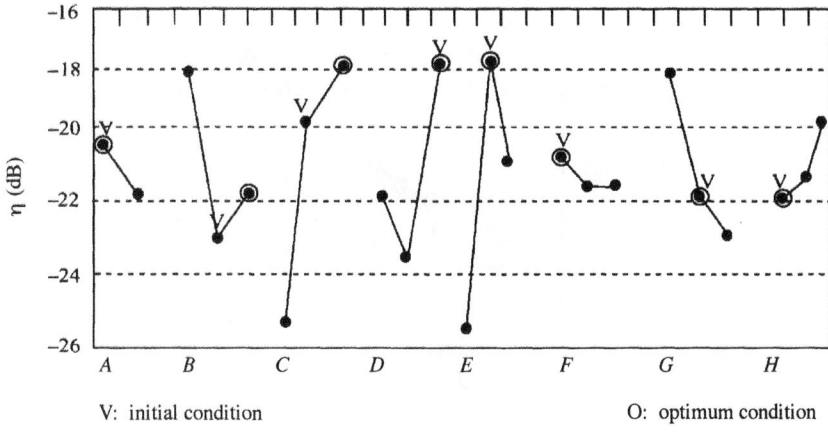

Figure 17.26
SN ratio of the temperature-measuring function

V: initial condition   O: optimum condition

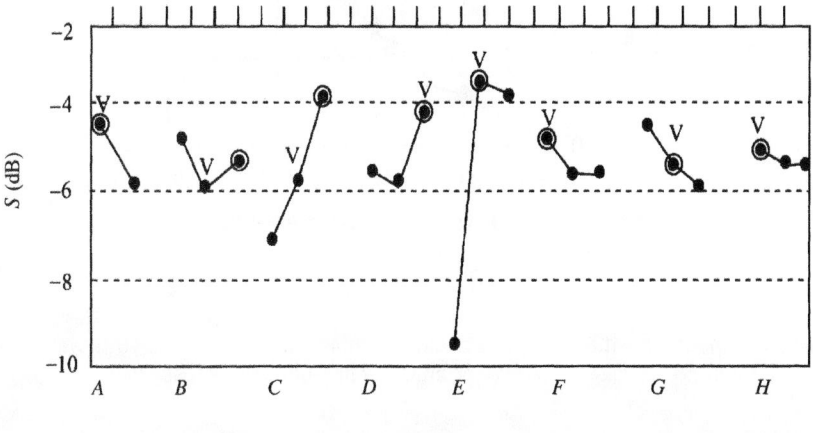

Figure 17.27
Sensitivity of the temperature-measuring function

V: initial condition   O: optimum condition

Table 17.8
Results of confirmation

	SN Ratio		SN Ratio	
	Estimation	**Confirmation**	**Estimation**	**Confirmation**
Initial condition	−16.39	−23.95	−2.39	−4.2
Optimum condition	−13.01	−20.87	0.12	−2.75
Gain	3.38	3.08	2.51	1.67

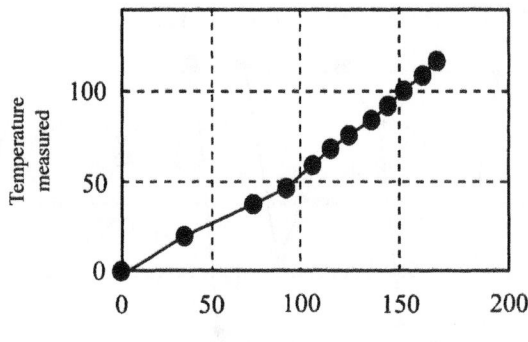

Figure 17.28
Input/output of initial
condition

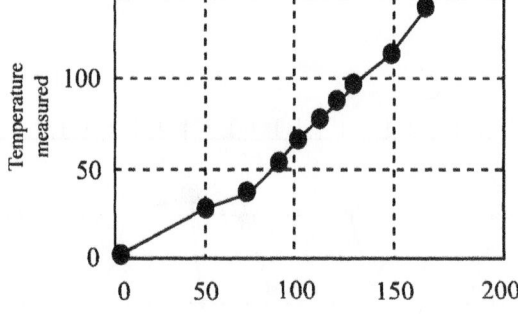

Figure 17.29
Input/output of
optimum condition

was continued and the sensitivity could be improved almost to 1, so it was concluded that no forecasting control or adjustment would be necessary.

References

1. Katsura Hino et al., 1996. Optimization of bonding condition of laminated thermosetting decorative sheets and steel sheets. Presented at the Quality Engineering Forum Symposium.

2. Eiji Okabayashi et al., 2000. Development of a new fusing system using four types of test pieces for four generic functions. *Journal of Quality Engineering Society*, Vol. 8, No. 1.

18 Robust Engineering: A Manager's Perspective

18.1. Introduction 377
Baseline Philosophy: Quality Loss Function 380
Noise Factors 381
18.2. Parameter Design 382
18.3. Tolerance Design 386
Conceptual Design, Parameter Design, Then Tolerance Design 387

18.1. Introduction

Robust design, a concept as familiar as it is misunderstood, will be clarified in this chapter. After we cover robust design, we discuss tolerance design, which is used to optimize tolerances at the lowest cost increase. It is important to note that robust design is not the same as tolerance design.

We conduct robust design optimization by following the famous two-step process created by Genichi Taguchi: (1) Minimize variability in the product or process (robust optimization), and (2) adjust the output to hit the target. In other words, first optimize performance to get the best out of the concept selected, then adjust the output to the target value to confirm whether all the requirements are met. The better the concept can perform, the greater our chances to meet all requirements. In the first step we try to kill many birds with one stone, that is, to meet many requirements by doing only one thing. How is that possible?

We start by identifying the ideal function, which will be determined by the basic physics of the system, be it a product or process. In either case, the design will be evaluated by the basic physics of the system. When evaluating a product or a manufacturing process, the ideal function is defined based on energy transformation from the input to the output. For example, for a car to go faster, the driver presses down on the gas pedal, and that energy is transformed to increased speed by sending gas through a fuel line to the engine, where it is burned, and finally to the wheels, which turn faster.

When designing a process, energy is not transformed, as in the design of a product, but information is. Take the invoicing process, for example. The supplier sends the company an invoice, and that information starts a chain of events that transforms the information into various forms of record keeping and results, finally, in a check being sent to the supplier.

In either case, we first define what the ideal function for that particular product or process would look like; then we seek a design that will minimize the variability of the transformation of energy or information, depending on what we are trying to optimize.

We concentrate on the transformation of energy or information because all problems, including defects, failures, and poor reliability, are symptoms of variability in the transformation of energy or information. By optimizing that transformation—taking out virtually all sources of "friction" or noise along the way—we strive to meet all the requirements at once. (We discuss this in greater detail later in this chapter.)

To understand fully Taguchi's revolutionary approach, let's first review how quality control has traditionally worked. Since virtually the advent of commerce, a "good" or acceptable product or process has been defined simply as one that meets the standards set by the company. But here's the critical weakness to the old way of thinking: It has always been assumed that *any* product or process that falls *anywhere* in the acceptable range is equal to any other that falls within that range.

Picture the old conveyer belt, where the products roll along the line one by one until they get to the end, where an inspector wearing goggles and a white coat looks at each one and tosses them into either the "acceptable" bin or the "reject" bin. In that case, there are no other distinctions made among the finished products, just "okay" or "bad."

If you were to ask that old-school inspector what separates the worst "okay" specimen from the best reject—in other words, the ones very close to the cutoff line—he'd probably say something like, "It's a hair's difference, but you've got to draw the line somewhere." But the inspector treats all acceptable samples the same: He just tosses them in the "okay" bin, and the same with the rejects. Even though he knows there are a million shades of gray in the output, he separates them all into black or white.

Now if you asked a typical consumer of that product if there was any difference between a sample that barely met the standards to make it into the "okay" bin and one that was perfect, she'd say, "Yes, absolutely. You can easily tell the difference between these two."

The difference between the inspector's and the customer's viewpoints can be clarified further with the following analogy: If both people were playing darts, the inspector would only notice whether or not the dart hit the dartboard, not caring if it landed near the edge of the board or right on the bull's-eye. But to the customer, there would be a world of difference between the dart that landed on the board's edge and the one that pierced the bull's-eye. Although she certainly wouldn't want any dart not good enough to hit the board, she would still greatly prefer the bull's-eye to the one just an inch inside the board's edge. The point is: With the old way of inspecting products, the manufacturer or service provider made no distinctions among acceptable outputs, but the consumer almost always did, which made the company out of step with the customer's observations and desires. The concept is illustrated in Figure 18.1.

The figure shows the result of 21 field-goal trials from 40 yards out by two field-goal kickers. Both kickers made all 21 successfully, but which place kicker would you want playing on your team? Easy, isn't it?

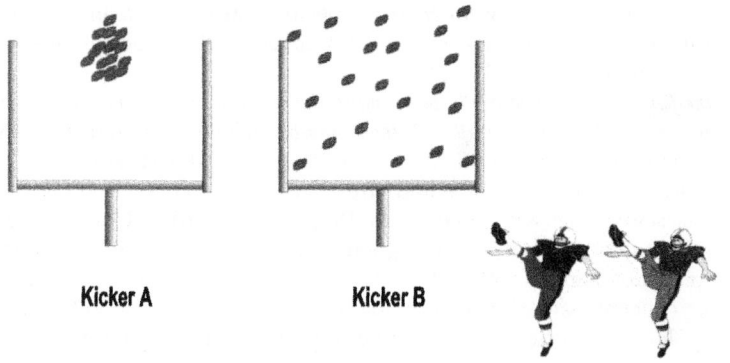

Kicker A   Kicker B

Figure 18.1
Two field-goal kickers

This dissonance between these two perspectives demonstrates that the traditional view of quality—"good enough!"—is not good enough for remaining competitive in the modern economy. Instead of just barely meeting the lowest possible specifications, we need to hit the bull's-eye. The way to do that is to replace the oversimplified over/under bar with a more sophisticated bull's-eye design, where the goal is not merely to make acceptable products, but to reduce the spread of darts around the target.

The same is also true on the other side of the mark. In the old system, once you meet the specification, that was that. No point going past it. But even if we're already doing a good job on a particular specification, we need to look into whether we can do it better and, if so, what it would cost. Would improving pay off?

You might wonder why a company should bother making already good designs into great designs. After all, there's no extra credit for exceeding specifications. Or is there?

Robust design requires you to free your employees—and your imaginations—to achieve the optimum performance by focusing on the energy/information transformation described earlier. This notion of having no ceiling is important, not just as a business concept, but psychologically as well. The IRS, of course, tells you how much to pay in taxes, and virtually no one ever pays extra. Most taxpayers do their best to pay as little as legally possible. Charities, on the other hand, never tell their donors what to pay—which might explain why Americans are by far the most generous citizens around the world in terms of charitable giving.

The point is simple: Don't give any employee, team, or project an upper limit. Let them optimize and *maximize* the design for robustness. See what's possible, and take advantage of the best performances you can produce! Let the sky be the limit and watch what your people can do! A limitless environment is a very inspiring place to work.

The next big question is: Once the energy/information transformation is optimized, is the design's performance greater than required? If so, you've got some decisions to make. Let's examine two extreme cases.

When the optimum performance exceeds the requirements, you have plenty of opportunities to reduce real cost. For example, you can use the added value in other ways, by using cheaper materials, increased tolerances, or by speeding up the process. The objective of robust design is to improve performance without increasing costs. Once you can achieve that, you can take advantage of the opportunities that cost reductions can create.

On the flip side, if the optimum performance comes in below the requirements, it's time to rethink the concept and come up with something better. The problem is that, in most corporate cultures, it is very difficult to abandon a concept because so many people have already spent so much time and effort on the project.

But this is where leadership comes in. Despite the heartbreak of letting an idea go, if it's not good enough, it's not good enough. So instead of spending good money on a doomed project and fighting fires later, it's best to cut your losses, reject the concept (salvaging the best ideas, if any), and move on to the next one, instead of locking yourself into a method of production that's never going to give you the results you want. Thus, it is extremely important to detect poor designs and reject them at the early stages of development.

Taguchi has built a model based on this concept that demonstrates the impact that variations from the target have on profits and costs. As the function of the product or process deviates from the target—either above or below it—the quality of the function is compromised. This in turn results in higher losses. The farther from the target, the greater the monetary losses will be.

Baseline Philosophy: Quality Loss Function

A tool called the quality loss function (QLF) is very helpful in estimating the overall loss resulting from function deviations from the target. Of course, to make the function more efficient, before you decide to minimize function variation, you need to be sure that the cost to do so is less than the cost would be to maintain the status quo. If you're losing more money from the variations than it would cost to reduce them, you have a very convincing case for taking action. If not, you're better off spending the money elsewhere.

We refer to the quality loss function again in Section 18.3. Because QLF is used to estimate the cost of functional variations, it is an excellent tool to assist in the decision making that evaluating the cost–performance trade-off requires. We discuss next how the QLF philosophy can motivate corporations to adopt the culture of robust design.

The quality loss function can be put to use for several tasks, including estimating the loss associated with one part of a product or process, the entire item, or even an entire population of products or services if, say, one is looking at something like distribution that affects a wide range of products.

I'll show you how this works mathematically, but don't be discouraged if you're not a math whiz. I'll also show how it works in everyday life, too, which is usually all you'll need to know.

Let's call the target t. To estimate the loss for a single product or part (or for a service or step, for that matter), we need to determine two things: (1) How far off a product or process is from the target $(y - t)$, and (2) how much that deviation is costing in dollars. We accomplish this by calculating the deviation $(y - t)$, squaring it, and then dividing the dollars lost by that figure. So it looks like this:

$$\frac{D \text{ (cost of loss in dollars)}}{(y - t)^2} = k \text{ (loss coefficient of process)}$$

Determining the first sum, the degree of deviation from the norm $(y - t)$ is relatively easy, squaring it presents no problems, but determining the second part of the equation, dollars lost, is much trickier. The reason is that to derive the loss coefficient, we must determine a typical cost of a specific deviation from the target. These two estimates, cost and deviation, will ultimately produce the loss coefficient of the process (k).

Now here's how it works in everyday life. For an easy example, let's take the old copier paper jam problem. A paper jam is simply a symptom of the variability of energy transformation within the copier. For a copy to be unacceptable, the paper displacement rate must not deviate too much but must stay within the realm of ($y - t$). When this variability is too great, a copier's performance becomes unacceptable. And when that happens, it will cost the company approximately $300, the cost of service, and lost productivity for each half-day it's out of operation for service.

Now it's time to see how this formula applies to a small number of products (or processes) with various deviations from the target. Those outputs that hit the target will cost the company no losses (apart from the original production costs), whereas those that lie farther and farther from the target cost the company more and more money.

One simple way to estimate the average loss per product is to total the individual offset deviations, square the sum, divide by the number of products, and finally, multiply the resulting figure by the loss coefficient (k). When you compare the average loss of the function to one with a "tighter" deviation, you will quickly see that the average loss of the second group would be much less. And that's the idea.

Having covered how to calculate deviation, loss, average loss, and the loss coefficient, now we need to ask the fundamental question: What creates the variation in our products and processes in the first place? Solve that puzzle, and we can adjust that variation as we see fit.

Noise Factors

The bugaboos that create the wiggles in the products and processes we create can be separated into the following general categories:

- ❏ Manufacturing, material, and assembly variations
- ❏ Environmental influences (not ecological, but atmospheric)
- ❏ Customer causes
- ❏ Deterioration, aging, and wear
- ❏ Neighboring subsystems

This list will become especially important to us when we look at parameter design for robust optimization, whose stated purpose is to minimize the system's sensitivity to these sources of variation. From here on, we will lump *all* these sources and their categories under the title of *noise*, meaning not just unwanted sound, but anything that prevents the product or process from functioning in a smooth, seamless way. Think of noise as the friction that gets in the way of perfect performance.

When teams confront a function beset with excessive variation caused by noise, the worst possible response is to ignore the problem—the slip-it-under-the-rug response. Needless to say, this never solves the problem, although it is a surprisingly common response.

As you might expect, more proactive teams usually respond by either attacking the sources of the noise, trying to buffer them, or compensating for the noise by other means. All these approaches can work to a degree, but they will almost always add to the costs.

Traditionally, companies have created new products and processes by the simple formula design–build–test, or, essentially, trial and error. This has its appeal, of course, but is ultimately time consuming, inefficient, and unimaginative. It's physically rigorous but intellectually lazy.

18.2. Parameter Design

Robust design takes a different approach. Instead of using the solutions listed above, which all kick in *after* the noise is discovered, parameter design works to eliminate the effect of noise *before* it occurs by making the function immune to possible sources of variation. It's the difference between prevention and cure, the latter being one of the biggest themes of design for six sigma. (Parameter design is another name for robust design, a design parameter optimization for robustness.)

We make the function immune to noise by identifying design factors we can control and exploiting those factors to minimize or eliminate the negative effects of any possible deviations—rather like finding a natural predator for a species that's harming crops and people. Instead of battling the species directly with pesticides and the like, it's more efficient to find a natural agent. The first step toward doing this is to discard the familiar approach to quality control, which really is a focus on failure, in favor of a new approach that focuses on success.

Instead of coming up with countless ways that a system might go wrong, analyzing potential failures, and applying a countermeasure for each, in parameter design we focus on the much smaller number of ways we can make things go right! It's much faster to think that way, and much more rewarding, too. Think of it as the world of scientists versus the world of engineers. It is the goal of scientists to understand the entire universe, inside and out. A noble goal, surely, but not a very efficient one. It is the engineer's goal simply to understand what he needs to understand to make the product or process he's working on work well. *We need to think like engineers, looking for solutions, not like pure scientists, looking for explanations for every potential problem.*

The usual quality control systems try to determine the symptoms of poor quality, track the rate of failure in the product or process, then attempt to find out what's wrong and how to fix it. It's a backward process: beginning with failure and tracing it back to how it occurred.

In parameter design we take a different tack: one that may seem a little foreign at first, but which is ultimately much more rewarding and effective. As discussed earlier, every product or process ultimately boils down to a system whereby energy is transferred from one thing to another to create that product or process. It's how electricity becomes a cool breeze pumping out of your air conditioner. (In the case of software or business processes, a system transforms information, not energy, and exactly the same optimization can be applied.)

In the parameter design approach, instead of analyzing failure modes of an air-conditioning unit, we measure and optimize the variability and efficiency of the energy transformation from the socket to the cool air pumping out of the unit.

In other words, we optimize the quality of energy transformation, as illustrated in Figure 18.2.

This forces us to define each intended function clearly so that we can reduce its variability and maximize its efficiency. In fact, that's another core issue of parameter design: the shift from focusing on what's wrong and how to fix it to focusing on what's right and how to maximize it. Mere debugging and bandaging are not effective.

To gain a deeper understanding of the distinctions between the old and new ways of thinking, it might be helpful to walk through an example. Let's look at the transfer case of a brand new four-wheel-drive truck. Now, as you probably know, the basic function of this system is as follows: The fuel system sends fuel to the engine, which turns it into active energy and sends it on to the transmission, which sends it on to the transfer case, whose job is to take that energy and distribute it to the front and rear axles for maximum traction and power. The transfer case, therefore, acts as the clearinghouse, or distribution center, for the car's energy.

Even with new transfer cases, common problems include audible noise, excessive vibration, excessive heat generation, poor driving feel, premature failure or breakdown, and poor reliability. When engineers see any of these conditions, they

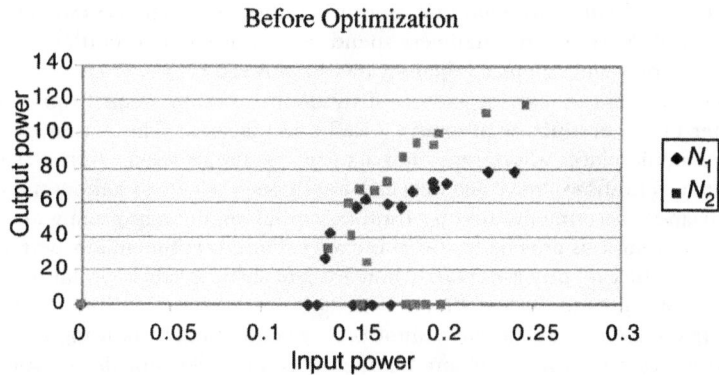

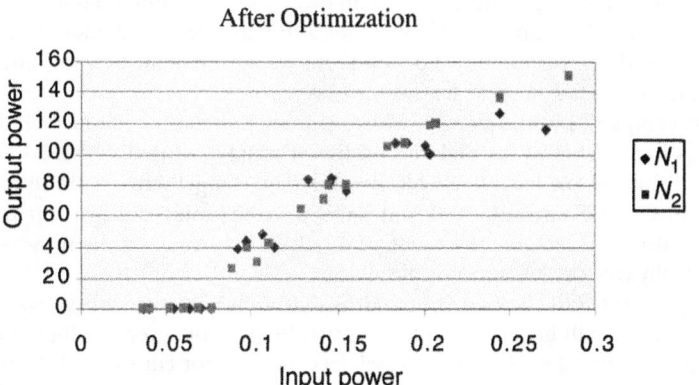

Figure 18.2
Robustness of energy transformation

traditionally have jumped right in to modify the transfer case's design to minimize the particular problem. The catch is, however, that often "fixing" one of these problems only makes another one worse. For example, we could reduce audible noise, only to find a dangerous increase in friction-generated heat.

It's like squeezing one end of a balloon only to see the other end expand, or quitting smoking only to see your weight increase. Using this approach, instead of eradicating the problem, we've only shifted the symptom of variability from one area to another, and have spent a lot of time, energy, and money in the process.

With parameter design, however, instead of trying to debug the transfer case bug by bug, which often results in us chasing our tails, we focus on reducing the variability of energy transformation, then maximizing the energy that goes through the transfer case cleanly. In other words, we shift our focus from defense to offense.

The theory goes like this: If we could create a perfect transfer case with zero energy loss, there would be no "wasted" energy necessary to create audible noise, heat, vibration, and so on. Sounds good, of course, but obviously building the perfect transfer case is still a pipe dream. But the thinking behind the perfect transfer case, however, can help us build a better one. Wouldn't it be better to try to achieve the perfect energy-efficient transfer case than to try to achieve perfection through endless debugging, putting out fire after fire in the hope of eliminating fires forever? As Ben Franklin said: "An ounce of prevention is worth a pound of cure." We try to build that prevention into the design. It's estimated that in a typical U.S. company, engineers spend 80% of their time putting out fires, not preventing them. Smart companies reverse this ratio.

Usually, the single biggest source of function variation stems from how the customer uses a product or process. (Recall noise factors.) The reason is simple: Labs are sterile places where sensible scientists test the product or process under reasonable conditions, but customers can use these products in a thousand different ways and environments, adding countless variables, including aging and wear. Virtually no one can anticipate the many ways that customers might be tempted to use a product or process. This is how we get warning labels on lawn mowers advising consumers not to use them on hedges.

But that's the real world. We cannot prevent customers from using their four-wheel-drive cars in just about any manner they wish. So how do we solve this problem?

Let's take a simple pair of scissors as an example. When designed well, as almost all of them are, they can cut regular paper and basic cloth well enough. But what can you do about customers who buy them to use on materials for which they were never intended, such as leather or plastic?

Most companies would do one of two things. Either they would include stern warnings in the owner's manual and on the product itself that the scissors are not intended for use on leather or plastic and that using them on those materials would render the warranty null and void, or companies can give up trying to educate customers, assume the worst, and bolster the design of the scissors so that they actually can cut leather and plastic.

The problem with the first approach is that such warnings only go so far; your company might still be found liable in court. In any case, even intelligent customers might be turned off by a pair of scissors that cannot cut through leather and plastic, even if they never intend to use theirs in that way. The problem with the second approach—making the scissors all but bulletproof—is that, for the vast majority of customers, the extra materials and joint strengthening would be over-kill and would raise the price of the product, even for people who will never need such additional force.

With parameter design, however, you don't need to resort to either unsatisfactory solution, because the method helps you create "perfect scissors" that require virtually no effort to cut almost any material. Instead of simply bolstering the device, parameter design streamlines the product to avoid the problems that arise when the product is being used on tough materials, in much the same way that offices solved their "paper problem" not by merely building more and bigger file cabinets, but by converting their information to microfilm, microfiche, and, finally, computer disks.

Making the scissors more efficient reduces the odds of damage and deterioration, and therefore effectively makes the scissors immune to the extremes of customer use variation without burdening the product with undue costs.

The same concept of parameter design for robust optimization can be applied to the design of a business transactional process. Let's take the efficiency of hospital service, for example. Even for a case like this, we can look at the system as an energy transformation.

Each patient visiting an emergency room (ER) represents the input to the system. Each of them has a different level of demands. One may require a simple diagnosis and a prescription; another may require immediate surgery. The total time spent by a patient in the hospital represents the output. Therefore, we can define the ideal function as the ideal relationship between the input demanded and the actual output. Then we want to optimize the system for robustness. We want the relationship between the input and output to have the least variability at the highest efficiency.

In other words, we want the design to address the number of beds, number of nursing staff, number of heath unit coordinators on staff, number of doctors on staff, pharmacy hours, in-house coverage, ER coordinator, dedicated x-ray services, private triage space, and so on. And we want the design to be the most robust against noise factors such as total number of patients visiting, time of patient visit, equipment down time, lab delay, private MD delay, absenteeism, and so on. In essence, we want the relationship between the inputs (the demands of each patient) and the outputs (the time spent on each patient) to have the smallest variability with the highest efficiency. Next, we formulate an experiment with this objective in mind which can be executed by computer simulation instead of more expensive, real-life models.

The eight key steps of parameter design are:

1. Define the scope. Define which system/subsystem you are optimizing.
2. Define the ideal function.
3. Develop a strategy for how you are going to reduce the effect of noises.
4. Determine design parameters and their alternative levels.
5. Conduct the test/simulation to obtain data.
6. Analyze the data using the SN ratio, a metric for robustness.
7. Predict the performance of optimum design.
8. Conduct a confirmation trial using the optimum design.

In summary, teams will learn how to apply the principles of Taguchi's parameter design to optimize the performance of a given system in a far more elegant fashion than just debugging or bolstering it would ever accomplish.

18.3. Tolerance Design

In parameter design we optimize the design for robustness by selecting design parameter values, which means defining the materials, configurations, and dimensions needed for the design. For a transfer case in a four-wheel-drive truck, for example, we define the type of gears needed, the gear material, the gear heat treatment method, the shaft diameter, and so on. For a hospital, we define the number of beds, pharmacy hours, and so on. In sum, in parameter design we define the nominal values that will determine the design.

The next step is tolerance design, in which we optimize our tolerances for maximum effect, which does not necessarily mean making them all as tight as they can be. What it does mean is making them tight where they need to be tight, and allowing looser ones where we can afford to have looser ones, thus maximizing the quality, efficiency, and thrift of our design.

For tolerance design optimization, we use the quality loss function to help us evaluate the effectiveness of changing dimensional or material tolerances. This allows us to see if our results are better or worse as we tweak a particular element up or down.

Let's start with tolerance design optimization. *Tolerancing* is a generic label often applied to any method of setting tolerances, be they tolerances for dimensions, materials, or time, in the case of a process.

Tolerance design means something more specific: a logical approach to establishing the appropriate tolerances based on their overall effect on system function (sensitivity) and what it costs to control them. As mentioned earlier, the key model employed in tolerance design is quality loss function. To say it another way, Tolerance design describes a specific approach to improving tolerances by tightening up the most critical tolerances (not all of them, in other words) at the lowest possible cost through QLF.

This requires us first to determine which tolerances have the greatest impact on system variability, which we accomplish by designing experiments using orthogonal arrays. These experiments are done by computer simulation (occasionally, by hardware). This allows us to prioritize our tolerances—to decide which changes reap the greatest rewards—and thereby helps us make wise decisions about the status of our various options, letting us know which ones we should tighten, loosen, or leave alone.

Think of it as a baseball team's batting order, and you're the manager. Your job is to maximize run production, and you do it by trying different players in different spots in the lineup. The key is isolating who helps and who does not. Substituting various players in the lineup and changing the order will give you the results you need to determine who works best in which position.

This brings us to the six steps in tolerance design:

1. Determine the quality loss function.

2. Design and run (or simulate) the experiment to determine the percent contribution of each tolerance.

3. Determine a tolerance upgrading action plan and estimate its cost for each of the high contributors.

4. Determine a tolerance degrading action plan for the no/low-contributing tolerances and estimate the cost reduction for using it.

5. Finalize your tolerance upgrading and degrading action plan based on the percentage of contribution, quality loss, and the cost.

6. Confirm the effectiveness of your plan.

Tolerance design will help teams meet one of the primary objectives of the program: developing a product or process with six sigma quality while keeping costs to a minimum. The steps of tolerance design are intended to help you and your team work through the process of establishing optimal tolerances for optimal effect.

The goal is not simply to tighten every standard, but to make more sophisticated decisions about tolerances. To clarify what we mean, let's consider a sports analogy. Billy Martin was a good baseball player and a great manager. He had his own off-field problems, but as a field general, he had no equal. One of the reasons he was so good was because he was smart enough, first, to see what kind of team he had, then to find a way to win with them, playing to their strengths and covering their weaknesses—unlike most coaches, who have only one approach that sometimes doesn't mesh with their players.

In the 1970s, when he was managing the Detroit Tigers, a big, slow team, he emphasized power: extra base hits and home runs. When he coached the Oakland A's a decade later, however, he realized that the team could never match Detroit's home-run power, but they were fast, so he switched his emphasis from big hits to base stealing, bunting, and hitting singles. In both places, he won division crowns, but with very different teams.

It's the same with tolerance design. We do not impose on the product or process what we think should happen. We look at what we have, surmise what improvements will obtain the best results, and test our theories. In Detroit, Martin didn't bother trying to make his team faster and to steal more bases because it wouldn't have worked. He made them focus on hitting even more home runs, and they did. In Oakland, he didn't make them lift weights and try to hit more homers, because they didn't have that ability. He made them get leaner and meaner and faster and steal even more bases. And that's why it worked: he played to his teams' strengths.

You don't want to spend any money at all to upgrade low-contributing tolerances. You want to reduce cost by taking advantage of these tolerances. You don't want to upgrade a high-contributing tolerance if it is too expensive. If the price is right, you will upgrade those high contributors. Tolerance design is all about balancing cost against performance and quality.

It is extremely important to follow the steps shown in Figure 18.3. One common problem is that people skip parameter design and conduct tolerance design. Be aware of the opportunities you are missing if you skip parameter design. By skipping parameter design, you are missing great opportunities for cost reduction. You may be getting the best possible performance by optimizing for robustness, but if the best is far better than required, there are plenty of opportunities to reduce cost. Further, you are missing the opportunity to find a bad concept, so that you can reject the bad concept at the early stage of product/process development. If

Conceptual Design, Parameter Design, Then Tolerance Design

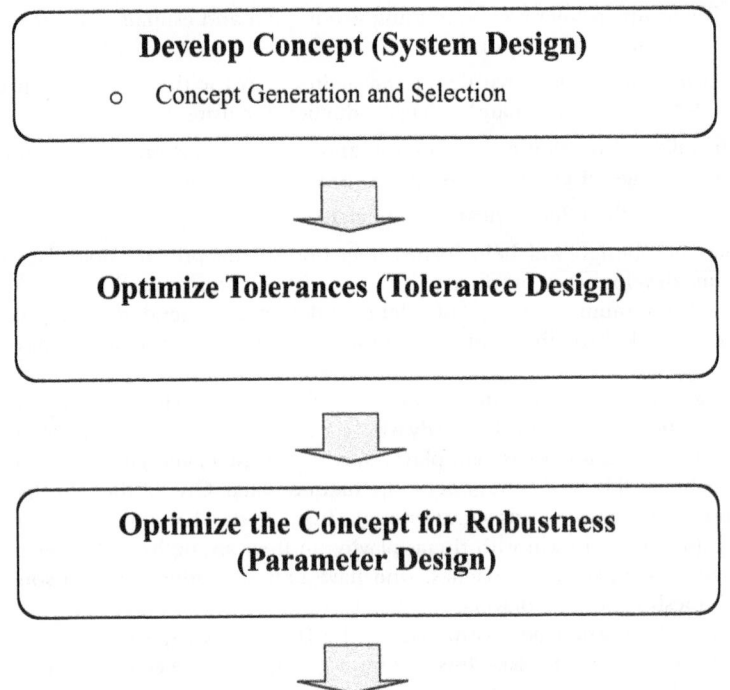

Figure 18.3
Steps in robust
engineering

the best concept you can come up with is not good enough, you have to change the concept.

The result of tolerance design on designs that have not been optimized is far different from the result of tolerance design after robust optimization has taken place. In other words, you end up tightening tolerances, which would have been unnecessary if the design had been optimized for robustness in the first place. Think of all firefighting activities your company is doing today. If the design were optimized, you would have fewer problems and the problems would be different. Hence, solutions would be different.

19 Implementation Strategies

19.1. Introduction   389
19.2. Robust Engineering Implementation   389
       Step 1: Management Commitment   390
       Step 2: Corporate Leader and the Corporate Team   390
       Step 3: Effective Communication   391
       Step 4: Education and Training   392
       Step 5: Integration Strategy   392
       Step 6: Bottom-Line Performance   393

19.1. Introduction

In this chapter we introduce the implementation processes of robust engineering into the corporate environment and provide complete and detailed strategies behind the process. In it we will also show compatibility and possible enhancements of any other quality program or process that an organization may already have in place. Some organizations absolutely thrive on change and embrace the challenges and opportunities that come with change and diversity. Those organizations will have the foresight and vision to see that the robust engineering implementation strategy can be a guide for them and ultimately, can help them to stay on the cutting edge and ahead of their competition.

19.2. Robust Engineering Implementation

The six critical components for a smooth implementation process for robust engineering are as follows:

1. Management commitment
2. Corporate leader and the corporate team
3. Effective vommunication
4. Education and training

5. Integration strategy

6. Bottom-line performance

It is extremely important that all of these elements be used. Additional favorable components may be present, of course, that complement corporate strengths and/ or enhance and build on less-developed areas of the organization. Companies that face major challenges and have the greatest needs for a working quality process will definitely have the most to gain from robust engineering.

Step 1: Management Commitment

Robust engineering is one of the most powerful tools for making higher-quality products with little or no waste, and making them faster too. However, as with any new quality process, without the full support of managers, the organization will not receive the full benefit of this program. Buy-in from top and middle management is essential. No matter how good or how efficient a program might be, it can be extremely difficult for personnel at the lower end of the organization if management is not committed to supporting it and making it work. It is essential that management fully understand all of the benefits of the process and see that the program is in no way a threat to their position or to their importance to the organization.

Today, any organization is dependent on the quality of its offerings. Management wants everything to be more efficient, faster, cheaper, and of higher quality. However, the only organizations that will achieve competitive status are those willing to invest in a more advanced and efficient methodology to achieve that quality.

When top management is not only supportive, but encouraging, the rest of the organization will follow suit, and momentum will be gained for the process. A negative attitude by the powers that be can cause the breakdown of the program.

Step 2: Corporate Leader and the Corporate Team

Once all levels of management have bought into the idea of implementing robust engineering methodology, it is imperative that someone be chosen to be the driver, the leader of the project. This person should be a manager, highly efficient as a leader, motivated, and able to motivate others. He or she should be someone whom others would be honored to follow. He or she should be dependable and committed to the goals of the program, be firm in delegation, very detail-oriented, and strong on follow-up. Clearly, it is in the best interest of the organization that the corporate leader be highly respected and have knowledge of the work of all departments. He or she must be able to assert authority and have strong project-management skills. In short, you are looking for a "champion." Top management must be certain that they have chosen the right person for the job.

At the start of the process, an open meeting between upper management, middle management, and the chosen corporate leader should take place. At this meeting, a complete understanding must be reached on all of the elements involved, and the tasks must be broken down to be more manageable and less overwhelming. If the methodology is not managed properly, the robust engineering project will not succeed.

The next most significant contributing factor to the success of the robust engineering implementation process is an empowered, multidisciplinary, cross-functional team. Since robust engineering implementation is an engineering project, it would be very beneficial to place as many engineers as possible on the team. The team must learn all of the methodology possible. The team should consist of representatives from every department and should be a mix of personalities and talents. It is also important to note, at this point, that there should be no cultural or organizational barriers that could cause the system to break down.

Each team member should have a clear understanding of the overall implementation process and of how his or her part contributes to the "big" picture. To this end, members of the team should divide and/or share all the tasks in the implementation according to responsibilities in the organization and personal abilities. Each team member should take full responsibility and be held accountable for his or her part in the process.

Defining the overall goals and the vision for the project will be the responsibility of a strategic planning team, and the project leader should be actively involved with the strategic planning process as well as with the project. It is also the responsibility of each member of the cross-functional team to support, acknowledge, and communicate his or her understanding of the project and its objectives throughout the entire organization. Success of the process requires control, discipline, and some type of benchmarking against competitors as it progresses.

Managing change is difficult in any organization, and under most circumstances, employees have a natural resistance to and feel threatened by change. A project of the magnitude of robust engineering requires extremely effective management of resources and people to assure its success. Remember, however, that anything new requires consistent encouragement and that mistakes and failures should be considered part of the process. The corporate leader must report back to management on the process at each phase. If he or she sees that something is not moving as planned, he or she must work through members of the team or upper management to keep the process moving in the right direction. Often, when a breakdown occurs, it can be linked directly to unclear or ineffective communication.

Step 3: Effective Communication

It is the responsibility of the cross-functional team to find ways to communicate with the entire organization. Every employee should receive the same messages, and the team must make sure that the meaning is clear to everyone. Robust engineering requires the organization to learn a new vocabulary, and it is the responsibility of the team to find interpretations that everyone can understand. The team members should consider what employees might ask about the process and then be prepared to answer those questions in language everyone will understand. Some of the questions that may come up follow:

❑ What is meant by *robustness?*

❑ What is meant by *optimization?*

❑ What is meant by *variability?*

❑ What is the robustness definition of failure?

❑ What is the operational meaning of prevention?

Of course, there are many other terms that must be defined. The task must be approached through well-thought-out, effective communication channels. Employees will react to what they perceive that a message says, so ambiguity must be avoided, and every base must be covered. The strategic planning team should put the necessary information into written form to help ensure that everyone is hearing and "getting" the same thing. Each cross-functional team member must see to it that his or her department members understand "what was said" as well as "what was meant." Every employee needs to be able to make decisions based on receiving the same information. Even the best of systems will fail if everyone is not pulling in the same direction, so clear, concise, and timely information is crucial.

It is vital to the success of the project that misconceptions and misinformation be dealt with before the process is started.

Step 4: Education and Training

The importance of training and education can never be overstated, especially when new projects or programs are being introduced into an organization. A lack of understanding by even one employee can cause chaos and poor performance. The robust engineering process requires every employee to have an understanding of the entire process and then have a complete understanding of how his or her part in the process contributes to the bottom line. To that end, some training in business terminology and sharing of profit and loss information should be undertaken for every employee. Training also can help to inspire and motivate employees to stretch their potential to achieve more. It is essential to a quality program such as robust engineering that everyone participate fully in all training. Engineers especially need to have a full understanding of the methodology and its potential rewards. But everyone, from top management to the lowest levels of the organization, must commit to focusing on the goals of the project and how they can affect the bottom line. Knowledge adds strength to any organization, and success breeds an environment conducive to happier workers.

Organizations the world over are discovering the value of looking at their organizations to find out what makes them work and what would make them work better. Through the development and use of an internal "expert" they are able to help their people work through changes, embrace diversity, stretch their potential, and optimize their productivity. These internal experts are corporate assets who communicate change throughout an organization and provide accurate information to everyone. Even when quality programs already exist, often the internal experts can work with the cross-functional team to communicate the right information in the right ways. These experts are also extremely efficient and helpful in spreading the goals and the purpose of robust engineering, as well as explaining the quality principles on which the methodology is built.

Step 5: Integration Strategy

Trying to implement robust engineering methods and create an environment where they can be seen as part of everyone's normal work activities is a challenge in itself. Many quality programs are in popular use at this time, and some view robust engineering as just another program. However, such programs as quality function deployment (QFD), Pugh analysis, failure modes and effects analysis (FMEA), test planning, and reliability analysis are much more effective, take less time, and give higher performance with measurable results when combined with robust engineering.

Robust engineering is a brand new approach to engineering thinking. The old way of thinking was "build it, test it, fix it." The robust engineering way is "optimize it, confirm it, verify it." The Robust engineering implementation process has revolutionized the engineering industry, and it is doing away with many of the more traditional tools and methods. Some companies, in part, have been able to change their perspectives on what constitutes failure and have been able to redefine reliability. With robust engineering, the emphasis is on variability.

When you make a product right in the first place, you don't have to make it again. Using the right materials and the right designs saves time, saves effort, saves waste, and increases profitability.

When one measures the results of new ways against the results of old ways, it is usually easy to tell which is better. When there are considerable differences, a closer look at the new way is certainly merited. Systems that implement robust engineering come out ahead on the bottom line. The real question is: Would you rather have your engineers spending their days putting out fires, or would you rather have them prevent the fires by finding and preventing the causes? What if you offered incentives to your workers, the ones who actually made it all happen? What if you motivated them by rewarding them for differences in the bottom line? Your employees will feel very important to the organization when they see how they have helped to enhance the bottom line, but they will feel sheer ecstasy when they are able to share in the increased profits.

Step 6: Bottom-Line Performance

Part VI

Mahalanobis–Taguchi System (MTS)

20 Mahalanobis–Taguchi System

20.1. Introduction   397
20.2. What Is the MTS?   398
20.3. Challenges Appropriate for the MTS   398
20.4. MTS Case Study: Liver Disease Diagnosis   401
20.5. Mahalanobis–Taguchi Gram–Schmidt Process   415
20.6. Gram–Schmidt's Orthogonalization Process   415
20.7. Calculation of MD Using the Gram–Schmidt Process   416
20.8. MTS/MTGS Method versus Other Methods   417
20.9. MTS/MTGS versus Artificial Neural Networks   417
20.10. MTS Applications   418
    Medical Diagnosis   418
    Manufacturing   418
    Fire Detection   419
    Earthquake Forecasting   419
    Weather Forecasting   419
    Automotive Air Bag Deployment   419
    Automotive Accident Avoidance Systems   419
    Business Applications   419
    Other Applications   420
20.11. Some Important Considerations   420

20.1. Introduction

Many people remember the RCA Victor Company, with its logo of a dog listening to "his master's voice." Dogs can easily recognize their masters' voices, but it is not easy for a machine to be created that recognizes its owner by voice recognition. Painting lovers can easily recognize the impressionist paintings of Vincent van Gogh. Can we develop a machine that will recognize van Gogh painting? Classical music lovers can distinguish the music composed by Johann Sebastian Bach from that of others. But even using the most advanced computers, it is not easy for human-made machines to recognize human voice, face, handwriting, or printing.

Human beings have the ability to recognize visual patterns because we have some 9 million optical sensing cells in our eyes. Assuming that the time to treat a captured image is 0.04 second, we can treat 225 million pieces of information in 1 second. We do not necessarily try to memorize such a vast amount of information, but our sensing cells are efficient and simplify and screen out unnecessary pieces of information. If we could develop a system that made it simple enough to use current personal computers for pattern recognition, there would be numerous applications in various areas.

20.2. What Is the MTS?

MTS is the acronym for the *Mahalanobis–Taguchi system*. P. C. Mahalanobis was a famous Indian statistician who established the Indian Statistical Institute. In 1930, Mahalanobis introduced a statistical tool called the *Mahalanobis distance* (MD), used to distinguish the pattern of a certain group from other groups. Mahalanobis used the Mahalanobis distance for an archaeological application to classify excavated bones and to make judgments as to which dinosaur a variety of sample bones belonged to. Another application was ethnological, to characterize differences among Asian races and tribes. The main objective of his application was to make statistical judgments to distinguish one group from another.

The Mahalanobis–Taguchi system is Taguchi's design of a systematic method for using the Mahalanobis distance. The objective of MTS is to develop and optimize a diagnostic system with a measurement scale of *abnormality*. In MTS, the signal-to-noise (SN) ratio is used to assess the effectiveness of a system. Moreover, Taguchi used the system not only for diagnosis but also for forecasting/prediction systems. Thus, MTS is used to develop and optimize a system of multivariable diagnosis, pattern recognition, and prediction of occurrence of particular events.

20.3. Challenges Appropriate for the MTS

Diagnosis or pattern recognition is not easy, since it is based on multivariable data consisting of both continuous and attribute data, with correlations among those variables. For example:

1. Physicians conduct diagnoses on patients; observe patients for symptoms, and look at patient data such as blood pressure, pulse, blood test results, and past data on the patient. A good doctor will come to a valid diagnostic result, and a poor one would misdiagnose the problem. Misdiagnosis can be very costly.

2. An inspection system must make a judgment as to whether a product or system can pass or fail based on multiple specifications. The same can be said for monitoring a complex manufacturing process.

3. It is difficult to evaluate and improve discrimination power among handwritten letters. There is no perfect handwriting, voice, fingerprint, or facial expression recognition system.

4. It is extremely difficult to predict earthquakes. How can we predict when and where a big one will occur?

To illustrate the difficulty of any of these, let's take a very simple example with two variables, x_1 and x_2. Let x_1 be the weight of an American male and x_2 be the height of an American male. Suppose that a weight is 310 pounds. If you look at this data, you may think that this person (John) is very heavy but not necessarily abnormal, since there are many healthy American males who weigh around 310 (Figure 20.1).

Suppose that John's height is 5 feet. If you look at his height by itself, you may think John is a very short person but not that there is necessarily anything abnormal about John, since there are many healthy American males whose height is around 5 feet (Figure 20.2).

On the other hand, if you look at both pieces of data on John, $x_1 = 310$ pounds and $x_2 = 5.0$ feet, you would think something was definitely out of the ordinary about John. You would think that he is not necessarily normal. This is because, as everyone knows, there is a correlation between weight and height. Suppose that we can define the "normal" group in terms of weight and height (Figure 20.3). That would have the shape of an ellipse because of the correlation between weight and height. We can clearly see that John is outside the normal group. The distance between the center of a normal group and a person under diagnosis is the Mahalanobis distance.

From this discussion, you can see the following points:

1. For an inspection system, two characteristics meeting the specification do not guarantee that it is a good product because of correlation (interaction).

2. It is easy to see the pattern when the number of variables, k, is equal to 2. What if $k = 5$, $k = 10$, $k = 20$, $k = 40$, $k = 120$, $k = 400$, or $k = 1000$? The greater the number of variables, the more difficult it is to see the pattern.

The steps to design and optimize the MTS are as follows:

❑ *Task 1:* Generation of normal space

    Step 1. Define the normal group.

    Step 2. Define k variables $(1 < k < 1000)$.

    Step 3. Gather data from the normal group on k variables with sample size n, $n >> k$.

    Step 4. Calculate the MD for each sample from the normal group.

❑ *Task 2:* Confirmation of discrimination power

    Step 5. Gather data on k variables for samples from outside the normal group; sample size $= r$.

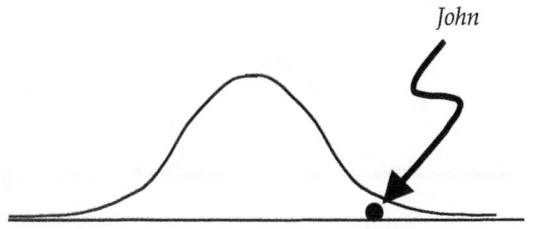

$x_1 = $ weight

Figure 20.1
John's weight

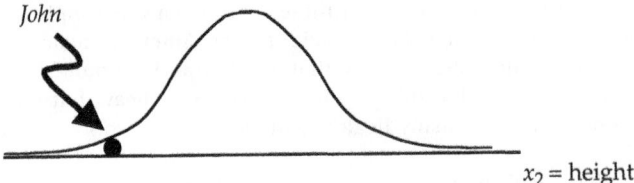

Figure 20.2
John's height

Step 6. Calculate the MD for each sample from outside the normal group.

Step 7. Evaluate the discrimination power.

❏ *Task 3:* Identification of critical variables and optimization

Step 8. Optimize the MTS. Evaluate critical variables. Optimize the number of variables. Define the optimum system.

There are two primary methods of computing the MD. One is to use an inverted matrix and the other is to use Gram–Schmidt orthogonal vectors, which we refer to as MTGS. First, a simple case study with inverted matrix method is illustrated. MTGS will be introduced later.

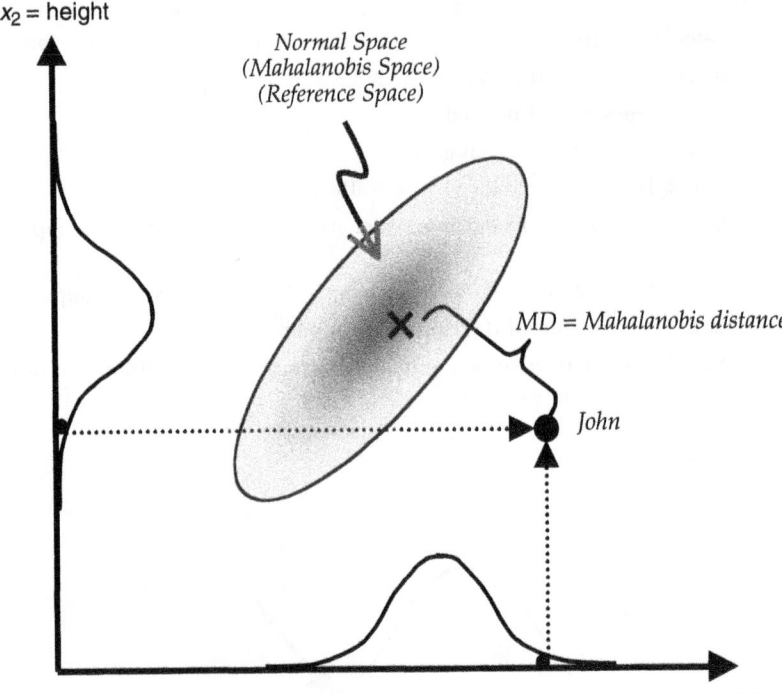

Figure 20.3
Both height and weight

20.4. MTS Case Study: Liver Disease Diagnosis*

Step 1. Define the normal group: people with a healthy liver ($n = 200$). Physicians specializing in liver disease identified 200 people who had healthy livers. They defined the normal group as any person with a healthy liver.

Step 2. Define the variables ($k = 15$). The following variables were selected:

x_1	Age	x_6	LDH	x_{11}	TG
x_2	Gender	x_7	ALP	x_{12}	PL
x_3	TP	x_8	GTP	x_{13}	C_r
x_4	Alb	x_9	LAP	x_{14}	BUN
x_5	ChE	x_{10}	Tch	x_{15}	UA

Physicians identified these 15 variables, 13 from blood test results. Variables can be continuous or discrete. In this case, gender was a discrete variable.

Step 3. Generate a database of a normal group for the variables selected. For a discrete variable with two classes, such as gender, simply assign a number for each class: say, 1 for male and 10 for female (Table 20.1).

Step 4a. Compute the mean and standard deviation for each variable and normalize the data. See Table 20.2.

$$Z = \frac{x - \bar{x}}{\sigma_x}$$

Step 4b. Compute the correlation matrix. See Table 20.3.

Step 4c. Compute the inverse matrix of the correlation matrix. See Table 20.4.

Step 4d. Compute MD for each sample from the normal group. Each person in the normal group is a vector of ($z_1, \ldots, z_{15}$). For each person, MD is given by the following equation and by Table 20.5 and Figure 20.4:

$$MD = \frac{1}{k} ZA^{-1}Z^T$$

$$= \frac{1}{k} (z_1, z_2, \ldots, z_{15}) \ (A^{-1}) \begin{pmatrix} z_1 \\ \vdots \\ z_{15} \end{pmatrix}$$

The distribution of Mahalanobis distances for samples from the normal group typically look like Figure 20.5, below, shown in Figure 20.6 as a curve. The average is 1.00. Note that it is very rare to have an MD value greater then 5.0; an MD for an abnormal sample should become much larger than 5.0; MD provides a measurement scale for abnormality.

---

*This case study is presented by courtesy of T. Kanetaka, Teishin Hospital, Tokyo.

Table 20.1
Data for normal group

Sample	Age x_1	Gender x_2	TP x_3	Alb x_4	ChE x_5	LDH x_6	ALP x_7	GTP x_8	LAP x_9	Tch x_{10}	TG x_{11}	PL x_{12}	Cr x_{13}	BUN x_{14}	UA x_{15}
1	59	10	7	4.2	0.7	190	7.7	12	275	220	140	228	0.8	13	3.8
2	51	1	7	4.4	0.98	217	6.3	23	340	225	87	227	1.1	15	4.6
3	53	1	6.7	4.2	0.83	220	9.2	16	278	220	67	222	0.8	17	3.2
4	52	1	6.6	4.3	0.9	204	4.6	15	289	171	59	175	0.8	13	3.8
5	52	1	6.7	3.9	0.97	198	5.2	13	312	192	51	203	1	14	4.1
6	56	1	7.2	4	0.93	188	6.1	16	304	216	86	213	1	13	4.2
...	...	...	...	...	...	...	...	...	...	...	...	...	...	...	...
198	35	10	7.3	4.3	0.7	203	6.1	19	343	165	59	178	1.3	14	6.4
199	35	10	7.1	4.4	0.86	190	4	36	358	190	155	200	1.4	19	5.7
200	53	1	6.8	4	0.88	225	8.2	18	330	189	124	203	1	17	4.3
Avg.	39.7	6.9	7.1	4.3	0.9	195.4	5.5	22.6	332.1	184.0	84.7	192.1	1.2	15.1	5.0
Sum	10.34	4.30	0.30	0.14	0.10	22.93	1.36	12.92	55.76	22.21	34.86	20.57	0.19	2.88	1.22

Table 20.2
Mean and standard deviation for normal group

Sample	Age z_1	Gender z_2	TP z_3	Alb z_4	ChE z_5	LDH z_6	ALP z_7	GTP z_8	LAP z_9	Tch z_{10}	TG z_{11}	PL z_{12}	Cr z_{13}	BUN z_{14}	UA z_{15}
1	1.87	0.73	-0.19	-0.66	-1.91	-0.24	1.59	-0.82	-1.02	1.62	1.59	1.75	-1.9	-0.74	-1.01
2	1.09	-1.36	-0.19	0.77	1	0.94	0.56	0.03	0.14	1.85	0.06	1.7	-0.33	-0.04	-0.35
3	1.29	-1.36	-1.2	-0.66	-0.56	1.07	2.7	-0.51	-0.97	1.62	-0.51	1.46	-1.9	0.65	-1.5
4	1.19	-1.36	-1.54	0.06	0.17	0.37	-0.69	-0.59	-0.77	-0.58	-0.74	-0.83	-1.9	-0.74	-1.01
5	1.19	-1.36	-1.2	-2.81	0.9	0.11	-0.25	-0.74	-0.36	0.36	-0.97	0.53	-0.85	-0.39	-0.76
6	1.58	-1.36	0.49	-2.09	0.48	-0.32	0.41	-0.51	-0.5	1.44	0.04	1.02	-0.85	-0.74	-0.68
⋮	⋮	⋮	⋮	⋮	⋮	⋮	⋮	⋮	⋮	⋮	⋮	⋮	⋮	⋮	⋮
198	-0.45	0.73	0.83	0.06	-1.91	0.33	0.41	-0.28	0.2	-0.85	-0.74	-0.68	0.73	-0.39	1.12
199	-0.45	0.73	0.15	0.77	-0.24	-0.24	-1.13	1.04	0.46	0.27	2.02	0.39	1.25	1.35	0.56
200	1.29	-1.36	-0.86	-2.09	-0.04	1.29	1.96	-0.35	-0.04	0.23	1.13	0.53	-0.85	0.65	-0.6
Avg.	0.0	0.0	0.0	0.0	0.0	0.0	0.0	0.0	0.0	0.0	0.0	0.0	0.0	0.0	0.0
Sum	1.00	1.00	1.00	1.00	1.00	1.00	1.00	1.00	1.00	1.00	1.00	1.00	1.00	1.00	1.00

Table 20.3
Correlation matrix

	z_1	z_2	z_3	z_4	z_5	z_6	z_7	z_8	z_9	z_{10}	z_{11}	z_{12}	z_{13}	z_{14}	z_{15}
z_1	1.0000	-0.2968	-0.2993	-0.4168	-0.0570	0.2084	0.3335	0.0234	-0.1119	0.2343	0.1405	0.2658	-0.2305	-0.0292	-0.2108
z_2	-0.2968	1.0000	0.0811	0.3874	0.1996	-0.1084	-0.0581	0.3638	0.3934	-0.1966	0.2707	-0.2189	0.6490	0.2675	0.5616
z_3	-0.2993	0.0811	1.0000	0.3966	0.0763	0.0454	-0.0149	0.1588	0.1762	0.0785	0.1225	0.0845	0.1376	-0.1167	0.2022
z_4	-0.4168	0.3874	0.3966	1.0000	0.1274	0.0139	-0.1341	0.1769	0.2266	-0.0864	0.0423	-0.0943	0.2767	0.0961	0.3684
z_5	-0.0570	0.1996	0.0763	0.1274	1.0000	0.1472	0.0264	0.3305	0.4005	0.2891	0.3198	0.2866	0.2535	0.0293	0.3500
z_6	0.2084	-0.1084	0.0454	0.0139	0.1472	1.0000	0.2230	0.2195	0.2285	0.2137	0.1051	0.2488	0.0666	0.0919	0.0514
z_7	0.3335	-0.0581	-0.0149	-0.1341	0.0264	0.2230	1.0000	0.1754	0.1615	0.1806	0.1523	0.1783	-0.0348	-0.0869	-0.0430
z_8	0.0234	0.3638	0.1588	0.1769	0.3305	0.2195	0.1754	1.0000	0.7297	0.2057	0.4462	0.2110	0.3851	0.0534	0.4663
z_9	-0.1119	0.3934	0.1762	0.2266	0.4005	0.2285	0.1615	0.7297	1.0000	0.1855	0.3378	0.1634	0.4610	0.0514	0.4427
z_{10}	0.2343	-0.1966	0.0785	-0.0864	0.2891	0.2137	0.1806	0.2057	0.1855	1.0000	0.3036	0.9334	0.0381	0.0173	0.1354
z_{11}	0.1405	0.2707	0.1225	0.0423	0.3198	0.1051	0.1523	0.4462	0.3378	0.3036	1.0000	0.3014	0.2553	-0.0288	0.3791
z_{12}	0.2658	-0.2189	0.0845	-0.0943	0.2866	0.2488	0.1783	0.2110	0.1634	0.9334	0.3014	1.0000	0.0204	0.0349	0.1191
z_{13}	-0.2305	0.6490	0.1376	0.2767	0.2535	0.0666	-0.0348	0.3851	0.4610	0.0381	0.2553	0.0204	1.0000	0.2485	0.5713
z_{14}	-0.0292	0.2675	-0.1167	0.0961	0.0293	0.0919	-0.0869	0.0534	0.0514	0.0173	-0.0288	0.0349	0.2485	1.0000	0.1447
z_{15}	-0.2108	0.5616	0.2022	0.3684	0.3500	0.0514	-0.0430	0.4663	0.4427	0.1354	0.3791	0.1191	0.5713	0.1447	1.0000

Table 20.4
Inverse matrix

	1	2	3	4	5	6	7	8	9	10	11	12	13	14	15
1	1.6383	0.1542	0.3455	0.3415	0.0986	-0.2063	-0.3587	-0.2414	0.2598	0.1133	-0.2578	-0.3422	0.1152	-0.0397	0.0843
2	0.1542	2.6720	0.2274	-0.4176	-0.0459	0.3648	0.1303	-0.1838	-0.1639	0.2613	-0.3976	0.4678	-1.0153	-0.3740	-0.5164
3	0.3455	0.2274	1.3519	-0.4626	0.0908	-0.0352	-0.0652	-0.0759	-0.0251	0.0999	-0.1414	-0.2343	-0.0860	0.1833	-0.0536
4	0.3415	-0.4176	-0.4626	1.5731	-0.0202	-0.1313	0.0640	0.0007	-0.0395	0.0292	0.1347	0.0027	0.1499	-0.0570	-0.2785
5	0.0986	-0.0459	0.0908	-0.0202	1.3626	-0.0757	0.0720	0.0646	-0.3733	-0.0687	-0.1840	-0.2407	0.0164	0.0470	-0.2076
6	-0.2063	0.3648	-0.0352	-0.1313	-0.0757	1.2503	-0.1651	-0.0903	-0.2222	0.2280	0.0225	-0.2908	-0.1468	-0.1634	0.0055
7	-0.3587	0.1303	-0.0652	0.0640	0.0720	-0.1651	1.2354	-0.0508	-0.1969	-0.1598	-0.0394	0.0640	0.0628	0.1115	0.1114
8	-0.2414	-0.1838	-0.0759	0.0007	0.0646	-0.0903	-0.0508	2.5327	-1.5365	0.3081	-0.3927	-0.3686	0.0939	0.0036	-0.3568
9	0.2598	-0.1639	-0.0251	-0.0395	-0.3733	-0.2222	-0.1969	-1.5365	2.6016	-0.4256	0.0977	0.3556	-0.3981	0.1034	0.0556
10	0.1133	0.2613	0.0999	0.0292	-0.0687	0.2280	-0.1598	0.3081	-0.4256	8.0089	-0.1736	-7.3840	-0.1133	0.0849	-0.1908
11	-0.2578	-0.3976	-0.1414	0.1347	-0.1840	0.0225	-0.0394	-0.3927	0.0977	-0.1736	1.5448	-0.1540	0.0449	0.1621	-0.2435
12	-0.3422	0.4678	-0.2343	0.0027	-0.2407	-0.2908	0.0640	-0.3686	0.3556	-7.3840	-0.1540	8.3166	-0.0523	-0.2711	-0.0348
13	0.1152	-1.0153	-0.0860	0.1499	0.0164	-0.1469	0.0628	0.0939	-0.3981	-0.1133	0.0449	-0.0523	2.1288	-0.1681	-0.4937
14	-0.0397	-0.3740	0.1833	-0.0570	0.0470	-0.1634	0.1115	0.0036	0.1034	0.0849	0.1621	-0.2711	-0.1681	1.1955	0.0172
15	0.0843	-0.5164	-0.0536	-0.2785	-0.2076	0.0055	0.1114	-0.3568	0.0556	-0.1908	-0.2435	-0.0348	-0.4937	0.0172	2.0420

Table 20.5
Mahalanobis distance for 200 samples from healthy group

2.11	0.61	1.13	0.75	0.82	0.74	0.57	0.86	0.64	1.20	0.85	0.69	0.66	0.99	0.76	0.52	0.81	0.77	1.29	1.16
0.58	0.95	0.64	1.15	0.67	0.72	0.83	0.89	0.69	1.22	0.64	0.88	1.04	1.19	1.25	0.72	1.09	0.47	2.77	1.55
1.41	1.11	0.43	0.66	2.63	0.74	1.12	1.04	0.93	1.25	1.52	0.69	1.69	0.72	0.72	0.41	1.81	1.06	0.95	0.59
0.70	1.02	0.90	0.76	0.82	1.00	0.90	0.56	2.59	0.70	0.90	1.09	0.52	0.88	1.32	0.63	1.64	0.98	1.39	0.53
0.95	0.86	1.18	0.84	0.50	0.92	0.96	0.94	0.82	0.91	1.75	0.37	0.65	0.46	0.81	0.45	2.76	1.47	0.49	0.76
0.82	0.58	1.07	0.97	3.34	0.84	0.61	1.65	0.82	1.88	0.96	1.13	1.58	0.78	1.63	0.94	0.73	0.51	0.64	0.88
1.99	0.78	0.91	1.26	2.10	1.35	0.79	0.51	0.72	0.97	1.04	0.59	0.61	0.79	1.45	0.83	2.21	0.69	0.66	1.19
1.06	0.97	0.80	0.62	1.08	0.61	0.61	0.81	0.79	0.52	0.83	0.49	0.68	1.62	1.03	0.98	0.88	1.11	0.80	0.64
1.35	0.78	1.67	0.61	0.95	0.67	0.86	1.08	0.78	0.90	0.88	0.71	0.74	1.37	0.88	0.97	1.24	0.80	0.44	1.18
1.70	1.30	0.45	1.46	2.05	1.20	1.25	0.88	1.20	0.71	0.57	0.55	1.87	1.20	1.03	0.69	0.90	0.77	0.75	0.91

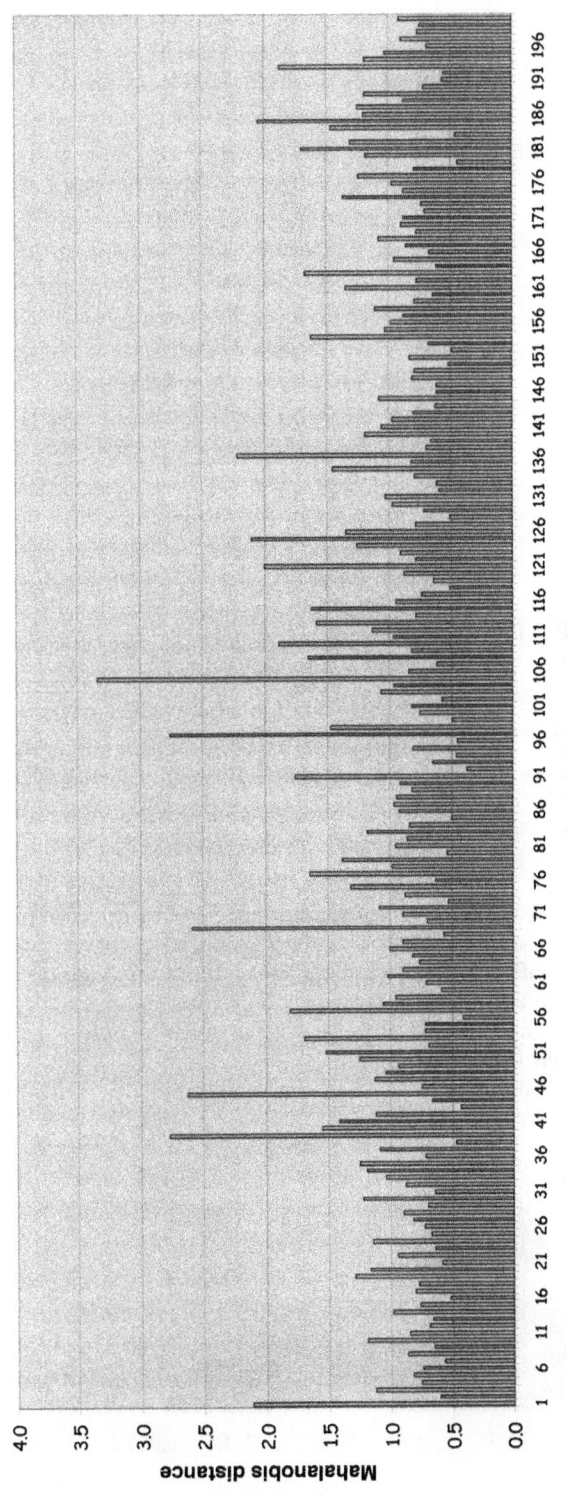

Figure 20.4
Sample showing Table 20.5 data

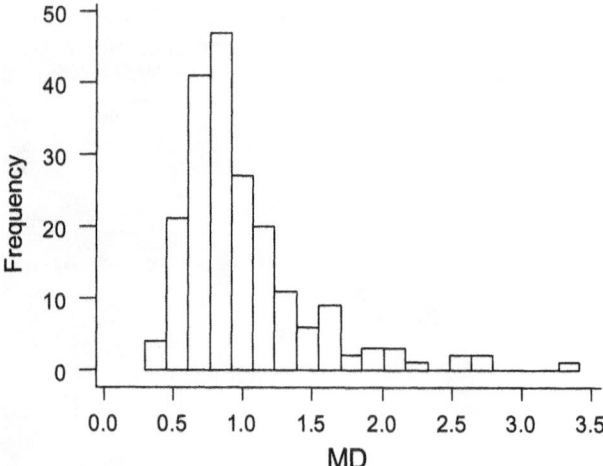

Figure 20.5
MD distribution

At this point it is important to recognize the following:

❑ The objective of MTS is to develop a measurement scale that measures abnormality. What we have just done is to define the zero point and one unit.

❑ MTS measures abnormality. Only the normal group has a population.

Now we will assess the discrimination power by gathering information.

Step 5. Gather data on k variables for samples from outside the normal group. See Table 20.6.

Step 6. Calculate the MD for each sample from outside the normal group. See Table 20.7.

Step 7. Evaluate The discrimination power. See Figure 20.7. Just by eyeballing Figure 20.7, one can see that the discrimination power is very good. In some studies it may be sufficient to have an MTS that can discriminate x percent of bad samples. For example, there is a situation called "no trouble found" for those samples that come back for warranty claims. In that situation, discriminating only 25% can be extremely beneficial.

After confirming a good discrimination, we are ready to optimize the MTS.

Figure 20.6
Distribution for Figure
20.6

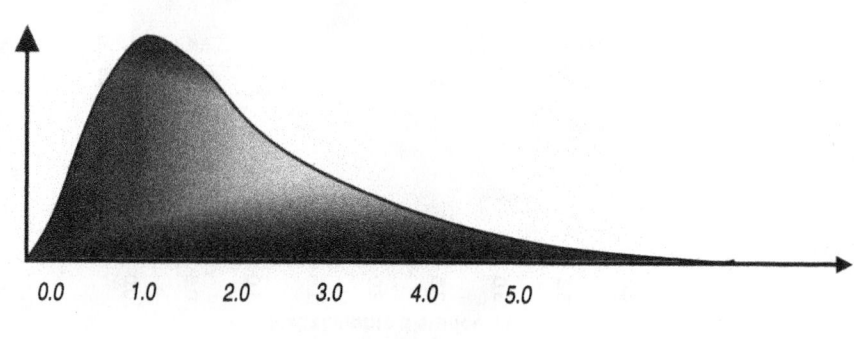

MD (Mahalanobis distance)

Table 20.6

k Variables for samples

Sample	Age x_1	Gender x_2	TP x_3	Alb x_4	ChE x_5	LDH x_6	ALP x_7	GTP x_8	LAP x_9	Tch x_{10}	TG x_{11}	PL x_{12}	Cr x_{13}	BUN x_{14}	UA x_{15}
1	30	10	8	4	0.59	336	13.7	192	856	144	138	199	1.6	16	3.5
2	46	10	7	3.3	0.48	365	25.8	274	1027	113	176	218	1.9	18	4.5
3	22	1	7.2	3.8	0.52	255	10.6	123	704	141	190	206	1.4	11	3.5
4	31	1	7	3.8	0.57	723	24.3	152	841	125	45	164	1.3	15	3.1
5	33	10	6.7	3.4	0.51	281	14.4	96	613	89	133	147	1.4	11	1.9
⋮	⋮	⋮	⋮	⋮	⋮	⋮	⋮	⋮	⋮	⋮	⋮	⋮	⋮	⋮	⋮
34	71	1	6.8	3.3	0.54	533	22.7	156	646	130	126	220	1.6	13	2.6
35	61	1	7.2	4	0.53	747	13.7	123	971	174	91	215	1.4	13	4.4
36	16	1	6.7	3.9	0.51	458	12.8	37	495	122	151	201	1.9	11	3.6

Table 20.7
Normalized data

Sample	Age z_1	Gender z_2	TP z_3	Alb z_4	ChE z_5	LDH z_6	ALP z_7	GTP z_8	LAP z_9	Tch z_{10}	TG z_{11}	PL z_{12}	Cr z_{13}	BUN z_{14}	UA z_{15}	MD Abnormal
1	-0.94	0.73	3.19	-2.09	-3.05	6.13	6.01	13.12	9.40	-1.80	1.53	0.34	2.30	0.31	-1.25	25.1
2	0.61	0.73	-0.19	-7.12	-4.20	7.39	14.93	19.47	12.46	-3.19	2.62	1.26	3.88	1.00	-0.43	66.5
3	-1.71	-1.36	0.49	-3.53	-3.78	2.60	3.73	7.78	6.67	-1.93	3.02	0.68	1.25	-1.43	-1.25	16.9
4	-0.84	-1.36	-0.19	-3.53	-3.26	23.01	13.82	10.02	9.13	-2.65	-1.14	-1.36	0.73	-0.04	-1.58	59.8
5	-0.65	0.73	-1.20	-6.40	-3.89	3.73	6.53	5.69	5.04	-4.28	1.38	-2.19	1.25	-1.43	-2.56	17.0
⋮	⋮	⋮	⋮	⋮	⋮			⋮	⋮	⋮		⋮	⋮	⋮	⋮	⋮
34	3.03	-1.36	-0.86	-7.12	-3.57	14.72	12.64	10.33	5.63	-2.43	1.18	1.36	2.30	-0.74	-1.99	41.8
35	2.06	-1.36	0.49	-2.09	-3.68	24.05	6.01	7.78	11.46	-0.45	0.18	1.12	1.25	-0.74	-0.52	52.8
36	-2.29	-1.36	-1.20	-2.81	-3.89	11.45	5.35	1.12	2.92	-2.79	1.90	0.43	3.88	-1.43	-1.17	25.0

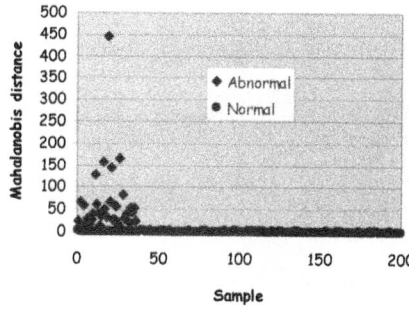
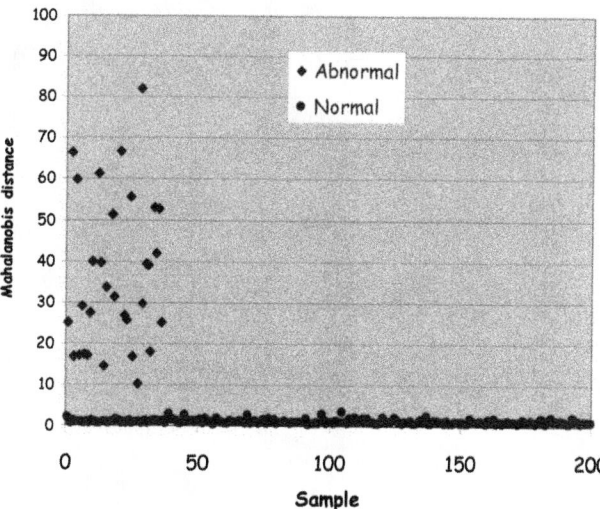

Figure 20.7
Discrimination power

Step 8. Optimize MTS.

Step 8.1. Assign variables to a two-level orthogonal array. Define level 1, to use variable, and level 2, not to use variable. Table 20.8 presents the two-level array. Recognize that run 1 of L_{16} is to use all 15 variables, and others are using different combinations of variables. The MTS is executed for all 16 runs, and data will be the MD for those 36 abnormal samples.

Step 8.2. Redo MTS for each run of the orthogonal array. First, calculate the MD for the abnormal sample as data and then use the SN ratio as an assessment criteria for discrimination power, as shown in Table 20.9. Recognize that the larger the MD values for abnormal samples, the better the discrimination is. In this example, number 4 shows very poor discrimination, and numbers 1, 7, and 16 show very good discrimination. The SN ratio for a larger-the-better response is used to assess the discrimination power. Ideally, a dynamic SN ratio where M is the true abnormality level and y is the computed MD value would be the ideal assessment.

Now we are ready to generate the response graph for the SN ratio (Figure 20.8).

Step 8.3. Make a response graph for the SN ratio. Evaluate how each variable contributes to discrimination power. Recognize that the higher the SN ratio, the better the discrimination power is. Notice that:

❑ Level 2 has a higher SN ratio for x_1, x_2, x_3, x_{14}, and x_{15}.

❑ Levels 1 and 2 have roughly the same performance for x_4, x_{11}, x_{12}, and x_{13}.

❑ Level 1 has higher SN ratios for x_5, x_6, x_7, x_8, x_9, and x_{10}.

From this we can conclude:

❑ x_1, x_2, x_3, x_{14}, and x_{15} are harming discrimination.

❑ x_4, x_{11}, x_{12}, and x_{13} do not contribute to discrimination.

❑ x_5, x_6, x_7, x_8, x_9, and x_{10} contributes greatly to discrimination.

Table 20.8
Two-level orthogonal array

L_{16}	X_1 1	X_2 2	X_3 3	X_4 4	X_5 5	X_6 6	X_7 7	X_8 8	X_9 9	X_{10} 10	X_{11} 11	X_{12} 12	X_{13} 13	X_{14} 14	X_{15} 15
1	1	1	1	1	1	1	1	1	1	1	1	1	1	1	1
2	1	1	1	1	1	1	1	2	2	2	2	2	2	2	2
3	1	1	1	2	2	2	2	1	1	1	1	2	2	2	2
4	1	1	1	2	2	2	2	2	2	2	2	1	1	1	1
5	1	2	2	1	1	2	2	1	1	2	2	1	1	2	2
6	1	2	2	1	1	2	2	2	2	1	1	2	2	1	1
7	1	2	2	2	2	1	1	1	1	2	2	2	2	1	1
8	1	2	2	2	2	1	1	2	2	1	1	1	1	2	2
9	2	1	2	1	2	1	2	1	2	1	2	1	2	1	2
10	2	1	2	1	2	1	2	2	1	2	1	2	1	2	1
11	2	1	2	2	1	2	1	1	2	1	2	2	1	2	1
12	2	1	2	2	1	2	1	2	1	2	1	1	2	1	2
13	2	2	1	1	2	2	1	1	2	2	1	1	2	2	1
14	2	2	1	1	2	2	1	2	1	1	2	2	1	1	2
15	2	2	1	2	1	1	2	1	2	2	1	2	1	1	2
16	2	2	1	2	1	1	2	2	1	1	2	1	2	2	1

L_{16}	X_1 1	X_2 2	X_3 3	X_4 4	X_5 5	X_6 6	X_7 7	X_8 8	X_9 9	X_{10} 10	X_{11} 11	X_{12} 12	X_{13} 13	X_{14} 14	X_{15} 15
1	Use	Use	Use	Use	Use	Use	Use	Use	Use	Use	Use	Use	Use	Use	Use
2	Use	Use	Use	Use	Use	Use	Use		Use	Use	Use	Use	Use	Use	
3	Use	Use	Use	Use	Use			Use	Use	Use	Use		Use		Use
4	Use	Use	Use					Use				Use			Use
5	Use		Use	Use	Use							Use			
6	Use			Use	Use	Use			Use	Use	Use		Use		Use
7	Use					Use	Use	Use	Use	Use	Use		Use		Use
8	Use					Use	Use	Use		Use		Use		Use	
9		Use		Use	Use	Use		Use	Use	Use	Use	Use	Use		Use
10		Use				Use		Use	Use	Use	Use	Use		Use	Use
11			Use	Use	Use		Use	Use	Use		Use	Use	Use	Use	Use
12							Use	Use	Use	Use		Use	Use	Use	
13			Use	Use							Use		Use	Use	
14			Use	Use	Use				Use	Use			Use	Use	
15		Use	Use	Use	Use	Use		Use		Use	Use	Use	Use	Use	
16			Use	Use	Use	Use		Use	Use	Use		Use			Use

Table 20.9
Response data

	1	2	3	4	5	6	7	8	9	10	. . .	34	35	36	SN Ratio
1	25.1	66.5	16.9	59.8	17.0	29.2	17.3	17.2	27.3	39.9	. . .	41.8	52.8	25.0	28.39
2	17.3	54.4	10.2	112.9	21.5	12.2	13.1	21.4	26.1	33.0	. . .	60.2	95.6	34.6	23.40
3	34.4	77.1	18.6	30.4	12.1	38.0	21.1	10.1	22.1	51.1	. . .	27.0	29.4	8.5	21.93
4	3.9	4.2	2.8	1.9	3.6	2.6	5.3	2.8	5.4	2.2	. . .	5.4	2.1	9.2	7.90
5	38.7	90.8	23.2	32.2	22.4	42.7	25.8	24.8	31.4	63.6	. . .	34.0	34.7	11.7	25.45
6	4.1	15.8	9.5	4.4	13.4	8.5	7.9	14.9	10.3	7.4	. . .	10.6	3.0	8.2	15.57
7	39.8	93.4	16.3	106.5	16.0	34.7	23.2	14.4	32.3	59.1	. . .	59.2	93.8	26.9	28.44
8	16.6	63.9	13.4	107.7	16.7	21.9	9.0	14.8	29.2	29.2	. . .	62.2	88.8	37.4	26.47
9	38.0	100.7	22.2	96.1	23.2	34.5	18.8	24.3	38.0	35.1	. . .	67.1	92.4	32.2	27.35
10	21.0	42.1	12.7	88.2	15.9	35.2	21.9	17.5	17.3	53.6	. . .	43.4	95.7	24.3	27.11
11	43.5	108.5	21.3	50.9	20.9	32.2	22.8	19.0	36.9	50.2	. . .	50.8	21.2	16.4	24.18
12	24.1	61.5	17.5	47.6	15.3	40.4	24.3	14.2	18.0	68.1	. . .	30.4	37.2	9.9	25.04
13	40.3	100.0	15.5	49.0	20.6	25.6	23.1	22.3	24.7	51.2	. . .	45.9	16.4	5.2	18.47
14	23.4	68.4	13.8	43.9	22.6	40.2	25.3	24.6	15.1	62.3	. . .	37.8	27.5	11.5	24.07
15	37.8	79.2	16.1	95.0	12.6	22.6	14.9	10.8	35.1	35.4	. . .	51.5	96.3	30.0	25.33
16	32.2	66.8	23.4	96.8	19.7	54.6	26.1	17.1	35.7	58.3	. . .	54.0	107.9	34.6	30.70

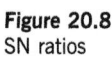

Figure 20.8
SN ratios

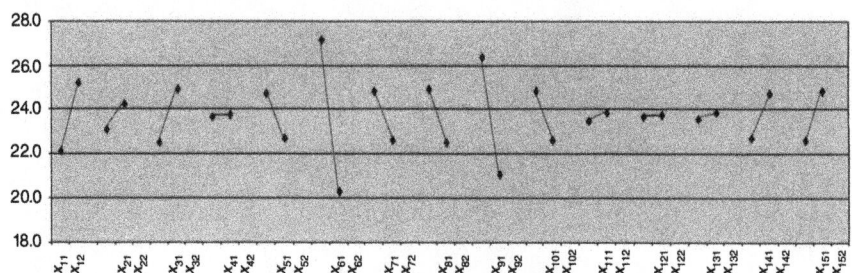

As of now our conclusion is as follows:

Optimization: Keep x_5, x_6, x_7, x_8, x_9, and x_{10}.

Discard all others.

As with all other Taguchi methods, we need to confirm this result.

Step 8.4. Confirm the optimum design. Now we evaluate the discrimination power under the optimum scheme. The result of confirmation using only x_5, x_6, x_7, x_8, x_9, and x_{10} is shown in Figure 20.9.

Conclusions: (1) discrimination has improved and (2) it only requires six variables. This new system has led to an improvement, as shown in Table 20.10. As can

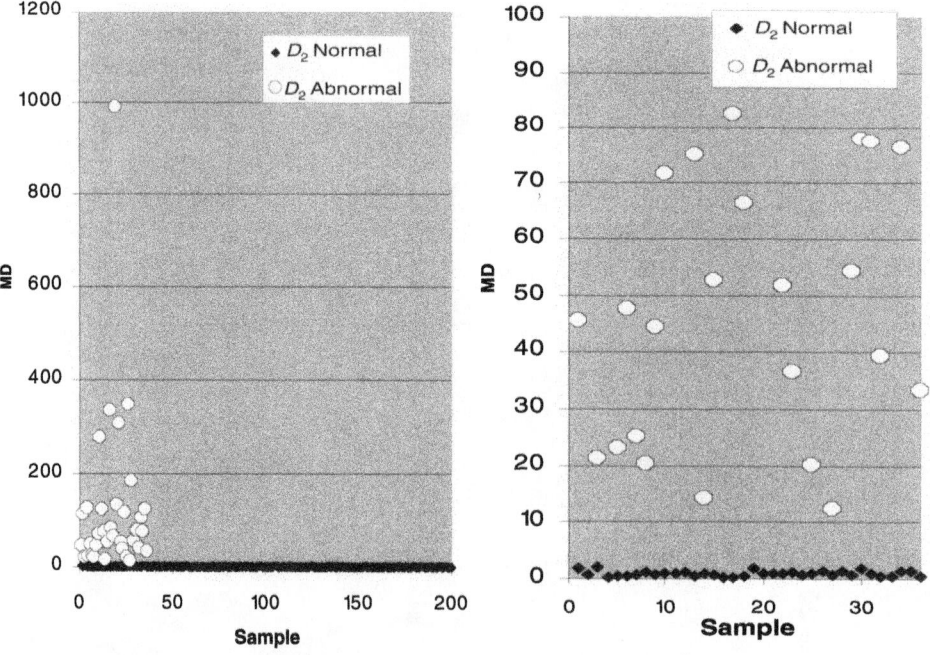

Figure 20.9
Result of confirmation

Table 20.10
Improvement in diagnosis of liver disease

	Traditional Method			Mahalanobis–Taguchi		
	Negative	Positive	Total	Negative	Positive	Total
Negative	28	51	79	63	16	79
Positive	1	15	16	1	15	16

be seen in Table 20.10, the current diagnosis method and Mahalanobis–Taguchi system have the same error rate for false negatives. The current diagnosis method has 65% false positives, requiring further testing, whereas the MTS has only a 20% false positive rate.

20.5. Mahalanobis–Taguchi Gram–Schmidt Process

There is also an alternative way of evaluating MD with the help of the Gram–Schmidt orthogonalization process (GSP). This process is described in the following section.

20.6. Gram–Schmidt's Orthogonalization Process

Given linearly independent vectors $Z_1, Z_2, \dots, Z_k$, there exist mutually perpendicular vectors $U_1, U_2, \dots, U_k$ with the same linear span.

The Gram–Schmidt vectors are constructed sequentially by setting (Figure 20.10)

$$U_1 = Z_1$$

$$U_2 = Z_2 - \frac{Z_2' U_1}{U_1' U_1} U_1$$

$$\vdots$$

$$U_k = Z_k - \frac{Z_k' U_1}{U_1' U_1} U_1 - \dots - \frac{Z_k' U_{k-1}}{U_{k-1}' U_{k-1}} U_{k-1}$$

where the prime denotes the transpose of a vector.

While calculating the MD using GSP, standardized values of the variables are used. Therefore, in the set of equations above, $Z_1, Z_2, \dots, Z_k$ correspond to

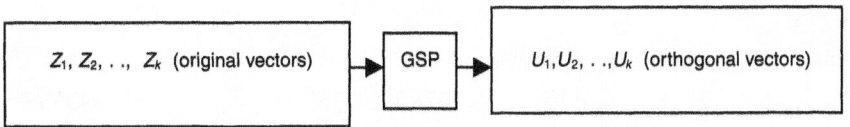

Figure 20.10
Gram–Schmidt's process

standardized values. From this set of equations it is clear that the transformation process depends largely on the first variable.

20.7. Calculation of MD Using the Gram–Schmidt Process

Let us suppose that we have a sample of size n and that each sample contains observations on k variables. After standardizing the variables, we will have a set of standardized vectors. Let these vectors be

$$Z_1 = (z_{11}, z_{12}, \dots, z_{1n})$$

$$Z_2 = (z_{21}, z_{22}, \dots, z_{2n})$$

$$\vdots$$

$$Z_k = (z_{k1}, z_{k2}, \dots, z_{kn})$$

After performing GSP, we have orthogonal vectors as follows:

$$U_1 = (u_{11}, u_{12}, \dots, u_{1n})$$

$$U_2 = (u_{21}, u_{22}, \dots, u_{2n})$$

$$\vdots$$

$$U_k = (u_{k1}, u_{k2}, \dots, u_{kn})$$

It can easily be followed that the mean of vectors $U_1, U_2, \dots, U_k$ is zero. Let $s_1, s_2, \dots, s_k$ be standard deviations of $U_1, U_2, \dots, U_k$, respectively. Since we have a sample of size n, there will be n different MDs. MD corresponding to the jth observation of the sample is computed using the equation

$$\text{MD}_j = \frac{1}{k} \left(\frac{u_{1j}^2}{s_1^2} + \frac{u_{2j}^2}{s_2^2} + \dots + \frac{u_{kj}^2}{s_k^2} \right)$$

where $j = 1, \dots, n$. The values of MD obtained from the inverted matrix method are approximately equal to each other. Table 20.11 below shows MD values by the inverted matrix method and the MTGS method from the preceding example.

Table 20.11
MD values using MTGS and MTS

(a) Healthy Group		
	MD Values (1,2, ... , 200)	Average MD
MTGS	2.105694, 0.605969, ... , 0.899975, 0.765149, 0.913744	0.99499991
MTS	2.105699, 0.605892, ... , 0.899972, 0.765146, 0.913695	0.99498413
(b) Abnormal Group		
	MD Values (1,2, ... , 36)	Average MD
MTGS	25.13597, 66.46272, ... , 41.84671, 52.76765, 24.98069	60.09035
MTS	25.13559, 66.4609, ... , 41.84529, 52.76777, 24.98046	60.08957

20.8. MTS/MTGS Method versus Other Methods

MTS/MTGS methods are intended to provide meaningful information based on patterns. These patterns are identified based on several characteristics. The methods are different from classical multivariate methods in the sense that the methods used in MTS/MTGS are data analytic (i.e., they do not require any assumptions on the distribution of input variables) rather than being on probability-based inference. Artificial neural networks (ANN) also use data analytic methods for pattern recognition problems.

MTS/MTGS methods are different from classical multivariate methods in the following ways:

1. In these methods the Mahalanobis distance is suitably scaled and used as a measure of severity of various conditions.
2. The conditions outside the normal space (i.e., abnormal conditions) are unique and do not constitute a separate population.

The objectives of these multivariate/pattern recognition techniques are implicit in MTS/MTGS methods. These are in addition to the primary objective of developing a multidimensional measurement scale. Like principal component analysis (PCA), MTS/MTGS methods can be used for dimensionality reduction. In fact, dimensionality reduction is done in terms of original variables. In PCA, although dimensionality reduction is done by calculating principal components, to calculate one principal component we need all of the original variables.

Like the discrimination and classification method and regression analysis, MTS/MTGS methods can be used to find the "normals" and "abnormals." Further, they can be used to measure the degree of abnormality. The degree of abnormality cannot be measured by using the discrimination and classification method because its objective is to classify observations into different groups. It is possible to measure the degree of abnormality with regression analysis, but the method of doing this is complex.

Like stepwise regression or a test for additional information, MTS/MTGS methods can be used to screen the variables. However, stepwise regression is a probabilistic approach. The method does not guarantee the best subset regression model. As it terminates with one final equation, an inexperienced analyst may conclude that he or she has found the optimal subset of the model.

Like multivariate charts, MTS/MTGS methods can be used to monitor and control various process conditions based on their severity. It is to be noted that in process control charts the degree of abnormality is judged with respect to probabilistic control limits, whereas in MTS/MTGS the degree of abnormality is judged with respect to a threshold obtained from QLF.

20.9. MTS/MTGS versus Artificial Neural Networks

As compared with artificial neural networks (ANNs), MTS/MTGS methods are similar in the sense that they are data analytic. Using ANNs, it is not as easy to achieve dimensionality reduction as in the case of MTS/MTGS methods. With ANNs, if a new pattern is added in the system, the weight value of the network will be changed because the ANN has to recognize this pattern along with old patterns. Also, randomization plays an important role because ANNs have a

tendency to forget old patterns. So patterns are randomized with ANNs. In MTS/
MTGS methods, if a new pattern is added, the MS corresponding to this pattern
has to be constructed and added to the system, and there is no need to randomize
the patterns. ANNs will not help in finding the relationship between input and
output. In MTS/MTGS methods, such relationships can be obtained. Another disadvantage of ANNs is setting the number of hidden layers, as this number would
affect ultimate results. There is no definite means to set the number of hidden
layers.

20.10. MTS Applications

MTS can be used for two major objectives: diagnosis (pattern recognition) and
forecasting. Following are some of the potential areas of application.

Medical Diagnosis   In a conventional checkup, diagnosis is made by the experience of a physician
along with the results of testing. There are usually many test items, and each item
has a range. A person will be judged, and if the tested result of one item falls
beyond the range, he or she will be sent to a secondary examiner, and a closer
examination will be made by increasing the number of check items.

In biochemical testing, normal values, in general, are determined arbitrarily by
the tester and test chemical manufacturers, and in extreme cases, textbook values
are used without modification.

Many of the test items have various levels of correlations. In the application of
MTS, test results from a healthy group are collected to construct a Mahalanobis
space (normal space or reference space) for diagnosis. This approach can be used
for all types of diagnosis. It can also be used to predict the time it will take a
patient to recover.

Perhaps the most exciting potential of application is in research on medicines
and medical treatments. Currently, a newly developed medicine is evaluated using
a double-blind test with two groups. One group is given the test drug, and the
other, a placebo. Not only does such a test require a large number of people, it is
inhumane for the group taking the placebo. By the use of MTS, the change in
Mahalanobis distance can detect the effect of the new drug or the new treatment
method, possibly using only one person in a short period of time.

Manufacturing   In manufacturing, pattern recognition is widely used. For example, many inspections are done by eyeball observation, such as the appearance of welded or soldered component parts. When the fraction defective in a production line is very
low, it is difficult for a worker to inspect 300 pieces an hour, and therefore easy
to overlook a defective. Moreover, meeting specifications one by one does not
guarantee that it is a good product because of correlations. Many successful studies
of MTS have been reported from the mechanical, electrical, and chemical
industries.

MTS is also very useful in monitoring a manufacturing process. For example,
a semiconductor process can be very complex and consists of many variables, such
as temperature, flow, and pressure. In that case, the normal group would be defined as the group existing when the process is producing perfect products.

In public buildings or hotels, fire alarm systems are installed by law. Perhaps most of us have experienced a false alarm: An alarm sounds but there is no fire, or a smoke detector sounds as a result of cigarette smoking or barbecuing. In the development of fire alarm systems, data such as temperature or the amount of smoke are collected from the situations without fire. These data are considered as a reference group for the construction of a Mahalanobis space.

Fire Detection

In the case of earthquakes, it is difficult to know what type of data should be collected. The method used for data collection depends on what type of forecasting is to be done. For example, we may try to forecast an earthquake one hour from now. But such forecasting may not be very useful except for being able to take such actions as shutting off the gas valve.

Earthquake Forecasting

Suppose that we are going to forecast an earthquake that will occur 24 to 48 hours from now. It is an interesting problem, and we must know in MTS that the data to be collected are under "normal" conditions. In medical checkups, the data are collected from the "healthy people." In fire detection, data are collected under "no fire" conditions. Similarly, the data "without earthquake" must be collected as a reference group. Weather bureaus have seismograph records for years. The amplitude of 48 hours without an earthquake is collected to construct a Mahalanobis space. Then the amplitudes during the 48 hours following an earthquake are collected and compared for the study of future forecasting.

MTS could be used for weather forecasting—not to discuss current forecasting methodology itself but to provide an approach that summarizes all types of existing data and to simplify the process by reducing information that is not contributing to a proper forecast.

Weather Forecasting

In designing air bags in a car, we want to avoid both types of error: an air bag actuated when there was no crash and an air bag that did not actuate at the moment of a crash. To construct a Mahalanobis space, acceleration at several locations in a car is recorded from time to time at an interval of a few thousandth of a second. Such data are collected at various driving conditions: driving on a very rough road, quick acceleration or quick stop, and so on. Those data are used with other sources of information, such as driving speed, to construct a Mahalanobis space. The computer in the air bag system is constantly calculating a Mahalanobis distance. When the distance becomes greater than a predicted threshold, the system sends a signal to actuate the air bag(s) before a crash. On the passenger side, you don't want to deploy the air bag when a child is sitting there or when an adult is sitting in the seat and bending over. Such MTS studies are currently being conducted within automotive companies.

Automotive Air Bag Deployment

Automotive manufacturers are developing sensors that detect a dangerous condition during driving. From various sensor outputs, we want to detect a situation immediately before an accident takes place so that the system can notify the driver or even take over the control of the car by braking, accelerating, and/or steering. Research is in the early stages.

Automotive Accident Avoidance Systems

Prediction is one of the most important areas in pattern recognition. In the business area, banks or credit card companies make predictions for loan or credit card

Business Applications

approval based on the application form or the information from credit companies. A Mahalanobis space can be formed from the existing good customers and the Mahalanobis distance used to predict creditworthy applicants.

Other Applications   MTS can be applied in many other areas, such as fingerprints, handwriting, or character or voice recognition. Although some products of this type are already in the market, they are far from mistake-free. The use of Mahalanobis distance could contribute to further improvement.

20.11. Some Important Considerations

The MTS basically provides a measurement system to assess abnormality. To define a measurement system, we need two parameters defined: the origin and a unit scale. For an MTS, the origin, zero, and one unit are defined by the normal group. The ideal function in measurement is y, the measured value, being proportional to M, the true value. Therefore, it is ideal to use a dynamic SN ratio where M is the true abnormality and y is the MD (or y is the square root the MD). The reciprocal of the SN ratio is the error variance of measurement. The procedure is to evaluate each sample in the abnormal group and assign a number that reflects the abnormality level. That number is used as its signal level. This evaluation and scoring must be done by a professional in the field (Figure 20.11.)

In a medical treatment center, for example, the Mahalanobis distance of a certain patient was calculated from medical test results before the patient was treated. After a certain period of treatment, another Mahalanobis distance was calculated from the new test results. If it is known that the larger the Mahalanobis distance, the more severe the illness, the value calculated from the new test should become smaller if the illness has improved. This is illustrated in Figure 20.12. This is a very important area in which MTS can be used for forecasting. From the relationship between the Mahalanobis distance and the time of treatment, a physician is able to predict when a patient will be able to leave the hospital.

In medical research, researching the effect of a new treatment or a new drug can be made using just one patient instead of relying on double-blind tests, where

Figure 20.11
Rating of abnormality
level

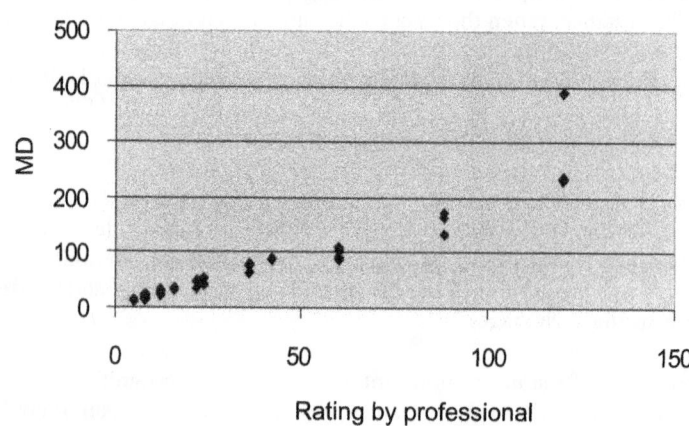

Rating by professional

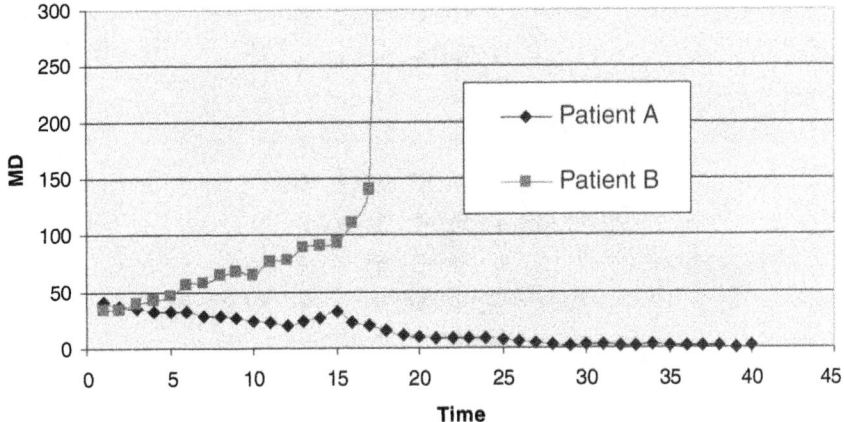

Figure 20.12
Monitoring abnormal
condition

thousands of people are studied; sometimes such a study takes years. By use of the
dynamic SN ratio, the trend can be observed in a short period of time. Such
applications have a huge potential for future research and would bring immeas-
urable benefits to society.

Of course, it is all right to discuss two types of errors, or chi-square distribution,
as a reference. But it is essential to realize that the objective of the MTS approach
is to use the SN ratio for estimation of the error of misjudgment and also for
forecasting.

Part VII

Software Testing and Application

21 Application of Taguchi Methods to Software System Testing

21.1. Introduction   425
21.2. Difficulty in System Testing   425
21.3. Typical Testing   426

21.1. Introduction

System testing is a relatively new area within the Taguchi methods. Systems under testing can be hardware and software or pure software: for example, printer operation, copy machine operation, automotive engine control system, automotive transmission control, automotive antilock brake (ABS) module, computer operating systems such as Windows XP, computer software such as Word and Excel, airline reservation systems, ATMs, and ticket vending machines. The objective of testing is to detect all failures before releasing the design. It is ideal to detect all failures and to fix them before design release. Obviously, you would rather not have customers detect failures in your system.

21.2. Difficulty in System Testing

The difficulty is that there are millions of combinations of customer usage commands. Let's consider a simple system such as printing, for which usage can include the following selections (the number of choices are shown in parentheses):

❏ Tray (5)

❏ Print range (5)

❏ Pages per sheet (6)

❏ Duplex options (4)

❏ Medium (10)

❏ Collate choice (2)

❏ Orientation (2)

❏ Scale to paper (6)

The total number of combinations with these basic parameters alone is

$$5 \times 5 \times 6 \times 4 \times 10 \times 2 \times 2 \times 6 = 144,000 \text{ combinations}$$

In addition to these parameters, there are yes/no choices for 11 print options. That would be $2^{11} = 2048$ combinations. Then the total combination will be

$$(2048)(144,000) = 294,912,000 \text{ combinations}$$

Suppose that testing takes 1 minute per combination and assuming that testing can be automated and can be performed 24 hours a day and 365 days per year, it will take 561 years to complete all combinations. If it takes 1 second per combination, it will take 9.35 years.

In addition to these parameters, there are several more user selections with variable input. One such command is the brightness setting, where user input is between −100 and 100 in increments of 1 unit. In other words, this variable has 201 choices; and there are several parameters like this. In other words, there exist literally billions of user input combinations, and it is impossible to complete them all.

21.3. Typical Testing

Testing people analyze the functionality of software and user operations. Then they select common operations that would go under testing. The number of combinations may vary depending on the complexity of the system. But it is typical that such testing requires several weeks to several months to complete, and such testing is not perfect in terms of area coverage or detection rate. Taguchi methods of software testing utilize an orthogonal array that will improve both area coverage and detection rate.

❐ Example

The procedure for system testing using Taguchi methods is explained below using a copy machine operation as an example.

Step 1: Design the test array. Determine the factors and levels. Factors are operation variables and levels are alternatives for operations. Assign them to an orthogonal array.

Factors and Levels. When an operation such as "paper tray selection" has, say, six options, you may just use three options out of six for the first iteration. In the next iteration you can take other levels. Another choice is to use all six options (a six-level factor). It is not difficult to modify an orthogonal array to accommodate many multilevel factors.

If the user gets to input a continuous number, such as "enlargement/shrinkage setting," where a choice can be any percentage between 25 and 256%, you can select some numbers. For example:

A: enlargement
$A_1 = 33\%$ $\qquad$ $A_2 = 100\%$ $\qquad$ $A_3 = 166\%$

A: enlargement
$A_1 = 25\%$ $\qquad$ $A_2 = 66\%$ $\qquad$ $A_3 = 100\%$ $\qquad$ $A_4 = 166\%$ $\qquad$ $A_5 = 216\%$

Basically, you can take as many levels as you like. The idea is to select levels to cover the operation range as much as possible. As you can see in the design of experiment section of this book, you may choose an orthogonal array to cover as many factors and as many levels as you like.

Orthogonal Array. Recall the notation of an orthogonal array (Figure 21.1). Following is a list of some standard orthogonal arrays:

$$L_8(2^7), L_{16}(2^{15}), L_{32}(2^{31}), L_{64}(2^{63})$$

$$L_9(3^4), L_{27}(3^{13}), L_{81}(3^{40})$$

$$L_{12}(2^{11}), L_{18}(2^1 \times 3^7), L_{36}(2^{11} \times 3^{12}), L_{54}(2^1 \times 3^{25})$$

$$L_{32}(2^1 \times 4^9), L_{62}(4^{21}), L_{25}(5^6), L_{50}(2^1 \times 5^{11})$$

Below is a list of some orthogonal arrays that you can generate from a standard array.

$L_8(4^1 \times 2^4)$

$L_{16}(4^1 \times 2^{12}) \, L_{16}(4^2 \times 2^9), L_{16}(4^3 \times 2^6), L_{16}(4^4 \times 2^3), L_{16}(4^5), L_{16}(8^1 \times 2^8)$

$L_{32}(4^1 \times 2^{28}), L_{32}(4^4 \times 2^{19}), L_{32}(4^8 \times 2^7), L_{32}(8^1 \times 2^{24}), L_{32}(8^1 \times 4^6 \times 2^{12})$

$L_{64}(4^1 \times 2^{60}), L_{64}(4^{12} \times 2^{27}), L_{64}(8^8 \times 2^7), L_{64}(8^4 \times 4^5 \times 2^{20}), L_{64}(8^9)$

$L_{27} \, (9^1 \times 3^9)$

$L_{81} \, (9^1 \times 3^{32}), L_{81}(9^6 \times 3^{16}), L_{81}(27^1 \times 3^{27}), L_{81}(9^{10})$

$L_{18}(6^1 \times 3^6)$

$L_{36}(6^1 \times 3^{13} \times 2^2)$

$L_{54}(6^1 \times 3^{24})$

For example, $L_{81}(9^6 \times 3^{16})$ can handle up to six nine-level factors and 16 three-level factors in orthogonal 81 runs.

In the design of experiments section of the book a technique called *dummy treatment* is described. This technique allows you to assign a factor that has fewer

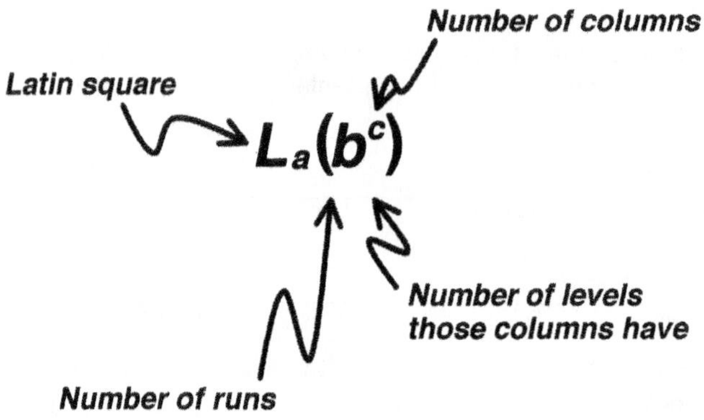

Figure 21.1
Orthogonal array
notation

levels than those of a column. Using the dummy treatment technique, you can assign:

❏ two-level factor to a three-level column

❏ two-level factor to a four-level column

❏ two-level factor to a five-level column   etc.

❏ three-level factor to a four-level column

❏ three-level factor to a five-level column

❏ three-level factor to a six-level column   etc.

❏ four-level factor to a five-level column

❏ four-level factor to a six-level column

❏ four-level factor to an eight-level column   etc.

❏ seven-level factor to a nine-level column   etc.

For example, using dummy treatment, $L_{81}(9^6 \times 3^{16})$ can handle:

❏ Five two-level factors

❏ Ten three-level factors

❏ One five-level factor

❏ Three seven-level factors

❏ Two eight-level factors

By having all these orthogonal arrays and dummy treatment techniques, you can assign just about any number of factors and levels. A Japanese company, Fuji Xerox, uses an L_{128} array for system testing. An L_{128} can handle just about any situation as long as the number of factors is on the order of tens, up to 80 or 90.

To return to our example of copy machine operation system, factors and levels are shown in Table 21.1. They can be assigned to an L_{18} ($2^1 \times 3^7$).

Table 21.1

Factors and levels

	Factor	Level 1	Level 2	Level 3
A	Staple	No	Yes	
B	Side	2 to 1	1 to 2	2 to 2
C	No. copies	3	20	50
D	No. pages	2	20	50
E	Paper tray	Tray 6	Tray 5 (LS)	Tray 3 (OHP)
F	Darkness	Normal	Light	Dark
G	Enlarge	100%	78%	128%
H	Execution	From PC	At machine	Memory

Step 2: Conduct system testing. Table 21.2 shows the L_{18} test array. Eighteen tests are conducted according to each row of the L_{18}. The response will simply be a 0–1 response where

$$y = \begin{cases} 0 & \text{if no problem} \\ 1 & \text{if problem} \end{cases}$$

It is okay to have some infeasible factor-level combinations. For example, when OHP is selected, users should not be able to perform copying "2 to 1 side" or "2 to 2 sides." In that case, the system should provide a proper response. If the system provides the proper output, the data is 0; the data is 1 otherwise.

Step 3: Construct two-way response tables showing the failure rate. Once all tests are run and 0–1 data corrected, a two-way response table is constructed for every two-factor combination. An $A \times B$ table, $A \times C$ table, and so on, up to an $G \times H$ table is then constructed. The entry of each combination is the sum of the 1's of the responses. The data are then converted into the percentage of failure (Table 21.3). For instance, since there are nine combinations of B_iC_j (for $i = 1, 2, 3$ and $j = 1, 2, 3$), there are two runs of B_2C_3 in an L_{18}. The result is two under B_2C_3, indicating 100% failure for B_2C_3. Similarly, A_1B_2 results in two failures in three runs, indicating a 66.7% failure rate.

In general, for a combination A_iB_j to generate 100% failure, the total number of failures must equal the size of the array ÷ the number of combinations A_iB_j. For this example, for A_1B_1 to become 100% failure, the total must be 3 (18/6 = 3).

Step 4: Investigate 100% combinations. By observing two-way tables, 100% failure occurred for B_2C_3, B_2F_3, B_2G_2, C_3G_2, and H_1F_3. Now we need to investigate those combinations for how failures occurred, and then fix the problem.

Table 21.2
L_{18} array

	A_1	A_2	B_1	B_2	B_3	C_1	C_2	C_3	D_1	D_2	D_3	E_1	E_2	E_3	F_1	F_2	F_3	G_1	G_2	G_3
B_1	0	0																		
B_2	2	1																		
B_3	0	0																		
C_1	0	0	0	0	0															
C_2	1	0	0	1	0															
C_3	1	1	0	2	0															
D_1	0	1	0	1	0	0	0	1												
D_2	1	0	0	1	0	0	1	0												
D_3	1	0	0	1	0	0	0	1												
E_1	1	0	0	1	0	0	0	1	0	0	1									
E_2	0	0	0	1	0	0	0	1	0	0	0									
E_3	1	1	0	1	0	0	1	0	1	1	0									
F_1	1	1	0	1	0	0	0	1	1	1	0	1	0	0						
F_2	0	0	0	0	0	0	0	0	0	0	0	0	1	0						
F_3	1	1	0	2	0	0	1	0	0	0	0	0	0	1						
G_1	1	0	0	1	0	0	1	0	0	1	0	0	0	0	0	0	1			
G_2	1	1	0	2	0	0	0	0	1	0	1	1	1	0	1	0	1			
G_3	0	0	0	0	0	0	0	1	0	0	0	0	0	0	0	0	0			
H_1	1	1	0	2	0	0	1	1	1	1	0	0	1	1	0	0	2	1	1	0
H_2	1	0	0	1	0	0	0	1	0	0	1	1	0	0	1	0	0	0	1	0
H_3	0	0	0	0	0	0	0	0	0	0	0	0	0	0	0	0	0	0	0	0

430

Table 21.2
(Continued)

	A 1	B 2	C 3	D 4	E 5	F 6	G 7	H 8	1 = Problem
1	1	1	1	1	1	1	1	1	0
2	1	1	2	2	2	2	2	2	0
3	1	1	3	3	3	3	3	3	0
4	1	2	1	1	2	2	3	3	0
5	1	2	2	2	3	3	1	1	1
6	1	2	3	3	1	1	2	2	1
7	1	3	1	2	1	3	2	3	0
8	1	3	2	3	2	1	3	1	0
9	1	3	3	1	3	2	1	2	0
10	2	1	1	3	3	2	2	1	0
11	2	1	2	1	1	3	3	2	0
12	2	1	3	2	2	1	1	3	0
13	2	2	1	2	3	1	3	2	0
14	2	2	2	3	1	2	1	3	0
15	2	2	3	1	2	3	2	1	1
16	2	3	1	3	2	3	1	2	0
17	2	3	2	1	3	1	2	3	0
18	2	3	3	2	1	2	3	1	0

Table 21.3
Percentage of failure

	A_1	A_2	B_1	B_2	B_3	C_1	C_2	C_3	D_1	D_2	D_3	E_1	E_2	E_3	F_1	F_2	F_3	G_1	G_2	G_3
B_1	0	0																		
B_2	67	33																		
B_3	0	0																		
C_1	0	0	0	0	0															
C_2	33	0	0	50	0															
C_3	33	33	0	**100**	0															
D_1	0	33	0	50	0	0	0	50												
D_2	33	0	0	50	0	0	0	50												
D_3	33	0	0	50	0	0	50	0												
E_1	33	0	0	50	0	0	0	50	0	0	50									
E_2	0	33	0	50	0	0	0	50	0	0	50									
E_3	33	0	0	50	0	0	50	0	0	50	0									
F_1	33	0	0	50	0	0	0	50	0	50	50	50	0	0						
F_2	0	0	0	0	0	0	0	0	0	0	0	0	50	50						
F_3	33	33	0	**100**	0	0	**100**	0	0	50	50	0	0	50						
G_1	33	0	0	50	0	0	50	50	0	50	50	0	50	50	0	0	50			
G_2	33	33	0	**100**	0	0	0	50	0	0	50	50	0	0	0	50	0			
G_3	0	0	0	0	0	0	0	0	0	0	0	0	50	0	0	0	0			
H_1	33	33	0	**100**	0	0	50	50	0	50	50	0	50	50	0	0	50	50	50	0
H_2	33	0	0	50	0	0	0	50	0	0	50	50	0	0	0	50	0	0	50	0
H_3	0	0	0	0	0	0	0	0	0	0	0	0	50	0	0	0	0	0	0	0

It is important to recognize that this system test method will lead us to where problems may reside, but it does not pinpoint the problem or provide solutions to solve problems.

Xerox, Seiko Epson, and ITT Defense Electronics are three companies that presented applications for this method in public conferences in the late 1990s. All three companies report that they achieved 400% improvement in both area coverage and detection rate. For more case studies, see Section 2 in this book.

Part VIII

On-Line Quality Engineering

22 Tolerancing and Quality Level

22.1. **Introduction**   437
  Design Quality and Production Quality   438
  Quality Problems before and after Production   438
  On-Line Approaches   438
22.2. **Product Cost Analysis**   440
22.3. **Responsibilities of the Production Division**   440
22.4. **Role of Production in Quality Evaluation**   441
22.5. **Determination of Tolerance**   442
22.6. **Quality Level at Factories for Nominal-the-Best Characteristics**   448
22.7. **Quality Level at Factories for Smaller-the-Better Characteristics**   450
22.8. **Quality Level at Factories for Larger-the-Better Characteristics**   451
  Reference   453

22.1. Introduction

The difference between parameter design and on-line quality engineering is that in parameter design, the parameters of the process are fixed at certain levels so that variability is reduced. Parameter design is used during manufacturing to vary manufacturing conditions and reduce variability. But in the on-line approach, a certain parameter level is varied from time to time based on the manufacturing situation to adjust the quality level to hit the target. In the case of a watch, for example, its error can be reduced by designing (purchasing) a more accurate watch—this is parameter design. Using the on-line approach, the inaccurate watch is used and errors are reduced by checking more frequently with the time signal and making adjustments to it.

Although it is important to improve working conditions and the method of operating machines and equipment on the job, such issues are off-line improvements in terms of production technology. Instead of such process improvement, here we deal with process control through checking, as well as process maintenance through prevention. In other words, general issues related to the design of process control of on-line real-time processing and production are covered in this chapter. It is important to control or automate an on-line production process based on this theory, using economic calculations.

Design Quality and Production Quality

It is important to discuss two types of quality: design and production. *Design quality* is represented by the properties contained in a product. Such properties are a part of the product and can be described in catalogs. The marketing department determines which quality items should be in the catalogs. Design quality includes the following four aspects, along with usage conditions and problem occurrence:

1. *Natural conditions:* temperature, humidity, atmospheric pressure, sunlight, hydraulic pressure, rain, snow, chemical durability, etc.
2. *Artificial conditions:* impact, centrifugal force, excess current, air pollution, power failure, chemical durability, etc.
3. *Individual differences or incorrect applications:* differences in individual ability, taste, physical conditions, etc.
4. *Indication of problems:* ease or difficulty of repair, disposability, trade in, etc.

Production quality represents deviations from design quality, such as variations due to raw materials, heat treatment, machine problems, or human error. Production quality is the responsibility of the production department.

Quality Problems before and after Production

The quality target must be selected by the R&D department based on feedback from the marketing department about customer desires and competitors products. After the development of a product, the marketing department evaluates the design quality, and the production department evaluates the manufacturing cost.

Generally, production quality problems are caused either by variability or mistakes, such as the variations in raw materials or heat treatment or problems with a machine. The production department is responsible not for deviation from the drawings but for deviation from design quality.

Since there is responsibility, there is freedom. It is all right for the production department to produce products that do not follow the drawings, but it must then be responsible for the claims received due to such deviations. Production departments are also responsible for claims received due to variations, even if the product was made within specifications.

Design quality also includes functions not included in catalogs because they are not advantageous for marketing. However, claims due to items included in catalogs but not in the design must be the responsibility of the marketing department. To summarize: Design quality is the responsibility of the marketing department, and production quality is the responsibility of the production department.

On-Line Approaches

Various concerns of on-line production are discussed below.

QUANTITATIVE EVALUATION OF PRODUCT QUALITY

Since defective products found on the line are not shipped, the problem of these defective products does not affect customers directly, but production costs increase. Product design engineers do not deal with process capability but work to design a product that works under various customers' conditions for a certain number of years. This is called *parameter design.* The point of parameter design for production is to provide a wide tolerance for manufacturing or to make it easy for production personnel to make the product.

In addition, production engineering should design a stable manufacturing process, to balance loss due to variations in cost. When a process is very stable, engineers should try to increase the production speed to reduce the manufacturing cost. When production speed increases, variation normally increases. An engineer should adjust the production speed to a point where the manufacturing cost and the loss due to variation balance. This is important because a product is normally sold at three or four times unit manufacturing cost (UMC). Therefore, any reduction in the UMC yields a tremendous benefit to the manufacturer in terms of profit.

DETERMINATION OF SPECIFICATIONS FOR MANUFACTURED PRODUCTS: TOLERANCING

Tolerancing is a method to quantitatively determine the specifications for transactions or a contract. A new and unique approach to tolerancing utilizes the safety factor, previously vague.

FEEDBACK CONTROL

In *feedback control*, product quality or manufacturing conditions are checked at intervals. Production is continued if the result is within a certain limit; otherwise, feedback control is conducted. The optimum design for such a feedback control system is explained below.

PROCESS DIAGNOSIS, ADJUSTMENT, AND MAINTENANCE DESIGN

We discuss the design of several preventive maintenance systems. In the case of soldering, for example, its apparent quality cannot be measured by using continuous variables. In the same way, it is difficult to design a gauge for a product with a complicated shape. The design of such process control systems using basic equations and various applications is illustrated later in the chapter.

FEEDFORWARD CONTROL

Feedback control involves measurement of a product's quality characteristics, then making changes to the process as required. Feedforward control involves process change *before* production so that the quality level can meet the target. In the manufacture of photographic film or iron, for example, the quality of incoming gelatin or iron ore is measured so that the mixing ratios or reacting conditions may be adjusted.

PRODUCTION CONTROL FOR THE QUALITY OF INDIVIDUAL PRODUCTS: INSPECTION

Using inspection, product quality characteristics are measured and compared with the target. A product whose quality level deviates from the target is either adjusted or discarded. This is inspection in a broad sense, where individual products are inspected.

22.2. Product Cost Analysis

Production costs can be shown by the following equation:

$$\text{(production costs)} = \text{(material costs)} + \text{(processing costs)}$$
$$+ \text{(control costs)} + \text{(pollution control costs)} \quad (22.1)$$

Control costs consist of both production control and quality control costs. The product design division is responsible for all the costs seen on the right-hand side of equation (22.1). When we say that a product designer is responsible for the four terms in equation (22.1), we mean that he or she is responsible for the design of a product so that it functions under various conditions for the duration of its design life (the length of time it is designed to last). In particular, the method of parameter design used in product design enables the stability of the product function to be improved while providing a wide range of tolerance. For example, doubling the stability means that the variability of an objective characteristic does not change even if the level of all sources of variability is doubled.

The production engineering division seeks to produce the initial characteristics of the product as designed, with minimum cost and uniform quality. The production engineering division is responsible for the sum of the second to fourth terms on the right-hand side of equation (22.1). In other words, they are responsible for designing stable production processes without increasing cost and within the tolerance of the initial characteristics given by the product design division.

However, no matter how stable a production process is, it will still generate defectives if process control is not conducted. Additionally, if production speed is increased to reduce cost by producing more product within tolerance limits, more loss will occur, due to increased variability. Basically, it is difficult to balance production costs and loss due to variability in quality, so an appropriate level of process control must be employed.

22.3. Responsibilities of the Production Division

Production line workers are responsible for the daily production of a product by using a given production process and specifications, that is, the quality characteristics that should be controlled at each step, as well as the major target values and control limits from technical and manufacturing standards. However, some process conditions are often overlooked, and therefore control limits may need to be adjusted through calibration.

Initially, the cost of a product is more important than its quality, since the price of a product results in the first loss to the consumer (customer). In this sense the balance between cost and quality is the balance between price and quality. Quality includes the following items: (1) operating cost, (2) loss due to functional variability, and (3) loss due to harmful quality items. The first and third items are estimated in the design stage, whereas item 2, operating cost, is usually evaluated under standard usage conditions. The second and third cost items are evaluated using a loss function, which is derived from the mean square of the difference from the ideal value. In daily production activities, on the other hand, economic loss due to the second and third items is evaluated using the mean square of the difference from the target value. The quality level is evaluated using the equation

$$L = \frac{A}{\Delta^2} \sigma^2 \qquad (22.2)$$

where A (dollars) is the loss when the initial characteristic value of a product does not satisfy the tolerance, Δ is the tolerance of the characteristic, and σ^2 is the mean square of the difference from the objective value of the initial characteristic.

If the cost is far more important than the quality and inventory losses, one has to use a number several times that of the real cost as the cost figure in the following calculations. According to Professor Trivus, the director of MIT's Advanced Technology Research Institute, Xerox Corporation used to set the product price at four times the UMC. Since the UMC does not include the costs of development, business operations, head-office overhead, and so on, corporations cannot generate profit if their product is not sold at a price several times higher than the UMC.

Production attempts to minimize the sum of process control costs and quality-related loss through process and product control. Process control costs are evaluated using real cost figures, although multiples of the figure are often used. If the cost is several times more important than the quality-related loss, one can take the cost figure times some number to use as the real cost. To balance the two types of unit costs necessary for control and the quality-related loss, we consider the following cost terms:

B: checking (measurement) cost
C: adjustment cost, to bring the process under control or to adjust the product

If the amount of control cost is far greater than the amount of quality-related loss, excessive control is indicated; alternatively, if quality-related losses dominate, a lack of control is indicated.

The main task of people on the production line is to control the process and product while machines do the actual processing. Therefore, we can also use quality engineering methods to determine the optimal number of workers on the production line.

22.4. Role of Production in Quality Evaluation

The manufacturing process is composed of the following six stages:

1. Product planning (to decide product function, price, and design life)
2. Product design (to design a product with functions determined by product planning)
3. Production process design (to design the production process necessary to produce a product)
4. Production
5. Sales
6. After-service

The production division is involved directly in stages 3 and 4. Its task is to produce products efficiently as specified by drawings and specifications, within quality standards.

Whereas production cost is obvious, a definition of product quality is not always clear. Many firms evaluate product quality using fraction defective; however, defectives do not cause problems for consumers because defectives are generally not delivered to them, Therefore, the production of defectives is a cost issue, not a quality issue. Quality is related to the loss incurred by consumers due to function variability, contamination, and so on, of the product as shipped. Therefore, the idea is to find a way to evaluate quality levels of accepted products. To conduct such an evaluation, it is necessary to determine product standards, especially tolerance.

Quality can be defined as "the loss that a product costs society after its shipment, not including the loss due to the function itself." Under such a definition, quality is divided roughly into two components: (1) loss due to function variability and (2) loss that is unrelated to function, such as loss due to side effects or poor fuel efficiency. Here we focus primarily on loss due to function variability.

Engineers in the United States and other countries have used the following two methods for tolerance design: (1) a method that uses the safety factor derived from past experience, and (2) a method that allocates tolerance based on reliability design. The Monte Carlo method, which obtains tolerance by generating a random vector, is used in the area of reliability design because other factors of variability also influence the function. In the following discussion, these methods are explained and some examples are given that compare them to using the loss function.

22.5. Determination of Tolerance

When it is desirable that an objective characteristic, *y*, be as close to a target value as possible, we call it a *nominal-the-best characteristic.*

❏ Example

In the power circuit of a TV set, 100-V ac input in a home should be converted into 115-V dc output. If the circuit generates 115-V of output constantly during its designed life, the power circuit has perfect functional quality. Thus, this power circuit is a subsystem with a nominal-the-best characteristic whose target value is 115-V.

Suppose that the target value is *m* and the actual output voltage is *y*. Variability of component parts in the circuit, deterioration, variability of input power source, and so on, create the difference $(y - m)$. Tolerance for the difference $(y - m)$ is determined by dividing the difference by the safety factor ϕ, derived from past experience. Suppose that the safety factor in this case is 10. When a 25% change in output voltage causes trouble, the tolerance is derived as follows:

$$\text{tolerance} = \frac{\text{functional limit}}{\text{safety factor}} = \frac{25}{10} = 2.5\% \qquad (22.3)$$

If the target value of output voltage is 115 V, the tolerance is

$$\Delta = (115)(0.025) = 2.9 \text{ V} \qquad (22.4)$$

Therefore, the factory standard of output voltage is determined as follows after rounding:

$$115 \pm 3 \text{ V} \qquad (22.5)$$

Let Δ_0 denote the functional limit, that is, deviation from the target value when the product ceases to function. At Δ_0, product function breaks down. In this case, other characteristics and usage conditions are assumed to be normal, and Δ_0 is derived as the deviation of y from m when the product stops functioning.

The assumption of normality for other conditions in the derivation of functional limits implies that standard conditions are assumed for the other conditions. The exact middle condition is where the standard value cuts the distribution in half, for example, median and mode, and it is often abbreviated as LD_{50} (lethal dosage), that is, there is a 50–50 chance of life (L) or death (D)—50% of people die if the quantity is less than the middle value and the other 50% die if the quantity is greater. Consequently, one test is usually sufficient to obtain the function limit.

The value of Δ_0 is one of the easiest values to obtain. For example, if the central value of engine ignition voltage is 20 kV, LD_{50} is the value when ignition fails after changing the voltage level. (The central value is determined by parameter design and not by tolerance design. It is better to have a function limit as large as possible in parameter design, but that is a product design problem and is not discussed here.) After reducing voltage, if ignition fails at 8 kV, -12 kV is the lower functional limit. On the other hand, the upper functional limit is obtained by raising the voltage until a problem develops. A high voltage generates problems such as corona discharge; let its limit be $+18$ kV. If the upper- and lower-side functional limits are different, different tolerances may be given using different safety factors. In such a case, the tolerance is indicated as follows:

$$+\Delta_1 \ m \ \ -\Delta_2 \qquad (22.6)$$

Upper and lower tolerance, Δ_1 and Δ_2, are obtained separately from respective functional limit Δ_{01} and Δ_{02} and safety factor ϕ_1 and ϕ_2:

$$\Delta_i = \frac{\Delta_{0i}}{\phi_i} \qquad (i = 1,2) \qquad (22.7)$$

The functional limit Δ_0 is usually derived through experiment and design calculations. However, derivation of the safety factor is not clear. Recently, some people have started using probability theory to determine tolerance [1]. The authors of this book firmly believe that the probability approach is inefficient and incorrect. The authors recommend the derivation of safety factors without using probability statistics. We suggest that safety factor ϕ be derived according to the following equation, which is becoming popular:

$$\phi = \sqrt{\frac{\text{average loss when falling outside the functional limit}}{\text{loss to the factory when falling outside the factory standard}}} \qquad (22.8)$$

Letting A_0 be the average loss when the characteristic value falls outside the functional limit, and A be the loss to the factory when the characteristic value falls outside the factory standard:

$$\phi = \sqrt{\frac{A_0}{A}} \qquad (22.9)$$

The reason for using this formula is as follows. Let y denote the characteristic value and m the target value. Consider the loss when y differs from m. Let $L(y)$ be the economic net loss when a product with characteristic value y is delivered to the market and used under various conditions. Assume that the total market size of the product is N and that $L_i(t,y)$ is the actual loss occurring on location. In the tth year after the product stops functioning, the loss jumps from zero to a substantial amount. With a given design life T, $L(y)$ gives the average loss over the entire usage period in the entire usage location; that is,

$$L(y) = \frac{1}{N} \sum_{i=1}^{N} \int_0^T L_i(t,y) \ dt \qquad (22.10)$$

Even if individual $L_i(t,y)$ is a discontinuous function, we can take the average usage condition over many individuals, and thus approximate it as a continuous function. We can conduct a Taylor expansion of $L(y)$ around target value m:

$$L(y) = L(m + y - m)$$

$$= L(m) + \frac{L'(m)}{1!} (y - m) + \frac{L'(m)}{2!} (y - m)^2 + \cdots \qquad (22.11)$$

In equation (22.11), the loss is zero when y is equal to m and $L'(m)$ is zero, since the loss is minimum at the target value. Thus, the first and second terms drop out and the third term becomes the first term. The cubic term should not be considered. By omitting the quartic term and above, $L(y)$ is approximated as follows:

$$L(y) = k(y - m)^2 \qquad (22.12)$$

The right-hand side of equation (22.12) is loss due to quality when the characteristic value deviates from the target value. In this equation, the only unknown is k, which can be obtained if we know the level of loss at one point where y is different from m. For that point we use the function limit Δ_0 (see Figure 22.1). Assuming that A dollars is the average loss when the product does not function in

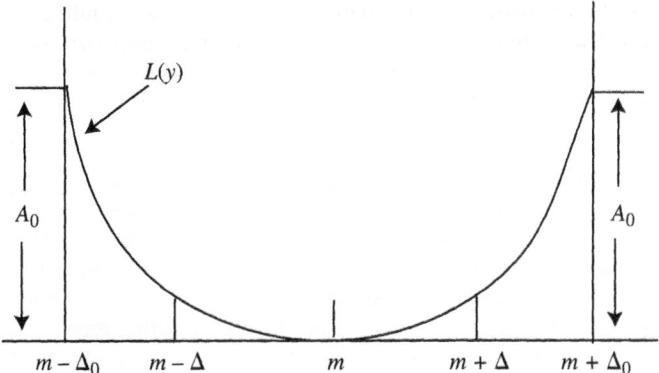

Figure 22.1
Loss function

the market, we substitute A_0 on the left-hand side of equation (22.12) and Δ_0 for $y - m$ on the right-hand side. Thus,

$$A_0 = k\Delta_0^2 \qquad (22.13)$$

Solving this for k gives us

$$k = \frac{A_0}{\Delta_0^2} \qquad (22.14)$$

Therefore, the loss function, $L(y)$, due to quality variability in the case of a nominal-the-best characteristic becomes

$$L(y) = \frac{A_0}{\Delta_0^2} (y - m)^2 \qquad (22.15)$$

As we can see from equation (22.10), the value of quality evaluation given by equation (22.15) is the average loss when the product with characteristic value y is used under various usage conditions, including a major problem such as an accident. Since it is not known under which condition the product is used, it is important to evaluate the average loss by assuming that the product is used under the entire range of usage conditions. Tolerance should be decided by taking the balance between average loss and cost.

To determine the safety factor, ϕ, we need to evaluate one more economic constant, the cost, A, when the product fails to meet factory standards. A product that does not meet these standards is called *defective*. A defective item needs to be repaired or scrapped. If it can be modified, the cost of modification is used for A. But in some cases the product may not even pass after modification. If the rate of acceptance is close to 10%, the product price is used as the cost of modification, A. If the rate of acceptance is lower, say q, A is defined as

$$A = \frac{\text{factory pine}}{q} \qquad (22.16)$$

Since the rate of acceptance cannot be determined before establishing the factor standards, the equation obtained by substituting the value of equation

(22.16) into the left-hand side of equation (22.15) becomes nonlinear and thus is difficult to solve. Generally, the method of sequential approximation is used to solve the problem.

❏ Example

Continuing the example introduced at the beginning of this section, the functional limit of the power circuit is assumed to be 25%, and the loss to the company when the product falls outside the function limit in the market, A_0, is assumed to be $300 (this includes repair cost and replacement cost during repair). We assume further that the power circuit, which does not satisfy factory standards, can be fixed by changing the level of resistance of the resistor within the circuit. Let the sum of labor and material cost for repair, A, be $1. The safety factor, ϕ, is determined as follows using equation (22.9).

$$\phi = \sqrt{\frac{A_0}{A}} = \sqrt{\frac{300}{1}} = 17.3 \qquad (22.17)$$

Since the safety factor is 17.3, the factory standard is

$$\Delta = \frac{\Delta_0}{\phi} = \frac{25}{17.3} = 1.45\% \qquad (22.18)$$

Therefore, with a target value of 115 V, the factory standard, Δ, is just 1.7 V:

$$115 \pm (115)(0.0145) \approx 115 \pm 1.7 \text{ V} \qquad (22.19)$$

Even if the functional limit in the market is 115 ± 29 V, the standard for the production division of the factory is 115 ± 1.7 V, and those products that do not satisfy this standard need to be repaired.

A characteristic value that is nonnegative and as small as possible is called a *smaller-the-better characteristic*. The safety factor for a smaller-the-better characteristic is found in equation (22.9). For example, if the fatal dose of a poisonous element (LD_{50}) is 8000 ppm, the loss, A_0, when someone has died due to negligence, for example, is obtained using the following equation:

A_0 = (national income) × average life (or average life expectancy)

= $20,000 × 77.5 years

= $1,550,000 \qquad (22.20)

The safety factor, ϕ, and factory tolerance, Δ, are determined from equation (22.9), assuming that the price of the product, A, is $3.

$$\phi = \sqrt{\frac{A_0}{A}} = \sqrt{\frac{1,550,000}{3}} = 719 \qquad (22.21)$$

Since the safety factor is 719, the factory standard is

$$\Delta = \frac{\Delta_0}{\phi} = \frac{8000}{719} = 11 \text{ ppm} \tag{22.22}$$

For a characteristic that is nonnegative and should be as large as possible, that is, the *larger-the-better characteristic*, the safety factor, ϕ, is again given by equation (22.9). For example, a pipe made of a certain resin breaks with 5000 kg$_f$ of external force, and the resulting loss, A_0, is \$300,000. Assume that the strength and cost of the pipe are proportional to the cross-sectional area and that their coefficients of proportion $b = 80$ (kg$_f$/mm^2) and $a = \$40$ per/mm^2. First, let us derive the optimal cross-sectional area.

The loss due to quality is handled as a smaller-the-better characteristic after taking the inverse. The sum, L, of the loss due to price and quality is

$$L = \text{price} + \text{quality} = ax + \frac{A_0 \Delta_0^2}{(bx)^2} \tag{22.23}$$

where x is the cross-sectional area.

To minimize the above, differentiate equation (22.23) and solve it by equating the derivative to zero. The following formula is derived:

$$x = \left(\frac{2A_0 \, \Delta_0^2}{ab^2}\right)^{1/3} \tag{22.24}$$

By substituting the parameter values, that is, $a = \$40$, $b = 80$ kg$_f$, $A_0 = \$300,000$, and $\Delta_0 = 5000$ kg$_f$, the optimal cross-sectional area x is derived as follows:

$$x = \left(\frac{(2)(300,000)(5000)^2}{(40)(80^2)}\right)^{1/3} = 388 \tag{22.25}$$

Therefore, the price is determined by multiplying the result for x in (22.25) by $a = \$40$:

$$A = (40)(388) = \$15,520 \tag{22.26}$$

The safety factor is derived by substituting $A = \$15,520$ and $A_0 = \$300,000$ into equation (22.9):

$$\phi = \sqrt{\frac{30}{1.55}} = 4.4 \tag{22.27}$$

Thus, the lower limit of the standard is 22 tons$_f$, that is, 4.4 times the strength corresponding to the function limit $\Delta_0 = 5$ tons$_f$.

The safety factor method can be applied to any component part and material tolerance in the same way. In this case you would use the point of LD$_{50}$, that is, the characteristic value when the objective characteristic stops functioning (the value of functional limit when the characteristic of other parts and usage conditions are at the standard levels), Δ_0 and A_0, as the loss incurred to the society. A is the loss incurred to the factory when the component part or raw material does not satisfy the delivery standard of the factory.

Therefore, the safety factor, ϕ, is given by the following equation in all cases of nominal-the-best, larger-the-better, and smaller-the-better characteristics:

$$\phi = \sqrt{\frac{A_0}{A}} \tag{22.28}$$

Further, the factory standard, Δ, is determined by the following:

$$\Delta = \begin{cases} \dfrac{\Delta_0}{\phi} & \text{nominal-the-best and smaller-the-better} & (22.29) \\[2mm] \phi\Delta_0 & \text{larger-the better} & (22.30) \end{cases}$$

22.6. Quality Level at Factories for Nominal-the-Best Characteristics

Production will accept the tolerances given by specifications and drawings and use it as the basis for quality-level evaluation. It is important to use the loss function for quality-level evaluation instead of fraction defective or rate of acceptance. In the case of nominal-the-best characteristics, the following equation is used:

$$L = \frac{A_0}{\Delta_0^2} \sigma^2 = \frac{A}{\Delta^2} \sigma^2 \tag{22.31}$$

where σ^2 is the mean square of the difference from the target value m of n data, $y_1, y_2, \dots, y_n$, or the mean-squared error, or variance.

$$\sigma^2 = \frac{1}{n} [(y_1 - m)^2 + (y_2 - m)^2 + \cdots + (y_n - m)^2] \tag{22.32}$$

❏ Example

Consider a plate glass product having a shipment cost of $3 per unit, with a dimensional tolerance, Δ, of 2.0 mm. Dimensional data for glass plates delivered from a factory minus the target value, m, are

```
0.3   0.6  -0.5  -0.2   0.0   1.0   1.2   0.8  -0.6   0.9   0.0
0.2   0.8   1.1  -0.5  -0.2   0.0   0.3   0.8   1.3
```

To obtain the loss due to variability, we must calculate the mean-squared error from the target value, Δ^2, and substitute it into the loss function. We refer to this quantity as the mean-squared error, σ^2 (variance):

$$\sigma^2 = \frac{1}{20} (0.3^2 + 0.6^2 + \cdots + 1.3^2) = 0.4795 \text{ mm}^2 \tag{22.33}$$

Substituting this into the loss function, equation (22.31), yields

$$L = \frac{A}{\Delta^2}\sigma^2 = \frac{3}{2^2}(0.4795) = 36 \text{ cents} \qquad (22.34)$$

The mean of the deviations of glass plate size from the target value is 0.365, indicating that the glass plates are a bit oversized on average. One can create an ANOVA table to observe this situation (see Chapter 29).

$$S_T = 0.3^2 + 0.6^2 + \cdots + 1.3^2 = 9.59 \qquad (f = 20) \quad (22.35)$$

$$S_m = \frac{(0.3 + 0.6 + \cdots + 1.3)^2}{20} = \frac{7.3^2}{20} = 2.66 \qquad (f = 1) \quad (22.36)$$

$$S_e = S_T - S_m = 9.59 - 2.66 = 6.93 \qquad (f = 19) \quad (22.37)$$

Variance, V, is obtained by dividing variation, S_e, by its degrees of freedom, f. Net variation, S', is derived by subtracting from variation, S, its error variance times the degrees of freedom. The rate of contribution, ρ, is derived by dividing net variation, S', by total variation, S_T. The calculations are illustrated in Table 22.1.

We can observe from Table 22.1 that the mean size is greater than the central value of the standard, thus increasing the mean-squared error and the loss due to variability. It is usually easy to adjust the mean to the target value in factories. Accordingly, if the mean is adjusted to the target value, the variance, σ^2, should be reduced to that of the error variance, V_e, in Table 22.1. In this case, the value of the loss function will be

$$L = \frac{3}{2^2}(0.365) = (36.0)(0.761) = 27.4 \text{ cents} \qquad (22.38)$$

Compared to equation (22.34), this means a quality improvement of

$$36.0 - 27.4 = 8.6 \text{ cents} \qquad (22.39)$$

per sheet of glass plate. If the monthly output is 100,000 sheets, quality improvement will be $8600 per month.

No special tool is required to adjust the mean to the target value. One needs to compare the size with the target value occasionally and cut the edge of the plate

Table 22.1
ANOVA table

Source	f	S	V	S'	ρ (%)
m	1	2.66	2.66	2.295	23.9
e	19	6.93	0.365	7.295	76.1
Total	20	9.59	0.4795	9.590	100.0

if it is oversized. By varying the interval of comparison with the measure, not only the mean but also the variance will change. This is the determination of calibration cycle in quality engineering.

22.7. Quality Level at Factories for Smaller-the-Better Characteristics

A smaller-the-better characteristic is nonnegative and its optimal value is zero. The following loss function, L, gives the quality level for smaller-the-better characteristics:

$$L = \frac{A_0}{\Delta_0^2} \sigma^2 \qquad (22.40)$$

where A_0 is the loss incurred by the company if the functional limit is exceeded, Δ_0 the functional limit, and σ^2 the mean-squared error, or variance.

In production factories, the following Δ and A are used instead of Δ_0 and A_0, respectively.

Δ: factory standard

A: loss incurred by the factory if the factory standard is not fulfilled

This is because Δ is obtained from

$$\Delta = \frac{\Delta_0}{\phi} \qquad (22.41)$$

and the following equality holds, as in equation (22.31):

$$\frac{A}{\Delta^2} = \frac{A_0}{\Delta_0^2} \qquad (22.42)$$

☐ Example

Suppose that the specification of roundness is less than 12 μm. The loss when the product is not acceptable, A, is 80 cents. Two machines, A_1 and A_2, are used in production, and the data for roundness of samples taken two per day for two weeks is summarized in Table 22.2. The unit is micrometers.

Table 22.2
Roundness of samples from two machines

A_1	0	5	4	2	3	1	7	6	8	4
	6	0	3	10	4	5	3	2	0	7
A_2	5	4	0	4	2	1	0	2	5	3
	2	1	3	0	2	4	1	6	2	1

We compare the quality level of two machines, A_1 and A_2. The characteristic value of this example is a smaller-the-better characteristic with an upper standard limit. Thus, equation (22.40) is the formula to use:

$$\Delta = 12 \ \mu m \tag{22.43}$$

$$A = 80 \ \text{cents} \tag{22.44}$$

Variance, σ^2, is calculated separately for A_1 and A_2. For A_1,

$$\sigma^2 = \tfrac{1}{20}(y_1^2 + y_2^2 + \cdots + y_{20}^2) = \tfrac{1}{20}(0^2 + 5^2 + 4^2 + \cdots + 7^2)$$

$$= 23.4 \ \mu m^2 \tag{22.45}$$

$$L = \frac{A}{\Delta^2}\sigma^2 = \frac{0.80}{12^2}(23.4) = 13 \ \text{cents} \tag{22.46}$$

For A_2,

$$\sigma^2 = \tfrac{1}{20}(5^2 + 4^2 + 0^2 + \cdots + 1^2) = 8.8 \ \mu m^2 \tag{22.47}$$

$$L = \frac{0.80}{12^2}(8.8) = 4.9 \ \text{cents} \tag{22.48}$$

Thus, compared to A_1, A_2 has a better quality level.

$$13.0 - 4.9 = 8.1 \ \text{cents} \tag{22.49}$$

or A_2 is

$$\frac{13.0}{4.9} = 2.7 \tag{22.50}$$

times better than A_1. If production is 2000 per day, the difference between the two machines will amount to $40,500 mil per year, assuming that the number of working days per year is 250.

22.8. Quality Level at Factories for Larger-the-Better Characteristics

A larger-the-better characteristic is a characteristic value that is nonnegative and whose most desirable value is infinity. For thermal efficiency, yield, and fraction acceptance, the maximum value is 1 (100%) even though the value never takes on a negative value and the larger value is better. Thus, characteristics that do not have a target value and whose number should be as large as possible (e.g., amplification factor, power, strength, harvest) are larger-the-better characteristics.

Suppose that y is a larger-the-better characteristic and that its loss function is

$$L(y) \qquad (0 \leq y \leq \infty) \tag{22.51}$$

$L(y)$ is expanded around $y = \infty$ using Maclaurin's expansion.

$$L(y) = L(\infty) + \frac{L'(\infty)}{1!}\frac{1}{y} + \frac{L''(\infty)}{2!}\frac{1}{y^2} + \cdots \qquad (22.52)$$

Loss is zero at $y = \infty$, and the following is immediate:

$$L(\infty) = 0 \qquad (22.53)$$

$$L'(\infty) = 0 \qquad (22.54)$$

Thus, by substituting the above into equation (22.52) and omitting the cubic terms and beyond, we get the following approximation:

$$L(y) = k\frac{1}{y^2} \qquad (22.55)$$

The proportion coefficient, k, can be determined by the following equation once the loss at a certain point, $y = y_0$, that is, $L(y_0)$, is known:

$$k = y_0^2\, L(y_0) \qquad (22.56)$$

For this point we can use the point when a problem actually occurs in the market, Δ_0, and derive k with loss A_0 at that point:

$$k = \Delta_0^2 A_0 \qquad (22.57)$$

Thus, the loss function becomes

$$L = \frac{\Delta_0^2 A_0}{y^2} \qquad (22.58)$$

Let $y_1, y_2, \ldots, y_n$ denote n observations. We can calculate the quality level per product by substituting the data into equation (22.58). Let L be the quality level per product. Then

$$L = A_0 \Delta_0^2 \sigma^2 \qquad (22.59)$$

where

$$\sigma^2 = \frac{1}{n}\left(\frac{1}{y_1^2} + \frac{1}{y_2^2} + \cdots + \frac{1}{y_n^2}\right) \qquad (22.60)$$

In other words, the inverse of a larger-the-better characteristic is approximated by a smaller-the-better characteristic.

❏ Example

For three-layered reinforced rubber hose, the adhesive strength between layers—K_1, the adhesion of rubber tube and reinforcing fiber, and K_2, the adhesion of reinforcing fiber and cover rubber—is very important. For both adhesive strengths, a lower standard, Δ, is set as 5.0 kg. If a defective product is scrapped, its loss is $5 per unit. Annual production is 120,000 units. Two different types of adhesive, A_1 (50 cents) and A_2 (60 cents), were used to produce eight test samples each,

Table 22.3
Adhesive strength

A_1	K_1	10.2	5.8	4.9	16.1	15.0	9.4	4.8	10.1
	K_2	14.6	19.7	5.0	4.7	16.8	4.5	4.0	16.5
A_2	K_1	7.6	13.7	7.0	12.8	11.8	13.7	14.8	10.4
	K_2	7.0	10.1	6.8	10.0	8.6	11.2	8.3	10.6

and the adhesive strength of K_1 and K_2 was checked. The data are given in Table 22.3.

Let us compare the quality level of A_1 and A_2. Since it is a larger-the-better characteristic, we want the mean to be large and the variability to be small. A measurement that takes both of these requirements into account is the *variance*, that is, the mean square of the inverse, given by equation (22.60). Variances for A_1 and A_2 are derived from equation (22.60) separately, and the loss function is obtained from equation (22.59). For A_1,

$$\sigma_1^2 = \frac{1}{16}\left(\frac{1}{0.2^2} + \frac{1}{5.8^2} + \cdots + \frac{1}{16.5^2}\right)$$

$$= 0.02284 \tag{22.61}$$

$$L_1 = A_0\Delta_0^2\sigma_1^2 = (5)(5^2)(0.02284)$$

$$= \$2.85 \tag{22.62}$$

For A_2,

$$\sigma_2^2 = \frac{1}{16}\left(\frac{1}{7.6^2} + \frac{1}{13.7^2} + \cdots + \frac{1}{10.6^2}\right)$$

$$= 0.01139 \tag{22.63}$$

$$L_2 = (5)(5^2)(0.01139) = \$1.42 \tag{22.64}$$

Therefore, even if the adhesion cost (the sum of adhesive and labor) is 10 cents higher in the case of A_2, the use of A_2 is better than the use of A_1.

The amount of savings per unit is

$$(0.50 + 2.85) - (0.60 + 1.42) = \$1.33 \tag{22.65}$$

In one year, this amounts to a savings of \$159,700.

Reference

1. Genichi Taguchi et al., 1994. *Quality Engineering Series 2: On-Line Production.* Tokyo: Japanese Standards Association and ASI Press.

23 Feedback Control Based on Product Characteristics

23.1. Introduction   454
23.2. System Design of Feedback Control Based on Quality Characteristics   454
23.3. Batch Production Process   459
23.4. Design of a Process Control Gauge (Using a Boundary Sample)   462

23.1. Introduction

Under feedback control, the characteristics of a product produced by a production process are checked, and if a difference between a characteristic and a target is greater than a certain level, the process is adjusted. If the difference is not so great, production is continued without adjustment. This type of control system can be applied without modification to cases where the process is controlled automatically. Even when a process is controlled automatically, routine control by human beings is necessary.

23.2. System Design of Feedback Control Based on Quality Characteristics

Many factories today use a system in which the characteristics of a product are checked and the process condition is controlled if the quality level is off-target. However, when it is costly to adjust process conditions, feedback about an entire process based on quality engineering is rarely used. The optimal design of a feedback controller to maintain quality is described below.

The following parameters are defined:

❑ *Specification of objective characteristic:* $m \pm \Delta$

❑ *Loss due to a defective:* A (dollars)

❑ *Checking (measuring) cost:* B (dollars)

❑ *Adjustment cost:* C (dollars)

❑ *Current checking interval:* n_0 (units)

❑ *Current mean adjustment interval (observed):* u_0 (units)

❏ *Current adjustment (control) limit: D_0*
❏ *Time lag of the checking method: l (units)*
❏ *Optimal checking interval: n (units)*
❏ *Estimate of mean adjustment interval: u (units)*

If checking is done by n-unit intervals at a cost of B dollars, the checking cost per product is given as

$$\text{checking cost} = \frac{B}{n} \quad \text{dollars} \tag{23.1}$$

The average checking interval depends on the individual process: in particular, the stability of the process given any adjustment limit, D. A production process changes in a complex manner due to various factors (e.g., external disturbances, tool wear, measurement error). Here the average adjustment interval, u, is assumed to be proportional to the squared adjustment limit, D. This assumption is an approximation since we can assume an optimal adjustment limit, D, and estimate the average adjustment interval, u, based on that assumption. We can validate this assumption later by comparing it with the data observed. Furthermore, the proportion coefficient, λ, is determined based on past performance. We first determine the current (initial) adjustment limit, D_0, then investigate the mean adjustment interval, u_0, under D_0, and then determine the proportion coefficient, λ, using the following equation:

$$\lambda = \frac{u_0}{D_0^2} \tag{23.2}$$

(Many Japanese firms use $D_0 = \Delta/3$, while many U.S. firms use $D_0 = D$. Which is more rational can be determined only through comparison with the actual optimal value, D.) After deciding on an optimal adjustment limit, D, the mean adjustment interval, u, is estimated by

$$u = u_0 \frac{D^2}{D_0^2} \tag{23.3}$$

Under this assumption, the checking adjustment costs are calculated based on the use of equations (23.1) and (23.3):

$$\frac{B}{n} + \frac{C}{u} = \frac{B}{n} + \frac{D_0^2 C}{u_0 D^2} \tag{23.4}$$

On the other hand, what would happen to the quality level if checking and adjustment were performed? Assuming the absence of checking error, the loss is given by

$$\frac{A}{\Delta^2} \left[\frac{D^2}{3} + \left(\frac{n+1}{2} + 1 \right) \frac{D^2}{u} \right] \tag{23.5}$$

The characteristic is considered to be distributed uniformly within the adjustment limit, $m \pm D$; therefore, $D^2/3$ is given by the variance of the characteristic value:

$$\frac{1}{2D} \int_{m-D}^{m+D} (y - m)^2 \, dy \tag{23.6}$$

When checking is done with interval n, and the last reading is within the adjustment limit but the current measurement is outside the adjustment limit, the average number of products outside the limit is roughly equal to $(n + 1)/2$. Although the mean of those falling outside the adjustment limit is greater than D, D is used as an approximation, since the difference is not great. Therefore, the variance when the measurement is outside the adjustment limit is obtained by dividing:

$$\left(\frac{n + 1}{2} + 1\right) D^2 \tag{23.7}$$

with mean adjustment interval u,

$$\left(\frac{n + 1}{2} + 1\right) \frac{D^2}{u} \tag{23.8}$$

Therefore, the quality level in terms of money is given by

$$\frac{A}{\Delta^2}\left[\frac{D^2}{3} + \left(\frac{n + 1}{2} + 1\right)\frac{D^2}{u}\right] \tag{23.9}$$

Total loss L, which is the sum of checking and adjustment costs [i.e., equation (23.4)] and the quality level [i.e., equation (23.9)], is

$$L = \frac{B}{n} + \frac{C}{u} + \frac{A}{\Delta^2}\left[\frac{D^2}{3} + \left(\frac{n + 1}{2} + 1\right)\frac{D^2}{u}\right] \tag{23.10}$$

After substituting equation (23.3), the loss, L, becomes

$$L = \frac{B}{n} + \frac{D_0^2 C}{u_0 D^2} + \frac{A}{\Delta^2}\left[\frac{D_0^2}{3} + \left(\frac{n_0 + 1}{2} + 1\right)\frac{D_0^2}{u_0}\right] \tag{23.11}$$

In particular, if n, u, and D are current n_0, u_0, and D_0, equation (23.11) becomes

$$L_0 = \frac{B}{n_0} + \frac{C}{u_0} + \frac{A}{\Delta^2}\left[\frac{D_0^2}{3} + \left(\frac{n_0 + 1}{2} + 1\right)\frac{D_0^2}{u_0}\right] \tag{23.12}$$

The optimal checking interval, n, and the optimal adjustment limit, D, are derived by differentiating equation (23.11) with respect to n and D and by solving them after equating them to zero:

$$\frac{dL}{dn} = -\frac{B}{n^2} + \frac{A}{\Delta^2}\left(\frac{1}{2}\right)\frac{D_0^2}{u_0} = 0 \tag{23.13}$$

Thus, the optimal checking interval is

$$n = \sqrt{\frac{2u_0 B}{A}\frac{\Delta}{D_0}} \tag{23.14}$$

The optimal adjustment limit D is found from

$$\frac{dL}{dD} = -2\frac{D_0^2 C}{u_0 D^3} + \frac{A}{\Delta^2}\left(\frac{2}{3}\right)D = 0 \tag{23.15}$$

$$D = \left(\frac{3C}{A}\frac{D_0^2}{u_0}\Delta^2\right)^{1/4} \tag{23.16}$$

In practice, the values of n and D can be rounded to convenient numbers. By substituting these numbers into equation (23.10), the loss, L, of the optimal system is derived. Here, u is estimated according to equation (23.3).

☐ Example

Consider the control of a certain component part dimension. The parameters are as follows.

- ☐ *Specification: $m \pm 15$ μm*
- ☐ *Loss due to a defective: $A = 80$ cents*
- ☐ *Checking cost: $B = \$1.50$*
- ☐ *Time lag: $I = 1$ unit*
- ☐ *Adjustment cost: $C = \$12$*
- ☐ *Current checking interval: $n_0 = 600$ units, once every two hours*
- ☐ *Current adjustment limit: $\pm D_0 = \pm 5$ μm*
- ☐ *Current mean adjustment interval: $u_0 = 1200$ units, twice a day*

(a) Derive the optimal checking interval, n, and the optimal adjustment limit, D, and estimate the extent of improvement from the current situation.

(b) Estimate the process capability index C_p.

(c) Assuming that it takes 3 minutes for checking and 15 minutes for adjustment, calculate the required worker-hours for this process to two digits after the decimal.

(a) We begin the derivation as follows:

$$n = \sqrt{\frac{2u_0 B}{A}} \frac{\Delta}{D_0} = \sqrt{\frac{(2)(1200)(1.50)}{0.80}} \left(\frac{15}{5}\right)$$

$$= 201 \rightarrow 200 \text{ units} \qquad (23.17)$$

$$D = \left(\frac{3C}{A} \frac{D_0^2}{u_0} \Delta^2\right)^{1/4} = \left[\frac{(3)(12)}{0.80} \left(\frac{5^2}{1200}\right) (15^2)\right]^{1/4}$$

$$= 3.8 \rightarrow 4.0 \text{ μm} \qquad (23.18)$$

Thus, loss function L of feedback control is

$$L = \frac{B}{n} + \frac{C}{u} + \frac{A}{\Delta^2}\left[\frac{D^2}{3} + \left(\frac{n+1}{2} + I\right)\frac{D^2}{u}\right] \qquad (23.19)$$

Substituting

$$u = u_0 \frac{D^2}{D_0^2} = (1200)\left(\frac{4^2}{5^2}\right)$$

$$= 768 \text{ units} \qquad (23.20)$$

into equation (23.19) yields

$$L = \frac{1.50}{200} + \frac{12}{768} + \frac{0.80}{15^2}\left[\frac{4^2}{3} + \left(\frac{201}{2} + I\right)\left(\frac{4^2}{768}\right)\right]$$

$$= 0.0075 + 1.0056 + 1.0090 + 0.75$$

$$= \$0.0496 \tag{23.21}$$

Under the current conditions,

$$L_0 = \frac{B}{n_0} + \frac{C}{u_0} + \frac{A}{\Delta^2}\left[\frac{D_0^2}{3} + \left(\frac{n_0 + 1}{2} + I\right)\frac{D_0^2}{u_0}\right]$$

$$= \frac{1.50}{600} + \frac{12}{1200} + \frac{0.80}{15^2}\left[\frac{5^2}{3} + \left(\frac{601}{2} + I\right)\left(\frac{5^2}{1200}\right)\right]$$

$$= 0.0025 + 0.01 + 0.0296 + 2.23$$

$$= \$0.0644 \tag{23.22}$$

Assuming that the total hours of operation per year are 2000, a production rate of 300 units per hour will result in a yearly improvement of

$$(0.0644 - 0.0496)\,(300)\,(2000) = \$8880 \tag{23.23}$$

(b) Although this demonstration of the improvement effect should be adequate to demonstrate the effectiveness of feedback control, the process capability index is derived below for reference. The square root of the term in brackets of in equation (23.19) is an estimate of the standard deviation, σ:

$$\sigma = \sqrt{\frac{D^2}{3} + \left(\frac{n+1}{2} + 1\right)\frac{D^2}{u}} \tag{23.24}$$

Since $\Delta = 15$ and the definition of the process capability index is [tolerance ÷ (6 × standard deviation)]

$$C_p = \frac{(2)(15)}{\sqrt{(4^2/3) + [(201/2) + 1]\,(4^2/768)}\,(6)} \approx 1.83 \tag{23.25}$$

The process capability index is about 1.8. This number deteriorates if there is measurement error. The C_p value of the current condition is

$$C_p = \frac{(2)(15)}{\sqrt{(5^2/3) + [(601/2) + 1]\,(5^2/1200)}\,(6)} = 1.31 \tag{23.26}$$

(c) Under the optimal checking interval, checking is done every 40 minutes; it takes 3 minutes for checking. Therefore, in one 8-hour day, there will be 12 checks, amounting to 36 minutes. The estimated adjustment interval is once every 768 pieces and

$$\frac{(300)(8)}{768} \approx 3.1 \tag{23.27}$$

times a day. It takes (15 minutes)(3.1) = 46.5 minutes per day. Assuming that work minutes per day per worker is 480 minutes, the number of personnel required is

$$\frac{36 + 46.5}{480} = 0.17 \text{ person} \tag{23.28}$$

Similar calculations are done for each process, and an allocation of processes per worker is determined. For example, one person can be in charge of seven processes (i.e., M_1, M_2, ... , M_7), and the checking interval is 200, 150, 180, 300, 100, 190, and 250, respectively. These can be rearranged with +30% flexibility to obtain the following schedule.

❏ *Once every 0.5 hour* (once every 150 pieces): M_1, M_2, M_3, M_5, M_6

❏ *Once every 1.0 hour* (once every 300 pieces): M_4, M_7

One should create an easily understandable checking system, and based on the system, the number of checking persons is determined.

23.3. Batch Production Process

Batch production is processing of a large number of products or a large volume of products at one time. If liquid, gas, or powder is flowing continuously in a pipe, it can be analyzed, as in Section 23.2, by setting the unit of production as 1 liter, and so on. On the other hand, if a film is deposited on 100 wafers at a time, the mean film thickness of the batch can be controlled, although one cannot control the film thickness among different wafers in the same batch. Control of the mean film thickness is done by estimating the mean of the entire batch and controlling process.

If many units are processed in one step, the unit of production is the number of units in a batch. Checking is done for one or several units from a batch to estimate the average of the batch. When this method is used to estimate the mean of all the products, we have to be concerned about the problem of estimation error. Let σ_m^2 be the measurement (estimated) error variance, and the loss function, L, be as follows:

$$L = \frac{B}{n} + \frac{C}{u} + \frac{A}{\Delta^2} \left[\frac{D^2}{3} + \left(\frac{n+1}{2} + 1 \right) \frac{D^2}{u} + \sigma_m^2 \right] \tag{23.29}$$

Measurement error in feedback control is counted as an independent error item.

❐ **Example**

In an injection molding process that produces 12 units in one shot, an average product dimension is estimated once every 100 shots and adjusted relative to its target value. It takes 1 hour of production to produce 100 shots, and the loss per defective is 30 cents. The tolerance is ± 120 μm; current adjustment limit is ± 50 μm; adjustment cost is \$18; mean adjustment interval is 800 shots; cost B for the estimation of average size is \$4; and time lag l is four shots.

(a) Derive the optimal checking interval, n, and optimal adjustment limit, D, and obtain the possible gain per shots and per year (200,000 shots/year).

(b) Suppose that the standard deviation of the estimation error of the current measuring methods is 15 μm. Compare this with a new measuring method whose estimation error is supposed to be one-third of the current error and whose cost is \$7 (\$3 more than the current cost). Which method is better, and by how much? Assume that the time lag is the same for both alternatives.

(c) The standard deviation between cavities is 6 μm. Obtain the process capability index when the process is controlled with the new measuring method.

(a) Since 12 units are produced at a time, 12 is treated as the unit of production. The same treatment will be used for a mass-produced chemical product (e.g., a 10-ton batch). A is the loss when all units in a batch are defective. Thus, the parameters can be summarized as follows:

❏ *Tolerance of the dimension:* $\Delta = 120$ μm
❏ *Loss due to defective:* $A = (30)(12) = \$3.60$
❏ *Cost of estimating the average dimension:* $B = \$4$
❏ *Adjusting cost:* $C = \$18$
❏ *Time lag:* $l = 4$ shots
❏ *Mean dimension checking interval:* $n_0 = 100$ shots
❏ *Current adjustment limit:* $\pm D_0 = \pm 50$ (μm)
❏ *Current mean adjustment interval:* $u_0 = 800$ shots

Derive the optimal checking interval, n, and optimal adjustment limit, D:

$$n = \sqrt{\frac{2u_0 B}{A} \frac{\Delta}{D}}$$

$$= \sqrt{\frac{(2)(800)(4)}{3.60} \left(\frac{120}{50}\right)}$$

$$= 101 \rightarrow 100 \text{ shots} \tag{23.30}$$

Thus, the current level of n is optimal:

$$D = \left(\frac{3C}{A} \frac{D_0^2}{u_0} \Delta^2\right)^{1/4} = \left[\frac{(3)(18)}{3.60} \left(\frac{50^2}{800}\right) (120^2)\right]^{1/4}$$

$$= 29 \rightarrow 30 \text{ μm} \tag{23.31}$$

Thus, the adjustment limit, 30, which is smaller than the current 50, is better. The loss under the current method, L_0, is

$$L_0 = \frac{B}{n_0} + \frac{C}{u_0} + \frac{A}{\Delta^2}\left[\frac{D_0^2}{3} + \left(\frac{n+1}{2}+1\right)\frac{D_0^2}{u_0}\right]$$

$$= \frac{4}{100} + \frac{18}{800} + \frac{3.60}{120^2}\left[\frac{50^2}{3} + \left(\frac{101}{2}+4\right)\left(\frac{50^2}{800}\right)\right]$$

$$= 0.04 + 0.0225 + 0.2083 + 4.26$$

$$= 31 \text{ cents} \qquad (23.32)$$

Under the optimal conditions, we have that

$$u = u_0\frac{D^2}{D_0^2} = (800)\left(\frac{30^2}{50^2}\right) = 288 \text{ shots} \qquad (23.33)$$

and the loss will be

$$L = \frac{4}{100} + \frac{18}{288} + \frac{3.60}{120^2}\left[\frac{30^2}{3} + \left(\frac{101}{2}+4\right)\left(\frac{30^2}{288}\right)\right]$$

$$= 0.04 + 0.0625 + 0.075 + 4.26$$

$$= 22 \text{ cents} \qquad (23.34)$$

Therefore, use of the optimal system results in an improvement over the current system by

$$(31 - 0.22)(100)(2000) = \$18,000 \qquad (23.35)$$

(b) For the current checking method, the loss due to measurement error (its standard deviation),

$$\frac{A}{\Delta^2}\sigma_m^2 = \frac{3.60}{120^2}(15^2) = 6 \text{ cents} \qquad (23.36)$$

is added to equation (23.34):

$$0.22 + 0.06 = 28 \text{ cents} \qquad (23.37)$$

On the other hand, the new measuring method costs \$7 rather than \$4, and thus it is necessary to derive the optimal checking interval, n, again:

$$n = \sqrt{\frac{(2)(800)(7)}{3.60}}\left(\frac{120}{50}\right)$$

$$= 134 \rightarrow 150 \text{ shots once every 1.5 hours} \qquad (23.38)$$

Therefore, the loss, L, is, by setting $\sigma_m = 15/3 = 5$ μm,

$$L = \frac{B}{n} + \frac{C}{u} + \frac{A}{\Delta^2}\left[\frac{D^2}{3} + \left(\frac{n+1}{2} + 1\right)\frac{D^2}{u} + \sigma_m^2\right]$$

$$= \frac{7}{150} + \frac{(18)}{288} + \frac{3.60}{120^2}\left[\frac{30^2}{3} + \left(\frac{151}{2} + 4\right)\left(\frac{30^2}{288}\right) + 5^2\right]$$

$$= 0.0467 + 0.0625 + 0.075 + 6.21 + 0.62$$

$$= 25 \text{ cents} \tag{23.39}$$

Because measuring accuracy is inferior using the current method, the new measuring method results in an improvement of

$$(0.28 - 0.25)(100)(2000) \approx \$6000 \tag{23.40}$$

(c) The on-line control of a production process does not allow control of variability among cavities created by one shot. Therefore, if σ_c is the standard deviation among cavities, the overall standard deviation, σ, is

$$\sigma = \sqrt{\frac{D^2}{3} + \left(\frac{n+1}{2} + 1\right)\frac{D^2}{u} + \sigma_m^2 + \sigma_c^2}$$

$$= \sqrt{\frac{30^2}{3} + \left(\frac{151}{2} + 4\right)\left(\frac{30^2}{288}\right) + 5^2 + 6^2}$$

$$= 24.7 \text{ μm} \tag{23.41}$$

Therefore, the process capability index, C_p, is

$$C_p = \frac{(2)(120)}{(6)(24.7)} = 1.6 \tag{23.42}$$

23.4. Design of a Process Control Gauge (Using a Boundary Sample)

When we are confronted with quality problems where continuous data are not available, control of process conditions and preventive maintenance become important. However, we cannot be 100% certain about such a process, even if we control process conditions as much as possible. Additionally, we cannot even be sure if it is rational to control all process conditions. This is analogous to a health checkup, where it is irrational to check for symptoms of all diseases, and thus we do not attempt to control problems that occur rarely. Even if continuous data are not given for certain quality items, it is possible in many cases to judge the quality level by a factor such as appearance. Thus, instead of attempting to control all process conditions, we should control major conditions only, together with the use of preventive maintenance based on quality data. The method of preventive maintenance can be applied to the preventive method for delivered products.

Suppose that tolerance, Δ, is specified for the appearance of a product. The appearance is currently controlled at D_0, using a boundary sample. Similar to the example in Section 23.2, parameters are defined as follows.

A: loss per defective

Δ: tolerance

B: checking cost

C: adjustment cost

n_0: current checking interval

u_0: current mean adjustment interval

D_0: current adjustment limit

l: time lag

The optimal checking interval, n, and optimal adjustment limit, D, are given by the following:

$$n = \sqrt{\frac{2u_0 B}{A} \frac{\Delta}{D_0}} \tag{23.43}$$

$$D = \left(\frac{3C}{A} \frac{D_0^2}{u_0} \Delta^2\right)^{1/4} \tag{23.44}$$

If pass/fail is determined using a boundary or a limit sample without an intermediate boundary sample, we have the following:

$$D_0 = \Delta \text{ (tolerance)} \tag{23.45}$$

We often want to conduct preventive maintenance by providing an appropriate limit sample before actually producing defectives. This is called the *design method for preventive maintenance* with inspection, described fully below.

When continuous data are not available, inspection is done with a boundary sample or a gauge. However, it is still possible to devise a method in which a boundary sample or a gauge midway between pass and fail is prepared so that process adjustment is required if a product exceeds the boundary sample. Suppose that the intermediate condition can be expressed quantitatively as a continuous value, and its ratio with the tolerance limit is ϕ, that is,

$$\phi = \frac{\text{value of intermediate condition}}{\Delta} = \frac{D}{\Delta} \tag{23.46}$$

Accordingly, the adjustment limit D can be expressed as

$$D = \phi\Delta \tag{23.47}$$

Since the current method is $D_0 = \Delta$, the optimal adjustment limit, D, and optimal checking interval, n, are determined by substituting $D_0 = \Delta$, and $u_0 = \bar{u}$ in the formula for the optimal adjustment limit, D:

$$D = \left(\frac{3C}{A} + \frac{D_0^2}{u_0} \Delta^2 \right)^{1/4}$$

$$= \left(\frac{3C}{A} \frac{\Delta^2}{\bar{u}} \Delta^2 \right)^{1/4}$$

$$= \left(\frac{3C}{A\bar{u}} \right)^{1/4} \Delta \tag{23.48}$$

From equation (23.47),

$$D = \phi\Delta \tag{23.49}$$

Thus, ϕ is given by

$$\phi = \left(\frac{3C}{A\bar{u}} \right)^{1/4} \tag{23.50}$$

Furthermore, the optimal checking interval, n, is

$$n = \sqrt{\frac{2u_0 B}{AD_0^2}} \, \Delta \tag{23.51}$$

By substituting

$$D_0 = \Delta$$

$$u_0 = \bar{u} \frac{D_0^2}{\Delta^2} = \bar{u} \frac{\Delta^2}{\Delta^2}$$

$$= \bar{u} \tag{23.52}$$

the following formula is obtained. The optimal checking interval, n, is determined independently from the adjustment limit, as shown by:

$$n = \sqrt{\frac{2\bar{u}B}{A\Delta^2}} \, \Delta = \sqrt{\frac{2\bar{u}B}{A}} \tag{23.53}$$

Furthermore, the loss function, L is obtained based on the formula for continuous values. That is,

$$L = \frac{B}{n} + \frac{C}{u} + \frac{A}{\Delta^2} \left[\frac{D^2}{3} + \left(\frac{n+1}{2} + 1 \right) \frac{D^2}{u} \right] \tag{23.54}$$

and by substituting equation (23.47) into (23.54).

$$L = \frac{B}{n} + \frac{C}{u} + \frac{A}{\Delta^2} \left[\frac{1}{3} (\phi\Delta)^2 + \left(\frac{n+1}{2} + 1 \right) \frac{(\phi\Delta)^2}{u} \right]$$

$$= \frac{B}{n} + \frac{C}{u} + A \left[\frac{\phi^2}{3} + \left(\frac{n+1}{2} + 1 \right) \frac{\phi^2}{u} \right] \tag{23.55}$$

❏ Example

Consider a characteristic value that cannot be measured easily, and product pass/fail is evaluated using a boundary sample. The loss, A, when the product is not accepted is $1.80; the mean failure interval, $\bar{u}$, is 2300 units; the adjustment cost, C is $120; the measuring cost, B, is $4; and the time lag, l, is 2 units.

In the following, we design the optimal preventive maintenance method. The parameters are:

$$A = \$1.80$$

$$B = \$4$$

$$C = \$120$$

$$\bar{u} = 2300 \text{ units}$$

$$l = 2 \text{ units}$$

From equation (23.50),

$$\phi = \left(\frac{3C}{A\bar{u}}\right)^{1/4} = \left(\frac{(3)(120)}{(1.80)(2300)}\right)^{1/4}$$

$$= 0.54 \rightarrow 0.5 \tag{23.56}$$

This is about one-half of the unacceptable boundary samples. In other words, optimal control should be done with a boundary sample that is half as bad as the unacceptable boundary sample. That level is D.

$$D = 0.5\Delta \tag{23.57}$$

Next, derive the optimal checking interval, n, and the loss function, L. From equation (23.53),

$$n = \sqrt{\frac{2\bar{u}B}{A}} = \sqrt{\frac{(2)(2300)(4)}{1.80}}$$

$$\approx 100 \text{ units} \tag{23.58}$$

This means that the optimal checking interval of inspection is 100. Loss does not vary much, even if we change the checking interval by 20 to 30% around the optimal value. In this case, if there are not sufficient staff, one can stretch the checking interval to 150.

Next is the loss function. Under current conditions,

$$L_0 = \frac{B}{n_0} + \frac{C}{u_0} + \frac{A}{\Delta^2}\left[\frac{D_0^2}{3} + \left(\frac{n_0 + 1}{2} + 1\right)\frac{D_0^2}{u_0}\right] \tag{23.59}$$

and since

$$u_0 = \bar{u} = 2300 \text{ units} \tag{23.60}$$

$$D_0 = \Delta \tag{23.61}$$

the current optimal checking interval, n_0, is

$$n_0 = \sqrt{\frac{2u_0 B}{AD_0^2}} \, \Delta = \sqrt{\frac{2\bar{u}B}{A\Delta^2}} \, \Delta$$

$$= \sqrt{\frac{2\bar{u}B}{A}} = \sqrt{\frac{(2)(2300)(4)}{1.80}}$$

$$\approx 100 \text{ units} \tag{23.62}$$

Therefore, the loss L of the current control method is

$$L_0 = \frac{B}{n_0} + \frac{C}{u_0} + \frac{A}{\Delta^2}\left[\frac{D_0^2}{3} + \left(\frac{n_0 + 1}{2} + 1\right)\frac{D_0^2}{u_0}\right]$$

$$= \frac{B}{n_0} + \frac{C}{\bar{u}_0} + \frac{A}{\Delta^2}\left[\frac{\Delta^2}{3} + \left(\frac{n_0 + 1}{2} + 1\right)\frac{\Delta^2}{\bar{u}}\right]$$

$$= \frac{B}{n_0} + \frac{A}{3} + \frac{n_0 + 1}{2}\frac{A}{\bar{u}} + \frac{C}{\bar{u}} + \frac{lA}{\bar{u}}$$

$$= \frac{4}{100} + \frac{1.80}{3} + \frac{101}{2}\left(\frac{1.80}{2300}\right) + \frac{120}{2300} + \frac{(2)(1.80)}{2300}$$

$$= 0.04 + 0.6 + 0.04 + 0.052 + 0.002$$

$$= 73 \text{ cents} \tag{23.63}$$

Furthermore, under optimal conditions,

$$L = \frac{B}{n} + \frac{C}{u} + \frac{A}{\Delta^2}\left[\frac{D^2}{3} + \left(\frac{n + 1}{2} + 1\right)\frac{D^2}{u}\right] \tag{23.64}$$

and A, B, C, n, and l are the same as in equation (23.63):

$$D = \phi\Delta \tag{23.65}$$

$$u = \bar{u}\frac{D^2}{\Delta^2} = \bar{u}\frac{(\phi\Delta)^2}{\Delta^2} = \bar{u}\phi^2 \tag{23.66}$$

Thus, by substituting into the loss function,

$$L = \frac{B}{n} + \frac{C}{u} + \frac{\phi^2 A}{3} + \frac{n + 1}{2}\frac{\phi^2 A}{u} + \frac{l\phi^2 A}{u} \tag{23.67}$$

the following,

$$u = \bar{u}\phi^2 = (2300)(0.5^2) = 575 \text{ units} \qquad (23.68)$$

$$\phi^2 A = (0.5^2)(1.80) = 45 \text{ cents} \qquad (23.69)$$

we obtain

$$L = \frac{4}{100} + \frac{120}{575} + \frac{0.45}{3} + \frac{101}{2}\left(\frac{0.45}{575}\right) + \frac{(2)(0.45)}{575}$$

$$= 0.04 + 0.21 + 0.15 + 0.04 + 0.002$$

$$= 44 \text{ cents} \qquad (23.70)$$

Compared with the current method, the optimal method results in a gain of

$$73 - 44 = 29 \text{ cents} \qquad (23.71)$$

If the yearly production were 500,000 units, there would be a savings of $146,500.

24 Feedback Control of a Process Condition

24.1. Introduction   468
24.2. Process Condition Control   468
24.3. Feedback Control System Design: Viscosity Control   469
24.4. Feedback Control System: Feedback Control of Temperature   471

24.1. Introduction

In this chapter the method for designing a feedback control system for changes in process conditions is explained. A case where changes in the process condition affect product quality is given. A case where deterioration of the process condition leads to stoppage of the process is discussed in later chapters.

24.2. Process Condition Control

Process condition control deals not only with product quality but also with process failure. This chapter deals with issues such as how long the checking interval should be and what the adjustment limit would be if a process condition such as temperature affects product quality.

The most important aspect of process condition control is the derivation of tolerance of the process condition D, that is, the functional limit for producing defectives. Since a change in the process condition by D causes defectives, the following equation is suggested for calculation of the proportion coefficient of the quality level:

$$k = \frac{\text{loss } A \text{ when the product is defective}}{\Delta^2} \tag{24.1}$$

Loss A in the numerator is the loss when a change in process condition creates defectives. A process condition is irrelevant to the concept of defect.

24.3. Feedback Control System Design: Viscosity Control

Consider a problem where process parameter x is controlled through feedback so as to achieve a certain level of a quality characteristic, y. Suppose that the tolerance of the objective characteristic, y, is $\pm\Delta_0$, and the loss when the product falls off the tolerance is A dollars. It is assumed that when process condition x changes by Δ, the objective characteristic y becomes unacceptable.

The parameters of process condition x is defined as follows:

❑ *Tolerance of x:* Δ (the tolerance of x that determines the pass/fail of objective characteristic y)

❑ *Measuring costs of x:* B (dollars)

❑ *Measuring time lag:* l (units)

❑ *Current adjustment limit of x:* C (dollars)

❑ *Current checking interval of x:* n_0 (units)

❑ *Current mean adjustment interval of x:* u_0 (units)

The optimal checking interval, n, and optimal adjustment limit, D, of process condition x are given by the following equations:

$$L_0 = \frac{B}{n_0} + \frac{C}{u_0} + \frac{A}{\Delta^2}\left[\frac{D_0^2}{3} + \left(\frac{n_0 + 1}{2} + l\right)\frac{D_0^2}{u_0}\right] \tag{24.2}$$

$$L = \frac{B}{n} + \frac{C}{u} + \frac{A}{\Delta^2}\left[\frac{D_0^2}{3} + \left(\frac{n + 1}{2} + l\right)\frac{D^2}{u}\right] \tag{24.3}$$

The estimated mean adjustment interval, u, is given by

$$u = u_0 \frac{D^2}{D_0^2} \tag{24.4}$$

❑ Example

Consider a process of continuous emulsion coating. The object of control is the viscosity of the emulsion. The standard of coating thickness is $m_0 + 8\ \mu m$ and the loss caused by failing to meet the standard is $\$3/m^2$. When the viscosity of the emulsion changes by 5.3 P, the coated film fails to meet the standard. The current checking interval, n_0, is $6000 m^2$ (once every 2 hours), the adjustment limit is 0.9 P, and the mean adjustment interval is $12,000 m^2$ (twice a day). Cost B of measuring x is \$2, and the adjustment cost of viscosity, C, is \$10. The time lag, l, of the measurement is $30 m^2$. Derive the optimal checking interval, n, optimal adjustment limit, D, and the profit per m^2 and per year (assume 2000-hour/year operation).

According to the assumptions, the tolerance Δ of viscosity x is 5.3. The parameters can be summarized as follows.

❏ *Tolerance: D = 5.3 P*
❏ *Loss due to defective: A = $3*
❏ *Measurement cost: B = $2*
❏ *Time lag for measurement: l = 30m^2*
❏ *Adjustment cost: C = $10*
❏ *Current checking interval: n_0 = 6000m^2*
❏ *Current adjustment limit: ±D_0 = 0.9 P*
❏ *Current mean adjustment interval: u_0 = 12,000m^2*

The current loss function, L_0, is

$$L_0 = \frac{B}{n_0} + \frac{C}{u_0} + \frac{A}{\Delta^2}\left[\frac{D_0^2}{3} + \left(\frac{n_0+1}{2} + 1\right)\frac{D_0^2}{u_0}\right]$$

$$= \frac{2}{6000} + \frac{10}{12,000} + \frac{3}{5.3^2}\left[\frac{0.9^2}{3} + \left(\frac{6001}{2} + 30\right)\left(\frac{0.9^2}{12,000}\right)\right]$$

$$= 0.00033 + 0.00083 + 0.02884 + 0.02185$$

$$= \$0.05185 \tag{24.5}$$

Optimal checking interval, n, and optimal adjustment limit, D, are

$$n = \sqrt{\frac{2u_0 B}{A}}\frac{\Delta}{D_0} = \sqrt{\frac{(2)(12,000)(2)}{3}}\left(\frac{5.3}{0.9}\right)$$

$$= 745 \rightarrow 1000 m^2 \tag{24.6}$$

$$D = \left(\frac{3C}{A}\frac{D_0^2}{u_0}\Delta^2\right)^{1/4} = \left[\frac{(3)(10)}{3}\left(\frac{0.9^2}{12,000}\right)(5.3^2)\right]^{1/4}$$

$$= 0.37 \rightarrow 0.4 \text{ P} \tag{24.7}$$

The expected mean adjustment interval, u, is

$$u = u_0\frac{D^2}{D_0^2} = (12,000)\left(\frac{0.4^2}{0.9^2}\right) = 2400 m^2 \tag{24.8}$$

and the loss function, L, is

$$L = \frac{B}{n} + \frac{C}{u} + \frac{A}{\Delta^2}\left[\frac{D^2}{3} + \left(\frac{n+1}{2} + 1\right)\frac{D^2}{u}\right]$$

$$= \frac{2}{1000} + \frac{10}{2,400} + \frac{3}{5.3^2}\left[\frac{0.4^2}{3} + \left(\frac{1001}{2} + 30\right)\left(\frac{0.4^2}{2400}\right)\right]$$

$$= 0.002 + 0.004 + 0.006 + 0.378 = 1.564 \text{ cents} \tag{24.9}$$

Thus, the yearly improvement is

$$(0.05 - 0.016)(3000)(2000) = \$204,000 \qquad (24.10)$$

24.4. Feedback Control System: Feedback Control of Temperature

☐ Example

Consider a heat process where objects are fed continuously under 24-hour operation. When the temperature changes by 40°C, the product becomes defective and
the unit loss is $3. Although the temperature is controlled automatically at all times,
a worker checks the temperature with a standard thermometer once every 4 hours.
The cost of checking, B, is $5; the current adjustment limit, D_0, is 5°C; the mean
adjustment interval is 6 hours; the adjustment cost is $5; the current adjustment
limit D_0 is 5°C; the mean adjustment interval is 6 hours; the adjustment cost is
$8; the number of units heat-treated in 1 hour is 300; the time lag of measurement
is 20 units; and the measurement error (standard deviation) of the standard thermometer is about 1.0°C.

   (a) Design an optimal temperature control system and calculate the yearly im-
       provement over the current system. Assume that the number of operation-
       days per year is 250.
   (b) If a new temperature control system is introduced, the annual cost will in-
       crease by $12,000, but the degree of temperature drift will be reduced to
       one-fifth of the existing level. Calculate the loss/gain of the new system.

Parameters are summarized as follows:

$$D = 40°C$$

$$A = \$3$$

$$B = \$5$$

$$C = \$18$$

$$n_0 = (3)(4) = 12$$

$$D_0 = 5°C$$

$$u_0 = (3)(6) = 18 \text{ units}$$

$$l = 20 \text{ units}$$

$$\sigma_s = 1.0°C$$

(a)] The loss function, L_0, of the current system is

$$L_0 = \frac{B}{n_0} + \frac{C}{u_0} + \frac{A}{\Delta^2}\left[\frac{D_0^2}{3} + \left(\frac{n_0+1}{2}+1\right)\frac{D_0^2}{u_0} + \sigma_s^2\right]$$

$$= \frac{5}{1200} + \frac{18}{1800} + \frac{3}{40^2}\left[\frac{5^2}{3} + \left(\frac{1201}{2}+20\right)\left(\frac{5^2}{1800}\right) + 1.0^2\right]$$

$$= 0.0042 + 0.0100 + 0.0156 + 0.0162 + 0.0478$$

$$= \$0.0478 \tag{24.11}$$

The optimal checking interval, n, is

$$n = \sqrt{\frac{2u_0 B}{A}}\frac{\Delta}{D_0} = \sqrt{\frac{(2)(1800)(5)}{3}}\left(\frac{40}{5}\right)$$

$$= 620 \rightarrow 600 \text{ units}\quad\text{(once in 2 hours)} \tag{24.12}$$

The optimal adjustment interval, D, is

$$D = \left(\frac{3C}{A}\frac{D_0^2}{u_0}\Delta^2\right)^{1/4} = \left(\frac{(3)(18)}{3}\frac{5^2}{1800}(40^2)\right)^{1/4}$$

$$= 4.5 \rightarrow 5.0°C \text{ (The current level is acceptable)} \tag{24.13}$$

Accordingly, the mean adjustment interval remains the same, and the new loss function is

$$L = \frac{5}{600} + \frac{18}{1800} + \frac{3}{40^2}\left[\frac{5^2}{3} + \left(\frac{601}{2}+20\right)\left(\frac{5^2}{1800}\right) + 1.0^2\right]$$

$$= 0.0083 + 0.0100 + 0.0156 + 0.0083 + 0.0019$$

$$= \$0.0442 \tag{24.14}$$

The yearly improvement is

$$(0.0479 - 0.0441)(300)(24)(250) = \$6840 \tag{24.15}$$

(b) The total production in one year is $(300)(24)(250) = 1,800,000$. Thus, the increase of cost per unit of production by introduction new equipment is

$$\frac{120}{180} = \$0.0067 \tag{24.16}$$

Reduction of drift to one-fifth of the current level can be interpreted as meaning that the mean adjustment interval becomes five times the current adjustment interval, or the standard deviation per unit of output is reduced to one-fifth.

Assuming that $u_0 = (1800)(5) = 9000$, optimal checking interval, n, and optimal adjustment limit, D, are

$$n = \sqrt{\frac{(2)(9000)(5)}{3}}\left(\frac{40}{5}\right)$$

$$= 1386 \rightarrow 1200 \text{ (once every 4 hours)}$$

$$D = \left[\frac{(3)(18)}{3}\left(\frac{5^2}{9000}\right)(40^2)\right]^{1/4}$$

$$= 3.0°C \tag{24.17}$$

Furthermore, the mean adjustment interval is estimated as

$$u = u_0 \times \frac{D^2}{D_0^2} = (9000)\left(\frac{3.0^2}{5^2}\right)$$

$$= 3240 \text{ units} \tag{24.18}$$

Thus,

$$L = \frac{5}{1200} + \frac{18}{3240} + \frac{3}{40^2}\left[\frac{3^2}{3} + \left(\frac{1201}{2} + 20\right)\left(\frac{3^2}{3240}\right) + 1.0^2\right]$$

$$= 0.0042 + 0.0056 + 0.0056 + 0.0032 + 0.0019$$

$$= \$0.0205 \tag{24.19}$$

Even if the increase in the cost given by equation (24.16) (i.e., \$0.0067) is added, the loss is \$0.0272. Therefore, compared to equation (24.14), a yearly improvement of \$30,420 can be expected:

$$(0.0441 - 0.0272)(300)(24)(250) \approx \$30,420) \tag{24.20}$$

25 Process Diagnosis and Adjustment

25.1. Introduction   474
25.2. Quality Control Systems and Their Cost during Production   474
25.3. Optimal Diagnostic Interval   477
25.4. Derivation of Optimal Number of Workers   479

25.1. Introduction

Some quality characteristics, such as soldering and electric welding, cannot be measured on a continuous scale; they either pass or fail. Although process conditions can be adjusted to control these characteristics, costs would increase significantly if we attempted to control all conditions. Frequently, we do not know what condition is causing problems. However, one cannot possibly anticipate and check for all potential future problems, such as arsenic contamination of milk or PCB (polychlorinated biphenyl) contamination of food oil, for example. For quality control testing to find unknown items, it would be necessary to design and conduct a performance test or animal test that would incorporate all possibilities. In this chapter we examine the design of a process control system when measurement technology is not available. The only possible test is to check whether or not a product functions or when go/no-go gauges are used. In this situation, the emphasis is on minimizing costs of control and not on potential loss due to the variability of products.

25.2. Quality Control Systems and Their Cost during Production

Let us consider a process control problem where the state of the process—normal or abnormal—is judged by inspecting the noncontinuous quality characteristics of the product. The unit of production will be determined as follows: In the case of parts and/or components, the unit is the number of units produced; in the case of liquid, power, gas, or electric wire, the unit of production is the quantity of output per second or per meter. It will be assumed that the product is checked within an interval of n units, and if the process is judged to be normal, production continues. However, if the process is judged to be abnormal, the process is brought back to its normal condition before continuing production. For example, health checks on human beings are done at certain intervals, and if there is any problem, the person will be treated and brought back to work after recovering to a normal healthy state. Our problem is to discover the most effective method for diagnosis and treatment of production processes.

The parameters are defined as follows:

- ❏ *A:* loss of producing a product under an abnormal process condition
- ❏ *B:* diagnostic cost (per diagnosis)
- ❏ *C:* adjustment cost, the cost of bringing an abnormal process back to its normal state, given as the sum of the loss due to process stoppage and the treatment cost, derived by (the cost of screening out defectives is included in some cases)

$$C = C' \text{ (loss due to process stoppage per unit of time)}$$
$$\times t \text{ (mean stoppage time due to process failure)}$$
$$+ C'' \text{ (direct adjustment cost such as labor cost for tool}$$
$$\text{replacement and the tool cost)}$$

- ❏ $\bar{u}$: mean failure interval, which is derived as

$$\bar{u} = \frac{\text{quantity of production in the period}}{\text{number of failures in the period}}$$

If the number of failures is zero from the start of production to the present, the following is used:

$$\bar{u} = 2(\text{quantity of production in the period})$$

- ❏ *l:* time lag, the number of output units produced between the time of sample taking and the time of process stoppage/adjustment if the process is diagnosed as abnormal (expressed in terms of the number of output units)

Therefore, *L*, the quality control cost per unit of production when process diagnosis and adjustment is performed using inspection interval, *n*, becomes the following:

L = (diagnostic cost per unit of production) + (loss of producing defectives between diagnoses using interval *n*) + (process adjustment cost) × (the possibility of process adjustment) + (loss due to time lag)

$$= \frac{B}{n} + \frac{n+1}{2} \times \frac{A}{\bar{u}} + \frac{C}{\bar{u}} + \frac{lA}{\bar{u}} \tag{25.1}$$

❐ Example

An automatic welding machine sometimes fails due to welding-head abrasion and carbon deposition. Currently, the welded spots of a product are inspected at

100-unit intervals. Tensile strength and welding surface are inspected to find non-welded or imperfectly welded product. If the machine fails, the product is scrapped and the loss per unit is 50 cents. Diagnosis cost B is $1.60. Although it does not take time to sample, inspection takes 4 minutes to perform. Thirty units can be produced during inspection. Therefore, the time lag, I, for this inspection method is 30 units. Furthermore, production in the past two months was 84,000 units and the number of failures during this period was 16.

The adjustment cost, C, when the machine fails is $31.70. It is derived in the following way: First, when the machine fails, it has to be stopped for about 20 minutes. The loss due to stoppage (which has to be covered by overtime work, etc.) is $47.10 per hour. The products were found to be acceptable at the previous inspection and remained so until the hundredth unit produced since the last inspection. A binary search procedure is used to screen the 100 units. If the fiftieth is checked and found to be acceptable, the first 50 products are sent to the next process or to the market. If the seventy-fifth, in the middle of the last 50, is found to be defective, the seventy-sixth to hundredth are scrapped. This process is continued to identify acceptable product. The search process is continued until no more than four items remain. They are scrapped, since further selection is uneconomical. This screening method requires inspection of five products; the cost is $8. Direct costs such as repair and replacement of the welding head, which can be done in 20 minutes, is $8. Thus, the total adjustment cost is $C = $31.70.

The parameters for this problem can be summarized as follows:

A: loss of producing a defective = 50 cents

B: diagnostic cost = $1.60

C: adjustment cost = $31.70

$\bar{u}$: mean failure interval = 84,000/16 = 5250 units

I: time lag = 30 units

n: diagnostic interval = 100 units

By substituting the above into equation (25.1), the following is derived:

$$L = \frac{B}{n} + \frac{n+1}{2}\frac{A}{\bar{u}} + \frac{C}{\bar{u}} + \frac{IA}{\bar{u}}$$

$$= \frac{1.60}{100} + \frac{101}{2}\left(\frac{0.50}{5250}\right) + \frac{31.70}{5250} + \frac{(30)\,(0.50)}{5250}$$

$$= 0.016 + 0.0048 + 0.006 + 0.0029 = \$0.0297 \qquad (25.2)$$

The quantity control cost for failure of the automatic welding machine is $0.0297 per unit of output when diagnosis/adjustment is done at 100-unit intervals. (*Note:* If we had a welding machine never known to fail, no diagnosis cost would be necessary and no welding machine adjustment would be necessary, since no defective products would ever be produced. Therefore, the value of L in equation (25.2) would be zero. In reality, such a machine does not exist, and even if it did, it would be very expensive.) If the interest and depreciation of the machine permit output higher than the current machine by 10 cents, introduction of a new machine would increase the cost by 7 cents per output.

It is surprising that the processing cost difference is only 3 cents between a machine that never fails or a tool whose life is infinite and a welding machine that fails eight times a month and has a limited life. Of course, if the diagnosis/adjustment method is irrational, the difference will be greater. Since the mean failure interval is once every three days, one may think that diagnosis can be done once a day. In such a case, the diagnostic interval n is 1500 units, and the quantity of daily production and quality control cost L becomes

$$L = \frac{1.60}{1500} + \frac{1501}{2}\left(\frac{0.50}{5250}\right) + \frac{31.70}{5250} + \frac{(30)(0.50)}{5250}$$

$$= 0.0011 + 0.0715 + 0.0060 + 0.0029$$

$$= \$0.0815 \tag{25.3}$$

This means an increase of loss due to lack of diagnosis by $0.0815 - 0.0297 = \$0.0518$ per unit of output over the current control method, which diagnoses the process at 100-unit intervals. This means an increase of cost due to an extended inspection interval by \$2176 per month.

On the other hand, if the inspection interval is reduced to $n = 50$, one-half of the current method, then

$$L = \frac{1.60}{50} + \frac{51}{2}\left(\frac{0.50}{5250}\right) + \frac{31.70}{5250} + \frac{(30)(0.50)}{5250}$$

$$= 0.032 + 0.0024 + 0.0060 + 0.0029$$

$$= \$0.0433 \tag{25.4}$$

In this case, the cost increases due to excessive diagnosis by \$0.0136 per product compared to the current diagnosis interval. What is the optimal diagnosis interval? This problem is addressed in the next section.

25.3. Optimal Diagnostic Interval

The three elements that are necessary to design an on-line quality control system for a production process are the process, the diagnostic method, and an adjustment method. The production process is characterized by loss A of producing a defective and mean failure interval $\bar{u}$, which are treated as system elements. The parameters of the diagnostic method are diagnostic cost, B, and time lag, l, and that of the adjustment method is adjustment cost, C. However, individual parameter values simply represent unorganized system elements. Improvement of the elements in a system is an engineering technology (specialized technology) issue and not an issue of quality and/or management engineering (universal technology).

What combines the process, diagnosis method, and adjustment method is diagnostic interval n. The determination of a diagnostic interval n is an issue of system design and not an issue of the technology in specialized fields. The main issue of quality improvement during production is to minimize the L, given by equation (25.1), as close to zero as possible. Improvement of the technology of specialized fields is not the only way to improve quality.

The cost of quality control given by equation (25.1) can be reduced through optimization of the diagnostic interval n. The optimal diagnostic interval is the interval that roughly equalizes the first term, the diagnostic cost per unit of output, and the second term, the loss of producing defectives between diagnoses in equation (25.1).

The optimal diagnostic interval, n, is given as

$$n = \sqrt{\frac{2(\bar{u} + l)B}{A - C/\bar{u}}} \qquad (25.5)$$

☐ Example

For the automatic welding machine example in Section 25.2,

$$n = \sqrt{\frac{(2)(5250 + 30)(1.60)}{0.50 - 31.70/5250}} = 185 \qquad (25.6)$$

The loss per unit product becomes [see equation (25.2)]

$$L = \frac{1.60}{185} + \frac{186}{2}\left(\frac{0.50}{5250}\right) + \frac{31.70}{5250} + \frac{(30)(0.50)}{5250}$$

$$= 0.0086 + 0.0088 + 0.0060 + 0.0029$$

$$= \$0.0263 \qquad (25.7)$$

Therefore, by altering the diagnostic interval from 100 to 185 units, we obtain a cost reduction of $0.0297 - 0.0263 = \$0.0034$ per unit. This translates into a cost reduction of $0.0034 \times 42,000 = \$143$ per month. Although this number appears to be small, if there are 200 machines of various types and we can improve the cost of each by a similar amount, the total cost improvement will be significant.

It does not matter if there is substantial estimation error in parameters A, B, C, $\bar{u}$, and l. This is because the value of L does not change much even if the value of these parameters changes by up to about 30%. For example, if loss A due to a defect is estimated to be 70 cents, which is nearly a 40% overestimation over 50 cents, then

$$n = \sqrt{\frac{(2)(5250 + 30)(1.60)}{0.70 - 31.70/5250}}$$

$$= 156 \text{ units} \qquad (25.8)$$

that is, a difference of about 15% from the correct answer, 185 units. Furthermore, quality control cost L in the case of $n = 156$ units is

$$L = \frac{1.60}{156} + \frac{157}{2}\left(\frac{0.50}{5250}\right) + \frac{31.70}{5250} + \frac{(30)\,(0.50)}{5250} = \$0.0266 \quad (25.9)$$

that is, a difference of $0.0003 per product compared to the optimal solution. Therefore, the optimal diagnostic interval can be anywhere between $\pm 20\%$ of 185 (i.e., between 150 and 220), depending on the circumstances of the diagnostic operation. For example, if there are not enough workers, choose 220 units; however, if there are surplus workers, choose 150 units.

Equations (25.1) and (25.5) are the basic formulas for process adjustment.

25.4. Derivation of Optimal Number of Workers

Various automatic machines do the actual processing in production processes, while humans do jobs such as transportation of material and parts, mounting, process diagnosis/adjustment, partial inspection, and tooling. The elimination of human labor has been achieved at a rapid pace: for example, by automation of transfer and mounting processes using conveyers and transfer machines and automatic mounting using automatic feeders. However, human beings are necessary for control systems, which involve process diagnosis, problem finding, and process adjustment. In other words, most of the work being conducted by humans in today's factories is process diagnosis and adjustment. This means that we cannot rationally determine the optimal number of workers in a factory without a diagnosis/adjustment system theory. The following simplified example illustrates this point.

❒ Example

LP records are produced by press machines (this is an example from the 1960s) in a factory that has 40 machines, and each machine produces one record every minute; weekly operation is 40 hours. Dust on the press machines and adjustment error sometimes create defective records. Process diagnosis/adjustment is currently conducted at 100-record intervals. The defective record, A, is $1.20; the cost per diagnosis is $8.00; diagnosis time lag I is 30 records; the mean failure interval of the press machine is 8000 records; adjustment time when a press machine breaks down is two hours on average; and adjustment cost C is $50. Derive the optimal diagnosis interval and the optimal number of workers necessary for the factory operation.

Since there are 40 press machines, the number of records produced in one week is (1 record)(60 minutes)(40 hours)(40 machines) = 96,000 records. Therefore, the number of diagnoses per week is 960 and total diagnosis time is

$$(960)(30 \text{ minutes}) = 28,800 \text{ minutes} = 480 \text{ hours} \qquad (25.10)$$

since the time required for diagnosis is equal to time lag l.

By dividing this by the weekly work hours of one worker, 40 hours, we obtain 12; in other words, 12 workers are necessary for diagnosis only. On the other hand, the number of failures per week is $96,000 \div 8000 = 12$. Thus, the time for repair and adjustment is

$$(2 \text{ hours})(12) = 24 \text{ hours} \qquad (25.11)$$

This is fewer than the weekly hours of one worker. Therefore, a total of 13 workers are currently working in this factory for diagnosis and adjustment.

From equation (25.5), the optimal diagnostic interval is

$$n = \sqrt{\frac{(2)(\bar{u} + l)B}{A - C/\bar{u}}} = \sqrt{\frac{(2)(8000 + 30)(8)}{1.20 - 50/8000}}$$

$$\approx 330 \text{ units} \qquad (25.12)$$

This implies that the diagnostic frequency can be increased by 3.3 times the current level. The number of diagnosing workers can be 1/3.3 of the current 12 workers, or 3.6. Thus, one worker should take 10 machines, checking one after another. It is not necessary to be overly concerned about the figure of 330 records. It can be 250 or 400 once in a while. A worker should diagnose one machine about every 30 minutes.

On the other hand, the frequency of failure will not change, even if the diagnostic interval is changed, meaning that one person is required for adjustment. Thus, five workers is more than enough for this factory. If a diagnosing worker also takes the responsibility for adjusting the machine that fails, he or she can take eight machines instead of 10, and the diagnostic interval can still be extended to 10 machines. This is because the diagnostic interval can be changed by 20 to 30%. Therefore, the maximum number of workers for diagnosis and adjustment of this press process is four or five. Using five workers, the number of defectives will increase by 3.3 times, and the total quality control cost, L, will be

$$L = \frac{B}{n} + \frac{n+1}{2}\frac{A}{\bar{u}} + \frac{C}{\bar{u}} + \frac{lA}{\bar{u}}$$

$$= \frac{8}{330} + \frac{331}{2}\left(\frac{1.20}{8000}\right) + \frac{50}{8000} + \frac{(30)(1.20)}{8000}$$

$$= 0.0242 + 0.0248 + 0.0062 + 0.0045$$

$$= \$0.0597 \qquad (25.13)$$

which is a cost improvement over the current quality control cost, L_0,

$$L_0 = \frac{8}{100} + \frac{101}{2}\left(\frac{1.20}{8000}\right) + \frac{50}{8000} + \frac{(30)(1.20)}{8000}$$

$$= 0.08 + 0.0076 + 0.0062 + 0.0045$$

$$= \$0.0983 \qquad (25.14)$$

by \$0.0386 per record. For one week, this will mean a profit from the quality control system by

$$(0.0386)(60 \text{ minutes})(40 \text{ hours})(40 \text{ machines}) \approx \$3700 \qquad (25.15)$$

If one year has 50 operation weeks, this is a rationalization of quality control by \$185,000 per year. Furthermore, the optimal defective ratio, p, of this factory is

$$p = \frac{n+1}{2} \times \frac{1}{\bar{u}} + \frac{1}{\bar{u}} = \frac{331}{2}\left(\frac{1}{8000}\right) + \frac{30}{8000}$$

$$= 0.024 = 2.4\% \qquad (25.16)$$

If the actual defective ratio is far greater than 2.4%, there is not enough diagnosis; however, if it is less than 2.4%, this means that there is excessive checking. Even the new diagnostic interval will incur a quality control cost of 6 cents per record; it is necessary to attempt to reduce the cost closer to zero.

Part IX

Experimental Regression

26 Parameter Estimation in Regression Equations

26.1. Introduction   485
26.2. Estimation of Operation Time and Least Squares Method   485
26.3. Experimental Regression Analysis   486
   Setting of Initial Values   488
   Selection of Orthogonal Array and Calculation   489
   Sequential Approximation   490
   Selection of Estimated Values   492

26.1. Introduction

Quite often we wish to express results or objectives by functions of causes or means in various specialized fields, such as natural science, social science, or engineering. If we cannot perform experiments to obtain results or objectives after arranging causes or means, that is, factors in a particular manner, we often estimate parameters of a regression function predicted by data obtained from research or observations. Although the least squares method is frequently used to estimate unknown parameters after a linear regression model is selected as a regression function, in most cases this procedure is not used appropriately and predicted regression equations do not function well from the standpoint of specialized engineering fields. Among several methods of avoiding this situation is experimental regression analysis.

26.2. Estimation of Operation Time and Least Squares Method

As a simple example, let's examine a problem of estimating standard operation times for office work. Since office work does not consist of several clear-cut portions and its transaction time has large variability, we need a considerable number of observations to estimate a standard operation time accurately after breaking total time down into many work elements. Now we define k types of major transaction times among various office works as $a_1, a_2, \ldots, a_k$ and the corresponding number of jobs to be completed as $x_1, x_2, \ldots, x_k$. The total operation time y_0 is expressed as follows:

$$y_0 = a_1 x_1 + a_2 x_2 + \cdots + a_k x_k \tag{26.1}$$

In actuality, since there are jobs other than the k types of work or slack times, we should add a constant a_0 to equation (26.1). Therefore, the total operation time y is

$$y = a_0 + a_1 x_1 + a_2 x_2 + \cdots + a_k x_k \tag{26.2}$$

Table 26.1 shows the incidence of seven different works $(x_1, x_2, \ldots, x_7)$ and corresponding operation time y (minutes) for each business office. For the sake of convenience, we take data for a certain day. However, to improve the adequacy of analysis, we sometimes sum up all data for one month. To estimate $a_0, a_1, a_2, \ldots, a_7$ of the following equation, using these data is our objective:

$$y = a_0 + a_1 x_1 + a_2 x_2 + \cdots + a_7 x_7 \tag{26.3}$$

This process is called *estimation of unknown parameters* $a_0, a_1, a_2, \ldots, a_7$.

An equation such as equation (26.2), termed a *single regression equation* or *linear regression equation*, takes y as the objective variable and $x_1, x_2, \ldots, x_7$ as the explanation variables with respect to unknowns $a_0, a_1, a_2, \ldots, a_7$.

As one of the estimation methods of unknowns in a linear regression equation, the multivariate analysis is known, which is based on the least squares method. By applying the least squares method to the data in Table 26.1, we obtain the following result:

$$y = 551.59 - 72.51 x_1 + 25.92 x_2 - 44.19 x_3 + 65.55 x_4$$
$$+ 6.58 x_5 + 24.02 x_6 + 11.35 x_7 \tag{26.4}$$

error variance $= V_e = 233{,}468.19$

Although the true values of $a_0, a_1, a_2, \ldots, a_7$ are unknown, the fact that a_1 are estimated as -72.51 or a_3 is estimated as -44.19 is obviously unreasonable because each of the values indicates the operation time for one job.

In the least squares method, the coefficients $a_1, a_2, \ldots, a_7$ are estimated in such a way that the coefficients may take a range of $(-\infty, \infty)$. It is well known that the estimation accuracy of $a_0, a_1, a_2, \ldots, a_7$ depends significantly on the correlation among $x_1, x_2, \ldots, x_7$. If a certain variable x_1 correlates strongly with other variables, the estimation accuracy of a_1 is decreased; otherwise, it is improved.

Actual data observed always correlate with each other. In this case, since the volume of all jobs is supposed to increase on a busy day, the data $x_1, x_2, \ldots, x_7$ tend to correlate with each other, reflecting actual situations. When the least squares method is applied to this case, the estimation obtained quite often digresses from the true standard time. In calculating appropriate coefficients, the least squares method is not effective.

26.3. Experimental Regression Analysis

To eliminate the drawbacks of the least squares method we need to use a new method that reflects on the following:

Table 26.1
Incidence of works and total operation time

No.	x_1	x_2	x_3	x_4	x_5	x_6	x_7	y
1	4	6	18	6	58	120	22	3667
2	2	2	16	12	74	85	31	3683
3	2	12	18	10	58	127	37	3681
4	0	2	3	4	82	39	14	2477
5	2	2	2	0	28	31	0	1944
6	2	0	0	0	26	27	12	414
7	0	2	6	8	52	31	14	1847
8	0	0	2	2	110	15	5	1689
9	0	2	6	6	86	31	11	2459
10	0	0	2	8	64	20	4	1934
11	0	0	2	6	84	15	1	2448
12	0	0	4	2	50	13	3	1228
13	0	0	2	4	72	12	2	1234
14	0	0	4	4	32	12	5	1240
15	0	0	2	0	40	7	0	648
16	0	0	2	8	70	13	2	1252
17	0	0	0	0	42	6	1	715
18	0	0	6	2	46	15	4	1309
19	0	2	4	4	84	62	22	3023
20	2	2	12	4	72	57	24	2433
21	2	2	6	6	86	45	0	2447
22	2	0	4	4	74	76	21	3210
23	0	2	2	4	58	24	2	1239
24	0	4	2	2	52	28	13	3635
25	0	2	4	4	94	22	6	1341
26	0	2	4	4	56	24	11	1930
27	0	0	4	8	56	18	4	1940
28	0	0	6	0	56	18	6	7565
29	0	0	2	0	12	3	0	639
30	0	0	2	0	36	4	2	1229
31	0	0	2	4	52	9	2	1239
32	0	0	2	4	94	22	6	1341
33	0	0	2	0	10	4	2	1245

Table 26.1
(*Continued*)

No.	x_1	x_2	x_3	x_4	x_5	x_6	x_7	y
34	0	0	0	0	48	4	0	748
35	0	0	0	4	36	4	0	1218
36	0	0	2	2	20	5	2	518
37	0	0	0	0	56	5	1	636
38	0	0	2	0	26	5	1	631
39	0	0	0	0	48	2	0	650
40	0	0	0	0	16	1	1	649
41	0	0	0	0	6	1	1	653
42	0	0	0	0	18	1	0	653
43	0	0	0	0	16	2	0	640

1. For unknowns (or coefficients), we set up a practically reasonable range and seek a suitable equation within the range. This is called *quadratic planning*.

2. To rationally seek reasonable coefficients, we narrow down the range of each coefficient using an orthogonal array.

3. When we determine coefficients based on narrowed-down results, we emphasize ideas of users of the equation or experts in specialized engineering fields.

As one of the methods satisfying these requirements, we describe the experimental regression analysis below. This is called *quadratic planning* in the Taguchi method.

Setting of Initial Values

First, by assuming how much time a_0, a_1, a_2, ... , a_7 affects the total operation time y when each piece of work increases by 1 unit (number of jobs or papers), or a rough range of the standard operation time required for each piece of work, we set up three levels within the range. More specifically:

1. After setting the minimum to level 1 and the maximum to level 3, we select the middle of the two values as level 2.

2. After setting the forecasting value to level 2, by creating a certain range before and after level 1 based on level 2's confidence, we establish level 2 − the range as levels 1 and 2 + the range as level 3.

Both of the methods above can be used. For example, we assume that work 1 takes approximately 2 hours on average and has a range of 30 minutes.

❑ *Level 1 of a_1*: a_{11} = 90 minutes
❑ *Level 2 of a_1*: a_{12} = 120 minutes
❑ *Level 3 of a_1*: a_{13} = 150 minutes

These are called *initial values of a_1*. Although the range should be as precise as possible, we do not need to be too sensitive to it. Table 26.2 shows the initial values

Table 26.2

Initial values of coefficients (minutes)

Coefficient	Level		
	1	2	3
a_1	90	120	150
a_2	0	60	120
a_3	0	30	60
a_4	0	30	60
a_5	0	10	20
a_6	0	20	40
a_7	0	20	40

for other work. Now, we do not set up a variable range of a_1 but calculate it using the following relationship:

$$a_0 = \bar{y} - a_1\bar{x}_1 - a_2\bar{x}_2 - \cdots - a_7\bar{x}_7 \tag{26.5}$$

If, as for level 2, all values of $a_0, a_1, a_2, \ldots, a_7$ follow the data in Table 26.1, the total operation time of business office 1, y, is calculated as follows:

Selection of Orthogonal Array and Calculation

$$y_1 = a_0 + a_{12}x_1 + a_{22}x_2 + \cdots + a_{72}x_7$$

$$= a_0 + (120)(4) + (60)(6) + \cdots + (20)(22)$$

$$= a_0 + 4980 \tag{26.6}$$

However, since the actual time is 3667 minutes, $\bar{y}_1 - y_1$ indicates the difference between the actual time and the estimate assuming $a_1 = 120$, $a_2 = 60$, ..., $a_7 = 20$ as a unit operation time. This difference involves assumptive errors. On the other hand, if each of the assumed coefficients a has little error, $\bar{y}_1 - y_1$ for each business office is not supposed to vary greatly. The magnitude of change can be evaluated by variation (a sum of squares of residuals, S_e). Therefore, we need to do the same calculation and comparison for other business offices. Now, we exclude a_0 in equation (26.3). In this case, $\bar{y}_1 - y_1$ involves not merely assumptive errors regarding $a_1 = 120$, $a_2 = 60$, ..., $a_7 = 20$ but also operation times other than $x_1, x_2, \ldots, x_7$.

When we have three assumed values for each of the unknowns $a_1, a_2, \ldots, a_7$, the eventual total number of combinations amounts to $3^7 = 2187$. However, we select a certain number of combinations from them. Which combination should be chosen depends on an orthogonal array.

By taking into account the number of unknown parameters, we select the size of an orthogonal array according to Table 26.3. The orthogonal arrays shown in Table 26.3 have no particular columns where inter-column interactions emerge. Since we can disperse interaction effects between parameters for the sum of squares of residuals S_e to various columns by using these arrays, we can perform sequential approximation even if there are no interactions.

Table 26.3
Selection of orthogonal array

Number of Unknown Parameters	Orthogonal Array to Be Used
2	3×3 two-dimensional layout
3–6	Three-level portion of L_{18} orthogonal array
7–13	Three-level portion of L_{36} orthogonal array
14–49	Three-level portion of L_{108} orthogonal array

Because the L_{18} orthogonal array has seven three-level columns, we can allocate up to seven parameters. However, if the actual calculation does not take long, we recommend that the L_{36} array be used in cases of seven parameters or more. For this case, we utilize the L_{36} array, where there are only seven unknown parameters. Table 26.4 illustrates the L_{36} array. In the table, each number from 1 to 13 in the top horizontal row shows a type of unknown, and each digit 1, 2, and 3 in the table itself indicates the unknown's level. Each row in the table represents a combination of levels: 36 combinations in case of an L_{36} orthogonal array.

Now, we assign each of the seven unknowns $a_1, a_2, \ldots, a_7$ to each column from 1 to 7. This process is called *assignment (layout) to an orthogonal array*. As a consequence, the numbers in rows 1 to 36 represent the following 36 equations:

$$\text{Row 1:} \quad y = 90x_1 + 0x_2 + \cdots + 0x_7$$

$$\text{Row 2:} \quad y = 120x_1 + 60x_2 + \cdots + 20x_7 \tag{26.7}$$

$$\vdots$$

$$\text{Row 36:} \quad y = 150x_1 + 60x_2 + \cdots + 20x_7$$

As a next step, we plug $x_1, x_2, \ldots, x_7$ of 43 business offices into these equations and calculate the following differences between the result calculated and its corresponding actual operation time:

$$e_{ij} = y_{ij} - y_j \quad (i = 1,2, \ldots, 36; j = 1,2, \ldots, 43) \tag{26.8}$$

Then we compute a sum of all of the differences and variation as follows:

$$T_1 = e_{11} + e_{12} + \cdots + e_{143} \tag{26.9}$$

$$S_1 = e_{11}^2 + e_{12}^2 + \cdots + e_{143}^2 - \frac{T_1^2}{43} \tag{26.10}$$

Table 26.5 summarizes the result for the first data in Table 26.2. However, the data in the first column of Table 26.5 do not indicate T itself but the averaged T, which is equivalent to the constant term a_0 in each equation.

Sequential Approximation   Using Table 26.5, we compute a level-by-level sum of S values for each of the unknown coefficients $a_1, a_2, \ldots, a_7$, as shown in Table 26.6. For each unknown we compare three different level-by-level sums of S. If all of them are almost the same, any of the three levels assumed for each coefficient leads to the same amount of

Table 26.4

L_{36} orthogonal array

No.	1	2	3	4	5	6	7	8	9	10	11	12	13
							Factor						
1	1	1	1	1	1	1	1	1	1	1	1	1	1
2	2	2	2	2	2	2	2	2	2	2	2	2	1
3	3	3	3	3	3	3	3	3	3	3	3	3	1
4	1	1	1	1	2	2	2	2	3	3	3	3	1
5	2	2	2	2	3	3	3	3	1	1	1	1	1
6	3	3	3	3	1	1	1	1	2	2	2	2	1
7	1	1	2	3	1	2	3	3	1	2	2	3	1
8	2	2	3	1	2	3	1	1	2	3	3	1	1
9	3	3	1	2	3	1	2	2	3	1	1	2	1
10	1	1	3	2	1	3	2	3	2	1	3	2	1
11	2	2	1	3	2	1	3	1	3	2	1	3	1
12	3	3	2	1	3	2	1	2	1	3	2	1	1
13	1	2	3	1	3	2	1	3	3	2	1	2	2
14	2	3	1	2	1	3	2	1	1	3	2	3	2
15	3	1	2	3	2	1	3	2	2	1	3	1	2
16	1	2	3	2	1	1	3	2	3	3	2	1	2
17	2	3	1	3	2	2	1	3	1	1	3	2	2
18	3	1	2	1	3	3	2	1	2	2	1	3	2
19	1	2	1	3	3	3	1	2	2	1	2	3	2
20	2	3	2	1	1	1	2	3	3	2	3	1	2
21	3	1	3	2	2	2	3	1	1	3	1	2	2
22	1	2	2	3	3	1	2	1	1	3	3	2	2
23	2	3	3	1	1	2	3	2	2	1	1	3	2
24	3	1	1	2	2	3	1	3	3	2	2	1	2
25	1	3	2	1	2	3	3	1	3	1	2	2	3
26	2	1	3	2	3	1	1	2	1	2	3	3	3
27	3	2	1	3	1	2	2	3	2	3	1	1	3
28	1	3	2	2	2	1	1	3	2	3	1	3	3
29	2	1	3	3	3	2	2	1	3	1	2	1	3
30	3	2	1	1	1	3	3	2	1	2	3	2	3
31	1	3	3	3	2	3	2	2	1	2	1	1	3
32	2	1	1	1	3	1	3	3	2	3	2	2	3

Table 26.4
(*Continued*)

No.	Factor												
	1	2	3	4	5	6	7	8	9	10	11	12	13
33	3	2	2	2	1	2	1	1	3	1	3	3	3
34	1	3	1	2	3	2	3	1	2	2	3	1	3
35	2	1	2	3	1	3	1	2	3	3	1	2	3
36	3	2	3	1	2	1	2	3	1	1	2	3	3

error. However, if S for level 2 is small while S values for levels 1 and 3 are large, the true value of the coefficient a_i will lie close to level 2. In this case, by resetting three levels in the proximity to level 2, we repeat the calculation. For practical calculation, we determine new level values using level-by-level sums of S values in Table 26.6 according to Table 26.7.

Now, for V_e shown in the determinant column of Table 26.7, we use error variance obtained by the least squares method. Indeed, coefficients calculated by the least squares method are unreliable; however, error variances can be considered trustworthy. For the sake of convenience, we recommend using 3 as the F-test criterion because a value of 2 to 4 has been regarded as the most appropriate in our experience.

For instance, coefficient a_1 has three levels 90, 120, and 150. Table 26.6 shows the following:

$$S(a_{11}) = 0.39111173 \times 10^9$$

$$S(a_{12}) = 0.34771989 \times 10^9$$

$$S(a_{13}) = 0.36822313 \times 10^9$$

The fact that the value of S at level 1 is the largest and the value at level 2 is the smallest indicates a V-shaped type (type V), in particular, pattern 3.

$$\frac{(1/r)\ (S_3 - S_2)}{V_e} = \frac{\frac{1}{12}(0.36822313 \times 10^9 - 0.34771989 \times 10^9)}{23{,}346.19}$$

$$= 87.8 > 3 \tag{26.11}$$

In this equation, we set r to 12 because both S_3 and S_2 are sums of 12 data.

Accordingly, the new levels for coefficient a_1 in the second round are established in the proximity to level 2. By proceeding with this process, we obtain the new level setting, as illustrated in Tables 26.7 and 26.8.

Selection of Estimated Values

Based on the newly selected levels, we repeat the same calculation as in the first round. In addition, after making the same judgment as that shown in Table 26.7, for coefficients not converging sufficiently, we set up the three new levels and move on to the third round. In this case, all coefficients have converged in the fifth round, as summarized Table 26.9.

Table 26.5

T and S in first round

No.	Averaged T	Variation S	Combination												
1	1,553.58140	36,740,548.5	1	1	1	1	1	1	1	1	1	1	1	1	1
2	143.581395	16,829,328.5	2	2	2	2	2	2	2	2	2	2	2	2	1
3	−1,266.41860	11,911,050.8	3	3	3	3	3	3	3	3	3	3	3	3	1
4	4,221.953488	10,251,155.9	1	1	1	1	2	2	2	2	3	3	3	3	1
5	−988.046512	81,015,535.9	2	2	2	2	3	3	3	3	1	1	1	1	1
6	996.837209	14,439,209.9	3	3	3	3	1	1	1	1	2	2	2	2	1
7	497.069767	18,175,430.8	1	1	2	3	1	2	3	3	1	2	2	3	1
8	−225.023256	45,653,675.0	2	2	3	1	2	3	1	1	2	3	3	1	1
9	158.697674	11,285,447.1	3	3	1	2	3	1	2	2	3	1	1	2	1
10	128.465116	46,156,958.7	1	1	3	2	1	3	2	3	2	1	3	2	1
11	512.186047	9,359,614.51	2	2	1	3	2	1	3	1	3	2	1	3	1
12	−209.906977	19,597,979.6	3	3	2	1	3	2	1	2	1	3	2	1	1
13	−232.930233	18,923,170.8	1	2	3	1	3	2	1	3	3	2	1	2	2
14	212.186047	45,252,574.5	2	3	1	2	1	3	2	1	1	3	2	3	2
15	451.488372	10,126,044.7	3	1	2	3	2	1	3	2	2	1	3	1	2
16	909.395349	13,132,844.3	1	2	3	2	1	1	3	2	3	3	2	1	2
17	231.255814	13,060,792.2	2	3	1	3	2	2	1	3	1	1	3	2	2
18	−709.906977	52,139,539.6	3	1	2	1	3	3	2	1	2	2	1	3	2
19	−692.00000	43,558,928.0	1	2	1	3	3	3	1	2	2	1	2	3	2
20	1,174.27907	16,310,512.7	2	3	2	1	1	1	2	3	3	2	3	1	2
21	−51.5348837	26,925,158.7	3	1	3	2	2	2	3	1	1	3	1	2	2
22	40.7906977	12,195,333.1	1	2	2	3	3	1	2	1	1	3	3	2	2
23	441.953488	31,986,195.9	2	3	3	1	1	2	3	2	2	1	1	3	2
24	−52.000000	27,263,368.0	3	1	1	2	2	3	1	3	3	2	2	1	2
25	−433.162791	71,126,909.9	1	3	2	1	2	3	3	1	3	1	2	2	3
26	209.395349	15,054,924.3	2	1	3	2	3	1	1	2	1	2	3	3	3
27	654.511628	13,870,504.7	3	2	1	3	1	2	2	3	2	3	1	1	3
28	717.534884	13,035,738.7	1	3	2	2	2	1	1	3	2	3	1	3	3
29	−508.279070	28,202,412.7	2	1	3	3	3	2	2	1	3	1	2	1	3
30	221.488372	48,144,624.7	3	2	1	1	1	3	3	2	1	2	3	2	3
31	−598.046512	78,524,235.9	1	3	3	3	2	3	2	2	1	2	1	1	3
32	254.511628	13,377,984.7	2	1	1	1	3	1	3	3	2	3	2	2	3

Table 26.5
(Continued)

No.	Averaged T	Variation S	Combination											
33	774.279070	13,789,392.7	3	2	2	2	1	2	1	1	3	1	3	3	3
34	−438.976744	29,290,475.0	1	3	1	2	3	2	3	1	2	2	3	1	3
35	264.976744	31,616,335.0	2	1	2	3	1	3	1	2	3	3	1	2	3
36	604.744186	11,531,352.2	3	2	3	1	2	1	2	3	1	1	2	3	3

As a reference, see Tables 26.10 and 26.11, showing averaged T and S values and level-by-level sums of S', respectively. When all coefficients are converged, we can use any of the three levels. However, without any special reasons to select a specific level, we should adopt level 2. Selecting level 2 for all of $a_1, a_2, \ldots, a_7$, we can express them as follows:

$$a_1 = 127.5 \pm 7.5 \text{ minutes}$$

$$a_2 = 30.0 \pm 15.0 \text{ minutes}$$

$$\vdots$$

$$a_7 = 15.0 \pm 5.0 \text{ minutes} \tag{26.12}$$

In addition, the constant term a_0 is calculated as

$$a_0 = \bar{y} - a_1\bar{x}_1 - a_2\bar{x}_2 - \cdots - a_7\bar{x}_7$$

$$= 1591.26 - (127.5)(0.4183) - (30.0)(1.02) \cdots - (15.0)(6.72)$$

$$= 573.52 \tag{26.13}$$

This value is equal to the value of averaged T in row 2 of Table 26.10 because all levels in the row are set to level 2.

Table 26.6
Level-by-level sum of S values[a]

Coefficient	Level 1	Level 2	Level 3
a_1	0.39111173E + 09	0.34771989E + 09	0.36822313E + 09
a_2	0.316029886E + 09	0.32800430E + 09	0.46302058E + 09
a_3	0.30145602E + 09	0.35595808E + 09	0.44964065E + 09
a_4	0.37578365E + 09	0.33903175E + 09	0.39223935E + 09
a_5	0.32961513E + 09	0.33368737E + 09	0.44375224E + 09
a_6	0.17658955E + 09	0.24090200E + 09	0.68956319E + 09
a_7	0.29273406E + 09	0.34254936E + 09	0.47177133E + 09

[a]$0.39111173E + 09 = 0.39111173 \times 10^9$.

Table 26.7
New setting of levels[a]

Plot of Variation S	Determinant	New Level 1'	2'	3'
I	(1) $\dfrac{(1/r)(S_3 - S_1)}{V_e} < F$	1	2	3
	(2) $\dfrac{(1/r)(S_2 - S_1)}{V_e} < F$	1	1.5	2
	(3) $\dfrac{(1/r)(S_2 - S_1)}{V_e} \geq F$	0.5 (1)	1 (1.5)	1.5 (2)
II	(1) $\dfrac{(1/r)(S_3 - S_2)}{V_e} < F$	1	2	3
	(2) $\dfrac{(1/r)(S_1 - S_2)}{V_e} < F$	1	1.5	2
	(3) $\dfrac{(1/r)(S_1 - S_2)}{V_e} \geq F$	1.5	2	2.5
III	Omitted	1	2	3
IV	Omitted	1	2	3
V	(1) $\dfrac{(1/r)(S_1 - S_3)}{V_e} < F$	1	2	3
	(2) $\dfrac{(1/r)(S_1 - S_2)}{V_e} < F$	2	2.5	3
	(3) $\dfrac{(1/r)(S_3 - S_2)}{V_e} \geq F$	1.5	2	2.5

Table 26.7
(Continued)

Plot of Variation S	Determinant	New Level		
		1'	2'	3'
VI	(1) $\dfrac{(1/r)(S_1 - S_3)}{V_e} < F$	1	2	3
	(2) $\dfrac{(1/r)(S_2 - S_3)}{V_e} < F$	2	2.5	3
	(3) $\dfrac{(1/r)(S_2 - S_3)}{V_e} \geq F$	2.5 (2)	3 (2.5)	3.5 (3)

[a] $r = N/3$, where N is the size of an orthogonal array; Level 1.5 means to create an intermediate level between levels 1 and 2; values in parentheses are used when newly determined values exceed the initial variable range.

Finally, we obtain the following regression equation:

$$y = 573.52 + 127.5x_1 + 30.0x_2 + \cdots + 15.0x_7 \qquad (S_e = 9279244.98) \qquad (26.14)$$

At the same time, $S_e = 9279244.98$ in (26.14) can be seen in variation S in row 2 of Table 26.10. Indeed, this value is somewhat larger than the variation of $S_e = 8171386.56$ obtained by the least squares method; however, this can be regarded as quite practical because the coefficients determined here do not involve any contradiction.

The regression equation obtained here is normally utilized for data other than those used for this regression equation: for example, for predicting future values or making a decision for other groups. Then it is important to evaluate the magnitude of errors by using different data. For instance, we can make the comparison

Table 26.8
Three levels for second round

Coefficient	Level		
	1	2	3
a_1	105	120	135
a_2	0	30	60
a_3	0	15	30
a_4	15	30	45
a_5	0	5	10
a_6	0	10	20
a_7	0	10	20

Table 26.9
Three levels for fifth round

Coefficient	Level		
	1	2	3
a_1	120.0	127.5	135.0
a_2	15.0	30.0	45.0
a_3	0.0	7.5	15.0
a_4	30.0	37.5	45.0
a_5	5.0	7.5	10.0
a_6	10.0	12.5	15.0
a_7	10.0	15.0	20.0

Table 26.10
T and S in fifth round

No.	Averaged T	Variation S	Combination												
1	864.976744	11,740,655.0	1	1	1	1	1	1	1	1	1	1	1	1	1
2	573.523256	9,279,244.98	2	2	2	2	2	2	2	2	2	2	2	1	
3	282.069767	11,255,440.8	3	3	3	3	3	3	3	3	3	3	3	1	
4	643.116279	9,535,102.42	1	1	1	1	2	2	2	2	3	3	3	3	1
5	351.662791	10,016,397.4	2	2	2	2	3	3	3	3	1	1	1	1	1
6	725.790698	9,962,633.12	3	3	3	3	1	1	1	1	2	2	2	2	1
7	661.895349	9,489,529.28	1	1	2	3	1	2	3	3	1	2	2	3	1
8	542.418605	9,702,168.47	2	2	3	1	2	3	1	1	2	3	3	1	1
9	516.255814	9,133,562.19	3	3	1	2	3	1	2	2	3	1	1	2	1
10	630.790698	9,877,123.12	1	1	3	2	1	3	2	3	2	1	3	2	1
11	604.627907	9,122,004.05	2	2	1	3	2	1	3	1	3	2	1	3	1
12	485.151163	9,326,132.77	3	3	2	1	3	2	1	2	1	3	2	1	1
13	479.395349	9,390,239.28	1	2	3	1	3	2	1	3	3	2	1	2	2
14	651.720930	9,654,002.65	2	3	1	2	1	3	2	1	1	3	2	3	2
15	589.453488	9,293,008.41	3	1	2	3	2	1	3	2	2	1	3	1	2
16	703.930233	9,779,070.79	1	2	3	2	1	1	3	2	3	3	2	1	2
17	595.441860	9,142,876.60	2	3	1	3	2	2	1	3	1	1	3	2	2
18	421.197674	9,549,964.57	3	1	2	1	3	3	2	1	2	2	1	3	2

Table 26.10
(*Continued*)

No.	Averaged T	Variation S	Combination											
19	425.674419	9,087,415.44	1	2	1	3	3	3	1	2	2	1	2	3	2
20	770.151163	10,164,835.3	2	3	2	1	1	1	2	3	3	2	3	1	2
21	524.744186	9,579,907.19	3	1	3	2	2	2	3	1	1	3	1	2	2
22	486.779070	9,120,635.15	1	2	2	3	3	1	2	1	1	3	3	2	2
23	648.116279	10,081,987.4	2	3	3	1	1	2	3	2	2	1	1	3	2
24	585.674419	9,342,025.44	3	1	1	2	2	3	1	3	3	2	2	1	2
25	490.383721	9,972,519.92	1	3	2	1	2	3	3	1	3	1	2	2	3
26	528.930233	9,688,640.79	2	1	3	2	3	1	1	2	1	2	3	3	3
27	701.255814	9,511,572.19	3	2	1	3	1	2	2	3	2	3	1	1	3
28	655.965116	9,647,941.20	1	3	2	2	2	1	1	3	2	3	1	3	3
29	410.558140	9,381,541.60	2	1	3	3	3	2	2	1	3	1	2	1	3
30	654.046512	9,862,355.91	3	2	1	1	1	3	3	2	1	2	3	2	3
31	449.162791	10,108,859.9	1	3	3	3	2	3	2	2	1	2	1	1	3
32	540.209302	9,418,093.12	2	1	1	1	3	1	3	3	2	3	2	2	3
33	731.197674	9,944,122.07	3	2	2	2	1	2	1	1	3	1	3	3	3
34	427.883721	9,123,062.42	1	3	1	2	3	2	3	1	2	2	3	1	3
35	664.918605	9,628,645.97	2	1	2	3	1	3	1	2	3	3	1	2	3
36	627.767442	9,686,485.67	3	2	3	1	2	1	2	3	1	1	2	3	3

Table 26.11
Level-by-level sum of S values

Coefficient	Level 1	Level 2	Level 3
a_1	0.11687215E + 09	0.11528044E + 09	0.11644715E + 09
a_2	0.11652424E + 09	0.11450165E + 09	0.11757385E + 09
a_3	0.11467273E + 09	0.11543298E + 09	0.11849404E + 09
a_4	0.11843048E + 09	0.11506510E + 09	0.11510416E + 09
a_5	0.11969653E + 09	0.11441208E + 09	0.11449113E + 09
a_6	0.11675750E + 09	0.11378532E + 09	0.11805692E + 09
a_7	0.11660350E + 09	0.11500287E + 09	0.11699338E + 09

Table 26.12

Comparison between least squares method and experimental regression analysis with different business office data[a]

No.	x_1	x_2	x_3	x_4	x_5	x_6	x_7	y	Least Squares Method	Experimental Regression
1	2	4	10	4	144	85	21	2980	596.37	651.02
2	0	0	4	10	120	28	5	2431	140.17	−127.48
3	0	2	2	4	42	19	1	1254	287.43	112.02
4	0	4	6	0	78	38	0	1815	19.38	−16.48
5	2	0	2	4	118	27	6	2452	−356.60	−145.98
6	0	0	4	4	56	14	5	1223	197.47	200.52
7	0	0	2	6	70	14	0	1234	441.32	279.52
							Sum of squares of residual:		819,226.319	592,554.321
							Variance:		117,032.331	84,650.617

[a] Variance = sum of square of residuals/7.

shown in Table 26.12. This table reveals that the mean sum of squares of deviations between the estimation and actual value (variance) calculated based on the least squares method is 1.38 times as large as that by the experimental regression analysis. This fact demonstrates that we cannot necessarily minimize the sum of squares of residuals for different business office data, whereas we can do so for the data to be used for coefficient estimation. This is considered to be caused the by uncertainty of parameters estimated using the least squares method. Then if volume of jobs such as x_1 and x_3 is increased in the future, the equation based on the least squares method will lose its practicality.

Part X

Design of Experiments

27 Introduction to Design of Experiments

27.1. What Is Design of Experiments?   503
27.2. Evaluation of Functionality   504
27.3. Reproducibility of Conclusions   504
27.4. Decomposition of Variation   505
      Reference   505

27.1. What Is Design of Experiments?

Traditional design of experiments was developed by Ronald A. Fisher of England in the 1920s for the rationalization of agricultural experimentation. The approach was later applied in medical and biological studies.

In biology, there is substantial variation in the output characteristics, which is caused by individuals. Therefore, the content of the traditional design of experiments consists primarily of ways to express individual variation and ways to separate such variation. These include blocking to separate conditional difference, randomization of experimental order, repetition of experiments, distribution, and test of significance, among others.

Design of experiments means to find the relationship between various factors (variables) and responses to them. It is therefore important to express their relationship precisely and efficiently by an equation. If the interactions between variables (factors) are significant, it is important to include the terms of interactions in the equation.

Traditional design of experiments is the study of response. To find an equation with a small error, many variables that might affect the output characteristic are listed and an experiment is planned. The equation is written as

$$y = f(M, x_1, x_2, \dots , x_n) \tag{27.1}$$

where y is the output characteristic, M the input signal, and $x_1, x_2, \dots , x_n$ the noise factors.

This approach is appropriate for scientific studies. But for engineering studies, where the objective is to develop a good product efficiently at low cost, the approaches are entirely different. In engineering, monetary issues are most important. But in science, there is no economical consideration.

There are basically two areas in quality engineering: off-line and on-line. The first relates to technology development, and the second is applied in daily production activities for process control or management.

There are three steps in off-line quality engineering: *system selection, parameter design,* and *tolerance design.* Selection of a system is a task for engineers working in a specialized engineering field. Parameter design is to improve the function of a product by varying the parameter levels in a system. The function of a system is expressed and evaluated by the SN ratio. Many design variables are listed up and assigned to an orthogonal array.

Tolerance design is to assign noise factors that affect variability to an orthogonal array and study the effects of every noise factor. Removal of noise factors by upgrading of components or raw materials is determined by the use of the loss function.

27.2. Evaluation of Functionality

For the evaluation of functionality, equation (27.1) is decomposed into two parts:

$$y = f(M, x_1, x_2, \dots, x_n)$$

$$= \beta M + [f(M, x_1, x_2, \dots, x_n) - \beta M] \tag{27.2}$$

where βM is the useful part and $[f(M, x_1, x_2, \dots, x_n) - \beta M]$ is the harmful part. The SN ratio is the ratio of the first term to the second. The larger the ratio, the better the functionality.

The basic difference between the traditional design of experiments and the design of experiments for quality engineering is that the SN ratio is used as an index for functionality instead of using a response. In the traditional design of experiments, there is no distinction between control factors and noise factors. Error is considered to vary at random, and its distribution is discussed. But in quality engineering, neither random error nor distribution are considered.

27.3. Reproducibility of Conclusions

In technology development or product design study, the reproducibility of conclusions is most important, since it is conducted at an early stage in the laboratory and the conclusions must be reproduced downstream.

Reproducibility of conclusions does not mean reproducibility under the same conditions but:

❏ Reproducibility of the results from a study conducted using test pieces or using a computer to simulate the actual product

❏ Reproducibility of the results from a small-scale experiment to a large-scale manufacturing process

❏ Reproducibility of the conclusions from limited conditions to various other conditions

How can one assure that the conclusions from a test piece study in a laboratory can be reproduced downstream, where the conditions are totally different? Since the conditions are different, the characteristic (output) to be measured will be different. Therefore, we have to give up reproducibility of the output characteristic. The characteristic will be adjusted or tuned at the second stage of parameter design. At the development stage, only improvement of functionality is sought. That is why the SN ratio is used in parameter design.

By use of the SN ratio, it is expected that the difference in the stability of a function can be reproduced. However, there is no guarantee of reproducibility. Therefore, a method to check reproducibility is needed. Orthogonal arrays are used for this purpose. This is discussed in a later chapter.

27.4. Decomposition of Variation [1]

In the design of experiments for quality engineering, the steps include (1) selection of characteristics, (2) layout of experiment, and (3) data analysis. Selection of the characteristics used for analysis is the most important and also the most difficult step.

Once the characteristics to be used are determined, the next step is the layout of the experiment. From many factors with different numbers of levels, we select the combinations of factor levels so that reliable conclusions may be obtained efficiently. Orthogonal arrays are used for the layout.

After the experiment is conducted and the data collected, data will be analyzed and conclusions drawn. The analysis includes decomposition of variation, calculation of factorial effects, estimation of the optimum condition, and many other factors, discussed in the following chapters.

Reference

1. Genichi Taguchi et al., 1973. *Design of Experiments*. Tokyo: Japanese Standards Association.

28 Fundamentals of Data Analysis

28.1. Introduction   506
28.2. Sum and Mean   506
28.3. Deviation   507
28.4. Variation and Variance   509

28.1. Introduction

In this chapter, the fundamentals of data analysis are illustrated by introducing the concepts of sum, mean, deviation, variation, and variance. This chapter is based on Genichi Taguchi et al., *Design of Experiments*. Tokyo: Japanese Standards Association, 1973.

28.2. Sum and Mean

The following data are the heights of 10 persons in centimeters:

163.2   171.6   156.3   159.2   160.0   158.7   167.5   175.4   172.8   153.5

Letting the total height of the 10 persons be T yeilds

$$T = 163.2 + 171.6 + \cdots + 153.5 = 1638.2 \qquad (28.1)$$

Letting the mean be $\bar{y}$, we have

$$\bar{y} = \frac{1}{10} T = \frac{1638.2}{10} = 163.82 \qquad (28.2)$$

To simplify the calculation, the sum and mean can be obtained for the data after subtracting a figure that is probably in the neighborhood of the mean. This number is called a *working mean*. Obviously, the numbers become smaller after subtracting the working mean from each side; thus the calculation is simplified. Also for simplification, data after subtracting a working mean could be multiplied by 10, or 100, or 0.1, and so on.

This procedure is called *data transformation*. The height data after subtracting a working mean of 160.0 cm are:

$$3.2 \quad 11.6 \quad -3.7 \quad -0.8 \quad 0.0 \quad -1.3 \quad 7.5 \quad 15.4 \quad 12.8 \quad -6.5$$

The sum T and mean $\bar{y}$ are now calculated as follows:

$$T = (160.0)(10) + [3.2 + 11.6 + \cdots + (-6.5)] = 1600 + 38.2 = 1638.2 \quad (28.3)$$

$$\bar{y} = 160.0 + \frac{38.2}{10} = 163.82 \quad (28.4)$$

Letting n measurements by $y_1, y_2, \ldots, y_n$, their sum, T, is defined as

$$T = y_1 + y_2 + \cdots + y_n \quad (28.5)$$

The mean, $\bar{y}$, is

$$\bar{y} = \frac{1}{n}(y_1 + y_2 + \cdots + y_n) \quad (28.6)$$

Letting the working mean be y_0, T and $\bar{y}$ may be written as

$$T = y_0 n + [(y_1 - y_0) + (y_2 - y_0) + \cdots + (y_n - y_0)] \quad (28.7)$$

$$\bar{y} = y_0 + \frac{1}{n}[(y_1 - y_0) + (y_2 - y_0) + \cdots + (y_n - y_0)] \quad (28.8)$$

28.3. Deviation

There are two types of deviations: deviation from an objective value and the deviation from the mean, $\bar{y}$. First, deviation from an objective value is explained. Let n measurements be $y_1, y_2, \ldots, y_n$. When there is a definite objective value, y_0, the differences of $y_1, y_2, \ldots, y_n$ from the objective value, y_0, are called the deviations from an objective value.

The following is an example to illustrate calculation of the deviation from an objective value. In order to produce carbon resistors valued at 10 kΩ, 12 resistors were produced for trial. Their resistance was measured to obtain the following values in kilohoms:

$$10.3 \quad 9.9 \quad 10.5 \quad 11.0 \quad 10.0 \quad 10.3 \quad 10.2 \quad 10.3 \quad 9.8 \quad 9.5 \quad 10.1 \quad 10.6$$

The deviations from an objective value of 10 kΩ are

$$0.3 \quad -0.1 \quad 0.5 \quad 1.0 \quad 0.0 \quad 0.3 \quad 0.2 \quad 0.3 \quad -0.2 \quad -0.5 \quad 0.1 \quad 0.6$$

An objective value is not always limited to the value that a person would assign arbitrarily. Sometimes standard values are used for comparison, such as nominal values (specifications or indications, such as 100 g per container), theoretical values (values calculated from theories or from a standard method used by a company), forecast values, and values concerning the product used by other companies or other countries.

Table 28.1 shows the results of measuring the electric current of an electric circuit when both the voltage (*E*: volts) and resistance (*R*: ohms) were varied.

From Ohm's law, the theoretical values for each voltage and resistance are calculated as 5.00, 1.00, 2.00, 0.75, and 1.50. The differences between each observational value, *y*, and the theoretical value, y_0, are shown in the right-hand column of the table.

When the heights of sixth-grade children in a primary school of a certain school district are measured, deviations from the mean height of sixth-grade children in the entire country are usually calculated. In this case, the mean value of the country is an objective value; therefore, deviations from the objective value must be calculated.

Next, let's discuss the deviation from a mean value. When there is neither an objective value nor a theoretical value, it is important to calculate the deviation from a mean value. For example, the average height of the 10 persons described earlier was $\bar{y} = 163.82$. Deviations from the mean are

$$163.2 - 163.82 = -0.62$$

$$171.6 - 163.82 = 7.78$$

$$156.3 - 163.82 = -7.52 \tag{28.9}$$

$$\vdots$$

$$153.5 - 163.82 = -10.32$$

The total of these deviations from the mean is equal to zero. In many cases, the word *deviation* signifies the deviation from an arithmetic mean. It is important to distinguish in each case whether a deviation is a deviation from a mean or from an objective value.

Again, using the 10 height data values, if these heights are data from 10 young men who live in a remote place, a deviation from an objective value should be calculated in addition to the deviation from the mean. The problem is to determine whether the mean value of the young men living in this remote place differs from the mean value of the entire country. Therefore, y_0, the deviation from the mean value of the country, is calculated. (*Note:* this is not the same as calculating the deviation from the mean or from an objective value.)

Table 28.1
Observational values *y* and theoretical values y_0

R	E	y	y_0	Deviation
10	50	5.10	5.00	0.10
20	20	0.98	1.00	−0.02
20	40	2.04	2.00	0.04
40	30	0.75	0.75	0.00
40	60	1.52	1.50	0.02

28.4. Variation and Variance

No matter whether a deviation is from a mean or from an objective value, it is plus, minus, or zero. When there are many deviations with different magnitudes, the magnitude as a whole is expressed by the sum of the deviations squared or by the mean of the sum squared.

The sum of the deviations squared is called *variation*, and its mean is called *variance*. In some books, the sum of the deviations squared is called the *sum of squares*. In using either variation or variance, the magnitude of the data change is obtained quantitatively.

Letting n measurements be y_1, y_2, ... , y_n and their objective value be y_0, the sum of squares of $(y_1 - y_0)$, $(y_2 - y_0)$, is called the *total sum squared* (or *total variation*), denoted by S_T:

$$S_T = (y_1 - y_0)^2 + (y_2 - y_0)^2 + \cdots + (y_n - y_0)^2 \qquad (28.10)$$

The number of independent squares in the total variation, S_T, is called the number of *degrees of freedom*. The number of degrees of freedom in equation (28.10) is therefore n.

The deviation of the resistance of carbon in Section 28.2 varied from -0.5 to $+1.0$. To express these different values by a simple figure, the sum of these deviations squared is calculated:

$$S_T = (0.3)^2 + (-0.1)^2 + (0.5)^2 + \cdots + (0.6)^2$$

$$= 0.09 + 0.01 + 0.25 + \cdots + 0.36$$

$$= 2.23 \qquad (28.11)$$

The deviations in equation (28.11) are from an objective value of 10 kΩ; its number of degrees of freedom is 12. The total variation, S_T, of Table 28.1 is

$$S_T = (0.10)^2 + (-0.02)^2 + (0.04)^2 + (0.00)^2 + (0.02)^2$$

$$= 0.0124 \qquad (28.12)$$

Its number of degrees of freedom is 5.

Let the mean of n measurements be $\bar{y}$. The sum of the deviation squared from the mean would correctly be called the sum of deviations squared from the mean, but it is usually called the *total variation* (or the *total sum squared*).

$$S_T = (y_1 - \bar{y})^2 + (y_2 - \bar{y})^2 + \cdots + (y_n - \bar{y})^2 \qquad (28.13)$$

There are n squares in equation (28.13); however, its number of degrees of freedom is $n - 1$. This is because there exist the following linear relationships among n deviations:

$$(y_1 - \bar{y}) + (y_2 - \bar{y}) + \cdots + (y_n - \bar{y}) = 0 \qquad (28.14)$$

Generally, the number of degrees of freedom for the sum of n number of squares is given by n minus the number of relational equations among the deviations before these deviations are squared.

There are n squares in the S_T equation (28.13), but there is a relational equation shown by (28.14); therefore, its number of degrees of freedom is $n - 1$.

When there are n measurements, the number of degrees of freedom of the sum of the deviations squared either from an objective value or from a theoretical value is n, and the number of degrees of freedom of the sum of the deviations squared from the mean, $\bar{y}$, is $n - 1$.

The sum of the deviations in equation (28.9) squared, denoted by S_T,

$$S_T = (-0.62)^2 + 7.78^2 + (-7.52)^2 + \cdots + (-10.32)^2$$

$$= 0.3844 + 60.5284 + 56.5504 + \cdots + 106.5204$$

$$= 514.9360 \tag{28.15}$$

is the sum of deviations (from the mean $\bar{y}$) squared with nine degrees of freedom.

To obtain a variation from the arithmetic mean, $\bar{y}$, it is easier to calculate by way of subtracting a working mean from $y_1, y_2, \ldots, y_n$, calculating their sum squared, then subtracting the correction factor:

$$S_T = (y_1 - \bar{y})^2 + (y_2 - \bar{y})^2$$

$$= (y_1')^2 + (y_2')^2 + \cdots$$

$$+ (y_n')^2 - \frac{(y_1' + y_2' + \cdots + y_n')^2}{n} \tag{28.16}$$

The measurements $y_1', y_2', \ldots, y_n'$ are the values subtracted by a working mean. It is relatively easy to prove equation (28.16). In equation (28.16),

$$CF = \frac{(y_1' + y_2' + \cdots + y_n')^2}{n} \tag{28.17}$$

is called the *correction factor*, usually denoted by CF or CT.

The total variation of the heights is calculated using equation (28.16) with a working mean of 160.0 cm:

$$S_T = \text{(sum of the data after subtracting the working mean squared)}$$

$$- \text{(sum of the data after subtracting the working mean)}^2$$

$$= (163.2 - 160.0)^2 + (171.6 - 160.0)^2 + \cdots + (153.5 - 160.0)^2$$

$$- \frac{[(163.2 - 160.0) + (171.6 - 160.0) + \cdots + (153.5 - 160.0)]^2}{10}$$

$$= 3.2^2 + 11.6^2 + \cdots + (-6.5)^2 - \frac{[3.2 + 11.6 + \cdots + (-6.5)]^2}{10}$$

$$= 660.32 - \frac{38.2^2}{10}$$

$$= 660.32 - 145.924$$

$$= 514.396 \tag{28.18}$$

Usually, a correction factor or a variation is calculated as far as the lowest unit after squaring its original data. In this case, the calculation is made to the second decimal place, so the third decimal place and the lower units are rounded.

$$S_T = 660.32 - 145.924 = 514.40 \qquad (f = 9) \tag{28.19}$$

The correction factor is

$$CF = \frac{38.2^2}{10} = 145.92 \tag{28.20}$$

Next, we discuss *variance*, the value of variation divided by the degrees of freedom:

$$\text{variance} = \frac{\text{variation}}{\text{degrees of freedom}} \tag{28.21}$$

Variance signifies the mean of deviation per unit degree of freedom. It really means the magnitude of error, or mistake, or dispersion, or change per unit. This measure corresponds to work or energy per unit time in physics.

Letting $y_1, y_2, \ldots, y_n$ be the results of n measurements, and letting an objective value be y_0, the variance of these measurements is

$$V = \frac{(y_1 - y_0)^2 + (y_2 - y_0)^2 + \cdots + (y_n - y_0)^2}{n} \tag{28.22}$$

The number of degrees of freedom of total variation S_T in equation (28.11) is 12; then its variance is

$$V = \frac{S_T}{\text{degrees of freedom}}$$

$$= \frac{2.23}{12}$$

$$= 0.186 \tag{28.23}$$

Sometimes the value above is called the *error variance* with 12 degrees of freedom. It is recommended that one carry out this calculation to one more decimal place. This is shown in equation (28.23), which is calculated to the third decimal place.

When the variation is calculated as the sum of the deviations squared from the arithmetical mean, as in the case of height, variance is obtained as variation divided by the degrees of freedom. Letting the arithmetical mean of $y_1, y_2, \ldots, y_n$ be $\bar{y}$, the variance, V, is then

$$V = \frac{(y_1 - \bar{y})^2 + (y_2 - \bar{y})^2 + \cdots + (y_n - \bar{y})^2}{(n - 1)}$$

$$= \frac{S_T}{n - 1} \tag{28.24}$$

In cases when there is an objective value of a theoretical value, deviations of each datum from the objective or from the theoretical value are calculated. For example, some carbon resistances were produced with an objective value of y_0 kilohms. In this case, $(y_1 - y_0), (y_2 - y_0), \ldots, (y_n - y_0)$ are discussed. Another example would be the differences between the time indicated by a certain watch and the actual time. The data would then show the time that the watch either gained or lost.

Letting the deviations from an objective or a theoretical value be $y_1, y_2, \ldots, y_n$, and the mean be $\bar{y}$,

$$\bar{y} = \frac{y_1 + y_2 + \cdots + y_n}{n} \tag{28.25}$$

This would then be called *deviation*, or more correctly, the *estimate of deviation*. When the initial observational values before subtracting an objective or a theoretical value are always larger than the objective value, deviation $\bar{y}$ in (28.25) would be a large positive value, or vice versa.

In the case of carbon resistance discussed earlier, the deviations from the objective value are

0.3   −0.1   0.5   1.0   1.0   0.0   0.3   0.2   0.3   −0.2   −0.5   0.1   0.6

The deviation $\bar{y}$ is

$$\bar{y} = \frac{1}{12}(0.3 - 0.1 + 0.5 + \cdots + 0.6) = 2.5$$

$$= 0.208 \tag{28.26}$$

It is also recommended that one calculate $\bar{y}$ to one or two more decimal places.

A deviation that shows a plus value indicates that the resistance values of carbon are generally larger than the objective value. To express the absolute magnitude of a deviation, no matter whether it is plus or minus, $\bar{y}$ is squared and then multiplied by n:

$$\text{magnitude of deviation} = n(\bar{y})^2$$

$$= n\frac{(y_1 + y_2 + \cdots + y_n^2)}{n} \tag{28.27}$$

This value signifies the variation of the mean, m, of $y_1, y_2, \ldots, y_n$, which are the deviations from an objective value or from a theoretical value.

The following decomposition is therefore derived:

$$y_1^2 + y_2^2 + \cdots + y_n^2 = \frac{(y_1 + y_2 + \cdots + y_n)^2}{n}$$

$$+ [(y_1 - \bar{y})^2 + (y_2 - \bar{y})^2 + \cdots + (y_n - \bar{y})^2] \tag{28.28}$$

The relational equation above is written as

sum of data squared = (variation of the mean of deviations)
                + (sum of the deviations from the mean squared)(28.29)

(*Note:* When the differences between an objective value or from a theoretical value are discussed, and letting these differences be $y_1, y_2, \ldots, y_n$, the variation of the mean of deviations must not be calculated from the data that were subtracted by a working mean. The variation of the mean of deviations must be calculated from the differences between the original data and either an objective value or a theoretical value.)

If the sum of the carbon resistances, S_T, in (28.11) is 2.23, then S_m, the variation of the mean (m), is

$$S_m = \frac{(0.3 - 0.1 + \cdots + 0.6)^2}{12}$$

$$= \frac{2.5^2}{12}$$

$$= 0.52 \tag{28.30}$$

Therefore, the equation

$$2.23 = 0.52 + (\text{sum of deviations from mean } \bar{y} \text{ squared}) \tag{28.31}$$

is calculated. The sum of deviations from the mean squared, S_e, is given as

$$S_e = S_T - S_m = 2.23 - 0.52 = 1.71 \tag{28.32}$$

S_e is called the *error variation*. The number of degrees of freedom of error variation is $12 - 1 = 11$.

Error variation divided by its degrees of freedom is denoted by V_e. Using the example of resistance,

$$V_e = \frac{S_e}{\text{degrees of freedom}} = \frac{1.71}{11} = 0.155 \tag{28.33}$$

where V_e would signify the extent of dispersion of individual carbon resistances, a large V_e value means a large difference between individual products. On the other hand, a large variation of mean, or a large S_m, shows that the mean value of the products is considerable apart from the objective value.

It is easy in many cases to reduce S_m, the variation of a mean, when it is large. In the case of carbon resistance, for example, it is possible to reduce the magnitude, either by prolonging the time of carbonation or by adjusting the cutting method for carbon.

It is the same for some other dimensional characteristics; such deviations are easily adjustable. However, many contrivances are required in order to reduce dispersion. In the case of height, the total variation, S_T, is calculated as the variation of deviations from the mean, since there is neither objective value nor theoretical value in height. Sometimes zero is taken as a theoretical value. In such case, let $y_1, y_2, \ldots, y_n$ be the differences from zero. The magnitude of deviations, S_m, is calculated from $y_1, y_2, \ldots, y_n$.

The relationship in equation (28.29) corresponds to a concept such as the electric current of the telephone, where direct current and alternating current are mixed. This would be expressed as follows:

$$\text{power of the total current} = (\text{square of the dc voltage}) + (\text{ac power})$$

$\bar{y}$ is the dc voltage (48 V in the case of a telephone), and the ac power is equal to the square of the ac voltage. It is very important to recognize that there is no additivity in voltage, but there is additivity in power, which is the square of voltage. Such decomposition is called *power spectrum analysis* in engineering.

Such concepts become easier to understand as one uses such calculations. In the case of telephone current, the dc voltage has nothing to do with the transmission of voice, whereas the ac voltage does.

When we discuss deviations from an objective or a theoretical value, the variation of mean (S_m) and the remaining part of variation are entirely different, both in technical meaning and in countermeasures that are to be taken. It is very important to distinguish between the two.

29

Introduction to Analysis of Variance

29.1. Introduction   515
29.2. Decomposition of Variation   515
29.3. Analysis-of-Variance Table   521

29.1. Introduction

In this chapter, linear equations of observational data with constant coefficients and the method of calculating their variations are explained. Analysis of variance is also introduced. This chapter is based on Genichi Taguchi et al., *Design of Experiments*. Tokyo: Japanese Standards Association, 1973.

29.2. Decomposition of Variation

Letting n observational values be $y_1, y_2, \ldots, y_n$, the equation for $y_1, y_2, \ldots, y_n$ with constant coefficients is called a *linear equation*. Let n coefficients be $c_1, c_2, \ldots, c_n$; also let the linear equation be L:

$$L = c_1 y_1 + c_2 y_2 + \cdots + c_n y_n \tag{29.1}$$

Some of the coefficients $c_1, c_2, \ldots, c_n$, could be zero, but assume that not all the coefficients are zero.

The sum of these coefficients squared is denoted by D, which is the number of units comprising the linear equation, L. D is

$$D = c_1^2 + c_2^2 + \cdots + c^{2n} \tag{29.2}$$

For example, the mean (deviation) of n measurements, denoted by $\bar{y}$, is

$$\bar{y} = \frac{y_1 + y_2 + \cdots + y_n}{n}$$

$$= \frac{1}{n} y_1 + \frac{1}{n} y_2 + \cdots + \frac{1}{n} y_n \tag{29.3}$$

Equation (29.3) is a linear equation in which all the coefficients are identical.

$$c_1 = c_2 = \cdots = c_n = \frac{1}{n} \qquad (29.4)$$

Its number of units is

$$D = \left(\frac{1}{n}\right)^2 n = \frac{1}{n} \qquad (29.5)$$

The square of a linear equation divided by its number of units forms a variation with one degree of freedom. Letting it be S_L, then

$$S_L = \frac{L^2}{c_1^2 + c_2^2 + \cdots + c_n^2} = \frac{L^2}{D} \qquad (29.6)$$

The sum of $y_1, y_2, \ldots, y_n$ is expressed by a linear equation with $c_1 = c_2 = \cdots = c_n = 1$, with n units. Variation S_L is written from equation (29.6) as

$$S_L = \frac{(y_1 + y_2 + \cdots + y_n)^2}{n} \qquad (29.7)$$

The variation corresponds to variation S_m, which is the correction factor. Following is an example to clarify linear equations and their variation.

□ Example 1

There are six Japanese and four Americans with the following heights in centimeters:

A_1 (Japanese): 158, 162, 155, 172, 160, 168

A_2 (American): 186, 172, 176, 180

The quality of total variation among the 10 persons is given by the sum of deviations (from the mean) squared. Since there is neither an objective value nor a theoretical value, a working mean of 170 is subtracted and the total variation among the 10 persons is calculated.

□ Deviations from the working mean for A_1: $-12, -8, -15, 2, -10, -2$
Total: -45
□ Deviations from the working mean for A_2: 16, 2 , 6, 10
Total: 34

Therefore,

$$CF = \frac{(-45 + 34)^2}{10} = \frac{(-11)^2}{10} = 12 \qquad (29.8)$$

$$S_T = (-12)^2 + (-8)^2 + \cdots + 10^2 - CF = 937 - 12 = 925 \qquad (29.9)$$

Observing the data above, one must understand that the major part of the total variation of heights among the 10 persons is caused by the difference between the Japanese and the Americans. How much of the total variation of heights among the 10 persons, which was calculated to be 925, was caused by the differences between the two nationalities?

Assume that everyone expresses the difference between the mean heights of the Japanese and the Americans as follows:

$$L = \text{(mean of Japanese)} - \text{(mean of Americans)}$$

$$= \left(170 + \frac{-45}{6}\right) - \left(170 + \frac{34}{4}\right) = \frac{-45}{6} - \frac{34}{4} \qquad (29.10)$$

Usually, the total of the Japanese (A_1) is denoted by the same symbol, A_1, and the total of the Americans (A_2), by A_2. Thus,

$$L = \frac{A_1}{6} - \frac{A_2}{4} \qquad (29.11)$$

Now let the heights of the 10 persons by $y_1, y_2, y_3, y_4, y_5, y_6$ (Japanese), y_7, y_8, y_9, y_{10} (Americans). The total variation, S_T, is the total variation of the 10 measurements of height $y_1, y_2, \dots , y_{10}$. Equations (29.10) and (29.11) are the linear equations of 10 observational values $y_1, y_2, \dots , y_{10}$ with constant coefficients.

Accordingly, the variation S_L for equation (29.11) is

$$S_A = S_L = \frac{(A_1/6 - A_2/4)^2}{(1/6)^2(6) + (-1/4)^2(4)} = \frac{(-45/6 - 34/4)^2}{1/6 + 1/4}$$

$$= \frac{[(4)(-45) - (6)(34)]^2}{(4^2)(6) + (-6)^2(4)} = \frac{(-384)^2}{240} = 614 \qquad (29.12)$$

It is seen that with the total variation among 10 heights, which in this case is 925, the difference between Japanese and Americans can be as much as 614. By subtracting 614 from 925, we then identify the magnitude of individual differences within the same nationality. This difference is called the *error variation*, denoted by S_e:

$$S_e = S_T - S_A = 925 - 614 = 311 \qquad (29.13)$$

The degrees of freedom are 9 for S_T, 1 for S_A, and $(9 - 1 = 8)$ for S_e. Therefore, the error variance V_e is

$$V_e = \frac{S_e}{8} = \frac{311}{8} = 38.9 \qquad (29.14)$$

Error variance shows the height differences per person within the same nationality.

The square-root value of error variance (V_e) is called the *standard deviation* of individual difference of height, usually denoted by s:

$$s = \sqrt{V_e} = \sqrt{38.9} = 6.24 \text{ cm} \qquad (29.15)$$

In this way, the total variation of height among the 10 persons, S_T, is decomposed or separated into the variation of the difference between the Japanese and the Americans and the variation of the individual differences.

$$S_T = S_A + S_e \tag{29.16}$$

$$925 = 614 + 311 \quad \text{(variation)}$$

$$9 = 1 + 8 \quad \text{(degrees of freedom)} \tag{29.17}$$

Since there are 10 persons, there must be 10 individual differences; otherwise, the data are contradictory. In S_e, however, there are only eight individual differences. To make the total of 10, one difference is included in correction factor CF, and one difference in S_A.

In general, one portion of the individual differences is included in S_A, which equaled 614. Therefore, the real differences between nations must be calculated by subtracting a portion of the individual differences (which is the error variance) from S_A. The result after subtracting the error variance is called the *pure variation* between Japanese and Americans, distinguished from the original variation using a prime:

$$S'_A = S_A - V_e = 614 - 38.9 = 575.1 \tag{29.18}$$

Another portion of the individual differences is also in the correction factor. But in the case of the variation in height, this portion does not provide useful information. However, it will be shown in Example 2 where the correction factor does indicate some useful information. For this reason, the total quantity of variation is considered to have nine degrees of freedom.

Accordingly, the pure variation of error, S'_e, is calculated by adding the error variance, V_e (which is the unit portion subtracted from S_A), to S_e:

$$S'_e = S_e + V_e = 311 + 38.9 = 349.9 \tag{29.19}$$

Then the following relationship is concluded:

$$S_T = S'_A + S'_e \tag{29.20}$$

$$925 = 575.1 + 349.9 \tag{29.21}$$

Each term of equation (29.20) is divided by S_T and then multiplied by 100. This calculation is called the *decomposition* of the degrees of contribution.

In this case, the following relationship is concluded:

$$100 = \rho_A + \rho_e$$

$$= \frac{575.1}{925}(100) + \frac{349.9}{925}(100) = 62.2 + 37.8 \tag{29.22}$$

Equation (29.22) shows that the total varying quantity of 10 heights is equal to 925; the differences between the Japanese and American nationalities comprises 62.2% of the cause, and the individual differences consitutes 37.8%

In this way, the decomposition of variation is necessary to calculate when each data value changes. This is to express the total change by a total variation and then calculate the degrees of contribution, which is the influence of individual causes among the total variation, S_T.

To decompose variations at ease, one must master the calculation of the analysis of variance, which, mathematically, is the expression of variation in quadratic form. However, the most important thing is the cause of the variation. Possible causes must first be suggested by a researcher in the specialized field being studied. Without knowledge of the field, one cannot solve the problem of analysis of variance, no matter how well he or she knows the mathematics of quadratic equations.

It is important for a research worker to become acquainted with the analysis of variance, to have the ability to decompose variations for any type of data, and to be able to quickly calculate the magnitude of the influence of the cause being considered. Next, an example using an objective or theoretical value is shown.

❐ Example 2

To improve the antiabrasive properties of a product, two versions were manufactured on a trial basis.

    A_1:   product with no addictive in the raw material

    A_2:   product additive with a particular in the raw material

Six test pieces were prepared from the two versions, and wear tests were carried out under identical conditions. The quantities of wear (mg) were as follows:

    A_1:   26, 18, 19, 21, 15, 29

    A_2:   15, 8, 14, 13, 16, 9

In a case such as abrasion, quantity of deterioration, or percent of deterioration, the objective value would be zero. Accordingly, the correction factor is calculated from the original data:

$$A_1 = 26 + 18 + \cdots + 29 = 128 \tag{29.23}$$

$$A_2 = 15 + 8 + \cdots + 9 = 75 \tag{29.24}$$

$$T = A_1 + A_2 = 203 \tag{29.25}$$

Therefore, the correction factor would be

$$CF = \frac{203^2}{12} = 3434 \tag{29.26}$$

When there is an objective value, the correction factor CF is also called the *variation of the general mean*, denoted by S_m. It signifies the magnitude of the average quantity of abrasion of all the data.

The difference between averages A_1 and A_2, denoted by L, is

$$L = \frac{A_1}{6} - \frac{A_2}{6} = \frac{A_1 - A_2}{6} \tag{29.27}$$

The variation of the differences between A_1 and A_2, or S_A, is calculated from the square of linear equation L divided by the sum of the coefficients of L squared.

$$S_A = S_L = \frac{L^2}{\text{sum of coefficients squared}}$$

$$= \frac{[(A_1 - A_2)/6]^2}{(1/6)^2(6) + (-1/6)^2(6)} = \frac{(A_1 - A_2)^2}{(1^2)(6) + (-1)^2(6)} = \frac{(A_1 - A_2)^2}{12}$$

$$= \frac{(128 - 75)^2}{12} = \frac{53^2}{12} = \frac{2809}{12} = 234 \tag{29.28}$$

The objective value in this case is zero, so the total variation is equal to the total sum squared.

$$S_T = 26^2 + 18^2 + \cdots + 9^2 = 3859 \tag{29.29}$$

Error variation is then

$$S_e = S_T - CF - S_A = 3859 - 3434 - 234 = 191 \tag{29.30}$$

Degrees of freedom are 12 for S_T, 1 for the correction factor (CF or general mean S_m), and $(12 - 2) = 10$ for the error variation S_e.

Accordingly, the decomposition of the variation as well as the degrees of freedom are concluded as follows:

$$S_T = S_m + S_A + S_e \tag{29.31}$$

$$3859 = 34.34 + 234 + 191 \quad \text{(variation)} \tag{29.32}$$

$$12 = 1 + 1 + 10 \quad \text{(degrees of freedom)} \tag{29.33}$$

Decomposition of the degrees of contribution is made as

$$S_m' = S_m - V_e = 3434 - \frac{S_e}{10} = 3434 - 19.1 = 3414.9 \tag{29.34}$$

$$S_A' = S_A - V_e = 234 - 19.1 = 214.9 \tag{29.35}$$

Also,

$$S_e' = S_e + 2V_e = 191 + (2)(19.1) = 229.2 \tag{37.36}$$

Degrees of contribution are

$$\rho_m = \frac{S'_m}{S_T} = \frac{3414.9}{3859} = 0.885 \qquad (29.37)$$

$$\rho = \frac{S'_A}{S_T} = \frac{214.9}{3859} = 0.056 \qquad (29.38)$$

$$\rho_e = \frac{S'_e}{S_T} = \frac{229.2}{3859} = 0.059 \qquad (29.39)$$

Therefore,

$$100 = \rho_m + \rho_A + \rho_e = 88.5 + 5.6 + 5.9\% \qquad (29.40)$$

29.3. Analysis-of-Variance Table

The results of the decomposing variation shown in Section 29.2 are arranged in an analysis of variance table. To affirm the existence of correction factor or the difference of A qualitatively, a *significance test* is made prior to calculation of the degrees of contribution in the traditional analysis of variance. In quality engineering, however, the significance is observed from the degrees of contribution. It is a basic rule to calculate the degrees of contribution only for those causes (called *significant factorial effects*) indicated by asterisks. However, we should not make light of insignificant factoral effects whose degrees of contribution are large. There is a high possibility that those factor effects will have a substantial influence on the result. An insignificant result is obtained in two situations: In one, the effect is really nonexistent; in the other, an effect does exist but there is insufficient evidence to affirm the significance associated with the small number of degrees of freedom of the error variance. Calculation of the degrees of contribution is shown

Table 29.1
ANOVA table

Source	Degrees of Freedom, f	Variation, S	Variance, V	Pure Variation, S'	Degrees of Contribution, ρ (%)
m	1	CF	V_m	S'_m	ρ_m
A	1	S_A	V_A	S'_A	ρ_A
e	$n-2$	S_e	V_e	S'_e	ρ_e
Total	n	S_T		S_T	100.0

Table 29.2
ANOVA table for Example 2

Source	Degrees of Freedom, f	Variation, S	Variance, V	Pure Variation, S'	Degrees of Contribution, ρ (%)
m	1	3434	3434	3414.9	88.5
A	1	234	234	214.9	5.6
e	10	191	19.1	229.2	5.9
Total	12	3859		3859.0	100.0

in Table 29.1, an *analysis of variance (ANOVA) table.* The symbols in the table are defined as follows:

$$V_m = \frac{CF}{1} = CF \tag{29.41}$$

$$V_A = \frac{S_A}{1} = S_A \tag{29.42}$$

$$S'_m = CF - V_e \tag{29.43}$$

$$S'_A = S_A - (\text{degrees of freedom for } A)V_e = S_A - V_e \tag{29.44}$$

$$S'_e = S_e + 2V_e \tag{29.45}$$

$$\rho_m = \frac{S'_m}{S_T} \times 100 \tag{29.46}$$

$$\rho_A = \frac{S'_A}{S_T} \times 100 \tag{29.47}$$

$$\rho_e = \frac{S'_e}{S_T} \times 100 \tag{29.48}$$

The calculating procedures above constitute the analysis of variance. The results for Example 2 are shown in Table 29.2.

30 One-Way Layout

30.1. Introduction   523
30.2. Equal Number of Repetitions   523
30.3. Number of Repetitions Is Not Equal   527

30.1. Introduction

One-way layout is a type of experiment with one factor with k conditions (or k levels). The number of repetitions of each level may be the same or may be different; they are shown as n_1, n_2, ... , n_k.

The experiments based on a k value of 2 were described in Chapter 29. In this chapter, experiments with k equal to or larger than 3 are described. For a more precise comparison, it is desirable that the number of repetitions in each level be the same. However, if this is not possible, the number of repetitions in each level can vary. This chapter is based on Genichi Taguchi et al., *Design of Experiments*. Tokyo: Japanese Standards Association, 1973.

30.2. Equal Number of Repetitions

In pain-killer experiments, if the dosage of a new pain killer were changed in three ways, such as

$$A_2 = 1 \text{ g/day}$$

$$A_3 = 2 \text{ g/day}$$

$$A_4 = 4 \text{ g/day}$$

and the result compared with the existing pain killer A_1 (the standard dosage), there would be four levels for factor A.

Suppose that an experiment for factor A were carried out with the same number of repetitions (n) for each of the a levels: A_1, A_2, ... , A_a. The data would be tabulated as shown in Table 30.1.

Table 30.1
Data with n repetitions

Level of A	Data	Total
A_1	$y_{11}\ y_{12}\ \cdots\ y_{1n}$	A_1
A_2	$y_{21}\ y_{22}\ \cdots\ y_{2n}$	A_2
$\vdots$	$\vdots$	$\vdots$
Total	$y_{a1}\ y_{a2}\ y_{an}$	A_a

Suppose that datum y_{ij} denotes the deviation from the objective value, y_0. Variations are calculated from $A_1, A_2, \ldots, A_a$.

$$S_m = \frac{(A_1 + A_2 + \cdots + A_a)^2}{an} \qquad (f = 1) \tag{30.1}$$

$$S_A = \frac{A_1^2 + A_2^2 + \cdots + A_a^2}{n} - S_m \qquad (f = a - 1) \tag{30.2}$$

$$S_T = y_{11}^2 + y_{12}^2 + \cdots + y_{an}^2 \qquad (f = an) \tag{30.3}$$

$$S_e = S_T - S_m - S_A \qquad [f = a(n - 1)] \tag{30.4}$$

The ANOVA table is shown in Table 30.2. The symbols in the table signify the following:

$$V_m = \frac{S_m}{1} = S_m \tag{30.5}$$

$$V_A = \frac{S_A}{a - 1} \tag{30.6}$$

$$V_e = \frac{S_e}{a(n - 1)} \tag{30.7}$$

$$S_m' = S_m - V_e \tag{30.8}$$

$$S_A' = S_A - (a - 1)V_e \tag{30.9}$$

$$S_e' = S_e + aV_e \tag{30.10}$$

$$\rho_m = \frac{S_m'}{S_T} \tag{30.11}$$

$$\rho_A = \frac{S_A'}{S_T} \tag{30.12}$$

$$\rho_e = \frac{S_e'}{S_T} \tag{30.13}$$

Table 30.2
ANOVA table

Source	f	S	V	S'	ρ (%)
m	1	S_m	V_m	S'_m	ρ_m
A	$a - 1$	S_A	V_A	S'_A	ρ_A
e	$a(n - 1)$	S_e	V_e	S'_e	ρ_e
Total	an	S_T		S_T	100.0

Estimation is made as follows: When A is insignificant, only the general mean is estimated in most cases. That is to determine that the effect of A is inconsequential; accordingly, S_A is pooled with S_e to get a new pooled error variance (using the same symbol, V_e):

$$V_e = \frac{S_A + S_e}{(a - 1) + a(n - 1)} = \frac{S_A + S_e}{an - 1} \qquad (30.14)$$

❐ Example

In a pinhole manufacturing process, the order of processing was changed in three levels. The experiment was carried out with 10 repetitions for each level. The roundness of pinholes were measured, as shown in Table 30.3.

A_1: (penetrate the lower hole) – (ream the lower hole)
– (ream) – (process the leading hole)

A_2: (process the lower hole to half-depth) – (ream the lower hole) – (ream)
– (process the leading hole)

A_3: (process the leading hole) – (process the lower hole)
– (ream the lower hole) – (ream)

Table 30.3
Data for roundness (μm)

Level	Data	Total
A_1	10, 15, 3, 18, 8, 4, 6, 10, 0, 13	87
A_2	12, 14, 5, 6, 4, 1, 11, 15, 7, 10	85
A_3	8, 2, 0, 4, 1, 6, 5, 3, 2, 4	35
Total		207

There is an objective value (i.e., zero), in this case, so the general mean, m, is tested:

$$S_m = \frac{207^2}{30} = 1428 \tag{30.15}$$

$$S_A = \frac{A_1^2 + A_2^2 + A_3^2}{10} - S_m$$

$$= \frac{87^2 + 85^2 + 35^2}{10} - 1428 = \frac{16{,}019}{10} - 1428$$

$$= 1602 - 1428 = 174 \tag{30.16}$$

$$S_T = 10^2 + 15^2 + 3^2 + \cdots + 4^2 = 2131 \tag{30.17}$$

$$S_m' = 1428 - 19.6 = 1408.4 \tag{30.18}$$

$$S_A' = 174 - (2)(19.6) = 134.8 \tag{30.19}$$

$$S_e' = 529 + (3)(19.6) = 587.8 \tag{30.20}$$

$$\rho_m = \frac{S_m'}{S_T} = \frac{1408.4}{2131} = 0.661 \tag{30.21}$$

$$\rho_A = \frac{134.8}{2131} = 0.063 \tag{30.22}$$

$$\rho_e = \frac{587.8}{2131} = 0.276 \tag{30.23}$$

The analysis of variance is summarized in Table 30.4.

Since the general mean, m, and A are significant, estimation is made for each level of A.

$$\bar{A}_1 = \frac{A_1}{n} = \frac{87}{10} = 8.7 \tag{30.24}$$

Table 30.4
ANOVA table

Source	f	S	V	S'	ρ (%)
m	1	1428	1428	1408.4	66.1
A	2	174	87	134.8	6.3
e	27	529	19.6	587.8	27.6
Total	30	2131		2131.0	100.0

Table 30.5
ANOVA table

Source	f	S	V	S'	ρ (%)
m	1	S_m	V_m	S'_m	ρ_m
A	$a-1$	S_A	V_A	S'_A	ρ_A
e	$n-a$	S_e	V_e	S'_e	ρ_e
Total	n	S_T		S_T	100.0

Similarly,

$$\bar{A}_2 = \frac{85}{10} = 8.5 \tag{30.25}$$

$$\bar{A}_3 = \frac{35}{10} = 3.5 \tag{30.26}$$

30.3. Number of Repetitions Is Not Equal

When the numbers or repetitions, $n_1, n_2, \dots , n_1$, for an a-level factor A, $A_1, A_2, \dots ,$ A_a are not equal, the following calculation is made:

S_T = sum of $(n_1 + n_2 + \cdots + n_a)$ individual values squared

$$(f = n_1 + n_2 + \cdots + n_a) \tag{30.27}$$

$$S_m = \frac{(\text{total})^2}{n_1 + n_2 + \cdots + n_a} \quad (f = 1) \tag{30.28}$$

$$S_A = \frac{A_1^2}{n_1} + \frac{A_2^2}{n_2} + \cdots + \frac{A_a^2}{n_a} - S_m \quad (f = a - 1) \tag{38.29}$$

$$S_e = S_T - S_m - S_A \quad (f = n_1 + n_2 + \cdots + n_a - a) \tag{30.30}$$

The analysis of variance is shown in Table 30.5. In the table,

$$n = n_1 + n_2 + \cdots + n_a$$

31 Decomposition to Components with a Unit Degree of Freedom

31.1. Introduction   528
31.2. Comparison and Its Variation   528
31.3. Linear Regression Equation   534
31.4. Application of Orthogonal Polynomials   542

31.1. Introduction

In previous chapters, the total sum of the squares was decomposed into the sum of the variations of a correction factor, the main effect of a factor, and the error. This decomposition can be extended by looking at what the researcher feels could be causing the variations; or the total sum of the squares can be decomposed into the sum of the variations of causes with a unit degree of freedom. The latter method is described in this chapter. This chapter is based on Genichi Taguchi et al., *Design of Experiments*. Tokyo: Japanese Standards Association, 1973.

31.2. Comparison and Its Variation

The analysis of variance for the example in Section 30.2 showed that A is significant. This meant that the roundness differed by the order of the various processes used in making the pinholes. Sometimes we would like to know whether the difference was caused by the difference between A_1 and A_2, or between A_2 and the mean of A_1 and A_2.

Everyone knows to compare A_1 and A_2 by

$$L_1 = \frac{A_1}{10} - \frac{A_2}{10}$$

(31.1)

However, A_3 and the mean of A_1 and A_1 would be compared by using the following linear equation:

$$L_2 = \frac{A_1 + A_2}{20} - \frac{A_3}{10}$$

(31.2)

L_1 and L_2 are linear equations; their sum of the coefficients for A_1, A_2, and A_3 is equal to zero either in L_1 or in L_2.

$$L_1: \quad \frac{1}{10} - \frac{1}{10} = 0 \tag{31.3}$$

$$L_2: \quad \frac{1}{20} + \frac{1}{20} - \frac{1}{10} = 0 \tag{31.4}$$

Assume that the sums A_1, A_2, ... , A_a each has the same number of data (b) making up their respective total.

In a linear equation with constant coefficients A_1, A_2, ... , A_a,

$$L = c_1 A_1 + c_2 A_2 + \cdots + c_a A_a \tag{31.5}$$

when the sum of the coefficients is equal to zero,

$$c_1 + c_2 + \cdots + c_a = 0 \tag{31.6}$$

then L is called either *contrast* or *comparison*.

As described previously, in a linear equation with constant coefficients,

$$L = c_1 A_1 + c_2 A_2 + \cdots + c_a A_a \tag{31.7}$$

the variation of L, or S_L, is given by

$$S_L \quad \frac{(c_1 A_1 + \cdots + c_a A_a)^2}{(c_1^2 + c_1^2 + \cdots + c_a^2) b} \tag{31.8}$$

where S_L has one degree of freedom. Calculation and estimation of a contrast are made in exactly the same way.

In two contrasts,

$$L_1 = c_1 A_1 + \cdots + c_a A_a \tag{31.9}$$

$$L_{21} = c_1' A_1 + \cdots + c_a' A_a \tag{31.10}$$

when their sum of products is equal to zero,

$$c_1 c_1' + c_2 c_2' + \cdots + c_a c_a' = 0 \tag{31.11}$$

L_1 and L_2 are *orthogonal.* When L_1 and L_2 are orthogonal, each of

$$S_{L_1} = \frac{L_1^2}{(c_1^2 + \cdots + c_a^2) b} \tag{31.12}$$

$$S_{L_2} = \frac{L_2^2}{(c_1'2 + \cdots + c_a'2) b} \tag{31.13}$$

is variation having one degree of freedom; each variation consists of one of the components in S_A. Therefore, when $(a - 1)$ comparisons, $L_1, L_2, ... , L_{(a-1)}$, are orthogonal to each other, the following equation is used:

$$S_A = S_{L_1} + S_{L_2} + \cdots + S_{L(a-1)} \qquad (31.14)$$

☐ Example 1

In an example of pinhole processing,

$$L_1 = \frac{A_1}{10} - \frac{A_2}{10} \qquad (31.15)$$

$$L_2 = \frac{A_1 + A_2}{20} - \frac{A_3}{10} \qquad (31.16)$$

The orthogonality between L_1 and L_2 is proven by

$$\left(\frac{1}{10}\right)\left(\frac{1}{20}\right) + \left(-\frac{1}{10}\right)\left(\frac{1}{20}\right) + 0\left(-\frac{1}{10}\right) = 0 \qquad (31.17)$$

Therefore, the following relation can be made:

$$S_A = S_L + S_{L_2} \qquad (31.18)$$

where S_{L1} and S_{L2} are calculated from equation (31.8) as

$$S_{L_1} = \frac{(A_1/10 - A_2/10)^2}{[(1/10)^2 + (-1/10)^2](10)} = \frac{(1/10)^2 (A_1 - A_2)^2}{(1/10)^2 [1^2 + (-1)^2](10)}$$

$$= \frac{(A_1 - A_2)^2}{20}$$

$$= \frac{(87 - 85)^2}{20}$$

$$= 0.2 \qquad (31.19)$$

$$S_{L_2} = \frac{[(A_1 + A_2)/20 - A_3/10)]^2}{[(1/20)^2 + (1/20)^2 + (-1/10)^2](10)}$$

$$= \frac{(1/20)^2 (A_1 + A_2 - 2A_3)^2}{(1/20)^2(6)(10)}$$

$$= 173.4 \qquad (31.20)$$

Thus, the magnitude of the variation in roundness, which is caused by A_1, A_2, and A_3, namely, S_A, is decomposed into the variation caused by the difference between A_1 and A_2, namely, S_{L_1}, and that variation caused by the difference between A_3 and the mean of A_1 and A_2, namely, S_{L_2}:

$$S_{L_1} + S_{L_2} = 0.2 + 173.4 = 174 = S_A \qquad (31.21)$$

◻ Example 2

We have the following four types of products:

A_1: foreign products

A_2: our company's products

A_3: domestic: α company's products

A_4: domestic: β company's products

Two, ten, six, and six products were sampled from each type, respectively, and a 300-hour continuous deterioration test was made. The percent of deterioration (Table 31.1) was determined as follows:

$$y = \frac{(\text{value after the test}) - (\text{initial value})}{\text{initial value}} \tag{31.22}$$

$$S_m = \frac{T^2}{n} = \frac{488^2}{24}$$

$$= 9923 \tag{31.23}$$

$$S_A = \frac{A_1^2}{2} + \frac{A_2^2}{10} + \frac{A_3^2}{6} + \frac{A_4^2}{6} - S_m$$

$$= \frac{26^2}{2} + \frac{175^2}{10} + \frac{147^2 + 140^2}{6} - 9923$$

$$= 346 \tag{31.24}$$

$$S_T = 12^2 + 14^2 + 20^2 + \cdots + 24^2$$

$$= 10{,}426 \tag{31.25}$$

$$S_e = S_T - S_m - S_A$$

$$= 10{,}426 - 9923 - 346$$

$$= 157 \tag{31.26}$$

The analysis of variance is shown in Table 31.2.

Instead of making an overall comparison among the four products, we usually want to make a more detailed comparison, such as:

L_1: difference between foreign and domestic products

L_2: difference between our company and the other domestic products

L_3: difference between the other domestic companies' products

Table 31.1
Percent of deterioration

Level	Data	Total
A_1	12, 14	26
A_2	20, 18, 19, 17, 15, 16, 13, 18, 22, 17	175
A_3	26, 19, 26, 28, 23, 25	147
A_4	24, 25, 18, 22, 27, 24	140
Total		488

The comparison above can be made by using the following linear equations:

$$L_1 = \frac{A_1}{2} - \frac{A_2 + A_3 + A_4}{22}$$

$$= \frac{26}{2} - \frac{462}{22}$$

$$= -8.0 \qquad (31.27)$$

$$L_2 = \frac{A_2}{10} - \frac{A_3 + A_4}{12}$$

$$= \frac{175}{10} - \frac{287}{12}$$

$$= -6.4 \qquad (31.28)$$

$$L_3 = \frac{A_3 - A_4}{6}$$

$$= \frac{147 - 140}{6}$$

$$= 1.2 \qquad (31.29)$$

Table 31.2
ANOVA table

Source	f	S	V	S'	ρ (%)
m	1	9,923	9,923	9,915	95.1
A	3	346	115.3	322	3.0
e	20	157	7.85	189	1.8
Total	24	10,426		10,426	100.0

These equations are orthogonal to each other, and also orthogonal with the general mean,

$$L_m = \frac{A_1 + A_2 + A_3 + A_4}{24} \qquad (31.30)$$

For example, the orthogonality between L_m and L_1 is proven by

$$\left(\frac{1}{2}\right)\left(\frac{1}{24}\right)(2) + \left(-\frac{1}{22}\right)\left(\frac{1}{24}\right)(22) = 0 \qquad (31.31)$$

where the sum of product of the coefficients is zero. The orthogonality between L_1 and L_2 is proven by

$$\left(\frac{1}{2}\right)(0)(2) + \left(-\frac{1}{22}\right)\left(\frac{1}{10}\right)(10) + \left(-\frac{1}{22}\right)\left(-\frac{1}{12}\right)(12) = 0$$
$$(31.32)$$

After calculating the variations of L_1, L_2, and L_3, we obtain the following decomposition:

$$S_A = S_{L_1} + S_{L_2} + S_{L_3} \qquad (31.33)$$

$$S_{L_1} = \frac{L_1^2}{\text{sum of the coefficients squared}}$$

$$= \frac{[A_1/2 - (A_2 + A_3 + A_4)/22]^2}{(1/2)^2(2) + (-1/22)^2(22)}$$

$$= \frac{[11A_1 - (A_2 + A_3 + A_4)]^2}{(11^2)(2) + (-1)^2(22)}$$

$$= \frac{(286 - 462)^2}{264}$$

$$= 117 \qquad (31.34)$$

$$S_{L_2} = \frac{[A_2/10 - (A_3 + A_4)/12]^2}{(1/10)^2(10) + (-1/12)^2(12)}$$

$$= \frac{[6A_2 - 5(A_3 + A_4)]^2}{(6^2)(10) + (-5)^2(12)}$$

$$= \frac{(1050 - 1435)^2}{660} = 225 \qquad (31.35)$$

$$S_{L_2} = \frac{(A_3/6 - A_4/6)^2}{(1/6)^2(6) + (-1/6)^2(6)}$$

$$= \frac{(A_3 - A_4)^2}{12}$$

$$= 4 \qquad (31.36)$$

An ANOVA table with the decomposition of A into three components with a unit degree of freedom is shown in Table 31.3.

From the analysis of variance, it has been determined that there is a significant difference between L_1 and L_2, but none between the other domestic companies.

When there is no significance, as in this case, it is better to pool the effect with the error to show the miscellaneous effect. Pooling the effect of L_3 with the error, the degrees of contribution would then be 1.9%.

Where the error variance, V_e, is calculated from the pooled error variation,

$$V_e = \frac{157 + 4}{21}$$

$$= 7.67 \tag{31.37}$$

number of units = sum of coefficients squared

$$= \left(\frac{1}{2}\right)^2 (2) + \left(-\frac{1}{22}\right)^2 (22)$$

$$= \frac{1}{2} + \frac{1}{22}$$

$$= \frac{12}{22}$$

$$= \frac{6}{11} \tag{31.38}$$

The other confidence intervals were calculated in the same way. Since L_3 of equation (31.29) is not significant, it is generally not estimated.

31.3. Linear Regression Equation

The tensile strength of a product was measured at different temperatures, x_1, x_2, ... , x_n, to get y_1, y_2, ... , y_n.

Table 31.3
ANOVA table

Source	f	S	V	S'	ρ (%)
m	1	9,923	9,923	9,915	95.1
A					
L_1	1	117	117	109	1.0
L_2	1	225	225	217	2.0
L_3	1	4	4		
e	20	157	7.85	185	1.8
Total	24	10,426		10,426	100.0

The relationship of the tensile strength, y, to temperature, x, is usually expressed by a linear function:

$$a + bx = y \qquad (31.39)$$

Then n observational values (x_1, y_1), (x_2, y_2), ... , (x_n, y_n), are put in equation (31.39):

$$a + bx_i = y_i \qquad (i = 1, 2, ... , n) \qquad (31.40)$$

Equation (31.40) is called an *observational equation*. When the n pairs of observational values are put in the equation, there are n simultaneous equations with two unknowns, a and b.

When the number of equations exceeds the number of unknowns, a solution that perfectly satisfies both of these equations does not exist; however, a solution that minimizes the differences between both sides of the equations can be obtained. That is, to find a and b would minimize the residual sum of the squares or the differences between the two sides:

$$S_e = (y_1 - a - bx_1)^2 + (y_2 - a - bx_2)^2 + \cdots + (y_n - a - bx_n)^2 \quad (31.41)$$

This solution was named by K. F. Gauss and is known as the *least squares method*. It is obtained by solving the following normal equations:

$$na + \left(\sum x_i\right)b = \sum y_i \qquad (31.42)$$

$$\left(\sum x_i\right)a + \left(\sum x_i^2\right)b = \sum x_i y_i \qquad (31.43)$$

where

$n =$ sum of coefficients squared of a in (31.40); the coefficients are equal to 1

$$= 1^2 + 1^2 + \cdots + 1^2 \qquad (31.44)$$

$\left(\sum x_i\right) =$ sum of products of the coefficients of a and b in (31.40)

$$= 1x_1 + 1x_2 + \cdots + 1x_n \qquad (31.45)$$

$\left(\sum y_i\right) =$ sum of products of the coefficients of a and observational values of y in (31.40)

$$= 1y_1 + 1y_2 + \cdots + 1y_n \qquad (31.46)$$

$\left(\sum x_i^2\right) =$ sum of the coefficients squared of b in (31.40)

$$= x_1^2 + x_2^2 + \cdots + x_n^2 \qquad (31.47)$$

$\sum x_i y_i =$ sum of products of the coefficients of b and observational values of y in (31.40)

$$= x_1 y_1 + x_2 y_2 + \cdots + x_n y_n \qquad (31.48)$$

It is highly desirable that the reader be able to write equations (39.42) and (31.43) at any time. Their memorization should not be difficult.

To solve the simultaneous equations (31.42) and (31.43), x_i is multiplied by both sides of (31.42). Also, n has to be multiplied by both sides of (31.43). After subtracting the same sides of the two equations from each other, term a disappears and term b is left.

$$\left[\left(\sum x_i\right)^2 - n\sum\sum x_i^2\right] b = \left(\sum x_i\right)\left(\sum y_i\right) - n\sum x_i y_i \tag{31.49}$$

From this,

$$b = \frac{n\sum x_i y_i - \left(\sum x_i\right)\left(\sum y_i\right)}{n\left(\sum x_i^2\right) - \left(\sum x_i\right)^2} \tag{31.50}$$

and then dividing the denominator and the numerator by n, respectively,

$$b = \frac{x_1 y_1 + \cdots + x_n y_n - [(x_1 + \cdots + x_n)(y_1 + \cdots + y_n)/n]}{x_1^2 + \cdots + x_n^2 - \dfrac{(x_1 + \cdots + x_n)^2}{n}}$$

$$= \frac{S_T(xy)}{S_T(xx)} \tag{31.51}$$

where $S_T(xx)$ is the total variation of the temperature, x:

$$S_T(xx) = x_1^2 + x_2^2 + \cdots + x_n^2 - \frac{(x_1 + x_2 + \cdots + x_n)^2}{n} \tag{31.52}$$

$S_T(xy)$ is called the *covariance* of x and y and is determined by

$$S_T(xy) = x_1 y_1 + x_2 y_2 + \cdots + x_n y_n$$
$$- \frac{(x_1 + x_2 + \cdots + x_n)(y_1 + y_2 + \cdots + y_n)}{n} \tag{31.53}$$

In the equation of covariance, the square term in the equation of variation is substituted for by the product of x and y.

Putting b of (31.50) into (31.42), a is obtained:

$$a = \frac{1}{n}\left[\sum y_i - \frac{S_T(xy)}{S_T(xx)}\left(\sum x_i\right)\right] \tag{31.54}$$

It is known from (31.54) that when

$$\sum x_i = 0 \tag{31.55}$$

then

$$a = \frac{y_1 + \cdots + y_n}{n} = \bar{y} \tag{31.56}$$

Thus, the estimation of the unknowns a and b becomes very simple. For this purpose, equation (31.39) may be expanded as follows:

$$m + b(x - \bar{x}) = y \tag{31.57}$$

where $\bar{x}$ is the mean of $x_1, x_2, \ldots, x_n$. Such an expansion is called an *orthogonal expansion*.

The orthogonal expansion has the following meaning: Either the general mean, m, or the linear coefficient, b, is estimated from the linear equation of y. Using equation (31.57), the two linear equations are orthogonal; therefore, the magnitudes of their influences are easily evaluated.

In the observational equation

$$m + b(x_i - \bar{x}) = y_i \qquad (i = 1, 2, \ldots, n) \tag{31.58}$$

the sum of the products of the unknowns m and b,

$$\sum (x_i - \bar{x}) = 0$$

becomes zero. Accordingly, the normal equations become

$$nm + 0b = \sum y_i$$

$$0m + \left[\sum (x_i - \bar{x})^2 \right] b = \sum (x_i - \bar{x}) y_i \tag{31.59}$$

Letting the estimates of m and b be $\hat{m}$ and $\hat{b}$,

$$\hat{m} = \frac{y_1 + y_2 + \cdots + y_n}{n} \tag{31.60}$$

$$\hat{b} = \frac{\sum (x_i - \bar{x}) y_i}{\sum (x_i - \bar{x})^2}$$

$$= \frac{S_T(xy)}{S_T(xx)} \tag{31.61}$$

Not only $\hat{m}$, but also $\hat{b}$, is a linear equation of $y_1, y_2, \ldots, y_n$.

$$c_1 = \frac{x_1 - \bar{x}}{S_T(xx)}$$

$$c_2 = \frac{x_2 - \bar{x}}{S_T(xx)}$$

$$\vdots$$

$$c_n = \frac{x_n - \bar{x}}{S_T(xx)} \tag{31.62}$$

The number of units of the sum of the coefficients squared is

$$c_1^2 + c_2^2 + \cdots + c_n^2 = \left[\frac{x_1 - \bar{x}}{S_T(xx)}\right]^2 + \left[\frac{x_2 - \bar{x}}{S_T(xx)}\right]^2 + \cdots + \left[\frac{x_n - \bar{x}}{S_T(xx)}\right]^2$$

$$= \frac{1}{[S_T(xx)]^2}\left[(x_1 - \bar{x})^2 + (x_2 - \bar{x})^2 + \cdots (x_n - \bar{x})^2\right]$$

$$= \frac{S_T(xx)}{[S_T(xx)]^2}$$

$$= \frac{1}{S_T(xx)} \tag{31.63}$$

The variations of m, b, and error are

$$S_m = CF = \frac{(y_1 + \cdots + y_n)^2}{n}$$

$$S_b = \frac{(b)^2}{\text{no. of units}}$$

$$= \frac{[S_T(xy)/S_T(xx)]^2}{1/S_T(xx)}$$

$$= \frac{S_T(xy)^2}{S_T(xx)} \tag{31.64}$$

$$S_e = y_1^2 + y_2^2 + \cdots + y_n^2 - S_m - S_b \tag{31.65}$$

The number of degrees of freedom is 1 for S_m or S_b, and $n - 2$ for S_e.

□ **Example**

To observe the change of tensile strength of a product, given a change in temperature, the tensile strength of two test pieces was measured at four different temperatures, with the following results (kg/mm²):

A_1 (0°C):   84.0,   85.2

A_2 (20°C):   77.2,   76.8

A_3 (40°C):   67.4,   68.6

A_4 (60°C):   58.2,   60.4

If there is no objective value, such as a given specification, the degrees of freedom of the total variation is 7, where the degree of freedom for the correction factor is not included.

Subtracting a working mean, 70.0, the data in Table 31.4 are obtained.

$$CF = \frac{17.8^2}{8} = 39.60 \tag{31.66}$$

$$S_T = 14.0^2 + 15.2 + \cdots + (9.6)^2 - CF = 765.24 - 39.60$$

$$= 725.64 \tag{31.67}$$

Assuming that tensile strength changes in the same way as the linear function of temperature, A, the observational equations will become

$$y = m + b(A - \bar{A}) \tag{31.68}$$

where A signifies temperature and $\bar{A}$ represents the mean value of the various temperature changes:

$$\bar{A} = \tfrac{1}{8}[(2)(0) + (2)(20) + (2)(40) + (2)(60)] = 30°C \tag{31.69}$$

In the linear equation, m is a constant and b is a coefficient indicating how much the tensile strength decreases with a 1°C temperature change. The actual observational values are put into equation (31.68), as follows:

$$m + b(0 - 30) = 84.0$$

$$m + b(0 - 30) = 85.2$$

$$m + b(20 - 30) = 77.2$$

$$m + b(20 - 30) = 76.8 \tag{31.70}$$

$$m + b(40 - 30) = 67.4$$

$$m + b(40 - 30) = 68.6$$

$$m + b(60 - 30) = 58.2$$

$$m + b(60 - 30) = 60.4$$

Table 31.4
Data after subtracting a working mean

Level	Data		Total
A_1	14.0,	15.2	29.2
A_2	7.2,	6.8	14.0
A_3	−2.6,	−1.4	−4.0
A_4	−11.8,	−9.6	−21.4
Total			17.8

The unknowns are m and b, and there are eight equations. Therefore, the least squares method is used to find m and b.

$$n = \text{sum of the coefficients squared of } m \qquad (31.71)$$
$$= 8$$

$$\left(\sum x_i \right) = \text{sum of the coefficients of } b$$
$$= (-30) + (-30) + (-10) + (-10) + 10 + 10 + 30 + 30$$
$$= 0 \qquad (31.72)$$

$$\left(\sum y_i \right) = \text{sum of products of } y \text{ and the coefficients of } m$$
$$= 84.0 + 85.2 + \cdots + 60.4$$
$$= (70.0)(8) + 17.8$$
$$= 577.8 \qquad (31.73)$$

$$\left(\sum x_i^2 \right) = \text{sum of the coefficients squared of } b$$
$$= (-30)^2 + (-30)^2 + (-10)^2 + (-10)^2 + 10^2 + 10^2 + 30^2 + 30^2$$
$$= 4000 \qquad (31.74)$$

$$\left(\sum x_i y_i \right) = \text{sum of products of } y \text{ and the coefficients of } b$$
$$= (-30)(84.0) + (-30)(85.2) + \cdots + (30)(58.2) + (30)(60.4)$$
$$= (-30)(169.2) + (-10)(154.0) + (10)(136.0) + (30)(118.6)$$
$$= -1698.0 \qquad (31.75)$$

The following simultaneous equations are then obtained:

$$8m + 0b = 577.8$$
$$0m + 4000b = -1698.0 \qquad (31.76)$$

Solving these yields

$$\hat{m} = \frac{577.8}{8}$$
$$= 72.22 \qquad (31.77)$$

$$\beta = \frac{-1698.0}{4000}$$

$$= -0.4245 \tag{31.78}$$

The orthogonality of these equations is proved as follows: Let the eight observational data be $y_1, y_2, \ldots, y_8$.

$$\hat{m} = \frac{y_1 + y_2 + \cdots + y_8}{8} \tag{31.79}$$

$$\beta = \frac{-30(y_1 + y_2) - 10(y_3 + y_4) + 10(y_5 + y_6) + 30(y_7 + y_8)}{4000}$$

$$= \frac{-3(y_1 + y_2) - (y_3 + y_4) + (y_5 + y_6) + 3(y_7 + y_8)}{400} \tag{31.80}$$

The sum of products of the corresponding coefficients of $\hat{m}$ and β is

$$2\left[\left(\frac{1}{8}\right)\left(\frac{-3}{400}\right) + \left(\frac{1}{8}\right)\left(\frac{-1}{400}\right) + \left(\frac{1}{8}\right)\left(\frac{3}{400}\right)\right] = 0 \tag{31.81}$$

The variation of m, S_m, is identical to the correction factor.

$$S_m = CF = \frac{577.8^2}{8}$$

$$= 41,731.60 \tag{31.82}$$

$$S_b = \frac{S_r(xy)^2}{S_r(xx)}$$

$$= \frac{(-1698.0)^2}{4000}$$

$$= 720.80 \tag{31.83}$$

The total sum of the observational values squared, S_T, is

$$S_T = 84.0^2 + 85.2^2 + \cdots + 60.4^2$$

$$= 42,457.24 \tag{31.84}$$

The error variation, S_e, is then

$$S_e = S_T - S_m - S_b$$

$$= 42,457.24 - 41,731.60 - 720.80$$

$$= 4.84 \tag{31.85}$$

The error variance, with 6 degrees of freedom, is

$$V_e = \frac{S_e}{6}$$

$$= \frac{4.84}{6}$$

$$= 0.807 \tag{31.86}$$

The pure variations of the general mean and the linear coefficient, b, are

$$S'_m = S_m - V_e$$

$$= 41,731.60 - 0.807$$

$$= 41,730.793 \tag{31.87}$$

$$S'_b = S_b - V_e$$

$$= 720.80 - 0.807$$

$$= 719.993 \tag{31.88}$$

Degrees of contributions are calculated as follows:

$$\rho_m = \frac{41,730.793}{42,457.24}$$

$$= 98.289\% \tag{31.89}$$

$$\rho_b = \frac{719.993}{42,457.24}$$

$$= 1.696\% \tag{31.90}$$

$$\rho_e = \frac{4.84 + (2)(0.807)}{42,457.24}$$

$$= 0.015\% \tag{31.91}$$

The analysis of variance is shown in Table 31.5.
The tensile strength y at temperature x is estimated as

$$y = \hat{m} + \hat{b}(x - \bar{x})$$

$$= 72.22 - 0.4245(x - 30) \tag{31.92}$$

31.4. Application of Orthogonal Polynomials

The type of contrast, or comparison, to cite is up to a researcher. The appropriate selection is crucial if the results achieved are to be based on practical and justifiable

Table 31.5
ANOVA table

Source	f	S	V	S'	ρ (%)
m	1	41,731.60	41,731.60	41,730.793	98.289
b	1	720.80	720.80	719.993	1.696
e	6	4.84	0.807	6.545	0.015
Total	8	42,457.24		42,457.240	100.000

comparisons rather than simply on the theoretical methodologies. Only a researcher or an engineer who is thoroughly knowledgeable of the products or conditions being investigated can know what comparison would be practical. However, the contrasts of orthogonal polynomials in linear, quadratic, ... , are often used. Let $A_1, A_2, ... , A_a$ denote the values of first, second, ... , a level, respectively.

Assuming that the levels of A are of equal intervals, the levels are expressed as

$$A_1 = A_1$$

$$A_2 = A_1 + h$$

$$A_3 = A_1 + 2h \tag{31.93}$$

$$\vdots$$

$$A_a = A_1 + (a - 1)h$$

From each level of $A_1, A_2, ... , A_a$, r data are taken. The sum of each level is denoted by $y_1, y_2, ... , y_a$, respectively. When A has levels with the same interval, an orthogonal polynomial, which is called the *orthogonal function* of P. L. Chebyshev, is generally used.

The expanded equation is

$$y = b_0 + b_1(A - \overline{A}) + b_2\left[(A - \overline{A})^2 - \frac{a^2 - 1}{12}h^2\right] + \cdots \tag{31.94}$$

where $\overline{A}$ is the mean of the levels of A:

$$\overline{A} = A_1 + \frac{a - 1}{2}h \tag{31.95}$$

The characteristics of the expansion above are that it attaches importance to the terms of the lower orders, such as a constant, a linear function, or even a quadratic function. First, the constant term b_0 is tried. If it does not fit well, a linear term is tried. If it still does not show linear tendency, the quadratic term is tried, and so on. When sums $y_1, y_2, ... , y_a$ were obtained from r data of the levels $A_1, A_2, ... , A_a$, respectively, the values are $b_0, b_1, ...$. Equation (31.94) will be obtained by solving the following observational equation with order a using the least squares method:

$$b_0 + b_1(A_1 - \overline{A}) + b_2\left[(A_1 - \overline{A})^2 - \frac{a^2 - 1}{12}h^2\right] + \cdots = \frac{y_1}{r}$$

$$\vdots \qquad (31.96)$$

$$b_0 = b_1(A_a - A) + b_2\left[(A_a - \overline{A})^2 - \frac{a^2 - 1}{12}h^2\right] + \cdots = \frac{y_a}{r}$$

The estimates of b_0, b_1, ... are denoted by b_0', b_0', b_0', ... :

$$b_0' = \frac{y_1 + \cdots + y_a}{ar} \qquad (31.97)$$

$$b_1' = \frac{(A_1 - \overline{A})y_1 + \cdots + (A_a - \overline{A})y_a}{[(A_1 - \overline{A})^2 + \cdots + (A_a - \overline{A})^2]r} \qquad (31.98)$$

It is easy to calculate b_0', since it is the mean of the total data. But it seems to be troublesome to obtain b_1', b_2', ... Actually, it is easy to estimate them by using Table 31.6, which shows the coefficients of orthogonal polynomials when $a = 3$ (three levels). b_1 and b_2 are linear and quadratic coefficients, respectively. These are given by

$$b_1' = \frac{W_1 A_1 + W_2 A_2 + W_3 A_3}{r(\lambda S)h} \qquad (31.99)$$

$$b_2' = \frac{W_1 A_1 + W_2 A_2 + W_3 A_3}{r(\lambda S)h^2} \qquad (31.100)$$

In the equations above, A_1, A_2, and A_3 are used instead of y_1, y_2, and y_3. For b_1, the values of W_1, W_2, and W_3 are -1, 0, and 1. Also, S is 2. b_1' is then

$$b_1' = \frac{-A_1 + A_3}{2rh} \qquad (31.101)$$

Similarly,

$$b_2' = \frac{A_1 - 2A_2 + A_3}{2rh^2} \qquad (31.102)$$

Table 31.6
Coefficients of orthogonal polynomial for three levels

Coefficients	b_1	b_2
W_1	-1	1
W_2	0	-2
W_3	1	1
$\lambda^2 S$	2	6
λS	2	2
S	2	$\frac{2}{3}$

The variations of b_1' and b_2' are denoted by S_{A1} and S_{Aq}, respectively. These are given by the squares of equations (31.101) and (31.102) divided by their numbers of units, respectively.

$$S_{A_l} = \frac{(-A_1 + A_3)^2}{2r} \tag{31.103}$$

$$S_{A_q} = \frac{(A_1 - 2A_2 + A_3)^2}{6r} \tag{31.104}$$

The denominator of equations (31.103) and (31.104) is $r(\lambda^2 S)$ ($\lambda^2 S$ is the sum of squares of the coefficients of W). The effective number of replication, n_e, is given by

$$b_1': \quad n_e = rSh^2 \tag{31.105}$$

$$b_2': \quad n_e = rSh^4 \tag{31.106}$$

In general, the ith coefficient b_i and its variation on S_{A_i} are

$$b_1' = \frac{W_1 A_1 + \cdots + W_a A_a}{r(\lambda S) h^i} \tag{31.107}$$

$$S_{A_i} = \frac{(W_1 A_1 + \cdots + W_a A_a)^2}{(W_1^2 + \cdots + W_a^2)r} \tag{31.108}$$

❑ Example

To observe the change of the tensile strength of a synthetic resin due to changes in temperature, the tensile strength of five test pieces was measured at $A_1 = 5°C$, $A_2 = 20°C$, and $A_3 = 35°C$, respectively, to get the results in Table 31.7. The main effect A is decomposed into linear, quadratic, and cubic components to find the correct order of polynomial to be used. From Table 31.8, find the coefficients at the number of levels $k = 4$.

Table 31.7
Tensile strength (kg/mm²)

Level	Data	Total
A_1 (5°C)	43, 47, 45, 43, 45	233
A_2 (20°C)	43, 41, 45, 41, 39	209
A_3 (35°C)	37, 36, 39, 40, 38	190
A_4 (50°C)	34, 32, 36, 35, 35	172

Table 31.8

Orthogonal polynomials with equal intervals

Coeff.	k = 2	k = 3		k = 4			k = 5			
	b_1	b_1	b_2	b_1	b_2	b_3	b_1	b_2	b_3	b_4
W_1	-1	-1	1	-3	1	-1	-2	2	-1	1
W_2	1	0	-2	-1	-1	3	-1	-1	2	-4
W_3		1	1	1	-1	-3	0	-2	0	6
W_4				3	1	1	1	-1	-2	-4
W_5							2	2	1	1
$\lambda'S$	2	2	6	20	4	20	10	14	10	70
λS	1	2	2	10	4	6	10	14	12	24
S	1/2	2	2/3	5	4	9/5	10	14	72/5	288/85
λ	2	1	8	2	1	10/3	1	1	5/6	85/12

Coeff.	k = 6					k = 7				
	b_1	b_2	b_3	b_4	b_5	b_1	b_2	b_3	b_4	b_5
W_1	-5	5	-5	1	-1	-3	5	-1	3	-1
W_2	-3	-1	7	-3	5	-2	0	1	-7	4
W_3	-1	-4	4	2	-10	-1	-3	1	1	-5
W_4	1	-4	-4	2	10	0	-4	0	6	0
W_5	3	-1	-7	-3	-5	1	-3	-1	1	5
W_6	5	5	5	1	1	2	0	-1	-7	-4
W_7						3	5	1	3	1
$\lambda'S$	70	84	180	28	252	28	84	6	154	84
λS	35	56	108	48	120	28	84	36	264	240
S	35/2	112/3	324/5	576/7	400/7	28	84	216	3,168/7	4,800/7
λ	2	3/2	5/3	7/12	21/10	1	1	1/6	7/12	7/20

Coeff.	k = 8					k = 9				
	b_1	b_2	b_3	b_4	b_5	b_1	b_2	b_3	b_4	b_5
W_1	-7	7	-7	7	-7	-4	28	-14	14	-4
W_2	-5	1	5	-13	23	-3	7	7	-21	11
W_3	-3	-3	7	-3	-17	-2	-8	13	-11	-4
W_4	-1	-5	3	9	-15	-1	-17	9	9	-9
W_5	1	-5	-3	9	15	0	-20	0	18	0
W_6	3	-3	-7	-3	17	1	-17	-9	9	9
W_7	5	1	-5	-13	-23	2	-8	-13	-11	4
W_8	7	7	7	7	7	3	7	-7	-21	-11
W_9						4	28	14	14	4
$\lambda'S$	168	168	264	616	2184	60	2772	990	2,002	468
λS	84	168	396	1,056	3,102	60	924	1,188	3,432	3,120
S	42	168	594	12,672/7	31,200/7	60	308	7,128/5	41,184/7	20,800
λ	2	1	2/3	12/7	10/7	1	3	5/6	7/12	3/20

Coeff.	k = 10					k = 11				
	b_1	b_2	b_3	b_4	b_5	b_1	b_2	b_3	b_4	b_5
W_1	-9	6	-42	18	-6	-5	15	-30	6	-3
W_2	-7	2	14	-22	14	-4	6	6	-6	6
W_3	-5	-1	35	-17	-1	-3	-1	22	-6	1
W_4	-3	-3	31	3	-11	-2	-6	23	-1	-4
W_5	-1	-4	12	18	-6	-1	-9	14	4	-4
W_6	1	-4	-12	18	6	0	-10	0	6	0
W_7	3	-3	-31	3	11	1	-9	-14	4	4
W_8	5	-1	-35	-17	1	2	-6	-23	-1	4
W_9	7	2	-14	-22	-14	3	-1	-22	-6	-1
W_{10}	9	6	42	18	6	4	6	-6	-6	-6

Table 31.8
Orthogonal polynomials with equal intervals (*Continued*)

Coeff.	k = 10					k = 11				
	b_1	b_2	b_3	b_4	b_5	b_1	b_2	b_3	b_4	b_5
W_{11}						5	15	30	6	3
$\lambda'S$	330	132	8,580	2,860	780	110	858	4,290	286	156
λS	165	264	5,148	6,864	7,800	110	858	5,148	3,432	6,240
S	165/2	528	15,444/5	82,368/5	78,000	110	858	30,886/5	41,184	249,600
λ	2	1/2	5/3	5/12	1/10	1	1	5/6	1/12	1/40

548

Coeff.	k = 12					k = 13				
	b_1	b_2	b_3	b_4	b_5	b_1	b_2	b_3	b_4	b_5
W_1	-11	55	-33	33	-33	-6	22	-11	99	-22
W_2	-9	25	3	-27	57	-5	11	0	-66	33
W_3	-7	1	21	-33	21	-4	2	6	-96	18
W_4	-5	-17	25	-13	-29	-3	-5	8	-54	-11
W_5	-3	-29	19	12	-44	-2	-10	7	11	-26
W_6	-1	-35	7	28	-20	-1	-13	4	64	-20
W_7	1	-35	-7	28	20	0	-14	0	84	0
W_8	3	-29	-19	12	44	1	-13	-4	64	20
W_9	5	-17	-25	-13	29	2	-10	-7	11	26
W_{10}	7	1	-21	-33	-21	3	-5	-8	-54	11
W_{11}	9	25	-3	-27	57	4	2	-6	-96	-18
W_{12}	11	55	33	33	33	5	11	0	-66	-33
W_{13}						6	22	11	99	22
$\lambda'S$	572	12,012	5,148	8,008	15,912	182	2,002	572	68,068	6,188
λS	286	4,004	7,722	27,456	106,080	182	2,002	3,432	116,688	106,080
S	143	4,004/3	11,583	658,944/7	707,200	182	2,002	20,592	1,400,256/7	12,729,600/7
λ	2	3	2/3	7/24	3/20	1	1	1/6	7/12	7/120

$$S_{A_l} = \frac{(W_1 A_1 + W_2 A_2 + W_2 A_2 + W_3 A_3 + W_4 A_4)^2}{r(\lambda^2 S)}$$

$$= \frac{[-3(223) - 209 + 190 + 3(172)]^2}{(5)(20)}$$

$$= \frac{(-172)^2}{100} = 296 \qquad\qquad (31.109)$$

$$S_{A_q} = \frac{(223 - 209 - 190 + 172)^2}{(5)(4)} = \frac{(-4)^2}{20} = 1 \qquad (31.110)$$

$$S_{A_c} = \frac{[-223 + 3(209) - 3(190) + 172]^2}{(5)(20)} = \frac{6^2}{100} = 0 \quad (31.111)$$

$$S_T = 43^2 + 47^2 + \cdots + 35^2 - CF = 348$$

$$S_e = S_T - S_{A_l} - S_{A_q} - S_{A_c}$$

$$= 51 \qquad\qquad (31.112)$$

It is known from Table 31.9 that only the linear term of A is significant; hence, the relationship between temperature, A, and tensile strength, y, can be deemed to be linear within the range of our experimental temperature changes. The estimate of the linear coefficient of temperature, b, is

$$b_1' = \frac{-3(223) - 209 + 190 + 3(172)}{r(\lambda S)h}$$

$$= \frac{-172}{(5)(10)(15)} = -0.23 \qquad\qquad (31.113)$$

Table 31.9
ANOVA table

Source	f	S	V	ρ (%)
A				
Linear	1	296	296	84.2
Quadratic	1	1	1	
Cubic	1	0	0	
e	16	51	3.2	
Total	19	348		100.0
(e)	(18)	(52)	(2.9)	(15.8)

The relationship between temperature and tensile strength is then

$$y = b_0' + b_1' (A - \bar{A})$$

$$= \frac{794}{20} - 0.23(A - 27.5)$$

$$= 39.70 - 0.23(A - 27.5) \qquad (31.114)$$

b_1: 1 *linear*

b_2: 2 *quadratic*

$$b = \frac{W_1 A_1 + \cdots + W_a A_a}{r(\lambda S)h^i} \qquad (31.115)$$

$$\mathrm{Var}(\hat{b}_i) = \frac{\sigma^2}{rSh^{2i}}$$

$$S_{b_i} = \frac{(W_1 A_1 + \cdots + W_a A_a)^2}{r(\lambda^2 S)}$$

32 Two-Way Layout

32.1. Introduction   552
32.2. Factors and Levels   552
32.3. Analysis of Variance   555

32.1. Introduction

Let us discuss how to determine a way of increasing the production yield of a chemical product. Many factors might affect yield of the product; in this case we only discuss an experiment designed for determining the catalyst quantity and synthesis temperature to obtain a high yield. This chapter is based on Genichi Taguchi et al., *Design of Experiments.* Tokyo: Japanese Standards Association, 1973.

32.2. Factors and Levels

Causes of a given result in an experiment are called *factors*. Factors such as temperature and catalyst quantity at the stage of synthesis are denoted by A and B, respectively. If the relationships between factor A, factor B, and yield are determined, we can decide the temperature and catalyst quantity that will result in a higher yield.

From our past experience, we can roughly estimate a temperature range that will give the highest yield: for example, a temperature range of 200 to 300°C. In addition, we may know that the quantity of a catalyst that will give us the highest yield is in the range 0.2 to 0.8%. Although such ranges are known roughly in many cases, the yield may vary substantially within the ranges above; for this reason we need an experiment that will determine precisely the best temperature and catalyst quantity.

To determine the relationships among yield, temperature, or catalyst quantity, the conditions of temperature and catalyst quantity are varied, and the resulting relationships are plotted in graphs. This particular type of experiment is usually carried out as follows: First, the ranges of temperature and catalyst quantity are fixed and two or three temperatures and catalyst quantities within the ranges are selected. For example, when the temperature range is between 200 and 300°C, the following temperatures may be selected: $A_1 = 200°C$, $A_2 = 225°C$, $A_3 = 250°C$, $A_4 = 275°C$, and $A_5 = 300°C$.

After the temperature range is fixed, the remaining problem is to decide the number of points within the range to be used. The number of points needed depends on the complexity of the curve of temperature, which is plotted against the yield. Like the experiment described above, three to five points are usually chosen because the curve of temperature and the yield normally has a mountainous shape or a simple smooth curve. If there is a possibility for a curve with two or more mountains, more temperature points must be chosen.

Suppose that we have the range 200 to 300°C and take five points. In this case we define the temperature at five different levels at equal intervals. In the same way, levels are chosen for the catalyst quantity factor. If it is known before the experiment begins that the catalyst quantity must be increased when the temperature is increased, the range of catalyst quantity should be adjusted for each temperature level.

This is not likely to happen in the relationship between temperature and catalyst quantity, but obviously does happen in the case of temperature and reaction time. This particular phenomenon often happens with various related factors of chemical reactions: for example, the reaction of hydrogen and oxygen to produce water,

$$2H_2 + O_2 = 2H_2O$$

where two molecules of hydrogen and one molecule of oxygen react. Consequently, the quantity of oxygen is in the neighborhood of half the chemical equivalent of hydrogen.

To find the effect caused by different flows per unit of time, suppose that the quantity of H_2 is varied as A_1, A_2, and A_3. The quantity of oxygen is then set at the following three levels:

B_1: theoretical amount (chemical equivalent)

B_2: 1.02 times B_1 (2% more than the theoretical amount)

B_3: 1.04 times B_1 (4% more than the theoretical amount)

Why are levels B_2 and B_3 necessary? To minimize the productivity loss caused by the amount of unreacted material, we want to investigate how much excess oxygen can be added to increase profitability maximally, since oxygen happens to be cheaper than hydrogen. (The purpose of setting three levels for the quantity of hydrogen is to investigate the effect of hydrogen flow.)

In this particular experiment, we assume that it is not necessary to forecast the temperature range for each distinct catalyst quantity; therefore, the range of catalyst quantity can be independent of temperature. Suppose that the range is 0.2 to 0.8% and that four levels are chosen at equal intervals. Thus, $B_1 = 0.2\%$,

$B_2 = 0.4\%$, $B_3 = 0.6\%$, and $B_4 = 0.8\%$. There are now 20 combinations of factors A and B as shown in Table 32.1.

The order of experimenting with each combination is determined by drawing lots. Random order is needed because other factors may also be influencing the yield. Uncontrollable variations of raw or subsidiary materials or the catalyst itself may influence the results. Errors in measurement of the yield may also create biased results. There will be differences in the experiments, even though the conditions are controlled as much as possible.

Suppose that the experiments were conducted and the yields obtained are those shown in Table 32.1. The following questions must now be answered:

Table 32.1
Experiment with a two-way layout

Experiment No.	Level of A	Level of B	Combination of A and B	Temperature (°C)	Catalyst Quantity (%)	Data in Yield (%)
					Actual Condition of Experiment	
1	1	1	A_1B_1	200	0.2	64
2	1	2	A_1B_2	200	0.4	65
3	1	3	A_1B_3	200	0.6	76
4	1	4	A_1B_4	200	0.8	64
5	2	1	A_2B_1	225	0.2	67
6	2	2	A_2B_2	225	0.4	81
7	2	3	A_2B_3	225	0.6	82
8	2	4	A_2B_4	225	0.8	91
9	3	1	A_3B_1	250	0.2	76
10	3	2	A_3B_2	250	0.4	81
11	3	3	A_3B_3	250	0.6	88
12	3	4	A_3B_4	250	0.8	90
13	4	1	A_4B_1	275	0.2	76
14	4	2	A_4B_2	275	0.4	84
15	4	3	A_4B_3	275	0.6	83
16	4	4	A_4B_4	275	0.8	92
17	5	1	A_5B_1	300	0.2	73
18	5	2	A_5B_2	300	0.4	80
19	5	3	A_5B_3	300	0.6	84
20	5	4	A_5B_4	300	0.8	91

1. How much influence against the yield will be caused when the temperature or catalyst quantity changes?

2. What is the appearance of the curve showing the relation of temperature and yield or catalyst quantity and yield?

3. Which temperature and what catalyst quantity will give the highest yield?

An investigation of these questions follows.

32.3. Analysis of Variance

First, the decomposition of variation has to be derived. We want to know how the yield changes when the temperature or catalyst quantity changes within the experimental range. In other words, we want to show the composite effects of temperature and catalyst quantity on yield and in such a way answer the three questions posed earlier.

If every experiment in Table 32.1 resulted in 64% yield, this would indicate that neither temperature nor the quantity of the catalyst affects yield within that particular factor range, or it would be shown that neither materials nor the measuring skill caused the variance.

However, the average yield is 64% instead of 100%; this is a big problem. This was caused by the factors that were kept fixed during the period of the experiment. That is, the average value of experimental data reflects the influence of nonvariable factors. The average might not have been 64% had the type of catalyst or quantity of materials in the reaction been changed.

The problem of discussing the influence of those fixed factors belongs to the technologists who work in this specialized field; it is normally excluded from the analysis of data. For this reason, the correction factor is

$$CF = \frac{(\text{total of all data})^2}{20} \qquad (32.1)$$

which is subtracted from the total sum of the squares to get the total variation, S_T; then S_T is decomposed. First, a working mean, 80, is subtracted from each result in Table 32.1. The data subtracted are shown in Table 32.2.

Table 32.2
Subtracted data

	B_1	B_2	B_3	B_4	Total
A_1	−16	−15	−4	−16	−51
A_2	−13	1	2	11	1
A_3	−4	1	8	10	15
A_4	−4	4	3	12	15
A_5	−7	0	4	11	8
Total	−44	−9	13	28	−12

To verify the accuracy of the table, the following totals are calculated: first, a total is made for each row and each column; next, the totals for both the rows and columns are brought to grand totals; these grand totals must then be equal. The correction factor is

$$CF = \frac{(-12)^2}{20} = 7 \qquad (f = 1) \tag{32.2}$$

The total amount of variation S_T is

$$S_T = (-16)^2 + (-15)^2 + \cdots + 11^2 - CF$$

$$= 1600 - 7 = 1593 \qquad (f = 19) \tag{32.3}$$

Next, calculate the effect of A, which is shown by S_A.

$$S_A = \tfrac{1}{4}[(-51)^2 + 1^2 + 15^2 + 15^2 + 8^2] - 7 = 772 \qquad (f = 4) \tag{32.4}$$

The following is the reason that S_A can be calculated by equation (32.4).

In Table 32.2, the effect of temperature, which is factor A, is included in each condition for each level of catalyst quantity: B_1, B_2, B_3, and B_4. Accordingly, the effect of A must be obtained from each level of B and the effects added together. This is especially important when the effects of temperature are not the same under different levels of catalyst quantity.

When the effects of temperature A are not the same under varying amounts of catalyst, it is known that interaction exists between temperature A and catalyst quantity B. This interaction is written as $A \times B$. Strictly speaking, interaction exists in any experiment. If someone does assume that the effect of temperature doesn't change when the quantity of the catalyst is changed, he or she will be incorrect. Following is a discussion of interaction and the practical way it should be handled.

The influence of temperature A differs with various quantities of catalyst B; hence, the curves showing the effect of temperature must be plotted separately for each catalyst quantity. But after the comparison of temperature and quantity of catalyst is finished, do we then have to investigate the effects of temperature plotted for various additives, types of catalyst, agitating conditions, the shapes of equipment, and so on?

Temperature and quantity of catalyst must be the only two factors among the many we intend to investigate. If so, we must consider the fact that the optimum conditions of temperature and quantity of the catalyst will differ according to different conditions of other factors, such as type of additive, quantity of additive, and others. In other words, there may be interactions between A, B, and other factors: C, D, ... The optimum conditions of A and B obtained from the curves of A and B may not be useful, since it may change when the type of additives changes.

This leads us to feel the need to plot the curves of temperature against all combinations of B, C, ... For this purpose we have to design a type of experiment to obtain the interactions between A and all the other factors. Hence, an experiment with only one factor, only two factors, or several factors would tend to be meaningless. Therefore, it would always be necessary to investigate the effects of all factors. If there are 10 factors on three levels, there are 59,049 different conditions.

If a research worker tried to obtain 59,049 curves of temperature A under different conditions of other factors in order to prepare a perfect report, this report could be 10,000 pages long. Clearly, no one can afford the time to observe 59,049 curves. We just want to know the range where the optimum temperature exists. Instead of getting 59,049 curves, the researchers should have conducted the next investigation.

Strictly speaking, the curves of temperature do differ under the conditions of other factors B, C, D, ... , but we must consider the average curve to be the representative curve of temperature A, and term it the *main effect* curve. Although curves do differ under the conditions of other factors; if the differences are not great, only a slight error will be caused when the optimum condition is determined from the main effect curve, which represents the effect of temperature A.

However, it is important to evaluate the extent of deviation from the main effect curve or the magnitude of the interaction. For this reason, the determination of interaction must be limited to the factors of concern to a research worker. It would be foolish to try to design an experiment for all interactions. It must be stressed that many interactions are not required; it is desirable that the magnitude of deviation be evaluated by the residual sum of squares.

When datum y_{ij} (the result of A_iB_j) is expressed by the composite effects of temperature A and quantity of catalyst B, we must find the extent of the error in such an expression. For this purpose we are going to determine the residual sum of squares, which includes interactions.

Since main effect A means the average curve of the effect of A, the means of the effects of A under conditions B_1, B_2, B_3, and B_4 are

$$\bar{A}_1 = \frac{A_1}{4} = \frac{-51}{4} = -12.75$$

$$\bar{A}_2 = \frac{A_2}{4} = \frac{1}{4} = 0.25$$

$$\bar{A}_3 = \frac{A_3}{4} = \frac{15}{4} = 3.75$$

$$\bar{A}_4 = \frac{A_4}{4} = \frac{15}{4} = 3.75$$

$$\bar{A}_5 = \frac{A_5}{4} = \frac{8}{4} = 2.00 \tag{32.5}$$

where A_1, A_2, A_3, A_4, and A_5 indicate the totals of the results under conditions A_1, A_2, A_3, A_4, and A_5.

Figure 32.1 shows the figures obtained after adding back the working mean, 80. The line of $\bar{T}$, which is 79.4, shows the average value of all experiments. Let a_1 be the difference between $\bar{A}_1$ and $\bar{T}$:

$$\hat{a}_1 = \frac{A_i}{4} - \frac{T}{20} \qquad (i = 1, 2, 3, 4, 5) \tag{32.6}$$

If all five points were on-line $\bar{T}$, everyone would agree that within this range there is no effect due to changes in temperature on yield. It is true, therefore,

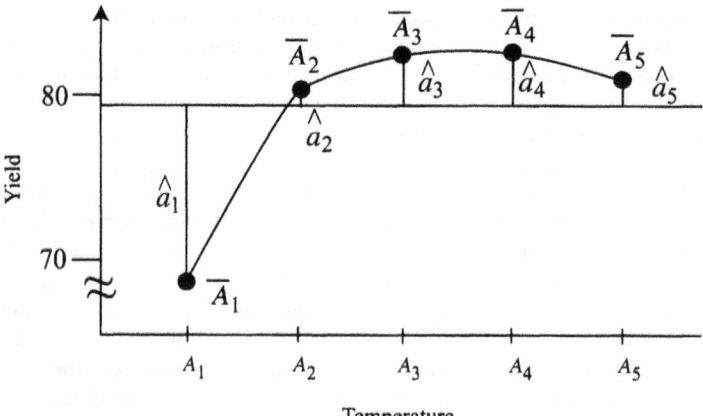

Figure 32.1
Effect of A

that the effect of temperature can be evaluated by making the following calculation:

$$\text{effect of } A = (\hat{a}_1)^2 + (\hat{a}_2)^2 + (\hat{a}_3)^2 + (\hat{a}_4)^2 + (\hat{a}_5)^2 \tag{32.7}$$

This is known as the *effect of the average curve*. Since there are four repetitions for each temperature, we can determine that four times the magnitude of equation (32.7) is the effect of temperature A, which constitutes part of the total variation of the data. Therefore, S_A, the effect of temperature A, is

$$S_A = 4[(\hat{a}_1)^2 + (\hat{a}_2)^2 + (\hat{a}_3)^2 + (\hat{a}_4)^2 + (\hat{a}_5)^2]$$

$$= 4\left[\left(\frac{A_1}{4} - \frac{T}{20}\right)^2 + \left(\frac{A_2}{4} - \frac{T}{20}\right)^2 + \cdots + \left(\frac{A_5}{4} - \frac{T}{20}\right)^2\right]$$

$$= 4\left[\frac{A_1^2}{16} - 2\frac{T}{20}\frac{A_1}{4} + \left(-\frac{T}{20}\right)^2\right] + \left[\frac{A_2^2}{16} - 2\frac{T}{20}\frac{A_2}{4} + \left(-\frac{T}{20}\right)^2\right]$$

$$+ \cdots + \left[\frac{A_5^2}{16} - 2\frac{T}{20}\frac{A_5}{4} + -\frac{T^2}{20}\right]$$

$$= \frac{A_1^2 + A_2^2 + A_3^2 + A_4^2 + A_5^2}{4}$$

$$- 2\frac{T}{20}\frac{A_1 + A_2 + A_3 + A_4 + A_5}{4}(4) + (4)(5)\left(\frac{T}{20}\right)^2 \tag{32.8}$$

$$= \frac{A_1^2 + A_2^2 + A_3^2 + A_4^2 + A_5^2}{4} - 2\frac{T^2}{20} + \frac{T^2}{20}$$

$$= \frac{A_1^2 + A_2^2 + A_3^2 + A_4^2 + A_5^2}{4} - \frac{T^2}{20} \tag{32.9}$$

To simplify equation (32.9), a working mean may be subtracted:

$$S_A = \frac{(-51)^2 + 1^2 + 15^2 + 15^2 + 8^2}{4} - \frac{(-12)^2}{20} = 772 \qquad (32.10)$$

In the calculation of S_A, $(A_1^2 + A_2^2 + \cdots + A_5^2)$ is divided by 4, since each of A_1, A_2, ... , A_5 is the total of four curves. Generally, in such calculations of variation, each squared value is divided by its *number of units*, which is the total of the square for each coefficient of a linear equation. Since A_1 is the sum of four observational values, A_1 is expressed by a linear equation with four observational values, where each value has a coefficient of 1. Therefore, A_1^2 must be divided by 4, the number of units of the linear equation, or the total of squares of each coefficient, 1. Similarly,

$$S_B = \frac{B_1^2 + B_2^2 + B_3^2 + B_4^2}{5} - \text{CF}$$

$$= \frac{(-44)^2 + (-9)^2 + 13^2 + 28^2}{5} - 7 = 587 \qquad (f = 3) \qquad (32.11)$$

All other factor-related effects (including interaction $A \times B$) are evaluated by subtracting S_A and S_B from S_T, the total variations of all data. The residual sum of the squares, S_e, is called the *error variation:*

$$S_e = S_T - S_A - S_B = 1593 - 772 - 587$$

$$= 234 \qquad (f = 19 - 4 - 3 = 12) \qquad (32.12)$$

There are four unknown items in S_A; its degrees of freedom is four. If all the items are negligible, or if the effect of temperature A is negligible, S_A will simply be about equal to $4V_e$.

There are 4 units of error variance in S_A. Therefore, the net effect of A, or the pure variation S_A, is estimated by

$$S_A' = S_A - 4V_e \qquad (32.13)$$

Therefore,

$$S_A' = 772 - (4)(19.5)$$

$$= 694.0 \qquad (32.14)$$

Similarly, the pure variation of B is

$$S_B' = S_B - 3V_e = 587 - (3)(19.5)$$

$$= 528.5 \qquad (32.15)$$

The total effect of all factors, including A and B but excluding the general mean, is shown by S_T, which is 1593. The magnitude of all factor-related effects, excluding A and B, is then

$$S_e' = (\text{total variation}) - (\text{pure variation of } A) - (\text{pure variation of } B)$$

$$= 1593 - 694.0 - 528.5 = 370.5 \qquad (32.16)$$

The yield results varied from 64 to 92%. The variation was caused by the effects of temperature, catalyst quantity, and other factors. How much of the total variation S_T was caused by temperature? This is calculated by dividing pure variation of A by S_T. If the percentage of variation caused by temperature is denoted by ρ_A, then

$$\rho_A = \frac{\text{pure variation of } A}{\text{total variation}} = \frac{S_A - 4V_e}{S_T}$$

$$= \frac{694}{1593} = 0.436 = 43.6\% \tag{32.17}$$

That is, of the yield variation ranging from 64 to 92%, in 43.5% of the cases, the change of temperature caused the variation in yield. Similarly,

$$\rho_B = \frac{528.5}{1593} = 33.2\% \tag{32.18}$$

$$\rho_e = \frac{370.5}{1593} = 23.2\% \tag{32.19}$$

There are similarities between the analysis of variance and spectrum analysis. The total amount of variance, S_T, corresponds to the total power of the wave in spectrum analysis. In the latter, total power is decomposed to the power of each component with different cycles. Similarly, in analysis of variance, total variance is decomposed to a cause system. The ratio of the power of each component to total power is determined as degrees of contribution and is denoted by a percentage. The "components" for the purpose of our investigation are factor A, factor B, and other factors that change without artificial control. In the design of experiments, the cause system is designed to be orthogonal. Accordingly, calculation of the analysis of variance is easy and straightforward (Table 32.3).

Degrees of contribution can be explained as follows: In the case of a wave, when the amplitude of oscillation doubles, its power increases four fold. When the power is reduced to half, its amplitude is $1/\sqrt{2}$. In Table 32.1, the minimum value is 64 and the maximum value is 92.

We found from the analysis above that temperature and the quantity of the catalyst do indeed influence yield. What we want to know next is the relationship between temperature and yield and the quantity of catalyst and yield.

Table 32.3
Analysis of variance

Source	f	S	V	E(V)	S'	ρ (%)
A	4	772	193	$\sigma^2 + 4\sigma_A^2$	694.0	43.6
B	3	587	196	$\sigma^2 + 5\sigma_B^2$	528.5	33.2
e	12	234	19.5	σ^2	370.5	23.2
Total	19	1593			1593	100.0

In the previous calculation we had four estimates of yield at each temperature (200, 225, 250, 275, and 300°C). A total estimate for each temperature results in −51, 1, 15, 15, and 8, respectively. These totals are divided by 4, and the working mean 80 is added.

$$200°C: \quad \frac{-51}{4} + 80 = 67.25$$

$$225°C: \quad \frac{1}{4} + 80 = 80.25$$

$$250°C: \quad \frac{15}{4} + 80 = 83.75$$

$$275°C: \quad \frac{8}{4} + 80 = 82.00 \tag{32.20}$$

Plotting these four points, a curve showing the relation of temperature and yield is obtained as shown in Figure 32.2. It is important to stress here that these values show the main effect of temperature and yield. The main effect curve shown in Figure 32.2 is the curve of the average values at B_1, B_2, B_3, and B_4.

Next, we want to find the temperature and quantity of catalyst that will contribute to an increase in yield. This procedure is actually the determination of optimum conditions. These conditions are normally found after observing the graphs in Figures 32.2 and Figure 32.3. The best temperature is the highest point of the curve. But the best temperature must result in the highest profit, and that may not necessarily mean the highest yield.

Actually, the heavy oil consumptions that may be necessary to maintain different temperatures are investigated first; these consumptions are then converted to the same cost per unit as that of the yield, and those units are then plotted against temperature. The temperature of the largest difference between yield and fuel cost is read from the abscissa, and it is then established as the operating standard.

From Figure 32.2 it is clear that the maximum value probably lies in the neighborhood of 250 to 275°C. If the value does not vary significantly, the lower temperature, 250°C, is the best temperature for our discussion. Similarly, the catalyst quantity that will give us the highest yield is determined from Figure 32.3. Let us assume that 0.8% is determined as the best quantity.

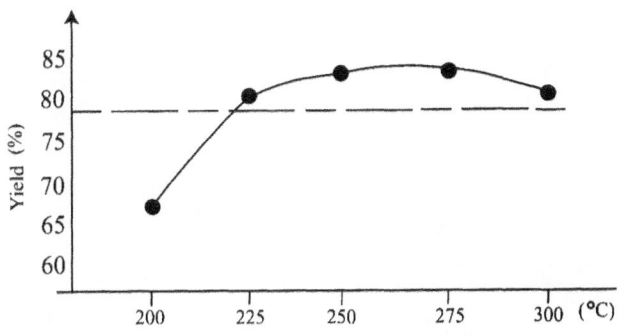

Figure 32.2
Relationship between temperature and yield

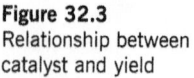

Figure 32.3
Relationship between
catalyst and yield

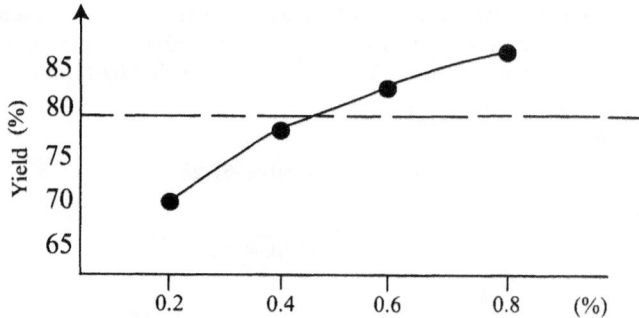

Our next step is to find the actual value for the yield when production is continued at 250°C using an 0.8% catalyst quantity. To do this, we need to calculate the process average, expressed by the following equation, where $\hat{\mu}$ is the estimate of process average of yield at A_3B_4:

$$\hat{\mu} = \text{(average of all experiments, } \overline{T})$$

$$+ \text{(yield increase above } \overline{T} \text{ when } A_3 \text{ is used)}$$

$$+ \text{(yield increase above } \overline{T} \text{ when } B_4 \text{ is used)}$$

$$= \overline{T} + (\overline{A}_3 - \overline{T}) + (\overline{B}_4 - \overline{T}) = \overline{A}_3 + \overline{B}_4 - \overline{T}$$

$$= 83.75 + 85.60 - 79.40 = 89.95 \tag{32.21}$$

33 Two-Way Layout with Decomposition

33.1. Introduction   563
33.2. One Continuous Variable   563
33.3. Two Continuous Variables   567

33.1. Introduction

When a factor in an experiment is a continuous variable, such as temperature or time, it is often necessary to know the linear or quadratic effects of the factor. For this purpose, the orthogonal polynomial equation is used. In this chapter, methods of analysis for one or two continuous variables in two-way layout experiments are shown. This chapter is based on Genichi Taguchi et al., *Design of Experiments*. Tokyo: Japanese Standards Association, 1973.

33.2. One Continuous Variable

To observe changes in the elongation of plastics, three plastics are prepared with three types of additives (A_1, A_2, and A_3), and four levels of temperature to be used in the experiment are as follows: $B_1 = -15°C$, $B_2 = 0°C$, $B_3 = 15°C$, $B_4 = 30°C$. Elongation is then measured for each combination of factors A and B. The results are shown in Table 33.1.

The most desirable data we want would be high elongation that will not change with temperature. This type of problem of wanting to produce a product whose characteristics do not change with temperature is very familiar to engineers and researchers. Temperature is considered an indicative factor. The indicative factor, in this case temperature, would indicate the temperature when the products are used by the customer. An indicative factor has levels like a control factor, which can be controlled in the stage of production to achieve optimum results. When an indicative factor is a condition such as temperature, humidity, or electric current, it is usually decomposed into linear and residual effects.

Table 33.1
Percent elongation

	B_1	B_2	B_3	B_4
A_1	15	31	47	62
A_2	11	20	37	45
A_3	34	42	49	54

The procedure for this analysis is as follows: First, subtract a working mean of 40 from all data (Table 33.2). The factorial effect, B, is decomposed into components of linear, quadratic, and cubic effects. For this purpose, coefficients of the four-level polynomial are used.

$$S_{B_l} = \frac{(W_1 B_1 + W_2 B_2 + W_3 B_3 + W_4 B_4)^2}{r(\lambda^2 S)}$$

$$= \frac{[(-3)(-60) - (1)(-27) + (1)(13) + (3)(41)]^2}{(3)(20)}$$

$$= \frac{343^2}{(3)(20)} = \frac{117,649}{60} = 1961 \tag{33.1}$$

$$S_{B_q} = \frac{(-60 + 27 - 13 + 41)^2}{(3)(4)} = \frac{(-5)^2}{12}$$

$$= \frac{25}{12} = 2 \tag{33.2}$$

$$S_{B_c} = \frac{[(-1)(-60) + (3)(-27) - (3)(13) + 41]^2}{(3)(20)}$$

$$= \frac{(-19)^2}{60} = \frac{361}{60} = 6.0 \tag{33.3}$$

Table 33.2
Data after subtracting working mean, 40

	B_1	B_2	B_3	B_4	Total
A_1	-25	-9	7	22	-5
A_2	-29	-20	-3	5	-47
A_3	-6	2	9	14	19
Total	-60	-27	13	41	-33

The main effect of B is calculated as

$$S_B = \frac{(-60)^2 + (-27)^2 + 13^2 + 41^2}{3} - \frac{(-33)^2}{12}$$

$$= \frac{3600 + 729 + 169 + 1681}{3} - 91$$

$$= \frac{6179}{3} - 19 = 1969 \qquad (f = 3) \tag{33.4}$$

The total of S_{B_l}, S_{B_q}, and S_{B_c} is calculated as

$$S_{B_l} + S_{B_q} + S_{BC} = 1961 + 2 + 6 = 1969 \tag{33.5}$$

which coincides with S_B.

When there are two factors, as in this example, it is also necessary to investigate their interaction, which signifies that the effect of one factor will change at different levels of the other factor. In our case it means discovering whether the effect of temperature (factor B) changes if a plastic is made from different additives.

The linear coefficient of temperature at A_1, denoted by $b_1(A_1)$, is calculated as

$$b_1(A_1) = \frac{(-3)(-25) - (1)(-9) + 7 + (3)(22)}{r(\lambda S)h}$$

$$= \frac{157}{(1)(10)(15)} = 1.05 \tag{33.6}$$

Linear coefficients of temperature at A_2 and A_3, denoted by $b_1(A_2)$ and $b_1(A_3)$, are calculated similarly. The variation of linear coefficients of B at different conditions of A, denoted by S_{AB_l}, is obtained as

$$S_{AB_l} = \frac{L(A_1)^2 + L(A_2)^2 + L(A_3)^2}{20} - \frac{[L(A_1) + L(A_2) + L(A_3)^2]}{(20)(3)} \tag{33.7}$$

where

$$L(A_1) = (-3)(-25) - (1)(-9) + 7 + (3)(22) = 157$$

$$L(A_2) = (-3)(-29) - (1)(-20) - 3 + (3)(5) = 119 \tag{33.8}$$

$$L(A_3) = (-3)(-6) - (1)(2) + 9 + (3)(14) = 67$$

Hence,

$$S_{AB_l} = \frac{157^2 + 119^2 + 67^2}{20} - \frac{(157 + 119 + 67)^2}{60}$$

$$= 2165 - 1961 = 204 \qquad (f = 2) \tag{33.9}$$

The effect of A is obtained from the totals of A_1, A_2, and A_3:

$$S_A = \frac{A_1^2 + A_2^2 + A_3^2}{4} - CF$$

$$= \frac{(-5)^2 + (-47)^2 + 19^2}{4} - \frac{(-33)^2}{12}$$

$$= \frac{2595}{4} - 91 = 558 \qquad (f = 2) \qquad (33.10)$$

The total variation, S_T, and error variation, S_e, are calculated.

$$S_T = (-25)^2 + (-9)^2 + \cdots + 14^2 - \frac{(-33)^2}{12}$$

$$= 2831 - 91 = 2740 \qquad (f = 11) \qquad (33.11)$$

$$S_e = S_T - S_A - S_B - S_{AB_l}$$

$$= 2740 - 558 - 1969 - 204$$

$$= 9 \qquad (f = 4) \qquad (33.12)$$

The analysis of variance is shown in Table 33.3.

B_q and B_c are not significant and are pooled with e. This composite is shown as (e). A, B_l, and AB_l are significant. The significance of A shows that elongation does change with different types of additives. The significance of B_l shows that the change of elongation due to temperature has a linear tendency. Since AB_l is significant in this case, the linear tendency changes due to the different types of additive.

Table 33.3
Analysis of variance

Source	f	S	V	S'	p (%)
A	2	558	279	552	20.1
B					
l	1	1961	1961	1958	71.5
q	1	2	2		
c	1	6	6		
AB_l	2	204	102	198	7.3
e	4	9	2.25		
(e)	(6)	(17)	(2.83)	32	1.1
Total	11	2740		2740	100.0

Next, an estimation is made against the significant effects. The main effect of A is calculated as

$$\overline{A}_1 = \frac{-5}{4} + 40 = 38.75 \tag{33.13}$$

$$\overline{A}_2 = \frac{-47}{4} + 40 + 2.06 = 28.25 \tag{33.14}$$

$$\overline{A}_3 = \frac{19}{4} + 40 \pm 2.06 = 44.75 \tag{33.15}$$

This shows that elongation of A_3 is the highest. Next, the change of elongation due to changes in temperature is calculated. B is significant, as is AB_l, so the linear coefficient at each level of A is calculated.

$$\hat{b}_1(A_1) = \frac{-3A_1B_1 - A_1B_2 + A_1B_3 + 3A_1B_4}{r(\lambda S)h}$$

$$= \frac{(-3)(-25) - (1)(-9) + 7 + (3)(22)}{(1)(10)(15)} = 1.05 \tag{33.16}$$

$$\hat{b}_1(A_2) = \frac{(-3)(-29) - (1)(-20) + (-3) + (3)(5)}{(1)(10)(15)} = 0.79 \tag{33.17}$$

$$\hat{b}_1(A_3) = \frac{(-3)(-6) - 2 + 9 + (3)(14)}{(1)(10)(15)} = 0.45 \tag{33.18}$$

From the results above, it is known that the change due to temperature is the smallest at A_3.

We can now conclude that the plastic-using additive, A_3, has the highest elongation, the least change of elongation, and the least change of elongation due to a change in temperature. If cost is not considered, A_3 is the best additive.

33.3. Two Continuous Variables

Now we describe an analysis with two continuous variable factors. In the production of 8% phosphor bronze (containing 0.45% phosphorus), an experiment was planned for the purpose of obtaining a material to substitute an alloy for the current material used in making a spring. To increase the tensile strength of the material, factor A, the extent of processing, and factor B, the annealing temperature, are analyzed. Each factor has four levels.

There are four types of material that can be used when making a spring. Their specifications are:

- ❏ Type 1: over 75 kg/cm^2
- ❏ Type 2: over 70 kg/cm^2
- ❏ Type 3: over 65 kg/cm^2
- ❏ Type 4: over 60 kg/cm^2

The levels of A and B are

Extent of processing, A (%): $A_1 = 30$, $A_2 = 40$, $A_3 = 50$, $A_4 = 60$

Annealing temperature, B (°C): $B_1 = 150$, $B_2 = 200$, $B_3 = 250$, $B_4 = 300$

Sixteen experiments were performed, combining the factors in random order.

In such experiments, two test pieces are usually chosen from each of the 16 combinations, and then their strengths are measured. Generally, r pieces are taken, so there are r data for each combination. But for the analysis of variance, averages are not adequate. Instead, the analysis must be done starting from the original data of r repetitions. In this example, let us assume that $r = 1$; the following problems are discussed:

1. What are the relationships between extent of processing (A), annealing temperature (B), and tensile strength, and how can we express them in the form of a polynomial?

2. What is the range of the extent of processing and annealing temperature that will satisfy the tensile strength specifications for each of the four products?

The observed data are shown in Table 33.4. From each item, subtract a working mean of 70 and multiply by 10 to get the data in Table 33.5.

To get an observational equation, variables A and B are expanded up to cubic terms, and the product of A and B is also calculated. Interaction may be considered as the product of the orthogonal polynomial of one variable.

$$
\begin{aligned}
y = {} & m + a_1(A - \bar{A}) + a_2\left[(A - \bar{A})^2 - \frac{a^2 - 1}{12}h_A^2\right] \\
& + a_3\left[(A - \bar{A})^3 - \frac{3a^2 - 7}{20}(A - \bar{A})h_A^2\right] + b_1(B - \bar{B}) \\
& + b_2\left[(B - \bar{B})^2 - \frac{b^2 - 1}{12}h_B^2\right] + b_3\left[(B - \bar{B})^3 - \frac{3b^2 - 7}{20}(B - \bar{B})h_B^2\right] \\
& + c_{11}(A - \bar{A})(B - \bar{B}) \qquad\qquad\qquad\qquad\qquad\qquad (33.19)
\end{aligned}
$$

where a and b are the numbers of levels of A and B, and h_A and h_B are the intervals of the levels of A and B.

The magnitude of each term is evaluated in the analysis of variance; then a polynomial is written only from the necessary terms. Letting the linear, quadratic,

Table 33.4
Tensile strength data (kg/cm²)

	B_1	B_2	B_3	B_4
A_1	64.9	62.6	61.1	59.2
A_2	69.1	70.1	66.8	63.6
A_3	76.1	74.0	71.3	67.2
A_4	82.9	80.0	76.0	72.3

Table 33.5
Supplementary table

	B_1	B_2	B_3	B_4	Total
A_1	−51	−74	−89	−108	−322
A_2	−9	1	−32	−64	−104
A_3	61	40	13	−28	86
A_4	129	100	60	23	312
Total	130	67	−48	−177	−28

and cubic terms of A be S_{a_l}, S_{A_q}, and S_{a_c}, and defining the term of the product of linear effects of A and B as $S_{A_l B_l}$, the calculations are then made as follows:

$$S_{A_l} = \frac{(-3A_1 - A_2 + A_3 + 3A_4)^2}{\gamma(\lambda^2 S)}$$

$$= \frac{[-3(-322) - 1(-104) + 86 + 3(312)]^2}{(4)(20)}$$

$$= \frac{(2092)^2}{80} = 54{,}706 \tag{33.20}$$

$$S_{A_q} = \frac{(A_1 - A_2 - A_3 + A_4)^2}{r(\lambda^2 S)}$$

$$= \frac{(-322 + 104 - 86 + 312)^2}{(4)(4)}$$

$$= \frac{8^2}{16} = 4 \tag{33.21}$$

$$S_{A_c} = \frac{(-A_1 + 3A_2 - 3A_3 + A_4)^2}{(4)(20)}$$

$$= \frac{[-(-322) + 3(-104) - 3(86) + 312]^2}{80}$$

$$= \frac{64^2}{80} = 51 \tag{33.22}$$

$$S_{B_l} = \frac{(-3B_1 - B_2 + B_3 + 3B_4)^2}{r(\lambda^2 S)}$$

$$= \frac{[-3(130) - (67) + (-48) + 3(-177)]^2}{(4)(20)}$$

$$= \frac{(-1036)^2}{80} = 13{,}416 \tag{33.23}$$

$$S_{B_q} = \frac{(130 - 67 + 48 - 177)^2}{(4)(4)}$$

$$= 272 \tag{33.24}$$

$$S_{B_q} = \frac{[-1(130) + 3(67) - 3(-48) + (-177)]^2}{(4)(20)}$$

$$= 18 \tag{33.25}$$

To calculate $S_{A_l B_l}$, comparisons of A_l at each level of B (L_1, L_2, L_3, and L_4) are written:

$$L_1 = (-3)(-51) + (-1)(-9) + (1)(61) + (3)(129) = 610$$

$$L_2 = (-3)(-74) + (-1)(1) + (1)(40) + (3)(100) = 561$$

$$L_3 = (-3)(-89) + (-1)(-32) + (1)(13) + (3)(60) = 492$$

$$L_4 = (-3)(-108) + (-1)(-64) + (1)(-28) + (3)(23) = 429 \tag{33.26}$$

The number of units of these levels is 20. The values of L_1, L_2, L_3, and L_4, divided by $r(\lambda S)h_A = (1)(10)(10) = 100$, are the increments of tensile strength when the extent of processing is increased by 1% for each annealing temperature. From the values of L_1, L_2, L_3, and L_4, it seems that the effect of processing decreases linearly when annealing temperature increases. $S_{A_l B_l}$, the extent of the processing effect, decreases linearly as the annealing temperature increases. This is calculated by multiplying the linear coefficients of the four levels: -3, -1, 1, and 3:

$$S_{A_l B_l} = \frac{(-3L_1 - L_2 + L_3 + 3L_4)^2}{[(-3)^2 + (-1)^2 + 1^2 + 3^2](20)}$$

$$= \frac{(-612)^2}{400} = 936 \tag{33.27}$$

The total variation is

$$S_T = (-51)^2 + (-74)^2 + \cdots + 23^2 - \frac{(-28)^2}{16}$$

$$= 69859 \tag{33.28}$$

Residual variation is obtained by subtracting the variations of factorial effects, which are being analyzed from the total variation, S_T.

$$S_e = S_T - (S_{A_l} + S_{A_q} + S_{A_r} + S_{B_l} + S_{B_q} + S_{B_c} + S_{A_l B_l})$$

$$= 69,859 - (54706 + 4 + 51 + 13416 + 272 + 18 + 936)$$

$$= 456 \tag{33.29}$$

The result of the calculation of S_A is checked by adding S_{A_l}, S_{A_q}, and S_{A_c}. It means that the differences between A_1, A_2, A_3, and A_4 are decomposed to the components of a polynomial to achieve more analytical meaning. If necessary, the terms $S_{A_q B_l}$ corresponding to $A^2 B$ can be calculated. If it were calculated, we would find it to be very small, since it is merely a component of S_e.

Table 33.6

Analysis of variance

Source	f	S	V	ρ (%)
A				
Linear	1	54,706	54,706	78.3
Quadratic	1	4	4	
Cubic	1	51	51	
B				
Linear	1	13,416	13,416	19.1
Quadratic	1	272	272	
Cubic	1	18	18	
A_1B_1	1	936	936	1.2
e	8	456	57.0	
Total	15	69,859		100.0
(e)	(12)	(801)	(66.7)	(1.4)

The result of the decomposition of variation and the analysis of variance are summarized in Table 33.6. The degree of contribution of A is calculated after pooling the insignificant effects, S_{A_q}, S_{A_c}, S_{B_q}, and S_{B_c}, with S_e.

$$\rho_A = \frac{S_A - V_e}{S_T} = \frac{54,706 - 66.7}{69,859} = 78.2\% \tag{33.30}$$

However, it is a general rule to pool insignificant factorial effects with error. In some special cases, insignificant factorial effects are not pooled, and in other cases, significant factorial effects are pooled. If the product of A and B is pooled with the residue, the degrees of contribution of error increases from 1.4% to 2.6%.

Now we can just consider four terms in the observational equation. These terms are the linear term of A, the linear term of B, the product of A and B, and the general mean. The observational equation is as follows:

$$y = m + a_1(A - \bar{A}) + b_1(B - \bar{B}) + c_{11}(A - \bar{A})(B - \bar{B}) \tag{33.31}$$

In this calculation, remember to convert the values of the four unknowns, m, a_1, b_1, and c_{11}, to their original units.

$$\hat{m} = \text{average of total} = \text{working mean} + \frac{1}{10}\left(\frac{T}{16}\right)$$

$$= 70.0 + \frac{1}{10}\left(\frac{-28}{16}\right) = 69.82 \tag{33.32}$$

$$\hat{a} = \frac{-3A_1 - A_2 + A_3 + 3A_4}{10r(\lambda S)h_A} = \frac{(-3)(-322) - (-104) + 86 + (3)(312)}{(10)(4)(10)(10)}$$

$$= 0.523 \tag{33.33}$$

In the denominator of equation (33.33), the first 10 is for the conversion to the original unit, since the original data were multiplied by 10; r is the number of replications; λS is obtained from the table of orthogonal polynomials; and h_A is the interval between the levels of A. Similarly,

$$\hat{b}_1 = \frac{-3B_1 - B_2 + B_3 + 3B_4}{10rSh_B} = \frac{-3(130) - 67 - 48 + (3)(-177)}{(10)(4)(10)(50)}$$

$$= -0.0518 \tag{33.34}$$

$$\hat{c}_{11} = \frac{-3L_1 - L_2 + L_3 + 3L_4}{10r(\lambda S)_A h_A (\lambda S)_B h_B} = \frac{(-3)(610) - (561) + 492 + (3)(429)}{(10)(1)(10)(10)(10)(50)}$$

$$= -0.00122 \tag{33.35}$$

The population mean shown as y in equation (33.31) is written as $\hat{\mu}$, since it is the estimation for any values of A and B:

$$\hat{\mu} = 69.82 + 0.523(A - 45) - 0.0518(B - 225)$$

$$- 0.00122(A - 45)(B - 225) \tag{33.36}$$

Equation (33.36) is the result of orthogonal expansion, so the estimates of the coefficient of each term are independent of each other.

34 Two-Way Layout with Repetition

34.1. Introduction   573
34.2. Same Number of Repetitions   573
34.3. Different Numbers of Repetitions   578

34.1 Introduction

In this chapter we describe experiments with repetitions. This chapter is based on Genichi Taguchi et al., *Design of Experiments*. Tokyo: Japanese Standards Association, 1973.

34.2. Same Number of Repetitions

The data in Table 34.1 show the height of bounding for two brands of golf balls: A_1 (Dunlop) and A_2 (Eagle). The bounding of two balls from each brand was measured at four temperatures: $B_1 = 0°C$, $B_2 = 10°C$, $B_3 = 20°C$, and $B_4 = 30°C$. After subtracting a working mean of 105.0 cm (Table 34.2a), the sums of repetitions, the sums of rows, and sums of columns are then calculated to get a supplementary table (Table 34.2b).

$$CF = \frac{27.0^2}{16}$$

$$= 45.56 \quad (f = 1) \tag{34.1}$$

$$S_T = (-6.0)^2 + 0.1^2 + \cdots + 11.0^2 - CF$$

$$= 833.50 \quad (f = 15) \tag{34.2}$$

Table 34.1
Bounding height of golf balls (cm)

	B_1	B_2	B_3	B_4
A_1	99.0	105.1	110.3	114.5
	98.2	104.6	112.8	116.1
A_2	96.1	101.6	109.8	117.1
	95.2	102.4	108.2	116.0

Next, the *variation between the combinations* of A and B, say S_{T_1}, is obtained. S_{T_1} can also be called either the *variation between experiments* (there are eight experimental combinations with different conditions), or the *variation between the primary units.*

$$S_{T_1} = \frac{1}{2} \left[(-12.8)^2 + (-0.3)^2 + \cdots + 23.1^2 \right] - \mathrm{CF}$$

$$= 826.04 \quad (f = 7) \tag{34.3}$$

The variation between experiments is decomposed into the following variations:

$$S_A: \text{main effect of } A$$

$$S_{B_l}: \text{linear effect of } B$$

$$S_{B_q}: \text{quadratic effect of } B$$

$$S_{B_c}: \text{cubic effect of } B$$

Table 34.2
Supplementary table

	B_1	B_2	B_3	B_4
A_1	-6.0	0.1	5.3	9.5
	-6.8	-0.4	7.8	11.1
A_2	-8.9	-3.4	4.8	12.1
	-9.8	-2.6	3.2	11.0

(a)

	B_1	B_2	B_3	B_4	Total
A_1	-12.8	-0.3	13.1	20.6	20.6
A_2	-18.7	-6.0	8.0	23.1	6.4
Total	-31.5	-6.3	21.1	43.7	27.0

(b)

S_{AB_l}: interaction between the linear effect of B and the levels of A (A_1 and A_2)

S_{e_l}: primary error (error between experiments)

$$S_A = \frac{1}{16}(20.6 - 6.4)^2$$

$$= 12.60 \quad (f = 1) \tag{34.4}$$

$$S_{B_l} = \frac{(-3B_1 - B_2 + B_3 + 3B_4)^2}{[(-3)^2 + (-1)^2 + 1^2 + 3^2](4)}$$

$$= \frac{[(-3)(31.5) - (1)(-6.3) + (1)(21.1) + (3)(43.7)]^2}{80}$$

$$= 800.11 \quad (f = 1) \tag{34.5}$$

$$S_{B_q} = \frac{[(1)(-31.5) + (-1)(-6.3) + (-1)(21.1) + (1)(43.7)]^2}{(4)(4)}$$

$$= 0.42 \quad (f = 1) \tag{34.6}$$

$$S_{B_c} = \frac{[(-1)(-31.5) + (3)(-6.3) + (-3)(21.1) + (1)(43.7)]^2}{(20)(4)}$$

$$= 0.61 \quad (f = 1) \tag{34.7}$$

Next, the comparisons of the linear effects of B for A_1 and A_2 are calculated:

$$L(A_1) = (-3)(-12.8) - (1)(-0.3) + (1)(13.1) + (3)(20.6)$$

$$= 113.6 \tag{34.8}$$

$$L(A_2) = (-3)(-18.7) - (1)(-6.0) + (1)(8.0) + (3)(23.1)$$

$$= 139.4$$

The number of units for $L(A_1)$ and $L(A_2)$ are both equal to

$$[(-3)^2 + (-1)^2 + 1^2 + 3^2](2) = 40$$

$$S_{AB_l} = \frac{L(A_1)^2 + L(A_2)^2}{40} - \frac{[L(A_1) + L(A)_2)]^2}{80}$$

$$= \frac{113.6^2 + 13.94^2}{40} - \frac{253.0^2}{80}$$

$$= 8.32 \quad (f = 1) \tag{34.9}$$

$$S_{e_l} = S_{T_1} - S_A - S_{B_l} - S_{B_q} - S_{B_c} - S_{AB_l}$$

$$= 3.98 \quad (f = 2) \tag{34.10}$$

Next, the *variation within repetitions* (also called the *variation of the secondary error*), namely S_{e_2}, is calculated:

$$S_{e_2} = \left[(-6.0)^2 + (-6.8)^2 - \frac{(-12.8)^2}{2}\right] + \cdots$$

$$+ \left(12.1^2 + 11.0^2 - \frac{23.1^2}{2}\right)$$

$$= (-6.0) + (-6.8)^2 + \cdots + 12.1^2 + 11.0^2 - CF$$

$$- \left[\frac{(-12.8)^2 + \cdots + 23.1^2}{2} - CF\right]$$

$$= S_T - S_{T_1}$$

$$= 833.50 - 826.04$$

$$= 7.46 \quad (f-8) \tag{34.11}$$

The analysis of variance is shown in Table 34.3. (*Note:* When e_1 is significant as tested against e_2, then B_q, B_c, and e_1 are pooled to get a new primary error variance, which is used to test A, B_l, and AB_l.)

Comparing the primary error against the secondary error, we find it to be insignificant. Hence, we deem that the error between experiments does not exist, so e_1 and e_2 are pooled.

The results from the analysis of variance enable the following conclusions to be made:

1. There is a slight difference between the two brands.
2. The bouncing height increases linearly according to a rise in temperature.
3. The bouncing height increases linearly according to a rise in temperature, but the trend differs slightly between A_1 and A_2.

Estimation is made as follows: The linear equations of B are formed for A_1 and A_2, respectively. For case A_1,

Table 34.3
ANOVA table

Source	f	S	V	S'	ρ (%)
A	1	12.60	12.60	11.56	1.4
B					
l	1	800.11	800.11	799.07	95.9
q	1	0.42	0.42		
c	1	0.61	0.61		
AB₁	1	8.32	8.32	7.28	0.9
E₁	2	3.98	1.99		
E₂	8	7.46	0.932		
(e)	(12)	(12.47)	(1.04)	15.59	1.8
Total	15	833.50		833.50	100.0

$$\hat{\mu} = \overline{A}_1 + b_1(B - \overline{B}) \qquad (34.12)$$

$$\overline{A}_1 = \text{working mean} + \frac{1}{8}A_1$$

$$= 105.0 + \frac{20.6}{8}$$

$$= 107.6 \qquad (34.13)$$

$$\hat{b}_1 = \frac{-3B_1 - B_2 + B_3 + 3B_4}{r(\lambda S)h}$$

$$= \frac{(-3)(-12.8) - (1)(-0.3) + (1)(13.1) + (3)(20.6)}{(2)(10)(10)}$$

$$= \frac{113.6}{200}$$

$$= 0.568 \qquad (34.14)$$

$$\overline{B} = \frac{1}{4}(0 + 10 + 20 + 30)$$

$$= 15 \qquad (34.15)$$

From the above, the relationship between the bouncing height of golf ball A_1 and temperature, B, is

$$\hat{\mu} = 107.6 + 0.568(B - 15) \qquad (34.16)$$

Case A_2 is similar to case A_1:

$$\overline{A}_2 = 105.0 + \frac{6.4}{8} = 105.8$$

$$\hat{b}_1 = \frac{139.4}{200} = 0.697$$

$$\hat{\mu} = 105.8 + 0.697(B - 15) \qquad (34.17)$$

The results above show that the effect of bouncing by temperature for A_2 (Eagle) is about 23% higher than A_1 (Dunlop).

It is said that professional golfers tend to have lower scores during cold weather if their golf balls are kept warm. The following is the calculation to show how much the distance will be increased if a golf ball is warmed to 20°C.

To illustrate, $B = 5°C$ and $B = 20°C$ are placed into equations (34.16) and (34.17). When $B = 5°C$,

$$\text{Dunlop:} \quad \hat{\mu} = 107.6 + (0.568)(5 - 15)$$

$$= 101.92 \qquad (34.18)$$

$$\text{Eagle:} \quad \hat{\mu} = 105.8 + 0.697(5 - 15)$$

$$= 98.83$$

When $B = 20°C$,

$$\text{Dunlop:} \quad \hat{\mu} = 107.6 + (0.568)(20 - 15)$$
$$= 110.44 \tag{34.19}$$
$$\text{Eagle:} \quad \hat{\mu} = 105.8 + (0.697)(20 - 15)$$
$$= 109.28$$

Hence, the increase in distance is

$$\text{Dunlop:} \quad (110.44 \div 101.92 - 1)(100) = 8.4\% \tag{34.20}$$
$$\text{Eagle:} \quad (109.28 \div 98.83 - 1)(100) = 10.6\%$$

34.3. Different Numbers of Repetitions

In this section, analysis with an unequal number of data for each combination of A and B is described. In the experiments shown in Section 34.2, it seldom happens that an unequal number of repetitions occurs; but it can happen, especially when the data are obtained from questionnaires.

Table 34.4 shows the tensile strength of a certain product from three different manufacturers:

A_1:   foreign product

A_2:   our company's products

A_3:   another domestic company's products

at four different temperatures: $B_1 = -30°C$, $B_2 = 0°C$, $B_3 = 30°C$, and $B_4 = 60°C$. The numbers of repetitions are 1 for A_1, 5 for A_2 (in A_2, 4 for B_3 and 3 for B_4), and 3 for A_3. A working mean of 80 has been subtracted. Now, how would we analyze these data?

First, calculate the mean for each combination. From the repeated data of each combination in Table 34.4, the mean is calculated to get Table 34.5. It is

Table 34.4
Tensile strength (kg/mm²)

	B_1	B_2	B_3	B_4
A_1	20	8	0	−9
A_2	22	12	−2	−12
	25	8	0	−14
	28	10	3	−13
	25	9	0	
	26	12		
A_3	17	6	−8	−20
	23	8	−6	−18
	20	4	−3	−22

Table 34.5

Table of means

	B_1	B_2	B_3	B_4	Total
A_1	20.0	8.0	0.0	−9.0	19.0
A_2	25.2	10.2	0.2	−13.0	22.6
A_3	20.0	6.0	−5.7	−20.0	0.3
Total	65.2	24.2	−5.5	−42.0	41.9

recommended that the means be calculated to at least one decimal place. Then do an analysis of variance using the means.

$$CF = \frac{41.9^2}{12} = 146.30 \tag{34.21}$$

$$S_A = \frac{19.0^2 + 22.6^2 + 0.3^2}{4} - CF = 71.66 \tag{34.22}$$

$$S_B = \frac{65.2^2 + 24.2^2 + \cdots + (-42.0)^2}{3} - CF \tag{34.23}$$

$$S_{T_1} = 20.2^2 + 8.0^2 + \cdots + (-20.0)^2 - CF = 2175.31 \tag{34.24}$$

S_{T_1} is the total variation between the combinations of A and B. From S_{T_1}, subtract S_A and S_B to get the variation, called the *interaction* between A and B and denoted by S_{AB}. S_A indicates the difference between the companies as a whole.

To make a more detailed comparison, the following contrasts are considered:

$$L_1(A) = \frac{A_1}{4} - \frac{A_2 + A_3}{8} \tag{34.25}$$

$$L_2(A) = \frac{A_2}{4} - \frac{A_3}{4} \tag{34.26}$$

L_1 shows the comparison between foreign and domestic products, and L_2 compares our company and another domestic company. Since L_1 and L_2 are orthogonal to each other, their variations, $S_{L_1(A)}$ and $S_{L_2(A)}$, are

$$S_A = S_{L_1}(A) + S_{L_2}(A) \tag{34.27}$$

where

$$S_{L_1}(A) = \frac{[L_1(A)]^2}{\text{sum of coefficients squared}} = \frac{2A_1 - A_2 - A_3}{(2^2)(4) + (-1)^2(8)}$$

$$= \frac{[(2)(19.0) - 22.6 - 0.3]^2}{24} = \frac{15.1^2}{24}$$

$$= 9.50 \tag{34.28}$$

$$S_{L_2}(A) = \frac{[L_2(A)]^2}{\text{sum of coefficients squared}} = \frac{(A_2 - A_3)^2}{(1^2)(4) + (-1)^2(4)}$$

$$= \frac{(22.6 - 0.3)^2}{8}$$

$$= 62.16 \tag{34.29}$$

Next, we want to know whether or not the relationship between temperature and tensile strength is expressed by a linear equation. The linear component of temperature B is given by

$$S_{B_l} = \frac{(-3B_1 - B_2 + B_3 + 3B_4)^2}{r(\lambda^2 S)}$$

$$= \frac{[(-3)(65.2) - 24.2 + (-5.5) + (3)(-42.0)]^2}{(3)(20)}$$

$$= 2056.86 \tag{34.30}$$

Let the variation of B, except its linear component, be B_{res}:

$$S_{B_{\text{res}}} = S_B - S_{B_l} = 2064.01 - 2056.86 = 7.15 \tag{34.31}$$

For the linear coefficients of temperature, their differences between foreign and domestic, or between our company and another domestic company, are found by testing $S_{L_1(A)B_l}$ and $S_{L_2(A)B_l}$, respectively.

$$S_{L_1}(A)B_l = \frac{[2L(A_1) - L(A_2) - L(A_3)]^2}{[2^2 + (-1)^2(2)][(-3)^2 + (-1)^2 + 1^2 + 3^2]}$$

$$= \frac{[(2)(-95.0) + 124.6 + 131.7]^2}{120}$$

$$= 36.63 \tag{34.32}$$

$$S_{L_2}(A)B_l = \frac{[L(A_2) - L(A_3)]^2}{[1^2 + (-1)^2][(-3)^2 + (-1)^2 + 1^2 + 3^2]}$$

$$= \frac{(-124.6 + 131.7)^2}{40}$$

$$= 1.26 \tag{34.33}$$

where

$$L(A_i) = -3A_iB_1 - A_iB_2 + A_iB_3 + 3A_iB_4 \tag{34.34}$$

The primary error variation, S_{e_1}, is obtained from S_{T_1} by subtracting all of the variations of the factorial effects that we have considered.

$$S_{e_1} = S_{T_1} - S_A - S_B - S_{L_1(A)B_l} - S_{L_2(A)B_l}$$

$$= 2175.31 - 71.66 - 2064.01 - 36.63 - 1.26$$

$$= 1.75 \quad (f = 4) \tag{34.35}$$

Next, the variation within repetitions, S_{e_2}, is calculated. The error variation of each combination of A and B in Table 34.4 is calculated such that the number of repetitions is different for each combination. These variations are summed and then divided by the harmonic mean of the numbers or repetitions. This is denoted by $\bar{r}$.

$$S_{e_2} = \frac{1}{r}[S_{11} \text{ (error variation within the repetitions of } A_1B_1) + \cdots$$

$$+ S_{34} \text{ (error variation within the repetions of } A_3B_4)]$$

$$= \frac{1}{r}\left\{\left(20 - \frac{20^2}{1}\right) + \cdots + \left[(-20)^2 + (-18)^2\right.\right.$$

$$\left.\left. + (-22)^2 - \frac{(-20 - 18 - 22)^2}{3}\right]\right\}$$

$$= \frac{1}{r}[\text{(sum of the individual data squared)}$$

$$- \text{(variation between the combinations of } A \text{ and } B)]$$

$$= \frac{1}{r}(93.02) \tag{34.36}$$

$$\frac{1}{r} = \frac{1}{\text{number of combinations of } A \text{ and } B}$$

$$\text{(sum of recipricals of the number of repetitions of } A_iB_j)$$

$$= \frac{1}{12}\left(\frac{1}{1} + \frac{1}{1} + \cdots + \frac{1}{3}\right) = \frac{1}{1.8997} \tag{34.37}$$

Putting equation (34.37) into equation (34.36) yields

$$S_{e_2} = 48.97 \tag{34.38}$$

Since S_A and S_B were obtained from the mean values, S_{e_2} is multiplied by $1/\bar{r}$. From the results above, the analysis of variance can be obtained, as shown in Table 34.6. In the table, S_T was obtained by

$$S_T = S_A + S_B + \cdots + S_{e_2} = 2224.28$$

The degrees of freedom for S_{e_2} are obtained by summing the degrees of freedom for the repetitions A_1B_1, A_1B_2, ... , A_3B_4 to be $0 + \cdots + 2 = 21$.

The analysis of variance shows that $L_{2(A)}$, B_1, and $L_{1(A)B_1}$ are significant. The significance of $L_{2(A)}$ means there is a significant difference between the two domestic companies. It has also been determined that the influence of temperature to tensile strength is linear; the extent to which temperature affects tensile strength differs between the foreign and domestic products.

Table 34.6
ANOVA table

Source	f	S	V	S'	ρ (%)
A					
$L_1(A)$	1	9.50	9.50[a]		
$L_2(A)$	1	62.16	62.16	59.793	2.69
B					
I	1	2056.86	2056.86	2054.493	92.37
Res	2	7.15	3.58[a]		
AB					
$L_1(A)B_1$	1	36.63	36.63	34.263	1.54
$L_2(A)B_1$	1	1.26	1.26[a]		
e_1	4	1.75	0.438[a]		
e_2	21	48.97	2.332[a]		(3.46)
e'	(29)	(68.63)	(2.367)	(75.731)	(3.40)
Total	32	2224.28		2224.28	100.00

[a] Pooled data.

In the case of foreign products, the influence per 1°C, or b_1, is given by

$$\hat{b} = \frac{L(A_1)}{r(\lambda S)h}$$

$$= \frac{(-3)(20) - 8.0 + 0.0 + (3)(-9.0)}{(1)(10)(30)}$$

$$= \frac{-95.0}{300}$$

$$= -0.317 \qquad\qquad (34.39)$$

Similarly, b_2 for domestic products is

$$\hat{b}_2 = \frac{L(A)_2 + L(A_3)}{r(\lambda S)h}$$

$$= \frac{-124.6 - 131.7}{(2)(10)(30)}$$

$$= \frac{-256.3}{600} = -0.427 \qquad\qquad (34.40)$$

Therefore, the tensile strength is expressed by the linear equation of temperature.

For foreign products,

$$\hat{\mu}_1 = \overline{A}_1 + b_1(B - \overline{B})$$
$$= 80 + 4.75 - (0.317)(B - 15)$$
$$= 84.75 - (0.317)(B - 15) \tag{34.41}$$

Our company's products:

$$\hat{\mu}_2 = \overline{A}_2 + b_2(B - \overline{B})$$
$$= 80 + 5.65 - (0.427)(B - 15)$$
$$= 85.65 - (0.427)(B - 15) \tag{34.42}$$

Another domestic company's products:

$$\hat{\mu}_3 = \overline{A}_3 + b_3(B - \overline{B})$$
$$= 80 + 0.075 - (0.427)(B - 15)$$
$$= 80.075 - (0.427)(B - 15) \tag{34.43}$$

35 Introduction to Orthogonal Arrays

35.1. Introduction   584
35.2. Quality Characteristics and Additivity   584
    Data Showing Physical Condition   584
    Noise Level and Additivity   586
    Data Showing Particle Size Distribution   587
35.3. Orthogonal Array L_{18} 589
35.4. Role of Orthogonal Arrays   594
35.5. Types of Orthogonal Arrays   596
    References   596

35.1. Introduction

Earlier, reproducibility of conclusions was discussed, and it was recommended that one use orthogonal arrays to check the existence of interactions, the cause of nonreproducibility. Followings are examples of poor-quality characteristics that cause interactions.

35.2. Quality Characteristics and Additivity [1]

Data Showing Physical Condition

At the stage of selecting a characteristic, it must be considered whether that characteristic is the most suitable one to use to achieve the purpose of the experiment.

A characteristic may not always exactly express the purpose itself. As an example, let's discuss an experiment for curing the physical condition of a patient. Since the data showing physical condition is directly related to the purpose of the experiment, a characteristic "condition" is selected. Another example of a characteristic condition would be data recorded under categories such as "delicious" or "unsavory."

Assume that there are three types of medicines: A, B, and C. First, medicine A was taken by the patient, then his or her physical condition became better. Such data can be expressed as:

584

A_1: Did not take A Bad condition

A_2: Took A Slightly better condition

Suppose that the results after taking medicines B and C were as follows:

B_1: Did not take B Bad condition

B_2: Took B Better condition

C_1: Did not take C Bad condition

C_2: Took C Slightly better condition

If, for the reason that the three medicines (A, B, and C) did the patient good, and if the patient became much better after taking the three medicines together, we say that there exists additivity in regard to the effects of the three medicines, A, B, and C. However, if the patient became worse or died after taking the three medicines together, we say that there is interaction (a large effect that reverses the results) between A, B, and C.

When there is interaction, it is generally useless to carry out the investigations in such a way that one factor (in this case, medicine) is introduced into the experiment followed by another factor.

For example, suppose that a patient is suffering from diabetes, which is caused by lack of insulin being produced by the pancreas. Also suppose that medicine A contains 80%, B contains 100%, and C contains 150% of the amount of insulin the patient requires. In this case, interaction does occur. Taking A, B, and C together would result in a 330% insulin content, which would definitely be fatal.

As far as physical condition is concerned, there is no additivity. But if the amount of insulin is measured, it becomes possible to estimate the result of taking A, B, and C together, but the combination can never be tested. If effect A is reversed by the other conditions, solely researching effect A would turn out to be meaningless. Accordingly, we have to conduct the experiment in regard to all the combinations of A, B, and C.

When there are three two-level factors, there are only eight combinations in total; however, when there are 10 factors to be investigated, 1024 combinations should be included in the experiment, which would be impossible.

If the amount of insulin is measured, there then exists additivity. That is, we can predict that physical condition becomes worse when A, B, and C are taken together. In other words, we do not need the data from all combinations. In this way, even when there are 10 two-level factors, we can estimate the results for all combinations by researching each individual factor, or by using orthogonal arrays (described in other chapters) and then finishing the research work by conducting only about 10 experiments.

"Physical condition" is not really an efficient and scientific characteristic, but "quantity of insulin" is. Research efficiency tends to drop if we cannot find a characteristic by which the results of the individual effect reappears or has consistent effects, no matter how the other conditions, variables, or causes vary. It is therefore most important from the viewpoint of research efficiency to find a characteristic with the additivity of effects in its widest sense.

Noise Level and Additivity

To measure noise level, it is usual to measure the magnitude of noise for an audible frequency range using "phon" or "decibel" as a unit. Generally, there are three types of factors affecting noise. Of these types, additivity exists for the factors of the following two types:

1. Factors that extinguish vibrating energy, the origin of noise.

2. Factors that do not affect vibrating energy itself, but convert the energy caused by vibration into a certain type of energy, such as heat energy. That is, the less vibrating source, the lower the noise level. Also, the more we convert the energy that causes noise into other types of energy, the lower the noise level.

The third type of factors imply:

3. Factors that change the frequency characteristics into frequency levels outside the audible range: for example, changing the shape or dimensions of component parts, changing the gap between component parts, changing surface conditions, or switching the relative positions of the various component parts.

Suppose that the frequency characteristic of a certain noise is as shown in Figure 35.1. There is a large resonance frequency at 500 cycles. It forms half of the total noise. There are two countermeasures, A and B, as follows:

1. *Countermeasure A:* converts resonance frequency from 500 cycles to 50,000 cycles. As a result, the noise level decreases by 50%.

2. *Countermeasure B:* converts resonance frequency from 500 cycles to 5 cycles. As a result, again the noise level decreases by 50%.

If the two countermeasures were instituted at the same time, the noise level might come back to the original magnitude.

Generally, most factors cited in noise research belong to the third type. If so, it is useless to use noise level as a characteristic. Unless frequency characteristics are researched with the conditions outside the audible range, it would turn out to be inefficient; therefore, researching all combinations is required to find the optimum condition.

In many cases, taking data such as the frequency of misoperation of a machine, the rate of error in a communications system or in a calculator, or the magnitude

Figure 35.1
Frequency and noise level

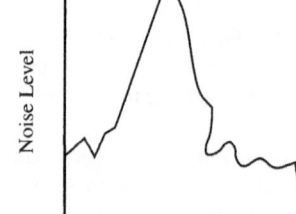

of electromagnetic waves when there are frequency changes does not make sense. These errors are not only the function of the signal-to-noise (SN) ratio, but may also be caused by the unbalance of a setting on the machine. In the majority of these cases, it is advisable to calculate the SN ratio, or it is necessary to find another method to analyze these data, which are called *classified attributes*.

Next we discuss an experiment concerned with a crushing process.

**Data Showing
Particle Size
Distribution**

☐ Example

The product becomes off grade when it is either too coarse or too fine; the fineness within the range of 15 through 50 mesh is desirable, it being most desirable to increase the yield. But additivity does not exist for yield or percentage in this case. One would not use such a characteristic for the purpose of finding the optimum condition efficiently.

Suppose that the present crushing condition is $A_1B_1C_1$. The resulting yields after changing A_1 to A_2, B_1 to B_2, and C_1 to C_2 are as follows:

$$A_1:\ 40\% \qquad B_1:\ 40\% \qquad C_1:\ 40\%$$

$$A_2:\ 80\% \qquad B_2:\ 90\% \qquad C_2:\ 40\%$$

The data signify that yield increases by 40% when crushing condition A changes from A_1 to A_2; the yield also increases by 50% when B is changed from B_1 to B_2, but yield does not change when C is changed from C_1 to C_2. Judging from these data, the optimum condition is either $A_2B_2C_1$ or $A_2B_2C_2$.

The yield of any one of the optimum conditions is expected to the nearly 100%. But if the crushing process is actually operated under condition $A_2B_2C_2$, and if a yield of less than 10% is obtained, the researcher would be puzzled. However, it

Table 35.1
Particle size distribution

	≤15	15–50	≥50
A_1	60	40	0
A_2	0	80	20
B_1	60	40	0
B_2	0	90	10
C_1	60	40	0
C_2	0	40	60

is very possible to obtain such a result. This is easy to understand from the particle distribution shown in Table 35.1.

At present operating conditions, a 40% low yield is obtained because the particles are too coarse. As to conditions C_1 and C_2, the percentage of particles in the range 15 to 50 mesh is equal, but factor C exerts the largest influence on particle distribution. Only by changing condition C from C_1 to C_2 do the particles become much finer. Furthermore, adding the effects of A and B, the yield of condition $A_2B_2C_2$ tends to be very small. Yield is therefore not a good characteristic for finding an optimum condition.

Even if the yield characteristic were used, the optimum condition could be obtained if all of the combinations were included in the experiment. In this case, the optimum condition is obtainable because there are only eight experiments ($2^3 = 8$). When there are only three two-level factors, the performance of eight experiments is not impractical. But when these factors are three and four levels, the total combinations increase to 27 and 64, respectively.

If there are 10 three-level factors, experiments with 59,049 combinations are required. Assuming that the experiments performed are 10 combinations a day, about 20 years would be required to test for all the combinations. Clearly, in order to find an optimum condition by researching one factor after another, or researching a few factors at one time, a characteristic must have additivity.

The example above is a case where yield becomes 100% when particle size is in a certain range. In such a case, it is nonsense to take yield as data. In addition, we need percentages of the coarser portions and the finer portions. If the three portions are shown, it is clear that condition $A_2B_2C_2$ is the worse condition without conducting any experiment.

There is no additivity with the yield data within the range 15 to 50 mesh, but there is additivity if the total (100%) is divided into three classes, called *classified variables*.

When a physical condition is expressed as good, average, and bad, this type of data would not be additive. The same holds true when yield is expressed as a percentage.

The yield in the case of a reversible chemical reaction, or the yield of a reaction that may be overreacted, would not be considered additive data. What type of scientific field would exist where "condition" or "balance" is seriously discussed, such as the color matching of dyes or paints, the condition of a machine or a furnace, taste, flavor, or physical condition? In such fields, no one could be as competent as an expert worker who has the experience of nearly all the combinations. So far a characteristic with additivity has not been discovered; we can never be out of the working level.

It is difficult in most cases to judge in a specialized technical field whether the characteristic is additive. Accordingly, we need a way to judge whether or not we are lost in a maze. Evaluation of experimental reliability is the evaluation of a characteristic in regard to its additivity. The importance of experiments using orthogonal arrays lies in the feasibility of such evaluations.

35.3. Orthogonal Array L_{18} [2]

In quality engineering, orthogonal arrays L_{12}, L_{18}, and L_{36} are generally used. L_{12} is used for two-level factors, L_{18} for three-level factors, and L_{36} for simulation. In this discussion, L_{18} is used for simplifying the explanation.

❑ Example [1]

An early application of robust design involved the optimization of a tile manufacturing process in Japan in the 1950s. In 1953, a tile manufacturing experiment was conducted by the Ina Seito Company. The flow of the manufacturing process is as follows:

raw material preparation: crushing and mixing → molding → calcining

→ glazing → calcining

The molded tiles are stacked in the carts and move slowly in a tunnel kiln as burners fire the tiles. The newly constructed kiln did not produce tiles with uniform dimensions. More than 40% of the outside tiles were out of specification. The inside tiles barely met the specification. It was obvious to the engineers at the plant that one of the causes of dimensional variation was the uneven temperature distribution.

The traditional way of improving quality is to remove the cause of variation. In this case it would mean redesigning the tunnel kiln to make temperature distribution more uniform, which was impossible because it was too costly. Instead of removing the cause of variation, the engineers decided to perform an experiment to find the formula for the tile materials that would produce consistently uniform tiles, regardless of their positions within the kiln.

In the study, control factors were assigned to orthogonal arrays, and seven positions in the kiln were assigned to the outer array as noise factors. The interactions between each control factor and the noise factors were studied to find the optimum condition. Today, the SN ratio is used to substitute for the tedious calculation to study the interactions between control and noise factors.

Following are the control factors and their levels:

A: amount of a certain material
  A_1 = 5.0%
  A_2 = 1.0% (current)
B: firmness of material
  B_1 = fine
  B_2 = (current)
  B_3 = coarse
C: amount of agalmatolite
  C_1 = less
  C_2 = (current)
  C_3 = new mix without additive

D: type of agalmatolite
     $D_1 = -0.0\%$
     $D_2 = 1.0\%$ (current)
     $D_3 =$
E: amount charged
     $E_1 =$ smaller
     $E_2 =$ current
     $E_3 =$ larger
F: amount of returned material
     $F_1 =$ less
     $F_2 =$ medium (current)
     $F_3 =$ more
G: amount of feldspar
     $G_1 = 7\%$
     $G_2 = 4\%$ (current)
     $G_3 = 0\%$
H: clay type
     $H_1 =$ only K-type
     $H_2 =$ half-and-half (current)
     $H_3 =$ only G-type

These were assigned to an orthogonal array L_{18} as the inner array, and seven positions in the kiln were assigned to the outer array. Dimensions of tiles for each combination between a control factor and the noise factor are shown in Table 35.2.

Data Analysis
As a quality characteristic, a nominal-the-best SN ratio is used for analysis. (For the SN ratio, see later chapters.) For experiment 1:

$$S_m = \frac{(10.18 + 10.18 + \cdots + 10.20)^2}{7} = 714.8782 \tag{35.1}$$

$$S_T = 10.18^2 + 10.18^2 + \cdots + 10.20^2 = 714.9236 \tag{35.2}$$

$$S_e = S_T - S_m = 714.9236 - 714.8782 = 0.0454 \tag{35.3}$$

$$V_e = \frac{S_e}{7-1} = \frac{0.0454}{6} = 0.00757 \tag{35.4}$$

$$\eta = 10 \log \frac{(1/n)(S_m - V_e)}{V_e}$$

$$= 10 \log \frac{7(714.8782 - 0.00757)}{0.00757}$$

$$= 41.31 \text{ dB} \tag{35.5}$$

The SN ratios of the other 17 runs are calculated similarly, as shown in Table 35.2.

Table 35.2
Layout and results

L_{18}	A 1	B 2	C 3	D 4	E 5	F 6	G 7	H 8	P_1	P_2	P_3	P_4	P_5	P_6	P_7	Mean	SN
1	1	1	1	1	1	1	1	1	10.18	10.18	10.12	10.06	10.02	9.98	10.20	10.11	41.31
2	1	1	2	2	2	2	2	2	10.03	10.01	9.98	9.96	9.91	9.89	10.12	9.99	42.19
3	1	1	3	3	3	3	3	3	9.81	9.78	9.74	9.74	9.71	9.68	9.87	9.76	43.65
4	1	2	1	1	2	2	3	3	10.09	10.08	10.07	9.99	9.92	9.88	10.14	10.02	40.36
5	1	2	2	2	3	3	1	1	10.06	10.05	10.05	9.89	9.85	9.78	10.12	9.97	37.74
6	1	2	3	3	1	1	2	2	10.20	10.19	10.18	10.17	10.14	10.13	10.22	10.18	50.03
7	1	3	1	2	1	3	2	3	9.91	9.88	9.88	9.84	9.82	9.80	9.93	9.87	46.34
8	1	3	2	3	2	1	3	1	10.32	10.28	10.25	10.20	10.18	10.18	10.36	10.25	43.21
9	1	3	3	1	3	2	1	2	10.04	10.02	10.01	9.98	9.95	9.89	10.11	10.00	43.13
10	2	1	1	3	3	2	2	1	10.00	9.98	9.93	9.80	9.77	9.70	10.15	9.90	35.99
11	2	1	2	1	1	3	3	2	9.97	9.97	9.91	9.88	9.87	9.85	10.05	9.93	42.88
12	2	1	3	2	2	1	1	3	10.06	9.94	9.90	9.88	9.80	9.72	10.12	9.92	37.05
13	2	2	1	2	3	1	3	2	10.15	10.08	10.04	9.98	9.91	9.90	10.22	10.04	38.46
14	2	2	2	3	1	2	1	3	9.91	9.87	9.86	9.87	9.85	9.80	10.02	9.88	43.15
15	2	2	3	1	2	3	2	1	10.02	10.00	9.95	9.92	9.78	9.71	10.06	9.92	37.70
16	2	3	1	3	2	3	1	2	10.08	10.00	9.99	9.95	9.92	9.85	10.14	9.99	40.23
17	2	3	2	1	3	1	2	3	10.07	10.02	9.89	9.89	9.85	9.76	10.19	9.95	36.60
18	2	3	3	2	1	2	3	1	10.10	10.08	10.05	9.99	9.97	9.95	10.12	10.04	43.48

Tables 35.3 and 35.4 show the response tables for the SN ratio and average. Figure 35.2 shows the response graphs of SN the ratio and average. For example:

$$\text{average SN ratio for } A_1 = \frac{41.31 + 42.19 + \cdots + 43.13}{9}$$

$$= 43.10 \tag{35.6}$$

$$\text{average SN ratio for } A_2 = \frac{35.99 + 42.88 + \cdots + 43.48}{9}$$

$$= 39.50 \tag{35.7}$$

$$\text{average SN ratio for } B_1 = \frac{41.31 + 42.19 + \cdots + 37.05}{6}$$

$$= 40.51 \tag{35.8}$$

Table 35.3
Response table: SN ratio

Level	A	B	C	D	E	F	G	H
1	43.10	0.51	40.45	40.33	44.53	41.11	40.44	39.90
2	39.50	41.24	40.96	40.88	40.12	41.38	41.47	42.82
3		42.16	42.51	42.71	39.26	41.42	42.00	41.19
Δ	3.60	1.65	2.06	2.38	5.27	0.31	1.57	2.92
Ranking	2	6	5	4	1	8	7	3

Optimization

The first thing in parameter design is to reduce variability or maximize the SN ratio. Select the optimum combination from the SN ratio response table as $A_1B_3C_3D_3E_1F_3G_3H_2$. Factors A, C, D, E, and H had a relatively strong impact on variability, whereas B, F, and G had a relatively weak impact. Therefore, A_1, C_3, E_1, and H_2 are definitely selected.

The second step is to adjust the mean. Adjusting the mean is easy by adjusting the dimension of the mold. In general, look for a control factor with a large impact on the mean and a minimum impact on variability.

Estimation and Confirmation

It is important to check the validity of experiments or the reproducibility of results. To do so, the SN ratios of the optimum conditions and initial condition are estimated using the additivity of factorial effects. Their difference, called *gain*, is then compared with the one calculated from the confirmatory experiments under optimum and initial conditions.

Additivity of factorial effects is simply the sum of the gains from the control factors. Every factorial effect may be used (added) to estimate the SN ratios of the optimum and initial conditions. But it is recommended that one exclude weak factorial effects from the prediction to avoid overestimates. In this example, factors B,

Table 35.4
Response table: Mean

Level	A	B	C	D	E	F	G	H
1	10.02	9.93	9.99	9.99	10.00	10.07	9.98	10.03
2	9.95	10.00	10.00	9.97	10.02	9.97	9.97	10.02
3		10.02	9.97	9.99	9.94	9.91	10.01	9.90
Δ	0.06	0.08	0.03	0.02	0.08	0.17	0.04	0.13
Ranking	5	3	7	8	3	1	6	2

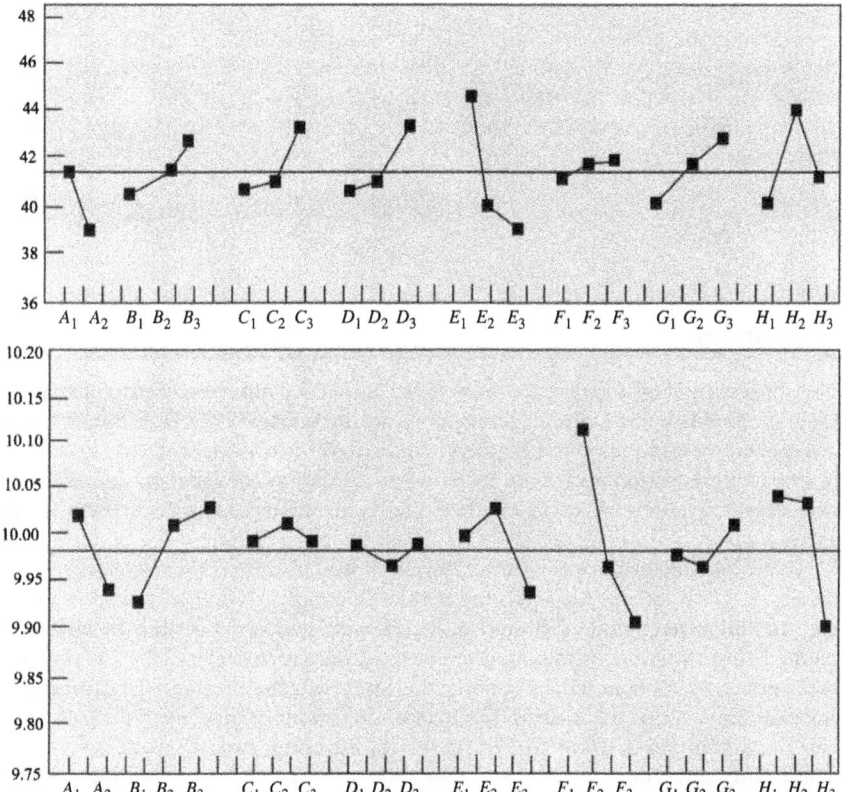

Figure 35.2
Response graphs

F, and G are excluded from the predictions. The SN ratios under the optimum and initial conditions, denoted by η_{opt} and $\eta_{initial}$, respectively, are predicted by

$$\eta_{opt} = \bar{T} + (\bar{A}_1 - \bar{T}) + (\bar{C}_3 - \bar{T}) + (\bar{D}_3 - \bar{T}) + (\bar{E}_1 - \bar{T}) + (\bar{H}_2 - \bar{T})$$

$$= \bar{A}_1 + \bar{C}_3 + \bar{D}_3 + \bar{E}_1 + \bar{H}_2 - 4\bar{T}$$

$$= 43.10 + 42.51 + 42.71 + 44.53 + 42.82 - (4)(41.30)$$

$$= 50.47 \text{ dB} \tag{35.9}$$

$$\eta_{initial} = \bar{T} + (\bar{A}_2 - \bar{T}) + (\bar{C}_2 - \bar{T}) + (\bar{D}_2 - \bar{T}) + (\bar{E}_2 - \bar{T}) + (\bar{H}_2 - \bar{T})$$

$$= \bar{A}_2 + \bar{C}_2 + \bar{D}_2 + \bar{E}_2 + \bar{H}_2 - 4\bar{T}$$

$$= 39.50 + 40.96 + 40.88 + 40.12 + 42.82 - (4)(41.30)$$

$$= 39.08 \text{ dB} \tag{35.10}$$

$$\text{gain predicted} = 50.47 - 39.08 = 11.39 \text{ dB} \tag{35.11}$$

Confirmatory experiments are conducted under the optimum and initial conditions. From these results, two SN ratios are calculated. Their difference is the gain confirmed. When the gain from prediction is close enough to the gain from the confirmatory experiments, it indicates there is additivity and that the conclusions are probably reproducible.

35.4. Role of Orthogonal Arrays

To estimate the eight factorial effects A, B, ... , H, it is not necessary to carry out the experiment by orthogonal arrays as shown in Section 35.3. Traditionally, one-factor-at-a-time experiments have been conducted. In this method, the condition of one factor is varied each time by fixing all conditions of other factors. Such an experiment is much easier to do than the experiments arranged by orthogonal arrays.

In the one-factor-by-one method, levels A_1 and A_2 are compared while other factors, B, C, ... , H, are fixed (to the first level, $B_1 C_1 D_1 E_1 F_1 G_1 H_1$). Generally speaking, the difference between A_1 and A_2 can be obtained very precisely by using this method. In orthogonal arrays, on the contrary, the average effect of A_1 and A_2 (or main effect A) is obtained by varying the levels of other factors. That means to calculate the average of A_1 from the data of the four experiments (No. 1, 2, to 9) and the average of A_2 from No. 10, 11, to 18. But using such a layout is very time consuming, since the conditions of many factors must be varied for each experiment; also, the data variation between experiments generally becomes larger compared with experiments by the one-factor-by-one method.

One way to look at it is that the comparison between single experimental figures in the one-factor-by-one method is less precise than the comparison between the average of four figures from the experiments that used orthogonal arrays. However, this advantage in orthogonal array design might be offset by the disadvantage caused by the tendency for increased variation when the levels of many factors vary from experiment to experiment. As a result, there is no guarantee of obtaining better precision. What, then, is the merit of recommending such time-consuming experiments as orthogonal arrays?

They are stressed because of the high reproducibility of factorial effects. In experiments with orthogonal arrays, the difference of the two levels, A_1 and A_2, is determined as the average effect, while the conditions of other factors vary.

If the influence of A_1 and A_2 on the experimental results is consistent while the conditions of other factors vary, the effect obtained from the experiments on orthogonal arrays tends to be significant. On the other hand, if the difference between A_1 and A_2 either reverses or varies greatly once the levels of other factors change, effect A tends to be insignificant.

If orthogonal arrays are used, a factor having a consistent effect, with other factors varying under different conditions, can be estimated. This means that a large factorial effect (or the order of the preferable levels) obtained from orthogonal arrays does not vary if there are some variations in the levels of other factors. The reliability of such factorial effects is therefore very good.

In experiments using the one-factor-by-one method, on the other hand, the difference between A_1 and A_2 is estimated under a certain constant condition of the other factors. No matter how precisely such an effect is estimated and how neatly the curve is plotted, the effect is correct only for the case where the levels of other factors are exactly identical to the condition that was fixed at the time of the experiment; there is no guarantee at all of obtaining a consistent factorial effect if other factorial conditions change. Accordingly, it is doubtful whether the results obtained from the experimental data of the one-factor-by-one method will be consistent if the researcher changes or if the raw material changes.

It is said that since orthogonal arrays have been used in experiments, the results of small-scale laboratory experiments have become adopted satisfactorily to actual manufacturing. That is, a factor with a consistent effect under the various conditions of other factors has a good possibility of reproducing its effect at a manufacturing scale.

The reason for using orthogonal arrays is not to reduce cost by improving the efficiency of experimentation. When orthogonal arrays such as L_{12}, L_{18} and L_{36} are used, the interactions between control factors are almost evenly distributed to other columns of the orthogonal arrays and confounded to various main effects. If the interactions between control factors are significant, the gain predicted under the optimum condition from the initial condition will become significantly different from the gain from confirmatory experiments.

Using the one-factor-at-a-time method, where all other factor levels are fixed, it is not certain that an effect will be consistent. In other words, the main effect may be different if the conditions of other factors change. So this method is good only when there are no interactions between control factors.

The method of assigning main factors only to an orthogonal array will be successful when there are no interactions. The chance of success is the same as the case of using the one-factor-at-a-time method. Then why is are orthogonal arrays used? It is to check the reproducibility of conclusions by conducting confirmatory experiments. Thus, the success or failure will be clear. If the gain predicted did not agree with the confirmed gain, it tells us that the experiment has failed and the optimum condition was not found.

It is the same as inspection. When a product passes an inspection, the inspection was wasted. Inspection has a value only when a defective is found. The defective product is either scrapped or repaired to prevent problems in the marketplace.

Similarly, an experiment using an orthogonal array is used to inspect a bad experiment. It is to prevent a wrongly designed product from being shipped to the market and causing problems. Therefore, when the gain predicted is reproduced in confirmatory experiments, the use of orthogonal arrays is wasted. The use of using orthogonal arrays is advantageous only when a gain was not reproduced.

To be successful in a study, it is necessary to find a quality characteristic with small interactions between control factors. It is believed in quality engineering that the dynamic SN ratio is based on the functionality of a product or a system. It is also believed that reproducibility using a dynamic SN ratio based on the generic function is superior to use of a nondynamic SN ratio.

35.5. Types of Orthogonal Arrays

There are many orthogonal arrays. $L_4(2^3)$, $L_8(2^7)$, $L_{12}(2^{11})$, $L_{16}(2^{15})$, or $L_{32}(2^{31})$ belong to two-level series; $L_9(3^4)$, $L_{27}(3^{13})$, or $L_{81}(3^{40})$ are three-level series; and $L_{18}(2 \times 3^7)$ or $L_{36}(2^3 \times 3^{13})$ are mixed-level arrays. In quality engineering, it is recommended that one use arrays such as L_{12}, L_{18}, and L_{36} because the interactions are almost evenly distributed to other columns, and there is no worry that an interaction confounds to a specific column or columns, thus leading to confusion.

References

1. Genichi Taguchi et al., 1973. *Design of Experiments.* Tokyo: Japanese Standards Association.
2. American Supplier Institute, Robust Design Workshop, 1998.

36 Layout of Orthogonal Arrays Using Linear Graphs

36.1. Introduction   597
36.2. Linear Graphs of Orthogonal Array L_8 597
       Linear Graph (1)   598
       Linear Graph (2)   598
36.3. Multilevel Arrangement   599
36.4. Dummy Treatment   605
36.5. Combination Design   606
       Reference   608

36.1. Introduction

Linear graphs have been developed by Taguchi for easy assignment of experiments. These graphs can be used to assign interactions between factors in an array to calculate interactions in the design of experiments. But in quality engineering, linear graphs are not used for assigning interactions but for special cases such as multilevel assignment. This chapter is based on Genichi Taguchi, *Design of Experiments*. Tokyo: Japanese Standards Association, 1973.

36.2. Linear Graphs of Orthogonal Array L_8

Before explaining the modifications of orthogonal arrays for experimental layout, it is necessary to know the column or columns where an interaction between columns is confounded. To simplify explanation, orthogonal array L_8 is used. Two linear graphs can be drawn for the L_8 table as shown in Figure 36.1.

Linear Graph (1)

Linear graph (1) in Figure 36.1 means that the interaction between columns 1 and 2 comes out to column 3, and the interaction between columns 1 and 4 comes out to column 5. Column 7 is shown as an independent point apart from the triangle.

(1)   (2)

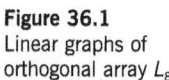

Figure 36.1
Linear graphs of
orthogonal array L_8

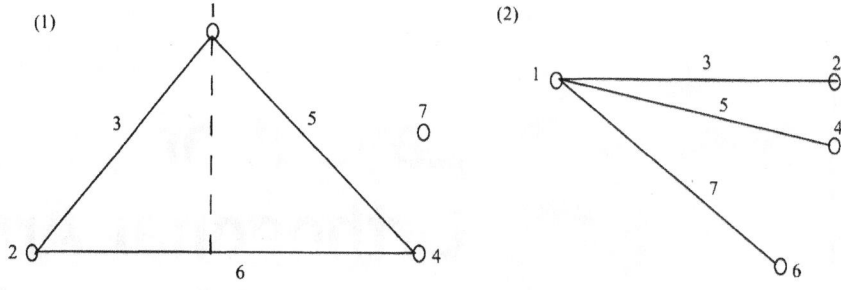

If there are four two-level factors, A, B, C, and D, and if A is assigned to column 1, B to column 2, C to column 4, and D to column 7, then interaction $A \times B$ is obtained from column 3, $A \times C$ from column 5, and $B \times C$ from column 6. Such assignment problems are easily solved using linear graph (1).

Linear Graph (2) Linear graph (2) is used for an experiment where the interactions between one particular factor and some other factors are important. This is illustrated in the following experiment, with two types of raw materials, A_1 and A_2; two annealing methods, B_1 and B_2; two temperatures, C_1 and C_2; and two treating times, D_1 and D_2.

It is expected that since B_1 and B_2 use different types of furnaces, their operations are quite different and the optimum temperature or optimum treating time might not be the same for B_1 as for B_2. On the other hand, it is expected that a better raw material must always be better for any annealing method. Accordingly, only the main effect is cited for factor A. Thus, we need to get the information of main effects A, B, C, D and interactions $B \times C$ and $B \times D$.

These requirements are shown in Figure 36.2. Such a layout is easily arranged using two lines out of the three radial lines of linear graph (2) shown in Figure 36.1.

The remaining columns, 6 and 7, are removed from the form, as shown in Figure 36.3, with the independent points as indicated. An independent point is used for the assignment of a main effect; factor A is assigned to any one of the remaining two points, since only the main effect is required for A. The layout is shown in Table 36.1.

Figure 36.2
Information required for
a linear graph

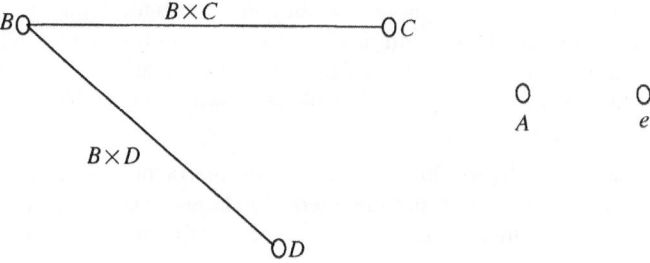

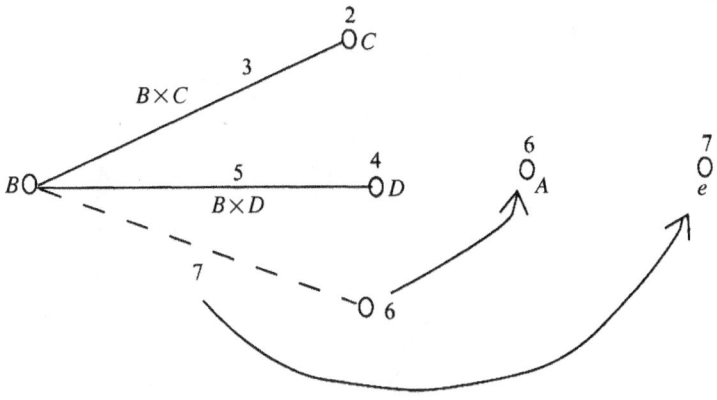

Figure 36.3
Layout of the experiment

Such a layout can also be arranged from linear graph (1). Remove column 6, which is the line connected by columns 2 and 4, indicate it as an independent point like column 7, then assign factor A, as shown in Figure 36.4.

36.3. Multilevel Arrangement

At the planning stage of an experiment, we sometimes want to include some multilevel factors. That is to arrange a four- or eight-level column in two-level series orthogonal arrays, or a nine-level column in three-level series orthogonal arrays. Using linear graphs, a four- or eight-level factor can be assigned from a two-level series orthogonal array.

Next, the method of arranging a four-level factor from orthogonal array L_8 is explained. Assume that A is a four-level factor, and B, C, and D are two-level factors.

Table 36.1
Layout of experiment

Column	B 1	C 2	BC 3	D 4	BD 5	A 6	e 7
1	1	1	1	1	1	1	1
2	1	1	1	2	2	2	2
3	1	2	2	1	1	2	2
4	1	2	2	2	2	1	1
5	2	1	2	1	2	1	2
6	2	1	2	2	1	2	1
7	2	2	1	1	2	2	1
8	2	2	1	2	1	1	2

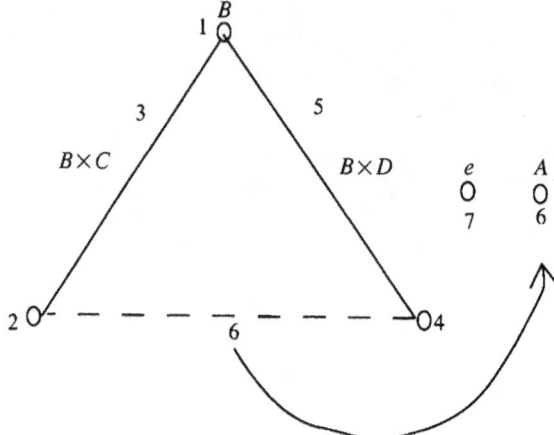

Figure 36.4
Layout using linear
graph (1)

To obtain these main effects from an L_8 array, it is necessary to prepare a four-level column in the array. For instance, line 3 with two points 1 and 2 in linear graph (1) is selected. Then three columns are removed from the linear graph and rearranged using a four-level column, as shown in Table 36.2. That is, from the two columns shown as the two points at both ends of the line (or any two of the three columns), four combinations (11, 12, 21, and 22) are obtained; these combinations are then substituted for 1, 2, 3, and 4, respectively, to form a four-level column. Then columns 1, 2, and 3 are removed from the table, as shown on the right side of Table 36.2.

Table 36.2
Preparation of orthogonal array $L_8(4 \times 2^4)$ from L_8 (2^7)

No.	$L_8(2^7)$							$L_8(4 \times 2^4)$				
	1	2	3	4	5	6	7	123	4	5	6	7
1	1	1	1(1)	1	1	1	1	1	1	1	1	1
2	1	1	1(1)	2	2	2	2	1	2	2	2	2
3	1	2	2(2)	1	1	2	2	2	1	1	2	2
4	1	2	2(2)	2	2	1	1	2	2	2	1	1
5	2	1	2(3)	1	2	1	2	3	1	2	1	2
6	2	1	2(3)	2	1	2	1	3	2	1	2	1
7	2	2	1(4)	1	2	2	1	4	1	2	2	1
8	2	2	1(4)	2	1	1	2	4	2	1	1	2

It is seen from the example above that for a four-level column, a line with points at each end is necessary, as shown in Figure 36.5. The figure shows a four-level column of three degrees of freedom replaced by three columns of one degree of freedom.

When a four-level column is arranged in a two-level series orthogonal array, the four-level column is substituted for the three columns of the array; the three columns consist of any two columns and the column of interaction between the two columns. These three columns are similar to those illustrated in the linear graph in Figure 36.5. Notice the graph has a line with two points at each end. To do this, pick up a line and the two points at the ends of the line, select any two points (normally select the two points at the ends) out of the three columns to form a new four-level column. Letting the four combinations of the two columns, 11, 12, 21, and 22, correspond to 1, 2, 3, and 4, respectively, a four-level column is then formed. Delete the three columns described above and then put the four-level column in the table, as shown in Table 36.2. To determine the degrees of freedom, the three columns with one degree of freedom each are replaced by a column with three degrees of freedom.

Next, the formation of an eight-level column is illustrated. For this purpose, a closed triangle (plus the line to represent the interaction between a point at the top and the baseline in the triangle) from a two-level series linear graph is used. These seven columns (i.e., three apexes, three bases, and one perpendicular line) with seven degrees of freedom in total are replaced by a eight-level column with seven degrees of freedom.

Figure 36.6 shows a triangle chosen from a linear graph of the L_{16} array. From apex 1, draw a perpendicular line to base 6; the line is found to be column 7 from Appendix A. There are eight combinations of 1 and 2 formed by the three columns (three apexes 1, 2, and 4). The combinations are 111, 112, 121, 122, 211, 212, 221, and 222. As in the case of the four-level arrangement, it is necessary to erase column 1 through 7 from the orthogonal array after inserting the eight-level column. Thus, the $L_{16}(2^{15})$ array is rearranged to be an $L_{16}(8 \times 2^8)$ table. The new orthogonal array is shown in Table 36.3.

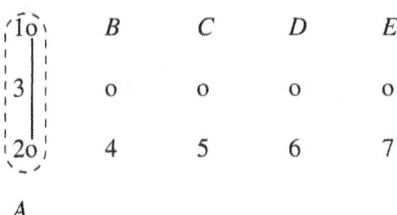

Figure 36.5
Layout with a four-level factor

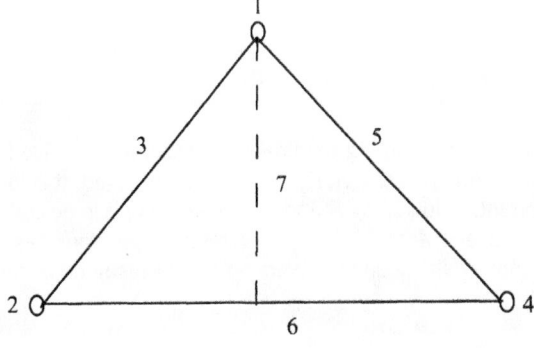

Figure 36.6
Linear graph from an
L_{16} array

Table 36.3
Orthogonal array $L_{16}(8 \times 2^8)$ from $L_{16}(2^{15})$

Column	1	2	4	A	1	2	4	A
	1	1	1	1	2	1	1	5
	1	1	2	2	2	1	2	6
	1	2	1	3	2	2	1	7
	1	2	2	4	2	2	2	8

Col. No.	1–7 1	8 2	9 3	10 4	11 5	12 6	13 7	14 8	15 9
1	1	1	1	1	1	1	1	1	1
2	1	2	2	2	2	2	2	2	2
3	2	1	1	1	1	2	2	2	2
4	2	2	2	2	2	1	1	1	1
5	3	1	1	2	2	1	1	2	2
6	3	2	2	1	1	2	2	1	1
7	4	1	1	2	2	2	2	1	1
8	4	2	2	1	1	1	1	2	2
9	5	1	2	1	2	1	2	1	2
10	5	2	1	2	1	2	1	2	1
11	6	1	2	1	2	2	1	2	1
12	6	2	1	2	1	1	2	1	2
13	7	1	2	2	1	1	2	2	1
14	7	2	1	1	2	2	1	1	2
15	8	1	2	2	1	2	1	1	2
16	8	2	1	1	2	1	2	2	1

☐ **Example**

To obtain information concerning the tires for passenger cars, five two-level factors from the manufacturing process, A, B, C, D, and E, are cited. It is judged that factor A is most important, followed by B and C. Accordingly, it is desirable to obtain the interactions $A \times B$ and $A \times C$. For the life test of tires, four passenger cars, R_1, R_2, R_3, and R_4, are available and there are no driving restrictions during the testing.

In other words, once the tires for the test are put on the cars, there is no restriction on the roads driven or the distance traveled, and the cars are allowed to run as usual. As a result, the driving conditions among these four cars must differ considerably.

On the other hand, it is known that the different positions of tires of a car cause different conditions; the condition of the rear tires is worse than the front and there is also a difference between the two front tires since a car is not symmetrical, a driver does not sit in the middle, the road conditions for both sides are not the same, and so on.

For these reasons, we must consider the difference among the four positions in a car: V_1: front right; V_2: front left; V_3: rear right; and V_4: rear left. Such influences should not interfere with the effects of the five factors and interactions $A \times B$ and $A \times C$.

The real purpose of the experiment is to obtain the information concerning factors A, B, C, D, and E, and not for R and V. The character of factors A, B, C, D, and E is different from that of factors R and V.

The purpose of factors A, B, C, D, and E, called *control factors,* is the selection of the optimum levels. There are four cars, R_1, R_2, R_3, and R_4. For each level of R, the number of tests is limited to four (four wheels). The factor that has restricted the number of tests for each level, in this case four, is called a *block factor.* From a technical viewpoint, the significance of the difference between levels in a block factor is not important. Therefore, the effect of such a factor is not useful for adjustment of the levels of other factors. The purpose of citing a block factor is merely to avoid its effect being mixed with control factors.

The factor, like the position of tire, V, is called an *indicative factor.* It is meaningless to find the best level among the levels of an indicative factor.

From the explanation of block factor or indicative factor, the information of R (cars) and V (positions of tires) are not needed; these factors are cited so to avoid mixing the control factors (A, B, C, D, and E).

This experiment is assigned as follows. Factors and degrees of freedom are listed in Table 36.4. All factorial effects belong to either two levels or four levels; the total degrees of freedom is 13; hence L_{16} (with 15 total degrees of freedom) is probably an appropriate one. (The selection of either a two- or three-level series orthogonal array depends on the majority of the number of levels of factors.)

R and V are both four-level factors. As described in Chapter 35, a four-level factor has to replace three two-level columns, which consist of a line and two points at the ends. Accordingly, the linear graph required in this experiment is shown in Figure 36.7.

There are six standard types of linear graphs given for an L_{16} array [1]. But normally, we cannot find an exactly identical linear graph to the one that is required in a particular experiment. What we must do is to examine the standard types of linear graphs, compare them with the linear graph required for our purpose, and select the one that might be most easily modified. Here, type 3 is selected, for example, and modified to get the linear graph we require. Figure 36.8 shows type 3.

Table 36.4
Distribution of degrees of freedom

Factorial Effect	Degrees of Freedom	Factorial Effect	Degrees of Freedom
A	1	A × B	1
B	1	A × C	1
C	1	R	3
D	1	V	3
E	1		

First, factorial effects A, B, C, $A \times B$, and $A \times C$ are assigned to the figure on the right side. Next, R and V are assigned as a line to the figure on the left side. To the remaining columns, factors D, E, and error are assigned. The layout is shown in Figure 36.9 and Table 36.5.

From the table, it is known that in experiment 10, the tire is manufactured under condition $A_1B_1C_2D_2E_1$; the tire thus manufactured is put on V_2, the front left wheel of car R_3. This is the case when only one tire is manufactured for each experiment number.

When four cars are used, two replications are performed for each experiment. When eight cars are used, the half portion, or 16 tires, are tested by cars R_1, R_2, R_3, and R_4, as shown in Figure 36.9. Then another half portion is tested using cars R_5, R_6, R_7, and R_8, which replace R_1, R_2, R_3, and R_4 as follows: R_5 for R_1, R_7 for R_3, R_6 for R_2, and R_8 for R_4.

Figure 36.7
Linear graph for experiment

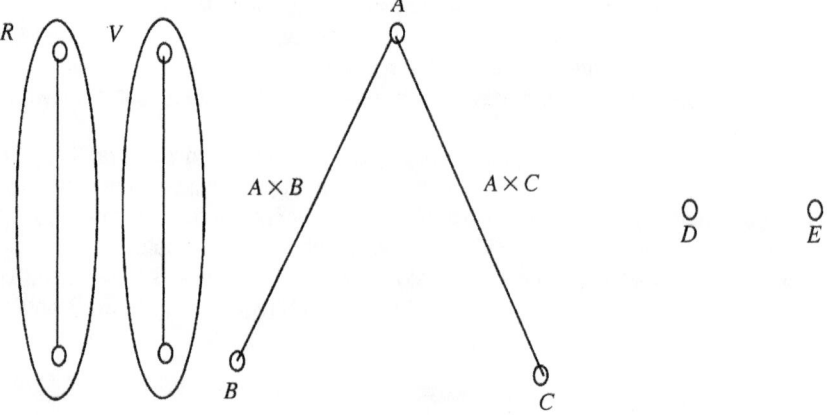

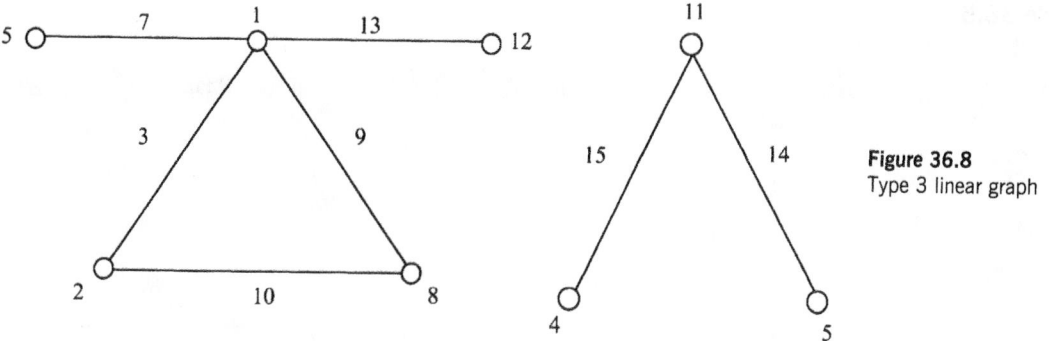

Figure 36.8
Type 3 linear graph

36.4. Dummy Treatment

To put a two-level factor A in a three-level series table, A is formally treated as a three-level factor: that is, to provide three levels for A and run one of the two actual levels twice. Usually, we select one level that is probably more important than the other, and let this level replicate. For instance, $A_1 = A_1$, $A_2 = A_2$, and $A_3 = A_1$. A_3 is formally treated as the third level of A, but actually it is A_1.

When a factor such as A is assigned to column 3 of an $L_9(3^4)$ array, the layout would look like the one shown in Table 36.6. In the table, 3 in the third column of the original L_9 array is rewritten as $1'$ to show that it is the dummy level of A.

The main effect A is calculated as

$$S_A = \frac{(A_1 + A_{1'})^2}{6r} + \frac{A_2^2}{3r} - \frac{(A_1 + A_1' + A_2)^2}{9r} \qquad (36.1)$$

or

$$S_A = \frac{(A_1 + A_1' - 2)^2}{18r} \qquad (36.2)$$

where r is the number of replication in orthogonal array L_9.

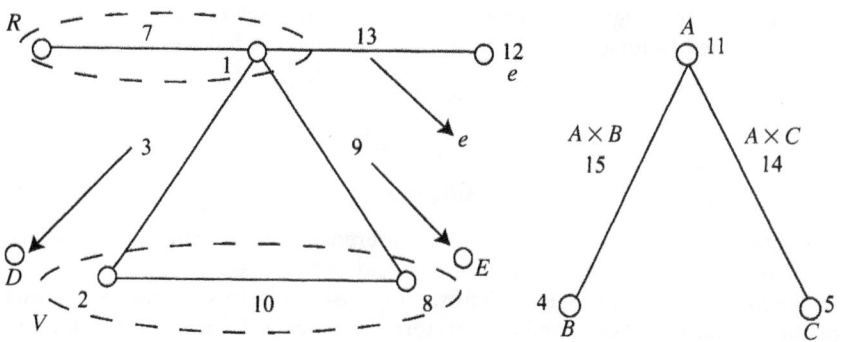

Figure 36.9
Layout of experiment

Table 36.5
Layout of experiment

No.	A	B	C	D	E	A × C	A × B	Car	Position	e	e
1	1	1	1	1	1	1	1	R_1	V_1	1	1
2	2	1	1	1	2	2	2	R_1	V_2	2	2
3	1	2	2	1	1	2	2	R_2	V_2	1	1
4	2	2	2	1	2	1	1	R_2	V_2	1	1
5	2	1	1	2	1	2	2	R_2	V_3	1	1
6	1	1	1	2	2	1	1	R_2	V_4	2	2
7	2	2	2	2	1	1	1	R_1	V_3	2	2
8	1	2	2	2	2	2	2	R_1	V_4	1	1
9	2	1	2	2	2	1	2	R_3	V_1	1	2
10	1	1	2	2	1	2	1	R_3	V_2	2	1
11	2	2	1	2	2	2	1	R_4	V_1	2	1
12	1	2	1	2	1	1	2	R_4	V_2	1	2
13	1	1	2	1	2	2	1	R_4	V_3	1	2
14	2	1	2	1	1	1	2	R_4	V_4	2	1
15	1	2	1	1	2	1	2	R_3	V_3	2	1
16	2	2	1	1	1	2	1	R_3	V_4	1	2
Column	11	4	5	3	9	14	15	1	2	1	1
								6	8	2	3
								7	10		

36.5. Combination Design

There is a technique to put two-level factors in a three-level series orthogonal array. When there are two factors with two levels, and if their interaction is not required, the combination design is used instead of the dummy treatment.

A combined factor (*AB*) with three levels is formed as follows:

$$(AB)_1 = A_1B_1$$

$$(AB)_2 = A_2B_1$$

$$(AB)_3 = A_1B_2$$

The main effect, *A*, is obtained from the difference of $(AB)_1$ and $(AB)_2$. Similarly, the main effect, *B*, is obtained from $(AB)_1$ and $(AB)_3$.

When there are 20 two-level factors, 10 three-level factors are formed by each of the two factors; these combined factors are assigned to an L_{27} array. Because

Table 36.6

Assignment of *A* to column 3

No.	1	2	3	4
		Level		
1	1	1	1	1
2	1	2	2	2
3	1	3	1′	3
4	2	1	2	3
5	2	2	1′	1
6	2	3	1	2
7	3	1	1′	2
8	3	2	1	3
9	3	3	2	1

there are so many two-level factors, it is better to assign these factors to an $L_{32}(2^{31})$ array. But for an experiment such as the $2^{10} \times 3^6$ type, it is more practical to combine each two for the 10 two-level factors to get five three-level combined factors and assign the other six three-level factors to an L_{27} array.

Also, if it is necessary to obtain the interactions between combined factor *AB* and the three-level factor *C*, *AB* and *C* are assigned to the two dots of a line to obtain interactions $A \times C$ and $B \times C$.

A three-level factor (*A*) and a two-level factor (*B*) are combined to form a combined four-level factor:

$$(AB)_1 = A_1 B_1$$

$$(AB)_2 = A_2 B_1$$

$$(AB)_3 = A_3 B_1$$

$$(AB)_4 = A_1 B_2$$

The variation of a combined three-level factor (*AB*) by two factors (*A*, *B*) of two levels is calculated as

$$S_{(AB)} = \frac{1}{r} [(AB)_1^2 + (AB)_2^2 + (AB)_3^2] - \text{CF} \qquad (36.3)$$

where *r* is the number of units of $A_1 B_1$, $A_2 B_1$, and $A_1 B_2$.

However, the following relationship does not hold, since *A* and *B* are not orthogonal:

$$S_{(AB)} = S_A + S_B \qquad (36.4)$$

If we want to separate A and B, calculate

$$S_A = \frac{[(AB)_1 - (AB)_2]^2}{2r} \tag{36.5}$$

$$S_B = \frac{[(AB)_1 - (AB)_3]^2}{2r} \tag{36.6}$$

and decompose $S_{(AB)}$ into

$$S_{(AB)} = S_A + S_{\text{res}} \tag{36.7}$$

or

$$S_{(AB)} = S_B + S'_{\text{res}} \tag{36.8}$$

In the analysis of variance table, $S_{(AB)}$ is listed inside the table, while S_A and S_B are listed outside.

Reference

1. ASI, 1987. *Orthogonal Arrays and Linear Graphs.* Livonia, Michigan: American Supplier Institute.

37

Incomplete Data

37.1. **Introduction   610**
37.2. **Treatment of Incomplete Data   611**
    Cases 1.1 to 1.4   611
    Case 2.1   611
    Case 2.2   611
    Case 2.3   612
    Cases 3.1 to 3.3   612
    Cases 4.1 to 4.3   613
    Case 4.4   613
    Case 4.5   613
37.3. **Sequential Approximation   613**
    References   616

37.1. Introduction

Sometimes during experimentation or simulation, data cannot be obtained as planned because of lost data, equipment problems, or extreme conditions. Sometimes, the SN ratios of a few runs of an orthogonal array cannot be calculated, for various reasons. In the treatment of missing data in the traditional design of experiments, no distinction is made between different types of incomplete data. In this chapter, treatments of such cases are described.

Although it is desirable to collect a complete set of data for analysis, it is not absolutely necessary. In parameter design, it is recommended that control-factor-level intervals be set wide enough so that a significant improvement in quality may be expected. Such determinations can only be made based on existing knowledge or past experiences. In product development, however, there is no existing knowledge to determine the range of control factor levels. To avoid generating undesirable results, engineers tend to conduct research within narrow ranges.

If control-factor-level intervals were set wide, the conditions of some control-factor-level combinations may become so extreme that there would be no product or no measurement: for example, an explosion in a chemical reaction. That's all right, because important information was produced from the result. It must be remembered that the objective of new product development is to produce good knowledge rather than producing a good product. The incomplete data in that case can be analyzed by a method called *sequential approximation*.

Various types and examples of incomplete data follow:

1. All data in one run (or a few runs) of an orthogonal array are missing.

   1.1 Samples to be measured are missing.

   1.2 Data sheets with the recorded results are misplaced.

   1.3 Some experimental runs were discontinued due to budget, time constraints, or job transfer.

   1.4 The raw materials or test pieces needed to complete the whole runs of experiments were insufficient.

2. The number of data points in one run are different from others.

   2.1 The number of signal factor levels in one run differs from those in the other runs.

   2.2 The number of repetitions in one run are different from other runs.

   2.3 All results in one noise factor level of one run are missing.

3. Part or all the results of one run is missing due to extreme conditions.

   3.1 The chemical reaction apparatus exploded.

   3.2 No current flowed in an electric circuit.

   3.3 No product was produced.

4. No data are missing, but calculation of the SN ratios results in values of either positive or negative infinity.

   4.1 When using the zero-point proportional, reference-point proportional, or linear equation, the error variance (V_e) is greater than the sensitivity (S_β).

   4.2 In the nominal-is-best case, the error variance (V_e) is greater than the sensitivity (S_m).

   4.3 In classified data with three classes or more, the error variance (V_e) is greater than the sensitivity (S_m).

   4.4 In the smaller-is-better case, all results are equal to zero.

   4.5 In the larger-is-better case, all results are equal to zero.

Levels of conditions must be within a certain range so that undesirable results are not obtained. Such determinations can only be made based on existing knowledge or past experience. In the case of new product development, however, there is no existing knowledge on which to base determination of the appropriate range of control factor levels within which a product or a system functions. To avoid generating undesirable results, engineers tend to conduct research within narrow ranges.

It is important to realize that the purpose of product development is not to produce good products but to generate useful knowledge. If all results from different experimental runs are close to each other, little knowledge can be gained. To get more knowledge about a new frontier, control-factor-level intervals need to be set wide enough to purposely produce some bad results, such as having explosions or cracked test pieces. Such situations are allowed at the R&D stage. In the case of an L_{18} orthogonal array experiment, there may be as many as several runs with incomplete data, but this is still good enough to draw valuable conclusions.

37.2. Treatment of Incomplete Data

To make analysis and optimization possible, the following treatments are suggested for the cases listed in Section 37.1.

Cases 1.1 to 1.4 are referred to as missing data in traditional experimental design books. The Fisher–Yates method is commonly used. The sequential approximation method is recommended.

Cases 1.1 to 1.4

An example of case 2.1 is when the number of signal factor level of run 4 is 3 but all other runs are 5. The SN ratio of run 4 is calculated as is and analyzed with the other SN ratios.

Case 2.1

For case 2.2, the number of repetitions under each signal factor are different, as shown in Table 37.1. In the case of the zero-point proportional equation, the SN ratio is calculated as follows:

Case 2.2

$$S_T = y_{11}^2 + y_{12}^2 + \cdots + y_{kr_k}^2 \qquad (f_T = r_1 + r_2 + \cdots + r_k) \qquad (37.1)$$

$$S_\beta = \frac{(M_1 y_1 + M_2 y_2 + \cdots + M_k y_k)^2}{r_1 M_1^2 + r_2 M_2^2 + \cdots + r_k M_k^2} \qquad (f_\beta = 1) \qquad (37.2)$$

$$S_e = S_T - S_\beta \qquad (f_e = f_T - 1) \qquad (37.3)$$

$$V_e = \frac{S_e}{f_e} \qquad (37.4)$$

$$\eta = 10 \log \frac{[1/(r_1 M_1^2 + r_2 M_2^2 + \cdots + r_k M_k^2)](S_\beta - V_e)}{V_e} \qquad (37.5)$$

Table 37.1
Incomplete data

Signal	M_1	M_2	$\cdots$	M_k
Repetition	y_{11} y_{12} $\vdots$ y_{1r_1}	y_{21} y_{22} y_{2r_2}	$\cdots$ $\cdots$ $\cdots$	y_{k1} y_{k2} y_{kr_k}
Total	y_1	y_2	$\cdots$	y_k

Table 37.2 is a simple numerical example [1]. For the data in the table:

$$S_T = 0.098^2 + 0.097^2 + \cdots + 0.488^2$$

$$= 2.002033 \qquad (f_T = 4 + 6 + 6 = 16) \tag{37.6}$$

$$S_\beta = \frac{[(0.1)(0.380) + (0.3)(1.750) + (0.5)(2.955)]^2}{(4)(0.1^2) + (6)(0.3^2) + (6)(0.5^2)} = \frac{2.0405^2}{2.08}$$

$$= 2.00175012 \qquad (f = 1) \tag{37.7}$$

$$S_e = 2.002033 - 2.00175012$$

$$= 0.0028288 \qquad (f = 16 - 1 = 15) \tag{37.8}$$

$$V_e = \frac{0.00028288}{15} = 0.00001886 \tag{37.9}$$

$$\eta = 10 \log \frac{(1/2.08)(2.00175012 - 0.00001886)}{0.00001886}$$

$$= 10 \log 51027.08$$

$$= 47.08 \text{ dB} \tag{37.10}$$

Case 2.3 An example of case 2.3 is when there are two compounded noise factor levels: the positive and negative extreme conditions. If one of these two extreme conditions is totally missing, do not use the results of another extreme condition to calculate the SN ratio. Instead, treat it as described for type 1.

Cases 3.1 to 3.3 Whether it's an explosion, lack of current, or lack of end product, each problem indicates that the condition is very bad. Use negative infinity as the SN ratio. It is important to distinguish this case from type 1, in which it is not known whether the condition is good or bad.

There are two ways to treat these cases: (1) classify the SN ratios, including positive and/or negative infinity, into several categories and analyze the results by accumulation analysis [2]; and (2) subtract 3 to 5 dB from the smallest SN ratio in the orthogonal array. Assign that number to the SN ratio for the missing run(s). Then follow with sequential approximation.

Table 37.2
Numerical example

Signal	$M_1 = 0.1$	$M_2 = 0.3$	$M_3 = 0.5$
	0.098	0.294	0.495
	0.097	0.288	0.493
Result	0.093	0.288	0.489
	0.092	0.296	0.495
		0.297	0.495
		0.287	0.488
Total	0.380	1.750	2.955

Cases 4.1 to 4.3 involve error variance being greater than the sensitivity. Use negative infinity for the SN ratio and treat these cases in the same way as case 3.1.

Cases 4.1 to 4.3

Case 4.4 is when all results are equal to zero for a smaller-is-better case. Use positive infinity as the SN ratio and classify the SN ratio into several categories and conduct accumulation analysis. Add 3 to 5 dB to the largest SN ratio in the orthogonal array. Assign that number to the SN ratio of the missing run(s), followed with sequential approximation.

Case 4.4

Case 4.5 is when all results are equal to zero for a larger-is-better case. Use negative infinity, and treat the problem in the same way as case 3.1.

Case 4.5

37.3. Sequential Approximation [3]

Sequential approximation is a method used to estimate the data that are supposed to exist. In the experiment for the amount of wear, assume that the data of experiment 3 in orthogonal array L_{12} are missing (Table 37.3). This is the case of missing data; the SN ratio for a smaller-the-better characteristic cannot be calculated.

Sequential approximation is made using the following procedure:

1. Place the average calculated from the existing SN ratio into the missing location; this is called the *zeroth approximation.*

Table 37.3
Missing data in experiment 3

Factor: No.\Column:	A 1	B 2	C 3	D 4	E 5	F 6	G 7	H 8	I 9	J 10	K 11	Amount of Wear (μm) N_1	N_2	η (dB)
1	1	1	1	1	1	1	1	1	1	1	1	39 19	9 10	−27.12
2	1	1	1	1	1	2	2	2	2	2	2	21 20	3 16	−24.42
3	1	1	2	2	2	1	1	1	2	2	2			
4	1	2	1	2	2	1	2	2	1	1	2	42 27	13 24	−29.08
5	1	2	2	1	2	2	1	2	1	2	1	37 35	9 29	−29.44
6	1	2	2	2	1	2	2	1	2	1	1	68 85	38 64	−36.38
7	2	1	2	2	1	1	2	2	1	2	1	21 10	5 2	−21.54
8	2	1	2	1	2	2	2	1	1	1	2	38 23	17 4	−27.55
9	2	1	1	2	2	2	1	2	2	1	1	30 34	30 24	−29.46
10	2	2	2	1	1	1	1	2	2	1	2	72 46	24 40	−33.75
11	2	2	1	2	1	2	1	1	1	2	2	10 6	1 2	−15.47
12	2	2	1	1	2	1	2	1	2	2	1	30 12	8 0	−24.42

2. Conduct response analysis and estimate the missing location using larger effects; call this the *first approximation*.

3. Repeat step 2 until the estimation converages.

The following is the estimation of experiment 3 using sequential approximation:

1. *Zeroth approximation.* The average of 11 pieces of data is −27.15 dB.

2. *ANOVA and estimation of experiment 3.* Put −27.15 dB into experiment 3, and the ANOVA table is used to find larger effects. Tables 37.4 and 37.5 show the ANOVA and supplementary tables.

From Table 37.4, larger effects are *A, C, I,* and *J.* They are used to estimate the condition of experiment 3: $A_1(B_1) C_2(D_2 E_2 F_1 G_1 H_1) I_2 J_2(K_2)$.

$$\hat{\mu}(\text{No. 3}) = \frac{-173.59 + (-175.81) + (-175.58) + (-142.44)}{6} - (3)\left(\frac{-325.78}{12}\right)$$

$$= -29.78 \text{ dB} \tag{37.11}$$

The first approximation is −29.78 dB.

Repeat step 2 using −29.78 dB. The following result is obtained:

$$\hat{\mu}(\text{No. 3}) = -30.88 \text{ dB} \tag{37.12}$$

Using the same procedure, the fifth approximation is obtained as −31.62 dB. This is close to the fourth approximation of −31.53, and calculation is discontinued. The actual result was −34.02; these are fairly close.

Table 37.4
ANOVA table using zeroth approximation

Factor	f	S	V
A	1	38.16	38.16
B	1	10.64	10.64°
C	1	55.64	55.64
D	1	4.84	4.84°
E	1	5.91	5.91°
F	1	0.01	0.01°
G	1	0.08	0.08°
H	1	7.68	7.68°
I	1	53.68	53.68
J	1	139.40	139.40
K	1	9.97	9.97
Total	11	326.02	

°: Pooled as error.

Table 37.5
Supplementary table

Factor	Level Total	Factor	Level Total
A_1	−173.59	G_1	−162.39
A_2	−152.19	G_2	163.39
B_1	−157.24	H_1	−158.09
B_2	−168.54	H_2	−167.69
C_1	−149.97	I_1	−150.20
C_2	−175.81	I_2	−175.58
D_1	−166.70	J_1	−183.34
D_2	−159.08	J_2	−142.44
E_1	−158.68	K_1	−168.36
E_2	−167.10	K_2	−157.42
F_1	−163.06		
F_2	−162.72	Total	−324.78

The calculation above is terminated at the fifth approximation using four factors: A, C, I, and J. Since the effect of E becomes larger and larger, the approximation could be better if E were included in the calculation.

Next, the optimum condition is estimated using the ANOVA and supplementary tables from the fifth approximation (Tables 37.6 and 37.7).

Table 37.6
Supplementary table using fifth approximation

Factor	Level Total	Factor	Level Total
A_1	−178.06	G_1	−166.86
A_2	−152.19	G_2	−163.39
B_1	−161.71	H_1	−162.56
B_2	−168.54	H_2	−167.69
C_1	−149.97	I_1	−150.20
C_2	−180.28	I_2	−180.05
D_1	−166.70	J_1	−183.34
D_2	−163.35	J_2	−146.92
E_1	−158.68	K_1	−168.36
E_2	−171.57	K_2	−161.89
F_1	−167.53		
F_2	−162.72	Total	−330.25

Table 37.7
ANOVA table using fifth approximation

Factor	f	S	V	ρ (%)
A	1	55.77	55.77	15.6
B	1	3.89	3.89°	
C	1	76.56	76.56	21.6
D	1	0.89	0.83°	
E	1	13.83	13.85	3.4
F	1	1.93	1.93°	
G	1	1.00	1.00°	
H	1	2.19	2.19°	
I	1	74.25	74.25	20.9
J	1	110.60	110.60	31.5
K	1	3.49	3.49°	
(e)	(6)	(13.33)	(2.22)	(7.0)
Total	11	344.36		100.0

°: Pooled as error.

The optimum condition is $A_2B_1C_1D_2E_1F_2G_2H_1I_1J_2K_2$. Using the effects of A, C, E, I, and J, we have

$$\hat{\mu} = \overline{A}_2 + \overline{C}_1 + \overline{E}_1 + \overline{I}_1 + \overline{J}_2 - 4\overline{T}$$

$$= -25.36 - 25.00 - 26.45 - 25.03 - 24.48 - (4)(-27.52)$$

$$= -16.24 \text{ dB} \tag{37.13}$$

Sequential approximation can be used when there is more than one missing experiment. But it cannot be used when all data under a certain level of a certain factor are missing.

References

1. Yuin Wu et al., 2000. *Taguchi Methods for Robust Design.* New York: ASME Press.
2. Genichi Taguchi, 1987. *System of Experimental Design.* Livonia, Michigan: Unipub/ American Supplier Institute.
3. Genichi Taguchi et al., 1993. *Quality Engineering Series,* Vol. 4. Tokyo: Japanese Standards Association.

38

Youden Squares

38.1. Introduction   617
38.2. Objective of Using Youden Squares   617
38.3. Calculations   620
38.4. Derivations   623
     Reference   625

38.1. Introduction

Youden squares are a type of layout used in design of experiments. A *Youden square* is a fraction of a *Latin square*. They are also called *incomplete Latin squares*. Table 38.1 shows a Latin square commonly seen in experimental design texts. Table 38.2 is another way of showing the contents of a Latin square using an L_9 orthogonal array, where only the first three columns of the array are shown.

Table 38.3 shows a Youden square. In the table, a portion of the layout from Table 38.1 is missing. Table 38.4 is another way of showing the contents of a Youden square.

From Tables 38.1 and 38.2, it is seen that all three factors, A, B, and C, are balanced (orthogonal), whereas in Tables 38.3 and 38.4, factors B and C are unbalanced. Although factors B and C are unbalanced in a Youden square, their effects can be calculated by solving the equations generated from the relationship shown in this layout.

38.2. Objective of Using Youden Squares

The objective of using Youden squares is to utilize an L_{18} orthogonal array efficiently to obtain more information from the same number of runs of the array. Normally, an L_{18} array is used to assign one two-level factor and seven three-level factors. By putting a Youden square in an L_{18} array, one more three-level factor can be assigned without adding extra experimental runs.

617

Table 38.1
Common Latin square

	B_1	B_2	B_3
A_1	C_1	C_2	C_3
A_2	C_2	C_3	C_1
A_3	C_3	C_1	C_2

Table 38.5 is a Youden square where an L_{18} array is assigned with an extra column; the ninth factor is denoted by C, as column 9. In the table, y_1', y_2', ... , y_{18}' are the outputs of 18 runs. Comparing Tables 38.4 and 38.5, the first two columns of the two arrays correspond to each other, with the exception that there are three repetitions in Table 38.5. For example, combination A_1B_1 appears once in Table 38.4, while the combination appears three times (numbers 1, 2, and 3) in Table 38.5. It is the same for other combinations.

It is known that the number of degrees of freedom means the amount of information available. There are 18 runs in an L_{18} array; its number of degrees of freedom is 17. When one two-level factor and seven three-level factors are assigned, the total number of degrees of freedom from this layout is 15. The difference, two degrees of freedom, is hidden inside. One of the ways to utilize these two degrees of freedom to obtain more information without having more runs is by substituting columns 1 and 2 with a six-level column to create a column that can assign a factor of up to six levels. This is called *multilevel assignment*. Another way of utilizing the two degrees of freedom is to assign a three-level column using a Youden square. Thus, 17 degrees of freedom are all utilized.

Table 38.2
Latin square using an orthogonal array

No.	A 1	B 2	C 3	y
1	1	1	1	y_1
2	1	2	2	y_2
3	1	3	3	y_3
4	2	1	2	y_4
5	2	2	3	y_5
6	2	3	1	y_6
7	3	1	3	y_7
8	3	2	1	y_8
9	3	3	2	y_9

Table 38.3
Youden square

	B_1	B_2	B_3
A_1	$C_1(y_1)$	$C_2(y_2)$	$C_3(y_3)$
A_2	$C_2(y_4)$	$C_3(y_5)$	$C_1(y_6)$

Table 38.4
Alternative Youden square depiction

No.	A 1	B 2	C 3	y
1	1	1	1	y_1
2	1	2	2	y_2
3	1	3	3	y_3
4	2	1	2	y_4
5	2	2	3	y_5
6	2	3	1	y_6

Table 38.5
Youden square

No.	A 1	B 2	C 9	D_3 3	⋯ ⋯	I 8	Results	Subtotal
1	1	1	1				y'_1	
2	1	1	1				y'_2	y_1
3	1	1	1				y'_3	
4	1	2	2				y'_4	
5	1	2	2				y'_5	y_2
6	1	2	2				y'_6	
7	1	3	3				y'_7	
8	1	3	3				y'_8	y_3
9	1	3	3				y'_9	
10	2	1	2				y'_{10}	
11	2	1	2				y'_{11}	y_4
12	2	1	2				y'_{12}	
13	2	2	3				y'_{13}	
14	2	2	3				y'_{14}	y_5
15	2	2	3				y'_{15}	
16	2	3	1				y'_{16}	
17	2	3	1				y'_{17}	y_6
18	2	3	1				y'_{18}	

38.3. Calculations

In Table 38.5, one two-level factor, A, and eight three-level factors, B, C, D, E, F, G, H, and I, were assigned. Eighteen outputs, y_1, y_2, ... , y_{18}, were obtained, and results, y_1', y_2', ... , y_{18}', were obtained. Calculation of main effects A, D, E, F, G, and I are exactly identical to regular calculation. Main effects B and C are calculated after calculating their level averages as follows:

$$y_1 = y_1' + y_2' + y_3'$$

$$y_2 = y_4' + y_5' + y_6'$$

$$\vdots$$

$$y_6 = y_{16}' + y_{17}' + y_{18}'$$

(38.1)

Let the effect of each factor level, A_1, A_2, B_1, B_2, B_3, C_1, C_2, and C_3, be denoted by a_1, a_2, b_1, b_2, b_3, c_1, c_2, and c_3, respectively, as shown in Figure 38.1.

For example, a_1 is the deviation of A_1 from the grand average, $\overline{T}$. As described previously, a_1 and a_2 can be determined as usual without considering the layout of Youden squares. The effects of factor levels of B and C are calculated as follows:

$$b_1 = \tfrac{1}{9}(y_1 - y_2 + y_4 - y_6)$$

$$b_2 = \tfrac{1}{9}(y_2 - y_3 - y_4 + y_5)$$

$$b_3 = \tfrac{1}{9}(-y_1 + y_3 - y_5 + y_6)$$

(38.2)

$$c_1 = \tfrac{1}{9}(y_1 - y_3 - y_4 + y_6)$$

$$c_2 = \tfrac{1}{9}(-y_1 + y_2 + y_4 - y_5)$$

$$c_3 = \tfrac{1}{9}(-y_2 + y_3 + y_5 - y_6)$$

(38.3)

❐ Example [1]

In a study of air turbulence in a conical air induction system (conical AIS) for automobile engines, the following control factors were studied:

A: honeycomb, two levels

B: large screen, three levels

C: filter medium, three levels

D: element part support, three levels

E: inlet geometry, three levels

F: gasket, three levels

G: inlet cover geometry, three levels

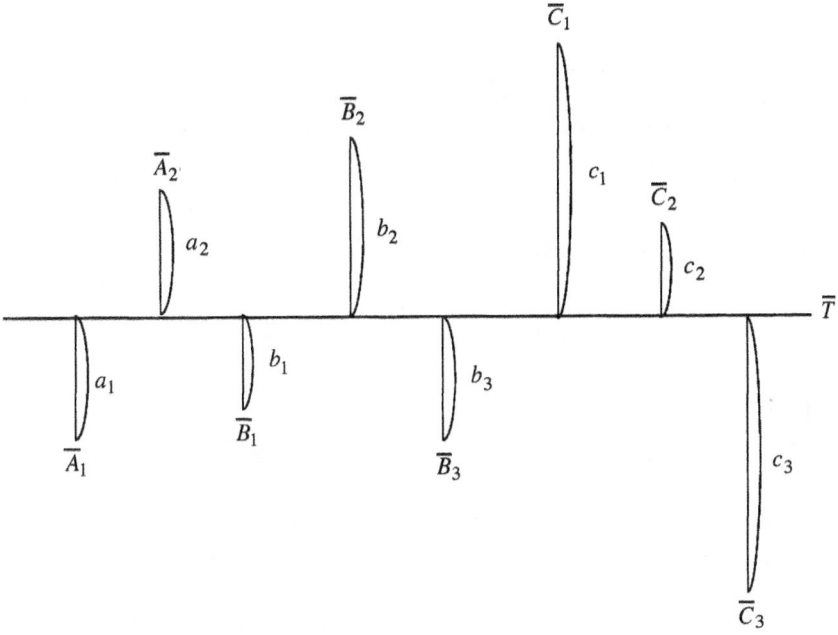

Figure 38.1
Main effects of A, B, and C

H: bypass screen, three levels

I: MAFS orientation to housing, three levels

A Youden square was used for assigning the factors above. The dynamic SN ratio and sensitivity were calculated from the experiment. Table 38.6 shows the layout and SN ratio.

Comparing Tables 38.5 and 38.6, it is seen that in Table 38.5, the combinations of columns 1, 2, and 9 are based on a Youden square, and factors A, B, and C are assigned. But in Table 38.6, the combinations of columns 1, 2, and 9 are still based on a Youden square, where factors A, B, and I are assigned.

The effect of each factor level of B and I is calculated as follows:

$$b_1 = \frac{y_1 - y_2 + y_4 - y_6}{9}$$

$$= \frac{9.05 - 8.77 + 8.88 - 8.23}{9} = 0.1033 \qquad (38.4)$$

$$b_2 = \frac{y_2 - y_3 - y_4 + y_5}{9}$$

$$= \frac{8.77 - 8.31 - 8.88 + 8.58}{9} = 0.0177 \qquad (38.5)$$

Table 38.6

Layout and results

| | Column | | | | | | | | | | |
| | 1 | 2 | 3 | 4 | 5 | 6 | 7 | 8 | 9 | | |
No.	A	B	C	D	E	F	G	H	I	SN	Subtotal
1	1	1	1	1	1	1	1	1	1	4.24	
2	1	1	2	2	2	2	2	2	1	2.21	$y_1 = 9.05$
3	1	1	3	3	3	3	3	3	1	2.60	
4	1	2	1	1	2	2	3	3	2	2.20	
5	1	2	2	2	3	3	1	1	2	4.01	$y_2 = 8.77$
6	1	2	3	3	1	1	2	2	2	2.56	
7	1	3	1	2	1	3	2	3	3	2.20	
8	1	3	2	3	2	1	3	1	3	3.87	$y_3 = 8.31$
9	1	3	3	1	3	2	1	2	3	2.24	
10	2	1	1	3	3	2	2	1	2	2.66	
11	2	1	2	1	1	3	3	2	2	2.66	$y_4 = 8.88$
12	2	1	3	2	2	1	1	3	2	2.31	
13	2	2	1	2	3	1	3	2	3	2.20	
14	2	2	2	3	1	2	1	3	3	2.35	$y_5 = 8.58$
15	2	2	3	1	2	3	2	1	3	4.03	
16	2	3	1	3	2	3	1	2	1	2.11	
17	2	3	2	1	3	1	2	3	1	2.16	$y_6 = 8.23$
18	2	3	3	2	1	2	3	1	1	3.96	

$$b_3 = \frac{-y_1 + y_3 - y_5 + y_6}{9}$$

$$= \frac{-9.05 + 8.31 - 8.58 + 8.23}{9} = -0.1211 \qquad (38.6)$$

$$i_1 = \frac{y_1 - y_3 - y_4 + y_6}{9}$$

$$= \frac{9.05 - 8.31 - 8.88 + 8.23}{9} = 0.01 \qquad (38.7)$$

$$i_2 = \frac{-y_1 + y_2 + y_4 - y_5}{9}$$

$$= \frac{-9.05 + 8.77 + 8.88 - 8.58}{9} = 0.0022 \qquad (38.8)$$

$$i_3 = \frac{-y_2 + y_3 + y_5 - y_6}{9}$$

$$= \frac{-8.77 + 8.31 + 8.58 - 8.23}{9} = -0.0122 \qquad (38.9)$$

The grand average, $\bar{T}$, is

$$\bar{T} = \frac{4.24 + 2.21 + \cdots + 3.96}{18} = 2.88 \qquad (38.10)$$

Therefore, the level averages of B and I are calculated as

$$\bar{B}_1 = \bar{T} + b_1 = 2.88 + 0.1033 = 2.98 \qquad (38.11)$$

$$\bar{B}_2 = \bar{T} + b_2 = 2.88 + 0.0177 = 2.90 \qquad (38.12)$$

$$\bar{B}_3 = \bar{T} + b_3 = 2.88 - 0.121 = 2.76 \qquad (38.13)$$

$$\bar{I}_1 = \bar{T} + i_1 = 2.88 + 0.01 = 2.89 \qquad (38.14)$$

$$\bar{I}_2 = \bar{T} + i_2 = 2.88 - 0.0022 = 2.88 \qquad (38.15)$$

$$\bar{I}_3 = \bar{T} + i_3 = 2.88 - 0.0122 = 2.87 \qquad (38.16)$$

Other level totals are calculated in the normal way. Table 38.7 shows the results of calculation.

38.4. Derivations

The derivations of equations (38.2) and (38.3) are illustrated. Table 38.8 is constructed from Table 38.4 using symbols a_1, a_2, b_1, b_2, b_3, c_1, c_2, and c_3, as described in Section 38.3. It is seen from the table that y_1 is the result of a_1, b_1, and c_1; y_2 is the result of a_1, b_2, and c_2; and so on. Therefore, the following equations can be written:

Table 38.7
Response table for SN ratio

	A	B	C	D	E	F	G	H	I
1	2.90	2.98	2.81	2.92	3.00	2.89	2.88	4.00	2.89
2	2.86	2.90	2.88	2.82	2.79	2.81	2.84	2.33	2.88
3		2.76	2.95	2.90	2.85	2.94	2.92	2.30	2.87

Table 38.8
Youden square

No.	A 1	B 2	C 3	y
1	a_1	b_1	c_1	y_1
2	a_1	b_2	c_2	y_2
3	a_1	b_3	c_3	y_3
4	a_2	b_1	c_2	y_4
5	a_2	b_2	c_3	y_5
6	a_2	b_3	c_1	y_6

$$y_1 = a_1 + b_1 + c_1 \tag{38.17}$$

$$y_2 = a_1 + b_2 + c_2 \tag{38.18}$$

$$y_3 = a_1 + b_3 + c_3 \tag{38.19}$$

$$y_4 = a_2 + b_1 + c_2 \tag{38.20}$$

$$y_5 = a_2 + b_2 + c_3 \tag{38.21}$$

$$y_6 = a_2 + b_3 + c_1 \tag{38.22}$$

In addition, the following three equations exist based on the definition of a_1, a_2, ... , c_3:

$$a_1 + a_2 = 0 \tag{38.23}$$

$$b_1 + b_2 + b_3 = 0 \tag{38.24}$$

$$c_1 + c_2 + c_3 = 0 \tag{38.25}$$

By solving these equations, b_1, b_2, b_3, c_1, c_2, and c_3 are calculated, as shown in equations (38.2) and (38.3).

For example, the effect of C_2 is derived as follows.

equation (38.17) − equation (38.18):

$$y_1 - y_2 = b_1 - b_2 + c_1 - c_2 \tag{38.26}$$

equation (38.20) − equation (38.21):

$$y_4 - y_5 = b_1 - b_2 + c_2 - c_3 \tag{38.27}$$

equation (38.26) − equation (38.27):

$$y_1 - y_2 - (y_4 - y_5) = c_1 - c_2 - (c_2 - c_3) \tag{38.28}$$

$$y_1 - y_2 - y_4 + y_5 = c_1 + c_3 - 2c_2 \tag{38.29}$$

From equation (38.25),

$$c_1 + c_2 + c_3 = 0 \tag{38.30}$$

Therefore, equation (38.29) can be written as

$$y_1 - y_2 - y_4 + y_5 = c_1 + c_3 + c_2 - 3c_2 \tag{38.31}$$

$$c_2 = \tfrac{1}{3}(-y_1 + y_2 + y_4 - y_5) \tag{38.32}$$

This is shown in equation (38.3).

The Youden square shown in Table 38.4 has six runs; each run (combination) has one repetition. When a Youden square is assigned in L_{18} arrays, as shown in Table 38.5, each combination of the Youden square has three repetitions. Therefore, the effect of each factor level of B and C must be divided by 9, as shown in equations (38.2) and (38.3), instead of 3 as in equation (38.31).

Reference

1. R. Khami, et al., 1994. Optimization of the conical air induction system's mass airflow sensor performance. Presented at the Taguchi Symposium.

Section 2

Application
(Case Studies)

Part I

Robust Engineering: Chemical Applications

Biochemistry (Cases 1–2)
Chemical Reaction (Cases 3–4)
Measurement (Cases 5–8)
Pharmacology (Case 9)
Separation (Cases 10–11)

Optimization of Bean Sprouting Conditions by Parameter Design

Abstract: In this study we observe bean sprout germination and growth as a change in weight and evaluate this change as a generic function of growth.

1. Introduction

For growing bean sprouts, we use a series of production processes where small beans are soaked in water and germinated and grown in a lightless environment. As shown in Figure 1, this process is divided into germination, growth, and decay periods. The major traits in each period are described below.

Germination Period

Although the raw material for bean sprouts is often believed to be soybeans or small beans, in actuality, they are specific types of small bean: mappe (ketsuru adzuki) and green gram. The former are imported from Thailand and Burma, and the latter mainly from China. Despite the fact that green gram has become mainstream, its price is three to five times as great as black mappe's. Because they grow under dry and hibernating conditions, they must be soaked in water and heated for germination. Germination has the important role of sprouting them and also increases the germination rate and sterilizes accompanying bacteria.

Growth Period

After germination, mappes absorb a considerable amount of water. Regarding three days before shipping as a growth period, we evaluate an SN ratio.

Decay Period

Although a plant needs photosynthesis after a growth period, bean sprouts normally decay and rot because no light is supplied in the production environment. Whether the preservation of bean sprouts is good or poor depends on the length of the decay period and the processing up to the decay period. On the basis of a criterion that cannot judge whether growth should be fast or slow, such as "a fast-growing bean sprout withers fast," or "a slowly growing bean sprout is immature and of poor quality," a proper growth period for bean sprouts has been considered approximately seven days. Additionally, a contradictory evaluation of appearance, one of which insists that completely grown bean sprouts are better because they grow "wide and long" and the other that growing young sprouts is better because they last longer, has been used. We believe that our generic function (in this case, a growth rate by weight is used) can evaluate the ideal growth, as shown in Figure 1, solving these contradictions. A state where no energy of bean sprouts are wasted for purposes other than growth is regarded as suitable for healthy bean sprouts.

2. Generic Function

Bean sprouts grow relying on absorption of water. This study observes a process of bean sprout germination and growth as a change in weight and evaluates this change as a generic function of growth. Figure 1 shows three states: actual, ideal, and optimal regarding the weight change of normal bean sprouts.

Although perpetual growth is ideal, in actuality, after using up nutrition, bean sprouts slow their

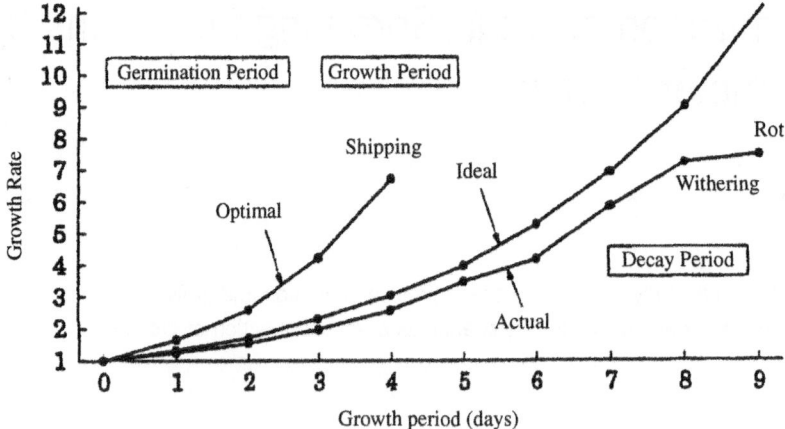

Figure 1
Growth curve of bean sprouts

growth and wither from around the seventh day because only water is supplied. This entire process, classified into germination, growth, and decay periods, cannot be discriminated perfectly. Then, for a period of water absorption and growth caused by germination, setting the initial weight of seeds to Y_0 and the grown weight after T hours to Y, we define the following as an ideal generic function:

$$Y = Y_0 e^{\beta T} \qquad (1)$$

By transforming equation (1), we obtain:

$$\ln(Y/Y_0) = \beta T \qquad (2)$$

Setting the left side of equation (2) to y, we have the following same zero-point proportional equation for time, as used in chemical reactions:

$$y = \beta T \qquad (3)$$

We use an exponential function in equation (1), based on the idea that a seed absorbs water and then divides its cells. However, whether this is the generic equation of a plant's growth is not clear. Therefore, if there are other equations to express the water absorption and growth period of a plant, we should make the most of them. Nevertheless, we do not need to use a complicated equation analyzing natural phenomena, even if it is regarded as rational to express the growth. What is most important is to predict the rationality of our idea

before putting it to a practical use. This is the quality engineering way of thinking.

1. *Selection of signal factor.* Considering the ideal function, we choose time as a signal factor. More specifically, with 24 hours as one day, we select five, six, and seven days as signal factor levels.

2. *Selection of noise factors.* Since simultaneous control of humidity and temperature is quite difficult, we focus on humidity, which is considered to be more easily regulated as an experimental noise factor.

   N_1: desiccator at the humidity of 60%

   N_2: desiccator at the humidity of 80%

Although we select humidity as a noise factor rather easily, we have a contradiction between the factor effects on SN ratio and sensitivity if we compute them after decomposing the two levels of the noise factor. Furthermore, under the condition of slightly high humidity, the gain in SN ratio is negative, whereas that in sensitivity is positive. This result seems to be of great significance because this result is consistent with our experience in growing bean sprouts, and the factor is likely to be one that can adjust our contradictory objectives that bean sprouts should have fast growth but delayed rot.

In addition, since according to the image of continuous water sprinkling in our operation, we might have set both humidity levels high in the experiment. As a result, it is believed that important conditions of water sprinkling selected as control factors become vague and thus lead to a lower contribution of control factors in our experiment. If we conduct another experiment, we should include humidity as a control factor and also add much drier conditions. In fact, the growth room for bean sprouts in our actual operation is not a high-humidity environment.

As an example, we calculate an SN ratio using the data for experiment 1 in the L_{18} orthogonal array.

Total variation:

$$S_T = 1.500^2 + 1.623^2 + 1.692^2 + 1.625^2$$
$$+ 1.697^2 + 1.758^2 = 16.359796 \quad (f = 6) \tag{4}$$

Effective divider:

$$\gamma = 5^2 + 6^2 + 7^2 = 110 \tag{5}$$

Linear equations:

$$L_1 = (1.500)(5) + (1.623)(6)$$
$$+ (1.692)(7) = 29.082 \tag{6}$$

$$L_2 = (1.626)(5) + (1.697)(6)$$
$$+ (1.758)(7) = 30.618 \tag{7}$$

Variation of proportional term:

$$S_\beta = \frac{(29.082 + 30.618)^2}{(2)(110)}$$
$$= 16.200409 \quad (f = 1) \tag{8}$$

Variation of differences between proportional terms:

$$S_{N\beta} = \frac{(29.082 - 30.618)^2}{(2)(110)}$$
$$= 0.010724 \quad (f = 1) \tag{9}$$

Error variation:

$$S_e = 16.359796 - 16.200409 - 0.010724$$
$$= 0.148627 \quad (f = 6) \tag{10}$$

Error variance:

$$V_e = \frac{0.148627}{4} = 0.037157 \tag{11}$$

Total error variance:

$$V_N = \frac{0.010724 + 0.148627}{1 + 4}$$
$$= 0.031870 \tag{12}$$

SN ratio:

$$\eta = 10 \log \frac{[1/(2)(110)]}{0.031870}$$
$$= -3.63 \text{ dB} \tag{13}$$

Sensitivity:

$$S = 10 \log \frac{1}{(2)(110)} (16.200409 - 0.031870)$$
$$= -11.34 \text{ dB} \tag{14}$$

3. Optimum Configuration and Confirmatory Experiment

To pursue an ideal function of growth, we select factors that are likely to affect growth and determine the experimental conditions. Since we study factors especially to determine a condition in the growth room in our experiment, from control factors we exclude germination conditions such as the water-soaking temperature. In each experiment we set up an identical germination condition because it influences the germination rate and sterilization effect of bacteria, which are not within the scope of this study. Table 1 shows the control factors and levels.

❏ *Type of seed* (A). Although this is not a condition of growth, selection of a seed (mappe) is regarded as essential to growth. Although black mappe is widely used, the demand for green gram (green bean) has started to soar because of its better growth, despite its being three to five times as high in price. We chose these two seed types for our experiment.

Table 1
Control factors and levels

	Control Factor	Level		
		1	2	3
A:	type of seed	Black mappe	Green gram	
B:	room temperature (°C)	18	24	30
C:	number of ethylene gas bathings	Morning (9 am)	Morning (9 am) Noon (12 pm)	Morning (9 am) Noon (12 pm) Evening (5 pm)
D:	concentration of ethylene gas (mL/L)	10	20	30
E:	sprinkling timing	Morning (9 am)	Morning (9 am) Noon (12 pm)	Morning (9 am) Noon (12 pm) Evening (5 pm)
F:	number of mist sprays (0.5 mL/spray)	1	2	3
G:	amount of mineral added to sprinkled water (%)	0	0.1	1.0

❑ *Number of ethylene gas bathings* (*C*). An ethylene gas, also called a plant hormone, is used for production of thick bean sprouts. In addition, it is believed that this is related to the aging and decay of bean sprouts. Furthermore, bean sprouts themselves are considered to generate an ethylene gas when they feel stress.

Figure 2 shows the response graphs for an L_{18} orthogonal array including each factor. Based on them, we select $A_2B_1C_3D_2E_1F_2G_1$ as the optimal configuration for the SN ratio. On the other hand, the counterpart for sensitivity is $A_1B_2C_1D_1E_2F_1G_1$.

The factors with the largest effects are $B_1D_2F_2$ for the SN ratio and A_1B_2 for sensitivity. Now, choosing

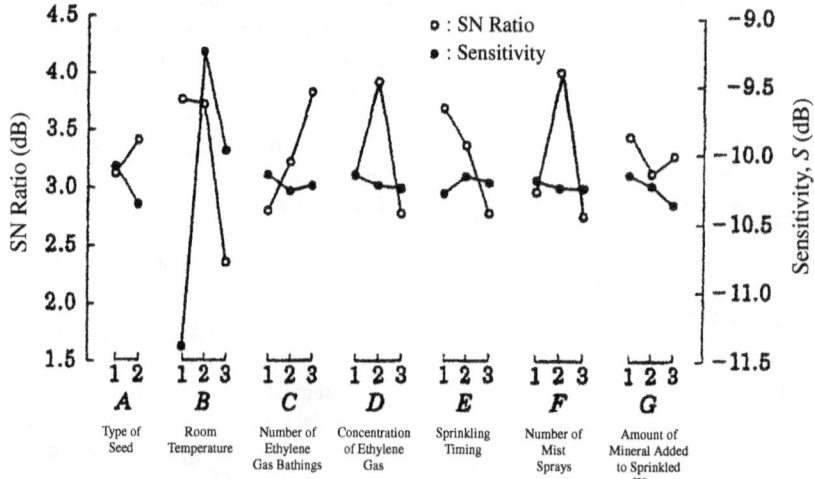

Figure 2
Response graphs

the following combinations for the optimal configurations, we estimate a sensitivity, SN ratio, and process average, respectively.

1. Optimization prioritizing growth rate (sensitivity)
2. Optimization prioritizing SN ratio
3. Optimization of SN ratio balancing growth rate (sensitivity)

For each combination, we estimate the process average. Table 2 shows the results. For the maximization of both characteristics, we can select combination 3. Table 3 summarizes the results of the confirmatory experiment.

Despite a difference from the estimated SN ratio, we conclude that fairly good reproducibility is obtained because of a small difference between the estimation and confirmation. On the other hand, we can attain gains of 2.20 and 2.56 dB in SN ratio and sensitivity, respectively, which are regarded as sufficient for our improvement in growth.

Our experiment cannot produce significant improvement in the SN ratio. Since there is a slight difference between estimation and confirmation in the SN ratio, we are somewhat skeptical of the additivity in the SN ratio. This is primarily because our selection of ranges for signal factor levels is not appropriate (i.e., we should use the growth range of an earlier stage). Other major reasons are as follows: (1) because humidity chosen as the noise factor is shifted too much to the side of high humidity, the sprinkling condition becomes ambiguous; (2) because of a long experimental period spanning approximately 20 weeks, changes in environmental conditions cannot be ignored; and (3) there is considerable variability in property among mappes.

Particularly for the variability in mappe properties mentioned above, to mitigate the variability we should have increased the number of mappes or sieved only appropriate sizes. In the case of handling organisms, it is difficult to select uniform samples because of large variability among individuals. Therefore, we need to plan an experimental method in such a way that individual variability is negligible.

The improvement in SN ratio leads to (1) smooth growth (decline or elimination of frequent occurrence in rot, prolonged period of preservation), and (2) improved controllability of a growth curve (controllability of a growth rate, shipping, or surplus and shortage in volume). In the meantime, the improvement in sensitivity brings the following: (1) a shortened period of growth, more accurate prediction of shipping day; and (2) reduction in production cost, streamlining of production operations, retrenchment of production equipment, more effective use of space for other products, mitigation of industrial waste. Shortening of a production period by improving sensitivity, S, leads to significant cost reduction together with high productivity.

In the meantime, among the unexpected significant benefits brought about by short-term growth or early shipping is a reduction in industrial waste. In producing bean sprouts, although little waste is caused by rottenness that occurs in normal production processes, we have had a considerable amount of waste due to excessive production by incorrect prediction of a shipped volume. The cost required to discard the waste (which is not a loss due to the discard of products themselves but the expense of turning the waste over to industrial waste disposal

Table 2
Estimation of optimal configuration

	Combination		
	1	**2**	**3**
SN ratio (dB)	4.02	5.30	5.26
Sensitivity (dB)	−9.24	−11.19	−9.24
β	0.325	0.276	0.345
Growth rate on seventh day	11.20	6.89	11.20

Table 3

Results of confirmation experiment for combination 3 (dB)

		Configuration		
		Optimal	Current	Gain
SN ratio	Estimation	5.26	—	—
	Confirmation	5.72	3.52	2.20
Sensitivity	Estimation	−9.24	—	—
	Confirmation	−8.93	−11.49	2.56

companies) sometimes amounts to several million yen monthly. Shortening a production period contributes greatly to improved prediction.

Similarly, higher production stability achieved through improvement in the SN ratio is among the benefits that is difficult to evaluate monetarily. Above all, a constant drop in the amount of rotten products is considered one of the most effective improvements, even though it is difficult to grasp.

The frequent occurrence of rotten products inside a factory is a serious problem. Although this does not take place often, when it happens, it is generally on an enormous scale. Proliferation of bacteria in a growth room with no sunlight and extremely high humidity and temperature is always of concern. Even if only a small number of bacteria proliferate, the appearance of the bean sprouts is damaged, so that not all of them can be shipped.

Although the proliferation of bacteria and rot are regarded as defects, we cannot succeed in im-

plementing an experiment focusing on defects, where we make evaluations based on the frequency of occurrence. Therefore, a decline in rotten products through an improved SN ratio is regarded as effective. In a study of the SN ratio, rotten product is not measured as a quality characteristic but is a good countermeasure for rotten products. (If we calculate experimentally the limitation that rotten products occur, we will be able to compute the cost using the loss function based on the on-site SN ratio at the production plant.)

Reference

Setsumi Yoshino, 1995. Optimization of bean sprouting condition by parameter design. *Quality Engineering,* Vol. 3, No. 2, pp. 17–22.

This case study is contributed by Setsumi Yoshino.

CASE 2

Optimization of Small Algae Production by Parameter Design

Abstract: By analyzing a proliferation curve for a production process for small algae on the basis of a dynamic characteristic, we applied an optimal configuration obtained from a small-scale, low-density cultivation experiment to large-scale, high-density cultivation. We had difficulty finding a really optimal configuration because of a time-lapsed change in initial conditions for control factors.

1. Introduction and Generic Function

Our objective in producing small algae was to maximize the volume of algae using the tube cultivation apparatus shown in Figure 1, which is considered promising as a closed-loop continuous cultivation device to solve technical problems such as contamination of bacteria. Since numerous factors affect the proliferation speed of a small alga, we attempted to solve these problems through a small-scale experiment, as an experiment using a large-scale continuous cultivation machine was not regarded as economical.

Ideally, the proliferation curve in the case of producing small algae should have a linear relationship between the logarithmized amount of algae and the cultivation period. However, in actuality, as a cultivation period increases, the proliferation curve's slope decreases, and in the end, the proliferation saturates, as shown in Figure 2. To enhance productivity, we need to cultivate the number of algae as densely as possible while retaining constant proliferation. That is, even in a high-density region, the proliferation curve must be linear. Therefore, considering that a generic function is a linear proliferation between a quantity of sunlight as input energy and the logarithm of the amount of algae (biomass), we studied the conditions for the most ideal proliferation curve. In other words, conditions

where both the SN ratio and sensitivity S of the proliferation curve are large are regarded as optimal.

2. Experimental Procedure

The factors affecting the proliferation speed of small algae include environment, cultivation apparatus, bacteria type, or medium. To study medium conditions for obtaining maximum biomass, we need to repeat experiments under the conditions for high-density cultivation. However, since the transparency deteriorates and light energy acts as a constraint factor, we had difficulty performing an experiment on the influence of a medium's ingredients. On the contrary, whereas it is easy to establish an optimal level of medium in the case of low-density cultivation, it is difficult to adopt the results for high-density cultivation. It was expected in this experiment, however, that the results of small-scale low-density cultivation would be applicable for large-scale cultivation.

Since there is a correlation between the amount of small algae produced (biomass, amount of dried cells per liter in the culture solution) and optical density (OD_{680}), the latter was considered as an alternative characteristic, for biomass. There are other substitutive characteristics, such as PCV (volume of cells in culture solution per liter) or cell

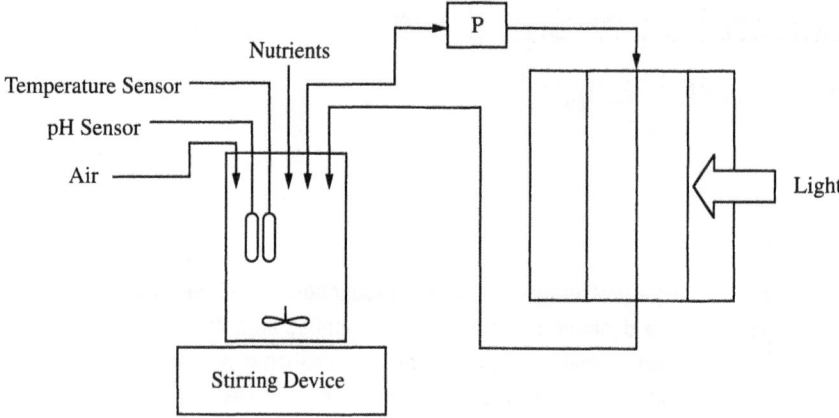

Figure 1
Tube cultivation apparatus

concentration (number of cells in culture solution per milliliter), but light absorption was used for its simplicity, small amount of sample required, and good precision of measurement.

As a signal factor, the number of cultivating days (one to nine days), which is the total amount of light, or input energy, was used. Light absorption at the same time of the day was measured. Artificial illumination with a constant amount of light was used.

As noise factors, environmental conditions (temperature, amount of light, amount of air, etc.) or the shape of apparatus may be considered. But these were not included because of trying to reduce

the scale of the experiment and also to minimize the error of algae propagation activity by starting each experiment under the same conditions. It is difficult in biological experimentation to include many noise factors. Therefore, it is important to select carefully both the factors and the scale of the experiment.

3. SN Ratio

To analyze the measurements of optical density using a linear relationship passing through the origin (zero-point proportional equation), we include their logarithmized values in Table 1. Now x_0 and x indicate optical density (OD_{680}) and y represents a logarithmized value.

$$y = 1000[\log(1000x) - \log(1000x_0)] \qquad (1)$$

For a logarithmized value (y), considering that a zero-point proportional $y = \beta M$ is ideal, we compute the SN ratio and sensitivity based on a dynamic characteristics as below. Next, we detail an example for experiment 1 (Table 1).

Total variation:

$$S_T = 298.20^2 + 577.40^2 + \cdots + 1318.98^2$$

$$= 8,434,868.8523 \qquad (f = 8) \qquad (2)$$

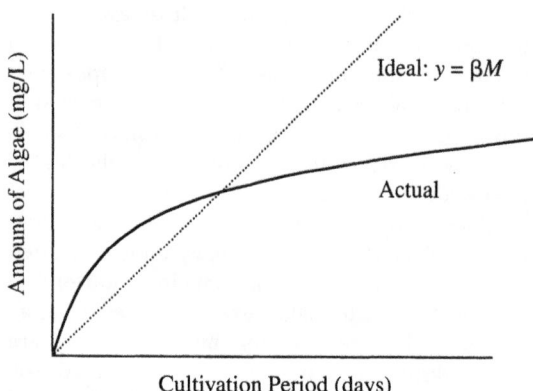

Figure 2
Proliferation curve

Table 1
Logarithmized measurement of optical density

No.	1	2	3	4	5	6	7	8	9
					Day				
1	0.0	298.2	577.4	833.7	1002.8	1127.6	1265.2	1303.8	1319.0
2	0.0	306.6	604.6	900.3	1089.9	1198.1	1285.8	1235.7	668.6
⋮	⋮	⋮	⋮	⋮	⋮	⋮	⋮	⋮	⋮
17	0.0	443.9	781.9	1020.9	1213.5	1314.4	1419.1	1497.5	1518.2
18	0.0	443.9	800.1	1089.9	1259.6	1291.6	1301.9	1340.6	1115.7

Effective divider:

$$\gamma = 1^2 + 2^2 + \cdots + 9^2 = 249 \qquad (3)$$

Variation of proportional term:

$$S_\beta =$$
$$\frac{[(1)(298.20) + (2)(577.40) + \cdots + (9)(1318.98)]^2}{249}$$

$$= 8,046,441.5662 \qquad (f = 1) \qquad (4)$$

Error variation:

$$S_e = 8,434,868.8523 - 8,046,441.5662$$
$$= 388,427.2861 \qquad (f = 7) \qquad (5)$$

Error variance:

$$V_e = \frac{388,427.2861}{7} = 55,489.6123 \qquad (6)$$

SN ratio:

$$\eta = 10 \log$$
$$\frac{(1/249)\ (8,046,441.5662 - 55,489.6123)}{55,489.6123}$$
$$= -2.38 \text{ dB} \qquad (7)$$

Sensitivity:

$$S = 10 \log \frac{1}{249}(8,046,441.5662 - 55,489.6123)$$
$$= 45.06 \text{ dB} \qquad (8)$$

4. Results of Experiment

Table 2 shows the control factors. Adding as inoculants small algae (*Nannnochrolopsis*) in 18 culture

Table 2
Control factors and levels

Control Factor	Level 1	2	3
A: illuminance (lx)	2000	4000	—
B: concentration of added seawater (%)	30	0	70
C: nitrogen coefficient	0.385	0.585	0.785
D: phosphorus coefficient	0.001	0.002	0.002
E: carbonic acid coefficient	0.010	0.016	0.022
F: coefficient of small-quantity ingredient	0.014	0.022	0.030
G: coefficient of added fertilizer	1.90	2.04	2.18

flasks that contain 300 mL of each type of medium assigned to the L_{18} orthogonal array, we began to cultivate them with two types of illuminance. The amount of air is fixed for all conditions. By sampling a small amount of culture solution at the same time of the day, we measured an optical density (OD_{680}). Based on measurements of an optical density, we computed the number of algae and added a required nutrient equivalent daily to that multiplied by each coefficient of the aforementioned nutrient-related control factors.

Figures 3 shows the response graphs of the SN ratio and sensitivity. The more the slope of a logarithmic proliferation inclines, the larger the dynamic sensitivity S becomes. The more linear and less varied the slope of a logarithmic proliferation becomes, the larger the dynamic SN ratio becomes. Therefore, picking up levels with a large sensitivity and SN ratio according to the response graphs, we determined an optimal configuration.

As a result, the optimal configuration selected from the response graphs was $A_2B_2C_1D_1E_3F_3G_1$. It was

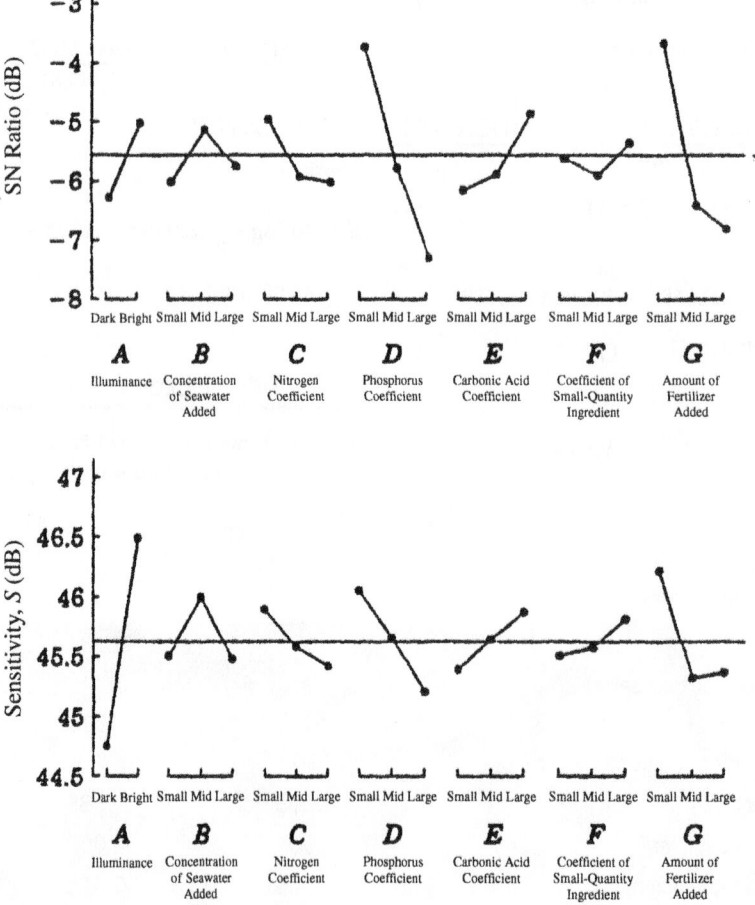

Figure 3
Response graphs for proliferation

found that the factor effects of phosphorus coefficient D and added fertilizer coefficient G were significant among all control factors. For factor F, because of its small effect on the SN ratio and sensitivity, we can select F_1, which requires only a small amount of chemical.

As the next step, to estimate the process average under the optimal configuration selected, $A_2B_2C_1D_1$ $E_3F_3G_1$ we calculated the SN ratio and sensitivity, S.

SN ratio:

$$\eta = -5.00 - 5.12 - 4.97 - 3.75 - 4.90 - 5.39$$
$$- 3.67 - (6)(-5.65)$$
$$= 1.10 \qquad (9)$$

Sensitivity:

$$S = 46.49 + 46.00 + 45.88 + 46.03 + 45.86$$
$$+ 45.81 + 46.20 - (6)(45.62)$$
$$= 48.55 \qquad (10)$$

On the other hand, the process average under the current configuration, $A_2B_2C_2D_2E_2F_2G_2$, is as follows.

SN ratio:

$$\eta = -5.00 - 5.12 - 5.94 - 5.82 - 5.90 - 5.96$$
$$- 6.44 - (6)(-5.65)$$
$$= -6.28 \qquad (11)$$

Sensitivity:

$$S = 46.49 + 46.00 + 45.57 + 45.64 + 45.63$$
$$+ 45.56 + 45.31 - (6)45.62$$
$$= 46.48 \qquad (12)$$

Table 3 compares the confirmatory experimental result and the estimate.

As a result of this confirmatory experiment, although there is little improvement in confirmatory experiment 1, fairly good reproducibility and significant improvement in the SN ratio and sensitivity were gained in confirmatory experiment 2, resting on a tube cultivation apparatus. In this case, the gains in SN ratio and sensitivity are 5.80 and 4.35 dB, respectively, which are quite consistent with the estimations. Consequently, our experiment implies that by analyzing a proliferation curve for a production process of small algae on the basis of a dynamic characteristic, we could apply an optimal configuration obtained from a small-scale, low-density cultivation experiment to large-scale, high-density cultivation.

In an experiment on a production process for algae, such as our study, even for the simple purpose of determining proper conditions for a medium, we had difficulty finding a really optimal configuration because of a time-lapsed change in initial conditions for control factors. For instance, focusing on illuminance selected as a control factor, we can see that the quantity of light that algae receives decreases gradually, as less external light reaches the inside of a culture solution, due to the proliferation of organisms. In addition, each nutrient diminishes accordingly, whereas the number of algae increases.

Two major solutions for this problem are (1) to devise a control system of maintaining these control factor conditions at constant levels, and (2) to determine an optimal configuration for the system by changing these control factors to noise factors.

Table 3
Estimation and confirmation of process average (dB)

| Configuration | SN Ratio | | | Sensitivity S | | |
| | Estimation | Confirmation | | Estimation | Confirmation | |
		1	2		1	2
Optimal	1.10	−2.81	2.72	48.55	43.85	47.37
Initial	−6.28	−3.08	−3.08	46.48	43.02	43.02
Gain	7.38	0.27	5.80	2.07	0.83	4.35

Although it is possible to maintain the conditions of control factors using current technology, it is costly and difficult to expect perfect control. Therefore, in the future, by attempting to use the control system to retain the same conditions to some extent, we might consider the control factors as noise factors in an economical study of an optimal configuration for biomass.

Reference

Makoto Kubonaga and Eri Fujimoto, 1993. Optimization of small algae production process by parameter design. *Quality Engineering*, Vol. 1, No. 5.

This case study is contributed by Makoto Kubonaga.

CASE 3

Optimization of Polymerization Reactions

Abstract: In this research we applied quality engineering principles to a synthetic plastics reaction known as a polymerization reaction, regarded as one of the slowest of all polymerization reactions, and a reaction with a wide molecular mass distribution, to enhance productivity and obtain a polymerized substance with a narrow molecular mass distribution.

1. Introduction

In manufacturing processes for chemical products, quality and cost depend strongly on a synthetic reaction condition. The reaction handled in this research, a synthetic reaction of resin called a *polymerization reaction*, is one of the slowest of all such reactions and has a wide molecular weight distribution. Although we had tried many times to improve productivity and molecular weight distribution, no significant achievement was obtained. In general cases of polymerization reaction, for a complete product we have conventionally measured its molecular weight distribution and analyzed the average molecular weight calculated. The objective of this study was to accomplish an unachieved goal by applying quality engineering to this polymerization reaction. Regarding the application of quality engineering to a chemical reaction, there is one case of doubling productivity by using a dynamic operating window for an organic synthesis reaction. However, the case contains some problems in reproducibility, such as having both peaks and V-shapes in a factor effect plot of the SN ratio.

To solve these problems, we have taken advantage of the ratio of the rate of main reaction β_1 and that of side reaction β_2: β_1/β_2. By doing this we obtained fairly good reproducibility. In our research, in reference to an organic synthesis reaction, we applied quality engineering to a polymerization reaction by improving a method of handling a polymerization degree distribution and calculating reaction speed β and the SN ratio.

2. Experimental Procedure

Objective

Since, in general, a polymer substance consists of various types of molecular weight, it has a particular molecular weight. A polymer in this experiment is an intermediate material used as a raw material in the following process. In this case, its molecular weight should be neither too large nor too small. Therefore, using the quality engineering method, we attempted to enhance productivity and obtain a polymerized substance with a narrow molecular weight. To enhance productivity, a high reaction speed and a stable reaction are both essential.

To obtain a polymerized substance with a narrow molecular weight distribution, the constitution of the resulting molecular weight should be high. Therefore, by classifying an ingredient not reaching a certain molecular weight M_n as an *underreacted substance*, one ranging from M_n to M_m as a *target substance*, and one exceeding M_m as an *overreacted substance*, we proceeded with our analysis.

Experimental Device

We heated material Y, solvent, and catalyst in a 1-L flask. After a certain time, we added material X to the flask continuously. Next came the reaction of X and Y. When the addition of X was complete, dehydration occurred and polymerization took place.

Table 1
Control factors and levels for the first L_{18} orthogonal array (%)

Column	Control Factor		Level 1	Level 2	Level 3
1	None		—	—	—
2	A:	amount of material X	96	100	106
3	B:	amount of agent 1	33	100	300
4	C:	amount of agent 2	85	100	115
5	D:	reaction speed	90	100	110
6	E:	maturation time	0	33	100
7	F:	material supply time	0	33	100
8	G:	continuously supplied material: X or Y	X	X	Y

Table 2
Control factors and levels for the second L_{18} orthogonal array (%)

Column	Control Factor		Level 1	Level 2	Level 3
1	C:	amount of agent 2	100	115	—
2	A:	amount of material X	92	96	100
3	B:	amount of agent 1	100	300	500
4	H:	agent type	1	2	3
5	I:	supply method of material	1	2	3
6	J:	supply method of agent 1	1	2	3
7	F:	material supply time	33	100	150
8	K:	material type of Y	1	2	3

Table 3
Change in constitution of yields at each reaction time (%)

	Reaction Time				
	0.3	1	2	3	4
Underreacted substance	97.1	30.9	20.7	19.1	12.3
Target substance	2.9	62.0	72.2	63.6	61.1
Overreacted substance	0	7.1	7.1	17.3	26.7
Total	100	100	100	100	100

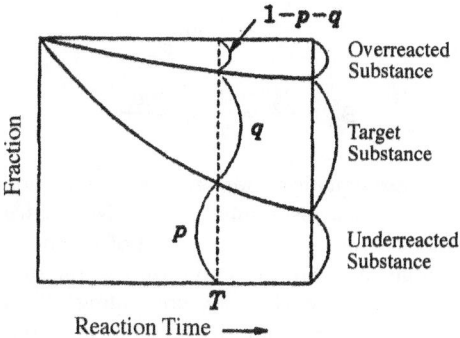

Figure 1
Dynamic operating window method

Layout of Orthogonal Array

In the first experiment, we selected seven control factors and allocated them to an L_{18} orthogonal array with a larger interval of level. Table 1 shows the factor and levels. After analyzing the experimental result of the first L_{18} experiment, we designed and conducted the second L_{18} experiment (Table 2).

Data Gathering for Analysis

We sampled the polymerizing solution in each of runs 1 to 18 allocated in the L_{18} orthogonal array at a certain interval. Then, using gel permeation chromatography (GPC), we gauged the constitution of a material and polymers.

As a specific example, Table 3 shows a change in the constitution at each reaction time in run 2 in the first L_{18} orthogonal array. The data for a molec-

ular mass and its distribution were computed as converted values of the standard polystyrene molecular mass.

3. Analysis by Dynamic Operating Window Method

Generic Function

Since a chemical reaction is an impact phenomenon between molecules, by setting the constitution of underreacted substances in a yielded substance to p, we defined the following equation as a generic function on the basis that reaction speed is proportional to density of material:

$$p = e^{-\beta T}$$

Taking the reciprocal of both sides and logarithmizing, we have

$$\ln(1/p) = \beta T$$

Setting $Y = \ln (1/p)$, we obtain the following zero-point proportional equation:

$$Y = \beta T$$

Now β is a coefficient of the reaction speed.

Concept of Operating Window

If we categorize a yield as underreacted, target, and overreacted based on the molecular mass, as a chemical reaction progresses, the constitution of the target substance rises (Figure 1).

Setting the constitution of an underreacted substance at the point of reaction time T to p and that of an overreacted substance to q, we can express that of an overreacted substance by $1 - p - q$. Since an ideal reaction is regarded as one with a faster reaction speed and higher constitution then a target substance (Figure 2), we should lower the constitution of an underreacted substance immediately and maintain that of an overreacted substance as low as possible.

Then, taking into account the following two reaction stages, we estimated a reaction speed for each stage β.

R_1: a material (molecular mass M_1, monomer) and polymers (molecular mass $M_2 - M_{n-1}$,

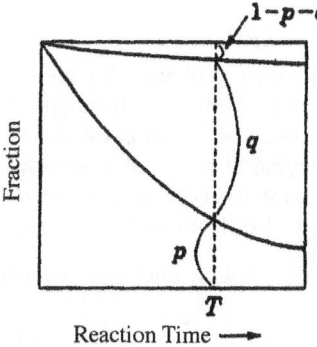

Figure 2
Reaction regarded as ideal

polymer, that is, a dimer consisting of two combined molecules) convert into target substances (molecular mass $M_n - M_{m-1}$) and over-reacted substances (molecular mass more than M_m):

material + polymer (M_2) + ⋯
   + polymer (M_{n-1}) (M_n + ⋯ + M_{m-1}) + M_m ⋯

We define the reaction speed at this stage as β_1.

R_2: A material (molecular mass M_1, monomer), polymers (molecular mass $M_2 - M_{n-1}$, polymer, that is, a dimer consisting of two combined molecules), and target substances (molecular mass $M_n - M_{m-1}$, monomer) convert into overreacted substances (molecular mass more than M_m):

material + polymer (M_2) + ⋯
   + polymer (M_{n-1}) + (M_n + ⋯ + M_{m-1}) M_m ⋯

We define the reaction speed at this stage as β_2.

A reaction speed in each stage is not a theoretical rate of a simple chemical reaction per se, however, but a rate in which the underreacted substances described above convert into a yield. The reason that we deal with the reaction in this manner is to prepare a mechanism of additivity. As a result, since R_1 indicates a yielding rate of target and over-reacted substances and R_2 represents that of an overreacted substance, our final objective is to obtain an optimal configuration to maximize the difference between two rates, R_1 and R_2.

Now for R_1 and R_2 we calculate the constitution of an underreacted substance at the point of time T_i, p_i and $p_i + q_i$, and compute rates of reaction β_{1i} and β_{2i}.

$$R_1:\quad p_i = e^{-\beta_{1i}/T_i} \qquad \ln \frac{1}{p_i} = \beta_{1i}T_i$$

$$R_2:\quad p_i + q_i = e^{-\beta_{2i}T_i} \qquad \ln \frac{1}{p_i + q_i} = \beta_{2i}T_i$$

In this case, the reaction speed to yield a target substance β_1 should be larger and the reaction speed to yield an overreacted substance β_2 should be smaller.

$$\beta_{1i} = \frac{Y_{1i}}{T_i} = \frac{1}{T_i}\ln\frac{1}{p_i}$$

$$\beta_{2i} = \frac{Y_{2i}}{T_i} = \frac{1}{T_i}\ln\frac{1}{p_i + q_i}$$

Using these equations, we can calculate β_{1i} and β_{2i} for each of K measurement points. Now, since we need to maximize a reaction speed of yielding a target substance, β_{1i} and to minimize that for an over-reacted substance, β_{2i}, by computing β_{1i} as a larger-the-better SN ratio η_1 and β_{2i} as a smaller-the-better SN ratio η_2, we find factors to maximize $\eta_1 + \eta_2 = \eta$. This is a dynamic operating window method.

$$\eta_1 = -10\log\left[\frac{1}{K}\left(\frac{1}{\beta_{11}^2} + \cdots + \frac{1}{\beta_{1K}^2}\right)\right]$$

$$\eta_2 = -10\log\left[\frac{1}{K}(\beta_{21}^2 + \cdots + \beta_{2K}^2)\right]$$

When a reaction speed is stable throughout a total reaction, we can analyze the reaction with only one β at a single observation point in a more efficient manner. If we use a single observation point, the calculation is

$$\eta = \eta_1 + \eta_2$$
$$= -10\log\frac{1}{\beta_1^2} - 10\log\beta_2^2 = 10\log\frac{\beta_1^2}{\beta_2^2}$$

As sensitivity S, we consider β_1 regarding productivity:

$$S = 10\log\beta_1^2$$

Example of Analysis

We now show the calculations for the experimental data for run 2 in the first L_{18} orthogonal array. The observation points of β_1 and β_2 can differ. Then, setting the observation points so as to estimate β_1 and β_2 most accurately at T_1 and T_2, respectively, we compute β_1 and β_2 from each single point.

Decreasing speed of an underreacted substance:

$$\beta_1 = \frac{1}{T_1}\ln\frac{1}{p_1} = \frac{1}{1}\ln\frac{1}{0.309} = 1.1757$$

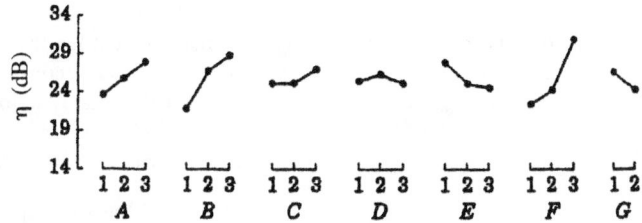

Figure 3
Factor effect plot of the SN ratio in the first L_{18} experiment

Yielding speed of an overreacted substance:

$$\beta_2 = \frac{1}{T_2} \ln \frac{1}{p_2 + q_2} = \frac{1}{4} \ln \frac{1}{0.123 + 0.611} = 0.0776$$

SN ratio:

$$\eta = 10 \log \frac{\beta_1^2}{\beta_2^2} = 10 \log \frac{(1.1757)^2}{(0.0776)^2} = 23.6105 \text{ dB}$$

Sensitivity:

$$S = 10 \log \beta_1^2 = 10 \log (1.1757)^2 = 1.4059 \text{ dB}$$

4. Result of the First L_{18} Experiment

Figures 3 and 4 show the factor effect plots of the SN ratio, η, and sensitivity, S, in the first L_{18} experiment. In addition, Table 4 shows the optimal levels for the SN ratio and sensitivity. By prioritizing sensitivity for factors whose optimal SN ratio and sensitivity are different, we determine factor levels in the second L_{18} experiment.

5. Result of the Second L_{18} Experiment

Considering the result of the first L_{18} experiment, we determined factors and levels for the second L_{18} experiment as follows. We have the following control factors:

E: We fixed its levels to zero at E_1 by focusing on productivity because the maturation time is zero under the optimal configuration.

D: We fixed its levels to D_2 by selecting the current level because of its small factor effect.

G: We fixed its levels to G_1 because X has turned out to be a good choice in terms of continuously supplied material.

For other factors, A, B, C, and F, in order to delve further into optimal levels, we changed each level. On the other hand, as new control factors, we deliberately selected H, I, J, and K. As for factor C, although we had tested as technically wide a range as possible in the first L_{18} experiment, we included it once again to associate it with the result in the first L_{18} experiment because the first column emptied in the layout of the second experiment.

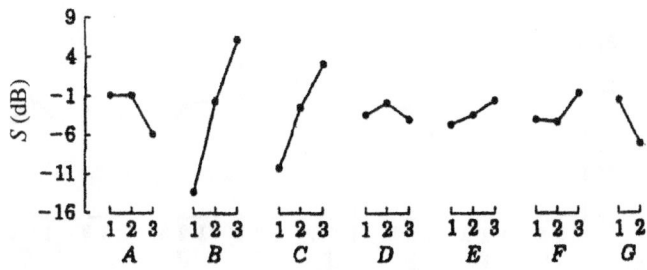

Figure 4
Factor effect plot of sensitivity in the first L_{18} experiment

Table 4

Optimal configuration in the first L_{18} experiment[a]

	Factor						
	A	B	C	D	E	F	G
SN ratio	3	3	(3)	(2)	1	3	(1)
Sensitivity	1	3	3	(2)	(3)	(3)	1

[a] Parentheses indicate small-effect level.

Figures 5 and 6 summarize the factor effect plots of the SN ratio, η, and sensitivity, S, in the second L_{18} experiment. According to each result for the SN ratio and sensitivity, we determined the optimal levels as shown in Table 5.

6. Results of Confirmatory Experiment

According to the first and second L_{18} experiments, we obtained less rugged results for the factor effects of the SN ratio and sensitivity. The implied high additivity of the SN ratio is based on reaction speed. Then we conducted a confirmatory experiment under each optimal configuration of the first and second L_{18} experiments. Tables 6 and 7 show the results of the first and second L_{18} confirmatory experiments, respectively.

For the first L_{18} experiment, we conducted a confirmatory experiment under the optimal configuration for each SN ratio and sensitivity value. The result of the first L_{18} experiment reveals that whereas we obtained fairly good reproducibility in

gain for the SN ratio, there was a small gap in gain for the sensitivity. Under the optimal configuration for sensitivity, we can confirm that the gain improved by 5.0 dB, the reaction speed almost doubled, and the reaction time approximately halved. Because sensitivity $S = 10 \log\beta_1^2$,

$$\text{Gain} = 10 \log \beta_{1\text{optimal}}^2 - 10 \log \beta_{1\text{current}}^2$$

$$= 10 \log \frac{\beta_{1\text{optimal}}^2}{\beta_{1\text{current}}^2}$$

On the other hand, since the gain is 5.0 dB, that is, the real (nondecimal) number is 3.1,

$$\frac{\beta_{1\text{optimal}}^2}{\beta_{1\text{current}}^2} = 3.1$$

$$\frac{\beta_{1\text{optimal}}}{\beta_{1\text{current}}} = 1.8$$

For the second L_{18} experiment, we conducted a confirmatory experiment under the optimal configuration determined by both the SN ratio and sensitivity. As a result of the second L_{18} confirmatory experiment, the gain in actual sensitivity is increased to 9.2 dB from that of the first L_{18} experiment. This is equivalent to a 2.9-fold improvement in reaction speed (i.e., the reaction speed is reduced by two-thirds).

On the other hand, if we compare two molecular mass distributions of polymerized substances under the current configuration (8720 dB) and the optimal configuration (8840 dB) by adjusting the average molecular mass to the target value, we can obtain almost the same result. That is, we cannot determine a way to narrow down the range of molecular mass distribution in this research.

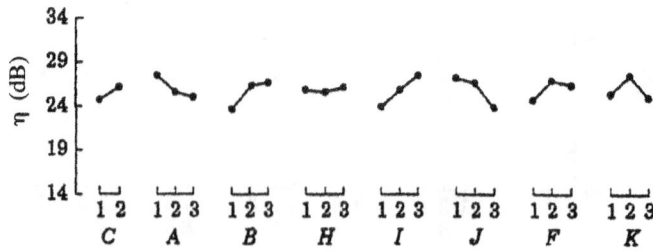

Figure 5
Factor effect plot of the SN ratio in the second L_{18} experiment

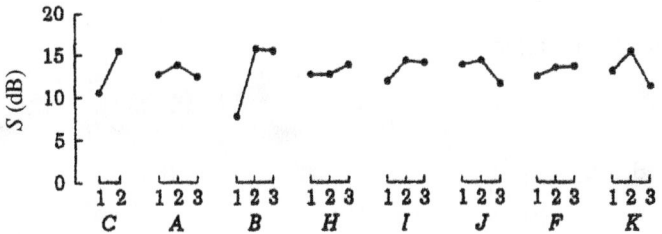

Figure 6
Factor effect plot of the sensitivity in the second L_{18} experiment

7. Conclusions

Experimental Achievements

The productivity improved dramatically because the reaction speed doubled under the optimal configuration compared to that under the current configuration. It was difficult to improve the molecular mass distribution; consequently, although the reaction speed was enhanced, the molecular mass distribution was not. However, improving the productivity threefold while retaining the same level of current quality (molecular mass distribution) was regarded as a major achievement.

Future Reearch

Since the reaction speed was increased despite little improvement in molecular mass distribution, our research can be regarded as a significant success from a company-wide standpoint. However, judging from the definition of SN ratio used, we expected

that the molecular mass distribution would also be improved. We explain considerations of improvement in molecular mass distribution as follows.

In the L_{18} experiments and their confirmatory experiments, the reaction speed β for estimation of the SN ratio is calculated by the data measured at the initial stage of reaction. Looking at the experimental data, we find that in the latter half of the experiment, the reaction saturates and deviates considerably downward from $Y = \beta T$ (particularly β_1), as shown in Figure 7. This is because a reverse or termination reaction occurred in the polymerization reaction. If we measure and compare molecular mass distributions in the initial stage of reaction, where such reactions are very minimal, we can observe improvement not only in the SN ratio but also in molecular mass distribution. Then the measurement point of molecular mass distribution in the confirmatory experiment is considered too late.

However, since the initial stage of reaction has small molecular masses as a whole, we need to

Table 5
Optimal configuration in the second L_{18} experiment[a]

	Factor							
	C	A	B	H	I	J	F	K
SN ratio	2	1	3	(3)	3	1	2	2
Sensitivity	2	(2)	2,3	(3)	2,3	1,2	(2)	2
Optimal level	2	1	3	3	3	1	2	2

[a] Parentheses indicate small effect level.

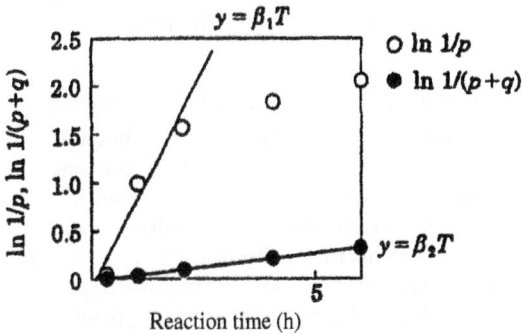

Figure 7
Example of lapse of reaction time

Table 6
Results of the first L_{18} confirmatory experiment

Configuration	Factors							SN Ratio		Sensitivity	
								Estimation	Confirmation	Estimation	Confirmation
Optimal η	A_3	B_3	C_3	D_3	E_1	F_3	G_1	38.7	34.8	10.8	9.9
Current	A_2	B_2	C_2	D_2	E_3	F_3	G_1	30.6	28.1	3.0	7.2
Gain								8.1	6.7	7.5	2.7
Optimal S	A_1	B_3	C_3	D_2	E_3	F_3	G_1	30.4	29.6	16.3	12.2
Current	A_2	B_2	C_2	D_2	E_3	F_3	G_1	30.6	28.1	3.0	7.2
Gain								−0.2	1.5	13.3	5.0

Table 7
Results of the second L_{18} confirmatory experiment

Configuration	Factors								SN Ratio		Sensitivity	
									Estimation	Confirmation	Estimation	Confirmation
Optimal η	C_2	A_1	B_3	H_3	I_3	J_1	F_2	K_2	33.1	35.5	19.2	16.4
Current	C_1	A_3	B_1	H_2	I_1	J_1	F_2	K_2	23.0	28.1	5.8	7.2
Gain									10.1	7.4	13.4	9.2

continue the polymerization reaction until the average molecular mass attains the target. Our confirmatory experiment demonstrates that after doing so, the molecular mass distribution is no longer improvable. Therefore, to improve molecular mass distribution, research on design of experiments and analytical methods, including consideration of a reverse reaction and consideration of a termination reaction, should be tackled in the future.

Achievement through Use of Quality Engineering

Experiments in production technology beginning at the beaker level in a laboratory using conventional development processes require a considerable amount of time. By evaluating a polymerization reaction with an SN ratio reflecting a generic function in a chemical reaction, we can improve reproducibility in factor effects and enhance productivity threefold, thereby obtaining a significant achievement in a short period of time.

References

Yoshikazu Mori, Koji Kimura, Akimasa Kume, and Takeo Nakajima, 1995. Optimization of synthesis conditions in chemical reaction (SN ratio analysis example no. 1). *Quality Engineering*, Vol. 3, No. 1.

Yoshikazu Mori, Koji Kimura, Akimasa Kume, and Takeo Nakajima, 1996. Optimization of synthesis conditions in chemical reaction (SN ratio analysis example no. 2). *Quality Engineering*, Vol. 4, No. 4.

1998 Report for NEDO (New Energy and Industrial Technology Development Organization), 1998. *Globally Standardized Research and Development: Study of Optimization Design and Evaluation Technique by Quality Engineering.* Tokyo: Japanese Standards Association.

This case study is contributed by Yoshikazu Mori, Kenji Ishikawa, Takeo Nakajima, Koya Yano, Masashi Matsuzaka, Akimasa Kume, and Hiroshi Yano.

CASE 4

Evaluation of Photographic Systems Using a Dynamic Operating Window

Abstract: The function of photographic image formation is represented by the characteristic curves of photosensitive materials. In color photography, variation in the characteristic curves of three colors [blue, green, and red (BGR)] affects color change. The purpose of this study was to use the chemical reaction function and compare it with the results from a traditional approach.

1. Evaluation of Color Negative Film by a Conventional Characteristic Curve

A characteristic curve as a total system of color film and its coating structure are shown in Figures 1 and 2, respectively.

Each photosensitive layer (BGR) has multiple layers, each of which consists of a mixture of several types of silver halide emulsion (subsystems). Figure 3 illustrates characteristic curves of three types of silver halide emulsion forming a green-sensitive layer. Then we need to adjust a total system to a linear gradation with ample latitude by combining these emulsions. Therefore, even if we set up an input/output relationship that regards a characteristic curve of commercial photosensitive material as ideal, we cannot evaluate an essential photographic function but assess only its objective function. This technical issue is a major motivation for introducing quality engineering to the photographic field.

As a typical example, an image transfer method used for a color photocopy machine is well known. In this case, the idea occurs when the density of the original image is proportional to that of the photocopied image with a slope equal to 1. For a photographic system, if we consider the function to be to transfer the subject's image into a photographic density, we can conduct a similar analysis.

Setting the permeability to T and the amount of light to E we consider the input/output relationship expressed by

$$1/T = \beta E \qquad (1)$$

$$\text{Log}(1/T) = \log E + \log \beta \qquad (2)$$

Logarithmizing both sides of the equation, similar to the case of a photocopy machine, we obtain the following relationship between exposure ($\log E$) and photographic density (more specifically, an omega-transformed value of permeability) with a slope of 1, as shown in Figure 4:

In a photographic system, there is no ideal relationship in this input/output system because the degree of slope (called *gradation* in photography) is an objective function determined by designers and is not appropriate in evaluating a generic function because photography density as an output depends on an amount of silver (or pigment) coming out in a development process.

If we regard a function of photography as forming the reaction of a latent image, a photographic system can also be considered as a system of converting light energy to latent image. Using a well-known theory of photosensitivity, we can derive the input/output relationship:

$$\log \frac{p}{1-p} = 2 \log E + A \qquad (3)$$

where p, E, and A are the reaction rate (a ratio of the amount of silver developed to the total amount), the quantity of light, and a constant respectively.

651

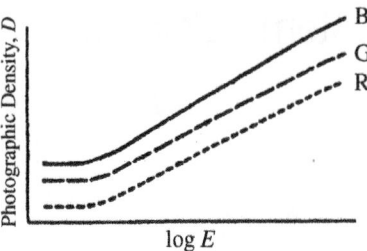

Figure 1
Characteristic curve of color negative film

In addition, considering that the photographic function is a developing reaction, we can regard the system as a system of amplifying a latent image in a development process. Since the ability to amplify the difference in the magnitude of the latent image is considered ideal, the following linear equation represents an input/output relationship (B is a constant):

$$\log \frac{p}{1-p} = \beta \log E + B \qquad (4)$$

In actuality, after analyzing the data using the foregoing two methods, we arrived at the same conclusion. This is because the slope cannot exceed 2, as shown in the latent-image-forming reaction.

Since these methods are tantamount to dealing with a photographic characteristic curve (log E and density), there is an advantage in expressing the variability in the characteristic curve. However, there are two problems in such an evaluation. The first problem is a considerable increase in the experimental effort. Figure 5 shows a characteristic curve for each development time in a subsystem using gradual exposure. The greater the quantity of light, the higher the corresponding photographic density. However, if the development time is lengthened, an unexposed area (i.e., a fogged area) develops.

Figure 6 illustrates the relationship of the SN ratio calculated using the aforementioned two methods for each development time with external disturbance given by the subsystem in Figure 5. This indicates a fluctuation in SN ratio in accordance with development time. Since each type of silver halide emulsion has a different development time when a peak emerges, we need a vast amount of experimental time simply to calculate this peak.

The second problem is one for which we probably cannot obtain a universally applicable result. Figure 7 shows the curve of E versus $\log[p/(1-p)]$ for a subsystem combining a basic condition and two external disturbances. Operators A and B have different regions for which to calculate an SN ratio, and eventually each SN ratio computed by each operator differs. So we cannot expect good reproduc-

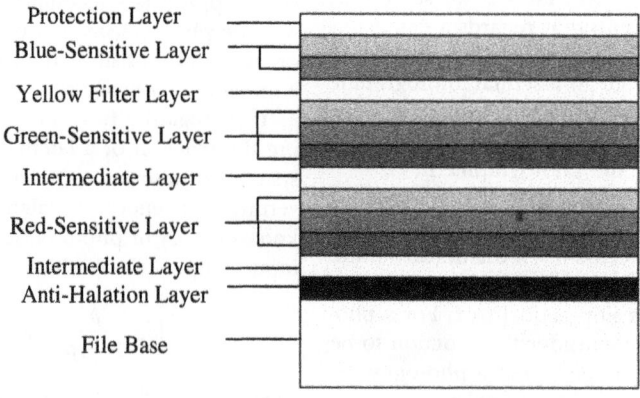

Silver Halide Particles

Figure 2
Structure of color negative film (cross section)

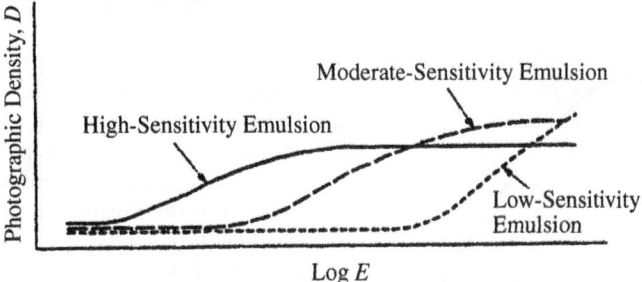

Figure 3
Characteristic curve of subsystem

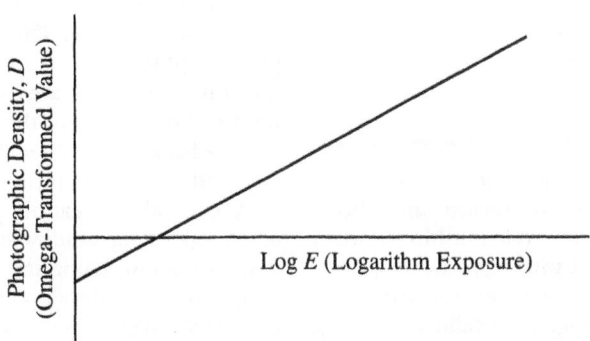

Figure 4
Input/output in the image transfer method

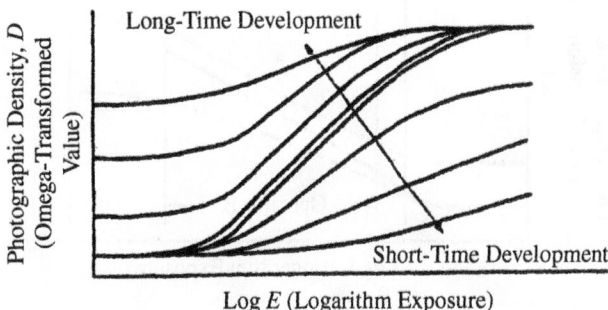

Figure 5
Fluctuation of characteristic curve for each development time

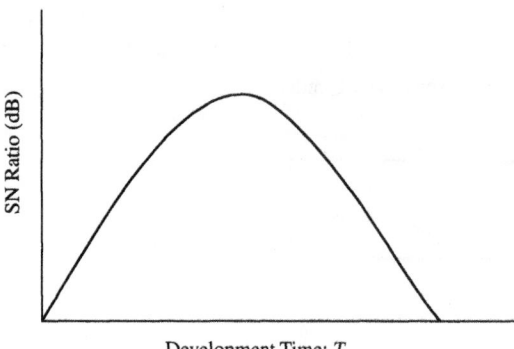

Figure 6
Development time and conventional SN ratio

silver halide used as the photosensitive material in a photograph forms a latent image in which several atoms of silver metal gather on a crystal of silver halide through a reaction activated by exposure. The latent image acts as a catalyst for the chemical reaction of development. In short, the existence of the latent image accelerates the developing reaction. Since the magnitude or number of latent images amplifies this action, exposure causes the developing speed to increase continuously. In a photographic system, by halting a developing reaction halfway and taking advantage of a difference between unexposed and exposed areas or of exposure in both areas as a difference in reaction speed, we form an image as a different amount of reaction.

As a matter of fact, as the total reaction speed increases, the reaction speed in an exposed area increases. However, an unexposed area's reaction speed also increases. Therefore, the function of a photographic system is to increase the developing speed in an exposed area without raising the speed (and without lowering the speed in an unexposed area). This is none other than the dynamic operating window in a chemical reaction.

A general calculation process based on a dynamic operating window in a photographic system is shown below. Using the characteristic curve of a sample for each development time in Figure 6, we converted image density into a reacted fraction of development, p. Since this system is a closed reaction system with a constant material used in chemical reactions, an ideal relationship between developing time and reacted fraction is identical to

ibility and generally practical results due to these differences.

2. Generic Function

In evaluating a photographic system using a dynamic operating window, we regard the photographic function to be the relationship between developing reaction speed with light. Development is a chemical reaction in which a microcrystal of silver halide is reduced to original metallic silver using a reducer, the developing chemical. Even unexposed photosensitive material will be developed sooner or later if the developing time is extended long enough. On the other hand, a microcrystal of

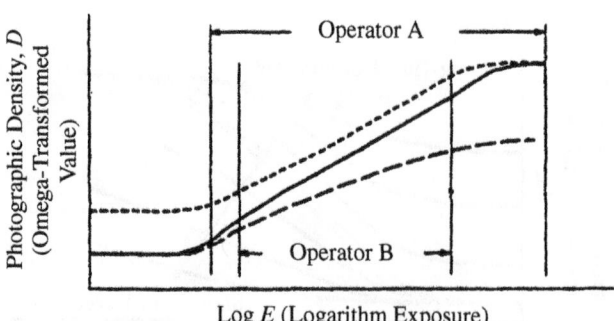

Figure 7
Signal range for calculating conventional SN ratios

the equation of the speed in a first-order reaction in the theory of reaction speed:

$$p(T) = 1 - e^{-\beta T} \tag{5}$$

where β is a constant of a developing reaction speed. Setting $y = \ln[1/(1 - p)]$, we can obtain a zero-point proportional equation (Figure 8):

$$y = \beta T \tag{6}$$

Defining exposures M_1, M_2, and M_3 and developing times T_1, T_2, ... , T_6 as signals, we calculated an SN ratio from Table 1.

Variation of differences between proportional terms due to exposure M:

$$S_{M\beta} = \frac{L_1^2 + L_2^2 + L_3^2)}{r - S_\beta} \tag{7}$$

Variation of proportional term due to exposure (fog) M_1:

$$S_{\beta M_1} = \frac{L_1^2}{r} \tag{8}$$

SN ratio:

$$\eta = 10 \log \frac{S_{M\beta} - 2V_e}{3r(V_e + S_{(M1)})} \tag{9}$$

Sensitivity:

$$S = 10 \log \frac{S_{M\beta} - 2V_e}{3r} \tag{10}$$

A key point in this calculation is that we handle a reaction at exposure M_1 [i.e., a reaction with no

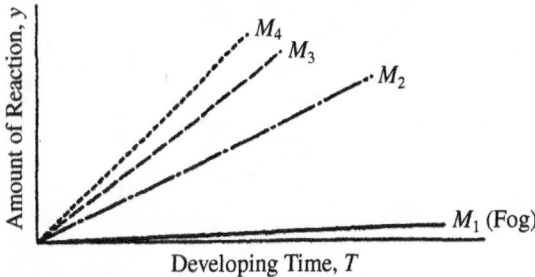

Figure 8
Input/output relationship for a dynamic operating window

exposure (fog)] as a noise and include it in the noise part of the SN ratio. This is because we expect that if light hits a film, the reaction will be faster; and if it does not, the reaction will be slower.

3. Analysis Example

As an example of further simplifying the dynamic operating window method for photography, we selected two levels for exposure, M (no exposure: M_1 = 0, maximum exposure: M_2). It is considered easy to alter the developing time in a photographic system. A photographic emulsion is made primarily through a two-stage process. The first stage is a process in which the crystal of silver halide is grown, and the second, one in which the sensitivity of the silver halide is chemically enhanced (or color-sensitized such that it can carry a preferred spectral sensitivity distribution). Then, taking up control factors from the silver halide development process, which determine the majority of fundamental characteristics of a photographic emulsion, we designed an experiment based on an L_{18} orthogonal array. Table 2 shows the control factors and levels. For the amount of exposure, we selected two levels, no exposure M_1 and maximum exposure, M_2. Ten levels were chosen for developing time T:

Level:	1	2	3	4	5	6	7	8	9	10
Time:	24	41	69	118	195	340	578	983	1670	2839

Figure 9 shows an example of raw data obtained via the above described above. This reveals that we could estimate a slope of β even if we had not picked up as many as 10 points. The following calculation is based on two points for a developing time, 69 and 195 seconds.

Linear equations:

$$L_1 = 69y_{11} + 195y_{12}$$

$$L_2 = 69y_{21} + 195y_{22} \tag{11}$$

Effective divider:

$$r = 69^2 + 195^2 = 42{,}786 \tag{12}$$

Table 1
Data for dynamic operating window

	Developing Time						Linear Equation
Exposure	T_1	T_2	T_3	T_4	T_5	T_6	
M_1 (= 0: fog)	y_{11}	y_{12}	y_{13}	y_{14}	y_{15}	y_{16}	L_1
M_2	y_{21}	y_{22}	y_{23}	y_{24}	y_{25}	y_{26}	L_2
M_3	y_{31}	y_{32}	y_{33}	y_{34}	y_{35}	y_{36}	L_3

Total variation:

$$S_T = y_{11}^2 + y_{12}^2 + y_{21}^2 + y_{22}^2 \quad (f = 4) \quad (13)$$

Variation of proportional term:

$$S_\beta = \frac{L_1 + L_2^2}{2}(42{,}786) \quad (f = 1) \quad (14)$$

Variation of differences between proportional terms due to exposure M:

$$S_{M\beta} = \frac{L_1^2 + L_2^2}{42{,}786} - S_\beta \quad (f = 1) \quad (15)$$

Variation of proportional terms due to exposure (fog) M_1:

$$S_{\beta M_1} = \frac{L_1^2}{42{,}786} \quad (f = 1) \quad (16)$$

Error variation:

$$S_e = S_T = (S_\beta + S_{M\beta}) \quad (f = 2) \quad (17)$$

Error variance:

$$V_e = \frac{S_e}{2} \quad (18)$$

Total error variance:

$$V_N = V_e + S_{\beta M_1} \quad (19)$$

Table 2
Control factors and levels

Control Factor	Level		
	1	2	3
A: density of ammonia	0	1.2	
B: pH	3	6	9
C: temperature (°C)	75	65	55
D: particle diameter of silver halide (μm)	0.55	0.65	0.75
E: silver ion index (p_{Ag})	6.5	8.1	9.1
F: constitution of iodine (mol %)	0	3	6
G: amount of additive agent A	10	5	0
H: surface treatment of silver halide	Oxidation	None	Reduction

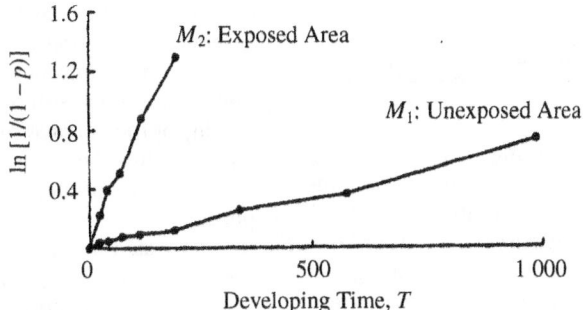

Figure 9
Actual plot for color negative film (dynamic operating window)

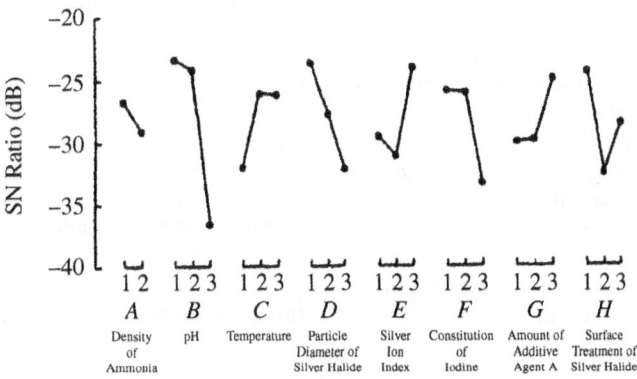

Figure 10
Response graphs of SN ratio by dynamic operating window

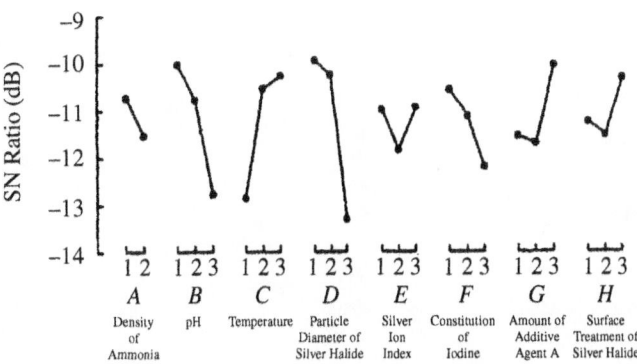

Figure 11
Response graphs of SN ratio by conventional procedure

SN ratio:

$$\eta = 10 \log \frac{[1/(2)(42{,}786)](S_{M\beta} - V_e)}{V_N} \qquad (20)$$

On the other hand, using the same L_{18} experimental configurations, we attempted to compute an SN ratio by the conventional method grounded on a characteristic curve as an input/output relationship. For the purpose of evaluating stability, in addition to the linearity of a characteristic curve, we allocated a compounded noise factor, including storage or development conditions, to the outer array and select exposure expressed by $\log E$ as a signal factor and the omega-transformed value of a reacted fraction as an output for an analysis using the zero-point proportional equation. Figures 10 and 11 show the response graphs of the SN ratio obtained from characteristic curves based on both the dynamic operating window and conventional (i.e., input/output of characteristic curve) procedures. Comparing the response graphs obtained by both a dynamic operating window and a conventional SN ratio, we can see that almost all of the plots are consistent. The SN ratio in the conventional procedure assesses the change of slope of gradation due to external disturbance ($S_{N\beta}$), magnitude of variability (S_N), and nonlinear portion (S_e) as

noises. Since we tried to avoid such problems, as the difference in developing conditions between experiments or the difference in signal factor ranges, quality features consistent with the magnitude of variability of the characteristic curve were evaluated. They can thus be considered to represent the stability of a photographic emulsion in this experiment.

In contrast, although the result derived from use of a dynamic operating Window never evaluates variations in quality features, such as the shape of a characteristic curve or external noise, it shows almost the same variation as that of the SN ratio of a characteristic curve. This implies that improvement in a photographic function based on developing speed eventually leads to stability of a characteristic curve, the objective quality characteristic.

Reference

Shoji Matsuzaka, Keisuke Tobita, Tomoyuki Nakayama, Nobuo Kubo, and Hideaki Haraga, 1998. Functional evaluation of photographic systems by dynamic operating window. *Quality Engineering*, Vol. 7, No. 5, pp. 31–40.

This case study is contributed by Shoji Matsuzaka.

CASE 5

Application of Dynamic Optimization in Ultra Trace Analysis

Abstract: Ultra trace analysis with physical and chemical methods is of considerable importance in various fields, including medicine, the food industry, and solid-state physics and electronics. For manufacturing electronic devices, ultrapure water has to be monitored for metallic contamination. Extreme demands exist related to the reliability and reproducibility of quantitative results at the lowest possible concentration level. With such data, interpretation of the linear response with the least variability and the highest sensitivity can be achieved. For proof, the lowest detectable concentration is calculated and a robust detection limit below 1 ppt has been confirmed for some elements.

1. Introduction

The analytical tool considered here is an inductively coupled plasma mass spectrometer (ICP-MS), used widely for the determination of metal impurities in aqueous solutions. The output signal of the measurement system responds to the concentration level of the metallic elements. For higher metal concentrations, the error portion of the signal can essentially be neglected, whereas for lower concentrations signal deviations become dominant. To avoid unwanted noise, ultrapure water sample preparation was carried out under flowbox conditions (class 10). The response (intensity/counts per time) is investigated for a variety of measurement conditions, equipment components, and sample preparation (10 control parameters and two noise parameters) as a function of the signal factor. Signal factor levels are ultrapure water solutions with calibrated metal concentrations from 2.5 to 100 ppt (= 10^{-9} g/L) and additional water samples for reference (blanks). This type of system, with its steadily increasing relative noise toward lower signal values, requires appropriate dynamic S/N ratio calculations. Analytical tools with more demanding requirements are used to elucidate process monitoring of metallic contaminations. Quantitative results must be reliable and reproducible from high metal concentrations to the lowest possible limit of detection (LOD). In the mass production of silicon wafers in semiconductor manufacturing, most metallic surface contaminations have to be below 1010 atoms/cm² to avoid malfunction of the final device, due to subsequent metal diffusion into the bulk.

In manufacturing, every process step is accompanied by or finishes with a water rinse. For this purpose, ultrapure water is prepared and used. Consequently, the final silicon wafer surface is only as clean (i.e., metal free) as the ultrapure water available. Therefore, water purity with respect to such common elements as Na, K, Cu, Fe, and others must be monitored routinely.

For stringent purity demands in semiconductor processing (metal concentration <2 to 10 ppt), easy methods (e.g., resistivity measurements) are not at all sufficient; the ion product of water at pH 7 corresponds to a salt concentration of less than 10 ppm. Instead, the Hewlett-Packard's ICP-MS 4500 analytical equipment is used, where the notation stands for the physical principle of an inductively coupled mass spectrometer.

2. Equipment

The starting point for the raw material water is drinking quality. The following salt contents are presented with typical concentrations beyond 10 ppm:

Salts > 10 ppm: Na^+, K^+, Ca^{2+}, Mg^{2+}

Salts < 10 ppm: Fe^{2+}. Sr^{2+}, Mn^{2+}, NH_4^+

Salts < 0.1 ppm: As^{3+}, Al^{3+}, Ba^{2+}, Cu^{2+}, Zn^{2+}, Pb^{2+}, Li^+, Rb^+

From this level any metal contamination has to be removed to ppt values for use in semiconductor cleaning. With numerous chemical and physical principles applied in multiple process steps (Figure 1), this can be achieved.

The ultrapure water output quality must be monitored (i.e., controlled routinely) to maintain the required low contamination level. For detection, liquid samples are fed into the ICP-MS 4500. Figure 2 shows the components of the inductively coupled plasma mass spectrometer. After ionization, the ions are focused and separated according to their charge/mass ratio, and the intensity is measured in counts per time.

Various equipment components can be tested combined with various measurement modes and methods of sample preparation. Altogether, 10 parameters have been defined, composed of

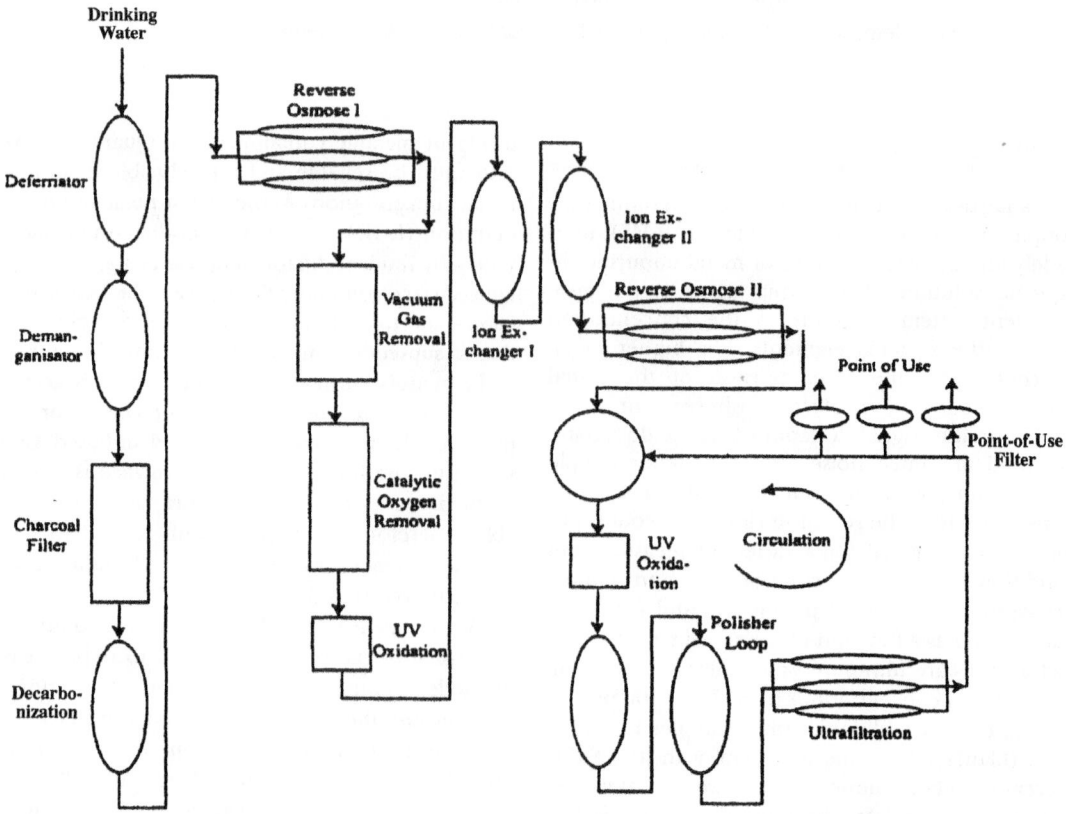

Figure 1
Ultrapure water supply system

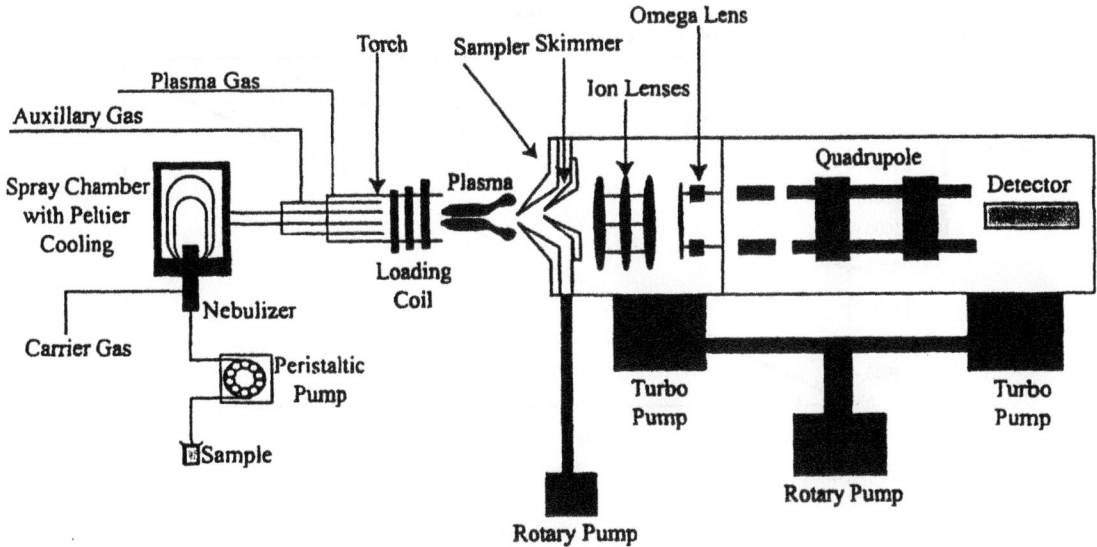

Figure 2
ICP-MS 4500 schematic

exchangeable hardware, measurement points and time, cleaning procedures, and sample handling.

3. Concept of Experiments

Mass production of semiconductor wafers requires a constant low-level ultrapure water supply. In principle, it is sufficient to prove the maintenance of a contamination level below an upper limit for the elements monitored. In common analytical practice, the measured sample is compared with a blank (sample without intentional content of elements to be detected). The information on the count rate of the blank alone does not allow one to calculate the lowest possible detection limit. Therefore, sample series with a calibrated content of elements are prepared and measured. Thus, a concentration range is covered and deviation of the signal output becomes visible. For physical reasons, the system is expected to behave linearly (plasma equilibrium), which means that the count rate (output) is proportional to the concentration of the elements. In designed experiments with 10 parameters, dynamic optimization enables one to reduce variation at the

highest equipment sensitivity with the goal of achieving the lowest possible limit of detection. The corresponding SN figure can be calculated according to the zero-point-proportional characteristic. However, this is valid only for a few elements that occur rarely in nature and usually not in contact with water (e.g., Li, Cs, Co, Pb). Of more interest are widespread elements such as salts (e.g., Na, K, Al, Fe). Here the relative detection error (noise) increases toward lower concentrations, and at a certain point the limit of detection is reached when the output signal of the element detected becomes too low and disappears in the system noise. Original count rates are plotted for two elements in Figure 3.

For this reason, appropriate SN figures with the smallest error have to be calculated. In most cases the dynamic SN ratio for a linear function (no origin) is used.

4. Experimental Design

Equipment parameters (hardware), measurement parameters (count sampling), and preparation steps total 10 parameters on two levels:

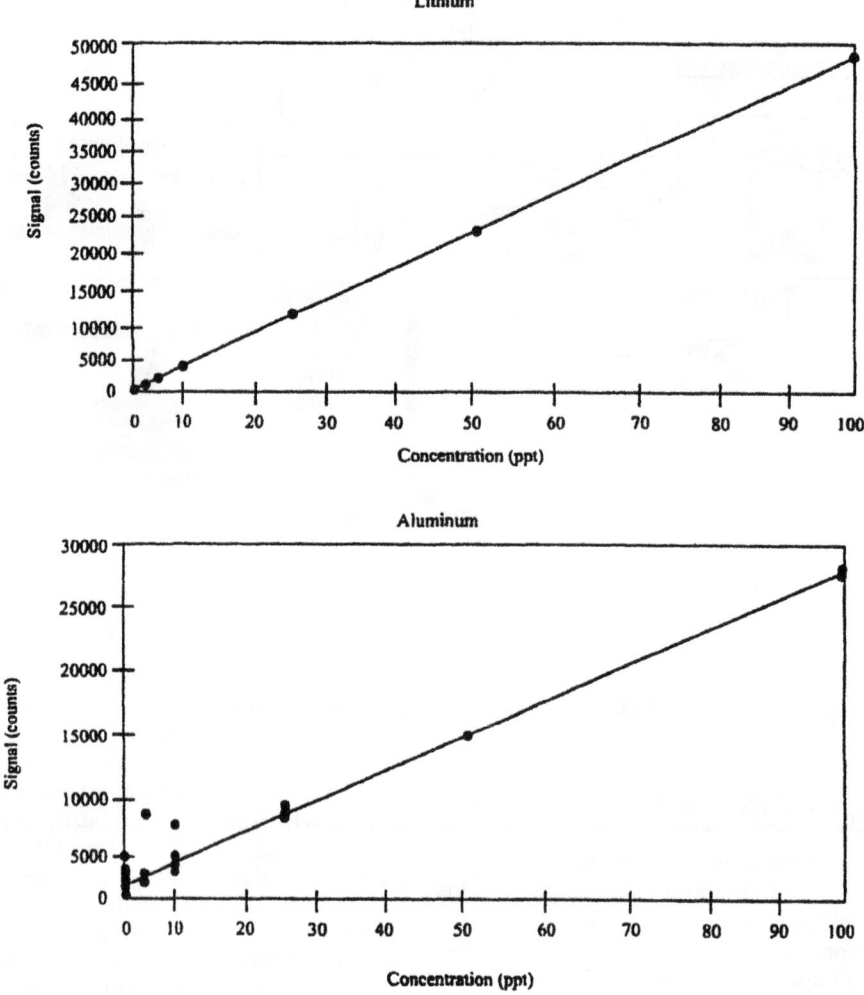

Figure 3
Element-specific count-rate deviations

A: component 1 (interface cone)

B: component 2 (plasma torch)

C: component 3 (spray chamber)

D: component 4 (nebulizer)

E: cleaning time

F: stabilization time

G: measurement points (sampling q/m)

H: integration time

I: spectrum scan

K: probe cleaning

L: error

The L_{12} layout has enough space for the parameters and additional error calculations. The calibration series are prepared with concentrations of (ppt).

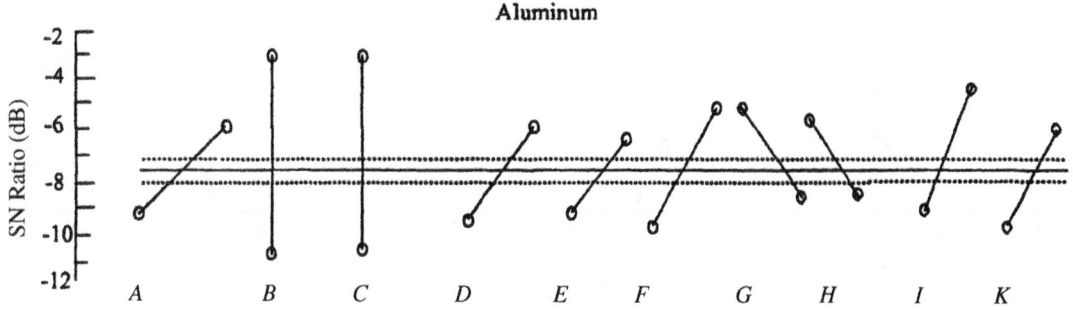

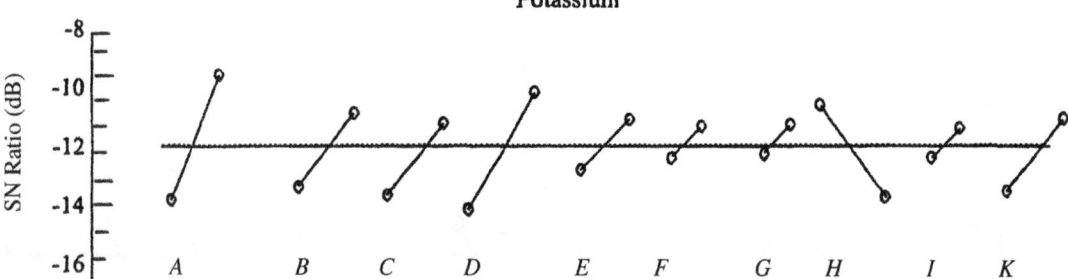

Figure 4
Results for aluminum and potassium

2.5   5   10   25   50   100   blank 1   blank 2

for two noise conditions (four combinations): long-term stability and dilution steps. In total, 12 experimental configurations (L_{12}) are carried out for eight samples with four replications (two noise factors) and one repetition; in each case 21 elements (under cold plasma conditions) are characterized, resulting in 16,128 calculations.

5. Data Analysis and Interpretation

System function is of type I: $Y = \beta M$ (M = signal factor, concentration of elements; β = slope) or of type II: $Y = \beta M + t$ (t = offset, sources of noise). The preparation conditions (long term/dilution) proved to be very stable and the SN calculations were performed for all noise conditions with the appropriate function. Surprisingly, the parameter effects for all elements were identical for the sensitivity of the count rate (slope); effects on the slope

deviation were element-specific. Results are presented for selected elements, classified for environmental occurrence: for example, Al and K compared with Li and Co (Figures 4 and 5).

SN Linear Function

In a coarse estimation, hardware parameters A to D (i.e., equipment components) have a greater influence than parameters E to K, which define measurement mode and preparation techniques. In addition, parameter settings for E to K have to be at the "low" level, which means short and quick measurements. For example, extended integration time does not improve the SN ratio of the output signal, because with increasing time and repeated cleaning procedures, the probability for contamination through the environment increases as well.

SN Zero-Point Proportional

Compared to the previous situation, the stronger effects are now from the software parameters, E to K,

Lithium

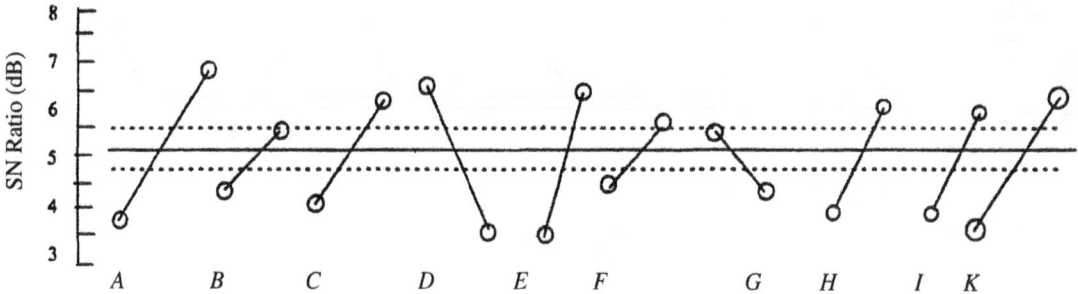

Cobalt

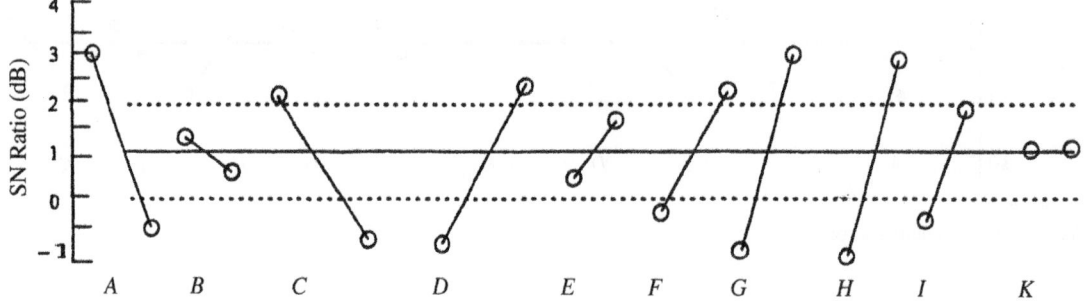

Figure 5
Results for lithium and cobalt

with a tendency toward increased integration time, higher measurement points, and enhanced cleaning procedures. Obviously, the risk of contamination from the environment is much less severe.

6. Prediction and Confirmation

Data analysis points out element-specific parameter settings due to individual occurrence in the environment and different ionization properties. If a specific element has to be detected at the lowest possible concentration, all parameters can be tuned element-specifically. More convenient is some type of average parameter setting that fits most important elements. Moreover, the interesting question is how low the detection limit appears for these pa-

rameter levels. Therefore, the real practical proof is reproducible detection at the lowest possible metal concentration predicted. For selected parameter settings, calculations are performed to predict the σ/β value of the linear function. This represents the tolerance (standard deviation) of element concentration; by convention three times this value is defined as the limit of detection. Results are presented in Table 1.

7. Conclusions

With appropriate tuned analytical equipment detection, limits below 1 ppt can be achieved for relevant metallic contaminations in ultrapure water:

Table 1
Predicted and confirmed detection limits (3σ) (ppt)

Element	Prediction	Confirmation	Element	Prediction	Confirmation
Ag	2	3	Li	0.6	1
Al	0.7	0.6	Mg	3	0.7
Ca	6	2	Mn	1.5	0.6
Co	2	0.6	Na	5	2
Cr	1	0.6	Ni	3	0.8
Cs	1	1	Pb	6	3
Cu	2	0.6	Rb	2	0.7
Fe	2	2	Sr	2	0.7
Ga	0.9	0.9	Ti	2	2
In	3	3	Zn	9	4
K	5	6			

0.1–0.2 ppt: Co

0.3–0.4 ppt: Al, Mg, Li, Cr, Mn, Rb, Sr, In, Cs

0.5–0.6 ppt: Fe, Ni, Ga, Ti, Pb

0.7–1.0 ppt: K, Zn, Ag, Cu, Na

Requirements in the semiconductor industry are thus fulfilled and can be guaranteed. Further improvements concern sampling techniques to benefit from this ultra-trace analytical capability. Future steps will lead toward new industrial cleaning procedures below the ppt level for the most important raw material: water.

References

Balazs, 1993. *Balazs Analytical Laboratory's Pure Water Specifications and Guidelines.* Austin, TX: Balazs Analytical Laboratory.

Hewlitt-Packard, 1994. *HP 4500 ChemStation and Application Handbook.* Palo Alto, CA: Hewlitt-Packard.

K. E. Jarvis, A. L. Gray, and R. S. Houk, 1992. *Handbook of Inductively Coupled Plasma Mass Spectrometry.* Glasgow: Blackie:

This case study is contributed by H. Rufer, R. Holzi, and L. Kota.

CASE 6

Evaluation of Component Separation Using a Dynamic Operating Window

Abstract: Liquid chromatographic (LC) analysis is an effective instrument for use in pharmaceutical analysis and pharmaceutical science. Various methods have been utilized in attempts to optimize LC analytical conditions. To obtain a high degree of separation, optimum conditions were determined by use of the dynamic SN ratio as an index for evaluation of the separation process. An L_{18} orthogonal array was used in this experiment. It was demonstrated that the SN ratio was an appropriate scale to indicate the separation of some peaks in an LC. Also, appropriate types of SN ratios to show the degree of separation in an LC are discussed.

1. Introduction

In a high-performance liquid chromatograph (HPLC), after moving components placed on liquid, called a *mobile phase*, we separated each component by taking advantage of its degree of adsorption for a solid phase, called a *carrier*, in the column. A flow rate indicates the mobile phase. The result of separation was recorded as a peak by sensing an optical deflection or a difference in the wavelength of a desorbed component in the liquid using a detector. Although components were separated according to a degree of adsorption, in actuality this distance is determined by the type and constitution of the mobile phase (primarily, organic solvent), pressure, temperature, and type of column (i.e., the solid phase's capability of adsorption and separation inside the column). Therefore, using parameter design we selected levels and types of parameters leading to an appropriate degree of separation from various variables.

As shown in Figure 1, in separating two peaks we used the separability, R_s, as a degree of separation of two components, which is defined as

$$R_S = \frac{2(t_{R_2} - t_{R_1})}{W_1 + W_2} \tag{1}$$

The larger R_s becomes, the better the degree of separation of two components. However, since there are numerous factors related to the separation of components' peaks, it really is difficult to conduct a simultaneous multiple-factor evaluation regarding a relationship between the separability and influential factors. In our study we attempted to evaluate the relationship quantitatively using the SN ratio in parameter design for the separation of components.

The HPLC method detects the peak value of a component. In this case, since the desorbing time of a component was unknown, we determiend that a reciprocal of a flow rate M in separation, $1/M$, was proportional to a desorbing time Y (time when a peak shows up), and clarified that the dynamic operating window method for components M_1^*, M_2^*, and M_3^* can be utilized with the SN ratio based on a desorbing time Y as an output. Yet we have come to the conclusion that it is more natural to choose a flow rate M per se in response to the function of the HPLC instead of selecting reciprocal $1/M$ as a

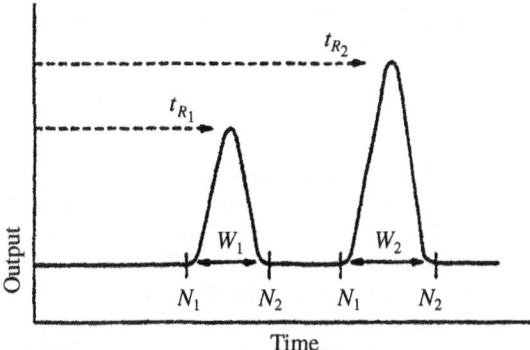

Figure 1
Correspondence of evaluation scale and separability

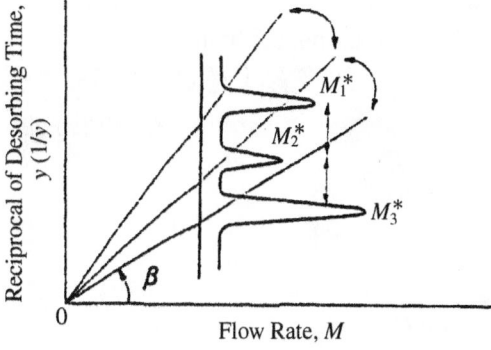

Figure 2
Chromatography and conceptual diagram of separation

signal factor, because we considered it rational to separate adsorbed components by flowing energy.

Then, setting a reciprocal of a desorbing time Y, $1/Y$, to y, we obtained

$$y = \beta M \qquad (2)$$

Based on this equation, our objective was to compute a variation of differences in proportional terms between components' M* values, denoted by $S_{M*\beta}$. As a component of a signal in the dynamic operating window method, we studied the following SN ratio η^*:

$$\eta^* = \frac{(\sigma_{M*\beta})^2}{\sigma_N^2} \qquad (3)$$

2. Quantification of Component Separation

The basic idea of the dynamic operating window method for component separation in a chemical reaction is shown in Figure 2. However, if we consider the principle of chromatography in the HPLC process, we let liquid flow in a column, and by doing so, desorb components in the column and flush the desorbed components using pressure. It is more reasonable to regard this phenomenon as a physical thrust of desorbed components by energy of liquid motion rather than as a chemical reaction. In other words, a flow rate is equivalent to energy that thrusts components by raising the flow rate of liquid,

thereby applying pressure. Since this thrust energy removes components, a component with weak adsorption is pushed off earlier than one with strong adsorption.

We adopted an extended version of a basic dynamic operating window method rather than a reaction rate for time in a chemical reaction. That is, we focused our analysis on improving the detectability of a difference of components in the case of multiple operating windows. As a noise factor, we selected rising and falling times before and after a peak in the amount of desorbed components, N_1 and N_2 (refer to Figure 1). We consider the following flow rates:

Signal:	M_1	M_2	M_3
Flow Rate (mL/min):	5.0	4.2	3.3

Using the data in the second column of the L_{18} orthogonal array shown in Table 1, we calculated the SN ratio following the procedure outlined below. The original value was a peak distance from the origin (mm), and 80 mm was equivalent to 1 minute. Therefore, we divided the original value by 80 into a desorbing time in minutes and took its reciprocal:

$$S_T = \left(\frac{1}{64^2} + \frac{1}{76^2} + \cdots + \frac{1}{145^2} \right) 80^2 = 12.8243 \quad (4)$$

$$r = 5.0^2 + 4.2^2 + 3.3^2 = 53.53 \qquad (5)$$

Table 1
Results of experiment 2 (mm)

Component		Flow Rate			Linear Equation
		M_1	M_2	M_3	
M_1^*	N_1	64	76	96	L_1
	N_2	73	85	107	L_2
M_2^*	N_1	84	98	124	L_3
	N_2	96	110	138	L_4
M_3^*	N_1	89	106	132	L_5
	N_2	93	116	145	L_6

$$L_1 = \left[(5.0)\left(\frac{1}{64}\right) + (4.2)\left(\frac{1}{76}\right) + (3.3)\left(\frac{1}{96}\right) \right] 80$$

$$= 13.4211 \tag{6}$$

By doing the same calculations, we obtain

$$L_2 = 11.8997$$

$$L_3 = 10.3195$$

$$L_4 = 9.1343$$

$$L_5 = 9.6642$$

$$L_6 = 9.0183$$

$$S_\beta = \frac{(L_1 + \cdots + L_6)^2}{(3)(2r)} = 12.5375 \tag{7}$$

$$S_{M*\beta} = \frac{(L_1 + L_2)^2 + (L_3 + L_4)^2 + (L_5 + L_6)^2}{2r} - S_\beta$$

$$= 0.2462 \tag{8}$$

$$S_{N\beta} = \frac{(L_1 + L_3 + L_5)^2 + (L_2 + L_4 + L_6)^2}{3r} - S_\beta$$

$$= 0.0350 \tag{9}$$

$$S_e = S_T - S_\beta - S_{N\beta} = 0.0056 \tag{10}$$

$$V_e = \frac{S_e}{14} = 0.0004 \tag{11}$$

$$V_N = \frac{S_{N\beta} + S_e}{14 + 1} = 0.0027 \tag{12}$$

$$\eta^* = 10 \log \frac{(1/2r)(S_{M*\beta} - 2V_e)}{V_N} = -0.71 \text{ dB} \tag{13}$$

$$S = 10 \log \frac{1}{(3)(2r)} (S_\beta - V_e) = -14.09 \text{ dB} \tag{14}$$

$$S^* = 10 \log \frac{1}{2r} (S_{M*\beta} - 2V_e) = -26.40 \text{ dB} \tag{15}$$

The SN ratio η^* represented the variability in the operating window's width between peaks M_1^*, M_2^*, and M_3^*. In other words, it indicated the stability of the differences among β_1, β_2, and β_3, each of which was the slope of each of M_1^*, M_2^*, and M_3^*, and $S_{M*\beta}$ is the signal's effect. Considering the separability for each of M_1^*, M_2^*, and M_3^*, the square root of an SN ratio represents the degree of separation. Therefore, improvement in the SN ratio leads to increased separability. Taking into account the definition of an SN ratio, an SN ratio is a scale not only for the degree of improvement in separability but also for a ratio of the variance between components to the variance within components. On the other hand, sensitivity S was used to accelerate the desorbing time. In addition, S^* was used to magnify an operating window's width, that is, to judge whether or not the gap among β values for M_1^*, M_2^*, and M_3^* is large. Separability was adjusted by tuning each parameter's sensitivity.

As illustrated in Table 2, we selected factors that were assumed to affect component separation and allocated them to an L_{18} orthogonal array. The levels of the control factors were selected according to the preliminary study of our current analytical con-

Table 2
Control factors and levels

	Control Factor	Level		
		1	2	3
A:	error			
B:	constitution of ethanol (%)	Low	Mid	High
C:	constitution of methanol (%)	Low	Mid	High
D:	constitution of acetonitrile (%)	Low	0	High
E:	constitution of chloroform (%)	Low	0	High
F:	temperature of column (°C)	Low	Mid	High
G:	error			
H:	error			

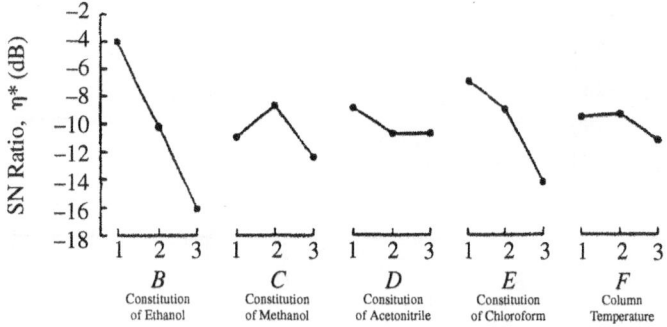

Figure 3
Factor effect plot of SN ratio η*

ditions. The components chosen in this research were three types of components included in nature resources. The second component was not easily separated from the third. In addition, because of its high constitution, the third tended to overlap with the others in terms of the peak. Therefore, it was difficult to gain high separability.

3. Optimal Configuration and Confirmatory Experiment

On the basis of the idea outlined above, we designed experiments on component separation and conducted them and a subsequent analysis.

Figures 3 to 5 show the response graphs; using them, we selected levels with large decibel values of SN ratio η* and sensitivities S and $S*$ to estimate the

optimal configuration. Primarily, we found that the factors largely affecting the SN ratio η* were factors B, C, D, and E, whereas factor F had little influence on the SN ratio. Secondarily, looking at the sensitivities S and $S*$, we noticed that they both became high at high SN ratio η* levels. Then we determined the combination $B_1C_1D_1E_2F_1$ by choosing a level with a high decibel value as an optimal level.

Although essentially, the second level should be selected for factor C, we picked up the first level for efficiency because there was little influence of the SN ratio. The reason for selecting the second level for factor E in most cases is that in a product development process, technical issues are prioritized. On the other hand, as the worst configuration, we selected $B_3C_3D_3E_3F_3$ to estimate the SN ratio η* and sensitivities S and $S*$.

We conducted a confirmatory experiment and compared the results with those under the worst

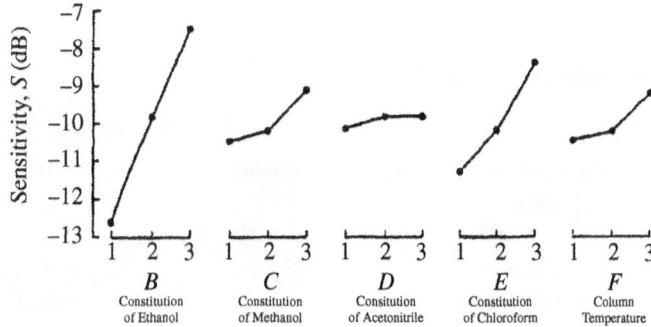

Figure 4
Factor effect plot of sensitivity S

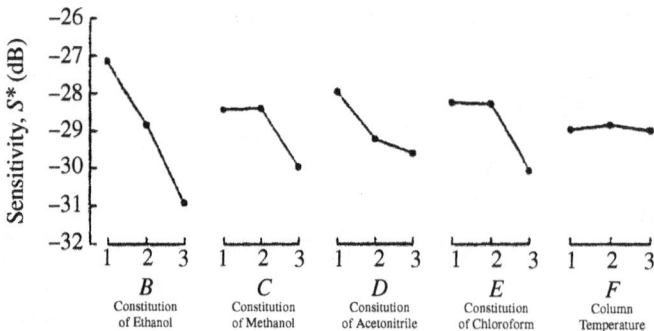

Figure 5
Factor effect plot of sensitivity S

configuration. As a result of the confirmatory experiment, we attained improvement in separation of peaks of M_1^*, M_2^*, and M_3^* (Table 3). Although separation was difficult in M_1^*, M_2^*, and M_3^*, by magnifying the operating window of three types of peaks and making an adjustment, we achieved sufficient separation. The remaining problem was poor reproducibility in gain. The response graphs explicitly revealed numerous interactions. Yet since reproducibility in the absolute value of the SN ratio under the worst configuration was poorer than that under the optimal configuration, we supposed that the worst configuration might have extremely poor separability. Despite inconsistent reproducibility in gain, by taking advantage of the adjustment factor (flow rate), our significant achievement was to reduce the analysis time from 10 to 20 minutes in the conventional process to only 1 minute.

The problem with our experimentation was that since the noise factors rely on the haphazard external disturbance of the variability in desorbing speed or desorbing time at the peak, the effects of the noise factor became less visible. The reason that the reproducibility in gain was poor was considered to be because the noise factor selected for our study was insufficient. Although recent analytical apparatuses have high accuracy and reproducibility, there remain some technical issues regarding inconsistent reproducibility in gain defined by quality engineering.

Although HPCL is well established as a means of evaluation, still needed is a comprehensive evaluation method to select an optimal configuration among various factors used for HPLC. In our traditional technique, there was no way other than use of a one-by-one method, based on our experience.

Table 3
Results of confirmatory experiment (dB)

Characteristic	Condition	Configuration		Gain
		Optimal	Current	
SN ratio η^*	Estimation	23.98	1.23	22.74
	Confirmation	28.96	−11.99	40.95
Sensitivity S	Estimation	0.43	−9.86	10.29
	Confirmation	−2.78	−8.20	5.42
Sensitivty S	Estimation	−11.82	−39.17	27.35
	Confirmation	−7.60	−50.19	42.59

Especially for the issues of stability and reproducibility in measurement, the only solution at present is to rely on a precision test. Additionally, for evaluation of separation, despite several proposed scales, comprehensive judgment has been limited. We see as our significant accomplishment the discovery that we can achieve not only a comprehensive judgment on separability but can also solve problems such as reproducibility and robustness by use of an SN-ratio-based evaluation method.

Reference

Kouya Yano, Hironori Kitzazki, and Yoshinobu Nakai, 1995. Evaluation of liquid chromatography analysis using quality engineering. *Quality Engineering*, Vol. 3, No. 4, pp. 41–48.

This case study is contributed by Kouya Yano.

CASE 7

Optimization of a Measuring Method for Granule Strength

Abstract: Our company produces granule products from herbal extracts. The granules are produced by a dry granulating process, including press molding, crushing, and classifying. The granules require a certain amount of hardness and fragility to meet the specifications and to make pills easier to swallow. Granules are difficult to dissolve if too hard and become powdery if too soft. In the current measuring method, granules are vibrated; then the powder is separated by screening. The amount of powder remaining on the screen is measured to indicate the hardness. This method is time consuming and the measurement error is high. This case reports the development of a new method with a short measurement time and high precision.

1. Introduction

Our research dealt with granules manufactured from solidified essence that is extracted from herbal medicines. Measurement of the granular strength were taken as follows. After sieving granules and removing fine powders, we forcibly abraded them with a shaker and then got rid of the abraded granules through sieving and suction. Measuring the remainder after suction, we defined the granular strength as

granular strength

$$= \frac{\text{sample weight} - \text{remainder weight})}{\text{sample weight}} \times 100\%$$

$$(1)$$

Figure 1 is a schematic of a new measuring instrument. The measuring method is as follows.

1. *Supply of a test sample.* We removed the lid on a test sample (of known weight), placed the sample on a sieve, and replaced the lid.

2. *Abrasion and classification.* By sucking the inside air from an exhaust outlet using a vacuum cleaner, we made air flow out of a rotating slit nozzle. Receiving this flowing-out air, the test sample hits the sieve frame and lid and is thus abraded. The abraded powders were removed from the exhaust outlet through suction.

3. *Measurement of the remainder.* After a certain period, we measured the weight of the sample remaining on the sieve.

This measurement instrument was considered to have two types of functions. The first function was the ratio of the amount of abrasion to the operation time: The shorter the operation time, the less the abrasion; the longer the time, the greater the abrasion. In terms of this function, by focusing not on hardness as a quality characteristic but on pulverization of granules as the measurement instrument's function, we evaluated the function. Considering a time-based change in the number of granules or their shapes, we supposed that like a chemical reaction, the amount of abrasion for the operation

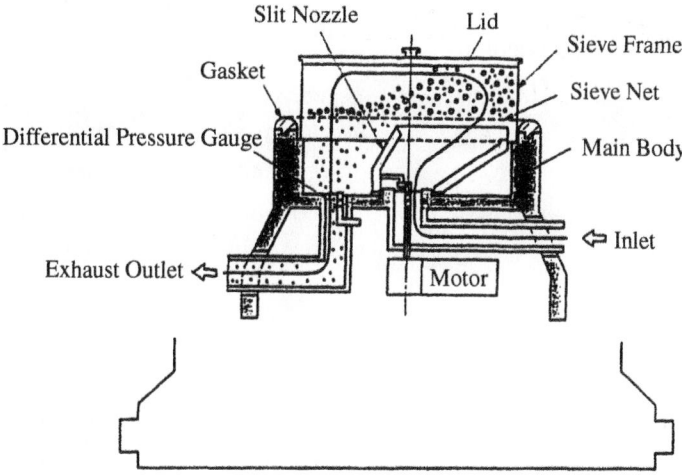

Figure 1
Schematic of new measurement instrument for granular strength

time followed not simple linearity but natural logarithmic proportionality.

The other function was the proportion of the amount of abrasion to the test sample's hardness. For a constant operation time, a test sample with low abrasive resistance (or high hardness) had a smaller amount of abrasion, whereas one with high resistance had more. The true value of hardness discussed here was unknown. However, since we obtained a relative difference by altering manufacturing conditions, we evaluated the difference as a signal.

2. SN Ratio

Considering the above as functions of a new measuring instrument, we developed a zero-point proportional equation based on two types of signal factors: operation time and abrasive resistance (hardness) of a test sample. Since it was difficult to collect and measure the abraded powders, we measured primarily the weight of a test sample (W_1), and secondarily, gauged the weight of the remainder left on the sieve net (W_2). The remaining rate was defined as

Table 1
Remaining fractions for experiment 1

Abrasive Resistance to Test Sample	Operation Time		
	T_1 (15 s)	T_2 (30 s)	T_3 (60 s)
M_1 (hard)	0.992	0.991	0.989
M_2 (medium)	0.990	0.988	0.986
M_3 (soft)	0.988	0.986	0.980

Table 2
Logarithmized values for experiment 1

Abrasive Resistance of Test Sample	Operation Time			Linear Equation
	T_1 (15 s)	T_2 (30 s)	T_3 (60 s)	
M_1 (hard)	0.00783	0.00944	0.01065	1.03961
M_2 (medium)	0.00984	0.01186	0.01389	1.33681
M_3 (soft)	0.01207	0.01388	0.02039	1.82062

$$\text{fraction remained} = \frac{W_2}{W_1} \qquad (2)$$

$$\frac{W_2}{W_1} = e^{-\beta T} \qquad (3)$$

The principle of abrasion was grounded on collision among granules; therefore, we focused on the same principle as that of a chemical reaction. Using this idea, we proposed that the remaining fraction of a test sample after an operation time of T seconds could be expressed by the following exponential equation of an operation time T:

$$\ln \frac{W_2}{W_1} = -\beta T \qquad (4)$$

$$y = -\ln \frac{W_2}{W_1} \qquad (5)$$

$$y = \beta T \qquad (6)$$

Table 3
Control factors and levels

Control Factor	Level		
	1	2	3
A: water content in granule	Low	Mid	High
B: humidity of sucked air (RH %)	$L_1 - 10$	L_1	$L_1 + 10$
$\quad A_1$	$M_1 - 10$	M_1	$M_1 + 10$
$\quad A_2$	$H_1 - 10$	H_1	$H_1 + 10$
$\quad A_3$			
C: mesh size (mm)	0.297	0.355	0.500
D: suction pressure (Pa)	L_2	$1.5 L_2$	$2 L_2$
$\quad F_1$	M_2	$1.5 M_2$	$2 M_2$
$\quad F_2$	H_2	$1.5 H_2$	$2 H_2$
$\quad F_3$			
E: lid height (mm)	57	42	27
F: amount of test sample	Low	Mid	High
G: distance between slit nozzle and sieve net	4	9	12

In short, this is identical to the idea of a chemical reaction.

As a test sample, we prepared granules of a certain size or larger obtained from an actual pre-treated product. In analyzing experimental data, by selecting a natural-logarithmized remaining fraction as output, we calculated an SN ratio and sensitivity based on the ideal relationship $y = \beta T$. Table 1 shows the remaining fraction for the data in the first row of an L_{18} orthogonal array. In addition, Table 2 summarizes the output computed through natural logarithmization of the remaining rates.

Total variation:

$$S_T = 0.00783^2 + 0.00944^2 + \cdots + 0.02039^2$$

$$= 0.001448 \qquad (f = 9) \tag{7}$$

Effective divider:

$$r = 15^2 + 30^2 + 60^2 = 4725 \tag{8}$$

Linear equations:

$$L_1 = (0.00783)(15) + (0.00944)(30) + (0.01065)(60) = 1.03961$$

$$L_2 = 1.33681, \qquad L_3 = 1.82062 \tag{9}$$

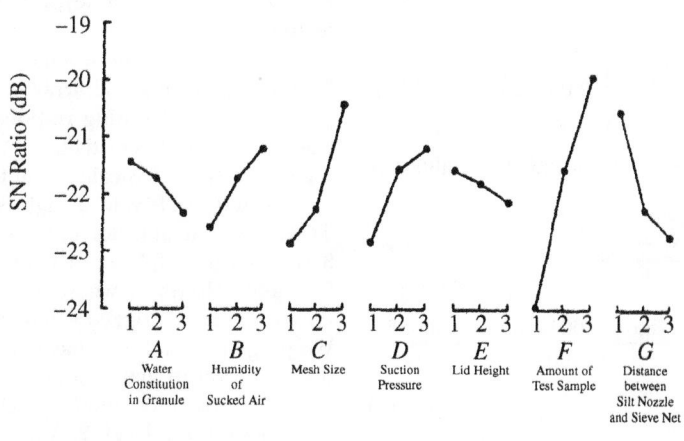

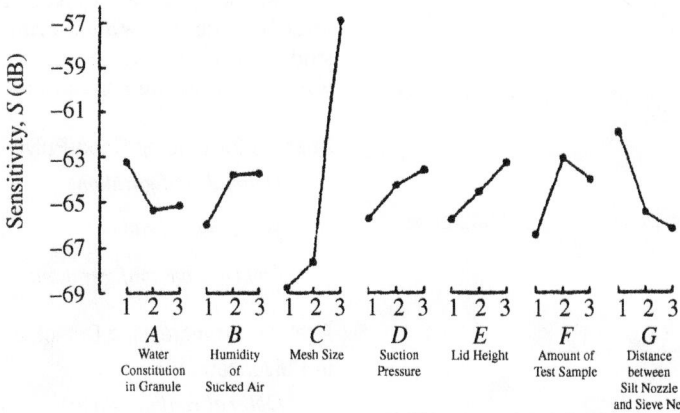

Figure 2
Response graphs for determining good pulverization conditions

Variation of proportional term:

$$S_\beta = \frac{(1.03961 + 1.33681 + 1.82062)^2}{(3)(4725)}$$

$$= 0.001243 \qquad (f = 1) \qquad (10)$$

Variation of differences between hardness between samples:

$$S_{M\beta} = \frac{1.03961^2 + 1.33681^2 + 1.82062^2}{4725} - 0.001243$$

$$= 0.000066 \qquad (f = 2) \qquad (11)$$

Error variation:

$$S_e = 0.001448 - 0.001243 - 0.000066$$

$$= 0.00014 \qquad (f = 6) \qquad (12)$$

Error variance:

$$V_e = \frac{0.00014}{6} = 0.000023 \qquad (13)$$

Analysis for Finding Good Pulverization Conditions

SN ratio:

$$\eta = 10 \log \frac{(1/3r)(S_\beta - V_e)}{V_e}$$

$$= 10 \log \frac{[1/(3)(4725)](0.001243 - 0.000023)}{0.000023}$$

$$= -24.32 \text{ dB} \qquad (14)$$

Sensitivity:

$$S = 10 \log \frac{1}{3r}(S_\beta - V_e)$$

$$= 10 \log \frac{1}{(3)(4725)}(0.001243 - 0.000023)$$

$$= -70.65 \text{ dB} \qquad (15)$$

Analysis for Increasing Detectability of Difference in Pulverization

SN ratio:

$$\eta' = 10 \log \frac{[1/(3r)(S_{M\beta} - V_e)}{V_e}$$

$$= 10 \log \frac{[1/(3)(4725)](0.000066 - 0.000023)}{0.000023}$$

$$= -38.90 \text{ dB} \qquad (16)$$

Sensitivity:

$$S = 10 \log (1/3r)(S_{M\beta} - V_e)$$

$$= 10 \log [1/(3)(4725)](0.000066 - 0.000023)$$

$$= -85.23 \text{ dB} \qquad (17)$$

3. Optimal Configuration and Confirmatory Experiment

As control factors, we selected A to G, shown in Table 3. In addition, for two pairs of factors that were supposed to have mutual interactions: between A (water content in granule) and B (humidity of sucked air) and between D (suction pressure) and F (amount of test sample), we used the sliding-level technique.

After allocating the seven control factors in Table 3 to an L_{18} orthogonal array, we conducted an experiment. The resulting response graphs of SN ratios (η, η^*) and sensitivities (S, S^*) are shown in Figures 2 and 3. Basically, from the response graphs we selected levels with a high SN ratio as optimal. Then, for control factor C, mesh size, we chose level 3 (mesh size = 0.500 mm), which had the highest SN ratio. However, when we used this level, the meshes became clogged when granules were sieved. We judged, therefore, that it would be difficult to use this level in actual production because proper abrasion would be unlikely. Thus, as an optimal level, we chose level 2, which had the second-highest SN ratio. In contrast, the lowest-SN-ratio level was set as the worst. Since there was no current level because a new measuring method was being studied, each of the second level of control factor was selected as the comparative level.

Analysis for Finding Good Pulverization Conditions

Optimal configuration: $A_1 B_3 C_2 D_3 E_1 F_3 G_1$

Worst configuration: $A_3 B_1 C_1 D_1 E_3 F_1 G_3$

Comparative configuration: $A_2 B_2 C_2 D_2 E_2 F_2 G_2$

Analysis for Increasing Detectability of Differences in Pulverization

Optimal configuration: $A_1 B_3 C_2 D_2 E_3 F_2 G_1$

Worst configuration: $A_2 B_1 C_1 D_1 E_2 F_1 G_3$

Comparative configuration: $A_2 B_2 C_2 D_2 E_2 F_2 G_2$

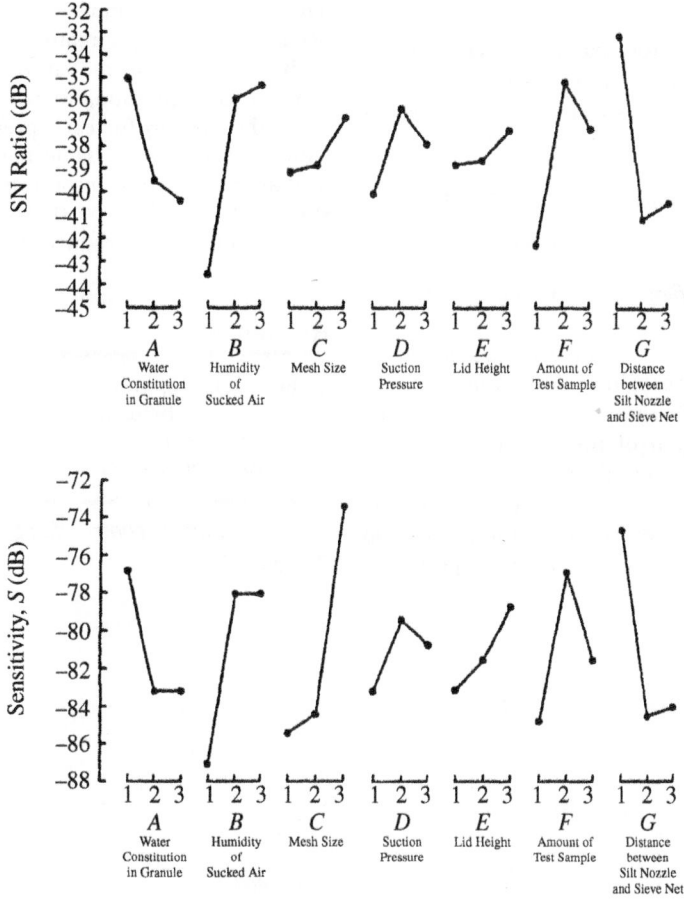

Figure 3
Response graphs for increasing detectability of differences between
materials

Table 4
Confirmation of SN ratio and sensitivity (dB)

Configuration	SN Ratio		Sensitivity	
	Estimation	Confirmation	Estimation	Confirmation
Optimal	−18.38	−13.13	−63.54	−56.23
Worst	−26.79	−21.54	−73.61	−75.63
Comparative	−22.09	−16.71	−67.05	−68.38
Gain between optimal and worst	8.41	8.41	10.07	19.40
Gain between optimal and comparative	3.71	3.58	3.51	12.15

Confirmatory Experiment

To conduct a confirmatory experiment, comprehensively judging all configurations obtained from the two types of analyses, we selected the following optimal, worst, and comparative configurations:

Optimal configuration:	$A_1 B_3 C_2 D_3 E_3 F_3 G_1$
Worst configuration:	$A_3 B_1 C_1 D_1 E_1 F_1 G_3$
Comparative configuration:	$A_2 B_2 C_2 D_2 E_2 F_2 G_2$

As a result of the confirmatory experiment, we calculated them as shown in Table 4. Comparing the estimation and confirmation, we could see that the reproducibility in gain of the SN ratio was fairly good, whereas that of sensitivity had a disparity of 9 dB. The reason was assumed to be that the sieving of test samples had been conducted improperly as a pretreatment in the confirmatory experiment.

Therefore, in the future we must sieve test samples accurately during pretreatment.

By adopting the measuring method under the optimal configuration, we could detect granular strength precisely for each prescription in parallel with maintaining the same measurement accuracy as that of the current procedure. We reduced measurement time to 1/30 that of the current level, thereby improving productivity.

Reference

Yoshiaki Ohishi, Toshihiro Ichida, and Shigetoshi Mochizuki, 1996. Optimization of the measuring method for granule fragility. *Quality Engineering*, Vol. 4, No. 5, pp. 39–46.

This case study is contributed by Yoshiaki Ohishi and Shigetoshi Mochizuki.

A Detection Method for Thermoresistant Bacteria

Abstract: In cultivating bacteria, time is regarded as a signal factor. Since a certain number of bacteria exist before starting, we used a reference-point proportional equation with the ninth hour from the start as a reference time. On the other hand, because the real number of bacteria was unknown, the numbers diluted to 1/5 and 1/25 were chosen as signal factors to be used for measurement. In other words, using dilution ratio as the signal, we calculated an SN ratio whose true values were unknown.

1. Introduction

Thermoresistant bacteria form heat-resisting spores in the body that survive in the food after cooking and rot heat-treated food. In order to detect a specific bacterium causing food poisoning, we developed a method of detecting poison produced by a bacterium or technique using an antigen–antibody reaction. However, since a heat-resistant bacterium is a common decomposing bacterium, a method of detecting cultivated bacteria is widely used. In this case, a key issue was to accelerate bacteria as quickly as possible and to detect them accurately. If the speed of occurrence of bacteria is slow, accuracy in detection is reduced. In contrast, poor accuracy of detection leads to a larger error in confirmation of occurrence. This relationship between a measuring method and a measured result cannot be separated in technological development, and each should be analyzed individually. Whereas cultivation of bacteria is considered a growth-related phenomenon, detection is a measurement- or SN ratio–related issue. In our study we dealt with both in a single experiment.

In cultivating bacteria, time is regarded as a signal factor (Figure 1). Since a certain number of bacteria exist before starting, we used a reference-point proportional equation with the ninth hour from the start as a reference time. On the other hand, because the real number of bacteria was unknown, the numbers diluted to 1/5 and 1/25 were chosen as signal factors to be used for measurement. In other words, using dilution ratio as the signal, we calculate an SN ratio whose true values were unknown (Figure 2). In addition, on the assumption that the pH of the food had already been adjusted to control microbes, it was considered as the noise factor:

N_1: diluted solution, pH 7

N_2: diluted solution, pH 3

Following our objective of study, we proceeded with the following data analysis:

1. To perform parameter design for bacteria cultivation, we calculated the SN ratio of the reference-point proportional equation using time as a signal factor (analysis 1).

2. By computing the SN ratio for bacteria detection at each point of time, we used bacteria dilution ratio as a signal factor (analysis 2).

Using the results obtained from both, we chose the optimal configuration for cultivation and detection methods for bacteria.

2. SN Ratio for Cultivation Conditions for Bacteria (Analysis 1)

As a data example, the data in the first row of an L_{18} orthogonal array are shown in Table 1. To

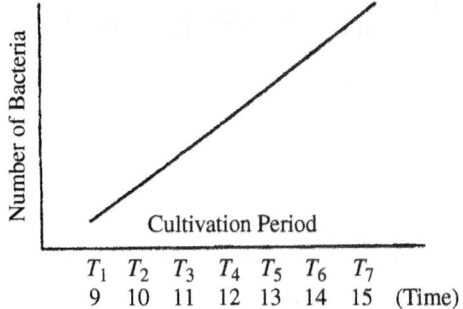

Figure 1
Cultivation of bacteria using time as the signal

compute the SN ratio using a reference-point proportional equation whose reference point is set to T_1, we created an auxiliary table (Table 2).

Total variation:

$$S_T = 79.5^2 + 147.5^2 + \cdots + 7.5^2 + 8.5^2$$

$$= 603,265 \tag{1}$$

Effective divider:

$$r = 0^2 + 1^2 + \cdots + 6^2 = 91 \tag{2}$$

Linear equation:

$$L_1 = (0)(79.5) + \cdots + (6)(346.5) = 6178.5$$

$$L_2 = 2965.5 \qquad L_3 = 2018.0$$

$$L_4 = 167.0 \qquad L_5 = 590.5 \qquad L_6 = 133.5 \tag{3}$$

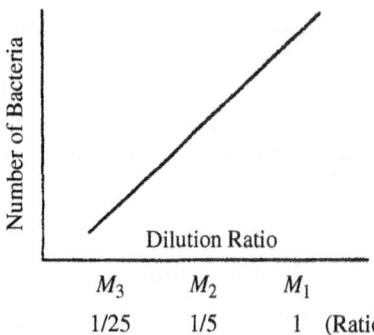

Figure 2
Measurement of bacteria using dilution of bacteria as the signal

Table 1
Measured data of number of bacteria in experiment 1

		M_1	M_2	M_3
T_1	N_1	496	119	11
	N_2	337	35	8
T_2	N_1	564	137	23
	N_2	425	50	12
T_3	N_1	612	153	27
	N_2	480	63	13
T_4	N_1	663	164	34
	N_2	525	76	14
T_5	N_1	710	171	40
	N_2	558	86	15
T_6	N_1	746	180	41
	N_2	578	92	17
T_7	N_1	763	186	41
	N_2	605	96	18

Table 2
Calibrated data for reference-point proportional equation

		M_1	M_2	M_3
0	N_1	79.5	42.0	1.5
	N_2	−79.5	−42.0	−1.5
1	N_1	147.5	60.0	13.5
	N_2	8.5	−27.0	2.5
2	N_1	195.5	76.0	17.5
	N_2	63.5	−14.0	3.5
3	N_1	246.5	87.0	24.5
	N_2	108.5	−1.0	4.5
4	N_1	293.5	94.0	30.5
	N_2	141.5	9.0	5.5
5	N_1	329.5	103.0	31.5
	N_2	161.5	15.0	7.5
6	N_1	346.5	109.0	31.5
	N_2	188.5	19.0	8.5
Linear equation	N_1	L_1	L_3	L_5
	N_2	L_2	L_4	L_6

Variation of proportional term:

$$S_\beta = \frac{(L_1 + L_2 + L_3 + L_4 + L_5 + L_6)^2}{6r}$$

$$= 266,071.08 \qquad (4)$$

Variation of difference between proportional terms by dilution:

$$S_{M\beta} = \frac{(L_1 + L_2)^2 + (L_3 + L_4)^2 + (L_5 + L_6)^2}{2r} - S_\beta$$

$$= 222,451.65 \qquad (5)$$

Variation of differences between proportional terms due to noise:

$$S_{N\beta} = \frac{(L_1 + L_3 + L_5)^2 + (L_2 + L_4 + L_6)^2}{3r} - S_\beta$$

$$= 55,826.82 \qquad (6)$$

Error variation:

$$S_e = S_T - S_\beta - S_{M\beta} - S_{N\beta} = 58,915.45 \qquad (7)$$

Error variance:

$$V_e = \frac{S_e}{38} = 1550.41 \qquad (8)$$

Total variance:

$$V_N = \frac{S_{N\beta} + S_e}{39} \times 2942.11 \qquad (9)$$

SN ratio:

$$\eta = 10 \log \frac{(S_\beta - V_e)/2r}{V_N} = -3.06 \qquad (10)$$

Sensitivity:

$$S = 10 \log \frac{S_\beta - V_e}{2r} = 31.62 \qquad (11)$$

3. SN Ratio for Measurement of Number of Bacteria (Analysis 2)

Using the data of time T_1 in Table 1, we calculated the SN ratio for using the bacteria dilution ratio. Table 3 summarizes each datum divided by 416.5, the average number of bacteria at M_1.

Table 3
Calibrated data

		1	0.2	0.04
T_1	N_1	1.191	0.286	0.026
	N_2	0.809	0.084	0.019

Total variation:

$$S_T = 1.191^2 + 0.286^2 + \cdots + 0.019^2 = 2.163 \qquad (12)$$

Effective divider:

$$r = 1^2 + 0.2^2 + 0.04^2 = 1.042 \qquad (13)$$

Linear equations:

$$L_1 = (1)(1.191) + \cdots + (0.04)(0.026) = 1.249$$

$$L_2 = 0.827 \qquad (14)$$

Variation of proportional term:

$$S_\beta = \frac{(L_1 + L_2)^2}{2r} = 2.068 \qquad (15)$$

Variation of difference between proportional terms:

$$S_{N\beta} = \frac{L_1^2 + L_2^2}{r} - S_\beta = 0.086 \qquad (16)$$

Error variation:

$$S_e = S_T - S_\beta - S_{N\beta} = 0.009 \qquad (17)$$

Error variance:

$$V_e = \frac{S_e}{4} = 0.002 \qquad (18)$$

Total variance:

$$V_N = \frac{S_{N\beta} + S_e}{5} = 0.019 \qquad (19)$$

SN ratio:

$$\eta = 10 \log \frac{(S_\beta - V_e)/2r}{V_N} = 17.216 \qquad (20)$$

Sensitivity:

$$S = 10 \log \frac{S_\beta - V_e}{2r} = -0.04 \qquad (21)$$

We also computed each value for T_2 to T_7 and then proceeded with the same calculation for experiments 2 to 18.

4. Optimization of Cultivation and Detection Methods and Confirmatory Experiment

Table 4 shows the control factors for the design of experiments on cultivation and detection of bacteria. Type of bacterium, A, was allocated as an indicative factor. For control factors, we chose type of dilution solution, B; type of medium, D; and amounts of elements E to H, which are assumed to facilitate growth of bacteria (added to medium). On the other hand, bench time C, measured in terms of the number of days, with bacteria maintained at 10°C after being heated up.

As below, we showed the response graphs for measurement of the number of bacteria (analysis 2). Figure 3 shows the response graphs for factor A only. Since analysis is conducted for each time level, there will be response graphs of T_1 to T_7 for other factors.

We obtained the following identical optimal configurations for analyses 1 and 2:

Analysis 1: $A_1B_3C_1D_3E_2F_1H_2$

Analysis 2: $A_1B_3C_1D_3E_2F_1H_2$

Under both optimal and current configurations, we performed a confirmatory experiment. The results are shown in Table 5.

5. Results of Experiment

According to the results of the confirmatory experiment shown in Table 5, we discovered that satisfactory reproducibility in the gain of the SN ratio cannot be obtained. However, since the trend toward increasing SN ratio with respect to time in the results of analysis 2 was similar to that of the estimation, we concluded that our experimental results were fairly reliable. On the other hand, a number of peaks and V-shapes in the response graphs for the control factors were regarded as problems to be solved in a future study.

Despite several remaining problems, the following improvements were achieved:

1. We reduced the experimentation time by two days because the bench time allocated as one of the control factors was not necessary. More-

Table 4
Control and indicative factors and levels

Factor		Level	
	1	2	3
A: type of bacterium (B. subtilis)	A	B	—
B: dilution solution	Sterile water	Phosphate buffer solution	Peptone–phosphate buffer solution
C: bench time (days)	0	1	2
D: type of medium	Standard	Glucose tryptone	Trypticase soy
E: amount of catalase (μg/mL)	0	50	100
F: amount of lysozyme (μg/mL)	0	0.01	0.1
G: amount of sodium pyruvate (%)	0.00	0.10	0.20
H: amount of alanine (μg/mL)	0	20	40

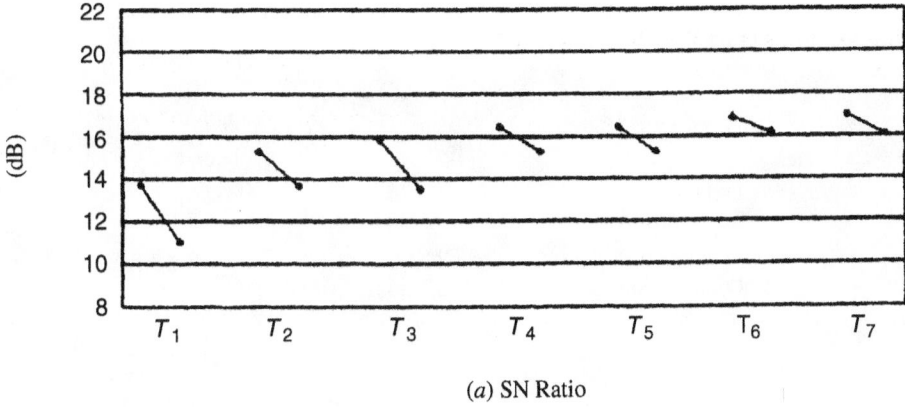

(*a*) SN Ratio

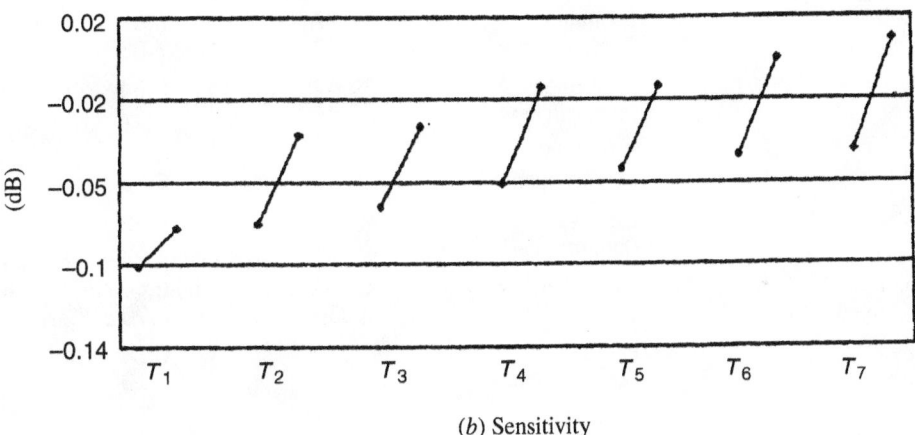

(*b*) Sensitivity

Figure 3
Analysis 2: Response graphs for factor *A*

over, whereas five days are used as a cultivation period in our current process, we found such a long-time cultivation not meaningful for experimentation. Therefore, we finished an inspection that normally takes one week in only two days.

2. The improvement above contributed not only to reducing the cost but also to taking countermeasures more swiftly. If we can detect the occurrence of bacteria sooner, we can more retrieve defective products immediately, and

thereby alleviate the risk of inflicting damage on customers.

3. For detectability of bacteria, we obtained improvement of 15 dB in the SN ratio (1/32 of the current variability) regarding a cultivation condition of bacteria when time was selected as a signal factor, whereas the bacteria dilution ratio was approximately 21 dB (1/122 of the current variability) after 15 hours.

The resulting economic benefits were computed as follows:

Table 5
Confirmatory experiment (dB)

Analysis 1			Configuration		
			Optimal	Current	Gain
SN ratio		Estimation	12.85	−14.01	26.87
		Confirmation	−2.53	−17.58	15.05
Sensitivity		Estimation	31.02	28.62	2.40
		Confirmation	30.84	23.30	7.54

Analysis 2			Configuration		
			Optimal	Current	Gain
Sn Ratio	T_1	Estimation	29.67	5.33	24.34
		Confirmation	20.01	7.61	12.40
	T_2	Estimation	37.77	5.08	32.69
		Confirmation	22.20	7.88	14.32
	T_3	Estimation	35.47	6.54	29.93
		Confirmation	23.84	7.87	16.44
	T_5	Estimation	40.66	6.33	34.33
		Confirmation	25.65	7.68	17.97
	T_6	Estimation	43.20	6.32	36.88
		Confirmation	26.75	7.69	19.05
	T_7	Estimation	42.77	6.40	36.37
		Confirmation	28.55	7.70	20.85
Sensitivity	T_1	Estimation	−0.19	−0.14	−0.05
		Confirmation	−0.07	−0.10	0.02
	T_2	Estimation	−0.13	−0.10	−0.03
		Confirmation	−0.04	−0.08	0.04
	T_3	Estimation	−0.12	−0.08	−0.04
		Confirmation	−0.03	−0.06	0.03
	T_4	Estimation	−0.10	−0.04	−0.05
		Confirmation	−0.03	−0.05	0.02
	T_5	Estimation	−0.08	−0.02	−0.06
		Confirmation	−0.02	−0.04	0.02
	T_6	Estimation	−0.07	−0.0	−0.07
		Confirmation	−0.02	−0.02	0.01
	T_7	Estimation	−0.06	−0.01	−0.05
		Confirmation	−0.01	−0.02	0.01

Streamlined Inspection due to Time Reduction

If we assume that

inspection cost = (sample cost per inspection)
+ (labor cost)
+ (machine running cost)

where the daily labor cost is 10,000 yen and the machine running cost is 155 yen, we obtain the following costs:

Current configuration:
$$235 + (10,000)(5) + (155)(5) = 51,010$$

Optimal configuration:
$$221 + (10,000)(2) + (155)(2) = 20,531$$

The cost benefit per inspection is 30,479 yen.

Risk Aversion due to Time Reduction

Now suppose that it takes two days from the point when the production of a product is completed to the point when it starts to be sold at a shop. If the optimally produced product contains a harmful bacterium, despite being in process of shipment, we can withdraw it shortly after checking the result of inspection. Then we assume that the loss in withdrawal costs 1 million yen. On the other hand, under the current configuration, the product reaches a consumer. In this case, we suppose that the loss to a consumer amounts to 150 million yen per person for the death, or 100,000 yen for hospitalization (as the number of victims is derived from a certain past record).

Current:
$$\frac{(180 \text{ persons} \times 100,000 \text{ yen}) + (3 \text{ persons} \times 150 \text{ million yen})}{183}$$

$$= 2,557,377 \text{ yen/product}$$

Optimal:
$$\frac{1 \text{ million yen}}{183}$$

$$= 5,264 \text{ yen/product}$$

Thanks to time reduction, we can slash the social loss by $2,557,377 - 5264 = 2,552,113$ yen/product.

Loss Function of Detectability of Bacteria

Now we assume that if 100,000 harmful bacteria exist in a certain product, a person who eats it will die, and the resulting loss totals 150 million yen (for loss of life). On the other hand, the loss due to discard of a product is 300 yen.

$$A_0 = 150 \text{ million yen}$$

$$A = 300 \text{ yen}$$

$$\Delta_0 = 100,000 \text{ products}$$

The tolerance is

$$\Delta = \sqrt{\frac{A}{A_0}} \, \Delta_0$$

$$= \sqrt{\frac{300}{150,000,000}} \, (100,000) = 141.42$$

Supposing that the current average number of bacteria is 100, we have the following loss function:

$$L = \frac{300}{(141.42)^2} \, (100^2) = 150 \text{ yen}$$

Converting the gain of 20.85 dB into an antilog number, we obtain 121.62. Then, the improvement achieved is $(150)(1/121.62) = 1.23$. Therefore, the current loss is 150 yen, whereas the optimal is 1 yen. As a result, we can reduce the loss by $150 - 1 = 149$ yen.

This study took only six days to complete. In our conventional process, we have had an enormous amount of time loss through reiterated trials and errors, and moreover, always doubted whether we could arrive at a conclusion. If research work can be completed within a short period by taking advantage of quality engineering, we believe that many achievements can be obtained for other inspection methods.

Reference

Eiko Mikami and Hiroshi Yano, 2001. Studies on the method for detection of thermoresistant bacteria. *Proceedings of the 9th Quality Engineering Symposium*, pp. 126–129.

This case study is contributed by Eiko Mikami and Hiroshi Yano.

Optimization of Model Ointment Prescriptions for In Vitro Percutaneous Permeation

Abstract: Evaluation of percutaneous permeation of a transdermal drug delivery system usually requires expensive and repeated animal experimentation, which also requires confirmation of a species difference between animals and human beings. In this study, considering these issues, we applied the quality engineering method to technological development in prescription optimization of an effective transdermal drug delivery system.

1. Introduction

In recent days, transdermal drug delivery that supplies medical essence through the skin has attracted a great deal of attention because this causes few side effects, such as gastrointestinal disorder, compared with a conventional drug delivery system via the digestive tract. The new method can be used with patients or aged people who have a feeble swallowing movement (movement of the muscles of the throat while swallowing food).

However, in general cases it is not easy for medicine to permeate the skin. Since normal skin is equipped with a barrier ability to block alien substances from the outside, penetration by medicine is also hampered. Therefore, to obtain sufficient permeation for medical treatment, several additives, including a drug absorption enhancer, need to be compounded. As a result, the prescription becomes more complex and a longer period of research on prescription optimization is required. Moreover, for the evaluation of percutaneous permeation of a transdermal drug delivery system (e.g., an ointment), animal experimentation is needed. This animal experimentation not only is economically inferior due to the recommendation that four or more repeated experiments be performed to eliminate individual differences but also requires confirmation of a species difference between animals and

human beings. In this study, considering these issues, we used the quality engineering method for the technological development of an effective transdermal drug delivery system.

2. Experimental Procedure

Among several existing evaluation methods of percutaneous permeation of a transdermal drug delivery system, we selected an in vitro percutaneous permeation experiment, the most inexpensive and handiest. Figure 1 shows the structure of a diffusion cell used in the in vitro percutaneous permeation experiment.

Between a donor cell (supplying medicine) and a receptor cell (receiving medicine), we place a horny layer of skin extracted from an animal on the side of the donor cell. Skin extracted from various types of animals was used for the horny layer. We started with an application of ointment on the skin extracted from the side of the donor cell. Since medical essence (a principal agent in medicine) that penetrates the skin is dissolved in the receptor solution, by measuring the amount of the principal agent in solution at each point in time, we calculated the cumulative amount of skin permeation per unit area.

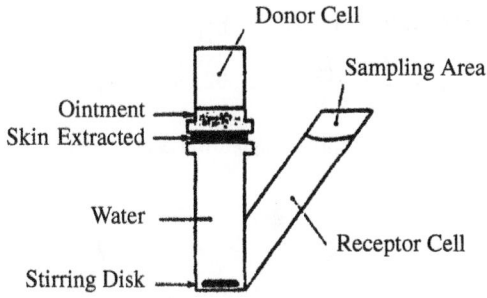

Figure 1
In vitro skin permeation experiment

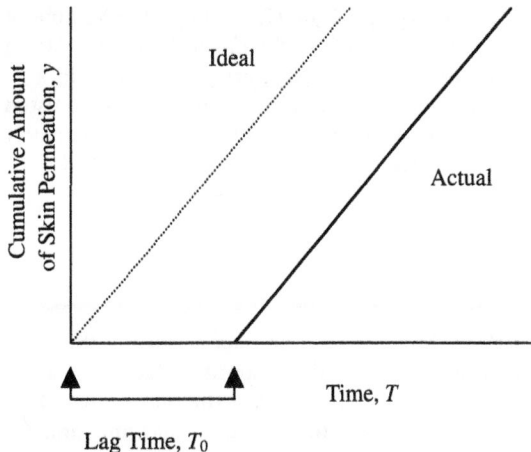

(Time to Saturation Density Inside Skin)

Figure 2
Relationship between cumulative amount of skin permeation (y) and time (T)

When optimal configuration is determined from the result of animal experimentation, the animal species difference comes into question, as is often the case with other experiments. If we could use human tissues, this problem would be solved easily. Yet for ethical reasons, it is difficult to do this in Japan. To solve this problem, we used an animal species difference as a noise factor. By doing so, we attempted to obtain an optimal prescription that could achieve skin permeation that is not only independent of the animal species difference but also superior for a human being with respect to the effects of barrier ability or absorption enhancer in skin permeation.

3. Signal and Noise Factors

Assuming that concentration of the principal agent included in the ointment that is applied to the donor cell is constant, the cumulative amount of skin permeation, y, normally changes with time, T (Figure 2). That is, while there occurs a lag time, T_0, because the rate of skin permeation is small due to the gentle slope of concentration of principal agents contained in the skin and receptor solution in the initial stage, as time elapses, the slope of a line corresponding to skin permeation speed becomes constant. The ideal is an ointment with a short lag time and high skin permeation speed.

Owing to this lag time, we can express the scientific phenomenon by $y = \beta(T - T_0)$. If this equation were selected as the generic function, we would spend most of our time pursuing this phenomenon.

Therefore, although it was unrealistic that lag time was equal to zero at this point in time, by assuming an ideal prescription where a principal agent permeates at a high and constant speed with no lag time shortly after being applied to the skin, we adopted a zero-point proportional equation, $y = \beta T$, as the generic function. Following this idea, we evaluated deviation from the ideal function using the SN ratio.

As levels of the noise factor, we selected an abdominal skin extracted from a rat (hairless rat, 8 weeks old, male) (N_1), which is considered normally to have higher skin permeation than that of a human and is highly subject to an absorption enhancer, and the back skin of a minipig (Yucatan micropig, 5 weeks old, female) (N_2), which is regarded to be analogous to a human's skin structure, less susceptible to an absorption enhancer, and has low permeation.

For N_1 (rat), after killing a rat in an ether atmosphere, we extracted the skin in its abdomen and used it for the in vitro skin permeation experiment. On the other hand, for N_2 (minipig), we obtained a commercial skin set and removed its under-skin fat after thawing it.

As a signal factor, we chose the sampling time of the receptor solution in the in vitro skin permeation experiment. Because of a great difference in skin permeation speed between N_1 and N_2, we adjusted the signal ranges of N_1 and N_2 to approximately equalize each amount of permeation (Table 1).

4. SN Ratio

Table 2 shows part of the data regarding the principal agent's cumulative amount of skin permeation per skin area for N_1 and N_2 as the output. Based on this result, we performed the following analysis (analysis 1).

Below is an example of calculation for experiment 2.

Total variation:

$$S_T = 26.5^2 + 152.8^2 + \cdots + 1138.7^2$$
$$= 2,461,900 \qquad (f = 9) \qquad (1)$$

Effective divider:

$$\gamma_1 = 2^2 + 4^2 + 6^2 + 8^2 + 10^2 = 220 \qquad (2)$$

$$\gamma_2 = 4^2 + 8^2 + 24^2 + 48^2 = 2960 \qquad (3)$$

Linear equations:

$$L_1 = (2)(26.5) + (4)(152.8) + \cdots + (10)(748.3)$$
$$= 142,892.2 \qquad (4)$$

$$L_2 = (4)(1.0) + (8)(13.4) + \cdots + (48)(1138.7)$$
$$= 65,503.9 \qquad (5)$$

Variation of proportional term:

$$S_\beta = \frac{(L_1 + L_2)^2}{\gamma_1 + \gamma_2} = 2,002,184 \qquad (f = 1) \qquad (6)$$

Variation of differences between proportional terms:

$$S_{N\beta} = \frac{(L_1 - L_2)^2}{\gamma_1 + \gamma_2} = 375,496 \qquad (f = 1) \qquad (7)$$

Error variation:

$$S_e = S_T - S_\beta - S_{N\beta} = 84,220 \ (f = 7) \qquad (8)$$

Error variance:

$$V_e = \frac{S_e}{7} = \frac{84,220}{7} = 12,031 \qquad (9)$$

SN ratio:

$$\eta = 10 \log \frac{[1/(\gamma_1 + \gamma_2)] \ (S_\beta - V_e)}{V_e}$$
$$= -12.84 \text{ dB} \qquad (10)$$

Sensitivity:

$$S = 10 \log \left[\frac{1}{\gamma_1 + \gamma_2} (S_\beta - V_e) \right]$$
$$= 27.97 \text{ dB} \qquad (11)$$

In this calculation we computed the SN ratio without including $N \times \beta$, corresponding to an animal species difference in the error variance. This is because the objective of our study was to obtain not a prescription with a small difference in permeation between a rat and minipig but an optimal prescription with superior permeation for a human being as well as for a rat and minipig. If an animal species difference was included in the error variance, we might select good levels for a rat at the sacrifice of a minipig, or vice versa. Furthermore, the effort of finding good levels for a human being might also be affected. By dealing with an animal species difference as an indicative factor without including $N \times \beta$ in the error variation, we could obtain an optimal prescription with excellent permeation for both a rat and a minipig and adjust the amount of principal agent and application time for humans. Therefore, we considered that a difference in per-

Table 1
Signal factors and levels (sampling point) (h)

Noise Factor	T_1	T_2	T_3	T_4	T_5
N_1	2	4	6	8	10
N_2	4	8	24	48	—

Table 2
Principal agent's cumulative amount of skin permeation per skin area ($\mu g/cm^2$)

Noise Factor	No.	T_1	T_2	T_3	T_4	T_5
N_1	1	0.0	0.0	2.3	4.6	7.7
	2	26.5	152.8	319.3	528.3	748.3
	⋮	⋮	⋮	⋮	⋮	⋮
	18	15.0	142.8	378.5	545.7	718.8
N_2	1	0.2	1.0	25.4	123.8	—
	2	1.0	13.4	447.4	1138.7	—
	⋮	⋮	⋮	⋮	⋮	⋮
	18	2.3	20.1	335.7	1059.7	—

meation due to an animal species difference was not a noise factor.

What is important in permeation is a large amount of permeation in a short time (i.e., a high SN ratio). According to Table 2, it is obvious that there is a significant difference in permeation per unit time between N_1 and N_2, that is, a considerable difference in sensitivity. In this case we should conduct an analysis using time as the output under the same amount of permeation. In addition, since a certain medical effect need was determined by the amount of medicine given, we attempted an analysis based on the cumulative amount of skin permeation as the signal, and its corresponding time as the output, that is, an analysis that swaps the input signal and the output characteristic in analysis 1 (analysis

2). The calculation procedure of analysis 2 was basically the same as that for analysis 1. However, since the amount of permeation differed largely from experiment to experiment, the range of a signal factor was determined by picking up an overlapping range between N_1 and N_2 for each experimental run. Thus, while the range of a signal factor for each column was different in analysis 2, the effective divider for each row, r, was common for N_1 and N_2.

5. Results of Experiment

The control factors are allocated as shown in Table 3. To determine an ointment prescription with

Table 3
Control factors and levels

	Control Factor	Level 1	2	3
A:	amount of sampled sollution (mL)	Mid	Large	—
B:	amount of applied ointment (mL)	Small	Mid	Large
C:	amount of principal agent (%)	Low	Mid	High
D:	amount of absorption enhancer 1 (%)	None	Mid	High
F:	amount of absorption enhancer 3 (%)	None	Mid	High
G:	amount of additive agent 1 (%)	Low	Mid	High
H:	amount of additive agent 2 (%)	Low	Mid	High

excellent permeation, we assigned each prescription ingredient to control factors C to H. Considering that the amounts of applied ointment or receptor solution affect the amount of skin permeation as well as the variability in an in vitro skip permeation experiment, we also allocated them to factors A and B.

Table 4 shows SN ratios and sensitivities. Now, for rows 5, 10, and 15, since we could not arrange a uniform prescription because suspension occurred in the concentrate of ointment, no experiment was implemented. Then, because we could not judge whether or not the absorption was good or poor, the process average was substituted into these rows. Figure 3 shows the response graphs of the SN ratio,

η, and sensitivity, S, for analysis 1. In analysis 2, the sensitivity shown in Figure 4 is the more desirable the smaller it is, since the output was time. This is contrary to analysis 1.

Based on the idea that the sensitivity is more important than the SN ratio to obtain an optimal prescription enabling a principal agent to be quickly absorbed, we determined the optimal configuration considering sensitivity a higher priority. In this case, since factor C, amount of principal agent, was an adjusting factor in the case of dealing with a human being, we selected the second level: the same as the initial configuration. According to analysis 1, the optimal configuration turned out to be $A_1B_2C_2D_2E_3 F_2G_1H_2$, whereas $A_2B_3C_2D_2E_3F_2G_1H_2$ was chosen from

Table 4
SN ratios and sensitivities (dB)

No.	A 1	B 2	C 3	D 4	E 5	F 6	G 7	H 8	SN Ratio	Sensitivity
1	1	1	1	1	1	1	1	1	−16.74	6.79
2	1	1	2	2	2	2	2	2	−12.84	27.96
3	1	1	3	3	3	3	3	3	−20.84	20.26
4	1	2	1	1	2	2	3	3	−20.59	3.75
5	1	2	2	2	3	3	1	1	−19.58[a]	17.64[a]
6	1	2	3	3	1	1	2	2	−20.04	10.38
7	1	3	1	2	1	3	2	3	−21.32	9.37
8	1	3	2	3	2	1	3	1	−22.12	25.67
9	1	3	3	1	3	2	1	2	−13.56	38.46
10	2	1	1	3	3	2	2	1	−19.58[a]	17.64[a]
11	2	1	2	1	1	3	3	2	−14.87	22.73
12	2	1	3	2	2	1	1	3	−23.29	24.52
13	2	2	1	2	3	1	3	2	−24.89	6.36
14	2	2	2	3	1	2	1	3	−19.65	17.59
15	2	2	3	1	2	3	2	1	−19.58[a]	17.64[a]
16	2	3	1	3	2	3	1	2	−16.78	17.33
17	2	3	2	1	3	1	2	3	−32.45	6.64
18	2	3	3	2	1	2	3	1	−13.79	27.25

[a] Process average was used.

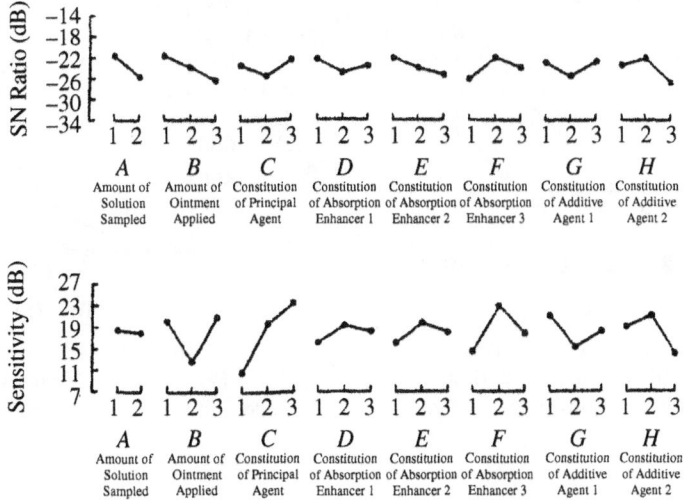

Figure 3
Analysis 1 of in vitro skin permeation experiment (signal: time)

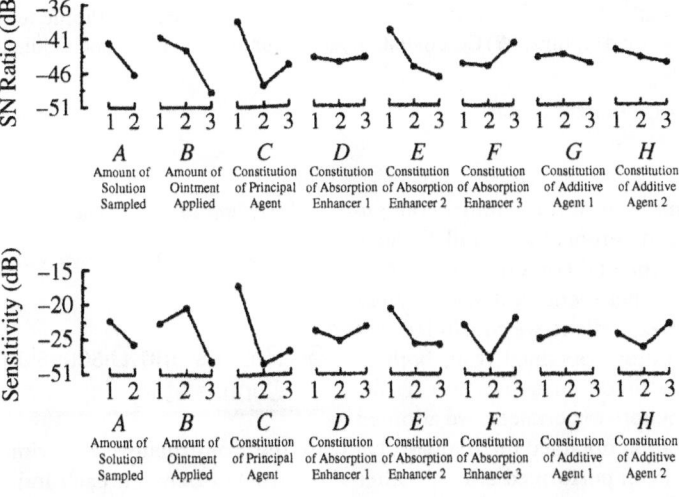

Figure 4
Analysis 2 of in vitro skin permeation experiment

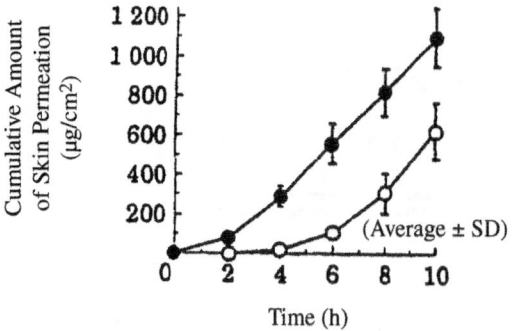

(a) Analysis 1 (Signal: Time)
Permeation Profile for N_1 (Rat)
under Initial (○) and Optimal (●) Configurations

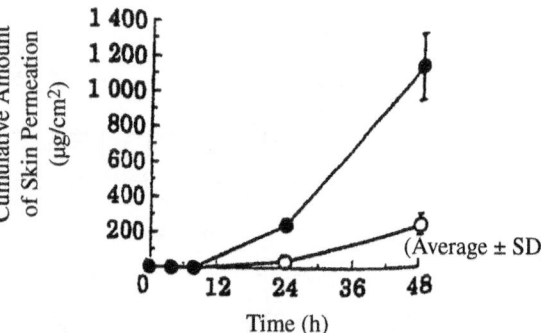

(b) Analysis 1 (Signal: Time)
Permeation Profile for N_2 (Minipig)
under Initial (○) and Optimal (●) Configurations

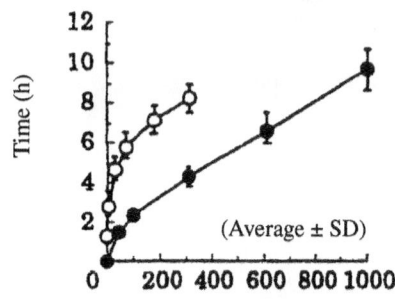

(c) Analysis 2 (Signal: Time)
Permeation Profile for N_1 (Rat)
under Initial (○) and Optimal (●) Configurations

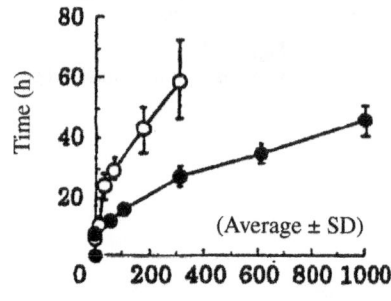

(d) Analysis 2 (Signal: Time)
Permeation Profile for N_2 (Minipig)
under Initial (○) and Optimal (●) Configurations

Figure 5
Graph for confirmatory experiment

analysis 2. Comparing both of the optimal configurations, we could see differences at A and E. Since factor A was an experimental condition in an in vitro skin permeation experiment, and also because E_1 and E_2 were nearly equivalent, we concluded that the optimal configurations obtained from both of the experiments were almost consistent as a result.

As for the confirmatory experiment, we adopted the optimal configuration obtained for analysis 1 because analysis 2 had been performed after all other experiments, including the confirmatory experiment, were completed.

Initial configuration: $A_1 B_3 C_2 D_2 E_1 F_1 G_1 H_1$

Optimal configuration: $A_1 B_3 C_2 D_2 E_2 F_2 G_2 H_2$

6. Results and Discussion of Confirmatory Experiment

In the confirmatory experiment, considering the differences between each individual animal, we repeated each experiment six times. Figure 5a and b show profiles of skin permeation for lapsed time un-

Table 5
Estimation and confirmation of SN ratio and sensitivity (dB)

Configuration	Analysis 1				Analysis 2			
	SN Ratio		Sensitivity		SN Ratio		Sensitivity	
	Estimation	Confirmation	Estimation	Confirmation	Estimation	Confirmation	Estimation	Confirmation
Initial	−18.9	−25.8	21.6	15.5	−45.9	−36.5	−28.4	−18.1
Optimal	−12.4	−15.5	36.0	27.9	−53.9	−46.1	−40.5	−30.2
Gain	6.5	10.3	14.5	12.4	−8.0	−9.6	−12.1	−12.1

der both initial and optimal configurations for N_1 and N_2.

Under the optimal configuration, we confirmed excellent skin permeation for both a rat and a minipig, which led to remarkable improvement compared with those under the initial configuration. Additionally, combining this result with that of analysis 2 and plotting each data point by setting the amount of permeation as the signal and time as the output characteristic, we obtained Figure 5c and d. These results reveal that a considerable amount of time was reduced when a certain amount of permeation is achieved.

Table 5 shows each pair of estimations and confirmations of the SN ratio, η, and sensitivity, S, for analyses 1 and 2. Now all factors are used for estimation. In analysis 2, since shorter permeation time was a characteristic (i.e., a smaller sensitivity was preferable), the estimated gain turns out negative. The reason that the estimated gain of the SN ratio becomes negative is that if an improvement in σ is relatively smaller than that of β because a decrease in β becomes larger than an improvement in σ, eventually the SN ratio (defined as β^2/σ^2) diminishes.

Comparing the estimated and measured values of the SN ratio and sensitivity, we found that both were nearly consistent in terms of the gain of sensitivity for both analyses 1 and 2; and for analysis 2, the reproducibility of the SN ratio also improved. In analysis 2 the gain in the SN ratio resulted in negative values for both estimation and confirmation. However, as the standard deviations in Figure 5c and d show, although the variability under the optimal configuration was less than or equal to that under the current, the improvement in sensitivity, which was larger than that of the variability, turned the gain of the SN ratio into negatives.

Although animal testing usually requires four or more repetitions because of individual differences, we confirmed that a single experiment allowed us to obtain good reproducibility, thereby streamlining the experimental process.

7. Conclusions

Using a characteristic of in vitro skin permeation, we have optimized prescription of ointment. Setting

a rat and minipig to noise factor levels and assuming that an animal species difference was an indicative factor, we calculated an SN ratio using V_e. As a result, we obtained an ideal, optimal prescription with a high skin permeation speed and short lag time.

Although we have shown two different types of analyses in this study, considering the quality engineering concept that a signal factor is a customer's demand in nature, we have realized that analysis 2 is more appropriate. In fact, reproducibility of the SN ratio is more improved in analysis 2 than in analysis 1, where time was selected as a signal factor.

In addition, despite omuitting the details in our research, response graphs based on a conventional SN ratio calculation method that uses V_N as an error variance have almost agreed with those calculated by V_e (Figure 5). This implies that even if there are significant noise conditions, such as a rat and a minipig, the resulting optimal configurations are not so different for the two types, and these optimal configurations are not susceptible to species differences. Therefore, we can predict that only a simple adjustment such as of principal agent, amount of applied medicine, or application time is needed to apply a transdermal drug delivery system effectively.

When optimizing an ointment prescription, not only skin permeation but also the problems of skin stimulation and stable storage need to be taken into consideration. Our research succeeded in reducing development cycle time drastically. In addition, for the characteristics obtained from the animal experiment, we confirmed that only one experiment is necessary for optimization, and consequently, achieved a considerable economic benefit. Therefore, we believe that the quality engineering method is applicable to prescription optimization of other types of drugs or other animal experiments.

Reference

Hatsuki Asahara, Kouya Yano, Yuichiro Ota, Shuichi Takeda, and Takashi Nakagawa, 1998. Optimization of model ointment prescription for the improvement of in vitro skin permeability *Quality Engineering*, Vol. 6, No. 2, pp. 65–73.

This case study is contributed by Hatsuki Asahara and Kouya Yano.

CASE 10

Use of a Dynamic Operating Window for Herbal Medicine Granulation

Abstract: This study applies the dynamic operating window analysis method in granulation technology. The original SN ratio, calculated from $M \times \beta$, was named the speed difference method, whereas the new one, calculated from the ratio of larger-the-better and smaller-the-better characteristics, was called the speed ratio method. In this study, parameter design was used for the granulation of herbal medicines.

1. Introduction

In 1994, Taguchi introduced the dynamic operating windows method for chemical reactions. Since then several applications have been reported. Later, a new analysis method was introduced to combine the larger-the-better and smaller-the-better characteristics, the former representing the main chemical reaction and the latter, side reactions. This new method was based on a practice that no noise factors are assigned for chemical reactions, although side reactions behave as noise to the main reaction.

2. Granulation

Granulation is a universal technology, used widely in various industries, such as fertilizer, food, catalysts, agricultural chemicals, and ceramics. In this study, parameter design was used for the granulation of herbal medicines. There are two types of granulation: dry and wet. Most of the granulation processes for medicines use the wet method. Since this study belongs to upstream technology development and no restrictions were to be followed, laboratory-scale equipment was used, without considering the scale of manufacturing. Instead, efficiency of development was considered the top priority.

Two types of granulation are used to obtain a uniform-particle-size product: either increasing or reducing the particle size. Most cases are of the former type. In fact, the latter is a crushing process and therefore is not described here. This study introduces a granulating process using the speed ratio method based on the concept that fine particles grow by contact in exponential fashion.

In the herbal medicine industry, efforts are being made to develop high-value-added products. Today, the traditional excellence of herbal medicines is reviewed, and further technological advancement through the use of quality engineering is expected. The development of medicines includes research departments to search for new substances; pharmacology departments to study the mechanism of medical effects, side effects, or the safety of medicines, and product planning departments. Once a product is designed and reaches the stage of production engineering, its engineering approaches are almost identical to those in other manufacturing industries.

In this case, evaluation of the granulating function is described from the quality engineering viewpoint. Granulation is an important process in the pharmaceutical industry to pretreat raw material or produce an objective product shape, such as granules. There are Japanese Pharmacopoeia regulations regarding particle-size distribution for medicines, and it is desirable to produce a medicine with uniform particle-size distribution and high yield.

3. Basic Reaction Theory

To correspond to chemical reactions from the viewpoint of quality engineering, an example of dissolution of a substance is used for explanation. When a reaction progresses,

$$aA + bB + \cdots \rightarrow pP + qQ + \cdots \qquad (1)$$

its reaction speed is given by

$$V = k[A]^a[B]^b \qquad (2)$$

The total of exponents of concentration of reacting substances A, B, ... , denoted by $a + b + \cdots = n$, is called *order*. V, the speed of A being dissolved and B increased, is expressed as

$$V = -\frac{dA}{dt} = \frac{dB}{dt} = k[A][B] \qquad (3)$$

It changes by the concentration of A and B.

There are different types of reactions: zeroth-, first-, and second-order reactions. Different order reactions are used depending on the subject being discussed. In quality engineering, only one type (e.g., first order) is considered. For example, let the concentration of substance A be denoted by [A]. In the case of a first-order reaction, the speed of A reacting to become P is given by

$$-\frac{d[A]}{dt} = k[A] \qquad (4)$$

Letting the initial concentration of A be $[A_0]$,

$$\ln[A] = -kt + \ln[A_0] \qquad (5)$$
$$[A] = [A_0]e^{-kt} \qquad (6)$$
$$-\ln\frac{[A]}{[A_0]} = kt \qquad (7)$$

Therefore, [A] decreases exponentially.

For second-order reactions, a reaction speed of A = B ··· P is

$$-\frac{d[A]}{dt} = -\frac{d[B]}{dt} = k[A][B] \qquad (8)$$

Letting the initial concentration of A and B be $[A_0]$ and $[B_0]$, respectively,

$$\frac{1}{[A_0] - [B_0]} \ln \frac{[B_0][A]}{[A_0][B]} = kt \qquad (9)$$

One can therefore derive a zero-point proportional equation of concentration and time for any first- or second-order equations, as shown in equations (7) and (9). The following discussion is based on a first-order reaction.

There are many parameters in a chemical reaction, such as pH, diffusion speed, or surface area. In scientific research, these parameters are put into equations. But in quality engineering, these are considered as noise factors because it would be ideal if the reaction were not affected by the effects of outer noise, such as temperature or pH. Since a reaction can be discussed at the molecular level, equations that describe phenomena exactly are not necessary.

Let's expand the concept of molecular collision. One mole of a substance includes 6.0231023 molecules. A generic function can be established to express a reaction between single molecules. Letting β be a constant, Y_0 the initial concentration (100%), Y the concentration after dissolving, and Y/Y_0 the fraction of reduction, the reaction is expressed by

$$\frac{d(Y/Y_0)}{dt} = \beta \left(1 - \frac{Y}{Y_0}\right) \qquad (10)$$

Equation (10) is described simply as the relationship between the fraction of reduction of concentration and time. Since the equation essentially expresses the function only, one may judge the equation to be insufficient. But that is the difference between science and engineering.

Let the left side of equation (15.11) be p_0 and solve the equation:

$$1 - \frac{Y}{Y_0} = e^{-\beta t} \qquad (11)$$

$$p = 1 - \frac{Y}{Y_0} \qquad (12)$$

$$-\ln p_0 = \beta t \qquad (13)$$

Rewriting $-\ln p_0$ as y_1, the following generic function is derived.

$$y_1 = \beta t \qquad (14)$$

4. Generic Function

Figure 1 shows the fluidized-bed granulating equipment used in this experiment. Powder is fed into the chamber and heated air is blown from the bottom to fluidize powder. While the particles continue to keep contact with each other, a liquid containing binder is sprayed to make the particles grow. The granulated particles sink by weight; the light particles flow up in the air and continue granulation. The process ends when the blowing air stops. Like the case of spray drying, it is difficult to balance heat energy and blowing energy by adjusting the temperature, liquid feeding speed, or air-blowing speed. Parameter design is an effective approach in these cases.

The growing mechanism in fluidized-bed granulation can be considered in the same way as a chemical reaction that follows exponential function. Letting the fraction of reduction of powder be y_2 and time be t,

$$y_2 = 1 - e^{-\beta t} \qquad (15)$$

Thus, it is ideal that the fine particles in the chamber decrease following equation (15).

Among many available granulating methods, a laboratory-scale fluidized-bed granulating device was selected using about 100 g of powder to conduct one run of the experiment. It was low in cost and easy to operate. The device was small and could be put on a table, as shown in Figure 1. Air was blown from the bottom of the chamber to cause the powder to mobilize and make contact. Binder was sprayed in a mist to keep the powder particles in contact, in which state they grow gradually. An important condition for granulation is to balance the energy of air blowing force and energy of heat to evaporate moisture. The conditions have been determined by experience and accumulated experimental results. In this study we tried to determine the conditions using parameter design.

5. Experimental Design

An L_{18} orthogonal array was used for the experiment. The factors and levels are given in Table 1. Particle size distribution was measured by the screening method shown in Figure 2.

6. Speed Ratio Method

Smaller-the-Better, Larger-the-Better, and Dynamic Operating Window

Since a chemical reaction occurs by collision, it is assumed that the collision in granulation follows equation (15). In the speed difference method of dynamic operating windows, the SN ratio, η^*, and sensitivity, S^*, are calculated from the $S_{M\beta}$ value of the principal and side reactions. The speed ratio method involves increasing the granulating speed, β_1, and reducing the growing speed of coarse particles, β_2. The former corresponds to p, the unreacted part, and the latter correspond to q, the

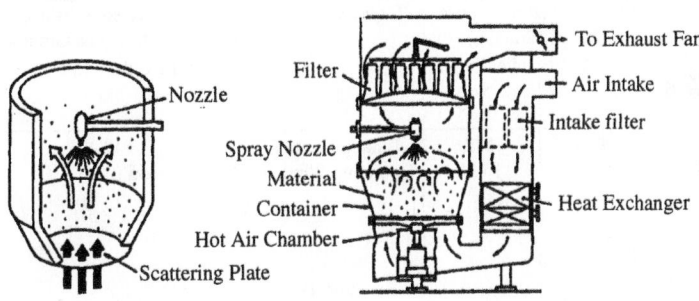

Figure 1
Fluidized-bed granulation equipment

Table 1
Factors and levels

Control Factor	Level 1	Level 2	Level 3
A: position of spray nozzle	Close	Far	
B: concentration of binder	Standard	Slightly high	High
C: amount of herbal extract supplied	Small	Standard	Large
D: air flow	Small	Standard	Large
E: nozzle spray pressure	Low	Standard	High
F: air temperature	Low	Standard	High
G: amount of fluidity assisting agent	Small	Standard	Large
H: flow of binding liquid	Small	Standard	Large

overreacted part in Figure 3. In other words, we tried to maximize the operating window, $1 - p - q$).

Since the generic function is expressed by $y_2 = 1 - e^{-\beta t}$:

Unreacted part: $q = e^{-\beta t}$

$$\beta_1: \text{ larger-the-better} \tag{16}$$

Overreacted part: $1 - p - e^{-\beta t}$

$$\beta_2: \text{ smaller-the-better} \tag{17}$$

Over No. 12 sieve (diameter 1400 mm):

overreacted (p), coarse

Between No. 12 and No. 42 sieve (diameter 1400 to 355 mm):

target $(1 - p - q)$

Under No. 42 sieve (diameter 355 mm):

underreacted (q), fine

According to product specification, the portion over a No. 12 sieve must be less than 5%, and the portion under a No. 42 sieve must be less than 15%.

Signal and Noise Factors

Three points of time, T, were set as signal factor levels to take samples. Among the three points, the two where reaction took place faster were selected

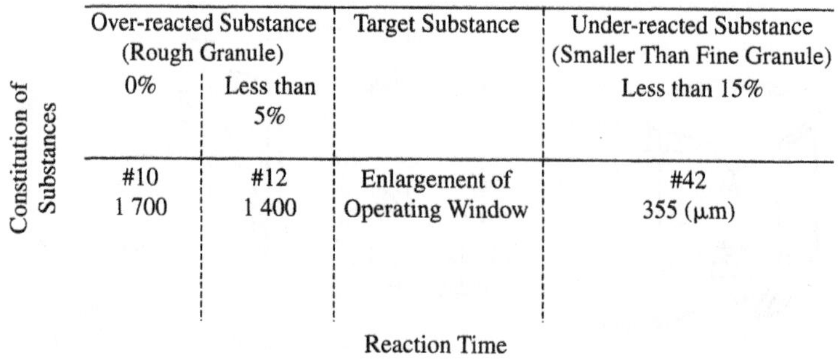
Over-reacted Substance (Rough Granule)
0%
Less than 5%
Target Substance
Under-reacted Substance (Smaller Than Fine Granule)
Less than 15%
Constitution of Substances
#10
1 700
#12
1 400
Enlargement of Operating Window
#42
355 (μm)
Reaction Time

Figure 2
Specification of particle size

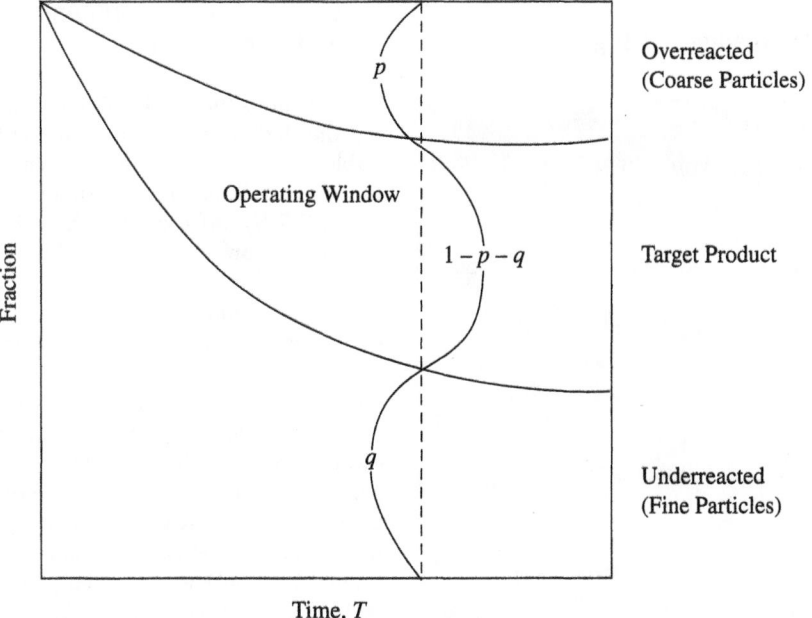

Figure 3
Dynamic operating window threshold (specification of particle size)

for calculation. This is because the fraction of the reacted part can be measured more precisely in the neighborhood of 50%. Noise factors were not assigned, considering the efficiency of the study. This was based on the thought that side reactions behave as noise.

7. SN Ratio and Sensitivity

Measuring the fractions of unreacted part (q) and overreacted part (p) at time T, reaction speeds β_1 and β_2 were calculated.

$$\beta_1 = \frac{1}{T} \ln \frac{1}{q} \tag{18}$$

$$\beta_2 = \frac{1}{T} \ln \frac{1}{1-p} \tag{19}$$

$$\eta_1 = -10 \log \frac{1}{\beta_1^2} \tag{20}$$

$$\eta_2 = -10 \log \beta_2^2 \tag{21}$$

The speed ratio method is to maximize $\eta_1 = \eta_1 + \eta_2$ as the SN ratio.

$$\textit{SN ratio:} \quad \eta = \eta_1 + \eta_2 \tag{22}$$

$$\textit{Sensitivity:} \quad S = 10 \log \beta_1^2 \tag{23}$$

Sensitivity, S, reflects reaction speed. Examples of calculation are shown in Tables 2 and 3 for cases demonstrating progress and no progress, respectively. As mentioned previously, two points were measured. The results in the tables are calculated from the average β^2 from the two points.

From Figures 4 and 5, the optimum configuration was determined as $A_2B_2C_1D_3E_3F_3G_3H_2$. As described later, A_1B_1 was used for the confirmatory experiment instead of A_2B_2. The initial condition was $A_2B_2C_2D_2E_2F_2G_2H_2$. Condition H_2 was selected as the optimum after taking sensitivity into consideration.

Table 2

Calculation of experiment 13: granulation progressed[a]

Reaction	Time	
	T_1 (25 min)	T_2 (40 min)
Overreacted substance p	0.0060	0.0099
Target substance $1 - p - q$	0.0930	0.0960
Underreacted substance q	0.9010	0.8941
	After Conversion	
	T_1	T_2
Overreacted substance p (β_2)	0.0002423	0.002488
Underreacted substance q (β_1)	0.0041715	0.0027974

[a] The results are $\eta_1 = -49.67$ dB, $\eta_2 = 72.20$ dB, $\eta = 22.53$ dB, and $S = -49.67$ dB.

8. Confirmatory Experiment and Improvement

Table 4 shows the results of optimum conditions using the data of one point. If the entire reaction is stable, analysis may be made from one point only. For the confirmatory experiment, condition A_1B_1 $C_1D_3E_3F_3G_3H_2$ was used, which is the optimum SN ratio condition selected from the smaller-the-better characteristic.

Under optimum conditions, the fraction of targeted particle size was high with a small fraction unreacted. The SN ratio was slightly lower, due to a larger overreaction portion (side reaction). In the confirmatory experiment, a gain of 20.86 dB was obtained, which is seen by a comparison of the particle-size distribution. As shown in Figure 6, nearly 80% was unreacted (i.e., fine particles) under the initial conditions. But under the optimum conditions, over 80% was targeted product.

In the confirmatory experiment, the SN ratio of the speed ratio increases, so that the reaction speed increases. But observing the particle-size distribution shown in Figure 7, it appears that the distributions of the two cases were not really very different; only the center of the distribution was shifting. In such a case, it is better to compare both after adjusting the center of distribution. But from the viewpoint of granulation, it looks as if the growth of particles is being hindered due to an im-

Table 3

Calculation of experiment 8: granulation did not progress

	Control Factor	Level		
		1	2	3
A:	position of spray nozzle	Close	Far	—
B:	density of binder	Standard	Slightly high	High
C:	amount of herbal essence supplied	Small	Standard	Large
D:	amount of air flow	Small	Standard	Large
E:	spraying pressure	Low	Standard	High
F:	temperature of heated air	Low	Standard	High
G:	amount of fluidity agent	Small	Standard	Large
H:	amount of binder supplied	Small	Standard	Large

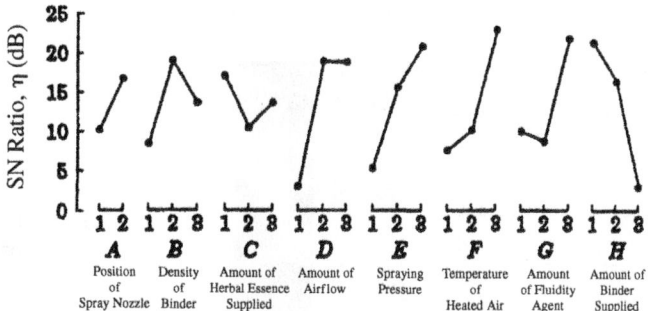

Figure 4
Response graphs of SN ratio

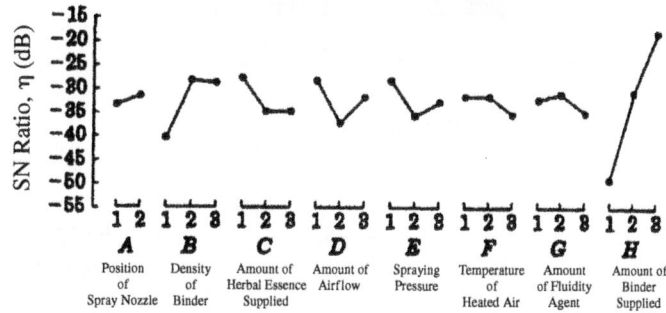

Figure 5
Response graphs of sensitivity

Table 4
Results of optimum configuration

Reaction	Time (37 min)
Overreacted substance p	0.0539
Target substance $1 - p - q$	0.7494
Underreacted substance q	0.1967
After Conversion	
Overreacted substance (β_1)	0.0014978
Underreacted substance (β_2)	0.0439504

[a] The results are $\eta_1 = -27.14$ dB, $\eta_2 = 56.49$ dB, $\eta = 29.35$ dB, and $s = -27.14$ dB.

balance of evaporation, heat, and airflow, so that it would be difficult to adjust the center to one functioning under optimum conditions, even spending a longer time. This is probably the reason that the SN ratios in Table 5 do not agree. In this regard, calculations were made to estimate the time necessary to move the center of distribution toward that under optimum conditions (the data for the initial conditions are given in Table 6):

$$\beta_{opt} = \frac{\beta_1 + \beta_2}{2} = 0.02272 \qquad (24)$$

$$\beta_{initial} = \frac{\beta_1 + \beta_2}{2} = 0.002545 \qquad (25)$$

$$y = \beta_{opt}\, T = \beta_{opt}\,(37) = 0.8406 \qquad (26)$$

The time corresponds to the center of the initial conditions:

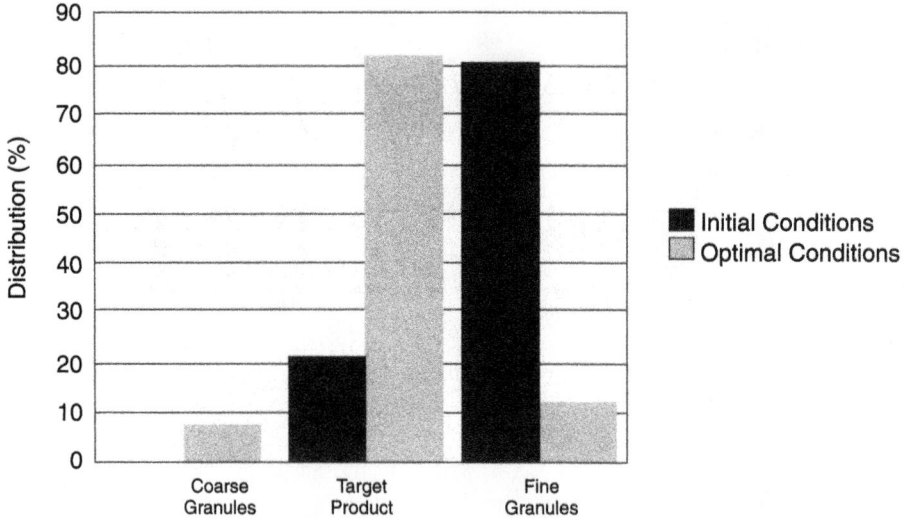

Figure 6
Granule-size distribution before and after optimization

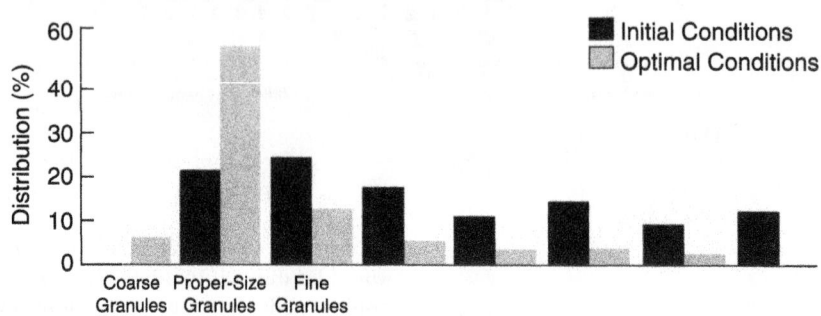

Figure 7
Improved granule-size distribution before and after optimization

$$0.8406 = 0.002545T$$

$$T = 330 \text{ min} \qquad (27)$$

The SN ratio calculated for the initial conditions is

Table 5
Gain of SN ratio from confirmatory experiment (dB)

	Configuration		
	Initial	Optimal	Gain
Estimation	13.78	41.11	27.33
Confirmation	8.49	29.35	20.86

$$\eta_{\text{initial}} = 10 \log \frac{\beta_1^2}{\beta_2^2} = 10 \log \frac{0.003701^2}{0.01388^2}$$

$$= 8.52 \text{ dB} \qquad (28)$$

It is therefore necessary to maintain 330 minutes in order to shift the center of distribution. It seems from past experience, however, that excess reaction

Table 6
Data on initial conditions

	Time		
Reaction	T_1 (10 min)	T_2 (20 min)	T_x (? min)
Overreacted substance p	0.0001	0.0385	0.3675
Target substance $1 - p - q$	0.0258	0.4671	0.3377
Underreacted substance q	0.9741	0.4944	0.2948

	After Conversion		
	T_1	T_2	Average
Overreacted substance p (β_2)	0.00001	0.001963	0.001388
Underreacted substance q (β_1)	0.002624	0.03522	0.003701

Table 7
SN ratio of the smaller-the-better characteristic (dB)

	Item[a]			
No.	Particle Size	Fluidity	Cohesion	SN Ratio
1	△	○	△	−4.7
2	△	△	△	−6.0
3	○	○	○	0.0
4	X	X	X	−9.5
5	△	○	○	−3.0
6	X	X	X	−9.5
7	△	X	X	−8.7
8	△	○	○	−3.0
9	○	X	△	−6.7
10	△	○	○	−3.0
11	○	△	X	−6.7
12	△	X	X	−8.6
13	○	○	○	0.0
14	X	X	X	−9.5
15	△	○	○	−3.0
16	△	△	△	−6.0
17	X	X	X	−9.5
18	△	△	△	−6.0

[a] If ○, score = 1. If △, score = 2. If X, score = 3.

time gives excess contact energy to particles, causing the crashing of particles and thus the production of fine particles. But it is interesting to see that the results calculated agree with those obtained by the speed ratio method.

9. Smaller-the-Better SN Ratio Using Scores

Because of time restrictions, an attempt was made to estimate the optimum conditions by eyeball observation of the results instead of using particle-size distribution. Qualitative scores were given by observing items such as particle size, fluidity, and cohesion, with the smaller-the-better SN ratio being used for evaluation:

$$\eta = -10 \log \sigma^2 \qquad (29)$$

This simple method is not quantitative. But to our surprise, the condition determined from this method was close to the optimum conditions using the speed ratio method. The items observed, such as particle size, fluidity, and cohesion, are phenomena in the process of granulation rather than objective characteristics. The dynamic operating window is, of course, a quantitative and more precise method.

For example, the SN ratio of row 1 is calculated as

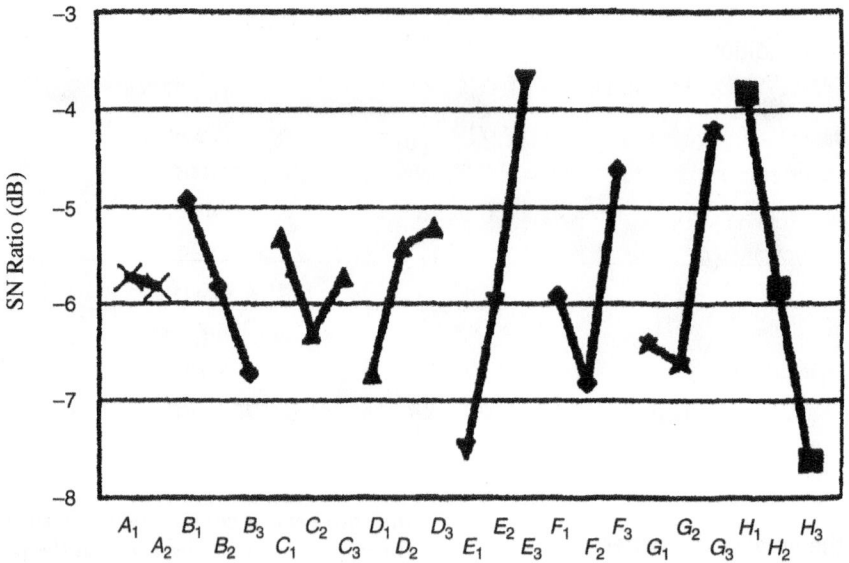

Figure 8
Response graphs of the smaller-the-better SN ratio

$$\eta = -10 \log \frac{2^2 + 1^2 + 2^2}{3} = -4.7 \text{ dB} \quad (30)$$

Table 7 shows the scores and SN ratios of 18 rows. The confirmatory experiment described in the preceding section was conducted using the optimum conditions of the smaller-the-better SN ratio of scored data instead of using the conditions selected from the speed ratio method: The former used condition A_1B_1 and the latter, A_2B_2. However, this is not a problem.

The optimum condition selected from the response graph of Figure 8 was $A_1B_1C_1D_3E_3F_3G_3H_2$, and the estimated score was about the same, showing good reproducibility. The SN ratios from the speed ratio method and the score were close to each other. It is probably because granulation becomes saturated as time passes, so the particle-size distribution after saturation is proportional to the initial granulating speed, β.

10. Conclusions

Use of the dynamic operating window in conjunction with the speed ratio method applied to gran-

ulation has been described in contrast with the existing dynamic operating window. It is a new concept, combining the smaller-the-better and larger-the-better characteristics instead of using $M\beta$ as the signal. The new approach is not only simpler but makes it possible to reduce the number of samples to be used. The existing method requires at least three data points for analysis, and in the case of incomplete data, sequential analysis has to be used for estimation.

In this granulation study, the reaction progressed rapidly and coagulated in some of the experimental runs, and size distribution could not be measured. But the possibility of analysis from even a single data point opens up critical areas in chemical reactions.

Reference

Kouya Yano, Masataka Shitara, Noriaki Nishiuchi, and Hideyuki Maruyama, 1998. Application of dynamic operating window for Kampo medicine granulation. *Quality Engineering*, Vol. 6, No. 5. pp. 60–68.

This case study is contributed by Kouya Yano.

CASE 11

Particle-Size Adjustment in a Fine Grinding Process for a Developer

Abstract: As a result of a parameter design based on an L_9 orthogonal array using easily changeable factors in a grinding process, we obtained sufficiently good results in the adjustability of particle size. However, since we could not achieve an adequate improvement in particle-size distribution, we tackled our second parameter design with an L_{18} orthogonal array, utilizing more control factors to improve both particle-size adjustment and distribution.

1. Introduction

A prototyping device used in the research stage is a combination of grinding and separating devices, which has the advantages of requiring less material for experiments and enabling us to arrange various factors easily. Using a parameter design based on an L_9 orthogonal array with easily changeable factors in the first experiment, we obtained a sufficiently good result in particle-size adjustability. However, since we could not achieve an adequate improvement in particle-size distribution, we tackled our second parameter design with an L_{18} orthogonal array by taking in more control factors to improve both particle-size adjustment and distribution.

Figure 1 shows the grinding and separating device utilized. At the supply area (1), roughly ground material (sample) is input, and subsequently is dispersed at the cyclone (2) and carried from the separating area to the grinding area. At the separating area, an airflow type of separating machine classifies particles of different diameters by taking advantage of a balance of centrifugal and suction forces. The material transported to the grinding area strikes an impact plate and is crushed into fine particles. In the separating area, only particles of a target size are sorted out; larger granules are reground in the grinding area. Thus, in the grinding and separating process, a granule is pulverized repeatedly until it matches the target size.

Although in our conventional process, we have focused on the separating conditions as the key point for improvement in granular diameter adjustment and distribution, we implemented experiments based primarily on grinding conditions by changing our approach drastically in this research. In addition, we arranged control factor conditions manually that cannot be prepared easily using our current machine, by using ready-made component parts, such as the impact plate, whose shape is regarded to largely affect diameter and distribution. Through reiterative preliminary experiments, we narrowed the control factor levels down to six types of shapes. More important, whereas we have traditionally considered grinding and separating as independent systems, we conducted experiments in this study by considering the two factors together.

2. Ideal Function and SN Ratio

The ideal function at the grinding area is that particle size varies linearly with little variability in grinding energy, which is an adjusting factor. Therefore, as shown in Figure 2, the reciprocal of grinding energy is considered as a signal factor such that the diameter of particles becomes zero when grinding energy approaches infinity, so we proceeded with our analysis based on zero-point proportional SN

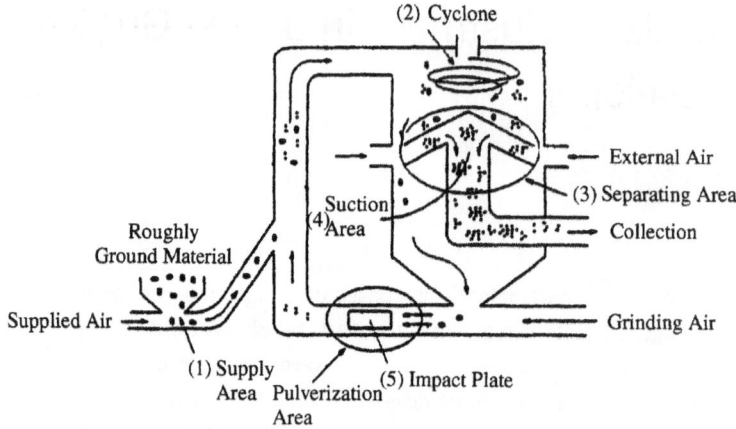

Figure 1
Schematic drawing of grinding and separating device

ratio. For the signal factor, grinding pressure that corresponds to grinding energy was used:

$$M_1: \frac{1}{\text{high pressure}}$$

$$M_2: \frac{1}{\text{medium pressure}}$$

$$M_3: \frac{1}{\text{low pressure}}$$

Choosing the hardness and weight of a sample as noise factors, we termed the level with a larger granular diameter after collection as N_1 (hard and light) and the one with a smaller diameter as N_2 (soft and heavy). The particle-size distribution of ground samples was measured using a Coulter counter, and the data were put in order in a histogram, with particle diameter on the horizontal axis and volumetric percentage on the vertical axis (Figure 3). We calculated the fraction of particles smaller than the average or target diameter from this histogram and used them as the characteristics for analysis in our conventional process. These characteristics are closely related to the quality of photocopy images and reliability. In addition,

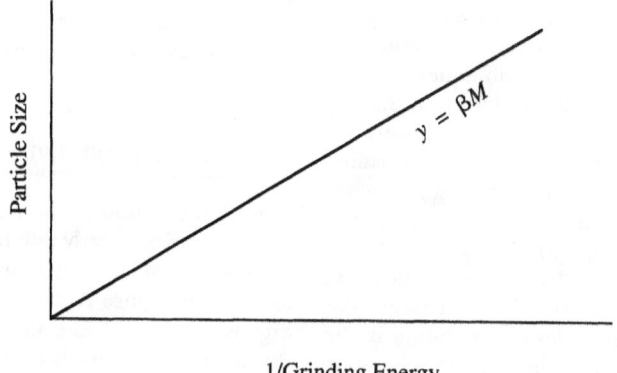

Figure 2
Relationship between grinding energy and particle size

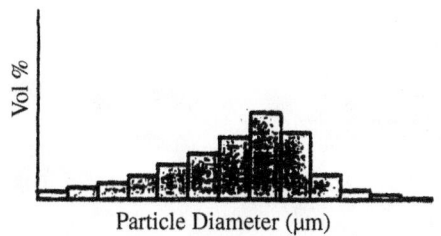

Figure 3
Particle-size distribution after grinding

production capacity and yield have to be considered, so we must make an overall judgment. However, it was quite difficult to satisfy all of them, so in most cases we had to compromise.

In this experiment we needed to compute an average y and variance V from the particle-size distribution data in Figure 3. Primarily, we tabulated the values of particle diameters $(x_1, \ldots, x_{14})$ and fractions $(p_1, \ldots, p_{14})$:

Measurement channel:

| 1 | 2 | 3 | 4 | ... | 11 | 12 | 13 | 14 |

Average particle diameter (μm):

x_1 x_2 x_3 x_4 ... x_{11} x_{12} x_{13} x_{14}

Fraction (%):

p_1 p_2 p_3 p_4 ... p_{11} p_{12} p_{13} p_{14}

After calculating the average (y) and variance (V) as

$$y = \frac{x_1 p_1 + x_2 p_2 + \cdots + x_{14} p_{14}}{100} \quad (1)$$

$$V = \frac{x_1^2 p_1 + x_2^2 p_2 + \cdots + x_{14}^2 p_{14}}{100} - y^2 \quad (2)$$

we obtained Table 1. Using these data, we computed a zero-point proportional SN ratio.

Total variation:

$$S_T = (y_{11})^2 + (y_{12})^2 + (y_{13})^2 + (y_{21})^2 + (y_{22})^2 + (y_{23})^2 \quad (3)$$

Effective divider:

$$\gamma = M_1^2 + M_2^2 + M_3^2 \quad (4)$$

Variation of proportional term:

$$S_\beta = \frac{(L_1 + L_2)^2}{2\gamma} \quad (5)$$

Variation of differences between proportional terms:

$$S_{N\beta} = \frac{(L_1 - L_2)^2}{2\gamma} \quad (6)$$

Error variation:

$$S_e = S_T - S_\beta - S_{N\beta} \quad (7)$$

Error variance:

$$V_e = \frac{S_e}{4} \quad (8)$$

Total error variance:

$$V_N = \frac{S_{N\beta}}{5} + S_e + \frac{V_{11} + V_{12} + V_{13} + V_{21} + V_{22} + V_{23}}{6} \quad (9)$$

Table 1
Average particle diameter (y) and variance (V) for each experiment

	M_1	M_2	M_3	Linear Equation
N_1	y_{11} V_{11}	y_{12} V_{12}	y_{13} V_{13}	$L_1 = M_1 y_{11} + M_2 y_{12} + M_3 y_{13}$
N_2	y_{21} V_{21}	y_{22} V_{22}	y_{23} V_{23}	$L_2 = M_1 y_{21} + M_2 y_{22} + M_3 y_{23}$

Table 2
Control factors and levels

	Control Factor	Level					
		1	2	3	4	5	6
A:	type of impact plate	Cone 1	Cone 2	Trapezoid 1	Trapezoid 2	Triangle 1	Triangle 2
B:	distance of impact plate	1.0	1.5	2.0			
C:	amount of material supplied	1.0	1.5	2.0			
D:	supply time	10	20	30			
E:	damper condition	10	25	40			
F:	separating area condition	20	18	16			
G:	particle diameter of supplied material	1.0	1.5	2.0			

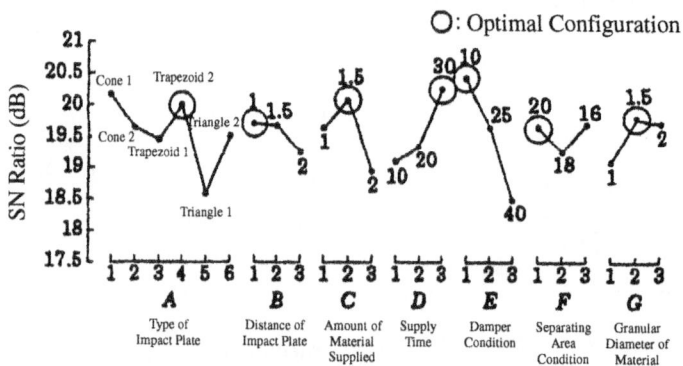

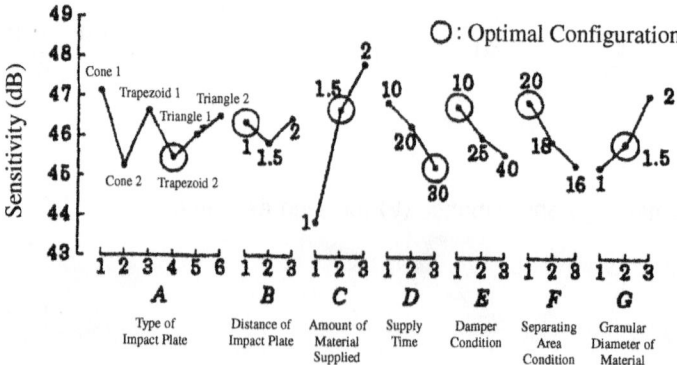

Figure 4
Response graphs

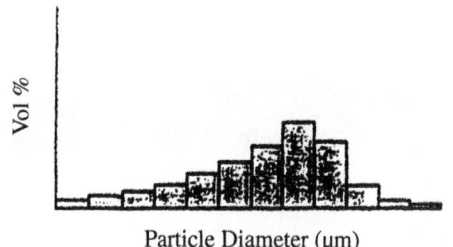

(*a*) Particle-size distribution under current condition

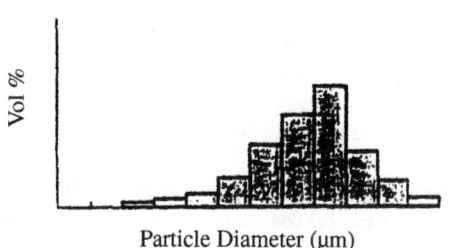

(*b*) Particle-size distribution under L_9 optimal condition

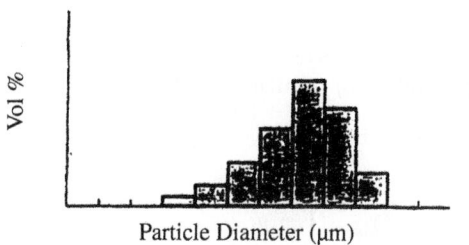

(*c*) Particle-size distribution under L_{18} optimal condition

Figure 5
Improvement in particle-size distribution

SN ratio:

$$\eta = 10 \log \frac{(1/2\gamma)(S_\beta - V_e)}{V_N} \tag{10}$$

Sensitivity:

$$S = 10 \log(1/2\gamma)(S_\beta - V_e) \tag{11}$$

As control factors, we set up A, B, C, D, E, F, and G, as summarized in Table 2.

3. Optimal Configuration and Confirmatory Experiment

Figures 4 and 5 show the response graphs of the SN ratio and sensitivity. Although the optimal configuration based on the SN ratio is $A_1 B_1 C_2 D_3 E_1 F_1 G_2$, we selected A_4 in place of A_1 because A_1's manufacturing conditions were poor and estimated the process average using A, C, D, and E.

Estimation of SN ratio

$$\eta = A_4 + C_2 + D_3 + E_1 - 3T = 22.15 \text{ dB} \tag{12}$$

Estimation of sensitivity:

$$S = A_4 + C_2 + F_1 + G_2 - 3T = 46.69 \text{ dB} \tag{13}$$

As a result of the confirmatory experiment, we can see that good reproducibility were achieved for the SN ratio and sensitivity, as summarized in Table 3. As for the current conditions, we estimated the gain using the optimal configuration in the first L_9 experiment. As the difference caused by N_1 and N_2 decreased, we adjusted the particle diameter by fine-adjusting with pressure. That is, with hardly any change in grinding or separating conditions, we

Table 3
Results of confirmatory experiment (dB)

Optimal Configuration	SN Ratio		Sensitivity	
	Estimation	Confirmation	Estimation	Confirmation
L_{18}	22.15	22.19	46.69	46.00
L_9	17.13	17.13	45.29	45.31
Gain	5.02	5.07	1.40	0.69

Table 4
Data for dynamic operating window

Signal Threshold	Error	Grinding Pressure			Linear Equation
		T_1	T_2	T_3	
M_1	N_1	y_{111}	y_{112}	y_{113}	L_1
	N_2	y_{121}	y_{122}	y_{123}	L_2
M_2	N_1	y_{211}	y_{212}	y_{213}	L_3
	N_2	y_{221}	y_{222}	y_{223}	L_4

converted samples of different sizes and weights into almost identical particles.

As shown in Figure 5, we can change the particle-size distribution under the L_9 optimal configuration into a narrower distribution. Consequently, not only a drastic improvement in yield but also the elimination of fine adjustment of particle diameter were attained. The proportions of small and large particles were reduced significantly under the optimal configuration as compared with those under the current and L_9 optimal configurations, and eventually we obtained an extremely sharp distribution, as illustrated in Figure 5c.

4. Analysis Using a Dynamic Operating Window

In the next analysis it was assumed that the relationship between grinding energy, T, and reduction of raw material speed follows equation (16.4), where Y and Y_0 denote the amount of after-ground product and before-ground material, respectively. The grinding efficiency should be highest when the grinding process follows the equation.

$$\frac{Y}{Y_0} = 1 - e^{-\beta T} \qquad (14)$$

When performing an analysis based on the dynamic operating window, an important key point is how to determine the portions corresponding to side reactions and underreactions in chemical reactions. Since we produce small particles from raw material in a grinding system, we can assume that a side-reacted part, p, is the part in the particle-size

distribution where the particle diameter is smaller than the targeted product, and unreacted part, q, is that where the particle diameter is larger than the targeted product. The objective product is in-between.

Although time is considered a preferable signal factor in a chemical reaction, we have not been able to measure grinding time, due to the technical limitations of this device. Therefore, similar to the situation in zero-point proportional analysis, we computed an SN ratio using grinding pressure as a signal factor.

Now the converted values of p and q were used for the following calculation of SN ratio:

M_1 (small particles):

$$y_1 = \ln \frac{1}{1 - p}, \qquad 1 - p = \frac{Y_1}{Y_0} \qquad (15)$$

M_2 (large particles):

$$y_2 = \ln \frac{1}{q}, \qquad q = \frac{Y_2}{Y_0} \qquad (16)$$

All data obtained are shown in Table 4. The SN ratio is computed as follows:

Total variation:

$$S_T = y_{111}^2 + y_{112}^2 + \cdots + y_{223}^2 \qquad (f = 12) \qquad (17)$$

Variation of proportional term:

$$S_\beta = \frac{(L_1 + \cdots + L_4)^2}{4\gamma} \qquad (f = 1) \qquad (18)$$

Variation of proportional terms due to M:

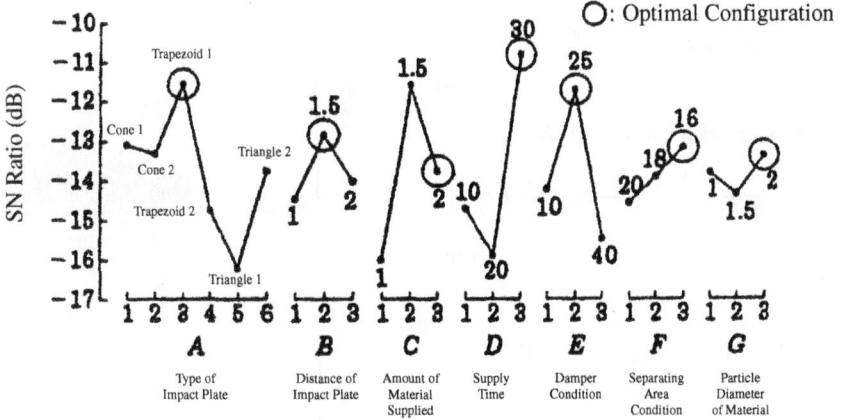

(a) SN Ratio, η, of Operating Window

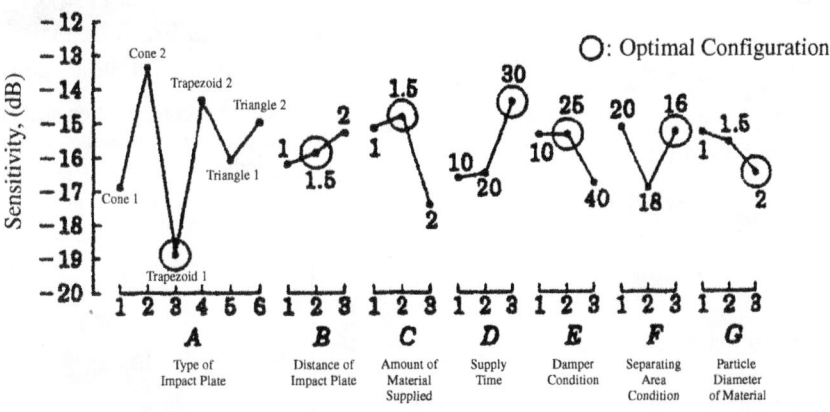

(b) Sensitivity, S*, in Operating Window

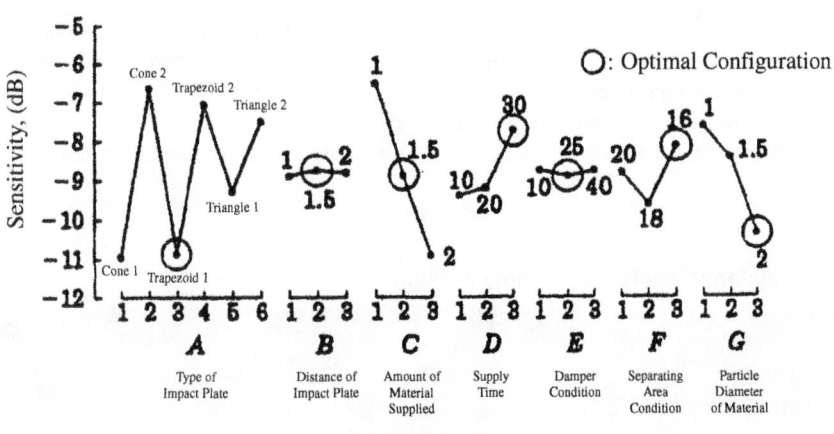

(c) Sensitivity, S

Figure 6
Response graphs of dynamic operating window

$$S^*_{M\beta} = \frac{(L_1 + L_2)^2 + (L_3 + L_4)^2}{2\gamma} - S_\beta \qquad (f = 1)$$

(19)

Variation of proportional terms due to N:

$$S^*_{N\beta} = \frac{(L_1 + L_3)^2 + (L_2 + L_4)^2}{2\gamma} - S_\beta \qquad (f = 1)$$

(20)

Error variation:

$$S_e = S_T - S_\beta - S^*_{M\beta} - S^*_{N\beta} \qquad (f = 9) \quad (21)$$

Error variance:

$$V_e = \frac{S_e}{9}$$

(22)

Total variance:

$$V_N = \frac{S_e + S^*_{N\beta}}{10}$$

(23)

SN ratio of operating window:

$$\eta = 10 \log \frac{(1/4\gamma)(S^*_{M\beta} - V_e)}{V_N}$$

(24)

Sensitivity:

$$S = 10 \log \frac{1}{4\gamma}(S_\beta - V_e)$$

(25)

Sensitivity of operating window:

$$S^* = 10 \log \frac{1}{4\gamma}(S^*_{M\beta} - V_e)$$

(26)

Figure 6 illustrates the response graphs for the SN ratio and sensitivity. Since factor levels with a high SN ratio are optimal, we prepared our estimates under the optimal configuration selected.

Particle Diameter (µm)

(a) Current configuration

Particle Diameter (µm)

(b) Optimal configuration analyzed by zero-point proportional equation with grinding pressure as signal factor

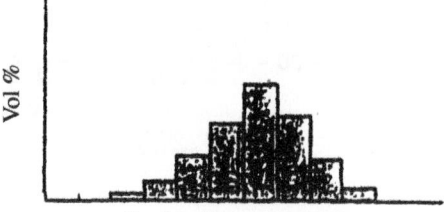

Particle Diameter (µm)

(c) Optimal configuration analyzed by dynamic operating window with grinding pressure as signal factor

Figure 7
Comparison of particle-size distribution

Table 5
Confirmation by dynamic operating window method

	Estimation	Confirmation
SN ratio of operating window	−9.97	−11.53
Sensitivity of operating window, S	−6.76	−5.75
Sensitivity of operating window, S^*	−11.36	−10.59

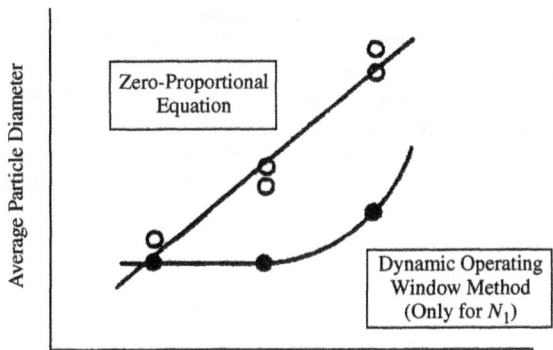

Figure 8
Relationship between grinding pressure and particle diameter

5. Optimal Configuration and Confirmatory Experiment

A confirmatory experiment under the optimal configuration could not be performed because only test samples for N_1 have remained. Then, using the optimal configuration grounded on the aforementioned zero-point proportional equation and the data for a granular diameter distribution in its confirmatory experiment, we computed the SN ratio and sensitivity and checked the reproducibility. The result is summarized in Table 5, which reveals that fairly good reproducibility of the SN ratio and sensitivity was obtained.

As a next step, using only test samples for N_1, we ground actual material under the optimal configuration. For all of the results under the initial, zero-point proportional–based optimal, and dynamic operating window–based optimal configurations, the particle diameter distribution for the N_1 sample is illustrated in Figure 7. Under each condition, the particle size distribution was different. But the figure shows that the conditions for the two optimal configurations (except the current configuration) can be used for mass production.

Next, we checked the particle diameter adjustability under the optimal configuration based on the dynamic operating window, as shown in Figure 8. Compared with that under the optimal configu-

ration using the zero-point proportional equation, adjustability using grinding pressure worsened.

Considering all of the above, we can summarize the dynamic operating window characteristics as follows:

1. When using the dynamic operating window method, the particle-size distribution was horizontally symmetrical, as shown in Figure 7c. Although this tendency was not regarded as problematic, but as rather good, we need to further separate the particles because of the broad distribution.

2. As compared to that under the optimal configuration using a zero-point proportional equation, the grinding efficiency improved significantly. In other words, if we grind under the same pressure, we can obtain a condition where we can make the particles smaller.

3. Despite the improvement noted above, the particle-size adjustability was poorer. This indicates that we achieved the results as defined in equation (14).

4. Our experiment proves that a dynamic operating window method can also be used in nonchemical reaction areas.

5. According to the results shown in Figure 8, grinding time should be chosen as a signal factor rather than using grinding pressure.

References

Hiroshi Shibano, Tomoharu Nishikawa, and Kouichi Takenaka, 1994. Establishment of particle size adjusting technique for a fine grinding process for developer. *Quality Engineering*, Vol. 2, No. 3, pp. 22–27.

Hiroshi Shibano, Tomoharu Nishikawa, and Kouichi Takemaka, 1997. Establishment of control technique for particle size in roles of fine grinding process for developer. *Quality Engineering*, Vol. 5, No. 1, pp. 36–42.

This case study is contributed by Hiroshi Shibano.

Part II

Robust Engineering: Electrical Applications

Circuits (Cases 12–15)
Electronic Devices (Cases 16–20)
Electrophoto (Cases 21–22)

CASE 12

Design for Amplifier Stabilization

Abstract: The authors have been studying some applications of the reproducibility of conclusions in electrical fields. One of the conclusions was that the characteristics of the measuring device, settings of the signal factor, and control factors from the circuit elements affected the SN ratio. In the study, a new amplifier was designed considering the foregoing points. Parameter design was conducted by varying two combinations of control factors.

1. Introduction

A circuit designed for our research is illustrated in Figure 1. This is a typical transistor amplification circuit whose function is expressed by $y = \beta M$, where M is the input voltage and y is the output voltage. We selected eight factors from resistance R, capacitor C, and inductor L as control factors and assigned them to an L_{18} orthogonal array. Among the control factors shown in Table 1, factors C, D, E, and F were determined in the design phase by considering the technical relationship of circuit characteristics. As for a signal factor of the input voltage of an amplification circuit, we allocated the values shown in Table 2 to each of four levels. In addition, we defined input frequency as an indicative factor and assigned three levels. As a noise factor, we selected environmental temperature for the circuit and set up two levels of room and high temperatures. The former is approximately 25°C, and the latter is heated up to around 80°C using a hair dryer.

2. Experimental Procedure

Using the RC oscillator depicted in Figure 2, we added a sinusoidal input signal to a circuit and measured the input voltage with a voltmeter as a signal factor. Setting the frequency of the input signal as an indicator, we proceeded with our experiment by

measuring the output voltage. Now, to simplify our measurement and analysis, we measured all data in root-mean-square form. Each resistance shown in Figure 2 was expressed as internal resistance, a typical characteristic of each measuring apparatus, and regarded as an important element in measuring an electronic circuit.

To evaluate how the difference in measurement conditions affected experimental results, we conducted the experiment under four distinct measurement conditions. We showed a few elements regarded as influential throughout the experimental process:

1. *Load resistance between a circuit and measurement apparatus.* An electric circuit designer attempts to keep measurement conditions constant by predetermining the input/output impedance of a circuit (representative values of circuitry characteristics) in reference to the measurement apparatus used. Because we can change the impedance by adding resistors, we usually measure a circuit using 50- or 75-Ω resistors. In our research we confirmed the effects on measurement by altering the characteristics of measurement apparatus with resistors added.

2. *Setup of signals reflecting the capable range of a circuit.* This research deals with a transistor amplification circuit. If the input voltage goes out of the capable range of a circuit, the output voltage tends to saturate, due to the

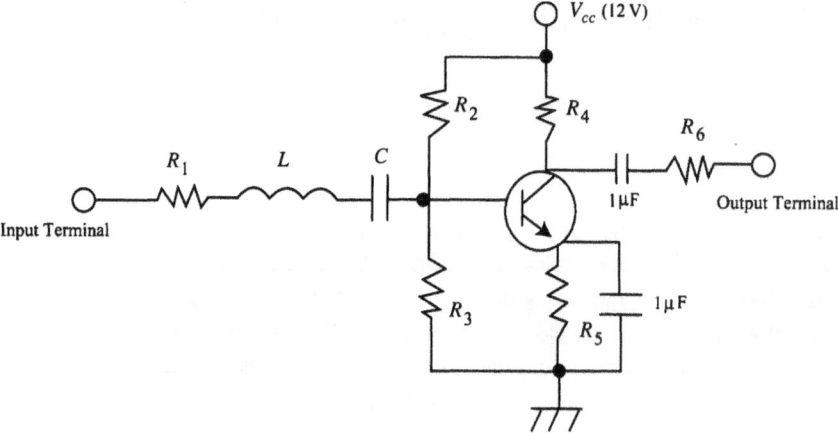

Figure 1
Transistor amplification circuit

characteristics of the circuit. Therefore, we postulated that selection of signal factors (input voltage) affects SN ratio and sensitivity and experimented on this presumption.

3. Experiment 1: Amplification Circuit Design without Considering the Effect on Measurement

Since in this experiment we focused on a case that we do not consider characteristic of measuring equipment, a circuit and measurement apparatus were connected directly, without resistors. A signal factor has four levels, as shown in Table 2. As for other factors, we used levels of Table 2. We discuss only the data from experiment 1, shown in Table 3, which is chosen from 18 experiments designed based on an L_{18} orthogonal array and explain the calculation procedure.

S_1, S_2, and S_3 represent the sensitivitys for each indicative factor (frequency of input signal).

Total variation:

$$S_T = 510^2 + 610^2 + \cdots + 3930^2 + 3810^2$$

$$= 125,390,500 \qquad (f = 24) \qquad (1)$$

Effective divider:

$$r = 5.0^2 + 10.0^2 + 40.0^2 = 2125 \qquad (2)$$

Linear equations:

$$L_{11} = (5.0)(510) + (10.0)(1020) + (20.0)(2010)$$
$$+ (40.0)(3690)$$

$$= 200,550$$

$$L_{21} = 217,250$$

$$L_{31} = 210,900$$

$$L_{12} = 200,700$$

$$L_{22} = 218,600$$

$$L_{32} = 212,550 \qquad (3)$$

Variation of proportional term:

$$S_\beta = \frac{(L_{11} + L_{21} + L_{31} + L_{12} + L_{22} + L_{32})^2}{6r}$$

$$= 124,626,376.7 \qquad (f = 1) \qquad (4)$$

Variation of proportional term difference of indicative factor:

$$S_{F\beta} = \frac{(L_{11} + L_{12})^2 + (L_{21} + L_{22})^2 + (L_{31} + L_{32})^2}{2r} - S_\beta$$

$$= 144,608.6 \qquad (f = 2) \qquad (5)$$

Table 1
Control factors

		Level		
Control Factor		1	2	3
A:	resistance R_1	0	6.8	—
B:	resistance R_2	10	22	33
C:	resistance ratio $R_3/(R_2 + R_3)$	0.1	0.2	0.4
D:	resistance R_5	$I_c \approx 1$ mA	$I_c \approx 3$ mA	$I_c \approx 5$ mA
E:	inductor $L(\mu H)$	10	22	56
F:	capacitor C	20	+0	−20
G:	resistance R_4	470	820	200
H:	resistance R_6	0	22	47

Variation of proportional term due to error factor:

$$S_{N(F)B} = \frac{L_{11}^2 + L_{21}^2 + L_{31}^2 + L_{12}^2 + L_{22}^2 + L_{32}^2}{r}$$
$$- S_\beta - S_{F\beta}$$
$$= 1074.7 \quad (f = 3) \tag{6}$$

Error variation:

$$S_e = S_T - S_\beta - S_{F\beta} - S_{N(F)\beta} = 618{,}440 \quad (f = 18) \tag{7}$$

Error variance:

$$V_e = \frac{S_e}{18} = 34{,}357.78 \quad (f = 18) \tag{8}$$

Total error variance:

$$V_N = \frac{S_e + S_{N(F)\beta}}{21} = 29{,}500.7 \quad (f = 21) \tag{9}$$

SN ratio:

$$\eta = 10 \log \frac{(1/6r)(S_\beta - V_e)}{V_N} = -4.80 \text{ dB} \tag{10}$$

Sensitivity:

$$S_1 = 10 \log \frac{1}{2r}(S_{\beta i} - V_e) = 39.50 \text{ dB} \tag{11}$$

$$S_{\beta i} = \frac{(L_{11} + L_{12})^2}{2r} \tag{12}$$

$$S_2 = 40.22 \quad S_3 = 39.96 \text{ dB} \tag{13}$$

Figure 3 shows the response graphs. Since we concentrated on reproducibility of the SN ratio in our study, only S_1 is posted in terms of sensitivity. Looking at the graphs, we find a remarkable V-shape in factors G and H. For G, we can judge that this is reasonable because this shape is caused by the order of levels (level 3 < level 1 < level 2), due

Table 2
Signal, indicative, and noise factors

			Level			
Factor			1	2	3	4
Signal factor	M:	input Voltage (mV)	5.0	10.0	20.0	40.0
Indicative factor	F:	frequency (kHz)	125.0	250.0	500.0	—
Noise factor	N:	temperature	Room	High	—	—

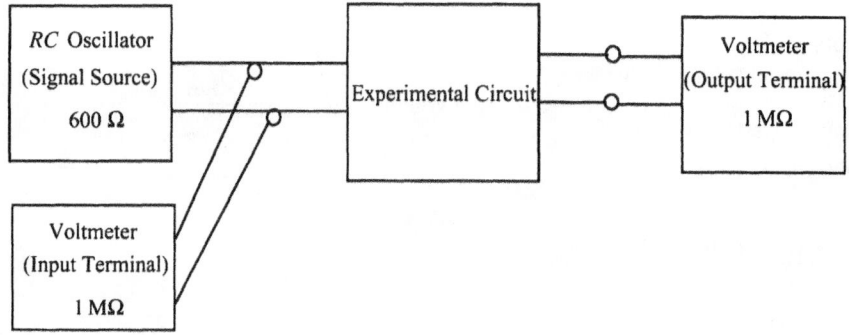

Figure 2
Measuring apparatus

to technical relationships with other factors. For H, this is because we did not take into account measurement characteristics at output terminals where a resistor as factor H was connected. The optimal configuration of SN ratio is $A_1B_1C_1D_3E_3F_2G_3H_1$. On the other hand, the current configuration is $A_1B_2C_2D_2E_2F_2G_2H_2$. Thus, we can see a considerable difference between the two.

Table 4 of the confirmatory experiment (shown in Table 3) reveals that we do not obtain sufficient reproducibility of gain. This is particularly because of the V-shape for factor H shown in the response graph. More specifically, we believed the reason to be that output measurement varies due to no resistance between a circuit and the measurement apparatus.

4. Experiment 2: Amplification Circuit Design Reflecting Effects on Measurement

In experiment 2, we adopted a common method to place a 50-Ω impedance between the input and output terminals of a circuit. As Figure 4 illustrates, we connected a 50-Ω resistor between the input and output terminals. This is called terminal resistance. Other conditions were exactly as in experiment 1, as was the analysis procedure. Table 5 shows the data of experiment 1 from the L_{18} orthogonal array. Due to the effect of the added terminal resistor, each measured datum was changed. This is because the circuit load changes, and eventually, measurement apparatus characteristics also vary. Although

Table 3
Measured data for experiment 1 (mV)

Noise Factor	Indicative Factor	M_1	M_2	M_3	M_4	Linear Equation
N_1	F_1	510	1020	2010	3690	L_{11}
	F_2	610	1180	2300	3910	L_{21}
	F_3	580	1140	2250	3790	L_{31}
N_2	F_1	520	1010	2000	3700	L_{12}
	F_2	700	1170	2310	3930	L_{22}
	F_3	590	1160	2280	3810	L_{32}

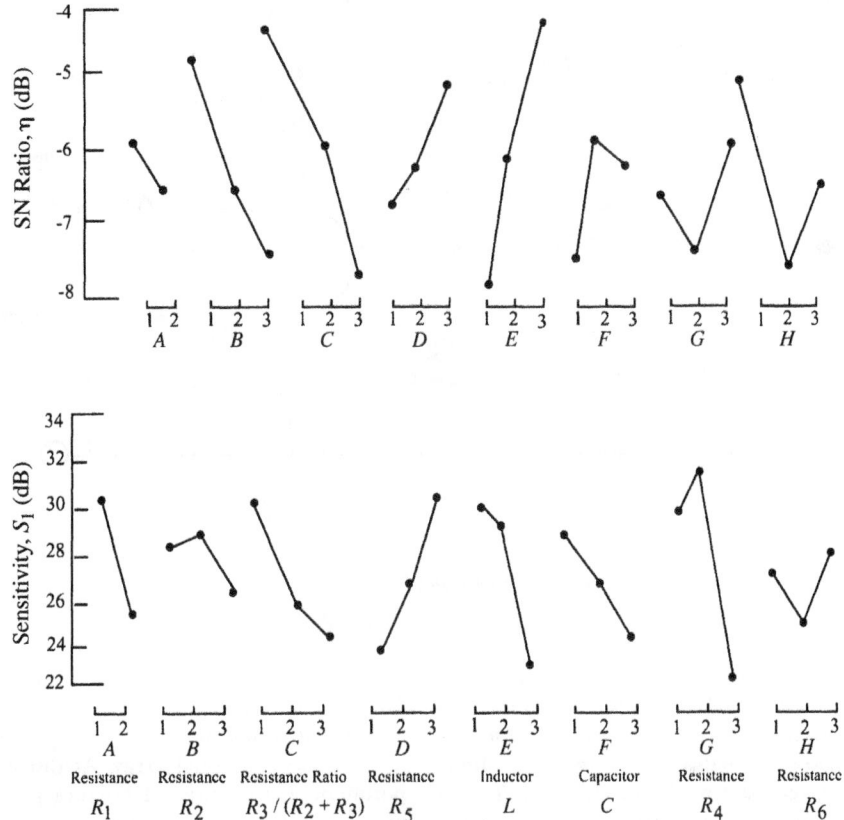

Figure 3
Response graphs of experiment 1

Table 4

Gain in SN ratio of confirmatory experiment for experiment 1 (dB)

SN Ratio	Configuration		
	Optimal	Current	Gain
Estimation	1.87	−8.09	9.96
Confirmation	−0.26	−6.88	6.62

we changed only terminal resistance in this experiment, the factor effect plot in Figure 5 shows characteristics similar to those of experiment 1 in Figure 3. Nevertheless, since the V-shape of factor H is somewhat less significant, the effect of terminal resistance was not negligible.

The optimal and current configurations obtained from the response graph in Figure 5 were the same as those of experiment 1. Table 6 shows the result of a confirmatory experiment. The fact that we did not obtain sufficient reproducibility is reasonable since the optimal and current configurations were unchanged. Although this experiment was performed on the assumption that insufficient reproducibility is caused by measurement apparatus characteristics, experiment 2 revealed that there are factors other than the characteristics affected by

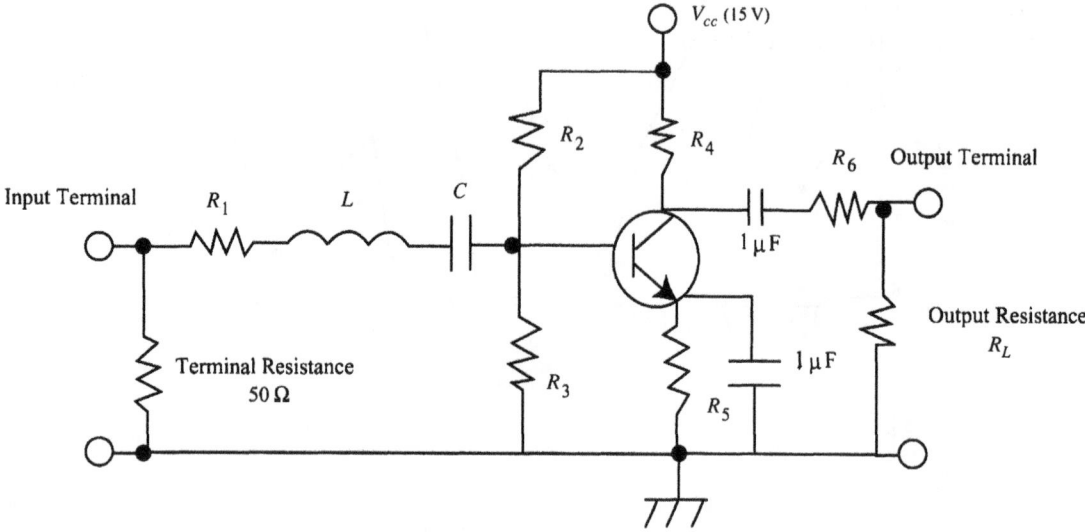

Figure 4
Terminal resistance-added amplification circuit for experiment 2

load resistance connected to a circuit. In sum, the reason is that a measured datum (output y) and signal (input M) whose relationship is supposed to satisfy the proportionality of $y = \beta M$ shows nonlinearity, as typified by M_4 in Figure 6. This is because we have added input voltage beyond a range where linearity of amplification capability is maintained. With this in mind, we conducted another experiment, experiment 3.

5. Experiment 3: Amplification Circuit Design Reflecting Characteristics Affecting Signals and Measurement

This experiment evaluated how appropriately a signal factor is set up and how much measurement apparatus terminal resistance affects SN ratio and sensitivity, both of which were regarded as technical problems from experiments 1 and 2. After checking the amplification capability range of a circuit, we selected three levels of signal factor from within the range of voltage (Table 7). For other factors we did not make any change and followed the setup shown in Table 2. Similar to experiment 2, we used a 50-Ω resistor as terminal resistance (Figure 4) because in

experiment 2 we confirmed that we should flatten the V-shape of factor H.

Table 8 shows the result of experiment 3 based on an L_{18} orthogonal array. As discussed with experiment 2, measured data were changed by terminal resistance. Since the number of data are different from those of experiments 1 and 2, the number of degrees of freedom was also altered in experiment 3 but is substantially equivalent to that of experiment 1. Figure 7 shows the response graph. Although these demonstrate trends similar to those of the former experiments, we can see that the V-shape of factor H is larger. This implies that there exist factors other than the effects of measurement apparatus characteristics that we have mentioned so far. The optimal configuration is $A_1B_1C_1D_3E_3F_2G_1H_1$. G's level is different from that of experiments 1 and 2.

On the other hand, the current configuration is $A_1B_2C_2D_2E_2F_2G_2H_2$, which is the same as that of experiments 1 and 2. Now we considered the reason that the trends of factors G and H changed. Factor G is a resistor connected to a resistor of factor H. Additionally, the resistor of factor H is connected to the terminal resistor near the output terminal. Then factor H is affected by three sources: factor G, output terminal resistance, and input voltage. On the

Table 5
Measured data of experiment 2 (mV)

Noise Factor	Indicative Factor	M_1	M_2	M_3	M_4	Linear Equation
N_1	F_1	53	109	212	399	L_{11}
	F_2	68	134	262	460	L_{21}
	F_3	74	142	280	468	L_{31}
N_2	F_1	55	110	219	410	L_{12}
	F_2	69	134	265	473	L_{22}
	F_3	75	144	282	485	L_{32}

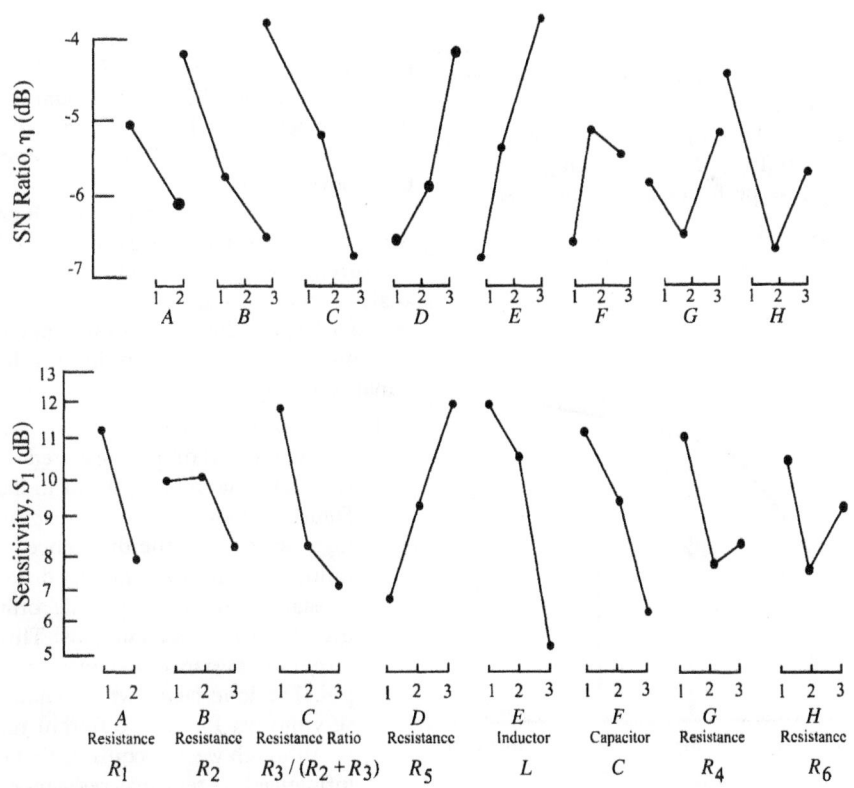

Figure 5
Response graphs of experiment 2

Table 6
Results of confirmatory experiment for experiment 2 (dB)

SN Ratio	Configuration		
	Optimal	Current	Gain
Estimation	0.58	−7.09	7.67
Confirmation	−1.42	−6.40	4.98

Table 7
Signal factors reflecting amplification capability

Signal Factor	M_1	M_2	M_3
Input Voltage (mV):	5.0	10.0	20.0

other hand, the sensitivities of experiment 3 are closer to those of experiment 2 than to those of experiment 1. This fact implies that a terminal resistor of 50 Ω as the circuit load affects the results significantly. As for reproducibility in the confirmatory experiment, we obtained the best results for all the experiments so far, as shown in Table 9. These results also prove that the insufficient SN ratios of experiments 1 and 2 were caused by signal factors.

However, interaction among Factors G and H, and terminal resistance, considered the other reason, are still unclear because they are quite complicated. Therefore, by keeping all the levels of a signal factor as shown in Table 7 and changing only terminal resistance, we performed experiment 4.

6. Experiment 4: Amplification Circuit Design Reflecting Effect of Terminal Resistance

To check the effect of terminal resistance added to a circuit by connecting a 50-Ω terminal resistor to the input terminals of a circuit (Figure 4), we set the impedance at to 50 Ω. At the same time, by adding a 50-Ω terminal resistor to the output terminals, we made the impedance equivalent to R_L. For R_L we set up four levels, and by using the three-level signal factor (input voltage) shown in Table 7, we performed an experiment. More specifically, we took measurements independently for each terminal resistance by following the same procedure as that of experiments 1 to 3. Bring at level 4 means no terminal resistance (Table 10), which was done by removing R_L in Figure 4.

In addition, to evaluate how much change of resistance affects voltage measurement, based on premeasured data, we implemented two types of analyses, one with an indicative factor of terminal resistance R_L and the other a two-signal analysis with the signal factor a product of input voltage and terminal resistance.

1. *Analysis of terminal resistance as an indicative factor.* First, based on premeasured data, we used terminal resistance R_L as an indicative factor. Figure 8 shows its response graph. Accordingly, we can see the differences for both SN ratio and sensitivity. Effects due to terminal resistance are in magnitude only, but each overall trend was analogous. This is because terminal resistance was proportional to output. The lone factor whose characteristic was very distinct for each different resistance was *G*. Although we can confirm that factor *G* was influenced by terminal resistance R_L, we cannot be sure of an interaction between input

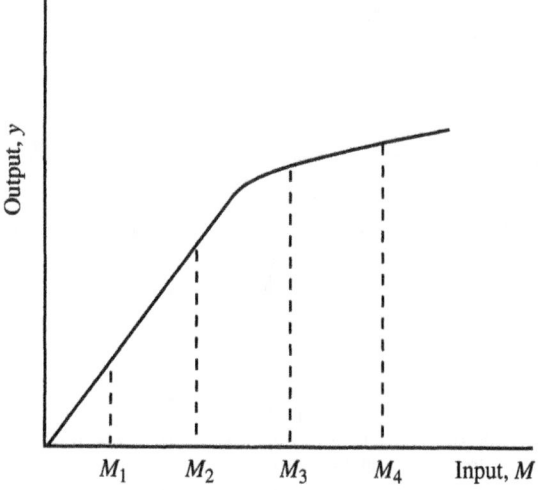

Figure 6
Nonlinearity of output characteristic for signal factor level

Table 8
Measured data of experiment 3 (mV)

Error Factor	Indicative Factor	M_1	M_2	M_3	Linear Equation
N_1	F_1	65	128	258	L_{11}
	F_2	88	175	348	L_{21}
	F_3	98	196	380	L_{31}
N_2	F_1	66	130	262	L_{12}
	F_2	87	170	340	L_{22}
	F_3	97	190	372	L_{32}

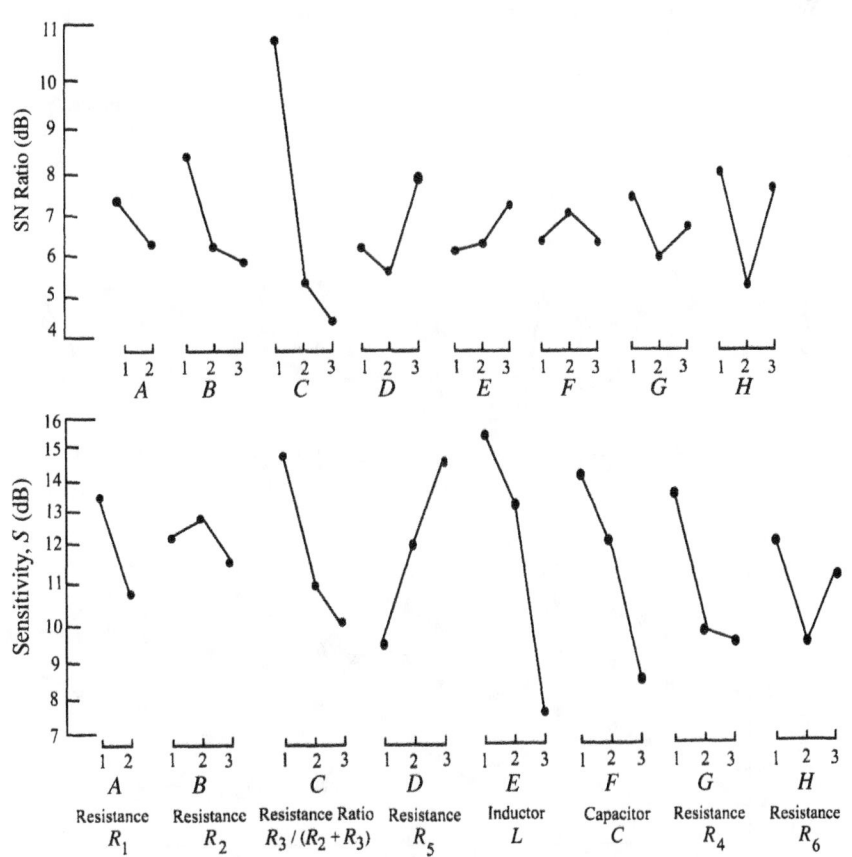

Figure 7
Response graphs of experiment 3

Table 9
Result of confirmatory experiment for experiment 3 (dB)

SN Ratio	Configuration		
	Optimal	Current	Gain
Estimation	16.55	4.19	12.36
Confirmation	17.07	7.36	9.71

Table 10
Levels of terminal resistance R_L

Level	1	2	3	4
Resistance, $R_L(\Omega)$	50	200	1000	∞

voltage and factor G/H and the terminal resistance R_L, as discussed before.

2. *Analysis with terminal resistance as a signal factor.* To clarify the mutual relationship, by combining all resistances of factors H and G and the terminal resistance into a single resistance, we

focused on this as a signal factor. When we set the combined resistance of factors H and G and the terminal resistance as a signal factor, we regarded this as two-signal factors, defining input voltage as M and terminal resistance as M^*. Due to the fact that both were proportional to the output voltage, we could combine them into a product of two signals MM^* and express its relationship with y as $y = \beta MM^*$. Although we calculated actual MM^* as

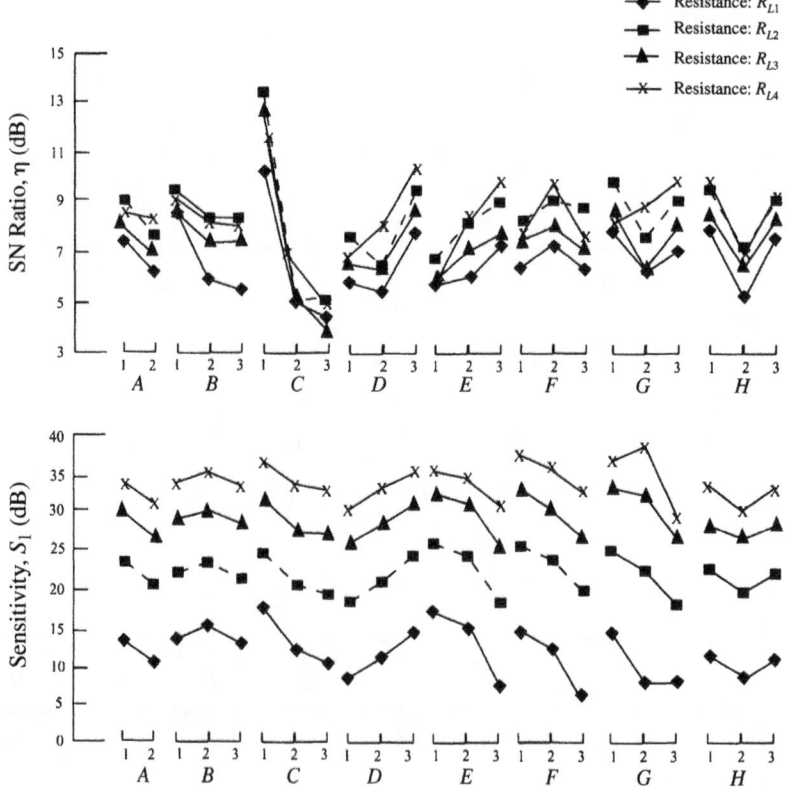

Figure 8
Response graphs for each terminal resistance of experiment 4

(signal factor of M) × (terminal resistance M^*), we needed to convert R_L into M^* because we did not have the equation $M^* = R_L$ in this case.

Figure 9 depicts a partial equivalent schematic of the terminal resistance in Figure 4. Setting each resistance at sections where V_{all} and V_{out} were added as R_{all} and R_{out}, respectively, we expressed M^* as shown in Figure 8.

$$R_{all} = \frac{1}{1/R_4 + 1/(R_6 + R_L)} \quad (14)$$

$$R_{out} = R_{all}\frac{R_L}{R_6 + R_L} \quad (15)$$

$$M^* = R_{out} \quad (16)$$

In sum, MM^* is a product of equation (16) and input voltage M.

After substituting a new signal for a product of two signals MM^*, we sorted them in increasing order (Table 11). Each subscript from M_1 to M_{12} in the table indicates a product of two level numbers, an input voltage level (three levels), and a terminal resistance level (four levels) (obviously, a maximum subscript is 3 × 4 = 12). Although the number of data was changed due to the increased number of

levels, the basic procedure for analyzing was the same as that of experiment 1 because the indicative and noise factors were identical. Thus, we do not describe it in detail.

Figure 10 shows the response graph for experiment 4. The optimal configuration is $A_1B_3C_2D_2E_2$ $F_1G_3H_1$. We can see that this is entirely different from those of the former experiments. The current configuration is $A_1B_2C_2D_2E_2F_2G_2H_2$, which is the same.

Looking at the optimal configuration, we find that it had a larger number of the same levels as the current configuration, but not as the previous experiments had. Also, in terms of the SN ratio, there were peaked and V-shaped characteristics in factors C, D, F, and G. Factor G had an especially steep V-shape.

Although factor G's characteristic was not unreasonable because its levels were not placed in regular order, we should pay attention to the fact that the V-shape of factor H, which was discussed for experiments 1 to 3, disappeared. However, since the number of peaks and V-shapes increased compared to the previous experiments, as a whole we believe that a change in terminal resistance affects the overall circuit to some extent. On the other hand, for sensitivity, there was no particular difference except

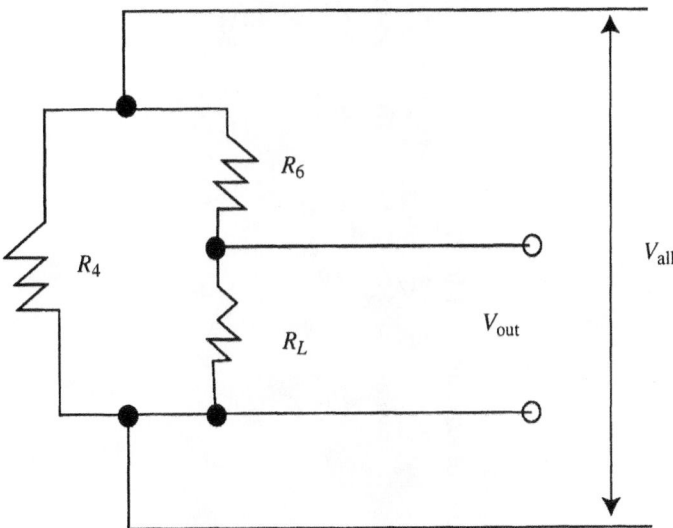

Figure 9
Schematic of combined resistances

Table 11
Results of experiment 4 (mV)

Noise	Frequency	M_1 226	M_2 452	M_3 702	M_4 904	M_5 1403	M_6 1599	M_7 2350	M_8 2806	M_9 3197	M_{10} 4700	M_{11} 6394	M_{12} 9400	Linear Equation
N_1	F_1	65	128	192	258	382	440	630	790	880	1270	1770	2540	L_{11}
	F_2	88	175	263	348	525	590	840	1040	1170	1680	2350	3310	L_{21}
	F_3	98	196	302	380	590	680	930	1160	1345	1820	2590	3500	L_{31}
N_2	F_1	66	130	190	262	380	438	620	785	870	1250	1770	2550	L_{12}
	F_2	87	170	258	340	510	570	820	1020	1150	1620	2300	3210	L_{22}
	F_3	97	190	288	372	580	670	910	1140	1310	1780	2550	3400	L_{32}

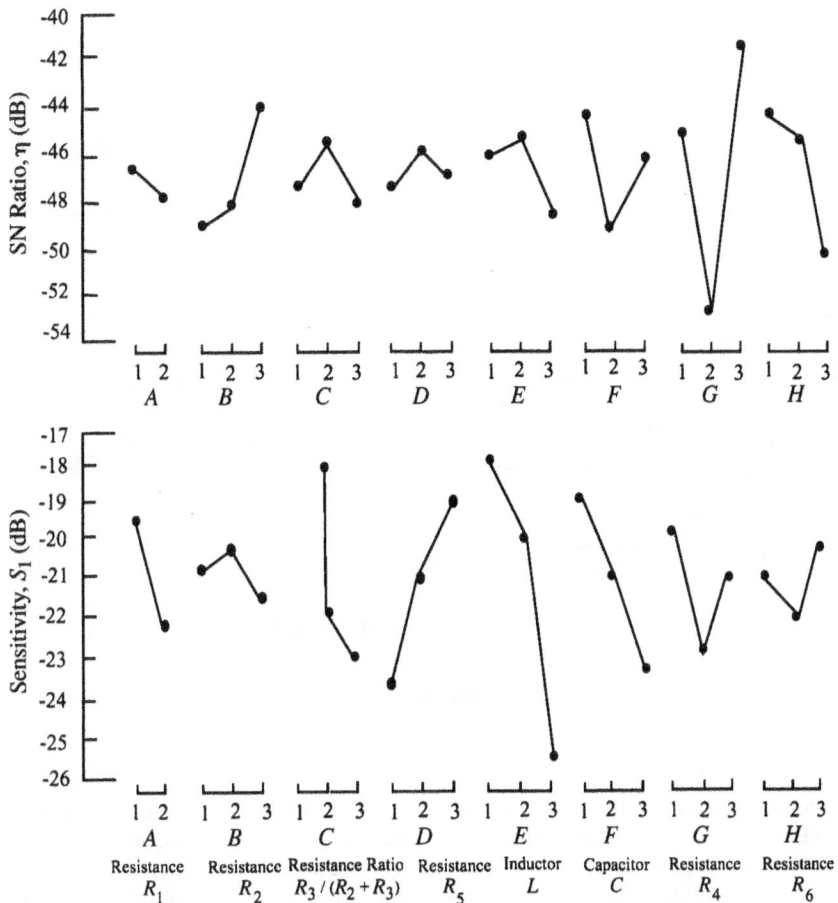

Figure 10
Response graph of experiment 4

7. Analysis of Experiment 4 and Confirmatory Experiment

By checking whether a circuit becomes nonlinear or saturated, as shown in Figure 6, we investigated the reason that many peaks and V-shapes appeared in the factor effect plot. We plotted graphs of $y =$ that some factors had a higher sensitivity value. Thus, we concluded that a change in terminal resistance does not have a significant impact on sensitivity.

βMM^* using the data regarding all 18 circuits, shown in Table 11. Then, as illustrated in Figure 11, we discovered irregularity in the graph of experiment 10 of the L_{18} orthogonal array. This graph was not only nonlinear but had a momentary voltage decrease. If the circuit becomes saturated, the slope may become less steep but the voltage cannot go down. This obviously demonstrates that a certain abnormality happened in measuring. Therefore, by altering the input voltage, input frequency, and terminal resistance in various steps, we found that without knowing the exact values, this irregularity occurs when a particular terminal resistance and input frequency are added to a circuit.

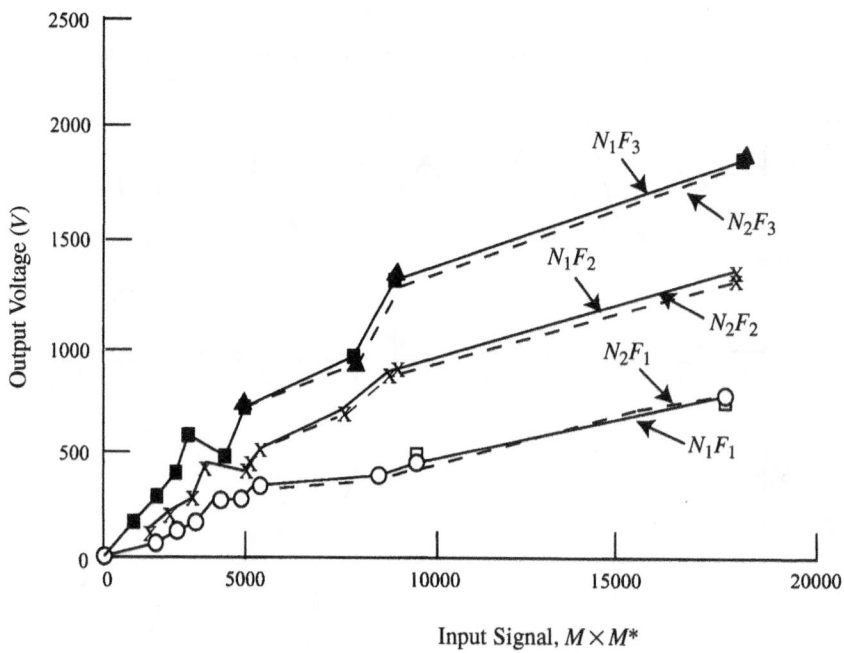

Figure 11
Irregularity unveiled by two-signal factor analysis

Then we conducted two confirmatory experiments to check reproducibility, the first an experiment that we performed using the optimal configuration of experiment 1 after removing the data causing this irregularity; the second was an experiment using the optimal configuration selected according to the response graphs of experiment 4, shown in Figure 8, for four different terminal resistances. (*Note:* For four different resistances, each optimal level was also different, but we chose a level that was selected most frequently as the optimal, to observe reproducibility.)

Confirmatory Experiment 1: Optimal Configuration When Irregular Part Was Removed

Since the combination of terminal resistance and a nominal factor level F_3 (500 Hz) causes irregularity, we removed F_3. We do not describe our analysis procedure because only two data combinations of $N_1 F_3$ and $N_2 F_3$ were eliminated from the measured data in Table 11.

As a result of analysis, we selected $A_1 B_1 C_1 D_3 E_3$ $F_2 G_3 H_1$ as the optimal configuration. This was the same as that of an experiment that does not make the setup range of four levels of a signal factor within the circuit capability in experiments 1 and 2. In short, this result shows that factor G was relatively easily affected, as discussed in the previous experiments. Using this configuration we conducted a confirmatory experiment and summarized the results in Table 12. We obtained the worst reproducibility in terms of gain as well as the optimal configuration.

Table 12

Result of confirmatory experiment 1 (dB)

SN Ratio	Configuration		
	Optimal	Current	Gain
Estimation	−34.02	−49.01	14.99
Confirmation	−39.93	−48.94	9.01

Table 13
Result of confirmatory experiment 2 (dB)

SN Ratio	Configuration		
	Optimal	Current	Gain
Estimation	−48.05	−50.99	2.94
Confirmation	−47.47	−52.45	4.98

Confirmatory Experiment 2: Optimal Configuration That Was Selected Most Often

This experiment used a combination of optimal levels that was selected most often according to the response graph in Figure 8. The optimal configuration was $A_1 B_1 C_1 D_3 E_3 F_2 G_1 H_1$. Using this configuration, we conducted a confirmatory experiment and summarized the results in Table 13. We have obtained good reproducibility.

Reference

Naoki Kawada, Yoshisige Kanemoto, Masao Yamamoto, and Hiroshi Yano, 1996. Design for the stabilization of an amplifier. *Quality Engineering,* Vol. 4, No. 3, pp. 69–81.

This case study is contributed by Naoki Kawada.

CASE 13

Parameter Design of Ceramic Oscillation Circuits

Abstract: In this study, parameters were designed using an orthogonal array to maintain stable oscillation even if the temperature and voltage vary.

1. Introduction

Oscillating frequency, accuracy, and oscillator type are different depending on usage. In general, as compared with a crystal oscillator, a ceramic oscillator tends to be used for a circuit that does not require less accuracy because of its lower price and accuracy (Figure 1). In this study we designed parameters to maintain stable oscillation even if the temperature and voltage vary.

2. Factorial Effects and Optimal Configuration

After selecting seven factors as control factors and two factors as error factors assigned to an outer array, we define their levels as shown in Table 1. Each factor was allocated to an L_{18} orthogonal array. After we implemented each of the 18 experiments assigned to the inner array under the nine conditions assigned to the outer array, we obtained oscillating frequency data as measured characteristics. In quality engineering it is recommended that a generic function be used. But at the time this experiment was conducted, the nominal-the-best characteristic was frequently used. In data development in quality engineering, the use of frequency was suggested for oscillating circuits. On the basic of these data, we computed SN ratios and sensitivities.

Variation of general mean:

$$S_m = \frac{\left(\sum_j y_{ij} \right)^2}{9} \qquad (f = 1) \qquad (1)$$

Error variation:

$$S_e = \sum_j y_{ij}^2 - S_m \qquad (f = 8) \qquad (2)$$

Error variance:

$$V_e = \frac{S_e}{8} \qquad (3)$$

$$\eta = 10 \log \frac{S_m - V_e}{V_e} \qquad (4)$$

$$S = 10 \log S_m \qquad (5)$$

Figure 2 depicts SN ratios and sensitivities of effective factors.

Although we calculated a constant by choosing a combination of levels to attain the maximum SN ratio, the eventual frequency fell below the target value. Thus, by adjusting it with the factors affecting sensitivity, we determined the optimal configuration.

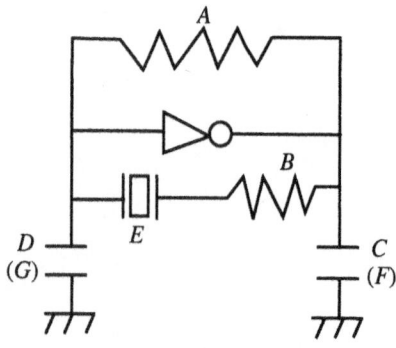

Figure 1
Ceramic oscillation circuit

Table 1
Factors and levels

Factor		Level		
		1	2	3
Control factors				
A:	resistance	200 kΩ	1 MΩ	4.7 MΩ
B:	resistance	100 Ω	1 kΩ	10 kΩ
C:	capacitance (C) (pF)	33	100	330
D:	capacitance (D) (pF)	33	100	330
E:	manufacturer of oscillator	M	N	K
F:	grade of capacitor (C)	Temperature compensating	Generic	—
G:	grade of capacitor (D)	Temperature compensating	Generic	—
Noise factors				
H:	environmental temperature (°C)	0	25	50
I:	voltage of power source (V)	−12	−15	−18

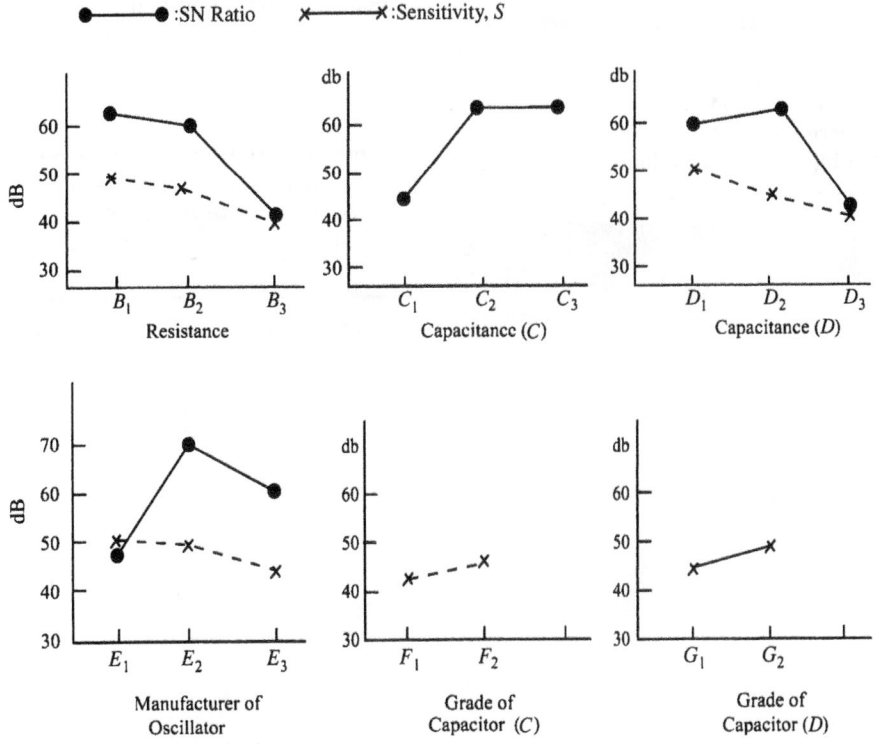

Figure 2
SN ratios and sensitivities of effective factors

Table 2
Confirmatory experiment result at optimal configuration (kHz)

Sample	0°C			25°C			50°C		
	−12 V	−15 V	−18 V	−12 V	−15 V	−18 V	−12 V	−15 V	−18 V
1	100.05	100.04	100.03	100.10	100.09	100.08	100.10	100.09	100.09
2	99.83	99.82	99.82	99.85	99.84	99.83	99.89	99.88	99.88

A: $A_1 = 220\ \Omega$ B: $B_1 = 100\ \Omega$

C: $C_3 = 330$ pF   D: $D_1 = 33$ pF

E: $E_2 = N$ F: $F_2 = $ generic

G: $G_2 = $ generic

3. Confirmatory Experiment

Based on the optimal configuration, we implemented a confirmatory experiment. Table 2 shows the result. We note that even if the power source's voltage and environmental temperature vary, the oscillation can remain stable. We obtained an optimal circuit successfully. In addition, we verified that the circuit could achieve good oscillation in the confirmatory experiment.

The following are η and S values for the confirmatory experiment:

Sample	η (dB)	S (dB)
1	80.80	49.55
2	80.04	49.53

We analyzed a measured characteristic as nominal-the-best. However, as a reference, by regarding it as a deviation from the target value of 100 kHz and using equations (6) and (7), we reanalyzed it according to the smaller-the-better procedure.

$$V_e = \frac{\sum_j (x_{ij} - 100)^2}{9} \qquad (6)$$

$$\eta = -10\log V_e \qquad (7)$$

In both cases we found that $B_1 D_1 E_2$ is optimal.

Reference

Kiyohiro Inoue and Genichi Taguchi, 1984. *Reliability Design Case Studies for New Product Development*, pp. 169–173. Tokyo: Japanese Standards Association.

This case study is contributed by Kiyohiro Inoue.

CASE 14

Evaluation of Electric Waveforms by Momentary Values

Abstract: In this report a new method of evaluating the performance of four-terminal circuits is proposed. In circuit design, sinusoidal waves are frequently used as the input signal to generate responses. From the viewpoint of research efficiency, responses are not good outputs to use. Here, pulses that include wave disorder are used as input instead of sinusoidal waves. One SN ratio was used to evaluate the robustness of four-terminal circuits. The results showed good reproducibility of the experimental conclusions.

1. Introduction

The most fundamental elements in an electric circuit are a capacitor, coil, and resistor, expressed as C, L, and R. These circuit elements represent the relationship between the current, i, and the terminal voltage, e. For instance, the resistance, R, whose unit is the ohm, is the coefficient of i to e. The coil's terminal voltage is proportional to the rate of current change, and its coefficient is denoted as L. Similarly, the coefficient of the capacitor, in which the current is proportional to the rate of voltage change, is represented as C.

In general, although these coefficients are designed on the assumption that they are constant in any case, they are not invariable in actuality. For example, they vary in accordance with environmental temperatures, frequencies, and magnitudes of input. Therefore, we design these coefficients such that an overall system can be robust. However, in most cases this robust design cannot be repeated.

2. Evaluation of the Functions of an Electronic Circuit

Although a coil and capacitor can be expressed in the form of a time-variable function because both

of them accumulate and emit energy, we normally handle them as time-constant functions, due to difficulty in measuring. In other words, by pruning their momentary values, we see only the average values.

This approach can improve measurement accuracy by canceling out noises through the averaging process; however, it hides their change rates because the smoothness of energy conversion according to the change rate is regarded as the essential criterion of the functionality of an electronic circuit and a key factor to secure repeatability. This holds true for the elements, as illustrated in Figure 1. Thus, as a new approach, we designed the parameters of a filter by using pulses as input signals. The reason that we chose pulses is that we concluded that the waveform of a pulse would be most appropriate to use to observe second-by-second changes of output energy to input energy. If this purpose could be satisfied, the observation of only a single cycle of a sine curve would be sufficient.

As a signal to be used for circuit evaluation, we usually consider a unit function. Response to the unit function, called *indicial response*, has been used to assess the pattern of the waveform qualitatively and to evaluate the quality characteristic of ramp-up time in case the high-frequency band used for a pulse amplifier or television display amplifier is

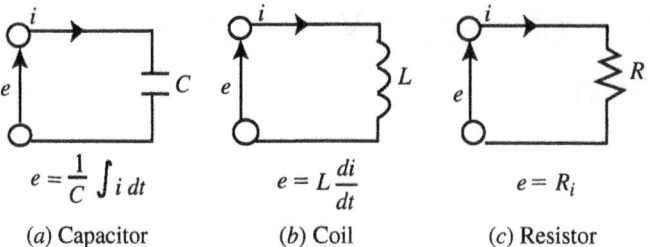

$$e = \frac{1}{C} \int i \, dt \qquad\qquad e = L \frac{di}{dt} \qquad\qquad e = R_i$$

(a) Capacitor   (b) Coil   (c) Resistor

Figure 1
Relationship between the voltage and current of typical elements

required. Yet these adjustments are made primarily to align ramp-up times to target values, not necessarily to stabilize functions.

3. Ideal Functions and Measurement Characteristics

By utilizing a pulse, which can be regarded as a repeated unit function, and computing a single SN ratio, we evaluated the robustness of a four-terminal circuit. The functionality of the circuit is (Figure 2) to convert input signals into output signals in accordance with functions tailored for each circuit. Although different circuits naturally have different output waveforms, a stable circuit should not have any disturbed patterns of waveform, even if condi-

tions such as input voltage, part characteristics, or environmental temperature alter it to some extent.

We selected a pulse voltage as an input signal and output voltage as the measurement characteristic. Next, we measured the input and output in a synchronized manner, and each measurement was taken for a certain time at a constant interval. As for the unit function, at $t < 0$, $e = 0$, and at $t \geq 0$, $e = 1$. Although basically we needed no signal data, actually we observed input data.

As illustrated in Figure 3, the generic function was to obtain output voltages proportional to input voltages. However, the important point is that both voltages were proportional to each other at every point in time, and this indicated the smoothness of energy conversion to the change rate. Although time was essentially regarded as a signal, we considered it as an indicative factor because we did not have any common methods to express all functions for various types of circuits in a proportional equation.

Figure 2
Examples of waveforms in four terminal circuits

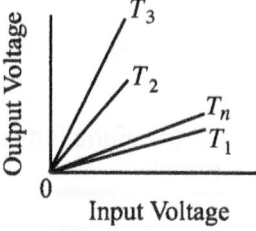

Figure 3
Input and output of an electronic circuit

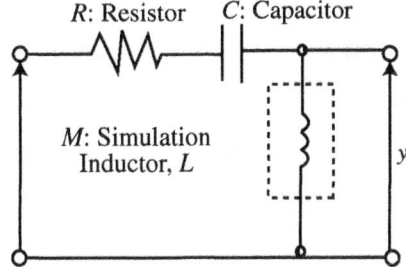

Figure 4
Filter circuit

4. Experimental Procedure

Next we considered an LC type of high-pass filter circuit (Figure 4). L is a simulation inductor consisting of two operational amplifiers, four resistors, and one capacitor. We measured the outputs by generating pulses with a function generator, then synchronizing the waveforms of input and output of a filter with a storage oscilloscope (Figure 5).

As control factors we assigned six elements chosen from resistances and capacitances in an L_{18} orthogonal array, since levels of the elements can be selected independently. The types and levels of control factors are given in Table 1. We set 250 and 500 mV as levels of the signal factor. For the time, t, as an indicative factor, we measured 15 points at a time interval of 8 μs between 8 and 120 μs. Finally, as a

Table 1
Control factors and levels

Control Factor		Level 1	Level 2	Level 3
A:	capacitor	0.1	0.22	—
B:	coil	−10	±0	+10
C:	coil	470	1000	2200
D:	coil	4.7	10	20
E:	coil	D(0.5)	D(1.0)	D(2.0)
F:	coil	D(0.5)	D(1.0)	D(2.0)
G:	resistor	430	820	1600
H:	A's type	H_1	H_2	H_3

noise factor, we picked up low and high environmental temperatures, N_1 and N_2.

5. SN Ratio and Sensitivity

Our analysis of the SN ratio and sensitivity was as follows. The data used for this analysis were based on experiment 1. Except that the levels of signal factors differ according to the levels of indicative factor and the calculation is somewhat cumbersome due to the number of levels, this analysis is the same

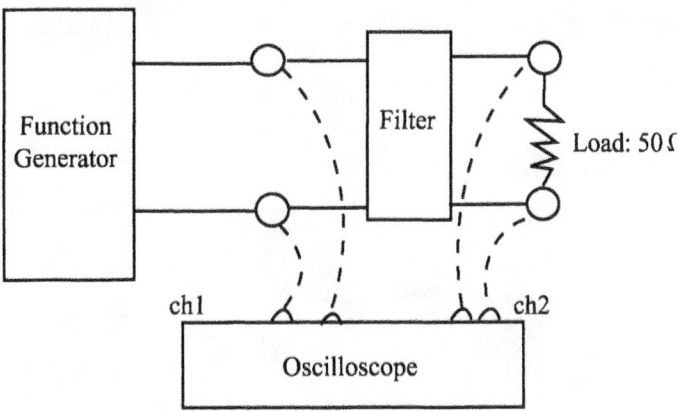

Figure 5
Measurement circuit for a filter circuit

as that of a proportional equation. Table 2 shows the data from experiment 1. The effective divider, r, the magnitude of input, is expressed as:

$$T_1 N_1 r_{11} = M_{111}^2 + M_{211}^2 \qquad (1)$$

$$T_1 N_2 r_{12} = M_{112}^2 + M_{212}^2 \qquad (2)$$

Total sum of effective dividers at a certain time:

$$r = r_{21} + r_{22} + \cdots + r_{215} \qquad (3)$$

Linear equation L is expressed as follows:

$$T_1 N_1 L_{11} = M_{111} y_{111} + M_{211} y_{211} \qquad (4)$$

$$T_1 N_2 L_{12} = M_{112} y_{112} + M_{212} y_{212} \qquad (5)$$

Total variation:

$$S_T = 284.2^2 + \cdots + 328.9^2 = 9,395,685.60$$

$$(f = 60) \qquad (6)$$

Variation of proportional term:

$$S_\beta = \frac{(L_{11} + \cdots + L_{152})^2}{r_{11} + \cdots + r_{152}}$$

$$= \frac{(4,474,591.41 + \cdots + 4,814,868.42)^2}{9,494,979.22}$$

$$= 9,088,388.92 \qquad (f = 1) \qquad (7)$$

Variation of proportional terms with respect to time difference:

Table 2
Results of experiment 1 (mV)

Time	Noise	Signal		Effective Divider	Linear Equation
		M_1	M_2		
T_1	N_1	M_{111} (263.2)	M_{211} (500.0)	r_{11} (319,274)	L_{11} (364,251)
		y_{111} (284.2)	y_{211} (578.9)		
	N_2	M_{112} (257.9)	M_{212} (513.2)	r_{12} (329,886)	L_{12} (386,643)
		y_{112} (294.7)	y_{212} (605.3)		
T_2	N_1	M_{121} (257.9)	M_{221} (500.0)	r_{21} (316,512)	L_{21} (356,195)
		y_{121} (284.2)	y_{221} (565.8)		
	N_2	M_{122} (257.9)	M_{222} (513.2)	r_{22} (329,887)	L_{22} (389,385)
		y_{122} (331.6)	y_{222} (592.1)		
⋮	⋮	⋮	⋮	⋮	⋮
T_3	N_1	M_{1151} (252.6)	M_{2151} (486.8)	r_{151} (300,781)	L_{151} (217,272)
		y_{1151} (226.3)	y_{2151} (328.9)		
	N_2	M_{1152} (252.6)	M_{2152} (500.0)	r_{152} (313,807)	L_{152} (224,266)
		y_{1152} (236.8)	y_{2152} (328.9)		

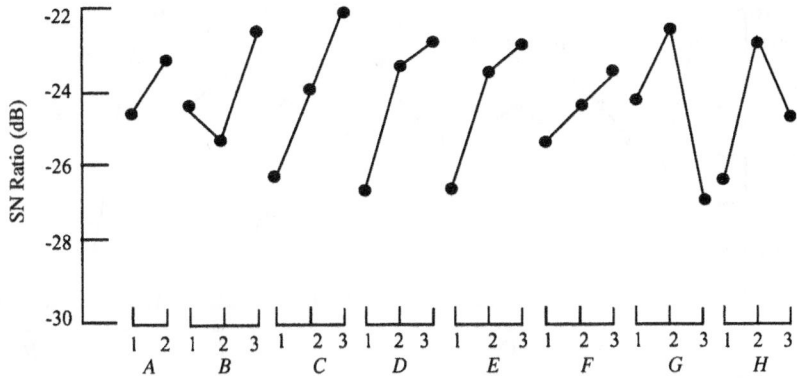

Figure 6
Response graph of SN ratio

$$S_{T\beta} = \frac{(L_{11} + L_{12})^2}{r_{11} + r_{12}} + \frac{(L_{12} + L_{22})^2}{r_{12} + r_{22}}$$
$$+ \cdots + \frac{(L_{151} + L_{152})^2}{r_{151} + r_{152}} - S_\beta$$

$$= 9{,}318{,}717.24 - 9{,}088{,}388.92$$

$$= 230{,}328.32 \qquad (f = 14) \qquad (8)$$

Variation due to noise factor:

$$S_{N\beta} = \frac{(L_{11} + L_{21} + \cdots + L_{151})^2}{r_{11} + r_{21} + \cdots + r_{151}}$$
$$+ \frac{(L_{12} + L_{22} + \cdots + L_{152})^2}{r_{12} + r_{22} + \cdots + r_{152}} - S_\beta$$

$$= \frac{4{,}747{,}591.41^2}{4{,}651{,}433.52} + \frac{4{,}814{,}868.42^2}{4{,}843{,}545.71}$$
$$- 9{,}088{,}388.92$$

$$= 2444.63 \qquad (f = 1) \qquad (9)$$

Error variation:

Table 3
Results of SN Ratio Confirmatory Experiment

Configuration	SN Ratio	
	Estimation	Confirmation
Optimal	−12.47	−17.69
Current	−19.69	−24.44
Gain	7.22	6.75

$$S_e = S_T - S_\beta - S_{T\beta} - S_{N\beta}$$

$$= 9{,}395{,}685.60 - 9{,}088{,}388.92$$
$$- 230{,}328.32 - 244.63$$

$$= 74{,}523.73 \qquad (f = 44) \qquad (10)$$

Error variance:

$$V_e = \frac{S_e}{44} = 1693.72 \qquad (11)$$

Total error variance:

$$V_N = \frac{S_{N\beta} + S_e}{45} = \frac{2444.63 + 523.73}{45}$$
$$= 1710.40 \qquad (12)$$

Whereas functionality was evaluated based on SN ratios of total data, sensitivity adjustment to target values was made based on time because time was regarded as an indicative factor in our case.

Function stability (SN ratio):

$$\eta = 10 \log \frac{[1/(r_{11} + \cdots + r_{152})](S_\beta - V_e)}{V_N}$$

$$= -32.52 \ (\text{dB}) \qquad (13)$$

$$S_i = 10 \log \beta_i^2 \qquad (i = 1, 2, 3, \ldots, 15) \qquad (14)$$

For $T = 1$,

$$\beta_1^2 = \frac{1}{r_{11} + r_{12}} \left[\frac{(L_{11} + L_{12})^2}{r_{11} + r_{12}} - V_e \right] \qquad (15)$$

For $T = 15$,

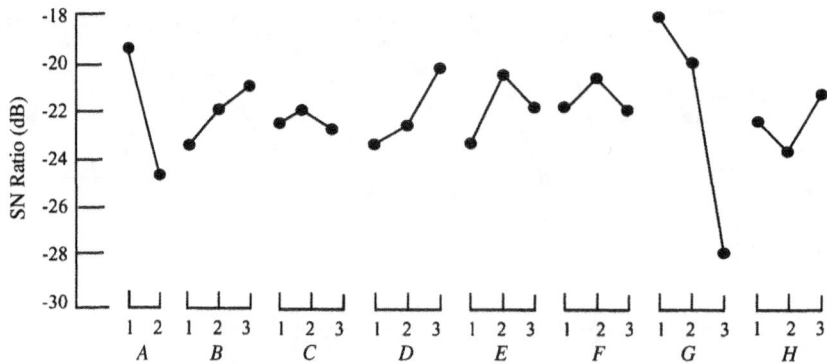

Figure 7
Response graph of sensitivity

$$\beta_{15}^2 = \frac{1}{r_{151} + r_{152}} \left[\frac{(L_{151} + L_{152})^2}{r_{151} + r_{152}} - V_e \right] \quad (16)$$

Although sensitivity adjustment is needed when each sensitivity at a certain time is different, we omitted it in this case.

6. Response Graphs and Confirmatory Experiment

Based on the SN ratios and sensitivities calculated so far, we drew the response graphs shown in Figure 6. Although, in fact, all factor effect diagrams for T_1 to T_{15} were required because sensitivity needs to be adjusted to target frequency characteristics by the least squares method, we omitted such fine tuning for now.

Looking at the factor effect diagram of the SN ratio, we found that factor H had a peak, but it was a capacitor type, which is a noncontinuous factor. Factors B and G were much less peaked. Now using the factor effect diagrams, we estimated optimal and current conditions. Taking the chart of the SN ratio into account, we knew that the optimal condition is regarded as $A_2B_3C_3D_3E_3F_3G_2H_2$, and the current condition is $A_2B_2C_2D_2E_2F_2G_2H_2$. On the basis of these two levels, we estimated and confirmed each SN ratio. Table 3 shows that the optimal condition is approximately 7 dB better than the current one. Additionally, the benefits of the estimation and confirmation coincide well. As a reference, we added the response graphs of sensitivity (Figure 7).

Reference

Yoshishige Kanemoto, Naoki Kawada, and Hiroshi Yano, 1998. New evaluation method for electronic circuit. *Quality Engineering*, Vol. 6, No. 3, pp. 57–61.

This case study is contributed by Yoshishige Kanemoto.

CASE 15

Robust Design for Frequency-Modulation Circuits

Abstract: In our research we verified highly repeatable experimentation (measurement characteristics) in regard to resonant circuits such as oscillator and modulation circuits, typical in the electrical engineering field.

1. Introduction

The majority of tasks requisite for electrical circuit design involve adjustment of quality characteristics to target values. More specifically, by implementing product confirmatory experiments called environmental or reliability tests, we have made a strong effort to match the quality characteristics to final specifications through the modification of part characteristics if the quality characteristics go beyond the standards. However, since this procedure is regarded only as "adjustment" and robustness cannot be built in, complaints in the marketplace regarding products have not been reduced significantly. This is the reason that parameter design in quality engineering is needed desperately. On the other hand, most conventional examples to which parameter design has been applied have not had sufficient repeatability because their designs are implemented based on measurement of powers such as output voltage. We assume that one of the most important characteristics related to experimentation is reproducibility. Even if a certain function is stabilized in a laboratory, if it cannot be reproduced under mass-production conditions, the experiments will be wasted.

2. Generic Function

As shown in Figure 1, the experimental circuit is designed based on a combination of a Corvit oscillator and a variable capacitor (voltage-variable capacitor). It consists of an inverter IC, ceramic

oscillator (C_3), two resistors (R_1 and R_2), two capacitors (C_1 and C_2), and two variable capacitors (C_{v1} and C_{v2}), whose capacitance varies according to the voltage.

As illustrated in Figure 2, in actual modulation, when an alternating current such as a human voice is given on a direct bias voltage, the output voltage alternates between the lower and upper voltages around the bias voltage. This alternating signal makes the terminal capacitance, and consequently the frequency, change.

Therefore, the proportional relationship between voltages imposed on the variable capacitors and the frequency is the function of the modulation circuit. We selected the voltage, V_p, of the variable capacitor as a signal factor, set the dc bias voltage to 2 V, and altered this voltage within the range of ± 1.0 V at an interval of 0.5 V equivalent to one level. In sum, as shown in Table 1, we laid out five levels.

The generic function of a frequency-modulation circuit is to maintain a relationship between a signal voltage, V_p, and a frequency deviation (difference between an initial and a deviated frequency; Figure 3). Obviously, the frequency deviation should not fluctuate in accordance with usage and environmental conditions. In our study, noise factors such as environmental temperature or time drift can be regarded as contained in the fluctuation of the power supply's voltage. After all, we chose only the power source's voltage as a noise factor, which ranges between ± 0.5 V, N_1 and N_2, around the power source's initial voltage.

Based on the design of experiments mentioned above, we collected measured data. As an example,

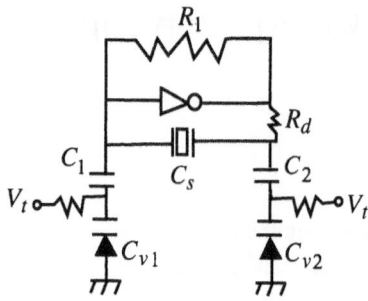

Figure 1
Experimental circuit

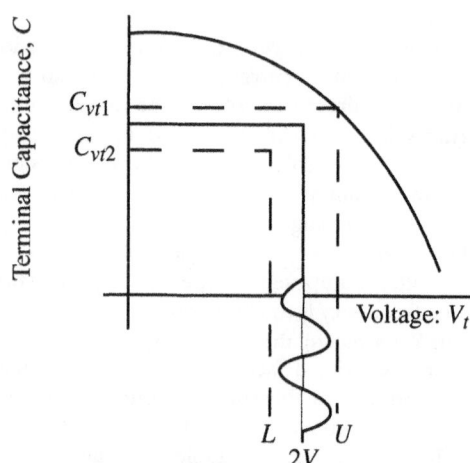

Figure 2
Frequency modulation voltage V_t and terminal capacitance

Table 1

Signal factors and levels (V)

	Signal				
	M_1	M_2	M_3	M_4	M_5
Absolute voltage	1.0	1.5	2.0	2.5	3.0
Signal voltage	−1.0	−0.5	±0	+0.5	+1.0

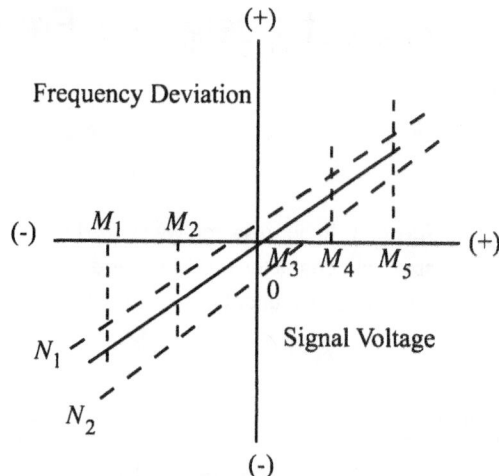

Figure 3
Generic function

raw and converted data of experiment 1 are shown in Table 2. Since under the initial state (initial power source's and signal voltages were 0 V) in experiment 1, the frequency (M_N) was 453,286; the converted frequency deviation was equal to the corresponding raw frequency minus the initial frequency, M_N. Next, using these data, we decomposed the total variation.

Total variation:

$$S_T = y_{11}^2 + y_{21}^2 + y_{12}^2 + y_{22}^2 + \cdots + y_{15}^2 + y_{25}^2$$

$$= (-558)^2 + (-542)^2 + \cdots + 661^2$$

$$= 1,753,035 \quad (1)$$

Linear equations:

$$L_1 = M_1 y_{11} + M_2 y_{12} + \cdots + M_4 y_{14} + M_5 y_{15}$$

$$= (-1.0)(-588) + \cdots + (1.0)(581) = 1453.5$$

$$L_2 = M_1 y_{21} + M_2 y_{22} + \cdots + M_4 y_{24} + M_5 y_{25}$$

$$= (-1.0)(-542) + \cdots + (1.0)(661) = 1496.0 \quad (2)$$

Effective divider:

$$r = M_1^2 + M_2^2 + M_3^2 + M_4^2 + M_5^2$$

$$= (1.0)^2 + (-0.5)^2 + 0.5^2 + 1.0^2 = 2.5 \quad (3)$$

Table 2
Results of experiment 1

	Signal				
Noise	M_1 (−1.0)	M_2 (−0.5)	M_3 (0.0)	M_4 (+0.5)	M_5 (+1.0)
Raw data					
N_1	452,698	452,971	453,246	453,540	453,867
N_2	452,744	453,022	453,304	453,608	453,947
Converted data					
N_1	−588	−315	−40	254	581
N_2	−542	−264	18	322	661

Variation of proportional term:

$$S_\beta = \frac{(L_1 + L_2)^2}{2\gamma}$$

$$= \frac{(1453.5 + 1496.0)^2}{(2)(2.5)(2.5)} \; 1{,}739{,}910.05 \quad (4)$$

Interaction between proportional term and noise:

$$S_{N\beta} = \frac{L_1^2 + L_2^2}{\gamma - S_\beta} = \frac{(1453.5 + 1496.0)^2}{2.5 - 1{,}739{,}910.05}$$

$$= 361.25 \quad (5)$$

Error variation:

$$S_e = S_T - (S_\beta + S_{N\beta})$$

$$= 1{,}753{,}035 - (1{,}739{,}910.05 + 361.25)$$

$$= 12{,}763.70 \quad (6)$$

Error variance:

$$\frac{V_e = S_e}{8} = 1595.46 \quad (7)$$

Total error variance:

Table 3
Control factors and levels

	Control Factor	Level		
		1	2	3
A:	ceramic oscillation element type	CSB 455	B456 F15	—
B:	resistance	680	820	1000
C:	resistance	3.3	4.7	5.6
D:	capacitance	560	680	820
E:	capacitance	Two levels below C_1	One level below C_1	Same as C_1
F:	capacitor type	Ceramic	Chip	Film
G:	capacitor type	Ceramic	Chip	Film
H:	power source voltage	3	5	7

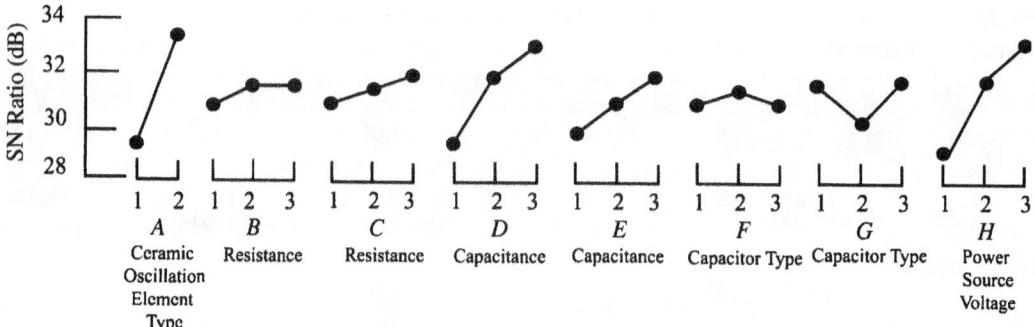

Figure 4
Response graph of SN ratio

$$V_N = \frac{S_e + S_{N\beta}}{9} = 1458.33 \qquad (8)$$

SN ratio:

$$\eta = 10 \log \frac{(1/2\gamma)(S_\beta - V_e)}{V_N} = 23.77 \text{ dB} \qquad (9)$$

Sensitivity:

$$S = 10 \log \frac{S_\beta - V_e}{2\gamma} = 55.41 \text{ dB} \qquad (10)$$

3. Results of Experiment

As control factors we selected seven constant values and types of parts, shown in Figure 1. Additionally, the voltage of the power voltage supplied to the inverter, IC, was chosen as another control factor. All the control factors were assigned to an L_{18} orthogonal array. Table 3 lists the control factors and levels.

Using the aforementioned calculation formulas, we analyzed all data from experiments 1 to 18 and created the response graph of SN ratio and sensitivity as Figure 4. The fact that the trend of factor G is V-shaped does not indicate the interaction between control factors because G, which represents a capacitor type, is not continuous.

4. Confirmatory Experiment and Analysis

Based on the factor effect plot, we defined optimal and current configurations and conducted confirmatory experiments. Although we selected the highest level for each control factor, we chose a chip capacitor as most practical for surfacing because there was no significant difference among all types of capacitors. In sum, the configurations are:

Optimal: $A_2 B_3 C_3 D_3 E_3 F_2 G_2 H_3$

Current: $A_1 B_2 C_2 D_2 E_2 F_2 G_2 H_2$

Table 4
Confirmatory experiment

Configuration	SN Ratio		Sensitivity	
	Estimation	Confirmation	Estimation	Confirmation
Optimal	37.53	37.27	60.25	60.32
Current	29.78	29.77	57.32	57.24
Gain	7.74	7.5	2.94	3.08

Table 4 shows the estimated values and results of confirmatory experiments. Both SN ratios and sensitivities have quite good repeatability. This fact proves that to measure frequencies as characteristics or response variables is effective enough to obtain good repeatability. On the other hand, the optimal configuration is 7.50 dB better than the current one and halves the functional variability, due to the environmental temperature and fluctuation of power source's voltage.

The fact that we obtained highly repeatable results can be regarded as epoch-making in that the results can certainly be reproduced in the marketplace in the mass-production phase. For instance, we have designed the stability of a filter by setting the output voltage as a characteristic value (normally the logarithm of the ratio of input to output) at several frequencies around the cutoff frequency.

However, because there is a radical change near the boundary between the pass and attenuation bands, the output voltage changes from 0 to 1 under a certain use and consequently does not function well as an addable characteristic. Our procedure can be applied to a wide range of devices, including digital circuits.

Reference

Yoshishige Kanemoto, 1998. Robust design for frequency modulation circuit. *Quality Engineering*, Vol. 6, No. 5, pp. 33–37.

This case study is contributed by Yoshishige Kanemoto.

CASE 16

Optimization of Blow-off Charge Measurement Systems

Abstract: Our objective was to optimize the measurement functionality of a toner charge measurement device to achieve stable measurement for the charged amount of toner in the developing solvent. Although we calculated for one type of developing solvent, if we apply this method to other types, we will be able to accomplish many more economic benefits.

1. Introduction

As a measurement device for the amount of toner charge, we selected blow-off charge measurement apparatus used widely because of its quick measurability and portability. Figure 1 depicts the magnified toner separator of the measurement system.

Our measurement procedure was as follows. Developing solvent consisting of toner and carrier was poured in a measurement cell that included a mesh at the bottom through which only the toner could pass. Then compressed air was blown into the cell, pushing the toner out while leaving the carrier, which has an equivalent but opposite-polarity charge to that of the separated toner. Finally, the amount of the charge is detected with an electrometer, and the polarity is read as opposite.

2. Generic Function

The blow-off charge measurement separates the toner by stirring the developing solvent with blown air. However, this process might alter a state of charge in the developing solvent to be measured and disturb actual measured values.

We considered it ideal that blown-air energy be used to separate the toner with as little disturbance of the developing solvent as possible. In this ideal state, we hoped to find a true charge because there is no unnecessary, subsequent variation of charge

caused by measurement. In this case it was essential to evaluate a dynamic process in which toner is gradually being separated as the energy introduced increases. Since the amount of toner contained in the measured developing solvent is limited, even if we prolonged the blown-air time, the separation of toner would not increase beyond a certain level and would arrive at a state in which a new charge was not detected. In other words, because this toner separation process is analogous to a chemical reaction, if we selected a value corresponding to the reaction rate as output, we could utilize a chemical reaction equation in our research.

Therefore, we chose airflow time, M, as the input and mass separation rate, Y/M^*, as the output (Figure 2):

$$
\begin{aligned}
\text{mass separation rate, } Y/M^* \\
= \frac{\text{separated mass of toner, } Y}{\text{total mass of toner, } M^*}
\end{aligned}
\quad (1)
$$

Now the total mass of toner, M^* could be measured by the mass density of toner in the developing solvent measured. On the other hand, the separated mass of toner, Y, is the total mass of toner separated from the beginning to the end of the blown air (Figure 3):

Additionally, we assumed that the mass separation rate, Y/M^*, is affected by the total amount of developing solvent measured. From our experience we assumed a relationship between blown-air time, M, and mass separation rate, Y/M^*.

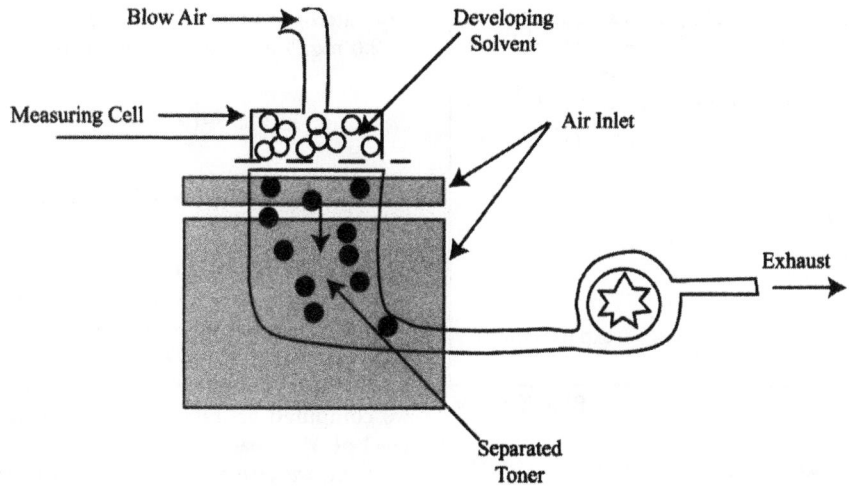

Figure 1
Toner separator

When there is a fixed mass density of toner, since the total mass of toner, M^*, is proportional to the total amount of developing solvent, the relationship of the mass separation rate, Y/M^*, to the blown-air time, M, and total mass of toner, M^*, was presumed to be

$$Y/M^* = 1 - \exp(-\beta MM^*) \qquad (2)$$

By taking a natural logarithm of both sides of equation (2), we obtained the following proportional equation through the origin:

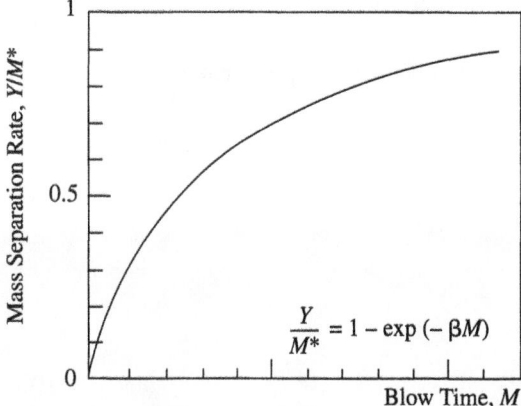

Figure 2
Blown-air time and mass separation rate

$$\ln[1 - (Y/M^*)] = -\beta MM^* \qquad (3)$$

Now, substituting $y = -\ln[1 - (y/M^*)]$, we have

$$y = -\ln[1 - (Y/M^*)]$$

$$y = \beta MM^* \qquad (4)$$

We selected blown-air time M, and total mass of toner, M^*, as signal factors. In this study we allocated the amount of developing solvent used to control factor B, which is discussed later. Because the total mass of toner M^* varies with the amount of developing solvent used and the mass density of toner, W, signal factor M^* and control factor B are not independent:

total mass of toner, M^* = (amount of developing
solvent used) × (mass
density of toner, W)

$$(5)$$

Then we made signal factor M^* and control factor B independent by setting nominal 90 and 110% values of factor BW as levels of total mass of toner, M^*. Signal and level factors are shown in Table 1.

On the other hand, to optimize the measurement apparatus by focusing on its separation function, we dealt with the toner charge amount as a noise factor, N. With its true values of levels unknown, we prepared two levels close to the mini-

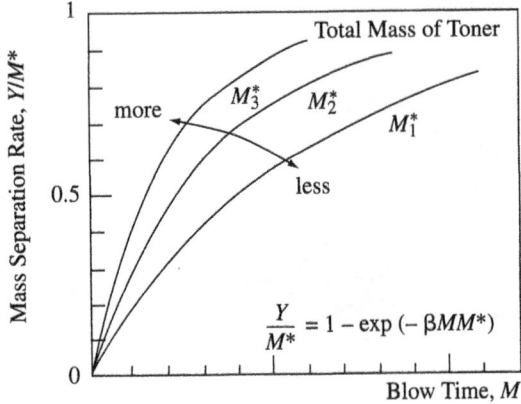

Figure 3
Total mass of toner and mass separation rate

mum and maximum limits of the amount of charge favorable for development.

3. SN Ratio and Sensitivity

As an example, we chose experiment 9. The converted data are shown in Table 2. First, we calculated signal factor M^*. Setting mass density of toner $W = 6$ wt % and using 898.0 mg as the amount of developing solvent, we computed the total mass of toner, M^*:

$$M_1^* = \text{amount of developing solvent used} \times W$$

$$= (898.0)(0.06) = 53.88 \qquad (6)$$

As a next step, we converted the mass of toner seperated, Y_{11}, into the output, y_{11}. The mass of toner

separated was obtained from the experiment as $Y_{11} = 2.0$ mg. We calculated the output, y_{11}, as follows:

$$y_{11} = -\ln\left(1 - \frac{Y_{11}}{M_1^*}\right)$$

$$= \ln\left(1 - \frac{2.0}{53.88}\right) = 0.0378 \qquad (7)$$

Since levels of signal factor M^* in Table 2 are representative values, each level of the signal factor depends on each value of output, y. The reason is that we cannot make each amount of developing solvent used completely identical. Needless to say, we computed SN ratios and sensitivities using each level of M^* measured.

Next, we give the calculation procedure for the SN ratio, η, and sensitivity, S.

Total variation:

$$S_T = y_{11}^2 + y_{12}^2 + \cdots + y_{33}^2,$$

$$= 0.3117 \qquad (f_T = 18) \qquad (8)$$

Effective divider:

$$r_i = (M_1M_i^*)^2 + (M_2M_i^*)^2 + (M_3M_i^*)^2 \qquad (9)$$

Linear equation:

$$L_i = (M_1M_i^*)y_{1i} + (M_2M_i^*)y_{2i} + (M_3M_i^*)y_{3i} \qquad (10)$$

Variation of proportional term:

$$S_\beta = \frac{(L_1 + L_2 + \cdots + L_3')^2}{r_1 + r_2 + \cdots + r_3'} = 0.2703 \qquad (f_\beta = 1) \qquad (11)$$

Variation between proportional terms:

Table 1
Signal and noise factors

		Level		
Factor		**1**	**2**	**3**
Signal factors				
M: blown-air time		1	2	3
M^*: total mass of toner		$BW - 10\%$	BW	$BW + 10\%$
Noise factor				
N: toner charge amount		High	Low	

Table 2
Logarithmized output data of experiment 9

Signal Factor, M*	Noise Factor, N	Signal Factor, M			Effective Divider, r	Linear Equation, L
		M_1	M_2	M_3		
54	N_1	0.00378	0.1358	0.1343	40,713	38.38
	N_2	0.1027	0.1570	0.2086	28,641	47.14
60	N_1	0.0479	0.1045	0.1074	50,528	34.81
	N_2	0.1236	0.1340	0.2000	35,092	49.63
66	N_1	0.0370	0.1001	0.1197	60,582	39.24
	N_2	0.0897	0.1379	0.2106	42,151	54.72

$$S_{N\beta} = \frac{(L_1 + L_2 + L_3)^2}{r_1 + r_2 + r_3} + \frac{(L_1' + L_2' + L_3')^2}{r_1' + r_2' + r_3'} - S_\beta$$

$$= 0.0297 \quad (f_{N\beta} = 1) \qquad (12)$$

Error variation:

$$S_e = S_T - S_\beta - S_{N\beta} = 0.0117 \quad (f_e = 16) \qquad (13)$$

Error variance:

$$V_e = \frac{S_e}{f_e} = \frac{0.0117}{16} = 0.00073 \qquad (14)$$

Combined error variance:

$$V_N = \frac{0.0297 + 0.0117}{1 + 16} = 0.00244 \qquad (15)$$

SN ratio:

$$\eta = 10 \log \frac{[1/(r_1 + r_2 + \cdots + r_3')](S_\beta - V_e)}{V_N}$$

$$= -33.7 \text{ dB} \qquad (16)$$

Sensitivity:

$$S = 10 \log \frac{1}{r_1 + r_2 + \cdots + r_3'} (S_\beta - V_e) = -59.8 \text{ dB} \qquad (17)$$

4. Optimal Configuration and Confirmatory Experiment

As controls we selected two factors evenly from each of the factors developing solvent, pressure, airflow,

Table 3
Control factors and levels

	Control Factor	Level		
		1	2	3
A:	Deviation of developing solvent	Yes	No	—
B:	Amount of developing solvent	Small	Mid	Large
C:	Primary pressure	Low	Mid	High
D:	Secondary pressure	Low	Mid	High
E:	Airflow amount	Low	Mid	Large
F:	Airflow length	Short	Mid	Long
G:	Air inlet shape	No. 1	No. 2	No. 3
H:	Air inlet opening	Closed	$\frac{1}{3}$ closed	Open

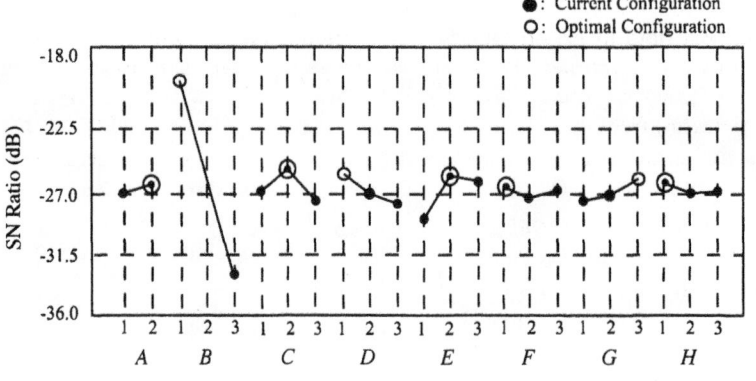

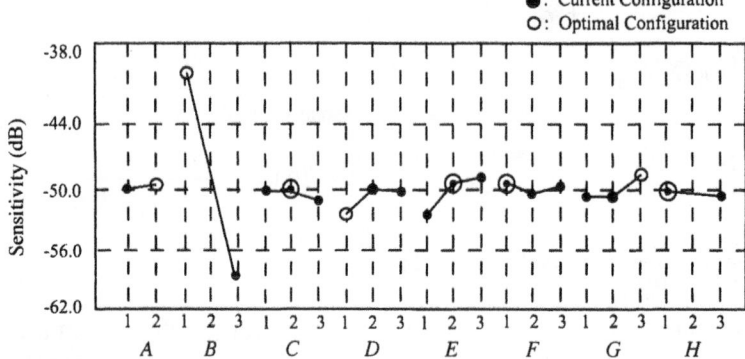

Figure 4
Response graphs

and apparatus design, and assigned all of them to an L_{18} orthogonal array (Table 3). Figure 4 shows factor effect plots for the SN ratio and sensitivity. According to these results, we selected the following configurations:

Current configuration: $A_1B_3C_2D_2E_2F_1G_2H_1$

Optimal configuration: $A_2B_1C_2D_1E_2F_1G_3H_1$

Table 4 shows the results of the confirmatory experiment. Consequently, we obtained high gains for

Table 4
Confirmatory experiment (dB)

Configuration	SN Ratio		Sensitivity	
	Estimation	Confirmation	Estimation	Confirmation
Current	−29.8	−28.5	−55.8	−54.3
Optimal	−12.4	−17.4	−38.0	−41.0
Gain	+17.4	+11.1	+17.8	+13.3

both SN ratio and sensitivity. Figure 5 reveals that the optimal configuration enables us to improve the separation efficiency of the toner. However, we could not obtain sufficient reproducibility of gain. This was probably because of the interaction between control factors C and D (pressure) and control factor E (airflow amount).

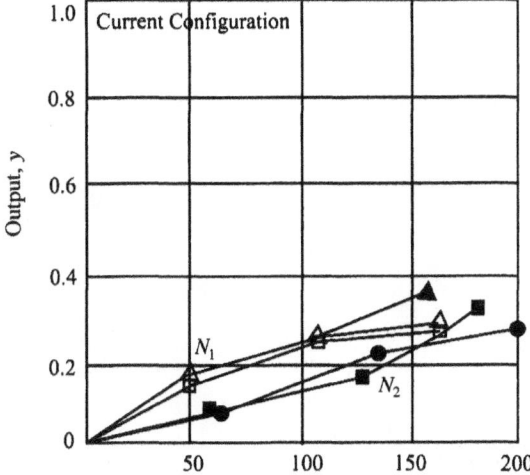

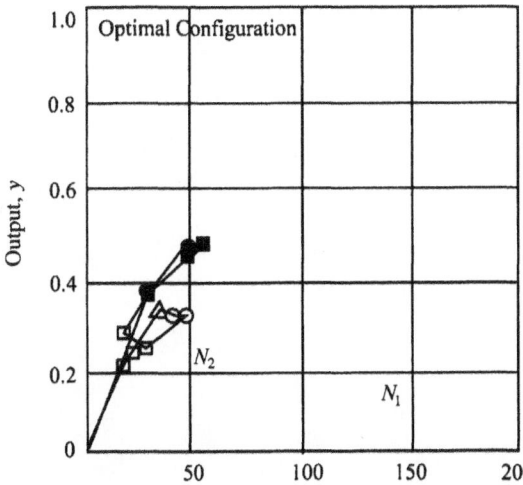

Confirmatory experiment (input/output)

To ascertain our achievement, we compared the optimal and current configurations using unit-mass charge amount q/m, expressed as

$$\frac{q}{m} = \frac{\text{charge amount detected, } q(\mu C)}{\text{separated mass of toner (g)}} \quad (18)$$

By fixing the blown-air time, we measured the unit-mass charge amount, q/m, for two types of developing solvent used as noise. Calculating the nominal-the-best SN ratio of amount of charge obtained, we found that the optimal configuration achieves approximately +11 dB more gain than that of the current configuraiton. In short, we demonstrated that in optimizing charge amount measurement apparatus, it is appropriate to focus on its function to separate toner instead of measuring the amount of charge.

From the results obtained so far, we calculated the economic benefit related to inspection on an assumption regarding the gains confirmed by measuring the amount of charge. Setting the lower functional limit of the developing solvent to 15 $\mu C/g$ and the upper limit to 30 $\mu C/g$, we have

$$\Delta_0 = 7.5 \ \mu C/g \quad (19)$$

Assuming that the developing solvent exceeded the functional limits as discarded, we regarded the cost per production lot as

$$A_0 = 1,000,000 \text{ yen (\$8000)} \quad (20)$$

Therefore, the loss coefficient, k, is as follows:

$$k = \frac{A_0}{\Delta_0^2} = 17,778 \text{ yen} \quad (21)$$

From the results of measurement of the amount of charge, we calculated the following values of σ:

Current configuration: $\quad \sigma_0^2 = \dfrac{S_0}{\eta_0} = 0.794 \quad (22)$

Optimal configuration: $\quad \sigma^2 = \dfrac{S}{\eta} = 0.047 \quad (23)$

Then the economic benefit per lot of developing solvent was computed as

$$\Delta L = (17,778)(0.794 - 0.047) = 13,280 \text{ yen} \quad (24)$$

Assuming a type of developing solvent whose production volume is 500 lots per year, we obtained

$$(13,280 \text{ yen})(500 \text{ lots}) = 6,640,000 \text{ yen/year} \quad (25)$$

an annual economic benefit of 6.64 million yen.

Although our calculations covered only one type of developing solvent, if we apply this method to other types, we will be able to accomplish many more economic benefits.

Reference

Kishio Tamura and Hiroyuki Takagiwa, 1999. Optimization of blow-off charge measurement system. *Quality Engineering*, Vol. 7, No. 2, pp. 45–52.

This case study is contributed by Kishio Tamura and Hiroyuki Takagiwa.

CASE 17

Evaluation of the Generic Function of Film Capacitors

Abstract: This research focuses on parameter design to clarify (1) a generic function to express a capacitor's functionality systematically, and (2) a practical measurement process to evaluate the generic function of a film capacitor.

1. Introduction

In applying quality engineering techniques to the functional evaluation of a capacitor, we used the following characteristics: (1) the charging characteristic by direct current (dc), (2) discharging characteristic by dc, (3) voltage and current characteristics by alternating current (ac), and (4) charging and discharging characteristics by a combination of dc and ac. Additionally, to apply it only to the technological development process to meet the following conditions was desirable: (a) that a large variation in SN ratios be obtained; (b) that a voltage similar to the one in practical use be set; (c) that a test sample not come up easily with missing data, even if the sample was defective; (d) that a time constant be easy to measure.

In applying quality engineering techniques to a film capacitor, we concluded that item 4 above was appropriate because we wished to evaluate energy loss of charge and discharge at a given dc stress.

On the other hand, since a capacitor stores and discharges energy proportional to charged voltage in the form of an electric charge, if we define capacitance as C (farads) and charged voltage as V, the electrical charge accumulated, Q (coulombs), is expressed as

$$Q = CV \qquad (1)$$

In our study we considered this relationship to be a basic function. We regard this as

$$y = \beta M \qquad (2)$$

2. Study of Measurement

Figure 1 illustrates a measurement circuit, and Figure 2 depicts the waveforms measured. Figure 2a indicates the waveform of a terminal voltage given to a capacitor, and 2b shows the waveform of a corresponding charging or discharging current. We divided a voltage waveform of one charge-to-discharge cycle (Figure 2a) by 10 equal time frames and set each of 10 voltage values (V_1 to V_{10}) at each point of time (T_1 to T_{10}) to 10 different levels. However, when we analyzed them, to assess the linearity of a waveform, we combined voltage values V_1 to V_5 at charging points of time T_1 to T_5 with signal factors M_6 to M_{10}. On the other hand, by subtracting V_5 from V_6 to V_{10} at discharging points of time T_6 to T_{10}, we created V_6^* to V_{10}^* and then related each of V_{10}^* to V_6^* with M_1 to M_5 in a reverse manner.

Since we established a relationship between terminal voltage and electric charge as the generic function, we needed to integrate the current over the time. Dividing the current waveform shown in Figure 2b at the same 10 points of time as those used for (1), we calculated an integral of current from zero (T_0) to each point of time (the accumulated area between a current waveform and the time axis) and set each integral to each of the electrical charges Q_1 to Q_{10}. In actuality, we measured Q_1 to Q_{10} in Figure 2b by reading a waveform from a digital oscilloscope, computing the area at each time frame divided into 10 equal lengths, and summing up all areas. When analyzed, for the same reason as in the case of signal factors, accumulated electrical

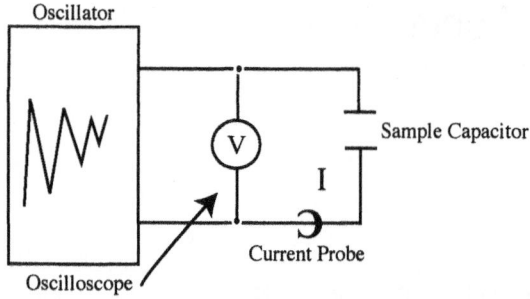

Figure 1
Measurement circuit

charges Q_{10} to Q_6 were combined with measured output values y_1 to y_5, and Q_1 to Q_5 correspond to y_6 to y_{10}.

We set a maximum voltage value to a three-level indicative factor and excluded their effects from errors. This is done because if a commercial capacitor

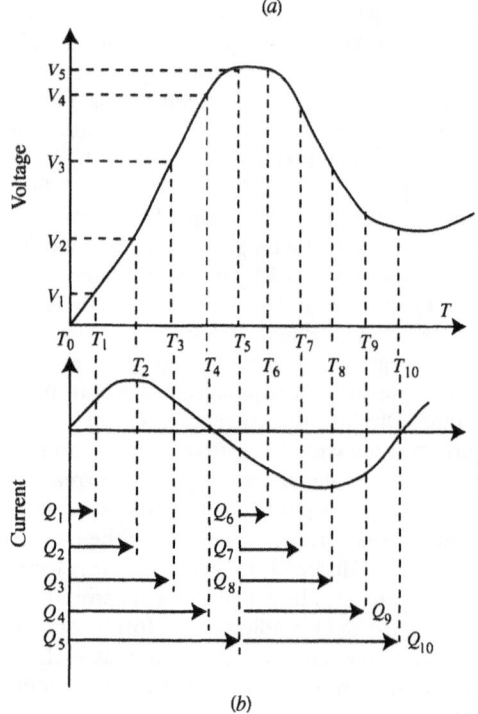

Figure 2
Waveform measured

has variations in the voltage designed, we often adjust them:

$$P_1: \quad 1 \text{ kV}$$
$$P_2: \quad 2 \text{ kV}$$
$$P_3: \quad 3 \text{ kV}$$

For a noise factor, we chose two levels:

$$N_1: \quad \text{predegradation state}$$
$$N_2: \quad \text{postdegradation state}$$

Our experiments are conducted based on an L_{18} orthogonal array with eight factors assigned to it (Table 1).

3. SN Ratio

Table 2 shows the data for experiment 1. We computed the SN ratios and sensitivities of experiment 1, as follows.

Total variation:

$$S_T = (-11.67)^2 + (-11.05)^2 + \cdots + 127.44^2$$
$$= 169,473.52 \qquad (f = 60) \qquad (3)$$

Six effective dividers:

$$r_1 = (-3.01)^2 + (-2.70)^2 + \cdots + 7.11^2 = 149.30$$
$$r_2 = 283.36 \qquad r_3 = 569.16 \qquad r_4 = 869.50$$
$$r_5 = 1097.47 \qquad r_6 = 1697.48 \qquad (4)$$

Six linear equations:

$$L_1 = (-3.01)(-11.67) + \cdots + (7.11)(50.98)$$
$$= 991.05$$
$$L_2 = 1264.93 \qquad L_3 = 3945.39 \qquad L_4 = 4558.54$$
$$L_5 = 7542.77 \qquad L_6 = 8726.45 \qquad (5)$$

Variation of proportional term:

$$S_\beta = \frac{(L_1 + L_2 + L_3 + L_4 + L_5 + L_6)^2}{r_1 + r_2 + r_3 + r_4 + r_5 + r_6}$$
$$= 160,059.15 \qquad (f = 1) \qquad (6)$$

Variation of proportional term to indicative factor:

Table 1
Control factors

	Control Factor	Levels
A:	conducting film material	2
B:	pre-treatment condition	3
C:	forming temperature	3
D:	forming pressure	3
E:	forming time	3
F:	posttreatment A	3
G:	posttreatment B	3
H:	impregnation condition	3

$$S_{P\beta} = \frac{(L_1 + L_2)^2}{r_1 + r_2} + \frac{(L_3 + L_4)^2}{r_3 + r_4} + \frac{(L_5 + L_6)^2}{r_5 + r_6} - S_\beta$$

$$= 197.16 \quad (f = 2) \tag{7}$$

Variation of proportional term to variability of sensitivity:

$$S_{P(N)\beta} = \frac{L_1^2}{r_1} + \frac{L_2^2}{r_2} + \frac{L_3^2}{r_3} + \frac{L_4^2}{r_4} + \frac{L_5^2}{r_5} + \frac{L_6^2}{r_6} - S_\beta - S_{P\beta}$$

$$= 4124.22 \quad (f = 3) \tag{8}$$

Error variation:

$$S_e = S_T - S_\beta - S_{P(N)\beta} = 5092.99 \tag{9}$$

Error variance:

$$V_e = \frac{S_e}{54} = 94.31 \tag{10}$$

Combined error variance:

$$V_N = \frac{S_{P(N)\beta} + S_e}{57} = 161.71 \tag{11}$$

SN ratio:

$$\eta = 10 \log \frac{(1/r)(S_\beta - V_e)}{V_N} = -6.74 \text{ dB} \tag{12}$$

Sensitivity:

$$S = 10 \log \frac{1}{r}(S_\beta - V_e) = 15.35 \text{ dB} \tag{13}$$

4. Response Graph and Confirmatory Experiment

The SN ratios and sensitivities of experiments 1 to 18 were calculated similarly. Figure 3 illustrates the corresponding response graph. Because the coefficient of proportionality, β, in the generic function is tantamount to the capacitor's capacitance, we selected the largest-value level as the optimal config-

Table 2
Results of experiment 1

			M_1	M_2	...	M_{10}	r/L
P_1	N_1	M	−3.01	−2.70	...	7.11	r_1
		Y	−11.67	−11.05	...	50.98	L_1
	N_2	M	−4.26	−3.63	...	8.36	r_2
		Y	−25.45	−28.03	...	39.38	L_2
P_2	N_1	M	−7.17	−6.86	...	13.74	r_3
		Y	−30.25	−28.03	...	105.81	L_3
	N_2	M	−7.48	6.55	...	15.30	r_4
		Y	−48.59	−42.38	...	81.84	L_4
P_3	N_1	M	−8.98	−8.04	...	21.99	r_6
		Y	−38.81	−36.13	...	147.50	L_5
	N_2	M	−8.98	−8.04	...	21.99	r_6
		Y	−38.31	−34.69	...	127.44	L_6

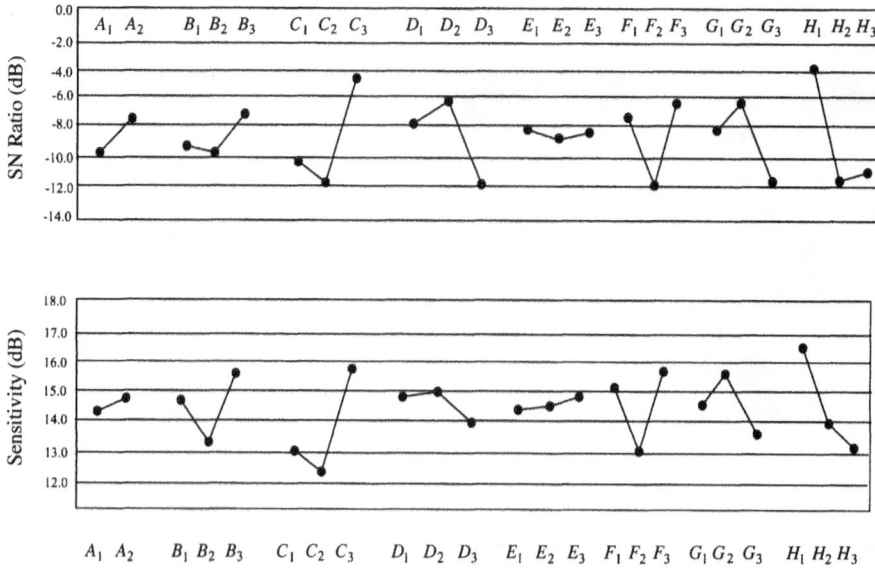

Figure 3
Response graphs

uration not only for the SN ratio but also for sensitivity. Consequently, the following configurations were chosen:

Optimal configuration: $A_2 B_3 C_3 D_2 E_1 F_3 G_2 H_1$

Current configuration: $A_1 B_1 C_2 D_2 E_1 F_1 G_1 H_1$

Based on the optimal and current configurations discussed above, we prepared prototypes and implemented the same measurement and analysis. Table 3 shows the results of the confirmatory experiment. As for the SN ratio, since the difference between estimation and confirmation at the optimal configuration is small, the reproducibility can be regarded

as sufficient. This was validated by the difference in the gains. However, the reproducibility of sensitivity was poor. Although we gathered that this might be due to different manufacturing conditions for prototypes that we used in the confirmatory experiment, we have not verified that assumption. Therefore, we continue to investigate this issue. Nevertheless, we do not consider it vital because there are some factors that easily control sensitivity.

Figure 4 compares the characteristics of the optimal and current configurations. Considering the gains in SN ratio and sensitivity, we can conclude a 12.5% improvement in capacitance at the optimal configuration, and a 75% reduction in variability at

Table 3
Results of confirmatory experiment (dB)

	SN Ratio		Sensitivity	
Configuration	Estimation	Confirmation	Estimation	Confirmation
Optimal	9.81	10.69	21.43	12.69
Current	−3.63	−2.41	15.83	11.66
Gain	13.44	13.10	5.60	1.03

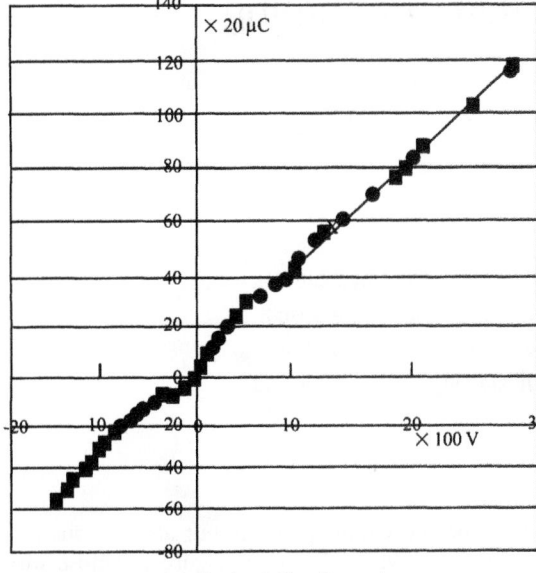

140
×20 µC
120
100
80
60
40
20
0
-20 -10   20 0   10   20   30
              ×100 V
-40
-60
-80

(a) Optimal Configuration

120
×20 µC
100
80
60
40
20
0
-20 -10 -20   10   20   30
                   ×100 V
-40
-60
-80

(b) Current Configuration

Figure 4
Characteristic of film capacitor

the optimal configuration. On the other hand, we noticed that it was unnecessary to study the indicative factor. We could conduct an equivalent experiment with only a noise factor by fixing the indicative factor at a maximum voltage (3 kV). By doing so, we could have scaled down the experiment by one-third.

Owing to the improved linearity of the generic function by parameter design, tolerable voltage, one of the important quality characteristics, was also ameliorated as follows:

Optimal configuration: no failures with all samples up to 5.00 kV

Current configuration: smoke-and-fire breakdowns with all samples at 4.25 kV

Reference

Yoichi Shirai, Tatsuru Okusada, and Hiroshi Yano, 1999. Evaluation of the Function for Film Capacitors. *Quality Engineering,* Vol. 7, No. 6, pp. 45–50.

This case study is contributed by Yoichi Shirai, Tatsuru Okusada, and Hiroshi Yano.

Parameter Design of Fine-Line Patterning for IC Fabrication

Abstract: We developed our research on the patterning process by including everything up to the actual process of forming metal circuits. More specifically, focusing on electrical characteristics as evaluation metrics, we investigated a circuit's ability to improve wire width stability and voltage and current characteristics simultaneously.

1. Introduction

To miniaturize the width of an electric wire is an essential technology not only for downsizing and high accumulation of integrated circuits (ICs) but also for performance enhancement of transistors. Currently, a 1.0-μm-wide wire is used as the smallest-width wire for a circuit on a printed circuit board. However, variability in the wire width causes fluctuation of a transistor's threshold voltage, and the ICs malfunction. Additionally, since the fluctuation of a transistor's threshold voltage is affected more by variability in width as the width decreases, to stabilize this is regarded as important in improving the degree of accumulation.

Until now we have focused especially on the patterning process in the IC manufacturing process, to improve the ability to pattern circuits from a mask to a photoresist using the line-and-space pattern of one-to-one wire width. Figure 1 shows the process that we researched. First, a photoresist applied onto a wafer was exposed to ultraviolet light through a mask. Only the exposed part of the photoresist was dissolved with solvent, and as a result, a pattern was formed. As a next step, after metal material was sputtered onto the patterned wafer, a metal pattern (circuit) was created through dissolution of the photoresist remaining.

On the other hand, considering that this experiment was also conducted to accumulate technical know-how that could be widely used, we laid out not only specific circuit patterns but also various patterns designed as shown in Figure 2, whose wire widths range from 0.4 to 1.2 μm in the middle and at the edge of the mask surface by grouping different types.

By utilizing simple patterns, we also tested the functionality of electrical characteristics of a pattern together with patternability in accordance with the wire width. We set the photoresist type, prebaking temperature, and developing time as control factors, assigned them to an L_{18} orthogonal array, and investigated the patternability. As a result, we found that exposure time, focusing depth, developing time, and descum time greatly affect patternability.

2. Evaluation Characteristics and Experiment

On the basis of our experimental results and experience, we selected four control factors and levels (Table 1). As a noise factor, we chose two pattern positions in the center and at the edge of the wafer. These factors were allocated to an L_9 orthogonal array. Although an L_{18} is recommended for this case in the field of quality engineering, we used an L_9 because we knew that there was no interaction among control factors. As evaluation characteristics, we selected the current and voltage characteristics of a circuit. More specifically, by setting the wire

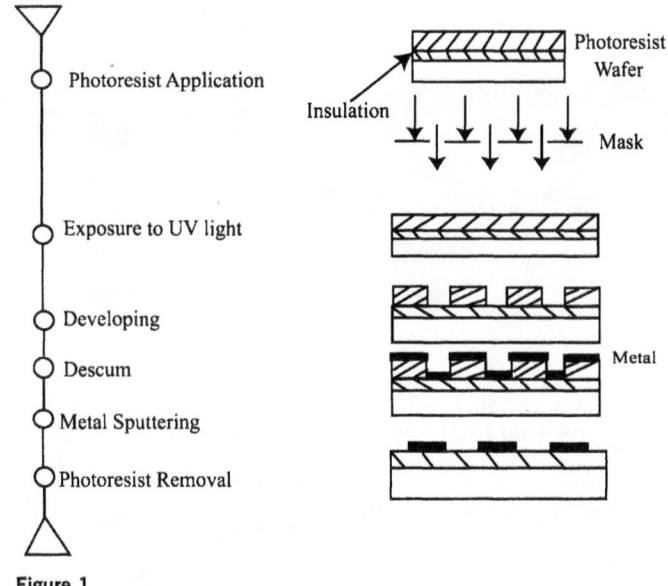

Figure 1
Patterning process

width as a signal factor M (with seven levels between 0.4 and 1.2 μm) and at a given voltage M^* (with seven levels between 0.5 and 4.0 V), we measured the current.

The resistance of a formed circuit is proportional to its length, L, and inversely proportional to its section area. On the other hand, the section area can be expressed by a wire width, M, and its height, t.

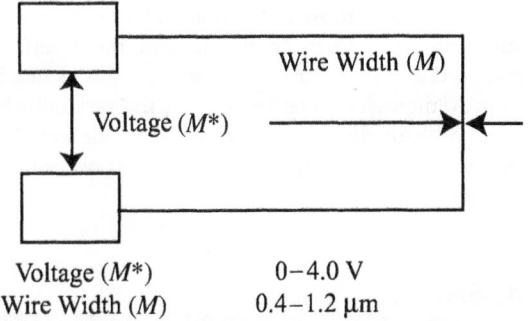

| Voltage (M^*) | 0–4.0 V |
| Wire Width (M) | 0.4–1.2 μm |

Figure 2
Mask pattern

In this experiment, since we fixed L and t, the resistance,

$$R = \frac{k}{m} \qquad \left(k = \frac{\rho L}{t}\right) \qquad (1)$$

Thus, the current y flowing in the circuit is expressed as follows:

$$y = \frac{M^*}{R} = \beta M M^* \qquad \left(\beta = \frac{1}{k}\right) \qquad (2)$$

The control factors were assigned to the inner L_9 orthogonal array, and the signal and error factors were allocated to the outer array. An example of the experimental data corresponding to the seventh row of the orthogonal array is illustrated in Table 2. Using these data, we calculated SN ratios and sensitivities as zero-point proportional dynamic characteristics through the origin by following the next procedure, as shown in Table 3:

$$S_T = 12.50^2 + 12.53^2 + 24.90^2$$
$$+ \cdots + 40.25^2 = 162{,}302 \qquad (f = 98) \qquad (3)$$

$$r = (2)\{[(1.2)(0.5)]^2 + [(1.2)(1.0)]^2$$
$$+ \cdots + [(0.4)(4.0)]^2\} = 336.35 \qquad (4)$$

Table 1

Control factors and levels

		Signal	
Control Factor	1	2	3
C: exposure time	Short	Middle	Long
D: focusing depth	Standard	Deep	Deeper
G: developing time	Short	Middle	Long
H: descum time	No	Yes	—

$$S_\beta = \frac{[(1.2)(0.5)(12.50 + 12.53) + \cdots + (0.4)(4.0)(41.98 + 40.25)]^2}{r}$$

$$= 160{,}667 \qquad (f = 1) \tag{5}$$

$$S_e = S_T - S_\beta = 162{,}302 - 160{,}667$$

$$= 1635 \qquad (f = 97) \tag{6}$$

$$V_e = \frac{S_e}{n-1} = \frac{1635}{98-1} = 16.86 \tag{7}$$

SN ratio:

$$\eta = 10 \log \frac{(1/r)(S_\beta - V_e)}{V_e}$$

$$= 10 \log \frac{(1/336.35)(160{,}667 - 16.86)}{16.86}$$

$$= 10 \log 28.3 = 14.52 \text{ dB} \tag{8}$$

Sensitivity:

$$S = 10 \log \frac{1}{r}(S_\beta - V_e)$$

$$= 10 \log \frac{1}{336.35}(160{,}667 - 16.86)$$

$$= 10 \log 477.63 = 26.79 \text{ dB} \tag{9}$$

3. Optimal Configuration

Based on the SN ratios and sensitivities illustrated in Table 3, the level-by-level computed averages of the SN ratio and sensitivity are as shown in Table 4 and in the factor effect diagram (Figure 3). According to Figure 3, the optimal configuration is

$C_3D_1G_3H_2$. We estimated the process average by selecting two effective factors, C and D.

SN ratio:

$$\eta = \bar{C}_3 + \bar{D}_1 - \bar{T} = 8.92 + 10.37 - 6.18$$

$$= 13.11 \text{ dB} \tag{10}$$

Since the control factors and levels for our experiment did not include the values for the current, configuration, we conducted an extra experiment for the current, whose results are shown at the bottom of Table 3. The SN ratio of the current configuration is

$$\eta_0 = 8.29 \text{ dB} \tag{11}$$

The optimal configuration, $C_3D_1G_3H_2$, matches that of experiment 7, whose SN ratio is

$$\eta = 14.52 \text{ dB} \tag{12}$$

We summarize these calculations in Table 5.

Under the optimal configuration, we improved the gain by 6.23 dB compared with the current configuration. The most influential factors were exposure time and focusing depth, and the levels of standard and long for each factor brought us much improvement. This may be because the variability in wire width or defects of wire would be increased if we made focusing depth deeper and exposure time shorter.

4. Sensitivity Analysis

The resistance of a formed circuit can be expressed as follows with circuit length, L, wire width, M, wire height, t, and resistivity, ρ:

Table 2
Data for experiment 7 (mA)

		M_1^* (0V)	M_2^* (0.5 V)	M_3^* (1.0 V)	M_4^* (1.5 V)	M_5^* (2.0 V)	M_6^* (2.5 V)	M_7^* (3.0 V)	M_8^* (4.0 V)
M_1 (1.2 μm)	Middle	0	12.50	24.90	37.12	49.10	60.71	71.91	92.87
	Edge	0	12.53	24.99	37.28	49.30	60.97	72.23	93.32
M_2 (1.0 μm)	Middle	0	10.74	21.41	31.95	42.28	52.34	62.07	80.47
	Edge	0	10.77	21.49	32.06	42.44	52.55	62.33	80.77
M_3 (0.8 μm)	Middle	0	9.210	18.38	27.42	36.32	45.01	53.45	69.44
	Edge	0	9.246	18.46	27.56	36.51	45.25	53.75	69.84
M_4 (0.7 μm)	Middle	0	8.525	17.00	25.36	33.61	41.65	49.49	64.39
	Edge	0	8.528	17.00	25.37	33.62	41.68	49.51	64.43
M_5 (0.6 μm)	Middle	0	7.746	15.46	23.09	30.61	37.96	45.14	58.82
	Edge	0	7.762	15.48	23.11	30.63	37.99	45.18	58.88
M_6 (0.5 μm)	Middle	0	6.864	13.71	20.48	27.15	33.70	40.11	52.36
	Edge	0	6.824	13.61	20.32	26.96	33.46	39.82	52.00
M_7 (0.4 μm)	Middle	0	5.461	10.91	16.30	21.65	26.90	32.05	41.98
	Edge	0	5.222	10.43	15.61	20.73	25.77	30.71	40.25

Table 3
Assignment and analysis of control factors (dB)

No.	Column and Factor				SN Ratio	Sensitivity
	1 C	2 D	3 G	4 H		
1	1	1	1	1	8.13	24.69
2	1	2	2	2	4.34	24.18
3	1	3	3	1'	−1.60	21.29
4	2	1	2	1'	8.46	25.02
5	2	2	3	1	8.46	25.23
6	2	3	1	2	1.15	25.39
7	3	1	3	2	14.52	26.79
8	3	2	1	1'	8.12	26.44
9	3	3	2	1	4.11	25.03
	Current configuration				8.29	25.22

Table 4
Level-by-level averages of SN ratio and sensitivity (dB)

Factor	SN Ratio			Sensitivity		
	1	2	3	1	2	3
C	3.62	6.01	8.92	23.39	25.21	26.10
D	10.37	6.96	1.22	25.50	25.28	23.92
G	5.80	5.64	7.11	25.50	24.76	24.44
H	5.94	6.67	—	24.62	25.45	—
Average		6.18			24.90	

$$R = \frac{\rho L}{tM} \tag{13}$$

When $\rho = 2.2 \times 10^{-5}\ \Omega \cdot$ mm, $L = 1.12$ mm, and $t = 0.0005$ mm, the resistance is

$$R = \frac{0.004928}{M} \tag{14}$$

If the wire width, $M = 1.0$ μm and the voltage, $M^* = 1.0$ V, the current is

$$y = \frac{M^*}{R} = \frac{1.0}{(0.004928)(0.001)}$$

$$= 0.02029\ \text{A} = 20.29\ \text{mA} \tag{15}$$

Therefore, the target value of the sensitivity is

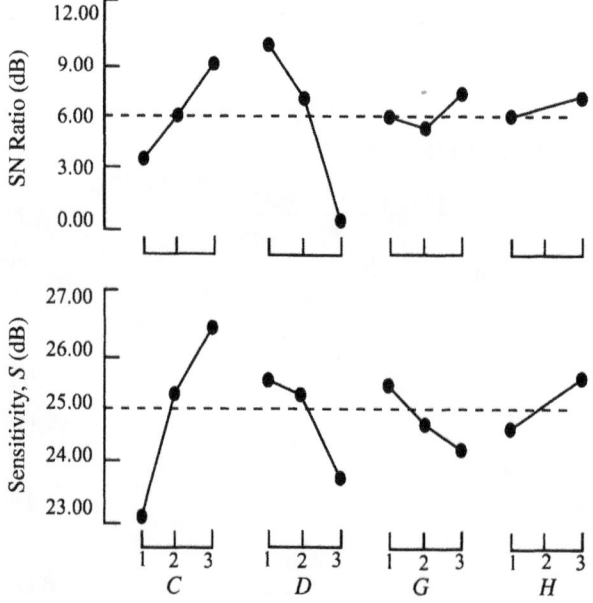

Figure 3
Response graphs

Table 5
SN ratio comparison of optimal and current configurations (dB)

	Configuration		
	Optimal	**Current**	**Gain**
Estimation	13.11	—	—
Configuration	14.52	8.29	6.23

$$S = 10 \log 20.29^2 = 26.15 \text{ dB} \qquad (16)$$

Under the optimal configuration the sensitivity is

$$S = \overline{C}_3 + \overline{D}_1 - \overline{T} = 26.10 + 25.50 - 24.90$$

$$= 26.70 \text{ dB} \qquad (17)$$

This reveals that the target and optimal values of sensitivity are consistent. Although we concluded that we do not need to adjust the sensitivity, when needed we can regulate using the exposure time, C, which affects sensitivity relatively more than the SN ratio does.

This study revealed that we can develop manufacturing conditions to form wires of fine width with little variability and confirm a fundamental principle for stabilizing voltage and current as basic functions of a circuit. This also demonstrates that it is possible to evaluate two signal factors of patternability β and functionality $\beta*$ at the same time, whereas traditionally each has been analyzed separately.

Reference

Yosuke Goda and Takeshi Fukazawa, 1993. Parameter design of fine line patterning for IC fabrication. *Quality Engineering*, Vol. 1, No. 2, pp. 29–34.

This case study is contributed by Takeshi Fukazawa and Yosuke Goda.

Minimizing Variation in Pot Core Transformer Processing

Abstract: This study addressed a transformer processing problem involving inductance changes. Due to inductance falling out of the ranges calculated, much time and expense were involved in an attempt to rectify the problem. In this L_{16} experiment it was determined that by adjusting the level settings of seven of these factors, average inductance could be reduced by 90%. Annual losses attributed to this product were reduced to 28% per power supply model.

1. Introduction

The objective of this study was to minimize inductance variation from transformer to transformer after processing. The inductance should be within calculated limits. A ferrite pot core transformer (Figure 1) was the subject of the experiment.

Manufacturing specifications are determined as follows:

$$L = N^2 \frac{A_L}{1 \times 10^6} \qquad (1)$$

where L is in millihenries, N is the number of primary turns, and A_L is in inductance factor from the core manufacturer's data sheets (specified with a tolerance). Initial out-of-tolerance inductance indicates an incorrect number of primary turns or incorrect core material. Large drops in inductance after processing indicate separated or cracked cores. This results in high magnetizing currents and flatter hysteresis loops. The present distribution of transformer inductance is shown in Figure 2, where 8.2% of the products are below the lower specification limit.

2. Factors and Experimental Layout

The brainstorming group consisted of representatives from quality, product design, and manufactur-

ing. A total of 12 people attended the brainstorming session, which lasted two hours. The transformer process flow diagram, with factors and measurements steps, is shown in Figure 3. The group originally discussed 14 possible factors, but this was reduced to nine controllable and two noise factors (Table 1). From Table 1 it was decided to select two orthogonal arrays, one for controllable factors and another for noise factors. An L_{16} orthogonal array was modified by using a linear graph (Figure 4) and multilevel arrangement to yield an L_{16} (8×2^8) array. A standard L_4 (2) orthogonal array was selected for the noise factors. Since available production time for the experiment was minimal, it was decided to process two transformers per experiment. The experiment layout is shown in Table 2.

3. Results of Experiment

A regular analysis was conducted using inner and outer orthogonal array concepts. This analysis identified the most significant factors and interactions between noise and control factors. The following equation was the basis for variation calculations:

$$S_T = \sum_{i=1}^{n} y_i^2 - \text{CF} \qquad (2)$$

Nominal-the-best SN ratio analysis was also done.

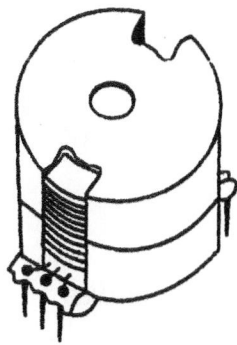

Figure 1
Pot core transformer

$$S_m = \frac{(9.44 + 10.21 + 9.54 + 9.73)^2}{4} = 378.70$$

$$V_e = \frac{\begin{array}{c}(9.44 - 9.73)^2 + (10.21 - 9.73)^2 \\ + (9.54 - 9.73)^2 + (9.73 - 9.73)^2\end{array}}{4 - 1} = 0.117$$

$$\eta = 10 \log \frac{1}{4}\left(\frac{378.70 - 0.117}{0.117}\right) = 29.08 \text{ dB}$$

This type of analysis is more sensitive than regular analysis because (among other advantages) it provides robustness against all sources of noise, not necessarily only the noise factors controlled. The SN ratio nominal-the-best equations for Table 2 are as follows:

$$\eta = 10 \log \frac{1}{n} \frac{S_m - V_e}{V_e} \qquad (3)$$

$$S_m = \frac{T^2}{n} \qquad (4)$$

$$V_e = \frac{\sum_{i=1}^{n} (y_i - \bar{y})^2}{n - 1} \qquad (5)$$

For run 1:

4. Optimization and Estimation

Optimum levels were determined from the response graphs (Figure 5) and Table 3. A trade-off between interaction with noise factors, main effects, and SN ratio was considered to determine how a confirmation experiment should be conducted. The SN ratio determines the best factor levels for robustness against noise factors. The higher the SN ratio, the more robust the level setting is to noise.

Regular analysis determined that the best factor levels were L_2, B_1, C_1, D_1, E_1, and F_2. Factor G had a conflict between interactions with noise. Level G_1 is optimum for robustness against noise factor Y; G_2 is optimum for robustness against noise factor X.

SN ratio analysis identified levels, L_2, L_4, L_6, B_2, C_1, D_2, F_1, G_2, H_1, and A_2 as best. Factor levels of L, B, D, and F were in conflict between regular analysis and SN ratio analysis. This conflict is not unusual, since SN ratio analysis considers the variability of the quality characteristic due to all sources of noise,

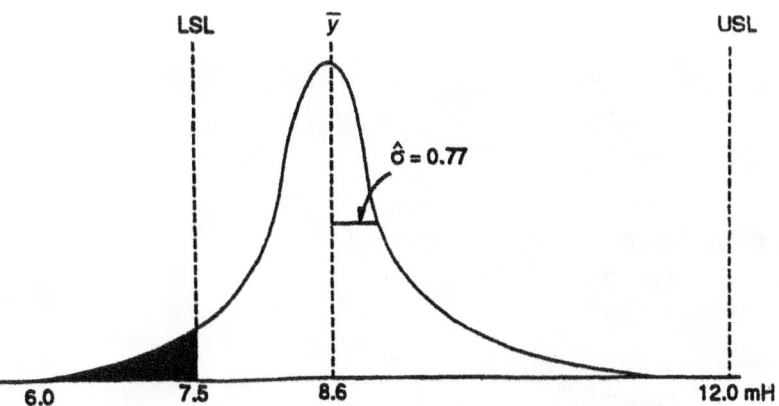

Figure 2
Present PS6501 auxiliary distribution curve

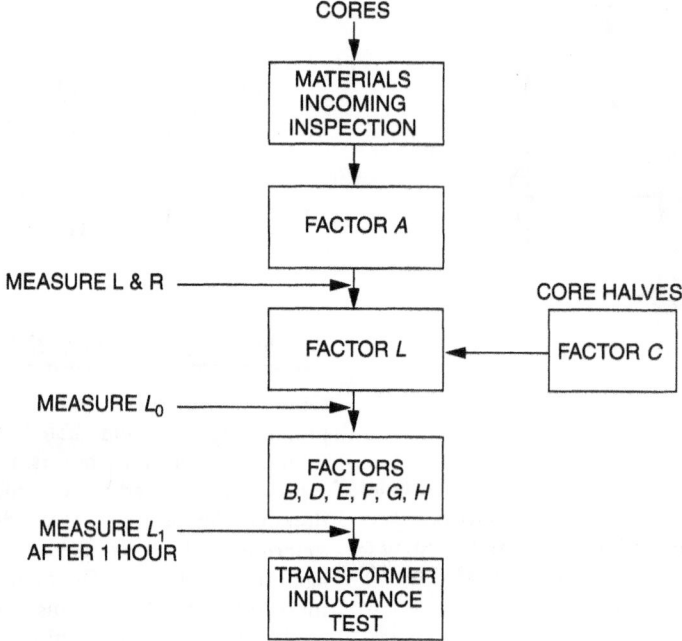

Figure 3
Transformer process

Table 1
Factors and levels

Factor	1	2	3	4	5	6	7	8
A	Level 1	Level 2						
B	Level 1	Level 2						
C	Level 1	Level 2						
D	Level 1	Level 2						
E	Level 1	Level 2						
F	Level 1	Level 2						
G	Level 1	Level 2						
H	Level 1	Level 2						
L	Level 1	Level 2	Level 3	Level 4	Level 5	Level 6	Level 7	Level 8

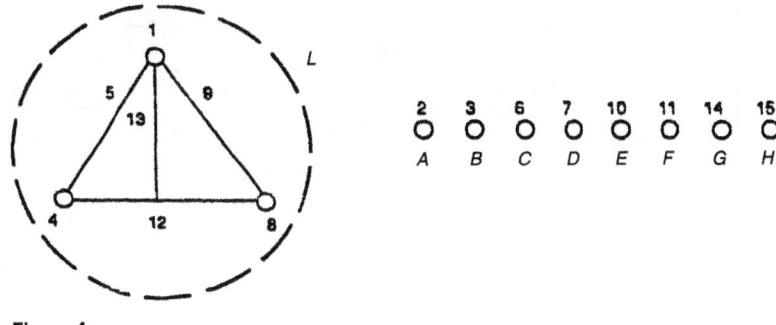

Figure 4
Linear Graph

Table 2
Orthogonal array

									Z	1	2	2	1
									Y	1	2	1	2
									X	1	1	2	2
									Exp.	1	2	3	4
Array	L	A	B	C	D	E	F	G	H	Data				T	SN[a]
1	1	1	1	1	1	1	1	1	1	9.44	10.21	9.54	9.73	38.92	29.08
2	2	1	1	1	1	2	2	2	2	9.07	9.68	8.82	8.84	36.41	11.42
3	3	1	1	2	2	1	1	2	2	8.41	7.23	8.87	8.17	32.68	21.42
4	4	1	1	2	2	2	2	1	1	10.20	10.48	10.62	11.08	42.38	29.20
5	1	2	2	2	2	2	2	2	2	9.56	8.39	8.85	7.87	34.67	21.64
6	2	2	2	2	2	1	1	1	1	9.08	9.18	9.30	8.94	36.50	35.59
7	3	2	2	1	1	2	2	1	1	9.30	8.11	9.43	9.04	35.88	23.55
8	4	2	2	1	1	1	1	2	2	9.72	9.83	9.92	9.85	39.32	41.40
9	5	1	2	1	2	1	2	1	2	9.10	8.88	9.43	10.08	37.49	25.07
10	6	1	2	1	2	2	1	2	1	9.63	9.77	9.90	9.73	39.03	38.82
11	7	1	2	2	1	1	2	2	1	9.94	9.17	10.40	9.15	38.66	23.96
12	8	1	2	2	1	2	1	1	2	9.63	7.85	9.52	7.87	34.87	18.87
13	5	1	2	2	1	2	1	2	1	10.11	8.52	9.84	7.83	36.30	18.46
14	6	2	1	2	1	1	2	1	2	9.89	10.65	10.19	10.71	41.44	28.49
15	7	2	1	1	2	2	1	1	2	10.20	9.87	10.87	10.73	41.67	27.00
16	8	2	1	1	2	1	2	2	1	8.72	8.94	9.14	8.91	35.71	34.24

[a] SN ratio nominal-the-best equations are given in Section 3.

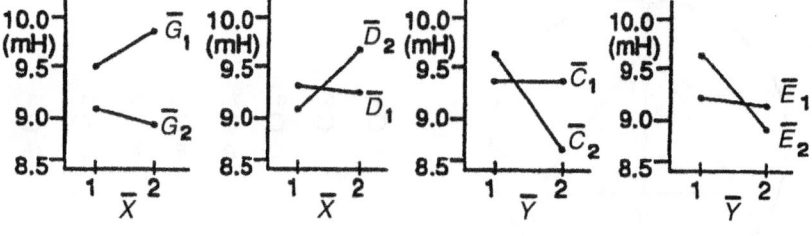

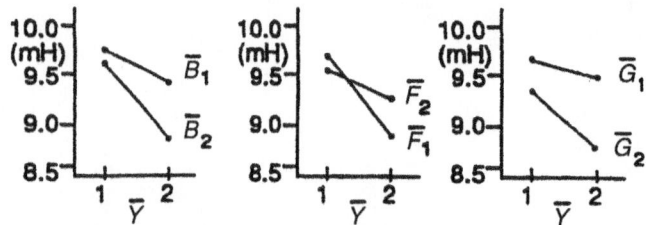

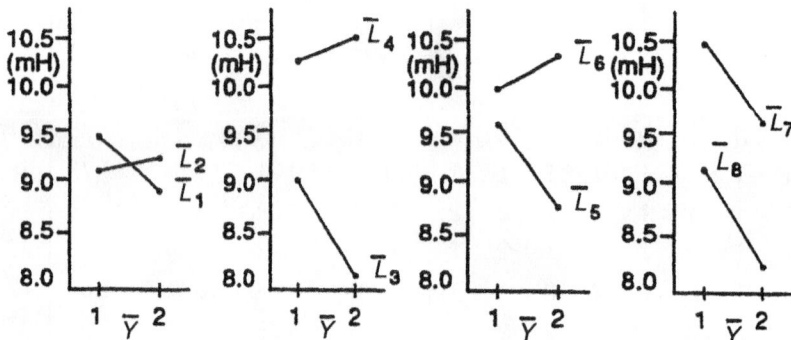

Figure 5
Response graphs

not just the noise factors controlled in the experiment. A confirmation experiment (Table 4) was designed to identify the optimum setting for the conflicting factors.

As a result of the confirmation experiment, optimum factor levels L_4, B_1, C_1, E_1, F_2, G_1, D_1, A_2, and H_1 were selected for production and cost reasons. The process average equation for each combination of noise factor is

$$\hat{u} = \overline{YL} + \overline{YB} + \overline{YC} + \overline{YE} + \overline{YF} + \overline{YG} + \overline{XD} \\ + \overline{XG} - \overline{G} - 5\overline{Y} - \overline{X} \quad (6)$$

The inductance results at optimum levels for each combination of noise factors is

$$\hat{u}x_1 y_1 = 10.01 \text{ mH}$$

$$\hat{u}x_1 y_2 = 11.40 \text{ mH}$$

Table 3
Summary table

		Regular Analysis					
		Interaction with Noise Factor				Main Effect	SN $-10 \log \eta$
		Y		X			
Factor		1	2	1	2		
L	1	9.35	9.05	—	—	9.20	6.11
	2	9.06	9.16	—	—	9.11	12.13
	3	9.00	8.14	—	—	8.57	3.86
	4	10.12	10.31	—	—	10.21	15.21
	5	9.62	8.83	—	—	9.22	2.47
	6	9.90	10.21	—	—	10.06	13.61
	7	10.35	9.73	—	—	10.04	5.45
	8	9.25	8.39	—	—	8.82	7.69
B	1	9.62	9.47	—	—	9.55	—
	2	9.54	8.98	—	—	9.26	—
C	1	9.51	9.51	—	—	9.51	11.25
	2	9.65	8.94	—	—	9.30	5.38
D	1	—	—	9.44	9.92	—	6.90
	2	—	—	9.23	9.53	—	9.72
E	1	9.44	9.35	—	—	—	10.48
	2	9.72	9.10	—	—	—	6.14
F	1	9.62	9.08	—	—	—	9.44
	2	9.54	9.37	—	—	—	7.18
G	1	9.73	9.59	9.51	9.82	9.66	—
	2	9.43	8.87	9.17	9.13	9.15	—
H	1	—	—	—	—	—	9.60
	2	—	—	—	—	—	7.02
A	1	—	—	—	—	—	—
	2	—	—	—	—	—	—

$$\hat{u}x_2 y_1 = 10.17 \text{ mH}$$

$$\hat{u}x_2 y_2 = 11.56 \text{ mH}$$

Table 4
Confirmation experiment

Exp.	L	C	E	D	H	A	B	F	G
				Level					
1	4	1	1	2	1	2	1	2	1
2	6	1	1	2	1	2	1	2	1
3	2	1	1	2	1	2	1	2	1
4	Best	1	1	2	1	2	2	1	2

5. Confirmation and Improvement in Quality

Average values obtained in the experiment were confirmed at the levels predicted. The mean inductance was influenced by the factors exactly as predicted.

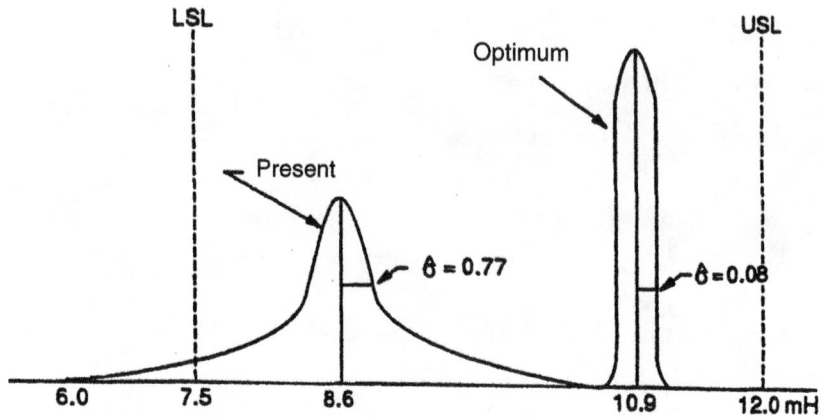

Figure 6
Comparison of transformer inductance distributions

The confirmation experiment variation was calculated from only two samples; therefore, more data should be obtained to determine the actual inductance distribution. The variation exhibited in the experiment falls within the range predicted in the analysis, although at the upper limit of the confidence interval.

Implementing the new factor levels will improve the capability of the process. A comparison of transformer inductance distributions is shown in Figure 6. Engineering and assembling time will not be wasted on questions regarding out-of-specification transformers. Scrap transformers and production delays will be eliminated.

Savings using the conventional method were calculated to be about 28% annually. This calculation considers yearly production volume, rejection percentage, cost of scrap, and engineering and assembly time. If costing information, specification limits, and target value are valid, savings calculated by the quality loss function (QLF) will be reflected in real life. In this case, the annual savings calculated by the QLF are 30.60%. After a more extensive confirmation/production run, the mean can be adjusted to the target value. This will increase savings due to minimizing loss due to quality to 99.67% annually (when the quality loss after optimization is compared to the quality loss before optimization).

This case study is contributed by Gerard Pfaff.

Optimization of the Back Contact of Power MOSFETs

Abstract: Since a large current flows in a power MOSFET for an automobile, its electric resistance (ON resistance) must be lowered to reduce the electric power loss. In our research we attempted to reduce the ON resistance by looking at the contact resistance of the back gate of a power MOSFET and optimizing the conditions for forming the back gate.

1. Introduction

In the development of automobile electronics, new technical issues, such as increasing space for control units or growing consumption energy, have emerged. To solve them, the power MOSFET, which is easy to obtain and is driven by low electrical power, is attracting much attention as a key next-generation device in the electronic field and is being developed as a reliable switching element to control large electric currents.

Since a large current flows in an automobile's power MOSFET, its electric resistance (ON resistance) must be lowered to reduce the electrical power loss. In our research we attempted to reduce the ON resistance by looking at the contact resistance of a back gate of a power MOSFET and optimizing the conditions for forming the back gate.

2. Structure of Back Contact

Figure 1 depicts the structure of a vertical-type power MOSFET. As illustrated in the figure, the current flows from the drain to the source. Figure 2 shows a structure of resistances.

To reduce the contact resistance between the back contact and silicon substrate, R_1, we optimize some factors, such as the condition for forming back contacts. This is because by abolishing the impurity doping and annealing processes instead of using the conventional method of contact-resistance reduction, we can both realize a thin silicon substrate, which has been difficult to manufacture to date, and lower the substrate resistance, R_2. It is said that if electronic devices become smaller in the future, the substrate resistance, R_2, will account for approximately 50% of the total resistance. Therefore, it should be possible to drastically reduce the ON resistance of a power MOSFET if the substrate resistance, can be lowered together with the contact resistance.

3. Fundamental Functions and Measurement Characteristics

The conventional technological development of back contacts has been dedicated to improvement of their quality and characteristics in terms of quality features such as back contact thickness and adhesive strength, which are measured by a peeling test. However, as a consequence of focusing on back contact's functions as a power MOSFET, by using voltage and current as measurement characteristics we established a forming technology for back contacts that have low resistance and maintain stable electrical characteristics for environmental fluctuations.

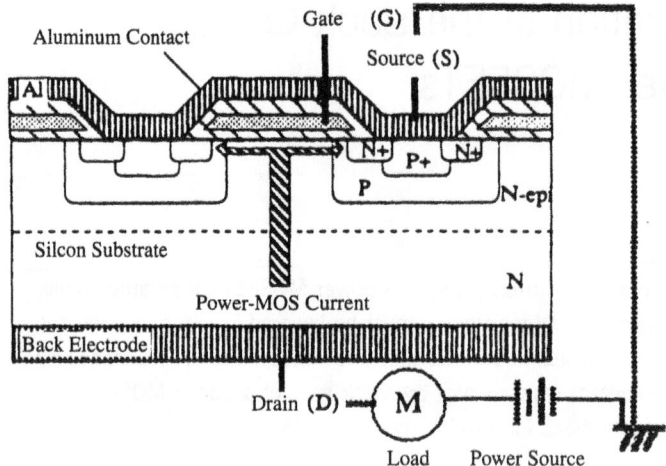

Figure 1
Structure of power MOSFET

We have also sought a way to reduce the contact resistance between the substrate and drain contact (back contact resistance, R_1) by concentrating on manufacturing processes. For measurement characteristics, we set different currents as signal levels (Table 1) and analyzed all data following the pro-cedure for dynamic characteristics by measuring voltage outputs at the back contact (Figure 3). Although this measurement included both the back contact resistance, R_1, and substrate resistance, R_2, we judged that it is reasonably possible to assess a fluctuation of the back contact resistance because it

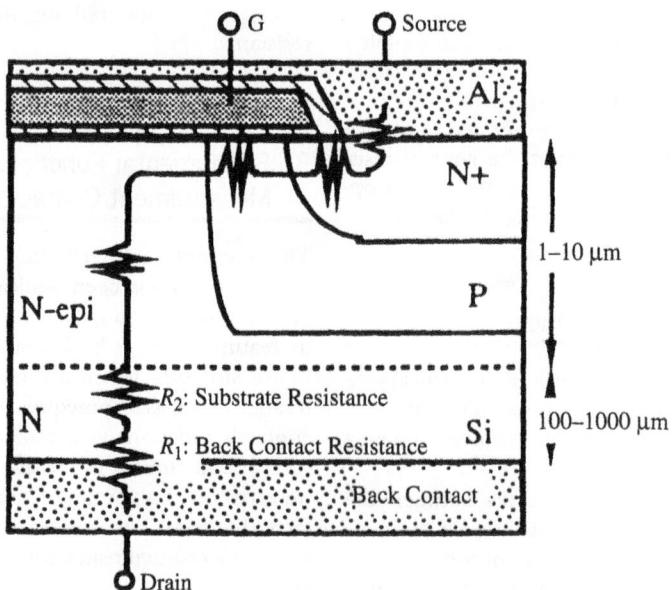

Figure 2
Structure of electric resistances

Table 1

Factors and levels

	Level		
Factor	1	2	3
Signal factor			
Current	0.5	1.0	1.5
Noise factors			
Environmental temperature	Low	Room	High
Heat cycle	Initial	1	2

is sufficiently larger than the substrate resistance in case the impurity density of the substrate is quite low. As noise factors, we selected environmental temperatures and heat cycles, as shown in Table 1.

4. SN Ratio and Sensitivity

Table 2 shows the data from experiment 1. Based on these data, we calculated SN ratios and sensitivity.

Total variation:

$$S_T = 428^2 + 523^2 + 597^2 + \cdots + 295^2$$
$$+ 375^2 + 430^2 = 4,761,567 \quad (f = 27) \quad (1)$$

Effective divider:

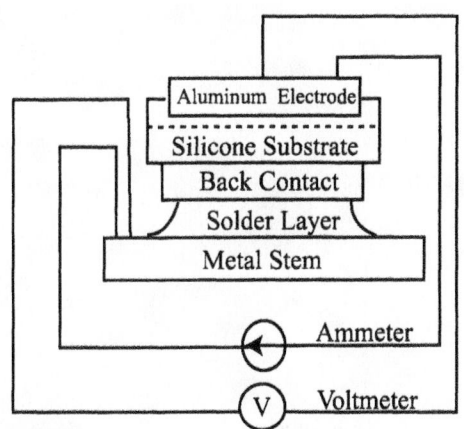

Figure 3

Voltage measurement

$$r = 0.5^2 + 1.0^2 + 1.5^2 = 3.50 \quad (2)$$

Linear equations:

$$L_1 = (0.5)(428) + (1.0)(523) + (1.5)(597)$$

$$= 1632.5$$

$$L_2 = (0.5)(295) + (1.0)(378) + (1.5)(435)$$

$$= 1178.0$$

$$L_3 = (0.5)(365) + (1.0)(440) + (1.5)(495)$$

$$= 1365.0$$

$$\vdots$$

$$L_9 = (0.5)(295) + (1.0)(375) + (1.5)(430)$$

$$= 1167.5 \quad (3)$$

Variation of proportional term:

$$S_\beta = \frac{(L_1 + L_2 + L_3 + \cdots + L_9)^2}{9r}$$

$$= \frac{(1632.5 + 1178.0 + \cdots + 1167.5)^2}{9r}$$

$$= 4,399,364.5710 \quad (f = 1) \quad (4)$$

Error variation:

$$S_e = S_T - S_\beta = 4,761,567 - 4,399,364.5710$$

$$= 362,202.4290 \quad (f = 26) \quad (5)$$

Error variance:

$$V_e = \frac{S_e}{26} = \frac{362,202.4290}{26} = 13,930.8627 \quad (6)$$

SN ratio:

Table 2
Voltage data

	Error Factor		Signal			Linear Equation
			M_1	M_2	M_3	
I_1:	low temperature	J_1 (initial)	428	523	597	L_1
		J_2 (cycle 1)	295	378	435	L_2
		J_3 (cycle 2)	365	440	495	L_3
I_2:	room temperature	J_1	385	485	565	L_4
		J_2	288	365	424	L_5
		J_3	295	371	425	L_6
I_3:	high temperature	J_1	372	474	548	L_7
		J_2	276	356	418	L_8
		J_3	295	375	430	L_9

$$\eta = 10 \log \frac{(1/9r)(S_\beta - V_e)}{V_e}$$

$$= 10 \log \frac{[1/(9)(3.50)]\left(\begin{array}{c}4{,}399{,}364.5710\\ - 13{,}930.8627\end{array}\right)}{13{,}930.8627}$$

$$= 10.00 \text{ dB} \tag{7}$$

Sensitivity:

$$S = 10 \log \frac{1}{9r}(S_\beta - V_e)$$

$$= 10 \log [1/(9)(3.50)]\left(\begin{array}{c}4{,}399{,}364.5710\\ - 13{,}930.8627\end{array}\right)$$

$$= 51.44 \text{ dB} \tag{8}$$

Table 3 shows selected control factors. The second level of H was assigned to a dummy. Judging from our technical knowledge and insight, we chose

Table 3
Factors and levels

	Control Factor	Level		
		1	2	3
A:	back contact metal type	1	2	—
B:	device temperature 1	20	150	350[a]
B':	sputtering temperature 1	Low	Mid	High
C:	treatment time 1 (min)	0[a]	1	5
D:	treatment time 2 (s)	0	20[a]	40
E:	treatment time 3 (min)	0[a]	1	5
F:	organic cleansing	None	IPA	Methanol
G:	treatment time 4 (min)	0	10[a]	20
H:	device temperature 2	Low	Low (dummy)	High[a]

[a] Current condition.

nine major factors supposed to affect back contact resistance significantly. These nine factors were assigned to an L_{18} orthogonal array (Table 4). This use of an L_{18} orthogonal array was so special that we added a new column, B', after the second column of B. By doing so, nine factors could be assigned. B' was not orthogonal to other columns because it is a column of interaction of A and B. Therefore, the independent effect could not be computed; however, for the columns of B and B', 75% of the total effects can be calculated with respect to the main effect. The analysis procedure is described later.

5. Design of Experiments and Results

Based on Table 4, we setup Table 5 of level-by-level averages of SN ratio and sensitivity. From this point

on, we explain primarily how to calculate the effects of factors assigned to columns 2 and $2'$. Except for this, the calculation procedure was exactly the same as the conventional procedure. While the averages of levels of factors $A, C, \ldots, H$ are computed in the conventional way, those of factors B and B' are computed as follows:

$$\overline{B_1} = \frac{\begin{array}{c} y_1 + y_2 + y_3) + (y_{16} + y_{17} + y_{18}) \\ - (y_{10} + y_{11} + y_{12}) - (y_7 + y_8 + y_9) \end{array}}{9} + \overline{T}$$

$$= \frac{\left[\begin{array}{c} (10.00 + 22.39 + 23.06) \\ + (11.80 + 9.64 + 7.75) \\ - (11.12 + 10.52 + 14.66) \\ - (5.27 + 5.17 + 17.67) \end{array}\right]}{9} + 11.47$$

$$= 13.72 \tag{9}$$

Table 4
Assignment of control factors, SN ratio, and sensitivity

Exp.	1 A	2 B	2' B'	3 C	4 D	5 E	6 F	7 G	8 H	SN Ratio	Sensitivity
1	1	1	1	1	1	1	1	1	1	10.00	51.44
2	1	1	1	2	2	2	2	2	1'	22.39	29.93
3	1	1	1	3	3	3	3	3	3	23.06	28.19
4	1	2	2	1	1	2	2	3	3	11.77	48.08
5	1	2	2	2	2	3	3	1	1	9.80	25.28
6	1	2	2	3	3	1	1	2	1'	9.68	42.23
7	1	3	3	1	2	1	3	2	3	5.27	38.19
8	1	3	3	2	3	2	1	3	1	5.17	38.46
9	1	3	3	3	1	3	2	1	1'	17.67	29.81
10	2	1	2	1	3	3	2	2	1	11.12	59.42
11	2	1	2	2	1	1	3	3	1'	10.52	61.12
12	2	1	2	3	2	2	1	1	3	14.66	52.72
13	2	2	3	1	2	3	1	3	1'	8.14	50.09
14	2	2	3	2	3	1	2	1	3	7.85	48.48
15	2	2	3	3	1	2	3	2	1	10.23	55.78
16	2	3	1	1	3	2	3	1	1'	11.80	44.94
17	2	3	1	2	1	3	1	2	3	9.64	59.55
18	2	3	1	3	2	1	2	3	1	7.75	48.55

Table 5

Level-by-level averages of SN ratio and sensitivity (dB)

	Control Factor	SN Ratio			Sensitivity		
		1	2	3	1	2	3
A:	back contact metal type	12.76	10.19	—	36.93	53.41	—
B:	device temperature 1	14.95	10.70	8.76	46.74	44.00	44.75
B':	sputtering temperature 1	13.72	9.90	10.79	43.18	47.94	44.38
C:	treatment time 1 (min)	9.68	10.89	13.84	47.81	43.80	42.88
D:	treatment time 2 (s)	11.64	11.33	11.45	50.96	40.91	43.62
E:	treatment time 3 (min)	8.51	12.67	13.24	48.45	44.99	42.06
F:	organic cleansing	9.55	13.09	11.78	49.08	44.05	42.37
G:	treatment time 4 (min)	11.96	11.39	11.07	42.11	47.63	45.75
H:	device temperature 2	11.19	—	12.04	44.75	—	45.99
Total average			11.47			45.17	

$$\bar{B_2'} = \frac{\left[\begin{array}{l} (y_4 + y_5 + y_6) + (y_{10} + y_{11} + y_{12}) \\ - (y_1 + y_2 + y_3) - (y_{13} + y_{14} + y_{15}) \end{array} \right]}{9} + \bar{T}$$

(10)

$$\bar{B_3'} = \frac{\left[\begin{array}{l} (y_7 + y_8 + y_9) + (y_{13} + y_{14} + y_{15}) \\ - (y_4 + y_5 + y_6) - (y_{16} + y_{17} + y_{18}) \end{array} \right]}{9} + \bar{T}$$

(11)

$$\bar{B_1} = \frac{\left[\begin{array}{l} (y_1 + y_2 + y_3) + (y_{10} + y_{11} + y_{12}) \\ - (y_{16} + y_{17} + y_{18}) - (y_4 + y_5 + y_6) \end{array} \right]}{9} + \bar{T}$$

$$= \frac{\left[\begin{array}{l} (10.00 + 22.39 + 23.06) \\ + (11.12 + 10.52 + 14.66) \\ - (11.80 + 9.64 + 7.75) \\ - (11.77 + 9.80 + 9.68) \end{array} \right]}{9} + 11.47$$

$$= 14.95$$

(12)

$$\bar{B_2} = \frac{\left[\begin{array}{l} (y_4 + y_5 + y_6) + (y_{13} + y_{14} + y_{15}) \\ - (y_{10} + y_{11} + y_{12}) - (y_7 + y_8 + y_9) \end{array} \right]}{9}$$
$$+ \bar{T}$$

(13)

$$\bar{B_3} = \frac{\left[\begin{array}{l} (y_7 + y_8 + y_9) + (y_{16} + y_{17} + y_{18}) \\ - (y_1 + y_2 + y_3) - (y_{13} + y_{14} + y_{15}) \end{array} \right]}{9} + \bar{T}$$

(14)

On the basis of Table 5, we created a factor effect diagram. H_1 is averaged together with the dummy level of H_2.

6. Analysis of Optimal Conditions and Confirmatory Experiment

To reduce the back contact resistance as well as to improve its stability, we should lower the sensitivity as much as possible. Looking at Figure 4, we noticed that for factors A, B', C, E, and G, we should select the levels that had a higher SN ratio because they were consistent with the levels that had a lower sensitivity. On the other hand, for factor B, whose tendency of SN ratio differed from that of sensitivity, we chose level 1 because it greatly affects SN ratio. For factors D, F, and H, by prioritizing the results of sensitivity, we selected levels 2, 3, and 1, respectively. The optimal condition was $A_1B_1B_1'C_3D_2E_3F_3G_1H_1$.

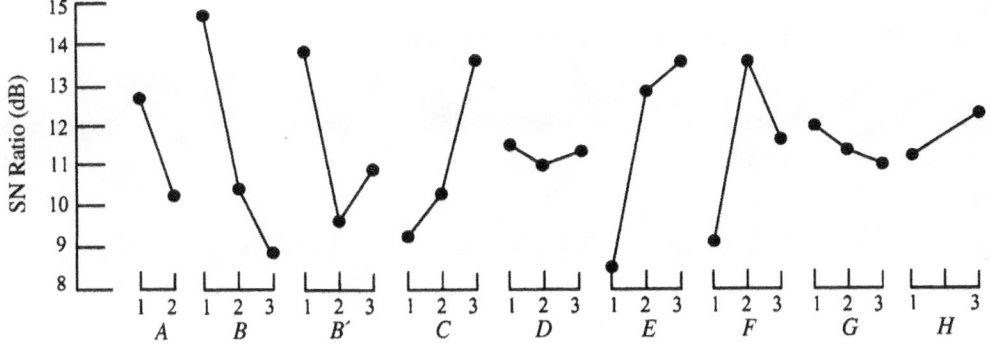

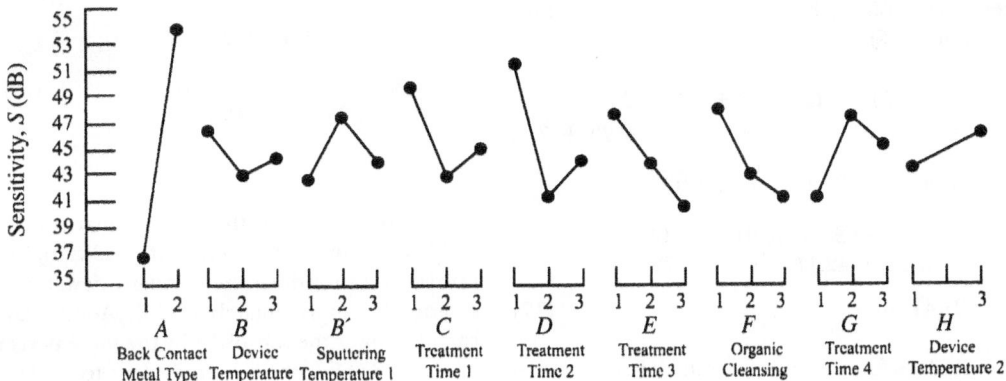

Figure 4
Response graph of SN ratio and sensitivity

Next, by using the five factors A, B, B', C, and E, whose differences between the SN ratio and total average were large, we estimated the SN ratio of the optimal condition selected above.

$$
\begin{aligned}
\eta &= (\overline{A_1} - \overline{T}) + (\overline{B_1} - \overline{T}) + (\overline{B'_1} - \overline{T}) \\
&\quad + (\overline{C_3} - \overline{T}) + (\overline{E_3} - T) + \overline{T} = \overline{A_1} + \overline{B_1} \\
&\quad + \overline{B'_1} + \overline{C_3} + \overline{E_3} - 4\overline{T} \\
&= 12.76 + 14.95 + 13.72 + 13.84 \\
&\quad + 13.24 - (4)(11.47) \\
&= 22.63 \text{ dB} \tag{15}
\end{aligned}
$$

On the other hand, we estimated the ratio of the current condition of $A_1 B_3 B'_3 C_1 D_2 E_1 F_3 G_2 H_3$ using A, B, B', C, and E.

$$
\begin{aligned}
\eta &= (\overline{A_1} - \overline{T}) + (\overline{B_3} - \overline{T}) + (\overline{B'_3} - \overline{T}) \\
&\quad + (\overline{C_1} - \overline{T}) + (\overline{E_1} - \overline{T}) - 4\overline{T} \\
&= \overline{A_1} + \overline{B_3} + \overline{B'_3} + \overline{C_1} + \overline{E_1} - 4\overline{T} \\
&= 12.76 + 8.76 + 10.79 + 9.68 \\
&\quad + 8.51 - (4)(11.47) \\
&= 4.62 \text{ dB} \tag{16}
\end{aligned}
$$

As a next step, to estimate the sensitivity of the

Table 6
SN ratio and sensitivity estimation from confirmatory experiments (dB)

	SN Ratio		Sensitivity	
Condition	Estimation	Confirmation	Estimation	Confirmation
Optimal	22.63	22.96	21.41	23.99
Current	4.62	4.77	39.25	40.35
Gain	18.01	18.19	-17.84	-16.36

optimal condition, we calculated the process average using the six factors A, C, D, E, F, and G, whose differences between the sensitivity and total average were large.

$$S = (\overline{A_1} - \overline{T}) + (\overline{C_3} - \overline{T}) + (\overline{D_2} - \overline{T})$$
$$+ (\overline{E_3} - \overline{T}) + (\overline{F_3} - T) + (\overline{G_1} - T) + \overline{T}$$

$$= \overline{A_1} + \overline{C_3} + \overline{D_2} + \overline{E_3} + \overline{F_3} + \overline{G_1} - 5\overline{T}$$

$$= 36.93 + 42.88 + 40.91 + 42.06$$
$$+ 42.37 + 42.11 - (5)(45.17)$$

$$= 21.41 \text{ dB} \tag{17}$$

Similarly, we computed the ratio of the current condition of $A_1B_3B_3'C_1D_2E_1F_3G_2H_3$ using A, C, D, E, F, and G.

$$S = (\overline{A_1} - \overline{T}) + (\overline{C_1} - \overline{T}) + (\overline{D_2} - \overline{T})$$
$$+ (\overline{E_1} - \overline{T}) + (\overline{F_3} - \overline{T}) + (\overline{G_2} - \overline{T}) + \overline{T}$$

$$= \overline{A_1} + \overline{C_1} + \overline{D_2} + \overline{E_1} + \overline{F_3} + \overline{G_2} - 5T$$

$$= 36.93 + 48.81 + 40.91 + 48.45 + 42.37$$
$$+ 47.63 - (5)(45.17)$$

$$= 39.25 \text{ dB} \tag{18}$$

This result showed that the optimal condition is 18.01 dB better that the current condition; in other words, we can reduce the standard deviation of resistance by approximately 87.5%. Additionally, we can also lower the sensitivity by the same percentage of about 87.5%, which is equivalent to 17.84 dB.

Under the estimated optimal and current conditions, we formed back contacts and conducted

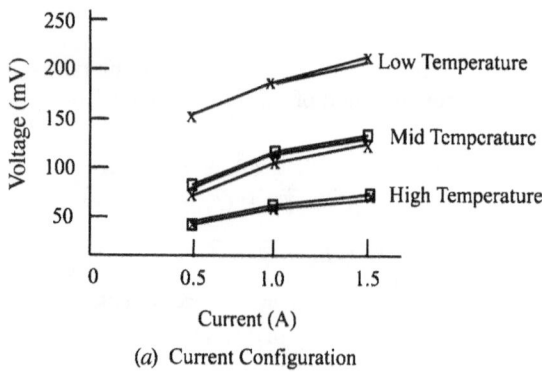

(a) Current Configuration

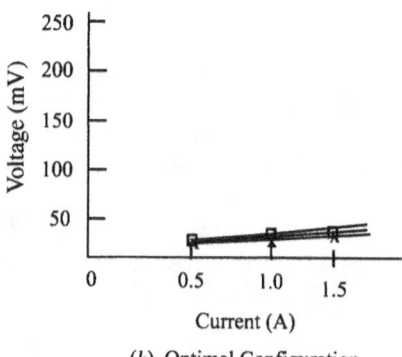

(b) Optimal Configuration

Figure 5
Confirmatory experiment results

confirmatory experiments. For the sake of convenience here, we omit details of the data measured and calculation procedure. The results are shown in Table 6.

Looking at the results, we see that both the SN ratio and sensitivity are consistent with the estimation and, in fact, the SN ratio was improved by 18.19 dB. This indicates that the standard deviation of resistance was lowered by approximately 87.5%. On the other hand, the sensitivity at the optimal condition was reduced by 16.26 dB, approximately 85.7% of the current resistance.

Figure 5 shows the results of the confirmatory experiment. We concluded that the back contact resistance is dramatically better stabilized for the fluctuations of environmental temperature selected as noise factors, whereas the optimal magnitude of resistance is considerably reduced compared to the current resistance.

Through our research, we developed a new manufacturing technology that achieves a back contact resistance equal to the current one without the impurity doping and heat treatment processes regarded as essential to reduce back contact resistances in conventional manufacturing. Furthermore, since abolishing these processes enabled us to make the thickness of a silicon substrate thinner and to lower its resistance, R_2, more reduction of ON resistance can be anticipated. For instance, halving the substrate thickness will lead to reducing its resistance by 50%. The 50% reduction in substrate resistance would be equivalent to a 25% enhancement of device performance were devices to become much smaller. For devices whose power loss is at the conventional level, we can shrink their chip size by approximately 25% and as a result, improve productivity and reduce production cost. In addition, the change in manufacturing conditions helps shorten production processes and solve process problems such as substrate defects. Consequently, 20% improvement in yield and 10% reduction in inspections can be achieved.

Reference

Koji Manabe, Takeyuki Koji, Shigeo Hoshino, and Akio Aoki, 1996. Characteristic improvement of back contact of power MOSFET. *Quality Engineering*, Vol. 4, No. 2, pp. 58–64.

This case study is contributed by Akio Aoki.

CASE 21

Development of High-Quality Developers for Electrophotography

Abstract: We have used Taguchi's methods in the development of developers and toners with good results. Here we report on the development of a high-quality developer. The developer formulation and production conditions were studied as control factors. The environmental conditions during the carrier and toner mixing stages were considered as noise factors. From this study, a developer demonstrating only a small fluctuation in triboelectrical charge to environmental change and deterioration was obtained.

1. Introduction

An electrophotography process consists of (1) electrical charge, (2) exposure, (3) development, (4) transfer/separation, and (5) fixing (Figure 1). In developing this electrophotography process, we utilized the magnetic brush development method with a two-component developer comprising toner and carrier (called simply *developer*). In magnetic brush development, the carrier, magnetic material, is held on a sleeve, and toner adheres to the carrier electrostatically. In the development process, toner is developed on photosensitive material, the carrier remains in the developing apparatus, and consequently, the same amount of toner as that developed is supplied from a stirring area.

The carrier charges the toner positively through friction. In this case, if the electrically charged amount of toner is small, overlap and spill will occur, due to toner scattering. On the other hand, if the amount of toner is large, insufficient image density occurs. Therefore, we need to stabilize the electrically charged toner to obtain a high-quality image.

However, because the charging force of the carrier decreases (durability) or the charged amount of carrier varies due to a change in environment such as temperature or humidity (environmental

dependability), a problem with the charged amount of toner also occurs in the developer. To solve this, it is necessary to stabilize the charging characteristics of the carrier providing electrical charge to the toner.

2. Signal Factors and Charging Characteristics

Carriers whose charging characteristics are stable satisfy the following two requirements at the same time. We evaluated these two requirements.

1. *Generic function.* The time of stirring (mixing) with toner is proportional to an integral of the electrically charged amount of toner. As shown in Figure 2a, stable chargeability of carrier is considered to increase the charged amount quickly in accordance with the time of stirring with toner and to maintain a constant amount afterward. To comprehend its input/output relationship in a dynamic manner, we integrated the charged amount with respect to time, as shown in Figure 2b. As a result, an ideal function is proportionality between the time of mixing with toner and a time-based integral of the charged amount of

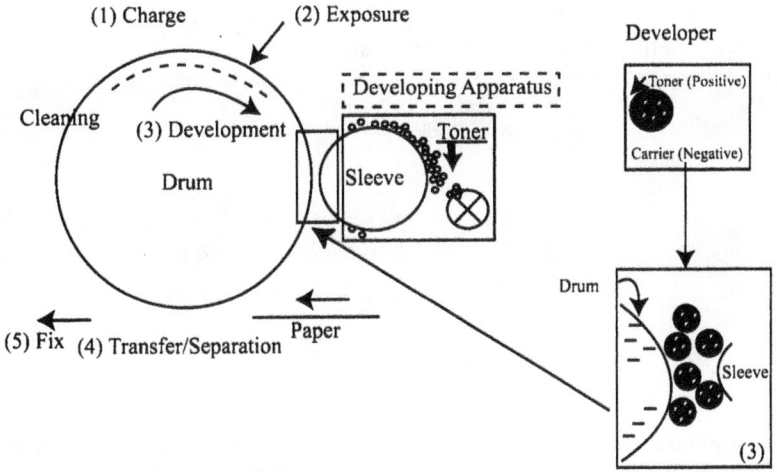

Figure 1
Electrophotography process

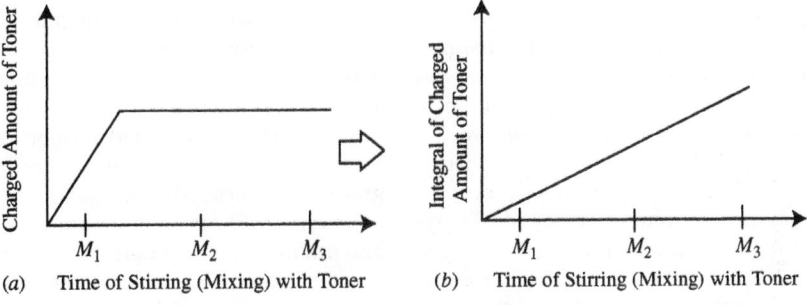

Figure 2
Generic function required for carrier

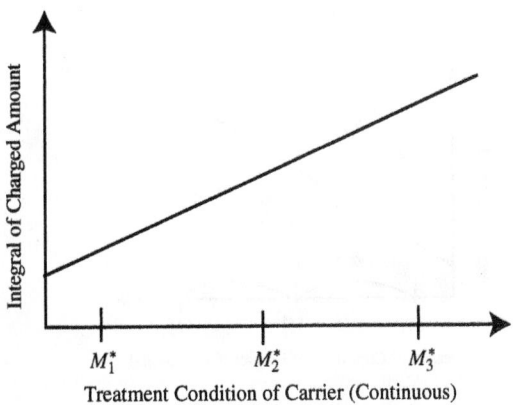

Figure 3
Adjusting the function required for the carrier

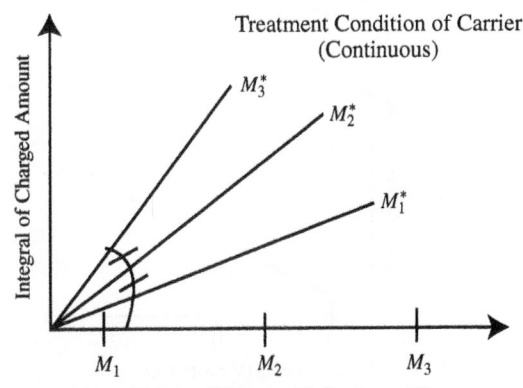

Figure 4
Function of the electrophotographic developer

Table 1
Output data of experiment 1 of an L_{18} orthogonal array

		1 (M_1)	10 (M_2)	20 (M_3)	Linear Equation	
M_1^*	N_1	12.9	275.7	596.7	L_1:	14,703.9
	N_2	10.5	189.1	373.1	L_2:	9,363.6
M_2^*	N_1	17.2	341.7	718.2	L_3:	17,798.2
	N_2	11.9	226.5	458.5	L_4:	11,446.9
M_3^*	N_1	19.8	391.5	818	L_5:	20,294.8
	N_2	14.2	269.3	549.3	L_6:	13,693.2

toner. Its disturbed linearity means instability of the amount of charge, which is represented by its slope.

2. *Adjusting factors.* The treatment condition of the carrier is proportional to an integral of the electrically charged amount of toner. The carrier has the characteristic that its charging force varies in the early phase of imaging but becomes constant after a certain period. Thus, by treating the carrier beforehand, we adjusted an initial charging force to a constant charging force. By doing so, we reduced the fluctuation in the amount charged. Therefore, to obtain a desirable amount charged from the beginning, we should have a linear relationship between the treatment condition of the carrier and an integral of the amount charged (Figure 3). Its deviation from linear-

ity indicates a fluctuation in the amount of toner charged in the initial stage of imaging.

As illustrated in Figure 4, the ideal function is the proportionality between the treatment condition of the carrier and an integral of the electrically charged amount of toner. In this case we evaluated it by setting the treatment condition (continuous) as an adjusting factor. To be brief, for this ideal function, (1) we evaluated proportionality between the treatment condition of the carrier and an integral of the electrically charged amount of toner by assessing an SN ratio that shows linearity, (2) and also evaluated a linear correlation between the treatment condition of the carrier and the charged amount of mixture of treated carrier and toner by assessing an SN* ratio that shows a linear characteristic of an equal interval.

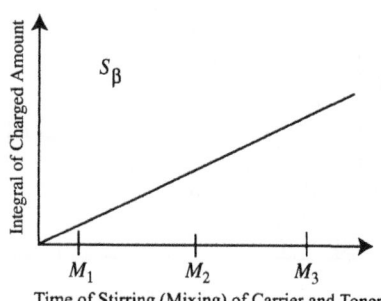

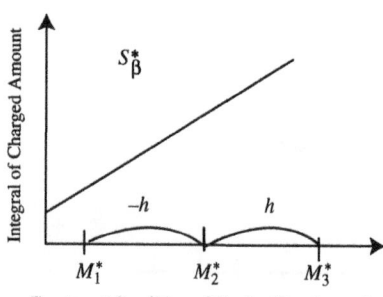

Figure 5
Developing function analysis

3. SN Ratio and Sensitivity

Our experiment was conducted in the following order: (1) by continuously changing the conditions, treat the carrier (M_1^*, M_2^*, M_3^*); (2) after treating the carrier in the toner, sample at a certain time interval (M_1, M_2, M_3); and (3) measure the sampled amount charged.

As noise factors, we selected the environment (the mixing environment with toner), and called low temperature and humidity (LL) N_1, and high temperature and humidity (HH) N_2.

Table 1 shows the output data of experiment 1 of an L_{18} orthogonal array. Additionally, we illustrate a basic process of simultaneous analysis of two signals in Figure 5.

Total variation:

$$S_T = 12.9^2 + \cdots + 549.3^2 = 2,699,075 \qquad (f = 18) \tag{1}$$

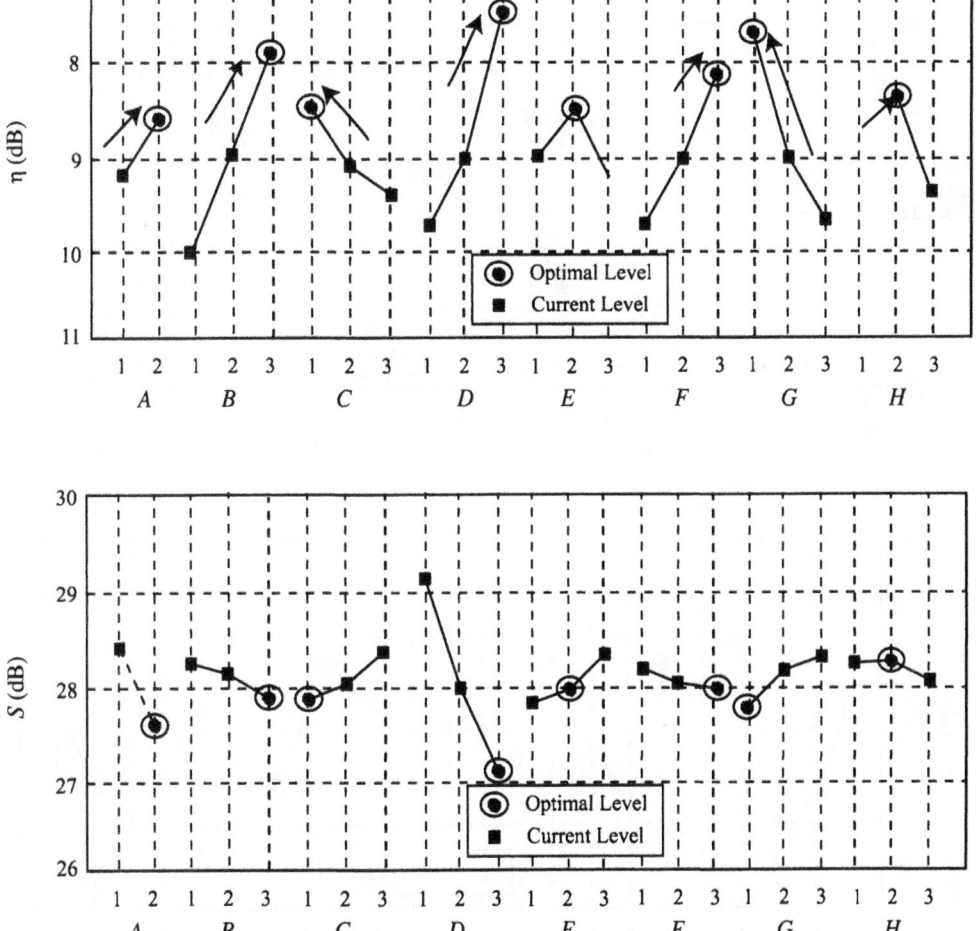

Figure 6
Response graphs of charging characteristic

Effective divider:

$$r = 1^2 + 10^2 + 20^2 = 501 \qquad (2)$$

Linear equations:

$$L_1 = (1)(12.9) + (10)(275.7) + (20)(596.7)$$

$$= 14,703.9 \qquad L_2 = 9363.6 \qquad L_3 = 17,798.2$$

$$L_4 = 11,446.9 \qquad L_5 = 20,294.8 \qquad L_6 = 13,693.2$$
$$(3)$$

Variation of proportional term:

$$S_\beta \text{(signal)}$$

$$= \frac{(L_1 + L_2 + \cdots + L_6)^2}{(2)(3)(M_1^2 + M_2^2 + M_3^2)}$$

$$= \frac{(14,703.9 + \cdots + 13,693.2)^2}{(2)(3)(1^2 + 10^2 + 20^2)}$$

$$= 2,535,417 \qquad (4)$$

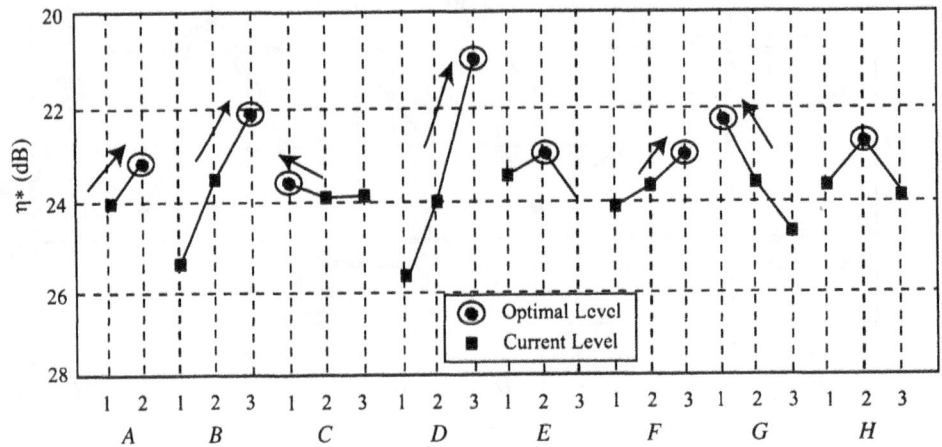

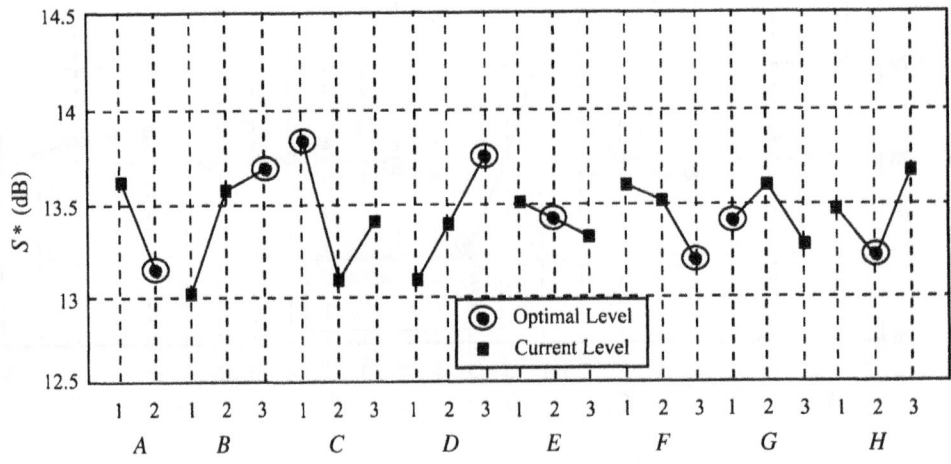

Figure 7
Response graphs of treatment condition

$S_{\beta *}$(adjusting)

$$= \frac{(-L_1 - L_2 + L_5 + L_6)^2}{(2)(2)(M_1^2 + M_2^2 + M_3^2)}$$

$$= \frac{(-14{,}703.9 - 9{,}363.2 + 20{,}294.8 + 13{,}693.2)^2}{(2)(2)(1^2 + 10^2 + 20^2)}$$

$$= 49{,}105 \tag{5}$$

Variation of proportional term difference:

$$S_{N\beta} = \frac{(L_1 + L_3 + L_5)^2 + (L_2 + L_4 + L_6)^2}{3r} - S_\beta$$

$$= 111{,}322 \tag{6}$$

Error variation:

$$S_e = S_T - S_\beta - S_{\beta *} - S_{N\beta} = 3231 \tag{7}$$

Error variance:

$$V_e = \frac{S_e}{15} = 215 \tag{8}$$

Total error variance:

$$V_N = \frac{S_{N\beta} + S_e}{16} = 7160 \tag{9}$$

SN ratio:

$$\eta = 10 \log \frac{[1/(2)(3r)](S_\beta - V_e)}{V_N} = -9.57 \text{ dB} \tag{10}$$

$$\eta^* = 10 \log \frac{[1/(2)(2r)](S_{\beta *} - V_e)}{V_N} = -24.96 \text{ dB} \tag{11}$$

Sensitivity:

$$S = 10 \log \frac{1}{(2)3r}(S_\beta - V_e) = 29.26 \text{ dB} \tag{12}$$

$$S^* = 10 \log \frac{1}{(2)2r}(S_{\beta *} - V_e) = 13.87 \text{ dB} \tag{13}$$

4. Response Graphs and Results of Confirmatory Experiment

Based on the results of the L_{18} experiment, we calculated process averages and created response graphs. Figure 6 shows charging characteristics, and Figure 7 shows treatment conditions. Comparing η and η^* with respect to each control factor in Figure 6, we can see that all levels that have the highest SN ratio are identical. On the other hand, Figure 7 reveals that sensitivities S and S^* have a different characteristic.

As shown in Table 2, after choosing two factors regarding carrier formulation and six factors regarding manufacturing conditions as control factors, we conducted an L_{18} (2×3^7) experiment.

Table 3 summarizes the result of estimated and confirmed gains of η and S at current and optimal configurations. According to this table, we find that we obtain sufficient gain for the SN ratio and good reproducibility of the SN ratio and sensitivity. Figure 8 shows the result of M_3^*, which achieves the target sensitivity. At the optimal configuration, we can greatly reduce its variability related to environment compared to the current configuration.

To ascertain if the optimal configuration selected in our study is the most effective in the actual machine, by changing our evaluation environment (LL and HH) we assessed the current and optimal configurations. Since we had a difference of 14 μC/g at the current configuration and 6 μC/g at the optimal configuration, the optimal configuration diminished the difference related to environment by more than 50%.

Table 2
Control factors of developer

	Control Factor	Level 1	2	3
A:	formulation	A_1	A_2	—
B:	manufacturing condition	B_1	B_2	B_3
C:	formulation	C_1	C_2	C_3
D:	manufacturing condition	D_1	D_2	D_3
E:	manufacturing condition	E_1	E_2	E_3
F:	manufacturing condition	F_1	F_2	F_3
G:	manufacturing condition	G_1	G_2	G_3
H:	manufacturing condition	H_1	H_2	H_3

Table 3
Estimation and confirmation of SN ratio and sensitivity of developer (dB)

Configuration	SN ratio		Sensitivity		SN ratio*		Sensitivity*	
	Estimation	Confirmation	Estimation	Confirmation	Estimation	Confirmation	Estimation	Confirmation
Optimal	−3.01	−0.30	26.08	27.68	−15.33	−14.90	13.78	13.08
Current	−9.33	−9.55	28.49	27.85	−24.36	−25.07	13.44	12.33
Gain	6.32	9.25	−2.41	−0.17	9.03	10.17	0.34	0.75

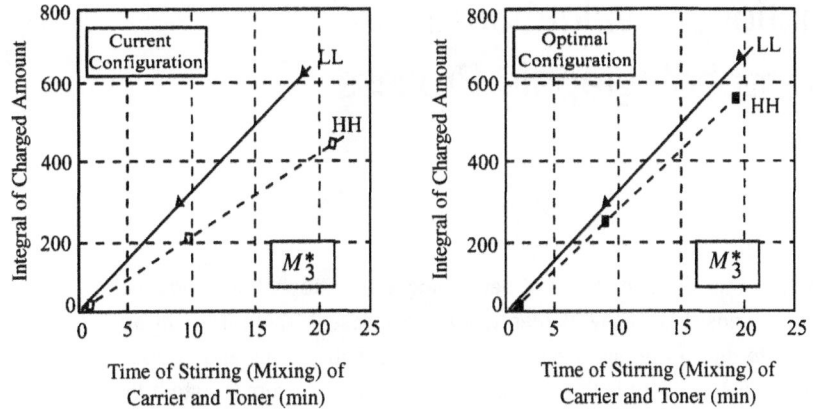

Figure 8
Results of confirmation experiment at current and optimal configurations

In addition, through durability (image stability) tests in the actual imaging, we confirmed that a new product reflecting the optimal configuration shows a more than fivefold durability compared to the current durability. Taking all the results into account, we proved that the optimal configuration obtained from the L_{18} experiment holds true for the actual machine.

5. Economic Calculation

Using the results of the experiment on actual imaging, we calculated the economic effect. While overlap and spill due to toner scattering occur when the amount charged falls below 10 $\mu C/g$, there is insufficient image density when it exceeds 40 $\mu C/g$. Thus, we can define functional limits as smaller than 10 $\mu C/g$ and greater than 40 $\mu C/g$ (range: 30 $\mu C/g$), respectively. The expected costs at functional limits were considered to be approximately 20,000 yen ($160), which includes both parts cost for replacement developer and service cost for repair personnel.

Thus, loss per photocopy machine can be calculated as follows:

$$L(\text{current}) = \frac{20,000 \text{ yen}}{(30/2)^2(14/2)^2}$$

$$= 4360 \text{ yen (difference due to environment: 14 } \mu C/g) \quad (14)$$

$$L(\text{optimal}) = \frac{20,000 \text{ yen}}{(30/2)^2(6/2)^2}$$

$$= 800 \text{ yen (difference due to environment: 6 } \mu C/g) \quad (15)$$

Based on this result, the difference in loss per product (ΔL) is

$$\Delta L = L(\text{optimal}) - L(\text{current}) = 3560 \text{ yen}$$

In other words, we can reduce loss per product by 3560 yen. Assuming that we sell 60,000 units annually, we can improve economic loss by 210,000 yen in a year.

Reference

Hiroyuki Kozuru, Yuji Marukawa, Kenji Yamane, and Tomoni Oshiba, 1999. Development of high quality and long life developer. *Quality Engineering*, Vol. 7, No. 2, pp. 60–66.

This case study is contributed by Hiroyuki Kozuru.

CASE 22

Functional Evaluation of an Electrophotographic Process

Abstract: In this study, methods of developing an electrophotographic process, which consist of many elements, were discussed. There are two approaches: to divide the entire system into several subsystems and evaluate each one individually, or is to evaluate the entire system. From the study we determinedthat the function of the entire system can be evaluated and part of the subsystem can be optimized at the same time.

1. Introduction

Electrophotography is widely used for plain-paper photocopy machines and laser printers. Figure 1 outlines a popular imaging device. As photosensitive material turns in the direction of the arrow, it is evenly charged on the surface, exposed by a laser beam modulated according to an image, developed by toner, transferred to transfer material such as paper, and the surface cleaned. Toner image transferred onto transfer material is fixed and carried out of a machine. The imaging device in Figure 1 can be regarded as consisting of functional subsystems such as electrical charge, exposure, development, transfer, and fixer.

In developing this device, two different approaches are possible: (1) organize a total system after optimizing each subsystem, or (2) evaluate the functionality of the entire system continuously from the beginning. The former can be ideal if there is no interaction among separated subsystems. However, we should consider that we have some interaction in case of a complex system such as an electrophotography device. In the following, we explain the method to integrate subsystems.

2. Generic Function

As shown in Figure 1, an imaging device is a system converting input image data into fixed toner images. Their relationship should be one-to-one and linear and controlled by units that are as small as possible in terms of adhesion. Once we achieve all of them, we can obtain both satisfactory gradation and vividness. Therefore, an ideal electrophotography system should reproduce dots of image data precisely with dots formed by toner. More specifically, an ideal characteristic has the following relationship between the square root of the number of image data dots, M, and the square root of the dot area of toner image, y:

Y (square root of dot area)

$= \beta M$ (square root of the number of image data (1)

Figure 2 shows this relationship.

Next we describe a parameter design based on the foregoing idea. Control factors can be selected from multiple subsystems or as common factors from such as developer. In this example, we assigned 12 factors to an L_{27} orthogonal array. As noise factors we chose factors that are special to electrophotography and significantly influential on results. We selected the following:

1. Paper type (popular noise factor, very influential on transfer and fixing)

2. Difference between the total and core areas of a dot (specific to electrophotography)

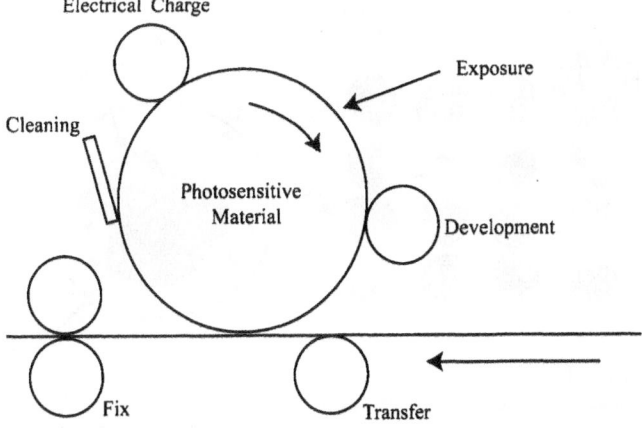

Figure 1
Electrophotography device

3. Maximum dot and minimum dot (specific to electrophotography)

As shown in Figure 3, quite often a dot formed on transfer material by toner is either separated or small portions are scattered around a large portion. As to factor 2, we regarded as noise the difference between the total sum of areas included in a dot, defined as one portion, and the area of its core area only. For factor 3 we set a dimensional difference between maximum and minimum dots to noise.

3. SN Ratio

After producing images using preset conditions, we measured magnified dot areas using image analysis software. The results are shown in Table 1. We surmised that the proportionality of input and output dot sizes would deviate if the beam diameter of the exposure light were large. This theory will be evaluated in the analysis. In short, we removed divergence from the proportionality, M_{res}, from a signal

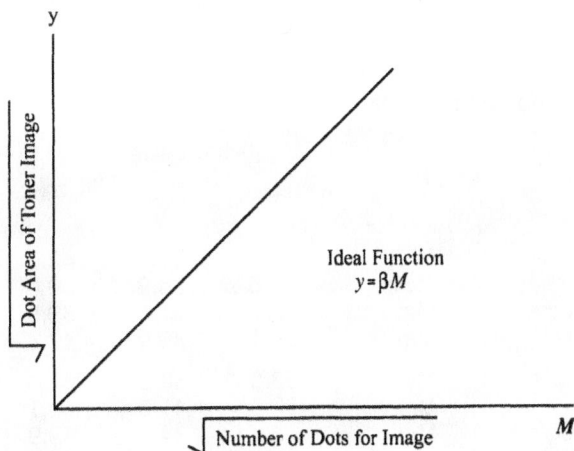

Figure 2
Generic function of imaging

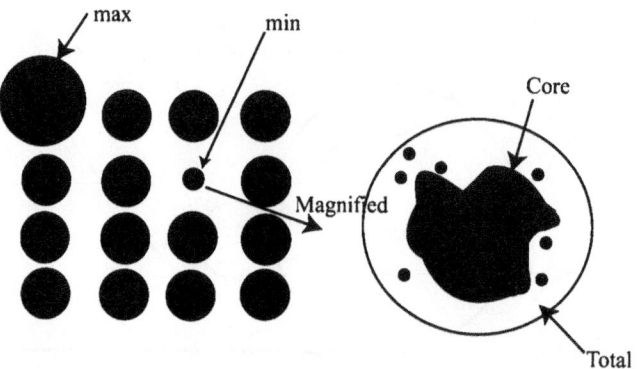

Figure 3
Noise factor in electrophotography image

effect, M. On the other hand, other variations, N, can be decomposed into the three error factors 1, 2, and 3 ($\beta \times$ paper type, $\beta \times$ gap, and $\beta \times$ size) and an error, e. Detailed calculations for the SN ratio and sensitivity were as follows.

Total variation:

$$S_T = 19.7^2 + 33.9^2 + \cdots + 54.6^2$$

$$= 54{,}167 \quad (f = 32) \quad (2)$$

Total error variance:

$$V_N = \frac{S_T - S_M}{28} = 13.82 \quad (3)$$

SN ratio:

$$\eta = 10 \log \frac{(1/8r)(S_\beta - V_e)}{V_N} = 12.02 \text{ dB} \quad (4)$$

Sensitivity:

$$S = 10 \log \frac{1}{8r}(S_\beta - V_e) = 23.43 \text{ dB} \quad (5)$$

4. Optimal Configuration and Confirmatory Experiment

Table 2 shows control factors assigned to an L_{27} orthogonal array. As a result of parameter design, we

Table 1
Example of square root of dot area of toner image

Paper Type	Gap	Min./Max.	Signal (Square Root of Number of Dots)				Linear Equation
			1	2	3	4	
Plain	Total	Min.	19.7	33.9	42.0	51.9	420.9
		Max.	23.5	35.9	49.0	56.7	469.0
	Core	Min.	13.7	31.4	39.9	49.9	395.9
		Max.	22.1	35.0	47.8	55.2	456.4
Overhead Projector	Total	Min.	23.5	37.0	47.4	53.4	453.3
		Max.	27.1	41.6	51.1	56.0	487.6
	Core	Min.	13.7	34.3	43.9	49.8	413.2
		Max.	22.5	39.3	48.8	54.6	465.9

Table 2
Control factors

	Control Factor	Level		
		1	2	3
Developer				
A:	ingredient content	A_1[a]	A_2	A_3
B:	ingredient content	B_1[a]	B_2	B_3
C:	ingredient type	C_1[a]	C_2	C_3
D:	ingredient content	D_1	D_2[a]	D_3
E:	manufacturing condition (quantity)	E_1	E_2[a]	E_3
Device				
F:	device condition (quantity)	F_1[a]	F_2	F_3
G:	device condition (quantity)	G_1	G_2[a]	G_3
H:	device condition (quantity)	H_1	H_2[a]	H_3
I:	device condition (quantity)	I_1	I_2[a]	I_3
J:	device condition (quantity)	J_1[a]	J_2	J_3
K:	device condition (quantity)	K_1[a]	K_2	K_3
L:	device condition (quantity)	L_1	L_2[a]	L_3
L:	device condition (type)			

[a] Current and optimal configurations.

Table 3
Result of confirmatory experiment

(a) Original Analysis Configuration	Estimation	Confirmation
Current	10.33	9.39
Optimal	16.61	13.21
Gain	6.28	3.82

(b) Reanalysis Configuration	Estimation	Confirmation
Current	12.94	10.80
Optimal	19.71	16.57
Gain	6.77	5.77

obtained factor effect plots (Figure 4). Based on the optimal and current configurations, we conducted a confirmatory experiment and obtained the results shown in Table 3a.

After we had investigated the reason for a difference between estimation and confirmation, we noticed that there was no control factor to improve the min./max. noise factor, so we proposed remov-

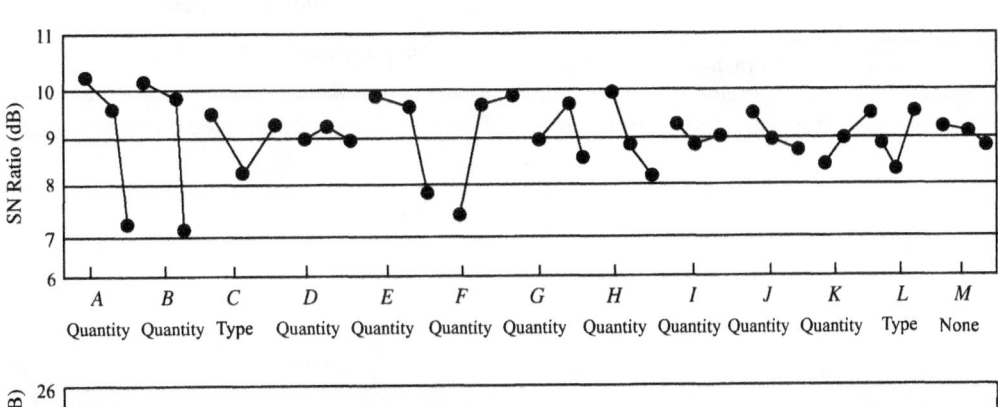

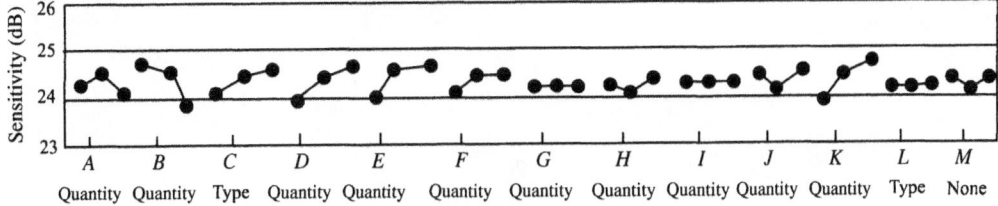

Figure 4
Response graphs of original analysis corresponding to Table 3a

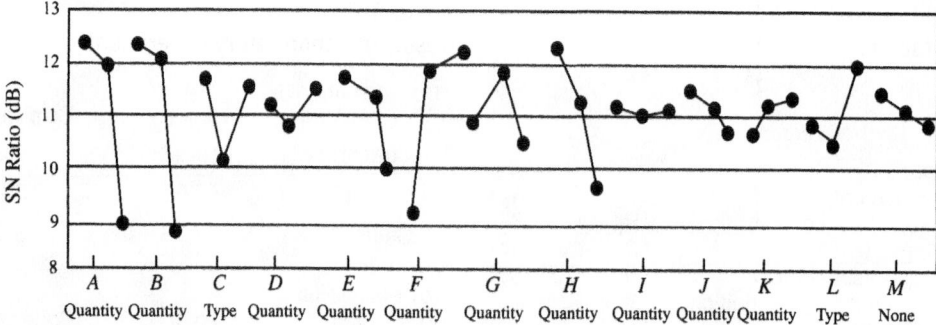

Figure 5
Response graphs of reanalysis corresponding to Table 3*b*

ing this contributing part. More specifically, in place of V_N as computed in equation (3), we used V'_N to calculate the SN ratio and sensitivity:

$$V'_N = \frac{S_{\beta \times error} + S_{\beta \times gap} + S_e}{27} = 8.43 \quad (6)$$

The eventual response graph was as shown in Figure 5, and the optimal configuration was the same as the original configuration shown in Table 3*a*. In the results of the confirmatory experiment shown in Table 3*b*, we see fairly good reproducibility of gain.

This procedure was applicable not only to the optimization of a total system but also to that of partial systems such as the development or transfer processes. As a consequence, we can ensure at an earlier stage that the total system is appropriate. One problem with this method is that we cannot use it without knowing the total system. Thus, when this method is used, we need a flexible application such as partial optimization determining the generic functions of subsystems.

Reference

Hisashi Shoji, Tsukasa Adachi, and Kohsuke Tsunashima, 2000. A study of functional evaluation for electrophotographic process. *Quality Engineering*, Vol. 8, No. 5, pp. 54–61.

This case study is contributed by Hisashi Shoji.

Part III

Robust Engineering: Mechanical Applications

Biomechanical (Case 23)

Machining (Cases 24–26)

Material Design (Cases 27–30)

Material Strength (Cases 31–33)

Measurement (Cases 34–35)

Processing (Cases 36–46)

Product Development (Cases 47–65)

Other (Case 66)

Biomechanical Comparison of Flexor Tendon Repairs

Abstract: The aim of this study was to determine the optimal method for flexor tendon repair. We used the Taguchi method of analysis to identify the strongest and most consistent repair. The optimum combination of variables was determined by the Taguchi method to be the augmented Becker technique. Five tendons repaired with the optimized combination were then tested and compared to the set of tendons repaired initially using the standard modified Kessler technique. The strength of the flexor tendon repair improved from an average of 17 N to 128 N with a standard deviation improving from 17% of the mean to 4% of the mean. Stiffness of the optimized repair technique improved from an average of 4.6 N/mm to 16.2 N/mm.

1. Introduction

Flexor tendon repair continues to evolve. Despite years of research by many investigators, the optimum suture technique remains elusive. As our understanding of tendon healing and biomechanics advances, new goals are set for ultimate motion. Historically, treatment involved delayed primary tendon grafting. Currently, primary repair is done followed by a postoperative active extension–passive flexion protocol [13]. There is now a trend toward primary repair followed by early *active* flexion, which aims to improve final results [24].

The premise is that early active flexion reduces adhesions and improves the ultimate range of motion and function [6]. Toward that end, investigators have tried to determine the strongest method of tendon repair [5,8,14,19,27,28,31]. Because of the many variables investigated, the various testing methods used, and the different focus of each study, a review of the literature does not provide the optimum combination of factors for the best repair technique.

The purpose of our experiment was to identify the factors that have the greatest effect on outcome variation and to determine the value of the factors that result in the most consistent outcomes Thus, the first goal of our study was to determine a method for tendon repair that provides greater strength combined with less outcome variability. To accomplish this, we used the Taguchi method of experimental design and analysis [16].

There are many examples in the medical field where the optimum method or process has not yet been determined because of the number of variables and the inherent limitations in patients of material for study. The second goal of our study was to evaluate the Taguchi method for future medical applications.

2. Materials and Methods

Material for Study

Flexor tendons were harvested from seven fresh-frozen human hands. Flexor pollicis longus, as well as index, middle, and ring finger flexor digitorum superficialis and flexor digitorum profundus tendons were each divided transversely into two sections of equal length. Comparison of proximal and distal sections have shown no substantial difference in tendon diameter or stiffness characteristics [3]. A total of 14 flexor tendon specimens were available per hand for testing. Tendon size is not under the control of the surgeon and was considered extraneous (so-called *noise*) according to the Taguchi method.

Static Testing

After each specific tendon repair, specimens were mounted in a servohydraulic mechanical testing machine (Instron, Canton, Massachusetts) for tension testing to failure. A jig was specially designed to grip the tendon ends so that minimal slippage would occur. The gripper was first tested with whole tendons and noted to withstand tensile loads of more than 1000 N before slippage was identified visually. Elongation was produced at a constant speed of 0.33 mm/s. A preload of 2.0 N was applied prior to testing to remove slack in the system. Ultimate strength was determined by the peak load recorded.

Real-time recording was made using an x–y plot for displacement–load analysis. This revealed that the initial portion of each curve was linear. Stiffness was determined by the slope of this initial curve. The load was determined at 2.0 mm displacement, and the result was divided by 2.0 to yield a stiffness of newtons per millimeter. We recognize that crosshead displacement for stiffness only provides an estimate of the true gap. During specimen testing we noted that the overwhelming majority of displacement was from the relatively weak and compliant repair. Increased stiffness is therefore reflective of less gap formation.

Variables under Study

Five control factors were studied: core tendon repair technique, core suture type, core suture size, epitenon technique, and distance from the repair site for core suture placement (Table 1).

Eight techniques of core tendon repair were examined, as shown in Figure 1: modified Kessler 1, consisting of a two-strand repair with a single knot within the repair site; modified Kessler 2, consisting of a two-strand repair with an epitenon-first repair with the core suture knot away from the repair site;

Table 1
Summary of factors under study

Control Factor	Levels[a]
Core suture technique	Modified Kessler 1 (MK1)
	Modified Kessler 2 (MK2)
	Double modified Kessler (Dbl. MK)
	Savage
	Lee
	Tsuge
	Tajima
	Augmented Becker (Aug. Beck)
Core suture material	Ethillon, Nurolon, Prolene, Mersilene
Core suture size	4-0, 3-0
Epitenon suture technique	None, volar only, simple, cross-stitch (cross)
Suture distance from cut edge (cm)	0.5, 1.0

[a] Modified Kessler 1, two-strand repair with a single knot within the repair site; modified Kessler 2, epitenon-first two-strand repair with a single knot outside the repair site; Lee, double-loop locking four-strand repair with two knots within the repair site; savage, six-strand repair with a single knot within the repair site; Tsuge, two-strand repair with a single knot outside the repair site; augmented Becker, four-strand repair with two knots outside the repair site; double modified Kessler, four-strand repair with a single not outside the repair site; Tajima, modified Kessler grasp with two knots within the repair site.

Tsuge, consisting of a two-strand repair with a single knot outside the repair site; Tajima, consisting of a two-strand repair with two knots within the repair site; Lee, consisting of four-strand repair with two knots within the repair site; double modified Kessler, consisting of a four-strand repair with the modified Kessler method repeated: a single suture with one knot outside the repair site being used to produce the four-strand core repair; augmented Becker, consisting of a four-strand with two knots outside the repair site; and Savage, consisting of a six-strand repair with a single knot within the repair site.

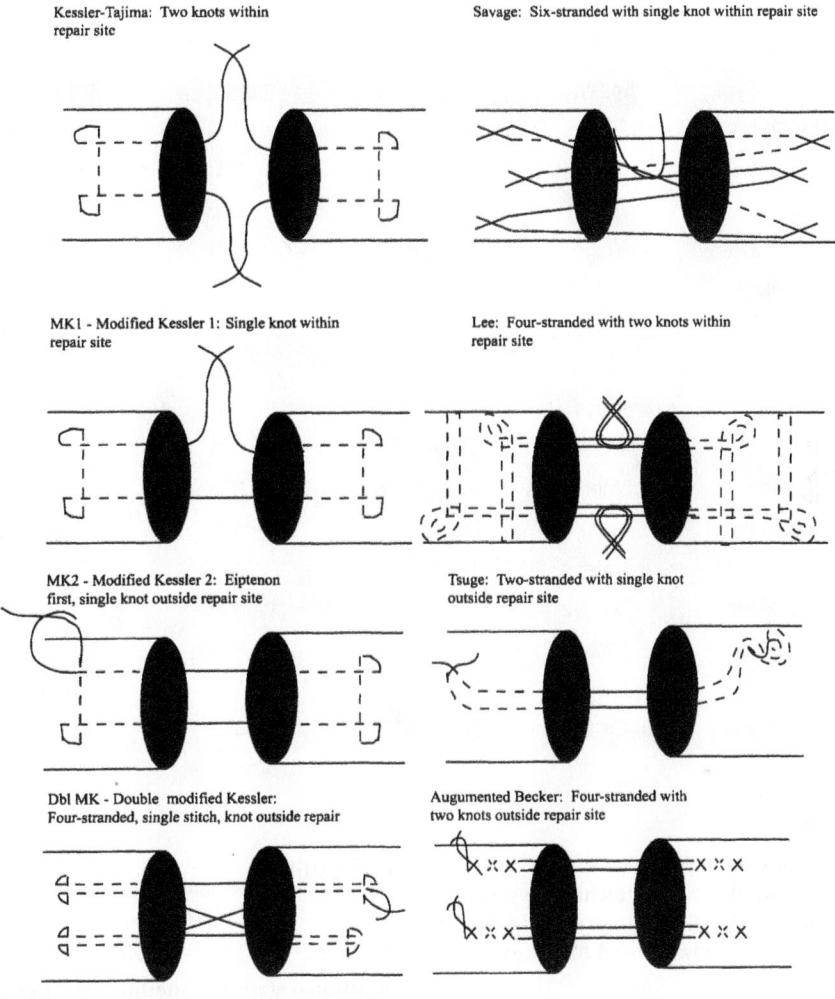

Figure 1
Eight repair techniques

Four core suture materials were examined: monofilament nylon (Ethilon), braided nylon (Nurolon), polypropylene monofilament (Prolene), and braided polyester (Mersilene) (Ethicon, Inc., Somerville, New Jersey). Two core suture sizes were examined: 4-0 and 3-0. Four epitenon techniques were examined: no epitenon, volar only, simple, and crossed. Suture distances of 0.5 and 1.0 cm were examined. (This refers to the distance from the repair site to the entry or exit and the transverse course of the core suture.)

All knots were three-throw square knots. All repairs were done by a single surgeon. All modified Kessler techniques used the locking-loop modification described by Pennington [17].

Experimental Design: Taguchi Method

The set of 16 experiments (Table 2) was repeated four times, for a total of 64 experiments. Within each set of 16 experiments, the order was randomized. In this case, 64 experiments gave a minimum

Table 2

Test matrix

No.	Suture Type	Epitenon	Technique	Core Size	Suture Distance (cm)
1	Ethilon	None	MK1	4-0	0.5
2	Ethilon	Volar	MK2	3-0	1.0
3	Nurolon	None	Lee	3-0	1.0
4	Nurolon	Volar	Savage	4-0	0.5
5	Nurolon	Simple	MK2	4-0	0.5
6	Nurolon	Cross	MK1	3-0	1.0
7	Ethilon	Simple	Savage	3-0	1.0
8	Ethilon	Cross	Lee	4-0	0.5
9	Prolene	None	Tsuge	4-0	1.0
10	Prolene	Volar	Aug. Becker	3-0	0.5
11	Mersilene	None	Dbl. MK	3-0	0.5
12	Mersilene	Volar	Tajima	4-0	1.0
13	Mersilene	Simple	Aug. Becker	4-0	1.0
14	Mersilene	Cross	Tsuge	3-0	0.5
15	Prolene	Simple	Tajima	3-0	0.5
16	Prolene	Cross	Dbl. MK	4-0	1.0

of eight repetitions of any particular variable. For example, the modified Kessler 1 technique was performed eight times, the Ethilon core was used 16 times, and the 4-0 suture was tested 32 times.

Initial and Confirmation Runs

To validate the method, the initial and the optimum repair techniques were compared. Initially, a set of eight flexor tendons was repaired using a standard method: a modified Kessler technique using a 4-0 Ethilon core suture placed 0.5 cm from the lacerated ends and a simple epitenon stitch with 6-0 Ethilon. At the completion of the 64 experiments, the optimum combination was determined using the Taguchi method. A set of five flexor tendons with this predicted optimum combination was then tested. The resulting mean and standard deviation were compared to the initial standard method to show the improvement that was obtained.

3. Results

SN Ratio

Traditional statistical methods use the mean to compare results. Using the standard deviation, one may then determine whether the difference between two groups is significant. With the Taguchi method, one uses a different statistic, the signal-to-noise (SN) ratio, to compare results.

In the Taguchi method, variables under study are divided into factors that either can be controlled (control factors) or factors which either cannot be controlled or are too expensive to control (noise factors). The greater the effect of the noise, the greater the inconsistency. The goal of the Taguchi method is to choose control factors that produce not only the desired result (such as stronger) but also to direct a process that is less sensitive to noise. Although noise cannot be eliminated, the effect of noise can be minimized. This produces a result that is not only stronger but is also less variable. Calculation of the SN ratio takes into account not only the mean but also the variation from one result to the next. Therefore, we can think of SN ratio analysis as being two-dimensional as opposed to regular analysis being only one-dimensional [16].

When the experimental goal is to maximize the outcome variable, the SN ratio in decibels is

$$\text{SN ratio} = -10 \log^{10} \frac{1/y_1^2 + \cdots + 1/y_n^2}{n}$$

where y is the strength or stiffness of each of the n repetitions of the experiments (here n was 4). The SN ratio is, in a sense, a combination of the mean and the variance. Mathematically, the ratio increases as the individual values become larger. Improved consistency or decreased variability between values also increases the ratio.

The SN ratios for tensile strength and stiffness for each of the 16 experimental combinations, with n equal to 4, were calculated (Table 3). We next calculated the SN ratios of each of the individual components of the five control factors. The values of these SN ratios are simply the average of the signal-to-noise ratios of all experiments containing that particular control factor component (Table 4). For example, the SN ratio for tensile strength considering the control factor suture and the component Ethilon is (Table 3, rows 1, 2, 7, 8)

SN ratio for Ethilon + $\frac{1}{4}$(25.8 + 30.4 + 37.2 + 31.2)
+ 31.2 dB

Results Using the Taguchi Method

Strength and stiffness were related only indirectly. Correlating the SN ratios of strength to those of stiffness yields a Pearson correlation coefficient of 0.51 ($p = 0.022$). This suggests that only about one-half of the variation in stiffness could be predicted by the strength. A higher ultimate strength therefore did not necessarily correlate with greater stiffness. Ultimate strength often occurred at more than 8.0 mm of displacement, whereas determination of stiffness was within the initial 2.0 mm of displacement. Diao et al. [3] also looked at strength and stiffness after tendon repair. Their group of tendons repaired with deep peripheral sutures failed catastrophically at 4 mm of displacement. Their group

repaired with superficial peripheral sutures failed gradually by a pattern of breaking and unwinding at 16 mm of displacement. Based on the biology of tendon healing, we would predict that a strong repair that allows a gap of 8 mm would not be successful. The performance of the repair should therefore be judged by studying both the strength and the stiffness.

A comparison of two SN ratios is less intuitive than a comparison of two means. Using the following formula, one may correlate decibel difference with percent difference:

$$\text{percent change in value} = (10^{x/20} - 1)(100)$$

where x is the change in SN ratio (in decibels). For example, a 1-dB difference in strength between factors is equivalent to a 12% difference in strength. A 10.0-dB difference is equivalent to a 215% difference.

Suture Technique

Of all the variables studied, suture technique had the greatest effect on both strength (change in ratio = 10.5 dB, maximum = 36.6 dB) and stiffness of the repair (change in ratio = 5.9 dB, maximum = 16.2 dB) (Table 4).

Core Suture

The SN ratios for the various core suture materials were similar, with the exception of Nurolon (Table 4). Several of the repairs with Nurolon failed because the knot untied. This did not occur with any of the other sutures. Three-throw square knots were useed throughout the study; four throws would be unrealistic given the bulk of the knot. This led to occasional low values, as reflected in the low SN ratio.

Epitenon

Moderate improvement with epitenon repair was obtained compared to no epitenon repair. Simple epitenon was 31.2 dB − 28.7 dB = 2.5 dB better than no epitenon repair (Table 4). This is approximately $10^{2.5/20}$ or 33% stronger.

Table 3
Tensile strength, stiffness, and SN ratios

	Suture Type	Epitenon	Technique	Core	Suture Distance (cm)	Tensile Strength for Each Repetition (N)				Stiffness for Each Repetition (N/mm)				Signal-to-Noise Ratio for 4 Repetitions (dB)	
						TS1	TS2	TS3	TS4	Stiff1	Stiff2	Stiff3	Stiff4	SN TS	SN Stiff
1	Ethilon	None	MK1	4-0	0.5	19	16	21	26	1.75	2.40	2.57	2.05	25.8	6.5
2	Ethilon	Volar	MK2	3-0	1.0	34	32	36	32	4.51	4.55	2.53	3.78	30.4	10.9
3	Nurolon	None	Lee	3-0	1.0	26	16	49	48	4.81	4.81	4.72	4.51	28.1	13.5
4	Nurolon	Volar	Savage	4-0	0.5	52	50	46	26	5.71	9.51	5.20	3.95	31.7	14.5
5	Nurolon	Simple	MK2	4-0	0.5	13	18	22	22	5.50	6.88	2.96	3.95	24.8	12.4
6	Nurolon	Cross	MK1	3-0	1.0	21	15	24	37	6.06	7.18	3.99	4.55	26.3	14.0
7	Ethilon	Simple	Savage	3-0	1.0	95	88	59	65	2.74	4.77	3.73	4.59	37.2	11.3
8	Ethilon	Cross	Lee	4-0	0.5	38	44	29	39	4.37	4.98	6.23	5.93	31.2	14.4
9	Prolene	None	Tsuge	4-0	1.0	21	19	25	24	2.48	2.66	2.01	2.53	27.0	7.5
10	Prolene	Volar	Aug. Becker	3-0	0.5	72	80	67	90	7.83	7.57	5.67	5.93	37.6	16.3
11	Mersilene	None	Dbl. MK	3-0	0.5	39	45	58	73	7.14	6.88	5.93	7.14	33.9	16.5
12	Mersilene	Volar	Tajima	4-0	1.0	14	27	27	25	5.24	4.64	6.66	3.52	26.3	13.3
13	Mersilene	Simple	Aug. Becker	4-0	1.0	53	70	66	60	5.89	7.65	5.28	7.22	35.7	16.0
14	Mersilene	Cross	Tsuge	3-0	0.5	21	33	20	21	7.14	7.5	7.78	8.82	27.0	17.8
15	Prolene	Simple	Tajima	3-0	0.5	21	26	27	19	8.86	6.49	3.35	5.84	27.1	14.1
16	Prolene	Cross	Dbl. MK	4-0	1.0	31	49	46	46	6.62	6.66	8.34	4.21	32.2	15.4

Table 4
SN ratios for strength and stiffness (dB)

	SN Ratio	
	Strength	**Stiffness**
I. Technique		
MK1	26.1	10.3
Taijima	26.7	13.9
Tsuge	27.0	12.7
MK2	27.6	11.7
Lee	29.7	14.0
Dbl MK	33.0	16.0
Savage	34.4	12.9
Aug. Becker	36.6	16.2
II. Core suture		
Nurolon	27.7	13.6
Mersilene	30.7	15.9
Prolene	30.9	13.3
Ethilon	31.1	10.8
III. Core size		
4-0	29.3	12.5
3-0	31.0	14.3
IV. Epitenon		
None	28.7	11.0
Cross	29.2	15.4
Simple	31.2	13.5
Volar	31.5	13.8
V. Suture distance (cm)		
0.5	29.9	14.1
1.0	30.4	12.7

Core Suture Size

The 3-0 Mersilene suture improved strength 1.7 dB compared to 4-0 Mersilene (31.0 dB − 29.3 dB). The stiffness improved by 1.8 dB (Table 4).

Suture Distance

Suture distance for core suture placement from the cut edge had the least effect on strength and stiffness of any variable tested. The failure mode of the core suture for repair at 0.5 cm showed more failures from suture pullout than from breakage. In repairs at 1.0 cm, only four of 32 (13%) repairs failed by pullout. With repairs at 0.5 cm, 11 of 32 (34%) failed by pullout. Pullout of the core suture is in a sense a premature failure in that the suture pulled out of the tendon before the suture itself failed. The

SN ratio shows that a smaller suture distance gives greater stiffness, an intuitively apparent result.

Optimum Combination of Factors

In choosing the optimum combination of factors, both strength and stiffness were considered. The optimum combination of variables was found to be an augmented Becker technique using 3-0 Mersilene core suture placed 0.75 cm from the cut edge with volar epitenon suture. As anticipated, this exact combination had not actually been tested in the Taguchi array. A set of tendons was then repaired with this optimized combination to confirm the results of the Taguchi analysis.

The results of the optimum combination tested confirmed the results predicted to be accurate. The initial standard combination was compared to the optimum combination, both predicted and tested (Table 5). Included are values for strength and stiffness. The low values for the series are given along with the percent decrease from the mean.

4. Discussion

The goal of flexor tendon repair is restoration of full motion of the finger. Historically, repair was followed by postoperative immobilization [10,11]. Healing of flexor tendons was thought to occur via an extrinsic process mediated by the flexor sheath [18]. This was logically thought to require immobilization of the tendon with necessary adhesion formation. Final motion of the digit was, not surprisingly, limited. Studies by Lundborg and Rank [12] and Gelberman et al. [6] provide evidence that the flexor tendon has an intrinsic repair capability. This revision of the understanding of the healing process gave impetus to the need to study postoperative motion protocols.

In a study of canine tendon healing, Gelberman et al. [6] found that tendon healing could occur without adhesion formation. In addition, mobilized tendons healed more rapidly than immobilized repairs and had greater ultimate strength [6]. In a clinical study, Kleinert et al. [9], using postoperative active extension and passive flexion, produced substantially improved results. In a subsequent clinical study, Strickland et al. [25] confirmed the benefits

Table 5

Comparison of standard and optimum combinations of variables

	Standard Method	Optimum Combination	
		Predicted	Tested
Mean tensile strength (N)	17.2 ± 2.9	94	128 ± 5.6
Low value (difference from mean)	13.1 (24%)		121 (5%)
Mean stiffness (N/mm)	4.6 ± 1.0	10	16.2 ± 5.8
SN ratio (dB)			
Tensile strength	24.4	40	42.1
Stiffness	12.7	20	22.8

of postoperative light-active rehabilitation after a four-strand core suture repair in zone II flexor tendon lacerations. The light-active mobilized group yielded 76% (19 of 25) excellent and good results. This was in comparison to 56% (14 of 25) excellent and good results compared to a previously reported group treated with passive motion after a two-strand core suture repair.

Prior to the healing tendon sharing the load, active flexion places increased demand on the repair. Currently, popular techniques are not sufficiently strong to withstand the forces associated with mobilization. Many factors have been studied to create a stronger repair. Most studies have focused on various core suture techniques [7,14,28,31]; some have investigated core suture materials [8,28,29] or epitenon techniques [15,29]. Comparisons and conclusions are difficult because these studies have involved a variety of tendon models (dog, rabbit, chicken, human), different size sutures for the same techniques, different techniques for testing, and have been both in vivo and in vitro studies.

Some investigators have focused on gap formation [1,22,23]. Logic dictates that a gap at the repair site will fill with fibrous tissue and lead to an inferior repair with regard to strength, stiffness, and tendon length. This will present clinically as decreased total active motion and an increased rupture rate. In a prospective clinical study using radiopaque markers, Seradge [22], found a direct correlation between the amount of elongation at the repair site and the incidence of secondary tenolysis. In contrast, Silverskiold et al. [23] found only a weak correlation between elongation and final interphalangeal joint motion.

Small and Colville [24] reported results of a prospective clinical study that used an early active flexion protocol in 98 patients. Using the modified Kessler technique with 4-0 Ethilon or 4-0 Monofil core suture and 6-0 Prolene epitenon suture, they had excellent or good results in 77% (90 of 117) of digits, but noted dehiscence in 9.4% (11 of 117) of digits. Cullen et al. [2] noted a rupture rate of 6.5% (two of 31) of digits. They used a modified Kessler technique using 3-0 Tycron core and 6-0 Prolene epitenon stitches.

To interpret expected demands on a tendon repair, it is important to examine the forces that the tendon may generate. Schuind et al. [21] measured in vivo forces using a specially designed device during carpal tunnel release. They found forces of up to 8.8 N in passive mobilization, up to 34.3 N in active unresisted finger motion, and up to 117 N in tip pinch. One would predict that in a digit with edema, forces even higher than 34.3 N may be generated in an early active motion protocol.

These studies on early active motion are encouraging [20]. They provide evidence that motion is improved following early active flexion. However, the higher rupture rate reflects the large load that is placed on the relatively weak repair.

Optimum Variables

In determining the optimum combination for repair, both strength and stiffness were considered

(Table 4). For each variable, such as technique or suture material, the ideal choice would be the variable that gives the maximum value for both strength and stiffness. When the maximums did not coincide, a rationale was provided for choosing the optimum variable.

For technique, the augmented Becker was chosen because it yielded the highest ratio for both strength and stiffness. For core suture type, Mersilene was chosen since it performed nearly the highest for strength and clearly the highest for stiffness; prolene would be the second choice. For suture size, 3-0 yielded both the highest strength and the highest stiffness; volar epitenon was chosen since strength was considered to be more important than stiffness. Distance from the end was the only continuous variable, and 0.75 cm was chosen to optimize both strength and stiffness. One may choose a combination of variables other than the optimum combination determined, but a confirmation experiment should be done for that combination to avoid the effect of unexpected interactions.

Suture technique was the most important variable studied (Table 4). As might be predicted, the two-strand techniques gave the lowest values. The six-strand Savage technique did not perform as well as expected. This may have resulted from the inability of each suture across the repair to share the load evenly. This would lead to earlier-than-expected failure because of overload on any one suture.

Schuind et al. [21] estimated the maximum force expected during active flexion to be 34.3 N. Substituting this value into the formula gives a value of 30.7 dB. This is an estimate of the minimum SN ratio required for tendon repair. None of the two-strand techniques was able to achieve this level (Table 4). An SN ratio of more than 31.0 dB was reached for both the double modified Kessler and the augmented Becker (Table 4). Both of these repair techniques should be strong enough to withstand early active flexion. In our opinion, the six-strand Savage technique is too difficult and time consuming to be widely accepted by surgeons.

Taking into account both strength and stiffness, Mersilene would be the first choice and Prolene the second choice (Table 4). Techniques that require suturing into as opposed to across or down the tendon fibers require more handling of the tendon.

Braided suture has more friction and tends to deform the tendon more than monofilament suture. Technical considerations may therefore lead to the choice of Prolene. Ethilon would have been chosen if strength alone was used for selection. Ethilon performed poorly with respect to stiffness and therefore is not the optimum choice when considering overall repair performance.

Increasing the core suture size substantially improved both strength and stiffness (Table 4), but not as much as would have been predicted from the material properties alone. Ethicon, Inc. [4] reports a suture strength for 4-0 Mersilene of 13.0 N and for 3-0 Mersilene of 18.0 N. This reflects an improvement in strength of 38% in going from 4-0 to 3-0 Mersilene. The Taguchi method showed that the 3-0 core suture improved strength 1.7 dB compared with 4-0 core suture (31.0 dB − 29.3 dB). This translated into a difference of approximately 22%.

Injury to flexor tendons outside zone II would certainly be better repaired with the large suture [26]. In zone II lacerations, additional studies would have to be made to evaluate the gliding characteristics before the 3-0 suture could be recommended because of the added bulk. There is a clinical precedent for using this size suture in zone II injuries. Cullen et al. [2] report on their results of early active motion in zone II repairs using 3-0 Tycron modified Kessler technique. They had 77% (24 of 31 digits) excellent or good results with a rupture rate of 6.5% (two of 31 digits).

The addition of a simple epitenon suture has been shown in previous studies to have an impact on strength [15,29]. Volar epitenon was chosen to test the clinical situation where suturing only the volar epitenon is technically feasible. Simple epitenon repair was chosen over volar only because performance was similar mechanically and the biology of tendon healing suggests that unexposed tendon leads to less scar production [30]. Simple epitenon was stronger than no epitenon by 2.5 dB, or approximately 33%. Wade et al. [29] found that adding a peripheral 6-0 polypropylene stitch improved strength by 12.7 N, with an ultimate strength of 31.3 N, a difference of 41%. The results here agree with Wade et al. that a peripheral stitch adds substantial strength to the repair.

Comparing modified Kessler 1 (core suture first) to modified Kessler 2 (epitenon suture first) repairs, the SN ratio for strength improved from 26.1 dB to 27.6 dB, respectively. This translates to an expected improvement in strength of $10^{1.5/20}$, or 19% improvement by suturing the epitenon first. This correlated well with the study by Papandrea et al. [15] using 26 matched canine tendons. They found an improvement of 22% when performing epitenon first repair using 4-0 braided polyester core suture and 6-0 braided polyester epitenon suture.

Standard versus Optimum Combination

The Taguchi method was used to identify a stronger and less variable repair method. The confirmation experiment determines whether the Taguchi method of study was accurate. Variation may be measured by the standard deviation, but in the spirit of attaining quality, the minimum result is also important. It is the occasional low value that is associated with the occasional failure.

The optimal flexor tendon repair improved in strength from 17.0 N to 128 N, with a low value going from 24% below the mean to 5% below the mean (Table 5). Anticipating stress on a repair of up to 35 N during unopposed active flexion, a repair that can resist 128 N of tension with minimal variation should give the surgeon enough confidence to begin an early postoperative active flexion protocol.

The increase in stiffness from 4.6 N/mm to 16.2 N/mm substantially reduces the gap under physiologic load. With the standard combination, a load of 34.0 N would exceed the expected strength of the repair, but if the repair remained intact, a gap of 7.4 mm is predicted. With the optimum combination, a maximum gap 2.1 mm is expected at 34.0 N of load.

Taguchi Method

The Taguchi method differs from traditional statistical methods by its focus on identifying a solution that is, in this instance, both a stronger and a less variable method of repair. The parameter used for optimization is the SN ratio, in which a low value is more heavily penalized by the formula used than a high value is rewarded. In other words, it is the low values that are associated with failure. Not only does the SN ratio help reduce variability, but it does so by identifying control factors that lead to those low

values. The Taguchi method identifies which factors provide the greatest contribution to variation and determines those settings or values that result in the least variability. However, the method does not allow easy analysis of interactions between factors.

The optimum technique for a flexor tendon repair shown in this study still may not satisfy many surgeons. It may, for example, be unacceptable because of time or effort to perform the augmented Becker technique. One may in fact decide that any four-strand repair technique is either technically unappealing, injurious to the tendon, or both. Inspection of the signal-to-noise ratio graph, however, allows one to realize that the two-strand techniques are simply too weak to allow for a reliable early active motion protocol. The clinical series by Small and Colville [24] and Cullen et al. [2] both used a two-strand repair technique followed by early active flexion. Small reported 77% (90 of 117 digits) excellent or good results with a dehiscence rate of 9.4% (11 of 117 digits). Cullen reported 77% (24 of 31 digits) excellent or good results with a rupture rate of 6.5% (two of 31 digits). The Taguchi method suggests that by using the optimum combination, a better range of motion and a smaller rupture rate are possible. The results show that only a four-strand technique is strong enough to perform immediate active flexion rehabilitation reliably.

References

1. H. Beckerand and M. Davidoff, 1977. Eliminating the gap in flexor tendon surgery. *Hand,* Vol. 9, pp. 306–311.

2. K. W. Cullen, P. Tolhurst, D. Lang, and R. Page, 1989. Flexor tendon repair in zone 2 followed by controlled active mobilization. *Journal of Hand Surgery,* Vol. 14-B, pp. 392–395.

3. E. Diao, J. S. Hariharan, O. Soejima, and J. C. Lotz, 1996. Effect of peripheral suture depth on strength of tendon repairs. *Journal of Hand Surgery,* Vol. 21-A, pp. 234–239.

4. Ethicon, 1983. *Ethicon Knot Tying Manual* Somerville, NJ: Ethicon, Inc., p. 39.

5. R. H. Gelberman, J. S. Vandeberg, G. N. Lundborg, and W. H. Akeson, 1983. Flexor tendon healing and restoration of the gliding surface. *Journal of Bone and Joint Surgery,* Vol. 65-A, pp. 70–80.

6. R. H. Gelberman and P. R. Manske, 1985. Factors influencing tendon adhesions. *Hand Clinics*, Vol. 1, pp. 35–42.

7. D. P. Greenwald, H. Z. Hong, and J. W. May, 1994. Mechanical analysis of tendon suture techniques. *Journal of Hand Surgery*, Vol. 19-A, pp. 642–647.

8. L. D. Ketchum, 1985. Suture materials and suture techniques used in tendon repair. *Hand Clinics*, Vol. 1, pp, 43–53.

9. H. E. Kleinert, J. E. Kutz, T. S. Ashbell, and E. Martinez, 1967. Primary repair of lacerated flexor tendons in "no man's land." *Journal of Bone and Joint Surgery*, Vol. 49-A, p. 577.

10. J. P. Leddy, 1993. Flexor tendons: acute injuries. In *Operative Hand Surgery*, D. Green (ed.), 3rd ed., Vol. 2. New York: Churchill Livingstone, pp. 1923–1851.

11. H. Lee, 1990. Double loop locking suture: a technique of tendon repair for early active mobilization. *Journal of Hand Surgery*, Vol. 15-A, pp. 945–958.

12. G. Lundborg and F. Rank, 1978. Experimental intrinsic healing of flexor tendons based upon synovial fluid nutrition. *Journal of Hand Surgery*, Vol. 3-A, pp. 21–31.

13. R. Meals, 1985. Flexor tendon injuries. *Journal of Bone and Joint Surgery*, Vol. 67-A, 817–821.

14. M. Noguchi, J. G. Seiler, R. H. Gelberman, R. A. Sofranko, and S. L. Y. Woo, 1993. In vitro biomechanical analysis of suture methods for flexor tendon repair. *Journal of Orthopedic Research*, Vol. 11, pp. 603–610.

15. R. Papandrea, W. H. Weitz, P. Shapiro, and B. Borden, 1995. Biomechanical and clinical evaluation of the epitenon-first technique of flexor tendon repair. *Journal of Hand Surgery*, Vol. 20-A, pp. 261–266.

16. G. S. Peace, 1993. *Taguchi Methods: A Hands-on Approach to Quality Engineering*. Reading, MA: Addison-Wesley, pp. 1–236 and 292–312.

17. D. G. Pennington, 1979. The locking loop tendon suture. *Plastic Reconstructive Surgery*, Vol. 63, pp. 648–652.

18. A. D. Potenza, 1963. Critical evaluation of flexor-tendon healing and adhesion formation within artificial digital sheaths. *Journal of Bone and Joint Surgery*, Vol. 45-A, 1217–1233.

19. W. E. Sander, 1992. Advantages of "epiteono first" suture placement technique in flexor tendon repair. *Clinical Orthopedics*, Vol. 280, pp. 198–199.

20. R. Savage, 1985. In vitro studies of a new method of flexor tendon repair. *Journal of Hand Surgery*, Vol. 10-B, pp. 135–141.

21. F. Schuind, M. Garcia-Elias, W. P. Cooney, and K. N. An, 1992. Flexor tendon forces: in vivo measurements. *Journal of Hand Surgery*, Vol. 17-A, pp. 291–298.

22. H. Seradge, 1983. Elongation of the repair configuration following flexor tendon repair. *Journal of Hand Surgery*, Vol. 8-A, pp. 182–185.

23. K. L. Silfverskiold, E. J. May, and A. H. Tornvall, 1992. Gap formation during controlled motion after flexor tendon repair in zone II: a prospective clinical study. *Journal of Hand Surgery*, Vol. 17-A, pp. 539–546.

24. J. C. Small and J. Colville, 1989. Early active mobilization following flexor tendon repair in zone 2. *Journal of Hand Surgery*, Vol. 14-B, pp. 383–391.

25. J. W. Strickland, N. M. Cannon, and K. H. Gettle, 1995. Early AROM program for zone II flexor tendon lacerations. *Proceedings of the Annual Meeting of the American Society of Surgery—Hand* (abstract).

26. J. W. Strickland and S. V. Glogovac, 1980. Digital function following flexor tendon repair in zone II: a comparison of immobilization and controlled passive motion techniques. *Journal of Hand Surgery*, Vol. 5-A, pp. 537–543.

27. K. Tsuge, Y. Ikuta, and Y. Matsuishi, 1997. Repair of flexor tendons by intratendinous tendon suture. *Journal of Hand Surgery*, Vol. 2-A, pp. 436–440.

28. J. R. Urbaniak, J. D. Cahill, and R. A. Mortenson, 1975. Tendon suturing methods: analysis of tensile strengths. *Symposium on Tendon Surgery in the Hand*. St. Louis, MO: Mosby, pp. 70–80.

29. P. J. Wade, I. F. Muir, and L. L. Hutcheon, 1986. Primary flexor tendon repair: the mechanical limitations of the modified Kessler technique. *Journal of Hand Surgery*, Vol. 11-B, pp. 71–76.

30. W. F. Wagner, C. Carroll, J. W. Strickland, D. A. Heck, and J. P. Toombs, 1994. A biomechanical comparison of techniques of flexor tendon repair. *Journal of Hand Surgery*, Vol. 19-A, pp. 979–983.

31. R. C. Wray and P. M. Weeks, 1980. Experimental comparison of techniques of tendon repair. *Journal of Hand Surgery*, Vol. 5-A, pp. 144–148.

This case study is contributed by Gordon Singer.

CASE 24

Optimization of Machining Conditions by Electrical Power

Abstract: Together with others, we reported research on parameter design using an L_{12} orthogonal array as a study of energy conversion at the initial stage to confirm additivity of measurements and effectiveness of energy evaluation. Test pieces of ferrous and copper materials were used. We obtained good reproducibility of gain using energy evaluation and satisfied quality characteristics as our final goal. However, due to the limited scale of our experiment, we could not investigate the details and left many issues behind in terms of machining efficiency and generality of research.

1. Functional Evaluation by Energy Conversion

To assess machining of stainless steel used for mass production, we conducted a practical experiment using an L_{18} orthogonal array. We surmised that there were certain technical issues because we have not been able to obtain satisfactory reproducibility, even though we have implemented several different analyses after encountering extremely poor reproducibility at first. Considering that there have been some problems with variability of energy during machine idling after referring to the research of Ford, which deals with energy evaluation during idle time, by adding electrical power during idling, we have analyzed the relationships among time, material removed, and electrical power by use of the SN ratio. For electrical power, we calculated the product of time and power as area so as to effectively reflect its variability. For a noise factor, we selected a difference between maximum and minimum electrical power. Using all of them, we computed SN ratios.

2. Generic Function

The objective of machining is to cut a product or part cost-effectively and accurately to realize a target shape. Therefore, machining engineers select optimal conditions by changing conditions of machines and tools used or cutting conditions such as cutting or feeding speeds, and measuring eventual dimensions and roughness of a product or part. In contrast, the objective of machining evaluation by energy is to assess general functions of machines and secure final quality characteristics (machining accuracy or surface roughness).

As an effective evaluation method of cutting, including machine performance, we can pick up a change between electrical power supplied to a machine and power used during cutting. In other words, we assumed that cutting efficiency can be assessed by the relationship between time consumed for cutting and electrical power consumed by a machine. We concluded that unsatisfactory precision of work is caused by inefficient consumption of energy for a target material amount to be removed, due to a factor such as unevenness of material, tool condition, or cutting setup. Generic function 1 is expressed as $y = \beta_1 T$ by the relationship between cutting time, T, and electrical power, y. In this case, the greater the SN ratio and sensitivity, the better. Generic function 2 is expressed as $\sqrt{y} = \beta_2 \sqrt{M}$, where the amount removed is M and the power is y. For this, less sensitivity and a higher SN ratio are desirable. Figures 1 and 2 illustrate these relation-

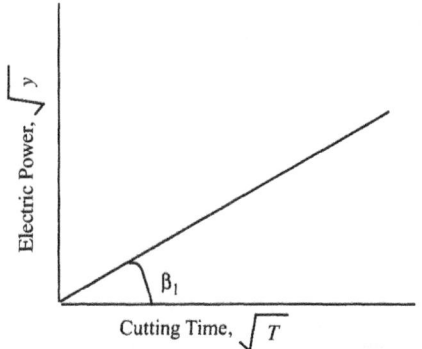

Figure 1
Generic function 1

ships. The reason that we take the square root of both sides of an equation is that in quality engineering, factors are dealt with as energy in decomposing total variation.

3. Measurement and Experimental Procedure

A common manual lathe was used for this experiment. A wattmeter connected to a three-phase distribution board located at a power source for the lathe measured effective power (W) consumed for cutting. Figure 3 depicts the shape of a test piece. Its material is SUS 304. Two grooves were added

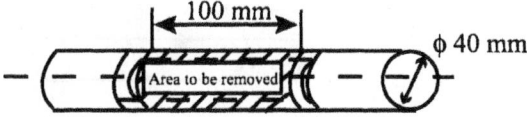

Figure 3
Shape of the test piece

beforehand at the start and end points to clarify them, and the length between them was held constant. Figure 4 outlines the cutting processes. We regarded the electrical power needed to run a lathe after a test piece is chucked on it as idling power. Subsequently, we measured each level of power while cutting the area removed three times. Figure 5 shows a magnified plot of fluctuation in power for cutting run 1. l_{before} indicates a fluctuation of power during idling and h represents a fluctuation while cutting material by feeding a tool. Once cutting is completed, fluctuation goes down to l_{after}. Because l_{before} and l_{after} show power during idling, only h indicates total electrical power that a machine consumes for cutting and idling. Therefore, subtracting l_{before} and l_{after} from h, we can obtain the actual power needed for cutting. Although we do not illustrate plots for other cutting conditions, they also showed great fluctuation. Additionally, the ratio of cutting power to idling power was small; in short, machining efficiency was regarded as poor. Then we concluded that we should evaluate the variability and instability of energy of a machine.

4. Design of Experiments

For generic function 1, as signal factors we selected each cumulative sum of 12 time intervals into which total time duration from start to finish of cutting was divided equally: T_1, T_2, T_3, ... , and T_{12}. We repeated three times cutting of the area removed. For generic function 2, by cutting the amount removed three times, we measured work for each cutting as a signal factor. Next, considering ease of measurement, we substituted change in the amount removed, M_1, M_2, and M_3, for mass removed per se. For both functions, as the output characteristic we selected the cumulative value of the electrical power, y, for signal at each factor level.

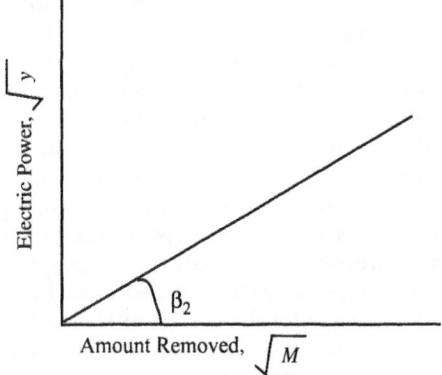

Figure 2
Generic function 2

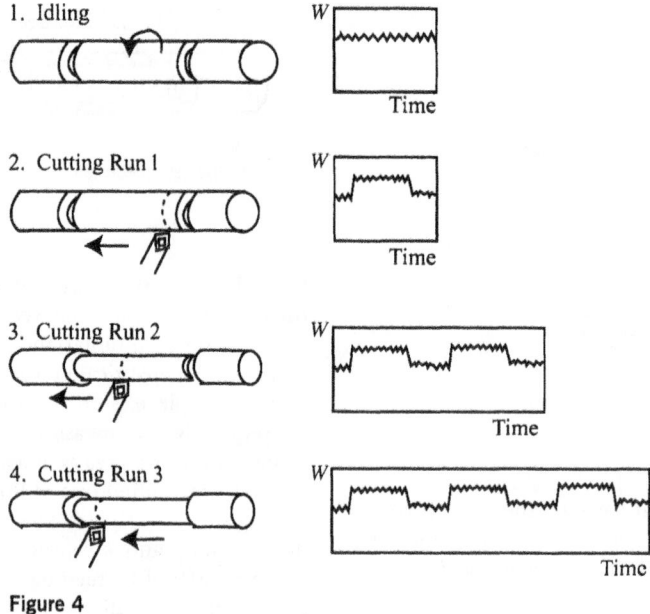

Figure 4
Cutting processes

Assuming that causes for errors affect the variability of power in a direct or indirect manner, we set a difference between maximum and minimum values of electrical power to a noise factor, which should be small. In addition, we selected a minimum value of power y_{min} as N_1 and a maximum y_{max} as N_2. Since the diameter of a test piece becomes smaller as we repeat cutting, a change in electrical power for each cutting run was also chosen as a noise factor.

As control factors, we picked various factors, such as machining setup and tool condition. We confirmed that a revolution during machining does not vary when measured with a tachometer. Control factors are summarized in Table 1.

Electricic Power Measurement Results

When a cumulative value of electrical power is used, its variability is sometimes hidden, due to its accumulation. To solve this problem we substituted a product or area of time and a minimum value W_{min} (or a maximum value W_{max}) for each divided time interval for the simple cumulative value (Figure 6).

For generic function 1, we calculated the maximum and minimum of electrical power, W, for each divide time interval (Figure 7). Table 2 shows a sample of the electrical power measured for cutting run 1. Idling power means before- or after-cutting power. Moreover, by including power during idling, we show the cumulative relationship between time, T, and electrical power, W, in Table 3 and Figure 8, which is a schematic of Table 3. For generic function 2, we computed the electrical power for each time duration from start to finish of cutting (Figure 9). Using electrical power during idling as a standard, we accumulated each area of power (Figure 10), which represent the data of experiment 1 of the L_{18} orthogonal array. The symbol P_0 indicates idling, and P_1, P_2, and P_3 indicate the cutting run number. T, M, and y are point of time, amount removed, and cumulative value of electrical power calculated as area, respectively. In addition, to evaluate the linearity of these data, we plotted Figure 11 for the change of electrical power during idling (Table 4), Figure 12 for the change of electrical power during cutting (Table 5), and for the change in electrical power versus mass removed. As a result, we can see the linearity for each case.

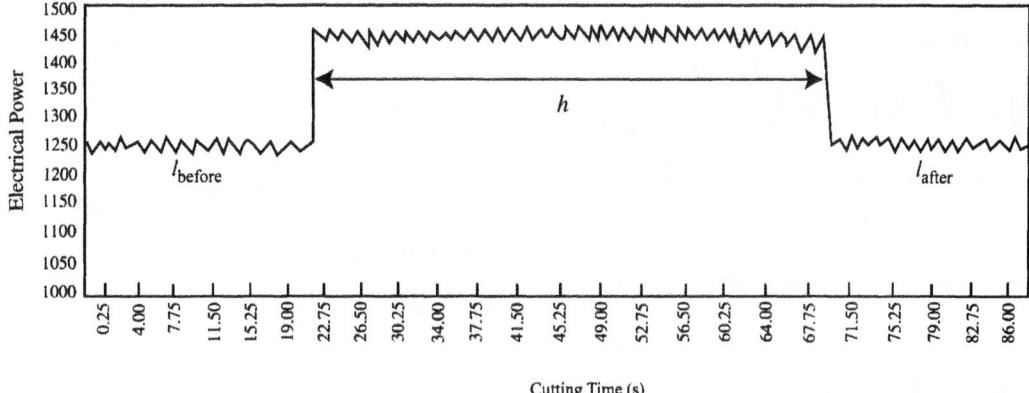

Figure 5
Fluctuation of electrical power at cutting run 1

Based on these results, we describe our calculation process in the following section.

5. SN Ratios and Response Graphs

SN Ratio for Generic Function: Time versus Electrical Power

By calculating the square root of each data point in Table 3, we obtained the converted data in Table 6. We computed $S_{M*\beta}$, which is the effect due to a difference between idling and cutting. For energy consumption, it should be smaller during idling and greater during cutting. By regarding this difference of effect as an effective portion of energy, we calculated the SN ratio.

Total variation:

$$S_T = 125.419^2 + 128.452^2 + \cdots + 349.428^2$$

$$= 4,574,898.032$$

Linear equations:

Table 1
Control factors and levels

	Control Factor	Level		
		1	2	3
A:	lubricant dilution (%)	Little	Mid	—
B:	depth of cut (mm)	0.5	1.0	2.0
C:	nose angle (deg)	Small	Mid	Large
D:	rake angle (deg)	Small	Mid	Large
E:	side cutting-edge angle (deg)	Small	Mid	Large
F:	tip face type	1	2	3
G:	revolutionary speed (rpm)	Slow	Mid	Fast
H:	feeding speed (mm/rev)	Slow	Mid	Fast

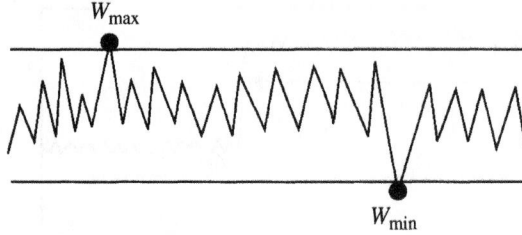

Figure 6
Minimum and maximum values of electrical power

$$L_1 = (3.32)(125.419) + (4.69)(177.370)$$
$$+ \cdots + (8.12)(304.877) = 8692.862$$

$$\vdots$$

$$L_{12} = 9940.218$$

Effective divider:

$$r = 3.32^2 + 4.69^2 + \cdots + 8.12^2 = 230.914$$

Variation of proportional terms:

$$S_\beta = \frac{(L_1 + \cdots + L_{12})^2}{(2)(3)(2)r} = 454{,}651{,}059$$

Variation of differences of proportional terms:

$$S_{N\beta} =$$
$$\frac{(L_1 + L_3 + \cdots + L_{11})^2 + (L_2 + L_4 + \cdots + L_{12})^2}{(2)3r}$$
$$- S_\beta = 225.049$$

$$S_{M\beta}^* =$$
$$\frac{(L_1 + L_2 + \cdots + L_6)^2 + (L_7 + L_8 + \cdots + L_{12})^2}{(3)2r}$$
$$- S_\beta = 27056211$$

$$S_{P\beta} =$$
$$\frac{(L_1 + L_2 + L_7 + L_8)^2 + \cdots + (L_5 + L_6 + L_{11} + L_{12})^2}{(2)2r}$$
$$- S_\beta = 1041.748$$

Error variation:

$$S_e = S_T - S_\beta - S_{N\beta} - S_{M^*\beta} - S_{P\beta} = 64.434$$

Error variance:

$$V_e = \frac{S_e}{67} = 0.962$$

Total error variance:

$$V_N = \frac{S_e + S_{N\beta} + S_{P\beta}}{70} = 19.018$$

SN ratio:

$$\eta = 10 \log \left[\frac{[1/(2)(3r)](S_{M^*\beta} - V_e)}{V_N} \right]$$
$$= -2.90 \text{ dB}$$

Sensitivity:

$$S = 10 \log \left[\frac{1}{(2)(3r)} (S_\beta - V_e) \right] = 32.15 \text{ dB}$$

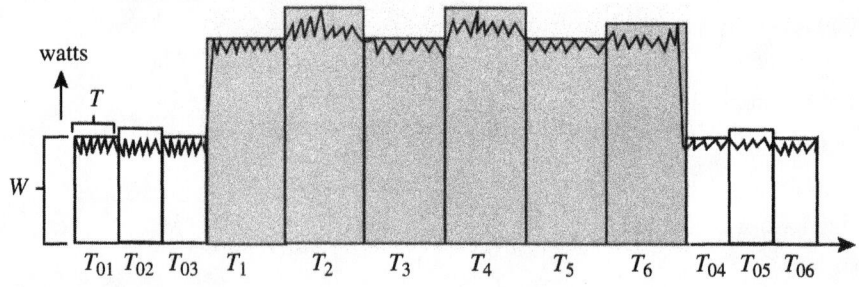

Figure 7
Electrical power for each time interval for cutting run 1 (generic function 1)

Table 2
Measured data for cutting run 1 (W) (generic function 1)

			T_{01} 13.00	T_{02} 13.00	T_{03} 13.00	T_{04} 13.00	T_{05} 13.00	T_{06} 13.00
$M_1^\star$: idling	N_1:	W_{min}	1430	1430	1420	1390	1390	1390
	N_2:	W_{max}	1500	1490	1480	1440	1440	1440
$M_2^\star$ cutting	N_1:	W_{min}	1960	1950	1950	1950	1940	1940
	N_2:	W_{max}	2030	2020	2010	2010	2000	2010

Table 3
Data of time versus electrical power

					T_1 11.00	T_2 22.00	T_3 33.00	T_4 44.00	T_5 55.00	T_6 66.00
$M_1^\star$: idling	P_{01}: cutting 1	N_1	y_{min}		15,730	31,460	47,080	62,370	77,660	92,950
		N_2	y_{max}		16,500	32,899	49,170	65,010	80,850	96,690
	P_{02}: cutting 2	N_1	y_{min}		15,180	30,360	45,650	60,500	75,350	90,200
		N_2	y_{max}		15,620	31,350	46,970	62,040	77,220	92,400
	P_{03}: cutting 3	N_1	y_{min}		14,740	29,840	44,330	58,960	73,590	88,000
		N_2	y_{max}		15,070	30,140	45,320	60,280	75,240	90,200
$M_2^\star$: cutting	P_1: cutting 1	N_1	y_{min}		21,560	43,010	64,460	85,910	107,250	128,590
		N_2	y_{max}		22,330	44,550	66,660	88,770	110,770	132,880
	P_2: cutting 2	N_1	y_{min}		20,680	41,360	61,930	82,390	102,850	23,420
		N_2	y_{max}		21,230	42,350	63,360	84,370	105,270	126,170
	P_3: cutting 3	N_1	y_{min}		19,910	39,930	59,840	79,750	99,550	119,350
		N_2	y_{max}		20,460	40,810	61,160	81,510	101,860	122,100

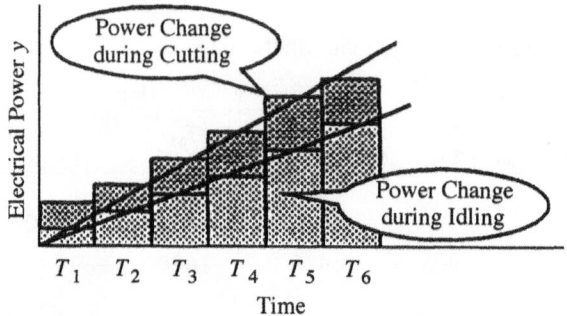

Figure 8
Change in electrical power during idling and cutting (generic function 1)

SN Ratio for Basic Function: Amount Removed versus Electric Power

By calculating the square root of each data point in Table 5, we obtained the converted data in Table 7. Next, using an average value of electrical power as a reference point, we converted the data from Table 7 into the reference-point proportional data in Table 8.

$$S_T = (-10.282^2) + 10.282^2 + \cdots + 389.986^2$$

$$= 509,577.5952$$

$$L_1 = (13.753)(158.498) + (18.912)(275.875)$$
$$+ \cdots + (22.505)(370.277) = 15,730.255$$

$$L_2 = 16,811.217$$

$$r = 13.753^2 + 18.912^2 + 22.505^2 = 1053.284$$

$$S_\beta = \frac{(L_1 + L_2)^2}{2r} = 502,688.4487$$

$$S_{N\beta} = \frac{L_1^2 + L_2^2}{r} - S_\beta = 554.6836$$

$$S_e = S_T - S_\beta - S_{N\beta} = 6334.4629$$

$$V_e = \frac{S_e}{6} = 1055.7438$$

$$V_N = \frac{S_e + S_{N\beta}}{7} = 984.1638$$

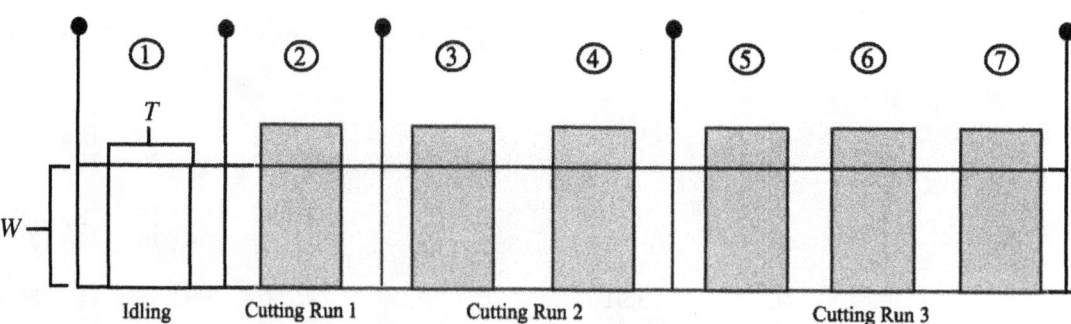

Figure 9
Electrical power for each cutting run, including the idling run (generic function 2)

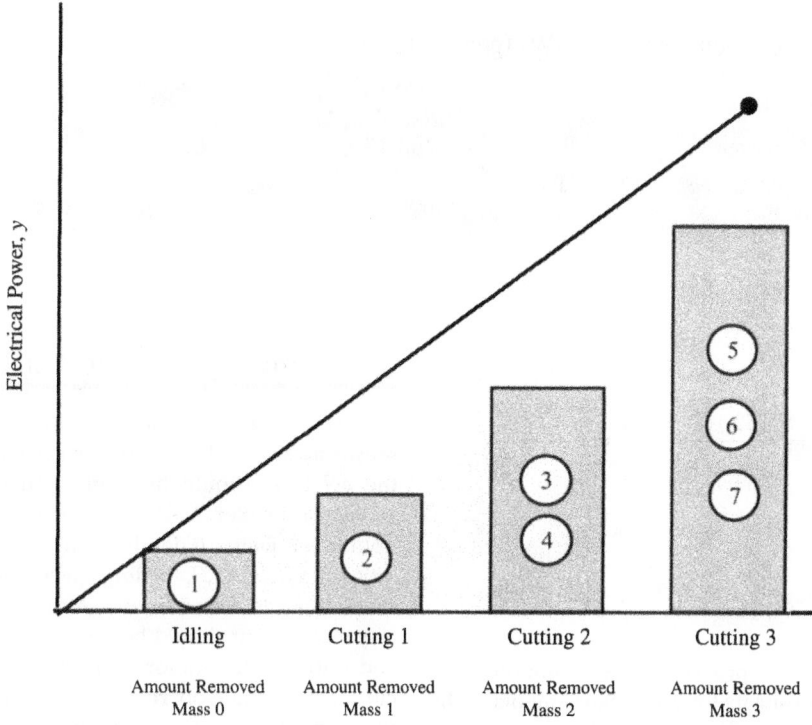

Figure 10
Time change in electrical power for each idling and cutting run (generic function 2)

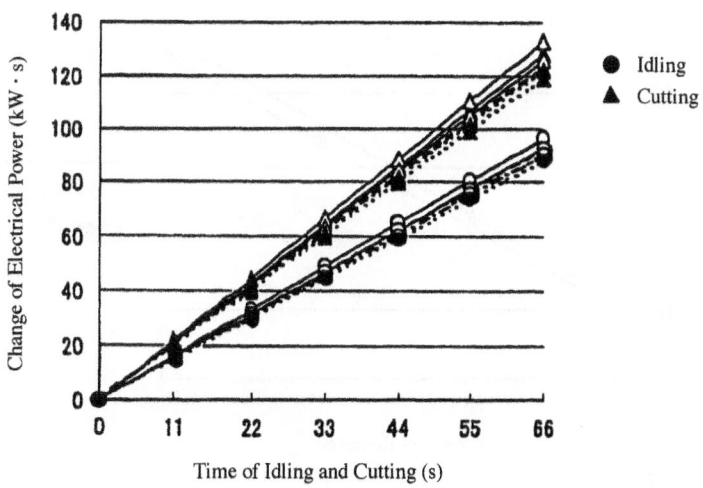

Figure 11
Change in electrical power during idling

Table 4

Measured data for each cutting run (W) (generic function 2)

M: Amount Removed (g):	P_0: Idling 0	P_1: Cutting Run 1 189.140	P_2: Cutting Run 2 168.510	P_3: Cutting Run 3 148.830
N_1 W_1 min.	1310	1940	1860	1800
N_2 W_1 max.	1500	2030	1930	1860

SN ratio:

$$\eta = 10 \log \frac{(1/2r)(S_\beta - V_e)}{V_N} = -6.16 \text{ dB}$$

Sensitivity:

$$S = 10 \log \left[\frac{1}{2r}(S_\beta - V_e) \right] = 23.77 \text{ dB}$$

Following these procedures, we computed the SN ratio and sensitivity for other experiments of the L_{18} orthogonal array. Figures 13 and 14 show a comparison of two generic functions for SN ratio and sensitivity, respectively.

6. Confirmatory Experiment and Analysis

While for generic function 1, both the SN ratio and sensitivity should be larger, for generic function 2, the SN ratio should be larger and the sensitivity should be smaller. Looking at each factor effect, we notice that factor B depth of cut, factor G, of revolution, and factor H, feeding speed, have a stronger effect than do other factors. Although a confirmatory experiment should be implemented at optimal and initial configurations for each function, by focusing on a trade-off relationship between generic functions 1 and 2 in our research, we selected $A_1 B_1 C_1 D_2 E_1 F_2 G_3 H_3$ as the optimal configuration and $A_2 B_3 C_2 D_2 E_2 F_2 G_1 H_1$ as the initial configuration. Table

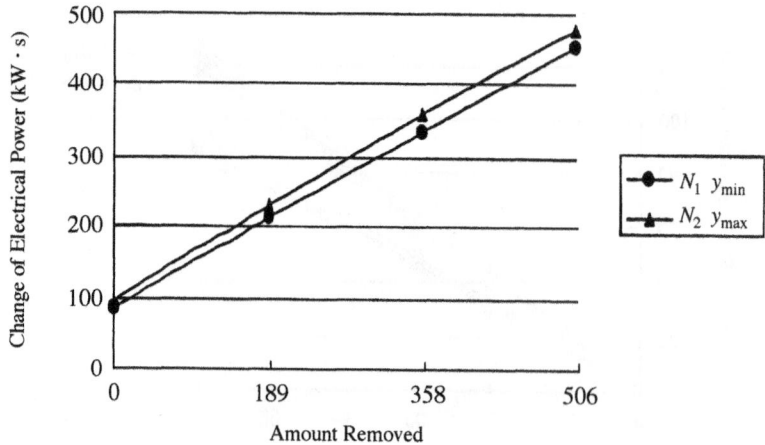

Figure 12
Change in electrical power during cutting

Table 5
Amount removed and electrical power (W) for each cutting run (generic function 2)

M: Amount Removed (g):	P_0: Idling 0	P_1: Cutting Run 1 189.140	P_2: Cutting Run 2 357.650	P_3: Cutting Run 3 506.480
N_1 y_{min}	86133	213,687.5	335,982.5	454,442.5
N_2 y_{max}	98625	232,097.5	358,995	481,290

9 shows the results. We believe that good reproducibility of gain is obtained.

7. Relationship between Energy Evaluation and Improvement in Dimension and Roughness

If we can obtain improvement effects at the optimal configuration–based energy evaluation, target quality characteristics such as machining dimensions or surface roughness should be improved. The quality of dimension means whether dimension y of a test piece is cut for each P, the number of cuts, without variability. Therefore, for dimension, the diameter of the test piece for each P is measured by a micrometer, and for roughness, average surface roughness is measured by a touch-probe surface roughness measuring instrument. Tables 10 and 11 show the measurement data for dimension and roughness. J indicates measurement points in the longitudinal direction of a test piece, X and Y represent measurement points in the radial direction,

Table 6
Converted data of time versus electrical power (postconversion)

				Time					
				T_1 3.32	T_2 4.69	T_3 5.74	T_4 6.63	T_5 7.42	T_6 8.12
M_1^*: idling	P_{01}: cutting run 1	N_1 y_{min}		125.419	177.370	216.979	249.740	278.675	304.877
		N_2 y_{max}		128.452	181.356	221.743	254.971	284.341	310.950
	P_{02}: cutting run 2	N_1 y_{min}		123.207	174.241	213.659	245.967	274.500	300.333
		N_2 y_{max}		124.980	177.059	216.726	249.078	277.885	303.974
	P_{03}: cutting run 3	N_1 y_{min}		121.408	171.697	210.547	242.817	271.275	296.648
		N_2 y_{max}		122.760	173.609	212.885	245.520	274.299	300.333
M_2^*: cutting	P_1: cutting run 1	N_1 y_{min}		146.833	207.389	253.5890	293.104	327.490	358.594
		N_2 y_{max}		149.432	211.069	258.186	297.943	332.821	364.527
	P_2: cutting run 2	N_1 y_{min}		143.805	203.372	248.857	287.037	320.702	351.312
		N_2 y_{max}		145.705	205.791	251.714	290.465	324.453	355.204
	P_3: cutting run 3	N_1 y_{min}		141.103	199.825	244.622	282.400	315.515	345.471
		N_2 y_{max}		143.038	202.015	247.305	285.500	319.155	349.428

Table 7
Converted data of amount removed versus electrical power (W) (postconversion)

M:	Amount Removed (g):	P_0: Idling 0	P_1: Cutting Run 1 13.753	P_2: Cutting Run 2 18.912	P_3: Cutting Run 3 22.506
N_1	y_{min}	293.483	462.263	574.640	674.042
N_2	y_{max}	314.046	481.765	599.162	693.751

Table 8
Data for reference-point proportional equation (W)

M:	Amount Removed (g):	P_0: Idling 0	P_1: Cutting Run 1 13.753	P_2: Cutting Run 2 18.912	P_3: Cutting Run 3 22.505
N_1	y_{min}	−10.22	158.498	275.875	370.277
N_2	y_{max}	10.282	178.000	295.397	389.986

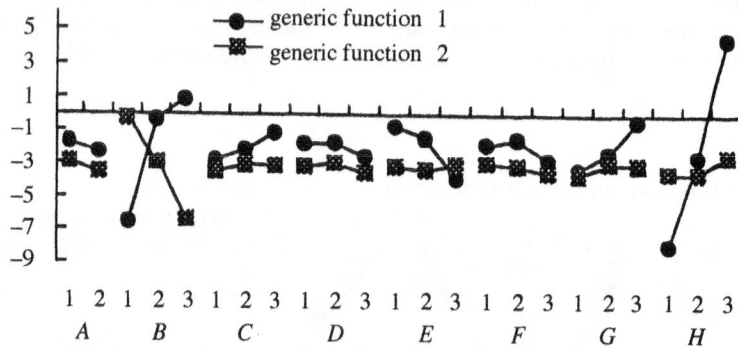

Figure 13
Response graphs of SN ratio of time versus electrical power

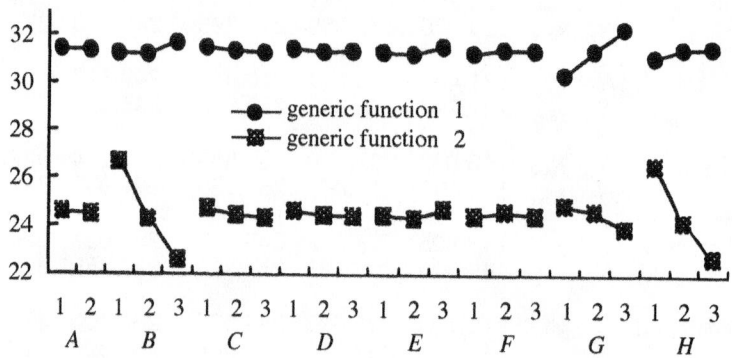

Figure 14
Response graphs of sensitivity of time versus electrical power

Table 9
Results of gain in confirmatory experiments

(a) Time vs. Electrical Power		Configuration		
		Optimal	Initial	Gain
SN ratio	Estimation	2.40	−5.53	7.93
	Confirmation	5.96	−4.92	10.88
Sensitivity	Estimation	32.40	30.36	2.04
	Confirmation	31.96	30.10	1.86

(b) Amount Removed vs. Electrical Power		Configuration		
		Optimal	Initial	Gain
SN ratio	Estimation	0.74	−7.25	7.99
	Confirmation	0.27	−6.93	7.20
Sensitivity	Estimation	24.56	24.97	−0.41
	Confirmation	23.98	24.87	−0.89

R represents repetition of measurement, and N indicates the number of test pieces machined. Table 12 shows the results calculated as a nominal-the-best characteristic. Consequently, we can confirm that we can estimate the final quality and machine products in a stable manner once we improve the cutting process based on energy evaluation.

8. Discussion and Conclusions

The reason that we have not been able to obtain good reproducibility of gain in the research on machining based on energy evaluation is that we have not assessed electrical power during idling (no loading) in a proper manner. As a result of combining

Table 10
Dimensional data at optimal configuration (mm)

			J_1		J_2		J_3		J_4	
			X	Y	X	Y	X	Y	X	Y
N_1	P_1	R_1	30.023	39.025	39.019	39.018	39.015	39.017	39.015	39.014
		R_2	39.031	39.031	39.029	39.029	39.019	39.027	39.030	39.027
	P_2	R_1	38.022	38.023	38.018	38.018	38.015	38.016	38.014	38.014
		R_2	38.020	38.020	38.021	38.019	38.012	38.020	38.022	38.020
	P_3	R_1	37.023	37.023	37.019	37.018	37.015	37.016	37.014	37.014
		R_2	37.023	37.024	37.024	37.022	37.015	37.021	37.025	37.021
N_2	P_1	R_1	39.032	39.025	39.032	39.033	39.027	39.026	39.026	39.025
		R_2	39.013	39.015	39.020	39.019	39.028	39.026	39.026	39.029
	P_2	R_1	38.052	38.045	38.050	38.050	38.045	38.044	38.043	38.043
		R_2	38.020	38.022	38.026	38.024	38.028	38.027	38.027	38.030
	P_3	R_1	37.038	37.027	37.034	37.034	37.029	37.028	37.026	37.026
		R_2	37.024	37.026	37.029	37.026	37.032	37.030	37.033	37.034

Table 11
Roughness data at optimal configuration (μm)

			J_1	J_2	J_3	J_4
N_1	P_1	R_1	2.500	2.460	2.450	2.450
		R_2	2.910	2.950	3.020	2.940
	P_2	R_1	2.500	2.450	2.500	2.530
		R_2	2.930	2.990	3.080	2.940
	P_3	R_1	2.240	2.500	2.480	2.500
		R_2	3.030	3.010	3.100	3.080
N_2	P_1	R_1	3.510	3.580	3.640	3.670
		R_2	1.980	2.030	2.060	2.120
	P_2	R_1	3.560	3.650	3.780	3.820
		R_2	2.211	2.360	2.240	2.780
	P_3	R_1	3.980	4.160	4.140	4.010
		R_2	2.487	2.512	2.604	2.564

Table 12
Gain of the SN ratio of dimension and roughness (dB)

| | Configuration | | | Improvement of |
	Optimal	Initial	Gain	Variance
SN ratio of dimension	72.86	60.20	12.66	1/18.45
SN ratio of roughness	13.64	11.98	1.66	1/1.47

data for electrical power with data for each generic function, we have obtained good reproducibility. In addition, we have proven that we can estimate final quality characteristics using the results.

As one of the analyses in this research, by using the difference between idling and cutting and regarding this difference as an effective amount of energy, we have calculated SN ratios. Next, looking at the relationship for $y = \beta_1 M$, β_1 should be greater; conversely, for $y = \beta_2 M$, β_2 should be smaller. The reason is that when $M = (\beta_1/\beta_2)\,T$, β_1/β_2 should be greater. Indeed, electrical power consumption seems great if β_1 is great; however, electrical power required for the same amount of machining can be smaller if β_2 is small. Thus, we conclude that electrical power during idling should be smaller, whereas that during cutting should be greater. These considerations are applicable to performance evaluation of a robot or other machines that have two functions, one during idling and one during loading.

Since a portion to be removed should be shaved uniformly at each microscopic area with even energy in cutting, we have proved that it reasonable to evaluate proportionality of energy with maximum and minimum values of electric power at each microscopic area.

Reference

Kazuhito Takahashi, Shinji Kousaka, Kiyoharu Hoshiya, Kouya Yano, Noriaki Nishiuchi, and Hiroshi Yano, 2000. Optimization of machining conditions through observation of electric power consumption. *Quality Engineering*, Vol. 8, No. 1, pp. 24–30.

This case study is contributed by Kazuhito Takahashi, Shinji Kousaka, Kiyoharu Hoshiya, Kouya Yano, Noriaki Nishiuchi, and Hiroshi Yano.

CASE 25

Development of Machining Technology for High-Performance Steel by Transformability

Abstract: In this research we focused on developing cutting technology for difficult-to-cut material. It is an urgent necessity to develop energy-efficient production engineering as well as fuel-efficient automobiles to conserve our energy resources. Among various technical items, improvement in surface roughness and tool life in a gear-cutting process using high-strength steel, which is quite difficult to cut, has been suggested. Instead of cutting the gear itself, test pieces were designed in such a way that the concept of transformability could be applied to develop machining technology.

1. Introduction

Because component parts used for power train or steering systems require high strength and durability, we have secured such requirements by cutting and carburizating low-carbon steel, which is easy to cut. However, since carburization takes approximately 10 hours and lowers productivity, currently, high-frequency heating is used to reduce process time drastically, to 1 minute. Yet high-frequency heating material shows 30 on the Rockwell C scale and is difficult to cut, whereas carburization material is quite easy to cut, with almost zero on the same scale. In this research we focused on developing cutting technology for difficult-to-cut material.

2. Shape to Be Machined

When we develop cutting technology, by paying too much attention to an actual product shape, we tend to digress from our original technological purpose. To avoid this situation and develop accurate and stable machining technology, a development method using the idea of transformability is extremely effective.

Applying the idea of transformability to our research, we pursued cutting conditions with the proportionality

$$y = \beta M \qquad (1)$$

where M is a signal factor of input data from a numerically controlled (NC) machine and y is the output of a product dimension corresponding to the input.

As a product shape to be machined, we used an easy-to-measure shape without sticking with a product shape to be manufactured, because our objective was technological development. In addition, to pursue cutting technology robust to the control shaft direction of a cutting machine and to simplify an experimental process by using analogous solid geometry, we selected the model detailed in Figure 1, which depicts the measured points a_1 to a_4, b_1 to b_4, and c_1 to c_4. The signal factors are defined as follows:

Signal factor:

$$M_1 \qquad M_2 \qquad \cdots \qquad M_{12} \qquad \cdots \qquad M_{66}$$

Linear distance:

$$a_1 - a_2 \quad a_1 - a_3 \quad \cdots \quad a_2 - a_3 \quad \cdots \quad c_3 - c_4$$

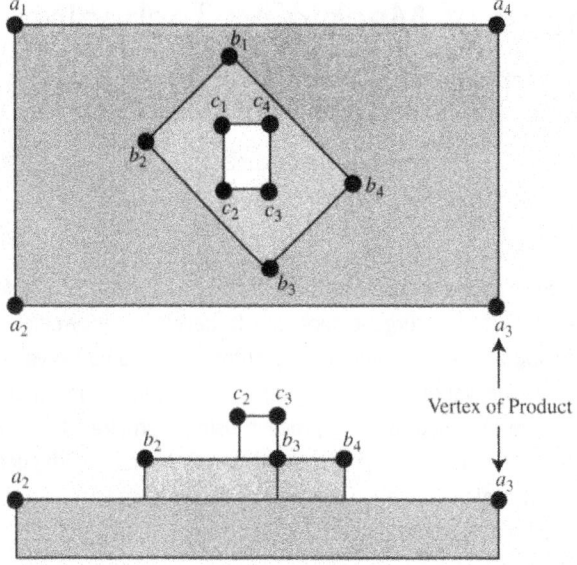

Figure 1
Test piece shape and measured points

A signal factor can be calculated from a coordinate of each vertex of the model shown in Figure 1. This is defined as a linear distance between all possible combinations of two vertices. Now since we have 12 vertices, the number of signal factor levels amounts to 66. By measuring a coordinate (X, Y, Z) of each vertex with a coordinate measuring machine, we computed a linear distance as output.

From our technical knowledge, we chose as control factors, eight factors that affect machining accuracy and tool life. For a noise factor, we picked material hardness and set up maximum and minimum values that were assumed to happen in the actual manufacturing processes. Table 1 summarizes the control and noise factors and levels selected.

3. Assignment of Factors and Levels to Orthogonal Array and Data Analysis

As illustrated in Table 2, each control factor was allocated to an L_{18} orthogonal array, and signal and noise factors were assigned to an outer array. Based on this, we conducted 18 experiments. Finally, we

obtained the results shown in Table 3. Using them, we implemented data analysis as follows.

Total variation:

$$S_T = 70.992^2 + 84.607^2 + \cdots + 10.955^2$$
$$= 150{,}021.740385 \quad (f = 132) \quad (2)$$

Linear equations:

$$L_1 = (71.000)(70.992) + (84.599)(84.607)$$
$$+ \cdots + (11.000)(10.958)$$
$$= 125{,}043.939335 \quad (3)$$

$$L_2 = (71.000)(70.991) + (84.599)(84.607)$$
$$+ \cdots + (11.000)(10.955)$$
$$= 125{,}040.360611 \quad (4)$$

Effective divider:

$$r = (2)(71.000^2 + 84.599^2 + \cdots + 11.000^2)$$
$$= 250{,}146.969846 \quad (5)$$

Variation of proportional term:

Table 1
Factors and levels

		Level		
	Factor	1	2	3
Control factors				
A:	cutting direction	Up	Down	—
B:	cutting speed (m/min)	Slow	Standard	Fast
C:	feeding speed (m/min)	Slow	Standard	Fast
D:	tool material	Soft	Standard	Hard
E:	tool rigidity	Low	Standard	High
F:	twisting angle (deg)	Small	Standard	Large
G:	rake angle (deg)	Small	Standard	Large
H:	depth of cut (mm)	Small	Standard	Large
Noise factor				
N:	material hardness	Soft	Hard	—

Table 2
Layout of control factors and results of analysis (dB)

No.	A	B	C	D	E	F	G	H	SN Ratio	Sensitivity
1	1	1	1	1	1	1	1	1	31.41	−0.0022
2	1	1	2	2	2	2	2	2	39.70	0.0058
3	1	1	3	3	3	3	3	3	39.68	0.0028
4	1	2	1	1	2	2	3	3	9.25	0.0730
5	1	2	2	2	3	3	1	1	44.56	−0.0001
6	1	2	3	3	1	1	2	2	42.02	0.0020
7	1	3	1	2	1	3	2	3	33.75	0.0057
8	1	3	2	3	2	1	3	1	44.59	0.0003
9	1	3	3	1	3	2	1	2	19.18	0.0114
10	2	1	1	3	3	2	2	1	42.80	0.0011
11	2	1	2	1	1	3	3	2	30.55	0.0145
12	2	1	3	2	2	1	1	3	26.41	0.0166
13	2	2	1	2	3	1	3	2	25.86	0.0148
14	2	2	2	3	1	2	1	3	35.24	0.0056
15	2	2	3	1	2	3	2	1	42.52	0.0022
16	2	3	1	3	2	3	1	2	41.01	−0.0009
17	2	3	2	1	3	1	2	3	2.63	0.1801
18	2	3	3	2	1	2	3	1	39.30	0.0025

Table 3

Results of experiment 1 (mm)

Noise Factor	Signal Factor			
	M_1 a_1–a_2 71.00	M_2 a_1–a_3 84.599	... ...	M_{66} c_3–c_4 11.000
N_1	70.992	84.607	...	10.958
N_2	70.991	84.607	...	10.955
Total	141.983	169.214	...	21.913

$$S_\beta = \cfrac{\cfrac{1}{250,146.969846}}{(125,043.939335 - 125,040.360611)^2}$$

$$= 250,021.645747 \qquad (f = 1) \qquad (6)$$

Variation due to proportional terms:

$$S_{N\beta} = \cfrac{\cfrac{1}{250,146.969846}}{(125,043.939335 - 125,040.360611)^2}$$

$$= 0.000051 \qquad (f = 1) \qquad (7)$$

Error variation:

$$S_e = 250,021.740385 - 250,021.645747 - 0.000051$$

$$= 0.094587 \qquad (f = 130) \qquad (8)$$

Error variance:

$$V_e = \frac{0.094587}{130} = 0.005944$$

$$= 0.000728 \qquad (9)$$

Total error variance:

$$V_N = \frac{0.094587 + 0.000051}{131}$$

$$= 0.000722 \qquad (10)$$

Now V_N is smaller than V_e, so we calculated the SN ratio using the equation below:

SN ratio:

$$\eta = 10 \log \cfrac{\cfrac{1}{250,146.969846}}{\left(\cfrac{250,021.645747 - 0.000728}{0.000728}\right)}$$

$$= 31.41 \text{ dB} \qquad (11)$$

Sensitivity:

$$S = 10 \log \frac{250,021.645747 - 0.000728}{250,146.969846}$$

$$= -0.0022 \text{ dB} \qquad (12)$$

We summarized these results in Table 2. Additionally, using these results, we created Table 4 as a level-by-level supplement table regarding the SN ratio and sensitivity. Further, we plotted the factor effects in Figure 2.

4. Estimation of Optimal Configuration and Prediction of Effects

Selecting levels for a higher SN ratio, we obtained the following optimal configuration: $A_1B_1C_3D_3E_1F_3$ G_2H_1. To estimate the SN ratio of the optimal configuration using factors B, D, F, and H, which have large differences from the average grand total of SN ratios, we calculated its process average as

$$\eta = B_1 + D_3 + F_3 + H_1 - 3T$$

$$= 35.09 + 40.89 + 38.68 + 40.86 - (3)(32.80)$$

$$= 57.24 \text{ dB} \qquad (13)$$

On the other hand, the estimation of the process average of the initial configuration $A_1B_2C_2D_2E_2F_2$ G_2H_2 was as follows:

$$\eta = B_2 + D_2 + F_2 + H_2 - 3T$$

$$= 33.24 + 34.93 + 30.91 + 33.05 - (3)(32.80)$$

$$= 33.73 \text{ dB} \qquad (14)$$

Comparing the two SN ratios, we noticed that we could obtain a gain of more than 23 dB from the initial configuration and reduce the variance to $1/220$.

Table 4
Supplementary table of the SN ratio and sensitivity (dB)

Control Factor	SN Ratio			Sensitivity		
	1	2	3	1	2	3
A: cutting direction	33.79	31.81	—	0.0110	0.0263	—
B: cutting speed	35.09	33.24	30.08	0.0065	0.0162	0.0332
C: feeding speed	30.68	32.88	34.85	0.0153	0.0344	0.0063
D: tool material	22.59	34.93	40.89	0.0465	0.0076	0.0018
E: tool rigidity	35.38	33.91	29.12	0.0047	0.0162	0.0350
F: twisting angle	29.82	30.91	38.68	0.0353	0.0166	0.0040
G: rake angle	32.97	33.90	31.54	0.0051	0.0328	0.0180
H: depth of cut	40.86	33.05	24.49	0.006	0.0079	0.0473
Average		32.80			0.0186	

Next, using factors B, D, E, F, and H, which have large differences from the average grand total of sensitivity, we estimated the sensitivity at the optimal configuration as

$$S = B_1 + D_3 + E_1 + F_3 + H_1 - 4T$$

$$= 0.0065 + 0.0018 + 0.0047 + 0.0040 + 0.0006 - (4)(0.0186)$$

$$= -0.0568 \text{ dB} \qquad (15)$$

Using the definition $S = 10 \log \beta^2$, we can compute the coefficient of proportionality: $\beta = 0.9935$. For the initial configuration, we estimated the following sensitivity:

$$S = B_2 + D_2 + E_2 + F_2 + H_2 - 4T$$

$$= 0.0162 + 0.0076 + 0.0162 + 0.0166 + 0.0079 - (4)(0.0186)$$

$$= -0.0099 \text{ dB} \qquad (16)$$

This value can be converted into $\beta = 0.9989$.

5. Confirmatory Experiment

We performed confirmatory experiments for both the optimal and initial configurations. Measured data and calculation procedures are tabulated in Table 5. We expressed sensitivity by the value of β. Consequently, we confirmed a dramatic improvement in machining accuracy, almost 20 dB, and at the same time verified that the gains and absolute values have high reproducibility. This 20 dB indicates that the standard deviation of accuracy decreases to $\frac{1}{10}$.

On the other hand, the sensitivity for the optimal configuration in the confirmatory experiment turns out to be $\beta = 0.9939$. Therefore, using the following equation, we can compute an NC input data, M, when actual products are produced:

$$M = \frac{y}{\beta} = \frac{y}{0.9939} = 1.0061y$$

where y is defined as product size.

6. Research Process

This example began with the following requests from a manager and engineer in charge of developing cutting technology: to coach them on the Taguchi method in developing cutting technology for high-strength steel; and to focus on two evaluation characteristics, surface roughness of a gear after it is machined and tool life.

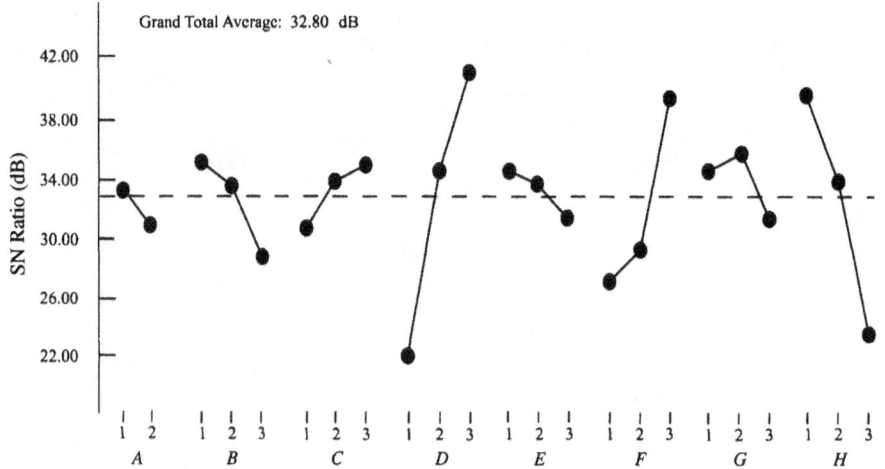

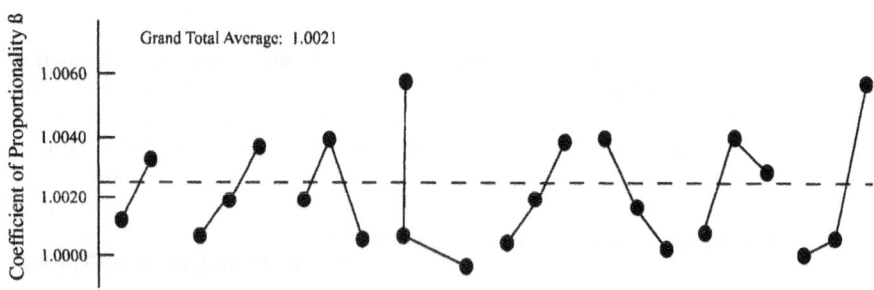

Figure 2
Response graphs of the SN ratio and coefficient of proportionality β

Table 5
Results of estimation and confirmatory experiment (dB)

Configuration	SN Ratio		Sensitivity	
	Estimation	Confirmation	Estimation	Confirmation
Optimal	57.24	54.09	0.9935	0.9939
Current	33.73	34.71	0.9989	0.9992
Gain	23.51	19.38	—	

Although their main goal was technological development, they were particularly interested in the gear as a product. Therefore, the original evaluation characteristics were quality characteristics (objective characteristics) of surface roughness and tool life through actual machining of the product.

Since it seemed that this plan would not be successful, I asked them various questions about previous technical trials. They confessed that they had not made any significant achievement in over one year of development before consulting with me. In addition, I answered their questions about why their efforts to date had failed, as follows: Their original evaluation characteristics, surface roughness and tool life, were related not to features to measure cutting functions but to superficial phenomena. Furtherm, even if material characteristic or material hardness were changed, development of a technology for smoothly cutting material would enable us to smooth surface roughness by improving surface evenness and prolonging tool life.

Although the engineer suggested measuring the machining and consumption powers, I recommended applying the idea of transformability instead because we wondered how we could measure them. Afterward, when they started to make a new experimental plan, they asked me to identify the key point in applying transformability. I suggested the following key points:

❏ The shape to be machined should be easy to measure.

❏ The shape to be machined can be assigned many signal factors with a wide range of levels.

❏ We should implement technological development using a test piece in place of a real product.

❏ The shape of the test price should have a geometry similar to that of the product so that it can also be analyzed.

❏ After completing generic technological development, we should apply the technology to a gear.

The underlying idea is that once we are able to cut a line smoothly, we will be able to cut a surface. In other words, we cannot cut a surface without being able to cut a line. As a result, we devised the model shape shown in Figure 3.

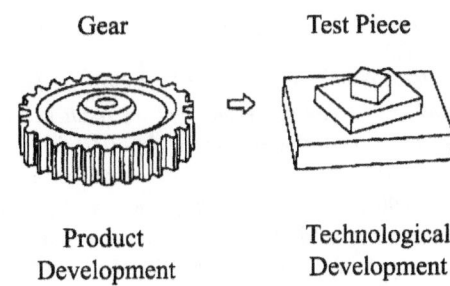

Gear   Test Piece

Product   Technological
Development   Development

Figure 3
Final product and test piece

Since the original plan focused mainly on a gear, the idea was to cut a gear directly in the actual production process and to use quality characteristics such as surface roughness and tool longevity as evaluation characteristics. Indeed, this type of procedure could be used to improve quality to some extent if we take steady steps for a long time; however, we would encounter the following serious problems:

❏ We could not survive in fierce competition because we would take too much time.

❏ A technology applicable only to a particular product demands that the same technological development be repeated each time a specification is changed.

Therefore, instead of a real product shape, we tackled a test piece to develop the cutting technology.

Nonetheless, we anticipate the following questions:

1. *Did you use machining lubricant in this experiment?* We did not use it because we assumed that if we succeeded in developing a technology to cut material smoothly without lubricant, we could cut even more smoothly in actual production processes.

2. *Looking at the experimental results in the orthogonal array, we see a great difference, ranging from 2.6 to 44.5 dB. Taking into account that the average grand total is 32.8 dB, don't the SN ratios of experiments 4 and 17 seem extraordinarily small?* We set control factor levels as well as selected material hardness as a noise factor within as large a range as possible. As a consequence, we made a difference between a good condition and a bad condition for a tool standout.

Therefore, we regarded this experiment as quite meaningful.

3. *Although you prove that a technological development has good results, can you expect that the result will hold true for a real product?* Because we obtained the result by using proportionality between various input signals regarding all directions *x, y,* and *z* and product sizes, we are confident that this method can work for any types of model shapes. Yet absolute values of the SN ratio of a test piece may not be consistent with those of an actual product. Even in this case, we ask readers to confirm the results because you can obtain good reproducibility of gain.

Reference

Kenzo Ueno, 1993. Machining technology development for high performance steel to which concept of transformability is applied. *Quality Engineering,* Vol. 1, No. 1, pp. 26–30.

This case study is contributed by Kenzo Ueno.

CASE 26

Transformability of Plastic Injection-Molded Gear

Abstract: To improve the process accuracy of small, precise plastic parts, by selecting after-molding part placement, die temperature, injection speed, and injection pressure as control factors, and die dimensions as a signal factor of transformability, we conducted an experiment. Since we usually adjust product size using molding conditions, we added holding pressure as an adjusting signal factor as well as the number of molding shots as a noise factor. In sum, we chose two signal factors. Furthermore, we proved that there are systematic differences in shrinkage rate due to positions of mold dimensions caused by product design. By separating the effect of shrinkage difference from noise terms through respective tuning, we improved the process accuracy.

1. Introduction

Small, precise plastic gears require an extremely rigorous tolerance between 5 and 20 μm. Furthermore, as the performance of a product equipped with gears improves, the tolerance also becomes more severe. Traditionally, in developing a plastic-molded part, we have repeated the following processes to determine mold dimensions:

1. Decide the plastic molding conditions based on previous technical experience.
2. Mold and machine the products.
3. Inspect the dimensions of the products.
4. Modify and adjust the mold.

For certain product shapes, we currently struggle to modify and adjust a die due to different shrinkage among different portions of a mold. To solve this, we need to clarify such technical relationships, thereby reducing product development cycle time, eliminating waste of resources, and improving product accuracy. In this experiment we applied the idea of transformability to assess our conventional, empirical method of determining molding conditions

and investigated the feasibility of improving molding accuracy.

2. Transformability Process

As signal factors, we chose mold dimensions to evaluate transformability and a parameter in the production process for adjusting. Transformability corresponds to the dimensions of a model gear (Figure 1). More specifically, M_1, M_2, M_3, M_4, M_5 and M_6 were selected, and in particular, each of M_1, M_2, M_3 and M_4 contains two directions of X and Y. In sum, since one model has six signal factor levels and one mold produces two pieces, $2 \times 6 = 12$ signal factors were set up in total. We chose holding pressure as a three-level adjusting signal factor.

On the other hand, for all control factors, we set the current factor levels to level 2. As a noise factor, we selected the number of plastic molding shots completed and set the third and twentieth shots to levels 1 and 2, respectively. They represented the noise at the initial and stable stages of the molding process. Table 1 summarizes signal and noise

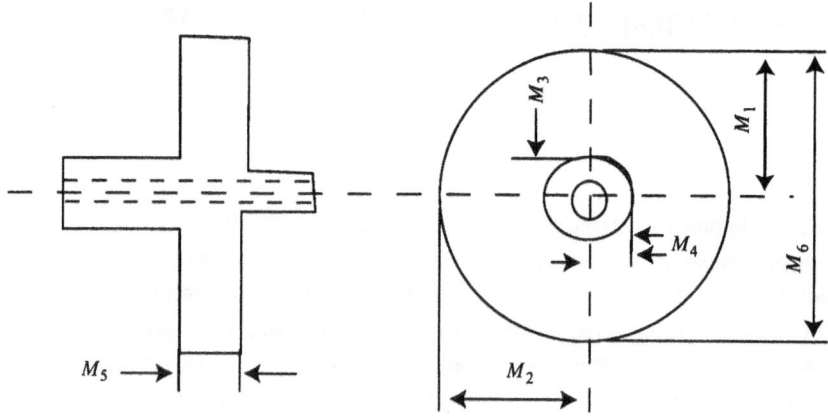

Figure 1
Section of model gear

factors. Dimensions corresponding to signals were chosen as measurement characteristics.

3. SN Ratio

We show some of the experimental data of the L_{18} orthogonal array in Table 2. Based on these, we proceeded with an analysis.

Total variation:

$$S_T = 9.782^2 + 9.786^2 + \cdots + 19.666^2 + 9.924^2$$
$$+ 9.921^2 + \cdots + 19.931^2$$

$$= 7049.764914 \qquad (f = 72) \qquad (1)$$

Table 1
Signal and noise factors

Factor	Level
Signal	
Transformability (mold dimension)	$M_1, M_2, M_3, M_4, M_5,$ $M_6, M_7, M_8, M_9, M_{10},$ M_{11}, M_{12}
Adjusting (holding pressure)	$M_1^* = 300, M_2^* = 550,$ $M_3^* = 800$ kgf/cm^2
Noise	
Number of shots	N_1, third shot; N_2, twentieth shot

Linear equations holding pressure corresponding to mold dimension and noise factor:

$$L_1 = (9.996)(9.782) + \cdots + (0.925)(0.900)$$
$$+ (20.298)(19.818)$$

$$= 1195.315475 \qquad (2)$$

$$L_2 = (9.996)(9.786) + \cdots + (0.925)(0.890)$$
$$+ (20.298)(19.820)$$

$$= 1195.845021 \qquad (3)$$

$$L_3 = (9.996)(9.819) + \cdots + (0.925)(0.900)$$
$$+ (20.298)(19.894)$$

$$= 1199.881184 \qquad (4)$$

$$L_4 = (9.996)(9.808) + \cdots + (0.925)(0.896)$$
$$+ (20.298)(19.892)$$

$$= 1199.350249 \qquad (5)$$

$$L_5 = (9.996)(9.832) + \cdots + (0.925)(0.910)$$
$$+ (20.298)(19.941)$$

$$= 1202.810795 \qquad (6)$$

$$L_6 = (9.996)(9.829) + \cdots + (0.925)(0.906)$$
$$+ (20.298)(19.931)$$

$$= 1202.271255 \qquad (7)$$

Effective divider:

Table 2
Example of one run of an L_{18} orthogonal array (mm)

Adjusting Factor	Noise Factor	Signal					
		M_1 9.996	M_2 9.989	M_3 1.018	M_4 1.026	M_5 0.978	M_6 20.004
M_1^*	N_1	9.782	9.768	0.903	0.905	0.955	19.545
	N_2	9.786	9.770	0.901	0.894	0.954	19.565
M_2^*	N_1	9.819	9.805	0.900	0.900	0.928	19.630
	N_2	9.808	9.808	0.896	0.900	0.959	19.616
M_3^*	N_1	9.832	9.836	0.908	0.892	0.977	19.680
	N_2	9.829	9.833	0.905	0.904	0.970	19.666

Adjusting Factor	Noise Factor	M_7 10.164	M_8 10.142	M_9 1.041	M_{10} 1.043	M_{11} 0.925	M_{12} 20.298	Linear Equation
M_1^*	N_1	9.924	9.901	0.883	0.864	0.900	19.818	L_1
	N_2	9.921	9.910	0.875	0.864	0.890	19.820	L_2
M_2^*	N_1	9.950	9.9934	0.880	0.857	0.900	19.894	L_3
	N_2	9.950	9.921	0.879	0.866	0.896	19.892	L_4
M_3^*	N_1	9.979	9.954	0.878	0.869	0.910	19.941	L_5
	N_2	9.970	9.964	0.873	0.869	0.906	19.931	L_6

$$r = 9.996^2 + \cdots + 20.004^2 + \cdots + 20.298^2$$
$$= 1224.108656 \qquad (8)$$

Variation of proportional term of transformability:

$$S_\beta = \frac{1}{6r}[(9.996)(58.856) + \cdots + (20.004)(117.70) \\ + (10.164)(59.694) + \cdots \\ + (20.298)(119.296)]^2$$
$$= 7049.325990 \qquad (f=1) \qquad (9)$$

Variation of the first-order term of adjustability, $S_{\beta*}$, can be calculated in a three-level orthogonal polynomial equation because M^* is orthogonal to M around M_2^*.

Variation of proportional term of adjustability:

$$S_{\beta*} = \frac{(-L_1 - L_2 + L_5 + L_6)^2}{(2)(2r)}$$
$$= 0.039582 \qquad (f=1) \qquad (10)$$

Variation of individual proportional term of transformability:

$$S_L = \frac{L_1^2 + L_2^2 + \cdots + L_5^2 + L_6^2}{r}$$
$$= 7049.366256 \qquad (f=6) \qquad (11)$$

Variation of proportional term due to noise:

$$S_{\beta N} = \frac{\left[\begin{array}{l}(9.996)(29.433) + \cdots + (20.004)(58.855) \\ + (10.164)(29.853) + \cdots + (20.298)(59.653)\end{array}\right]^2}{(3)(9.996^2 + \cdots + 20.004^2 + 10.164^2 + \cdots + 20.298^2)} \\ + \frac{\left[\begin{array}{l}(9.996)(29.423) + \cdots + (20.004)(58.847) \\ + (10.164)(29.841) + \cdots + (20.298)(59.643)\end{array}\right]^2}{(3)(9.996^2 + \cdots + 20.004^2 + 10.164^2 + \cdots + 20.298^2)} \\ - S_\beta$$
$$= 0.000039 \qquad (12)$$

Residual variation due to individual proportional term:

$$S_{res} = S_L - S_\beta - S_{\beta*} - S_{\beta N}$$
$$= 0.000645 \qquad (f=3) \qquad (13)$$

Table 3
ANOVA table of one run of the L_{18} orthogonal array

	Level	f	S	V
β:	proportional term	1	7049.325990	7049.325990
βM':	positional noise	1	0.377766	0.377766
βN:	compounded noise	1	0.000039	0.000039[a]
β*:	proportional term	1	0.039582	0.039582
res:	residual	3	0.000645	0.000215[a]
e:	error	65	0.020853	0.000321[a]
e':	error (after pooling factors indicated by [a])	69	0.021537	0.000312
Total		72	7049.764914	

[a] Factors to be pooled.

Now, looking at these experimental data in detail, we notice that contraction rates for die dimensions corresponding to signal factors M_3, M_4, M_9, and M_{10} have a tendentious difference compared to other dimensions, because of mold structure, including gate position or thickness. Since we have a similar tendency for other experiments of the L_{18} orthogonal array, by substituting M_1' for M_1, M_2, M_5, M_6, M_7, M_8, M_{11}, and M_{12}, and M_2' for M_3, M_4, M_9, and M_{10}, we calculated the variation of interaction between M's and β and removed this from error variation.

$$S_{\beta M'} = \text{(variation of proportional term for factors}$$
$$\text{other than } M_3, M_4, M_9, \text{ and } M_{10})$$
$$+ \text{(variation of proportional term for}$$
$$M_3, M_4, M_9, \text{ and } M_{10}) - S_\beta$$

$$= \frac{[(9.996)(58.856) + \cdots + (20.298)(119.296)]^2}{(6)(9.996^2 + \cdots + 20.298^2)}$$
$$+ \frac{[(1.018)(5.413) + \cdots + (1.043)(5.189)]^2}{(6)(1.018^2 + \cdots + 1.043^2)}$$
$$- S_\beta$$

$$= \frac{(7173.532160)^2}{(6)(1219.848126)} + \frac{(21.941819)^2}{(6)(4.26053)} - S_\beta$$

$$= 7030.870285 + 18.833471 - S_\beta$$

$$= 0.377766 \quad (f = 1) \tag{14}$$

Error variation:

$$S_e = S_T - S_\beta - S_{\beta M'} - S_{\beta N} - S_{\beta *} - S_{\text{res}}$$

$$= 0.020853 \quad (f = 65) \tag{15}$$

To summarize the above, we show Table 3 for ANOVA (analysis of variance). Based on this result, we compute the SN ratio and sensitivity.

SN ratio of transformability:

$$\eta = \frac{[1/(6)(1224.108656)]}{0.000312}(7049.325990 - 0.000312)$$

$$= 3076.25$$

$$10 \log \eta = 34.88 \text{ dB} \tag{16}$$

Since the range of level of holding pressure, 250 kgf/cm^2, does not have any significance as an absolute value when its SN ratio is calculated, we set $h = 1$.

SN ratio of adjustability:

$$\eta^* = \frac{[1/(2)(2)(1224.108656)(1^2)]}{0.000312}(0.039582 - 0.000312)$$

$$= 0.02571$$

$$10 \log \eta^* = -15.90 \text{ dB} \tag{17}$$

For the sensitivity of transformability, S, we calculated the sensitivity, S_2, of signal factors M_3, M_4,

M_9, and M_{10}, and the S_1 values of other signal factor levels, respectively, because we found a different shrinkage rate for a different portion of a model gear. In short, for the data in Table 3, we can calculate each sensitivity according to each variation of proportional term in equation (14), as follows.

Sensitivity of dimensions M_1':

$$S_1 = \frac{1}{(6)(1219.848126)}(7030.870285 - 0.000154) = 0.960621$$

$$10 \log S_1 = -0.17 \text{ dB} \tag{18}$$

Sensitivity of dimensions M_2':

$$S_2 = \frac{1}{(6)(4.26053)}(18.833471 - 0.000787) = 0.736711$$

$$10 \log S_2 = -1.33 \text{ dB} \tag{19}$$

4. Factors and Levels

We designed a model gear for our molding experiment. Table 4 shows control factor and levels. As control factors, we selected mold temperature (for both fixed and movable), cylinder temperature, injection speed, injection pressure, and cooling time from molding conditions. In addition, after-molding part placement, which is believed to affect the di-

mensional accuracy of small, precise parts, was also chosen as one of the control factors. According to the sliding-level method, we related the low level of the fixed mold to the low level of the movable mold as level 2, which is the same temperature as that of the fixed mold, and set the low level $\pm 5°C$ to levels 1 and 3, respectively. The control factors were assigned to an L_{18} as the inner array. The signal and noise factors were assigned to the outside.

Following the foregoing procedure, we can compute each SN ratio of each experiment in the L_{18} orthogonal array as shown in Table 5. Table 6 shows the level-by-level SN ratios, and Figure 2 plots the factor effects.

5. Estimation of Molding Conditions and Confirmatory Experiment

According to the results obtained thus far, the control factors affecting transformability to a large degree are B, D, E, F, and G. However, B, E, and G are either peaked or V-shaped. Since control factor B represents a difference between the upper and lower mold temperatures, B_2 was set to the condition with no temperature difference, and others to a configuration with a certain difference. Judging from Figure 3, showing A's factor effect for each of B's levels, we can see that all B factors except B_2 are regarded as unstable because their effects on a

Table 4
Control and noise factors and levels

Factor		Level 1	Level 2	Level 3
A:	mold temperature (fixed die)	Low (A_1)	High (A_2)	—
B:	mold temperature (movable die)	$A_1 - 5$ / $A_2 - 5$	A_1 / A_2	$A_1 + 5$ / $A_2 + 5$
C:	cylinder temperature	Low	Current	High
D:	injection speed	Slow	Current	Fast
E:	injection pressure	Low	Current	High
F:	cooling time	Short	Current	Long
G:	part placement	I	II	III

Table 5
SN ratios of the L_{18} orthogonal array (dB)

No.	Transformability	Adjustability	No.	Transformability	Adjustability
1	34.88	−15.90	10	33.53	−24.26
2	34.38	−17.82	11	36.07	−14.38
3	32.73	−13.78	12	30.07	−21.60
4	34.96	−15.87	13	35.46	−14.81
5	34.69	−16.41	14	33.39	−22.37
6	35.57	−21.92	15	36.29	−14.55
7	35.38	−14.82	16	33.17	−22.37
8	28.97	−31.51	17	35.49	−16.22
9	36.54	−14.91	18	31.14	−32.88

model gear are not constant due to a temperature difference. This is the reason that B has a peak in the plot. Although we believed that control factor E assumes no peaked shape, we suggest that this may be caused by certain interactions. We should reexamine this phenomenon in the future. Since control factor G is associated with part placement and is not continuous, it can become peaked. Thus, the best level of G, G_2, demonstrates that our current part placement is best. The adjustability SN ratio shows a tendency similar to that of transformability. Finally, as the optimal configuration, we selected the combination of $A_1B_2C_1D_1E_3F_3G_2$ because we judged that B and E should be excluded in calculating the

SN ratio, due to their instability. The SN ratio at the optimal configuration was calculated as follows.

Transformability:

$$\mu = 35.70 + 34.72 + 35.11 - (2)(34.06)$$

$$= 37.41 \text{ dB} \tag{20}$$

Adjustability:

$$\mu^* = -15.42 - 16.05 - 18.26 + (2)(19.28)$$

$$= -11.17 \text{ dB} \tag{21}$$

The SN ratio at the current configuration of A_1B_2 $C_2D_2E_2F_2G_2$ was calculated as follows.

Table 6
Average SN ratios by level (dB)

		Transformability			Adjustability		
	Factor	1	2	3	1	2	3
A:	mold temperature (fixed die)	34.23	33.87	—	−18.18	−20.38	—
B:	mold temperature (movable die)	33.66	35.06	33.45	−18.07	−17.66	−22.12
C:	cylinder temperature	34.56	33.83	33.77	−18.12	−19.79	−19.94
D:	injection speed	35.70	33.57	32.89	−15.42	−19.72	−22.70
E:	injection pressure	34.40	33.02	34.74	−20.50	−20.62	−16.73
F:	cooling time	33.46	33.99	34.72	−20.44	−21.35	−16.05
G:	part placement	33.84	35.11	33.22	−19.04	−18.26	−20.54

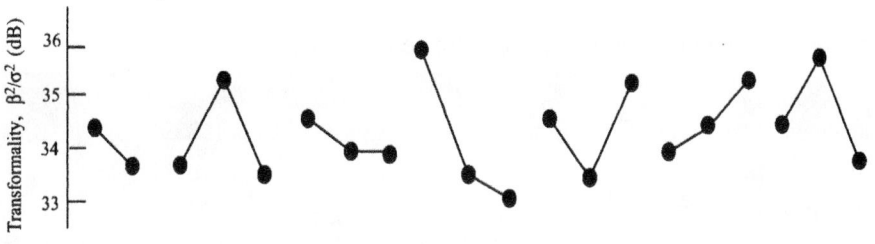

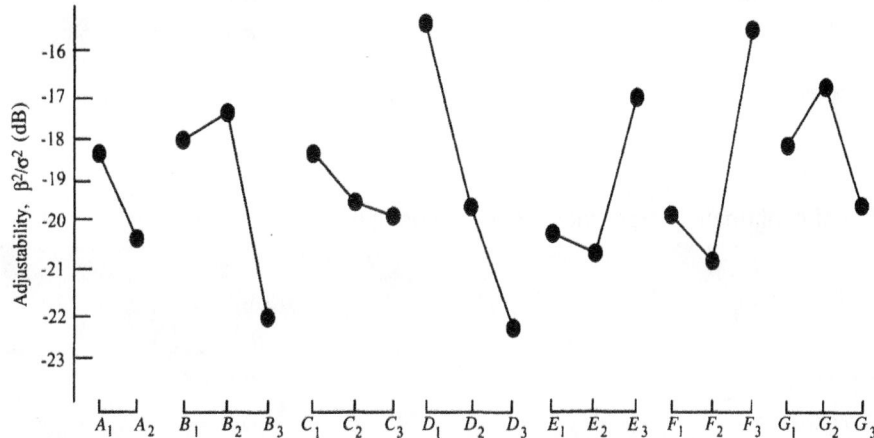

Figure 2
Response graphs

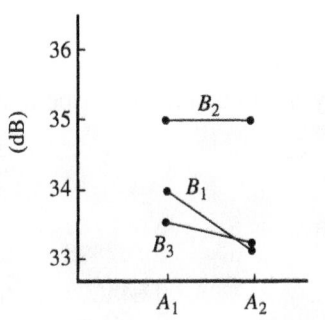

Figure 3
Interaction between A and B

Transformability:

$$\mu = 33.57 + 33.99 + 35.11 - (2)(34.06)$$

$$= 34.55 \text{ dB} \tag{22}$$

Adjustability:

$$\mu^* = -19.72 - 21.35 - 18.26 - (2)(19.28)$$

$$= -20.77 \text{ dB} \tag{23}$$

As a result, we can obtain $(37.41 - 34.55) = 2.86$ dB and $(-11.17 + 20.77) = 9.60$ dB as the gains of transformability and adjustability.

Although we chose the third and twentieth shots as error factor levels, we found that there was only a small fluctuation between them because our gear model quickly became stable after being molded, due to its small dimension. Indeed, our original setup of signal factor levels were not wide enough;

Table 7
ANOVA table of confirmatory experiment (optimal configuration)

	Level	f	S	V
β:	proportional term	1	7075.810398	7075.810398
βM':	positional noise	1	0.318380	0.318380
βN:	compounded noise	1	0.000202	0.000202[a]
β*:	first-order term	1	0.038291	0.038291
res:	residual	4	0.000442	0.000111[a]
e:	error	64	0.013529	0.000211[a]
e':	error (after pooling factors indicated by [a])	69	0.014173	0.000205
Total		72	7076.181243	

[a] Factors to be pooled.

Table 8
ANOVA table of confirmatory experiment (current configuration)

	Level	f	S	V
β:	proportional term	1	7065.759455	7065.759455
βM':	positional noise	1	0.311792	0.311792
βN:	compounded noise	1	0.000128	0.000128[a]
β*:	first-order term	1	0.044723	0.044723
res:	residual	4	0.002015	0.000504[a]
e:	error	64	0.017974	0.000281[a]
e':	error (after pooling factors indicated by [a])	69	0.020117	0.000292
Total		72	7066.136083	

[a] Factors to be pooled.

Table 9
Estimation and confirmation (dB)

	Configuration		
	Optimal	Current	Gain
Transformability			
Estimation	37.41	34.55	2.86
Confirmation	36.72	35.18	1.54
Adjustability			
Estimation	−11.17	−20.77	9.6
Confirmation	−16.08	−16.84	0.76

Table 10
Sensitivity S for confirmatory experiment

		Configuration	
		Current	Optimal
Transformability	S_1	0.962783	0.964159
	S_2	0.757898	0.757105
Adjustability	S_1	0.044507	0.031529
	S_2	0.000166	0.000152

however, by prioritizing transformability, we performed a confirmatory experiment based on the optimal configuration $A_1B_2C_1D_1E_3F_3G_2$ and current configuration $A_1B_2C_2D_2E_2F_2G_2$. Tables 7 and 8 show the ANOVA tables, and we summarize the SN ratios in Table 9.

Although we obtained fairly good reproducibility of transformability, for adjustability we concluded that we should examine the reason that its reproducibility was not satisfactory. Since the shrinkage of different dimensions in a piece is different, sensitivity from M_3, M_4, M_9 and M_{10}, and also that from other dimensions, were calculated from the confirmatory experiment, as shown in Table 10. No significant differences were found.

Reference

Takayoshi Matsunaga and Shinji Hanada, 1992. *Technology Development of Transformability*. Quality Engineering Application Series. Tokyo: Japanese Standards Association, pp. 83–95.

This case study is contributed by Takayoshi Matsunaga and Shinji Hanada.

Optimization of a Felt-Resist Paste Formula Used in Partial Felting

Abstract: In this research, by taking advantage of other easily obtainable paste agents in place of PNIPAM and minimizing the use of costly PVME, we attempted to apply the partial felting technique to jacquard cloth, which requires a variety of threads and fabrics. After optimization, a paste formula was developed successfully, resulting in a one-third cost reduction.

1. Introduction

Partial felting is a forming process based on felting shrinkage where by using a paste made primarily of heat-sensitive polymers as an antifelting agent, after printing a predetermined pattern onto wool cloth, it is felted above the temperature at which heat-sensitive polymers become insoluble. Through this process, the part not printed shrinks, whereas the part printed is not felted, and as a result, both felted and unfelted parts remain, creating an expressive appearance including rugged surface areas and various fluffs.

What felting means here is a phenomenon unique to wool (knitted) cloth, where because of cuticles on the surface of wool, its fibers are interwoven with each other, the wool shrinks in an irreversible manner, and subsequently, its texture becomes thicker and finer. This is similar to the case where a sweater made of wool shrinks after being washed and can never be returned to its original size. Unlike other common substances, heat-sensitive polymers are soluble in cold water but insoluble in hot water. PVME [poly(vinyl methyl ether)], especially, used in our study, is a very heat-sensitive polymer that quite acutely repeats the reversible states of becoming dissolved and clouded around 35°C.

Another heat-sensitive polymer, is a mixture of PNIPAM ([poly(*N*-isopropylacrylamide)] with

PVME for partial felting. However, since PNIPAM has been obtainable only for experimental purposes and PVME is costly as a fiber-forming agent, it has been difficult for us to put this technique into practice. Therefore, in this research, by taking advantage of other easily obtainable paste agents in place of PNIPAM and minimizing the use of costly PVME, we attempted to apply the partial felting technique to jacquard cloth, which requires a variety of threads and fabrics. At the same time, we considered not only the anti-felting ability but also sufficient dyeablity.

2. Generic Function and Measurement Characteristics

We defined as ideal the state where, even if felted, the part printed by antifelting paste neither shrank nor warped: in other words, remained unchanged. Now we set the distance between each pair of vertices of the printed part to the input and the distance between the identical pair after felting to the output. The ideal is to have a zero-point proportional relationship with a slope of 1 between them.

As measurement characteristics, we used the before- and after-felted distances for 12 levels of each pair of vertices based on one unequilateral quadrilateral-shaped and two different triangle-shaped test pieces on which antifelting paste was printed. As a noise factor, considering that we apply the partial felting technique to jacquard cloth consisting of various types of threads and fabrics, we selected mousseline, which is made with a single-filament plain weave, and serge, which is made with two-up and two-down twill weaves, as test cloth. In addition, since actual products require different levels of felting shrinkage, we defined the felting time as a compound noise factor whose two levels are 50 and 70 min (Table 1).

On the other hand, this antifelting starch needs to have sufficient dyeability as well as antifelting capability because we wished to expand the number of products to which we can apply our partial felting technique. As for a measurement characteristic of dyeability, we use the K/S value that is derived from the theory proposed by Kubelka and Munk in 1913 and computed by the spectral reflectance of a dyed material. Ideally, the K/S value is proportional to the volume of dye included in test cloth. According to the orthogonal array, we mixed antifelting paste with dye with 1.0, 0.4, and 0.1% for each of the deep, intermediate, and light colors and printed each of them with the dies (Figure 1). The generic function of dyeability is expressed by a reference-point proportional equation having an origin where the line (showing the relationship between dye concentration in paste and K/S value) passes through the point of K/S value where the antifelt paste containing no dye is printed. Since the ability of anti-felting and dyeability were different, we performed an analysis of each function separately and determined the optimal configuration by combining results.

3. SN Ratio and Sensitivity

Table 2 summarizes the data of experiment 1 in an L_{18} orthogonal array. Using them, we explain com-

putation of the SN ratio and sensitivity. In this study we used dynamic characteristics of transformability.

Total variation:

$$S_T = 123.0^2 + \cdots + 83.4^2$$
$$= 493,403.9 \quad (f = 24) \quad (1)$$

Effective divider:

$$r = 127.5^2 + \cdots + 90.5^2 = 276,021.8 \quad (2)$$

Linear equation:

$$L_1 = (127.5)(123.0) + \cdots + (90.5)(88.0)$$
$$= 269,879.3$$
$$L_2 = (127.5)(118.9) + \cdots + (90.5)(83.4)$$
$$= 251,651.5 \quad (3)$$

Variation of proportional terms:

$$S_\beta = \frac{(L_1 + L_2)^2}{2r} = 492,704.6 \quad (f = 1) \quad (4)$$

Variation of differences of proportional terms:

$$S_{N\beta} = \frac{(L_1 - L_2)^2}{2r} = 601.86 \quad (f = 1) \quad (5)$$

Error variation:

$$S_e = S_T - S_\beta - S_{N\beta} = 97.44 \quad (f = 22) \quad (6)$$

Error variance:

$$V_e = \frac{S_e}{22} = 4.43 \quad (7)$$

Total error variance after pooling:

$$V_N = \frac{S_{N\beta} + S_e}{23} = 30.40 \quad (8)$$

SN ratio:

$$\eta = 10 \log \frac{(1/2r)(S_\beta - V_e)}{V_N} = -15.32 \text{ dB} \quad (9)$$

Sensitivity:

$$S = 10 \log \frac{1}{2r}(S_\beta - V_e) = -0.49 \text{ dB} \quad (10)$$

Table 1
Noise factors

	N_1	N_2
Cloth type	Mousseline	Serge
Felting time (min)	50	70

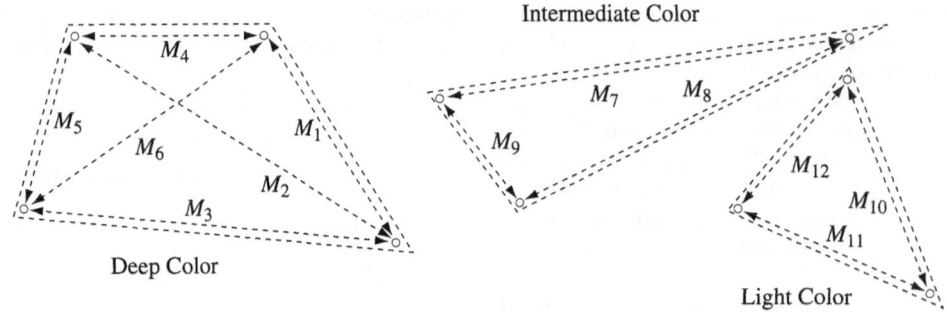

Figure 1
Signal factors

4. Optimal Configuration and Results of Confirmatory Experiment

As control factors, as shown in Table 3, we selected eight from the factors needed both for antifelting and for dyeability and assigned them to an L_{18} orthogonal array. Figure 2 illustrates the response graphs obtained from the SN ratio and sensitivity for each experiment based on the L_{18} orthogonal array. At the same time, as for dyeability, we calculated the SN ratios and sensitivities (but we omit them here). By focusing on the results regarding antifelting, we determined the optimal and worst configurations and confirmed them with confirmatory experiments. The results are summarized in Table 4.

These results prove that we obtained good reproducibility for both SN ratio and sensitivity. If we can recognize visually the difference between felted and unfelted portions in the partial felting process, we can evaluate quality using a six-point of inspection, then fine-tune to satisfy SN ratios and sensitivities. To this end, we adjusted the amount of heat-sensitive polymers used such that the total cost became minimal. For experiments 3, 6, 10, 14, and 16, all of whose visually inspected values are judged to be less than 2 (i.e., sufficient antifelting ability for partial felting), their average SN ratio and sensitivity are computed as -12.94 dB and -0.40 dB, respectively. The optimal configuration still allows these values to have a 0.66-dB slack for the SN ratio and a 0.04-dB slack for sensitivity. By reducing costly PVME with these slacks, we can slash the amount used from that at the optimal configuration by 20% in the SN ratios and by 27% in sensitivity. Computing the final cost of paste in accordance with the smaller reduction of 20%, we can arrive at approximately 850 yen/kg. As compared with approximately 2600 yen/kg when two types of heat-sensitive polymers are mixed in a conventional manner, this

Table 2
Experimental data (mm)

	M_1 (127.5)	M_2 (207.0)	M_3 (194.0)	M_4 (101.0)	M_5 (92.5)	M_6 (150.5)	Visual Judgment
N_1	123.0	201.0	188.0	99.0	90.0	146.6	5
N_2	118.9	187.4	173.0	89.6	86.0	136.0	5
	M_7 (230.5)	M_8 (204.5)	M_9 (71.0)	M_{10} (126.5)	M_{11} (117.0)	M_{12} (90.5)	Visual Judgment
N_1	228.0	202.5	69.0	123.5	113.5	88.0	3, 4
N_2	210.8	184.8	68.0	119.9	108.9	83.4	5, 5

Table 3
Control factors and levels

		Level		
	Control Factor	1	2	3
A:	mutual solubility improvement	LN: Type-1	YH: Type-2	—
B:	print-dyeing paste	SM 4000	SH 4000	SH 4000
C:	paste concentration	Low	Mid	High
D:	PVME concentration	Low	Mid	High
E:	osmotic agent concentration	0	Mid	High
F:	uric acid concentration	0	Mid	High
G:	pH adjustment agent	None	Malic acid	Ammonium sulfate
H:	dry temperature	High	Low	(Low)

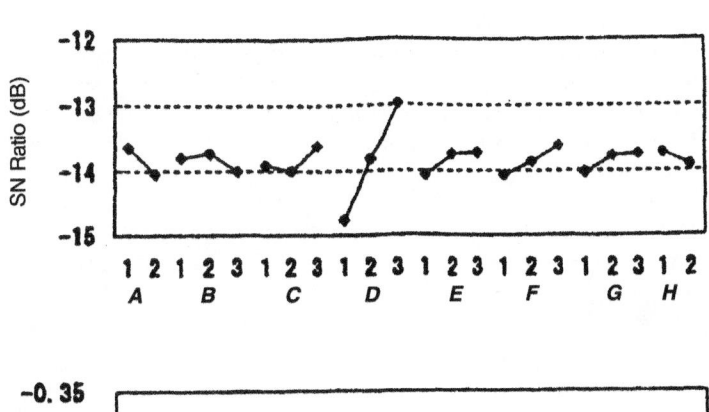

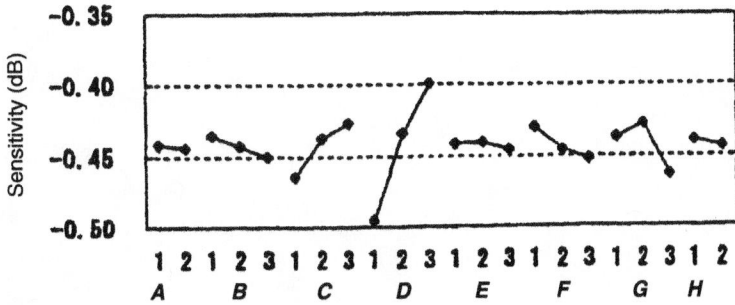

Figure 2
Response graphs

Table 4
Results of confirmatory experiment (dB)

Configuration	SN Ratio		Sensitivity	
	Estimation	Confirmation	Estimation	Confirmation
Optimal	−11.90	−12.28	−0.37	−0.36
Worst	−15.87	−16.20	−0.51	−0.51
Gain	3.97	3.92	0.14	0.15

cost reduction is one-third. Additionally, although the paste used currently often contaminates unprinted portions when mixed with dye during felting, the optimal paste does not cause the same problem, as long as the dyeing color ranges between light and intermediate, because we consider its dyeability as well. Considering this advantage, together with the cost benefit, we can obtain a greater economic benefit.

Reference

Katsuaki Ishii, 2001. Optimization of a felting resist paste formula used in the partial felting process. *Quality Engineering*, Vol. 9, No. 1, pp. 51–58.

This case study is contributed by Katsuaki Ishii.

Development of Friction Material for Automatic Transmissions

Abstract: In the conventional development process, friction material development has been made in conjunction with the development of automatic transmissions so that the performance of friction material may meet the requirements for ease of use. In this research, besides developing an automatic transmission, we attempted to develop a friction material that has high functionality.

1. Introduction

Figure 1 illustrates the structure of an automatic transmission. At the portions that are circled, friction materials are required for smooth clutching and releasing actions to meet targeted performance goals such as reduction of shift shock or vibration and improvement in fuel efficiency.

2. Generic Function

Traditionally, we have evaluated friction materials by using primarily the μ–V characteristic (representing the coefficient of dynamic friction corresponding to a relative rotating speed between a friction material and its contact surface; Figure 2). The μ–V characteristic need not have a negative slope to prevent shift shock or vibration. However, it tends to become negative. Thus, through many repeated experiments, engineers have made final decisions by looking at μ–V characteristics.

Fundamentally, the friction force follows Coulomb's law, expressed by $F = \mu W$, where F is the friction force, W the vertical load, and μ the friction coefficient. Taking into consideration the fact that a friction material inside an automatic transmission is used in a rotational direction, a commonly used experimental device (Figure 3) was utilized for our

study. As a generic function, we defined the proportionality of $y = \beta M$ by setting surface pressure to an input signal, M (equal to load during experiments), and torque to an output, y.

For signal factor levels, we determined them by considering loads applied to friction materials used at various parts of an automatic transmission. The rotating speed is required to evaluate the slopes of μ–V characteristics, which we desire to be flat. However, because of external factors such as complicated part structure in an automatic transmission, the slopes are expected to be slightly positive. So we assigned the rotating speed, M^*, to the outer array to assess the slope of the μ–V characteristic. As noise factors, we picked up some factors from what most affect the μ–V characteristic (Table 1). Knowing from our experience that the characteristic does not change uniformly until it arrives at the deteriorating phase, we took up fit-in conditions to stabilize the characteristic from the initial phase up to the actual use. As for setting of deteriorating conditions, we selected load, which needs a short time for evaluation from the preliminary experimental phase.

3. SN Ratio

Based on the format shown in Table 2, we obtained measurement data. To add the slope of μ–V char-

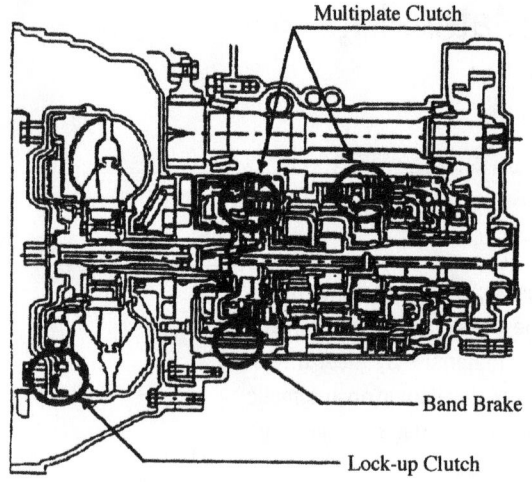

Figure 1
Structure of automatic transmission (sectional view)

acteristics to our evaluation, we prepared the summary table of linear equations illustrated in Table 3 and proceeded with our analysis based on the next procedure. Figure 4 indicates measured data under a certain noise condition. With respect to β, we can replot these data in Figure 5 by setting the rotating speed velocity, M^*, to the horizontal axis. Next we defined the slope in Figure 5 as β*. That β* becomes positive is equivalent to a positive slope of a μ–V characteristic. $S_{\beta*}$ corresponds to the variation in the first-order element among the variations $S_{M^*\beta}$ due to the relative rotating speed M^*. Orthogonal polynomial equations are used to decompose these data.

Total variation:

$$S_T = y_{111}^2 + y_{112}^2 + \cdots + y_{135}^2 + \cdots$$
$$+ y_{611}^2 + \cdots + y_{635}^2$$

$$= 1032.4750 \quad (f = 90) \quad (1)$$

Effective divider:

$$r = M_1^2 + M_2^2 + M_3^2 = 3.50 \quad (2)$$

Variation of proportional term:

$$S_\beta = \frac{(L_1 + L_2 + L_3 + L_4 + L_5 + L_6)^2}{(6)5r}$$

$$= 1029.1453 \ (f = 1) \quad (3)$$

Variation of differences of proportional terms due to rotating speed:

$$S_{M^*\beta} = \frac{K_1^2 + K_2^2 + \cdots + K_5^2}{6r} - S_\beta$$

$$= 0.0319 \quad (f = 4) \quad (4)$$

Variation of proportional term due to rotating speed:

$$S_{\beta*} = \frac{(W_1 K_1 + W_2 K_2 + \cdots + W_5 K_5)^2}{J(6r)}$$

$$= 0.0030 \quad (f = 1) \quad (5)$$

where W_i is the coefficient for an orthogonal polynomial equation. $W_i = X_i - \overline{X}$ and $J = \Sigma \ w_i^2$.

Variation of differences of residual terms:

$$S_{res \times \beta} = S_{M^*\beta} - S_{\beta*} = 0.0290 \quad (f = 3) \quad (6)$$

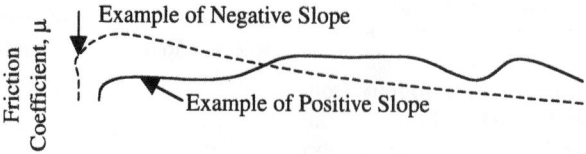

Figure 2
Typical μ–V characteristic

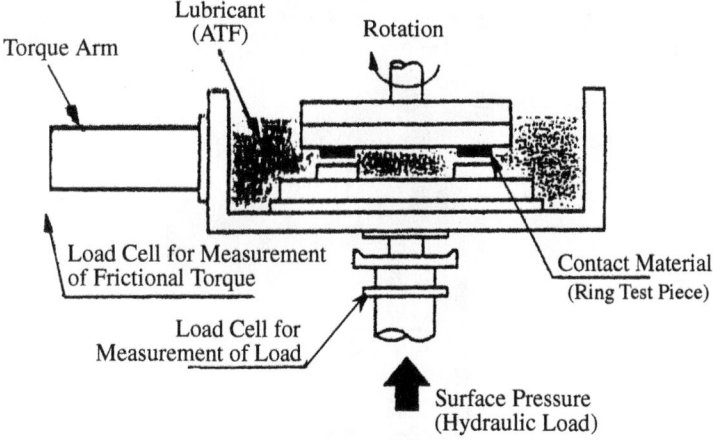

Figure 3
Experimental device

Error variation:

$$S_e = S_T - S_\beta - S_{M*\beta} = 3.2977 \quad (f = 85) \quad (7)$$

Error variance:

$$V_e = \frac{S_e}{85} = 0.0388 \quad (8)$$

Total error variance:

$$V_N = \frac{0.0290 + 3.2977}{88} = 0.0378 \quad (9)$$

SN ratio:

$$\eta = 10 \log \frac{(1/30r)(S_\beta - V_e)}{V_N} = 24.14 \text{ dB} \quad (10)$$

Sensitivity:

$$S = 10 \log \frac{1}{30r}(S_\beta - V_e) = 9.91 \text{ dB} \quad (11)$$

Slope:

$$\beta* = \frac{W_1 K_1 + W_2 K_2 + \cdots + W_5 K_5}{J \times 6 \times r} = 0.00004 \quad (12)$$

Table 1
Signal and noise factors

Factor	Level				
Signal factors Main signal M: surface pressure	Low	Mid	High		
Subsignal $M*$: relative rotating speed	M_1^*	M_2^*	M_3^*	M_4^*	M_5^*
Noise factors O: ATF oil temperature P: fit-in, deterioration	Low Initial	High Fit-in	— After Degradation	— —	— —

Table 2
Examples of torque data

Noise Condition (Oil Temperature)			Signal Factor (Surface Pressure) Relative Rotating Speed (rpm)									
			M_1					M_2		$\cdots$	M_3	
			M_1^*	M_2^*	M_3^*	M_4^*	M_5^*	M_1^*	M_2^*	$\cdots$	M_4	M_5^*
Low	N_1	Initial	y_{111}	y_{112}	y_{113}	y_{114}	y_{115}	y_{121}	y_{122}	$\cdots$	y_{134}	y_{135}
	N_2	Fit-in										
	N_3	Deterioration										
High	N_4	Initial										
	N_5	Fit-in							y_{622}			
	N_6	Deterioration	y_{611}	y_{612}	y_{613}	y_{614}	y_{615}	y_{621}	y_{634}	$\cdots$	y_{635}	

844

Table 3
Summary of linear equations

| | Linear Equation for Each Rotating Speed | | | | | Linear Equation for Each Error Condition | |
	M_1^*	M_2^*	M_3^*	M_4^*	M_5^*		
N_1	L_{11}	L_{12}	L_{13}	L_{14}	L_{15}	55.0480	L_1
N_2	L_{21}	L_{22}	L_{23}	L_{24}	L_{25}	56.8663	L_2
N_3	L_{31}	L_{32}	L_{33}	L_{34}	L_{35}	57.6040	L_3
N_4	L_{41}	L_{42}	L_{43}	L_{44}	L_{45}	51.7542	L_4
N_5	L_{51}	L_{52}	L_{53}	L_{54}	L_{55}	52.3932	L_5
N_6	L_{61}	L_{62}	L_{63}	L_{64}	L_{65}	55.0596	L_6
	K_1	K_2	K_3	K_4	K_5		
Subtotal	65.6614	66.4108	65.6962	65.6730	65.2838		

4. Results of Analysis

Figure 6 summarizes the response graphs of the analysis of results obtained by the aforementioned analysis procedure.

5. Optimal Configuration and Results of Confirmatory Experiment

To determine the optimal configuration, we performed the parameter design after selecting the control factors enumerated in Table 4. In selecting each level for the optimal configuration, we took into account the slope β^* and cost issue as well as SN ratios. Table 5 shows the estimations and confirmatory experimental results. We can see that both the SN ratio and sensitivity have a high reproducibility, and the SN ratio, particularly, is improved by 3.65 dB. Now the sensitivity level was regarded as sufficient for the automatic transmission system. Although we believed that the slope β^* was good enough because it was approximately equivalent to that under the initial configuration, a better result was obtained. In addition, a confirmatory experiment was conducted using the oil of the initial development stage. We can see that an approximate 4.5 dB improvement in the SN ratio and a flat characteristic of β^* were realized. We confirmed that by selecting the optimal specifications, we could obtain a better tendency even if the property of the oil is

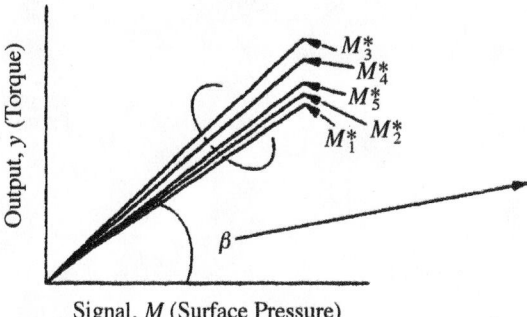

Signal, M (Surface Pressure)

Figure 4
Typical measured data

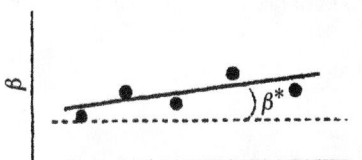

Relative Rotating Speed, M^* (rpm)

Figure 5
Relationship between relative rotating speed and β^*

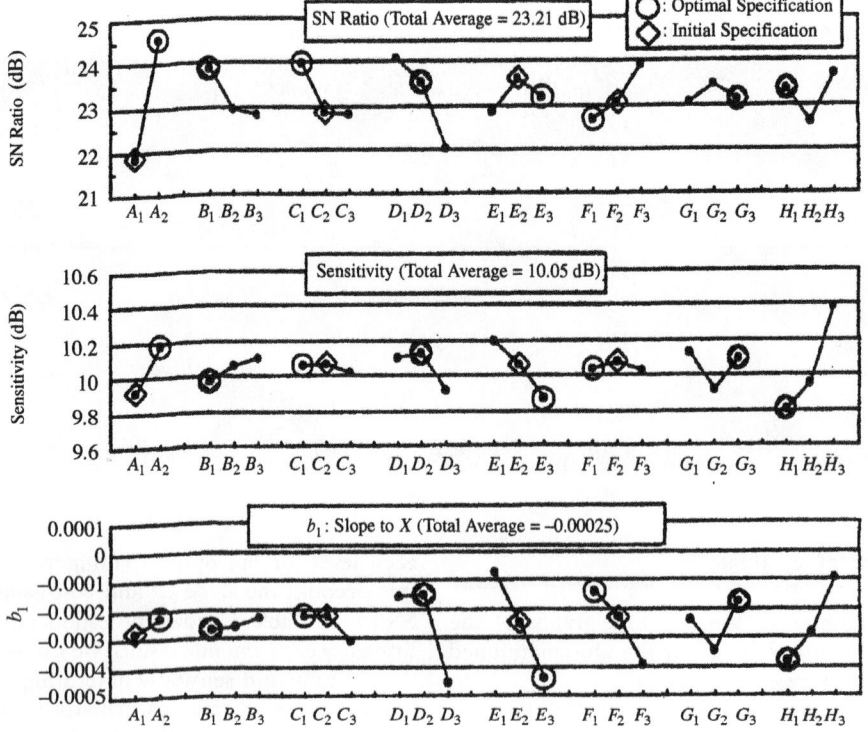

Figure 6
Response graphs

Table 4
Control factors and levels

		Level		
Control Factor		**1**	**2**	**3**
A:	fiber diameter	Small	Large	—
B:	ratio of abrasion material to adjustment material	Small	Mid	Large
C:	amount of fiber	Small	Mid	Large
D:	fiber ratio	Small	Mid	Large
E:	amount of resin	Small	Mid	Large
F:	compression rate	Small	Mid	Large
G:	plate thickness	Small	Mid	Large
H:	surface treatment	None	1	2

Table 5

Results of confirmatory experiment

Configuration	SN Ratio, η (dB)		Sensitivitiy, S (dB)		Slope, β^*	
	Estimation	Confirmation	Estimation	Confirmation	Estimation	Confirmation
Optimal	25.80	27.29	9.82	9.74	−0.00024	−0.00017
Initial	22.72	23.64	9.79	9.65	−0.00018	−0.00026
Gain	3.08	3.65	0.03	0.09	−0.00006	0.00009

different. More important results from oil development are expected.

Reference

Makoto Maeda, Nobutaka Chiba, Tomoyuki Daikuhara, and Tetsuya Ishitani, 2000. Development of wet friction: material for automatic transmission. *Quality Engineering*, Vol. 8, No. 4, pp. 39–46.

This case study is contributed by Makoto Maeda, Nobutaka Chiba, Tomoyuki Daikuhara, and Tetsuya Ishitani.

CASE 29

Parameter Design for a Foundry Process Using Green Sand

Abstract: Because of its low cost and easy recycling, green sand casting (casting method using green sand) is used in many cast products. Since the caking strength of green sand tends to vary widely when mixed and it fluctuates over time after being mixed, it is considered difficult to control on actual production lines. Therefore, by applying parameter design we attempted to determine an optimal manufacturing condition such that we can improve its caking strength after kneading and make it more robust to time-lapsed change.

1. Introduction

The conventional evaluation method has used the compression strength that is measured by a cylindrical sample filled firmly with green sand. However, since some green sand is brittle even if it has a high compression strength, we considered not only the strength but also displacement up to a breaking point as important elements. Thus, after preparing a sample of green sand filled in an upper-and-lower-split cylinder (Figure 1), we measured the relationship between the displacement and load by pulling the sample vertically. As a generic function, we assumed that there exists a proportionality of $y = \beta M$ between the displacement, M, and the load, y, according to Hooke's law.

As a signal factor, we selected the displacement, M, when performing a tension experiment; and for the factor levels, we picked up the next five, $M = 0.02, 0.04, 0.06, 0.08,$ and 0.10 mm. Considering the fact that a stoppage time at a production line strongly affects the caking strength, as a noise factor we chose a lapsed time after mixture of green sand with two levels, $N_1 = 0$ min and $N_2 = 60$ min.

2. SN Ratio and Sensitivity

Based on the result of the pull test, we calculated using the SN ratio and sensitivity zero-point proportional equations. Using the data of experiment 1, in Table 1 we show the calculation procedure of SN ratio and sensitivity.

Total variation:

$$S_T = 28.3^2 + 47.9^2 + \cdots + 2.4^2 + 1.5^2$$
$$= 6880.46 \qquad (f = 10) \qquad (1)$$

Linear equations:

$$L_1 = (0.02)(28.3) + (0.04)(47.9)$$
$$+ \cdots + (0.10)(8.3) = 7.168$$

$$L_2 = (0.02)(34.2) + (0.04)(18.1)$$
$$+ \cdots + (0.10)(1.5) = 2.002 \qquad (2)$$

Effective divider:

$$r = 0.02^2 + 0.04^2 + \cdots + 0.10^2 = 0.022 \qquad (3)$$

Variation of proportional term:

$$S_\beta = \frac{(L_1 + L_2)^2}{2r} = 1911.111 \qquad (f = 1) \qquad (4)$$

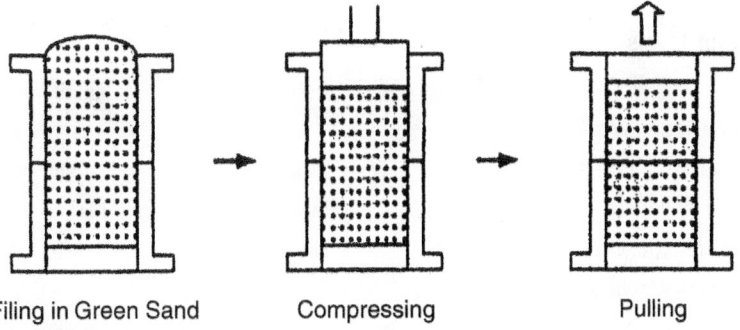

Filing in Green Sand   Compressing   Pulling

Figure 1
Method for measurement of caking strength

Variation of differences of proportional term:

$$S_{N\beta} = \frac{L_1^2 + L_2^2}{r} - S_\beta = 606.536 \qquad (f = 1) \qquad (5)$$

Error variation:

$$S_e = S_T - S_\beta - S_{N\beta} = 4362.819 \qquad (f = 8) \qquad (6)$$

Error variance:

$$V_e = \frac{S_e}{8} = 545.352 \qquad (7)$$

Total error variance:

$$V_N = \frac{S_{N\beta} + S_e}{9} = 552.151 \qquad (8)$$

SN ratio:

$$\eta = 10 \log \frac{(1/2r)(S_\beta - V_e)}{V_N} = 17.50 \text{ dB} \qquad (9)$$

Sensitivity:

$$S = 10 \log \frac{1}{2r}(S_\beta - V_e) = 44.92 \text{ dB} \qquad (10)$$

3. Optimal Configuration and Results of Confirmatory Experiment

As shown in Table 2, we selected as control factors, based on our technical knowledge, eight factors that significantly influence caking strength. Figure 2 illustrates the response graphs.

Considering the response graphs of the SN ratio and sensitivity, we determined that the optimal configuration is $A_1B_3C_3D_3E_1F_1G_1H_1$. To verify the estimation, we performed confirmatory experiments both at the optimal and at the initial configurations. Table 3 shows estimations of the SN ratio and sensitivity and results of the confirmatory experiments. The fact that with respect to both SN ratio and sensitivity, the gains obtained in the confirmatory

Table 1
Results of pull test (mm)

	Signal Factor (Displacement of Green Sand)				
Noise Factor	M_1 (0.02)	M_2 (0.04)	M_3 (0.06)	M_4 (0.08)	M_5 (0.10)
N_1 (0 min)	28.3	47.9	44.4	14.8	8.3
N_2 (60 min)	34.2	18.1	4.2	2.4	1.5

Table 2
Control factors and levels

	Control Factor	Level		
		1	2	3
A:	amount of water before kneading	Small	Large	—
B:	CB value after kneading	Low	Mid	High
C:	amount of bonding agent	No	Small	Large
D:	amount of surface stabilizing agent	No	Small	Large
E:	amount of micropowder sand	No	Small	Large
F:	amount of coal powder	No	Small	Large
G:	amount of new sand	Small	Mid	Large
H:	maturation time	Short	Mid	Long

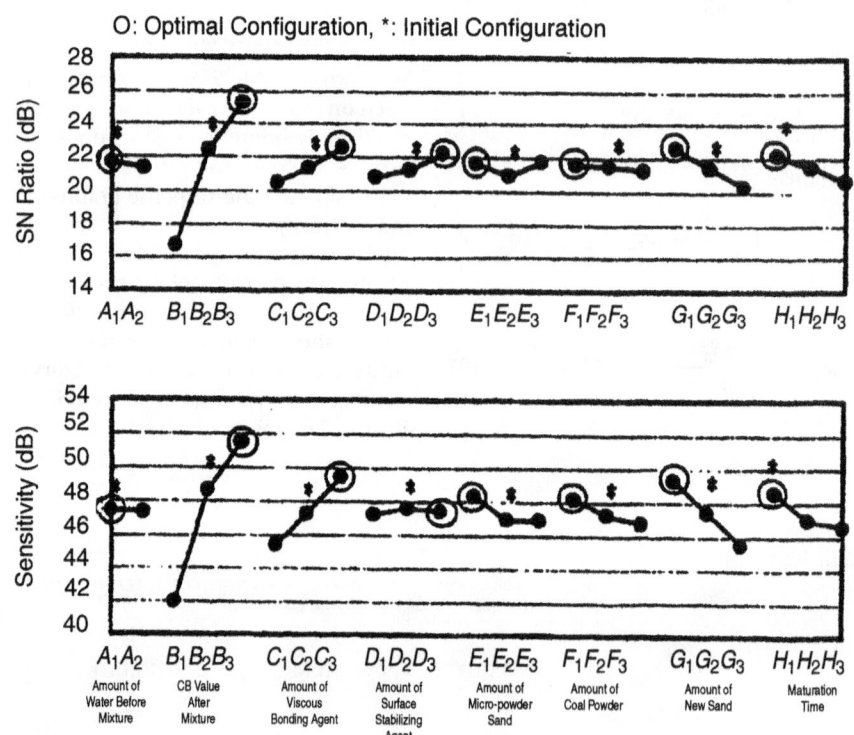

Figure 2
Response graphs

Table 3
SN ratios and sensitivity and results of experiments

	SN		Sensitivity	
	Prediction	**Confirmation**	**Prediction**	**Confirmation**
Optimum	27.61	33.89	55.55	55.96
Baseline	22.20	29.13	48.51	49.55
Gain	5.41	4.76	7.04	6.41

experiments were proven to be almost the same as the estimations reveals the validity of the experiments.

In addition, we confirmed the effects on quality characteristics. Currently, as a quality characteristic to evaluate the viscous bonding strength of green sand, we used the jolt toughness (JT) test, which can be implemented easily on production lines. As Figure 3 illustrates, this test measures how many impacts are needed to break a test sample, with repeated impacts on the sample, firmly filled with green sand.

Table 4 summarizes the JT values measured at both the optimal and initial configurations. For the JT value right after mixing, we obtained a result of 64 times at the optimal configuration rather than

Table 4
Jolt toughness values

Configuration	Right after Kneading (N_1)	60 min after Kneading (N_2)
Optimal	64	48
Initial	33	16

33 at the initial configuration, thereby improving the viscous caking strength by 94%. On the other hand, for the decreased JT value 60 minutes after mixing, there was only a 25% drop, from 64 times to 48 times at the optimal configuration, whereas a 52% fall from 33 times to 16 times took place at the initial configuration. According to this result in the confirmatory experiments for the objective quality characteristic, we can see that it is possible to improve the caking strength of green sand by changing the configuration from initial to optimal.

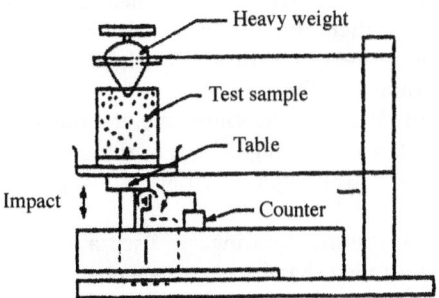

Figure 3
Jolt toughness test

Reference

Yuji Hori and Satomi Shinoda, 1998. Parameter design on a foundry process using green sand. *Quality Engineering*, Vol. 6, No. 1, pp. 54–59.

This case study is contributed by Yuji Hori.

Development of Functional Material by Plasma Spraying

Abstract: New materials composed of metal and ceramic have many interesting and useful properties in the mechanical, electrical, and chemical fields. Recently, a great deal of attention has been paid to the potential utilization of these properties. In this study, the forming process for such material was researched using low-pressure plasma spraying equipment, including two independent metal and ceramic powder supplying devices and a plasma jet flame. Using quality engineering approaches, the generic function of plasma spraying was used for evaluation. After optimization, metal and ceramic materials can be sprayed under the same conditions. Also, it is possible to produce a sprayed deposit layer and metal/ceramic mixing ratio for both dispersed-type and inclined-type products.

1. Introduction

Major technical problems in developing a functional material by plasma spraying were as follows:

1. The conditions for spraying metal are different from those for ceramic because they have extremely different heat characteristics.
2. Metal and ceramics are so different in specific gravity that they cannot comingle.

To tackle these issues, by utilizing two separate powder supply systems connected to a single decompressed plasma thermal spraying device, and by supplying metal and ceramics simultaneously to a plasma heat source, we created conditions that can develop a compound thin film.

2. Generic Function and Measurement Characteristic

Since we wanted to spray two materials that possess different properties, such as heat characteristics or specific gravity, in the first place we considered it important to create conditions where thin-film coating of any material can be achieved. Therefore, we regarded it as a generic function that when we greatly altered a mixture ratio of metal and ceramics and threw them into the same plasma jet, the film amount formed (thickness and weight), y, would be proportional to the number of reciprocal spraying motions, M, and at the same time, the film creation speed, β, would be high. That is, the generic function is $y = \beta M$. Figure 1 shows our experimental device.

When thermal spraying, we used a atmospherically controlled chamber, moving a powder-supplied thermal spraying gun horizontally and moving test pieces vertically until the spray count reaches the predetermined number. As a signal factor, we chose the number of reciprocal motions of a test piece and set each of its three levels to 1, 3, and 5 times. In addition, as a noise factor, we picked the supply ratio of metal powder (nickel) and ceramics powder (alumina) and took two levels of 7:3 and 3:7 in volume. The powder supply amount was kept con-

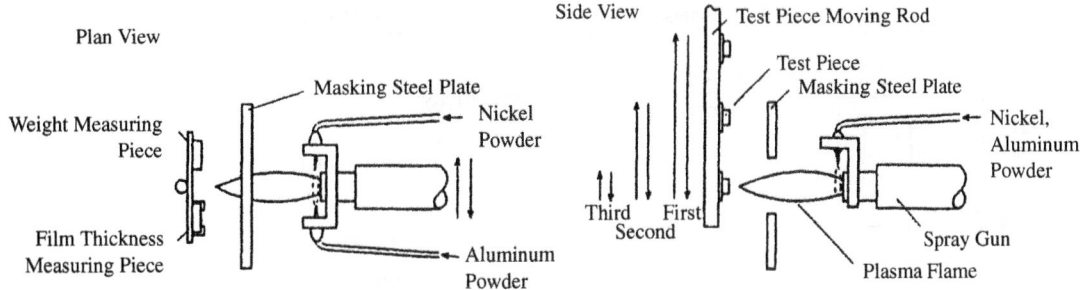

Figure 1
Experimental device

stant at 30 g/min. Finally, we selected two different characteristics, film thickness and film weight. Table 1 summarizes all signal factor noise factors and characteristics chosen in this experiment. See Table and Figure 2 for the thicknesses to be measured, where I_1, I_2, and I_3 and J_1, J_2, and J_3 indicate measurement positions.

3. SN Ratio and Sensitivity

Tables 2 and 3 illustrate the results of thickness and weight measurements. Assuming that the number of reciprocal sprays (signal factor M) and thickness and weight of a thin film (y) are expressed by the zero-point proportional equation ($y = \beta M$), we proceeded with our analysis based on the dynamic SN ratio.

Thickness Experiment
Total variation:

$$S_T = 78^2 + 80^2 + \cdots + 269^2$$
$$= 3{,}081{,}538 \qquad (f = 54) \qquad (1)$$

Effective divider:

$$r = 1^2 + 3^2 + 5^2 = 35 \qquad (2)$$

Linear equations:

$$L_1 = (1)(714) + (3)(2336) + (5)(3575) = 25{,}597$$
$$L_2 = 17{,}667 \qquad (3)$$

Variation of proportional terms:

$$S_\beta = \frac{(L_1 + L_2)^2}{(9)(2r)} = 2{,}971{,}069.3 \qquad (f = 1) \qquad (4)$$

Variation of differences of proportional terms:

Table 1
Factors and characteristics

		Level		
Factor		**1**	**2**	**3**
Signal factor Number of reciprocating sprays		1	3	5
Noise factor Compound ratio (in volume) Ni vs. alumina		7:3	3:7	—
Characteristic		Thickness Weight		—

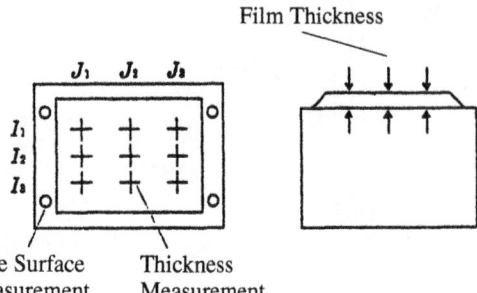

Base Surface   Thickness
Measurement   Measurement

Figure 2
Thicknesses to be measured

$$S_{N\beta} = \frac{L_1^2 + L_2^2}{9r} - S_\beta = 99,817.30 \qquad (f = 1) \qquad (5)$$

Error variation:

$$S_e = S_T - S_\beta - S_{N\beta} = 10,651.34 \qquad (f = 52) \qquad (6)$$

Error variance:

$$V_e = \frac{S_e}{52} = 204.83 \qquad (7)$$

Total error variance:

$$V_N = \frac{S_{N\beta} + S_e}{1 + 52} = 2084.31 \qquad (8)$$

SN ratio:

$$\eta = \frac{[1/(9)(2r)](S_\beta - V_e)}{V_N} = 2.26 \quad (3.55 \text{ dB}) \qquad (9)$$

Sensitivity:

$$S = \frac{1}{(9)(2r)}(S_\beta - V_e) = 4715.66 \quad (36.74 \text{ dB}) \qquad (10)$$

Weight Experiment
Total variation:

$$S_T = 0.363^2 + 1.0292^2 + \cdots + 1.1260^2$$

$$= 59,201,730 \qquad (f = 6) \qquad (11)$$

Linear equations:

$$L_1 = (1)(0.3630) + (3)(1.0292) + (5)(1.7196)$$

$$= 12.0486$$

$$L_2 = 7.8716 \qquad (12)$$

Variation of proportional terms:

$$S_\beta = \frac{(L_1 + L_2)^2}{2r} = 5.6687766 \qquad (f = 1) \qquad (13)$$

Variation of differences of proportional terms:

$$S_{N\beta} = \frac{L_1^2 + L_2^2}{r} - S_\beta = 0.2492475 \qquad (f = 1) \qquad (14)$$

Error variation:

$$S_e = S_T - S_\beta - S_{N\beta} = 0.0021488 \qquad (f = 4) \qquad (15)$$

Error variance:

Table 2
Data for 16 of the L_{18} orthogonal array (μm)

		M_1 (First Round Trip)				M_2 (Third Round Trip)				M_3 (Fifth Round Trip)			
		J_1	J_2	J_3	Total	J_1	J_2	J_3	Total	J_1	J_2	J_3	Total
N_1	l_1	78	80	79	237	282	272	274	828	434	419	401	1254
	l_2	75	81	81	237	270	263	257	790	408	400	385	1193
	l_3	78	82	80	240	240	241	237	718	385	375	368	1128
	Total	231	243	240	714	792	776	768	2336	1227	1194	1154	3575
N_2	l_1	48	47	51	146	174	167	160	501	293	292	273	858
	l_2	56	53	52	161	175	178	172	525	294	281	282	857
	l_3	51	49	47	147	160	160	155	475	288	270	269	827
	Total	155	149	150	454	509	505	487	1501	875	843	824	2542

Table 3
Data for experiment 16 of the L_{18} orthogonal array (g)

	M_1 (First Round Trip)	M_2 (Third Round Trip)	M_3 (Fifth Round Trip)
N_1	0.3630	1.0292	1.7196
N_2	0.1842	0.6858	1.1260

$$V_e = \frac{S_e}{4} = 0.0005372 \qquad (16)$$

Total error variance:

$$V_N = \frac{S_{N\beta} + S_e}{1 + 4} = 0.0502792 \qquad (17)$$

SN ratio:

$$\eta = \frac{(1/2r)(S_\beta - V_e)}{V_N} = 1.6105014 \ (2.07 \ dB) \qquad (18)$$

Sensitivity:

$$S = \frac{1}{2r}(S_\beta - V_e) = 0.0809748 \quad (-10.92 \ dB) \qquad (19)$$

4. Optimal Conditions and Confirmatory Experiment

Table 4 illustrates control factors for this study. Additionally, Figures 3 and 4 show response graphs of thickness and weight.

To confirm the reproducibility of our experimental results, we performed a confirmatory experiment on a combination of the optimal and worst conditions. In selecting the conditions, we used a combination of SN ratios for weight experiments because both thickness and weight experiments have almost identical tendencies of factor effects; moreover, the large-effect levels in the thickness experiments were consistent with those in the weight experiments. The results are shown in Table 5, which indicates that both estimation and confirma-

Table 4
Control factors and levels

Control Factor		Level 1	Level 2	Level 3
A:	secondary gas type	Hydrogen	Helium	—
B:	electric power (kW)	25	35	45
C:	current/voltage (A/V)	12	15	18
D:	decompression degree (torr)	50	200	400
E:	spraying distance (relative proportion to standard frame length)	0.8	1	1.2
F:	moving speed of spray gun (m/min)	6	13	24
G:	average particle diameter of metal powder (μm)	6	30	60
H:	average particle diameter of ceramics powder (relative proportion to G)	0.1	0.5	1

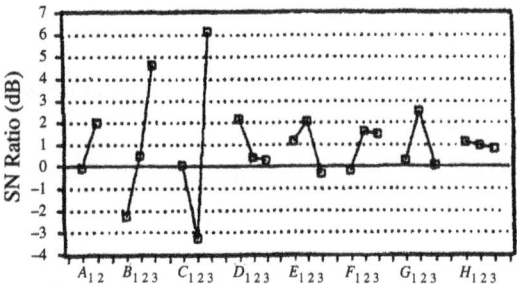

Figure 3
Response graphs of thickness experiment

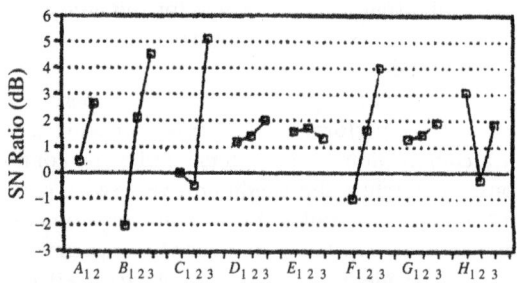
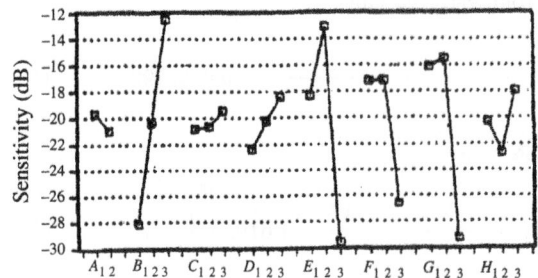

Figure 4
Response graphs of weight experiment

Table 5
Estimation and confirmatory experiment results

| | | Configuration | | |
		Optimal	Worst	Gain
Thickness	Estimation	9.83	−6.49	16.32
	Confirmation	12.70	−3.04	15.74
Weight	Estimation	12.16	−8.48	20.64
	Confirmation	15.12	−1.61	16.73

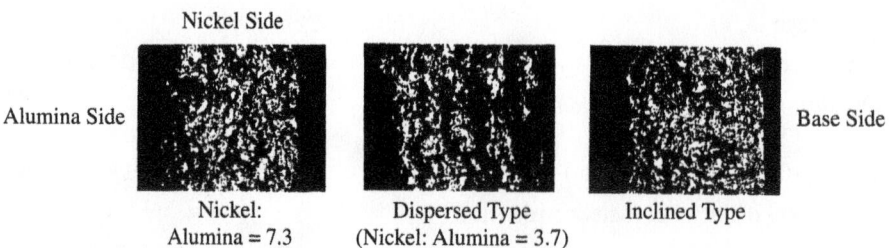

Nickel Side

Alumina Side   Base Side

Nickel: Dispersed Type   Inclined Type
Alumina = 7.3   (Nickel: Alumina = 3.7)

Figure 5
Dispersed and inclined films

tion are fairly consistent and that good reproducibility exists.

Optimal condition: $A_2B_3C_3D_3E_2F_3G_2H_1$

Worst condition: $A_1B_1C_2D_1E_3F_1G_3H_2$

By taking advantage of the optimal plasma-spraying condition obtained in this experiment, we formed both dispersed and inclined plasma-sprayed thin films whose sectional structures (of only thin films) are shown in Figure 5. The right-hand side indicates a base, and the white and dark portions show nickel and alumina, respectively. Based on these, we can see that nickel and alumina are distributed evenly even though their supply ratios are different, and that the distribution rate is close to the supply ratio. On the other hand, for the in-

clined thermal-sprayed thin film, we gradually alter the supply ratio from the nickel-abundant state to the alumina-abundant state, from right to left in the figure. These results enabled us to deal with metal and ceramics under the same conditions and to develop compounded functional materials by plasma spraying.

Reference

Kazumoto Fujita, Takayoshi Matsumaga, and Satoshi Horibe, 1994. Development of functional materials by spraying process. *Quality Engineering*, Vol. 2, No. 1, pp. 22–29.

This case study is contributed by Kazutomo Fujita.

CASE 31

Optimization of Two-Piece Gear Brazing Conditions

Abstract: Bonding is indispensable in automobile manufacturing for a wide variety of component parts. Today, brazing is not used often because of its insufficient strength and variability, although the process is very suitable for mass production, The improvement in the brazing process toward higher strength would make the process much more applicable in manufacturing. Here we introduce a two-piece gear brazing experiment to find a robust condition for stable, high-strength joints.

1. Introduction

A two-piece gear consists of two parts. After charging a circular brazing material in a groove of gear 2, we press-fit gear 1 to gear 2 and braze them together in a furnace. Figure 1 outlines the brazing process.

Good brazing that can result in constant strength needs to have the state in which the brazing material is permeated all over the brazing surface. In other words, it should ideally have the same characteristics as that of a one-piece gear. As shown in Figure 2, when pressure is applied, a one-piece gear is supposed to have proportionality between the load and the displacement, following Hooke's law within the elastic range. Therefore, to establish the proportionality between the load, y, and the displacement, M, regardless of gear size, we select the proportionality between them and the proportionality between the load and the brazing area, M^*, as the generic function of brazing.

As signal factors, the displacement, M, and brazing area, M^*, were chosen according to the strength standard and preliminary experimental results, as illustrated in Table 1. On the other hand, as a noise factor, temperature (Table 1) was chosen because the variability in temperature inside the brazing furnace greatly affects brazing permeation.

The experiment was performed according to the pulling strength test shown in Figure 3, and the test piece used had a three-level sectional area with a constant length of a and b (Figure 4) such that the combined area of its vertical and horizontal brazing areas was proportional to the gear diameter, d, resulting in the same amount of brazing material in a unit area.

2. SN Ratio and Sensitivity

According to the measurements obtained by means of the strength test, by using the displacement curve we observed each load for each displacement. Table 2 shows the data of experiment 1 in an L_{18} orthogonal array. Based on these data, we calculated the SN ratio and sensitivity. The table shows the computation procedure.

Total variation:

$$S_T = 0.00^2 + 0.13^2 + \cdots + 7.33^2 + 0.00^2$$

$$= 99.3 \qquad (f = 24) \qquad (1)$$

Effective divider:

$$r = [(0.1)(554)]^2 + [(0.1)(679)]^2 + (0.1)(804)]^2$$
$$+ \cdots + [(0.4)(804)]^2$$

$$= 423{,}980 \qquad (2)$$

Variation of proportional terms:

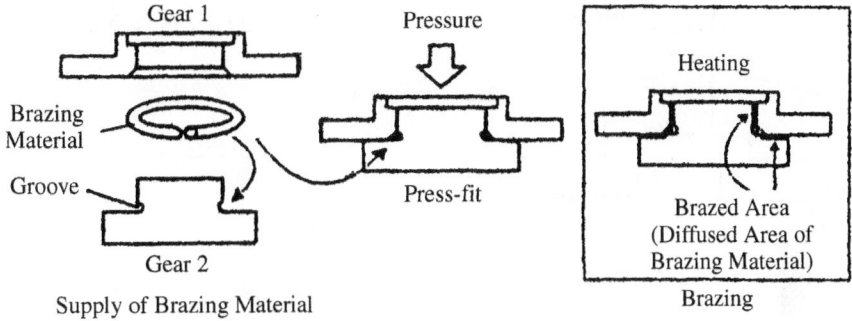

Figure 1
Gear brazing process

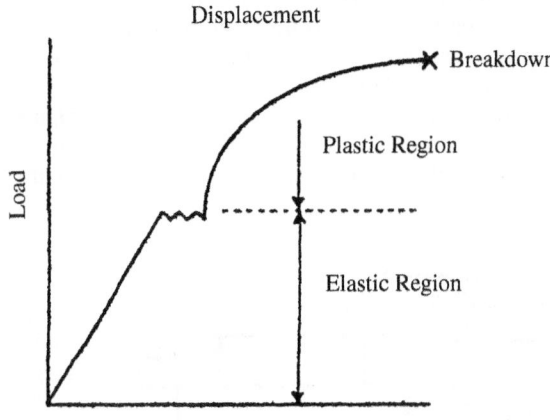

Figure 2
Material property of a one-piece gear

$$S_\beta = \frac{1}{2r} \begin{bmatrix} (0.1)(554)(0.00 + 0.13) \\ + (0.1)(679)(0.1 + 0.005) \\ + \cdots + (0.4)(679) \\ (0.00 + 0.00) + (0.4)(804) \\ (7.33 + 0.00) \end{bmatrix}$$

$$= 24.54 \qquad (f = 1) \tag{3}$$

Variation of differences of proportional terms:

$$S_{N\beta} = \frac{1}{r} \begin{bmatrix} (0.1)(554)(0.00) \\ + \cdots + (0.4)(804)(733)^2 \\ + (0.1)(554)(0.13) \\ + \cdots + (0.4)(804)(0.00)^2 \end{bmatrix} - 24.54$$

$$= 22.61 \qquad (f = 1) \tag{4}$$

Error variation:

Table 1
Signal and noise factors and levels

	Level			
Factor	**1**	**2**	**3**	**4**
Signal factors				
Displacement, M (mm)	0.1	0.2	0.3	0.4
Brazing area, M^* (mm²)	554	679	804	—
Noise factor				
Furnace temperature, N	High	Low	—	—

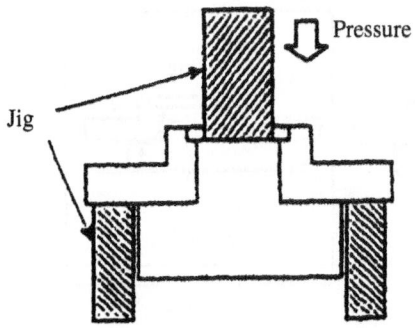

Figure 3
Testing method for a two-piece gear (pulling strength test)

$$S_e = S_T - S_{N\beta} - S_\beta = 52.15 \quad (f = 22) \quad (5)$$

Error variance:

$$V_e = \frac{S_e}{22} = 2.370 \quad (6)$$

Total error variance after pooling:

$$V_N = \frac{S_{N\beta} + S_e}{23} = 3.25 \quad (7)$$

SN ratio:

$$\eta = 10 \log \frac{(1/2r)(S_\beta - V_e)}{V_N} = -50.95 \text{ dB} \quad (8)$$

Sensitivity:

$$S = 10 \log \frac{1}{2r}(S_\beta - V_e) = -45.83 \text{ dB} \quad (9)$$

3. Optimization and Results of Confirmatory Experiment

Table 3 shows the control factors selected. After choosing factors regarding material and shape that affect the permeation of brazing material, we assigned each of the initial conditions to level 2 and varied it with a large range as levels 1 and 3. Figure 5 summarizes the response graphs. In addition, we performed confirmatory experiments at both the

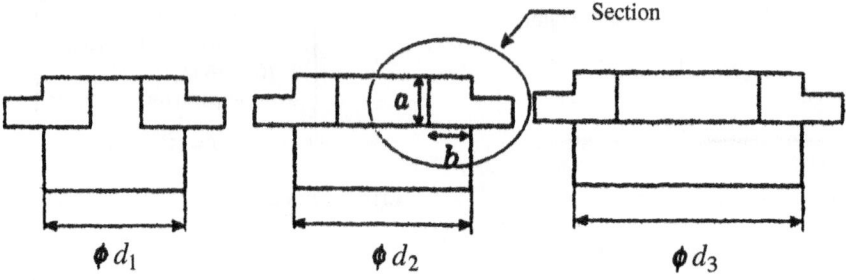

Figure 4
Test piece shape of a two-piece gear

Table 2
Strength test measurements

	M_1 (0.1 mm)			M_2 (0.2 mm)			M_3 (0.3 mm)			M_4 (0.4 mm)		
	M_1^*	M_2^*	M_3^*	M_1^*	M_2^*	M_3^*	M_1^*	M_2^*	M_3^*	M_1^*	M_2^*	M_3^*
N_1	0.00	0.10	2.08	0.00	0.23	3.55	0.00	0.25	5.33	0.00	0.00	7.33
N_2	0.13	0.05	0.00	0.15	0.08	0.00	0.20	0.10	0.00	0.00	0.00	0.00

Table 3
Control factors and levels

	Control Factor	Level		
		1	2	3
A:	flux supply position	Bottom	Top	—
B:	flux concentration	Thin	Med.	Concentrate
C:	amount of flux	Less	Med.	More
D:	amount of brazing material	Less	Med.	More
E:	amount of melting-point-lowering material	Less	Med.	More
F:	roughness of contacting surface	Fine	Med.	Rough
G:	tighting amount	Small	Med.	Large
H:	pressure	Low	Med.	High

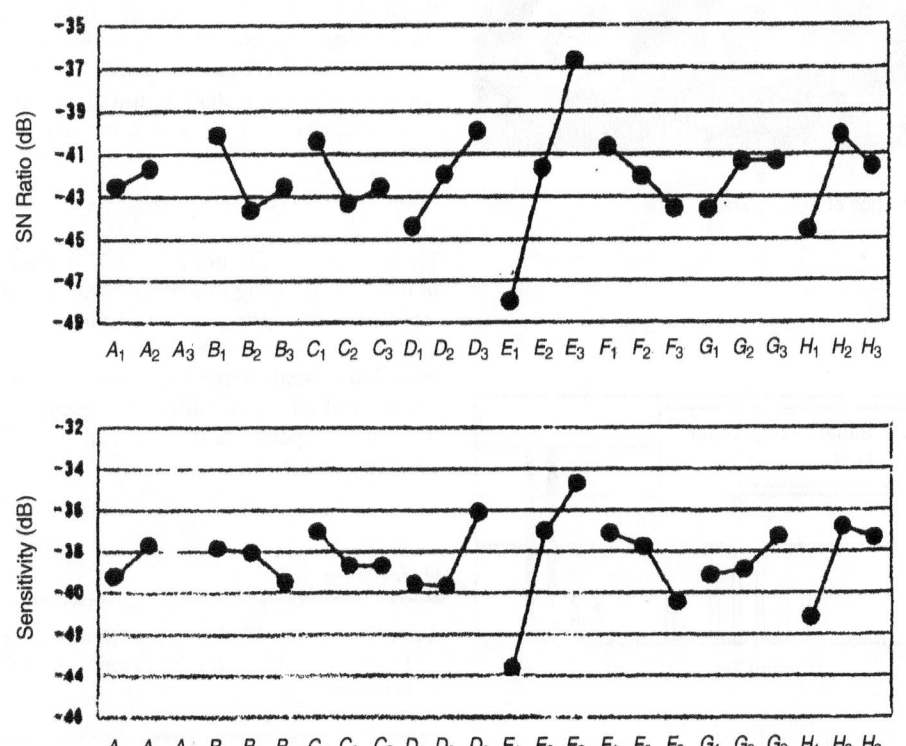

Figure 5
Response graphs

Table 4
Estimation and confirmation of SN ratio and sensitivity (dB)

Configuration	SN Ratio		Sensitivity	
	Estimation	Confirmation	Estimation	Confirmation
Optimal	−34.5	−36.0	−32.4	−30.7
Initial	−41.6	−43.4	−38.3	−34.6
Gain	7.1	7.4	5.9	3.9

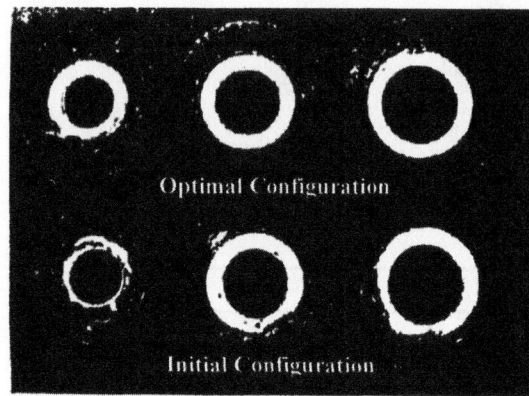

Figure 6
Ruptured section of brazing area

optimal and initial conditions. The results are outlined in Table 4.

For the SN ratio, a gain of 7.4 dB and good reproducibility were obtained. For the sensitivity, a somewhat smaller gain of 3.9 dB but sufficient reproducibility were earned. These results are confirmed in a photograph of the ruptured section (Figure 6), which indicates that brazing material is diffused all over the circumference, regardless of the brazing area size under the optimal configuration, whereas there exists a portion with no brazing material diffused under the initial configuration.

Converting 7.45 dB into a nondecibel number, we have

$$\mu^* = 10^{0.745} = 5.55 \qquad (10)$$

Then the error variance of $1/5.55$ is the initial configuration. Applying this result to an actual product, we obtain the histogram in Figure 7, based on a torsion test to evaluate the original function of a gear. This reveals that we can improve both the variability and strength under the optimal configuration as compared with those under the initial configuration.

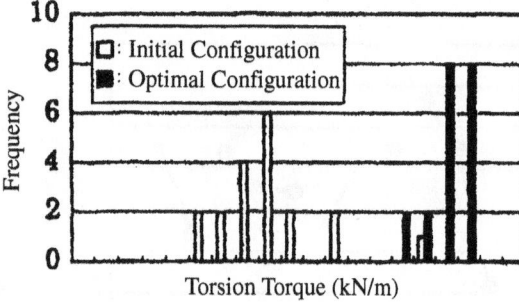

Figure 7
Result of torsion test

Reference

Akira Hashimoto, 1997. Optimization of two-piece gear brazing condition. *Quality Engineering*, Vol. 5, No. 3, pp. 47–54.

This case study is contributed by Akira Hashimoto.

Optimization of Resistance Welding Conditions for Electronic Components

Abstract: In the manufacture of automobile switches using electronic components such as light-emitting diodes or resistors, a good electrical connection between the component and the terminal is essential. Soldering has long been used to make this connection, but resistance welding is becoming more and more popular, due to easy automation and the shorter time required for the process. Currently, evaluation of resistance welding relies on a strength test; therefore, evaluation cannot be made without destroying the delicate wire in the component. In this study, evaluation is made by the generic function for welding current–voltage characteristics rather than by strength.

1. Introduction

We often need to use resistance welding to join leads in electronic parts with electrodes such as a light-emitting diode. Resistance welding is a technology used to combine metal materials by heating and the application of pressure, with resistance heat generated by electric current supplied between the metal materials. Among some of the major evaluation methods of resistance welding conditions for electronic parts are strength tests such as tension and peeling tests, and cross-sectional observation tests. In general, because of the thin diameter of leads in electronic parts, if we perform a strength test on a fixed lead, it often breaks at the point of the lead whose strength is lower than that of a joint. In this case, while the strength at a joint can be regarded as sufficient, we cannot assess how well the joint is welded. Taking the objective of resistance welding into account, we know that conductivity needs to be secured between a lead and a jointed electrode. Since good conductivity implies a good joint, to satisfy our objective we should evaluate electrical conductivity by measuring electrical characteristics for resistance-welding electronic leads.

2. Evaluation Method

Since the objective function of a joint is to secure the conductivity between a lead and an electrode, we selected current and voltage characteristics as the generic functions of a joint. Setting a supplied current to a signal factor, M, and a voltage difference to an output, y, we proceeded with the analysis using a zero-point proportional equation. If the joint condition between a lead and an electrode is poor, we assume that the resistance of the joint increases and that the slope of the input and output relationship flattens.

On the other hand, if the entire joint is uniform in condition, we are supposed to measure the same current and voltage characteristic no matter which part of the joint we pick up. That is, a small fluctuation in sensitivity among different measurement positions at the joint indicates high uniformity of resistance welding. Thus, as noise factors, we selected four joints and measured the voltage drop to assess uniformity. In addition, to find welding conditions to satisfy current and voltage characteristics, even if the electrode becomes somewhat contaminated, we also chose electrode contamination as another noise factor. Figure 1 outlines the measure-

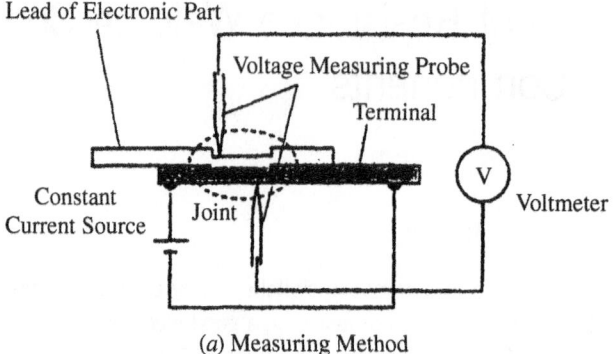

(a) Measuring Method

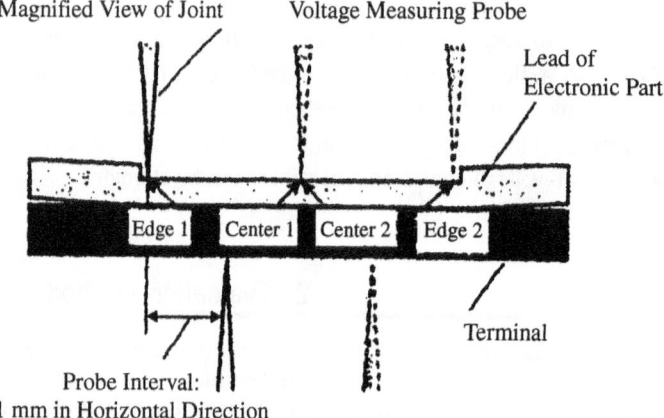

(b) Measurement Location

Figure 1
Measuring method of current and voltage characteristics

ment method and positions. Because of the tiny resistance at the joint, to measure voltage drop properly, we selected three levels of current, 5, 10, and 15 A, which is a wider range than that of a normal tolerable current in electronic parts. Additionally, we prepared a special measurement device to achieve the accuracy of locating the voltage measuring probe.

3. SN Ratio

Using test pieces for each welding condition laid out in an L_{18} orthogonal array, we measured the current and voltage characteristics. Table 1 gives one example of the experimental data. Based on this, we

show the calculations for the SN ratio and sensitivity.

Total variation:

$$S_T = 0.020^2 + 0.029^2 + \cdots + 0.101^2 + 0.058^2$$

$$= 0.091721 \qquad (f = 24) \qquad (1)$$

Linear equations:

$$L_{11} = (5)(0.020) + (10)(0.041) + (20)(0.093)$$

$$= 2.370$$

$$L_{12} = 3.045$$

$$\vdots$$

$$L_{24} = 1560 \qquad (2)$$

Table 1
Example of voltage drop at welding joint (mV)

Noise Factor		Signal Factor			
Contamination of Electrode	Measurement Position	M_1 5 A	M_2 10 A	M_3 20 A	Linear Equation
l_1: contaminated	J_1: edge 1	0.020	0.041	0.093	L_{11}
	J_2: center 1	0.029	0.058	0.116	L_{12}
	J_3: center 2	0.024	0.050	0.105	L_{13}
	J_4: edge 2	0.018	0.037	0.070	L_{14}
l_2: not contaminated	J_1	0.020	0.040	0.086	L_{21}
	J_2	0.027	0.055	0.108	L_{22}
	J_3	0.024	0.049	0.101	L_{23}
	J_4	0.016	0.032	0.058	L_{24}

Effective divider:

$$r = 5^2 + 10^2 + 20^2 = 525 \qquad (3)$$

Variation of proportional term:

$$S_\beta = \frac{(2.370 + 3.045 + \cdots + 2.630 + 1.560)^2}{(8)(525)}$$

$$= 0.088229 \qquad (f = 1) \qquad (4)$$

Variation of differences of proportional terms due to contamination of electrodes:

$$S_{l\times\beta} = \frac{(2.370 + \cdots + 1.860)^2 + \cdots + (2.220 + \cdots + 1.560)^2}{(4)(525)} - 0.088229$$

$$= 0.000130 \qquad (f = 1) \qquad (5)$$

Variation of differences of proportional terms due to measurement positions:

Table 2
Control factors and levels

Control Factor	Level 1	Level 2	Level 3
A: error column	1	2	—
B: electrode material	B_1	B_2	B_2'
C: electrode diameter	Small	Large	Large
D: upslope time	Short	Mid	Long
E: electricity supplying time	Short	Mid	Long
F: welding current	Small	Mid	Large
G: applied pressure	Small	Mid	Large
H: downslope time	Short	Mid	Long

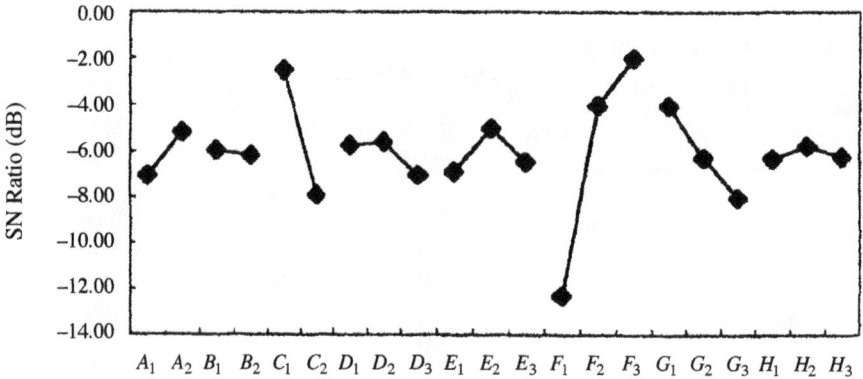

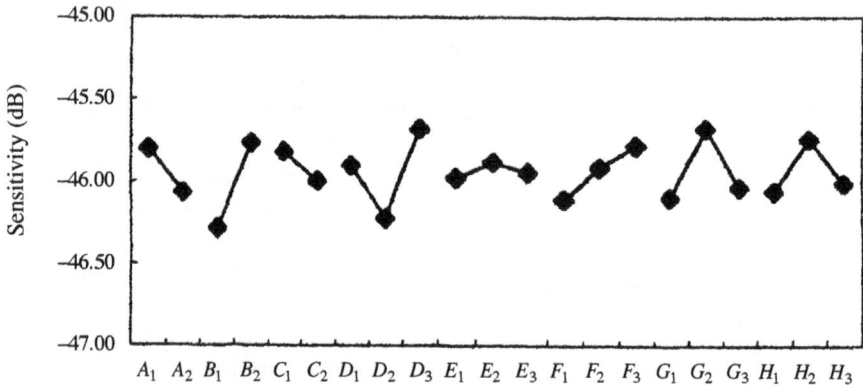

Figure 2
Response graphs

Table 3
Results of confirmatory experiment (dB)

Configuration	SN Ratio		Sensitivity	
	Estimation	Confirmation	Estimation	Confirmation
Worst	−16.02	−13.22	−46.27	−47.05
Optimal	3.68	1.95	−45.83	−45.68
Gain	19.70	15.17	0.44	1.37

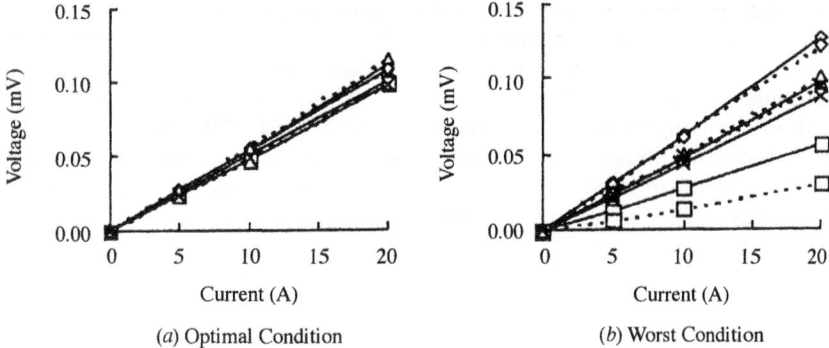

Figure 3
Characteristics of current and voltage in confirmatory experiment

(a) Optimal Condition (b) Worst Condition

$$S_{J\times\beta} = \frac{(2.370 + 2.220)^2 + \cdots + (1.860 + 1.560)^2}{(2)(525)}$$

$$- 0.088229$$

$$= 0.003275 \quad (f = 3) \tag{6}$$

Error variation:

$$S_e = 0.091721 - 0.088229 - 0.000130 - 0.003275$$

$$= 0.000087 \quad (f = 19) \tag{7}$$

Error variance:

$$V_e = \frac{0.000087}{19} = 0.000005 \tag{8}$$

Total error variance

$$V_N = \frac{S_{I\beta} + S_{J\beta} + S_e}{23} = 0.000152 \tag{9}$$

SN ratio:

$$\eta = 10 \log \frac{[1/(8)(525)](0.088229 - 0.000005)}{0.000152}$$

$$= -8.59 \text{ dB} \tag{10}$$

Sensitivity:

$$S = 10 \log \frac{1}{(8)(525)} (0.088229 - 0.000005)$$

$$= -46.78 \text{ dB} \tag{11}$$

4. Optimization and Result of Confirmatory Experiment

Table 2 shows the control factors and levels chosen for the experiment. A is an error column. Since control factor B, electrode material, and control factor C, electrode diameter, have two levels, their third levels are assigned as a dummy level. The levels of control factors E, electricity supply time, and F, welding current, are dealt with as sliding levels around D, the upslope time, such that the welding energy in each experiment becomes neither extremely high nor extremely low.

Figure 2 shows the response graphs obtained from the experiments based the orthogonal array. As the optimal configuration, we selected the combination of $B_1 C_1 D_2 E_2 F_3 G_1 H_2$ by chosing each level with a high SN ratio. To estimate the gain, we performed a confirmatory experiment based on factors C, F, and G with large effects. Table 3 summarizes estimations of the SN ratio and sensitivity and the experimental results. In addition, Figure 3 shows the characteristics of current and voltage in the confirmatory experiment.

According to the results in the confirmatory experiment, we can see that fairly good reproducibility was obtained in the gain in the SN ratio. The characteristics of current and voltage in the confirmatory experiment demonstrate that the viability in measurement position was obviously reduced under the optimal configuration compared to that under the worst configuration. As a result of observing the sections of the joints, we have confirmed that all of them have good joint conditions. Additionally, a normal strength test under the optimal configuration used as a double-check has revealed that rupture at the lead, which indicates sufficient strength at the joint, was observed. Although this evaluation method is not easy to implement in terms of measurement, it is regarded as applicable to other types of joints.

Reference

Takakichi Tochibora, 1999. Optimization of resistance welding conditions for electronic components. *Quality Engineering*, Vol. 7, No. 3, pp. 52–58.

This case study is contributed by Takakichi Tochibora.

CASE 33

Tile Manufacturing Using Industrial Waste

Abstract: In this study we performed experiments based on the generic function of tile manufacture, forming a tile with evenly distributed density with the intention of saving on material in order to avoid waste. As a signal factor and output, we selected weight in the air and weight in the water and measured the proportionality of an underwater weight to an in-the-air weight.

1. Generic Function

We filled a press die evenly with material, that is, formed a tile with evenly distributed density. Then, as a signal factor and output, we selected a weight in the air and a weight in the water, respectively, and measured the proportionality of underwater weight to in-the-air weight. To measure the underwater weight, we attempted to use underwater weighing. Although it was not easy to calculate the underwater weight of a absorptive tile, we were able to do so through our technical know-how.

In actuality, we split a formed tile into several pieces, as shown in Figure 1a, and defined $y = \beta M$ as the ideal relationship equation of the in-the-air weight, M, and the underwater weight, y, according to Figure 1b. In addition, since we know that to measure the geometric dimensions of a formed tile and to judge its appearance are relatively easy, for the purpose of fundamental research we checked the transformability of its shape and appearance. As illustrated in Figure 2, setting the dimensions of a die to a signal factor, M, and each corresponding dimensions of a formed tile to y, we defined $y = \beta M$ as the ideal relationship equation. A caliper was used for measurement.

On the other hand, since a tile that is not easily separated from the base plate after being baked tends to have a poor surface, we performed a smaller-the-better analysis by setting the value of an easily separated tile to 0, that of a not-easily separated tile to 100, and that of a tile in between to 50. For noise factors, we select the following four:

1. Baking temperature, N (two levels)
2. Major row materials maintaining conditions, N^* (two levels)
3. Drying condition of mixture of sand and water, N^{**} (two levels)
4. Tile thickness, N^{***} (two levels)

As an analysis example, we showed the analysis of in-the-air versus underwater weights. Table 1 shows the data examples and the calculation process.

Total variation:

$$S_T = 5.40^2 + 2.81^2 + \cdots + 0.48^2 = 468.8553$$
$$(f = 80) \tag{1}$$

Effective divider:

$$r_1 = 12.91^2 + 6.87^2 + 3.81^2 + 1.78^2$$
$$+ 1.73^2 = 234.5424 \tag{2}$$

Total of effective dividers:

$$r = r_1 + r_2 + \cdots + r_{16} = 3080.1503 \tag{3}$$

Variation of proportional term:

$$S_\beta = \frac{(L_1 + L_2 + \cdots + L_{16})^2}{r} = 459.9931$$
$$(f = 1) \tag{4}$$

Variation of sensitivity:

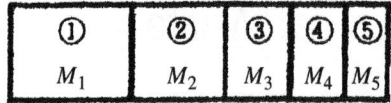

(a) Division of baked tile

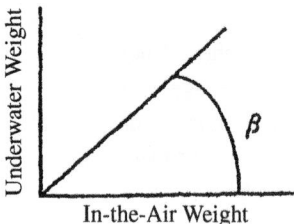

(b) Relationship between in-the-air weight
and underwater weight

Figure 1
Proportionality of divided tile

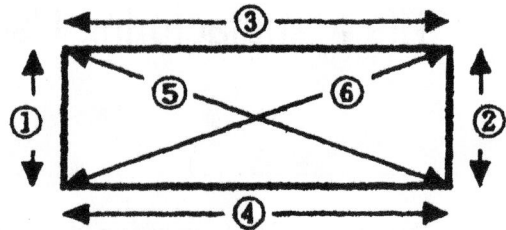

Figure 2
Measurement of tile's dimensions

Table 1
Data examples of in-the-air and underwater weights (experiment 1)

					M_1	M_2	M_3	M_4	M_5
N_1	N_1^*	N_1^{**}	N_1^{***}	In-the-air weight	12.91	6.87	3.81	1.78	1.73
				Underwater weight	5.40	2.81	1.59	0.76	0.72
			N_2^{***}	In-the-air weight	10.61	4.58	2.72	1.48	1.25
				Underwater weight	4.40	1.87	1.13	0.62	0.58
		N_2^{**}	N_1^{***}	In-the-air weight	12.85	6.62	3.23	2.04	1.53
				Underwater weight	5.86	3.00	1.44	0.92	0.70
			N_2^{***}	In-the-air weight	11.06	6.74	3.52	1.56	1.29
				Underwater weight	4.98	3.08	1.59	0.72	0.60
	N_2^*	N_1^{**}	N_1^{***}	In-the-air weight	11.92	7.15	3.89	2.20	1.36
				Underwater weight	3.65	2.12	1.15	0.69	0.56
			N_2^{***}	In-the-air weight	10.24	5.43	3.05	1.80	1.13
				Underwater weight	3.04	1.56	0.92	0.48	0.33
N_2	N_2^*	N_1^{**}		In-the-air weight	3.69	2.13	0.97	0.68	0.44
			N_2^{***}	In-the-air weight	10.64	5.69	2.58	1.56	1.01
				Underwater weight	3.70	1.93	0.99	0.67	0.38
		N_2^{**}	N_1^{***}	In-the-air weight	12.17	6.88	3.27	1.85	1.60
				Underwater weight	4.33	2.39	1.19	0.70	0.62
			N_2^{***}	In-the-air weight	9.98	5.24	2.43	1.45	1.30
				Underwater weight	3.73	1.88	0.92	0.51	0.48

Table 2

Control factors and levels

		Level		
Control Factor		**1**	**2**	**3**
A:	pulverizing time	Long	Short	—
B:	pressing pressure	Small	Mid	Large
C:	clay 1 (%)	5	10	15
D:	waste 1 (%)	0	3	6
E:	waste 2 (%)	0	3	6
F:	waste 3 (%)	10	15	20
G:	waste 4 (%)	10	15	20
H:	clay 2 (%)	0	2	4

$$S_{\beta N} = \frac{(L_1 + \cdots + L_8)^2}{r_1 + \cdots + r_8} + \frac{(L_9 + \cdots + L_{16})^2}{r_9 + \cdots + r_{16}}$$

$$= 2.1038 \qquad (f = 1) \tag{5}$$

$$S_{\beta N*} = \frac{\begin{array}{c}(L_1 + L_2 + L_3 + L_4 \\ + L_9 + L_{10} + L_{11} + L_{12})^2\end{array}}{\begin{array}{c}r_1 + r_2 + r_3 + r_4 \\ + r_9 + r_{10} + r_{11} + r_{12}\end{array}}$$
$$+ \frac{(L_5 + \cdots + L_{16})^2}{r_5 + \cdots + r_{16}} - S_\beta$$

$$= 6.8997 \qquad (f = 1) \tag{6}$$

Similarly,

$$S_{\beta N**} = 1.6786 \qquad (f = 1) \tag{7}$$

$$S_{\beta N***} = 0.0035 \qquad (f = 1) \tag{8}$$

Error variation:

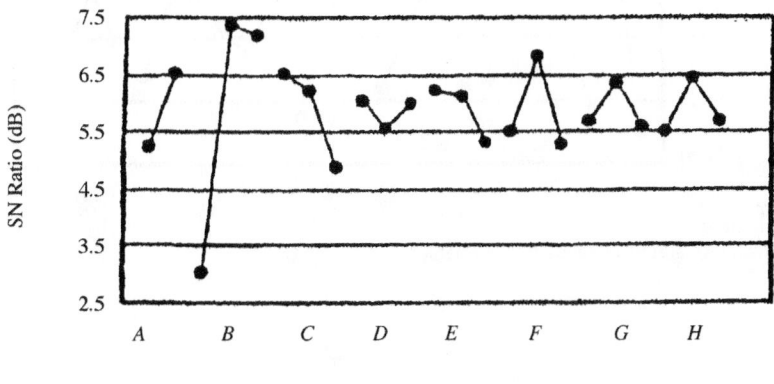

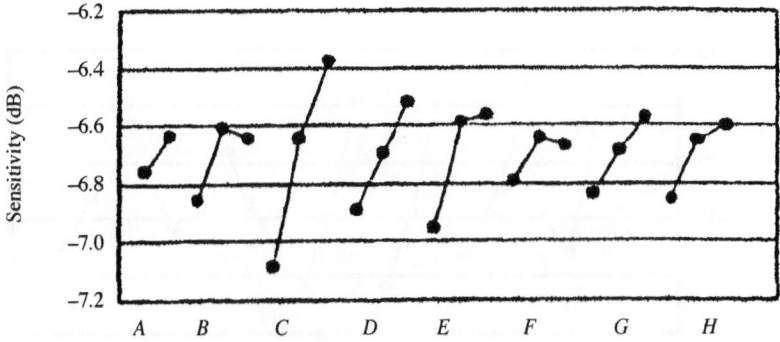

Figure 3
Response graphs of in-the-air versus underwater weights

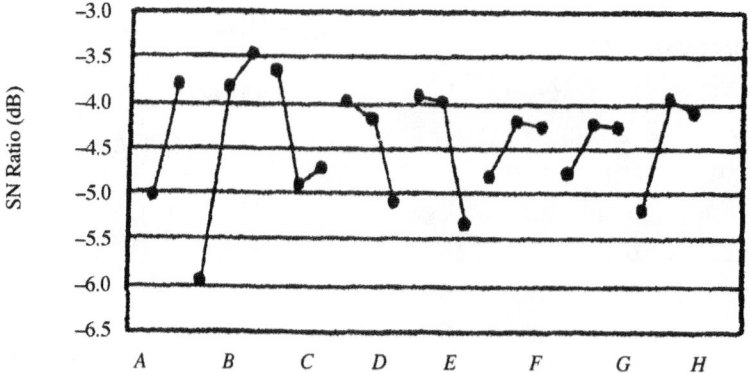

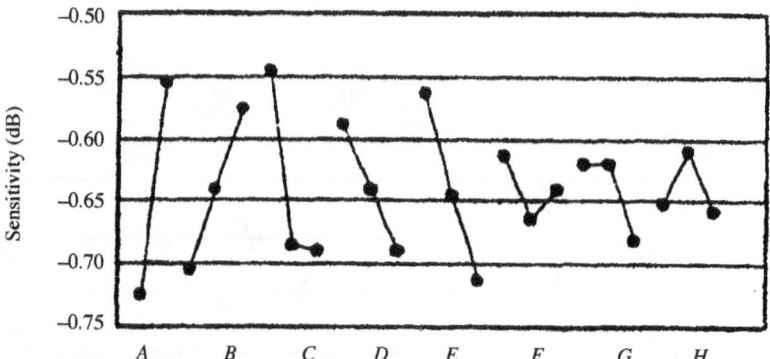

Figure 4
Response graphs of shape transformability

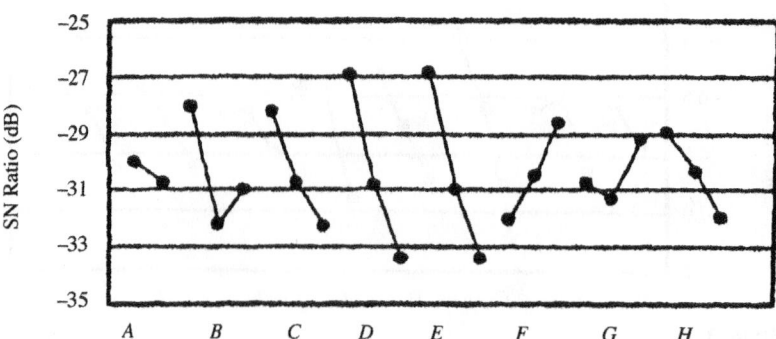

Figure 5
Response graph of separability

Table 3
Optimal and compared configurations

	Factor							
	A	B	C	D	E	F	G	H
In-the-air weight Underwater weight	2	2 3	1	(1)	1	(2)	(2)	(2)
Transformality of shape	2	3	1	(1)	1	(2)	(2)	(2)
Appearance	(1)	1	1	1	1	3	(3)	1
Optimal configuration	2	3	1	(1)	1	(3)	(3)	(2)
Compared configuration	1	1	2	(2)	3	(2)	(3)	(1)

$$S_e = S_T - S_\beta - S_{\beta N} - S_{\beta N*} - S_{\beta N**} - S_{\beta N***}$$
$$= 0.2641 \quad (f = 75) \tag{9}$$

Error variance:

$$V_e = \frac{S_e}{75} = 0.0035 \tag{10}$$

Compounded error variance:

$$V_N = \frac{S_e + S_{\beta N} + S_{\beta N*} + S_{\beta N**} + S_{\beta N***})}{79}$$
$$= 0.1122 \tag{11}$$

SN ratio:

$$\eta = \frac{(1/r)(S_\beta - V_e)}{V_N} = 1.3312 = 1.24 \text{ dB} \tag{12}$$

Sensitivity:

$$S = \frac{1}{r}(S_\beta - V_e) = 0.1493 = -8.26 \text{ dB} \tag{13}$$

2. Control Factors and Experimental Results

Table 2 shows the control factors selected for our study. Now we assign to an L_{18} orthogonal array factors related primarily to material mixture. For major industrial wastes, we mix them using the following equation:

$$100 - C - D - E - F - G - H \text{ (\%)}$$

Figure 3 shows the response graphs of the SN ratio and sensitivity for underwater versus in-the-air weights. When the filling density is high, the sensitivity becomes large. Figure 4 shows the response graph of the transformability of shape. Now, for the analysis of shape, among 18 experiments in total, the tiles used in experiments 6, 12, and 14 cannot be separated from the base plate. Therefore, we handled the columns of these three experiments as missing values, and by substituting the averages of the SN ratio and sensitivity as a provisional value, we proceeded with the analysis of sequential ap-

Table 4
SN ratio and sensitivity for in-the-air versus underwater weights

Configuration	SN Ratio		Sensitivity	
	Estimation	Confirmation	Estimation	Confirmation
Optimal	8.83	8.63	−7.21	−6.87
Compared	2.17	4.48	−6.40	−6.57
Gain	6.66	4.15		

Table 5
SN ratio and sensitivity for transformability of shape

Configuration	SN Ratio		Sensitivity	
	Estimation	Confirmation	Estimation	Confirmation
Optimal	−1.58	−2.26	−0.32	−0.33
Compared	−7.66	−7.44	−0.84	−0.96
Gain	6.08	5.18		

proximation. A large value of sensitivity, S, indicates a small contraction rate. Figure 5 shows the response graph of appearance (the smaller-the-better characteristic).

Considering all results obtained so far, we concluded that the tendencies of the data related both to weighing and to transformability are almost consistent. Additionally, for separability, because the contraction rate becomes smaller as the filling density gets smaller in terms of the sensitivity, S, we can say that there is some consistency.

3. Optimal Configuration and Confirmatory Experiment

Based on the aforementioned results, we determined the combination of optimal conditions by first prioritizing the in-the-air versus underwater weights, and next, by referring to the transformability of shape and appearance (Table 3).

Tables 4 and 5 summarize SN ratios and sensitivities, at both the optimal and compared configura-

tions. Now we computed the estimations using four factors, A, B, C, and E. Since the estimations and confirmatory results are fairly consistent, we can conclude that there is good reproducibility.

This experiment enabled us to mix a smaller amount of virgin materials (clay 1 and clay 2) and a larger amount of waste. Considering that the cost of industrial waste treatment is soaring and we are scheduled to use approximately 2800 tons of such tile materials on a yearly basis, we are expecting to contribute more to the disposal of industrial wastes and to the preservation of our environment and the Earth.

Reference

Toshichika Hirano, Kenichi Mizuno, Kouji Sekiguchi, and Takayoshi Matsunaga, 2000. Make tile of industry rejection by quality engineering. *Quality Engineering*, Vol. 8, No. 1, pp. 31–37.

This case study is contributed by Toshichika Hirano and Takayoshi Matsunaga.

Development of an Electrophotographic Toner Charging Function Measuring System

Abstract: Through this research we developed a measuring technology for toner particles ahead of designing functions for a developer and a developing device. Here we describe a two-ingredient developer as a typical example. This is a developer that supplies a target amount of charge to a toner particle by mixing a charged particle and a toner particle and charging the toner with friction.

1. Introduction

An imaging method in electrophotography used for photocopy machines or printers is to develop an electrostatic latent image that is transferred onto photosensitive material to obtain an objective image using charged toner particles. Therefore, in designing electrophotographic developer to reproduce an electrostatic latent image accurately, control of the amount of charge of toner particles is a major technical problem. To solve this, a measuring technology for a charging function focusing on individual toner particles, which can be used to design a charging function for toner particles, needs to be established. Since we have realized the importance of developing a measuring technology through this research, we developed the technology ahead of designing functions for the developer and the developing device.

Here, as a typical example, we describe two-ingredient developer. This is developer that supplies a target amount of charge to a toner particle by mixing a charged particle (called a *carrier particle*) and a toner particle and charging the toner with friction. Figure 1 shows an outline of electrophotography. A toner particle is required to have a function, particularly in a developing area. This is a process using a toner's characteristic of being charged.

In a developing area, a toner particle, the particle to be developed (Figure 2), is transferred onto photosensitive material and developed in accordance with the strength of an electric field formed by imaging information between a developing sleeve and the photosensitive material. That is, development is a system of controlling toner particles by the strength of an electric field. Now, since it is expected that an individual toner particle is transferred to a target position and developed according to the strength of an electric field, a constant amount of charge in a toner particle is considered ideal. In other words, the ideal function of a toner particle is that each toner particle has an identical amount of charge. To proceed with the functional design of a toner particle, we need to measure the amount of charge each toner particle and to evaluate the uniformity. In the following sections we discuss optimization of a measuring system.

2. Measuring System and Generic Function

As a measurement of an amount of charge in an individual toner particle, we selected a measuring system based on the E-Spart method, codeveloped by the University of Arkansas and Hosokawa Micron Corporation as a method of measuring the

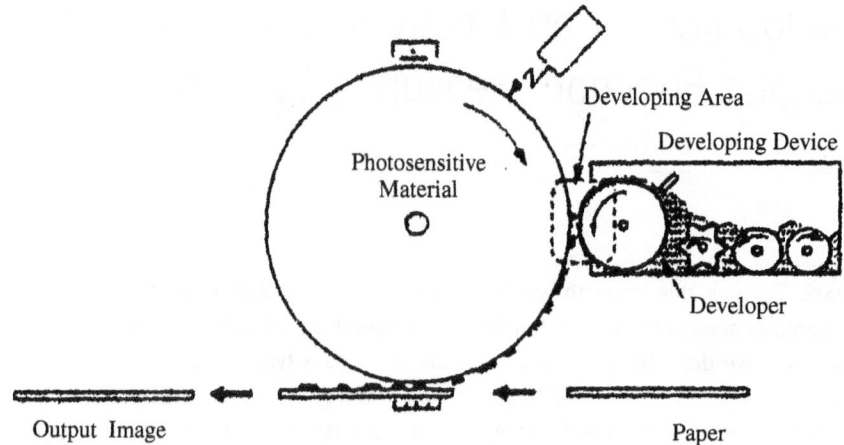

Figure 1
Electrophotography process

distribution of an amount of charge. The measuring system for an amount of charge is composed of a particle supplier to provide toner particles as our objective to be measured, an air blower to remove particles, and a charged amount detector to sense the movement of particles (Figure 3). At the charged amount detector, particles pass through an electric field generated by a pair of electrodes. At the same time, because of an air vibration field formed between the electrodes, together with the

electric field, each particle runs through with oscillation. By detecting this particle movement using a laser beam, we can simultaneously measure the amount of charge and the diameter of each particle.

In general, the generic function of a measuring system is regarded as a proportionality between the true value, M, and the measurement, y. However, since it is realistically difficult to prepare the true amount of charge as a signal factor, we need to arrange an alternative signal factor for the true

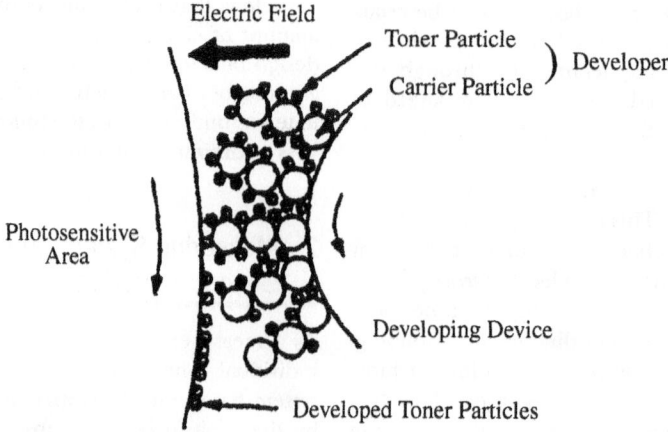

Figure 2
Movement of toner particles in developing area

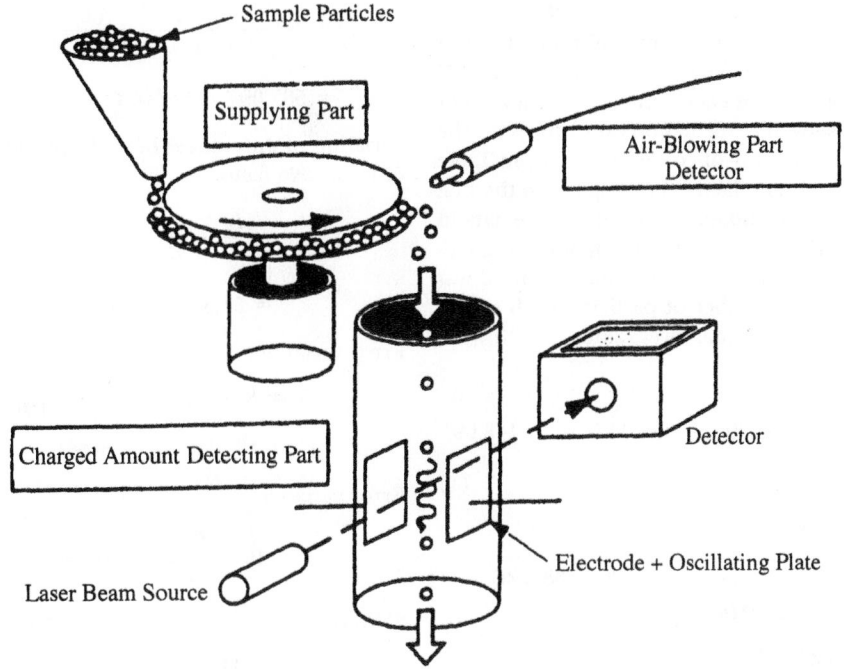

Figure 3
Measuring system for amount of charge

amount. Then, emphasizing cost and simplicity in this research, we studied whether it is possible to conduct the functional design of a charged amount measuring system by taking advantage of existing toner particles with a sufficiently good track record as a signal factor. When we use an actual product as a signal factor level, the reliability of the factor level comes into question. As a result of our thoroughness, an amount of charge per surface area was considered the most stable. Therefore, setting the sum of surface areas of all toner particles to an input, M, and the sum of their charge amounts to an output, y, we defined their zero-point proportional relationship as a generic function. In this case, the surface area of each toner particle was calculated by its diameter on the assumption that it was a perfect sphere.

To select a surface area as a signal factor, we used three different levels with regard to the surface area of a particle:

1. *Signal factor, M:* sum of surface areas of particles (three levels). Because of the difficulty of

checking the surface area of a particle during measurement, by taking a measurement for approximately 30, 60, and 90 toner particles, we converted each particle diameter measured into a surface area as a signal factor.

2. *Indicative factor, M*:* average amount of charge of a particle (two levels). In this study, as an indicative factor, we chose an average amount of charge of a toner particle with two levels. To cover a practical range of charge amounts, we chose two levels, one for a sufficiently high amount and the other for a sufficiently low amount. As a matter of course, the true value of each amount of charge was unknown.

Since no noise factor is handled here, we performed a split-type analysis.

3. SN Ratio and Sensitivity

Because no error factor was chosen, we could not provide the degrees of freedom of the error through normal calculations of the SN ratio and sensitivity. Thus, we used a split analysis method in our research. This provides a way to calculate an SN ratio and sensitivity for each control factor level. For instance, we show the calculation procedures for the SN ratio and sensitivity of control factor A_1 in Table 1. In this experiment, factor A is assigned to the first column of an L_{18} orthogonal array using the data in experiments 1 through 9, where A_1 was assigned. Now since a surface area as a signal factor M was computed from a number of particles, each experiment had different signal level values.

Total variation:

$$S_T = 216.43^2 + 385.09^2 + \cdots + 112.96^2 + 148.09^3$$

$$= 2{,}896{,}492 \qquad (f = 54) \tag{1}$$

Linear equations:

$$L_1 = (7750)(216.43) + (13{,}269)(385.09)$$
$$\quad + (18{,}957)(498.32)$$
$$= 16{,}233{,}744$$

$$L_1' = (8440)(95.41) + (12{,}522)(135.99)$$
$$\quad + (17{,}875)(173.65)$$
$$= 5{,}612{,}121$$

$$\vdots$$

$$L_9' = (7996)(65.42) + (12{,}243)(112.96)$$
$$\quad + (16{,}032)(148.09)$$
$$= 4{,}280{,}246 \tag{2}$$

Effective divider:

$$r_1 = 7750^2 + 13{,}269^2 + 18{,}957^2 = 595{,}496{,}710$$

$$r_1' = 8440^2 + 12{,}522^2 + 17{,}875^2 = 547{,}549{,}709$$

$$\vdots$$

$$r_9' = 7996^2 + 12{,}243^2 + 16{,}032^2 = 470{,}852{,}089 \tag{3}$$

Variation of proportional term:

$$S_\beta = \frac{(L_1 + L_1' + \cdots + L_9')^2}{r_1 + r_1' + \cdots + r_9'} = 2{,}269{,}031 \quad (f = 1) \tag{4}$$

Variation of M*χβ between rows:

$$S_{M*\beta} = \frac{(L_1 + \cdots + L_9)^2}{r_1 + \cdots + r_9} + \frac{(L_1' + \cdots + L_9')}{r_1' + \cdots + r_9'} - S_\beta$$

$$= 582{,}097 \qquad (f = 1) \tag{5}$$

Variation of differences between proportional terms of indicative factor:

$$S_{\text{row}(M*\beta)} = \frac{L_1}{r_1} + \frac{L_1'}{r_1'} + \cdots + \frac{L_9'}{r_9'} - S_\beta - S_{M*\beta}$$

$$= 31{,}846 \qquad (f = 16) \tag{6}$$

Error variation:

$$S_e = S_T - S_\beta \; S_{M*\beta} - S_{\text{row}(M*\beta)} - S$$

$$= 13{,}518 \qquad (f = 36) \tag{7}$$

Error variance:

$$V_e = \frac{S_e}{36} = \frac{13{,}518}{36} = 376 \tag{8}$$

SN ratio:

$$\eta_{A1} = 10 \log \frac{[1/(r_1 + r_1' + \cdots + r_9')](S_\beta - V_e)}{V_e}$$

$$= -61.2 \text{ dB} \tag{9}$$

Sensitivity:

$$S_{A1} = 10 \log \frac{1}{r_1 + r_1' + \cdots + r_9'} (S_\beta - V_e)$$

$$= -35.5 \text{ dB} \tag{10}$$

For other control factor levels, we computed the SN ratios and sensitivities by following the same procedure.

4. Optimal Configuration and Confirmatory Experiment

Using an L_{18} orthogonal array, we selected as control factors the eight factors shown in Table 2. Figure 4 shows the response graphs.

When an optimal configuration was selected, we prioritized levels with a high SN ratio. However, for some factor levels, we selected those that provided easier handling of equipment. We cannot calculate

Table 1
Data in each experiment under control factor A_1 (M, μm^2; y, fC)

No.	Experimental Data					Effective Divider	Linear Equation
1	M_1^* (large charge amount)	M	7,750	13,269	18,957	r_1 (595,496,710)	L_1 (16,233,744)
		y	216.43	385.09	498.32		
	M_2^* (small charge amount)	M'	8,440	12,522	17,875	r_1' (547,549,709)	L_1' (5,612,121)
		y'	95.41	135.99	173.65		
⋮	⋮					⋮	⋮
9	M_1^* (large charge amount)	M	5,387	12,337	18,259	r_9 (363,931,139)	L_9 (9,442,478)
		y	140.38	301.42	455.37		
	M_2^* (small charge amount)	M'	7,996	12,243	16,032	r_9' (470,852,089)	L_9' (4,280,248)
		y'	65.42	112.96	148.09		

the process average obtained through split-type analysis. To estimate gains, by adding up each gain for each control factor, we need to compute a sum of gains directly. As shown in Table 3, according to the confirmatory experiment, we achieved fairly good reproducibility in gain. Using an optimized measuring system, we can expect that our design of the charging function of toner particles is sufficiently streamlined.

Next, we calculated the cost/benefit ratio on the assumption that the total cost/benefit advantage earned through our measuring technology could be attributed to an improved inspection process. First, the functional limit of a charge amount based on the current inspection method Δ_0 is

$$\Delta_0 = 7.5 \ \mu C/g \qquad (11)$$

When the amount of charge exceeded the func-

Table 2
Control factors and levels

Control Factor		Level		
		1	2[a]	3
A:	particle supplier condition 1	Standard − 5	Standard	—
B:	particle supplier condition 2	6	9	12
C:	air-supplying condition 1	Standard − 5	Standard	Standard + 5
D:	air-supplying condition 2	0.2	0.3	0.4
E:	air-supplying condition 3	1	3	5
F:	particle supplier condition 3	5	20	35
G:	detector condition 1	2	3	4
H:	detector condition 2	5	10	15

[a] Current level.

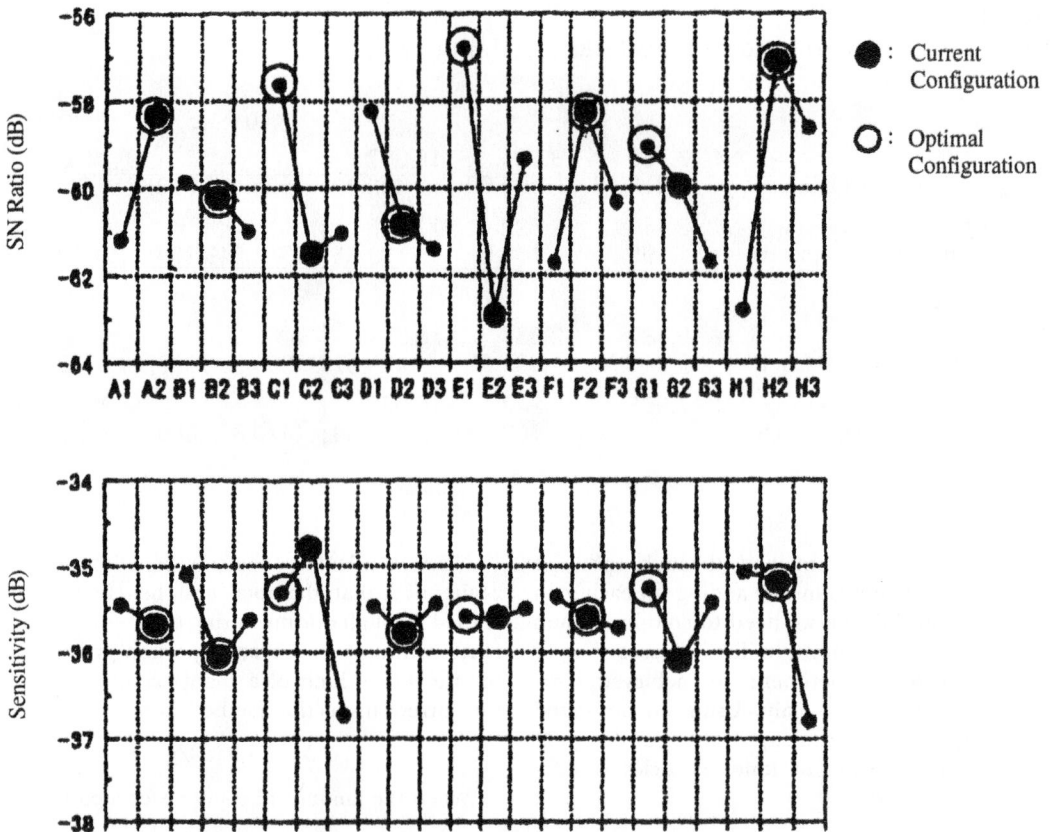

Figure 4
Response graphs

Table 3
Confirmation of SN ratio and sensitivity (dB)

Configuration	SN Ratio		Sensitivity	
	Estimation	Confirmation	Estimation	Confirmation
Current	—	−57.7	—	−36.5
Optimal	Value at current configuration, +10.8	−48.7	Value at current configuration, +0.4	−37.7
Gain	+10.8	+9.0	+0.4	−1.2

tional limit, the production lot of developer was discarded. Therefore, the cost of developer per se results in a loss, A_0. The loss function, L_0, is as follows:

$$A_0 = 100 \text{ million yen/lot} \tag{12}$$

$$L_0 = \frac{A_0}{\Delta_0^2} \sigma_0^2 = \frac{100 \text{ million yen}}{7.5^2} \sigma_0^2$$

$$= 17{,}778\sigma_0^2 \tag{13}$$

Assuming that the sensitivity can be adjusted, we obtained the following from the confirmatory experiment:

$$\frac{\sigma_{\text{optimal}}^2}{\sigma_{\text{current}}^2} \approx 0.0956 \tag{14}$$

Since we know that the current inspection of amounts of charge has the error variance, $\sigma_0^2 =$ approximately 0.8, the cost benefit per lot is as follows:

$$\Delta L = L_{\text{current}} - L_{\text{optimal}}$$

$$= (17{,}778)(0.8)(1 - 0.0956) \approx 12{,}863 \text{ yen/lot} \tag{15}$$

Now, setting the annual production volume of developer to 500 lots, the annual cost benefit is

$$(12{,}863 \text{ yen/lot})(500 \text{ lots/year})$$
$$= \text{approximately } 6{,}430{,}000 \text{ yen/year} \tag{16}$$

Reference

Kishio Tamura and Hiroyuki Takagiwa, 1999. Optimization of a charge measuring system for electrophotographic toner focused on each individual particle. *Quality Engineering*, Vol. 7, No. 5, pp. 47–54.

This case study is contributed by Kishio Tamura.

CASE 35

Clear Vision by Robust Design

Abstract: Clear vision is as the angle of the steering wheel while a vehicle is being driven in a straight line. Since the process of measuring CV introduces measurement variability and to ensure that the setting process is stable, the CV angle is audited on several vehicles at each shift and corrective action is taken if the CV average is outside SPC limits. A Ford gauge repeatability study indicated that gauge measurement error was approximately 100% of the tolerance of the CV specification. Due to the large measurement error, corrective actions were taken too often, not often enough, or incorrectly. This study used methods developed by Taguchi to reduce the gauge error by developing a process robust to noise factors that are present during the measurement process.

1. Introduction

Clear vision (CV) is the perceived angle of the steering wheel while a vehicle is being driven in a straight line (Figure 1). The perception of this angle is influenced strongly by the design of the steering wheel hub as well as by the design of the instrument panel. Clear vision for each vehicle is set in conjunction with the vehicle's alignment in the assembly plant at a near-zero setting determined by a customer acceptance study. Since the process of measuring CV introduces measurement variability to ensure that the setting process is stable, the CV angle is audited on several vehicles at each shift using the following audit process:

1. A CV audit tool is mounted to the steering wheel. The tool measures the inclination (i.e., rotation) angle of the wheel.

2. The vehicle is driven along a straight path while the CV tool collects and stores data.

3. At the completion of the drive, the CV tool algorithm calculates the average CV angle, and corrective action is taken if the CV average is outside SPC limits.

Currently, CV is audited by driving a vehicle along a straight path (Figure 2). The driver mounts a gauge on the steering wheel which measures the inclination (rotation) of the steering wheel and stores the data. When the drive is complete, the average CV reading for the drive is calculated using the gauge.

Quality Concerns

Historical gauge repeatability and reproducibility metrics indicated that gauge measurement error was approximately 100% of the tolerance of the CV specification. Due to the large measurement error in the facility, corrective actions were taken too often, not often enough, or incorrectly.

Measurement studies conducted in two different facilities indicated errors of 95 and 108%, respectively. These errors indicated that the facilities were not capable of accurate tracking of their own performance setting of clear vision. Most important, without the ability to track their own progress accurately, the facility was unable to respond properly to CV setting problems. Improving the robustness of the measurement system would reduce the variability of the audit measurements and allow the facilities to monitor and correct the setting process more accurately when required. This would lead to

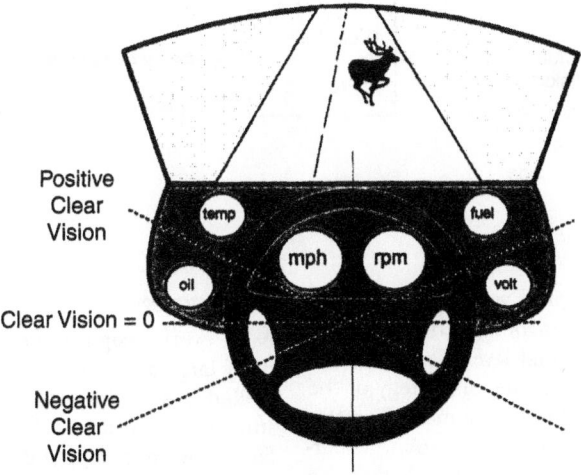

Figure 1
Clear vision: driver's view

improved customer satisfaction and reduce CV warranty cost.

2. Parameter Design

Ideal Function

The objective of this study was to reduce the measurement error of the CV tool from 100% of the clear vision tolerance. The ideal function (Figure 3)

of the CV audit process is to produce an accurate and repeatable measurement of the actual steering wheel angle using the CV audit tool. The purpose of this tool is to indicate a CV angle response that reflects the actual clear vision angle condition.

The signal, actual clear vision angle, was set at two levels. The first signal, M_1, was the CV value of the vehicle as it was built at the assembly plant. M_2 was achieved by removing the steering wheel, rotating it clockwise approximately 15° from the original position, and reassembling it to the column.

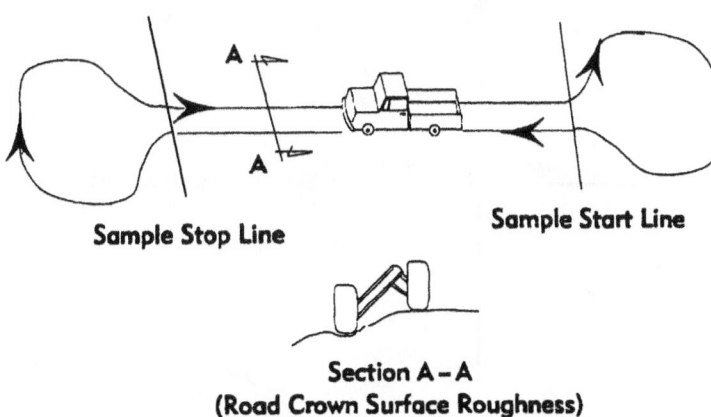

Sample Stop Line   **Sample Start Line**

Section A–A
(Road Crown Surface Roughness)

Figure 2
Testing diagram

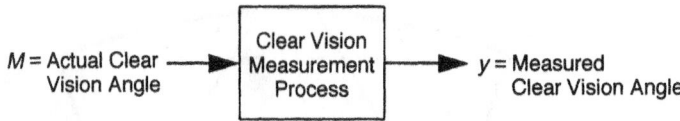

Figure 3
Ideal function

In traditional engineering, we do not typically change the signal to look at the linearity of the response to the changing signal level. Testing at different signal levels is a fundamental strategy of the robust design method, and testing at more than two signal levels is desirable. However, the spline (steering wheel attachment) was designed such that the allowable attachment rotation was in 15° increments. If the steering wheel was moved two notches in either direction, the total degree of movement would have been outside the limitations of the measurement tool.

The true value for the M_1 condition was not known, but the change in signal, M_2, was known. Also, the midpoint between positive and negative M was not zero. Therefore, reference-point proportional analysis was recommended as being most appropriate (Figure 4). (See the Appendix for the original data analysis process.)

Noise Strategy

As in every design, noise factors can induce a relatively large amount of variability and cannot be controlled easily. The noise strategy selected for this study included driver skill and driver weight. These factors were compounded as N_1 = skilled, lightweight driver and N_2 = unskilled, heavyweight driver.

Experimental Layout

The experimental layout was completed in two phases. The first phase was the execution of parameter design to determine the best combination of control factors to reduce the measurement system variability. An L_{18} orthogonal array (Figure 5) was used to define the control factor configuration for each of the 18 tests. The experiment was conducted at two signal levels, M_1 and M_2. The control factors

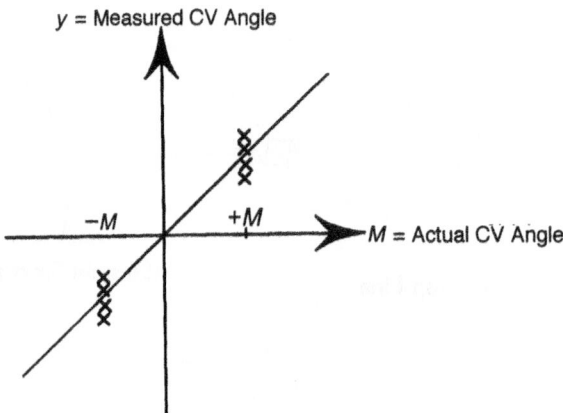

Figure 4
Measured CV angle versus actual CV angle: reference-point proportional

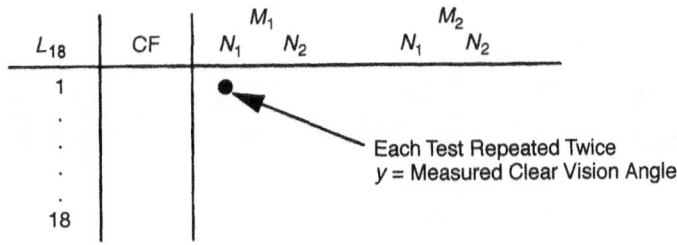

Figure 5
Experimental layout

were road service (A), navigational signs (B), route type (C), drive distance (D), drive speed (E), tool mounting (F), steering wheel tilt (G), and algorithm (H). Table 1 defines the control factor levels. Test results are shown in Table 2. Signal-to-noise (SN) and beta values were calculated for each configuration based on four responses. Response plots are shown in Figures 6 and 7.

The current process (baseline configuration) was $A_1 B_1 C_2 D_3 E_1 F_2 G_3 H_1$. The optimum configuration selected was $A_1 B_1 C_1 D_3 E_2 F_1 G_3 H_1$. The optimum combination has an SN ratio of -6.06 dB; the baseline method yielded a -16.42 dB ratio. The improvement predicted was 10.36 dB, which corresponds to a 70% reduction in variability. Time constraints did not allow a confirmation run at both signal levels. Instead, a gauge repeatability and reproducibility study which measures the standard deviation at one signal level was conducted using the optimum configuration. Results (Table 3) indicated that measurement error was reduced to 39% versus the

predicted error of 30% (based on a 70% reduction in variability).

Sensitivity Analysis

The second phase of this experiment was a sensitivity analysis conducted by performing CV audit drives with configurations of vehicle factors and operators, as specified by an L_{16} orthogonal array. The sensitivity analysis was undertaken to determine how different vehicles and operator factors biased the actual CV setting. The percent contribution of each factor to the total system variability was calculated.

Completion of this analysis showed that the CV audit measurement was affected significantly by operator weight, caster split values, and steering system hysteresis. Since the audit driver cannot affect the caster split values or steering system hysteresis, it was decided to confirm only the effect of differences in operator weight. By eliminating differences in operator weights, the measurement system error was reduced further from 39% to 32%.

Table 1
Control factors and levels

	Factor							
Level	A	B	C	D	E	F	G	H
1	Smooth	Line	Loop	50 ft	10 mph	BHND WHL	Bottom	Current
2	Rough	Diam. sign	Modified	150 ft	20 mph	FRT WHL	Center	New 1
3	—	Goalpost	Modified	300 ft	30 mph	BHND WHL	Top	New 2

Table 2
Raw data, SN ratio, and beta values

	$M_1 = -1$				$M_2 = 1$					
	N_1	N_1	N_2	N_2	N_1	N_1	N_2	N_2	SN	β
1	-2.80	-5.90	-7.30	-5.10	10.70	9.20	11.10	9.40	15.38	17.27
2	-6.30	-4.10	-5.60	-5.50	9.90	10.40	10.10	11.40	15.83	22.87
3	-4.90	-5.80	-7.40	-6.20	12.20	9.80	9.10	9.10	16.13	19.04
4	-2.10	-6.40	-1.60	-3.80	11.10	9.90	10.90	17.80	15.90	11.44
5	-5.10	-5.80	-4.60	-3.40	8.80	10.60	11.10	12.20	15.40	18.93
6	-7.20	-7.40	-6.40	-7.20	8.80	9.10	9.00	8.60	15.93	30.13
7	-9.10	-9.50	-6.20	-11.60	13.00	4.60	12.10	17.80	20.98	10.95
8	-6.50	-6.70	-6.00	-5.80	8.00	8.80	9.40	9.50	15.18	25.48
9	-0.40	-1.00	-8.30	7.30	15.70	11.50	17.10	17.80	11.98	6.44
10	-5.90	-4.60	-5.10	-5.40	9.40	9.70	10.50	9.00	14.90	25.01
11	-6.10	-6.90	-5.70	-6.00	7.30	7.80	8.80	8.10	14.18	24.86
12	-5.30	-5.30	-6.20	-6.80	9.70	8.20	10.00	10.30	15.45	22.28
13	-5.50	-4.70	-3.70	-1.10	7.00	11.20	13.90	11.90	14.75	12.52
14	-6.50	-7.60	-5.60	-7.90	10.20	13.60	11.50	2.90	16.45	10.66
15	-4.30	-4.60	-3.40	-1.60	9.50	8.50	11.40	12.90	14.05	15.41
16	-5.30	-5.50	-5.70	-6.10	8.80	10.30	10.00	9.80	15.38	26.42
17	-7.30	-5.00	17.80	5.40	10.20	11.30	17.80	17.80	11.55	-0.80
18	-7.90	-6.50	-5.40	-5.40	8.50	8.50	9.00	10.00	15.30	20.89

Deviation of SN and β

$$SN = 10 \log_{10} \frac{(1/r)(S_\beta - V_e)}{V_N} = 10 \log_{10} \frac{\beta_2}{\sigma^2}$$

$$\beta = \sqrt{\frac{1}{r}(S_\beta - V_e)}$$

In the following equations, m is the number of signal levels, r_0 the number of repetitions at each noise level, and n the number of noise levels.

$$M'_i = M_i - M_{\text{ref}} \qquad i = 1, 2, 3, \dots, m$$

$$y'_{ijk} = y_{ijk} - Y_{\text{ref}} \qquad i = 1, 2, 3, \dots, m;$$

$$j = 1, 2, 3, \dots, n; \quad k = 1, 2, 3, \dots, r_0$$

$$Y_{ij} = \sum_{k=1}^{r_0} y'_{ijk} \qquad i = 1, 2, 3, \dots, m;$$

$$j = 1, 2, 3, \dots, n$$

$$L_j = \sum_{i=1}^{k} M'^2_i$$

$$r = nr_0 \sum_{i=1}^{k} M'^2_i$$

$$S_\beta = \frac{1}{r} \left(\sum_{j=1}^{n} L_j \right)^2 \qquad (f_\beta = 1)$$

$$S_N = \sum_{j=1}^{n} \frac{L_j^2}{r/n} - S_\beta \qquad (f_N = n - 1)$$

$$S_T = \sum_{k=1}^{r_0} \sum_{j=1}^{n} \sum_{i=1}^{m} (y'_{ijk})^2 \qquad (f_T = nkr_0)$$

$$S_e = S_T - S_N - S_\beta \qquad [f_e = n(kr_0 - 1)]$$

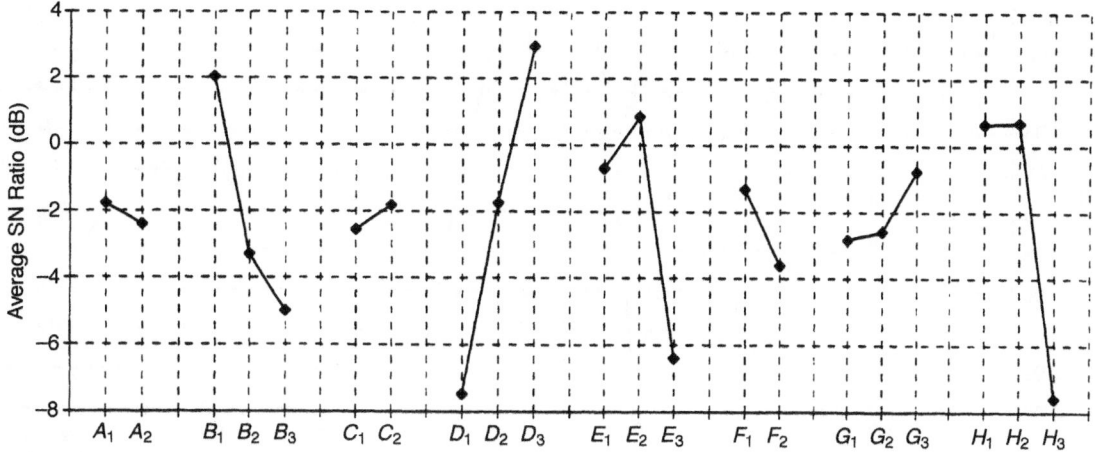

Figure 6
SN response plot

$$V_e = \frac{S_e}{f_e}$$

$$V'_N = \frac{S_e + S_N}{f_e + f_N}$$

Sample Calculations

We have $m = 2$, $n = 2$, $r_0 = 2$, and $M_{\text{ref}} = M_1$.

$M_1 = M_1 - M_1 = 0$

$M_2 = M_2 - M_1 = 15$

Y_{ref} = average response at M_1
$= \frac{1}{4}[(-28) + (-5.9) + (-73) + (-5.1)]$
$= -5.275$

$Y_{11} = (-2.8 + 5.275) + (-5.9 + 5.275) = 1.85$

$Y_{12} = (10.70 + 5.275) + (9.20 + 5.275) = 30.75$

$Y_{21} = (-7.30 + 5.275) + (-5.10 + 5.275) = -18.5$

$Y_{22} = (11.10 + 5.275) + (9.40 + 5.275) = 31.05$

$L_1 = 0(1.85) + 15(30.45) = 456.75$

$L_2 = 0(-1.85) + 15(31.05) = 465.75$

$r = (2)(2)(0^2 + 15^2) = 900$

$S_\beta = \frac{1}{900}(456.75 + 465.75)^2 = 945.5625$
$(f_{\beta=1})$

$$S_N = \frac{(456.75)^2}{(\frac{1}{2})(900)} + \frac{(465.75)^2}{(\frac{1}{2})(900)} - 945.5625$$
$$= 0.09 \qquad (f_N = 1)$$

$S_T = (-2.8 + 5.275)^2 + (-5.9 + 5.275)^2 + \cdots$
$\quad + (9.4 + 5.275)^2 = 958.87 \qquad (f_T = 8)$

$S_e = 958.87 - 0.09 - 945.5625 = 13.2175$
$\quad (f_e = 6)$

$$V_e = \frac{13.2175}{6} = 2.202916667$$

$$V_N = \frac{13.2175 + 0.09}{7} = 1.901071429$$

$\beta = \sqrt{\frac{1}{900}(945.5625 - 2.202916667)}$
$\quad = 1.023805311$

$$\text{SN} = 10 \log \frac{(1.023805311)^2}{1.901071429} = -2.585636787$$
$$= -2.59$$

Appendix

The original analysis sets the value for M_1 at -1 and M_2 at $+1$. See Figures 1 to 5 and Tables 1 and 2. Note C_2 and C_3 had similar levels and were grouped in analysis. The same is true for F_1 and F_3.

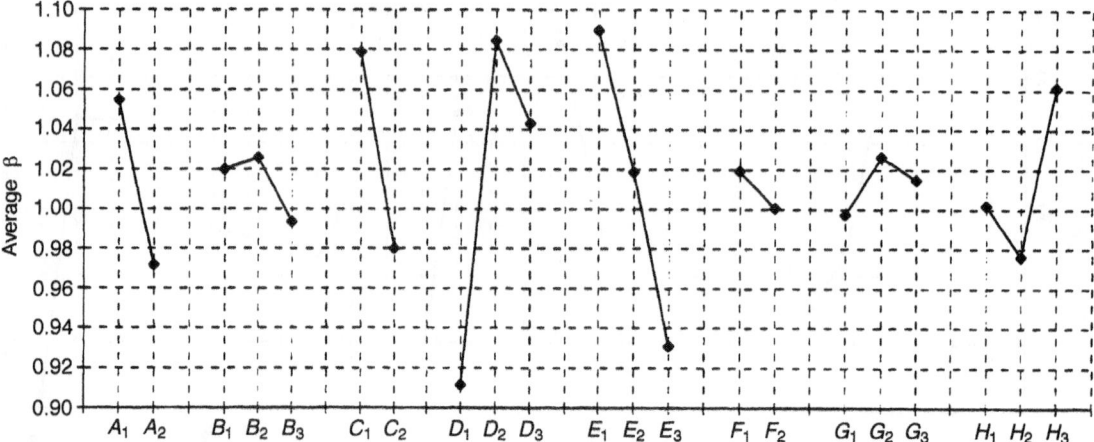

Figure 7
Beta response plot

The current process (baseline configuration) was $A_1B_1C_2D_3E_1F_2G_3H_1$. The optimum configuration selected was $A_1B_1C_1D_3E_2F_1G_3H_1$. Although the optimum configuration for level A is A_1 (smooth), A_2 (rough) is acceptable. Allowing a facility to use its current condition (either smooth or rough) will result in a cost savings.

The optimum combination has an SN ratio of 34.54 dB; the baseline method yielded a 24.19 dB ratio. The improvement predicted was 10.35 dB, which corresponds to a 70% reduction in variability. Time constraints did not allow a confirmation run at both signal levels. Instead, a gauge repeatability and reproducability study, which measures standard deviation at one signal level, was conducted using the optimum configuration. Results (Table A1) in-

dicated that measurement error was reduced to 39%, versus the predicted error of 30% (based on a 70% reduction in variability).

Derivation of SN and β

$$SN = 10 \log_{10} \frac{\frac{1}{4}(S_\beta - V_e)}{V_N} = 10 \log_{10} \frac{\beta^2}{\sigma^2}$$

$$\beta = \sqrt{\frac{S_\beta}{2}}$$

$$S_\beta = \frac{Y_1^2}{4} + \frac{Y_2^2}{4} - \frac{7'^2}{8} \qquad V_N = \frac{S_N}{f_N}$$

$$Y_1 = \sum_{i=1}^{4} Y_i \qquad Y_2 = \sum_{i=5}^{8} Y_i$$

Table 3
Values predicted and confirmed

	SN	Std. Dev. Relative to Spec. (%)	
		Predicted	Confirmed
Baseline	−16.42	100	100
Optimum	−6.06	30	39
Gain	10.36	70	61

Table A1
Values predicted and confirmed (at one signal level only)

| | Std. Dev. Relative to Spec. (%) | | |
	SN	Predicted	Confirmed
Baseline	24.19	100	100
Optimum	34.54	30	39
Gain	10.35	70	61

$$T_2 = (Y_1 + Y_2)^2$$

$$S_T = \sum_{i=1}^{8} Y_i^2 - \frac{T^2}{8} \qquad S_N = S_T - S_\beta$$

Sample Calculations
$$Y_1 = (-2.8 - 5.9 - 7.3 - 5.1) = -21.1$$
$$Y_2 = (10.7 + 9.2 + 11.1 + 9.4) = 40.4$$

$$T^2 = (-21.1 + 40.4)^2 = 372.49$$

$$S_T = [-2.8^2 + (-5.9^2) + (-7.3^2)$$
$$+ (-5.1^2) + 10.7^2 + 9.2^2$$
$$+ 11.1^2 + 9.4^2] - \frac{372.49}{8} = 486.09$$

$$S_\beta = \frac{(-21.2)^2}{4} + \frac{40.4^2}{4} - \frac{372.49}{8} = 472.78$$

$$S_N = 486.09 - 472.78 = 13.31$$

$$V_N = \frac{13.31}{6} = 2.22$$

$$S/N = 10 \log_{10} \frac{\frac{1}{4}(472.78 - 2.45)}{2.22} = 17.24$$

$$\beta = \sqrt{\frac{472.78}{2}} = 15.38$$

This case study is contributed by Ellen Barnes, Eric W. Crowley, L. Dean Ho, Lori L. Pugh, and Brian C. Shepard.

Optimization of Adhesion Condition of Resin Board and Copper Plate

Abstract: Electronic component parts installed in resin board require shape stability, flatness, or heat radiation with good heat conductivity. To evaluate adhesion conditions, peel tests have been used traditionally. For other tests, the occurrence of bubbles is inspected by ultrasonic detection. But these methods are not easy to quantify. In this study the characteristics of a capacitor at the contact surface was used as the generic function with good reproducibility.

1. Introduction

Electronic parts installed on resin boards (e.g., motherboards) need shape stability, flatness, or heat radiation, depending on how they will be used. To this end, we need to fix a resin plate onto a more rigid plate made of material that has good heat conductivity. Therefore, a product combining a resin board and a copper plate is used. As a typical method of evaluating good adhesion, we have conventionally used the peel test to judge good or poor adhesion. This method is still regarded as being in most common use to indicate adhesion conditions per se and as the easiest to use to control an adhesion process.

In the meantime, other methods exist, such as observation of the occurrence of bubbles displayed in a monitor by an ultrasonic detection technique. However, since they are not easy to quantify, we use these methods primarily for inspection. The adhesion structure studied in our research is shown in Figure 1. Using adhesive, we glued a copper plate onto a resin board that had a metal pattern on the surface. The most critical issue was that a conventional adhesion strength test could not be used because the plastic board used is a thin film. In the peel test, after adhesion the film tends to break down when either the resin board or copper plate

is fixed. Therefore, an optimal evaluation method was needed to judge adhesion condition.

2. Generic Function

Taking into consideration the function of a copper plate as a heat radiator, we would usually evaluate the relationship between electricity consumption and temperature on the copper side. However, since errors are often generated due to the attachment of the copper to a plastic board, in this research we focused on stable adhesion quality and defined a capacitance characteristic on the adhesion area as a generic function.

Looking at the fact that the copper plate and metal pattern shown in Figure 1 clamp the adhesive and plastic board, we note that this structure forms a capacitor. Although we have attempted to assess $Q = CV$ as a generic function, we cannot prepare a proper measuring instrument for this. Therefore, we took advantage of a capacitance characteristic $C = \varepsilon S/d$ (where C is the capacitance, ε the permittivity, S the surface area of an electrode, and d the distance between electrodes; Figure 2). By regarding a small change in capacitance (a high SN ratio) as a good condition both before and after an accelerated test that places a high-temperature

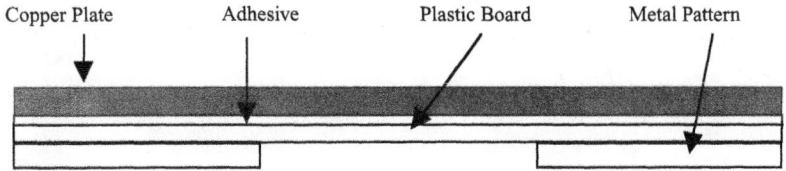

Figure 1
Joint area of resin board and copper plate

stress, we evaluated the characteristic. Since the change in permitivity is constant for the temperature change before and after the accelerated test, we assumed that it does not affect the capacitance.

Signal Factor

Assuming the sizes of electronic parts in a product, we selected an area of the metal pattern on the side of the resin board as a signal factor. As shown in Figure 3, we prepared resin boards with small (M_1), medium (M_2), and large (M_3) metal pattern areas for the evaluation. The outer surrounding area shown in Figure 3 indicates a metal pattern.

Noise Factor

High-temperature stress before and after the accelerated test were selected as noise factor levels:

N_1: shortly after adhesion (after adhesive solidifies)

N_2: after accelerated test (three sets of water absorption and reflow at a temperature of 245°C)

As the experimental device in the accelerated test, we used our reflow furnace to melt solder on terminals where motherboards were mounted.

3. SN Ratio

Table 1 shows the calculations for the SN ratio and sensitivity.

Total variation:

$$S_T = 44.14^2 + 40.33^2 + \cdots + 103.88^2 = 42{,}972.83 \tag{1}$$

Effective divider:

$$r = 69.25^2 + \cdots + 213.75^2 = 70{,}506.88 \tag{2}$$

Linear equations:

$$L_1 = (69.25)(44.14) + \cdots + (213.75)(128.51)$$
$$= 42{,}783.85$$
$$L_2 = (69.25)(40.33) + \cdots + (213.75)(103.88)$$
$$= 34{,}581.00 \tag{3}$$

Variation of proportional term:

$$S_\beta = \frac{(L_1 + L_2)^2}{2r} = 42{,}444.94 \qquad (f = 1) \tag{4}$$

Variation of differences between proportional terms:

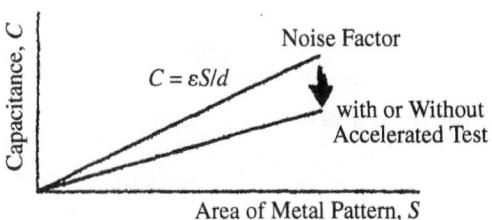

Figure 2
Generic function of resin board and copper plate

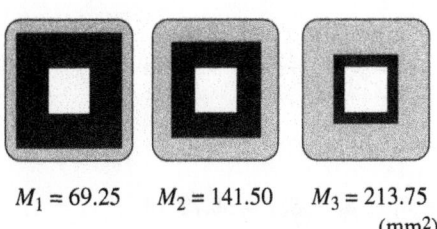

$M_1 = 69.25 \qquad M_2 = 141.50 \qquad M_3 = 213.75$
(mm²)

Figure 3
Signal factor for adhesion (metal pattern area)

Table 1
Capacitance data (pF)

No.	M_1 (69.25)		M_2 (141.50)		M_3 (213.75)	
	N_1	N_2	N_1	N_2	N_1	N_2
1	44.14	40.33	86.63	67.73	128.51	103.88

Table 2
Control factors and levels

	Control Factor	Level		
		1	2	3
A:	storage condition of copper plate	Normal temperature and humidity	High temperature and humidity	—
B:	storage condition of plastic board	Dry	High temperature and humidity	Normal temperature and humidity
C:	adhesion load	Small	Mid	Large
D:	adhesion temperature	Low	Mid	High
E:	adhesion time	Short	Mid	Long
F:	leave-as-is time	Short	Mid	Long

Table 3
Confirmatory results (dB)

Configuration	SN Ratio		Sensitivity	
	Estimation	Confirmation	Estimation	Confirmation
Optimal	−11.97	−13.74	−3.89	−4.50
Current	−19.72	−19.17	−4.79	−4.66
Gain	7.75	5.43	0.89	0.16

$$S_{N\beta} = \frac{(L_1 - L_2)^2}{2r} = 477.165 \qquad (f = 1) \quad (5)$$

$$V_e = \frac{S_e}{4} = 12.6822 \qquad (7)$$

Error variation:

Total error variance:

$$S_e = S_T - S_\beta - S_{N\beta} = 50.72878 \qquad (f = 4) \quad (6)$$

$$V_N = \frac{S_{N\beta} + S_e}{5} = 105.5788 \qquad (8)$$

Error variance:

SN ratio:

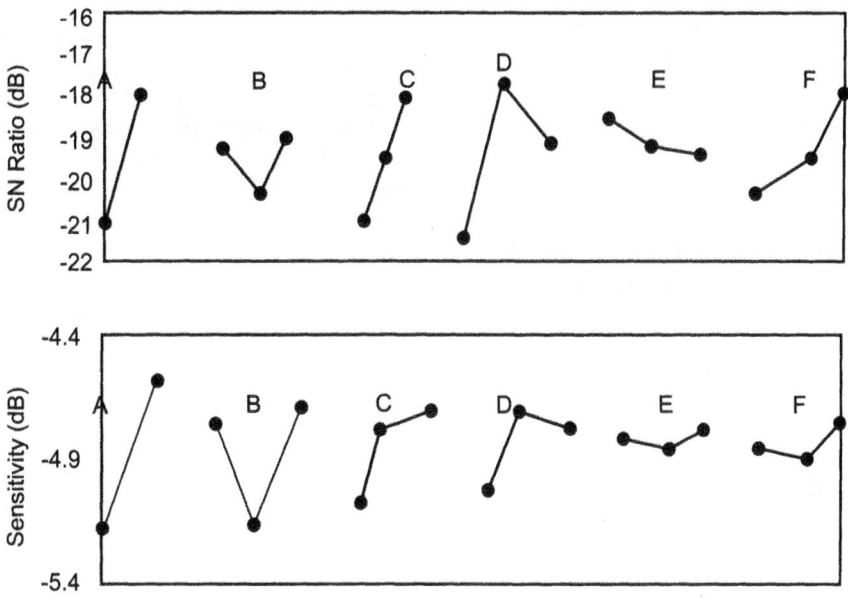

Figure 4
Response graphs

$$\eta = 10 \log \frac{(1/2r)(S_\beta - V_e)}{V_N} = -25.4514 \text{ dB} \quad (9)$$

Sensitivity:

$$S = 10 \log \frac{1}{2r}(S_\beta - V_e) = -5.21565 \text{ dB} \quad (10)$$

4. Optimization Configuration and Results of Confirmatory Experiment

As control factors, we selected factors that can be controlled during adhesion (Table 2), such as pressure- or heat-related parameters, storage conditions of adhesives, or as-is conditions. Table 3 shows the results of the confirmatory experiment under the optimal and current conditions. This table and the response graphs in Figure 4 reveal that the gain in SN ratio has 70% reproducibility, which is regarded as somewhat poor. Whereas under the current configuration the estimation and confirmation are extremely consistent, under the optimal config-

uration, the confirmation is somewhat smaller. One possible reason is that the condition of high temperature and humidity as the optimal level of factor A tends to cause errors, because this condition approximates that at the point when the adhesive solidifies. To improve reproducibility, it seems that we need to separate temperature and humidity when evaluating.

Based on several assumptions, we calculated the cost/benefit ratio obtained from our functionality optimization. Assuming a change in capacitance when there is a peeling problem at the adhesion area, caused by voids, we computed the following loss function:

$$\sigma^2_{\text{optimal}} = \frac{\beta^2_{\text{optimal}}}{10^{\eta_{\text{optimal}}/10}} = 6.426 \quad (11)$$

$$\sigma^2_{\text{current}} = \frac{\beta^2_{\text{current}}}{10^{\eta_{\text{current}}/10}} = 31.12 \quad (12)$$

Now, setting $A_0 = 2000$ yen and $\Delta = 11$ pF (assumed value), we have

$$L_{\text{optimal}} = \frac{A_0}{\Delta^2 \sigma^2_{\text{optimal}}} = 106.2 \qquad (13)$$

$$L_{\text{current}} = \frac{A_0}{\Delta^2 \sigma^2_{\text{current}}} = 514.4 \qquad (14)$$

In sum, a loss of 408.2 yen/piece is eliminated through this optimization. In the case of monthly production volume of 10,000 units, the cost/benefit ratio amounts to 4,082,000 yen/month.

Reference

Yugo Koyama, Kazunori Sakurai, and Atsushi Kato, 2000. Optimization of adhesion condition of resin board and copper plate. *Proceedings of the 8th Quality Engineering Symposium,* pp. 262–265.

This case study is contributed by Yugo Koyama.

CASE 37

Optimization of a Wave Soldering Process

Abstract: In optimizing manufacturing conditions for soldering, quality characteristics such as bridges or no-solder joints are traditionally measured for evaluation. By using the characteristics of current and voltage as an ultimate objective function of soldering, we optimized manufacturing conditions in the automated soldering process.

1. Introduction

For optimization of manufacturing conditions in the automated soldering process, it is important to make the process robust not only to the products of current-level density but also to those with high density, to arrange conditions for manufacturing various types of motherboards without adjustment. The process studied is outlined in Figure 1.

In optimizing manufacturing conditions for soldering, quality characteristics such as bridges or no-solder joints are traditionally measured for evaluation. Soldering is used to join a land pattern printed on the board to a lead of an electronic part (Figure 2). Defects called *no-solder* indicate that the two parts are not joined completely; *bridge* is the name given to the defect of unwanted joints.

In one print board, some portions need considerable energy for soldering and others need less energy. This manufacturing process provides the same amount of energy to each portion. Our technical objective is to solder circuit boards evenly no matter how much energy is needed. In addition, ideally, the amount of current flowing in solder is proportional not only to its area but also to the corresponding voltage. Therefore, by using the characteristics of current and voltage as an ultimate objective function of soldering, we optimize manufacturing conditions in the automated soldering process.

2. SN Ratio for Current and Voltage Characteristics

For calculating the SN ratio (Table 1), we set up a sectional area of solder, M^*, to a three-level signal factor for a production-related signal, and voltage to a four-level signal factor for a soldering-function-related signal. We measured continuously the current corresponding to the incrementing voltage.

As the noise factor, we separated the factor levels to those needing a lot of energy for soldering and those needing little, then compounded them. For experimentation, we designed and fabricated test pieces. A test piece had a certain cross-sectional area and a compounded noise factor condition. Based on the data in Table 2, we proceeded with our calculation of the SN ratio for the voltage and current characteristics as follows.

Total variation

$$S_T = 18^2 + 16^2 + \cdots + 178^2$$

$$= 241{,}073 \qquad (f = 24) \tag{1}$$

Linear equations:

$$L_1 = (5)(15)(18) + \cdots + (20)(62)(192)$$

$$= 796{,}540 \tag{2}$$

$$L_2 = 737{,}770 \tag{3}$$

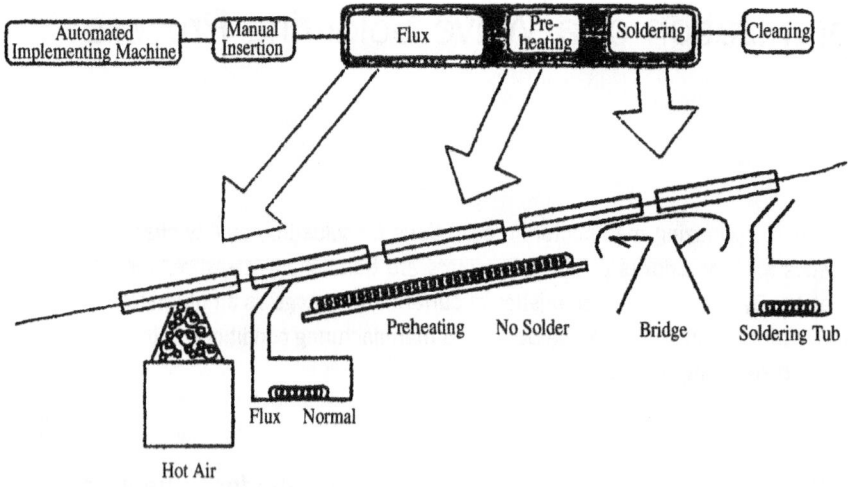

Preheating   No Solder   Bridge   Soldering Tub

Flux   Normal

Hot Air

Figure 1
Automated soldering process

Effective divider:

$$r = [(5)(15)]^2 + [(5)(50)]^2 + [(5)(62)]^2 + \cdots + [(20)(62)]^2$$

$$= 4{,}926{,}750 \qquad (4)$$

Variation of proportional term:

$$S_\beta = \frac{(L_1 + L_2)^2}{2r} = 238{,}910.76 \qquad (f_\beta = 1) \quad (5)$$

Variation due to proportional terms:

$$S_{N\beta} = \frac{L_1^2 + L_2^2}{r} - S_\beta = 350.53 \qquad (f_{N\beta} = 1) \quad (6)$$

Error variation:

$$S_e = S_T - S_\beta - S_{N\beta} = 1811.71 \qquad (f_e = 22) \quad (7)$$

Error variance:

$$V_e = \frac{S_e}{f_e} = \frac{1811.71}{22} = 82.35 \qquad (8)$$

Total error variance:

$$V_N = \frac{S_{N\beta} + S_e}{1 + 22} = \frac{350.53 + 1811.71}{23} = 94.01 \quad (9)$$

SN ratio:

Lead of Electronic Part

Print Board

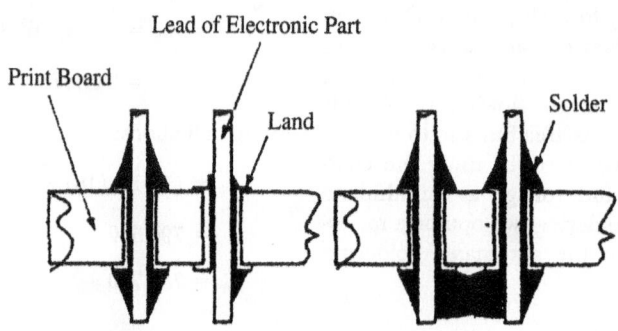

Land

Solder

Figure 2
No-solder and bridge

Table 1
Signal and noise factors

	Level			
Factor	**1**	**2**	**3**	**4**
Signal factors	15	50	62	—
Sectional area M^* ($\times 10^{-1}$ mm²)				
Voltage M (mV)	5	10	15	20
Noise factor				
Energy demand	Much N_1		Little N_2	
Pattern interval of motherboard	Narrow		Wide	
Number of motherboard layers	Many		One	

Table 2
Current data (dB $\times$ 10^{-2} A)

Sectional Area:		M_1^* (15)		M_2^* (50)		M_3^* (62)	
	Noise:	N_1	N_2	N_1	N_2	N_1	N_2
Voltage	M_1 (5)	18	16	44	42	61	55
	M_2 (10)	33	28	84	79	118	105
	M_3 (15)	44	38	116	112	163	145
	M_4 (20)	52	47	143	138	192	178

Table 3
Control factors and levels

	Control Factor	Level		
		1	**2**	**3**
A:	amount of flux air flow	1.5	Current	—
B:	specific gravity of flux	−0.01	Current	+0.01
C:	opening of flux air knife	Fully closed	Half closed	Fully open
D:	distance between solder and board	Close	Mid	Far
E:	wave height	Low	Mid	High
F:	solder temperature	Low	Mid	High
G:	preheating temperature	Low	Mid	High
H:	conveyor speed	Slow	Mid	Fast

$$\eta = 10 \log \frac{(1/2r)(S_\beta - V_e)}{V_N} = -35.89 \text{ dB} \quad (10)$$

Sensitivity:

$$S = 10 \log \frac{1}{2r}(S_\beta - V_e) = -16.16 \text{ dB} \quad (11)$$

3. Optimal Configuration and Confirmatory Experiment

As control factors we chose several production-related variables (Table 3), and allocated them to an L_{18} orthogonal array. Since three control factors, solder temperature, preheating temperature, and conveyor speed, are energy-related factors, mutual interaction between them is assumed to exist. In other words, if we set the levels of these factors independently, we cannot provide adequate energy to each factor–level combination (Figure 3). Thus, as a result of studying levels along energy contours, we used a sliding-level method (Figure 4 and Table 4). According to the result in the L_{18} orthogonal array, we calculated an SN ratio and sensitivity for each experiment. Their response graphs are shown in Figure 5.

Table 5 shows the estimates of the SN ratio under the current and optimal configurations. The fact that the gains in the confirmatory experiment were consistent with those estimated demonstrates success in our experimental process. Next, we calculated the gains obtained by the loss function. Assuming that the resulting social loss when the function reaches its limitation, A_0, is 50,000 yen, we have

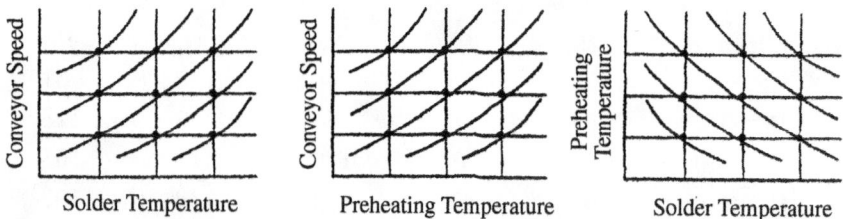

Figure 3
Energy balance by combination of solder temperature, conveyor speed, and preheating temperature

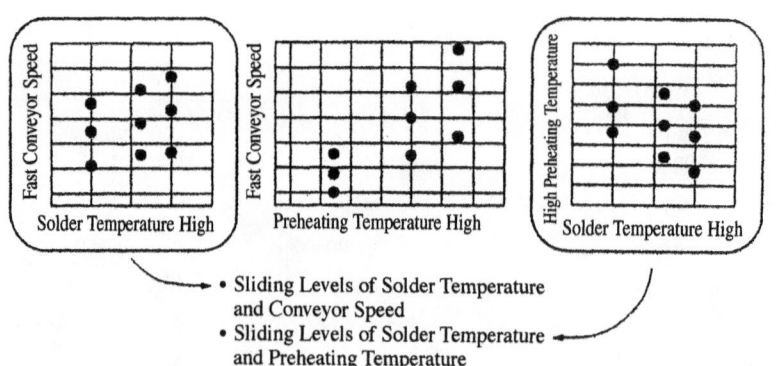

Figure 4
Selection of values at factor levels for soldering by sliding-level method

Table 4
Relative factor levels for solder temperature

Solder Temperature	Conveyor Speed			Preheating Temperature		
	G_1 Slow	G_2 Mid	G_3 Fast	G_1 Low	G_2 Mid	G_3 High
F_1: low	0.63	0.9	1.125	90	120	160
F_2: mid	0.7	1.0	1.25	70	100	130
F_3: high	0.742	1.06	1.325	40	90	120

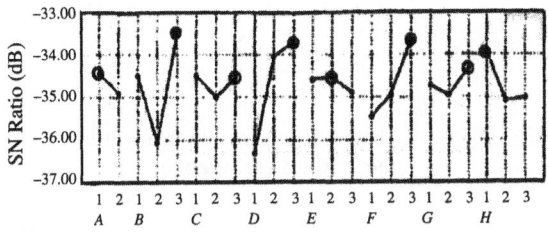 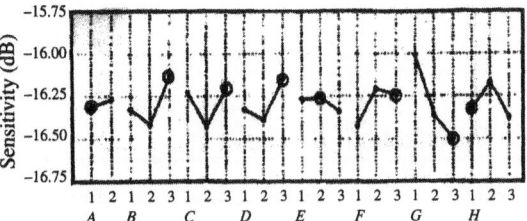

Figure 5
Response graphs

$$L_0 = k\sigma^2 = \frac{50,000}{(60.2^2)(2951.9)(3000)}$$

$$= 122.2 \text{ million yen} \qquad (12)$$

$$L_1 = k\sigma_1^2 = \frac{50,000}{(60.2^2)(935.4)(3000)}$$

$$= 38.7 \text{ million yen} \qquad (13)$$

Thus, our improvement will result in a benefit of $L_0 - L_1 = 83.5$ million yen/month.

Table 5
Results of confirmatory experiment

	Configuration		
	Optimal	Current	Gain
Estimation	−30.70	−36.09	5.39
Confirmation	−29.71	−34.70	4.99

Reference

Shinichi Kazashi and Isamu Miyazaki, 1993. Process optimization of wave soldering by parameter design. *Quality Engineering,* Vol. 1, No. 3, pp. 22–32.

This case study is contributed by Shinichi Kazashi.

CASE 38

Optimization of Casting Conditions for Camshafts by Simulation

Abstract: We attempted to optimize the casting process of a camshaft by taking advantage of parameter design using simulation, analyzing how we prevent gas from being dragged in a casting die when filling it with molten iron.

1. Introduction

A camshaft, one of the most vital parts of an automobile engine, is commonly made of cast iron. If there are casting defects on the machined surfaces of frictional areas such as a cam, functional deterioration of an engine takes place, thereby causing quality assurance cost. To reduce this, we need to eliminate casting defects in essential formed areas of a camshaft.

A camshaft is cast by pouring cast iron melted at a high temperature, called *molten iron*, into a cavity of a die called a *shell mold*. This casting method and flow of molten iron are shown in Figure 1. Molten iron poured from a pour basin is gradually charged from a gate located at the bottom to a hollow area for a camshaft. Since casting conditions such as constituents or the temperature of molten iron are varied within a certain range under control, we need to design a casting method robust to variability in these conditions.

2. Casting Function and Evaluation of Casting

If we eliminate gas from entering molten iron when poured, casting defects in formed areas can be prevented. To do so, ideally, no turbulence in molten iron flowing through the gate should occur. The Reynolds number, Re, expressed as an index, indicates the degree of fluid turbulence:

$$\text{Re} = V\frac{d}{\nu} \qquad (1)$$

where V is the flow velocity of molten iron (m/s), d the diameter of the gate section (m), and ν the coefficient of dynamic viscosity (m²/s). The Reynolds number of molten iron flowing through the gate can be obtained by the flow speed of molten iron computed by simulation.

Considering that a smaller Reynolds number indicates less fluid turbulence, ideally, we should lower it as much as possible. However, if it is extraordinarily small, other casting defects could be caused. Therefore, based on our prior experience with similar products, a Reynolds number of 4800 is considered ideal. So we evaluate the casting function by using a Reynolds number of molten fluid flowing through the gate as a nominal-the-best measurement characteristic.

Since analyzing time is regarded as an issue when casting simulation is implemented, we improved our analysis workflow beforehand. Primarily, we attempted to highly streamline the analysis by optimizing the element size to be divided in a three-dimensional model. In creating the model, we used a three-dimensional CAD system called I-DEAS and shortened the modification time of the model by taking advantage of the parametric geometry controls in the history editing function.

900

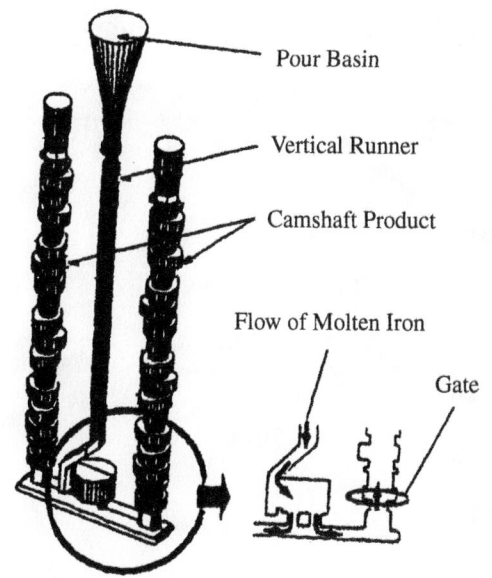

Pour Basin

Vertical Runner

Camshaft Product

Flow of Molten Iron

Gate

Figure 1
Method and flow of camshaft casting

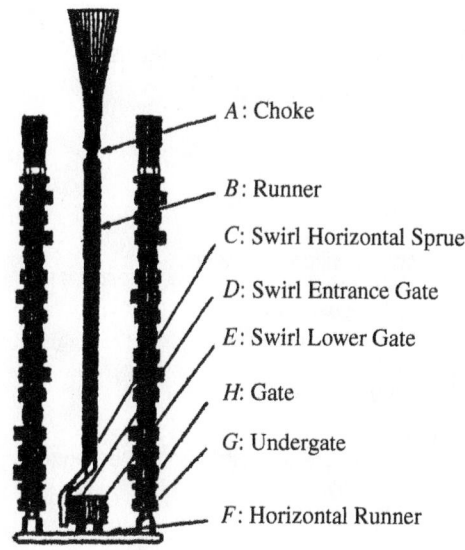

A: Choke

B: Runner

C: Swirl Horizontal Sprue

D: Swirl Entrance Gate

E: Swirl Lower Gate

H: Gate

G: Undergate

F: Horizontal Runner

Figure 2
Control factors

3. Control and Noise Factors

Figure 2 shows the control factors selected from sectional dimensions (areas) of principal parts, A to H, which were considered to greatly affect the flow speed of molten iron flowing through the gate (Figure 1). Each level chosen for simulation is shown in Table 1. Considering that the variability in casting needs to be included in simulation and there is an asymmetric characteristic in the right and left runners, we selected a temperature of molten iron, I, charging rate of molten iron, J, and side of a runner used (right or left), K, in Table 1 as noise factors. To evaluate turbulence due to the anisotropy of a product, we chose five levels only for the filling rate of molten iron.

4. SN Ratio and Sensitivity of Reynolds Number

After allocating the control and noise factors to an L_{18} orthogonal array, we conducted casting simulation and computed Reynolds numbers. Table 2 shows the experimental data for experiment 1. Us-

ing these, we calculated the SN ratio and sensitivity as follows.

Total variation:

$$S_T = 5749^2 + 5900^2 + \cdots + 3104^2 + 3879^2$$
$$= 545{,}917{,}217 \qquad (f = 20) \qquad (2)$$

Variation of general mean:

$$S_m = \frac{(5749 + 5900 + \cdots + 3104 + 3879)^2}{20}$$
$$= 531{,}738{,}281 \qquad (f = 1) \qquad (3)$$

Error variation:

$$S_e = S_T - S_m = 545{,}917{,}217 - 531{,}738{,}281$$
$$= 14{,}178{,}936 \qquad (f = 19) \qquad (4)$$

Error variance:

$$V_e = \frac{S_e}{19} = \frac{14{,}178{,}936}{19} = 746{,}260 \qquad (5)$$

SN ratio:

Table 1
Control and noise factors and levels

		Level				
	Factor	1	2	3	4	5
Control factors						
A:	choke	Down	Std.[a]	—	—	—
B:	vertical runner	Down	Std.[a]	Up	—	—
C:	swirl horizontal runner	Down	Std.[a]	Up	—	—
D:	swirl entrance gate	Down	Std.	Up[a]	—	—
E:	swirl lower gate	Down	Std.[a]	Up	—	—
F:	horizontal runner	Down	Std.[a]	Up	—	—
G:	undergate	Down	Std.[a]	Up	—	—
H:	gate	Down	Std.[a]	Up	—	—
Noise factors						
I:	temperature of molten iron when poured	Down	Std.[a]	—	—	—
J:	filling rate of molten iron (%)	35	40	45	50	55
K:	side of a runner	Left	Right	—	—	—

[a] Initial level.

Table 2
Analysis results for casting of camshaft (experiment 1)

		K_1 (35%)	K_2 (40%)	K_3 (45%)	K_4 (50%)	K_5 (55%)
I_1:	J_1 (left)	5749	5900	4722	4552	4070
lower temperature of molten iron	J_1 (right)	5732	5728	5484	4967	4712
I_2:	J_1 (left)	6162	6127	6298	5138	5062
standard temperature of molten iron	J_2 (right)	6069	5278	4392	3104	3879

$$\eta = 10 \log \frac{\frac{1}{20}(S_m - V_e)}{V_e}$$

$$= 10 \log \frac{\frac{1}{20}(531{,}738{,}281 - 746{,}260)}{746{,}260}$$

$$= 15.51 \text{ dB} \tag{6}$$

Sensitivity:

$$S = 10 \log \frac{1}{20}(S_m - V_e)$$

$$= 10 \log \frac{1}{20}(531{,}738{,}281 - 746{,}260)$$

$$= 74.24 \text{ dB} \tag{7}$$

5. Optimal Configuration and Confirmation

Figure 3 shows the response graphs for the SN ratio and sensitivity.

To make our casting method robust to the variability in casting conditions and prevent casting defects, we needed not only to mitigate the variability of Reynolds numbers, but also to lower the absolute values. Therefore, considering that we should determine the optimal configuration based on a criterion of a higher SN ratio and lower sensitivity, we selected the combination $A_2 B_2 C_3 D_3 E_2 F_2 G_1 H_2$. Using estimations of the SN ratio and sensitivity under this configuration, we estimated the gains. Then, to ver-

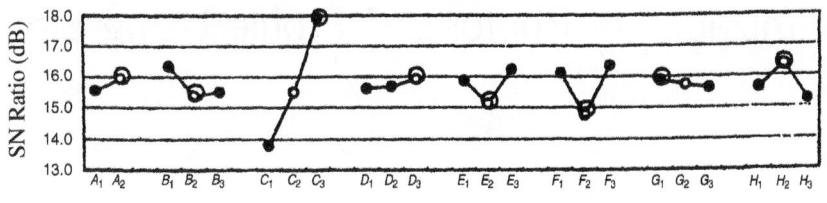

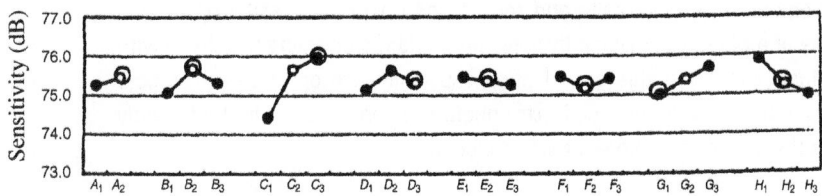

Figure 3
Response graphs

Table 3
Results of confirmatory experiment (dB)

Configuration	SN Ratio		Sensitivity	
	Estimation	Confirmation	Estimation	Confirmation
Optimal	17.10	18.09	75.67	75.27
Current	14.52	12.95	75.73	75.78
Gain	2.58	5.15	−0.06	−0.51

ify the validity of these estimations, we conducted a confirmatory experiment under the optimal configuration.

Table 3 shows the results of the confirmatory experiment. This reveals that an improvement of approximately 5 dB is expected, whereas the gains in SN ratio do not have good reproducibility. The actual casting method reflecting the optimal configuration had reduced the occurrence rate of casting defects, and the casting cost was reduced by 34%. As a side effect, we can anticipate a shorter development cycle time by applying this engineering process to preliminary studies before mass production of a new product.

Reference

Kazuyuki Shiino and Yasuhiro Fukumoto, 2001. Optimization of casting conditions for camshafts. *Quality Engineering*, Vol. 9, No. 4, pp. 68–73.

This case study is contributed by Kazuyuki Shiino.

Optimization of Photoresist Profile Using Simulation

Abstract: In this optimization case, our objective was to shorten the product development cycle drastically and reduce the operating cost by optimizing stability of a photoresist shape through simulation, confirming the shape with an actual pattern. In this optimization especially, we optimized a process dealing with relatively thick (2.2 μm) photoresist on a metal film that greatly affects the shape of photoresist when etched.

1. Introduction

Micromachining used in semiconductor manufacturing is conducted through a repeated process of photomasking, photomask patterning, and etching (Figure 1). In the photomask patterning process (photo process), it is essential to control not only dimensions but also cross-sectional photoresist shapes for stabilizing etching shapes. This is considered a key technology in semiconductor manufacturing that introduces advanced micromachining technologies year after year. Figure 1 shows a case of positive photoresist in which photosensitive parts are removed in alkaline solution.

To optimize conditions for stabilizing dimensions of photoresist and controlling its design shape, we made the most of a simulation technology. As simulation software, we used Prolith/2, supplied by Finle Technologies, Inc. For exposure computation, this software utilizes a theoretical calculation method based on an optical and physical model, and for development computation, a method using actual measurements of the developer dissolution rate. Therefore, we could more realistically analyze the after-development photoresist shape with respect to its dimensions and also perform nonlinear analyses. Although to create and measure a photoresist shape requires a considerable amount of skill, the simulation enabled us to confirm a shape and dimensions at each position instantaneously.

2. Simulation

In the photomask patterning process using exposure, a pattern of light (the width of a light shield) generated by a mask (light shield) is transferred onto photoresist. In the case of the positive photoresist used for this research, photoresist in the area that does not receive any light remains after development. In this case, the ideal situation is that a 1-μm-wide pattern in design is transferred as 1 μm and a 0.5-μm-wide pattern as 0.5 μm.

Therefore, in this study, we regarded it as a generic function that a dimension of the pattern transferred is proportional to that of a mask. As the ideal input/output relationship, we considered the proportionality between a dimension of a mask, M, and the counterpart of a pattern, y ($y = \beta M$) (Figure 2).

A dimension of a mask pattern was chosen as a signal factor, M, that has three different values, 0.6, 0.8, and 1.0 μm. We defined as an ideal condition that the photoresist shape becomes rectangular and does not change at any position in the direction of the photoresist's thickness. Then, as a noise factor, we selected a dimension of photoresist, setting the dimension of a boundary between the photoresist and the metal film below to L_{bot}, the dimension of photoresist at the top to L_{top}, and that in the middle to L_{mid} (Figure 3). When the shape is to be controlled separately, it was analyzed as an indicative factor. Eight factors that affect dimension and shape

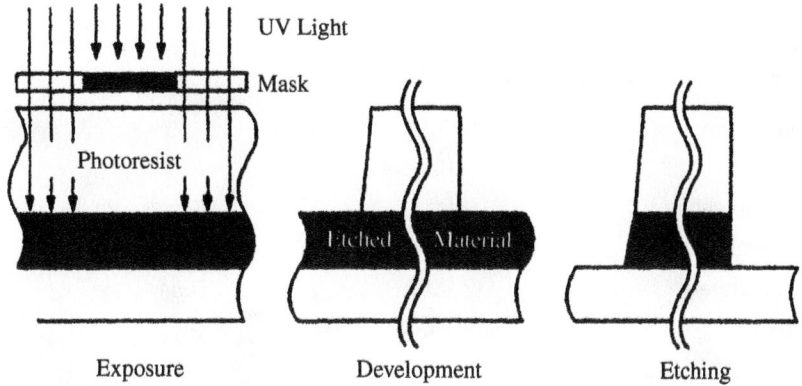

Figure 1
Photoresist and etched shapes

in the process were listed as control factors (Table 1).

For the amount of light (F), we used a sliding-level technique for each experiment. We calculated beforehand the amount of light needed to form a 0.6-μm pattern for a mask of the same size and selected the quantity as the optimal amount of light (EOP) for each experiment. Then we defined factor levels for the amount of light (F) as $\pm10\%$ of the optimal amount.

3. Parameter Design with Position in Thickness Direction as Noise Factor

When using a position in the direction of thickness (vertical direction) as the noise factor (Table 2), the SN ratio is calculated as follows.

Total variation:

$$S_T = 0.325^2 + 0.568^2 + 0.783^2 + \cdots + 1.041^2$$
$$= 4.670 \quad (f = 9) \tag{1}$$

Effective divider:

$$r = 0.6^2 + 0.8^2 + 1.0^2 = 2.0 \tag{2}$$

Linear equations:

$$L_1 = (0.6)(0.325) + (0.8)(0.568) + (1.0)(0.783)$$
$$= 1.432 \tag{3}$$

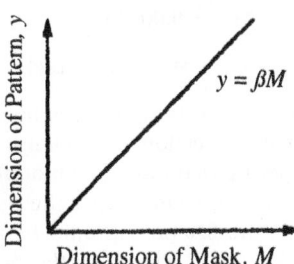

Figure 2
Generic function

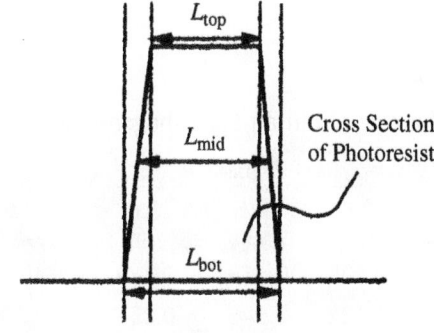

Figure 3
Dimensions checked

Table 1
Control factors and levels

		Level		
	Control Factor	1	2	3
A:	photoresist type	Current	New	—
B:	prebaking temperature (°C)	80	90	100
C:	mask bias (μm)	−0.05	0	+0.05
D:	numerical aperture	0.50	0.54	0.57
E:	focus (μm)	−0.3	0	+0.3
F:	exposure time (ms)	−10%	EOP	+10%
G:	baking temperature (°C)	100	110	120
H:	development time	60	90	120

$$L_2 = 1.636$$

$$L_3 = 2.135 \tag{4}$$

Variation of proportional term:

$$S_\beta = \frac{(1.432 + 1.636 + 2.135)^2}{(3)(2.0)} = \frac{27.07}{6}$$

$$= 4.513 \quad (f = 1) \tag{5}$$

Variation of differences between proportional terms:

$$S_{N\beta} = \frac{1.432^2 + 1.636^2 + 2.135^2}{2.0} - 4.513$$

$$= 0.131 \quad (f = 2) \tag{6}$$

Error variation:

$$S_e = 4.670 - 4.513 - 0.131$$

$$= 0.026 \quad (f = 6) \tag{7}$$

Error variance:

$$V_e = \frac{0.026}{6} = 0.00438 \tag{8}$$

Total error variance:

$$V_N = \frac{4.670 - 4.513}{8} = 0.0196 \tag{9}$$

SN ratio:

$$\eta = \frac{[1/(3)(2.0)](4.513 - 0.00438)}{0.0196}$$

$$= 38.30 \ (15.83 \ \text{dB}) \tag{10}$$

Sensitivity:

$$S = \frac{1}{(3)(2.0)} (4.513 - 0.00438)$$

$$= 0.751 \ (-1.241 \ \text{dB}) \tag{11}$$

Table 2
Data for one run in an L_{18} orthogonal array

	Signal			
Noise	0.6 μm	0.8 μm	1.0 μm	Linear Equation
N_1	0.325	0.568	0.783	L_1
N_2	0.422	0.645	0.867	L_2
N_3	0.694	0.8477	1.041	L_3

Figure 4 shows the response graphs when using a position in the direction of thickness as a noise factor. Considering that only the amount of light is used as an adjusting factor under the optimal configuration and the SN ratio should be prioritized, we selected $A_2 B_1 C_3 D_3 E_2 F^* G_3 H_1$ as the optimal configuration. In the response graphs, O and Δ indicate the optimal and current levels, respectively. Since

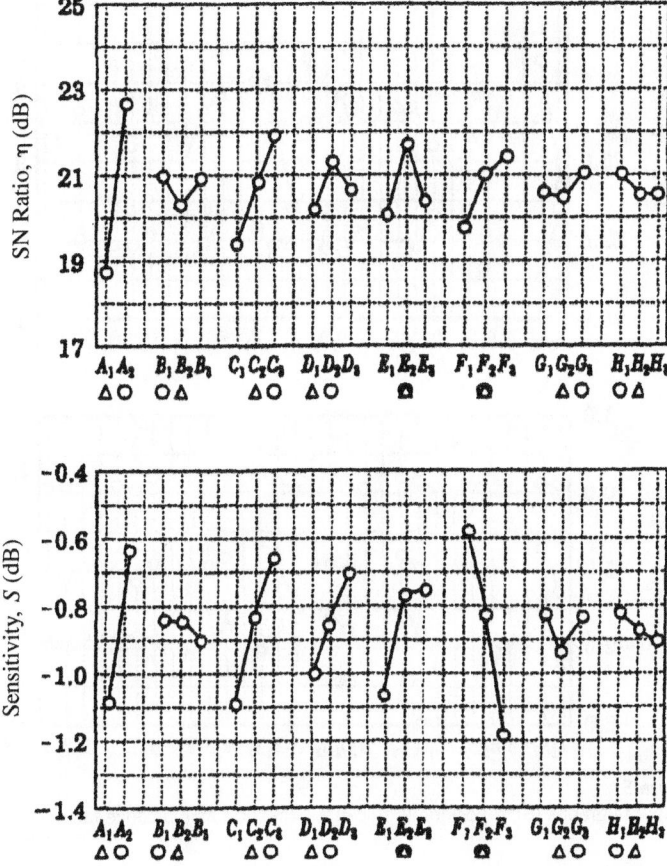

Figure 4
Response graphs using position in thickness direction as a noise factor

the SN ratio depends greatly on the type of photoresist, we can improve the SN ratio by overexposing photoresist to light under the conditions of a positive mask bias.

4. Parameter Design with Position in Thickness Direction as Indicative Factor

To control the shape (taper angle), we also performed an analysis using each measurement point as an indicative factor. Since the variation in the differences between proportional terms $S_{N\beta}$ in the preceding section is dealt with as an indicative factor,

we did not use it for calculating the SN ratio. In this case, as adjusting sensitivities, we computed sensitivities S_1 to S_3, corresponding to the dimension at each measurement point.

SN ratio:

$$\eta^* = \frac{(1/3r)(S_\beta - V_e)}{V_e}$$

$$= \frac{[1/(3)(2.0)](4.513 - 0.00438)}{0.00438}$$

$$= 171.5 \quad (22.34 \text{ dB}) \tag{12}$$

Variation of proportional term for each indicative factor, N_1, N_2, and N_3:

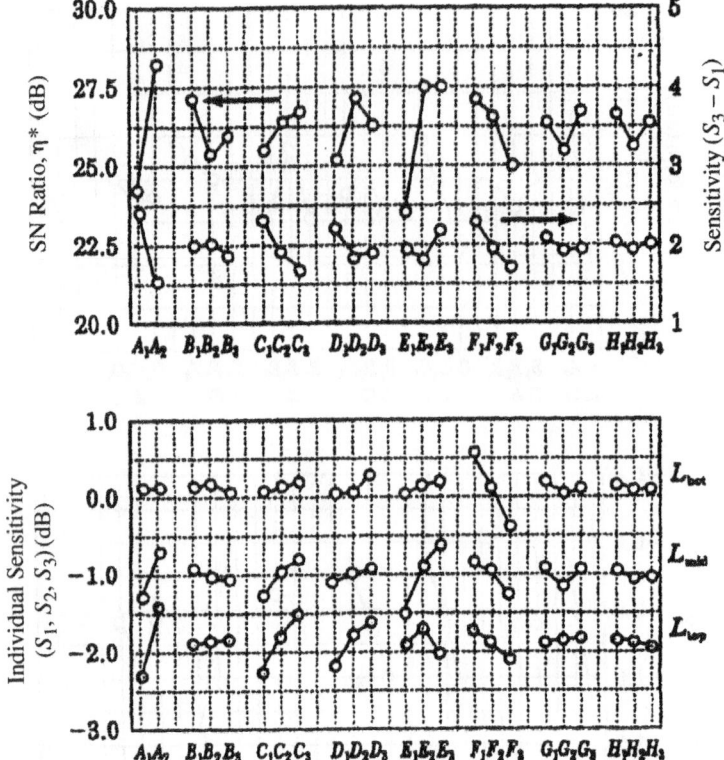

Figure 5
Response graphs using position in thickness direction as indicative factor

$$S_{\beta 1} = \frac{L_1^2}{r} = \frac{1.432^2}{2.0} = 1.026$$

$$S_{\beta 2} = \frac{L_2^2}{r} = 1.338$$

$$S_{\beta 3} = \frac{L_3^2}{r} = 2.279 \qquad (13)$$

Sensitivity:

$$S_1 = \frac{1}{r}(S_{\beta 1} - V_e) = 0.511 \ (-2.92 \text{ dB})$$

$$S_2 = 0.667 \ (-1.76 \text{ dB})$$

$$S_3 = 1.137 \ (0.056 \text{ dB}) \qquad (14)$$

Now, to make the analysis of the taper shape more

understandable, we calculated $S_3 - S_1$ as the sensitivity:

$$S_3 - S_1 = 0.056 - (-2.92) = 3.48 \qquad (15)$$

Figure 5 shows the response graphs when using a measurement point in the direction of thickness as an indicative factor.

A smaller sensitivity, $S_3 - S_1$, used to control the taper angle, indicates a more vertical shape. Shape improvement was confirmed from both the SN ratio and sensitivity, $S_3 - S_1$. $S_3 - S_1$ can be used to tune the shape. The change in the amount of light, F, gives the opposite tendency in SN ratio and sensitivity, $S_3 - S_1$, thus improving both the SN ratio and the shape. To control taper shape, we can solve the orthogonal polynomial equation for sensitivity at each position and make adjustments.

Table 3
Confirmation of SN ratio and sensitivity (dB)

	SN Ratio		Sensitivity	
Configuration	Estimation	Confirmation	Estimation	Confirmation
Optimal	26.63	26.73	−0.19	−0.27
Current	18.76	19.37	−1.13	−1.10
Gain	7.87	7.36	0.94	0.83

5. Confirmatory Calculation and Comparison with Actual Pattern for Photoresist

Estimating the SN ratios and sensitivities under the optimal and current configurations, we confirmed the reproducibility in gain based on the results obtained from our resimulation (Table 3).

Optimal configuration: $A_2 B_1 C_3 D_2 E_2 F_2 G_3 H_1$

Current configuration: $A_1 B_2 C_2 D_1 E_2 F_2 G_2 H_2$

As a result, we confirmed high reproducibility in gain: 93.8% for SN ratio and 88.8% for sensitivity.

Figure 6 illustrates the relationships between the dimension of a mask and the dimension at each position of the actual photoresist. According to these charts, we can see that under the current configuration, there is a large variability in the direction of

thickness, and that the smaller the dimensions of a mask become, the more the bottom dimension is enlarged and the top, diminished. In contrast, under the optimal configuration, this tendency is improved significantly. This result implies improvement in the phenomenon whereby the pattern of photoresist crashes as the dimension of a pattern becomes smaller.

The cross-sectional shapes under the optimal and current configurations when using simulation and in actual patterns are compared in Figure 7. For both cases we used a 0.6-μm-wide pattern. Using a scanning electron microscope we measured the actual pattern. The shapes confirmed by simulation under the optimal and current configurations were sufficiently consistent with those of actual patterns. When using each measurement as an error, the reproducibility in gain of the nominal-the-best SN ratio is also consistent (Table 4).

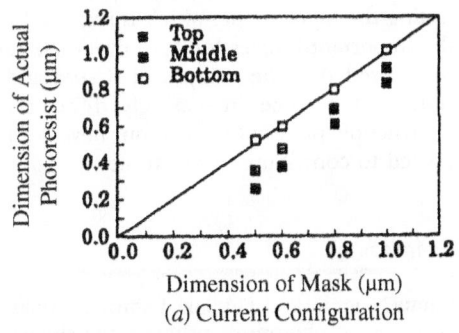

(a) Current Configuration

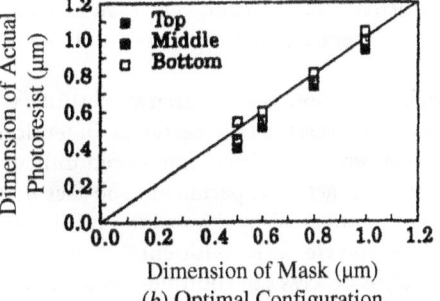

(b) Optimal Configuration

*Data at 0.5 μm are plotted as a reference

Figure 6
Comparison of dimensions before and after optimization

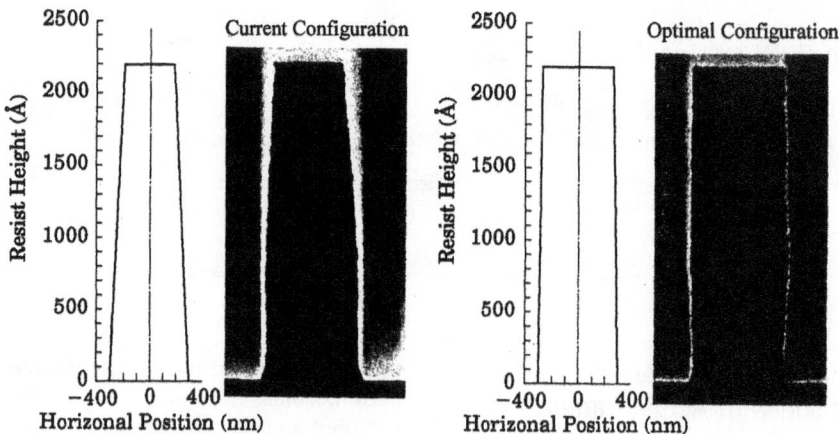

Figure 7
Comparison of shapes of photoresist before and after optimization

Table 4
Reproducibility of shapes (dB)

	Reproducibility: SN Ratio		
	Optimal	Current	Gain
Actual cross-sectional shape	29.59	16.80	12.79
Simulation	27.55	15.59	11.96

Since the simulation technique in this research used as developer model parameters, parameters calculated from data for dissolution rates of photoresist in actual development, our analysis based on dynamic characteristics reflected the actual situation.

Through simulation, our research successfully visualizes and optimizes the cross-sectional dimension of photoresist, which has been considered difficult to measure in an actual experiment. The fact that both the shape obtained by simulation and the actual pattern shape are fairly consistent demonstrates that our analysis process can contribute much to optimizing the photoresist shape and variability in dimensions. In addition, our new method takes only two days, whereas a conventional method required a few months. Consequently, we can shorten the analytical cycle time dramatically.

On the other hand, by selecting the dimension in the direction of thickness as an indicative factor in our analysis, we can obtain parameters for adjusting the taper shape, which is regarded as one of the important indices for optimization. Although it is believed that the analysis of shapes and dimensions will become increasingly difficult in more microscopic pattern forming, our new method is expected to contribute to shorter-term optimization.

Reference

Isamu Namose and Fumiaki Ushiyama, 2000. Optimization of photoresist profile using simulation. *Proceedings of the 8th Quality Engineering Symposium*, pp. 302–305.

This case study is contributed by Isamu Namose.

CASE 40

Optimization of a Deep-Drawing Process

Abstract: In this study we attempt to reduce production cost for deep-drawn products by minimizing the number of processes through optimization of manufacturing conditions for deep drawing.

1. Introduction

In general, a manufacturing process to form from sheet metal a cylindrical container with a bottom is called a *drawing process,* (Figure 1). In most cases, a deep-drawn product is rarely formed in a single process but is stamped with the required depth in multiple processes. To reduce production cost for deep-drawn products, we minimize the number of processes through optimization of manufacturing conditions for deep drawing.

A drawing defect (i.e., a crack) is caused by partial strain concentration following uneven stretch of a material under an inappropriate forming condition. Ideal stamping is considered as forming any shape with uniform stretch and no cracks in the material. In a deep-drawing process, we obtain the required depth through multiple drawing processes. In the processes following process 2, the rate of drawing is regarded to decrease, compared with that in process 1, under the influence of work hardening. Therefore, since we need to evaluate deep drawing in the processes after process 2, as a signal factor the number of drawings was selected. At each drawing, we measured material thickness and defined a logarithmized value of the ratio of the before and after drawing thickness as an output, y. That is, setting the number of drawings to a signal factor, M, and a logarithmized value of the initial thickness, Y_0, and after-drawing thickness, Y, to the following output, we regarded a zero-point proportional equation of y and βM as a generic function:

$$y = -\log \frac{Y}{Y_0} \qquad (1)$$

In addition, since a product should be formed with uniform thickness, we selected a measurement of product thickness, N (five points) as a noise factor to determine a condition with minimal variability.

2. SN Ratio and Sensitivity

Table 1 shows the data of experiment 1 in the L_{18} orthogonal array. The calculations for the SN ratio and sensitivity follow.

Total variation:

$$S_T = 0.021^2 + \cdots + 0.061^2 = 0.025325 \qquad (f = 15) \qquad (2)$$

Effective divider:

$$r = 1^2 + 2^2 + 3^2 = 14 \qquad (3)$$

Linear equations:

$$L_1 = (1)(0.021) + (2)(0.029) + (3)(0.057)$$
$$= 0.250$$
$$L_2 = (1)(0.021) + (2)(0.026) + (3)(0.058)$$
$$= 0.247$$
$$L_5 = (1)(0.023) + (2)(0.027) + (3)(0.061)$$
$$= 0.260 \qquad (4)$$

Variation of proportional term:

$$S_\beta = \frac{(L_1 + \cdots + L_5)^2}{5r} = 0.024816 \qquad (f = 1) \qquad (5)$$

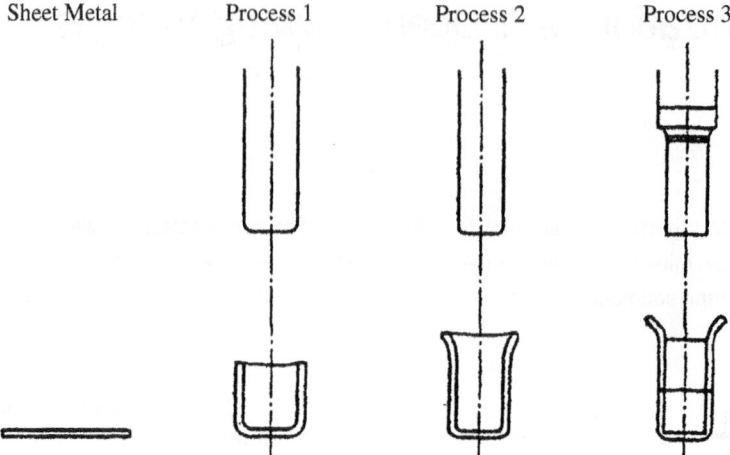

Figure 1
Deep-drawing process

Variation of differences between proportional terms:

$$S_{N\beta} = \frac{L_1^2 + \cdots + L_5^2}{r} - S_\beta = 0.000079 \quad (f = 4) \quad (6)$$

Error variation:

$$S_e = S_T - S_\beta - S_{N\beta} = 0.00043 \quad (f = 10) \quad (7)$$

Error variance:

$$V_e = \frac{S_e}{10} = 0.000043 \quad (8)$$

Total error variance:

$$V_N = \frac{S_{N\beta} + S_e}{14} = 0.000036 \quad (9)$$

SN ratio:

$$\eta = 10 \log \frac{(1/5r)(S_\beta - V_e)}{V_N} = 9.88 \text{ dB} \quad (10)$$

Sensitivity:

Table 1
Measured data of experiment 1 in the L_{18} orthogonal array (converted)

Noise Factor	Signal Factor			Linear Equation
	M_1 1	M_2 2	M_3 3	
N_1	0.02	0.03	0.06	L_1
N_2	0.02	0.03	0.06	L_2
N_3	0.02	0.04	0.06	L_3
N_4	0.02	0.03	0.06	L_4
N_5	0.02	0.03	0.06	L_5

$$S = 10 \log \frac{1}{5r} (S_\beta - V_e) = -34.51 \text{ dB} \quad (11)$$

3. Optimization Configuration and Results of Confirmatory Experiment

As control factors, we selected the seven factors shown in Table 2 and allocated them to an L_{18} orthogonal array. Figure 2 shows the response graphs. Based on these graphs, we determined $B_1 C_3 D_1 E_3 F_1 G_2 H_3$ as the optimal configuration. On the other hand, as a comparative configuration, we chose $B_2 C_1 D_3 E_2 F_3 G_3 H_1$, leading to the worst SN ratio. For these optimal and comparative configurations, a confirmatory experiment was conducted. Table 3 shows estimations of the SN ratio and sensitivity. Considering all of the above, we concluded that the gains in SN ratio and sensitivity have good reproducibility.

The 15.23 dB gain obtained from the confirmatory experiment was converted into an antilog value of 33.37, which implies that we can reduce the variability to 1/33.37 (1/5.58) of the initial value. In actuality, the standard deviation under the optimal configuration is decreased to 1/7.07 of the counterpart under the comparative configuration. Because of large gains and good reproducibility, reduction in the number of drawing processes can be expected under the optimal configuration.

On the other hand, sensitivity, S, indicates the degree of influence for displacement of thickness at the bottom of a product. As the sensitivity becomes larger, the rate of change becomes larger and the thickness at the bottom becomes smaller, causing cracks more often. Therefore, a smaller value of sensitivity is advantageous to drawing.

4. Economic Benefits under the Optimal Configuration Based on the Loss Function

This product has a much larger tolerance in dimensions compared with the actual standard deviation. Thus, even if we minimize the variability and lower the standard deviation, we cannot expect much benefit through the loss improvements as a whole. In contrast, even if the variability increases to some extent, less production cost through reduction in the number of drawing processes can mitigate the total loss. Then we reduce the production cost by integrating drawing processes under the optimal configuration. By applying the optimal configuration and unifying processes, we lowered the number of processes by 40%. Consequently, we arrived at considerably less production cost.

The production cost for one deep-drawn product is 8.5 yen. All products that exceed the tolerance have been discarded. The loss that occurs to a

Table 2
Control factors and levels

	Control Factor	Level		
		1	2	3
A:	error	—	—	—
B:	punch type	A	B	C
C:	dice type	A	B	C
D:	clearance	Small	Mid	Large
E:	blank holder force	Small	Mid	Large
F:	lubricant type	Company A	Company B	Company C
G:	knockout pressure	Small	Mid	Large
H:	material type	A	B	C

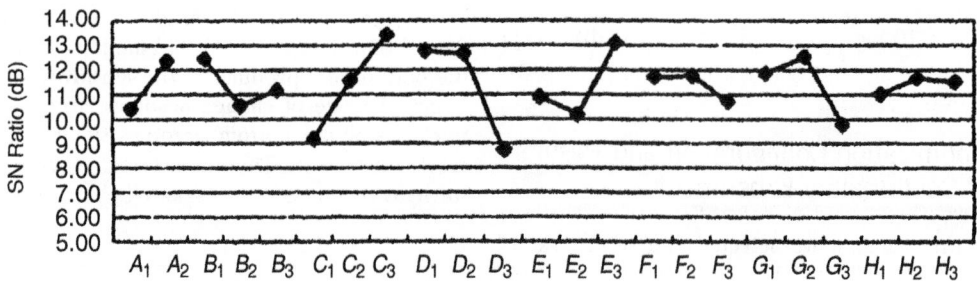

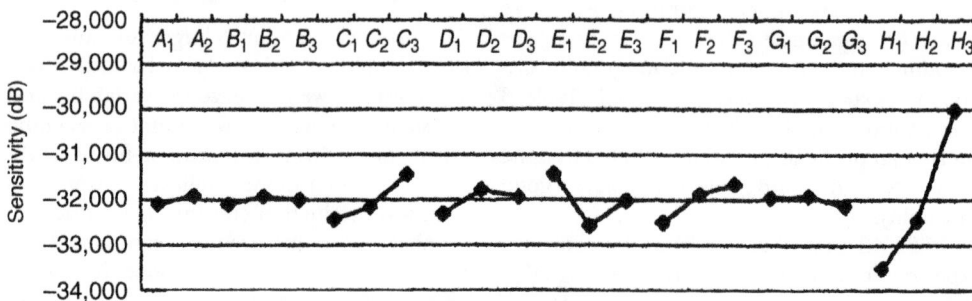

Figure 2
Response graphs

customer in this case is defined as A_0 yen. The production cost that has to date amounted to 8.5 yen/product can be reduced to 4 yen under the optimal configuration. The improvement is calculated as follows.

Loss function:

$$L = \frac{A_0}{\Delta^2} \sigma^2 + \text{product cost} \qquad (12)$$

Under the current configuration:

$$L = \frac{8.5}{0.2^2} (0.0032^2) + 8.5 = 8.502 \qquad (13)$$

Under the optimal configuration:

$$L = \frac{8.5}{0.2^2} (0.0068^2) + 4 = 4.01 \qquad (14)$$

Table 3
Estimation and confirmation of SN ratio and sensitivity

Configuration	SN Ratio		Sensitivity	
	Estimation	Confirmation	Estimation	Confirmation
Optimal	19.13	18.63	−34.47	−30.26
Current	2.17	3.40	−30.26	−30.79
Gain	16.96	15.23	−4.21	−1.41

Thus, the improvement expected per product is 8.502 yen − 4.01 yen = 4.49 yen. As we produce 30,000 units/day and 600,000 units/year, we have the following benefit:

$$(4.49 \text{ yen}) (600,000 \text{ units}) (12 \text{ months})$$
$$= 32,400,000 \text{ yen}$$

As a result of applying the optimal configuration, we can improve approximately 30 million yen on a yearly basis through streamlined production with no increased variability.

Reference

Satoru Shiratsukayama, Yoshihiro Shimada, and Yasumitsu Kawasumi, 2001. Optimization of deep drawing process. *Quality Engineering*, Vol. 9, No. 2, pp. 42–47.

This case study is contributed by Satoru Shiratsukayama.

Robust Technology Development of an Encapsulation Process

Abstract: The purpose of this experiment was to determine the optimum process parameters to use to reduce encapsulation height variation for the enhanced plastic ball grid array (EPBGA) package. Optimum process parameters for the EPBGA encapsulation process require encapsulation height standard deviations of less than 1.5 and on target, satisfying a capacity requirement of 150 units per hour. The experiment's results show that the encapsulation height variation is reduced by increasing the gel time, decreasing the temperature for the dispensing and gelling stages, utilizing an effective dispensing pattern, and dispensing a flow control dam ring of a specified height. The standard process settings for the EPBGA encapsulation process yielded an SN ratio of 24.33 dB. The experiment's optimum process settings achieve an SN ratio of 27.31 dB. The robust technology development experiment increased the SN ratio by 2.98 dB (27.31 dB − 24.33 dB).

1. Introduction

Robust technology development of an encapsulation process was conducted at LSI Assembly Engineering's Research and Development Facility, located in Fremont, California. The experiment focused on minimizing encapsulation height variation for the single-tier cavity-down EPBGA electronic package. The encapsulation process is a critical process for ensuring surface mountability [1].

Objectives

Determining optimum process parameters to reduce encapsulation height variation for the EPBGA package was the main purpose of this experiment. Optimum process parameters for the EPBGA encapsulation process require (1) encapsulation height standard deviations of less than 1.5 and on target, and (2) satisfying a capacity requirement of 150

units per hour. The function of the EPBGA encapsulation process is to dispense a flow control dam ring and a glob-top liquid epoxy (encapsulant) over the silicon die and delicate bonding wires [3]. Encapsulation offers several basic functions for an electronic package: (1) filling the cavity to create a seal, (2) adding mechanical strength to the package, and (3) providing good moisture resistance.

Flow Control Dam (Dam Ring) The dam ring is a high-viscosity liquid damming material that supports the glob-top epoxy. The thixotropic property of the dam ring epoxy enables the dam ring "to thin upon isothermal agitation and thicken upon subsequent rest" [2]. The dam ring is dispensed in between the solder ball pads and the bonding fingers. The distance between the inner array of solder ball pads and the middle bonding fingers is 0.040 in., requiring accurate placement of the dam ring. Figure 1

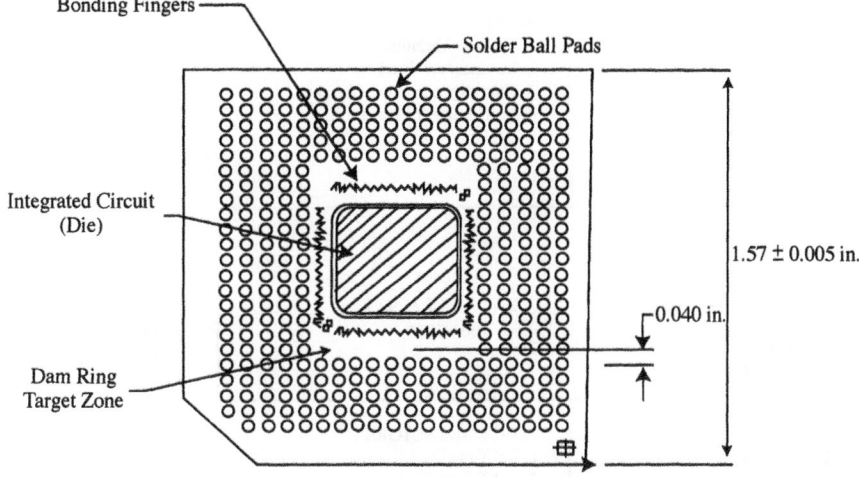

Figure 1
Top view of a single-tier cavity-down electronic process

shows a top view of an EPBGA single-tier cavity-down electronic package.

Glob-Top Encapsulant The glob-top encapsulant is a low-viscosity, low-stress liquid epoxy that offers excellent moisture resistance. The encapsulant exhibits a high flow rate and requires a dispensed thixotropic dam for containment [4]. Figure 2 shows an electronic package with an integrated circuit (silicon die), dam ring, and encapsulant.

2. Design of Experiment

Parameter Diagram

A parameter diagram (Figure 3) represents the EPBGA encapsulation process. Noise, signal, and controllable factors are inputs to the EPBGA encapsulation process. The output of the EPBGA encapsulation process is called the *response* (encapsulation height).

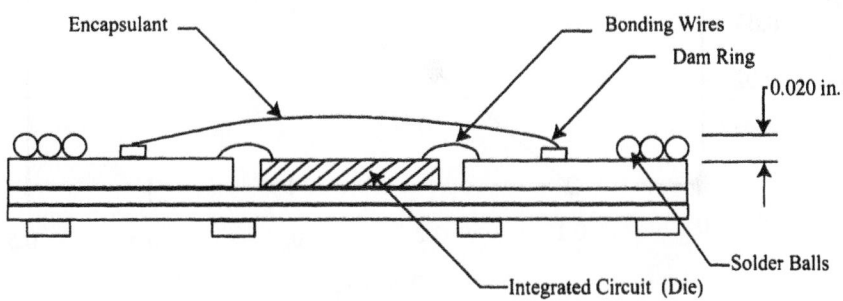

Figure 2
Side view of encapsulant process

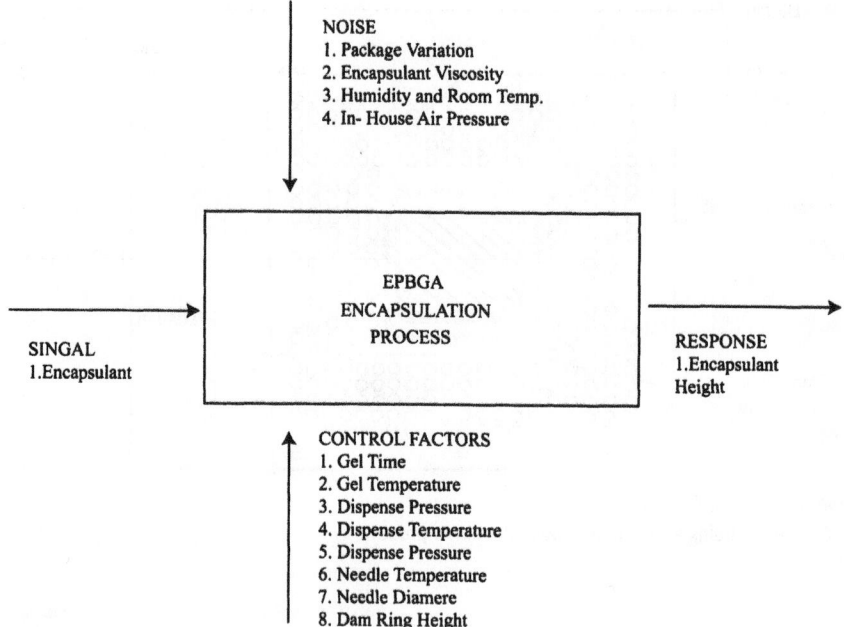

Figure 3
EPBGA encapsulation process parameter design

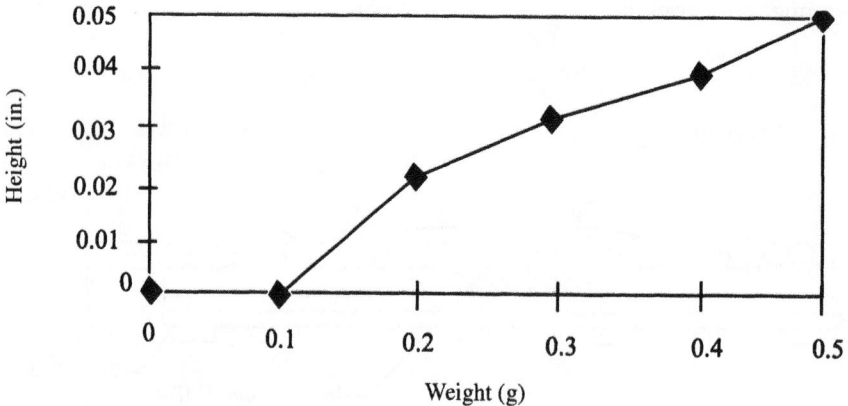

Figure 4
Ideal function between encapsulation weight and height

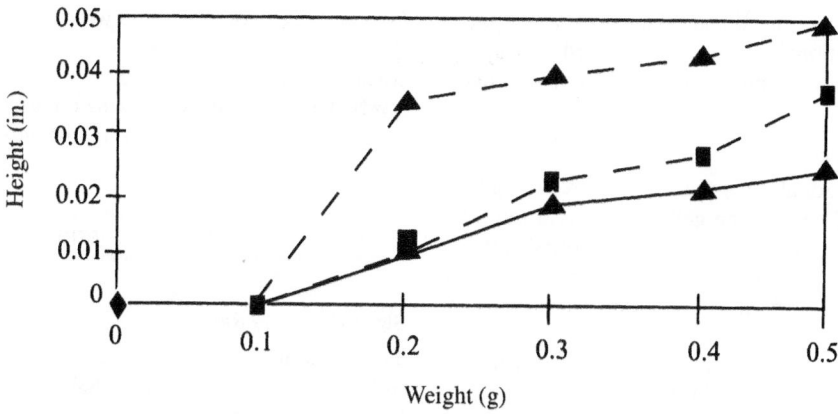

Figure 5
Actual function between encapsulation weight and height

❑ *Noise factors* are process parameters whose settings are difficult or expensive to control.

❑ *Controllable factors* are parameters that can be specified by the designer.

❑ *Signal* is the parameter set by the operator to achieve the intended value for the response.

❑ *Response* is the measured output for the EPBGA encapsulation process.

Cured encapsulation height is measured by utilizing a mechanical height indicator and a designer fixture. The fixture contains five measuring holes and four zeroing holes located on the top plate to ensure repeatable encapsulation height measurements. The drop indicator is zeroed on the substrate and is then moved on top of the cured encapsulant for a digital display representing encapsulation height [6].

Ideal Function

The ideal function of the encapsulation process is a linear function between the encapsulation weight and height (Figure 4). A correlation exists between the weight and height: Increasing the weight increases the height [5].

The EPBGA encapsulation process is sensitive to the following noise factors:

1. *Package-to-package variation.* The volume of the cavity could be different for every package, due to variations in the die attach material.

2. *Encapsulation viscosity variation.* The viscosity decreases when heat is applied and increases over time.

Figure 5 represents an undesirable function between encapsulation weight and height. The experiment considered several process parameters to

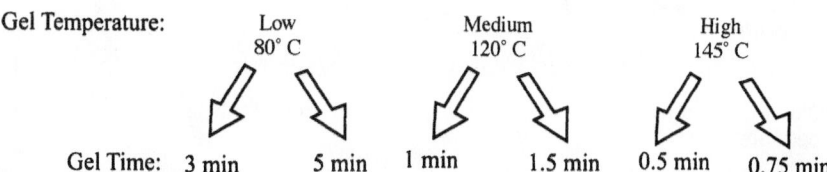

Figure 6
Sliding level settings

make this function linear (Figure 4). The zero-point-proportional SN ratio was utilized. A large SN ratio decreases the encapsulation height variance.

Layout of Experiment

An L_{18} orthogonal array was selected. A sliding level was utilized between the gelling time and temperature to enlarge the range of experimental settings. The low and high gelling times were paired with low, medium, and high gel temperature settings, (Figure 6).

The sliding level was incorporated into run 1 of the inner array, as seen in Table 1. During the experiment the encapsulant's viscosity (noise factor) was controlled by the thaw time in hours prior to dispensing. Controlling the low-viscosity level, the thaw time was 1.5 hours. Controlling the high-viscosity level, the thaw time totaled 6 hours. Two encapsulation viscosity levels (in centipoise) were incorporated into the experiment:

Low (0–6 hours):   40,600

High (6–12 hours):   61,200

Figure 7 shows an L_{18} inner array and an outer array with signal and noise factors. The signal was assigned to three levels of encapsulation weight.

Low and high viscosity levels were incorporated into the experiment as noise factors. The inner array contains eight control factors with two and three levels. The outer array contains the experiment's responses (540 encapsulation height measurements).

3. Experimental Results and Analysis

Signal-to-Noise Ratio

The dynamic zero-point-proportional SN ratio was utilized for experimental analysis. The general idea behind the SN ratio is

$$\text{SN ratio} = \frac{\text{useful energy}}{\text{not useful energy}}$$

$$= 10 \log_{10} \frac{\beta^2}{\sigma^2}$$

An SN ratio was calculated for each of the 18 experimental runs. Choosing the highest SN ratio decreases the encapsulation height variation and increases the linearity. Increasing the linearity between the encapsulation weight and height will help target the encapsulation height. Refer to Table 2 for the experiment's SN ratio and linear coefficient per

Table 1
Control factors and levels for the inner array

Run	Control Factor	Level		
		1	2	3
1	Gel time	(0.5-GT, GT, GT + 2.0)	(0.75 − GT, GT, GT + 3.5)	—
2	Gel temperature	GT − 40	GT	GT + 25
3	Dispense pressure	DP − 4	DP	DP + 3
4	Dispense temperature	DT − 15	DT	DT + 50
5	Dispense pattern	Pattern A	Pattern B	Pattern C
6	Needle temperature	NT	NT + 10	NT + 20
7	Needle diameter	ND − 0.010	ND	ND + 0.010
8	Dam ring height	DR − 0.003	DR	DR + 0.003

INNER ARRAY									OUTER ARRAY						
FACTORS									Signal:	LOW		MED		HIGH	
Run	A	B	C	D	E	F	G	H	Noise:	LOW	HIGH	LOW	HIGH	LOW	HIGH
1	1	1	1	1	1	1	1	1		1...5	1...5	1...5	1...5	1...5	1...5
2	1	1	2	2	2	2	2	2							
3	1	1	3	3	3	3	3	3							
4	1	2	1	1	2	2	3	3							
5	1	2	2	2	3	3	1	1							
6	1	2	3	3	1	1	2	2		30 Encapsulation Height					
7	1	3	1	2	1	3	2	3		Measurements per Run					
8	1	3	2	3	2	1	3	1		Totaling ($18 \times 6 \times 5$)					
9	1	3	3	1	3	2	1	2		540 Measurements					
10	2	1	1	3	3	2	2	1							
11	2	1	2	1	1	3	3	2							
12	2	1	3	2	2	1	1	3							
13	2	2	1	2	3	1	3	2							
14	2	2	2	3	1	2	1	3							
15	2	2	3	1	2	3	2	1							
16	2	3	1	3	2	3	1	2							
17	2	3	2	1	3	1	2	3							
18	2	3	3	2	1	2	3	1							

Figure 7
Experimental design and layout

Table 2
SN ratio and linear coefficient per run

Run	SN Ratio (dB)	Linear Coefficient
1	24.20	33.77
2	25.58	32.64
3	20.84	34.45
4	25.11	31.02
5	22.17	24.13
6	24.98	31.42
7	23.73	30.76
8	24.33	32.19
9	24.85	32.52
10	26.59	34.48
11	27.37	30.60
12	26.38	29.77
13	26.48	31.24
14	24.23	33.92
15	24.28	33.00
16	24.87	31.06
17	24.69	31.79
18	25.45	31.21

run. Prior to this experiment the standard process settings for the EPBGA encapsulation process yielded an SN ratio of 24.33 dB. (Figure 8).

Level Averages for SN Ratio

Tables 3 and 4 show the average SN ratio and linear coefficient for the control factors and their level responses.

Choice Table

Control factors and their level settings that showed a strong effect on the SN ratio are shown in Table 5. The linear coefficient is not represented in the table and would only be useful in this application for targeting a specified encapsulation height.

Choice Table Process Settings

1. Increasing the gelling time (level II) yielded the highest SN ratio and decreased the throughput from 150 units/h to 60.

2. Reducing the gelling temperature (level I) minimized the encapsulation height variation and eliminated vapors from the encapsulation gelling stage.

3. A particular dispensing pattern (level II) provided an even distribution of encapsulant over the die and cavity.

4. Building the dispensed dam ring at a specified height (level II) reduced the encapsulation height variation.

5. Using a needle heater appeared to have a negligible effect on reducing encapsulation height variation.

Trade-off Process Settings Due to the capacity requirement, trade-off settings were created based on the experiment's results. The trade-off process settings increased the throughput from 60 units/h to 150 units/h.

1. Gel time was reduced from the level II setting in order to maintain acceptable throughput levels.

2. The preheat, dispense, and gel temperatures were increased to compensate for the shorter gel time.

4. Confirmation

Prior to the confirmation run, factors containing strong effects (a large SN ratio) from the choice table were placed into the following SN ratio prediction equation. (T_{av} of SN ratio = 24.79):

$$
\begin{aligned}
\text{SN ratio}_{\text{prediction}} = T &= (\text{GTI}_2 - T) + (\text{DH}_2 - T) \\
&\quad + (\text{TT}_1 - T) + (\text{GTE}_1 - T) \\
&= 24.79 + (25.59 - 24.79) \\
&\quad + (25.69 - 24.79) + (25.18 \\
&\quad - 24.79) + (25.16 - 24.79) \\
&= 27.32
\end{aligned}
$$

The process factors with strong effects included gel

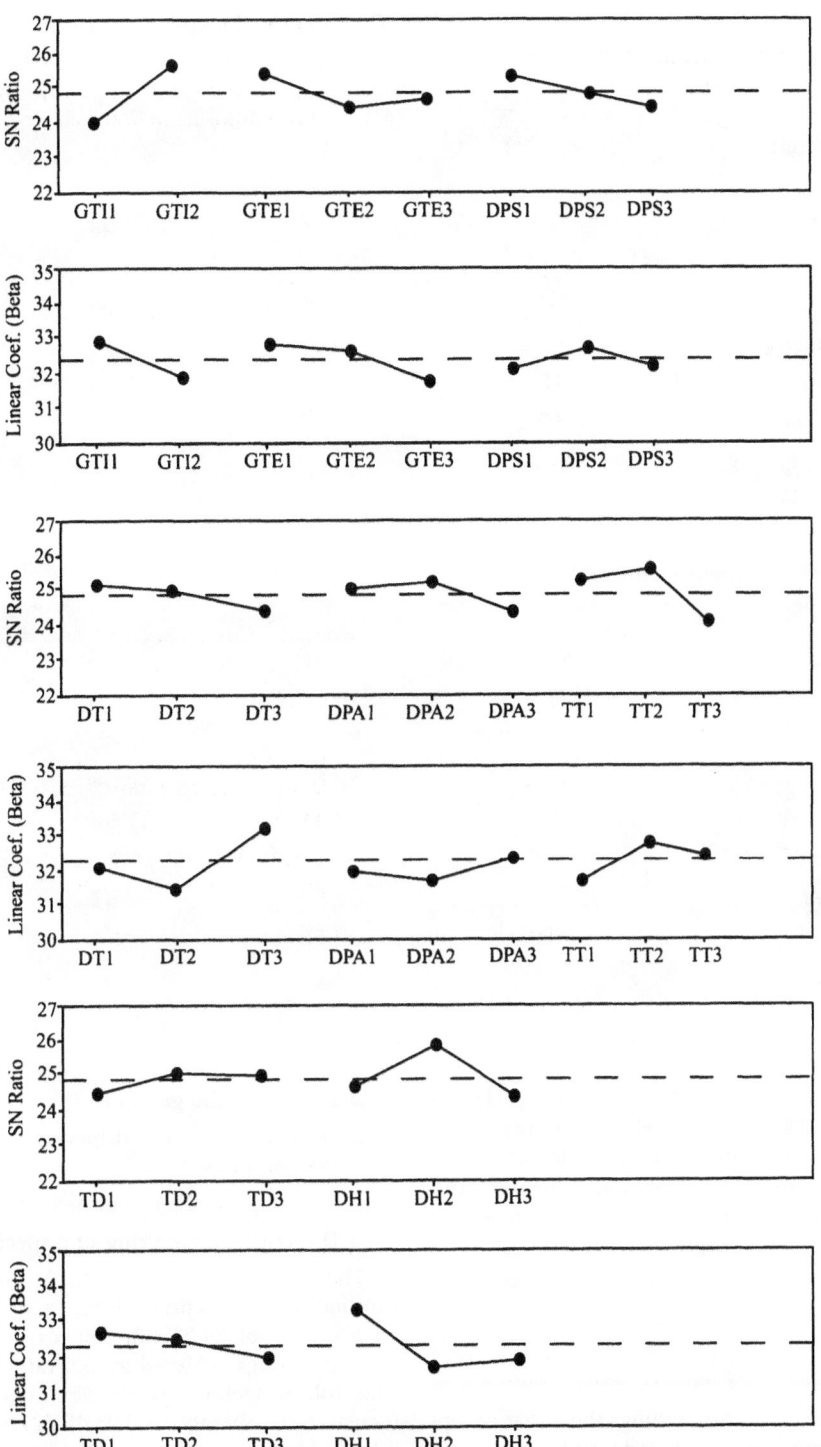

Figure 8
Level average for SN ratios and linear coefficients

Table 3
Level averages for SN ratio (dB)

Control Factor	Symbol	Level		
		1	2	3
Gel time	GTI	23.98	25.59	—
Gel temperature	GTE	25.16	24.54	24.46
Dispense pressure	DPS	25.17	24.73	24.46
Dispense temperature	DT	25.09	24.96	24.31
Dispense pattern	DPA	25.00	25.09	24.27
Needle heater	TT	25.18	25.30	23.88
Needle diameter	TD	24.40	24.90	24.90
Dam ring height	DH	24.50	25.60	24.16

Table 4
Level averages for linear coefficient

Control Factor	Symbol	Level		
		1	2	3
Gel time	GTI	32.54	31.89	—
Gel temperature	GTE	32.61	32.45	31.58
Dispense pressure	DPS	32.05	32.54	32.06
Dispense temperature	DT	32.11	31.62	32.91
Dispense pattern	DPA	31.94	31.61	33.10
Needle heater	TT	31.69	32.63	32.33
Needle diameter	TD	32.58	32.34	31.78
Dam ring height	DH	33.13	31.57	31.95

time (GTI), dispense DAM ring height (DH), tip temperature (TT), and the tel temperature (GT). Based on the SN –ratio prediction calculation, the SN ratio from the choice table and the SN ratio from the tradeoff settings, no large interactions were present.

5. Summary

The experiment's process settings show that encapsulation height variation was reduced by:

❑ Increasing the gel time
❑ Decreasing the dispense and gelling temperatures
❑ Utilizing an effective dispensing pattern
❑ Dispensing a dam ring of a specified height

The standard process settings at the EPBGA encapsulation process prior to this experiment yielded an SN ratio of 24.33 dB. The experiment's final process settings achieved an SN ratio of 27.31 dB. The robust technology development experiment increased the SN ratio by 2.98 dB (27.31 dB − 24.33 dB) (Table 6).

Table 5
Control factors and levels

Control Factor	Code	SN Ratio[a]	Chosen Process Level	Reason for Selecting the Process Control Settings
Gel time	GTI	*	2	Highest SN ratio
Gel temperature	GTE	*	1	Highest SN ratio and less vapor
Dispense pressure	DPS	—	—	Fine tuning on-line for targeting
Dispense temperature	DT	—	—	Utilize previous DOE data
Dispense pattern	DP	—	2	Good encapsulant coverage
Tip temperature	TT	*	1	Satisfies the capacity requirement
Needle diameter	TD	—	2	Adequate diameter for encapsulation quantity
Dam ring	DH	*	2	Highest SN ratio

[a] Choice has effect (*) or no effect (—).

Table 6
SN ratios for various settings

Setting	SN Ratio$_{prediction}$ (dB)	SN Ratio (dB)	Units/h
Standard prior to experiment	—	24.33	150
Choice table	27.32	31.85	60
Trade off	—	27.31	150

References

1. Fred W. Billmeyer, Jr., 1984. *Textbook of Polymer Science*. Hoboken, NJ: Wiley.

2. Hal F. Brinson, 1990. *Engineered Materials Handbook*, Vol. 3, Adhesives and Sealant. Materials Park, OH: ASM International.

3. Marilyn Hwan, 1996. Robust technology development: an explanation with examples. Presented at the ASI/Western Product Development Symposium, Pomona, CA, November 6–8.

4. Brian Lynch, 1992. Asymtek automatic encapsulation coating for COT packages. *LSI Logic*, May.

5. Madhav S. Phadke, 1989. *Quality Engineering Using Robust Design*. Upper Saddle River, N.J.: Prentice Hall.

6. Felipe Sumagaysay and Brent Bacher, Repeat a fixture. U.S. patent pending, November 22, 1995.

This case study is contributed by Brent Bacher.

CASE 42

Gas-Arc Stud Weld Process Parameter Optimization Using Robust Design

Abstract: This paper describes a parameter design experiment utilized to optimize a gas-arc stud welding process. The desired outcome of the project was to determine the optimum process parameter levels leading to minimized variation and maximized stud weld strength across a wide array of real-world processing conditions. The L_{18} orthogonal array experiment employed a zero-point proportional dynamic SN ratio, with signal levels varying across runs. The optimized process parameters yielded a confirmed SN ratio improvement of 6.4 dB, thus reducing the system's energy transformation variability by 52%. Implementation of the optimized process parameters is expected to result in significant cost savings, due to the elimination of a secondary welding operation and a significant reduction in downstream stud strength failures.

1. Introduction

Welding Process

Stud welding is a general term for joining a metal stud to a metal base material. In *gas-arc stud welding,* the base of the stud is joined to the base material by heating both parts with an arc drawn between them and then bringing the parts together under pressure.

The typical steps in gas-arc stud welding (Figure 1) are as follows:

1. The stud is loaded into the chuck and the stud gun is positioned properly for welding by inserting the spark shield into the welding fixture bushing until the stud base contacts the workpiece.

2. The stud gun's trigger is depressed, starting the automatic welding cycle. Immediately upon triggering the weld cycle, the original atmosphere within the spark shield is purged with a controlled gas mixture. Upon purge completion the solenoid coil within the gun body becomes energized, concurrently lifting the stud off the workpiece and creating an arc.

3. The base of the stud and surface of the workpiece are melted by the arc.

4 Upon completion of the preset arc period, the welding current is shut off automatically. This deenergizes the solenoid coil, causing the gun's main spring to plunge the stud into the molten pool of the workpiece, completing the weld.

5. The controlled gas mixture continues to be applied for a specific time after deenergization to ensure the proper gas atmosphere as the weld solidifies.

6. The stud gun is removed, yielding the finished product.

Problems with the Process

Case Corporation's Racine tractor plant had been experiencing an unacceptably high level of

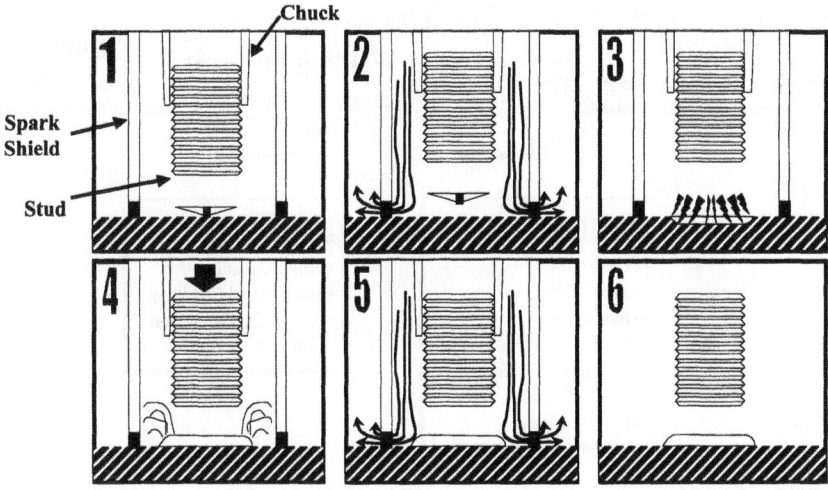

Figure 1
Gas-arc stud welding process

threaded metal studs breaking off during the tractor assembly process. Stud weld strength variation was identified as one of the primary causes of assembly breakage. To ensure adequate strength in the short term, the weld department began reinforcing each stud by adding an additional manual tack-weld step. This additional step added more cost to the process and only masked the underlying variation; therefore, an improved long-term solution was needed.

The problem-solving team decided that its first step in determining a long-term solution was to optimize the controllable parameters (control factors) of the gas-arc stud weld process using robust design methodology.

Desired Outcome

The desired outcome of the project was to determine the optimum process parameter levels, leading to minimized variation and maximized stud weld strength across a wide array of real-world processing conditions.

2. Experimental Procedure

The study was performed using studs 10 mm in diameter and 28 mm in long, of low carbon content,

with a 1.5-mm thread, and 50,000 psi tensile yield strength, and HSLA ROPS steel base material.

Engineered System

The stud welding process can be broken down into an engineering system with three types of input variables and one output (response) variable (Figure 2). Each variable type is discussed in detail below.

Signal Factor

The signal factor is typically the primary energy input into the engineering system. For a gas-arc stud welding system the signal factor (M) is defined as

$$M = M_I M_T M_V$$

where M is the energy input into the system, M_I the average welding current as set on the welding power supply, M_T the welding time (arc period) as set on the welding power supply, and M_V the average peak voltage during welding as determined by the power supply's automatic feedback loop. The power supply is programmed to maintain a constant current; therefore, the voltage is adjusted automatically based on Ohm's law ($I = VR$). Thus, voltage could not be set; instead, it had to be measured for each trial.

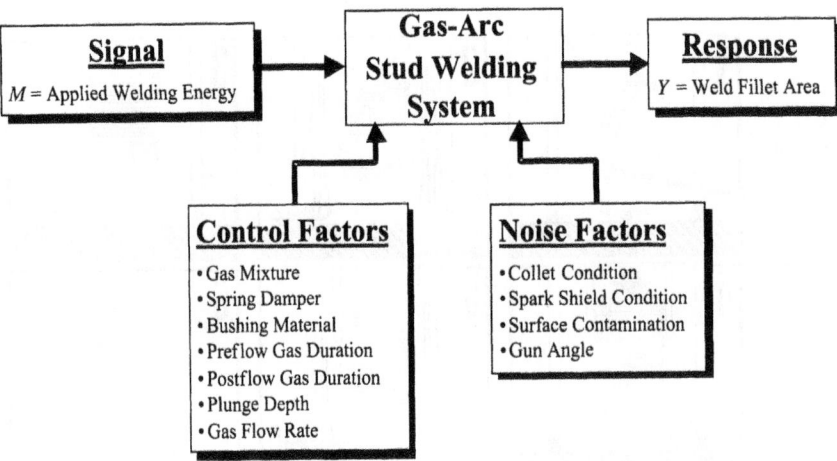

Figure 2
Engineered system

Studs were welded at nine different signal levels for each experiment run (Table 1). However, due to the inherent variation in the welding power supply, the actual values varied somewhat from the set point. Therefore, an oscilloscope was used to record actual welding current, welding time, and peak voltage for each stud welded. The actual energy input (M_i) was then calculated for each trial based on the oscilloscope readings.

Table 1
Signal factor settings

Trial	M_i: Average Current (A)	M_r: Welding Time (s)
M_1 and M_2	560	0.35
M_3 and M_4	560	0.40
M_5 and M_6	560	0.45
M_7 and M_8	700	0.35
M_9 and M_{10}	700	0.40
M_{11} and M_{12}	700	0.45
M_{13} and M_{14}	840	0.35
M_{15} and M_{16}	840	0.40
M_{17} and M_{18}	840	0.45

Response Factor

The response factor (Y) is the measurable intended output of the engineered system. The use of maximum applied torque prior to stud breakage as the response factor was considered. This approach would work well if weld strength is less than the stud material yield strength; however, it would not give an accurate indication if weld strength equaled or exceeded the stud material yield strength. In addition, it could not detect instances of excessive applied energy. Therefore, a more suitable response factor had to be found.

Based on discussions with several welding experts, it was determined that measuring the annular weld fillet area at the base of the stud would provide a reasonable estimate of the amount of energy applied to make a weld. Therefore, annular weld fillet area was chosen as the response factor (Y).

Ideal Function

In an ideal stud welding system, 100% of the welding input energy (signal) would be efficiently transformed into the weld, bonding the stud and base material. However, under real-world conditions, the efficiency of the energy transformation will vary due to the presence of noise factors. The goal of robust design is to minimize the effect of noise factors and to maximize the efficiency of the energy transformation.

The ideal function for the stud welding system is given by

$$y = \beta M$$

where y is the weld fillet area, M the welding energy, and β the slope of the best-fit line between y and M.

Noise Strategy

Noise factors cause variability in the energy transformation of the engineered system. These are factors that are difficult, impossible, or too expensive to control. Robust design strives to make the engineered system insensitive to these noise factors.

Based on brainstorming, the project team determined that the following noise factors were the most important: collet condition, spark shield condition, stud gun angle, and workpiece surface contamination. A compound noise strategy was chosen to simulate mild (N_1) and severe (N_2) noise conditions. Table 2 shows the noise factor settings for each of the compound noise levels.

Control Factors

Control factors are the parameters that can be specified by process designers. The project team's brainstorming efforts yielded the following critical factors: gas mixture, spring damper, bushing material, preflow gas duration, postflow gas duration, plunge depth, and gas flow rate.

In addition, the team felt that arc length was an important factor. However, it was excluded from the experiment because changing the arc length would increase arc resistance and thus interact significantly with the feedback control loop of the welding power supply. Therefore, based on existing engineering knowledge, the arc length was held constant at 3/

32 in. for the entire experiment. Level settings for each of the control factors are shown in Table 3.

Orthogonal Array

The use of an orthogonal array allows for efficient sampling of the multidimensional design space. An L_{18} (2×3^7) orthogonal array was chosen for this experiment. The L_{18} array is shown under the inner array portion of Table 4. Each run within the inner array was repeated for each signal and noise level, as shown in the outer array portion of the table. The experiment design required a total of 324 stud welds (18 runs $\times$ 9 signal levels $\times$ 2 noise levels).

3. Results and Data Analysis

The annular weld fillet area of each stud was estimated using the methodology summarized in Figure 3. In addition, each stud was stressed to failure using a torque testing apparatus that closely simulated assembly forces. Torque testing results are presented in Table 5. Figure 4 provides a graphical comparison of annular weld fillet area versus maximum torque for all 324 welded studs. The comparison shows that the fillet area provides a good indication of the relative probability that a stud will survive the assembly torque upper specification limit (15 ft-lb for 10-mm studs).

Out of 224 studs with a fillet area larger than 34 mm^2, only one stud (run 10, M_7, N_1) failed below the assembly torque upper specification limit. This stud exhibited a large hollow fillet (>5 mm wide) on one side of the stud base and no fillet on the other side. Therefore, fillet width consistency also needs to be considered when predicting a stud's ability to survive assembly torque requirements.

Table 2
Compound noise factor settings

Noise Factor	N_1: Mild Noise	N_2: Severe Noise
Collet condition	New	Near end of useful life
Spark shield condition	New	Near end of useful life
Stud gun angle	Perpendicular	Leaning against bushing
Surface contamination	Clean part	Oil-coated part

Table 3
Control factors and levels

	Control Factor	Level 1	Level 2	Level 3
A:	gas mixture	90% argon–10% CO$_2$	98% argon–2% O$_2$	—
B:	spring damper	None	Weak	Mid
C:	bushing material	Plastic	Steel	Brass
D:	preflow gas duration(s)	0.50	0.75	1.0
E:	postflow gas duration(s)	0.25	0.50	1.0
F:	plunge depth (in.)	0.040	0.070	0.100
G*:	dummy factor (result of eliminating arc length from experiment)	—	—	—
H:	gas flow rate (ft^3/h)	10	20	30

Additionally, Figure 4 shows that the transition point between weld failure and stud material failure occurs near 20 ft-lb of applied torque.

SN Ratio

The signal-to-noise ratio (SN or η) is the best indicator of system robustness. As this ratio increases, the engineered system becomes less sensitive to noise factor influences, and the efficiency of the energy transformation increases. The goal of SN analysis is to determine the set of control factor settings that maximize the SN.

Zero-Point Proportional SN Ratio

The experiment was designed to treat the response factor (fillet area) as a zero-point proportional dynamic characteristic. The formula for a zero-point proportional SN ratio is

$$\eta = 10 \log \frac{(1/r)(S_\beta - V_e)}{V_e}$$

where $V_e = S/(k-1)$. The magnitude of the energy input (r) can be calculated as

$$r = \sum_{i=1}^{k} M_i^2$$

where M_i is the welding energy input and k is the number of input levels per run. The variation caused by the linear effect of β (S_β) can be determined as

$$S_\beta = \frac{1}{r} \left(\sum_{i=1}^{k} M_i y_i \right)^2$$

where y_i is the fillet area. To calculate the error variance (V_e), the total variation sum of squares (S_T) and error variation (S_e) need to be calculated as

$$S_T = \sum_{i=1}^{k} y_i^2$$

The error variance can now be calculated:

$$S_e = S_T - S_\beta$$

Calculation Example

The methodology used to determine the SN ratio for each experimental run is demonstrated using data from run 3. Table 6 shows the energy input (M) and fillet area (y) data for each trial of run 3. The calculations used to determine the SN ratio (η) for run 3 are shown below.

$$r = M_1^2 + M_2^2 + \cdots + M_{18}^2 = 6232^2$$
$$+ 6611^2 + \cdots + 15365^2$$
$$= 2{,}158{,}878{,}662$$

Table 4
Experiment array

	Inner Array								Outer Array																	
	Control Factor Level								M_1		M_2		M_3		M_4		M_5		M_6		M_7		M_8		M_9	
Run	A	B	C	D	E	F	G	H	N_1	N_2	N_1	N_2	N_1	N_2	N_1	N_2	N_1	N_2	N_1	N_2	N_1	N_2	N_1	N_2	N_1	N_2
1	1	1	1	1	1	1	1	1																		
2	1	1	2	2	2	2	2	2																		
3	1	1	3	3	3	3	3	3																		
4	1	2	1	1	2	2	3	3																		
5	1	2	2	2	3	3	1	1																		
6	1	2	3	3	1	1	2	2																		
7	1	3	1	2	1	3	2	3																		
8	1	3	2	3	2	1	3	1																		
9	1	3	3	1	3	2	1	2																		
10	2	1	1	3	3	2	2	1																		
11	2	1	2	1	1	3	3	2																		
12	2	1	3	2	2	1	1	3																		
13	2	2	1	2	3	1	3	2																		
14	2	2	2	3	1	2	1	3																		
15	2	2	3	1	2	3	2	1																		
16	2	3	1	3	2	3	1	2																		
17	2	3	2	1	3	1	2	3																		
18	2	3	3	2	1	2	3	1																		

$$S_\beta = \frac{1}{r}(M_1 y_1 + M_2 y_2 + \cdots + M_{18} y_{18})^2$$

$$= \frac{1}{2,158,878,662}[(6232)(16.48) \\ + (6611)(27.93) \\ + \cdots + (15,365)(93.61)]^2$$

$$= 62,058.0$$

$$S_T = y_1^2 + y_2^2 + \cdots + y_{18}^2 = 16.48^2 \\ + 27.93^2 + \cdots + 93.61^2$$

$$= 73,397.8$$

$$S_e = S_T - S_\beta = 73,397.8 - 62,058.0$$

$$= 11,339.8$$

$$V_e = \frac{S_e}{k-1} = \frac{11,339.8}{18-1} = 667.0$$

$$\eta = 10 \log \frac{[1/(2.16)(10^9)](62,058.0 - 11,339.8)}{667.0}$$

$$= -73.70 \text{ dB}$$

The same methodology was used to calculate SN ratios for all 18 runs.

Response Analysis

The calculated SN ratio and beta values for each experimental run are displayed in Table 7. The average SN ratio for each control factor level is shown

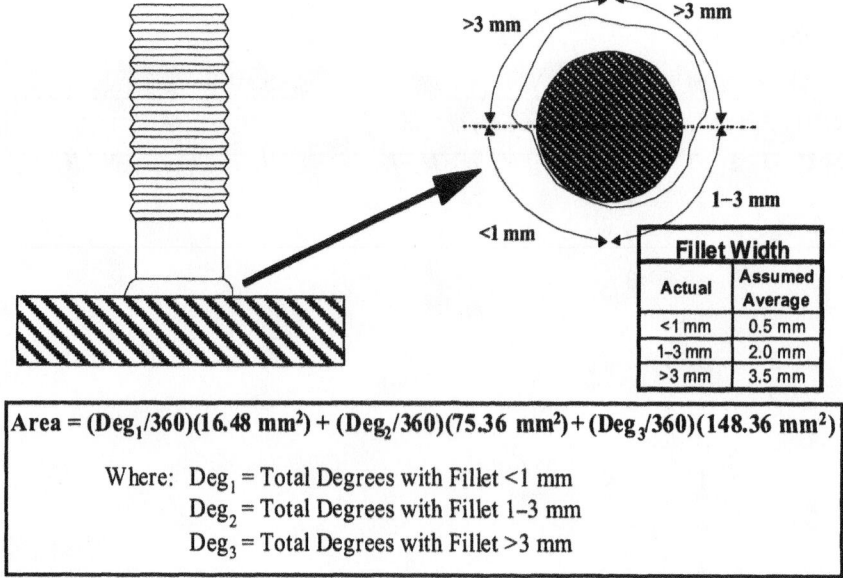

Figure 3
Annular weld fillet area calculation

in Table 8. The data from the table are shown graphically in Figure 5. It can be seen that the spring damper (B), bushing material (C), and gas flow rate (H) have the largest effect on the SN ratio. It should also be noted that the dummy factor (G) had the lowest impact on the SN ratio, thus providing an indication that interactions between main effects are minimal.

Optimal Control Factor Settings

The optimal control factor settings were determined by selecting the combination of levels that maximize the SN ratio. Table 9 shows the optimum control factor levels and current process control factor levels predicted. The SN ratios predicted and gain expected are also presented in the table. Calculations for the values predicted are shown below.

optimal SN ratio predicted
$= B_2 + C_1 + H_2 - 2(\text{av. } \eta)$
$= (-67.76 \text{ dB}) + (-68.30 \text{ dB}) + (-68.36 \text{ dB})$
$\quad - 2(-69.48 \text{ dB})$
$= -65.46 \text{ dB}$

current SN ratio
$= B_1 + C_2 + H_1 - 2(\text{av. } \eta)$
$= (-71.59 \text{ dB}) + (-69.38 \text{ dB}) + (-70.68 \text{ dB})$
$\quad - 2(-69.48 \text{ dB})$
$= -72.69 \text{ dB}$

4. Beta Analysis

The value of β was calculated so that the squares of the distances between the right and left sides of the ideal function equation were minimized (least squares method). The calculation method is

$$\beta = \frac{1}{r}\left(\sum_{i=1}^{k} M_i y_i\right)$$

The β values calculated for each experimental run are displayed in Table 8. The average β values for each control factor level are shown in Table 10. The data from Table 10 are shown graphically in Figure 6. It can be seen that the spring damper (B) has the largest impact on β.

Table 5
Maximum torque prior to stud failure

	Maximum Torque (ft-lb)																	
	M_1		M_2		M_3		M_4		M_5		M_6		M_7		M_8		M_9	
Run	N_1	N_2	N_1	N_2	N_1	N_2	N_1	N_2	N_1	N_2	N_1	N_2	N_1	N_2	N_1	N_2	N_1	N_2
1	29.9	51.6	38.0	38.9	24.0	23.7	34.0	31.6	43.9	25.0	53.5	27.4	43.7	31.2	51.8	33.8	51.0	37.2
2	54.4	40.0	28.0	36.7	49.3	33.6	31.8	27.2	46.2	33.4	49.9	39.2	52.8	40.4	45.7	28.9	53.2	35.8
3	12.5	21.4	15.4	33.0	12.5	32.1	26.5	32.7	35.4	38.7	41.4	31.5	44.2	30.0	38.1	29.7	52.5	36.1
4	36.8	35.4	45.2	32.1	37.6	42.9	27.0	44.1	26.2	38.4	39.2	33.1	36.1	35.7	37.8	39.7	43.9	38.8
5	39.4	33.6	36.8	36.3	37.2	26.9	41.5	31.3	43.8	39.0	10.0	31.5	49.2	35.1	40.7	35.9	43.1	37.6
6	44.5	43.6	41.0	38.0	45.5	34.4	44.8	31.6	50.3	29.6	50.6	40.8	50.4	43.9	44.3	39.6	44.3	43.5
7	34.1	38.8	22.5	16.7	22.2	4.2	33.7	41.3	36.3	36.1	37.8	34.7	38.0	41.7	38.9	34.8	38.0	37.2
8	39.9	34.7	35.9	32.1	31.1	4.0	25.5	27.2	43.1	33.8	46.6	34.2	45.9	42.7	54.9	37.1	45.9	36.0
9	36.2	26.3	26.2	35.1	38.2	34.5	33.8	32.8	27.0	31.0	26.0	30.0	38.5	28.4	35.3	25.5	43.1	29.9
10	33.1	28.4	38.7	28.1	36.6	7.5	35.0	26.8	34.0	22.9	39.2	34.0	10.0	26.3	36.0	30.8	39.6	38.7
11	22.8	31.5	18.8	32.8	31.2	38.5	35.6	39.5	34.2	32.9	39.3	17.5	38.6	33.5	37.5	31.0	41.3	35.5
12	10.0	29.0	32.4	27.2	32.9	36.9	38.5	32.9	35.9	23.7	29.8	32.9	27.3	34.1	32.2	36.6	39.9	40.8
13	37.5	33.7	35.2	33.8	41.9	37.4	38.1	35.9	36.7	32.9	40.3	35.9	32.4	37.1	39.2	38.2	36.3	35.2
14	25.9	31.2	8.5	34.1	33.5	34.6	33.3	33.5	33.0	31.7	29.4	31.2	32.4	34.8	33.3	35.5	29.8	35.8
15	8.5	35.2	27.6	31.4	31.7	33.1	32.4	32.1	31.7	29.9	37.8	26.1	37.3	36.1	38.5	33.5	34.2	29.2
16	31.2	34.8	29.1	37.3	32.2	36.3	39.2	39.6	36.6	40.4	38.2	31.2	34.1	36.3	36.9	38.7	36.3	40.2
17	10.0	33.7	36.8	35.2	16.7	32.4	35.1	32.4	29.8	24.9	37.2	35.0	39.2	29.6	37.7	26.8	36.6	28.0
18	10.0	32.0	31.7	30.9	31.8	32.9	38.4	36.5	13.2	38.2	31.5	28.5	36.1	37.0	33.3	31.7	36.8	36.7

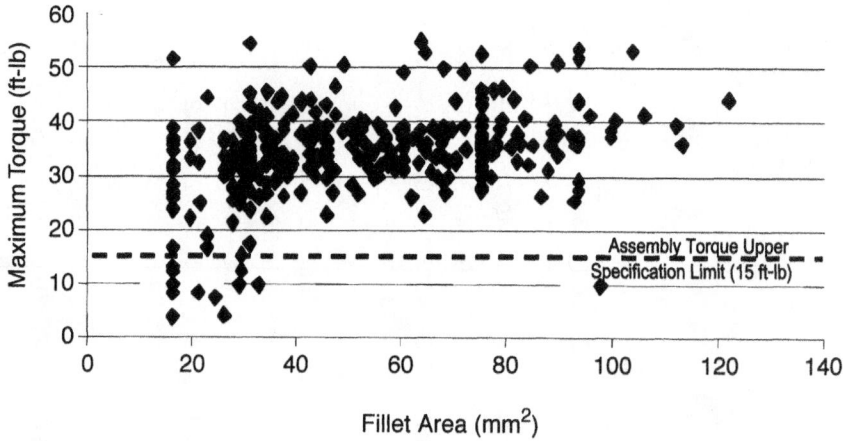

Figure 4
Weld fillet area versus maximum torque prior to stud failure

Table 6
Data for run 3

Trial	Actual Current (A)	Actual Time (s)	Voltage (V)	Energy, M (J)	Area, y Fillet (mm²)
M_1	584.2	0.35	30.5	6,232	16.48
M_2	566.1	0.35	33.3	6,611	27.93
M_3	575.1	0.39	31.8	7,139	29.56
M_4	560.1	0.38	35.5	7,585	34.47
M_5	583.5	0.43	31.0	7,786	29.56
M_6	563.2	0.43	34.8	8,460	56.45
M_7	737.3	0.35	38.5	9,935	16.48
M_8	760.4	0.35	33.7	8,958	43.43
M_9	731.9	0.38	39.1	10,875	78.04
M_{10}	756.6	0.39	34.3	10,185	29.56
M_{11}	745.2	0.30	37.2	8,305	105.78
M_{12}	761.6	0.44	33.3	11,112	37.74
M_{13}	885.5	0.35	42.1	13,060	121.93
M_{14}	904.7	0.35	39.7	12,655	38.46
M_{15}	887.1	0.39	42.0	14,517	89.55
M_{16}	909.8	0.39	38.9	13,894	75.36
M_{17}	884.5	0.44	42.4	16,482	75.36
M_{18}	909.9	0.43	39.0	15,365	93.61

Table 7
SN ratio and beta values for each run

	Control Factor									
Run	A	B	C	D	E	F	G	H	η	β
1	1	1	1	1	1	1	1	1	−71.74	0.00542
2	1	1	2	2	2	2	2	2	−69.56	0.00544
3	1	1	3	3	3	3	3	3	−73.70	0.00536
4	1	2	1	1	2	2	3	3	−64.27	0.00476
5	1	2	2	2	3	3	1	1	−67.56	0.00478
6	1	2	3	3	1	1	2	2	−68.07	0.00438
7	1	3	1	2	1	3	2	3	−68.34	0.00397
8	1	3	2	3	2	1	3	1	−69.63	0.00382
9	1	3	3	1	3	2	1	2	−68.69	0.00420
10	2	1	1	3	3	2	2	1	−72.39	0.00522
11	2	1	2	1	1	3	3	2	−70.76	0.00504
12	2	1	3	2	2	1	1	3	−71.40	0.00540
13	2	2	1	2	3	1	3	2	−65.87	0.00450
14	2	2	2	3	1	2	1	3	−70.45	0.00417
15	2	2	3	1	2	3	2	1	−70.34	0.00426
16	2	3	1	3	2	3	1	2	−67.19	0.00421
17	2	3	2	1	3	1	2	3	−68.31	0.00407
18	2	3	3	2	1	2	3	1	−72.45	0.00347
								Average:	−69.48	0.00458

Table 8
SN ratio response table

	Control Factor							
Level	A	B	C	D	E	F	G*	H
1	−69.06	−71.59	−68.30	−69.02	−70.30	−69.17	−69.51	−70.69
2	−69.91	−67.76	−69.38	−69.20	−68.73	−69.64	−69.50	−68.36
3	—	−69.10	−70.78	−70.24	−69.42	−69.65	−69.45	−69.41
Δ	0.84	3.83	2.48	1.22	1.57	0.48	0.06	2.33

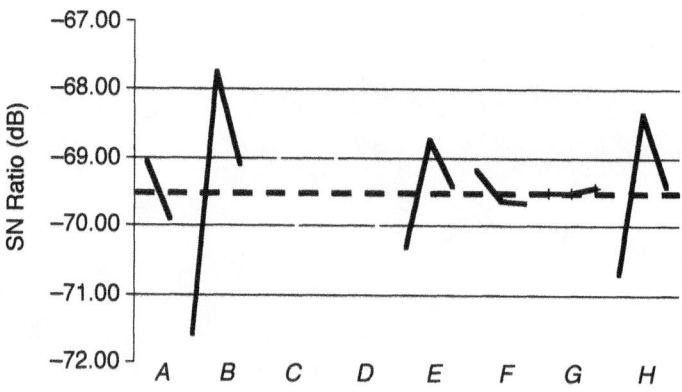

Figure 5
SN ratio effect plot

Table 9
Optimal control factor settings and SN ratio predictions

Configuration	Gas Mixture	Spring Damper	Bushing Material	Preflow Gas Duration (s)	Postflow Gas Duration (s)	Plunge Depth (in.)	Gas Flow Rate (ft³/h)	SN Ratio Predicted (dB)
Optimal	90/10	Weak	Plastic	0.50	0.50	0.040	20	−65.46
Current	98/2	None	Steel	0.50	0.50	0.070	10	−72.69
							Gain predicted:	7.23

Table 10
Beta response

Level	Control Factor							
	A	B	C	D	E	F	G*	H
1	0.00468	0.00531	0.00468	0.00463	0.00441	0.00460	0.00470	0.00450
2	0.00448	0.00448	0.00455	0.00459	0.00465	0.00454	0.00456	0.00463
3	—	0.00396	0.00451	0.00453	0.00469	0.00460	0.00449	0.00462
Δ	0.00020	0.00136	0.00017	0.00010	0.00028	0.00006	0.00020	0.00013

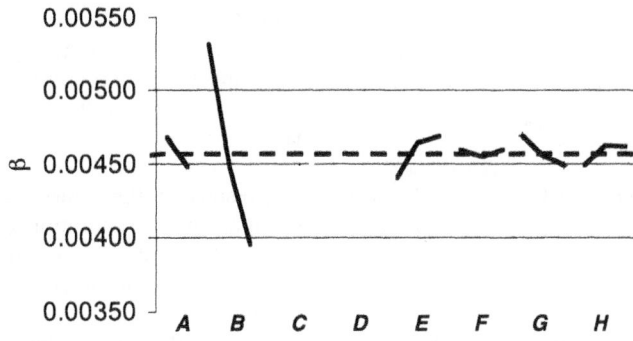

Figure 6
Beta effect plot

Table 11
Predicted versus confirmed results

	Predicted		Confirmed	
	SN Ratio (dB)	β	SN Ratio (dB)	β
Optimal	−65.46	0.00458	−66.85	0.00420
Current	−72.69	0.00520	−73.23	0.00518
Gain	7.23		6.38	

Beta Predictions

In the production environment, the input energy will be set to an optimum level and held constant. Therefore, β is relatively unimportant in this experiment. However, a comparison of predicted β values to confirmed values will provide increased insight into the validity of the experimental results. Therefore, β values were predicted using the optimal control factor levels as identified in the SN ratio analysis. Beta predictions are as follows:

Optimal: 0.00458

Current: 0.00520

Calculations for the values predicted are:

optimal SN ratio predicted
$$= B_2 + C_1 + H_2 - 2(\text{av. } \eta)$$
$$= 0.00448 + 0.00463 + 0.00463 - (2)(0.00458)$$
$$= 0.00458$$

current SN ratio
$$= B_1 + C_2 + H_1 - 2(\text{av. } \eta)$$
$$= 0.00531 + 0.00455 + 0.00450 - (2)(0.00458)$$
$$= 0.00520$$

5. Confirmation Test

The confirmation test was conducted in three steps:

1. Testing at *current* control factor settings across all 18 energy input levels:

   Control factor settings: $A_1 B_1 C_2 D_1 E_1 F_2 H_1$

2. Testing at *optimal* control factor settings predicted across all 18 energy input levels:

   Control factor settings: $A_1 B_2 C_1 D_2 E_2 F_1 H_2$

3. Testing at *optimal* control factor settings predicted at the *optimum* energy input level:

Table 12

Constant-set-point versus full-energy input range

Confirmed Optimal	SN Ratio (dB)	β
Full energy input range	−66.85	0.00420
Constant set point	−66.86	0.00475

Control factor settings: $A_1 B_2 C_1 D_2 E_2 F_1 H_2$

Signal factor settings: 840 A, 0.35 s

Each confirmation test was conducted in exactly the same manner as each run in the original L_{18} experimental array. SN ratio results from the initial two steps of the confirmation run are compared to their predicted values in Table 11. The confirmed results compare favorably to the values predicted.

The 6.38-dB gain between current and optimal conditions translates into a 52.1% reduction in energy transformation variability, as shown by the following calculation:

$$\text{variability reduction} = 1 - 0.5^{\text{gain}/6} = 1 - 0.5^{6.38/6}$$
$$= 52.1\%$$

Based on weld fillet area, torque testing, and general visual inspection results of studs welded under optimal parameter settings, the team chose 840 A and 0.35 s as the most desirable signal factor settings. Therefore, a third confirmation run was performed at a constant 840 A, 0.35 s set point. Even

though the input energy set point was held constant, the actual energy input varied from stud to stud due to the inherent variability of the power supply. The one positive effect of this inherent variability was that it enabled calculation of a SN ratio and β for the run. Table 12 shows a comparison of the results of the constant-set-point run and the original optimal condition confirmation run utilizing the full input energy range. It can be observed that the results of the runs were virtually identical.

Figure 7 shows the torque testing results from studs welded under optimal control factor settings and the constant input energy setpoint of 840 A, 0.35 s. For all studs in this run, the weld yield strength was greater than the stud material yield strength.

The average observed maximum torque prior to failure (39.6 ft-lb) is 2.6 times larger than the assembly torque upper specification limit. The optimal process data exhibit a state of statistical control; therefore the process capability index (C_p) can be calculated as follows:

$$C_p = \frac{\text{average max. torque} - \text{assembly torque upper spec. limit}}{\sigma}$$

where σ is the estimated standard deviation (mRbar/D_2 method). Therefore, for the optimal process,

$$C_p = \frac{39.6 - 15.0}{(3)(2.72)}$$
$$= 9.03$$

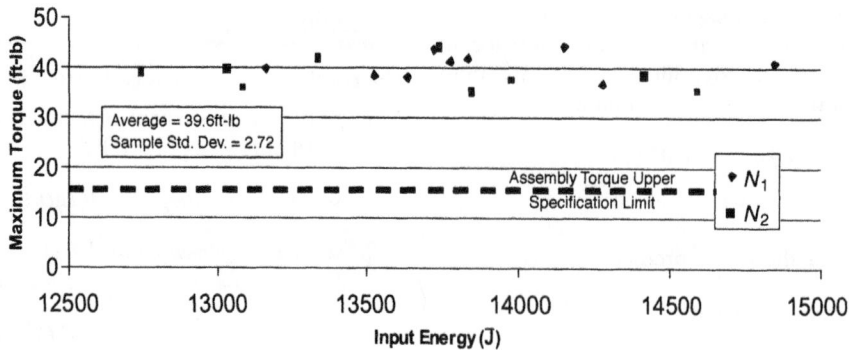

Figure 7

Maximum torque prior to stud failure: Optimal process

6. Conclusions

Process parameter optimization utilizing robust design methodology clearly resulted in a substantial improvement in the gas-arc welding process. The optimal combination of control factor settings was determined, resulting in a 52.1% process variability reduction and an average stud weld strength of 2.6 times the level required by the assembly operation. Of the seven control factors tested, only two factors had been set at their optimal level. Additionally, the preexperiment welding energy input settings (700 A, 0.4 s) were found to be suboptimal. Implementation of the optimized process parameters is expected to result in significant cost savings (greater than $30,000), due to the elimination of a secondary welding operation and a significant reduction in downstream stud strength failures.

References

American Supplier Institute, 1998. *Robust Design Using Taguchi Methods: Workshop Manual.* Livonia, MI: American Supplier Institute, pp. I1–II36.

American Welding Society, 1993. *ANSI/AWS C5.4-93: Recommended Practices for Stud Welding.* Miami, FL: American Welding Society.

Yuin Wu, 1999. *Robust Design Using Taguchi Methods.* Livonia, MI: American Supplier Institute, pp. 47–51.

This case study is contributed by Jeff Stankiewicz.

CASE 43

Optimization of Molding Conditions of Thick-Walled Products

Abstract: In an injection molding process, products of a thickness over 5 mm are called thick-walled products. Many of these products have problems such as sink, void, or abnormal shrinkage. To avoid such defects, molding time and cooling time have to be extended, resulting in a long overall molding cycle time. To improve the process, the generic function was considered where material must be filled uniformly to any spot inside the mold. After optimization, the molding cycle time was reduced and the quality loss was cut to one-third.

1. Introduction

Traditionally, the generic function of injection molding has been considered transformability to a mold or the capability of molding a product proportional to the mold shape. However, the resin caster discussed in this study is required to have sufficient strength because it is used as a cart carrying baggage. In addition, since internal voids occur quite often because the caster has a considerably thick wall, instead of the concept of transformability we selected as a generic function even filling of the inside of a mold with resin, that is, uniformity of density (specific gravity).

As a measuring method for analyzing density uniformity, we used a specific gravity measurement based on an underwater weighing method. A molded product was cut in five pieces (Figure 1). We chose in-the-air weight as a signal factor, M, and underwater weight as the output, y.

More specifically, splitting up a molded product and measuring both in-the-air weight (M) and underwater weight (y), we set the data to an ideal relationship equation (Figure 2).

The mixing ratio of recycled material was used as a noise factor. It affects the fluidity of resin

material. At level 1 the mixing ratio is 1, and at level 2 it is 50. The control factors are listed in Table 1.

2. SN Ratio

The measured data are listed in Table 2. The data analysis procedure was as follows.

Total variation:

$$S_T = 19.1^2 + 22.3^2 + \cdots + 82.8^2 + 176.6^2$$

$$= 82{,}758.13 \qquad (f = 10) \tag{1}$$

Effective divider:

$$r_1 = 22.2^2 + 25.1^2 + \cdots + 202.8^2 = 54{,}151.02$$

$$r_2 = 23.3^2 + 26.0^2 + \cdots + 200.9^2 = 52{,}852.51 \tag{2}$$

Linear equations:

$$L_1 = (22.2)(19.1) + \cdots + (202.8)(178.5)$$

$$= 47{,}668.88$$

$$L_2 = (23.3)(20.2) + \cdots + (200.9)(176.6)$$

$$= 46{,}434.04 \tag{3}$$

Variation of proportional term:

$$S_\beta = \frac{(L_1 + L_2)^2}{r_1 + r_2} = 82{,}757.64 \qquad (f = 1) \quad (4)$$

Variation of proportional terms due to noise:

$$S_{N\beta} = \frac{L_1^2}{r_1} + \frac{L_2^2}{r_2} - S_\beta = \frac{47{,}668.88^2}{5451.02} + \frac{46{,}434.04^2}{52{,}852.51}$$
$$- 82{,}757.64 = 0.08 \qquad (f = 1) \quad (5)$$

Error variation:

$$S_e = S_T - S_\beta - S_{\beta N} = 82{,}758.13 - 82{,}757.64$$
$$- 0.08 = 0.41 \qquad (f = 8) \quad (6)$$

Error variance:

$$V_e = \frac{S_e}{8} = \frac{0.41}{8} = 0.05 \quad (7)$$

Total error variance after pooling:

$$V_N = \frac{S_e + S_{\beta N}}{n - 1} = \frac{0.41 + 0.08}{9} = 0.05 \quad (8)$$

SN ratio:

$$\eta = 10 \log \frac{[1/(r_1 + r_2)](S_\beta - V_e)}{V_N} = 11.89 \text{ dB} \quad (9)$$

Sensitivity:

$$S = 10 \log \frac{1}{r_1 + r_2} (S_\beta - V_e) = -1.11 \text{ dB} \quad (10)$$

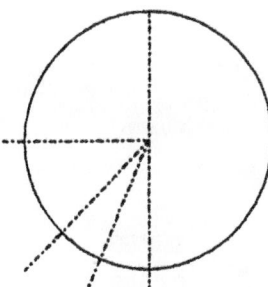

Figure 1
Division of thick-walled molded product

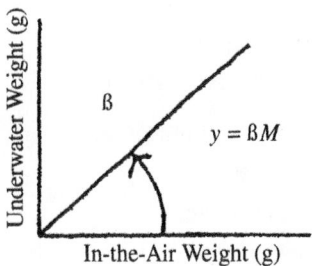

Figure 2
Ideal relationship

3. Optimal Condition and Confirmatory Experiment

The injection molding conditions were listed as control factors (Table 3). Figure 3 shows the shape of molded product. Nylon plastic was used as the resin material. Control factors were assigned to an L_{18} orthogonal array and the noise factor to the outer array. Eighteen experiments were conducted twice.

Due to poor setup conditions in experiments 1, 4, 11, and 17, the resin was not fully charged in the mold, so no products were obtained from the mold. Because of no data available, these conditions were treated as incomplete data. Analyses were made using the following two methods:

1. Find the worst SN ratio from experiment 1 to experiment 18; experiment 8 is the worst (-6.5). Double the value (-13). Put this value into experiments 1, 4, 11, and 17 to calculate the averages.

2. Ignore the missing values to calculate level averages. For example, the average of A_1 is calculated as $A_1 = $ (total of experiments 1 to 9 except experiments 1 and 4) $\div$ 7.

Based on these analyses, we created the response graphs in Figure 4.

The optimal configuration is estimated as follows:

1. Select levels with less missing data for each factor.

2. If the numbers of missing data are the same, select the level with a high SN ratio for the

Table 1
Control factors and levels

Control Factor	Level		
	1	2	3
A: injection pressure (%)	Low	High	—
B: resin temperature (°C)	Low	Mid	High
C: mold temperature (°C)	B − 200	B − 190	B − 180
D: injection speed V_1 (%)	Low	Mid	High
E: injection speed V_2 (%)	<D	D	<D
F: injection speed V_3 (%)	E	<F_1	<F_2
G: condition holding pressure (%) × time (s)	Low × long	Mid × mid	High × short
H: cooling time (s)	Short	Mid	High

Table 2
Example of measured data

Noise Factor [Mixing Ratio of Recycled Material (wt %)]		Signal Factor (M) and Output (y) (g)				
0	M	22.2	25.1	50.3	96.8	202.8
	y	19.1	22.3	44.3	85.3	178.5
50	M	23.3	26.0	48.4	94.5	200.9
	y	20.2	23.0	42.6	82.8	176.6

Table 3
Estimation and confirmation (dB)

	SN Ratio		Sensitivity	
	Estimation	Confirmation	Estimation	Confirmation
Optimal configuration 1	20.1662	14.2205	−1.0344	−1.1160
Optimal configuration 2	—	14.5307	—	−1.1170
Comparison	6.4157	9.8173	−1.1193	−1.0925
Gain 1	13.7505	4.4032	0.0850	−0.0235
Gain 2	—	4.7134	—	−0.0245

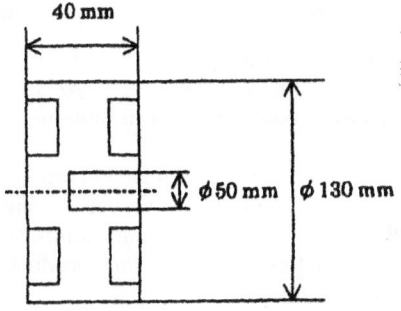

Figure 3
Shape of molded product

case when factor effects are estimated ignoring missing data.

Based on this procedure, we chose the following optimal configuration: $A_2B_2C_3D_2E_3F_3G_2H_2$. Under this configuration, the SN ratio is estimated as

$$\eta = 20.1662 \text{ dB}$$

Next, for the current configuration, $A_1B_2C_2D_1E_2F_3G_2H_2$, we calculated the SN ratio as

$$\eta = 6.4157 \text{ dB}$$

Then the improvements in gain under the optimal configuration were computed as:

gain in SN ratio for the current configuration
$$= 20.1662 - 6.4157 = 13.7505 \text{ dB}$$

Although we conducted a confirmatory experiment under the optimal configuration estimated above, assuming that a significant reduction of molding time cannot be expected and the level range of cooling time, H, is set up too wide, for factors A to G, we defined the experimental results themselves as optimal levels. For factor H, we regarded an intermediate value between levels 1 and 2 (10 minutes less than the current cooling time) for another optimal calculation. This table reveals that we obtained fairly good reproducibility. Additionally, since optimal configuration 2, which was added to shorten molding time (or to reduce molding cost), has a higher SN ratio by 0.3 dB than the estimated optimal configuration 1 does, we concluded that it is possible to shorten the molding time under optimal configuration 2.

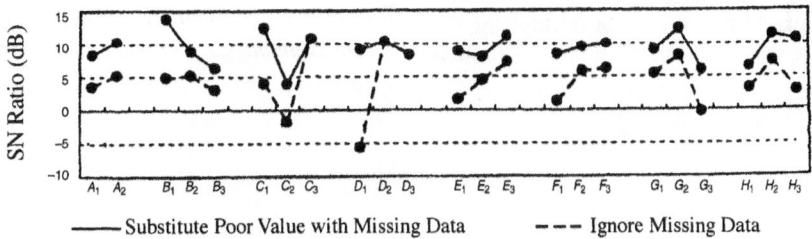

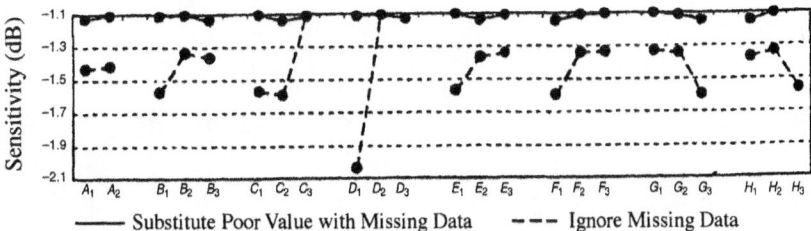

Figure 4
Response graphs

4. Economic Effect of Improvement

Now we computed a loss according to the effects obtained from this experiment.

$$\text{loss function, } L = \frac{A}{\Delta^2 \sigma^2}$$

where A (loss when a defective product is produced) is assumed to be 400 yen/unit, Δ (tolerance in the process) is assumed to be 0.1, and σ (variance) is computed by the result of the confirmatory experiment. That is, converting $\eta = \beta^2/\sigma^2$, we obtain $\sigma^2 = \beta^2/\eta$.

The loss function for the case of the reference configuration is

$$L_1 = \frac{400}{1^2}(0.081) = 3244 \text{ yen/unit}$$

In contrast, the loss function for the optimal configuration is

$$L_2 = \frac{400}{1^2}(0.0272) = 1088 \text{ yen/unit}$$

Therefore, we can reduce the loss by two-thirds.

On the other hand, considering that the molding time is reduced by 10 s/unit, by multiplying this by the unit cost, we can estimate the monetary benefit as

benefit by shortened time = 10 yen/unit

where 10/3600 is the shortened time (hours) and 3600 is the hourly molding cost (yen/hour). As a result, a 10-yen cost reduction in molding for each product is expected.

Converting this monetary benefit into a yearly amount, we obtain the following benefit due to the loss reduction in reference to our monthly production volume of 1000 units or annual production volume of 12,000 units:

$$(3244 - 1088)(12,000) = 25,872,000 \text{ yen/year}$$

At the same time, the monetary benefit due to the shortened molding time is

$$(10)(12,000) = 120,000 \text{ yen/year}$$

In the end, we can anticipate 25,992,000 yen/year.

Reference

Shinji Hanada, Takayoshi Matsunaga, Hirokazu Sakou, and Yoshihisa Takano, 1999. Optimization of molding condition for thick walled mold. *Quality Engineering*, Vol. 7, No. 6, pp. 56–62.

This case study is contributed by Shinji Hanada, Takayoshi Matsunaga, Hirokazu Sakou, and Yoshihisa Takano.

Quality Improvement of an Electrodeposited Process for Magnet Production

Abstract: In a conventional development process of electrodeposition, improvement has been focused mainly on pinholes as one of the quality characteristics. However, in our research, we regarded a function of coating formation in the electrodeposition process as an input and output of energy. By reducing the variability in coating thickness on each side and minimizing the difference in coating thickness in both sides of a magnet, we aimed to minimize pinholes as a quality characteristic.

1. Introduction

The magnet for motors is electrodeposited to prevent it from releasing dust and from becoming rusty. There are coating methods other than electrodeposition: for example, spraying and plating. One of the major advantages of electrodeposition is its good "throwing power," which enables us to obtain uniform coating thickness even on a complicated shape. However, since the rare-earth-bonded magnet used is compression molded from metal powder, it has plenty of hollow holes inside. Therefore, when its coating is baked, gas frequently comes out of the inside, leading to the development of tiny pinholes, ranging in diameter from a few micrometers to a few dozen on the coat.

To electrodeposit a magnet, water-soluble resin made primarily of epoxy resin is used. After charging the magnet negatively and the electrode positively by applying voltage in its solution, we can obtain water electrolysis, which raises the pH of the solution, with hydrogen generated around the magnet. On the other hand, since the coating material in the solution is charged positively, it is attracted by the magnet and consequently condensed and educed on the surface of the magnet due to alkali.

The coating thus formed on the surface has many minute holes generated by O_2 gas. Because of current flowing through these holes, new coatings are reduced continuously. This is regarded as a growing coating process. The educed coatings per se have high resistance. As the coatings develop, the number of holes for O_2 gas decreases and a smaller amount of current flows. Therefore, since the current tends to flow from a high-resistance area to a low-resistance area, it forms new coatings in areas where coatings are not well developed. This phenomenon, throwing power, is peculiar to electrodeposition.

2. Generic Function and Measurement Characteristics

In a conventional electrodeposition development process, improvement has focused mainly on pinholes as one quality characteristic. However, in our research, we regarded a function of coating formation in the electrodeposition process as an input and output of energy.

Most magnets are ring-shaped and have electrode terminals on the outer circumference. Therefore, electrical current flows less easily in the inner circumference. In addition, because of the circulation of the coating solution, the inner side is

Table 1
Signal and noise factors

Signal Factor:	M_1	M_2	M_3
Coulomb value	2	3	4
Noise Factor:	N_1	N_2	
1: measured side	Outer side	Inner side	
2: deterioration of solution	Yes	No	
3. coating thickness	Maximum	Minimum	

disadvantageous. That is, coating forms differently on the outer and inner circumferences. As a result, on the inner side, where a thinner coating develops, a defect related to pinholes occurs more often than on the outer side. By reducing the variability in coating thickness on each side and minimizing the difference in coating thickness on both sides of a magnet, we aimed to minimize pinholes as a quality characteristic.

As Table 1 shows, after setting a Coulomb value as an integral of current to a signal factor, we conducted an analysis based on dynamic characteristics by measuring coating thickness with a β-ray thickness measuring instrument. As the noise factor, we used a compounded factor consisting of coating thickness on both the outer and inner sides of a

magnet, maximum and minimum thickness on each side, and deterioration of solution. Figure 1 outlines the experimentation on electrodeposition. A beaker was considered an easy experimental device as a downscale of the mass-production line.

Setting up two ring-shaped magnets at the same time, we measured each. Next, using magnets that were longer than their outer diameter, we attempted to simulate the situation of our mass-production line (producing tubular magnets).

3. Calculation of SN Ratio

Table 2 shows data examples of experiment 1 in an L_{18} orthogonal array. Using these data we computed the SN ratios and sensitivities as follows.

Total variation:

$$S_T = 9.5^2 + 9.3^2 + \cdots + 13.6^2 + 14.5^2$$

$$= 2188.01 \quad (f = 12) \quad (1)$$

Linear equations:

$$L_1 = (2)(9.5 + 9.3) + \cdots + (4)(21.1 + 20.6)$$

$$= 291.7$$

$$L_2 = (2)(6.7 + 6.7) + \cdots + (4)(13.6 + 14.5)$$

$$= 204.0 \quad (2)$$

Effective divider:

$$r = 2^2 + 3^2 + 4^2 = 29 \quad (3)$$

Variation of proportional term:

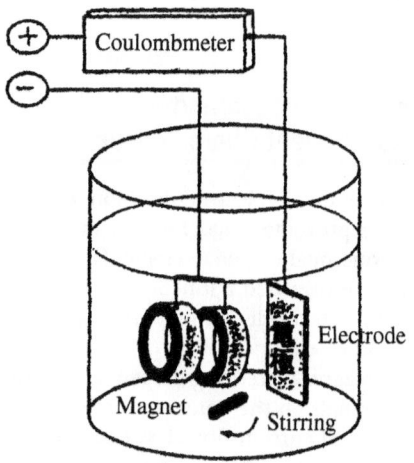

Figure 1
Experimental device

Table 2
Data of experiment 1 on coating thickness
(μm)

	M_1 (2)		M_2 (3)		M_3 (4)	
N_1	9.5	9.3	14.6	14.5	21.1	20.6
N_2	6.7	6.7	10.7	10.9	13.6	14.5

$$S_\beta = \frac{(L_1 + L_2)^2}{(2)(2r)} = \frac{(291.7 + 204.0)^2}{(2)(2)(29)}$$

$$= 21{,}18.26 \qquad (f = 1) \qquad (4)$$

Variation of differences between proportional terms:

$$S_{N\beta} = \frac{L_1^2 + L_2^2}{2r} - S_\beta$$

$$= \frac{291.7^2 + 204.0^2}{(2)(29)} - 2118.26$$

$$= 66.31 \qquad (f = 1) \qquad (5)$$

Error variation:

$$S_e = S_T - S_\beta - S_{N\beta} = 2188.01 - 2118.26 - 66.31$$

$$= 3.44 \qquad (f = 10) \qquad (6)$$

Error variance:

$$V_e = \frac{S_e}{(2)(2)(3) - 2} = \frac{3.44}{10} = 0.34 \qquad (7)$$

Total error variance:

$$V_N = \frac{S_T - S_\beta}{(2)(2)(3) - 1}$$

$$= \frac{2188.01 - 2118.26}{11} = 6.34 \qquad (8)$$

SN ratio:

$$\eta = 10 \log \frac{(1/4r)(S_\beta - V_e)}{V_N}$$

$$= 10 \log \frac{[1/(4)(29)](2118.26 - 0.34)}{6.34}$$

$$= 4.59 \text{ dB} \qquad (9)$$

Sensitivity:

$$S = 10 \log \frac{1}{4r}(S_\beta - V_e)$$

$$= 10 \log \frac{1}{(4)(29)}(2118.26 - 0.34)$$

$$= 12.61 \text{ dB} \qquad (10)$$

Table 3
Control factors and levels

			Level		
	Signal		1	2	3
A:	distance between electrodes		Far	Close	—
B:	temperature		Low	Mid[a]	High
e:	—		—	—	—
C:	NV value		Small	Mid[a]	Large
D:	amount of ash		Small	Mid[a]	Large
E:	amount of solvent		Small	Mid[a]	Large
F:	flow of solution		Small	Mid	High
G:	voltage (V)		115	175	235

[a] Current level.

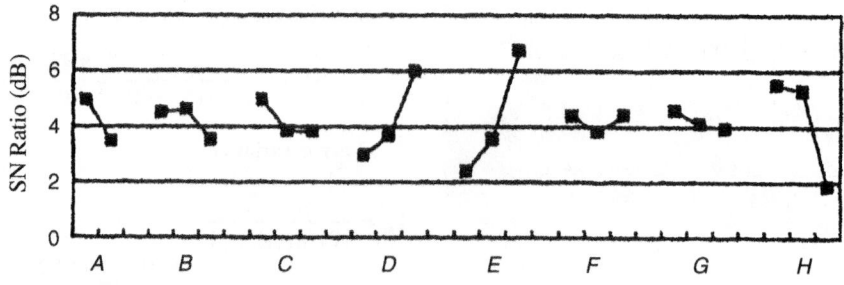

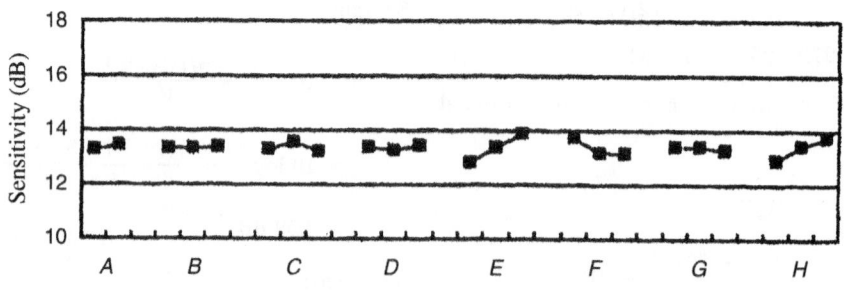

Figure 2
Response graphs

4. Optimal Configuration and Confirmatory Experiment

Table 3 shows the control factors selected. All seven factors were selected based on our existing technical knowledge. Figure 2 shows the response graphs for the SN ratio and sensitivity. Table 4 and Figure 3 show estimations of the SN ratio and sensitivity and results obtained from the confirmatory experiment.

Looking at these results, we can see good reproducibility in gains. In addition, comparing the re-

sults with our technical knowledge, many of them are regarded as quite reasonable. While the linearity between a Coulomb value and coating thickness had been fairly good under the current configuration, the SN ratio was improved under the optimal configuration. Because of a reduced difference in coating thickness between the outer and inner circumferences, pinhole defects are expected to decrease. On the other hand, the sensitivity is never improved. Since this research focused more on quality improvement without major changes in the

Table 4
Results of estimation and confirmatory experiment

Configuration	SN Ratio		Sensitivity	
	Estimation	Confirmation	Estimation	Confirmation
Optimal	8.49	11.70	13.95	14.28
Current	3.01	7.42	13.27	14.16
Gain	5.48	4.28	0.68	0.12

Current Configuration

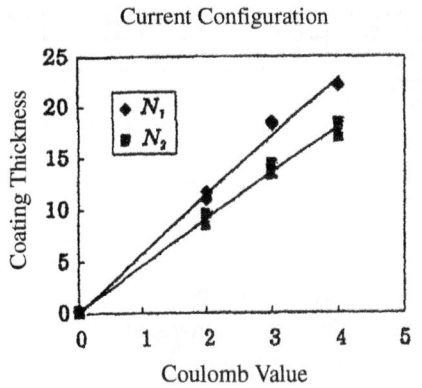

Optimal Configuration

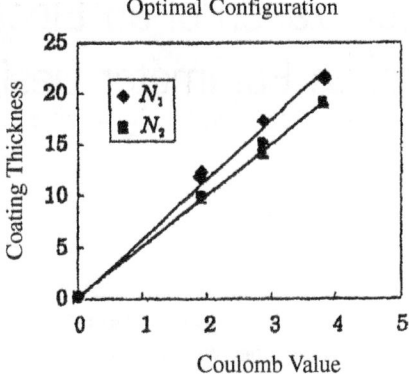

Figure 3
Results of confirmatory experiment

existing production line, we have not chosen factors that influence sensitivity significantly.

As a result of deploying the optimal configuration obtained in this research, we have succeeded in slashing the number of pinhole-related defects remarkably, by 90%. On the other hand, by stabilizing the coating thickness, we have also enhanced the dimensional accuracy of products.

Reference

Hayato Shirai and Yukihiko Shiobara, 1999. Quality improvement of electro-deposited magnets. *Quality Engineering*, Vol. 7, No. 1, pp. 38–42.

This case study is contributed by Hayato Shirai.

Optimization of an Electrical Encapsulation Process through Parameter Design

Abstract: A process used to generate an encapsulant layer, which provides electrical isolation between elements on the outer surface of a night-vision image-intensifier tube, was optimized using Taguchi's L_{18} orthogonal array and SN ratio. This parameter design effort was launched in response to the need to optimize an encapsulation process using a lower processing time and temperature to cure the encapsulant. The SN ratio of the engineered encapsulation system was increased 7.2 dB compared to baseline conditions, and β, the slope of the ideal function, was reduced from 0.6298 to 0.1985, or 68%. As a result of this process improvement effort, a new encapsulant was qualified using a process temperature reduced by 20% and process cycle time reduced by 40%.

1. Introduction

We used Taguchi's dynamic parameter design approach to optimize an electrical encapsulation process. Here we present a parameter design effort focused on reducing the process cycle time and curing temperature while maintaining the same reliability performance for the image sensor module used in the night-viewing product lines at ITT Night Vision.

Figure 1 shows an F5050 system that has two sensor modules (Figure 2), one located behind each objective lens opposite the eyepiece. At the heart of each module is an image intensifier tube (Figure 3), whose outer surface has exposed electrical elements that must be covered to provide environmental isolation.

In addition, the material covering the tube's surface must isolate the exposed metallic elements electrically. The space allowed for the encapsulant material is very small, to reduce sensor module weight and size, increasing the challenge of introducing a new encapsulant material. A paramount constraint on this engineered system is temperature exposure. The image tube cannot be exposed to temperatures greater than 95°C to maintain the tube's ability to amplify low levels of input light uniformly and consistently over time (stable luminous gain).

2. Background and Objectives

ITT customers require robust products that endure environmental extremes. This robustness is measured using a test involving thermal excursions while simultaneously exposing the viewing device and image sensor module to a wide range of relative humidity. Image sensor module reliability was defined at the customer's quality level as the ability of the device to maintain luminous gain over time. (Luminous gain is the light-amplifying power of the image sensor module.)

The historical approach for improving the reliability of the product and image sensor module was to evaluate design and process changes while ex-

Figure 1
F5050 night vision goggle

Figure 2
Cross section of sensor module

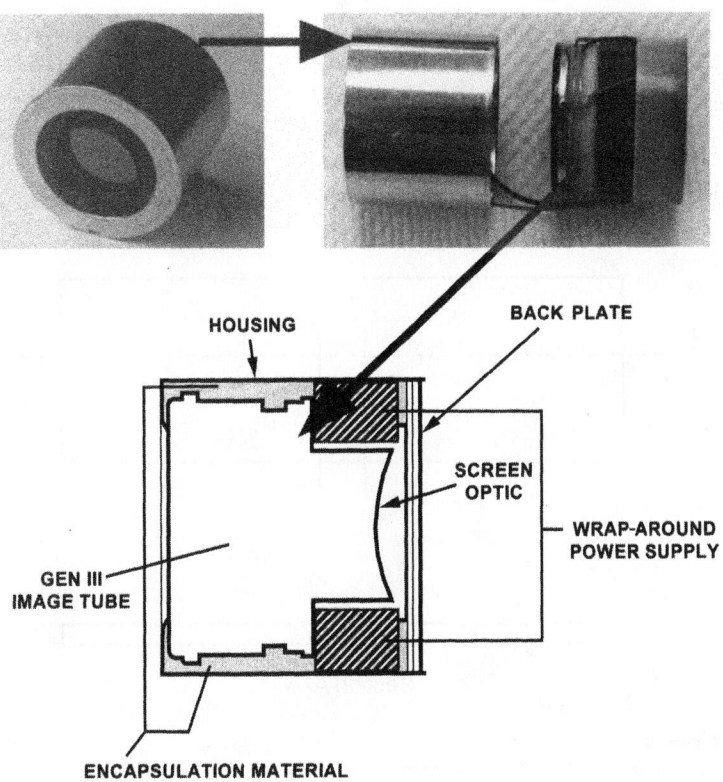

HOUSING

BACK PLATE

SCREEN
OPTIC

WRAP-AROUND
POWER SUPPLY

GEN III
IMAGE TUBE

ENCAPSULATION MATERIAL

Figure 3
Cross section of image-intensifier tube

posing these devices to a customer's environmental test. These evaluations were typically single factorial in nature, and the quality characteristic measured was number of cycles until the device's luminous gain degraded past a certain value or the total percent of gain degradation. For this experiment, a sample coupon was used to investigate the ideal function of the encapsulant in order to develop the encapsulating/insulating technology independent of the product on which it was used.

At the engineering quality level, the quality characteristic used was interface leakage current measured at several voltage potentials applied between an electrode on a sample coupon, not the image sensor module itself.

Figure 4 shows a representation of the coupon used to gather the data needed to generate the dynamic SN ratio and slope. This metric was used to select the best levels of the controllable factors in the encapsulating/insulating engineered system. The coupon consisted of a ceramic substrate with a set of evenly spaced electrode lines deposited on its surface. The approach was to minimize the slope of the ideal function, thus reducing the effect of the noise factors while reducing leakage current proportional to applied voltage.

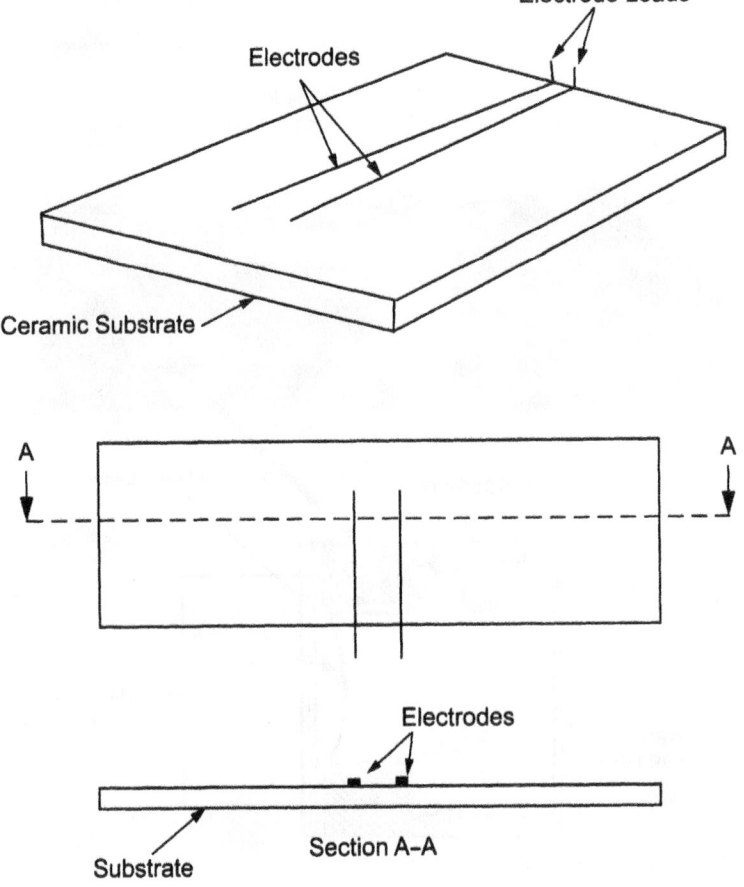

Figure 4
Test coupon to simulate the product

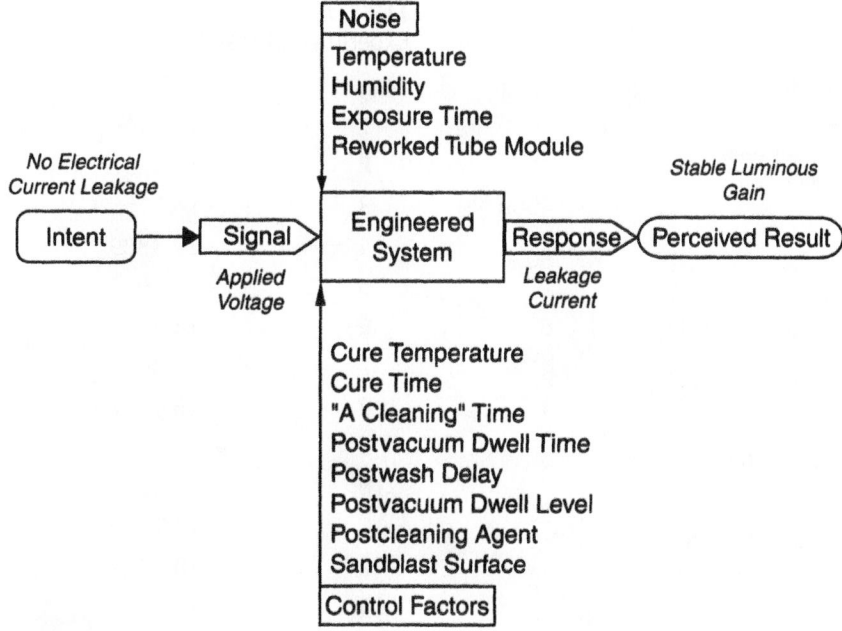

Figure 5
Encapsulation system

3. Parameter Design
Experimental Approach

Figure 5 presents the encapsulating/isolating engineered system. The signal factor is applied voltage. The response is the resultant current flow between the metallic elements. Figure 6 presents the ideal function for the system. The compound noise factor was thermal/humidity exposure, performed cyclically over many days, as well as the condition of the coupon simulating a reworking situation. The electrical data were collected before and after this cycling. The data were analyzed using the zero-point proportional dynamic SN ratio using applied voltage as the signal factor. The objective was to minimize the slope, β. For the ideal function of voltage versus current, the slope is inversely proportional to resistance, and for this system, the objective was to maximize resistance in order to increase the electrical isolation properties of the encapsulant system.

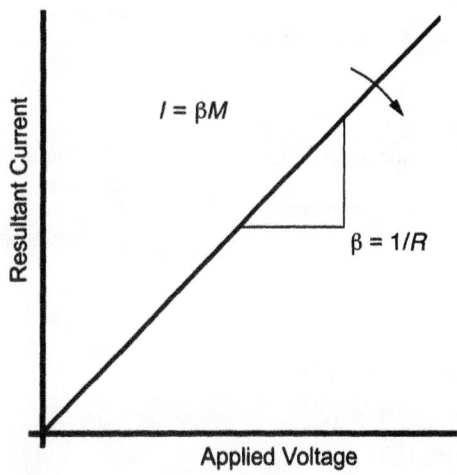

Figure 6
Ideal function

Table 1
Factors and levels

Factor		Level
Signal factors		
M:	voltage between electrodes (V)	1: 2000
		2: 4000
		3: 6000
		4: 8000
Controllable factors		
A:	cure temperature (°C)	1: 60
		2: 80
B:	cure time (h)	1: 2
		2: 4
		3: 6
C:	"A cleaning" time (min)	1: 5
		2: Std.
		3: 30
D:	postvacuum dwell time (min)	1: 0
		2: 5
		3: 10
E:	postwash delay (h)	1: <0.5
		2: 12
		3: 24
F:	postvacuum dwell level (torr)	1: 250m
		2: 2
		3: 200
G:	postcleaning agent	1: IPA
		2: Std.
		3: Acetone
H:	sandblast surface	1: Light
		2: Std.
		3: Heavy
Noise factors		
X:	humidity/temperature	1: Before
		2: After
Y:	coupon condition	1: New
		2: Reworked

Table 3
Experimental results

Row	SNR (dB)	Slope, β
1	−16.5406	1.9895
2	−15.6297	0.7347
3	−14.5655	0.6423
4	−15.3339	0.928
5	−14.9945	0.6889
6	−18.344	0.7758
7	−15.4606	0.9989
8	−15.1447	0.8003
9	−15.77	0.2967
10	−14.9883	0.4587
11	−15.7399	0.1666
12	−16.3414	0.5862
13	−18.0843	0.4627
14	−15.2836	0.3669
15	−17.7041	0.5584
16	−15.9816	0.1713
17	−16.9162	0.8485
18	−14.4645	0.2468

4. Experimental Layout

Table 1 lists the factors and levels used in the L_{18} experimental layout. Thirty-six coupons were fabricated, half of which were treated to represent one of the noise factors called "reworked." Each was encapsulated according to the orthogonal combinations using the process developed for sensor module

Table 2
Baseline raw data

2 kV				4 kV				6 kV				8 kV	
Rework		New		Rework		New		Rework		New		Rework	
Before	After	Before	After	Before	After	Before	After	Before	After	Before	After	Before	After
0.041	1.59	0.056	5.05	0.131	3.61	0.118	8.01	0.252	6.02	0.205	10.93	1.25	9.13

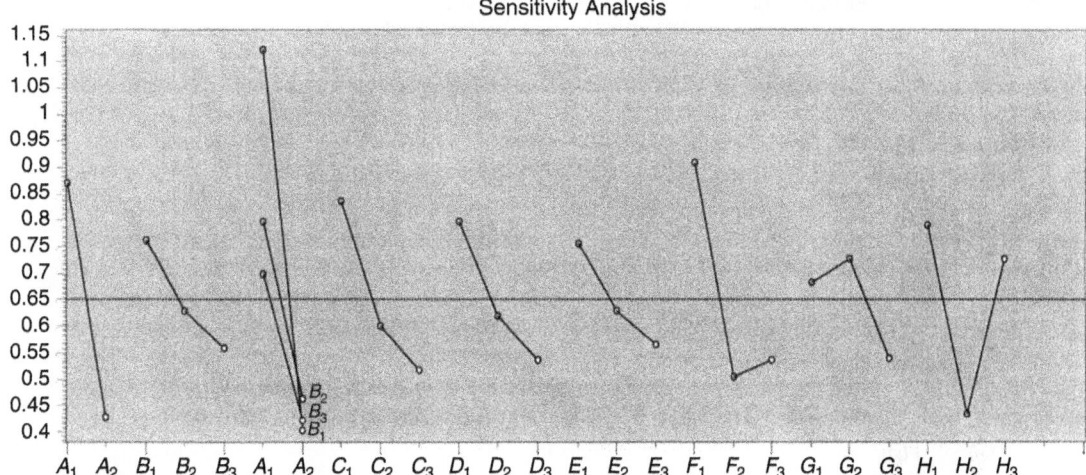

Figure 7
Average response graph for β

fabrication. Once these were encapsulated, all the "before" electrical measurements were obtained, and then the coupons were exposed to repeated cycles of temperature and humidity. Then each was tested again for the electrical characteristics, and the data collection sheets were completed.

5. Experimental Results and Analysis

The electrical current measurements made on the 36 coupons for each voltage potential before and after thermal/humidity exposure are too numerous to present here. The SN ratio and β were calculated for the 18 combinations of control factors and used

Table 4
Process average and confirmed values

Condition	Process Average Estimate		Confirmed Value	
	η (dB)	β	η (dB)	β
Baseline	—	—	−14.61	0.6298
Optimum	−15.12	0.1256	−7.42	0.1985
Gain	—	—	7.19	0.4313

as the system response for analysis in the orthogonal array.

Process Baseline

Before the parameter design experiment began, the existing encapsulation system was baselined by producing a coupon under standard conditions and making the measurements shown in Table 2. The baseline SN ratio is −14.61 and the baseline β value is 0.6298.

η and β ANOVA Tables and Response Graphs

Table 3 presents the η and β values that resulted from the L_{18} experiment. These values were generated using data from the zero-point proportional dynamic SN ratio equation. Of particular importance in Table 3 is the large range in β values. It was the objective of the study to find control factors that influence β, adjusting their levels to *minimize* β. Figure 7 presents the factor-level average response graphs for $10 \log \beta^2$. The best levels focusing on minimizing β are A_2, B_1, C_3, D_3, E_3, F_2, G_3, H_2.

6. Confirmation

Table 4 presents the process average estimates and confirmation results for η and β. Figure 8 compares

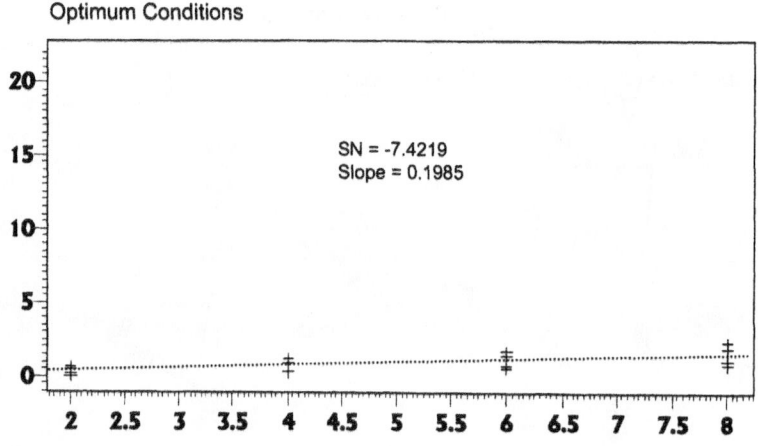

Figure 8
Optimum versus standard process conditions

the reduced slope confirmation to the standard process baseline, showing the improvements in both SNR and slope.

7. Conclusions

As a result of the parameter design effort focused on the electrical encapsulation engineered system, the SN ratio increased 7.2 db while decreasing the slope, β, 68% from 0.6298 to 0.1985. This translated into a 20% reduction in processing temperature and a 40% reduction in cycle time. Of far more importance is the improvement in encapsulation technology because the parameter design effort was performed on coupons and focused on improving the function of the system independent of product application.

This case study is contributed by Joe Barker, William Finn, Lapthe Flora, and Ron Ward.

Development of Plastic Injection Molding Technology by Transformability

Abstract: Because the selection of molding conditions greatly influences the dimensional accuracy of a product in the process of plastic injection molding, we inject, then cool melted plastic material in the mold. We used the concept of transformability to develop the plastic injection molding technology to secure stable dimensional accuracy.

1. Introduction

CFRP is material made of a mixture of superengineering plastics and carbon fibers. Compared with materials used for plastic injection molding, this has a much higher melting point and becomes carbonized and solidified at temperatures 50 to 60° higher than the melting point. Therefore, since a conventional plastic injection machine cannot mold this material, by adjusting the shapes of the nozzle and screw in the preliminary experiment, we confirmed factors and their ranges regarding possible molding conditions.

As the shape to be molded, we selected an easily measurable shape for both mold and product without considering actual product shapes to be manufactured in the future, because our objective was technological development of plastic injection molding. On the other hand, taking into account a shape with which we can analyze the proportionality (resemblance relationship), we prepared the shape shown in Figure 1 by using a changeable mold. As a reference, we note the die structure in Figure 2. The measurements are as follows:

> *Bottom Stage:*   *ABCDE* (Points)
>
> *Third Stage:*   *FGHIJK* (Points)

The signal factors are defined in Table 1.

Using a coordinate measuring machine as measurements, we read each set of x, y, and z coordinates for both the mold and the product. Then we defined the linear distance for every pair of measurements in the mold as a signal factor.

As control factors, we selected from the factors confirmed in the preliminary experiment eight molding conditions believed to greatly affect moldability and dimensions. For a noise factor, we chose the number of shots whose levels are the second and fifth shots. Table 2 summarizes all factor levels of the control and noise factors.

2. SN Ratio of Transformability Calculated from the Entire Product Shape

After assigning the control factors selected to an L_{18} orthogonal array and the signal and error factors to the outer array, we performed 18 experiments to calculate the distance data shown in Table 3.

Total variation:

$$S_T = 34.668^2 + \cdots + 35.877^2$$
$$+ 34.672^2 + \cdots + 35.885^2$$
$$= 179,318.477815 \quad (f = 110) \quad (1)$$

Linear equation for the second shot:

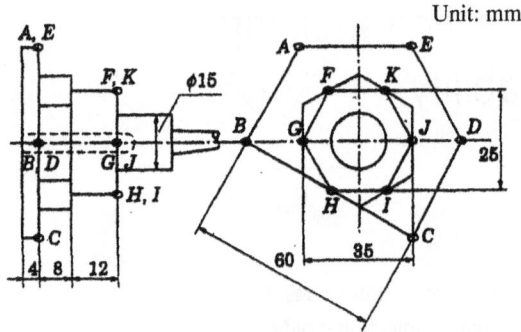

Figure 1
Shape of molded product and measurements

$$L_1 = (34.729)(34.668) + (60.114)(59.996)$$
$$+ \cdots + (36.316)(35.877)$$
$$= 90,229.763098 \qquad (2)$$

Linear equation for the fifth shot:

$$L_2 = 90,265.630586$$

Effective divider:

$$r = (2)(34.729^2 + 60.114^2 + \cdots + 36.316^2)$$
$$= 181,682.363768 \qquad (3)$$

Variation of proportional term:

$$S_{N\beta} = \frac{1}{181,682.363768} (90,229.763098$$
$$+ 90,265.630586)^2$$
$$= 179,316.178332 \qquad (f = 1) \qquad (4)$$

Variation of proportional term due to differences between molding shots:

$$S_N = \frac{1}{181,682.363768} (90,229.763098$$
$$- 90,265.630586)^2$$
$$= 0.007108 \qquad (f = 1) \qquad (5)$$

Error variation:

$$S_e = 179,318.477815 - 179,316.178332$$
$$- 0.007108$$
$$= 2.292402 \qquad (f = 108) \qquad (6)$$

Error variance:

$$V_e = \frac{2.292402}{108} = 0.021226 \qquad (7)$$

Total error variance:

$$V_N = \frac{2.292402 + 0.007108}{109} = 0.021096 \qquad (8)$$

SN ratio:

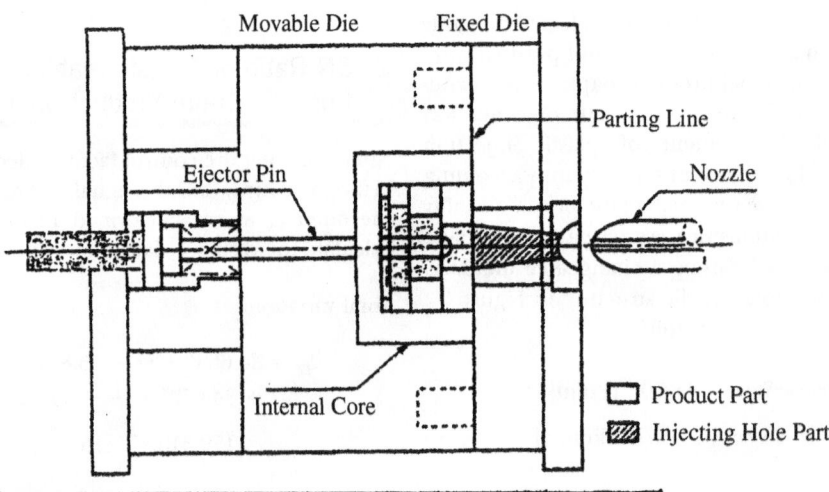

Figure 2
Model mold used for injection molding of CFRP

Table 1
Signal factors

Planar Distance on Bottom Stage				Planar Distance on Third Stage				Distance Connecting Points on Bottom and Third Stages			
A–B	A–C	...	D–E	F–G	F–H	...	J–K	A–F	A–G	...	E–K
M_1	M_2	...	M_{10}	M_{11}	M_{12}	...	M_{25}	M_{31}	M_{32}	...	M_{60}

$$\eta = 10 \log \frac{\dfrac{1}{181,682.363768}(179,316.178332 - 0.021226)}{0.021096}$$

$$= 16.70 \text{ dB} \tag{9}$$

Sensitivity:

$$S(\beta) = 10 \log \frac{1}{181,682.363768}$$
$$(179,316.178332 - 0.021226)$$

$$= -0.0569 \text{ dB} \tag{10}$$

3. Optimal Configuration and Prediction of Effects

Selecting levels with a high SN ratio from the response graphs shown in Figure 3, we can estimate the following optimal configuration: $A_2B_1C_3D_1E_3$-$F_3G_2H_1$. Now if the difference between the initial and optimal conditions for control factor C, nozzle temperature, is small, we fix the factor level to level 2 because we wish to leave the temperature at the lower level. As a result, we have:

Optimal configuration: $A_2B_1C_2D_1E_3F_3G_2H_1$

Initial configuration: $A_1B_2C_2D_2E_2F_2G_2H_2$

When preparing a test piece for the confirmatory experiment, we set up the following factor levels after determining that we recheck the factor effects of factors E and H, both of which have a V-shaped effect, with the lowest value at level 2. Since for cooling time, H, we have better productivity at a shorter time, we selected two levels, H_1 and H_2. For the switching position of the holding pressure, E, we reconfirmed all levels, E_1, E_2, and E_3. By combining the levels of E and H in reference to each optimal level estimated before, we performed a

Table 2
Control factors and levels[a]

Factor	Level		
	1	2	3
Control Factors			
A: cylinder temperature (°C)	Low[a]	Mid	—
B: mold temperature (°C)	Low	Mid[a]	High
C: nozzle temperature (°C)	Low	Mid[a]	High
D: injection speed (%)	5	8[a]	15
E: switching position of holding pressure (mm)	5	10[a]	15
F: holding pressure (%)	70	85[a]	99
G: holding time (s)	15	30[a]	45
H: cooling time (s)	40	110[a]	180
Noise factor			
N: number of shots	Second shot	Fifth shot	—

[a] Initial conditions.

Table 3
Distance data of molded product for transformability (experiment 1) (mm)

Noise Factor	Dimensions of Mold (Signal Factor, Number of Levels: 55)							
	M_1 34.729	M_2 60.114	...	M_{11} 14.447	M_{12} 25.018	...	M_{31} 28.541	...	M_{60} 36.316
N_1	34.668	59.966	...	14.294	24.769	...	28.104	...	35.877
N_2	34.672	59.970	...	14.302	24.780	...	28.137	...	35.885
Total	79.340	119.936	...	28.596	49.549	...	56.241	...	71.762

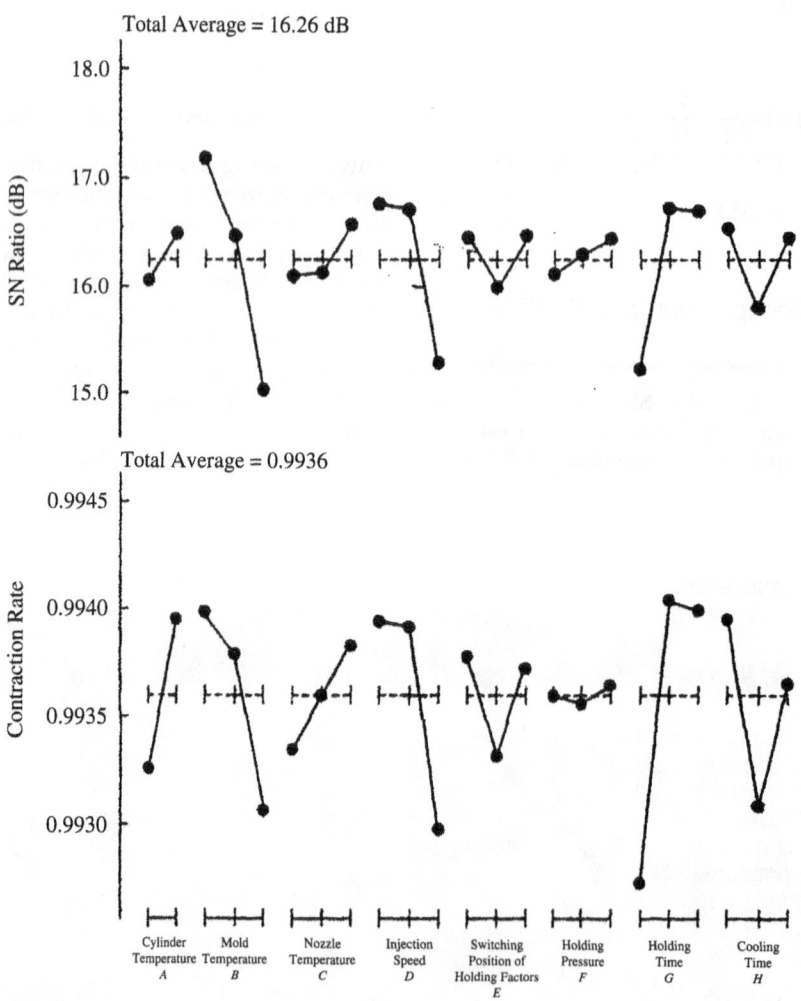

Figure 3
Response graphs

confirmatory experiment and summarize the analyzed results in Table 4.

Based on these results, we selected the following as the optimal configuration: $A_2B_1C_2D_1E_3F_3G_2H_1$. The confirmations under the optimal and initial configurations are tabulated in Table 5. These results demonstrate that the SN ratio and its gain are about the same as were estimated; that is, both have high reproducibility. Even so, their absolute values did not satisfy our targets completely. We could not obtain any remarkable improvement for the gain, possibly because it already approached the optimal value in the preliminary experiment.

The reason for the small SN ratio was probably related to the fact that shrinkage has a direction because we mixed plastic material with carbon fibers. By studying each portion or direction of a test piece under the optimal configuration, we computed an SN ratio and shrinkage for each of the following three analyses:

1. *Planar bottom stage:* analysis using planar distance dimensions on the bottom stage
2. *Planar top stage:* analysis using planar distance dimensions on the third stage
3. *Vertical direction:* analysis using distance dimensions between the bottom and third stages

The results summarized in Table 6 imply that the differences between shrinkage generated in each pair of portions or directions lower the SN ratio of the whole. Next, we computed the SN ratio when changing the shrinkage for each portion.

4. Analysis Procedure

Essentially, the reason that we applied the concept of transformability to injection molding was to de-termine an optimal configuration where the shrinkage remains constant, thereby designing a mold that allows us to manufacture a product with high accuracy. Nevertheless, in actuality, we modified the dimensions of the mold through tuning processes after calculating the shrinkage for each portion.

Table 7 shows the decomposition of $S_{\beta \times \text{portion}}$, variation caused by different portions. Only S_e is treated as the error. For the calculation procedure, we rewrote the distance data obtained from the test piece prepared for the optimal configuration in Table 8. Based on them, we proceeded with the calculation as follows.

Analysis of Planar Bottom Stage

Total variation:

$$S_{T1} = 34.705^2 + 34.697^2 + \cdots + 34.690^2$$
$$= 57{,}884.917281 \qquad (f = 20) \qquad (11)$$

Linear equation:

$$L_1 = (34.729)(69.402) + (60.114)(120.149)$$
$$+ \cdots + (34.729)(69.383)$$
$$= 57{,}931.241963 \qquad (12)$$

Effective divider:

$$r_1 = (2)(34.729^2 + 60.114^2 + \cdots + 34.729^2)$$
$$= 57{,}977.605121 \qquad (13)$$

Variation of proportional term:

$$S_{\beta 1} = \frac{L_1^2}{r_1} = \frac{57{,}931.241963^2}{57{,}977.605121}$$
$$= 57{,}884.915880 \qquad (f = 1) \qquad (14)$$

Error variation:

$$S_{e1} = S_{T1} - S_{\beta 1} = 57{,}884.917281$$
$$- 57{,}884.915880$$
$$= 0.001401 \qquad (f = 19) \qquad (15)$$

Error variance:

$$V_{e1} = \frac{S_{e1}}{19} = \frac{0.001401}{19} = 0.0000738 \qquad (16)$$

Shrinkage:

Table 4

Results of confirmatory experiment (dB)

	SN Ratio, η		Sensitivity, $S(\beta)$	
	H_1	H_2	H_1	H_2
E_1	17.16	18.32	−0.0478	−0.0305
E_2	18.76	17.88	−0.0345	−0.0340
E_3	19.19	18.63	−0.0320	−0.0346

Table 5
Results of estimation and confirmatory experiment (dB)

Configuration	SN Ratio		Shrinkage	
	Estimation	Confirmation	Estimation	Confirmation
Optimal	18.52	19.19	0.995	0.996
Initial	16.39	16.44	0.993	0.994
Gain	2.13	2.75		

$$\beta_1^{\cdot} = \sqrt{\frac{1}{r_1}(S_{\beta 1} - V_e)}$$

$$= \sqrt{\frac{1}{57{,}977.605121}(57{,}884.915880 - 0.0000738)}$$

$$= 0.999 \qquad (17)$$

Similar analyses are made for planar top stage and vertical direction.

Overall Analysis
Total variation:

$$S_T = S_{T1} + S_{T2} + S_{T3}$$

$$= 57{,}884.917281 + 14{,}868.183649 + 10{,}8094.440286$$

$$= 180{,}847.541216 \qquad (f = 110) \qquad (18)$$

Table 6
SN ratio and shrinkage for each portion of product

	SN Ratio (dB)	Shrinkage
Planar bottom stage	41.31	0.999
Planar top stage	38.83	0.995
Vertical direction	31.01	0.995

Linear equation:

$$L = L_1 + L_2 + L_3$$

$$= 57{,}931.241963 + 14{,}944.716946 + 108{,}637.586755$$

$$= 181{,}513.545664 \qquad (19)$$

Effective divider:

$$r = r_1 + r_2 + r_3$$

$$= 57{,}977.605121 + 15{,}021.647989 + 109{,}183.504276$$

$$= 182{,}182.757386 \qquad (20)$$

Variation of proportional term:

$$S_\beta = \frac{1}{r}(L_1 + L_2 + L_3)^2$$

$$= \frac{1}{182{,}182.757386}\left(\begin{array}{c}57{,}931.241962\\+14{,}944.716946\\+108{,}637.586755\end{array}\right)^2$$

$$= 180{,}846.792157 \qquad (21)$$

Variation of proportional terms between portions:

$$S_{\beta\times\text{proportion}} = \frac{L_1^2}{r_1} + \frac{L_2^2}{r_2} + \frac{L_3^2}{r_3} - S_\beta$$

$$= \frac{57{,}931.241963^2}{57{,}977.605121} + \frac{14{,}944.716946^2}{15{,}021.647989}$$
$$+ \frac{108{,}637.586755^2}{109{,}183.504276} - 180{,}846.792157$$

$$= 0.702439 \qquad (f = 2) \qquad (22)$$

Table 7
ANOVA table separating portion-by-portion effects

Factor	f	S
β	1	S_β
$\beta_1, \beta_2, \beta_3$	2	$S_{\beta \times \text{portion}}$
e	$n-3$	S_e
Total	n	S_T

Error variation:

$$S_e = S_T - S_\beta - S_{\beta \times \text{proportion}}$$

$$= 180{,}847.541216 - 180{,}846.792157 \quad (23)$$
$$- 0.702439$$

$$= 0.046620 \ (f = 107)$$

Error variance:

$$V_e = \frac{S_e}{107} = \frac{0.046620}{107} = 0.0004357 \quad (24)$$

SN ratio:

$$\eta = 10 \log \frac{(1/r)(S_\beta - V_e)}{V_e}$$

$$= 10 \log \frac{(1/182{,}182.757386)}{(180{,}846.792157 - 0.0004357)}{0.0004357}$$

$$= 33.58 \text{ dB} \quad (25)$$

In Table 9, we put together the data calculated for the SN ratio and shrinkage and those computed for a test piece under the initial configuration. According to these results, we can observe that differences in shrinkage among different portions of a mold contribute greatly to the total SN ratio.

To date, we have used a constant shrinkage figure of 0.998 based on technical information from material suppliers. However, by applying the concept of transformability, we have clarified that there exists a difference in shrinkage among different portions. To reflect this result, we should design a mold and manufacture products to meet dimensional requirements by using a different shrinkage value for each portion of a mold.

Table 8
Distance data under optimal configuration

	Noise Factor	M_1 34.729	M_2 60.114	...	M_{10} 34.729
Planar bottom stage	N_1	34.705	60.075	...	34.693
	N_2	34.697	60.074	...	34.690
	Total	69.402	120.149	...	69.383
Planar top stage		M_{11} 14.447	M_{12} 25.016	...	M_{25} 14.450
	N_1	14.347	24.873	...	14.368
	N_2	14.357	24.887	...	14.364
	Total	28.704	49.760	...	28.732
Vertical direction		M_{31} 28.541	M_{32} 36.277	...	M_{60} 36.316
	N_1	28.313	36.032	...	36.082
	N_2	28.315	36.024	...	36.122
	Total	56.628	72.056	...	72.204

Table 9
Results of confirmatory experiment (dB

		Shrinkage		
Configuration	SN Ratio (dB)	Planar Bottom Stage	Planar Top Stage	Vertical Direction
Optimal	33.58	0.999	0.995	0.995
Initial	28.19	0.999	0.993	0.993
Gain	5.39			

In addition, although several production trials have been needed to obtain even a target level under the initial configuration in the conventional trial-and-error method of selecting molding conditions to eliminate internal defects, by taking advantage of our developed process base on the concept of transformability, we can confirm a better molding condition, including internal uniformity, in just a single experiment, and at the same time, earn a gain of approximately 5 dB.

Reference

Tamkai Asakawa and Kenzo Ueno, 1992. *Technology Development for Transformability*. Quality Engineering Application Series. Tokyo: Japanese Standards Association, pp. 61–82.

This case study is contributed by Tamaki Asakawa and Kenzo Ueno.

Stability Design of Shutter Mechanisms of Single-Use Cameras by Simulation

Abstract: In general, for a shutter mechanism of a single-use camera, a single-plate reciprocal opening mechanism of the guillotine type is adopted that controls the amount of light onto a photosensitive film by opening and closing the cover plate. Predetermination of the shutter speed is regarded as one of the most difficult performances to predict. In terms of the initial design of each dimension and specification, for efficient and objective design of a stable mechanism in a limited period, we conducted parameter design using simulation in the early design stage.

1. Introduction

Since we open and close the cover plate (which we subsequently call a *sector*) by pressing a shutter button that releases a compressed coil spring's force, there is no transitional control process for the series of actions (Figure 1). Therefore, it is difficult to predict a shutter speed in the early design phase, and if it causes estimated design constants or part dimensions to exceed their upper and lower limits, we need to go back to the prototype confirmation stage and modify the initial design.

In our conventional design procedure, based on data created by three-dimensional computer-aided design systems, we performed kinematic simulation. In this case, after allocating the initial value of each design parameter using our technical feeling and experience, we repeated simulation runs and fine tunings of parameters. According to the design values determined through this process, we prototyped an actual model and matched actual values with target values by the final adjustment. However, this was considered inefficient because we needed to modify the initial design if the amount of modification exceeded the limit predicted.

2. Input and Output of Shutter Mechanism

If we rely on simulation systems, we cannot handle environmental effects such as conditions regarding actual uses or time-lapsed deterioration. In this study, it was assumed that environmental effects appear as design factor errors, including part dimensions and physical properties. Therefore, errors were associated with control factors for simulation. It was considered ideal that there be no variation from the output under standard conditions. That is, a design whose output is insensitive to noise fluctuations can be regarded as robust to environmental deterioration. The three-dimensional model used for simulation is shown in Figure 2.

Method 1: Dynamic SN Ratio

This is a method of evaluating a sector's stability for motion time of a shutter mechanism by the SN ratio, which is based on the idea proposed in quality engineering. Although we can regard time as a signal and sector's position as an output value, we cannot deal with sector's motion as a dynamic characteristic because a sector has reciprocal motion (Figure 3a). To solve this problem, we

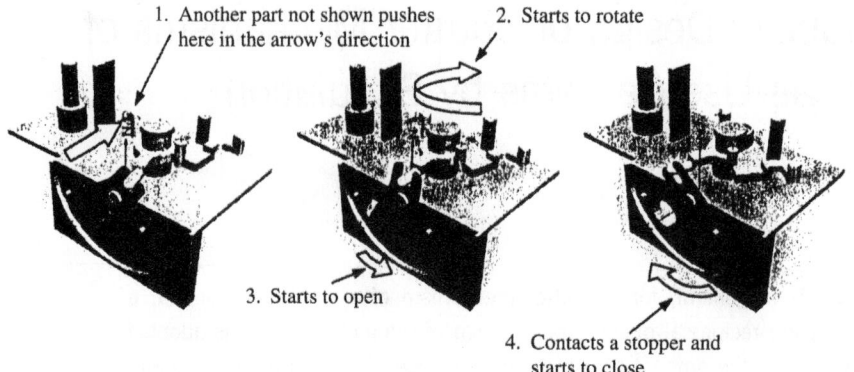

Figure 1
Shutter mechanism

substituted a cumulative value of motion angle that increases monotonously (Figure 3b).

Method 2: Standard SN Ratio

In method 1, because we obtained input/output relationships, as illustrated in Figures 4 and 5, we cannot see any direct relationship to the system. Thus, by setting an output value of the sector's position under the standard condition as signal, we considered a dynamic characteristic based on simulation results of the sector's position under noise conditions as outputs.

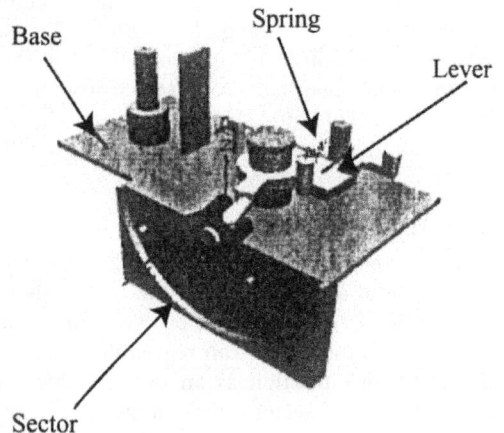

Figure 2
Three-dimensional model and part names

3. Parameter Design Using Method 1

After altering each control factor level by 5 to 10% around a standard condition, we compounded noise into two levels, positive side and negative side. By adding a standard condition to these two levels, we prepared three levels. We assigned eight factors, such as part dimensions, located positions, and spring constant to an L_{18} orthogonal array (Table 1).

Setup

We defined a motion time of a shutter mechanism as a signal factor, T, and a cumulative value of the sector's motion angle as an output value, y (Table 2).

Dynamic SN Ratio

We calculated the SN ratio by following the steps below. Because the system is nonlinear, V_e, which includes a nonlinear effect, is excluded from calculation of the SN ratio and sensitivity. Only the effect of $N \times \beta$ is used as error.

$$S_T = y_1^2 + y_2^2 + \cdots + y_{25}^2 \qquad (f = 15) \qquad (1)$$

$$r = M_1^2 + M_2^2 + \cdots + M_5^2 \qquad (2)$$

$$L_1 = y_{11}M_1 + y_{12}M_2 + \cdots + y_{15}M_5$$

$$L_2 = y_{21}M_1 + y_{22}M_2 + \cdots + y_{25}M_5 \qquad (3)$$

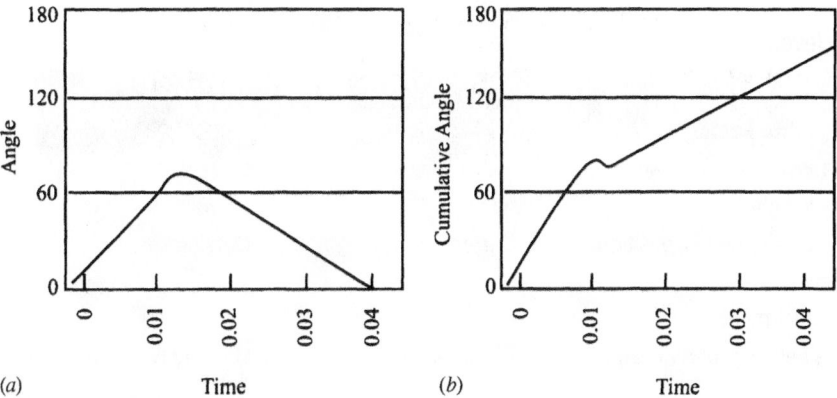

(a)   Time   (b)   Time

Figure 3
Data conversion of shutter angle

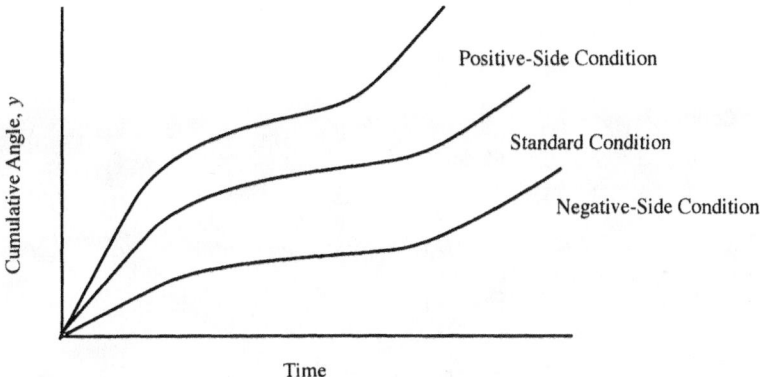

Time

Figure 4
Input/output relationship in method 1

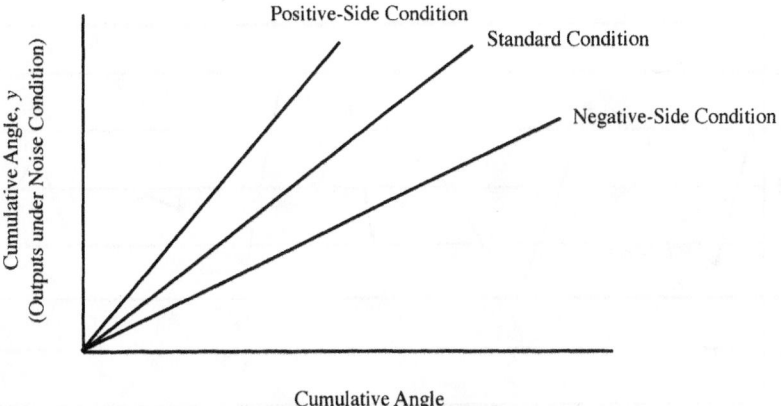

Figure 5
Input/output relationship in method 2

Table 1
Factors and levels

	Control Factor	Level		
		1	2	3
A:	part A's weight	×0.5	Current	—
B:	part B's weight	×0.5	Current	×2.0
C:	part B's gravity center position	Current	Outside	More outside
D:	spring constant	Current	×1.3	×1.5
E:	spring attachment angle	A	B	C
F:	distance between parts C and D	Inside gravity center path	On gravity center path	Current
G:	part A's stroke	Short	Current	Long
H:	motion input	Weak	Current	Strong

Table 2
Data format

No.	A	B	C	D	E	F	G	H	Error		T_1	T_2	T_3	T_4	T_5
										Signal Factor				
1	1	1	1	1	1	1	1	1	N_1:	standard condition	y_{11}	y_{12}	y_{13}	y_{14}	y_1
									N_2:	positive-side condition	y_{21}	y_{22}	y_{23}	y_{24}	y_{25}
									N_3:	negative-side condition	y_{31}	y_{32}	y_{33}	y_{34}	y_{35}
.	.	.	.	.	.	.	.	.		—	—	—	—	—
.	.	.	.	.	.	.	.	.						
18	2	3	3	2	1	2	3	1	N_1:	standard condition					
									N_2:	positive-side condition					
									N_3:	negative-side condition					

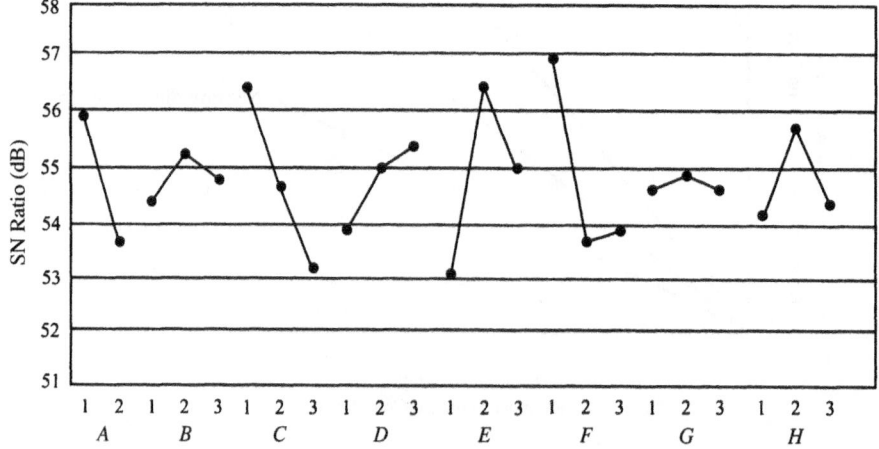

Figure 6
Response graphs of method 1

Table 3
SN ratio in confirmatory experiment dB

| | Condition | | |
	Current	Optimal	Gain
Estimation	53.26	63.08	9.82
Confirmation	51.72	56.07	4.35

$$S_\beta = \frac{(L_1 + L_2 + L_3)^2}{3r} \quad (f = 1) \qquad (4)$$

$$S_{N\beta} = \frac{L_1^2 + L_2^2 + L_3^2}{r} - S_\beta \quad (f = 2) \qquad (5)$$

$$S_e = S_T - S_\beta - S_{N\beta} \quad (f = 12) \qquad (6)$$

$$V_e = \frac{S_e}{12} \qquad (7)$$

$$V_N = \frac{S_{N\beta}}{2} \qquad (8)$$

$$\eta = 10 \log \frac{(1/3r)(S_\beta - V_e)}{V_N} \qquad (9)$$

Optimal Condition and Confirmatory Experiment
See Figure 6 and Table 3 for results of the confirmatory experiment.

4. Estimation of Gains and Results of Confirmatory Experiment

Figure 6 shows the response graphs. The current and optimal conditions are as follows:

Current Condition: $A_2 B_2 C_1 D_1 E_1 F_3 G_2 H_2$

Optimal Condition: $A_1 B_2 C_1 D_3 E_2 F_1 G_2 H_2$

The reproducibility is shown in Table 3; see also Figures 7 and 8.

Although we expected almost 10 dB of gain under the optimal condition as compared to that under standard conditions, our confirmatory experiment revealed that we can obtain only 4.3-dB improvement with poor reproducibility.

5. Parameter Design Using Method 2

We used the standard SN ratio in method 2 to compare with method 1.

Setup
We defined a simulation output (cumulative value of the sector's motion angle) under standard conditions and a simulation output (cumulative value of the sector's motion angle) under noise conditions as an output value, y (Table 4).

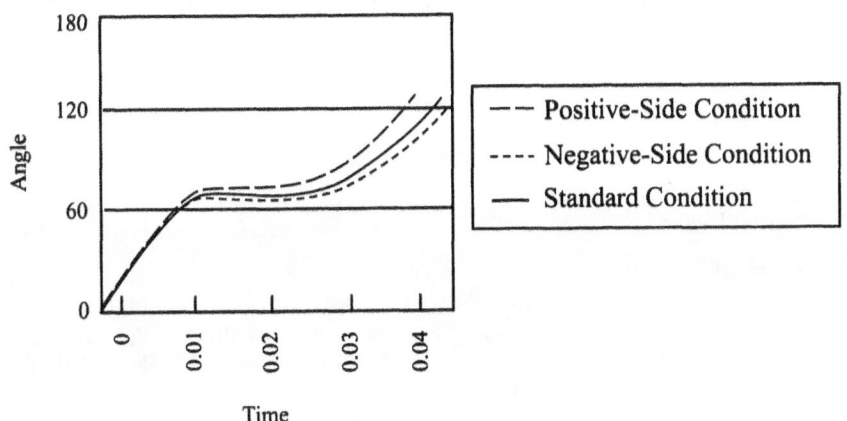

Figure 7
Angle change using method 1

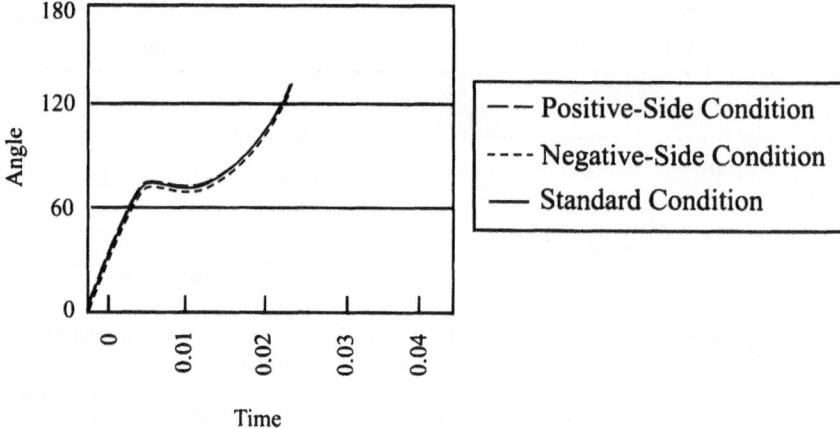

Figure 8
Angle under optimal condition in method 1

Standard SN Ratio
The calculations are as follows:

$$S_T = y_{11}^2 + y_{12}^2 + \cdots + y_{25}^2 \qquad (f = 10) \qquad (10)$$

$$r = M_1^2 + M_2^2 + \cdots + M_5^2 \qquad (11)$$

$$L_1 = y_{11}M_1 + y_{12}M_2 + \cdots + y_{15}M_5$$
$$L_2 = y_{21}M_1 + y_{22}M_2 + \cdots + y_{25}M_5 \qquad (12)$$

$$S_\beta = \frac{(L_1 + L_2)^2}{2r} \qquad (f = 1) \qquad (13)$$

$$S_{N\beta} = \frac{L_1^2 + L_2^2}{r} - S_\beta \qquad (f = 1) \qquad (14)$$

$$S_e = S_T - S_\beta - S_{N\beta} \qquad (f = 8) \qquad (15)$$

$$V_e = \frac{S_e}{8} \qquad (16)$$

$$V_N = \frac{S_{N\beta} + S_e}{9} \qquad (17)$$

$$\eta = 10 \log \frac{(1/2r)(S_\beta - V_e)}{V_N} \qquad (18)$$

Table 4
Data format

No.	Control Factor										
	A	B	C	D	E	F	G	H			
1	1	1	1	1	1	1	1	1	Signal	N: standard condition	M_1 M_2 M_3 M_4 M_5
									Output	N_1: positive-side condition	y_{11} y_{12} y_{13} y_{14} y_{15}
										N_2: negative-side condition	y_{21} y_{22} y_{23} y_{24} y_{25}
⋮											
18	2	3	3	2	1	2	3	1	Signal	N: standard condition	
									Output	N_1: positive-side condition	
										N_2: negative-side condition	

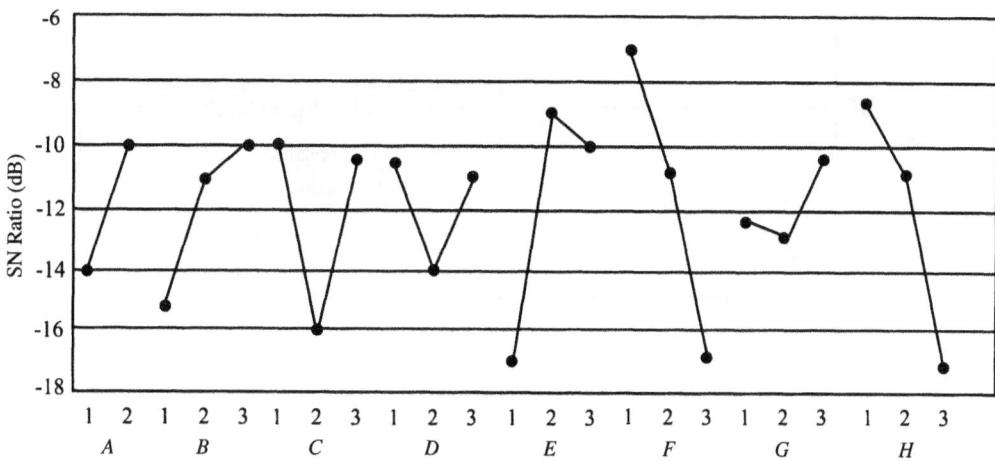

Figure 9
Response graphs of method 2

Optimal Condition and Confirmatory Experiment

The response graph of method 2 (Figure 9) was different from that in method 1 and we notice that there are fewer continuous-value factors that have a steep peak or V-shape in method 2 than in method 1. The current and optimal conditions are:

> *Current condition:* $A_2 B_2 C_1 D_1 E_1 F_3 G_2 H_2$
>
> *Optimal condition:* $A_2 B_3 C_1 D_1 E_2 F_1 G_3 H_1$

The reproducibility is shown in Table 5.

Comparing the gains of the two methods, both have poor reproducibility. In method 2, the gain is almost 20 dB from the estimate but only 13 dB from the confirmation. However, we concluded that a larger amount of improvement can be expected.

Table 5
SN ratio in confirmatory experiment (dB)

	Condition		
	Current	Optimal	Gain
Estimation	−13.49	6.49	19.98
Confirmation	−20.36	−7.65	12.71

The optimal condition under method 2 is shown in Figure 10.

6. Simulation Using Standard SN Ratio

This study predicates applying quality engineering to the simulation process in the uppermost stream of development, aiming for quality improvement. In simulation, environmental effects cannot be calculated. Under such circumstances, method 2 can bring more universal and satisfactory results than can method 1. This procedure can be regarded to have high generality and practicality because we can define input and output regardless of simulation contents or types. For mechanism designers, this is considered the most powerful tool to use to reduce technical risks in later phases because they can objectively evaluate unknown designs before actual products take shape.

In this example, since the sector's motion angle (Figure 3a) does not have a target curve, we have not made any adjustment tuning to a target value. When such tuning is needed, after securing functional stability as discussed in this example, we should make an adjustment by using simulation. This will enable us to develop products in a more efficient manner.

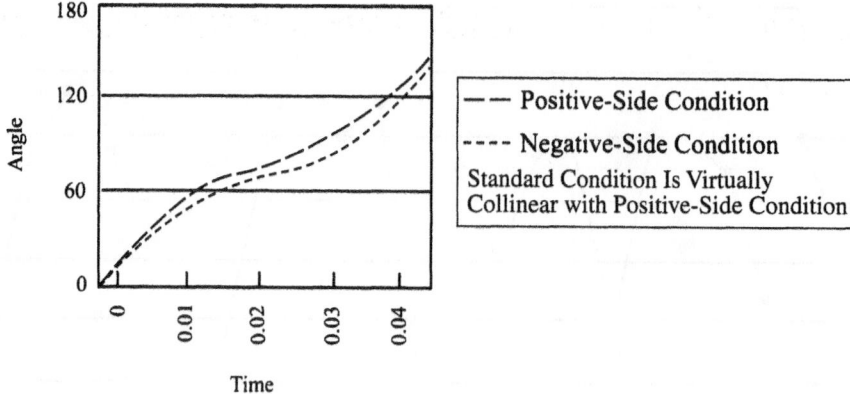

Figure 10
Angle change under optimal condition using method 2

Reference

Shuri Mizoguchi, 2001. Optimization of parameters in shutter mechanism designing. *Proceedings of the 9th Quality Engineering Symposium,* pp. 230–233.

This case study is contributed by Shuri Mizoguchi.

CASE 48

Optimization of a Clutch Disk Torsional Damping System Design

Abstract: During the development of a new clutch disk damping system and after three months of hard work, the engineering team got stuck with unsatisfactory system performance (rattle noise) despite the help of a computer simulation software. The only alternative seemed to be using a more expensive damping concept, but that option would increase part cost by $200 plus development cost. The team then decided to try Taguchi's robust engineering optimization approach. In just two weeks we got a gain of 4.6 dB for the SN ratio. That allowed us to keep the system proposed initially, thus avoiding the cost increase and keeping the development schedule on track. Perhaps even more important, we can now apply our engineering knowledge more effectively and make better use of the computer simulator, which results in a faster and more reliable development process.

1. Introduction

More and more attention is being paid to audible noise caused by torsional vibrations in the engine–clutch–transmission–rear axle system (the drive train). On the one hand, modern automotive design is producing stronger mechanical excitation due to more powerful engines; on the other hand, transmissions that have been optimized for weight and efficiency are highly sensitive to excitation. This situation brings a number of comfort problems, including rattle noise.

Many sources of excitation must be considered related to torsional vibrations. The main source is engine irregularity resulting from the ignition cycle. This basic excitation in the motor vehicle drive train is the focus of this study. Irregular ignition or even misfiring can also lead to serious vibration problems, but those factors were not included in this work because they can usually be eliminated by optimizing the ignition itself. The main source of excitation for torsional vibrations is the fact that engine ignition is not a continuous process but rather is discrete. That is also the primary cause of low-speed gear rattle. The basic principles involved in this phenomenon are described below.

Torque at the crankshaft is a periodic function of time. The torque curve is determined by the gas forces in the cylinder, by the geometry of the crank drive, and by the acceleration moments of the crank drive as it changes mass moments of inertia during one rotation. Significant mass forces occur at higher speeds only, where they can even become dominant. Because gear rattle occurs primarily at low engine speeds, inertial forces are negligible for our observations. That yields the very simple torque relationship shown in Figure 1.

According to that equation, the time-dependent torque acting on the crankshaft is the product of piston displacement, a geometric factor dependent on the angle of the crankshaft and the pressure curve of the cylinder. The graph shown at the bottom of Figure 1 illustrates that principle for one cylinder. In a six-cylinder engine, that curve is repeated three times per revolution.

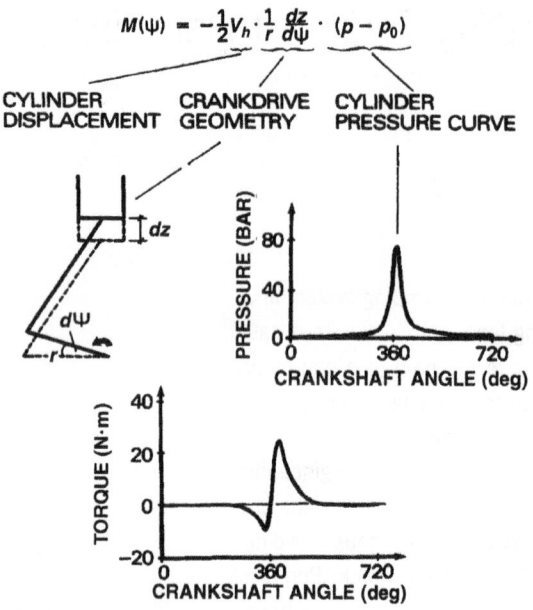

Figure 1
Engine excitation due to gas forces

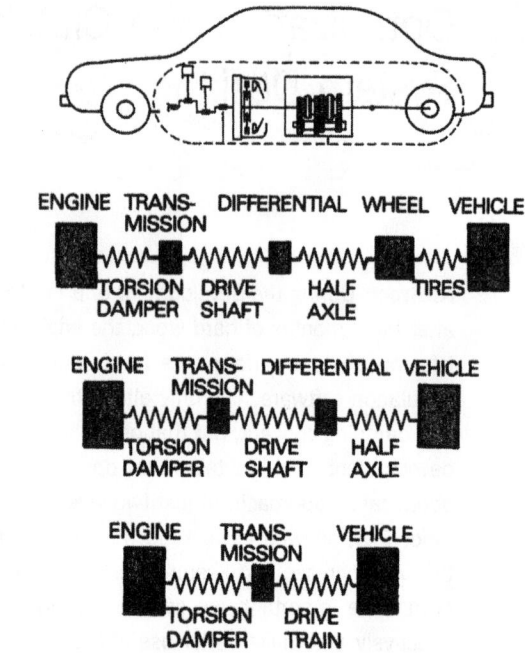

Figure 2
Analytical model for vehicle drive train vibrations

Peak torque magnitude (M) by itself does not reveal anything about engine irregularity. This value is based on the mass moment of inertia, J_m, of the crankshaft, together with components such as the flywheel and the clutch, which are attached to it rigidly. The angular acceleration, $\dot{\Omega}_m$, is determined using the equation

$$M = J_m \dot{\Omega}_m$$

That factor exhibits an almost constant amplitude at the low speeds in which we are interested. The curve for the angular velocity or the speed as a function of time can be plotted using integration:

$$\Omega_m = \int \dot{\Omega}_m \, dt$$

A vehicle drive train represents a torsional vibration system. It can be described roughly as a chain of rotating inertia and torsion springs. We can assume that we are dealing with a linear chain (Figure 2). For purposes of clarity, we will try to get by with as few different rotating inertia and torsion springs as possible. This type of simple model provides adequate information for many vibration problems. Of

course, it is important to emphasize that it cannot be used to study all drive train problems.

Figure 3 shows the possible vibration modes for a simple three-mass vibration system. In the case of rattle noise, the transmission vibrates at high amplitudes. Natural frequencies of 40 to 80 Hz are typical. This vibration mode occurs in the case of gear rattle noise.

Whenever engine irregularity excites the drive train vibration system, resonance can occur if the excitation frequency equals the natural frequency. Because engine excitation is not sinusoidal, additional high-frequency components must occur in the excitation, but their intensity generally decreases with increased frequency. Rattle vibration is excited primarily by the third and sixth orders in a six-stroke engine. Associated resonance speeds typically occur in the approximate range 700 to 2000 rpm. The lower the gear selected, the higher those speeds.

In theory, there are several options for modifying the vibration performance of a vehicle drive train in order to reduce undesirable torsional vibrations.

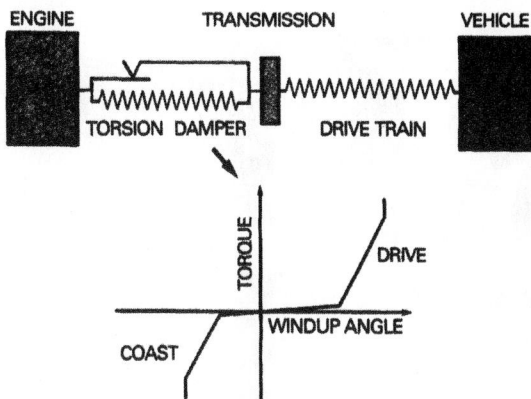

Figure 3
Torsion damper characteristic curve in analytical model

Unfortunately, however, compelling engineering considerations prevent us from changing most of the components making up the drive train. Therefore, with the exception of the flywheel mass, it is hardly possible to change the mass moments of inertia involved in the system, nor is it readily feasible to alter the spring rates (stiffness) of the tires, half axles, and drive shafts. Consequently, the only actual option left to target for positive change is the connection between the engine and the transmission.

An appropriate damped, torsionally elastic coupling (torsional damper) can be used to shift the resonance speeds and damp resonance amplitudes. Generally, gear rattle in the drive, coast, and idle modes requires different torsional elasticity and damping characteristics. Consequently, most torsion dampers include precisely tuned multistage characteristic curves, with each stage optimized individually for its respective load ranges.

Based on a calculated example, vibration performance can be influenced by changing torsional spring rate and damping. Each section of the matrix graph shows the vibration amplitude of the engine and the transmission speed plotted as a function of the mean engine speed. Engine speed is plotted as a solid line, and transmission speed is plotted as a dashed line. Each section was based on the calculation for a torsional damper with different frictional hysteresis and torsional spring rate.

We note that vibration isolation does not occur until we reach the speed range above natural frequency in the 2000 rpm range. If we reduce the torsional spring rate, we can shift the resonance to lower speeds and introduce the vibration isolation range sooner. At very low torsional spring rates of about 1 N · m/deg, the vibration amplitude also decreases significantly in the resonance range. With that design, the very flexible coupling passes hardly any vibrations at all through to the rest of the drive train. This would constitute the ideal torsion damper.

Unfortunately, it is impossible to accommodate this type of flat spring rate in the space available for installing the clutch. It is impossible to achieve vibration insulation in drive mode over the engine speed range using an elastic coupling such as a conventional torsion damper. As a result, torsion damper tuning always represents a compromise. Low hysteresis results in high resonance amplitudes at low speeds and good vibration insulation at high speeds, whereas high hysteresis leads to rigid performance. If a satisfactory compromise proves impossible, other vibration damping procedures must be adopted.

By using traditional, knowledge-guided one-factor-at-a-time experiments using computer simulation, we have tried successive design–prototype–test cycles. But after three months we still did not find an acceptable solution for the aforementioned compromise, and the design release deadline was too close to venture major changes. Then one of the team members suggested the application of robust engineering in the simulation analysis, whose results we now report.

2. Experimental Procedure

Scope of Experiment

Figure 4 shows the engineering system under study. Its parts are spring set and friction components. System frontiers are, at the input, the engine flywheel, and at the output, the input transmission shaft.

System Function

The objective function is to minimize the angular variation from the engine while transferring torque through the system. The basic function (work) is to

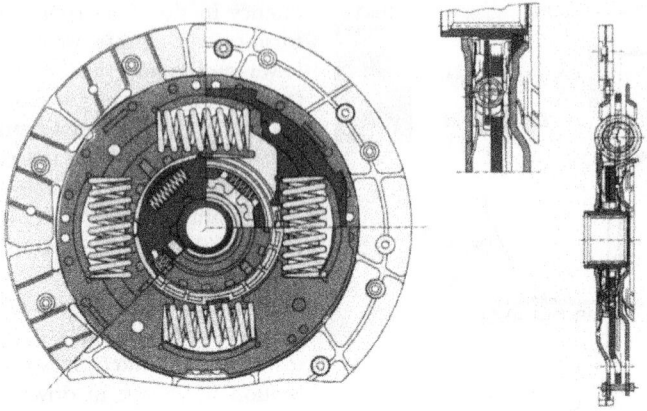

Figure 4
Damping system

attenuate angular variation. A P-diagram and ideal function are shown in Figure 5.

Since the function of the damping system is to attenuate the angular variation generated by the engine, we defined as the signal factor, M, the angular variation in the flywheel, and as the response, y, the attenuated angular variation, which ideally should be equal to M. Graphically, the ideal function is a straight line through zero, with unity angular coefficient (Figure 6).

The following relationship holds for the system:

$$y = \Delta - M$$

where y is the attenuated angular variation, Δ the angular variation at the output, and M the angular variation at the input. If we want the ideal behavior $y = M$, Δ must be zero. Therefore, because of this one-to-one relationship between y (dynamic characteristic) and Δ (smaller-the-better characteristic), for the sake of easier analysis, we decided to work on the minimization of Δ. We found that approach acceptable in this case, because one can see that Δ is the single and only functional symptom of y and is measured in the same unit scale.

Our response y becomes Δ, the angular variation at the output of the damping system (which, by the

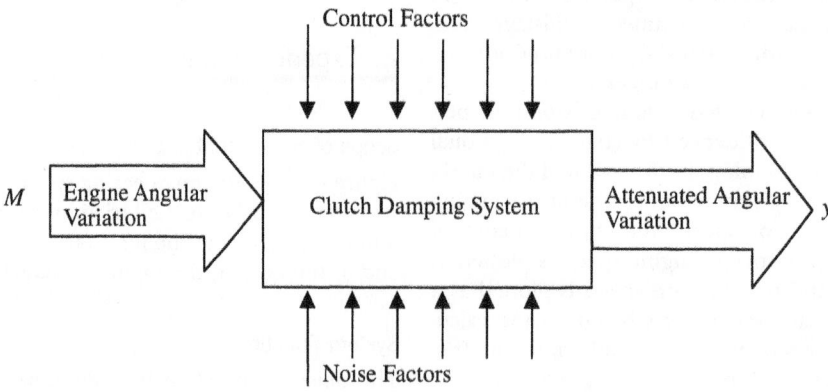

Figure 5
P-diagram for the engineering system

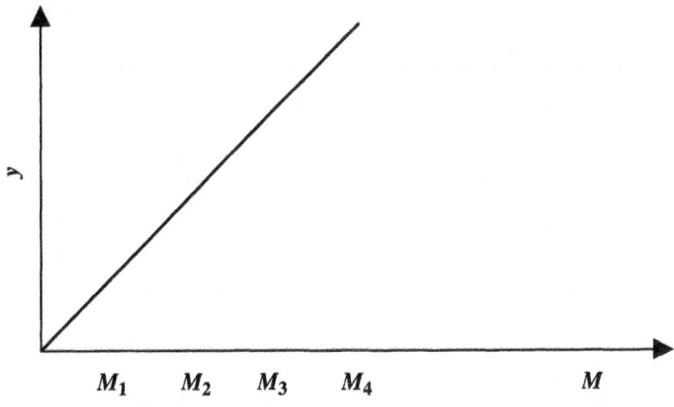

Figure 6
Ideal function

way, is the software response) and is analyzed as a smaller-the-better characteristic because it represents the angular variation that was not attenuated by the system and was released at the output (whose ideal value is zero).

Signal and Noise Factor Strategies

The signal, M, is a function of the engine speed; therefore, the software was fed with a curve of angular variation from 500 to 2500 rpm. To analyze the software output, we took the response y (Δ) in incremental steps of 50 rpm, starting at 750 rpm (engine idle speed). For each run of the orthogonal array, a smaller-the-better SN ratio was calculated from the 10 highest peak values of the response curve, in selected points of interest in the rpm range. (See the example in Figure 7 and Table 1.)

Besides sweeping the input signal, M (which for smaller-the-better calculations can also be regarded as noise), we simulated additional noise factors related to deterioration and variability existing in the manufacturing environment (Table 2).

Control Factors

The control factors are as follows:

❑ *Elastic constant* (K) *of the spring* (A). Six levels are included in order to explore the design space as much as possible.

❑ *Angle at the beginning of the second friction level* (B). In the damping system under study, it is

possible to use two friction levels along the total work angle. The control factor, B, is of great importance to identify the range of work angle for each friction level, or even to answer the following question: Do we need two friction levels? Therefore, we considered the maximum work angle of 20° divided into three levels.

❑ *Second friction.* As described above, the system under study allows the use of two friction levels along its range of 20°. To identify the value to be adopted, three wide (although feasible) levels were defined, from low to high.

❑ *First friction.* The basic requirement for the friction system to work is that the first level must be smaller than the second one, so the three levels for factor D were defined as a percentage of factor C.

These control factors were assigned to an orthogonal L_{18} array (Table 3).

3. Experimental Data and Analysis Using a Smaller-the-Better SN Ratio

For each combination of L_{18}, we obtained a group of data composed by the highest 10 peaks, under each noise condition N_1 and N_2 (Table 4). See Tables 5 to 7 for data analysis.

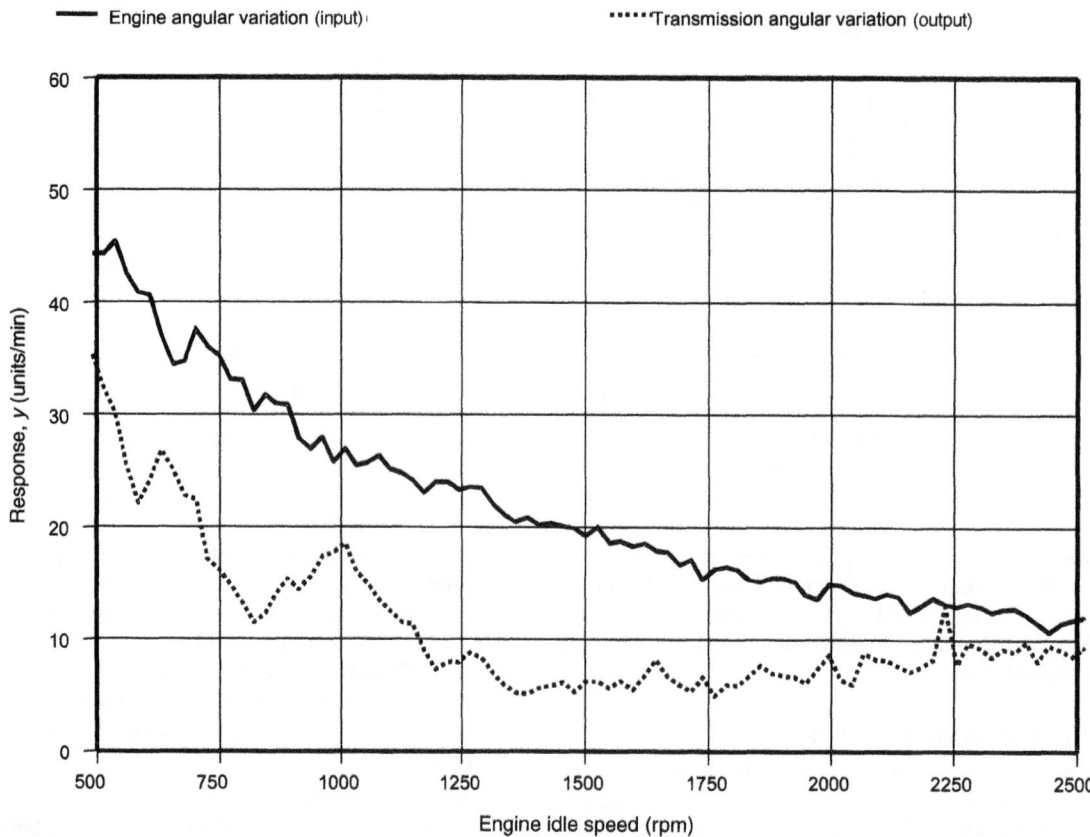

Figure 7
Example of simulator response

Table 1
Example of response for each experimental run

Run	Signal (rpm)		Response, y (units/min)
25	M_7:	1050	16.5
27	M_6:	1000	18
28	M_5:	950	17.5
31	M_4:	900	16
⋮			

4. Results

The SN results (in decibels) are:

Initial configuration:	-23.46
Optimum configuration $(A_6B_3C_1D_1)$:	-18.58
Confirmed by simulation:	-18.90
Gain:	4.56

As expected, due to the good correlation between the simulator and reality, the SN gain was confirmed subsequently in vehicle testing. Figures 8 and 9 show the positive impact obtained in the damping system performance by comparing the initial situation (Figure 8) with the optimized design (Figure 9).

Table 2
Compound noise

	Noise Factor	
Compound Noise[a]	N$_1$ (y high)	N$_2$ (y low)
Variation of K$_{spring}$	Control fator A value, minus X% fatigue	Control factor A vaue, plus X% manufacturing variation
Variation of K$_{vehicle}$	Min.	Max.

[a] K, elastic constant.

With that optimization, the subjective evaluation rating improved from 4.0 (rated as failure by all customers) to 5.5 (rated as disturbing by some customers). That improvement was enough to keep us from having to adopt a more complex and expensive damping system that would add $200 to the system cost. To achieve the development target grade of 7.0 (satisfactory to noticeable by all customers, but not causing disturbance), we decided to increase the mass of the engine flywheel by 20%, with minimum impact on piece cost. Those actions combined were effective, and the development schedule was kept on track.

5. Conclusions

In the process of conventional design using simulation, the control factors were being analyzed separately, one at a time. Although the computer analysis was fast, the responses that we obtained were far from optimum. Thus, after vehicle prototype testing, we had to get back to a new cycle of simulation, prototype construction, and evaluation in the vehicle, which increased the time and cost of development.

With the use of the robust engineering methodology, the technology under study (clutch damping system) could be optimized quickly using the simulator and there was no need to build more than one physical prototype. In that way, besides system performance quality improvement, we were able to achieve a faster and less expensive development cycle.

This robust engineering study also allowed us to maintain the low-complexity system, thus avoiding an apparently unavoidable cost increase that would occur by using a more sophisticated damper. We

Table 3
Control factors and levels

	Control Factor			
Level	A: Spring's K	B: Angle to Second (deg) Friction Starts	C: Second Friction (N)	D: First Friction (% of C)
1	4.6	0	3	25
2	9.2	7	15	50
3	13.8	14	27	75
4	18.4			
5	23			
6	27.6			

Table 4

L_{18} orthogonal array and responses under noise extremes

No.	Factor				y at:	
	A	B	C	D	N_1	N_2
1	4.6	0°	3	0.75	120 95 60 37 36 26 22.5 22.5 21 21	105 65 60 39 28 26.5 21 18.5 18 16.5
2	4.6	7°	15	7.5	49 44 34 29 26 26 26 23 23 23	41.5 41 30.5 27 26 24.5 24 24 22 21
3	4.6	14°	27	20.25	30 29.5 29 29 27 26 23.5 23.5 23 21.5	29.5 29 25.5 25 24.5 22.5 22.5 21 21 20.5
4	9.2	0°	3	1.5	97 76 62 38 34 26.5 22.5 21 18.5 14.5	87 66 54 39 28 27 21 18 13 11
5	9.2	7°	15	11.25	49 44.5 33 30 26 26 25 23.5 23 21.5	42 41 31 26 26 24 24 23.5 22.5 20.5
6	9.2	14°	27	6.75	30.5 29.5 29 27.5 26 25.5 23.5 23 22 21.5	28.5 25.5 25.5 24.5 22.5 22.5 21 21 20.5 20
7	13.8	0°	15	3.75	49.5 44 33 29 26.5 26 23 23 21.5 21.5	23.5 23 22 19.5 19 18 15 15 14 14
8	13.8	7°	27	13.5	30 29.5 29 27 25.5 25 23 23 22 19.5	26 24.5 23.5 22.5 21 20 19 18 17.5 17
9	13.8	14°	3	2.25	60 54 50.5 37 36 21 15 11.5 10.5 10	13.5 11 11 10.5 8.5 7.5 5.5 5.5 5 5
10	18.4	0°	27	20.25	30.5 27 25 23.5 21.5 20 19 18.5 17 16.5	28.5 26 24.5 22.5 20.5 20 19.5 19 16.5 16
11	18.4	7°	3	0.75	12 10 9.5 6.5 6 4.5 4.5 4 3.5 3.5	12.5 11.5 10 7 6.5 5 5 4 4 3.5
12	18.4	14°	15	7.5	21 20 20 18 15 13.5 13.5 13.5 13.5 13.5	28 25.5 22.5 20.5 19.5 18 16.5 16 15.5 15
13	23	0°	15	11.25	26.5 23 23 20.5 18.5 18 15.5 14 14 13.5	24 23 22 20 19.5 18 16 15 14.5 14.5
14	23	7°	27	6.75	30 27 24 23.5 21 19.5 18.5 17 17 16.5	28.5 25.5 22.5 20.5 19.5 19 18 16.5 16 16
15	23	14°	3	1.5	12 10 9 6 5.5 4 4 4 3.5 3.5	13.5 11.5 9.5 6.5 6.5 4.5 4 3.5 3.5 2.5
16	27.6	0°	27	13.5	30 27 25 23 21 20 19 18.5 18 17	28.5 26 24.5 22 20.5 20.5 19.5 18.5 17 16
17	27.6	7°	3	2.25	14.5 13 12 8.5 8 6 5.5 4.5 4 4	15.5 15 13 9 8 6.5 5.5 5.5 4.5 4
18	27.6	14°	15	3.75	15 13.5 13 9 8 6.5 5.5 5 4.5 4	15.5 14.5 13.5 9.5 9.5 7 6.5 5.5 5 4.5

Table 5
Control factors and calculated values of SN and average (smaller-the-better)

No.	Factor A 1	B 3	C 4	D 5	SN (dB)	Average
1	1	1	1	1	−34.66	42.93
2	1	2	2	2	−29.85	29.23
3	1	3	3	3	−28.30	25.15
4	2	1	1	2	−33.52	38.70
5	2	2	2	3	−29.83	29.10
6	2	3	3	1	−28.07	24.48
7	3	1	2	1	−28.40	24.00
8	3	2	3	2	−27.62	23.13
9	3	3	1	3	−28.52	19.43
10	4	1	3	3	−27.06	21.58
11	4	2	1	1	−17.50	6.65
12	4	3	2	2	−25.51	17.93
13	5	1	2	3	−25.82	18.65
14	5	2	3	1	−26.76	20.80
15	5	3	1	2	−17.30	6.35
16	6	1	3	2	−27.04	21.58
17	6	2	1	3	−19.51	8.33
18	6	3	2	1	−19.85	8.75
Grand averages:					−26.40	21.48

Table 6
Response table for the SN ratio

No.	Factor A	B	C	D
1	−30.94	−29.42	−25.17	−25.87
2	−30.47	−25.18	−26.54	−26.81
3	−28.18	−24.59	−27.48	−26.51
4	−23.35			
5	−23.29			
6	−22.13			
Diff.	8.80	4.82	2.31	0.94

Table 7
Response table for the mean

No.	Factor A	B	C	D
1	32.43	27.90	20.40	21.27
2	30.76	19.54	21.28	22.82
3	22.18	17.01	22.78	20.37
4	15.38			
5	15.27			
6	12.88			
Diff.	19.55	10.89	2.39	2.45

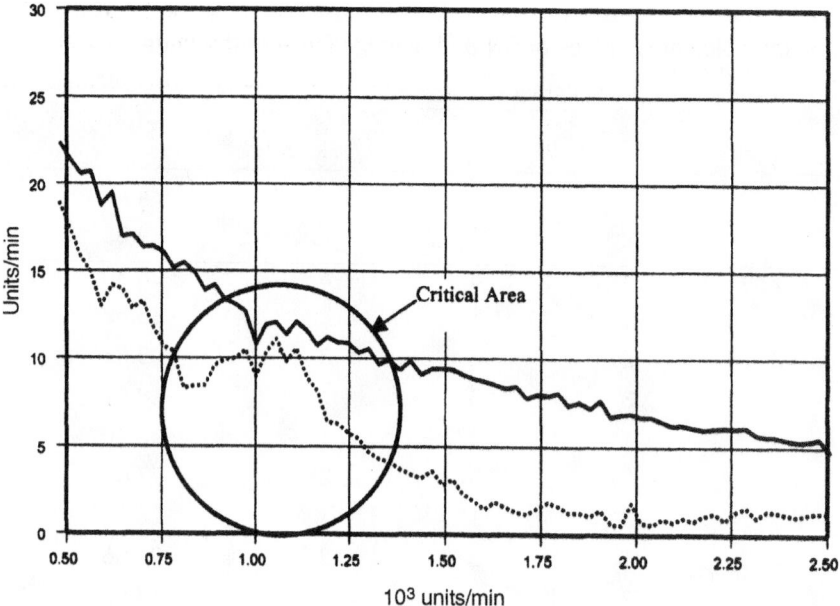

Figure 8
Damping system performance before study

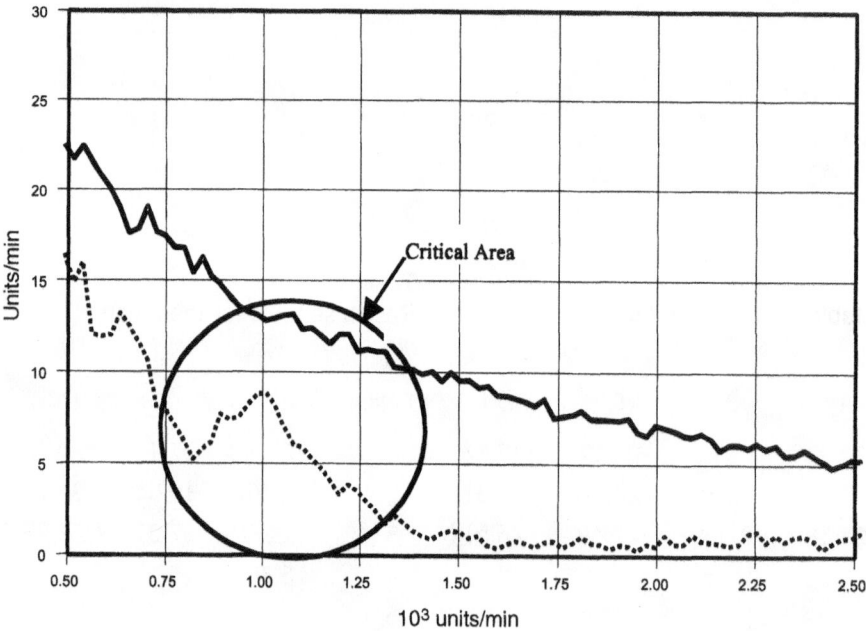

Figure 9
Damping system performance after study

also got rid of a critical development schedule pressure, because a solution that at first seemed impossible could quickly be identified.

Perhaps even more important, from now on we can more effectively apply our engineering knowledge and make better use of the computer simula-

tor, which should result in faster and more reliable development cycles in the future.

This case study is contributed by Luiz H. Riedel, Claudio Castro, Nélio de Lucia, and Eduardo Moura.

CASE 49

Direct-Injection Diesel Injector Optimization

Abstract: Delphi Automotive Systems is entering the direct-injection diesel business, which requires a significant shift in technologies from the current diesel injection approach. The injector itself is the key to mastering this goal, and its high sensitivity to sources of variation makes robust engineering a valuable approach to optimization. A robust engineering dynamic experiment based on a numerical model of the injector allowed us to achieve (1) a 4.46-dB improvement in the SN ratio through parameter design, representing a reduction of about 40% in the variability of the quantity injected; (2) an improvement of about 26% in the manufacturing process end-of-line first-time quality (percentage of good injectors at the end of the line) predicted; and (3) the generation of tolerance design-based charts to support manufacturing tolerance decisions and reduce cost further. This robust engineering case study shows that cost-effective and informative robust engineering projects can be conducted using good simulation models, with hardware being used to confirm results.

1. Introduction

The common rail direct-injection diesel fuel system is an important technology for Delphi. For this reason, a product and process engineering team, dedicated to the design and implementation of the best possible common rail system with respect to both product and process, was set up at the European Technical Center. The direct-injection diesel common rail system is comprised of the following core components of the engine management system:

❏ Injectors

❏ High-pressure pump

❏ Fuel rail and tubes

❏ High-pressure sensor

❏ Electronic control module

❏ Pressure control system

The main challenge with diesel common rail systems as opposed to indirect diesel injection systems is the continuous high operating pressures. Current common rail systems are designed to operate with pressure levels of 1350 to 1600 bar.

Figure 1 shows a sample injector and its key variability sources. The injector is the most critical and the most complex element of the system. A diesel common rail system will not be successful if the injectors are not "world class" in term of quality and reliability.

Problem

The injector complexity is due to very tight manufacturing tolerances and challenging customer requirements for quantity injected. These issues are confirmed by the problems reported by our competitors. Some of them experienced very high scrap levels in production. An alternative approach is to build quality into the product at an early stage of design.

Injector operation is affected by the following sources of variability: (1) other elements of the engine and (2) the management system.

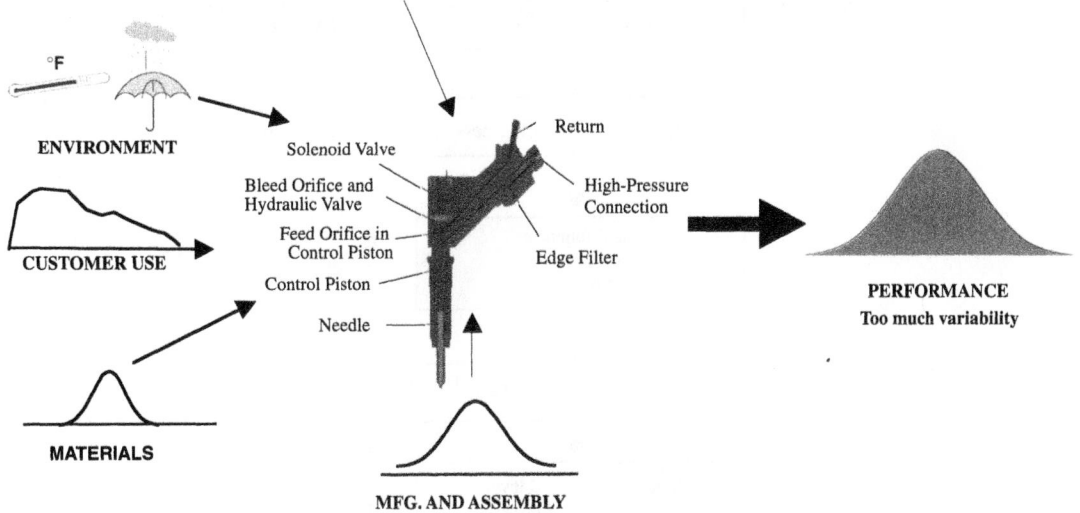

Figure 1
Injector and key variability sources

Objectives and Approach to Optimization

The optimization process followed the flowchart in Figure 2. A simulation model was developed, improved, and used to perform the orthogonal array experiments. Because of the very high confidence level in the simulation model (see Figure 7), we decided to use hardware only for the confirmation runs.

The main deliverables assigned to this optimization process were:

❏ Reduced part-to-part and shot-to-shot variation in quantity injected

❏ Part-to-part variation among several injectors

❏ Shot-to-shot variation from one injection to the next within the same injector

❏ Decreased sensitivity to manufacturing variation and ability to reduce cost by increasing component tolerances as appropriate

❏ Graphical tools to increase understanding of downstream quality drivers in the design

The following labeling is used in the experiment:

A: control factor for parameter design

A': control factor variation considered as a noise factor

A'': control factor included in the tolerance design

2. Simulation Model Robustness

A direct-injection diesel injector for a common rail system is a complex component with high-precision parts and very demanding specifications. To simulate such a component means representing the physical transient interactions of a coupled system, including a magnetic actuator, fluid flows at very high pressure, possibly with cavitation phenomena, and moving mechanical parts. The typical time scale for operation is a few microseconds to a few milliseconds.

Because pressure wave propagation phenomena are very important in the accurate representation of injector operation, the simulation code cannot be limited to the injector itself but must include the common rail and the connection pipe from the rail to the injector (Figure 3)

The rail and the connection line are standard hydraulic elements in which the model calculates wave propagation using classical methods for solving wave equations. The internal structure of the injector is illustrated in Figure 4. We can distinguish two hydraulic circuits and one moving part (plus the mobile part of the control valve). The first hydraulic circuit feeds the control volume at the common rail high pressure through a calibrated orifice. The pressure, P_C, in the control volume is controlled by

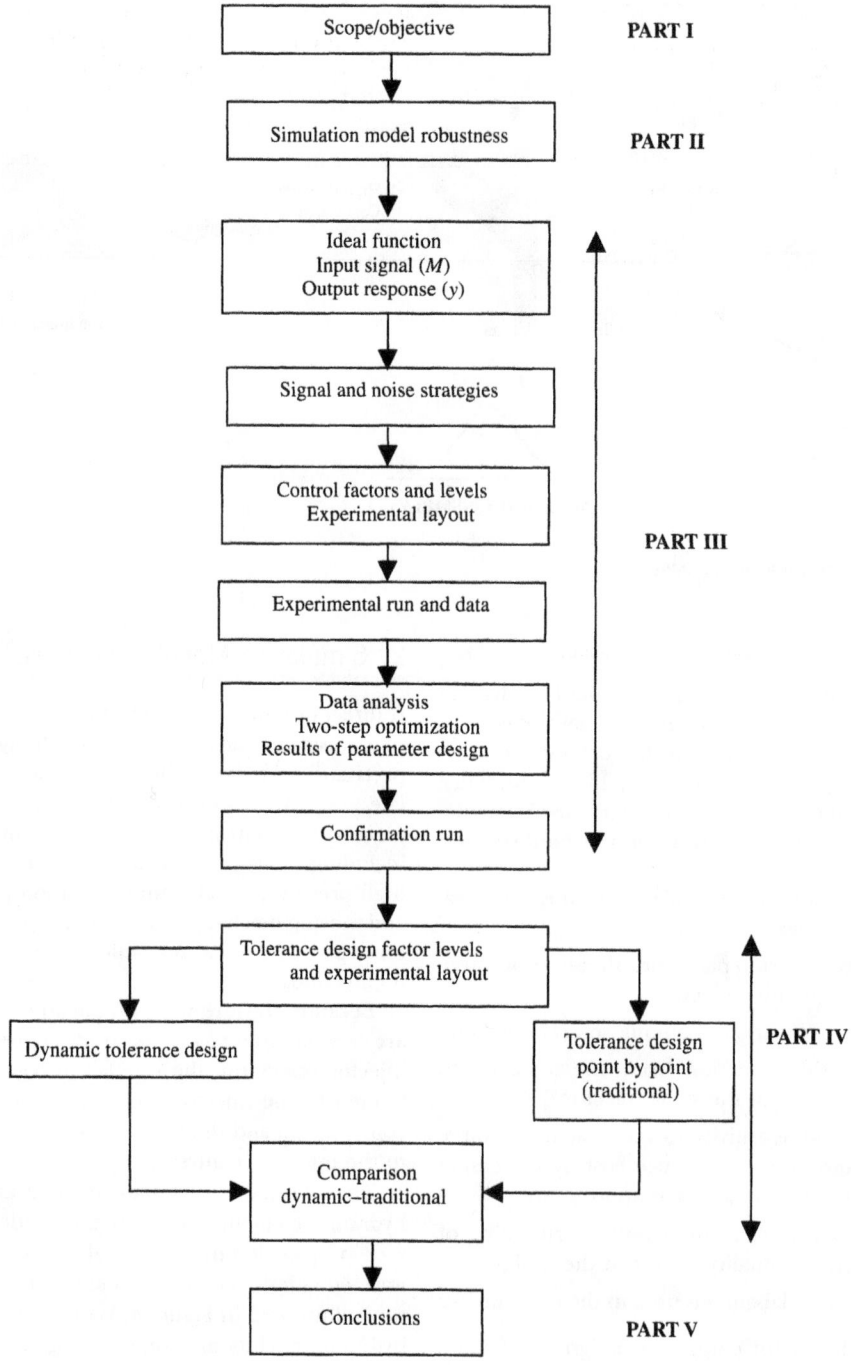

Figure 2
Optimization process

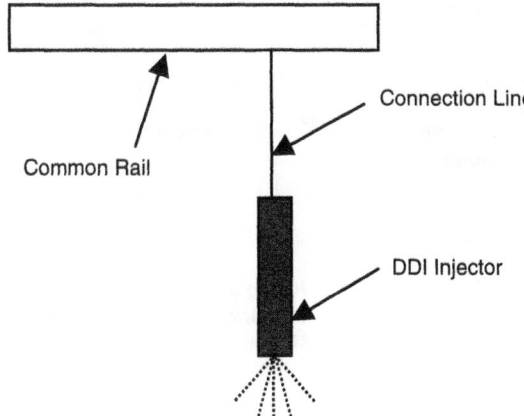

Figure 3
Modeled system

activating the electromechanical control valve and bleeding off a small amount of fluid. The duration of electrical activation, called the *pulse width*, is calculated by the engine control module (ECM), depending on driver demand, rail pressure, and engine torque needs. The other hydraulic circuit feeds the nozzle volume at a pressure, P_N, that remains close to common rail pressure. Using internal sealing, the area stressed by the pressure is larger on the control volume side than on the nozzle side. Thus, as long as P_C and P_N remain equal, the needle is pushed against the nozzle seat and the injector is closed. To start an injection event, the control valve is activated, which lowers the pressure in the control chamber until the force balance on the needle changes sign. Then the needle moves up, and injection occurs. To close the injector, the control valve is closed and the pressure builds up again in the control chamber until the force balance on the needle changes sign again and the needle moves down and closes.

The structure of the mathematical model of the injector can be deduced from its physical structure (Figure 5). Three submodels are coupled:

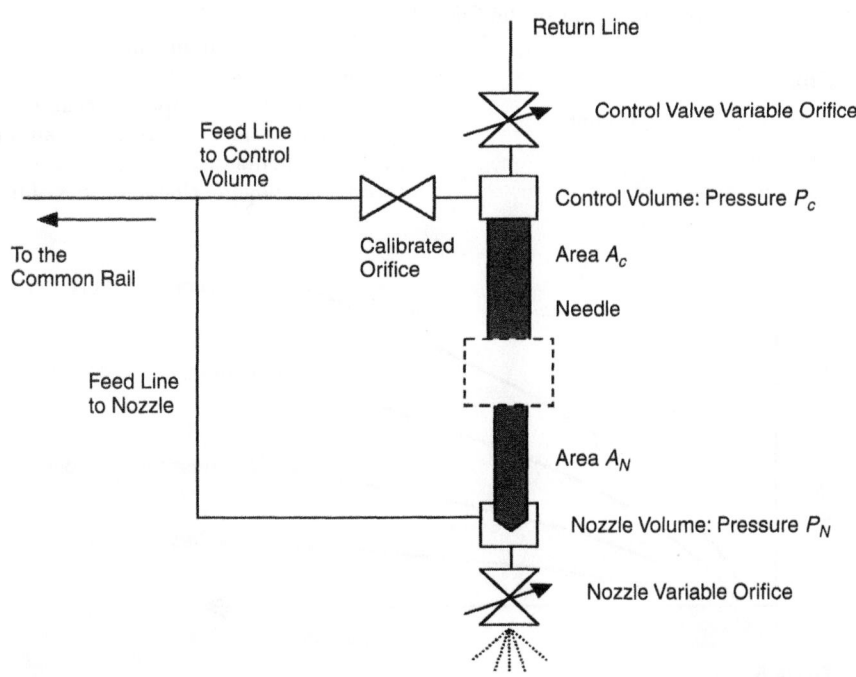

Figure 4
Injector

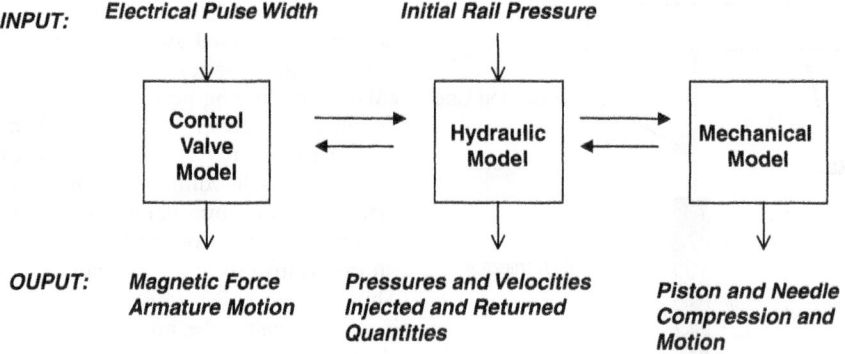

Figure 5
Mathematical model

1. Electromechanical control valve

2. Hydraulic circuits

3. Mechanical moving parts

Model inputs are as follows:

❏ Duration of electrical activation of electromagnetic actuator

❏ Rail pressure

Model outputs include time variation of the following elements:

❏ Magnetic force

❏ Fuel pressure and velocity in lines

❏ Flows through orifices

❏ Displacement of moving parts

The behavior of a direct-injection diesel injector can be characterized by a *mapping*, which gives the fuel quantity injected versus pulse width and rail pressure (Figure 6). Given the initial rail pressure and the pulse width, the model is expected to predict accurately the quantity of fuel that should be injected.

Approach to Optimization

The difficulty in building and then optimizing such a model is that some features have a very strong effect on the results. Imprecision about their model representation can affect results in an unacceptable way. Experimental investigations have been used to thoroughly study the characteristics of internal flows

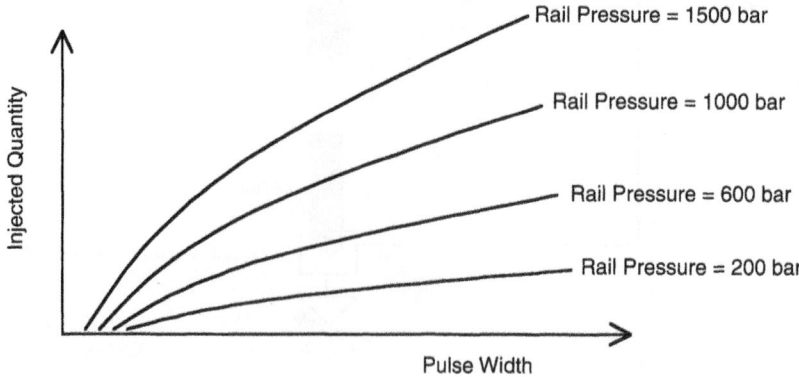

Figure 6
Quantity injected versus pulse width for various rail pressure levels

in key parts such as the control valve and the nozzle, for instance. From these experiments, empirical equations have been fitted and then implemented in the model. Finally, the transient nature of the flows is also a problem because most of the experiments are done statically.

In the end, after implementing the most realistic physical representation of variables, based either on experiments or on theoretical calculation, injector model optimization is a feedback loop between model results and experimentally observable injection parameters, such as injected quantity, injection timing, and control valve displacement. Engineering judgment is then necessary to assess the cause of any discrepancy and to improve or even change the physical representation of variables shown to be inadequate.

Results

Figure 7 shows the correlation between model and experiment for quantity injected, for rail pressures and pulse widths covering the full range of injector operation, and for three different injectors with different settings of control factor values. The agreement between model and hardware is good over the full range of operation. Figure 7 is the basis of the high confidence level we have in the model outputs.

3. Parameter Design

Ideal Function

The direct-injection diesel injector has several performance characteristics:

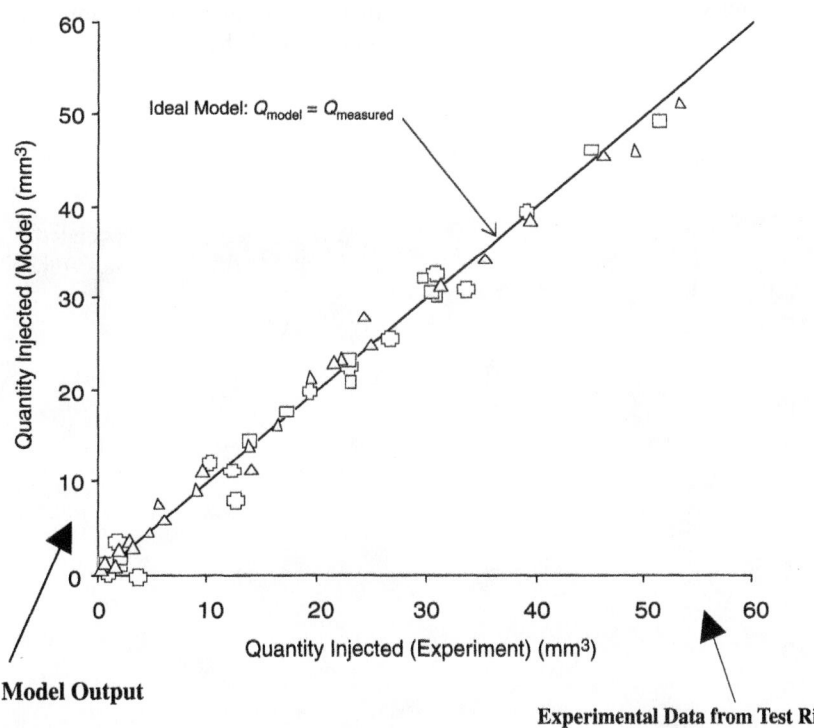

Figure 7
Comparison model data versus experimental data for various rail pressure, pulse width, and injector control factor settings ($\square$, reference; Δ, control factors A, B, G, E, and F changed; O, control factors A, B, G, E, and I changed)

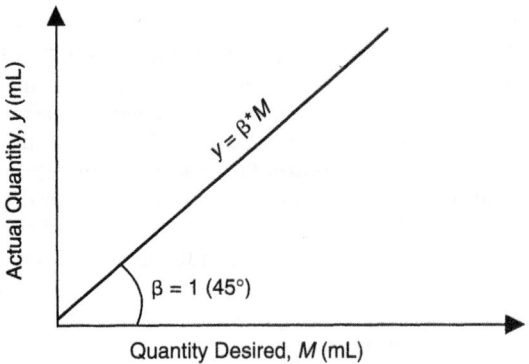

Figure 8
Ideal function

❑ Recirculated quantity
❑ Total opening time
❑ Delay of main injection
❑ Quantity injected

In the traditional approach, these performance characteristics would require individual optimizations and trade-off based on experimental results. In such a case, the power of a carefully chosen ideal function is invaluable. For an injection system, the customer requirement can be expressed as the quantity of fuel required for good combustion at a given operating point. By considering the customer need as a signal, we can compute through the simulation model, under noise conditions, what the actual injector will deliver. This last value is the *y*-axis of the ideal function. A perfect injector will have a slope of 1, with the actual quantity equal to the value desired (Figure 8). A design that exhibits the ideal function shown in Figure 8 is likely to be acceptable for any of the performance characteristics listed above.

Signal and Noise Strategies

Five levels of injected quantity desired were selected based on customer specifications: 2, 5, 15, 25, and 40 mL. Table 1 shows the impact of the noise factors

Table 1
Noise factors

Noise Factor	Level	Level	Relationship with Quantity Injected	Compounded Noise Level 1 (N_1)	Compounded Noise Level 2 (N_2)	Rationale[a]
B'	−10	+10	+	1	2	Mfg.
C'	−5	+5	+	1	2	Age
D'	−2	+2	+	1	2	Mfg., Age
E'	−3	+3	−	2	1	Mfg., Age
F'	−3	+3	−	2	1	Age
G'	−3	+3	+	1	2	Mfg., Age
H'	−4	+4	−	2	1	Age
I'	−30	+30	+	1	2	Mfg.
J'	−3	+3	+	1	2	Age
L	0.6	1	+	1	2	Mfg., Age
M	−10	+10	+	1	2	System
N	0	0.03	+	1	2	Mfg.
O	2	5	+	1	2	Mfg.

[a] Mfg., manufacturing variation; Age, aging; System, influence of other components.

Table 2
Control factors and levels (%)[a]

Factor	Level 1	Level 2	Level 3
A	−16.6	X	+16.6
B	−11.1[b]	X	+11.1
C	−16.6	X	+16.6
D	−11.1[b]	X	+11.1
E	−28.6	X	+28.6[b]
F	−18.5	X	+18.5
G	−18.2	X	+22.7[b]
H	+33.3	X	+66.6
I	−11.1	X	+11.1[b]
J	−8.62	X	+6.9[b]
K	−2.3	X	+2.3[b]

[a] X is a reference value. An L_{27} (3^{13}) orthogonal array was used to perform the experiment. Two columns were unused (see Figure 9).
[b] Current design level.

on the quantity injected. A (+) indicates that an increase in the noise factor will increase the injected quantity (direct relationship). A (−) indicates that a decrease in the noise factor will increase the injected quantity (inverse relationship).

The compounded noise level 1 (N_1) groups all the noise factor levels that have the effect of reducing the quantity injected. The compounded noise level 2 (N_2) groups the noise factor levels that have the effect of increasing the quantity injected.

B', C' ... , J' are noise factors obtained by considering variation (from manufacturing or other sources) of the corresponding control factors. L, M, N, and O are noise factors not derived from control factors. We expect these noises to be present during product use.

Control Factors and Levels

Table 2 shows control factors and levels for the experiment. Table 3 shows the experimental layout.

Data Analysis and Two-Step Optimization

The data were analyzed using the dynamic SN ratio:

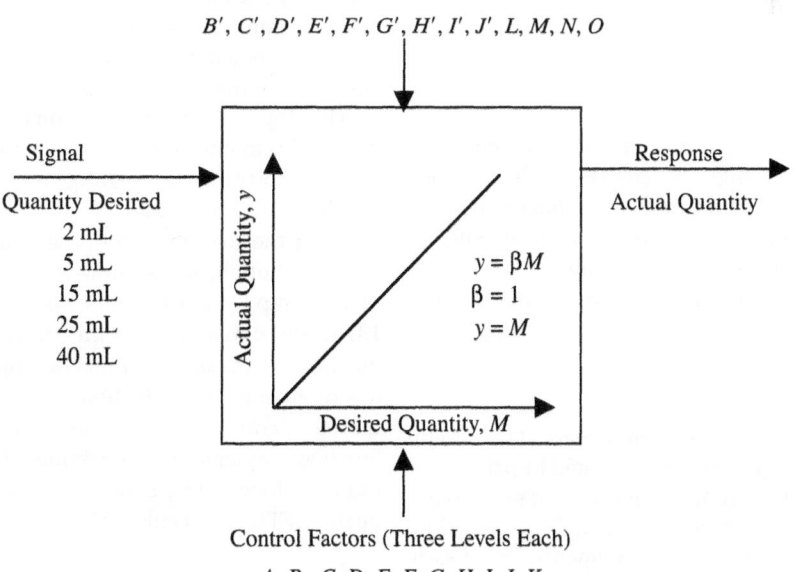

Noise Factors (Compounded to Two Levels)

$B', C', D', E', F', G', H', I', J', L, M, N, O$

Signal

Quantity Desired
2 mL
5 mL
15 mL
25 mL
40 mL

Actual Quantity, y

$y = \beta M$
$\beta = 1$
$y = M$

Desired Quantity, M

Response

Actual Quantity

Control Factors (Three Levels Each)

$A, B , C, D, E, F, G, H, I, J, K$

Figure 9
Parameter design

Table 3
Experimental layout

No.	A	B	C	D	E	F	G	H	I	J	K	12	13	2 mL		5 mL		15 mL		25 mL		40 mL		S/N	β
			Control Factor Array											N_1	N_2	N_1	N_2	N_1	N_2	N_1	N_2	N_1	N_2		
1	1	1	1	1	1	1	1	1	1	1	1	1	1												
2	1	1	1	1	2	2	2	2	2	2	2	2	2												
3	1	1	1	1	3	3	3	3	3	3	3	3	3												
⋮																									
25	3	3	2	1	1	3	2	3	2	1	2	1	3												
26	3	3	2	1	2	1	3	1	3	2	3	2	1												
27	3	3	2	1	3	2	1	2	1	3	1	3	2												

$$\eta = 10 \log \frac{S_\beta - V_e}{rV_e}$$

where S_β is the sum of squares of distance between zero and the least squares best-fit line (forced through zero) for each data point, V_e the mean square (variance), and r the effective divider. See Figure 10 for the SN and sensitivity analyses. Note the SN Y-axis values are all negative and expressed in decibels in the figure. See Tables 4 and 5 for further results.

Confirmation

The model in Table 6 confirms the expected improvement with a slight difference. Neither the optimum nor the current design combinations were part of the control factor orthogonal array. The initial hardware testing on the optimized configuration showed promising results. See Figure 11 for results.

Results

B and C are control valve parameters. The design change on these parameters suggested by parameter design is likely to improve injected quantity, part-to-part, and shot-to-shot variation. The simulation model confirms these improvements. Hardware confirmation is ongoing. A, D, E, G, H, I, J are hydraulic parameters. Implementing the changes

suggested by parameter design will decrease our sensitivity to most of the noise factors and to pressure fluctuations in the system.

An improvement in SN ratio can be translated directly into a reduction in variability:

$$\text{variability improvement} = (\tfrac{1}{2})^{\text{gain}/6} \times \text{initial variability}$$

where gain is the SN ratio gain in decibels; in our case, the model confirmed gain is 4.46 dB (Table 6). We are making an almost 40% reduction in variability from the initial design.

The slope of the ideal function is distributed normally. The tolerance on the slope can be obtained from the tolerances at each signal point of injection (Table 7 and Figure 12).

From the tolerances on β, and assuming a normal distribution, we can draw a curve to relate end-of-line scrap level or first-time quality (FTQ) (Figure 13) to the deviation from the target, β. This is not meant to reject the loss function approach. Rather, it is to quantify the reduction in part rejection (and therefore cost avoidance) at the end of the assembly line if we implement the recommendations from parameter design. The end-of-assembly line first-time quality (FTQ) is calculated by the formula

$$\text{FTQ} = \frac{\text{number of injectors within tolerances}}{\text{total number of injectors produced}}$$

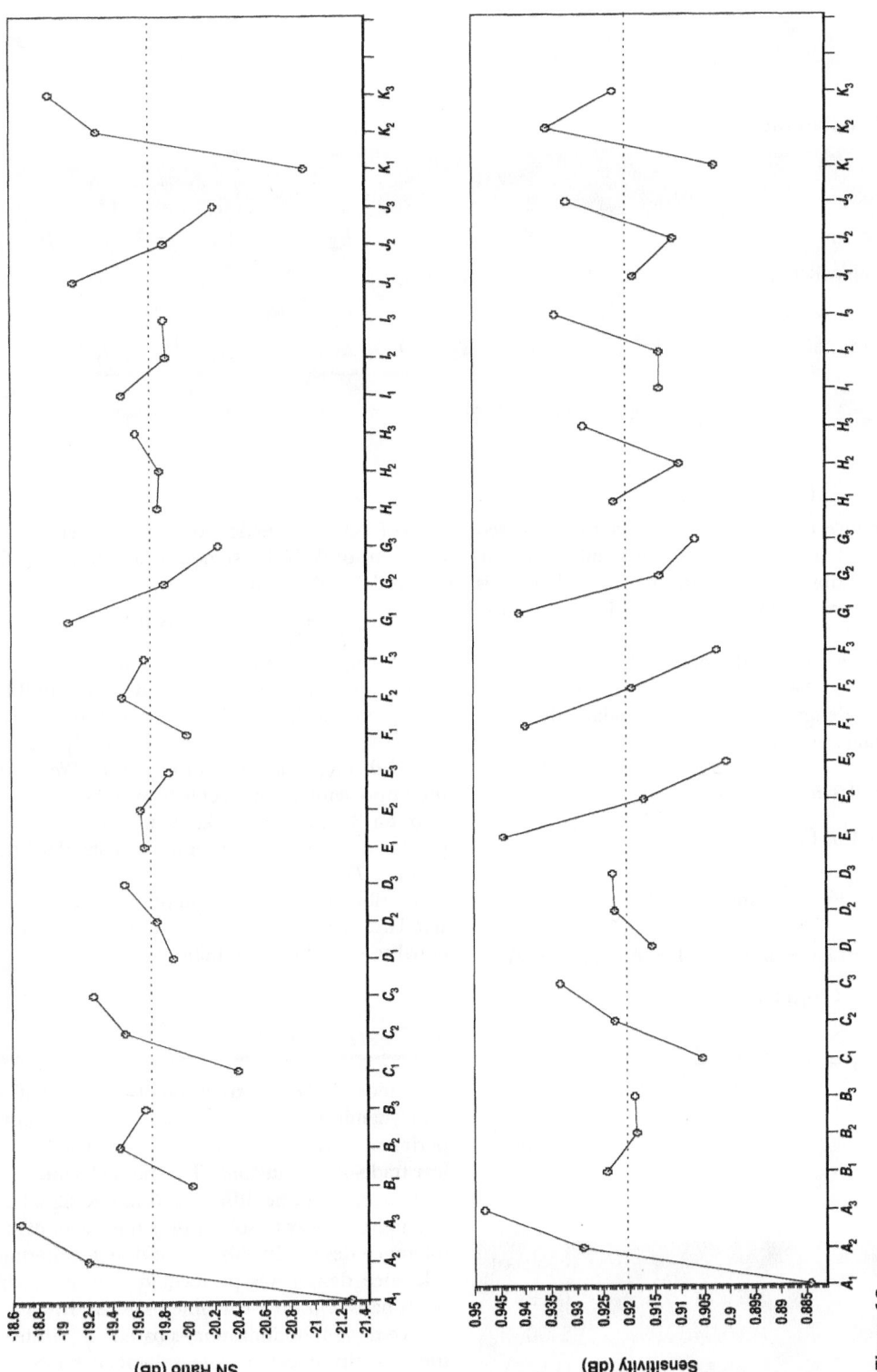

Figure 10
SN ratio and sensitivity plots

993

Table 4
Two-step optimization

	Factor										
	A	**B**	**C**	**D**	**E**	**F**	**G**	**H**	**I**	**J**	**K**
Initial design	A_2	B_1	C_1	D_1	E_3	F_2	G_3	H_2	I_3	J_3	K_3
SN maximization (first step)	A_3	B_2	C_3	D_3	$E_?$	F_2	G_1	H_3	I_1	J_1	K_3
Adjustment for beta (second step)					E_1						
Optimized design	A_3	B_2	C_3	D_3	E_1	F_2	G_1	H_3	I_1	J_1	K_3

Design change required ⎵ (A–E)   Design change required ⎵ (G–J)

The figure identifies FTQ. If β is the observed slope, $\mu = 1$ is the mean, z is the number of standard deviations away from the mean, and $\phi(z)$ is the standard cumulative normal distribution function.

In Figure 13, σ is the observed standard deviation on β, σ_i the standard deviation on β before parameter design, and σ_{opt} the standard deviation after parameter design.

$$\beta_{min} = 0.945$$

$$\beta_{max} = 1.054$$

$$Z_2 = \frac{\beta_{max} - 1}{\sigma} \quad \text{and} \quad Z_1 = \frac{\beta_{min} - 1}{\sigma}$$

$$FTQ = \phi(Z_2) - \phi(Z_1) = [1 - \phi(Z_1)] - \phi(Z_1)$$

$$= 1 - 2\phi(Z_1)$$

$$= 1 - 2\phi\left(\frac{\beta_{max} - 1}{\sigma}\right)$$

Table 5
Prediction

Design	SN (dB)	β (Slope)
Initial	−20.7118	0.8941
Optimized	−15.0439	1.0062
Gain	+5.66	0.1121

So FTQ is a function of the observed standard deviation of β. If the standard deviation of β from the initial design is σ_i,

$$\sigma_{opt} = \sqrt{0.6\sigma_i^2} = 0.77\sigma_i$$

considering the 40% reduction in variability from parameter design, the standard deviation of β from the optimized design after parameter design is the predicted end-of-line fraction of good parts is obtained through the curve in Figure 14. We simulate the improvement on FTQ with two different values of σ_i on β. It is hard to know the current value of σ_i, but we know that the optimized standard deviation is 0.77 σ_i.

Possible FTQ improvement in percent with a standard deviation on β from the initial design equal to σ_i is shown in Table 8.

4. Tolerance Design

Tolerance design is traditionally conducted with nondynamic responses. With a dynamic robust experiment, this is not optimal and can result in endless trade-off discussions. The ranked sensitivity to tolerances might be different from one signal point to another. A way to solve this is the use of dynamic tolerance design. In this robust design experiment, tolerance design was performed both on a signal point basis and on a dynamic basis. In dynamic tolerance design, optimization is based on β instead of the quantity injected. The tolerance on β is obtained by considering the tolerances at each signal

Table 6
Model confirmation

| Design | Predicted | | Confirmed | | | |
| | SN (dB) | β | Model | | Hardware | |
			SN (dB)	β	SN	β
Initial	−20.7118	0.8941	−20.2644	0.9150	Ongoing	Ongoing
Optimum	−15.0439	1.0062	−15.8048	0.9752	Ongoing	Ongoing
Gain	5.66	0.1121	4.46	0.06		

point and drawing a best-fit line. This ensures a more comprehensive tolerance design. (Figure 15). To illustrate, we perform tolerance design with the two approaches and compare them in Tables 9 and 10.

Signal Point by Signal Point Tolerance Design

Factors and Experimental Layout We used a three-level tolerance design with 13 parameters. Therefore, we will once again use an L_{27} array. The signal point–based tolerance design is performed using the data in the corresponding signal point column. Analysis of variance (ANOVA) for each injection point is shown in Table 11. The percent contribution is shown in Figure 16 (injection point 1: 2 mL desired). (The ANOVA was performed at each signal point.)

Loss Function For the signal point tolerance design, the loss function was nominal-the-best on the quantity injected. The calculations in Table 12 were performed for each signal point. In the table, $\%r$ is the percent contribution of the design parameter to the injected quantity variability. L_{tc} is the total current loss in dollars in the Taguchi loss function. σ_{pc} is the design parameter current standard deviation or its current tolerance level. σ_t is the current total output standard deviation, considering the variation of all design parameters. σ_p is a variable representing the standard deviation of a design parameter. L_{cp} is the current fraction of the total loss caused by the corresponding parameter. The higher the contribution percentage ($\%r$), the higher the L_{cp} value for a given design parameter. FTQ is the first-time

quality (end of assembly line). L_p is the loss caused by a design parameter given any value of the parameter standard deviation σ_p (L_p is a function of σ_p). See also Figure 17 and Table 13. K is the proportional constant of the loss function.

$$K = \frac{A_0}{\Delta_0^2} = 9.18C$$

A_0 = base average injector cost in dollars = $\$C$

Δ_0 = specification on quantity injected = 0.33 mL at 1σ

V_T = current total variance from ANOVA = 0.7542
$L_{tc} = KV_T = \$6.9235C$

Figure 18 shows the loss/cost as a function of each parameter.

Dynamic Tolerance Design

The control factor settings are the same as in the signal point tolerance design. In this case the response becomes the β from each combination. We use a nominal-the-best approach with the target β = 1 (ideal function). Results are shown in Table 14.

Dynamic Analysis of Variance β replaces the quantity injected. The target for β is 1 in this case. The boundaries for injector first-time quality correspond here to the upper and lower lines that we can draw on each signal point based on customer specifications (see Figure 12). The ANOVA is given in Table 15 and the percent contribution in Figure 19.

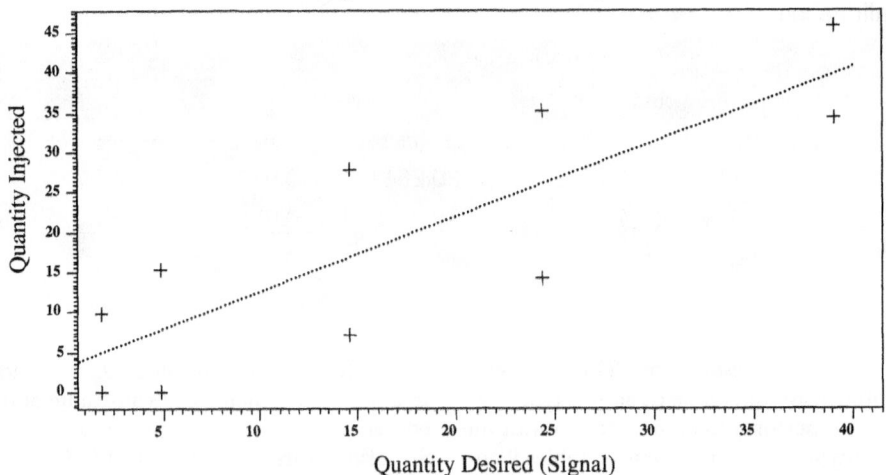

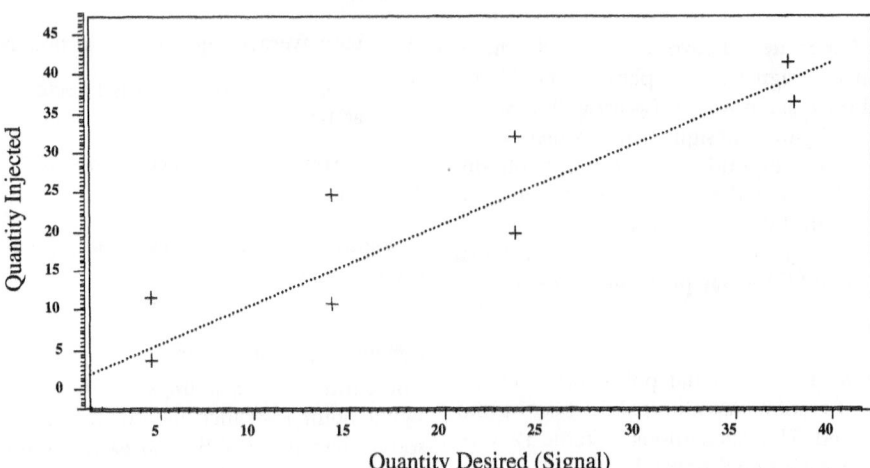

Figure 11
Graphical results of initial design compared to optimized design

Dynamic Loss Function The current loss distribution is shown in Table 16 and the dynamic tolerance design in Figure 20.

Dynamic characteristic: slope β. This is the same as signal point by signal point tolerance design. The response is β in this case; $\%r$ is the percent contribution of the design parameter on the β variability;

and K is the proportional constant of the loss function.

$$K = \frac{A_0}{\Delta_0^2} = 2500°C$$

A_0 = average injector cost in dollars = C

Table 7
Slope specifications

Quantity Desired	Minimum Quantity Acceptable (at 3σ)	Target	Maximum Quantity Acceptable (at 3σ)
2	1	2	3
25	21.5	25	28.5
40	36.5	40	43.5
β (slope)	0.945 (at 3σ)	1	1.054 (at 3σ)

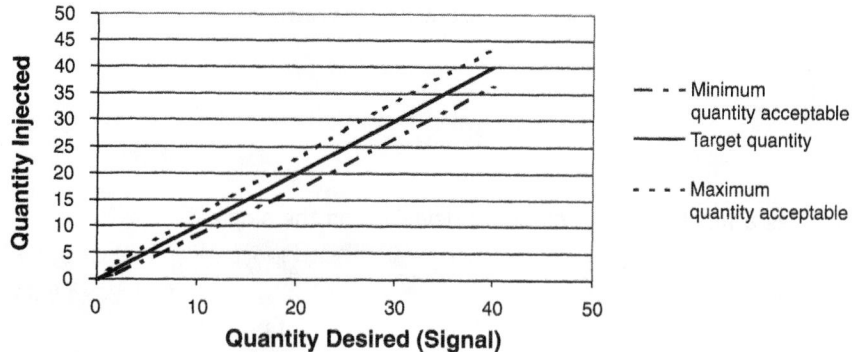

Figure 12
Slope tolerance

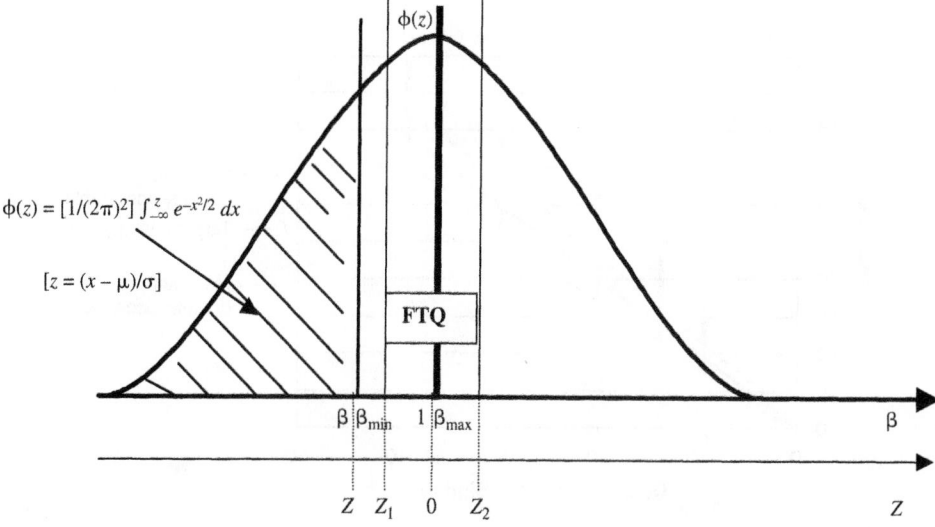

$$\phi(z) = [1/(2\pi)^2] \int_{-\infty}^{z} e^{-x^2/2}\, dx$$

$$[z = (x - \mu)/\sigma]$$

Figure 13
FTQ representation

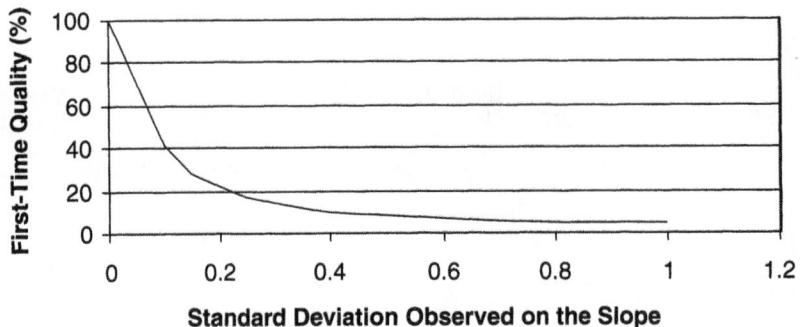

Figure 14
End-of-line FTQ as a function of the standard deviation observed on β

Table 8
End-of-line FTQ as a function of the standard deviation on the slope

σ_i	σ_{opt}	FTQ$_i$ (%)	FTQ$_{opt}$ (%)	ΔFTQ (%)	Gain (on FTQ %) Initial to Optimized (%)
0.10	0.076	41	52	11	26
0.05	0.038	72	84	12	16

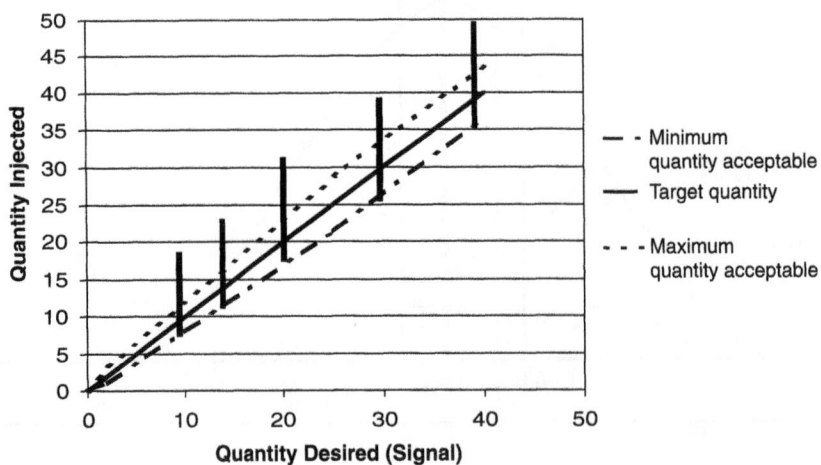

Figure 15
Tolerance design: signal point versus dynamic

Table 9

Factors and levels for tolerance design[a]

Factor	Level 1 $X - \sqrt{3/2}\,\sigma$	Level 2 X	Level 3 $X + \sqrt{3/2}\,\sigma$	Current σ (%)
A''				3.810
B''				3.704
C''				2.381
D''				1.333
E''				0.010
F''				0.004
G''				0.002
H''				2.667
I''				6.250
J''				0.002
K''				0.114
L				0.022
M				2.632

[a] X is the optimized level of each design parameter; σ is the current tolerance of each design parameter.

Table 10

Tolerance design experimental layout

No.	Tolerance Factor A''	B''	C''	D''	E''	F''	G''	H''	I''	J''	K''	L	M	Signal 2 mL	5 mL	15 mL	25 mL	40 mL
1	1	1	1	1	1	1	1	1	1	1	1	1	1					
2	1	1	1	1	2	2	2	2	2	2	2	2	2					
3	1	1	1	1	3	3	3	3	3	3	3	3	3					
⋮																		
25	3	3	2	1	1	3	2	3	2	1	2	1	3					
26	3	3	2	1	2	1	3	1	3	2	3	2	1					
27	3	3	2	1	3	2	1	2	1	3	1	3	2					

Table 11
ANOVA for signal point 1

Source	d.f.	S	V	F	S	ρ
A''	2	2.6413	1.3206	15.9851	2.4760	12.63
B''	2	4.3410	2.1705	26.2721	4.1758	21.30
C''	2	0.0672	0.0336	—	—	—
D''	2	2.2450	1.1225	13.5872	2.0798	10.61
E''	2	0.3953	0.1977	—	—	—
F''	2	0.9158	0.4579	5.5426	0.7506	2.83
G''	2	1.0480	0.5240	6.3427	0.8828	4.50
H''	2	0.0332	0.0166	—	—	—
I''	2	0.6072	0.3036	3.6750	0.4420	2.25
J''	2	0.5165	0.2582	3.1257	0.3512	1.79
K''	2	2.0346	1.0173	12.3134	1.8693	9.53
L	2	4.2696	2.1348	25.8403	4.1044	20.93
M	2	0.4940	0.2470	2.9899	0.3288	1.68
e_1						
(e)	6	0.4957	0.0826	—	2.1480	10.95
Total	26	19.6087	0.7542			

(e) is pooled error.

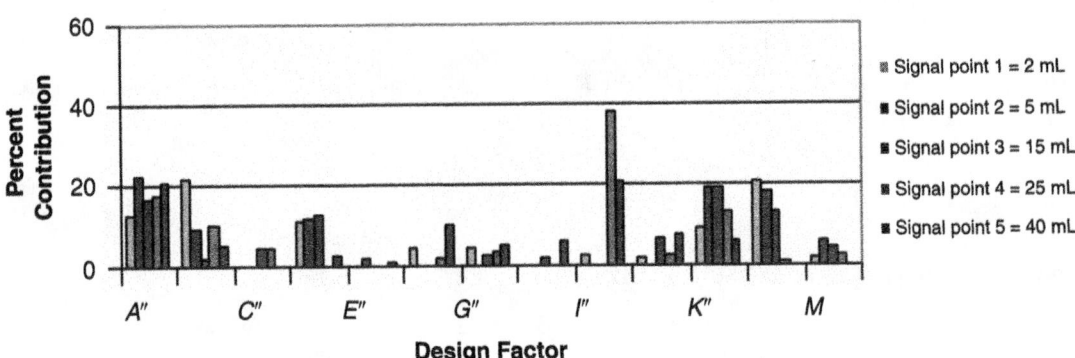

Figure 16
Percent contribution of design factors to variability per signal point

Table 12
Percent contribution of design factors to variability (signal point 1: 2 mL desired)

Factor	r (%)	L_{tc} (c)	Current, σ_{pc}	L_{cp} (c)
A''	12.63	6.9235	3.81	0.8744381
B''	21.30	6.9235	3.70	1.474706
C''	0	6.9235	2.38	0
D''	10.61	6.9235	1.33	0.7345834
E''	0	6.9235	0.01	0
F''	3.83	6.9235	0.00	0.2651701
G''	4.5	6.9235	0.00	0.3115575
H''	0	6.9235	2.67	0
I''	2.25	6.9235	6.25	0.1557788
J''	1.79	6.9235	0.00	0.1239307
K''	9.53	6.9235	0.11	0.6598096
L	20.93	6.9235	0.02	1.449089
M	1.68	6.9235	2.63	0.1163148

Δ_0 = specification on β = 0.02 at 1σ

V_T = current total variance from ANOVA = 0.0012

$L_{tc} = KV_T = \$3C$

Comparison of dynamic tolerance design and signal point tolerance design. The total loss with dynamic tolerance design was almost half of the total loss when tolerance design was conducted signal point by signal point. We do not have a precise explanation for this finding. The percent contributions of the

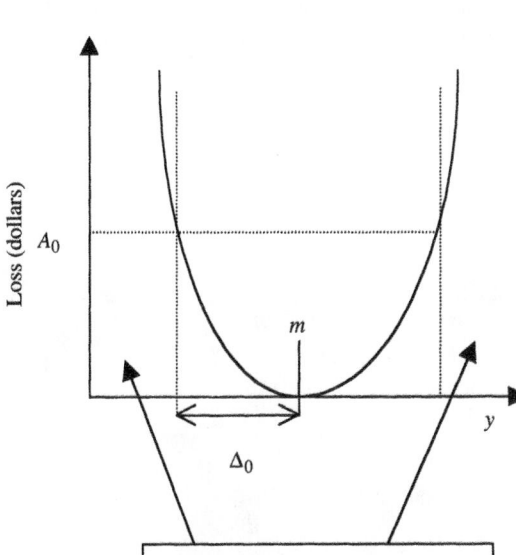

Figure 17
Taguchi loss function

Table 13
Summary of current total loss for some signal points

Signal Point (mL)	Current Total Loss
2	$6.9235C
25	$0.5541C
40	$0.5541C

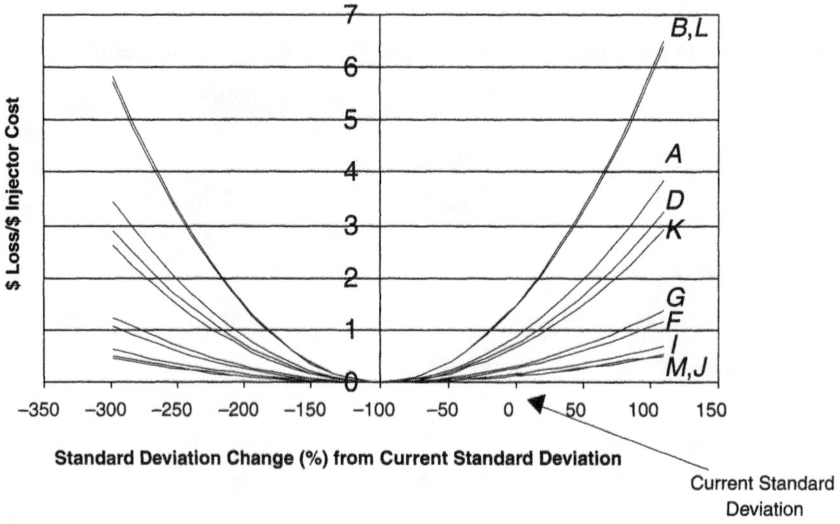

Figure 18
Chart to support manufacturing tolerance decision making (signal point 1: 2 mL)

design parameters at each signal point (see Figure 16) were difficult to use for decision making, due to the complexity of trade-off. We found it easier to consider the global percent contributions obtained from dynamic tolerance design (see Figure 19).

5. Conclusions

A gain of 4.46 dB in the SN ratio was realized in the modeled performance of the direct-injection diesel injector. This gain represents a reduction of about

Table 14
Experimental layout for dynamic tolerance design

	Tolerance Factor													
No.	*A"*	*B"*	*C"*	*D"*	*E"*	*F"*	*G"*	*H"*	*I"*	*J"*	*K"*	*L*	*M*	
1	1	1	1	1	1	1	1	1	1	1	1	1	1	β_1
2	1	1	1	1	2	2	2	2	2	2	2	2	2	β_2
3	1	1	1	1	3	3	3	3	3	3	3	3	3	β_3
⋮														⋮
25	3	3	2	1	1	3	2	3	2	1	2	1	3	β_{25}
26	3	3	2	1	2	1	3	1	3	2	3	2	1	β_{26}
27	3	3	2	1	3	2	1	2	1	3	1	3	2	β_{27}

Table 15
Dynamic tolerance design ANOVA

Source	d.f.	S	V	F	S'	ρ
A″	2	0.0117	0.0059	35.4725	0.0114	35.50
B″	2	0.0010	0.0005	2.9348	0.0006	1.99
C″	2	0.0066	0.0033	20.1535	0.0063	19.73
D″	2	0.0001	0.0000	—	—	—
E″	2	0.0006	0.0003	—	—	—
F″	2	0.0002	0.0001	—	—	—
G″	2	0.0000	0.0000	—	—	—
H″	2	0.0005	0.0003	—	—	—
I″	2	0.0029	0.0015	8.9340	0.0026	8.17
J″	2	0.0038	0.0019	11.5096	0.0035	10.82
K″	2	0.0005	0.0003	—	—	—
L	2	0.0004	0.0002	—	—	—
M	2	0.0037	0.0018	11.0911	0.0033	10.39
e_1						
e_2						
(e)	14	0.0023	0.0002	—	0.0043	13.39
Total	26	0.0320	0.0012			

(e) is pooled error.

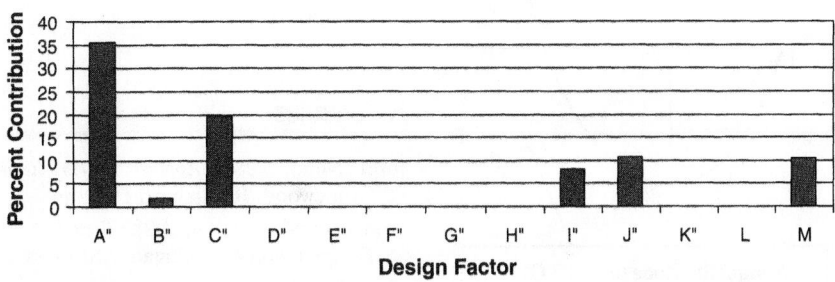

Figure 19
Percent contribution of design factors on the slope (dynamic tolerance design)

Table 16
Total current loss distribution among parameters

Factor	r (%)	L_{tc} (c)	L_{cp} (c)
A″	35.5	3	1.065
B″	1.99	3	0.0597
C″	19.73	3	0.5919
D″	0	3	0
E″	0	3	0
F″	0	3	0
G″	0	3	0
H″	0	3	0
I″	8.17	3	0.2451
J″	10.82	3	0.3246
K″	0	3	0
L	0	3	0
M	10.39	3	0.3117

40% in the variability in the quantity injected. We anticipate that the variability reduction will translate to a 16 to 26% increase of the fraction of parts inside tolerances for quantities injected at the end of the manufacturing line.

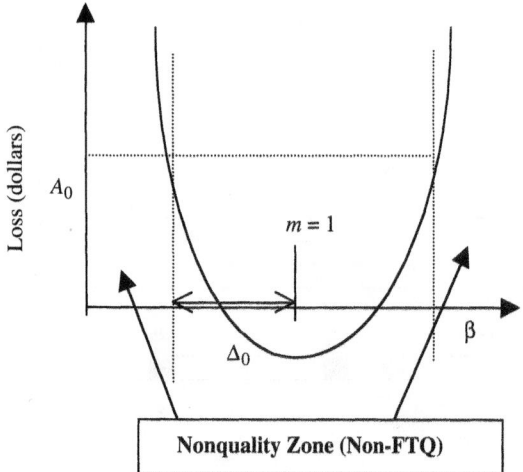

Figure 20
Taguchi loss function for dynamic tolerance design

Several charts were developed to support manufacturing process tolerance decision making. As information on tolerance upgrade or degrade cost is made available, the charts will be used to reduce cost further by considering larger tolerances where appropriate.

Recommendations

We recommend the use of dynamic tolerance design for dynamic responses. Optimization at individual signal points may not always lead to a clear and consistent design recommendation. This case study proposes a solution for a nominal–the-best situation for tolerance design. A standardized approach needs to be developed to cover dynamic tolerance design.

Hardware cost is one of the roadblocks to Taguchi robust design implementation. The development of good simulation models will help overcome this roadblock while increasing the amount of good information that one can extract from the projects. With a good model, we can conduct parameter design and tolerance design with a great number of control factors. We are also able to simulate control factor manufacturing variation as noise factors. The ability to consider manufacturing variation is an important advantage compared to full hardware-based robust engineering. Hardware confirmation should be conducted to ascertain the results and should not lead to a surprise if a good robustness level is achieved up front on the simulation model.

Acknowledgments We would like to acknowledge Dr. Jean Botti, Dr. Martin Knopf, Aparicio Gomez, Mike Seino, Carilee Cole, Giulio Ricci from Delphi Automotive Systems and Shin Taguchi and Alan Wu from the American Supplier Institute for their technical support.

References

John Neter, 1983. *Applied Linear Regression Models.* Homewood, IL: Richard D. Irwin.

Yuin Wu and Alan Wu, 1998. *Taguchi Methods for Robust Design.* Livonia, Michigan; ASI Press.

This case study is contributed by Desire Djomani, Pierre Barthelet, and Michael Holbrook.

CASE 50

Optimization of Disk Blade Mobile Cutters

Abstract: In this study we attempted initially to consider the function as input/output of energy, as we studied in machining. However, we could not grasp the difference in a motor's power consumption when a cutter is moved without and with cutting. Then we conducted an experiment using cutting resistance as output. As a consequence, we succeeded in optimizing each parameter in the cutter.

1. Introduction

Figure 1 shows the structure of a disk blade mobile cutter. We move a carriage with a disk blade along a fixed blade with an edge on one side by using a motor and cut a label placed on the fixed blade. The carriage has a shaft for rotating the disk blade and compressed spring to press the disk blade onto the fixed blade.

A good label cutter is considered a cutter that can cut a label with less energy regardless of width or material the label is made from. Therefore, the generic function of label cutting is defined as proportionality between cutting resistance of a cutter and width of a cutout label with a cutting range in accordance with the edge's movement. In other words, if cutting length and resistance are set to an input signal, M, and output, y, respectively, the equation of $y = \beta M$ holds true. In addition, sensitivity, β, should be smaller.

For measurement of cutting resistance, load imposed on the carriage from start to end of cutting is measured in the cutting direction (X-direction in Figure 1) and the direction perpendicular to it (Y-direction in Figure 1). As shown in Figure 2, the cutting resistance when the carriage cuts a label is calculated by subtracting the load with no label from the total load and integrating it within the range of a label width. Then, by measuring both cutting resistances X and Y in the cutting and per-

pendicular directions, respectively, and calculating the following value,

$$y = \sqrt{X^2 + Y^2} \tag{1}$$

we define y.

As noise factors, we chose edge deterioration, N, and carriage moving direction, N'. For the former, we deteriorated the edge at an increasing speed by cutting a polish tape with it. Additionally, as an indicative factor, we selected P_1 for difficult-to-cut paper due to its thickness and P_2 for difficult-to-cut cloth due to its softness.

2. SN Ratio

Table 1 shows the data of cutting resistance for row 1 in the L_{18} orthogonal array. The following is a calculation method for the SN ratio and sensitivity.

Total variation:

$$S_T = 7.71^2 + 9.47^2 + \cdots + 24.44^2$$

$$= 3274.8 \qquad (f = 12) \tag{2}$$

Effective divider:

$$r = 30^2 + 67^2 + 100^2 = 15,389 \tag{3}$$

Linear equation:

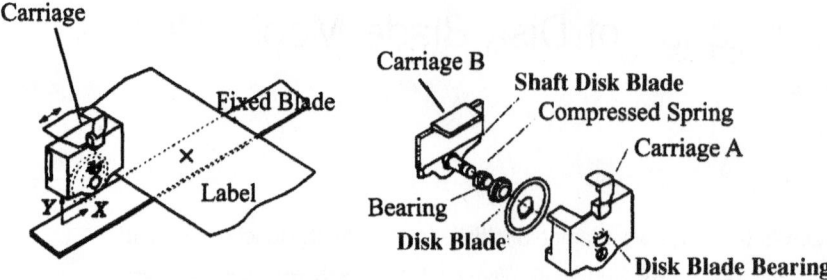

Figure 1
Structure of a disk blade mobile cutter

$$L_1 = (7.71)(30) + (15.21) + (67)(21.69)(100)$$

$$= 3419.95$$

$$L_2 = 3422.28$$

$$L_3 = 3583.16$$

$$L_4 = 3732.99 \tag{4}$$

Variation of proportional term:

$$S_\beta = \frac{(L_1 + L_2 + L_3 + L_4)^2}{4r} = 3256.55 \quad (f = 1) \tag{5}$$

Variation of differences of proportional terms $S_{N\beta}$ for edge deterioration:

$$S_{N\beta} = \frac{(L_1 + L_2)^2 + (L_3 + L_4)^2}{2r} - S_\beta$$

$$= 3.65 \quad (f = 1) \tag{6}$$

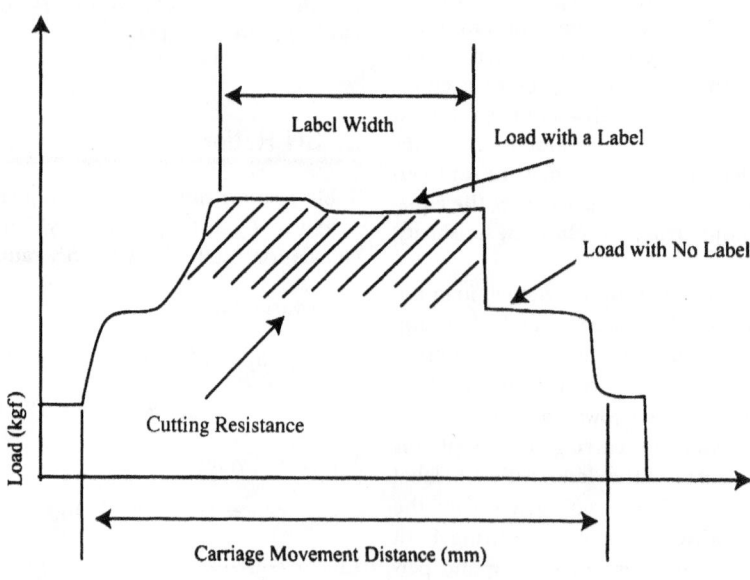

Figure 2
Cutting resistance generated during carriage movement

Table 1
Cutting resistance data for row 1 in L_{18} orthogonal array[a]

			M_1 30 mm	M_2 67 mm	M_3 100 mm	Linear Equation
P_1	N_1	N'_1	7.71	15.21	21.69	L_1
		N'_2	9.47	14.52	21.65	L_2
	N_2	N'_1	8.10	15.28	23.16	L_3
		N'_2	8.67	15.35	24.44	L_4
P_2	N_1	N'_1	1.01	1.31	1.88	L_5
		N'_2	2.11	1.44	1.31	L_6
	N_2	N'_1	11.69	18.10	25.83	L_7
		N'_2	12.96	19.52	29.64	L_8

[a] M, signal factor; N, edge deterioration; N', carriage movement direction; P_1, thick paper; P_2, cloth.

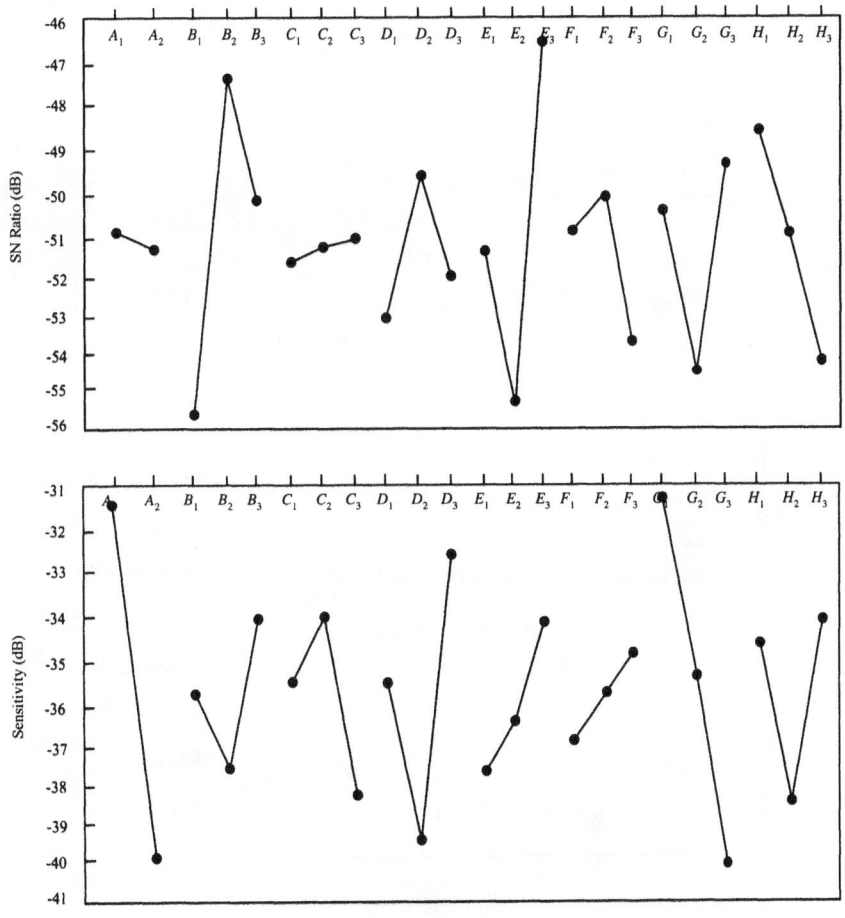

Figure 3
Response graphs

Table 2
Control factors and levels (dB)

	Control Factor	Level 1	Level 2	Level 3
A:	pressing position on carriage	A_1[a]	A_2	—
B:	rake angle of disk blade	Small[a]	Mid	Large
C:	edge width of disk blade	Large[a]	Mid	Small
D:	material of disk blalde	D_1[a]	D_2	D_3
E:	rake angle of fixed blade	E_1[a]	E_2	E_3
F:	material of fixed blade	Low	Same as current[a]	High
G:	pressing force	Small[a]	Mid	Large
H:	bearing position on carriage	H_1[a]	H_2	H_3

[a]Current condition.

Table 3
Confirmatory experimental results (dB)

	SN Ratio		Sensitivity	
Condition	Estimation	Confirmation	Estimation	Confirmation
Optimal	−45.26	−37.94	−47.89	−48.60
Current	−54.06	−52.86	−26.80	−31.67
Gain	8.80	14.92	−21.09	−16.93

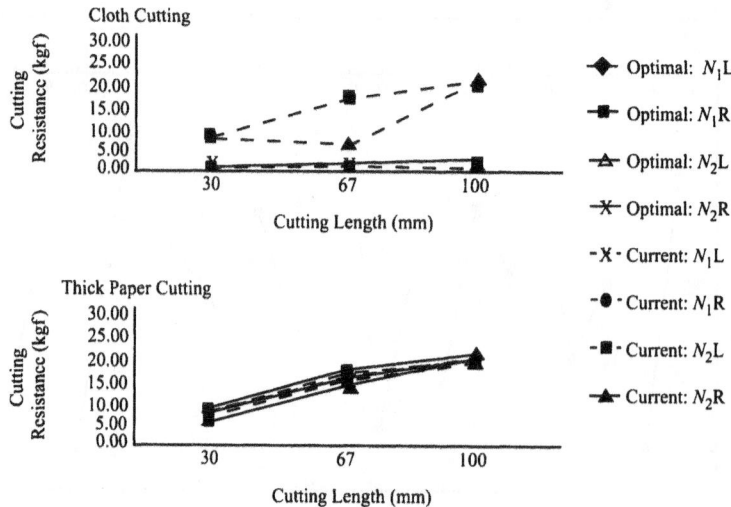

Figure 4
Measured data in confirmatory experiment

Table 4

Confirmatory experimental results

	Number of Cutting Polish Tape until Cutter Becomes Unable to Cut Standard Cloth					
Condition	0	50	100	150	200	250
Current	$\longrightarrow$					
Optimal	$\xrightarrow{\hspace{6cm}}$					

Variation of differences of proportional terms $S_{N\beta}$ for carriage movement direction:

$$S_{N\beta} = \frac{(L_1 + L_3)^2 + (L_2 + L_4)^2}{2r} - S_\beta$$

$$= 0.38 \qquad (f = 1) \qquad (7)$$

Error variation:

$$S_e = S_T - S_\beta - S_{N\beta} - S_{N'\beta} = 14.30 \qquad (f = 9) \quad (8)$$

Error variance:

$$V_e = \frac{S_e}{9} = 1.59 \qquad (9)$$

Total error variance:

$$V_N = \frac{S_{N\beta} + S_{N'\beta} + S_e}{11} = 1.67 \qquad (10)$$

Table 5

Evaluation of cutting performance

	Current Condition		Optimal Condition	
Material	N_1	N_2	N_1	N_2
Thick paper	○	○	○	○
Standard cloth	○	×	○	○
Soft cloth	×	—	○	×
Extremely soft film	×	—	○	×

ᵃ○, can be cut; ×, cannot be cut.

SN ratio:

$$\eta = 10 \log \frac{(1/4r)(S_\beta - V_e)}{V_N} = -14.98 \text{ dB} \qquad (11)$$

Sensitivity:

$$S = 10 \log \frac{1}{4r}(S_\beta - V_e) = -12.77 \text{ dB} \qquad (12)$$

3. Optimal Condition and Confirmatory Experimental Results

Figure 3 illustrates the response graphs of the SN ratio and sensitivity. As shown in Table 2, we chose control factors supposed to contribute much to cutting. By calculating the SN ratio and sensitivity for thick paper and cloth as an indicative factor and considering all aspects of SN ratios, sensitivities, cost, and ease of application to real products, we chose $A_2B_2C_3D_2E_1F_1G_3H_2$ as the optimal condition. After implementing a confirmatory experiment under optimal and current conditions, we obtained the results shown in Table 3, which reveals an insufficient reproducibility for gain.

Figure 4 shows the data for cutting resistance when cloth and thick paper are cut under the current and optimal conditions in the confirmatory experiment. As compared to the current condition, both the variability and magnitude of cutting resistance are reduced. In addition, Tables 4 and 5 demonstrate the robustness in basic cutting performances.

Figure 4 shows the comparison of durability for cutting a standard cloth under the current and optimal conditions. To evaluate edge deterioration, we attempted to cut standard cloth after cutting a polish tape a certain number of times. As a result, under current conditions, the cutter became unable to cut the standard cloth after cutting the polish tape approximately 25 times, whereas the number was 250 under the optimal condition. On the other hand, Table 5 highlights that extremely soft cloth and film, both of which cannot be cut under the current condition, can be cut under the optimal condition, N_1 (initial condition).

These results prove that we can improve a cutter's durability, achieve a wider range of objectives to be cut, and increase versatility of a cutter under the optimal condition.

In this study we attempted initially to consider the function as input/output of energy, as we studied in machining. However, we could not grasp the difference in a motor's power consumption when a cutter is moved without and with cutting. Then we gave up an experiment based on input/output of energy and conducted an experiment using cutting resistance as output. As a consequence, we succeeded in optimizing each parameter in the cutter.

Reference

Genji Oshino, Isamu Suzuki, Motohisa Ono, and Hideyuki Morita, 2001. Robustness of circle-blade cutter. *Quality Engineering*, Vol. 9, No. 1, pp. 37–44.

This case study is contributed by Genji Oshino.

CASE 51

D-VHS Tape Travel Stability

Abstract: In the case where a capstan and head cylinder rotate and a tape travels in an ideal manner under a proper condition, the output waveform of a D-VHS tape is determined theoretically. Thus, we supposed that we could evaluate the travel stability of a tape by assessing this difference between ideal and actual waveforms. First focusing on time assigned to the horizontal axis of the output waveform plot, we considered the relationship between the time the output reaches the peak point and the ideal time to reach as the generic function.

1. Introduction

What controls travel of a VTR tape is a pinch roller and capstan (Figure 1). Therefore, the generic function is regarded as travel distance of a tape for the number of revolutions of a capstan (Figure 2). However, since the travel stability of the tape used for this study is a few dozens of micrometers and too small compared to the travel distance, we cannot evaluate it. By focusing on a shorter time interval, we considered some generic functions that can be used to evaluate stability at a minute level of a few dozen micrometers. Finally, we selected output waveform during tape playing.

Figure 3 shows the relationship between a signal track and head when a tape travels, and Figure 4 represents the output waveform. If a head traces a record track, output proportional to an area traced is produced.

When a capstan and head cylinder rotate and a tape travels in an ideal manner under proper conditions, the output waveform is determined theoretically. Thus, we supposed that we could evaluate the travel stability of a tape by assessing this difference between ideal and actual waveforms.

First focusing on time assigned to the horizontal axis of the output waveform plot, we considered the relationship between the time the output reaches the peak point and the ideal time to reach as the generic function. While the VTR head traverses the tape one time, the output waveform has six peaks. By selecting 10 out of 12 peak points (maximum, minimum), we set them to signal factors. As control factors, eight items enumerated in Table 1 were chosen. As noise factors, we selected the following three:

1. *Start and end of tape winding.* Any type of tape should travel in a stable manner. In this study the force applied to a tape at the start and at the end of tape winding was considered as a noise factor.

2. *Head.* Since a VTR has two heads, P_1 and P_2, their phases are shifted. Because travel of a tape should be stabilized for both heads, we chose head as a noise.

3. *Positions of head cylinder and tape.* Since a head traces a tape in a moment (1/30 s), we cannot evaluate VTR's travel stability for this short time interval. Therefore, we assessed it while a tape travels more than half circumference of a head cylinder (while it traces 100 times).

2. SN Ratio

We show the data for experiment 4 as an example in Table 2.

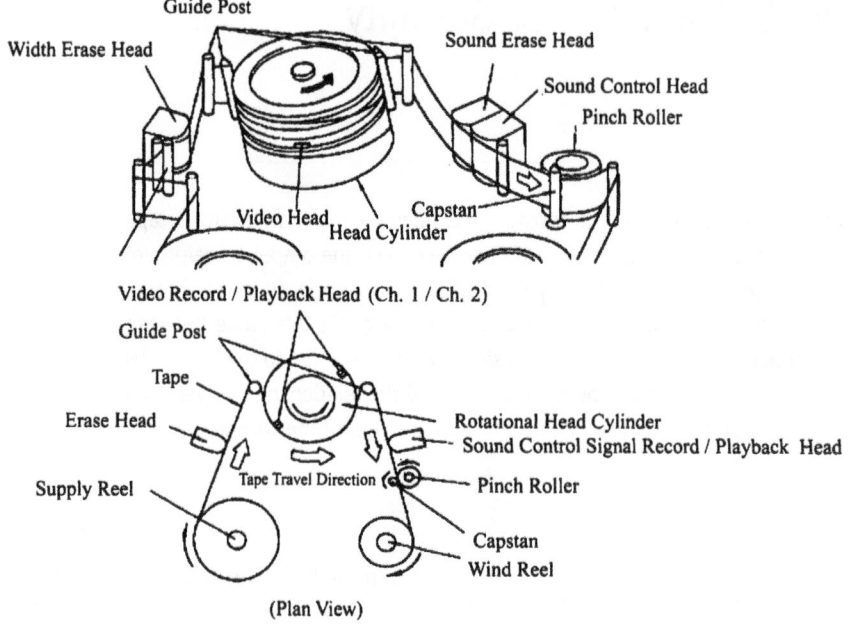

Figure 1
VTR mechanism

Total variation:

$$S_T = 25^2 + \cdots + 308^2 + \cdots + 320^2$$

$$= 153{,}751{,}460 \qquad (f = 4000) \qquad (1)$$

Linear equation:

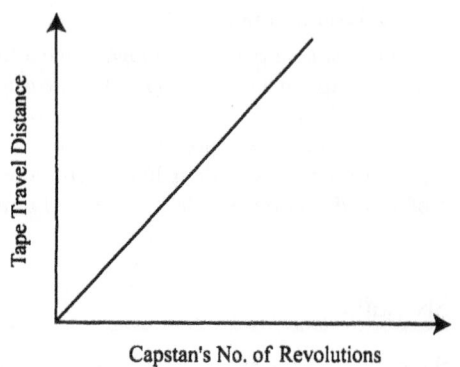

Tape Travel Distance

Capstan's No. of Revolutions

Figure 2
Generic function

$$L_1 = (32.1)(25) + \cdots + (320.5)(308)$$

$$= 377{,}307.7 \qquad (2)$$

Similar calculations are continued up to L_{400}.

Effective divider:

$$r = 32.1^2 + \cdots + 320.5^2 = 395504.6 \qquad (3)$$

Variation of proportional terms:

$$S_\beta = \frac{1}{400r}(L_1 + \cdots + L_{400})^2$$

$$= 153{,}671{,}134.7 \qquad (f = 1) \qquad (4)$$

$$S_{O\beta} = \frac{(L + \cdots + L_{200})^2 + (L_{201} + L_{400})^2}{200r} - S_\beta$$

$$= 19{,}468 \qquad (5)$$

$$S_{P\beta} = \frac{\begin{array}{c}(L_1 + \cdots + L_{100} + L_{201} + \cdots + L_{300})^2 \\ + (L_{101} + \cdots + L_{200} + L_{301} + \cdots + L_{400})^2\end{array}}{200r}$$

$$= 1964 \qquad (f = 1) \qquad (6)$$

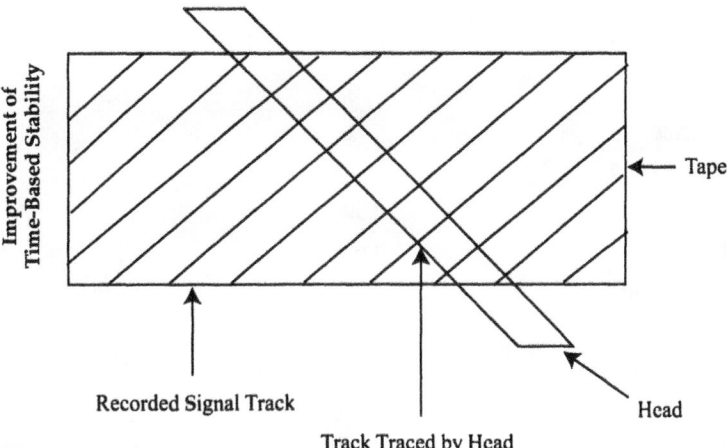

Figure 3
Relationship between signal track and head

$$S_{Q\beta} = \frac{(L_1 + \cdots + L_{301})^2 + (L_2 + L_{302})^2 + \cdots}{4r} - S_\beta$$

$$= 5584 \quad (f = 99) \tag{7}$$

$$S_e = S_T - S_\beta - S_{O\beta} - S_{P\beta} - S_{Q\beta}$$

$$= 53{,}309 \quad (f = 3898) \tag{8}$$

Error variance:

$$V_e = \frac{S_e}{3898} = 13.676 \tag{9}$$

Magnitude of noise:

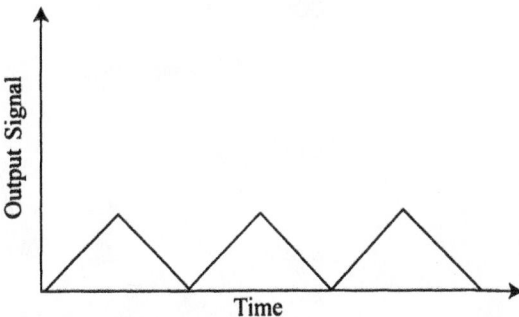

Figure 4
Output waveform during tape traveling

$$V_N = \frac{S_{O\beta} + S_{P\beta} + S_{Q\beta} + S_e}{3999} = 20.086 \tag{10}$$

SN ratio:

$$\eta = 10 \log \frac{(1/400r)(S_\beta - V_e)}{V_N} = -13.16 \text{ dB} \tag{11}$$

Sensitivity:

$$S = 10 \log \frac{S_\beta - V_e}{400r} = -0.126 \text{ dB} \tag{12}$$

3. Optimal Condition and Confirmatory Experiment

In the experiment, tapes had been damaged under 10 conditions. Since we could not analyze the conditions where a tape becomes damaged, from the SN ratio and sensitivity of the remaining eight conditions, we estimated optimal levels using the sequential approximation method. Table 3 shows the experimental results. From the results shown in the table we selected $A_2 B_3 C_1 D_1 E_3 F_2 G_2 H_2$ as the optimal condition by taking into account the number of NGs.

Using the result calculated through sequential approximation of the data that were obtained even when a tape was damaged, we also estimated the

Table 1
Control factors and levels

Control Factor	Level		
	1	2	3
A: stick-out length of head	Low	Current	—
B: height A of tape guide	Low	Current	High
C: height B of tape guide	Low	Current	High
D: height C of tape guide	Low	Current	High
E: reel clutch torque	Small	Current	Large
F: angle D of tape guide	Small	Current	Large
G: angle E of tape guide	Small	Current	Large
H: angle F of tape guide	Small	Current	Large

optimal condition (for experiments 1 and 8, we obtained no data because of the serious damage of a tape). This estimation showed the same results as the case using no-damage data only.

To confirm the effect under the optimal condition, we conducted a confirmatory experiment. The result is shown in Table 4 (for sensitivity, we do not use the data as judging criteria because there is no significant difference around 0 dB). From this we

noticed that poor reproducibility was obtained with a small gain.

4. Reproducibility

Control factor D, which is likely to have an interaction, has to be determined by the level of control

Table 2
Raw data for experiment 4 of the L_{18} orthogonal array (bits[a])

Error Factor			Measurement			
Tape O	Head P	Position Q	M_1 32.1	...	M_{10} 320.5	Linear Equation
Start	P_1	1	25		308	L_1
		...	...	...	...	...
		100	28		313	L_{100}
	P_2	1	29		316	L_{101}
		...	...	...	...	...
		100	26		313	L_{200}
End	P_1	1	32		317	L_{201}
		...	...	...	...	...
		100	32		323	L_{300}
	P_2	1	28		315	L_{301}
		...	...	...	...	...
		100	32		320	L_{400}

[a]1 bit = 4×10^{-5} s.

Table 3
SN ratio of control factor levels for tape traveling and "damage" data[a]

Control Factor	Level					
	1	NG	2	NG	3	NG
A	−13.04	6	**−13.60**	4	—	—
B	−13.87	4	−14.05	4	**−12.96**	2
C	**−13.46**	1	−13.70	4	−12.98	5
D	**−13.38**	2	−13.94	4	−13.15	4
E	−13.75	4	−13.15	4	**−13.47**	2
F	−13.88	4	**−13.09**	3	−13.55	3
G	−13.06	4	**−12.96**	3	−14.23	3
H	−13.15	5	−13.95	2	**−12.96**	3

[a]NG, not good. Boldface represents optimum conditions selected.

factor C. Next, selecting factors B and E, considered to contribute much to gain among the remaining control factors, we reset the levels. We did not select other control factors because their optimal levels were identical to the ones under the current condition. In contrast, control factor I, which was not chosen in the previous experiment, was added.

Because of four factors to be investigated, we used an L_9 orthogonal array for the reexperiment. Table 5 illustrates control factors for the experiment. Based on the reexperimental result, we plotted the factor effects in Figure 5. This plot implies that the optimal condition was estimated to be $B_2C_3E_3I_2$. Table 6 shows the results of the confirmatory experiment. Although there might exist some problems because the graph has many V-shapes and peaks, we obtained relatively good reproducibility.

As a reference, we calculated the eventual economic effect. The loss is expressed by $L = (A/\Delta^2)\sigma^2$. Now A is the loss when the travel exceeds a tolerance, Δ the tolerance, and σ the variance. As an example, we supposed that when the travel exceeds 30 μm, the VTR cannot display any picture. As a result, the customer has $A = 10,000$ yen of repair cost.

Taking into account the fact that the difference of the travel between under the current and optimal conditions is approximately 2.4 μm, we obtained the following economic effect per unit:

$$L_{\text{current}} - L_{\text{opt}} = \frac{A}{\Delta^2} (\sigma^2_{\text{current}} - \sigma^2_{\text{opt}})$$

$$= \frac{10,000}{30^2} (2.4^2) = 64 \text{ yen} \tag{13}$$

Table 4
Confirmatory experimental results for the L_{18} orthogonal array

	SN Ratio	
Configuration	Estimation	Confirmation
Optimal: $A_2B_3C_1D_1E_3F_2G_2H_2$	−6.52	−11.06
Current: $A_2B_2C_2D_2E_2F_2G_2H_2$	−12.50	−12.41
Gain	5.98	1.35

Table 5

Control factors for the L_9 orthogonal array (reexperiment)

		Level		
Control Factor		**1**	**2**	**3**
B:	height A of table guide	Current	—	+
C:	height B of table guide	—	Current	+
E:	reel clutch torque	—	Current	+
I:	tension A	—	Current	+

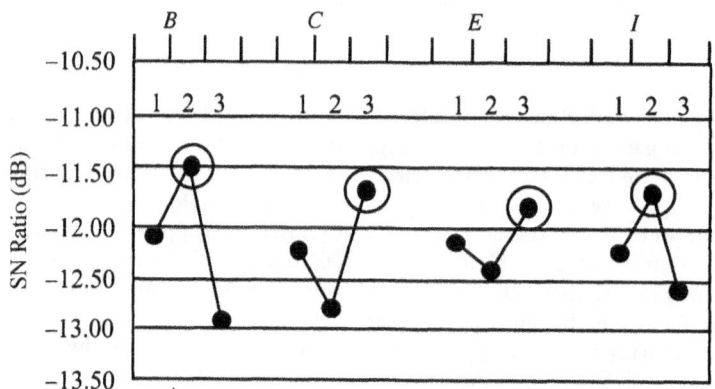

Figure 5
Response graphs of the SN ratio for the L_9 orthogonal array

Table 6

Confirmatory experimental result for the L_9 orthogonal array (economic effect) (dB)

	SN		Sensitivity	
Condition	**Estimation**	**Confirmation**	**Estimation**	**Confirmation**
Optimal	−10.17	−10.38	−0.22	−0.01
Current	−12.17	−12.41	−0.06	−0.06
Gain	2.00	2.03	−0.16	−0.05

Since the annual production volume of D-VHS is about 100,000, the total annual economic effect amounts to 6.4 million yen. Moreover, if this result is applied to all VTRs, we expect a larger-scale economic effect because the total annual production volume is over 4 million.

Reference

Hideaki Sawada and Yoshiyuki Togo, 2001. An investigation on the stability of tape run for D-VHS. *Quality Engineering*, Vol. 9, No. 2, pp. 63–70.

This case study is contributed by Hideaki Sawada and Yoshiyuki Togo.

CASE 52

Functionality Evaluation of Spindles

Abstract: In the development of spindles for machining, many quality characteristics, such as temperature rise of bearings, vibration noise, or deformation, have traditionally been studied, spending a long time for evaluation. Recently, the evaluation of machining by electricity was developed, which makes the job easier. This study was conducted from the energy transformation viewpoint, aiming for the stability of machining and reducing development cycle time.

1. Introduction

Although in most studies of machine tools, we evaluate performance by conducting actual machining, essentially it is important to change the design of machine tools. However, since this type of study can be implemented only by machine manufacturers, we tentatively conduct research on a main spindle as one of the principal elements in a machine tool. Figure 1 depicts a main spindle in a machining center. A tool is to be attached at the main spindle end and a motor is connected to its other end. The main spindle is required of stable performance for various tools and cutting methods under the condition ranging from low to high numbers of revolutions.

In general, manufacturers and users stress quality of a product for evaluating performance. In the traditional study of high-speed machining, temperature rise at bearings is the most focused on, and quality characteristics such as vibration, noise, and deformation are measured separately, thereby leading to an enormous amount of labor hours and cost. But to secure its stability in the development phase and shorten the development cycle from the standpoint of energy conversion, we studied the evaluation method grounded on functionality as well as quality characteristics.

2. Generic Function and Measurement Characteristic

A good main spindle is regarded as what can maintain a smooth revolution for instructed numbers of revolutions ranging from low to high speed and generate less heat and energy loss caused by vibration and noise. Setting a square root of spinning time to a signal and a square root of cumulative electric power to output (Figure 2), we considered the generic function, $\sqrt{y} = \beta\sqrt{T}$. On the other hand, since the main spindle is driven by the motor, we defined as the on-state the revolution when the main spindle is connected to the motor and as the off-state the revolution made only by the motor, and set the difference between the on- and off-states to the SN ratio. Figure 3 summarizes the transition of the measurements, and Figure 4 magnifies a part of them. Details are as follows:

❑ *Characteristics:* cumulative electric power when the main spindle is spinning, $\sqrt{y}$.

❑ *Signal factor:* revolution time, $\sqrt{T}$.

❑ *Noise factor:* since the number of revolutions changes all the time, we select the revolution-increasing phase (N_1) and decreasing phase (N_2). In addition, we chose the minimum

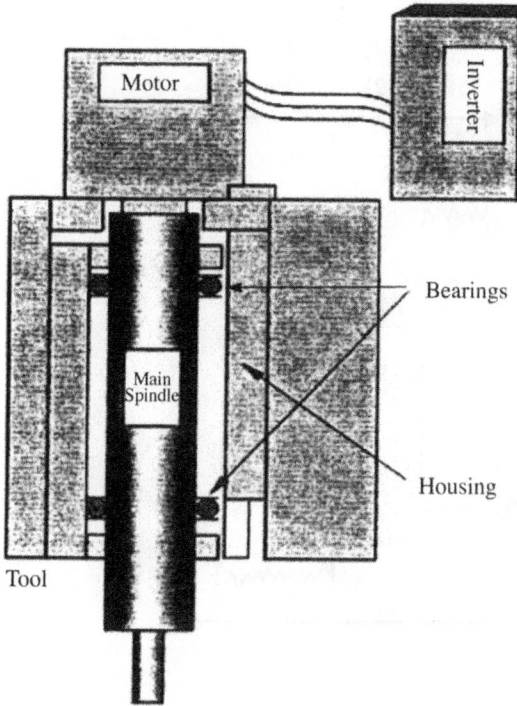

Figure 1
Main spindle in machining center

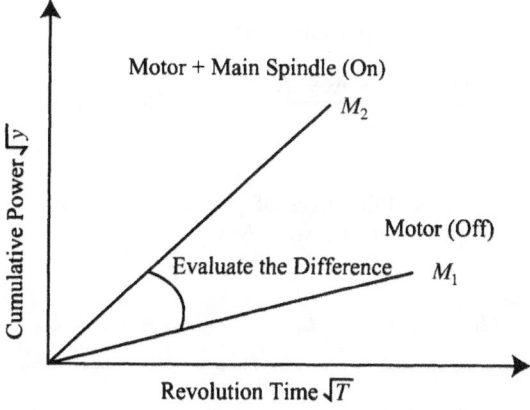

Figure 2
Generic function

value of electric power (N_1') and maximum value (N_{2r}') at each number of revolutions.

For the number of revolutions (k indicates 1000 revolutions), three values, 8k, 12k, and 16k per minute, were allocated to the outer array as the second signal. As control factors, we selected eight factors, such as type of bearing, part dimension, part shape, or cooling method, and assigned them to an L_{18} as the inner array.

3. Experimental Procedure and Calculation of the SN Ratio

Only for the main spindle shown in Figure 1, we prepared the experimental device. Using this, we measured momentary voltage by an electric power meter connected to the secondary terminal of the inverter control board to drive the motor. Two measurements were taken, one for revolution only by the motor, denoted by M_1^*, and the other for revolution by both motor and spindle, denoted by M_2^*. Table 1 shows the results of experiment 1. All data in Table 1 are expressed in the unit of a square root of cumulative electric power. The reason that we take a square root of time and cumulative electric power is that the original data is energy. That is, we followed the principle of quality engineering that squared data should become energy so that we can obtain energy through decomposition of variations. Based on these, by regarding the SN ratio, η^*, and sensitivity, S^*, of the effect difference between M_1^* obtained only by the motor only (off-state) and M_2^* by both the motor and main spindle (on-state) as the effective portion of the energy of the main spindle's revolution, we proceeded with calculations as follows.

Total variation:

$$S_T = 0.410^2 + 0.578^2 + \cdots + 2.945^2$$

$$= 257.62 \qquad (f = 120) \qquad (1)$$

Effective divider:

$$r = 5.477^2 + 7.746^2 + \cdots + 12.247^2 = 450.00 \qquad (2)$$

Linear equations:

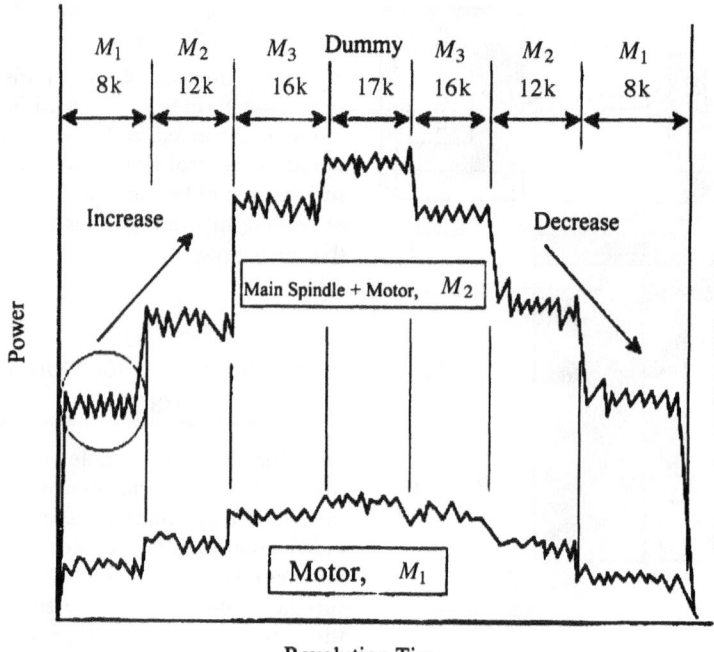

Figure 3
Transition of electric power at on- and off-states when main spindle is spinning

$$L_1 = (0.410)(5.477) + \cdots + (0.913)(12.247)$$

$$= 32.566$$

$$\vdots$$

$$L_{24} = 107.842 \qquad (3)$$

Variation of proportional term:

$$S_\beta = \frac{(L_1 + L_2 + \cdots + L_{24})^2}{24r} = 219.13 \qquad (f = 1) \qquad (4)$$

Variation of differences of proportional terms regarding number of revolutions:

$$S_{M\beta}$$
$$= \frac{\begin{array}{c}(L_1 + L_2 + L_3 + L_4 + L_{13} + L_{14} + L_{15} + L_{16})^2 \\ + (L_5 + \cdots + L_{20})^2 + (L_9 + \cdots + L_{24})^2\end{array}}{8r}$$
$$- S_\beta = 12.00990 \qquad (f = 2) \qquad (5)$$

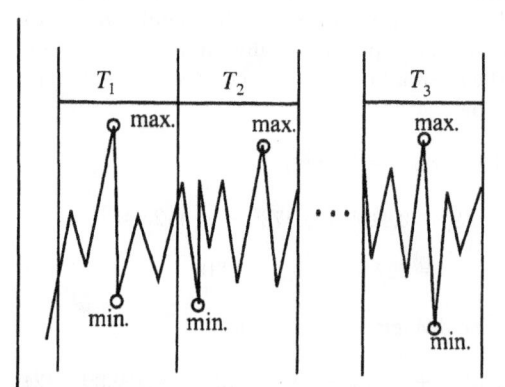

Figure 4
Data selection of minimum and maximum electric power values when main spindle is spinning

Variation of differences of proportional terms regarding motor only and motor and main spindle:

Table 1
Square root data of cumulative electric power

Time: Revolutions:		T_1 5.477	T_2 7.746	T_3 9.487	T_4 10.954	T_5 12.247	Linear Equation	
M_1^* (motor only)	8k increase max.	0.410	0.578	0.708	0.816	0.913	33.566	L_1
	min.	0.394	0.556	0.681	0.786	0.879	32.305	L_2
	8k decrease max.	0.403	0.575	0.707	0.818	0.915	33.546	L_3
	min.	0.396	0.560	0.686	0.791	0.883	32.488	L_4
	12k increase max.	0.529	0.748	0.917	1.058	1.177	43.399	L_5
	min.	0.515	0.728	0.893	1.023	1.138	42.080	L_6
	12k decrease max.	0.529	0.749	0.917	1.058	1.181	43.452	L_7
	min.	0.515	0.726	0.891	1.028	1.147	42.206	L_8
	16k increase max.	0.670	0.948	1.160	1.338	1.496	54.991	L_9
	min.	0.644	0.915	1.119	1.291	1.443	53.051	L_{10}
	16k decrease max.	0.674	0.951	1.167	1.349	1.511	55.409	L_{11}
	min.	0.650	0.923	1.136	1.313	1.471	53.883	L_{12}
M_2^* (motor and main spindle)	8k increase max.	0.904	1.216	1.461	1.673	1.855	69.279	L_{13}
	min.	0.733	1.057	1.303	1.506	1.683	61.680	L_{14}
	8k decrease max.	0.642	0.914	1.121	1.297	1.453	53.230	L_{15}
	min.	0.629	0.887	1.087	1.256	1.402	51.557	L_{16}
	12k increase max.	1.351	1.866	2.132	2.344	2.535	98.801	L_{17}
	min.	1.211	1.588	1.852	2.077	2.280	87.186	L_{18}
	12k decrease max.	0.937	1.325	1.623	1.876	2.094	76.994	L_{19}
	min.	0.911	1.292	1.581	1.826	2.038	74.964	L_{20}
	16k increase max.	1.407	1.981	2.422	2.794	3.119	114.838	L_{21}
	min.	1.330	1.896	2.331	2.690	3.009	110.398	L_{22}
	16k decrease max.	1.345	1.907	2.343	2.711	3.035	111.231	L_{23}
	min.	1.304	1.847	2.269	2.629	2.945	107.842	L_{24}

$$S_{M*\beta} = \frac{(L_1 + L_2 + L_3 + \cdots + L_{12})^2 + (L_{13} + L_{14} + \cdots + L_{24})^2}{12r} - S_\beta$$

$$= 22.92859 \qquad (f = 1) \qquad (6)$$

Variation of differences of proportional terms regarding increasing and decreasing phases:

$$S_{N\beta} = \frac{(L_1 + L_2 + L_5 + L_6 + \cdots + L_{21} + L_{22})^2 + (L_3 + L_4 + L_7 + L_8 + \cdots + L_{23} + L_{24})^2}{12r}$$

$$- S_\beta = 0.38847 \qquad (f = 1) \qquad (7)$$

Variation of differences of proportional terms regarding min and max:

$$S_{N'\beta} = \frac{(L_1 + L_3 + L_5 + \cdots + L_{21} + L_{23})^2 + (L_2 + L_4 + L_6 + \cdots + L_{22} + L_{24})^2}{12r} - S_\beta$$

$$= 0.14159 \qquad (f = 1) \qquad (8)$$

Variation of differences of proportional terms

regarding rpm and motor/(motor and spindle) combinations:

$$S_{MM*\beta} = \frac{1}{4r}$$

$$\left[\begin{array}{l} (L_1 + L_2 + L_3 + L_4)^2 + (L_5 + L_6 + L_7 + L_8)^2 \\ + (L_9 + \cdots + L_{12})^2 + \cdots + (L_{17} + \cdots + L_{20})^2 \\ + (L_{21} + \cdots + L_{24})^2 \end{array} \right]$$

$$- S_\beta - S_{M*\beta} - S_{M\beta} = 2.10615 \qquad (f = 2) \qquad (9)$$

Error variation:

$$S_e = S_T - S_\beta - S_{M*\beta} - S_{M\beta} - S_{N'\beta} - S_{N\beta} - S_{MM*\beta}$$

$$= 0.91277 \qquad (f = 112) \qquad (10)$$

Error variance:

$$V_e = \frac{S_e}{112} = 0.00815 \qquad (11)$$

Total error variance:

Table 2
Control factors and levels

	Control Factor	Level		
		1	2	3
A:	bearing type	A_1	A_2	—
B:	housing dimension	Small	Mid	Large
C:	main spindle dimension	Small	Mid	Large
D:	gap seat dimension	Small	Mid	Large
E:	gap seat shape	E_1	E_2	E_3
F:	tightening dimension	Small	Mid	Large
G:	bearing shape	Small	Mid	Large
H:	cooling method	H_1	H_2	H_3

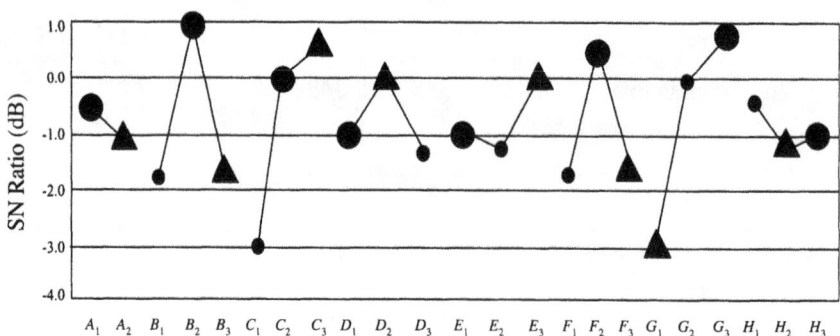

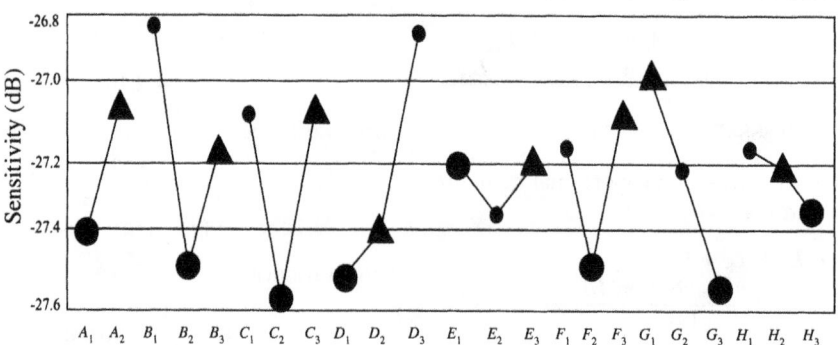

Figure 5
Factor effects plot

Table 3
Confirmatory experimental results (dB)

Condition	SN Ratio		Sensitivity	
	Estimation	Confirmation	Estimation	Confirmation
Optimal	4.33	1.20	−28.80	−28.25
Current	−2.71	−3.23	−26.55	−27.54
Gain	7.04	4.43	−2.25	−0.71

$$V_N = \frac{S_{N^*\beta} + S_{N\beta} + S_e}{114} = 0.012656 \quad (12)$$

SN ratio:

$$\eta^* = 10 \log \frac{(1/24r)(S_{M^*\beta} - V_e)}{V_N} = -7.76 \text{ dB} \quad (13)$$

Sensitivity:

$$S^* = 10 \log \frac{1}{24} (S_{M^*\beta} - V_e) = -26.73 \text{ dB} \quad (14)$$

4. Optimal Condition and Confirmatory Experiment

As control factors, we selected eight factors such as bearing type, part dimension, part shape, and cool-ing method and assigned them to an L_{18} orthogonal array (Table 2). Based on the results above, we obtained the response graphs shown in Figure 5. Despite many peaks and V-shapes, by focusing on levels that have a higher SN ratio and a lower sensitivity value, we selected each factor level.

Optimal condition: $A_1B_2C_2D_1E_1F_2G_3H_3$

In addition, we chose the comparative conditions by taking into account the workability or compatibility of parts to maximize the gain.

Comparative condition: $A_2B_3C_3D_2E_3F_3G_1H_2$

Table 3 shows the estimations and confirmations. The measurements of the main spindle's temperature as our initial objective are illustrated in Figure 6. As we aimed to do at the beginning, we were able to reduce the temperature at the bearings of the

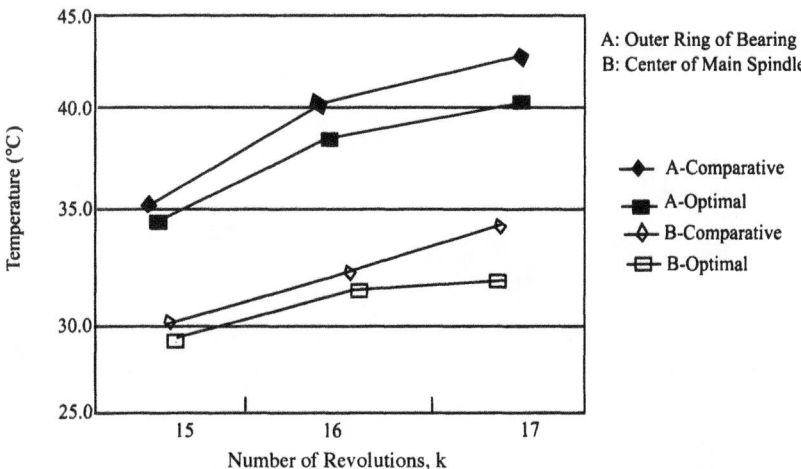

Figure 6
Temperature change of main spindle

main spindle. By measuring the electric power of the motor and evaluating its functionality, we found the optimal condition, which can be confirmed based on the quality characteristics.

Although it has long been believed that we should continue to do experiments on a main spindle until it breaks down, in our study, no breakdown happened. Therefore, we were able to drastically reduce the time necessary for evaluation.

Reference

Machio Tamamura, Makoto Uemura, Katsutoshi Matsuura, and Hiroshi Yano, 2000. Study of evaluation method of spindle. *Proceedings of the 8th Quality Engineering Symposium*, pp. 34–37.

This case study is contributed by Machio Tamamura and Hiroshi Yano.

CASE 53

Improving Minivan Rear Window Latching

Abstract: In this project the authors built up an ADAMS computer model to simulate the closing/opening motions of the rear window glass latch of a minivan. This project was initially formulated to solve a high-closing/opening-effort problem with an initial latch design. However, in this study, closing/opening efforts are considered to be downstream quality characteristics and an upstream quality characteristic, input/output energy efficiency, was chosen as the surrogate. The objective and ideal functions of the latch system were also developed to evaluate the input/output relationship of the latch system. A design constraint of self-locking energy was also taken into consideration. Five control factors and one noise factor were selected to optimize the associated dynamic SN ratio and to meet the design constraint of self-locking energy through an L_{18} design of experiment (DOE) matrix. The improvements in the latch closing/opening efforts were validated and confirmed through prototypes.

1. Introduction

This project was initiated in October 1997. At that time, the project was at its early development stage; as a result, there was still much design freedom allowed for optimization of the latch system. The basic design of the latch is a four-bar linkage with a detent inside the handle to lock or unlock the latch (Figure 1).

2. Objective and Basic Functions

Initial clinic study showed that the opening/closing effort of the initial design was too high (around 110 N versus the target of 50 N) and quite sensitive to tolerance variations in the latch system. However, after thorough consideration, opening/closing effects were considered to be downstream quality characteristics; thus, they were not treated as the objective measurement in the study. Instead of the

opening/closing efforts, the authors focused on the objective function of the latch system, to keep the window fully latched against the window seal strips. In other words, the basic function of a latch system is to convert latching (closing) energy into the sealing energy of weather strips so as to keep wind, audible noises, water, and so on, from coming into the vehicle. Simply put, the purpose of this case study was to enhance and smooth out the energy transformation of the latch system.

As mentioned above, the basic function of a latch system is to convert the closing energy of a latch into the sealing energy of window weather strips. Thus, the input is the energy required to close the latch, and the output would be the energy stored in the compressed weather strips. From this description, we defined the basic function of the latch (Figure 2).

In an ideal condition, all input energy would be converted in the sealing energy of weather strips and no energy would be wasted. However, under real working conditions, there would be numerous

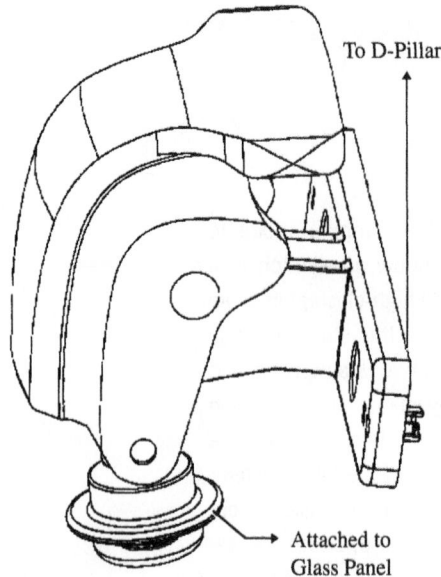

Figure 1
Rear window latch of a minivan

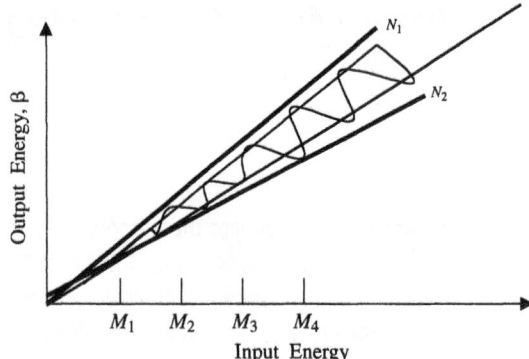

Figure 3
Real function

3. Design Constraints

In addition to its basic function, this latch has a design constraint, that is, it needs a certain amount of self-locking energy to prevent itself from unlocking. There is no strict requirement for this self-locking energy. However, the maximum opening effort will be strongly correlated to this self-locking energy (Figure 4).

4. Control and Noise Factors

The goal of a robust engineering study is to find out a good combination of control factors to make the

noise factors that take away some input energy and convert it into wear, deterioration, vibration, rattle, and so on, in the latch system. The purpose of the robust engineering study was to maximize the energy efficiency and to reduce the variation of the basic function (Figure 3). To achieve this purpose, the authors maximized the dynamic SN ratio of zero-point proportional type.

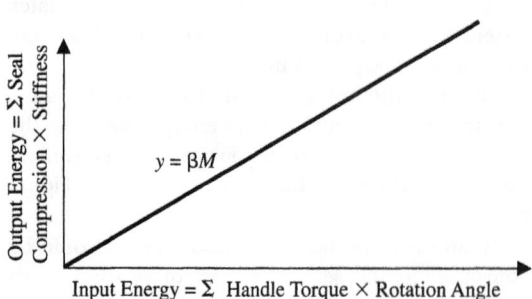

Figure 2
Ideal basic function

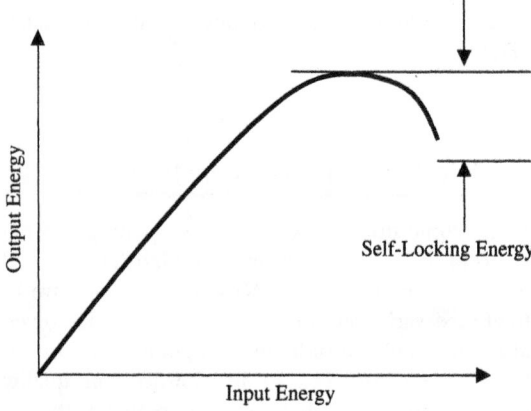

Figure 4
Design constraint of self-locking energy

target system insensitive to the noise factors selected and also to make the mean (i.e., average) output response meet its predetermined target value. In the study, the dimensions of the four-bar linkage of the latch were considered control factors and applied to achieve the robust engineering objective above. For convenience in computer simulations, the authors chose the handle length and the nominal values of the *x*- and *y*-coordinates of two key points, *A* and *B*, as control factors. In addition, the internal friction coefficient of the latch was considered to be a noise factor, *N*. Details of control and noise factors are shown in Table 1.

5. Dynamic SN Ratio

The following equations were applied to calculate the dynamic SN ratio for the real function of Figure 3. In the calculation, k is 4 and r_0 is 2.

$$y_{ij} = \beta M_i \quad (i = 1, \ldots, k, \quad j = 1, \ldots, r_0) \quad (1)$$

$$S_\beta = \frac{\left(\sum Y_i M_i\right)^2}{r_0 \sum M_i^2} \qquad Y_i = \sum y_{ij} \quad (2)$$

$$S_T = \sum\sum y_{ij}^2 \quad (3)$$

$$\beta = \frac{\sum Y_i M_i}{r_0 \sum M_i^2} \quad (4)$$

$$\sigma^2 = V_e = \frac{S_T - S_\beta}{kr_0 - 1} \quad (5)$$

$$SN = 10 \log \left(\frac{\beta^2}{\sigma^2}\right) \quad (6)$$

6. Simulation Data and Analysis

Next, the input/output energy data in the L_{18} DOE matrix of Table 2 were generated through the ADAMS/AVIEW model. The data were analyzed through General Motors' DEXPERT system. The main effects charts of the five control factors are shown in Figures 5 and 6.

In addition to the main effects charts, the ANOVA tables for SN, β, and the self-locking energy are given in Tables 3, 4, and 5.

Using the main effects charts and ANOVA tables above, the authors chose a good combination of the five control factors to maximize the SN ratio and to keep the self-locking energy at a reasonable level. Based on trade-off among SN ratios, β, and

Table 1
Control and noise factors

Factor	Abbreviation	Level 1	Level 2	Level 3
Control factors				
Handle length	C	Low	High	—
x-coordinates of A	A_x	Low	Mid	High
y-coordinates of A	A_y	Low	Mid	High
x-coordinates of B	B_x	Low	Mid	High
y-coordinates of B	B_y	Low	Mid	High
Noise factor				
Friction coefficient of the latch	N	Low	High	—

Table 2
L_{18} DOE matrix[a]

Run	Factor					M_1 10,000 Energy Units		20,000 Energy Units		30,000 Energy Units		40,000 Energy Units		β	SN Ratio	Average Self-Locking Energy	Seal Energy at Locking Position
	C	A_1	A_2	B_1	B_2	N_1	N_2	N_1	N_2	N_1	N_2	N_1	N_2				
1	1	3	2	3	1	5,679	5,925	12,469	12,407	20,617	20,493	−10,000	0	0.241143	−95.9907	37,839	48,519
2	2	1	3	3	2	3,750	3,846	7,211	7,115	10,250	10,769	13,519	13,461	0.345375	−59.6395	10,115	41,676
3	1	1	2	2	3	3,600	3,900	7,360	7,400	11,300	11,060	15,000	15,020	0.373633	−50.2029	7,730	48,515
4	1	2	1	3	3	3,452	3,571	7,095	7,142	11,056	11,000	15,000	15,119	0.370235	−56.1026	809	59,286
5	1	1	1	1	1	4,167	4,214	8,367	8,761	13,381	13,310	18,095	18,476	0.448323	−58.1531	6,125	55,000
6	2	1	2	2	3	4,660	4,620	8,760	8,780	11,700	11,820	15,100	15,020	0.392333	−65.1758	7,620	48,293
7	2	3	1	2	2	5,625	5,630	11,188	11,313	18,750	19,312	0	9	0.284072	−90.7171	35,812	34,255
8	1	2	2	1	2	5,125	5,313	11,250	11,312	18,563	18,125	38,000	36,750	0.774377	−76.3228	37,906	12,096
9	2	2	1	3	3	3,928	3,952	7,809	7,857	12,071	12,119	16,404	16,476	0.405503	−54.0386	1,761	23,886
10	2	1	1	1	1	4,190	4,690	8,405	8,571	13,286	13,452	18,404	18,571	0.451577	−59.2203	6,023	55,353
...																	

[a] Per unit of input or output energy = 1.74533 × 10⁻⁵ J.

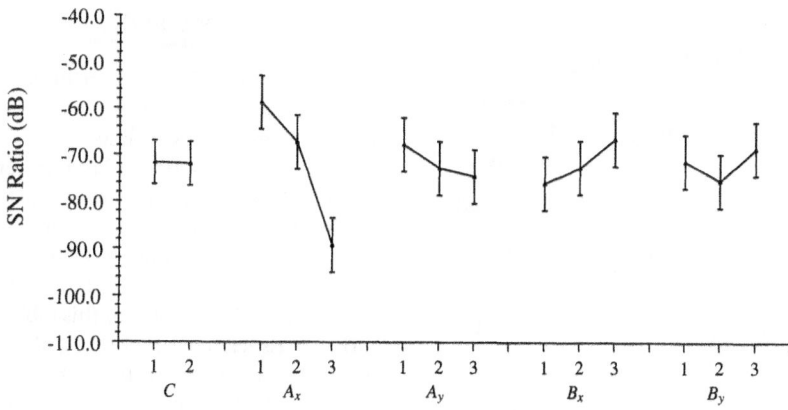

Figure 5
Main effects on the SN ratio

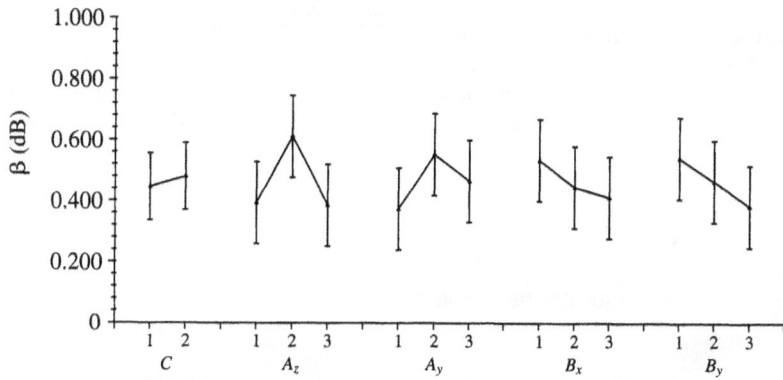

Figure 6
Main effects on beta (energy efficiency)

Table 3
ANOVA for the SN ratio

Source	d.f.	Percent Contribution
C (handle length)	1	0.00
A_x	2	78.50
A_y	2	2.07
B_x	2	5.26
B_y	2	0.57
Error	8	13.60
Total	17	100.00

Table 4
ANOVA for beta

Source	d.f.	Percent Contribution
C (handle length)	1	0.00
A_x	2	31.40
A_y	2	11.10
B_x	2	1.45
B_y	2	6.80
Error	8	49.25
Total	17	100.00

Table 5
ANOVA for the self-locking energy

Source	d.f.	Percent Contribution
C (handle length)	1	0.02
A_x	2	59.40
A_y	2	13.30
B_x	2	9.71
B_y	2	17.00
Error	8	0.57
Total	17	100.00

self-locking energy, the authors determined that the optimal design is C = level 1, A_x = 1, A_y = 2, B_x = 1, B_y = 1. Table 6 is a comparison of an initial (2,3,3,3,3) and an optimal (1,1,2,1,1) design.

7. Confirmations and Validations

Five prototypes of new latch design were made to validate the results of the computer simulation. Since it is extremely difficult to measure the input and output energy of the latch system, the authors applied downstream quality characteristics such as maximum opening/closing efforts and a water-leak test to conduct the validation tests. The maximum closing/opening efforts and water-leak test results are shown in Table 7. From this table the latch effort and the associated variation in the optimal design are seen to be better than the initial design. In other words, the new latch design meets customer requirements and also exhibits less variation.

8. Conclusions

In this project, the authors developed an objective and ideal function for the rear window latch of minivans. Next, the authors optimized five control fac-

Table 6
Comparison between initial and optimal designs

	Design		
	Initial	Optimal	Improvement
SN radio (dB)	−86.44	−63.56	−22.88
β (energy efficiency)	0.2715	0.6116	0.3410
Total closing energy (J)	1.30	1.28	0.02
Sealing energy at locking position (J)	0.68	0.84	0.16
Self-locking energy (opening energy) (J)	0.47	0.33	0.14

Table 7
Validation test results

	Design	
	Initial	Optimal
Water leak	Questionable	Pass
Maximum opening/closing effort (N)	95.70 ± 31.05	44.73 ± 10.52

tors to maximize the SN ratio while maintaining a reasonable amount of self-locking energy for the latch. An optimal design was achieved and the energy efficiency improved. Consequently, the mean value and variation of the maximum opening/closing effort have been improved by 53 and 66%, individually. As a result, the latch effort met the customer requirements and is more robust against the variation caused by assembly and usage conditions.

Acknowledgments The authors would like to thank Ron Nieman of Exterior Design, MLCG, and Jeffrey Flaugher of LOF for their input and efforts in this project, and Mike Wright of VSAS, MLCG, General Motors, for his guidance in using the DOE features of ADAMS/AVIEW. Thanks are also due to Tony Sajdak of MDI for his technical support of ADAMS/AVIEW. Finally, the authors wish to thank Dan Drahushak for his support in making this project a reality.

References

Genichi Taguchi, 1993. *Taguchi Methods: Design Of Experiment.* Quality Engineering Series, Vol. 4. Livonia, Michigan: ASI Press, pp. 34–38.

Kenzo Ueno and Shih-Chung Tsai (translator), 1997. *Company-Wide Implementations of Robust-Technology Development.* New York: ASME Press.

This case study is contributed by Shih-Chung Tsai and Manohar B. Motwani.

CASE 54

Linear Proportional Purge Solenoids

Abstract: The purpose of an evaporative emission control system is to control fuel vapors that accumulate in a gas tank and carbon canisters. Due to strict Environmental Protection Agency regulations on the control of fuel vapor and to customer requirements, the need for a new technology became apparent. This new solenoid offers a better continuous proportional flow with a high degree of flow controllability. To eliminate the sensitivity of the current design on all forms of variation, this design was selected for robust engineering methods. Using math-based models, the subfunctions were improved from 30 to 60%. This study shows that decomposition of the functional block diagram is a powerful tool in the design process using computer simulation model in the early stage of concept development.

1. Introduction

The principal components of an evaporative emissions control system are a carbon canister, tank pressure transducer, rollover orifice, liquid separator, and the purge solenoid, a valve that functions to meter fuel vapor from the fuel tank and a carbon canister to the intake manifold (Figure 1). The linear proportional purge solenoid project was initiated to develop a linear purge solenoid that offers a continuous proportional flow with a high degree of flow controllability.

This particular solenoid design proposes to meet or exceed the existing flow and noise specification and has the potential to replace or compete with comparable products currently in use in the marketplace.

2. Problem Statement

The objective of using robust engineering methods in the design of the linear proportional purge solenoid was to reduce part-to-part flow and opening point variation and to minimize the effects of the following sources of variability: (1) environment (temperature, atmospheric pressure), (2) materials variation, (3) manufacturing and assembly variation, and (4) operating conditions (vacuum).

3. Objectives and Approach to Optimization

The objective of the study was to optimize the design through a validated simulation model. The success factors assigned to the optimization process were as follows:

1. *Proportional purge flow.* The flow is controlled by a current level that is maintained using a high-frequency pulse-width-modulated (PWM) signal (Figure 2). Flow is proportional to the percentage of duty cycle at the PWM signal (Figure 3). A current driver controlling the solenoid is needed.

2. *Vacuum independent flow.* Under operating conditions, the vacuum effect on the flow has to be minimized.

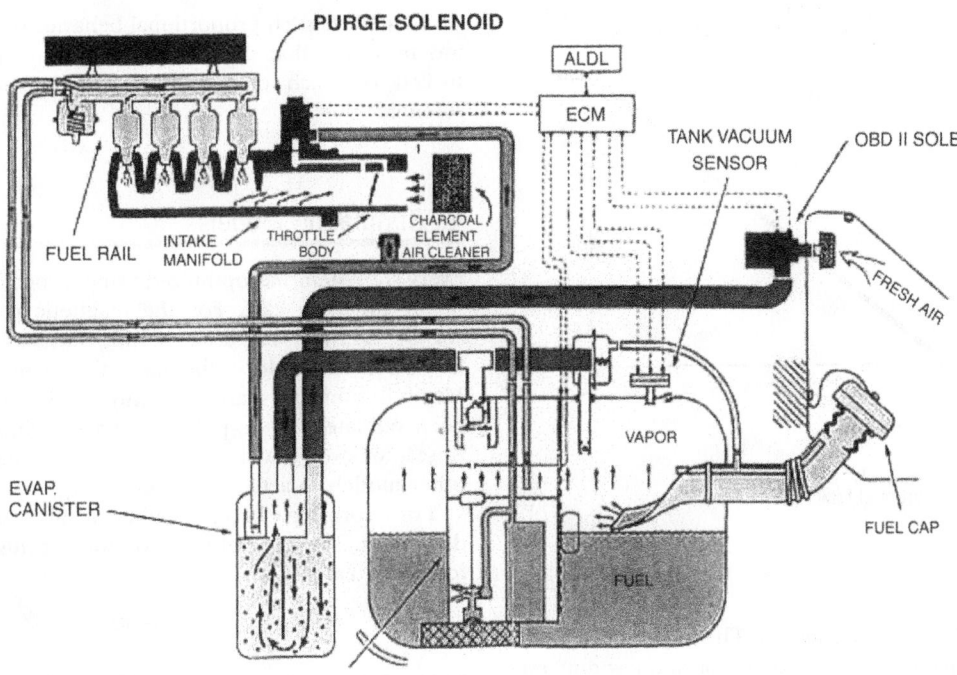

Figure 1
System hardware

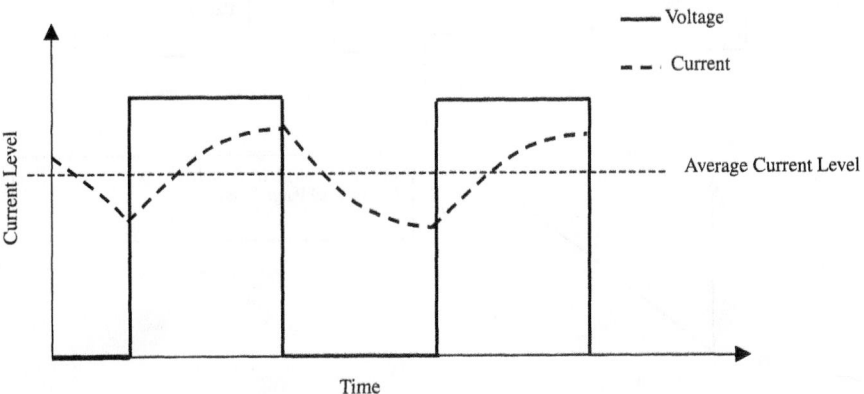

Figure 2
Current level maintained using high-frequency PWM signal

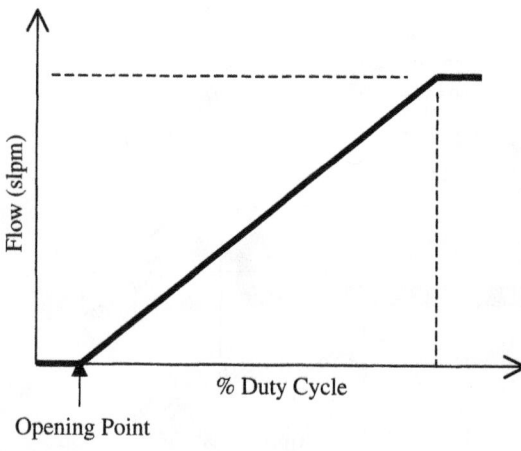

Figure 3
Linear proportional flow

3. *Opening point stability.* The solenoid has to control the flow of fuel vapor at a low duty cycle percentage.

4. *Low part-to-part variation.* Reduction of part-to-part variation of flow and opening point is required.

To achieve such proportional behavior of flow, it was necessary that the energy conversion be controlled on each of the three functions of the solenoid.

4. Simulation Models

Every subsystem was optimized using a specific simulator for each case. For the magnetic package, Ansoft Maxwell was used, solving an axisimetric finite-element model of the magnetic circuit. Ansoft Maxwell simulators are currently used and have been validated in the past. The error estimate for Ansoft Maxwell is about 1.5%, based on comparing other models to actual test data.

For a spring package, a mathematical model was developed. The general equation for a spring-mass-forced system is

$$m\ddot{x} + c\dot{x} + kx = F(t) \qquad (1)$$

where m is the mass, c the damping coefficient, k the spring constant, and F the excitation force. From the definition of a linear proportional purge solenoid, the magnetic force is not a function of the travel; therefore, the deexcitation force is not a

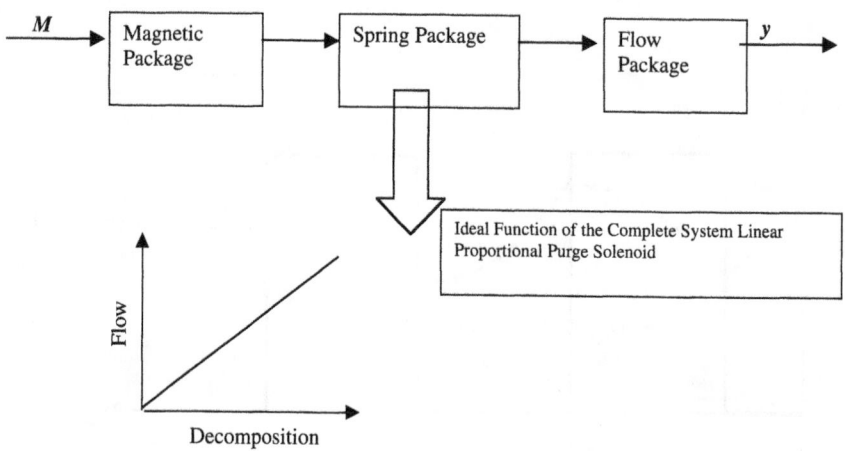

Figure 4
Functional decomposition into three subsystems

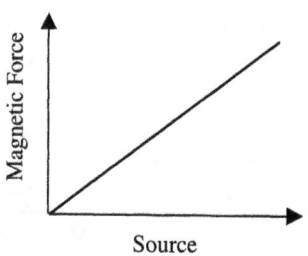

Figure 5
Ideal function

Magnetic package: force is
proportional to the number of
amp-turns (source)

function of the travel. The solution was obtained using the numerical method of finite differences.

The Taylor series for the position is given as

$$X_{i+1} = x_i + \frac{dx}{dt}\Delta t + \frac{d^2x}{dt^2}\frac{\Delta t^2}{2} + \cdots + \frac{d^nx}{dx^n}\frac{(\Delta x)^n}{n!} \quad (2)$$

$$X_{i-1} = x_i - \frac{dx}{dt}\Delta t + \frac{d^2x}{dt^2}\frac{\Delta t^2}{2} + \cdots + (-1)^n\frac{d^nx}{dx^n}\frac{(\Delta x)^n}{n!} \quad (3)$$

Equations (2) and (3) can be written as

$$\frac{dx}{dt} = \frac{x_{i+1} - x_i}{\Delta t} \quad (4)$$

$$\frac{d^2x}{dt^2} = \frac{x_{i+1} - 2x_i + x_{i-1}}{\Delta t^2} \quad (5)$$

Placing equations (4) and (5) in equation (1), the position (through the time) is given as

$$x_{i+1} = \frac{F + (2x_i - x_{i-1})(m/\Delta t^2) + (c/\Delta t)\,x_i}{m/\Delta t^2 + c/\Delta t + k} \quad (6)$$

For the flow package, Star CD was used, solving an axial finite-element model of the pneumatic valve. Star CD simulator is currently used and has been validated in the past. The numerical error estimated from the solver is around 3%.

5. Functional Decomposition

To control the energy conversion of the solenoid, the design optimization process was made using functional decomposition. The study focused on three different packages or subsystems. The three subsystems were optimized separately. They were connected linearly (Figure 4) to complete an ideal function of the system level of the solenoid. Each engineered system uses energy transformation to

Table 1
Control factors and levels[a]

Factor	Level 1	Level 2	Level 3
A	Low[a]	Average	High
B	Low	Average[a]	High
C	Low	Average[a]	High
D	Low	Average[a]	High
E	1	2	3[a]
F	1	2	3[a]
G	1	2	3[a]
H	1	2	3[a]
I	Low[a]	Average	High
J	Low	Average[a]	High
K	Low	Average[a]	High
L	Low	Average[a]	High
M	Low	Average[a]	High

[a] Current design level.

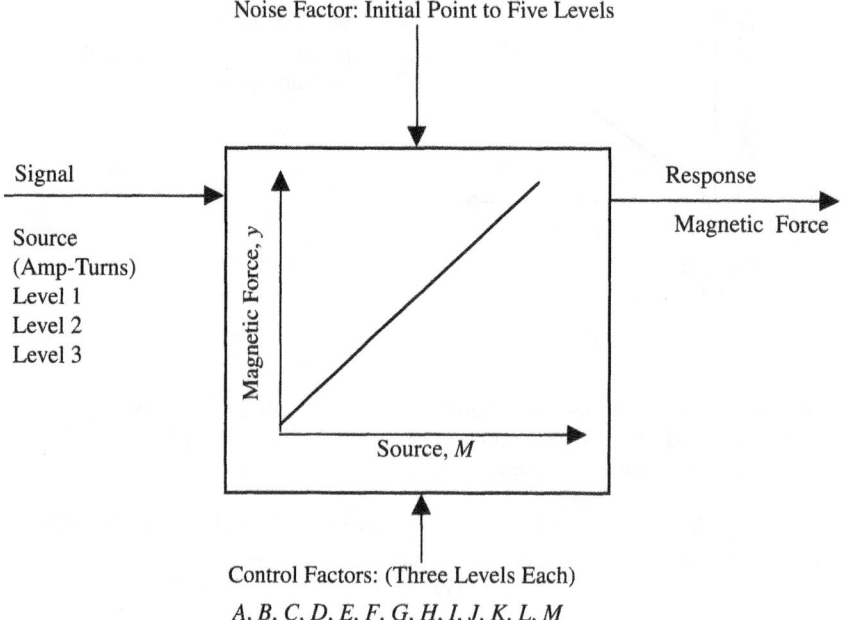

Figure 6
Parameter design

convert input energy into specific output energy and they are connected linearly (dynamic function).

6. Parameter Design

Ideal Function for the Magnetic Package

The function of a magnetic package is to provide a magnetic force to the plunger movement. For this package the input signal is an average electrical current level that is controlled by the percent of the duty cycle and the frequency (high-frequency PWM signal).

The design of the magnetic package for the linear proportional solenoid was based on (1) the magnitude of the magnetic force being proportional to the source (amp-turns), and (2) the magnetic force being independent of the position of the plunger.

According to the two points mentioned above, the ideal function for this package is as shown in Figure 5.

The first robust design experiment to be studied is the magnetic force (newtons) out of the magnetic package.

Signal and Noise Strategies Three levels of source (amp-turns) were selected (Table 1). These values represent the complete range of the electrical current under operation conditions. The initial point or position of the plunger was treated as a noise because the magnetic force should obtain its linear behavior insensitive to the effects of the travel setting, and the magnitude of the magnetic force is independent of the plunger position. Five levels of the initial point were selected.

The parameter diagram, illustrating the relationship of the control factors, noise factors, signal factor, and the response to the subsystem of the

Table 2
Experiment layout

No.	A	B	C	D	E	F	G	H	I	J	K	L	M	1					2					3					SN	β
						Control Factor													**Signals and Initial Points**											
														1	2	3	4	5	1	2	3	4	5	1	2	3	4	5		
1	1	1	1	1	1	1	1	1	1	1	1	1	1																	
2	2	2	2	2	2	2	2	2	2	2	2	2	1																	
3	3	3	3	3	3	3	3	3	3	3	3	3	1																	
4	1	1	1	1	2	2	2	2	3	3	3	3	1																	
5	2	2	2	2	3	3	3	3	1	1	1	1	1																	
6	3	3	3	3	1	1	1	1	2	2	2	2	1																	
7	1	1	2	3	1	2	3	3	2	2	2	3	1																	
8	2	2	3	1	2	3	1	2	3	3	3	1	1																	
9	3	3	1	2	3	1	2	2	1	1	1	2	1																	
10	1	1	3	2	1	3	2	3	3	3	3	2	2																	
11	2	2	1	3	2	1	3	1	2	2	2	3	2																	
12	3	3	2	1	3	2	1	2	3	3	1	1	1																	
13	1	2	3	1	3	2	1	1	1	2	3	2	2																	
14	2	3	1	2	1	3	2	2	2	3	1	3	2																	
15	3	1	2	3	2	1	3	3	3	1	2	1	2																	
16	1	2	3	2	2	1	3	2	3	3	2	1	2																	
17	2	3	1	3	3	2	1	3	1	1	3	2	2																	
18	3	1	2	1	1	3	2	1	2	2	1	3	2																	
19	1	2	1	3	3	3	1	2	2	1	2	3	2																	

Table 2
(*Continued*)

No.	Control Factor													Signals and Initial Points																SN	β
	A	B	C	D	E	F	G	H	I	J	K	L	M	1					2					3							
														1	2	3	4	5	1	2	3	4	5	1	2	3	4	5		
20	2	3	3	1	1	1	2	3	3	2	3	1	2																	
21	3	1	3	2	2	2	3	1	1	3	1	2	2																	
22	1	2	2	3	3	1	2	1	1	3	3	2	2																	
23	2	3	3	1	1	2	3	2	2	1	1	3	2																	
24	3	1	1	2	2	3	1	3	3	2	2	1	2																	
25	1	3	2	1	2	3	3	1	3	1	2	2	3																	
26	2	1	3	2	3	1	1	2	1	2	3	3	3																	
27	3	2	1	3	1	2	2	3	2	3	1	1	3																	
28	1	3	2	2	2	1	1	3	2	3	1	3	3																	
29	2	1	3	3	3	2	2	1	3	1	2	1	3																	
30	3	2	1	1	1	3	3	2	1	2	3	2	3																	
31	1	3	3	3	2	3	2	2	1	2	1	1	3																	
32	2	1	1	1	3	1	3	3	2	3	2	2	3																	
33	3	2	2	2	1	2	1	1	3	1	3	3	3																	
34	1	3	1	2	2	2	1	1	2	2	3	1	3																	
35	2	1	2	3	3	3	2	2	3	3	1	2	3																	
36	3	2	3	1	1	1	2	3	1	1	2	3	3																	

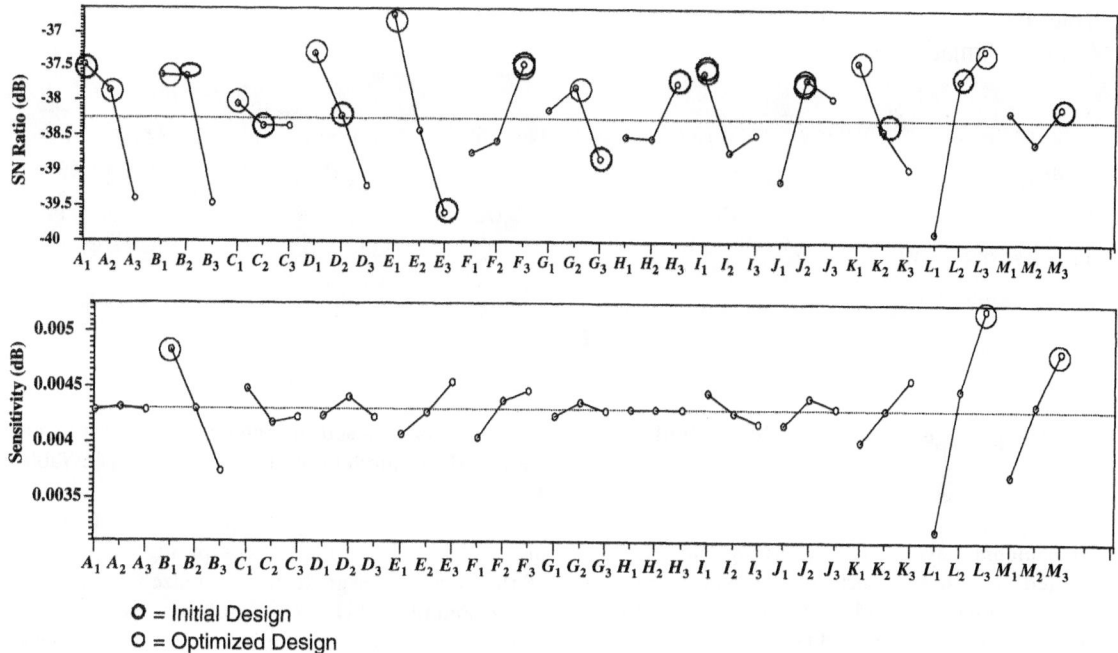

O = Initial Design
O = Optimized Design

Figure 7
SN ratio and sensitivity plots

Table 3
Two-step optimization

	Control Factor												
	A	B	C	D	E	F	G	H	I	J	K	L	M
Original	1	2	2	2	3	3	3	3	1	2	2	2	3
SN ratio (first step)	1	1	1	1	1	3	2	3	1	2	1	3	3
Sensitivity (second step)		1										3	3
Optimized	2	1	1	1	1	3	2	3	1	2	1	3	3

Table 4
Model confirmation

| | SN | | Sensitivity | |
Condition	Prediction	Confirmation	Prediction	Confirmation
Optimum	−29.1	−32.2	0.0048	0.0049
Current	−35.65	−38.02	0.0041	0.0041
Improvement (dB)	6.55	5.82	0.0007	0.0008
Variation reduction (%)	53	49	17	19

magnetic package, is shown in Table 1 and Figure 6.

Experimental Layout An L_{36} orthogonal array was selected and used to perform the experiment (Table 2), so that we could study three levels of each control factor using only 13 columns.

Data Analysis and Two-Step Optimization The data were analyzed using the dynamic SN ratio:

$$\eta = 10 \log \frac{S_\beta - V_e}{rV_e} \qquad (7)$$

where S_β is the sum of the squares of distance between zero and the least squares best-fit line (forced through zero) for each data point, V_e the mean square (variance), and r the sum of squares of the signals. (*Note:* The same dynamic SN ratio cited earlier was used in each experiment.)

Figure 7 illustrates the main effects of each control factor at the three different sources. The effect of a factor level was given by the deviation of the response from the overall mean due to the factor level. The main effects plots show which factor levels are best for increasing the SN ratio. They should give the best response with the smallest effect due to noise.

Control factor settings were selected to maximize the SN ratio, minimizing the sensitivity to the noise factor (initial point) under study, providing a uniform linearity regardless of the travel position. The

optimum nominal settings selected are shown in Table 3 and the model confirmation is given in Table 4.

In robust design experiments, the level of improvement is determined by comparing the SN ratio of the current design to the optimized design. An improvement in SN ratio signifies a reduction in variability. In this case the gain (SN ratio in decibels) is 5.82 dB; this represents a reduction in the variation about 49% from the original design using the following formula:

$$\text{variability improvement} = (\tfrac{1}{2})^{\text{Gain}/6} \times \text{initial variability} \qquad (8)$$

Results A significant improvement in the linearity at different initial points of the solenoid was achieved. As can be seen in Figure 8, about 50% of the linearity was gained between the original and the optimized designs. The magnetic force was increased considerably. The β value increased slightly at the beginning of the excitation of the solenoid. Therefore, we expect an earlier opening point.

Ideal Function for the Spring Package

A spring element's function is to regulate the plunger position and movement. For this package the input was the magnetic force, with results shown in Figure 9. The design of the spring element for the linear proportional solenoid is based on the fact that the travel is proportional to the magnetic force

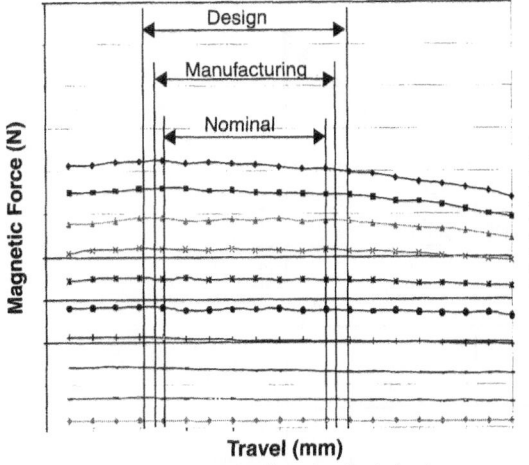

Before Robust Engineering Optimization
(0.7 mm Linear Nominal Travel)

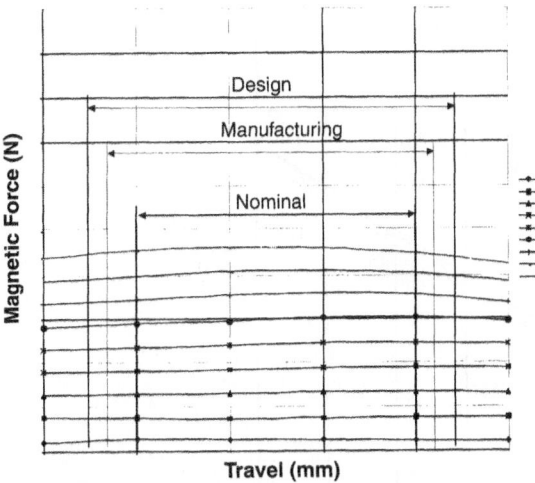

After Robust Engineering Optimization
(1.5 mm Linear Nominal Travel)

Figure 8
Performance of the magnetic package before and after
robust engineering

(main excitation). The ideal function for this package is shown in Figure 10. The second robust design experiment we conducted was on the travel (millimeters) of the spring package.

Signal and Noise Strategies Three levels of the magnetic force (newtons) were selected: force 1, force 2, and force 3, which represent the total range of the magnetic force on operation conditions. The model (vacuum/area) was the principal source that cannot be controlled during normal operation of the solenoid and was treated as a noise factor for this experiment. This strategy reduced the mechanical response variation due to the pneumatic forces generated by the orifice diameter variation and vacuum variation.

Control Factors and Levels Table 5 gives the layout for control factors and levels. The parameter diagram, illustrating the relationship of the control factors, noise factors, signal factor, and the response to the subsystem of the spring package, is shown in Figure 11.

Experimental Layout An L_{18} orthogonal array was selected and used to perform the experiment (Figure 12), so that we could study two levels of one control factor and three levels of the three other control factors.

Data Analysis and Two-Step Optimization The data were analyzed using the dynamic SN ratio with equation (7).

Figure 13 illustrates the main effects of each control factor at the three sources. The effect of a factor level is given by the deviation of the response from the overall mean due to the factor level. The main effects plots show which factor levels are best for increasing the SN ratio. They should give the best response with the smallest effect due to noise.

Control factor settings were selected to maximize the SN ratio, minimizing the sensitivity to the noise factor (vacuum/area) under study, providing a uniform linearity of travel regardless of the pneumatic force. The optimum nominal settings selected are shown in Table 6.

Confirmation and Improvement in the SN Ratio Table 7 shows the predictions and confirmatory data. In robust design experiments, the level of improvement is determined by comparing the SN ratio of the current design to the optimized design.

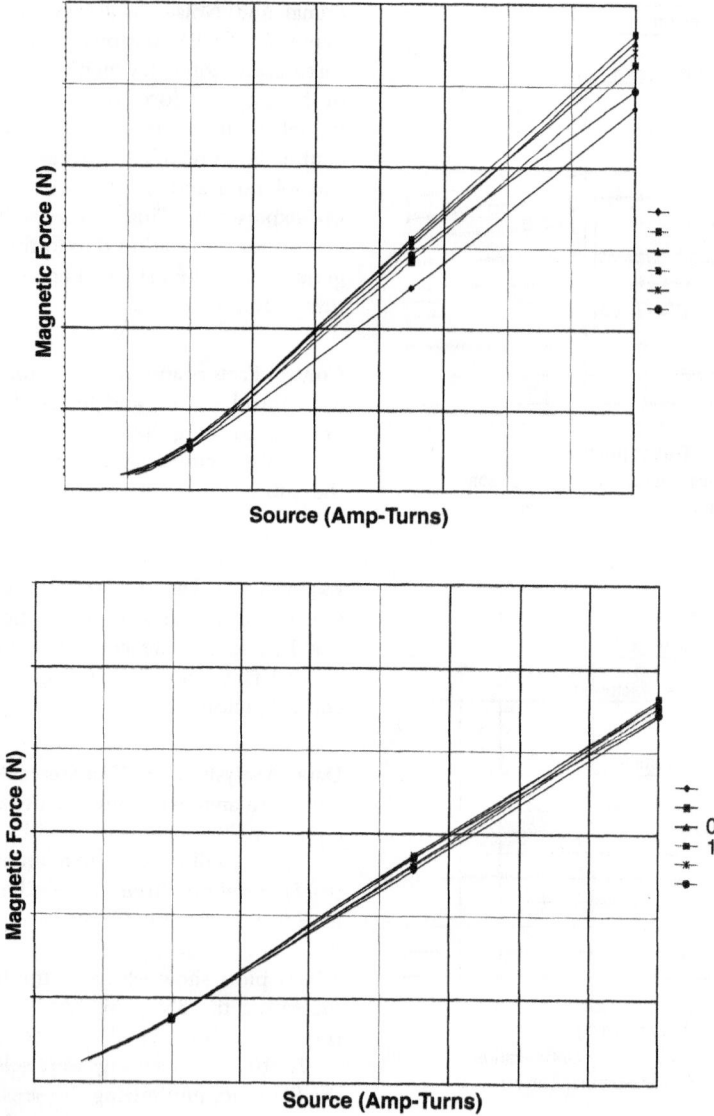

Figure 9
Ideal function

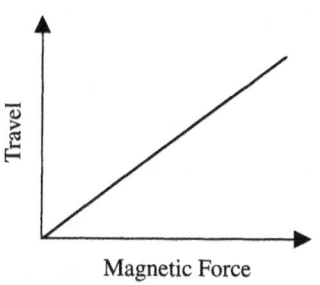

Spring package: travel is proportional to the magnetic force

Figure 10
Ideal function

An improvement in SN ratio signifies a reduction in variability. In this case the gain (SN ratio in decibels) is 3.62 dB; this represents a reduction in the variation by about 34% from the original design using equation (8). Performance improvement results are shown in Figure 14.

A linear excitation force was obtained earlier by optimizing the magnetic package combined with the spring package. The offset between the curves in Figure 14 is a function of the effects of the vacuum force compared to the magnetic and spring forces. This offset could be reduced by making the resultant of forces less sensible to vacuum variation.

Ideal Function for the Flow Package

The purpose of the flow package is to control flow through the solenoid. For this package the input is the position of the plunger (travel). The design of the magnetic package for linear proportional solenoid is based on flow being proportional to the

movement of the mobile part of the valve. The ideal function for this package is shown in Figure 15. The third robust design experiment to be studied is the flow in standard liters per minute (slpm) out of the flow package.

Signal and Noise Strategies Four levels of travel (millimeters) were selected. The flow of linear proportional purge solenoid should not be unstable at a low duty cycle and should have a high hysteresis, due to the effects of varying environmental factors. High and low temperature and vacuum conditions lead to poor flow performance. Therefore, the temperature and vacuum were treated as a noise factor (two levels: low and high). They are shown in Table 8. The parameter diagram, illustrating the relationship of the control factors, noise factors, signal factor, and the response to the subsystem of the flow package is shown in Figure 16.

Experimental Layout An L_{18} orthogonal array was selected and used to perform the experiment (Figure 17), so that we could study two levels of one control factor and three levels of the four other control factors.

Data Analysis and Two-Step Optimization The data were analyzed using the dynamic SN ratio with equation (7). Figure 18 illustrates the main effects of each control factor at the three different sources. The effect of a factor level is given by the deviation of the response from the overall mean due to the factor level. The main effect plots show which factor

Table 5
Control factors levels[a]

Factor	Level		
	1	2	3
A	With[a]	Without	—
B	Low	Average	High[a]
C	Low[a]	Average	High
D	Low[a]	Average	High

[a] Current design level.

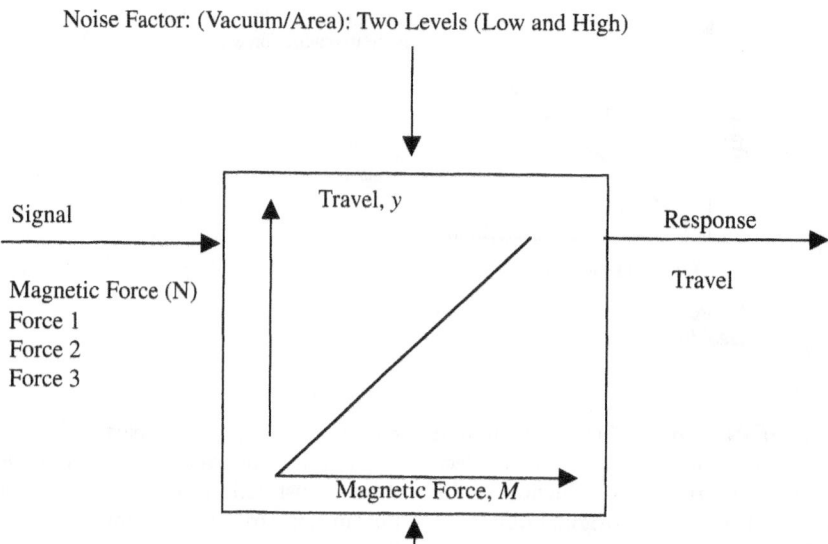

Noise Factor: (Vacuum/Area): Two Levels (Low and High)

Signal

Magnetic Force (N)
Force 1
Force 2
Force 3

Travel, y

Magnetic Force, M

Response

Travel

Control Factors: One Factor at Two Levels and Three Factors at Three Levels Each

A, B, C, D

Figure 11
Parameter design

Control Factor Array				Magnetic Force	1		2		3		SN	β
				Vacuum / Area	Low	High	Low	High	Low	High		
Run	A	B	C	D									
1	1	1	1	1									
2	1	1	2	2									
3	1	1	3	3									
4	1	2	1	1									
5	1	2	2	2									
6	1	2	3	3									
7	1	3	1	2									
8	1	3	2	3									
9	1	3	3	1			DATA						
10	2	1	1	3									
11	2	1	2	1									
12	2	1	3	2									
13	2	2	1	2									
14	2	2	2	3									
15	2	2	3	1									
16	2	3	1	3									
17	2	3	2	1									
18	2	3	3	2									

Figure 12
Experimental layout

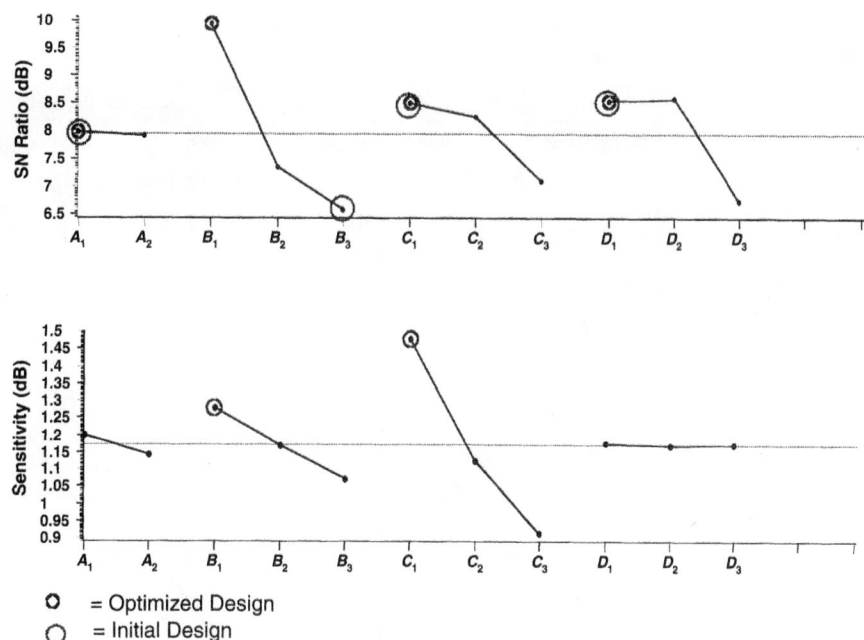

O = Optimized Design
O = Initial Design

Figure 13
SN ratio and sensitivity plots

levels are best for increasing the SN ratio. They should give the best response with the smallest effect due to noise.

Control factor settings were selected to maximize the SN ratio, minimizing the sensitivity to the noise factor (vacuum/area) under study, providing a uniform linearity of travel regardless of the pneumatic force. The optimum nominal settings selected are

Table 6
Two-step optimization

	Control Factor			
	A	**B**	**C**	**D**
Initial design	A_1	B_3	C_1	D_1
SN ratio (first step)	A_1	B_1	C_1	D_1
Sensitivity (second step)			C_1	
Optimized	A_1	B_1	C_1	D_1

shown in Table 9. Confirmation and improvement in the SN ratio are shown in Table 10.

In robust design experiments, the level of improvement is determined by comparing the SN ratio of the current design to the optimized design. An improvement in SN ratio signifies a reduction in variability. In this case the gain (SN ratio) is 8.13 dB; this represents a reduction in variation by about 61% from the original design using equation (8). Performance improvement results are shown in Figure 19.

The full flow or maximum flow has been improved and has become insensitive to the effect of the vacuum and temperature level, in such a manner reducing the flow variation at different vacuums that can exist in actual conditions.

7. Summary

The largest contributors to the SN ratio for the magnetic package were, in order of importance, E, L, A,

Table 7
Model confirmation

	SN		Sensitivity	
Condition	Prediction	Confirmation	Prediction	Confirmation
Optimum	11.11	10.66	1.62	1.63
Current	7.73	7.04	1.37	1.37
Improvement (dB)	3.38	3.62	0.21	0.26
Variation reduction (%)	32.3	34	15	19

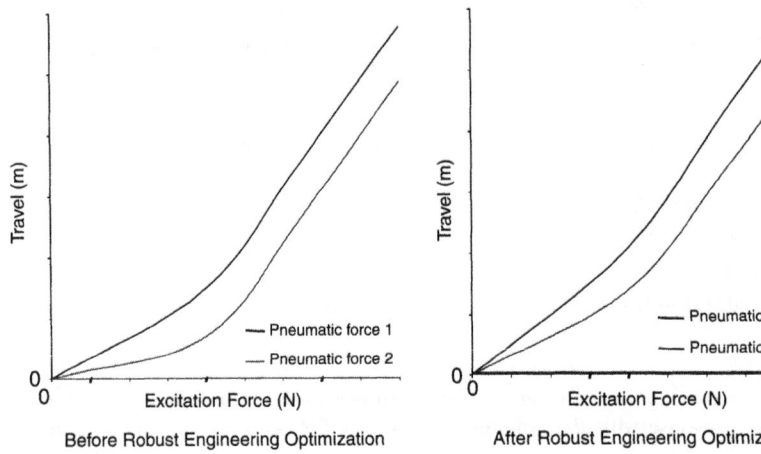

Figure 14
Improvement results

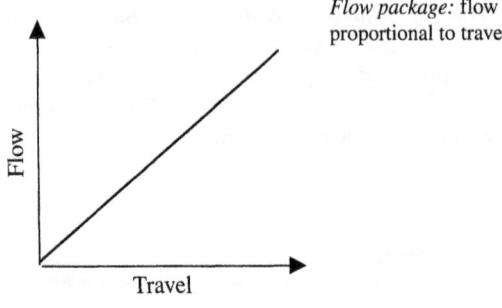

Figure 15
Ideal function

Flow package: flow is proportional to travel

Table 8
Control factors and levels[a]

	Level		
Factor	1	2	3
A	Small[a]	Medium	Large
B	Low	Average[a]	High[a]
C	Low[a]	Average	High
D	Low	Average[a]	High
E	Small[a]	Medium	Large

[a] Current design level.

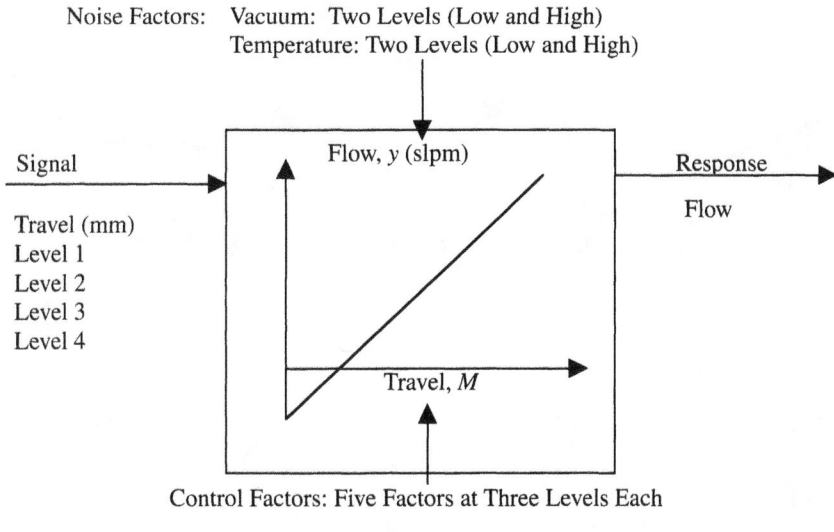

Noise Factors: Vacuum: Two Levels (Low and High)
Temperature: Two Levels (Low and High)

Signal

Travel (mm)
Level 1
Level 2
Level 3
Level 4

Flow, y (slpm)

Travel, M

Response

Flow

Control Factors: Five Factors at Three Levels Each

A, B, C, D, E

Figure 16
Parameter design

No.	A	B	C	D	E	Travel	Level 1				Level 2				Level 3				Level 4				SN	β
						Temp.	L	H	L	H	L	H	L	H	L	H	L	H	L	H	L	H		
						Vacuum	L	L	H	H	L	L	H	H	L	L	H	H	L	L	H	H		
1	1	1	1	1	1																			
2	1	2	2	2	2																			
3	1	3	3	3	3																			
4	2	1	1	2	2																			
5	2	2	2	3	3																			
6	2	3	3	1	1																			
7	3	1	2	1	3																			
8	3	2	3	2	1						DATA												
9	3	3	1	3	2																			
10	1	1	3	3	2																			
11	1	2	1	1	3																			
12	1	3	2	2	1																			
13	2	1	2	3	1																			
14	2	2	3	1	2																			
15	2	3	1	2	3																			
16	3	1	3	2	3																			
17	3	2	1	3	1																			
18	3	3	2	1	2																			

Figure 17
Experimental layout

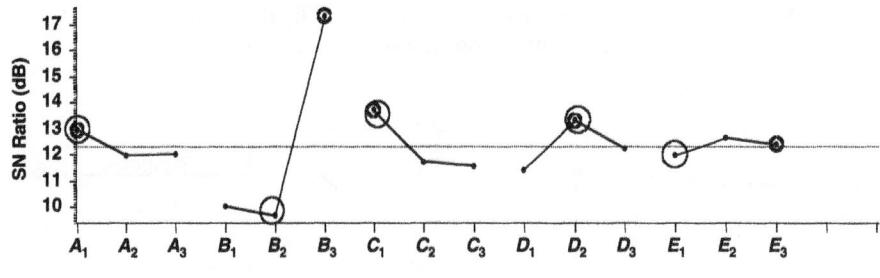

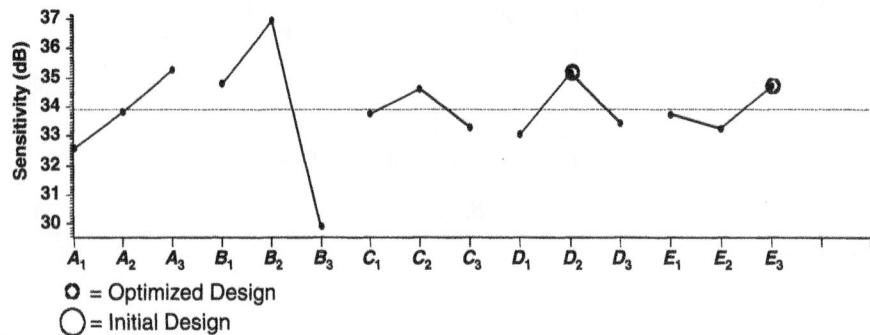

O = Optimized Design

○ = Initial Design

Figure 18
SN ratio and sensitivity plots

D, B, and K. C contributed little to the SN ratio. The optimized control factor settings for the magnetic package are L, M, and B. A trade-off for the control factor A between the levels A_1 and A_2 was made for manufacturing concerns and cost.

The largest contributors to the SN ratio for the spring package were, in order of importance, B and

D. A contributed little to the SN ratio. The optimized control factor settings for the spring package is C.

The largest contributors to the SN ratio for the flow package were, in order of importance, B, C, and D. A contributed little to the SN ratio. The optimized control factor settings for the magnetic package is B.

Table 9
Two-step optimization

	Control Factor				
	A	B	C	D	E
Original	A_1	B_2	C_1	D_2	E_1
SN ratio (first step)	A_1	B_3	C_1	D_2	E_2
Sensitivity ratio (second step)				D_2	E_3
Optimized	A_1	B_3	C_1	D_2	E_3

8. Conclusions

The linear purge solenoid was broken down into subsystems. Decoupling subassemblies and components from the system allows more efficient and manageable experiments. The β values can be tuned for optimum system performance when the subassemblies and components are reassembled into the system. In this case verification and confirmation of the system level with the optimized condition of each subsystem will be pursued when the hardware becomes available.

Table 10
Model confirmation

Condition	SN		Sensitivity	
	Prediction	Confirmation	Prediction	Confirmation
Optimum	20.37	21.57	30.64	32.52
Current	12.31	13.44	29.32	29.61
Improvement (dB)	8.06	8.13	1.32	2.91
Variation reduction (%)	60.5	61	4.5	10

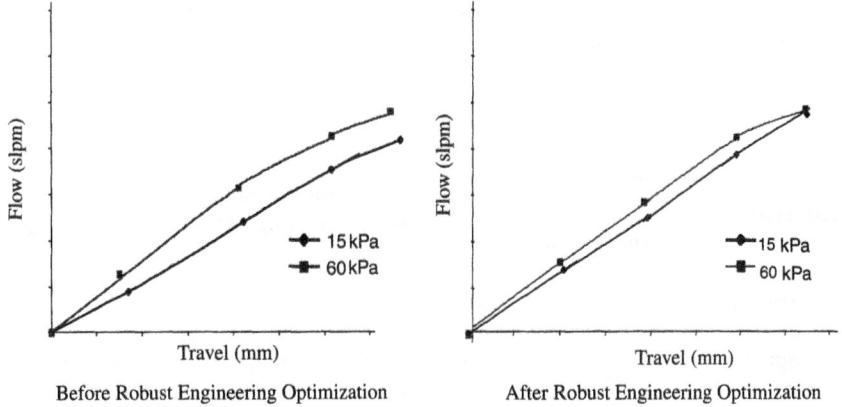

Figure 19
Performance of the flow package before and after robust engineering

The functional block diagram is a powerful tool in the design process. We should consider the confirmation run for best-case and worst-case conditions to prove the exactness of the simulation configuration and calculation. Robust engineering as part of the design concept defines the critical features on the design concepts at the early stages of development and allows for faster definition of the best concept.

Acknowledgments We would like to acknowledge Mark Shost, Sanjay Gupta, Lorenzo Rodriguez, and Dr. Amiyo Basu from Delphi Automotive Systems, and Alan Wu and Shin Taguchi from the American Supplier Institute for their technical support.

Reference

Steven C. Chapar and Raymond P. Canale, 1988. *Métodos Numericos para Ingenieros*. New York: McGraw-Hill Mexico.

This case study is contributed by Conrado Carrillo, Michael Holbrook, and Jean-François Pelka.

CASE 55

Optimization of a Linear Actuator Using Simulation

Abstract: Since the structure of magnetic circuits is advanced and complicated, it is difficult to design magnetic circuits without finite-element method (FEM) analysis. In our research, by integrating FEM magnetic field analysis and quality engineering, we designed parameters of a magnetic circuit to secure its robustness through the evaluation of reproducibility not only in simulation but also in an actual device. Additionally, we found that reducing cost and development cycle time was possible using this process.

1. Introduction

Figure 1 depicts a circuit of a magnetic drive system. The magnetic circuit used for this research was cylindrical and horizontally symmetric. Therefore, when electric current is not applied, static thrust is balanced in the center position with magnetic force generated by magnets. On the other hand, once electric current is applied, one-directional thrust is produced by the imbalance of force caused by magnetic flux around a coil. The direction of thrust can be controlled by the direction of current. As illustrated in Figure 2, it is ideal that static thrust, y, be proportional to input current, M, as well as a large amount of thrust by a small input of current.

In our research, by using static simulation software for magnetic drive, which was developed in-house, instead of conducting an actual experiment, we implemented parameter design. After selecting an optimal configuration, we confirmed reproducibility in simulation and in an actual device.

2. Design of Experiments for Simulation

Based on the parameters in Figure 3, we created Figure 4, a Y-type cause-and-effect diagram for design of experiment.

Signal Factors and Characteristic

According to the definition of an ideal state, we selected current (M_1, M_2, M_3) as a signal factor and static thrust as a characteristic.

Control Factors

We chose design parameters associated with a magnetic drive structure as control factors and set levels for each factor (Table 1).

Noise Factors

We set dimensional variability of design parameters as a substitute for stresses of an actual device (Figure 4). More specifically, for each design parameter, we set two levels, one a certain variability above a nominal design parameter and the other below a nominal. As a next step, by conducting a preliminary experiment with an L_{12} orthogonal array (Table 2), we studied a magnitude of effects on the characteristic and a trend for each level of each factor (from Table 3). Using the data of the preliminary experiment, we calculated factor effects in Table 4 and plotted them in Figure 5.

Indicative Factors

Essentially, stroke affects the conversion efficiency rate of static thrust for electric current and it is de-

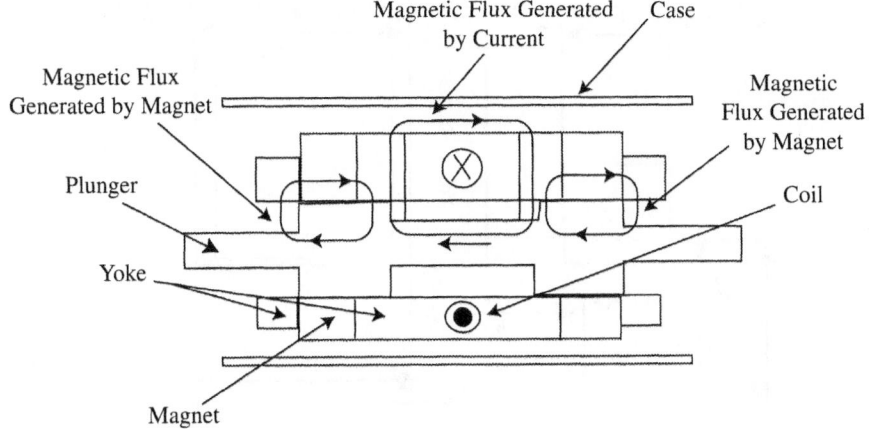

Figure 1
Magnetic drive system

sirable to stabilize its input/output relationship for each stroke. Thus, we set up five levels (k_1 to k_5) as an indicative factor (Table 5).

Considering a dynamic movement, we hoped the slope of the difference between each stroke would be flat. Yet since it can be adjusted afterward, we prioritized the system's functionality and planned to confirm the slope as a reference later.

Finally, we assigned all factors to an L_{18} orthogonal array (Table 6). Control factors are allocated to the inner array and signal, compounded noise, and indicative factors are laid out in the outer array.

3. SN Ratio

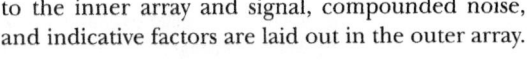

Based on the L_{18} orthogonal array, we obtained the data shown in Table 7. Using the zero-point proportional equation, we analyzed them using the following calculation procedure.

Total variation:

$$S_T = \sum y_{ij}^2 \qquad (f = 30) \qquad (1)$$

Effective divider:

$$r = \sum M_j^2 \qquad (2)$$

Linear equations:

$$L_1 = M_1 y_{11} + M_2 y_{12} + M_3 y_{13}$$

$$\vdots$$

$$L_{10} = M_1 y_{101} + M_2 y_{102} + M_3 y_{103} \qquad (3)$$

Slope:

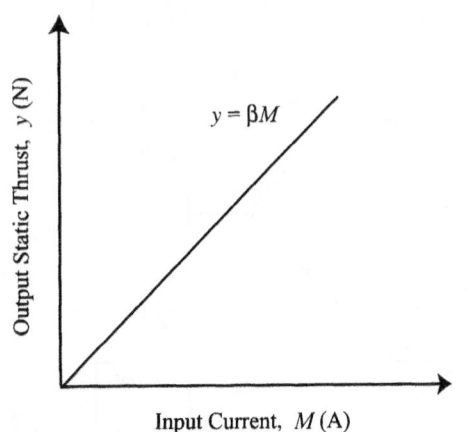

Figure 2
Ideal state

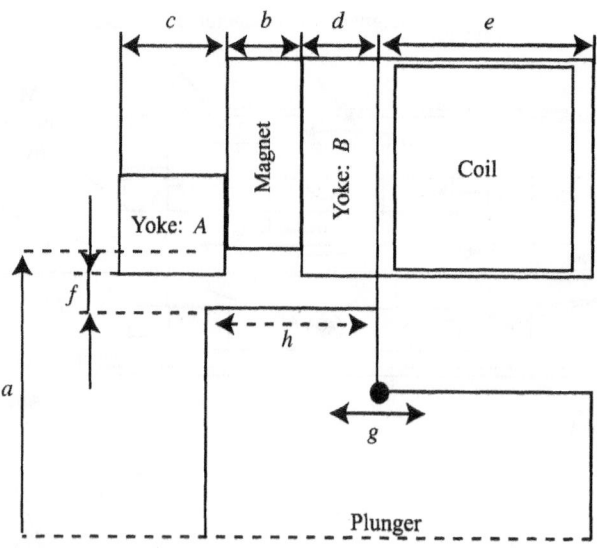

a: Internal Diameter of Magnet   e: Width of Bobbin
b: Width of Magnet   f: Air Gap
c: Width of Yoke A   g: Position of Convex Portion of Plunger
d: Width of Yoke B   h: Width of Convex Portion of Plunger

Figure 3
Magnetic drive and design parameters

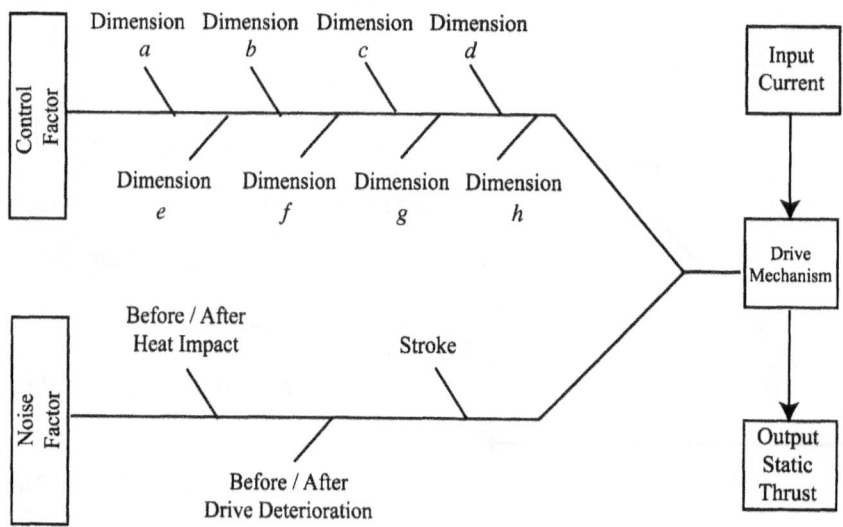

Figure 4
Y-type cause-and-effect diagram

Table 1
Control factors and levels

	Control Factor	Level 1	Level 2	Level 3
A:	dimension e	A_1	A_2	—
B:	dimension $b + c + d$	B_1	B_2	B_3
C:	(dimension $c + d$)/factor B	C_1	C_2	C_3
D:	(dimension d)/(dimension $c + d$)	D_1	D_2	D_3
E:	dimension a	E_1	E_2	E_3
F:	dimension f	F_1	F_2	F_3
G:	dimension g	G_1	G_2	G_3
H:	dimension h/(factor B)	H_1	H_2	H_3

Table 2
Layout and results of L_{12} orthogonal array

No.	P	Q	R	S	T	U	V	W	Data
1	1	1	1	1	1	1	1	1	2.95
2	1	1	1	1	1	2	2	2	3.12
⋮	⋮	⋮	⋮	⋮	⋮	⋮	⋮	⋮	⋮
11	2	2	1	2	1	2	1	1	3.85
12	2	2	1	1	2	1	2	1	2.80

Table 3
Factor levels of preliminary experiment

	Control Factor	Positive Side 1	Negative Side 2
P:	dimension e	$e + \alpha e$	$e - \alpha e$
Q:	dimension c	$c + \alpha c$	$c - \alpha c$
R:	dimension d	$d + \alpha d$	$d - \alpha d$
S:	dimension a	$a + \alpha a$	$a - \alpha a$
T:	dimension b	$b + \alpha b$	$b - \alpha b$
U:	dimension f	$f + \alpha f$	$f - \alpha f$
V:	dimension g	$g + \alpha g$	$g - \alpha g$
W:	dimension h	$h + \alpha h$	$h - \alpha h$

$$\beta_m = \frac{L_m + L_{m+5}}{2r} \tag{4}$$

Variation of proportional term:

$$S_\beta = \frac{\left(\sum L_i\right)^2}{10r} \quad (f = 1) \tag{5}$$

Variation of proportional term due to stroke:

$$S_{K\beta} = \frac{1}{2r}\left[\begin{array}{l}(L_1 + L_6)^2 + (L_2 + L_7)^2 \\ + (L_3 + L_8)^2 + (L_4 + L_9)^2 \\ + (L_5 + L_{10})^2\end{array}\right] - S_\beta \quad (f = 4) \tag{6}$$

Variation of proportional term due to compounded noise:

Table 4
Factor effects of preliminary experimental data

No.	\multicolumn{8}{c}{Factor}							
	P	*Q*	*R*	*S*	*T*	*U*	*V*	*W*
1	3.17	3.17	3.18	3.04	3.19	2.90	3.29	3.30
2	3.18	3.19	3.18	3.32	3.17	3.46	3.07	3.06

$$S_{N\beta} = \frac{1}{5r}\left[(L_1 + \cdots + L_5)^2 + (L_6 + \cdots + L_{10})^2\right]$$
$$- S_\beta \quad (f = 1) \tag{7}$$

Error variation:

$$S_e = S_T - S_\beta - S_{K\beta} - S_{N\beta} \quad (f = 24) \tag{8}$$

Error variance:

$$V_e = \frac{S_e}{f_e} \tag{9}$$

Total error variation:

$$S_N = S_T - S_\beta - S_{K\beta} \quad (f = 25) \tag{10}$$

Total error variance:

$$V_N = \frac{S_N}{25} \tag{11}$$

SN ratio:

$$\eta = 10 \log \frac{(1/10r)(S_\beta - V_e)}{V_N} \text{ dB} \tag{12}$$

Sensitivity:

$$S = 10 \log \frac{1}{10r}(S_\beta - V_e) \text{ dB} \tag{13}$$

SN Ratio of Slope β (Flatness)

Using β_1 to β_5, we computed nominal-the-best SN ratio as follows:

$$\eta_\beta = 10 \log \frac{\frac{1}{5}(S_m - V_{N'})}{V_{N'}} \text{ dB} \tag{14}$$

4. Optimal Condition and Confirmatory Experiment

SN ratios and sensitivity for each experiment are summarized in Table 8 and the factor effects are

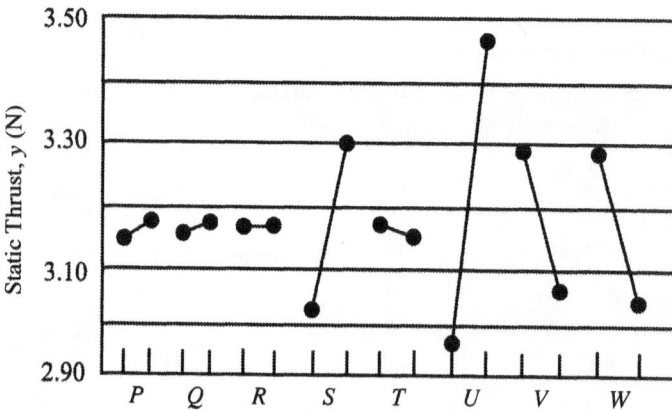

Figure 5
Response graphs for preliminary experiment

Table 5
Compounded noise factors and levels

Control Factor		Positive Side N_1	Negative Side N_2
S:	dimension a	$\alpha + \alpha a$	$\alpha - \alpha a$
U:	dimension f	$f + \alpha f$	$f - \alpha f$
V:	dimension g	$g - \alpha g$	$g + \alpha g$
W:	dimension h	$h - \alpha h$	$h + \alpha h$

illustrated in Figure 6. According to the response graphs, we determined the optimal configuration by selecting levels leading to greater SN ratios and sensitivities.

Optimal configuration: $A_2B_3C_3D_2E_1F_2G_2H_2$

Current configuration: $A_1B_2C_2D_2E_2F_2G_2H_2$

According to the magnitude of factor effects, we estimated SN ratio by factors B, C, D, F, and H and sensitivity by factors B, C, F, G, and H at the optimal and current configurations selected. Also, as shown in Table 9, we confirmed reproducibility through confirmatory experiments for the optimal and current configurations. Whereas we obtained good reproducibility of the SN ratio, we did not for sensitivity. However, we considered this not so critical because of the large gain in sensitivity.

5. Confirmation of Simulated Results Using Actual Device

Using the optimal and current configurations obtained from simulation, we prepare an actual device equipped with actual condition of use and stress (before/after thermal shock and continuous operation). As a result we obtained a large amount of gain (Table 10). Despite different absolute values of SN ratios and sensitivities, we found a correlation between the results in simulation and the actual device, which enabled us to ensure robustness in simulation. Figure 7 shows actual data at the optimal and current configurations.

On the other hand, the fact that a difference between each slope of stroke was also improved by approximately 3 dB proves that we do not have any problems in dealing with dynamic movement. Since we confirmed that simulation enables us to ensure robustness and reproducibility, by implementing product development in a small scale for the same series of products without prototyping, we can expect to reduce development labor hours and innovate our development process.

The expected benefits include elimination of the prototyping process (cut development cost in half) and a shortened development cycle time (cut labor hours for development in half). Although we focused on static simulation in this research, we will attempt to apply this technique to dynamic simulation or other mechanisms to innovate our development process and, consequently, standardize our de-

Table 6
Layout of an L_{18} orthogonal array

No.	1 A	2 B	...	8 H	N_1 M_1 k_1–k_5	M_2 k_1–k_5	M_3 k_1–k_5	N_2 M_1 k_1–k_5	M_2 k_1–k_5	M_3 k_1–k_5
1	1	1	...	1	...	...	...	...	...	...
2	1	1	...	2	...	...	...	...	...	...
⋮	⋮	⋮	...	⋮						
17	2	3	...	3	...	...	...	...	...	...
18	2	3	...	1	...	...	...	...	...	...

Table 7
Results of experiments

Noise	Stroke	M_1	M_2	M_3	L	β
N_1	k_1	y_{11}	y_{12}	y_{13}	L_1	β_1
	k_2	y_{21}	y_{22}	y_{23}	L_2	β_2
	k_3	y_{31}	y_{32}	y_{33}	L_3	β_3
	k_4	y_{41}	y_{42}	y_{43}	L_4	β_4
	k_5	y_{51}	y_{52}	y_{53}	L_5	β_5
N_2	k_1	y_{61}	y_{62}	y_{63}	L_6	
	k_2	y_{71}	y_{72}	y_{73}	L_7	
	k_3	y_{81}	y_{82}	y_{83}	L_8	
	k_4	y_{91}	y_{92}	y_{93}	L_9	
	k_5	y_{101}	y_{102}	y_{103}	L_{10}	

Table 8
Results of SN ratio and sensitivity analysis

No.	SN Ratio	Sensitivity	SN Ratio of Slope
1	17.97	3.55	30.10
2	25.79	8.72	18.55
3	21.42	7.49	17.15
4	18.51	8.54	14.96
5	17.28	3.15	26.28
6	28.60	9.67	24.28
7	28.09	8.71	22.38
8	23.08	9.51	25.23
9	37.14	8.15	23.12
10	23.65	5.83	25.10
11	20.33	7.71	15.33
12	29.83	7.71	24.57
13	29.41	10.20	14.10
14	29.66	7.74	26.29
15	25.02	7.26	20.82
16	21.37	4.76	22.53
17	21.10	8.56	21.36
18	30.88	9.68	17.28
Average	24.95	7.61	21.64

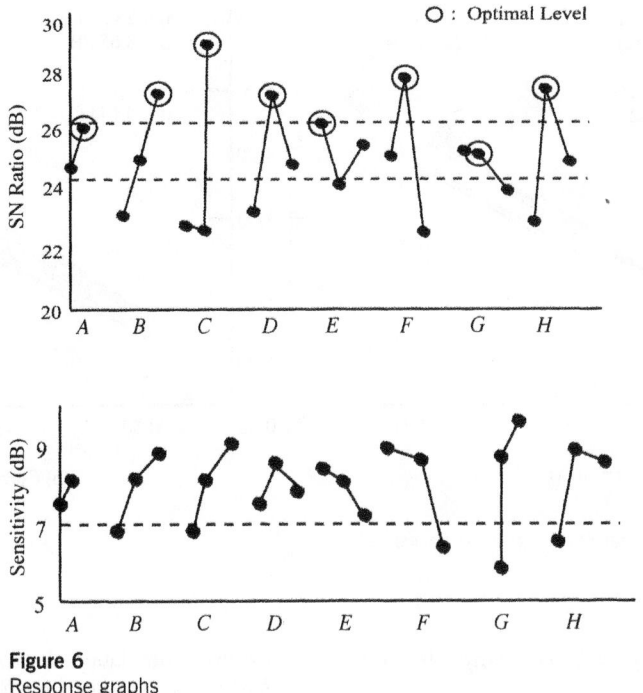

Figure 6
Response graphs

Table 9
Estimation and confirmation of gain

	Estimation		Confirmation	
Condition	SN Ratio	Sensitivity	SN Ratio	Sensitivity
Optimal	37.54	10.56	34.08	9.01
Current	29.41	9.33	27.82	9.22
Gain	8.14	1.23	6.16	−0.021

Table 10
Results confirmed by actual experimental device

Condition	SN Ratio	Sensitivity	SN Ratio of Slope
Optimal	36.43	8.97	24.22
Current	25.50	8.65	21.15
Gain	10.93	0.32	3.07

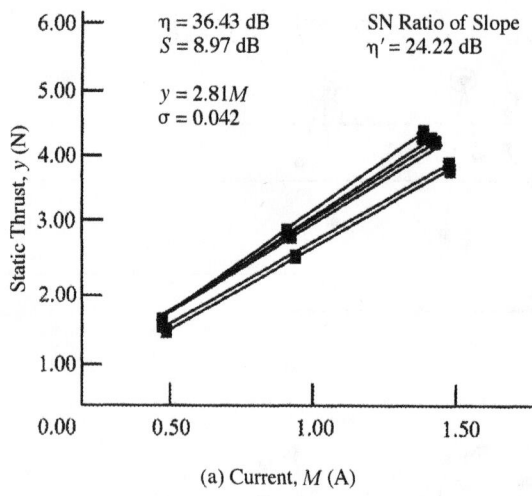

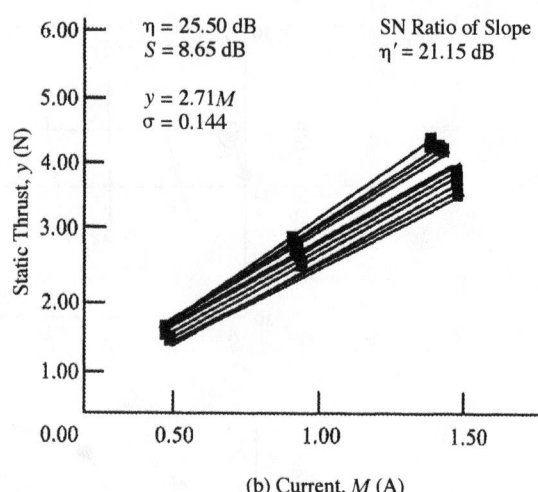

(a) Current, M (A)

(b) Current, M (A)

Figure 7
Confirmation of simulated results using actual device

sign process and deploy it companywide in the future.

Reference

Tetsuo Kimura, Hidekazu Yabuuchi, Hiroki Inoue, and Yoshitaka Ichii. 2001. Optimization of linear actua-tor using simulation. *Proceedings of the 2001 Quality Engineering Symposium*, pp. 246–249.

This case study is contributed by Tetsuo Kimura, Hidekazu Yabuuchi, Hiroki Inoue, and Yoshitaka Ichii.

CASE 56

Functionality Evaluation of Articulated Robots

Abstract: In this study we assessed robot functionality. There are still a number of problems to be solved for obtaining a clear-cut conclusion; however, we have determined that we may be able to evaluate robot functionality by measuring electricity consumption of an articulated robot.

1. Introduction

Many multijoint robots have the function of grasping and moving an object located at an arbitrary position with perfect freedom, as human arms do. Robots used primarily for product assembly are called industrial robots. Each arm of an articulated robot moves in accordance with each joint angle changed by a motor and aligns its end effecter to a target. That is, the movement of a robot's end effecter is controlled by an indicated motor's angle. We used the vertical articulated robot shown in Figures 1 and 2 and instructed it with angular values so that it could move by recognizing both angular and positional values.

Using Cartesian coordinates (X, Y, Z) (Figure 1), and joint coordinates (J_1, ... , J_5) (Figure 2), we can express the position of this robot. In this case, the locating point of this robot is the center point of the flange at the end of the robot's arm (point P in Figure 1), whose coordinates are indicated on the controller.

2. Robot Arm's Displacement

For the robot used for our study, we instructed successive positions of the end effecter so that it could move according to our target path. There were two major instruction methods: one a method of giving digital information related to positions, the other a method by which an operator instructs a robot as to target positions by moving the robot with controller buttons. For each arm's rotational angle as a signal, M, we adopted the latter method. On the other hand, we measured end effecter's displacement as a characteristic, y, by using a coordinate measuring machine after the robot moved automatically by following the program code. Using these signals and characteristics, we evaluated the robot's functionality.

In this study, to obtained the characteristic, y, we evaluated the locating performance of the end effecter as follows. First, we attached a pen at point P (center point) on the flange in such a way that it was parallel with the flange (Figure 3), of the experimental device. Second, a three-dimensional wall consisting of three plates was set up within the range of the robot arm. Third, a piece of paper was attached on each plate. Fourth, by manipulating the robot arm, we plotted four points on paper with the pen attached at point P of the robot (Figure 4). While we plotted points A, B, and C by manipulating the robot with the controller buttons, point D was plotted by automated movement of the robot based on the program code. After connecting points A and B with a line, we defined the line as the Y-axis of the robot. Then we set point C as the origin for locating. Since it was difficult to measure points located by the robot based on the coordinate origin of the robot, we set up another point as the origin (in Figure 4, point C). Therefore, point D was regarded as a located point that was translated with a certain amount of displacement from point C by the robot arm.

Since a located point is expressed in the Cartesian coordinated system (in this case, Y and Z

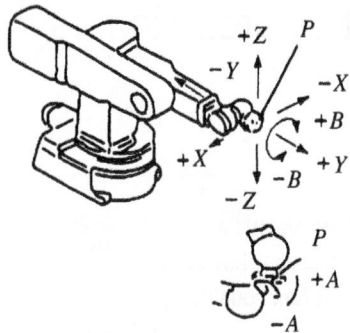

When we perform motion A in the Cartesian coordinate system, the position of the center point on the flange (point P) never changes position, as shown on the right, but the robot's posture changes.

Figure 1
Robot's movement in Cartesian coordinate system

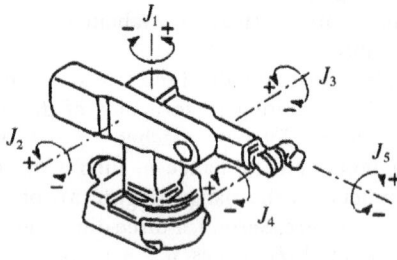

Figure 2
Robot's movement in joint coordinate system

coordinates), by transforming its coordinates through geometric calculation, we computed the position of the end effecter relative to the rotational center of the arm. Next, we detail the calculation procedure.

Since point P did not coincide with the pen's center point (Figure 5), when we transformed the Cartesian coordinates to the rotational coordinates, we defined the angle between the horizontal line and the line connecting the pen's center and the rotational center of joint J_4 as α_4 when the flange was set up in a vertical orientation. Next, we defined the length from the pen's center to the rotational center of joint J_4 as L_4. As illustrated in Figure 6, if we set the Y-axis directional displacement to GH_Y and the Z-axis directional displacement to GH_Z between points G and H, we obtained the following equations:

$$GH_Y = L_4 \cos \alpha_4 - L_4 \sin \alpha_4 \qquad (1)$$

$$GH_Z = L_4 \sin \alpha_4 - L_4 \cos \alpha_4 \qquad (2)$$

Using these two equations, we can calculate L_4 and α_4 as follows:

$$L_4 = \sqrt{\frac{GH_y^2 - GH_Z^2}{2}} \qquad (3)$$

$$\alpha_4 = \cos^{-1} \frac{GH_y + GH_Z}{2L_4} \qquad (4)$$

Focusing on two arms moved with rotation of

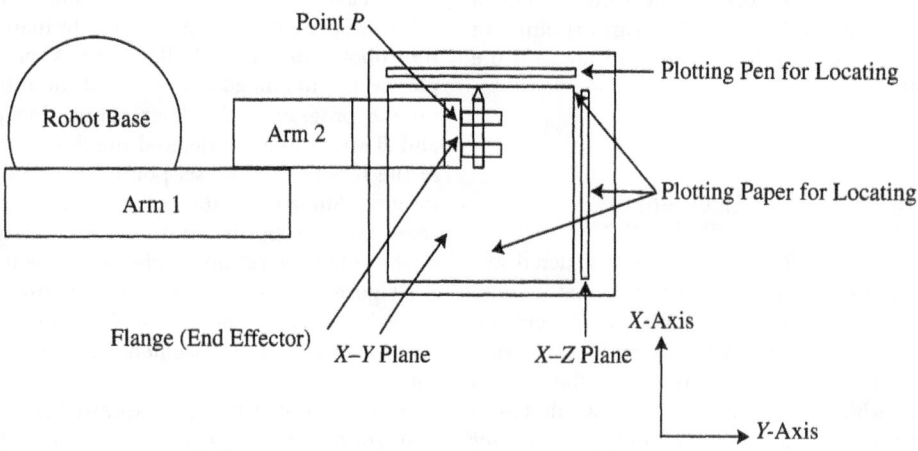

Figure 3
Positions of robot and box (wall)

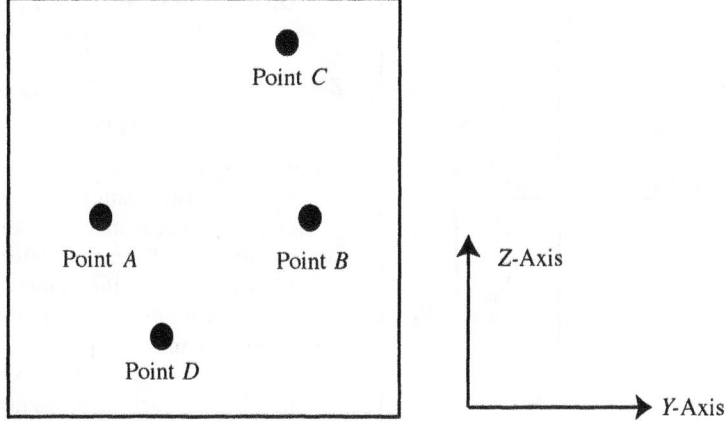

Figure 4
Plotting sequence for displacement of robot arm

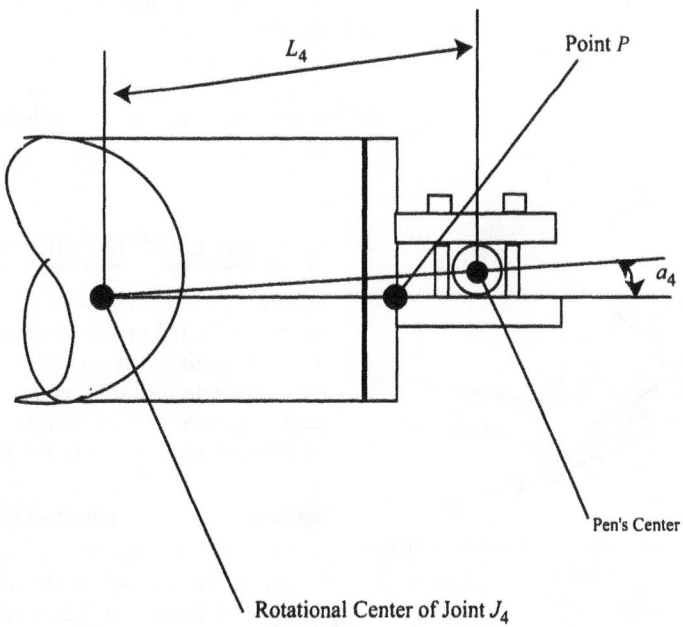

Figure 5
Relationship among pen's center point, point P, and rotational center of
joint J_4

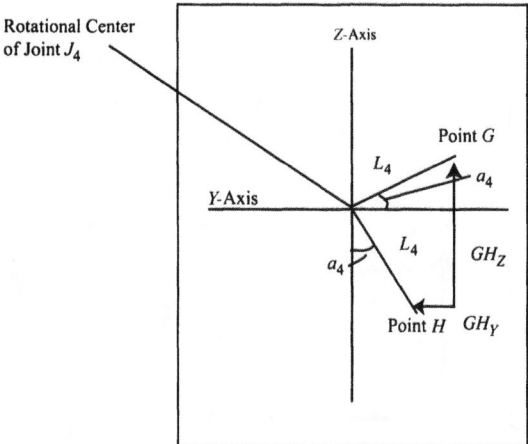

Figure 6
Relationship between pen's center point and rotational center of joint J_4

joint J_2, we set the length of arm 1 to L_2 and its angle to θ_2, the length of arm 2 to L_3 and its angle to θ_3 (Figure 7). In addition, defining the coordinates of the origin for locating as (O_Y, O_Z), we had the following equation:

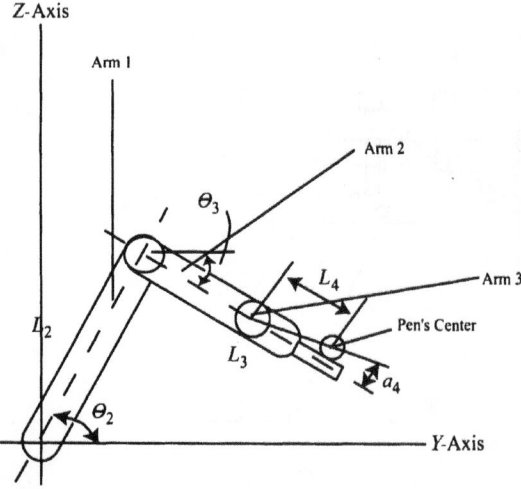

Figure 7
Relationship between pen's center point and rotational center

$$Y + O_Y = L_2 \cos \theta_2 + (L_3 + L_4 \cos \alpha_4)\cos \theta_3$$
$$- L_4 \sin \alpha_4 \sin \theta_3 \qquad (5)$$

$$Z + O_Z = L_2 \cos \theta_2 + (L_3 + L_4 \cos \alpha_4)\sin \theta_3$$
$$- L_4 \sin \alpha_4 \cos \theta_3 \qquad (6)$$

Using the equations addressed above, we performed coordinate transformation.

Next, let's discuss measurement and the modification method of errors. After adhering point-plotted papers on the table of a coordinate measuring machine, we measured coordinates of points using a microscope. Our study of measurement errors in the coordinate measuring machine by using die-cast parts with unknown dimensions revealed that the level of errors was approximately 10 mm. Since 10 mm is equivalent to 0.001° in terms of the rotational angle of joint J_2 and the minimum instruction unit of the robot is 0.01°, the errors in the coordinate measuring machine were considered much smaller than those in the robot.

Nevertheless, because the pen held by the robot plots points on the paper in our evaluation method, we should consider the following effects related to the robot arm's locating: (1) effect caused by the slant of paper attached to the table of the coordinate measuring machine, and (2) effect caused by the slant of a box (walls) onto which paper is attached. If we measure pertinent slant angles beforehand and perform geometric coordinate transformation, we can eliminate the foregoing effects.

3. Robot Locating Performance

According to the measurement method discussed so far, we evaluated errors in robot locating. However, it was difficult to set up control factors because we used only one robot. Thus, substituting factors originally regarded as error factors for indicative factors, we studied additivity of factor effects.

Setup of Signal and Indicative Factors

Defining the angular change setting values of arms 1 and 2, M and M' as signals, and the angular changes of arms 1 and 2, y and y' for automated movement based the program code as characteristics, we proceeded with the experiment.

Table 1
Signal factors regarding arm's angle (deg)

	Level		
Factor	1	2	3
Angular change setting value of joint J_2	−5	−10	−15
Angular change setting value of joint J_3	−5	−10	−15

$$y = \beta M \qquad (7)$$

$$y = \beta' M' \qquad (8)$$

In Table 1 we tabulated signal factors and levels. In the table, M corresponds to the setting values of joint J_2 while M' accords with the setting values of joint J_3. In addition, although we could not widen ranges of signal factor levels due to the constraint of our experimental device, we judged them to be sufficient for our evaluation. All indicative factors are shown in Table 2. These factors are allocated to columns 2, 3, and 4 in an L_{18} orthogonal array.

Indicative factor A, arm's posture, was selected so that we could investigate the difference between two cases where arms 1 and 2 start to move from a folded posture and from an extended posture. As shown in Figure 8, if we begin to move the arm from a folded state, we set up a point closer to the robot position as the locating origin. If we begin to move the arm from an extended state, we set up a point far from the robot position as the locating point. Next, indicative factor B, acceleration, indicates the magnitude of acceleration until the velocity of the robot arm reaches a constant value from zero, when it is manipulated automatically by the program code. Indicative factor C, constant velocity, represents the magnitude of velocity after the velocity of

the robot arm becomes constant by following the program code. Three levels of constant velocity, 6, 12, and 18%, represent mean the ratio of a selected velocity to the maximum movement velocity of the robot arm.

Calculation of SN Ratio and Sensitivity

For this study we used an L_{18} orthogonal array. Since procedures for analyzing data for joints J_2 and J_3 were identical, and additionally, a procedure for analyzing data in each row was also the same, as one typical example, we show the analysis of data in the first row in the case of joint J_2 in Table 3. In the table, M_1, M_2, and M_3 represent values when angular values at joint J_2 are set to −5°, −10°, and −15°, respectively. Similarly, M_1', M_2', and M_3' indicate values when angular values at joint J_3 are set to −5°, −10°, and −15°, respectively.

When the robot used for this study locates a target position, there is a technical constraint that we cannot move only one of joint J_2 or J_3. Therefore, setting the angular value at joint J_2 to a signal, we can regard joint J_3's movement caused by joint J_2's movement as an error. Similarly, defining the angular value at joint J_3 as a signal, we can consider joint J_2's movement triggered by joint J_3 as an error.

Table 2
Indicative factors

	Level		
Factor	1	2	3
A: arm's posture	Folded	Intermediate	Extended
B: acceleration	Low	Mid	High
C: velocity (constant)	6%	12%	18%

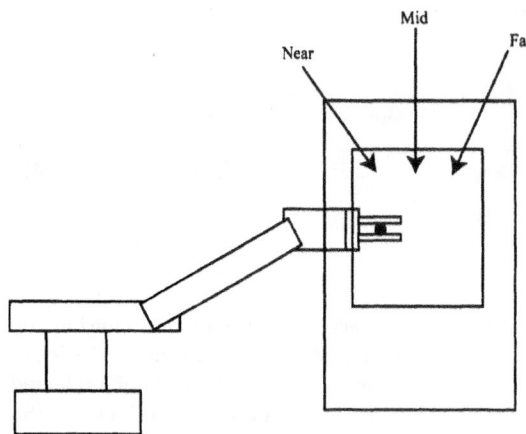

Figure 8
Position of origin for robot locating

Total variation:

$$S_T = (-4.98)^2 + (-9.83)^2 + \cdots + (-14.83)^2$$

$$= 51{,}030.1378 \qquad (f = 9) \tag{9}$$

Effective divider:

$$r = (-5)^2 + (-10)^2 + (-15)^2 = 350 \tag{10}$$

Linear equation:

$$L_1 = (-5)(-4.98) + (-10)(-9.83)$$
$$\quad + (-15)(-14.87)$$

$$= 346.250$$

$$L_2 = 348.950$$

$$L_3 = 344.800 \tag{11}$$

Variation of proportional term:

$$S_\beta = \frac{(L_1 + L_2 + L_3)^2}{3r} = 1030.0952 \qquad (f = 1) \tag{12}$$

Variation of proportional terms by joint J_3's movement:

$$S_{\beta\beta} = \frac{L_1^2 + L_2^2 + L_3^2}{r} - S_\beta = 0.0253 \qquad (f = 2) \tag{13}$$

Error variation:

$$S_e = S_T - S_\beta - S_{\beta\beta} = 0.0172 \qquad (f = 6) \tag{14}$$

Error variance:

$$V_e = \frac{S_e}{6} = 0.0029 \qquad (f = 6) \tag{15}$$

Total error variance:

$$V_N = \frac{S_{\beta\beta} + S_e}{7} = 0.0061 \qquad (f = 7) \tag{16}$$

SN ratio:

$$\eta = 10 \log \frac{(1/3r)(S_\beta - V_e)}{V_N} = 26.85 \text{ dB} \tag{17}$$

Sensitivity:

$$S = 10 \log \frac{1}{3r}(S_\beta - V_e) = 4.69 \text{ dB} \tag{18}$$

Analyzed Result

The response graphs of SN ratio and sensitivity regarding joint J_2 are shown in Figures 9 and 10, respectively, and those regarding joint J_3 are shown in Figures 11 and 12. Looking at the SN ratio factor effect plots, we can see that factor C (constant velocity) greatly affects the locating performance.

Table 3
Measurement data for $A_1B_1C_1$ (deg)

	M_1 (−5°)	M_2 (−10°)	M_3 (−15°)	Linear Equation
M_1' (−5°)	−4.98	−9.83	−14.87	L_1
M_2' (−10°)	−4.96	−9.96	−14.97	L_2
M_3' (−15°)	−4.85	−9.81	−14.83	L_3

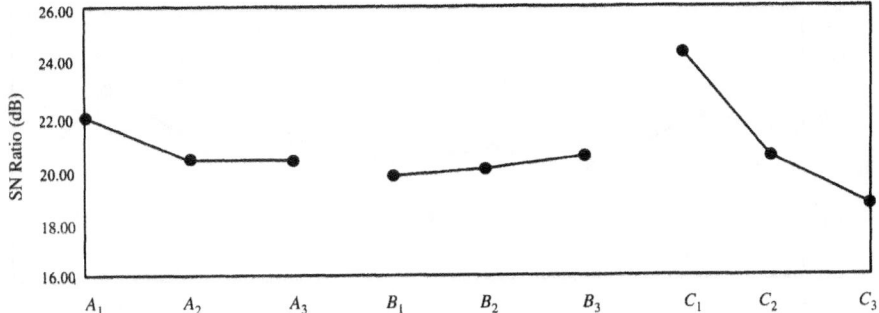

Figure 9
SN ratio for joint J_2

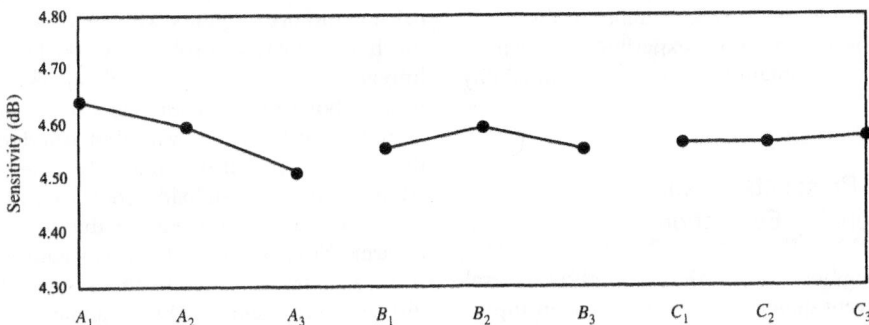

Figure 10
Sensitivity for joint J_2

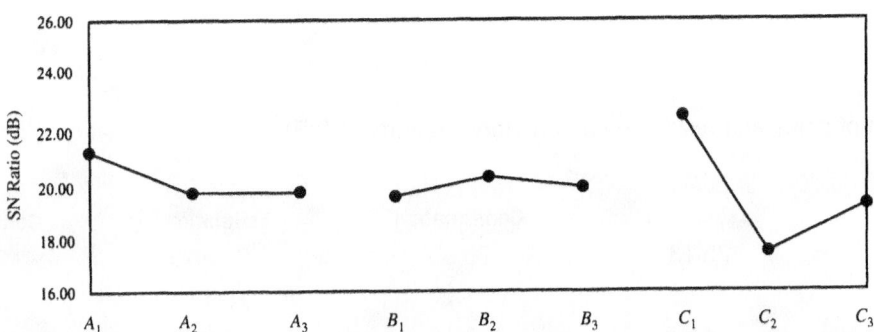

Figure 11
SN ratio for joint J_3

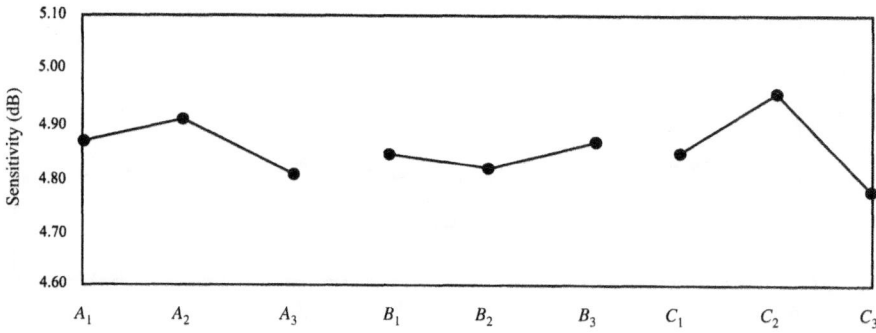

Figure 12
Sensitivity for joint J_3

According to the SN ratio factor effect plots, we notice that the optimal condition is $A_1B_2C_1$, whereas the initial condition is $A_2B_2C_2$. Table 4 shows the results in the confirmatory experiment, which reveals that we can obtain fairly good reproducibility.

4. Future Prospects in Robot Functionality Evaluation

In quality engineering, a variety of methods of evaluating different generic functions based on input/output of energy are proposed. As one application of the methods, we assessed robot functionality. Indeed, a number of problems remain to be solved before reaching a clear-cut conclusion; however, we may be able to evaluate robot functionality by measuring electricity consumption of an articulated robot.

At first, we considered the energy consumed by a robot. Many articulated robots move their arms by driving motors at joints with electric power. This mechanism enables a robot to work. Taking this fact into account, we wished to evaluate its functionality using robot arm's work and electricity consumption. However, work carried by a robot arm does not have the same weight all the time. Then we turned our attention to a free-of-load robot arm, as a robot is considered to consume energy due to the weight of its arm. Now, because of the constant weight of a robot arm, we evaluated its functionality based on a displacement element (i.e., angular value at each joint).

Considering all of the above, we should study its functionality by setting an instruction (angular) value at each joint to a signal factor and an amount of electricity consumed by a robot to a characteristic. In addition, to compare this evaluation method with the one based on "signal = instruction value

Table 4
Estimation of gains and results in confirmatory experiment (dB)

Condition	Joint J_2		Joint J_3	
	Estimation	Confirmation	Estimation	Confirmation
Optimal	25.13	25.33	24.25	23.81
Initial	20.69	23.17	18.44	19.28
Gain	4.44	2.16	5.81	4.53

Table 5
Indicative factors

	Control Factor	Level		
		1	2	3
A:	teaching method	Controller	Digit	—
B:	arms used	J_2, J_3	J_2, J_4	J_3, J_4
C:	posture	Folded	Intermediate	Extended
D:	weight (kg)	None	1	3
E:	acceleration	Low	Mid	High
F:	velocity	Low	Mid	High
G:	deceleration	Low	Mid	High

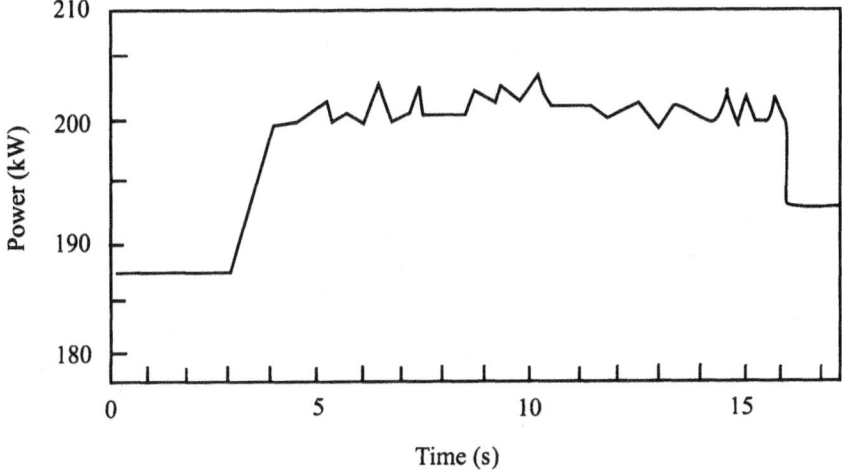

Figure 13
Electric power data

(angular change)" and "characteristic = angular change at a joint," we measured angular changes at joints as eventual movements as well as electricity consumption and defined them as characteristics.

The signal factor levels are as follows:

Level:	1	2	3
Joint instruction value (deg):	5	10	15

Allocating the indicative factors enumerated in Ta-

ble 5, we performed an experiment. Figure 13 indicates the electricity data measured. For analysis of electricity consumption and locating values, we followed the functionality evaluation based on the joint angle discussed before.

Judging from the result of SN ratios measured in electric power, we found that factor F has a considerable effect on electric power. For locating, factors A, D, and G have a strong influence. Consequently, the fact that functionality evaluation method using

electric power can be applied to an articulated robot demonstrates the validity of functionality evaluation method based on energy.

Reference

Takatoshi Imai, Naoki Kawada, Masayoshi Koike, Akira
Sugiyama, and Hiroshi Yano, 1997. A study on the

measurement of the generic function of articulated
robots. *Quality Engineering*, Vol. 5, No. 6, pp. 30–37.

*This case study is contributed by Naoki Kawada and
Masayoshi Koike.*

New Ultraminiature KMS Tact Switch Optimization

Abstract: A robust design effort was initiated using Taguchi's dynamic parameter design approach. An L_{18} orthogonal array was used to evaluate the control factors with respect to the signal factor (dome travel), the response (actuating force), and the noise factors (process and material variations and number of operational cycles). The numerical experiments were carried out by using computer simulations only. The signal-to-noise ratio increased 48.7 dB as a result of this optimization. The analysis and selection of the best system parameters confirmed the process estimates and resulted in a design with dramatically improved mechanical and reliability performance.

1. Introduction

The newest multifunction switch at ITT Cannon, the KMS switch, was designed specifically to meet phone market requirements. This switch is a new type of ultraminiature tact switch (Figure 1).

An engineering team was created to address the switch design and to improve its mechanical and electrical performances. A parameter design approach was selected using the method of maximizing the SN ratio and of optimizing the forces and the tactile feel.

The essence of the switch is the K2000 dome (Figure 2). The dome provides the spring force, the tactile feel, and the electrical contact. A properly designed dome functions in a unique manner: As the dome is compressed, a point is reached where the actuating force actually declines and goes through a minimum value. The person actuating the switch feels the change in force, so it is called a *tactile feel*. The tactile feel is usually quantified by the ratio of actuating force to forward force (Figure 3).

2. Background

The KMS switch uses a very small dome with the following specifications:

- Ultraminiature dimensions: 1.7×2.8 mm
- Actuating force, F_a
- Forward force, F_{ra}
- Maximum tactile feel, given by the ratio $(F_a - F_{ra})/F_a$ (see Figure 3)
- Operation life of 300 K_{op}/min.

3. Objectives

The objective of the parameter design effort using Taguchi's approach was to determine a set of design parameters that would result in the largest number of operating cycles while keeping a good tactile feel and a constant actuating force for the life of the product. The function of the switch is to give a tactile feel at a predetermined actuating force and to transmit an electrical current when it is actuated during the life of the product. This can be described using the force–deflection (F/D) curve (Figure 3). The ideal switch would exhibit an F/D curve that would not change during the life of the product.

The analytical objective for Taguchi's parameter design approach is to maximize the SN ratio. The SN ratio is a metric that measures the switch

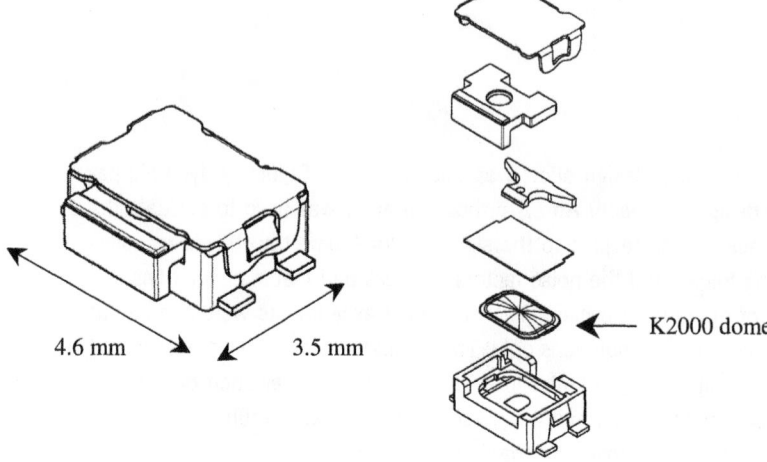

Figure 1
KMS tact switch

function when exposed to external factors, called noise factors, which may affect that performance. As the effect of these noise factors increases, the SN ratio decreases. The performance of the switch is measured while being exposed to these noise factors for various combinations of design parameters, using an orthogonal array to provide a balanced treatment of these control factors to the noise factors. The performance characteristics measured were the typical points of the F/D curve and specifically the actuating force F_a (M_1) and the forward force F_{ra} (M_2).

4. Finite-Element Analysis (Computer Simulations)

All the experiments in this study are numerical experiments that were conducted by using computer simulation (finite-element analysis). A different F/D curve was calculated for each combination of factors. The aging effect (due to the number of operations) corresponding to each dome design was obtained from the calculated maximum stress level by using a modelization of the Wöhler curve (Figure 4).

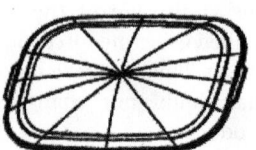 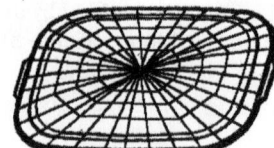

Figure 2
K2000 dome

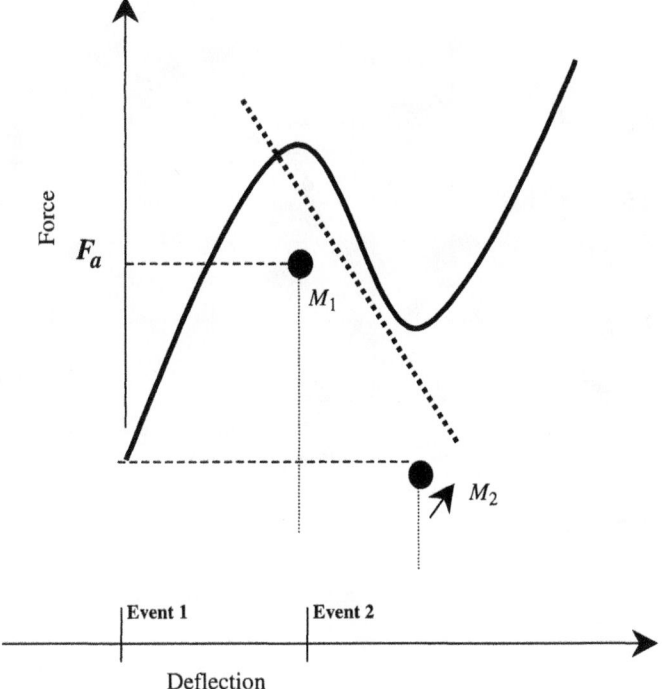

Figure 3
F/D curve for electromechanical switch

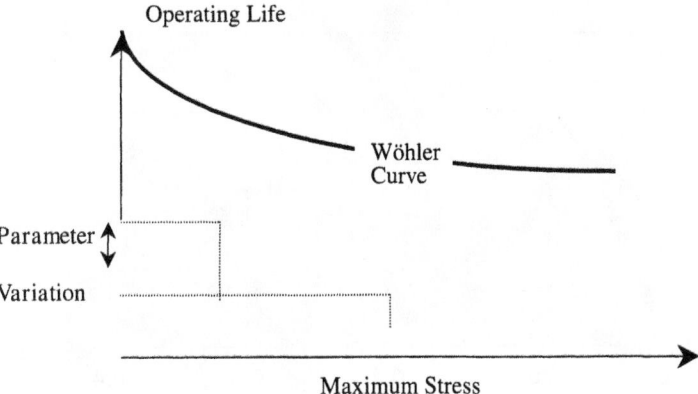

Figure 4
Operating life obtained from the Wöhler curve

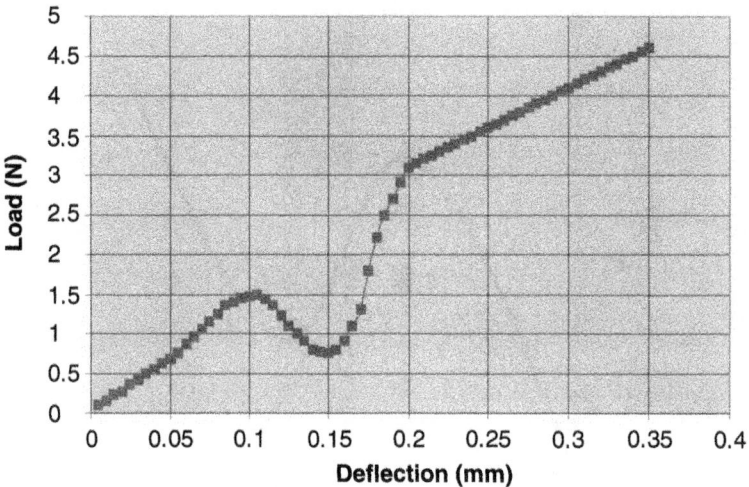

Figure 5
Ideal F/D curve

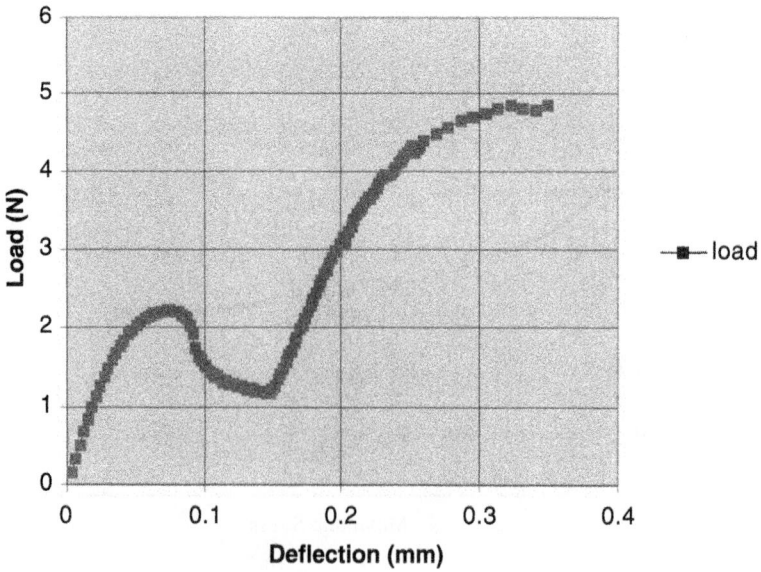

Figure 6
Simulated F/D curve (corresponding to one design)

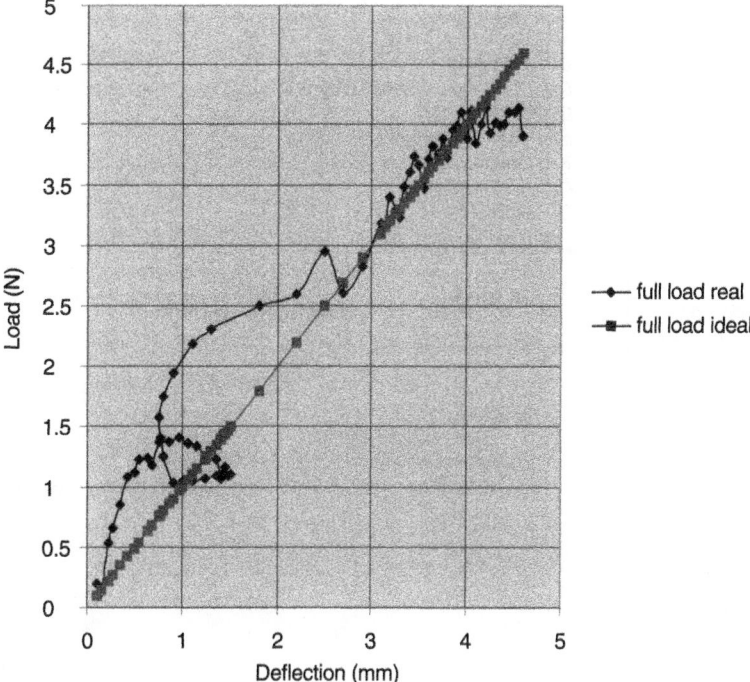

Figure 7
Simulated F/D curve versus the ideal switch F/D curve

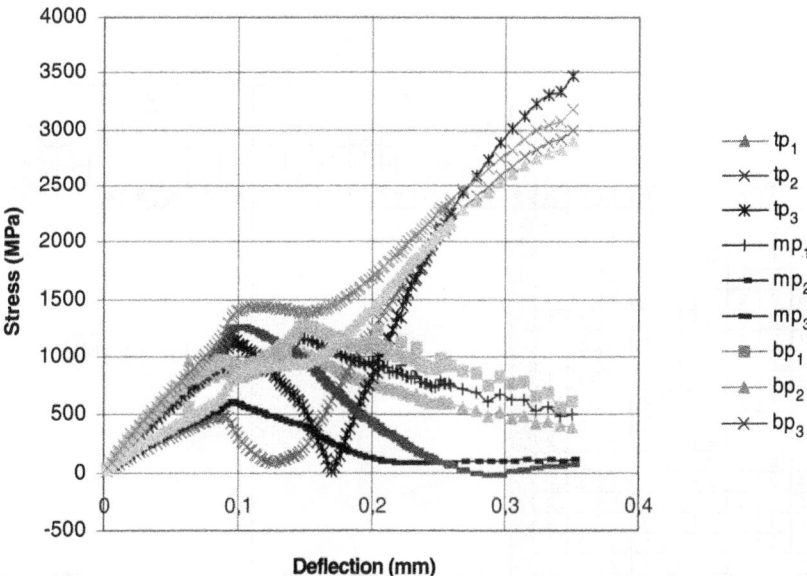

Figure 8
Stress curves at different points of the dome (corresponding to one dome design)

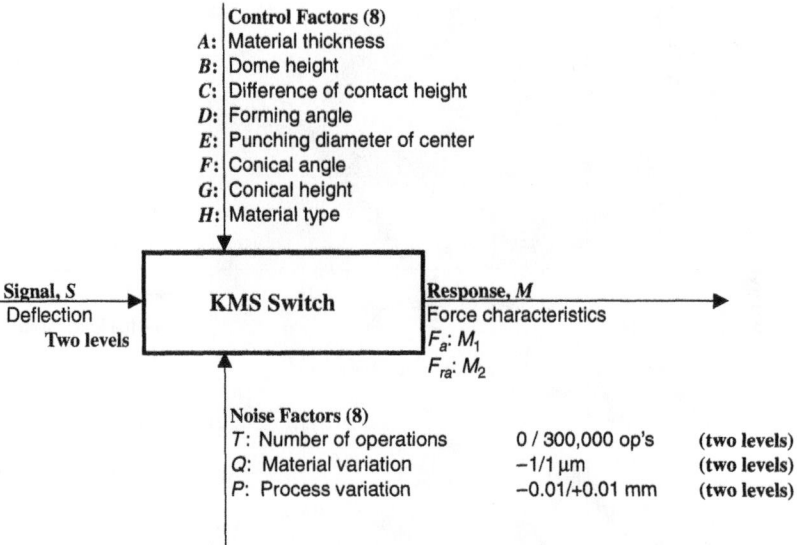

Figure 9
P-diagram

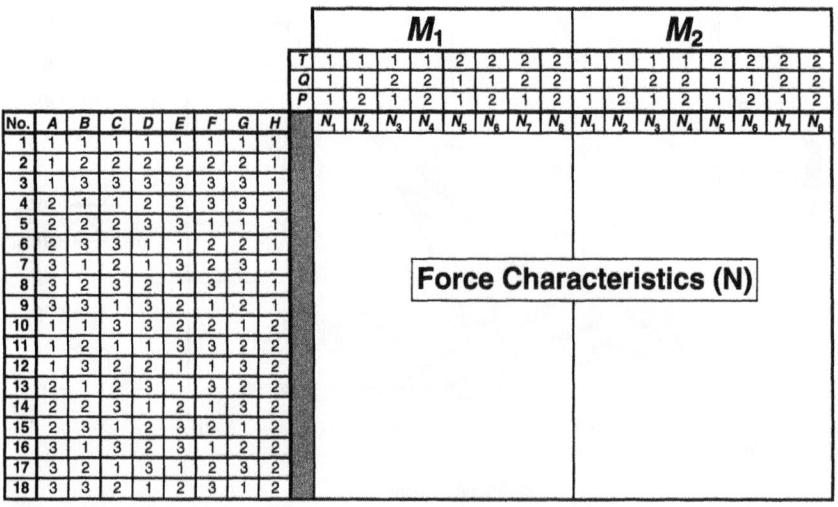

Figure 10
Experimental layout showing L_{18} and outer group

Table 1
Control factors and levels

	Control Factor	Level		
		1	2	3
A:	material thickness (μm)	30	35	38
B:	dome height (mm)	0.12	0.15	0.2
C:	difference of contact height (mm)	0	0.05	0.1
D:	forming profile (three-dimensional)	1	2	3
E:	punching diameter/depth of center area (mm)	0/0	0.25/0.025	0.5/0.05
F:	conical angle (deg)	12.5	25	40
G:	conical height (mm)	0.02	0.05	0.1
H:	material type	1	2	

We obtain the following charts:

❏ The ideal F/D function (Figure 5)

❏ One example of F/D curve obtained by computer simulation (Figure 6)

❏ The F/D curve of Figure 6 versus the ideal F/D curve (Figure 7)

❏ One example of stress curves at different points versus dome deflection (Figure 8)

5. P-Diagram

The switch product, generally called an engineered system, can be described by a visual method called a P-diagram. This diagram indicates the relationship between parameter types and the input and outputs of the switch system. The dome deflection is the

Table 2
Noise factors and levels

	Noise Factor	Level	
		1	2
N:	material variation (μm)	−1	1
P:	process variation	−0.01	0.01
T:	number of operations (K_{op})	0	300

signal. Figure 9 presents the P-diagram for the KMS switch system. The control factors are selectable by the designer, but the noise factors are not, except when they are controlled to evaluate their impact on the system. The typical response points are, respectively, the actuating force F_a (M_1) and the forward force F_{ra} (M_2).

6. Experimental Layout

The factors shown in Figure 9 were assigned to an experimental layout consisting of an L_{18} orthogonal array and a full-factorial combination of the signal and noise factors (Figure 10).

The factors and their associated levels are described in Table 1. The control factors were assigned to the columns of the orthogonal array with the coefficients in the column representing the level of the corresponding factor. The noise factors were assigned to the outer, fully orthogonal group (Table 2).

7. Results and Analysis

Figure 11 presents the results from the computer simulations done by finite element analysis. For

No.	M_1 (F_a: actuating force)								M_2 (F_{ra}: forward force)								S_M	S_N	V_M	V_N	S_T	S_{NxM}	V_N	SN	10 log V_M	10 log V_m
	1	1	1	2	1	1	1	2	2	1	2	1	2	1	2	2										
	N_1	N_2	N_3	N_4	N_5	N_6	N_7	N_8	N_1	N_2	N_3	N_4	N_5	N_6	N_7	N_8										
1	1.40	1.60	1.95	2.20	1.27	1.41	1.44	1.86	1.10	1.30	1.00	1.20	1.01	1.16	0.83	1.06	1.25	0.53	1.25	0.08	2.15	0.366	0.06	12.90	0.96	14.72
2	2.10	2.20	2.75	3.15	1.14	1.18	0.95	0.90	1.60	2.00	1.55	1.90	0.94	1.10	0.71	0.71	0.93	6.41	0.93	0.92	8.13	0.79	0.51	2.57	-0.32	15.87
3	2.95	3.80	3.85	4.40	0.50	0.50	0.50	0.50	2.90	3.80	2.95	3.60	0.50	0.50	0.50	0.50	0.19	38.11	0.19	5.44	38.83	0.53	2.76	-11.59	-7.18	18.13
4	3.60	4.00	4.00	4.20	2.52	2.78	2.08	2.35	3.61	4.01	4.01	4.21	2.53	2.79	2.09	2.36	0.00	10.15	0.00	1.45	10.15	0.00	0.73	-32.58	-33.98	22.13
5	3.90	4.50	5.40	6.00	1.44	1.50	1.85	1.88	2.20	2.30	2.40	2.20	0.97	0.95	1.02	0.93	11.38	22.56	11.38	3.22	39.19	5.26	1.99	7.58	10.56	19.87
6	4.30	4.95	5.35	6.00	0.50	0.50	0.50	0.50	2.10	2.90	4.90	5.35	0.50	0.50	0.50	0.50	1.79	70.98	1.79	10.14	75.82	3.04	5.29	-4.71	2.53	19.97
7	6.00	6.60	6.00	7.00	2.84	2.94	2.15	2.13	6.01	6.61	6.01	7.01	2.85	2.95	2.16	2.14	0.00	63.01	0.00	9.00	63.01	0.00	4.50	-40.51	-33.98	25.03
8	3.20	3.75	4.35	4.40	0.50	0.50	0.50	0.50	2.70	3.20	2.60	2.55	0.89	0.82	0.84	0.50	1.35	32.76	1.35	4.68	36.28	2.17	2.49	-2.66	1.31	17.72
9	6.90	7.80	7.70	8.50	1.14	0.87	0.91	0.79	4.40	6.95	7.20	8.20	0.86	0.82	0.79	0.76	1.25	172.00	1.25	24.57	175.69	2.44	12.46	-10.00	0.96	24.09
10	1.80	2.10	2.15	2.40	0.86	0.82	0.91	0.79	1.80	2.10	1.65	2.20	0.86	0.82	0.79	0.76	0.05	6.07	0.05	0.87	6.22	0.11	0.44	-9.85	-13.40	15.12
11	1.90	2.05	2.70	3.00	1.20	1.28	1.71	1.25	1.90	2.00	1.70	2.05	1.20	1.25	1.16	0.97	0.51	3.88	0.51	0.55	5.02	0.63	0.32	2.01	-2.91	16.69
12	4.20	4.80	4.50	5.15	2.17	1.58	0.90	0.73	3.30	0.95	3.95	4.50	1.76	0.61	0.85	0.70	3.43	36.34	3.43	5.19	45.06	5.30	2.97	0.62	5.35	20.14
13	4.00	4.30	4.00	4.60	1.73	1.64	1.38	0.71	4.01	4.31	4.01	4.61	1.74	1.65	1.39	0.72	0.00	34.58	0.00	4.94	34.58	0.00	2.47	-37.91	-33.98	20.98
14	3.40	3.90	4.60	5.10	1.08	0.93	0.50	0.50	2.30	2.80	4.00	4.50	0.86	0.79	0.50	0.50	0.88	43.96	0.88	6.28	45.56	0.72	3.19	-5.58	-0.54	19.15
15	8.00	6.10	6.05	7.40	1.63	0.58	0.50	0.50	4.00	5.00	3.20	1.50	1.03	0.50	0.50	0.50	13.05	83.53	13.05	11.93	113.78	17.20	7.19	2.59	11.16	21.38
16	3.10	3.70	3.00	3.45	0.76	0.58	1.13	0.50	3.10	3.70	2.95	3.40	0.76	0.58	1.11	0.50	0.00	27.32	0.00	3.90	27.32	0.00	1.95	-33.92	-31.02	18.15
17	5.40	6.00	4.70	6.00	2.22	2.15	0.92	1.19	5.41	6.01	4.71	6.01	2.23	2.16	0.93	1.20	0.00	65.97	0.00	9.42	65.97	0.00	4.71	-40.71	-33.98	23.11
18	5.30	6.00	6.35	7.00	0.50	0.50	0.50	0.50	4.90	2.40	4.80	1.90	0.50	0.50	0.50	0.50	7.09	77.38	7.09	11.05	98.14	13.68	6.50	0.37	8.51	20.56

Figure 11
Numerical results

Table 3
ANOVA for the SN ratio

Source	Pool	d.f.	S	V	F	S'	R
A		2	1286.0658	643.0329	80.2653	1270.0431	23.84
B		2	1413.2788	706.6394	88.2049	1397.2561	26.22
C		2	0.5300	0.2650			
D		2	376.9884	188.4942	23.5284	360.9657	6.77
E		2	36.5609	18.2804			
F		2	380.5481	190.2740	23.7506	364.5254	6.84
G		2	1718.4728	859.2364	107.2525	1702.4501	31.95
H		1	104.5913	104.5913	13.0554	96.5800	1.81
HA		2	10.9772	5.4886			
e_1							
e_2							
(e)		6	48.0681	8.0113		136.1928	2.56
Total		17	5328.0132	313.4125			

(e) is pooled error.

each line of the design of experiment, the ANOVA decomposition can be carried out as follows:

Source	d.f.	S	V
m	1	S_m	$V_m = S_m$
M	1	S_M	$V_M = S_M$
N	7	S_N	$V_N = S_N/7$
MN	7	S_{MN}	$V_{MN} = S_{MN}/7$
	16		

To optimize the system, Taguchi recommended that we use the following data transformations. This is a two-step optimization process:

1. *Signal-to-noise ratio:*

$$SN = 10 \log \frac{V_M}{V_{N'}}$$

where $V_{N'} = (S_N + S_{MN})/14$.

2. *Sensitivity:*

$$S_M = 10 \log V_M$$

to maximize the delta of the force difference that plays on the tactile feel $(F_a - F_{ra})/F_a$

$$S = 10 \log V_m$$

to optimize the actuation force (F_a)

The SN ratio had to be maximized first. The sensitivity, S_M, had to be maximized next, and last, the S parameter had to be optimized to focus on the nominal force value. The data transformations resulted in the ANOVA tables and response graphs shown in Tables 3 to 5 and Figures 12 to 14.

1. *Signal-to-noise:* $10 \log(V_M/V_N)$ transformation (Table 3 and Figure 12)
2. *Sensitivity 1:* $S_M = 10 \log V_M$ transformation (Table 4 and Figure 13)
3. *Sensitivity 2:* $S = 10 \log V_m$ transformation (Table 5 and Figure 14)

Table 6 summarizes the results and the best level choice.

8. Confirmation

Table 7 presents a comparison of process estimates calculated from the numerical results and the confirmation values obtained from actual switch under

Table 4
ANOVA for $10 \log V_M$

Source	Pool	d.f.	S	V	F	S'	r
A		2	424.8138	212.4069	9.4948	380.0722	7.73
B		2	2461.2643	1230.6322	55.0106	2416.5227	49.13
C		2	16.8809	8.4404			
D		2	223.2589	111.6294	4.9900	178.5172	3.63
E		2	33.0715	16.5358			
F		2	328.7046	164.3523	7.3467	283.9630	5.77
G		2	1323.5088	661.7544	29.5812	1278.7672	26.00
H		1	55.7912	55.7912			
HA		2	50.8520	25.4260			
e_1							
e_2							
(e)		7	156.5957	22.3708		380.3037	7.73
Total		17	4918.1461	289.3027			

(e) is pooled error.

Table 5
ANOVA for $10 \log V_M$

Source	Pool	d.f.	S	V	F	S'	r
A		2	73.9262	36.9631	39.6632	72.0624	48.30
B		2	12.2648	6.1324	6.5804	10.4010	6.97
C		2	21.9876	10.9938	11.7969	20.1238	13.49
D		2	3.4679	1.7339			
E		2	0.6887	0.3444			
F		2	2.0802	1.0401			
G		2	28.8232	14.4116	15.4643	26.9594	18.07
H		1	0.2866	0.2866			
HA		2	5.6751	2.8376	3.0448	3.8113	2.55
e_1							
e_2							
(e)		7	6.5235	0.9319		15.8427	10.62
Total		17	149.2005	8.7765			

(e) is pooled error.

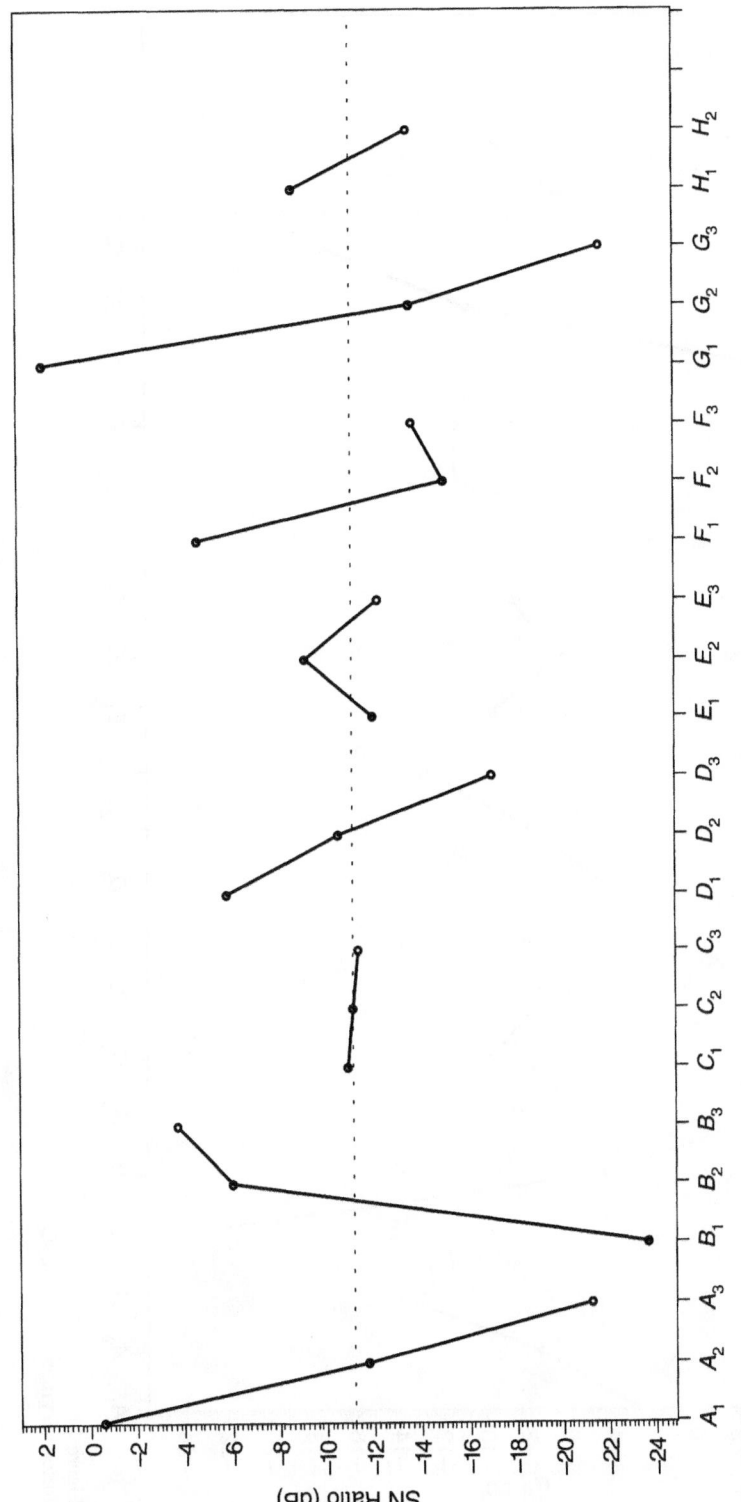

Figure 12
Response graph for the SN ratio

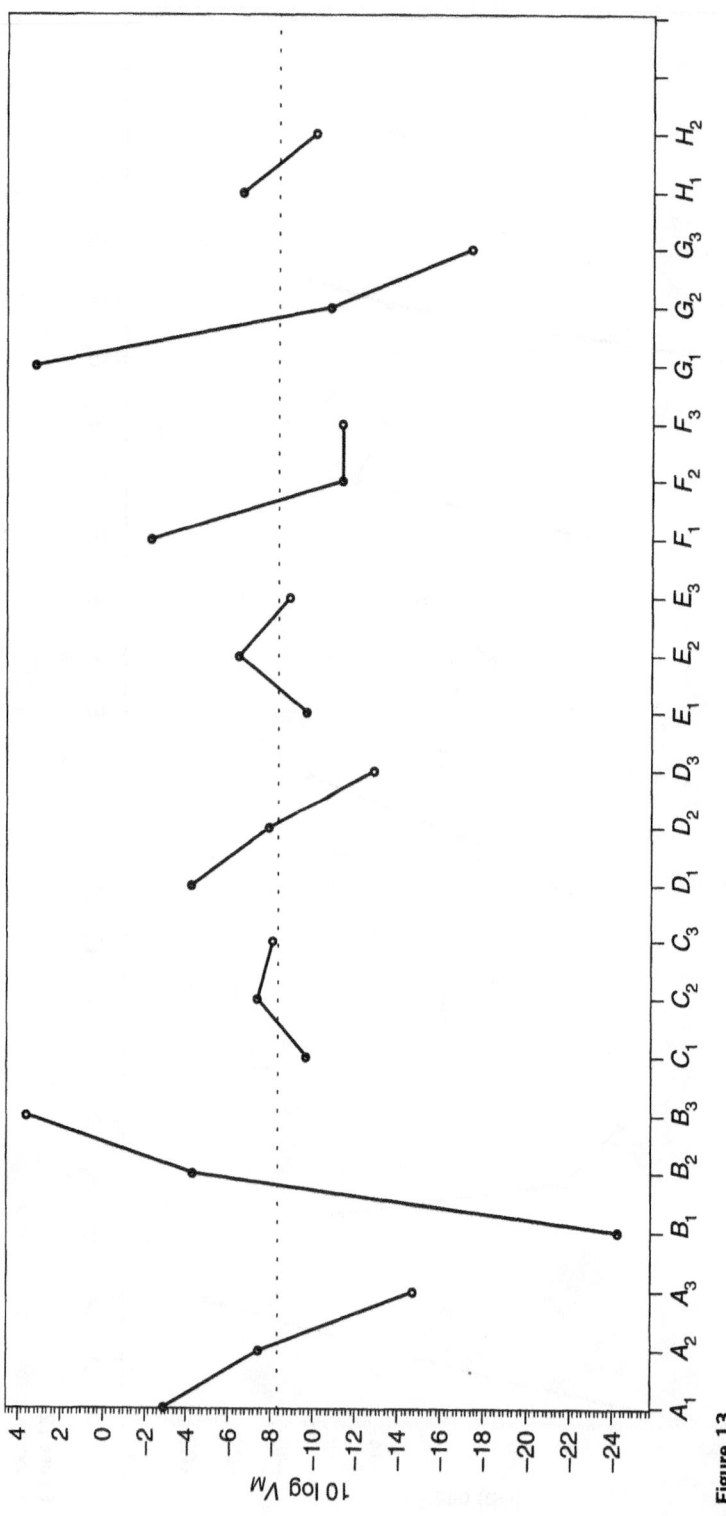

Figure 13
Response graph for 10 log V_M

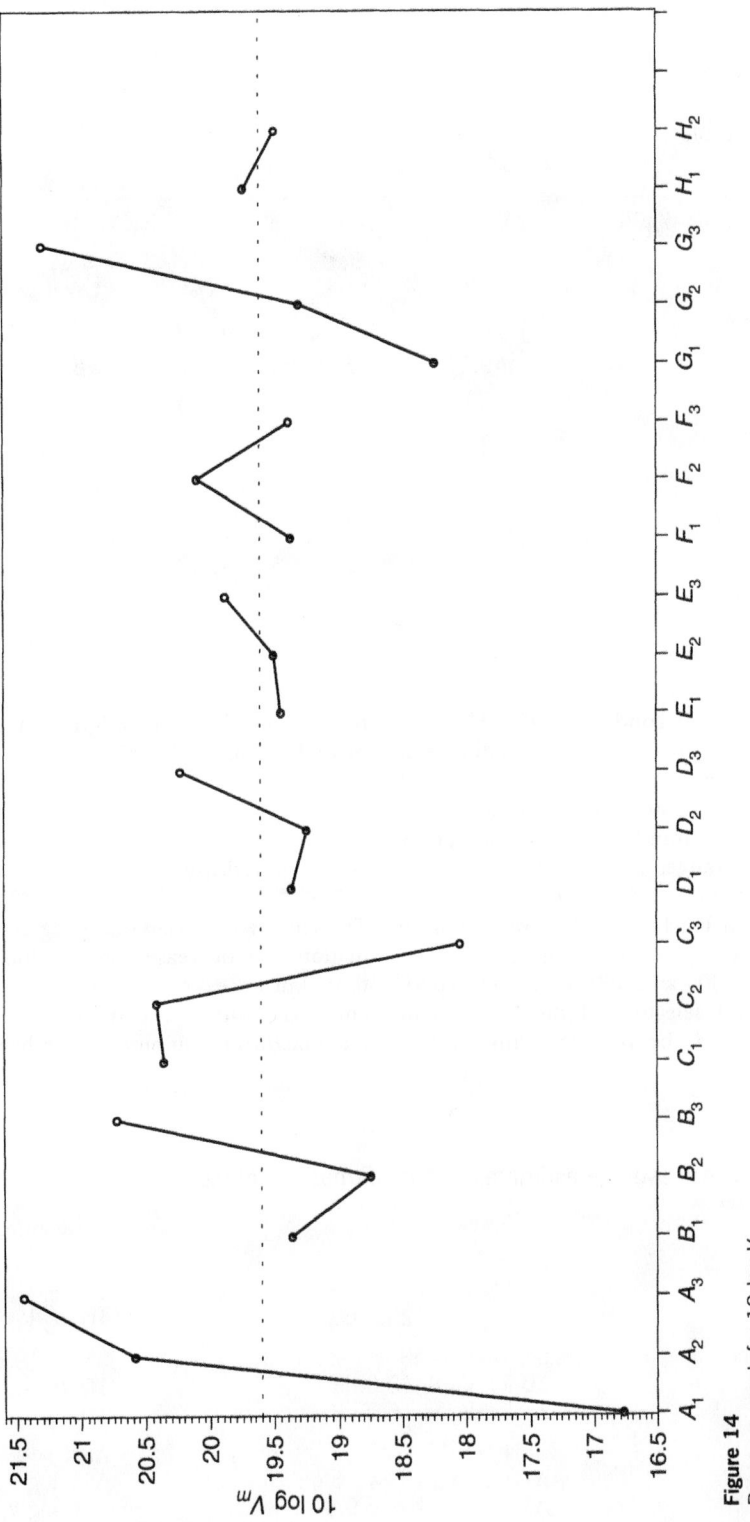

Figure 14
Response graph for 10 log V_m

Table 6

Best-level-choice table

	SN						
	$10 \log (V_M/V_{N'})$		$10 \log V_M$		$10 \log V_M$		
Factor	r (%)	Best Level	r (%)	Best Level	r (%)	Best Level	Best Choice
A	23.84	A_1	7.73	A_1	48.30	A_3	A_1
B	26.22	B_3	49.13	B_3	6.97	B_3	B_3
C					13.49	C_2	C_2
D	6.77	D_1	3.63	D_1			D_1
E							E_1
F	6.84	F_1	5.77	F_1			F_1
G	31.95	G_1	26.00	G_1	18.07	G_3	G_1
H	1.81	H_1					H_1

standard and optimized conditions. The SN ratio and $10 \log V_M$ (difference between actuating and forward forces) increased, respectively, by 48.7 and 37.1 dB as the result of study. In addition, the average force calculated from S_m is 1.67 N, that is, fully in the specification range.

We made prototypes starting from the results of this study. Different batches of KMS switches have been tested, and very good operating life has been obtained (between 300 and 500 K_{op}). The experimental analysis and selection of the best system parameters confirmed the process estimates and

resulted in a switch design with dramatically improved mechanical performances.

9. Conclusions

This study was carried out using only computer simulations, to decrease the development costs and time, but an experimental confirmation was made and was positive. The designs for the switch and manufacturing equipment have been made accord-

Table 7

Comparison of process average estimates and confirmation values

Parameter Set	Process Estimates	Confirmation Values
Initial product	SN = −20.4 dB $10 \log V_M$ = −8.5 dB $10 \log V_m$ = 23.7 dB	SN = −20.4 dB $10 \log V_M$ = −8.5 dB $10 \log V_m$ = 23.7 dB
Optimized product	SN = 33.0 dB $10 \log V_M$ = 32.0 dB $10 \log V_m$ = 16.8 dB	SN = 28.3 dB $10 \log V_M$ = 28.6 dB $10 \log V_m$ = 16.5 dB
Improvement	ΔSN = +53.4 dB $\Delta 10 \log V_M$ = +40.5 dB $\Delta 10 \log V_m$ = −6.9 dB	ΔSN = +48.7 dB $\Delta 10 \log V_M$ = +37.1 dB $\Delta 10 \log V_m$ = −7.2 dB

ingly as a result of the confirmation of this parameter design effort. The right key characteristics have been reached directly (actuating force, tactile feel, and operating life higher than 300 K_{op}). The switch robustness has greatly improved, resulting in no problem of actuating forces and tactile feel in the switch along its operational life.

There was no large change in dome shape, but the SN ratio increased 48.7 dB as a result of this optimization. The analysis and selection of the best system parameters confirmed the process estimates and resulted in a design with dramatically improved mechanical and reliability performances.

References

S. Rochon and P. Bouysses, October 1997. Optimization of the MSB series switch using Taguchi's parameter design method. Presented at the ITT Defense and Electronics Taguchi Symposium '97, Washington, DC.

S. Rochon and P. Bouysses, October 1999. Optimization of the PROXIMA rotary switch using Taguchi's parameter design method. Presented at the ITT Defense and Electronics Taguchi Symposium '99, Washington, DC.

S. Rochon and T. Burnel, October 1995. A three-step method based on Taguchi design of experiments to optimize robustness and reliability of ultraminiature SMT tact switches: the top actuated KSR series. Presented at the ITT Defense and Electronics Taguchi Symposium '95, Washington, DC.

S. Rochon, T. Burnel, and P. Bouysses, October 1999. Improvement of the operating life of the TPA multifunction switch using Taguchi's parameter design method. Presented at the ITT Defense and Electronics Taguchi Symposium '99, Washington, DC.

Genichi Taguchi, 1993. *Taguchi on Robust Technology Development.* New York: ASME, Press.

Genichi Taguchi, Shin Taguchi, and Subir Chowdhury. 1999. *Robust Engineering.* New York: McGraw-Hill.

This case study is contributed by Sylvain Rochon and Peter Wilcox.

Optimization of an Electrical Connector Insulator Contact Housing

Abstract: A computer simulation was used to optimize an electrical connector product design and associated plastic injection molding process. Taguchi's parameter optimization method, in conjunction with finite-element computer simulation, was used to eliminate the need to fabricate expensive tooling and molded parts, leading to faster, more robust product/process designs.

To function correctly the electrical connector must be flat and free of dimensional distortion, to allow reliable insertion of a credit card and proper electrical connection to the embedded computer chip. Three approaches were studied and compared to determine which provides the best set of product/process design parameter values. Smaller-the-better Z-axis deflection, dynamic Z-axis position, and dynamic mold dimension versus part dimension approaches were compared with respect to the resulting part flatness (Z-axis deflection).

1. Introduction

A new generation of CCM02 electrical connectors has been developed within Cannon for use with Smart Cards, a credit card with an embedded computer chip capable of direct communication with a host computer. The connector insulator part (Figure 1) is produced using plastic injection molding technology. For the connector to function reliably, the insulator part needs to be exceptionally flat for surface mounting, allowing the Smart Card to be inserted and interfaced electrically with the contacts mounted in the insulator.

This study presents a comparison between three approaches based on Taguchi's parameter design methods used to optimize the electrical connector product and associated plastic injection molding process. The three approaches were studied and compared to determine which provides the best set of product/process design parameter values. Smaller-the-better Z-axis deflection, dynamic Z-axis

position, and dynamic mold dimension versus part dimension approaches were compared. A computer simulation was used to eliminate the need to fabricate expensive tooling and molded parts. Figure 2 shows the finite-element analysis node mesh used in the computer simulation, which modeled the geometrical position and stress of each node after molding the part.

2. Objective

The objective of the robust engineering effort was to select the best product and process parameter value set which maximizes the function of the injection molding process and results in electrical connector parts that are flat and distortion free. The goal of the study was to explore different robust engineering approaches [i.e., signal-to-noise (SN) ratios] to determine which one resulted in the best product/process parameter configuration.

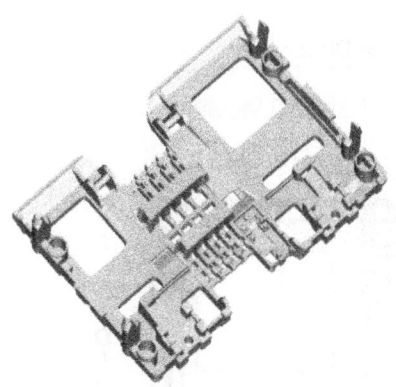

Figure 1
CCM02 MKII electrical connector insulator part

Figure 2
FEA mesh

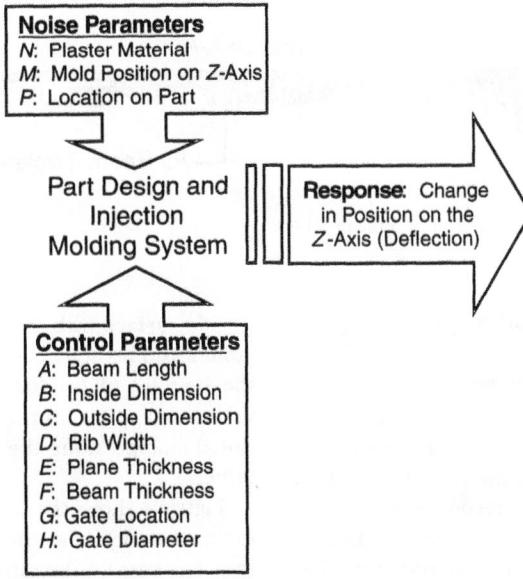

Figure 3
P-diagram for the smaller-the-best approach

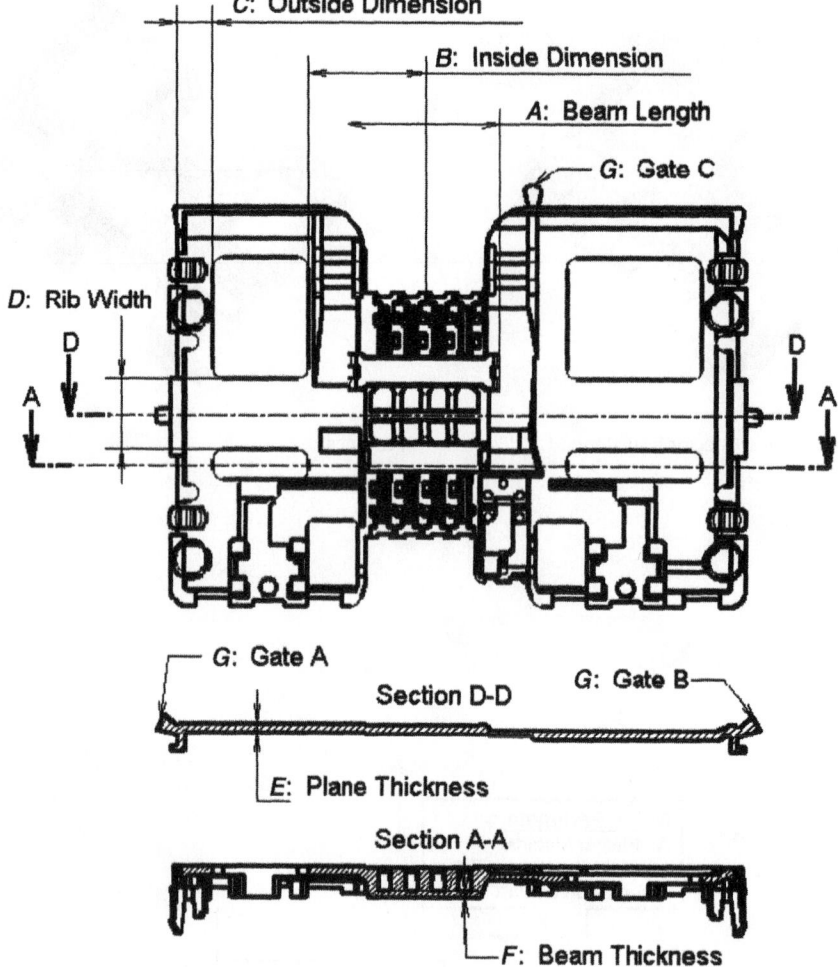

Figure 4
Electrical connector insulator and factor labels

3. Approach 1: Smaller-the-Best Deflection on the Z-Axis

After brainstorming the system and comparing past analyses on injection-molded parts of previous versions, the parameter diagram shown in Figure 3 was developed. The static system response was dimensional change along the Z-axis from a reference plane parallel to the flat plane of the part. This plane would be analogous to the part lying flat on a table surface with the Z-axis perpendicular to the table top. The objective of this approach was to find the best set of product/process design parameter values that results in the smallest amount of dimensional change along the Z-axis, zero being the ideal value.

Figure 4 shows the control factors associated with the part design and process parameters. These parameters remain the same for the three approaches discussed in this document. The control parameter

Table 1
Factor levels for the smaller-the-best approach

Factor		Level											
		1	2	3	4	5	6	7	8	9	10	11	12
Control factors													
A:	beam length (mm)	10	12										
B:	inside dimension (mm)	11.3	13.6	15.9									
C:	outside dimension (mm)	3.38	5.68	7.98									
D:	rib width (mm)	0	7	14									
E:	plane thickness (mm)	0.6	0.88	1.25									
F:	beam thickness (mm)	0.6	0.88	1.25									
G:	gate location	A	B	C									
H:	gate diameter (mm)	0.4	0.8	1.2									
Noise factors													
N:	plastic material	A	B										
M:	Z-axis mold position (mm)	0.44	1.04	1.34	1.64	1.96	2.23	2.63	3.04	3.38			
P:	location on part (node no.)	1	2	3	4	5	6	7	8	9	11	11	12

values used in an $L_{18}(2 \times 3^7)$ orthogonal array to explore the parameter value design space are presented in Table 1. The noise parameter values, also shown in Table 1, were exposed to the L_{18} using a full-factorial layout.

Table 2 presents the L_{18} orthogonal array with the column assignments used in this computer-simulated parameter optimization study. In addition, the smaller-the-best SN ratios calculated from each row of the data are shown. The raw data values are not presented, due to the large amount of numbers generated for this layout.

This experimental layout was used for all three of the approaches discussed in this document. Figure 5 presents the control-factor-level average graphs for approach 1: smaller-the-best, indicating that the optimum set of control parameter values is $A_1B_1C_2D_3E_3F_2G_1H_2$. Confirmation studies were conducted with this set of control parameter values as well as those of the other two optimization approaches. These are presented later.

4. Approach 2: Dynamic Zero-Point Proportional Z-Axis Mold Position versus Part Position

This alternative approach's objective was to select the optimum set of control parameter values that

Table 2
Orthogonal layout for the control factors

OA Row	Control Factor								Smaller-the-Best SN (dB)
	A	B	C	D	E	F	G	H	
1	1	1	1	1	1	1	1	1	10.30
2	1	1	2	2	2	2	2	2	14.24
3	1	1	3	3	3	3	3	3	15.27
4	1	2	1	1	2	2	3	3	9.54
5	1	2	2	2	3	3	1	1	15.43
6	1	2	3	3	1	1	2	2	9.49
7	1	3	1	2	1	3	2	3	9.01
8	1	3	2	3	2	1	3	1	12.04
9	1	3	3	1	3	2	1	2	17.75
10	2	1	1	3	3	2	2	1	17.08
11	2	1	2	1	1	3	3	2	8.84
12	2	1	3	2	2	1	1	3	12.03
13	2	2	1	2	3	1	3	2	13.38
14	2	2	2	3	1	2	1	3	10.30
15	2	2	3	1	2	3	2	1	12.53
16	2	3	1	3	2	3	1	2	13.94
17	2	3	2	1	3	1	2	3	17.30
18	2	3	3	2	1	2	3	1	7.50

maximizes the correlation between the signal factor Z-Axis mold position and the system's response Z-axis part position. The metric selected was the zero-point proportional dynamic SN ratio, and the goal was to select control parameter values levels that maximize this SN ratio and result in a slope as close to 1.00 as possible. This would result in a system that produces a part exactly the shape and size of the mold. The parameter diagram for this approach is shown in Figure 6. Figure 7 shows the ideal function associated with this P-diagram.

Table 3 presents the control, noise, and signal factor levels used in this approach. To accommodate the noise factor location on the Part, 12 locations were selected whose mesh nodes contained the same nine Z-axis coordinates used as signal factor values. In this approach, the noise factor Z-axis mold

position of the smaller-the-best approach became the signal factor for the zero-point proportional dynamic approach.

Figure 8 shows the electrical insulator part finite-element node mesh, indicating the positions used for the noise factor location on the part. Table 4 presents the orthogonal layout used, along with the resulting zero-point proportional SN ratio and slope, β.

Figure 9 shows the control-factor-level average graphs for the zero-point proportional SN ratio for the Z-axis position approach, using the values shown in Table 4. Figure 10 presents the sensitivity or slope-factor-level average graphs for the data values shown in Table 4, β.

From these graphs, the best set of control parameter values is $A_1 B_1 C_2 D_3 E_3 F_2 G_1 H_2$. Confirmation stud-

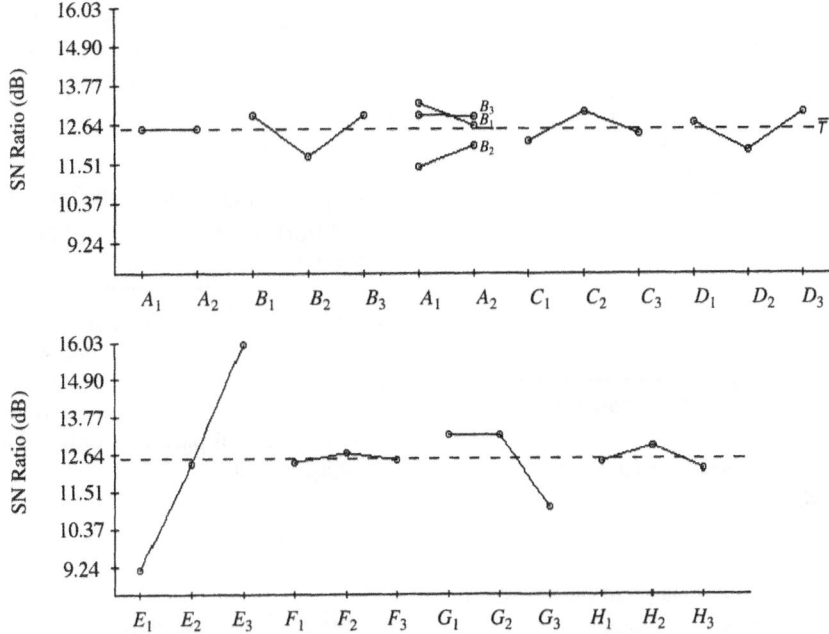

Figure 5
Smaller-the-best control-factor-level average graphs

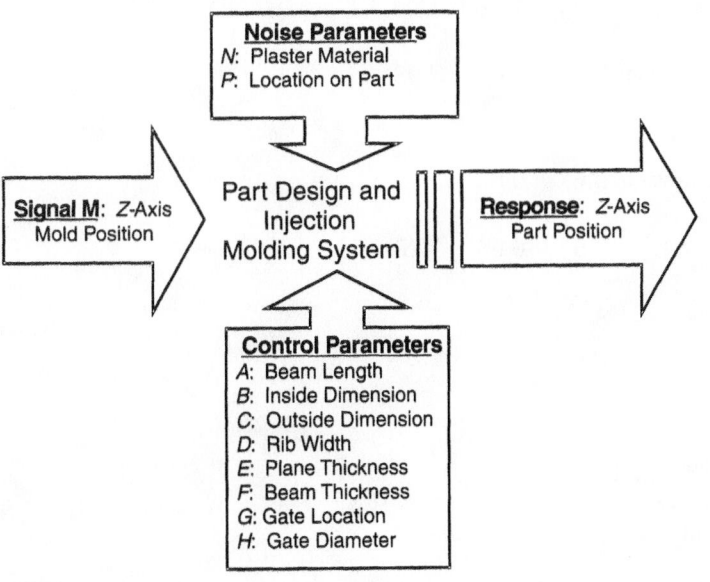

Figure 6
P-diagram for the dynamic Z-axis position approach

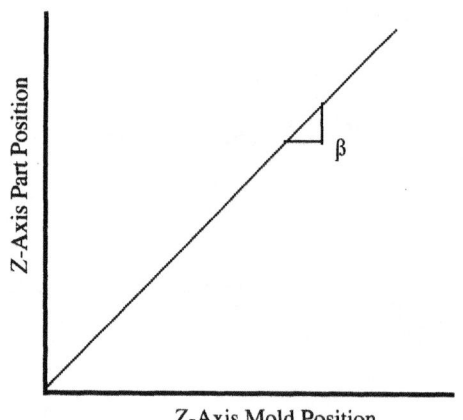

Figure 7
Ideal function for the Z-axis position approach

ies were conducted on this set of control factor levels, the results of which are presented along with the other approaches later.

5. Approach 3: Dynamic Zero-Point Proportional Mold Dimension versus Part Dimension, All Axes

The objective of this approach was to select the optimum set of control parameter values that maximizes the correlation between the signal factor, mold dimension, and the system's response, part dimension, using all part axes rather than just the Z-axis of approach 2.

Table 3
Factors levels for the Z-axis position approach

Factor		Level										
		1	2	3	4	5	6	7	8	9	10	11	12
Control factors													
A:	beam length (mm)	10	12										
B:	inside dimension (mm)	11.3	13.6	15.9									
C:	outside dimension (mm)	3.38	5.68	7.98									
D:	rib width (mm)	0	7	14									
E:	plane thickness (mm)	0.6	0.88	1.25									
F:	beam thickness (mm)	0.6	0.88	1.25									
G:	gate location	A	B	C									
H:	gate diameter (mm)	0.4	0.8	1.2									
Noise factors													
N:	plastic material	A	B										
P:	location on part (node no.)	1	2	3	4	5	6	7	8	9	11	11	12
Signal factor													
M:	Z-axis mold position (mm)	0.44	1.04	1.34	1.64	1.96	2.23	2.63	3.04	3.38			

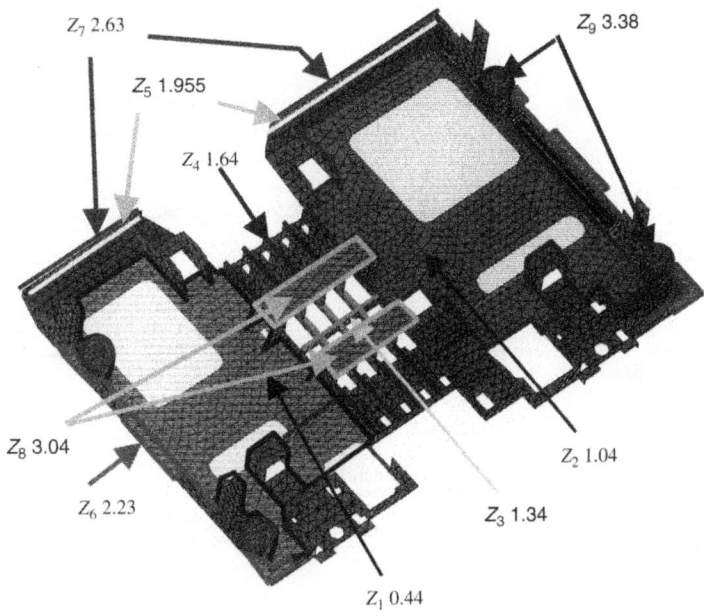

Figure 8
Finite-element node mesh for the Z-axis position approach

Table 4
Orthogonal layout and results for the Z-axis position approach

OA Row	Control Factor								Zero-Point SN	Slope, β
	A	B	C	D	E	F	G	H		
1	1	1	1	1	1	1	1	1	10.22	0.969
2	1	1	2	2	2	2	2	2	14.24	0.981
3	1	1	3	3	3	3	3	3	15.22	0.990
4	1	2	1	1	2	2	3	3	9.43	0.967
5	1	2	2	2	3	3	1	1	15.44	0.984
6	1	2	3	3	1	1	2	2	9.36	0.969
7	1	3	1	2	1	3	2	3	8.89	0.961
8	1	3	2	3	2	1	3	1	11.96	0.983
9	1	3	3	1	3	2	1	2	17.87	0.985
10	2	1	1	3	3	2	2	1	17.12	0.986
11	2	1	2	1	1	3	3	2	8.69	0.970
12	2	1	3	2	2	1	1	3	11.97	0.978
13	2	2	1	2	3	1	3	2	13.35	0.982
14	2	2	2	3	1	2	1	3	10.22	0.969
15	2	2	3	1	2	3	2	1	12.49	0.978
16	2	3	1	3	2	3	1	2	13.98	0.977
17	2	3	2	1	3	1	2	3	17.42	0.984
18	2	3	3	2	1	2	3	1	7.32	0.971

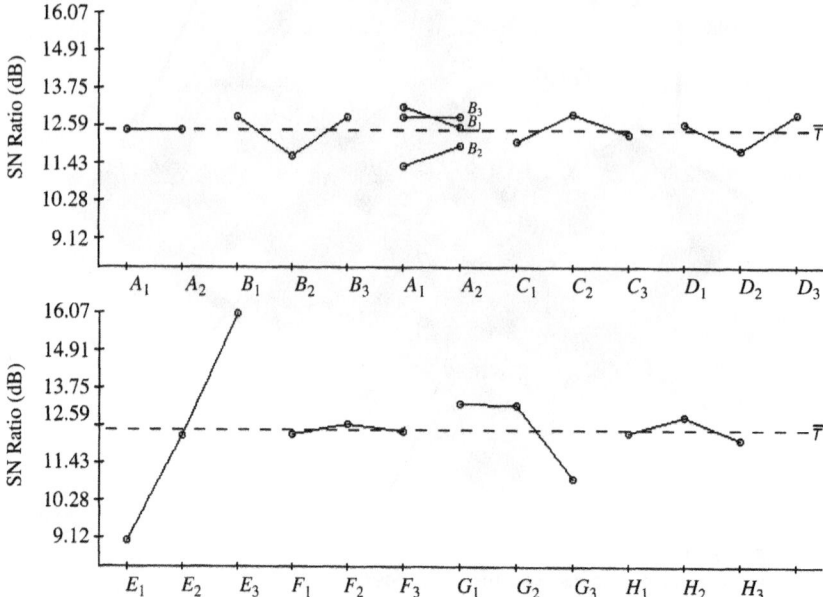

Figure 9
Control-factor-level average graphs for zero-point proportional SN ratio

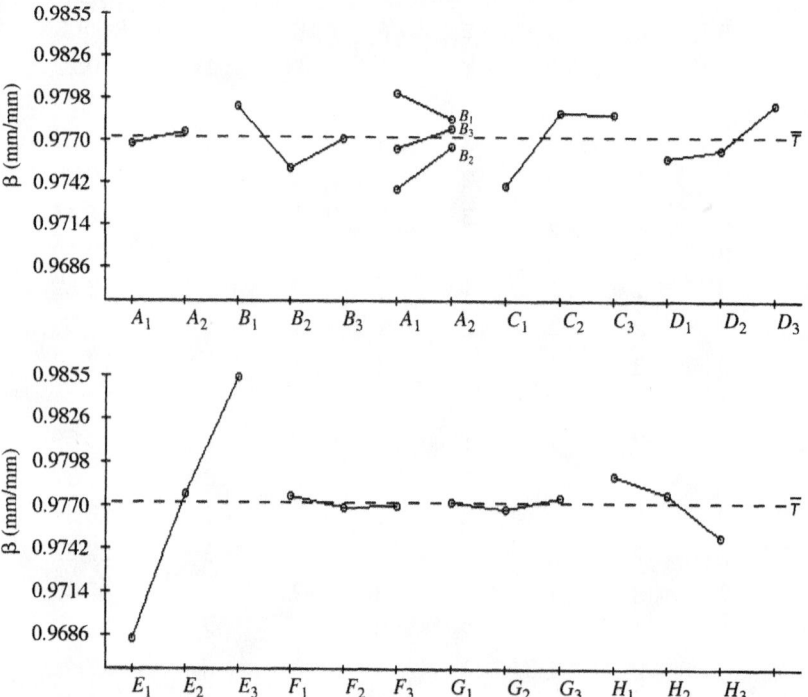

Figure 10
Control-factor-level average graphs for zero-point proportional sensitivity, β

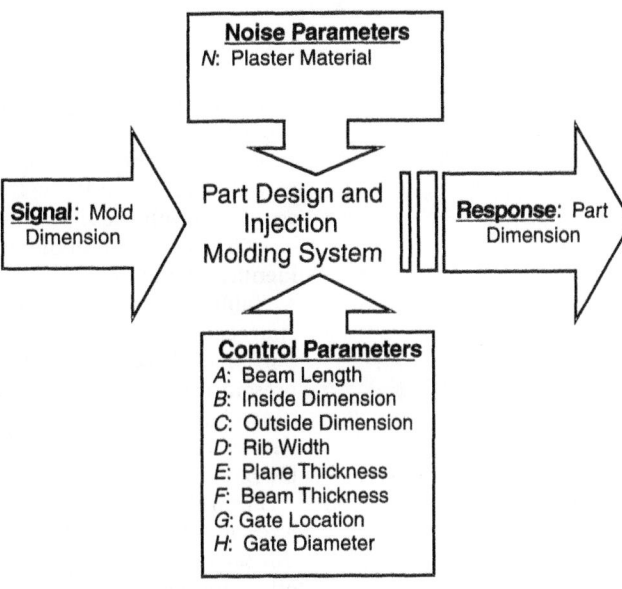

Figure 11
Parameter diagram for the all-axes dimension approach

Table 5
Factors and levels for the all-axes dimension approach

	Factor	1	2	3	4	5	6	7	8	9	10	11	12
							Level						
Control factors													
A:	beam length (mm)	10	12										
B:	inside dimension (mm)	11.3	13.6	15.9									
C:	outside dimension (mm)	3.38	5.68	7.98									
D:	rib width (mm)	0	7	14									
E:	plane thickness (mm)	0.6	0.88	1.25									
F:	beam thickness (mm)	0.6	0.88	1.25									
G:	gate location	A	B	C									
H:	gate diameter (mm)	0.4	0.8	1.2									
Noise factors													
N:	plastic material	A	B										
Signal factor													
M:	mold dimension (mm)	0.48 2.94	0.60 3.10	0.65 6.42	0.80 20.0	0.90 35.1	1.07 55.4	1.20	1.35	1.80	2.04	2.20	2.65

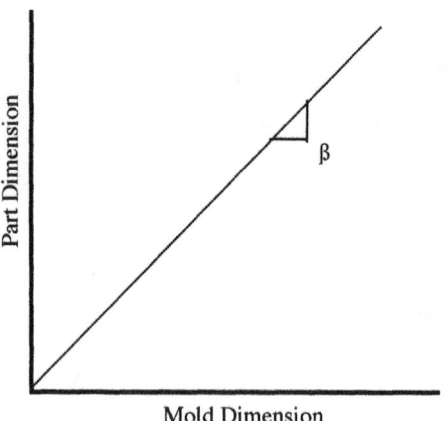

Figure 12
Ideal function for the all-axes dimension approach

Figure 11 presents the P diagram for this approach, and Table 5 the list of factors and levels.

Figure 12 shows the ideal function for this approach, and Figure 13 the finite-element node mesh associated with the signal factor levels (18). A slope of 1.00 would result in the mold and the part having identical dimensions.

Table 6 presents the orthogonal layout for this approach, along with the resulting zero-point proportional SN ratio and slope, β. Figures 14 and 15 present the zero-point SN and sensitivity (slope, β) control-factor-level average graphs for the dynamic all-axes dimension approach using the data in Table 6.

From these graphs, the best set of control factor levels is $A_2 B_3 C_2 D_1 E_3 F_2 G_1 H_1$. Verification studies using this set were done and are reported along with the other approaches later.

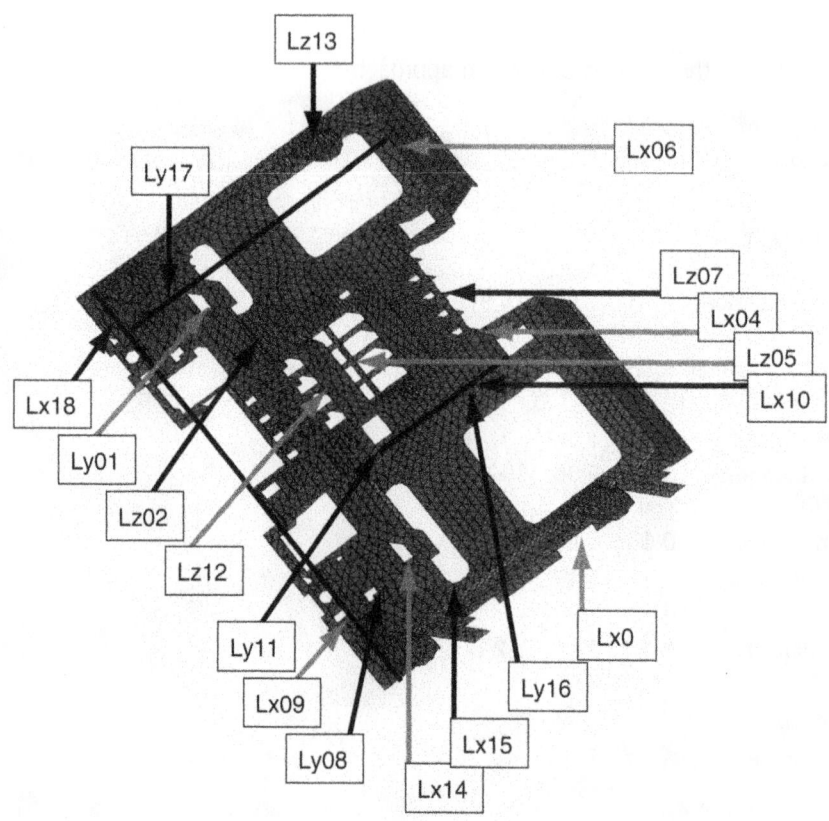

Figure 13
FE node mesh for the all-axes dimension approach

Table 6
Orthogonal layout and results for the all-axes approach

OA Row	Control Factor								Zero-Point SN	Slope, β
	A	B	C	D	E	F	G	H		
1	1	1	1	1	1	1	1	1	32.32	0.985
2	1	1	2	2	2	2	2	2	32.80	0.983
3	1	1	3	3	3	3	3	3	34.17	0.982
4	1	2	1	1	2	2	3	3	33.41	0.985
5	1	2	2	2	3	3	1	1	38.93	0.982
6	1	2	3	3	1	1	2	2	26.07	0.984
7	1	3	1	2	1	3	2	3	26.45	0.984
8	1	3	2	3	2	1	3	1	33.23	0.984
9	1	3	3	1	3	2	1	2	38.23	0.981
10	2	1	1	3	3	2	2	1	30.88	0.982
11	2	1	2	1	1	3	3	2	29.71	0.986
12	2	1	3	2	2	1	1	3	33.15	0.984
13	2	2	1	2	3	1	3	2	30.31	0.985
14	2	2	2	3	1	2	1	3	35.31	0.982
15	2	2	3	1	2	3	2	1	33.95	0.983
16	2	3	1	3	2	3	1	2	36.13	0.981
17	2	3	2	1	3	1	2	3	32.40	0.983
18	2	3	3	2	1	2	3	1	32.21	0.983

6. Verification and Comparison

Table 7 presents the predicted (assuming complete additivity in the L_{18} orthogonal array) and verified (computer simulation confirmed) SN and slope, β, values for the three approaches studied. In addition, a Z-axis deflection value was calculated using the computer simulation model for the three approaches.

Figures 16 and 17 show the computer simulation results of the Z-axis deflection calculated by using the Z_{max} minus Z_{min} deflection for material A added to Z_{max} minus Z_{min} for material B. Figure 16 is the same for approach 1 (smaller-the-best) and approach 2 (Z-axis position zero-point proportional) because these analyses resulted in the selection of the same optimum set of control parameter values.

Figure 17 shows Approach 3 (all-axes dimension zero-point proportional). Figure 18 shows the worst-case image from the L_{18}.

7. Conclusions

1. All the computer simulation confirmed values are within the confidence interval predicted.

2. Approach 3, all-axes dimension zero-point proportional, results in the largest undesirable deflection on the electrical connector insulator part in the Z-axis.

3. Approach 1, smaller-the-best, and approach 2, Z-axis position zero-point proportional, results

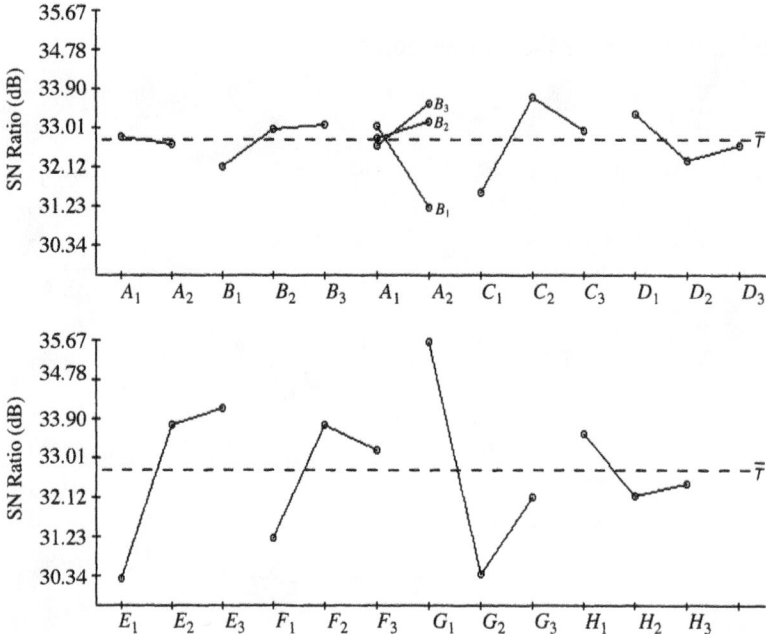

Figure 14
Control-factor-level average graphs

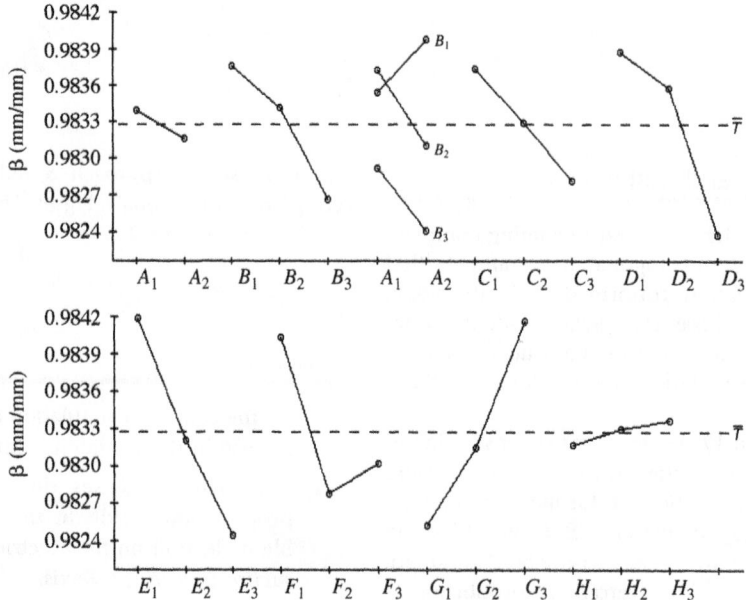

Figure 15
Control-factor-level average graphs zero-point slope, β, all-axes dimension

Table 7
Confirmation predictions and verification

Parameter Design Approach	Gate	Predicted		Verified (Simulation Confirmed)		Calculated Deflection from Mold Reference Z-Axis
		SN (dB)	β	SN (dB)	β	Delta (mm)
Smaller-the-best deflection on z-axis	G_1	18.99 ± 1.53	—	17.76	—	1.03
	G_2	18.28 ± 1.53	—	19.30	—	0.93
Zero-point proportional, z-axis position	G_1	18.80 ± 1.49	0.991 ± 0.004	17.86	0.986	1.03
	G_2	18.77 ± 1.49	0.990 ± 0.004	19.33	0.992	0.93
Zero-point proportional, all axes dimension	G_1	40.76 ± 4.42	0.981 ± 0.001	35.08	0.983	1.37
	G_2	35.50 ± 4.42	0.982 ± 0.001	31.61	0.984	1.17

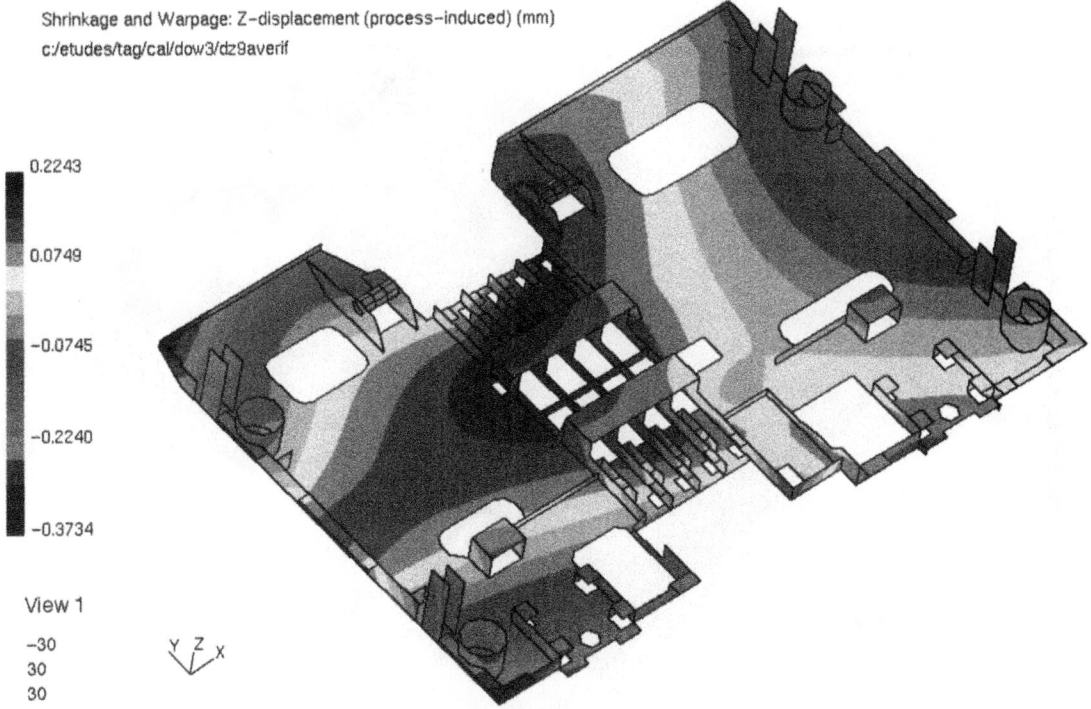

Shrinkage and Warpage: Z-displacement (process–induced) (mm)
c:/etudes/tag/cal/dow3/dz9averif

0.2243

0.0749

−0.0745

−0.2240

−0.3734

View 1

−30
30
30

Figure 16
Verification of approach 1 (smaller-the-best Z-axis deflection) and approach 2 (Z-axis position zero-point proportional)

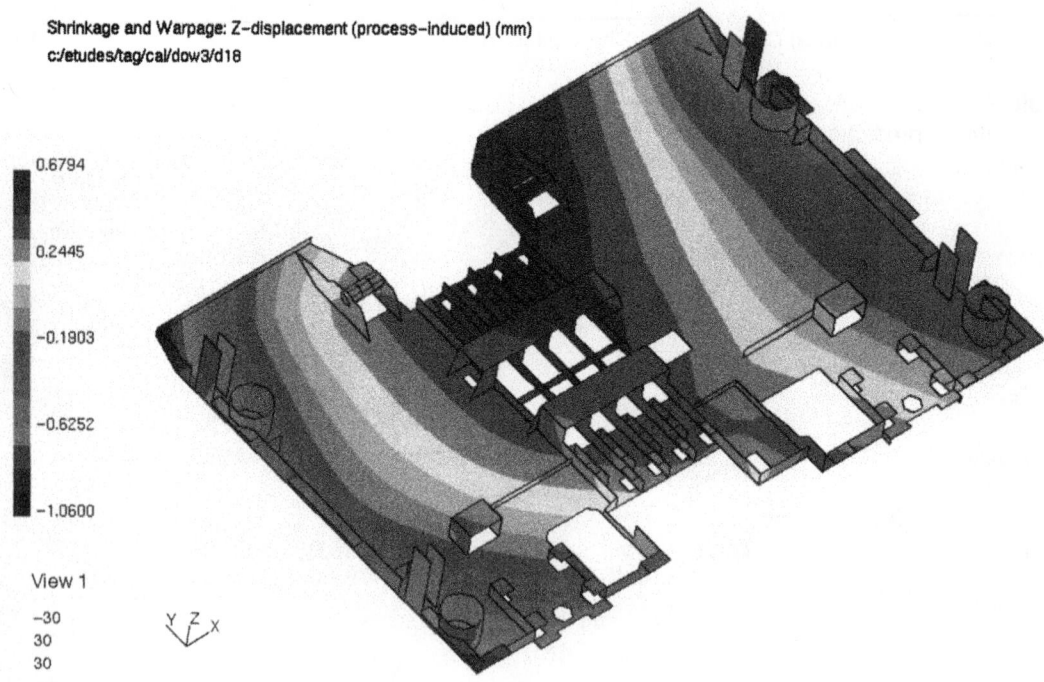

Shrinkage and Warpage: Z–displacement (process–induced) (mm)
c:/etudes/tag/cal/dow3/d18

0.6794
0.2445
−0.1903
−0.6252
−1.0600

View 1
−30
30
30

Y Z X

Figure 17
Verification of approach 3 (all-axes dimension zero-point proportional)

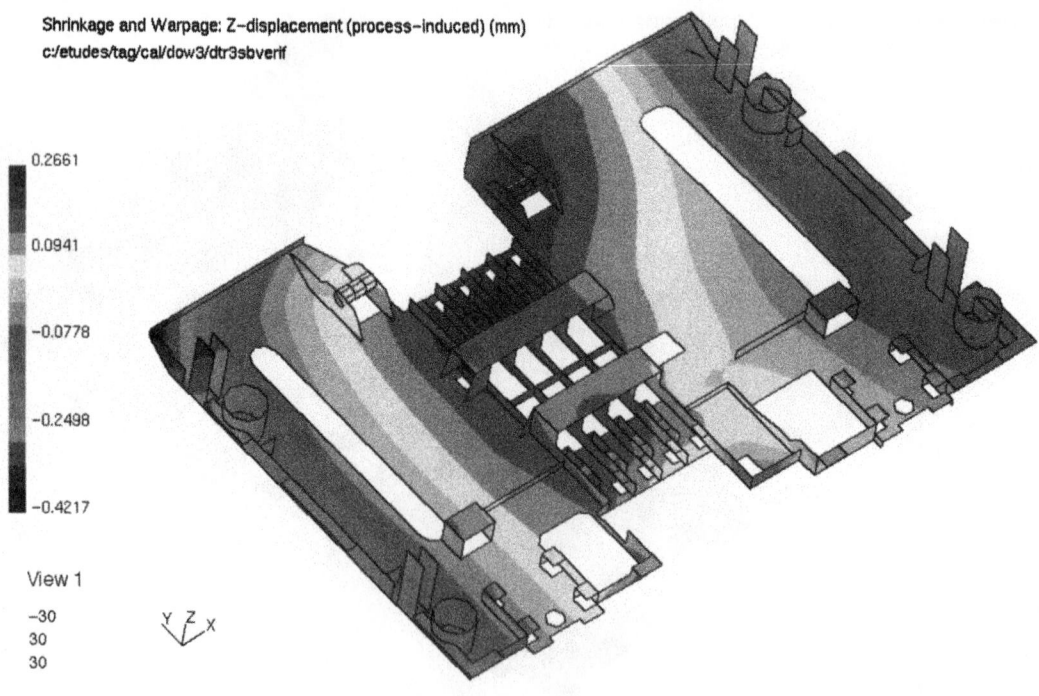

Shrinkage and Warpage: Z–displacement (process–induced) (mm)
c:/etudes/tag/cal/dow3/dtr3sbverif

0.2661
0.0941
−0.0778
−0.2498
−0.4217

View 1
−30
30
30

Y Z X

Figure 18
L_{18} worst case

1098

are the same with the lowest associated Z-axis deflection.

4. Taguchi's parameter optimization methodology, coupled with computer simulation, provides a powerful tool for optimizing plastic part and tool design as well as the associated injection molding process. In addition, these coupled tools reduced product and process design time and tooling cost by eliminating the need to build prototypes with injection molding tools of various designs.

5. Applying the results of this study to the CCM02 MKII electrical connector contact housing part and tool resulted in less than 0.15 mm of Z-axis deformation for hardware produced using the optimum control parameter value set. This was less that the 1.0-mm average confirmed using the finite-element simulation software.

This case study is contributed by Marc Bland.

CASE 59

Airflow Noise Reduction of Intercoolers

Abstract: Trying to solve the problem of airflow noise motivated us to apply the quality engineering technique. However, we attempted to measure noise, but the ideal function was discussed to improve the uniformity of airflow in the intercooler. As a result, the optimum parameter setting not only solved the noise problem but also improved the cooling system function and reduced the cost.

1. Introduction

As shown in Figure 1, an intercooler (I/C) is placed in a suction air path between a turbocharger (T/C) and engine. Although two cooling methods (air and water) exist, the former is more widely used because its structure is simple, its capacity is easy to increase, and it needs no maintenance.

Figure 2 outlines the structure of an air-cooled I/C. Compressed and heated up by a T/C, sucked air is cooled down when it passes through tubes inside the I/C. At this point, if there is a large amount of resistance against the airflow traveling each tube, the cooling performance deteriorates, and at the same time, the charging efficiency of air into engine cylinders decreases. In addition, if imbalance in airflow occurs in some portions, the cooling performance worsens, and in some cases, a noise problem called airflow noise takes place.

Trying to solve the problem of airflow noise motivated us to apply the quality engineering technique. We attempted not only to take measures for this noise problem but also to establish a design technique applicable to future products and to reduce airflow noise by improving the function of an air-cooled I/C.

2. SN Ratio

The function of an I/C is to cool air, which is compressed and heated up by a compressor of a turbo-charger while it passes through tubes inside the I/C. Therefore, to improve the charging efficiency and cooling performance, we needed to reduce resistance against air passing through each tube and to equalize its velocity at any position of the tubes.

Since we can theoretically calculate the airflow velocity traveling inside the I/C, we regarded as the ideal state of the I/C system function that the airflow velocity inside the I/C, y, is proportional to the theoretical velocity, M (i.e., $y = \beta M$), and the velocity is equal at each position.

Thus, in this study we computed airflow velocity from the revolution of I/C and airflow for each condition and evaluated its relationship with actual velocity. For theoretical velocity as a signal factor, denoted by M (m/s), we determined its level by considering the entire range of a car-driving condition.

As noise factors, we picked the position inside the I/C and fluctuation in airflow. For the former we chose two levels, denoted by I_1 and I_2, the nearest (upper) and farthest (lower) positions to the intercooler's inlet. This is because no velocity difference at both positions was considered to indicate that the velocity is equalized all over the I/C. For the latter, maximum and minimum airflows, denoted by J_1 and J_2, respectively, were measured (Table 1).

Judging from our technical experience and knowledge, as control factors we selected eight factors, including dimension and shape, which were regarded to greatly affect the airflow velocity inside the I/C. Table 2 shows the control factor and level selected.

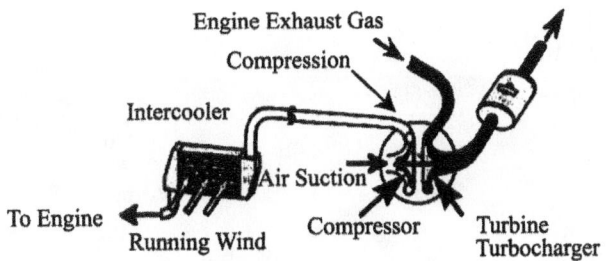

Figure 1
Turbocharger and intercooler

Using an L_{18} orthogonal array, we assigned control factors to its inner array and signal and error factors to its outer array for experiments 1 to 18. Table 2 shows the data. Based on these data, we proceed with our analysis for computing SN ratio and sensitivity (Figure 3) using the following calculations.

Total variation:

$$S_T = 3.4^2 + 5.6^2 + 6.7^2 + \cdots + 34.4^2$$

$$= 11{,}009.41 \qquad (f = 28) \qquad (1)$$

Effective divider:

$$r = 29.0^2 + 30.1^2 + 40.0^2 + \cdots + 58.1^2$$

$$= 26{,}007.54 \qquad (2)$$

Linear equations:

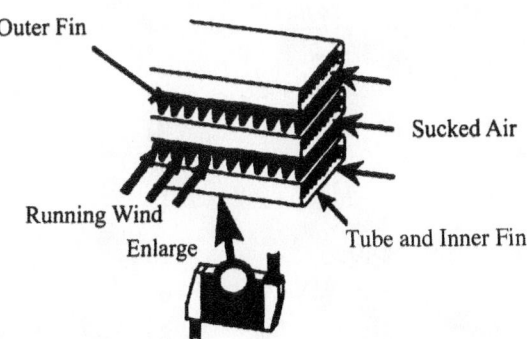

Figure 2
Structure of intercooler

$$L_1 = (29.0)(3.4) + (30.1)(5.6) + (40.0)(6.7)$$
$$+ \cdots + (58.1)(12.2) = 2539.58$$

$$L_2 = 2003.07$$

$$L_3 = 8772.67$$

$$L_4 = (29.0)(13.6) + (30.1)(18.3) + (40.0)(20.0)$$
$$+ \cdots + (58.1)(34.4) = 7340.69 \qquad (3)$$

Variation of proportional term:

$$S_\beta = \frac{(L_1 + L_2 + L_3 + L_4)^2}{4r}$$

$$= 8202.828 \qquad (f = 1) \qquad (4)$$

Variation of proportional terms due to noise:

$$S_{N\beta} = \frac{L_1^2 + L_2^2 + L_3^2 + L_4^2}{r} - S_\beta$$

$$= 2663.807 \qquad (f = 3) \qquad (5)$$

Error variation:

$$S_e = S_T - S_\beta - S_{N\beta} = 142.775 \qquad (f = 24) \qquad (6)$$

Error variance:

$$V_e = \frac{S_e}{24} = 5.9490 \qquad (7)$$

Total error variance:

$$V_N = \frac{S_T - S_\beta}{27} = 103.9475 \qquad (8)$$

SN ratio:

Table 1
Measured data of airflow (m/s)

Number of Revolutions of T/C:		6		8			10		
Amount of Airflow:		6	6	8	10	6	8	10
M: Theoretical Velocity (m/s):		M_1 29.0	M_2 30.1	M_3 40.0	M_4 46.1	M_5 41.3	M_6 49.5	M_7 58.1
l_1: upper	J_1: max.	3.4	5.6	6.7	9.0	8.5	10.7	12.2
	J_2: min.	2.7	4.4	5.4	7.0	6.5	8.4	9.8
l_2: lower	J_1: max.	15.4	21.8	24.0	26.3	31.3	36.0	41.7
	J_2: min.	13.6	18.3	20.0	21.8	26.8	30.0	34.4

$$\eta = 10 \log \frac{(1/4r)\,(S_\beta - V_e)}{V_N} = -28.19 \text{ dB} \quad (9)$$

Sensitivity:

$$S = 10 \log \frac{1}{4r}\,(S_\beta - V_e) = -8.02 \text{ dB} \quad (10)$$

Under the estimated optimal and current conditions, we prototype I/Cs and conduct a confirmatory experiment. For the sake of simplicity, we do not detail the measured data and calculation procedure. The results are summarized in Table 3 and Figure 4.

Looking at the confirmatory experimental results, we can see that reproducibility was good for the SN ratio. Under the optimal condition, we improved the SN ratio by 8.77 dB compared to that under the current conditions. This is equivalent to reduction of variability in airflow velocity by two-thirds. On the other hand, the sensitivities under both conditions did not differ significantly. As a result, we notice that we can drastically reduce the difference in airflow velocity at each position without changing the average flow (Figure 4).

3. Confirmation of Improvement

Figure 5 shows the result for improvement confirmed based on measurement of airflow noise under the current and optimal conditions.

As the figure shows, under the optimal condition, we can reduce the noise level by 6 dB at the

Table 2
Control factors and levels[a]

	Control Factor	Level		
		1	2	3
A:	inlet tank length (mm)	Standard[a]	Standard + 20	—
B:	tube thickness (mm)	5*	7	9
C:	inlet tank shape	1	2[a]	3
D:	inflow direction of inlet tank	1[a]	2[a]	3[a]
E:	tube length	120	140[a]	160
F:	tube end shape (deg)	0	30[a]	60
G:	inner fin length (mm)	Standard −7.5	Standard[a]	Standard + 7.5
H:	inner diameter of inlet tube (mm)	Φ45	Φ55[a]	Φ65[a]

[a] Current level.

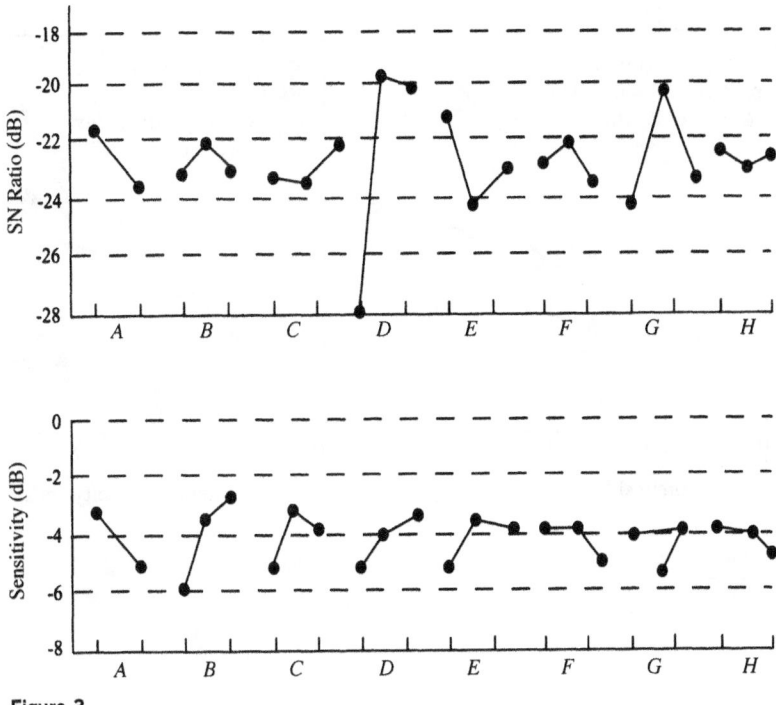

Figure 3
Response graphs

driving condition where a maximum airflow noise is generated in contrast to that under the current condition. This level can be regarded as satisfactory for our design target.

Similarly, we confirm the improvement of cooling performance and show the results in Figure 6. The temperature shown in the figure indicates the decrease in the temperature of airflow per unit length of tube. The larger the value becomes, the higher the cooling performance.

Under the optimal condition, the cooling performance was improved by 20% as compared to that under the current condition. This performance improvement represents that the total efficiency of the engine was also ameliorated by 2%.

In addition, if the cooling performance required for a certain I/C was the same as before, we could reduce the size of the I/C, thereby leading not only to cost reduction but also to higher flexibility in the layout of an engine room and easier development.

Table 3
Estimation and confirmation of SN ratio and sensitivity (dB)

Condition	SN Ratio		Sensitivity	
	Estimation	Confirmation	Estimation	Confirmation
Optimal	−19.07	−19.74	−4.33	−4.22
Current	−26.55	−28.51	−5.76	−5.15
Gain	7.48	8.77	1.43	0.93

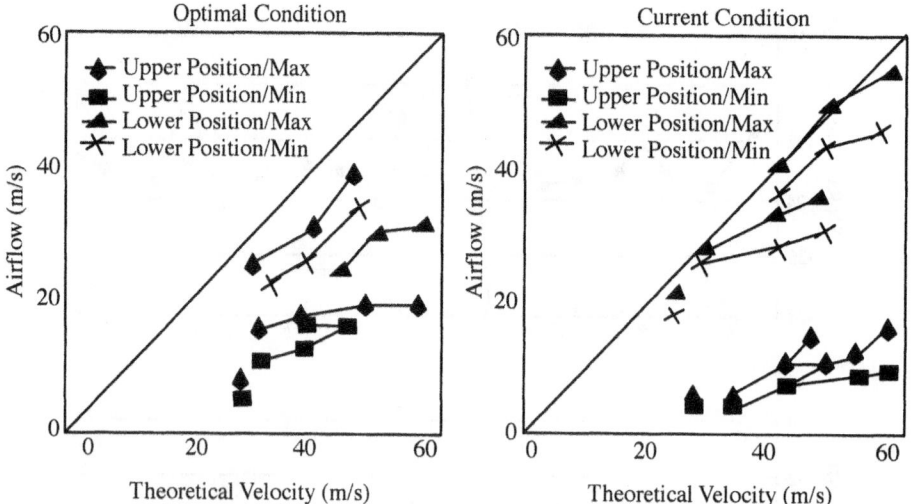

Figure 4
Airflow in confirmatory experiment

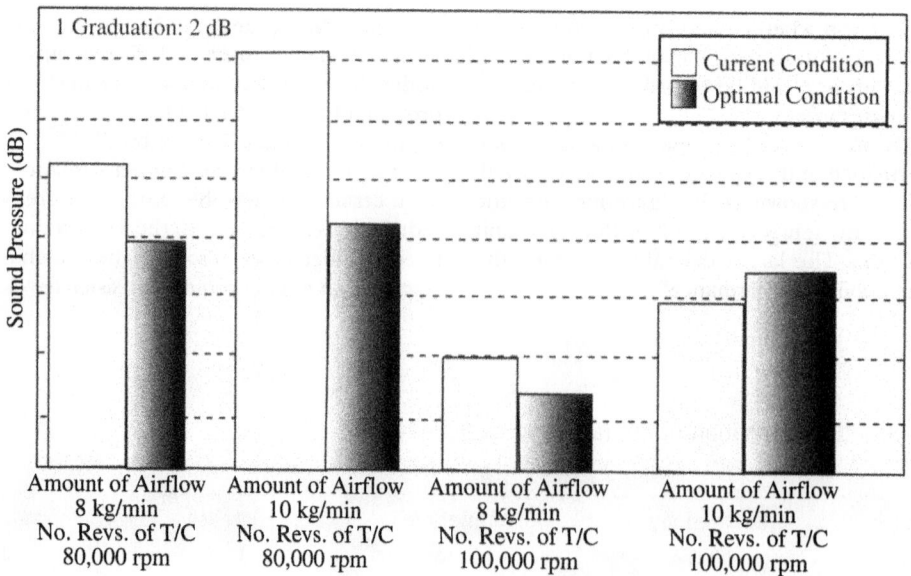

Figure 5
Confirmatory result for intercooler airflow noise

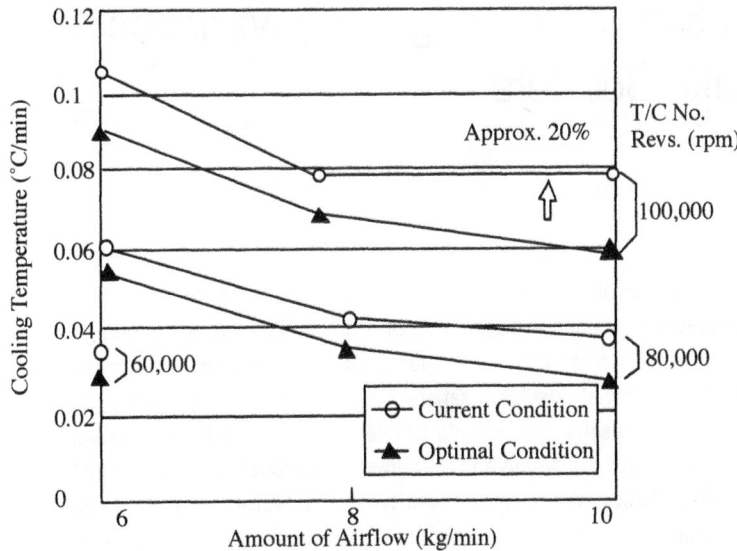

Figure 6
Confirmatory result for cooling performance

Reference

Hisahiko Sano, Masahiko Watanabe, Satoshi Fujiwava, and Kenji Kurihara, 1998. Air flow noise reduction of inter-cooler system. *Quality Engineering*, Vol. 6, No. 1, pp. 48–53.

This case study is contributed by Hisahiko Sano, Masahiko Watanabe, Satoshi Fujiwara, and Kenji Kurihara.

Reduction of Boosting Force Variation of Brake Boosters

Abstract: A brake booster is a component of an automotive brake system whose function is to boost the pedal force of a driver. The hydraulic force of each cylinder in the brake system is proportional to the foot force. However, the defect rate has not always been satisfactory, due to relatively large boost force variations. To resolve this chronic problem, the robust design method was utilized to improve the design and manufacturing processes. In the robust design experiment, five control factors and two noise factors were allocated in an L_{12} orthogonal array and the dynamic signal-to-noise (SN) ratios were calculated. Through use of the robust design method, we improved the rolled throughput yield dramatically, and the optimum conditions found will be adopted in the design and manufacturing processes.

1. Introduction

The brake booster for the automotive brake system under study, located between the brake pedal and master cylinder, boosts a driver's pedal force to generate enough braking force to stop a vehicle (Figure 1). It uses the vacuum pump system (the intake manifold or built-in vacuum pump) of the engine to boost the pedal force, and that's why it is called *vacuum servo* or *master vac*. To check the brake booster quality, booster forces at two given input forces were inspected to see whether or not they met the target values (Figures 2 and 3). However, unacceptably large variations were found in the booster forces. Therefore, it was decided to use robust design to solve this chronic problem.

2. Background and Objectives

The brake booster forces are generated through three different phases (Figure 3). In the first phase,

no force is generated and the input stroke in this phase is regarded as "lost travel." As the input force rises into the second phase, a sudden stepwise booster force, a *jump-in force*, is generated. After the jump-in, the booster force is increased linearly as the input force increases up to the knee point, where the booster force increase rate is suddenly reduced. In the last phase, the booster force increase continues at a reduced rate. *Boost ratios,* ratios of output forces to input forces, jump-in force, and knee point are the major design criteria of the brake booster.

The objective of this study was to reduce the high defect rate of the brake booster by reducing the booster force variations in the second phase, which had been a major chronic problem. To reduce the defect rate, we decided to use the robust design technique to improve the design as well as the manufacturing processes. Since booster manufacturing began about 20 some years ago, various approaches had been adopted earlier to solve this problem. However, no significant improvement had been made until the robust design technique was utilized.

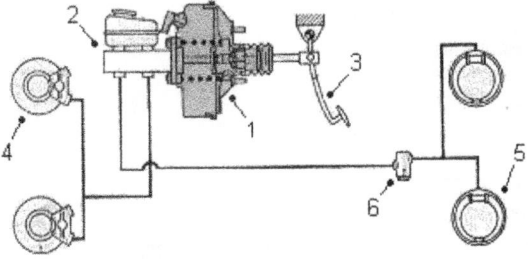

1. Booster
2. Master Cylinder
3. Brake Pedal
4. Caliper
5. Drum
6. Pressure Limit Valve

Figure 1
Automotive brake system

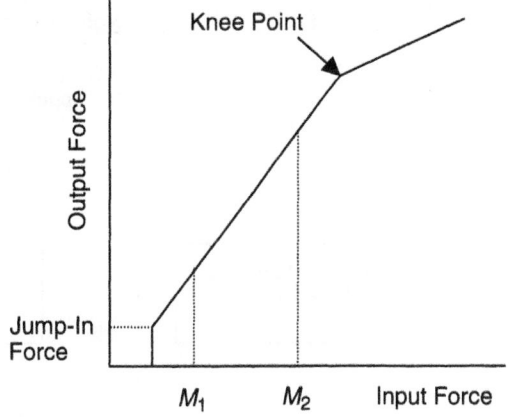

Figure 3
Output characteristics

3. Robust Design Formulation

Several design and manufacturing parameters were considered and a few noise factors were studied, and finally, five control factors and three noise factors were selected, as shown in the P-diagram (Figure 4).

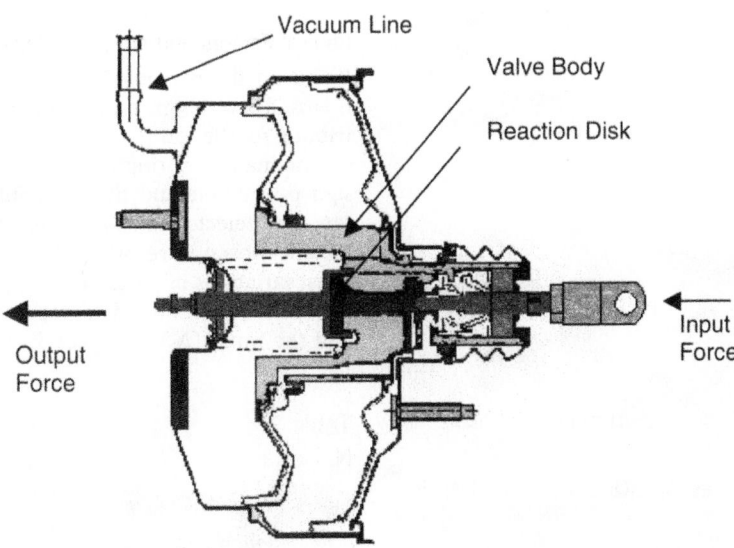

Figure 2
Brake booster

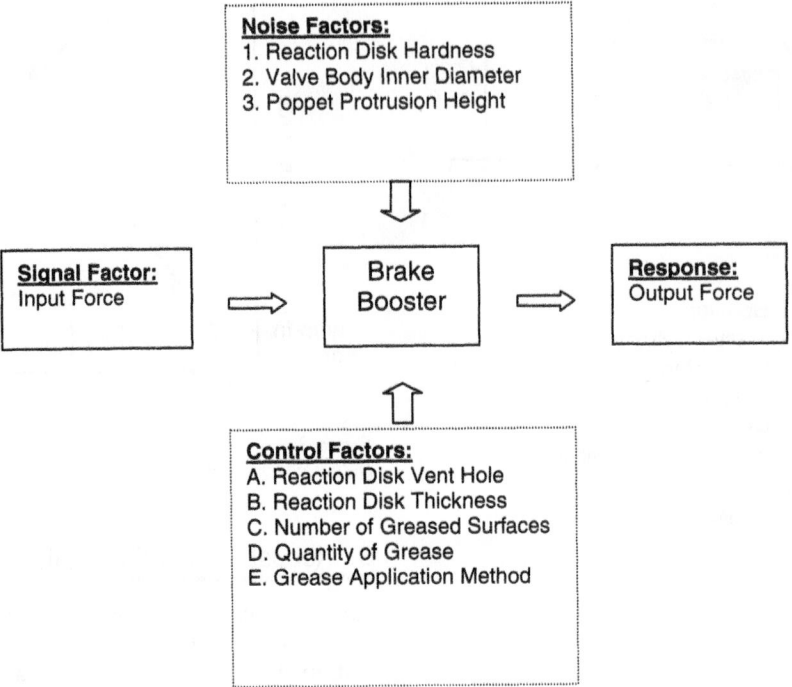

Figure 4
P-diagram

Control Factors and Noise Strategy

Based on the results of several preliminary experiments, a few design parameters were found to contribute to the output force variations as well as several manufacturing processes. Therefore, two design parameters and three manufacturing parameters were selected as control factors (Table 1). Noise factors chosen were primarily geometrical and material variations of parts from suppliers. Three noise factors were combined to represent the best (N_1) and the worst (N_2) noise conditions (Table 2).

Table 1
Control factors

	Control Factor	Level 1	Level 2
A:	reaction disk vent hole	Yes	No
B:	reaction disk thickness	Current	Thicker
C:	number of greased surface	One side	Both sides
D:	quantity of grease	Current	More
E:	grease application method	Current	Better

Table 2
Noise factors

Noise Factor	N_1	N_2
1. Reaction disk hardness	Soft	Hard
2. Valve body inner diameter	Large	Small
3. Poppet protrusion height	Small	Large

Table 3
L_{12} orthogonal array

No.	A	B	C	D	E	F	G	H	I	J	K	M_1		M_2	
												N_1	N_2	N_1	N_2
1	1	1	1	1	1	1	1	1	1	1	1				
2	1	1	1	1	1	2	2	2	2	2	2				
3	1	1	2	2	2	1	1	1	2	2	2				
4	1	2	1	2	2	1	2	2	1	1	2				
5	1	2	2	1	2	2	1	2	1	2	1				
6	1	2	2	2	1	2	2	1	2	1	1				
7	2	1	2	2	1	1	2	2	1	2	1				
8	2	1	2	1	2	2	2	1	1	1	2				
9	2	1	1	2	2	2	1	2	2	1	1				
10	2	2	2	1	1	1	1	2	2	1	2				
11	2	2	1	2	1	2	1	1	1	2	2				
12	2	2	1	1	2	1	2	1	2	2	1				

Figure 5
Booster performance tester

Table 4
SN ratio response

	A	B	C	D	E
1	−2.11	−1.31	−2.55	−0.65	−1.32
2	0.40	−0.41	0.84	−1.06	−0.39
Δ	2.51216	0.89757	3.38785	0.40884	0.92659
Rank	2	4	1	5	3

Table 5
β response

	A	B	C	D	E
1	7.86959	7.80028	7.92845	7.85619	7.8032
2	7.76598	7.83529	7.70712	7.77938	7.83237
Δ	0.10361	0.03501	0.22132	0.07681	0.02917
Rank	2	4	1	3	5

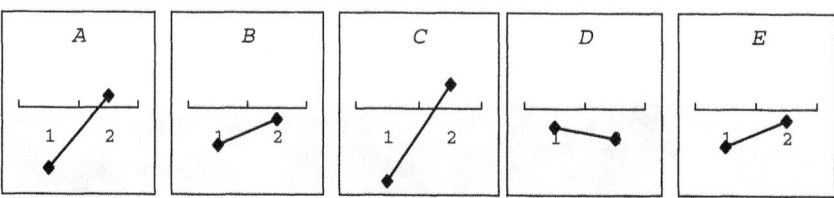

Figure 6
SN ratio response

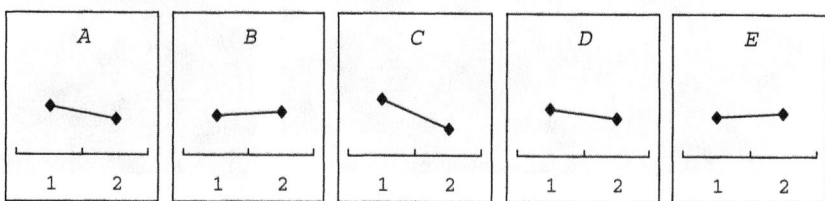

Figure 7
β response

Table 6
SN ratio predicted for a derived optimal condition

Condition	SN Predicted (dB)	Control Factors
Current	−2.00	$A_2B_1C_1D_1E_1$
Optimal	2.31	$A_2B_1C_2D_1E_2$
Gain	4.31	

Orthogonal Array

Since five control factors were identified, the L_{12} orthogonal array was selected for experiments. These control and noise factors were allocated, as shown in the Table 3.

Ideal Function

The following ideal function was used for this study, which can easily be derived from Figure 3:

$$y = \beta M \tag{1}$$

where y is the output force and M is the input force.

Test Equipment

A booster performance tester was used to measure input and output forces of the booster (Figure 5). It supplies vacuum to the booster just like the vacuum pump system of the engine of a vehicle. Its measurement system both records input/output forces and calculates some performance parameters, such as jump-in force, boost ratio, lost travel, and so on.

4. Results and Analysis

The signal-to-noise (SN) ratios (η) and sensitivities (β) were calculated using the following equations:

$$\eta = 10 \log \frac{(1/r_0)(S_\beta - V_e)}{r V_e} \tag{2}$$

$$\beta = \sqrt{\frac{1}{r_0}(S_\beta - V_e)} \tag{3}$$

where r is the sum of squares of the signals, r_0 the

number of linear equations, S_β the variation caused by β, and V_e the error variance.

The response tables (Tables 4 and 5) and response graphs (Figures 6 and 7) summarize our results. The results show that control factors A, C, and E had large effects on the SN ratio. Factor B was shown to have a minor influence on the variation but a large effect on cost. The SN ratio gain of the optimum condition was predicted to be 4.31 dB (Table 6).

5. Confirmation Test

A confirmation test revealed good reproducibility:

Gain predicted (dB): 4.31

Gain confirmed (dB): 5.78

Reproducibility (%): 134

The ultimate object of the project was, however, improvement of the productivity of booster manufacturing processes. To verify the optimized condition in manufacturing processes, a few hundred units were mass produced, and substantial improvement in the rolled throughput yield was realized.

6. Conclusions

The robust design method was used to solve a chronic brake booster problem. Using this highly efficient optimization process, considerable improvement in reducing the defect rate due to booster force variation was achieved. It would also be very helpful in enhancing productivity.

References

Genichi Taguchi, Subir Chowdhury, and Shin Taguchi, 2000. *Robust Engineering*. New York: McGraw-Hill.

Yuin Wu and Alan Wu, 2000. *Taguchi Methods for Robust Design*. New York: ASME Press; *ASI 18th Annual Taguchi Methods Symposium Proceedings*, Dearborn, MI.

This case study is contributed by Myoung-June Kim, Jun-Gyu Song, Chung-Keun Lee, Kyu-Yeol Ryu, and Ki L. Chang.

Reduction of Chattering Noise in Series 47 Feeder Valves

Abstract: Field reports of *chattering* inspired us to apply the Taguchi robust design technique to enhance the design of the Series 47 feeder valve to make it robust against field noise conditions. Chattering manifests itself when the feeder valve is slightly open or slightly closed. A system harmonic is created, resulting in a loud, rolling noise through the heating pipes. This noise is considered by customers to be a significant nuisance and must be minimized or eliminated.

Instead of measuring chattering noise, the proportionality among flow rate, flow area, and inlet pressure was considered as the ideal function. The experiment analyzed the level effects of six control factors and three noise factors. The zero-point proportional equation was used for the analysis. The inlet pressure to the feeder valve and the cross-sectional flow area were utilized as dynamic signals. The flow rate through the valve was used as the response.

Experimental results showed a SN ratio gain of 4.63 dB over our present design. This indicates the potential for a significant improvement in the chattering problem.

1. Introduction

Series 47 and Series 51 mechanical water feeder and low-water cutoff combinations, as well as Series 101A electric water feeders, are essential to our product mix. All three series incorporate the same feeder valve. The valve cartridge that controls the water flow is shown in Figure 1. This valve handles cold water feed at inlet pressures up to 150 psig. This project was undertaken to correct a field condition that is related to the valve, commonly referred to as *chattering*.

2. Background

To increase its reliability and life and to make it more installer-friendly, the feeder valve was rede-

signed in 1995. The new design incorporates a plastic (Noryl) replaceable cartridge. Valve shutoff is achieved when the moving poppet inside the cartridge is locked into place mechanically by an external stem that exerts force on the diaphragm. The configuration of the cartridge within the valve is shown in Figure 2. Internal components of the cartridge are shown in Figure 3.

Upon its introduction to the market, we began receiving complaints of valve chattering in 47-LWCO series. The 51-LWCO series received almost no complaints. As the Series 47 is used in residential boilers and the Series 51 in industrial boilers, we surmised that the high complaint level on the Series 47 was due to by homeowners' sensitivity to noise. The 101A feeder was unaffected because it has a quick-shutoff mechanism. The chattering could not

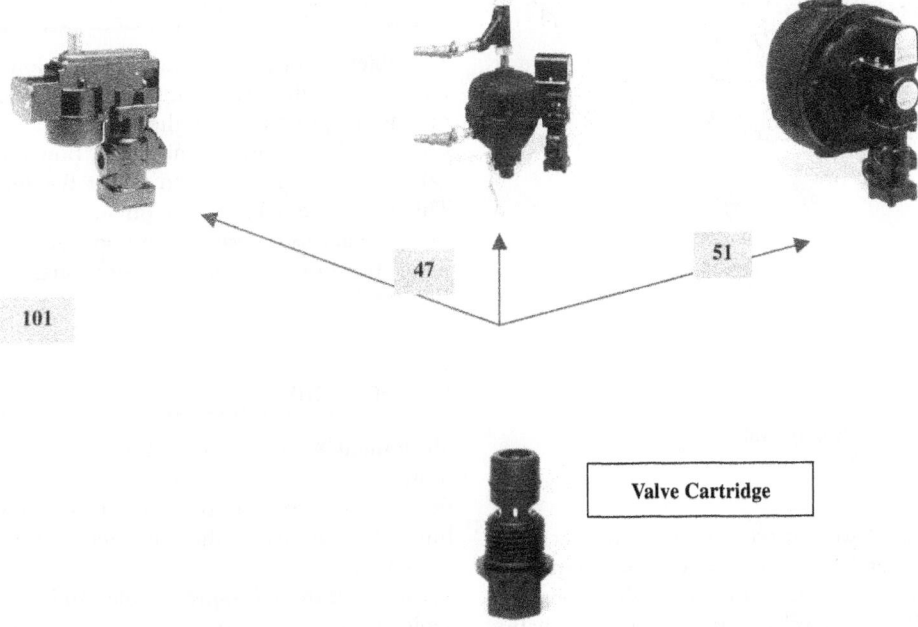

Figure 1
Feeder valve

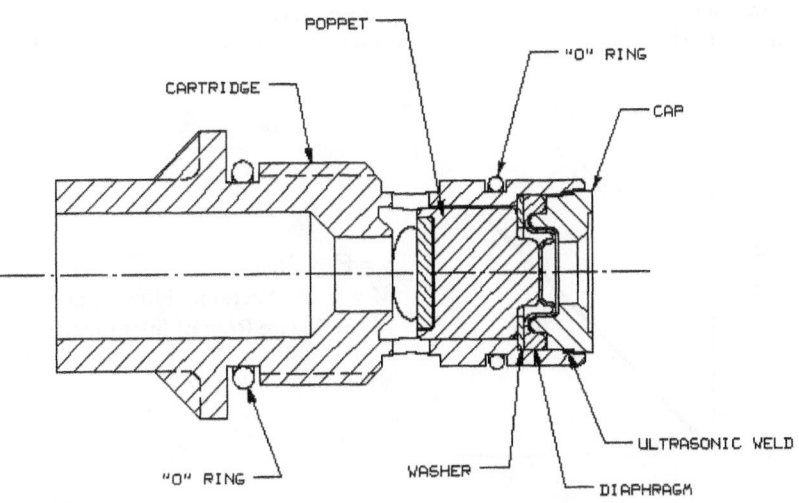

Figure 2
Cartridge within the valve

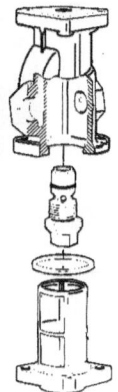

Figure 3
Internal components of the valve

be corrolated with specific system dynamics, nor could it be replicated completely in the engineering laboratory. Nevertheless, it was clear that under certain conditions the valve would cause an oscillation in the heating pipes. The resulting vibration and noise had customers, as well as manufacturer's representatives, calling for a solution.

Earlier attempts at solving the chattering noise did not follow a designed experiment methodology. Two one-factor-at-a-time solutions, in particular the quadring and the round poppet, caused ancillary difficulties and had to be reversed later. One conclusive fact was uncovered: The chattering occurred when the valve was slightly open, either right at opening or right before shutting off.

3. Objectives

The objective of this product optimization was two-fold: (1) to reduce variability under field noise conditions (i.e., to maximize the SN ratio), and (2) to optimize the design for minimum flow rate of 2.0 gal/min at 40 psig (i.e., to adjust the sensitivity). This would be achieved by obtaining a consistent flow rate across the valve in proportion to the inlet water pressure and the cross-sectional flow area (Figure 4).

4. The Team

McDonnell & Miller selected a team of personnel to improve the robustness of the feeder valve under field noise conditions. The team was cross-functional, to ensure that all interests were represented and all resources accessed. The team consisted of the following people: Amjed Shafique, senior product support engineer (team leader); Azi Feifel, senior quality engineer; Nestor Galimba, quality technician; Vladimir Ulyanov, senior laboratory technician; Chris Nichols, assistant laboratory technician; and Greg Roder, product specialist.

5. Experimental Layout

Figure 5 is a P-diagram that shows the control, noise, and signal factors. Factor definitions follow. Factor levels are shown in Table 1.

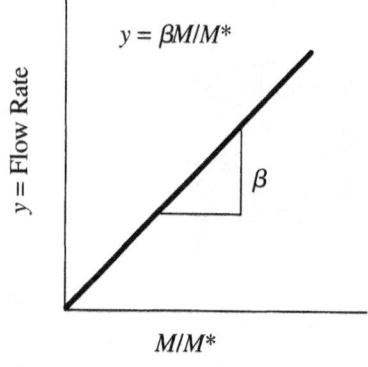

$$y = \beta M/M^*$$

y = Flow Rate
M = Cross-Sectional Flow Area
M^* = Square Root of Inlet Pressure

Figure 4
Ideal function

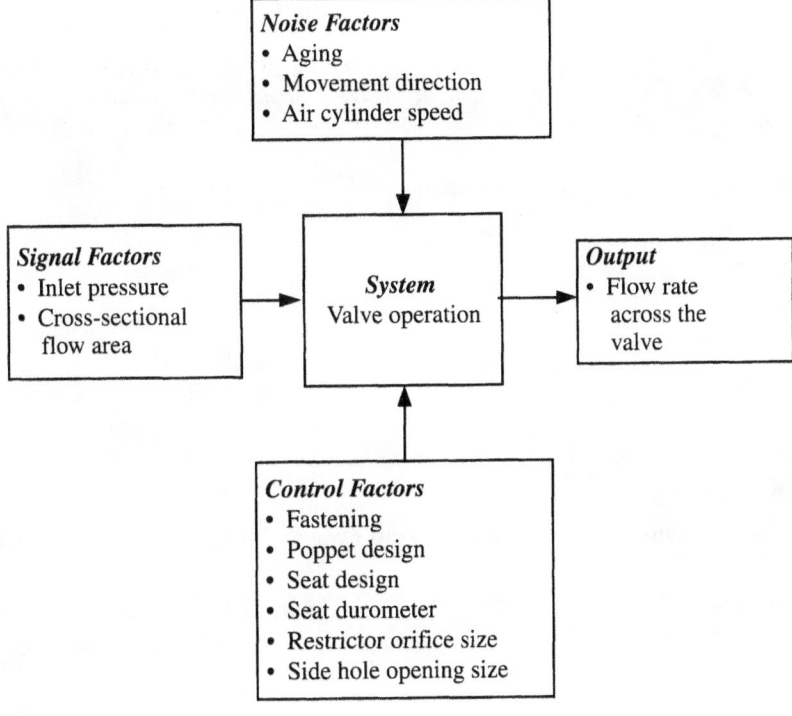

Figure 5
P-diagram

Control Factors

❏ *Fastening:* the cartridge poppet, either mechanically fastened to the diaphragm or unfastened, and allowed to move freely in the cylinder (see Figure 3 for a graphical depiction)

❏ *Poppet design:* the shape of the poppet, determining the amount of flow around the poppet

❏ *Seat Design:* the relative shape of the EPDM rubber seat on the poppet that is forced on the inlet orifice and shuts off the water flow

❏ *Durometer:* the hardness of the EPDM rubber seat on the shutoff end of the poppet

❏ *Restrictor size:* the diameter of the hole in the restrictor plug inserted into the hex end of the cartridge (the restrictor acts as a pressure-reducing mechanism for the inlet water feed)

❏ *Cartridge side hole:* the slots around the neck of the cartridge through which the water flows when the valve is open

Noise Factors

❏ *Aging:* controls whether or not the cartridge had been in use (the effect of age is recognized by the lasting impression of the inlet orifice on the poppet seat)

❏ *Air cylinder speed:* reflects the rate at which force is applied to the valve as it is shutting and opening

❏ *Movement direction:* controls whether the response is being measured as the valve is opening or closing

Dynamic Signal Factors

❏ *Inlet pressure:* the water pressure as it enters the water feeder valve. The outlet of the valve is exposed to the atmospheric pressure during testing, which is why we used inlet pressure instead of pressure drop. It is typically determined by the system pressure of the locale.

Table 1
Factor and levels

		Level		
Factors		**1**	**2**	**3**
Control factors				
A:	fastening	Not fastened	Fastened	
B:	poppet design	Hex	Square	
C:	seat design	Flat	Conical	Spherical
D:	durometer	80	60	90
E:	restrictor size (in.)	$\frac{1}{2}$	$\frac{1}{8}$	$\frac{1}{16}$
F:	cartridge side hole size (in.)	0.187	0.210	
Noise factors				
N:	aging	New	Aged	
FR:	air cylinder speed (in./sec)	0.002	0.001	
DR:	air cylinder direction	Down (as valve closing)	Up (as valve opening)	
Signal factors				
M:	inlet pressure (psig)	40	150	80
M*:	cross-sectional flow area	Flow area calculated as valve closing or opening.		

❑ *Cross-sectional flow area:* the cross-sectional area between the seat and the inlet orifice as the valve is opening or shutting. This is calculated from the shutoff force.

$$\frac{(\text{maximum shutoff force}) - (\text{current shutoff force})}{(\text{maximum shutoff force})} \times 100$$

Cross-Sectional Flow Area Calculations

❑ Maximum valve shutoff force occurs when the valve is held fully shut. At this point, 0% of the cross-sectional flow area is open, and flow of water through the valve has ceased.

❑ Minimum valve shutoff force occurs when the valve is fully open. At this point 100% of the cross-sectional flow area is open, and maximum flow through the valve occurs.

❑ Current valve shutoff force refers to the force exerted on the valve at any relative cross-sectional flow area between 100%, fully open, and 0%, fully closed. This value is variable as the valve opens and shuts and is recorded by the data collection system twice per second.

❑ The actual cross-sectional flow area percentage at any point along the spectrum is interpolated and calculated as

6. Test Setup

The test setup is shown in Figures 6 and 7. Important notes:

❑ Inlet feed pressure, one of the dynamic signals, and air cylinder speed, the rate at which the force gauge closes and opens the valve, were regulated as part of the setup.

❑ Flow rate, the output response, and force to close or open the valve, used to calculate the cross-sectional flow area, were taken in real time from a flow meter and force gauge, respectively.

❑ As the valve opens and shuts, the flow rate and shutoff force were taken simultaneously at the rate of twice per second. The data were downloaded electronically directly into a computer spreadsheet.

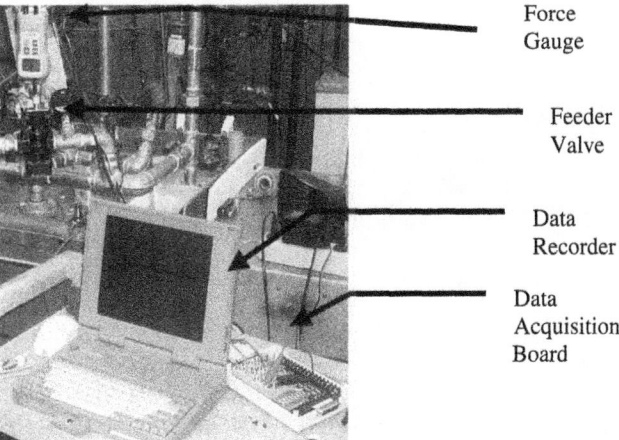

Figure 6
Test setup 1

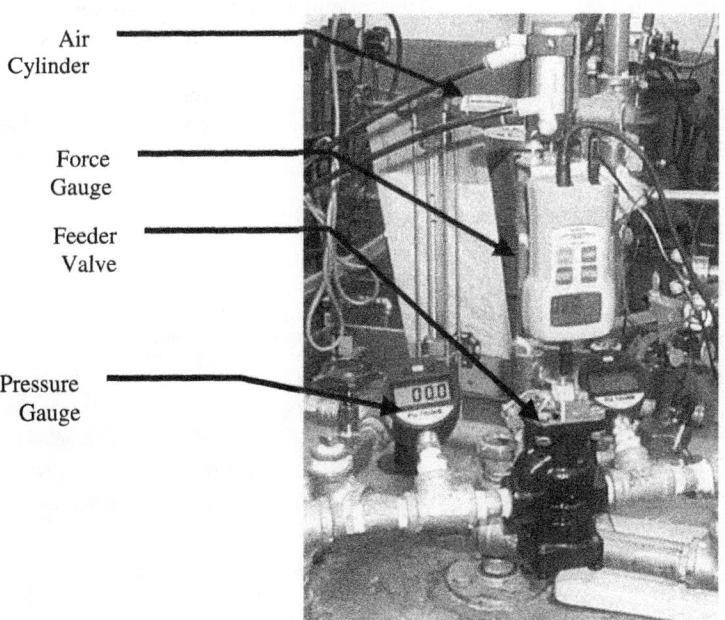

Figure 7
Test setup 2

Table 2
SN response

	Factor					
Level	A: Fastening	B: Poppet Design	C: Seat Design	D: Durometer	E: Restrictor Size	F: Side Hole Size
1	−6.05	−5.76	−5.90	−6.12	−7.35	−5.64
2	−5.12	−5.25	−5.93	−5.24	−4.87	−5.48
3			−4.94	−5.39	−4.55	
Delta	0.93	0.51	0.99	0.88	2.80	0.15
Ranking	3	5	2	4	1	6

7. Experimental Results and Analysis

A dynamic L_{18} orthogonal array experiment was performed to optimize the valve design. Data resulting from the experiment across the signal and noise factors are presented in the appendix for this case. As this experiment and the response were dynamic, the equations used to calculate the SN ratio and slope (β) were as follows:

$$SN = \eta_{dB} = 10 \log \frac{(1/r_0)(S_\beta - V_e)}{V_e}$$

$$\beta = \sqrt{\frac{1}{r_0}(S_\beta - V_e)}$$

The SN ratio and sensitivity are given in Tables 2 and 3, and the SN response graph is shown in Figure 8.

The tables show that factor E, the restrictor size, and then factor C, the seat design, had the greatest effect on the variability in the flow rate. Factor A, fastening, was also shown to have some effect. The graph shows the greatest SN differentials at different levels for factors E and C, with a smaller differential for factor A. Factors B, D, and F showed no significant differential in SN ratios at the various levels.

Factor E, restrictor size, had by far the greatest impact on the valve flow rate. Factor C, seat design, also showed less impact, with fastening (factor A)

Table 3
Sensitivity response

	Factor					
Level	A: Fastening	B: Poppet Design	C: Seat Design	D: Durometer	E: Restrictor Size	F: Side Hole Size
1	0.4665	0.4702	0.4734	0.4788	0.8957	0.4800
2	0.4856	0.4877	0.5046	0.4764	0.4075	0.4680
3			0.4501	0.4728	0.1248	
Delta	0.0574	0.0751	0.0342	0.0635	0.7134	0.0455
Ranking	4	2	6	3	1	5

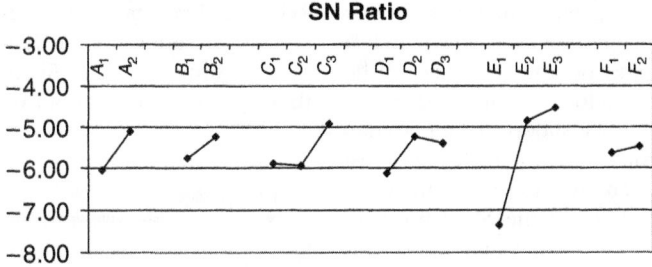

Figure 8
Response graph

shown to have less influence yet. Factor B, poppet design, showed no effect at all, with factors D and F, seat durometer and side hole size, showing little effect.

We applied the two-step optimization process to reduce flow rate variability and to meet minimum flow rate requirements. The optimum design will give us an average flow rate of 2.25 gal/min. The control factor levels of the optimum design are as follows:

A_2: fastened

B_1: hex

C_3: spherical

D_2: 60

E_2: 0.125 in.

F_1: 0.187 in.

8. Confirmation

Based on the robust design experimentation, design optimization would predictably result in an SN ratio

gain of 5.90 dB. A confirmation run at the optimal design resulted in a 4.63-dB gain. This is not perfect; nevertheless, it shows good reproducibility and confirms the success of the experiment. Results are shown in Table 4.

9. Conclusions

To solve the water feeder valve chattering problem, a dynamic Taguchi L_{18} experiment was performed. The experiment analyzed the level effects of six control factors and three noise factors. The zero-point proportional equation was used for the analysis. The study employed a double signal, inlet pressure into the feeder and cross-sectional flow area, with flow rate through the valve used as the response.

Experimental results showed a predictive SN ratio gain of 5.90 dB. This indicates that design optimization and a significant improvement in the chattering problem could be achieved. A confirmation study using the optimal design factors resulted in a 4.63-dB gain.

Table 4
Results of confirmatory run

Design	SN	Gain	Beta	Gain
Initial production	−8.87	—	0.88	—
Optimum prediction	−2.97	5.90	0.46	−0.42
Optimum confirmed	−4.25	4.63	0.37	−0.51

The Series 47 chattering experiment was the first Taguchi robust design experiment for McDonnell & Miller. We put the robust design technique to the test by applying it to a product that has been in the field for several years and has shown a sporadic chattering problem from the beginning. Utilizing robust design techniques we were able to predict the quality of the optimum design via the SN ratio gain. Based on the 4.63-db gain, we felt that this design needs to be tested in the field for customer approval. At the same time, we started working on a new cartridge that will eliminate altogether the free-moving part in the cartridge. With the traditional design–test–build method we could not pinpoint control factors that would make the product robust against field noise conditions as explicitly as we were able to do using the Taguchi robust design method.

Acknowledgments The author expresses his sincere appreciation to Shin Taguchi of American Supplier Institute and Tim Reed of ITT Industries for their guidance throughout the experiment.

References

J. L. Lyons, 1982. *Lyons Valve Designer's Handbook.* New York: Van Nostrand Reinhold.

R. L. Panton, 1984. *Incompressible Flow.* New York: Wiley.

G. H. Pearson, 1972. *Valve Design.* New York: Pitman.

Genichi Taguchi, Shin Taguchi, and Subir Chowdhury, 1999. *Robust Engineering.* New York: McGraw-Hill.

Yuin Wu and Alan Wu, 1998. *Taguchi Methods for Robust Design.* Livonia, Michigan: ASI Press.

This case study is contributed by Amjed Shafique.

Appendix

			Inner Array							
No.	A: Fastening	B: Poppet Design	C: Seat Design	D: Durometer	e	E: Restrictor Size	F: Cartridge Side Hole Size	e	SN Ratio	Sensitivity
1	Not fastened	Hex	Flat	80		0.500	0.187		−8.27	0.8794
2	Not fastened	Hex	Conical	60		0.125	0.210		−4.00	0.3998
3	Not fastened	Hex	Spherical	90		0.063	0.187[a]		−1.95	0.1214
4	Not fastened	Square	Flat	80		0.125	0.187[a]		−6.62	0.4158
5	Not fastened	Square	Conical	60		0.063	0.187		−2.94	0.1318
6	Not fastened	Square	Spherical	90		0.500	0.210		−5.01	0.8376
7	Not fastened	Hex[a]	Flat	60		0.063	0.210		−4.57	0.1228
8	Not fastened	Hex[a]	Conical	90		0.500	0.187[a]		−9.58	0.9315
9	Not fastened	Hex[a]	Spherical	80		0.125	0.187		−2.73	0.3582
10	Fastened	Hex	Flat	90		0.125	0.210		−3.72	0.3540
11	Fastened	Hex	Conical	80		0.063	0.187[a]		−2.34	0.1257
12	Fastened	Hex	Spherical	60		0.500	0.187		−4.46	0.8111
13	Fastened	Square	Flat	60		0.500	0.187[a]		−5.48	0.9460
14	Fastened	Square	Conical	90		0.125	0.187		−3.54	0.4702
15	Fastened	Square	Spherical	80		0.063	0.210		−1.95	0.4702
16	Fastened	Hex[a]	Flat	90		0.063	0.187		−2.66	0.1223
17	Fastened	Hex[a]	Conical	80		0.500	0.210		−7.66	0.9668
18	Fastened	Hex[a]	Spherical	60		0.125	0.187[a]		−2.30	0.4471
Average									−5.59	0.4760

[a] Dummy treated.

CASE 62

Optimal Design for a Small DC Motor

Abstract: In automobiles, dc motors are used for power windows, wipers, and other purposes. To improve quality and reduce cost, development was conducted together with the supplier of motors. In the beginning, the main objective was trying to reduce audible noise. However, the functionality, including power consumption, was evaluated through energy transformation. As a result, efficiency was improved substantially, nose level was reduced, and motor downsizing was achieved.

1. Function of a DC Motor

A dc motor has a mechanism that converts electric power (electricity, I, × voltage, V) to motive power ($2 \times \pi \times$ no. revolutions, n, × torque, T), and their relationship is expressed as follows:

$$2\pi n T = \beta I E \qquad \text{watts} \qquad (1)$$

Now β corresponds to energy conversion efficiency, and equation (1) represents a technical means per se. On the other hand, since the requirement of a motor as a product is motive power to drive a target object, if we cannot obtain the power, the motor does not work. Therefore, we regard the motive power required for a product as a signal and the required motive power with less electric power as the dc motor's function. This relationship is expressed as

$$I E = \beta \times 2\pi n T \qquad \text{watts} \qquad (2)$$

Now β is considered the magnitude of electric power when a certain amount of motive power is generated, defined as *electric power consumption rate*. Thus, as shown in Figure 1, enhancing the stability for noise and reducing the electric consumption rate, β, in equation (2) are equivalent to functional improvement, and we evaluate an objective by using a means.

2. Experimental Procedure and Measurement Characteristics

In our experiment we measured the number of revolutions and internal current at intervals of 0.1 second for 180 seconds by applying a certain amount of voltage when one of three levels of torque, which simulate actual target objects, was applied to the motor. Therefore, 1800 data were recorded for each load condition by a measurement machine. However, since, judging from the total lapse of time, we concluded that we could elucidate the overall trend by using only three points of time: the initial point, the point at 90 seconds, and the point at 180 seconds. We selected only 10 data at the time interval of 1 second from the 10-second-lapsed point in each time frame, thereby obtaining a total of 30 data for each load condition. For experiment 1, Figure 2 shows the transition of electric consumption and subsequent relationship of input and output. The deviation from linearity in this input/output relationship represents nonlinear elements in the change of current for a change in electric power consumption and load torque as well as error elements when the function is evaluated. By using a motor's heat as the noise effect in a functionality evaluation, we reduced our analysis period by 50% from that under conventional conditions, by assigning temperature or deterioration to an outer array in an orthogonal array.

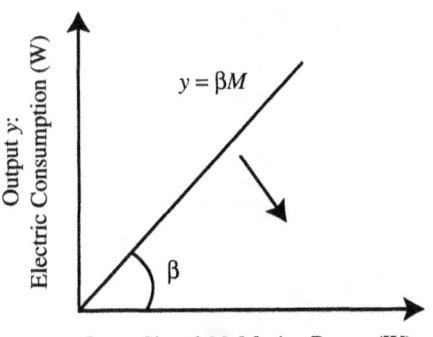

Figure 1
DC motor's function

3. SN Ratio and Sensitivity

Table 1 shows the data for experiment 1. We computed the motive power as a signal factor by multiplying by 2π the product of the measured number of revolutions, n, and the load torque, T. In addition, the power consumption as an output was equivalent to the product of the measured internal current of the motor, I, and the applied voltage, E. Since both the signal and output measured in the unit of power (watts) are characteristics related to energy, we analyzed their square roots by considering the additivity of factor effects obtained through decomposition of their variations.

Using the data of experiment 1 shown in Table 1, we detailed the calculation procedure of SN ratio and sensitivity as shown below.

Total variation:

$$S_T = 8.56^2 + 8.50^2 + \cdots + 10.14^2 + 10.06^2$$

$$= 7413.58 \qquad (f = 90) \qquad (3)$$

Effective divider:

$$r = 3.95^2 + 3.98^2 + \cdots + 4.70^2 + 4.73^2 = 1854.37 \qquad (4)$$

Linear equation:

$$L = (3.95)(8.56) + (3.98)(8.50)$$
$$+ \cdots + (4.70)(10.14) + (4.73)(10.06)$$
$$= 3701.03 \qquad (5)$$

Variation of proportional term:

$$S_\beta = \frac{L^2}{r} = \frac{3701.03^2}{1854.37} = 7386.65 \qquad (f = 1) \qquad (6)$$

Error variation:

$$S_e = S_T - S_\beta = 7413.58 - 7386.65$$
$$= 26.93 \qquad (f = 89) \qquad (7)$$

Error variance:

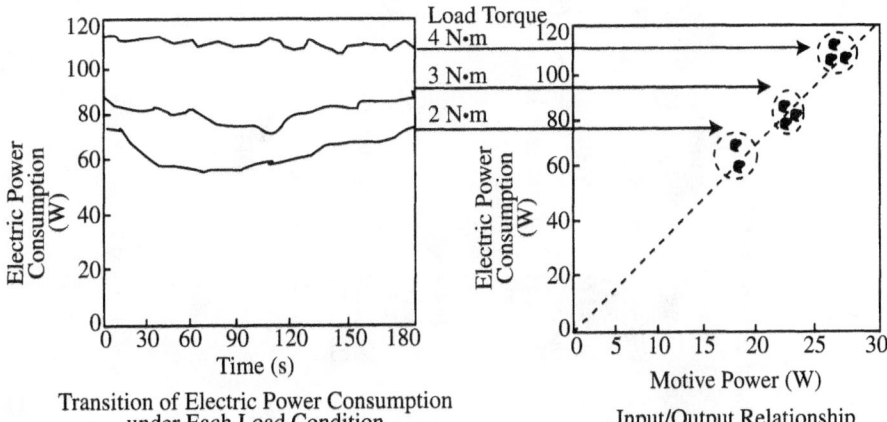

Transition of Electric Power Consumption
under Each Load Condition

Input/Output Relationship

Figure 2
Input/output relationship

Table 1
Results of experiment 1

Load (Nm)	Data Type[a]	Measurement Point of Time	Fluctuation at Each Measurement Point of Time									
2	Signal M	Initial	3.95	3.98	3.98	3.97	3.98	3.97	4.01	4.00	3.99	4.01
	Output y		8.56	8.50	8.42	8.35	8.32	8.43	8.38	8.31	8.26	8.23
	Signal M	90-s-lapsed	4.10	4.10	4.09	4.10	4.11	4.09	4.11	4.10	4.11	4.09
	Output y		7.68	7.61	7.80	7.72	7.66	7.70	7.75	7.69	7.56	7.76
	Signal M	180-s-lapsed	4.09	4.13	4.11	4.11	4.13	4.12	4.12	4.11	4.13	4.13
	Output y		7.64	7.61	7.61	7.59	7.68	7.58	7.51	7.69	7.55	7.56
3	Signal M	Initial	4.61	4.62	4.63	4.67	4.65	4.65	4.66	4.64	4.64	4.65
	Output y		9.28	9.31	9.28	9.25	9.15	9.15	9.16	9.17	9.18	9.12
	Signal M	90-s-lapsed	4.75	4.73	4.73	4.73	4.76	4.78	4.76	4.76	4.74	4.75
	Output y		8.70	8.76	8.76	8.67	8.72	8.73	8.70	8.74	8.74	8.73
	Signal M	180-s-lapsed	4.64	4.65	4.66	4.64	4.64	4.67	4.65	4.63	4.66	4.69
	Output y		8.98	9.01	8.90	8.93	8.96	8.80	8.80	8.91	8.92	8.88
4	Signal M	Initial	4.90	4.91	4.93	4.95	4.95	4.96	4.97	4.96	4.94	4.96
	Output y		10.57	10.50	10.43	10.38	10.36	10.37	10.40	10.39	10.37	10.37
	Signal M	90-s-lapsed	4.81	4.76	4.76	4.82	4.88	4.85	4.82	4.83	4.84	4.90
	Output y		10.28	10.33	10.28	10.11	10.05	10.19	10.16	10.19	10.04	10.00
	Signal M	180-s-lapsed	4.73	4.73	4.80	4.72	4.64	4.68	4.77	4.70	4.70	4.73
	Output y		10.09	10.00	10.01	10.19	10.29	10.11	10.01	10.20	10.14	10.06

[a] Signal M is $\sqrt{2\pi n T}$ and output y is $\sqrt{IV}$.

$$V_e = \frac{S_e}{89} = \frac{26.93}{89} = 0.3026 \qquad (8)$$

SN ratio:

$$\eta = 10 \log \frac{(1/r)(S_\beta - V_e)}{V_e}$$

$$= 10 \log \frac{(1/1854.37)(7386.65 - 0.3026)}{0.3026}$$

$$= 11.19 \text{ dB} \qquad (9)$$

Sensitivity:

$$S = 10 \log \frac{1}{r}(S_\beta - V_e)$$

$$= 10 \log \frac{1}{1854.37}(7386.65 - 0.3026)$$

$$= 6.00 \text{ dB} \qquad (10)$$

4. Optimal Condition and Confirmatory Experiment

As control factors we selected eight factors (Table 2) and assigned them to an L_{18} orthogonal array. For each experiment in the L_{18} orthogonal array, we calculated the SN ratio and sensitivity using the calculation method discussed in the preceding section. Based on the result, we computed the level-by-level average SN ratio and sensitivity and plotted the factor effects (Figure 3).

The optimal condition determined by the response graph of the SN ratio is shown together with the current condition as follows:

Optimal condition:　$A_1 B_2 C_3 D_1 E_2 F_1 G_3 H_3$

Current condition:　$A_1 B_2 C_1 D_3 E_2 F_1 G_1 H_1$

In Table 3 we show an estimate of the process averages of the SN ratio and sensitivity under both optimal and current conditions, together with the result of the confirmatory experiment. As a result, we confirmed good reproducibility for both the SN ratio and sensitivity. Additionally, by assessing a benchmarked product regarded as comparable to our target, we confirmed that the optimal condition is equivalent to its level.

Next, we illustrated the plot of the input/output relationship under each condition in Figure 4. Looking at the figure we can see that we can minimize the fluctuation in power consumption and reduce the power consumption rate β by 17% under current conditions and by 11% for the benchmarked product. The reduction in power

Table 2
Factors and levels

		Level		
Factor		**1**	**2**	**3**
Control factors				
A: fixing method for part A		Current	Rigid	—
B: thickness of part B		Small	Mid	Large
C: shape of part B		1	2	3
D: width of part D		Small	Mid	Large
E: shape of part E		1	2	3
F: inner radius of part F		Small	Mid	Large
G: shape of part G		1	2	3
H: thickness of part G		Small	Mid	Large
Signal factor				
M: motive power ($2\pi nT$)		Product of load torque and number of revolutions		
Noise factor				
I: time		Initial	90-s-lapsed	180-s-lapsed

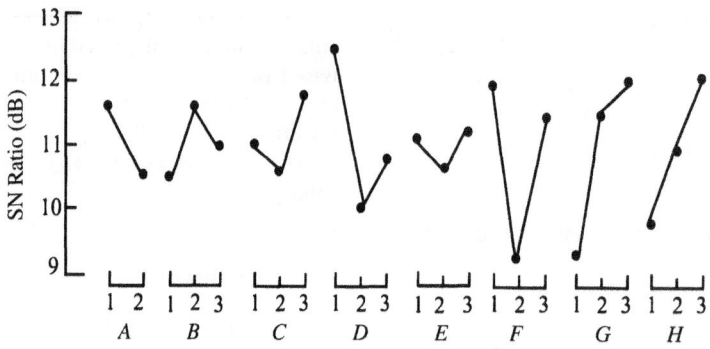

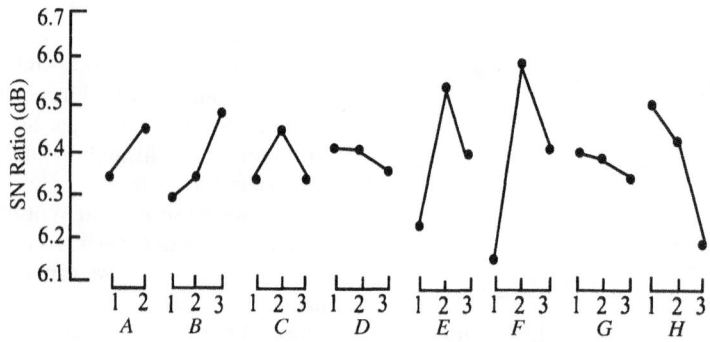

Figure 3
Response graphs

consumption enhances the motive power gained from the same power and increases the efficiency, β, by 4.3% for those under the current condition and by 2.7% for those of the benchmarked product. Owing to this achievement, we improved the motive power obtained from a car battery by 20% com-

pared with that under current conditions. In addition, we can downsize components such a magnet or yoke without losing the current level of the motive power and, moreover, can slash material and other costs by several dozen million yen on a yearly basis.

Table 3
SN ratio and sensitivity in confirmatory experiment (dB)

Condition	SN Ratio		Sensitivity	
	Estimation	Confirmation	Estimation	Confirmation
Optimal	16.88	16.43	5.99	6.11
Current	10.06	11.73	6.31	6.93
Gain	6.82	4.70	−0.32	−0.82
Benchmark	—	16.77	—	6.62

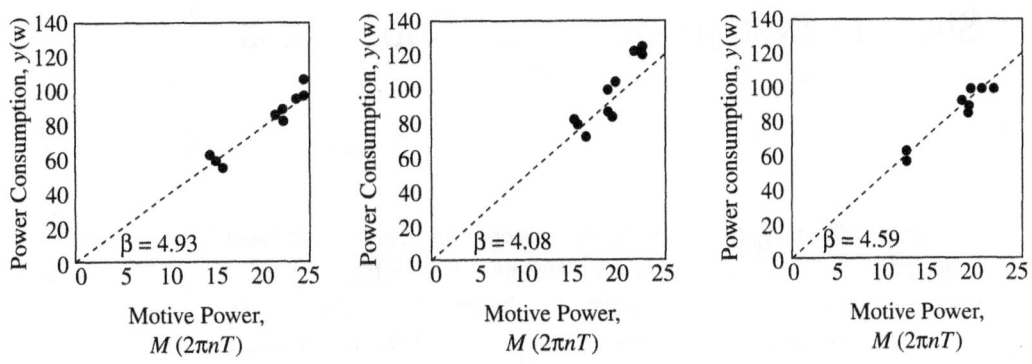

Figure 4
Results of confirmatory experiment

Reference

Kanya Nara, Kenji Ishitsubo, Mitsushiro Shimura, Akira Harigaya, Naoto Harada, and Hideki Rikanji, 2000. Parameter design for a small dc motor. *Proceedings of the 8th Quality Engineering Symposium*, pp. 194–197.

This case study is contributed by Kanya Nara.

CASE 63

Steering System On-Center Robustness

Abstract: Traditionally, on-center steering response for a vehicle has been evaluated primarily through testing performed on actual prototype vehicles. A good on-center steering system is such that the driver can perceive small changes in road reaction forces through the steering wheel. This allows a driver to respond to slight changes in the vehicle heading. This is typically referred to as *on-center road feel*. A good on-center steering system will also have high overall stiffness, to achieve a narrow center and quick response.

The purpose of this study was to determine the steering system parameter values that optimize the robustness of vehicle on-center steering performance. A full vehicle ADAMS dynamic simulation model was used to generate all data required. The study employs the latest robust design methodology, incorporating the ideal function and Taguchi's dynamic signal-to-noise (SN) ratio.

1. Introduction

The goal of this project was to determine the vehicle steering and suspension system combination that optimizes the robustness of on-center steering performance. An ADAMS dynamic simulation computer model of a light truck vehicle was used to generate the data necessary for evaluation of the design. The model consisted of a detailed representation of the steering system combined with a full vehicle model. To establish the validity of the ADAMS model, the steering system and vehicle responses predicted were correlated with on-center road test data from a baseline vehicle prior to performing this study.

Many steering systems fail to achieve good on-center steering system performance, due partly to large amounts of lash and friction. A steering system with good road feel at high speeds must have maximum stiffness and minimum friction. These parameters are especially important in the on-center region, so that the driver can detect and respond to extremely small changes in road reaction forces resulting from road camber, bumps, wind, and so on, so as to maintain a constant heading.

There are limits as to what information from the road and vehicle the driver should be subjected to at the steering wheel. For higher-frequency noise, the steering system should function as a low-pass band filter that suppresses noise above 10 Hz from the road or vehicle but allows lower-frequency noise, below 3 Hz, through to the driver, so that the driver can respond appropriately and make corrections to his or her heading without having to rely solely on visual indicators, such as waiting for the vehicle to drift sufficiently far from the center of a traffic lane.

A steering system should also have linear on-center sensitivity, so that for small steering wheel angles, an increase in the steering wheel angle results in a proportional increase in the lateral acceleration experienced by the vehicle. The steering sensitivity level desired for a particular vehicle depends on the targeted image for that vehicle within the market.

Another important consideration in steering system on-center performance is the amount of dead

Figure 1
ADAMS full vehicle model

band in the system. This is a result of the hysteresis in the response of the vehicle to steering wheel inputs. Typically, smaller dead bands are preferred, but they cannot be eliminated entirely, due to a large contribution from the overall vehicle dynamics. The hysteresis is a function of the peak lateral acceleration, since larger accelerations will excite more nonlinear vehicle and tire dynamics.

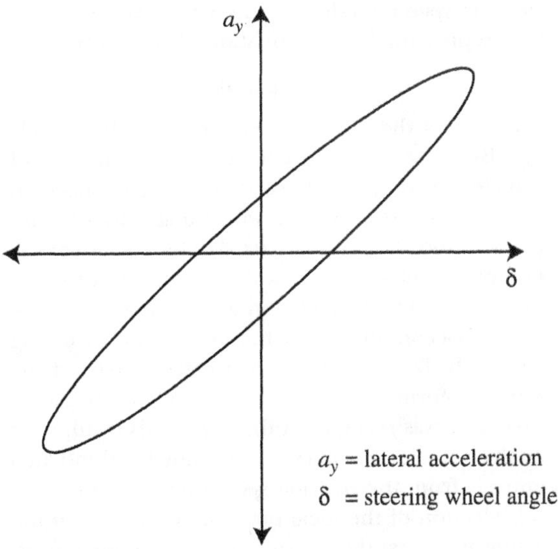

2. Description of the Model

The steering system model, with over 90 degrees of freedom, is highly detailed in its description of individual steering system components. It was coupled with a full vehicle ADAMS model developed by Ford Light Truck Engineering (Figure 1).

Inputs to the steering system are at the steering wheel, and the outputs are reflected in the lateral acceleration response of the vehicle to the steering wheel input angle. For the on-center test simulated in this study, a sinusoidal steering wheel input at 0.2 Hz was imposed with a constant amplitude designed to generate an approximate 0.2-g peak lateral acceleration at a forward vehicle speed of 60 mph. A low level of lateral acceleration was chosen to reduce the nonlinear effects of vehicle dynamics on the on-center performance, so that the steering system performance could be isolated and evaluated.

Outputs from the simulation were characterized by plotting the vehicle lateral acceleration versus the steering wheel angle (Figure 2). The speed of the vehicle was varied, as in an ordinary on-center test, while the steering wheel angle amplitude was held

a_y = lateral acceleration
δ = steering wheel angle

Figure 2
Typical data for on-center a_y versus δ cross plot

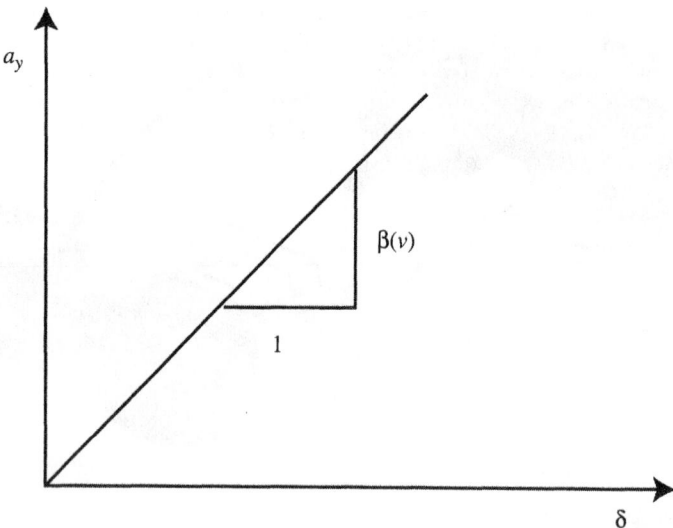

Figure 3
Ideal function for on-center steering

constant throughout the simulation. Results from actual vehicle on-center tests are typically characterized by the features defining the shape of the curve in cross plots of input and output variables. Among the significant indicators are the slope of the loop, which is a measure of hysteresis in vehicle-handling performance.

Steering sensitivity is a measure of the responsiveness of the vehicle to steering wheel inputs, usually tailored to fit a specific target value that depends on the vehicle image desired. As an example, sporty vehicles would typically be targeted to have a higher steering sensitivity than a family sedan would. Steering sensitivity is essentially a nominal-the-best target.

Hysteresis at low levels of peak lateral acceleration is an indicator of the lash and friction in the steering system, as well as the dynamic contribution due to the suspension and tires. At higher levels of vehicle lateral acceleration, hysteresis increases due to vehicle roll dynamics in the suspension and other nonlinearities in the tires and steering. Hysteresis can be thought of as a lag in the response of the vehicle to steering system inputs. Increased lag in the steering system response results in more steering dead band, which is undesirable. In this sense, hysteresis may be considered as a smaller-the-better target.

3. Ideal Function

Much discussion and effort went into defining the ideal function for the case study. After careful consideration of many possible alternatives, it was decided that the ideal function characterizing the steering system model on-center behavior would be best represented by an equation of the form

$$a_y = \beta(v)\delta$$

where a_y is the lateral acceleration of the vehicle (g), $\beta(v)$ the sensitivity coefficient as a function of vehicle speed (g/deg), (v is the vehicle speed in mph), and δ the steering wheel angle (deg). This formulation, shown in Figure 3, allowed for consideration of vehicles with both constant and varying understeer. The formulation also assumed that hysteresis is a condition to be minimized in the steering system. In fact, hysteresis was treated as part of the noise in formulating the system's SN ratio. Nonetheless, it was recognized that hysteresis is inherent in the vehicle dynamics and cannot be eliminated entirely from the steering system performance.

Selection of the ideal function was based on the assumption that the steering system open-loop function is as shown in Figure 3, where steering wheel angle (δ) is the driver input and lateral acceleration (a_y) of the vehicle is the output.

The ideal function selected for this study was derived from the results of an on-center steering test. The steering wheel angle versus lateral acceleration cross plot shown in Figure 2 is representative of a typical curve obtained from a vehicle on-center test. It can be seen that the ideal function shown in Figure 3 is a representation of the overall slope of the loop, but without hysteresis.

The response of the system in Figure 4 can be affected by environmental and other factors outside the direct control of the design engineer. Such factors are referred to as noise factors. Examples of noise factors are tire inflation pressure, tire wear, and weight distribution. It is necessary to design the system to achieve the most consistent input/output response regardless of the operating conditions experienced by the customer. Figure 5 shows the overall system diagram, indicating the noise factors and control factors considered in this study.

4. Noise Strategy

Several noise factors are shown in Figure 5. However, it is not necessary that all noise factors be included in the analysis. Typically, if a system is robust against one noise factor, it will be robust against other noise factors having a similar effect on the system. Therefore, the noise factor strategy involves selecting a subset of noise factors from all possible noise factors affecting the system. The noise factors selected should be those factors that tend to excite the greatest variation in the system response.

The steering system response to noise has two significant modes, one being the change in slope of the ideal function (sensitivity), the other a change in the size of the hysteresis loop. Therefore, it is necessary to group the noise factors into subsets, each of which excites one or the other mode of response to variations in noise levels. For example, some noise levels may expand the hysteresis loop

with little or no effect on the steering sensitivity. These will be referred to as type P noise. Other noise factors may affect the slope of the loop, but not the width, thus affecting only steering sensitivity and not hysteresis, which will be referred to as Type Q noise. Any noise factors found to influence both modes simultaneously are not acceptable as part of the noise strategy, since these will impose interactions in the effects of noise on the system.

A study was conducted wherein each noise factor was varied individually between high and low levels using the vehicle steering system. Each noise factor was ranked according to its effect on the system response for both noise-induced response modes (i.e., hysteresis and/or sensitivity effects).

It was decided to select the two most significant noise factors affecting each of the two modes of noise-induced system response variability. Although column lash appeared second in significance in terms of the sensitivity variability, it also exhibited a strong interaction with the hysteresis mode. Thus, column lash was not a good candidate for the noise strategy, since it excited both noise modes simultaneously, which precluded separation of the effects of the two types of noise-induced modes on the system. Therefore, the noise strategy was as shown in Table 1, where the associated noise factor levels for each noise level (N_1 to N_4) are also indicated.

5. Case Study Results

The analysis was performed using an L_{27} array to accommodate three levels of each control factor and was conducted for three speeds: 45, 60, and 75 mph. Each ADAMS run generated response curves similar to those in Figure 2, where the lateral acceleration of the vehicle was plotted against steering wheel angle. From these curves, a least squares linear fit was made (passing through the origin), using points on both the top and bottom halves of the

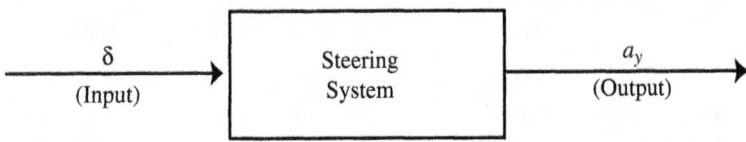

Figure 4
Steering system function

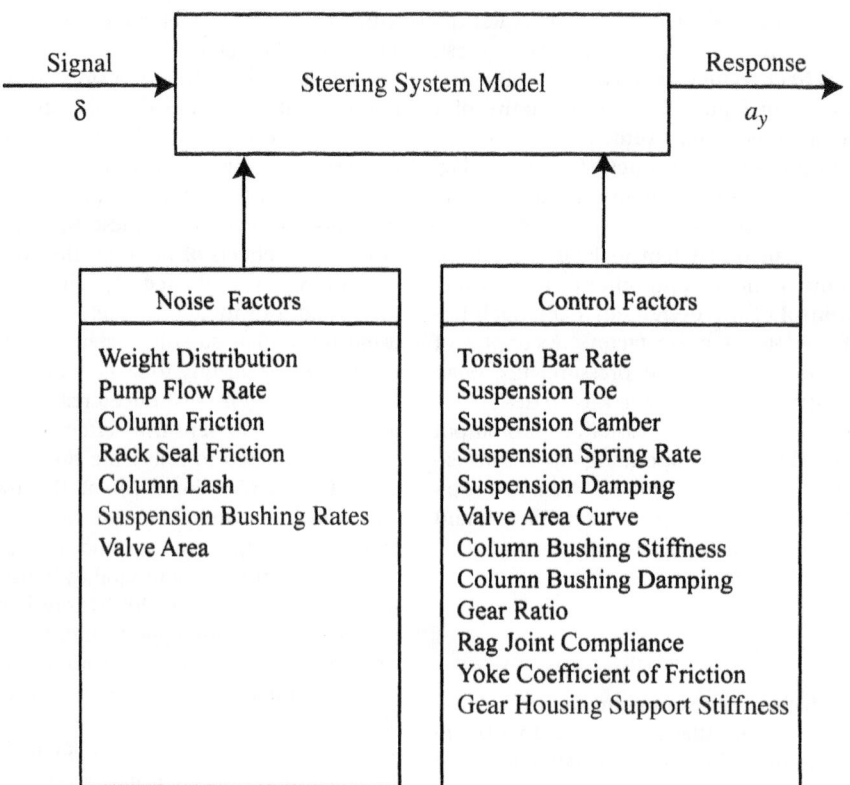

Figure 5
Engineering system for the steering system model

Table 1
Noise strategy

Noise Factor	Noise Level			
	N_1 P_1, Q_1	N_2 P_1, Q_2	N_3 P_2, Q_1	N_4 P_2, Q_2
Rack friction	Low	Low	High	High
Column friction	Low	Low	High	High
Weight distribution	High	Low	High	Low
Valve areas	High	Low	High	Low

loop. This line represents the ideal function of the steering system. The closer the actual data correspond with the straight-line approximation, the better the robustness of the system. Figure 6 illustrates the best-fit approach.

6. Analysis

For each run in the L_{27} array used in this experiment, an evaluation of the SN ratio can be made, using the relationship

$$SN = 10 \log \left(\frac{\beta^2}{\sigma^2} \right)$$

where, SN is the signal-to-noise ratio (dB), β the sensitivity (g/deg), and σ^2 the variance (g^2). Although the SN ratio is a concept used extensively in control theory, it has application to general systems analysis as well. For the SN ratio to be maximized, the variance must be low, which indicates that the system response is consistent over a wide range of inputs and that its sensitivity to noise effects is small.

A robust system is one that results consistently in the corresponding ideal output signal response, as determined by the system ideal function (Figure 3), regardless of the noise factor levels in the surrounding environment. Keeping this definition in mind, it is apparent how the SN ratio can be used as an effective measure of system robustness.

In analyzing the SN data resulting from the simulations, it was decided first to perform separate robustness analysis for each of the three simulation velocities: in effect, with the assumption that the on-center robustness of the steering system is independent of vehicle speed. If this analysis were to produce varying results at different vehicle speeds, a more sophisticated approach to handling the data would be required to reconcile the results from varying speeds.

The SN ratio for each of the runs in the L_{27} array was computed over the signal levels (steering wheel angle inputs) and noise factor levels (N_1 to N_4) for each run. Next, an average was computed for each of the three levels for every control factor. A similar calculation was performed for the sensitivity (β) value.

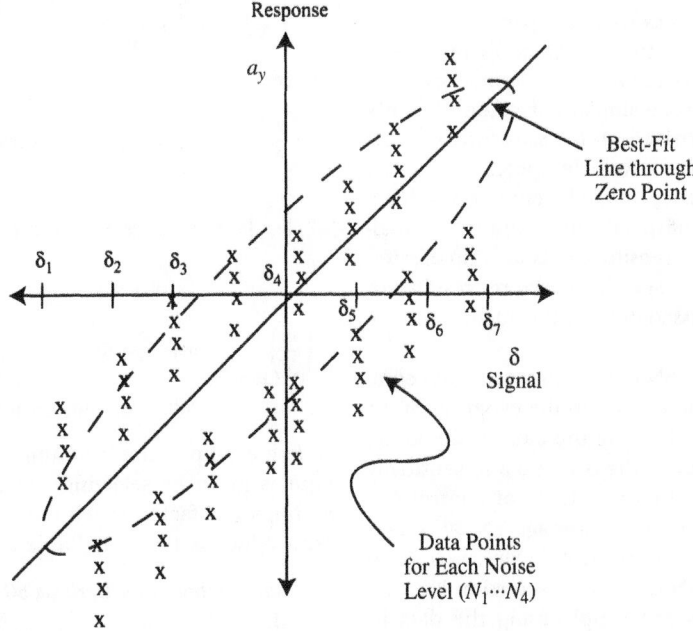

Figure 6
Best-fit approximation

With the average SN ratios, the optimum configuration was selected for maximizing the robustness of the system by choosing the level of each control factor with the highest SN ratio. An estimate of the overall system SN ratio was then obtained by summing the improvements due to each optimal control factor setting from the overall average of SN ratios for all 27 runs.

7. Simulation Results and Analysis of Data

A procedure for automating successive ADAMS simulation runs was developed and executed using a series of input data sets for each configuration (324 total) and command scripts to queue the analyses. Results were postprocessed automatically to obtain tabular data for the steering wheel angle versus vehicle lateral acceleration curve using seven steering wheel angle values, from -15 to $+15°$ to define the upper and lower halves of the curve. The tabular data were then transferred into a spreadsheet that performed the curve fit (Figure 6) and least squares regression analysis, to compute the SN ratio and sensitivity values. As mentioned earlier, each of the three vehicle speeds was treated separately.

Results showed that the change in SN ratio from one run to the next were fairly consistent, regardless of vehicle speed. Also, a similar correlation was observed for the sensitivity, with the sensitivity increasing proportionally with vehicle speed. Thus, the signal quality (SN ratio) was independent of vehicle speed for the range of speed considered in this analysis, and the relative sensitivity was also unaffected by changes in vehicle speed. Therefore, results are presented and discussed only for the 60-mph vehicle speed simulations.

Table 2 shows the SN ratios and sensitivities computed for each of the 27 runs in the experiment for 60-mph vehicle speed. Using the data in the table, it is possible to calculate the SN ratio and sensitivity average for each of the three levels of control factors. Tables 3 and 4 show the average SN ratios and sensitivity values for the control factor levels at 60 mph. The average SN ratios for the torsion bar rate control factor levels at 60 mph, using the data in Table 1, were computed as follows.

Level 1:
$$\left(\frac{S}{N}\right)_{av} = \frac{1}{9} - 18.6 - 18.4 - 18.3 - 18.9 - 18.9$$
$$- 18.9 - 19.8 - 19.9 - 19.9) = -19.06$$

Level 2:
$$\left(\frac{S}{N}\right)_{av} = \frac{1}{9}(-20.0 - 20.0 - 20.2 - 18.4 - 18.4$$
$$- 18.2 - 18.5 - 18.7 - 18.6) = -19.02$$

Level 3:
$$\left(\frac{S}{N}\right)_{av} = \frac{1}{9}(-18.9 - 19.1 - 19.0 - 19.8 - 19.6$$
$$- 19.9 - 18.4 - 18.2 - 18.3) = -19.01$$

A chart comparing the effect on the SN ratios for the various control factor levels is presented in Figure 7 for the 60-mph simulations, and a similar plot for the sensitivity is shown in Figure 8.

From the individual control factor average SN ratio values shown in Figure 7 it is possible to predict the SN ratio for the optimum configuration, using the equation

$$\left(\frac{S}{N}\right)_{opt} = \left(\frac{S}{N}\right)_{av} + \sum_n \left[\left(\frac{S}{N}\right)_{n,opt} - \left(\frac{S}{N}\right)_{av}\right]$$

where

$\left(\frac{S}{N}\right)_{opt}$ = optimum configuration SN ratio

$\left(\frac{S}{N}\right)_{av}$ = average SN ratio for all runs in the L_{27} array

$\left(\frac{S}{N}\right)_{n,opt}$ = average SN ratio for the nth control factor at the optimum level

For example, the optimum configuration at 60 mph is given by selecting the following levels for each control factor, which represent the highest SN ratio values as shown in Table 3.

Optimum control factor levels at 60 mph:

A_3 B_3 C_1 D_1 E_2 F_1 G_3
H_3 I_2 J_3 K_1 L_3

Table 2
Summary of results for the L_{27} array at 60 mph

| | Factor[a] | | | | | | | | | | | | | |
Run	A	B	C	D	E	F	G	H	I	J	K	L	SN	β
1	1	1	1	1	1	1	1	1	1	1	1	1	−18.6	0.724
2	1	1	1	1	2	2	2	2	2	2	2	2	−18.4	0.730
3	1	1	1	1	3	3	3	3	3	3	3	3	−18.3	0.740
4	1	2	2	2	1	1	1	2	2	2	3	3	−18.9	0.749
5	1	2	2	2	2	2	2	2	3	3	1	1	−18.9	0.802
6	1	2	2	2	3	3	3	1	1	1	2	2	−18.9	0.755
7	1	3	3	3	1	1	1	3	3	3	2	2	−19.8	0.802
8	1	3	3	3	2	2	2	1	1	1	3	3	−19.9	0.754
9	1	3	3	3	3	3	3	2	2	2	1	1	−19.9	0.803
10	2	1	2	3	1	2	3	1	2	3	1	2	−20.0	0.669
11	2	1	2	3	2	3	1	2	3	1	2	3	−20.0	0.669
12	2	1	2	3	3	1	2	3	1	2	3	1	−20.2	0.729
13	2	2	3	1	1	2	3	2	3	1	3	1	−18.4	0.818
14	2	2	3	1	2	3	1	3	1	2	1	2	−18.4	0.775
15	2	2	3	1	3	1	2	1	2	3	2	3	−18.2	0.741
16	2	3	1	2	1	2	3	3	1	2	2	3	−18.5	0.826
17	2	3	1	2	2	3	1	1	2	3	3	1	−18.7	0.852
18	2	3	1	2	3	1	2	2	3	1	1	2	−18.6	0.830
19	3	1	3	2	1	3	2	1	3	2	1	3	−18.9	0.668
20	3	1	3	2	2	1	3	2	1	3	2	1	−19.1	0.730
21	3	1	3	2	3	2	1	3	2	1	3	2	−19.0	0.722
22	3	2	1	3	1	3	2	2	1	3	3	2	−19.8	0.785
23	3	2	1	3	2	1	3	3	2	1	1	3	−19.6	0.762
24	3	2	1	3	3	2	1	1	3	2	2	1	−19.9	0.791
25	3	3	2	1	1	3	2	3	2	1	2	1	−18.4	0.852
26	3	3	2	1	2	1	3	1	3	2	3	2	−18.2	0.808
27	3	3	2	1	3	2	1	2	1	3	1	3	−18.3	0.771

[a] A, torsion bar rate; B, gear ratio; C, valve curve; D, yoke friction; E, column bearing radial stiffness; F, column bearing radial damping; G, rag join torsional stiffness; H, gear housing support radial stiffness; I, suspension torsion bar rate; J, shock rate; K, toe alignment; L, camber alignment.

Table 3
SN ratios for control factor levels at 60 mph

Level	Factor											
	A	B	C	D	E	F	G	H	I	J	K	L
1	−19.06	−19.15	−18.93	−18.35	−19.03	−19.02	−19.08	−19.04	−19.07	−19.04	−19.02	−19.11
2	−19.02	−19.00	−19.09	−18.84	−19.03	−19.03	−19.03	−19.03	−19.01	−19.03	−19.03	−19.02
3	−19.01	−18.93	−19.07	−19.89	−19.03	−18.98	−19.02	−19.01	−19.02	−19.03	−19.03	−18.95
Delta	0.05	0.22	0.16	1.54	0.00	0.01	0.10	0.02	0.06	0.02	0.01	0.16

Table 4
Sensitivities for control factor levels at 60 mph

	Factor											
Level	A	B	C	D	E	F	G	H	I	J	K	L
1	0.762	0.709	0.782	0.773	0.766	0.764	0.762	0.751	0.761	0.765	0.756	0.789
2	0.768	0.775	0.756	0.771	0.765	0.765	0.766	0.765	0.764	0.764	0.766	0.764
3	0.765	0.811	0.757	0.752	0.765	0.767	0.768	0.779	0.770	0.766	0.773	0.742
Delta	0.006	0.102	0.026	0.021	0.001	0.003	0.006	0.028	0.009	0.002	0.017	0.047

The average SN ratio for all 27 runs at 60 mph can be computed by averaging the SN ratios computed for the three levels of any control factor in Table 3, since these values represent averages of nine runs each, so that

$$\left(\frac{S}{N}\right)_{av} = \frac{-19.06 - 19.02 - 19.01}{3} = -19.03$$

Thus, the optimum SN ratio at 60 mph is

$$\begin{aligned}\left(\frac{S}{N}\right)_{opt} &= -19.03 + (-19.01 + 19.03) \\ &+ (-18.93 + 19.03) + (-18.93 + 19.03) \\ &+ (-18.35 + 19.03) + (-19.03 + 19.03) \\ &+ (-19.02 + 19.03) + (-18.98 + 19.03) \\ &+ (-19.02 + 19.03) + (-19.01 + 19.03) \\ &+ (-19.02 + 19.03) + (-19.02 + 19.03) \\ &+ (-18.95 + 19.03) \\ &= -17.94\end{aligned}$$

The optimum control factor configuration at each vehicle speed was predicted. Similarly, predictions were also made for the worst-case configuration, having control factor-level settings corresponding to the lowest SN ratios as well as for the nominal configuration. The SN ratios predicted were verified by performing simulation at the control factor-level settings corresponding to the configuration predicted (optimum, nominal, and worst-case) and performing regression analysis to determine the actual SN ratio. Table 5 compares the results of the confirmation runs for all three cases: optimum, worst-case, and nominal. Values predicted were in good agreement with actual values, indicating that there are negligible interactions among the control factors.

Once the optimum control factor settings were determined to optimize system robustness, it was necessary to determine which control factors could be used to tune the system sensitivity to the target value desired. It can be seen in Figure 8 that the gear ratio has by far the greatest effect on the sensitivity, with several other parameters, such as camber, having appreciable effects as well. In using the gear ratio or camber as control factors to tune the steering system, there will be a slight effect on the steering system robustness. Camber changes will also affect tire wear characteristics, making it a poor choice for steering sensitivity adjustment. Among the control factors that had no appreciable effect on the steering system robustness (SN ratio) and that exhibited an effect on the sensitivity were the gear housing support radial stiffness and toe alignment angle.

Thus, the gear ratio, toe alignment, and gear housing support stiffness are good candidates as control factors available for tuning the system sensitivity. However, the gear ratio is probably the most practical, as changes to the gear housing support stiffness will also have an effect on the vehicle understeer characteristics. The toe alignment effect on steering sensitivity is in a rather narrow range, and there is evidence of toe alignment changes having an increasing effect on the steering system robustness at higher speeds, indicating that at very high speeds, toe effects may influence system robustness appreciably.

One other important observation from this study relates to the control factors that had no appreciable effect on the steering system robustness and sensitivity: including the column bearing radial

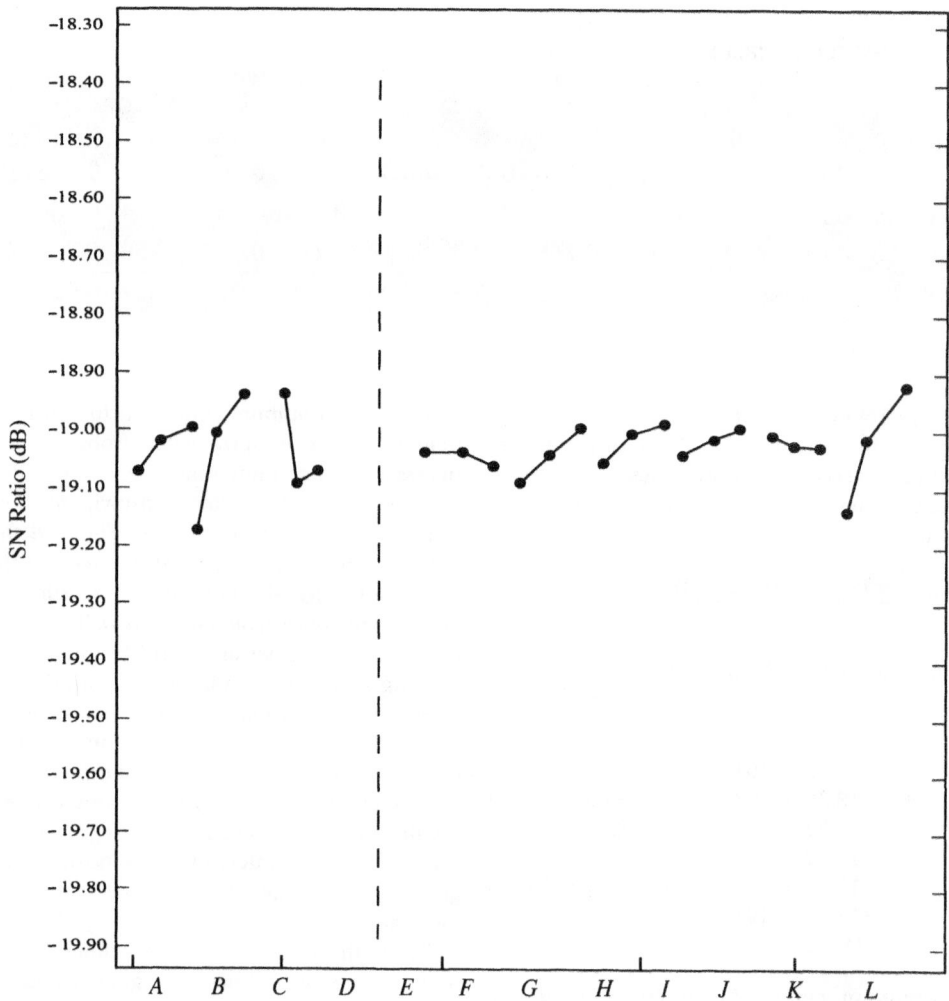

Figure 7
SN ratio comparison for control factor levels at 60 mph

stiffness, column bearing radial damping, and shock rate. These factors can be set to any level without affecting steering system robustness or sensitivity, and they can therefore be eliminated from consideration in any future on-center steering system robustness studies.

8. Conclusions

The close correlation between predicted and actual SN ratios and sensitivities for the optimum, nominal, and worst-case configurations demonstrate that the Taguchi robustness system analysis approach is accurate. Close correlation also indicates that there is little interaction between the control factors.

Important control factors for optimizing the steering system on-center robustness include the yoke friction, torsion bar rate, valve curve, gear ratio, rag joint torsional stiffness, camber angle, and suspension torsion bar rate. Other control factors, such as column bearing radial stiffness and damping, gear housing support radial stiffness, shock rate, and toe alignment angle had no effect on the

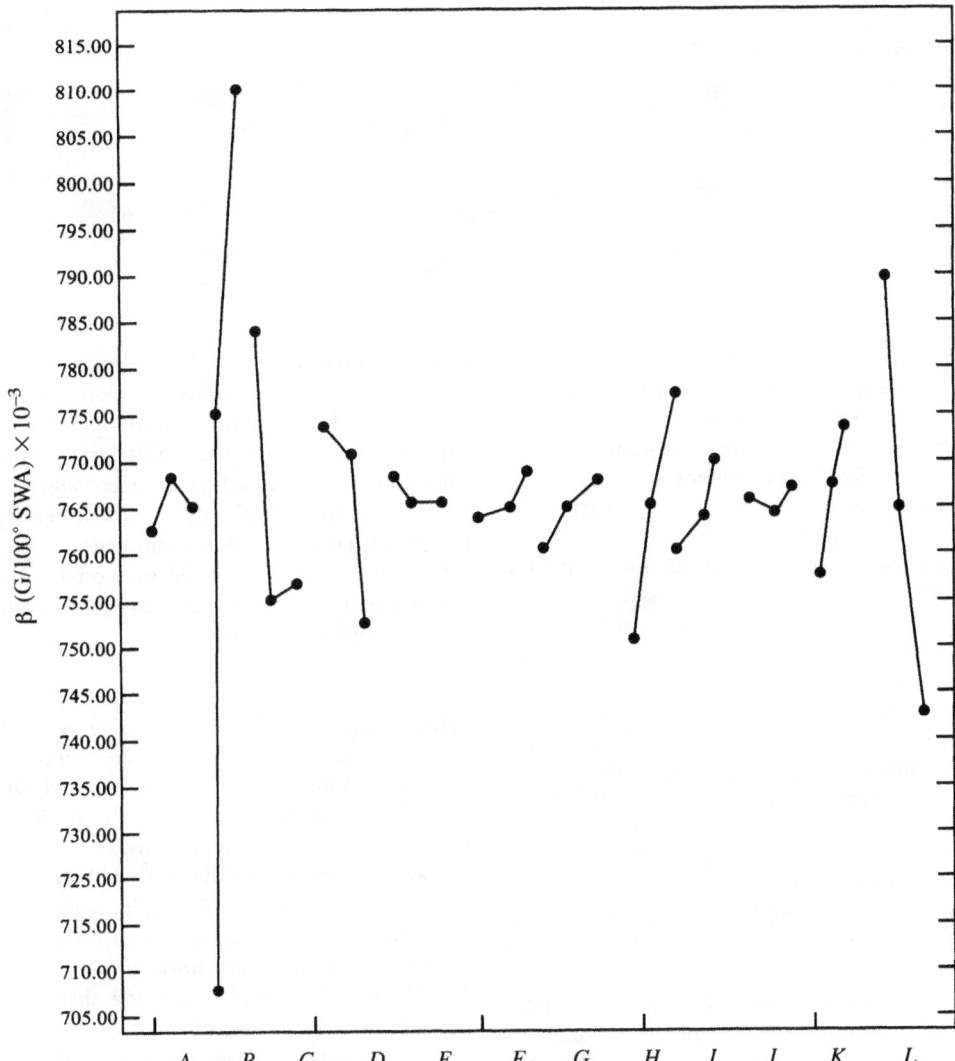

Figure 8
Sensitivity (β) comparison for control factor levels at 60 mph

SN ratio and can be set at any level desired without affecting steering system robustness. Future studies can narrow down control factors, eliminating those with little or no effect on signal quality in order to reduce the size of the problem.

In evaluating the optimum control factor levels, it is important to keep in mind other steering system considerations that may be affected by changes in the design. For example, the torsion bar rate in-

crease will result in greater steering effort, which may be unacceptable. Also, yoke friction and rag joint stiffness changes may reduce the noise-handling characteristics of the steering system, requiring further design changes, such as incorporating a vibration damper into the steering system.

The SN ratio change from nominal to optimum is on the order of 1.6 dB for all speeds, which translates into an improvement of about 17%. The

Table 5
Predicted versus actual SN ratios (dB)

Configuration	45 mph		60 mph		75 mph	
	Predicted	Actual	Predicted	Actual	Predicted	Actual
Optimum	−17.75	−17.80	−17.94	−18.01	−17.85	−17.95
Nominal	−19.47	−19.41	−19.65	−19.58	−19.57	−19.52
Worst-Case	−19.94	−19.90	−20.33	−20.27	−20.31	−20.31

improvement in signal quality is especially apparent in Table 4, where the current nominal level is much closer to the worst-case configuration than to the optimum. Subjectively, a driver can sometimes perceive even a 0.25-dB improvement.

In general, control factors influence steering sensitivity and steering system signal quality (SN ratio) in the same way over the range of speed studied. Thus, future on-center steering system studies can be performed without the need to consider several speeds, thereby reducing overall simulation requirements.

Gear ratio and camber are the most important control factors in determining the sensitivity of the steering system, among those investigated in this analysis. The gear housing support stiffness is a good candidate for adjusting vehicle steering sensitivity within a limited range without affecting steering system on-center robustness. However, it may have an effect on understeer and other handling characteristics of the vehicle that must be considered.

Since friction plays such a dominant role in determining signal quality, it is important to study in further detail sources of friction in the rack, particularly the yoke, gear, and seal friction. Efforts are currently under way to study yoke and seal frictional effects, with the intent to incorporate more sophisticated gear friction models into the vehicle simulation model.

The main benefits of using the dynamic simulation techniques outlined in this study are reduced product development cost and time, achieved by a reduction in actual vehicle development test requirements. In addition, the development engineer can study many more possible design combinations than can be accomplished by exercising even the most ambitious vehicle development testing programs. Dynamic computer simulation techniques offer a much more versatile method by which to determine the best system configuration for product performance and quality.

Acknowledgments   We would like to thank the following additional case study team members from the Ford Motor Company, Car Product Development, for their contributions to the project: Chris D. Clark, Paul Hurley, Phil Lenius, and Bernard Los. We would also like to thank Don Tandy and the people from the Vehicle Dynamics Section, Chassis Engineering, Light Truck Engineering, at Ford Motor Company for their work in developing the ADAMS vehicle model that made this case study possible. Finally, we would like to thank Genichi Taguchi and Shin Taguchi of the American Supplier Institute and John King and John Coleman at the Ford Design Institute for their guidance and assistance with this project.

This case study is contributed by V. G. Lenses and P. Jayakumar.

CASE 64

Improvement in the Taste of Omelets

Abstract: In this study, to improve the typical physical taste of an omelet, texture, that is, the "mouth feel" when an omelet is chewed, we undertook technological development based on quality engineering by using a mechanical measurement method instead of a sensory test.

1. Taste Measurement Method

The taste of an omelet, which accounts for 70 to 80% of the eating experience, is not chemical but physical. In this study, in order to improve the texture (i.e., the way the mouth feels when an omelet is eaten), we attempted technological development based on quality engineering by using a mechanical measurement method instead of a sensory test.

A typical tasty omelet is a handmade omelet, cooked at home or in a restaurant kitchen. Assuming such an omelet to be ideal and at the same time, targeting popular commercial omelets, we proceeded with our study.

Among major mechanical measurement methods for texture are compression, shearing, cutting, and tension tests. Using these we can evaluate texture from many aspects. In addition, when we eat an omelet, we can perceive deformation, cracks, fragmentation, and so on, in the mouth. We adopted a plunge-type measurement device with a hollow plunger tube (Figure 1).

Primarily, we measured and compared "handmade," "regarded as tasty," and "regarded as worst" omelets (Figure 2). The thickness of the plunger's wall is 0.25 mm. This experiment revealed that for the elastic region (area A), a handmade or regarded-as-tasty omelet has a gentle slope and small rupture load.

On the other hand, for the stage of a plunger's plunge (area B) after the omelet's surface is broken, or after area A, the handmade or regarded-as-tasty omelet has a small plunge load and relatively flat curve. This area is considered to represent textures such as a "feel of chewing" or "feel of crushing."

2. Calculation of SN Ratio and Sensitivity Using Texture Data

For area A in Figure 2 we selected displacement M (0.05, 0.10, ... , 0.25 cm) as a signal factor. For area B, instead of setting up signal factors, we took advantage of a nominal-the-best characteristic. We set cooling time in hours to a noise factor ($N_1 = 24$, $N_2 = 48$), the wall thickness of the plunger in centimeters as a measurement jig was chosen as an indicative factor ($P_1 = 0.25$, $P_2 = 0.50$, $P_3 = 0.75$).

On the basis of the load–displacement curve obtained from the plunge test, for the analysis of area A we read a load value for each signal. For area B we read minimum and maximum loads in the stage of plunge after the surface was broken. As a result, we obtained the data shown in Tables 1 and 2.

Calculation for Area A (Experiment 1)

Since the wall thickness of the plunger is an indicative factor, we did not include it as a noise factor. However, its interaction with the thickness was regarded as noise.

Total variation:

$$S_T = 9194^2 + \cdots + 63{,}439^2$$

$$= 60{,}665{,}521{,}314 \qquad (f = 30) \qquad (1)$$

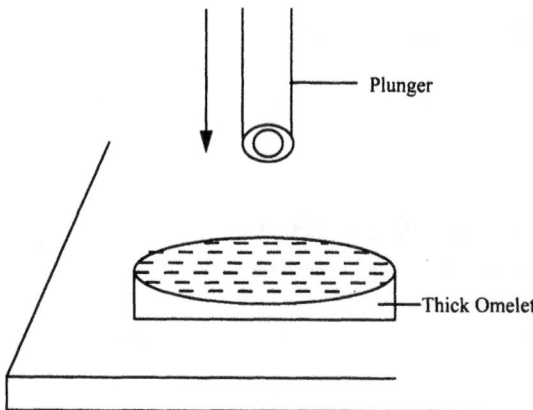

Figure 1
Measurement method for texture of omelet

Effective divider:

$$r = 0.05^2 + \cdots + 0.25^2 = 0.1375 \qquad (2)$$

Linear equation:

$$L_1 = (0.05)(9194) + \cdots + (0.25)(68,710)$$

$$= 36,384$$

$$\vdots$$

$$L_6 = (0.05)(3371) + \cdots + (0.25)(63,439)$$

$$= 30,662 \qquad (3)$$

Variation of proportional terms:

$$S_\beta = \frac{(L_1 + L_2 + \cdots + L_6)^2}{6r}$$

$$= 59,129,381,251 \qquad (f = 1) \qquad (4)$$

Variation of differences of proportional terms:

$$S_{N\beta} = \frac{(L_1 + L_3 + L_5)^2 + (L_2 + L_4 + L_6)^2}{3r} - S_\beta$$

$$= 13,926,068 \qquad (f = 1) \qquad (5)$$

$$S_{P\beta} = \frac{(L_1 + L_2)^2 + (L_3 + L_4)^2 + (L_5 + L_6)^2}{2r} - S_\beta$$

$$= 574,278,389 \qquad (f = 2) \qquad (6)$$

Error variation:

$$S_e = S_T - S_\beta - S_{N\beta} - S_{P\beta}$$

$$= 947,935,606 \qquad (f = 26) \qquad (7)$$

Error variance:

$$V_e = \frac{S_e}{26} = 36,459,062 \qquad (8)$$

Total error variance:

$$V_N = \frac{S_{N\beta} + S_e}{27} = 35,624,506 \qquad (9)$$

SN ratio:

$$\eta_1 = 10 \log \frac{(1/6r)(S_\beta - V_e)}{V_N} = 33.03 \text{ dB} \qquad (10)$$

Sensitivity:

$$S_1 = 10 \log \frac{1}{6r}(S_\beta - V_e) = 108.55 \text{ dB} \qquad (11)$$

Calculation for Area B (Experiment 1)

Again for this case, since the variation between each plunger naturally exists, we did not include it as a noise factor.

Total variation:

$$S_T = 102,728^2 + \cdots + 150,599^2$$

$$= 221,531,535,262 \qquad (f = 12) \qquad (12)$$

Variation of general mean:

$$S_m = \frac{(102,728 + \cdots + 150,599)^2}{(3)(4)}$$

$$= 213,712,627,107 \qquad (f = 1) \qquad (13)$$

y's in Table 2:

$$y_1 = 102,728 + \cdots + 126,572 = 451,551$$

$$y_2 = 130,004 + \cdots + 139,750 = 520,629$$

$$y_3 = 159,609 + \cdots + 150,599 = 629,242 \qquad (14)$$

Variation between each plunger:

$$S_p = \frac{y_1^2 + y_2^2 + y_3^2}{4} - S_m = 4,011,867,234 \qquad (f = 2)$$

$$\qquad (15)$$

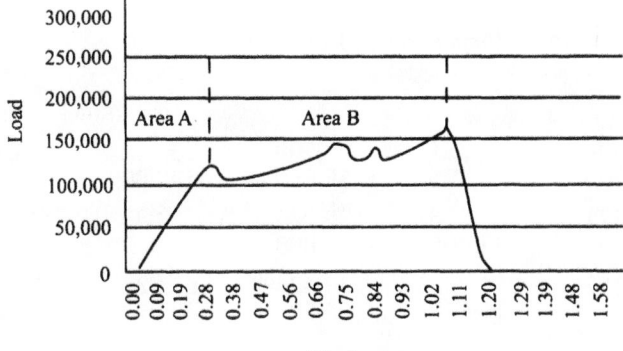

(a) Worst (estimation): 0.25 cm (24 hours after made)

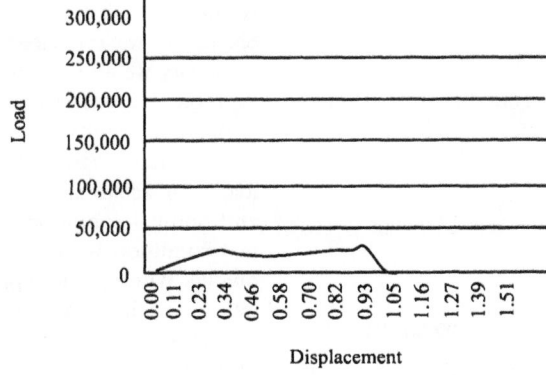

(b) Tayaki hondama: 0.25 cm (handmade)

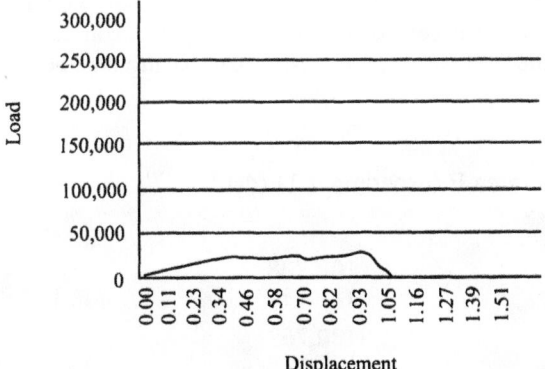

(c) Double omelet by Company A: 0.25 cm (rumored to be tasty)

Figure 2
Comparative measurement of omelet

Table 1
Measured texture data for area A (experiment 1) (dyn)

		M_1	M_2	M_3	M_4	M_5	Linear Equation
P_1	N_1	9,194	23,721	38,309	53,142	68,710	L_1
	N_2	9,133	24,395	38,860	53,080	67,484	L_2
P_2	N_1	7,171	22,066	38,983	55,839	72,572	L_3
	N_2	15,323	33,405	49,893	65,952	81,888	L_4
P_3	N_1	4,107	14,098	32,424	52,038	70,978	L_5
	N_2	3,371	11,830	28,379	45,970	63,439	L_6

Error variation:

$$S_e = S_T - S_m - S_p = 3{,}807{,}040{,}921 \qquad (f = 9) \tag{16}$$

Error variance:

$$V_e = \frac{S_e}{9} = 423{,}004{,}547 \tag{17}$$

SN ratio:

$$\eta_2 = 10 \log \frac{\frac{1}{12}(S_m - V_e)}{V_e} = 16.23 \text{ dB} \tag{18}$$

Sensitivity:

$$S_2 = 10 \log \tfrac{1}{12}(S_m - V_e) = 102.50 \text{ dB} \tag{19}$$

3. Optimal Condition and Confirmatory Experiment

As shown in Table 3, we selected control factors from the mixing, baking, and sterilization processes.

Figure 3 shows the response graphs for the SN ratio and sensitivity. Although there exist V-shapes and peaks for some particular factors in the SN ratio response graph for areas A and B, by adding both SN ratios, we evaluated each factor. Similarly, for sensitivity, we added both values.

Table 4 shows estimations and confirmatory experimental results for the SN ratio and sensitivity. In addition, Figure 4 illustrates a part of the load–displacement curve under current (level 2) and optimal conditions. The curve under the optimal condition is similar to that for the handmade omelet. Whereas the reproducibility for the SN ratio was poor, that for sensitivity was fairly good. Although the reason for this tendency is not clear, we note that there exist haphazard causes for the trends related to noise and indicative factors according to the data in Tables 1 and 2. Therefore, for example, an analysis of each categorized data might increase the reliability of error variance.

While we conducted this experiment with a constant baking time because we could not change the

Table 2
Measured texture data for area B (experiment 1) (dyn)

		P_1	P_2	P_3
N_1	min.	102,728	130,004	159,609
	max.	121,484	158,015	181,613
N_2	min.	100,767	92,860	137,421
	max.	126,572	139,750	150,599
Total		y_1	y_2	y_3

Table 3
Control factors and levels

	Control Factor	Level 1	Level 2	Level 3
A:	mixing during baking	Yes	No	—
B:	egg stirring time (min)	0	15	30
C:	ratio of frozen egg (%)	0	50	100
D:	amount of starch (%)	0	3	5
E:	amount of water (%)	0	10	20
F:	initial temperature (pot)	Low	Mid	High
G*: flow of gas	when F_1	Low	Mid	High
	when F_2	Low	Mid	High
	when F_3	Low	Mid	High
H:	sterilization time (h)	0	0.5	1

*Sliding-level technique is used for defining levels of G.

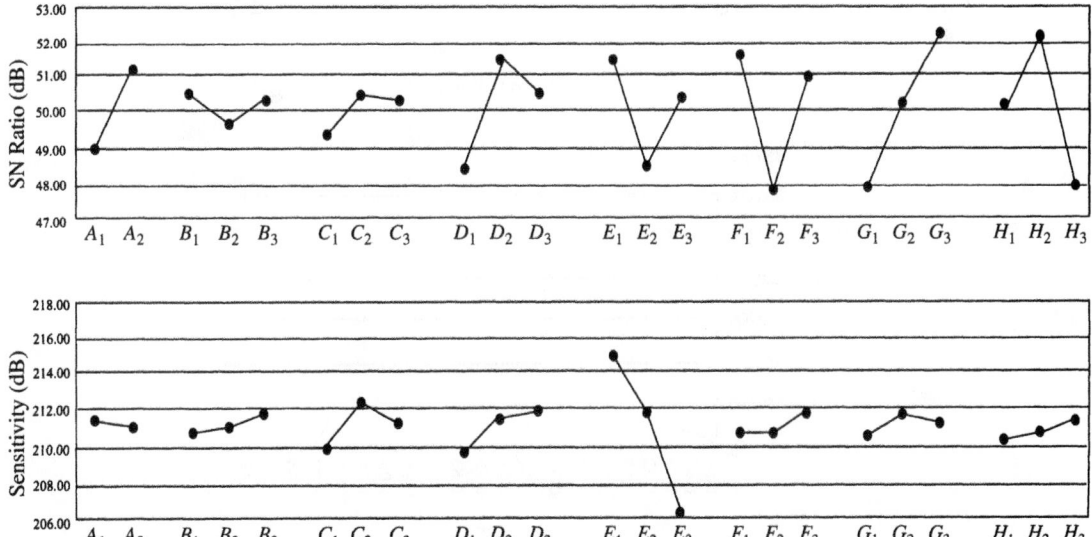

Figure 3
Response graphs

Table 4
Estimation and confirmation of gain (dB)

| Condition | SN Ratio | | Sensitivity | |
	Estimation	Confirmation	Estimation	Confirmation
Optimal	51.10	51.90	213.13	212.41
Initial	58.08	54.60	204.84	203.35
Gain	6.98	2.70	−8.29	−9.06

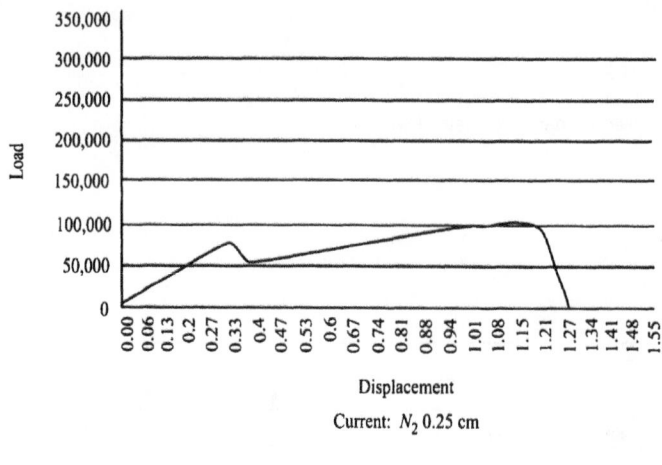

Current: N_2 0.25 cm

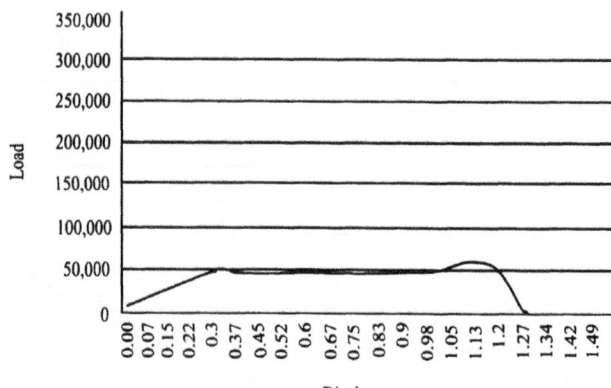

Optimal: N_2 0.25 cm

Figure 4
Measured data under current and optimal conditions

production speed materially in our manufacturing process, in the future we intend to attempt to reduce the number of V-shapes and peaks and improve the production speed, regarded as one of the objectives of quality engineering, by implementing an experiment based on heat energy.

Reference

Toshio Kanatsuki, 2000. Improvement of the taste of thick-baked eggs. *Quality Engineering*, Vol. 8, No. 4, pp. 23–30.

This case study is contributed by Toshio Kanetsuki.

CASE 65

Wiper System Chatter Reduction

Abstract: The wiper arm/blade forms an integral part of a windshield wiper system. *Chatter* occurs when the wiper blade does not take the proper set and "skips" across the windshield during operation, causing both acoustic and wipe quality deterioration, which significantly affects customer satisfaction. Instead of measuring wiper quality deterioration, a robust design approach was used by evaluating the actual time that a blade takes to reach a fixed point on the windshield at a given cycle. That means that the focus was on measuring the functionality of the system and not the end characteristics. Nine control factors were used under all categories of noise factors. Eighteen design configurations were tested in this study.

1. Introduction

Windshield Wiper System

A windshield wiper system is composed of four major subsystems: (1) the wiper motor, (2) the wiper linkage assembly, (3) the pivot towers, and (4) the wiper arm/blade combination. The wiper motor provides driving power to the system. The wiper linkage assembly (usually, a series- or parallel-coupled four-bar mechanism) is a mechanical transmission configuration to transfer the driving power from the motor to the pivots and to change the rotary motion of the motor to an oscillating motion at the pivots. The pivot towers provide a firm anchor from which the pivots operate, holding the wiper pivots in proper orientation with the windshield surface. When the arm/blade combinations are attached to the pivot and the system is cycled, the blades clear two arc-shaped wipe patterns on the windshield.

The customer expectations for a good windshield wiper system are:

- ❏ Clear vision under a wide variety of operating conditions
- ❏ Multiple levels of wiper speed
- ❏ Uniform wiper pattern
- ❏ Quiet under all weather conditions
- ❏ Long life and high reliability

Definition of Chatter

Chatter occurs when a wiper blade does not take the proper set and skips across the windshield during operation, causing both acoustic and wipe quality deterioration, which significantly affect customer satisfaction. From the results of previous studies it was determined that the components that contribute most to chatter are the arms and blades. This study was conducted to understand more thoroughly the specific design factors relating to arm/blade combinations that affect system performance.

2. Description of the Experiment

Engineered System

Control and noise factors were selected based on knowledge of chatter. The engineered system is shown in Figure 1.

Ideal Function

The ideal function for this study was based on the following hypothesis: For an ideal system, the actual

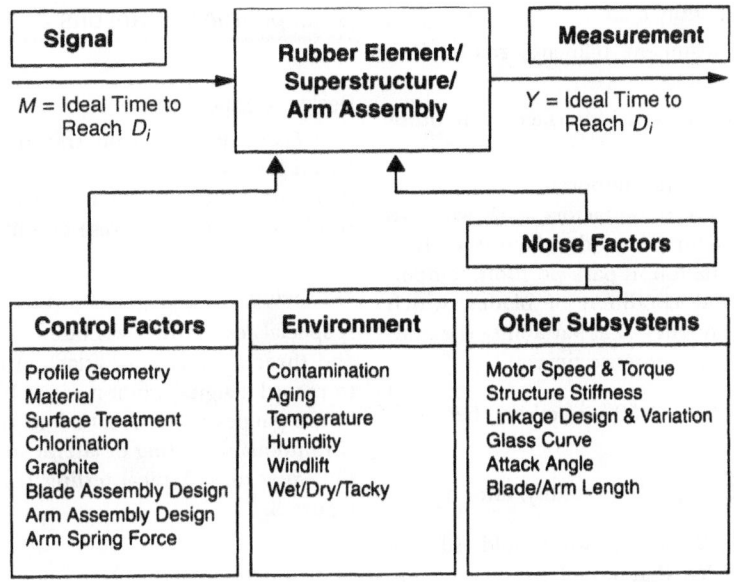

Figure 1
Engineered system

time for the wiper blades to reach a fixed point on the windshield during a cycle should be the same as the theoretical time (ideal time) for which the system was designed. Under the presence of the noise factors, the actual time will differ from the theoretical time. Furthermore, the actual time of a robust system under varying noise conditions should have less variation and be closer to the theoretical

time. Therefore, the system ideal function is given by

$$Y_i = \beta M_i$$

where Y_i is the actual time it takes the blade to reach a fixed point on the windshield at the ith cycle, M_i the theoretical time for the blade to reach a fixed point on the windshield at the ith cycle, and $\beta = i/$ motor rpm. The ideal system function is shown in Figure 2.

Three rpm values were used in the experiment: 40, 55, and 70. In general, due to the noise effects, the actual time would always be longer than the theoretical time determined by the design (i.e., $b > 1$, and $b = 1$ would be the ideal case).

In this experiment, other measurements were also observed: (1) the lateral load on the wiper arm and (2) the normal load on the wiper arm.

Noise Strategy

In general, noises can be categorized as follows:

1. Piece-to-piece variation (manufacturer, material variation, etc).

2. Change in dimension due to wear, fatigue, and so on

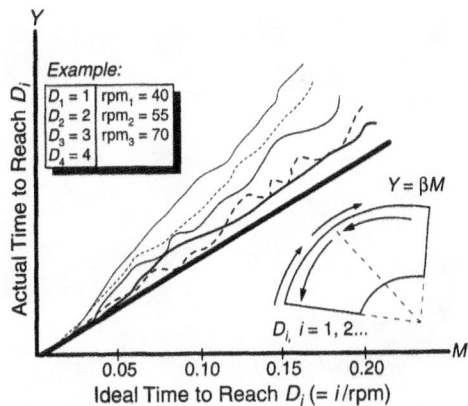

Figure 2
Ideal function

3. Customer use/duty cycle

4. External environment (climate, road, wind, temperature, etc).

5. Internal environment (interface with other body systems)

In this experiment, the important noise factors were grouped into two noise factors, each with two levels: (1) Noise factor S included factors such as wiper angular orientation in park position, temperature, humidity, and condition of blades, which would produce a tendency to ac variability (i.e., variation about the mean response time):

$S_1 = -30$ at park/200°C/50% humidity/before aging

$S_2 = 10$ at park/20°C/90% humidity/after aging

(2) Noise factor T is the windshield surface condition, wet or dry, that would produce a tendency to dc variability in the response time (i.e., shifting the mean):

$$T_1 = \text{wet surface condition}$$

$$T_2 = \text{dry surface condition}$$

The Control Factors

The control factors and their levels (Table 1) were selected through team brainstorming.

3. Experimental Results and Data Analysis

Design of Experiment

The L_{18} orthogonal array shown in Table 2 was selected for the design of experiment. It may be seen that under each run there are 12 tests at different rpm's and combined noise conditions.

Test Setup

A special test fixture was built for this experiment, and three sensors were attached to the windshield to record a signal when a wiper blade passes them. Strain gauges were attached to the wiper arm for continuous recording of lateral and normal loads to the wiper arm. Typical testing results are shown in Figure 3.

Test Results

In this study, only the analysis of time measurement data is presented. Under each test, the system was run for a certain period of operating time, and the actual accumulated operating times for the blade to pass a fixed point were recorded against the ideal operating time, determined by the nominal rpm setting of the constant-rpm motor. A testing data plot for this measurement is shown in Figure 4.

The regression analyses were conducted for those data and the slopes of the regression lines

Table 1
Control factors and levels

	Control Factor	1	2	3
			Level	
A:	arm lateral rigidity design	1	2	3
B:	superstructure rigidity	Median	High	Low
C:	vertebra shape	Straight	Concave	Convex
D:	spring force	Low	High	—
E:	profile geometry	1	2	New design
F:	rubber material	I	II	III
G:	graphite	Current	Higher graphite	None
H:	chlorination	Median	High	Low
I:	attachment method	Current clip	Modified	Twin screw

Table 2
Design of experiment

Run	D	A	B	C	E	F	G	H	I	rpm 1				rpm 2				rpm 3			
										S_1		S_2		S_1		S_2		S_1		S_2	
			Factor							Wet	Dry	Wet	Dry	Wet	Dry	Wet	Dry	Wet	Dry	Wet	Dry
1	1	1	1	1	1	1	1	1	1												
2	1	1	2	2	2	2	2	2	1												
3	1	1	3	3	3	3	3	3	1												
4	1	2	1	1	2	2	3	3	2												
5	1	2	2	2	3	3	1	1	2												
6	1	2	3	3	1	1	2	2	2												
7	1	3	1	2	1	3	2	3	3												
8	1	3	2	3	2	1	3	1	3												
9	1	3	3	1	3	2	1	2	3												
10	2	1	2	3	3	2	2	1	2												
11	2	1	2	1	1	3	3	2	2												
12	2	1	3	2	2	1	1	3	2												
13	2	2	1	2	3	1	3	2	3												
14	2	2	2	3	1	2	1	3	3												
15	2	2	3	1	2	3	2	1	3												
16	2	3	1	3	2	3	1	2	1												
17	2	3	2	1	3	1	2	3	1												
18	2	3	3	2	1	2	3	1	1												

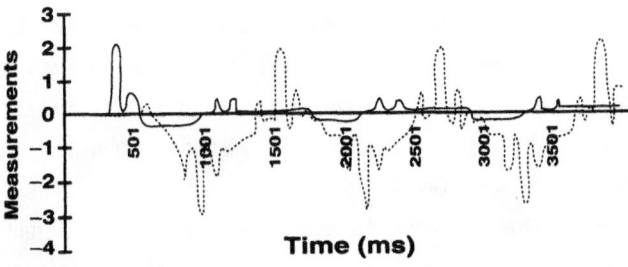

Figure 3
Plot of a typical testing data

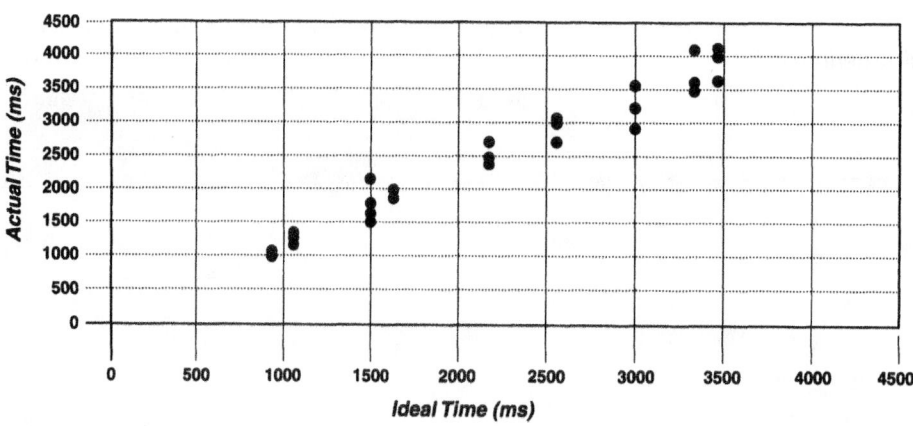

Figure 4
Typical testing data plot

Table 3
SN ratio and β value for each run

					Factor						
Run	D	A	B	C	E	F	G	H	I	SN	β
1	Low	Design 1	Median	Straight	Geo 1	Material I	Current	Median	Current	−34.27	1.078
2	Low	Design 1	High	concave	Geo 2	Material II	Higher	High	Current	−34.41	1.089
3	Low	Design 2	Low	Convex	New	Material III	None	Low	Current	−44.27	1.133
4	Low	Design 2	Median	Straight	Geo 2	Material II	None	Low	Modified	−37.27	1.096
5	Low	Design 3	High	Concave	New	Material III	Current	Median	Modified	−30.81	1.073
6	Low	Design 3	Low	Convex	Geo 1	Material I	Higher	High	Modified	−32.80	1.093
7	Low	Design 3	Median	Concave	Geo 1	Material III	Higher	Low	Twin screw	−34.65	1.076
8	low	Design 3	High	Convex	Geo 2	Material I	None	Median	Twin screw	−38.89	1.094
9	Low	Design 3	Low	Straight	New	Material II	Current	High	Twin screw	−33.39	1.65
10	High	Design 1	Median	Convex	New	Material II	Higher	Median	Modified	−30.46	1.063
11	High	Design 1	High	Straight	Geo 1	Material III	None	High	Modified	−35.24	1.079
12	High	Design 1	Low	Concave	Geo 2	Material I	Current	Low	Modified	−42.59	1.096
13	High	Design 2	Median	Concave	New	Material I	None	High	Twin screw	−30.54	1.075
14	High	Design 2	High	Convex	Geo 1	Material II	Current	Low	Twin screw	−32.52	1.068
15	High	Design 2	Low	Straight	Geo 2	Material III	Higher	Median	Twin screw	−32.70	1.065
16	High	Design 3	Median	Convex	Geo 2	Material III	Current	High	Current	−31.00	1.065
17	High	Design 3	High	Straight	New	Material I	Higher	Low	Current	−33.67	1.075
18	High	Design 3	Low	Concave	Geo 1	Material II	None	Median	Current	−34.18	1.079

were estimated; the mean square was used to estimate the variation. In Table 3, the SN ratio and β values are evaluated by

$$\text{SN ratio} = 10 \log_{10} \frac{\beta^2}{\sigma^2}$$

where β is the slope of regression line (forced through zero) and σ is the square root of the mean-squared deviation.

4. Data Analysis and the Optimal Condition

The SN ratios and β values given in the last two columns of Table 3 were analyzed further to determine the significance of the control factors and to select the best levels.

SN Ratio

A plot of the SN ratio against the control factors and their levels is shown in Figure 5. From the plot it can be seen that different levels of arm rigidity, superstructure rigidity, graphite, and chlorination yield quite different results, whereas load distribution has a relatively small effect on the SN ratio.

Beta Value

From Figure 6 it may be observed that in contrast to the plot of the SN ratio, where the differences in the impact of the control factor levels are large, differences in the impact of the control factor levels on the β value are relatively small.

Optimal Condition

The optimal condition was determined using two-step optimization, in which selection of the control factor combination was carried out by (1) maximizing the SN ratio and (2) having a β value close to 1.00.

Since the objective was to have minimum variability in motion and minimum difference between actual and theoretical times for the wiper blades to reach a fixed point on the windshield in each cycle, maximizing the SN ratio was the top priority.

Table 4 gives the optimal condition and baseline design. The values predicted for the SN ratio and for β are also shown in Table 4. It may be seen that the SN ratio of the optimal condition has an approximate 10-dB increase over the baseline.

The predicted values shown in Table 4 were calculated as follows.

SN ratio of optimal
$$= -[34.647 + (32.77 - 34.647)$$
$$+ (33.03 - 34.647) + (34.42 - 34.647)$$
$$+ (33.66 - 34.647) + (33.86 - 34.647)$$
$$+ (33.70 - 34.647) + (33.11 - 34.647)$$
$$+ (33.55 - 34.647) + (33.78 - 34.647)]$$
$$= -24.71$$

SN ratio of baseline
$$= - [34.647 + (36.87 - 34.647)$$
$$+ (33.03 - 34.647) + (34.42 - 34.647)$$
$$+ (35.64 - 34.647) + (33.94 - 34.647)$$
$$+ (35.46 - 34.647) + (34.10 - 34.647)$$
$$+ (33.55 - 34.647) + (35.30 - 34.647)]$$
$$= -35.13$$

β value of optimal
$$= [1.08249 + (1.0783 - 1.08249)$$
$$+ (1.0792 - 1.08249) + (1.0763 - 1.08249)$$
$$+ (1.0765 - 1.08249) + (1.0807 \ 1.08249)$$
$$+ (1.0767 - 1.08249) + (1.U768 - 1.08249)$$
$$+ (1.0752 - 1.08249) + (1.0737 - 1.08249)]$$
$$= 1.033$$

β value of baseline
$$= [1.08249 + (1.0898 - 1.08249)$$
$$+ (1.0792 - 1.08249) + (1.0763 - 1.08249)$$
$$+ (1.0885 - 1.08249) + (1.0788 - 1.08249)$$
$$+ (1.0852 - 1.08249) + (1.0780 - 1.08249)$$
$$+ (1.0752 - 1.08249) + (1.0904 - 1.08249)]$$
$$= 1.082$$

5. Confirmation Tests and Results

The confirmation test plan is as follows:

Baseline configuration: $A_1B_1C_1D_1E_1F_1G_1H_1I_1$

Optimal configuration: $A_2B_1C_1D_2E_3F_2G_2H_2I_3$

See Table 5.

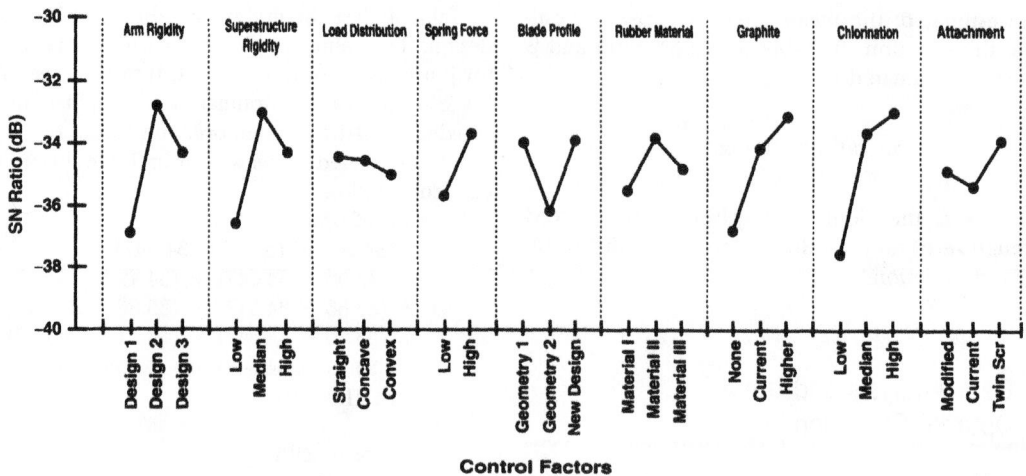

Figure 5
SN ratio chart

6. Conclusions

This robust design study for a windshield wiper system indicates that the chlorination, graphite, arm rigidity, and superstructure rigidity have significant impacts on the optimal wiper system. Load distribution on the blade has a minimal influence. As a

result, low friction and high rigidity of the wiper arm and blade will lead to a more robust windshield wiper system.

Acknowledgments The authors wish to thank John King, John Coleman, Dr. Ken Morman, Henry Kopickpo, and Mark Garaseia of Ford Motor Company,

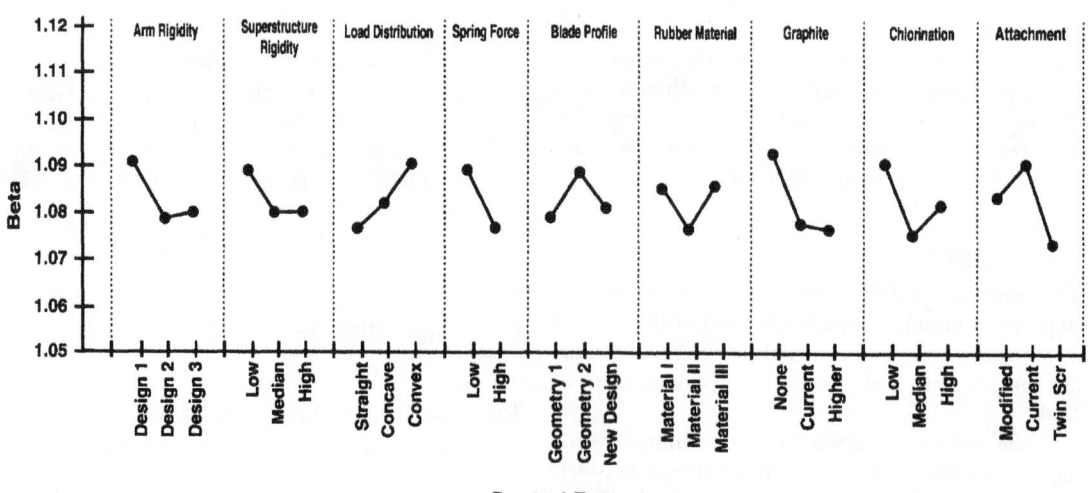

Figure 6
β value chart

Table 4
Optimal conditions and prediction

	Arm Rigidity	Super-structure	Load Distribution	Spring Force	Blade Profile	Rubber Material	Graphite	Chlorination	Attachment	SN	β
Optimal	Design 2	Median	Straight	High	New design	Material 2	Higher	Median	Twin screw	−24.71	1.033
Baseline	Design 1	Median	Straight	Low	Geo 1	Material 1	Current	Median	Current	−35.13	1.082

Table 5
Confirmation test results

Condition	Predicted		Confirmed	
	SN Ratio	β	SN Ratio	β
Optimal	−24.71	1.033	−25.41	1.011
Baseline	−35.13	1.082	−36.81	1.01
Gain	10.42	—	11.4	—

and Brian Thomas and Don Hutter for their support during the course of this work.

This case study is contributed by Michael Deng, Dingjun Li, and Wesley Szpunar.

Fabrication Line Capacity Planning Using a Robust Design Dynamic Model

Abstract: We needed to expand a fabrication facility to increase production capacity. To clearly understand the various expansion alternatives, we evaluated the performance of fabrication line configurations according to how closely they followed the ideal dynamic relationship:

$$Y = (\text{lot start factor} + \text{workweek factor} \times M_2)M_1$$

where Y is the product ship rate (wafers/day), M_1 the lot start rate (lot starts/week), and M_2 the scheduled workweek (workdays/week).

We set up fabrication line simulations according to an L_{36} ($2^{11} \times 3^{12}$) inner array for configuration factors. We identified noise factors to evaluate robustness to workforce fluctuations, equipment availability, and operation protocols, and then selected worst- and best-case combinations. We simulated each combination of line configuration factors twice at each noise factor case for three workweek schedules and five lotstart rates. We determined an optimum line configuration by comparing ANOVA results for the lot start factor sensitivity and signal-to-noise (SN) ratio, and the workweek factor sensitivity and SN ratio calculated for each inner array row. We ran confirmation simulations for the optimum and a second (selected) configuration; the results agreed with linear predictions.

1. Introduction

We established a fabrication facility for accelerometer sensors, evolving from pilot line to moderate production levels. We now required a substantial increase in production capacity. To help define the equipment and workforce additions needed to maintain the target production levels, we used a simulation software package to model the fabrication line and robust design techniques to find an optimum stable configuration.

We developed a model of the present fabrication line and validated its behavior against our existing production history. We then developed a baseline expansion configuration using conventional capacity planning methods with our line production history data, tempered by space and budget constraints. We included the baseline as the initial (all level 1) control factor combination in this experiment.

Monitoring the stability of the sensor fabrication line (or a faithful simulation) involves tracking work-in-progress and scrap profiles, cycle and queue times, and other parameters. We defined an ideal function for the line that allowed more efficient evaluation of stability and capacity by focusing on one response parameter.

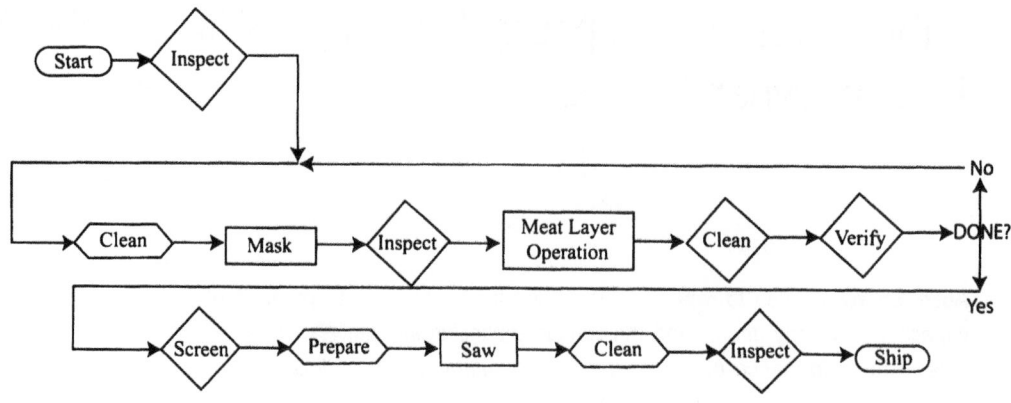

Figure 1
Flow diagram

2. Fabrication Line Model

The fabrication line model was constructed using Taylor II for Windows 4.20 [1], a menu-driven discrete-event simulation package. The Taylor II software features include predefined basic element types, integrated runtime animation, user-defined functions, and batch run mode.

The model has the following basic characteristics:

❏ The model was expanded from that implemented in an earlier study [2].

❏ Each in-line piece of equipment was represented; each task performed with the equipment was defined.

❏ Cycle times and variations were derived from operation standards and production history data.

❏ Conservative line loss (scrap) was modeled.

❏ Unit yield per wafer was considered constant.

❏ Automated operations were modeled in three parts: operator-assisted load, untended operation, and operator-assisted unload.

❏ Extra preparation time was included at the beginning of each process module.

❏ No mask step rework was included.

❏ End-of-shift protocols, operator task sharing, and maintenance events were not modeled directly. For simplicity, these were included as workforce and equipment availability variations combined into the outer array noise.

The process flow for the simulation model (Figure 1) was derived from the product traveler (the formal step-by-step definition of the process sequence). The accelerometer sensor fabrication process resembles semiconductor wafer fabrication in that many workstations (e.g., cleaning, masking, and inspection) are visited multiple times by a product wafer as it is processed to completion. The product wafers start in lots processed together until the structures are fully formed. They are then regrouped into smaller batches for the remaining "tail-end" processing.

The key response parameter of a fabrication line is the number of products shipped. The number of products shipped will ideally be proportional to the number of product lot starts for at least light-to-moderate production levels. The degree to which the product ship rate deviates from this proportionality defines the basic line stability. The onset of line saturation, when further increases in the rate of lot starts do not increase the rate of products shipped, defines line capacity. Therefore, we selected the lot start rate as the primary signal factor (M_1) for the fabrication ideal function.

We selected the strategy of changing the workweek schedule to adjust line capacity while keeping workforce and equipment constant. We included days per workweek as the second signal factor (M_2) to evaluate its effectiveness and sensitivity.

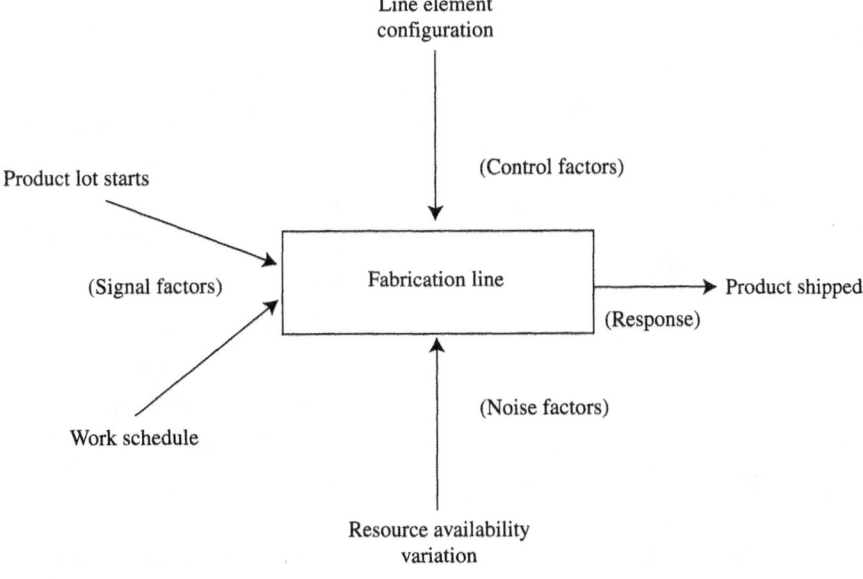

Figure 2
P-diagram

3. Design of Experiment

Figure 2 shows the P-diagram for the fabrication line ideal function. The zero-point proportional double dynamic model was selected to establish the product ship rate dependence on lot start rate (start none–ship none). The effect of varying the sched-uled workweek was included as a modifier to the lot start rate coefficient in the form

$$Y = \beta_1 + \beta_2 \ (M_2 - M_{20}) M_1$$

where Y is the product ship rate (wafers/day), M_1 the lot start rate (lots/week), M_2 the scheduled workweek (workdays/week), M_{20} the nominal

Table 1
Signal and noise factors

			Level	
Factor		**Description**	**Value**	**Code**
M_1:	lot start rate (lots/week)	Number of product wafer lots started per week	2 3 4 5 6	2 3 4 5 6
M_2:	scheduled workweek (days/week)	Number of days per week scheduled for production shifts	5 6 7	−1 0 1
N:	noise	Combined variation in the availability of inspection stations, steppers, saws, and operators	Worst Best	−1 1

Table 2
Control factors

	Factor	Level 1	Level 2	Level 3	Description
A:	meas	Baseline	Added	—	Number of verification measurement stations
B:	spcr1	Baseline	Added	—	Number of spcr1 stations
C:	spcr2	Baseline	Added	—	Number of spcr2 station
D:	etch 1	Baseline	Added	—	Number of etch1 stations
E:	etch2	Baseline	Added	—	Number of etch2 stations
F:	snsr1	Baseline	Added	—	Number of snsr1 stations
G:	snsr2	Baseline	Added	—	Number of snsr2 stations
H:	PC_cln	Baseline	Added	—	Number of pc cleaning baths
I:	D_cln	Baseline	Added	—	Number of d cleaning baths
J:	B_cln	Baseline	Added	—	Number of b cleaning baths
K:	oven	Baseline	Added	—	Number of curing/drying ovens
L:	metal_dep	A	B	C	Number and type of metal deposition systems
M:	coater	1+	2	3	Number of front-end coaters
N:	process	Baseline	Improved	Improved HD	Production process improvement strategies
O:	stepper	1	1+	2	Number of active steppers
P:	develop	1	2	3	Number of develop baths
Q:	descum	2	3	4	Number of descum units
R:	screen	2/1	2+/1+	3/2	Number of screening stations
S:	saw	1	2	3	Number of saws
T:	batch	s	m	1	Tail-end batch size
U:	staff	Base	Base+1	Base+2	Workforce staffing level
V:	empty				
W:	empty				

workweek (6 workdays/week), and β_1 and β_2 are sensitivity coefficients.

The noise cases were generated by combining values for the number of inspection stations, specific equipment availability, process module preparation time, and absenteeism, which were all "lean" for the worst case and all "augmented" for the best case. Each noise case was run twice to sample two simulation random number generator sequences.

The signal and noise factors are described in Table 1. Fully crossing the two signal factors with the noise cases required that 60 simulations be evaluated for each line configuration.

The inner array factors are described in Table 2. Control factors A to K, M, and O to S represent fabrication line elements viewed as determiners of production capacity. They were evaluated with baseline values as level 1 and augmented values as level 2 (and 3). Three strategies involving number and type of metal deposition systems used in the process were represented in factor L. Two process improvement strategies were compared with the current

Table 3

Inner array (L_{36}) response summary

No.	A	B	C	D	E	F	G	H	I	J	K	L	M	N	O	P	Q	R	S	T	U	V	W	β_1	SN$_1$ (dB)	β_2	SN$_2$ (dB)
1	1	1	1	1	1	1	1	1	1	1	1	1	1	1	1	1	1	1	1	1	1	1	1	1.314	−3.98	0.186	−20.95
2	1	1	1	1	1	1	1	1	1	1	2	2	2	2	2	2	2	2	2	2	2	2	2	1.580	0.22	0.239	−16.38
3	1	1	1	1	1	1	1	1	1	1	3	3	3	3	3	3	3	3	3	3	3	3	3	1.602	1.48	0.220	−15.75
4	1	1	1	1	1	1	2	2	2	2	1	1	1	1	2	2	2	3	3	3	3	3	3	1.629	2.50	0.212	−15.21
5	1	1	1	1	1	1	2	2	2	2	2	2	2	2	3	3	3	1	1	1	3	3	3	1.329	−3.11	0.220	−18.72
6	1	1	1	1	1	2	2	2	2	2	3	3	3	3	1	1	1	2	2	2	3	3	3	1.472	−1.03	0.227	−17.27
7	1	1	2	2	2	1	1	1	2	2	1	1	2	3	1	2	3	3	1	2	2	2	3	1.73	0.39	0.254	−15.97
8	1	1	2	2	2	1	1	1	2	2	2	2	3	1	2	3	1	1	2	3	3	3	1	1.584	0.58	0.219	−16.59
9	1	1	2	2	2	1	1	1	2	2	3	3	1	2	3	1	2	2	3	1	1	1	2	1.265	−2.02	0.220	−17.22
10	1	2	1	2	2	1	2	2	1	1	1	1	3	2	1	3	2	3	2	1	3	3	2	1.300	−2.10	0.213	−17.80
11	1	2	1	2	2	1	2	2	1	1	2	2	1	3	2	1	3	1	3	2	1	1	3	1.406	−1.07	0.226	−16.96
12	1	2	1	2	2	1	2	2	1	1	3	3	2	1	3	2	1	2	1	3	2	2	1	1.811	2.06	0.247	−15.25
13	1	2	2	1	2	2	1	2	1	2	1	2	3	3	1	1	3	2	1	3	1	2	2	1.430	0.06	0.228	−15.88
14	1	2	2	1	2	2	1	2	1	2	2	3	1	1	2	2	1	3	2	1	2	3	3	1.679	−0.64	0.264	−16.70
15	1	2	2	1	2	2	1	2	1	2	3	1	2	2	3	3	2	1	3	2	3	1	1	1.311	−2.29	0.214	−18.02
16	1	2	2	2	1	2	2	1	2	1	1	2	3	1	3	2	3	2	3	1	2	2	1	1.574	2.46	0.219	−14.66
17	1	2	2	2	1	2	2	1	2	1	2	3	1	2	1	3	1	3	1	2	3	1	2	1.278	−4.16	0.233	−18.94
18	1	2	2	2	1	2	2	1	2	1	3	1	2	3	2	1	2	1	2	3	1	2	3	1.562	0.03	0.237	−16.34

Production Ship Rate

Table 3 (Continued)

No.												Factor														Production Ship Rate		
	A	B	C	D	E	F	G	H	I	J	K	L	M	N	O	P	Q	R	S	T	U	V	W	β_1	SN_1 (dB)	β_2	SN_2 (dB)
19	2	1	2	2	1	1	2	2	1	2	1	2	2	1	3	3	3	1	2	2	1	2	3	1.379	−1.97	0.199	−18.80
20	2	1	2	2	1	1	2	2	1	2	1	2	3	3	1	2	1	2	3	3	2	3	1	1.486	0.76	0.219	−15.86
21	2	1	2	2	1	1	2	2	1	2	1	3	1	2	2	3	2	3	1	1	3	1	2	1.638	−2.43	0.240	−19.12
22	2	1	2	2	1	2	1	1	2	1	2	1	2	3	3	2	3	2	1	1	3	3	2	1.755	0.54	0.244	−16.59
23	2	1	2	2	1	2	1	1	2	1	2	2	3	3	1	1	1	2	2	1	1	3	3	1.300	−2.67	0.235	−17.53
24	2	1	2	2	1	2	1	1	2	1	2	3	1	1	2	2	2	3	3	3	2	1	1	1.394	−3.47	0.239	−18.79
25	2	1	2	1	2	2	2	2	1	1	3	1	1	2	2	3	3	1	3	2	1	2	2	1.238	−2.15	0.197	−18.09
26	2	1	2	1	2	2	2	2	1	1	3	2	2	3	3	1	1	2	1	3	2	3	3	1.476	−3.00	0.260	−18.08
27	2	1	2	1	2	2	2	2	1	1	3	3	3	1	1	2	2	3	2	1	3	1	1	1.573	−2.75	0.260	−18.39
28	2	2	1	2	2	2	2	1	2	1	1	1	1	3	2	2	3	1	3	3	1	3	3	1.592	−2.36	0.249	−18.48
29	2	2	1	2	2	2	2	1	2	1	1	2	2	1	3	3	1	2	1	1	2	1	1	1.180	−3.04	0.200	−18.46
30	2	2	1	2	2	2	2	1	2	1	1	3	3	2	1	1	2	3	2	2	3	2	2	1.717	0.67	0.258	−15.79
31	2	2	1	2	2	1	1	2	1	2	2	1	2	3	3	3	2	1	2	1	1	1	1	1.618	−1.33	0.233	−18.15
32	2	2	1	2	2	1	1	2	1	2	2	2	3	1	1	1	3	2	3	2	2	2	2	1.752	2.52	0.218	−15.59
33	2	2	1	2	2	1	1	2	1	2	2	3	1	2	2	2	1	3	1	3	3	3	3	1.214	−2.84	0.208	−18.16
34	2	2	1	1	1	2	1	1	1	2	1	1	2	3	2	3	2	3	1	2	3	1	1	1.504	−2.82	0.241	−18.763
35	2	2	1	1	1	2	1	1	1	2	1	2	3	1	3	1	3	1	2	3	1	2	2	1.484	−2.14	0.220	−18.74
36	2	2	1	1	1	2	1	1	1	2	1	3	1	2	1	2	1	2	3	1	2	3	3	1.295	−3.54	0.203	−19.62

baseline process using factor N. Factor T was used to evaluate the effect of small, medium, and large batch sizes in the tail-end process on-line capacity and stability. Overall workforce staffing level was varied by factor U. This was done both to evaluate fabrication line sensitivity and to examine the simulation model sensitivity to workforce variations.

We selected an L_{36} orthogonal array [36] to accommodate the 11 two-level and 10 three-level inner array factors, with column assignments as indicated in Table 2.

4. Experimental Procedure

All simulations and analyses were run using Pentium-class personal computers. Experiment definitions and results were performed using Quattro Pro for Windows 5.0 [4], and the model simulations were created and executed using Taylor II software.

We used a previously developed spreadsheet template with associated macro programs to generate the simulation model control factor combinations according to the L_{36} inner array. Each of these 36

combinations was implemented as a separate model by modifying the baseline (row 1) model.

We imposed the outer array conditions using a Taylor II batch run file to set the signal and noise factor levels at the start of each simulation. The 12 outer array conditions for each level of lot start rate (M_1) were simulated and the results extracted as a group. The results for all 60 outer array runs were stored in five separate files for later analysis.

5. Results

The five data files for each inner array condition were combined into one spreadsheet and the dynamic model sensitivities (β_1 and β_2) and signal-to-noise (SN) ratios (SN_1 and SN_2) were computed using the method presented in Section 7 of the Robust Design Workshop handout [5]. The β's and SN ratios calculated for all of the inner array conditions are shown in Table 3. The data for the conditions with best and worst SN_1 are shown in Figure 3 with the dynamic model lines.

The average factorial effects for each performance parameter were calculated and are shown

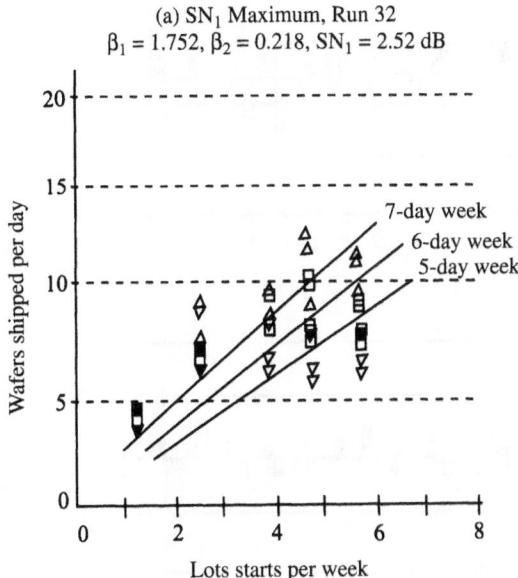

(a) SN_1 Maximum, Run 32
$\beta_1 = 1.752$, $\beta_2 = 0.218$, $SN_1 = 2.52$ dB

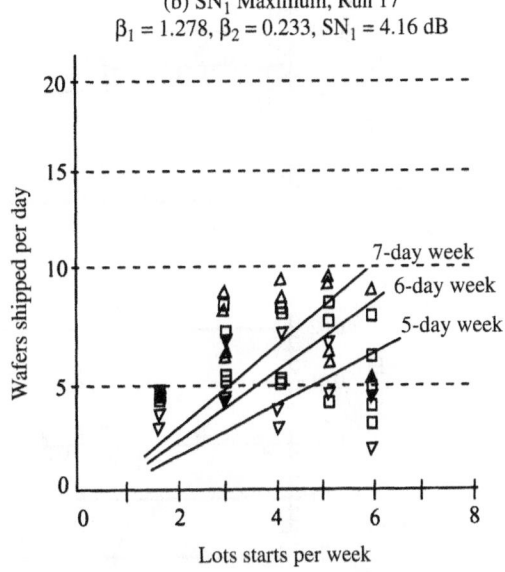

(b) SN_1 Maximum, Run 17
$\beta_1 = 1.278$, $\beta_2 = 0.233$, $SN_1 = 4.16$ dB

Figure 3
Best- and worst-case inner array conditions

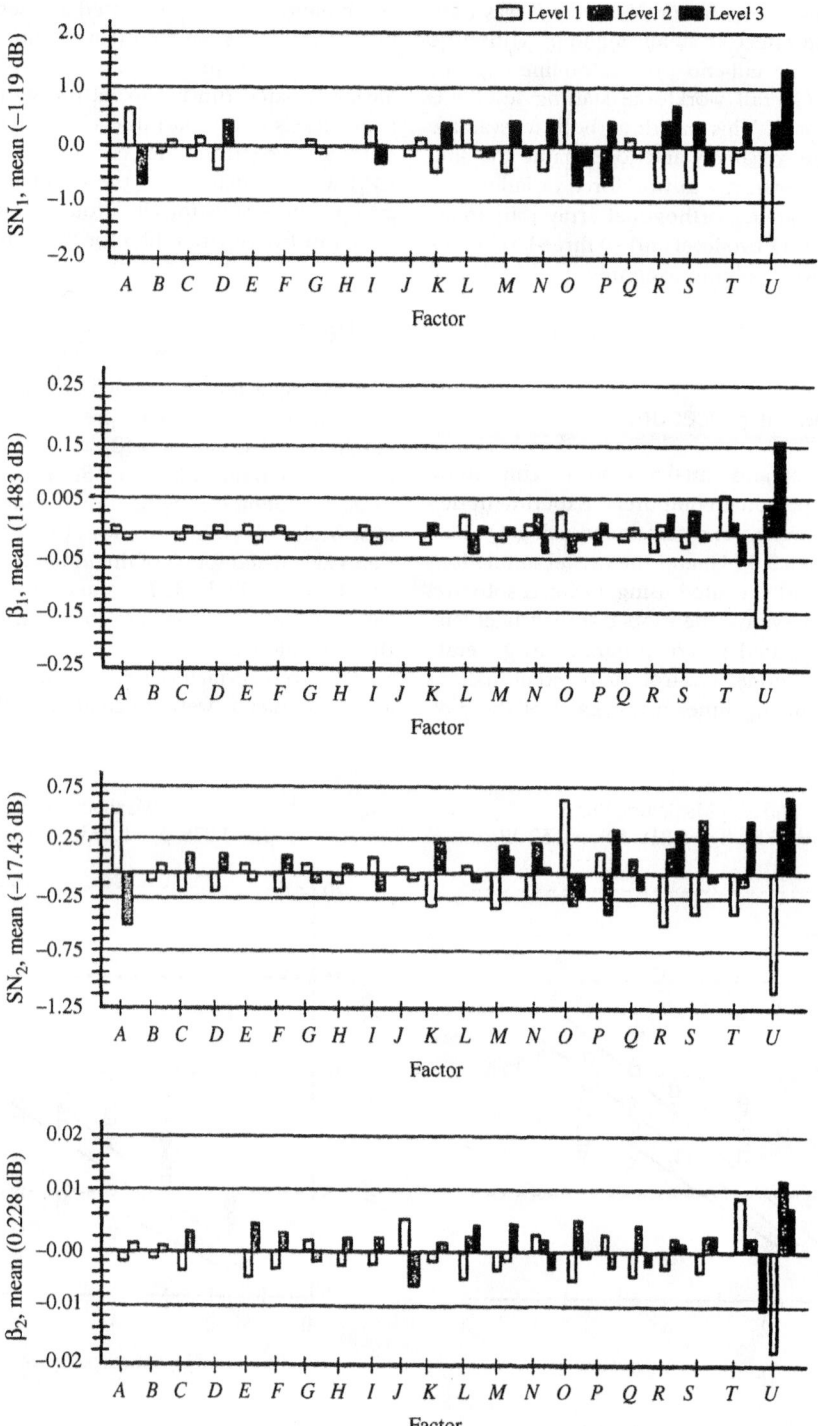

Figure 4
Average factorial effects

Table 4
Summary of inner array response ANOVAs

| | Factor | Response Parameter[a] (%) | | | | Optimization Decision | | |
		SN_1	β_1	SN_2	β_2		Choice	Reason
A:	meas	8.4 (1)	—	1.6%	—	A_1:	baseline	SN 1 and 2
B:	spcr1	—	—	—	—	B_1:	baseline	No effect
C:	spcr2	—	—	1.5 (2)	3.7 (2)	C_2:	added	Beta 2
D:	etch1	0.8 (2)	—	1.2 (2)	—	D_2:	added	SN 1 and 2
E:	etch2	1.7 (1)	0.4 (1)	—	5.3 (2)	E_2:	added	Beta 2
F:	snsr1	—	—	—	1.5 (2)	$F2$:	added	Beta 2
G:	snsr2	—	—	—	—	G_1:	baseline	No effect
H:	PC–cln	—	—	—	—	H_1:	baseline	No effect
I:	D_cln	—	0.7 (1)	—	—	I_1:	baseline	Beta 1
J:	B_cln	—	—	—	2.7 (1)	J_1:	baseline	Beta 2
K:	oven	1.8 (2)	0.8 (2)	3.2 (2)	—	K_2:	added	SN 1 and 2
L:	metal_dep	—	0.6 (1)	—	2.3 (3)	L_3:	c	Beta 2
M:	coater	—	—	—	1.8 (3)	M_3:	3	Beta 2
N:	process	—	1.1 (2)	—	—	N_2:	improved	Beta
O:	stepper	10.5 (1)	1.1 (1)	5.2 (1)	6.3 (2)	O_1:	1	SN 1 and 2
P:	develop	1.7 (3)	0.7 (3)	2.9 (3)	—	P_2:	3	SN 1 and 2
Q:	descum	—	—	—	3.2 (2)	Q_2:	3	Beta 2
R:	screen	4.4 (3)	1.4 (3)	5.9 (3)	—	R_3:	3/2	SN 1 and 2
S:	saw	2.5 (2)	1.3 (2)	3.9 (2)	0.9 (3)	S_2:	2	SN 1 and 2
T:	batch	—	11.2 (1)	3.9 (3)	16.7 (1)	T_1:	s	Beta 1 and 2
U:	staff	38.8 (3)	75.9 (3)	27.6 (3)	39.3 (2)	U_3:	base +2	Scaling factor
	Pooled error	29.3	4.9	33.0	16.2			

[a] Numbers in parentheses represent the level that maximizes response.

in Figure 4. An analysis of variance (ANOVA) was performed for each parameter; the results are summarized in Table 4. We used the ANOVA results to identify the influential factors and the factorial effects analysis to choose the levels for the optimum factor combination detailed in Table 4.

We modified the optimum into a second configuration based on fabrication line operational concerns. Factor A (number of verification measurement stations) had been an occasional bottleneck in the past, so we changed to level A_2 ("added"). The largest gain in performance due to factor R (number of screening stations) is between R_1 and R_2; R_3 shows less improvement over R_2. Since R_2 ("2+/1+") demands less equipment and less space, we selected that level. Because the product

Table 5
Confirmation run results summary

	Factor																					Product Ship Rate			
Run	A	B	C	D	E	F	G	H	I	J	K	L	M	N	O	P	Q	R	S	T	U	β_1	SN_1 (dB)	β_2	SN_2 (dB)
37	1	1	2	2	2	2	1	1	1	1	2	3	3	2	1	3	2	3	2	1	3	1.845	3.32	0.267	−14.25
																						1.831	2.29	0.234	−15.57
38	2	1	2	2	2	1	1	1	1	1	2	3	3	3	1	3	2	2	2	1	3	1.788	1.93	0.267	−15.42
																						1.867	2.52	0.240	−15.29

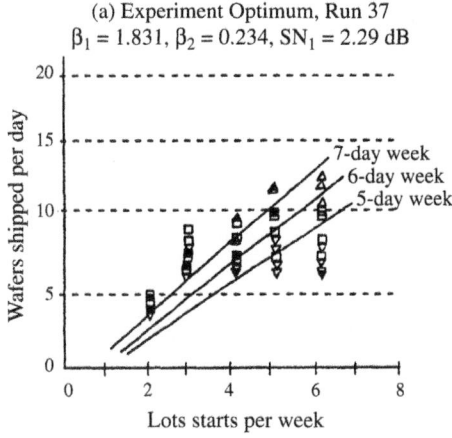

(a) Experiment Optimum, Run 37
$\beta_1 = 1.831$, $\beta_2 = 0.234$, $SN_1 = 2.29$ dB

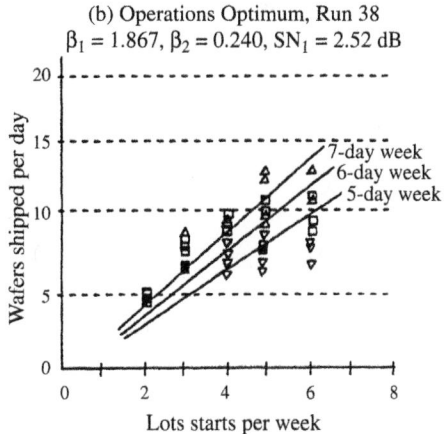

(b) Operations Optimum, Run 38
$\beta_1 = 1.867$, $\beta_2 = 0.240$, $SN_1 = 2.52$ dB

Figure 5
Optimum conditions run results

ship rates spanned by this study were not high enough to meet long-term requirements comfortably, we changed factor M (production process improvement strategies) to M_3 ("improved HD"), which includes provision for a greater number of units per wafer.

The linear predictions and confirmation run results for both the experiment optimum and the modified operations optimum configurations are shown in Table 5, with the factor-level differences shaded. The run data are shown plotted in Figure 5 with the dynamic model lines. Both configurations gave results close to those predicted and with less difference between the two configurations than predicted.

6. Summary and Conclusions

We evaluated the sensitivity of the accelerometer sensor fabrication line expansion to 21 different factors by using an L_{36} orthogonal array to direct the configuration of a validated simulation model. We developed a double-signal dynamic model of the fabrication line ideal function that greatly clarifies the assessment of line capacity and stability. We are now using these results to verify and refine the fabrication line expansion planning and to guide capacity forecasting.

Acknowledgments This study was performed in support of Sensor Fabrication Facility of TRW AEN, California Operations. I would like to thank A. Arrignton for his encouragement and support and R. Bilyak, J. Kidd, and T. Roth for their support throughout the long evolution of this project. I have used the pronoun "we" in this study not to dodge responsibility for the results, but rather, to reflect that without the operators, engineers, and managers of the Sensor Fabrication Facility, there would have been no process to simulate.

References

1. *Taylor II for Windows 4.2*, 1997. Orem, UT: F&H Simulations.

2. L. Farey, 1994. TRW/TED accelerometer wafer process production facility; manufacturing simulation. TRW internal report for R. Bilyak.

3. G. Taguchi and S. Konishi, 1987. *Orthogonal Arrays and Linear Graphs*. Dearborn, MI: ASI Press.

4. *Quattro Pro for Windows 5.0*, 1993. Scotts Valley, CA: Borland International.

5. ASI, 1997. Robust design using Taguchi methods. Handout for the Robust Design Workshop.

This case study is contributed by R. K. Ellis.

Part IV

Mahalanobis–Taguchi System (MTS)

Human Performance (Cases 67–68)
Inspection (Cases 69–73)
Medical Diagnosis (Cases 74–78)
Product (Cases 79–80)

CASE 67

Prediction of Programming Ability from a Questionnaire Using the MTS

Abstract: In this research, by taking advantage of the Mahalanobis–Taguchi system (MTS), we evaluated the ability of respondents to write software, based on their answers on questionnaires.

1. Objective of Experiment

To collect the necessary data, we asked 83 testees to answer a questionnaire and create a program. The 83 testees broadly covered programmers, business-people, undergraduate students, and college students, all with some knowledge of programming. The questionnaire consisted of 56 questions related to programming and each testee was asked to respond from the following seven-point scale: "positively no (−3)," "no (−2)," "somewhat no (−1)," "yes or no (0)," "somewhat yes (+1)," "yes (+2)," and "positively yes (+3)." The program that each testee worked on was a simple mask calculation (BASE + BALL = GAMES: each alphabet corresponds to a single digit and the same alphabet needs to have the same digit). That is, we believed the results would show that people who were able to write such a program would answer in a certain way. Our model is shown in Figure 1.

2. Calculation of Base Space

After checking over all the programs that the testees created, we defined those who created an easily readable program without bugs as "persons with a sense of programming." Among the 83 testees there were only 17 persons with a sense of programming. Using the remaining 66 persons with no sense of programming, we formed a base space. The total number of items was 60.

Figure 2 shows a histogram of Mahalanobis distances for all data. Give an appropriate threshold, we can completely separate "persons with a sense of programming" and "those with no sense of programming." According to this result, we concluded that a Mahalanobis distance can be applied to questionnaire data.

However, when we judged whether or not a person has a sense of programming, the probability of making a correct judgment is significant. That is, we need to ascertain the reliability of the base space that we have formed.

Provided that we have numerous data, for example, 300 normal data, using 285 normal data to create a base space, we compute the distance from the base space for each of the remaining 15 data. As long as the distances of these 15 sets of data are distributed around 1, the base space created turns out to be reliable. If the distance becomes larger, we conclude that the number of data needed to construct the base space is lacking or that the items selected are inappropriate.

Yet, due to the insufficient number of data for creation of the base space in this study, even if only five pieces of data are removed from the base space, the reliability of the space tends to be lowered dramatically. Therefore, by excluding only one data point from the base space, we created a base space without the one point excluded and calculated the distance of the point excluded. As a next step, we removed another data point and restored the point excluded previously in the base space. Again, creating a new base space without one piece of data,

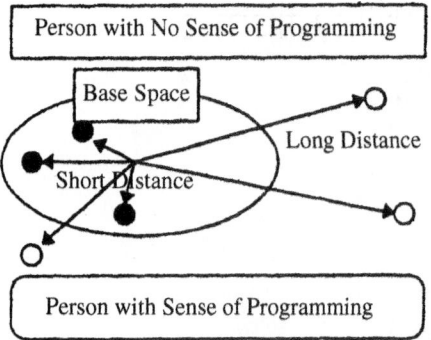

Figure 1
Base space for programming ability

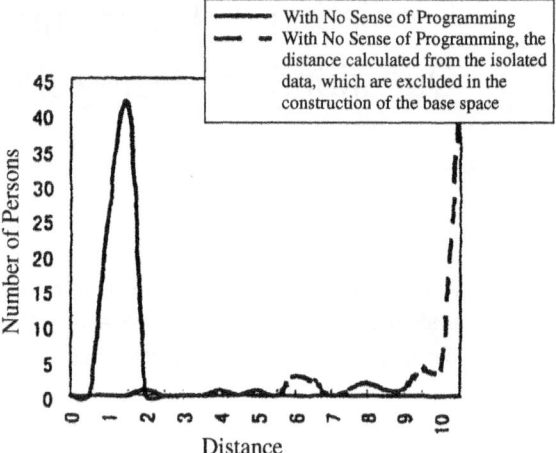

Figure 3
Data in base space for reliability evaluation

we computed the distance for the data currently removed. By reiterating this process, we evaluated the reliability of the standard space.

Figure 3 shows the results obtained using the reliability evaluation method described above. The result reveals that the current base space cannot be used for a judgment because "persons with no sense of programming" is also distant from the base space. Such a base space cannot be used for judgment, primarily because of the small number of data.

3. Selection of Items

There are some reasons that the distance for "a person with no sense of programming" resulted in a

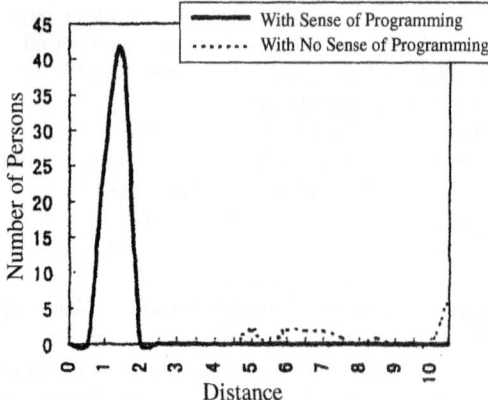

Figure 2
Distances from base space for "persons with no sense of programming"

large value. When there are many items in a base space, unnecessary items for a judgment are sometimes included in the space. By identifying these items, we attempted to exclude them from the base space.

We allocated each item as a factor to an orthogonal array with two levels. Since we now had 60 items, an L_{64} orthogonal array was selected. For each item, level 1 was used to create a base space. Without using level 2 for the base space, from each row in the L_{64} orthogonal array, we formed one base space for "persons no sense of programming." For the resulting base space, we computed distances for abnormal data or "persons with a sense of programming." Since these abnormal data should be distant from the base space, we utilized a larger-the-better SN ratio.

$$\eta = -10 \log \frac{1}{n}\left(\sum_{i=1}^{n} \frac{1}{D_i^2}\right) \qquad (1)$$

Based on the response graphs, we checked over effective and ineffective items to decide whether a certain person had a sense or no sense of programming. The items at level 1 were used to create a base space. Now, because the items at level 2 were not used for the base space, an item with a declining tendency from left to right was considered effective for a judgment, whereas an item with a contrary trend was regarded to affect judgment negatively.

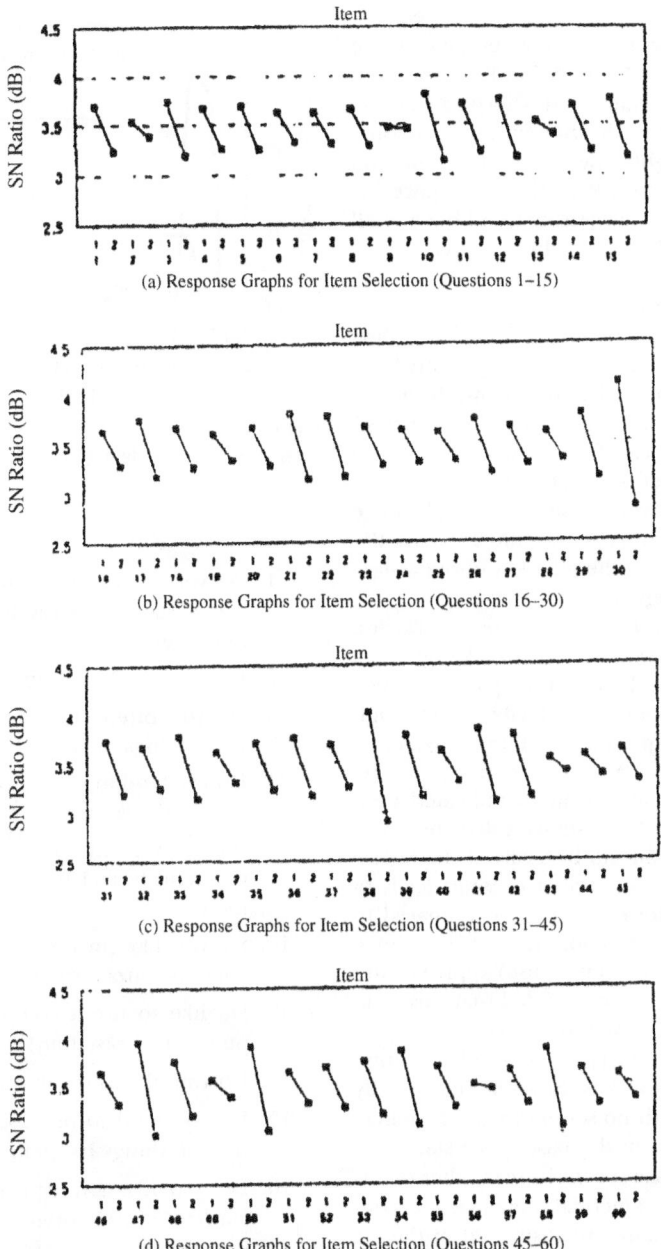

(a) Response Graphs for Item Selection (Questions 1–15)

(b) Response Graphs for Item Selection (Questions 16–30)

(c) Response Graphs for Item Selection (Questions 31–45)

(d) Response Graphs for Item Selection (Questions 45–60)

Figure 4
Response graphs for item selection

Figure 4 demonstrates that all of the items affected a judgment positively. Now, by eliminating the least effective item for question 9, we selected effective items once again. In this second item selection, only question 13 showed an increasing trend from left to right. Then, excluding question 13, we returned question 9 to the base space because it did not have an increasing trend but a small effect (in fact, question 9 will be removed in the later item selection because it did show a rising trend from left to right).

In the third and later item selections, every time we had at least one item with a rising trend from left to right we eliminated it (or them) from the base space and selected other items. Nevertheless, once the total number of items was reduced to about 50, all of the items turned out to be effective or with a decreasing trend from left to right once again. At this point in time, fixing the 20 items, the larger ones, and keeping them fixed without assigning them to an orthogonal array, but assigning the remaining 30 data to an L_{32} orthogonal array, we selected effective items. In sum, by changing an orthogonal array used, we attempted to seek items that showed poor reproducibility. Through this process, we expected to find only high-reproducibility items. Nevertheless, since we removed items, even if all the items indicated were more or less effective, the distance calculated from abnormal data was reduced. In terms of the SN ratio, the reduction in the number of items leads to a smaller SN ratio. However, since as compared to this diminution of the SN ratio, the average distance outside the base space (with no sense) approximates that in the base space, we concluded that this analysis procedure was improved as a whole.

Through this procedure, we reduced the number of items from 60 to 33 as shown in Figure 5. By doing so, "persons with no sense of programming" who were excluded from the base space started to overlap 25% of the base space. To make them overlap more, we needed to increase the data. The contents of the questionnaire are abstracted in the list below. All of the questions were composed of items that some software- or programming-related books regard as important for programmers and were considered essential from our empirical standpoint. The list includes only items that have high reproducibility and are associated with a sense of programming.

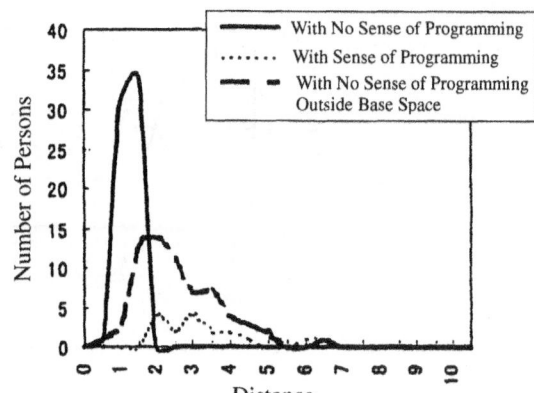

Figure 5
After-item-selection data

1. Do you tend to be distracted from your job if a certain job's result will come out after one week?

2. Do you mind your appearance?

3. Do you often clean up things around you (belongings or files in a hard disk)?

4. Do you tend to be absorbed without minding time while you are doing favorite things?

5. When everyday jobs change drastically, do you tend to feel happy about responding to them?

10. Do you like puzzles (jigsaw puzzles, Rubik's cube, or puzzle rings)?

12. Do like to use a computer (personal computer or workstation)?

14. Do you like to speak publicly?

15. Do you tend to be superstitious when unfavorable things happen?

16. Do you like any type of programming regardless of its contents?

17. Do you often play TV games?

19. Are you in the forefront of fashion?

20. Do you like to explain with a chart when asked by others?

21. Do you have beliefs?

22. Do you like to handcraft something?

Table 1
Average and standard deviation for each effective question

Question	With No Sense		With Sense		Difference between Averages
	Average	Standard Deviation	Average	Standard Deviation	
1	0.17	1.61	0.18	1.51	0.01
2	0.67	1.42	0.29	1.40	0.37
3	−0.05	1.55	0.12	1.76	0.16
4	2.12	0.92	1.76	1.09	0.36
5	0.41	1.53	−0.35	1.11	0.76
10	0.58	1.56	0.65	1.90	0.07
12	1.29	1.41	1.53	1.37	0.24
14	−0.47	1.60	−1.24	1.39	0.77
16	0.32	1.63	1.35	0.86	1.03
17	0.32	1.97	−0.06	2.14	0.38
19	−0.24	1.48	−0.18	1.55	0.07
20	0.53	1.25	0.29	1.53	0.24
21	0.77	1.41	0.40	1.22	0.37
22	1.44	1.45	1.14	1.50	0.30
23	0.85	1.42	0.29	1.36	0.55
24	0.41	1.93	0.53	1.81	0.12
26	0.39	1.50	0.76	1.52	0.37
27	−0.14	1.82	0.41	1.87	0.55
30	−0.08	1.71	−0.41	1.97	0.34
31	0.31	1.76	0.29	1.76	0.02
35	−1.09	1.57	−0.76	1.68	0.33
37	1.33	1.40	1.18	1.29	0.16
38	−0.03	1.69	−0.41	1.28	0.38
39	0.65	1.45	0.29	1.45	0.36
41	0.33	1.27	0.35	1.54	0.02
47	0.75	1.19	0.24	1.52	0.52
50	0.67	1.14	0.29	1.05	0.37
51	−0.63	1.77	0.53	1.87	1.15
54	0.19	1.87	−0.53	1.94	0.72
55	0.19	1.58	0.18	1.38	0.01
56	−0.16	1.60	−0.06	1.78	0.10
58	1.12	0.33	1.00	0.00	0.12

Table 2
Distances from base space using data classified by type of program

Class	Loop Type	Logic Type and Other	Outside Loop Type
0	0	0	0
0.5	0	0	0
1	28	0	0
1.5	38	0	0
2	0	0	0
2.5	0	0	0
3	0	0	1
3.5	0	1	0
4	0	0	0
4.5	0	0	0
5	0	0	1
5.5	0	0	1
6	0	0	1
6.5	0	0	1
More than 7	0	16	61
Total	66	17	66

Table 3
Distances from base space using after-item-selection data classified by type of program

Class	Loop Type	Logic Type and Other	Outside Loop Type
0	0	0	0
0.5	0	0	0
1	37	0	0
1.5	29	2	3
2	0	0	4
2.5	0	3	9
3	0	1	5
3.5	0	6	16
4	0	1	2
4.5	0	1	6
5	0	0	4
5.5	0	0	3
6	0	0	1
6.5	0	1	4
More than 7	0	2	9
Total	66	17	66

23. Are you worried about things that have gone wrong with you?

24. Are you willing to watch TV sports programs?

26. Do you like to read specialized books?

27. Can you continue to do simple work with patience?

30. Do you like to draw pictures?

31. Do you tend to be annoyed by your surroundings while pondering something?

35. Do you sometimes forget to button up your clothes?

37. Are you curious about the mechanisms of machines?

38. Do you mind if your name is spoken incorrectly?

39. Do you tend to pursue perfection in your job?

41. Do you tend to take action as soon as you make up your mind?

47. Do you feel that you are different from others?

50. Do you tend to think about things that are coming up when doing a job?

51. Can you wake up early in the morning?

54. Do you tend to get lost when visiting an unknown town?

55. Do you make more mistakes in spelling or in filling out a form?

56. Do you dislike having to socialize with others, for example, at a party?

58. Gender

Table 1 shows the distribution of the data for "persons with a sense of programming" and "those with no sense of programming" in the questionnaire. However, for any question, there is no significant difference in sense of programming. In other words, we cannot make a judgment by using only individual questions. Once a certain number of questions are grouped, we distinguished "persons with a sense of programming" from "those with no sense of programming."

4. Base Space Classified by Type of Program

We attempted to create a base space using another classification, categorizing the data by a type of program. The programs that we used as a sample can be roughly classified into three types. *Loop type* is a program that changes numbers from 0 to 9 in a loop process and searches for an answer on a trial-and-error basis in the mask calculation. In this research, the largest number of testees, 66, took up this method. Despite its wide applicability, it takes a lot of time to arrive at the answer. *Logic type* finds out an answer according to the logic that "if A + B = C, then. . . ." Although this method is not broadly applicable, we can use it to arrive at an answer quickly. Finally, *other* is a category for complicated programs that cannot be classified in the aforementioned two groups. In fact, some are excellent programs, whereas some do not take shape as a program.

Assuming that "if a program is written based on the same idea as those of other persons, they can also easily debug the program," by using the data for loop type we created a base space to seek a group with the same way of thinking in program-

ming. The total number of data in the base space turned out to be identical to that in the case of classifying the data by a sense of programming.

Table 2 summarizes the histogram of the before-item-selection data (Figure 5), and Table 3 shows that of the after-item-selection data. Even after the items were selected, no judging ability can be seen in the base space based on the classification by type of program. This fact demonstrates that even using data from the same questionnaire, if we opt to use a different setup of a base space or a different classification of data, we can never make a judgment. In sum, the initial classification is regarded as the most crucial.

References

Kei Takada, Muneo Takahasi, Narushi Yamanouchi, and Hiroshi Yano, 1997. The study on evaluation of capability and errors in programming. *Quality Engineering*, Vol. 5, No. 6, pp. 38–44.

Kei Takada, Kazuhito Takahashi, and Hiroshi Yano, 1999. The study of reappearance of experiment in evaluation of programmer's ability. *Quality Engineering*, Vol. 6, No. 1, pp. 39–47.

Kei Takada, Kazuhito Takahashi, and Hiroshi Yano, 1999. A prediction on ability of programming from questionnaire using MTS method. *Quality Engineering*, Vol. 7, No. 1, pp. 65–72.

Muneo Takahashi, Kei Takada, Narushi Yamanouchi, and Hiroshi Yano, 1996. A basic study on the evaluation of software error detecting capability. *Quality Engineering*, Vol. 4, No. 3, pp. 45–53.

This case study is contributed by Kei Takada, Kazuhito Takahashi, and Hiroshi Yano.

CASE 68

Technique for the Evaluation of Programming Ability Based on the MTS

Abstract: In our research, by advancing Takada's research [1] to some extent, we set up persons creating a normal program in a base space, judged directions of two abnormal data groups using the Mahalanobis–Taguchi system (MTS) method with Schmidt's orthogonalization [2], and proposed a method of assessing a programmer's ability with data from questionnaires about lifestyle-related matters.

1. Introduction

In a software development process, a programmer's ability is regarded as an important human factor that affects software quality directly. Currently, many companies have adopted a programmer aptitude test as a method of evaluating a programmer's potential. However, in fact, approximately 30% of them are not making the best of the programmer aptitude test [3]. The main reason is its reliability problem: that the test result is not relevant to a programmer's ability or that the criteria for judgment of aptitude are not clear. Therefore, establishment of an effective method of evaluating a programmer's ability has been desired; in fact, many attempts have been made. Among them is Takada's and other researchers' evaluation method of programming ability using the MTS method.

Using two types of programs, mask calculation and sorting programs, we gathered data for programming ability. Table 1 shows the difference between the two programs. Lifestyle-related data were collected through questionnaires, which consist of 56 questions and the four items of age, gender, programming experience, and computer experience (from now on, this questionnaire's item is simply called an "item").

Combining the data for sorting and mask calculation programs, we obtained 62 sets of data. By adding the data used in Takada's research, the total number of data amounts to 153.

2. Setup of Normal and Abnormal Data

In the study by Takada and others, the people who do not have a high level of programming ability were considered as the base space, since the people with high ability were considered as abnormal. But since the people with low ability are also abnormal, the base space was defined as the people who write normal programs. It is a fact that the programs produced by the people with high ability are generally difficult to use and do not fit well in a large system.

A program created by a testee was judged by two examiners (Suzuki and Takada). A testee who created a normal program was classified as a normal datum, or a datum belonging to a base space, whereas testees making a good and poor programs were categorized as abnormal data. It can be seen from Table 2 that there are not many people in the "good program" category, indicating the difficulty of using these people to construct a base space.

3. Calculation of Mahalanobis Distance Using Schmidt's Orthogonal Variable

A Schmidt orthogonal expansion's calculation process is shown below. Using measurements contained in a base space, we calculated a mean and standard deviation for each item and computed normalized

Table 1
Experimental difference between two programs

	Sorting Program	Mask Calculation Program
Testee	Student	Businessperson and student
Programming time	Limited	Unlimited
Programming language	C	Any language

variables, $X_1, X_2, \ldots, X_{60}$. Then, with $X_1, X_2, \ldots, X_{60}$, we created orthogonalized variables, $x_1, x_2, \ldots, x_{60}$.

$$x_1 = X_1 \tag{1}$$

$$x_2 = X_2 - b_{21}x_1 \tag{2}$$

$$\vdots$$

$$x_{60} = X_{60} - b_{601}x_1 - b_{602}x_2 - \cdots - b_{6059}x_{59} \tag{3}$$

Next we set each variance of $x_1, x_2, \ldots, x_{60}$ to $V_1, V_2, \ldots, V_{60}$. The normalized orthogonal variables, $Y_1, Y_2, \ldots, Y_{60}$, are expressed as follows:

$$Y_1 = \frac{x_1}{\sqrt{V_1}} \tag{4}$$

$$Y_2 = \frac{x_2}{\sqrt{V_2}} \tag{5}$$

$$\vdots$$

$$Y_{60} = \frac{x_{60}}{\sqrt{V_{60}}} \tag{6}$$

Table 3 summarizes the calculation results above.

Next, for measurements of testees creating good and poor programs, we computed $Y_1, Y_2, \ldots, Y_{60}$, as shown in Table 4. Then the Mahalanobis distance D^2 was computed by $Y_1, Y_2, \ldots, Y_{60}$:

$$D^2 = \frac{1}{60}(Y_1^2 + Y_2^2 + \cdots + Y_{60}^2) \tag{7}$$

Table 5 shows the calculation results of Mahalanobis Distances for the base space and abnormal data.

4. Selection of Items for Measurement Ability Using the Larger-the-Better Characteristic

As a next step, we judged the direction of a distance and set up each item's order. It is important in using the Schmidt method to determine the order of items based on their importance as well as on cost, but these were unknown because each item in our research was a question. Next, selecting items by a larger-the-better characteristic based on an orthogonal array, after creating response graphs, we placed all items in descending order with respect to the magnitude of a factor effect. (Level 1 was used for creating a standard space. Since level 2 was not used for the purpose, items with decreasing value from left to right are effective for judgment.) Figures 1 and 2 show the results. The former is the item-selection response graph for testees creating a good program, and the latter is that for those creating a poor one.

According to Figure 1, we note that items 28, 2, 20, and 46 are effective for testees creating a good program. In contrast, Figure 2 reveals that effective items for those creating a poor program are 29, 32, 36, and 30. These results highlight that there is an

Table 2
Classification of normal and abnormal data for software programming

	Normal Data, Normal Program	Abnormal Data	
		Good Program	Poor Program
Judgment criterion	A program after removing abnormal data	A program that contains no bugs and understandable to others	A program that contains fatal bugs
Number of Data	102	14	23

Table 3
Database for the base space

	Item 1	Item 2	...	Item 60
Measurement	2	−1		5
	3	1	...	6
	⋮	⋮		⋮
	−2	−1		7
Average	0.24	0.66	...	6.42
Standard deviation	1.72	1.37	...	4.77
Normalized variable X	1.02	−1.21		−0.98
	1.60	0.25	...	0.54
	⋮	⋮		⋮
	−1.30	−1.21		0.33
Regression coefficient b	—	0.12	...	−0.10
				−0.12
				⋮
				0.70
Variance V	1	0.99	...	0.10
Normalized orthogonal variable Y	1.02	−1.34		−1.20
	1.60	0.06	...	0.04
	⋮	⋮		⋮
	−1.30	−1.07		1.64

Table 4
Normalized orthogonal variables of abnormal data

	Item 1	Item 2	...	Item 60
Measurement				
Good program	−2	0		5
	2	0	...	6
	⋮	⋮		⋮
	0	−2		7
Poor program	0	0		2
	1	2	...	3
	⋮	⋮		⋮
	1	1		7
Orthogonal variable				
Good program	−1.30	−0.33		5.18
	1.02	−0.61		−3.98
	⋮	⋮	...	⋮
	−0.14	−1.94		−5.62
Poor program	−0.14	−0.47		−2.50
	0.44	0.93	...	1.94
	⋮	⋮		⋮
	0.44	0.20		5.01

obvious difference between those creating good and poor programs. By performing a Schmidt orthogonal expansion on the data for those creating good and poor programs, we rearranged each item's order.

5. Determination of Item Order by Normalized Orthogonal Variables

In accordance with the procedure below, using orthogonal normalized variables Y_1, Y_2, ... , Y_{60}, we computed larger-the-better SN ratios η_1, η_2, ... , η_{60} for A_1, A_2, ... , A_{60}, and rearranged each item's order such that we could eliminate items leading to a smaller SN ratio:

$$A_1 = \text{only } Y_1 \text{ is used}$$

$$A_2 = Y_1 \text{ and } Y_2 \text{ are used}$$

$$\vdots$$

$$A_k = Y_1, Y_2, ... , Y_{60} \text{ are used}$$

Then we detailed calculation of the SN ratio for testees creating a good program (refer to Table 6) as follows.

SN ratio for A_1:

$$\eta_1 = -10 \log \frac{1}{14} \left(\frac{1}{1.20^2} + \frac{1}{1.20^2} + \cdots + \frac{1}{4.99^2} \right)$$

$$= -5.90 \text{ dB} \tag{8}$$

SN ratio for A_2:

$$\eta_2 = -10 \log \frac{1}{14} \left(\frac{1}{0.86^2} + \frac{1}{1.38^2} + \cdots + \frac{1}{3.57^2} \right)$$

$$= -5.21 \text{ dB} \tag{9}$$

SN ratio for A_{60}:

$$\eta_{60} = -10 \log \frac{1}{14} \left(\frac{1}{2.18^2} + \frac{1}{2.52^2} + \cdots + \frac{1}{1.54^2} \right)$$

$$= 4.07 \text{ dB} \tag{10}$$

Figure 3 shows the relationship between items and SN ratios when using the data for testees creating a good program, and the relationship for those creating a poor program. Up to this point, calculations were made using software from the Oken Company.

Table 5
Mahalanobis distance of abnormal data calculated from a base space constructed by the group of people who wrote normal programs

Distance	Base Space	Good Program	Poor Program
0	0	0	0
0.5	2	0	0
1	49	0	1
1.5	51	1	2
2	0	2	1
2.5	0	4	5
3	0	4	3
3.5	0	0	4
4	0	0	4
4.5	0	1	1
5	0	1	2
5.5	0	0	0
6	0	0	0
6.5	0	1	0
7	0	0	0
7.5	0	0	0
8	0	0	0
8.5	0	0	0
9	0	0	0
9.5	0	0	0
10	0	0	0
Total	102	14	23

6. Directional Judgment by Schmidt Orthogonal Expansion

By taking advantage of the rearranged data, we made a judgment on the direction of abnormal data as to whether the abnormal data were abnormal in a favorable or unfavorable direction: whether abnormally good or abnormally bad. Since we did not have enough data for the base space, it was decided

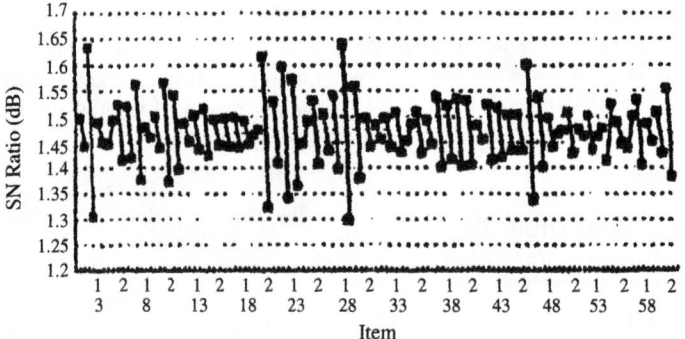

Figure 1
Response graphs for item selection (for testees creating a good program)

to judge one side at a time rather than judging the sides at the same time. As a first step, we made a separate directional judgment for each piece of the data for those creating good and poor programs.

After each type was evaluated and classified into two levels, we set up the levels of a signal, m, as follows:

m_1: most favorable = 2 (8 testees matched)

m_2: favorable = 1 (6 testees matched)

m_3: unfavorable = −1 (8 testees matched)

m_4: most unfavorable = −2 (15 testees matched)

As a next step, we calculated the normalized orthogonal variables Y's for each item for those

creating good and poor programs. Because the number of items was 60, we calculated Y_1, Y_2, ... , Y_{60}. We rearranged the data for those creating a good program and computed the normalized orthogonal variables for abnormal data (Table 7).

Note the following calculation example of β for the first item after rearranged for those creating a good program: For the first item we computed the average of Y_1's, denoted by $\overline{Y}_1$ for the data belonging to each of m_1 and then computed the same for m_2, m_3, and m_4. Then, for signals, setting the true value as m and the output normalized orthogonal variables as Y, we applied a zero-proportional equation $Y = \beta m$. Computing each $\overline{Y}_1$ belonging to each of m_1, m_2, m_3, and m_4, we estimated β_1 by the least squares method.

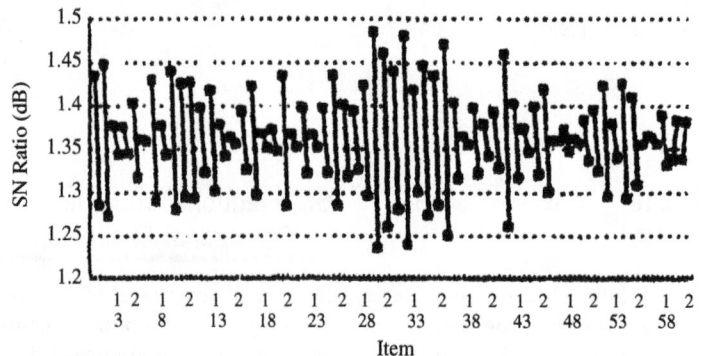

Figure 2
Response graphs for item selection (for testees creating a poor program)

Table 6
Rearranging data for testees creating a good program

No.	A_1	A_2	...	A_{60}
1	1.20	0.86	...	2.18
2	1.20	1.38	...	2.52
⋮	⋮	⋮	...	⋮
8	0.69	1.94	...	1.53
9	1.16	0.96	...	2.03
10	0.22	0.18	...	1.72
⋮	⋮	⋮	...	⋮
14	4.99	3.57	...	1.54

$\overline{Y}_1$ for data belonging to m_1:

$$\overline{Y}_{1m1} = \frac{-1.20 + \cdots + 0.69}{8} = -0.38 \quad (11)$$

$\overline{Y}_1$ for data belonging to m_2:

$$\overline{Y}_{1m2} = \frac{1.16 + \cdots + (-4.99)}{6} = -0.49 \quad (12)$$

$\overline{Y}$ for data belonging to m_3:

$$\overline{Y}_{1m3} = \frac{-0.73 + \cdots + 0.22}{8} = -0.02 \quad (13)$$

$\overline{Y}_1$ for data belonging to m_4:

$$\overline{Y}_{1m4} = \frac{-0.26 + \cdots + 1.16}{15} = 0.34 \quad (14)$$

$$\beta_1 = \frac{m_1 \overline{Y}_{1m1} + m_2 \overline{Y}_{1m2} + m_3 \overline{Y}_{1m3} + m_4 \overline{Y}_{1m4}}{m_1^2 + m_2^2 + m_3^2 + m_4^2}$$

$$= -0.19 \quad (15)$$

In a similar manner, we computed β_2, β_3, ... , β_{60}. Then, for Y_1, Y_2, ... , Y_{60} (abnormal data), whose directions we attempted to judge, for example, if we define the direction of the first item as M_1, we know the direction of M_1 by using the relationship $Y_1 = \beta_1 M_1$ because all of β_1, β_2, ... , β_{60} are known. Similarly, determining M_2, ... , M_{60} for Y_2, ... , Y_{60} with β_2, ... , β_{60}, we calculated the sum of M's as the direction of the data's distances, M. Since we defined the signals for testees creating a good program as $+2$ and $+1$, and those for testees creating a poor program as -2 and -1, if testees creating a good program have a positive direction and testees creating a poor program have a negative direction, our judgment method can be regarded to work properly. Table 8 shows the judgment result for both of the cases of using the data for testees creating good or poor programs.

Table 8 reveals that in the case of using the data for those creating a good program, 8 of 14 testees creating a good program and 12 of 23 creating a poor program were judged correctly. On the other hand, in the case of using the data for testees creating a poor program, 8 of 14 creating a good program and 13 of 23 creating a poor program were

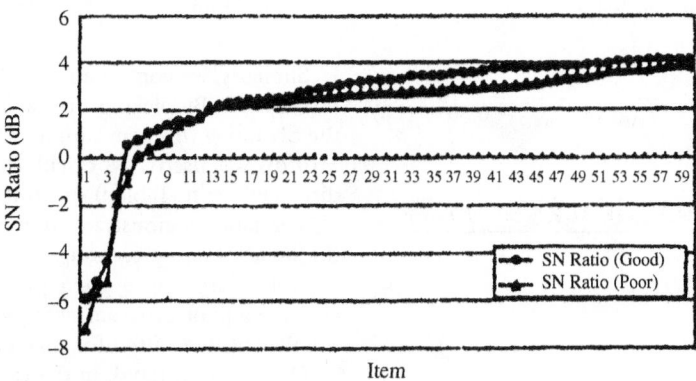

Figure 3
Rearrangement of items by Schmidt orthogonal expansion

Table 7
Normalized orthogonal variables of abnormal data

	No.	m Level	Normalized Orthogonal Variable			
			Y_1	Y_2	...	Y_{60}
Testees creating good program	1	m_1	−1.20	−0.35	...	−2.84
	2	m_1	−1.20	−0.35	...	−5.38
	⋮	⋮	⋮	⋮	...	⋮
	8	m_1	0.69	−2.04	...	−0.98
	9	m_2	1.16	1.59	...	−0.85
	10	m_2	0.22	−1.98	...	2.87
	⋮	⋮	⋮	⋮	...	⋮
	14	m_2	−4.99	−1.39	...	0.92
Testees creating poor program	1	m_2	−0.73	1.07	...	1.71
	2	m_3	0.22	−1.98	...	0.53
	⋮	⋮	⋮	⋮	...	⋮
	15	m_3	0.22	0.23	...	2.06
	16	m_4	−0.26	−0.46	...	−3.67
	17	m_4	0.22	0.96	...	−1.69
	⋮	⋮	⋮	⋮	...	⋮
	23	m_4	1.16	0.85	...	−2.19

judged correctly. However, the data do not necessarily indicate an accurate judgment.

Item Selection for Programming Ability Evaluation

To improve the accuracy in judgment, by using the SN ratio, we evaluated item selection for programming ability evaluation whether or not outputs for signals from −2 to +2 were stable (the linearity of $Y = \beta M$). For the first item obtained after rearranging the data for testees creating a good program, we calculated the SN ratio as below.

Total variation:

$$S_T = \overline{Y_{1M1}}^2 + \overline{Y_{1M2}}^2 + \overline{Y_{1M3}}^2 \, \text{pl} \, \overline{Y_{1M4}}^2 = 0.50 \quad (16)$$

Variation of β:

$$S_\beta = \frac{(M_1\overline{Y_{1M1}} + M_2\overline{Y_{1M2}} + M_3\overline{Y_{1M3}} + M_4\overline{Y_{1M4}})^2}{r}$$

$$= 0.36 \quad (17)$$

Effective divider:

$$r = M_1^2 + M_2^2 + M_3^2 + M_4^2 = 10 \quad (18)$$

Error variation:

$$S_e = S_T - S_\beta = 0.14 \quad (19)$$

Error variance:

$$V_e = \frac{S_e}{f} = 0.05 \quad (f = 3) \quad (20)$$

SN ratio:

$$\eta = 10 \log \frac{(1/r)(S_\beta - V_e)}{V_e} = -1.62 \text{ dB} \quad (21)$$

Similarly, we computed the SN ratios for items 2 through 60. By picking up stable items from all of the SN ratios (no item with an SN ratio of less than −10 dB was used), we calculated the directions of distances again. Table 9 shows the result.

The table demonstrates that in the case of using the data for those creating a good program, 12 of 14 testees creating a good program and 17 of 23 creating a poor program were judged correctly. The number of items used for this analysis was 17.

On the other hand, in the case of using the data for testees creating a poor program, for 12 of 14

Table 8
Directional judgment using all items

Good Program No.	M for Good Program	M for Poor Program	Poor Program No.	M for Good Program	M for Poor Program
1	2125.61	780.12	1	12.51	391.49
2	3973.87	−527.55	2	118.96	−10.84
3	−4842.26	428.87	3	7785.43	316.39
4	−1027.02	140.29	4	−2097.13	−1702.27
5	−3437.98	897.78	5	−3453.95	1185.03
6	−3258.63	631.64	6	−2755.32	−1036.88
7	5805.01	−594.37	7	−1602.08	966.77
8	5062.20	−753.75	8	−1070.22	−731.54
9	−2984.79	−1713.75	9	7412.83	192.16
10	4531.27	464.39	10	−2571.95	713.33
11	1095.78	−321.21	11	1006.82	1890.32
12	4209.38	1084.77	12	905.96	−216.42
13	−1011.24	−670.23	13	6814.81	−2039.88
14	2580.05	170.22	14	1184.24	−143.26
			15	−2647.86	−1052.03
			16	5393.82	121.56
			17	589.65	−964.47
			18	3003.96	−248.19
			19	−350.94	385.61
			20	456.86	10.69
			21	3193.06	−71.63
			22	−3413.29	253.36
			23	−1104.85	−287.06

creating a good program and 15 of 23 creating a poor program, were correctly judged. The number of items used was 11. Consequently, these results prove that the accuracy in judgment was improved compared with the former analysis.

After obtaining the results that single-directional judgment produced satisfactory accuracy, we attempted two-directional judgment. More specifi-

cally, after combining the data for testees creating good and poor programs into a single set of abnormal data, we selected items by using an orthogonal array, rearranged them according to the response graphs shown in Figure 4, and reordered them by a Schmidt orthogonal expansion (figure 5) such that items leading to a smaller SN ratio can be eliminated.

Table 9
Directional judgment using only stable items

Good Program No.	M for Good Program	M for Poor Program	Poor Program No.	M for Good Program	M for Poor Program
1	11.92	48.48	1	−60.26	−19.46
2	0.64	6.49	2	39.77	4.19
3	104.51	72.78	3	−80.13	−38.91
4	10.62	13.21	4	−10.76	−33.91
5	95.44	7.22	5	−28.88	−4.31
6	−11.47	14.04	6	14.21	7.94
7	16.24	24.18	7	26.12	18.67
8	34.97	5.35	8	−88.30	−25.71
9	29.00	−31.36	9	−44.31	19.67
10	37.79	22.85	10	−72.51	−53.59
11	40.22	32.68	11	−16.22	−3.78
12	26.22	−1.93	12	−24.20	−25.94
13	33.57	14.21	13	−13.18	−17.04
14	87.14	61.13	14	−52.91	3.27
			15	−12.71	2.45
			16	10.68	11.67
			17	32.93	−12.76
			18	−66.87	−93.26
			19	−59.76	−33.73
			20	−61.35	−22.68
			21	−15.74	−16.04
			22	74.36	10.28
			23	−38.94	−26.52

Then, based on the rearranged order of the items, we made a directional judgment focused on the SN ratio, indicating a linear trend. Table 10 shows the result of directional judgment on abnormal data.

Table 10 shows that 13 of 14 testees creating a good program and 19 of 23 creating a poor program were judged correctly. Eleven items were used for this analysis. Although initially, we attempted to make a separate directional judgment on each group of testees creating good and poor programs, because of the small number of data used, the two groups' combined double-directional judgment resulted in higher accuracy.

7. Results

Observing the Mahalanobis distances of normal (base space) and abnormal data, it was found that

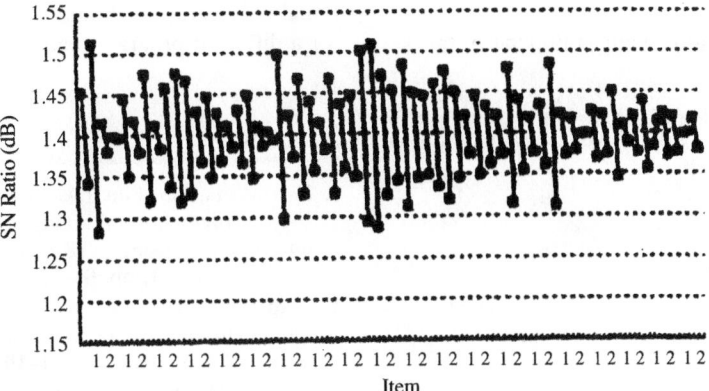

Figure 4
Response graphs for item selection

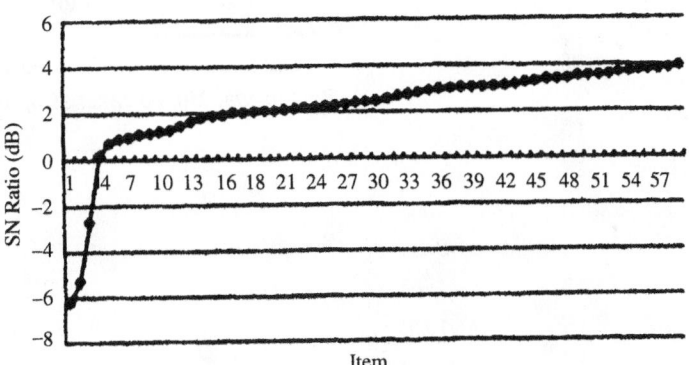

Figure 5
Rearrangement of items by Schmidt orthogonal expansion

the center of distribution of each case was separate; therefore, it is possible to distinguish between normal and abnormal data. Regarding the directional judgment for abnormal data, its accuracy was poor if all items were used. As an alternative approach, the SN ratio was calculated to observe the linearity of $Y = \beta M$. Then those items with good linearity were used. It means that the items useful for discrimination of normal/abnormal data were not necessarily useful for the judgment of direction unless they have linearity on $y = \beta m$. As a result, judging accuracy was improved and the direction of all abnormal data was judged correctly for almost all the data.

In our experiment, two types of abnormal data were used. By taking advantage of a Schmidt orthogonal expansion, with direction expressed by a positive or negative sign, we judged successfully whether the abnormal data were abnormal in a favorable or unfavorable direction. According to this result, after creating a base space with data for persons creating a normal program, we can assess each programmer's ability by comparing abnormal data for persons creating good and poor programs with the base space. The remaining issue is to form a more stable base space consisting of a larger number of data. Additionally, while we classified a signal, m, into four groups, to improve accuracy in direc-

Table 10
Directional judgment using abnormal data

Good Program No.	Direction of Distance M	Poor Program No.	Direction of Distance M
1	53.17	1	−40.63
2	18.92	2	9.31
3	55.64	3	−45.33
4	3.11	4	−18.44
5	28.49	5	−23.91
6	7.04	6	19.25
7	17.66	7	−7.10
8	12.49	8	−38.26
9	−29.90	9	17.12
10	15.67	10	−35.41
11	23.18	11	−5.86
12	0.47	12	−29.15
13	7.77	13	−12.16
14	60.18	14	9.74
		15	4.83
		16	−3.24
		17	−4.79
		18	−61.08
		19	−29.90
		20	−11.70
		21	−16.68
		22	−3.69
		23	−22.33

tional judgment, we should set up a separate signal for a different programmer.

References

1. Kei Takada, Kazuhito Takahashi, and Hiroshi Yano, 1999. A prediction on ability of programming from questionnaire using MTS method. *Quality Engineering*, Vol. 7, No. 1, pp. 65–72.
2. Genichi Taguchi, 1997. Application of Schmidt's orthogonalization in Mahalanobis space. *Quality Engineering*, Vol. 5, No. 6, pp. 11–15.
3. Hiroshi Miyeno. 1991. *Aptitude Inspection of Programmers and SE.* Tokyo: Ohmsha, pp. 105–106.
4. Takayuki Suzuki, Kei Takada, Muneo Takahashi, and Hiroshi Yano, 2000. A technique for the evaluation of programming ability based on MTS. *Quality Engineering*, Vol. 8, No. 4, pp. 57–64.

This case study is contributed by Takayuki Suzuki, Kei Takada, Muneo Takahashi, and Hiroshi Yano.

CASE 69

Application of Mahalanobis Distance for the Automatic Inspection of Solder Joints

Abstract: In order to solve a problem with an appearance inspection of soldering using lasers, we applied the Mahalanobis–Taguchi system (MTS) to a soldering fine-pitch quad flat package (QFP) implementation process where erroneous judgment occurs quite often, and attempted to study (1) the selection of effective inspection logic and (2) the possibility of inspection using measurement results.

1. Introduction

Figure 1 illustrates the appearance of a QFP. The QFP studied in our research has 304 pins with a 0.5-mm pitch and is reflow soldered using tin–lead alloy solder paste onto a copper-foil circuit pattern on a glass epoxy board.

NLB-5000, laser inspection apparatus developed by Nagoya Electric Words Co. Ltd., was used. Figure 2 outlines this apparatus's basic principle of inspection. An He–Ne laser, following a prescribed route, is swept over a soldered area of an inspection subject. The solder forms a specific fillet shape whose angle changes the reflection angle of the laser. Since several receptor cells are located separately and independently, we can quantitatively capture the solder's shape according to changes in a laser swept over the solder's surface. Therefore, by using this inspection apparatus, we can grasp the entire shape, from the copper-foil pattern's end to the QFP lead's end in the form of sequential data for a resolution of the laser sweeping pitch. Calculating these sequential data in accordance with a particular inspection logic and comparing them with the judging standard, we can make a judgment as to a good or defective solder.

An example of sequential data is shown in Table 1. From left to right in the table, each sequence represents a periodic change of reflection angle

when a laser is swept. A blank in the middle indicates the tip of a lead. Each numeral signifies an allocated value that corresponds uniquely to the angle of a solder's fillet. The larger it becomes, the more gentle the fillet's slope becomes, and conversely, the smaller, the steeper. The most gentle slope close to a flat surface is denoted by 6; the steepest is represented by 2. Therefore, a downward trend in a sequence indicates that a solder's fillet retains a concave shape. By analyzing the numerical order and magnitude of a sequence, we attempted to organize an appropriate inspection logic.

Figure 3 shows the cross section of a soldered area on a QFP lead, and Figure 4 shows good and defective soldering.

2. Collection of Normal and Abnormal Data

In an appearance inspection process of soldering in a mass production plant, based on a judgment result shown by an automated appearance inspection machine, a final inspector rechecks a product with an inspection tool such as a magnifying glass. When finding a true defect, he or she repairs it and passes it on to the next process.

From 11 QFP leads judged as defectives that had to be rechecked, selected from the boards produced

1189

Figure 1
Appearance of QFP implementation

Table 1
Example of sequential data

No.	Data
1	6666666666554443332 34466666
2	6666666666554433322 34446666
⋮	⋮
n	6666666666554433332 33466666

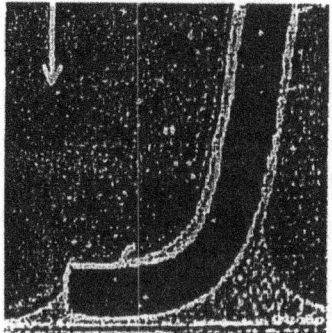

Figure 3
Cross section of soldering on QFP lead

in one week and from randomly choosing 2670 QFP leads judged as normal, we measured a solder shape using automated laser appearance inspection apparatus. The pitch for sweeping a laser from the tip of a pattern to that of a lead was set at 20 μm.

In our research, since we assumed that normality and abnormality are discriminated properly, we verified the validity of discriminability by the bonding strength, which is the generic function required for soldering. Since the function required is the bonding strength after a product is shipped to market, bonding strength should be measured after a tem-

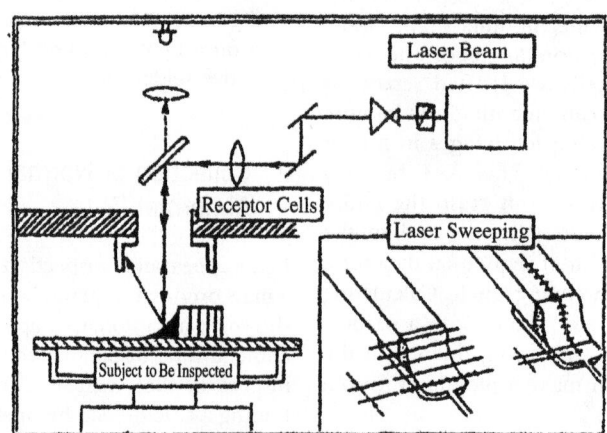

Figure 2
Basic principle of laser appearance inspection of soldering

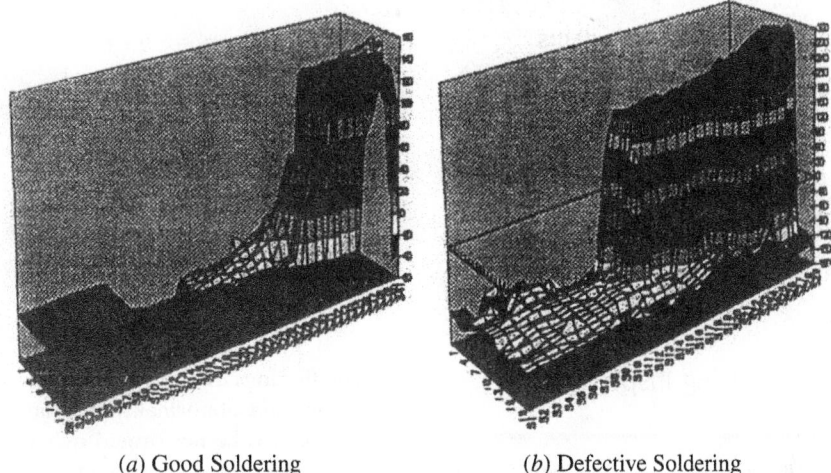

(*a*) Good Soldering   (*b*) Defective Soldering

Figure 4
Good and defective soldering

perature cycle test. But due to time constraints, we substituted initial bonding strength. As for a measuring method, after obtaining data for a soldering shape, using a pull tester supplied by Rhesca Co. Ltd., we measured the destructive tensile strength in the direction perpendicular to the surface of a circuit pattern on a board. For normal products, we measured 50 products randomly. All of the measurement results are shown in Figure 5. The clear-cut difference in bonding strength between normal and abnormal products proves good discriminability.

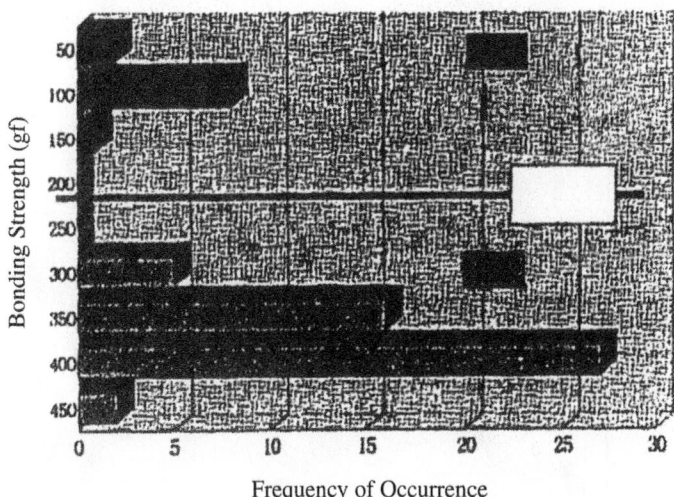

Frequency of Occurrence

Figure 5
Bonding strength of normal and abnormal units

Table 2
Classification of inspection logic items

Category	Frequency	Used or Not Used for Standard Space
1. Item for base space	77	Use
2. Item with variance of zero	49	Do not use
3. Item with different value for normality and abnormality	3	Do not use
4. Item with character data mixed up	1	Do not use

3. MTS Method Using Inspection Logic Terms

After verifying 130 types of inspection logic from the data obtained, Y_{ij}'s, we categorized them into four groups and, as a result, used 77 items to create a base space (Table 2).

The findings that the three items can distinguish normality and abnormality indicate that a proper inspection can be performed only with them. However, considering that we need to develop an inspection method with more logic in order to enhance the validity of inspection for various defects, we decided to leave them for further study.

Table 3
Mahalanobis distances of normal and abnormal data

	Normal Data			Abnormal Data, Distance
Class	Frequency	Class	Frequency	
0	0	5.5	4	37.09
0.5	267	6	1	28.24
1	1626	6.5	2	100.39
1.5	423	7	9	37.91
2	175	7.5	2	31.96
2.5	79	8	1	89.38
3	36	8.5	2	89.75
3.5	21	9	2	29.49
4	11	9.5	0	36.64
4.5	8	10	0	49.19
5	5	Next classes	5	25.17
Distances in next classes	29.20 / 10.12 / 11.92 / 12.69 / 10.39			

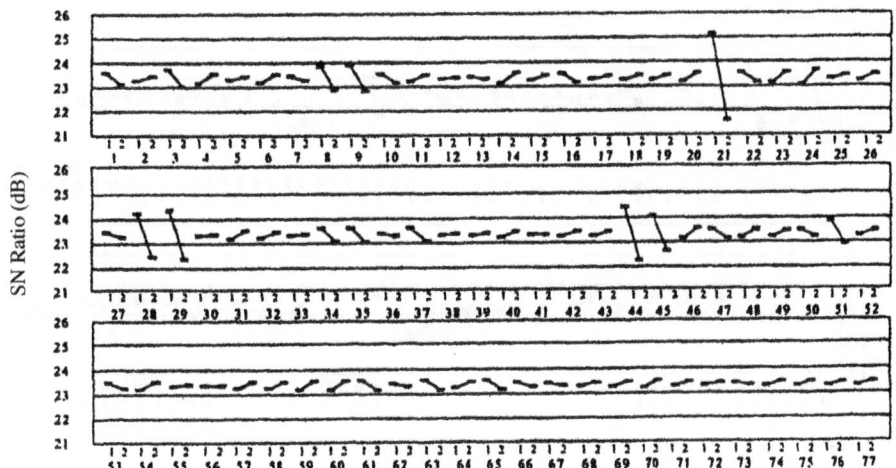

Figure 6
Response graphs of the SN ratio for inspection logic item

Table 3 shows the distribution of the Mahalanobis distances calculated. This table reveals high discriminability of normal data (proper products) and abnormal data (defective products) by the use of the 77 pieces of data as the base space. In the table, the bottom part in a longer middle column shows the distribution of higher classes.

In an actual inspection process, a smaller number of inspection logic items is desirable for a high-performance inspection, due to the shortened

Table 4
Inspection logic items after selection

No.	Content of Inspection Logic
8	Convexity of tip in pad
9	Average height of solder
21	Degree of diffused reflection at the back of the swept laser
28	Circuit pattern's flatness 1
29	Circuit pattern's flatness 2
44	Roundness 1 of solder shape
45	Roundness 2 of solder shape
51	Convexity of solder shape

inspection time and simpler maintenance. Next, we verify whether we were able to select appropriate inspection logic items. The following is our verification procedure. Because of using the aforementioned 77 items, we took advantage of an L_{128} orthogonal array and allocated two levels, "use of an item" and "no use of an item," to this array to create a standard space.

As a next step, forming a base space for each row according to the orthogonal array, we calculated the distance for each abnormal data item from the corresponding base space. Defining as $D_1, D_2, \ldots, D_n$ the square root of the Mahalanobis distance for each abnormal data in each row, we performed an evaluation with a larger-the-better SN ratio. For all 77 inspection logic items, Figure 6 shows the response graphs with the two levels "use of an item" and "no use of an item."

Each lower number below the x-axis in Figure 6 indicates each logic item, and each upper number represents use or no use (i.e., 1 for use of an item and 2 for no use of an item). Therefore, an item with a decreasing value from left to right was regarded as effective in an actual inspection, whereas an item with a contrary trend was considered to blunt the inspection accuracy (discriminability of normality and abnormality).

Analyzing these results with the criterion of a more-than-1-dB difference in the SN ratio, we se-

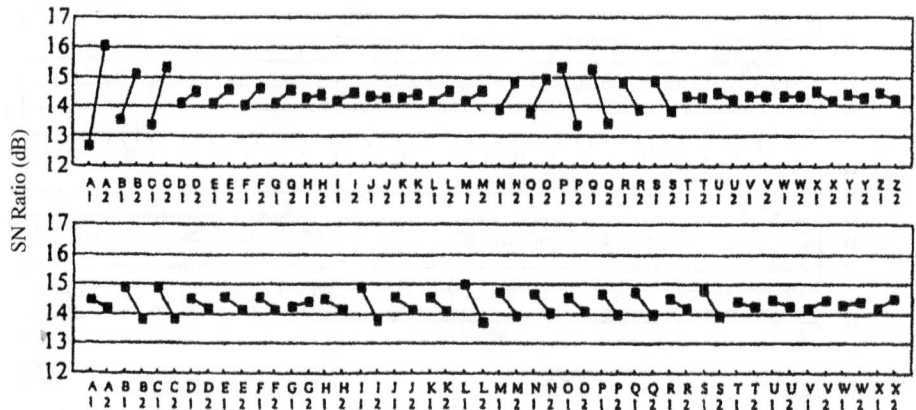

Figure 7
Response graphs of SN ratio by laser reflection data

lected extremely effective inspection logic items (Table 4). Since these items were selected from the 11 abnormal data obtained for our research, they do not reflect a whole diversity of defects that take place in an inspection process of a mass production line. However, together with three inspection logic items (height and length of a solder's fillet and convexity of a solder's shape) for category 3 in Table 2,

they can be regarded as very important inspection logic items to identify abnormality.

Furthermore, using 30 inspection logics that are not necessarily considered to have no effect in Figure 6 (with decreasing value from left to right), we created a new base space and selected effective inspection logic items (selection 2) in a manner similar to the prior selection (selection 1). As a result,

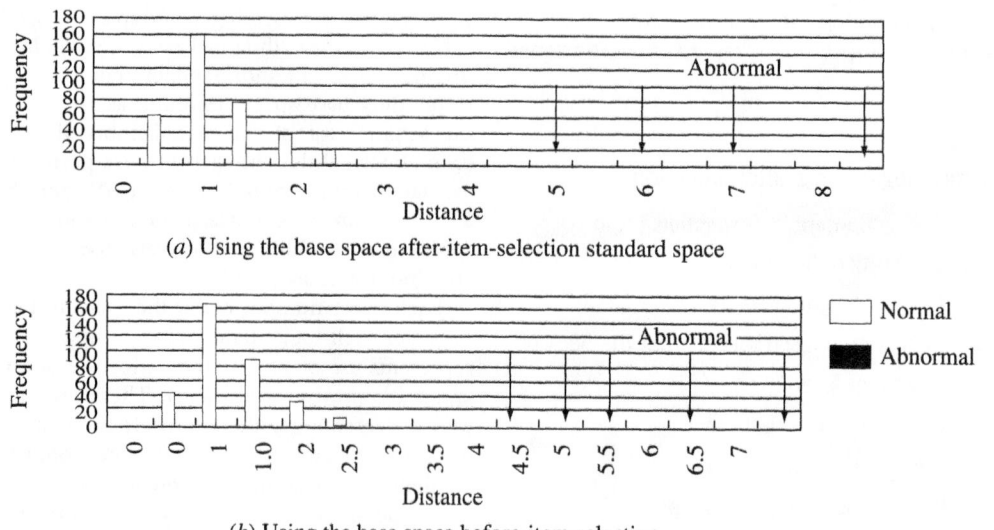

(a) Using the base space after-item-selection standard space

(b) Using the base space before-item selection

Figure 8
Mahalanobis distances for normal and abnormal units

interestingly, some inspection logic that has been judged as extremely effective in selection 1 (e.g., inspection logic 28 and 29: circuit pattern's flatness) turned out to be ineffective.

4. MTS Method Using Reflection Data

Similar to inspection logic items, we obtained 350 normal and 9 abnormal sets of data for laser reflection in a mass production line. Each set of data consisted of 50 items with a 20-μm pitch in a total sweep length of 1000 μm. Using an L_{64} orthogonal array, items were selected in the same way as in the previous analysis. Figure 7 shows the response graphs.

These results demonstrate that laser reflection data as raw data also enabled us to select effective items. We created a base space using only 31 selected items (with decreasing value from left to right in Figure 7). Figure 8a shows the result of distances calculated for normal and abnormal data, and Figure 8b indicates the distances for normal and ab-

normal data in the base space formed with all 50 items before item selection.

According to these results, we can see that while the data in the after-item-selection base space are distributed somewhat broadly, the distance between the normal and abnormal data were sufficient for discrimination. Thus, we concluded that we could discriminate normality and abnormality by Mahalanobis distances using a base space created only with laser reflection data.

Reference

Akira Makabe, Kei Takada, and Hiroshi Yano, 1998. An application of the Mahalanobis distance for solder joint appearance automatic inspection. Presented at the Quality Engineering Forum Symposium, 1998. Japanese Standards Association.

This case study is contributed by Akira Makabe, Kei Takada, and Hiroshi Yano.

Application of the MTS to Thermal Ink Jet Image Quality Inspection

Abstract: Efficient conversion of electrical and thermal energy into robust ink drop generation (and subsequent delivery) to a wide variety of substrates requires target directionality, drop velocity, drop volume, image darkness, and so on, over a wide range of signal and noise space. One hundred percent inspection is justified based on loss function estimates, considering inspection cost, defective loss, and quality loss after shipment. The Mahalanobis–Taguchi system (MTS) was applied to the final inspection of image quality characteristics for a number of reasons: (1) to summarize multivariate results with a continuous number expressing distance from the centroid of "good" ones; (2) to reduce measurement cost through identification and removal of measurements with low discrimination power; (3) to balance misclassification cost for both type I and type II errors; and (4) to develop a case study from which practitioners and subject-matter experts in training could learn the MTS.

1. Introduction

Classification of thermal ink jet printheads as either acceptable (for shipment customers) or rejectable (for scrap) is a decision with two possible errors. Type I error occurs when a printhead is declared conforming when it is okay. Type II error occurs when a printhead is declared conforming when actually it is not. Type I error means that scrap costs may be high or orders cannot be filled due to low manufacturing process yield. Type II error means that the quality loss after shipment may be high when customers discover they cannot obtain promised print quality, even with printhead priming and other driver countermeasures. Over time, type II errors may be substantially more costly to Xerox than type I errors. Consequently, engineers are forced to consider a balance between shipping bad printheads and scrapping good ones.

A Mahalanobis space was constructed using multivariate image quality data from 144 good-print-quality printheads. The data were used to create a Mahalanobis space. Nonsequential serial numbers were used from several months of production. In addition, the Mahalanobis distance, D_2, of 45 nonconforming printheads were calculated using the aforementioned Mahalanobis space. The feasibility of the technique to better discriminate good from bad printheads was demonstrated. In addition, from these results, the contribution rate of each measurement characteristic to the Mahalanobis distance was calculated using L_{64} and analysis of variance (ANOVA) methods. The minimum set of measurement characteristics was then selected to reduce measurement costs and to speed up the analysis process. The accelerating pace of printhead fabrication, pulled by customer demand, has put considerable stress on the manufacturing system (which worked quite well for lower quantities of printheads). Reduction of these inspection data to one continuous Mahalanobis number and reducing measures, which have very low discrimination power, will speed up the inspection and disposition process.

2. Camera Inspection System

Assembled printheads were print-tested using 100% inspection. During the test, power was applied to the device, ink nozzles were primed, and a variety of patterns of drops were ejected from the linear array of nozzles. A standard Xerox paper was the medium, which received the various patterns of drops. A tricolor line scan camera captured the images of the print pattern automatically and sent the image data to a machine vision processor. The vision processor controlled the camera and used video data to place templates and gather statistics on the printed patterns. A host PC initiated inspection and configuration processes, created reports, and provided a user interface for the operators.

Certain camera calibration steps were done before testing, including white balance (to compensate for uneven light distribution over the image plane and the spatial nonuniformity of the sensitivity of the camera's sensors). Other calibrations included printed pattern measurement and alignment (to center the image in the camera's field of view), focusing and camera pixel-size calibration, focusing the light guides so that the maximum amount of light was in the camera's field of view, and calibration of color dot size so that dots of different colors were considered to be of the same size.

Dot aspect ratio, the ratio of a dot's height to its width, was one of the many characteristics captured by the camera system. It is shown in Figure 1 for magenta ink. A maximum aspect ratio is allowed, to control certain image quality problems.

A second characteristic captured by the camera system was dot diameter. The diameter is an equivalent diameter calculated for noncircular dots. For each color, the mean and standard deviation of both repeated dots and same jets were calculated. Upper and lower tolerances were applied to the dot size. Missing dot counts (or missing jets counts) were tabulated. Dot misallocation (directionality) results were summarized as both x and y error variances from least squares fitting routines of multiple dot patterns. If any single criteria limit is violated, the printhead is scrapped. Inspection results are generally reported in terms of pass/fail percentages considering decomposition by dot, by jet, and by color for each printhead. All of this existing inspection process does not consider any correlation among response characteristics in the decision making.

Figure 2 shows a hypothetical plot of aspect ratio versus dot diameter. With correlation, a printhead is classified as abnormal; with no correlation, a printhead is classified as normal. It is clear that a point labeled as abnormal (because it is outside the ellipse) would be considered normal if each characteristic were considered in isolation. It is at +1.4 from the mean of the spot diameter and 0.7 from the mean of the aspect ratio. Because of the correlation between the characteristics measured, one can say that this printhead has too large a spot diameter for its aspect ratio, and therefore it should be classified as abnormal.

3. Mahalanobis Distance

The quantity D^2 is called the Mahalanobis distance from the measured featured vector x to the mean vector mx, where Cov x is the covariance matrix x. The future vector represents the vector-measured characteristics for a given printhead in this study. The question is: How far is a given printhead from the database of good ones? If the distance is greater than some threshold, it is classified as defective. If the distance is less than some threshold, it is classified as no different from the population of good ones. It can be shown that the surfaces on which D^2 is constant are ellipsoids that are centered about

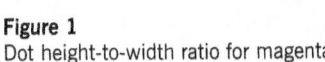

Aspect ratio 1.2   Aspect ratio 1.0

Figure 1
Dot height-to-width ratio for magenta

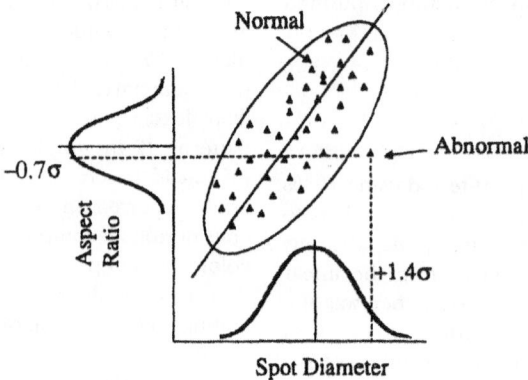

Figure 2
Hypothetical plot of aspect ratio versus dot diameter

the mean mx. In the special case where the measurements are correlated and the variances in all directions are the same, these spheres and the Mahalanobis distance become equivalent to the Euclidean distance.

The Mahalanobis distance, D^2, was calculated as:

$$D^2 = (x - \mu_x) C_x^{-1} (x - \mu_x)^{\mathrm{T}} \qquad (1)$$

where x is an n-dimensional vector of measured characteristics for a given printhead. The population of good printheads was characterized by a mean feature vector $m = (m_1, m_2, \ldots, m_{144})$. C^1 is the inverse of the variance–covariance matrix, σ_{ij}, for the measured characteristics x, where

$$\mathrm{Cov}(x_i x_j) = \frac{1}{m} \sum_1^m (x_{i\alpha} - m_i)(x_{je} - m_j) \qquad (2)$$

For $i = j$ this becomes the usual equation for variance. D^2 is the generalized distance between the αth product and the mean feature vector for the population of good printheads.

The covariance of two measures captures their tendency to vary together (i.e., to co-vary). Where the variance is the average of squared deviation of a measure from its mean, the covariance is the average of the products of the deviations of measured values from their means. To be more precise, consider measure I and measure j. Let $\{x(1,I), x(2,I), \ldots, x(n,I)\}$ be a set of n examples of characteristic I, and le$\{x(1,j), x(2,j), \ldots, x(n,j)\}$ be a corresponding

set of n examples of characteristic j. [That is, $x(k,I)$ and $x(k,j)$ are measures of the same pattern, pattern k.] Similarly, let $m(i)$ be the mean of feature I and $m(j)$ be the mean of feature j. Then the covariance of features I and j is defined by $c(i,j) = \{[x(1,i) - m(i)]\,[x(1,j) - m(j)] + \cdots + [x(n,i) - m(i)]\,[x(n,j) - m(j)]\}/(n - 1)$.

The covariance has several important properties:

❑ If measure i and measure j tend to increase together, then $c(i,j) > 0$

❑ If measure i tends to decrease when feature j increases, then $c(i,j) < 0$.

❑ If features i and j are independent, then $c(i,j) = 0$.

❑ $|c(i,j)| \leq s(i)s(j)$, where $s(i)$ is the standard deviation of feature i.

❑ $c(i,i) = s(i)^2 = v(i)$.

Thus, the covariance $C(i,j)$ is a number between $-s(i)s(j)$ and $+s(i)s(j)$ that means dependence between measures i and j, with $c(i,j) = 0$ if there is no dependence. All of the covariances $c(i,j)$ can be collected together into a covariance matrix as shown below:

$c(1,1)$	$c(1,2)$	$\cdots$	$c(1,d)$
$c(2,1)$	$c(2,2)$	$\cdots$	$c(2,d)$
$\vdots$	$\vdots$		$\vdots$
$c(d,1)$	$c(d,2)$	$\cdots$	$c(d,d)$

This matrix provides us with a way to measure distance that is invariant to linear transformations of the data. The covariance among the responses is a valuable additional piece of information characterizing the se of good printheads. For this study, the covariance matrix (among all of the responses for the good printhead data) was calculated. The inspection and rationalization of these results can provide very important insights into the performance of the printheads. For example, the spot size performance for the cyan ink is highly correlated with the spot size performance for the magenta ink, and moderately correlated with the spot size for the cyan ink. None of these three measures is correlated with the black ink spot size performance. This should make sense to the technologist.

Mahalanobis distance calculations were performed using SAS/I ML with custom macros. Fifty-eight different image-quality summary measures were considered, approximately 15 for each ink color. All 144 good printhead data were standardized according to

$$Y_{ij} = (X_{ij} - m_i)\sigma_i$$

This scaling creates a standardized distance of characteristic X to mean, m_x, relative to the standard deviation, σ_x. This scaling removes the effects of different units for the various characteristics. All measures were continuous data in this case, but different units for the various characteristics. All measures were continuous data in this case, although other types of data could be considered. Scaling is a particular example of a linear transformation. This standardization is also useful since it reduces the variance–covariance matrix to the better-known correlation matrix $R = \rho_{ij}$. It can be shown that the Mahalanobis distance for the xth product is given by

$$D_\alpha^2 = XR^{-1}X^T \tag{3}$$

This is the form used to calculate D^2 by the software code.

The use of the Mahalanobis metric removes several of the limitations of the Euclidean metric for calculating the distance between the printhead and the population of good ones:

1. It accounts automatically for the scaling of the coordinate axes.

2. It corrects for correlation between the different measures.

3. It can provide curved as well as linear decision boundaries.

4. Mahalanobis Distance Results

The Mahalanobis distance was calculated for each good printhead using the inverse correlation matrix, and values for D^2 ranged from 0.5 to 2.5. Singular matrix problems on inversion indicated very high correlation between several of the measured characteristics, and the elimination of one of them was required.

The average D^2 value was 1.000 for the good ones, as expected. (These distances are in units of standard deviations from the centroid of the multivariate good printhead data.) From these results an origin and unit amount were established. The inverse correlation matrix from the good ones, in turn, was used to calculate the Mahalanobis distance D^2 for each of the 44 bad printheads. It was expected that the D^2 values would be considerably larger than D^2 values for the good printheads. As a rule of thumb, a value below 4.0 for the bad ones usually indicated a misclassification with type 1 error. These numbers ranged from 2.6 to 179.4.

The Mahalanobis distance identifies the range of variation of the bad ones, which is much more informative than a simple good or bad binary classification system. It is a simple matter to go back through the individual printhead data to see why the really bad ones have much higher D^2 values. Knowing the level of badness enables continuous improvement activities.

If the distribution of the many measured characteristics is multivariate Gaussian, the D^2 follow a chi-square distribution with k degrees of freedom. Probabilities of membership in a defective classification can be determined. Again, as a rule of thumb, those printheads with D^2 values greater than about 3.0 to 4.0 can be classified as not belonging to the good group of printheads with fairly high confidence.

The Mahalanobis–Taguchi system allows you to distinguish normal and abnormal printhead conditions as long as you have data for the normal condition. Only the normal condition produces a uniform state, and infinite abnormal conditions exist. To collect data for all those abnormal conditions produced the Mahalanobis space. For new data we calculated the Mahalanobis distance, and by judging whether it exceeds the normal range, we can distinguish between normal and abnormal printhead conditions.

The traditional way is to discriminate which of multiple preexisting categories the actual device belongs to. For printheads, where one has to decide if it is good or not, a large volume of defective condition data and normal condition data are measured beforehand, and a database is constructed with both. For newly measured data (for which you do not know whether it is a good one or bad one), you calculate which group the data belongs to.

However, in reality, there are a multitude of image-quality problems. So if information about various types of images is not obtained beforehand, printheads cannot be distinguished. Because the traditional discrimination method requires this information, it is not practical.

5. Results

A $d \times d$ correlation matrix of the 58 observed characteristics is partially shown in Table 1. These correlations make looking at individual measured characteristics problematic for classifying printheads. When there is correlation between characteristics, diagnosis cannot be made from single characteristics. Table 2 shows the calculated D^2 values for the 145 good printheads. Table 3 shows the Mahalanobis distance for the 44 bad printheads, using the inverse correlation matrix from the good printheads.

Figure 3 shows the Mahalanobis distances D^2 for both the good and bad printheads. A wide range of D^2 values were observed for the nonconformant printheads. Note that the good ones are quire uniform. This is usually the case. There are one or two cases of printheads classified as bad but more likely part of the good ones. The threshold value for D^2 was determined by economic considerations. Misclassification countermeasure costs are usually considered.

6. Measurement System Cost Reduction

Among the 58 different measured characteristics in this example, there were a number that were doing most of the discrimination work. An L_{64} orthogonal array was used to identify those characteristics. In this case, the orthogonal array was used as a decision tool. One measured characteristic was assigned to each column, using 58 of the 63 available columns. L_{64} is a two-level design with 63 available degrees of freedom for assignment. A value of $+1$ in any cell of the standard array layout indicated that the measured characteristic was considered, while a value of -1 in any given cell meant that the measured characteristic was not considered. In row 1, for example, 27 of the 58 measures were considered since there were 27 positive ones. Using just those 27 columns, the analysis was run to generate a Mahalanobis space with the good printheads, and Mahalanobis distances D^2 for each of the 44 bad printheads was calculated. From these 27 responses, and 44 bad printheads, a single SN ratio (larger-the-better characteristic) was calculated according to

$$\eta = -10 \log \frac{1}{m} \left(\frac{1}{D_1^2} + \frac{1}{D_2^2} + \cdots + \frac{1}{D_m^2} \right) \quad (4)$$

where summation is from $m = 1$ to $m = 44$.

This same procedure was followed for the remaining L_{64} rows. A larger-the-better form was used, since the intent was to amplify the discrimination power by selecting a subset of measures. This in turn would reduce cost and time to process inspection data. The results are shown in Table 4 for each of the L_{64} rows. The SN ratios ranged from -16.6 dB to -4.7 dB. Row 64 of the L_{64} included use of each of the 58 measures, and consequently had the highest positive SN ratio.

An ANOVA procedure was conducted on these 64 SN ratios to decompose the total variation into constituent parts. Table 5 shows the large effects of a handful of measures (e.g., measure C_2, C_8, C_{15}, C_{17}, C_{18}, C_{19}, C_{29}, C_{43}, C_{57}), and the small effects of many of the others. This suggests that not all these measures are adding value to the decision to scrap or ship printheads. Much of the time, the measurements are adding little or nothing to the decision. Reducing the number of measures substantially

Table 1
Print quality inspection correlation matrix (partial for 58 variables)

1.00	−0.24	0.24	−0.25	0.12	−0.15	−0.03	−0.24	0.34	0.24	0.37	0.44	0.42
−0.20	−0.10	0.18	0.18	−0.29	0.30	−0.27	0.05	0.07	0.03	−0.08	0.29	0.38
0.25	−0.05	−0.15	0.21	0.45	−0.17	−0.05	−0.01	0.01	−0.00	0.02	0.27	0.24
0.7	0.25	0.04	−0.13	0.24	0.03	−0.19	0.07	−0.18	−0.03	0.04	−0.02	0.06
−0.04	0.20	0.16	0.06	−0.16	0.27							
−0.24	1.00	0.08	0.03	0.19	0.13	0.25	−0.18	−0.42	0.18	0.12	−0.40	−0.32
0.03	−0.24	−0.15	−0.31	0.97	0.06	0.14	−0.26	0.16	−0.25	0.11	0.02	−0.41
−0.28	0.05	−0.24	−0.21	−0.11	0.94	0.01	0.06	0.02	0.06	0.01	−0.06	0.07
−0.35	−0.29	0.27	−0.28	−0.20	−0.29	0.83	−0.15	0.14	−0.19	0.23	−0.31	0.07
0.08	−0.37	−0.28	−0.02	−0.25	−0.22							
0.24	0.08	1.00	−0.53	0.44	−0.21	−0.11	−0.35	−0.02	0.21	0.22	0.02	−0.01
0.16	−0.10	−0.06	−0.18	0.08	0.29	−0.35	−0.09	0.07	−0.15	−0.30	0.05	−0.03
−0.05	0.14	−0.16	−0.05	0.13	0.15	0.08	0.08	−0.22	0.08	−0.12	0.26	0.09
−0.07	−0.11	0.29	−0.15	−0.02	−0.17	0.04	0.21	−0.32	−0.18	0.28	−0.18	0.21
0.06	−0.18	−0.07	−0.03	−0.19	−0.02							
−0.25	0.03	−0.53	1.00	−0.20	−0.02	−0.23	0.29	−0.10	−0.34	−0.33	−0.26	−0.16
0.04	0.24	−0.09	−0.03	0.05	−0.28	0.82	0.00	−0.09	−0.13	0.14	−0.11	−0.20
−0.09	−0.01	0.28	−0.10	−0.13	0.00	−0.15	0.29	0.37	−0.17	−0.04	−0.34	−0.05
0.11	−0.02	−0.10	0.24	−0.12	0.03	0.11	−0.28	0.72	0.04	−0.23	0.06	−0.27
−0.01	0.02	0.01	0.09	0.31	−0.12							
0.12	0.19	0.44	−0.20	1.00	−0.26	−0.05	−0.08	−0.21	0.31	0.17	−0.13	0.02
−0.01	−0.06	−0.09	−0.14	0.15	0.28	−0.08	−0.05	−0.05	−0.15	−0.05	0.11	−0.22
−0.16	0.05	−0.07	−0.12	0.13	0.25	−0.06	0.13	0.03	−0.13	0.08	0.17	0.11
−0.08	−0.01	0.19	−0.06	−0.12	−0.10	0.22	−0.07	−0.06	−0.12	0.07	−0.08	0.12
0.05	−0.12	−0.10	0.07	−0.08	−0.12							
−0.15	0.13	−0.21	−0.02	−0.26	1.00	0.48	0.20	−0.16	−0.07	0.01	−0.25	−0.23
−0.04	0.07	−0.02	−0.01	0.11	0.04	−0.08	0.03	0.67	0.38	0.19	0.10	−0.18
−0.09	0.06	0.11	−0.05	−0.25	0.10	0.04	−0.50	−0.02	0.53	0.14	0.01	0.03
−0.20	−0.08	−0.09	0.11	−0.07	0.10	0.01	0.12	−0.28	0.12	0.14	0.12	−0.05
−0.08	−0.10	−0.01	0.01	0.12	−0.08							
−0.03	0.25	−0.11	−0.23	−0.05	0.48	1.00	0.11	−0.10	0.31	0.26	−0.11	−0.10
−0.12	−0.10	−0.01	−0.11	0.22	0.08	−0.11	−0.12	0.14	0.17	0.28	0.02	−0.11
−0.04	0.03	−0.06	−0.06	−0.09	0.23	0.00	−0.14	−0.14	0.10	0.25	0.07	0.03
−0.07	−0.04	0.05	−0.07	−0.04	−0.05	0.28	−0.18	−0.15	−0.04	−0.04	0.07	0.13
−0.01	0.02	0.02	0.03	−0.07	−0.05							

reduces the time and cost of inspection and dispositioning.

Using equation (4), a larger-the-better type of SN ratio was calculated from each of the L_{64} combinations, since we were looking for ways to amplify Mahalanobis distance by selecting just a few measures out of the total of 58 (Tables 6 and 7).

7. Conclusions

The Mahalanobis–Taguchi system was used to analyze image-quality data collected during final inspection of thermal ink jet printheads. A Mahalanobis space was constructed using results from a variety of printheads classified as good by all criteria,

Table 2
Print quality inspection Mahalanobis distances for 145 good printheads

0.9788539	0.7648473	0.8365425	1.1541308	1.5237813	0.6360767	1.9261662
0.6253195	1.8759381	0.6151459	0.8409623	2.3004483	0.8254882	0.9997392
0.8141697	1.0554864	1.3502383	1.4477616	1.575254	0.6408639	0.9636183
0.6377904	1.1194686	0.7462556	1.0316818	1.8069662	1.9241475	1.0319332
0.9899644	1.2182822	1.2182822	1.1331171	0.8370348	2.2826932	0.9207974
1.1467803	1.0811343	1.005008	0.9170464	1.2295005	0.5892113	0.4927686
0.8277777	0.9806735	0.9319074	0.9434869	0.8671445	0.8750086	0.8219719
1.2870564	0.6186984	1.5087154	1.0080104	1.343378	1.3982634	1.338509
0.5055908	0.9222672	1.2616672	2.0557088	0.4774843	0.7116405	0.4109008
0.9073186	1.0098421	0.610628	1.8020641	0.7664054	0.5304222	1.22455
0.6284057	1.407896	0.9961997	0.6662852	0.8032502	1.2031408	0.9997437
1.2198565	1.0614666	0.5896821	0.4663961	1.63111	0.7763159	1.0746786
0.744278	0.5047632	1.0467326	1.0163124	0.9345957	0.6408224	0.5880229
1.3450012	0.7046379	1.4418307	0.7789877	1.0798978	0.6764585	0.8649927
0.4805313	0.4805313	1.0123116	0.7207266	1.3288415	0.6955458	0.9248074
1.2550094	0.9059842	1.1129645	0.7871329	0.6158813	0.762891	0.7765336
0.5245411	0.5759779	1.0465223	1.2335352	1.6694431	0.9200819	1.4373301
0.8082051	0.9922023	0.9889715	0.5822106	1.5744132	0.5348402	1.0790753
0.6256058	0.8307524	0.8303376	1.3077433	0.6317183	1.2935359	0.509095
1.2105074	1.1197155	0.7580503	0.9266635	0.6304593		

Self-check: 58.00000

Table 3
Mahalanobis distances for 44 bad printheads

22.267015	6.3039776	47.516228	86.911572	15.757659	47.675354	9.2641579
2.5930287	14.052908	11.691446	15.090696	13.894336	8.6390361	179.43874
20.041785	10.695167	29.33736	13.926274	34.2429	9.3282624	44.442126
35.30109	18.388437	55.787193	6.0059889	63.12819	16.813476	45.355212
78.914301	29.200082	42.105292	28.056818	46.802471	121.26875	8.1542113
33.730109	9.3395192	6.777362	22.493329	16.492573	4.6645559	26.666635
22.46681	153.76889					

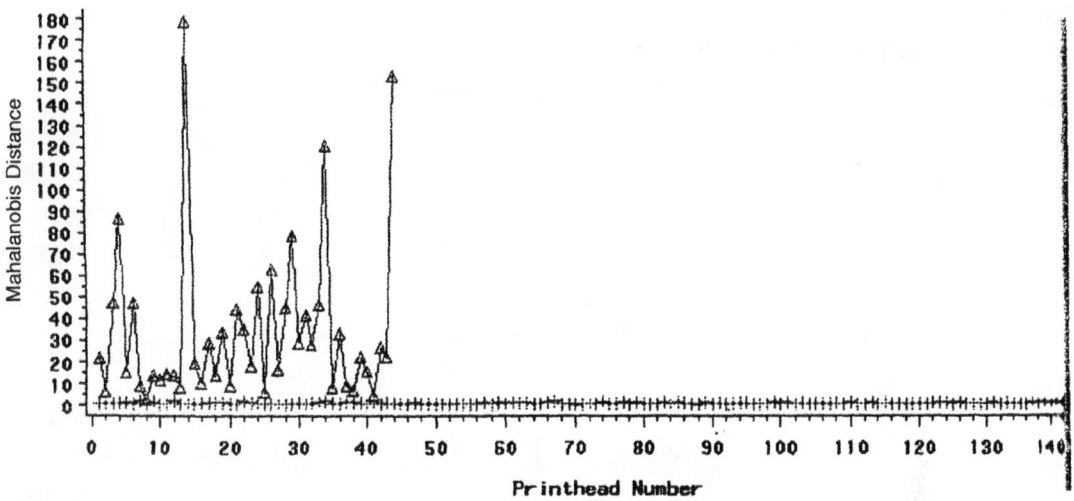

Figure 3
Mahalanobis distance versus printhead number for both passing and failing printheads

establishing an origin and unit number. Mahalanobis distance for nonconforming printheads was calculated, demonstrating the ability to discriminate normal and abnormal printheads. Among the 58 product image quality attributes investigated, only a fraction of the total were doing the work of discriminating good from bad for actual production, identifying the opportunity to reduce measurement and overhead cost for the factories. Misclassification of printheads was checked, and as expected, scrapping good printheads was more prevalent than shipping bad printheads. The scrap cost was approximately one-tenth the cost for type II error, so naturally one would be more likely to scrap good ones.

Other opportunities to apply the Mahalanobis–Taguchi system are being explored for asset recovery. For example, when copiers and printers are returned from customer accounts, recovery of motors, solenoids, clutches, lamps, and so on, that are still classified as good can be reused in new or refurbished machines. Other opportunities include image segmentation algorithm development, pattern recognition to prevent counterfeit copying of currencies, outlier detection in data analysis, system test improvement, diagnosis and control of manufacturing processes, chemical spectral analysis,

specification-free disposition, assembly process diagnosis, Kanji symbol recognition, magnetic ink check reading, and others.

The Mahalanobis–Taguchi system considers the following issues:

❑ Whether or not a newly manufactured or remanufactured product (or subsystem or component) is a member of the population of previously made good products (scrap/rework/disposition decision)

❑ How to characterize the degree of badness of recently made product with one scalar number derived from a multivariate set of measures

❑ How to define the contribution of each feature measured to the overall distance from the good population (cause detection)

❑ How to eliminate measures that do not help discriminate good from bad products (cost reduction due to misclassification)

❑ Use of loss function minimization for balancing false positive and false negative errors

❑ Optimization of multivariate data to move all output to the good classification without adding cost

Table 4

SN ratios for each L_{64} combination

Obs.	Expt.	η	Obs.	Expt.	η
1	1	−16.1455	33	33	−137804
2	2	−15.1823	34	34	−15.8187
3	3	−13.4513	35	35	−11.5528
4	4	−14.7418	36	36	−14.3503
5	5	−13.0265	37	37	−14.3373
6	6	−13.9980	38	38	−14.3891
7	7	−14.8928	39	39	−14.0671
8	8	−15.1808	40	40	−14.8759
9	9	−14.3932	41	41	−15.3622
10	10	−15.7381	42	42	−15.4632
11	11	−16.0573	43	43	−12.5654
12	12	−13.2949	44	44	−13.9813
13	13	−11.3572	45	45	−16.3452
14	14	−12.5372	46	46	−13.0218
15	15	−14.0662	47	47	−15.3362
16	16	−15.9616	48	48	−15.8292
17	17	−10.4232	49	49	−12.2162
18	18	−12.4296	50	50	−14.5625
19	19	−14.1015	51	51	−15.3548
20	20	−11.5102	52	52	−16.6430
21	21	−15.1508	53	53	−13.9844
22	22	−14.3844	54	54	−8.9751
23	23	−12.2187	55	55	−8.3045
24	24	−15.3981	56	56	−13.8523
25	25	−13.3026	57	57	−11.0553
26	26	−9.4273	58	58	−14.9251
27	27	−12.2470	59	59	−15.4196
28	28	−15.1932	60	60	−11.7635
29	29	−14.4695	61	61	−9.8751
30	30	−14.4875	62	62	−14.5514
31	31	−14.8804	63	63	−13.5409
32	32	−14.6857	64	64	−4.7039

Table 5
Mahalanobis distances for a sample of L_{64}

			Row 1 of L_{64}			
3.3948335	0.7672923	3.7212996	7.5160489	2.1628283	0.4735541	0.9902953
0.8921714	0.8890575	0.4336029	0.7916304	3.1689748	0.3589341	15.351063
1.6113871	0.8156152	5.1049158	3.53441	0.4765632	0.6596119	0.5178774
7.8548821	0.6789542	2.3212435	1.1780486	0.620615	4.3913711	30.419276
1.3079055	1.5259677	0.5979884	0.9019846	4.0599383	6.7014764	0.35931
0.6227532	3.3683819	1.3396863	5.3121466	0.7085592	0.7498931	11.770036
1.7869636	2.6817723					

			Row 2 of L_{64}			
2.1538859	0.5691297	5.0493192	6.0833953	2.26442	1.1433147	1.1618087
0.8776536	0.743007	0.861313	0.5233098	2.0991015	1.2277685	14.323771
3.5200479	0.2022022	4.8885268	2.3330918	0.8417254	1.1887293	0.7514819
4.1954716	6.0096008	1.0382971	3.080755	1.5613432	2.0078949	3.9968699
0.9804545	2.4407172	0.5850997	1.1164188	1.5777033	7.4701672	0.5262845
0.7533587	1.1707115	1.3342356	2.7278281	1.0099527	0.8908078	12.579455
2.772986	7.0534854					

			Row 3 of L_{64}			
2.432585	1.8731996	6.2922649	8.802438	2.2404416	1.631539	1.9070269
1.0529453	1.2410093	1.0671635	4.211971	2.7976614	0.7359613	6.9516176
1.2869917	3.3668795	4.4021798	5.6621817	1.9823413	2.2946022	0.7033499
14.94987	1.6268848	2.0984149	1.7826083	1.5005829	7.4753245	25.570186
0.9018103	7.125921	2.3458211	1.0736915	6.0013726	31.945727	3.6939773
3.2325816	2.7765327	0.5454912	3.371364	1.5793315	1.2316633	10.05149
2.5569832	2.9596748					

			Row 64 of L_{64}			
22.267015	6.3039776	47.516228	86.911572	15.757659	47.675354	9.2641579
2.5930287	14.052908	11.691446	15.090696	13.894336	8.6390361	179.43874
20.041785	10.695767	29.33736	13.926274	34.2429	9.3282624	44.442126
35.30109	18.388437	55.787193	6.0059889	63.12819	16.813476	45.355212
78.914301	29.200082	42.105292	28.056818	46.802471	121.26875	8.1542113
33.730109	9.3395192	6.777362	22.493329	16.492573	4.6645559	26.666635
22.46681	153.76889					

Table 6
ANOVA procedure for L_{64} results

Dependent variable: η, larger-the-better form					
Source	d.f.	Sum of Squares	Mean Square	F Value	Pr > F
Model	58	275.93384768	4.75748013	1.06	0.5408
Error	5	22.47309837	4.49461967		
Corrected total	63	298.40694604			
	R^2	C.V.	Root MSE		η Mean
	0.924690	−15.50422	2.12005181		−13.67403198

Table 7
ANOVA table

Source	d.f.	ANOVA SS	Mean Square	F Value	Pr > F
C_1	1	2.86055676	2.86055676	0.64	0.4612
C_2	1	34.60858396	34.60858396	7.70	0.0391
C_3	1	0.187724428	0.18724428	0.04	0.8463
C_4	1	1.49095175	1.49095175	0.33	0.5896
C_5	1	0.37607339	0.37607339	0.08	0.7840
C_6	1	1.14916741	1.14916741	0.26	0.6346
C_7	1	0.16314465	0.16314465	0.04	0.8564
C_8	1	29.49021893	29.49021893	6.56	0.0506
C_9	1	0.82834518	0.82834518	0.18	0.6856
C_{10}	1	8.20578845	8.20578845	1.83	0.2346
C_{11}	1	0.19192667	0.19192667	0.04	0.8444
C_{12}	1	0.11836547	0.11836547	0.03	0.8774
C_{13}	1	0.34995540	0.3499540	0.08	0.7914
C_{14}	1	2.07436008	2.07436008	0.46	0.5271
C_{15}	1	33.60580077	33.630580077	7.48	0.0411
C_{16}	1	7.51486220	7.51486220	1.67	0.2525
C_{17}	1	13.34268884	13.34268884	2.97	0.1455
C_{18}	1	9.47454201	9.47454201	2.11	0.2062
C_{19}	1	10.53510713	10.53510713	2.34	0.1863
C_{20}	1	1.69292927	1.69292927	0.38	0.5662
C_{21}	1	1.27346455	1.27346455	0.28	0.6173
C_{22}	1	0.00896986	0.00896986	0.00	0.9661
C_{23}	1	0.37507340	0.37507340	0.08	0.7843
C_{24}	1	0.70398097	0.70398097	0.16	0.7086
C_{25}	1	0.13738739	0.13738739	0.03	0.8681
C_{26}	1	0.00175670	0.00175670	0.00	0.9850
C_{27}	1	0.09980698	0.9980698	0.02	0.8874
C_{28}	1	0.01056163	0.01056163	0.00	0.9632
C_{29}	1	9.74085337	9.74085337	2.17	0.2010
C_{30}	1	0.68426748	0.68426748	0.15	0.7125
C_{31}	1	0.05612707	0.05612707	0.01	0.9154
C_{32}	1	0.26460629	0.26460629	0.06	0.8179
C_{33}	1	2.35115887	2.35115887	0.52	0.5019
C_{34}	1	5.05833213	5.05833213	1.13	0.3373

Table 7 (*Continued*)

Source	d.f.	ANOVA SS	Mean Square	F Value	Pr > F
C_{35}	1	0.04174250	0.04174250	0.01	0.9270
C_{36}	1	0.69712660	0.69712660	0.16	0.7099
C_{37}	1	0.00644011	0.00644011	0.00	0.9713
C_{38}	1	0.17791141	0.17791141	0.04	0.8501
C_{39}	1	0.21159149	0.21159149	0.05	0.8368
C_{40}	1	0.00395108	0.00395108	0.00	0.9775
C_{41}	1	0.06935294	0.06935294	0.02	0.9060
C_{42}	1	0.03615277	0.03615277	0.01	0.9320
C_{43}	1	32.02214448	32.02214448	7.12	0.0444
C_{44}	1	0.93469126	0.93469126	0.21	0.6675
C_{45}	1	0.01268770	0.01268770	0.00	0.9597
C_{46}	1	1.39043263	1.39043263	0.31	0.6020
C_{47}	1	2.01474985	2.01474985	0.45	0.5328
C_{48}	1	5.24140378	5.24140378	1.17	0.3295
C_{49}	1	0.03621724	0.03621724	0.01	0.9320
C_{50}	1	0.31732921	0.31732921	0.07	0.8011
C_{51}	1	0.61788448	0.61788448	0.14	0.7260
C_{52}	1	5.03501292	5.03501292	1.12	0.3383
C_{53}	1	0.00618522	0.00618522	0.00	0.9718
C_{54}	1	0.38732998	0.38732998	0.09	0.7809
C_{55}	1	0.91071239	0.91071239	0.20	0.6715
C_{56}	1	0.00739509	0.00739509	0.00	0.9692
C_{57}	1	46.41832533	46.41832533	10.33	0.0236
C_{58}	1	0.31011793	0.31011793	0.07	0.8033

❏ Definition of the probability of membership in a certain classification category

❏ Temporal change and control of a manufacturing process

❏ Getting practical use from advanced statistical concepts combined with robust design ideas

References

T. Kamoshita, K. Tabata, H. Okano, K. Takahashi, and H. Yano, 1996. Optimization of multi-dimensional information system using Mahalanobis distance: the case of a fire alarm system. *Journal of Quality Engineering Forum*, Vol. 4, No. 3.

Charles Naselli, 1996. Mahalanobis distance: a statistical inspection tool. Presented at the ITT Symposium.

Shin Taguchi. Private communication

A. White, K. Harrison, and R. Altavela. Private communication (IJFF).

This case study is contributed by Louis LaVallee.

CASE 71

Detector Switch Characterization Using the MTS

Abstract: This study confirmed the feasibility and improved discrimination of the multivariable Mahalanobis–Taguchi system (MTS) approach to detect and quantify the parameters specified. Based on the specified switch parameters, an MTS study was carried out with both good parts and bad production parts in order to select and to quantify the useful parameters that would be used for specifying and for checking the products at the lowest cost. Future evaluations will increase the sample size and the number of variables considered to improve the results. Implementation of this approach allows early detection of product performance (enabling shortened testing), detailed evaluation of product, and the potential to comprehend bias introduced by test conditions.

1. Introduction

The primary switch product types manufactured at Dole are tact, key, coding, rotary switches, and smart card connectors, designed for the communication, automotive, consumer, and industrial market. The KSM6 switch (Figure 1) was designed specifically to meet automotive market requirements as a switch to detect the portion of the ignition key in a high-end car model.

An engineering team was created to address the switch design and to improve its mechanical and electrical performances. The specification was defined with the customer according to the constraints given by the application. It was decided to use lots of parameters to characterize the product.

2. Background

The KSM6 is a detector switch with the following parameters (Figure 2): (1) force characteristics, (2) travel characteristics, (3), hysteresis performances between the ON and OFF curves, and (4) noise and bounces characteristics.

3. Objectives

As far as the KSM6 switch is concerned, we selected quite a lot of specified parameters necessary to guarantee both the quality and reliability of this product. Indeed, 19 parameters were chosen. The analytical objective for the Mahalanobis–Taguchi system approach was to reduce the number of parameters specified and to validate the characteristics according to the 19 parameters selected for the KSM6 product.

4. Experiment

The measurements were conducted in the ITT lab in Dole by using the *F/D method* (force–deflection electrical and mechanical measurements). It dealt with a traction/compression machine that enables it to establish the evolution curves of the component force according to the travel applied, thanks to an actuator. A connection enabled us to obtain the electric state of the product according to the same travel in the same way.

The F/D curve gave the points necessary to establish the mechanical and electrical characteristics

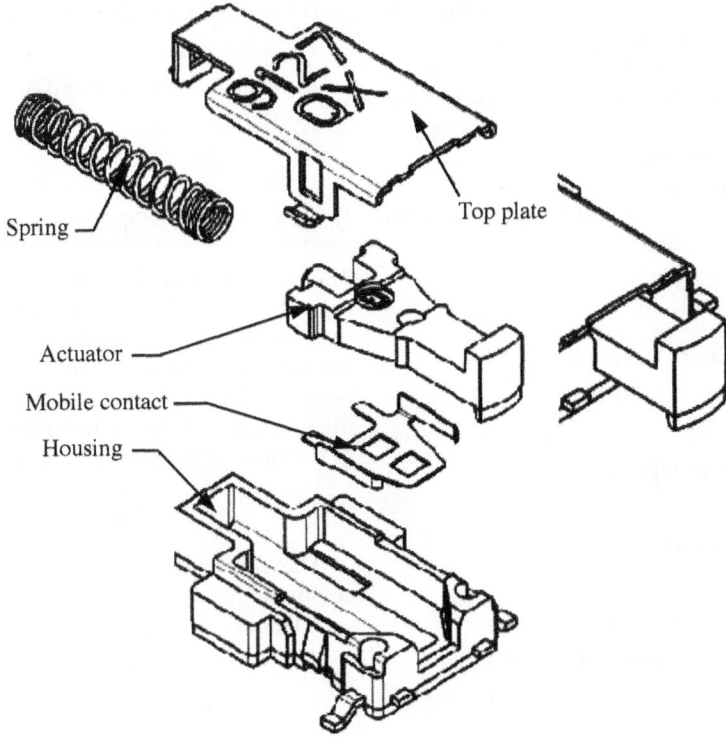

Figure 1
Three-dimensional and exploded assembly views of the KSM6 tact switch

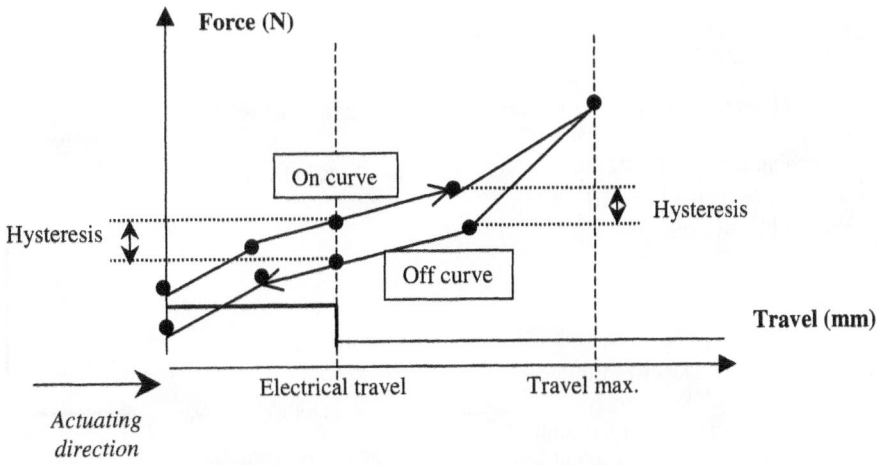

Figure 2
Typical force–deflection curve of the KSM6 switch

of the product. These characteristics allowed the validation of a product according to the specification (see Figure 3). This evaluation was based on switches coming from the assembly line. There were two groups: (1) good switches and (2) scrapped switches (rejected because of one or more parameters out the specification).

Following are the 19 specified parameters selected for the KSM6 switch:

1. Contact preload (N)
2. Fa (N)
3. Electrical travel ON (mm)
4. Electrical travel OFF (mm)
5. Mechanical travel (mm)
6. Electrical force ON (N)
7. Electrical force OFF (N)
8. Return force (N)
9. Force (at 1.85 mm)
10. Return force (at 1.85 mm)
11. Delta preload force/return force (N)
12. Delta electrical force ON/elect. force OFF (N)
13. Delta forces at 1.85 mm (N)
14. Noise beginning ON curve
15. Noise beginning OFF curve

16. Noise total ON curve
17. Noise total OFF curve
18. Contact resistance (mΩ)
19. Bounces (ms)

5. Mahalanobis Distance Calculations

The purpose of the MTS evaluation is to detect signal behavior outside the reference group. Existing data for the 19 characteristics of interest were organized for the 80 reference switches. The data were normalized for this group (Table 1) by considering the mean and standard deviation of this population of switches for each variable of interest:

$$Z_i = x_i - \frac{\bar{x}_i}{\sigma_i} \qquad (1)$$

The correlation matrix was then calculated to comprehend all 19 variables and their respective correlations:

$$\mathbf{R} = \begin{bmatrix} 1 & r_{12} & \cdots & r_{1k} \\ r_{21} & 1 & \cdots & r_{2k} \\ \vdots & \vdots & \cdots & \vdots \\ r_{k1} & r_{k2} & \cdots & 1 \end{bmatrix} \qquad (2)$$

$$r_{ij} = \frac{\sum x_{i1} x_{j1}}{n} \qquad (1: 1,2, \ldots , n)$$

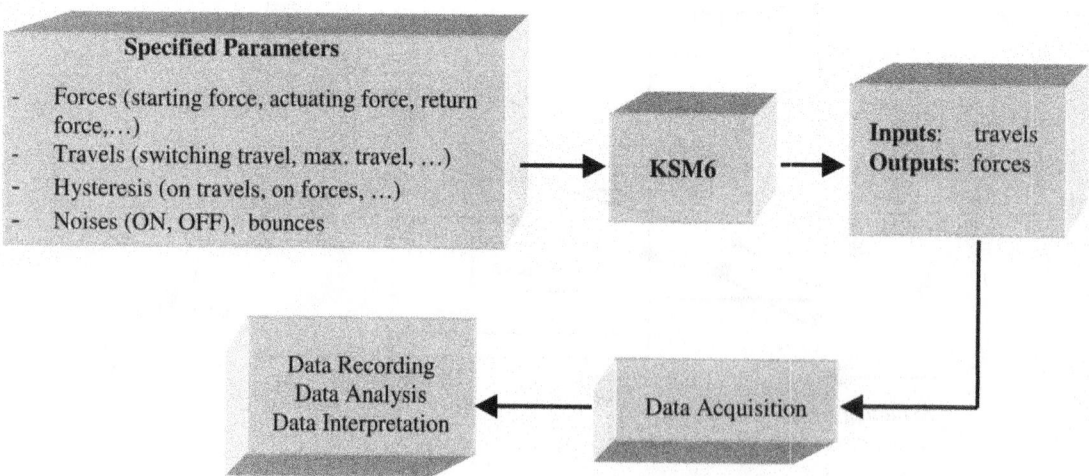

Figure 3
System, subsystems, and components

Table 1
Reference group output data normalization

	Reference Data					Normalized Data				
No.	Variable 1 X_1	Variable 2 X_2	Variable 3 X_3	...	Variable 19 X_{19}	Variable 1 Z_1	Variable 2 Z_2	Variable 3 Z_3	...	Variable 19 Z_{19}
1										
2										
3										
4										
5										
6										
7										
8										
...										
80										
Mean	$\overline{X}_1$	$\overline{X}_2$	$\overline{X}_3$	...	$\overline{X}_{19}$	0.0	0.0	0.0	...	0.0
St. Dev.	σ_1	σ_2	σ_3	...	σ_{19}	1.0	1.0	1.0	...	1.0

Table 2
Correlation matrix results for the reference group

		PRL 1	FA 2	CE ON 3	CE OFF 4	OM 5	FOE ON 6	FOE OFF 7	FR 8	FM ON 9	FM OFF 10	HYST1 11	HYST2 12	HYST3 13	NB OFF 14	NB ON 15	NT ON 16	NT OFF 17	RC 18	BOUN 19
PRL	1	1.000	0.812	0.511	0.477	-0.099	0.839	0.811	-0.281	0.814	0.797	0.716	0.510	0.587	-0.133	-0.130	-0.030	-0.025	-0.011	0.042
FA	2	0.812	1.000	0.770	0.734	-0.063	0.968	0.942	-0.733	0.983	0.961	0.869	0.569	0.753	-0.057	-0.236	-0.100	-0.096	-0.012	-0.002
CE ON	3	0.511	0.770	1.000	0.961	0.072	0.807	0.831	-0.755	0.745	0.747	0.729	0.280	0.616	-0.042	-0.277	-0.009	-0.158	-0.035	0.081
CE OFF	4	0.477	0.734	1.961	1.000	0.014	0.758	0.812	-0.736	0.719	0.715	0.711	0.210	0.577	-0.019	-0.231	-0.025	-0.171	0.012	0.054
CM	5	-0.099	-0.063	0.072	0.014	1.000	-0.056	-0.030	-0.025	-0.143	-0.151	-0.001	-0.114	-0.204	0.006	0.179	0.112	0.007	0.041	-0.067
FCE ON	6	0.839	0.968	0.807	0.758	-0.056	1.000	0.937	-0.668	0.956	0.935	0.846	0.602	0.766	-0.072	-0.290	-0.083	-0.107	-0.017	0.056
FCE OFF	7	0.811	0.942	0.831	0.812	-0.030	0.937	1.000	-0.650	0.935	0.948	0.807	0.345	0.646	-0.102	-0.164	-0.025	-0.107	0.012	0.020
FR	8	-0.281	-0.733	-0.755	-0.736	-0.025	-0.668	-0.650	1.000	-0.705	-0.705	-0.792	-0.419	-0.577	-0.085	0.196	0.127	0.127	-0.105	0.044
FM ON	9	0.814	0.983	0.745	0.719	-0.143	0.956	0.935	-0.705	1.000	0.957	0.851	0.544	0.753	-0.054	-0.214	-0.091	-0.096	0.002	0.016
FM OFF	10	0.797	0.961	0.747	0.715	-0.151	0.935	0.948	-0.705	0.957	1.000	0.822	0.486	0.654	-0.100	-0.167	-0.074	-0.085	0.008	0.030
HYST1	11	0.716	0.869	0.729	0.711	-0.001	0.846	0.807	-0.792	0.851	0.822	1.000	0.549	0.674	0.009	-0.298	-0.066	-0.083	0.092	0.005
HYST2	12	0.510	0.569	0.280	0.210	-0.114	0.602	0.345	-0.419	0.544	0.486	0.549	1.000	0.689	0.004	-0.254	-0.192	-0.068	-0.038	0.017
HYST3	13	0.587	0.753	0.616	0.577	-0.204	0.766	0.646	-0.577	0.753	0.654	0.674	0.689	1.000	-0.026	-0.330	-0.187	-0.124	0.050	0.004
NB OFF	14	-0.133	-0.057	-0.042	-0.019	0.006	-0.072	-0.102	-0.085	-0.054	-0.100	0.009	0.004	-0.026	1.000	0.038	0.073	0.309	-0.080	-0.007
NB ON	15	-0.130	-0.236	-0.277	-0.231	0.179	-1.290	-0.164	0.196	-0.214	-0.167	-0.298	-0.254	-0.330	0.038	1.000	0.039	0.009	-0.028	-0.176
NT ON	16	-0.030	-0.100	-0.009	-0.025	0.112	-0.083	-0.025	0.127	-0.091	-0.074	-0.066	-0.192	-0.187	0.073	0.039	1.000	-0.009	-0.122	0.050
NT OFF	17	-0.025	-0.096	-0.158	-0.171	0.007	-0.107	-0.107	0.127	-0.096	-0.085	-0.083	-0.068	-0.124	0.309	0.009	-0.009	1.000	-0.033	0.010
RC	18	-0.011	-0.012	-0.035	0.012	0.041	-0.017	0.012	-0.105	0.002	0.008	0.092	-0.038	0.050	-0.080	-0.028	-0.122	-0.033	1.000	-0.008
BOUN	19	0.042	-0.002	0.081	0.054	-0.067	0.056	0.020	0.044	0.016	0.030	0.005	0.017	0.004	-0.007	-0.176	0.050	0.010	-0.008	1.000

Upon review of the correlation matrix (Table 2), it is clear that correlation between parameters exists. For this reason, the application of the multivariable Mahalanobis–Taguchi system approach makes sense because no single characteristic can describe the output fully.

The inverse of the matrix was then calculated:

$$\mathbf{R}^{-1} = \begin{bmatrix} a_{11} & a_{12} & \cdots & a_{1k} \\ a_{21} & a_{22} & \cdots & a_{2k} \\ \vdots & \vdots & & \vdots \\ a_{k1} & a_{k2} & \cdots & a_{kk} \end{bmatrix} \quad (3)$$

and finally the Mahalanobis distance:

$$\mathrm{MD} = \frac{1}{k}ZR^{-1}Z^{\mathrm{T}} \quad (4)$$

where k is the number of characteristics, Z the 1×19 normalized data vector, R^{-1} the 19×19 inverse correlation matrix, and Z^{T} the transposed vector (19×1).

This completes the calculations of the normal group. All reference samples had MD distances of less than 2 (Table 3).

TMD for the abnormal samples was then calculated. Again the data were normalized, but now the mean and standard deviations of the reference group were considered. The inverse correlation matrix of the reference group solved previously was

Table 3
Mahalanobis distance values for the reference and abnormal groups

Sample	Good Parts	Rejected Parts
1	0.45588274	3.34905375
2	1.11503408	2.40756945
3	0.17740621	4.13031615
4	0.67630344	2.44623681
5	0.51367029	1.70684039
6	0.91088082	3.62376137
7	0.47617251	1.8801606
8	0.48574861	2.74470151
9	0.61893043	4.58774521

Table 3 (*Continued*)

Sample	Good Parts	Rejected Parts
10	1.27624221	2.46283229
11	0.91560766	
12	0.73373554	
13	0.4391565	
14	0.37539039	
15	0.91071876	
16	0.29173633	
17	0.28862911	
18	0.40312754	
19	0.46821194	
20	0.29330727	
⋮	⋮	
60	0.24485985	
61	0.45349242	
62	0.22177811	
63	0.29027765	
64	0.08698667	
65	0.15542392	
66	0.31067779	
67	0.09868037	
68	0.34478916	
69	1.23338162	
70	0.45290798	
71	0.29085425	
72	0.76586855	
73	0.3832427	
74	1.15630344	
75	0.70401821	
76	0.15559801	
77	0.29566716	
78	0.81947543	
79	0.35900551	
80	2.58171136	

also used. The resulting MDs of the abnormal samples are summarized in Table 3.

6. Discussion

As is evident in the MDs of the abnormal samples, tremendous discrimination between good and bad switches was accomplished (Figure 4). To reduce data-processing complexity, it is desirable to consider fewer characteristics and eliminate those not contributing to product discrimination. Four out of 19 characteristics were selected as very important, and these characteristics were used all of the time and were not considered for screening. The other 15 characteristics were assigned to an L_{16} array (Table 4).

All these 15 characteristics were considered at two levels. Level 1 used the variable to calculate the Mahalanobis distance, and level 2 did not use the variable to calculate the MD. Reconsideration of both the reference group and abnormal group MD was made for each run. The experiment design and results are shown in Table 4.

From these runs, SN ratios and mean responses were calculated for the main effects of each variable. As the goal was to improve discrimination, larger MDs were preferred and the larger-the-better SN ratio was used:

$$SN = \eta = -10 \log \left(\frac{1}{n} \sum_{i=1}^{n} \frac{1}{y_i^2} \right) \tag{5}$$

The data transformations gave the results shown in the response charts and ANOVA tables in xc-Figures and 6 and Table 5. Variables A, C, F, G, I, J, and O are shown to have little contribution to the SN ratio and could be considered for elimination. This would reduce the MD calculation to 12 characteristics. Some of the variables rejected contribute significantly to the mean, but as a whole, the effects of the factors on the mean compensate mutually to obtain a small variation to the mean (ca. 7%), which is fully acceptable. So we can confirm the choice of eliminated characteristics.

7. Confirmation

The confirmation method consists in doing the same MTS calculations by taking off the nonsignificant factors. By selecting only the significant factors, 12 out of 19 in our case, we created a new MTS graph (Figure 7). MD values for the normal and abnormal samples are shown in Table 6. The optimization evaluation gives very good results. Indeed, we can confirm that there is still a very good selection between the bad and the good pieces, even though we eliminated seven nonsignificant parameters.

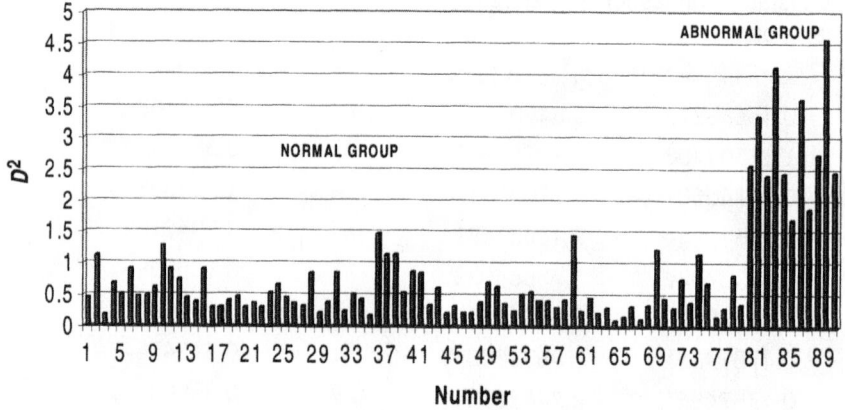

Figure 4
Mahalanobis distance for normal and abnormal groups

Table 4
MDs calculated for abnormal group within the L_{16} orthogonal array

No.	A	B	C	D	E	F	G	H	I	J	K	L	M	N	O	1	2	3	4	5	6	7	8	9	10
			Factor																**MD for 10 Abnormal Samples**						
1	1	1	1	1	1	1	1	1	1	1	1	1	1	1	1	3.349	2.408	4.13	2.446	1.707	3.624	1.88	2.745	4.588	2.463
2	1	1	1	1	1	1	1	2	2	2	2	2	2	2	2	5.066	3.555	3.507	2.848	2.09	4.867	2.506	0.235	7.818	0.726
3	1	1	1	2	2	2	2	1	1	1	1	2	2	2	2	4.615	1.247	1.826	3.466	1.209	2.64	2.28	2.403	7.784	0.805
4	1	1	1	2	2	2	2	2	2	2	2	1	1	1	1	4.213	1.19	2.264	3.437	1.548	2.866	2.633	3.762	7.788	3.278
5	1	2	2	1	1	2	2	1	1	2	2	1	1	2	2	5.198	1.873	6.523	3.057	1.443	1.094	1.203	0.918	7.835	0.56
6	1	2	2	1	1	2	2	2	2	1	1	2	2	1	1	4.404	1.541	2.829	2.241	0.796	1.089	1.021	4.061	7.822	3.894
7	1	2	2	2	2	1	1	1	1	2	2	2	2	1	1	4.492	2.8	1.141	2.645	2.029	3.73	1.035	2.965	7.799	3.23
8	1	2	2	2	2	1	1	2	2	1	1	1	1	2	2	4.321	2.906	1.658	3.11	2.435	4.467	1.204	2.898	7.791	0.586
9	2	1	2	1	2	1	2	1	2	1	2	1	2	1	2	4.842	3.027	4.105	3.384	1.983	1.426	2.37	2.825	7.81	3.474
10	2	1	2	1	2	1	2	2	1	2	1	2	1	2	1	4.183	2.941	3.128	1.898	1.318	1.366	2.146	3.102	7.849	0.617
11	2	1	2	2	1	2	1	1	2	1	2	2	1	2	1	5.135	0.668	1.337	2.097	1.548	5.294	2.334	1.094	7.8	0.911
12	2	1	2	2	1	2	1	2	1	2	1	1	2	1	2	5.262	1.117	2.499	3.1	2.092	4.619	2.166	3.782	7.804	3.809
13	2	2	1	1	2	2	1	1	2	2	1	1	2	2	1	4.878	1.251	4.353	2.739	1.657	4.367	2.037	2.648	7.826	0.552
14	2	2	1	1	2	2	1	2	1	1	2	2	1	1	2	4.119	1.045	3.491	1.422	1.475	4.627	1.534	3.672	7.794	3.371
15	2	2	1	2	1	1	2	1	2	2	1	2	1	1	2	4.621	2.465	1.894	1.644	1.415	2.481	1.483	4.436	7.794	3.705
16	2	2	1	2	1	1	2	2	1	1	2	1	2	2	1	4.383	2.847	2.668	2.976	2.037	2.78	1.962	0.504	7.793	0.531

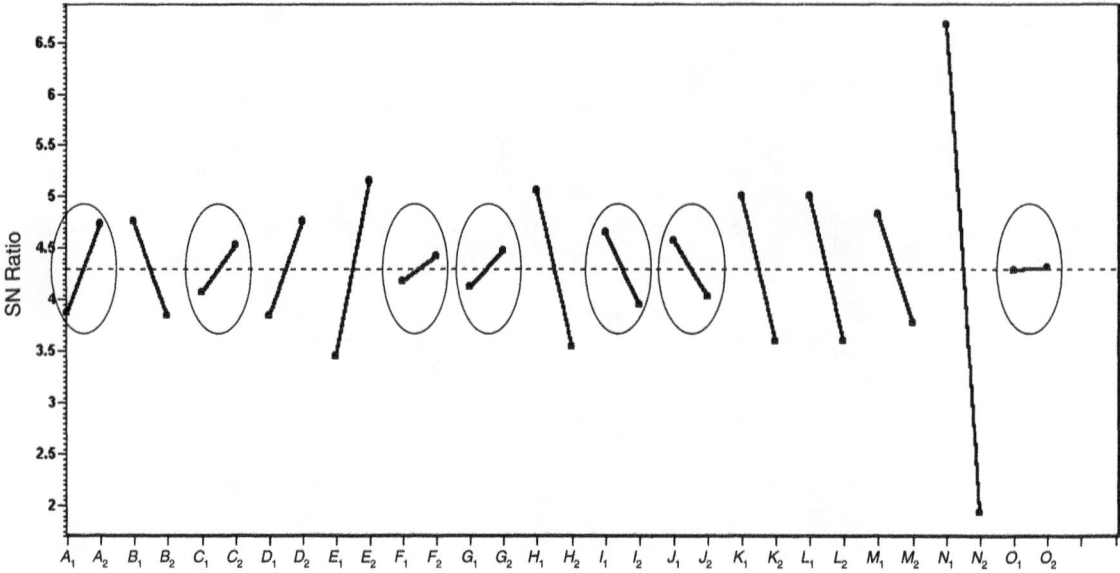

Figure 5
Response graph for the SN ratio

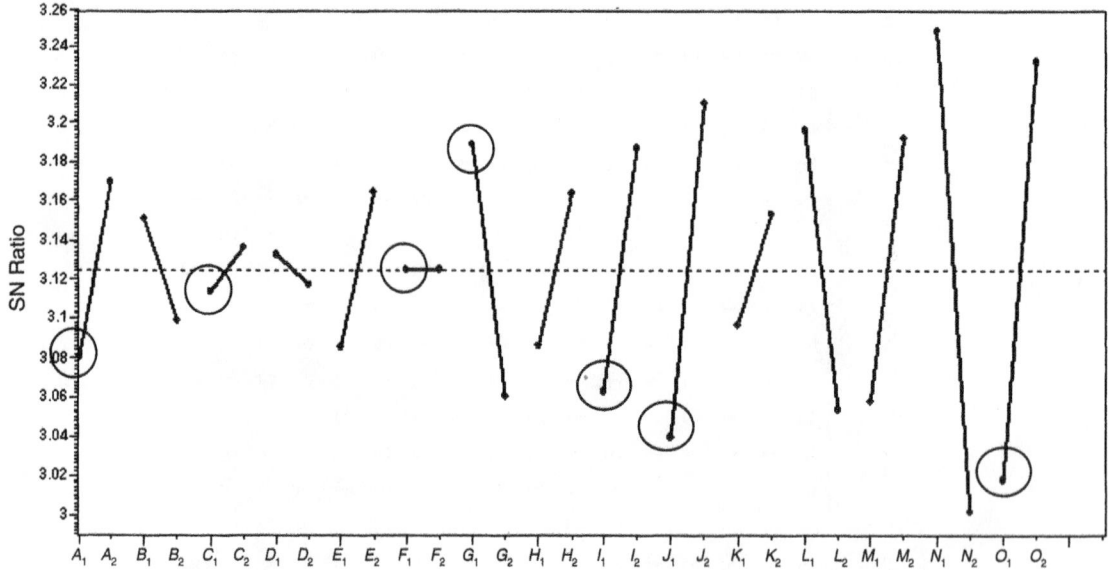

Figure 6
Response graph for the mean values

Table 5
ANOVA for the SN Ratio

Source	d.f.	S	V	F	S'	r
A	1	3.064	3.064			
B	1	3.3156	3.3156	2.9981	2.2097	1.53
C	1	0.8761	0.8761			
D	1	3.2996	3.2996	2.9836	2.1937	1.52
E	1	11.447	11.447	10.3507	10.341	7.15
F	1	0.2492	0.2492			
G	1	0.4806	0.4806			
H	1	9.0515	9.0515	8.1847	7.9456	5.49
I	1	1.9429	1.9429			
J	1	1.1272	1.1272			
K	1	7.9037	7.9037	7.1468	6.7978	4.7
L	1	7.9029	7.9029	7.146	6.797	4.7
M	1	4.342	4.342	3.9262	3.2361	2.24
N	1	89.7075	89.7075	81.1166	88.6016	61.23
O	1	0.0013	0.0013			
e_1						
e_2						
(e)	7	7.7414	1.1059		16.5886	11.46
Total	15	144.7111	9.6474			

(e) is pooled error.

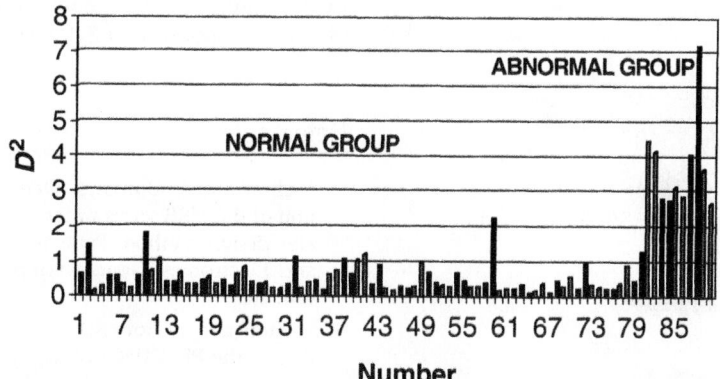

Figure 7
Mahalanobis distance for normal and abnormal groups

Table 6

Mahalanobis distance values for the reference and abnormal groups

Sample	Good Parts	Rejected Parts
1	0.58527515	4.4383654
2	1.42813712	4.15092412
3	0.15154057	2.82153352
4	0.29781518	2.75523643
5	0.54822524	3.12827267
6	0.5647138	2.88094718
7	0.32688929	4.01820264
8	0.24148032	7.18625986
9	0.5559233	3.62863953
10	1.80277538	2.67755672
11	0.70160571	
12	1.05331414	
13	0.36788692	
14	0.35625463	
15	0.64054606	
16	0.33201852	
17	0.31591459	
18	0.43327776	
19	0.53893526	
20	0.33842311	
⋮	⋮	
60	0.18403453	
61	0.22158889	
62	0.2094948	
63	0.3233402	
64	0.08915155	
65	0.16588856	
66	0.37570123	
67	0.11199467	
68	0.46124969	
69	0.28254858	
70	0.57035309	

Table 6 (*Continued*)

Sample	Good Parts	Rejected Parts
71	0.2190487	
72	0.95366322	
73	0.32950224	
74	0.26453747	
75	0.22518346	
76	0.20727962	
77	0.35644774	
78	0.90913625	
79	0.44455372	
80	1.26760236	

8. Conclusions

The feasibility of applying the MTS approach has been demonstrated. In our present case we used it to characterize and to select the parameters specified for a detector switch. Indeed, thanks to this method, we were able to keep 12 specific parameters out of the 19 initially selected. The confirmation of the discrimination between the good and the bad switches without using the nonselected parameters was very good.

The MTS method was very helpful to eliminate the seven parameters from the specification and realize a significant reduction of the checking costs. In our case, we could reduce these costs by 37%, which corresponds to a $200,000 yearly profit.

References

S. Rochon and P. Bouysses, October 1997. Optimization of the MSB series switch using Taguchi's parameter design method. Presented at the ITT Defense and Electronics Taguchi Symposium '97, Washington, DC.

S. Rochon and P. Bouysses, October 1999. Optimization of the PROXIMA rotary switch using Taguchi's parameter design method. Presented at the ITT Defense and Electronics Taguchi Symposium '99, Washington, DC.

S. Rochon and T. Burnel, October 1995. A three-step method based on Taguchi design of experiments to optimize robustness and reliability of ultra miniature SMT tact switches: the top actuated KSR series. Presented at the ITT Defense and Electronics Taguchi Symposium '95, Washington, DC.

S. Rochon and T. Burnel, October 2000. Optimization of the KMS ultraminiature switch using Taguchi's parameter design method. Presented at the ITT Defense and Electronics Taguchi Symposium 2000, Washington, DC.

S. Rochon, T. Burnel, and P. Bouysses, October 1999. Improvement of the operating life of the TPA multifunction switch using Taguchi's parameter design method. Presented at the ITT Defense and electronics Taguchi Symposium '99, Washington, DC.

S. Surface and J. Oliver II, 2001. Exhaust sensor output characterization using MTS. Presented at the 18th ASI Taguchi Methods Symposium.

G Taguchi, 1993. *Taguchi on Robust Technology Development*. New York: ASME.

G. Taguchi, S. Chowdhury, and Y. Wu, 2001. *The Mahalanobis–Taguchi System*. New York: McGraw Hill.

G. Taguchi, S. Chowdhury, and S. Taguchi, 1999. *Robust Engineering*. New York: McGraw-Hill.

This case study is contributed by Sylvain Rochon.

CASE 72

Exhaust Sensor Output Characterization Using the MTS

Abstract: The Mahalanobis–Taguchi system (MTS) evaluation described here considers the change in exhaust sensor signal performance resulting from an accelerated engine test environment. This study confirmed the feasibility and improved discrimination of the multivariable MTS approach to detect and quantify even small changes in signal output response. Future evaluations will increase the sample size and number of variables considered to verify the results. Implementation of this approach allows early detection of product performance shifts (enabling shortened testing), detailed evaluation of product design changes, and the potential to comprehend bias introduced by test conditions.

1. Introduction

Delphi Automotive Systems manufactures exhaust oxygen sensors for engine management feedback control. The stoichiometric switching sensor is located in the exhaust stream and reacts to rich and lean exhaust conditions. The sensor output signal (0 to 1 V) must maintain a consistent response throughout its life to ensure robust operation and allow tight engine calibrations that minimize tailpipe emissions.

Test Engineering at Delphi performs a variety of accelerated test schedules to expose the sensor realistically to representative vehicle conditions. Sensor performance measurements are conducted to monitor the sensor output characteristics throughout its life. Characterizing the sensor performance is often accomplished by recording and analyzing the sensor output voltage under a range of controlled exhaust conditions.

As emission control standards become more stringent and sensor technology improves to meet these demands, the testing community needs to improve its techniques to describe product perform-

ance accurately. The multivariable MTS evaluation presented here considers change in the stoichiometric exhaust oxygen sensor signal performance resulting from an accelerated test environment.

2. Sensor Aging Responses

The exhaust oxygen sensor is expected to perform in a high-temperature environment with exposures to water, road salt and dirt, engine vibration, and exhaust-borne contaminants with minimal change in performance from the manufacturing line to the end of vehicle life, which can exceed 150,000 miles. However, decreasing development cycles do not permit the accumulation of extensive vehicle mileage, so accelerated durability cycles have been developed in the test laboratory. These test schedules simulate the thermal, chemical, environmental, and mechanical demands that would be experienced on a vehicle.

A new sensor and an aged sensor will respond differently to these exposures based on product design combinations that include the electrodes, coatings, and package design selections. Test Engineering evaluates these design combinations by exposing the product to various accelerated durability tests and reports the sensor response. Figure 1 shows different sensor output responses after exposure to a few of these durability tests.

3. Sensor Performance Testing

Various methods exist to evaluate sensor performance, including electrical checks, flow bench tests using single or blended gases, and engine dynamometers. Engine dynamometers create a realistic exhaust gas stream as typical of a vehicle and with proper engine control can create a wide variety of stable engine running conditions.

One of the engine dynamometer performance tests is an open-loop perturbation test where the test sensor reacts to alternating rich and lean air/fuel mixtures about stoichiometry. These air/fuel ratio perturbations can be conducted at different frequencies and amplitudes and under different exhaust gas temperatures. From the simple output waveform, the measured signal is analyzed to derive more than 100 characteristics. The most descriptive characteristics were chosen for this preliminary evaluation. Figure 2 is a schematic of the system considered. The data used to support this analysis were available from previous traditional studies.

4. Experiment

Traditional methods of sensor performance analyses consider key characteristics of interest to customers and ensure that product specifications are met. Product development teams, however, want to understand not just time to failure but also the initiation and rate of signal degradation. Sensor output characteristics must indicate this response change over time.

The goals of this evaluation were to determine whether discrimination among aged test samples could be achieved, whether the analysis could comprehend both product and test setup variations, and whether it would provide a tool capable of detecting early sensor signal change. The MTS method used here generates a numerical comparison of a reference group to an alternative group to detect levels of abnormality. The method also identifies the key factors associated with these differences.

Definition of Groups

This evaluation was based on sensors with differing levels of aging. Twenty-six oxygen sensors were chosen as the reference (normal) group. These sensors had no aging and were of the same product design with similar fresh performance. These sensors were characterized in two groups with similar test conditions.

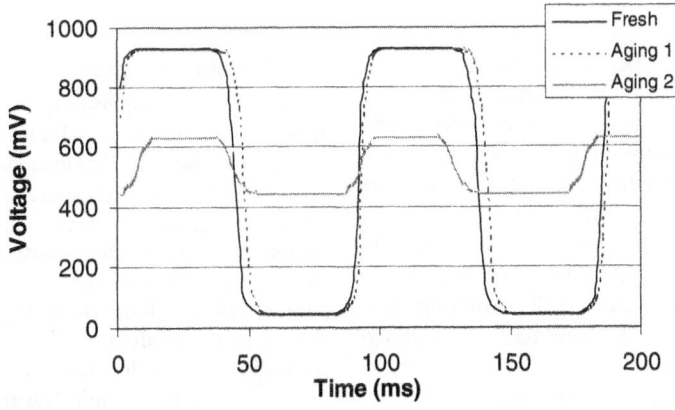

Figure 1
Sensor output for different sensors after various aging tests

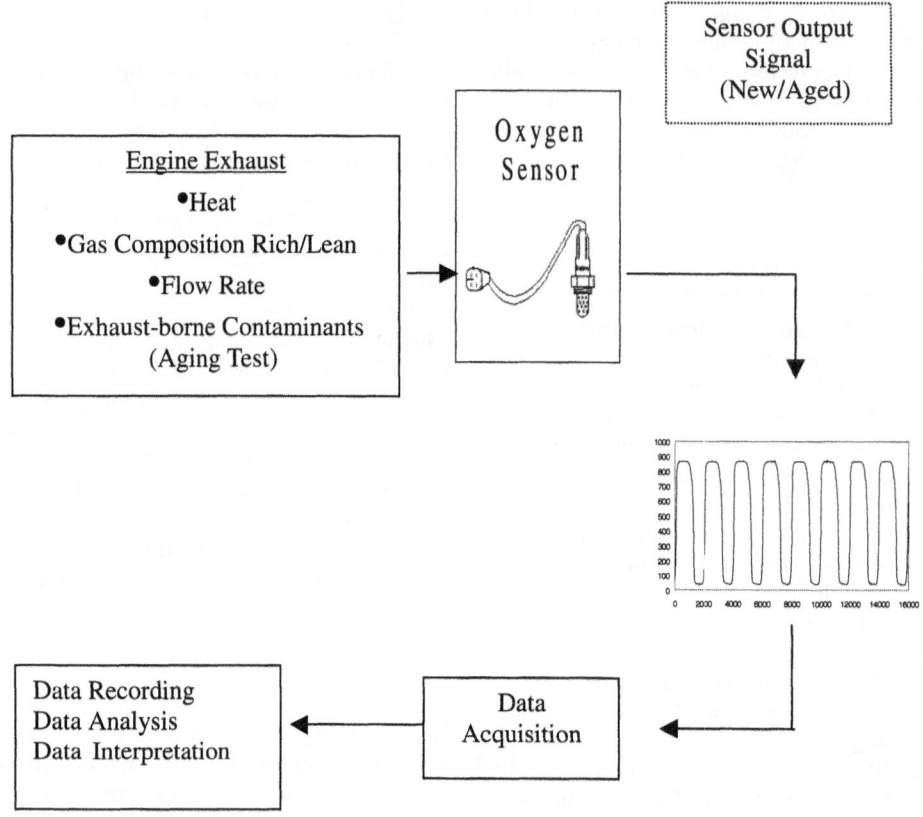

Figure 2
System, subsystems, and components

Next, the abnormal population was selected. A total of nine sensors were selected based on end-of-test postaging performance. The sensors had completed the full exposure of a highly accelerated engine-aging environment. Six of the sensors showed excellent posttest performance with little to no signal degradation ("aged"). Three of the abnormal sensors selected showed noticeable degradation, although they were still switching and functional ("degraded"). Representative voltage signals are shown in Figure 3. The "fresh" belonged to the reference group, and the "aged" and "degraded" were in the abnormal group.

Definition of Characteristics

As discussed, the open-loop engine-based performance test generates an exhaust stream to which the test sensors respond. Traditional sensor output characteristics that are often specified include maximum voltage, minimum voltage, voltage amplitude, response time in the lean-to-rich direction, and response time in the rich-to-lean direction. These parameters are denoted in Figure 4 and were included in the evaluation. One test-related parameter indicating the location of the test sensor (nest position) was also included, as multiple sensors were tested simultaneously. Considering the reference

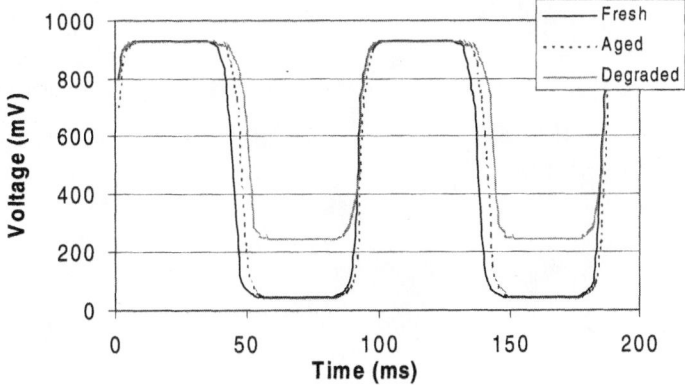

Figure 3
Sensor output voltage traces before and after aging

group sample size of 26, only nine additional characteristics were selected, for a total of 15 (referred to as factors *A* to *O*). The other characteristics, although not defined here, comprise a best attempt at variables that describe the waveform.

5. Mahalanobis Distance

The purpose of an MTS evaluation is to detect signal behavior outside the reference group. Existing

data for the 15 characteristics of interest were organized for the 26 reference (nonaged) sensors. The data were normalized for this group (Table 1) by considering the mean and standard deviation of this population for each variable of interest:

$$Z_i = \frac{x_i - \overline{x_i}}{\sigma_i} \qquad (1)$$

The correlation matrix was then calculated to comprehend all 15 variables and their respective correlations:

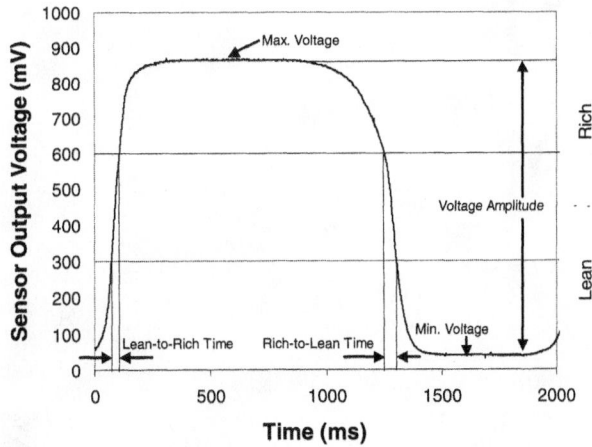

Figure 4
Sensor output parameters during the engine perturbation test

Table 1

Reference group output data normalization

	Reference Data					Normalized Data				
No.	Variable 1 X_1	Variable 2 X_2	Variable 3 X_3	...	Variable 15 X_{15}	Variable 1 Z_1	Variable 2 Z_2	Variable 3 Z_3	...	Variable 15 Z_{15}
1										
2										
3										
4										
5										
6										
7										
8										
...										
26										
Mean	$\bar{X}_1$	$\bar{X}_2$	$\bar{X}_3$	...	$\bar{X}_{15}$	0.0	0.0	0.0	...	0.0
Std. Dev.	σ_1	σ_2	σ_3	...	σ_{15}	1.0	1.0	1.0	...	1.0

Table 2
Correlation matrix results for the reference group

		A 1	B 2	C 3	V_{min} 4	V_{max} 5	V_{ampl} 6	LR_{Time} 7	RL_{Time} 8	I 9	J 10	K 11	L 12	M 13	Position 14	O 15
A	1	1.000	0.956	0.679	0.696	0.710	-0.096	0.116	0.732	0.933	0.772	-0.478	-0.043	0.597	0.388	0.005
B	2	0.956	1.000	0.636	0.542	0.631	0.042	0.135	0.611	0.897	0.597	-0.304	-0.056	0.618	0.389	-0.046
C	3	0.679	0.636	1.000	0.553	0.305	-0.480	-0.123	0.456	0.659	0.599	-0.572	-0.326	0.074	0.218	0.144
V_{min}	4	0.696	0.542	0.553	1.000	0.815	-0.446	0.208	0.439	0.812	0.968	-0.538	-0.364	0.048	0.236	0.231
V_{max}	5	0.710	0.631	0.305	0.815	1.000	0.156	-0.156	0.393	0.850	0.796	-0.112	-0.295	0.241	0.414	0.244
V_{ampl}	6	-0.096	0.042	-0.480	-0.446	0.156	1.000	0.119	-0.150	0.075	-0.422	0.743	0.167	0.286	0.230	-0.025
LR_{Time}	7	0.116	0.135	-0.123	0.208	-0.156	0.119	1.000	0.174	-0.090	-0.052	-0.366	0.899	0.719	0.281	-0.948
RL_{Time}	8	0.732	0.611	0.456	0.439	0.393	-0.150	0.174	1.000	0.544	0.597	-0.581	0.213	0.587	0.281	-0.005
I	9	0.933	0.897	0.659	0.812	0.850	0.075	-0.090	0.544	1.000	0.826	-0.337	-0.206	0.358	0.353	0.167
J	10	0.772	0.597	0.599	0.968	0.796	-0.422	-0.052	0.597	0.826	1.000	-0.660	-0.206	0.200	0.291	0.119
K	11	-0.478	-0.304	-0.572	-0.538	-0.112	0.743	-0.366	-0.581	-0.337	-0.660	1.000	-0.251	-0.282	-0.093	0.326
L	12	-0.043	-0.056	-0.326	-0.364	-0.295	0.167	0.899	0.213	-0.304	-0.206	-0.251	1.000	0.717	-0.241	-0.822
M	13	0.597	0.618	0.074	0.048	0.241	0.286	0.719	0.587	0.358	0.200	-0.282	0.717	1.000	0.453	-0.561
Position	14	0.388	0.389	0.218	0.236	0.414	0.230	0.281	0.281	0.353	0.291	-0.093	0.241	0.453	1.000	-0.225
O	15	0.005	-0.046	0.144	0.231	0.244	-0.026	-0.948	-0.005	0.167	0.119	0.326	-0.822	-0.561	-0.225	1.000

Table 3

Mahalanobis distances for the reference and abnormal groups

No.	SN	0 hours Normal Group MD	Aged Abnormal Group MD
1	24,720	0.9	14.1
2	24,716	1.2	15.7
3	24,730	0.8	9.6
4	24,719	0.6	14.3
5	24,728	1.4	15.1
6	24,723	1.6	5.3
7	22,963	0.8	79,651.8
8	23,013	1.5	86,771.7
9	23,073	1.1	84,128.8
10	24,673	1.4	
11	24,700	0.8	
12	24,694	0.9	
13	24,696	1.1	
14	24,701	1.2	
15	24,697	1.1	
16	24,633	0.6	
17	24,634	1.0	
18	24,635	1.1	
19	24,636	1.0	
20	24,637	0.6	
21	24,593	1.1	
22	24,595	0.5	
23	24,598	0.7	
24	24,599	1.2	
25	24,602	1.0	
26	24,603	0.7	

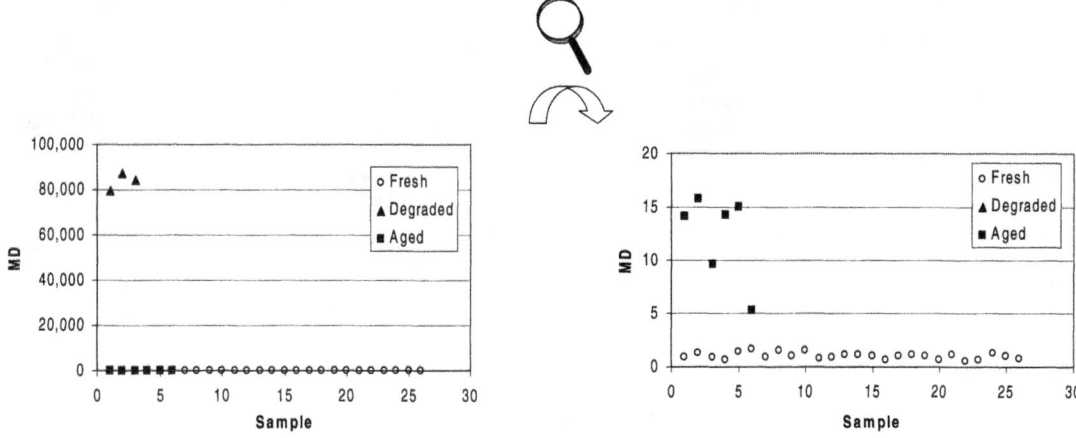

Figure 5
Power of discrimination for the reference to abnormal groups

$$R = \begin{bmatrix} 1 & r_{12} & \cdots & r_{1k} \\ r_{21} & 1 & \cdots & r_{2k} \\ \vdots & \vdots & & \vdots \\ r_{k1} & r_{k2} & \cdots & 1 \end{bmatrix} \tag{2}$$

$$r_{ij} = \frac{\sum x_{il} x_{jl}}{n} \ (l = 1, 2, \ldots, n)$$

Upon review of the correlation matrix (Table 2), it is clear that a correlation exists between parameters. For this reason, application of the multivariable MTS approach makes sense because no single characteristic can describe the output fully.

The inverse of the matrix was then calculated [equation (3)] and finally, the Mahalanobis distance [equation (4)], denoted by MD This completes the calculations for the normal group. All reference samples had MD distances of less than 2 (Table 3):

$$R^{-1} = \begin{bmatrix} a_{11} & a_{12} & \cdots & a_{1k} \\ a_{21} & a_{22} & \cdots & a_{2k} \\ \vdots & \vdots & & \cdots \\ a_{k1} & a_{k2} & \cdots & a_{kk} \end{bmatrix} \tag{3}$$

$$MD = \frac{1}{k} ZR^{-1}Z^{T} \tag{4}$$

where k is the number of characteristics, Z is the 1×15 normalized data vector, R^{-1} is the 15×15 inverse correlation matrix, and Z^T is the transposed vector (15×1).

The MD values for the abnormal samples were then calculated. Again the data are normalized, but now the mean and standard deviations of the reference group were considered. The inverse correlation matrix of the reference group solved previously was also used. The resulting MD values of the abnormal samples were calculated and are summarized in Table 3. As evident in the MD values of the abnormal samples, clear discrimination between fresh and degraded performance was made possible (Figure 5). Additionally, complete discrimination of the aged samples from the fresh samples was seen. Although the MD value calculated is nondirectional and does not indicate "goodness" or "badness," it does indicate differences from normal. A consistent signal (low MD over time) is one goal for sensor output.

Importantly, this performance separation is not apparent in the traditional one-variable-at-a-time approach. For this type of aging test, typical performance metrics that are recognized to change are V_{min} and RL_{Time}. As shown in Figure 6, this independent variable consideration would merely allow detection of the degraded sensors with no clear discrimination of the aged sensors.

6. Selection of Characteristics

Optimization with L_{16} Orthogonal Array

To reduce data processing complexity, it is desirable to consider fewer characteristics and eliminate those

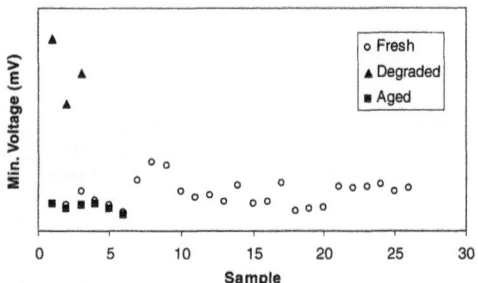

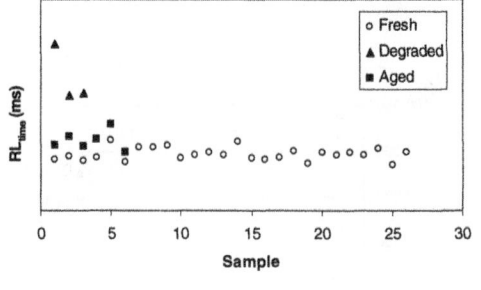

Figure 6
Discrimination not evident with traditional single variable approach

not contributing to product discrimination. An L_{16} orthogonal array was used for this purpose (Table 4).

All 15 characteristics were considered at two levels. Level 1 used the variable to calculate the Mahalanobis distance, and level 2 did not use the variable to calculate the MD. Reconsideration of both the reference group and abnormal group MD was made for each run. The experiment design and results are shown in Table 4.

From these runs, SN ratios and mean responses were calculated for the main effects of each variable. As the goal was to improve discrimination, larger MDs were preferred and the larger-the-better SN ratio was used:

$$\eta = -10 \log \left(\frac{1}{n} \sum_{i=1}^{n} \frac{1}{y_i^2} \right) \qquad (5)$$

Response charts and tables are shown in Figure 7 and Table 5.

Variables C, F, I, and J are shown to have little contribution to the SN ratio and could be considered for elimination. This would reduce the MD calculation to 11 characteristics. All variables, however, contributed positively to the mean.

Confirmation

A confirmation run with 11 variables (eliminating the low-SN contributors) showed reduced discrimination. However, these variables all contribute significantly to the mean (Figure 7) and therefore cannot be eliminated. This conclusion is somewhat indicated within the results of the L_{16} array (Table

4). Run 1, which considered all the variables, had by far the largest calculated MD compared to any other run, which considered only seven variables each. Therefore, all of the 15 variables initially selected should be used to maximize discrimination.

Discussion

Although the optimization evaluation may seem disappointing, as no variables can be eliminated, it is not an unlikely conclusion, as there are over 100 variables output in an effort to characterize the waveform. This initial evaluation only considers 15 "best guess" candidates, and more than likely, other important variables are still to be identified. The MTS method does, however, confirm the value of combining the influence of many variables to interpret a change in response, as compared to considering one variable or even a few variables at a time.

The ranking of the "traditional" metrics, factors D and H, used to detect signal degradation, are ranked high, but not the highest in terms of contribution (Table 5). The influence of factor N, the nest position during test, is seen to have a low influence on variability and a low contribution to the mean, as is desired.

7. Conclusions

The feasibility of use of the MTS multivariable approach has been demonstrated. The combination of 15 variables from the sensor output waveform allowed much improved discrimination compared to

Table 4
MDs calculated for abnormal group within the L_{16} orthogonal array

No.	Factor															MD for Nine Abnormal Samples								
	A	B	C	D	E	F	G	H	I	J	K	L	M	N	O	1	2	3	4	5	6	7	8	9
1	1	1	1	1	1	1	1	1	1	1	1	1	1	1	1	14.1	15.7	9.6	14.3	15.1	5.3	79,651.8	86,771.7	84,128.8
2	1	1	1	1	1	1	1	2	2	2	2	2	2	2	2	1.6	2.2	1.5	2.2	6.5	1.0	9,722.0	11,988.6	11,349.3
3	1	1	1	2	2	2	2	1	1	1	1	2	2	2	2	2.9	5.2	1.8	3.3	8.0	1.1	25,731.9	32,363.5	30,961.9
4	1	1	1	2	2	2	2	2	2	2	2	1	1	1	1	2.4	2.8	1.6	3.0	9.5	1.5	6,300.9	7,283.8	7,102.0
5	1	2	2	1	1	2	2	1	1	2	2	1	1	2	2	2.5	2.8	1.0	2.9	7.3	2.6	25,535.4	28,754.8	27,635.1
6	1	2	2	1	1	2	2	2	2	1	1	2	2	1	1	4.0	3.0	3.3	2.0	5.1	1.5	1,560.3	1,538.7	1,526.0
7	1	2	2	2	2	1	1	1	1	2	2	2	2	1	1	1.8	2.5	1.2	2.3	6.1	1.1	9,156.7	10,742.5	10,330.4
8	1	2	2	2	2	1	1	2	2	1	1	1	1	2	2	1.6	1.8	1.3	1.1	5.4	3.1	11,568.6	13,508.6	12,824.1
9	2	1	2	1	2	1	2	1	2	1	2	1	2	1	2	2.0	3.3	1.5	3.1	5.6	1.7	151.6	30.2	49.2
10	2	1	2	1	2	1	2	2	1	2	1	2	1	2	1	6.4	1.9	2.3	4.8	9.5	3.1	279.1	37.0	57.8
11	2	1	2	2	1	2	1	1	2	1	2	2	1	2	1	3.4	3.8	1.6	3.4	6.9	1.9	348.6	25.7	35.4
12	2	1	2	2	1	2	1	2	1	2	1	1	2	1	2	7.4	1.7	2.6	4.7	7.3	1.7	287.0	33.4	47.5
13	2	2	1	1	2	2	1	1	2	2	1	1	2	2	1	4.0	4.9	2.0	3.9	10.5	2.6	379.9	44.9	59.5
14	2	2	1	1	2	2	1	2	1	1	2	2	1	1	2	0.9	1.1	0.8	0.6	4.9	1.7	74.1	12.3	25.3
15	2	2	1	2	1	1	2	1	2	2	1	2	1	1	2	1.2	3.0	1.0	2.3	7.9	2.4	161.1	35.1	53.1
16	2	2	1	2	1	1	2	2	1	1	2	1	2	2	1	0.5	2.3	0.8	0.7	7.8	2.2	131.4	35.3	49.3

SN Ratio Analysis

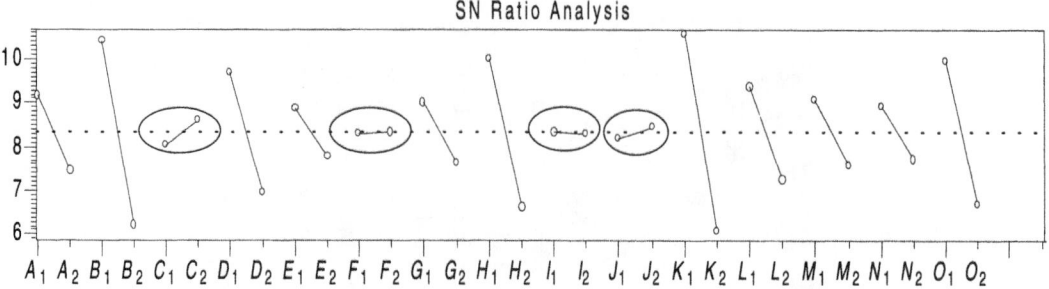

Mean Analysis

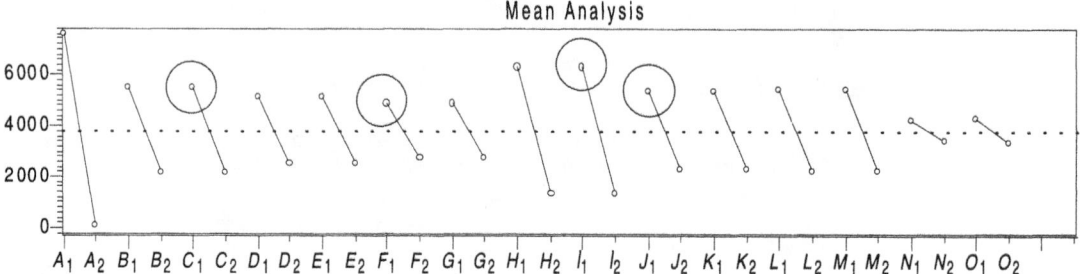

Figure 7
SN and means response charts

the traditional one-variable-at-a-time approach. The method also identified some alternative variables that contributed more to discrimination than did the traditional "favorites."

MTS allows discrimination of even very subtle differences due to aging, thereby allowing detailed feedback on product performance. If continued studies support these preliminary findings, the excellent discrimination will allow product optimization based on significantly shorter tests. Full-length test exposures could then be confined to the product confirmation and validation stages. Robust engineering evaluations for sensor product optimization are ongoing. By applying MTS to characterize sensor performance, the MD can confirm improvements in aging response over time (Figure 8).

Future Evaluations

Although this project proved excellent feasibility in terms of its approach to discriminate performance,

much remains to be done to extend the study and consider other variables for better understanding and implementation. The number of sensors in the reference group limited this study. It would be desirable to increase this sample size to allow consideration of many more output characteristics. Additionally, alternative test parameters, which may influence sensor output response during the test, should be considered. Future evaluations that consider more test parameters should ideally comprehend any bias introduced by the test, thereby identifying true product differences.

Implications of Study

Many ideas have evolved as a result of this study. The first key finding is related to detecting small changes in signal output through MTS. With this detection method, shortened tests (to save time and money) should suffice to understand aging trends and optimize designs. Further, rather than supplying our

Table 5
SN and means response tables

	A	B	C	V_{min}[a] D	V_{max} E	V_{mpl} F	LR_{Time} G	RL_{Time}[a] H	I	J	K	L	M	N	O
							SN Response								**Position**
Level 1	9.2	10.4	8.0	9.7	8.9	8.3	9.0	10.0	8.4	8.2	10.6	9.4	9.1	8.9	9.9
Level 2	7.5	6.2	8.6	7.0	7.8	8.4	7.6	6.7	8.3	8.5	6.1	7.3	7.6	7.7	6.7
Delta	1.7	4.2	0.6	2.7	1.1	0.1	1.4	3.3	0.1	0.3	4.5	2.1	1.5	1.2	3.2
Rank	7	2	12	5	11	15	9	3	14	13	1	6	8	10	4
							Means Response								
Level 1	7614	5486	5481	5161	5161	4903	4907	6297	6292	5324	5331	5452	5450	4267	4275
Level 2	36	2165	2170	2490	2489	2748	2743	1354	1358	2327	2319	2198	2200	3383	3376
Delta	7578	3321	3311	2671	2672	2156	2164	4943	4934	2997	3012	3254	3250	884	899
Rank	1	4	5	11	10	13	12	2	3	9	8	6	7	15	14

[a] V_{min} and RL_{time} are traditional metrics for performance characterization.

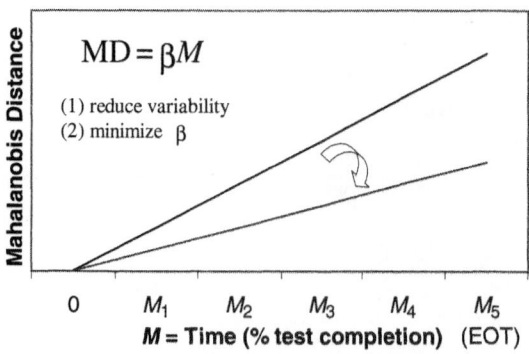

$$MD = \beta M$$

(1) reduce variability
(2) minimize β

Mahalanobis Distance

0   M_1 M_2 M_3 M_4 M_5
M = Time (% test completion) (EOT)

Figure 8
Used in conjunction with product optimization, MD can
confirm improvements in aging response over time

calculations, demonstrating the power of appropriate data analysis.

Acknowledgments We would like to recognize
Mike Holbrook and Kimberly Pope of Delphi for
sharing their expertise in Taguchi methods and
MTS studies; also Bob Stack of Delphi for recognizing the role that these tools can play and allowing
us time to devote efforts applying them; and finally,
thanks to Shin Taguchi, of American Supplier Institute, who enthusiastically teaches and encourages
the application of these methods at Delphi. Application of these techniques at Delphi will assure
product and marketplace leadership.

Reference

S. Teshima, T. Bando, and D. Jin, 1997. A research of
defect detection using the Mahalanobis–Taguchi
system method. *Journal of Quality Engineering Forum*,
Vol. 5, No. 5, pp. 169–180.

*This case study is contributed by Stephanie C.
Surface and James W. Oliver II.*

product engineers with more than 100 data variables related to a waveform, the MD distance could
help them make decisions during product development evaluations.

This study used existing data with no new experimental tests needed. The study points out the potential of its application, as no new test procedures
are required, with the exception of additional data

CASE 73

Defect Detection Using the MTS

Abstract: In our research, selecting differential and integral characteristics calculated by a pattern as feature values, we take advantage of the Mahalanobis–Taguchi system (MTS), which creates a Mahalanobis space with normal products (base data) to study the visual inspection process. In this research, using disks judged as "normal" by inspectors, we created a base space.

1. Introduction

Since an inspector responsible for a visual inspection is considered to have many subconscious inspection standards in order to realize an automated inspection process, we need to build up a system of integrating multidimensional information. On the other hand, for the automation of an inspection process, necessary multidimensional information, or a group of characteristics, is not necessarily identical to that held by an inspector. That is, although characteristics used by an inspector are unknown in some cases, even characteristics easily handled by a computer can be utilized successfully if they lead to the same results as those determined by a visual inspection.

In our research, selecting differential and integral characteristics calculated by a pattern as feature values, we take advantage of the MTS method, which creates a Mahalanobis space with normal products (base data). In this research, using disks judged as "normal" by inspectors, we created a base space.

To obtain a highly accurate image at low cost, we adopted a method of capturing an image with a linear CCD (change-coupled device) camera while rotating a disk at constant speed (Figure 1). The captured image is input into a personal computer in real time by way of a special image input board. By doing so we can obtain extremely thin and long image data, that is, (width of a disk) × (circular

length for one rotation), when the disk is turned around in one rotation. The starting point for capture in a rotational direction was set arbitrarily.

2. Differential and Integral Characteristics

An image datum was represented by color concentration. Looking at the fluctuation in concentration of a single charge-coupled device in a linear CCD camera, we can see that a waveform such as that shown in Figure 2 is obtained when a disk is turned one rotation. Its horizontal and vertical axes indicate rotational position and gradation, respectively. Now, since there are a few dozen charge-coupled devices, we can capture the same number of different waveforms as that of the devices.

The waveform shown in part (a), which exemplifies that of a normal disk, demonstrates that concentration peaks appear regularly around a constant concentration level. In addition, in (b) and (c) we show examples of waveforms for defective disks. More specifically, (b) represents the case for a disk with adhesive, and (c) indicates the case for a disk with frictional material peeled off. Comparing those with the normal pattern in Figure 2, we can see a significant disturbance in a waveform of (c), whereas a fluctuation in gradation of (b) due to adhesive stuck is not distinct. However, a certain type of disturbance is involved in its waveform despite its subtlety.

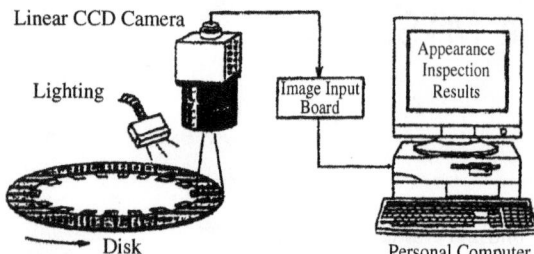

Figure 1
Appearance inspection system for disks

A key point in our research was to deal with image data of a disk as a group of waveforms. By recognizing a disturbance in a waveform, we attempted to detect a defect. As a feature value for a waveform pattern, we calculated the following differential and integral characteristics, explained below.

Next, we defined a waveform for one unit (a waveform consisting of pixels for one rotation) as $Y(t)$. As Figure 3 illustrates, we drew p parallel lines to the t-axis at an equal interval along the Y-axis. Counting the number of intersections between each line and the waveform, $Y(t)$, as per each line, we set this as a differential characteristic. In addition, calculating the sum of all intervals between intersections, we defined it as an integral characteristic.

The differential characteristic can be regarded to indicate a frequency of fluctuations in a waveform (i.e., information equivalent to the frequency of a waveform). Since we can obtain a frequency for each amplitude, the distribution of a frequency in the direction of an amplitude can be known. On the other hand, the integral characteristic indicates an amount of occupation for each amplitude. Both the differential and integral characteristics were considered to cover a frequency and amplitude of a waveform and useful information regarding them. In addition, since the differential and integral characteristics can be calculated more simply than the traditional characteristics such as a frequency, we can expect a faster processing speed.

Furthermore, in addition to the differential and integral characteristics, in this research we added four characteristics that are regarded to be effective for a disk appearance inspection.

3. Preparation of Base Space

From clutch disks that we confirmed had no defect we captured a total of 1000 images. One image con-

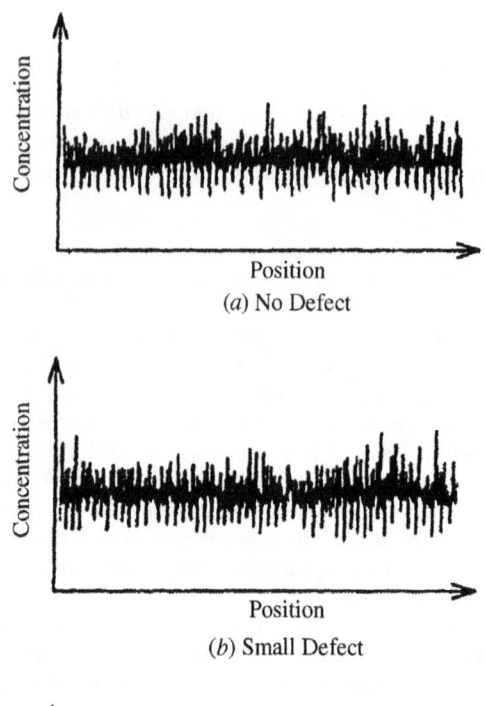

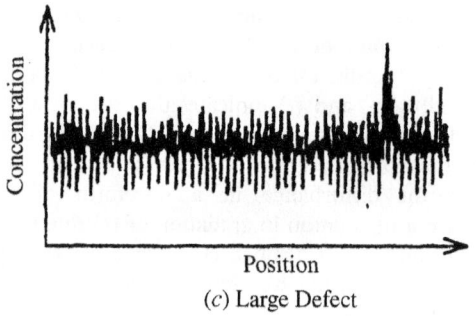

Figure 2
Waveform of disk image

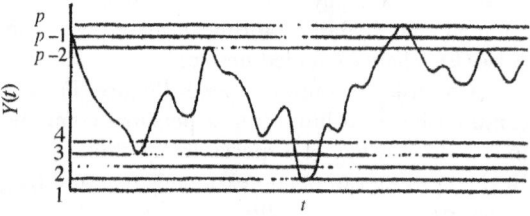

Figure 3
Differential and integral characteristics

Table 1
Feature values in the wave pattern of a disk[a]

	Characteristic					
P	Differential	Integral	A	B	C	D
1	18	5989	75	1318	54	66
2	46	5971	68	1098	82	66
3	72	5944	65	828	92	65
⋮						
37	8	4	1	1	40	1593
38	8	4	1	1	40	1593
39	4	2	1	1	1525	2552
40	2	1	1	1	2552	3447

[a] Row numbers correspond to numbers on the vertical axis of Figure 3.

sists of 36 waveforms. Then we computed 240 feature values, including the differential and integral characteristics for each waveform. However, after excluding characteristics that do not obviously contribute to image recognition, we finally obtained 160 feature values (items). Therefore, using 1000×160 base data, we computed 36 correlation matrices. A typical example of feature values that are not con-

tributing to image recognition is a value with a quite large standard deviation, which can be any value.

As inspection data, we measured a certain number of data from nondefective and defective disks. Table 1 shows an example of feature values.

Figure 4 shows examples of the recognition results. In this plot the horizontal and vertical axes indicate a radial position on a disk and the

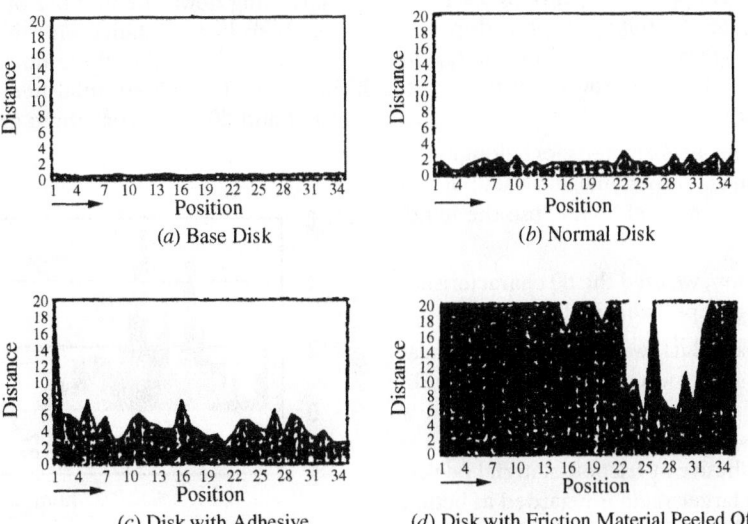

(a) Base Disk

(b) Normal Disk

(c) Disk with Adhesive

(d) Disk with Friction Material Peeled Off

Figure 4
Distances of disk and wave patterns

Mahalanobis distance, respectively. In (a), representing base data, all Mahalanobis distances show a small value. Part (b) represents the distances of a nondefective disk; almost all lie below 2. For (c), for a disk with a small amount of adhesive stuck, many data exceed the value of 4. This cannot be regarded as normal. In (d) for a disk with friction material peeled off, many high peaks can be seen on the left side of the plot. This implies that relatively large defects exist on the corresponding positions. In addition, as a result of recognizing various types of disks, we observed that most of the recognition results were consistent with results via a human visual inspection.

4. Item Selection Using Orthogonal Array L_{64}

One hundred sixty items were used for recognition thus far. Now, to improve processing time and reduce measurement cost, we picked up items that were effective for recognition. To this end, using an L_{64} orthogonal array, we attempted to select items.

Primarily, from all of the 160 items used in the prior experiment, we selected 40 items that are considered essential for recognition without being selected. Secondly, after dividing the remaining 120 items into three groups of 40 items by each attribute, according to the following procedure, we performed three experiments based on an L_{64} orthogonal array to verbatim narrow down the items.

For selection 1:

❑ We allocate the 40 characteristics belonging to group 1 to an L_{64} orthogonal array, setting "no use of an item" to level 1 and "use the item" to level 2.

❑ For recognition, we used the 80 characteristics included in groups 2 and 3, 40 in each group.

❑ Using the base data, we calculated a correlation matrix under each experimental condition. Then, in each experiment we compute a Mahalanobis distance for each of 10 defective disks with adhesive or friction material peeled off. Since its larger value is regarded as better, we calculated a larger-the-better SN ratio with the following equation:

$$\eta = -10 \log \frac{1/D_1^2 + 1/D_2^2 + \cdots + 1/D_m^2}{m} \quad \text{dB}$$

(1)

where m indicates the total number of data.

❑ A response graph was created. Figure 5 represents the difference between the sum of SN ratios for level 2 and that of SN ratios for level 1. Because level 2 corresponds to the case of "use the item (characteristic)," a factor with a larger difference is regarded to have a greater effect on defect recognition.

❑ Twenty items among 40 with a larger factor effect were selected.

For selections 2 and 3:

❑ After assigning each of the 40 characteristics belonging to groups 2 or 3, respectively, to an L_{64} orthogonal array, we performed a similar experiment.

❑ Similar to group 1, for each group we selected 20 larger items that were effective for recognition.

Because we split up the 120 items into three groups of 40 items each, any interactions among groups were ignored. However, since all groups had initially been classified according to their nature, we considered that there were no significant interactions.

Narrowing down the number of items on a step-by-step basis in accordance with the foregoing procedure, we finally reduced the number of the items from 160 to 100: 40 essential items without being selected and 20 each from three groups. As a next

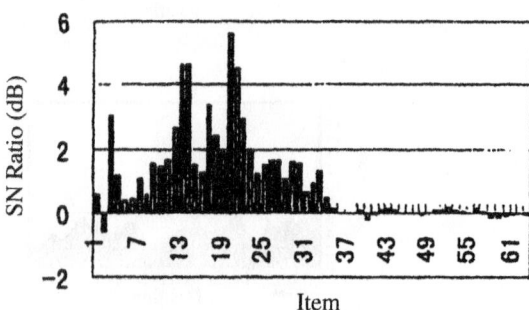

Figure 5
Response graphs for item selection

step, we conducted a confirmatory experiment using the surviving 100 items. Figure 6, which shows the recognition result for a disk with adhesive, highlights the difference in the SN ratio between the two cases of using the 160 and 100 feature values. The shaded part above a fluctuating line represents an increment in the Mahalanobis distance caused by the items selected. Basically, a Mahalanobis distance for a defect is expected to emerge as a larger value.

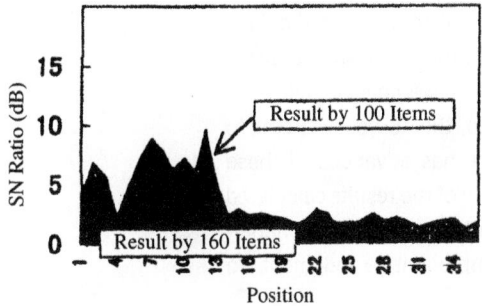

Figure 6
Recognition result by selected items

Since the Mahalanobis distances for defective areas in this case were increased by the items selected, we can conclude that the detectability of defects is enhanced. This is supposedly because we have picked up only items that were effective for defect recognition, thereby mitigating noises in recognition. On the other hand, no increase in Mahalanobis distance can be seen in the normal areas. Since the base space is created from the data for a normal disk, its Mahalanobis distance lies around 1.

Our research demonstrates that by narrowing down the original items, we can enhance the capability of recognition with less calculation.

Reference

Shoichi Teshima, Tomonori Bando, and Dan Jin, 1997. A research of defect detection using the Mahalanobis–Taguchi system. *Quality Engineering*, Vol. 5, No. 5, pp. 38–45.

This case study is contributed by Shoichi Teshima, Tomonori Bando, and Dan Jin.

Application of Mahalanobis Distance to the Measurement of Drug Efficacy

Abstract: Despite various evaluation methods of drug efficacy designed to date, we do not yet have a single definitive technique. It is widely known that there are few cases in which the data follow a normal distribution; particularly in the clinical field, there is no guarantee that the data follow such a distribution. Moreover, medical data for patients tend to have considerable variability compared with those for healthy people. Also, the number of items to be studied has been increasing as medical science has advanced. If these data are analyzed as raw response data, the reliability of the results calculated is considered to deteriorate. In these contexts we attempted to use a Mahalanobis distance (D^2), which is regarded as a comprehensive judgment, to assess drug efficacy.

1. Introduction

In general, few drugs are effective for chronic hepatitis. Even Interferon, which has lately been said to be a cure for hepatitis C, has only a 40 to 50% efficacy rate. For hepatitis B the rate is believed to be much smaller. Moreover, Interferon causes significantly strong sideeffects, and at the same time, cannot be applied to all patients suffering from hepatitis. In our research we use Azelastine, an antiallergy medicine. We examined two groups: (1) healthy persons and (2) patients with chronic hepatitis.

Healthy Persons (Normal Group)

We examined 200 people, who took a biochemical test consisting of 16 examination items (described later), and from the diagnostic history at Tokyo Teishin Hospital, these people were diagnosed as being in good health with no disease. Although initially, we wished to increase the number of examinees, because of the limited capacity of the computer used at that time, we had to be content with only 200 people.

Patients with Chronic Active Hepatitis

The patients with chronic active hepatitis to be studied were 32 males who had visited the Department of Digestive Systems at Tokyo Teishin Hospital over three years and whose medical checkup data had been stored. The active period of chronic hepatitis does not represent a stable condition but indicates a stage when many liver cells die. Using more popular terminologies, we can view it as a period with high GOT or GPT values (see below).

- ❏ *Contrast group.* This group consisted of 16 patients with chronic active hepatitis who had taken an orginary three-year course of therapy. No Interferon had been prescribed for them.

- ❏ *Azelastine group.* This group comprised 16 patients with chronic active hepatitis who had been medicated with Azelastine for one year and who were now in the second year of therapy.

Items Studied

The healthy persons and patients with chronic active hepatitis had already had the following 16 blood tests: total protein (TP), albumin (Alb), A/G (albumin/globulin) ratio, cholinesterase (ChE), glutamate oxaloacetate transaminase (GOT), glutamate pyruvate transaminase (GPT), lactate dehydrogenase (LDH), alkaline phosphatase (ALP), γ-glutamyltranspeptidase (γ-GTP), leucine aminopeptidase (LAP), total cholesterol (TCh), triglyceride (TG), phospholipases (PL), creatinine (Cr), blood urea nitrogen (BUN), and uric acid (UA).

In addition, assuming the age and gender of the examinees to be relevant and that no extra cost was needed for data collection, we added both to the 16 examination items.

Analysis Method

The Mahalanobis distance D^2 was not convenient to analyze as raw data. We divided D^2 by the degrees of freedom, 18. The average of the resulting values for the normal group was equal to 1. Next, by logarithmizing the value above and multiplying it by 10, we obtained the value used in our analysis. Then the data for the ith patient were computed as follows:

$$Y_i = 10 \log \frac{D_i^2}{18}$$

This value is called a *decibel value*.

Table 1

Data for normal group

Age:	No. 1 59 Male	No. 2 51 Female	No. 3 38 Male	No. 4 56 Female
TP (g/dL)	6.5	7.0	7.1	7.2
Alb (g/dL)	3.7	4.4	4.3	4.0
A/G	1.32	1.69	1.54	1.25
ChE (ΔpH)	0.70	0.98	0.92	0.93
GOT (IU/L)	12	21	16	22
GPT (IU/L)	6	18	15	16
LDH (GU/dL)	190	247	152	188
ALP (KAU/dL)	7.7	6.3	4.8	6.1
γ-GTP (IU/L)	12	23	40	16
LAP (U/dL)	275	340	355	304
TCh (mg/dL)	235	225	177	216
TG (mg/dL)	140	87	93	86
PL (mg/dL)	238	227	185	213
Cr (mg/dL)	0.8	1.1	1.4	1.0
BUN (mg/dL)	1.3	15	15	13
UA (mg/dL)	3.8	4.6	4.6	4.2

2. Analysis Results

From here on, we detail the actual data and analysis results.

Data for the Normal Group

Table 1 shows a part of the data for 200 healthy persons.

Data for Patients with Chronic Active Hepatitis

Since the total data for patients with chronic active hepatitis were composed of 32 examinees, 36 months, and 18 items, we show only a part of them in Table 2.

Mahalanobis Distance

Since D^2 for the normal group comprises variances with 18 degrees of freedom and is considered rela-

tively small, we do not show the actual data here. For the contrast and Azelastine groups for chronic active hepatitis, we show a part of the data of D^2 in Table 3. however.

Performing an analysis of variance for the decibel values (Y's) and calculating the estimations us-

Table 2
Data for patients with chronic active hepatitis[a]

	1 M	2 M	3 M	4 M
TP (mg/dL)	7.0	7.9	7.9	7.8
Alb (mg/dL)	3.6	4.0	3.6	4.1
A/G	1.06	1.03	0.97	1.08
ChE (IU/L)	303	275	312	297
GOT (IU/L)	178	168	160	174
GPT (IU/L)	225	228	218	202
LDH (IU/L)	446	429	454	467
ALP (IU/L)	152	146	146	162
γ-GTP (IU/L)	27	31	28	30
LAP (IU/dL)	59	69	64	68
TCh (mg/dL)	186	185	181	197
TG (mg/dL)	153	148	100	157
PL (mg/dL)	193	222	198	216
Cr (mg/dL)	0.7	0.7	0.7	0.7
BUN (mg/dL)	11	12	11	15
UA (mg/dL)	5.9	0.7	5.6	5.5

[a]Case 17 in Azelastine group: 47-year-old male.

Table 3
Mahalanobis distance for patients with chronic active hepatitis

Item	No. 17	No. 18	No. 19	No. 20
1	1554.0	1873.6	1216.5	1418.6
2	1741.7	2869.2	2550.1	1164.8
3	1531.9	1412.8	1190.5	1239.7
4	1248.8	1877.5	1656.3	903.1
5	1871.2	1756.8	2558.0	533.3
6	1362.9	1881.7	2100.2	896.3
7	2346.0	2346.3	2808.8	623.7
8	2489.1	1195.0	2207.3	710.8
9	2492.7	790.5	1843.8	790.3
10	3380.7	990.1	367.7	918.0
11	3196.8	1416.1	1784.8	862.9
12	2815.7	1301.7	2398.7	811.8
13	2191.2	131.2	473.6	739.5
14	2218.3	979.4	204.1	690.8
15	1930.3	1306.5	560.2	884.2
16	1787.3	1662.5	157.2	332.8
17	2572.1	1348.3	178.1	393.9
18	1893.3	1650.7	498.1	604.6
19	1298.3	1223.9	738.0	520.2
20	1508.3	1329.2	948.3	969.2
21	1634.1	1180.4	1175.2	274.6
22	1785.5	1106.5	942.8	339.6
23	1546.5	560.4	366.6	269.9
24	1284.4	782.4	1049.5	56.6
25	2037.8	1286.7	1652.9	59.2
26	2294.8	588.8	432.9	84.1

ing them, we arrived at Figure 1. For the sake of simplicity, we do not indicate the 95% confidence limits.

Looking at Figure 1, we can see that the Azelastine group shows a greater distance from the normal group than the contrast group'does. Considered not so good as a comparative test of group 2, this fact is likely to reflect the effects of a special drug on patients with poor values in the clinical test or with serious conditions.

In addition, for both the contrast and Azelastine groups, the Mahalanobis distance is on the increase in the first year. However, for the second year, the distance is on a downward trend, due partly to the effect of more active therapy. One of the most remarkable changes is that the Azelastine group's distance ends at approximately the same place as that of the contrast group.

In the third year, although the contrast group's distance increases temporarily, it tends to return back to the initial state. In contrast, for the Azelastine group, the distance continues to decline and finally assimilates with that of the contrast groups.

Judgment by the Linear Trend

We notice no relationship among all data on the whole. Although various types of analysis methods have been attempted to date, we detail the linear trend analysis as a typical example.

Table 4 shows the average of each case, the corresponding value of the linear equation, and the value that we obtain by dividing it by the total of squares of coefficients (which is equivalent to a monthly slope for the linear trend). Performing an analysis of variance for these linear trend values, we obtain Table 5. Based on this result, we calculated the estimations shown in Table 6. From these we can conclude that since a Mahalanobis distance is considered a distance from the gravity center of a multidimensional space, if the distance for a certain patient from the normal group increases, the degree of his or her disease is regarded to increase. On the contrary, as the distance decreases, the patient becomes better and closer to the state of a normal person. As a whole, as compared with the contrast group for chronic active patients, the Azelastine group obviously has a good linear trend. However, this trend does not necessarily hold true for all patients. Yet if we look at the values beyond the 95% confidence limits, the values of the linear trend reveals that the patients are getting better. For example, for the linear trend for the contrast group, the average is -19.9572 and the 95% confidence interval is -202.137 ± 162.2232. On the other hand, the average for the Azelastine group is -115.5413 and its confidence interval is -297.7217 ± 66.6391. Therefore, patients 17, 19, 20, and 32, those who take the linear trend value below the 95% confidence interval, are judged to be getting

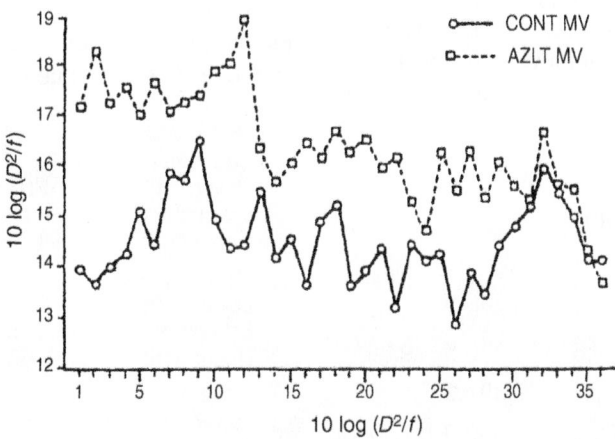

Figure 1
Plots for contrast and Azelastine groups

Table 4
Linear trend of 10 log(D^2/f) (for one month)

	Contrast Group			Azelastine Group	
CN	L	L/D	CN	L	L/D
1	−58.3	−0.0065	17	−419.8	−0.0470
2	+51.3	+0.0057	18	−121.8	−0.0136
3	−67.9	−0.0076	19	−249.0	−0.0279
4	+87.7	+0.0098	20	−289.6	−0.0324
5	−116.4	−0.0130	21	−189.8	−0.0213
6	−46.9	−0.0053	22	−123.5	−0.0138
7	−27.1	−0.0030	23	−82.5	−0.0092
8	+69.2	+0.0078	24	−185.7	−0.0208
9	−68.6	−0.0077	25	−48.9	−0.0055
10	+48.2	+0.0054	26	+160.5	+0.0180
11	+33.9	−0.0038	27	−22.3	−0.0025
12	−144.45	−0.0162	28	−27.2	−0.0030
13	+122.0	−0.0137	29	+75.5	+0.0085
14	−182.8	−0.0205	30	−72.3	−0.0081
15	−2.1	−0.0002	31	−48.8	−0.0055
16	−17.1	−0.0019	32	−203.5	−0.0228

healthy. In addition, in the Azelastine group, there is no patient whose disease is aggravated with a value beyond the 95% confidence interval of the contrast group. Furthermore, in the Azelastine group, there is no one with a value below the 95% confidence interval of the contrast group. When di-

Table 5
ANOVA for linear trend (L)

Source	f	S	V	F_0
L	1	146,878.7	146,878.7	19.87**
C	1	73,090.5	73,090.5	9.89**
Within C_1	15	110,878.7	7,391.9	
Within C_2	15	300,000.6	20,000.0	2.71*
Total	32	630,848.5		

Table 6
Estimations of linear trend and that divided by D and 95% confidence limit

	L	L/D
Contrast group		
Lower limit	−202.1376	−0.022646
Estimation	−19.9572	−0.002236
Upper limit	+162.2232	+0.018174
Azelastine group		
Lower limit	−297.7217	−0.033354
Estimation	−115.5413	−0.012944
Upper limit	+66.6391	+0.007466
Average		
Lower limit	−242.8962	−0.027212
Estimation	−67.7492	−0.007590
Upper limit	−107.3977	+0.012032

viding the value above by the number of units (total of squares of coefficients), the result is completely similar, even in terms of F_0 for an analysis of variance.

dividual patient data, no distinct relationship can be seen. As a final judgment, we conclude that Azelastine can contribute quite effectively to treatment of many patients with chronic active hepatitis.

3. Conclusions

As a result of comparing an Azelastine-based therapy with common therapies on patients with chronic active hepatitis by taking advantage of the Mahalanobis distance, we conclude that the Azelastine group obviously has a better trend health than that of the contrast group. However, looking at in-

Reference

Tatsuji Kanetaka, 1992. Application of Mahalanobis distance to measurement of drug efficiency. *Standardization and Quality Control*, Vol. 45, No. 10, pp. 40–44.

This case study is contributed by Tatsuji Kanetaka.

Use of Mahalanobis Distance in Medical Diagnosis

Abstract: This case deals with use of the Mahalanobis–Taguchi system (MTS) and multivariate analysis together with a special medical examination focusing on liver disease. More specifically, as a medical application of statistics, 16 blood biochemical data, ages, and genders were analyzed to evaluate examinees' health conditions. This is the first research case study of the MTS method used in an application of Mahalanobis distance (D^2) to a multivariate analysis.

1. Introduction

Generally speaking, when analyzing multiple variables, we should create a good database above all. A good database is generally homogeneous. In the case of biochemical data for human beings, while data for healthy people have little variability and distribute homogeneously, those for patients suffering diseases vary widely and cannot be viewed as homogeneous.

For these reasons it is desirable to prepare a database by using examination results of healthy people (normal persons). However, when attempting to diagnose a specific disease, unless necessary examination items to diagnose the disease are included, we cannot make use of the database. In a periodic medical checkup, due to constraints of budget, personnel, or time, few examination items are selected for diagnosis. Even in a complete physical examination, only common items are checked. Therefore, if the data for a regular medical checkup or complete physical examination are used as a normal contrast, we sometimes lack examination items. Then it is quite difficult to obtain enough data on normal persons. However, when we determine a standard value in certain examination facilities, we can obtain healthy people's data. Yet it is not indisputable whether judgment of health conditions is made accurately enough. In this research, as our

database, we capitalized on good biochemical data for normal persons, which we obtained by chance.

2. Database and the Base Space

For the people whose examination data can be obtained, after examining their disease histories in medical checkups at Tokyo Teishin Hospital over the last year, we selected 200 persons regarded to have no disease. Although we initially attempted to obtain data for 1000 persons, due to our limited computer capacity in the 1980s, we decided to perform an analysis using the data for only 200 persons. If a larger-scale computer were available and more people could be involved, it would be better to analyze a greater number.

Based on the performance of the automated analyzer used, we determined 18 examination items, including the following 16 biochemical data plus age and gender: total protein (TP), albumin (Alb), A/G (albumin/globulin) ratio, cholinesterase (ChE), glutamate oxaloacetate transaminase (GOT), glutamate pyruvate transaminase (GPT), lactate dehydrogenase (LDH), alkaline phosphatase (ALP), γ-glutamyltranspeptidase (γ-GTP), leucine aminopeptidase (LAP), total cholesterol (TCh), triglyceride (TG), phospholipases (PL), creatinine

Table 1

Age, gender, and biochemical data of normal persons

No.	Gender	Age	TP	Alb	A/G	ChE	GOT	GPT	LDH	γ-GTP	LAP	TCh	TG	PL	Cr	BUN	UA
1	Male	59	6.5	3.7	1.32	0.70	12	6	190	12	275	235	140	238	0.8	13	3.8
2	Female	51	7.0	4.4	1.69	0.98	21	18	247	23	340	225	87	227	1.1	15	4.5
3	Female	53	6.7	4.2	1.68	0.83	18	8	220	16	278	230	67	232	0.8	17	3.2
4	Female	52	6.6	4.3	1.87	0.90	12	6	244	15	289	171	59	175	0.8	13	3.8
5	Female	52	6.7	3.9	1.39	0.97	13	7	198	13	312	192	51	203	1.0	14	4.1
6	Female	56	7.2	4.0	1.25	0.93	22	16	188	16	304	216	86	213	1.0	13	4.2
7	Female	51	7.1	4.0	1.29	0.88	16	9	187	13	272	235	96	251	1.0	14	3.5
8	Female	50	6.6	3.6	1.20	0.71	14	5	190	12	270	149	57	165	0.9	14	3.4
9	Female	41	6.9	4.3	1.65	0.81	16	12	195	13	319	160	55	175	1.0	12	3.0
10	Female	48	7.0	4.2	1.50	0.93	21	19	230	35	411	197	110	213	1.0	8	5.2

Table 2
Normalized data of normal persons

No.	Age	Gender	TP	Alb	A/G	ChE	GOT	GPT	LDH	ALP	γ-GTP	LAP	TCh	TG	PL	Cr	BUN	UA
191	−1.42	+0.73	−0.34	−0.26	−0.08	−1.18	−0.44	−0.65	−0.06	−0.21	−0.57	+0.59	−1.24	−0.38	−0.72	+0.20	−0.39	−0.24
192	−1.32	+0.73	−0.34	+0.51	+0.94	+0.69	+0.90	−0.65	−0.92	−0.21	−0.57	+0.24	−1.36	−0.38	−1.44	−0.33	+0.23	+0.40
193	−1.32	+0.73	+1.48	+1.27	+0.04	+1.21	+1.16	+1.29	−0.41	+0.25	−0.26	+0.03	−1.09	+2.15	−1.15	+0.20	+1.16	−1.10
194	−1.90	+0.73	+1.48	+1.65	+0.49	−1.21	−0.67	−1.09	−0.68	+0.71	−0.88	+1.04	−1.75	−0.55	−1.80	+0.73	−1.32	+0.96
195	−1.23	+0.73	−0.65	+0.51	+1.32	+0.38	−1.36	−0.87	−0.51	−0.34	−0.72	−0.93	−1.24	−1.22	−1.11	+0.20	−1.63	+0.08
196	−1.03	−1.36	−0.65	+0.13	+0.75	−0.87	−1.13	−1.30	−1.87	+0.01	−0.41	+0.56	−1.30	−1.36	−1.11	+0.73	−0.39	+0.48
197	+0.80	+0.73	+0.04	−0.26	−0.36	+0.48	+1.39	+1.51	+1.57	+0.86	+2.81	+1.99	+0.25	−0.14	+0.35	+0.20	−0.39	+1.12
198	+0.45	+0.73	−0.34	−2.16	−2.05	−1.91	−0.67	−0.44	−2.18	+0.45	−0.26	+0.21	−1.45	−0.74	−0.33	+0.73	−0.39	+1.12
199	−0.45	+0.73	+0.26	+0.89	+0.78	−0.24	−0.47	+0.43	−0.24	−0.93	+1.04	+0.47	+0.31	+1.85	+0.38	+1.25	+1.18	+0.56
200	+1.29	−1.36	−0.65	−1.40	−1.11	+0.04	−0.93	−0.93	+0.96	+1.83	−0.34	−0.02	+1.79	+1.02	+1.39	−0.85	−0.54	−0.55
Total	0.000	0.000	0.000	0.000	0.000	0.000	0.000	0.000	0.000	0.000	0.000	0.000	0.000	0.000	0.000	0.000	0.000	0.000
Ave.	0.000	0.000	0.000	0.000	0.000	0.000	0.000	0.000	0.000	0.000	0.000	0.000	0.000	0.000	0.000	0.000	0.000	0.000

Table 3
D^2/f and $10 \log(D^2/f)$ of normal persons

No.	D^2/f	$10 \log(D^2/f)$	No.	D^2/f	$10 \log(D^2/f)$	No.	D^2/f	$10 \log(D^2/f)$
1	1.989	+2.987	36	0.513	−2.897	71	0.722	−1.413
2	0.609	−2.153	37	0.530	−2.761	72	1.165	+0.66
3	1.104	+0.428	38	0.567	−2.465	73	0.927	−0.328
4	0.747	−1.267	39	0.675	−1.704	74	0.946	−0.242
5	0.571	−2.432	40	0.746	−1.273	75	1.186	+0.741
6	0.893	−0.491	41	1.587	+2.006	76	0.557	−2.545
7	0.779	−1.087	42	0.859	−0.662	77	1.47	+1.674
8	0.872	−0.593	43	0.391	−4.053	78	1.244	+0.947
9	0.587	−2.316	44	0.616	−2.102	79	1.719	+2.352
10	1.274	+1.053	45	0.724	−1.400	80	0.440	−3.567
11	0.698	−1.564	46	0.813	−0.901	81	0.999	−0.005
12	0.596	−2.251	47	1.027	+0.117	82	0.707	−1.505
13	0.623	−2.053	48	1.107	+0.443	83	1.128	+0.524
14	1.249	+0.965	49	0.907	−0.422	84	0.713	−1.468
15	0.673	−1.722	50	1.773	+2.486	85	0.755	−1.219
16	0.482	−3.174	51	1.129	+0.527	86	1.086	+0.357
17	1.144	+0.586	52	0.634	−1.979	87	0.842	−0.749
18	0.529	−2.766	53	0.786	−1.047	88	0.972	−0.124
19	1.131	+0.535	54	1.368	+1.360	89	0.555	−2.560
20	1.299	+1.135	55	1.184	+0.733	90	0.761	−1.183
21	0.341368	−4.668	56	0.324	−4.900	91	1.90	+2.801
22	1.028	+0.118	57	1.412	+1.499	92	0.576	−2.395
23	0.588	−2.33	58	1.039	+0.165	93	0.775	−1.106
24	1.776	+2.495	59	0.782	+1.068	94	0.417	−3.803
25	0.657	−1.825	60	1.140	+0.570	95	1.682	+2.746
26	0.600	−2.222	61	1.213	+0.839	96	0.571	−2.433
27	0.720	−1.427	62	0.629	−2.015	97	2.251	+3.523
28	1.084	+0.349	63	1.135	+0.548	98	1.195	+0.775
29	3.442	+5.368	64	0.652	−1.858	99	0.678	−1.690
30	1.082	+0.343	65	0.811	−0.910	100	1.286	+1.093
31	0.960	−0.177	66	1.609	+2.06			
32	0.521	−2.831	67	0.905	−0.436			
33	1.076	+0.319	68	0.999	−0.003			
34	1.629	+2.120	69	2.264	+3.549			
35	0.896	−0.475	70	0.978	−0.098			

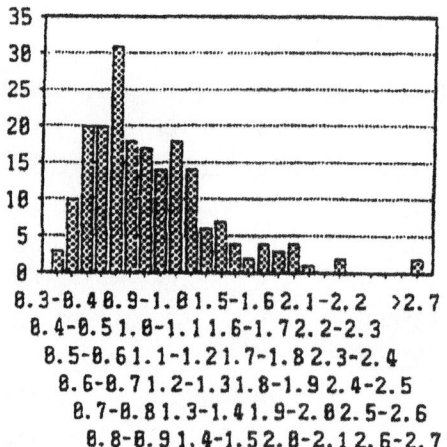

8.3-8.4 8.9-1.8 1.5-1.6 2.1-2.2 >2.7
8.4-8.5 1.8-1.1 1.6-1.7 2.2-2.3
8.5-8.6 1.1-1.2 1.7-1.8 2.3-2.4
8.6-8.7 1.2-1.3 1.8-1.9 2.4-2.5
8.7-8.8 1.3-1.4 1.9-2.8 2.5-2.6
8.8-8.9 1.4-1.5 2.8-2.1 2.6-2.7

Figure 1
D^2/f of normal persons

(Cr), blood urea nitrogen (BUN), and uric acid (UA).

Since age and gender do not require extra cost and money for investigation and the biochemical data are associated with them, we considered it better to study the biochemical data along with age and gender. As the n value for age, considering that each datum is logarithmized, we set male as 10 and female as 1. Table 1 shows a part of the data.

Strictly speaking, although these data approximately follow the normal distribution, the normalization process facilitates the succeeding step. Setting the mean of the ith person's jth examination

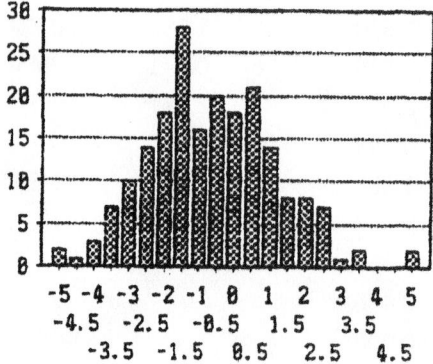

Figure 2
$10 \log(D^2/f)$ of normal persons

item to m_j, the standard deviation to σ_j, and the data from the ith person's jth examination item to X_{ij}, we calculated the normalized data:

$$Y_{ij} = \frac{X_{ij} - m_j}{\sigma_j} \qquad (1)$$

Table 2 shows a part of the normalized data. The total sum and mean for each item result in zero, as shown in the table.

Although we can proceed with the analysis with D^2, now we use the value of D^2 divided by a degree of freedom, f, and its logarithm multiplied by 10:

$$Z = 10 \log \frac{D^2}{f} \qquad (2)$$

Table 3 shows a part of D^2/f' and $10 \log(D^2/f')$. In addition, Figures 1 and 2 illustrate their distributions. Although neither distribution is completely homogeneous because the number of data are relatively small, it is considered homogeneous enough to analyze.

3. Data for Examinees Taking Special Medical Checkups

The examinees for a special medical checkup are basically clerical workers 35 years of age or older, 45 males and 50 females.

Collection of Data

The examinees were tested in the afternoon without diet restrictions. If even a single piece of data is beyond a target limit, the examinee takes the second test on another day with no breakfast.

Diagnosis of Examinees

1. If all data were within the target limit in the first test, also taking into account other data through a medical examination by interview or other checkups, we diagnosed the examinee as being within a normal limit (WNL).

2. If there were no data beyond the target limit in the second test, although a certain number of data were out of limit in the first test, we judged the examinee as WNL if the data beyond the limit were caused by the intake of

Table 4
Data of special medical examination

No.	Age	Gender	TP	Alb	A/G	ChE	GOT	GPT	LDH	ALP	γ-GTP	LAP	TCh	TG	PL	Cr	BUN	UA
1	35	Male	7.0	4.0	1.33	0.67	17	9	144	4.1	19	265	151	53	183	1.5	1.4	5.7
2	42	Female	7.5	4.2	1.27	0.67	15	10	187	5.4	14	315	202	148	218	1.3	14	5.6
3	41	Female	7.1	4.1	1.37	0.64	19	15	295	11.4	13	230	191	384	270	1.3	16	3.5
4	52	Female	6.6	3.9	1.44	0.88	14	10	243	5.8	10	289	185	96	218	1.3	13	4.7
5	53	Female	7.4	4.2	1.31	0.77	16	15	204	6.1	44	408	283	344	300	1.4	14	5.3
6	45	Female	7.3	4.3	1.43	0.81	15	14	196	5.4	15	312	169	70	181	0.9	11	3.3
7	41	Female	6.7	4.0	1.48	0.74	9	6	178	4.5	7	273	135	98	160	1.0	15	3.1
8	41	Female	6.7	3.7	1.23	0.82	16	12	180	6.3	13	269	214	56	218	1.1	14	4.4
9	48	Female	6.7	3.8	1.31	0.73	26	5	119	6.3	5	251	176	69	190	0.9	14	3.6
10	51	Female	7.1	4.3	1.54	0.99	12	9	220	6.1	7	305	305	114	203	1.2	15	4.7

Table 5
D^2/f and $10 \log(D^2/f)$ of special medical examination

No.	D^2/f	$10 \log(D^2/f)$	No.	D^2/f	$10 \log(D^2/f)$	No.	D^2/f	$10 \log(D^2/f)$
1	1.894	+4.933	21	3.114	+4.933	41	1.594	+2.025
2	1.320	+1.207	22	3.280	+5.159	42	1.872	+2.723
3	18.591	+12.683	23	3.797	+5.794	43	2.366	+3.741
4	2.160	+3.345	24	5.305	+7.670	44	0.959	−0.181
5	5.648	+7.519	25	1.537	+1.867	45	4.609	+8.813
6	0.630	−2.006	26	4.611	+6.638	46	4.274	+6.308
7	1.075	+0.314	27	2.856	+4.557	47	5.847	+7.670
8	0.846	−0.725	28	1.706	+2.320	48	1.653	+2.184
9	2.956	+4.70	29	4.848	+6.855	49	7.325	+8.648
10	0.815	−0.888	30	5.498	+7.402	50	2.076	+3.172

alcohol the previous night or meals taken before the test, and there was no necessity of considering liver disease, the examinee was diagnosed as normal.

3. If there were data out of the limit even in the second test, the examinee takes a precise examination later.

4. Among the people who took precise examinations, those who were judged to have certain signs that indicated the potential for a specific liver ailment, and thus to whom a warning should be given for future health care, were diagnosed as slightly abnormal (SAB). On the other hand, those who did not need to be warned were diagnosed as WNL.

5. People who evidenced a fatty liver due to obesity, excessive intake of sugar, a fatty liver due to diabetes, slight liver disease due to a gallstone or other abnormal organs, or to be an asymptomatic carrier of hepatitis B were judged to be in the category of slight liver disease (SLI).

6. Those with a chronic liver disease due to hepatitis B or C virus, cirrhosis, or alcoholic liver disease were diagnosed as having a liver disease (LVD).

Data for Examinees

Table 4 shows a part of data for examinees who took a special medical examination. Normalizing these data using equation (1), we calculated D^2 by using the inverse matrix. Table 5 shows a part of D^2/f and $10 \log(D^2/f)$

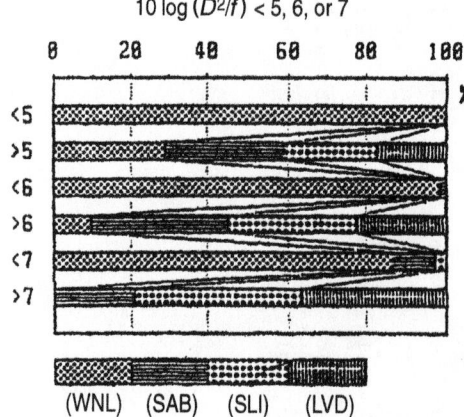

Figure 3
Final diagnosis distribution change under different thresholds

4. Analysis and Results

Now we detail the analysis of the D values calculated. We studied what value should be selected as a

Table 6
Threshold for $10 \log(D^2/f)$ and final diagnosis (estimation and 95% confidence interval)

Threshold	95% Confidence Interval	$10 \log(D^2/f)$ Less Than Threshold				$10 \log(D^2/f)$ Greater Than Threshold			
		WNL	SAB	SLI	LVD	WNL	SAB	SLI	LVD
5	Lower limit	98.19	$-\infty$	$-\infty$	$-\infty$	19.31	20.07	15.72	10.64
	Estimation	100.00	0.00	0.00	0.00	29.27	30.27	24.39	17.07
	Upper limit	$+\infty$	1.81	1.81	1.81	41.71	42.87	35.86	26.25
6	Lower limit	97.43	0.94	$-\infty$	$-\infty$	6.05	24.86	22.27	14.93
	Estimation	98.44	1.56	0.00	0.00	9.68	35.48	32.26	22.58
	Upper limit	99.05	2.57	1.74	1.74	0.1512	47.76	44.19	32.65
7	Lower limit	79.54	6.48	1.57	$-\infty$	$-\infty$	13.58	30.00	25.58
	Estimation	86.84	10.53	2.63	0.00	0.00	21.05	42.11	36.84
	Upper limit	91.65	16.05	4.38	1.77	1.77	31.16	55.25	49.75

threshold for $10 \log(D^2/f)$ in order to obtain a good result.

Figure 3 and Table 6 go along with the following analyses.

1. *Threshold* = 5. If 5 is selected as the threshold, all cases with a value less than or equal to 5 result in WNL with no abnormal cases. However, since 29.27% of the cases with a value greater than or equal to 5 (or from 19.31 to 41.71% in a 95% confidence interval) belong to WNL, the indication is that unnecessary retesting or a precise examination was conducted.

2. *Threshold* = 6. Compared to the case of threshold = 5, the proportion of WNL among cases diagnosed as abnormal decreased to 9.68% (6.05 to 15.12%). Yet among cases with a value below 6 diagnosed as normal, 1.56% (0.94 to 2.57%) of abnormal cases were included. However, these abnormal cases were not con-

sidered to be a serious problem because all of them had only a slight abnormality.

3. *Threshold* = 7. If the threshold is set to 7, the possibility of mistakenly diagnosing a normal person for an abnormal one is eliminated. However, since 10.53% of slight abnormality and 2.63% (1.57 to 4.38%) of slight liver disease are mingled, this threshold cannot be regarded as correct.

Biochemical Data within and beyond Target Limit and Final Diagnosis

If the judgment based on biochemical data (within the target limit or not) affects our final diagnosis of slight liver disease, it can be seen from Table 7 and Figure 4 that 6.67% (3.44 to 12.54%) are included in the cases judged as normal. In addition, the fact that 65% (48.05 to 78.85%) of the cases diagnosed as abnormal are WNL, in fact, demonstrates that the

Table 7
Biochemical data and final diagnosis

95% Confidence Interval	Within Target Limits				Beyond Target Limits			
	WNL	SAB	SLI	LVD	WNL	SAB	SLI	LVD
Lower limit	87.46	$-\infty$	3.44	$-\infty$	48.05	8.08	5.94	4.56
Estimation	93.33	0.00	6.67	0.00	65.00	15.00	11.25	8.75
Upper limit	96.56	2.09	12.54	2.09	78.85	26.16	20.29	16.14

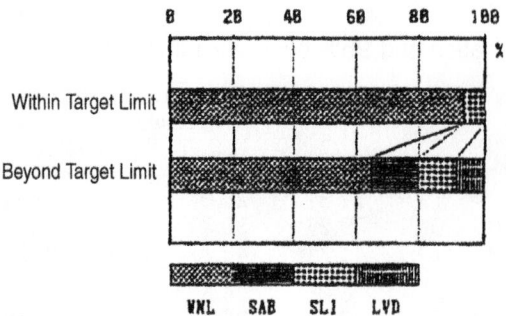

Figure 4
Within and beyond target limits and final diagnosis

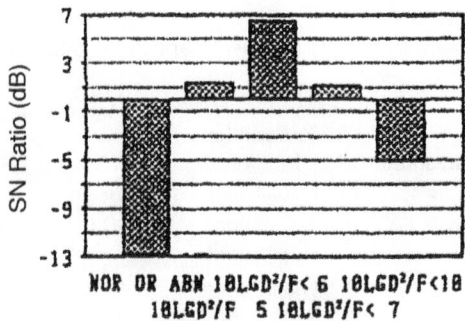

Within Normal Limit or Others

Figure 6
Within normal limits or others

method based on biochemical data cannot be used for diagnosis. Since those doctors who can judge cases with several data outside a target limit as normal are, in most cases, quite experienced, it is extremely difficult to make a similar judgment.

This is one of the reasons that we attempted to use a somewhat bothersome calculation such as D^2. It is quite easy to imagine the feeling of persons who are asked to take a retest or precise examination in a medical checkup. It is possible that some of them could have high blood pressure or a stomach ulcer due to mental stress or have an attack of angina. Thus, this study was not necessarily a waste of time, labor, and budget.

Next we studied how a limited number of examination items influences the result. Figure 5 shows the result obtained by only the five items of GOT, GPT, γ-GTP, TCh, and TG prescribed by the

Industrial Safety and Health Law. Of the cases diagnosed as normal, 5.55% (2.71 to 10.94%) and 2.68% (1.23 to 5.35%) were actually SAB and SLI, respectively. Additionally, 55.93% (37.57 to 72.80 percent) of the cases judged as abnormal were categorized as WNL. Considering these results, we cannot say that constraint in the number of examination items leads to better judgment.

Determination of Data within and beyond the Target Limit

Since we were dealing with a medical checkup, we studied the relationships among the final diagnosis (of whether data were within or beyond a normal

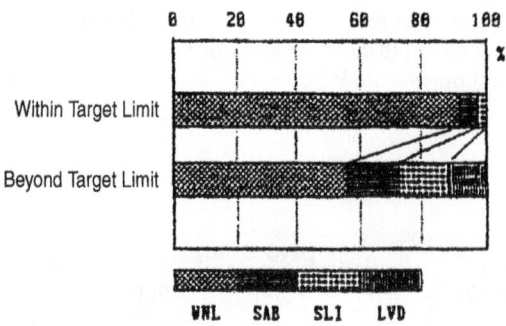

Figure 5
Within and beyond target limits using five test items and final diagnosis

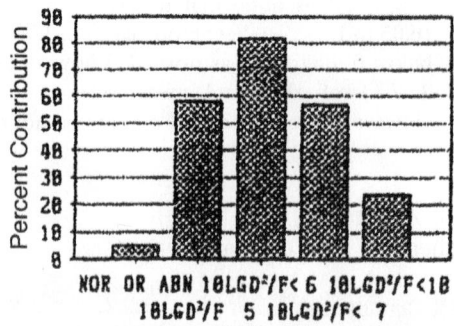

Within Normal Limit or Others

Figure 7
Percent contribution for biochemical data limits and 10 log(D^2/f) threshold

Table 8

Data within or beyond target limit versus threshold for 10 log(D^2/f) versus final diagnosis (estimation and 95% confidence interval)

Final Diagnosis	95% Confidence Interval	Within or Beyond Target Limits		10 log(D^2/f) = 5		10 log(D^2/f) = 6		10 log(D^2/f) = 7		10 log(D^2/f) = 10	
		Within Target Limit	Beyond Target Limit	Less Than Threshold	Greater Than Threshold	Less Than Threshold	Greater Than Threshold	Less Than Threshold	Greater Than Threshold	Less Than Threshold	Greater Than Threshold
Within normal limit	Lower limit	87.28	47.65	98.32	20.47	97.87	7.26	80.32	-∞	63.54	-∞
	Estimation	93.33	65.00	1.00	29.27	98.44	9.68	86.84	0.00	76.74	0.00
	Upper limit	96.62	79.12	+∞	39.95	98.85	12.79	91.43	1.69	88.20	1.97
Beyond normal limit	Lower limit	3.38	20.88	-∞	60.05	1.15	87.21	8.67	98.31	13.80	98.03
	Estimation	6.67	35.00	0.00	70.73	1.56	90.32	13.16	100.00	23.26	100.00
	Upper limit	12.72	52.35	1.68	79.53	2.13	92.74	19.68	+∞	36.46	+∞
Contribution		5.0331		57.87		81.66		56.90		23.82	
SN ratio		-12.76		1.38		6.49		1.21		-5.05	

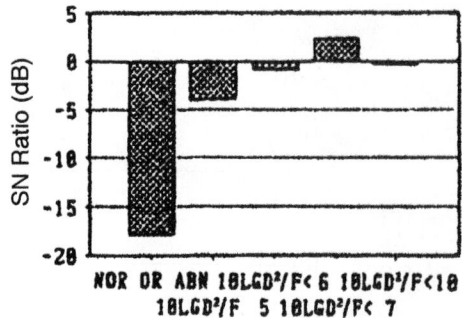

No Liver Disease or Liver Disease

Figure 8
SN ratio for liver disease, biochemical data limits, and 10 $\log(D^2/f)$ threshold

limit), biochemical data within or beyond a target limit, and a threshold of 10 $\log(D^2/f)$, as shown in Figures 6 and 7 and Table 8.

1. *Biochemical data within or beyond the target limit.* While 93.33% of the cases within the target limit reasonably belong to WNL, 6.67% (3.38 to 12.72%) of them are categorized as abnormal. On the other hand, 65% (47.65 to 79.12%) of the cases judged as abnormal are WNL. The percent contribution computed results in 5.03, and the SN ratio was 12.76. Therefore, we do not regard this as a recommendable method.

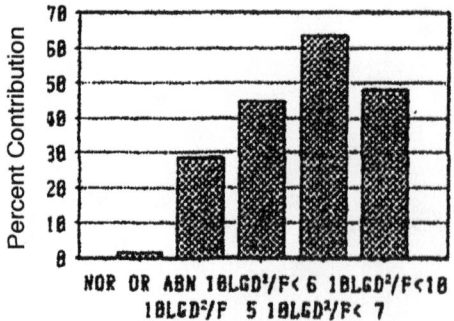

No Liver Disease or Liver Disease

Figure 9
Percent contribution for liver disease, biochemical data limits, and 10(D^2/f) threshold

2. *Threshold for* 10 $\log(D^2/f)$ *set to* 5. All of the cases are WNL (i.e., no abnormality is mingled with the cases with a value less than or equal to the threshold). In contrast, the fact that 29.27% (20.47 to 39.95%) of the cases with a value greater than the threshold are diagnosed as WNL was regarded as a problem. A higher percent contribution and SN ratio, 57.87% and 1.38, respectively, were obtained.

3. *Threshold for* 10 $\log(D^2/f)$ *set to* 6. While 98.44% of the cases with a value below the threshold are WNL, 1.56% (1.15 to 2.13%) of slight abnormality cases were mixed with them. On the other hand, only 9.68% of the cases with a value greater than or equal to the threshold were WNL, which may be regarded as tolerable. The resulting contribution of 81.67% and SN ratio of 6.49 are a good value. In actuality, we decided that we should use this method.

4. *Threshold for* 10 $\log(D^2/f)$ *set to* 7. Although no normal case was mingled with the cases with a value greater than the threshold, 23.26% (13.80 to 36.46%) of the abnormality was mixed with the cases with a value less than the threshold. We consider this impractical. In fact, the contribution and SN ratio are 23.82% and −5.05, respectively, both of which are poor.

Relationship with Liver Disease

Despite slight digression from our main focus, we looked at the relationship among the final diagnosis of liver disease, biochemical data within or beyond a target limit, and threshold for 10 $\log(D^2/f)$. Because we dealt with an issue with a different objective, the resulting contribution and SN ratio are somewhat poor (Figures 8 and 9 and Table 9).

1. *Biochemical data within or beyond the target limit.* While 6.67% (2.66 to 15.76%) of the cases categorized as normal based on biochemical data are mixed up with the cases of liver disease, 80% (60.44 to 91.62%) of the cases categorized as abnormal do not have a liver disease based on the final diagnosis. The contribution is 1.61%, and the SN ratio turns out to be 17.87. Both values indicate a poor classification.

Table 9

Data within or beyond target limit versus threshold for $10 \log(D^2/f)$ versus final diagnosis of liver disease

Final Diagnosis	95% Confidence Interval	Within or Beyond Target Limits		$10 \log(D^2/f) = 5$		$10 \log(D^2/f) = 6$		$10 \log(D^2/f) = 10$		$10 \log(D^2/f) = 10$	
		Within Target Limit	Beyond Target Limit	Less Than Threshold	Greater Than Threshold	Less Than Threshold	Greater Than Threshold	Less Than Threshold	Greater Than Threshold	Less Than Threshold	Greater Than Threshold
With liver disease	Lower limit	84.24	60.44	97.65	38.44	97.86	28.61	95.37	12.92	83.01	$-\infty$
	Estimation	93.33	80.00	100.00	58.54	100.00	45.16	97.37	21.05	90.70	0.00
	Upper limit	97.34	91.62	$+\infty$	76.15	$+\infty$	62.85	98.52	32.40	95.11	2.08
With no liver disease	Lower limit	2.66	8.38	$-\infty$	23.85	$-\infty$	37.15	1.48	67.60	4.89	97.92
	Estimation	6.67	20.00	0.00	41.46	0.00	54.84	2.63	78.95	9.30	100.00
	Upper limit	15.76	39.56	2.35	61.56	2.14	71.39	4.63	87.08	16.99	$+\infty$
Contribution		0.0161		28.71		45.10		63.42		48.02	
SN ratio		−17.86		−3.95		−8.89		2.39		−0.34	

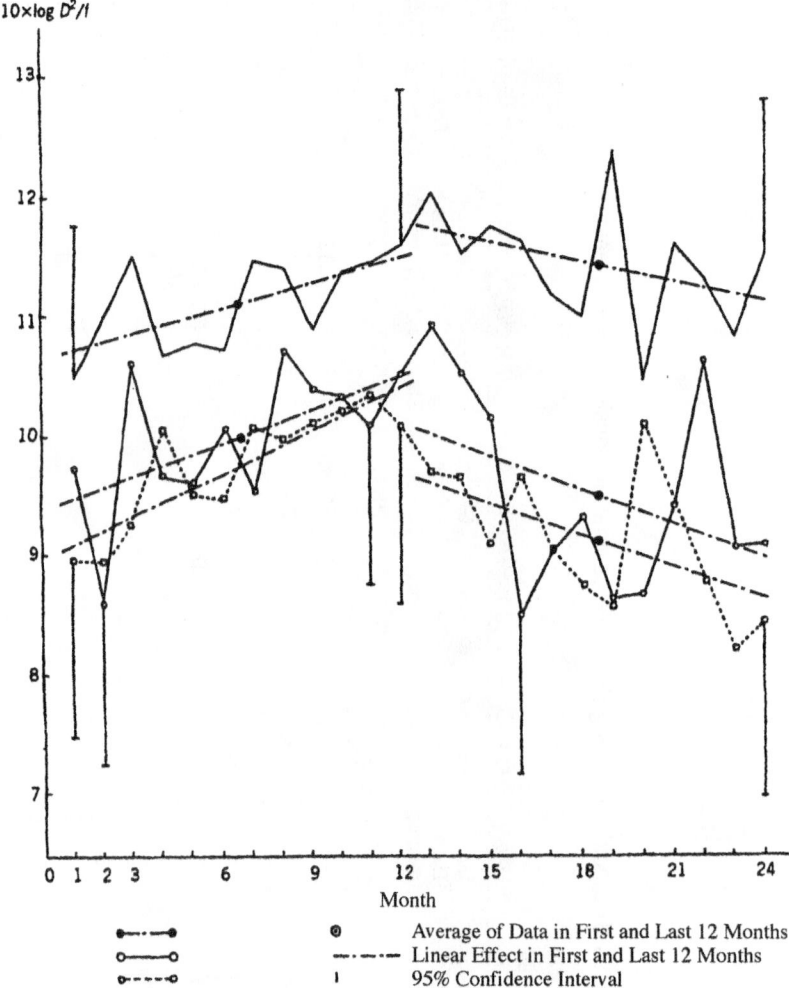

Figure 10
Transition of chronic active hepatitis

2. *Threshold for* $10 \log(D^2/f)$ *set to* 5. No case of liver disease was mingled with cases with a value below threshold. In contrast, 58.54% (38.44 to 76.15%) of the cases with a value greater than the threshold actually have no liver disease. The contribution and SN ratio result in 28.71% and −3.96, respectively, both of which are not good enough.

3. *Threshold for* $10 \log(D^2/f)$ *set to* 6. No case of liver disease was mixed up with cases with a

value below threshold. However, 45.16% (28.61 to 62.85%) of the cases with a value greater than threshold had no liver disease. The resulting contribution and SN ratio were computed as 45.10% and −0.89, respectively.

4. *Threshold for* $10 \log(D^2/f)$ *set to* 7. Of the cases with a value below threshold, 2.63% (1.48 to 4.63%) had liver disease. In addition, 21.05% (12.92 to 32.40%) of the cases with a value above threshold had no liver disease. We ob-

tained the best contribution of 63.42% and SN ratio of 2.39 among all for this threshold. Since our main objective was judgment of health conditions rather than diagnosis of whether or not there was liver disease, such moderate results seem reasonable enough.

5. *Threshold for* 10 $\log(D^2/f)$ *set to* 10. While 9.30% (4.89 to 16.99%) of the cases with a value below threshold had liver disease, the cases with a value greater than threshold had no case of liver disease. The resulting contribution and SN ratio were 48.02% and −0.34.

As can be seen, when we diagnose liver disease, the correctness of the judgment deteriorates slightly. To improve our judgment, we need to use a different concept in building when using the same database.

5. Another Application of Mahalanobis Distance

To date, we have attempted to apply the D^2 value to the medical field by judging whether or not certain data belong to a certain group, as Mahalanobis originally intended in his research. Here, our discussion is based on the idea that D^2 is regarded as the distance from the center of gravity of the group forming a database.

Let's take the case of people with a certain disease: for example, patients with chronic active hepatitis. Defining their distance from normal people as D^2, we can see that their distance from the gravity center of the normal group is equivalent to how serious their disease is. Since a decreasing D^2 value indicates proximity to a normal group when we keep track of the condition of patients suffering chronic active hepatitis, their health is expected to improve. On the contrary, when D^2 increases gradually, their disease is considered to be aggravated because the distance from the normal group changes by increments.

Figure 10 shows the 18-item data measured for 24 months for 21 patients with chronic active hepatitis: eight for group 1, seven for group 2, and six for group 3. Each of the three groups received a different treatment in the last 12 months.

Looking at D^2 and the linear effect line for each group, we can see that whereas health degraded during the first 12 months, for the last 12 months, the D^2 value decreased, due to the active therapy (i.e., health was improved). This implies that a time-based analysis of D^2 enables us to judge the transition of data. That is, in the medical science field, we can make a judgment on medical efficacy.

Although the Mahalanobis distance was complicated and impractical as developed by Mahalanobis, today, when computer technology is widely used, even a microcomputer can easily calculate the distance. As a result, it is regarded as one of the most broadly applicable techniques for multivariate analysis.

References

Tatsuji Kanetaka, 1987. An application of Mahalanobis distance: an example of chronic liver disease during active period. *Standardization and Quality Control,* Vol. 40, No. 11, pp. 46–54.

Tatsuji Kanetaka, 1997. An application of Mahalanobis distance to diagnosis of medical examination. *Quality Engineering,* Vol. 5, No. 2, pp. 35–44.

This case study is contributed by Tatsuji Kanetaka.

CASE 76

Prediction of Urinary Continence Recovery among Patients with Brain Disease Using the Mahalanobis Distance

Abstract: We created an experiment to see how accurately we could predict the rate of independent urination after four weeks using a Mahalanobis distance after one week.

1. Introduction

For our research, we collected data on 232 patients with brain disease, which we classified into four groups, each of which was a possible combination of whether urinary continence was attained by a patient at the end of the first or fourth week after he or she suffered brain trauma (Table 1). Group I consisted of 150 patients who had recovered continence at the end of the first and fourth weeks, respectively. Group II contains only one patient who had recovered continence at the end of the first week but had not at the end of the fourth week. Eighty-one patients who had not recovered continence at the end of the first week belong to groups III and IV. Among them, 30 patients who had recovered at the end of the fourth week were classified as group III, and the remaining 51 patients who could not recover were categorized as group IV.

The most interesting groups in our pattern recognition were groups III and IV, because if we can predict precisely whether a patient who has not been able to recover continence after one week will recover it after four weeks, we can use such information to establish a nursing procedure for each patient.

A Mahalanobis space was constructed using the data of the first week from group III: The patients in this group did not recover continence at the end of the first week but recovered by the end of the fourth week.

As a next step, using after-one-week data for groups III and IV, we computed the Mahalanobis distances. With the central point of a resulting Mahalanobis space defined as a reference point, we predicted the possibility of each patient's continence recovery at the end of the fourth week. Table 2 lists the items.

Items

Twelve factors, including the six diseases shown in Table 2, were used to calculate Mahalanobis distances. Disturbance of consciousness and urinary continence were, according to their degrees, classified into the following categories and quantified as an ordinal number:

1. Disturbance of consciousness 1 (eye-opening level): 4 points

2. Disturbance of consciousness 2 (utterance level): 5 points

3. Disturbance of consciousness 3 (movement level): 6 points

6. Urinary continence level after one week: 8 points

For each of the above, a larger number indicates a better condition. Since 8 points were added to a state of complete continence recovery, all other possible scores from 1 to 7 indicated incomplete recovery.

Table 1
Classification of patients by pattern in starting time of urinal continence recovery

Group	Starting Time of Urinal Continence Recovery[a]		Number of Patients
	End of First Week	End of Fourth Week	
I	+	+	150
II[b]	+	−	1
III	−	+	30
IV	−	−	51
		Total	232

[a] +, Continence recovered; −, continence not recovered.
[b] The data in group II can be viewed as exceptional because the patient was operated on again.

Binary Decomposition of Categorical Variables

We defined a male as 0 and a female as 1. This conversion was regarded as a binary expression for categorical variables. Six types of diseases can be expressed by five binary variables (Table 3). Although we can drop any single item, item 10 was removed. For example, when a patient with disease 1 (head injury, operation group) is considered, item 7 is set to 1, whereas items 8 to 12 are set to 0. On the other hand, for a patient with disease 4 (cerebral aneurysm, operation group), 0 was assigned to all items 7 to 12 because item 10 was excluded. For patients with other diseases, we allocated 0 or 1 in a similar manner to that for disease 1.

2. Prediction of Continence Recovery after Four Weeks

To see how accurately we could predict the rate of independent urination after four weeks by using a Mahalanobis distance after one week, we created Figures 1 to 6. The X-axis denotes a Mahalanobis distance after one week, and its calculation is expressed as

$$y = 10 \log_{10} \frac{D^2}{f}$$

Table 2
Items used for calculation of Mahalanobis distance

Factor No.	Factor
1	Disturbance of consciousness 1 (eye-opening level)
2	Disturbance of consciousness 2 (utterance level)
3	Disturbance of consciousness 3 (movement level)
4[a]	Gender (male = 0, female = 1)
5	Age
6	Urinary continence level after one week
7[a]	Disease 1 (head injury, operation group)
8[a]	Disease 2 (head injury, nonoperation group)
9[a]	Disease 3 (brain tumor, operation group)
10[a]	Disease 4 (cerebral aneurysm, operation group)
11[a]	Disease 5 (brain infarction, nonoperation group)
12[a]	Disease 6 (brain hemorrhage, operation and nonoperation group)

[a] Categorical variable.

Table 3
Binary expression for six types of diseases

	(7)	(8)	(9)	(11)	(12)
Patient with disease 1	1	0	0	0	0
Patient with disease 2	0	1	0	0	0
Patient with disease 3	0	0	1	0	0
Patient with disease 4	0	0	0	0	0
Patient with disease 5	0	0	0	1	0
Patient with disease 6	0	0	0	0	0

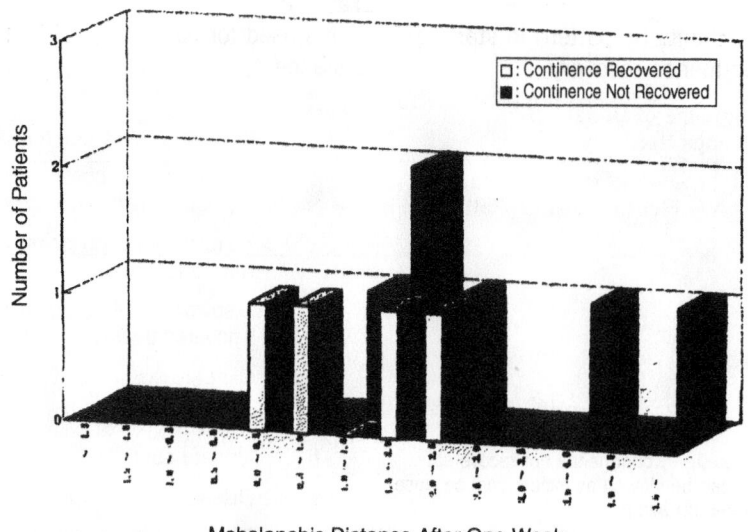

Figure 1
Mahalanobis distances after the end of the first and fourth weeks and the progress
of urinary continence recovery for disease 1 (head injury, operation group)

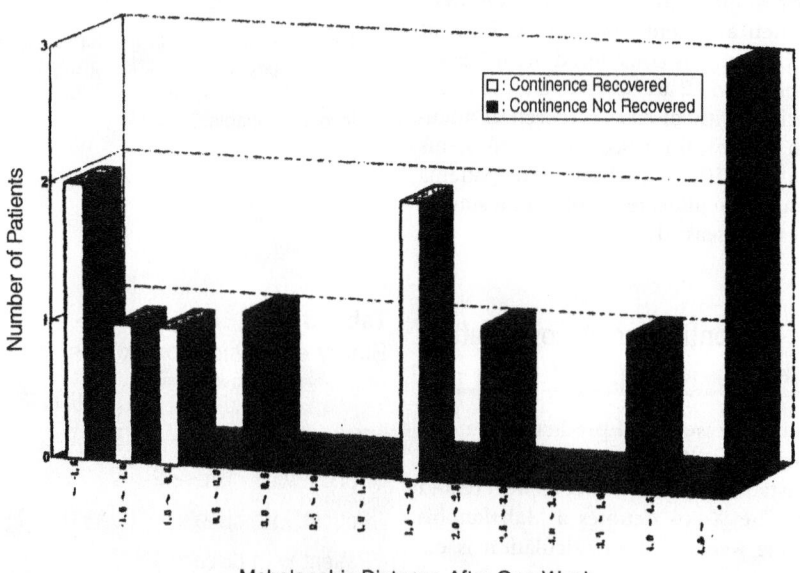

Figure 2
Mahalanobis distances after the end of the first and fourth weeks and the progress of
urinary continence recovery for disease 2 (head injury, nonoperation group)

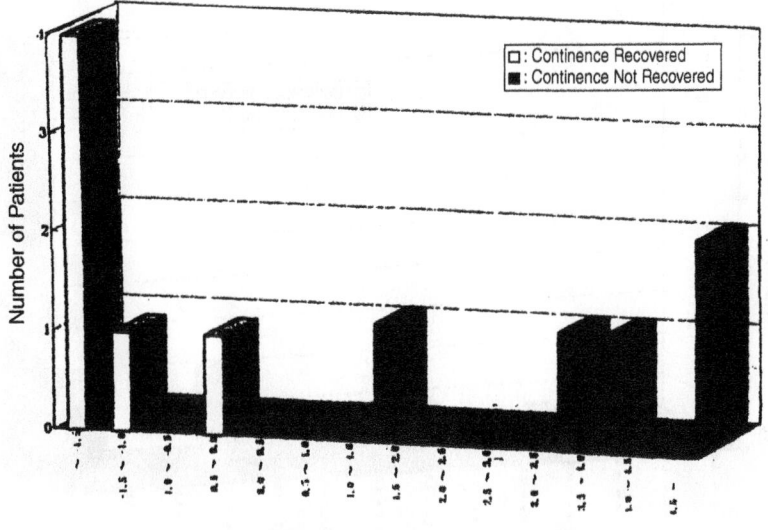

Figure 3
Mahalanobis distances after the end of the first and fourth weeks and the progress of
urinary continence recovery for disease 3 (brain tumor, operation group)

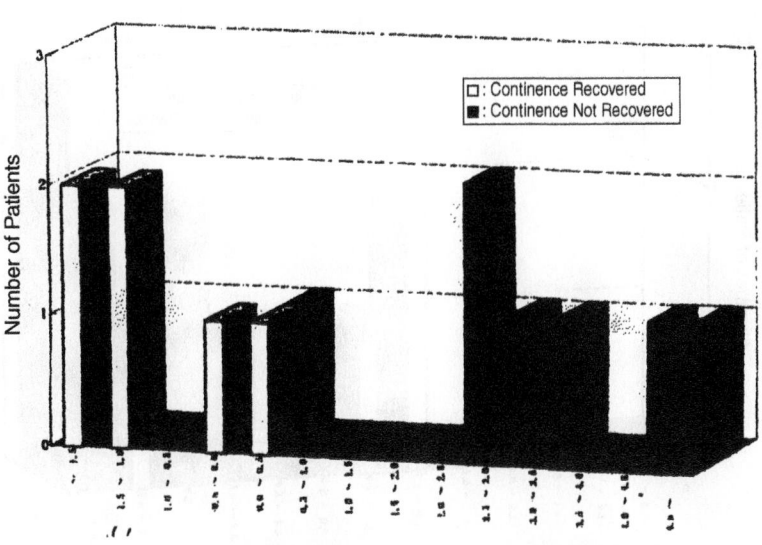

Figure 4
Mahalanobis distances after the end of the first and fourth weeks and the progress of
urinary continence recovery for disease 4 (cerebral aneurysm, operation group)

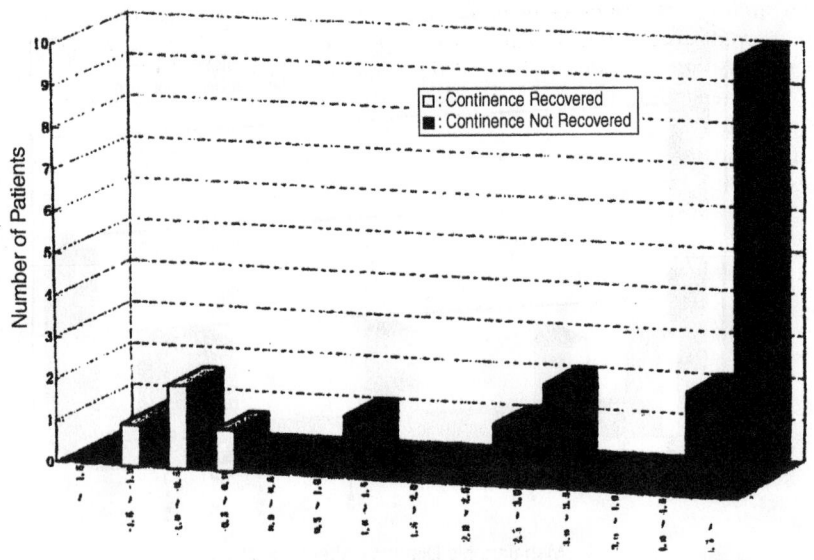

Figure 5
Mahalanobis distances after the end of the first and fourth weeks and the progress of urinary continence recovery for disease 5 (brain infection, nonoperation group)

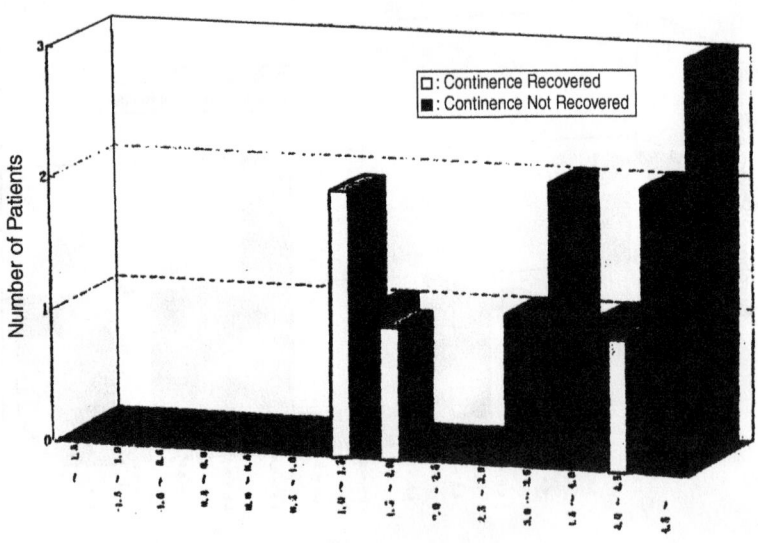

Figure 6
Mahalanobis distances after the end of the first and fourth weeks and the progress of urinary continence recovery for disease 6 (brain hemorrhage, operation and nonoperation groups)

where the Y-axis indicates the number of patients. "Continence recovered" means that a patient can urinate independently at the end of the fourth week, whereas "continence not recovered" signifies that he or she cannot do so.

Our research revealed that for disease 3 (brain tumor, operation group) and disease 5 (brain infarction, nonoperation group), we can predict perfectly from the Mahalanobis distance at the end of the first week the recovery after the fourth week. In addition, for disease 4 (cerebral aneurysm, operation group), we can foresee the progress relatively accurately.

3. Selection of Variables by Orthogonal Array

The number of items used to create a Mahalanobis space was 11, from items 1 to 12 except item 11. However, when handling a larger number of items, the data collection cost could become enormous. How to reduce the cost and make an accurate prediction was the next issue. Using an L_{12} orthogonal array, we assigned items as shown in Table 4.

The levels used are (1) select an item, and (2) do not select the item. Numbers arranged in verti-

cally are experiment numbers. For example, in experiment 1 we calculated a Mahalanobis distance with all 11 factors, and in experiment 2 computed MD with five variables, such as factors 3, 4, 6, 8, and 12.

The right side of Table 4 shows the SN ratio and sensitivity. Let the Mahalanobis distances of 30 patients who belong to group III, the patients who recovered continence at the end of the fourth week, be $y_{i,1}, y_{i,2}, \ldots, y_{i,30}$. Also let the Mahalanobis distances of 50 patients who belong to group IV, the patients who could not recover continence, be $y_{i,31}, y_{i,32}, \ldots, y_{i,81}$. The SN ratio and sensitivity were defined as follows. We evaluated the certainty of judgment by the SN ratio and the distance between two groups by sensitivity S:

$$\eta = 10 \log_{10} \frac{(1/r)(S_G - V_e)}{V_e} \quad (1)$$

$$S = \overline{G}_4 - \overline{G}_3 \quad (2)$$

where G_3 and G_4 are the averages of Mahalanobis distances belonging to groups III and IV, respectively.

The harmonic mean is expressed as

$$\frac{1}{r} = \frac{1}{2}\left(\frac{1}{30} + \frac{1}{51}\right) \quad (3)$$

Table 4
SN ratio sensitivity

No.	1	2	3	4	5	6	7	8	9	10	11	η (dB)	S
1	1	1	1	1	1	1	1	1	1	1	1	1.47	4.92
2	1	1	1	1	1	2	2	2	2	2	2	−2.70	4.78
3	1	1	2	2	2	1	1	1	2	2	2	−6.61	2.66
4	1	2	1	2	2	1	2	2	1	1	2	−5.07	3.30
5	1	2	2	1	2	2	1	2	1	2	1	−6.91	2.07
6	1	2	2	2	1	2	2	1	2	1	1	−4.54	2.97
7	2	1	2	2	1	1	2	2	1	2	1	−4.54	3.17
8	2	1	2	1	2	2	2	1	1	1	2	−5.88	2.35
9	2	1	1	2	2	2	1	2	2	1	1	−3.58	4.82
10	2	2	2	1	1	1	1	2	2	1	2	−7.82	2.02
11	2	2	1	2	1	2	1	1	1	2	2	−10.35	2.20
12	2	2	1	1	2	1	2	1	2	2	1	−3.87	4.42

Total variation, S_T, general mean, S_m, variation between groups, S_G, error variation, S_e, and error variance, V_e, were computed as follows:

$$S_T = y_1^2 + y_2^2 + \cdots + y_{81}^2 \qquad (f = 81) \quad (4)$$

$$S_m = \frac{(y_1 + y_2 + \cdots + y_{81})^2}{81} \qquad (f = 1) \quad (5)$$

$$S_G = \frac{(y_1^2 + y_2^2 + \cdots + y_{30}^2)}{30}$$
$$+ \frac{(y_{31}^2 + y_{32}^2 + \cdots + y_{81}^2)}{51} \qquad (f = 1) \quad (6)$$

$$S_e = S_T - S_m - S_G \qquad (f = 79) \quad (7)$$

$$V_e = \frac{S_e}{79} \qquad (8)$$

Using the equations above, we calculated SN ratios, η, and sensitivities, S, and summarized them in Table 4. Based on the SN ratios and sensitivities shown in the table, we created the response graphs shown in Figure 7, where 1 through 12 indicated above the X-axis represent item numbers and 2 and 1 below the axis indicate the orthogonal array level. Level 1 denotes use of an item, whereas level 2 indicates no use of the item.

The upper part of the plot shows the response graphs for the sensitivity, S. As for the sensitivity S of items 7, 8, and 9, level 1, representing use of a variable, is smaller than level 2, showing negative effects. For other items, use of the items is more effective. The lower part of the plot shows the response graphs for the SN ratio, η, which reveals that use of a variable is better that no use for all items except items 7 and 9.

4. Optimal Selection and Results

Figure 8 shows the relationship between Mahalanobis distances after one week and continence recovery after four weeks when using all items for all diseases. In this case, the SN ratio, η, and sensitivity, S, result in 1.47 dB and 4.92. In contrast, according to the result in the preceding section, Figure 9 illustrates the relationship between the Mahalanobis distances after one week and the recovery after four weeks, when using all items except items 7, 8, and 9. The resulting SN ratio, η, and sensitivity, S, are computed as 0.40 dB and 5.30.

Excluding the three items yields a better sensitivity, S, but a 1.07 dB lower SN ratio than when using all factors. A comparison of Figures 8 and 9 demonstrates that taking all items can lead to better discriminability.

The items excluded, 7, 8, and 9, are some of the items of diseases expressed as binary variables. This example highlights that when we select some cate-

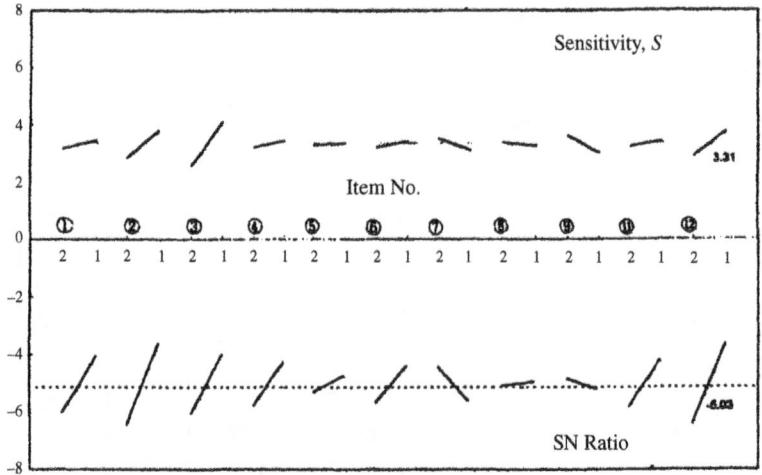

Figure 7
Response graphs for item selection

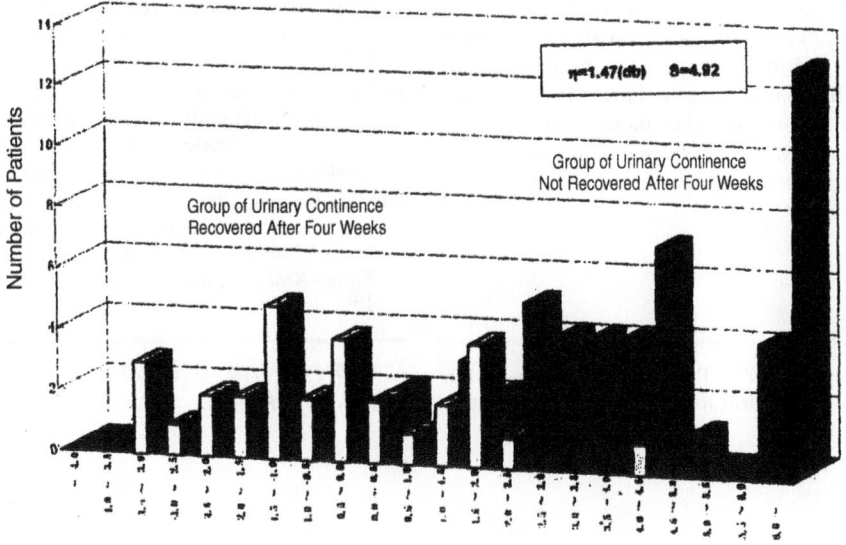

Figure 8
Mahalanobis distances after the end of the first and fourth weeks and the progress in urinary
continence recovery for all diseases, including all items

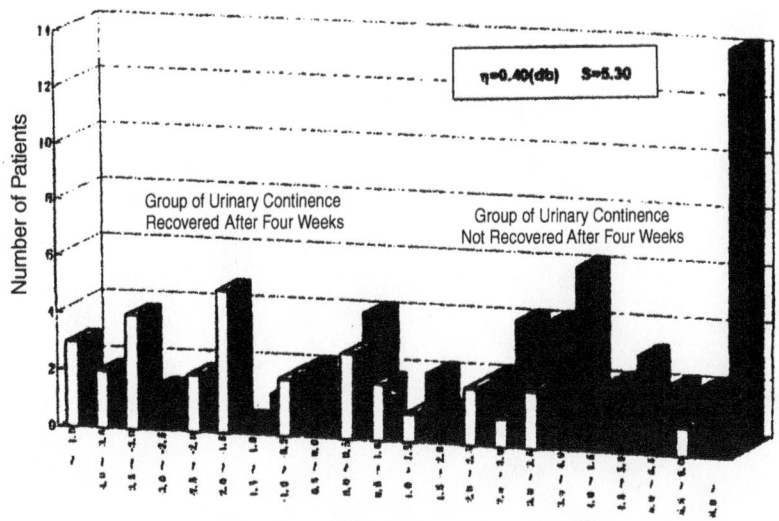

Figure 9
Mahalanobis distances after the end of the first and fourth weeks and the progress in
urinary continence recovery for all diseases when items were selected

gorical variables, it is regarded as inappropriate to exclude some of the variables from all the variables. If a selection cannot be averted, we should carefully exclude each variable in turn. This is because each binary variable contains other factors' effects.

In our research, using all the factors was chosen as the optimal selection for creating a Mahalanobis space.

References

Yoshiko Hasegawa, 1995. Pattern recognition by Mahalanobis: application to medical field—prediction of urinary continence recovery among patients with brain disease. *Standardization and Quality Control*, Vol. 48, No. 11, pp. 40–45.

Yoshiko Hasegawa and Michiyo Kojima, 1997. Prediction of urinary continence recovery among patients with brain disease using Mahalanobis distance. *Quality Engineering*, Vol. 5, No. 5, pp. 55–63.

Michiyo Kojima, 1986. Basic research on prediction of urinary continence recovery among patients with brain disease using conscious urinary diagram. *Kango Kenkyu/Igaku Shoin*. Vol. 19, No. 2, pp. 161–189.

This case study is contributed by Yoshiko Hasegawa and Michiyo Kojima.

Mahalanobis Distance Application for Health Examination and Treatment of Missing Data

Abstract: When a Mahalanobis space is formed, there occasionally exist missing data in a group of healthy people. If these are not treated properly, the base point and unit distance cannot be determined accurately. In this study, a method to supplement the missing data was studied.

1. Introduction

In studies utilizing the Mahalanobis–Taguchi system (MTS), missing data are sometimes generated, caused by an inability to calculate a Mahalanobis distance. For example, when the inverse matrix of a correlation matrix cannot be calculated due to its multiple collinearity, and if such data are left untreated, most computer software automatically treats them as zero. As a result, calculation of an inverse matrix becomes possible, but such results are meaningless.

As a solution to this problem, the following two methods are proposed:

1. If it is possible to collect enough information with no missing data for normal people, we can create a Mahalanobis space with them.

2. Using only data for items other than those with missing data, we form a Mahalanobis space.

Method 2 is considered to be intricate and impractical because we need to recreate a Mahalanobis space for each combination of items with no missing data. In addition, although theoretically, method 2 does not require any countermeasures for missing data, accuracy in judgment could be lowered because of the decreased number of examination items.

Since we could secure no missing data for 354 normal persons in our research, we created a Mahalanobis space by adopting method 1. The main

reason that the data with no missing measurements were used is that(1) we evaluated discriminability by using the data with no missing measurements, and (2) after generating missing measurements intentionally and randomly, we took effective measures and assessed the new discriminability.

Therefore, the objective of our research was to obtain a guideline for taking effective measures for missing data such that a Mahalanobis distance can be calculated accurately for a certain examinee if his or her medical checkup data were missing.

2. Collection and Sorting of Data

As our analysis data we used the data with no missing measurements in blood test items, which had been measured at the periodic medical checkups for 1377 over-40-year-old persons working at two offices of company A for three years, from February 1992 to March 1995.

Since company A's offices are distributed all over the nation, medical checkups, data processing, and final judgments are implemented at multiple medical examination facilities. If we analyzed all of the medical checkup data measured at the multiple facilities, the variability among the facilities could affect the result as a disturbance factor. Therefore, in our research we only used the data from the examination site run by company B.

Company B's comprehensive judgment on a medical checkup has the eight categories described in Table 1.

In our research, a "normal" person was defined as a person judged in the comprehensive judgment to be A_1 or A_2. Eventually, 354 people were selected as normal persons. On the other hand, for an "abnormal" person, diagnosed as C_1 or C_2, 221 people were chosen. People judged as G_1 or G_2 would have needed to retake the test or have a thorough reexamination and are not included in our study.

3. Medical Checkup Method

Using the data for 354 people matching the definition of "normal," we formed a Mahalanobis space. What is important here is which items (characteristics) to select in creating a Mahalanobis space. In general, since the results in a clinical test are largely affected by gender or age, by adding these two to the 25 items for a blood test shown in Table 2 (i.e., using a total of 27 items), we created a Mahalanobis space. In the gender category, we set data for a male to 0 and for a female to 1 and then treated the data similarly to other data.

If a Mahalanobis distance is below a "certain value," we judged it as normal and categorized it as "no guidance and precise examination are needed." If above a certain value, we categorized it as "guidance or no precise examination is needed" or "therapy is needed." To this end, we set up a threshold.

A threshold should be determined by doctors from the viewpoint of defining how far a Mahalanobis distance is positioned from a group of normal people. However, since at a medical checkup we cannot compare the health condition for each examinee with the corresponding Mahalanobis distance, we attempted to determine the threshold by using type I and type II error.

In the case of determining a threshold by the type of error, medical experts should calculate from an economic standpoint such that losses due to type I and type II error are balanced. However, how to determine the threshold using both types of error is a difficult matter. Since the objective of our research was to assess discriminability through a method of taking measures for missing data, by assuming for the sake of simplicity that the losses caused by both errors were equal, we determined the threshold.

4. Simulation of Missing Measurements and Countermeasures

To perform a simulation of measures for missing data in the medical checkup, we set up a model to generate missing data randomly for each data item, consisting of 1377 checkups for a total of 25 blood test items (Table 3). Now we defined 1, 5, 10, 20, and 30% as five levels of missing data. Following are procedures for missing-measurement simulation.

Table 1
Definition of normal and abnormal persons in comprehensive judgment

Definition	Comprehensive Judgment		Frequency	Proportion (%)
Normal persons	A_1:	normal	59	4.3
	A_2:	healthy with comments	295	21.4
	B_1:	observation needed	345	25.1
	B_2:	under observation	6	0.4
Abnormal persons	C_1:	therapy needed	56	4.1
	C_2:	under therapy	165	12.0
	G_1:	reexamination needed	15	1.1
	G_2:	precise reexamination needed	436	31.7
		Total	1377	100.1

Table 2
Blood test items used for our study

No.	Test Item	Abbreviation
1	White blood cell count	WBC
2	Red blood cell count[a]	RBC
3	Hemoglobin count[a]	Hb
4	Hematocrit count	Hct
5	Aspartate amino transferase[a]	AST
6	Alanine amino transferase[a]	ALT
7	Alkaline phosphatase	ALP
8	Glutamyltranspeptidase[a]	γ-GTP
9	Lactatdehydrogenase	LDT
10	Total bilirubin	TB
11	Thymol turbidity test	TTT
12	Zinc turbidity test	ZTT
13	Total protein	TP
14	Albumin	Alb
15	α_2-Globulins	α_2-Gl
16	β-Globulins	β-Gl
17	γ-Globulins	γ-Gl
18	Amylase	AMY
19	Total cholesterol[a]	TC
20	Triglyceride[a]	TG
21	High-density lipoprotein cholesterol	HDL-C
22	Fasting blood sugar levels	FBS
23	Blood urea nitrogen	BUN
24	Creatinine	Cr
25	Uric acid	UA

[a] Requisite. Indicates examination items that the Industrial Safety and Health Regulations (Article 44) required to be checked during a periodical medical checkup.

Table 3
Simulation model of missing data[a]

Checkup	Examination Item				
	1	2	3	...	25
1		•			•
2	•			•	
3		•			
⋮	•				•
		•		•	
1377		•			•

[a] A dot indicates missing data.

Procedure 1

Regarding the proportion of missing data as a parameter, we calculated the total number of data missing. However, since the items age and gender are not often missed, we excluded both.

> Total number of missing data
> = (total number of checkups)
>   × (number of blood test items)
>   × (proportion of missing data)

Procedure 2

We generated two types of random numbers:

R_1: Determine a checkup number with missing data j from all 1377 checkups.

R_2: Determine an item number with missing data i from all 25 blood test items.

If we had already generated missing data for a checkup number, j, and item number, i, we repeated procedure 2.

Procedure 3

Missing data for a checkup number, j, and item number, i, is complemented with a *complementary value*, described later.

Procedure 4

Counting the number of missing data generated, we repeated procedures 2 to 4 until the number counted reached the required total number of missing data calculated in procedure 1.

In the case of taking countermeasures for missing data, in procedure 3 we complemented a missing measurement for a checkup number, j, and item number, i, with the following three types of averages as a complementary value:

1. *Average from all examinees.* When there are missing measurements, they are quite often complemented with an average of all data because of its handiness. Our research also studied this method.

2. *Complement with item-by-item average from comprehensive judgment.* When a comprehensive judgment has been made by company B, it was expected that by complementing the data with an average from an examination item stratified by a comprehensive judgment, we could improve the discriminability more than by using an item-by-item average for all examinees.

3. *Complement with item-by-item average from normal persons.* Because a group of normal persons was regarded as homogeneous, an item-by-item average from normal persons was expected to be of significance, stable, and reliable.

This strategy is based on our supposition that examination items in which even a group of abnormal persons has missing data are likely to have almost the same values as those for a group of normal persons. Table 4 shows item-by-item averages from normal persons.

5. Results of Missing Data Calculation

Classification of Missing Data Calculation

Among the three types of countermeasures for missing data, we detailed the discriminative result in the case of complement with an item-by-item average from normal persons. Next, we compared the discriminative result in the case of leaving missing data and filling them automatically with a value of zero. For the evaluation of discriminability, an average of Mahalanobis distances for each comprehensive judgment, contribution, ρ, and SN ratio, η, were used.

Table 4

Item-by-item average from normal persons
(units omitted)

Item	Examination Item	Average
1	WBC	5.85
2	RBC	476.7
3	Hb	14.93
4	Hct	44.74
5	AST	19.6
6	ALT	14.9
7	ALP	104.2
8	γ-GTP	27.3
9	LDT	343.3
10	TB	0.84
11	TTT	1.11
12	ZTT	7.27
13	TP	7.37
14	Alb	68.61
15	α_2-Gl	8.03
16	β-Gl	7.84
17	γ-Gl	12.53
18	AMY	156.9
19	TC	201.9
20	TG	94.9
21	HDL-C	58.5
22	FBS	92.6
23	BUN	16.09
24	Cr	0.87
25	UA	5.56
Number of items		354

Result of Discriminability in Case of Complementing with Item-by-Item Average from Normal Persons

Proportion of Missing Data and Fluctuation of Average of Mahalanobis Distances for Each Comprehensive Judgment Table 5 shows the fluctuation of the average of Mahalanobis distance for each comprehensive judgment when we change the proportion of the missing data in the case where they are complemented by item-by-item averages from normal persons.

When the proportion of missing data is 10% the averages of Mahalanobis distances for the two groups of normal persons, A_1 and A_2, were 1.27 and 1.60, respectively, both of which are regarded as relatively small. On the other hand, those for the two groups of abnormal persons, C_1 and C_2, were 16.11 and 6.01, respectively, which are viewed as large. Therefore, we can discriminate between normal and abnormal persons by the averages of Mahalanobis distances.

Even when the proportion of missing data is 20 or 30%, the averages of Mahalanobis distances for A_1 and A_2 are obviously smaller than those for C_1 and C_2. As a result, normal and abnormal persons can be discriminated.

Proportion of Missing Data and Change in Discriminability A 2×2 table (Table 6) summarizes the results of discriminability based on thresholds selected on the basis that a type I error occurs as often as a type II error does, in the case of complementing missing data with item-by-item averages from normal persons.

1. *Proportion of missing data is 1%.* When the threshold is 1.28, the occurrences of type I and type II error are 16.95% and 16.78%, respectively. The contribution, ρ, results in 43%. Since the SN ratio, η, is reduced by 0.220 dB from −1.077 dB when the proportion of missing data is zero, to −1.297 dB, the resulting discriminability decreases slightly, to 95.06% of that in the case of no missing data.

2. *Proportion of missing data is 5%.* If the threshold is set to 1.35, the occurrences of type I and type II error are 20.92 and 19.00%, respectively, whereas the contribution is 35%. The SN ratio declines by 1.605 dB, from −1.077 dB when the proportion of missing data is zero to −2.682 dB. The eventual discriminability is then lowered to 69.01% of that when there is no missing data.

3. *Proportion of missing data is 10%.* If the threshold is set to 1.55, type I and type II error take place at the possibilities of 20.92 and 19.00%,

Table 5
Average of Mahalanobis distances in case of complement with item-by-item averages from normal persons

Percent of Missing Measurements	Comprehensive Judgment							
	A_1	A_2	B_1	B_2	C_1	C_2	G_1	G_2
0	0.92	1.02	2.41	1.35	16.16	5.08	9.28	4.62
1	0.92	1.05	2.51	1.35	16.29	5.12	9.32	4.67
5	1.03	1.22	2.76	1.85	16.26	5.35	9.33	4.89
10	1.27	1.60	2.86	2.26	16.11	6.01	9.29	5.72
20	1.85	1.98	3.51	1.62	16.65	5.36	8.70	5.78
30	1.97	2.38	4.11	1.72	17.02	5.72	8.27	5.62

Table 6
Result of discriminability in case of complement with item-by-item averages for normal persons

Percent of Missing Data	Comprehensive Judgment	Mahalanobis Distance		Total	Error	ρ	η (dB)
		$D^2 \leqslant 1.26$	$D^2 > 1.26$				
0	$A_1 - A_2$	297	57	354	16.10	0.44	−1.077
	$C_1 - C_2$	37	184	221	16.74		
		$D^2 \leqslant 1.28$	$D^2 > 1.28$				
1	$A_1 - A_2$	294	60	354	16.95	0.43	−1.297
	$C_1 - C_2$	37	184	221	16.74		
		$D^2 \leqslant 1.35$	$D^2 > 1.35$				
5	$A_1 - A_2$	281	73	354	20.62	0.35	−2.682
	$C_1 - C_2$	42	179	221	19.00		
		$D^2 \leqslant 1.55$	$D^2 > 1.55$				
10	$A_1 - A_2$	283	71	354	20.06	0.36	−2.545
	$C_1 - C_2$	42	179	221	19.00		
		$D^2 \leqslant 1.80$	$D^2 > 1.80$				
20	$A_1 - A_2$	249	105	354	29.66	0.17	−6.766
	$C_1 - C_2$	61	160	221	27.60		
		$D^2 \leqslant 2.05$	$D^2 > 2.05$				
30	$A_1 - A_2$	248	106	354	29.94	0.15	−7.504
	$C_1 - C_2$	67	154	221	30.32		

respectively. Perhaps because of an uneven generation of random numbers, the resulting contribution is 35% and the SN ratio rises to −2.545 dB, up slightly from that when ρ is 5%. At the same time, since the SN ratio drops by 1.468 dB, from −1.077 dB when the proportion of missing data is zero to −2.545 dB, the discriminability diminishes to 71.32% of that when there is no missing data.

In addition, when the proportions of missing data are 20 and 30%, the SN ratios are −6.766 and −7.504 dB. These results reveal that when the proportion of missing data exceed 10%, the discriminability tends to deteriorate drastically.

Relationship between Proportion of Missing Data and Threshold

To compare the discriminability for each method of taking measures for missing data, we studied the relationship between the proportion of missing data and threshold. Figure 1 shows changes in thresholds for an increase in the number of missing data in the case where the occurrence of type I error is almost the same as that of type II error.

In the case where the proportion of missing data is 0%, if the threshold is set to 1.26, the occurrences of type I and type II error almost match. In the case of leaving missing data alone without complementing (we define this case as case 1), the threshold increases to 1.50 when the proportion of missing data is 1%, and to 6.00 when that is 5%, in a diverging manner.

On the other hand, in any case of taking measures for missing data, that is, complement with item-by-item averages from all examinees (labeled case 2), complement with item-by-item averages from comprehensive judgment (labeled case 3), and complement with item-by-item averages from normal persons (labeled case 4), the threshold increases gradually as the proportion of missing data rises.

These results demonstrates that even if we use any complementary method, the threshold becomes more stable than that in the case of leaving missing data as they are (case 1). Therefore, it is regarded as inappropriate to keep missing data intact. In addition, the threshold in the case of complement with item-by-item averages from normal persons (case 4) becomes smaller than those in the cases of complement with item-by-item averages from all

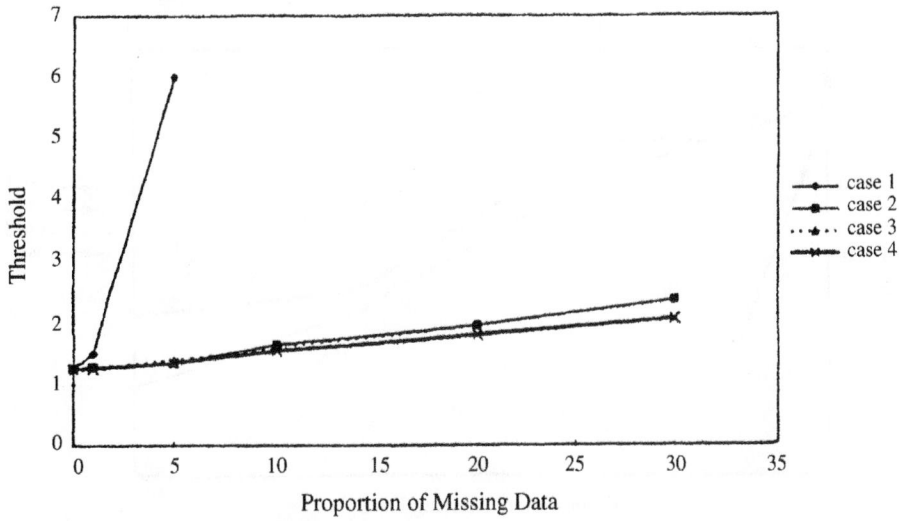

Figure 1
Proportion of missing data for each case of countermeasure and change in threshold

examinees (case 2) or with item-by-item averages from comprehensive judgment (case 3).

Relationship between Proportion of Missing Measurements and SN Ratio

Now, to compare the discriminability for each countermeasure for missing data, we studied the relationship between the proportion of missing data and the SN ratio. Figure 2 illustrates how the SN ratio varies for each measure for missing measurements as their proportion increases.

When the proportion of missing measurements is 0%, the resulting SN ratio is −1.077 dB. When missing data are kept intact (case 1), when the proportions are 1 and 5%, the corresponding SN ratios plummet drastically, to −3.548 and −10.523 dB. However, for the complement with item-by-item averages from all examinees (case 2), that of complement with item-by-item averages from each comprehensive judgment (case 3), and that of complement with item-by-item averages from normal persons (case 4), all of the SN ratios gradually decrease in a manner similar to the way the number of missing data rises.

These results indicate that no matter what complementary method is used, the SN ratio becomes more stable than when leaving missing data as they are (case 1). Comparing the discriminabilities by the SN ratio for the three complementary methods described above, complement with item-by-item averages from comprehensive judgment (case 3) is regarded as the best.

For example, when the proportion of missing measurements is 10%, the SN ratio is computed as −3.563, −2.009, or −2.545 dB for each case of complement with item-by-item averages from all examinees (case 2), complement with item-by-item averages from comprehensive judgment (case 3), and complement with item-by-item averages from normal persons (case 4). Since a larger SN ratio signifies superior discriminability, simply judging from all of the SN ratios, we can view the case of complement with item-by-item averages from comprehensive judgment (case 3) as the best.

Yet the case of complement with item-by-item averages for each comprehensive judgment (case 3) involves a problem when assuming that a comprehensive judgment can be made despite the existence of missing data. Since cases other than A_1 and A_2 in terms of a comprehensive judgment were not regarded as homogeneous, an item-by-item average from comprehensive judgment does not necessarily have a meaning. Therefore, we concluded that complement with item-by-item averages from comprehensive judgment (case 3) is not a proper method.

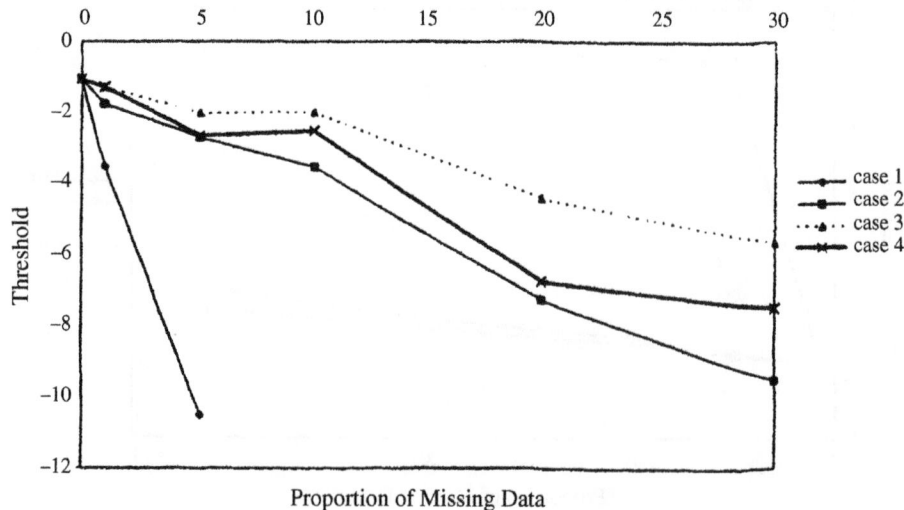

Figure 2
Proportion of missing data for each case of countermeasure and change in SN ratio

On the other hand, we compared cases of complement with item-by-item averages from all examinees (case 2) and complement with item-by-item averages from normal persons (case 4). The SN ratio for complement with item-by-item averages from normal persons (case 4) shows a better value (i.e., better discriminability) than complement with item-by-item averages for normal persons (case 4) does.

Since an item-by-item average for all examinees denotes a mean value for all normal, abnormal, and even other-type persons, its stability depends greatly on the constitution of data collected. In contrast, because a group of normal persons is homogeneous, an item-by-item average from normal persons can be considered stable.

Considering all of the above, as the most appropriate method of complementing missing data for medical checkups, we selected the complement with item-by-item averages from normal persons (case 4).

6. Countermeasures for Missing Data

Missing Data Occurrence and Simulation Model

In our research, we performed a simulation based on a model that generates missing measurements equally and randomly for any examination item. However, the actual occurrence of missing data for each examination item for medical checkups is not uniform but is rather low for the particular requisite items prescribed by Industrial Safety and Health Regulations (Table 2). This implies that important items tend to have a smaller number of missing data.

In addition, when we complemented missing measurements with item-by-item averages from normal persons (case 4), because an item-by-item average from normal persons was used even for abnormal persons' missing data, the threshold resulted in a smaller value than for complementing with other types of averages (cases 2 and 3). The reason is likely to relate to our setup of a simulation model that randomly generates missing data for the 25 blood test items.

Taking all of the above into account, we can observe that by setting up a simulation model that creates missing data in accordance with an item-by-item proportion of actual missing data, we can properly

complement missing data with an item-by-item average from normal persons.

Setup of Threshold

For the sake of convenience, our research determined thresholds on the assumption that type I and type II error have the same loss. Although a threshold should essentially be computed from the loss function, it is not easy to determine a threshold that balances future losses of type I and type II error. Thus, focusing on our objective of taking effective measures for missing data for medical checkups, on the assumption that both errors' losses are equal, we determined thresholds. Since the issue of threshold selection is regarded as quite crucial, we continue to work on this issue.

7. Conclusions

In this study of a medical checkup method based on a Mahalanobis distance, we studied a calculation method for estimating a Mahalanobis distance accurately for each examinee in the case where there were missing data for medical checkups, and finally, obtained the following results.

1. When we kept missing data intact, and eventually most computer software complemented them automatically with a value of zero, even if the proportion of missing data was only 1 or 5%, the average of the Mahalanobis distances for each comprehensive judgment diverges, preventing us from discriminating between normal and abnormal persons. Therefore, it was regarded as inappropriate to leave missing data untreated.

2. The method of complementing missing data with item-by-item averages from all examinees has better discriminability than that of keeping missing data intact. However, different constitution of data collected often causes instability of averages.

3. Discriminability by the method of complementing missing data with item-by-item averages from comprehensive judgment is viewed as the best of all from the perspective of the

SN ratio. However, a group of abnormal persons is so heterogeneous that its average hardly has significance. Thus, if the method of complementing missing data with item-by-item averages from comprehensive judgment is used for cases other than A_1 and A_2, its use is not considered appropriate.

4. In terms of the SN ratio, discriminability by the method of complementing missing data with item-by-item averages from normal persons turns out to be better than the case of complement with item-by-item averages from all examinees. The reason is that because a group of normal persons is homogeneous enough that its average is regarded as meaningful, and because missing data are quite unlikely to occur in essential examination items that are supposed to distinguish between normality and abnormality, viewing missing data as normal values is not considered to distort the actuality.

5. Taking into consideration all of these matters, we judged that the method of complementing missing measurements with item-by-item averages from normal persons was the best of the measures for missing measurements.

The greatest significance in our study is that we have successfully elucidated the possibility of retaining medical checkup discriminability resting on the Mahalanobis distance with a minimal loss of information by complementing with item-by-item averages from normal persons when missing data for medical checkups. The result gained from our research is believed to contribute to the rationalization of a medical checkup: that is, reduction in medical expenses through the elimination of excessive detailed examinations, mitigation of the time and economic cost to examinees, alleviation of examinee anxiety, or assistance with information or instructions from doctors, nurses, or health workers.

Pattern recognition using a Mahalanobis distance can be applied not only to a medical checkup but also to a comprehensive judgment in many fields. Our research could contribute to the problem of missing measurements in any application.

References

Yoshiko Hasegawa, 1997. Mahalanobis distance application for health examination and treatment of missing data: 1. The case with no missing data. *Quality Engineering*, Vol. 5, No. 5, pp. 46–54.

Yoshiko Hasegawa, 1997. Mahalanobis distance application for health examination and treatment of missing data: 2. The case with no treatment and substituted zeros for missing data. *Quality Engineering*, Vol. 5, No. 6, pp. 45–52.

This case study is contributed by Yoshiko Hasegawa.

CASE 78

Forecasting Future Health from Existing Medical Examination Results Using the MTS

Abstract: For our research, we used medical checkup data stored for three years by a certain company. Among 5000 examinees, there were several hundred sets of data with no missing measurements. In this case, since the determination of a threshold between A and B was quite difficult, a different doctor sometimes classified a certain examinee into a different category, A or B. It was the aim of this study to improve the certainty of such judgment.

1. Introduction

The medical checkup items can be classified roughly into the following 19:

1. Diseases under medical treatment (maximum of three types for each patient)
2. Diseases experienced in the past (maximum of three types for each patient)
3. Diseases that family members had or have (categorized by grandparents, parents, and brothers)
4. Preferences (cigarette, alcohol)
5. Subjective symptoms
6. Time since meals
7. Physical measurements (height, weight, degree of obesity)
8. Eyesight
9. Blood pressure
10. Urinalysis
11. Symptoms of other senses
12. Hearing ability
13. Chest x-ray
14. Electrocardiogram
15. Blood (19 items)
16. Photo of eyeground
17. Additional blood test items
18. Gender
19. Age

A doctor's overall judgment should be added to a medical diagnosis, but there is difficulty when there are too many examinees. In this study, a doctor analyzes a medical checkup list (blood work, electrocardiogram, chest x-ray, eyesight, medical checkup interview, blood pressure, etc.), then categorizes an examinee in one of four categories:

A: normal

B: observation needed

C: treatment needed or under treatment

D: detailed examination needed

In this case, since the determination of a threshold between A and B is quite difficult, a different doctor sometimes classifies a certain examinee into a different category, A or B. It was our aim in this study to improve the certainty of such judgment.

Characteristics such as gender are quantified into numerals as a category data, such as "male = 1" or "female = 2." Age or biochemical data are used as they are. Items with the same value among all data are excluded (e.g., all of the examinees have "1" or "normal" for hearing ability because they are

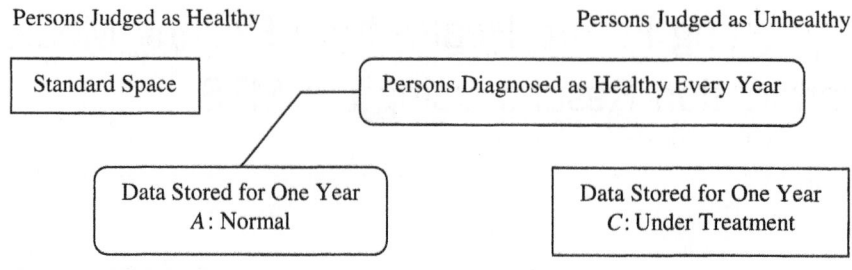

Persons Judged as Healthy

Persons Judged as Unhealthy

Standard Space

Persons Diagnosed as Healthy Every Year

Data Stored for One Year
A: Normal

Data Stored for One Year
C: Under Treatment

Figure 1
Base space I for one-year-healthy persons

healthy). In addition, items that lead to multicollinearity are also omitted. Eventually, the original nearly 100 items were reduced to 66.

2. Preparation of the Base Space

The following base spaces were prepared to observe the judgment being affected by different base space preparation.

Base Space I

Primarily, we investigated whether a medical checkup judgment could be made using only the one-year healthy persons' data (685 persons diagnosed as A and 735 as C). According to Figure 1, we formed base space I with the data for 685 persons judged as A and studied whether the data could be discriminated from those for 735 under-treatment patients identified as C. Furthermore, to improve discriminability, we selected examination items.

Base Space II

With the data for two-year-healthy persons, we looked for further improvement in accuracy of discriminability. According to Figure 2, we created base space II with the data for two-year-healthy 159 per-

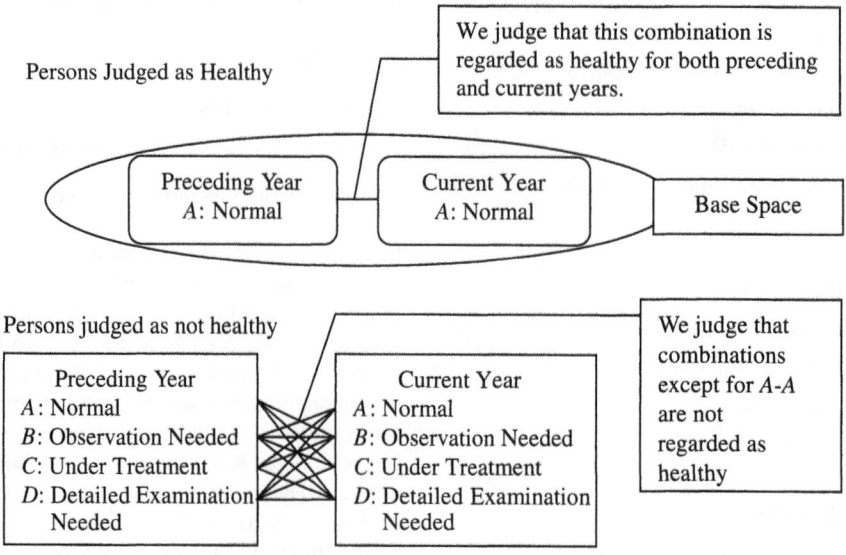

Persons Judged as Healthy

We judge that this combination is regarded as healthy for both preceding and current years.

Preceding Year
A: Normal

Current Year
A: Normal

Base Space

Persons judged as not healthy

Preceding Year
A: Normal
B: Observation Needed
C: Under Treatment
D: Detailed Examination
Needed

Current Year
A: Normal
B: Observation Needed
C: Under Treatment
D: Detailed Examination
Needed

We judge that combinations except for A-A are not regarded as healthy

Figure 2
Base space II for two-year-healthy persons

sons diagnosed as *A* and calculated the distances for 159 persons diagnosed as *A* and the ones for 37 persons as *C*. Since the result within one year was treated as one data point, from a two-year-healthy person diagnosed as *A*, we obtained two data points. Since missing data leads to lower accuracy in judgment, we did not use situations with missing data.

Base Space III

After studying discriminability using the Mahalanobis–Taguchi system, we investigated predictability, which characterizes a Mahalanobis distance. That is, we studied the possibilities not only of judging whether a certain person is currently healthy but also foreseeing whether he or she will be healthy in the next year based on the current medical checkup data.

First, combining the two-year medical checkup data, we created base space III based on Figure 3. The persons who were diagnosed as category *A* this year were defined as healthy persons. Base space III was prepared from the preceding year's data for these healthy persons. In other words, a person cat-

egorized as healthy for the current year could be healthy or unhealthy the preceding year. Since the medical checkup data for the preceding year were used, only one-year data are available, although two-year data (for the preceding and current years) are available. No matter what category they belonged to in the preceding year, those who are judged as unhealthy are viewed as *C* in the current year. Through this analysis it is possible to predict what type of base space should be used to predict a healthy person in the following year.

Base Space IV

From the perspective of reliability in prediction, prediction with three-year-long data is more accurate than that with one- or two-year data. Whereas one- or two-year data do not reflect time-series elements, with three-year data, we studied the time-lapsed change for medical checkup data. As a matter of course, the number of data covering all measurements in three-consecutive-year medical checkups is rather limited. We formed base space IV with data only for the preceding two years from 120

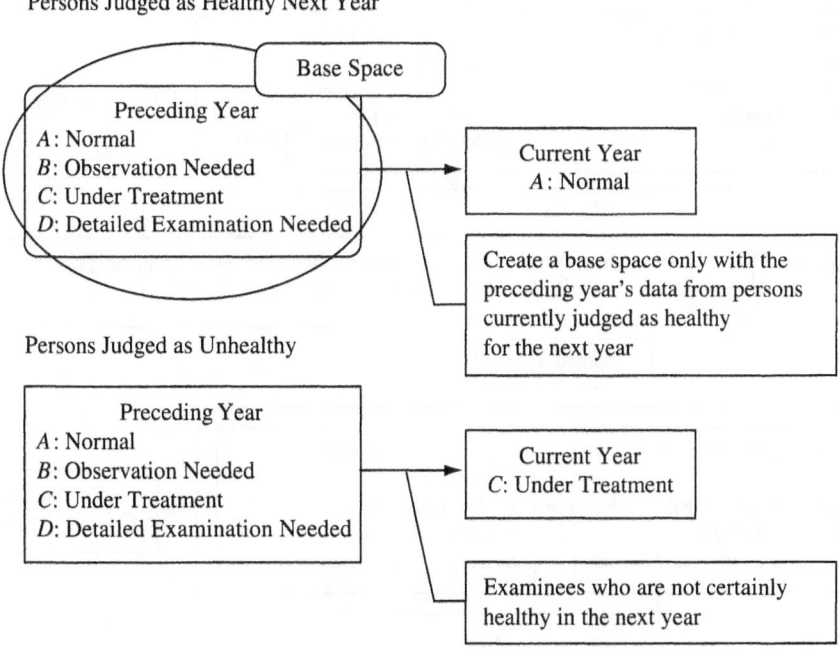

Figure 3
Base space III: predicting current year's health condition using two-year data

patients. The number of medical checkup items was approximately 200 as a two-year time-based item. However, because the number of datapoints was quite small, considering the stability of the base space, we reduced it to 101. Base space IV was prepared according to Figure 4. We also studied base space IV after item selection. If the data were judged as *A* for the current year, it does not matter what category they are diagnosed before. In addition, those who were diagnosed as unhealthy for the current year were regarded as *C*-diagnosed examinees, no matter what data they had for the preceding two years.

3. Selection of Examination Items

After having predicted and judging a health condition by using a Mahalanobis distance, we distin-

guished between necessary and unnecessary items for prediction and judgment based on the base space to reduce measurement cost. That is, without lowering predictability and discriminability grounded on a Mahalanobis distance, we selected essential examination items.

To design parameters of all items, we set "use of an item for creating a standard space" to level 1 and "no use of an item for creating a standard space" to level 2. Because 100 items were used for our medical checkup, an L_{128} orthogonal array was selected.

It is all right to leave some columns empty. Since the first row has only one item allocated, we created a base space using all items. For the second row, about half the row is assigned a 1, so we formed a base space using approximately half of all the items. Similarly for the remaining rows, we created base spaces, and finally, obtained 128 different spaces.

As a next step, for the 128 base spaces created thus far, we computed a distance for each of the

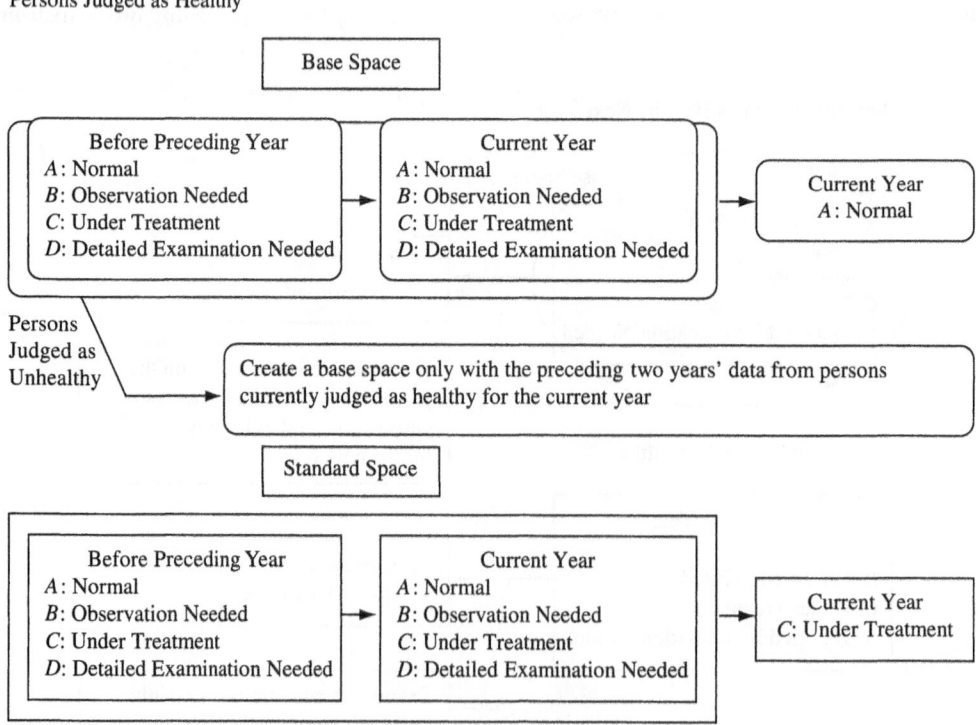

Figure 4
Base space IV: predicting current year's health condition with three-year-healthy persons' data

data that were expected to be distant (abnormal data). The greater the number of abnormal data, the more desirable. However, under some circumstances, we cannot collect a sufficient number of abnormal data. In this case, even a few data are regarded as applicable. We defined the distance for each abnormal data item, which was calculated from the base space for the first row, as $D_1, D_2, \ldots, D_n$. Now, since the abnormal data should be distant from the normal space, we used a larger-the-better SN ratio for evaluation:

$$\eta = -10 \log \frac{1}{n} \left(\sum_{i=1}^{n} \frac{1}{D_i^2} \right)$$

To select necessary items, we calculated the average for each level and created the response graphs where an item whose value decreases from left to right was regarded as essential, whereas one with a contrary trend was considered irrelevant to a judgment.

4. Results of Analysis

The possibility of judging health examinations using base spaces prepared in different ways was studied.

Prediction by Base Space I

Using only the one-year data, we investigated whether or not we could make a medical checkup judgment. The data have no missing data and consist of those for 685 persons judged as healthy (A) and 735 persons judged as unhealthy (C), with 109 items in total. In the hope of improving discriminability, we selected the following necessary items for a judgment by excluding unnecessary ones:

- ❏ *Blood test:* GOT, GPT, ALP, γ-GTP, T.Bil, ZTT, TG, T.Chol, HDL, UA, BS, HbA$_1$c, WBC, RBC, Hb, Ht, MCHC
- ❏ *Medical checkup interview:* gender, age, height, diseases experienced, diseases cured
- ❏ *Blood pressure:* maximum, minimum
- ❏ *Others:* Chest x-ray electrocardiogram

Since the discriminability deteriorates if we iterate item selection over four times, by creating base space I with items that came up with the highest discriminability without reaching a downward trend

in discriminability, we computed the distance for the base space and summarized the results in Table 1. This table reveals that there is an overlap area between judgments A and C, which leads to inability of judgment, even though the number of data points is quite large.

Prediction by Base Space II

Second, we investigated whether we could make a judgment using the previous two-year data. According to Figure 2, we created base space II with the 67-item data for 159 persons judged as healthy. Figure 5 and Table 2 show the results of the Mahalanobis distances for 159 persons judged as healthy (A) and 37 persons under treatment (C). These results demonstrate that if a certain examinee has a Mahalanobis distance less than or equal to 2.0 from base space II, he or she is diagnosed as healthy. On the other hand, if the distance becomes greater than 2.0, he or she can be judged as unhealthy or possibly suffering some disease even without a doctor's diagnosis. However, for judgment C discussed here, the one-year data for those who were judged as C in two years were used. For people who were diagnosed as A in the preceding year but as C in the current year, it was difficult to make an accurate

Table 1

Distribution of one-year data's distances from base space I after item selection

Data Class	A	C	Data Class	A	C
0	0	0	6	0	11
0.5	17	0	6.5	0	7
1	356	42	7	0	9
1.5	261	121	7.5	0	4
2	49	141	8	0	7
3	1	71	8.5	0	6
3.5	0	44	9	0	4
4	0	21	9.5	0	5
4.5	0	21	10	0	3
5.5	0	17	Next class	0	77

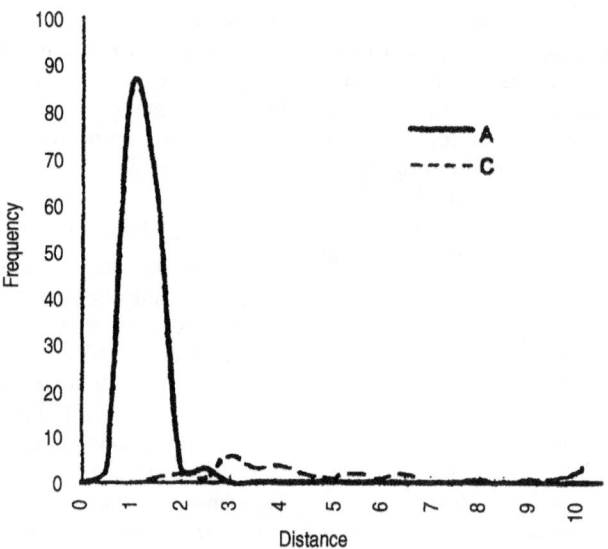

Figure 5
Distances from base space II using two-year data

Table 2
Distribution of distances from base space II for next year using two-year data

Data Class	A	C	Data Class	A	C
0	0	0	6	0	1
0.5	3	0	6.5	0	2
1	85	0	7	0	0
1.5	65	1	7.5	0	0
2	3	2	8	0	1
2.5	3	1	8.5	0	0
3	0	6	9	0	1
3.5	0	3	9.5	0	1
4	0	4	10	0	2
4.5	0	2	Next class	0	7
5	0	2			
5.5	0	2			
			Total	159	37

judgment on their health conditions because their data were distributed between judgments A and C, shown in Figure 2.

Prediction by Base Space III

Using the two-year data, we studied predictability. The data used for base space III were composed of the complete 100-item data in the preceding year for 336 persons judged as healthy (A) and in the previous year for 137 persons under treatment (C). Table 3 shows the distances from the base space for judgments A and C. Although we could not clearly separate A and C, despite being able to make a judgment on whether a certain person was currently in good health or not, when we selected 1.5 as the threshold, the accuracy in predictability was regarded as approximately 80%.

Prediction by Base Space IV

We investigated predictability based on the time-series data. Using the three-year data for 120 persons regarded as healthy, we formed base space IV. While the number of the two-year time-series items as medical checkup items was approximately 200, considering the stability of base space IV, we nar-

Table 3
Distribution of distances from base space III using two-year data for the prediction of next year

Data Class	A	C	A(%)	C(%)
0	0	0	100	0
0.5	40	0	100	0
1	140	7	88	5
1.5	112	14	46	15
2	39	24	13	33
2.5	4	12	1	42
3	0	18	0	55
3.5	1	18	0	68
4	0	8	0	74
4.5	0	12	0	82
5	0	7	0	88
5.5	0	6	0	92
6	0	1	0	93
6.5	0	2	0	94
7	0	0	0	94
7.5	0	0	0	94
8	0	0	0	94
8.5	0	1	0	95
9	0	0	0	95
9.5	0	0	0	95
10	0	1	0	96
Next class	0	6	0	100
Total	336	137		

Table 4
Distribution of distances from base space IV from three-year data before item selection

Data Class	A	C	A'	A (%)	C (%)	A' (%)
0	0	0	0	100.0	0.0	100.0
0.5	0	0	0	100.0	0.0	100.0
1	52	0	0	100.0	0.0	100.0
1.5	68	0	0	56.7	0.0	100.0
2	0	0	1	0.0	0.0	100.0
2.5	0	0	0	0.0	0.0	98.8
3	0	1	1	0.0	2.6	98.8
3.5	0	0	7	0.0	2.6	97.6
4	0	0	4	0.0	2.6	89.4
4.5	0	1	4	0.0	5.3	84.7
5	0	1	4	0.0	7.9	80.0
5.5	0	1	7	0.0	10.5	75.3
6	0	1	6	0.0	13.2	67.1
6.5	0	1	4	0.0	15.8	60.0
7	0	1	7	0.0	18.4	55.3
7.5	0	2	1	0.0	23.7	47.1
8	0	0	4	0.0	23.7	45.9
8.5	0	0	4	0.0	23.7	41.2
9	0	2	6	0.0	28.9	36.5
9.5	0	0	4	0.0	28.9	29.4
10	0	0	1	0.0	28.9	24.7
Next class	0	27	20	0.0	100.0	23.5
Total	120	38	85			

rowed them down to 101 because the total number of data points was small. Table 4 shows the results of judgments A and C. Those below 1.5 were judged as A, whereas the ones above 3.0 were seen as C. The two classes are separated perfectly. A' indicates the case where, after excluding one person's data from the 120 persons' judged as A and creating base space IV using only the remaining 119 persons' data, we calculated the distance using the data for this extended person, from base space IV generated from the 119 persons' data. Similarly, we calculated the distances for the 85 persons' data judged as A by using such a base space. We can see that some data judged as A' deviated from base space IV. The reason is that because of a small number of data in base space IV, the space becomes so sensitive that a part of the data judged as A' are not separated from those in judgment C. Therefore, through item selection, by eliminating ineffective items for a judgment on abnormal data, we reduced the number of items to 60. Then, using 119 of the 120 normal data, we created base space IV'. The distance of the one data point excluded was computed using the base space formed with the 119 data points. Next, removing another data point from the 120 data, we created base space IV' with the remaining 119 data once again. And then the distance for the one data item removed was calculated. Iterating this calculation process for all 120 data, we arrived at Figure 6 and Table 5, which reveal that judgment A' becomes closer to judgment A. Therefore, we came to the conclusion that considerable improvement in accuracy was obtained.

5. Selection of Examination Items by a Larger-the-Better SN Ratio

To improve prediction accuracy, through item selection with the two-year data, we excluded items that are not closely related to prediction and formed base space III' once again. While the number of the data is still 336, the number of the items is cut down from 100 to 66. Based on the data, we created response graphs for item selection. The distances

from the base space after item selection are shown in Table 6 and Figure 7.

That is, with the preceding year's data for those judged as A in the current year, we created a base space. Then we computed the distances using this base space for persons judged as A and C. Setting the threshold to 1.5 and classifying all the distances D's by this criterion, we summarize them in Table 7. According to this table, we can see that the proportion of accurate prediction of judgment A results in 90%, whereas the proportion of wrong prediction of judgment C, which are judged as A, is only 15%. As a result, these numbers prove that our predictability was fairly high.

6. Discussion

In our first analysis, we studied base space I generated from the one-year data for persons judged as healthy and created a base space with items that maximized discriminability after item selection. However, despite a large number of data, including those for 685 persons judged as A and 735 as C, an overlap area in the distribution has occurred and caused insufficient discriminability.

In the case where we took advantage of a Mahalanobis distance from base space II, which were formed by the two-year data for persons diagnosed as healthy, judgments A and C can be distinguished.

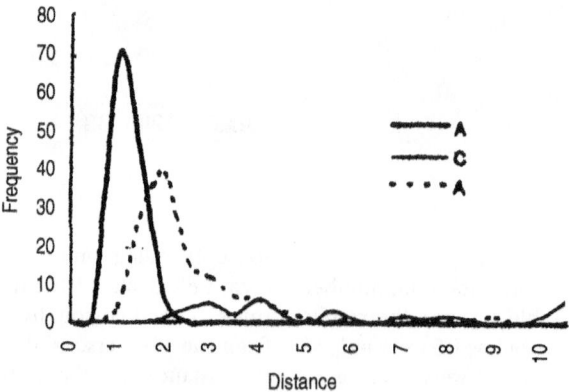

Figure 6
Distances from base space IV' using three-year data after item selection

Table 5
Distribution of distances from base space IV' using three-year data after item selection

Data Class	A	C	A'	A (%)	C (%)	A' (%)
0	0	0	0	100.0	0.0	100.0
0.5	1	0	0	100.0	0.0	100.0
1	69	0	4	99.2	0.0	100.0
1.5	44	0	27	41.7	0.0	96.7
2	6	1	38	5.0	2.6	74.2
2.5	0	4	15	0.0	13.2	42.5
3	0	5	12	0.0	26.3	30.0
3.5	0	2	7	0.0	31.6	20.0
4	0	6	6	0.0	47.4	14.2
4.5	0	2	3	0.0	52.6	9.2
5	0	0	1	0.0	52.6	6.7
5.5	0	3	0	0.0	60.5	5.8
6	0	1	1	0.0	63.2	5.8
6.5	0	0	0	0.0	63.2	5.0
7	0	2	2	0.0	68.4	5.0
7.5	0	1	0	0.0	71.1	3.3
8	0	2	1	0.0	76.3	3.3
8.5	0	1	1	0.0	78.9	2.5
9	0	0	2	0.0	78.9	1.7
9.5	0	0	0	0.0	78.9	0.0
10	0	2	0	0.0	84.2	0.0
Next class	0	6	0	0.0	100.0	0.0
Total	120	38	120			

Table 6
Distribution of distances from base space III' using two-year data after item selection

Data Class	A	C	A (%)	C (%)
0	0	0	100	0
0.5	15	0	100	0
1	170	3	88	4
1.5	117	15	45	15
2	31	28	10	35
2.5	3	25	1	53
3	0	17	0	66
3.5	0	12	0	74
4	0	10	0	82
4.5	0	3	0	84
5	0	6	0	88
5.5	0	3	0	91
6	0	2	0	92
6.5	0	0	0	82
7	0	2	0	83
7.5	0	1	0	84
8	0	0	0	94
8.5	0	0	0	94
9	0	0	0	94
9.5	0	1	0	95
10	0	0	0	95
Next class	0	7	0	100
Total	338	137		

If the threshold for a Mahalanobis distance is set to 2.0, we can predict that the data below the threshold will be judged as healthy, whereas those above the threshold will be judged as unhealthy. Consequently, we may make a medical checkup judgment without a doctor's diagnosis. As a next step, using the preceding year's data of those judged as A in the current year, we calculated the distances from base space III to classify A and C. As a result, in the case of choosing 1.5 as the threshold, we have been able to obtain predictability of approximately 80%.

Since we have confirmed that judgment and prediction can be attained by using a Mahalanobis distance to reduce measurement cost without worsening predictability and discriminability, we selected necessary and unnecessary items for creating

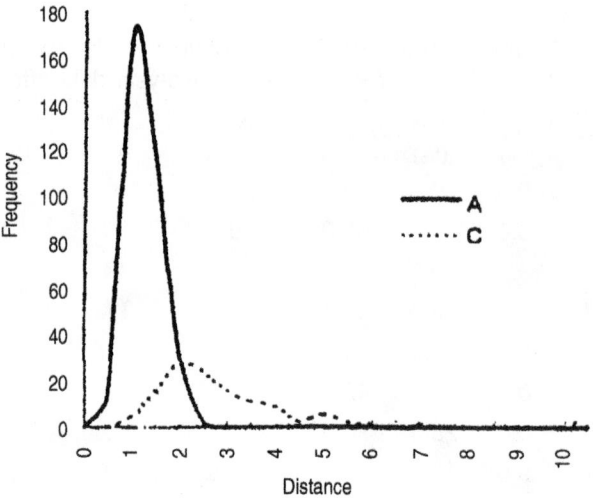

Figure 7
Distances from base space III′ using two-year data after item
selection

a standard space. After item selection, utilizing the preceding year's data for persons diagnosed as A in the current year, we created base space III′ and calculated the distances for judgments A and C. In this case, if 1.5 is chosen as the threshold and all of the distances D's are categorized, the proportion of the data judged accurately as A amounts to 90%, whereas the proportion of the data diagnosed erroneously as A even though they belong to C is only 15%. As a result, we have obtained fairly good predictability.

On the other hand, three-year data are obviously more effective than two-year data to improve the accuracy in prediction of health conditions. Thus,

we have investigated whether we can raise data reliability by including time-series data in the calculation. However, since there are numerous changes in examination items for a medical checkup at a company because of budgetary issues, job rotations, or employees' ages, the number of data that can cover all medical checkups for three years in a row without missing data is very limited. Although we have created a base space by using the three-year data for 120 persons, eventually, the sum of the items has decreased because the number of data turned out to be small. Therefore, considering the stability of base space IV, we reduced the number of items to 101. As a result, we have been able to separate judgments A and C completely. But the situation occurred where the data that should be judged as A′ were distant from a base space. That is, a small number of data in the base space resulted in the space becoming sensitive, so judgments A′ and C overlapped.

In most cases, to confirm whether abnormal data's distances become distant after creating a base space, we make sure that normal data's distances that are not used to construct the base space lie around 1.0. In this case we faced two typical phenomena, an overlap between abnormal data and the base space and an increase in normal data's dis-

Table 7

Prediction of medical checkup for threshold = 1.5 (%)

Expected Judgment for Next Year	Threshold		
	<1.5	>1.5	Total
A	90	10	100
C	15	85	100

tances. In our research, since the distances of the abnormal data have also increased, distance calculated from the normal data that are not used to construct the base space are not close to 1.0. One of the possible reasons is that the excessive number of items in the base space has led to a difficult judgment. Therefore, it was considered that we should select only items effective to discriminate abnormal data. Yet the most important issue is how many items should be eliminated. Removal of too many items will blur the standard space. We left 60 items in our study. Through the item selection addressed above, we eventually obtained fairly good improvement in prediction accuracy, even though the normal data that are not used for the base space were somewhat distant from the base space. By increasing the number of data, we can expect to realize a highly accurate prediction in the future. In addition, all of the items sieved through the item selection are regarded as essential from the medical viewpoint. By improving the accuracy in item selection with an increased number of data points, we can obtain a proof of medically vital items, thereby narrowing down examination items and streamlining a medical checkup process.

7. Conclusions

According to the results discussed above, we confirmed primarily that a Mahalanobis distance is highly applicable to medical checkup data. For the prediction of health conditions for the next year, despite its incompleteness, it was concluded that we obtained relatively good results. Additionally, although we have attempted to perform an analysis of the three-year data, we considered that a larger number of data are needed to improve the prediction accuracy.

We summarize all of the above as follows:

1. Although we cannot fully separate healthy and unhealthy persons by using Mahalanobis distances based on the one-year medical checkup data despite a large amount of data used, if the two-year data are used, we can distinguish both groups accurately.

2. If the two-year data are used and necessary items are selected, it is possible to predict the current year's health conditions for most people on the basis of the preceding year's data. Furthermore, if the three-year data are used, we could improve the predictability drastically.

3. The effective examination items sieved through item selection cover all the items that are regarded as essential for a daily medical consultation.

4. The MTS method used in this research not only is effective for improvement of a medical checkup system but will also be applicable to the diagnosis and medical treatment of diseases if the methodology is advanced further.

References

Yoshiko Hasegawa, 1997. Mahalanobis distance application for health examination and treatment of missing data: III. The case with no missing data. *Quality Engineering*, Vol. 5, No. 5, pp. 46–54.

Yoshiko Hasegawa, 1997. Mahalanobis distance application for health examination and treatment of missing data: II. A study on the case for missing data, *Quality Engineering*, Vol. 5, No. 6, pp. 45–52.

Yoshiko Hasegawa and Michiyo Kojima, 1994. Prediction of urinary continence recovery among patients with brain disease using pattern recognition. *Standardization and Quality Control*, Vol. 47, No. 3, pp. 38–44.

Yoshiko Hasegawa and Michiyo Kojima. 1997. "Prediction of urinary continence recovery among patients with brain disease Using Mahalanobis Distance," *Quality Engineering*, Vol. 5, No. 5, p. 55–63.

Tatsuji Kanetaka, 1997. An application of Mahalanobis distance to diagnosis of medical examination. *Quality Engineering*, Vol. 5, No. 2, pp. 35–44.

Hisato Nakajima, Kei Takada, Hiroshi Yano, Yuka Shibamoto, Ichiro Takagi, Masayoshi Yamauchi, and Gotaro Toda, 1998. Forecasting of the future health from existing medical examination results using Mahalanobis–Taguchi system. *Quality Engineering*, Vol. 7, No. 4, pp. 49–57.

This case study is contributed by Hisato Nakajima, Kei Takada, Hiroshi Yano, Yuka Shibamoto, Ichiro Takagi, Masayoshi Yamauchi, and Gotaro Toda.

CASE 79

Character Recognition Using the Mahalanobis Distance

Abstract: Today, the demand for electronic document processing is becoming stronger and stronger. However, technologies to recognize handwriting have not been well developed. In this study, improvement in handwriting recognition, to recognize a pattern or multidimensional information, was studied using the Mahalanobis–Taguchi system (MTS). It has been confirmed that the use of MTS is effective for multidimensional information.

1. Introduction

In order for a computer to recognize a handwriting character by taking advantage of a multidimensional information process, we create a base space. Considering that the base space needs to recognize various types of commonly used characters, we collected as many writings for a particular character but different shapes as possible by asking a number of testees to hand-write characters. Each testee writes down a *hiragana* character, or Japanese syllabary, on a 50 mm by 50 mm piece of paper as a sample character. The total number of writings amounted to approximately 3000. Figure 1 shows a sample character.

A sample character is saved as graphical data according to the procedure shown in Figure 2, and its features are extracted. The graphical data are divided into $n \times n$ cells (Figure 3). If a part of the character lies on a certain cell, the cell is set to 1. Otherwise, the cell is set to 0. This process is regarded as binarization of the graphical data. Since the number of binarized cells is $n \times n$ without special treatment, if n is equal to 50, the number adds up to 2500. In such a case it is quite difficult to calculate a 2500×2500 inverse matrix with a personal computer, which is required to create a base space. Therefore, by extracting the character's features, we reduce the number of binarized cells.

For the 25×25 binary data for the character shown in Figure 4, we performed feature extraction in accordance with the following three procedures:

1. *Integral method I.* In each row and column, we summed up the number of consecutive cells, including empty cells, from the cell with a character data item (or 1) to the one with a character data item.

2. *Integral method II.* In each row and column, we simply counted the number of cells with a character data item.

3. *Differential method.* In each row and column, we counted the number of groups of a single cell or consecutive cells with a character data item.

The rightmost columns and bottom rows in Figure 4 show the values calculated by the three types of feature extraction methods above.

2. Procedure for Character Recognition

Figure 5 summarizes the procedure, starting from the collection of a handwriting character through the extraction of features and creation of a base space, to recognition of a character.

Without any special modification, handwriting characters were gathered from as many people as possible. The characters collected were read with a scanner and saved as graphical data. These graphical data were divided into $n \times n$ cells, converted into 0/1 data, and their features were extracted.

Figure 1
Handwriting the character "ah"

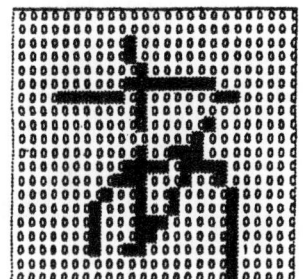

Figure 3
Binarization of handwriting character in Figure 1

Setting the features extracted to our basic data, we determined a base space and computed the Mahalanobis distance for each of the characters collected. As a nest step, we gathered new characters irrelevant to creation of the base space. For these new characters, we calculated the Mahalanobis distances and performed character recognition for their target characters.

3. Preparation of Base Space for Handwriting Character

Here we focused only on the case of using the character "ah" as a base space. Preparing four character groups: the character "ah" used as the base space, another "ah" (not used for the standard space), and "oh" and "nu," which look similar to "ah." We calculated the Mahalanobis distances for the standard space created by the character "ah" from the first group. Figure 6 shows the distribution of the Mahalanobis distance for each character. Using Figure

6 we investigated the probability of recognizing an unknown character. The horizontal and vertical axes indicate a Mahalanobis distance and frequency, respectively. The Mahalanobis distances for the character used for the base space are distributed on both sides of 1, which is their average. The distances for the second group of the character "ah" are almost fully contained in the base space. On the other hand, although both of the distributions of the distances for "oh" and "nu" to some extent overlap the base space around 1.5, we are likely to discriminate them from the character "ah" as different characters. Ideally, the second group of "ah," which was not used to construct the base space, should overlap the first group of "ah," which was used to construct the base space.

Although the number of "ah" characters used to create a base space had initially been set to 1500, by increasing the number to 3000 we sharpened the distribution and made the discrimination easier. On

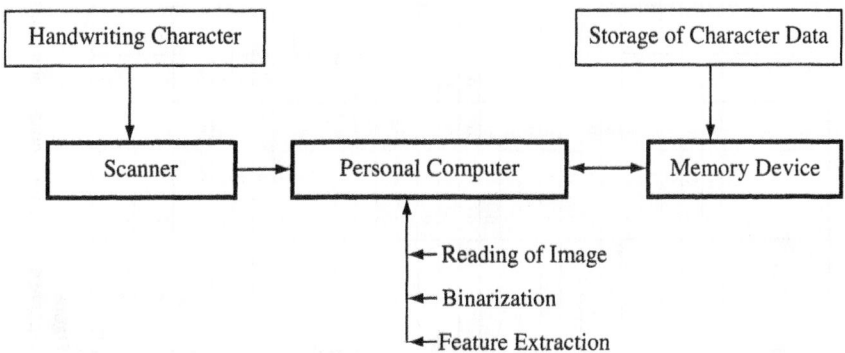

Figure 2
Reading of handwriting characters and extraction of features

																										Integral Method I	Integral Method II	Differential Method
	0	0	0	0	0	0	0	0	0	0	0	1	0	0	0	0	0	0	0	0	0	0	0	0	0	1	1	1
	0	0	0	0	0	0	0	0	0	0	0	1	0	0	0	0	0	0	0	0	0	0	0	0	0	1	1	1
	0	0	0	0	0	0	0	0	0	0	0	1	0	0	0	0	0	0	0	0	0	0	0	0	0	1	1	1
	0	0	0	0	0	0	0	0	0	1	1	1	1	1	1	1	1	1	0	0	0	0	0	0	0	9	9	1
	1	1	1	1	1	1	1	1	1	1	1	0	0	0	0	0	0	1	1	1	1	1	1	1	1	25	19	2
	0	0	0	0	0	0	0	0	0	0	1	0	0	0	0	0	0	0	0	0	0	0	0	0	0	1	1	1
	0	0	0	0	0	0	0	0	0	0	1	0	0	0	0	0	0	0	0	0	0	0	0	0	0	1	1	1
	0	0	0	0	0	0	0	0	0	0	1	0	0	0	0	0	0	0	0	0	0	0	0	0	0	1	1	1
	0	0	0	0	0	0	0	0	0	0	1	0	0	0	0	0	0	0	0	0	1	0	0	0	0	11	2	2
	0	0	0	0	0	0	0	0	0	0	1	0	1	1	1	0	0	0	0	1	1	0	0	0	0	11	6	3
	0	0	0	0	0	0	0	0	0	0	1	1	1	0	1	1	1	0	1	1	0	0	0	0	0	10	8	3
	0	0	0	0	0	0	0	0	1	1	1	1	0	0	0	0	1	1	1	0	0	0	0	0	0	11	7	2
	0	0	0	0	0	0	1	1	1	0	0	1	0	0	0	0	0	1	1	1	1	0	0	0	0	15	8	3
	0	0	0	0	0	1	1	0	0	0	0	1	0	0	0	0	0	1	0	0	1	1	0	0	0	17	6	4
	0	0	0	0	1	1	0	0	0	0	0	1	0	0	0	0	1	1	0	0	0	1	1	0	0	19	7	4
	0	0	0	1	1	0	0	0	0	0	0	1	0	0	0	0	1	0	0	0	0	0	1	0	0	20	5	4
	0	0	1	1	0	0	0	0	0	0	0	1	0	0	0	0	1	0	0	0	0	0	1	1	0	22	6	4
	0	0	1	1	0	0	0	0	0	0	0	1	0	0	0	1	1	0	0	0	0	0	0	1	0	22	6	4
	0	1	1	1	0	0	0	0	0	0	0	1	1	0	1	1	0	0	0	0	0	0	1	1	0	23	9	4
	0	1	0	0	0	0	0	0	0	0	0	0	1	1	1	0	0	0	0	0	0	0	1	0	0	22	5	3
	0	1	1	0	0	0	0	0	0	0	0	0	1	1	0	0	0	0	0	0	0	1	1	0	0	22	6	3
	0	1	1	1	0	0	0	0	0	0	0	1	1	0	0	0	0	0	0	0	1	1	0	0	0	21	7	3
	0	0	1	1	1	1	0	0	0	0	1	1	1	0	0	0	0	0	1	1	1	0	0	0	0	19	10	3
	0	0	0	0	0	1	1	1	1	1	1	0	0	0	0	0	0	1	1	0	0	0	0	0	0	14	8	2
	0	0	0	0	0	0	0	1	1	0	0	0	0	0	0	0	0	0	0	0	0	0	0	0	0	2	2	1
Integral Method I	1	18	19	19	19	20	20	21	21	21	21	23	20	18	17	16	15	21	20	19	19	18	17	15	1			
Integral Method II	1	5	7	7	4	5	4	4	5	4	11	15	8	4	5	4	7	7	6	5	7	5	7	4	1			
Differential Method	1	2	3	3	3	3	3	3	3	3	2	3	2	3	3	3	3	3	3	4	4	3	3	2	1			

Figure 4
Japanese character whose features are extracted in 25 × 25 cells

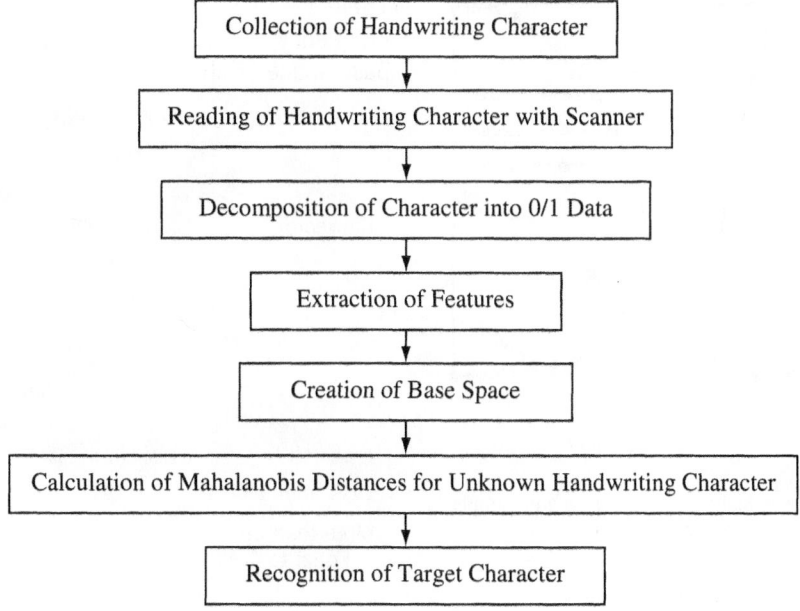

Figure 5
Procedure for handwriting character recognition

the other hand, the total number of "oh," "nu," and "ah" that were not used for the base space is 100.

To clarify the recognition rate of target characters according to Figure 6, for the character "ah" that is not used for the base space, we plotted the relationship (1 − cumulative frequency) versus Mahalanobis distance, whereas for "oh" and "nu" the relationship of cumulative frequency versus Mahalanobis distance is drawn. Both of the results are shown in Figure 7.

If the Mahalanobis distance of a target character is less than 1, we can judge it as a character "ah" without fail. However, when it becomes more than 1, we cannot reject the probability that the target character should be "oh" or "nu," which look similar to "ah." On the other hand, if the Mahalanobis distance of the target character exceeds 3, the likelihood that the character is "ah" diminishes to almost zero. For characters that do not resemble "ah" in the slightest, we do not experience difficulty with recognition.

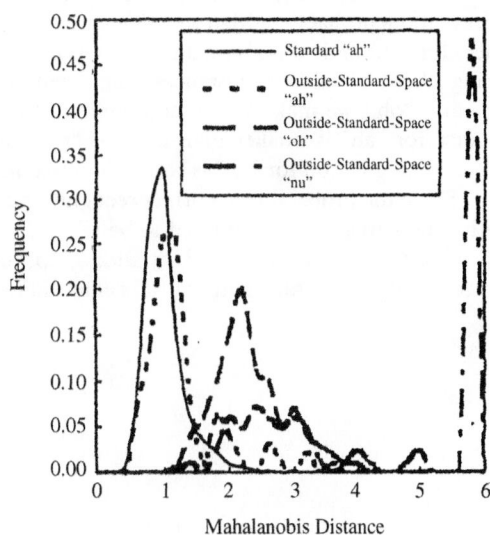

Figure 6
Distribution of characters by the Mahalanobis distance

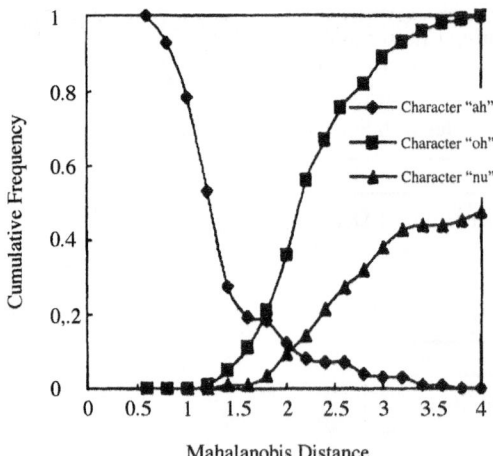

Figure 7
Relationship between Mahalanobis distance and cumulative frequency using base space for "ah"

Table 1
Frequency of occurrence of target character for base space of character "ah"

(a) Occurrences of "ah" and "oh"

	Mahalanobis Distance	
Character	Less Than 1.5	More Than or Equal to 1.5
"ah"	0.8	0.2
"oh"	0.1	0.9

(b) Occurrences of Combinations of "ah" and "oh"

	"ah"	
"oh"	Less Than 1.5	More Than or equal to 1.5
Less than 1.5	0.08	0.02
More than or Equal to 1.5	0.72	0.18

4. Conclusions

Table 1 shows the frequencies of occurrence for "ah," "oh," and "nu" when the Mahalanobis distance = 1.5. Using (a) of Table 1, we obtain (b) for the character recognition rate, which indicates 0.72 as the probability of correct recognition and 0.02 as that of incorrect recognition. The hatched areas represent the probabilities that the Mahalanobis distances for "ah" and "oh" above and below 1.5. Comparing the Mahalanobis distances calculated for "ah" and "oh," we make a correct judgment if the distance for "ah" is smaller than that for "oh," an incorrect judgment if for "ah" is larger than that for "oh." Since the probability of correct recognition in the hatched areas can be assumed to be more than half the total probability, or 0.13, by adding up this correct recognition rate and the aforementioned 0.72, we obtain over 90% as the overall recognition rate.

Reference

Takashi Kamoshita, Ken-ichi Okumura, Kazuhito Takahashi, Masao Masumura, and Hiroshi Yano, 1998. Character recognition using Mahalanobis distance. *Quality Engineering*, Vol. 6, No. 4, pp. 39–45.

This case study is contributed by Takashi Kamoshita, Ken-ichi Okumura, Kazuhito Takahashi, Masao Masumura, and Hiroshi Yano.

Printed Letter Inspection Technique Using the MTS

Abstract: There are many products, such as computers or mobile telephones, whose operation buttons or keys are printed with letters. The quality of these printed letters is inspected visually to find defects. However, it is difficult to clearly determine the quality or defects because of defect types. This problem has not been solved. It is therefore necessary to develop an automatic inspection method for quantification of quality. In this study the Mahalanobis–Taguchi System (MTS) was used for this purpose. The experimental results agreed well with the results of human inspection.

1. Introduction

Since in recent years the number of cellular phones has increased rapidly, not only are size and weight reduction under active research but also the phone's appearance. More attractive, easy-to-understand printed characters and labels are highly sought after.

A keypad character is normally integrated into a keypad. While the size of a character changes to some degree with the type of cellular phone, the digits from "0" to "9" and symbols like "*" and "#," which were studied in our research, are approximately 3 mm × 2 mm with a linewidth of 0.4 mm. On the other hand, Kana (Japanese syllabaries) or alphabetical characters are about 1 mm × 1 mm with a linewidth of approximately 0.16 mm. While fade resistance is required for these characters, for characters on keypads, a material and printing method that make a back light pass through are used such that a cellular phone can be used in a dark environment.

The keypad characters are printed in white on the blacklike background. Various defects are caused by a variety of errors in a printing method. Figure 1 shows a typical example. While Figure 1a is judged as a proper character, Figure 1b, having a chipped area on the left, is judged as a defect. In fact, a part of the line is obviously concave. As long as this defect lies within a range of 10% of a linewidth, it does not stand out much. However, if it exceeds 30%, a keypad is judged as defective because the defect gives a poor impression on a character. In addition to this, there are other types of defects, such as extreme swelling of a line, extremely closed round area of an "8," or additional unnecessary dots in a free space on a keypad.

2. Image-Capturing System

In order to capture a highly accurate image of a character, we organized an image-capturing system with a CCD (charge-coupled device) camera, image input board, and personal computer. Figure 2a shows a photo of an image-capturing device, and Figure 2b outlines the overall system. The CCD camera used for our research can capture an image with 1000 × 1000 pixels at a time. Using this image-capturing system, we can capture a 3 mm × 2 mm character as an image of 150 × 100 pixels and a 0.4-mm linewidth as that of 20 pixels.

Because the light and shade of a captured image are represented in a 256-level scale, we binarize the image data such that only part of a character can be extracted properly. By setting up an appropriate threshold, we obtained a clear-cut binary image. Furthermore, we performed a noise-reduction

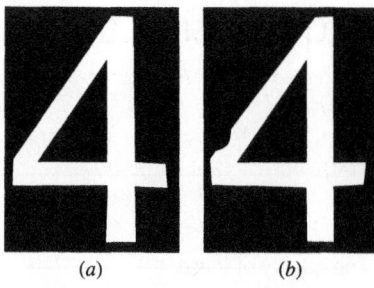

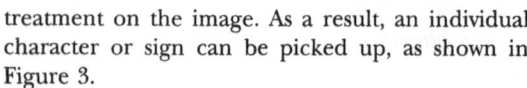

Figure 1
Printed character

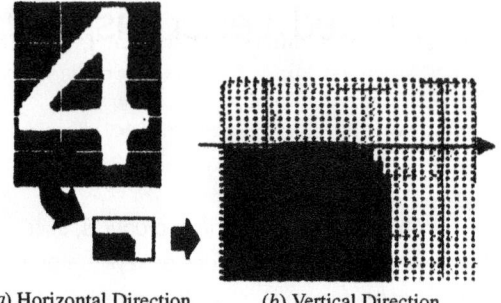

(a) Horizontal Direction   (b) Vertical Direction

Figure 3
Extraction of feature values

treatment on the image. As a result, an individual character or sign can be picked up, as shown in Figure 3.

3. Base Space

A judgment on whether a certain character is proper or defective is made in accordance with the inspection standard. However, as discussed above, it is quite difficult to prescribe detailed inspection standards for all defect types, such as a chipped area or swelling, or all defect cases on a character. In addition, even if a printed character is too bold or fine on the whole, when it is well balanced, we may

judge it as good. The overall shape of a character may be considered as one of the judgment factors. Then the final judgment depends greatly on a human experience.

Therefore, assuming that the inspector's judgment is correct, we gather proper and defective characters. Two inspectors participated in the experiment. The results judged by the inspectors as correct were defined as standard data. The proper characters collected were 200 sets of digits from "0" to "9" and the symbols "#" and "*".

To use the MTS method we extracted some multidimensional feature values that are considered necessary to judge whether a certain character pat-

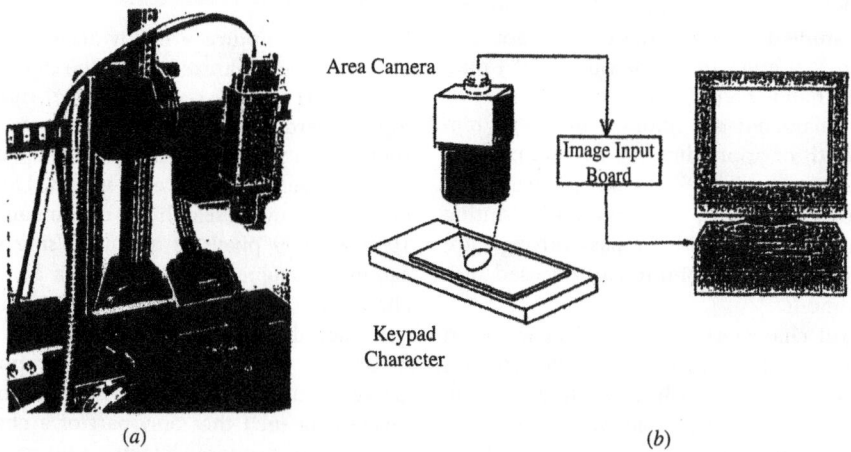

Area Camera

Image Input Board

Keypad Character

(a)   (b)

Figure 2
Character inspection system

tern is proper or defective. In our research we used as a feature value differential and integral characteristics which are applied to character recognition.

A character obtained as a binary image is, for example, represented in a pixel image 150 in height × 100 in width. Since the extract of feature values from this original image results in a considerable number of feature values, we divided the image into appropriate-size unit regions. For example, we split them up into many unit regions, each with 30 × 30 pixels. These were divided into 5 × 4 unit regions. Then, for each unit region, we calculated the following feature values.

1. For 30 lines in a horizontal direction, we selected the number of areas that change from white to black, or vice versa, as a differential characteristic. In addition, the sum of lengths of areas occupied with a character data was chosen as an integral characteristic.

2. Similarly, for 30 lines in a vertical direction, we computed differential and integral characteristics.

Thus, the total number of feature values adds up to 120, or the sum of all differential and integral characteristics. This workflow and an example of feature values are shown in Figure 3 and Table 1, respectively.

4. Results of Inspection

After creating software that covers the aforementioned workflow, ranging from image capturing to character extraction, character division into unit regions, extraction of feature values, generation of base space, and calculation of Mahalanobis distances, we performed an experiment. For the division into unit regions, the software automatically split up an image into unit regions so that each consists of 30 × 30 pixels. Therefore, the number of unit regions varies with the size of a character or sign. For instance, when the size of a character is 150 pixels in height × 100 pixels in width, its image is divided into 5 × 4 unit regions, whereas it is split up into 4 × 3 unit regions in the case of a character of 120 pixels in height × 80 pixels.

Figure 4 shows an example of displaying an inspection result. A binarized character is indicated on the left of the screen. On the right, a Mahalanobis distance for each unit region is shown. In addition, if a Mahalanobis distance exceeds a predetermined threshold, the corresponding area in the character on the left lights up. On the other hand, for all Mahalanobis distances surpassing 1000, 1000 is displayed.

As discussed before, using 200 sets of characters that have been preselected as proper by the two in-

Table 1
Feature values extracted from a character

	Differential Characteristic	Integral Characteristic		Differential Characteristic	Integral Characteristic
1	0	30	1	2	15
2	0	30	2	3	16
3	0	30	3	3	15
⋮	⋮	⋮	⋮	⋮	⋮
27	1	15	27	2	27
28	1	14	28	1	26
29	2	14	29	2	27
30	1	12	30	2	27

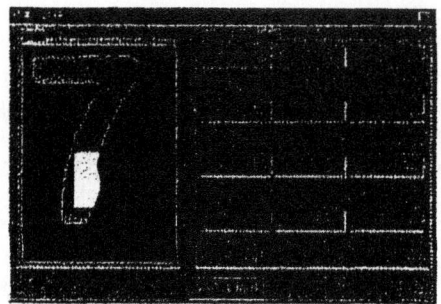

Figure 4
On-screen inspection result for character recognition by MTS

spectors, we created a Mahalanobis space for each unit region of each character or sign. Then, with 100 sets of characters, including both proper and defective, we performed an inspection experiment.

In the example shown in Figure 4, we can see a swelling in the lower part of a digit "7" whose Mahalanobis distance is computed as 10.337. All other regions have a smaller distance. Figure 5 indicates other examples together with a table of Mahalanobis distances. In the figure a small swelling can be found in the bottom left part of the symbol "#."

In fact, the corresponding Mahalanobis distance amounts to 10.555. Although one of the right-hand regions shows the value of 3.815, because the threshold is set to 4, the region is judged as proper on the display. For the digit "0," due to dots scattered in the bottom-left and top-right regions, both of the Mahalanobis distances exceed 1000. The reason for displaying these enormous values is that there is no dot in the standard data.

As a next step, in Table 2 we show the relationship between an inspector's judgment and a Mahalanobis distance based on the inspection of 1000 unit regions. Table 2 indicates the correspondence between proper-or-defective judgment by one inspector and the Mahalanobis distance, irrespective of a judgment based on a Mahalanobis distance. Based on this table, we can see that, for example, if we select 4 as the threshold for a proper-or-defective judgment [(a) of Table 2], the error of judging a defect as proper results in $22\% = (4 + 14)/81$, and the contrary error of judging a proper outcome as defective turns out to be $23\% = (22 + 36)/249$. When setting the threshold to 3 [(b) of Table 2], the error of judging a defect as proper is reduced to 5%, whereas the opposite error of confusing a proper character with a defect is increased.

0.009	1.006	0.582	0.000
1.471	0.914	0.984	0.237
0.784	1.189	0.205	3.815
10.555	0.668	0.488	0.117
0.200	0.398	0.223	0.000

0.789	0.523	1000.000
0.719	0.491	0.533
5.318	0.958	0.336
1000.000	0.411	0.183

Figure 5
Inspection results

Table 2
Comparison of inspector's judgment and Mahalanobis distance

	Inspector's Judgment		
Mahalanobis Distance	Proper	Defective	
$0 \leq MD < 1$	16	0	
$1 \leq MD < 2$	78	4	
$2 \leq MD < 3$	62	0	← (a)
$3 \leq MD < 4$	35	14	← (b)
$4 \leq MD < 10$	22	19	
$10 \leq MD$	36	44	
Total	249	81	

5. Conclusions

As shown in Table 2, since the Mahalanobis distances indicate a clear-cut trend for a judgment on whether a certain character is proper or defective, the evaluation method grounded on the MTS method in our research can be regarded as valid to detect a defective character. Despite the fact that this method needs to be improved further because it sometimes judges a proper character as a defective, we have successfully demonstrated that our conventional evaluation of quality in character printing, which has long relied heavily on human senses, can be quantified.

The following are possible reasons that a proper character is judged as defective:

1. Since the resolution of character image capturing (a character of 3 mm × 2 mm is captured by an image of 150 × 100 pixels) is regarded as insufficient, higher resolution is needed. Although 200 sets of proper characters were used in this research as the data for creating a Mahalanobis space, the number is still insufficient to take in patterns that are tolerable as a proper character. At this time, we believe that the second problem is more significant. Considering that the research on character recognition deals with 3000 handwriting characters for a standard space, we need to collect at least 1000 characters as the standard data.

2. The inspection processing time in our study, which ranges from extraction of feature values to calculation of a Mahalanobis distance, has resulted in less than 0.1 second. In addition, preprocessing time for data processing and binarization of a captured image is needed. However, since our current visual inspection requires 10 to 15 seconds for one sheet of keypads, we can expect to allocate a sufficient amount of preprocessing time.

In this research we have not sieved necessary essential feature values. On reason is that because a character is divided into unit regions, it is quite difficult to analyze comprehensively the priority of each of 120 differential and integral characteristics defined in each unit region, and the other is that 120 feature values do not lead to an enormous amount of calculation.

Reference

Shoichi Teshima, Dan Jin, Tomonori Bando, Hisashi Toyoshima, and Hiroshi Kubo, 2000. Development of printed letter inspection technique using Mahalanobis–Taguchi system. *Quality Engineering*, Vol. 8, No. 2, pp. 68–74.

This case study is contributed by Shoichi Teshima, Dan Jin, Tomonori Bando, Hisashi Toyoshima, and Hiroshi Kubo.

Part V

Software Testing and Application

Algorithms (Cases 81–82)
Computer Systems (Case 83)
Software (Cases 84–87)

Optimization of a Diesel Engine Software Control Strategy

Abstract: A zero-point-proportional dynamic SN ratio was used to quantify vibration and tracking accuracy under six driving conditions, which represented noise factors. An L_{18} orthogonal array explored combinations of six software strategy control factors associated with controlling fuel delivery to the engine. The result was a 4- to 10-dB improvement in vibration reduction, resulting in virtual elimination of the hitching condition. As a result of this effort, an $8 million warranty problem was eliminated. The robust design methodology developed in this application may be used for a variety of applications to optimize similar feedback control strategies.

1. Introduction

What makes a problem difficult? Suppose that you are assigned to work on a situation where (1) the phenomenon is relatively rare; (2) the phenomenon involves not only the entire drive train hardware and software of a vehicle, but specific road conditions are required to initiate the phenomenon; (3) even if all conditions are present, the phenomenon is difficult to reproduce; and (4) if a vehicle is disassembled and then reassembled with the same parts, the phenomenon may disappear completely!

For many years, various automobile manufacturers have occasionally experienced a phenomenon like this associated with slow oscillation of vehicle rpm under steady pedal position (ringing) or cruise-control conditions (hitching). Someone driving a vehicle would describe hitching as an unexpected bucking or surging of the vehicle with the cruise control engaged, especially under load (as in towing). Engineers define hitching as a vehicle in speed-control mode with engine speed variation of more than 50 rpm (peak to peak) at a frequency below 16 Hz.

A multifunction team with representatives from several areas of three different companies was brought together to address this issue. Their approaches were more numerous than the team members and included strategies ranging from studies of hardware variation to process FMEAs and dynamic system modeling. The situation was resolved using TRIZ and robust design. The fact that these methods worked effectively and efficiently in a complex and difficult situation is a testament to their power, especially when used in tandem.

TRIZ, a methodology for systemic innovation, is named for a Russian acronym meaning "theory of inventive problem solving." Anticipatory failure determination (AFD), created by Boris Zlotin and Alla Zusman of Ideation, is the use of TRIZ to anticipate failures and determine root cause. Working with Vladimir Proseanic and Svetlana Visnepolschi of Ideation, Dmitry Tananko of Ford applied TRIZ AFD to the hitching problem. Their results, published in a case study presented at the Second Annual Altshuller Institute for TRIZ Studies Conference, found that resources existed in the system to support seven possible hypotheses associated with hitching. By focusing on system conditions and circumstances associated with the phenomenon, they narrowed the possibilities to one probable hypothesis, instability in the controlling system.

By instrumenting a vehicle displaying the hitching phenomenon, Tananko was able to produce the plot shown in Figure 1. This plot of the three main signals of the control system (actual rpm, filtered rpm, and MF_DES, a command signal) verified the AFD hypothesis by showing the command signal out of phase with filtered rpm when the vehicle was kept at constant speed in cruise-control mode.

Actual rpm is out of phase with the command signal because of delays associated with mass inertia. In addition, the filtered rpm is delayed from the actual rpm because of the time it takes for the filtering calculation. The specific combination of these delays, a characteristic of the unified control system coupled with individual characteristics of the drive train hardware, produces the hitching phenomena. The solution lies in using Taguchi's techniques to make the software/hardware system robust.

2. System Description

A simple schematic of the controlling system is shown in Figure 2. The mph set point is determined by the accelerator pedal position or cruise-control setting. Depending on a number of parameters, such as vehicle load, road grade, and ambient temperature, the control system calculates the amount of fuel to be delivered for each engine cycle as well as other fuel delivery parameters. Accordingly, the engine generates a certain amount of torque, resulting in acceleration/deceleration of the vehicle. The feedback loop parameters and the speed sensor parameters must be set at appropriate values to achieve smooth vehicle behavior with no hitching/ringing.

3. P-Diagram

The parameters studied in this project are given in the P-diagram shown in Figure 3.

4. Noise Factors

Different driving profiles constitute important noise factors because they cause major changes to the load on the engine. The following six noise levels were used in this experiment:

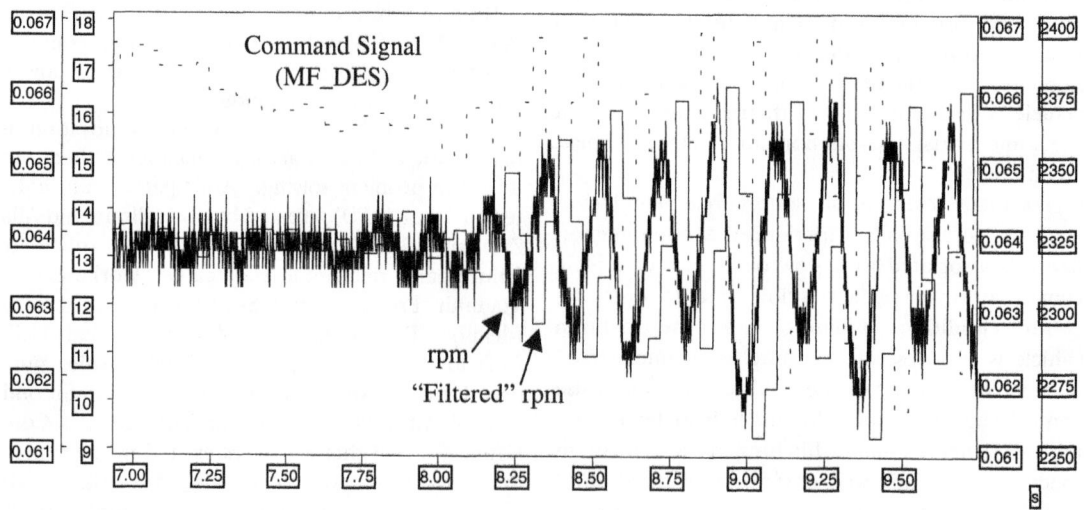

Figure 1
Hitching phenomena

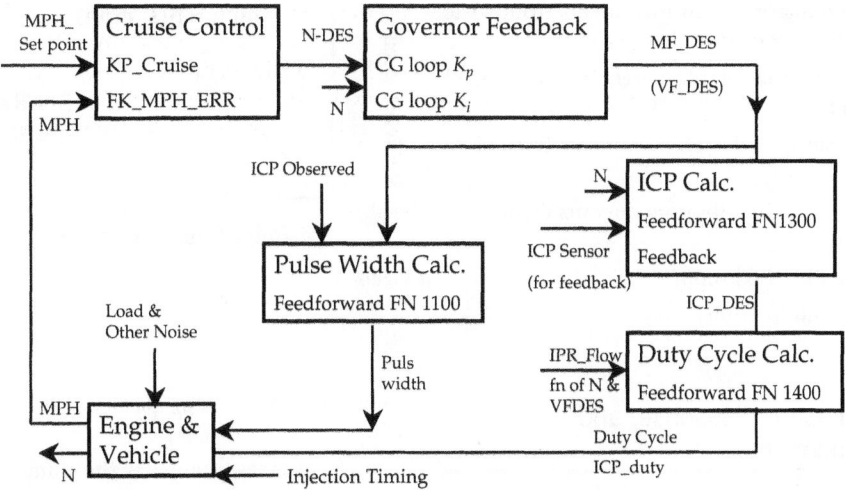

Figure 2
Simplified functional flow

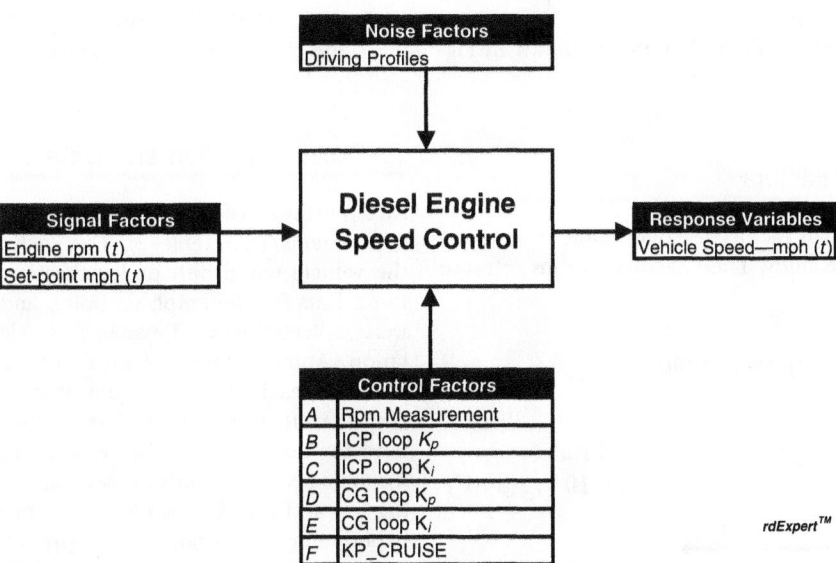

Figure 3
P-diagram

1. Accelerating in 1-mph increments from 47 to 56 mph

2. Accelerating in 1-mph increments from 57 to 65 mph

3. Decelerating in 1-mph increments from 65 to 57 mph

4. Decelerating in 1-mph increments from 56 to 47 mph

5. Rolling hill at 65 mph

6. Rolling hill 57 mph

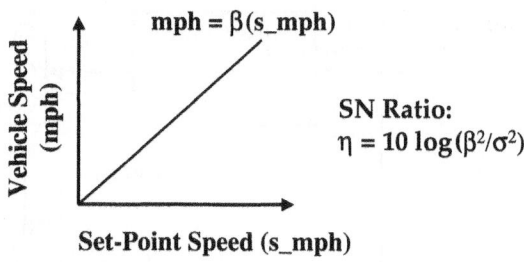

$$mph = \beta(s\_mph)$$

SN Ratio:
$$\eta = 10 \log(\beta^2/\sigma^2)$$

Figure 5
Ideal function 2: tracking

5. Signal Factor, Response, and Ideal Function

There would be no vibration or hitching or ringing if the vehicle speed (mph) were directly proportional to the engine speed (rpm) at every instant of time. Of course, the gear ratio was constant over the time period considered. Thus, the ideal function selected was zero-point-proportional with scaled engine rpm as the signal and vehicle speed (mph) as the response (Figure 4). The scale depends on the gear ratio and tire type.

While eliminating hitching, it is also important to have good tracking between the set-point mph and the actual mph. We need another ideal function and corresponding SN ratio, as shown in Figure 5.

6. Control Factors

The six control factors listed in Table 1 were selected for the study. These factors, various software speed control strategy parameters, are described below:

A: Rpm measurement is the number of consecutive measurements over which the rotational speed is averaged for estimating rpm.

B: ICP loop K_p is the proportional constant for the ICP loop.

C: ICP loop K_i is the integral constant for the ICP loop.

D: CG loop K_p is the proportional constant for the governor feedback.

E: CG loop K_i is the integral constant for the governor feedback.

F: KP_CRUISE is the proportional constant for the cruise-control feedback loop.

7. Experiment Plan and Data

An L_{18} orthogonal array was used for conducting the experiments (see Table 2). For each experiment, the vehicle was driven under the six noise conditions. Data for rpm, mph set point, and actual mph were collected using Tananko's vehicle instrumentation. About 1 minute's worth of data were collected for each noise condition. Plots of scaled rpm (signal factor) versus actual mph (response) were used for calculation of the zero-point-proportional dynamic SN ratios. Plots for two experiments, showing low and high values for the SN ratio in the L_{18} experiment [corresponding to pronounced hitching (experiment 6) and minimal hitching (experiment 5)], are shown in Figure 6a and b, respectively. The corresponding SN ratios were −1.8 and 11.8. This is an empirical validation that the SN ratio is

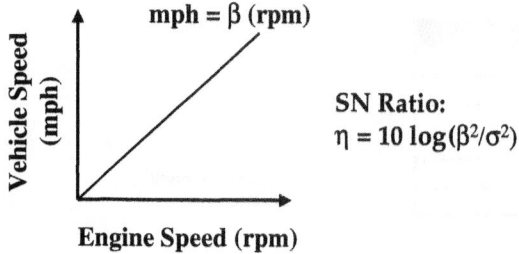

$$mph = \beta \ (rpm)$$

SN Ratio:
$$\eta = 10 \log(\beta^2/\sigma^2)$$

Figure 4
Ideal function 1: hitching

Table 1
Control factors and levels

	Control Factor	No. of Levels	Level		
			1	2	3
A:	Rpm measurement	2	6 teeth	12 teeth	
B:	ICP loop K_p	3	0.0005	0.0010	0.0015
C:	ICP loop K_i	3	0.0002	0.0007	0.0012
D:	CG loop K_p	3	0.8[a]	fn[a]	1.2[a]
E:	CG loop K_i	3	0.027	0.032	0.037
F:	KP_CRUISE	3	0	0.5[a]	[a]

[a] Current level.

Table 2
Control factor orthogonal array[a]

No.	A: Col. 1 rpm Measurement	B: Col. 2 ICP loop K_p	C: Col. 3 ICP loop K_i	D: Col. 4 CG loop K_p	E: Col. 5 CG loop K_i	F: Col. 6 KP_CRUISE
1	(1) 6 teeth	(1) 0.0005	(1) 0.0002	(1) 0.8[b]	(1) 0.027	(1) 0
2	(1) 6 teeth	(1) 0.0005	(2) 0.0007	(2) fn[b]	(2) 0.032	(2) 0.5[b]
3	(1) 6 teeth	(1) 0.0005	(3) 00012	(3) 1.2[b]	(3) 0.037	(3) [b]
4	(1) 6 teeth	(2) 0.0010	(1) 0.0002	(1) 0.8[b]	(2) 0.032	(2) 0.5[b]
5	(1) 6 teeth	(2) 0.0010	(2) 0.0007	(2) fn[b]	(3) 0.037	(3) [b]
6	(1) 6 teeth	(2) 0.0010	(3) 0.0012	(3) 1.2[b]	(1) 0.027	(1) 0
7	(1) 6 teeth	(3) 0.0015	(1) 0.0002	(2) fn[b]	(1) 0.027	(3) [b]
8	(1) 6 teeth	(3) 0.0015	(2) 0.0007	(3) 1.2[b]	(2) 0.032	(1) 0
9	(1) 6 teeth	(3) 0.0015	(3) 00012	(1) 0.8[b]	(3) 0.037	(2) 0.5[b]
10	(2) 12 teeth	(1) 0.0005	(1) 0.0002	(3) 1.2[b]	(3) 0.037	(2) 0.5[b]
11	(2) 12 teeth	(1) 0.0005	(2) 0.0007	(1) 0.8[b]	(1) 0.027	(3) [b]
12	(2) 12 teeth	(1) 0.0005	(3) 00012	(2) fn[b]	(2) 0.032	(1) 0
13	(2) 12 teeth	(2) 0.0010	(1) 0.0002	(2) fn[b]	(3) 0.037	(1) 0
14	(2) 12 teeth	(2) 0.0010	(2) 0.0007	(3) 1.2[b]	(1) 0.027	(2) 0.5[b]
15	(2) 12 teeth	(2) 0.0010	(3) 00012	(1) 0.8[b]	(2) 0.032	(3) [b]
16	(2) 12 teeth	(3) 0.0015	(1) 0.0002	(3) 1.2[b]	(2) 0.032	(3) [b]
17	(2) 12 teeth	(3) 0.0015	(2) 0.0007	(1) 0.8[b]	(3) 0.037	(1) 0[b]
18	(2) 12 teeth	(3) 0.0015	(3) 00012	(2) fn[b]	(1) 0.027	(2) 0.5[b]

[a] Level in parentheses.
[b] Current condition.

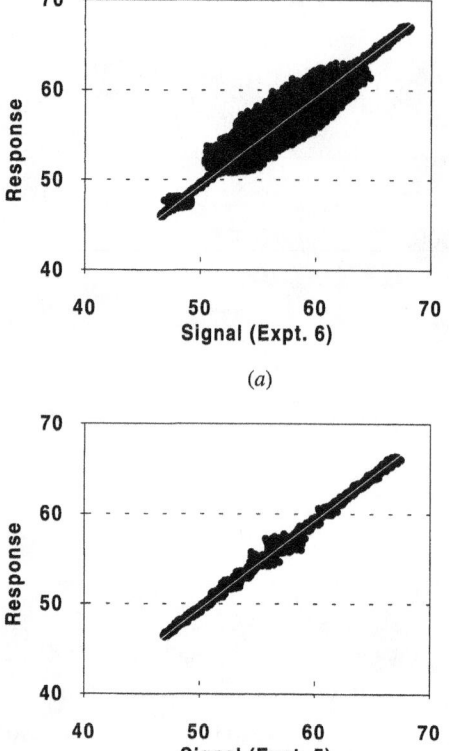

(a)

(b)

Figure 6
Data plots for hitching ideal function

capable of quantifying hitching. See Table 3 for the SN ratio from each run of L_{18}.

8. Factor Effects

Data from the L_{18} experiment were analyzed using *rdExpert* software developed by Phadke Associates, Inc. The control factor orthogonal array is given in the appendix to the case. The signal-to-noise (SN) ratio for each factor level is shown in Figure 7. From the analysis shown in the figure, the most important factors are A, D, and F.

1. Factor A is the number of teeth in the flywheel associated with rpm calculations. The more teeth used in the calculation, the longer the time associated with an rpm measurement and the greater the smoothing of the rpm measure. Level 2, or more teeth, gives a higher SN ratio, leading to reduced hitching.

2. Factor D is CG loop K_p, a software constant associated with gain in the governor loop. Here level 1, representing a decrease in the current function, is better.

3. Factor F is KP_CRUISE, a software constant in the cruise control strategy associated with gain. Level 3, maintaining the current value for this function, is best, although level 2 would also be acceptable.

Confirmation experiments using these factors were then conducted. Predicted values and observed values were computed for the best levels of factors, the worst levels of factors, and the vehicle baseline (original) levels of factors.

$$Best: A_2, B_3, C_2, D_1, E_2, F_3$$

$$Worst: A_1, B_1, C_3, D_3, E_1, F_1$$

$$Baseline: A_1, B_2, C_1, D_2, E_2, F_3$$

The results are shown in Table 4. We have shown the SN ratios separately for noise conditions 1–4 and 5–6 to be able to ascertain that the hitching problem is resolved under the two very different driving conditions. As can be seen in this table, there was very good agreement between the predicted and observed SN ratios under the foregoing conditions.

The confirmatory experiment plot of rpm versus mph for the best factor combination is shown in Figure 8. This plot clearly supports the conclusions reached by the SN ratio analysis.

An additional SN ratio analysis of the mph set point versus vehicle speed (mph) was done to evaluate ability of the speed control software to track the set-point speed accurately. The factor effects for the tracking ideal function are shown in Figure 9. Only factor F, KP_CRUISE, is important for tracking. Furthermore, the direction of improvement for the tracking ideal function is the same as that for the hitching ideal function. Thus, a compromise is not needed. The confirmation results for the tracking ideal function are also given in Table 2.

Table 3
SN ratios

No.	Hitching SN		Tracking SN	
	Noises 1–4	Noises 5–6	Noises 1–4	Noises 5–6
1	2.035	9.821	3.631	−2.996
2	11.078	4.569	11.091	7.800
3	4.188	4.126	9.332	9.701
4	15.077	7.766	11.545	8.256
5	11.799	3.908	12.429	9.233
6	−1.793	3.001	3.415	−1.390
7	9.798	4.484	11.841	9.793
8	5.309	6.212	4.392	−2.053
9	8.987	8.640	9.618	9.324
10	13.763	12.885	10.569	10.267
11	18.550	18.680	12.036	14.106
12	2.538	15.337	1.826	−3.128
13	0.929	16.065	1.492	−1.200
14	9.022	9.501	8.688	7.856
15	18.171	18.008	11.260	14.804
16	11.734	11.823	12.031	11.852
17	18.394	16.338	7.000	−1.881
18	13.774	17.485	9.943	9.142
Average	9.631	10.480	8.452	6.083

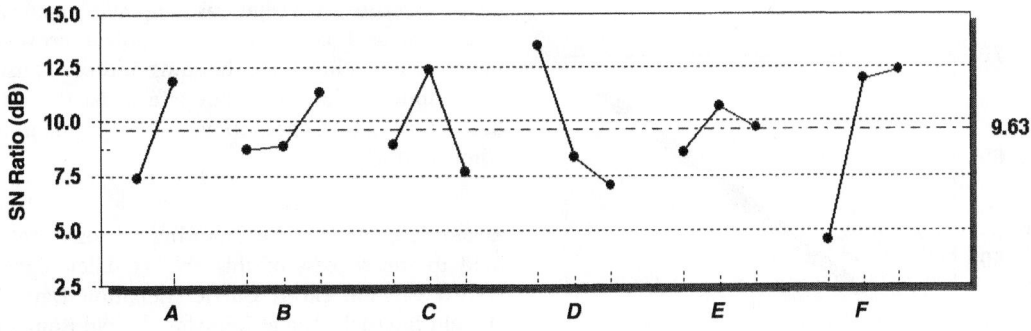

F Value	5.9	0.8	2.3	4.6	0.4	7.5
%SS	13.6	3.9	10.7	21.3	2.0	34.6

Figure 7
Factor effects for ideal function 1: hitching

Table 4

Results of confirmatory experiment

| | Ideal Function 1: Hitching | | Ideal Function 2: Tracking | |
	Noise Conditions 1–4	Noise Conditions 5–6	Noise Conditions 1–4	Noise Conditions 5–6
Best				
Observed	18.44	19.01	11.80	15.39
Predicted	21.25	17.85	12.31	12.37
Worst				
Observed	−0.04	6.45	4.04	−1.56
Predicted	−2.26	3.28	2.74	−2.89
Baseline				
Observed	14.88	9.56	12.86	10.31
Predicted	8.08	5.66	11.55	10.8

9. Further Improvements

The factor effect plots of Figures 7 and 9 indicate that improvements beyond the confirmation experiment can be achieved by exploring beyond level A_2 for factor A, below level D_1 for factor D, and beyond level F_3 for factor F. These extrapolations were subsequently tested and validated.

10. Conclusions

The team now knew how to eliminate hitching completely. Many members of this team had been working on this problem for quite some time. They believed it to be a very difficult problem that most likely would never be solved. The results of this study surprised some team members and made them believers in the robust design approach. In the words of one of the team members, "When we ran that confirmation experiment and there was no hitching, my jaw just dropped. I couldn't believe it. I thought for sure this would not work. But now I am telling all my friends about it and I intend to use this approach again in future situations."

After conducting only one L_{18} experiment, the team gained tremendous insights into the hitching phenomenon and how to avoid it. They understood on a root-cause level what was happening, made adjustments, and conducted a complete prove-out program that eliminated hitching without causing other undesirable vehicle side effects. As a result of this effort, an $8 million warranty problem was eliminated.

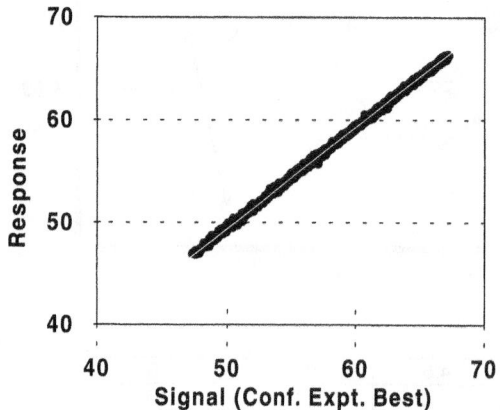

Figure 8
Plot of ideal function 1 (hitching) with best factor combination

Acknowledgments The following persons contributed to the success of this project: Ellen Barnes, Harish Chawla, David Currie, Leighton Davis Jr., Donald Ignasiak, Tracie Johnson, Arnold Komberg, Chris Kwasniewicz, Bob McCliment, Carl Swanson, Dmitry Tananko, Laura Terzes, and Luong-Dave Tieu at Ford Motor Company; Dan Henriksen and William C. Rudhman at International Truck Company; and David Bowden and Don Henderson at Visteon Corporation. The authors thank Dr. Carol Vale especially for her valuable comments in editing

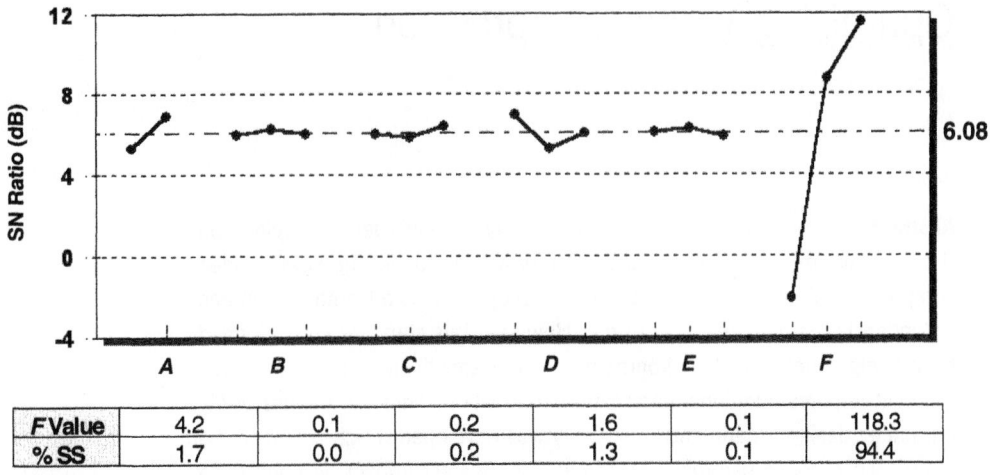

	A	B	C	D	E	F
F Value	4.2	0.1	0.2	1.6	0.1	118.3
% SS	1.7	0.0	0.2	1.3	0.1	94.4

Figure 9
Factor effects for the tracking ideal function

this manuscript. The data for this case were analyzed by using the *rdExpert* software developed by Phadke Associates, Inc. *rdExpert* is a trademark of Phadke Associates, Inc.

References

M. S. Phadke, 1989. *Quality Engineering Using Robust Design*, Prentice Hall, Englewood Cliffs, NJ.

V. Proseanic, D. Tananko, and S. Visnepolschi, 2000. The experience of the anticipatory failure determination (AFD) method applied to hitching/ringing problems. *Proceedings of the 2nd Annual Altshuller Institute TRIZ Conference*, Nashua, NH, pp. 119–126.

Genichi Taguchi, 1987. *System of Experimental Design*, edited by Don Clausing. New York: Unipub/Krass International, Vols. 1 and 2, 1987.

This case study is contributed by Larry R. Smith and Madhav S. Phadke.

CASE 82

Optimizing Video Compression

Abstract: Recent advances in Internet telephony and high-definition television (HDTV) have made real-time video compression a reality. The H.263 video compression standard investigated by this project takes advantage of image redundancies in both space and time. However, this standard was designed for a benign commercial environment. Using it effectively in a military environment requires additional parameter optimization. Through a series of L_{36} parameter design experiments, a strategy was developed which identified the H.263 video codec parameters that had the greatest impact on both video quality and transmitted bit rate. Using Taguchi experiment techniques, an optimal arrangement of control parameters were developed and tested.

1. Introduction

ITT Industries Aerospace/Communications Division, (IIN/ACD) located in Clifton, New Jersey, is a leader in the design and manufacturing of secure communications systems for military and government applications. The video codec investigated in this study is currently in use for the HMT and SUO programs.

Source Coding Algorithm

The H.263 video code is the ITU standard for very low bit rate video compression that makes it a prime candidate for use in a bandlimited military environment. Currently, it is used primarily for video-conferencing over analog telephone lines. Tests performed by the MPEG-4 committee in February 1996 determined the H.263 video quality outperformed all other algorithms optimized for very low bit rates.

The basic form of the H.263 encoder is shown in Figure 1. The video input to the codec consists of series of noninterlaced video frames occurring approximately 30 times per second. Images are encoded using a single luminance (intensity) and two chrominance (color difference) components. This format is commonly known as the Y, U, V format. Each pixel is sampled using an 8-bit linear quantizer, as defined in CCIR Recommendation 601. According to this standard, four pixels of the Y component are sampled and transmitted for every pixel of U and V (e.g., 4:1:1 format). This format was chosen by the ITU because human vision is more sensitive to luminance information than color information. Five standard image formats are supported by the H.263 codec: sub-QCIF, QCIF, CIF, 4CIF and 16CIF. For this investigation, the codec was operated using the QCIF format (e.g., 176 × 144 pixels/picture).

Images in the compressed video sequence are classified as either I-frames or P-frames. Intraframes (I-frames) do not rely on any previous spatial information and are encoded in a manner similar to still image compression. This processing results in a moderate amount of compression. Predicted frames (P-frames) are encoded based on spatial differences from the previous frame and allow a greater amount of compression to take place.

Both I and P frames are divided into 16 × 16 pixel luminance and 8 × 8 pixel chrominance macroblocks. H.263 uses motion estimation techniques to predict the pixel values of the target macroblock from the values of a reference macroblock in the

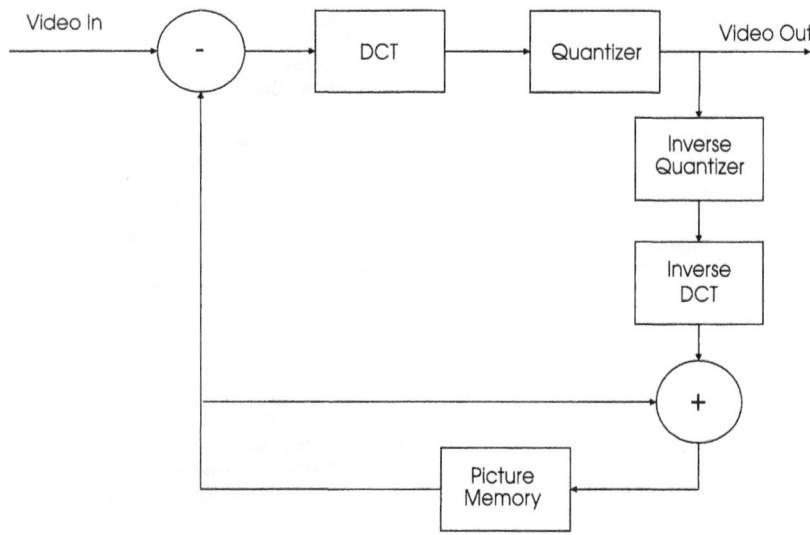

Figure 1
H.263 video codec system

previous frame. The best-fit prediction is found, usually as the result of an optimized search algorithm, and its location is transmitted as a motion vector that indicates its change from the target macroblock location. The macroblock is therefore encoded using two pieces of information, its motion vector and a prediction error.

Each macroblock undergoes a frequency-domain transform by using a discrete cosine transform (DCT). The DCT is a two-dimensional transform that maps values from the spatial domain into the frequency domain. For interframe macroblocks, the prediction error is calculated and transformed, while for intraframe macroblocks the pixel values themselves are transformed. The DCT coefficients are then quantized, mapping them into discrete steps. The step size, or Q-factor, remains constant throughout the video transmission and can range from 1 to 31. Because quantization introduces loss, higher Q values introduce more loss than lower Q values. Finally, the quantized coefficients are entropy encoded and transmitted to the decoder along with the motion vectors and other relevant formatting information.

The main purpose of the picture memory is to provide a reference image for the construction of P-frames. It accomplishes this by decoding the resid-uals of the transmitted video image through an inverse quantizer and DCT. It then adds these residuals with the previously stored image in the picture memory to construct the updated reference image necessary for P-frame difference processing.

In addition to I-frames and P-frames, four negotiable coding options are available for improved video performance (can be used together or separately):

1. *Unrestricted motion vectors.* For the default operation, motion vectors are restricted so that all pixels fall within the coded frame area. In unrestricted motion vector operation this restriction is removed and motion vectors are allowed to point outside the frame area.

2. *Syntax-based arithmetic coding.* For default operation, the H.263 codec using a variable-length encoding/decoding scheme for packing the compressed video into the bit stream. In this mode all variable-length encoding/decoding operations are replaced with arithmetic encoding/decoding techniques.

3. *Advanced prediction.* For the default H.263 codec, one motion vector is allocated per macroblock on a 16 × 16 pixel block size. As an

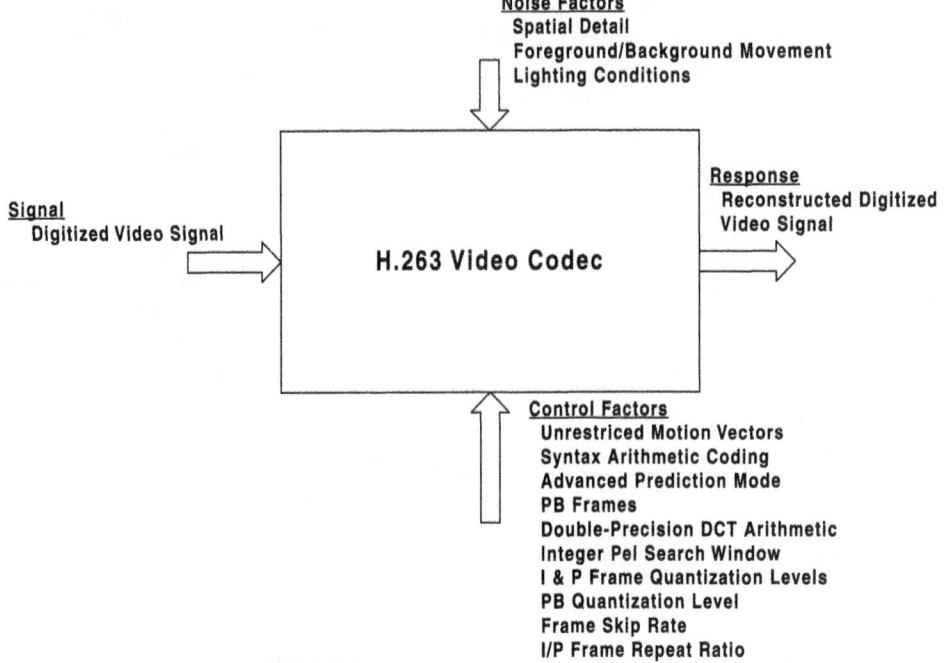

Figure 2
Parameter diagram

option, under advanced prediction mode, 8 × 8 vectors may be used.

4. *PB-frames.* Default operation of the H.263 video code only allows I and P frames. Under this option a PB frame is allowed, which consists of two pictures being coded as one unit. This PB-frame is predicted from a previously decode P-picture and a B-picture that is bi-directionally decoded.

2. Objectives

The objective of this experiment is to determine the sensitivity of the H.263 video codec parameters with respect to overall video quality and to identify the parameters that will best increase video quality without sacrificing the compressed video bit rate.

Parameter Diagram

To facilitate the robust design process, a parameter diagram of the system was generated. The parameter diagram is shown in Figure 2. The system noise and control parameters are as follows:

Noise Factors

Spatial detail: The amount of spatial detail in the image

Foreground motion: the amount of motion present in the primary or foreground object in the video scene

Background motion: the amount of motion present in the secondary or background objects in the video scene

Lighting conditions: the quality and direction of lighting in the video scene

Control Factors

Unrestricted motion vectors. Normal operation of the H.263 codec does not allow motion vectors to point "outside" the image areas. This option relieves this restriction and allows unrestricted motion outside the image area.

Syntax arithmetic coding. Normal operation of the H.263 codec uses variable-length coding/decoding for image transmission. Under this option, an arithmetic coding technique replaces all of the variable-length encoding/decoding operations.

Advanced prediction mode. For the H.263 encoder, one motion vector is allocated per macroblock on a 16×16 pixel block size. Under this option, 8×8 vectors are allowed.

PB frame. This allows the H.263 codec to encode pictures as PB-frames. A PB-frame consists of two pictures being codes as one unit. The PB-frame consists of a P-picture, which is predicted from a previously decoded P-picture, and a B-picture, which is bidirectionally decoded from past and future P-pictures.

Double-precision DCT. This option forces the decoder to use double-precision floating-point arithmetic in calculating the inverse DCT.

Integer pel search. This parameter controls the limit of the macroblock search window when calculating the motion vectors. The window size can range from 2 through 15.

I and P frame quantization. This parameter controls the quantization level for I-frames and P-frames. These parameters are set independently and do not vary during the course of a transmission. The quantization levels range from 2 through 30, with lower values indicating a greater number of quantization steps.

Frame skip rate. This parameter controls the amount of frames dropped prior to entering the encoder. These dropped frames are not encoded or used in any of the subsequent video analysis. This parameter can range from 1 thorugh 3.

PB quantization. This parameter controls the quantization level for the PB frames. It can range from 2 through 5.

I/P ratio. This parameter controls the number of P-frames transmitted between successive I-frames. It can range from 5 through 40.

Signal Factor

Digitized video image. The signal factor is the pixel valus for the luminance and two chrominance components of the video signal. The signal is measured on a frame-by-frame basis, and a composite SNR is calculated from these measurements.

Response Variable

Reconstructed video image. The response variable is thed pixel values for the luminance and two chrominance components of the reconstructed video signal.

Ideal Function

The effects of the control parameters on video performance were studied using an ideal function that measured the intensities of the reference video image pixels versus the uncompressed, reconstructed H.263 video image pixels for each of the video components (e.g., luminance and the two chrominance components; Figure 3). This ideal function was developed under this project and measures the overall video quality in an objective manner. Previously developed image quality metrics were based largely on the subjective techniques of a juried evaluation system. The ideal function has many advantages over a perceptive evaluation system. First, the measurement is repeatable. Given an image, the ideal function will always return the same measurement for image quality. Under a juried system, the image quality metric can vary considerably, depending on the makeup and composition of the jury. Second, the ideal function takes less time to score than the juried system. Typically, a video sequence of over 1000 frames took only 10 minutes of postprocessing analysis to calculate the quality metric. Third, the ideal function examines and scores each image in the sequence independent of the preceding image. This is an important distinction from a juried system, in which the human jurors' vision can "integrate" variations from frame to frame and miss some of the losses in image quality.

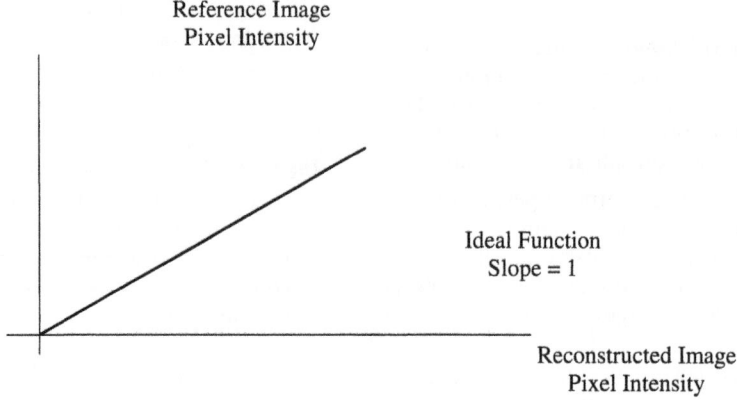

Figure 3
Image quality ideal function

For the ideal function, a perfect match between the input reference and the reconstructed video signals would result in a 45-degree straight line originating at the intersection of the x–y axis. However, since H.263 is a lossy compression scheme, the reconstructed video will deviate somewhat from the reference video, resulting in a degradation of video quality. This deviation increases with increasing video compression.

The SNR is calculated using the incoming video sequence as a reference. Since the incoming video is sampled at 30 frames/s and the video codec drops some of these video frames, the reconstructed video will be replayed at a much lower frame rate. Typically, the image decoder will decompress and display only 10 frames/second, resulting in a loss of video images. To account for the missing video images, the decoder was modified to update the display at a rate of 30 frames/s. "Missing" uncom-

pressed images were displayed using the previous frames decompressed image. This resulted in an effective video display rate equal to the incoming video frame rate. The SN ratios in the experiments were calculated using the effective video signal against the reference video signal at a rate of 30 frames/s.

3. Experimental Layout

Through a series of brainstorming sessions, noise factors and control factors were identified and levels assigned to these parameters, as described below.

Noise Factor Levels

Levels for the noise factors are shown in Table 1. Since there were a limited number of noise factors, a full factorial noise source was created, which used two video sequences. The first video sequence, called *claire,* is a "talking head" type of video with a low amount of both foreground and background movement and spatial details. The lighting condition for this sequence is low and does not vary. The second sequence, called *car phone,* contains a high amount of spatial detail and movement.

Control Factor Levels

Table 2 indicates the control factors, their mnemonics, and the levels used in this experiment.

Table 1
Noise factors and levels

| Noise Factor | Level | |
	1	2
Spatial detail	Low	High
Lighting variation	Low	Medium
Movement	Low	High

Table 2
Control factors and levels

	Control Factor	Mnemonic	Level 1	Level 2	Level 3
A:	unrestricted motion vector	UVecFlag	0(off)	1(on)	
B:	syntax arithmetic coding	SCodFlag	0(off)	1(on)	
C:	advanced prediction	AFlag	0(off)	1(on)	
D:	PB frames	PBFlag	0(off)	1(on)	
E:	double precision DCT	DCTFlag	0(off)	1(on)	
F:	pel search window size	IPelWin	2	7	15
G:	I-frame quantization	IQP	2	10	30
H:	P-frame quantization	PQP	2	10	30
I:	frame skip rate	FSkip	2	3	4
J:	PB quantization	DBQuant	0	1	3
K:	I/P ratio	IPRatio	4/1	19/1	39/1

Experimental Design

Based on the identified control factors, an L_{36} orthogonal array was chosen for the test configuration for the Taguchi experiments. Since there are five two-level factors and six three-level factors, a $2^{11} \times 3^{12}$, an L_{36} array was chosen to hold the parameters and was modified as follows. The first five columns in the standard L_{36} array were used to house the two-level control factors, and columns 12 through 17 were used to house the three-level control factors. Columns 6 through 11 and 18 through 23 were not used in the experimental design. All subsequent calculations were modified as appropriate to take into account the missing columns in the experiment design.

Each set of L_{36} Taguchi experiments was repeated six times, once for each of three video components (Y, U, V) and once for each of the two video scenes (car phone and claire), for a grand total of 216 experiments. In each experiment, the pixel intensity levels and the compression bit rates were measured and recorded. For each set of experiments, a data reduction and slope and SN ratio analysis was performed. Finally, a composite SN ratio analysis was performed for overall video quality and for the bit rate.

The final orthogonal array with the control factor levels is shown in the Table 3.

4. Experimental Results and Analyses

Derivation of Best-Fit Line (Sensitivity)

A data reduction analysis effort was performed on each set of collected data to obtain insight into the relationship between the reference video signal and the reconstructed video signal. The data from each of the Taguchi experiments was fitted to a linear relationship, which passed through the origin, using standard regression analysis. This regression produced a sensitivity (slope) analysis for the design.

Derivation of Composite SN Ratios

After deriving the best-fit sensitivity for each of the experiments, a composite SN ratio was calculated. In forming the composite SN ratio, we have taken into account two observations:

1. The sensitivities of the Y, U, and V components can be adjusted independently.

Table 3
Control factors L_{36} orthogonal array

No.	A	B	C	D	E	F	G	H	I	J	K
1	0	0	0	0	0	1	2	2	2	0	5
2	0	0	0	0	0	7	10	10	3	1	20
3	0	0	0	0	0	15	30	30	4	3	40
4	0	0	0	0	0	1	2	2	2	1	20
5	0	0	0	0	0	7	10	10	3	3	40
6	0	0	0	0	0	15	30	30	4	0	5
7	0	0	1	1	1	1	2	10	4	0	20
8	0	0	1	1	1	7	10	30	2	1	40
9	0	0	1	1	1	15	30	2	3	3	5
10	0	1	0	1	1	1	2	30	3	0	40
11	0	1	0	1	1	7	10	2	4	1	5
12	0	1	0	1	1	15	30	10	2	3	20
13	0	1	1	0	1	1	10	30	2	3	20
14	0	1	1	0	1	7	30	2	3	0	40
15	0	1	1	0	1	15	2	10	4	1	5
16	0	1	1	1	0	1	10	30	3	0	5
17	0	1	1	1	0	7	30	2	4	1	20
18	0	1	1	1	0	15	2	10	2	3	40
19	1	0	1	1	0	1	10	2	4	3	40
20	1	0	1	1	0	7	30	10	2	0	5
21	1	0	1	1	0	15	2	30	3	1	20
22	1	0	1	0	1	1	10	10	4	3	5
23	1	0	1	0	1	7	30	30	2	0	20
24	1	0	1	0	1	15	2	2	3	1	40
25	1	0	0	1	1	1	30	10	2	1	40
26	1	0	0	1	1	7	2	30	3	3	5
27	1	0	0	1	1	15	10	2	4	0	20
28	1	1	1	0	0	1	30	10	3	1	5
29	1	1	1	0	0	7	2	30	4	3	20
30	1	1	1	0	0	15	10	2	2	0	40
31	1	1	0	1	0	1	30	30	4	1	40
32	1	1	0	1	0	7	2	2	2	3	5
33	1	1	0	1	0	15	10	10	3	0	20
34	1	1	0	0	1	1	30	2	3	3	20
35	1	1	0	0	1	7	2	10	4	0	40
36	1	1	0	0	1	15	10	30	2	1	5

Table 4
Y, U, V, composite SN ratio, and BPS results

No.	SNR (dB)				
	Y	U	V	Composite	BPS
1	−17.4	−6.4	−3.8	−15.7	55.0
2	−18.9	−11.1	−9.5	−17.4	44.7
3	−21.0	−14.8	−12.4	−19.7	38.3
4	−17.3	−6.4	−3.9	−15.7	54.1
5	−17.6	−10.4	−9.4	−16.2	43.4
6	−21.6	−14.9	−12.2	−20.2	41.9
7	−19.9	−10.4	−8.4	−18.3	46.1
8	−19.5	−12.5	−10.5	−18.0	41.1
9	−19.2	−10.2	−7.5	−17.6	52.8
10	−19.5	−10.7	−10.0	−17.9	41.9
11	−19.6	−8.7	−6.4	−18.0	51.9
12	−18.6	−11.8	−9.7	−17.2	45.2
13	−19.8	−12.1	−10.2	−18.3	42.2
14	−16.5	−5.3	−4.6	−14.8	51.4
15	−19.7	−8.8	−6.5	−18.1	48.9
16	−19.8	−11.5	−9.6	−18.3	44.8
17	−19.6	−8.6	−6.0	−18.0	51.4
18	−17.8	−9.7	−8.1	−16.3	45.6
19	−18.5	−6.6	−5.2	−16.8	51.0
20	−19.1	−12.5	−10.3	−17.7	47.1
21	−19.9	−11.0	−9.0	−18.3	44.7
22	−20.0	−11.3	−9.3	−18.4	46.0
23	−20.6	−14.8	−12.1	−19.2	41.3
24	−16.3	−4.2	−4.0	−14.7	51.6
25	−18.2	−11.4	−9.7	−16.8	44.9
26	−19.5	−8.8	−6.2	−17.9	50.0
27	−19.5	−8.0	−5.5	−17.8	51.7
28	−19.8	−12.6	−10.3	−18.4	45.9
29	−20.7	−11.3	−9.0	−19.1	43.5
30	−16.9	−6.2	−4.2	−15.2	53.0
31	−21.1	−14.8	−12.3	−19.7	38.1
32	−17.4	−6.4	−3.8	−15.7	54.3
33	−18.9	−11.1	−9.5	−17.4	44.3

Table 4
(Continued)

No.	SNR (dB)				BPS
	Y	U	V	Composite	
34	−18.3	−7.7	−5.4	−16.6	52.4
35	−19.1	−9.9	−8.6	−17.5	43.9
36	−19.0	−11.2	−9.2	−17.5	45.9
Avg.	−19.1	−10.1	−8.1	−17.5	47.0
Min.	−21.6	−14.9	−12.4	−14.7	55.0
Max.	−16.3	−4.2	−3.8	−20.2	38.1
Range	5.3	10.6	8.5	5.5	16.9

2. The sensitivities are static and cannot be changed, depending on the type of image being compressed because the image type is a noise factor.

For each component (Y, U, and V), a straight line with zero intercept was fitted to the scatter plot of the pixel values of the reconstructed image versus the original image. This slope is referred to as β_i, where $i = Y$, U, or Y. Similarly, the variance around this fitted line is σ_i, where $i = Y$, U, or V.

The composite variance was defined as the weighted average of the Y, U, and V component variances, which were normalized for the individual slopes (e.g., make each slope equal to 1) and the number of data points for each component. The composite variance is given by

$$\sigma^2_{comp} = \frac{4\sigma_Y^2/\beta_Y^2 + \sigma_U^2/\beta_U^2 + \sigma_V^2/\beta_V^2}{6}$$

and the composite SN ratio is given by

$$\eta = 10 \log \frac{1}{\sigma^2_{comp}}$$

where η is the SN ratio and σ_{comp} is the composite variance.

Composite Bit-Rate SN Ratio

The second component measured was the bit rate of the compressed video bit stream (BPS). For each experiment the bit rate of the compressed signal was measured and recorded. A composite bit-rate SN ratio was calculated using the relationship

$$BPS = 10 \log \Gamma$$

where BPS is the composite bit rate SN ratio and Γ is the compressed video stream bit-rate (bits/s).

Experimental Results

The results of the Taguchi experiments are shown in Table 4, which lists the results of the SN calculations for the Y, U, V and composite video quality metric and the compressed video stream bit rate (BPS).

Factor Effects Plots

Based on the results of the Taguchi experiments and data reduction efforts, factor effects plots were generated for the composite image quality and BPS SNR. These results are shown in Figures 4 and 5.

5. Factor Effects Summary

Since the main objective of this project was to optimize video quality, an examination of the composite video quality and BPS factor effects plot is appropriate. An examination of Figures 4 and 2 indicates that only four of the 11 control factors significantly affect image quality and compression: I-frame quantization level (IQP), P-frame quantization level (PQP), frame skip rate (FSkip), and the I/P frame ratio (I/P ratio). The results of an ANOVA analysis (Table 5) reveals that these four factors account for over 92% of the total sum of

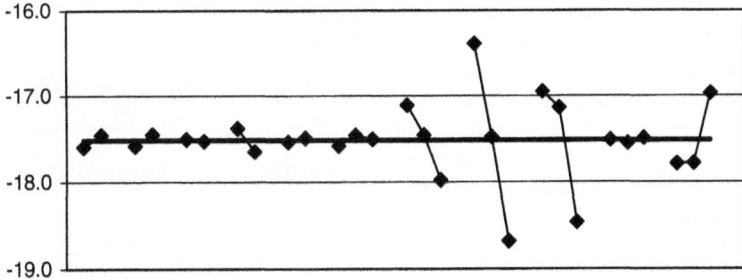

Figure 4
Composite image quality factor effects

squares for video quality and 94% of the total sum of squares for the BPS data.

The effect of the P-frame quantization parameter, PQP, is the strongest among all the control factors. For I-frame and P-frame quantization, the factor effects plot indicate that the smaller index values signify more levels of quantization, thereby increasing image quality. However, this increase in image quality does not come without an addition cost, as indicated from the BPS factor effects plot. That is, an increase in the number of quantization levels increases both the image quality and the compressed video bit stream. That is, for these parameters, an increase in image quality leads to a significant increase in bandwidth. A cost–benefit analysis is needed to arrive at the best values for these parameters. The factor effects plot shows that PQP has a greater effect on image quality than IQP. This is because the P-frame quantizer operates on the residuals of pixel intensities, while the I-frame quantizer operates on absolute pixel intensities.

Frame skip (FSkip) has the second-largest effect on image quality, with level 1 being the most desirable. This level implies that only alternating frames should be compressed and transmitted. Level 2 provides only a small degradation in image quality, while level 3, which corresponds to transmitting every fourth frame, causes a significant drop in image quality. An examination of the BPS factor effects plot indicates that the effect of FSkip on BPS is small, signifying that very little needs to be spent, in terms of bandwidth, to achieve significant gains in image quality.

A further examination of the slope of the FSkip parameter suggests an alternative scheme in which no frames are skipped, and the video codec compresses every image. If the factor effects plots were extrapolated from their present values, the image quality factor effects plot suggests a significant increase in image quality, while the BPS factor effects plot suggests a small increase in bandwidth because of it low slope.

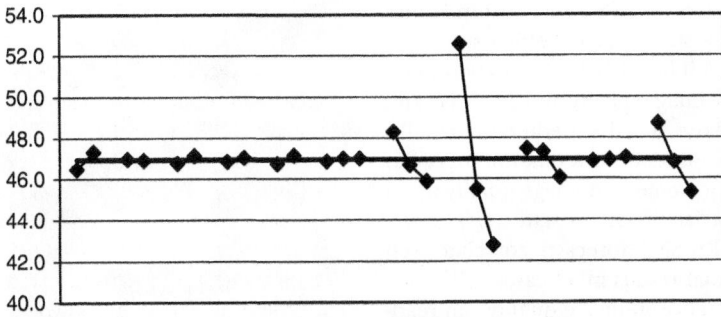

Figure 5
BPS factor effects

The data also suggest that larger values of I/P ratio are more desirable. It should be noted that this is the only win–win factor (i.e., larger values of the I/P ratio reduce BPS and at the same time increase image quality). Larger values of the I/P ratio should be investigated to confirm that this trend is accurate. However, I believe that there comes a point of diminishing returns, especially in noisy environments. Transmission errors may have a major interaction with the I/P ratio. Therefore, this factor should be studied when a simulation with transmission errors becomes available.

6. Video Parameter Design and Confirmation

To study the effects of the four major contributors to the SN ratio performance (e.g., IQP, PQP, FSkip, and I/P ratio), an L_9 confirmation experiment was run. In this experiment, the following factors were set for all nine experiments: A_1, B_1 C_1, D_1, E_1, F_2, and J_2. Also, the confirmation experiment used the same video sequences as the L_{36} experiment. In addition, the control factors were set as shown in Table 6.

Table 7 shows the results for each of the nine experiments plus the baseline experiment. This table shows both experimental results and the forecast for all nine experiments, plus the baseline configuration. Also included is an "optimal" image quality experiment, which consisted of factors A_1, B_1, C_1, D_1, E_1, F_2, G_1, H_1, I_1, J_2, and K_3.

7. Discussion

The factor effects plots for BPS are shown in Figure 6. This plot shows a similar factor effect sensitivity behavior, as was exhibited by the L_{36} experiment. The predictions for image quality and BPS SN ratio were confirmed by the actual experimental results. The experimental results confirm that the users can expect a 2.5-dB improvement in image quality when using the optimum factor effects settings. The image quality and BPS SNR forecasts correlate well with the experimental results in all cases.

Finally, the difference in image quality can readily be seen in Figure 7. This figure shows one image

Table 5
ANOVA results

Factor	Composite SN Ratio as % Sum of Squares	BPS SN Ratio as % Sum of Squares
UVecFlag	0.0	0.3
SCodFlag	0.3	0.0
AFlag	0.0	0.2
PBFlag	1.1	0.1
DCTFlag	0.0	0.2
IPelWin	0.1	0.0
IQP	7.3	4.7
PQP	50.3	78.8
FSkip	26.0	1.9
DBQuant	0.0	0.0
IPRatio	8.5	8.9

Table 6
Confirmation experiment orthogonal array

No.	A (IQP)	B (PQP)	C (FSkip)	D (IPRatio)
1	2	2	2	5
2	2	10	3	20
3	2	30	4	40
4	10	2	3	40
5	10	10	4	5
6	10	30	2	20
7	30	2	4	20
8	30	10	2	40
9	30	30	3	5
B	10	10	3	20

Table 7
Confirmation experiment results summary

	Composite SNR (dB)		BPS SNR (dB)	
No.	**Measured**	**Forecast**	**Measured**	**Forecast**
1	−15.7	−15.7	55.9	55.3
2	−17.2	−17.0	46.6	46.2
3	−18.9	−18.7	41.0	40.8
4	−14.7	−15.4	50.1	50.2
5	−18.5	−18.6	47.0	45.2
6	−18.3	−18.3	43.4	42.0
7	−17.9	−18.1	52.9	49.6
8	−16.8	−16.8	45.1	42.5
9	−19.8	−19.0	43.7	43.0
Optimum	−14.5	−14.9	53.0	52.8
Baseline	−17.6	−17.3	44.8	44.6

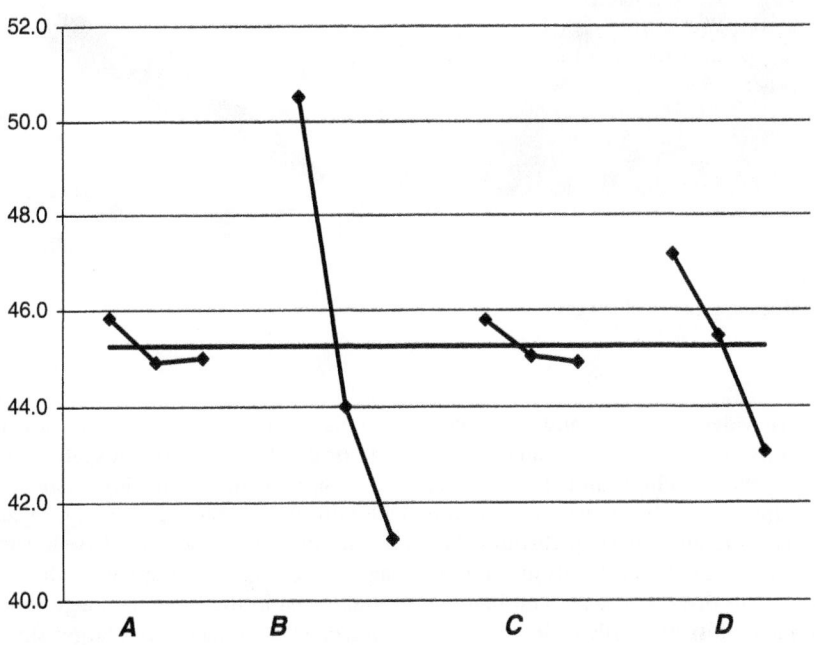

Figure 6
BPS factors effects

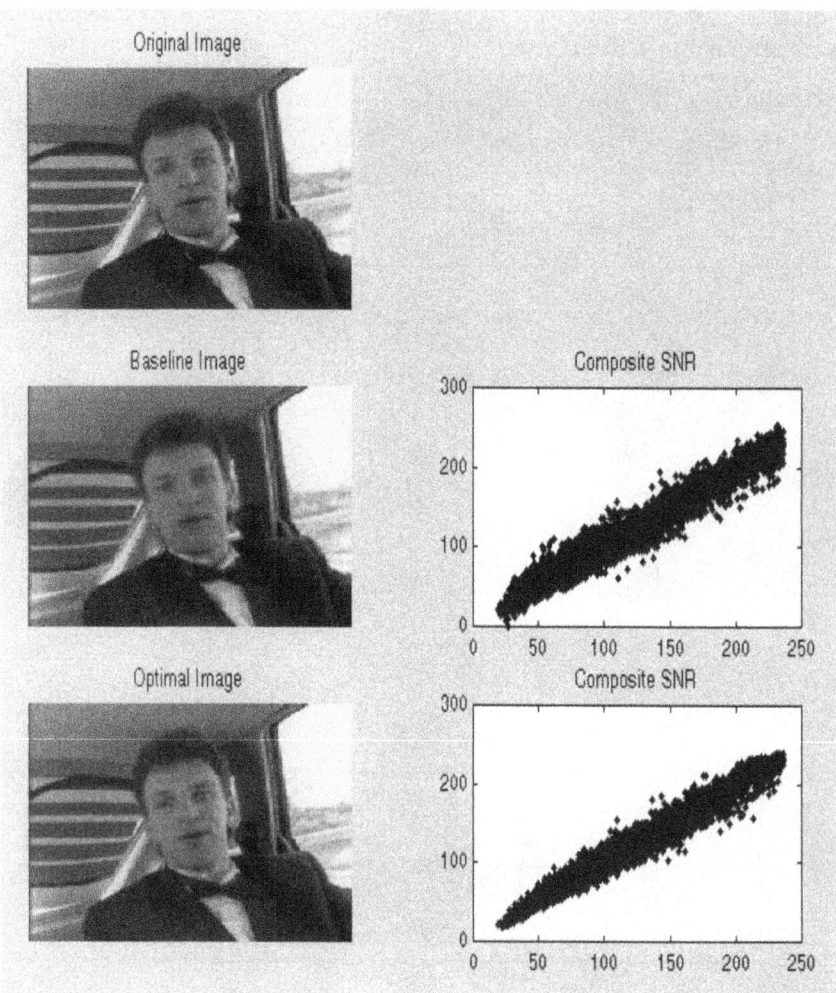

Figure 7
Verification experiment images

from the original video sequence and the reconstructed images from two of the validation experiments (e.g., the baseline and the optimal experiments). This figure shows that the baseline image suffers a considerable image quality loss. This can readily be seen in the facial details in the baseline image versus the original image. The baseline image suffers quality loss primarily in loss of facial details. In the optimal image sequence, the facial

details are much improved and are very similar to the original image. Loss of spatial details can result in a significant quality loss, especially in combat identification systems. The figure also shows the ideal function for both the baseline and optimal images. Once again, it is apparent that there is a signal quality loss in the baseline image versus the optimal image. This shows up as a larger signal variance on the composite SNR graphs for these images.

8. Conclusions

In conclusion, the objective video scoring methods developed under this project, in conjunction with robust design techniques and Taguchi methods, have shown to be an unbiased, cost-effective approach to parameter design for the H.263 video codec. Using these techniques, an improvement of 2.3 dB was observed over the baseline approach in terms of the image quality metric derived. The analysis in this paper demonstrates the importance of applying robust design techniques early in the design process.

Additionally, the image quality SN ratio metric developed in this project was independent of any compression technologies and can be applied to a wide range of still and motion video codecs. Through the use of an objective quality metric, the time and costs involved with evaluating and quantifying video codecs have been reduced significantly.

Acknowledgments The author expresses his sincere thanks to Sessan Pejan of Sarnoff Research Labs, Princeton, New Jersey, for his help and expertise in video processing.

Reference

ITU-T, 1996. *Video Coding for Low Bit Rate Communications*, ITU-T Recommendation H.263.

This case study is contributed by Edward Wojciechowski and Madhav S. Phadke.

Robust Optimization of a Real-Time Operating System Using Parameter Design

Abstract: In this study, a series of L_8 parameter design experiments provide the strategy for selecting the real-time operating system (RTOS) features that will provide optimal performance. Since these parameters were selected specifically for the radio controller architecture and used a realistic test bed, the final product will behave in a predictable and optimal fashion that would not be possible without using Taguchi techniques. This application of Taguchi techniques is not unique to a radio controller but rather can be used with any RTOS and architecture combination. These experiments can form the template for any software engineer to enhance RTOS performance for other applications.

1. Introduction

ITT Industries A/CD is a world leader in the design and manufacture of tactical wireless communications systems. ITT was awarded a major design contract for a very high frequency (VHF) radio. This radio requires a complex controller. As a result of a trade study, the CMX real-time operating system (RTOS) was selected as the scheduler for the radio controller. The CMX RTOS provides many real-time control structure options. Selecting these structures based on objective criteria early in the design phase is critical to the success of this real-time embedded application.

2. Background

Although the RTOS is a commercial off-the-shelf (COTS) product, there are still many features and options to select to achieve an optimal result. The features and options selected are highly dependent on the system architecture. For this reason, a param-

eter design was undertaken to optimize the selection of these factors.

The VHF radio combines a collection of independent concurrent processes that run using preemptive priority scheduling. In a real-time system, events enter the system through the hardware interrupt. When a hardware interrupt occurs, the software is interrupted automatically. The software services this interrupt through the use of an interrupt service routine (ISR). Once the interrupt is serviced, program control is returned to the processing point prior to the interrupt. The software architecture will rely on the communication between various tasks and ISRs to control the radio.

RTOS vendors typically provide a set of generic communications structures. These structures are then combined to form systems and architectures. Vendors strive to provide a varied set of functions to increase market dominance. Three common communications requirements are (1) ISR to task, (2) task to task, and (3) task to hardware. The system architecture requires these three mechanisms. In each case, there are two operating system constructs for achieving acceptable interprocessor communication. Each is implemented differently and will therefore affect the system performance differently. To choose the correct constructs, a parameter design is required. Another major concern is the speed at which the processor operates. The faster the processor runs, the more power it draws. It is also critical that the RTOS behave predictably over all processor speeds. Therefore, it is important to choose an experiment that combines the different interprocess communications structures and processor speed.

Figure 1 shows how events are processed in the software architecture. The VHF radio accepts commands and traffic events. Each type of event enters the system via an ISR. This ISR is a low-level assembly language routine that is invoked when a hardware interrupt register state changes. The event, having been registered, must be passed to the tasks that will process the message. There are four differ-

ent types of message handlers: traffic, background commands, immediate commands, and posttraffic commands; each interacts and behaves differently. Background commands are handled when traffic is not communicating actively with the hardware. As the name implies, posttraffic commands wait until traffic is complete before running. Immediate commands are special tasks that terminate the current traffic and then execute.

Comparing the radio controller architectural requirements to RTOS features results in the following:

Communication Requirement	*RTOS Capability Meeting Requirement*
ISR to task	Mailbox versus queue
Task to task	Mailbox versus events
Task to hardware	Semaphore versus resource

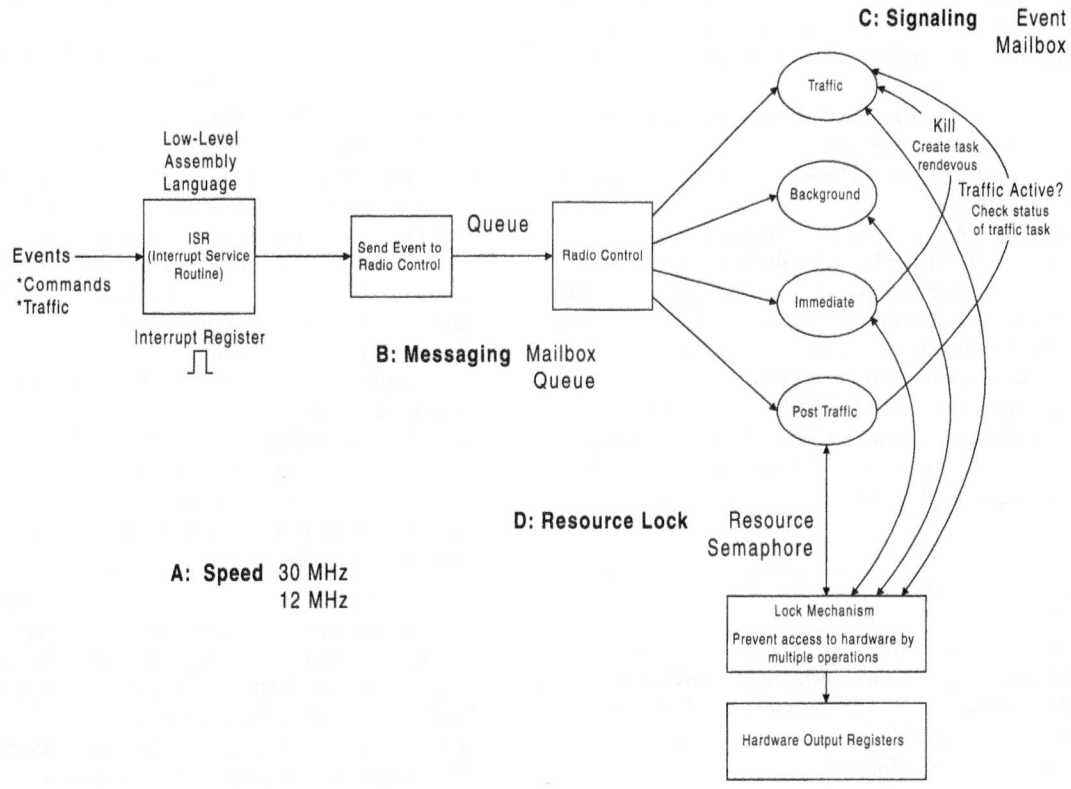

Figure 1
Event processing in the controller architecture

As part of the design phase, functional flows were constructed for every radio control and traffic operation. These functional flows focused on how the architecture processed commands. The ultimate test of the architecture is to handle each functional flow efficiently. Figure 2 shows an example of the volume functional flow.

There is a direct correlation between the functional flow and the architecture. The ISR can be seen as the input mechanism that starts a thread. The queue is represented by the queue column on the functional flow. It queues up requests going to radio control. The signaling mechanism in the architecture is shown in the queue request. Volume is a benign event that is performed as a background process by radio control. The resource lock is shown in the radio control column of the thread. It indicates where the hardware must be locked to prevent hardware contention. If contention were allowed to occur, the radio would lock up, thus failing catastrophically. Each of the primary communications mechanisms is identified clearly in the flow: messaging, signaling, and resource lock. The architecture shown in Figure 1 implements these mechanisms.

Since events are constantly being passed to radio control at a variable rate, they must be queued so that they are not lost. The RTOS provides two mechanisms for passing these events to radio control: mailboxes and queues. The various processing tasks must communicate between themselves to send terminate messages or wait for traffic to be complete. Interprocess communication can be achieved using mailboxes or events. Since each of the processing tasks must have exclusive access to hardware output registers, a lock must be implemented so that there is never simultaneous access to the output registers by more than one task. This can be accomplished by using semaphores or by defining a resource.

3. Objectives

The objective of this experiment was to determine the correct parameters, which will give maximum RTOS performance for the embedded application. The parameter diagram is given in Figure 3. The *noise factors* are as follows:

❑ *Interrupt mix:* The mixture of different types of interrupts entering the system

❑ *Interrupt rate:* The rate at which interrupts are entering the system

The control factors are as follows:

❑ *Processor speed.* This is the speed at which the processor is set. The radio design is trading off the performance at two speeds, 12.8 and 25.6 MHz (12- and 30-MHz oscillators are available and will be used for the parameter design)

❑ *Messaging.* The system architecture is based on the passing of messages between tasks. Requests enter the system via interrupts which are transformed into message requests, which are acted upon by the system. The two types of messaging schemes being investigated are mailboxes and queues. *Mailboxes* allow a task to wait for a message returning a pointer to the message passed. Any task can fill a mailbox through the use of a mailbox ID. Mailboxes are implemented as a first in, first out. Mailboxes can have multiple ownership. *Queues* are circular first in, first out and last in, first out with up to 32k slots. Slot sizes remain constant and are fixed at compile time. There are many manipulation functions for queues.

❑ *Signaling.* There are several signals, which must be implemented in the radio controller. An example is for an event to send a terminate signal to the current-running traffic thread. The two methods for efficiently sending messages to tasks are events and mailboxes. *Events* use a 16-bit mask to be used to interrupt a task or allow a task to wait for a specified event or condition to be present. It should be noted that this limits the design to 16 signals per task. *Mailboxes* can be used to allow a task to wait for and then check the message when it is received. This allows more information to be passed.

❑ *Resource locking.* A key problem with previous controllers was the inability to block competing operations from seizing a resource. This is a particular problem between the controller and the digital signal processors (DSPs). If a second command is sent to the DSPs before the current one is acted upon, the system can, and often will, lock up. By using a resource locking mechanism, this detrimental behavior

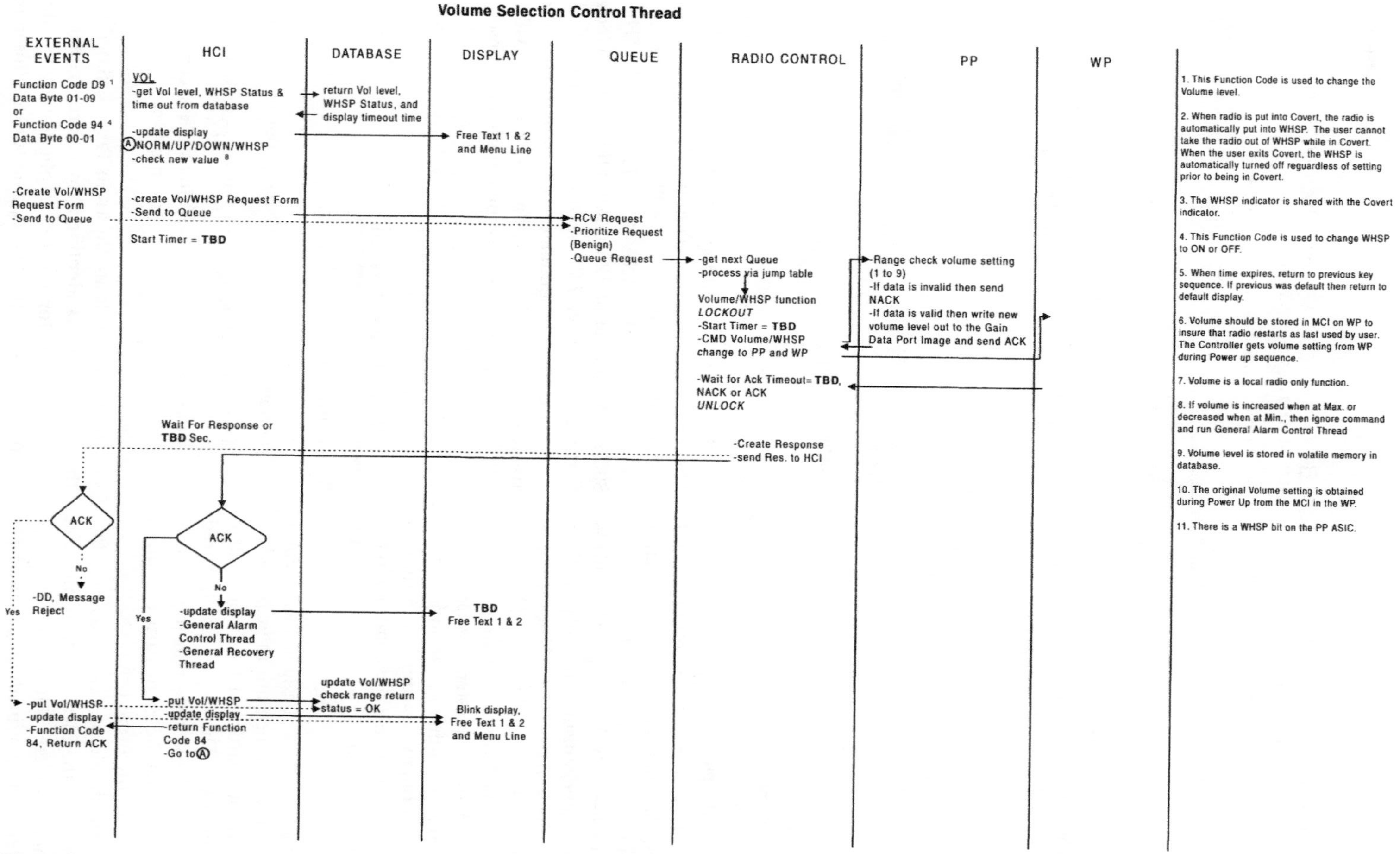

Volume Selection Control Thread

1. This Function Code is used to change the Volume level.

2. When radio is put into Covert, the radio is automatically put into WHSP. The user cannot take the radio out of WHSP while in Covert. When the user exits Covert, the WHSP is automatically turned off regaurdless of setting prior to being in Covert.

3. The WHSP indicator is shared with the Covert indicator.

4. This Function Code is used to change WHSP to ON or OFF.

5. When time expires, return to previous key sequence. If previous was default then return to default display.

6. Volume should be stored in MCI on WP to insure that radio restarts as last used by user. The Controller gets volume setting from WP during Power up sequence.

7. Volume is a local radio only function.

8. If volume is increased when at Max. or decreased when at Min., then ignore command and run General Alarm Control Thread.

9. Volume level is stored in volatile memory in database.

10. The original Volume setting is obtained during Power Up from the MCI in the WP.

11. There is a WHSP bit on the PP ASIC.

WP

PP

Range check volume setting (1 to 9)
- If data is invalid then send NACK
- If data is valid then write new volume level out to the Gain Data Port Image and send ACK

RADIO CONTROL

- get next Queue
- process via jump table
Volume/WHSP function
LOCKOUT
- Start Timer = **TBD**
- CMD Volume/WHSP change to PP and WP
- Wait for Ack Timeout= **TBD**, NACK or ACK *UNLOCK*

- Create Response
- send Res. to HCI

QUEUE

RCV Request
Prioritze Request (Benign)
- Queue Request

DISPLAY

Free Text 1 & 2 and Menu Line

TBD
Free Text 1 & 2

Blink display,
Free Text 1 & 2 and Menu Line

DATABASE

return Vol level, WHSP Status, and display timeout time

update Vol/WHSP
check range return
status = OK

HCI

VOL
- get Vol level, WHSP Status & time out from database
④*NORM/UP/DOWN/WHSP*
- update display
- check new value [8]

- create Vol/WHSP Request Form
- Send to Queue

Start Timer = **TBD**

Wait For Response or TBD Sec.

ACK — No

Yes

- update display
- General Alarm Control Thread
- General Recovery Thread

- put Vol/WHSP
- update display
- return Function Code 84
- Go to ⑩

EXTERNAL EVENTS

Function Code D9 [1]
Data Byte 01-09
or
Function Code 94 [4]
Data Byte 00-01

- Create Vol/WHSP Request Form
- Send to Queue

ACK — No

Yes

- DD, Message Reject

- put Vol/WHSP
- update display
- Function Code 84, Return ACK

Figure 2
Volume functional flow

1327

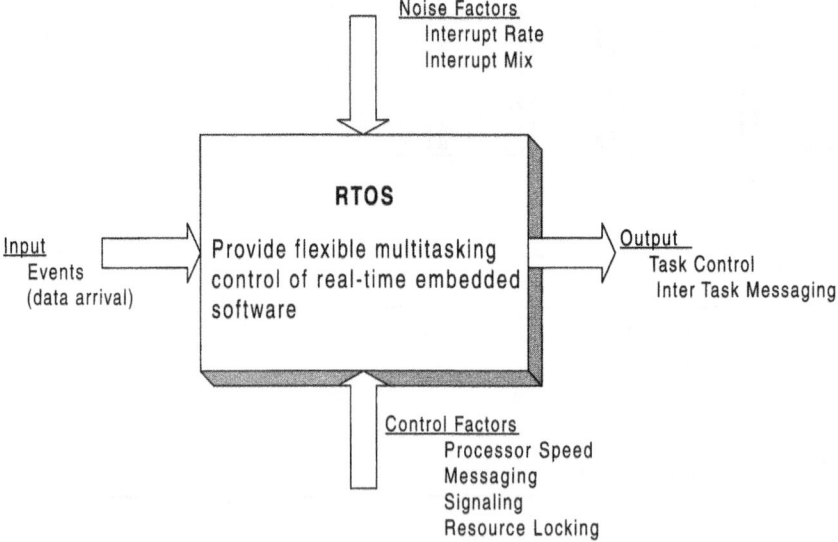

Figure 3
RTOS parameters

can be avoided. The two types of resource locking being investigated are resources and semaphores. *Resources* are RTOS commands that allow a resource to be locked, preventing simultaneous use by competing tasks. This is a binary-state mechanism that can provide a lock and unlock state. The highest-priority task requesting a resource is granted the resource. The locked procedure then inherits the priority of the requester. *Semaphores* are implemented as counting semaphores, which can be used as binary semaphores. Although more flexible because they are counting in nature, our design needs to prevent all secondary access.

The input signal is a series of events. Events enter the system as interrupts, which are then parsed to waiting tasks to execute particular traffic or control threads. The input is t_{ISR}. The *output* was measured as a series of times: (1) the time that the event is queued, (2) the time that the event starts execution and (3) the time that the event completes. Ideally the system should process messages as fast as possible. Therefore, the ideal system is one that has no latency and completes work in the minimum time. The smaller-the-better method was used for the analysis.

All tests were run using CMX RTOS version 5.2. The RTOS was run on a Pentium PC with a serial Nohau Emulator Model 51XA-PC and a companion trace card EMUL51XA-PC/IETR 128-25. The 80C51XA was run at 12 and 30 MHz. The HiTech C compiler version 7.73 was used to compile and link the code. The emulator was configured for a 16-bit bus and 1-Mg code/data. The default CPU wait state of four or five instruction cycles was used.

4. Experimental Layout

Test conditions are as follows:

Average case:	7 times per sec
Worst case with margin:	30 times per second
Failure point:	N times per second (no idle time)

Control factor levels are shown in Table 1. Based on the control factors identified, an L_8 orthogonal array was chosen for the test configuration for the Taguchi experiments. The L_8 2^4 array was chosen to hold the four two-level factors (Table 2).

Table 1

Control factors and levels

Control Factor	Level	
	1	2
Speed (MHz)	12	30
Messaging	Mailbox	Queue
Signaling	Event	Mailbox
Resource lock	Resource	Semaphore

5. Experimental Results and Analysis

The data were analyzed using a smaller-the-better method. The signal-to-noise (SN) ratio was computed using

$$SN = -10 \log \frac{\sum (t_{end} - t_{ISR})^2}{N}$$

The SN ratio quantifies the composite latency of all message types that the system is expected to handle. The delta time includes from the time the event enters the system until all processing is completed and the thread ends. This includes the time that the event is in the queue, waiting on other events to obtain a resource lock, and the time required to communicate with other tasks (i.e., terminate or wait message times).

Timing measurements were made of individual messages. The times captured were time entering the system, t_{ISR}; time message queued, t_{queue}; time processing starts, t_{start}; and time the processing ended, t_{end}.

The time delta that proved most useful was $t_{end} - t_{ISR}$, because it includes the time significant for all factors. The time between entering the system and being queued was in all cases insignificant, proving that the RTOS was handling this efficiently. The time between a message being queued and starting processing focuses on factor B, the type of queuing. Similarly, the time between starting and ending processing focuses on factors C and D. These times were used to validate the conclusions that were made. Therefore, the overall time delta was used since each factor is then represented by the experiment.

These measurements were made using message rates of 7 and 30 times per second. This proved to be the most valuable data collected, since individual threads as well as composite characteristics could be examined. Table 3 lists the results of the SN ratio calculations for 7 and 30 events per second.

Based on the results of the Taguchi experiments and the data reduction, the following factor effects plots were calculated for 7 and 30 events per second. The components of the factor effects plots are obtained by averaging the SN ratio values for each factor as specified in Figure 4. Examining the factors relating to speed, A_1 and A_2 are the averages of experiments 1 to 4 and 5 to 8, respectively.

Table 2

L_8 design experiment

No.	Speed	Messaging	Signaling	Resource lock
1	12	Mailbox	Event	Resource
2	12	Mailbox	Semaphore	Semaphore
3	12	Queue	Event	Semaphore
4	12	Queue	Semaphore	Resource
5	30	Mailbox	Event	Semaphore
6	30	Mailbox	Semaphore	Resource
7	30	Queue	Event	Resource
8	30	Queue	Semaphore	Semaphore

Table 3

SN ratio for $t_{end} - t_{ISR}$

No.	SN 7	SN 30
1	−27.7	−27.7
2	−27.7	−27.7
3	−32.0	−32.3
4	−32.0	−32.6
5	−21.7	−21.8
6	−21.8	−21.8
7	−28.7	−28.4
8	−28.5	−28.3

consequential. It appears the RTOS treats each of these methods in a way that guarantees similar performance. These factors can be chosen for ease of design and maintenance.

It is surprising that factor B is more significant than A. It was expected that clock speed was going to show strong dominance over the other factors. What this experiment showed is that the choice of how messages are passed is even more significant to performance than is speed. Choosing the wrong messaging type would totally eliminate the performance gained by a faster processor. Without the experiment, this parameter would have been chosen out of convenience, and significant performance might have been sacrificed.

Factor Effects Summary

The overall objective of this parameter design was to optimize RTOS performance. Examining Figure 4 shows almost identical results for both 7 and 30 events per second. Choosing the correct parameters for 30 events per second will yield a design that is robust over the entire region of performance expected.

The factor effects plot shows that processor speed and messaging are the dominant factors and very comparable in amplitude. Therefore, choosing 30 MHz and using mailboxes for messaging will provide the optimal design.

Further examination of the parameter design shows that signaling and resource locking were in-

6. Parameter Design Confirmation

Since parameters C and D are nondominant factors that can be chosen for other design considerations rather than for system optimization, the confirmation can be performed with the data already collected as part of the experiment. Choosing A_2 (30 MHz) and B_1 (mailboxes), one would expect to see a little more than a 10-dB performance increase over choosing A_1 (12 MHz) and B_2 (queues). Table 4 identifies the two scenarios.

As expected, the confirmation results were extremely consistent with the results expected. Since A and B are the dominant factors and C and D show little effect, the internal confirmation yielded positive confirmation that the factors cho-

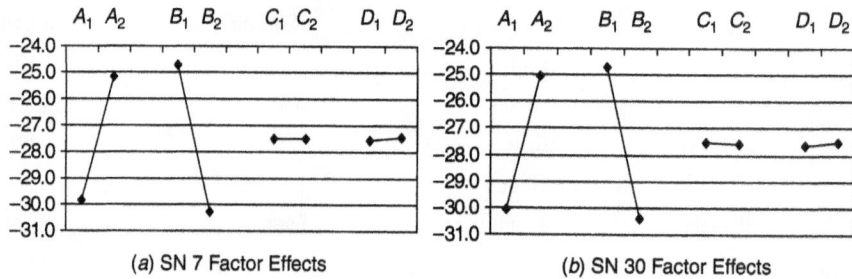

(a) SN 7 Factor Effects   (b) SN 30 Factor Effects

Figure 4

Factor effects for $t_{end} - t_{ISR}$

Table 4

Design confirmation using factors *A* and *B*

Confirmation Check	7 Events per Second	30 Events per Second
A_2 (30 MHz) and B_1 (mailboxes)	−21.7	−21.8
A_1 (12 MHz) and B_2 (queues)	−32.0	−32.5
Delta	10.3	10.7

sen are the correct factors to optimize the system properly.

7. Performance Analysis/System Stress Test

Flexibility, modifications, and engineering changes characterize the nature of software engineering systems. Software is commonly seen as easily changed compared to hardware, which requires radical modification. As a result, software is seen as the answer to increased product life, longevity, and product enhancement. The life cycle of software is often measured in months. Our parameter design focused on the current expected needs of the VHF radio. Optimum RTOS parameters were chosen that provide excellent performance within the expected range and include design margin.

It is the job of the software engineer to answer additional questions that affect the quality and products being designed. These questions aim at the heart of the reason that products become obsolete and require major upgrades prematurely. Margin is the ability of a system to accept change. Far too often, margin is measured as a static number early in the design phase of a project. Tolerances are projected and the margin is set. This answer does not take into account the long-term changes that a system is likely to see toward the end of its life or the enhanced performance required to penetrate new markets.

Additional information needs to be gathered. How will the system perform when stressed to its limits? Where does the system fail? Figure 5 shows how this system will perform past the expected limit of 30 events per second. The system was set to use mailboxes, events, and resources and run at 12 and 30 MHz. The message rate was increased steadily. The real simple syndication (RSS) of the event time was computed and plotted versus the message rate. Clearly shown in the figure is the fact that 12 MHz will lead to a much shorter product life since the performance cannot be increased significantly, and the steep failure curve. However, 30 MHz shows a system that can almost triple its target event rate and still maintain acceptable performance. The shallow assent of the event processing time RSS indicates that the system will behave more robustly. The 30-MHz system will degrade gracefully rather than catastrophically.

Initially, it was expected that the two graphs would have an identical shape displaced by the 2.5 times speed increase. This is not the case because the real-time interrupts controlling RTOS scheduling operate at a constant rate of 1 millisecond. Since the timer rate is constant, it has a larger effect on the slower clock rate than on the faster clock. This

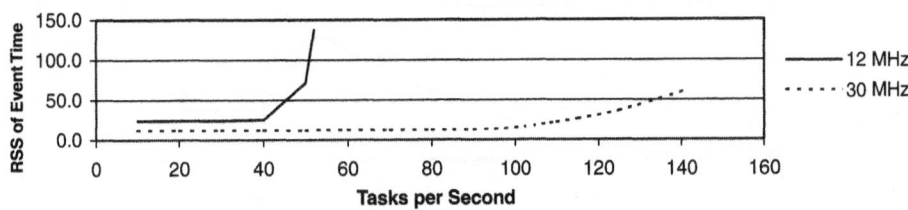

Figure 5

Performance versus event rate

is because the percentage of the processor that is required to process the timer for the 12 MHz is greater than for the 30 MHz. Therefore, when the speed is increased to 30 MHz, not only does the processing power increase but the percentage of the processor available to execute the threads also increases. Ultimately, this provides for a more robust system capable of performing more work to the user at 30 MHz than at 12 MHz.

Using this information, the software design engineer is equipped with one more piece of information from which a cost–benefit analysis can be performed. The cost of the additional part and power must be compared objectively against the benefit to the user in longevity of the radio and performance. The RSS of the event time is significant because, as it increases, the performance degrades to a point where the system is unacceptable. The user will become aware of this when the event time increases past 50 ms.

8. Quality Loss Function

The performance analysis solely looks at performance but fails to account for the cost. The quality loss function provides the software engineer with a way to evaluate objectively the cost to the user of the various clock speeds (Figure 6). The user wants a long-lasting radio that will meet the growing needs of the armed forces. If the radio ceases to meet military needs for expanding services, it will have to be

replaced. This could result in a negative customer perception of ITT's radios. Additionally, this might result in customer inconvenience and damaged reputation for the company, which opens the door to increased competition, affecting market share. In addition, there is the financial impact of upgrading the processor to meet the requirements. All of these factors affect quality loss. Taguchi summarized this by saying: "The quality of a product is the (minimum) loss imparted by the product to the society from the time the product is shipped."

Consider the traditional way of viewing quality costs. If the radio meets the specifications that were provided, it has the desired quality. This view does not take into account that a product specification is merely a summary of the known requirements and uses and cannot begin to capture the future military needs. The quality loss function quantifies the quality loss in a formal manner.

The quality loss function is expressed as

$$\text{loss} = \frac{A_0}{\Delta_0^2} t^2 = 0.18 t^2$$

where A_0 is the cost of replacing the controller module and Δ_0 is the response time that 50% of users would find unacceptable. In this case the quality loss function represents the cost of the quality loss as the response time grows. This occurs because the user becomes dissatisfied with the product as the response time increases. Table 5 shows the quality loss for the current system. The message traffic and re-

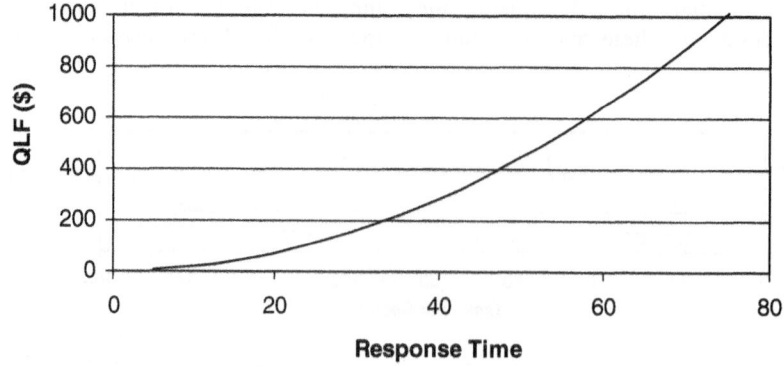

Figure 6
Quality loss function

Table 5

Current system

	Events per Second			
	7 (50%)	10 (40%)	20 (10%)	Quality Loss
12 MHz	24	24	24	$103.68
30 MHz	12	12	12	25.92

sulting loss expected were computed. Table 6 provides the same information for a future system with anticipated growth in processing functions. The quality loss is computed by summing the quality loss at each event rate times the percentage at that rate. As an example, the quality loss of the future system at 12 MHz is

$$\$241.29 = (0.5)[(0.18)(24^2)] + (0.4)[(0.18)(35^2)] + (0.1)[(0.18)(75^2)]$$

These data provide another view of the cost–benefit trade-off:

Current system: $77.76

Future system: $215.37

from which we can evaluate the present and future effectiveness of changing the processor speed. Clearly, if we look only at the current system, $77.76 becomes the cost breakpoint for upgrading the clock to 30 MHz. Upgrading becomes much more attractive when looking at the cost benefit for the future system. The future system provides a $215.27 cost breakpoint, which is a significant percentage of the total controller subsystem cost. The quality loss

Table 6

Future system

	Events per Second			
	20 (50%)	40 (40%)	50 (10%)	Quality Loss
12 MHz	24	35	75	$241.29
30 MHz	12	12	12	25.92

of the future system makes a compelling argument for upgrading to a 30-MHz processor now. This will return significant dividends in customer satisfaction and long-term market share.

It is important to note that had queues been chosen, there would have been a significant quality loss. Although it costs nothing to change to mailboxes, it results in a tremendous gain when amortized over 20,000 or 30,000 units.

9. Conclusions

In conclusion, applying robust design techniques and Taguchi methods to the selection of RTOS features for the controller architecture has yielded an optimal selection of key communication structures. These techniques allowed for an optimal design using objective methods for parameter selection. In addition to gaining valuable insight into the RTOS internal operation, several key observations were made and problems discovered early in the design cycle. This study demonstrates that robust design techniques and Taguchi methods can be applied to software engineering successfully to yield optimal results and replace intuition and guesswork. This technique can be applied to all types of operating environments. Although this application used an embedded RTOS, it was applied successfully to the UNIX operating system.

Taguchi's techniques have not only allowed for selection of optimum performance parameters for the VHF radio but have provided a tool for expanding the functionality of this radio and future product lines. The performance analysis and the accumulation analysis provided valuable insight into the robustness of future enhancements. Looking at the quality loss function, a designer can objectively predict how this system will behave under increased workloads. The accumulation analysis shows that the parameter selected will behave optimally even at the system failure point. This gives added confidence that the correct parameters have been chosen not only for this immediate project but also for future projects.

Additionally, it was learned that the RTOS exhibits catastrophic failures in two ways when the queues become full. One type of failure is when requests continue to be queued without the system performing any work. This is observed by the idle time increasing with tasks only being queued and not executed. The other type of failure is when the queue fills and wraps around, causing requests to become corrupted. By performing these experiments during the design phase, these failures can be avoided long before code and integration.

Finally, it was observed that when events have properties that couple them, as seen when one event waits on the result of another event to complete, the dominance of processor speed diminishes at the slower event rates. This confirms that the system would respond well to processor speed throttling at low event rates.

Using these optimization techniques resulted in a 10.7-dB improvement in system response time by using the optimal parameter set. Using optimal parameters reduces the message latency from 40.6 ms to 12.1 ms, a decrease of 28.5 ms. A performance analysis and quality loss function showed that there are significant cost and performance benefits in choosing 30 MHz over 12 MHz for both the current system and future systems. The quality loss for the current and future systems is $78 and $215, respectively, since the added cost of changing to a 30-MHz processor is minor, as compared to the quality loss it is a highly desirable modification. Finally, getting the equivalent power of a 30-MHz processor, with speed throttling during idle, will require only a 1% power increase with the same board area.

This case study is contributed by Howard S. Forstrom.

CASE 84

Evaluation of Capability and Error in Programming

Abstract: In assuring software quality, there are few evaluation methods with established grounds. Previously, we have been using a trial-and-error process. One of the reasons is that unlike the development of hardware that machines produce, that of software involves many human beings who are regarded to have a relatively large number of uncertain factors. Therefore, the Sensuous Value Measurement Committee was founded in the Sensory Test Study Group of the Japanese Union of Scientists and Engineers, which has been dealing with the application of quality engineering to the foregoing issues. This research, as part of the achievement, focuses on applying quality engineering to a production line, including human beings with many uncertain factors.

1. Introduction

Because of diversity in production lines of software, it is quite difficult to perform all evaluations at the same time. Figure 1 summarizes the flow of a software production line. We focused on software production that can be relatively easily evaluated. Figure 2 shows the details of software production. By application of the standard SN ratio, we can obtain a good result [4]. On the basis of the experiment, delving more deeply into this issue, we attempted to evaluate the relationship between capability in programming (coding) and errors.

2. SN Ratio for Error Occurrence in Software Production

We created software from specifications. That is, correctness of software in a coding process is our focus. First, we handed specifications and a flowchart to a testee and asked him or her to write a program. When the testee had completed a program that was satisfactory to him or her (or whose framework is completed), he or she saved the program (data 1). Next, compiling this with a compiler for the first time, the testee debugged it by himself or herself. After this program was checked by an examiner, a program with almost no bugs (data 2) was obtained as a final output. Then, by comparing data 1 and 2, we recognized a difference of code between them as an error:

Line number in program: 1   2   3 $\cdots$ N

Representation of correct data: y_1 y_2 y_3 $\cdots$ y_n

$y = 1$ for a correct line and 0 for an incorrect line.

The analysis procedure (calculation of the SN ratio) is as follows [2]. We compute the fraction of the number of correct codes in data 1 to that of lines in data 2, n:

$$p = \frac{y_1 + y_2 + \cdots + y_n}{n} \tag{1}$$

Because y takes up only 0 and 1, the total variation is

$$S_T = y_1^2 + y_2^2 + \cdots + y_n^2 = np \tag{2}$$

According to the variation in equation (1), the effect by the signal factor results in

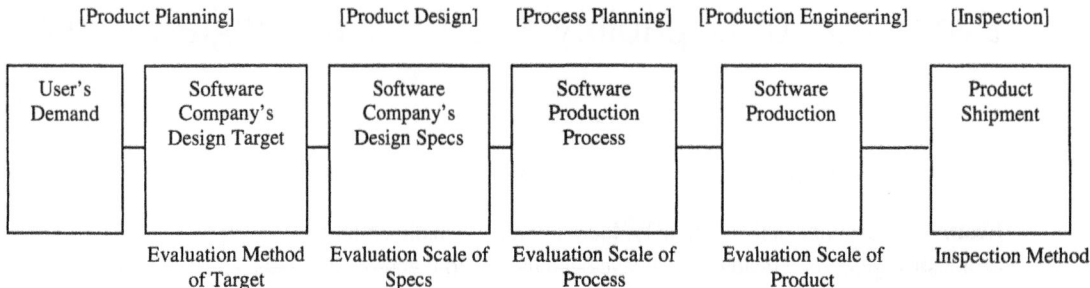

Figure 1
Work flow of software production

$$S_p = \frac{p^2}{\text{total of squares of coefficients}}$$

$$= \frac{[(y_1 + y_2 + \cdots + y_n/n]^2}{(1/n)^2 + (1/n)^2 + \cdots + (1/n)^2}$$

$$= \frac{(y_1 + y_2 + \cdots + y_n)^2}{n} = np^2 = S_m \quad (3)$$

Then the error variation

$$S_e = S_T - S_\beta = np - np^2 = np(1 - p) \quad (4)$$

An SN ratio for 0/1 data is expressed by a ratio of signal factor variation to error variation. To obtain additivity of measurement in the analysis, we expressed the SN ratio as a decibel value:

$$\eta = 10 \log \frac{S_\ell}{S_e} = 10 \log \frac{np^2}{np(1 - p)}$$

$$= 10 \log \frac{p}{1 - p}$$

$$= -10 \log \left(\frac{1}{p} - 1\right) \quad \text{dB} \quad (5)$$

3. Evaluation Method of Type-by-Type Error with SN Ratio

As a next step, we devised a method of evaluating errors of various types [1]. First, all errors were classified into four categories (Table 1). *Mistake* is an error that can be corrected by a testee if others advise him or her. *Unknown* is an error caused by a testee's ignorance of coding due to insufficient experience in programming or the C language. That is, in this case, even if others give advice, the testee cannot correct the error.

We allocated these error types to an L_8 orthogonal array (Table 2) We defined as level 1 a case of using an error type, and no use as level 2, and at the same time, evaluated an error's significance. The right side of Table 2 shows the number of errors.

As below, we show an example of the calculation for row 2 of the L_8 orthogonal array. We set each number of errors for A', B', C', and D' to 2, 15, 3, and 7, respectively, and the number of lines to 86. Since we use level 1 for the analysis, the only nec-

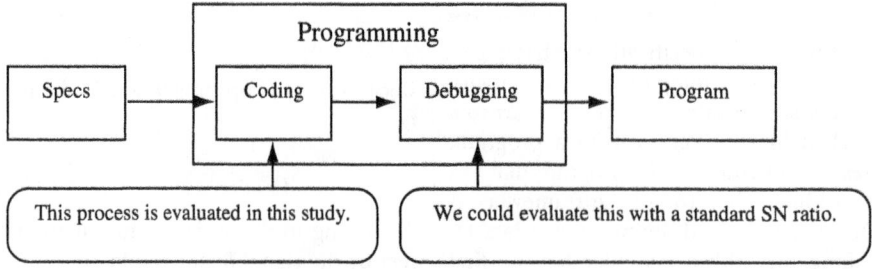

Figure 2
Details of software production

Table 1
Error types and levels

Factor	Level 1	Level 2
Logical error		
A': Mistake	Use	No use
B': Unknown	Use	No use
Syntax error		
C': Mistake	Use	No use
D': Unknown	Use	No use

essary data are A', B', and C; D' is not used for the analysis of row 2 because its number of levels is two.

The sum of errors for experiment 1 in the L_8 orthogonal array is

$$A' + B'' + C' = 2 + 15 + 3 = 20$$

Therefore, the number of lines where correct codes are written is

(total number of lines − number of errors for D')
− number of errors = 59

Then the proportion of correct codes, p, is computed as

$$p = \frac{59}{86 - 7} = 0.747 \tag{6}$$

In this case, because we disregard D' itself, it should be excluded from the total number of lines.

Total variation:

$$S_T = np = (79)(0.747) = 79.0 \tag{7}$$

Effect by a signal factor:

$$S_p = np^2 = (79)(0.747^2) = 32.91 \tag{8}$$

Error variation:

$$S_e = S_T - S_p = 79.0 - 32.91 = 46.09 \tag{9}$$

Thus,

$$\eta = 10 \log \frac{S_p}{S_e} = -10 \log \left(\frac{1}{p} - 1 \right)$$

$$= 4.70 \text{ dB} \tag{10}$$

Using these results we can evaluate the capability of software production for each type of error.

4. Capability Evaluation of Software Production

With an L_{12} orthogonal array, we studied an evaluation method of capability of software production. As factors regarding an instructional method and evaluation, 10 types of factors were selected (Table 3). The C language is used and the program consists of about 100 steps. As testees, six engineering students who have had 10 hours of training in the C language were chosen. Next, we tested these testees according to the levels allocated in the L_{12} orthogonal array.

Table 2
Error types and number of errors assigned to L_8

No.	A'	B'	C'	D'	e	e	e	A'	B'	C'	D'	Total
1	1	1	1	1	1	1	1	2	15	3	7	27
2	1	1	1	2	2	2	2	2	15	3		20
3	1	2	2	1	1	2	2	2			7	9
4	1	2	2	2	2	1	1	2				2
5	2	1	2	1	2	1	2		15		7	22
6	2	1	2	2	1	2	1		15			15
7	2	2	1	1	2	2	1			3	7	10
8	2	2	1	2	1	1	2			3		3

Table 3
Factors and levels

		Level	
Factor		**1**	**2**
A:	type of program	Business	Mathematical
B:	instructional method	Lecture	Software
C:	logical technique	F.C.	PAD
D:	period before deadline (days)	3	2
E:	coding guideline	Yes	No
F:	ability in Japanese language	High	Normal
G:	ability for office work	High	Normal
H:	number of functions	Small	Large
I:	intellectual ability	High	Normal
J:	mathematical ability	High	Normal

Because we were dealing with six testees, dividing the L_{12} orthogonal array into the upper half (1 to 6) and the lower half (7 to 12), we first performed the upper-half experiment using the six testees. Then we performed experiments on the lower half conducted on the same testees. That is, each testee needs to write two programs (business and mathematical programs). In this case, although a learning effect was considered to influence them, it hardly has an effect because this level of programming had been completed a few times prior to the training.

The following is an explanation of the factors selected.

❏ *A: type of program.* Business and mathematical programs were prepared for the experiment.

❏ *B: instructional method.* A testee may have been trained in the C language by a teacher or may have self-studied it using instructional software.

❏ *C: logical technique.* A schematic of a program's flow was given along with specifications.

❏ *D: deadline.* A testee was required to hand in a program within two or three days after the specifications were given.

❏ *E: coding guideline.* This material states ways of thinking in programming.

❏ *F: ability in Japanese language.* The ability to understand Japanese words and relevant concepts and to utilize them effectively is judged.

❏ *G: ability for office work.* The ability to comprehend words, documents, or bills accurately and in detail is judged.

❏ *H: number of functions.* As long as the difficulty level of each program was not changed, there are two types of specifications, minimal and redundant, in which a testee writes a program.

❏ *I: intellectual ability.* The ability to understand explanations, instructions, or principles is judged.

❏ *J: mathematical ability.* The ability to solve mathematical problems is judged.

For human abilities (*F, G, I,* and *J*), we examined the testees using the general job aptitude test issued by the Labor Ministry. After checking each testee's ability, we classified them into high- and low-ability groups.

5. Experimental Results

We computed interactions between the factors selected and error types. More specifically, an L_8 or-

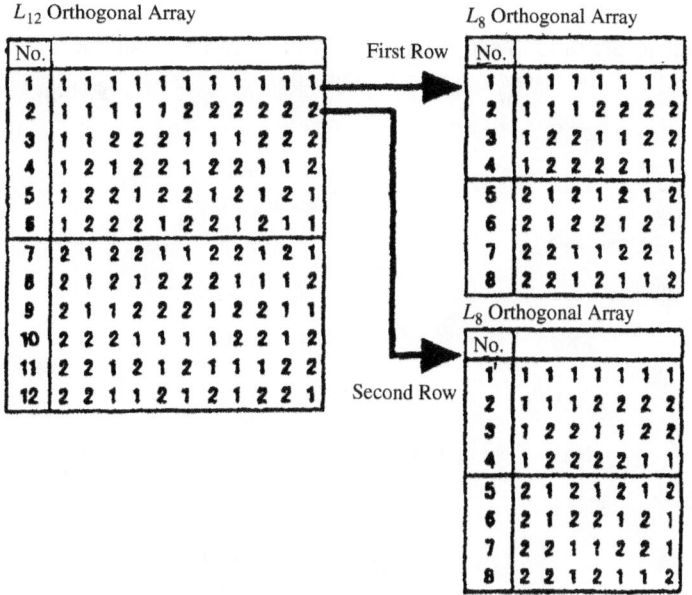

Figure 3
Layout of direct-product experiment using an L_{12} orthogonal array to evaluate the capability of software production and an L_8 orthogonal array to evaluate error types

thogonal array (outer array for error types) was allocated to each row in an L_{12} orthogonal array (inner array for combinations of all factors), as shown in Figure 3. Table 4 shows the data obtained from the experiment, and Table 5 shows the result computed through an analysis of variance with respect to the SN ratio. Boldfaced entries in Table 5 indicate a large factor effect.

Next we inquired into the main effect of the factor shown in Figure 4. The factor with the greatest main effect is the number of functions (H). A small number of functions leads to a small number of

Table 4
Number of errors obtained from experiment

L_8	L_{12}											
	1	2	3	4	5	6	7	8	9	10	11	12
A'	2	7	3	0	6	6	4	4	6	4	4	5
B'	15	51	7	75	50	13	40	12	160	47	2	35
C'	3	2	57	8	2	0	2	0	6	1	10	5
D'	7	12	0	3	2	0	15	4	12	12	2	0
Total number of lines	86	96	266	113	146	60	104	84	300	94	103	116

Table 5
ANOVA with respect to SN ratio based on capability of software production and error types[a] (dB)

Factor	f	S	Factor	f	S	Factor	f	S
A	1	22.9	A × A'	1	11.0	A × C'	1	12.8
B	1	41.8	B × A'	1	1.5	B × C'	1	0.0
C	**1**	**178.5**	C × A'	1	90.5	C × C'	1	109.5
D	1	31.1	D × A'	1	2.4	D × C'	1	70.3
E	1	77.2	E × A'	1	85.2	E × C'	1	79.7
F	1	106.8	**F × A'**	**1**	**141.3**	**F × C'**	**1**	**218.4**
G	1	17.5	G × A'	1	47.5	**G × C**	**1**	**146.7**
H	**1**	**929.8**	H × A'	1	123.1	H × C'	1	11.4
I	1	33.7	I × A'	1	50.1	I × C	1	21.4
J	1	22.9	J × A'	1	27.5	J × C'	1	79.2
A'	1	66.4	A × B'	1	29.6	A × D'	1	7.6
B'	**1**	**2408.7**	B × B'	1	44.4	B × D'	1	8.1
C'	1	6.5	C × B'	1	23.5	C × D'	1	0.1
D'	**1**	**236.7**	D × B'	1	1.5	D × D'	1	9.1
e	41	938.4	E × B'	1	48.1	E × D'	1	0.1
(e)	86	(2466.8)	F × B'	1	21.1	F × D'	1	0.9
ST	95	6980.9	**G × B'**	**1**	**147.1**	G × D'	1	9.7
			H × B'	1	77.0	H × D'	1	14.7
			I × B'	1	0.6	I × D'	1	32.2
			J × B'	1	39.3	J × D'	1	17.8

[a] Parentheses indicate pooled error.
$V_e = 938.4/41 = 22.9$. $(V_e) = (2466.8)/86 = (28.7)$.

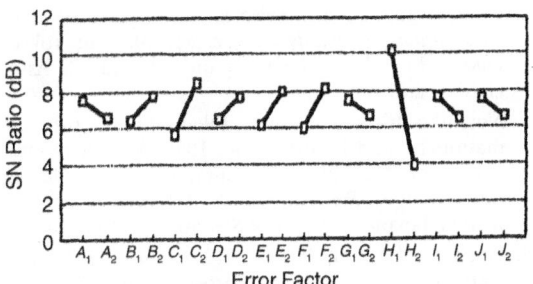

Figure 4
Response graphs of the SN ratio for capability of software production

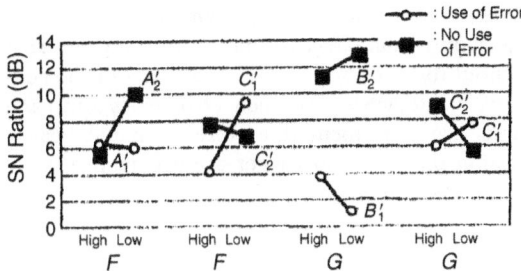

Figure 6
Interaction between SN ratios for capability of software production and SN ratios for error type

errors and subsequently, to a high SN ratio. In addition, logical technique (C) and ability in the Japanese language (F) have a large effect. Since factors other than these have a small effect, good reproducibility cannot be expected. On the other hand, since we observed no significant effect for intellectual ability, even in the experiment on error-finding ability, this factor's effect in programming was considered trivial.

In the response graphs for error types shown in Figure 5, level 1 represents error used for analysis, and level 2, when it was not used. Since the SN ratio when not using error is high, all the errors in this study should be included in an analysis. Unknown error in logical error (B') turned out to be particularly important.

Furthermore, according to the analysis of variance table in Figure 6, while some factors generate significant interactions with A', B', and C', interactions with D' had only trivial interactions with other factors, but D' per se was relatively large compared

with other error types. By combining large-interaction errors into one error and making a different classification, we are able to categorize errors into one with clear interactions and one without interactions in the next experiment.

It was observed that an interaction occurs more frequently in the factors regarding ability in the Japanese language and ability for office work. Considering that a similar result was obtained in a confirmatory experiment on error finding [3], we supposed that the reason was a problem relating to the ability test itself.

6. Conclusions

As a result of classifying capability of software production by each error type and evaluating it with an SN ratio for the number of errors, we concluded that our new approach was highly likely to evaluate the true capability of software production. We are currently performing a confirmatory experiment on its reliability.

Since students have been used as testees in this study, we cannot necessarily say that our results would hold true for actual programmers. Therefore, we need to test this further in the future.

The second problem is that a commercial ability test has been used to quantify each programmer's ability in our experiment. Whether this ability test is itself reproducible is questionable. An alternative method of ability evaluation needs to be developed.

On the other hand, while we have classified errors into four groups, we should prepare a different error classification method in accordance with an

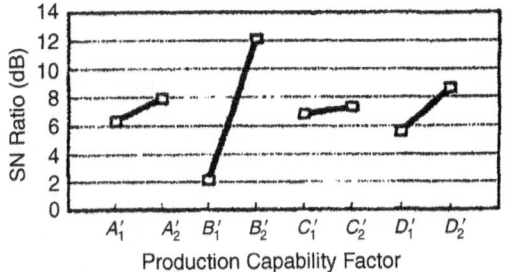

Figure 5
Response graphs of the SN ratio for error type

experimental objective. For example, if we wish to check only main effects, we need to classify the error without the occurrence of interaction. On the contrary, if interactions between error types and each factor are more focused, a more mechanical classification that tends to trigger the occurrence of interactions is necessary.

References

1. Kei Takada, Muneo Takahashi, Narushi Yamanouchi, and Hiroshi Yano, 1997. The study of evaluation of capability and errors in programming. *Quality Engineering*, Vol. 5, No. 6, pp. 38–44.

2. Genichi Taguchi and Seiso Konishi, 1988. *Signal-to-Noise Ratio for Quality Evaluation*, Vol. 3 in Tokyo: *Quality Engineering Series*. Japanese Standards Association, and ASI Press.

3. Norihiko Ishitani, Muneo Takahashi, Narushi Yamanouchi, and Hiroshi Yano, 1996. A study of evaluation of capability of programmers considering time changes. *Proceedings of the 26th Sensory Test Symposium,* Japanese Union of Scientists and Engineers.

4. Norihiko Ishitani, Muneo Takahashi, Kei Takada, Narushi Yamanouchi, and Hiroshi Yano, 1996. A basic study on the evaluation of software error detecting capability. *Quality Engineering*, Vol. 4, No. 3, pp. 45–53.

This case study is contributed by Kei Takada, Muneo Takahashi, Narushi Yamanouchi, and Hiroshi Yano.

CASE 85

Evaluation of Programmer Ability in Software Production

Abstract: This experiment was conducted to evaluate two types of ability: software error finding and program production. As a first step, we asked each testee to debug a program that included intentional errors. Then we evaluate his or her ability on the basis of the number of errors corrected and those not corrected. Because the results generated from a third party cannot be used to confirm experimental results and it is difficult to set the levels of human conditions, enough people need to be provided for a confirmatory experiment.

1. Evaluation of Software Error-Finding Ability

We asked testees to debug a program that had intentional errors. The results were two types of output data (Table 1). The types of error found can be expressed by a 0/1 value. A mistake can occur when no error is judged as an error. It is essential to judge an error as an error and also to judge no error as no error. Since Genichi Taguchi has proposed a method using the standard SN ratio for this type of case, we attempted to use this method.

We show the calculation process of the standard SN ratio as below. We set the total number of lines to n, the number of correct lines to n_0, that of lines including errors to n_1, that of correct lines judged as correct to n_{00}, that of correct lines judged as incorrect to n_{01}, that of incorrect lines judged as correct to n_{10}, and that of incorrect lines judged as incorrect to n_{00}. The input/output results are shown in Table 2.

Now the fraction of mistakes is

$$p = \frac{n_{01}}{n_0} \qquad (1)$$

$$q = \frac{n_{10}}{n_1} \qquad (2)$$

Using these, we calculated the standard SN ratio, η_0:

$$p_0 = \frac{1}{1 + \sqrt{[(1/p) - 1][(1/q) - 1]}} \qquad (3)$$

$$\eta = -10 \log \left[\frac{1}{(1 - 2p_0)^2} - 1 \right] \quad \text{dB} \qquad (4)$$

Choosing an L_{12} orthogonal array, we picked the 11 factors shown in Table 3. As a programming language, we used C and selected six scientific and engineering students with no knowledge of C. The number programs to be debugged was approximately 30. As a program type, a logical calculation program was chosen.

Each factor is defined as follows:

❏ *A: instruction time:* time needed for training in the C language

❏ *B: instructional method:* lecture by an instructor or individual study via software

❏ *C: number of reviewers:* review by a single person or by two people

❏ *D: review time:* time for reviewing

❏ *E: checklist:* whether or not to use a checklist

❏ *F, G, I: individual ability:* aptitude test (general job aptitude test issued by the Labor Ministry)

Table 1
Results of software error-finding experiment

	Output	
Input	Judged as Correct	Judged as Incorrect
Correct	O	×
Incorrect (bug)	×	O

examining testee's ability in the Japanese language, business ability, and intellectual ability, and allocation of the results to an L_{12} orthogonal array

❑ H: *target value:* because of errors included in a program intentionally, whether or not to inform testees of the number of errors to be found

❑ J: *work intermission:* whether or not to give the testee a 10-minute pause to do other work that has no relation to the review work

❑ K: *number of errors:* number of errors included intentionally in a program

Using the control factors above, we performed the following experiment. First, we provided 4-hour-long instruction in the C language to six testees to conduct the experiments of six combinations as numbers 1 to 7 of L_{12}. Then we provided another four hours of training to the same testees (in total, eight hours of instruction) and conducted an experiment based on combinations of numbers 7 to 12 in the L_{12} orthogonal array. However, because each testee was dealing with three characteristics at the same time, it was difficult to allocate them in-

Table 3
Factors and levels for experiments on error finding

	Factor	Level	
		1	2
A:	instruction time (hours)	4	2
B:	instructional method	Lectured	Self-taught
C:	number of reviewers	1	2
D:	review time (min)	20	30
E:	checklist	Yes	No
F:	ability in Japanese language	High	Normal
G:	ability for office work	High	Normal
H:	target	Yes	No
I:	intellectual ability	High	Normal
J:	work intermission	Yes	No
K:	number of errors	3	6

dependently. So by using a sequential approximation method, we estimated and calibrated SN ratios.

As a result of comparing the result obtained from this experiment and our prior experience, we could see that the trends in almost all of the control factors were understood satisfactorily. For instance, longer instruction time gave better results, and a review by two persons was more favorable than a review by one person. Therefore, we believe that the analysis method in our study contributed to evaluating factors related to error-finding capability.

2. Reproducibility of Results for Software Error-Finding Capability

As mentioned before, there are two types of errors: inability to find errors intentionally included in a program, and mistakenly considering a correct code to be an error.

While quality engineering requires a confirmatory experiment for reproducibility in gain under

Table 2
Input/output for two types of mistakes

	Output		
Input	0 (Correct)	1 (Incorrect)	Total
0 (correct)	n_{00}	n_{01}	n_0
1 (incorrect)	n_{10}	n_{11}	n_1
Total	r_0	r_1	n

an optimal configuration, we decided that instead of observing reproducibility under an optimal configuration, the trends of response graphs should be confirmed. Using new testees and a new program, we performed an experiment similar to our previous one.

However, to improve experimental reliability, we reiterated this new experiment based on a new program and repeated it three times (R_1, R_2, and R_3). Because this was a confirmatory experiment of the earlier one, we again selected six testees, an L_{12} orthogonal array, and 11 control factors (Table 3).

Following the conditions noted above, we obtained each datum. Figure 1 superimposes the response graphs of the first experiment on the confirmatory experiment. While the factor effects of ability in the Japanese language, business ability, and instruction time are still questionable, for other factors we see sufficient reproducibility. One of the possible reasons of poor reproducibility for ability in the Japanese language and business ability is that the reliability of the commercial aptitude test used to measure the two abilities is questionable. However, since we obtained fairly good reproducibility by changing testees and program, the reproducibility in this experiment can be regarded as sufficient.

3. Evaluation of Programming Capability

The previous experiment was conducted as follows. We handed specifications and a flowchart to a testee

and asked him or her to write a program. When the testee had completed a program that was satisfactory to him or her, he or she saved the program. Next, compiling this with a compiler for the first time, the testee debugged it by himself or herself. After this program was checked by an examiner, a program with almost no bugs was obtained as a final output. Then, by comparing the two programs, we recognized a difference in code between them as an error.

In addition, we evaluated the data using an SN ratio for each error type. First, we classified errors into each type and allocated all of the types to a new orthogonal array (outer factor). As level 1, "use of error" was selected, whereas "no use of error" was level 2. By assigning these outer factors with the control factors (inner factors), thereby obtaining a direct product layout, we analyzed interactions between the control factors and noise types. Consequently, we had a total of 144 (12 × 12) SN ratios. The errors were classified as follows:

- ❏ A': insufficient understanding of specs, or eventual lack of functions
- ❏ B': a change in a program caused through introduction of a new idea
- ❏ C': functions, but incomplete
- ❏ D': inability of grasping variables and keeping track of data changes
- ❏ E': errors in a specific area despite correct codes in other areas (with no program framework)

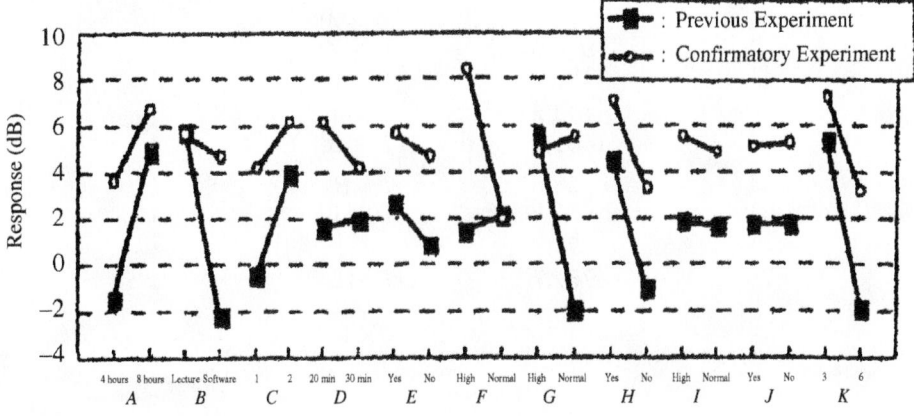

Figure 1

Response graphs for previous and confirmatory experiments

Table 4
Factors and levels for experiment and programming

		Level	
Factor		**1**	**2**
A:	type of program	Business	Mathematical
B:	instructional method	Lectured	Self-taught
C:	logical technique	F.C.	PAD
D:	period before deadline (days)	3	2
E:	coding guideline	Yes	No
F:	ability in Japanese language	High	Normal
G:	ability for office work	High	Normal
H:	number of specs	Small	Large
I:	intellectual ability	High	Normal
J:	mathematical ability	High	Normal

❏ F': errors in a specific area despite correct codes in other areas (executable but incorrect program)

❏ G': correct idea but no knowledge of syntax

❏ H': inability to write codes

❏ I': addition of { }'s in accordance with an increase in program codes

The data are collected as follows:

Line number in program: 1 2 3 ⋯ n

Representation of Correct Data: y_1 y_2 y_3 ⋯ y_n

where $y = 1$ for a correct line and $y = 0$ for incorrect line.

Table 5
ANOVA for experiment on programming

Factor	f	S	Factor	f	S
E	1	270.2	$G \times A'$	1	45.4
H	1	972.0	$H \times A'$	1	44.0
J	1	152.2	$I \times A'$	1	45.7
A'	1	124.4	$J \times A'$	1	44.8
B'	1	1461.1	$E \times B'$	1	319.7
$A \times A'$	1	58.6	$I \times B'$	1	115.4
$B \times A'$	1	51.8	$D \times C'$	1	93.8
$D \times A'$	1	67.9	$E \times C'$	1	69.4
e	76	767.7	$F \times C'$	1	59.2
			$G \times C'$	1	62.0
			$I \times C'$	1	73.9
			Total 95		4899.2

$V_e = 767.7/76 = 10.1$.

Table 6

Factors and levels for confirmatory experiment

		Level	
		1	**2**
A:	type of program	With objective	With objective
B:	instructional method	Lectured	Self-taught
C:	specs	With flowchart	With no flowchart
D:	period before deadline (days)	4	2
E:	test pattern	Yes	No
F:	ability in Japanese language	High	Normal
G:	ability for office work	High	Normal
H:	number of specs	Small	Large
I:	intellectual ability	High	Normal
J:	mathematical ability	High	Normal

We used the following calculation process. The fraction of the number of correct codes to that of lines, p, is computed as follows:

$$p = \frac{y_1 + y_2 + \cdots + y_n}{n} \qquad (5)$$

An SN ratio for 0/1 data is expressed by a ratio of signal factor variation and error variation. To include additivity of measurement in the analysis, we expressed the SN ratio as a decibel value:

$$\eta = 10 \log \frac{S_p}{S_e} = -10 \log \left(\frac{1}{p} - 1 \right) \qquad \text{dB} \qquad (6)$$

An L_{12} orthogonal array was used for this experiment. As control factors, 10 types of factors were selected (Table 4). As an ability test, a commercial aptitude test was prepared. The C language is used and the program consisted of about 100 steps. As testees, six engineering students who had taken 10 hours of training in the C language were chosen. We tested them according to the levels allocated in the L_{12} orthogonal array.

The following is an explanation of the factors selected:

❑ A: *type of program:* type of program created by a testee

❑ B: *instructional method:* lecture on the C language from a teacher, or the use of instructional software

❑ C: *logical technique:* schematic of a program's flow

❑ D: *period before deadline:* requirement that program be handed in within two or three days after specifications are given

❑ E: *coding guideline:* material that states ways of thinking in programming

❑ H: *number of specs:* as long as the difficulty level of each program is not changed, minimal and

Table 7

Classification of errors in confirmatory experiment

	Level	
Factor	**1**	**2**
Logical error		
A': mistake	Use	No use
B': unknown	Use	No use
Syntax error		
C': mistake	Use	No use
D': unknown	Use	No use

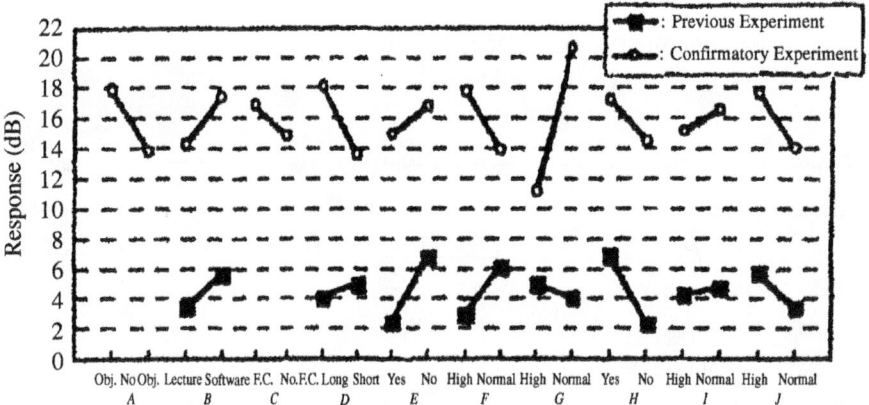

Figure 2
Comparison of response graphs between previous and confirmatory experiments

redundant specifications on which a testee was to write a program

❏ *F, G, I,* and *J: individual ability:* similar to abilities measured in the experiment on evaluation of error finding, plus mathematical ability

Since this experiment also tested six testees, using the same testees, we performed a separate experiment for the upper and lower halves of the orthogonal array. The results reveal that main effects are brought about by human abilities and number of specs. In addition, in terms of interactions between factors and errors, *A'*, *B'*, or the error of *C'* demonstrated a relatively large effect. That is, it was assumed that these errors were affected by individual differences (Tables 4 and 5).

4. Reproducibility of Results for Programming Capability Evaluation

This experiment was also performed under almost the same conditions. The only difference was addition of repetitions (R_1 and R_2) and a change in some control factors. Because of the repetitions added, considering the increased workload on the testees, we prepared two 50-line programs in this experiment instead of one 100-line program.

We describe the changed factors as follows (see Table 6):

❏ *A: type of program.* Because of using a different program, we could not follow the classification in the previous experiment. Rather, we focused on whether or not the objective of a program was clarified.

❏ *C: specs.* Since we did not have enough time to instruct the testees in two types of flow-

Table 8
ANOVA for experiment on programming

Factor	f	S	Factor	f	S
A	1	425.5	A × A'	1	655.6
B	1	266.2	G × A'	1	327.1
D	1	474.5	A × B'	1	820.0
F	1	369.4	G × B'	1	398.5
G	1	2063.1	A × C'	1	409.5
K	1	319.4	A × D'	1	211.7
A'	1	1952.3	C × D'	1	229.0
D'	1	911.2	D × D'	1	382.9
e	77	3061.8	H × D'	1	417.0
			K × D'	1	215.7
			Total	95	13,910.3

$V_e = 3061.8/77 = 39.8.$

Table 9
Gain obtained from experiments on error finding and programming

	Experiment on Error Finding		Experiment on Programming	
	Previous	**Confirmatory**	**Previous**	**Confirmatory**
All factors				
Optimal	22.6	17.6	13.3	31.0
Worst	−19.1	−7.3	0.0	0.5
Gain	41.6	24.9	13.4	30.5
All factors except				
Japanese language, office work, and time				
Optimal	18.4	13.0	12.9	22.2
Worst	−14.9	−2.7	0.4	9.4
Gain	12.6	19.0	12.6	12.8

charts, in this experiment we considered whether a flowchart was included in the specs.

❑ *D: period before deadline.* Because there is little difference between two and three days, we changed the periods to two and four days so that the difference would be greater.

❑ *E: test pattern.* The existence and nonexistence of a test pattern were compared.

Since there were approximately 50 lines in one program in the confirmatory experiment, we cannot classify the errors due to the small number as we have done in the preceding experiment. Additionally, because a small number of lines mitigates differences among individuals, we proceeded with the analysis following the categorization of errors shown in Table 7.

Figure 2 illustrates the response graphs. For the factors in the preceding experiment, we indicated only the factors that were the same as, or similar to, the corresponding ones in the confirmatory experiment. As a result, we can see that many of the factors have good reproducibility. On the other hand, it is quite reasonable that ability in the Japanese language, business ability, and period before deadline (time) still have poor reproducibility because all of them show the same conclusions in the experiment on error-finding capability. Table 8 lists the ANOVA for the experiment, and Figure 2 compares the response graphs.

5. Confirmation of Gains in Experiments on Error Finding and Programming

Table 9 shows calculation of the gains. While the upper half represents a normal calculation of gain, the lower indicates a computation excluding ability of Japanese language, business ability, and time, all of which had poor reproducibility. Because of no current configuration in our study, we defined a difference between SN ratios under the optimal and worst configurations. When using all the control factors, the reproducibility was extremely poor, but we obtained fairly good reproducibility when some factors were excluded.

According to the two experiments discussed thus far, we can see that effects caused by interactions need to be considered in an analysis because of significant interactions among control factors (differences among individuals) in the case of dealing with a 100-line program, whereas we can perform an analysis with no consideration of interactions when a 30- to-50-line program is used. On the other hand, since the commercial aptitude test used for evaluating human ability factors is problematic in terms of reliability, we need to seek a substitute.

Yet since we obtained sufficient reproducibility in an operational process including human factors, it would be reasonable to say that our analysis method was effective for the evaluation of programming capability.

Reference

Kei Takada, Muneo Takahashi, Narushi Yamanouchi, and Hiroshi Yano, 1998. The study of reappearance of experiment in evaluation of programmer's ability of software. *Quality Engineering*, Vol. 6, No. 1, pp. 39–47.

This case study is contributed by Kei Takada, Muneo Takahashi, Narushi Yamanouchi, and Hiroshi Yano.

Robust Testing of Electronic Warfare Systems

Abstract: The process of development testing for electronic warfare (EW) systems can be both expensive and time consuming, since the number of individual radar emitter types and modes is large. Robust engineering methods have been applied to development testing of EW systems whereby the confirmed test cycle time has been decreased significantly, by a factor of 4 to 1, while providing equivalent diagnostics of system performance. An L_{18} array has effectively demonstrated the ability to test more system functionality in significantly less time than do conventional methods.

1. Introduction

The development of EW systems includes testing and evaluation of new systems against a wide and varied range of radar emitter types, taking a significant amount of time and effort. To minimize development time while improving product quality and performance, robust engineering methods were implemented. The objective was to reduce EW system test time for a large quantity of emitters by defining a reduced set of orthogonal emitter types.

2. Experimental Approach

Most quality engineering methods employ control factors to determine the sensitivity of system performance to various design parameters. In this situation the approach was used to evaluate a system's performance against a wide variety of simulated field conditions (noise factors), in a time-efficient manner. A secondary goal is to determine system deficiencies in order to develop a more robust system design. A third result of the task was to yield a single measure of system performance (mean signal-to-noise ratio) to track hardware and software improvements. Since the objective of this task was to permit detection, noise factors were selected in lieu

of control factors to define the orthogonal array selected.

The purpose of an EW system is to search for various radar systems (emitters) in its operational environment, determine their location, and classify function by measuring various radiated parameters and received characteristics. If a radar site is considered hostile, effective measures are taken by the EW system to mitigate emitter function. This is illustrated in Figure 1. As an example, airport radar systems may be searching for aircraft, whereas a hostile emitter site may be seeking to launch and control armaments. The determination of radar type (airport, hostile, etc.) and radar mode (search, acquisition, track) is therefore a very critical function. The scope of this project was limited to assessing EW system performance to identify a radar system and its mode correctly.

The electronic warfare system makes these critical determinations by measuring various characteristics of a radar's transmitted signal. Measuring the signal amplitude from two or more EW system receiving antennas permits the determination of radar position by determining the angle of arrival (AOA). Measuring parameters transmitted from the radar, such as the frequency, pulse, scan type, and amplitude, help determine the type and mode of the radar system. These parameters become noise factors for an orthogonal array.

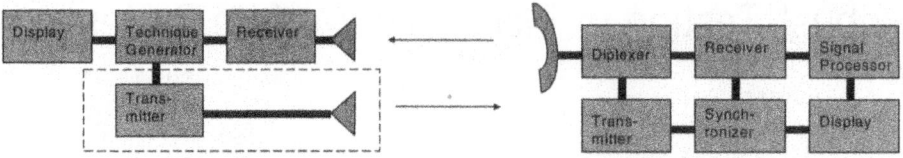

Electronic Warfare (EW) System   **Radar System (Emitter)**

Function: **Function (Modes):**
 Identification of Emitter   **Search for Targets**
 Mode of Emitter   **Acquisition of Targets**
 Location of Emitter   **Tracking of Targets**

Figure 1
Electronic warfare scenario

The challenge for this project was to select a signal (limited group of emitters) that represents the functional variation of the entire ensemble of radar types. The transfer function is to identify each emitter in the signal set without error. The Taguchi problem type was selected to be "maximum the best," given that correct identification of the emitter is considered to be a correct signal, and any error in the system output is considered noise. In this case the system must detect all signals without radar type or mode identification error.

It is recognized that the EW system will need to be tested against the entire group of radar systems. However, since software upgrades are made on a nearly weekly basis, the 16 hours required to evaluate the entire ensemble of emitter types becomes expensive and time consuming. The goal of this project was therefore to cut the weekly 16 hours of test time by at least one-half while maintaining test integrity.

3. P-Diagram

Figure 2 illustrates the P-diagram for this project. The input signal is the domain of emitters as described by the L_{18} array. The output signal is correct identification of each radar emitter and mode to which the EW system is subjected. The control factors consist of the EW system hardware and software configuration that is tested in a particular system configuration. The system configuration is variable depending on project phase due to improvements in hardware and software.

Noise factors are those that define the variable domain of conditions to which the EW system is subjected. This includes the variation in emitter characteristics over the domain of all emitter types; the variation in emitter position, including angle of arrival and signal amplitude; the variation in emitter mode; and the background noise.

4. Selection of Orthogonal Array

Selection of array size was based on the determination of one two-level factor and five three-level factors. (Table 1).

The L_{18} array was selected because it has more than 14 degrees of freedom and will accommodate one two-level factor and up to seven three-level factors. Selection of an L_{18} array will also yield growth for two three-level factors, as noted in Table 1. The next issue in developing this project was selection of 18 emitters from a group on the order of 100 in size.

5. Selection of Robust Emitter

Selection of emitters started with the characterization of each radar system in the total domain in terms of the noise factors and parameters selected. The orthogonal array is based on the noise factors shown in Table 1. Hence, each of the 18 experiments defined by the array would consist of a different set of emitter characteristics. Table 2 illus-

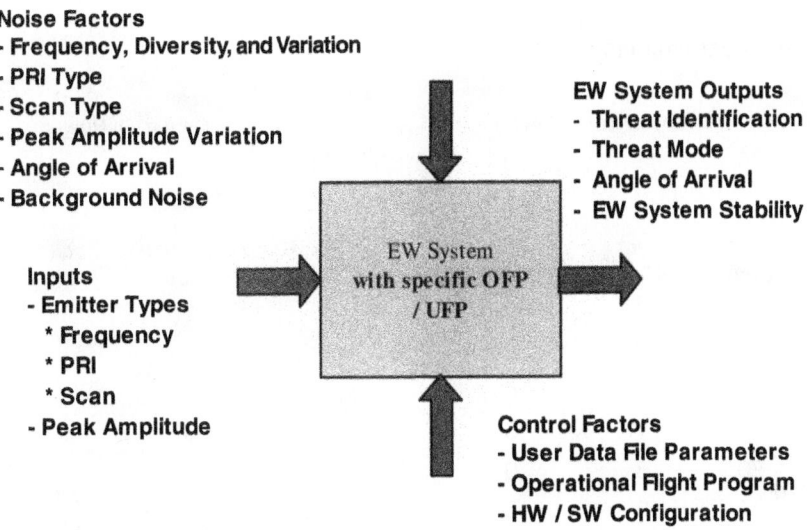

Noise Factors
- **Frequency, Diversity, and Variation**
- **PRI Type**
- **Scan Type**
- **Peak Amplitude Variation**
- **Angle of Arrival**
- **Background Noise**

EW System Outputs
- **Threat Identification**
- **Threat Mode**
- **Angle of Arrival**
- **EW System Stability**

Inputs
- **Emitter Types**
  * **Frequency**
  * **PRI**
  * **Scan**
- **Peak Amplitude**

EW System
with specific OFP
/ UFP

Control Factors
- **User Data File Parameters**
- **Operational Flight Program**
- **HW / SW Configuration**

Figure 2
P-diagram

Table 1
Selection of array

Factor	Level 1	Level 2	Level 3	d.f.
Overall mean	—	—	—	1
Frequency diversity	Single	Multiple	—	1
Frequency	Low	Mid	High	2
PRI type	CW	Stable	Agile	2
Scan type	Steady	Circular	Raster	2
Peak power	Nominal	Medium	Low	2
AOA[a]	Boresite	Offset 1	Offset 2	2
Illumination	Short	Medium	100 %	2
Background[a]	None	Low density	High density	2
			Total	16

[a] Growth factor.

Table 2
Selection of 18 robust emitters

Experiment	Item	Mean Distance	RF Diversity	RF	PRI Type	Scan Type	Peak Power	Illumination
2	76	1	Single	High	Stable	Circular		Medium
			Single	High	Stable	Circular		Short

	Item	Item Distance	RF Diversity	RF	PRI Type	Scan Type	Peak Power	Illumination
	66	4	0	1	1	1		1
	70	2	0	0	1	0		1
	74	2	0	1	0	0		1
	76	1	0	0	0	0		1
	80	3	0	0	1	1		1
	82	3	0	1	1	1		0
	83	2	0	0	0	1		1
	84	2	0	0	0	1		1
	88	4	1	1	1	0		1
	90	3	0	0	1	1		1

trates the selection process for experiment 2. The principal characteristics were single RF frequency, high RF frequency, stable PRI, circular scan, and short illumination time. The difficulty in choosing emitters from the ensemble available was that only a few provided a perfect match to the Taguchi-defined set. This problem was handled by calculating the shortest distance for each emitter in the ensemble against the 18 experiments defined.

Table 2 shows that for experiment 2, the shortest distance exists for item 76, with the only incorrect parameter being short illumination time, hence a distance of 1. Item 76 was therefore selected for experiment 2. Peak power was adjusted for each experiment and therefore was not considered in the selection process. Note that "0" indicates an exact correlation to the desired parameter, and "1" indicates no match. All experiments were selected in the same manner.

Table 3 presents the entire set of 18 test emitters. Several experiments match the orthogonal requirement exactly with an item distance of zero. Several emitters, such as item 76 used in experiment 2, are not quite orthogonal. That is, the distance is other than zero, either 1 or 2. The correct or desired parameter is located in the center of each parameter box, while the parameter exhibited by the emitter is located underneath. The count of valid parameter matches and incorrect matches are summed at the bottom of each factor column. The important point to be made here is that an orthogonal array selected for the purpose of cause detection need not be exactly orthogonal to be effective.

6. Signal-to-Noise Ratio

The ideal transfer function for this problem is to identify each emitter and its mode without error over the domain of emitter types. Each emitter was tested and evaluated to determine whether its radar type, radar mode, and angle of arrival were correct. Any incorrect output was assigned a count of 1 per incorrect output. Since there may be subjective considerations during this evaluation, the operator also

Table 3

L_{18} array for robust testing of EW systems

| Experiment | Item | Distance Item (Mean) | Expert Array Laboratory Performance: Column Factor and Parameter Assigned | | | | | | | | Observation m_1 (Defects) |
			A RF Diversity	B RF	C PRI Type	D Scan Type	E Peak Power	F	G Illumination	H	
1	10	0	Single	High	CW	Steady	Nominal	F_1	100%	H_1	1.000
2	76	1	Single	High	Stable	Circular	Medium	F_2	1 Medium Short	H_2	0.000
3	80	0	Single	High	Agile	Raster	Low	F_3	Short	H_3	6.000
4	61	1	Single	Mid	CW	Steady	Medium	F_2	1 Short 100%	H_3	0.000
5	12	1	Single	Mid	Stable	1 Circular Steady	Low	F_3	100%	H_1	4.000
6	70	2	Single	1 Mid High	Agile	Raster	Nominal	F_1	1 Medium Short	H_2	0.000
7	33	0	Single	Low	CW	Circular	Nominal	F_3	Medium	H_3	2.000
8	147	0	Single	Low	Stable	Raster	Medium	F_1	Short	H_1	0.000
9	141	1	Single	1 Low Mid	Agile	Steady	Low	F_2	100%	H_2	2.000
10	82	2	1 Multiple Single	1 High Mid	CW	Raster	Low	F_2	Medium	H_1	3.000
11	5	2	Multiple	1 High Low	Stable	1 Steady Circular	Nominal	F_3	Short	H_2	2.000
12	89	2	Multiple	1 High Mid	Agile	Circular	Medium	F_1	1 100% Short	H_3	4.000

Table 3
(*Continued*)

Experiment	Item	Distance Item (Mean)	A RF Diversity	B RF	C PRI Type	D Scan Type	E Peak Power	F	G Illumination	H	Observation m_1 (Defects)
13	65	0	Multiple	Mid	CW	Circular	Low	F_1	Short	H_2	2.000
14	69	1	1 Multiple Single	Mild	Stable	Raster Conical	Nominal	F_2	100%	H_3	0.000
15	149	2	Multiple	Mild	Agile	1 Steady Raster	Medium	F_3	1 Medium Short	H_1	4.000
16	32	1	1 Multiple Single	Low	CW	Raster Conical	Medium	F_3	100%	H_2	1.000
17	134	1	Multiple	Low	Stable	Steady	Low	F_1	1 Medium 100%	H_3	0.000
18	155	1	1 Multiple Single	Low	Agile	Circular	Nominal	F_2	Short	H_1	0.000
Iteam mean		1.00									
Total fault analysis Valid (mean)		15.75	14	13	18	15	18	18	12	18	
Incorrect (mean)		2.26	4	5	0	3	0	0	6	0	

1356

places a "color" valuation on the experiment in the form of green, yellow, or red (a defect count of 0, 1, or 2, respectively). Green is a stable signal with no variations or anomaly, yellow may indicate some concern regarding timeliness or stability of the identification, and red indicates a more serious malfunction, such as multiple emitter types identified for the one emitter tested. Typical emitter performance evaluation was graded as shown in Table 4.

The signal-to-noise (SN) ratio for this experiment was calculated using the relationship

$$SN = -20 \log\left(\frac{\text{total defects}}{n + 1}\right)$$

Since the problem type is "maximum-the-best", the SN ratio for a perfect system is 0 dB. Any error in performance yields a negative SN ratio. The relationship including "defects plus one" was used since the log (0) was undefined and log (1) was zero. The value for n is 18 to compute the entire array SN ratio, while n may vary from mean parameter count (6 or 9) because certain emitters were not perfect orthogonal fits. As an example, the RF diversity factor is segmented with $n = 5$ for the parameter "multiple" and $n = 13$ for the parameter "single."

A sample calculation is shown here for computing the SN ratio for "RF diversity." Since this is a two-level factor, an orthogonal array would contain nine entries for the parameter "single" and nine entries for the parameter "multiple." Since only five of the experiments contain the parameter "multiple," only those parameters are used for the calculation. These include experiments 11, 12, 13, 15, and 17. Hence,

Table 4

Calculation of emitter performance

| Experiment | Defects | | | | |
(Emitter)	Type	Mode	AOA	Color	Total
1	0	0	1	0	1
2	0	0	0	0	0
3	2	2	0	2	6

$$SN = 20 \log\left(\frac{M_{11} + M_{12} + M_{13} + M_{15} + M_{17}}{5 + 1}\right)$$

$$SN = 20 \log\left(\frac{2 + 4 + 2 + 4 + 0}{5 + 1}\right)$$

$$= -10.6 \text{ dB}$$

For the parameter "single,"

$$SN = 20 \log$$

$$\left(\frac{\begin{array}{c} M_1 + M_2 + M_3 + M_4 \\ + M_5 + M_6 + M_7 + M_8 + M_9 \\ + M_{10} + M_{14} + M_{16} + M_{18} \end{array}}{13} + 1\right)$$

$$SN = 20 \log$$

$$\left(\frac{\begin{array}{c} 1 + 0 + 6 + 0 + 4 + 0 + 2 \\ + 0 + 2 + 3 + 0 + 1 + 0 \end{array}}{13} + 1\right)$$

$$= -7.8 \text{ dB}$$

7. Confirmation

The reduction in test time surpassed the goal of a 50% decrease in weekly test time. The 18 emitters can be tested in only four hours, a reduction of 4:1 in effort. This is quite significant given that the sensitivity to test results did not yield any additional system weakness when the monthly test sequence was conducted. The monthly test sequence extended over a 16-hour period, with a number of emitters remaining to be evaluated.

Table 5 shows test results during several weeks of development but does not represent the final system configuration. Several important results can be observed from this table:

1. The overall SN ratio increased from -9.4 dB to -5.0 dB. This yields a steady improvement for each system upgrade, which can be used to track system performance.

2. From the first test conducted, the system demonstrated a sensitivity to input signal peak power. This was significant because peak power is one parameter that is not normally evaluated in the conventional sequence, due

Table 5
System performance summary

Factor/ Parameter	Test					
	1	2	3	4	5	6
Mean SN ratio (dB)	−9.4	−8.2	−8.7	−6.9	−5.8	−5.0
RF diversity						
Single	−8.9	−5.0	−7.8	−5.0	−5.3	−3.3
Multiple	−10.6	−13.3	−10.6	−10.6	−6.9	−8.3
RF frequency						
Low	−8.8	−2.0	−8.8	−4.9	−3.5	0.0
Mid	−10.9	−9.5	−10.6	−8.8	−7.5	−8.8
High	−7.4	−9.1	−5.3	−5.3	−4.4	−0.0
PRI type						
CW	−9.5	−2.5	−8.0	−4.4	−1.3	−6.0
Stable	−8.2	−9.5	−6.0	−8.0	−8.0	−2.5
Agile	−10.0	−10.5	−11.3	−8.0	−6.7	−6.0
Scan type						
Steady	−13.3	−5.1	−7.6	−6.0	−6.8	−5.1
Circular	−7.6	−11.6	−8.3	−9.5	−7.6	−6.8
Raster	−7.0	−7.0	−9.6	−5.5	−3.5	−3.5
Peak power						
Nominal	−4.4	−6.7	−5.3	−5.3	−1.3	0.0
Medium	−8.5	−10.9	−8.0	−6.7	−7.4	−4.4
Low	−13.0	−6.0	−11.7	−8.5	−7.4	−8.5

to the large quantity of emitters tested. Test 2 shows an improvement in the mean SN ratio not consistent in sequence with the remainder of the group. Upon investigation it was determined that correct peak power levels were not maintained for that group. It was therefore concluded that the Taguchi method identified, early in the development process, a system-sensitive parameter that would not normally have been recognized.

3. The Taguchi method provides a perspective of system performance perhaps not readily evident when testing a group of independent emitter types. That is the correlation of EW system sensitivity to the specific noise factors selected.

4. Additional tests may be added to the L_{18} array to provide an added level of robustness to the project. As noted in Table 1, the emitter angle

of arrival and background noise have been noted as noise factors but not yet implemented. Angle of arrival testing will consist of placing the emitter at three different angular positions from the EW system. Background noise has been defined to simulate unknown emitter types.

8. Conclusions

An L_{18} orthogonal array was employed successfully to yield a robust test methodology for EW systems. The robust group of 18 emitter types, coupled with operationally based noise factors, has surpassed the capability established by testing each radar type individually. Analysis of the L_{18} test results compared with test results from the entire domain of emitters demonstrates that the L_{18} can find more system sensitivities than testing the entire ensemble of emitters in a conventional manner.

What is more, the L_{18} array used contained experiments (emitters) that were not quite orthogonal, thereby permitting a selection of emitters from a standard group. This eliminated the time and effort necessary to generate simulations for a new, al-

beit not real emitter set. The L_{18} orthogonal array has effectively demonstrated the capability to test more system functionality in significantly less time than is possible using conventional methods.

This case study is contributed by Stan Goldstein and Tom Ulrich.

Streamlining of Debugging Software Using an Orthogonal Array

Abstract: In the development process of software, debugging before shipping a product is one of the most important and time-consuming processes. Where bugs are found by users after shipment, not only the software per se but also its company's reputation will be damaged. On the other hand, thanks to the widely spreading Internet technology, even if software contains bugs, it is now easy to distribute bug-fix software to users in the market through the Internet. Possibly because of this trend, the issue of whether there are software bugs for a personal computer seems to have become of less interest lately. However, it is still difficult to correct bugs after shipping in the case of software installed in hardware. Thus, we need to establish a method of removing as many bugs as possible within a limited period before shipment. In this study, a debugging method using an orthogonal array is illustrated.

1. Introduction

No effective debugging method has been developed to date. Since a debugging procedure or the range depends completely on each debugging engineer, a similar product might be of different quality in some cases. In addition, because it is so labor intensive, debugging is costly. Thus, a more efficient method of finding bugs could lead to significant cost reduction.

In the context addressed above, Genichi Taguchi has proposed an effective debugging method in two articles titled "Evaluation of objective function for signal factor" [1, 2] and "Evaluation of signal factor and functionality for software" [3]. In reference to these, we detail the results of our experiment on debugging using an orthogonal array.

Since this methodology is a technique for evaluating an objective function for multiple signals, it is regarded as related not to quality engineering but rather to the design of experiments, in a sense.

The method that we propose suggests that we allocate items selected by users (signal factors) to an L_{18} or L_{36} orthogonal array, run software in accordance with a combination of signal factors in each row of the orthogonal array, and judge, using 0 and 1, whether or not an output is normal [4]. Subsequently, using the output obtained, we calculated a variance or interaction to identify which combination of factors was most likely to cause bugs. Through these steps we can find almost all bugs caused by each combination of factors, and we only need to check for the remaining bugs based on single factor changes.

2. Debugging Experiment Using Orthogonal Array

Using the β version of our company's software, we performed an experiment grounded on an orthogonal array. This is because the β version contains numerous bugs, whose existence has been recognized. Therefore, the effectiveness of this experiment can easily be confirmed. However, since bugs detected cannot be corrected, we cannot check whether or not the trend regarding the number of bugs is decreasing.

As signal factors we selected eight items that can frequently be set up by users, allocating them to an L_{18} orthogonal array. When a signal factor has four or more levels—for example, continuous values ranging from 0 to 100—we selected 0, 50, and 100. When dealing with a factor that can be selected, such as patterns 1 to 5, three of the levels that are most commonly used by users were selected. Once we assigned these factors to an orthogonal array, we noticed that there were quite a few two-level factors. In this case we allocated a dummy level to level 3. (Because of our company's confidentiality rule, we have left out the details about signal factors and levels.)

For the output, we used a rule of normal = 0 and abnormal = 1, based on whether the result was what we wanted. In some cases, "no output" is the right output. Therefore, normal or abnormal is determined by referring to the specifications. Signal factors and levels are shown in Table 1.

3. Analysis of Bugs

From the results of Table 2, we created approximate two-way tables for all combinations. The upper left part of Table 3 shows the number of each combination of A and B: A_1B_1, A_1B_2, A_1B_3, A_2B_1, A_2B_2, and A_2B_3. Similarly, we created a whole table for all combinations. A part where many bugs occur on one side of this table was regarded as a location with bugs.

Looking at the overall result, we can see that bugs occur at H_3. After investigation it was found that bugs do not occur in the on-factor test of H, but occur by its combination with G_1 ($= G_1'$, the same level because of the dummy treatment used) and B_1 or B_2. Since B_3 is a factor level whose selection blocks us from choosing (or annuls) factor levels of H and has interactions among signal factors,

Table 1
Signal factors and levels

	1	2	3
A	A_1	A_2	—
B	B_1	B_2	B_3
C	C_1	C_2	C_3
D	D_1	D_2	D_3
E	E_1	E_2	E_2'
F	F_1	F_2	F_1'
G	G_1	G_2	G_1'
H	H_1	H_2	H_3

Table 2
L_{18} orthogonal array and output

No.	A	B	C	D	E	F	G	H	Ouput
1	1	1	1	1	1	1	1	1	0
2	1	1	2	2	2	2	2	2	0
3	1	1	3	3	2'	1'	1'	3	1
4	1	2	1	1	2	2	1'	3	1
5	1	2	2	2	2'	1'	1	1	0
6	1	2	3	3	1	1	2	2	0
7	1	3	1	2	1	1'	2	3	0
8	1	3	2	3	2	1	1'	1	0
9	1	3	3	1	2'	2	1	2	0
10	2	1	1	3	2'	2	2	1	0
11	2	1	2	1	1	1'	1'	2	0
12	2	1	3	2	2	1	1	3	1
13	2	2	1	2	2'	1	1'	2	0
14	2	2	2	3	1	2	1	3	1
15	2	2	3	1	2	1'	2	1	0
16	2	3	1	3	2	1'	1	2	0
17	2	3	2	1	2'	1	2	3	0
18	2	3	3	2	1	2	1'	1	0

Table 3
Binary table created from L_{18} orthogonal array

	B_1	B_2	B_3	C_1	C_2	C_3	D_1	D_2	D_3	E_1	E_2	E_3	F_1	F_2	F_3	G_1	G_2	G_3	H_1	H_2	H_3
A_1	1	1	0	1	0	1	1	0	1	0	1	1	0	1	1	0	0	2	0	0	2
A_2	1	1	0	0	1	1	0	1	1	1	1	0	1	1	0	2	0	0	0	0	2
B_1				0	0	2	0	1	1	0	1	1	1	0	1	1	0	1	0	0	2
B_2				1	1	0	1	0	1	1	1	0	0	2	0	1	0	1	0	0	2
B_3				0	0	0	0	0	0	0	0	0	0	0	0	0	0	0	0	0	0
C_1							1	0	0	0	1	0	0	1	0	0	0	1	0	0	1
C_2							0	0	1	1	0	0	0	1	0	1	0	0	0	0	1
C_3							0	1	1	0	1	1	1	0	1	1	0	1	0	0	2
D_1										0	1	0	0	1	0	0	0	1	0	0	1
D_2										0	1	0	1	0	0	1	0	0	0	0	1
D_3										1	0	1	0	1	1	1	0	1	0	0	2
E_1													0	1	0	1	0	0	0	0	1
E_2													1	1	0	1	0	1	0	0	2
E_2'													0	0	1	0	0	1	0	0	1
F_1																1	0	0	0	0	1
F_2																1	0	1	0	0	2
F_1'																0	0	1	0	0	1
G_1																			0	0	2
G_2																			0	0	0
G_1'																			0	0	2

it was considered as the reason that this result was obtained.

Now the calculated variance and interaction were as follows.

Variation between A and B combinations:

$$S_{AB} = \frac{1^2 + 1^2 + 0^2 + 1^2 + 1^2 + 0^2}{3} - \frac{4^2}{18}$$

$$= 0.44 \qquad (f = 5) \qquad (1)$$

Variation of A:

$$S_A = \frac{2^2 + 2^2}{9} - \frac{4^2}{18} = 0.00 \qquad (f = 1) \qquad (2)$$

Variation of B:

$$S_B = \frac{2^2 + 2^2 + 0^2}{6} - \frac{4^2}{18} = 0.44 \qquad (f = 2) \qquad (3)$$

A summary of all the main effects is shown in Table 4.

Variation of A, B interaction:

$$S_{A \times B} = S_{AB} - S_A - S_B$$

$$= 0.44 - 0.00 - 0.44$$

$$= 0.00 \qquad (f = 2) \qquad (4)$$

Table 4
Main effect

Factor	Main Effect[a]
A	0.0
B	**0.44**
C	0.11
D	0.11
E	0.03
F	0.03
G	**0.44**
H	**1.77**

[a] Boldface numbers indicate significant factorial effect.

As the next step, we divided the combinational effect, S_{AB}, and interaction effect, $S_{A \times B}$, by each corresponding degree of freedom:

$$\text{combinational effect} = \frac{S_{AB}}{5} = 0.09 \qquad (5)$$

$$\text{interaction effect} = \frac{S_{A \times B}}{2} = 0.00 \qquad (6)$$

Now, since these results are computed from our approximate two-way tables, if the occurrence of bugs is infrequent, as in this example, we should consider such results as a clue for debugging. When there are more bugs or a large-scale orthogonal array is used, we need to use these values for finding bug locations.

4. Bug Identification

Finally, we succeeded in finding bugs by taking advantage of each combination of factors (Table 5). As below, using the method as described, the bugs can be found from an observation of specific combinations. Following are the differences between our current debugging process and the method using an orthogonal array:

1. Efficiency of finding bugs
   a. *Current process.* What can be found through numerous tests are mainly independent bugs. To find bugs caused by a combination of factors, we need to perform many repeated tests.
   b. *Orthogonal array.* Through a small number of experiments, we can find independent bugs and bugs generated by a combination of factors. However, for a multiple-level factor, we need to conduct one-factor tests later.

2. Combination of signal factors
   a. *Current process.* We tend to check only where the bug may exist and unconsciously neglect the combinations that users probably do not use.
   b. *Orthogonal array.* This method is regarded as systematic. Through nonsubjective combinations that do not include debug engineers' presuppositions, well-balanced and broadband checkup can be performed.

3. Labor required
   a. *Current process.* After preparing a several-dozen-page check sheet, we have to investigate all of its checkpoints.
   b. *Orthogonal array.* The only task that we need to do is to determine signal factors and levels. Each combination is generated automatically. The number of checkups required is much smaller, considering the number of signal factors.

4. Location of bugs
   a. *Current process.* Since we need to change only a single parameter for each test, we can easily notice whether or not changed items or parameters involve bugs.
   b. *Orthogonal array.* Locations of bugs are identified by looking at the numbers after the analysis.

5. Judgment of bugs or normal outputs
   a. *Current process.* We can easily judge whether a certain output is normal or abnormal only by looking at one factor changed for the test.
   b. *Orthogonal array.* Since we need to check the validity for all signal factors for each

Table 5
Combinational and interaction effects

Factor	Combination	Interaction
AB	0.09	0.00
AC	0.09	0.17
AD	0.09	0.17
AE	0.09	0.25
AF	0.04	0.00
AG	0.15	0.00
AH	0.36	0.00
BC	0.26	0.39
BD	0.14	0.14
BE	0.17	0.19
BF	0.42	0.78
BG	0.22	0.11
BH	0.39	0.22
CD	0.14	0.22
CE	0.17	0.36
CF	0.22	0.44
CG	0.12	0.03
CH	0.26	0.06
DE	0.07	0.11
DF	0.12	0.19
DG	0.12	0.03
DH	0.26	0.11
EF	0.12	0.22
EG	0.16	0.01
EH	0.23	0.01
FG	0.20	0.06
FH	0.42	0.11
GH	0.62	0.44

output, it is considered cumbersome in some cases.

6. When there are combinational interactions among signal factors

   a. *Current process.* Nothing in particular.

   b. *Orthogonal array.* We cannot perform an experiment

following combinations determined in an orthogonal array.

Although several problems remain before we can conduct actual tests, we believe that through use of our method, the debugging process can be streamlined. In addition, since this method can be employed relatively easily by users, they can assess newly developed software in terms of bugs. In fact, as a result of applying this method to software developed by outside companies, we have found a certain number of bugs. From now on, we will attempt to incorporate this method into our software development process.

References

1. Genichi Taguchi, 1999. Evaluation of objective function for signal factor (1). *Standardization and Quality Control,* Vol. 52, No. 3, pp. 62–68.
2. Genichi Taguchi, 1999. Evaluation of objective function for signal factor (2). *Standardization and Quality Control,* Vol. 52, No. 4, pp. 97–103.
3. Genichi Taguchi, 1999. Evaluation of signal factor and functionality for software. *Standardization and Quality Control,* Vol. 52, No. 6, pp. 68–74.
4. Kei Takada, Masaru Uchikawa, Kazuhiro Kajimoto, and Jun-ichi Deguchi, 2000. Efficient debugging of a software using an orthogonal array. *Quality Engineering,* Vol. 8, No. 1, pp. 60–69.

This case study is contributed by Kei Takada, Masaru Uchikawa, Kazuhiro Kajimoto, and Jun-ichi Deguchi.

Part VI

On-Line Quality Engineering

On-Line (Cases 88–92)

CASE 88

Application of On-Line Quality Engineering to the Automobile Manufacturing Process

Abstract: In this study we apply on-line quality engineering (On-QE) to a sampling inspection process for a transmission case after the machining process has been completed.

1. Conventional Approach to Quality in Automobile Manufacturing

In our conventional procedure, we inspected one sample every 50 to 350 parts, according to an interval predetermined by the nature of the characteristics being inspected. By prioritizing operational efficiency, we have made a judgment as to whether a work lies within or beyond a tolerance without process management based on process stability factors such as trends or process capability indexes, C_{pk}. In addition, once the measurement cycle is set up based on engineers' technical experience before a mass production phase, it has been modified very little except through design changes or when there are significant quality problems.

In other words, our conventional management is based on the mindset that "work within tolerances never causes a loss." A problem with this approach is that we have not really known whether working within a tolerance is close to the target, lies on the verge of the limits, or has an increasing/decreasing trend. When a product unit was confirmed to be out of tolerance, we had to measure all the completed products again to try to determine the point at which products beyond tolerance limits had begun to be machined.

If we adjust a target according to a trend, we can prevent a product beyond tolerance from occurring. On the other hand, if a process is stable, with no dispersion and deviation, the measurement cycle should be lengthened. Otherwise, it needs to be shortened. However, conventionally, we have quite often taken an excessive number of measurements, even if a process is stabilized, or an insufficient number of measurements even when unstable.

2. Obstacle in Application to an Automobile Production Line

To perform an On-QE analysis with current data, we collected actual data regarding 18 major characteristics in a process. Table 1 reveals the fact that obvious imbalance between control cost and quality loss leads to a large loss under current conditions, especially in that the quality loss accounts for the majority of the total loss. On the other hand, the optimal configuration improves the balance of loss, thereby reducing the proportion of quality loss to 50%. In addition, by adopting a calculated optimal measurement interval, adjustment interval, and adjustable limit, we could compress the total loss to 1/33 and gain a considerable improvement.

However, judging from our current facilities capability, we realized that the optimal adjustable limit calculated contains infeasible matters such as a 1-μm-level adjustment. Therefore, by determining an optimal adjustable limit, as long as it is practical under the current conditions, we had to recalculate an optimal adjustment interval and measurement interval. As shown in Figure 1, although we needed

Table 1

Evaluation of balance between control cost and quality loss[a]

	Current Condition		Optimal Condition	
	Monetary Value[b]	Proportion to Total Loss (Balance) (%)	Monetary Value[b]	Proportion to Total Loss (Balance) (%)
Control cost				
Measurement cost (1)	5	0.5	3	10.0
Adjustment cost (2)	1	0.1	12	40.0
Quality loss (3), (4), (5)	994	99.4	15	50.0
Total loss [sum of (1)–(5)]	1000	100.0	30	100.0

[a](1) Measurement cost in process, (2) adjustment cost in process, (3) loss within adjustable limit, (4) loss beyond adjustable limit, (5) loss due to measurement error.
[b] The monetary value is estimated by assuming that the total loss = 100.

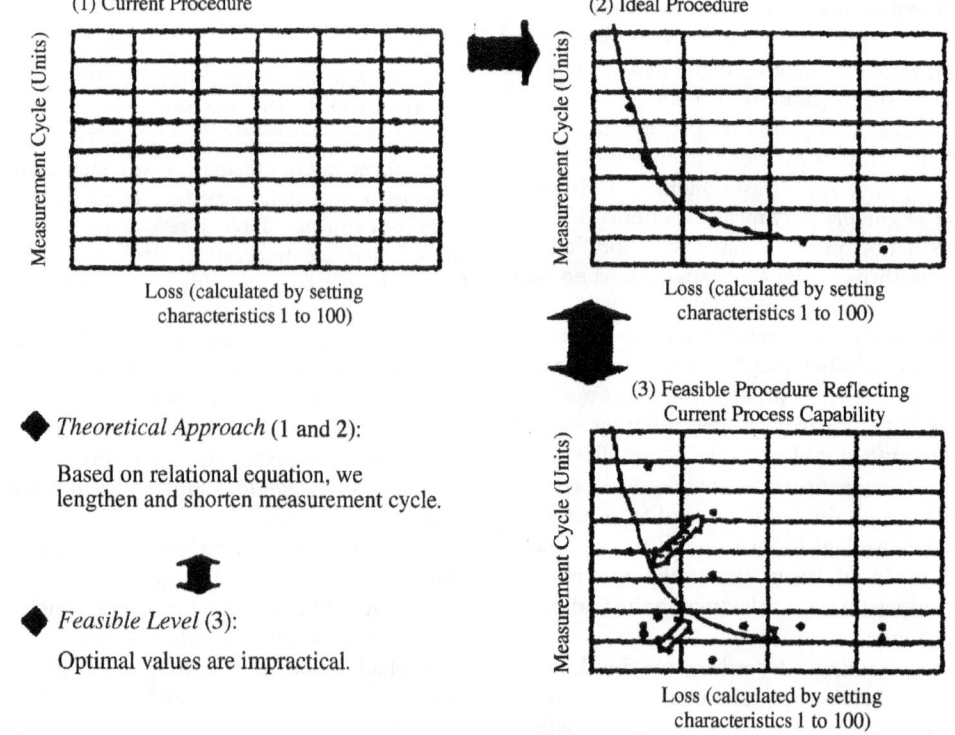

Figure 1
Limitation of application of analytical result to actual manufacturing line

to lengthen or shorten the fixed measurement interval (1) by following an ideal curve (2), the measurement cycle based on the current feasible adjustable limit follows (3).

As a next step, we analyzed the relationship between the process capability indexes C_p and C_{pk}, which were regarded as a basis in our conventional process control and loss. To convince management people, we had to find a connection between the On-QE method and our traditional management method. By doing this analysis, we confirmed that "a larger loss leads to smaller C_p and C_{pk}, and a smaller loss leads to larger C_p and C_{pk}" (Figure 2). In sum, our conventional approach to improving a process capability index is consistent with the new approach to reducing a loss.

To obtain the full understanding and cooperation of people in manufacturing lines for the use of On-QE, we improved processes with both the loss function and process capability indexes for the time being.

3. Design of System for Applying On-QE to an Automobile Production Line

Although we have successfully clarified the expected benefit earned thus far through the use of On-QE, the following three issues exist for its actual application to our manufacturing processes:

1. Since data analysis of the four-month-long results unveiled instability in the manufacturing processes, we were urged to build up a controlled condition (control limit: $\pm\Delta/3$), which is regarded as a precondition for applying the On-QE. More specifically, it was necessary to design a system for facilitating this corrective activity or a business system for effectively collection and analysis measurements with immediate feedback to operations.

2. Because we lack a system for managing and accumulating actual data, including time for a measurement or adjustment operation, we have not been able to understand actual data such as producer's loss, adjustment interval, time lag, or measurement/adjustment cost. Because of this, we need to build up a system to improve data accuracy by gathering and incorporating data efficiently into the calculation of loss.

3. Some of the theoretical optimal adjustable limits and the like are not feasible because of the technical capability of current machinery or tools. For instance, even if an optimal adjustable limit is calculated to be 1 μm, current machinery is only accurate to 5 μm. By identifying all these technical limitations, we need to prepare an operational rule reflecting the actual feasibility of our theoretical values.

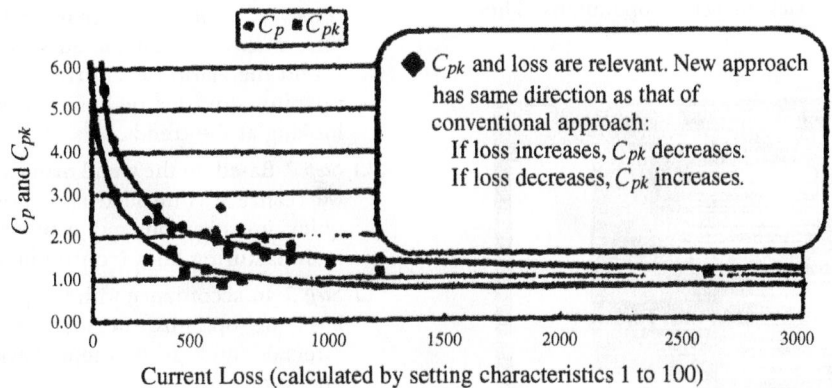

Current Loss (calculated by setting characteristics 1 to 100)

Figure 2
Relationship between quality loss and C_p and C_{pk} in the manufacturing process

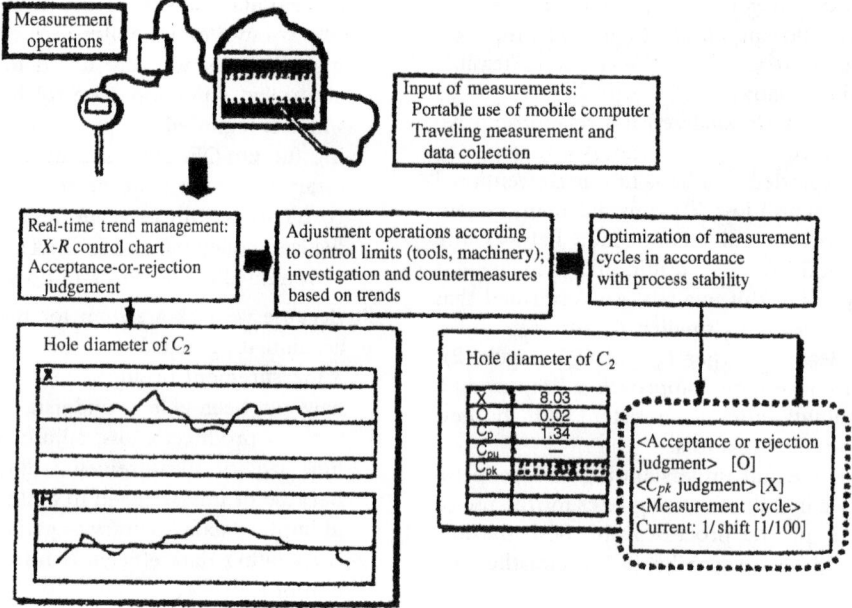

Figure 3
Work flow based on real-time quality data trend management and analysis system

For the foregoing issues, we took the following technical measures. First, although we have conventionally stored measurements on paper, they have not been fully used for analytical tools such as a control chart. Therefore, using the following three-step process, we built up a business system for efficient gathering and analysis of measurement data and feedback to actual operations. Then, in accordance with each level of process stability, we applied On-QE to each characteristic.

❑ *Step 1.* We developed and introduced a real-time quality data trend management and analysis system (QTS), shown in Figures 3 to 5. More specifically, immediately after collecting sampling test data in real time, we create or renew an $\bar{X}$–R control chart and at the same time make a judgment on control limits or sound the alarm for an irregular trend. Each operator does his or her own operations by looking at the trends.

❑ *Step 2.* Based on the trend management above, we realize a controlled condition through a quick investigation and countermeasures in manufacturing lines (control limit: $\pm \Delta/3$).

❑ *Step 3.* In accordance with the process stability level, we optimize measurement/adjustment intervals and feed them back to actual operations.

In addition, for reliable feedback of plant-administrative and production engineering issues that are clearly defined through our new activity im-

Figure 4
QTS: real-time display of control charts

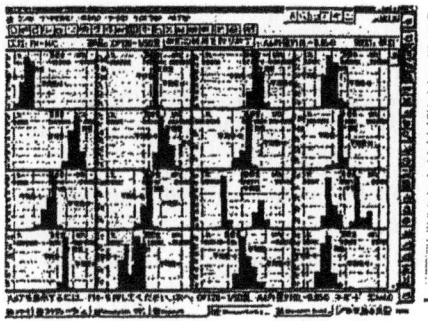

Figure 5
QTS: display of histograms for all characteristics

plemented in an efficient and intensive manner, we organize a promotion group and determine each role together with plant administrative and production engineering departments.

By making the most of periodic data related to measurements or tool changes, which are gathered by the QTS, we calculate a measurement/adjustment interval and time lag. By multiplying them by cost indexes, we convert them into a measurement/adjustment cost such that a loss can be computed later (parameters in the On-QE: C, D, u, and l). As for application procedures and the applicability of

theoretical optimal values, we create business rules reflecting operational constraints or efficiency in the promotion meetings.

4. Achievements and Future Considerations

Based on analysis of the actual data for the 18 major characteristics, we defined four different levels representing process stability (from good to bad, each figure in parentheses indicates the number of characteristics for each level):

- ❑ *Level 3* (7). Trend management with a control limit of $\pm \Delta / 3$ can be implemented because of process stability.

- ❑ *Level 2* (1). Plant administration needs to be improved because of significant deviation (intentionally biased adjustment of tools when changed).

- ❑ *Level 1* (5). A process needs to be improved because of the large variability.

- ❑ *Level 0* (5). A process needs to be investigated, analyzed, and corrected immediately because of process instability and various trends (e.g.,

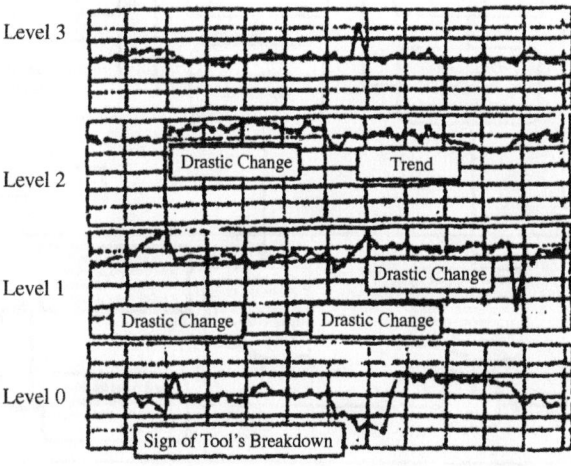

Figure 6
Example of X control chart for each process stability

insufficient adjustment of tools when changed or neglect of adjustment of tools when a part beyond control limits is found).

(See Figure 6.) Accordingly, we prepared a flowchart of process improvement and began to implement a systematic activity (Figure 7).

Through this activity, we obtained the following benefits:

1. The loss by process management in the current manufacturing lines was clarified, and drastic improvement in process stability was obtained (Figure 8). Consequently, we reduced problems dramatically in the process,

thereby diminishing quality problems (Figure 9). In addition, by prolonging the measurement cycle (from 1/50 in our conventional process to 1/100 or 1/350), we achieved a 55% reduction in labor hours (Figure 10).

2. As a basis for a systematic activity, we established a business work flow that can improve quality continuously.

3. By clarifying the gap between the optimal measurement/adjustment cycle calculated and limitations in its use in the current processes and technical issues obtained through tracking of countermeasures and results, we

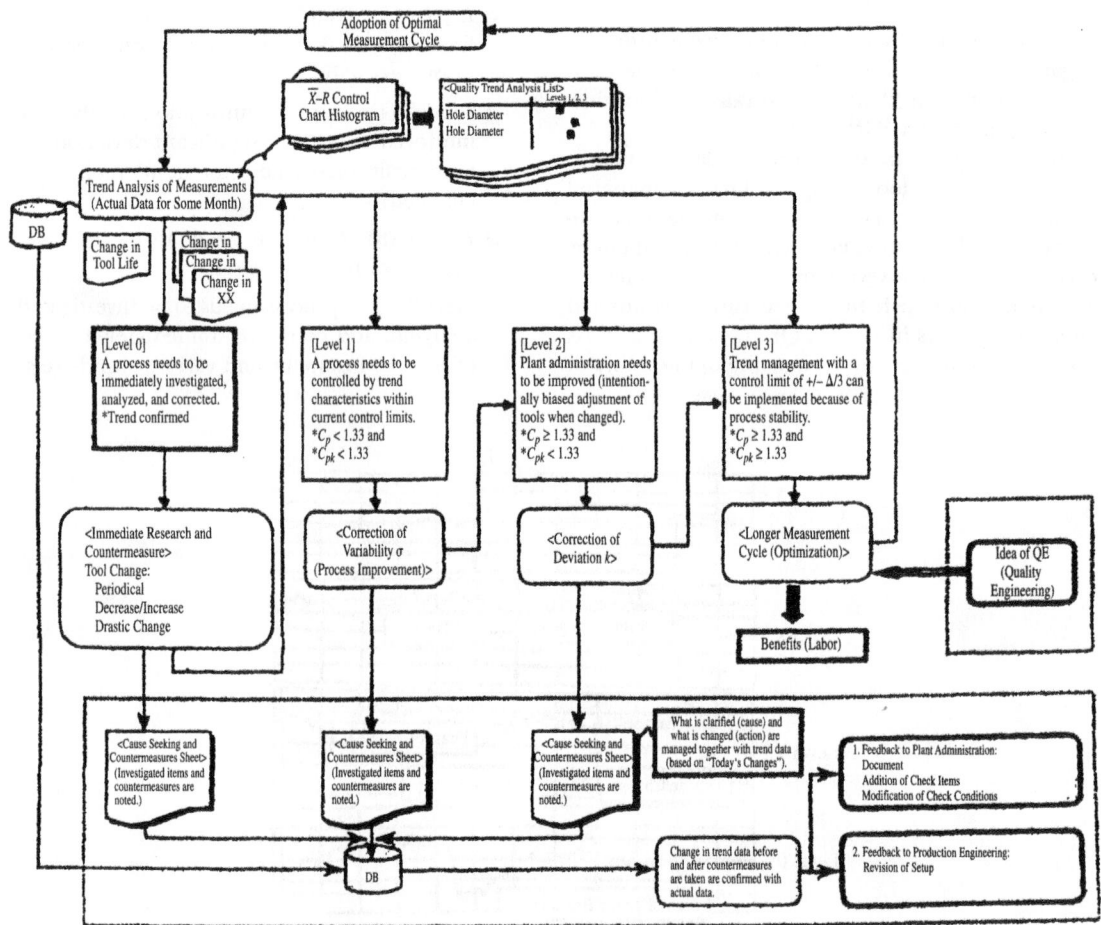

Figure 7
Flowchart of plant administration/process improvement based on process stability focused on trend values

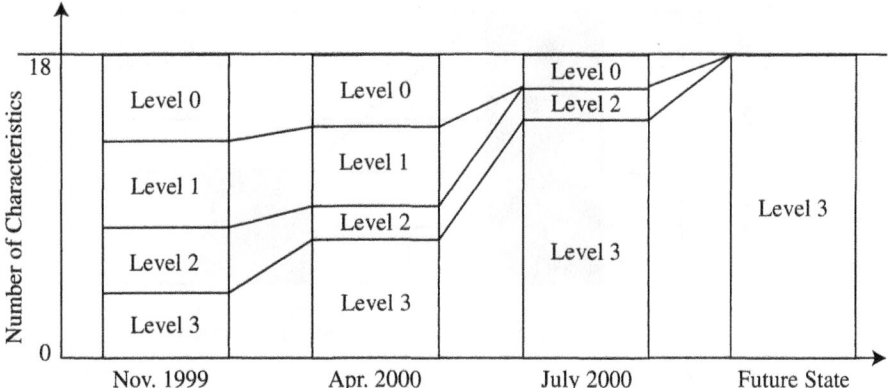

Figure 8
Transition of process stability for characteristics in 1/50 sampling test

established standards in measurement design in the mass production preparation phase.

The introduction to a new idea of the On-QE focus on a loss greatly affects conventional quality management, such as process management based on C_p and C_{pk}, or efficiency-oriented measurement design and operations that rest on a sampling test method in production engineering departments and plants. Particularly, in order to obtain an optimal value identified through calculation with the loss function, more resources, such as technology, labor hours, or management are needed for measurement or adjustment operations. Therefore, two

key factors are that new technologies are developed from the two perspectives of accuracy and efficiency in measurement and that manufacturing management can essentially understand the loss function.

Therefore, when introducing a new manufacturing line, we need to design processes that enable us to implement precision measurement or adjustment operations while utilizing On-QE in the future. On the other hand, to obtain management's understanding, it is quite important to accumulate examples that make all participants feel involved: for example, the cost of tools or wasted materials or savings from reduced labor hours. In short, by

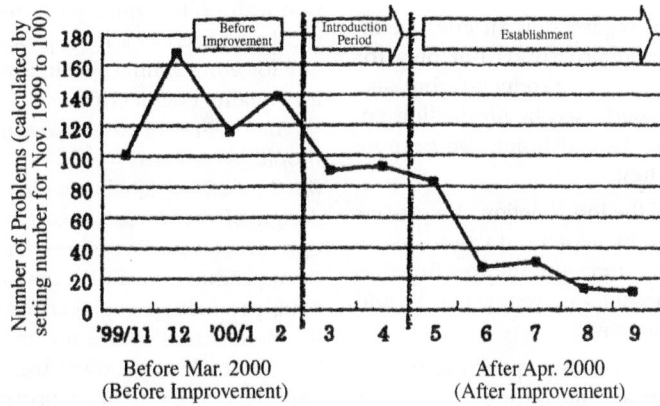

Figure 9
Transition of quality problems

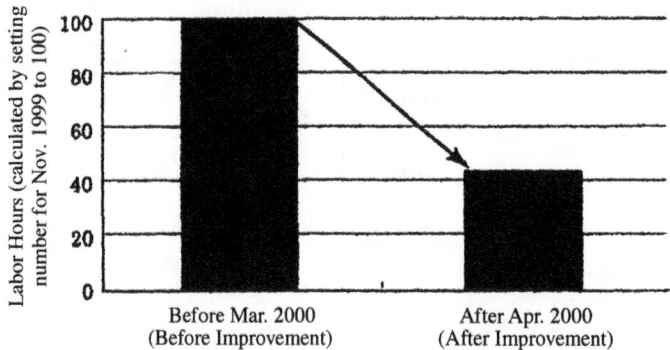

Figure 10
Labor reduction for sampling test (On-QE-applied seven characteristics)

5. Effective Application of On-QE

When applying a new idea or methodology to a production line, we must understand a new activity's significance by explaining the differences from and connection with a conventional procedure. In this case, to foster examples that can make all participants feel comfortable with the benefits is a key to success. In our study we achieved a longer measurement cycle that is closely related to actual operations. Through our application of On-QE to the current lines, we realized one of the most practical uses of On-QE.

The loss function is regarded as an effective index that should be used proactively not only for process improvement in mass production but also for process planning conducted by production engineering departments. In conclusion, we propose the following approaches:

1. Make the best of the loss function as a process evaluation index in mass production. As we have done in this study, use the loss function to analyze and optimize current processes and measuring instruments.

2. As an index to determine specifications for optimal processes and measuring instruments in mass production lines, utilize the loss function. If data obtained in technical trials are used, we can set up optimal methods of taking measurements, making adjustments, or calibrating measuring instruments according to a tolerance and loss for each characteristic: that is, proactively make the most of the loss function not only to improve current lines but also to design new lines. More specifically, fix specifications of machinery and measuring instruments that can be adjusted in accordance with On-QE after mass production begins or to design measurement processes.

6. Approach to Machining Systems for Developing Production Technologies

On-QE has proven its merits when used in association with off-line quality engineering. In this case it can be utilized as a mass production evaluation index for a machining system that is developed by off-line quality engineering.

An off-line quality engineering–applied system should be assessed on the basis of On-QE. In other words, if off-line quality engineering achieves robust design of the system, it can stabilize mass production processes. Therefore, On-QE analysis during mass production can become a yardstick for evaluating production engineering. On the other hand, if new technical issues are identified, as long as a business cycle feeds them back to production engineering development or process or machinery design, we can steadfastly enhance our technological level.

Reference

Yoshito Ida and Norihisa Adachi, 2001. Progress of on-line quality engineering. *Standardization and Quality Control*, Vol. 54, No. 5, pp. 24–31.

This case study is contributed by Yoshito Ida and Norihisa Adachi.

CASE 89

Design of Preventive Maintenance of a Bucket Elevator through Simultaneous Use of Periodic Maintenance and Checkup

Abstract: To solve the problem of apparent failure on a casting machine, we studied three different types of preventive maintenance using on-line quality engineering. We discuss one of them, for which an optimal maintenance method was attempted.

1. Introduction

Preventive maintenance is a scheduled maintenance method to prevent failures during use of an item and to maintain availability of the item. A key issue on a practical basis is how to conduct optimal maintenance.

When operating a casting machine (Figure 1), we suffered considerable damage because of an "apparent" failure caused when a roller of a chain (Figure 2b), used for a bucket elevator (Figure 2a), was worn out and suddenly tore apart. Like a tray elevator, a bucket elevator transports a work by hanging it on a chain. An *apparent failure* is defined as a case where a machine stops when an automated measurement system fails.

To prevent the same failures from occurring again, referring to the history of repairs in the maintenance record book, we designed an optimal maintenance method in accordance with the following procedures:

1. We understand the current situation correctly.

2. Using on-line quality engineering, we analyze preventive maintenance from the viewpoint of the following three types and select one optimal maintenance method (Figure 3):

    a. Periodic checking

    b. Periodic maintenance

    c. Simultaneous use of periodic checking and maintenance

3. Using the loss function, we study how much the optimal maintenance method can improve the current and determine the final optimal method.

2. Current Maintenance of Bucket Elevator

To determine the optimal maintenance method, we began by examining the current maintenance method.

Mean Time between Failures

For the current maintenance of chains, we checked the wear and fracture in a roller once a month, and with these data we predicted machinery failures and planned a maintenance schedule. To understand the failure modes that have occurred, using recent repair records for the bucket elevator, we rearranged 19 consecutive failure data (Table 1) and estimated the mean time between failures:

mean time between failures

$$= \frac{\text{total of operational days}}{\text{number of stops}}$$

$$= \frac{(3.5)(2) + (5.3)(5) + \cdots + (11.0)(3)}{19} = \frac{130.4}{19}$$

$$= 6.9 \text{ months} \tag{1}$$

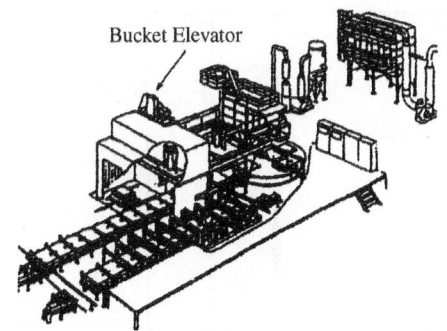

Figure 1
Vpro casting machine

As a result, we came up with approximately seven months, or 147 days, as the total number of operational days in a month is 21.

Arrangement of Parameters

The following are necessary parameters for the design of a preventive maintenance method.

- ❏ *Mean time between failures:* $\overline{u} = 147$
- ❏ *Loss due to stoppage when a machine goes out of order:* $A = 3,600,000$ yen
- ❏ *Checkup cost:* $B = 500$ yen
- ❏ *Current maintenance cost:* $C = 870,000$ yen
- ❏ *Periodic maintenance cost:* $C' = 670,000$ yen
- ❏ *Functional limit:* $\Delta_0 = 90\%$

❏ *Current checkup interval:* $n_0 = 21$ days
❏ *Current maintainable limit:* $D_0 = 80\%$

D_0 and Δ_0 indicate the amount of roller wear. A usable limit (functional limit Δ_0) that a manufacturer defines is the time when "a roller has a hole or cracks due to wear and tear" or "a bush has a hole due to wear and tear." Wear and tear occurs because of wear between each link plate and interference between the side of a roller and the inner surface of a link plate (Figure 4). If the amount of wear exceeds one-third of the thickness of a link plate, the possibility of insufficient chain strength needs to be assessed. In our experience, we set as Δ_0 90% of a period until a hole appears due to wear and defined D_0 as a point in time slightly before the functional limit. (We did not set up a control for the case where wear exceeds one-third of the regular thickness.)

We considered the difference between the current maintenance cost, C, and the periodic maintenance cost, C'. When implementing periodic maintenance, we need only C'. However, when the current checking method is used, because we need to check the machinery due to possible abrupt failures, a loss attributed to extra labor is required, in addition to the periodical maintenance cost, C'.

The checkup cost was estimated to be 500 yen, which is much lower than other costs. The reason is that only 5 minutes is required for this checkup because it requires only a visual examination of dimensional wear of representative rollers once a month.

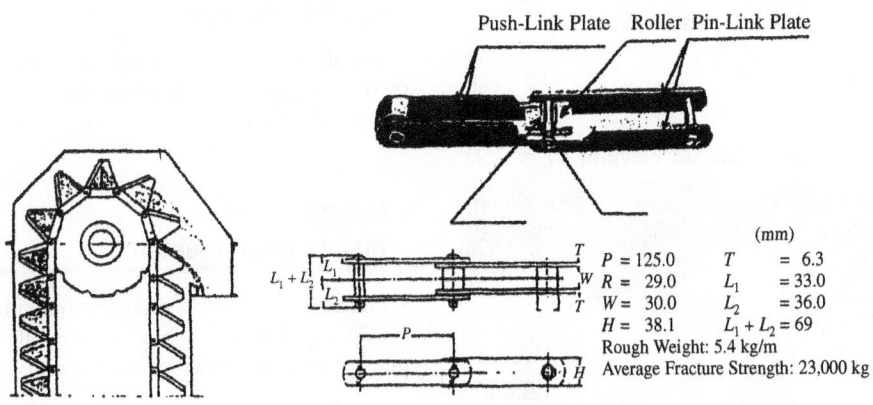

Push-Link Plate   Roller   Pin-Link Plate

(mm)

$P = 125.0$	$T = 6.3$
$R = 29.0$	$L_1 = 33.0$
$W = 30.0$	$L_2 = 36.0$
$H = 38.1$	$L_1 + L_2 = 69$

Rough Weight: 5.4 kg/m
Average Fracture Strength: 23,000 kg

(*a*) Suspension Elevator   (*b*) Chain Mechanism and Dimensions

Figure 2
Vertical bucket elevator

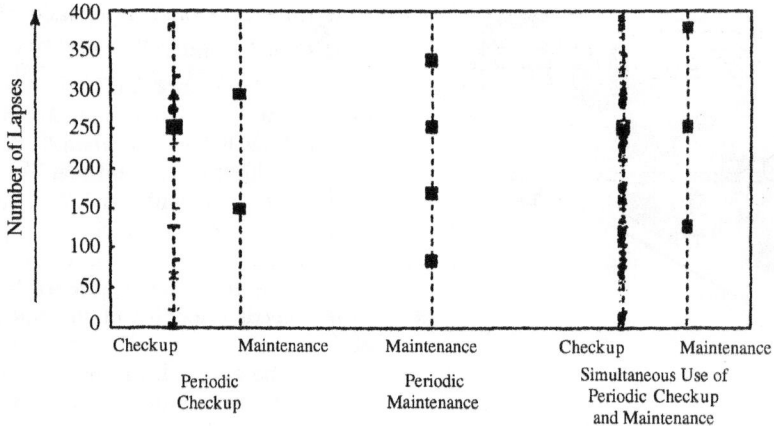

Figure 3
Types of preventive maintenance

Additionally, this casting machine recycles and reuses casting sand. After taking out a cast product, the remaining cast sand is transported by a bucket elevator and put into a sand hopper after it is cooled down. The operational time of this bucket elevator is generally constant, so we estimated not per product but per day.

Current Evaluation

The current management method is based on an ex post facto process in which a part is replaced by a new part when it has broken. In this case, no diagnosis is attempted. Using the idea of operations management, we calculated the loss function, L_0, The maintenance cost, C, was computed as

repair cost,

$$C = \text{(loss by machinery stoppage)} \\ + \text{(current maintenance cost)}$$
$$= 3{,}600{,}000 + 870{,}000 = 4{,}470{,}000 \text{ yen} \qquad (2)$$

$$L_0 = \text{(checkup cost)} \\ + \text{(loss by machinery stoppage)} \\ + \text{(repair cost)} + \text{(time lag loss)}$$

$$= \frac{B}{n_0} + \frac{n_0 + 1}{2}\frac{A}{u} + \frac{C}{u} + \frac{lA}{u}$$

$$= 0 + 0 + \frac{4{,}470{,}000}{147} + 0$$

$$= 30{,}408.2 \text{ yen/days} \qquad (3)$$

Therefore, the annual total of losses amounts to 7.66 million yen/year. [The symbol l in equation (3) indicates the time lag.]

3. Optimal Preventive Maintenance Method

To choose an optimal maintenance method, we used the loss function to analyze the following three types of preventive maintenance methods.

Using Only Periodic Checkups

We calculated an optimal periodic checkup interval, n, and maintenance limit, D, that minimize equa-

Table 1
Maintenance data for bucket elevator ($n = 19$)

Months until Next Replacement	Cumulative Frequency	Cumulative Percentage
3.5	2	10.5
5.3	7	36.8
6.0	10	52.6
7.1	13	68.4
8.2	16	84.2
11.0	19	100.0

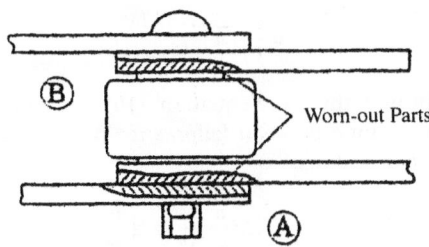

Figure 4
Wear and tear of link plate

tion (3). The optimal periodic checkup interval n is computed as

$$n = \sqrt{\frac{2B}{A}}\, \bar{u} = \sqrt{\frac{(2)(500)}{3,600,000}}\,(147) = 2.45 \quad (4)$$

Next, we calculated an optimal maintenance limit, D. Assuming the time to reach the functional limit, Δ_0, is proportional to the square of the limit, the average maintenance interval, u, for a certain maintenance limit, D, is calculated by

$$u = u_0 \frac{D^2}{D_0^2} \quad (5)$$

Setting a mean time between failures with no maintenance to $\bar{u}$, we can predict a failure rate according to equation (5):

$$\bar{u} = u_0 \frac{\Delta_0^2}{D_0^2} \quad (6)$$

Then the optimal maintainable limit, D, is

$$D = \left(\frac{3C}{A}\frac{\bar{u}}{u_0}\,D_0^2\Delta_0^2\right)^{1/4} \quad (7)$$

Substituting equation (6) into (7), we have

$$D = \left(\frac{3C}{A}\right)^{1/4}\Delta_0 \quad (8)$$

The optimal maintainable limit D is computed as follows:

$$D = \left(\frac{3C}{A}\right)^{1/4}\Delta_0$$

$$= \left[\frac{(3)(870,000)}{3,600,000}\right]^{1/4}(90) = 83.04\% \quad (9)$$

Therefore, we can see that a periodic checkup

every three days and a maintainable limit of 83% are optimal. As a next step, we calculated the loss function L_1 under optimal conditions:

$L_1 =$ (checkup cost) + (maintenance cost) + (loss by machinery stoppage due to maintainable limit) + (loss by machinery stoppage due to periodical checkup)

$$= \frac{B}{n} + \frac{C}{u} + \frac{A}{u}\Delta_0^2\left(\frac{D^2}{3} + \frac{n}{2}\frac{D^2}{u}\right)$$

$$= \frac{500}{3} + \frac{870,000}{125}$$

$$+ \frac{3,600,000}{(147)(90^2)}\left[\frac{83^2}{3} + \frac{3}{2}\left(\frac{83^2}{125}\right)\right]$$

$$= 166.7 + 6960 + 6942.8 + 249.9$$

$$= 14{,}319.4 \text{ yen/day} \quad (10)$$

So the annual losses total 3.608 million yen. The mean time between failures for the optimal periodic checkup interval is computed as

$$u = \bar{u}\frac{D_0^2}{\Delta_0^2} = 147\left(\frac{83^2}{90^2}\right) = 125 \text{ days} \quad (11)$$

Using Only Periodic Maintenance

The loss function for periodic maintenance L_2 is determined by a periodic maintenance cost and loss by failures occurring within a maintenance interval, calculated by the following equation:

$L_2 =$ (periodical maintenance cost) + (loss by machinery stop page)

$$= \frac{C'}{u'} + \frac{A}{2\bar{u}^2}u'$$

$$= \frac{670,000}{84} + \frac{3,600,000}{(2)(147^2)}(84)$$

$$= 7976.2 + 6997.1$$

$$= 14{,}973.3 \text{ yen/day} \quad (12)$$

This amounts to 3.773 million yen/year. Next, the optimal periodic maintenance interval u' was computed:

$$u' = \sqrt{\frac{2C'}{A}}\,\bar{u}$$

$$= \sqrt{\frac{(2)(670,000)}{3,600,000}}\,(147) = 89.7 \qquad (13)$$

Since 89.7 days is equivalent to 4.3 months, by rounding it off to a simple number, we set the optimal periodic maintenance interval to 4.0 months (84 days).

Using Both Periodic Checkups and Maintenance

The loss function, L_3, was calculated by the equation

L_3 = (periodic maintenance cost) + (checkup cost)
    + (maintenance cost) + (loss by machinery stoppage due to maintainable limit) + (loss by machinery stoppage due to periodic checkup)

$$= \frac{C'}{u'} + \left[\frac{B}{n} + \frac{C}{u} + \frac{A}{u^*\Delta_0^2}\left(\frac{D^3}{3} + \frac{n}{2}\frac{D^2}{u}\right)\right]$$

$$= \frac{670,000}{126} + \left\{\frac{500}{6} + \frac{870,000}{291} + \frac{3,600,000}{(343)(90^2)}\right.$$

$$\left.\left[\frac{83^2}{3} + \frac{6}{2}\left(\frac{83^2}{291}\right)\right]\right\}$$

$$= 5317.5 + (83.3 + 2989.7 + 2975.5 + 92.0)$$

$$= 5317.5 + 6140.5 = 11,458.0 \text{ yen/day} \qquad (14)$$

On a yearly basis, this is equal to 2.887 million yen. The optimal periodical checking interval, u', was estimated as follows. If periodical replacement is made at time u' hours, the mean time to the functional limit, Δ_0, turns out to be $\bar{u}$ with no maintenance. If we consider the case where maintenance of periodic replacement is implemented for u' hours, the periodic maintenance cost, C' yen, and mean time between failures, u^*, are estimated as

$$\text{maintenance cost } C' = \frac{C}{u} \qquad (15)$$

Then

$$u^* = 2\bar{u}\,\frac{\Delta_0^2}{(D')^2} \qquad (16)$$

$(D')^2$ is a root mean square of periodic maintenance characteristic differences from their standard values:

$$u' = \bar{u}\,\frac{(D')^2}{\Delta_0^2} \qquad (17)$$

Substituting this into equation (16), we obtained the mean time between failures for periodic maintenance as follows:

$$u^* = 2\bar{u}\Delta_0^2\frac{\bar{u}}{u'}\frac{1}{\Delta_0^2} = 2\frac{\bar{u}^2}{u'} \qquad (18)$$

Using this equation, the loss function when periodic replacement is implemented every u' hours is estimated as

$$L = \frac{C'}{u'} + \frac{u'}{2\bar{u}^2}\left(\sqrt{2AB} + \sqrt{\frac{4}{3}AC}\right) \qquad (19)$$

Partially differentiating this equation with respect to u', we set both sides to zero:

$$-\frac{C'}{(u')^2} + \frac{u'}{2\bar{u}^2}\left(\sqrt{2AB} + \sqrt{\frac{4}{3}AC}\right) = 0 \qquad (20)$$

Transforming this equation, we calculated the optimal periodic maintenance interval as

$$u' = \sqrt{\frac{2C'}{\sqrt{2AB} + \sqrt{\frac{4}{3}AC}}}\,\bar{u}$$

$$= \sqrt{\frac{(2)(670,000)}{\sqrt{(2)(3,600,000)(500)} + 870,000}}\,(147) \qquad (21)$$

$$= 117.3 \text{ days} \qquad (21)$$

For the sake of simplicity, we rounded 117.3 days up to six months (126 days). In this case the mean time between failures u^* is

$$u^* = \frac{2\bar{u}^2}{u'} = \frac{(2)(147^2)}{(21)(6)} = 343 \text{ days} \qquad (22)$$

This is approximately 1.4 years. With equations (21) and (22), we computed the optimal periodic checkup interval, n, and maintainable limit, D. The optimal periodic checkup interval, n, was estimated as

$$n = \sqrt{\frac{2B}{A}}\,u^* = \sqrt{\frac{(2)(500)}{3,600,000}}\,(343) = 5.7 \text{ days} \qquad (23)$$

5.7 days is rounded up to 1 week. Thus, we can perform one checkup a week.

    The optimal maintainable limit D is calculated as

Table 2

Cost comparison among all maintenance methods

| Maintenance Method | Cost (1000 yen) | | | | | | |
	Normalization	Checkup	Maintenance	Periodic Maintenance	Machinery Stop	Total	Proportion (%)
Current management (operation management)	7660					7660	100.0
Preventive Maintenance							
Only periodic checkup		42	1754		1812	3608	47.1
Only periodic maintenance				2010	1763	3773	49.2
Both periodic checkup and maintenance		21	753	1340	773	2887	37.7

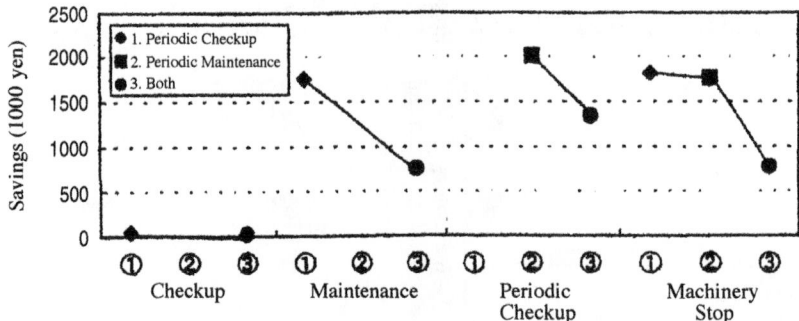

Figure 5
Effects of preventive maintenance methods

$$D = \left(\frac{3C}{A}\right)^{1/4} \Delta_0 = \left[\frac{(3)(870,000)}{3,600,000}\right]^{1/4} \quad (90)$$

$$= 83.04\% \qquad (24)$$

Now the mean time between failures turns out to be

$$u = u^* \frac{D^2}{\Delta_0^2} = 343\left(\frac{83^2}{90^2}\right) = 291 \text{ days} \quad (25)$$

4. Conclusions

Table 2 and Figure 5 show the cost calculation results for each type of maintenance method. Looking at the ex post facto maintenance method, we can see that the repair cost amounts to 7.66 million yen. Among all the maintenance methods, simultaneous use of the periodic checkup and maintenance methods were regarded as being most effective in overall cost and the best balanced. Considering all of the above, we come to the following conclusions:

1. As a maintenance method for chains, we perform periodic maintenance and checkups simultaneously. The following are the principal management concerns:

    a. *Optimal periodic maintenance interval:* once in six months

    b. *Optimal checking interval:* once a week

    c. *Maintainable limit:* 83%

2. By taking advantage of both periodic maintenance and checkups, we can obtain an annual improvement of 4.8 million yen compared with our current checkup method.

3. As a result of implementing the optimal maintenance method, we have confirmed that there has been no sudden failure, and the wear of rollers and bushes in a chain has been found quite close to the maintainable limit.

While through our calculations we have concluded that simultaneous use of periodic maintenance and checkups leads to the minimum overall loss, we plan to study the following three improvements to reduce the loss further:

1. *Improved checkup procedure* (improvement in B). Because the checkup cost is low, by enhancing the accuracy we can balance this and other costs and decrease the overall loss.

2. *Improved maintenance method* (improvement in C and C'). We will review the current labor requirements and operations.

3. *Improved wear resistance* (improvement in $\bar{u}$). We changed the current material to a wear-resistant material and decreased the pressure at a bearing through increased chain size.

Reference

Tadao Kondo, 1996. Design of preventive maintenance: combination of periodic maintenance and regular check of bucket elevator. *Quality Engineering*, Vol. 4, No. 6, pp. 30–35.

This case study is contributed by Tadao Kondo.

Feedback Control by Quality Characteristics

Abstract: This study demonstrates the use of on-line quality engineering at an early stage. It is an interesting case study for many reasons. On-line quality engineering has not been fully utilized.

1. Introduction

In off-line quality engineering, since we consider a generic function, signal factor, control factor, or noise factor and optimize design parameters with an orthogonal array, we can easily and intuitively follow the PDCA (plan–do–check–action) cycle. But in on-line quality engineering, to derive a loss function we need to scrutinize and clarify measurement or adjustment cost through on-site research at production lines. Figuratively speaking, on-line quality engineering requires untiring and low-key activities as groundwork, but only a company that wrestles actively with both off-line and on-line quality engineering can survive fierce competition.

Many companies, including ours, are striving consistently to reduce waste in direct and indirect workforce departments on a daily basis. However, we have been urged to innovate not only with new products and processes but also with existing ones by making the most of quality engineering techniques. So to promote on-line quality management in our company, we hold training seminars directed to managers and supervisors in the production division. The objective of the seminars is to understand the basic concepts and use of the loss function as a first step to improving quality and reducing cost.

For building a case study as a typical example for members in our in-house training seminar, we picked the following case: "Feedback Control by Quality Characteristics in Machining Component K" under the slogan "Do it first."

2. Current Feedback Control of Component K

Component K is machined in an integrated production line consisting of eight processes and nine quality characteristics. Among them we selected dimension H as associated with more than 40 subsequent processes for each component, required to be within a predetermined tolerance in all the processes. Since we had been told to attempt feedback control under a stable process condition rather than a warm-up process condition, we began to improve process stability and take measurements (including periodical fluctuations) as a grass-roots activity.

Through this research, analysis, and improvement, with the cooperation of the manufacturing and production engineering departments, many problems, such as treatment of removed material, dimensional adjustment procedures, and or measuring methods have come up. To make dimensions range within adjustable limits, we selected the idea of "fast, low-cost, and safe" as our basic policy.

Parameters

As a result of investigating the processes relevant to component K, we obtained the following parameters, which are currently controlled (some of them are denoted by a sign because of their confidentiality).

❑ *Design standard:* $m \pm 30$ μm; tolerance: $\Delta = 30$ μm

❑ *Loss by defect:* $A = 60$ yen (disposal cost per component at the point of completing component K)

- *Measurement cost:* $B_0 = 1572$ yen (estimated by the measurement time for component K)
- *Current checking interval:* $n_0 = 300$ components (one component per box is measured)
- *Current adjustable limit:* $D_0 = 25$ μm
- *Time lag:* $l_0 = 50$ components
- *Current average adjustment interval:* $U_0 = 19,560$ components (harmonic mean of all adjustment intervals regarding dimension H)
- *Current average adjustment cost:* $C_0 = 5370$ yen (weighted mean of all adjustment intervals regarding dimension H)
- *Measurement error variance:* $\sigma_{m0}^2 = 10^2$ μm^2 (large error due to poor precision of the special measurement instrument)

Current Loss Function

Based on our investigation into the current processes, we computed the current loss function as

$$L_0 = \frac{B_0}{n_0} + \frac{C_0}{U_0}$$
$$+ \frac{A}{\Delta^3} + \left[\frac{D_0^2}{3} + \left(\frac{n_0 + 1}{2} + l_0 \right) \frac{D_0^2}{U_0} + \sigma_{m0}^2 \right]$$

$$= 5.24 + 0.27 + 13.89 + 0.43 + 6.67$$

$$= 26.50 \text{ yen} \tag{1}$$

We see that the current loss amounts to 26.5 yen, and Figure 1 is a bar chart representing its content. This chart reveals that the loss inside the adjustable limit accounts for 52%.

3. Optimal System

To minimize the current loss, L_0, we designed an optimal control system. Since each process is controlled by quantitative characteristics and the adjustment method of quality characteristic H is clearly defined, using a quality management system based on feedback control, we attempted to optimize the processes.

Calculation of Loss L_1 by Feedback Control

While maintaining our current process conditions, we studied the optimization of processes using only the on-line approach. We computed the optimal measurement interval, n_1, and optimal adjustable limit, D_1:

$$n_1 = \sqrt{\frac{2U_0 B_0}{A} \frac{\Delta}{D_0}} \tag{2}$$

$$= 1215 \text{ components}$$
$$\rightarrow 1250 \text{ components every 5 hours}$$

$$D_1 = \left(\frac{3 C_0 D_0^2 \Delta_0^2}{A u_0} \right)^{1/4} = 9.4 \text{ μm}$$
$$\rightarrow 10 \text{ μm} \tag{3}$$

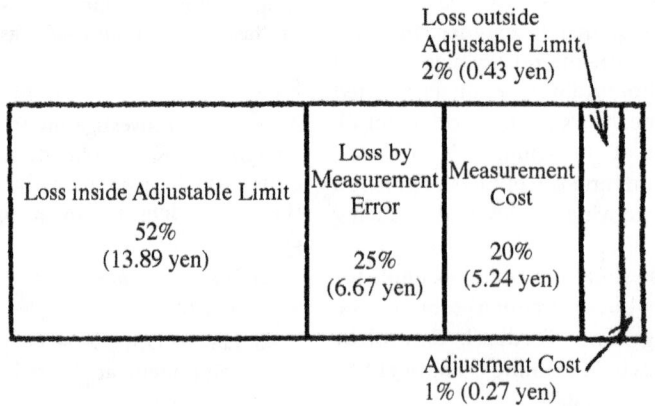

Loss outside Adjustable Limit
2% (0.43 yen)

| Loss inside Adjustable Limit 52% (13.89 yen) | Loss by Measurement Error 25% (6.67 yen) | Measurement Cost 20% (5.24 yen) | |

Adjustment Cost
1% (0.27 yen)

Figure 1
Current loss, L_0

Once the optimal measurement interval and adjustable limit were determined, we could predict the average adjustment interval (U_1) under the condition

$$U_1 = U_0 \frac{D_1^2}{D_0^2} = 19{,}560 \left(\frac{10^2}{25^2} \right)$$

$$= 3130 \text{ components} \qquad (4)$$

The loss by feedback control under the optimal configuration can be calculated by substituting the newly computed optimal measurement interval, n_1, adjustable limit, D_1, and average adjustment interval, U_1, into equation (1) as follows:

$$L_1 = \frac{B_0}{n_1} + \frac{C_0}{U_1} + \frac{A}{\Delta^2}$$

$$+ \left[\frac{D_1^2}{3} + \left(\frac{n+1}{2} + l_0 \right) \frac{D_1^2}{U_1} + \sigma_{m_0}^2 \right]$$

$$= 1.26 + 1.72 + 2.22 + 1.44 + 6.67$$

$$= 13.31 \text{ components} \qquad (5)$$

Comparison between L_0 and L_1 by Analysis Chart

Using a bar chart, we compared the current loss, L_0, and optimal loss, L_1 (Figure 2).

Problems in an Optimal System

By adopting the optimal condition, we obtained a reduction in total loss from 26.50 yen to 13.31 yen, an almost 50% improvement. However, the optimal configuration still involves the following three problems: (1) the measurement loss still accounts for the majority, 50%; (2) both the measurement cost (9%) and adjustment cost (13%) need to be reduced; (3) the technical feasibility of the optimal adjustable limit of $D_1 = 10 \ \mu m$ to reduce the loss is still unknown.

4. Further Improvement from the Optimal Configuration

Next, we took measures for further improvement. We studied technical problems lurking in the optimal system from the standpoint of some specific technologies.

Loss by Measurement Error

The loss due to the measurement error causes 50% of the total loss after optimization. Therefore, we took the following improvement measures.

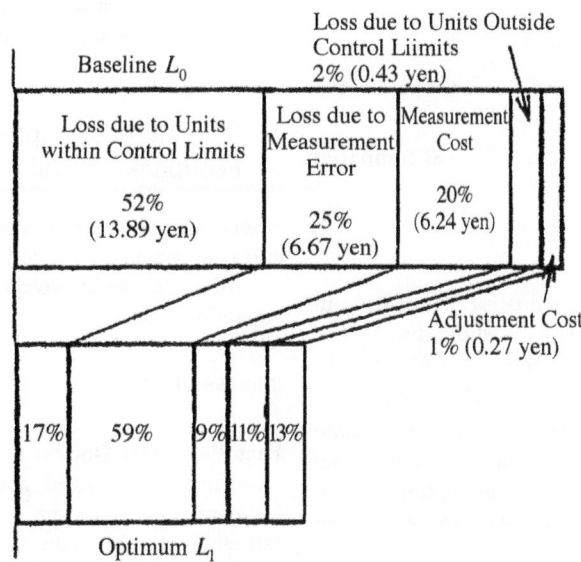

Figure 2
Comparison of loss between current and optimal conditions

Detailed Measures

❑ Renewal of the current measuring instrument, jigs, and tools

❑ Preparation of measurement standards

After adding two 5-mm peaks on the raw material of component K, we used it as our actual standard. In that way we could calibrate environmental errors such as temperature in machining processes.

Estimation of Measurement Error Variance (σ_{m1}^2) after Taking Measures Toward Improvement
Throughout the improvement process, we can ameliorate each parameter and obtain the following new values:

❑ *Checking interval:* $n_m = 1$ day (once before starting operation)

❑ *Calibration limit:* $D_m = 3$ μm (checked using the actual standard)

❑ *Average calibration interval:* $U_m = 264$ days (estimated to be calibrated after a year)

❑ *Time lag:* $l_m = 0$ (checked when a machine is stopped before starting operation)

❑ *Standard error variance:* $\sigma_s \approx 0$

Then the measurement error variance, σ_{m1}^2, can be estimated with the following equation for the error variance:

$$\sigma_{m1}^2 = \frac{D_m^2}{3} + \left(\frac{n_m}{2} + l_m \right) \frac{D_m^2}{u_m} + \sigma_s^2 \qquad (6)$$

Consequently, we can reduce it to 17% as compared with the current 10 μm.

Measurement Cost

The measurement cost accounts for 9% of the entire loss. Since there are 40 places whose dimensions were equal to leading the dimension of H to a long measurement time, we studied the following items:

1. Introduction to three-dimensional coordinate measuring or a fast measuring machine. Because of the long measurement time, transportation cost, and machinery cost, we cannot adopt it.

2. Automation of data processing in the current measuring instrument. As a result, we could reduce measurement time by two-thirds.

Expressing these items as measurement cost, we obtained $B_1 = 524$ yen.

Adjustment Cost

As a result of investigating current adjustment methods, we found that three methods among the current seven involve time-consuming procedures. So we modified these adjustment methods. The modifications can reduce the adjustment time, and we finally came up with the following adjustment cost: $C_1 = 4110$ yen (estimated by C_0).

Loss within Adjustable Limit

The optimization turned out to improve the loss inside the adjustable limit drastically. However, since the new adjustable limit was decreased to $D_1 = 10$ μm, we needed to secure $1/2.5$ times as much as the current 25 μm. To ensure this adjustable limit, we adopted the following technical measures:

❑ Adjustment of standards for peaks 1 to 5 in the data (input) in the machining program

❑ Adjustment in the balance of both the right and left sides in the peak cutter

We investigated modification of the optimal adjustable limit D_1. It turns out to be difficult to secure $D_1 = 10$ μm for all dimension H's, even if we assume the technical measures above. Therefore, we set up $D_1^* = 15$ μm on a practical basis.

5. Estimation of Improvement of Feedback Control

After taking the aforementioned improvement measures, we obtained the following parameters. Then we estimated the improvements in loss and process capability index.

Improved Loss

Each Parameter Obtained after Improvement The following are changed parameters after technical improvement measures. Except for them, all parameters are the same as those for the optimal configuration.

❑ *Measurement cost:* $B_1 = 524$ yen

❑ *Adjustment cost:* $C_1 = 4110$ yen

❑ *Adjustable limit:* $D_1^* = 15$ μm (on a practical basis)

❑ *Time lag:* $l_1 = 17$ components (through reduction of B_1 by two-thirds)

❑ *Measurement error variance:* $\sigma_{m1}^2 = 3.0$ μm^2

Using these parameters, we estimated the optimal measurement interval, n_1^*, and average adjustment interval, U_1^*.

$$n_1^* = \sqrt{\frac{2U_0 B_1}{A} \frac{\Delta}{D_0}}$$

$$= 701 \text{ components}$$
$$\rightarrow 750 \text{ components every 3 hours} \quad (7)$$

Under these conditions we predicted the average adjustment interval:

$$U_1^* = U_0 \frac{D_1^{*2}}{D_0^2} = 19{,}560 \left(\frac{10^2}{25^2}\right)$$

$$= 7042 \text{ components} \quad (8)$$

Estimation of Loss

$$L_2 = \frac{B_1}{n_1^*} + \frac{C_1}{U_1^*} + \frac{A}{\Delta^2}$$
$$+ \left[\frac{D_1^{*2}}{3} + \left(\frac{n_1^* + 1}{2} + l_0\right) \frac{D_1^{*2}}{U_1} + \sigma_{m0}^2\right]$$

$$= 0.70 + 0.58 + 5.00 + 0.84 + 0.20$$

$$= 7.32 \text{ yen} \quad (9)$$

Comparison of Improved Benefits For the improved benefits discussed above, we compared losses under current conditions, optimal current conditions, and after improvement. Figure 3 is an analytical chart for losses L_0, L_1, and L_2. Comparing the loss function under current conditions and the practical loss function after improvement, we can save 19.18 yen per component, or 10.1 million yen annually. However, even after improvement, the loss inside the adjustable limit accounts for a large portion of the 68%, so further improvement is required.

Process Capability Index

Comparing process capability indexes under both current and optimal conditions, we obtained the following improvement:

current $C_p = \dfrac{2\Delta}{6\sigma_0}$

$$= \frac{2\Delta}{6\sqrt{\dfrac{D_0^2}{3} + \left(\dfrac{n_0 + 1}{2} + l_0\right)\dfrac{D_0^2}{U_0} + \sigma_{m0}^2}}$$

$$= 0.56$$

optimal $C_p = \dfrac{2\Delta}{6\sigma_2}$

$$= \frac{2\Delta}{6\sqrt{\dfrac{D_1^{*2}}{3} + \left(\dfrac{n_1^* + 1}{2} + l_1\right)\dfrac{D_1^{*2}}{U_1} + \sigma_{m1}^2}}$$

$$= 1.06$$

Thus, an almost twofold improvement can be expected.

6. Notes on Feedback Control Activities

In our case, selecting one of the nine quality characteristics, we attempted to reduce cost by using the loss function. For other quality characteristics, by taking full advantage of each specific technology, we can build an optimal system for the entire machine through cost reduction based on a loss function. At this point we need to take notice of the following. For measurement costs, we should consider a feasible and efficient setup for reduction of measurement time and optimal measurement interval for all characteristics, including the use of specialized measurement technicians.

As for adjustment cost and inside- and outside-adjustable limit costs, the key point is to establish component production processes with less variability as well as those that are fast, low-cost, and safe, through use of off-line quality engineering. The resulting benefits through these improvements are unveiled gradually in terms of workforce reduction or enhanced company credibility. From the viewpoint of management, the most vital issue is to estimate how much the reduction in labor cost, users' complaints, or repair cost contributed to cost reduction. That is, instead of stressing workforce reduction after improvement, we should be able to

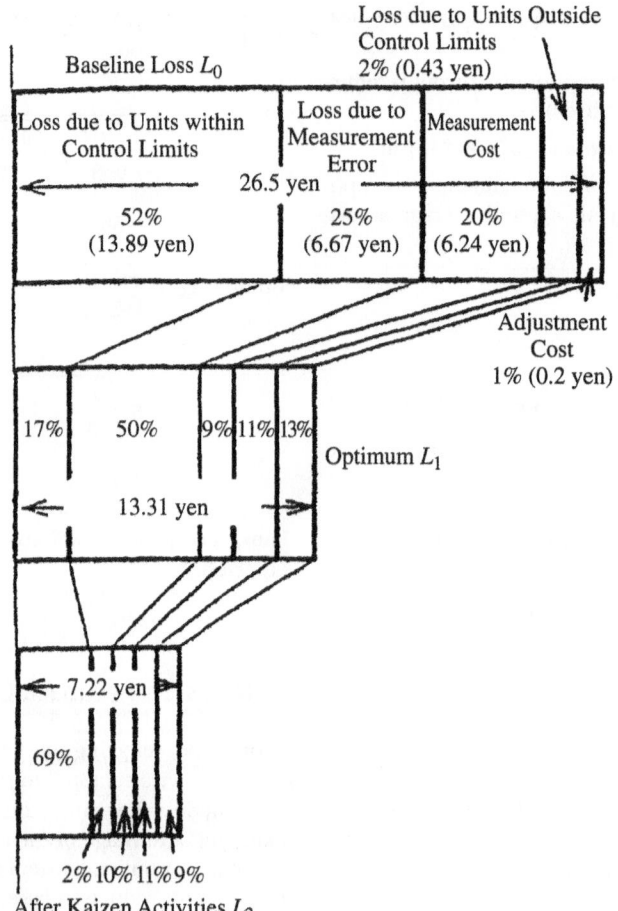

Figure 3
Cost and loss reduction through on-line quality engineering

reflect the financial benefit that accrues through the use of quality engineering.

We strongly hope that a number of case studies focusing on off-line and on-line quality engineering in a reciprocal manner will be reported in the future at conferences held by associations all over the country, including the Quality Engineering Study Group and the Quality Engineering Symposium.

Reference

Toshio Kanetsuki, 1996. An application of on-line quality engineering. *Quality Engineering*, Vol. 4, No. 1, pp. 13–17.

This case study is contributed by Toshio Kanetsuki.

Control of Mechanical Component Parts in a Manufacturing Process

Abstract: To evaluate actual dimensional control practices in a manufacturing line, we analyzed them using on-line quality engineering. With the cooperation of several companies, we sampled mechanical component parts that were actually produced in manufacturing lines in accordance with each line's checking interval, and remeasured their dimensions. At the same time, by obtaining actual operation data from companies we attempted to identify inherent problems using on-line quality engineering.

1. Introduction

The following are products that we used and their dimensions. The dimensions that we used for our study are boldfaced.

- ❏ *Metal plate* (*thickness*): **2.0** × 10.0 × 20.0 mm
- ❏ *Plastic tube A* (*height*): φ19.5 × **8.255** mm
- ❏ *Plastic tube B* (*height*): φ19.3 × **10** mm
- ❏ *Metal cap* (*inside diameter*): φ**29.55** × 3.9 mm

Using on-line quality engineering, we performed several studies regarding these products and dimensions. Control constants (parameters) related to the analysis were researched by the suppliers.

2. Actual and Loss Function–Based Tolerances

Table 1 compares actual tolerances in process and optimal tolerances, Δ, calculated by the following equation:

$$\Delta = \sqrt{\frac{A}{A_0}}\, \Delta_0 \qquad (1)$$

where A is the producer's loss, A_0 the average consumer's loss, and Δ_0 the consumer's functional limit.

For the metal cap, we can see a remarkable difference between the calculated optimal and current tolerances. Our investigation revealed that the reason was that the tolerance was arbitrarily set to ±50 μm for a functional limit of ±100 μm. In addition, the difficulty in the measurement of the metal cap's inner diameter was also related. In this type of case, where there is a considerable difference between the current and optimal situations, tolerance design was implemented improperly.

3. Actual Control of Measuring Instrument and Evaluation by the Loss Function

Investigating measurement control in actual processes and performing an analysis according to JIS Z 9090:1991 *Measurement: General Rules for Calibration System*, we determined problems with our current control system. As line 1 of Table 2 shows, we used a measuring instrument identical to those introduced by each supplier.

Optimal Checking and Correction Interval

Table 2 summarizes the management parameters. On the basis of the current controlling conditions shown in lines 2–8, using the following equations

Table 1
Comparison of tolerances of mechanical component parts (nominal-the-best characteristics)

	Metal Plate (Thickness)	Plastic Tube A (Height)	Plastic Tube B (Height)	Metal Cap (Inside Diameter)
1. Producer's loss, A (yen/unit)	15	5	5	5
2. Functional limit, Δ_0 (μm)	15	40	400	100
3. Average loss, A_0 (yen/unit)	500	25	1000	400
4. Calculated optimal tolerance, Δ (μm)	2.6	18	28	11
5. Current tolerance (μm)	3	25	35	50

derived in Appendix 1 of JIS Z 9090, we estimated losses related to the optimal controlling condition in line 13 and the losses in lines 9–12 and 16–19 of the same table.

Next, we set up the following equations where A is the producer's loss, B the checking cost, C the correction cost, D_0 the current correctable limit, n_0 the current checking interval, u_0 the current correction interval, Δ the current tolerance, and σ_0^2 the error variance of the standard used for calibration.

Loss:

$$L_0 = \frac{B}{n_0} + \frac{C}{u_0} + \frac{A}{\Delta_0}\left(\frac{D_0^2}{3} + \frac{n_0}{2}\frac{D_0^2}{u_0} + \sigma_s^2\right) \quad (2)$$

Optimal checking interval:

$$n = \sqrt{\frac{2u_0 B}{A}\frac{\Delta}{D_0}} \quad (3)$$

Optimal correctable limit:

$$D = \left(\frac{3C}{A}\frac{D_0^2}{u_0}\Delta^2\right)^{1/4} \quad (4)$$

Since the optimal correction interval can be calculated after the optimal correction limit is clarified, we computed the estimation of the optimal correction interval by using the following equation:

$$u = u_0\frac{D^2}{D_0^2} \quad (5)$$

Our current control procedure was formed through empirical rules. However, as compared with the optimal conditions, it cannot balance the control cost

(checking cost plus correction cost) and quality loss, thereby inflating the total loss. In short, this highlights how empirical rules are ambiguous.

Estimation of Measurement Error

1. *Error in using measuring instrument.* In reference to Appendix 2 of JIS Z 9090, considering the measurement environment in the actual processes, we experimentally estimated the magnitude of error in using a measuring instrument, σ^2 (Table 2, line 20).

2. *Equation for calibration and error of calibration.* Referring to Appendix 3 of JIS Z 9090, we selected a reference-point proportional equation as the calibration equation and calculated the estimated magnitude of error in calibration operation, σ_c^2 (Table 2, line 21).

3. *Error of the standard for calibration.* Since a commercial metal gauge is used as the standard for calibration, we defined its nominal error in the specification as the estimated magnitude of the standard's error, σ_0^2 (Table 2, line 22).

4. *Overall error in measuring operation.* The estimated magnitude of error variance in measuring operation σ_T^2 is expressed by the sum of the error variances 1 to 3 (Table 2, line 23):

$$\hat{\sigma}_T^2 = \hat{\sigma}^2 + \hat{\sigma}_c^2 + \hat{\sigma}_0^2 \quad (6)$$

Now, as the estimated range of error, we defined the double value of σ_T (Table 2, line 24), which is regarded as an error corresponding to uncertainty. We can see that the error for the metal cap is re-

Table 2
Comparison in control of measuring instrument used for mechanical component parts

	Metal Plate (Thickness)	Plastic Tube A (Height)	Plastic Tube B (Height)	Metal Cap (Inside Diameter)
1. Measuring instrument Used	Snap micrometer	Low-measuring force micrometer	Low-measuring force micrometer	Caliper
2. Producer's loss (yen/unit)	15	5	5	5
3. Current tolerance (μm)	3	25	70	50
4. Current checking interval (unit)	600	23,000	17,280	4800
5. Current correctable limit (μm)	0.5	5	5	10
6. Current correction interval (unit)	3,500	23,000	17,280	4800
7. Checking cost (yen)	100	200	200	200
8. Correction cost (yen)	200	800	800	400
9. Current checking cost per unit (yen/unit)	0.1667	0.0087	0.116	0.0417
10. Current correction cost per unit (yen/unit)	0.0571	0.0348	0.0463	0.0833
11. Current quality loss per unit (yen/unit)	0.1746	0.1667	0.0417	0.0850
12. Current total loss per unit (yen/unit)	0.3984	0.2102	0.0995	0.2100
13. Optimal checking interval (unit)	1300	6780	11,760	4340
14. Optimal correctible limit (μm)	0.4	4	6	13
15. Estimated optimal correction interval (unit)	2250	16,610	28,800	7510
16. Optimal checking cost per unit (yen/unit)	0.0772	0.0295	0.017	0.0461
17. Optimal correction cost per unit (yen/unit)	0.0891	0.0482	0.0278	0.0532
18. Optimal quality loss per unit (yen/unit)	0.1662	0.0776	0.0448	0.0993
19. Optimal total loss per unit (yen/unit)	0.3325	0.1553	0.0896	0.1987
20. Estimated error variance in use (μm²)	0.09	1.05	8.68	103.24
21. Estimated calibration error variance (μm²)	0.01	1.61	5.71	104.68
22. Estimated error variance of standard (μm²)	0.0004	0.0004	0.0441	1
23. Estimated error variance in measuring operation (μm²)	0.10	2.66	14.44	208.92
24. Range of error (2σ) (μm)	0.6	3	8	29

markably large. Considering that it exceeds the estimated optimal tolerance, we should use a more
accurate measuring instrument in place of a caliper,
which has a large measurement error. However, a
part can be judged as proper in a functional inspection process when it can be assembled or inserted, even if its error is larger in an actual process,
indicating that dimensional inspection is not greatly
emphasized. In other words, the manufacturer does
not trust tolerance to a great extent.

As for a control practice, because the optimal
checking (measurement) interval and other values
are calculated numbers, we should adjust them into
values that can be managed more easily. For instance, looking at the measuring instrument control
of plastic tube A, we can see that one checkup is
conducted every day because the current checkup
and correction interval is 23,000 units, and at the
same time, 23,000 units are produced daily using
this process. Under optimal conditions, these numbers are estimated as 6780 for checking and 16,610
for correction. Thus, since these numbers are not
convenient for checking operations, we changed
them slightly to 6710-unit (7-hour) and 16,290-unit
(17-hour) intervals. Furthermore, with careful consideration of the loss, we can also change them to
7670-unit (8-hour) and 15,330-unit (16-hour) intervals by taking into account that three 8-hour shifts
are normally used in a day.

4. Actual Process Control and Evaluation by the Loss Function

As we had earlier, we investigated manufacturing
process control of the products studied and compared the parameters with the optimal parameters
based on the loss function. Using the following
equations, we computed the optimal conditions and
summarized them (Table 3). The magnitude of er-
ror variance, σ_m^2, was not added to the quality loss.
Additionally, as in the measuring instrument control, the value estimated was used as an optimal adjustment interval.

Each value is defined as follows: A, producer's
loss, B, measurement cost, C, adjustment cost, D_0,
current adjustable limit, l, time lag, n_0, current
checking interval, u_0, current adjustment interval, Δ,
current tolerance, and σ_m^2, error variance of
measurement.

Loss:

$$L_0 = \frac{B}{n_0} + \frac{C}{u_0} + \frac{A}{\Delta_0}\left[\frac{D_0^2}{3} + \left(\frac{n_0+1}{2} + l\right)\frac{D_0^2}{u_0} + \sigma_m^2\right] \tag{7}$$

Optimal checking interval:

$$n = \sqrt{\frac{2u_0 B}{A}\frac{\Delta}{D_0}} \tag{8}$$

Optimal adjustable limit:

$$D = \left(\frac{3C}{A}\frac{D_0^2}{u_0}\Delta^2\right)^{1/4} \tag{9}$$

Estimated optimal adjustment limit:

$$u = u_0 \frac{D^2}{D_0^2} \tag{10}$$

As we can see by measuring instrument control,
the control cost (measurement cost plus adjustment
cost) and quality loss are better balanced and the
total loss is reduced under optimal conditions. The
fact that there is a particularly large difference between the current and optimal losses indicates a lax-
ity in management.

5. Evaluation of Product's Error in the Loss Function

After sampling products from the actual processes
for 30 checkup intervals, we computed the losses for
them using the loss function and summarized the
results in Table 4.

$$L = \frac{A}{\Delta^2}\sigma^2 = \frac{A}{\Delta_0}\frac{1}{n}\sum_i^n (y_i - m)^2 \tag{11}$$

In addition, we decomposed the error in equation (11) into a factor of deviation from a target
value and dispersion factor. Using the resulting
equation, we calculated the split-up losses:

$$L = \frac{A}{\Delta^2}\frac{1}{n}\left[n(\bar{y} - m)^2 + \sum_i^n (y_i - \bar{y})^2\right] \tag{12}$$

It is commonly known that the deviation in process
can easily be corrected. With the computed loss
by deviation, we can estimate the benefit due to
correction.

Table 3
Comparison in process control of mechanical component parts

	Metal Plate (Thickness)	Plastic Tube A (Height)	Plastic Tube B (Height)	Metal Cap (Inside Diameter)
1. Producer's loss (yen/unit)	15	5	5	5
2. Current tolerance (μm)	3	25	70	50
3. Current checking interval (units)	600	23,000	5760	4800
4. Current adjustable limit (μm)	1	5	25	50
5. Current adjustment interval (units)	2400	23,000	17,280	177,600
6. Time lag	300	240	120	80
7. Measurement cost (yen)	200	700	800	200
8. Adjustment cost (yen)	1200	1500	1600	20,000
9. Current measurement cost per unit (yen/unit)	0.3333	0.0304	0.1389	0.0417
10. Current adjustment cost per unit (yen/unit)	0.5000	0.0652	0.0926	0.1126
11. Current quality loss per unit (yen/unit)	0.9706	0.1688	0.3233	1.7365
12. Current total loss per unit (yen/unit)	1.8059	0.2644	0.5548	1.8908
13. Optimal checking interval (units)	760	12,690	6580	3770
14. Optimal adjustable limit (μm)	1	5	20	25
15. Estimated optimal correction interval (units)	2280	22,750	11,400	46,170
16. Optimal measurement cost per unit (yen/unit)	0.2635	0.0552	0.1215	0.0531
17. Optimal adjustment cost per unit (yen/unit)	0.5271	0.0659	0.1403	0.4332
18. Optimal quality loss per unit (yen/unit)	0.9993	0.1232	0.2662	0.4886
19. Optimal total loss per unit (yen/unit)	1.7898	0.2443	0.5281	0.9748

Table 4
Summary of quality losses in process control of mechanical component parts

	Metal Plate (Thickness)	Plastic Tube A (Height)	Plastic Tube B (Height)	Metal Cap (Inside Diameter)
1. Quality loss	0.227	1.6074	0.1469	0.5713
2. σ^2 (μm)	0.14	200.9	144.0	285.7
3. σ (μm)	0.37	14.2	12.0	16.9
4. Loss due to deviation	0.219	0.4128	0.0416	0.2385
5. Loss due to dispersion	0.008	1.1946	0.1053	0.3328

The magnitude of error discussed here should be the sum of the error in the measuring operation in Table 2 and the sum of the error in process control in Table 3, yet Table 4 reveals that it is not equivalent to the sum. Scrutinizing this phenomenon in depth, we found that the operator sometimes remeasures a part or takes another action after discarding a part when the measured result falls in an abnormal range.

Since parameters handled in on-line quality engineering have not often been considered to date, it has been quite difficult to help companies to understand the process. Therefore, we have spent a great deal of time investigating actual parameters. However, considering that many problems of current control practices have been identified through a series of analyses, we believe that these parameters should be managed as a matter of course. This example symbolizes vagueness in many current conventional management methods.

6. Conclusions

Through a comparative investigation of actual processes and optimal results based on the ideas of on-line quality engineering, we have shown how to utilize this management tool.

Reference

Yoshitaka Sugiyama, Kazuhito Takahashi, and Hiroshi Yano, 1996. A study on the on-line quality engineering for mechanical component parts manufacturing processes. *Quality Engineering*, Vol. 4, No. 2, pp. 28–38.

This case study is contributed by Yoshitaka Sugiyama.

CASE 92

Semiconductor Rectifier Manufacturing by On-Line Quality Engineering

Abstract: In quality engineering, the optimum controlling conditions are determined by balancing the cost of control and the cost of quality. This approach is effective and widely applicable in all manufacturing processes. This study reports a successful case that was applied to a manufacturing process for semiconductor rectifiers. In the process, the temperature-measuring interval was determined such that the cost was reduced while the quality was maintained, which contributed to customer satisfaction.

1. Introduction

Where lead and pellets are soldered in a semiconductor manufacturing process, we place a product in a heat-treating furnace. Figure 1 outlines a semiconductor manufacturing process. To control the furnace we take a periodic measurement with another temperature recording instrument (Figure 2). However, while the adjustable limits are set the same as in the manufacturing specifications, we have never done a thorough economical study of the checking interval because it had always been determined according to past experience and results. To verify its validity, we applied the concept of feedback control of process conditions from quality engineering.

2. Selection of Parameters for Mounting Process Conditions

By investigating actual situations in the mounting process in our semiconductor manufacturing line, we selected parameters for the case where there was feedback control.

1. *Measurement cost: B* yen. Fifteen minutes is needed to measure the heat-treating furnace.

We have

$$\frac{2400 \text{ yen (hourly labor cost)}}{60 \text{ minutes}} \times 15 \text{ minutes}$$
$$= 600 \text{ yen} \tag{1}$$

2. *Measurement interval: n* units. The hourly number of heat-treating products is 170,000, and they are measured every eight hours. Thus,

$$(170,000)(8) = 1,360,000 \text{ units} \tag{2}$$

3. *Adjustment cost: C* yen. Adjustment cost is based on hourly labor cost. Therefore, considering that one hour is required for adjustment, the hourly labor cost is 2,400 yen, and the adjustment needs another measurement, we obtain

$$2400 \text{ yen} + 600 \text{ yen} = 3000 \text{ yen} \tag{3}$$

4. *Mean adjustment interval: u* units. Over the past year, out-of-tolerance situations occurred approximately once every 100 days after a long-term consecutive holiday (e.g., at the beginning of the year, in the "Golden Week" from April 29 through May 5, in the summertime Bon Festival holidays).Thus, the mean adjustment interval, *u*, is estimated as follows:

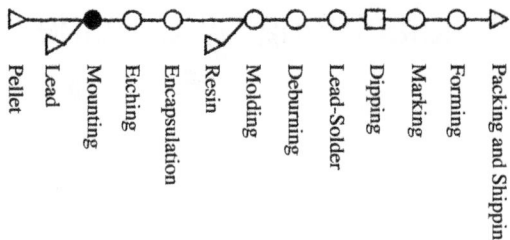

Figure 1
Quality control process chart for semiconductor manufacturing

170,000 units (hourly production volume)
× 2400 hours = 408,000,000 (units)   (4)

5. *Rejected product handling cost: A yen.* We discard rejected products after screening them. Therefore, we need the sum of processing cost, material cost, and screening cost up to the mounting process. In this case, this costs 3 yen per rectifier.

6. *Tolerance of process condition's temperature: Δ.* The temperature range that satisfies the tolerance of the objective characteristic of a semiconductor rectifier $\pm \Delta_0$ is $m \pm 10°C$. Thus, the tolerance of process condition's temperature is 10°C.

7. *Adjustable limit: D_0.* The adjustable limit is 5°C).

8. *Measurement time lag: l.* Since the measurement time lag is represented by the number of products manufactured during measurement of the heat-treating furnace, it amounts to 5900 units.

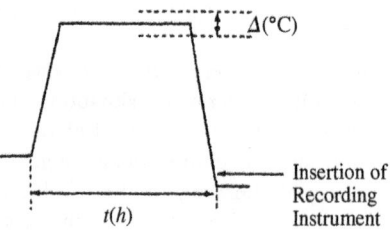

Figure 2
Temperature recording of heat-treating furnace in mounting process

9. *Measurement error variance: σ_m^2.* For the error variance of the temperature-recording instrument, using the formula in JIS Z 8403:1996, *Quality Characteristics of Products: General Rules for Determination of Specific Values*, we computed the value. As the parameters of the measurement error variance, we selected once a year as the checking interval, drift of the measuring instrument within the depreciation period of five years as the correctable limit, D, and 0.5°C as the error of the standard, σ_s:

$$\sigma_m^2 = \frac{D^2}{3} + \frac{n}{2}\frac{D^2}{u} + \sigma_s^2$$

$$= \frac{2^2}{3} + \frac{979,200,000}{2}\left(\frac{2^2}{4,896,000,000}\right) + 0.5^2$$

$$= 1.98°C \qquad (5)$$

Necessary values for calculating an optimal control system are $A = 3$ yen, $B = 600$ yen, $C = 3000$ yen, $D = 5°C$, $n = 1,360,000$ units, $l = 5900$ units, $u = 408,000,000$ units, $\Delta = 10°C$, $\sigma_m^2 = 1.98°C$.

3. Optimization Calculation for Mounting Process

We computed process management parameters according to the definitions in on-line quality engineering.

Current overall loss:

$$L_0 = \frac{B}{n_0} + \frac{C}{u_0} + \frac{A}{\Delta^2}\left[\frac{D_0^2}{3} + \left(\frac{n_0+1}{2} + l\right)\frac{D_0^2}{u_0} + \sigma_m^2\right]$$

$$= \frac{600}{1,360,000} + \frac{3000}{408,000,000} + \frac{3}{10^2}$$

$$\left\{\frac{5^2}{3} + \left[\frac{1,360,000+1}{2} + 5900\right.\right.$$

$$\left.\left.\left(\frac{5^2}{408,000,000}\right) + 1.96^2\right]\right\}$$

$$= 0.000441 + 0.000007 + 0.25$$
$$+ 0.001261 + 0.117612$$

$$= 0.369 \text{ yen} \qquad (6)$$

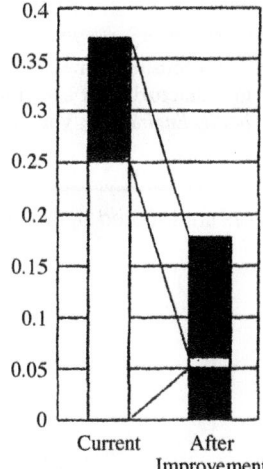

Figure 3
Overall loss for each product in mounting process

Optimal measurement interval:

$$n = \sqrt{\frac{(2)(408,000,000)(600)}{3}}\left(\frac{10}{5}\right)$$

$$= 807,960 \rightarrow 805,000 \text{ units}$$
$$\text{(once per 5 hours)} \qquad (7)$$

Optimal adjustable limit:

$$D = \left[\frac{(3)(3000)}{3}\left(\frac{5^2}{408,000,000}\right)(10^2)\right]^{1/4} = 0.37°C \qquad (8)$$

Mean adjustment interval:

$$u = 408,000,000\left(\frac{0.37^2}{5^2}\right) = 2,234,208 \text{ units} \qquad (9)$$

The result reveals that the optimal adjustable limit, D, is approximately $\frac{1}{10}$ of the current adjustable limit. Yet, due to unavailability of a comparable high-precision measuring instrument that can meet this level, we decided to set up $D = 3°C$ as a provisional intermediate point between the optimal and current limits.

Using $D = 3°C$, we determined an improvement measurement and estimated the overall loss:

$$L = \frac{600}{800,000} + \frac{3000}{2,234,208} + \frac{3}{10^2}$$
$$\left[\frac{3^2}{3}\left(\frac{800,000 + 1}{2} + 5900\right)\right.$$
$$\left.\left(\frac{3^2}{2,234,208}\right) + 1.98^2\right]$$

$$= 0.00075 + 0.00134 + 0.009$$
$$+ 0.0495 + 0.117612$$

$$= 0.178 \text{ yen} \qquad (10)$$

The difference between the current and overall losses is

$$L_0 - L = 0.369 - 0.178 = 0.19 \text{ yen} \qquad (11)$$

The improvement amounts to 0.19 yen per product (Figure 3 and Table 1). On a yearly basis, this turns out to be 18,605 million yen = (0.19)(170,000) (24)(240).

The result that the current loss is beyond the correctable limit and the standard errors are computed to be a considerable amount proves objectively how ambiguous our empirical management values are. In contrast, under the optimal configuration, using a high-precision measuring

Table 1
Current losses versus improvements

	Current	After Improvement
Measurement error variance	0.117612	0.117612
Loss beyond adjustable limit	0.00126	0.00075
Loss within adjustable limit	0.25	0.009
Adjustment loss	0.000006	0.00134
Measurement loss	0.000441	0.04905

instrument, we would be able to obtain an improvement of 0.25 yen per product, which is equal to approximately 88 million yen on a yearly basis. We need to improve or develop the measuring instrument in-house, since such an instrument is not currently available in the marketplace.

Reference

Nobuhiro Ishino and Mitsuo Ogawa, 1999. Study of online QE for manufacturing of the "rectifiers" semiconductor. *Quality Engineering*, Vol. 7, No. 6, pp. 51–55.

This case study is contributed by Nobuhiro Ishino.

Part VII

Miscellaneous

Miscellaneous (Cases 93–94)

Estimation of Working Hours in Software Development

Abstract: This study involves the estimation of total working hours in software development. The objective was to determine the coefficients of the items used for the calculation of total working time. This method was developed by Genichi Taguchi and is called *experimental regression*.

1. Introduction

In many cases, office computer sales companies receive an order for software as well as one for hardware. Since at this point in time, functional software is not well developed, the gap between the contract fee and actual expense quite often tends to be significant. Therefore, if the accuracy in rough estimation (also called the *initial estimation*) of working hours based on tangible information at the point of making a contract were to be improved, we could expect a considerable benefit from the viewpoint of sales.

On the other hand, once a contract is completed, detailed functional design is started. After the functional design is completed, actual development is under way. In this phase, to establish a development organization, development schedule, and productivity indexes, higher accuracy in labor-hour estimation than the one for the initial estimation is required. Thus, it is regarded as practically significant to review the estimated number of development working hours after completion of the contract. This is called working-hour estimation for actual development.

2. Estimation of Actual Development Working Hours

For approximately 200 software products that have been sold by manufacturers and suppliers of office computers, we requested development engineers to answer a questionnaire that contains the following items:

1. *Number of files:* only files for system design are included; work and program files are excluded
2. *Number of items:* total number of items in a file
3. *Number of data input programs:* number of programs for data input or voucher issuance
4. *Number of table creation programs:* number of programs for documentation after data processing
5. *Number of renewal programs:* number of programs for balance renewal, repeated data processing, and simple data conversion
6. *Number of calculation programs:* number of programs for master data renewal through daily transaction and calculation
7. *Number of COBOL-based programs:* number of COBOL-based programs
8. *Number of all files used:* number of files accessed from all programs
9. *Use of existing designs:* designs of existing similar systems are used or not
10. *Work experience as software engineer (SE):* SE in charge of system design has work experience in developing similar systems or not

11. *Familiarity with operating systems as SE:* SE in charge of system design has knowledge about or work experience in operating systems that developed system rests upon

12. *Number of subsystems:* transaction consisting of approximately 10 documents related to sales, purchase, and inventory, etc.

13. *Use of special devices:* use of devices other than printer, cathode-ray tube, keyboards, disk, and floppy disk

14. *Development manpower:* total time needed for system development

Assumed Functional Equation and Initial Values

In the case of estimating actual development working hours after understanding the entire system, items 1 to 13 are considered explanatory variables, whereas item 14 is regarded as an objective variable. That is, assuming that the development labor hours result in zero when all items are zero, the equation can be expressed as follows:

$$y = a_1 x_1 + a_2 x_2 + \cdots + a_{13} x_{13} \qquad (1)$$

This is a linear proportional regression equation.

Although 200 questionnaires were collected, by excluding missing and incomplete answers or exceptions, such as "a development cycle was doubled because it was redesigned after its full operation," the number of effective questionnaire results was 54. After obtaining a rough calculation result and using actual data, we realized that some cases have large gaps between actual data and estimations. To determine whether there are special reasons for this gap, by asking those who answered questionnaires about development situations in more detail and excluding items with a clear reason for the large gap, we still had 37 useful questionnaires. Several of the major special reasons were incomplete data, doubled development time because of redesign after starting, and too-specific orders from customers. But the majority of the cause was due to legacy availability and use of existing systems, and inability to calculate actual development working hours due to lack of records to trace back overtime work, and so on. For

Table 1
Initial values of coefficients in equation of actual labor-hour estimation

i	Variable x_i	Initial Value of Coefficient a_i for Level:		
		1	2	3
1	Number of files	0	3	6
2	Number of items	0	1	2
3	Number of data input programs	0	10	20
4	Number of table creation programs	0	3	6
5	Number of renewal programs	0	2	4
6	Number of calculation programs	0	5	10
7	Number of COBOL-based programs	0	2	4
8	Number of all files used	0	1	2
9	Use of existing designs	−300	−150	0
10	Work experience as SE	0	25	50
11	Familiarity with operating system as SE	0	20	40
12	Number of subsystems	0	15	30
13	Use of special devices	0	30	60

these reasons, we estimated parameters with 37 cases. According to practical assumptions by relevant engineers, initial values were set as shown in Table 1.

Converged Value and Analysis

Table 2 shows the three levels of coefficients after the sixth convergence calculation. Table 3 illustrates a part of the deviations and sums of residuals squared after the sixth convergence calculation.

If the values at level 2 after the sixth calculation are used, the resulting regression equation is expressed as

$$y = 3.188x_1 + 0.312x_2 - 5.625x_3 + 3.188x_4$$
$$+ 1.75x_5 + 5.312x_6 + 2.125x_7 + 0.469x_8$$
$$- 159.375x_9 + 25.0x_{10} + 20.0x_{11} + 15.938x_{12}$$
$$+ 58.125x_{13} \tag{2}$$

Setting the total of squares of actual working hours to S_T, we obtain the following value:

$$S_T = 12,734,085 \tag{3}$$

On the other hand, according to Table 3, the sum of the squared residuals, S_e, turns out to be

$$S_e = 681,703 \tag{4}$$

Therefore, the present contribution ρ (%) is

$$\rho = \left(1 - \frac{S_e}{S_T}\right)(100) = 94.6\% \tag{5}$$

Now, since a linear proportional regression equation was used, we defined a ratio to the total sum of squares as the contribution. In addition, the standard deviation of errors, σ, is

$$\sigma = \sqrt{\frac{S_e}{n}} = 135.73 \tag{6}$$

As a reference, we show the estimation equation based on a least squares method as

Table 2
Converged values after sixth calculation

Coefficient	Level		
	1	2	3
a_1	3.0	3.1875	3.375
a_2	0.28125	0.3125	0.34375
a_3	5.0	5.625	6.25
a_4	3.0	3.1875	3.375
a_5	1.5	1.75	2.0
a_6	5.0	5.3125	5.625
a_7	2.0	2.125	2.25
a_8	0.4375	0.46875	0.5
a_9	−168.75	−159.375	−150.0
a_{10}	23.4375	25.0	26.5625
a_{11}	18.75	20.0	21.25
a_{12}	15.0	15.9375	16.875
a_{13}	56.25	58.125	60.0

Table 3
Part of the deviations and sum of residuals squared

No.	Deviation	Sum of Residuals Squared	Combination												
1	43.3114871	734,657.989	1	1	1	1	1	1	1	1	1	1	1	1	1
2	0.460980390	681,703.051	2	2	2	2	2	2	2	2	2	2	2	2	1
3	−42.3895264	780,436.115	3	3	3	3	3	3	3	3	3	3	3	3	1
4	15.9880074	694,443.380	1	1	1	1	2	2	2	2	3	3	3	3	1
5	−3.93175610	667,459.864	2	2	2	2	3	3	3	3	1	1	1	1	1
⋮	⋮	⋮							⋮						
36	−8.90810745	678,690.578	3	2	3	1	2	1	2	3	1	1	2	3	3

$$y = 7.919x_1 + 0.00234x_2 - 7.654x_3$$
$$- 5.712x_4 - 4.1021x_5 + 0.984x_6$$
$$+ 4.346x_7 + 2.8637x_8 - 224.3517x_9$$
$$+ 92.6744x_{10} - 140.3562x_{11} + 9.1052x_{12}$$
$$+ 130.9149x_{13} \qquad (7)$$

Although the error variation, contribution, and standard deviation of errors are 311,112.77, 97.6%, and 113.72, respectively, we considered this equation impractical because it contains unrealistic coefficients.

Figure 1 illustrates the correspondence between the estimation and actual result when using equation (2).

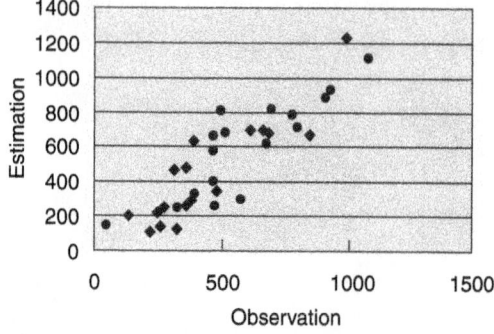

Figure 1
Estimation by observation and result using equation (2)

3. Rough Estimation

Comparing with the actual development working-hour estimation, the initial estimation at the time of receiving order is discussed below.

Selection of Items
As we stated at the beginning, we needed to determine a contract fee when receiving an order for software. However, most items were not determined at this point. More specifically, the only items that could be clarified were the (1) number of files, (2) number of table creation programs, (3) number of subsystems, and (4) number of special devices. Next, we studied whether we could roughly estimate the development working hours using only these four items.

Setup of Initial Values
A linear proportional regression equation based on the four items selected is assumed. Since all factors other than these four are represented by only these four, the initial values of the coefficients are set with wide intervals (Table 4). Now, the suffixes of the coefficients follow those in the estimation equation of actual development working hours.

Convergence Result and Evaluation
Because the number of unknowns is four, we used an L_{18} orthogonal array. The sequential approximation converged after six trials (Table 5).

Table 4

Initial values of coefficients for initial estimation equation

i	Variable x_i	Initial Value of Coefficient a_i at Level: 1	2	3
1	Number of files	0	15	30
4	Number of table creation programs	0	15	30
12	Number of subsystems	0	50	100
13	Use of special devices	0	50	100

Next we compared coefficients of common items in the equation for estimation of actual development working hours and one for initial estimation. In Table 6, we compare the values at level 2.

While the coefficients for number of subsystems and use of special devices are not so different, those for number of files and number of table creation programs are very distinct. The reason is that many items that are included from this analysis largely affect number of files and number of table creation programs. In other words, there is a correlation between the excluded items and each of number of files and number of table creation programs.

Table 5

Converged values of initial estimation equation after six trials

Coefficient	Level 1	2	3
a_1	15.0	15.46875	15.9375
a_4	11.25	11.71875	12.1875
a_{12}	12.5	14.0625	15.625
a_{13}	43.75	50.0	56.25

Table 6

Coefficients of common items in two equations

Coefficient	Development Working Hours Actual	Estimated
Number of files	3.1875	15.46875
Number of table creation programs	3.1875	11.71875
Number of subsystems	15.9375	14.0625
Use of special devices	58.125	50.0

As the coefficients of the initial estimation equation, we adopted the values at level 2. The sum of squared residuals S_e results in 977,692.517 when the values at level 2 are used. The percent contribution was 92.4% and the standard deviation of errors was 162.55. Then, this is approximately 1.2 times as much as 135.73, the counterpart for the estimation equation of actual development working hours based on 13 items.

In a conventional process, most companies have determined a contract fee according to only the numbers of different screen displays and documents. Thus, we calculated a regression equation based on them. Even in the case of using a general linear regression equation computed by the least squares method that minimizes errors [$y = 216.68 + 16.9$ (number of documents) $+ 9.82$ (number of screen displays)], the resulting standard deviation of errors amounts to 256.33, which is 1.6 times that for the initial estimation equation and about twice as much as that for the actual estimation equation. In addition, since the regression equation grounded on actual labor-hour results was not used, the resulting standard deviation of errors would be much larger. Therefore, our initial estimation equation is worthwhile to adopt.

This case study is contributed by Yoshiko Yokoyama.

CASE 94

Applications of Linear and Nonlinear Regression Equations for Engineering

Abstract: Experiments are normally conducted following the layout of design of experiments. But in many cases, data have existed without following an experimental layout. These data are not orthogonal between factors. For such cases, experimental regression analysis can be used as a powrful tool to establish an equation that will explain the relationship between result and factors.

This study concerns linear and nonlinear cases using experimental regression analysis. The former includes an example of chemical reaction and the case using partial data from an orthogonal array, which were not used in the previous analysis to establish an equation. The latter includes the estimation of the volume of trees and the starting temperature of room heaters.

1. L_{16} Experiment with Supplemental Characteristics

To improve the yield of a chemical product, an experiment was planned including factors A and B with four levels and factors C, D, E, F, G, H with two levels, assigned to an L_{16} orthogonal array. Beside those factors, an uncontrollable factor, x, which probably affects yield, the objective characteristic, was observed. Therefore, the magnitude of x was observed beside yield, y, for each experiment. The layout and results of this experiment are shown in Table 1.

Estimation of Factorial Effects Using Experimental Regression Analysis

Data with supplementary characteristics are generally analyzed using the analysis of covariance. But similar to regression analysis, it often occurs with some problems. In addition, calculation is tedious when there are many supplementary characteristics. In an experimental regression analysis, the following equation is used:

$$y = m + z_2 x_1 + a_3 x_2 + a_4 x_3 + b_2 x_4 + b_3 x_5 + b_4 x_6$$
$$+ c_2 x_7 + d_2 x_8 + e_2 x_8 + e_2 x_9 + f_2 x_{10}$$
$$+ g_2 x_{11} + h_2 x_{12} + ax \tag{1}$$

After estimating parameters, the factorial effects can be calculated. The variables are set as follows and shown in Table 2:

x_1: If level 2 is selected for A, x is set to 1, others are set to 0.

x_2: If level 3 is selected for A, x is set to 1, others are set to 0.

x_3: If level 4 is selected for A, x is set to 1, others are set to 0.

x_4: If level 2 is selected for B, x is set to 1, others are set to 0.

x_5: If level 3 is selected for B, x is set to 1, others are set to 0.

x_6: If level 4 is selected for B, x is set to 1, others are set to 0.

Table 1

Layout and results of experiment

	Column												
Row	A 123	B 48–12	C 5	D 6	e 7	e 9	E 10	F 11	G 13	H 14	e 15	x	y
1	1	1	1	1	1	1	1	1	1	1	1	138	54
2	1	2	1	1	1	2	2	2	2	2	2	120	55
3	1	3	2	2	2	1	1	1	2	2	2	95	24
4	1	4	2	2	2	2	2	2	1	1	1	95	41
5	2	1	1	2	2	1	2	2	1	2	2	105	66
6	2	2	1	2	2	2	1	1	2	1	1	90	49
7	2	3	2	1	1	1	2	2	2	1	1	129	58
8	2	4	2	1	1	2	1	1	1	2	2	130	96
9	3	1	2	1	2	2	2	2	2	1	2	130	63
10	3	2	2	1	2	1	2	1	1	2	1	124	73
11	3	3	1	2	1	2	1	2	1	2	1	102	31
12	3	4	1	2	1	1	2	1	2	1	2	95	34
13	4	1	2	2	1	2	2	1	2	2	1	104	91
14	4	2	2	2	1	1	1	2	1	1	2	123	54
15	4	3	1	1	2	2	2	1	1	1	2	119	50
16	4	4	1	1	2	1	1	2	2	2	1	95	61

x_7:   If level 2 is selected for C, x is set to 1, others are set to 0.

x_8:   If level 2 is selected for D, x is set to 1, others are set to 0.

x_9:   If level 2 is selected for E, x is set to 1, others are set to 0.

x_{10}:   If level 2 is selected for F, x is set to 1, others are set to 0.

x_{11}:   If level 2 is selected for G, x is set to 1, others are set to 0.

x_{12}:   If level 2 is selected for H, x is set to 1, others are set to 0.

x: Supplementary characteristic.

Following are definitions for $m, a_2, \dots, h_2$, and a:

m: estimate when all variables are at the first level

a_2:   estimate of $\overline{A}_2 - \overline{A}_1$

a_3:   estimate of $\overline{A}_3 - \overline{A}_1$

a_4:   estimate of $\overline{A}_4 - \overline{A}_1$

b_2:   estimate of $\overline{B}_2 - \overline{B}_1$

b_3:   estimate of $\overline{B}_3 - \overline{B}_1$

b_4:   estimate of $\overline{B}_4 - \overline{B}_1$

c_2:   estimate of $\overline{C}_2 - \overline{C}_1$

d_2:   estimate of $\overline{D}_2 - \overline{D}_1$

e_2:   estimate of $\overline{E}_2 - \overline{E}_1$

f_2:   estimate of $\overline{F}_2 - \overline{F}_1$

Table 2
Integer variables

Row	x_1	x_2	x_3	x_4	x_5	x_6	x_7	x_8	x_9	x_{10}	x_{11}	x_{12}	x	y
1	0	0	0	0	0	0	0	0	0	0	0	0	138	54
2	0	0	0	1	0	0	0	0	1	1	1	1	120	55
3	0	0	0	0	0	0	1	1	0	0	1	1	95	24
4	0	0	0	0	0	1	1	1	1	1	0	0	95	41
5	1	0	0	0	0	0	0	1	1	1	0	1	105	66
6	1	0	0	1	0	0	0	1	0	0	1	0	90	49
7	1	0	0	0	0	0	1	0	1	1	1	0	129	58
8	1	0	0	0	0	1	1	0	0	0	0	1	130	96
9	0	1	0	0	0	0	1	0	0	1	1	0	130	63
10	0	1	0	1	0	0	1	0	1	0	0	1	124	73
11	0	1	0	0	0	0	0	1	0	1	0	1	102	31
12	0	1	0	0	0	1	0	1	1	0	1	0	95	34
13	0	0	1	0	0	0	1	1	1	0	1	1	104	91
14	0	0	1	1	0	0	1	1	0	1	0	0	123	54
15	0	0	1	0	0	0	0	0	1	0	0	0	119	50
16	0	0	1	0	0	1	0	0	0	1	1	1	95	61

g_2: estimate of $\overline{G}_2 - \overline{G}_1$

h_2: estimate of $\overline{H}_2 - \overline{H}_1$

a: change of yield, y, by unit change of x

Table 3 shows the initial and converged values at the fifth iteration. The initial level intervals were set wide. These can be narrower if engineering knowledge is utilized. Using the second levels of the fifth iteration yields

$$m = \overline{y} - (a_2 x_1 + a_3 x_2 + \cdots + h_2 x_{12} + a\overline{x}) = -26 \tag{2}$$

Comparison was made as shown in Table 4. For verification it was desirable to use the data that were not used for the estimation of parameters. But in this case, all 16 data points are used for verification. The average of differences squared was calculated as 1412.60. This corresponds to error variance.

2. Partial Data from an Experiment Using an Orthogonal Array

In an experiment using an L_{18} array, two-level factor A was assigned to the first column and three-level factors B, C, D, E, and F to columns 2 to 6, respectively. The smaller-the-better characteristic was used, calculated from the noise factors assigned to the outer array. Factors A, D, and F are significant. Using these factors to estimate the SN ratio under the optimum configuration, the SN ratio was -36.97 dB, and that under current conditions was -38.61. There is an improvement of about 1.6 dB.

The estimate under current conditions was close to the result of number 1, where all factors are at their first level (Table 5). The SN ratios of number 6 or 8 are better than the one under the optimum, suggesting that the SN ratio under the optimum might be better. Although the SN ratio analysis of

Table 3
Initial values and values after convergence

Parameter	Initial Level			Level after Fifth Iteration		
	1	2	3	1	2	3
a_2	0.0	16.0	32.0	16.0	24.0	32.0
a_3	0.0	16.0	32.0	0.0	8.0	16.0
a_4	0.0	16.0	32.0	16.0	24.0	32.0
b_2	−32.0	−16.0	0.0	−16.0	−8.0	0.0
b_3	−32.0	−16.0	0.0	−32.0	−16.0	0.0
b_4	−8.0	0.0	8.0	−8.0	0.0	8.0
c_2	8.0	0.0	8.0	−8.0	0.0	8.0
d_2	−8.0	0.0	8.0	−8.0	0.0	8.0
e_2	−8.0	0.0	8.0	−8.0	0.0	8.0
f_2	−8.0	0.0	8.0	−8.0	0.0	8.0
g_2	−8.0	0.0	8.0	−8.0	0.0	8.0
h_2	−8.0	0.0	8.0	4.0	6.0	8.0
a	0.0	0.8	1.6	0.4	0.6	0.8

L_{18} is not shown here, the confidence interval was quite wide from the analysis of variance. In such a case it is natural that the researcher wants to have a better estimate.

It was noted from observations that there is a big difference between the SN ratios from numbers 1 to 9 and those from numbers 10 to 18. In addition, the SN ratios under A_2 are close to each other. In other words, the effects of B, C, D, E, F, and G under A_2 are insignificant. But it is not clear whether there are problems in the level setting for those factors or whether they are truly insignificant. Since the effects of factors B to G are different at different levels of A, which indicates the existence of interactions, the error calculated may become unsatisfactorily large.

Since the results under A_1 were better, it may be that the effects of B to G can be estimated from the results under A_1. As seen in the upper half of L_{18}, the array of columns 2, 3, 6, and 7 are identical to the L_9 array. Even for the case when there are other factors assigned to other columns in the upper half of L_{18}, if their effects were small, analysis could be made from the upper half, although it is not a general practice.

If we want to estimate the effects from the upper half of L_{18}, it is necessary that the factors be orthogonal to each other. To analyze the data that are not orthogonal, experimental regression analysis is useful.

Preparation of Data for Regression Analysis

Table 5 shows the SN ratios of experiments 1 to 9. To analyze the effects of B through G, those factors were replaced with integer-type variables and the following equation was used:

$$y = m + b_2 x_1 + b_3 x_2 + c_2 x_3 + c_3 x_4 + d_2 x_5 + d_3 x_6 + e_2 x_7 + e_3 x_8 + f_2 x_9 + f_3 x_{10} \tag{3}$$

Variables were set as follows (Table 6):

x_1: If level 2 is selected for B, x_1 is set to 1, others are set to 0.

Table 4
Comparison between estimation and
observation

No.	Observation	Estimation	(Est.) − (Obs.)
1	54.0	52.6	−1.4
2	55.0	41.4	−13.6
3	24.0	39.4	15.4
4	41.0	34.4	−5.6
5	66.0	67.4	1.4
6	49.0	33.4	−15.6
7	58.0	73.0	15.0
8	96.0	93.4	−2.6
9	63.0	49.4	−13.6
10	73.0	75.0	2.0
11	31.0	42.2	11.2
12	34.0	34.4	1.4
13	91.0	75.0	−16.0
14	54.0	54.6	0.6
15	50.0	77.0	27.0
16	61.0	55.4	−5.6

x_2: If level 3 is selected for B, x is set to 1, others are set to 0.

x_3: If level 2 is selected for C, x is set to 1, others are set to 0.

x_4: If level 3 is selected for C, x is set to 1, others are set to 0.

x_5: If level 2 is selected for D, x is set to 1, others are set to 0.

x_6: If level 3 is selected for D, x is set to 1, others are set to 0.

x_7: If level 2 is selected for E, x is set to 1, others are set to 0.

x_8: If level 3 is selected for E, x is set to 1, others are set to 0.

x_9: If level 2 is selected for F, x is set to 1, others are set to 0.

x_{10}: If level 3 is selected for F, x is set to 1, others are set to 0.

The coefficients show the following contents:

Table 5
SN ratios

Row	A 1	B 2	C 3	D 4	E 5	F 6	e 7	e 8	dB
1	1	1	1	1	1	1	1	1	−39.79
2	1	1	2	2	2	2	2	2	−42.74
3	1	1	3	3	3	3	3	3	−46.52
4	1	2	1	1	2	2	3	3	−44.33
5	1	2	2	2	3	3	1	1	−46.65
6	1	2	3	3	1	1	2	2	−34.92
7	1	3	1	2	1	3	2	3	−43.75
8	1	3	2	3	2	1	3	1	−32.81
9	1	3	3	1	3	2	1	2	−40.98

Table 6
Placement of factors in Table 5 as integer-type variables

B 2	C 3	D 4	E 5	F 6	x_1 B_2	x_2 B_3	x_3 C_2	x_4 C_3	x_5 D_2	x_6 D_3	x_7 E_2	x_8 E_3	x_9 F_2	x_{10} F_3	y
1	1	1	1	1	0	0	0	0	0	0	0	0	0	0	−39.79
1	2	2	2	2	0	0	1	0	1	0	1	0	1	0	−42.74
1	3	3	3	3	0	0	0	1	0	1	0	1	0	1	−46.52
2	1	1	2	2	1	0	0	0	0	0	1	0	1	0	−44.33
2	2	2	3	3	1	0	1	0	1	0	0	1	0	1	−46.55
2	3	3	1	1	1	0	0	1	0	1	0	0	0	0	−34.92
3	1	2	1	3	0	1	0	0	1	0	0	0	0	1	−43.75
3	2	3	2	1	0	1	1	0	0	1	1	0	0	0	−32.81
3	3	1	3	2	0	1	0	1	0	0	0	1	1	0	−40.98

m: estimate of $(A_1)B_1C_1D_1E_1F_1G_1$

b_2: estimate of $\bar{B}_2 - \bar{B}_1$

b_3: estimate of $\bar{B}_3 - \bar{B}_1$

c_2: estimate of $\bar{C}_2 - \bar{C}_1$

c_3: estimate of $\bar{C}_3 - \bar{C}_1$

d_2: estimate of $\bar{D}_2 - \bar{D}_1$

d_3: estimate of $\bar{D}_3 - \bar{D}_1$

e_2: estimate of $\bar{E}_2 - \bar{E}_1$

e_3: estimate of $\bar{E}_3 - \bar{E}_1$

f_2: estimate of $\bar{F}_2 - \bar{F}_1$

f_3: estimate of $\bar{F}_3 - \bar{F}_1$

Ranges of Initial Values
The initial values of $b_2, \ldots, f_3$ are determined based on engineering knowledge (Table 7). Setting the results of level 2 of the eighth iteration,

$$m = y - (b_2x_1 + b_3x_2 \cdots = h_3x_{10}) = -38.71 \quad (4)$$

y is calculated as

$$y = -38.71 + x_1 + 2.5x_2 + x_3 + 1.5x_4 + x_5 + x_6 + x_7 - x_8 - 7x_9 - 9x_{10} \quad (5)$$

Table 8 shows the comparison by putting the values in Table 7. The error variance was calculated as 0.64. Figure 1 shows the response graphs where the effect is the gain from level 1.

Estimation of the Optimal Configuration
Since the optimum configuration is $B_3C_3D_2(D_3)E_2F_1$,

$$y = -38.71 = x_1 + 2.5x_2 + x_3 + 1.5x_4 + x_5 + x_6 + x_7 - x_8 - 7x_9 - 9x_{10} \quad (6)$$

where $x_1 = 0$, $x_2 = 1$, $x_3 = 0$, $x_4 = 1$, $x_5 = 1$, $x_6 = 0$, $x_7 = 1$, $x_8 = 0$, $x_9 = 0$, and $x_{10} = 0$.

3. Other Applications

In the section above, only the optimum configuration was estimated. But it is possible to estimate any combination. Sometimes there are incomplete (missing) data in an orthogonal experiment, and sequential analysis is conducted followed by analysis of variance. Even in such a case, the estimation of missing combination(s) can be made by calculating

Table 7
Initial and converged values of parameters

	Level					
	Initial Three levels			Conveyed Eighth Iteration		
Coefficient	1	2	3	1	2	3
$b_2: \bar{B}_2 - \bar{B}_1$	0	2	4	0	1	2
$b_3: \bar{B}_3 - \bar{B}_1$	0	2	4	2	2.5	3
$c_2: \bar{C}_2 - \bar{C}_1$	−2	0	2	0	1	2
$c_3: \bar{C}_3 - \bar{C}_1$	−2	0	2	1	1.5	2
$d_2: \bar{D}_2 - \bar{D}_1$	−2	0	2	0	1	2
$d_3: \bar{D}_3 - \bar{D}_1$	−2	0	2	0	1	2
$e_2: \bar{E}_2 - \bar{E}_1$	−2	0	2	0	1	2
$e_3: \bar{E}_3 - \bar{E}_1$	−2	0	2	−2	−1	0
$f_2: \bar{F}_2 - \bar{F}_1$	−8	−4	0	−8	−7	−6
$f_3: \bar{F}_3 - \bar{F}_1$	−16	−8	0	−10	−9	−8

the coefficients using experimental regression analysis from the rest of the data.

In the experiments of manufacturing areas, data are collected based on the layout of experimental design. But in some cases, we want to utilize the existing data from observation prior to experimentation. In most cases, there is no orthogonality between factors. For such cases, experimental regression analysis can be used as a powerful tool.

4. Estimation of the Volume of Trees

The volume of a tree is estimated from chest-high diameter, D, and height, H. The following equation is generally used:

$$y = a_1 D^{a_2} H^{a_3} \qquad (7)$$

If the coefficients a_1, a_2, and a_3 are known, the volume of a tree is estimated from chest height and tree height.

To determine a_1, a_2, and a_3, 52 trees were cut down and the values, D and H, and volume, y, were measured (Table 9).

Although it is possible to conduct the logarithmic transformation, then estimate the coefficients by linear regression analysis, nonlinear equation (7) is used without transformation in this case.

Sequential Approximation

Since D and H are diameter and height, respectively, it looks like D is close to the power of 2, and H is around 1. In other words, a_2 is around 2 and a_3 is

Table 8
Comparison between estimation and observation

No.	Observation	Estimation	(Est.) − (Obs.)
1	−39.79	−38.71	1.08
2	−42.74	−42.71	0.03
3	−46.52	−46.21	0.31
4	−44.33	−43.71	0.62
5	−46.55	−45.71	0.84
6	−34.92	−35.21	−0.29
7	−43.75	−44.21	−0.46
8	−32.81	−33.21	−0.40
9	−40.98	−42.71	−1.73

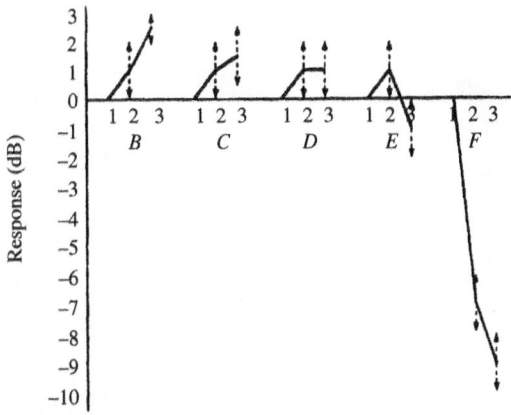

Figure 1
Response graphs of partial data

Table 10
Initial coefficients

	1	2	3
a_1	0.00005	0.00010	0.00015
a_2	1.5	2.0	2.5
a_3	0.5	1.0	1.5

around 1. Considering the units of D being in centimeters, H being in meters, and y being in square meters, the unit of a_1 is probably around 0.0001. The initial values are set as shown in Table 10.

Converged Values and Their Evaluation

After conducting sequential approximation, it converged after nine iterations to obtain the results shown in Table 11. Volume, y, is given by the following equation using the values of the second level:

$$y = 0.0001096 D^{1.839} H^{0.8398} \qquad (8)$$

The total of residuals squared, S_e, is 0.02423, and the standard deviation is 0.021.

Next, the average of residual error variation was calculated from the data in Table 12, the observa-

Table 9
Observational results for trees

No.	D (cm)	H (m)	y (m³)
1	36.3	22.50	1.161
2	24.5	18.83	0.461
3	26.3	17.35	0.535
4	29.5	19.22	0.701
⋮	⋮	⋮	⋮
52	14.5	14.37	0.124

tions of 10 trees that were not used in the estimation of parameters. It was calculated from the average of the differences between estimation and observation squared.

$$V = \frac{1}{10[(-0.003)^2 + 0.060^2 + \cdots + 0.009^2]}$$

$$= 0.00554 \qquad (9)$$

The standard deviation was 0.023, showing good reproducibility of conclusions for the data that were not used for the equation of estimation, shown in Tables 11, 12, and 13.

Comparison with Logarithmic Transformation

As described above, equation (7) can be converted into a linear equation by logarithmic transformation:

$$\log y = \log a_1 + a_2 \log D + a_3 \log H \qquad (10)$$

If a_2 and a_3 are known, $\log a_1$ is calculated as a constant. The following equation, which was obtained from the least squares method, does not have any technical contradictions and is considered as one of the solutions.

Table 11
Converged coefficients after ninth iteration

	1	2	3
a_1	0.0001094	0.0001096	0.0001098
a_2	1.8379	1.8398	1.8418
a_3	0.8387	0.8398	0.8418

Table 12
Data for verification and estimated values

No.	D (cm)	H (m)	y (m³)	ŷ (m³)	ŷ − y
1	46.3	21.90	1.701	1.698	−0.003
2	25.8	20.65	0.491	0.551	0.060
3	27.7	17.60	0.566	0.549	−0.017
4	23.6	20.05	0.473	0.456	−0.017
5	21.2	14.50	0.272	0.285	0.013
6	20.2	17.90	0.337	0.312	−0.025
7	18.1	17.33	0.264	0.248	−0.016
8	14.8	13.51	0.128	0.139	0.011
9	14.4	13.83	0.126	0.135	0.011
10	11.8	12.11	0.075	0.083	0.008

$$\log y = -4.08817 + 1.87475 \log D + 0.902218 \log H \tag{11}$$

A verification was made similarly to Table 12 using this equation, and the average squared error was calculated as

$$V = \frac{1}{10(0.052^2 + 0.064^2 + \cdots + 0.004^2)}$$

$$= 0.000864 \tag{12}$$

This is slightly larger than the result from equation (9).

Table 13
Data for verification and estimated values

No.	D (cm)	H (m)	y (m³)	ŷ (m³)	ŷ − y
1	46.3	21.90	1.701	1.753	0.052
2	24.8	20.65	0.491	0.555	0.064
3	27.7	17.60	0.566	0.549	−0.017
4	23.6	20.05	0.473	0.458	−0.015
5	21.2	14.50	0.272	0.279	0.007
6	20.2	17.90	0.337	0.309	−0.028
7	18.1	17.33	0.264	0.244	−0.020
8	14.8	13.51	0.128	0.134	0.006
9	14.4	13.83	0.126	0.130	0.006
10	11.8	12.11	0.075	0.079	0.004

Comparing the far-right columns in Tables 12 and 13, it can be seen that there is a big difference between the two tables on tree 1. The volume of this tree is larger than the volume of others. Generally, the absolute error becomes larger after logarithmic transformation. In the case of estimating the volume of the trees in forests, a 3% error of larger trees with a volume of such as 1.5 m^3 is more important than a 10% error of smaller trees with a volume of 0.2 m^3. Therefore, using logarithmic transformation to estimate a_2 or a_3, the weight of larger trees must be taken into consideration.

When using equation (7), the absolute error is not affected by volume, even when weight is not considered. It is therefore recommended that logarithmic transformation not be used when the loss caused by estimation error is affected by the absolute error.

5. Starting Temperature of Room Heaters

Most office workers begin their day at 9 a.m., and it is therefore desirable that the temperature at 9 a.m. be at the targeted room temperature in winter. To make this possible, the heater (or air conditioner in summer) must start 30 or 60 minutes earlier. Although the same period of time is used for preheating, overheating or underheating occur due to the outside temperature, sunshine, the length of time operated on the previous day, and so on, so the temperature often does not hit the target at the time that one wishes to start using the room.

Let the temperature to start heating be θ_n and the temperature difference between outside temperature and design temperature (0°C for Tokyo and −1°C for Yokohama) to start heating be $\Delta\theta_A$. The room temperature, T, hours later is given from heat transfer theory by

$$\theta_r = \theta_n + (1 - e^{-\alpha t})k(1 + \beta\,\Delta\theta_A) \qquad (13)$$

where α is the coefficient of transfer, k the constant (usually set to 4.5 based of past experiments), and β the calibration coefficient for outer temperature (usually set to 0.031). The preheating time, T_0, with a target room temperature of θ_0 is given from equation (13) as

$$T_0 = \frac{1}{\alpha \ln\{1 - [\theta_0 - \theta_n / k(1 + \beta\Delta\theta_A)]\}} \qquad (14)$$

The values of α, β, and k are different for different building structures and other conditions. It is important to determine those values for the development of energy-saving equipment.

Results of Observation and Estimation for Coefficients of Function

An air-conditioning control equipment manufacturer collected the data from some buildings to estimate the coefficients mentioned in the preceding section. During the 26-day period, observation was made on the day of the week, climate, temperature difference between outer and design target ($\Delta\theta_A$), room temperature when the heater started (θ_n), the time from starting (T), and current room temperature (θ_r). Measurements were done three times a day. Some of the results are shown in Table 14.

To estimate α, β and k in equation (13), some of the data in Table 14 were converted into integer-type variables, to obtain 11 variables in total. Regarding the week of the day, there are holidays beside Sunday; therefore, it is better to set the variable as follows (Table 15):

x_1: Second day from holiday is set to 1, others are set to 0.

x_2: Third day from holiday is set to 1, others are set to 0.

x_3: Fourth day from holiday is set to 1, others are set to 0.

x_4: Fifth day from holiday is set to 1, other are set to 0.

x_5: Sixth day from holiday is set to 1, others are set to 0.

For example, x_1 to x_5 are equal to 0 for the day right after a holiday. For climate,

x_6: Cloudy is set to 1, others are set to 0.

x_7: Snow is set to 1, others are set to 0.

Therefore, x_6 and x_7 are equal to zero for a fine day. The variables of other items are

Table 14
Results of observation

No.	Day[a]	Climate	$\Delta\theta_A$	θ_r	T (hours)	θ_n
1	Tuesday[b]	Cloudy	4.0	20.0	1.00	18.5
2	—	—	—	21.7	2.00	—
3	—	—	—	22.5	3.00	—
4	Wednesday	Snow	0.0	21.0	1.00	19.5
5	—	—	—	22.5	2.00	—
6	—	—	—	23.0	3.00	—
7	Thursday	Cloudy	0.0	20.5	0.17	20.0
8	—	—	—	22.0	0.33	—
9	—	—	—	23.0	0.50	—
⋮	⋮	⋮	⋮	⋮	⋮	⋮
76	Monday[b]	Clear	2.5	22.0	0.25	21.5
77	—	—	—	22.8	0.50	—
78	—	—	—	23.3	0.75	—

[a]—, observation same as above since made on the same day.
[b]Day next to a holiday when the heater was not operated.

Table 15
Variables, including integer types for room temperature

No.	1	2	3	4	5	6	7	8	9	10	11
1	0	0	0	0	0	1	0	4.0	20.0	1.00	18.5
2	0	0	0	0	0	1	0	4.0	21.7	2.00	18.5
3	0	0	0	0	0	1	0	4.0	22.5	3.00	18.5
4	1	0	0	0	0	0	1	0.0	21.0	1.00	19.5
5	1	0	0	0	0	0	1	0.0	22.5	2.0	19.5
6	1	0	0	0	0	0	1	0.0	23.0	3.00	19.5
⋮						⋮					
76	0	0	0	0	1	0	0	2.5	22.0	0.25	21.5
77	0	0	0	0	1	0	0	2.5	22.8	0.50	21.5
78	0	0	0	0	1	0	0	2.5	23.3	0.78	21.5

Table 16
Initial and convergent values of parameters for room temperature

| | Level | | | | | |
| | Initial Values | | | After 16th Iteration | | |
	1	2	3	1	2	3
a_1	0	0.8	1.6	0.3	0.4	0.5
a_2	0	0.8	1.6	0.7	0.75	0.8
a_3	0	0.8	1.6	0.7	0.75	0.8
a_4	0	0.8	1.6	0.4	0.5	0.6
a_5	0	0.8	1.6	0.4	0.45	0.5
a_6	−0.8	−0.4	0	−0.275	−0.2625	−0.25
a_7	−0.8	−0.4	0	−0.4	−0.35	−0.3
a_8	0.4	0.8	1.4	0.7	0.725	0.75
a_9	4	8	12	5.0	5.25	5.5
a_{10}	0	0.016	0.032	0.0	0.001	0.002

$$x_8: \quad \Delta\theta_A$$

$$x_9: \quad \theta_r (= y)$$

$$x_{10}: \quad T$$

$$x_{11}: \quad \theta_n$$

α, the heat transfer coefficient, seemingly has a small value if the heater was not operated the preceding day. β is the coefficient to calibrate the outer temperature: a constant for a particular building. k is a proportional constant for temperature rise during time passage, it is affected by climate.

The equations for α and k are set as

$$\alpha = a_8(1 + a_1x_1 + a_2x_2 + a_3x_3 \\ + a_4x_4 + a_5x_5) \tag{15}$$

$$k = a_9(1 + a_6x_6 + a_7x_7) \tag{16}$$

Variable a_{10} represents β. Determinating α, β, and k therefore involves determining 10 variables, a_1 to a_{10}.

Letting a_8 be the value of α the day after a holiday:

$$\alpha \text{ (2 days after a holiday)} = a_8(1 + a_1)$$

$$\alpha \text{ (3 days after a holiday)} = a_8(1 + a_2)$$

$$\vdots$$

$$\alpha \text{ (6 days after a holiday)} = a_8(1 + a_5)$$

Letting a_9 represent k for a fine day, then

$$k(\text{cloudy}) = a_9(1 + a_6)$$

$$k(\text{snow}) = a_9(1 + a_7)$$

Parameter Estimation by Experimental Regression Analysis
Letting $\theta_r(x_9)$ be the objective variable, y:

$$y = x_{11} + (1 - e^{\alpha x_{10}})k(1 + \alpha_{10}x_8) \tag{17}$$

where

$$\alpha = a_8(1 + a_1x_1 + a_2x_2 + a_3x_3 + a_4x_4 + a_5x_5) \tag{18}$$

$$k = a_9(1 + a_6x_6 + a_7x_7) \tag{19}$$

Table 17

Deviations and total residuals squared after the sixth iteration

No.	Deviation	Total Residuals Squared	Combination												
1	0.296194191	33.0726635	1	1	1	1	1	1	1	1	1	1	1	1	1
2	0.881743940E-01	28.4722355	2	2	2	2	2	2	2	2	2	2	2	2	1
3	−0.132098118	31.0850840	3	3	3	3	3	3	3	3	3	3	3	3	1
4	0.372788513E-01	29.2814816	1	1	1	1	2	2	2	2	3	3	3	3	1
⋮		⋮						⋮							
35	−0.146478391E-01	27.7087209	2	1	2	3	1	3	1	2	3	3	1	2	3
36	0.149078921	29.7143803	3	2	3	1	2	1	2	3	1	1	2	3	3

The initial parameter values and the ones after convergence are shown in Table 16.

For sequential approximation, the data of twenty-three out of twenty-six days were used. Some of the deviations and the total residuals squared are shown in Table 17. Using the second level, the values of a, b, and k are shown in Table 18.

The proportional coefficient of heat transfer (α) for the day after a holiday is about 30% smaller than on other days. The result that coefficients three and four days after a holiday are larger than those five

and six days after a holiday does not make sense. Probably, only a comparison between the day next to a holiday and other days has actual meaning. The calibration coefficient for outer temperature, β, is 0.001, showing that the effect of outer temperature is small.

The constant for temperature rise due to time passage, k, is affected by climate. The value is about 70% of that on a fine day. On a bad weather day, the temperature becomes lower, with other conditions unchanged.

Table 18

Values of α, β, and k

Parameter	x	Relation to Unknown Parameter a_1	Value
α	Day after holiday	a_8	0.725
	2 days after holiday x_1	$a_8(1 + a_1)$	0.725 (1 + 0.4) = 1.015
	3 days after holiday x_2	$a_8(1 + a_2)$	0.725 (1 + 0.75) = 1.26875
	4 days after holiday x_3	$a_8(1 + a_3)$	0.725 (1 + 0.75) = 1.26875
	5 days after holiday x_4	$a_8(1 + a_4)$	0.725 (1 + 0.5) = 1.0875
	6 days after holiday x_5	$a_8(1 + a_5)$	0.725 (1 + 0.45) = 1.05125
β		a_{10}	0.001
k	Clear	a_9	5.25
	Cloudy	$a_9 (1 + a_6)$	5.25 (1 − 0.2625) = 3.871875
	Snow	$a_9 (1 + a_7)$	5.25 (1 − 0.35) = 3.4125

Estimation of Preheating Time

Preheating time (T_0) can be estimated from the target temperature (θ_0) by putting the parameters above into equation (14).

There might be a tendency to use equation (14) as the regression equation since what we need is the time to preheat, T_0. However, what was measured in this study was room temperature against the time after heating started. In other words, time passage is the cause and room temperature is the effect. What can be used for adjustment is time, and the result is room temperature. It must be noted as a general rule that a regression equation is supposed to express a result, y, by the cause, x.

This case study is contributed by Genichi Taguchi.

Taguchi's Methods versus Other Quality Philosophies

39 Quality Management in Japan

39.1. History of Quality Management in Japan   1423
  Dawn of Quality Management   1423
  Period Following World War II   1424
  Evolution from SQC to TQC   1426
  Diffusion of Quality Management Activity   1427
  Stagnation of Quality Management in the Second Half of the 1990s   1428
39.2. Quality Management Techniques and Development   1428
  Seven QC Tools and the QC Story   1428
  Quality Function Deployment to Incorporate the Customers' Needs   1430
  Design of Experiments   1431
  Quality Engineering   1435
  References   1440

39.1. History of Quality Management in Japan

The idea that the quality of a product should be managed was born in Japan a long time ago. We can see some evidence that in making stone tools in the New Stone Age, our ancestors managed quality to obtain the same performance around 10,000 years ago. In the Jomon Period, a few centuries before Christ, earthenware were unified in terms of material and manufacturing method. In the second half of the Yayoi Period, in the second or third century, division of work (specialization) occurred in making earthenware, agricultural implements, weapons, and ritual tools. After the Tumulus Period, in the fourth or fifth century, the governing organization summoned specialized engineers from the Asian continent and had them instruct Japanese engineers. For the geographical reason that Japan is a small country consisting of many archipelagoes, this method of convening leaders from technologically advanced foreign countries has been continued until the twentieth century.

Japan did not document these techniques before the twentieth century. The major reason for this is that numerous technologies were imported from overseas, and superb technologies, such as the manufacture of Japanese swords, were inherited through an apprenticeship system. In Japan, technological inheritance

Dawn of Quality Management

often relies more on "learn, steal, and realize" than on "instruct or communicate." In fields other than manufacturing [e.g., Judo, Kendo (swordsmanship), Kado (flower arrangement), or Sado (tea ceremony)], Japanese leaders give detailed instructions to apprentices and make them learn the reasons for themselves. This is regarded as one of the most typical instructional methods in Japanese culture. Another instructional method relied on masters leaving know-how to descendants in the next generation (called "ogisho"). These were considered not modern commentaries or manuals but difficult-to-comprehend documents that could be understood only through repeated discipline. Therefore, even if a superb technology was born, it was not spread widely and did not boost the total technological level. Even after the Meiji Era arrived in 1868 after the Edo Shogunate was toppled, the major methods of technological innovation were introduced through foreign documents or importation of engineers.

In the area of scientific quality management, quite a few technologies were introduced. In 1931, when Shewhart completed *Economic Control of Quality of Manufactured Product*, the Shewhart control chart began to be used in the manufacturing process for light bulbs. It was in 1935 when the design of experiments, pioneered by Fisher, was first utilized in Japan. Even under the militaristic regime during World War II, which suppressed research activities, translation of bibliographies by Carl Pearson was continued and a Japanese-unique sampling inspection chart was invented by Toshio Kitagawa. By the end of the war, statistical methods were being applied to production processes in some military plants [1,2]. Nevertheless, as a whole, quality management levels in manufacturing products, except for craft products such as pottery or lacquerware, was not so high; therefore, "if it's cheap, it's bad" symbolized prewar Japanese products.

Period Following World War II

Many books have already detailed quality management in Japan after the war. In 1946, the GHQ of the Allied Forces summoned experts in statistical research from the United States to study required resources and telecommunication network systems needed for reviving Japan. The investigation significantly stimulated Japanese statisticians and quality management researchers. In 1950, W. Edwards Deming, a U.S. specialist in sampling theory and statistical quality management, instructed Japanese quality management researchers in statistical quality management. His key advice was: (1) product development and production is an endless cycle (shown in Figure 39.1), and by repeating this cycle, we can advance our products and corporate activity to produce them (this cycle is called the *Deming cycle* in Japan); and (2) statistical procedures can be applied to various stages in corporate management. After his instruction in Japan, various companies put his advice into practice.

Based in part on money, Deming donated from royalties on the textbook used for his lectures, the Deming Prize was founded in 1951 [3]; it is still awarded to companies that accomplish significant results through strenuous quality management. The first companies to win the prize in 1951 were Yahata Steel, Fuji Steel, Showa Denko, and Tanabe Seiyaku. Since the Deming Prize played an essential role in developing Japan's industry worldwide, it was regarded as "the Messiah" in Japan's industries.

Between the end of the war and the 1960s, statistical quality management methods, developed primarily in Europe and the United States, were studied vigorously and subsequently adopted by many companies and formulated as the Japan Industry Standards (JIS). In this period, companies focused mainly on introducing

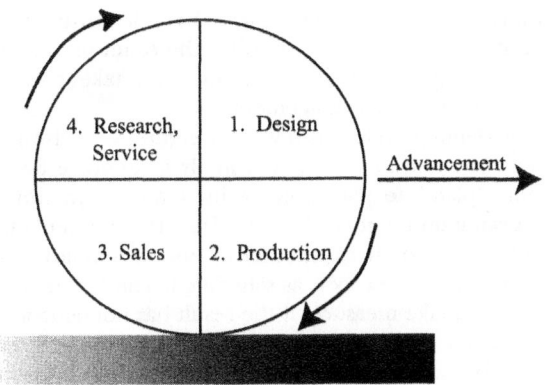

Figure 39.1
Deming cycle

statistical methods such as the Shewhart control chart, statistical testing, sampling inspection, and Fischer's design of experiments. It is often dubbed the *SQC* (statistical quality control) *period.*

While the Shewhart control chart [4] (Figure 39.2) contributed considerably to the production process, it brought about another result. The fundamental managerial concept behind the control chart—that we should recognize that any type

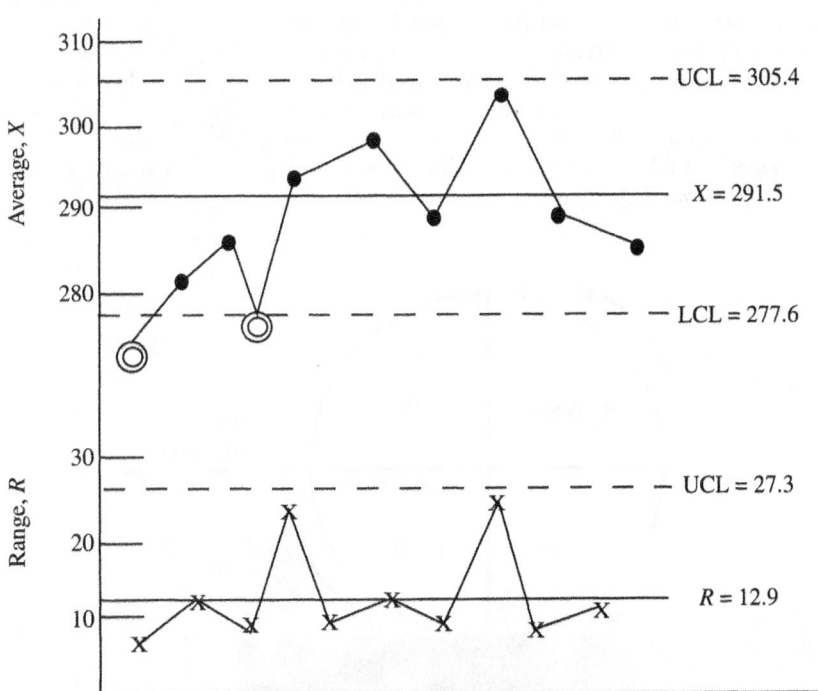

Figure 39.2
X-R control chart

of job always has variability, define a normal result using past data, determine control limits for an abnormal state, monitor the result once an abnormality happens, start an investigation, find the cause, and take corrective measures—contributed much to general management.

Based on the Deming cycle illustrated in Figure 39.1, the management cycle (*PDCA cycle*) was invented and used repeatedly to improve business. The "P" in PDCA represents "plan," to plan how we implement a job and what results we attain before working on the job. The "D" "Do," to fulfill the job as planned. The "C" represents "check," to check whether or not we have accomplished the same result or followed the same process as scheduled. The "A" represents "action," to analyze the cause and take measures if the result has not been achieved. Since this cycle is applicable not only to management and improvement of the production process but also to many other kinds of businesses, job improvement based on quality management has started to spread. The PDCA cycle (Figure 39.3) has been emphasized as a key method in TQC (total quality control), in line with Shewhart's concept of management.

In 1954, J. M. Juran, an advisor in the American Management Association at that time, was invited to Japan by Japanese quality management researchers. In Juran's lectures and instructions, statistical methods as one of the pillars of modern quality management were not mentioned. On the contrary, managerial aspects of quality management were stressed. Although this concept was implemented by Feigenhaum in the United States, it stood out as total quality control (TQC) in the second half of the 1960s in Japan.

Evolution from SQC to TQC

In the 1960s, trade in products such as automobiles, machinery, and electric appliances, all of which had been protected by the Japanese government during Japan's postwar rehabilitation period, was liberalized. Although quality management activity was reinforced to make the products more competitive, it was necessary to enhance products' attractiveness as well as their quality. Therefore, all departments from product planning to sales, or all hierarchical people from top management to manufacturing operators, were encouraged to participate in quality management activity, and at the same time, interdepartmental activity was reinforced. This was TQC (later changed to TQM).

Figure 39.3
PDCA cycle

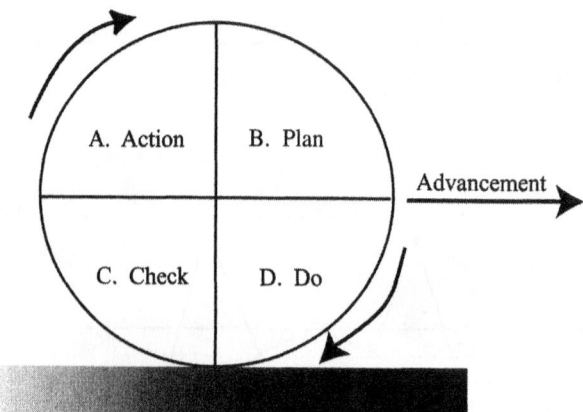

TQC stresses not only presentation of the PDCA as top management's policy and deployment of the policy to middle management in charge of a core of corporate administration but also companywide promotion of job improvement. To filter the job improvement activity down to bottom-level employees, the *QC circle* was introduced and subsequently spread nationwide. The QC circle, founded by the Japanese Union of Scientists and Engineers in 1962, contributed tremendously to quality improvement.

The other characteristic of TQC was to synthesize quality management procedures by introducing quality assurance and reliability management, which had been pioneered in the United States. The concept of quality assurance, including compensation to customers, was introduced to solve the problems caused by huge complicated public systems or their failures. The mileage assurance for automobiles was a typical example. Since indemnification incurred by system failures diminishes corporate profit, each company was urged to reduce system failures. As a result, quality assurance systems were established and reliability management was introduced in many companies. For example, reliability management was used in the Apollo program developed by NASA at that time. In Japan it was introduced in large-scale public works such as the Shinkansen bullet train program by National Railway (currently, JR) and the satellite program by the National Space Development Agency (NASDA), the automotive industry, or private joint enterprises with U.S. companies. The reliability management program manages what tasks should be completed, when they should be implemented, which departments should take responsibility, and how the problems should be solved from the product planning stage to use by consumers. Reliability management techniques such as FMEA, FTA, and design review were introduced in the 1970s and 1980s.

At the same time, *quality function deployment* (QFD) was developed by Shigeru Mizuno and Yoji Akao and adopted immediately by various companies. The quality table had been tried since 1966 and was proposed in the official book published in 1978 [5]. QFD and quality engineering (to be detailed later) are the two major powerful quality management techniques that were born in Japan and spread worldwide.

Since to improve and assure product quality, improvement and assurance of material and part quality were required, quality management and assurance were expanded to suppliers of materials and parts. This trend diffused from assembly manufacturers to part suppliers, from part suppliers to material suppliers, and then to all industries. As a consequence, many Japanese products are accepted worldwide. To illustrate this, one of every four automobiles being driven in the United States is produced by Japanese manufacturers. Ironically, self-regulation of exportation of automobiles to the United States was continued for a long time because of complaints by the U.S. automotive industry. Sony and Panasonic brand products are sold throughout the world; and the electric scoreboard in Yankee Stadium, known as the U.S. baseball hall of fame, was replaced by a Japanese product.

Since quality management activity evolved into the TQC and contributed greatly to the prosperity of the entire country, led by the manufacturing industry, TQC became known as a crucial means of managing a corporation in Japan. In the 1970s, nonmanufacturing industries such as the building, electric power, sales, and service industries adopted TQC. For the building industry, Takenaka Corp. and Sekisui Chemical in 1979, Kajima Corp. in 1982, Shimizu Corp. in 1983, Hazama

Diffusion of Quality Management Activity

in 1986, and Maeda in 1989 challenged and won the Deming Prize. Kansai Electric Power in the electric power industry won the prize in 1984, and Joban Hawaiian Center in the leisure industry was awarded the prize in 1988. Since the 1980s, all types of industries have introduced quality management activities. Bank and distribution industries, book-selling industries (e.g., Yaesu Book Center), the restaurant industry (e.g., Ringer Hut), and even public agencies such as prefectural and municipal offices have attempted to adopt quality management.

As TQM spread over other industries, its original aspect that quality problems should be solved based on data faded somewhat; in contrast, managerial aspects, including policy management, started to be emphasized. As a result, the concept of statistical methods, which had been playing a key role, was lowered in importance. In fact, the percentage of statistical topics released in the research forum of the Japanese Society for Quality Control has declined to a quarter of the total topics presented previously. In the meantime, the seven new QC tools, dealing with language information in the 1980s, and the seven product planning tools, handling planning issues in the 1990s, were newly invented. Both tools reflect the movement of quality management to nonmanufacturing industries.

Stagnation of Quality Management in the Second Half of the 1990s

When the Japanese economy, led by the manufacturing industry, reached a peak in the latter half of the 1980s, a number of companies attained top-rank status in the world. Also at that time, the income of the Japanese people became the highest in the world. As a consequence, a "bubble economy" was born whereby, instead of making a profit through strenuous manufacturing activity (or value added), we invested surplus money in land, stocks, golf courses, leisure facilities, or paintings to make money in the short term. The value of major products and services was determined only by supply and demand. The key players in the bubble economy were the financial industry, such as securities companies and banks, the real estate industry, the construction industry, and the leisure industry, all of which introduced TQM later or not at all. Since the "bubble" broke in 1992 and the prices of stocks and land slumped, these industries have been suffering. Unlike the manufacturing industry, these industries have blocked foreign companies from making inroads into the Japanese market.

The tragedy was that many manufacturers rode the wave of the bubble economy and invested a vast amount of money in stocks and land to make a short-term profit. These manufacturers suffered tremendous losses, and consequently, lost time and money they could have used to maintain quality management activities. Since the mid-1990s, the Japanese quality management activity has lost momentum. In fact, the quality management movement itself has sometimes been regarded as a bubble, just at the time that quality management is desperately needed.

39.2. Quality Management Techniques and Development

Quality engineering, with design of experiments as a main driver in developing it, is covered in this section.

Seven QC Tools and the QC Story

The first step in quality management is to solve quality problems quickly and thus to improve quality. Most quality management techniques were invented and developed for this purpose. While high-level statistical methods such as control charts, statistical tests, multivariate analysis, or design of experiments are utilized in quality management, in most cases, one of the simpler methods, often dubbed the *seven QC tools*, are used. These tools are easy to understand and use. (The number 7 has strong connotations in Japan and was used purposefully.) The seven tools are as follows: (1) control chart, (2) Pareto chart, (3) histogram, (4) graphs, (5) cause-and-effect diagram, (6) scatter diagram, and (7) check sheet.

As the quality management movement spread nationwide, when and how to utilize it to solve quality problems was questioned by many. Therefore, manuals called *QC story* were compiled so that anyone could easily take advantage of the methods. QC story was originally invented by Komatsu, a Japanese manufacturer of bulldozers that had manuals written for quality improvement giving examples of their use of quality circles. In the QC story, the process of quality problem solving or quality improvement consists of the following eight steps, and each step clarifies its objective and the QC tools to be used:

Step 1: Setup of the item to be improved (reason for selection of the item)

❑ *Objective:* Determine the problems or item to be improved.

❑ *Means:* Compare the objective and current problem and clarify the gap. Define the importance of the problem and the priority.

❑ *QC tools to be used:* check sheet, control chart, Pareto chart, histogram.

Step 2: Grasp of current problem

❑ *Objective:* Pick up a key point for improvement.

❑ *Means:* Convert the problem into a magnitude of variability. Find "good" and "bad" and clarify where variability to be eliminated lies by comparing both. This is a key point for improvement.

❑ *QC tools to be used:* check sheet, histogram, Pareto chart, graphs.

Step 3: Planning of improvement schedule

❑ *Objective:* Plan a schedule to make an efficient improvement. At the same time, establish an improvement target.

❑ *Means:* Draw up a schedule to investigate the cause of variability leading to "good" and "bad," take measures to turn "bad" into "good," and confirm the effect. Also, prove whether the objective is attained if "bad" is changed into "good."

❑ *QC tools to be used:* None.

Step 4: Analysis of causes

❑ *Objective:* Identify the cause to generate the variability of "good" and "bad."

❑ *Means:* (1) Select candidates generate the variability of "good" and "bad"; (2) specify the differences that cause the variability of "good" and "bad" using data analysis; (3) replace the differences with technical meanings.

❑ *QC tools to be used:* check sheet, cause-and-effect diagram, graphs, Scatter Diagram.

Step 5: Consideration and implementation of measures

❑ *Objective:* Consider the conditions to produce only "good."

❑ *Means:* Take measures for the cause to generate the variability of "good" and "bad."

❑ *QC tools to be used:* None.

Step 6: Confirmation of effects

❏ *Objective:* Confirm the effects of measures.

❏ *Means:* Confirm whether to produce only "good" and how to attain the objective using data analysis.

❏ *QC tools to be used:* Tools used in step 2 (to compare situations before and after improvement).

Step 7: Standardization, systematization, and prevention

❏ *Objective:* Consider the measures to prevent the same problem from occurring again.

❏ *Means:* Understand the causes in terms of methodology and system and implement standardization or systematization to prevent them.

❏ *QC tools to be used:* None.

Step 8: Review of completed process

❏ *Objective:* Learn from experience to improve the problem-solving process.

❏ *Means:* Wrap up good and bad points in the completed process and improvements.

❏ *QC tools to be used:* None.

Quality Function Deployment to Incorporate the Customers' Needs

After several trials and errors in the second half of the 1960s, *quality function deployment* was proposed in the Japanese quality management field in 1978 [5]. Proposed by Shigeru Mizuno and Yoji Akao, quality function deployment is a method of reflecting customers' voices in product design and clarifying production management items to satisfy them. After it was released, this method was adopted in the development process by many corporations.

Quality function deployment is implemented in accordance with Figure 39.4. In the first step, by allocating customers' voices in the rows of a matrix and product quality characteristics in the columns, we determine corresponding product quality

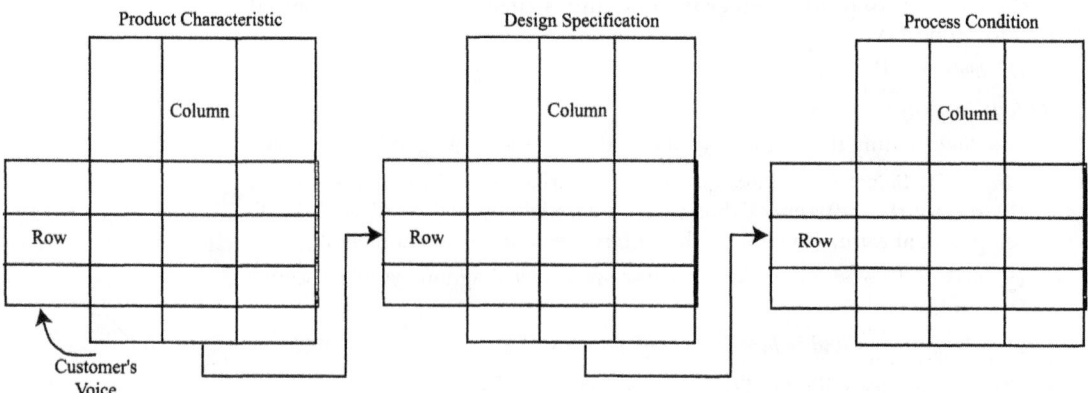

Figure 39.4
Structure of quality function deployment

characteristics to prevent customers' voices from being overlooked. In the second step, by assigning product quality characteristics in the rows of a matrix and characteristics of a subsystem or part in the columns, we determine corresponding characteristics of a subsystem or part to satisfy product quality characteristics completely. This process is launched at the stage of determining process conditions to attain part characteristics and maintenance procedures to product performances.

In short, quality function deployment is a method of putting information in order. Even if we use this method, no new information is generated. However, it still helps us significantly. Human beings tend not to pay attention to the importance of information because they lack knowledge of others' experiences. Even our own experiences tend to fade away as time goes by. Sometimes miscommunications happen. For all these reasons, if we arrange information using quality function deployment and all people in charge of development take advantage of this information, the possibility of higher customer satisfaction will be increased.

As for statistical methods, the principles of component analysis, factor analysis, and cluster analysis were used more widely to put customers' voices and product quality characteristics in order. In addition, the multiregression analysis was adopted by many users to comprehend the relationship between customers' voices and product quality characteristics. All of them are categorized in multivariate analysis [6].

Quality function deployment was introduced vigorously by U.S. quality management specialists between the latter half of the 1980s and the 1990s as a competitive tool. One well-known example in the United States was Ford's development of a new car model, the Taurus.

Design of Experiments

The design of experiments dates back to research on an experimental method of improving barley conducted by R. A. Fisher working for the Agricultural Laboratory in Rosenstadt in the suburbs of London. Used widely for quality improvement in Japan, primarily after World War II, this method contributed greatly to quality improvement in Japanese products.

BEGINNING OF THE APPLICATION OF DESIGN OF EXPERIMENTS IN JAPAN

In 1925, Fisher's *Statistical Methods for Research Workers* was released [7]. Ten years later, the Faculty of Agriculture of the Hokkaido University led the way in applying Fisher's design of experiments (experimentation using a small number of samples) to agricultural production. In 1939, Motosaburo Masuyama applied this technique to medical science.

APPLICATION OF DESIGN OF EXPERIMENTS TO INDUSTRIAL PRODUCTION

The first experiment using an orthogonal array in the scene of industrial production was performed during World War II in the United States to improve the yield in penicillin production. Shortly after the war, Morinaga Pharmaceutical conducted an experiment under the direction of Motosaburo Masuyama. Thereafter, the design of experiments started to be used on a regular basis in postwar Japan. Motosaburo Masuyama, Genichi Taguchi, and other experts in the design of experiments, as well as applied statisticians, helped various industries conduct a great number of experiments. The number of applications of the design of experiments increased in line with a boom in quality management activity in Japan, thereby boosting Japan as one of the top-ranked countries in applying the technique. It was not unusual that operators themselves studied the design of experiments and performed experiments.

After the mid-1950s, the number of universities or colleges introducing the design of experiments in their curriculums rose and contributed to engineers being familiar with the technique. In addition, nonprofit organizations offered lectures on the design of experiments to industrial engineers, which contributed considerably to promoting the technique. In this sense, the Japanese Union of Scientists and Japanese Standards Association were outstanding. To date, several thousands of company engineers have attended high-level lectures about the design of experiments given by both organizations above. As a result, the number of companies applying the technique has increased annually. In Europe and the United States, experts in statistical methods often instruct engineers in-house. Engineers graduating from mechanical, electrical, or chemical engineering departments learn the design of experiments and apply it to their own fields. Because of this, Japan boasts of being number 1 in terms of the number of experimental applications.

With respect to the number of researchers using the design of experiments, Japan reached the top in the 1970s. Among these researchers, Genichi Taguchi is one of the most famous in the world. Although he is known as a researcher and promoter of quality engineering, he is still considered a great expert in the design of experiments. The layout design of experiments, such as the level-labeled orthogonal array (called Taguchi's orthogonal array), assignment of interactions using a linear graph, multilevel assignment, dummy-level method, pseudofactor method, and the data analysis method, such as accumulation analysis and 0/1 data analysis, are Taguchi's major contributions to the design of experiments.

The ability to use the design of experiments to comprehend the causes of quality problems contributed much to quality improvement of Japanese products after World War II. This technique is an essential method of quality improvement for engineers. Yet in most cases, the design of experiments was used not in the stage of design optimization but in the phase of cause analysis after quality problems occurred [9,10].

STUDY OF ORTHOGONAL ARRAYS AND ORTHOGONAL EXPERIMENTS
One of the most typical applications of the design of experiments is experiments using an orthogonal array to assign all experiments systematically. The concept of experimentation using an orthogonal array dates back to Fisher. Due to the technical difficulty he experienced interpreting high-order interactions, he studied experimental layout leaving them out [7]. He based his experiments on the Latin square (also called the Greco Latin square). Thereafter, engineers the world over studied the orthogonal Latin square and orthogonal array Consequently, one research report by a U.S. team was released in 1946 [8]. Although full-scale research on an orthogonal array began in Japan after the war, much research was related to the characteristics of an orthogonal array and application methods based on them, whereas the U.S. study focused on how to create an orthogonal array and its mathematical aspects. A typical example was Taguchi's research in the 1950s and 1960s. Taguchi replaced elements labeled -1 and 1 or -1, 0, and 1 in an orthogonal array by 1, 2, and 3, representing the level of a factor for an experiment, and rearranged the array by reordering its columns in increasing order. At

the same time, by taking advantage of the research on columns where interactions emerge, he assigned interactions called *linear graphs* and recommended multilevel assignment to improve ease of use. In his books, he illustrated actual applications to explain how to use the method. His development of easy-to-use orthogonal arrays, linear graphs to allocate interactions and multiple levels, and demonstration of actual applications boosted the popularity of the orthogonal arrays in Japan. Sample arrays are shown in Table 39.1 and Figure 39.5.

While he assigned interactions to linear graphs, Taguchi had a negative opinion about selecting interactions in actual experiments. He insisted that he had devised linear graphs to allocate interactions easily only in response to many engineers' keen interest, but that we should not pursue interactions. He also maintained that since an orthogonal array was used not for the purpose of partial experimentation but for checking the reproducibility of factor effects, main effects should be allocated to an orthogonal array. As a result, other researchers who were using Fisher's design of experiments stood strongly against Taguchi's ideas, and this controversy continued until the 1980s. This dispute derived from the essential difference in objectives between Taguchi's and Fisher's experiments. Whereas Taguchi emphasized design optimization, Fisher looked at "quantification of contribution for each factor." Because this controversy was caused partially by Taguchi's avoidance of the conditional "in case of," after his concept was systematized as quality engineering in the mid-1980s, the dispute gradually calmed down.

In addition to Taguchi's linear graphs, other assignment techniques to deftly allocate factor interactions were studied. For example, to vie with Western research, techniques such as the resolution III, IV, and V were released and applied to actual experiments.

BEGINNINGS OF QUALITY ENGINEERING

In 1953, an advanced experiment was performed at Ina Seito (currently known as INAX) located in Aichi prefecture in the middle of Japan. Masao Ina described this experiment in the book *Quality Control.* In addition, Taguchi's book on the

Table 39.1

Taguchi's orthogonal array (L_8)

No.	Column						
	1	2	3	4	5	6	7
1	1	1	1	1	1	1	1
2	1	1	1	2	2	2	2
3	1	2	2	1	1	2	2
4	1	2	2	2	2	1	1
5	2	1	2	1	2	1	2
6	2	1	2	2	1	2	1
7	2	2	1	1	2	2	1
8	2	2	1	2	1	1	2

Figure 39.5
Taguchi's linear graphs
(L_8)

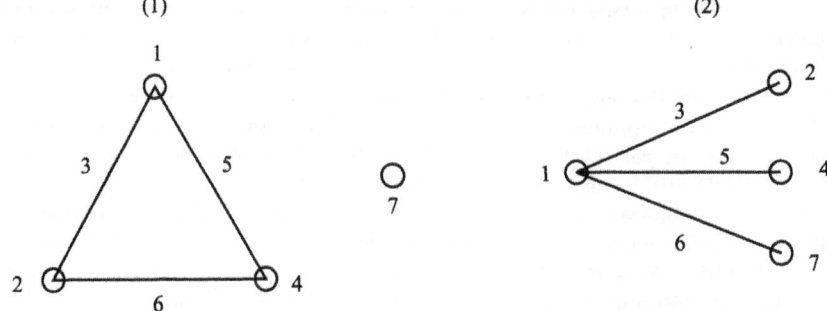

design of experiments described the experiment, after which much attention was paid by some Japanese engineers.

Seito had baked tiles using an expensive tunnel kiln. Because of the large positional variability in the tunnel's inside temperature, they had faced much variation in dimensions, glosses, and warping after baking. Although a common countermeasure in quality management was to reduce the temperature variability, they modified the mixture of materials (design) instead. As a consequence, once a certain mixture was fixed and a certain "secret medicine" was added at 0.5% by weight, tiles baked evenly and variability in dimensions, glosses, and warps was drastically reduced. Since Seito kept this experiment confidential afterward, even Taguchi, who cooperated in the experiment, did not introduce the details.

This experiment was very significant . At that time, there had been two known methods of improving quality: (1) to find the causes of variability and to manage them, or (2) a feedback (or feedforward) method of making repeated adjustments to eliminate eventual variability. The Seito experiment shed light on a third option: "attenuation of causal effects."

At this time Taguchi began to dedicate much of his research to how to apply the attenuation of causal effects to other engineering fields. His efforts became known as parameter design (robust design) in the latter half of the 1970s.

CLASSIFICATION OF FACTORS

In terms of experimental purpose, Taguchi's design of experiments greatly differed from Fisher's, although the former was founded on the latter. Many of the experiments undertaken by Taguchi focused not on a natural scientific concept of calculating factor-by-factor contributions for variability in product characteristics but on an idea of stabilizing product characteristics for variability, such as electrical power fluctuation or lapse of time. This fundamental difference between their purposes stems from the fact that Fisher worked for the Agricultural Laboratory and thus dealt with nature, whereas Taguchi worked for the Telecommunication Laboratory of the Nippon Telegraph and Telephone Public Corporation (currently NTT), a technological laboratory. Therefore, what derived from this technological purpose was factor classification. Taguchi took into account stability and reliability in actual production and use, such as production variation, consumers' use, and environmental conditions. Although these factors were called *indicative factors* at that time, they are now called *noise factors* (or *error factors*). While noise factors affect values of product characteristics, engineers cannot set up or fix their values at a desired level. Factors that engineers can select and fix at their disposal are dubbed *control* factors. Each is handled differently in layout of experiments and *analysis*.

Quality engineering has been developing since the Seito experiment in 1953 and is still under development. Although some engineers became interested in the experiment after release of the report in 1959, a limited number of engineers understood it well. Since many of the leaders of TQM had mastered Fischer's design of experiments, quality engineering was not accepted as being in the mainstream of TQM. Not until the late 1980s did quality management experts began to understand the significance of quality engineering. Thereafter, younger researchers in quality management and applied statistics started to evaluate quality engineering, and consequently, it gained acceptance, especially after presentation of a paper on parameter design offered in *Technometrics* by the American Statistical Association in 1992 [11].

Although quality engineering was seen as a method of incorporating quality in the design stage of a new product in the 1980s, it developed into a method of incorporating quality in the technological development stage. One of the main reasons that it started to be widely accepted in Japan is that the Quality Engineering Society was founded in the spring of 1993.

DEVELOPMENT INTO PARAMETER DESIGN (ROBUST DESIGN)

After the Seito experiment in 1953, Taguchi developed "attenuation of causal effects" into "parameter design" by helping to conduct experiments at several companies. His motivation was to apply the design of experiments to incorporation of quality at the initial design stage instead of quantitatively analyzing causes of quality problems after they occurred. He thought that a method of "attenuation of causal effects" must be used to incorporate quality. Taking into account that the mainstream in quality management at that time was into problem solving, Taguchi's idea can be seen not as quality management but as technological improvement. This was the main reason that some engineers began to pay attention to it. Figure 39.6 shows the debug model.

Initially, after conditions regarding the environment or use were assigned to the outer array of an orthogonal array and control factors (design parameters) were allocated (direct product experiment), whether there existed interactions

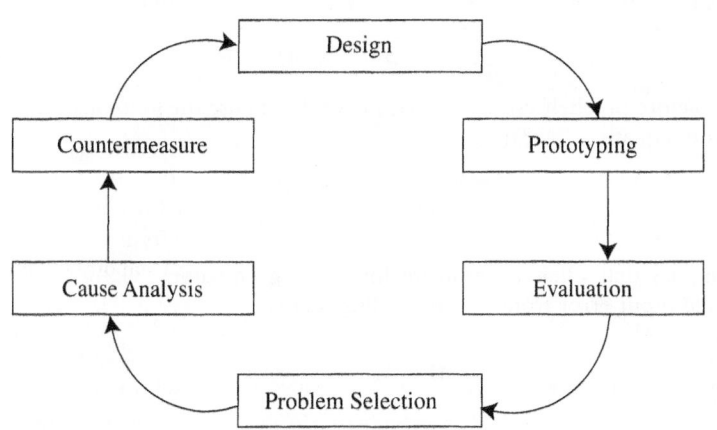

Figure 39.6
Development using
debug model

between control and indicative factors was analyzed, and control factor levels then could be stabilized as part of a procedure of determining design parameters to attenuate causal effects.

Questioning the concept of *repetition* (accidental error), which played a key role in Fischer's design of experiments, Taguchi conducted experiments by forcibly changing levels of factors leading to errors (indicative factors). Although this reinforced the parameter design, it generated mathematical difficulty for the SN ratio, as detailed below. An example of a direct product experiment is shown in Table 39.2.

At the same time, Taguchi was aware that the magnitudes of errors varied in accordance with design conditions. Therefore, he proposed that the magnitudes of errors be constant under all design conditions. Also, he doubted the feasibility of a statistical test based on homogeneity of variance. Thus, he questioned the concept of discount coefficient, which is not detailed in this book, and the SN ratio.

BIRTH OF THE SN RATIO

Research was done on the accuracy of measuring instruments in the 1960s and 1970s, where the scale of an SN ratio represented the intensity of signal to noise. Noises such as variability in measuring environment conditions or measurement methods altered output values and degraded measurement accuracy. Ideally, we would expect that output, y, is proportional to objective value, M, and has no errors. Thus, an ideal condition can be expressed as:

$$y = \beta M \qquad (39.1)$$

A measurement instrument is an apparatus to estimate input based on output. Since we cannot be aware of what types of errors are added when measuring, we estimate input using equation (39.1), indicating the ideal condition with no error assumed. That is, we can express input M by transforming equation (39.1) as follows:

$$M = \frac{y}{\beta} \qquad (39.2)$$

Considering that an actual measurement instrument is affected by various noises $e(M)$, we rewrite equation (39.1) as follows:

$$y = \beta M + e(M) \qquad (39.3)$$

Thus, to more precisely estimate input M, we should use the following transformed version of equation (39.3):

$$M = \frac{y - e(M)}{\beta} \qquad (39.4)$$

This implies that when we estimate input using equation (39.2), the following estimated input error from equation (39.2) exists:

$$M - M = \frac{-e(M)}{\beta} \qquad (39.5)$$

Table 39.2
Example of a direct product experiment

No.	Column											Distance
	A 11	B 4	C 5	D 3	E 9	A×C 14	A×B 15	R(Car) 1,6,7	V(Position) 2,8,10	e 12	e 13	$K_1\ K_2\ K_3$
1	1	1	1	1	1	1	1	1	1	1	1	
2	2	1	1	1	2	2	2	1	2	2	2	
3	1	2	2	1	1	2	2	2	1	2	2	
4	2	2	2	1	2	1	1	2	2	1	1	
5	2	1	1	2	1	2	2	2	3	1	1	
6	1	1	1	2	2	1	1	2	4	2	2	
7	2	2	2	2	1	1	1	1	3	2	2	
8	1	2	2	2	2	2	2	1	4	1	1	
9	2	1	2	2	2	1	2	3	1	1	2	
10	1	1	2	2	1	2	1	3	2	2	1	
11	2	2	1	2	2	2	1	4	1	2	1	
12	1	2	1	2	1	1	2	4	2	1	2	
13	1	1	2	1	2	2	1	4	3	1	2	
14	2	1	2	1	1	1	2	4	4	2	1	
15	1	2	1	1	2	1	2	3	3	2	1	
16	2	2	1	1	1	2	1	3	4	1	2	

1437

The estimated input error above does not indicate the total value. To estimate this, the following mean sum of squares is often calculated:

$$\frac{\Sigma[-e(M)/\beta]^2}{n-1} \tag{39.6}$$

Substituting the following σ^2 into equation (39.6),

$$\sigma^2 = \frac{\Sigma[-e(M)]^2}{n-1} \tag{39.7}$$

we obtain

$$\frac{\sigma^2}{\beta^2} \tag{39.8}$$

This equation (39.8) becomes better as the error increases. Thus, to convert this into an inverse scale, we take its reciprocal:

$$\frac{\beta^2}{\sigma^2} \tag{39.9}$$

The error computed by equation (39.9) is called the *SN ratio error* in the field of measurement.

When we analyze data using the SN ratio, we often make use of the following scale, called the *decibel value of the SN ratio* by taking a logarithm of both sides of the following equation because the equation has an unwieldy form of ratio:

$$10 \log \frac{\beta^2}{\sigma^2} \tag{39.10}$$

The reason that we multiply the logarithm by 10 is that we wish to unify this form with the SN ratio used to indicate the magnitude of noises of electric appliances and to transform original data into more handy ones when analyzed.

In the initial parameter design introduced earlier, we determined optimal levels using plot graphs of interactions between control and noise factors. However, as the number of control and noise factors increases, the number of graphs also rises. For this reason it is painstaking to find the best level for each control factor in actual analysis. For example, even if the effect of a certain noise diminishes at a certain control factor level, another noise quite often strengthens its effect. In this case we cannot easily judge which control factor levels are best. Since the SN ratio is a scale representing an entire effect of all noises, we can determine the best level if we simply select control factor levels leading to the largest SN ratio, which is regarded as quite a handy scale.

Nevertheless, it took years until this SN ratio was understood completely and applied to practical uses. One of the reasons is that equation (39.10) per se has been difficult for ordinary engineers to comprehend. The other is an issue of its statistical background. That is, since it was not known what type of statistical value the SN ratio in equation (9) was supposed to be, any method of significance test and confidence interval calculation of the SN ratio could be introduced when it was invented. Although the fact that if the error is an accidental error (repeated error), noncentral chi-square distribution is applicable was proved afterward, the issue was not easily solved. Because Taguchi only proposed calculating the SN ratio by adding noises to a denominator in evaluating the accuracy of measurement instruments, he did not show what statistical distribution's value equation (39.9) should become. Therefore, many statisticians believe it to be a difficult-to-use metric, and moreover, some of them are skeptical of it. Although this issue is not settled perfectly, it has been proved through accumulated practical applications that we can successfully make an improvement using the SN ratio. In quality engineering, the SN ratio detailed thus far is applied not only for measurement instruments but also for common systems with input and output, currently called the *dynamic SN ratio.*

SYSTEMATIZATION OF PARAMETER DESIGN

After Taguchi succeeded in reducing variability effects of used parts on output voltage of a electric circuit at Nippon Denso (currently known as Denso) and Nippon Electric (currently known as NEC) in the late 1970s, he took advantage of his successes by using quality engineering methods in other corporations. Since in an electric circuit, theoretical formulas expressing output voltage values by constants of used parts was well established in most cases, not many prototypes were needed to easily show that parts' variability effects for output voltage change according to parameter values. In indicating output voltage variability, a scale called *nominal-the-best* was used. This is expressed by equation (39.9) and equivalent to a reciprocal square of variation coefficient that is often used in quality management:

$$-10 \log \frac{\sigma^2}{m^2} \tag{39.11}$$

Since the examples in the parameter design of an electric circuit were extremely easy to understand, an increasing number of companies introduced Parameter design in their development process in the 1980s. One typical company was Fuji Xerox, a joint corporation founded by Xerox Corp. of the United States and Fuji Film of Japan. In 1985 the company implemented parameter design and prescribed parameter design procedures for new product development. Other companies gradually began to adopt the technique as a method of improving product design [13].

In those days, systematization of parameter design was advanced and templates for procedures were also created so that anyone could use the method. Additionally, by taking advantage of examples of an electric circuit, Taguchi showed not merely a parameter design procedure but also tactics to incorporate quality in product design: *two-step optimization.* This consists of the following two steps: (1) to determine design parameters in order to enhance the SN ratio (i.e., to reduce noise effects), and (2) to make adjustments to a target value using design parameters that can change average output and are unrelated to SN ratio. Since two-step optimization can reduce noise effects at first, we can expect to minimize the incidence of quality troubles caused by noises. Consequently, we can reallocate personnel from following up on quality problems to productive design tasks. This change can be appealing to managers in a development department.

TRANSITION OF PARAMETER DESIGN FROM PRODUCT DEVELOPMENT TO TECHNOLOGICAL DEVELOPMENT

If we call the 1980s a period of parameter design in *product design,* quality engineering can then be dubbed a period of parameter design in *technological development.* Thus, if we attempt to use parameter design, we cannot take advantage of nominal-the-best parameters as product characteristics. However, if we can incorporate quality in technologies ahead of product planning, quality problems can be minimized and labor hours can be dramatically reduced in the development phase. In addition, if we can make technologies robust to noises, they can be applied to a variety of products.

Two-step optimization implies that we make technologies robust to noise in the first step, and once each product's target is fixed, we make adjustment to each target in the next step. Then what do we measure for optimization in the phase of technological development? The answer is the dynamic SN ratio. By regarding a technology as an input/output system, such as a measurement instrument, we optimize the technology using the dynamic SN ratio, regardless of an individual product's target. At first, Taguchi explained this idea by using the example of plastic injection molding [12]. After 1990, he devoted much of his time to research on what to select as input and output in each technological field. He announced ideas of input and output in various technologies and called each of them a *generic technological function.*

The Quality Engineering Society founded in Japan in the spring of 1993 is regarded as an academic society to study this generic function. As of March 2001, 1835 people are members of this society, and over 70 researchers attended the Research Report Forum in 2000. In fact, most of the research is focused on technological input and output (generic functions). Moreover, some companies have started to report their quantitative results obtained in actual engineering jobs according to quality engineering.

If two-step optimization is regarded as a *tactical* approach, parameter design in the stage of technological development is considered a *strategic* approach. Since more and more people have become aware that the parameter design can contribute much to streamlining the product development process, an increasing number of corporations such as photocopy machine or printer manufacturers are introducing the method companywide.

MAHALANOBIS DISTANCE

Since the late 1990s, Taguchi has begun to propose applying multiple dimensional distance (commonly known as *Mahalanobis distance*) advocated by Mahalanobis, an Indian statistician. Indeed, this is not a means of quality design; however it is applicable to a wide variety of applications, such as inspection, prediction, or pattern recognition. Due to space limitation, we do not cover it here, but refer the reader to Part VI in Section 2 of this handbook.

References

1. Takanori Yoneyama, 1979. *Talk about Quality Management.* Tokyo: Japan Union of Scientists and Engineers.
2. Various Authors, 1977. *Quality Management Handbook,* 2nd ed. Tokyo: Japanese Standards Association.
3. Deming Prize Committee, *A Guide to the Deming Prize.*
4. W. A. Shewhart and W. E. Deming, 1960. *Fundamental Concept of Quality Management.* Tokyo: Iwanami Shoten.

5. Shigeru Mizuno, Yoji Akao, et al., 1978. *Quality Function Deployment: Approach to Company-Wide Quality Management*. Tokyo: Japan Union of Scientists and Engineers.

6. Don Clausing, 1994. *Total Quality Development*. New York: ASME Press.

7. R. A. Fisher. 1951. *Design of Experiments*. London: Oliver & Boyd.

8. R. L. Placket and J. P. Burman, 1946. The Design of optimum multi-factorial experiments. *Biometrika*, Vol. 33.

9. Genichi Taguchi, 1957. *Design of Experiments*, Vols. 1 and 2. Tokyo: Maruzen.

10. Genichi Taguchi, 1976. *Design of Experiments*, Vols. 1 and 2, 3rd ed. Tokyo: Maruzen.

11. Vijayan Nair, 1992. Taguchi's parameter design: a panel discussion. *Technometrics*, Vol. 32, No. 2.

12. Genichi Taguchi et al., 1992. *Technological Development of Transformability*. Tokyo: Japanese Standards Association.

13. Genichi Taguchi et al., 1994. *Quality Engineering for Technological Development*. Tokyo: Japanese Standards Association.

40 Deming and Taguchi's Quality Engineering

40.1. Introduction   1442
40.2. Brief Biography of William Edwards Deming   1442
40.3. Deming's Key Quality Principles   1444
    System of Profound Knowledge   1444
    Deming's 14 Points for Management   1445
    Deming's Seven Deadly Diseases   1446
    Special Cause versus Common Cause   1446
    Deming's Cycle for Continuous Improvement   1447
40.4. Complementing Philosophies   1447
    References   1448

40.1. Introduction

Many modern quality engineering and quality management philosophies originated in the United States throughout the twentieth century. Several people played key roles in the development and implementation of these revolutionary methods for managing organizations. Among the most influential of these people was William Edwards Deming. His philosophies on quality broadened the scope of quality engineering from a tactical nature to a strategic management way of life.

40.2. Brief Biography of William Edwards Deming

William Edwards Deming was born on October 14, 1900, to William Albert Deming and Pluma Irene Edwards in Sioux City, Iowa [1]. Seven years later, his family moved to Wyoming after receiving a land grant. The 40 acres of land given to the Deming family was useless for farming, making life very hard for Edwards and his two siblings. The family's first home was nothing more than a tarpaper shack.

Pluma Irene and William Albert Deming were well educated for the time. Pluma had studied music in San Francisco, and William Albert had studied mathematics and law. Both parents emphasized the importance of education and encouraged their children to focus on their studies. Although Deming was going to school, he began picking up odd jobs to support the family when he was 12.

After graduating from high school in 1917, Deming enrolled in classes at the University of Wyoming in Laramie. Fortunately, the university did not have a tuition fee. Deming studied for four years, graduating with a bachelor of science degree in physics. He stayed at the university in Laramie for an additional year, teaching engineering and taking additional classes in mathematics. In 1918 he accepted a teaching position for two years at the Colorado School of Mines. He returned briefly to Wyoming in 1922 and married Agnes Bell.

Deming continued his formal education at the University of Colorado in Boulder, earning a master's degree in mathematics and physics in 1924. With the support and recommendation from one of his professors, he was accepted to Yale University as a doctoral candidate with a scholarship and part-time teaching position during the school year. During the summers, he traveled to Chicago to conduct research on telephone transmitters at the Western Electric Hawthorne Plant. While working at the Hawthorne Plant, Deming met and was mentored by Walter Shewhart, the pioneer of statistical process control. After completing his dissertation in 1928 on the packing of nucleons in helium atoms, Deming was awarded a Ph.D. degree in mathematical physics.

Although Deming found employment with the U.S. Department of Agriculture working in the Fixed Nitrogen Research Laboratory, times became difficult. In 1930, he was left a widower, raising his daughter alone. He remarried two years later. Lola Elizabeth Shupe remained his wife for 52 years until her death in 1984.

In 1933, Deming became head of the Mathematics and Statistics Department at the Graduate School of the U.S. Department of Agriculture. During his tenure, he developed basic research on sampling. Although his scientific papers focused on physics, his work involved the application of statistical methods, which extended into other disciplines.

During the 1930s, the U.S. Census Bureau began working on the implementation of statistical sampling as it related to demographics. Traditional full-count methods were tedious. In many cases, census figures were outdated before they were calculated. Improved methods were needed. The Census Bureau named Deming the head of mathematics and advisor on sampling in 1939. While in this position, Deming began applying Shewhart's statistical control methods to non-manufacturing applications. He led the development of sampling procedures, which are now used throughout the world for current information on employment, housing, trade, and production.

After World War II, Deming established his private consulting practice and developed the 14 points for management, the seven deadly diseases, the Deming cycle for continuous improvement, and the system of profound knowledge. In 1946 Deming led the formation of the American Society for Quality Control. As a consultant to the War Department, he visited Japan for the first time in 1947 to aid in the 1951 census. He returned numerous times throughout the 1950s as a teacher and consultant to Japanese industry, through the Union of Japanese Scientists and Engineers. In 1960, the Emperor of Japan recognized Deming with the Second Order Medal of the Sacred Treasure for the improvement of the Japanese economy through the statistical control of quality.

Although Deming is best known for his work in Japan, he served as a consultant to many other countries, including Mexico, India, Greece, Turkey, Taiwan,

Argentine, and France. He was also a member of the United Nations Subcommission on Statistical Sampling from 1947 to 1952.

Although Deming was awarded the Shewhart Medal in 1955 from the American Society for Quality Control, most businesses in the United States ignored his methods until the 1980s. On June 15, 1987, President Reagan presented Deming with the National Medal of Technology. After 40 years, his own country publicly acknowledged the significance of his work. Deming continued consulting until his death in 1993.

40.3. Deming's Key Quality Principles

While working as a consultant after World War II, Deming developed five key quality principles. Unlike those of his predecessor and mentor, Shewhart, these principles are strategic in nature, focusing on management. The principles include (1) the system of profound knowledge, (2) the 14 points for management, (3) the seven deadly diseases, (4) applications of common and special causes in management, and (5) the Deming cycle for continuous improvement. Each of these concepts is discussed in the following sections.

System of Profound Knowledge

Deming advocated a radical change in management philosophy based on his observations in applying statistical quality principles to common business practices. To make this transformation, people must understand four key elements of change, which he termed *profound knowledge.*

> The layout of profound knowledge appears here in four parts, all related to each other: appreciation for a system, knowledge about variation, theory of knowledge, and psychology.
>
> One need not be eminent in any part nor in all four parts in order to understand it and to apply it. The 14 points for management in industry, education, and government follow naturally as application of this outside knowledge, for transformation from the present style of Western management to one of optimization.
>
> The various segments of the system of profound knowledge proposed here cannot be separated. They interact with each other. Thus, knowledge of psychology is incomplete without knowledge of variation. [2, p. 93]

Deming realized that most scientific management philosophies did not take into account variation in people. To implement the changes he advocated, management needs to understand the psychology of its workers and the variation in psychology among them individually.

> A leader of transformation, and managers involved, need to learn the psychology of individuals, the psychology of a group, the psychology of society, and the psychology of change.
>
> Some understanding of variation, including appreciation of a stable system, and some understanding of special causes and common causes of variation are essential for management of a system, including management of people. [2, p. 95]

To summarize what is needed to implement the profound system of knowledge within an organization, Deming developed the 14 points for management [3]. These key management guidelines can be applied to any organization, regardless of size or product produced.

Deming's 14 Points for Management

1. Create consistency of purpose toward improvement of product and service, with the aim to become competitive and to stay in business and to provide jobs.

2. Adopt a new philosophy. We are in a new economic age. Western management must awaken to the challenge, must learn their responsibilities, and take on leadership for change.

3. Cease dependence on inspection to achieve quality. Eliminate the need for inspection on a mass basis by building quality into the product in the first place.

4. End the practice of awarding business on the basis of price tag. Instead, minimize total cost. Move toward a single supplier for any one item, on a long-term relationship of loyalty and trust.

5. Improve constantly and forever the system of production and service, to improve quality and productivity, and thus constantly decrease cost.

6. Institute training on the job.

7. Institute leadership. The aim of supervision should be to help people and machines and gadgets to do a better job. Supervision of management is in need of overhaul, as well as supervision of production workers.

8. Drive out fear, so that everyone may work effectively for the company.

9. Break down barriers between departments. People in research, design, sales, and production must work as a team to foresee problems of production and in use that may be encountered with the product or service.

10. Eliminate slogans, exhortations, and targets for the workforce by asking for zero defects and new levels of productivity. Such exhortations only create adversarial relationships, as the bulk of the cause of low quality and low productivity belongs to the system and thus lies beyond the power of the workforce.

11. Eliminate work standards (quotas) on the factory floor. Substitute leadership. Eliminate management by objectives. Eliminate management by numbers, numerical goals. Substitute leadership.

12. Remove barriers that rob the hourly workers of their right to pride of workmanship. The responsibility of supervisors must be to change the sheer number of quality. Remove barriers that rob people in management and in engineering of their right to pride of workmanship. This means, inter alia, abolishment of annual or merit rating and of management by objectives.

13. Institute a vigorous program of education and self-improvement.

14. Put everybody in the company to work to accomplish the transformation. The transformation is everybody's job.

The 14 points are the basis for transformation of American industry. It will not suffice merely to solve problems, big or little. Adoption and action on the 14 points are a signal that the management intends to stay in business and aim to protect investors and jobs. Such a system formed the basis for lessons for top management in Japan in 1950 and in subsequent years. [3, p. 23]

Although the 14 points of management are numbered, they are not listed in any specific priority. Each of the points must be implemented within an organization to achieve the highest level of performance.

Deming's Seven Deadly Diseases

Just as the 14 points for management define what an organization must do to implement the system of profound knowledge, Deming identified seven points that inhibit its achievement [3]:

1. Lack of consistency of purpose to plan a product and service that will have a market, keep the company in business, and provide jobs

2. Emphasis on short-term profits: short-term thinking (just the opposite of consistency of purpose to stay in business), fed by fear of unfriendly takeover, and by push from bankers and owners of dividends

3. Evaluation of performance, merit rating, or annual review

4. Mobility of management; job hopping

5 Management by use only of visible figures, with little or no consideration of figures that are unknown and unknowable

6. Excessive medical costs

7. Excessive costs of liability, swelled by lawyers who work on contingency fees

Although some of these "diseases" can be avoided by following the 14 points for management, two sins in particular are currently unavoidable. Organizations currently exist in a culture in which litigation and medical costs are in many cases beyond their control.

Special Cause versus Common Cause

Shewhart developed the concepts of special cause (intermittent) and common (inherent) cause variation. In the late 1940s and throughout the 1950s, Deming showed that these principles can be applied not just to manufacturing, but to any system. With this in mind, management can make the following mistakes [2]: (1) to react to an outcome as if it came from special cause, when in reality it came from common causes of variation; and (2) to treat an outcome as if it came from common causes of variation, when in reality it came from a special cause.

Deming continually noted that management often punishes employees for errors and mistakes that are caused inherently by the system. The worker has no control but is responsible for the problem. No matter how much an employee is lectured or disciplined, improvement will not occur unless the system is improved. Improvement of the system is the responsibility of management.

Management must be able to distinguish between problems caused by common cause and special cause variation. Another destructive behavior in which management often engages is solving every problem as though it had a special cause. This approach to problem solving drains resources, leading down a path of frustration.

So obvious, so fruitless. The vice president of a huge concern told me that he has a strict schedule of inspection of final product. To my question about how they use the data came the answer: "The data are in the computer. The computer provides a record and description of every defect found. Our engineers never stop until they find the cause of every defect."

Why was it, he wondered, that the level of defective tubes had stayed relatively stable, around $4 1/2$ to $5 1/2$ percent, for two years? My answer, The engineers were confusing common cause with special causes. Every fault was to them a special cause, to track down, discover, and eliminate. They were trying to find the causes of ups and downs in a stable system, making things worse, defeating their purpose. [2, pp. 181–182]

Based on Shewhart's steps in a dynamic scientific process for acquiring knowledge, the Deming cycle for continuous improvement was developed as a general problem-solving methodology for any product or process, regardless of level [2].

Deming's Cycle for Continuous Improvement

Step 1: Plan. Somebody has the idea for improvement of a product or of a process. This is the 0-th stage, embedded in Step 1. It leads to a plan for a test, comparison, experiment. Step 1 is the foundation of the whole cycle. A hasty start may be ineffective, costly, and frustrating. People have a weakness to short-circuit this step. They do not wait to get into motion, to be active, to look busy, move into step 2.

Step 2: Do. Carry out the test, comparison, or experiment, preferably on a small scale, according to the layout decided in Step 1.

Step 3: Study. Study the results. Do they correspond with hopes and expectations? If not, what went wrong? Maybe we tricked ourselves in the first place, and should make a fresh start.

Step 4: Act. Adopt the change. or abandon it. or run through the cycle again, possibly under different environmental conditions, different materials, different people, different rules. [2, pp. 131–133]

40.4. Complementing Philosophies

When comparing Taguchi methods to Deming's teachings, one finds that the philosophies complement each other in several key areas. Deming stated relentlessly that management needed to change the system and reduce common cause variation in order to realize significant quality improvements. Moreover, he concurs with Taguchi's teachings regarding quality loss and the deficient philosophy of quality by conformance, as presented in the following quote:

The fallacy of zero defects. There is obviously something wrong when a measured characteristic barely inside a specification is declared to be conforming; outside it is declared to be non-conforming. The supposition that everything is all right inside the specification and all wrong outside does not correspond to this world.

A better description of the world is the Taguchi Loss Function in which there is minimum loss at the nominal value, and an ever-increasing loss with departure either way from the nominal value. [3, p. 141]

Although Deming challenged management with improving the system, he never offered management a way to achieve reductions in a system's common cause variation, whether it be a product or process. Taguchi methods can be used to address this missing element of Deming's teachings by applying robust parameter design. The optimization of an ideal function maximizes its signal-to-noise ratio, reducing common cause variation of a given system. Furthermore, robust parameter design is consistent with Deming's cycle of continuous improvement.

Both Taguchi and Deming recognize that the largest gains in quality occur upstream in a product's life cycle. During the initial stages of product design and process development, the engineer has the most freedom, as no parameters have been specified. However, as a product or process is developed, the number of undetermined design parameters declines while noise factors begin to emerge. This concept was elegantly conveyed by Deming in *Out of the Crisis:* "A theme that appears over and over in this book is that quality must be built in at the design stage. It may be too late once plans on their way. There must be continual improvement in test methods, and ever better understanding of the customer's needs and of the way he uses and misuses the product" [3, p. 49].

By applying Taguchi methods as early as possible in the life cycle of a product or process, the effects of noise on a system can be curtailed. The result is a system with minimized common cause variation at launch.

References

1. R. Aguayo, 1990. *Dr. Deming: The American Who Taught the Japanese about Quality.* New York: Simon & Schuster.

2. W. E. Deming, 1994. *The New Economics.* Cambridge, MA: MIT Press.

3. W. E. Deming, 1982. *Out of the Crisis.* Cambridge, MA: MIT Press.

This chapter is contributed by Ed Vinarcik.

41 Enhancing Robust Design with the Aid of TRIZ and Axiomatic Design

41.1. Introduction   1449
41.2. Review of the Taguchi Method, TRIZ, and Axiomatic Design   1451
Taguchi Method   1451
TRIZ   1452
Axiomatic Design   1454
Comparison of Disciplines   1456
41.3. Design Response Structural Analysis: Identification of System Output Response   1456
41.4. Examples   1462
41.5. Limitations of the Proposed Approach   1465
41.6. Further Research   1466
41.7. Conclusions   1468
References   1468

41.1. Introduction

The importance and benefits of performing robust parameter design advocated by G. Taguchi are well known [1,2]. Many people are familiar with Taguchi's robust parameter design with such terminologies as orthogonal array, signal-to-noise ratio, and control and noise factors. However, a generally ignored but most important task for a successful robust parameter design project is to select an appropriate system output characteristic.

The identification of a proper output characteristic is a key step to having a higher success rate for robust design projects. To identify a proper output characteristic, Taguchi suggests the following guidelines [2,3]:

❏ Identify the ideal function or ideal input/output relationship for the product or process. The quality characteristic should be related directly to the energy transfer associated with the basic mechanism of the product or process.

❏ Select quality characteristics that are continuous variables, as far as possible.

❏ Select additive quality characteristics.

❑ Quality characteristics should be complete. They should cover all dimensions of the ideal function or input/output relationship.

❑ Quality characteristics should be easy to measure.

According to Taguchi, it is important to avoid using quality symptoms such as reliability data, warranty information, scrap, and percent defective in the late product development cycle and manufacturing environment as the output characteristic. But improving a symptom may not be helpful in improving the robustness of a system's ability to deliver its functions, which is really the key objective of a robust design project. Understanding a system function, especially the basic function, is the key for robust technology development [1]. Defining the ideal state of the basic function, called the *ideal function*, is the centerpiece for identifying output characteristics.

The reason for using an energy-related system output response, according to the discussion of Pahl and Beitz [8] and Hubka and Eder [9], is due to the fact that an engineering system is always designed for delivering its basic function. To deliver its basic function, at least one of the three types of transformation must be used (Figure 41.1).

1. *Energy:* mechanical, thermal, electrical, chemical; also force, current, heat, and so on

2. *Material:* liquid, gas; also raw material, end product, component

3. *Signal:* information, data, display, magnitude

For example, in a machining process as an engineering system, the ideal relationship between output and input should be that the output dimensions are exactly the same as the dimension intended. This type of transformation system is the material transformation, since energy transformation is a very important type of transformation and there are many similarities in using these three types of transformation to identify the appropriate output characteristic. Without loss of generality, energy transformations are used as examples throughout this chapter.

Some of the published literature and articles point out that an energy-related characteristic is very helpful to identify proper quality characteristic and should be considered. Nair and Vigayan [4] cite Phadke's discussion, finding system output response that meets all of these guidelines is sometimes difficult or simply not possible with the technical know-how of the engineers involved. In general, it will be quite challenging to identify system output responses that will meet all of these criteria. Taguchi acknowledges this fact and states that the use of Taguchi methods will be inefficient to the certain extent if these guidelines are not satisfied. Revelle et al. [5] cite Shin Taguchi, Verdun, and Wu's work and points out that the selection of system output response that properly reflects the engineering function of a product or process is the most important and perhaps the most difficult task of the quality engineer. The choice of an energy-related system output response is

Figure 41.1
Technical system: conversion of energy, material, and signals

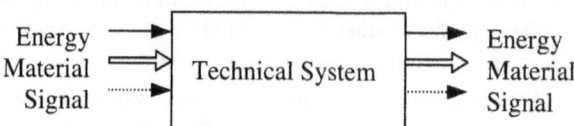

vital to ensure that the system output response is monotonic. According to Box and Draper [6], the monotonicity property requires that effects of control factor be both linear and additive. Based on Box and Draper's study, Wasserman [7] concludes that from a response surface methodology perspective, the requirement of the monotonicity property is equivalent to an assumption that the true functional response is purely additive and linear. The reconciliation of Taguchi's viewpoint is possible based on the assumption that energy-related characteristics are used to ensure that interactions are minimal.

Therefore, identification of the key transformation process is very important to understanding and identifying the ideal functions of the engineering system. By the choice of a good functional output, there is a good chance of avoiding interactions [2,3]. Without interactions, there is additivity or consistency or reproducibility. Laboratory experiments will be reproduced and research efficiency improved.

However, the foregoing guidelines for selecting an appropriate output characteristic are still very conceptual, and their implementation is highly dependent on the project leader's personal experience. There is very little literature that shows how a system output response can be designed and selected in a systematic fashion.

In this chapter we address these shortcomings. With an emphasis on robustness at the early stages of the product development, the proposed methodology will integrate the concept of Taguchi methods with the aid of TRIZ and axiomatic design principles. The proposed methodology has the following three mechanisms: (1) definition and identification of different system architectures, inputs/outputs, and the ideal function for each of the system/subsystem elements; (2) systematic attempts to facilitate a design that is insensitive to various variations caused by inherent functional interactions or user conditions; and (3) bridging the gap between robust conceptual design and robust parameter design through proper identification and selection of a system/subsystem output response.

41.2. Review of the Taguchi Method, TRIZ, and Axiomatic Design

Taguchi Method

Robust design using the Taguchi method is an efficient and systematic methodology that applies statistical experimental design for improving product and manufacturing process design. Genichi Taguchi's development of robust design is a great engineering achievement [10]. By 1990, concurrent engineering was becoming widespread in U.S. industry. It brought great improvements. However, pioneers such as Ford and Xerox realized that more was needed. Robust design, especially, needed to be practiced widely throughout the development of new products and processes.

Taguchi essentially uses the conventional statistical tools, but he simplifies them by identifying a set of stringent guidelines with an energy transformation model–focused engineering system for experimental layout and analysis of results. Taguchi used and promoted statistical techniques for quality from an engineer's perspective rather than that of a statistician.

As Taguchi's ideas have become more widespread, more and more design engineers use Taguchi's methodology in their everyday lives. Due to the growing popularity of robust design methods, more and more quality and engineering professionals have shifted their quality paradigm from defect inspecting and problem solving to designing quality and reliability into products or processes.

Taguchi's approach to design emphasizes continuous improvement and encompasses different aspects of the design process grouped into three main stages:

1. *System design.* This corresponds broadly to conceptual design in the generalized model of the design process. System design is the conceptual design stage in which scientific and engineering expertise is applied to develop new and original technologies. Robust design using the Taguchi method does not focus on the system design stage.

2. *Parameter design.* Parameter design is the stage at which a selected concept is optimized. Many variables can affect a system's function. The variables need to be characterized from an engineering viewpoint. The goals of parameter design are to (a) find that combination of control factors settings that allow the system to achieve its ideal function, and (b) remain insensitive to those variables that cannot be controlled. Parameter design provides opportunities to reduce the product and manufacturing costs.

3. *Tolerance design.* Although generally considered to be part of the detailed design stage, Taguchi views this as a distinct stage to be used when sufficiently small variability cannot be achieved within a parameter design. Initially, tolerances are usually taken to be fairly wide because tight tolerances often incur high supplier or manufacturing costs. Tolerance design can be used to identify those tolerances that, when tightened, produce the most substantial improvement in performance.

Taguchi offers more than techniques of experimental design and analysis. He has a complete and integrated system to develop specifications, engineer the design to specifications, and manufacture the product to specifications. The essence of Taguchi's approach to quality by design is this simple principle: Instead of trying to eliminate or reduce the causes for product performance variability, adjust the design of the product so that it can be insensitive to the effects of uncontrolled (noise) variation. The losses incurred to society by the poor product design are quantified using what Taguchi calls a *loss function,* which is assumed to be quadratic in nature. The five principles of Taguchi's methods are:

1. Select the proper system output response.
2. Measure the function using the SN ratio.
3. Take advantage of interactions between control and noise factors.
4. Use orthogonal arrays.
5. Apply two-step optimization.

TRIZ   TRIZ is a Russian acronym that stands for the *theory of inventive problem solving,* originated by Genrikn Altshuller [18]. How can the time required to invent be reduced? How can a process be structured to enhance breakthrough thinking? It was Altshuller's quest to facilitate the resolution of difficult inventive problems and pass the process for this facilitation on to other people. In trying to answer these questions, Altshuller realized how difficult it is for scientists to think outside their fields of reference, because that involves thinking with a different technology or "language." In the course of the study of some 400,000 inventions as depicted in patent descriptions, Altshuller noticed a consistent approach used by the best inventors to solve problems. At the heart of the best solutions, as described by the patents, existed an engineering conflict or contradiction. The best inventions consistently solved conflicts without compromise. Upon closer examination and classification of innovative solutions, natural patterns of solutions started to emerge. Altshuller discovered that when an engineering system was reduced to reveal the essential system contradictions, inventive solutions eliminated the contradictions completely. Furthermore, Altshuller noticed that the same inventive solutions appeared repeatedly at different points in time and in different places.

SUBSTANCE-FIELD ANALYSIS

Substance-field (S-F) analysis is a TRIZ analytical tool for modeling problems related to existing technological system. *Substance field* is a model of a minimal, functioning and controllable technical system [11]. Every system is created to perform some functions. The desired function is the output from an object or substance (S_1), caused by another object (S_2) with the help of some means (types of energy, F). The general term *substance* has been used in the classical TRIZ literature to refer to some object. Substances are objects of any level of complexity. They can be single items or complex systems. The action or means of accomplishing the action is called a *field*. Within the database of patents, there are 76 standard substance-field solutions that permit the quick modeling of simple structures for analysis. If there is a problem with an existing system and any of the three elements is missing, substance-field analysis indicates where the model requires completion and offers directions for innovative thinking. In short, S-F analysis is a technique used to model an engineering problem. S-F analysis looks at the interaction between substances and fields (energy) to describe the situation in a common language. In cases where the engineering system is not performing adequately, the S-F model leads the problem solver to standard solutions to help converge on an improvement. There are four steps to follow in making the substance-field model [12]:

1. Identify the elements.
2. Construct the model.
3. Consider solutions from the 76 standard solutions.
4. Develop a concept to support the solution.

OTHER TRIZ TOOLS, STRATEGIES, AND METHODS

The TRIZ innovative process consists of two parts: the *analytical stage* and the *synthesis stage*. A basic description of some of the instruments/tools is as follows:

1. *Ideality concept.* Every system performs functions that generate useful and harmful effects. *Useful effects* are the desirable functions of the system; *harmful effects* are the undesirable effects of the system. When solving problems, one of the goals is to maximize the useful functions of a system. The ideality concept has two main purposes. First, it is a law that all engineering systems evolve to increasing degrees of ideality. Second, it tries to get the problem solver to conceptualize perfection and helps him or her to break out of psychological inertia or paradigms.

2. *ARIZ.* This algorithm of inventive problem solving is a noncomputational algorithm that helps the problem solver take a situation that does not have

obvious contradictions and answer a series of questions to reveal the contradictions to make it suitable for TRIZ. There are four main steps in ARIZ.

3. *Contradiction table.* This is one of Altshuller's earliest TRIZ tools to aid inventors. It shows how to deal with 1263 common engineering contradictions (e.g., when improving one parameter, another is degraded).

4. *Inventive principles.* These are the principles in the contradiction table. There are 40 main principles and approximately 50 subprinciples. These are proposed solution pathways or methods of dealing with or eliminating engineering contradictions between parameters.

5. *Separation principles.* This technique has been used with great success to deal with physical contradictions. The most common separation principles can take place in space, time, or scale.

6. *Laws of evolution of engineering systems.* Altshuller found through his study of patents that engineering systems evolve according to patterns. When we understand these patterns or laws and compare them to our engineering system, we can predict and accelerate the advancement of our products.

7. *Functional analysis and trimming.* This technique is helpful in defining the problem and improving ideality or value of the system. The functions of a system are identified and analyzed with the intent of increasing the value of the product by eliminating parts while keeping the functions. Functionality is maximized and cost is minimized.

Axiomatic Design   Design is attained by the interactions between the goal of the designer and the method used to achieve the goal. The goal of the design is always proposed in the functional domain, and the method of achieving the goal is proposed in the physical domain. Design process is the mapping or assigning relationship between the domains for all the levels of design.

Axiomatic design is a principle-based design method focused on the concept of domains. The primary goal of axiomatic design is to establish a systematic foundation for design activity by two fundamental axioms and a set of implementation methods [13]. The two axioms are:

❑ *Axiom 1: Independence Axiom.* Maintain the independence of functional requirements.

❑ *Axiom 2: Information Axiom.* Minimize the information content in design.

In the axiomatic approach, design is modeled as a mapping process between a set of functional requirements (FRs) in the functional domain and a set of design parameters (DPs) in the physical domain. This mapping process is represented by the design equation:

$$FR = [A]DP \tag{41.1}$$

where

$$A_{ij} = \frac{\partial FR_i}{\partial DP_j} \tag{41.2}$$

Suh defines an *uncoupled design* as a design whose A matrix can be arranged as a diagonal matrix by an appropriate ordering of the FRs and DPs. He defines a *decoupled design* as a design whose A matrix can be arranged as a triangular matrix by an appropriate ordering of FRs and DPs. He defines a *coupled design* as a design whose A matrix cannot be arranged as a triangular or diagonal matrix by an appropriate ordering of the FRs and DPs. The categories of design based on the structure of the design matrix are shown is Figure 41.2.

The first axiom advocates that for a good design, the DPs should be chosen so that only one DP satisfies each FR. Thus, the number of FRs and DPs is equal. The best design has a strict one-to-one relationship between FRs and DPs. This is uncoupled design. If DP influences the FR, this element is nonzero; otherwise, it is zero. The independence axiom is satisfied for uncoupled design matrix [A] having all nonzero elements on its diagonal, indicating that the FRs are completely independent. However, complete uncoupling may not be easy to accomplish in a complex world, where interactions of factors are common. Designs where FRs are satisfied by more than one DP are acceptable, as long as the design matrix [A] is a triangle; that is, the nonzero elements occur in a triangular pattern either above or below the diagonal. This is decoupled design. A decoupled design also satisfies the independence axiom provided that the DPs are specified in sequence such that each FR is ultimately controlled by a unique DP. Any other formation of the design matrix that cannot be transformed into a triangular formations represents a coupled design, indicating the dependence of the FRs. Therefore, the design is unacceptable, according to axiomatic design.

The information axiom provides a means of evaluating the quality of designs, thus facilitating selection among available design alternatives. This is accomplished by comparing the information content of the several designs in terms of their respective probabilities of satisfying the FRs sucessfully. Information content is defined in terms of *entropy*, which is expressed as the logarithm of the inverse of the probability of success, p:

$$I = \log_2 \frac{1}{p} \tag{41.3}$$

In the simple case of uniform probability distribution, equation (3) can be written as

$$I = \log_2 \frac{\text{system range}}{\text{common range}} \tag{41.4}$$

where *system range* is the capability of the current system, given in terms of tolerances; *common range* refers to the amount of overlap between the design range and the system capability; and *design range* is the acceptable range associated with the DP specified by the designer. If a set of events is statistically independent, the

$$\begin{pmatrix} \times & \circ & \circ \\ \circ & \times & \circ \\ \circ & \circ & \times \end{pmatrix} \qquad \begin{pmatrix} \times & \circ & \circ \\ \times & \times & \circ \\ \times & \times & \times \end{pmatrix} \qquad \begin{pmatrix} \times & \times & \times \\ \times & \times & \times \\ \times & \times & \times \end{pmatrix}$$

Uncoupled **Decoupled** **Coupled**

Figure 41.2
Structure of the design matrix

probability of the union of the events is the product of the probabilities of the individual events.

Comparison of Disciplines

The purpose of such a comparison is to point out the strength and focuses of different contemporary disciplines, such as axiomatic design (Suh), robust design (Taguchi), and TRIZ (Altshuller).

A product can be divided into functionally oriented operating systems. Function is a key word and the basic need for describing our product, behavior. Regardless of what method is to be used to facilitate a design, they all have to start with an understanding of functions. However, what is the definition of function? How is the function defined in these disciplines? Understanding the specific meanings of function (or the definition of function) within each of these disciplines could help us to take advantage of tools to improve design efficiency and effectiveness.

According to *Webster's* dictionary, function has three basic explanations: (1) the acts or operations expected of a person or thing, (2) the action for which a person or thing is specially fitted or used, or (3) to operate in the proper or expected manner. Generally, people would agree that a function describes what a thing does and that it can be expressed as the combination of noun and verb: for example, creating a seal or sending an e-mail.

In axiomatic design, function is defined as desired output that is the same as the original definition. However, the importance of functional requirements is not identified in the axiomatic design framework. There are no guidelines or termination criteria for functional requirement decomposition. Functional requirements are treated as equally important, which is not necessary practical and feasible.

In robust design, the definition of function has the same general meaning but with more meaning in terms of ideal function, which is concerned about what fundamental things a system is supposed to do so that the energy can be transferred smoothly. For example, how can a seal be formed effectively? What is the basic function of an engineered seal system? Therefore, the definition of function in robust design using the Taguchi method may best be defined as energy transformation.

In TRIZ methodology, the definition of function also has the same general meaning, with negative thinking in terms of physical contradictions. Altshuller sought to deliver all system functions simultaneously with maximization of existing resources.

Table 41.1 shows a comparison of axiomatic design, TRIZ, and Robust Design; Table 41.2 shows the comparison using design axioms. Based on the comparisons, we can see that these three disciplines have their own focuses. They complement each other. The strengths and weakness are summarized in Table 41.3.

41.3. Design Response Structural Analysis: Identification of System Output Response

Any system output response is in one of the forms of energy, material, or signal. If the energy-related system output response can help to reduce the interactive effects of design parameter to minimal for the purpose of design optimization, we can better find a way of converting non-energy-related system output response

Table 41.1
Comparison of axiomatic design, TRIZ, and robust design

Approach	Function Focus	Best to Be Applied	Thought Process	Emphasis
Axiomatic design	Desired output	System structure and foundation in conceptual design	Positive thinking; how a design can be created perfectly; how a design is immune	Mapping from functional requirements to design requirement
TRIZ	Basic function	System structure and foundation in conceptual design	Negative thinking; start with conflicts or contradictions; how a contradiction can be resolved	Attacking on contradictions; start with design parameter, then back to functional requirements
Robust design	Energy transformation	Given specific technology optimization or a given structure or concept design optimization	Within a given structure or design, how an engineered system can be optimized to desensitize the side effects of uncontrollable conditions	Effective application of engineering strategies; identify ideal function (ideal relationship); start with a proper system output response, then maximize the system's useful function

Table 41.2
Comparison using design axioms

Approach	Independence Axiom	Information Axiom
Axiomatic design	Maintain the independence of the functional requirements.	Minimize the information.
TRIZ	Eliminate technical or physical contractions (maintaining independence of parameters).	Apply the concept of ideality.
Robust design	Identify ideal function; select proper system output characteristic and control factors to promote the additivity of effects of parameters.	Maximize the signal-to-noise (SN) ratio.

Table 41.3
Summary of strengths and weaknesses of axiomatic design, TRIZ, and robust design

Approach	Strengths	Weaknesses
Axiomatic design	It provides a good structural foundation for system (concept design. Design axioms are a strong referent. Domains are well defined. Quantitative models exist for coupled, uncoupled, and decoupled design.	Customer attributes are vague. "Zigzagging" between domains is lengthy. Information content is difficult to apply.
TRIZ	Conflict domain, physical contradiction, and its elimination target functional requirements and design parameters more precisely.	It is difficult to work on large, complicated systems. There is no customer attributes process.
Robust design using Taguchi methods	It improves the robustness of basic technology. It provides more depth of understanding of a given technology or a system's functional behavior. Within the domain of given design parameters, the side effects of uncontrollable (noise) factors can be desensitized through the optimization of levels of control factors.	There is no process for system (concept) design. It is limited to a given concept design. It is a black-box approach.

to an energy-related system output response. Instead of searching blindly for an energy-related system output response based on an empirical approach or experience, it is necessary to develop an energy-related system output response. With respect to a technical system, any technical system consists of three minimal numbers of elements: two substances (objects) and a field (energy) [11]. A substance can be modified as a result of direct action performed by another substance. Having the same thought process, a system output response can also be modified as a result of direct action performed by another substance, which can be used as an input signal from the perspective of Taguchi methods. The substance field analysis concept furnishes a clue to the direction of developing a system output response.

Consider as an example a product improvement task in which the plastic molding strength has to be improved to withstand a certain force. The objective function in this case is to improve the strength. What is the output response in this case? Many people would agree that the characteristic (output response) of push force (force to break the molding) could be the one (Figure 41.3). The concern about using push force as the output characteristic may be summarized as follows:

❑ It is difficult to understand the structure of material such as bubbles or voids.

❑ It is a destructive test.

❑ It is hard to take the advantage of a signal factor in a robust design experiment. In other words, it is difficult to understand the input and output relationship in this engineered system.

In an evaluation of the functional behavior of a system, failure modes are only symptoms. The evaluation of that will not provide insight on how to improve the system. Therefore, the push force characteristic is not a good system output response in this case. How can we have a proper characteristic instead of using push force to evaluate the strength?

Let us analyze the problem and its solution in detail. First, as the conditions of the problem suggest, nothing else can be selected to evaluate the strength: the direct response of the engineered system is out of consideration. Therefore, a new system output response should be created.

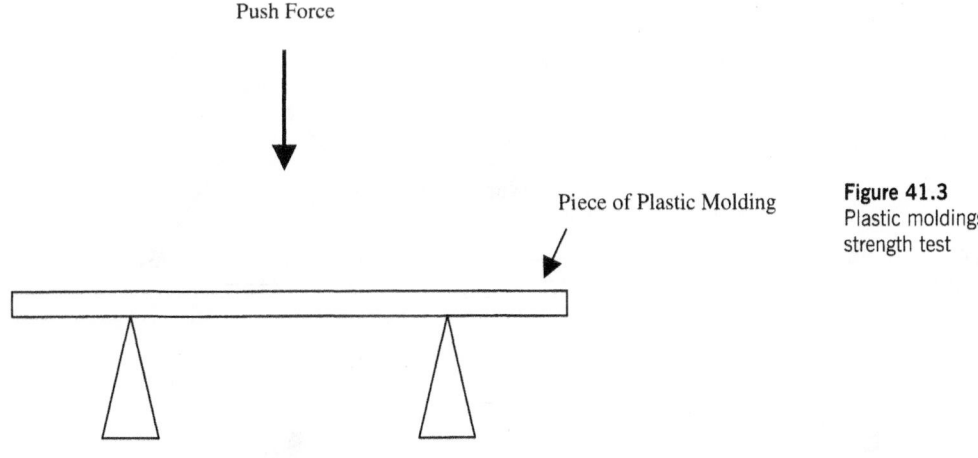

Push Force

Piece of Plastic Molding

Figure 41.3
Plastic moldings
strength test

In Figure 41.4, there is one substance (a piece of plastic molding) at the beginning; in the end there are two substances (a piece of plastic molding and a push bar) and a force field and the bent (not broken) piece plastic molding. We use the following symbols to represent the initial situation:

S_1:   straight piece of plastic molding

S_2:   bent piece of plastic molding

S_3:   push bar

The result is represented by

F: push force

Let us now look at how the system works. Mechanical field ($F_{\text{push force}}$) acts on push bar S_2, which, in turn, acts on the piece of plastic molding (S_1). As a result, S_1 is deformed (bent) to S_1'. Graphically, the operation can be represented as in Figure 41.4.

Until now, can we see the alternative system output characteristic? Can the S_1 be used to evaluate system behavior instead of push force? Let's validate this idea: Can the evaluation system work if we take off any of the substance? No, the system will fall apart and cease to apply the force to the piece of plastic molding. Does this mean that the evaluation system's operation is secured by the presence of all of its three elements? Yes. This follows from the main principle of materialism: Substance can be modified only by material factors [i.e., by matter or energy (field)]. With respect to technical systems, substance can be modified only as a result of direct action performed by another substance (e.g., impact-mechanical field) or by another substance. S_1' is modified from S_1 and is the output due to system input force of $F_{\text{push force}}$. The characteristic S_1' is closer to the structure of plastic molding than the push force.

According to Hubka and Eder [9], to obtain a certain result (i.e., an intended function); various phenomena are linked together into an action chain in such a way that an input quantity is converted into an output quantity. This chain describes the mode of action of a technical system. The mode of action describes

Figure 41.4
System diagram

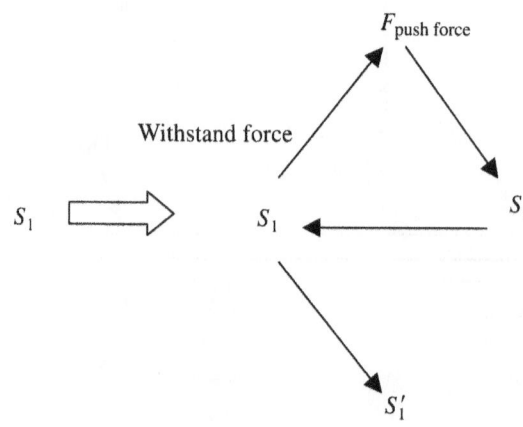

the way in which the inputs to a technical system are converted into its output effects. The mode of action is characterized by the technical system internal conversion of inputs of material, energy, and information. The output effect is achieved as the output of an action process (through an action chain) internal to the technical system in which the input measures are converted into the effects (output measures) of that technical system. The action process is a direct consequence of the structure of the technical system. Every technical system has a purpose, which is intended to exert some desired effects on the objects being changed in a transformation process. The behavior of any technical system is closely related to its structure.

As a consequence, the S_1' (the bent S_1) in terms of displacement (bent distance) is a better system output response (Figure 41.5). As a matter of fact, the displacement of S_1' is proportional to the push force, which enhances effectiveness of the efforts of robust parameter design. A robust parameter design case study has been developed successfully using the output characteristic of displacement in an automotive company [14,15].

In a robust design approach using the Taguchi method, the displacement, M, can also be used as an input signal. The spring force, Y, within the elastic limit, can be used as system output response. The displacement is an input signal, M. The ideal function will be given by

$$Y = \beta M \tag{41.5}$$

Y will be measured over the range of displacement (Figure 41.6). The signal-to-noise (SN) ratio will be optimized in the space of noise factors such as environment and aging.

Identification of system output response using S-field models sheds light on the essence of transformation of engineered systems and allows one to use universal technical or engineering terminology rather than customer perceptions, such as percent of failures, good or bad, to evaluate the system's behavior. The key idea is how the material, information, and energy are formed or transferred.

Searching for system output response based on S-field model analysis presents a general formula that shows the direction of identifying the possible system output characteristic. This direction depends heavily on the design intent of the system.

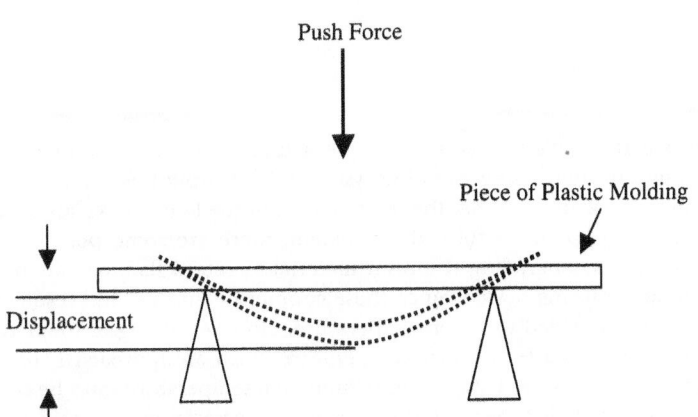

Figure 41.5
Better system output response

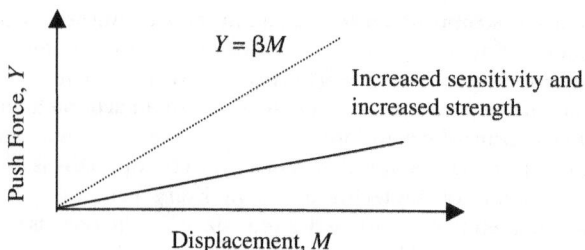

Figure 41.6
Range of displacement

Consider the example above: Introducing a substance or a field will profoundly change the process of identifying the system output response.

Gathering expert knowledge about the engineered system and various components in the product and how they affect one another is of the most importance if the identification of system output response is going to be more effective.

There are several rules for identifying system output response using S-field synthesis. Since we are interested in identifying proper system output characteristics in this chapter, our goal is to develop some principles for identification of system output response using S-field analysis.

- ❏ *Rule 1: Substance-Field Model Development for System Output Response.* If there is an output characteristic that is not easy to measure or not proper to reflect the system design intent, and the conditions do not contain any limitations on the introduction of substances and fields, the output characteristic can be identified through synthesizing a system output response–based S-field: The output characteristic is subjected to the action of a physical field that produces the necessary corresponding physical effects in the engineered system.

- ❏ *Rule 2: Change in Scope or Boundary of a Technical System.* If the conditions contain limitations on the existing system output response, the alternative output response has to be identified by synthesizing an S-field using external environment as the system output response. Changing the scope or the boundary of the technical system can help to identify a proper system response.

41.4. Examples

To illustrate this, let's use as an example a case study on a temperature-rising problem in a printer light-generating system [16]. During the development stage of a printer, it was noticed that the temperature in the light source area was much higher than expected. To solve this problem, there are some possible countermeasures, such as upgrading the resin material to retard flammability or adding a certain heat-resisting device. Since these countermeasures would result in a cost increase, it was decided to lower temperature. However, trying to lower temperature creates the need to measure temperature. Such an approach is not recommended, for two reasons. First, the environmental temperature must be controlled during experimentation. Second, the selection of material must consider all three aspects of heat transfer: conduction, radiation, and convection. It would take a long time to do.

In the system in this example, there are two subsystems: S_1 lamp (light-generating system) and S_2, fan (cooling system). The heat (field) in this system must be reduced. Since the heat energy is created by S_1 (lamp) and the cooling energy by S_2 (fan), the S-field system diagram may be drawn as in Figure 41.7.

The constraints for problem solving in this example are that (1) S_1 cannot be changed, (2) temperature is not preferred to measure the heat accumulated around the system, and (3) an rpm meter gauge is not available. What else can be measured to evaluate the status of temperature? Obviously, the rotation of the fan to remove the air surrounding the heat source could be another way of improving temperature condition. To improve the rotation of the fan, the rpm value has to be measured. The ideal situation is: "The air speed surrounding the heat source changes proportionally to the fan rotation. The sensitivity must also high." The modified S-field is shown is Figure 41.8.

However, as stated in the constraints, measuring rpm is not possible at that time, unfortunately. What can we do now? According to rule 2, we may have to change the scope or the boundary of the technical system. Can we find something that is not related to temperature directly? Of course, our goal is still to find a way of measuring heat for the purpose of achieving lower temperature if possible. Can we use motor voltage to measure the temperature indirectly? Let's validate this idea. Voltage is the input energy to drive a motor. The rpm of a fan, as the result of motor rotation, is probably proportional to motor voltage. Therefore, the ideal situation can be redefined as "the air speed surrounding the heat source is proportional to motor voltage with high sensitivity." The further-modified S-field is shown in Figure 41.9. Based on robust design, the ideal relationship between motor voltage and air speed may be shown in Figure 41.10.

Technical systems display numerous internal and external connections, both with subsystems (components of each technical system), systems of a higher rank, and the environment. Each technical system can be represented as a sum of S-field. The tendency is to increase the number of S-fields in a technical system with consideration of the chain of action mode as necessary.

❑ *Rule 3*. Efficiency of system output response–based S-field analysis can be improved by transforming one of the parts of the system output response–based S-field into an independently controllable system output response–based S-field, thus forming a chain of system output–based S-field analysis.

The graphical view of the chain of the system output response–based S-field is shown in Figure 41.11.

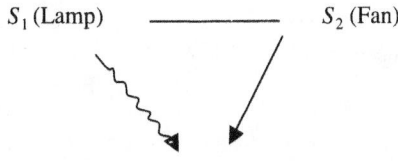

S_1 (Lamp) ——————— S_2 (Fan)

F_1 Temperature (Heat)

Figure 41.7
Harmful side effect

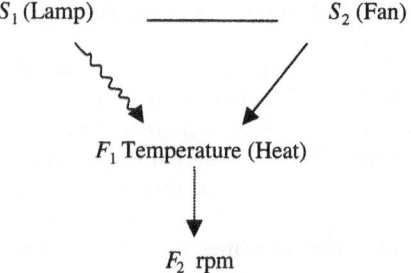

Figure 41.8
Rotated S-field

❏ *Rule 4: Chain of Action and Effect for System Output Response.* If an output characteristic is conflicting with another output characteristic in terms of the same design parameters, it is necessary to improve the efficiency by introducing a substance or a sub S-field and consider the chain of action in a technical system.

Rules 3 and 4 are often used together to identify a proper system output characteristic. For example, in a mechanical crimped product case study [17], pull strength force and voltage drop have to be optimized simultaneously (Figure 41.12). But the optimized design parameters are not the same with respect to the two different system output responses. Obviously, something may have to be compromised, unfortunately.

The reliability of complex electrical distribution systems can be dramatically affected through problems in the connecting elements of wires to the terminal in this case study. Minimum voltage drop is the design intent, and maximum pull strength is required for the long-term reliability concerns.

In this example, the pull strength is created by crimping force (F_1) acting on wire (S_1) and terminal (S_2). The S-field system diagram may be expressed as shown in Figure 41.13.

The pull strength, F_2, is not a good system output response for two reasons: first, pull strength has to be compromised by voltage drop. Second, the pull strength does not take the long-term reliability into consideration in terms of gas holes, void, and so on. According to rule 4, we could introduce an output response and consider the chain of action modes and the chain of effects. What effect can

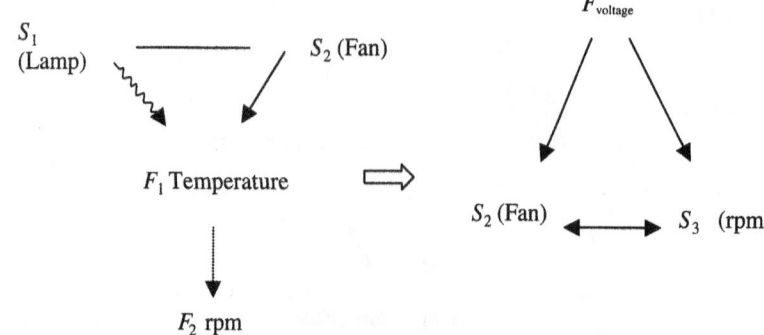

Figure 41.9
Changed boundary of
the technical system

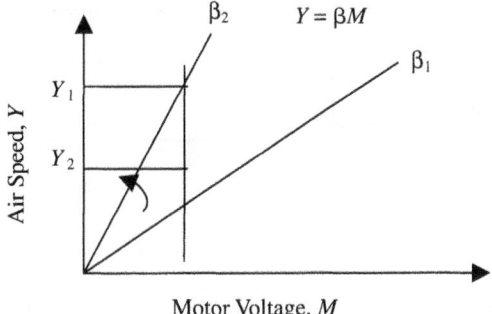

Figure 41.10
Relationship between voltage and speed

we find before the effect of pull strength is formed? When we crimp the wires and terminal, the wires and terminal are compressed into a certain form. Such a form can be measured by compactness. Can the compactness be used as a system output response? Let's validate this idea. The compactness is formed before the pull strength, and the compactness takes the gas holes and voids into consideration. What is the relationship between the compactness and the pull strength? The data show that the compactness is strongly related to the pull strength and the voltage drop. Therefore, the compactness could be used as a system output characteristic. The S-field diagram can be modified as shown in Figure 41.14.

The identification system output response using substance-field analysis is based on the law of energy transformation and the law of energy conductivity. Selecting a proper system output response using S-field analysis is one of the approaches based on the energy transformation thought process. Any technical system consists of three elements: two substances and a field. The identification system output response using substance-field analysis furnishes a clue to the direction of identifying a system output response for the purpose of conducting robust parameter design through a dynamic approach. This approach is very helpful when it is not clear how an object or a system, especially in the process of identifying a system output response, is related to the energy transformation for the purpose of design optimization.

41.5 Limitations of the Proposed Approach

Searching for a proper system output characteristic through the system output response based–S-field model, we often look at the technical system at only one

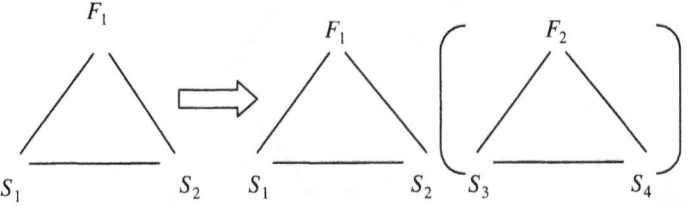

Figure 41.11
System output response

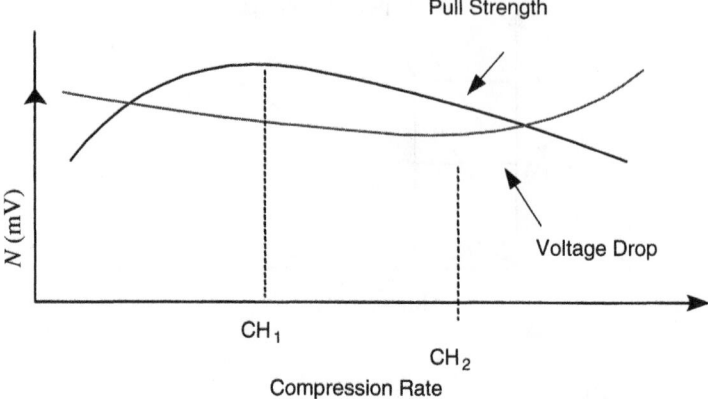

Figure 41.12
Pull strength and
voltage drop

level. In a more complex system, it is difficult to identify a proper system output response without looking into the structure of the system design. A thorough understanding of the design intent is essential for finding a way to identify a truly engineering-related output response.

41.6. Further Research

One interesting topic might be to investigate how the framework of axiomatic design could be used to improve the limitations of identifying system output response using substance-field analysis. Of course, we would like to investigate a way of bridging the gap between the conceptual design and parameter design so that the up-front robustness thinking and testability can be emphasized. Design through an axiomatic approach is attained by interactions between the goal of the designer and the method used to achieve the goal. The goal of the design is always proposed in the functional domain, and the method of achieving the goal is proposed in the physical domain. The design process is the mapping or assigning relationship between the domains for all levels of design.

As the functional requirements become diverse, satisfying the requirements becomes more difficult. Therefore, concentrating on the functional requirements for the given stage or level of the design process is necessary. A design or a problem

Figure 41.13
S-field diagram

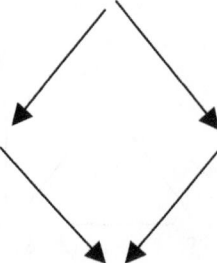

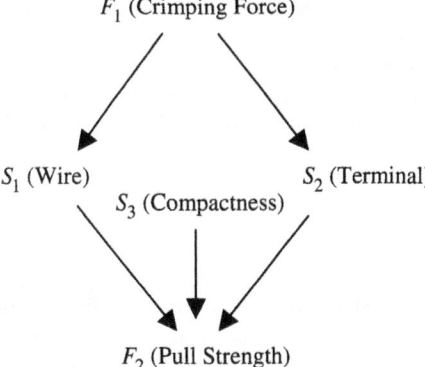

Figure 41.14
Modified S-field diagram

with many variables is very complicated. To prioritize the tasks and the proper focus, it is necessary to sort the primary and secondary functional requirements and handle each functional requirement according to its importance. For the purpose of design evaluation and optimization, it is essential to select a proper system output response to evaluate and understand an engineered system or a product's functional behavior. Such system output characteristics (responses) should be related to basic functions. A *basic function* is a function to transfer material, energy, and information from the input of the system to the output of the system. Obviously, the basic function of a product or process technology is related to its capability (highest probability) to transform input to output in terms of material, information, and energy.

Functional requirements are included in the functional domain. The designer should thoroughly understand problems in the functional domain and should not limit possible selections without a special reason. Clearly defining the problem is closely related to defining the functional requirements. On the other hand, the designer should select the design elements in the physical domain by specifying the functional requirements. Selecting a system output response characteristic is closely related to the physical domain to reflect how material, information, and energy are transferred smoothly from input to output in the technical system.

According axiomatic design principles, the essence of the design process lies in the hierarchies. The designer begins with functional requirements (top-down approach); and because of the different priorities of all the functional requirements, the designer can categorize all the functional requirements into different hierarchies. The important point in this process is that the functional requirements must be satisfied with specific design parameters. As it goes to the lower level, more details should be considered. This can be a very effective way of considering all details of the design. The functional requirements of the higher level must be satisfied through the appropriate design parameters in order for the lower-level functional requirements to be satisfied.

By using an axiomatic approach, the ideas in the initial stages of the design can be brought to bear in a scientific way once the design zigzagging mappings have been completed according to the design axiom. To evaluate the system's functional behavior, of course, a key system output response has to be identified. The lower level of functional requirement in the axiomatic design framework is not necessarily the best system output response for the purpose of system evaluation. But the lower level of functional requirement is certainly the proper starting point to identify or develop a proper system output characteristic. Additional creativity in the design can be induced when going through this task. The bottom-up approach is necessary to identify a system output response based on a result of zigzagging mapping.

41.7. Conclusions

In this chapter we suggest an approach for identifying a proper system output response using substance-field analysis along with analysis of the chain of action mode. The approach presented consists of four elements: (1) system output response–focused substance-field model development, (2) change in the scope or boundary of a technical system, (3) efficiency of system output response–focused substance-field model, and, (4) chain of action and effect for system output response.

The law of energy transformation and the law of energy conductivity guide the identification of system output response using substance-field analysis. One of the biggest advantages of using this approach is that the signal factor will come with the system output response identified. With the proper identification of signal factor and system output response, the chance of using dynamic robust design will be increased. Of course, the effectiveness of the robust parameter design will be improved.

Compared with other approaches to the identification of system output response, the approach presented in this chapter provides specific and detailed directions not only to search for but also to create an energy-related system output response. The approach has been applied successfully to several challenging case studies at some automotive companies. The findings from the case studies motivated the researchers to bridge the gap between the robust conceptual design and the robust parameter design.

References

1. G. Taguchi, 1993. *Taguchi on Robust Technology Development: Bring Quality Engineering Upstream.* New York: ASME Press.

2. G. Taguchi, 1987. *System of Experimental Design.* White Plains, NY: Unipub/Kraus International Publications.

3. M. S. Phadke, 1989. *Quality Engineering Using Robust Design.* Englewood Cliffs, NJ: Prentice Hall.

4. Vijayan Nair, 1992. Taguchi's parameter design: a panel discussion. *Technometrics,* Vol. 32, No. 2.

5. J. B. Revelle, J. W. Moran, and C. A. Cox, 1998. *The QFD Handbook.* New York: Wiley.

6. G. E. Box and N. R. Draper, 1988. *Empirical Model-Building and Response Surface Methodology.* New York: Wiley.

7. G. S. Wasserman, 1997–1998. The use of energy-related characteristics in robust product design. *Quality Engineering,* Vol. 10, No. 2, pp. 213–222.

8. G. Pahl and W. Beitz, 1988. *Engineering Design: A Systematic Approach.* New York: Springer-Verlag.

9. V. Hubka and W. E. Eder, 1984. *Theory of Technical System: A Total Concept Theory for Engineering Design.* New York: Springer-Verlag.

10. D. Clausing. 1998. Product development, robust design, and education. Presented at the 4th Annual Total Product Development Symposium, American Supplier Institute.

11. Y. Salamatov, 1999. *TRIZ: The Right Solution at the Right Time.* The Netherlands: Insytec.

12. J. Terninko, A. Zusman, and B. Zlotin, 1996. *Step-by-Step TRIZ: Creating Innovative Solution Concepts.* Nottingham, NH: Responsible Management.

13. N. P. Suh, 1990. *The Principles of Design.* New York: Oxford University Press.

14. Improvement of an aero-craft material casting process. Presented at the 1995 Quality Engineering Forum.

15. M. Hu, 1997. Reduction of product development cycle time. Presented at the 3rd Annual International Total Product Development Symposium, American Supplier Institute.

16. A research on the temperature rising problem for a printer light generating system. Presented at the 1996 Quality Engineering Forum.

17. M. Hu, A. Meder, and M. Boston, 1999. Mechanical crimping process improvement using robust design techniques. *Robust Engineering: Proceedings of the 17th Annual Taguchi Methods Symposium,* Cambridge, MA.

18. G. Altshuller, translated by L. Shulyak, 1996. *And Suddenly the Inventor Appeared—TRIZ: The Theory of Inventive Problem Solving.* Worcester, MA: Technical Innovation Center.

This chapter is contributed by Matthew Hu, Kai Yang, and Shin Taguchi.

42 Testing and Quality Engineering

42.1. Introduction   1470
42.2. Personnel   1470
42.3. Preparation   1472
42.4. Procedures   1473
42.5. Product   1474
42.6. Analysis   1475
42.7. Simulation   1475
42.8. Other Considerations   1476
External Test Facilities   1476
Documentation   1477
42.9. Conclusions   1477
Reference   1477

42.1. Introduction

Successful testing is a crucial component of any robust design project. Even if all other related activities are executed flawlessly, poorly conducted testing and evaluation can result in suboptimal results or failure of the project. Despite its critical nature, execution of the experimental design is sometimes considered a "black box" process. Successful execution of a robust design experiment depends on attention to the four "P's" of testing: personnel, preparation, procedure, and product.

42.2. Personnel

Numerous quality methodologies have received considerable attention in the past decade. Robust design, total quality management, six sigma . . . to many engineers these are perceived as nothing more than the latest in a never-ending stream of quick-fix fads adopted by management. Each technique has its proponents, appropriate applications, and case history; each also has its skeptics, misapplications, and anecdotal failures. The interaction between these philosophies and the existing corporate culture of the adopting organization can either lead to empowering synergies or nigh-impenetrable barriers to success. Because of the team-oriented interdisciplinary nature of modern engineering, successful testing in a robust design context cannot ignore the psychology of the people involved in the process.

This issue is of particular importance in organizations that give development responsibility to one functional area and testing and/or validation duties to another. It must also be addressed if the predominant engineering mindset is focused on design-test-fix and testing to targets (or "bogeys"). To obtain full participation and commitment from the team, one must eliminate misconceptions and ensure that each team member has an appropriate level of understanding of robust engineering techniques. It is not necessary for each team member (whether from design, quality, or testing activities) to be an expert in robust design principles; however, each should be familiar enough with the methodology to be able to contribute and apply his or her knowledge and experience to the project.

One positive outcome of the recent attention that quality management has received is that most people in engineering and testing activities are more attentive to quality issues and are somewhat familiar with the associated terms, principles, and methodologies. However, some of this knowledge may be secondhand and not entirely accurate. Although robust design and engineering techniques (Taguchi methods) have received considerable attention in the United States, particularly in the past decade, many engineers and technicians do not have a full understanding of the underlying principles.

One common misconception is that quality is "tested in." This mindset is often found within organizations that separate design and testing activities and use the inefficient design–test–fix cycle to develop products. It is essential to convince engineers in this type of environment that quality is primarily a function of their designs and not something added as an afterthought by testing. Once this hurdle is overcome, one can demonstrate how robust engineering practices can assist them in their design activities.

Another frequent belief is that robust engineering is "just about design of experiments." Although classical and Taguchi experimental design techniques have some similarities, the philosophical differences are considerable. It is important to educate people clinging to this point of view about the underlying conceptual differences between the two methodologies.

Some people may be skeptical of any test plan other than one-factor-at-a-time evaluation. The best way to illustrate the power of robust engineering principles to adherents of this outlook is to share appropriate case studies with them. The ever-growing published body of work available about robust engineering applications should yield a number of suitable illustrations that can be used to educate the team.

Finally, "quality is conformance to specifications" is a mantra that will survive well into the twenty-first century, despite its inherent flaws. People with this type of "goalpost" mentality believe that all product within specification limits is of equal quality; an explanation of the loss function and appropriate case study examples should eliminate this misconception.

Some of the most critical members of the team are those who will be performing the tests. The engineers and/or technicians responsible for conducting the experiment and collecting data must be included in the planning process. Many technicians, particularly within "chimney" organizations that segregate testing and design activities, have a wealth of knowledge about test procedures, instrument capabilities, past design failures, and other institutional knowledge that may not be readily available to the team from other sources. Fostering an inclusive relationship with these people can yield enormous benefits and improve the quality of the test results.

42.3. Preparation

Few oversights will limit the success of a quality engineering project more than that of improper scoping. The focus of the project, whether it is on a component, subsystem, system, or entire product, must be defined clearly at the onset. The development of P-diagrams, ideal functions, and experimental designs is dependent on the boundaries specified by the team. The robust engineering process is also heavily influenced by the nature of the issue: research and development activities validating new technologies will have different needs and requirements than will design engineers creating a product for an end user.

Once the region of interest within the design space has been identified, the team must define the ideal function to be studied and construct an appropriate P-diagram. Existing documents can provide a wealth of information of use in this development: warranty information, failure mode effects analyses (FMEAs), design guides, specifications, and test results from similar products can contribute to the development of a comprehensive P-diagram. It is difficult to develop an exhaustive list of error states, noise factors, and control factors without examining this type of supporting information. Published case studies and technical papers may also yield insights into similar products or technologies.

Although the primary focus during this exercise must be the operating environment and the customer's use of the product, some attention should be paid to the capabilities of the testing infrastructure to be used. Although it is important to specify an ideal function that is energy-based (to minimize the possibility of interactions between control factors), it does little good to select one that cannot be measured readily. Input from the engineer or technician who will perform the trials can assist the team in developing a feasible concept. Many enabling test technologies have been developed (such as miniature sensors, noncontact measuring techniques, and digital data collection and analysis), and the people who use them are best able to accurately communicate their capabilities and availability. If the input signal or output response cannot be measured easily, appropriate surrogates must be selected. These alternative phenomena should be as closely coupled to the property of interest as possible; second- and third-order effects should be avoided whenever possible.

Once the ideal function has been specified and the P-diagram has been constructed, the team can begin development of the experimental design. The control factors to be studied should be selected based on the team's engineering knowledge and data from the sources listed above. Techniques such as Altshuller's theory of inventive problem solving (TRIZ/TIPS), structured inventive thinking (SIT), Pugh's total design, or Suh's axiomatic design can be employed in the development of design alternatives and the specification of control factor levels.

Some consideration should be given to resource constraints at this time. Test length, cost, and facility and/or instrument availability may restrict the magnitude of the overall experiment. It is important to consider these feasibility issues without sacrificing parameters that the team feels are critical. If a meaningful test plan cannot be developed within these limits, it may be necessary to request additional resources or to select an alternative test method.

Testing personnel should be involved in the selection of signal and noise factors. Their insights and knowledge of available test capabilities and facilities can facilitate this phase of test plan construction. Analysis of customer usage profiles and warranty data will also help the team compound appropriate noise factors. Once this phase is complete, the experimental design should be complete, with signal and noise factors identified and the inner and outer arrays established.

42.4. Procedures

As the team develops the experimental design, it must match its expectations with the capabilities of the test facility. Some robust design case studies showcase experiments done on a tabletop with virtually no test equipment; others describe studies that required elaborate measurement devices. Sophistication is not a guarantee of success; appropriate instrumentation maximizes the probability of a favorable outcome.

As the team considers the test procedures to be used, its members should discuss the following questions:

❑ What ancillary uses are anticipated for the data?

❑ What measurements will be taken?

❑ What is the desired frequency of measurements?

❑ What measurement system will be used?

❑ What test procedure is appropriate?

When establishing the test plan, it is important to consider what other uses may be appropriate for the data collected. For example, the development of useful analytical models depends on correlation with observations and measurements of actual performance. Data obtained by the team in the course of the robust engineering process may be useful in the development of analytical tools that can speed future design of similar components or systems. Therefore, collection of data not directly related to the control and noise factors being studied may be desirable; if this unrelated information can be recorded without disruption of the main experiment, it may be cost-effective to do so.

Once the superset of data has been defined, the team must determine what measurements will be taken. The test plan should clearly define what data are to be collected; it may be useful to develop a check sheet or data collection form to simplify and standardize the recording of information. Parameters such as technician identity, ambient temperature, relative humidity, and measurement time should be considered for inclusion if they are relevant to the experiment.

Measurement frequency must also be established. If the intent of the study is to examine the functions of components that do not degrade appreciably, simple one-time measurements may be appropriate (e.g., airflow through a variety of duct configurations). However, if the system is expected to degrade over its service life, time and/or cycle expectations should be included in the experimental design as noise factors, and the measurement interval must be specified. Some consideration must also be given to the capabilities of the test facility. Will the test run continuously? When are personnel available to make measurements? What data can be captured automatically? The answers to these questions will affect the test plan.

The levels of precision and accuracy necessary for successful completion of the experiments must be also be established. Simple A-to-B performance comparisons may be more tolerant of measurement error than detailed experiments that will also be used to develop analytical models of system behavior. These requirements will drive selection of the measurement technique. It may be necessary to conduct a gauge repeatability and reproducibility study before using an instrument; at the very least, the team should review its calibration records and verify its correct operation by testing a reference standard. It is critical that the team understand the uncertainty that the measurement system will contribute and assure that it is sufficiently small so as not to influence the analysis.

The final consideration is the selection or development of a test procedure. A large number of test methods have been published by organizations such as the American Society for Testing and Materials (ASTM), the Society of Automotive Engineers (SAE), and the Institute for Electrical and Electronics Engineers (IEEE); procedures released by government agencies and included within military specifications also may be relevant. A careful review of available methods will provide the team with guidelines for selecting or establishing an appropriate test technique.

A detailed written test procedure should be provided to the testing activity; it should be an independent document that summarizes all of the test requirements specified by the team. Not only does a complete well-written test procedure serve to eliminate ambiguities, but it also will become an important part of the project's documentation. It must clearly specify what data are to be collected, their method of collection, the frequency of measurement, and any other special instructions. Appropriate images or schematics should be included to clearly illustrate test fixture setups and relevant features of the component or system. Ideally, any person not intimately familiar with the test plan and development of this experiment should be able to conduct the measurements successfully without any instructions other than the written procedure. It may be worthwhile to have a peer review of the document by a person unfamiliar with the process; this will help to identify any assumptions made by the writer that should be defined clearly in the document.

42.5. Product

Once test planning is complete and the procedures have been established, the team must procure the materials to be tested. A number of questions must be addressed:

❑ Are surrogate parts appropriate?

❑ What adjacent components are required?

❑ Is the testing destructive?

The nature of the study and the noise factors involved will determine whether surrogate parts are suitable or design-intent components must be fabricated. For example, if the study is attempting to measure airflow in a duct, a proxy may be appropriate. An existing duct may be reshaped to reflect the desired configurations specified by the experimental design. However, care must be taken not to introduce unintended noise. In this example, if the reshaping process introduced surface finish changes or burrs that affected the airflow, this testing would not accurately reflect the performance of a normally manufactured component. It may be more appropriate to conduct the tests using samples made by stereolithography or three-dimensional printing processes. This would normalize the noise induced, affecting each sample equally and allow the testing to reflect more accurately the impact of each control factor.

Adjacent components may also be required for the testing. Use of nearby components or systems may simplify fixturing and improve the simulation of the test samples' operating environment. If the surrounding items generate heat or electromagnetic interference, for example, omitting them may cause the test to reflect actual performance of the components in service inaccurately. In the case of software testing, processor loading and memory consumption of other executing programs may influence the results. Although the project scope may be limited to a certain subsystem or component, the team must remain aware of those systems with which it interfaces.

If the testing is destructive, the team must consider how to minimize unwanted variability. Certain types of piece-to-piece variation can be compounded into the noise factors; if this is not practical, every effort must be made to make the test samples as uniform as possible to assure test outcomes are a result of changes to control factors and are not due to component variation.

42.6. Analysis

Once testing is complete, the team must analyze the results. Some postprocessing of test data may be necessary: for example, frequency analysis of recorded sounds. All data should be subjected to identical processing to assure that no spurious effects are introduced by these procedures.

Commercial products are available for the calculation of signal-to-noise ratios, main effects, and analysis of variance (ANOVA), or the team may elect to compute these values themselves using spreadsheets or other computational aids. Once the analysis is complete, the team will conduct confirmation and verification of the optimal configuration.

42.7. Simulation

Although they are not considered to be testing in the traditional sense, analytical models and computer-aided engineering (CAE) have their place in the robust engineering process.

If a simulation or model has been correlated adequately to real-world performance and behavior, it can be used to conduct the tests required during the optimization phase of robust engineering. An obvious advantage of this approach is that the use of computational modeling minimizes the costs associated with fabricating test samples, and typically reduces the time to conduct the tests. The sophistication of commercially available modeling software continues to increase, and the power and accuracy of these methods will continue to improve. However, the use of virtual testing does require the engineering team to exercise increased caution and diligence. In *Engineering and the Mind's Eye* [1], Eugene S. Ferguson wrote: "Engineers need to be continually reminded that nearly all engineering failures result from faulty judgments rather than faulty calculations." This adage is particularly appropriate to engineers attempting to use modeling and simulation.

For example, a robust engineering team may need to evaluate the response of a metal part fabricated by stamping. Their analytical model may yield incorrect results if the sheet metal's nominal thickness and material properties are used. To be more accurate, the simulation should reflect the thinning and potential anisotropy of the material due to the stamping operation. Other important factors that might influence the quality of the results are heat treatment, burrs and surface finish, and other artifacts of the fabrication process. In many ways, testing on actual components is more tolerant of oversights and erroneous assumptions; real specimens require fabrication and have intrinsic properties that can be inspected and scrutinized by the team. Virtual test samples are totally dependent on the quality of the parameters entered into the simulation, and it may be difficult to detect errors. For this reason it is important for the team to be just as rigorous in the development of a virtual test plan; all assumptions and boundary conditions should be documented. If no members of the team have experience with the manufacturing of the components tested, a consultant should be queried to ensure that any relevant manufacturing artifacts are considered.

In addition to monitoring the inputs to the model, the team should have some familiarity with the mechanisms it uses to derive its results. Teams that treat the analytical model as a "black box" do so at their peril. Many parameters within modeling software have default values specified that may or may not be appropriate for the team's simulation. Also, whether the software in use is proprietary or available commercially, the possibility of errors in the code does exist. For these reasons, the team should have an intuitive understanding of the experiment and the expected output of the model; this will help to identify unreasonable results due to simulation or input error.

Once the optimization phase has been completed by CAE methods, the team should proceed with actual testing during the confirmation and verification phases. The use of real components during theses steps in the robust engineering process will assure that any errors or oversights inherent in the virtual testing will be detected.

42.8. Other Considerations

External Test Facilities   If the testing is conducted by an external service provider, a number of issues must be considered that do not apply to internal facilities. First, it may be necessary to obtain a nondisclosure or other privacy agreement from the test agency. Many standard contracts include this requirement, but the team should take appropriate steps to assure that any results of their work are retained as a competitive advantage by their organization and not shared with competitors. Second, resource availability and test timing may be less flexible, due to competing demands from other customers. Since the team's influence over the facility's schedule may be limited, it is important that test specimens and other materials be delivered in advance of all scheduling milestones. Finally, extra care must be taken in the development of the test procedure. Technicians and engineers at general test facilities may not be intimately familiar with jargon or nonstandard terminology used by the team's organization. It is important that the team clearly communicate its requirements and expectations.

The team should maintain adequate documentation of its project. Meeting minutes, P-diagrams, supporting engineering information, test plans, analysis, and final outcomes should be compiled into a master record maintained by the responsible engineer. This information should become part of the organization's institutional knowledge system; it can also serve as a basis for publication (if appropriate). **Documentation**

Because these documents may ultimately be distributed to an audience wider than the team originally intended, some care should be taken in their development. In particular, engineering documents are subpoenable, and therefore terms that may confuse a jury of nonengineers should be avoided. Inflammatory language and jargon should be minimized; the results of the project should be characterized fairly, in plain language whenever possible. This will minimize the possibility of the content being misconstrued in the event of litigation.

42.9. Conclusions

Testing is the heart of the robust engineering process. It is this fulfillment of the experimental design that allows subsequent analysis to determine optimal configurations and confirm and validate those conclusions. If testing resources are appropriately managed and utilized, a robust engineering team maximizes its probability of success.

Reference

1. Eugene S. Ferguson, 1992. *Engineering and the Mind's Eye.* Cambridge, MA: MIT Press.

This chapter is contributed by Michael J. Vinarcik.

43 Total Product Development and Taguchi's Quality Engineering

43.1. Introduction   1478
43.2. The Challenge   1478
43.3. Model of New Product Development   1479
43.4. Event Thinking in New Product Development   1480
43.5. Pattern Thinking: Big Ideas Make Big Differences   1481
43.6. Taguchi's System of Quality Engineering   1485
43.7. Structure Thinking in New Product Development   1486
43.8. Synergistic Product Development   1490
43.9. Conclusions   1491
    References   1491

43.1. Introduction

Quality in product development began with attempts to inspect quality into products or services in the process domain, the design domain, or the customer domain. The evolution of quality involves a significant change in thinking from reacting to inspection events to utilizing process patterns in engineering and manufacturing to build quality into the product. The use of pattern- and structure-level tools, such as the Taguchi system of quality engineering, TRIZ, and axiomatic design, provide a foundation for world-class product development.

43.2. The Challenge

Although the challenge is worldwide and exists in every industry, perhaps nowhere is the challenge so readily and clearly visible as in the U.S. automotive industry. The oil embargo of 1973 was a shocking introduction to the energy crisis, with consumers experiencing scarce gasoline and fuel oil and correspondingly high prices. In 1980, with a hostage crisis in Iran and soaring domestic gas prices, U.S. purchasers suddenly shifted to small cars and trucks, catching U.S. automakers

1478

with bulging inventories of full-sized carryover models. These automobile manufacturers found themselves in a full-blown economic recession, with consumers purchasing significant numbers of smaller, more-fuel-efficient, imported automobiles. Consumers discovered that these imported vehicles also had superior quality, and the challenge began.

From 1975 to 1979, total auto sales in the United States were 51.8 million, with U.S.-made vehicles accounting for 42.3 million. From 1985 to 1989, 53.3 million vehicles were sold in the United States. Compared to 1975–1979, U.S.-made vehicles dropped 10%, to 38.1 million, of which 3.3 million were from U.S. plants built by importers, while imported vehicles rose 60%, to 15.2 million [1]. In 1984, Louis Ross of Ford Motor Company said [2]: "Imports in 1984 took 26 percent of the U.S. car market. So here in our home market domestic auto manufacturers—and their supply base—are playing in the World Series, and in order to score we have to be able to meet or beat the world's best in quality, in cost, and in productivity, and in product content."

This challenge continues today. So far, much of the work in improving quality and productivity has been in implementing lean manufacturing and assembly systems. The player's scorecard is market share [3]. Following is a summary of the U.S. automotive market share (%):

	1988	1995	2001
GM	36	33	28
Ford	25	26	23
DaimlerChrysler	15	15	14
Japanese	20	23	27
European	2	2	4
Korean	2	1	4

The challenge will be won by the companies that can build quality and productivity into the design (product and process) during the product development process.

43.3. Model of New Product Development

To understand how quality may be built into the design during product development, consider the model shown in Figure 43.1. George Box, professor emeritus at the University of Wisconsin, has observed: "All models are wrong, but some are useful." So although this model does not match reality completely, it is useful.

This model was developed based on the ideas of Nam Suh [4] and Peter Senge [5], both of MIT. Nam Suh created the domain model of product development, with the belief that great products or services are a result of proper mappings between various domains. The first domain involves understanding the customer and defining desired customer attributes. These customer attributes are then mapped from the customer domain to functional requirements in the functional

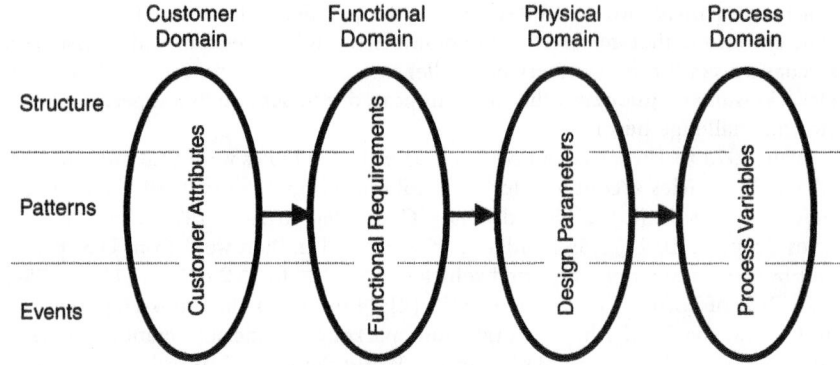

Figure 43.1
Senge's levels of
thinking overlaid on
Suh's domain model of
product development

domain. Functional requirements are then mapped to design parameters in the physical domain. Finally, design parameters are mapped to process variables in the process domain. The better or more complete the mappings between domains, the better the resulting product or service.

Peter Senge defined different levels of thinking based on events, patterns, and structure. *Event thinking* is at the lowest level. Here, one simply waits for something to happen, then reacts to the result. *Pattern thinking* involves monitoring process trends, then managing these trends before an unpleasant event occurs. *Structural thinking* deals with establishing the fundamental structure of the system so that the process patterns can work more effectively. Figure 43.1, an overlay of Peter Senge's patterns of thinking onto Nam Suh's domain model of product development, is useful in understanding the history of quality in new product development.

43.4. Event Thinking in New Product Development

After World War II, the primary way of assuring quality to customers was inspection after the process domain. Parts were produced, and then these parts were checked to see whether they were good enough to ship. If a part was not good, an *event* had occurred, resulting in rework or scrap and problem solving. Event thinking also occurred in the physical domain. Many engineers simply threw a design together and then tested it, expecting the design to fail. The failure of a design verification test is an event that the engineers answered with a sequence of build–test–fix cycles.

Build–test–fix is actually a method used today by many designers to "inspect" quality into the product or service. Supposedly, the design will be "Band-Aided" enough so that it will function properly before the product or service reaches the customer. Otherwise, the inevitable result is consumer complaints and warranty cost in the customer domain.

Unfortunately, many companies today still depend on event thinking to assure quality to customers. These companies learn about customers through analysis of warranty cost, try to assure design integrity via batteries of expensive reliability tests, and rely on checks after assembly to assure that the product is good enough to ship. At a company like GE, the cost of quality associated with event thinking in 1996 was estimated to be as high as $10 billion each year in scrap, reworking of parts, correction of transactional errors, inefficiencies, and lost productivity [6]. Problem-solving opportunities associated with event thinking in product development are illustrated in Figure 43.2.

43.5. Pattern Thinking: Big Ideas Make Big Differences

Serious pattern-level thinking was introduced to product development when W. Edwards Deming, Joseph M. Juran, and others were invited to Japan shortly after World War II. In 1950, Ichiro Ishikawa, president of the Union of Japanese Scientists and Engineers, arranged for Deming to meet with 21 top management executives of Japanese industry and lecture about quality.

Deming began by introducing some ideas he had learned from Walter Shewhart, specifically the plan–do–study–act cycle of learning and statistical process control (SPC). SPC, a pattern-level quality method in the process domain, focuses on patterns or trends in process data so that the process can be adjusted before an inspection event occurs. When Japanese companies began to implement SPC, quality improved dramatically. For the first time, product or design engineers knew that the parts they had designed were being manufactured according to print.

Use of pattern-level thinking in the other design domains began when Kaoru Ishikawa, known for Ishikawa diagrams and formalization of quality circles, noticed that even though parts were being made to print, customers were still unhappy with the products [7]. Specifications and tolerance limits were stated in the drawings. Measurements and chemical analyses were being performed. Standards existed for all these things and the standards were being met, but these standards were created without regard to what the customer wanted.

Ishikawa wrote: "When I ask the designer what is a good car, what is a good refrigerator, and what is a good synthetic fiber, most of them cannot answer. It is obvious that they cannot produce good products." You simply cannot design a good product or service if you do not know what "good" means to a customer. Ishikawa encouraged people to think at the pattern level in the customer domain

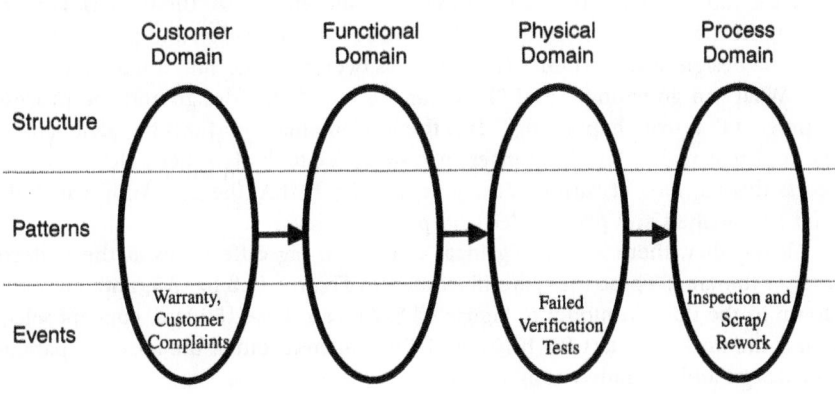

Figure 43.2
Problem-solving opportunities associated with event thinking in the various design domains

instead of just reacting to a warranty event. He said that if you don't know what a good product is, ask your customers. Customers will give you what Ishikawa called the *true quality characteristics*.

The problem with true quality characteristics, or customer attributes (CAs), is that the designer often cannot use them directly. For example, a customer may want the steering of an automobile to be "comfortable." An engineer cannot write on a drawing, "Make the steering comfortable." The engineer must find substitute quality characteristics, dimensions, or characteristics of the design that are correlated with customer desires but have meaning to an engineer [engineering characteristics (ECs)].

Therefore, Ishikawa said that the designer must create a map that moves from the "world of the customer" to the "world of the designer." He used a tree diagram to create such a map and called these maps *quality tables*. The Kobe Shipyard of Mitsubishi Heavy Industries created the first quality table in 1972. Once the quality table was completed, Ishikawa felt that the designer had a customer-driven definition of a good product or service. This definition of quality could then be deployed into the product development activity. Thus, quality function deployment (QFD) was born [8].

In subsequent years, about 120 different quality tools and methods were created at the pattern level for designers to manage product development process trends, so that inspection "events" would become more of a nonevent. Some of the most popular and powerful of these tools include:

- ❏ Failure mode and effects analysis (FMEA) for both the product and process domains
- ❏ Taguchi's quality engineering, including the methods of parameter design (for the product and process domains) and tolerance design (for the product domain)
- ❏ Design for assembly (DFA) and design for manufacturing (DFM), which improve the mapping from the product to the process domain
- ❏ System engineering, value analysis (VA), and value engineering (VE) in the functional domain

Rather than mentioning over 120 tools, it is easier and more effective to consider the fundamental ideas behind the tools and emphasize these ideas. For example, the main idea behind the tool of FMEA for both design and process is that after the design of the product (process) has been roughed out, it makes sense to ask "What can go wrong?" and "How can we modify the design and the process to prevent that from happening?" It is the implementation of this big idea behind the tool that makes big differences in results. Six tools have been developed to access this big idea in various situations. Besides FMEA, the best known include fault tree analysis and process decision program charts.

All together, there are 15 big ideas that make big differences at the pattern level of thinking. These ideas are illustrated in Figures 43.3 to 43.7, and a few are shown in the domain model of Figure 43.8. Among these 15 ideas, concept selection is the work of Stuart Pugh [9], and Taguchi contributed the ideas on parameter design and tolerance design.

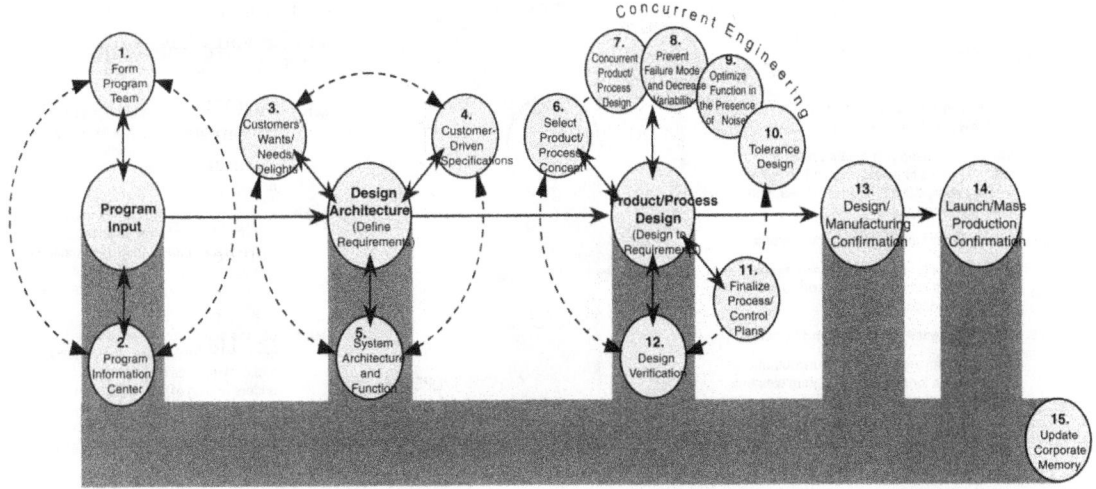

Figure 43.3
Pattern-level big ideas that make big differences

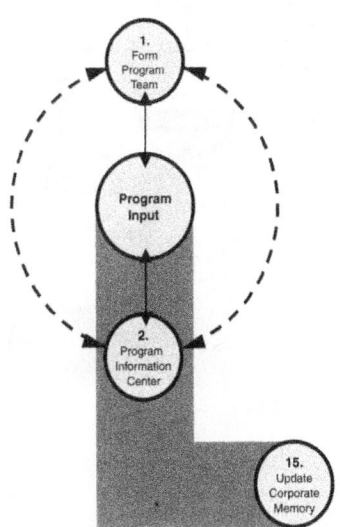

Purpose

Step 1. People Work as a Program Team, Share a Vision

Establish and maintain a program development team (for both product and process) that has a shared vision.

Step 2. Create a Program Information Center

Establish and maintain a program information center and understand program, social, and institutional lessons learned.

Breakthrough Idea

Step 1. High-performance teams emerge and grow through systematic application of learning organization and team disciplines to foster synergy, shared direction, interrelationships, and a balance between intrinsic and extrinsic motivation factors.

Step 2. Create an information environment with networks to foster program/product knowledge via: prior lessons learned (corporate memory), best practices (institutional, technical, and social), and inter- and intradisciplinary communication and collaboration.

Figure 43.4
Pattern-level big ideas associated with program input

Purpose

Step 3. Establish/Prioritize Customers' Wants/Needs/Delights

Identify customers and create opportunities for team members to establish/prioritize customer wants, needs, delights, usage profiles, and demographics.

Step 4. Derive Customer-Driven Specifications

Translate customer, corporate, and regulatory requirements into product/process specifications and engineering/test plans.

Step 5. Define System Architecture and Function

Define system architecture, inputs/outputs, and ideal function for each of the system elements, and identify interfaces.

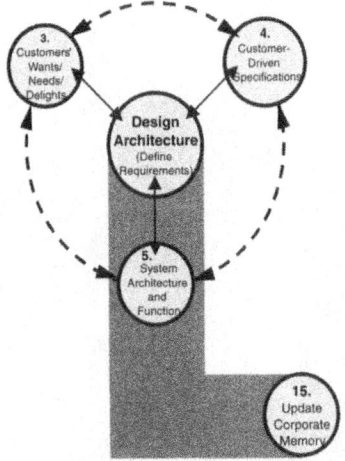

Breakthrough Idea

Step 3. Foster intense customer engagement to identify base expectations as well as distinctive opportunities which differentiate and characterize a winning product.

Step 4. Establish the foundation (maximum potential) for customer satisfaction by systematically translating the customer definition of a "good" product into engineering language and competitive targets.

Step 5. Lay the foundation for analytical optimization of function, cost, quality, and performance by gaining understanding of how the system and system elements function ideally, and by gaining understanding of the interfaces and interactions between functional system elements.

Figure 43.5
Pattern-level big ideas associated with design architecture

Purpose

Step 6. Select Product/Process Concept

Create/establish alternative product design and manufacturing process concepts and derive enhanced alternatives for development.

Step 7. Concurrent Product and Process Design

Design and model product and process concurrently using low-cost tolerances and inexpensive materials.

Step 8. Prevent Failure Modes and Decrease Variability

Improve product and process through reduction of failure modes and variability.

Step 9. Optimize Function in the Presence of Noise

Optimize product and manufacturing/assembly process functions by testing in the presence of anticipated sources of variation, or noise.

Step 10. Tolerance Design

Selectively tighten tolerances and upgrade materials to achieve desired performance (with cost/benefit trade-offs). Identify key characteristics for manufacturing control and variability reduction.

Step 11. Finalize Process/Control Plans

Finalize process and establish tooling, gauges, and control plans.

Step 12. Design Verification

Integrate and verify design and manufacturing process functions with production-like hardware/software.

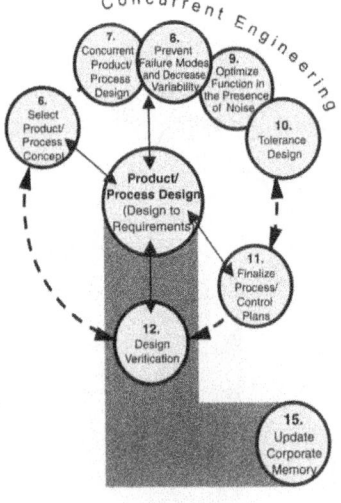

Breakthrough Idea

Step 6. Derive a concept to meet or exceed customer expectations through systematic exploration of many alternatives.

Step 7. Achieve superior performance through simultaneous integration of engineering, manufacturing, and delivery functions.

Step 8. Improve product and process by asking: "What can go wrong?" and "Where can variation come from?" Revise design and process to prevent occurrence and reduce variation.

Step 9. Improve performance against customer targets, up front in the development process, by adjusting controllable parameters to minimize deviations from the intended/ideal function.

Step 10. Achieve functional targets at lowest cost by selectively tightening tolerances and upgrading materials only where necessary. Demonstrated customer-sensitive characteristics are chosen for ongoing variation reduction.

Step 11. The manufacturing and assembly processes, tooling, gauges, and control plans are appropriately designed to control and reduce variation in characteristics that influence customer satisfaction.

Step 12. Improve quality and reduce time-to-market by enabling a single prototype build/test/fix cycle.

Figure 43.6
Pattern-level big ideas associated with product/process design

Purpose

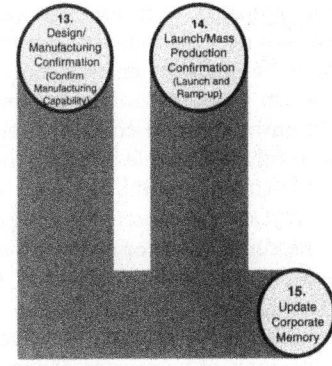

Step 13. Design/Manufacturing Confirmation

Confirm manufacturing and assembly process capability to achieve design intent.

Step 14. Launch/Mass Production Confirmation

Launch the product, ramp-up, and confirm that mass production delivers function, cost, quality, and performance objectives.

Step 15. Update Corporate Memory

Update corporate knowledge database with technical, institutional, and social lessons learned.

Breakthrough Idea

Step 13. Enable rapid, smooth confirmation of production and assembly operations with minimal refinements.

Step 14. Implementation of robust processes and up-front launch planning, with just-in-time training, will promote a smooth launch and rapid ramp-up to production speed. Final confirmation of deliverables enables team learning with improved understanding of cause/effect relationships.

Step 15. Retain whats been learned. Create and maintain capability for locating, collecting, and synthesizing data and information into profound knowledge. Communicate in a timely, user-friendly manner all technical, institutional, and social lessons learned.

Figure 43.7
Pattern-level big ideas associated with prelaunch and launch

43.6. Taguchi's System of Quality Engineering

One of the most significant achievements associated with designing quality into a product is Taguchi's system of quality engineering [10]. As shown in Figure 43.8, this system spans the physical and process domains and consists of:

❑ Parameter design in the physical domain

❑ Tolerance design in the physical domain

❑ Process parameter design in the process domain

❑ On-line quality control in the process domain

Use of these methods can make improvements in quality by factors of 10 and also improve both cost and productivity. Taguchi's intent is to produce the highest-quality product at the lowest possible cost. The big idea behind parameter design is the concept of hitting the initial product (and/or process) design with "noise"

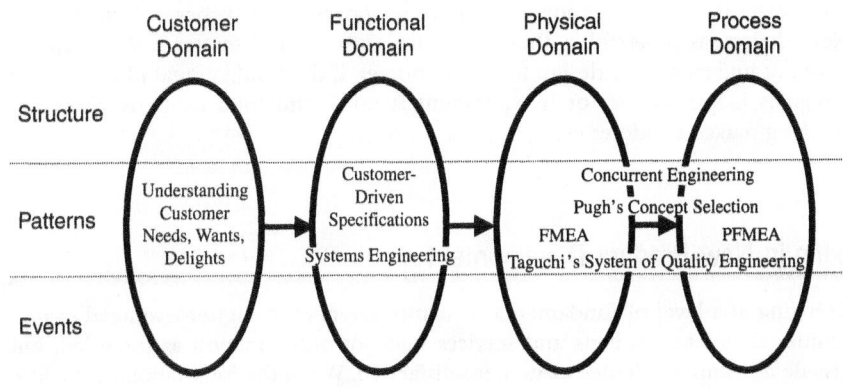

Figure 43.8
Examples of pattern thinking for problem prevention in various design domains

and finding a combination of factors that can be adjusted in the design systematically to make its functional performance insensitive to noise. Here noise is defined as variation the engineer either cannot control or chooses not to control, but which may affect product (and/or process) performance.

For example, environmental conditions are a noise. An automotive engineer cannot control whether a vehicle will be required to start in the freezing arctic cold or in a very hot desert, or whether the humidity will be very dry or very moist, but the vehicle must start and perform in all conditions. User variation is a noise. The driver may be conservative or extremely aggressive, yet the vehicle must function as intended. Less extreme variation in environment or operators may also affect process performance. Variation in material and/or part characteristics is also noise. So is functional deterioration over time.

Prior to the creation of parameter design, the best an engineer could do was to understand what is important to the customer in terms of product and process characteristics, find the target or set point, and work to reduce variation around this target point. With parameter design, one can make the variation in a characteristic insensitive from the customer point of view and open up tolerances to achieve high quality at low cost. This idea is a significant improvement to the product development process—so much so that if your competition is doing this and you are not, you will not be able to compete.

Similarly, tolerance design is a big idea that can improve quality and reduce cost. Too often, tolerances are chosen because "We have always done it that way" or "This is the standard tolerance for this situation." Taguchi's huge contribution involves using a designed experiment (orthogonal arrays) to understand the cause-and-effect relationships between tolerances and/or material choices and functional performance. With this knowledge, large tolerances and low-cost materials may be used for characteristics whose variation does not affect customers, achieving low cost. When customers are sensitive to a characteristic's variation, the characteristic can then be processed to reduce this variation with a tolerance that meets functional targets, thus achieving high quality. Of course, this exercise should be done after parameter design, because the number of customer-sensitive characteristics will then be reduced.

On-line quality control involves checking and managing manufacturing and assembly process trends in the most efficient and practical way possible so that subsequent process inspection events will not occur. One must not underestimate the impact that Taguchi's ideas can have on product and process performance. Nevertheless, as powerful as these ideas are, they will not matter if the engineer does not understand or design for the customer, if the fundamental physics of the design is inappropriate, or if equipment is not maintained properly. Structure thinking makes a difference.

43.7. Structure Thinking in New Product Development

Thinking at a level of fundamental structure offers even higher-leveraged opportunities to create products and services that not only function as intended, but also deliver unprecedented customer satisfaction. When the foundational structure of design is properly established, the methods at the pattern level are much more effective. When pattern-level methods work well, the event outcomes become world-class.

In the evolution of quality, two very powerful design methods have emerged at the structural level: axiomatic design and TRIZ (see Figure 43.9).

Axiomatic design is the result of work by Nam Suh [11]. In the late 1970s, he asked himself the question: Can the field of design be scientific? Suh wanted to establish a theoretical foundation for design by identifying the underlying principles or axioms that every great design must have in common. He knew that he could never prove that these principles were true, but could he find a set of principles for which no one could find a counterexample? After a decade of work, two principles emerged. From these two principles, theorems and corollaries could be derived that together form a scientific basis for the structure of great design.

The first principle that Suh discovered was the principle of independence. Consider Suh's domain model shown in Figure 43.1. Mapping between domains represents a mapping of "what's" to "how's." The *independence axiom* states that in great designs, the "how's" are chosen such that the "what's" maintain independence; that is, design parameters must be chosen in such a way that functional requirements maintain independence. For example, consider the water faucet designs shown in Figure 43.10.

The functional requirements for a water faucet are to control the flow rate and the water temperature. As functional requirements, these are independent by definition. A design that obeys the independence axiom will maintain this independence.

The faucet on the left of Figure 43.10 has two design parameters: a hot water knob and a cold water knob. What is the relationship between these design parameters and the functional requirements? When the hot water knob is turned, temperature is affected, but so is flow. Turning of the cold water knob also affects temperature and flow. Therefore, this design is coupled and the independence of the functional requirements has not been maintained. If a consumer has optimized flow rate, then turns one of the knobs to optimize temperature, the flow rate is changed and is no longer optimal. Designs of this type can eventually satisfy customers only by iterating between the two design parameters.

Consider the design on the right of Figure 43.10. This faucet has one handle and the design parameters are to lift the handle to adjust flow and to move the

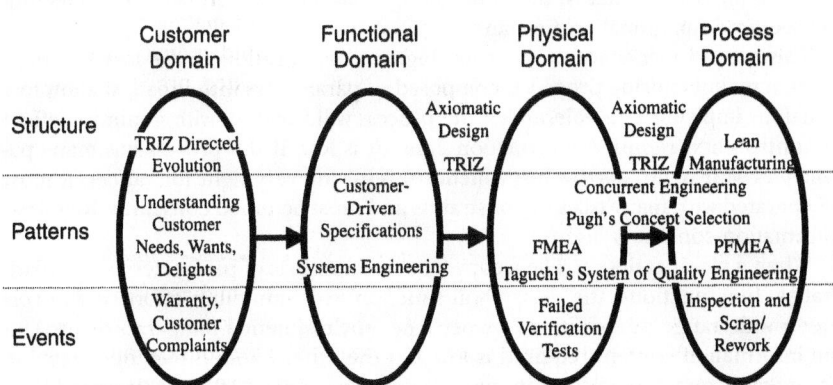

Figure 43.9
Quality evolution in the various design domains

Figure 43.10
Water faucet example

Functions for water faucet:
FR_1 = control flow rate FR_2 = control temperature

Design1:
DP_1 = hot water knob
DP_2 = cold water knob

Design 2:
DP_1 = handle lifting
DP_2 = handle moving side to side

handle from side to side to adjust temperature. In this design, adjusting temperature does not affect flow, and adjusting flow does not affect temperature. From an axiomatic design point of view, this design is superior because it maintains the independence of the functional requirements.

Imagine what happens when a designer is working in a situation with a dozen or more functional requirements. If the design is coupled, optimization of one function may adversely affect several other functions. When these functions are fixed, the original function no longer works well. The designer is always tuning and "Band-Aiding" such a design and the customer will never be completely happy. However, if the design is created in such a way that each of the functional requirements is handled independently by the design parameters, each function of the design can easily be optimized with pattern-level tools. Taguchi's additive model allows us to reach the same conclusion. Functional independence (Suh) or the absence of control factor interactions (Taguchi's additive model) provides the opportunity to design and manufacture products that will attenuate all forms of unwanted variation, including variation due to conditions of customer use.

The independence axiom can be used to evaluate how good a design will be when the design is still on paper! But suppose that you have two design alternatives that both follow the independence axiom. Which one is better? Suh's second principle states that the better design will minimize the information content necessary for implementation; that is, the better design has a higher probability of meeting the required functional performance [4].

This second principle is related to Taguchi's application of the loss function. When a manufacturing process is composed of parameters with broad, shallow loss functions implying wide tolerances, the process will function with minimum effort or controls and required information content is low. If the process has many parameters with steep, narrow loss functions, implying very tight tolerances, it must be operated with many rules or constraints and must be tuned constantly. Required information content is high.

When a product design, like a copy machine, consists of parameters with broad, shallow loss functions, the design will function with minimum effort or controls over a wide range of conditions (paper type, environmental temperature, etc.) so the information content required is low. On the other hand, if the copy machine has many parameters with steep, narrow loss functions, it must be operated with many rules or constraints, has a very high downtime, and must be tuned continually by a repairman. Required information content is high. Designs that have a solid axiomatic foundation simply work better than designs that do not.

Suppose the designer cannot find a set of design parameters that maintain the independence of all of the functional requirements. In this situation, improving one function typically degrades another. An example in automotive steering is steering "road feel" and "parking efforts." When the steering efforts are high, the customer experiences good "road feel." However, high efforts can make it difficult for customers to park. Adjusting efforts to make the vehicle "easy to park" will result in degraded "road feel." A typical approach to resolve this situation is compromise—trade off customer functionality and hope for the best.

Another example of compensation was in the early introduction of front-wheel-drive vehicles. These vehicles tended to steer to the right as the accelerator was compressed. The driver had to compensate by applying a negative steering wheel angle. This approach is clearly undesirable here and must also be avoided in the design stage if we wish performance to stay on target.

This is where TRIZ is most helpful. TRIZ, a Russian acronym for *theory of inventive problem solving*, is the result of over 45 years of research by Genrich Altshuller and colleagues [12]. Altshuller hated compromise. He called the situation where functions oppose each other contradictions and developed a methodology where design teams could systematically innovate and find design parameters that resolved contradictions, thereby creating "win–win" functional situations.

The methodology began by identifying all possible contradictions that existed in patent databases and identifying how these contradictions were resolved. Altshuller found that only a few particular principles of resolution have ever been widely used to resolve certain pairs of functional contradictions. For example, suppose that the functions of weight and reliability contradict. When the design is changed to improve reliability, weight increases. When weight is decreased, reliability degrades. Altshuller found that for the most part, there are only four principles that have been commonly used to resolve this contradiction [12]. He created a matrix of contradictions and resolution principles and used this information to guide design teams so that they could brainstorm in areas that are likely to lead to "win–win" solutions. Altshuller also believed in minimizing information content; this he called the *principle of ideality*. Later, TRIZ was expanded to include an entire database of innovation techniques, including the study of system evolution.

Altshuller found that systems tend to evolve along specific laws and lines of evolution. By studying system evolution for the past and present (for the super-system, system, and subsystem), a designer can identify the current stage of system evolution. By applying laws and lines of evolution, design teams can predict what the next developments of the system will be. This is a huge competitive advantage. Companies that operate at the event level obtain information about the customer through warranty data. Companies at the pattern level interact with customers and find out what customers believe is important today. No customer can tell the designer what will be important tomorrow. At the structural level, TRIZ-directed evolution predicts what will be important to customers tomorrow!

So if TRIZ works at the structural foundation of design, why does the name imply problem solving? The answer is simply that higher-level thinking can always be used as a methodology to solve lower-level problems. The fact that problems exist in the event realm means that the original design process had serious flaws—pattern or structural work that should have been done is either missing or incomplete. The designer can always go back and complete this work at any time. This is why completing work associated with pattern or structural tools can quickly lead to problem resolution. A good example of this is six sigma, which addresses problems created by event-level thinking, but the methodology of six sigma utilizes pattern-level tools such as FMEA.

43.8. Synergistic Product Development

The world has yet to see what can happen when all the big ideas associated with structure, pattern, and event thinking are implemented iteratively and synergistically. As product development technology moves in this direction, consumers can look forward to truly robust, reliable products and services—the big ideas are applicable to both—with unprecedented levels of customer satisfaction.

Great designs must start with a solid foundation. This means using TRIZ-directed evolution to understand how the system is evolving. In addition, principles of axiomatic design and TRIZ must be used to assure that an adequate architectural structure is established so that design parameters are chosen such that functional requirements maintain independence.

With this foundation, the tools at the pattern level of thinking will work so much better. With a proper understanding of system evolution, the results of consumer research can be interpreted correctly so that the planned product or service is something that consumers will want. Using structural principles of design and TRIZ, system engineering can be deployed effectively by Suh's method of zigzagging between the functional and physical domains, creating tree diagrams of functions and related design parameters. Now the stage is set to use quality engineering.

How does this up-front structural work make it easier to utilize Taguchi's system? First, one of the things engineers struggle with in parameter design is the question of what to measure. An effective measure is one that is related to the underlying physics and energy of the system. Doing the up-front structural work will make engineers better understand the fundamental physics of the system, which is extremely useful in selecting effective performance measures for optimization.

Second, if the foundation of the design is poor, it is impossible to optimize all the functions of the system effectively. However, if the design parameters are chosen such that the functional requirements maintain independence, different control factors may be used to optimize different functions, so that parameter design optimization can be effective and efficient. Efforts to get the structural foundation of the design right must first be done up-front.

How does quality engineering enable activities at the event level of thinking to go well? Imagine a process that is not robust, that is, a process sensitive to noise. Such a process rarely runs well for long intervals. Parts are OK, OK, OK, and then suddenly a part is out of specification. Something changed, a noise hit the process, and now tuning is needed to bring the parts back to specification. Operators scramble with adjustments until everything is running well again. Then, suddenly, the parts are out of print in the other direction of the specification. Such a process never works well and requires a great deal of management attention and energy. A robust process does not care about changes in noise, and parts track on target.

Imagine a product that is robust—a product that always performs close to target, no matter the circumstances. Such products will cause customers to brag about this performance to their neighbors. Deming used to say that the best thing we could do is to make products (or services) that cause customers to brag to their friends and neighbors. When it comes time to purchase another product, not only do they come back to buy yours, but they bring friends and neighbors with them. The industry challenge will finally be won by those companies that integrate and synergistically implement the big ideas that make big differences into the way they design products and services. People will go out of their way to find such products and services.

43.9. Conclusions

Quality tools and methods have evolved utilizing three stages of thinking (event, pattern, and structure) across four domains associated with product development. Taguchi's system of quality engineering plays a central role at the pattern or process level of thinking. Implementation of axiomatic design and TRIZ at the structure level will enhance the effectiveness of pattern-level processes. When the pattern-level processes work to their capability, the customer-related events are world-class. Companies that wish to accelerate development of their own quality programs can utilize the concepts (big ideas) explained in this chapter to understand their current level of evolution and to implement focused actions that can move them quickly past their competition.

ACKNOWLEDGMENT
The authors would like to thank Dr. Carol Vale of the Ford Motor Company for her assistance in editing this document and making the content more readable.

References

1. *Ward's Automotive Yearbook, 1979, 1985, 1990.* Southfield, MI: Ward's Communications.
2. *Ward's Automotive Yearbook, 1985.* Southfield, MI: Ward's Communications.
3. *Time,* January 14, 2002, p. 43.
4. Nam P. Suh, 2001. *Axiomatic Design: Advances and Applications.* New York: Oxford University Press.
5. Peter M. Senge, 1990. *The Fifth Discipline.* New York: Doubleday.
6. M. Harry and R. Schroeder, 2000. *Six Sigma.* New York: Doubleday.
7. K. Ishikawa, 1988. Quality Analysis. *ASQC Conference Transactions—Philadelphia,* pp. 423–429.
8. Dan Clausing and John R. Hauser, 1998. *House of Quality.* Cambridge: Harvard Business School Press.
9. Stuart Pugh, 1991. *Total Design.* Reading, MA: Addison-Wesley.
10. Genichi Taguchi, 1993. *Taguchi on Robust Technology Development: Bringing Quality Engineering Upstream.* New York: ASME Press.
11. Nam P. Suh. 1990. *The Principles of Design.* New York: Oxford University Press.
12. G. Altshuller. 1998. *40 Principles: TRIZ Keys to Technical Innovation.* Worchester, MA: Technical Innovation Center.

This chapter is contributed by Larry R. Smith and Kenneth M. Ragsdell.

44 Role of Taguchi Methods in Design for Six Sigma

44.1. **Introduction**   **1492**

44.2. **Taguchi Methods as the Foundation of DFSS**   **1493**
Overview of DFSS (IDDOV)   1494
Overview of Quality Engineering   1496
Quality Management of the Manufacturing Process   1498
Linkage between Quality Engineering and IDDOV   1498

44.3. **Aligning the NPD Process with DFSS**   **1498**
Synthesized Case Study   1499
Developing Linkages between NPD and IDDOV   1501

44.4. **Technical Benefits of IDDOV**   **1507**
Range of Variation and Sigma Level   1510
Concept/Technology Set Evaluation   1512
Failure Rate   1514
Innovation   1515

44.5. **Financial Benefits of IDDOV**   **1515**
Warranty Cost/Service Cost   1516
Product Cost   1516
Loss to Society   1516

44.6. **Taming the Intersection of Six Sigma and Robust Engineering**   **1517**
Case Study: Robust DMAIC   1519

44.7. **Conclusions**   **1520**
References   **1520**

44.1. Introduction

Quality engineering (also called Taguchi methods) has enjoyed legendary successes in many Japanese corporations over the past five decades and dramatic successes in a growing number of U.S. corporations over the past two decades. In 1999, the American Supplier Institute Consulting Group [1] introduced the IDDOV formulation of DFSS that builds on Taguchi methods to create an even stronger methodology that is helping a growing number of corporations deliver dramatically stronger products.

Quality engineering pursues both off-line and on-line quality. *Off-line quality engineering* (also called *robust engineering*) provides the core methodologies upon which the first four phases of IDDOV are built: identify project, define requirements, develop concept, and optimize design. On-line quality engineering provides some unique and powerful methods for establishing and maintaining the quality of manufacturing processes within the verify and launch phase of IDDOV.

Since the total set of robust engineering [2,3] methodologies is fully incorporated within IDDOV, robust engineering case studies double as IDDOV case studies. However, the converse does not necessarily hold since IDDOV contains elements that are not encompassed within robust engineering.

Thousands of robust engineering case studies on products ranging from automobiles to medical equipment and consumer electronics produced by U.S. manufacturing corporations have reported, on average, a gain in the signal-to-noise ratio of 6 dB. In terms of product performance metrics such as failure rate, a gain of 6 dB translates into a factor of 2 to 4 (200 to 400%) reduction in failure rate and in warranty cost. A sample of 16 case studies with an average gain of 7 dB is provided in the book *Robust Engineering: Learn How to Boost Quality While Reducing Cost and Time to Market* by Taguchi et al. [2]. These 16 case studies cover a broad range of applications, including acceleration of the growth rate of bean sprouts; improving electronic warfare and communications algorithms; and improving the performance of mechanical, electrical, and electronic systems and components in many different industries. Yuin Wu and Alan Wu provide a "deep dive" into the subject in their book, *Taguchi Methods for Robust Design*. [3]. Subir Chowdhury provides an exceptionally readable introduction to robust engineering and the whole of DFSS in his book, *Design for Six Sigma: The Revolutionary Process for Achieving Extraordinary Profits* [1].

Any manufacturing corporation in any country can rapidly enjoy significant gains by pervasively implementing *design for six sigma* (DFSS), founded on robust engineering. Just as six sigma has helped many corporations enjoy unprecedented financial gains through improvements in their processes, DFSS has helped a growing number of corporations achieve significant, additional financial gains due to improvements in engineered products delivered to customers and design or re-design of business, engineering, manufacturing, and service processes.

The next five sections explore in sequence:

❏ Taguchi methods as the foundation of DFSS

❏ Overview of DFSS (IDDOV)

❏ Overview of quality engineering

❏ Linkage between quality engineering and IDDOV

❏ Aligning the NPD process with DFSS

❏ Technical benefits of IDDOV

❏ Financial benefits of IDDOV

❏ Taming the intersection of six sigma and robust engineering

44.2. Taguchi Methods as the Foundation of DFSS

Taguchi methods, design for six sigma, and *new product development* (NPD) processes work together. NPD leverages DFSS, which in turn leverages Taguchi methods to achieve competitive leadership. NPD guides a corporation's product development activities. Design for six sigma (DFSS) is carefully designed to support a corporation's new product development process. The three methodologies form a natural hierarchy: (1) new product development (NPD) process, (2) design for six sigma (IDDOV), and (3) Taguchi methods (quality engineering).

The relationship between Taguchi methods and design for six sigma is developed to set the stage for developing the relationship between DFSS and a typical new product development process. A revised new product development process is presented that leverages the unique strengths of DFSS and Taguchi methods.

Overview of DFSS (IDDOV)

American Supplier Institute Consulting Group's formulation of DFSS [1] is structured into five phases:

1. *Identify project.* Select project, refine its scope, develop project plan, and form high-powered team.

2. *Define requirements.* Understand customer requirements and translate them into technical requirements using quality function deployment.

3. *Develop concept.* Generate concept alternatives using Pugh concept generation and TRIZ and select the best concept using Pugh concept selection methods; conduct FMEA [1]. The develop concept may be conducted at several levels, starting with the system architecture for the entire product. Then concepts are developed for the various system elements as needed.

4. *Optimize design.* Optimize technology set and concept design using robust optimization, parameter design, and tolerance design. Optimization is conducted at the system element levels.

5. *Verify and launch.* Finalize manufacturing process design using Taguchi's online quality engineering, conduct prototype cycle and pilot run, ramp-up to full production, and launch product.

The first two phases, identify project and define requirements, focus on getting the right product. The last two phases, optimize design and verify and launch, focus on getting the product right.

The middle phase, develop concept, is the bridge between getting the right product and getting the product right. Bridging across external customer requirements and internal technical requirements makes development of conceptual designs a difficult and critical step in the product development process. As the bridge between upstream and downstream requirements, conceptual designs should creatively respond to upstream requirements developed through the quality function deployment (QFD) process and downstream engineering, manufacturing, and service requirements that do not necessarily flow out of QFD. The central technical requirement that the selected concept and technology set should optimize well to provide good robustness (high signal-to-noise ratio) leads to the potential loopback from optimization to concept development indicated below.

The five phases, IDDOV, are segmented into 20 steps.

Identify Project

1. Refine charter and scope.
2. Develop project plan (plan, do, check, act).
3. Form high-powered team.

Define Requirements (*QFD*)

4. Understand customer requirements (QFD phase I).

5. Build house of quality (QFD phase I).

Develop Concept (*Pugh, TRIZ, and FMEA*)

6. Generate concepts:

    a. Pugh concept generation and creativity toolkit.

    b. TRIZ (theory of inventive problem solving).

7. Select concept (Pugh concept selection process).

    a. Conduct first run of evaluation matrix.

    b. Conduct confirmation run.

    c. Conduct controlled convergence.

8. Conduct FMEA.

Optimize Design (*Taguchi methods*)

9. Develop design planning matrix (QFD phase II).

10. Optimize (concept) design (step 1 of two-step optimization).

11. Adjust to target (step 2 of two-step optimization).

12. Conduct tolerance design.

13. Develop process planning matrix (QFD phase III).

14. Optimize process design.

Verify and Launch

15. Develop operations planning matrix (QFD Phase IV).

16. Finalize operations and service processes.

17. Conduct prototype cycle.

18. Conduct pilot run.

19. Launch, ramp-up, and confirm full production.

20. Track and improve field performance.

Stuart Pugh [4], creator of Pugh concept generation and selection methods, emphasizes the importance of concept design with the statement: "The wrong choice of concept in a given design situation can rarely, if ever, be recouped by brilliant detailed design." The ability to evaluate robustness of concept designs/ technology sets provides the means to avoid the "wrong choice of concept."

The loopback from optimize design to the develop concept phase can become a critical iteration since a concept design that does not optimize well cannot be recovered later in the process. The only feasible solution is to create a new concept that does optimize well. This is a strong and troublesome statement. The prospect of discarding a concept is far easier said than done. Whether the concept is newly created or an existing product in the field, it is very difficult to recognize and act on the recognition that a concept or embedded technology set or both should be discarded. A concept/technology set represents a significant investment in time and resources that nearly always appears easier to fix than to replace, especially in the absence of a definitive way to differentiate good from bad. The notion of robustness provides the conclusive technical differentiator between good and poor concepts.

The 20-step process outlined above is used for creating new products and processes. The develop concept and optimize design phases combine to provide powerful find and fix firefighting methods and tools. Many of the case studies that contribute to the average gain of 6 dB involve improvement of existing concepts. IDDOV is

❏ An engineering methodology that supports a new product development process

❏ Based on the world's best practices, many of which were not introduced into the Western world until the 1980s and are not yet adequately included in university engineering curricula

❏ An important implementation of Taguchi's Quality Engineering

❏ An engineering process developed by engineers rather than a statistically based process like that commonly employed in Six Sigma

IDDOV emphasizes relatively new methods and tools such as quality function deployment, Pugh concept development, TRIZ, and robust engineering, as described by Subir Chowdhury [1]. These newer, powerful methodologies are relatively familiar within technical communities; however, they are not implemented consistently within Western world NPD processes.

IDDOV is *not* a completely new product development process (NPD process). It is designed as an augmentation to an NPD process. As an augmentation, IDDOV does not encompass all of the elements of mechanical, electrical, chemical, software, and other engineering disciplines necessary to deliver products. IDDOV does not even contain a phase that relates to detailed product and process design of NPD.

Overview of Quality Engineering

Quality engineering (called Taguchi methods in the United States) consists of off-line and on-line quality engineering. (Off-line quality engineering is called robust engineering in the West.)

The purpose of robust engineering is to complement traditional engineering methods with robust methods to increase the competitiveness of new products by reducing their cost and improving their quality, starting with research and development prior to specific product applications. A central focus of robust engineering is robust optimization of conceptual designs and technology sets.

Quality engineering is applicable to all aspects of R&D, product development, and manufacturing process development: *system/concept design, parameter design, and tolerance design.* The three phases are common for technology, product, and manufacturing applications.

Off-line quality engineering (robust engineering) encompasses technology development and product design, which is characterized in more detail.

Technology Development (*R&D Prior to Knowledge about Requirements*)

❏ *System/concept design.* Select the best technology set from all possible technology alternatives that might be able to perform the objective function.

❏ *Parameter design.* Determine the optimal values of the parameters that affect the performance (*robustness*) of the technology set selected.

❏ *Tolerance design.* Find the optimal trade-off between the cost of quality loss due to variation of the objective functions and the cost of technology set materials and processes.

Product Design (*Creation and Optimization of Concepts/Technology Sets*)

❑ *System/concept design. Select the best system/concept design and technology set from all possible alternatives that can perform the objective function.* System/concept design retains the common meaning of developing innovative conceptual designs that respond to customer, company, and regulatory requirements. Concept design is conducted at various levels in a complex product. At the highest level, system design is the process of partitioning the product into an architectural hierarchy of system elements typically identified by terms such as subsystems, modules, components, and parts. Software implementations involve a corresponding hierarchy of levels. Once system architecture is defined, a conceptual design and technology set for each system element is developed. System design and concept design are frequently used interchangeably, since both terms refer to the process of stratifying a larger entity into its constituent elements.

❑ *Parameter design. Determine the optimal values of the parameters that affect the performance (robustness and target) of the selected concept/technology set.* Parameter design is the process of identifying and determining the values of design parameters that minimize the sensitivity of the design to sources of variation (i.e., sources of problems, including environmental conditions, customer usage conditions, wear of parts, manufacturing variations, etc.). Parameter design (*robust optimization*) is normally conducted on lower-level system elements that perform a single primary function. Two-step optimization facilitates optimization of concept designs and technology sets. Step 1 maximizes the functional robustness of the product or process concept design. Step 2 adjusts the product or process concept to target without affecting robustness significantly.

❑ *Tolerance design. Find the optimal trade-off between the cost of quality loss due to variation of the objective functions and the cost of components and technology set materials.* Tolerance design is the process of balancing product cost and quality losses (using the quality loss function). Application of the process is often biased to minimize product cost without significantly sacrificing the cost related to the customer's perspective of product performance, quality, or reliability. The gains realized through robust optimization typically exceed customer needs. Tolerance design identifies elements of the product that can be implemented with lower-cost materials, components, and parts without significantly affecting product performance, quality, or reliability. Tolerance design also works the other way around when the need is to upgrade product elements to reduce field cost.

Tolerance specification depends on the information developed during tolerance design. Tolerance specification is the process of deciding on the tolerance limits that will be entered on drawings and used for quality control of manufacturing and suppliers. Tolerance specification is typically combined with tolerance design.

On-line quality engineering addresses design and quality management of the manufacturing process.

Manufacturing Process Design

❑ *System/concept design.* Select the best production processes from all possible alternatives.

❏ *Parameter design.* Determine the optimal values of the parameters that affect the performance (*robustness and target*) of the production processes selected.

❏ *Tolerance design.* Determine tolerance specifications for the parameters that affect the quality and cost of the production processes.

Quality Management of the Manufacturing Process

Taguchi's on-line quality engineering differs from the typical American process of controlling the manufacturing process against tolerance specification limits. Taguchi's methodology focuses on maintaining the process close to target rather than within control limits. Taguchi observes [5]: "Right now, many American industries use very sophisticated equipment to measure the objective characteristics of their products. However, because some quality managers of these companies do not interrupt the production process as long as their products are within control limits, the objective characteristics of their products are usually uniformly distributed, not normally distributed."

It is clear from studies and common sense that products with all parameters set close to targets are more robust under actual customer usage conditions than products with some portion of parameters set close to control limits. Characteristics set close to control limits are sometimes suggestively called *latent defects* since they can rapidly become very real defects (*or premature failures*) under customer usage.

Linkage between Quality Engineering and IDDOV

The American Supplier Institute Consulting Group (ASI CG) structure of DFSS expands the system/concept design phase of off-line quality engineering into three phases: identify project, define requirements, and develop concept. Parameter design and tolerance design are aggregated under "optimize design." On-line quality engineering pertains to the verify and launch phase of IDDOV. Off-line quality engineering pertains to the following areas:

❏ *System/concept design.* Identify project, define requirements and concept.

❏ *Parameter design.* Optimize design.

❏ *Tolerance design.* Perform on-line quality engineering, verify and launch.

The correlation among the three phases of quality engineering and the five phases of DFSS suggest that IDDOV can be regarded as an implementation of quality engineering. The relationship is shown in Figure 44.1.

44.3. Aligning the NPD Process with DFSS

While IDDOV supports rather than replaces a new product development process, it does introduce new engineering capabilities that suggest modifications to the NPD process. As might be expected from the emphasis on robustness above, the

Figure 44.1
Aligning QE and DFSS

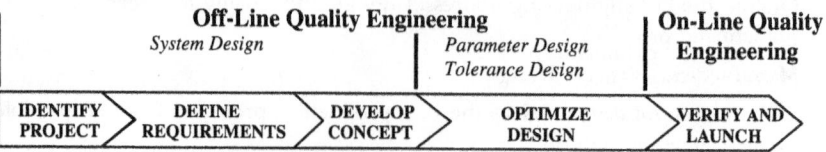

refined process will be identified as the new robust product development process (NRPD process).

Some of the problems addressed by the new process are illustrated in the style of a case study that is actually an amalgamation of case studies conducted during the 1990s and early 2000s in several large manufacturing corporations. The products ranged across automotive, office equipment, computer, medical diagnostic equipment, and chemical industries. The synthesized case study sets the stage for developing the NRPD process.

This study is derived from experience with several large corporations. Each study of an existing NPD process was conducted as background for improving the corporation's capability to deliver winning products consistently. The NPD processes in the study tended to focus more on management of the development process than on management of the engineering process. The NPD processes were universally designed to coordinate the myriad of details related to the development of complex products guided by a plan that contained the types of information identified in the reporter's credo of 5W's1H: who, what, when, where, why, how. The process starts with *why* the product is needed for the business as described in a business case document. The remaining four W's address information about *what* needs to be done *where* by *when* and by *whom. Where* identifies physical locations, organizations, and suppliers. *When* identifies inch-stones and milestones in the plan. *How* refers to the engineering activities, the weakest element in the process.

Synthesized Case Study

Managers and engineers, through an elegant software program, could readily access any element of the NPD process. Drop-down menus identified the engineering methodologies and tools that should be used at any point in the process. The 5W's1H and "go/kill phase gate" criteria and process were also available in drop-down menus. A red–yellow–green traffic light evaluation was used on all projects within the program and accumulated up to indicate the status of the total program.

Initial management presentations explaining their in-place NPD processes seem very orderly and, in some cases, very impressive. However, interviews and focus group–style discussions revealed a surprisingly high level of chaos in the engineering process. A seemingly endless series of build–test–fix cycles were conducted throughout the development process–build and test to see if the product met requirements and specifications, and fix the shortfalls, over and over and over.

Engineers and managers alike identified the popular phase gate model that defines every phase gate as a go/kill decision [6] as a major source of problems. If the model (*hardware and/or simulation*) designed and constructed for a build–test–fix cycle in a particular phase of NDP did not meet specifications to the degree required by the phase gate criteria, the team members feared that the program would be killed during the next phase gate review. The continuous series of go/kill tests throughout the development process put the kind of pressure on the team that destroyed confidence and fostered "look good" rather than "do good" behaviors. For example, the teams often struggled to appear as if they were meeting requirements and specifications just before each phase gate review. What better way to look good than to build and show off a sophisticated-looking model that appeared to meet specifications. The new demonstration model was often wasted effort, since its purpose was to help the team to get through the Phase Gate rather than to advance the engineering process.

One senior manager proudly summed up the situation as follows: "Ninety percent of the work gets done in the last 10 percent of the time before the next phase gate review." Others appeared to share his perspective that phase gates were effective fear factors. Whether in engineering or manufacturing, team members "tinkered" until they adjusted parameters to meet specifications, often just barely within specification limits.

In engineering, the spec may be the target value and allowable variation of functional performance that meets customer requirements. Tinkering to meet specs involved changing values of a selected design parameter and one more build–test–fix cycle. In manufacturing, the spec is typically the plus–minus tolerance specification of physical dimensions. Tinkering to meet specs involved adjustment, rework, scrap, and one more build–test–fix cycle.

Generally, manufacturing engineers seemed to understand better than the product engineers that the process of squeezing parameter values just inside "goalpost" specifications led to fragile products. However, their job was defined as meeting specifications, not striving for the centerline target. The conflict between their job description and their knowledge about what was right was a source of deep but silent frustration.

The engineering team members did not share a similar understanding about "just within specs." Their constant daily work was build–test–fix various levels of prototype models in an effort to meet requirements and specifications. That was literally the definition of their job, what they had spent their entire careers doing. The product engineers revealed no recognition that just within specs was even more damaging in engineering than in manufacturing. Anywhere inside the specified limits of allowable variation was regarded as an acceptable target without regard to the obvious fact that expected variation around an off-center target would quickly migrate beyond specification limits.

Neither managers nor engineers were aware of studies (and common sense) that make it clear that products with all parameters set close to targets are more robust under actual customer usage conditions than products with some portion of parameters set close to specification limits, whether in engineering or manufacturing.

Optimization of designs was regarded as just one of the many engineering activities conducted in the later detailed product and process design phases. Different teams within the same corporation conducted optimization in different ways, if at all. Optimization was lightly regarded as "a good thing to do if time allowed."

The degree of chaos was very high in the repetitious build–test–fix efforts to meet specifications under the pressure of go/kill style phase gates. Over half of the product engineers had never enjoyed the opportunity to launch a product into the market. Employee satisfaction, morale, and loyalty were trashed. A bunker mentality had developed as concerns about personal survival overwhelmed the desire to satisfy customers.

Conclusions

The engineering activities within NPD are strongly influenced by the specifics of the engineering methodologies that are used. Most in-place NPD processes do not encompass the newer methodologies, such as robust engineering or IDDOV.

Product development involves two very different activities: (1) concept development and (2) detailed product and process design. These activities are usually further partitioned into four to six phases linked by Phase Gates, as shown in Figure 44.2. The new product development (NPD) process shown in the figure is representative of the many variants. The phase–phase gate construct is a broadly accepted model for NPD [6].

A new product development process is effectively characterized through four elements.

1. *Phase–phase gate structure:* high-level structure, as illustrated in Figure 46.2
2. *Phase processes:* management and engineering activities within each phase
3. *Phase gate process:* activities supporting transitions between phases
4. *Project review process:* steps and style of project review process

PHASE–PHASE GATE STRUCTURE

The relationship of IDDOV and a typical NDP process is indicated in Figure 44.3. The dashed lines indicate the relationship of IDDOV and NPD phases. The stars represent phase gates. The phase gate indicated by the double star and solid line delineates completion of concept development activities and the beginning of detailed product and process design activities. As mentioned previously, the first four phases of DFSS are crowded into preconcept and concept phases of NPD.

While IDDOV can improve the capability to deliver competitive products significantly using the existing NPD process, the alignment with IDDOV is at best awkward. The new engineering methodologies contained within IDDOV suggest changes to the phase–phase gate structure of development processes.

Popular NPD processes simply do not capture the benefits that derive from the relatively new capability to optimize concept designs and technology sets. This important new capability provides competitive benefits that must be captured to enjoy leadership. NPD needs to be modified to capture the benefits of contemporary best practice engineering methodologies. Progressive corporations that refined their NPD process to capture the benefits of IDDOV over the last several years are benefiting from improved competitiveness.

Two-step optimization creates the opportunity to conduct robust optimization (step 1) during concept development and then set the system to target (step 2) during detailed product and process design. Taguchi introduced the notion of optimizing concepts and technology sets in the early 1990s [6]. He states (p. 90): "Research done in research departments should be applied to actual mass production processes and should take into account all kinds of operating conditions. Therefore, research departments should focus on technology that is related to the functional robustness of the processes and products, and not just to meeting customers' requirements. Bell Laboratories called this type of research 'two-step design methods.' In two-step design methods, the functional robustness of product

Concept Development ‖ *Detailed Product and Process Design*

| PRE-CONCEPT | CONCEPT | DESIGN AND DEVELOPMENT | VERIFY AND VALIDATE | RAMP-UP AND LAUNCH |

Figure 44.2
Phases of a typical new product development process

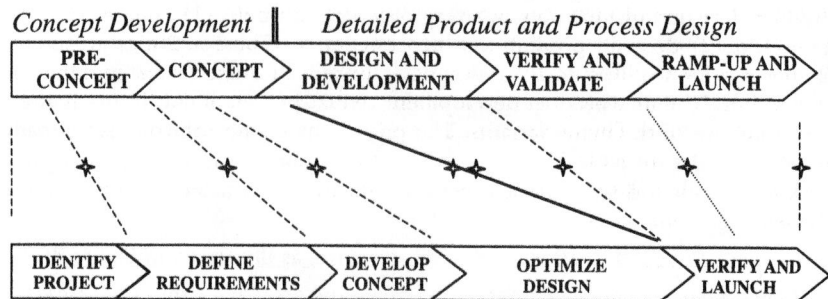

Figure 44.3
Linkage between IDDOV
and NPD

and processes is first maximized and then these product and processes are tuned
to meet customers' requirements."

This quote correctly suggests that two-step optimization can be brought all the
way forward to R&D activities that precede NPD concept development activities.
A refined new product development process explicitly identifies optimization up-
front in the development process. Figure 44.4 depicts one possible structure for a
refined NPD process supported by IDDOV that will serve as the foundation for
building a winning development process.

Elevating optimization to appear explicitly in what is now properly identified
as the new robust product development (NRPD) process is an important refine-
ment. Optimization serves somewhat different purposes in its two appearances in
the NRPD phases. Some subsystems may involve new concept designs and/or new
technology sets, while other subsystems may involve carryover designs from precedent products.

New concept designs should be optimized (step 1 of the two-step optimization
process) in the concept optimization phase and adjusted to target (step 2) either
in the concept optimization phase or the optimization, design, and development
phase.

Candidate carryover designs should be assessed for potential robustness prior to
accepting them as elements of the concept design. Most carryover designs can be
improved by conducting robust optimization experiments. Such carryover designs
might be both optimized and set to target in the optimization, design, and development phase.

The differences between NPD and NRPD depicted in Figures 44.3 and 44.4,
respectively, have huge implications. As noted earlier, optimization is normally re-

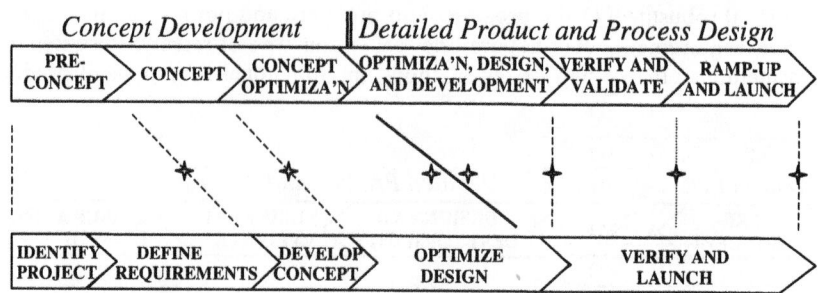

Figure 44.4
Optimization cited
explicitly in NRPD

garded as just one of many functions performed during detailed design that are too numerous to be identified in the high level of NPD phases. Elevating optimization to appear explicitly in the phase structure of NRPD is a big change. Pulling optimization forward to become part of the concept development activities is an even bigger change. The ability to optimize concepts and technology sets early in the development process streamlines the NRPD engineering process to significantly reduce the risk of innovation, enhance engineering excellence, reduce likelihood of canceling a development program, reduce the number of build–test–fix cycles needed, shorten development schedules, and reduce development, product, and warranty costs.

Figures 44.3 and 44.4 provide a reasonable representation of the linkages between IDDOV, NPD, and NRPD, respectively, for relatively simple products or for subsystems within a more complex product. The figures provide a good representation for suppliers of functional subsystems to original equipment manufacturers that deliver complex products. However, a different representation is needed for complex products.

The development of complex products usually involves a number of IDDOV projects in various points within the NRPD process, as depicted in Figure 44.5.

The activities in the verify and launch phase of IDDOV differ between the different applications. In the concept development phases, a lowercase v indicates the "conduct confirmation run" step of the robust optimization process. In the detailed design and development phases, V references all of the activities necessary to verify and validate product and process design, ramp-up production, and launch the product. In the verify and validate phase of NRPD, the activities indicated by V? are situational, as suggested by the question mark. Complexity determines how to align IDDOV projects with the product or subsystem under development.

PHASE PROCESSES

Xerox and numerous other corporations have encompassed off-line quality engineering (robust engineering) in their NPD [7]. In this section, the activities within each of the phases of NRPD are described briefly. The full significance of the NRPD process may not become evident until the compatible phase gate processes are introduced in "Phase Gate Processes."

In the first three phases of the NRPD process (preconcept, concept, concept optimization), the team focuses on getting the right product concept up-front for a flawless GOOD START transition from concept development activities to detailed product and process design. In the three detailed product and process design

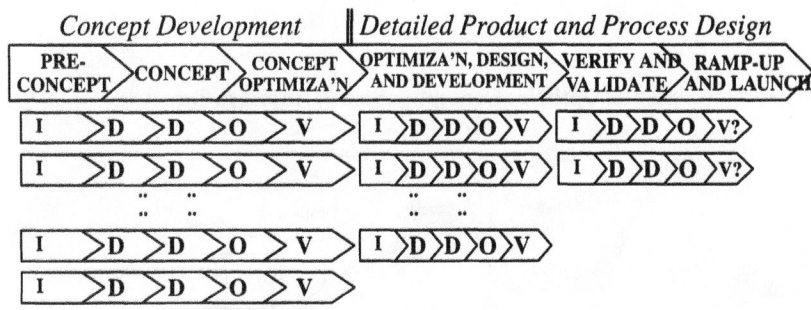

Figure 44.5
IDDOV projects conducted in different phases of NRPD for a complex product

phases (optimization, design, and development; verify and validate; ramp-up and launch), the team focuses on getting the product right.

1. *Preconcept.* Understand customer requirements and market potential, translate customer, company, and regulatory requirements into technical requirements (e.g., QFD), develop product description and business case.

2. *Concept.* Develop product (system) concept design that meets high-level requirements. Develop concepts/technology sets and identify carryover designs for all subsystems in the architectural hierarchy. Develop optimization plans.

3. *Concept optimization.* Two-step optimization dramatically simplifies the new product development process by separating optimization of technology sets and concepts from the challenges of meeting product specifications (Figure 44.6).

The first step of two-step optimization provides the opportunity first to optimize concept/technology set without regard to detailed product specifications. The optimized concept/technology set is subsequently set to product specific specifications in the second step.

1. Optimize the concept/technology set (determine the values of control factors to maximize the SN ratio).

2. Adjust the system to meet the specifications (select the adjustment factor that does not strongly affect the SN ratio and use it to set the system to target specifications).

For new concept designs and technology sets, step 1 is best performed in the concept optimization phase, and step 2 may be conducted in either the concept optimization phase or in the optimization, design, and development phase, depending on circumstances, such as knowledge of target applications. For subsystems, technologies, and concepts carried over from precedent products, optimization might be conducted in either the optimize concept phase or in the optimization, design, and development phase.

If the carryover subsystem has previously been optimized using robust optimization, it may only be necessary to conduct step 2 to set the subsystem to the new product specifications.

1. *Optimization, design.* Conduct detailed design, and optimize designs as needed.

2. *Development.* Conduct modeling, simulation, and bench tests; finalize product and process designs; and release drawings: build and test manufacturing intent prototypes.

Step 1: Optimize Concept   Step 2: Adjust to Target

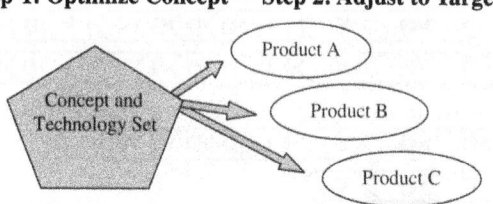

Figure 44.6
Optimized concept and technology set directed to different targets

3. *Verify and validate.* Finalize manufacturing process design including quality control. Taguchi's on-line quality engineering can usefully be applied in conducting this task. Evaluate manufacturing intent prototypes against performance, quality, reliability, and durability requirements under simulated customer usage conditions to verify product function and user friendliness. Conduct a pilot run to validate manufacturing and service processes.

4. *Ramp-up and launch.* Ramp-up production and launch product

PHASE GATE PROCESS

The phase gate and project review processes are closely linked. The project review process is broken out because it can be used at numerous points during the development process beyond the formal phase gates. The phase gate processes need to be designed to be compatible with the engineering and management processes. The ability to establish the technical sturdiness of the combination of a conceptual design and its technology set through robust optimization early in the development process opens up the opportunity to streamline the phase gate processes. The ability to verify the robustness of a concept and technology set decreases the likelihood of unanticipated downstream technical problems dramatically. Hence, the need for demoralizing, downstream go/kill[6]-style phase gates is largely eliminated. Proactive help/go-style phase gates become more effective in the context of early optimization. The intent of help/go-style phase gates is to help teams accelerate progress by resolving issues and demonstrating readiness to move forward into the next phase. Forward-looking preparation for entry into the next phase creates a more positive environment than backward-looking inquisitional evaluations against exit criteria.

The only go/(not-ready-yet or kill) phase gate resides between concept and technology activities and detailed product and process design activities, as indicated in Figures 44.4 and 44.5. In many corporations, the creative advanced development activities are conducted in different organizations by different teams from product development activities. The two-organization model makes the transition from conceptual development activities to detailed product and process design activities a significant event with the commitment of major resources. In the early 1980s, Xerox identified this event as GOOD START.

The primary criteria for a GOOD START are (1) business readiness and (2) technical readiness. *Business readiness* requires that completed business plans project financial attractiveness. *Technical readiness* requires that a strong product concept that meets upstream customer requirements and downstream technical requirements has been selected, and that technical feasibility based on mature technologies has been demonstrated.

The ability to optimize concepts and technology sets prior to transitioning to detailed product and process design supports the effective implementation of the GOOD START discipline. Robust optimization provides solid evidence about the strengths of concept designs and technology sets. This is an enormously important new capability. The engineering and management teams can know whether a new technology or a new concept design will be robust under actual customer usage conditions very early in the development process, before making the major resource commitment necessary to support detailed product and process design.

In addition, robust optimization assessment methods can be used to evaluate potential carryover designs prior to making reuse versus new concept design decisions. Robust optimization helps to remove uncertainties about technical readiness prior to GOOD START.

The GOOD START model features a strong management commitment to take the product to market with only two acceptable "showstoppers": (1) an unanticipated competitive action that makes the product nonfeasible in the marketplace, and (2) technical problems that cannot be recovered within the window of market opportunity.

A disciplined GOOD START with a hard commitment to go to market reduces the likelihood of killing a program for financial convenience, such as trying to fix this quarter's profit problems, cost overruns, staffing shortfalls, and a litany of other excuses.

The GOOD START model promotes continuity of effort, confidence, winning attitudes, and engineering excellence, in contrast to models that treat every phase gate as a go/kill decision. [6] When GOOD START is defined as the only go/(not-ready-yet or kill) phase gate, management and program/project team members are strongly motivated to get the potential program right up-front. The rigorous up-front process required to get through the GOOD START phase gate dramatically increases the probability that the program team will launch an outstanding product successfully.

PROJECT REVIEW PROCESS

A strong PROJECT REVIEW process fosters engineering excellence by ensuring excellence of execution of every step of the development process. A three-step project review process supports excellence of execution: (1) self-inspection, (2) peer (design) review, and (3) management review. The review process helps to ensure that all of the actions necessary to prepare for GOOD START have been executed properly. The project review process continues to ensure excellence of execution throughout the development process.

Team self-assessment should be facilitated by a team member. The facilitator role can be rotated among team members. Team members can assess their own work or ask for other team members to assess their work. The self-assessment should consider:

- ❏ Degree of excellence in carrying out process steps
- ❏ Did team complete their work or run out of time?
- ❏ Things done right and things gone wrong
- ❏ Ideas for improvement
- ❏ Were results world class? (While assessment should focus on execution of process, results, when available, provide a strong indication of excellence of execution. *In-process checks are leading indicators. Results are lagging indicators.*)

Document the assessment as a text document and presentation slides suitable for use as input to the expert peer review team. *Expert peer review (design review)* should be facilitated by an external person if possible. Reviewers should be external, neutral, and competent/expert in the field. The review team usually consists of two to four members. They should be reminded that their job is to help improve the process and the outcomes, not to criticize. The review team should assess the self-assessment, coherence of team, and external factors, including suppliers, customers, and management.

The output of the review should be submitted to the team leader, black belt, or master black belt. Document the assessment as a text document and presentation slides suitable for use as input for the management review.

Management reviews should be motivational and helpful. Makeup of the review committee might include top executives, involved managers, champions, and, when appropriate, supplier and customer representatives (i.e., the extended team).

Chair management reviews with a supportive style:

❑ Avoid inquisitional, backward-looking questions.

❑ Ask "How can I help?", forward-looking questions.

❑ Ask what, not who.

❑ Be a friend, not a command-and-control general.

❑ Drive out fear. *Fear stifles clear thinking, innovation, and bold actions.*

❑ Knock down barriers, remove obstacles, solve problems.

❑ Recognize mistakes as progress. *Suggest, only partially in jest: "Make mistakes as rapidly as possible to get them out of the way as quickly as possible."*

"Out-of-the-box" suggestions:

❑ Discourage dry runs. *Grant trust to the people, not to the system.*

❑ Encourage "chalk talks" on white boards and flipcharts.

Foster a passion for excellence. Seek to understand the quality of events. Do the team members feel that they were encouraged and given the time to pursue excellence in performing each step of the design process? Did team members perform a self-assessment and a peer review to ensure excellence prior to this management review?

A properly structured NRPD supported by IDDOV, GOOD START, and project review management processes creates a powerful development and engineering environment. Highly motivated teams with winning attitudes use the new processes and methodologies to deliver winning products that help a corporation profitably grow market share and revenue.

44.4. Technical Benefits of IDDOV

The success of quality engineering is legendary in Japan and is recognized increasingly in Western countries. This article was written at a time when six sigma was consuming management attention. Six sigma enjoys enormous success in improving existing processes to yield quick financial benefits. As a relative of six sigma, design for six sigma (IDDOV) is gaining recognition as the means for creating innovative, low-cost, trouble-free products on significantly shorter schedules. In addition, robust optimization, the "O" in IDDOV, is gaining recognition as a powerful problem-solving methodology for developing robust solutions that prevent the reappearance of problems.

The purposes of the three methodologies might be summarized as follows:

❑ Six sigma is an improvement process.

❑ Robust optimization is an improvement and prevention process.

❑ IDDOV is a creation, innovation, improvement, and prevention process.

In *six sigma,* improvement is achieved by digging down (*analyze*) into the details of an existing process or product to find and then eliminate the cause of the problem (*improve*). It is a powerful find-and-fix process that is helping many corporations to achieve huge performance and financial gains. Analyze and improve are two of the phases of the six sigma DMAIC process: define, measure, analyze, improve, control.

A difficulty with six sigma–style problem-solving processes, especially when applied to product problems, is that whacking down one problem frequently causes another problem to pop up. This good–bad dichotomy is somewhat pejoratively characterized as "whack-a-mole engineering." The game consists of a number of holes in a board with a mole's head sticking up through one of the holes. When one mole is hit on the head, another mole pops up out of another hole.

A number of actual case studies illustrate the equivalent engineering game. In one study, a team worked for many months to reduce the audible noise of a drive belt. But another problem popped up; the already short life of the belt was further shortened by a factor of 2. In another study, a team reduced the audible noise due to piston slap in a diesel engine. Problems of performance and fuel use popped up.

In *robust optimization,* improvement means something very different from its meaning in six sigma. Rather than seeking to eliminate an identified problem such as the audible noise in the aforementioned drive belt, robust optimization seeks to maximize the performance of the intended function of the product or process system. In the drive belt case study, robust optimization simultaneously eliminated the audible noise and doubled the life of the belt [7]. In the diesel engine example, optimizing engine efficiency dramatically reduced audible noise while improving performance and fuel use.

A simple explanation for working on the intended function rather than the problems (e.g., audible noise) has proved useful in helping nonexperts begin to accept the principle of signal-to-noise ratio based on energy thinking. Think of functions (intended and unintended) as energy transformations. By conservation of energy, maximizing the energy in the intended function (signal) minimizes the energy available (noise) to cause all unintended functions (all problems), not just audible noise. Hence, symptomatic problems such as inefficiency, noise, vibration, aging, wear, and heat are all improved by one set of robust optimization experiments to maximize signal-to-noise ratio. The tedious task of digging to find root causes of a symptomatic problem is bypassed.

Old AM radios provide a familiar example of signal and noise contributions to performance. Signal is the intended music. Static is the unintended noise. The total energy received by the radio is the sum of the energy that produces music, the signal, and the energy that produces static, the noise. Increasing the portion of total energy that produces music clearly reduces the portion of the total energy that is available to produce static. In an ideal radio, the total energy would go into producing music, leaving no energy to cause static. The function of the radio would be ideal. No static would be present to interfere with the music.

In robust engineering, if a function is performed perfectly, it is called the *ideal function.* Of course, the ideal function does not exist in our world. Try as we might, perpetual-motion machines do not exist. However, the measure of how far the value of the actual function differs from the value of the ideal function is a useful measure of robustness. It tells us about the relative volumes of music and static.

The objective of robust optimization is to maximize the relative volumes of music and static. This process moves the actual function as close as possible to the ideal function. The measure of robustness is the *signal-to-noise ratio,* the ratio of the energy in the intended function (signal) to the energy in the unintended functions (noise).

Assume that the ideal system response, y, is a linear relation to an M. Then the ideal function is $y = \beta M$. Now assume that the actual response to M is some function of M together with various system parameters, $x_1, x_2, \ldots, x_n$, then $y = f(M, x_1, x_2, \ldots, x_n)$. The actual function may be rearranged into two parts, the ideal function (useful part) and the deviation from the ideal function (harmful part) by adding and subtracting the ideal function, βM:

$$
\begin{array}{cc}
\text{Ideal} & \text{Deviation from} \\
\text{Function} & \text{Ideal Function}
\end{array}
$$

$$
y = f(M, x_1, x_2, \ldots, x_n) = \beta M + [f(M, x_1, x_2, \ldots, x_n) - \beta M]
$$

The ideal function represents all the radio signal energy going into the music with no energy available to cause static. The deviation from ideal is the portion of energy in the actual function that goes into causing static. Robust optimization is a methodology for maximizing the portion of energy that goes into the ideal function, which in turn minimizes the energy available to cause deviation from the ideal function. Referencing the functions above, robust optimization is the process of minimizing the deviation from the ideal function by finding the values of controllable x_i's that move the actual function, $f(M, x_1, x_2, \ldots, x_n)$, as close as possible to the ideal function, βM. When the value of the actual function is close to the value of the ideal function, the system is robust.

Some of the x_i's are usually not controllable, such as environmental and usage conditions, variations in materials and part dimensions, and deterioration factors such wear and aging. Uncontrollable factors are called *noise factors*. When the value of the actual function remains close to the values of the ideal function for all anticipated values of the noise factors, the system is robust in the presence of noise. Such a system is insensitive to sources of variation.

Taguchi uses the terms *useful part* and *harmful part* for ideal function and deviation from ideal function, respectively. The harmful part is the portion of energy that goes into causing problems. These terms provide the rationale for working on the useful part, the intended function of making music. Maximizing the energy in the useful part minimizes the energy available to go into the harmful part. The harmful part can cause multiple problems. Maximizing the actual function minimizes the harmful part that causes problems. Robust optimization is truly both an improvement and a prevention methodology. All moles are kept down permanently with a single whack.

In *IDDOV,* the methodologies for the creation of products and processes are built around the robust optimization methodology to formulate an engineering process that supports new product development processes described previously. In addition to being a creation, improvement, and prevention methodology, the develop concept phase of IDDOV provides the maximum opportunity for innovation.

IDDOV supports NRPD to deliver *"better, faster, cheaper"* benefits, including:

❏ Innovative, low-cost, reliable products that many customers around the globe purchase in preference to competitive offerings

❏ Low warranty cost and delighted customers

❏ Low development cost and reduced time to market

❏ Improved employee satisfaction and loyalty

❏ Growth of profit, revenue, and market share

The signal-to-noise (SN) ratio is, of course, the single measure of robustness, and gain in the SN ratio is the single measure of improvement. The SN ratio is an abstract measure with little meaning until it is related to *better, faster, cheaper* benefits.

In this section concerning technical benefits, the metrics selected to illustrate the relation between gain in SN to better, faster types of benefits include (1) range of variation and sigma level (better), (2) concept/technology set evaluation (better and faster), (3) failure rate (better), and (4) innovation (better and faster). *Cheaper* benefits are addressed appropriately in the financial benefits section.

Range of Variation and Sigma Level   A pictorial representation illustrating improvement in a manufacturing process achieved by conducting a robust optimization experiment is shown in Figure 44.7. A typical gain of 6 dB is assumed. The reduction in the range of variation depicted is given by the formula in the figure, $\sigma_{OP}/\sigma_{BL} = (\frac{1}{2})^{gain/6}$. When the gain equals 6 dB, the formula reduces to $6\sigma_{OP} = \frac{1}{2}\sigma_{BL}$ (a factor of 2 reduction in the range of variation).

The sigma level is related to the number of standard deviations (σ's) that fit within the control limits. The factor of 2 increase in sigma level from 3.5σ to 7σ is correlated with the increase in SN ratio along the vertical axis. The additional improvement to 8.5σ assumes that a 1.5σ shift existed that was eliminated by optimization.

The initial sigma level of 3.5σ was arbitrarily chosen as roughly representative of current manufacturing industry performance. The corresponding baseline SN ratio of 24 dB is also chosen arbitrarily, for illustrative purposes. For a troublesome

Figure 44.7
Improvement due to robust optimization

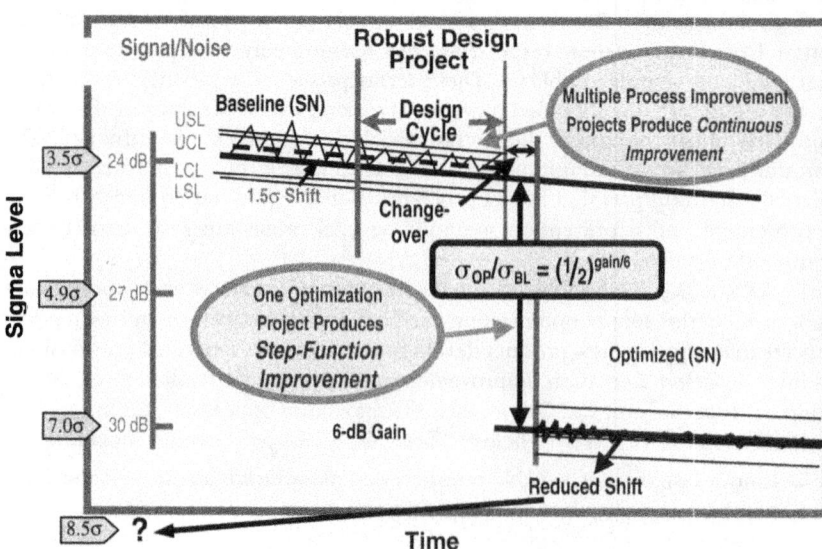

element of a system, a more representative choice for the initial sigma level might be 2σ; then the sigma level would increase to 4σ for a gain of 6 dB.

Dealing with the famous 1.5σ shift can be troublesome. Perhaps the shift was not present prior to optimization. Does it apply to distributions of variation in products as well as processes? Six sigma programs often assume that the 1.5σ shift is always present. This assumption does not work in the context of Taguchi methods.

MANUFACTURING PROCESS PERSPECTIVE

Both the similarities and the differences between Figure 44.7 and the Juran trilogy [8] shown in Figure 44.8 are striking. A small difference is the choice of metrics. The vertical axis in the Juran trilogy is the cost of poor quality, which, of course, is related directly to the SN and sigma level (added to figure) used in Figure 44.7. The most striking difference is the slight tilt in the control charts in Figure 44.7 that results from continuous improvement actions. As the range of variation is reduced over time, the sigma level increases to create the slight tilt.

Juran's trilogy contains three steps for improving quality:

1. *Quality planning.* Establish plans to improve quality to meet requirements.

2. *Quality control.* Get process in control before trying to improve it.

3. *Quality improvement.* Make improvement and maintain process control at new levels.

The Juran trilogy represents one action in a sequence of continuous improvement actions. The axis of the control chart remains horizontal between improvement actions. The 1.2σ improvement indicated in Figure 4.8 is normally regarded as a large improvement for a manufacturing process but a small step compared to the gain typically achieved using robust optimization. The tilted lines in Figure 44.7 represent a series of Juran trilogies (i.e., the continuous improvement that results from a sequence of small step-function improvements smoothed into lines).

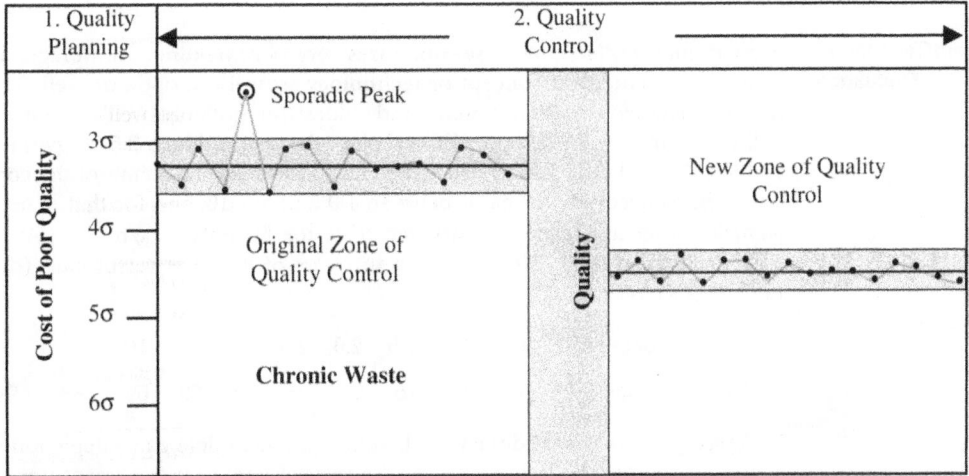

Figure 44.8
Juran's quality trilogy with representative sigma levels added (Adapted from Joseph M. Juran and A. Blanton Godfrey, *Juran's Quality Handbook*, McGraw-Hill, New York, 1999, Section 2.7.)

PRODUCT PERSPECTIVE

The part-to-part variation in a manufacturing process is one of three sources of variation. There are two other sources of variation in a product: internal and external noises. *Internal noise* includes aging, wear, vibration, heat, and so on; *external noise* includes environmental conditions and actual usage profiles.

The before-and-after variation depicted in Figure 44.7 can be regarded as representing a product in the presence of the three sources of variation: part-to-part variations, internal noise, and external noise. The depicted control charts are reinterpreted as tolerance charts that show variations in the context of specification limits (the manufacturing control limits are ignored since they are not appropriate to tolerance charts). Assume that the graphical representation of variation is made with data taken from an instrumented product operated under simulated usage conditions, as many manufacturers do as part of their internal evaluation processes.

The tolerance limits are determined by the degree of variation in performance that causes a user to take corrective action. Two types of failures are identified: functional failure (tow truck to service center) and performance degradation failure (drive to service center).

The slope of the tolerance charts relates to changes in the range of variation over time. The failure rate should change in the same way as range of variation does. The negative slope of the tolerance chart indicates improvement, a positive slope indicates degradation, and a zero slope indicates no change over time.

The objective of robust engineering and IDDOV is to maximize the robustness. Robustness is defined by Taguchi as "the state where the technology, product, or process performance is minimally sensitive to factors causing variability (either in manufacturing or user's environment) and aging at the lowest unit manufacturing cost."

With this definition, the slope of the tolerance charts should remain near zero, or horizontal, in the user's environment, with minimal variation in product performance (well within the upper and lower tolerance limits), as shown for the optimized control chart shown in the *lower* right-hand corner of Figure 44.7.

Concept/Technology Set Evaluation

A good concept/technology set optimizes very well, yielding, on average, a 6-dB gain. A poor choice of concept or technology set will not optimize well. The distinction between "optimizes well" and "does not optimize well" is not a well-defined boundary. Rough guidelines suggest that more than a 2.5-dB gain is good and less than a 1.0-dB gain is bad. *Good* and *bad* refer to the quality of the concept. Judgment is necessary for gains between 1.0 and 2.5 dB, and for that matter, any gain. Relations are readily constructed from the formula, $\sigma_{OP}/\sigma_{BL} = (\frac{1}{2})^{gain/6}$, to help make judgments. The reduction in range of variation versus gain (dB) for $(\frac{1}{2})^{gain/6}$ is as follows:

Gain (dB):	0.1	1.0	1.5	2.0	2.5	3.0	6.0	10.0	18.0	24.0
Range reduction (%):	1	11	16	21	25	29	50	69	88	94

Range of variation provides a useful, although incomplete visual depiction of the degree of robustness. In Figure 44.9, the vertical goal posts represent upper and lower tolerance limits. The separation between the goalposts, USL − LSL, is the design tolerance around some target $\pm\Delta$, where 2Δ = USL − LSL. The chart

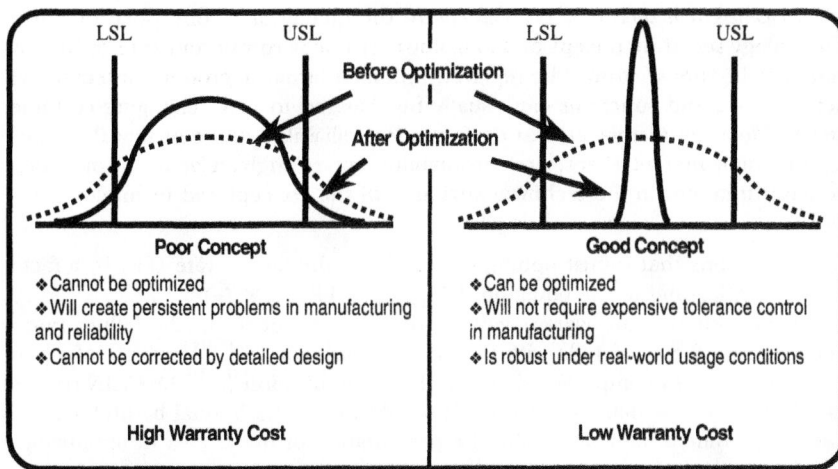

Figure 44.9
Poor concepts cannot
be recouped

shows variation compared to tolerance specifications. The distributions illustrated may represent product or process variations.

Robustness is a valuable metric for detecting poor concepts or technology sets. A concept/technology set that does not optimize well is probably a poor choice of either the conceptual design or the technology set.

Poor concepts or technology sets are the sources of chronic problems that reappear program after program to provide permanent employment for expensive armies of firefighters. Since neither poor concepts nor technology sets can be recouped by brilliant detailed design or firefighting, the only solution is to throw out the old design and invest in the development of a new concept or technology set that does optimize well. Throwing out a design is always a difficult decision with potentially major near-term financial consequences.

Most corporate environments, accidentally or intentionally, place higher priority on *urgency* about near-term cost avoidance than on *importance* about longer-term financial benefits. Costs are immediate and certain, while benefits are longer-term and uncertain. Near-term cost pressures to salvage an existing concept or technology set tend to outweigh potential, longer-term financial benefits of reduced warranty cost and improved customer satisfaction. *The urgent always pushes out the important.* The actual costs of fruitless and endless efforts to salvage unsalvageable concepts are usually not very visible.

In addition to plaguing current field performance, bad concepts get carried over into next-generation offerings to plague future field performance. As development schedule and cost pressures increase, so do the pressures to carry over precedent designs into the next development program. Development engineers like to create designs; that is, after all, what they do. In addition, downstream firefighters usually persist with the belief that they can fix the problems with the current concept or technology set; that is, after all, what they do. In the presence of conflicting inputs from well-intended and respected development and firefighting engineers, together with the absence of objective measures about the potential quality of a concept or technology set, managers have little choice beyond doing whatever seems necessary to fix and maintain the current concept and technology set.

Robustness is a conclusive discriminator of the quality of a conceptual design or technology set. If a concept or technology set is not robust and cannot be optimized to be robust, it must be replaced with an alternative concept or technology set. Features and functions can usually be changed to meet changing customer needs. However, factors such as performance, reliability, and cost that determine growth rate, market share, and profitability are strongly related to robustness, which is inherent in the technical attributes of the concept and technology set.

Failure Rate   Taguchi asserts that robust optimization reduces the failure rate (FR) by a factor of $(\frac{1}{2})^{\text{gain}/3}$, leading to the equation, $\text{FR}_{\text{optimized}} = \text{FR}_{\text{initial}} \times (\frac{1}{2})^{\text{gain}/3}$.

Again, assuming an average gain in the SN ratio of 6 dB, the failure rate is reduced by a factor of 4. As with the range of variation, the failure rate of a poor concept may not be improved significantly by optimization (i.e., the GAIN realized is substantially less than 1.5 dB). In Figure 44.10, the traditional bathtub curve is used to provide a familiar graphical representation of the impact of optimizing a good concept design.

Many teams build in a safety factor by using the more conservative gain/6 rather than the gain/3 relation to project improvements in failure rates. In Figure 44.10, the more conservative gain/6 relation is used, which yields a factor of 2 reduction in failure rate rather than the more aggressive gain/3 relation, which would yield a factor of four improvement. Whether to use gain/3 or gain/6 to make failure rate predictions is a matter of judgment.

The $(\frac{1}{2})^{\text{gain}/3}$ or $(\frac{1}{2})^{\text{gain}/6}$ factor applies to the failure rate, not initial quality, reliability, or durability, which are identified in Figure 44.10 with portions of the bathtub curves. Reliability is calculated from the failure rate. Condra [9] defines reliability as "quality over time," a definition that Juran repeats in his handbook [8]. For a constant failure rate typical of electronic systems, reliability, $R(t) = e^{-\text{FR} \cdot t}$.

For failure rates that change over time, FR(t), typical of mechanical systems and certain electronic components, reliability, $R(t)$, changes with time according to the formula $R(t) = \exp[-\int \text{FR}(t) \, dt]$, where the integral range is 0 to t. Reliability predictions are typically made using the Weibull function.

In either case, changes in the failure rate [normally called the *hazard rate, h(t)*, in reliability engineering] are related by an exponential function to changes in reliability. The brief discussion about reliability is presented only for the purpose of clarifying how the failure-rate bathtub curves in Figure 44.10 relate to Taguchi's $(\frac{1}{2})^{\text{gain}/3}$ factor.

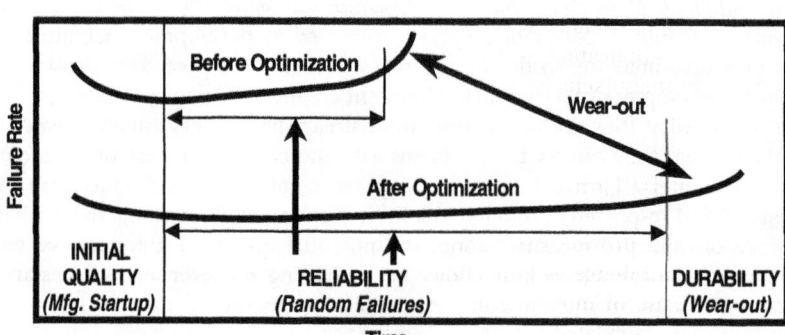

Figure 44.10
Good concepts optimized to deliver improved initial quality, reliability, and durability

Robust optimization of a good concept is the most effective way to improve reliability. Reliability engineering is more about predicting reliability from data collected from measurements of portions of the product than it is about improving reliability. Condra [9] provides an early application of Taguchi methods as the best way to improve reliability in the context of reliability engineering.

Find-and-fix improvement efforts may be constrained by major sunk investments in tooling, and so on, that can severely limit the selection of control factors. If control factors with a major impact on robustness are not available, the amount of improvement may be disappointing. This said, the average gain realized, even with the constraints of "fixing" existing products, appears to be around the average of 6 dB. Perhaps the aforementioned build–test–fix process left too many parameters just inside the goalpost specs so that even constrained optimization experiments yielded significant gains in the SN ratio. Surely even better performance could be achieved by conducting robust optimization experiments during the development process.

R&D labs are seldom equipped or funded to fully evaluate and maturate promising **Innovation** innovations to levels suitable for product applications. R&D teams tend to work around these constraints by tying emerging technologies and innovations to product development programs. This high-risk practice contributes to schedule slips and canceled programs. Understandably, after several disappointments, management teams become reticent to take innovations to market. Over time, the rate of innovation slows to a snail's pace.

Robust optimization fosters innovation, a benefit of robust engineering that is seldom heralded. Taguchi methods do not cause the word *innovation* to spring to mind. Nevertheless, the ability to conduct robust optimization experiments on concepts and technology sets in R&D or during the concept phases of NRPD prior to GOOD START significantly reduces the risk of implementing innovations into a development program.

Robust optimization can change the way that R&D is conducted and provide a way out of the "innovation paralysis" trap. [6] The major challenge for most R&D teams is in devising a noise strategy that represents real-world conditions that may not be familiar to the teams. Technology sets and concepts can be optimized efficiently without regard to product specific requirements. If an innovative concept/technology set optimizes well, it can be taken to market with minimal risk. Successes breed confidence, and confidence will again open the spigots of innovation.

Progressive corporations are beginning to move application of robust optimization upstream into R&D laboratories to help accelerate the rate of innovation. With increasing upstream applications, corporations are beginning to take advantage of the full application range of Taguchi methods from research through engineering and manufacturing to products already in the market.

44.5. Financial Benefits of IDDOV

Taguchi methods provide new ways to both reduce cost and make financial projections with confidence. A critical factor propelling the popularity of six sigma is the consistent use of money as the measure of success. The financial impact of every project is calculated. In typical applications of six sigma such as reducing scrap and rework in manufacturing, the calculations are straightforward using cost of quality methods: cost of (poor) internal quality, cost of (poor) external quality, and cost of lost opportunity.

Major impacts of DFSS are derived from product improvements. Relating product design changes to financial benefits can be very difficult. For example, projections of reduction in warranty cost that result from elimination or reduction of root causes depend on the difficult task of relating design changes to reductions in the frequency of occurrence (failure rate) of the problem. In contrast, robust optimization provides an explicit formula that relates gain in signal-to-noise ratio to failure rate, which is readily converted to warranty cost.

The intent of this section is to indicate the types of financial benefits that can be calculated with confidence, with occasional pointers about how to carry out the easier calculations. It is not intended as a complete treatise on making financial projections.

Three types of financial benefits are described: (1) warranty/cost service, (2) product cost, and (3) loss to society (balance between internal and external cost).

Warranty Cost/ Service Cost Calculations of field cost start with the now familiar relation for failure rate. To illustrate a reasonable assumption, the failure rate formula is adjusted by taking an exponential factor between gain/6 and gain/3 (e.g., gain/4.5):

$$FR_{new} = FR_{initial} \times (\tfrac{1}{2})^{gain/4.5}$$

This relationship provides the basis for making financial projections of service cost. Service cost is inclusive of warranty cost and customer cost.

Define annual service cost = (no. failures/year) $\times$ (average cost per failure)

Then

$$\text{annual service cost}_{new} = \text{annual service cost}_{initial} \times (\tfrac{1}{2})^{gain/4.5}$$

Assuming an initial annual service cost of \$500, a gain of 6 dB reduces the annual service cost to \$200:

$$\text{annual service cost}_{new} = \$500 \times (\tfrac{1}{2})^{6/4.5} = \$200$$

This simple means of projecting changes in service cost (warranty and customer cost) provides better estimates than more complex alternative methods.

Product Cost Taguchi suggests that when the gain in SN ratio is sufficiently large, half of the gain should be allocated to reduce cost. Consider, for example, an automotive rotor/brake pad subsystem. Optimization motivated by need to reduce audible noise yielded a 20% improvement in performance in addition to eliminating the audible noise problem. The 20% performance improvement was captured as a 20% weight reduction that translated into about a 12% cost reduction.

Loss to Society Minimize loss to society using tolerance design to balance internal company cost and external field cost. External field cost may be warranty cost or customer cost. For example, starting with current internal cost, balance company and customer costs by conducting tolerance design using the quality loss function. The methodology is too extensive to present here. Tolerance design is developed elsewhere in this handbook.

Taguchi suggests starting with the lowest-cost materials and parts as reasonable and upgrade only as necessary to minimize loss to society. This approach counters the engineer's tendency to overspecify materials and parts because it is "safe," albeit, potentially expensive. Again, tolerance design provides the means to determine what upgrades are needed.

44.6. Taming the Intersection of Six Sigma and Robust Engineering

Six sigma was born in Motorola as a structured methodology for improving existing manufacturing processes. Over the years, application of six sigma has been extended successfully to business and service processes. Mikel Harry is broadly recognized as the primary force behind popularizing six sigma [10]. He originally defined eight phases: recognize, define, measure, analyze, improve, control, standardize, and integrate with measure, analyze, improve, and control (MAIC) as the core.

Five phases have become the most frequently used structure for six sigma:

1. *Define.* Define project scope and plan.
2. *Measure.* Conduct measurement system evaluation, measure defect levels, understand customer needs.
3. *Analyze.* Dig, dig, dig in search of "root causes."
4. *Improve.* Eliminate root causes.
5. *Control.* Hold the gains through process control.

For more detail about six sigma in an easy-to-read, entertaining style, see *The Power of Six Sigma* [11]. The six sigma DMAIC process is usefully represented as a block arrow diagram to indicate the temporal sequence of activities.

DMAIC, IDDOV, and NPD block arrow diagrams represent temporal sequences of activities. However, the logic of the DMAIC process is fundamentally different. DMAIC consists of a sequence of activities for digging vertically down into the details of some portion of an existing product or process to find and eliminate the causes.

DMAIC and IDDOV intersect NPD differently. An unconventional representation of the relationship between DMAIC and NPD is depicted in Figure 44.11. DMAIC is oriented vertically to reflect its application for digging down into lower levels of an existing product to find and correct causes of problems. Define,

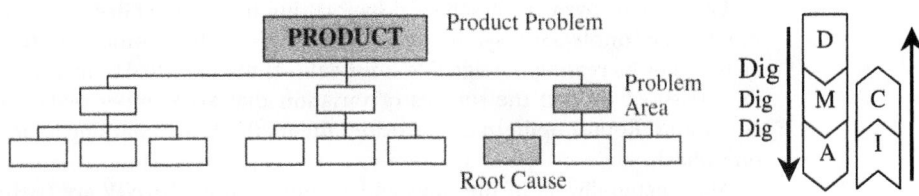

Figure 44.11
System architecture block diagram

measure, and analyze (DMA) flow vertically downward in search of root causes. *Improve* (*I*) flows vertically upward as improvements are made at lower levels and verified at higher system levels. *Control* (*C*) refers to the processes put in place to hold the gains.

The unconventional representation is intended to bring to mind logical rather than temporal relationships. DMAIC is used to find and fix problems in existing processes and products. Product applicability ranges from later portions of NPD when first-instance prototypes begin to exist and extend through the entire life cycle of customer use.

Complex products are typically represented as a hierarchy of system elements in two-dimensional block diagrams, as depicted in Figure 44.11. Define, measure, and analyze are applied to dig vertically down into increasing detail of a system to identify the causes of problems in existing processes. A hypothetical path from the product level to causes is indicated by the shaded system elements.

Once the cause is located, the methods within the improve phase are used to eliminate the cause. The process of verifying the improvement proceeds sequentially upward through the higher-level system elements to the product level. Then the methods within the control phase are utilized to hold the gains.

This unconventional representation of the relationship between DMAIC and NPD grew out of frustration from unsuccessful efforts to explain the differences between IDDOV and DMAIC to candidate DFSS black belts and to six sigma master black belts. The master black belts with the deepest knowledge of six sigma had the greatest difficulty breaking their "fixation" on the DMAIC paradigm. The accepted explanations that IDDOV is a creation process and DMAIC is an improvement process did not quell the continuing stream of questions that could extend for weeks. Analogies, different explanations, and alternative pictures did not seem to help. Finally, the vertical representation of DMAIC created "aha's" of understanding at first sight.

DMAIC, founded on statistical processes used for controlling and improving manufacturing processes, has proved to be a very effective methodology for improving business, engineering, manufacturing, and service processes. Its effectiveness as a find-and-fix methodology for product problems is arguably less spectacular.

Finding and fixing the causes of product problems is a process of working on the symptoms of a fundamental weakness. Two types of disappointments often occur, identified here as type C and type WAM. Type C is the "chronic" recurrence of the same problem in program after program. Type WAM is the previously identified "whack-a-mole" engineering process of whacking down one problem only to have another problem pop up (e.g., fix audible noise but shorten life). Both types of problems have a common origin: lack of robustness.

Lasting improvement is achieved by focusing on the intended function, not by hiding the unpleasant symptoms of dysfunction. Improving the function is achieved by increasing the signal-to-noise ratio of the intended function to increase its robustness against the sources of variation that arise under real-world usage conditions. Robust optimization provides the methods for whacking all moles with one whack.

More generally, all of the tools and methods within IDDOV are useful in the DMAIC process when applied to product improvement. The following case study illustrates the use of a portion of the IDDOV tool set in a DMAIC-like find-and-fix process. The case study involves a type C chronic, recurring problem. Because the source of the problem was found to be in the software, the choice of control factors was not constrained by sunk investments in tooling. The freedom to select the appropriate control factors facilitated a complete resolution of the problem.

The case study entitled "Optimization of a Diesel Engine Software Control Strategy" by Larry R. Smith and Madhav S. Phadke [12] was presented at the American Supplier Institute Consulting Group, 19th Annual Taguchi Symposium, held in 2002. Genichi Taguchi selected the paper for the Taguchi Award as the symposium's best case study.

Case Study: Robust DMAIC

The case study addresses a long-standing, intermittent problem in diesel engine power train control system called *hitching* under cruise control and *ringing* in idle mode. Hitching and ringing refer to oscillations in engine rpm. The intermittent character and relatively rare occurrence of the problem had confounded engineers in numerous previous attempts to solve the problem.

The essence of the six sigma style case study was the unusual use of engineering methodologies in a DMAIC-like find-and-fix process to eliminate completely a long-standing and expensive problem that many previous efforts failed to fix. TRIZ was used to analyze the problem and identify software as the location of the cause. Robust optimization was used to improve performance by eliminating the problem permanently. Elimination of the hitching/ringing problem saved $8 million in warranty cost.

A fascinating element in the case study that is not emphasized in the paper is the method used to develop a noise strategy. Several engineers had the patience to drive several trucks for extended periods under a variety of conditions. Different driving profiles constitute important noise factors because they cause major changes to the load on the engine. The noise factors were selected from driving conditions that the team of drivers eventually found that would initiate hitching. Six noise factors were identified that related to acceleration and deceleration in 1-mph increments for different speed bands and rolling hills at different speeds.

The case study illustrates the potential power of an engineering-oriented DMAIC process for improving existing products. The notion is straightforward. The DMAIC process works best when engineering methods are used to improve existing engineered products and statistical methods are used to improve existing statistical processes. The heart of the new robust six sigma process would be robust engineering just as it is in IDDOV. The new process would have the power to change find and fix, fix, fix a chronic, reoccurring problem into find and eradicate the problem and prevent its reoccurrence, provided that the selection of control factors was not overly constrained.

In the conclusions section, the authors state:

Many members of this team had been working on this problem for quite some time. They believed it to be a very difficult problem that most likely would never be solved. The results of this study surprised some team members and made them believers in the Robust Design approach. In the words of one of the team members, "When we ran that confirmation experiment and there was no hitching, my jaw just dropped. I couldn't believe it. I thought for sure this would not work. But now I am telling all my friends about it and I intend to use this approach again in future situations."

44.7. Conclusions

The dramatic benefits of robust optimization and IDDOV are characterized through quantification of improvements in terms of familiar metrics such as sigma level, product failure rate, and financial projections assuming the average gain in an SN ratio of 6 dB.

In addition, four new developments are introduced:

1. The IDDOV formulation of design for six sigma is shown to be a powerful implementation of Taguchi's quality engineering.
2. A new NPD process is described that takes advantage of IDDOV and two-step optimization.
3. An explicit formula for projecting reduction in warranty cost due to optimization is presented.
4. Innovation is fostered by the ability to optimize concepts and technology sets to reduce risk prior to a GOOD START commitment in NPD.

Finally, the tantalizing notion of creating a new robust DMAIC process focused on improving existing engineered products is introduced. Although this chapter focused only on product applications, nearly all of the notions and methodologies discussed are equally applicable to business, engineering, manufacturing, and service processes.

ACKNOWLEDGMENTS

The incredibly talented members of the ASI CG family made this chapter possible. I am especially indebted to my friends and colleagues, Subir Chowdhury, Shin Taguchi, Alan Wu, and Jim Wilkins for support, suggestions, and corrections. We are all indebted to Genichi Taguchi for his continuing stream of many outstanding contributions focused on helping corporations and engineers reduce the loss to society. I am also indebted to my good friend, Larry R. Smith, for continuing support over many years and for reading this manuscript and making valuable suggestions.

References

1. Subir Chowdhury, 2002. *Design for Six Sigma: The Revolutionary Process for Achieving Extraordinary Profits.* Chicago: Dearborn.
2. Genichi Taguchi, Subir Chowdhury, and Shin Taguchi, 2000. *Robust Engineering: Learn How to Boost Quality While Reducing Cost and Time to Market.* New York: McGraw-Hill.
3. Yuin Wu and Alan Wu, 2000. *Taguchi Methods for Robust Design.* New York: ASME Press.
4. Stuart Pugh, 1990. *Total Design: Integrated Methods for Successful Product Engineering.* Harlow, Essex, England: Addison-Wesley.
5. Genichi Taguchi, 1993. *Taguchi on Robust Technology Development: Bringing Quality Engineering Upstream.* New York: ASME Press.
6. Robert G. Cooper, 1993. *Winning at New Product Development: Accelerating the Process from Idea to Launch.* Reading, MA: Addison-Wesley.

7. How to bring out better products faster. Industrial Management & Technology issue of *Fortune*, November 23, 1998.

8. Joseph M. Juran and A. Blanton Godfrey, 1998. *Juran's Quality Handbook*, 5th ed. New York: McGraw-Hill p. 2.7.

9. Lloyd W. Condra, 1993. *Reliability Improvement with Design of Experiments*. New York: Marcel Dekker.

10. Mikel Harry and Richard Schroeder, 2000. *Six Sigma: The Breakthrough Management Strategy Revolutionizing the World's Top Corporations*. New York: Doubleday.

11. Subir Chowdhury, 2001. *The Power of Six Sigma: An Inspiring Tale of How Six Sigma Is Transforming the Way We Work*. Chicago: Dearborn.

12. Larry R. Smith and Madhav S. Phadke, 2002. Optimization of a diesel engine software control strategy. *Proceedings of the 19th Annual Taguchi Symposium*, sponsored by the American Supplier Institute Consulting Group.

This chapter is contributed by Barry Bebb.

Appendixes

Appendix A

Orthogonal Arrays and Linear Graphs: Tools for Quality Engineering*

Introduction   1526
Orthogonal Arrays   1526
   Assigning Interactions between Two Columns   1527
   Some General Considerations   1527
Linear Graphs   1527
   Linear Graph Symbols   1527
Suggested Steps in Designing Experiments   1528
General Considerations   1529

Arrays $L_4(2^3)$ 1530
$L_8(2^7)$ 1530
$L_{12}(2^{11})$ 1531
$L_{16}(2^{15})$ 1532
$L_{32}(2^{31})$ 1536
$L_{64}(2^{63})$ 1550
$L_9(3^4)$ 1564
$L_{18}(2^1 \times 3^7)$ 1564
$L_{27}(3^{13})$ 1565
$L_{54}(2^1 \times 3^{25})$ 1568
$L_{81}(3^{40})$ 1570
$L_{16}(4^5)$ 1586
$L_{32}(2^1 \times 4^9)$ 1587
$L_{64}(4^{21})$ 1588
$L_{25}(5^6)$ 1591
$L_{50}(2^1 \times 5^{11})$ 1592
$L_{36}(2^{11} \times 3^{12})$ 1593

*This material is from the appendixes of Genichi Taguchi's *System of Experimental Design*
(White Plains, NY: Unipub/American Supplier Institute, 1987).

$L_{36}(2^3 \times 3^{13})$ 1594

$L_9'(2^{21})$ 1595

$L'_{27}(3^{22})$ 1596

Introduction

This appendix has been produced to provide a convenient and easily accessible reference source for students and practitioners of quality engineering as developed by Dr. Genichi Taguchi. It contains the most commonly used orthogonal arrays and applicable linear graphs. The specific arrays and graphs were compiled by Dr. Taguchi and Shozo Konishi, and the American Supplier Institute wishes to acknowledge the importance and value of their contribution to quality engineering.

Experience has shown that students and practitioners in plants need a separate and convenient collection of the arrays and graphs to save time when developing experiments. The American Supplier Institute has, therefore, produced this appendix with the invaluable assistance of Professor Yuin Wu, Shin Taguchi, and Dr. Duane Orlowski.

Orthogonal Arrays

In design of experiments, orthogonal means "balanced," "separable" or "not mixed." A major feature of Dr. Taguchi's utilization of orthogonal arrays is the flexibility and capability for assigning a number of variables to them. An even more important feature, however, is the "reproducibility" or "repeatability" of the conclusions drawn from small scale experiments, in research and development work for product/process design, as well as in actual production and in the field.

Orthogonal arrays have been in use for many years, but their application by Dr. Taguchi has some special characteristics. At first glance, Dr. Taguchi's methods seem to be nothing more than an application of Fractional Factorials. In his approach to quality engineering, however, the primary goal is the optimization of product/process design for minimal sensitivity to "noise."

In Dr. Taguchi's methodology, the main role of an orthogonal array is to permit engineers to evaluate a product design with respect to "robustness" against "noise," and the cost involved. It is, in effect, an inspection device to prevent a "poor design" from going "downstream."

Quality engineering, according to Dr. Taguchi, focuses on the contribution of factor levels to robustness. In contrast with pure research, quality engineering applications in industry do not look for cause and effect relationships. Such a detailed approach is usually not cost-effective. Rather, good engineers should be able to involve their skills and experience in designing experiments by selecting characteristics with minimal interactions in order to focus on pure main effects.

Dr. Taguchi stresses the need for a confirmatory experiment. If there are strong interactions, a confirmatory experiment will show them. If predicted results are not confirmed, the experiment must be re-developed.

The convention for naming arrays is $L_a(b^c)$ where a is the number of experimental runs, b the number of levels of each factor, and c the number of columns in the array.

Arrays can have factors with many levels, although two and three level factors are most commonly encountered. An L_8 array, for example, can handle seven factors at two levels each, under eight experimental conditions.

L_{12}, L_{18}, L_{36}, and L_{54} arrays are among a group of specially designed arrays that enable the practitioner to focus on main effects. Such an approach helps to increase the efficiency and reproducibility of small scale experimentation as developed by Dr. Taguchi.

As noted above, Dr. Taguchi recommends that experiments be designed with emphasis on main effects, to the extent possible. Because there will be situations in which interactions must be analyzed, Dr. Taguchi has devised specific triangular tables for certain arrays to help practitioners assign factors and interactions in a non-arbitrary method to avoid possible confounding.

Assigning Interactions between Two Columns

There are six such triangular matrices or tables in this appendix for arrays $L_{16}(2^{15})$, $L_{32}(2^{31})$, $L_{64}(2^{63})$, $L_{27}(3^{13})$, $L_{81}(3^{40})$, and $L_{64}(4^{21})$. These triangular tables have been placed on pages facing their respective arrays for convenient reference.

The method for using them is as follows. Let's use array $L_{16}(2^{15})$ and its triangular table as an example. When assigning an interaction between two factors, designated A and B, factor A is assigned to column 1 of the array, and factor B to column 4. Their interaction effect is assigned to column 5. Note the triangular table. Where row (1) and column 4 intersect is the number 5. This indicates that the interaction $A \times B$ is assigned to column 5. For factors assigned to columns 5 and 4, their interaction would be assigned to column 1.

❑ As conditions permit, focus on main effects.

❑ The most frequently used arrays are: L_{16}, L_{18}, L_{27}, and L_{12}.

Some General Considerations

Linear Graphs

One of Dr. Taguchi's contributions to the use of orthogonal arrays in the design of experiments is the linear graph. These graphs are pictorial equivalents of triangular tables or matrices that represent the assignment of interactions between all columns of an array. Linear graphs are, thus, a graphic tool to facilitate the often complicated assignment of factors and interactions, factors with different numbers of levels, or complicated experiments such as pseudo factor design to an orthogonal array.

A linear graph is used as follows:

1. Factors are assigned to points of the graph.

2. An interaction between two factors is assigned to the line segment connecting the two respective points.

The special symbols and the groups they represent are used for "split unit" design. For all arrays except $L_{32}(2^{31})$ and $L_{64}(2^{63})$, we use the following designations:

Linear Graph Symbols

○	Group 1
◎	Group 2
◉	Group 3
●	Group 4

For $L_{32}(2^{31})$ and $L_{64}(2^{63})$, we use the following designations:

○	Groups 1 and 2
◎	Group 3
◉	Group 4
●	Group 5

Suggested Steps in Designing Experiments

1. Define the problem.

   _____ Express a clear statement of the problem to be solved.

2. Determine the objective.

   _____ Identify output characteristics (preferably measurable and with good additivity).

   _____ Determine the method of measurement (may require a separate experiment).

3. Brainstorm.

   _____ Group factors into control factors and noise factors.

   _____ Determine levels and values for factors.

4. Design the experiment.

   _____ Select the appropriate orthogonal arrays for control factors.

   _____ Assign control factors (and interactions) to orthogonal array columns.

   _____ Select an outer array for noise factors and assign factors to columns.

5. Conduct the experiment and collect the data.

6. Analyze the data by:

Regular Analysis	SN Analysis
Average response tables	SN response tables
Average response graphs	SN response graphs
ANOVA	SN ANOVA

7. Interpret results

   _____ Select optimum levels of control factors. (For nominal-the-best use mean response anlaysis in conjunction with SN analysis.)

   _____ Predict results for the optimal condition.

8. ALWAYS–ALWAYS–ALWAYS run a confirmatory experiment to verify predicted results.

   _____ If results are not confirmed or are otherwise unsatisfactory, additional experiments may be required.

General Considerations

- ❏ It is generally preferable to consider as many factors (rather than many interactions) as economically feasible for the initial screening.
- ❏ Remember that L_{12} and L_{18} arrays are highly recommended, because interactions are distributed uniformly to all columns.

$L_4 (2^3)$

No.	1	2	3
1	1	1	1
2	1	2	2
3	2	1	2
4	2	2	1
	a	b	a b
Group	1	2	

(1) 1 o——3——2 ⊚

$L_8 (2^7)$

No.	1	2	3	4	5	6	7
1	1	1	1	1	1	1	1
2	1	1	1	2	2	2	2
3	1	2	2	1	1	2	2
4	1	2	2	2	2	1	1
5	2	1	2	1	2	1	2
6	2	1	2	2	1	2	1
7	2	2	1	1	2	2	1
8	2	2	1	2	1	1	2
	a	b	a b	c	a c	b c	a b c
Group	1	2			3		

(1)   (2)

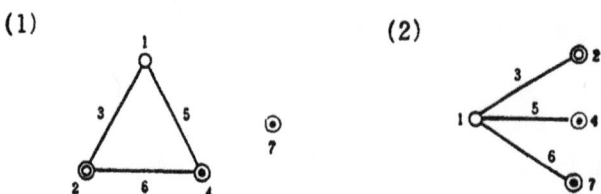

$$L_{12} (2^{11})$$

No.	1	2	3	4	5	6	7	8	9	10	11
1	1	1	1	1	1	1	1	1	1	1	1
2	1	1	1	1	1	2	2	2	2	2	2
3	1	1	2	2	2	1	1	1	2	2	2
4	1	2	1	2	2	1	2	2	1	1	2
5	1	2	2	1	2	2	1	2	1	2	1
6	1	2	2	2	1	2	2	1	2	1	1
7	2	1	2	2	1	1	2	2	1	2	1
8	2	1	2	1	2	2	2	1	1	1	2
9	2	1	1	2	2	2	1	2	2	1	1
10	2	2	2	1	1	1	1	2	2	1	2
11	2	2	1	2	1	2	1	1	1	2	2
12	2	2	1	1	2	1	2	1	2	2	1
Group	1					2					

The $L_{12} (2^{11})$ is a specially designed array, in that interactions are distributed more or less uniformly to all columns. Note that there is no linear graph for this array. It should not be used to analyze interactions. The advantage of this design is its capability to investigate 11 main effects, making it a highly recommended array.

$L_{16} (2^{15})$

No.	1	2	3	4	5	6	7	8	9	10	11	12	13	14	15
1	1	1	1	1	1	1	1	1	1	1	1	1	1	1	1
2	1	1	1	1	1	1	1	2	2	2	2	2	2	2	2
3	1	1	1	2	2	2	2	1	1	1	1	2	2	2	2
4	1	1	1	2	2	2	2	2	2	2	2	1	1	1	1
5	1	2	2	1	1	2	2	1	1	2	2	1	1	2	2
6	1	2	2	1	1	2	2	2	2	1	1	2	2	1	1
7	1	2	2	2	2	1	1	1	1	2	2	2	2	1	1
8	1	2	2	2	2	1	1	2	2	1	1	1	1	2	2
9	2	1	2	1	2	1	2	1	2	1	2	1	2	1	2
10	2	1	2	1	2	1	2	2	1	2	1	2	1	2	1
11	2	1	2	2	1	2	1	1	2	1	2	2	1	2	1
12	2	1	2	2	1	2	1	2	1	2	1	1	2	1	2
13	2	2	1	1	2	2	1	1	2	2	1	1	2	2	1
14	2	2	1	1	2	2	1	2	1	1	2	2	1	1	2
15	2	2	1	2	1	1	2	1	2	2	1	2	1	1	2
16	2	2	1	2	1	1	2	2	1	1	2	1	2	2	1
	a	b	a	c	a	b	a	d	a	b	a	c	a	b	a
			b		c	c	b		d	d	b	d	c	c	b
							c				d		d	d	c
															d
Group	1	2		3				4							

$L_{16} (2^{15})$ Interactions between Two Columns

	1	2	3	4	5	6	7	8	9	10	11	12	13	14	15
	(1)	3	2	5	4	7	6	9	8	11	10	13	12	15	14
		(2)	1	6	7	4	5	10	11	8	9	14	15	12	13
			(3)	7	6	5	4	11	10	9	8	15	14	13	12
				(4)	1	2	3	12	13	14	15	8	9	10	11
					(5)	3	2	13	12	15	14	9	8	11	10
						(6)	1	14	15	12	13	10	11	8	9
							(7)	15	14	13	12	11	10	9	8
								(8)	1	2	3	4	5	6	7
									(9)	3	2	5	4	7	6
										(10)	1	6	7	4	5
											(11)	7	6	5	4
												(12)	1	2	3
													(13)	3	2
														(14)	1

Linear Graphs for L_{16} (2^{15})

(1)

a

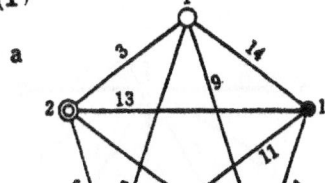

(2)

a

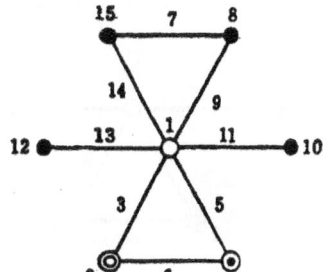

b

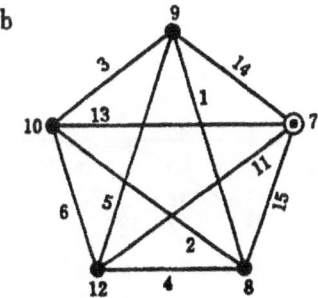

b

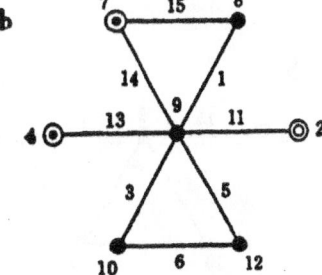

c

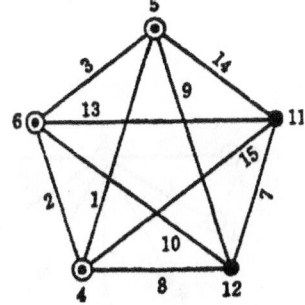

c

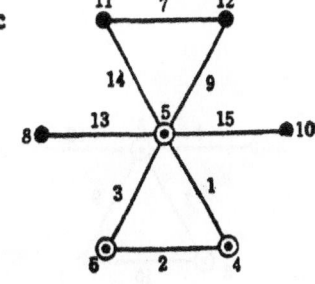

$$L_{16}\ (2^{15})\ (Continued)$$

(3)   **(4)**

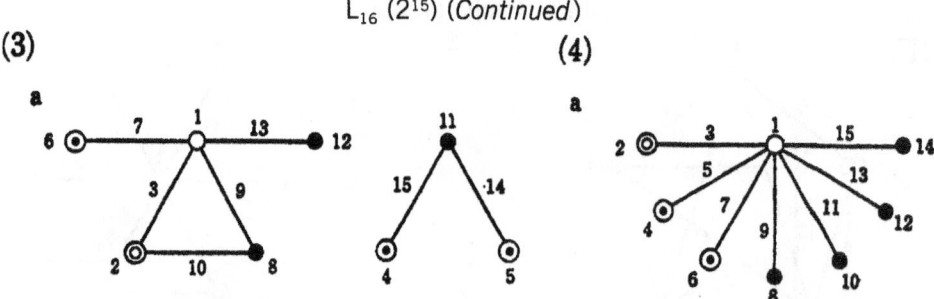

a   **a**

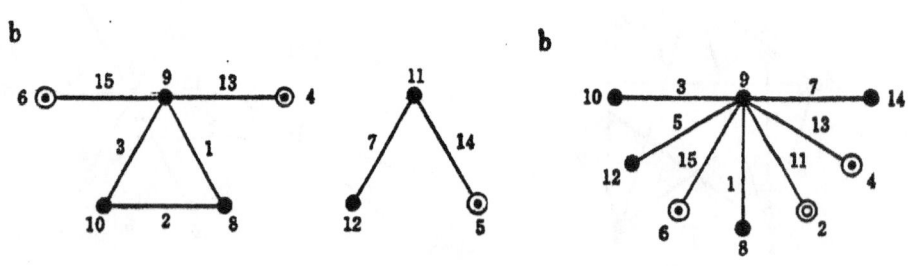

b   **b**

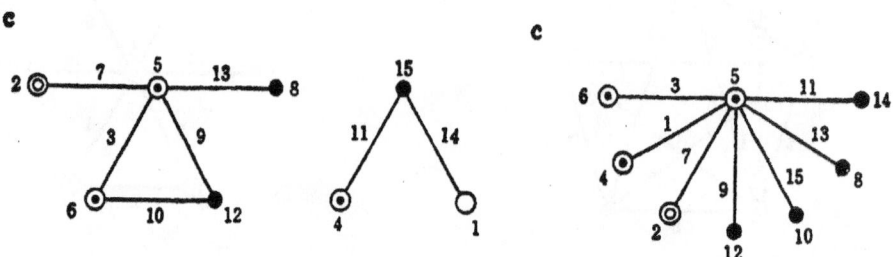

c   **c**

$$L_{16} \ (2^{15}) \ (Continued)$$
1.   2ⁿ

(5)

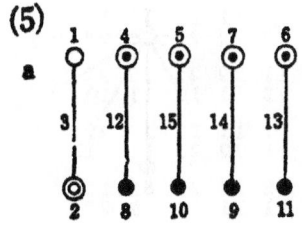

(6)

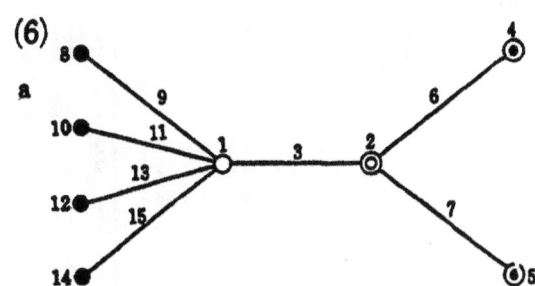

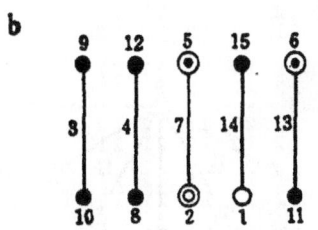

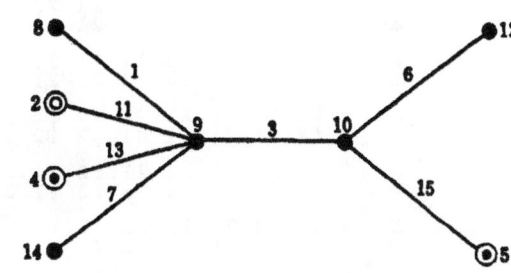

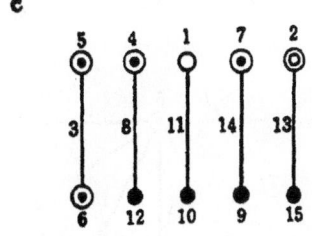

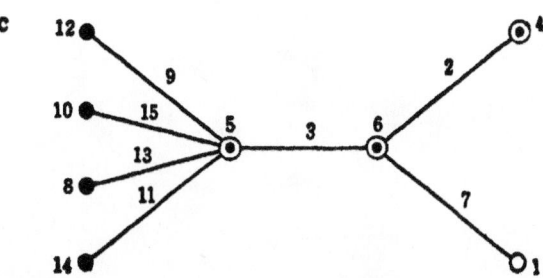

$L_{32} (2^{31})$

No.	1	2 3	4 5	6 7	8 9	1011	1213	1415	1617	1819	2021	2223	2425	2627	2829	3031
1	1	11	11	11	11	11	11	11	11	11	11	11	11	11	11	11
2	1	11	11	11	11	11	11	11	22	22	22	22	22	22	22	22
3	1	11	11	11	22	22	22	22	11	11	11	11	22	22	22	22
4	1	11	11	11	22	22	22	22	22	22	22	22	11	11	11	11
5	1	11	22	22	11	11	22	22	11	11	22	22	11	11	22	22
6	1	11	22	22	11	11	22	22	22	22	11	11	22	22	11	11
7	1	11	22	22	22	22	11	11	11	11	22	22	22	22	11	11
8	1	11	22	22	22	22	11	11	22	22	11	11	11	11	22	22
9	1	22	11	22	11	22	11	22	11	22	11	22	11	22	11	22
10	1	22	11	22	11	22	11	22	22	11	22	11	22	11	22	11
11	1	22	11	22	22	11	22	11	11	22	11	22	22	11	22	11
12	1	22	11	22	22	11	22	11	22	11	22	11	11	22	11	22
13	1	22	22	11	11	22	22	11	11	22	22	11	11	22	22	11
14	1	22	22	11	11	22	22	11	22	11	11	22	22	11	11	22
15	1	22	22	11	22	11	11	22	11	22	22	11	22	11	11	22
16	1	22	22	11	22	11	11	22	22	11	11	22	11	22	22	11
17	2	12	12	12	12	12	12	12	12	12	12	12	12	12	12	12
18	2	12	12	12	12	12	12	12	21	21	21	21	21	21	21	21
19	2	12	12	12	21	21	21	21	12	12	12	12	21	21	21	21
20	2	12	12	12	21	21	21	21	21	21	21	21	12	12	12	12
21	2	12	21	21	12	12	21	21	12	12	21	21	12	12	21	21
22	2	12	21	21	12	12	21	21	21	21	12	12	21	21	12	12
23	2	12	21	21	21	21	12	12	12	12	21	21	21	21	12	12
24	2	12	21	21	21	21	12	12	21	21	12	12	12	12	21	21
25	2	21	12	21	12	21	12	21	12	21	12	21	12	21	12	21
26	2	21	12	21	12	21	12	21	21	12	21	12	21	12	21	12
27	2	21	12	21	21	12	21	12	12	21	12	21	21	12	21	12
28	2	21	12	21	21	12	21	12	21	12	21	12	12	21	12	21
29	2	21	21	12	12	21	21	12	12	21	21	12	12	21	21	12
30	2	21	21	12	12	21	21	12	21	12	12	21	21	12	12	21
31	2	21	21	12	21	12	12	21	12	21	21	12	21	12	12	21
32	2	21	21	12	21	12	12	21	21	12	12	21	12	21	21	12
	a	ba	ca	ba	da	ba	ca	ba	ea	ba	ca	ba	da	ba	ca	ba
		b	c	cb	d	db	dc	cb	e	eb	ec	cb	ed	db	dc	cb
				c			d	dc		e	ec	ec		ed	ed	dc
								d				e			e	ed
																e
Group	1	2	3		4				5							

$$L_{32} (2^{31}) \ (Continued)$$

Interactions between Two Columns

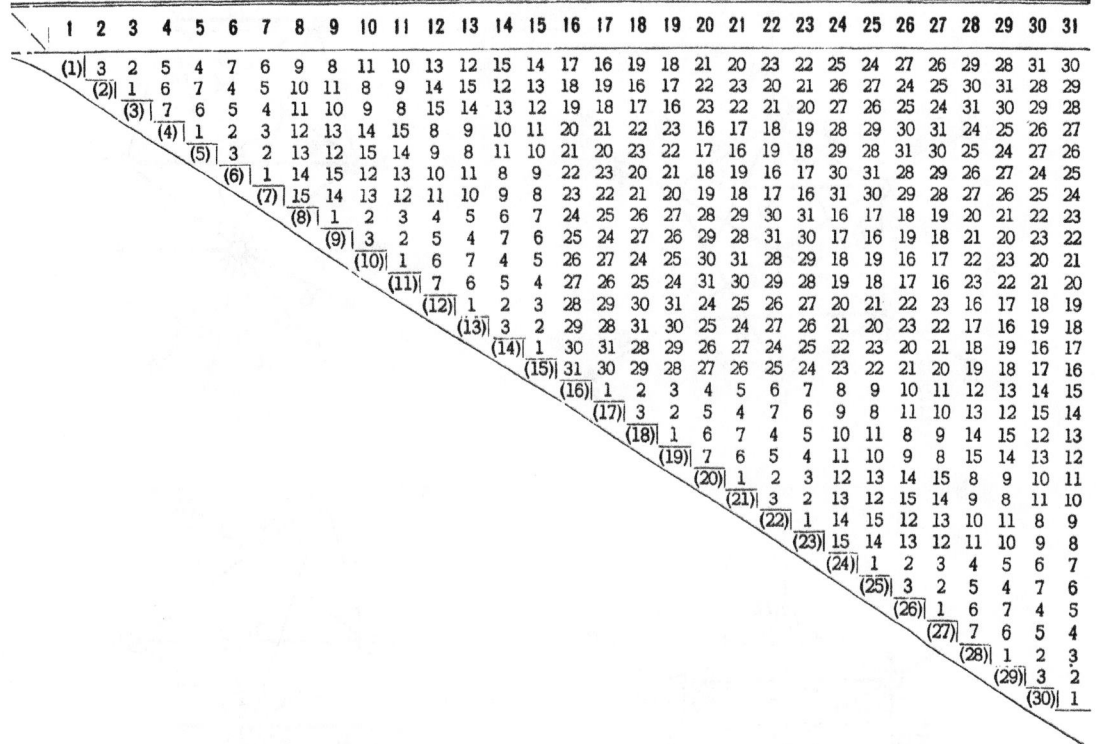

	1	2	3	4	5	6	7	8	9	10	11	12	13	14	15	16	17	18	19	20	21	22	23	24	25	26	27	28	29	30	31
(1)		3	2	5	4	7	6	9	8	11	10	13	12	15	14	17	16	19	18	21	20	23	22	25	24	27	26	29	28	31	30
(2)			1	6	7	4	5	10	11	8	9	14	15	12	13	18	19	16	17	22	23	20	21	26	27	24	25	30	31	28	29
(3)				7	6	5	4	11	10	9	8	15	14	13	12	19	18	17	16	23	22	21	20	27	26	25	24	31	30	29	28
(4)					1	2	3	12	13	14	15	8	9	10	11	20	21	22	23	16	17	18	19	28	29	30	31	24	25	26	27
(5)						3	2	13	12	15	14	9	8	11	10	21	20	23	22	17	16	19	18	29	28	31	30	25	24	27	26
(6)							1	14	15	12	13	10	11	8	9	22	23	20	21	18	19	16	17	30	31	28	29	26	27	24	25
(7)								15	14	13	12	11	10	9	8	23	22	21	20	19	18	17	16	31	30	29	28	27	26	25	24
(8)									1	2	3	4	5	6	7	24	25	26	27	28	29	30	31	16	17	18	19	20	21	22	23
(9)										3	2	5	4	7	6	25	24	27	26	29	28	31	30	17	16	19	18	21	20	23	22
(10)											1	6	7	4	5	26	27	24	25	30	31	28	29	18	19	16	17	22	23	20	21
(11)												7	6	5	4	27	26	25	24	31	30	29	28	19	18	17	16	23	22	21	20
(12)													1	2	3	28	29	30	31	24	25	26	27	20	21	22	23	16	17	18	19
(13)														3	2	29	28	31	30	25	24	27	26	21	20	23	22	17	16	19	18
(14)															1	30	31	28	29	26	27	24	25	22	23	20	21	18	19	16	17
(15)																31	30	29	28	27	26	25	24	23	22	21	20	19	18	17	16
(16)																	1	2	3	4	5	6	7	8	9	10	11	12	13	14	15
(17)																		3	2	5	4	7	6	9	8	11	10	13	12	15	14
(18)																			1	6	7	4	5	10	11	8	9	14	15	12	13
(19)																				7	6	5	4	11	10	9	8	15	14	13	12
(20)																					1	2	3	12	13	14	15	8	9	10	11
(21)																						3	2	13	12	15	14	9	8	11	10
(22)																							1	14	15	12	13	10	11	8	9
(23)																								15	14	13	12	11	10	9	8
(24)																									1	2	3	4	5	6	7
(25)																										3	2	5	4	7	6
(26)																											1	6	7	4	5
(27)																												7	6	5	4
(28)																													1	2	3
(29)																														3	2
(30)																															1

L_{32} (2^{31}) *(Continued)*

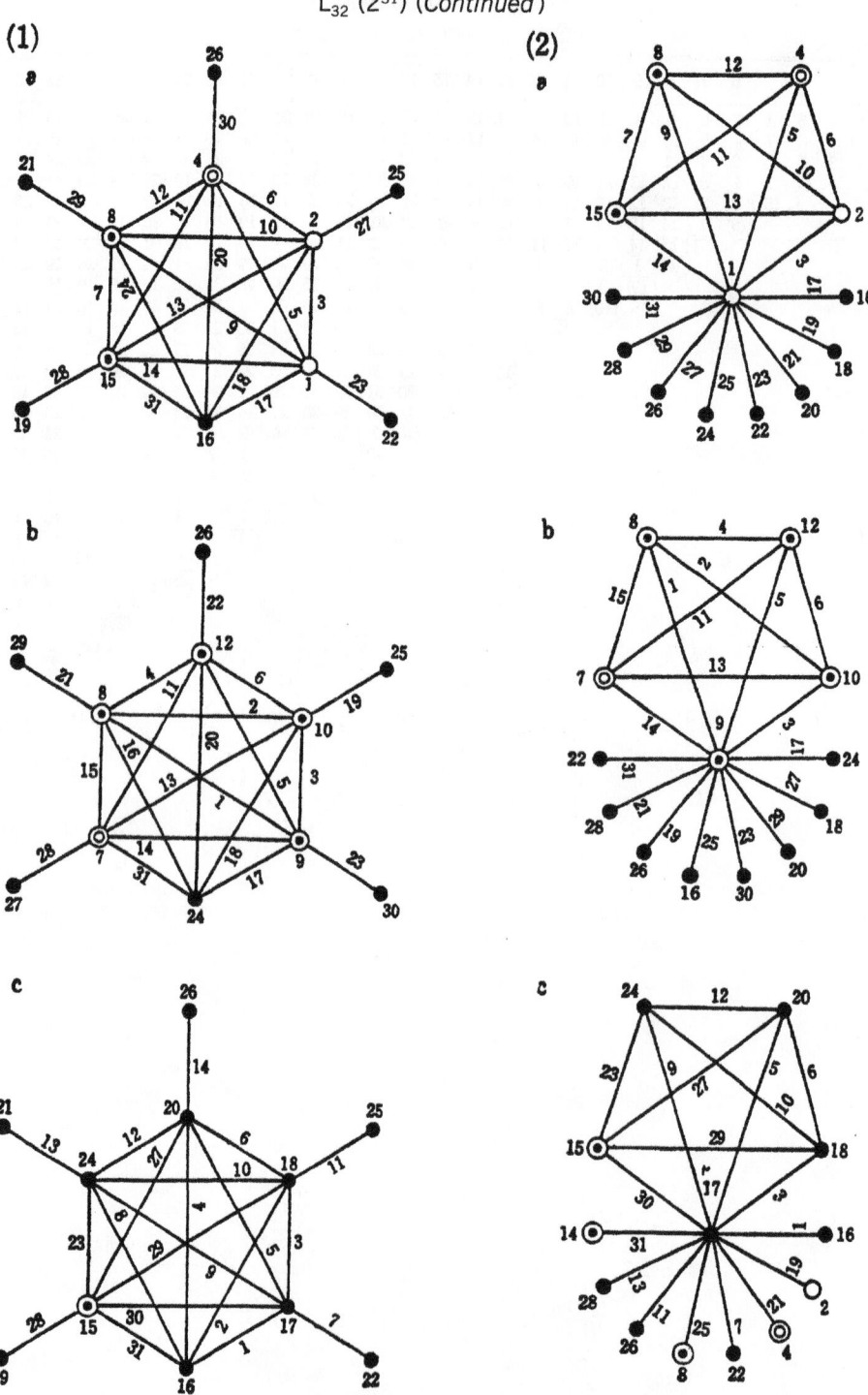

L_{32} (2^{31}) (Continued)

(3)

a

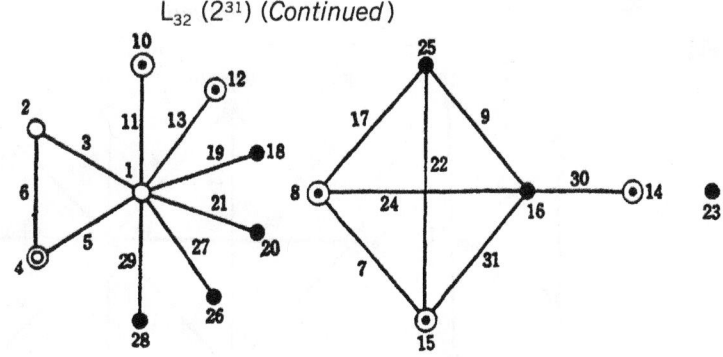

b

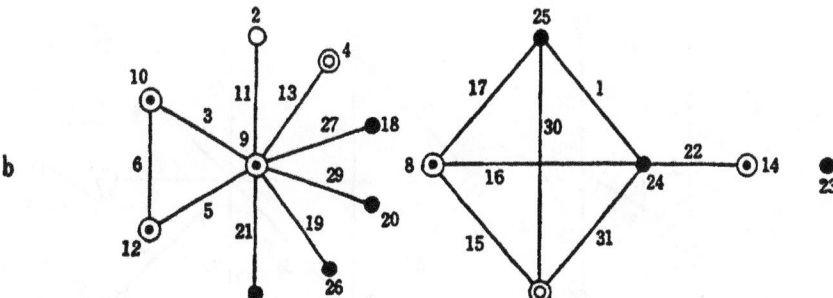

c

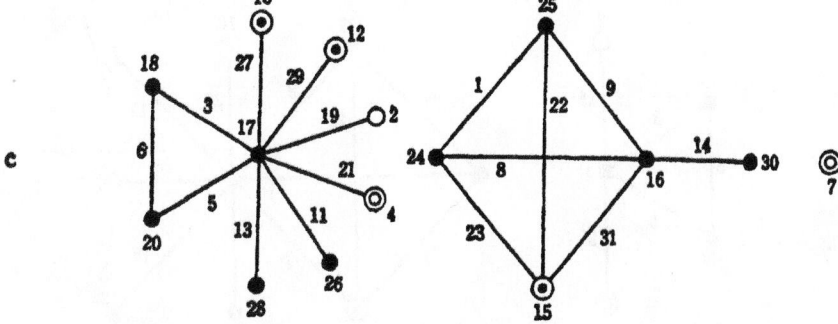

L_{32} (2^{31}) (*Continued*)

(4)

a

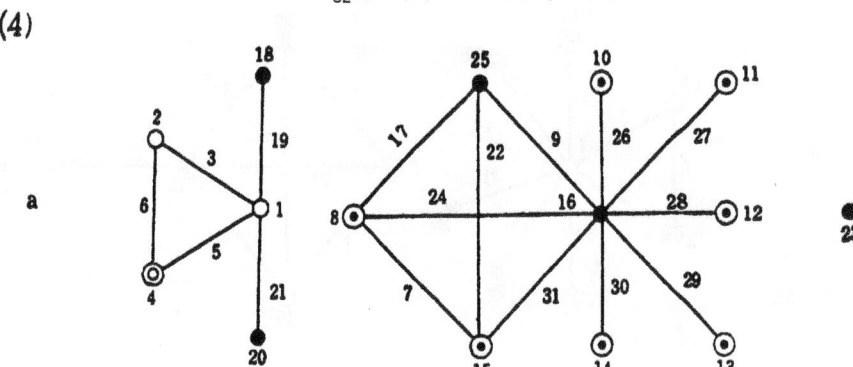

b

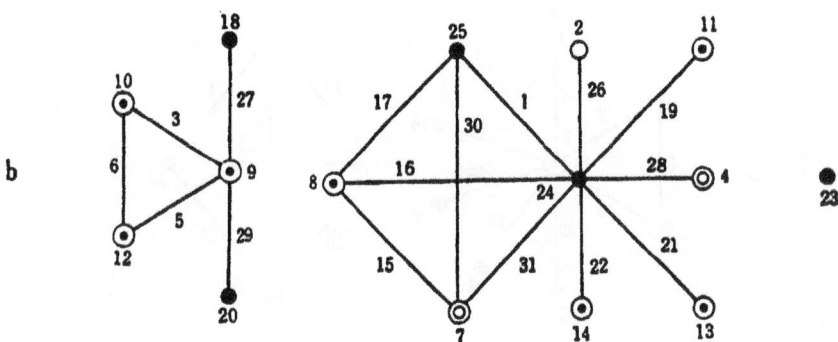

c

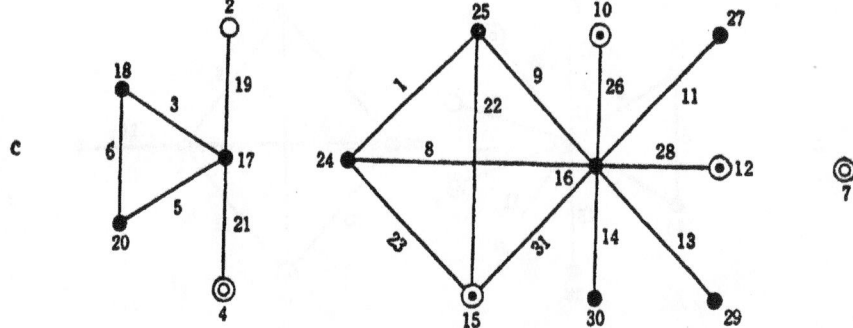

L_{32} (2^{31}) (*Continued*)

(5)

a

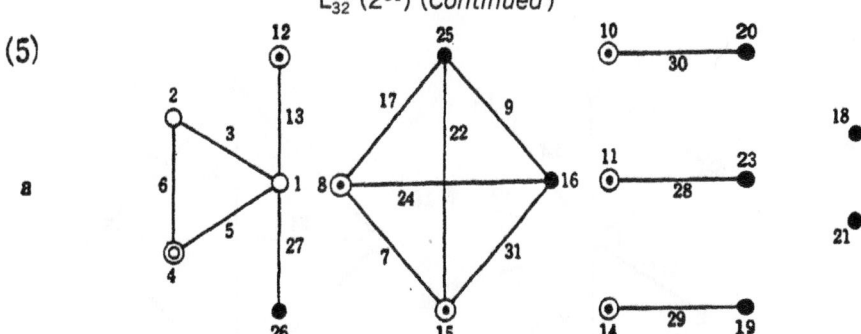

b

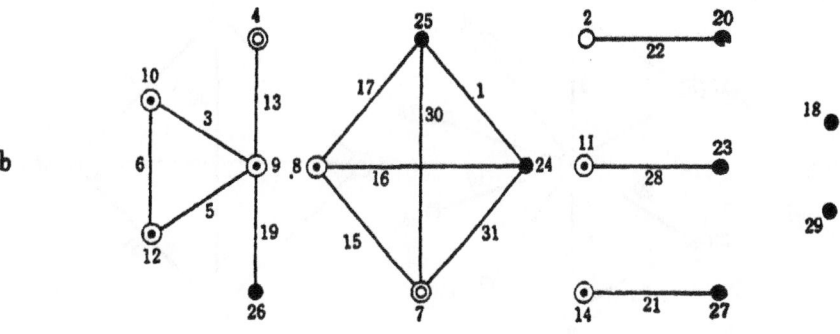

c

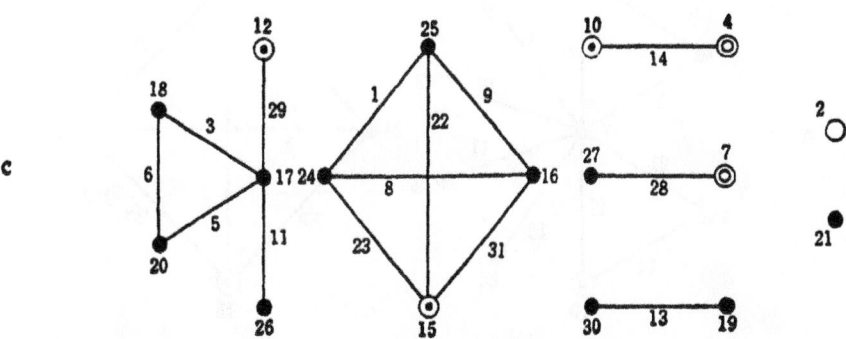

$$L_{32} \ (2^{31}) \ (\textit{Continued})$$

(6)

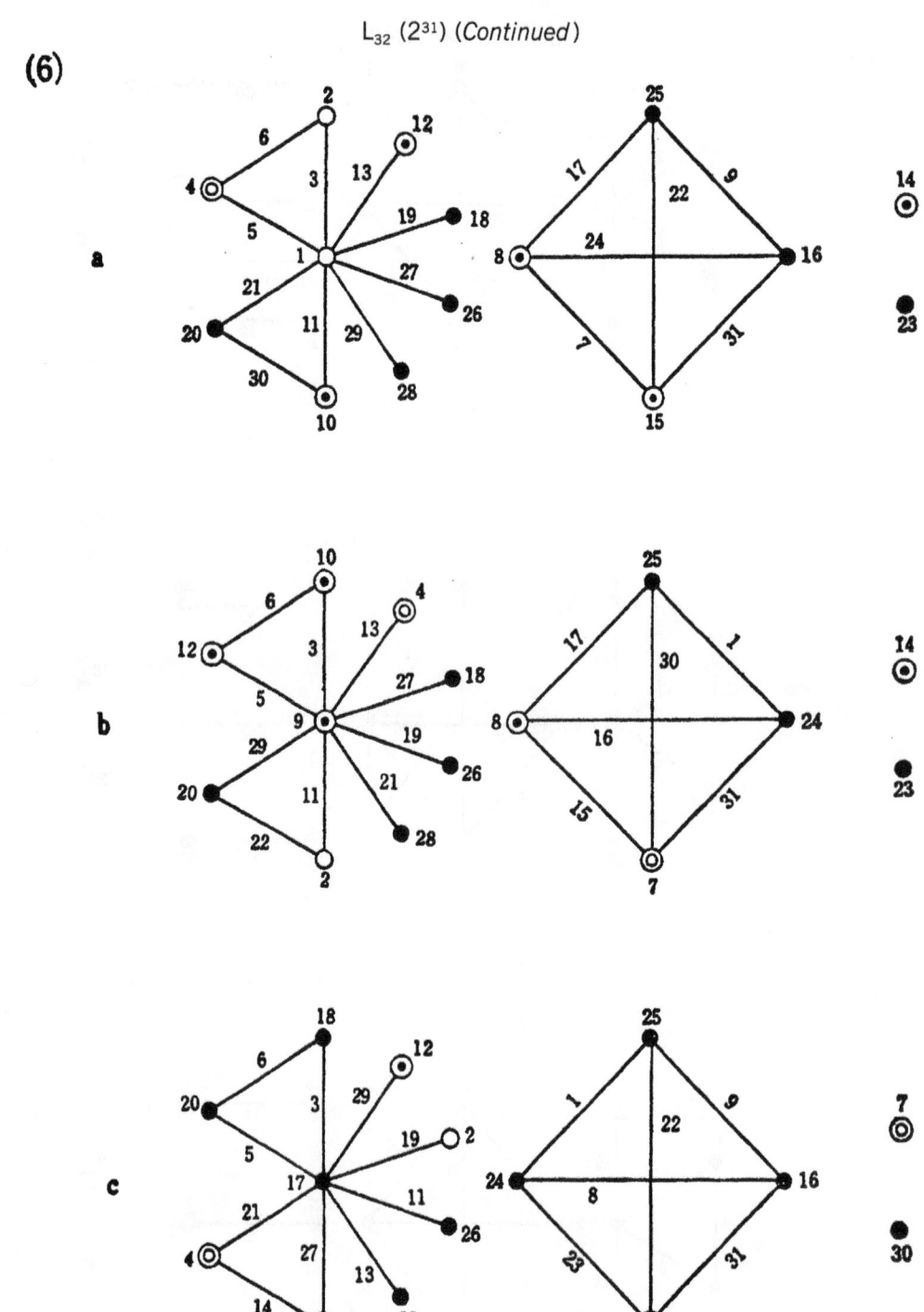

L_{32} (2^{31}) *(Continued)*

(7)

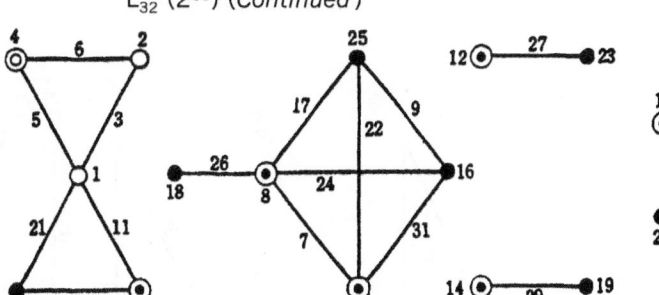

a

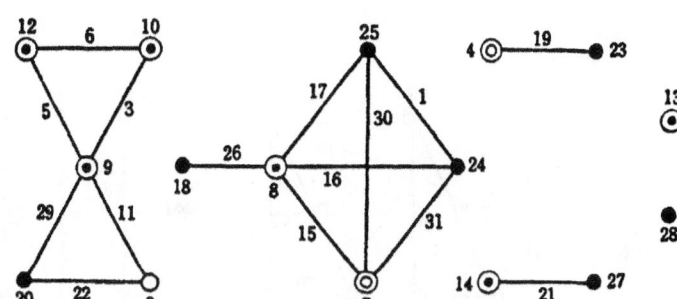

b

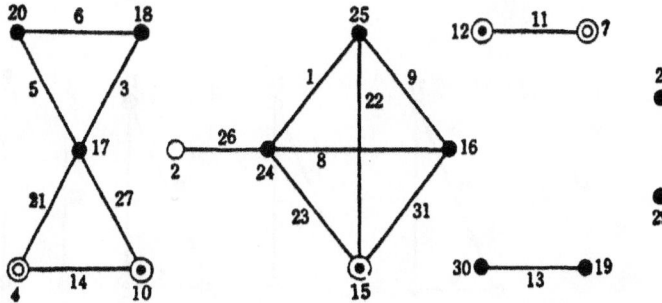

c

L_{32} (2^{31}) (*Continued*)

(8)

a

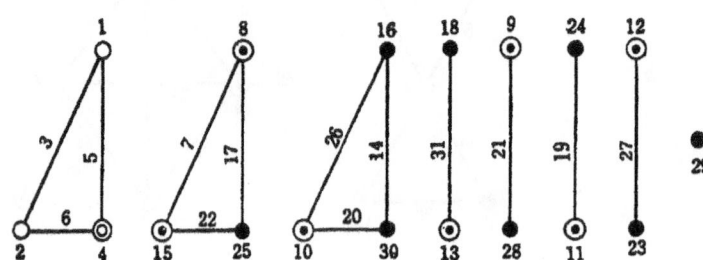

b

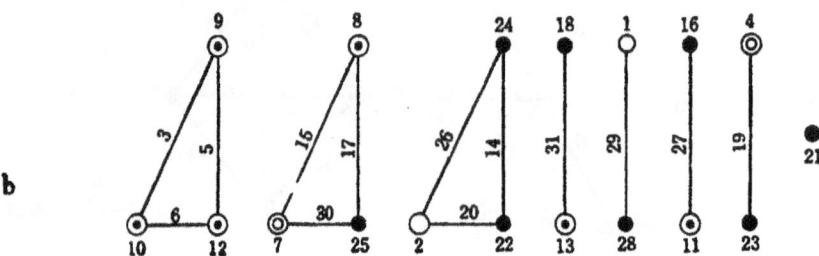

c

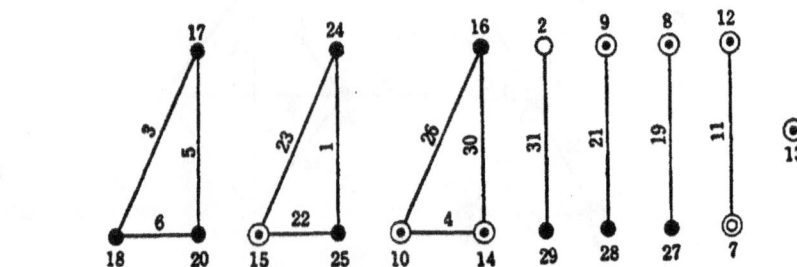

L_{32} (2^{31}) (*Continued*)

(9)

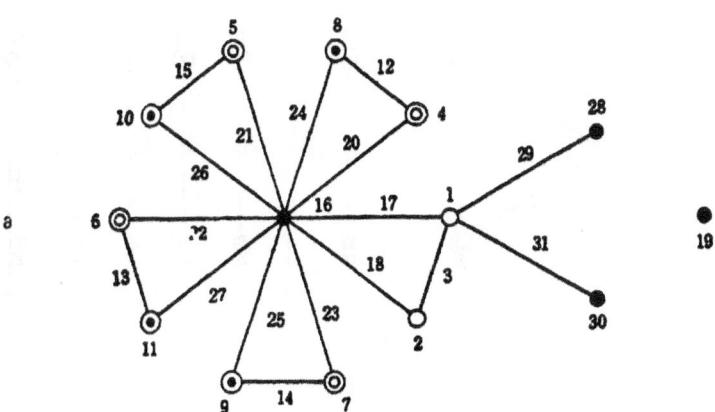

a

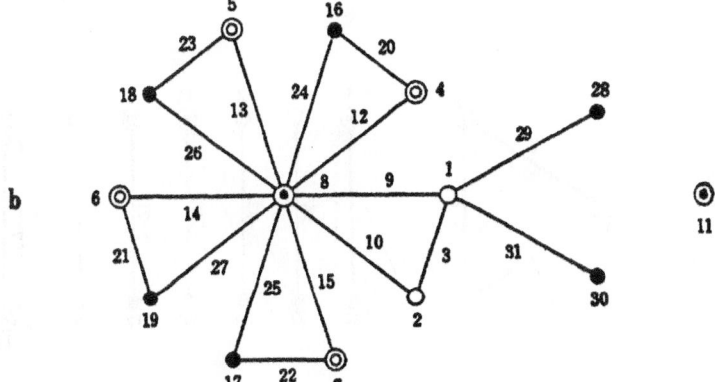

b

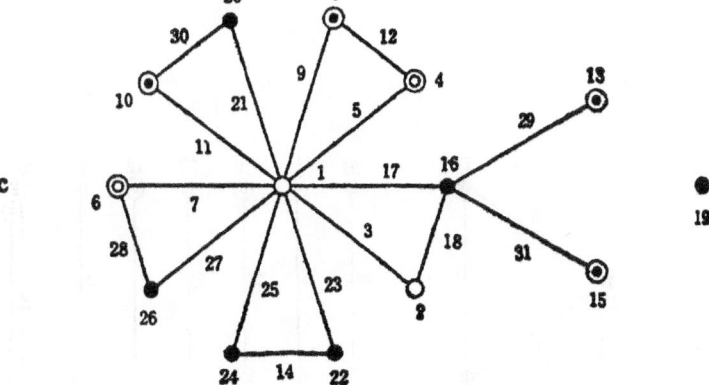

c

$$L_{32} (2^{31}) \text{ (Continued)}$$

(10)

a

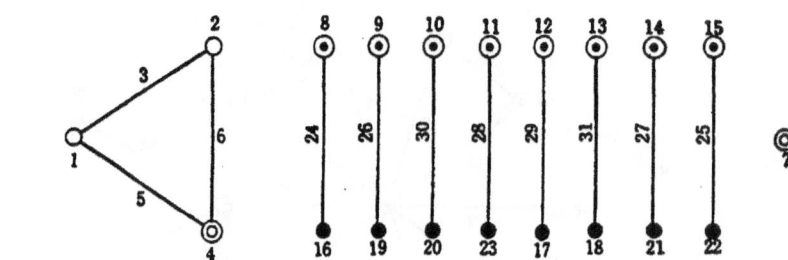

b

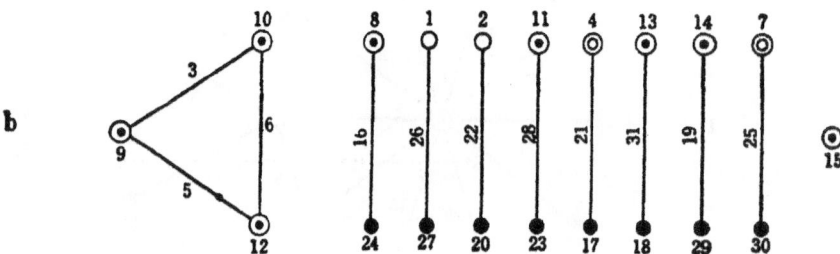

c

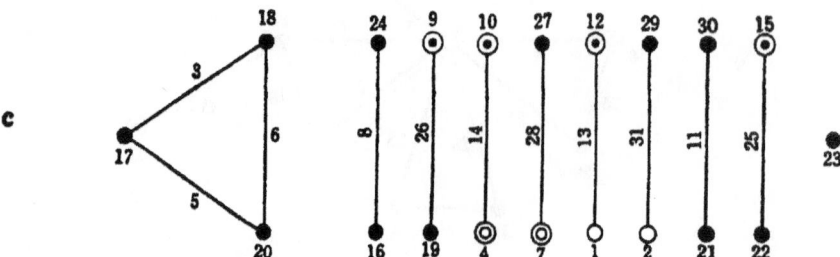

L_{32} (2^{31}) (*Continued*)

(11)

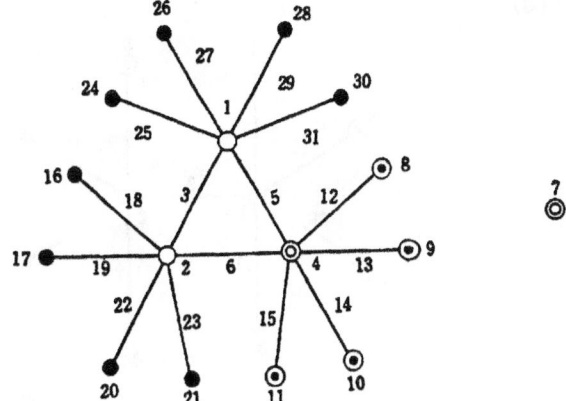

a

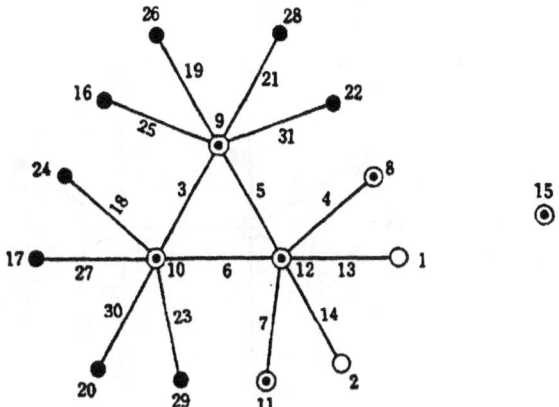

b

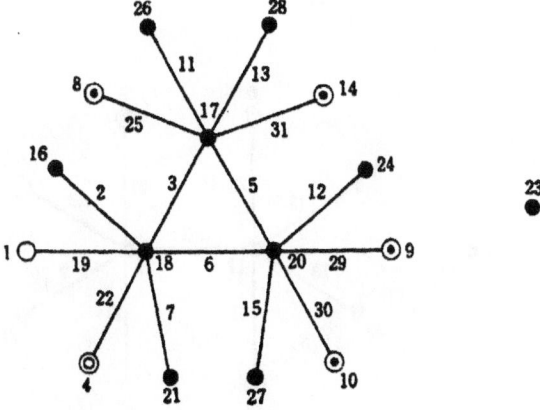

c

L_{32} (2^{31}) *(Continued)*

(12)

a

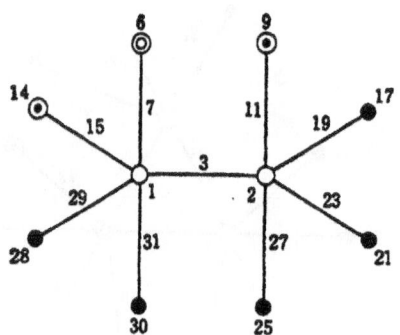
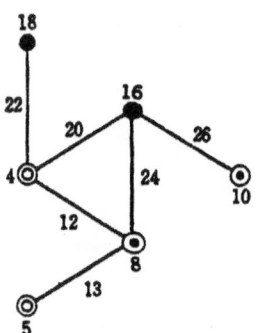

b

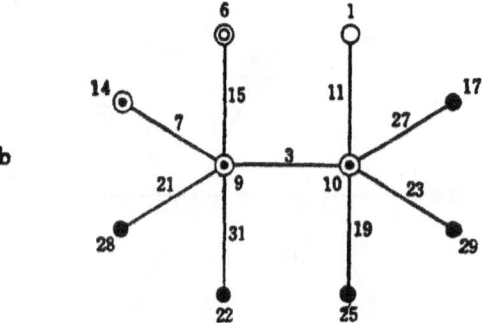
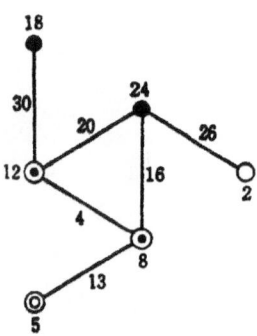

c

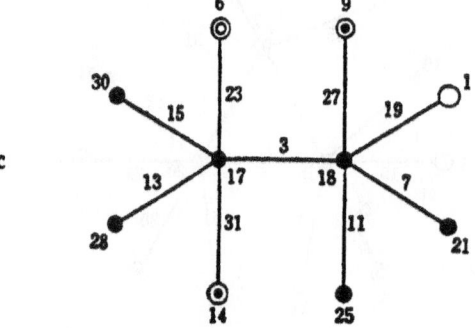
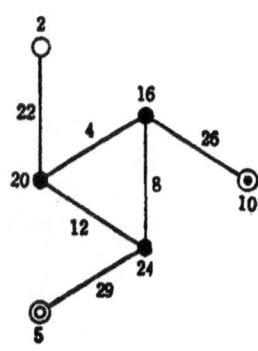

L_{32} (2^{31}) *(Continued)*

(13)

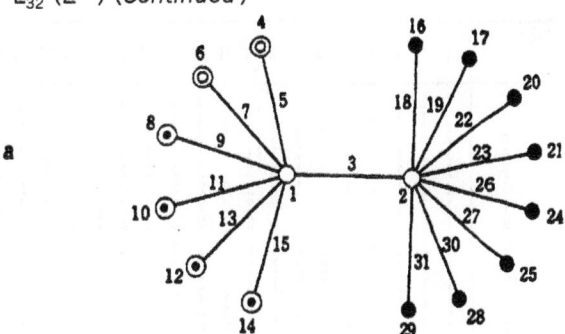

a

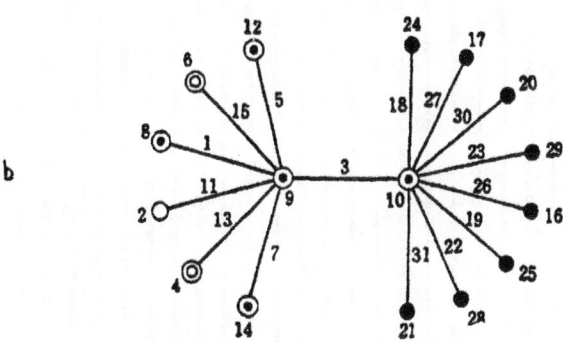

b

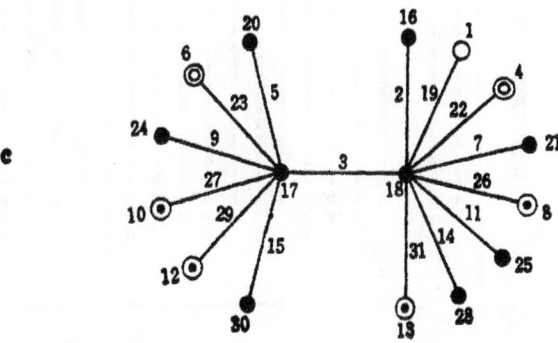

c

$$L_{64}\ (2^{63})$$

No.	1	2	3	4	5	6	7	8	9	10	11	12	13	14	15	16	17	18	19	20	21	22	23	24	25	26	27	28	29	30	31
1	1	1	1	1	1	1	1	1	1	1	1	1	1	1	1	1	1	1	1	1	1	1	1	1	1	1	1	1	1	1	1
2	1	1	1	1	1	1	1	1	1	1	1	1	1	1	1	2	2	2	2	2	2	2	2	2	2	2	2	2	2	2	2
3	1	1	1	1	1	1	1	1	1	1	1	1	1	1	1	1	1	1	1	1	1	1	1	1	1	1	1	1	1	1	1
4	1	1	1	1	1	1	1	1	1	1	1	1	1	1	1	2	2	2	2	2	2	2	2	2	2	2	2	2	2	2	2
5	1	1	1	1	1	1	1	2	2	2	2	2	2	2	2	1	1	1	1	1	1	1	1	2	2	2	2	2	2	2	2
6	1	1	1	1	1	1	1	2	2	2	2	2	2	2	2	2	2	2	2	2	2	2	2	1	1	1	1	1	1	1	1
7	1	1	1	1	1	1	1	2	2	2	2	2	2	2	2	1	1	1	1	1	1	1	1	2	2	2	2	2	2	2	2
8	1	1	1	1	1	1	1	2	2	2	2	2	2	2	2	2	2	2	2	2	2	2	2	1	1	1	1	1	1	1	1
9	1	1	1	2	2	2	2	1	1	1	1	2	2	2	2	1	1	1	1	2	2	2	2	1	1	1	1	2	2	2	2
10	1	1	1	2	2	2	2	1	1	1	1	2	2	2	2	2	2	2	2	1	1	1	1	2	2	2	2	1	1	1	1
11	1	1	1	2	2	2	2	1	1	1	1	2	2	2	2	1	1	1	1	2	2	2	2	1	1	1	1	2	2	2	2
12	1	1	1	2	2	2	2	1	1	1	1	2	2	2	2	2	2	2	2	1	1	1	1	2	2	2	2	1	1	1	1
13	1	1	1	2	2	2	2	2	2	2	2	1	1	1	1	1	1	1	1	2	2	2	2	2	2	2	2	1	1	1	1
14	1	1	1	2	2	2	2	2	2	2	2	1	1	1	1	2	2	2	2	1	1	1	1	1	1	1	1	2	2	2	2
15	1	1	1	2	2	2	2	2	2	2	2	1	1	1	1	1	1	1	1	2	2	2	2	2	2	2	2	1	1	1	1
16	1	1	1	2	2	2	2	2	2	2	2	1	1	1	1	2	2	2	2	1	1	1	1	1	1	1	1	2	2	2	2
17	1	2	2	1	1	2	2	1	1	2	2	1	1	2	2	1	1	2	2	1	1	2	2	1	1	2	2	1	1	2	2
18	1	2	2	1	1	2	2	1	1	2	2	1	1	2	2	2	2	1	1	2	2	1	1	2	2	1	1	2	2	1	1
19	1	2	2	1	1	2	2	1	1	2	2	1	1	2	2	1	1	2	2	1	1	2	2	1	1	2	2	1	1	2	2
20	1	2	2	1	1	2	2	1	1	2	2	1	1	2	2	2	2	1	1	2	2	1	1	2	2	1	1	2	2	1	1
21	1	2	2	1	1	2	2	2	2	1	1	2	2	1	1	1	1	2	2	1	1	2	2	2	2	1	1	2	2	1	1
22	1	2	2	1	1	2	2	2	2	1	1	2	2	1	1	2	2	1	1	2	2	1	1	1	1	2	2	1	1	2	2
23	1	2	2	1	1	2	2	2	2	1	1	2	2	1	1	1	1	2	2	1	1	2	2	2	2	1	1	2	2	1	1
24	1	2	2	1	1	2	2	2	2	1	1	2	2	1	1	2	2	1	1	2	2	1	1	1	1	2	2	1	1	2	2
25	1	2	2	2	2	1	1	1	1	2	2	2	2	1	1	1	1	2	2	2	2	1	1	1	1	2	2	2	2	1	1
26	1	2	2	2	2	1	1	1	1	2	2	2	2	1	1	2	2	1	1	1	1	2	2	2	2	1	1	1	1	2	2
27	1	2	2	2	2	1	1	1	1	2	2	2	2	1	1	1	1	2	2	2	2	1	1	1	1	2	2	2	2	1	1
28	1	2	2	2	2	1	1	1	1	2	2	2	2	1	1	2	2	1	1	1	1	2	2	2	2	1	1	1	1	2	2
29	1	2	2	2	2	1	1	2	2	1	1	1	1	2	2	1	1	2	2	2	2	1	1	2	2	1	1	1	1	2	2
30	1	2	2	2	2	1	1	2	2	1	1	1	1	2	2	2	2	1	1	1	1	2	2	1	1	2	2	2	2	1	1
31	1	2	2	2	2	1	1	2	2	1	1	1	1	2	2	1	1	2	2	2	2	1	1	2	2	1	1	1	1	2	2
32	1	2	2	2	2	1	1	2	2	1	1	1	1	2	2	2	2	1	1	1	1	2	2	1	1	2	2	2	2	1	1
33	2	1	2	1	2	1	2	1	2	1	2	1	2	1	2	1	2	1	2	1	2	1	2	1	2	1	2	1	2	1	2
34	2	1	2	1	2	1	2	1	2	1	2	1	2	1	2	2	1	2	1	2	1	2	1	2	1	2	1	2	1	2	1
35	2	1	2	1	2	1	2	1	2	1	2	1	2	1	2	1	2	1	2	1	2	1	2	1	2	1	2	1	2	1	2
36	2	1	2	1	2	1	2	1	2	1	2	1	2	1	2	2	1	2	1	2	1	2	1	2	1	2	1	2	1	2	1
37	2	1	2	1	2	1	2	2	1	2	1	2	1	2	1	1	2	1	2	1	2	1	2	2	1	2	1	2	1	2	1
38	2	1	2	1	2	1	2	2	1	2	1	2	1	2	1	2	1	2	1	2	1	2	1	1	2	1	2	1	2	1	2
39	2	1	2	1	2	1	2	2	1	2	1	2	1	2	1	1	2	1	2	1	2	1	2	2	1	2	1	2	1	2	1
40	2	1	2	1	2	1	2	2	1	2	1	2	1	2	1	2	1	2	1	2	1	2	1	1	2	1	2	1	2	1	2
41	2	1	2	2	1	2	1	1	2	1	2	2	1	2	1	1	2	1	2	2	1	2	1	1	2	1	2	2	1	2	1
42	2	1	2	2	1	2	1	1	2	1	2	2	1	2	1	2	1	2	1	1	2	1	2	2	1	2	1	1	2	1	2
43	2	1	2	2	1	2	1	1	2	1	2	2	1	2	1	1	2	1	2	2	1	2	1	1	2	1	2	2	1	2	1
44	2	1	2	2	1	2	1	1	2	1	2	2	1	2	1	2	1	2	1	1	2	1	2	2	1	2	1	1	2	1	2
45	2	1	2	2	1	2	1	2	1	2	1	1	2	1	2	1	2	1	2	2	1	2	1	1	2	1	2	1	2	1	2
46	2	1	2	2	1	2	1	2	1	2	1	1	2	1	2	2	1	2	1	1	2	1	2	1	2	1	2	2	1	2	1
47	2	1	2	2	1	2	1	2	1	2	1	1	2	1	2	1	2	1	2	2	1	2	1	1	2	1	2	1	2	1	2
48	2	1	2	2	1	2	1	2	1	2	1	1	2	1	2	2	1	2	1	1	2	1	2	1	2	1	2	2	1	2	1
49	2	2	1	1	2	2	1	1	2	2	1	1	2	2	1	1	2	2	1	1	2	2	1	1	2	2	1	1	2	2	1
50	2	2	1	1	2	2	1	1	2	2	1	1	2	2	1	2	1	1	2	2	1	1	2	2	1	1	2	2	1	1	2
51	2	2	1	1	2	2	1	1	2	2	1	1	2	2	1	1	2	2	1	1	2	2	1	1	2	2	1	1	2	2	1
52	2	2	1	1	2	2	1	1	2	2	1	1	2	2	1	2	1	1	2	2	1	1	2	2	1	1	2	2	1	1	2
53	2	2	1	1	2	2	1	2	1	1	2	2	1	1	2	1	2	2	1	1	2	2	1	2	1	1	2	2	1	1	2
54	2	2	1	1	2	2	1	2	1	1	2	2	1	1	2	2	1	1	2	2	1	1	2	1	2	2	1	1	2	2	1
55	2	2	1	1	2	2	1	2	1	1	2	2	1	1	2	1	2	2	1	1	2	2	1	2	1	1	2	2	1	1	2
56	2	2	1	1	2	2	1	2	1	1	2	2	1	1	2	2	1	1	2	2	1	1	2	1	2	2	1	1	2	2	1
57	2	2	1	2	1	1	2	1	2	2	1	2	1	1	2	1	2	2	1	2	1	1	2	1	2	2	1	2	1	1	2
58	2	2	1	2	1	1	2	1	2	2	1	2	1	1	2	2	1	1	2	1	2	2	1	2	1	1	2	1	2	2	1
59	2	2	1	2	1	1	2	1	2	2	1	2	1	1	2	1	2	2	1	2	1	1	2	1	2	2	1	2	1	1	2
60	2	2	1	2	1	1	2	1	2	2	1	2	1	1	2	2	1	1	2	1	2	2	1	2	1	1	2	1	2	2	1
61	2	2	1	2	1	1	2	2	1	1	2	1	2	2	1	1	2	2	1	2	1	1	2	2	1	1	2	1	2	2	1
62	2	2	1	2	1	1	2	2	1	1	2	1	2	2	1	2	1	1	2	1	2	2	1	1	2	2	1	1	2	1	2
63	2	2	1	2	1	1	2	2	1	1	2	1	2	2	1	1	2	2	1	2	1	1	2	2	1	1	2	1	2	2	1
64	2	2	1	2	1	1	2	2	1	1	2	1	2	2	1	2	1	1	2	1	2	2	1	1	2	2	1	1	2	1	2
	a	b	ab	c	ac	bc	abc	d	ad	bd	abd	cd	acd	bcd	abcd	e	ae	be	abe	ce	ace	bce	abce	de	ade	bde	abde	cde	acde	bcde	abcde

Group: Group 1 = column 1; Group 2 = columns 2–3; Group 3 = columns 4–7; Group 4 = columns 8–15; Group 5 = columns 16–31.

L$_{64}$ (2^{63}) (Continued)

32	33	34	35	36	37	38	39	40	41	42	43	44	45	46	47	48	49	50	51	52	53	54	55	56	57	58	59	60	61	62	63
1	1	1	1	1	1	1	1	1	1	1	1	1	1	1	1	1	1	1	1	1	1	1	1	1	1	1	1	1	1	1	1
2	2	2	2	2	2	2	2	2	2	2	2	2	2	2	2	2	2	2	2	2	2	2	2	2	2	2	2	2	2	2	2
1	1	1	1	1	1	1	1	1	1	1	1	1	1	1	1	2	2	2	2	2	2	2	2	2	2	2	2	2	2	2	2
2	2	2	2	2	2	2	2	2	2	2	2	2	2	2	2	1	1	1	1	1	1	1	1	1	1	1	1	1	1	1	1
1	1	1	1	1	1	1	1	2	2	2	2	2	2	2	2	1	1	1	1	1	1	1	1	2	2	2	2	2	2	2	2
2	2	2	2	2	2	2	2	1	1	1	1	1	1	1	1	2	2	2	2	2	2	2	2	1	1	1	1	1	1	1	1
1	1	1	1	1	1	1	1	2	2	2	2	2	2	2	2	2	2	2	2	2	2	2	2	1	1	1	1	1	1	1	1
2	2	2	2	2	2	2	2	1	1	1	1	1	1	1	1	1	1	1	1	1	1	1	1	2	2	2	2	2	2	2	2
1	1	1	1	2	2	2	2	1	1	1	1	2	2	2	2	1	1	1	1	2	2	2	2	1	1	1	1	2	2	2	2
2	2	2	2	1	1	1	1	2	2	2	2	1	1	1	1	2	2	2	2	1	1	1	1	2	2	2	2	1	1	1	1
1	1	1	1	2	2	2	2	1	1	1	1	2	2	2	2	2	2	2	2	1	1	1	1	2	2	2	2	1	1	1	1
2	2	2	2	1	1	1	1	2	2	2	2	1	1	1	1	1	1	1	1	2	2	2	2	1	1	1	1	2	2	2	2
1	1	1	1	2	2	2	2	2	2	2	2	1	1	1	1	1	1	1	1	2	2	2	2	2	2	2	2	1	1	1	1
2	2	2	2	1	1	1	1	1	1	1	1	2	2	2	2	2	2	2	2	1	1	1	1	1	1	1	1	2	2	2	2
1	1	1	1	2	2	2	2	2	2	2	2	1	1	1	1	2	2	2	2	1	1	1	1	1	1	1	1	2	2	2	2
2	2	2	2	1	1	1	1	1	1	1	1	2	2	2	2	1	1	1	1	2	2	2	2	2	2	2	2	1	1	1	1
1	1	2	2	1	1	2	2	1	1	2	2	1	1	2	2	1	1	2	2	1	1	2	2	1	1	2	2	1	1	2	2
2	2	1	1	2	2	1	1	2	2	1	1	2	2	1	1	2	2	1	1	2	2	1	1	2	2	1	1	2	2	1	1
1	1	2	2	1	1	2	2	1	1	2	2	1	1	2	2	2	2	1	1	2	2	1	1	2	2	1	1	2	2	1	1
2	2	1	1	2	2	1	1	2	2	1	1	2	2	1	1	1	1	2	2	1	1	2	2	1	1	2	2	1	1	2	2
1	1	2	2	1	1	2	2	2	2	1	1	2	2	1	1	1	1	2	2	1	1	2	2	2	2	1	1	2	2	1	1
2	2	1	1	2	2	1	1	1	1	2	2	1	1	2	2	2	2	1	1	2	2	1	1	1	1	2	2	1	1	2	2
1	1	2	2	1	1	2	2	2	2	1	1	2	2	1	1	2	2	1	1	2	2	1	1	1	1	2	2	1	1	2	2
2	2	1	1	2	2	1	1	1	1	2	2	1	1	2	2	1	1	2	2	1	1	2	2	2	2	1	1	2	2	1	1
1	1	2	2	2	2	1	1	1	1	2	2	2	2	1	1	1	1	2	2	2	2	1	1	1	1	2	2	2	2	1	1
2	2	1	1	1	1	2	2	2	2	1	1	1	1	2	2	2	2	1	1	1	1	2	2	2	2	1	1	1	1	2	2
1	1	2	2	2	2	1	1	1	1	2	2	2	2	1	1	2	2	1	1	1	1	2	2	2	2	1	1	1	1	2	2
2	2	1	1	1	1	2	2	2	2	1	1	1	1	2	2	1	1	2	2	2	2	1	1	1	1	2	2	2	2	1	1
1	1	2	2	2	2	1	1	2	2	1	1	1	1	2	2	1	1	2	2	2	2	1	1	2	2	1	1	1	1	2	2
2	2	1	1	1	1	2	2	1	1	2	2	2	2	1	1	2	2	1	1	1	1	2	2	1	1	2	2	2	2	1	1
1	1	2	2	2	2	1	1	2	2	1	1	1	1	2	2	2	2	1	1	1	1	2	2	1	1	2	2	2	2	1	1
2	2	1	1	1	1	2	2	1	1	2	2	2	2	1	1	1	1	2	2	2	2	1	1	2	2	1	1	1	1	2	2
1	2	1	2	1	2	1	2	1	2	1	2	1	2	1	2	1	2	1	2	1	2	1	2	1	2	1	2	1	2	1	2
2	1	2	1	2	1	2	1	2	1	2	1	2	1	2	1	2	1	2	1	2	1	2	1	2	1	2	1	2	1	2	1
1	2	1	2	1	2	1	2	1	2	1	2	1	2	1	2	2	1	2	1	2	1	2	1	2	1	2	1	2	1	2	1
2	1	2	1	2	1	2	1	2	1	2	1	2	1	2	1	1	2	1	2	1	2	1	2	1	2	1	2	1	2	1	2
1	2	1	2	1	2	1	2	2	1	2	1	2	1	2	1	1	2	1	2	1	2	1	2	2	1	2	1	2	1	2	1
2	1	2	1	2	1	2	1	1	2	1	2	1	2	1	2	2	1	2	1	2	1	2	1	1	2	1	2	1	2	1	2
1	2	1	2	1	2	1	2	2	1	2	1	2	1	2	1	2	1	2	1	2	1	2	1	1	2	1	2	1	2	1	2
2	1	2	1	2	1	2	1	1	2	1	2	1	2	1	2	1	2	1	2	1	2	1	2	2	1	2	1	2	1	2	1
1	2	1	2	2	1	2	1	1	2	1	2	2	1	2	1	1	2	1	2	2	1	2	1	1	2	1	2	2	1	2	1
2	1	2	1	1	2	1	2	2	1	2	1	1	2	1	2	2	1	2	1	1	2	1	2	2	1	2	1	1	2	1	2
1	2	1	2	2	1	2	1	1	2	1	2	2	1	2	1	2	1	2	1	1	2	1	2	2	1	2	1	1	2	1	2
2	1	2	1	1	2	1	2	2	1	2	1	1	2	1	2	1	2	1	2	2	1	2	1	1	2	1	2	2	1	2	1
1	2	1	2	2	1	2	1	2	1	2	1	1	2	1	2	1	2	1	2	2	1	2	1	2	1	2	1	1	2	1	2
2	1	2	1	1	2	1	2	1	2	1	2	2	1	2	1	2	1	2	1	1	2	1	2	1	2	1	2	2	1	2	1
1	2	1	2	2	1	2	1	2	1	2	1	1	2	1	2	2	1	2	1	1	2	1	2	1	2	1	2	2	1	2	1
2	1	2	1	1	2	1	2	1	2	1	2	2	1	2	1	1	2	1	2	2	1	2	1	2	1	2	1	1	2	1	2
1	2	2	1	1	2	2	1	1	2	2	1	1	2	2	1	1	2	2	1	1	2	2	1	1	2	2	1	1	2	2	1
2	1	1	2	2	1	1	2	2	1	1	2	2	1	1	2	2	1	1	2	2	1	1	2	2	1	1	2	2	1	1	2
1	2	2	1	1	2	2	1	1	2	2	1	1	2	2	1	2	1	1	2	2	1	1	2	2	1	1	2	2	1	1	2
2	1	1	2	2	1	1	2	2	1	1	2	2	1	1	2	1	2	2	1	1	2	2	1	1	2	2	1	1	2	2	1
1	2	2	1	1	2	2	1	2	1	1	2	2	1	1	2	1	2	2	1	1	2	2	1	2	1	1	2	2	1	1	2
2	1	1	2	2	1	1	2	1	2	2	1	1	2	2	1	2	1	1	2	2	1	1	2	1	2	2	1	1	2	2	1
1	2	2	1	1	2	2	1	2	1	1	2	2	1	1	2	2	1	1	2	2	1	1	2	1	2	2	1	1	2	2	1
2	1	1	2	2	1	1	2	1	2	2	1	1	2	2	1	1	2	2	1	1	2	2	1	2	1	1	2	2	1	1	2
1	2	2	1	2	1	1	2	1	2	2	1	2	1	1	2	1	2	2	1	2	1	1	2	1	2	2	1	2	1	1	2
2	1	1	2	1	2	2	1	2	1	1	2	1	2	2	1	2	1	1	2	1	2	2	1	2	1	1	2	1	2	2	1
1	2	2	1	2	1	1	2	1	2	2	1	2	1	1	2	2	1	1	2	1	2	2	1	2	1	1	2	1	2	2	1
2	1	1	2	1	2	2	1	2	1	1	2	1	2	2	1	1	2	2	1	2	1	1	2	1	2	2	1	2	1	1	2
1	2	2	1	2	1	1	2	2	1	1	2	1	2	2	1	1	2	2	1	2	1	1	2	2	1	1	2	1	2	2	1
2	1	1	2	1	2	2	1	1	2	2	1	2	1	1	2	2	1	1	2	1	2	2	1	1	2	2	1	2	1	1	2
1	2	2	1	2	1	1	2	2	1	1	2	1	2	2	1	2	1	1	2	1	2	2	1	1	2	2	1	2	1	1	2
2	1	1	2	1	2	2	1	1	2	2	1	2	1	1	2	1	2	2	1	2	1	1	2	2	1	1	2	1	2	2	1
f	a	b	a	c	a	b	a	d	a	b	a	c	a	b	a	e	a	b	a	c	a	b	a	d	a	b	a	c	a	b	a
	f	f	b	f	c	c	b	f	d	d	b	d	c	c	b	f	e	e	b	e	c	c	b	e	d	d	b	d	c	c	b
			f		f	f	c		f	f	d	f	d	d	c		f	f	e	f	e	e	c	f	e	e	d	e	d	d	c
							f				f		f	f	d				f		f	f	e		f	f	e	f	e	e	d
															f								f				f		f	f	e
																															f

6

$$L_{64} \, (2^{63}) \ (\textit{Continued})$$
Interactions between Two Columns

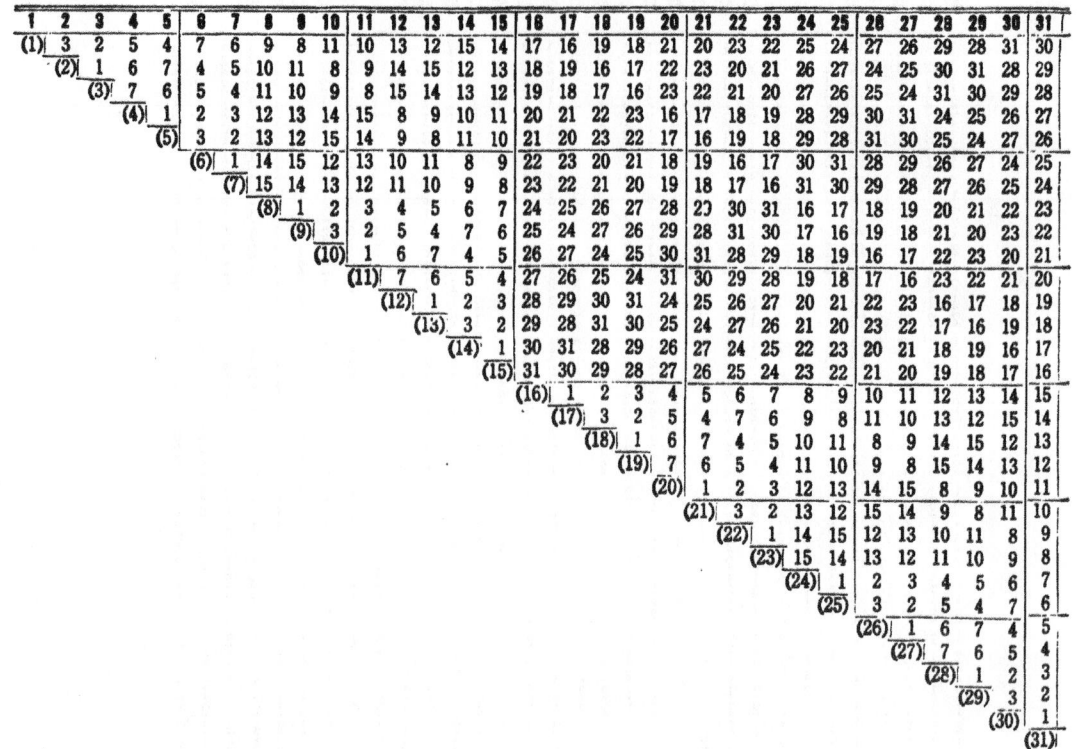

1	2	3	4	5	6	7	8	9	10	11	12	13	14	15	16	17	18	19	20	21	22	23	24	25	26	27	28	29	30	31
(1)	3	2	5	4	7	6	9	8	11	10	13	12	15	14	17	16	19	18	21	20	23	22	25	24	27	26	29	28	31	30
	(2)	1	6	7	4	5	10	11	8	9	14	15	12	13	18	19	16	17	22	23	20	21	26	27	24	25	30	31	28	29
		(3)	7	6	5	4	11	10	9	8	15	14	13	12	19	18	17	16	23	22	21	20	27	26	25	24	31	30	29	28
			(4)	1	2	3	12	13	14	15	8	9	10	11	20	21	22	23	16	17	18	19	28	29	30	31	24	25	26	27
				(5)	3	2	13	12	15	14	9	8	11	10	21	20	23	22	17	16	19	18	29	28	31	30	25	24	27	26
					(6)	1	14	15	12	13	10	11	8	9	22	23	20	21	18	19	16	17	30	31	28	29	26	27	24	25
						(7)	15	14	13	12	11	10	9	8	23	22	21	20	19	18	17	16	31	30	29	28	27	26	25	24
							(8)	1	2	3	4	5	6	7	24	25	26	27	28	29	30	31	16	17	18	19	20	21	22	23
								(9)	3	2	5	4	7	6	25	24	27	26	29	28	31	30	17	16	19	18	21	20	23	22
									(10)	1	6	7	4	5	26	27	24	25	30	31	28	29	18	19	16	17	22	23	20	21
										(11)	7	6	5	4	27	26	25	24	31	30	29	28	19	18	17	16	23	22	21	20
											(12)	1	2	3	28	29	30	31	24	25	26	27	20	21	22	23	16	17	18	19
												(13)	3	2	29	28	31	30	25	24	27	26	21	20	23	22	17	16	19	18
													(14)	1	30	31	28	29	26	27	24	25	22	23	20	21	18	19	16	17
														(15)	31	30	29	28	27	26	25	24	23	22	21	20	19	18	17	16
															(16)	1	2	3	4	5	6	7	8	9	10	11	12	13	14	15
																(17)	3	2	5	4	7	6	9	8	11	10	13	12	15	14
																	(18)	1	6	7	4	5	10	11	8	9	14	15	12	13
																		(19)	7	6	5	4	11	10	9	8	15	14	13	12
																			(20)	1	2	3	12	13	14	15	8	9	10	11
																				(21)	3	2	13	12	15	14	9	8	11	10
																					(22)	1	14	15	12	13	10	11	8	9
																						(23)	15	14	13	12	11	10	9	8
																							(24)	1	2	3	4	5	6	7
																								(25)	3	2	5	4	7	6
																									(26)	1	6	7	4	5
																										(27)	7	6	5	4
																											(28)	1	2	3
																												(29)	3	2
																													(30)	1
																														(31)

L_{64} (2^{63}) (*Continued*)
Interactions between Two Columns

32	33	34	35	36	37	38	39	40	41	42	43	44	45	46	47	48	49	50	51	52	53	54	55	56	57	58	59	60	61
33	32	35	34	37	36	39	38	41	40	43	42	45	44	47	46	49	48	51	50	53	52	55	54	57	56	59	58	61	60
34	35	32	33	38	39	36	37	42	43	40	41	46	47	44	45	50	51	48	49	54	55	52	53	58	59	56	57	62	63
35	34	33	32	39	38	37	36	43	42	41	40	47	46	45	44	51	50	49	48	55	54	53	52	59	58	57	56	63	62
36	37	38	39	32	33	34	35	44	45	46	47	40	41	42	43	52	53	54	55	48	49	50	51	60	61	62	63	56	57
37	36	39	38	33	32	35	34	45	44	47	46	41	40	43	42	53	52	55	54	49	48	51	50	61	60	63	62	57	56
38	39	36	37	34	35	32	33	46	47	44	45	42	43	40	41	54	55	52	53	50	51	48	49	62	63	60	61	58	59
39	38	37	36	35	34	33	32	47	46	45	44	43	42	41	40	55	54	53	52	51	50	49	48	63	62	61	60	59	58
40	41	42	43	44	45	46	47	32	33	34	35	36	37	38	39	56	57	58	59	60	61	62	63	48	49	50	51	52	53
41	40	43	42	45	44	47	46	33	32	35	34	37	36	39	38	57	56	59	58	61	60	63	62	49	48	51	50	53	52
42	43	40	41	46	47	44	45	34	35	32	33	38	39	36	37	58	59	56	57	62	63	60	61	50	51	48	49	54	55
43	42	41	40	47	46	45	44	35	34	33	32	39	38	37	36	59	58	57	56	63	62	61	60	51	50	49	48	55	54
44	45	46	47	40	41	42	43	36	37	38	39	32	33	34	35	60	61	62	63	56	57	58	59	52	53	54	55	48	49
45	44	47	46	41	40	43	42	37	36	39	38	33	32	35	34	61	60	63	62	57	56	59	58	53	52	55	54	49	48
46	47	44	45	42	43	40	41	38	39	36	37	34	35	32	33	62	63	60	61	58	59	56	57	54	55	52	53	50	51
47	46	45	44	43	42	41	40	39	38	37	36	35	34	33	32	63	62	61	60	59	58	57	56	55	54	53	52	51	50
48	49	50	51	52	53	54	55	56	57	58	59	60	61	62	63	32	33	34	35	36	37	38	39	40	41	42	43	44	45
49	48	51	50	53	52	55	54	57	56	59	58	61	60	63	62	33	32	35	34	37	36	39	38	41	40	43	42	45	44
50	51	48	49	54	55	52	53	58	59	56	57	62	63	60	61	34	35	32	33	38	39	36	37	42	43	40	41	46	47
51	50	49	48	55	54	53	52	59	58	57	56	63	62	61	60	35	34	33	32	39	38	37	36	43	42	41	40	47	46
52	53	54	55	48	49	50	51	60	61	62	63	56	57	58	59	36	37	38	39	32	33	34	35	44	45	46	47	40	41
53	52	55	54	49	48	51	50	61	60	63	62	57	56	59	58	37	36	39	38	33	32	35	34	45	44	47	46	41	40
54	55	52	53	50	51	48	49	62	63	60	61	58	59	56	57	38	39	36	37	34	35	32	33	46	47	44	45	42	43
55	54	53	52	51	50	49	48	63	62	61	60	59	58	57	56	39	38	37	36	35	34	33	32	47	46	45	44	43	42
56	57	58	59	60	61	62	63	48	49	50	51	52	53	54	55	40	41	42	43	44	45	46	47	32	33	34	35	36	37
57	56	59	58	61	60	63	62	49	48	51	50	53	52	55	54	41	40	43	42	45	44	47	46	33	32	35	34	37	36
58	59	56	57	62	63	60	61	50	51	48	49	54	55	52	53	42	43	40	41	46	47	44	45	34	35	32	33	38	39
59	58	57	56	63	62	61	60	51	50	49	48	55	54	53	52	43	42	41	40	47	46	45	44	35	34	33	32	39	38
60	61	62	63	56	57	58	59	52	53	54	55	48	49	50	51	44	45	46	47	40	41	42	43	36	37	38	39	32	33
61	60	63	62	57	56	59	58	53	52	55	54	49	48	51	50	45	44	47	46	41	40	43	42	37	36	39	38	33	32
62	63	60	61	58	59	56	57	54	55	52	53	50	51	48	49	46	47	44	45	42	43	40	41	38	39	36	37	34	35
63	62	61	60	59	58	57	56	55	54	53	52	51	50	49	48	47	46	45	44	43	42	41	40	39	38	37	36	35	34
(32)	1	2	3	4	5	6	7	8	9	10	11	12	13	14	15	16	17	18	19	20	21	22	23	24	25	26	27	28	29
	(33)	3	2	5	4	7	6	9	8	11	10	13	12	15	14	17	16	19	18	21	20	23	22	25	24	27	26	29	28
		(34)	1	6	7	4	5	10	11	8	9	14	15	12	13	18	19	16	17	22	23	20	21	26	27	24	25	30	31
			(35)	7	6	5	4	11	10	9	8	15	14	13	12	19	18	17	16	23	22	21	20	27	26	25	24	31	30
				(36)	1	2	3	12	13	14	15	8	9	10	11	20	21	22	23	16	17	18	19	28	29	30	31	24	25
					(37)	3	2	13	12	15	14	9	8	11	10	21	20	23	22	17	16	19	18	29	28	31	30	25	24
						(38)	1	14	15	12	13	10	11	8	9	22	23	20	21	18	19	16	17	30	31	28	29	26	27
							(39)	15	14	13	12	11	10	9	8	23	22	21	20	19	18	17	16	31	30	29	28	27	26
								(40)	1	2	3	4	5	6	7	24	25	26	27	28	29	30	31	16	17	18	19	20	21
									(41)	3	2	5	4	7	6	25	24	27	26	29	28	31	30	17	16	19	18	21	20
										(42)	1	6	7	4	5	26	27	24	25	30	31	28	29	18	19	16	17	22	23
											(43)	7	6	5	4	27	26	25	24	31	30	29	28	19	18	17	16	23	22
												(44)	1	2	3	28	29	30	31	24	25	26	27	20	21	22	23	16	17
													(45)	3	2	29	28	31	30	25	24	27	26	21	20	23	22	17	16
														(46)	1	30	31	28	29	26	27	24	25	22	23	20	21	18	19
															(47)	31	30	29	28	27	26	25	24	23	22	21	20	19	18
																(48)	1	2	3	4	5	6	7	8	9	10	11	12	13
																	(49)	3	2	5	4	7	6	9	8	11	10	13	12
																		(50)	1	6	7	4	5	10	11	8	9	14	15
																			(51)	7	6	5	4	11	10	9	8	15	14
																				(52)	1	2	3	12	13	14	15	8	9
																					(53)	3	2	13	12	15	14	9	8
																						(54)	1	14	15	12	13	10	11
																							(55)	15	14	13	12	11	10
																								(56)	1	2	3	4	5
																									(57)	3	2	5	4
																										(58)	1	6	7
																											(59)	7	6
																												(60)	1
																													(61)

L_{64} (2^{63}) (*Continued*)

(1)

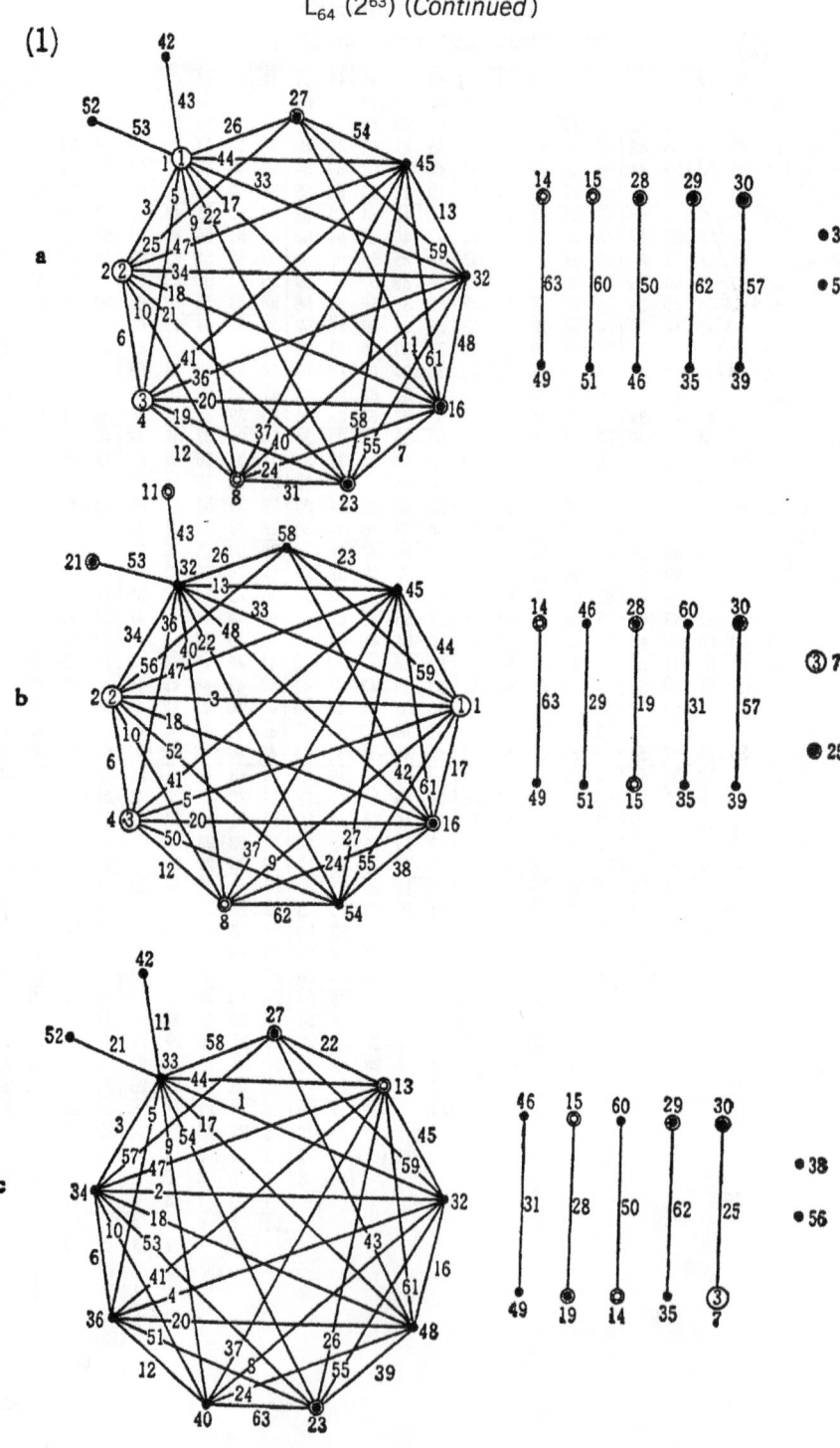

L_{64} (2^{63}) (*Continued*)

(2)

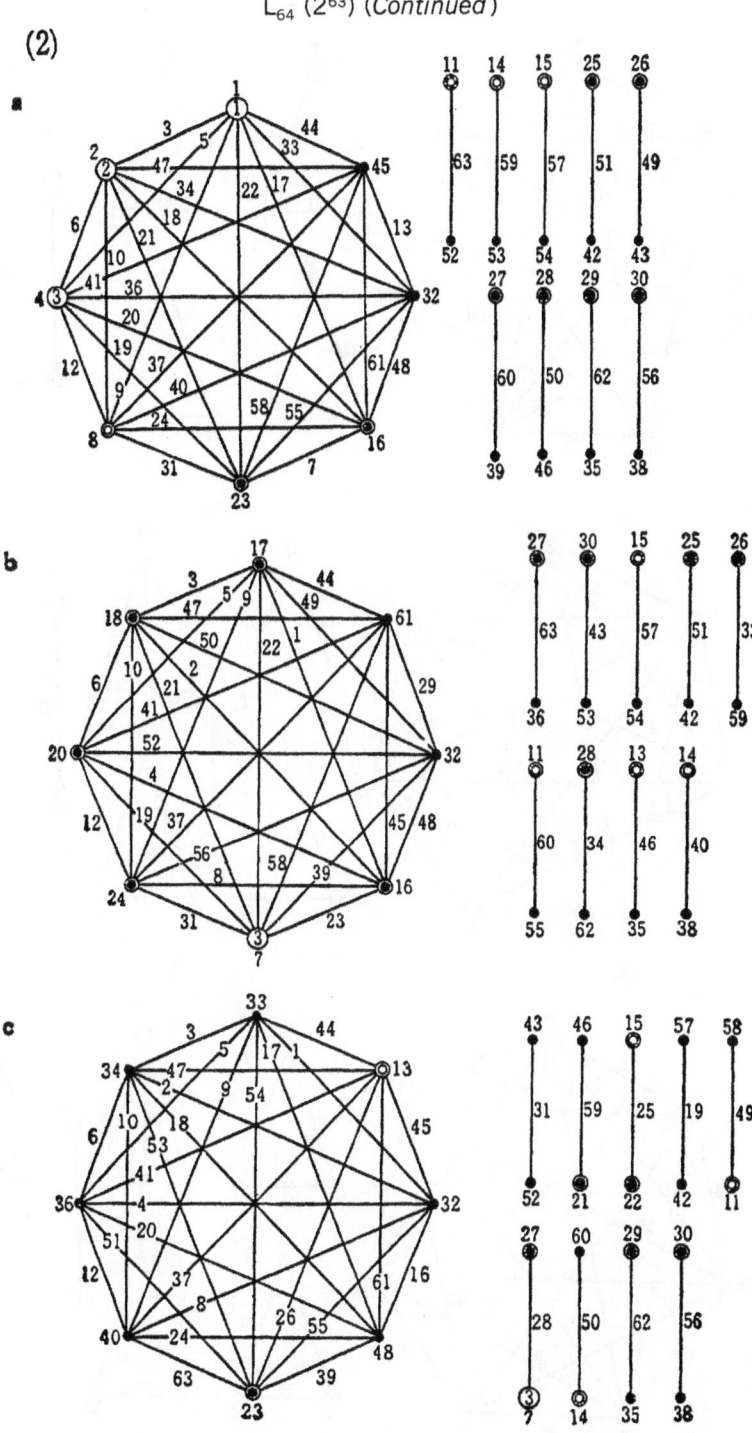

L_{64} (2^{63}) (*Continued*)

(3)

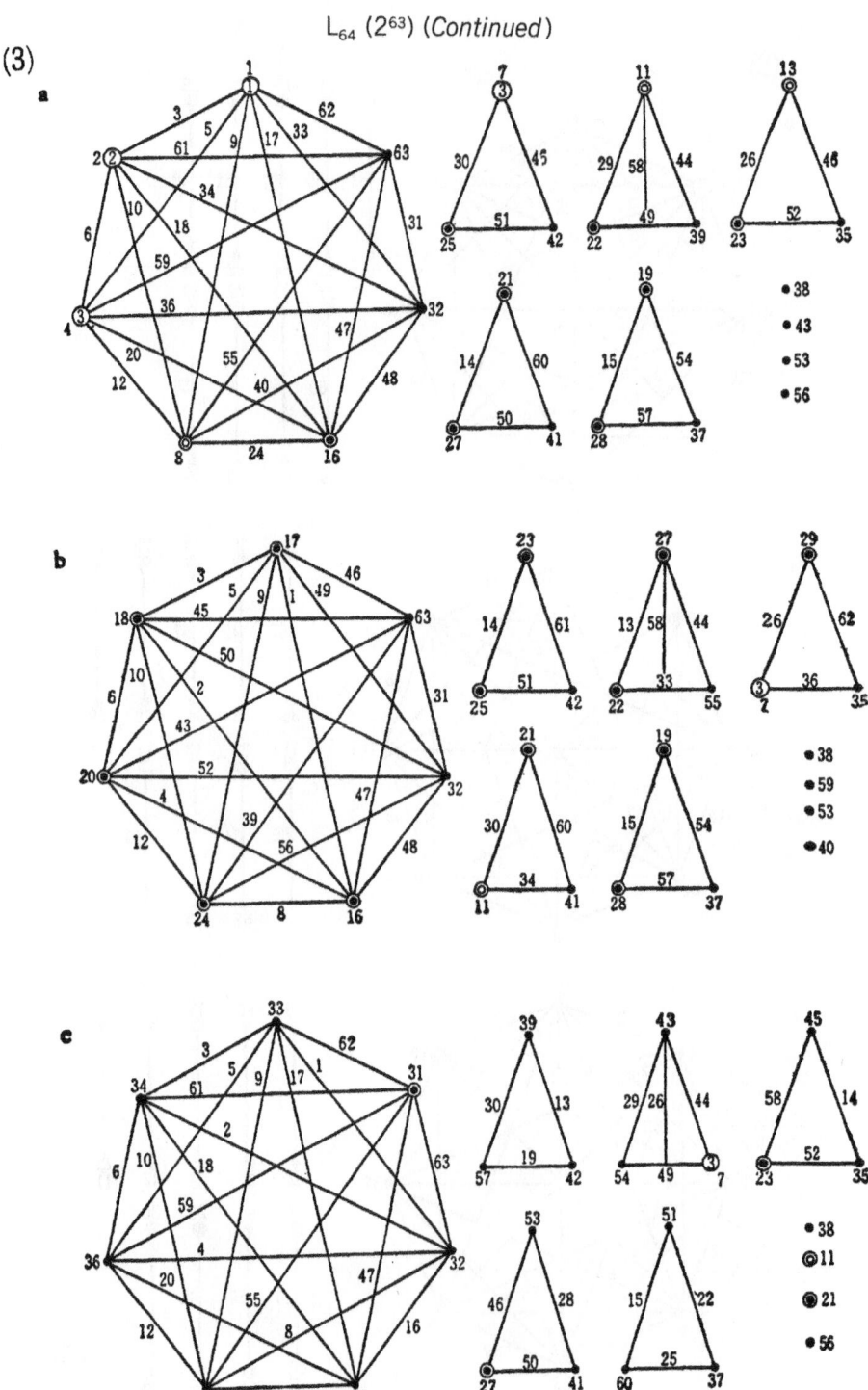

L_{64} (2^{63}) (*Continued*)

(4)

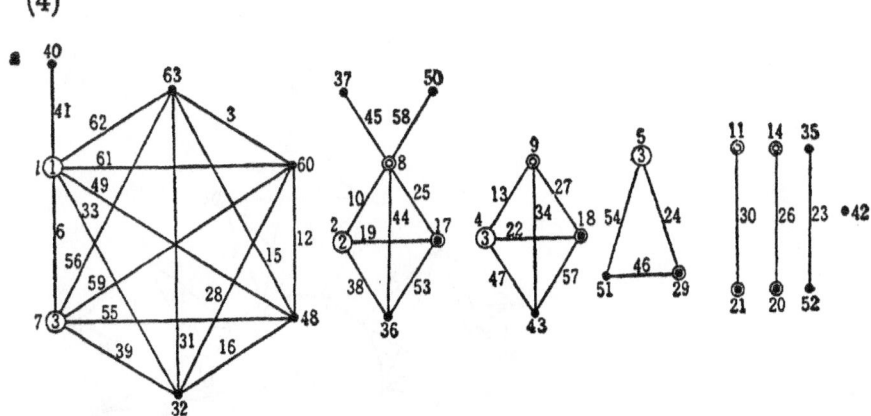

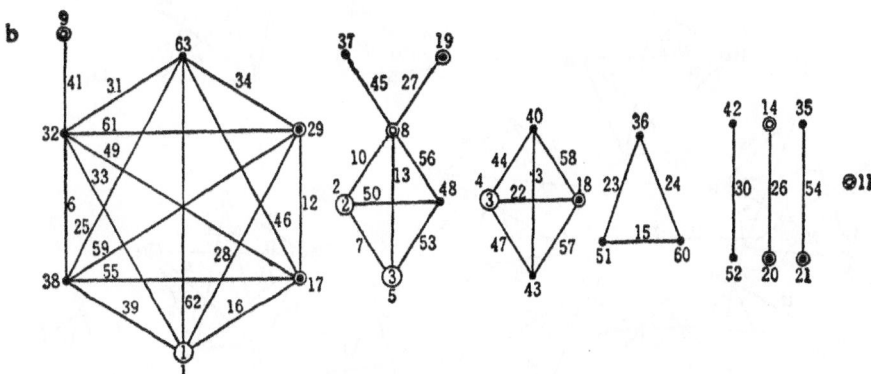

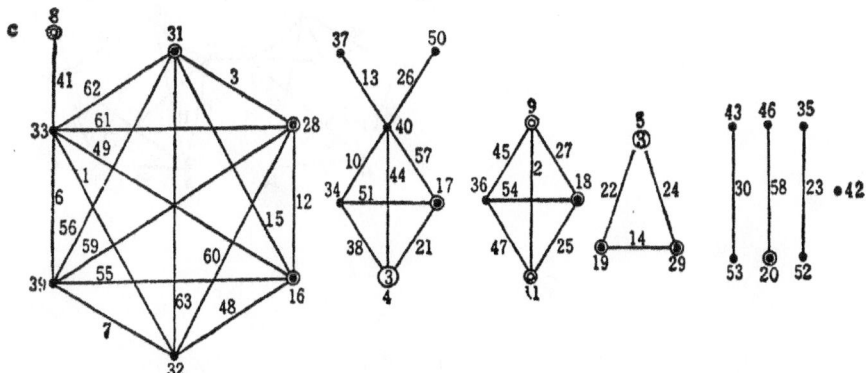

L_{64} (2^{63}) (*Continued*)

(5)

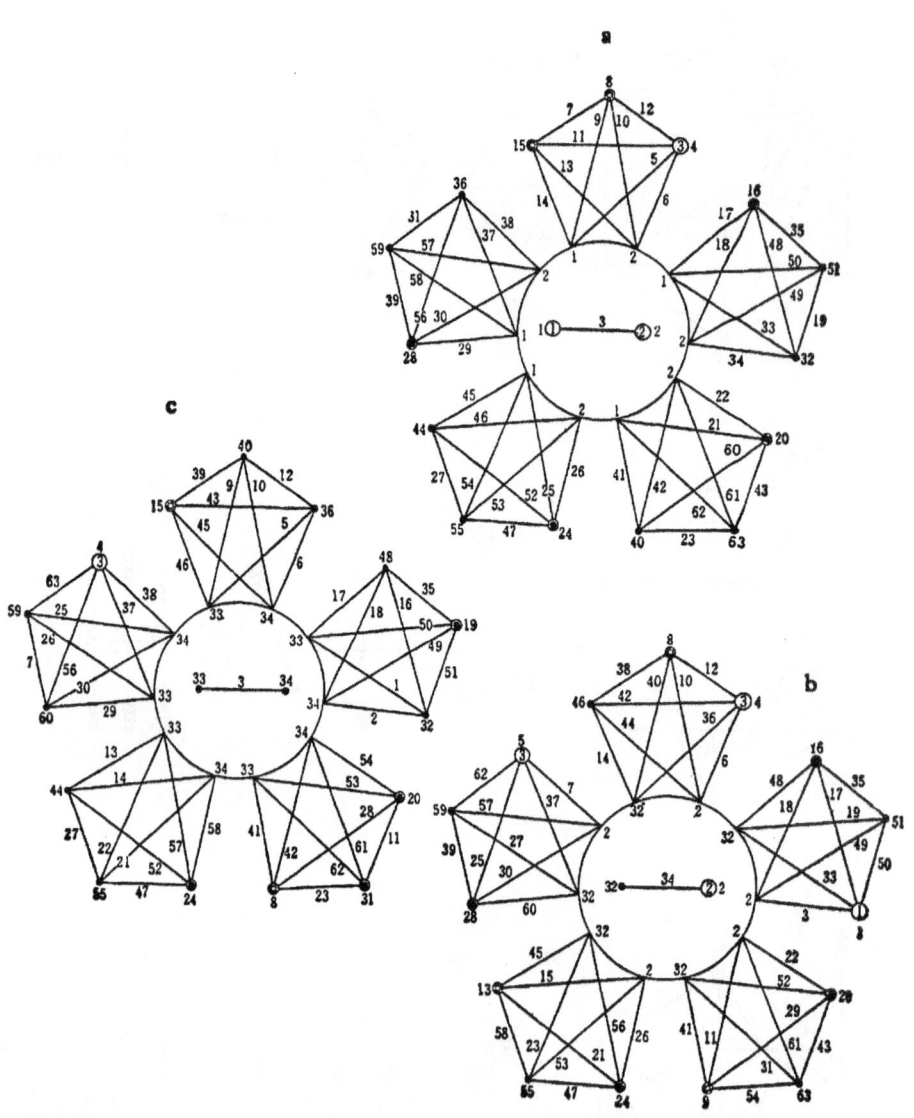

L_{64} (2^{63}) (Continued)

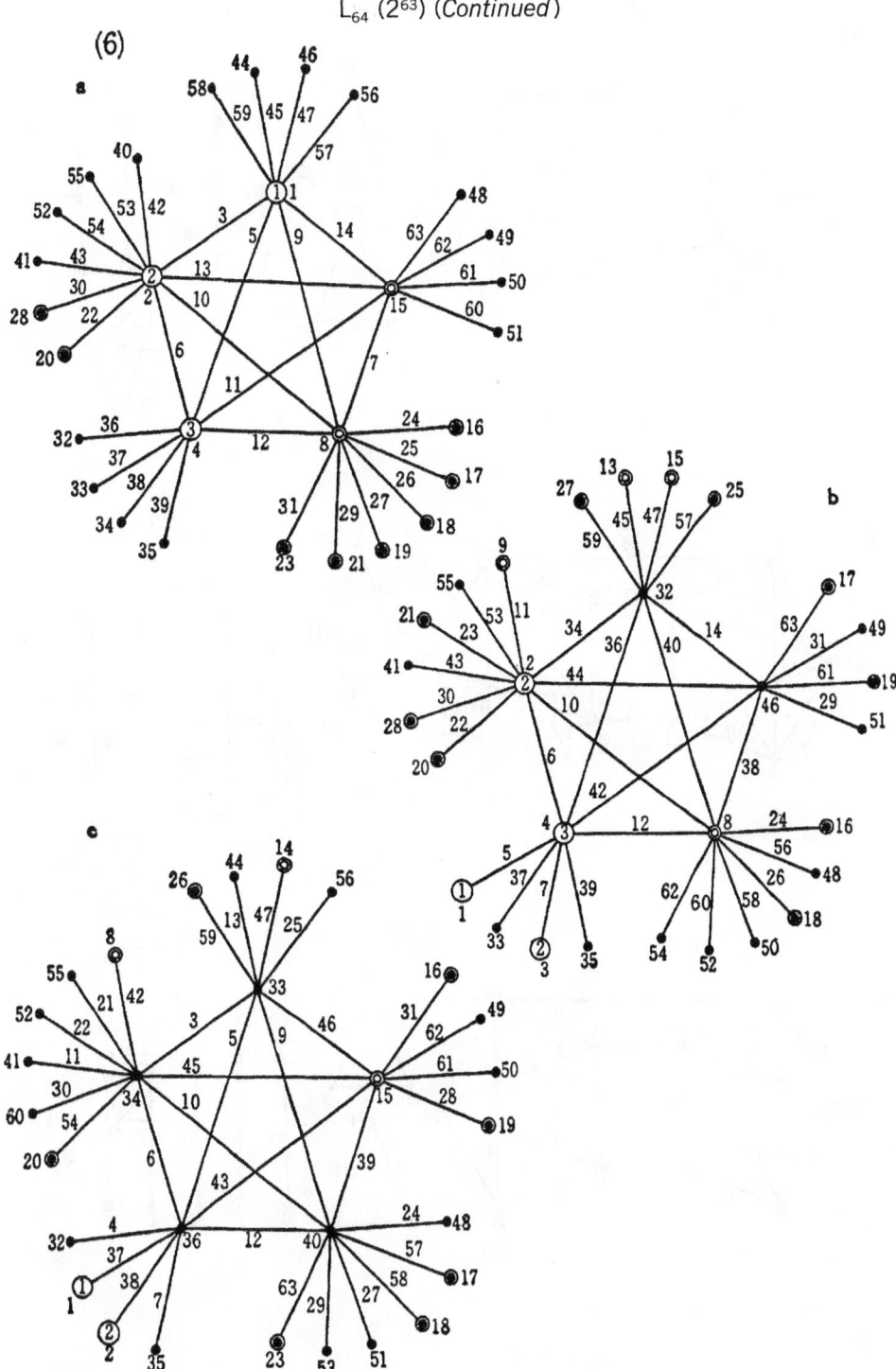

$$L_{64} \ (2^{63}) \ (Continued)$$

(7)

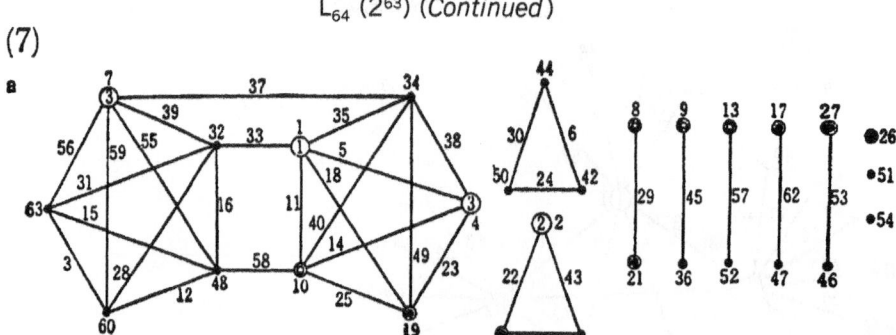

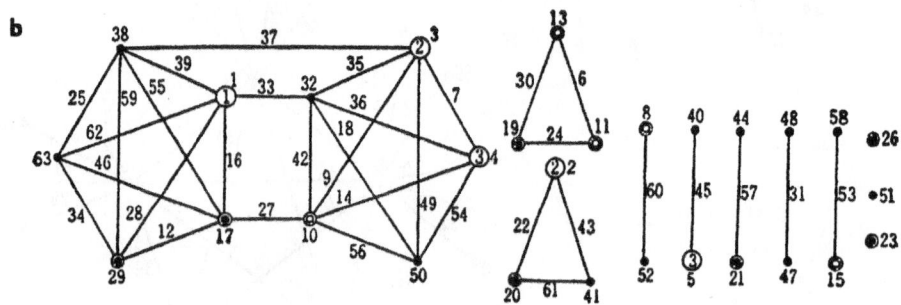

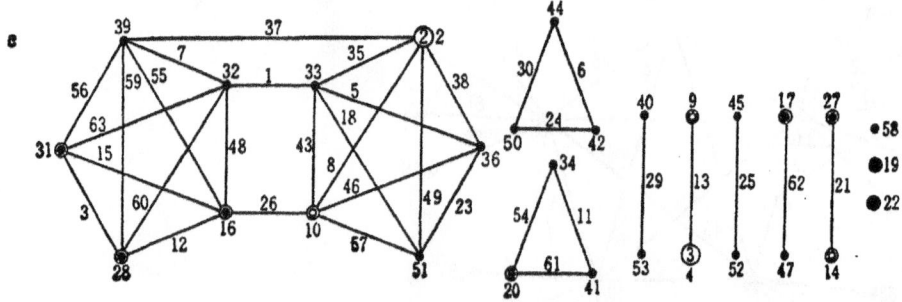

L_{64} (2^{63}) *(Continued)*

(8)

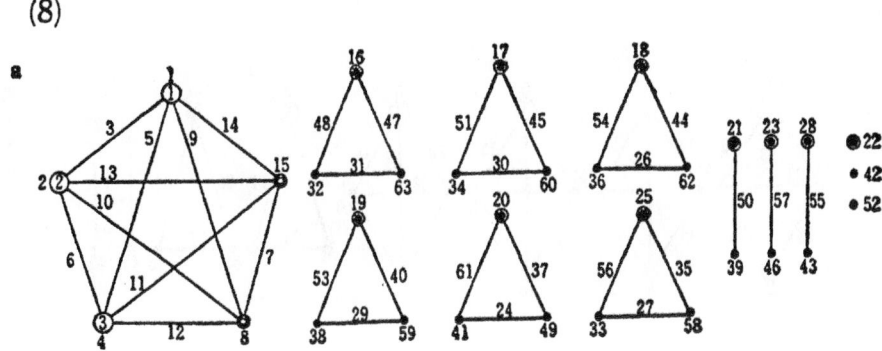

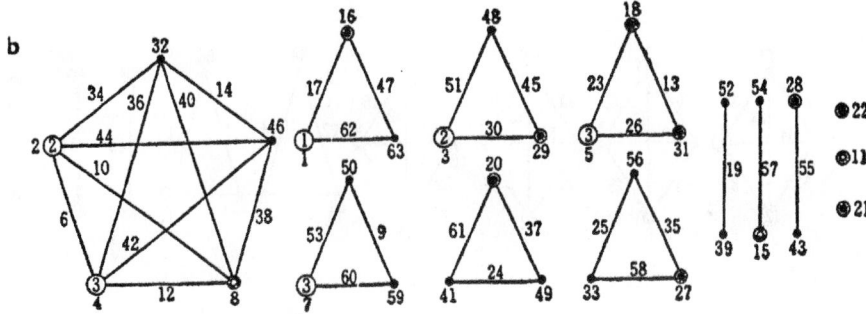

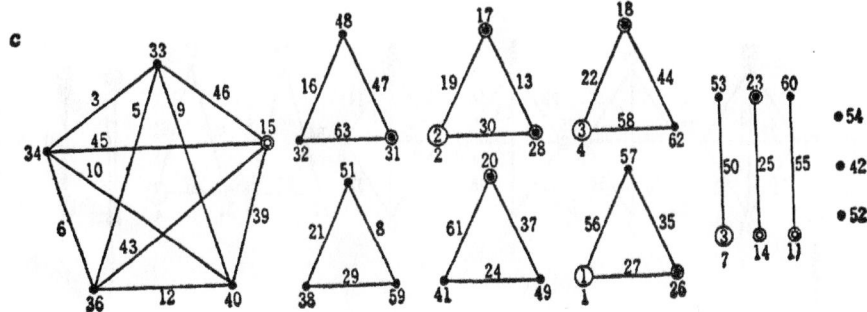

$$L_{64} (2^{63}) (Continued)$$

(9)

(a)

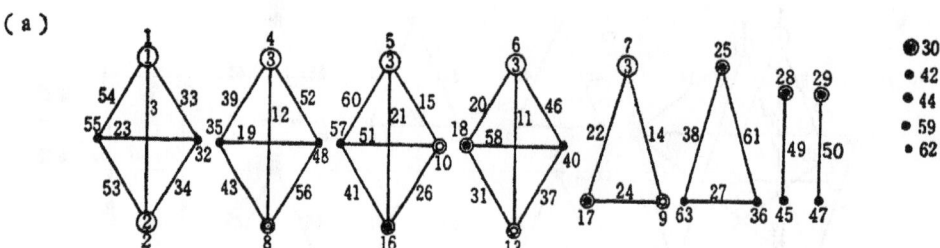

(b)

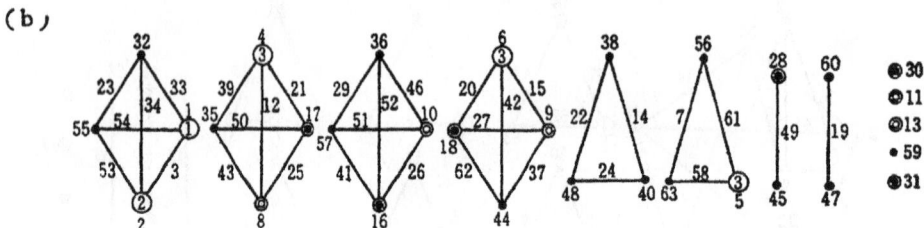

(c)

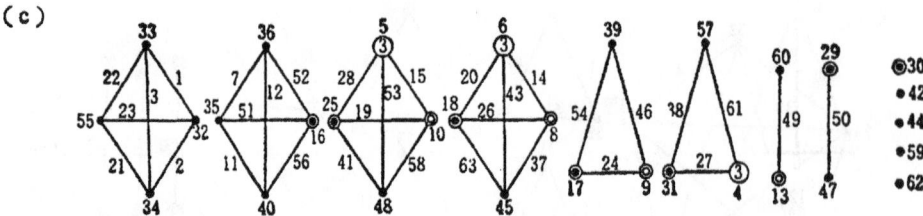

L_{64} (2^{63}) (*Continued*)

(10)

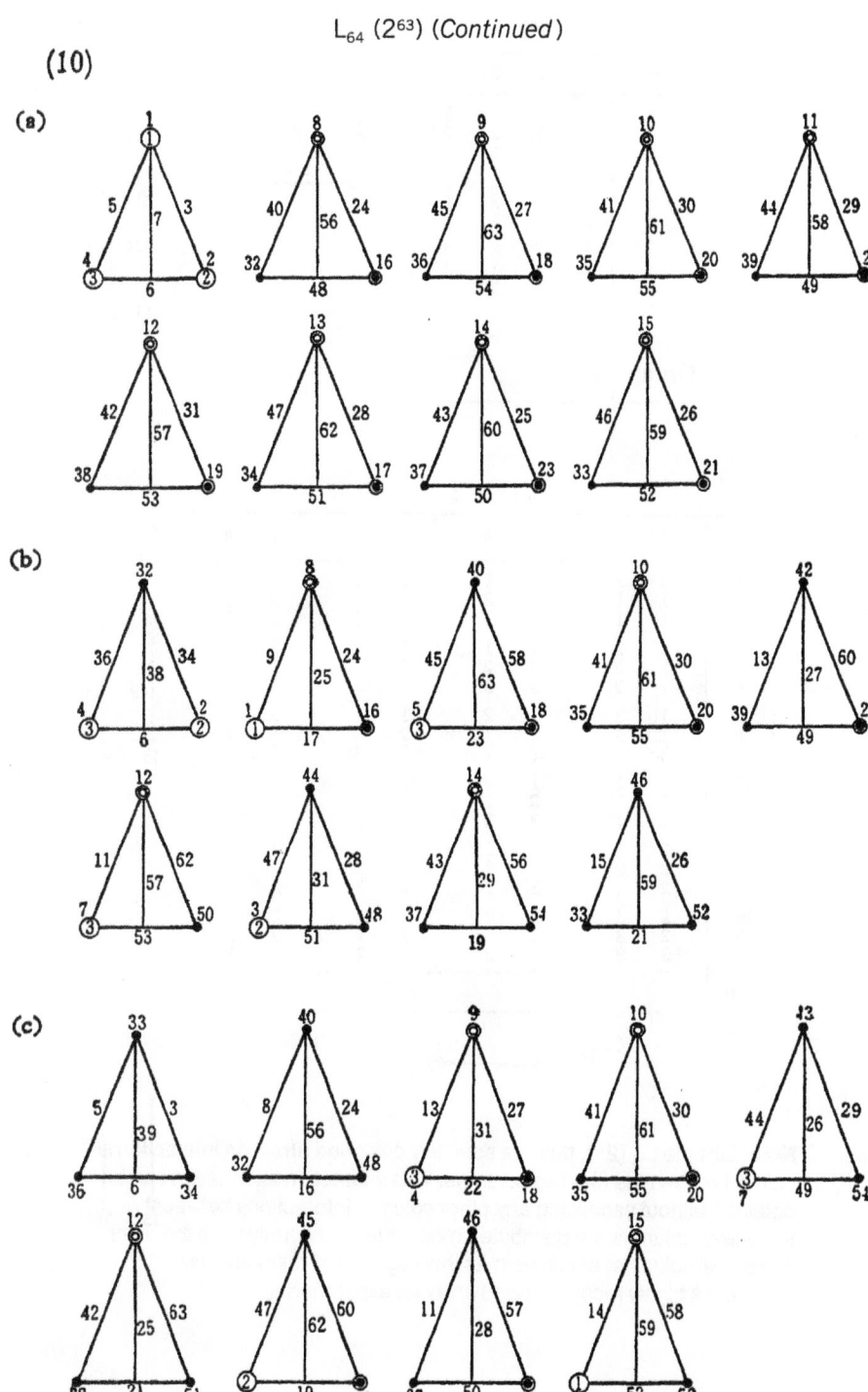

$L_9 (3^4)$

No.	1	2	3	4
1	1	1	1	1
2	1	2	2	2
3	1	3	3	3
4	2	1	2	3
5	2	2	3	1
6	2	3	1	2
7	3	1	3	2
8	3	2	1	3
9	3	3	2	1
	a	b	a b	a b²
Group	1	2		

(1)

$L_{18} (2^1 \times 3^7)$

No.	1	2	3	4	5	6	7	8
1	1	1	1	1	1	1	1	1
2	1	1	2	2	2	2	2	2
3	1	1	3	3	3	3	3	3
4	1	2	1	1	2	2	3	3
5	1	2	2	2	3	3	1	1
6	1	2	3	3	1	1	2	2
7	1	3	1	2	1	3	2	3
8	1	3	2	3	2	1	3	1
9	1	3	3	1	3	2	1	2
10	2	1	1	3	3	2	2	1
11	2	1	2	1	1	3	3	2
12	2	1	3	2	2	1	1	3
13	2	2	1	2	3	1	3	2
14	2	2	2	3	1	2	1	3
15	2	2	3	1	2	3	2	1
16	2	3	1	3	2	3	1	2
17	2	3	2	1	3	1	2	3
18	2	3	3	2	1	2	3	1
Group	1	2	3					

(1)

Note: Like the $L_{12} (2^{11})$, this is a specially designed array. An interaction is built in between the first two columns. This interaction information can be obtained without sacrificing any other column. Interactions between three-level columns are distributed more or less uniformly to all the other three-level columns, which permits investigation of main effects. Thus, it is a highly recommended array for experiments.

$L_{27}(3^{13})$

No.	1	2	3	4	5	6	7	8	9	10	11	12	13
1	1	1	1	1	1	1	1	1	1	1	1	1	1
2	1	1	1	1	2	2	2	2	2	2	2	2	2
3	1	1	1	1	3	3	3	3	3	3	3	3	3
4	1	2	2	2	1	1	1	2	2	2	3	3	3
5	1	2	2	2	2	2	2	3	3	3	1	1	1
6	1	2	2	2	3	3	3	1	1	1	2	2	2
7	1	3	3	3	1	1	1	3	3	3	2	2	2
8	1	3	3	3	2	2	2	1	1	1	3	3	3
9	1	3	3	3	3	3	3	2	2	2	1	1	1
10	2	1	2	3	1	2	3	1	2	3	1	2	3
11	2	1	2	3	2	3	1	2	3	1	2	3	1
12	2	1	2	3	3	1	2	3	1	2	3	1	2
13	2	2	3	1	1	2	3	2	3	1	3	1	2
14	2	2	3	1	2	3	1	3	1	2	1	2	3
15	2	2	3	1	3	1	2	1	2	3	2	3	1
16	2	3	1	2	1	2	3	3	1	2	2	3	1
17	2	3	1	2	2	3	1	1	2	3	3	1	2
18	2	3	1	2	3	1	2	2	3	1	1	2	3
19	3	1	3	2	1	3	2	1	3	2	1	3	2
20	3	1	3	2	2	1	3	2	1	3	2	1	3
21	3	1	3	2	3	2	1	3	2	1	3	2	1
22	3	2	1	3	1	3	2	2	1	3	3	2	1
23	3	2	1	3	2	1	3	3	2	1	1	3	2
24	3	2	1	3	3	2	1	1	3	2	2	1	3
25	3	3	2	1	1	3	2	3	2	1	2	1	3
26	3	3	2	1	2	1	3	1	3	2	3	2	1
27	3	3	2	1	3	2	1	2	1	3	1	3	2
	a	b	a	a	c	a	a	b	a	a	b	a	a
			b	b^2		c	c^2	c	b	b^2	c^3	b^2	b
									c	c^3		c	c^2
Group	1	2			3								

L_{27} (3^{13}) (*Continued*)   Interactions between Two Columns

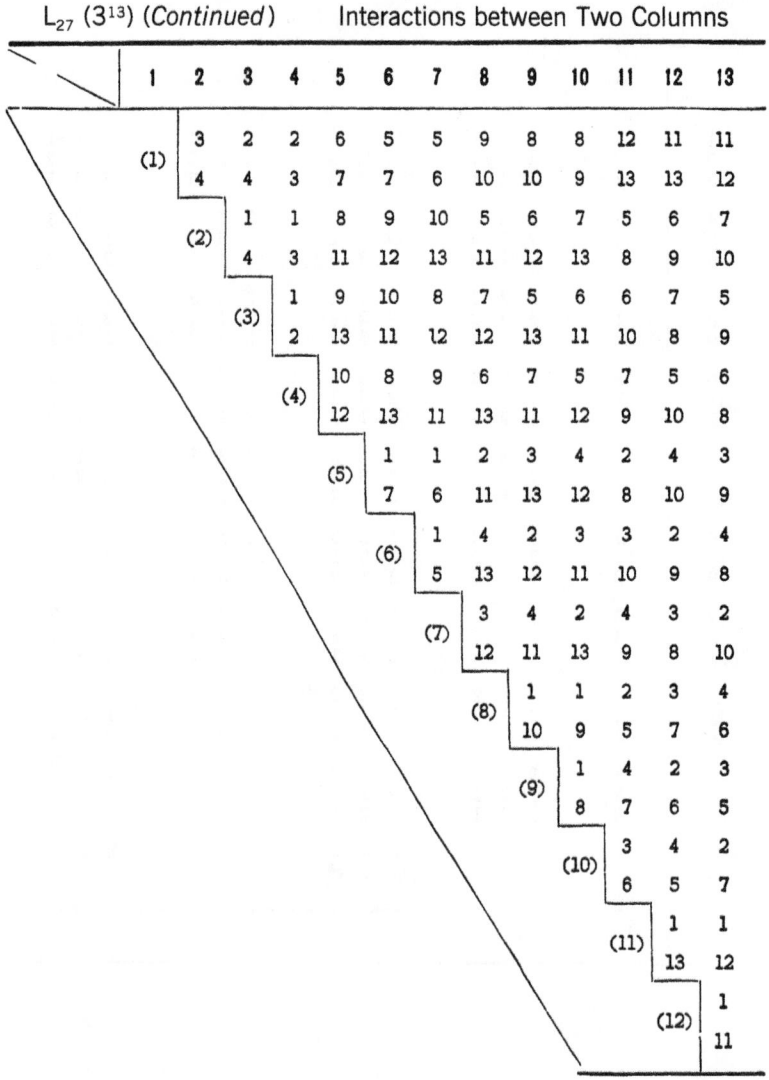

	1	2	3	4	5	6	7	8	9	10	11	12	13
(1)	3	2	2	6	5	5	9	8	8	12	11	11
	4	4	3	7	7	6	10	10	9	13	13	12
(2)		1	1	8	9	10	5	6	7	5	6	7
		4	3	11	12	13	11	12	13	8	9	10
(3)			1	9	10	8	7	5	6	6	7	5
			2	13	11	12	12	13	11	10	8	9
(4)				10	8	9	6	7	5	7	5	6
				12	13	11	13	11	12	9	10	8
(5)					1	1	2	3	4	2	4	3
					7	6	11	13	12	8	10	9
(6)						1	4	2	3	3	2	4
						5	13	12	11	10	9	8
(7)							3	4	2	4	3	2
							12	11	13	9	8	10
(8)								1	1	2	3	4
								10	9	5	7	6
(9)									1	4	2	3
									8	7	6	5
(10)										3	4	2
										6	5	7
(11)											1	1
											13	12
(12)												1
												11

L_{27} (3^{13}) (*Continued*)

(1)

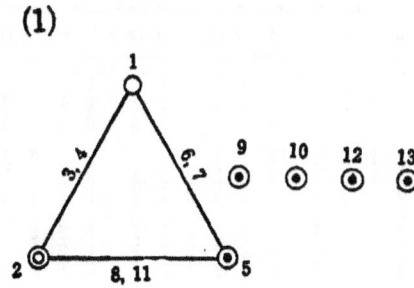

(2)

a

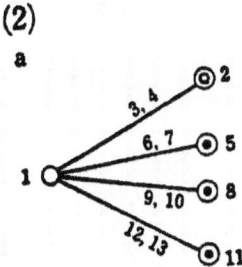

b

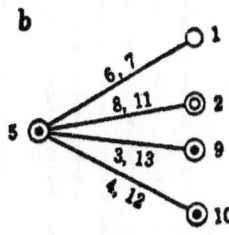

$$L_{54} \ (2^1 \times 3^{25})$$

No.	1	2	3	4	5	6	7	8	9	10	11	12	13	14	15	16	17	18	19	20	21	22	23	24	25	26
1	1	1	1	1	1	1	1	1	1	1	1	1	1	1	2	2	2	2	2	2	1	1	1	1	1	1
2	1	1	1	1	1	1	1	1	2	2	2	2	2	2	2	2	2	2	2	2	2	2	2	2	2	2
3	1	1	1	1	1	1	1	1	3	3	3	3	3	3	3	3	3	3	3	3	3	3	3	3	3	3
4	1	1	2	2	2	2	2	2	1	1	1	1	1	1	2	3	2	3	2	3	2	3	2	3	2	3
5	1	1	2	2	2	2	2	2	2	2	2	2	2	2	3	1	3	1	3	1	3	1	3	1	3	1
6	1	1	2	2	2	2	2	2	3	3	3	3	3	3	1	2	1	2	1	2	1	2	1	2	1	2
7	1	1	3	3	3	3	3	3	1	1	1	1	1	1	3	2	3	2	3	2	3	2	3	2	3	2
8	1	1	3	3	3	3	3	3	2	2	2	2	2	2	1	3	1	3	1	3	1	3	1	3	1	3
9	1	1	3	3	3	3	3	3	3	3	3	3	3	3	2	1	2	1	2	1	2	1	2	1	2	1
10	1	2	1	1	2	2	3	3	1	1	2	2	3	3	1	1	1	2	2	3	2	3	2	3	2	3
11	1	2	1	1	2	2	3	3	2	2	3	3	1	1	2	2	2	3	3	1	3	1	3	1	3	1
12	1	2	1	1	2	2	3	3	3	3	1	1	2	2	3	3	3	1	1	2	1	2	2	1	2	1
13	1	2	2	2	3	3	1	1	1	1	2	2	3	3	2	3	2	3	3	2	3	2	1	1	1	1
14	1	2	2	2	3	3	1	1	2	2	3	3	1	1	3	1	3	1	1	3	1	3	2	2	2	2
15	1	2	2	2	3	3	1	1	3	3	1	1	2	2	1	2	1	2	2	1	2	1	3	3	3	3
16	1	2	3	3	1	1	2	2	1	1	2	2	3	3	3	2	3	2	1	1	1	1	2	3	2	3
17	1	2	3	3	1	1	2	2	2	2	3	3	1	1	1	3	1	3	2	2	2	2	3	1	3	1
18	1	2	3	3	1	1	2	2	3	3	1	1	2	2	2	1	2	1	3	3	3	3	1	2	1	2
19	1	3	1	2	1	3	2	3	1	2	1	3	2	3	1	1	2	3	1	1	3	2	2	3	2	1
20	1	3	1	2	1	3	2	3	2	3	2	1	3	1	2	2	3	1	2	2	1	3	3	1	1	3
21	1	3	1	2	1	3	2	3	3	1	3	2	1	2	3	3	1	2	3	3	2	1	1	2	2	1
22	1	3	2	3	2	1	3	1	1	2	1	3	2	3	2	3	3	2	2	3	1	1	3	2	1	1
23	1	3	2	3	2	1	3	1	2	3	2	1	3	1	3	1	1	3	3	1	2	2	1	3	2	2
24	1	3	2	3	2	1	3	1	3	1	3	2	1	2	1	2	2	1	1	2	3	3	2	1	3	3
25	1	3	3	1	3	2	1	2	1	2	1	3	2	3	3	2	1	1	3	2	2	3	1	1	2	3
26	1	3	3	1	3	2	1	2	2	3	2	1	3	1	1	3	2	2	1	3	3	1	2	2	3	1
27	1	3	3	1	3	2	1	2	3	1	3	2	1	2	2	1	3	3	2	1	1	2	3	3	1	2
28	2	1	1	3	3	2	2	1	1	3	3	2	2	1	1	1	3	2	3	1	2	3	2	3	1	1
29	2	1	1	3	3	2	2	1	2	1	1	3	3	2	2	2	1	3	1	3	3	1	3	1	2	2
30	2	1	1	3	3	2	2	1	3	2	2	1	1	3	3	3	2	1	2	1	1	2	1	2	3	3
31	2	1	2	1	1	3	3	2	1	3	3	2	2	1	2	3	1	1	1	1	3	2	3	2	2	3
32	2	1	2	1	1	3	3	2	2	1	1	3	3	2	3	1	2	2	2	2	1	3	1	3	3	1
33	2	1	2	1	1	3	3	2	3	2	2	1	1	3	1	2	3	3	3	3	2	1	2	1	1	2
34	2	1	3	2	2	1	1	3	1	3	3	2	2	1	3	2	2	3	2	3	1	1	1	1	3	2
35	2	1	3	2	2	1	1	3	2	1	1	3	3	2	1	3	3	1	3	1	2	2	2	2	1	3
36	2	1	3	2	2	1	1	3	3	2	2	1	1	3	2	1	1	2	1	2	3	3	3	3	2	1
37	2	2	1	2	3	1	3	2	1	2	3	1	3	2	1	1	2	3	3	2	1	1	3	2	2	3
38	2	2	1	2	3	1	3	2	2	3	1	2	1	3	2	2	3	1	1	3	2	2	1	3	3	1
39	2	2	1	2	3	1	3	2	3	1	2	3	2	1	3	3	1	2	2	1	3	3	2	1	1	2
40	2	2	2	3	1	2	1	3	1	2	3	1	3	2	2	3	3	2	1	1	2	3	1	1	3	2
41	2	2	2	3	1	2	1	3	2	3	1	2	1	3	3	1	1	3	2	2	3	1	2	2	1	3
42	2	2	2	3	1	2	1	3	3	1	2	3	2	1	1	2	2	1	3	3	1	2	3	3	2	1
43	2	2	3	1	2	3	2	1	1	2	3	1	3	2	3	2	1	1	2	3	3	2	2	3	1	1
44	2	2	3	1	2	3	2	1	2	3	1	2	1	3	1	3	2	2	3	1	1	3	3	1	2	2
45	2	2	3	1	2	3	2	1	3	1	2	3	2	1	2	1	3	3	1	2	2	1	1	2	3	3
46	2	3	1	3	2	3	1	2	1	3	2	3	1	2	1	1	3	2	2	3	3	2	1	1	2	3
47	2	3	1	3	2	3	1	2	2	1	3	1	2	3	2	2	1	3	3	1	1	3	2	2	3	1
48	2	3	1	3	2	3	1	2	3	2	1	2	3	1	3	3	2	1	1	2	2	1	3	3	1	2
49	2	3	2	1	3	1	2	3	1	3	2	3	1	2	2	3	1	1	3	2	1	1	2	3	3	2
50	2	3	2	1	3	1	2	3	2	1	3	1	2	3	3	1	2	2	1	3	2	2	3	1	1	3
51	2	3	2	1	3	1	2	3	3	2	1	2	3	1	1	2	3	3	2	1	3	3	1	2	2	1
52	2	3	3	2	1	2	3	1	1	3	2	3	1	2	3	2	2	3	1	1	2	3	3	2	1	1
53	2	3	3	2	1	2	3	1	2	1	3	1	2	3	1	3	3	1	2	2	3	1	1	3	2	2
54	2	3	3	2	1	2	3	1	3	2	1	2	3	1	2	1	1	2	3	3	1	2	2	1	3	3
Group	1	2	3						4																	

$$L_{54} \ (2^1 \times 3^{25}) \ (\textit{Continued})$$

(1)

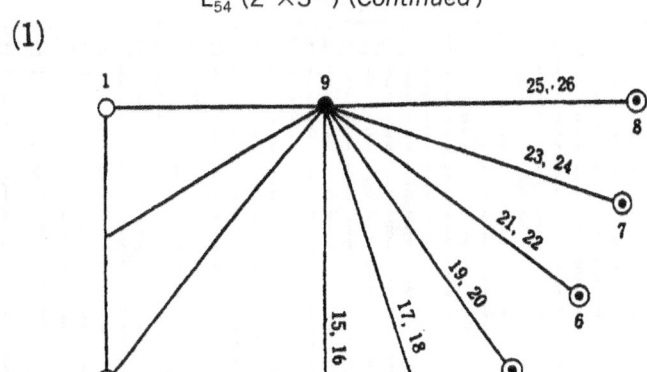

Notes:

1. Column 1 has two levels.

2. An interaction has been built in between Columns 1 and 2 without the need to sacrifice another column.

3. The two-way interactions between Columns 1 and 9, 2 and 9, and the 3-way interaction between Columns 1, 2, and 9 are distributed more or less uniformly to Columns 10-14.

   The interactions between Columns 1 and 9; 2 and 9; and 1, 2, and 9 can be obtained from 2-way and 3-way tables respectively.

4. Also, the interaction between the 6-level factor (created by combining Columns 1 and 2) and Column 9 can be obtained when leaving Columns 10-14 empty.

L_{81} (3^{40})

No.	1	2	3	4	5	6	7	8	9	10	11	12	13	14	15	16	17	18	19	20	21	22	23	24	25	26	27	28	29	30	31	32	33	34	35	36	37	38	39	40
1	1	1	1	1	1	1	1	1	1	1	1	1	1	1	1	1	1	1	1	1	1	1	1	1	1	1	1	1	1	1	1	1	1	1	1	1	1	1	1	1
2	1	1	1	1	1	1	1	1	1	1	1	1	1	2	2	2	2	2	2	2	2	2	2	2	2	2	2	2	2	2	2	2	2	2	2	2	2	2	2	2
3	1	1	1	1	1	1	1	1	1	1	1	1	1	3	3	3	3	3	3	3	3	3	3	3	3	3	3	3	3	3	3	3	3	3	3	3	3	3	3	3
4	1	1	1	1	2	2	2	2	2	2	2	2	2	1	1	1	1	1	1	1	1	1	2	2	2	2	2	2	2	2	2	3	3	3	3	3	3	3	3	3
5	1	1	1	1	2	2	2	2	2	2	2	2	2	2	2	2	2	2	2	2	2	2	3	3	3	3	3	3	3	3	3	1	1	1	1	1	1	1	1	1
6	1	1	1	1	2	2	2	2	2	2	2	2	2	3	3	3	3	3	3	3	3	3	1	1	1	1	1	1	1	1	1	2	2	2	2	2	2	2	2	2
7	1	1	1	1	3	3	3	3	3	3	3	3	3	1	1	1	1	1	1	1	1	1	3	3	3	3	3	3	3	3	3	2	2	2	2	2	2	2	2	2
8	1	1	1	1	3	3	3	3	3	3	3	3	3	2	2	2	2	2	2	2	2	2	1	1	1	1	1	1	1	1	1	3	3	3	3	3	3	3	3	3
9	1	1	1	1	3	3	3	3	3	3	3	3	3	3	3	3	3	3	3	3	3	3	2	2	2	2	2	2	2	2	2	1	1	1	1	1	1	1	1	1
10	1	2	2	2	1	1	1	2	2	2	3	3	3	1	1	1	2	2	2	3	3	3	1	1	1	2	2	2	3	3	3	1	1	1	2	2	2	3	3	3
11	1	2	2	2	1	1	1	2	2	2	3	3	3	2	2	2	3	3	3	1	1	1	2	2	2	3	3	3	1	1	1	2	2	2	3	3	3	1	1	1
12	1	2	2	2	1	1	1	2	2	2	3	3	3	3	3	3	1	1	1	2	2	2	3	3	3	1	1	1	2	2	2	3	3	3	1	1	1	2	2	2
13	1	2	2	2	2	2	2	3	3	3	1	1	1	1	1	1	2	2	2	3	3	3	2	2	2	3	3	3	1	1	1	3	3	3	1	1	1	2	2	2
14	1	2	2	2	2	2	2	3	3	3	1	1	1	2	2	2	3	3	3	1	1	1	3	3	3	1	1	1	2	2	2	1	1	1	2	2	2	3	3	3
15	1	2	2	2	2	2	2	3	3	3	1	1	1	3	3	3	1	1	1	2	2	2	1	1	1	2	2	2	3	3	3	2	2	2	3	3	3	1	1	1
16	1	2	2	2	3	3	3	1	1	1	2	2	2	1	1	1	2	2	2	3	3	3	3	3	3	1	1	1	2	2	2	2	2	2	3	3	3	1	1	1
17	1	2	2	2	3	3	3	1	1	1	2	2	2	2	2	2	3	3	3	1	1	1	1	1	1	2	2	2	3	3	3	3	3	3	1	1	1	2	2	2
18	1	2	2	2	3	3	3	1	1	1	2	2	2	3	3	3	1	1	1	2	2	2	2	2	2	3	3	3	1	1	1	1	1	1	2	2	2	3	3	3
19	1	3	3	3	1	1	1	3	3	3	2	2	2	1	1	1	3	3	3	2	2	2	1	1	1	3	3	3	2	2	2	1	1	1	3	3	3	2	2	2
20	1	3	3	3	1	1	1	3	3	3	2	2	2	2	2	2	1	1	1	3	3	3	2	2	2	1	1	1	3	3	3	2	2	2	1	1	1	3	3	3
21	1	3	3	3	1	1	1	3	3	3	2	2	2	3	3	3	2	2	2	1	1	1	3	3	3	2	2	2	1	1	1	3	3	3	2	2	2	1	1	1
22	1	3	3	3	2	2	2	1	1	1	3	3	3	1	1	1	3	3	3	2	2	2	2	2	2	1	1	1	3	3	3	3	3	3	2	2	2	1	1	1
23	1	3	3	3	2	2	2	1	1	1	3	3	3	2	2	2	1	1	1	3	3	3	3	3	3	2	2	2	1	1	1	1	1	1	3	3	3	2	2	2
24	1	3	3	3	2	2	2	1	1	1	3	3	3	3	3	3	2	2	2	1	1	1	1	1	1	3	3	3	2	2	2	2	2	2	1	1	1	3	3	3
25	1	3	3	3	3	3	3	2	2	2	1	1	1	1	1	1	3	3	3	2	2	2	3	3	3	2	2	2	1	1	1	2	2	2	1	1	1	3	3	3
26	1	3	3	3	3	3	3	2	2	2	1	1	1	2	2	2	1	1	1	3	3	3	1	1	1	3	3	3	2	2	2	3	3	3	2	2	2	1	1	1
27	1	3	3	3	3	3	3	2	2	2	1	1	1	3	3	3	2	2	2	1	1	1	2	2	2	1	1	1	3	3	3	1	1	1	3	3	3	2	2	2
28	2	1	2	3	1	2	3	1	2	3	1	2	3	1	2	3	1	2	3	1	2	3	1	2	3	1	2	3	1	2	3	1	2	3	1	2	3	1	2	3
29	2	1	2	3	1	2	3	1	2	3	1	2	3	2	3	1	2	3	1	2	3	1	2	3	1	2	3	1	2	3	1	2	3	1	2	3	1	2	3	1
30	2	1	2	3	1	2	3	1	2	3	1	2	3	3	1	2	3	1	2	3	1	2	3	1	2	3	1	2	3	1	2	3	1	2	3	1	2	3	1	2
31	2	1	2	3	2	3	1	2	3	1	2	3	1	1	2	3	1	2	3	1	2	3	2	3	1	2	3	1	2	3	1	3	1	2	3	1	2	3	1	2
32	2	1	2	3	2	3	1	2	3	1	2	3	1	2	3	1	2	3	1	2	3	1	3	1	2	3	1	2	3	1	2	1	2	3	1	2	3	1	2	3
33	2	1	2	3	2	3	1	2	3	1	2	3	1	3	1	2	3	1	2	3	1	2	1	2	3	1	2	3	1	2	3	2	3	1	2	3	1	2	3	1
34	2	1	2	3	3	1	2	3	1	2	3	1	2	1	2	3	1	2	3	1	2	3	3	1	2	3	1	2	3	1	2	2	3	1	2	3	1	2	3	1
35	2	1	2	3	3	1	2	3	1	2	3	1	2	2	3	1	2	3	1	2	3	1	1	2	3	1	2	3	1	2	3	3	1	2	3	1	2	3	1	2
36	2	1	2	3	3	1	2	3	1	2	3	1	2	3	1	2	3	1	2	3	1	2	2	3	1	2	3	1	2	3	1	1	2	3	1	2	3	1	2	3
37	2	2	3	1	1	2	3	2	3	1	3	1	2	1	2	3	2	3	1	3	1	2	1	2	3	2	3	1	3	1	2	1	2	3	2	3	1	3	1	2
38	2	2	3	1	1	2	3	2	3	1	3	1	2	2	3	1	3	1	2	1	2	3	2	3	1	3	1	2	1	2	3	2	3	1	3	1	2	1	2	3
39	2	2	3	1	1	2	3	2	3	1	3	1	2	3	1	2	1	2	3	2	3	1	3	1	2	1	2	3	2	3	1	3	1	2	1	2	3	2	3	1
40	2	2	3	1	2	3	1	3	1	2	1	2	3	1	2	3	2	3	1	3	1	2	2	3	1	3	1	2	1	2	3	3	1	2	1	2	3	2	3	1
41	2	2	3	1	2	3	1	3	1	2	1	2	3	2	3	1	3	1	2	1	2	3	3	1	2	1	2	3	2	3	1	1	2	3	2	3	1	3	1	2
42	2	2	3	1	2	3	1	3	1	2	1	2	3	3	1	2	1	2	3	2	3	1	1	2	3	2	3	1	3	1	2	2	3	1	3	1	2	1	2	3
43	2	2	3	1	3	1	2	1	2	3	2	3	1	1	2	3	2	3	1	3	1	2	3	1	2	1	2	3	2	3	1	2	3	1	3	1	2	1	2	3
44	2	2	3	1	3	1	2	1	2	3	2	3	1	2	3	1	3	1	2	1	2	3	1	2	3	2	3	1	3	1	2	3	1	2	1	2	3	2	3	1
45	2	2	3	1	3	1	2	1	2	3	2	3	1	3	1	2	1	2	3	2	3	1	2	3	1	3	1	2	1	2	3	1	2	3	2	3	1	3	1	2
46	2	3	1	2	1	2	3	3	1	2	2	3	1	1	2	3	3	1	2	2	3	1	1	2	3	3	1	2	2	3	1	1	2	3	3	1	2	2	3	1
47	2	3	1	2	1	2	3	3	1	2	2	3	1	2	3	1	1	2	3	3	1	2	2	3	1	1	2	3	3	1	2	2	3	1	1	2	3	3	1	2
48	2	3	1	2	1	2	3	3	1	2	2	3	1	3	1	2	2	3	1	1	2	3	3	1	2	2	3	1	1	2	3	3	1	2	2	3	1	1	2	3
49	2	3	1	2	2	3	1	1	2	3	3	1	2	1	2	3	3	1	2	2	3	1	2	3	1	1	2	3	3	1	2	3	1	2	2	3	1	1	2	3
50	2	3	1	2	2	3	1	1	2	3	3	1	2	2	3	1	1	2	3	3	1	2	3	1	2	2	3	1	1	2	3	1	2	3	3	1	2	2	3	1
51	2	3	1	2	2	3	1	1	2	3	3	1	2	3	1	2	2	3	1	1	2	3	1	2	3	3	1	2	2	3	1	2	3	1	1	2	3	3	1	2
52	2	3	1	2	3	1	2	2	3	1	1	2	3	1	2	3	3	1	2	2	3	1	3	1	2	2	3	1	1	2	3	2	3	1	1	2	3	3	1	2
53	2	3	1	2	3	1	2	2	3	1	1	2	3	2	3	1	1	2	3	3	1	2	1	2	3	3	1	2	2	3	1	3	1	2	2	3	1	1	2	3
54	2	3	1	2	3	1	2	2	3	1	1	2	3	3	1	2	2	3	1	1	2	3	2	3	1	1	2	3	3	1	2	1	2	3	3	1	2	2	3	1
55	3	1	3	2	1	3	2	1	3	2	1	3	2	1	3	2	1	3	2	1	3	2	1	3	2	1	3	2	1	3	2	1	3	2	1	3	2	1	3	2
56	3	1	3	2	1	3	2	1	3	2	1	3	2	2	1	3	2	1	3	2	1	3	2	1	3	2	1	3	2	1	3	2	1	3	2	1	3	2	1	3
57	3	1	3	2	1	3	2	1	3	2	1	3	2	3	2	1	3	2	1	3	2	1	3	2	1	3	2	1	3	2	1	3	2	1	3	2	1	3	2	1
58	3	1	3	2	2	1	3	2	1	3	2	1	3	1	3	2	1	3	2	1	3	2	2	1	3	2	1	3	2	1	3	3	2	1	3	2	1	3	2	1
59	3	1	3	2	2	1	3	2	1	3	2	1	3	2	1	3	2	1	3	2	1	3	3	2	1	3	2	1	3	2	1	1	3	2	1	3	2	1	3	2
60	3	1	3	2	2	1	3	2	1	3	2	1	3	3	2	1	3	2	1	3	2	1	1	3	2	1	3	2	1	3	2	2	1	3	2	1	3	2	1	3
61	3	1	3	2	3	2	1	3	2	1	3	2	1	1	3	2	1	3	2	1	3	2	3	2	1	3	2	1	3	2	1	2	1	3	2	1	3	2	1	3
62	3	1	3	2	3	2	1	3	2	1	3	2	1	2	1	3	2	1	3	2	1	3	1	3	2	1	3	2	1	3	2	3	2	1	3	2	1	3	2	1
63	3	1	3	2	3	2	1	3	2	1	3	2	1	3	2	1	3	2	1	3	2	1	2	1	3	2	1	3	2	1	3	1	3	2	1	3	2	1	3	2
64	3	2	1	3	1	3	2	2	1	3	3	2	1	1	3	2	2	1	3	3	2	1	1	3	2	2	1	3	3	2	1	1	3	2	2	1	3	3	2	1
65	3	2	1	3	1	3	2	2	1	3	3	2	1	2	1	3	3	2	1	1	3	2	2	1	3	3	2	1	1	3	2	2	1	3	3	2	1	1	3	2
66	3	2	1	3	1	3	2	2	1	3	3	2	1	3	2	1	1	3	2	2	1	3	3	2	1	1	3	2	2	1	3	3	2	1	1	3	2	2	1	3
67	3	2	1	3	2	1	3	3	2	1	1	3	2	1	3	2	2	1	3	3	2	1	2	1	3	3	2	1	1	3	2	3	2	1	1	3	2	2	1	3
68	3	2	1	3	2	1	3	3	2	1	1	3	2	2	1	3	3	2	1	1	3	2	3	2	1	1	3	2	2	1	3	1	3	2	2	1	3	3	2	1
69	3	2	1	3	2	1	3	3	2	1	1	3	2	3	2	1	1	3	2	2	1	3	1	3	2	2	1	3	3	2	1	2	1	3	3	2	1	1	3	2
70	3	2	1	3	3	2	1	1	3	2	2	1	3	1	3	2	2	1	3	3	2	1	3	2	1	1	3	2	2	1	3	2	1	3	3	2	1	1	3	2
71	3	2	1	3	3	2	1	1	3	2	2	1	3	2	1	3	3	2	1	1	3	2	1	3	2	2	1	3	3	2	1	3	2	1	1	3	2	2	1	3
72	3	2	1	3	3	2	1	1	3	2	2	1	3	3	2	1	1	3	2	2	1	3	2	1	3	3	2	1	1	3	2	1	3	2	2	1	3	3	2	1
73	3	3	2	1	1	3	2	3	2	1	2	1	3	1	3	2	3	2	1	2	1	3	1	3	2	3	2	1	2	1	3	1	3	2	3	2	1	2	1	3
74	3	3	2	1	1	3	2	3	2	1	2	1	3	2	1	3	1	3	2	3	2	1	2	1	3	1	3	2	3	2	1	2	1	3	1	3	2	3	2	1
75	3	3	2	1	1	3	2	3	2	1	2	1	3	3	2	1	2	1	3	1	3	2	3	2	1	2	1	3	1	3	2	3	2	1	2	1	3	1	3	2
76	3	3	2	1	2	1	3	1	3	2	3	2	1	1	3	2	3	2	1	2	1	3	2	1	3	1	3	2	3	2	1	3	2	1	2	1	3	1	3	2
77	3	3	2	1	2	1	3	1	3	2	3	2	1	2	1	3	1	3	2	3	2	1	3	2	1	2	1	3	1	3	2	1	3	2	3	2	1	2	1	3
78	3	3	2	1	2	1	3	1	3	2	3	2	1	3	2	1	2	1	3	1	3	2	1	3	2	3	2	1	2	1	3	2	1	3	1	3	2	3	2	1
79	3	3	2	1	3	2	1	2	1	3	1	3	2	1	3	2	3	2	1	2	1	3	3	2	1	2	1	3	1	3	2	2	1	3	1	3	2	3	2	1
80	3	3	2	1	3	2	1	2	1	3	1	3	2	2	1	3	1	3	2	3	2	1	1	3	2	3	2	1	2	1	3	3	2	1	2	1	3	1	3	2
81	3	3	2	1	3	2	1	2	1	3	1	3	2	3	2	1	2	1	3	1	3	2	2	1	3	1	3	2	3	2	1	1	3	2	3	2	1	2	1	3
	a	b	a b	a b²	c	a c	a c²	b c	a b c	a b² c²	b c²	a b² c	a b c²	d	a d	a d²	b d	a b d	a b² d²	b d²	a b² d	a b d²	c d	a c d	a c² d²	b c d	a b c d	a b² c² d²	b c² d²	a b² c d	a b c² d²	c d²	a c² d	a c d²	b c² d	a b c² d	a b² c d²	b c d²	a b² c² d	a b c d²
Group	1	2	2	2	3	3	3	3	3	3	3	3	3	4	4	4	4	4	4	4	4	4	4	4	4	4	4	4	4	4	4	4	4	4	4	4	4	4	4	4

L_{81} (3^{40}) *(Continued)* Interactions between Two Columns

	14	15	16	17	18	19	20	21	22	23	24	25	26	27	28	29	30	31	32	33	34	35	36	37	38	39	40
(1)	15	14	14	18	17	17	21	20	20	24	23	23	27	26	26	30	29	29	33	32	32	36	35	35	39	38	38
	16	16	15	19	19	18	22	22	21	25	25	24	28	28	27	31	31	30	34	34	33	37	37	36	40	40	39
(2)	17	18	19	14	15	16	14	15	16	26	27	28	23	24	25	23	24	25	35	36	37	32	33	34	32	33	34
	20	21	22	20	21	22	17	18	19	29	30	31	26	27	28	38	39	40	38	39	40	35	36	37	35	36	37
(3)	18	19	17	16	14	15	15	16	14	27	28	26	25	23	24	24	25	23	36	37	35	34	32	33	33	34	32
	22	20	21	21	22	20	19	17	18	30	31	29	28	26	27	40	38	39	40	38	37	35	34	32	34	32	36
(4)	19	17	18	15	16	14	16	14	15	28	26	27	24	25	23	25	23	24	37	35	36	33	34	32	34	32	33
	21	22	20	22	20	21	18	19	17	31	29	30	27	28	26	39	40	38	38	39	36	37	35	37	35	37	35
(5)	23	24	25	26	27	28	29	30	31	14	15	16	17	18	19	20	21	22	14	15	16	17	18	19	20	21	22
	32	33	34	35	36	37	38	39	40	23	24	25	26	27	28	29	30	31	23	24	25	26	27	28	29	30	31
(6)	24	25	23	27	28	26	30	31	29	16	14	15	19	17	18	22	20	21	15	16	14	18	19	17	21	22	20
	34	32	33	37	35	36	40	38	39	25	23	24	28	26	27	31	29	30	24	25	23	27	28	26	30	31	29
(7)	25	23	24	28	26	27	31	29	30	15	16	14	18	19	17	21	22	20	16	14	15	19	17	18	22	20	21
	33	34	32	36	37	35	39	40	38	24	25	23	27	28	26	30	31	29	25	23	27	28	26	30	31	29	16
(8)	26	27	28	29	30	31	23	24	25	20	21	22	14	15	16	17	18	19	17	18	19	20	21	22	14	15	16
	38	39	40	32	33	34	35	36	37	29	30	31	23	24	25	26	27	28	26	27	28	29	30	31	23	24	25
(9)	27	28	26	30	31	29	24	25	23	22	20	21	16	14	15	19	17	18	18	19	17	21	22	20	15	16	14
	40	38	39	34	32	33	37	35	36	31	29	30	25	23	24	28	26	27	28	26	27	31	29	30	25	23	24
(10)	28	26	27	31	29	30	25	23	24	21	22	20	15	16	14	18	19	17	19	17	18	22	20	15	16	14	15
	39	40	38	33	34	32	36	37	35	37	35	36	40	38	39	34	32	33	30	31	29	24	25	23	27	28	26
(11)	29	30	31	23	24	25	26	27	28	17	18	19	20	21	22	14	15	16	20	21	22	14	15	16	17	18	19
	35	36	37	38	39	40	32	33	34	35	36	37	38	39	40	32	33	34	29	30	31	23	24	25	26	27	28
(12)	30	31	29	24	25	23	27	28	26	19	17	18	22	20	21	16	14	15	21	22	20	15	16	14	19	17	18
	37	35	36	40	38	39	34	32	33	39	40	38	37	35	36	33	34	32	33	37	35	28	26	27	31	29	30
(13)	31	29	30	25	23	24	28	26	27	18	19	17	21	22	20	15	16	14	22	20	21	16	14	15	19	17	18
	36	37	35	39	40	38	33	34	32	40	38	39	34	32	33	37	35	36	27	28	26	31	29	30	25	24	23
(14)		16	15	20	22	21	17	19	18	23	38	40	39	35	37	36	23	25	24	29	31	30	26	28	27	31	30
		1	1	2	3	4	2	4	3	5	6	7	8	9	10	11	12	13	5	7	6	11	13	12	8	10	9
(15)			14	22	21	20	19	18	17	34	23	36	35	37	36	38	37	36	35	24	23	31	30	29	26	27	26
			1	4	2	3	3	2	4	7	5	6	10	8	9	13	11	12	6	5	7	12	11	13	9	8	10
(16)				21	20	22	18	17	19	33	32	34	39	38	40	36	35	37	24	23	25	30	29	31	27	26	28
				3	4	2	4	3	2	11	12	13	5	6	7	8	9	10	9	5	7	6	11	13	12	8	10
(17)					19	18	14	16	15	38	40	39	35	37	36	32	34	33	29	31	30	26	28	27	23	25	24
					1	2	3	4	2	13	11	12	7	5	6	10	8	9	6	5	7	12	11	13	9	8	10
(18)						17	16	15	14	40	39	38	37	36	35	34	33	32	31	30	29	28	27	26	25	24	23
						3	4	2	3	11	12	13	11	6	7	5	9	10	8	9	7	6	5	13	12	11	19
(19)							15	14	16	39	38	40	36	35	37	33	32	34	30	29	31	27	26	28	24	23	25
							8	9	10	11	12	13	5	6	7	9	5	7	6	13	12	11	9	8	10		
(20)								22	21	35	37	36	32	34	33	38	40	39	26	28	27	23	25	24	29	31	30
								1	10	8	9	13	11	12	7	5	6	12	11	13	9	8	10	6	5	7	
(21)									20	37	36	35	34	33	32	40	39	38	27	26	25	24	23	31	30	29	
									20	37	36	35	13	11	12	6	7	5	13	12	11	10	9	8	7	6	5
(22)										36	35	37	33	32	34	39	38	40	27	26	28	23	25	30	29	31	
										25	24	29	31	30	26	28	27	14	16	15	20	22	21	17	19	18	
(23)											23	31	30	26	28	27	14	16	15	20	22	21	17	19	18		
											3	4	2	3	3	2	4	7	5	6	10	8	9	13	11	12	
(24)												30	29	28	27	26	16	15	14	22	21	20	19	18	17		
												1	4	2	3	3	2	4	7	5	9	10	8	12	13	11	
(25)													29	31	27	26	28	15	14	16	21	20	22	18	17	19	
													3	4	2	4	3	2	6	7	5	9	10	8	12	13	11
(26)														28	27	23	25	24	20	22	21	17	19	18	14	16	15
														1	2	3	4	2	11	12	13	5	6	7	8	9	10
(27)															26	25	24	23	22	21	20	19	18	17	16	15	14
															1	4	2	3	13	11	12	11	6	7	5	9	10
(28)																24	23	25	21	20	22	18	17	19	15	14	16
																3	4	2	12	13	11	6	7	5	9	10	8
(29)																	31	30	17	19	18	14	16	15	20	22	21
																	1	1	8	9	10	11	12	13	5	6	7
(30)																		29	19	18	17	16	15	14	22	21	20
																		1	10	8	9	13	11	12	7	5	6
(31)																			18	17	19	15	14	16	21	20	22
																			1	2	3	4	2	3	3	2	4
(32)																				34	33	38	40	39	35	37	36
																				1	4	2	3	3	2	4	3
(33)																					32	40	39	38	37	36	35
																					3	4	2	4	3	2	4
(34)																						39	38	40	36	35	37
																						1	2	3	4	2	3
(35)																							37	36	32	34	33
																							1	4	2	3	3
(36)																								35	34	33	32
																								3	4	2	3
(37)																									33	32	34
																									3	4	2
(38)																										40	39
																										1	1
(39)																											38

Note: For interactions among Columns 1-13, see the Triangular Table for array L_{27} (3^{13}).

L_{81} (3^{40}) (*Continued*)

(1)

a

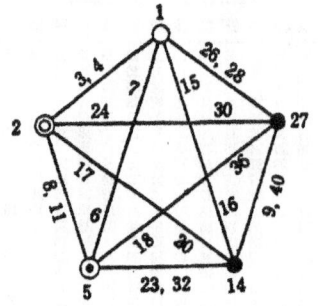
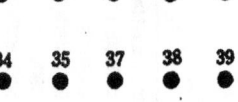

10	12	13	19	21
⊙	⊙	⊙	●	●
22	25	29	31	33
●	●	●	●	●
34	35	37	38	39
●	●	●	●	●

b

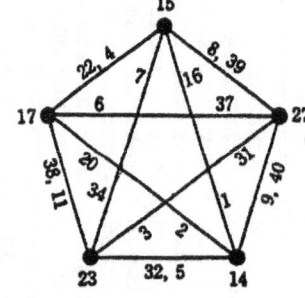

28	30	36	19	21
●	●	●	●	●
18	25	29	13	33
●	●	●	⊙	●
24	35	12	26	10
●	●	⊙	●	⊙

c

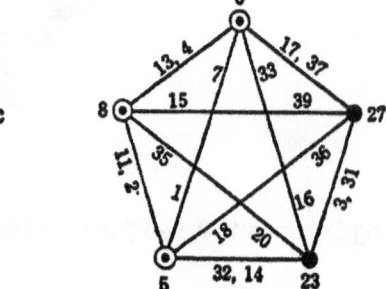
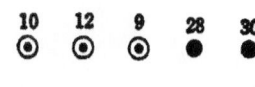

10	12	9	28	30
⊙	⊙	⊙	●	●
40	25	29	22	24
●	●	●	●	●
34	26	19	38	21
●	●	●	●	●

L_{81} (3^{40}) *(Continued)*

(2)

a

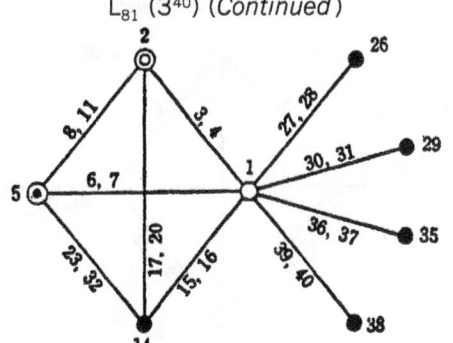

b

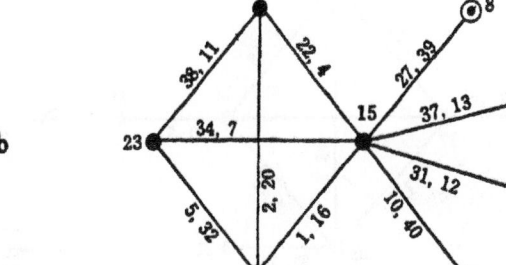

c

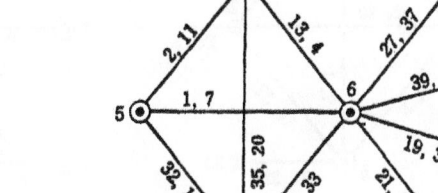

L_{81} (3^{40}) *(Continued)*

(3)

a

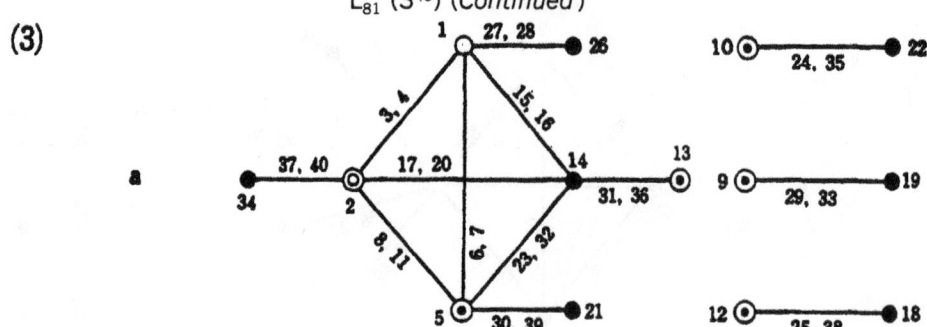

b

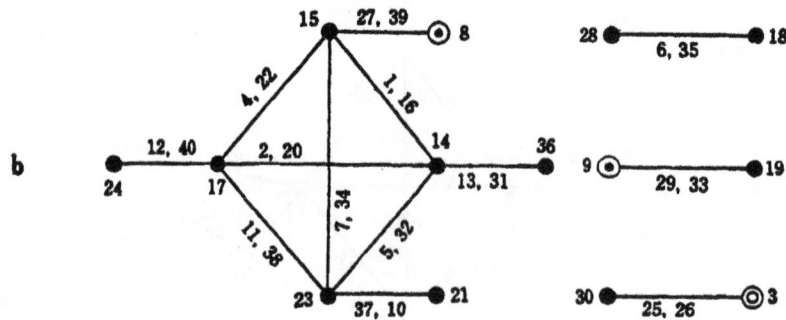

c

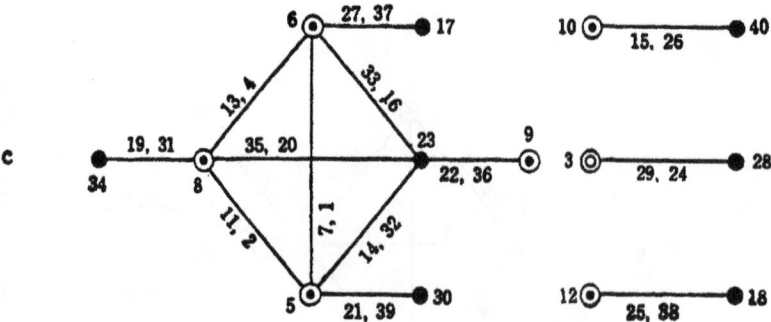

L_{81} (3^{40}) *(Continued)*

(4)

a

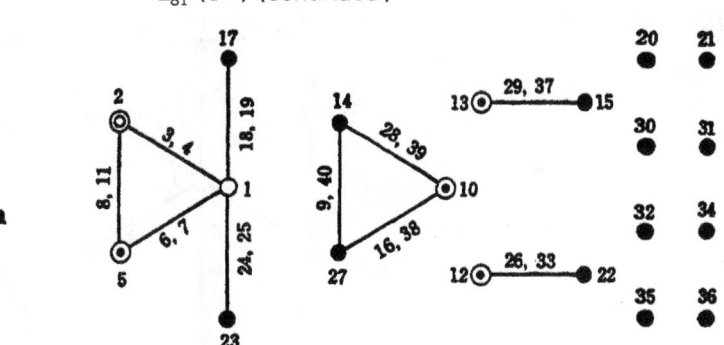

b

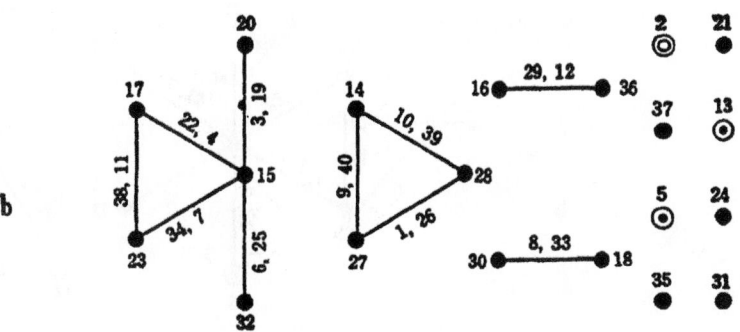

c

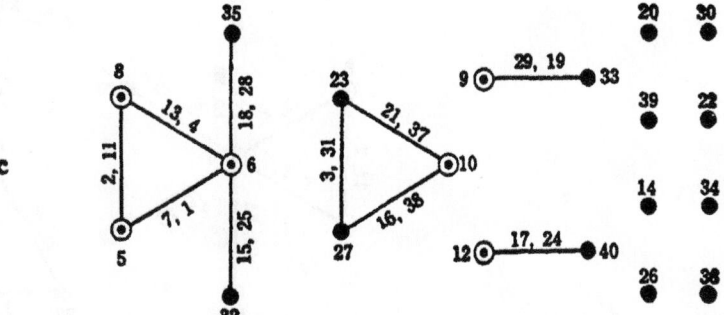

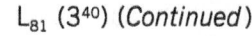

L_{81} (3^{40}) (*Continued*)

(5)

a

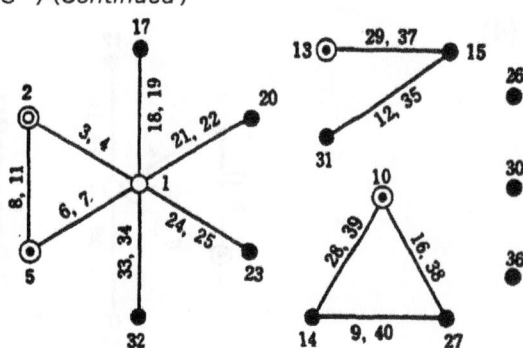

b

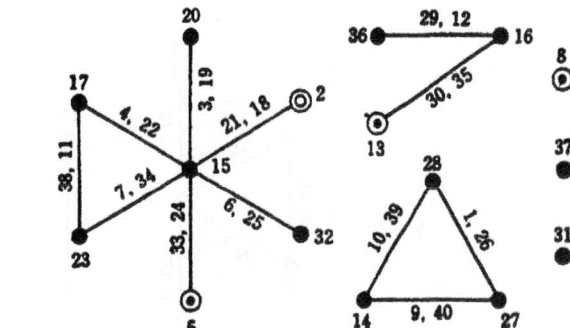

c

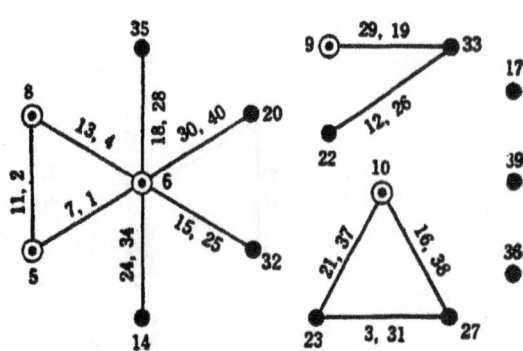

L_{81} (3^{40}) (Continued)

(6)

a

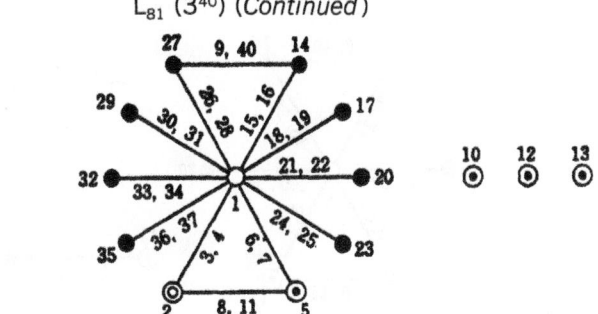

b

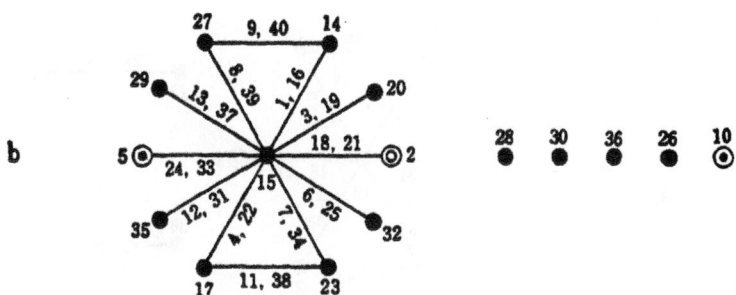

c

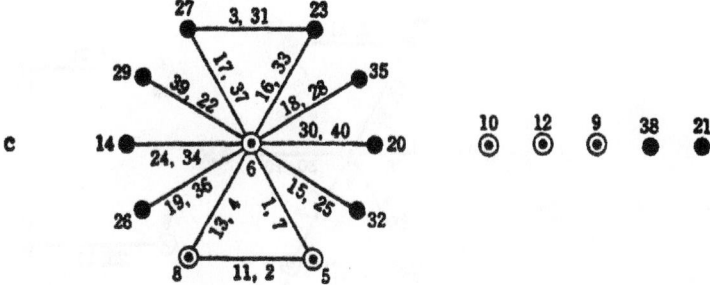

L_{81} (3^{40}) *(Continued)*

(7)

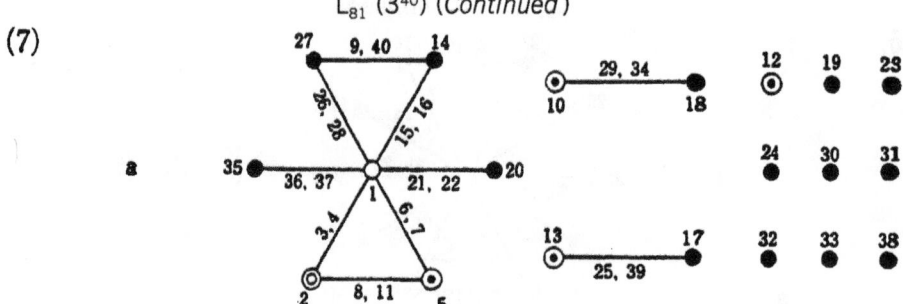

a

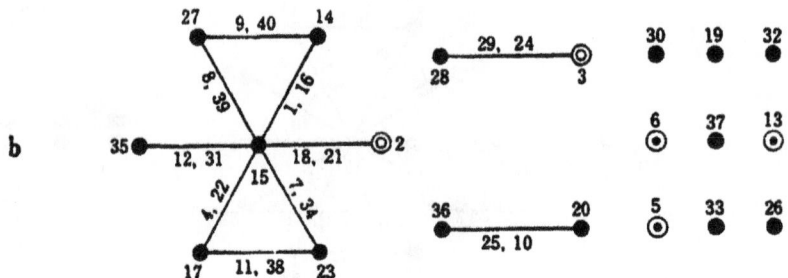

b

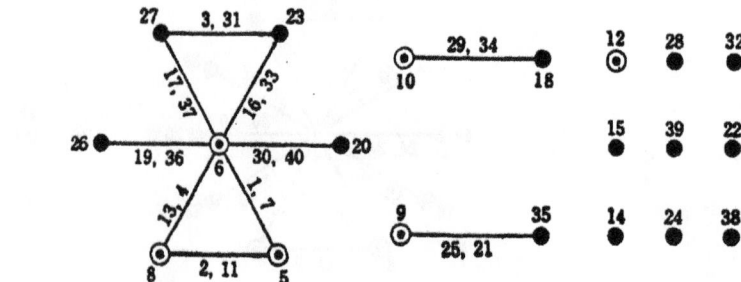

c

L_{81} (3^{40}) *(Continued)*

(8)

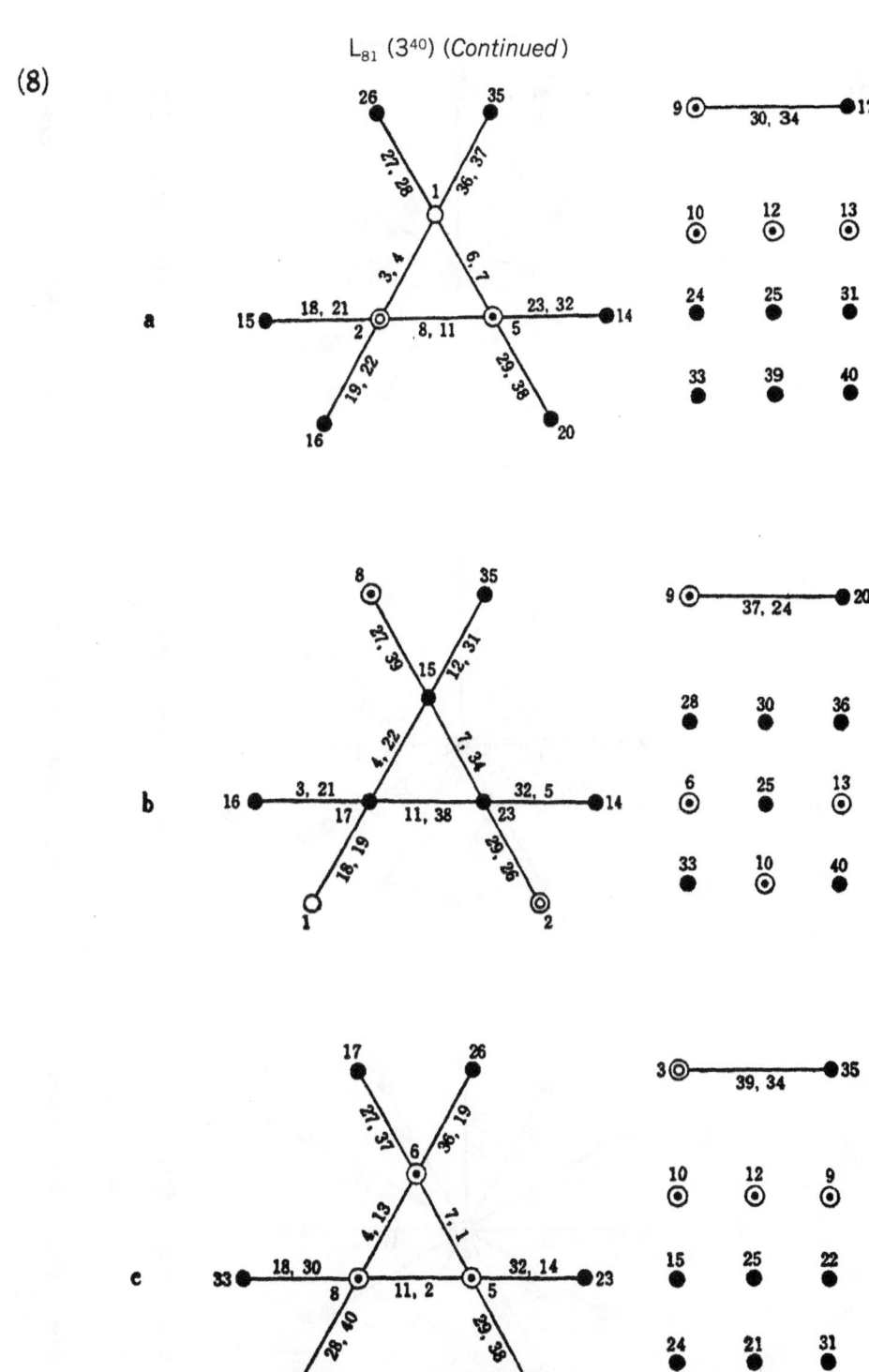

L_{81} (3^{40}) (*Continued*)

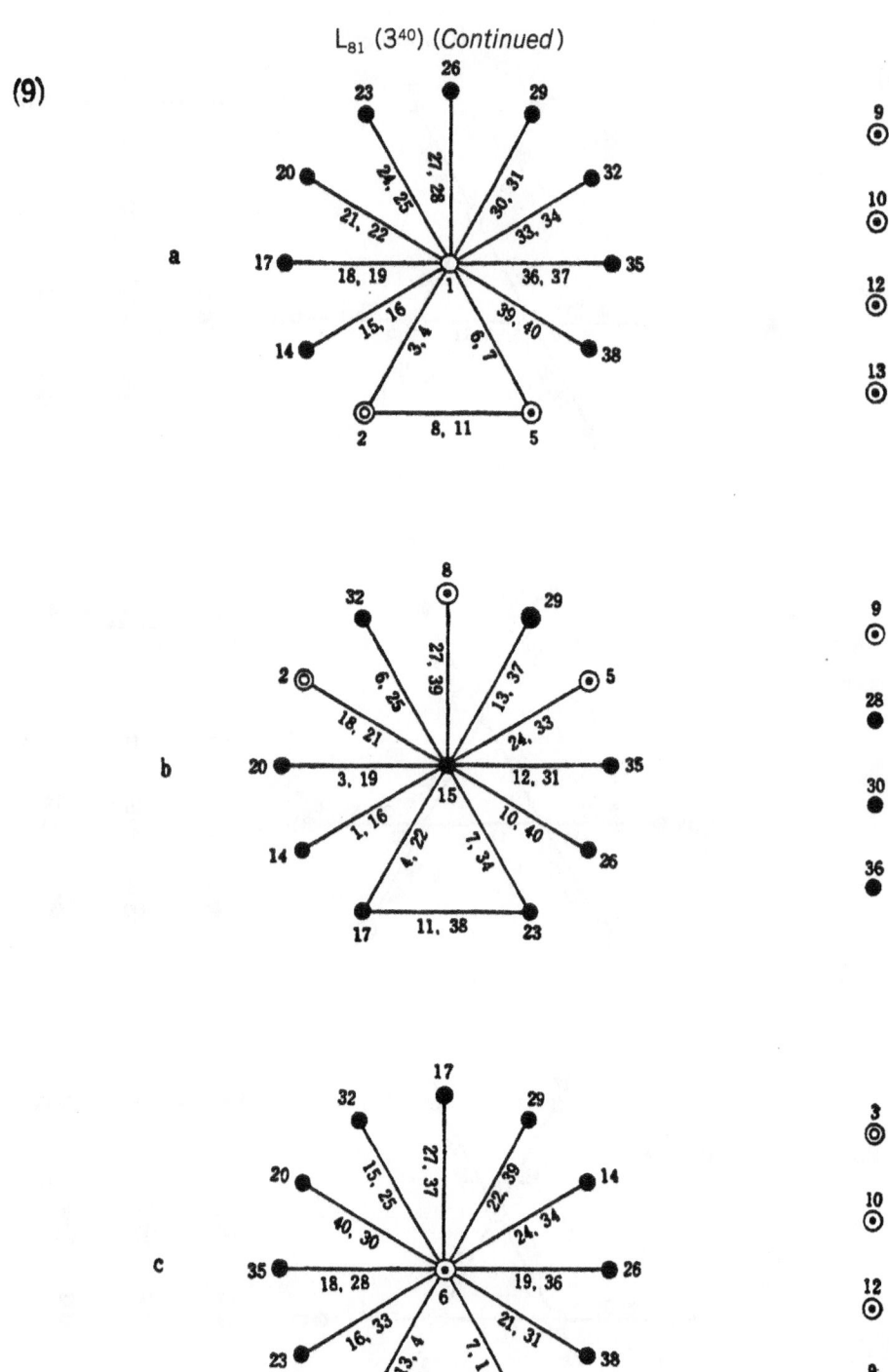

L_{81} (3^{40}) (*Continued*)

(10)

a

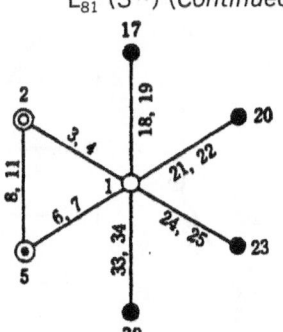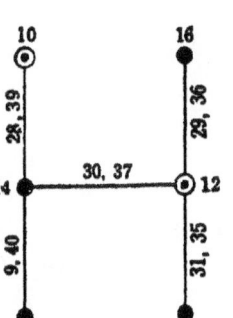

b

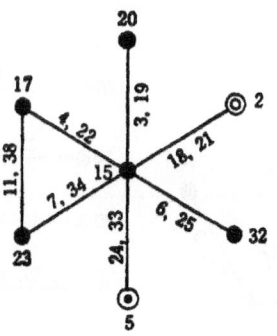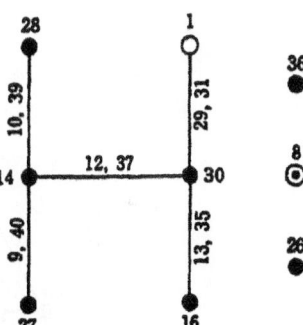

c

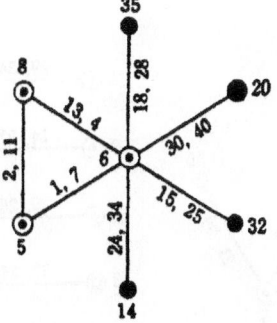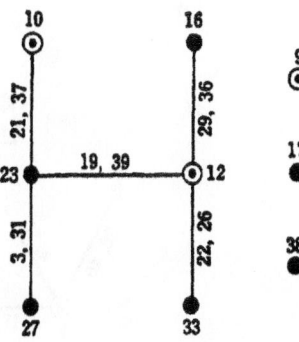

$$L_{81} \ (3^{40}) \ (Continued)$$

(11)

a

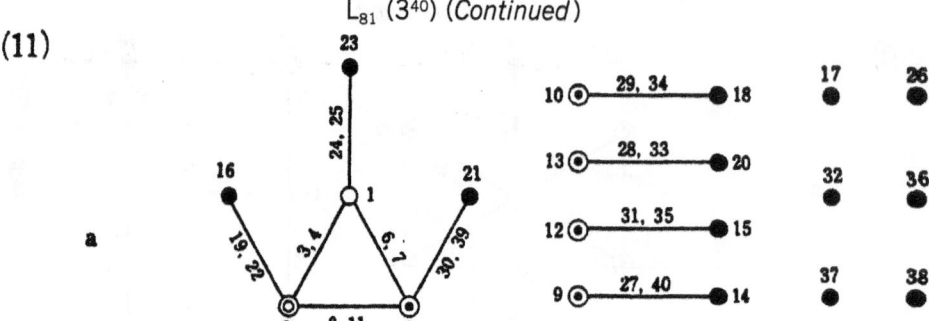

b

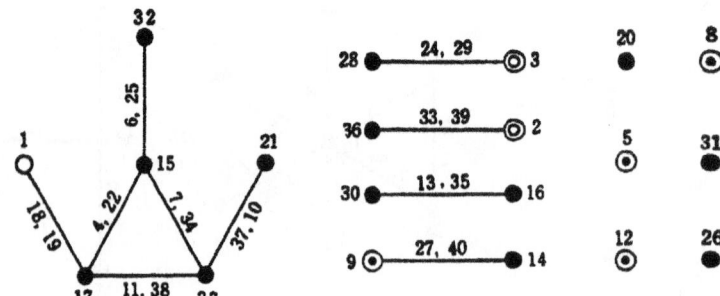

c

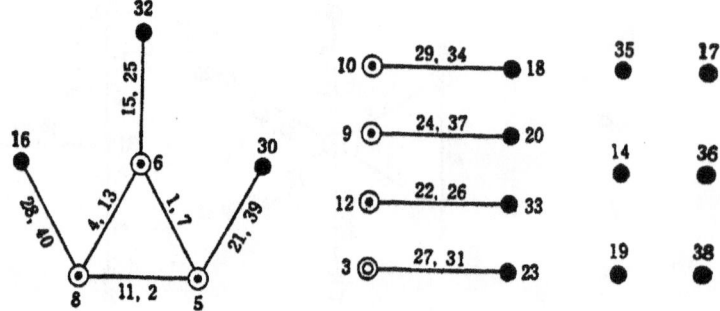

L_{81} (3^{40}) (Continued)

(12)

a

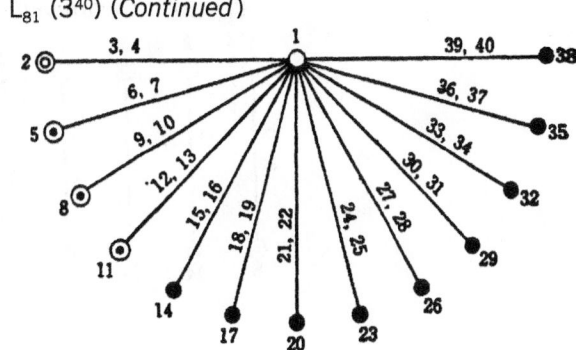

b

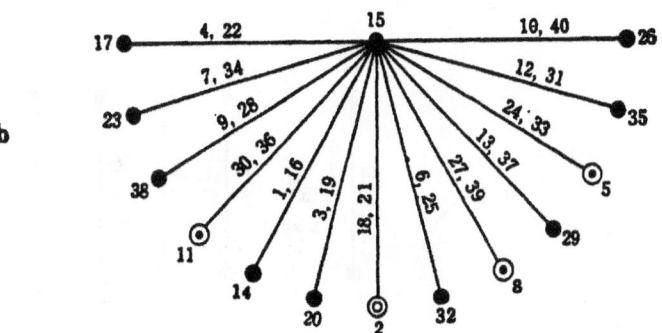

c

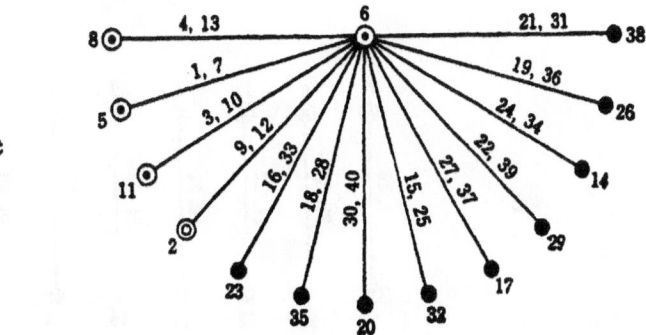

L_{81} (3^{40}) (*Continued*)

(13)

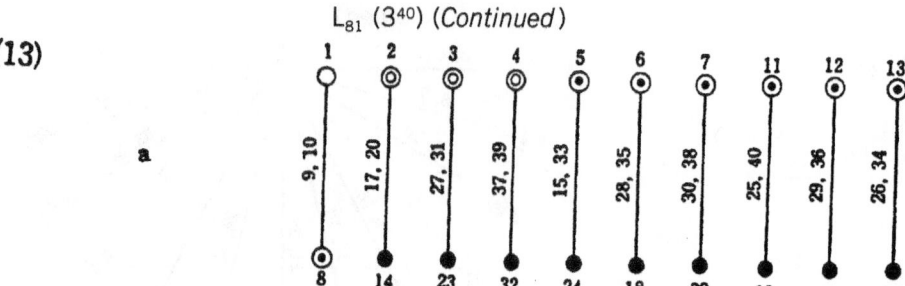

a

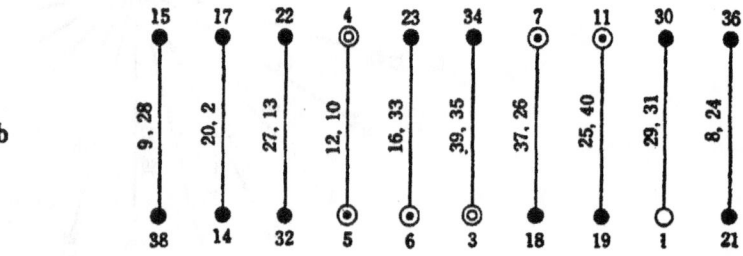

b

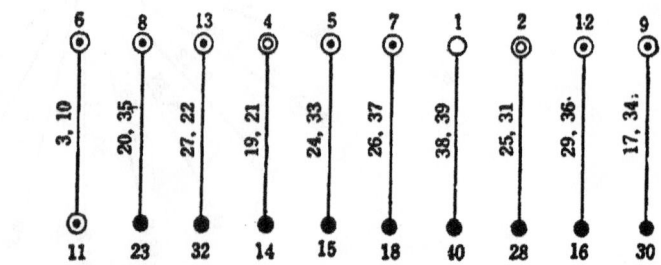

c

L_{81} (3^{40}) (Continued)

(14)

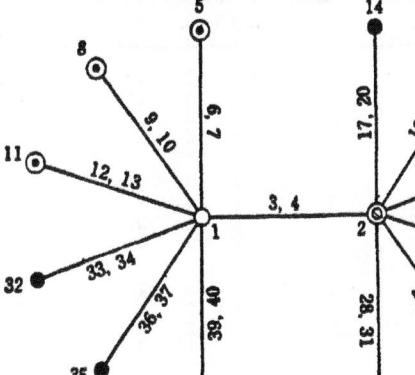

a

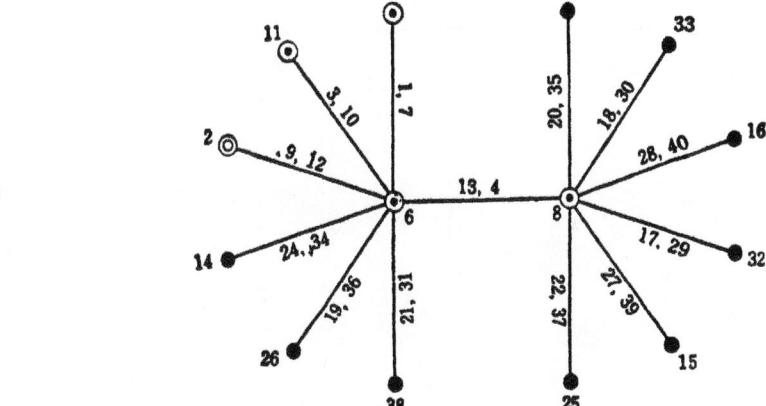

b

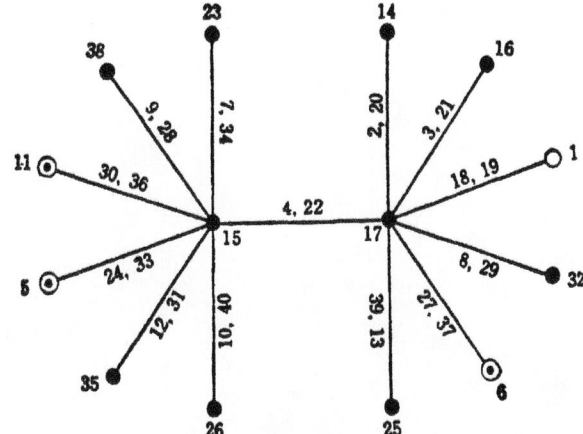

c

$L_{16} (4^5)$

No.	1	2	3	4	5
1	1	1	1	1	1
2	1	2	2	2	2
3	1	3	3	3	3
4	1	4	4	4	4
5	2	1	2	3	4
6	2	2	1	4	3
7	2	3	4	1	2
8	2	4	3	2	1
9	3	1	3	4	2
10	3	2	4	3	1
11	3	3	1	2	4
12	3	4	2	1	3
13	4	1	4	2	3
14	4	2	3	1	4
15	4	3	2	4	1
16	4	4	1	3	2
Group	1	2			

(1)

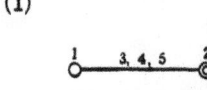

L_{32} $(2^1 \times 4^9)$

No.	1	2	3	4	5	6	7	8	9	10
1	1	1	1	1	1	1	1	1	1	1
2	1	1	2	2	2	2	2	2	2	2
3	1	1	3	3	3	3	3	3	3	3
4	1	1	4	4	4	4	4	4	4	4
5	1	2	1	1	2	2	3	3	4	4
6	1	2	2	2	1	1	4	4	3	3
7	1	2	3	3	4	4	1	1	2	2
8	1	2	4	4	3	3	2	2	1	1
9	1	3	1	2	3	4	1	2	3	4
10	1	3	2	1	4	3	2	1	4	3
11	1	3	3	4	1	2	3	4	1	2
12	1	3	4	3	2	1	4	3	2	1
13	1	4	1	2	4	3	3	4	2	1
14	1	4	2	1	3	4	4	3	1	2
15	1	4	3	4	2	1	1	2	4	3
16	1	4	4	3	1	2	2	1	3	4
17	2	1	1	4	1	4	2	3	2	3
18	2	1	2	3	2	3	1	4	1	4
19	2	1	3	2	3	2	4	1	4	1
20	2	1	4	1	4	1	3	2	3	2
21	2	2	1	4	2	3	4	1	3	2
22	2	2	2	3	1	4	3	2	4	1
23	2	2	3	2	4	1	2	3	1	4
24	2	2	4	1	3	2	1	4	2	3
25	2	3	1	3	3	1	2	4	4	2
26	2	3	2	4	4	2	1	3	3	1
27	2	3	3	1	1	3	4	2	2	4
28	2	3	4	2	2	4	3	1	1	3
29	2	4	1	3	4	2	4	2	1	3
30	2	4	2	4	3	1	3	1	2	4
31	2	4	3	1	2	4	2	4	3	1
32	2	4	4	2	1	3	1	3	4	2
Group	1	2	3							

(1)

Notes:

1. The interactions between 1 and 2 can be obtained without sacrificing any other column(s).

2. A two-way interaction between any of the four-level columns is distributed more or less uniformly to the four-level columns.

$$L_{64} (4^{21})$$

No.	1	2	3	4	5	6	7	8	9	10	11	12	13	14	15	16	17	18	19	20	21
1	1	1	1	1	1	1	1	1	1	1	1	1	1	1	1	1	1	1	1	1	1
2	1	1	1	1	1	2	2	2	2	2	2	2	2	2	2	2	2	2	2	2	2
3	1	1	1	1	1	3	3	3	3	3	3	3	3	3	3	3	3	3	3	3	3
4	1	1	1	1	1	4	4	4	4	4	4	4	4	4	4	4	4	4	4	4	4
5	1	2	2	2	2	1	1	1	1	2	2	2	2	3	3	3	3	4	4	4	4
6	1	2	2	2	2	2	2	2	2	1	1	1	1	4	4	4	4	3	3	3	3
7	1	2	2	2	2	3	3	3	3	4	4	4	4	1	1	1	1	2	2	2	2
8	1	2	2	2	2	4	4	4	4	3	3	3	3	2	2	2	2	1	1	1	1
9	1	3	3	3	3	1	1	1	1	3	3	3	3	4	4	4	4	2	2	2	2
10	1	3	3	3	3	2	2	2	2	4	4	4	4	3	3	3	3	1	1	1	1
11	1	3	3	3	3	3	3	3	3	1	1	1	1	2	2	2	2	4	4	4	4
12	1	3	3	3	3	4	4	4	4	2	2	2	2	1	1	1	1	3	3	3	3
13	1	4	4	4	4	1	1	1	1	4	4	4	4	2	2	2	2	3	3	3	3
14	1	4	4	4	4	2	2	2	2	3	3	3	3	1	1	1	1	4	4	4	4
15	1	4	4	4	4	3	3	3	3	2	2	2	2	4	4	4	4	1	1	1	1
16	1	4	4	4	4	4	4	4	4	1	1	1	1	3	3	3	3	2	2	2	2
17	2	1	2	3	4	1	2	3	4	1	2	3	4	1	2	3	4	1	2	3	4
18	2	1	2	3	4	2	1	4	3	2	1	4	3	2	1	4	3	2	1	4	3
19	2	1	2	3	4	3	4	1	2	3	4	1	2	3	4	1	2	3	4	1	2
20	2	1	2	3	4	4	3	2	1	4	3	2	1	4	3	2	1	4	3	2	1
21	2	2	1	4	3	1	2	3	4	2	1	4	3	3	4	1	2	4	3	2	1
22	2	2	1	4	3	2	1	4	3	1	2	3	4	4	3	2	1	3	4	1	2
23	2	2	1	4	3	3	4	1	2	4	3	2	1	1	2	3	4	2	1	4	3
24	2	2	1	4	3	4	3	2	1	3	4	1	2	2	1	4	3	1	2	3	4
25	2	3	4	1	2	1	2	3	4	3	4	1	2	4	3	2	1	2	1	4	3
26	2	3	4	1	2	2	1	4	3	4	3	2	1	3	4	1	2	1	2	3	4
27	2	3	4	1	2	3	4	1	2	1	2	3	4	2	1	4	3	4	3	2	1
28	2	3	4	1	2	4	3	2	1	2	1	4	3	1	2	3	4	3	4	1	2
29	2	4	3	2	1	1	2	3	4	4	3	2	1	2	1	4	3	3	4	1	2
30	2	4	3	2	1	2	1	4	3	3	4	1	2	1	2	3	4	4	3	2	1
31	2	4	3	2	1	3	4	1	2	2	1	4	3	4	3	2	1	1	2	3	4
32	2	4	3	2	1	4	3	2	1	1	2	3	4	3	4	1	2	2	1	4	3
33	3	1	3	4	2	1	3	4	2	1	3	4	2	1	3	4	2	1	3	4	2
34	3	1	3	4	2	2	4	3	1	2	4	3	1	2	4	3	1	2	4	3	1
35	3	1	3	4	2	3	1	2	4	3	1	2	4	3	1	2	4	3	1	2	4
36	3	1	3	4	2	4	2	1	3	4	2	1	3	4	2	1	3	4	2	1	3
37	3	2	4	3	1	1	3	4	2	2	4	3	1	3	1	2	4	4	2	1	3
38	3	2	4	3	1	2	4	3	1	1	3	4	2	4	2	1	3	3	1	2	4
39	3	2	4	3	1	3	1	2	4	4	2	1	3	1	3	4	2	2	4	3	1
40	3	2	4	3	1	4	2	1	3	3	1	2	4	2	4	3	1	1	3	4	2
41	3	3	1	2	4	1	3	4	2	3	1	2	4	4	2	1	3	2	4	3	1
42	3	3	1	2	4	2	4	3	1	4	2	1	3	3	1	2	4	1	3	4	2
43	3	3	1	2	4	3	1	2	4	1	3	4	2	2	4	3	1	4	2	1	3
44	3	3	1	2	4	4	2	1	3	2	4	3	1	1	3	4	2	3	1	2	4
45	3	4	2	1	3	1	3	4	2	4	2	1	3	2	4	3	1	3	1	2	4
46	3	4	2	1	3	2	4	3	1	3	1	2	4	1	3	4	2	4	2	1	3
47	3	4	2	1	3	3	1	2	4	2	4	3	1	4	2	1	3	1	3	4	2
48	3	4	2	1	3	4	2	1	3	1	3	4	2	3	1	2	4	2	4	3	1
49	4	1	4	2	3	1	4	2	3	1	4	2	3	1	4	2	3	1	4	2	3
50	4	1	4	2	3	2	3	1	4	2	3	1	4	2	3	1	4	2	3	1	4
51	4	1	4	2	3	3	2	4	1	3	2	4	1	3	2	4	1	3	2	4	1
52	4	1	4	2	3	4	1	3	2	4	1	3	2	4	1	3	2	4	1	3	2
53	4	2	3	1	4	1	4	2	3	2	3	1	4	3	2	4	1	4	1	3	2
54	4	2	3	1	4	2	3	1	4	1	4	2	3	4	1	3	2	3	2	4	1
55	4	2	3	1	4	3	2	4	1	4	1	3	2	1	4	2	3	2	3	1	4
56	4	2	3	1	4	4	1	3	2	3	2	4	1	2	3	1	4	1	4	2	3
57	4	3	2	4	1	1	4	2	3	3	2	4	1	4	1	3	2	2	3	1	4
58	4	3	2	4	1	2	3	1	4	4	1	3	2	3	2	4	1	1	4	2	3
59	4	3	2	4	1	3	2	4	1	1	4	2	3	2	3	1	4	4	1	3	2
60	4	3	2	4	1	4	1	3	2	2	3	1	4	1	4	2	3	3	2	4	1
61	4	4	1	3	2	1	4	2	3	4	1	3	2	2	3	1	4	3	2	4	1
62	4	4	1	3	2	2	3	1	4	3	2	4	1	1	4	2	3	4	1	3	2
63	4	4	1	3	2	3	2	4	1	2	3	1	4	4	1	3	2	1	4	2	3
64	4	4	1	3	2	4	1	3	2	1	4	2	3	3	2	4	1	2	3	1	4
Group	1		2										3								

L_{64} (4^{21}) (*Continued*)   Interactions between Two Columns

	1	2	3	4	5	6	7	8	9	10	11	12	13	14	15	16	17	18	19	20	21
(1)		3 4 5	2 4 5	2 3 5	2 3 4	7 8 9	6 8 9	6 7 9	6 7 8	11 12 13	10 12 13	10 11 13	10 11 12	15 16 17	14 16 17	14 15 17	14 15 16	19 20 21	18 20 21	18 19 21	18 19 20
		(2)	1 4 5	1 3 5	1 3 4	10 14 18	11 15 19	12 16 20	13 17 21	6 14 18	7 15 19	8 16 20	9 17 21	6 10 18	7 11 19	8 12 20	9 13 21	6 10 14	7 11 15	8 12 16	9 13 17
			(3)	1 2 5	1 2 4	11 16 21	10 17 20	13 14 19	12 15 18	7 17 20	6 16 21	9 15 18	8 14 19	8 13 19	9 12 18	6 11 21	7 10 20	9 12 16	8 13 17	7 10 14	6 11 15
				(4)	1 2 3	12 17 19	13 16 18	10 15 21	11 14 20	8 15 21	9 14 20	6 17 19	7 16 18	9 11 20	8 10 21	7 13 18	6 12 19	7 13 16	6 12 17	9 11 14	8 10 15
					(5)	13 15 20	12 14 21	11 17 18	10 16 19	9 16 19	8 17 18	7 14 21	6 15 20	7 12 21	6 13 20	9 10 19	8 11 18	8 11 17	9 10 16	6 13 15	7 12 14
						(6)	1 8 9	1 7 9	1 7 8	2 14 18	3 16 21	4 17 19	5 15 20	2 10 18	5 13 20	3 11 21	4 12 19	2 10 14	4 12 17	5 13 15	3 11 16
							(7)	1 6 9	1 6 8	3 17 20	2 15 19	5 14 21	4 16 18	2 12 21	4 11 19	3 13 18	5 10 20	5 13 16	3 11 15	2 10 17	4 12 14
								(8)	1 6 7	4 15 21	5 17 18	2 16 20	3 14 19	3 13 19	4 10 21	2 12 20	5 11 18	5 11 17	3 13 14	2 12 16	4 10 15
									(9)	5 16 19	4 14 20	3 15 18	2 17 21	4 11 20	3 12 18	5 10 19	2 13 21	3 12 15	5 10 16	4 11 14	2 13 17
										(10)	1 12 13	1 11 13	1 11 12	2 6 18	4 8 21	5 9 19	3 7 20	6 14 20	9 16 14	7 17 16	8 15 17
											(11)	1 10 13	1 10 12	4 9 20	2 7 19	3 6 21	5 8 18	5 8 17	2 7 15	4 15 14	3 14 16
												(12)	1 10 11	5 7 21	3 9 18	2 8 20	4 6 19	3 9 15	5 6 17	4 8 16	2 7 14
													(13)	3 8 19	5 6 20	4 7 18	2 9 21	4 7 16	3 8 14	5 6 15	2 9 17
														(14)	1 16 17	1 15 17	1 15 16	2 6 10	3 8 13	4 9 11	5 7 12
															(15)	1 14 17	1 14 16	3 9 12	2 7 11	5 6 13	4 8 10
																(16)	1 14 15	4 7 13	5 9 10	2 8 12	3 6 11
																	(17)	5 8 11	4 6 12	3 7 10	2 9 13
																		(18)	1 20 21	1 19 21	1 19 20
																			(19)	1 18 21	1 18 20
																				(20)	1 18 19

L_{64} (4^{21}) (*Continued*)

(1)

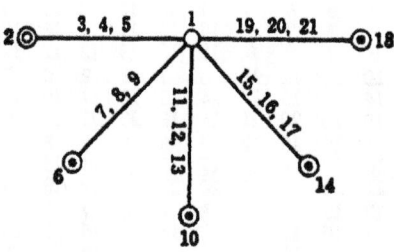

(2)

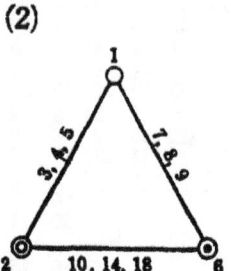 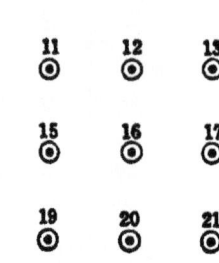

$$L_{25} (5^6)$$

No.	1	2	3	4	5	6
1	1	1	1	1	1	1
2	1	2	2	2	2	2
3	1	3	3	3	3	3
4	1	4	4	4	4	4
5	1	5	5	5	5	5
6	2	1	2	3	4	5
7	2	2	3	4	5	1
8	2	3	4	5	1	2
9	2	4	5	1	2	3
10	2	5	1	2	3	4
11	3	1	3	5	2	4
12	3	2	4	1	3	5
13	3	3	5	2	4	1
14	3	4	1	3	5	2
15	3	5	2	4	1	3
16	4	1	4	2	5	3
17	4	2	5	3	1	4
18	4	3	1	4	2	5
19	4	4	2	5	3	1
20	4	5	3	1	4	2
21	5	1	5	4	3	2
22	5	2	1	5	4	3
23	5	3	2	1	5	4
24	5	4	3	2	1	5
25	5	5	4	3	2	1
	a	b	ab	ab^2	ab^3	ab^4
Group	1	2				

(1)

L_{50} ($2^1 \times 5^{11}$)

No.	1	2	3	4	5	6	7	8	9	10	11	12
1	1	1	1	1	1	1	1	1	1	1	1	1
2	1	1	2	2	2	2	2	2	2	2	2	2
3	1	1	3	3	3	3	3	3	3	3	3	3
4	1	1	4	4	4	4	4	4	4	4	4	4
5	1	1	5	5	5	5	5	5	5	5	5	5
6	1	2	1	2	3	4	5	1	2	3	4	5
7	1	2	2	3	4	5	1	2	3	4	5	1
8	1	2	3	4	5	1	2	3	4	5	1	2
9	1	2	4	5	1	2	3	4	5	1	2	3
10	1	2	5	1	2	3	4	5	1	2	3	4
11	1	3	1	3	5	2	4	4	1	3	5	2
12	1	3	2	4	1	3	5	5	2	4	1	3
13	1	3	3	5	2	4	1	1	3	5	2	4
14	1	3	4	1	3	5	2	2	4	1	3	5
15	1	3	5	2	4	1	3	3	5	2	4	1
16	1	4	1	4	2	5	3	5	3	1	4	2
17	1	4	2	5	3	1	4	1	4	2	5	3
18	1	4	3	1	4	2	5	2	5	3	1	4
19	1	4	4	2	5	3	1	3	1	4	2	5
20	1	4	5	3	1	4	2	4	2	5	3	1
21	1	5	1	5	4	3	2	4	3	2	1	5
22	1	5	2	1	5	4	3	5	4	3	2	1
23	1	5	3	2	1	5	4	1	5	4	3	2
24	1	5	4	3	2	1	5	2	1	5	4	3
25	1	5	5	4	3	2	1	3	2	1	5	4
26	2	1	1	1	4	5	4	3	2	5	2	3
27	2	1	2	2	5	1	5	4	3	1	3	4
28	2	1	3	3	1	2	1	5	4	2	4	5
29	2	1	4	4	2	3	2	1	5	3	5	1
30	2	1	5	5	3	4	3	2	1	4	1	2
31	2	2	1	2	1	3	3	2	4	5	5	4
32	2	2	2	3	2	4	4	3	5	1	1	5
33	2	2	3	4	3	5	5	4	1	2	2	1
34	2	2	4	5	4	1	1	5	2	3	3	2
35	2	2	5	1	5	2	2	1	3	4	4	3
36	2	3	1	3	3	1	2	5	5	4	2	4
37	2	3	2	4	4	2	3	1	1	5	3	5
38	2	3	3	5	5	3	4	2	2	1	4	1
39	2	3	4	1	1	4	5	3	3	2	5	2
40	2	3	5	2	2	5	1	4	4	3	1	3
41	2	4	1	4	5	4	1	2	5	2	3	3
42	2	4	2	5	1	5	2	3	1	3	4	4
43	2	4	3	1	2	1	3	4	2	4	5	5
44	2	4	4	2	3	2	4	5	3	5	1	1
45	2	4	5	3	4	3	5	1	4	1	2	2
46	2	5	1	5	2	2	5	3	4	4	3	1
47	2	5	2	1	3	3	1	4	5	5	4	2
48	2	5	3	2	4	4	2	5	1	1	5	3
49	2	5	4	3	5	5	3	1	2	2	1	4
50	2	5	5	4	1	1	4	2	3	3	2	5
Group	1	2					3					

(1)

Note: Interaction between Columns 1 and 2 can be obtained without sacrificing another column.

L_{36} ($2^{11} \times 3^{12}$) See note below for L_{36} ($2^3 \times 3^{13}$)

No.	1	2	3	4	5	6	7	8	9	10	11	12	13	14	15	16	17	18	19	20	21	22	23	1'	2'	3'	4'
1	1	1	1	1	1	1	1	1	1	1	1	1	1	1	1	1	1	1	1	1	1	1	1	1	1	1	1
2	1	1	1	1	1	1	1	1	1	1	1	2	2	2	2	2	2	2	2	2	2	2	2	1	1	1	1
3	1	1	1	1	1	1	1	1	1	1	1	3	3	3	3	3	3	3	3	3	3	3	3	1	1	1	1
4	1	1	1	1	1	2	2	2	2	2	2	1	1	1	1	2	2	2	2	3	3	3	3	1	2	2	1
5	1	1	1	1	1	2	2	2	2	2	2	2	2	2	2	3	3	3	3	1	1	1	1	1	2	2	1
6	1	1	1	1	1	2	2	2	2	2	2	3	3	3	3	1	1	1	1	2	2	2	2	1	2	2	1
7	1	1	2	2	2	1	1	1	2	2	2	1	1	2	3	1	2	3	3	1	2	2	3	2	1	2	1
8	1	1	2	2	2	1	1	1	2	2	2	2	2	3	1	2	3	1	1	2	3	3	1	2	1	2	1
9	1	1	2	2	2	1	1	1	2	2	2	3	3	1	2	3	1	2	2	3	1	1	2	2	1	2	1
10	1	2	1	2	2	1	2	2	1	1	2	1	1	3	2	1	3	2	3	2	1	3	2	2	2	1	1
11	1	2	1	2	2	1	2	2	1	1	2	2	2	1	3	2	1	3	1	3	2	1	3	2	2	1	1
12	1	2	1	2	2	1	2	2	1	1	2	3	3	2	1	3	2	1	2	1	3	2	1	2	2	1	1
13	1	2	2	1	2	2	1	2	1	2	1	1	2	3	1	3	2	1	3	3	2	1	2	1	1	1	2
14	1	2	2	1	2	2	1	2	1	2	1	2	3	1	2	1	3	2	1	1	3	2	3	1	1	1	2
15	1	2	2	1	2	2	1	2	1	2	1	3	1	2	3	2	1	3	2	2	1	3	1	1	1	1	2
16	1	2	2	2	1	2	2	1	2	1	1	1	2	3	2	1	1	3	2	3	3	2	1	1	2	2	2
17	1	2	2	2	1	2	2	1	2	1	1	2	3	1	3	2	2	1	3	1	1	3	2	1	2	2	2
18	1	2	2	2	1	2	2	1	2	1	1	3	1	2	1	3	3	2	1	2	2	1	3	1	2	2	2
19	2	1	2	2	1	1	2	2	1	2	1	1	2	1	3	3	3	1	2	2	1	2	3	2	1	2	2
20	2	1	2	2	1	1	2	2	1	2	1	2	3	2	1	1	1	2	3	3	2	3	1	2	1	2	2
21	2	1	2	2	1	1	2	2	1	2	1	3	1	3	2	2	2	3	1	1	3	1	2	2	1	2	2
22	2	1	2	1	2	2	2	1	1	1	2	1	2	2	3	3	1	2	1	1	3	3	2	2	2	1	2
23	2	1	2	1	2	2	2	1	1	1	2	2	3	3	1	1	2	3	2	2	1	1	3	2	2	1	2
24	2	1	2	1	2	2	2	1	1	1	2	3	1	1	2	2	3	1	3	3	2	2	1	2	2	1	2
25	2	1	1	2	2	2	1	2	1	1	2	1	3	2	1	2	3	3	1	3	1	3	2	1	1	1	3
26	2	1	1	2	2	2	1	2	1	1	2	2	1	3	2	3	1	1	2	1	2	1	3	1	1	1	3
27	2	1	1	2	2	2	1	2	1	1	2	3	2	1	3	1	2	2	3	2	3	2	1	1	1	1	3
28	2	2	2	1	1	1	1	2	2	1	2	1	3	2	2	2	1	1	3	2	3	1	3	1	2	2	3
29	2	2	2	1	1	1	1	2	2	1	2	2	1	3	3	3	2	2	1	3	1	2	1	1	2	2	3
30	2	2	2	1	1	1	1	2	2	1	2	3	2	1	1	1	3	3	2	1	2	3	2	1	2	2	3
31	2	2	1	2	1	2	1	1	2	2	1	1	3	3	3	2	3	2	2	1	2	1	1	2	1	2	3
32	2	2	1	2	1	2	1	1	2	2	1	2	1	1	1	3	1	3	3	2	3	2	2	2	1	2	3
33	2	2	1	2	1	2	1	1	2	2	1	3	2	2	2	1	2	1	1	3	1	3	3	2	1	2	3
34	2	2	1	1	2	1	2	1	2	2	1	1	3	1	2	3	2	3	1	2	2	3	1	2	2	1	3
35	2	2	1	1	2	1	2	1	2	2	1	2	1	2	3	1	3	1	2	3	3	1	2	2	2	1	3
36	2	2	1	1	2	1	2	1	2	2	1	3	2	3	1	2	1	2	3	1	1	2	3	2	2	1	3
Group	1	2										3															

Note:

1. Replacing columns 1-11 with columns 1'-4' gives L_{36} ($2^3 \times 3^{13}$).

2. In L_{36} ($2^{11} \times 3^{12}$) array interactions are not orthogonal to other columns. Therefore do not use this array to obtain interactions.

3. Linear graphs for L_{36} ($2^3 \times 3^{13}$) are shown on the next page.

$$L_{36} \ (2^3 \times 3^{13})$$

(1)

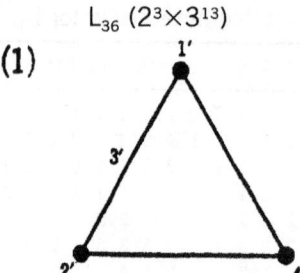

Note: 1'x4' and 2'x4' can be determined without sacrificing other columns.

(2)

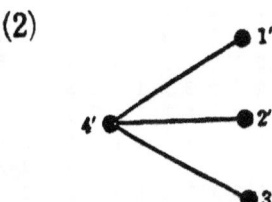

Note: These interactions can be determined without sacrificing other columns.

L_9' (2^{21}) Partially Orthogonal

No.	1	2	3	4	5	6	7	8	9	10	11	12	13	14	15	16	17	18	19	20	21
1	1	1	1	1	1	1	1	1	1	1	1	1	1	1	1	1	1	1	1	1	1
2	1	1	1	1	1	1	1	1	1	2	2	2	2	2	2	2	2	2	2	2	2
3	1	1	1	2	2	2	2	2	2	1	1	1	1	2	1	2	2	2	2	2	2
4	1	2	2	1	1	2	1	2	2	1	2	1	1	2	2	2	2	1	1	2	2
5	1	2	2	1	2	1	1	2	2	2	1	1	2	1	1	1	1	2	2	2	1
6	1	2	2	2	1	1	2	1	1	1	1	2	1	2	1	2	1	2	2	1	2
7	2	1	2	1	2	1	2	1	2	1	1	2	2	1	2	2	1	2	1	2	2
8	2	1	2	1	1	2	2	1	2	2	1	1	1	2	1	1	2	1	2	1	2
9	2	1	2	2	1	1	1	2	1	1	2	1	2	1	1	2	2	1	2	2	1

Note: You can not get an interaction from this array, which is considered almost or partially orthogonal.

$$L'_{27} \ (3^{22})$$

No.	1	2	3	4	5	6	7	8	9	10	11	12	13	14	15	16	17	18	19	20	21	22
1	1	1	1	1	1	1	1	1	1	1	1	1	1	3	3	3	2	2	2	1	1	1
2	1	1	1	1	2	2	2	2	2	2	2	2	2	3	3	3	3	3	3	2	2	2
3	1	1	1	1	3	3	3	3	3	3	3	3	3	1	1	1	2	2	2	2	2	2
4	1	2	2	2	1	1	1	2	2	2	3	3	3	2	2	2	3	3	3	1	1	1
5	1	2	2	2	2	2	2	3	3	3	1	1	1	1	1	1	3	3	3	3	3	3
6	1	2	2	2	3	3	3	1	1	1	2	2	2	1	1	1	1	1	1	1	1	1
7	1	3	3	3	1	1	1	3	3	3	2	2	2	2	2	2	2	2	2	3	3	3
8	1	3	3	3	2	2	2	1	1	1	3	3	3	3	3	3	1	1	1	3	3	3
9	1	3	3	3	3	3	3	2	2	2	1	1	1	2	2	2	1	1	1	2	2	2
10	2	1	2	3	1	2	3	1	2	3	1	2	3	1	2	3	1	2	3	1	2	3
11	2	1	2	3	2	3	1	2	3	1	2	3	1	3	1	2	2	3	1	3	1	2
12	2	1	2	3	3	1	2	3	1	2	3	1	2	2	3	1	3	1	2	2	3	1
13	2	2	3	1	1	2	3	2	3	1	3	1	2	2	3	1	2	3	1	2	3	1
14	2	2	3	1	2	3	1	3	1	2	1	2	3	3	1	2	3	1	2	3	1	2
15	2	2	3	1	3	1	2	1	2	3	2	3	1	1	2	3	1	2	3	1	2	3
16	2	3	1	2	1	2	3	3	1	2	2	3	1	1	3	2	2	1	3	1	3	2
17	2	3	1	2	2	3	1	1	2	3	3	1	2	2	1	3	3	2	1	2	1	3
18	2	3	1	2	3	1	2	2	3	1	1	2	3	3	2	1	1	3	2	3	2	1
19	3	1	3	2	1	3	2	1	3	2	1	3	2	2	1	3	1	3	2	2	1	3
20	3	1	3	2	2	1	3	2	1	3	2	1	3	3	2	1	2	1	3	3	2	1
21	3	1	3	2	3	2	1	3	2	1	3	2	1	1	3	2	3	2	1	1	3	2
22	3	2	1	3	1	3	2	2	1	3	3	2	1	3	2	1	1	3	2	3	2	1
23	3	2	1	3	2	1	3	3	2	1	1	3	2	1	3	2	2	1	3	1	3	2
24	3	2	1	3	3	2	1	1	3	2	2	1	3	2	1	3	3	2	1	2	1	3
25	3	3	2	1	1	3	2	3	2	1	2	1	3	2	1	3	2	1	3	3	2	1
26	3	3	2	1	2	1	3	1	3	2	3	2	1	3	2	1	3	2	1	1	3	2
27	3	3	2	1	3	2	1	2	1	3	1	3	2	1	3	2	1	3	2	2	1	3

Groups: 1   2   3

Notes:

1. Columns 1, 11-13 are orthogonal to each other.

2. Columns 2-10 are orthogonal to each other and to Columns 1, 11-13. Although Columns 2-10 are not orthogonal to Columns 14-22, they are so near to being orthogonal to them that in some instances (where expediency is desired and permissible), they could be treated as orthogonal.

3. Columns 14-22 are orthogonal to each other and to Columns 1, 11-13.

4. The columns are arranged in three groups (1, 2, 3) plus a "mixed" group which is a combination of the first three groups. Because of this lay-out, care must be exercised when using this array for a Split Unit design.

L'_{27} (3^{22}) (*Continued*)

(1)

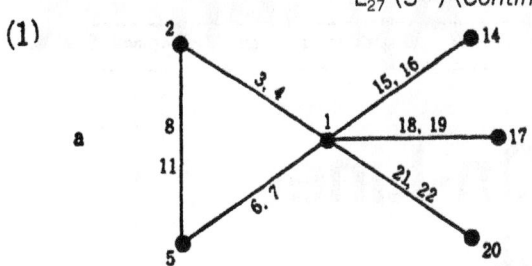 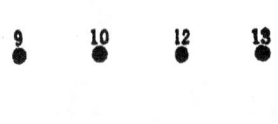

a

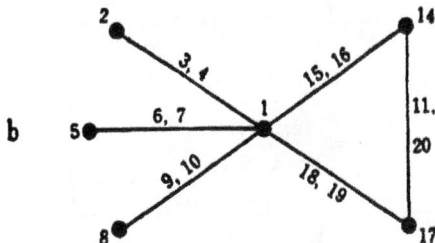 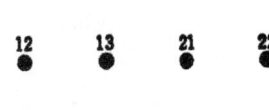

b

(2)

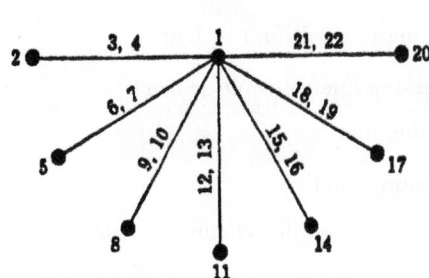

Appendix B

Equations for On-Line Process Control

Loss Functions and Equations

$$L_0 = \frac{B}{n_0} + \frac{C}{u_0} + \frac{A}{\Delta^2}\left[\frac{D_0^2}{3} + \left(\frac{n+1}{2} + l\right)\frac{D_0^2}{u_0} + \sigma_m^2\right]$$

$$L = \frac{B}{n} + \frac{C}{u} + \frac{A}{\Delta^2}\left[\frac{D^2}{3} + \left(\frac{n+1}{2} + l\right)\frac{D^2}{u} + \sigma_m^2\right]$$

where

$$n = \sqrt{\frac{2u_0 B}{A}}\,\frac{\Delta}{D_0}$$

$$D = \left(\frac{3C}{A}\frac{D_0^2}{u_0}\Delta^2\right)^{1/4}$$

$$u = u_0\frac{D^2}{D_0^2}$$

Parameters

Δ: tolerance of objective characteristics

A: loss of a defective (yen)

B: checking cost (yen)

C: adjustment cost (yen)

n_0: current checking interval (unit or batch)

n: optimum checking interval (unit or batch)

D_0: current adjusting limit

D: optimum adjusting limit

u_0: current mean adjusting interval (unit or batch)

u: forecasted mean adjusting interval after optimization (unit or batch)

l: time lag of checking method (unit or batch)

σ_m: measurement error in standard deviation

L_0:   current loss function (yen/unit or batch)   **Remarks**

L: optimum loss function (yen/unit or batch)

Feedback Control by Quality Characteristics (Using Limit Samples or Gauges; Chapter 24)

$$L_0 = \frac{B}{n_0} + \frac{C}{u_0} + \frac{A}{\Delta^2}\left[\frac{D_0^2}{3} + \left(\frac{n_0 + 1}{2} + l\right)\frac{D_0^2}{u_0}\right]$$

Loss Functions and Equations

$$L = \frac{B}{n} + \frac{C}{u} + \frac{A}{\Delta^2}\left[\frac{D^2}{3} + \left(\frac{n + 1}{2} + l\right)\frac{D^2}{u}\right]$$

$$= \frac{B}{n} + \frac{C}{u} + A\left[\frac{\Psi^2}{3}\left(\frac{n + 1}{2} + l\right)\frac{\Psi^2}{u}\right]$$

where

$$n = \sqrt{\frac{2u_0 B}{A}}\frac{\Delta}{D_0} = \sqrt{\frac{2\bar{u}B}{A}} \quad \left.\begin{array}{c} \end{array}\right\} \begin{array}{l}\text{Set } u_0 = \bar{u} \\ \text{when } D_0 = \Delta\end{array}$$

$$D = \left(\frac{3C}{A\bar{u}}\right)^{1/4} \quad \Delta = \Psi\Delta$$

$$\Psi = \left(\frac{3C}{A\bar{u}}\right)^{1/4}, \; u = \bar{u}\Psi^2$$

Δ: tolerance of objective characteristics   **Parameters**

A: loss per defective (yen)

B: checking cost (yen)

C: adjustment cost (yen)

n_0:   current checking interval (units)

n: optimum checking interval (units)

D_0:   current adjusting limit

D: optimum adjustment limit

u_0:   current mean adjustment interval (units)

$\bar{u}$:   mean failure interval (units)

u: mean adjustment interval after optimization (units)

Ψ: ratio of tolerance Δ and optimum adjustment limit D ($\Psi = D/\Delta$)

l: time lag of checking (units)

Remarks L_0: current loss function (yen/unit)

L: loss function after optimization (yen/unit)

Feedback Control of Process Conditions (for Continuous Variables; Chapter 25)

Loss Functions and Equations

$$L_0 = \frac{B}{n_0} + \frac{C}{u_0} + \frac{A}{\Delta^2}\left[\frac{D_0^2}{3} + \left(\frac{n+1}{2} + l\right)\frac{D_0^2}{u_0} + \sigma_m^2\right]$$

$$L = \frac{B}{n} + \frac{C}{u} + \frac{A}{\Delta^2}\left[\frac{D^2}{3} + \left(\frac{n+1}{2} + l\right)\frac{D^2}{u} + \sigma_m^2\right]$$

where

$$n = \sqrt{\frac{2u_0 B}{A}\frac{\Delta}{D_0}}$$

$$D = \left(\frac{3C}{A}\frac{D_0^2}{u_0}\Delta^2\right)^{1/4}$$

$$u = u_0\frac{D^2}{D_0^2}$$

Parameters Δ: limit of process condition (x) when objective characteristic exceeds tolerance

A: loss when objective characteristic exceeds tolerance (yen)

B: unit checking cost of process condition (yen)

C: adjustment cost of process condition (yen)

n_0: current checking interval of process condition (x) (units)

n: optimum checking interval of process condition (x) (units)

D_0: current adjustment limit of process condition (x)

D: optimum adjustment limit of process condition (x)

u_0: current mean adjustment interval of process condition (x) (units)

u: optimum mean adjustment interval (forecast value) of process condition (x)

l: time lag of checking process condition (x)

σ_m: standard deviation of measurement error of process condition (x)

L_0:   current loss function (yen/unit)

L: loss function after optimization (yen/unit)

Process Diagnosis and Adjustment (Basic Equations; Chapter 26)

$$L_0 = \frac{B}{n_0} + \frac{n_0 + 1}{2} \frac{A}{\bar{u}} + \frac{C}{\bar{u}} + \frac{lA}{\bar{u}}$$

$$L = \frac{B}{n} + \frac{n + 1}{2} \frac{A}{\bar{u}} + \frac{C}{\bar{u}} + \frac{lA}{\bar{u}}$$

where

$$n = \sqrt{\frac{2(\bar{u} + l)B}{A - C/\bar{u}}}$$

When $\bar{u} \gg l$ and $A \gg C/\bar{u}$,

$$n \approx \sqrt{\frac{2\bar{u}B}{A}}$$

A: loss of producing unit product under abnormal process condition (yen)

B: unit diagnosis cost (yen)

C: adjustment cost (loss to bring abnormal process condition back to normal, total of process stoppage loss and treatment cost, including screening cost) (yen) [$C = C'$ (process stoppage loss) $\times t$ (mean stoppage time) $+ C''$ direct adjustment cost)]

n_0:   current process diagnosis interval (units)

n: optimum process diagnosis interval (units)

$\bar{u}$:   mean failure interval (units) [(production of a certain period) $\div$ (number of failures that occurred during the period); when the number of failures is equal to zero since startup, $\bar{u} = 2 \times$ (production during the period)]

L: time lag (unit) [when a process is diagnosed as abnormal, the number of products produced from the time the product was processed to the time the process was stopped]

L_0:   current loss function (yen/unit)

L: loss function after optimization (yen/unit)

Points for improvement:

1. Prolong $\bar{u}$: Introduction of preventive maintenance or using long-life tools.
2. Reduce A: Improve defective treatment methods.
3. Reduce l: Improve diagnosis methods, improve diagnosis point.
4. Reduce C: Introduction of spare machines or spare molds.

Appendix C

Orthogonal Arrays and Linear Graphs for Chapter 38

$L_4(2^3)$

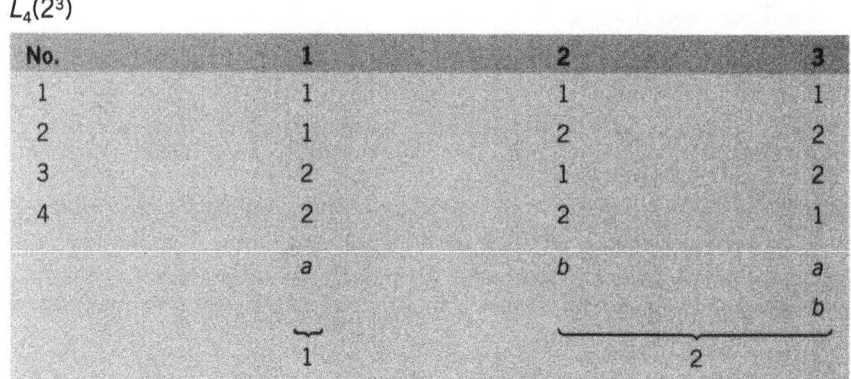

No.	1	2	3
1	1	1	1
2	1	2	2
3	2	1	2
4	2	2	1
	a	b	a
			b
	1	2	

$L_8(2^7)$

No.	1	2	3	4	5	6	7
1	1	1	1	1	1	1	1
2	1	1	1	2	2	2	2
3	1	2	2	1	1	2	2
4	1	2	2	2	2	1	1
5	2	1	2	1	2	1	2
6	2	1	2	2	1	2	1
7	2	2	1	1	2	2	1
8	2	2	1	2	1	1	2
	a	b	a	c	a	b	a
		b			c	c	b
							c
	1	2			3		

Interaction between Two Columns

1	2	3	4	5	6	7
(1)	3	2	5	4	7	6
	(2)	1	6	7	4	5
		(3)	7	6	5	4
			(4)	1	2	3
				(5)	3	2
					(6)	1
						(7)

(1)

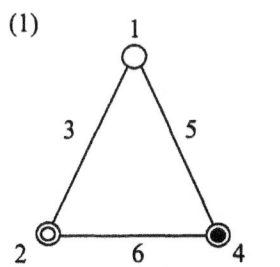

(2)

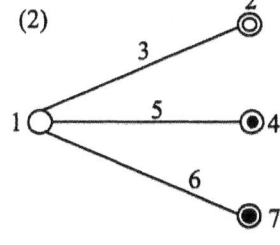

$L_{12}(2^{11})$

No.	1	2	3	4	5	6	7	8	9	10	11
1	1	1	1	1	1	1	1	1	1	1	1
2	1	1	1	1	1	2	2	2	2	2	2
3	1	1	2	2	2	1	1	1	2	2	2
4	1	2	1	2	2	1	2	2	1	1	2
5	1	2	2	1	2	2	1	2	1	2	1
6	1	2	2	2	1	2	2	1	2	1	1
7	2	1	2	2	1	1	2	2	1	2	1
8	2	1	2	1	2	2	2	1	1	1	2
9	2	1	1	2	2	2	1	2	2	1	1
10	2	2	2	1	1	1	1	2	2	1	2
11	2	2	1	2	1	2	1	1	1	2	2
12	2	2	1	1	2	1	2	1	2	2	1

       1   2

$L_{16}(2^{15})$

No.	1	2	3	4	5	6	7	8	9	10	11	12	13	14	15
1	1	1	1	1	1	1	1	1	1	1	1	1	1	1	1
2	1	1	1	1	1	1	1	2	2	2	2	2	2	2	2
3	1	1	1	2	2	2	2	1	1	1	1	2	2	2	2
4	1	1	1	2	2	2	2	2	2	2	2	1	1	1	1
5	1	2	2	1	1	2	2	1	1	2	2	1	1	2	2
6	1	2	2	1	1	2	2	2	2	1	1	2	2	1	1
7	1	2	2	2	2	1	1	1	1	2	2	2	2	1	1
8	1	2	2	2	2	1	1	2	2	1	1	1	1	2	2
9	2	1	2	1	2	1	2	1	2	1	2	1	2	1	2
10	2	1	2	1	2	1	2	2	1	2	1	2	1	2	1
11	2	1	2	2	1	2	1	1	2	1	2	2	1	2	1
12	2	1	2	2	1	2	1	2	1	2	1	1	2	1	2
13	2	2	1	1	2	2	1	1	2	2	1	1	2	2	1
14	2	2	1	1	2	2	1	2	1	1	2	2	1	1	2
15	2	2	1	2	1	1	2	1	2	2	1	2	1	1	2
16	2	2	1	2	1	1	2	2	1	1	2	1	2	2	1
	a	b	a	c	a	b	a	d	a	b	a	c	a	b	a
		b		c	c	b		d	d	d	b	d	c	c	b
					c						d		d	d	c
															d

| | 1 | | 2 | | | 3 | | | | | 4 | | | | |

Interaction between Two Columns

1	2	3	4	5	6	7	8	9	10	11	12	13	14	15
(1)	3	2	5	4	7	6	9	8	11	10	13	12	15	14
	(2)	1	6	7	4	5	10	11	8	9	14	15	12	13
		(3)	7	6	5	4	11	10	9	8	15	14	13	12
			(4)	1	2	3	12	13	14	15	8	9	10	11
				(5)	3	2	13	12	15	14	9	8	11	10
					(6)	1	14	15	12	13	10	11	8	9
						(7)	15	14	13	12	11	10	9	8
							(8)	1	2	3	4	5	6	7
								(9)	3	2	5	4	7	6
									(10)	1	6	7	4	5
										(11)	7	6	5	4
											(12)	1	2	3
												(13)	3	2
													(14)	1

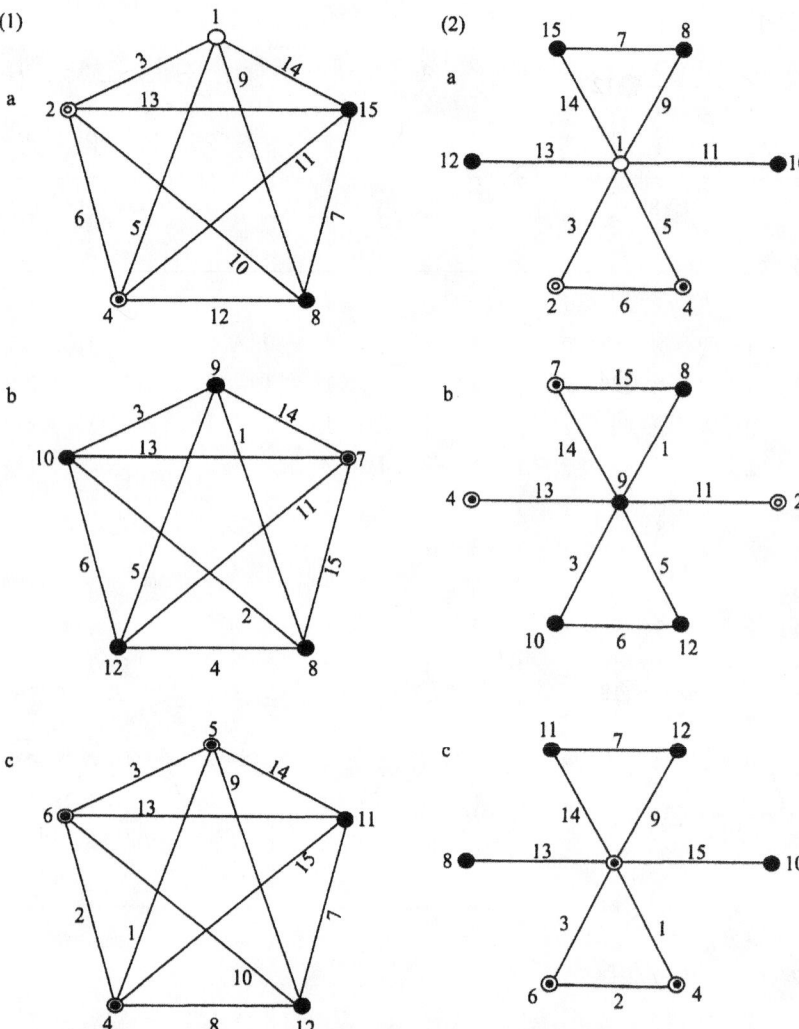

(3)

a

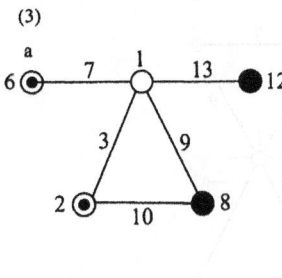

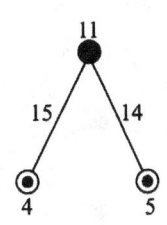

(4)

a

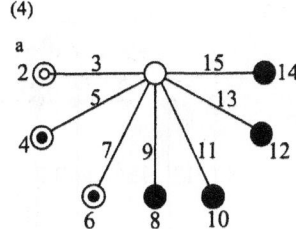

b

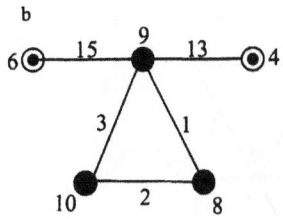

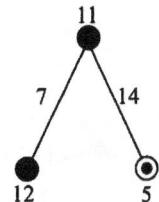

b

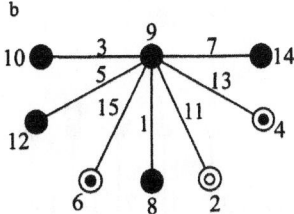

c

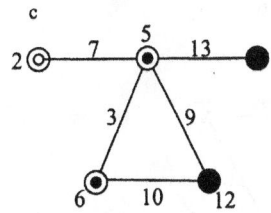

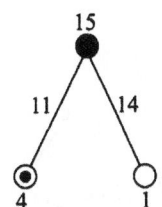

c

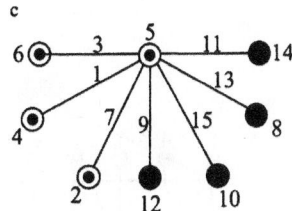

(5)

(6)

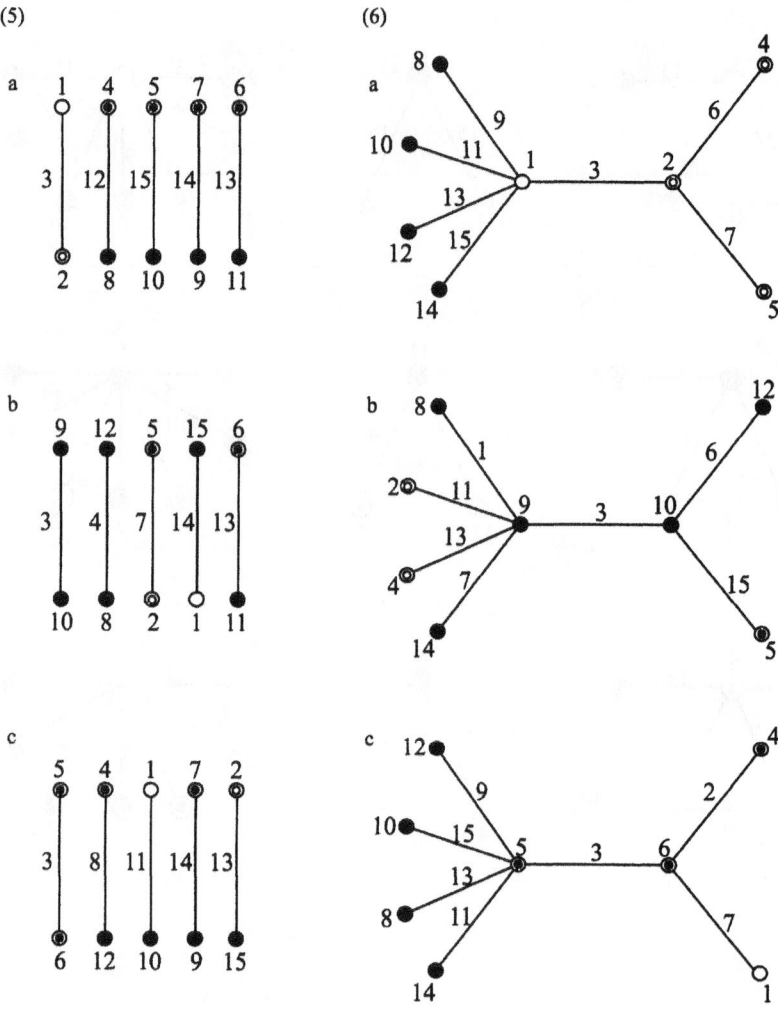

$L_9(3^4)$

No.	1	2	3	4
1	1	1	1	1
2	1	2	2	2
4	2	1	2	3
5	2	2	3	1
6	2	3	1	2
7	3	1	3	2
8	3	2	1	3
9	3	3	2	1
	a	b	a	a
			b	b^2
	1		2	

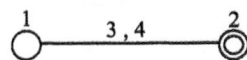

$L_{18}(2^1 \times 3^7)$

No.	1	2	3	4	5	6	7	8
1	1	1	1	1	1	1	1	1
2	1	1	2	2	2	2	2	2
3	1	1	3	3	3	3	3	3
4	1	2	1	1	2	2	3	3
5	1	2	2	2	3	3	1	1
6	1	2	3	3	1	1	2	2
7	1	3	1	2	1	3	2	3
8	1	3	2	3	2	1	3	1
9	1	3	3	1	3	2	1	2
10	2	1	1	3	3	2	2	1
11	2	1	2	1	1	3	3	2
12	2	1	3	2	2	1	1	3
13	2	2	1	2	3	1	3	2
14	2	2	2	3	1	2	1	3
15	2	2	3	1	2	3	2	1
16	2	3	1	3	2	3	1	2
17	2	3	2	1	3	1	2	3
18	2	3	3	2	1	2	3	1
	1	2				3		

(1)

$L_{27}(3^{13})$

No.	1	2	3	4	5	6	7	8	9	10	11	12	13
1	1	1	1	1	1	1	1	1	1	1	1	1	1
2	1	1	1	1	2	2	2	2	2	2	2	2	2
3	1	1	1	1	3	3	3	3	3	3	3	3	3
4	1	2	2	2	1	1	1	2	2	2	3	3	3
5	1	2	2	2	2	2	2	3	3	3	1	1	1
6	1	2	2	2	3	3	3	1	1	1	2	2	2
7	1	3	3	3	1	1	1	3	3	3	2	2	2
8	1	3	3	3	2	2	2	1	1	1	3	3	3
9	1	3	3	3	3	3	3	2	2	2	1	1	1
10	2	1	2	3	1	2	3	1	2	3	1	2	3
11	2	1	2	3	2	3	1	2	3	1	2	3	1
12	2	1	2	3	3	1	2	3	1	2	3	1	2
13	2	2	3	1	1	2	3	2	3	1	3	1	2
14	2	2	3	1	2	3	1	3	1	2	1	2	3
15	2	2	3	1	3	1	2	1	2	3	2	3	1
16	2	3	1	2	1	2	3	3	1	2	2	3	1
17	2	3	1	2	2	3	1	1	2	3	3	1	2
18	2	3	1	2	3	1	2	2	3	1	1	2	3
19	3	1	3	2	1	3	2	1	3	2	1	3	2
20	3	1	3	2	2	1	3	2	1	3	2	1	3
21	3	1	3	2	3	2	1	3	2	1	3	2	1
22	3	2	1	3	1	3	2	2	1	3	3	2	1
23	3	2	1	3	2	1	3	3	2	1	1	3	2
24	3	2	1	3	3	2	1	1	3	2	2	1	3
25	3	3	2	1	1	3	2	3	2	1	2	1	3
26	3	3	2	1	2	1	3	1	3	2	3	2	1
27	3	3	2	1	3	2	1	2	1	3	1	3	2
	a	b	a	a	c	a	a	b	a	a	b	a	a
		b	b^2		c	c^2	c	b	b^2	c^2	b^2	b	
									c	c^2		c	c^2

$\underbrace{\qquad}_{1}$ $\underbrace{\qquad\qquad}_{2}$ $\underbrace{\qquad\qquad\qquad\qquad\qquad\qquad}_{3}$

$L_{27}(3^{13})$ (Continued)

Interaction between Two Columns

1	2	3	4	5	6	7	8	9	10	11	12	13
(1)	3 4	2 4	2 3	6 7	5 7	5 6	9 10	8 10	8 9	12 13	11 13	11 12
	(2)	1 4	1 3	8 11	9 12	10 13	5 11	6 12	7 13	5 8	6 9	7 10
		(3)	1 2	9 13	10 11	8 12	7 12	5 13	6 11	6 10	7 8	5 9
			(4)	10 12	8 13	9 11	6 13	7 11	5 12	7 9	5 10	6 8
				(5)	1 7	1 6	2 11	3 13	4 12	2 8	4 10	3 9
					(6)	1 5	4 13	2 12	3 11	3 10	2 9	4 8
						(7)	3 12	4 11	2 13	4 9	3 8	2 10
							(8)	1 10	1 9	2 5	3 7	4 6
								(9)	1 8	4 7	2 6	3 5
									(10)	3 6	4 5	2 7
										(11)	1 13	1 12
											(12)	1 11

(1)

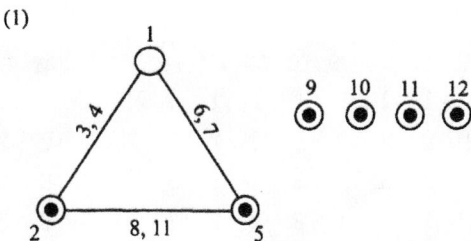

(2)

a

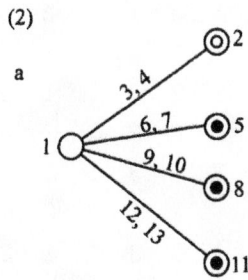

b

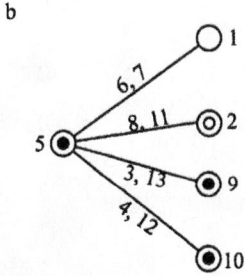

$L_{36}(2^{11}\times3^{12})^a$

No.	1	2	3	4	5	6	7	8	9	10	11	12	13	14	15	16	17	18	19	20	21	22	23	1'	2'	3'	4'
1	1	1	1	1	1	1	1	1	1	1	1	1	1	1	1	1	1	1	1	1	1	1	1	1	1	1	1
2	1	1	1	1	1	1	1	1	1	1	1	2	2	2	2	2	2	2	2	2	2	2	2	1	1	1	1
3	1	1	1	1	1	1	1	1	1	1	1	3	3	3	3	3	3	3	3	3	3	3	3	1	1	1	1
4	1	1	1	1	1	2	2	2	2	2	2	1	1	1	1	2	2	2	2	3	3	3	3	1	2	2	1
5	1	1	1	1	1	2	2	2	2	2	2	2	2	2	2	3	3	3	3	1	1	1	1	1	2	2	1
6	1	1	1	1	1	2	2	2	2	2	2	3	3	3	3	1	1	1	1	2	2	2	2	1	2	2	1
7	1	1	2	2	2	1	1	1	2	2	2	1	1	2	3	1	2	3	3	1	2	2	3	2	1	2	1
8	1	1	2	2	2	1	1	1	2	2	2	2	2	3	1	2	3	1	1	2	3	3	1	2	1	2	1
9	1	1	2	2	2	1	1	1	2	2	2	3	3	1	2	3	1	2	2	3	1	1	2	2	1	2	1
10	1	2	1	2	2	1	2	2	1	1	2	1	1	3	2	1	3	2	3	2	1	3	2	2	2	1	1
11	1	2	1	2	2	1	2	2	1	1	2	2	2	1	3	2	1	3	1	3	2	1	3	2	2	1	1
12	1	2	1	2	2	1	2	2	1	1	2	3	3	2	1	3	2	1	2	1	3	2	1	2	2	1	1
13	1	2	2	1	2	2	1	2	1	2	1	1	2	3	1	3	2	1	3	3	2	1	2	1	1	1	2
14	1	2	2	1	2	2	1	2	1	2	1	2	3	1	2	1	3	2	1	1	3	2	3	1	1	1	2
15	1	2	2	1	2	2	1	2	1	2	1	3	1	2	3	2	1	3	2	2	1	3	1	1	1	1	2
16	1	2	2	2	1	2	2	1	2	1	1	1	2	3	2	1	1	3	2	3	3	2	1	1	2	2	2
17	1	2	2	2	1	2	2	1	2	1	1	2	3	1	3	2	2	1	3	1	1	3	2	1	2	2	2
18	1	2	2	2	1	2	2	1	2	1	1	3	1	2	1	3	3	2	1	2	2	1	3	1	2	2	2
19	2	1	2	2	1	1	2	2	1	2	1	1	2	1	3	3	3	1	2	2	1	2	3	2	1	2	2
20	2	1	2	2	1	1	2	2	1	2	1	2	3	2	1	1	1	2	3	3	2	3	1	2	1	2	2
21	2	1	2	2	1	1	2	2	1	2	1	3	1	3	2	2	2	3	1	1	3	1	2	2	1	2	2
22	2	1	2	1	2	2	2	1	1	1	2	1	2	2	3	3	1	2	1	1	3	3	2	2	2	1	2
23	2	1	2	1	2	2	2	1	1	1	2	2	3	3	1	1	2	3	2	2	1	1	3	2	2	1	2
24	2	1	2	1	2	2	2	1	1	1	2	3	1	1	2	2	3	1	3	3	2	2	1	2	2	1	2
25	2	1	1	2	2	2	1	2	2	1	1	1	3	2	1	2	3	3	1	3	1	2	2	1	1	1	3
26	2	1	1	2	2	2	1	2	2	1	1	2	1	3	2	3	1	1	2	1	2	3	3	1	1	1	3
27	2	1	1	2	2	2	1	2	2	1	1	3	2	1	3	1	2	2	3	2	3	1	1	1	1	1	3
28	2	2	2	1	1	1	1	2	2	1	2	1	3	2	2	2	1	1	3	2	3	1	3	1	2	2	3
29	2	2	2	1	1	1	1	2	2	1	2	2	1	3	3	3	2	2	1	3	1	2	1	1	2	2	3
30	2	2	2	1	1	1	1	2	2	1	2	3	2	1	1	1	3	3	2	1	2	3	2	1	2	2	3
31	2	2	1	2	1	2	1	1	1	2	2	1	3	3	3	2	3	2	2	1	2	1	1	2	1	2	3
32	2	2	1	2	1	2	1	1	1	2	2	2	1	1	1	3	1	3	3	2	3	2	2	2	1	2	3

$L_{36}(2^{11} \times 3^{12})^a$ *(Continued)*

No.	1	2	3	4	5	6	7	8	9	10	11	12	13	14	15	16	17	18	19	20	21	22	23	1'	2'	3'	4'
33	2	2	1	2	1	2	1	1	1	2	2	3	2	2	2	1	2	1	1	3	1	3	3	2	1	2	3
34	2	2	1	1	2	1	2	1	2	2	1	1	3	1	2	3	2	3	1	2	2	3	1	2	2	1	3
35	2	2	1	1	2	1	2	1	2	2	1	2	1	2	3	1	3	1	2	3	3	1	2	2	2	1	3
36	2	2	1	1	2	1	2	1	2	2	1	3	2	3	1	2	1	2	3	1	1	2	3	2	2	1	3

| | 1 | | 2 | | | | 3 | | |

[a] By introducing columns 1', 2', 3', and 4' in place of columns 1, 2, ... , 11, one obtains $L_{36}(2^3 \times 3^{13})$.

$L_{54}(2^1 \times 3^{25})$

No.	1	2	3	4	5	6	7	8	9	10	11	12	13	14	15	16	17	18	19	20	21	22	23	24	25	26
1	1	1	1	1	1	1	1	1	1	1	1	1	1	1	1	1	1	1	1	1	1	1	1	1	1	1
2	1	1	1	1	1	1	1	1	2	2	2	2	2	2	2	2	2	2	2	2	2	2	2	2	2	2
3	1	1	1	1	1	1	1	1	3	3	3	3	3	3	3	3	3	3	3	3	3	3	3	3	3	3
4	1	1	2	2	2	2	2	2	1	1	1	1	1	1	2	3	2	3	2	3	2	3	2	3	2	3
5	1	1	2	2	2	2	2	2	2	2	2	2	2	2	3	1	3	1	3	1	3	1	3	1	3	1
6	1	1	2	2	2	2	2	2	3	3	3	3	3	3	1	2	1	2	1	2	1	2	1	2	1	2
7	1	1	3	3	3	3	3	3	1	1	1	1	1	1	3	2	3	2	3	2	3	2	3	2	3	2
8	1	1	3	3	3	3	3	3	2	2	2	2	2	2	1	3	1	3	1	3	1	3	1	3	1	3
9	1	1	3	3	3	3	3	3	3	3	3	3	3	3	2	1	2	1	2	1	2	1	2	1	2	1
10	1	2	1	1	2	2	3	3	1	1	2	2	3	3	1	1	1	1	2	3	2	3	3	2	3	2
11	1	2	1	1	2	2	3	3	2	2	3	3	1	1	2	2	2	2	3	1	3	1	1	3	1	3
12	1	2	1	1	2	2	3	3	3	3	1	1	2	2	3	3	3	3	1	2	1	2	2	1	2	1
13	1	2	2	2	3	3	1	1	1	1	2	2	3	3	2	3	2	3	3	2	3	2	1	1	1	1
14	1	2	2	2	3	3	1	1	2	2	3	3	1	1	3	1	3	1	1	3	1	3	2	2	2	2
15	1	2	2	2	3	3	1	1	3	3	1	1	2	2	1	2	1	2	2	1	2	1	3	3	3	3
16	1	2	3	3	1	1	2	2	1	1	2	2	3	3	3	2	3	2	1	1	1	1	2	3	2	3
17	1	2	3	3	1	1	2	2	2	2	3	3	1	1	1	3	1	3	2	2	2	2	3	1	3	1
18	1	2	3	3	1	1	2	2	3	3	1	1	2	2	2	1	2	1	3	3	3	3	1	2	1	2
19	1	3	1	2	1	3	2	3	1	2	1	3	2	3	1	1	2	3	1	1	3	2	2	3	3	2
20	1	3	1	2	1	3	2	3	2	3	2	1	3	1	2	2	3	1	2	2	1	3	3	1	1	3
21	1	3	1	2	1	3	2	3	3	1	3	2	1	2	3	3	1	2	3	3	2	1	1	2	2	1
22	1	3	2	3	2	1	3	1	1	2	1	3	2	3	2	3	3	2	2	3	1	1	3	2	1	1
23	1	3	2	3	2	1	3	1	2	3	2	1	3	1	3	1	1	3	3	1	2	2	1	3	2	2
24	1	3	2	3	2	1	3	1	3	1	3	2	1	2	1	2	2	1	1	2	3	3	2	1	3	3
25	1	3	3	1	3	2	1	2	1	2	1	3	2	3	3	2	1	1	3	2	2	3	1	1	2	3
26	1	3	3	1	3	2	1	2	2	3	2	1	3	1	1	3	2	2	1	3	3	1	2	2	3	1
27	1	3	3	1	3	2	1	2	3	1	3	2	1	2	2	1	3	3	2	1	1	2	3	3	1	2
28	2	1	1	3	3	2	2	1	1	3	3	2	2	1	1	1	3	2	3	2	2	3	2	3	1	1
29	2	1	1	3	3	2	2	1	2	1	1	3	3	2	2	1	3	1	3	3	1	3	1	2	2	1
30	2	1	1	3	3	2	2	1	3	2	2	1	1	3	3	2	1	2	1	1	2	1	2	3	3	3
31	2	1	2	1	1	3	3	2	1	3	3	2	2	1	2	3	1	1	1	3	2	3	2	2	3	1
32	2	1	2	1	1	3	3	2	2	1	1	3	3	2	3	1	2	2	2	2	1	3	1	3	3	1
33	2	1	2	1	1	3	3	2	3	2	2	1	1	3	1	2	3	3	3	3	2	1	2	1	1	2

$L_{54}(2^1 \times 3^{25})$ (Continued)

No.	1	2	3	4	5	6	7	8	9	10	11	12	13	14	15	16	17	18	19	20	21	22	23	24	25	26
34	2	1	3	2	2	1	1	3	1	3	3	2	2	1	3	2	2	3	2	3	1	1	1	1	3	2
35	2	1	3	2	2	1	1	3	2	1	1	3	3	2	1	3	3	1	3	1	2	2	2	2	1	3
36	2	1	3	2	2	1	1	3	3	2	2	1	1	3	2	1	1	2	1	2	3	3	3	3	2	1
37	2	2	1	2	3	1	3	2	1	2	3	1	3	2	1	1	2	3	3	2	1	1	3	2	2	3
38	2	2	1	2	3	1	3	2	2	3	1	2	1	3	2	2	3	1	1	3	2	2	1	3	3	1
39	2	2	1	2	3	1	3	2	3	1	2	3	2	1	3	3	1	2	2	1	3	3	2	1	1	2
40	2	2	2	3	1	2	1	3	1	2	3	1	3	2	2	3	3	2	1	1	2	3	1	1	3	2
41	2	2	2	3	1	2	1	3	2	3	1	2	1	3	3	1	1	3	2	2	3	1	2	2	1	3
42	2	2	2	3	1	2	1	3	3	1	2	3	2	1	1	2	2	1	3	3	1	2	3	3	2	1
43	2	2	3	1	2	3	2	1	1	2	3	1	3	2	3	2	1	1	2	3	3	2	2	3	1	1
44	2	2	3	1	2	3	2	1	2	3	1	2	1	3	1	3	2	2	3	1	1	3	3	1	2	2
45	2	2	3	1	2	3	2	1	3	1	2	3	2	1	2	1	3	3	1	2	2	1	1	2	3	3
46	2	3	1	3	2	3	1	2	1	3	2	3	1	2	1	1	3	2	2	3	3	2	1	1	2	3
47	2	3	1	3	2	3	1	2	2	1	3	1	2	3	2	2	1	3	3	1	1	3	2	2	3	1
48	2	3	1	3	2	3	1	2	3	2	1	2	3	1	3	3	2	1	1	2	2	1	3	3	1	2
49	2	3	2	1	3	1	2	3	1	3	2	3	1	2	2	3	1	1	3	2	1	1	2	3	3	2
50	2	3	2	1	3	1	2	3	2	1	3	1	2	3	3	1	2	2	1	3	2	2	3	1	1	3
51	2	3	2	1	3	1	2	3	3	2	1	2	3	1	1	2	3	3	2	1	3	3	1	2	2	1
52	2	3	3	2	1	2	3	1	1	3	2	3	1	2	3	2	2	3	1	1	2	3	3	2	1	1
53	2	3	3	2	1	2	3	1	2	1	3	1	2	3	1	3	3	1	2	2	3	1	1	3	2	2
54	2	3	3	2	1	2	3	1	3	2	1	2	3	1	2	1	1	2	3	3	1	2	2	1	3	3

```
        1   2      3              4
```

Glossary

additivity A concept relating to the independence of factors. The effect of additive factors occurs in the same direction (i.e., they do not interact).

adjusting factor A signal factor used to adjust the output.

analysis of variance (ANOVA) Analysis of the impacts of variances caused by factors. ANOVA is performed after the decomposition of the total sum of squares.

array A table of rows and columns used to make the combinations of an experiment.

base space A normal group selected by specialists to construct Mahalanobis space.

CAMPS Computer-aided measurement and process design and control system.

classified attribute The type of quality characteristic that is divided into discrete classes rather than being measured on a continuous scale.

compounding of noise factors A strategy for reducing the number of experimental runs by combining noise factors into a single output for experimentation.

confirmation experiment A follow-up experiment run under the conditions defined as optimum by a previous experiment. The confirmation experiment is intended to verify the experimental predictions.

confounding A condition in which experimental information on variables cannot be separated. The information becomes intermingled with other information.

control factor A product or process parameter whose values can be selected and controlled by the design or manufacturing engineer.

data analysis The process performed to determine the best factor levels for reducing variability and adjusting the average response toward the target value.

decomposition of variation The decomposition of collected data into the sources of variation.

degrees of freedom The number of independent squares associated with a given factor (usually, the number of factor levels minus 1).

design of experiments Experimental and analysis methods to construct mathematical models related to the response to the variables.

direct product design A layout designed to determine the interaction between any control factor assigned to an orthogonal array and any noise factor assigned to the outer array.

distribution A way of describing the output of a common-cause system variation in which individual values are not predictable but in which the outcomes as a group form a pattern that can be described in terms of its location, spread, and shape. Location is commonly expressed by the terms of the mean or median, and spread is commonly expressed in terms of the standard deviation or the range of a sample. Shape involves many characteristics, such as symmetry and peakedness. These are often summarized by using the name of a common distribution, such as the normal, binomial, or Poisson.

downstream quality Also called customer quality, such as car noise, vibration, or product life. It is the type of quality easily noticed by the customer. This is the type of quality to avoid using for quality improvement.

dummy treatment A method used to modify an orthogonal array to accommodate a factor with fewer levels.

dynamic characteristic A characteristic that expresses the functionality, transformation, and adjustability of the output of a system or subsystem.

dynamic operating window The gap between the total reaction rate and the side-reaction rate in chemical reactions. The greater the gap, the more desirable the result.

environmental noise factors Sources of variability due to environmental conditions when a product is used.

error sums of squares The total of the sums of squares that are considered as residual.

expected value of a variance The mean of an infinite number of variances estimated from collected data.

factor A parameter or variable that may affect product or process performance.

factorial effect The effect of a factor or an interaction or both.

feedforward control A way to compensate for the variability in a process or product by checking noise factor levels and sending signals that change an adjustment factor.

finite element method A computer-aided method used in many areas of engineering and physics to determine an approximate solution to problems that are continuous in nature, such as in the study of stress, strain, heat transfer, or electrical fields.

fractional factorial layout An experimental design that consists of a fraction of all factor-level combinations.

generic function The relationship between the objective output and the means as the signal that engineers use in order to generate the objective output.

go/no-go specifications The traditional approach to quality, which states that a specification part, component, or assembly that lies between upper and lower specifications meets quality standards.

hi–low table A table showing the magnitude of effects in descending order.

ideal function The ideal relationship under the standard condition between output and the signal by the users.

indicative factor A factor that has a technical meaning but has nothing to do with the selection of the best level.

inner array A layout or orthogonal array for the control factors selected for experimentation or simulation.

interaction A condition in which the impact of a factor or factors on a quality characteristic changes depending on the level of another factor (i.e., the interdependent effect of effect of two or more factors).

larger-the-better characteristic The type of performance parameter that gives improved characteristic performance as the value *of* the parameter increases (e.g., tensile strength, power, etc.) This type of characteristic belongs to the category of quality characteristics that has infinity as the ideal value and has no negative value.

least squares method A method to express the results by a function causes using existing data in such a way that the residual error of the function is minimized.

linear equation The equation showing the case when the input/output regression line does not pass through the origin.

linearity A measure of the straightness of a response plot. Also the extent to which a measuring instrument's response is proportional to the measured quantity.

loss function *See* quality loss function.

Mahalanobis distance A statistical tool introduced by P. C. Mahalanobis in 1930 to distinguish the pattern of one group from others.

Mahalanobis–Gram–Schmidt process (MTGS) One of the processes used to compute a Mahalanobis distance.

Mahalanobis space (MS) A normal space, such as a healthy group in a medical study.

Mahalanobis–Taguchi system (MTS) A measuring method of recognizing patterns using a Mahalanobis distance.

manufacturing tolerance The assessment of tolerance prior to shipping. The tolerance manufacturing tolerance is usually tighter than the consumer tolerance.

mean The average value of some variable.

mean-squared deviation (MSD) A measure of variability around the mean or target value.

mean-squared error variance The variance considered as the error.

mean sum of squares The sum of squares per unit degree of freedom.

measurement accuracy The difference between the average result of a measurement with a particular instrument and the true value of the quantity being measured.

measurement error The difference between the actual value and measured value of a measured quantity.

measurement precision The extent to which a repeated measurement gives the same result. Variations may arise from the inherent capabilities of the instrument, from changes in operating condition, and so on. *See also* repeatability *and* reproducibility.

midstream quality Also called *specified quality*, such as dimension or specification. This is the type of quality to avoid using for quality improvement.

multiple correlation Correlation by multiple variables.

noise factor Any uncontrollable factor that causes product quality to vary. There are three types of noise: (1) noise due to external causes (e.g., temperature, humidity, operator, vibration, etc.); (2) noise due to internal causes (e.g., wear, deterioration, etc.); and (3) noise due to part-to-part or product-to-product variation.

nominal-the-best characteristic The type of performance characteristic parameter that has an attainable target or nominal value (e.g., length, voltage, etc.).

nondynamic operating window A gap between the maximum and the minimum functional limits. A wide gap means a more robust function.

number of units The total of the square of the coefficients in a linear equation. It is used to calculate the sum of squares.

objective function The relationship between the objective characteristic and the signal used by the users.

off-line quality control Activities that use design of experiments or simulation to optimize product and process designs. These activities include system design, parameter design, and tolerance design.

on-line quality control Activities that occur at the manufacturing phase and include use of the quality loss function to determine the optimum inspection interval, control limits, and so on. On-line quality control is used to maintain the optimization gained through off-line quality control.

origin quality Also called *functional quality*. This is the type of quality used to improve the robustness of product functions.

orthogonal array A matrix of numbers arranged in rows and columns in such a way that each pair of columns is orthogonal to each other. When used in an experiment, each row represents the state of the factors in a given experiment. Each column represents a specific factor or condition that can be changed from experiment to experiment. The array is called orthogonal because the effects of the various factors in the experimental results can be separated from each other.

orthogonal polynomial equation An equation using orthogonal polynomial expansion.

outer array A layout or orthogonal array for the noise factors selected for experimentation.

parameter design The second of three design stages. During parameter design, the nominal values of critical dimensions and characteristics are established to optimize performance at low cost.

P-diagram A diagram that shows control, noise, and signal factors along with output response.

percent contribution The pure sum of squares divided by the total sum of squares to express the impact of a factorial effect.

point calibration Calibration of a specific point in measurement.

pooled error variance The error variance calculated by pooling the smaller factorial effects.

preliminary experiment An experiment conducted with only noise factors to determine direction and tendencies of the effect of these noise factors. The results of the preliminary experiment are used to compound noise factors.

pure (net) error sum of squares The sum of squares after adding the error terms originally included in the regular sums of squares of factorial effects.

pure (net) sum of squares The sum of squares of a factorial effect after subtracting the error portion.

quality The loss imparted by a product to society from the time the product is shipped.

quality characteristic A characteristic of a product or process that defines product or process quality. The quality characteristic measures the degree of conformity to some known standard.

quality engineering A series of approaches to predict and prevent the problems that might occur in the marketplace after a product is sold and used by the customer under various environmental and applied conditions for the duration of designed product life.

quality function deployment (QFD) The process by which customer feedback is analyzed and results are incorporated into product design. The QFD process is often referred to as determining the "voice of the customer."

quality loss function (QLF) A parabolic approximation of the quality loss that occurs when a quality characteristic deviates from its best or target value. The QLF is expressed in monetary units: the cost of deviating from the target increases quadratically the farther it moves from the target. The formula used to compute the QLF depends on the type of quality characteristic being used.

reference-point proportional equation The equation showing the case when a signal level is used as a reference input.

repeatability The variation in repeated measurements of a particular object with a particular instrument by a single operator.

reproducibility The state whereby the conclusions drawn from small-scale laboratory experiments will be valid under actual manufacturing and usage conditions (i.e., consistent and desirable).

response factor The output of a system or the result of a performance.

response table A table showing level averages of factors.

robust design A process within quality engineering of making a product or process insensitive to variability without removing the sources.

robust technology development An approach to maximize the functionality of a group of products at the earliest stage, such as at research-and-development stage, and to minimize overall product development cycle time.

robustness The condition used to describe a product or process design that functions with limited variability despite diverse and changing environmental conditions, wear, or, component-to-component variation. A product or process is robust when it has limited or reduced functional variation, even in the presence of noise.

sensitivity The magnitude of the output per input shown by the slope of the input/output relationship.

sensitivity analysis Analysis performed to determine the mean values of experimental runs used when the means are widely dispersed.

sequential approximation A method of providing figures to fill the open spots in an incomplete set of data to make data analysis possible.

signal factor A factor used to adjust the output.

signal-to-noise (SN) ratio Any of a group of special equations that are used in experimental design to find the optimum factor-level settings that will create a robust product or process. The SN ratio originated in the communications field, in which it represented the power of the signal over the power of the noise. In Taguchi methods usage, it represents the ratio of the mean (signal) to the standard deviation (noise). The formula used to compute the SN ratio depends on the type of quality characteristic being used.

signal-to-noise (SN) ratio analysis Analysis performed to determine the factor levels required to reduce variability and achieve the ideal function of a characteristic.

six sigma A quality program developed by Motorola Corporation with a detailed set of quality processes and tools.

slope calibration Calibration of slope in measurement.

smaller-the-better characteristic The type of performance characteristic parameter that has zero as the best value (e.g., wear, deterioration, etc.). This type of characteristic belongs to the category of quality characteristics that has zero as the best value and has no negative value.

speed difference method The SN ratio of a dynamic operating window calculated from the difference of two quality modes.

speed ratio method The SN ratio of a dynamic operating window calculated from the ratio of two quality modes.

split-type analysis A method to determine SN ratios without noise factors.

standard deviation A measure of the spread of the process output or the spread of a sampling statistic from the process (e.g., of subgroup averages). Standard deviation is denoted by the Greek letter σ (sigma).

standard rate of mistake The rate of mistake when type I and type II mistakes are adjusted to be equal.

standard SN ratio The SN ratio calculated from the standard rate of mistake.

subsystem A system to consist as part of the total system.

system design The first of three design stages. During system design, scientific and engineering knowledge is applied to produce a functional prototype design. This prototype is used to define the initial settings of a product or process design characteristic.

Taguchi methods Methods originated by Genichi Taguchi such as quality engineering, MTS, or experimental regression analysis.

tolerance design The third of three design stages. Tolerance design is applied if the design is not acceptable at its optimum level following parameter design. During tolerance design, more costly materials or processes with tighter tolerances are considered.

transformation The function of transforming the input signal, such as mold dimension or CNC machine input program, into the output, such as product dimension.

tuning factor A signal factor used to adjust the output.

two-step optimization An approach used in parameter design by maximizing robustness (SN ratio), first finding the optimum control factor set points to minimize sensitivity to noise and then adjusting the mean response to target.

upstream quality Also called *robust quality*. This is the type of quality used to improve the robustness of a specific product.

variability   The property of exhibiting variation (i.e., changes or differences).

variance   The mean-squared deviation (MSD) from the mean. The sum of squares divided by the degrees of freedom.

variation   The inevitable differences among individual outputs of a process.

variation of a factor   The sum of squares of a factorial effect.

Youden square   An incomplete Latin square.

zero-point proportional equation   The equation showing the case when the input/output regression line passes through the origin. The equation most frequently used in dynamic characteristics.

Bibliography

Genichi Taguchi's Publications

	Year	Title	Note	Publisher
1	1951	*Statistical Methods for Product Life Test and Estimation of Product Life*		Science Promotion Co. (Kagaku-shinkou-sha)
2	1956	*How to Set Tolerances (Assembly and Others)*		JSA (Japanese Standards Association); (Nihon-kikaku Kyoukai)
3	1957	*Design of Experiment*, Vol. 1		Maruzen
4	1958	*Design of Experiment*, Vol. 2		Maruzen
5	1959	*How to Plan Experiment with Orthogonal Array*	Coauthored	JUSE (Japanese Union of Scientists and Engineers)
6	1962	*Quality Control Handbook*	Coauthored	JSA
7	1962	*Production Control and Management*	Coauthored	JSA
8	1962	*Mathematical Table for Design of Experiment*		Maruzen
9	1962	*Modern Statistics Dictionary*	Coauthored	Toyo Economics Shinpou (Toyo Keizai Sninpou Sha)
10	1966	*Statistical Analysis*		Maruzen
11	1966	*Department Assessment and Appraisal System*		JSA
12	1967	*Process Control*	Coauthored	JSA
13	1969	*Quality Control*		JSA
14	1971	*Design of Processes and Systems*	Coauthored	JSA
15	1972	*Quality Assessment*	Coauthored	JSA
16	1972	*S/N Ratio Manual for Testing and Measurement for Performance Assessment*	Coauthored	JSA
17	1972	*Design of Experiment*, Vol. 1, 2nd edition		Maruzen Co.
18	1972	*Design of Experiment*, Vol. 2, 2nd edition		Maruzen Co.

	Year	Title	Note	Publisher
19	1972	*Statistical Analysis,* New Edition		Maruzen Co.
20	1972	*Signal-to-Noise Ratio Manual*	Coauthored	JSA
21	1973	*Design of Experiment Text*	Coauthored	JSA
22	1975	*Business Data Analysis*	Coauthored	JSA
23	1976	*Design of Experiment,* Vol. 1, 3rd edition		Maruzen Co.
24	1977	*Design of Experiment,* Vol. 2, 3rd edition		Maruzen Co.
25	1977	*Quality Control Text for Executives: Quality Planning and Economics*		JSA
26	1979	Management Engineering Series No. 18, *Design of Experiment*		JSA
27	1979	*Assessment of Measurement Error and Improvement: New Statistical Method with S/N Ratio*	Coauthored	JSA
28	1980	*Design of Experiment for Metrological Engineering*		Corona
29	1980	Management Engineering Series No. 5, *Statistical Analysis*	Coauthored	JSA
30	1981	*On-Line Quality Control during Production*		JSA
31	1982	*Measurement Control for Executives and Management*	Coauthored	JSA
32	1982	*Off-line Quality Control for Design and Development Engineers*		JSA
33	1983	*Measurement Control Manual for Quality Control*	Coauthored	JSA
34	1984	*Parameter Design for New Product Development*	Coauthored	JSA
35	1984	*How to Set Tolerance and Specifications*		JSA
36	1984	*Reliability Design for New Product Development: Case Studies*	Coauthored	JSA
37	1986	*Introduction to Quality Engineering* (English)		APO (Asian Productivity Organization)
38	1987	*Systems of Experimental Design* (English, Translation of DoE, Vols. 1 and 2)		Unipub
39	1988	*Quality Engineering for New Product Design and Development*		JSA
40	1988	*Quality Control Handbook,* New Edition	Coauthored	JSA
41	1988	Quality Engineering Lecture Series, Vol. 1, *Quality Engineering for New Product Development*		JSA
42	1988	Quality Engineering Lecture Series, Vol. 3, *S/N Ratio for Quality Assessment*		JSA

	Year	Title	Note	Publisher
43	1988	Quality Engineering Lecture Series, Vol. 4, *Design of Experiment for Quality Design*		JSA
44	1988	*Design of Measurement System*	Chief of Publication Committee	JSA
45	1988	*Application of Design of Measurement System*	Coauthored	JSA
46	1989	Quality Engineering Lecture Series, Vol. 2, *Quality Engineering for Manufacturing Process and its Control*		JSA
		Quality Engineering Lecture Series, Vols. 1–7	Chief of Publication Committee	JSA
47	1989	*Quality Engineering in Production Systems*	Coauthored	McGraw-Hill
48	1991	*Taguchi Methods: History of Its Advancement*		San-nou University
49	1992	Quality Engineering Application Series, *Technology Development by Transformability*	Coauthored	JSA
50	1992	JIS How-to Series, *Measuring System Calibration Manual*		JSA
51	1992	*Taguchi, on Robust Technology Development* (English)		ASME
52	1992	*Measurement Control Handbook*		Corona Co.
53	1994	*Quality Engineering for Technology Development*		JSA
54	1994	Quality Engineering Application Series, *Semiconductor Process Technology*	Coauthored	JSA
55	1994	*Taguchi Methods: On-line Production*		ASI
56	1995	*Quality Engineering for Executives*		Quality Month Committee
57	1996	*Management of Technology Development*		JSA
58	1999	*Mathematics for Quality Engineering*		JSA
59	1999	Quality Engineering Application Series, *Technology Development in Chemical, Pharmaceutical and Biological Systems*	Coauthored	JSA
60	1999	*Taguchi Methods: My Way of Thinking*		Keizaikai
61	1999	*Robust Engineering*	Coauthored	McGraw-Hill
62	2000	*Performance Assessment for Robust Design*		JSA
63	2000	Quality Engineering Application Series, *Technology Development in Electric and Electronics Systems*	Coauthored	JSA

	Year	Title	Note	Publisher
64	2000	*Mahalanobis–Taguchi System*	Coauthored	McGraw-Hill
65	2002	*The Mahalanobis–Taguchi Strategy*	Coauthored	John Wiley & Sons
66	2002	*Quality Engineering Application Series, Technology Development Using Mahalanobis–Taguchi Systems*	Coauthored	JSA
67	2002	*Paradoxical Strategy for Technology*	Coauthored	JSA

Index

A

Abnormality:
  measurement scale for, 398. *See also*
    Mahalanobis–Taguchi System
  methods for identifying, 417
Active dynamic SN ratio, 132–133
Active signal factors, 39, 97
Active SN ratios, 236
Additive model (Taguchi), 1488
Additivity:
  of factorial effects, 592
  and noise level, 586
  with particle size data, 588
  and physical condition, 585
Adhesion condition of resin board and copper
  plate, 890–894
Adjustability, 225
Adjusting factors, 60. *See also* Control factors
Adjustment, 60, 319. *See also* Tuning
  basic equations for, 1601
  in feedback control:
    based on product characteristics, 455–459
    of process condition, 468–473
  linearity in, 227–228
  in on-line quality engineering, 116
  process, 439
Adjustment cost, 441
Advanced Machining Tool Technology
  Promotion Association, 137
Airflow noise reduction of intercoolers, 1100–
  1105
Akao, Yoji, 1427, 1430
Algorithms case studies (software):
  optimization of a diesel engine software
    control strategy, 1301–1309

  video compression optimization, 1310–1323
Altshuller, Genrich, 1452–1453, 1489
American Society for Quality Control, 1443,
  1444
American Statistical Association, 1435
American Supplier Institute (ASI), 134, 144,
  158, 160, 162–164, 168, 1494
American Supplier Institute Consulting Group
  (ASI CG), 1498
Amplifier stabilization, design for, 717–731
Analysis of variance (ANOVA), 515–522
  and decomposition of variation, 515–521
  and spectrum analysis, 560
  table, 521–522
  in tolerance design, 342–346
  for two-way layout, 555–562
Analysis stage, 40
Analytical models, 1475–1476
Analytical stage (TRIZ), 1453
ANNs, *see* Artificial neural networks
ANOVA, *see* Analysis of variance
Aoyama Gakuin University, 155
Approximation:
  sequential, 610, 613–616
    in experimental regression analysis, 490,
      492–496
    with incomplete data, 613–616
  zero, 613
ARIZ, 1453–1454
Articulated robots, functionality evaluation of,
  1059–1068
Artificial environment, 14
Artificial neural networks (ANNs), 417–418
The Asahi, 181

ASI, *see* American Supplier Institute

ASI CG (American Supplier Institute Consulting Group), 1498

Assignment (layout) to an orthogonal array, 490

Attributes:
classified, 234, 587
signal-to-noise ratios for, 234–236, 290–308
value of continuous variables vs., 239
customer, 1482

Attribute data, disadvantages of using, 234–235

Automated process management, 31–33

Automatic inspection of solder joints, Mahalanobis distance in, 1189–1195

Automatic transmissions, friction material for, 841–847

Automotive industry, 7–8, 419. *See also specific companies*

Automotive industry case studies:
airflow noise reduction of intercoolers, 1100–1105
brake boosters, reduction of boosting force variation of, 1106–1111
casting conditions for camshafts, simulation of, 900–903
clear vision by robust design, 882–889
clutch disk torsional damping system design, 973–983
DC motor, optimal design for, 1122–1127
diesel engine software control strategy, 1301–1309
direct-injection diesel injector optimization, 984–1004
exhaust sensor output characterization with MTS, 1220–1232
friction material for automatic transmissions, 841–847
linear proportional purge solenoids, 1032–1049
minivan rear window latching improvement, 1025–1031
on-line quality engineering in, 1367–1375
product development challenge in, 1478–1479
steering system on-center robustness, 1128–1140
two-piece gear brazing conditions, optimization of, 858–862
wiper system chatter reduction, 1148–1156

Axiomatic design, 1454–1458
comparison of TRIZ, robust design, and, 1454–1458
function in, 1456
and structure thinking, 1487–1489

B

Baba, Ikuo, 140

Banks, 10, 11

Base space, 143

Batch production, 459

Bean sprouting conditions, optimization of, 631–636

Bebb, Barry, 158

Becker technique, 795

Bell Laboratories, 41, 57, 60, 154–157, 1501

Benchmarking, 10, 30, 96

Between-product noise, 176, 319

Biochemistry case studies:
bean sprouting conditions, optimization of, 631–636
small algae production, optimization of, 637–642

Biomechanical case study, 795–805

Block factors, 603

Blow-off charge measurement systems, optimization of, 746–752

Bobcat Division, Ingersoll-Rand, 161

Boosting force variation in brake boosters, reduction of, 1106–1111

Bossidy, Larry, 161

Box, George, 1479

Brain disease, urinary continence recovery in patients with, 1258–1266

Brake boosters, reduction of boosting force variation of, 1106–1111

Brazing conditions for two-piece gear, 858–862

Bucket elevator, preventive maintenance design for, 1376–1382

Build-test-fix cycles, 1480

Business applications of MTS, 419–420

Business readiness, 1505

C

CAE, *see* Computer-aided engineering

Calibration, 116, 177
linearity in, 227–228
point, 247–250
reference-point, 247–250
slope, 228, 247

zero-point, 228

Cameras, single-use, stability design of shutter mechanisms for, 965–972

CAMPS (Computer-Aided Measurement and Process Design and Control System), 140

Camshafts, simulation of casting conditions for, 900–903

Cannon, 1084

CAs (customer attributes), 1482

Casting conditions for camshafts, simulation of, 900–903

Cause, special vs. common, 1446–1447

Cause-and-effect diagrams, 1429

Cause parameters, 340. *See also* Low-rank characteristics

Central Japan Quality Control Association (CJQCA), 148, 155

Central Japan Quality Management Association, 144

Ceramic oscillation circuits, parameter design of, 732–734

CF, *see* Correction factor

Champion, corporate, 390

Change management, 390–391

Character recognition, Mahalanobis distance in, 1288–1292

Chebyshev, P. C., 543

Chebyshev's orthogonal polynomial equation, 254–257, 543

Checking interval:
  feedback control based on product characteristics, 455–459
  feedback control of process condition, 468–473

Check sheets, 1429

Checkup interval, 117

Chemical applications. *See also* Chemical reactions
  for on and off conditions, 83–85
  pharmacology, 686–694

Chemical applications case studies, 629–714
  biochemistry:
    bean sprouting conditions, optimization of, 631–636
    small algae production, optimization of, 637–642
  dynamic operating window:
    component separation, evaluation of, 666–671

herbal medicine granulation, use of, 695–704
  photographic systems, evaluation of, 651–658
  measurement:
    dynamic operating window, evaluation of component separation with, 666–671
    granule strength, optimization of measuring method for, 672–678
    thermoresistant bacteria, detection method for, 679–685
    ultra trace analysis, dynamic optimization in, 659–665
  separation:
    developer, particle-size adjustment in fine grinding process for, 705–714
    herbal medicine granulation, use of dynamic operating window for, 695–704
  in vitro percutaneous permeation, optimization of model ointment prescriptions for, 686–694

Chemical engineering, Taguchi method in, 85–93
  evaluation of images, 88–91
  functionality, such as granulation or polymerization distribution, 91
  function of an engine, 86–87
  general chemical reactions, 87–88
  separation system, 91–93

Chemical reactions:
  basic function of, 296–298
  general, 87–88
  ideal functions of, 230–232
  and pharmacy, 142
  with side reactions:
    reaction speed difference method, 298–301
    reaction speed ratio method, 301–308
  without side reactions, 296–298

Chemical reactions case studies:
  evaluation of photographic systems with dynamic operating window, 651–658
  polymerization reactions, optimization of, 643–650

Chowdhury, Subir, 168, 1493, 1496

Circuits case studies:
  design for amplifier stabilization, 717–731
  evaluation of electric waveforms by momentary values, 735–740

Circuits case studies (*continued*)
  parameter design of ceramic oscillation
      circuits, 732–734
  robust design for frequency-modulation
      circuits, 741–745
CJQCA, *see* Central Japan Quality Control
    Association
Clarion, 133, 141
Classified attributes, 587
  signal-to-noise ratios for, 234–236, 290–308
    for chemical reactions without side
        reactions, 296–298
    for chemical reactions with side reactions,
        298–308
    for two classes, one type of mistake, 291–
        293
    for two classes, two types of mistake, 293–
        296
    three-or-more-class, 234
    two-class, 234
  value of continuous variables vs., 239
Clausing, Don, 158
Clear vision by robust design, 882–889
Clutch disk torsional damping system design
    optimization, 973–983
Common cause, 1446–1447
Communication (in robust engineering
    implementation), 391–392
Comparison, 529, 542–543
Computer-aided engineering (CAE), 1475–1476
Computer-Aided Measurement and Process
    Design and Control System (CAMPS), 140
Computer systems case study, 1324–1334. *See
    also* Software testing and application case
    studies
Concepts:
  creation of, 39
  optimization of, 1497, 1512–1514
  selection of, 26–28, 40
Concept design, 1496, 1497. *See also* System
    design
Concept optimization phase, 1503, 1504
Concept phase, 1503, 1504
Concurrent engineering, 276, 1451
Conformance to specifications, quality as, 171
Consumer loss, quality level and, 17–18
Continuous improvement cycle, 1447
Continuous variables:
  signal-to-noise ratios for, 234–236, 239–289

double signals, 275–284
estimation of error, 271–275
with known signal factor level interval,
    256–260
with known signal factor level ratio, 262–
    263
linear equation, 250–254
linear equation using tabular display of
    orthogonal polynomial equation, 254–
    256
with no noise, 284–289
with no signal factor, 264–271
reference-point proportional equation,
    247–250
split analysis, 284–289
with unknown true values of signal factor
    levels, 264
when signal factor levels can be set up,
    260–262
zero-point proportional equation, 241–247
  in two-way layout experiments, 563–572
  value of classified attributes vs., 239
Contradiction table, 1454
Contrast, 529, 542
Contribution, 107
Control charts, 1429
Control costs, 17
Control factors, 30, 603, 1435. *See also* Adjusting
    factors; Indicative factors
  avoiding interactions between, 290, 356
  definition of, 74
  interactions of noise factors and, 227
  optimum levels for, 212
  in two-stage optimization, 40, 41
Copper, separation system for, 91–93
Copper plate, adhesion condition of resin
    board and, 890–894
Corporate leader/corporate team, 390–391
Corporate strategy, R&D investment in, 26
Correction factor (CF, CT), 510–511
Correction (in on-line quality engineering), 116
Cost(s):
  adjustment, 441
  balance of quality and, 18–22, 440–441
  control, 17
  and DFSS, 1516–1517
  for loss of quality, 171
  process control, 441
  production, 17, 442

of quality control systems during production, 474–477

and quality level, 18–22

and quality loss function, 380

unit manufacturing cost, 18, 121, 439, 441

Coupled design, 1455

Covariance, 536

Credit, 10–11

Criminal noises, 15–16

Cross-functional team, 390–392

CT, *see* Correction factor

Customers' tolerance, 182

Customer attributes (CAs), 1482

Customer loss, 193

Customer quality, *see* Downstream quality

Customer satisfaction, 134

Cycle for continuous improvement, 1447

Cycle time reduction, 39–40, 228

D

Data analysis, 506–514

  deviation, 507–508

  with incomplete data, 609–616

    sequential approximation, 613–616

    treatment, 611–613

  sum and mean, 506–507

  variation and variance, 509–514

Data transformation, 507

DC motor, optimal design for, 1122–1127

Debugging software, streamlining of, 1360–1364

Debug model, 1435

Decibel value of the SN ratio, 1438

Decomposition, 64. *See also* Decomposition of variation

  of degrees of contribution, 520–521

  of degrees of freedom, 518

  of indicative factors, 563–565

  two-way layout with, 563–572

    one continuous variable, 563–567

    two continuous variables, 567–572

Decomposition of variation, 505, 513

  and analysis of variance, 515–521

  to components with unit degree of freedom, 528–551

    comparison and its variation, 528–534

    linear regression equation, 534–543

    orthogonal polynomials, application of, 542–551

  between experiments, 574–575

Decoupled design, 1455

Deep-drawing process, optimization of, 911–915

Defect detection, MTS in, 1233–1237

Defective items, 445

Degrees of contribution:

  calculation of, 521–522

  decomposition of, 520–521

  explanation of, 560

Degrees of freedom, 509–510, 618

  decomposition of, 518

  decomposition of components with, 528–551

    comparison and its variation, 528–534

    linear regression equation, 534–543

    orthogonal polynomials, application of, 542–551

Delphi, 161

Delphi Automotive Systems, 984, 1220

Deming, Pluma Irene, 1442–1443

Deming, William Albert, 1442–1443

Deming, William Edwards, 1424, 1442–1448

  biographical background, 1442–1444

  continuous improvement cycle, 1447

  as father of quality in Japan, 157

  at Ford Motor Company, 157

  14 points for management, 1445–1446

  on making excellent products, 1491

  and pattern-level thinking, 1481

  quality principles of, 1444–1447

  seven deadly diseases, 1446

  Taguchi methods vs. teachings of, 1447–1448

Deming cycle, 1424, 1426, 1443

Deming Prize, 1424, 1428

Derivation(s):

  of quality loss function, 440

  of safety factors, 443–448

  with Youden squares, 623–625

Design. *See also specific topics*

  interaction between signals/noises and, 57

  management for quality engineering in, 30–31

Design-build/simulate-test-fix cycle, 164

Design constants:

  selection of, 40

  tuning as alteration of, 57

Design for assembly (DFA), 1482

Design for manufacturing (DFM), 1482

Design for six sigma (DFSS, IDDOV), 161–163, 1492–1520

  aligning NPD process with, 1498–1507

Design for six sigma (*continued*)
  financial benefits, 1515–1517
  IDDOV structure, 1494–1496
  and intersection of six sigma and robust
      engineering, 1517–1519
  and NPD, 1501–1507
  and quality engineering, 1498
  Taguchi methods as foundation for, 1493–
      1494
  technical benefits, 1507–1515
Design for Six Sigma (Subir Chowdhury), 1493
Design life, 57
Design method for preventive maintenance,
    463–467
Design of experiments (DoE), 503–505
  decomposition of variation, 505
  definition of, 503–504
  development of, 154
  at Ford Motor Company, 158
  functionality, evaluation of, 504
  in Japan, 1431–1435
  objective of, 178, 356
  quality engineering vs., 356
  and reproducibility, 504–505
  in United States, 159
Design of Experiments (Genichi Taguchi), 131,
    515, 523, 552, 563, 573, 597
Design process:
  analysis stage of, 40
  stages of, 40
  synthesis stage of, 40
  tools for, 30
Design quality, 438
Design quality test, 38
Design review, 1427
Design-test-fix cycle, 1471
Detector switch characterization with MTS,
    1208–1219
Deteriorating characteristics, 205–207
Developer:
  high-quality, development of, 780–787
  particle-size adjustment in fine grinding
      process for, 705–714
Deviation, 507–508
  estimate of, 512
  magnitude of, 509
  from main effect curve, 557
  from mean, 507, 508
  from objective value, 507–508

  standard, 517
  in variance, 511–514
  in variation, 509
DFA (design for assembly), 1482
DFM (design for manufacturing), 1482
DFSS, *see* Design for six sigma
Diagnosis. *See also* Medical diagnosis
  basic equations for, 1601
  MTS for, 398
  to prevent recalls, 33–38
  process, 439, 477–479
Diagnosis interval, 117
Diesel engine software control strategy, 1301–
    1309
Direct-injection diesel injector optimization,
    984–1004
Direct product design, 227
Direct product layout, 323
Discrimination and classification method, 417
Disk blade mobile cutters, optimization of,
    1005–1010
Distribution of tolerance, 204–205
DMAIC (six sigma), 1517–1519
Documentation of testing, 1477
DoE, *see* Design of experiments
Dole, 1208
Domain model of product development, 1479–
    1480
Double signals, 275–284
Downstream quality, 269, 354, 355
Drug efficacy, Mahalanobis distance in
      measurement of, 1238–1243
Dummy treatment, 427, 428, 605
D-VHS tape travel stability, 1011–1017
Dynamic operating window method, 142
  for chemical reactions with side reactions,
      298–308
  for component separation evaluation, 666–
      671
  for herbal medicine granulation, 695–704
  for photographic systems evaluation, 651–658
Dynamic SN ratio, 139, 225
  nondynamic vs., 139, 225
  for origin quality, 354–355
  in U.S. applications, 160
  uses of, 139, 236

E

Earthquake forecasting, MTS in, 419

ECL, *see* Electrical Communication Laboratory

Economy:
  direct relationship of SN ratio and, 225
  and productivity, 5–7

ECs (engineering characteristics), 1482

Education (in robust engineering
    implementation), 392

Effective power, 61

Effect of the average curve, 558–559

Elderly:
  improving standard of living for, 12
  predicting health of, 15

Electrical applications case studies, 715–792
  circuits:
    amplifier stabilization, design for, 717–731
    ceramic oscillation circuits, parameter
        design of, 732–734
    electric waveforms, evaluation of, 735–740
    frequency-modulation circuits, robust
        design for, 741–745
  electronics devices:
    back contact of power MOSFETs,
        optimization of, 771–779
    blow-off charge measurement systems,
        optimization of, 746–752
    fine-line patterning for IC fabrication,
        parameter design of, 758–763
    generic function of film capacitors,
        evaluation of, 753–757
    pot core transformer processing,
        minimizing variation in, 764–770
  electrophotography:
    functional evaluation of process, 788–792
    high-quality developer, development of,
        780–787

Electrical Communication Laboratory (ECL),
    154–156

Electrical connector insulator contact housing,
    optimization of, 1084–1099

Electrical encapsulants:
  ideal function for, 234
  optimization of process, 950–956
  robust technology development of process,
      916–925

Electric waveforms, momentary value in
    evaluation of, 735–740

Electro-deposited process for magnet
    production, 945–949

Electronic and electrical engineering, Taguchi
    method in, 59–73

functionality evaluation of system using
    power, 61–65
quality engineering of system using
    frequency, 65–73

Electronic components:
  adhesion condition of resin board and
      copper plate, 890–894
  on and off conditions, 83–85
  resistance welding conditions for, 863–868

Electronic devices case studies:
  back contact of power MOSFETs,
      optimization of, 771–779
  blow-off charge measurement systems,
      optimization of, 746–752
  fine-line patterning for IC fabrication,
      parameter design of, 758–763
  generic function of film capacitors,
      evaluation of, 753–757
  variation in pot core transformer processing,
      minimizing, 764–770

Electronic warfare systems, robust testing of,
    1351–1359

Electrophotography case studies:
  functional evaluation of an
      electrophotographic process, 788–792
  high-quality developer, 780–787
  toner charging function measuring system,
      875–881

Emission control systems, 1032–1049

Employment, productivity and, 6

Engines:
  function of, 86–87
  idle quality, ideal function of, 230

Engineering and the Mind's Eye (Eugene S.
    Ferguson), 1476

Engineering characteristics (ECs), 1482

Engineering quality, 10

Engineering systems, laws of evolution of, 1454

Enhanced plastic ball grid array (EPBGA)
    package, 916–925

Environmental noises, 14–15

EPBGA package, *see* Enhanced plastic ball grid
    array package

Equalizer design, ideal function in, 233

Error, human, 13–14

Error factors, 1434, 1436

Error variance, 240, 420, 511

Error variation, 513, 517, 518, 559

Estimated values, 492, 494–499

Estimate of deviation, 512
Estimation of error, 271–275
Europe, paradigm shift in, 29
Evaluation technologies, 26
Event thinking, 1480–1481, 1490
Evolution of engineering systems, laws of, 1454
Exchange rates, 7
Exhaust sensor output characterization with
    MTS, 1220–1232
Experimental regression analysis, 149, 484, 488–
    499
  estimated values, selection of, 492, 494–499
  initial values, setting of, 488–489
  orthogonal array, selection and calculation of,
    489–494
  parameter estimation with, 484, 488–499
  sequential approximation, 490, 492–496
Experimentation:
  confirmatory, 361
  development of experimental design, 1472–
    1473
  orthogonal array L_{18} for, 357
  successful use of orthogonal arrays for, 358
Experts, internal, 392
External noise, 1512
External test facilities, 1476–1477

F

Fabrication line capacity planning using a
    robust design dynamic model, 1157–1167
Factors. *See also specific types, e.g.,* Control factors
  classification of, 1434–1435
  definition of, 552
Failure mode and effects analysis (FMEA), 392,
    1427, 1482
Failure rates, 1514–1515
FDI (Ford Design Institute), 161
Feedback control, 32, 79, 439
  based on process condition, 468–473
    temperature control, 471–473
    viscosity control, 469–470
  based on product characteristics, 454–467
    batch production process, 459–462
    process control gauge design, 462–467
    system design, 454–459
  designing system of, 118–120
  measurement error in, 459
  by quality characteristics, 1383–1388
Feeder valves, reduction of chattering noise in,
    1112–1121

Feedforward control, 32, 79, 439
Felt-resist paste formula, optimization of, 836–
    840
FEM (finite-element method) analysis, 1050
Ferguson, Eugene S., 1476
Film capacitors, generic function of, 753–757
Financial system product design, 10–11
Find-and-fix improvement efforts, 1515
Fine-line patterning for IC fabrication, 758–763
Finite-element method (FEM) analysis, 1050
Fire detection, MTS in, 419
Firefighting, 159, 167, 318
Fire prevention, 167, 168
Fisher, Ronald A., 503, 1424, 1431–1434
Fixed targets, signal-to-noise ratios for, 236
Flexible manufacturing systems (FMSs), 12
Flexor tendon repairs, biomechanical
    comparison of, 795–805
Flex Technology, 160
FMEA, *see* Failure mode and effects analysis
FMSs (flexible manufacturing systems), 12
Ford Design Institute (FDI), 161
Ford Motor Company, 16, 81, 157–158, 160,
    161, 229, 230, 1431
Ford Supplier Institute (FSI), 157, 158
Forecasting. *See also* Prediction
  of earthquakes, 419
  of future health, 1277–1287
  MTS for, 398
  of weather, 419
Foundry process using green sand, 848–851
Four "P's" of testing, 1470
  personnel, 1470–1472
  preparation, 1472–1473
  procedures, 1473–1474
  product, 1474–1475
14 points for management, 1445–1446
Freedoms, individual, 9, 11
Frequency modulation, 67–69, 741–745
Friction material for automatic transmissions,
    841–847
FSI, *see* Ford Supplier Institute
Fuel delivery system, ideal function in, 229–230
Fuji Film, 24, 135, 1439
Fujimoto, Kenji, 84
Fuji Steel, 1424
Fuji Xerox, 149, 158, 428, 1439
Function(s):
  definitions of, 1456

in design, 1456
generic, 8
  definition of, 85, 229, 355
  determining evaluation scale of, 148
  identification of, 360
  machining, 80–83
  objective function vs., in robust technology
      development, 355
  optimization of, 352
  robustness of, 352
  selecting, 313–314
  SN ratio for, 133
  technological, 1440
ideal, 60, 229–234
  actual function vs., 1508–1509
  based on signal-to-noise ratios, 228–234
  chemical reactions, 230–232
  definition of, 1450, 1508
  with double signals, 279
  electrical encapsulant, 234
  engine idle quality, 230
  equalizer design, 233
  fuel delivery system, 229–230
  grinding process, 232
  harmful part of, 1509
  identifying, 377–378
  injection molding process, 229
  intercoolers, 232
  low-pass filters, 233–234
  machining, 229
  magnetooptical disk, 233
  power MOSFET, 234
  printer ink, 234
  specifying, 1472
  transparent conducting thin films, 233
  useful part of, 1509
  voltage-controlled oscillators, 234
  wave soldering, 233
  welding, 230
  wiper systems, 230
loss, 1452, 1488
modulation, 67–69
objective, 10
  definition of, 229, 355
  and design of product quality, 26
  generic function vs., 355
  standard conditions for, 40
  of subsystems, 313
  transformability as, 140

tuning as, 57
on and off, 138
orthogonal, 543
phase modulation, 68–72
quality loss function, 133–134, 171–179, 380
  classification of quality characteristics, 180
  derivation of, 440
  equations for, 1598–1601
  justice-related aspect of, 175–176
  for larger-the-better characteristic, 188, 191
  for nominal-the-best characteristic, 180–
      186, 190, 445
  and process capability, 173–176
  quadratic representation of, 174
  quality aspect of, 176
  for smaller-the-better characteristic, 186–
      188, 190
  and steps in product design, 176–179
  for tolerance design optimization, 386
in TRIZ, 1456
Functional analysis and trimming (TRIZ), 1454
Functional evaluation:
  of electrophotographic process, 788–792
  SN ratio for, 133
Functional independence, 1488
Functionality(-ies):
  evaluation, functionality, 28–30, 147, 504
    of articulated robots, 1059–1068
    of spindles, 1018–1024
    standardization draft for, 148
    in trading, 148
  improvement of, 27
  and system design, 314–317
Functionality design, 11, 73
Functional limit, 443
Functional quality, *see* Origin quality
Functional risk, 11
Functional tolerance, 193
Function limit, 196–201

G

Gain, 79
Gas-arc stud weld process parameter
    optimization, 926–939
Gauss, K. F., 535
GDP, *see* Gross domestic product
General Electric (GE), 161, 1481
Generic function(s), 8
  definition of, 85, 229, 355

Generic function(s) (*continued*)
  determining evaluation scale of, 148
  identification of, 360
  machining, 80–83
  objective function vs., in robust technology
      development, 355
  optimization of, 352
  robustness of, 352
  selecting, 313–314
  SN ratio for, 133
  technological, 1440
Generic technology, 26, 39, 59
Global Standard Creation Research and
    Development Program, 148
GNP (gross national product), 6
GOOD START model, 1505–1507
Gram–Schmidt orthogonalization process
    (GSP), 400, 415–418. *See also* Mahalanobis–
    Taguchi Gram–Schmidt process
Granule strength, optimization of measuring
    method for, 672–678
Graphs, 1429. *See also* Linear graphs
Greco Latin square, 1432
Green sand, foundry process using, 848–851
Grinding process:
  for developer, particle-size adjustment in,
      705–714
  ideal function of, 232
Gross domestic product (GDP), 11, 12
Gross national product (GNP), 6
GSP, *see* Gram–Schmidt orthogonalization
    process

H

Hackers, 15
Handwriting recognition, 1288–1292
Hara, Kazuhiko, 141
Hardware design, downstream reproducibility
    in, 58–59
Harmful effects (TRIZ), 1453
Harmful part (ideal functions), 1509
Harry, Mikel, 161, 1517
Hazama, 1427
Health. *See also* Medical diagnosis
  of elderly, 12, 15
  forecasting, 1277–1287
  MT method for representing, 102–104
Herbal medicine granulation, dynamic
    operating window method for, 695–704

Hermitian form, 62–64. *See also* Decomposition
    of variation
Hewlett-Packard, 27–29, 73
Hicks, Wayland, 158
Higashihara, Kazuyuki, 84
High-performance liquid chromatograph
    (HPLC), 666–671
High-performance steel, machining technology
    for, 819–826
High-rank characteristics, 201
  definition of, 340
  relationship between low-rank characteristics
      and, 203–204
  tolerance for, 205
Histograms, 1429
History of quality engineering, 127–168
  in Japan, 127–152
    application of simulation, 144–147
    chemical reaction and pharmacy, 142
    conceptual transition of SN ratio, 131–133
    design of experiments, 1431–1435
    development of MTS, 142–143
    electric characteristics, SN ratio in, 140–
        141
    evaluation using electric power in
        machining, 138–139
    machining, SN ratio in, 135–137
    on-line quality engineering and loss
        function, 133–135
    origin of quality engineering, 127–131
    QFD for customers' needs, 1430–1431
    seven QC tools and QC story, 1428–1430
    since 1950s, 1435–1440
    transformability, SN ratio of, 139–140
  in United States, 153–168
    from 1980 to 1984, 156–158
    from 1984 to 1992, 158–160
    from 1992 to 2000, 161–164
    from 2000 to present, 164–168
    and history of Taguchi's work, 153–156
Holmes, Maurice, 158
Hosokawa, Tetsuo, 27, 28
House of quality, 164–166
HPLC, *see* High-performance liquid
    chromatograph
Human errors, 13–14
Human performance case studies:
  evaluation of programming ability, 1178–1188
  prediction of programming ability, 1171–1177

I

ICP-MS (inductively coupled plasma mass spectrometer), 659
IDDOV, *see* Design for six sigma
Ideal function(s), 60, 229–234
actual function vs., 1508–1509
based on signal-to-noise ratios, 228–234
chemical reactions, 230–232
definition of, 1450, 1508
with double signals, 279
electrical encapsulant, 234
engine idle quality, 230
equalizer design, 233
fuel delivery system, 229–230
grinding process, 232
harmful part of, 1509
identifying, 377–378
injection molding process, 229
intercoolers, 232
low-pass filters, 233–234
machining, 229
magnetooptical disk, 233
power MOSFET, 234
printer ink, 234
specifying, 1472
transparent conducting thin films, 233
useful part of, 1509
voltage-controlled oscillators, 234
wave soldering, 233
welding, 230
wiper systems, 230
Ideality:
concept of, 1453
principle of, 1489
Ideation, 1301
IHI, *see* Ishikawajima–Harima Heavy Industries
IIN/ACD, *see* ITT Industries Aerospace/
Communications Division
Image, definition of, 88
IMEP (indicated mean effective pressure), 230
Implementation of robust engineering, 389–393
and bottom-line performance, 393
corporate leader/corporate team in, 390–391
critical components in, 389–390
education and training in, 392
effective communication in, 391–392
and integration strategy, 392–393
management commitment in, 390
Ina, Masao, 1433

Inao, Takeshi, 131
Ina Seito Company, 155, 589, 1433, 1434
Incomplete data, 609–616
sequential approximation with, 613–616
treatment of, 611–613
Incomplete Latin squares, 617. *See also* Youden squares
Independence axiom, 1454, 1487–1488
Indian Statistical Institute, 155, 398
Indicated mean effective pressure (IMEP), 230
Indicative factors, 30, 68, 563, 603, 1434, 1436. *See also* Control factors
Individual freedoms, 9, 11
Inductively coupled plasma mass spectrometer (ICP-MS), 659
Industrial Revolution, 9. *See also* Second Industrial Revolution
Industrial waste, tile manufacturing using, 869–874
Information axiom, 1454
Information technology, 10–11
Ingersoll-Rand, 161
Initial values (experimental regression analysis), 488–489
Injection molding process. *See also* Plastic injection molding
ideal function in, 229
thick-walled products, 940–944
Ink, printer:
ideal function for, 234
MTS in thermal ink jet image quality inspection, 1196–1207
Inner noise, 176, 319
Inner orthogonal arrays, 322–323
Input, signal-to-noise ratios based on, 236–237
Inspection, traditional approach to, 378–379
Inspection case studies:
automatic inspection of solder joints, Mahalanobis distance in, 1189–1195
defect detection, 1233–1237
detector switch characterization, 1208–1219
exhaust sensor output characterization, 1220–1232
printed letter inspection technique, 1293–1297
thermal ink jet image quality inspection, 1196–1207
Inspection design, 32–33, 80, 463–467
Inspection interval, 117

Integration strategy (robust engineering), 392–393

Intention, signal-to-noise ratios based on, 236

Interaction(s), 579
  and choice of functional output, 1451
  software diagnosis using, 98–102

Intercoolers:
  airflow noise reduction of, 1100–1105
  ideal function of, 232

Internal experts, 392

Internal noise, 1512

International Society for Robust Engineering Professionals, 168

Inventive principles, 1454

Inverted matrix (for MD computation), 400

In vitro percutaneous permeation, model ointment prescriptions for, 686–694

Ishikawa, Ichiro, 1481

Ishikawa, Kaoru, 1481, 1482

Ishikawajima–Harima Heavy Industries (IHI), 83–85, 137, 138, 141, 148

Isuzu Motors, 131–132

ITT, 161, 162

ITT Cannon, 144, 1069

ITT Defense Electronics, 433

ITT Industries Aerospace/Communications Division (IIN/ACD), 1310, 1324

J

Japan:
  fire loss in, 14
  functional evaluation of systems in, 11
  history of quality engineering in, 127–152, 1428–1440
    application of simulation, 144–147
    chemical reaction and pharmacy, 142
    conceptual transition of SN ratio, 131–133
    design of experiments, 1431–1435
    development of MTS, 142–143
    electric characteristics, SN ratio in, 140–141
    electric power in machining, evaluation using, 138–139
    machining, SN ratio in, 135–137
    on-line quality engineering and loss function, 133–135
    origin of quality engineering, 127–131
    QFD for customers' needs, 1430–1431
    seven QC tools and QC story, 1428–1430
    since 1950s, 1435–1440
    transformability, SN ratio of, 139–140
  quality management in, 1423–1440
    design of experiments, 1431–1435
    history of, 1423–1428
    quality engineering, 1435–1440
    quality function deployment, 1430–1431
  R&D investment in, 26
  standard of living and income in, 7
  statistical process control in, 1481

Japanese Society for Quality Control, 1428

Japanese Standards Association (JSA), 148–149, 155, 1432

Japanese Union of Scientists, 1432

Japanese Union of Scientists and Engineers (JUSE), 157, 1335, 1427, 1443, 1481

Japan Industry Standards (JIS), 196, 1424

Japan Society for Seikigakkai, 136

Japan Society of Plastic Technology, 140

JIS, *see* Japan Industry Standards

Joban Hawaiian Center, 1428

JR (Japan Railroad), 199

JSA, *see* Japanese Standards Association

Juran, J. M., 1426, 1481

Juran's trilogy, 1511

JUSE, *see* Japanese Union of Scientists and Engineers

K

Kajima Corp., 1427

Kakar, Rague, 156

Kanemoto, Yoshishige, 133, 141

Kanetaka, Tatsuji, 104–106, 143

Kansai Electric Power, 1428

Kawamura, Masanobu, 131

Kessler technique, 795

Kitagawa, Toshio, 1424

Kodak, 161

Kokai, Tomi, 153

Komatsu, 1429

Konica, 133

L

Larger-the-better characteristic, 17, 180
  quality level at factories for, 451–453
  quality level calculation for, 19, 21–22
  quality loss function for, 188, 191
  safety factor for, 447
  tolerance design for, 210–211

Larger-the-better SN ratio, 95, 139

Latent defects, 1498
Latin squares, 617, 1432
Laws of evolution of engineering systems, 1454
Layout to an orthogonal array, 490
LC (liquid chromatographic) analysis, 666
Leadership for robust engineering, 390–391
Least squares, 60, 535
  for estimation of operation time, 485–486
  parameter estimation with, 485–489
LIMDOW disk, 27, 28
Limits, 379
Linear actuator optimization using simulation, 1050–1058
Linear equations, 240, 241
  number of units in, 559
  for signal-to-noise ratios for continuous variables, 250–256
  zero-point proportionality as replacement for, 137
Linear graphs, 597–608, 1433, 1527–1528
  combination design, 606–608
  dummy treatment, 605
  for L_4, 1530
  for L_8, 597–600, 602, 604, 1530, 1603
  for L_9, 1564, 1609
  for L_{16}, 1532–1535, 1586, 1606–1608
  for L_{18}, 1564, 1610
  for L_{25}, 1591
  for L_{27}, 1567, 1613
  for L'_{27}, 1597
  for L_{32}, 1538–1549, 1587
  for L_{36}, 1594
  for L_{50}, 1592
  for L_{59}, 1568
  for L_{64}, 1554–1563, 1590
  for L_{81}, 1572–1585
  for multilevel factors, 599–606
Linearity (SN ratios), 225, 226
  in adjustment and calibration, 227–228
  dynamic, 236
Linear proportional purge solenoids, 1032–1049
Linear regression equations, 1406–1419
Liquid chromatographic (LC) analysis, 666
Liver disease diagnosis case studies, 401–415, 1244–1257
Loss:
  definition of, 171
  due to evil effects, 18
  due to functional variability, 18–19, 440–442
  due to harmful quality items, 440–441
  to society, 1516–1517
Loss function, 1452, 1488. *See also* Quality loss function
Low-pass filters, ideal function for, 233–234
Low-rank characteristics. *See also* Parameter design; Tolerance design
  definition of, 340
  deterioration of, 205–207
  determination of, 201–204
  and high-rank characteristics, 203–204
  specification tolerancing with known value of, 201–204

M

MACH, *see* Minolta advanced cylindrical heater
Machine and Tool, 136
Machine Technology, 136
Machining, 80–83
  analysis of cutting performance, 81, 82
  ideal function, 229
  improvement of work efficiency, 82–83
Machining case studies:
  development of machining technology for high-performance steel by transformability, 819–826
  optimization of machining conditions by electrical power, 806–818
  spindles, functionality evaluation of, 1018–1024
  transformability of plastic injection-molded gear, 827–835
Maeda, 1428
Magnetic circuits, optimization of linear actuator using simulation, 1050–1058
Magnetooptical disk, ideal function for, 233
Magnet production, electro-deposited process for, 945–949
Mahalanobis, P. C., 103, 155, 398, 1440
Mahalanobis distance (MD), 102–103, 114–116, 398, 1440
  applications of, 398
  in automatic inspection of solder joints, 1189–1195
  in character recognition, 1288–1292
  computing methods for, 400
  Gram–Schmidt process for calculation of, 416
  for health examination and treatment of missing data, 1267–1276

Mahalanobis distance (*continued*)
  in measurement of drug efficacy, 1238–1243
  in medical diagnosis, 1244–1257
  in prediction:
    of earthquake activity, 14
    of urinary continence recovery in patients
        with brain disease, 1258–1266
  and Taguchi method, 114–116
Mahalanobis space, 114–116
  in predicting health of elderly, 15
  in prediction of earthquake activity, 14
  and Taguchi method, 114–116
Mahalanobis–Taguchi Gram–Schmidt (MTGS)
    process, 415–418
Mahalanobis–Taguchi (MT) method, 102–109
  general pattern recognition and evaluation
      procedure, 108–109
  and Mahalanobis distance/space, 114–116
  for medical diagnosis, 104–108
Mahalanobis–Taguchi System (MTS), 102–103,
    109–116, 397–421
  applications, 418–420
  in automotive industry:
    accident avoidance systems, 419
    air bag deployment, 419
  base space creation in, 143
  in business, 419–420
  definition of, 398
  discriminant analysis vs., 143
  early criticism/misunderstanding of, 142–143
  in earthquake forecasting, 419
  in fire detection, 419
  general pattern recognition and evaluation
      procedure, 109–114
  and Gram–Schmidt orthogonalization
      process, 415–418
  liver disease diagnosis case study, 401–415
  and Mahalanobis distance/space, 114–116
  in manufacturing, 418
  in medical diagnosis, 418
  other methods vs., 417–418
  in prediction, 15, 419
  sample applications of, 398–400
  for study of elderly health, 12
  in weather forecasting, 419
Mahalanobis–Taguchi System (MTS) case
    studies, 1169–1297
  human performance:
    evaluation of programming ability, 1178–
        1188

    prediction of programming ability, 1171–
        1177
  inspection:
    automatic inspection of solder joints,
        Mahalanobis distance in, 1189–1195
    defect detection, 1233–1237
    detector switch characterization, 1208–1219
    exhaust sensor output characterization,
        1220–1232
    thermal ink jet image quality inspection,
        1196–1207
  medical diagnosis:
    future health, forecasting of, 1277–1287
    health examination and treatment of
        missing data, 1267–1276
    liver disease, 401–415
    measurement of drug efficacy, Mahalanobis
        distance for, 1238–1243
    urinary continence recovery among
        patients with brain disease, prediction
        of, 1258–1266
    use of Mahalanobis distance in, 1244–1257,
        1267–1276
  products:
    character recognition, Mahalanobis
        distance for, 1288–1292
    printed letter inspection technique, 1293–
        1297
Main effect curve, 557
Maintenance design, 439. *See also* Preventive
    maintenance
Maintenance system design, 33
Management:
  commitment to robust engineering by, 390
  general role of, 23
  for leadership in robust engineering, 390–
      391
  major role of, 10
  in manufacturing, 16–22, 120–123
    consumer loss, evaluation of quality level
        and, 17–18
    cost vs. quality level, 18–22
    at Ford Motor Company, 16
    tolerance, determination of, 16–17
  for quality engineering, 25–38
    automated process management, 31–33
    design process, 30–31
    functionality, evaluation of, 28–30
    recalls, diagnosis to prevent, 33–38

research and development, 25–28, 39–40
and robust engineering, 377–388
  parameter design, 382–386
  tolerance design, 386–388
strategic planning in, 13
Management by total results, 149
Management reviews, 1507
Manufacturing. *See also* Manufacturing processes
management in, 16–22
  balance of cost and quality level, 18–22
  determining tolerance, 16–17
  evaluation of quality level and consumer
    loss, 17–18
  at Ford Motor Company, 16
MTS in, 418
of semiconductor rectifier, 1395–1398
Manufacturing case studies. *See also* Processing
  case studies
control of mechanical component parts,
  1389–1394
on-line quality engineering in automobile
  industry, 1367–1375
semiconductor rectifier manufacturing, 1395–
  1398
Manufacturing processes:
automated, 31–33
on-line quality engineering for design of,
  1497–1498
quality management of, 1498
stages of, 441
Manufacturing tolerance, 194–195
Market noises, 13
Market quality, items in, 18
Martin, Billy, 387
Masuyama, Motosaburo, 1431
Material design case studies:
felt-resist paste formula used in partial
  felting, optimization of, 836–840
foundry process using green sand, parameter
  design for, 848–851
friction material for automatic transmissions,
  841–847
functional material by plasma spraying, 852–
  857
Material strength case studies:
resistance welding conditions for electronic
  components, optimization of, 863–868
tile manufacturing using industrial waste,
  869–874

two-piece gear brazing conditions,
  optimization of, 858–862
Matsuura Machinery Corp., 138
Maxell, Hitachi, 27
McDonnell & Miller Company, 330–337
MD, *see* Mahalanobis distance
Mean, 506–507
deviation from, 507, 508
as variance, 509
variation from, 510
working, 506–507
Mean-squared deviation (MSD), 183–184
Measurement. *See also* Measurement systems
of abnormality, 398
on continuous scale vs. pass/fail, 474
establishing procedures for, 1473–1474
in on-line quality engineering, 116
technologies for, 26
Measurement case studies:
clear vision by robust design, 882–889
component separation using a dynamic
  operating window, 666–671
drug efficacy, Mahalanobis distance in
  measurement of, 1238–1243
electrophotographic toner charging function
  measuring system, 875–881
granule strength, optimization of measuring
  method for, 672–678
thermoresistant bacteria, detection method
  for, 679–685
ultra trace analysis, dynamic optimization in,
  659–665
Measurement error, 459
Measurement interval, 117
Measurement systems:
efficiency of, 228
for electrophotographic toner charging
  function, 875–881
in MTS, 420. *See also* Mahalanobis–Taguchi
  System
SN ratio evaluation of, 225
types of calibration in, 247
Mechanical applications case studies, 793–1167
control of mechanical component parts in
  manufacturing, 1389–1394
fabrication line capacity planning, 1157–
  1167
flexor tendon repairs, biomechanical
  comparison of, 795–805

Mechanical applications case studies (*continued*)
machining:
electric power, optimization of machining conditions by, 806–818
high-performance steel by transformability, technology for, 819–826
plastic injection-molded gear, transformability of, 827–835
material design:
automatic transmissions, friction material for, 841–847
felt-resist paste formula used in partial felting, optimization of, 836–840
foundry process using green sand, 848–851
plasma spraying, development of functional material by, 852–857
material strength:
resistance welding conditions for electronic components, optimization of, 863–868
tile manufacturing using industrial waste, 869–874
two-piece gear brazing conditions, optimization of, 858–862
measurement:
clear vision of steering wheel by robust design, 882–889
electrophotographic toner charging function measuring system, development of, 875–881
processing:
adhesion condition of resin board and copper plate, optimization of, 890–894
casting conditions for camshafts, optimization of, 900–903
deep-drawing process, optimization of, 911–915
electrical encapsulation process, optimization of, 950–956
electro-deposited process for magnet production, quality improvement of, 945–949
encapsulation process, development of, 916–925
gas-arc stud weld process parameter optimization, 926–939
molding conditions of thick-walled products, optimization of, 940–944
photoresist profile, optimization of, 904–910

plastic injection molding technology, development of, 957–964
wave soldering process, optimization of, 895–899
product development:
airflow noise reduction of intercoolers, 1100–1105
articulated robots, functionality evaluation of, 1059–1068
brake boosters, reduction of boosting force variation of, 1106–1111
chattering noise in Series 47 feeder valves, reduction of, 1112–1121
clutch disk torsional damping system design, optimization of, 973–983
DC motor, optimal design for, 1122–1127
direct-injection diesel injector optimization, 984–1004
disk blade mobile cutters, optimization of, 1005–1010
D-VHS tape travel stability, 1011–1017
electrical connector insulator contact housing, optimization of, 1084–1099
linear actuator using simulation, 1050–1058
linear proportional purge solenoids, 1032–1049
minivan rear window latching, improvement of, 1025–1031
omelets, improvement in the taste of, 1141–1147
shutter mechanisms of single-use cameras by simulation, 965–972
spindles, functionality evaluation of, 1018–1024
steering system on-center robustness, 1128–1140
ultraminiature KMS tact switch optimization, 1069–1083
wiper system chatter reduction, 1148–1156
Mechanical engineering, Taguchi method in, 73–85
conventional meaning of robust design, 73–74
functionality design method, 74–79
generic function of machining, 80–83
problem solving and quality engineering, 79–80
signal and output, 80

when on and off conditions exist, 83–85

Medical diagnosis:
  Mahalanobis–Taguchi method for, 104–108
  MTS in, 418

Medical diagnosis case studies:
  future health, forecasting of, 1277–1287
  health examination and treatment of missing data, Mahalanobis distance for, 1267–1276
  liver disease, 401–415, 1244–1257
  measurement of drug efficacy, Mahalanobis distance for, 1238–1243
  urinary continence recovery among patients with brain disease, prediction of, 1258–1266
  use of Mahalanobis distance in, 1244–1257

Medical treatment and efficacy experimentation, Taguchi method in, 93–97

Metrology Management Association, 134, 140–141

Midstream quality (specified quality), 354, 355

Minivan rear window latching improvement, 1025–1031

Minolta advanced cylindrical heater (MACH), 361–376
  functions of, 362, 363
  heat generating function, 365, 367–370
  peel-off function, 367, 369, 371–373
  power supply function, 362–366
  temperature measuring function, 369, 374–376
  traditional fixing system vs., 361–362

Minolta Camera Company, 232

Mitsubishi Heavy Industries, 1482

Mitsubishi Research Institute, 26

Mixed-level orthogonal arrays, 596

Miyamoto, Katsumi, 136

Mizuno, Shigeru, 1427, 1430

Modification, 177, 319

Modulation function, 67–69

Momentary values, evaluation of electric waveforms by, 735–740

Monotonicity property, 1451

Monte Carlo method, 442

Moore, Willie, 16, 158

Morinaga Pharmaceutical, 1431

Moritomo, Sadao, 136

Motorola, 1517

MSD, *see* Mean-squared deviation

MTGS process, *see* Mahalanobis–Taguchi Gram–Schmidt process

MT method, *see* Mahalanobis–Taguchi method

MTS, *see* Mahalanobis–Taguchi System

Multilevel assignment, 618, 1433

Multiple factors, tolerance design for, 212–220

Multiple targets, signal-to-noise ratios for, 236

Multivariate charts, MTS/MTGS vs., 417

Musashino Telecommunication Laboratory, 127–129

My Way of Thinking (Genichi Taguchi), 127–129

N

Nachi-Fujikoshi Corp., 137, 138, 141

NASA, 1427

NASDA (National Space Development Agency), 1427

National Railway, 81, 83, 1427

National Research Laboratory of Metrology, 140

National Space Development Agency (NASDA), 1427

Natural environment, 14

Natural logarithms, 142

New product development (NPD), 1498–1507
  and design for six sigma, 1498–1507
  and IDDOV, 1496, 1501–1507
  leveraging of DFSS by, 1493–1494
  phase gate process, 1505–1506
  phase-phase gate structure, 1501–1503
  phase processes, 1503–1505
  project review process, 1506–1507
  structure thinking in, 1486–1490
  Suh/Senge model of, 1479–1480
  synthesized case study, 1499–1500

New robust product development (NRPD), 1499, 1502–1503, 1507

Night-vision image-intensifier tube, 950

Nikkei Mechanical, 27, 28

Nikon Corp., 27, 28

Nippon Denso, 135, 157, 1439

Nippon Electric, 1439

Nippon Telegraph and Telephone Public Corporation (NTT), 15, 73, 127–129, 199

Noise(s), 13. *See also* Signal-to-noise ratio
  between-product, 319
  criminal, 15–16
  definition of, 319, 381
  environmental, 14–15

Noise(s) (*continued*)
external, 1512
human errors, 13–14
inner, 319
interaction between design and, 57
internal, 1512
outer, 319
parameter design for minimization of, 319–320
types of, 319
and types of variability, 224
Noise factors, 29, 30, 381–382, 1434
compounded, 40–41, 360–361
indicative, 30
interactions of control factors and, 227
in robust design, 353
testing personnel in selection of, 1473
in tolerance design, 208–210
true, 30
in two-stage optimization, 40, 41
variation caused by repetitions, 242
Noise level, additivity for, 586–587
Noise variables, 176
Nominal system parameters, 40
Nominal-the-best characteristic, 17, 180
definition of, 442
quality level at factories for, 448–450
quality level calculation for, 19–21
quality loss function for, 180–186, 190, 445
tolerance design for, 208–210
Nominal-the-best scale, 1439
Nominal-the-best SN ratio, 139
Nomura Research Institute, 26
Nondynamic characteristics, 236
Nondynamic SN ratio, 139, 225, 264–271
dynamic vs., 139, 225
larger-the-better applications, 269
nominal-the-best applications, 265–268
operating window, 269–271
smaller-the-better applications, 268
for upstream quality, 243
uses of, 236
Nonlinear regression equations, 1406–1419
Normalization, 103
Normal state, 196
NPD, *see* New product development
NRPD, *see* New robust product development
NTT, *see* Nippon Telegraph and Telephone Public Corporation
Number of workers, optimal, 469–471

O

Objective function(s), 10
definition of, 229, 355
and design of product quality, 26
generic function vs., 355
standard conditions for, 40
of subsystems, 313
transformability as, 140
tuning as, 57
Objective value, deviation from, 507–508
Observational equation, 535
Off functions, 138
Off-line quality engineering, 135, 504, 1496–1497. *See also* Robust engineering
for product design, 1497
role of, 1493
steps in, 504
for technology development, 1496
Ohm's law, 508
Oken, 143
Oki Electric Industry, 44–55, 144
Omega transformation, 292–293
Omelets, improvement in taste of, 1141–1147
On and off conditions, 83–85
One-factor-at-a-time approach, 291, 357
One-factor-by-one method, orthogonal arrays vs., 594–595
One-way layout, 523–527
with equal number of repetitions, 523–527
with unequal number of repetitions, 527
On functions, 138
On-line process control:
equations for, 1598–1601
feedback control:
of process conditions, 1600–1601
by quality characteristics, 1598–1600
process diagnosis and adjustment, 1601
On-line quality engineering (OQE), 504, 1497–1498
applications of, 135
definition of, 14, 79, 116
feedback control:
process condition, based on, 468–473
product characteristics, based on, 454–467
major concept in, 135
for manufacturing process design, 1497–1498
origin of field, 135
parameter design vs., 437
process condition, feedback control based on, 468–473

temperature control, 471–473
viscosity control, 469–470
process diagnosis and adjustment, 474–481
  and cost of quality control systems during
    production, 474–477
  diagnostic interval, optimal, 477–479
  number of workers, optimal, 469–471
product characteristics, feedback control
    based on, 454–467
  batch production process, 459–462
  process control gauge design, 462–467
  system design, 454–459
role of, 1493
Taguchi method in, 116–123
tolerancing and quality level in, 437–453
  determination of tolerance, 442–448
  larger-the-better characteristics, quality level
    at factories for, 451–453
  nominal-the-best characteristics, quality
    level at factories for, 448–450
  product cost analysis, 440
  production division responsibilities, 440–
    441
  related issues, 437–439
  and role of production in quality
    evaluation, 441–442
  smaller-the-better characteristics, quality
    level at factories for, 450–451
On-line quality engineering case studies, 1365–
    1398
  automobile manufacturing process,
    application to, 1367–1375
  feedback control by quality characteristics,
    1383–1388
  manufacturing process, control of mechanical
    component parts in, 1389–1394
  preventive maintenance of bucket elevator,
    1376–1382
  semiconductor rectifier manufacturing, 1395–
    1398
Op-amp, 27
Operating cost, 18, 440
Operating window:
  dynamic, 142, 298–308
    for chemical reactions with side reactions,
      298–308
    for component separation evaluation, 666–
      671
    for herbal medicine granulation, 695–704

for photographic systems evaluation, 651–
      658
  nondynamic, 269–271
Optimal diagnostic interval, 477–479
Optimization. *See also specific case studies*
  of concept/technology sets, 1497, 1512–1514
  evaluation technique for, 148
  of generic functions, 352
  in research and development, 1502, 1515
  robust, 1497, 1501, 1502. *See also* parameter
    design
    improvement in, 1508
    and innovation, 1515
    for reliability improvement, 1515
  in robust technology development, 361
  stage, 40–55
    example of, 44–55
    formula of orthogonal expansion, 43–44
    orthogonal expansion, 40–41
    testing method and data analysis, 42–43
    variability improvement, 41–55
  of technology, 1497, 1512–1514
  of tolerance design, 386
  two-stage:
    control factors in, 40, 41
    noise factors in, 40, 41
  two-step, 178, 353–354, 377, 1502. *See also*
    Parameter design
    concept step in, 1504
    definition of, 60
    in new product development, 1501–1502
    in product design, 1439, 1497
    in robust technology development, 353–
      354
    in technological development, 1440
Optimum performance, exceeding, 379–380
OQE, *see* On-line quality engineering
Origin quality (functional quality), 354–355
Orthogonal arrays, 30, 356–358, 584–596, 1526–
    1527
  assignment (layout) to, 490
  inner, 322–323
  Japanese experimentation using, 1432–1433
  L_4, 1530, 1602
  L_8, 597–600, 602, 604, 1530, 1603
  L_9, 1564, 1609
  L'_9, 1595
  L_{12}, 589, 1531, 1604
  L_{16}, 1532, 1586, 1605

Orthogonal arrays (*continued*)
  L_{18}, 74, 589–594, 1564, 1610
  L_{25}, 1591
  L_{27}, 1565–1566, 1611–1612
  L'_{27}, 1596
  L_{32}, 1536–1537, 1587
  L_{36}, 491–492, 589, 1593, 1614–1617
  L_{50}, 1592
  L_{54}, 1568
  L_{64}, 1550–1553, 1588–1589
  L_{81}, 1570–1571
  layout of signal factors in, 97–98
  linear graphs of, *see* Linear graphs
  mixed, 286–289
  objective of using, 357
  one-factor-by-one method vs., 594–595
  outer, 322–323
  quality characteristics and additivity, 584–588
  for reproducibility assessment, 30
  split analysis for:
    mixed, 286–289
    two-level, 284–286
  two-level, 284–286
  types of, 596
Orthogonal expansion:
  formula of, 43–44
  meaning of, 537
  for standard SN ratios and tuning robustness,
    40–41
Orthogonal function, 543
Orthogonal polynomial equation, 254–257
Outer noise, 176, 319
Outer orthogonal arrays, 73, 322–323
Out of the Crisis (W. Edwards Deming), 1448
Output, signal-to-noise ratios based on, 236–237
Output characteristics, 1449–1450

P

Panasonic, 1427
Paradigm shift, 29
Parameter design, 318–338, 1439–1440. *See also*
    Two-step optimization
  applied to semiconductors, 141
  control-factor-level intervals in, 609
  definition of, 40, 177, 314, 1452
  evaluation for, 39
  history of, 140–141
  initial example of, 140
  key steps of, 387

  management perspective on, 382–386
  for manufacturing process design, 1498
  for noise minimization, 319–320
  objective of, 284, 438–439
  on-line quality engineering vs., 437
  optimum control factor levels in, 212
  origin of, 1435–1436
  primary uses of, 356
  for product design, 18, 177–178, 1497
  purpose of, 504
  and quality loss function, 177–178
  research on, 177
  and robust design, 382, 1435, 1452. *See also*
    Robust design
  in robust engineering, 382–386
  for selected system, 40
  in simulation, 144
  systematization of, 1439
  systems for, 39
  for technology development, 1496
  transition to technological development,
    1439–1440
  of water feeder valve, 330–338
  of Wheatstone Bridge, 320–330, 349–350
Parameter Design for New Product Development
    (Genichi Taguchi), 140–141
Parameter estimation:
  with experimental regression analysis, 484,
    488–499
  with least squares method, 485–489
Pareto charts, 1429
Particle size distribution, data showing, 587–588
Pass/fail criteria, specification limits vs., 181–
    182
Passive dynamic SN ratio, 132
Passive signal factors, 39, 97
Passive SN ratios, 236
Patterning:
  fine-line patterning for IC fabrication, 758–
    763
  transformability, patterning, 141
Pattern recognition, 397–398
  design of general procedure for, 108–114
  in earthquake-proof design, 14
  in manufacturing, 418
  with MT method, 108–109
  with MTS, 109–114, 398
  MTS/MTGS vs. other techniques for, 417
Pattern thinking, 1480–1485, 1490

PCA (principal component analysis), 417
PDCA (plan, do, check, action) cycle, 1426, 1427
P-diagrams, 1472
Pearson, Carl, 1424
Performance, 393. *See also* Functionality(-ies)
Personnel (for testing), 1470–1472
Peterson, Don, 157
Phadke, Madhave, 156, 1519
Pharmacology case studies:
   drug efficacy, Mahalanobis distance in measurement of, 1238–1243
   optimization of model ointment prescriptions for in vitro percutaneous permeation, 686–694
Phase gate process, 1505–1506
Phase modulation function, 68–72
Phase-phase gate structure, 1501–1503
Phase processes, 1503–1505
Photography:
   evaluation of photographic systems with dynamic operating window, 651–658
   photographic film, 28, 88–91
   stability design of shutter mechanisms of single-use cameras, 965–972
Photoresist profile using simulation, 904–910
Physical condition:
   additivity with, 585
   data showing, 584–585
Physics, quality engineering vs., 31
Plan, do, check, action cycle, 1426, 1427
Plan-do-study-act cycle, 1481
Plasma spraying, development of functional material by, 852–857
Plastic injection molding:
   development of technology by transformability, 857–864, 957–964
   electrical connector insulator contact housing, optimization of, 1084–1099
   gear, transformability of, 827–835
Point calibration, 247–250
Pollution:
   monitoring, 24
   post-World War II, 23–24
   and quality improvement, 7–9
Polymerization reactions, optimization of, 643–650
Pot core transformer processing, variation in, 764–770

Power factor, 61
Power MOSFET:
   ideal function for, 234
   optimizing back contact of, 771–779
Power spectrum analysis, 513
Preconcept phase, 1503, 1504
Prediction. *See also* Forecasting
   MTS for, 398
   of urinary continence recovery in patients with brain disease, 1258–1266
Premature failures, 1498
Preparation (for testing), 1472–1473
Preventive maintenance, 79
   for bucket elevator, simultaneous use of periodic maintenance and checkup, 1376–1382
   design method for, 463–467
Price:
   and production quality, 439
   and total cost, 121
   and unit manufacturing cost, 18, 439, 441
Primary error variation, 574–576, 580
Princeton University, 155
Principal component analysis (PCA), 417
Principle of ideality, 1489
Printed letter inspection, MTS for, 1293–1297
Procedures, testing, 1473–1474
Process capability, quality loss function and, 173–176
Process capability index, 173
Process condition control, 468–473
Process control, 32, 441
Process design:
   robust, 58
   steps in, 314. *See also specific steps*
   value of SN ratios in, 227
Process development, SN ratio and reduction of cycle time, 228
Process diagnosis and adjustment, 474–481
   and cost of quality control systems during production, 474–477
   optimal diagnostic interval, 477–479
   optimal number of workers, 469–471
Processing case studies:
   adhesion condition of resin board and copper plate, optimization of, 890–894
   casting conditions for camshafts by simulation, 900–903
   deep-drawing process, optimization of, 911–915

Processing case studies (*continued*)
electro-deposited process for magnet production, improvement of, 945–949
encapsulation process:
optimization of, 950–956
robust technology development of, 916–925
gas-arc stud weld process parameter optimization, 926–939
molding conditions of thick-walled products, optimization of, 940–944
photoresist profile using simulation, 904–910
plastic injection molding technology by transformability, 957–964
wave soldering process, optimization of, 895–899
Process maintenance design, 32
Product(s):
product planning:
definition of, 108
design life in, 57
and production engineering, 12–13
product quality, 10, 171
definition of, 442
design of, 26
quantitative evaluation of, 32, 438–439
standards for, 32
product species, 171
for testing, 1474–1475
Product case studies:
character recognition, Mahalanobis distance for, 1288–1292
printed letter inspection technique, 1293–1297
Product costs, 1516
Product cost analysis, 440
Product design:
direct, 227
for financial systems, 10–11
focus of, 18
off-line quality engineering for, 1497
orthogonal array for modification in, 74
paradigm shift for, 359–360
parameter design in, 177–178. *See also* Parameter design
quality loss function and steps in, 176–179
robust, 58
robustness in, 318
steps in, 314. *See also specific steps*

system design in, 176–177. *See also* System design
tolerance design in, 179. *See also* Tolerance design
value of SN ratios in, 227
Product development, 1478–1491. *See also* New product development; Robust technology development
control-factor-level intervals in, 609–610
event thinking in, 1480–1481
paradigm shift for, 359–360
pattern thinking in, 1481–1485
product vs. engineering quality in, 10
with robust engineering, 167
SN ratio and reduction of cycle time, 228
structure thinking in, 1486–1490
Suh/Senge model of, 1479–1480
synergistic, 1490–1491
Taguchi system for, 1485–1486
Product development case studies:
airflow noise reduction of intercoolers, 1100–1105
articulated robots, functionality evaluation of, 1059–1068
brake boosters, reduction of boosting force variation of, 1106–1111
chattering noise in Series 47 feeder valves, reduction of, 1112–1121
clutch disk torsional damping system design, optimization of, 973–983
DC motor, optimal design for, 1122–1127
direct-injection diesel injector optimization, 984–1004
disk blade mobile cutters, optimization of, 1005–1010
D-VHS tape travel stability, 1011–1017
electrical connector insulator contact housing, optimization of, 1084–1099
linear actuator, simulation for optimization of, 1050–1058
linear proportional purge solenoids, 1032–1049
minivan rear window latching, improvement in, 1025–1031
new ultraminiature KMS tact switch optimization, 1069–1083
omelets, improvement in the taste of, 1141–1147
spindles, functionality evaluation of, 1018–1024

stability design of shutter mechanisms of single-use cameras by simulation, 965–972

steering system on-center robustness, 1128–1140

wiper system chatter reduction, 1148–1156

Production control, 439

Production control costs, 17

Production cost, 17, 18, 442

Production division responsibilities:
  in quality evaluation, 441–442
  for tolerancing in testing, 440–441

Production engineering, 12–13
  for reduction of corporate costs, 18
  responsibilities of, 17

Production processes, 18

Production quality, 438

Productivity:
  definition of, 11
  developing, 10–13
  and economy, 5–7
  improvement in, 6–7
  and individual freedom, 9
  and quality, 7–8

Profound knowledge, 1444

Programming ability:
  evaluation of, 1343–1350
  evaluation of capability and error in, 1335–1342
  MTS evaluation technique for, 1178–1188
  questionnaire for prediction of, 1171–1177

Project leaders, 391

Project review process, 1505–1507

Proportional equation:
  reference-point, 247–250
  zero-point, 241–247

Proseanic, Vladimir, 1301

Prototypes, 318

Pugh, Stuart, 164, 1482, 1495

Pugh analysis, 392

Pugh concept, 1495

Pure variation, 518, 559

Q

QC circle, 1427

QCRGs, *see* Quality Control Research Groups

QC story, 1428–1430

QE, *see* Quality engineering

QFD, *see* Quality function deployment

QFP, *see* Quad flat package

QLF, *see* Quality loss function

Quad flat package (QFP), 1189, 1190

Quadratic planning, 488

Quality. *See also specific topics*
  as conformance to specifications, 171
  and corporate organization, 10
  and cost, 16, 440–441
  definition of, 171, 314, 442
  Deming's key principles of, 1444–1447
  design, 438
  downstream, 354, 355
  engineering, 10
  expressing, through loss function, 133–134
  four levels of, 354
  as generic term, 57
  measurement of, 138
  midstream, 354, 355
  origin, 354–355
  product, 10, 171
  in product design for financial systems, 10–11
  production, 438
  and productivity, 7–8
  risks to, 13–16
  Taguchi's definition of, 171, 193
  and taxation, 8–9
  as "tested in," 1471
  types of, 10
  upstream, 354, 355

Quality assurance department, 23–24, 38

Quality assurance (in Japan), 1427. *See also* Quality control

Quality characteristics, 16–17. *See also specific types*
  and additivity, 584–588
  classification of, 180
  downstream, 269
  high-rank, 340
  larger-the-better, 17
  low-rank, 340
  nominal-the-best, 17
  quality loss function and classification of, 180
  smaller-the-better, 17
  true, 1482

Quality control:
  cost of, during production, 474–477
  paradigm shift for, 359–360
  specification limits vs., 182
  traditional method of, 378–379

Quality control costs, 17
Quality Control (Ina Seito), 1433, 1434
Quality control in process, 116. *See also* On-line
   quality engineering
Quality Control Research Groups (QCRGs),
   148–149
Quality Engineering for Technological Development
   (Genichi Taguchi), 133, 147
Quality engineering (QE). *See also* Robust
   design; Taguchi method
  conventional research and design vs., 77, 78
  definition of, 314
  and design for six sigma, 1498
  at Ford Motor Company, 16
  functionality design in, 11
  generic (essential) functions in, 8
  history of, 127–168
    in Japan, 127–152, 1428–1440
    in United States, 153–168
  in Japan, 1427, 1435–1440
  main objective of, 23
  management for, 25–38
    automated process management, 31–33
    in design process, 30–31
    diagnosis to prevent recalls, 33–38
    evaluation of functionality, 28–30
    management role in research and
      development, 25–28, 39–40
  off-line, 1496–1497. *See also* Off-line quality
    engineering
  on-line, 1497–1498. *See also* On-line quality
    engineering
  physics vs., 31
  in research and development, 13
  research and development strategy in, 39–55
    cycle time reduction, 39–40
    stage optimization, 40–55
  responsibilities of, 23–24
  as robust design, 58
  standardization of, 148
  Taguchi method vs., 58
  timeline of, 129–130
  transition and development of, 135
  tuning in, 57–58
  U.S. Taguchi method vs., 58
Quality Engineering Series, 133
Quality Engineering Society (Japan), 147, 149,
   161, 162, 168, 1435, 1440
Quality function deployment (QFD), 164, 392
  and DFSS process, 1494

  in Japan, 1427, 1430–1431
  origin of, 1482
Quality improvement, 7–9
Quality level:
  and consumer loss, 17–18
  and cost, 18–22
  for larger-the-better characteristic, 19, 21–22,
    451–453
  for nominal-the-best characteristic, 19–21,
    448–450
  for smaller-the-better characteristic, 19, 21,
    450–451
  in tolerance design, 437–453
Quality loss function (QLF), 133–134, 171–179,
   380
  and classification of quality characteristics,
    180
  derivation of, 440
  equations for, 1598–1601
  justice-related aspect of, 175–176
  for larger-the-better characteristic, 188, 191
  for nominal-the-best characteristic, 180–186,
    190, 445
  and process capability, 173–176
  quadratic representation of, 174
  quality aspect of, 176
  for smaller-the-better characteristic, 186–188,
    190
  and steps in product design, 176–179
    parameter design, 177–178
    system design, 176–177
    tolerance design, 179
  for tolerance design optimization, 386
Quality management:
  design of experiments, 1431–1435
  history of, 1423–1428
  in Japan, 1423–1440
  of manufacturing processes, 1498
  quality engineering, 1435–1440
  quality function deployment, 1430–1431
  seven QC tools and QC story, 1428–1430
Quality programs, 392
Quality tables, 1482
Quality target, 438

R

Radio waves, 65
RCA Victor Company, 397
R&D, *see* Research and development

Reactive power, 61
Reagan, Ronald, 1444
Real-time operating system optimization, 1324–1334
Recalls, prevention of, 33–38
Reference-point calibration, 247–250
Reference-point proportional equation, 240, 241, 247–250
Regression analysis:
  experimental, 149, 484, 486, 488–499
    estimated values, selection of, 492, 494–499
    initial values, setting of, 488–489
    orthogonal array, selection and calculation of, 489–494
    parameter estimation with, 484, 488–499
    sequential approximation, 490, 492–496
  MTS/MTGS vs., 417
Regression equations, parameter estimation in, 485–489
Reliability, 40. *See also* Functionality
Reliability analysis, 392
Reliability and Maintainability 2000 Program (U.S. Air Force), 159
Reliability design, Monte Carlo method in, 442
Reliability management, 1427
Repeatability, definition of, 224
Repetition, 1436
Reproducibility, 504–505
  definition of, 224
  in design of experiments, 504–505
  evaluation of, 30–31
  for hardware design, 58–59
  meaning of term, 356–357
  with orthogonal arrays, 594
Research and development (R&D), 6
  application of quality engineering to, 1496
  in banks, 11
  cycle time reduction, 39–40
  functionality prediction in, 28–30
  management of, 10, 25–28, 39–40
  need for, in Japan, 7
  quality engineering strategy in, 39–55
    cycle time reduction, 39–40
    stage optimization, 40–55
  robust optimization in, 1515
  stage optimization, 40–55
    orthogonal expansion for standard SN ratios and tuning robustness, 40–41
    variability improvement, 41–55

  strategy in, 13
  two-step optimization in, 1502
research and development (R&D):
  management role in, 39–40
Research Institute of Electrical Communication, 57
Resin board, adhesion condition of copper plate and, 890–894
Resistance welding conditions for electronic components, 863–868
Resource constraints, 1472
Responses, 228, 503
Response tables, 361
Ringer Hut, 1428
Risk(s):
  countermeasures against, 13–16
  functional, 11
  to quality, 13–16
  as signals vs. noises, 13
Robots, articulated, 1059–1068
Robust design. *See also* Quality engineering; Robust engineering; Robust technology development
  conventional meaning of, 73–74
  function in, 1456
  noise conditions in, 353
  objective of, 58
  origin of, 154, 155
  as parameter design, 382. *See also* Parameter design
  quality engineering as, 58
  Taguchi method for, 1451–1452
  as term, 57
  TRIZ and axiomatic design for enhancement of, 1449–1468
    comparison of disciplines, 1451–1458
    further research, possibilities for, 1466–1468
    limitations of proposed approach, 1465–1466
    system output response, identification of, 1456, 1459–1465
  two-step process for, 377
  using Taguchi methods, 1451–1452
  value of SN ratios in, 227
Robust engineering. *See also* Off-line quality engineering
  development of, in U.S., 161, 162
  implementation strategies for, 389–393

Robust engineering (*continued*)
   and bottom-line performance, 393
   corporate leader/corporate team in, 390–391
   critical components in, 389–390
   education and training in, 392
   effective communication in, 391–392
   and integration strategy, 392–393
   management commitment in, 390
  institutionalization of, 164–168
  manager's perspective on, 377–388
   parameter design, 382–386
   tolerance design, 386–388
  misconceptions about, 1471
  parameter design in, 382–386
  product development with, 167
  purpose of, 1496
  and six sigma, 1517–1519
  steps in, 387, 388
  thinking approach in, 392
  tolerance design in, 386–388
Robust engineering case studies:
  chemical applications, 629–714
   biochemistry, 631–642
   chemical reaction, 643–658
   measurement, 659–685
   pharmacology, 686–694
   separation, 695–714
  electrical applications, 715–792
   circuits, 717–745
   electronics devices, 746–779
   electrophoto, 780–792
  mechanical applications, 793–1167
   biomechanical, 795–805
   fabrication line capacity planning using a robust design dynamic model, 1157–1167
   machining, 806–835
   material design, 836–857
   material strength, 858–874
   measurement, 875–889
   processing, 890–964
   product development, 965–1156
Robust Engineering (Genichi Taguchi, Subir Chowdhury, and Shin Taguchi), 1493
Robust optimization, 1497, 1501, 1502. *See also* parameter design
  improvement in, 1508
  and innovation, 1515
  for reliability improvement, 1515
  in research and development, 1515
Robust quality, *see* Upstream quality
Robust technology development, 352–376
  advantages of, 358–359
  encapsulation process, 916–925
  guidelines for application of, 359–361
  key concepts in, 353
  Minolta advanced cylindrical heater, 361–376
  objective function vs. generic function, 355
  orthogonal array, 356–358
  selection of what to measure, 354–355
  SN ratio, 228, 355–356
  two-step optimization, 353–354
Rolls-Royce, 33–35
Ross, Louis, 1479
Rules of Dimensional Tolerances of a Plastic Part (JIS K-7109), 140

S

Safety design, 199–201
Safety factor(s), 32
  derivation of, 443–448
  equation for, 197–198
  for larger-the-better characteristic, 447
  for smaller-the-better characteristic, 446
  in specification tolerancing, 195–201
Scatter diagrams, 1429
Scoping of project, 1472
Second Industrial Revolution, 5–24
  automated management operations in, 31–32
  elements of productivity in, 5–9
  and management in manufacturing, 16–22
  productivity development in, 10–13
  and quality assurance departments, 23–24
  risks to quality in, 13–16
Seed technology, 164
Seiki, Sanjo, 137
Seiko Epson, 433
Sekisui Chemical, 1427
Semiconductor rectifier manufacturing, 1395–1398
Senge, Peter, 1479, 1480
Sensitivity (SN ratios), 225, 226, 236, 360
Separation case studies:
  herbal medicine granulation, dynamic operating window for, 695–704
  particle-size adjustment in a fine grinding process for a developer, 705–714

Separation principles, 1454
Sequential approximation, 610, 613–616
  in experimental regression analysis, 490, 492–496
  with incomplete data, 613–616
Service costs, 1516
Seven deadly diseases (Deming), 1446
Seven QC tools, 1428–1430
S-F analysis, *see* Substance-field analysis
S-field analysis, 1461–1468
Sharp Corp., 141
Shewhart, Walter, 155, 1424, 1443, 1446, 1481
Shewhart control charts, 1424–1426
Shimizu Corp., 1427
Shoemaker, Anne, 156
Showa Denko, 1424
Sigma level, 1510, 1511
Signals, 13, 57
Signal factors, 29–30
  active vs. passive, 29, 97
  for degree of disease, 143
  in orthogonal array, 97–98
  in software testing, 97–98
  testing personnel in selection of, 1473
Signal-to-noise (SN) ratio, 13, 223–237
  active, 236
  active dynamic, 132–133
  based on input and output, 236–237
  based on intention, 236
  benefits to using, 225, 227–228
  classifications of, 234–237
  for classified attributes, 234–236, 290–308
    chemical reactions without side reactions, 296–298
    chemical reactions with side reactions, 298–308
    two classes, one type of mistake, 291–293
    two classes, two types of mistake, 293–296
  conceptual transition of, 131–133
  for continuous variables, 234–236, 239–289
    double signals, 275–284
    estimation of error, 271–275
    with known signal factor level interval, 256–260
    with known signal factor level ratio, 262–263
    linear equation, 250–254
    linear equation using tabular display of orthogonal polynomial equation, 254–256
    with no noise, 284–289
    with no signal factor, 264–271
    reference-point proportional equation, 247–250
    split analysis, 284–289
    with unknown true values of signal factor levels, 264
    when signal factor levels can be set up, 260–262
    zero-point proportional equation, 241–247
  correspondence course on, 132
  decibel value of, 1438
  definition of, 224, 225, 314
  determining evaluation scale of, 148
  differences in (gain), 79
  direct relationship of economy and, 225
  dynamic, 139, 225
    nondynamic vs., 139, 225
    for origin quality, 354–355
    in U.S. applications, 160
    uses of, 139, 236
  in electric characteristics, 140–141
  elements of, 224–226
  fixed vs. multiple targets, 236
  functional evaluation, 133
  generic function, 133
  ideal functions based on, 228–234
  improving, 41–55
  larger-the-better, 95, 139
  in machining, 135–137
  in measurement systems evaluation, 225
  in metrology field, 132
  in MTS, 398
  new method of calculating, 144–147
  nominal-the-best, 139
  nondynamic, 139, 225, 264–271
    dynamic vs., 139, 225
    larger-the-better applications, 269
    nominal-the-best applications, 265–268
    operating window, 269–271
    smaller-the-better applications, 268
    for upstream quality, 243
    uses of, 236
  origin of, 1436, 1438–1439
  orthogonal expansion for, 40–41
  passive, 236
  passive dynamic, 132
  in process design, 227
  in product design, 227

Signal-to-noise (SN) ratio (*continued*)
  for reactions without side reactions, 296–298
  for reactions with side reactions, 298–308
  and reduction of process/product
    development cycle time, 228
  and reproducibility, 505
  in robust design, 227
  in robust technology development, 355–356
  selection of, 240
  for shape retention, 140
  in simulations, 144
  smaller-the-better, 95, 139
  standard, orthogonal expansion and tuning
    robustness for, 40–41
  in technology development, 225
  and traditional approach for variability, 224
  of transformability, 139–140
Significance test, 521
Significant factorial effects, 521
Simulation(s), 1475–1476
  casting conditions for camshafts, 900–903
  design, 40, 72–73
  linear actuator, optimization of, 1050–1058
  orthogonal array L_{36} for, 357
  parameter design in, 144
  photoresist profile, optimization of, 904–910
  SN ratio applied to, 144
  stability design of shutter mechanisms of
    single-use cameras, 965–972
  technological development using, 148
Simultaneous engineering, 276
Single-use cameras, stability design of shutter
  mechanisms of, 965–972
Six sigma, 161, 162. *See also* Design for six
  sigma
  improvement in, 1508
  and robust engineering, 1517–1519
  thinking levels used in, 1490
  and whack-a-mole engineering, 1508
Slope calibration, 228, 247
Small algae production, optimization of, 637–
  642
Smaller-the-better characteristic, 17, 180
  quality level at factories for, 450–451
  quality level calculation for, 19, 21
  quality loss function for, 186–188, 190
  safety factor for, 446
  tolerance design for, 208–210
Smaller-the-better SN ratio, 95, 139

Smart Cards, 1084
Smith, Larry R., 1519
SN ratio, *see* Signal-to-noise ratio
SN Ratio Manual (Genichi Taguchi), 131
Software:
  simulation, 1475–1476
  for test results analysis, 1475
Software testing, 425–433
  main difficulty in, 425–426
  Taguchi method in, 97–102, 425–433
    layout of signal factors in orthogonal array,
      97–98
    software diagnosis using interaction, 98–
      102
    system decomposition, 102
    two types of signal factor and software, 97
  typical procedures for, 426–433
Software testing and application case studies,
  1299–1364
  algorithms:
    diesel engine software control strategy,
      optimization of, 1301–1309
    video compression, optimization of, 1310–
      1323
  debugging, streamlining of, 1360–1364
  electronic warfare systems, robust testing of,
    1351–1359
  prediction of programming ability from
    questionnaire, 1171–1177
  programmer ability in software production,
    evaluation of, 1343–1350
  programming, evaluation of capability and
    error in, 1335–1342
  real-time operating system, robust
    optimization of, 1324–1334
Soldering:
  automatic inspection of joints, Mahalanobis
    distance in, 1189–1195
  optimization of wave soldering process, 895–
    899
Sony, 134, 1427
SPC, *see* Statistical process control
Special cause, 1446–1447
Specifications, conformance to, 171
Specification limits:
  pass/fail criteria vs., 181–182
  quality control vs., 182
Specification tolerancing, 192–207
  deteriorating characteristics in, 205–207

distribution of tolerance, 204–205

safety factor in, 195–201

when central value of low-rank characteristics has been determined, 201–204

Specified quality, *see* Midstream quality

Spectrum analysis, ANOVA and, 560

Speed difference method, 142, 299

Speed ratio method, 142, 302

Spindles, functionality evaluation of, 1018–1024

Split analysis, 284–289

SQC, *see* Statistical quality control

Stability, definition of, 224

Stage optimization, 40–55

  example of, 44–55

  formula of orthogonal expansion, 43–44

  orthogonal expansion, 40–41

  testing method and data analysis, 42–43

  variability improvement, 41–55

Standards, quality, 32

Standard deviation, 517

Standardization and Quality Control, 133, 147, 156

Standardization and Quality Management, 147

Standard of living, 7, 12

Statistical Methods for Research Workers (Ronald A. Fisher), 1431

Statistical process control (SPC), 135, 1481. *See also* On-line quality engineering

Statistical quality control (SQC), 157, 1425, 1426

Steel, high-performance, machining technology for, 819–826

Steering system on-center robustness, 1128–1140

Stepwise regression, MTS/MTGS vs., 417

Strategic planning, 13

Strategic planning team, 391, 392

Strategy, 13

  definition of, 23

  for R&D, 25, 26

  Sun Tzu on the Art of War as, 25

  technological, 39

Structure thinking, 1480, 1486–1490

Substance field model, 1453

Substance-field (S-F) analysis, 1453, 1459, 1462

Subsystems, 313

Suh, Nam P., 1454–1456, 1479–1480, 1487

Suh/Senge model, 1479–1480

Sullivan, Larry, 157

Sum, 506–507

Sumitomo Electric Industries, 141

Sum of squares, 509

Sun Tzu on the Art of War, 25

Synergistic product development, 1490–1491

Synthesis stage, 40, 1453

System(s), 102

  creation of, 39

  optimization of, 30

  of profound knowledge, 1444

  selection of, 26–28, 40, 176–177, 313

System design, 313–317, 387–388, 504

  definition of, 313, 1452

  and functionality of selected system, 314–317

  manufacturing process design, 1497

  product design, 176–177, 1497

  and quality loss function, 176–177

  robust design, 1452

  and selection of system, 313

  technology development, 1496

System engineering, 1482

System output responses, 1449–1451, 1456, 1459–1465

System testing, 425. *See also* Software testing

T

Tactics, 13

  management systems as, 23

  technology as, 25

Taguchi, Fusaji, 153

Taguchi, Genichi. *See also* Taguchi method

  on American quality management, 1498

  at Bell Labs, 156–157

  biography of, 153–156

  on bridging MTS methods, 143

  and cells in studying main/side effects, 142

  and cost vs. quality, 138

  and design of experiments, 149, 1431, 1432, 1434–1436

  and development of quality engineering, 127–131

  at Ford Motor Company, 157

  foresight of, 131

  on function in plastic injection molding, 140

  history of work, 153–156

  honors received by, 162

  impact of work, 148

  at Konica, 133

  and Mahalanobis distance, 1440

  and management by total results, 149

Taguchi, Genichi (*continued*)
  and MTS application to fire alarm, 143
  and natural logarithms, 142
  and on/off functions, 138
  and origin of parameter design, 1434
  and output characteristic identification,
    1449–1450
  and parameter design, 141
  and pattern level of thinking, 1482
  publications by, 1625–1628
  in Quality Engineering Society, 161
  quality engineering's reliance on, 144
  and relationships between time/work and
    electric power, 137
  and selection of interactions for experiments,
    1433
  semiannual U.S. visits by, 160
  and sensor welding study, 141
  and signal and noise factors as consumer
    conditions, 138
  and simulation, 144
  and two-step optimization, 1439, 1440
  at Xerox, 158
Taguchi, Shin, 157, 158, 1493
Taguchi method, 56–133. *See also* Quality
  engineering
  in chemical engineering, 85–93
    evaluation of images, 88–91
    functionality such as granulation or
      polymerization distribution, 91
    function of an engine, 86–87
    general chemical reactions, 87–88
    separation system, 91–93
  and Deming's teachings, 1447–1448
  in design for six sigma, 1493–1494
  in electronic and electrical engineering, 59–
    73
    functionality evaluation of system using
      power, 61–65
    quality engineering of system using
      frequency, 65–73
  essence of, 58
  five principles of, 1452
  and leveraging of DFSS, 1493–1494
  and Mahalanobis distance/space, 114–116
  and Mahalanobis–Taguchi method, 102–109
  and Mahalanobis–Taguchi System, 102–103,
    109–116

  in mechanical engineering, 73–85
    conventional meaning of robust design,
      73–74
    functionality design method, 74–79
    generic function of machining, 80–83
    problem solving and quality engineering,
      79–80
    signal and output, 80
    when on and off conditions exist, 83–85
  in medical treatment and efficacy
    experimentation, 93–97
  misconceptions of, 163, 164
  naming of, 158
  in on-line quality engineering, 116–123
  in product development, 1485–1486
  quality engineering vs., 58
  for robust design, 1451–1452
  in software testing, 97–102, 425–433
    layout of signal factors in orthogonal array,
      97–98
    software diagnosis using interaction, 98–
      102
    system decomposition, 102
    two types of signal factor and software, 97
Taguchi Methods for Robust Design (Yuin Wu and
  Alan Wu), 1493
Taguchi paradigm, *see* Taguchi method
Taguchi quality engineering, 1482. *See also*
  Taguchi method
Takahaski, Kazuhito, 84
Takenaka Corp., 1427
Takeyama, Hidehiko, 136
Tanabe Seiyaku, 1424
Tananko, Dmitry, 1301
Target value/curve, 60–61
Taxation:
  and quality, 8–9
  for quality improvement, 8–9
Teams:
  familiarity with robust design methods, 1471
  for robust engineering implementation, 390–
    392
  self-assessment of, 1506
  and strategic planning, 391, 392
  testing considerations for, *see* Testing
Technical Committee of the Quality
  Engineering Society, 149
Technical readiness, 1505

Technological Development in Electric and Electronic Engineering (Kazuhiko Hara), 141
Technological development methodology, 147
Technological Development of Transformability (Ikuo Baba), 140
Technological strategy, 39. *See also* Generic technology
Technology(-ies):
  generic, 26
  optimization of, 1497, 1512–1514
  as tactics, 25
Technology development. *See also* Robust technology development
  dynamic SN ratios in, 225
  off-line quality engineering for, 1496
  transition of parameter design to, 1439–1440
  using simulation, 148
Technometrics, 1435
Temperature control, 471–473
Test for additional information, MTS/MTGS vs., 417
Testing, 1470–1477
  analysis of results, 1475
  design quality test, 38
  during development, 360
  documentation of, 1477
  by external service providers, 1476–1477
  four "P's" of, 1470
  measurement and evaluation technologies, 26
  personnel for, 1470–1472
  preparation for, 1472–1473
  procedures for, 1473–1474
  product for, 1474–1475
  simulation, 1475–1476
  of software, *see* Software testing
  of systems, 425
  tolerancing in, 437–453
    determination of tolerance, 442–448
    issues related to, 437–439
    product cost analysis, 440
    production division responsibilities for, 440–441
    quality level at factories for larger-the-better characteristics, 451–453
    quality level at factories for nominal-the-best characteristics, 448–450
    quality level at factories for smaller-the-better characteristics, 450–451

role of production in quality evaluation, 441–442
Test planning, 392
Theory of inventive problem solving, *see* TRIZ
Thermal ink jet image quality, 1196–1207
Thermoresistant bacteria, detection method for, 679–685
Thermotherapy, 96, 97
Thick-walled products, optimization of molding conditions for, 940–944
Thinking, Senge's levels of, 1480
  event thinking, 1480–1481
  pattern thinking, 1480–1485
  structural thinking, 1480, 1486–1490
Three-level orthogonal arrays, 596
Three-or-more-class classified attributes, 234
Tile manufacturing using industrial waste, 869–874
Toagosei and Tsumura & Co., 133, 142
Toa Gosei Chemical Company, 232
Tokumaru, Soya, 132
Tolerance, 192–193
  definition of, 193
  determination of, 16–17, 442–448
  functional, 193
  manufacturing, 194–195
Tolerance design, 208–220, 340–351. *See also* Specification tolerancing
  after parameter design, 318
  analysis of variance in, 342–346
  definition of, 79, 340
  for larger-the-better characteristics, 210–211
  management perspective on, 386–388
  for manufacturing process design, 1498
  for multiple factors, 212–220
  for nominal-the-best characteristics, 208–210
  optimization of, 386
  for product design, 179, 1497
  purpose of, 504
  quality and cost balanced by, 16
  and quality characteristics, 16–17
  and quality loss function, 179
  in robust design, 1452
  in robust engineering, 386–388
  for smaller-the-better characteristics, 208–210
  steps in, 212–220, 386–387
  for technology development, 1496

Tolerance design (*continued*)
two methods for, 442
of Wheatstone bridge, 346–351
Tolerance specification, 1497. *See also*
Specification tolerancing
Tolerancing, 439
Toner charge measurement, 746–752
Tools:
design, 30
preparation of, 39
Tosco and Sampo Chemical, 133
Total output, 314
Total quality control (TQC), 1426–1428
Total quality management (TQM), 1426, 1428,
1435
Total social productivity, 11–12
Total sum squared, 509
Total variation, 509
Toyota Motor Company, 12, 26, 135, 155
TQC, *see* Total quality control
TQM, *see* Total quality management
Trading, functionality evaluation in, 148
Training (in robust engineering
implementation), 392
Transformability, 229
machining technology for high-performance
steel by, 819–826
as objective function, 140
origin of, 137
of plastic injection-molded gear, 827–835
plastic injection molding technology
development by, 957–964
Transmitting wave, stability of, 65–67
Transparent conducting thin films, ideal
function for, 233
TRIZ (theory of inventive problem solving),
1452–1454, 1456–1458
comparison of axiomatic design, robust
design, and, 1456
function in, 1456
and pattern-level thinking, 1490
and structure thinking, 1489–1490
True noise factors, 30
True quality characteristics, 1482
Tuning, 57–58, 319
cause-and-effect relationship in, 41
orthogonal expansion for robustness in, 40–
41
in quality engineering, 57–58

in two-stage optimization, 60
Two-class classified attributes, 234
Two-level control factors, 357
Two-level orthogonal arrays, 596
Two-piece gear brazing conditions, optimization
of, 858–862
Two-stage optimization, 40–41
Two-step optimization, 178, 353–354, 377. *See
also* Parameter design
concept step in, 1504
definition of, 60
in new product development, 1501–1502
in product design, 1439, 1497
in robust technology development, 353–354
in technological development, 1440
Two-way layout, 552–562
analysis of variance for, 555–562
with decomposition, 563–572
one continuous variable, 563–567
two continuous variables, 567–572
factors and levels in, 552–555
with repetitions, 573–583
different number of, 578–583
same number of, 573–578

U

Uehara, Yoshito, 136
Ueno, Kenzo, 139, 140
Ultraminiature KMS tact switch optimization,
1069–1083
Ultra trace analysis, dynamic optimization in,
659–665
UMC, *see* Unit manufacturing cost
Uncoupled design, 1454, 1455
Unemployment:
and productivity improvement, 6
and quality improvement, 8
Union of Japanese Scientists and Engineers, *see*
Japanese Union of Scientists and Engineers
United States:
history of quality engineering in, 153–168
from 1980 to 1984, 156–158
from 1984 to 1992, 158–160
from 1992 to 2000, 161–164
from 2000 to present, 164–168
and history of Taguchi's work, 153–156
paradigm shift in, 29
productivity improvement in, 6
quality engineering in, 58

quality targets in, 16
U.S. Air Force, 159
U.S. Census Bureau, 1443
U.S. Department of Agriculture, 1443
U.S. Department of Defense, 159
Unit manufacturing cost (UMC), 18, 121, 439, 441
University of Electro-Communications, 133, 137, 138, 141, 143
Upstream quality (robust quality), 354, 355
Urinary continence recovery in patients with brain disease, prediction of, 1258–1266
Useful effects (TRIZ), 1453
Useful part (ideal functions), 1509
User friendly design, 14
Utilization of interaction between control and noise factors, *see* Parameter design
Utilization of nonlinearity, *see* Parameter design

V

Values, objective, 507–508
Value analysis (VA), 1482
Value engineering (VE), 1482
Variability, 225, 226
    dynamic, 236
    improving, 41–55
    nondynamic, 236
    reducing, 40
    repeatability as, 224
    reproducibility as, 224
    SN ratios and traditional approach for, 224
    stability as, 224
Variables:
    continuous:
        signal-to-noise ratios for, 234–236, 239–289
        in two-way layout experiments, 563–572
        value of classified attributes vs., 239
    noise, 176
Variance, 509–514. *See also* Analysis of variance
    definition of, 509
    error, 511. *See also* Error variance
Variation, 509–514
    decomposition of, *see* Decomposition of variation
    definition of, 509
    error, 513, 517, 518, 559
    between experiments, 574
    of primary error, 574–576
    between primary units, 574

pure, 518, 559
range of, 1510–1513
within repetitions, 575–576, 581
of the secondary error, 575–576, 581
total, 509
Variation of the general mean, 520
VE (value engineering), 1482
Video compression, optimization of, 1310
Virtual testing, 1475–1476
Viscosity control, 469–470
Vision, project, 391
Visnepolschi, Svetlana, 1301
Visual pattern recognition, 397–398
Voltage-controlled oscillators, ideal function for, 234

W

Wakabayaski, Kimihiro, 137
Warranty costs, 1516
Water feeder valve, parameter design of, 330–338
Wave soldering:
    ideal function for, 233
    optimization of process, 895–899
Weather forecasting, MTS in, 419
Welch, Jack, 161
Welding:
    gas-arc stud weld process parameter optimization, 926–939
    ideal function of, 230
    resistance welding conditions for electronic components, 863–868
Whack-a-mole engineering, 1508
Whack-a-mole research, 73, 74
Wheatstone bridge:
    parameter design of, 320–330, 349–350
    purpose of, 320
    tolerance design of, 346–351
Window latching, minivans, 1025–1031
Wiper systems:
    chatter reduction in, 1148–1156
    ideal function of, 230
Working hours, estimation of, 1401–1405
Working mean, 506–507
Wu, Alan, 1493
Wu, Yuin, 157–158, 1493

X

Xerox Corporation, 18, 158, 161, 433, 441, 1439, 1503, 1505

Y

Yaesu Book Center, 1428
Yahata Steel, 1424
Yano, Hiroshi, 132–133, 136, 140, 161
Yano Laboratory, 133
Yield, 235, 293
Yokogawa Hewlett-Packard, 141, 313
Yoshino, Fushimi, 142
Youden squares, 617–625
    calculations with, 620–623
        derivations with, 623–625
        objective of using, 617–619

Z

Zero approximation, 613
Zero operating window, 269, 270
Zero-point calibration, 228
Zero-point proportional equation, 240–247
Zero-point proportionality, 137
Zlotin, Boris, 1301
Zusman, Alla, 1301